D1156362

THE
RF AND
MICROWAVE
HANDBOOK

The Electrical Engineering Handbook Series

Series Editor
Richard C. Dorf
University of California, Davis

Titles Included in the Series

The Avionics Handbook, Cary R. Spitzer

The Biomedical Engineering Handbook, 2nd Edition, Joseph D. Bronzino

The Circuits and Filters Handbook, Wai-Kai Chen

The Communications Handbook, Jerry D. Gibson

The Control Handbook, William S. Levine

The Digital Signal Processing Handbook, Vijay K. Madisetti & Douglas Williams

The Electrical Engineering Handbook, 2nd Edition, Richard C. Dorf

The Electric Power Engineering Handbook, Leo L. Grigsby

The Electronics Handbook, Jerry C. Whitaker

The Engineering Handbook, Richard C. Dorf

The Handbook of Formulas and Tables for Signal Processing, Alexander D. Poularikas

The Industrial Electronics Handbook, J. David Irwin

The Measurement, Instrumentation, and Sensors Handbook, John G. Webster

The Mechanical Systems Design Handbook, Osita D.I. Nwokah

The RF and Microwave Handbook, Mike Golio

The Mobile Communications Handbook, 2nd Edition, Jerry D. Gibson

The Ocean Engineering Handbook, Ferial El-Hawary

The Technology Management Handbook, Richard C. Dorf

The Transforms and Applications Handbook, 2nd Edition, Alexander D. Poularikas

The VLSI Handbook, Wai-Kai Chen

The Mechatronics Handbook, Robert H. Bishop

Forthcoming Titles

The Communications Handbook, 2nd Edition, Jerry D. Gibson

The Circuits and Filters Handbook, 2nd Edition, Wai-Kai Chen

THE

RF AND

MICROWAVE

HANDBOOK

Editor-in-Chief

MIKE GOLIO

Director, RF/Power Design Center
Rockwell Collins, Inc.

CRC Press
Boca Raton   London   New York   Washington, D.C.

111.26

10/81/c

Library of Congress Cataloging-in-Publication Data

The RF and microwave handbook / editor-in-chief, Mike Golio.
      p. cm.—(The electrical engineering handbook series)
    Includes bibliographical references and index.
    ISBN 0-8493-8592-X (alk. paper)
    1. Microwave circuits. 2.Radio circuits. 3. Wireless communication systems. I. Golio,
John Michael, 1954- II. Series.

TK7876 .R493 2000
621.381′32—dc21
                                                                      00-052885

This book contains information obtained from authentic and highly regarded sources. Reprinted material is quoted with permission, and sources are indicated. A wide variety of references are listed. Reasonable efforts have been made to publish reliable data and information, but the author and the publisher cannot assume responsibility for the validity of all materials or for the consequences of their use.

Neither this book nor any part may be reproduced or transmitted in any form or by any means, electronic or mechanical, including photocopying, microfilming, and recording, or by any information storage or retrieval system, without prior permission in writing from the publisher.

All rights reserved. Authorization to photocopy items for internal or personal use, or the personal or internal use of specific clients, may be granted by CRC Press LLC, provided that $.50 per page photocopied is paid directly to Copyright clearance Center, 222 Rosewood Drive, Danvers, MA 01923 USA. The fee code for users of the Transactional Reporting Service is ISBN 0-8493-8592-X/01/$0.00+$.1.50. The fee is subject to change without notice. For organizations that have been granted a photocopy license by the CCC, a separate system of payment has been arranged.

The consent of CRC Press LLC does not extend to copying for general distribution, for promotion, for creating new works, or for resale. Specific permission must be obtained in writing from CRC Press LLC for such copying.

Direct all inquiries to CRC Press LLC, 2000 N.W. Corporate Blvd., Boca Raton, Florida 33431.

Trademark Notice: Product or corporate names may be trademarks or registered trademarks, and are used only for identification and explanation, without intent to infringe.

Visit the CRC Press Web site at www.crcpress.com

© 2001 by CRC Press LLC

No claim to original U.S. Government works
International Standard Book Number 0-8493-8592-X
Library of Congress Card Number 00-052885
Printed in the United States of America   2 3 4 5 6 7 8 9 0
Printed on acid-free paper

Preface

The field of microwave engineering has undergone a radical transformation in recent years as the Defense and Government work effort that once dominated the industry is now exceeded by commercial wireless efforts. The explosive growth of commercial wireless markets has not only altered the Defense/non-Defense work balance, but also brought about significant shifts in the perspective and focus of working microwave and RF engineers. Engineering emphasis has changed from optimum performance design to design for manufacturing, from one-of-a-kind parts to high volume production, from performance at all cost to minimum cost with acceptable performance, from widest possible bandwidth to regulated narrow band, etc. Even engineers and researchers working in the traditional high performance, low volume microwave markets have seen a shift in emphasis of their work as every effort is made to reduce cost through the re-use of commercial-off-the-shelf (COTS) parts. Although the physics and mathematics of microwave and RF engineering is the same, the job of the typical microwave engineer has changed dramatically. The modern microwave and RF engineer is expected to be knowledgeable of customer expectations, market trends, manufacturing technologies, and factory models to a degree that is unprecedented in the history of RF/microwave engineering. Unfortunately, the 40+ years of close association of the microwave industry solely with Defense/Government agencies has left a legacy of microwave literature that is narrowly focused on high performance, low volume applications and is deficient in many areas of critical importance today. This handbook comes at a time when there is an emerging interest in a broader range of microwave material with more balance than that which has been previously available.

The purpose of *The RF and Microwave Handbook* is to provide a single-volume comprehensive reference for modern microwave and RF engineers. The articles that comprise the handbook provide important information for practicing engineers in industry, government, and academia. The intended audience also includes microwave and other electrical engineers requiring information outside of their area of expertise as well as managers, marketers, and technical support workers who need better understanding of the fields driving and affected by their decisions.

The book is organized into nine chapters, with all but the first chapter consisting of several articles on related topics. The first chapter consists of a single introductory article that provides important definitions and spectral information. Two appendices containing useful information for practicing microwave engineers are also included. By design, there is some overlap of information presented in some of the chapters. This allows the reader to investigate a topic without needing to make numerous references to other portions of the handbook. A complete table of contents is presented at the front of the book to provide the reader with a means of locating desired information as easily and rapidly as possible. Finally, all of the articles provide the reader with additional references to related expert literature.

Acknowledgments

Developing a handbook like this one is a big job — much bigger than I originally anticipated. This handbook would simply never have been completed if it were not for the efforts of the managing editor, Janet Golio. Her focus, persistence, software expertise, and organizational skills were essential to the project. Her efforts can be measured in the nearly 10,000 pieces of email correspondence she handled, the 100+ telephone conversations she initiated, or the tracking of approximately 80 articles from initial contact to commitment to receipt to review through final modification and submission. Yet all of these metrics combined do not completely capture the full magnitude of her contribution to this handbook. I cannot offer enough gratitude to compensate for the many long evenings she spent on this project.

I am also significantly indebted to the Handbook Editorial Board. This Board contributed to every phase of the handbook development. Their efforts are reflected in the organization and outline of the material, selection and recruitment of authors, article contributions, and review of the articles. Their labors were essential to the project and I am happy to acknowledge their help.

And, of course, I must thank the handbook professionals at CRC Press. Richard Dorf, Editor for the CRC Electrical Engineering Handbook series, and Ron Powers, President of Book Publishing, were instrumental in getting this project started. Special thanks is extended to Nora Konopka, Acquisitions Editor, who has worked most closely with the project during article development and has been more patient and encouraging than I deserve. Finally, Helena Redshaw, Production Manager, has taken the stacks of manuscripts, disks and CDs, identified and added the missing bits and pieces, and turned them into a book. Thanks also to all the CRC staff that I have not had the pleasure to work closely with, but who have contributed to this effort.

Editor-in-Chief

Mike Golio received his B.S.E.E. degree from the University of Illinois in 1976. He worked designing microwave oscillators and amplifiers for electronic counter-measure applications for two years before returning to school to complete his M.S.E.E. and Ph.D. degrees at North Carolina State University in 1980 and 1983, respectively. After graduate school, he worked as an Assistant Professor of Electrical Engineering before returning to industry. He spent five years developing MMIC circuits, tools, and techniques for microwave satellite applications prior to participating in the establishment of a GaAs fabrication facility. The GaAs facility was constructed to manufacture devices and circuits to serve commercial wireless markets — including cellular telephone, pager, and wireless LAN businesses. His role shifted from individual technical contributor to Group Manager during this time. He is currently the Director of the Rockwell Collins RF/Power Design Center tasked with identifying, acquiring, developing, and implementing emerging RF and power conversion technologies. The Center provides design expertise in the radio frequency (RF), microwave, and millimeter wave domain. This includes receivers, transmitters, antennas, frequency control circuitry, and power supplies for commercial and military avionics applications.

Dr. Golio's research has resulted in 15 patents and over 200 professional presentations and publications. His technical publications include four book chapters, one book, and one software package. He was selected to serve as the IEEE Microwave Theory and Techniques-Society (MTT-S) Distinguished Microwave Lecturer from 1998 to 2000. He is currently co-editor of the *IEEE Microwave Magazine,* to which he contributes a regular column. He serves as a member of the MTT-S Administrative Committee, has organized several professional workshops and panel sessions, and served as a member of the Technical Program Committee for several professional conferences. He was elected Fellow of the IEEE in 1996.

Managing Editor

Janet R. Golio received her B. S. in Electrical Engineering Summa Cum Laude from North Carolina State University in 1984. Prior to that she was employed by IBM. Between 1984 and 1996 she worked at Motorola in Arizona. Since 1996, she has been at Rockwell Collins in Cedar Rapids, IA. She holds one patent and has written four technical papers. When not working, she actively pursues her interests in archaeology, trick roping, and country western dancing. She is also the author of young adult books, *A Present From the Past* and *A Puzzle From the Past*.

Advisory Board

Peter A. Blakey obtained a B.A. in Physics from the University of Oxford in 1972, a Ph.D. in Electronic Engineering from University College London in 1976, and an MBA from the University of Michigan in 1989. Between 1972 and 1983 he worked on the theory and simulation of microwave semiconductor devices and circuits, with an emphasis on the impact of non-local transport phenomena. Between 1983 and 1995 he worked on the development, application, and commercialization of technology CAD software for silicon VLSI applications. Since 1995, he has been the manager of Predictive Engineering in the GaAs Technology Department at Motorola.

Lawrence P. Dunleavy received the B.S.E.E. degree from Michigan Technological University in 1982, and the M.S.E.E. and Ph.D. degrees in 1984 and 1988, respectively, from the University of Michigan. He has worked in industry for E-Systems and Hughes Aircraft Company and was a Howard Hughes Doctoral Fellow. In 1990 he joined the Electrical Engineering Department at the University of South Florida, where he now holds the title of Associate Professor. From August 1997 to to August 1998 he enjoyed a 1-year sabbatical research appointment with the Microwave Metrology Group of the National Institute of Standards and Technology (NIST) in Boulder, CO. His current research interests are in the area of accurate microwave and millimeter-wave measurements, measurement-based active and passive component modeling, MMIC design, and wireless systems characterization and CAD. Dr. Dunleavy is a Senior Member of IEEE, is very active in the IEEE MTT Society, and has been an ARFTG member since 1986.
Dr. Dunleavy has served as local host and technical program chair for ARFTG, as MTT/ARFTG liaison, and as conference chair for various conferences. He has also served ARFTG as the Short Course Coordinator from 1995 to 1998, as Director from 1998 to present, as Chair of the Education Committee from 1997 to present, and as a member of the Executive Committee from 1998 to present.

Jack East received his B.S.E., M.S., and Ph.D. degrees from the University of Michigan. He is presently with the Solid State Electronics Laboratory at the University of Michigan conducting research in the areas of high speed microwave device design, fabrication, and experimental characterization of solid state microwave devices, nonlinear device and circuit modeling for communications circuits and low energy electronics, and THz technology.

Patrick J. Fay is an Assistant Professor in the Department of Electrical Engineering at the University of Notre Dame. Dr. Fay received the B.S. degree in electrical engineering from the University of Notre Dame in 1991, and the M.S. and Ph.D. degrees in electrical engineering from the University of Illinois at Urbana-Champaign in 1993 and 1996, respectively. From 1992 to 1994, he was a National Science Foundation Graduate Research Fellow. After graduation, Dr. Fay served as a visiting assistant professor in the Department of Electrical and Computer Engineering at the University of Illinois at Urbana-Champaign in 1996 and 1997.

Dr. Fay joined the University of Notre Dame as an assistant professor in 1997. His educational activities have included the development of an advanced undergraduate laboratory course in microwave circuit design and characterization, and graduate courses in optoelectronic devices and electronic device characterization. His research interests include the design, fabrication, and characterization of microwave and high-speed optoelectronic devices and circuits for ultra-fast analog and digital signal processing, wireless communications, and fiber optic telecommunications. His research also includes the development and use of micromachining techniques for the fabrication of active and passive microwave components.

Dr. Fay is a member of the IEEE Microwave Theory and Techniques Society, the Solid-State Circuits Society, and the Electron Device Society. He was awarded the Department of Electrical Engineering's IEEE Outstanding Teacher Award in 1998–1999, and has served as a consultant for the Division of Electrical and Communications Systems at the National Science Foundation.

David Halchin received the Ph.D. degree in electrical engineering from North Carolina State University in 1987. From 1987 to 1989 he was an Assistant Professor in the ECE Department of Santa Clara University in Santa Clara, CA. His main interests in teaching and research were device modeling and microwave design. From 1989 until 1996 he was with Motorola Corporate Research and SPS in Phoenix, AZ. While at Motorola, he worked on a wide variety of projects. These included microwave and millimeter wave characterization of MESFETs and PHEMTs, modeling of PHEMTs and MESFETs, load pull measurements, device (MESFETs and PHEMTs) engineering and process development. From 1996 until 1997 he worked for Rockwell International in Newbury Park, CA as the head of a team tasked to develop and implement into production a linear MESFET and PHEMT for wireless PA applications. He rejoined Motorola SPS in 1997 for 1 year and again worked with the PHEMT/MESFET device development team in Phoenix. In 1998 he joined RF Micro Devices in Greensboro, NC working as a PA designer and a group leader in the area of WCDMA applications. He is a member of the Technical Program Committee for the GaAs IC Symposium, a reviewer for Transactions on Electron Devices and a member of the Editorial Board for Microwave Theory and Techniques and an IEEE member. He has 3 issued U.S. patents and several pending.

Roger B. Marks received his B.A. in Physics in 1980 from Princeton University and his Ph.D. in Applied Physics in 1988 from Yale University. Following a postdoctoral appointment at the Delft University of Technology (The Netherlands), he began a professional career with U.S. Commerce Department's National Institute of Standards and Technology (NIST) in Boulder, CO.

In 1998, Dr. Marks initiated an effort to standardize fixed broadband wireless access within the IEEE 802 LAN/MAN Standards Committee. This effort led to creation, in March 2000, of the IEEE 802.16 Working Group on Broadband Wireless Access. With Dr. Marks as Chair, 802.16 has grown to include three projects and hundreds of participants from over 100 companies.

A Fellow of the IEEE and the author of over 70 journal and conference publications, Dr. Marks has received many awards, including an IEEE Technical Field Award (the 1995 Morris E. Leeds Award). He has served in numerous professional capacities. In the IEEE Microwave Theory and Techniques Society, he served on the Adminstrative Committee and as Distinguished Microwave Lecturer. He developed the IEEE Radio and Wireless Conference (RAWCON) and chaired it from 1996 through 1999. He served as Vice Chair of the 1997 IEEE MTT-S International Microwave Symposium.

Dr. Marks lives in Denver, CO with his wife, Robbie Bravman Marks, and their children, Dan and Amy. His drumming can be heard on the RAWCON theme song and in a recording by the band Los Lantzmun.

Alfy Riddle operates Macallan Consulting, which specializes in solving RF and microwave circuit design problems as well as selling Mathematica-based design software. Dr. Riddle also teaches a course in electronic noise at Santa Clara University. He received his Ph.D. from North Carolina State University in 1986 for work on oscillator noise. After spending several years employed in the electronics industry, he began Macallan Consulting in 1989. When he is not working, he can be found on the tennis courts, hiking in the Sierras, taking pictures with an old Leica M3, or making and playing Irish whistles.

Robert J. Trew serves as Director of Research for Department of Defense (DoD). He has management oversight responsibility for the $1.2 billion yearly basic research programs of the Military Services and Defense Agencies. Dr. Trew has extensive experience in industry, academia, and government. He received the Ph.D. degree in electrical engineering from the University of Michigan in 1975. His industrial experience includes employment at General Motors Corp. in Michigan and Watkins-Johnson Company in California. In addition to his present DoD assignment, he worked for the U.S. Army Research Office as a program manager for five years. Dr. Trew was on the faculty of North Carolina State University where he served as Professor of Electrical and Computer Engineering from 1976 to 1993. From 1993 to 1997 he was George S. Dively Professor of Engineering and Chair of the Department of Electrical Engineering and Applied Physics at Case Western Reserve University. He was a Visiting Professor at the University of Duisburg in Germany in 1985. Dr. Trew is a Fellow of the IEEE and serves on the Microwave Theory and Techniques Society ADCOM. He is currently serving as Chair of the Publications Committee and was the Editor-in-Chief of the *IEEE Transactions on Microwave Theory and Techniques* from 1995 to 1997. He is currently co-editor of the new *IEEE Microwave Magazine*. Dr. Trew was the recipient of the 1998 IEEE MTT Society Distinguished Educator Award. He received the 1992 Alcoa Foundation Distinguished Engineering Research Award and a 1991 Distinguished Scholarly Achievement Award from NCSU. He was a recipient of an IEEE Third Millennium Medal award, and is currently an IEEE Microwave Distinguished Lecturer. Dr. Trew has 3 patents and has published more than 140 technical articles, 13 book chapters, and given over 250 scientific and technical presentations.

Contributors

David Anderson
Maury Microwave Corporation
Ontario, Canada

Carl Andren
Intersil
Palm Bay, Florida

Saf Asghar
Advanced Micro Devices, Inc.
Austin, Texas

Avram Bar-Cohen
University of Minnesota
Minneapolis, Minnesota

James L. Bartlett
Rockwell Collins
Cedar Rapids, Iowa

Melvin L. Belcher, Jr.
Georgia Tech
Smyrna, Georgia

Peter A. Blakey
Motorola
Tempe, Arizona

Mark Bloom
Motorola
Tempe, Arizona

Nicholas E. Buris
Motorola
Schaumburg, Illinois

Prashant Chavarkar
CREE Lighting Company
Goleta, California

John C. Cowles
Analog Devices–Northwest
   Laboratories
Beaverton, Oregon

Walter R. Curtice
W.R. Curtice Consulting
Princeton Junction, New Jersey

W.R. Deal
Malibu Networks, Inc.
Calabasas, California

Lawrence P. Dunleavy
University of South Florida
Tampa, Florida

Jack East
University of Michigan
Ann Arbor, Michigan

Stuart D. Edwards
Conway Stuart Medical Inc.
Sunnyvale, California

K.F. Etzold
IBM Thomas J. Watson Research
   Center
Yorktown Heights, New York

Leland M. Farrer
Cadence Design Systems, Inc.
Sunnyvale, California

John Fakatselis
Intersil
Palm Bay, Florida

Patrick Fay
University of Notre Dame
Notre Dame, Indiana

S. Jerry Fiedziuszko
Space Systems/LORAL
Palo Alto, California

Paul G. Flikkema
Northern Arizona University
Flagstaff, Arizona

Karl J. Geisler
University of Minnesota
Minneapolis, Minnesota

Ian C. Gifford
M/A-COM, Inc.
Lowell, Massachusetts

Mike Golio
Rockwell Collins
Cedar Rapids, Iowa

Madhu S. Gupta
San Diego State University
San Diego, California

Ramesh K. Gupta
COMSTAT Laboratories
Clarksburg, Maryland

R.E. Ham
Consulting Engineer
Austin, Texas

Mike Harris
Georgia Tech Research Institute
Atlanta, Georgia

Robert D. Hayes
RDH Incorporated
Marietta, Georgia

T. Itoh
University of California
Los Angeles, California

Daniel E. Jenkins
Dell Computer Corp.
Austin, Texas

Nils V. Jespersen
Lockheed Martin Space Electronics
   and Communications
Reston, Virginia

Christopher Jones
M/A-COM TycoElectronics
Lowell, Massachusetts

J. Stevenson Kenney
Georgia Institute of Technology
Atlanta, Georgia

Ron Kielmeyer
Motorola
Scottsdale, Arizona

Allan D. Kraus
Allan D. Kraus Associates
Beachwood, Ohio

Andy D. Kucar
4U Communications Research, Inc.
Ottawa, Ontario, Canada

Jakub Kucera
Infineon Technologies
Munich, Germany

Jean-Pierre Lanteri
M/A-COM TycoElectronics
Lowell, Massachusetts

Michael Lightner
University of Colorado
Boulder, Colorado

William Liu
Texas Instruments
Dallas, Texas

Urs Lott
Acter AG
Zurich, Switzerland

Leonard MacEachern
Carleton University
Ottawa, Ontario, Canada

John R. Mahon
M/A-COM TycoElectronics
Lowell, Massachusetts

Michael E. Majerus
Motorola
Tempe, Arizona

Donald C. Malocha
University of Central Florida
Orlando, Florida

Tajinder Manku
University of Waterloo
Waterloo, Ontario, Canada

Brent A. McDonald
Dell Computer Corp.
Austin, Texas

Umesh K. Mishra
University of California
Santa Barbara, California

Karen E. Moore
Motorola
Tempe, Arizona

Charles Nelson
California State University
Sacramento, California

Josh T. Nessmith
Georgia Tech
Smyrna, Georgia

Robert Newgard
Rockwell Collins
Cedar Rapids, Iowa

John M. Osepchuk
Full Spectrum Consulting
Concord, Massachusetts

Anthony E. Parker
Macquarie University
Sydney, Australia

Anthony M. Pavio
Motorola
Tempe, Arizona

Jeanne S. Pavio
Motorola
Phoenix, Arizona

Jim Paviol
Intersil
Palm Bay, Florida

Michael Pecht
University of Maryland
College Park, Maryland

Benjamin B. Peterson
U.S. Coast Guard Academy
New London, Connecticut

Ronald C. Petersen
Lucent Technologies Inc./Bell Labs
Murray Hill, New Jersey

Brian Petry
3Com Corporation
San Diego, California

Y. Qian
University of California
Los Angeles, California

Vesna Radisic
HRL Laboratories, LLC
Malibu, California

Arun Ramakrishnan
University of Maryland
College Park, Maryland

James G. Rathmell
University of Sydney
Sydney, Australia

Alfy Riddle
Macallan Consulting
Milpitas, California

Arye Rosen
Drexel University
Philadelphia, Pennsylvania

Harel D. Rosen
UMDNJ/Robert Wood Johnson
    Medical School
New Brunswick, New Jersey

Matthew N.O. Sadiku
Avaya, Inc.
Holmdel, New Jersey

George K. Schoneman
Rockwell Collins
Cedar Rapids, Iowa

Jonathan B. Scott
Agilent Technologies
Santa Rosa, California

Warren L. Seely
Motorola
Scottsdale, Arizona

John F. Sevic
UltraRF, Inc.
Sunnyvale, California

Michael S. Shur
Renssalaer Polytechnic Institute
Troy, New York

Thomas M. Siep
Texas Instruments
Dallas, Texas

Richard V. Snyder
RS Microwave
Butler, New Jersey

Jan Stake
Chalmers University
Goteborg, Sweden

Wayne E. Stark
University of Michigan
Ann Arbor, Michigan

Joseph Staudinger
Motorola
Tempe, Arizona

Michael B. Steer
North Carolina State University
Raleigh, North Carolina

Daniel G. Swanson, Jr.
Bartley R.F. Systems
Amesbury, Massachusetts

Toby Syrus
University of Maryland
College Park, Maryland

Manos M. Tentzeris
Georgia Institute of Technology
Atlanta, Georgia

Robert J. Trew
U.S. Department of Defense
Arlington, Virginia

Karl R. Varian
Raytheon
Dallas, Texas

John P. Wendler
M/A–Com Components Business
  Unit
Lowell, Massachusetts

James B. West
Rockwell Collins
Cedar Rapids, Iowa

James C. Wiltse
Georgia Tech
Atlanta, Georgia

Jerry C. Whitaker
Technical Press
Morgan Hill, California

Contents

1

Introduction

Patrick Fay

University of Notre Dame

1.1   Overview of Microwave and RF Engineering

Modern microwave and radio frequency (RF) engineering is an exciting and dynamic field, due in large part to the symbiosis between recent advances in modern electronic device technology and the current explosion in demand for voice, data, and video communication capacity. Prior to this revolution in communications, microwave technology was the nearly exclusive domain of the defense industry; the recent and dramatic increase in demand for communication systems for such applications as wireless paging, mobile telephony, broadcast video, and tethered as well as untethered computer networks is revolutionizing the industry. These systems are being employed across a broad range of environments including corporate offices, industrial and manufacturing facilities, and infrastructure for municipalities, as well as private homes. The diversity of applications and operational environments has led, through the accompanying high production volumes, to tremendous advances in cost-efficient manufacturing capabilities of microwave and RF products. This, in turn, has lowered the implementation cost of a host of new and cost-effective wireless as well as wired RF and microwave services. Inexpensive handheld GPS navigational aids, automotive collision-avoidance radar, and widely available broadband digital service access are among these. Microwave technology is naturally suited for these emerging applications in communications and sensing, since the high operational frequencies permit both large numbers of independent channels for the wide variety of uses envisioned as well as significant available bandwidth per channel for high speed communication.

Loosely speaking, the fields of microwave and RF engineering together encompass the design and implementation of electronic systems utilizing frequencies in the electromagnetic spectrum from approximately 300 kHz to over 100 GHz. The term "RF" engineering is typically used to refer to circuits and systems having frequencies in the range from approximately 300 kHz at the low end to between 300 MHz and 1 GHz at the high end. The term "microwave engineering", meanwhile, is used rather loosely to refer to design and implementation of electronic systems with operating frequencies in the range of from 300 MHz to 1 GHz on the low end to upward of 100 GHz. Figure 1.1 illustrates schematically the electromagnetic spectrum from audio frequencies through cosmic rays. The RF frequency spectrum covers the medium frequency (MF), high frequency (HF), and very high frequency (VHF) bands, while the microwave portion of the electromagnetic spectrum extends from the upper edge of the VHF frequency range to just below the THz radiation and far-infrared optical frequencies (approximately 0.3 THz and above). The wavelength of free-space radiation for frequencies in the RF frequency range is from approximately 1 m (at 300 MHz) to 1 km (at 300 kHz), while those of the microwave range extend from 1 m to the vicinity of 1 mm (corresponding to 300 GHz) and below.

0-8493-8592-X/01/$0.00+$.50
© 2001 by CRC Press LLC

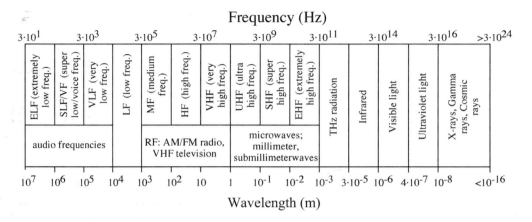

FIGURE 1.1 Electromagnetic frequency spectrum and associated wavelengths.

The boundary between "RF" and "microwave" design is both somewhat indistinct as well as one that is continually shifting as device technologies and design methodologies advance. This is due to implicit connotations that have come to be associated with the terms "RF" and "microwave" as the field has developed. In addition to the distinction based on the frequency ranges discussed previously, the fields of RF and microwave engineering are also often distinguished by other system features as well. For example, the particular active and passive devices used, the system applications pursued, and the design techniques and overall mindset employed all play a role in defining the fields of microwave and RF engineering. These connotations within the popular meaning of microwave and RF engineering arise fundamentally from the frequencies employed, but often not in a direct or absolute sense. For example, because advances in technology often considerably improve the high frequency performance of electronic devices, the correlation between particular types of electronic devices and particular frequency ranges is a fluid one. Similarly, new system concepts and designs are reshaping the applications landscape, with mass market designs utilizing ever higher frequencies rapidly breaking down conventional notions of microwave-frequency systems as serving "niche" markets.

The most fundamental characteristic that distinguishes RF engineering from microwave engineering is directly related to the frequency (and thus the wavelength) of the electronic signals being processed. For low-frequency and RF circuits (with a few special exceptions such as antennae), the signal wavelength is much larger than the size of the electronic system and circuit components. In contrast, for a microwave system the sizes of typical electronic components are often comparable (i.e., within approximately 1 order of magnitude) to the signal wavelength. This gives rise to a reasonable working definition of the two design areas based on the underlying approximations used in design. Since in conventional RF design, the circuit components and interconnections are generally small compared to a wavelength, they can be modeled as lumped elements for which Kirchoff's voltage and current laws apply at every instant in time. Parasitic inductances and capacitances are incorporated to accurately model the frequency dependencies, but these quantities can, to good approximation, be treated as lumped elements. For microwave frequencies, however, the finite propagation velocity of electromagnetic waves can no longer be neglected. For these frequencies, the time delay associated with signal propagation from one end of a component to the other is an appreciable fraction of the signal period, and thus lumped-element descriptions are no longer adequate to describe the electrical behavior. A distributed-element model is required to accurately capture the electrical behavior, in contrast to RF models. The time delay associated with finite wave propagation velocity that gives rise to the distributed circuit effects is a distinguishing feature of the mindset of microwave engineering. An alternative viewpoint is based on the observation that microwave engineering lies in a "middle ground" between traditional low-frequency electronics and optics, as shown in Fig. 1.1. As a consequence of RF, microwaves, and optics simply being different regimes of the same electromagnetic phenomena, there is a gradual transition between these regimes. The continuity of these regimes results in constant reevaluation of the appropriate design strategies and trade-offs as device and circuit technology advances. For example, miniaturization of active and passive components often increases the frequencies at which lumped-element circuit models are sufficiently accurate, since by reducing component dimensions, the time delay for propagation through a component is proportionally reduced. As a consequence, lumped-element components at "microwave" frequencies are becoming increasingly common in systems previously based on distributed elements due to significant advances in miniaturization, even though the operational frequencies remain unchanged. Component and circuit miniaturization also leads to tighter packing of interconnects and components, potentially introducing new parasitic coupling and distributed-element effects into circuits that could previously be treated using lumped-element RF models.

The comparable scales of components and signal wavelengths has other implications for the designer as well, since neither the ray-tracing approach from optics nor the lumped-element approach from RF circuit design are valid in this middle ground. In this regard, microwave engineering can also be considered to be "applied electromagnetic engineering," as the design of guided-wave structures such as waveguides and transmission lines, transitions between different types of transmission lines, and antennae all require analysis and control of the underlying electromagnetic fields. The design of these guided-wave structures is treated in detail in Section 9.3.

The distinction between RF and microwave engineering is further blurred by the trend of increasing commercialization and consumerization of systems using what have been traditionally considered to be microwave frequencies. Traditional microwave engineering, with its historically military applications, has long been focused on delivering performance at any cost. As a consequence, special-purpose devices intended solely for use in high performance microwave systems and often with somewhat narrow ranges of applicability were developed to achieve the required performance. With continuing advances in silicon microelectronics, including SiGe heterojunction bipolar transistors (HBTs) and conventional scaled CMOS, microwave frequency systems can now be reasonably implemented using the same devices as conventional low-frequency baseband electronics. In addition, the commercialization of low-cost III-V compound semiconductor electronics, including ion-implanted metal-semiconductor field-effect transistors (MESFETs), pseudomorphic high electron mobility transistors (PHEMTs), and III-V HBTs, has dramatically decreased the cost of including these elements in high-volume consumer systems. This convergence, with silicon microelectronics moving ever higher in frequency into the microwave spectrum from the low-frequency side and compound semiconductors declining in price for the middle of the frequency range, blurs the distinction between "microwave" and "RF" engineering, since "microwave" functions can now be realized with "mainstream" low-cost electronics. This is accompanied by a shift from physically large, low-integration-level hybrid implementations to highly integrated solutions based on monolithic microwave integrated circuits (MMICs). This shift has a dramatic effect not only on the design of systems and components, but also on the manufacturing technology and economics of production and implementation as well. A more complete discussion of the active device and integration technologies that make this progression possible is included in Chapter 7.

Aside from these defining characteristics of RF and microwave systems, a number of physical effects that are negligible at lower frequencies become increasingly important at high frequencies. Two of these effects are the skin effect and radiation losses. The skin effect is caused by the finite penetration depth of an electromagnetic field into conducting material. This effect is a function of frequency; the depth of penetration is given by $\delta_s = \frac{1}{\sqrt{\omega f \mu \sigma}}$, where μ is the permeability, f is the frequency, and σ is the conductivity of the material. As the expression indicates, δ_s decreases with increasing frequency, and so the electromagnetic fields are confined to regions increasingly near the surface as the frequency increases. This results in the microwave currents flowing exclusively along the surface of the conductor, significantly increasing the effective resistance (and thus the loss) of metallic interconnects. Further discussion of this topic can be found in Sections 9.2 and 9.6.1. Radiation losses also become increasingly important as the signal wavelengths approach the component and interconnect dimensions. For conductors and other components of comparable size to the signal wavelengths, standing waves caused by reflection of the electromagnetic waves from the boundaries of the component can greatly enhance the radiation of electromagnetic energy. These standing waves can be easily established either intentionally (in the case of antennae and resonant structures) or unintentionally (in the case of abrupt transitions, poor circuit layout, or other imperfections). Careful attention to transmission line geometry, placement relative to other components, transmission lines, and ground planes, as well as circuit packaging is essential for avoiding excessive signal attenuation and unintended coupling due to radiative effects.

A further distinction in the practice of RF and microwave engineering from conventional electronics is the methodology of testing. Due to the high frequencies involved, the capacitance and standing-wave effects associated with test cables and the parasitic capacitance of conventional test probes make the use of conventional low-frequency circuit characterization techniques impractical. Although advanced measurement techniques such as electro-optic sampling can sometimes be employed to circumvent these difficulties, in general the loading effect of measurement equipment poses significant measurement challenges for debugging and analyzing circuit performance, especially for nodes at the interior of the circuit under test. In addition, for circuits employing dielectric or hollow guided-wave structures, voltage and current often cannot be uniquely defined. Even for structures in which voltage and current are well-defined, practical difficulties associated with accurately measuring such high-frequency signals make this difficult. Furthermore, since a DC-coupled time-domain measurement of a microwave signal would have an extremely wide noise bandwidth, the sensitivity of the measurement would be inadequate. For these reasons, components and low-level subsystems are characterized using specialized techniques. These approaches are treated in detail in Chapter 4.

1.2   Frequency Band Definitions

The field of microwave and RF engineering is driven by applications — originally for military purposes such as radar and, more recently, for commercial, scientific, and consumer applications. As a consequence of this diverse applications base, microwave terminology and frequency band designations are not entirely standardized, with various standards bodies, corporations, and other interested parties all contributing to the collective terminology of microwave engineering. Figure 1.2 shows graphically some of the most common frequency band designations, with their approximate upper and lower bounds. As can be seen, some care must be exercised in the use of the "standard" letter designations; substantial differences in the definitions of these bands exist in the literature and in practice. Diagonal hashing at the ends of the frequency bands in Fig. 1.2 indicates variations in the definitions by different groups and authors; dark regions in the bars indicate frequencies for which there appears to be widespread agreement in the literature. The double-ended arrows appearing above some of the bands indicate the IEEE definitions for these bands. Two distinct definitions of K-band are in use; the first of these defines the band as the range from 18 GHz to approximately 26.5 GHz, while the other definition extends from 10.9 to 36 GHz. Both of these definitions are illustrated in Fig. 1.2. Similarly, L-band has two overlapping frequency range definitions; this gives rise to the large "variation" regions shown in Fig. 1.2. In addition, some care must be taken with these letter designations, as the IEEE and U.S. military specifications both define an L-band, but with very different frequencies. The IEEE L-band resides at the low end of the microwave spectrum, while the military definition of L-band is from 40 to 60 GHz. The IEEE designations (L–W) are presently used widely in practice and the technical literature, with the newer U.S. military designations (A–N) having not yet gained widespread popularity outside of the military community.

1.3   Applications

The field of microwave engineering is currently experiencing a radical transformation. Historically, the field has been driven by applications requiring the utmost in performance with little concern for cost or manufacturability. These systems have been primarily for military applications, where performance at nearly any cost could be justified. The current transformation of the field involves a dramatic shift from defense applications to those driven by the commercial and consumer sector, with an attendant shift in

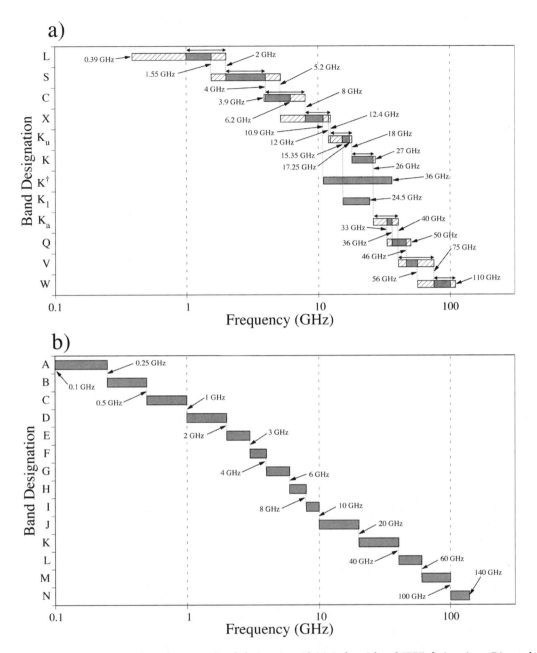

FIGURE 1.2 Microwave and RF frequency band designations.[1-5] (a) Industrial and IEEE designations. Diagonal hashing indicates variation in the definitions found in literature; dark regions in the bars indicate frequencies for which there is widespread agreement. Double-ended arrows appearing on bands indicate the current IEEE definitions for these bands where they exist, and K[†] denotes an alternative definition for K-band found in Reference 5. (b) U.S. military frequency band designations.[1-3]

focus from design for performance to design for manufacturability. This transformation also entails a shift from small production volumes to mass production for the commercial market, and from a focus on performance without regard to cost to a focus on minimum cost while maintaining acceptable performance. For wireless applications, an additional shift from broadband systems to systems having very tightly regulated spectral characteristics also accompanies this transformation.

For many years the driving application of microwave technology was military radar. The small wavelength of microwaves permits the realization of narrowly focused beams to be achieved with antennas small enough to be practically steered, resulting in adequate resolution of target location. Long-distance terrestrial communications for telephony as well as satellite uplink and downlink for voice and video were among the first commercially viable applications of microwave technology. These commercial communications applications were successful because microwave-frequency carriers (f_c) offer the possibility of very wide absolute signal bandwidths (Δf) while still maintaining relatively narrow fractional bandwidths (i.e., $\Delta f/f_c$). This allows many more voice and data channels to be accommodated than would be possible with lower-frequency carriers or baseband transmission.

Among the current host of emerging applications, many are based largely on this same principle, namely, the need to transmit more and more data at high speed, and thus the need for many communication channels with wide bandwidths. Wireless communication of voice and data, both to and from individual users as well as from users and central offices in aggregate, wired communication including coaxial cable systems for video distribution and broadband digital access, fiber-optic communication systems for long- and short-haul telecommunication, and hybrid systems such as hybrid fiber-coax systems are all poised to take advantage of the wide bandwidths and consequently high data carrying capacity of microwave-frequency electronic systems. In addition to the explosion in both diversity and capability of microwave-frequency communication systems, radar systems continue to be of importance with non-military and non-navigational applications including radar systems for automotive collision avoidance and weather and atmospheric sensing becoming increasingly widespread.

In addition to these "traditional" microwave applications, other fields of electronics are increasingly encroaching into the microwave frequency range. Examples include wired data networks based on coaxial cable or twisted-pair transmission lines with bit rates of over 1 Gb/s, fiber-optic communication systems with data rates well in excess of 10 Gb/s, and inexpensive personal computers and other digital systems with clock rates of over 1 GHz. The continuing advances in the speed and capability of conventional microelectronics is pushing traditional circuit design ever further into the microwave frequency regime. These trends promise to both invigorate and reshape the field of microwave engineering in new and exciting ways.

References

1. Collin, R. E., *Foundations for Microwave Engineering*, McGraw-Hill, New York, 1992, 2.
2. Harsany, S. C., *Principles of Microwave Technology*, Prentice-Hall, Upper Saddle River, NJ, 1997, 5.
3. Laverghetta, T. S., *Modern Microwave Measurements and Techniques*, Artech House, Norwood, MA, 1988, 479.
4. Rizzi, P. A., *Microwave Engineering*, Prentice-Hall, Englewood Cliffs, NJ, 1988, 1.
5. *Reference Data for Radio Engineers*, ITT Corp., New York, 1975.

2

Microwave and RF Product Applications

Paul G. Flikkema
Northern Arizona University

Andy D. Kucar
4U Communications Research, Inc.

Brian Petry
3Com Corporation

Saf Asghar
Advanced Micro Devices, Inc.

Jim Paviol
Intersil

Carl Andren
Intersil

John Fakatselis
Intersil

Thomas M. Siep
Texas Instruments

Ian C. Gifford
M/A-Com, Inc.

Ramesh K. Gupta
COMSAT Laboratories

Nils V. Jespersen
Lockheed Martin Space Electronics and Communications

Benjamin B. Peterson
U.S. Coast Guard Academy

James L. Bartlett
Rockwell Collins

James C. Wiltse
Georgia Tech

Melvin L. Belcher, Jr.
Georgia Tech

0-8493-8592-X/01/$0.00+$.50
© 2001 by CRC Press LLC

Josh T. Nessmith
Georgia Tech

Robert D. Hayes
RDH Incorporated

Madhu S. Gupta
San Diego State University

Arye Rosen
Drexel University

Harel D. Rosen
*UMDNJ/Robert Wood Johnson
Medical School*

Stuart D. Edwards
Conway Stuart Medical, Inc.

2.1   Cellular Mobile Telephony

Paul G. Flikkema

The goal of modern cellular mobile telephone systems is to provide services to telephone users as efficiently as possible. In the past, this definition would have been restricted to mobile users. However, the cost of wireless infrastructure is less than wired infrastructure in new telephone service markets. Thus, wireless mobile telephony technology is being adapted to provide in-home telephone service, the so-called wireless local loop (WLL). Indeed, it appears that wireless telephony will become dominant over traditional wired access worldwide.

The objective of this section is to familiarize the RF/microwave engineer with the concepts and terminology of cellular mobile telephony ("cellular"), or mobile wireless networks. A capsule history and a summary form the two bookends of the section. In between, we start with the cellular concept and the basics of mobile wireless networks. Then we take a look at some of the standardization issues for cellular systems. Following that, we cover the focus of the standards battles: channel access methods. We then take a look at some of the basic aspects of cellular important to RF/microwave engineers: first, modulation, diversity, and spread spectrum; then coding, interleaving, and time diversity; and finally nonlinear channels. Before wrapping up, we take a glimpse at a topic of growing importance: antenna array technology.

A Brief History

Mobile telephone service was inaugurated in the U.S. in 1947 with six radio channels available per city. This evolved into the manual Mobile Telephone System (MTS) used in the 1950s and 1960s. The year 1964 brought the Improved MTS (IMTS) systems with eight channels per city with — finally — no telephone operator required. Later, the capacity was more than doubled to 18. Most importantly, the IMTS introduced narrowband frequency modulation (NBFM) technology. The first cellular service was introduced in 1983, called AMPS (Advanced Mobile Phone Service). Cities were covered by cells averaging about 1 km in radius, each serviced by a base station. This system used the 900 MHz frequency band still in use for mobile telephony. The cellular architecture allowed frequency reuse, dramatically increasing capacity to a maximum of 832 channels per cell.

The age of digital, or second-generation, cellular did not arrive until 1995 with the introduction of the IS-54 TDMA service and the competing IS-95 CDMA service. In 1996–1997, the U.S. Federal Communications Commission auctioned licenses for mobile telephony in most U.S. markets in the so-called PCS (Personal Communication System) bands at 1.9 GHz. These systems use a variety of standards, including TDMA, CDMA, and the GSM TDMA standard that originated in Europe. Outside the U.S., a similar evolution has occurred, with GSM deployed in Europe and the PDC (Personal Digital Cellular) system in Japan. In other countries there has been a pitched competition between all systems. While not succeeding in the U.S., so-called low-tier systems have been popular in Europe and Japan. These systems are less robust to channel variations and are therefore targeted to pedestrian use. The European system is called DECT (Digital European Cordless Telephony) and the Japanese system is called PHS (Personal Handyphone System).

Third-generation (or 3G) mobile telephone service will be rolled out in the 2001–2002 time frame. These services will be offered in the context of a long-lived standardization effort recently renamed IMT-2000 (International Mobile Telecommunications–2000) under the auspices of the Radio Communications Standardization Sector of the International Telecommunications Union (ITU-R; see http://www.itu.int). Key goals of IMT-2000 are:[13]

1. Use of a common frequency band over the globe.
2. Worldwide roaming capability.
3. Transmission rates higher than second-generation systems to handle new data-over-cellular applications.

Another goal is to provide the capability to offer asymmetric rates, so that the subscriber can download data much faster than he can send it.

Finally, it is hoped that an architecture can be deployed that will allow hardware, software, and network commonality among services for a range of environments, such as those for vehicular, pedestrian, and fixed (nonmoving) subscribers. While also aiming for worldwide access and roaming, the main technical thrust of 3G systems will be to provide high-speed wireless data services, including 144 Kbps service to subscribers in moving vehicles, 384 Kbps to pedestrian users, 2 Mbps to indoor users, and service via satellites (where the other services do not reach) at up to 32 Kbps for mobile, hand-held terminals.

The Cellular Concept

At first glance, a logical method to provide radio-based communication service to a metropolitan area is a single, centrally located antenna. However, radio-frequency spectrum is a limited commodity, and regulatory agencies, in order to meet the needs of a vast number of applications, have further limited the amount of RF spectrum for mobile telephony. The limited amount of allocated spectrum forced designers to adopt the cellular approach: using multiple antennas (base stations) to cover a geographic area, each base station covers a roughly circular area called a cell. Figure 2.1 shows how a large region can be split into seven smaller cells (approximated by hexagons). This allows different base stations to use the same frequencies for communication links as long as they are separated by a sufficient distance. This is known as frequency reuse, and allows thousands of mobile telephone users in a metropolitan area to share far fewer channels.

There is a second important aspect to the cellular concept. With each base station covering a smaller area, the mobile phones need less transmit power to reach any base station (and thus be connected with the telephone network). This is a major advantage, since, with battery size and weight a major impediment to miniaturization, the importance of reducing power consumption of mobile phones is difficult to overestimate.

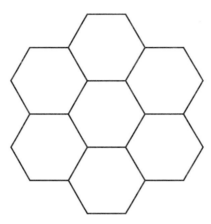

FIGURE 2.1 A region divided into cells. While normally the base stations are placed at the center of the cells, it is also possible to use edge-excited cells where base stations are placed at vertices.

If two mobile units are connected to their respective base stations at the same frequency (or more generally, channel), interference between them, called *co-channel interference*, can result. Thus, there is a trade-off between frequency reuse and signal quality, and a great deal of effort has resulted in frequency assignment techniques that balance this trade-off. They are based on the idea of clustering: taking the available set of channels, allocating them in chunks to each cell, and arranging the cells into geographically local clusters. Figure 2.2 shows how clusters of seven cells (each allocated one of seven mutually exclusive

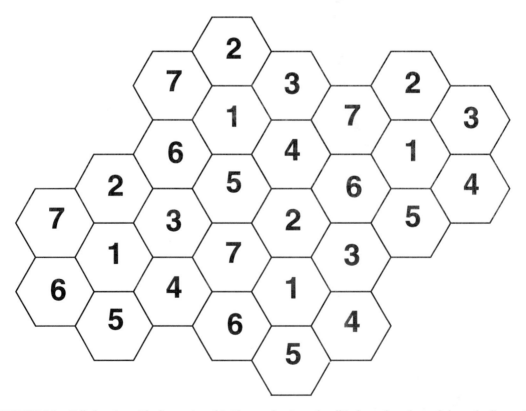

FIGURE 2.2 Cell planning with cluster size of 7. The number in each cell indexes the subset of channels allocated to the cell. Other cluster sizes, such as 4, 7, or 12 can be used.

channel subsets) are used to cover a large region; note that the arrangement of clusters maximizes the reuse distance — the distance between any two cells using the same frequency subset. Increasing the reuse distance has the effect of reducing co-channel interference.

Although it might seem attractive to make the cells smaller and smaller, there are diminishing returns. First, smaller cell sizes increase the need for management of mobile users as they move about. In other words, smaller cell sizes require more handoffs, where the network must transfer users between base stations. Another constraint is antenna location, which is often limited by available space and esthetics. Fortunately, both problems can be overcome by technology. Greater handoff rates can be handled by increases in processing speed, and creative antenna placement techniques (such as on lamp posts or sides of buildings) are allowing higher base station densities.

Another issue is evolution: how can a cellular system grow with demand? Two methods have been successful. The first is cell splitting: by dividing a cell into several cells (and adjusting the reuse pattern), a cellular service provider can increase its capacity in high-demand areas. The second is sectoring: instead of a single omnidirectional antenna covering a cell, a typical approach is to sectorize the cell into N_S regions, each served by an antenna that covers an angular span of $2\pi/N_S$ ($N_S = 3$ is typical). Note that both approaches increase handoff rates and thus require concurrent upgrading of network management. Later we will describe smart antennas, the logical extension to sectorization.

Networks for Mobile Telephony

A communication network that carries only voice — even a digital one — is relatively simple. Other than the usual digital communication system functions, such as channel coding, modulation, and synchronization, all that is required is call setup and takedown. However, current and future digital mobile telephony networks are expected to carry digital data traffic as well.

Data traffic is by nature computer-to-computer, and requires that the network have an infrastructure that supports everything from the application (such as Web browsing) to the actual transfer of bits. The data is normally organized into chunks called packets (instead of streams as in voice), and requires a much higher level of reliability than digitized voice signals. These two properties imply that the network must also label the packets, and manage the detection and retransmission of packets that are received in error. It is important to note that packet retransmission, while required for data to guarantee fidelity, is not possible for voice because it would introduce delays that would be intolerable in a human conversation.

Other functions that a modern digital network must perform include encryption and decryption (for data security) and source coding and decoding. The latter functions minimize the amount of the channel resource (in essence, bandwidth) needed for transferring the information. For voice networks this involves the design of voice codecs (coder/decoders) that not only digitize voice signals, but strip out the redundant information in them. In addition to all the functions that wired networks provide, wireless networks with mobile users must also provide mobility management functions that keep track of calls as subscribers move from cell to cell.

The various network functions are organized into *layers* to rationalize the network design and to ease internetworking, or the transfer of data between networks.[11] RF/microwave engineering is part of the *physical layer* that is responsible for carrying the data over the wireless medium.

Standards and Standardization Efforts

The cellular industry is, if anything, dense with lingo and acronyms. Here we try to make sense of at least some of the important and hopefully longer-lived terminology.

Worldwide, most of the cellular services are offered in two frequency bands: 900 and 1900 MHz. In each of the two bands, the exact spectrum allocated to terrestrial mobile services varies from country to country. In the U.S. cellular services are in the 800 to 900 MHz band, while similar services are in the 800 to 980 MHz band in Europe under the name GSM900. (GSM900 combines in one name a radio communication standard — GSM, or Global System for Mobile Communications — and the frequency band in which it is used. We will describe the radio communication, or *air interface,* standards later). In the mid-1990s, the U.S. allocated spectrum for PCS (Personal Communication Services) from 1850 to 2000 MHz; while many thought PCS would be different from cellular, they have converged and are interchangeable from the customer's perspective. Similarly, in Europe GSM1800 describes cellular services offered using the 1700 to 1880 MHz band.

The 1992 World Administrative Radio Conference (WARC '92) allocated spectrum for third-generation mobile radio in the 1885 to 1980 and 2110 to 2160 MHz bands. The ITU-Rs IMT-2000 standardization initiative adopted these bands for terrestrial mobile services. Note that the IMT-2000 effort is an umbrella that includes both terrestrial and satellite-based services — the latter for areas where terrestrial services are unavailable.

Please note that all figures here are approximate and subject to change in future WARCs; please consult References 2 and 13 for details.

The cellular *air interface* standards are designed to allow different manufacturers to develop both base station and subscriber (mobile user handset) equipment. The air interface standards are generally different for the downlink (base station to handset) and uplink (handset to base station). This reflects the asymmetry of resources available: the handsets are clearly constrained in terms of power consumption and antenna size, so that the downlink standards imply sophisticated transmitter design, while the uplink standards emphasize transmitter simplicity and advanced receive-side algorithms. The air interface standards address channel access protocols as well as traditional communication link design parameters such as modulation and coding. These issues are taken up in the following sections.

Channel Access

In a cellular system, a fixed amount of RF spectrum must somehow be shared among thousands of simultaneous phone conversations or data links. *Channel access* is about (1) dividing the allocated RF spectrum into pieces and (2) allocating the pieces to conversations/links in an efficient way.

The easiest channel access method to envision is FDMA (Frequency Division Multiple Access), where each link is allocated a sub-band (i.e., a specific carrier frequency; see Fig. 2.3). This is exactly the access method used by first generation (analog) cellular systems. The second generation of cellular brought two newer channel access methods that were enabled by progress in digital process technology. One is TDMA (Time Division Multiple Access), wherein time is divided into *frames,* and links are given short *time slots* in each frame (Fig. 2.4). FDMA and TDMA can be seen as time/frequency duals, in that FDMA subdivides the band into narrow sub-bands in the frequency domain, while TDMA subdivides time into slots, during which a link (within a cell) uses the entire allocated bandwidth.

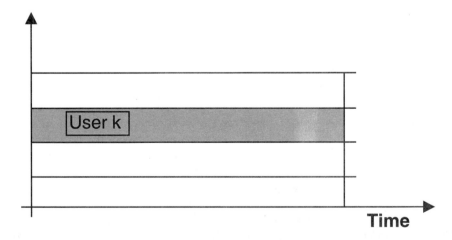

FIGURE 2.3 FDMA depicted on the time-frequency plane, with users assigned carrier frequencies, or channels. Not shown are guard bands between the channels to prevent interference between users' signals.

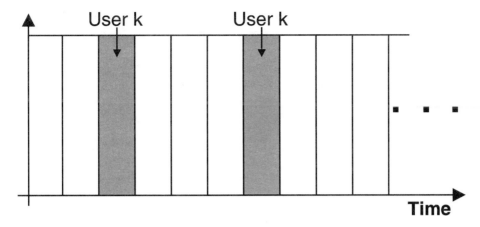

FIGURE 2.4　Depiction of TDMA on the time-frequency plane. Users are assigned time slots within a frame. Guard times (not shown) are needed between slots to compensate for timing inaccuracies.

The second generation of cellular also brought CDMA (Code Division Multiple Access). In CDMA, all active links simultaneously use the entire allocated spectrum, but sophisticated codes are used that allow the signals to be separated in the receiver.[1] We will describe CDMA in more depth later.

It should be noted that both TDMA- and CDMA-based cellular systems also implicitly employ FDMA, although this is rarely mentioned. The reason is that the cellular bands are divided into smaller bands (a form of FDMA), and both TDMA and CDMA are used within these sub-bands.

In the U.S., the TDMA and CDMA standards are referred to by different acronyms. The TDMA standard originally was called IS-54, but with enhancements became IS-136. The CDMA standard was called IS-95, and has been re-christened as cdmaOne by its originator, Qualcomm. These standards were created under the auspices of the Telecommunications Industry Association (TIA) and the Electronic Industries Alliance (EIA).

In Europe, the second generation brought digital technology in the form of the GSM standard, which used TDMA. (The GSM acronym originally referred to Group Special Mobile, but was updated to capture its move to worldwide markets.) Japan also chose TDMA in its first digital offering, called PDC (Personal Digital Cellular).

The three multiple access approaches use different signal properties (frequency, time, or code) to allow the distinguishing of multiple signals. How do they compare? In the main, as we move from FDMA to TDMA to CDMA (in order of their technological development), complexity is transferred from the RF section to the digital section of the transmitters and receivers. The evolution of multiple access techniques has tracked the rapid evolution of digital processing technology as the latter has become cheaper and faster. For example, while FDMA requires a tunable RF section, both TDMA and CDMA need only a fixed-frequency front end. CDMA relieves one requirement of TDMA — strict synchronization among the various transmitters — but introduces a stronger requirement for synchronization of the receiver to the received signal. In addition, the properties of the CDMA signal provide a natural means to exploit the multipath nature of the digital signal for improved performance. However, these advantages come at the cost of massive increases in the capability of digital hardware. Luckily, Moore's Law (i.e., that processing power roughly doubles every 18 months at similar cost) still remains in effect as of the turn of the century, and the amount of processing power that will be used in the digital phones in the 21st century will be unimaginable to the architects of the analog systems developed in the 1970s.

---

[1]It is fashionable to depict CDMA graphically using a "code dimension" that is orthogonal to the time-frequency plane, but this is an unfortunate misrepresentation. Like any signals, CDMA signals exist (in fact, overlap) in the time-frequency plane, but have correlation-based properties that allow them to be distinguished.

Modulation

The general purpose of modulation is to transform an information-bearing message signal into a related signal that is suitable for efficient transmission over a communication channel. In *analog modulation*, this is a relatively simple process: the information-bearing analog (or continuous-time) signal is used to alter a parameter (normally, the amplitude, frequency, or phase) of a sinusoidal signal (or carrier, the signal carrying the information). For example, in the NBFM modulation used in the AMPS system, the voice signal alters the frequency content of the modulated signal in a straightforward manner.

The purpose of *digital modulation* is to convert an information-bearing discrete-time symbol sequence into a continuous-time waveform. Digital modulation is easier to analyze than analog modulation, but more difficult to describe and implement.

Modulation in Digital Communication

Before digital modulation of the data in the transmitter, there are several processing steps that must be applied to the original message signal to obtain the discrete-time symbol sequence. A continuous-time message signal, such as the voice signal in telephony, is converted to digital form by sampling, quantization, and source coding. *Sampling* converts the original continuous-time waveform into discrete-time format, and *quantization* approximates each sample of the discrete-time signal using one of a finite number of levels. Then *source coding* jointly performs two functions: it strips redundancy out of the signal and converts it to a discrete-time sequence of symbols.

What if the original signal is already in discrete-time (sampled format), such as a computer file? In this case, no sampling or quantization is needed, but source coding is still used to remove redundancy.

Between source coding and modulation is a step critical to the efficiency of digital communications: channel coding. This is discussed later; it suffices for now to know that it converts the discrete-time sequence of symbols from the source coder into another (better) discrete-time symbol sequence for input to the modulator. Following modulation, the signal is upconverted, filtered (if required), and amplified in RF electronics before being sent to the antenna. All the steps described are shown in block-diagram form in Fig. 2.5. In the receiver, the signal from the antenna, following filtering (again, if required), is amplified and downconverted prior to demodulation, channel decoding, and source decoding (see Fig. 2.5).

What is the nature of the digital modulation process? The discrete-time symbol sequence from the channel coder is really a string of symbols (letters) from a *finite* alphabet. For example, in *binary* digital modulation, the input symbols are 0's and 1's. The modulator output converts those symbols into one of a *finite* set of waveforms that can be optimized for the channel.

While it is the finite set of waveforms that distinguishes digital modulation from analog modulation, that difference is only one manifestation of the entire paradigm of *digital communication*. In a good digital communication design, the source coder, channel coder, and modulator all work together to maximize the efficient use of the communication channel; even two of the three are not enough for good performance.

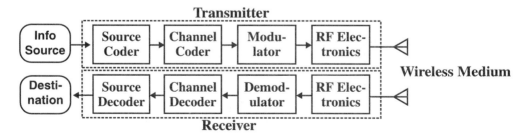

FIGURE 2.5 Communication system block diagram for wireless communication. In the wireless medium, multipath propagation and interference can be introduced. For system modeling purposes, the two blocks of RF electronics are combined with the wireless medium to form the wireless channel — a channel that distorts the signal, and adds noise and interference. The other blocks are designed to maximize the system performance for the channel.

Selection of Digital Modulation Formats

There are several (often conflicting) criteria for selection of a modulation scheme. They are:

- BER (bit error rate) performance
  - in wireless, particularly in cellular mobile channels, the scheme must operate under conditions of severe fading
  - cellular architectures imply co-channel interference
  - Typically, a BER of 10^{-2} or better is required for voice telephony, and 10^{-5} or better is required for data.
- Spectral (or bandwidth) efficiency (measured in bits/s/Hz)
- Power efficiency (especially for hand-held/mobile terminals)
- Implementation complexity and cost

In the U.S. cellular market, complexity is of special importance: with the number of standards growing, many handsets are now dual- and triple-mode; for example, a phone might have both GSM and 3G capability. While some hardware can be shared, multimode handsets clearly place additional constraints on the allowable complexity for each mode.

Classification of Digital Modulation Schemes

Broadly, modulation techniques can be classified into two categories.

Linear methods include schemes that use combinations of amplitude and phase modulation of a pulse stream. They have higher spectral efficiencies than constant-envelope methods (see the following), but must use more-expensive (or less efficient) linear amplifiers to maintain performance and to limit out-of-band emissions.

Examples of linear modulation schemes include PSK (phase-shift keying) and QAM (quadrature amplitude modulation). QAM can be viewed as a generalization of PSK in that both the amplitude and the phase of the modulated waveform are altered in response to the input symbols.

Constant-envelope methods are more complicated to describe, but usually are sophisticated methods based on frequency modulation. Their key characteristic is a constant envelope (resulting in a constant instantaneous signal power) regardless of the source symbol stream. They allow use of less expensive amplification and/or higher amplification efficiencies (e.g., running amplifiers in the nonlinear region), at the expense of out-of-band emissions. Historically, they are limited to spectral efficiencies of about 1 bit/s/Hz.

Examples of constant envelope methods include FSK (frequency-shift keying) and more sophisticated methods such as MSK and GMSK (these will be described shortly). These methods can be thought of as digital (finite alphabet) FM in that the spectrum of the output signal is varied according to the input symbol stream.

The spectral occupancy of a modulated signal (per channel) is roughly

$$S_O = B + 2\Delta f,$$

where B is the bandwidth occupied by signal power spectrum and Δf is the maximum one-way carrier frequency drift.[2] We can express the bandwidth

$$B = \frac{R_d}{\epsilon},$$

where R_d is the channel data rate (in bits/s) and ϵ is the spectral efficiency (in bits/s/Hz). Combining, we obtain

$$S_O = \frac{R_d}{\epsilon} + 2\Delta f.$$

---

[2]This drift can be caused by oscillator instability or Doppler due to channel time variations.

Thus, to minimize spectral occupancy (thus maximizing capacity in number of users) we can:

1. Reduce R_d by lowering the source coding rate (implying more complexity or lower fidelity), or
2. Improve the spectral efficiency of the modulation (implying higher complexity), or
3. Improve the transmitter/receiver oscillators (at greater cost).

Modulation, Up/Downconversion, and Demodulation

To transmit a string of binary information symbols (or bits — zeros and ones), $\{b_0, b_1, b_2, \ldots\}$, we can represent a 1 by a positive-valued pulse of amplitude one, and a 0 by a negative pulse of the sample amplitude. This mapping from the bit value at time n, b_n, to amplitude a_n can be accomplished using

$$a_n = 2b_n - 1.$$

To complete the definition, we define a pulse of unit amplitude with start time of zero and stop time of T as $p_T(t)$. Then the modulated signal can be efficiently written as

$$u(t) = \sum_n a_n p_T(t - nT).$$

This signal is at baseband — centered at zero frequency — and is therefore unsuitable for wireless communication media. However, this signal can be upconverted to a desired RF by mixing with a sinusoid to get the passband signal

$$x(t) = u(t)\cos(2\pi f_c t) = \cos(2\pi f_c t)\sum_n a_n p_T(t - nT),$$

where f_c is the carrier frequency.

Multiplying a sinusoid by ± 1 is identical to changing its phase between 0 and π radians, so we have

$$x(t) = \cos\left(2\pi f_c t + \sum_n d_n p_T(t - nT)\right),$$

where we assign $d_n = 0$ when $a_n = -1$ and $d_n = \pi$ when $a_n = 1$. This equation shows that we are simply shifting the phase of the carrier between two different values: this is BPSK (binary phase-shift keying).

Why not use more than two phase values? In fact, four are ordinarily used for better efficiency: pairs of bits are mapped to four different phase values, 0, $\pm\pi/2$, and π. For example, the CDMA standards employ this scheme, known as quaternary PSK (QPSK).

In general, the baseband signal will be complex-valued, which leads to the general form of upconversion from baseband to passband:

$$x(t) = \sqrt{2}\,\Re\left\{u(t)e^{j2\pi f_c t}\right\},$$

where the $\sqrt{2}$ factor is simply to maintain a consistency in measurement of signal power between passband and baseband. The motivation of using the baseband representation of a signal is twofold: first, it retains the amplitude and phase of the passband signal, and is thus independent of any particular carrier frequency; second, it provides the basis for modern baseband receiver implementations that use high-speed digital signal processing. The baseband representation is also known as the *complex envelope* representation.

BPSK and QPSK are linear modulation methods; in contrast, FSK is a constant-envelope modulation scheme. For binary FSK (BFSK), there are two possible signal pulses, given at baseband by

$$u_0\left(t\right) = A e^{-j\pi \Delta ft} p_T\left(t\right), \quad u_1\left(t\right) = A e^{j\pi \Delta ft} p_T\left(t\right),$$

where A is the amplitude. Notice that we have two (complex) tones separated by Δf. MSK (minimum-shift keying) and GMSK (Gaussian prefiltered MSK) are special forms of FSK that provide greater spectral efficiency at the cost of higher implementation efficiency. The GSM standard and its next-generation version, currently known as EDGE (for Enhanced Data Rates for Global Evolution), use GMSK.

At the receiver, the RF signal is amplified and downconverted with appropriate filtering to remove interference and noise. The downconverted signal is then passed to the demodulator, whose function is to detect (guess in an optimum way) what symbol stream was transmitted. Following demodulation (also referred to as detection), the symbol stream is sent to subsequent processing steps (channel decoding and source decoding) before delivery to the destination.

At this point it is typical to consider the BERs and spectral efficiencies of various digital modulation formats, modulator and demodulator designs, and the performance of different detection strategies for mobile cellular channels. This is beyond the scope of this section, and we direct the reader to a good book on digital communications (e.g., References 1, 4, 6, 7, 8) for more information.

Diversity, Spread Spectrum, and CDMA

A mobile wireless channel causes the transmitted signal to arrive at the receiver via a number of paths due to reflections from objects in the environment. If the channel is linear (including transmit and receive amplifiers), a simple modeling approach for this multipath channel is to assume that it is specular, i.e., each path results in a specific amplitude, time delay, and phase change. If the channel is also at least approximately time-invariant, its impulse response under these conditions can be expressed as[3]

$$h\left(t\right) = \sum_{\lambda=0}^{\Lambda} \alpha_\lambda e^{j\theta_\lambda} \delta\left(t - \tau_\lambda\right),$$

where α_λ, τ_λ, and θ_λ are, respectively, the amplitude, time delay, and phase for the λ-th path.

Let the transmitted signal be

$$s\left(t\right) = \sum_n a_n f_n\left(t\right),$$

a sequence of pulses $f_n(t)$ each modulated by a transmitted symbol a_n at a symbol rate of $1/T$. When transmitted via a specular multipath channel with Λ paths, the received signal — found by the convolution of the transmitted signal and the channel impulse response — is

$$y\left(t\right) = \sum_{\lambda=0}^{\Lambda} \alpha_\lambda e^{j\theta_\lambda} s\left(t - \tau_\lambda\right).$$

For simplicity, consider sending only three symbols a_{-1}, a_0, a_1. Then the received signal becomes

$$y\left(t\right) = \sum_{n=-1}^{1} a_n \sum_{\lambda=0}^{\Lambda} \alpha_\lambda e^{j\theta_\lambda} f_n\left(t - \tau_\lambda\right).$$

---

[3]Here $\delta(t)$ denotes the Dirac delta function.

Two effects may result: fading and intersymbol interference. *Fading* occurs when superimposed replicas of the same symbol pulse nullify each other due to phase differences. *Intersymbol interference (ISI)* is caused by the convolutive mixing of the adjacent symbols in the channels. Fading and ISI may occur individually or together depending on the channel parameters and the symbol rate T^{-1} of the transmitted signal.

Let us consider in more detail the case where the channel *delay spread* is a significant fraction of T, i.e., τ_Λ is close to, but smaller than T. In this case, we can have both fading and ISI, which, if left untreated, can severely compromise the reliability of the communication link. Direct-sequence spread-spectrum (DS/SS) signaling is a technique that mitigates these problems by using clever designs for the pulses $f_n(t)$. These pulse designs are wide bandwidth (hence "spread spectrum"), and the extra bandwidth is used to endow them with properties that allow the receiver to separate the symbol replicas.

Suppose we have a two-path channel, and consider the received signal for symbol a_0. Then the DS/SS receiver separates the two replicas

$$\alpha_0 e^{j\theta_0} a_p f_0\left(t - \tau_0\right), \quad \alpha_1 e^{j\theta_1} a_0 f_0\left(t - \tau_1\right).$$

Then each replica is adjusted in phase by multiplying it by $e^{-j\theta\lambda}$, $\lambda = 0, 1$ yielding (since $zz^* = |z|^2$)

$$\alpha_0 a_0 f\left(t - \tau_0\right), \quad \alpha_1 a_0 f\left(t - \tau_1\right).$$

Now all that remains is to delay the first replica by $\tau_1 - \tau_0$ so they line up in time, and sum them, which gives

$$\left(\alpha_0 + \alpha_1\right) a_0 f\left(t - \tau_1\right).$$

Thus DS/SS can turn the multipath channel to advantage — instead of interfering with each other, the two replicas are now added constructively. This *multipath combining* exploits the received signal's inherent *multipath diversity*, and is the basic idea behind the technology of RAKE reception[4] used in the CDMA digital cellular telephony standards.

It is important to note that this is the key idea behind all strategies for multipath fading channels: we somehow exploit the redundancy, or *diversity* of the channel (recall the multiple paths). In this case, we used the properties of DS/SS signaling to effectively split the problematic two-path channel into two benign one-path channels. Multipath diversity can also be viewed in the frequency domain, and is in effect a form of *frequency diversity*. As we will see later, frequency diversity can be used in conjunction with other forms of diversity afforded by wireless channels, including time diversity and antenna diversity.

CDMA takes the spread spectrum idea and extends it to the separation of signals from multiple *transmitters*. To see this, suppose M transmitters are sending signals simultaneously, and assume for simplicity that we have a single-path channel. Let the complex (magnitude/phase) gain for channel m be denoted by $\beta^{(m)}$. Finally, the transmitters use different spread-spectrum pulses, denoted by $f^{(m)}(t)$. If we just consider the zeroth transmitted symbols from each transmitter, we have the received signal

$$y\left(t\right) = \sum_{m=1}^{M} \beta^{(m)} a_0^{(m)} f^{(m)}\left(t - t_m\right)$$

where the time offset t_m indicates that the pulses do not necessarily arrive at the same time.

---

[4]The RAKE nomenclature can be traced to the block diagram representation of such a receiver — it is reminiscent of a garden rake.

The above equation represents a complicated mixture of the signals from multiple transmitters. If narrowband pulses are used, they would be extremely difficult — probably impossible — to separate. However, if the pulses are spread-spectrum, then the receiver can use algorithms to separate them from each other, and successfully demodulate the transmitted symbols. Of course, these ideas can be extended to many transmitters sending long strings of symbols over multipath channels.

Why is it called CDMA? It turns out that the special properties of the signal pulses $f^{(m)}(t)$ for each user (transmitter) m derive from high-speed *codes* consisting of periodic sequences of chips $c_k^{(m)}$ that modulate chip waveforms $\varphi(t)$. One way to envision it is to think of $\varphi(t)$ as a rectangular pulse of duration $T_c = T/N$. The pulse waveform for user m can then be written

$$f^{(m)}(t) = \sum_{k=0}^{N-1} c_k^{(m)} \varphi(t - kT_c).$$

The fact that we can separate the signals means that we are performing code-division multiple access — dividing up the channel resource by using codes. Recall that in FDMA this is done by allocating frequency bands, and in TDMA, time slots. The pulse waveforms in CDMA are designed so that many users' signals occupy the entire bandwidth simultaneously, yet can still be separated in the receiver. The signal-separating capability of CDMA is extremely important, and can extend beyond separating desired signals within a cell. For example, the IS-95 CDMA standard uses spread-spectrum pulse designs that enable the receiver to reject a substantial amount of co-channel interference (interference due to signals in other cells). This gives the IS-95 system (as well as its proposed 3G descendants) its well-known property of universal frequency reuse.

The advantages of DS/SS signals derive from what are called their *deterministic correlation* properties. For an arbitrary periodic sequence $\{c_k^{(m)}\}$, the deterministic *autocorrelation* is defined as

$$\phi^{(m)}(i) = \frac{1}{N} \sum_{k=0}^{N-1} c_k^{(m)} c_{k+i}^{(m)},$$

where i denotes the relative shift between two replicas of the sequence. If $\{c_k^{(m)}\}$ is a direct-sequence spreading code, then

$$\phi^{(m)}(i) \approx \begin{cases} 1, & i = 0 \\ 0, & 1 < |i| < N. \end{cases}$$

This "thumbtack" autocorrelation implies that relative shifts of the sequence can be separated from each other. Noting that each chip is a fraction of a symbol duration, we see that multipath replicas of a symbol pulse can be separated even if their arrival times at the receiver differ by less than a symbol duration.

CDMA signal sets also exhibit special deterministic *cross-correlation* properties. Two spreading codes $\{c_k^{(l)}\}$, $\{c_k^{(m)}\}$ of a CDMA signal set have the cross-correlation property

$$\phi^{(l,m)}(i) = \frac{1}{N} \sum_{k=0}^{N-1} c_k^{(l)} c_{k+i}^{(m)} \approx \begin{cases} 1, & l = m, i = 0, \\ 0, & l = m, 0 < |i| < N, \\ 0, & l \neq m. \end{cases}$$

Thus, we have a set of sequences with zero cross-correlations and "thumbtack" autocorrelations. (Note that this includes the earlier autocorrelation as a special case.) The basic idea of demodulation for CDMA is as follows: if the signal from user m is desired, the incoming received signal — a mixture of multiple transmitted signals — is correlated against $\{c_k^{(m)}\}$. Thus multiple replicas of a symbol from user m can be separated, delayed, and then combined, while all other users' signals (i.e., where $l \neq m$) are suppressed by the correlation.

Details of these properties, their consequences in demodulation, and descriptions of specific code designs can be found in References 3, 4, 7, and 10.

Channel Coding, Interleaving, and Time Diversity

As we have mentioned, channel coding is a transmitter function that is performed after source coding, but before modulation. The basic idea of channel coding is to introduce highly structured redundancy into the signal that will allow the receiver to easily detect or correct errors introduced in the transmission of the signal.

Channel coding is fundamental to the success of modern wireless communication. It can be considered the cornerstone of digital communication, since, without coding, it would not be possible to approach the fundamental limits established by Shannon's information theory.[9,12]

The easiest type of channel codes to understand are *block codes:* a sequence of input symbols of length k is transformed into a code sequence (codeword) of length $n > k$. Codes are often identified by their rate R, where $R = k/n \leq 1$. Generally, codes with a lower rate are more powerful. Almost all block codes are *linear,* meaning that the sum of two codewords is another codeword. By enforcing this linear structure, coding theorists have found it easier to find codes that not only have good performance, but have reasonably simple decoding algorithms as well.

In wireless systems, *convolutional codes* are very popular. Instead of blocking the input stream into length-k sequences and encoding each one independently, convolutional coders are finite-state sequential machines. Therefore they have memory, so that a particular output symbol is determined by a contiguous sequence of input symbols. For example, a rate-1/2 convolutional coder outputs two code symbols for each information symbol that arrives at its input. Normally, these codes are also linear.

Error-correcting codes have enough power so that errors can actually be corrected in the receiver. Systems that use these codes are called *forward error-control (FEC)* systems. *Error-detecting codes* are simpler, but less effective: they can tell *whether* an error has occurred, but not where the error is located in the received sequence, so it cannot be corrected.

Error-detecting codes can be useful when it is possible for the receiver to request retransmission of a corrupted codeword. Systems that employ this type of feedback are called *ARQ,* or Automatic Repeat-reQuest systems.

As we have seen, the fading in cellular systems is due to multipath. Of course, as the mobile unit and other objects in the environment move, the physical structure of the channel changes with time, causing the fading of the channel to vary with time. However, this fading process tends to be slow relative to the symbol rate, so a long string of coded symbols can be subjected to a deep channel fade. In other words, the fading from one symbol to the next will be highly correlated. Thus, the fades can cause a large string of demodulation (detection) errors, or an *error burst.* Thus, fading channels are often described from the point of view of coding as *burst-error channels.*

Most well-known block and convolutional codes are best suited to random errors, that is, errors that occur in an uncorrelated fashion and thus tend to occur as isolated single errors. While there have been a number of codes designed to correct burst errors, the theory of random error-correcting codes is so well developed that designers have often chosen to use these codes in concert with a method to "randomize" error bursts.

This randomization method, called *interleaving,* rearranges, or scrambles, the coded symbols in order to minimize this correlation so that errors are isolated and distributed across a number of codewords. Thus, a modest random-error correcting code can be combined with interleaving that is inserted between the channel coder and the modulator to shuffle the symbols of the codewords. Then, in the receiver, the de-interleaver is placed between the demodulator and the decoder is reassemble the codewords for decoding.

We note that a well-designed coding/interleaving system does more than redistribute errors for easy correction: it also exploits *time diversity*. In our discussion of spread-spectrum and CDMA, we saw how the DS/SS signal exploits the frequency diversity of the wireless channel via its multipath redundancy. Here, the redundancy added by channel coding/interleaving is designed so that, in addition to the usual performance increase due to just the code — the *coding gain* — there is also a benefit to distributing the redundancy in such a way that exploits the time variation of the channel, yielding a *time diversity gain*.

In this era of digital data over wireless, high link reliability is required. This is in spite of the fact that most wireless links have a raw bit error rate (BER) on the order of 1 in 1000. Clearly, we would like to see an error rate of 1 in 10^{12} or better. How is this astounding improvement achieved? The following two-level approach has proved successful. The first level employs FEC to correct a large percentage of the errors. This code is used in tandem with a powerful error-detecting algorithm to find the rare errors that the FEC cannot find and correct. This combined FEC/ARQ approach limits the amount of feedback to an acceptable level while still achieving the necessary reliability.

Nonlinear Channels

Amplifiers are more power-efficient if they are driven closer to saturation than if they are kept within their linear regions. Unfortunately, nonlinearities that occur as saturation is approached lead to *spectral spreading* of the signal. This can be illustrated by observing that an instantaneous (or memoryless) nonlinearity can be approximated by a polynomial. For example, a quadratic term effectively squares the signal; for a sinusoidal input this leads to double-frequency terms.

A more sophisticated perspective comes from noting that the nonlinear amplification can distort the symbol pulse shape, expanding the spectrum of the pulse. Nonlinearities of this type are said to cause AM/AM distortion. Amplifiers can also exhibit AM/PM conversion, where the output phase of a sinusoid is shifted by different amounts depending on its input power — a serious problem for PSK-type modulations.

A great deal of effort has gone into finding transmitter designs that allow more efficient amplifier operation. For example, constant-envelope modulation schemes are insensitive to nonlinearities, and signaling schemes that reduce the peak-to-average power ratio (PAPR) of the signal allow higher levels. Finally, methods to linearize amplifiers at higher efficiencies are receiving considerable attention.

Modeling and simulating nonlinear effects on system performance is a nontrivial task. AM/AM and AM/PM distortions are functions of frequency, so if wideband amplifier characterization is required, a family of curves is necessary. Even then the actual wideband response is only approximated, since these systems are limited in bandwidth and thus have memory. More accurate results in this case can be obtained using Volterra series expansions, or numerical solutions to nonlinear differential equations. Sophisticated approaches are becoming increasingly important in cellular as supported data rates move higher and higher. More information can be found in References 1 and 5 and the references therein.

Antenna Arrays

We have seen earlier how sectorized antennas can be used to increase system performance. They are one of the most economical forms of multielement antenna systems, and can be used to reduce interference or to increase user capacity. A second use of multielement systems is to exploit the *spatial diversity* of the wireless channel. Spatial diversity approaches assume that the received antenna elements are immersed in a signal field whose strength varies strongly with position due to a superposition of multipath signals arriving via various directions. The resulting element signal strengths are assumed to be at least somewhat statistically uncorrelated. This spatial uncorrelatedness is analogous to the uncorrelatedness over time or frequency that is exploited in mobile channels.

One of the simplest approaches is to use multiple (normally omnidirectional in azimuth) antenna elements at the receiver, and choose the one with the highest signal-to-noise ratio. More sophisticated schemes combine — rather than select just one of — the element signals to further improve the signal-to-noise ratio at the cost of higher receiver complexity. These approaches date from the 1950s, and do not take into account other interfering mobile units. These latter schemes are often grouped under the category of *antenna diversity* approaches.

More recently, a number of proposals for systems that combine error-control coding mechanisms with multiple elements have been made under the name of *space-time coding.* One of the main contributions of these efforts has been the recognition that multiple-element transmit antennas can, under certain conditions, dramatically increase the link capacity.

Another approach, beamforming or phased-array antennas, is also positioned to play a role in future systems under the new moniker *smart antennas.* Space-time coding and smart antenna methods can be seen as two approaches to exploiting the capabilities of multiple-input/multiple-output (MIMO) systems. However, in contrast to space-time coding approaches, strong inter-element correlation based on the direction of arrival of plane waves is assumed in smart antennas. The basic idea of smart antennas is to employ an array of antenna elements connected to an amplitude- and phase-shifting network to adaptively tune (steer electronically) the antenna pattern based on the geographic placement of mobile units. Much of the groundwork for smart antenna systems was laid in the 1950s in military radar research. The ultimate goal of these approaches can be stated as follows: to track individual mobile units with optimized antenna patterns that maximize performance (by maximizing the ratio of the signal to the sum of interference and noise) minimize power consumption at the mobile unit, and optimize the capacity of the cellular system. One can conjecture that the ultimate solution to this problem will be a class of techniques that involve joint design of channel coding, modulation, and antenna array processing in an optimum fashion.

Summary

Almost all wireless networks are distinguished by the characteristic of a shared channel resource, and this is in fact the key difference between wireless and wired networks. Another important difference between wired and wireless channels is the presence of multipath in the latter, which makes diversity possible. What is it that distinguishes cellular from other wireless services and systems? First, it historically has been designed for mobile telephone users, and has been optimized for carrying human voice. This has led to the following key traits of cellular:

- efficient use of spectrum via the cellular concept;
- system designs, including channel access mechanisms, that efficiently handle large numbers of uniform — i.e., voice — links; and
- difficult channels: user mobility causes fast variations in channel signal characteristics compared with other wireless applications such as wireless local area networks.

We close by mentioning two apparent trends. First, as we mentioned at the outset of this article, wireless local loop services, where home telephone subscribers use wireless phones — and the "last mile" is wireless rather than copper — are a new application for mobile wireless technology. Secondly, at this time there is a great deal of effort to make next-generation cellular systems useful for data networking in addition to voice. Certainly, the amount of data traffic on these networks will grow. However, one of the largest questions for the next ten years is whether mobile wireless will win the growing data market, or if new data-oriented wireless networks will come to dominate.

References

1. Zeng, M., Annamalai, A., and Bhargava, V. K., Recent advances in cellular wireless communications, *IEEE Communications Magazine,* 37, 9, 128–138, September 1999.
2. Walrand, J., and Varaiya, P., *High-Performance Communication Networks,* Morgan Kaufman, San Francisco, CA, 1996.
3. Chaudhury, P., Mohr, W., and Onoe, S., The 3GPP proposal for IMT-2000, *IEEE Communications Magazine,* 37 (12), 72–81, December 1999.

4. Anderson, J. B., *Digital Transmission Engineering*, IEEE Press, Piscataway, NJ, 1999.

5. Lee, E. A., and Messerschmitt, D. G., *Digital Communication*, Kluwer Academic, second edition, 1994.

6. Proakis, J. G., *Digital Communications*, 3rd ed., McGraw-Hill, New York, 1995.

7. Haykin, S., *Communication Systems*, Wiley, New York, 1994.

8. Proakis, J. G., and Salehi, M., *Communication Systems Engineering*, Prentice-Hall, Englewood Cliffs, NJ, 1994.

9. Flikkema, P., Introduction to spread spectrum for wireless communication: a signal processing perspective, *IEEE Spectrum Processing Magazine*, 14, 3, 26–36, May 1997.

10. Viterbi, A. J., *CDMA: Principles of Spread Spectrum Communication*, Addison-Wesley, 1995.

11. Shannon, C. E., Communication in the presence of noise, *Proceedings of the IRE*, 37, 1,10–21, January 1949.

12. Wyner, A. D., and Shamai (Shitz), S., Introduction to "Communication in the presence of noise" by C. E. Shannon, *Proceedings of the IEEE*, 86, 2, 442–446, February 1998. Reprinted in the *Proceedings of the IEEE*, vol. 86, no. 2, February 1998, 447–457.

13. Jeruchim, M. C., Balaban, P., and Shanmugan, K. S., *Simulation of Communication Systems*, Plenum, New York, 1992.

2.2 Nomadic Communications

Andy D. Kucar

Nomadic peoples of desert oases, tropical jungles, steppes, tundras, and polar regions have shown a limited interest in mobile radio communications, at the displeasure of some urbanite investors in mobile radio communications. The focus of this contribution with a delegated title *Nomadic Communications* is on terrestrial and satellite mobile radio communications used by urbanites while roaming urban canyons or golf courses, and by suburbanites who, every morning, assemble their sport utility vehicles and drive to urban jungles hunting for jobs. The habits and traffic patterns of these users are important parameters in the analysis and optimization of any mobile radio communications system. The mobile radio communications systems addressed in this contribution and illustrated in Fig. 2.6 include:

1. the first generation *analog cellular mobile radio systems* such as North American AMPS, Japanese MCS, Scandinavian NMT, and British TACS. These systems use analog voice data and frequency modulation (FM) for the transmission of voice, and coded digital data and a frequency shift keying (FSK) modulation scheme for the transmission of control information. Conceived and designed in the 1970s, these systems were optimized for vehicle-based services such as police and ambulances operating at possibly high vehicle speeds. The first generation *analog cordless telephones* include CT0 and CT1 cordless telephone systems, which were intended for use in the household environment;

2. the second generation *digital cellular and personal mobile radio systems* such as *Global System for Mobile Communications* (GSM), *Digital* AMPS > IS–54/136, DCS 1800/1900, and *Personal Digital Cellular* (PDC), all Time Division Multiple Access (TDMA), and IS–95 *spread spectrum Code Division Multiple Access* (CDMA) systems. All mentioned systems employ digital data for both voice and control purposes. The second generation *digital cordless telephony systems* include CT2, CT2Plus, CT3, *Digital Enhanced Cordless Telephone* (DECT), and *Personal Handyphone System* (PHS); *wireless data mobile radio systems* such as ARDIS, RAM, TETRA, *Cellular Digital Packet Data* (CDPD), IEEE 802.11 *Wireless Local Area Network* (WLAN), and recently announced Bluetooth; there are also projects known as *Personal Communication Network* (PCN), *Personal Communications Systems* (PCS) and FPLMTS > UMTS > IMT–2000 > 3G, where 3G stands for the third generation systems. The second generation systems also include *satellite mobile radio systems* such as INMARSAT, OmniTRACS, MSAT, AUSSAT, Iridium, Globalstar, and ORBCOMM.

After a brief prologue and historical overview, technical issues such as the repertoire of systems and services, the airwaves management, the operating environment, service quality, network issues and cell

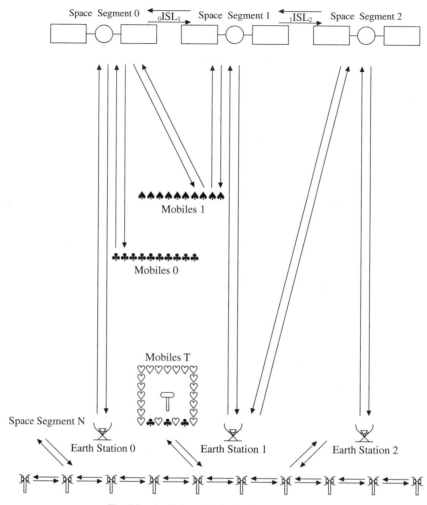

FIGURE 2.6   A Model of Fixed and Mobile, Satellite and Terrestrial Systems.

size, channel coding and modulation, speech coding, diversity, multiple broadcasting (FDMB, TDMB, CDMB), and multiple access (FDMA, TDMA, CDMA) are briefly discussed.

Many existing mobile radio communications systems collect some form of information on network behavior, users' positions, etc., with the purpose of enhancing the performance of communications, improving handover procedures and increasing the system capacity. Coarse positioning is usually achieved inherently, while more precise positioning and *navigation* can be achieved by employing LORAN-C and/or GPS, GLONASS, WAAS signals, or some other means, at an additional, usually modest, increase in cost and complexity.

Prologue

Mobile radio systems provide their users with opportunities to travel freely within the service area while being able to communicate with any telephone, fax, data modem, and electronic mail subscriber anywhere in the world; to determine their own positions; to track the precious cargo; to improve the management of fleets of vehicles and the distribution of goods; to improve traffic safety; to provide vital communication links during emergencies, search and rescue operations, to browse their favorites Websites, etc. These

TABLE 2.1 Glossary of Terms

AMPS	Advanced Mobile Phone Service
ASIC	Application Specific Integrated Circuits
BER	Bit Error Rate
CAD	Computer Aided Design
CB	Citizen Band (mobile radio)
CDMA	Spread spectrum Code Division Multiple Access
CEPT	Conference of European Postal and Telecommunications (Administrations)
CT	Cordless Telephony
DOC	Department of Communications (in Canada)
DSP	Digital Signal Processing
FCC	Federal Communications Commission (in USA)
FDMA	Frequency Division Multiple Access
FPLMTS	Future Public Land Mobile Telecommunications Systems
GDSS	Global Distress Safety System
GOES	Geostationary Operational Environmental Satellites
GPS	Global Positioning System
GSM	Groupe Spécial Mobile (now Global System for Mobile communications)
ISDN	Integrated Service Digital Network
ITU	International Telecommunications Union
MOS	Mean Opinion Score
MMIC	Microwave Monolithic Integrated Circuits
NMC	Network Management Center
NMT	Nordic Mobile Telephone (system)
PCN	Personal Communications Networks
PCS	Personal Communications Systems
PSTN	Public Switched Telephone Network
SARSAT	Search And Rescue Satellite Aided Tracking system
SERES	SEarch and REscue Satellite
TACS	Total Access Communication System
TDMA	Time Division Multiple Access
WAAS	Wide Area Augmentation System
WARC	World Administrative Radio Conference
WRC	World Radiocommunications Conference

Source: 4U Communications Research Inc., 2000.06.10~00:09, Updated: 2000.05.03

tieless (*wireless, cordless*) communications, the exchange of information, and the determination of position, course, and distance traveled are made possibly by the unique property of the radio to employ an *aerial* (*antenna*) for radiating and receiving electromagnetic waves. When the user's radio antenna is stationary over a prolonged period of time, the term *fixed radio* is used. A radio transceiver capable of being carried or moved around, but stationary when in operation, is called a *portable radio.* A radio transceiver capable of being carried and used, by a vehicle or by a person on the move, is called *mobile radio, personal and/or handheld device.* Individual radio users may communicate directly, or via one or more intermediaries, which may be *passive radio repeater(s), base station(s),* or *switch(es).* When all intermediaries are located on the Earth, the terms *terrestrial radio system* and *radio system* have been used. When at least one intermediary is a satellite borne, the terms *satellite radio system* and *satellite system* have been used. According to the location of a user, the terms *land, maritime, aeronautical, space,* and *deep-space radio systems* have been used. The second unique property of all terrestrial and satellite radio systems is that they share the same natural resource — the *airways* (*frequency bands* and *space*).

Recent developments in microwave monolithic integrated circuit (MMIC), application specific integrated circuit (ASIC), analog/digital signal processing (A/DSP), and battery technology, supported by computer-aided design (CAD) and robotics manufacturing allow the viable implementation of miniature radio transceivers at radio frequencies as high as 6 GHz, i.e., at wavelengths as short as about 5 cm. Up to these frequencies additional spectra have been assigned to mobile services; corresponding shorter wavelengths allow a viable implementation of adaptive antennas necessary for improvement of the quality of transmission and spatial frequency spectrum efficiency. The continuous flux of market forces (excited by the possibilities of a myriad of new services and great profits), international and domestic standard forces (who manage a common natural resource — the airwaves), and technology forces (capable of creating viable products) acted harmoniously and created a broad choice of communications (voice and data), information, and navigation systems, which propelled the explosive growth of mobile radio services for travelers.

A Glimpse of History

Late in the 19th century, Heinrich Rudolf Hertz, Nikola Tesla, Alexander Popov, Edouard Branly, Oliver Lodge, Jagadis Chandra Bose, Guglielmo Marconi, Adolphus Slaby, and other engineers and scientists experimented with the transmission and reception of electromagnetic waves. In 1898 Tesla made a demonstration in Madison Square Garden of a radio remote controlled boat; later the same year Marconi established the first wireless ship-to-shore telegraph link with the royal yacht Osborne. These events are now accepted as the birth of the mobile radio. Since that time, mobile radio communications have provided safe navigation for ships and airplanes, saved many lives, dispatched diverse fleets of vehicles, won many battles, generated many new businesses, etc.

Satellite mobile radio systems launched in the seventies and early eighties use ultrahigh frequency (UHF) bands around 400 MHz and around 1.5 GHz for communications and navigation services.

In the fifties and sixties, numerous private mobile radio networks, citizen band (CB) mobile radio, ham operator mobile radio, and portable home radio telephones used diverse types and brands of radio equipment and chunks of airwaves located anywhere in the frequency band from near 30 MHz to 3 GHz. Then, in the seventies, Ericsson introduced the *Nordic Mobile Telephone* (NMT) system, and AT&T Bell Laboratories introduced *Advanced Mobile Phone Service* (AMPS). The impact of these two *public land mobile telecommunication systems* on the standardization and prospects of mobile radio communications may be compared with the impact of Apple and IBM on the personal computer industry. In Europe systems like AMPS competed with NMT systems; in the rest of the world, AMPS, backed by Bell Laboratories' reputation for technical excellence and the clout of AT&T, became *de facto* and *de jure* the technical standard (British TACS and Japanese MCS–L1 are based on). In 1982, the Conference of European Postal and Telecommunications Administrations (CEPT) established Groupe Spécial Mobile (GSM) with the mandate to define future Pan-European cellular radio standards. On January 1, 1984, during the phase of explosive growth of *AMPS* and similar cellular mobile radio communications systems and services, came the divestiture (breakup) of AT&T.

Present and Future Trends

Based on the solid foundation established in 1970s the buildup of mobile radio systems and services at the end of the second millennium is continuing at an annual rate higher than 20%, worldwide. Terrestrial mobile radio systems offer analog and digital voice and low- to medium-rate data services compatible with existing public switching telephone networks in scope, but with poorer voice quality and lower data throughput. Wireless mobile radio data networks are expected to offer data rates as high as a few Mbit/s in the near future and even more in the portable environment.

Equipment miniaturization and price are important constraints on the systems providing these services. In the early fifties, mobile radio equipment used a considerable amount of a car's trunk space and challenged the capacity of car's alternator/battery source while in transmit mode. Today, the pocket-size, ≈100-gram handheld cellular radio telephone, manual and battery charger excluded, provides a few hours of talk capacity and dozens of hours in the standby mode. The average cost of the least expensive models of battery powered cellular mobile radio telephones has dropped proportionally and has broken the $100 U.S. barrier. However, one should take the price and growth numbers with a grain of salt, since some prices and growth itself might be subsidized. Many customers appear to have more than one telephone, at least during the promotion period, while they cannot use more than one telephone at the same time.

These facts need to be taken into consideration while estimating growth, capacity, and efficiency of recent wireless mobile radio systems.

Mobile satellite systems are expanding in many directions: large and powerful single unit geostationary systems; medium-sized, low orbit multi-satellite systems; and small-sized, and low orbit multi-satellite systems, launched from a plane, see [Kucar, 1992], [Del Re, 1995]. Recently, some financial uncertainties experienced by a few technically advanced LEO satellite systems, operational and planned, slowed down explosive growth in this area. Satellite mobile radio systems currently offer analog and digital voice, low to medium rate data, radio determination, and global distress safety services for travelers.

During the last five years numerous new digital radio systems for mobile users have been deployed. Presently, users in many countries have been offered between 5 and 10 different mobile radio communications systems to choose from. There already exists radio units capable of operating on two or more different systems using the same frequency band or even using a few different frequency bands. Overviews of mobile radio communications systems and related technical issues can be found in [Davis, 1984], [Cox, 1987], [Mahmoud, 1989], [Kucar, 1991], [Rhee, 1991], [Steele, 1992], [Chuang, 1993], [Cox, 1995], [Kucar, 1991], [Cimini, March 1999], [Mitola, 1999], [Cimini, July 1999], [Ariyavisitakul, 1999], [Cimini, November 1999], [Oppermann, 1999] and [Oppermann, 2000].

Repertoire of Systems and Services

The variety of services offered to travelers essentially consists of information in analog and/or digital form. Although most of today's traffic consists of analog or digital voice transmitted by analog frequency modulation FM (or phase modulation PM), or digital quadrature amplitude modulation (QAM) schemes, digital signaling, and a combination of analog and digital traffic, might provide superior frequency reuse capacity, processing, and network interconnectivity. By using a powerful and affordable microprocessor and digital signal processing chips, a myriad of different services particularly well suited to the needs of people on the move could be realized economically. A brief description of a few elementary systems/services currently available to travelers follows. Some of these elementary services can be combined within the mobile radio units for a marginal increase in the cost and complexity with respect to the cost of a single service system; for example, a mobile radio communications system can include a positioning receiver, digital map, Web browser, etc.

Terrestrial systems. In a terrestrial mobile radio network labeled Mobiles T in Fig. 2.6, a repeater was usually located at the nearest summit offering maximum service area coverage. As the number of users increased, the available frequency spectrum became unable to handle the increase traffic, and a need for frequency reuse arose. The service area was split into many subareas called cells, and the term *cellular radio* was born. The frequency reuse offers an increased overall system capacity, while the smaller cell size can offer an increased service quality, but at the expense of increased complexity of the user's terminal and network infrastructure. The trade-offs between real estate complexity, and implementation dynamics dictate the shape and the size of the cellular network. Increase in the overall capacity calls for new frequency spectra, smaller cells, which requires reconfiguration of existing base station locations; this is usually not possible in many circumstances, which leads to suboptimal solutions and even less efficient use of the frequency spectrum.

The *satellite systems* shown in Fig. 2.6 employ one or more satellites to serve as base station(s) and/or repeater(s) in a mobile radio network. The position of satellites relative to the service area is of crucial importance for the coverage, service quality, price, and complexity of the overall network. When a satellite encompass the Earth in 12-hour, 24-hour etc. periods, the term *geosynchronous orbit* has been used. An orbit inclined with respect to the equatorial plane is called an *inclined orbit*; an orbit with an inclination of about 90° is called a *polar orbit*. A circular geosynchronous 24-hour orbit in the equatorial plane (0° inclination) is known as the *geostationary orbit* (GSO), since from any point on the surface of the Earth, the satellite appears to be stationary; this orbit is particularly suitable for land mobile services at low latitudes, and for maritime and aeronautical services at latitudes of < |80|°. Systems that use geostationary satellites include INMARSAT, MSAT, and AUSSAT. An elliptical geosynchronous orbit with the inclination angle of 63.4° is known as *Tundra orbit*. An elliptical 12-hour orbit with the inclination angle of 63.4° is known as *Molniya orbit*. Both Tundra and Molniya orbits have been selected for coverage of the Northern latitudes and the area around the North Pole — for users at those latitudes the satellites appear to wander around the zenith for a prolonged period of time. The coverage of a particular region (*regional coverage*), and the whole globe (*global coverage*), can be provided by different constellations of satellites including ones in inclined and polar orbits. For example, inclined circular orbit constellations have been used by GPS (18 to 24 satellites, 55 to 63° inclination), Globalstar (48 satellites, 47° inclination), and Iridium (66 satellites, 90° inclination — polar orbits) system. All three systems provide global coverage. ORBCOM system employs Pegasus launchable low-orbit satellites to provide uninterrupted coverage of the Earth below ±60° latitudes, and an intermittent, but frequent coverage over the polar regions.

Satellite antenna systems can have one (*single beam global system*) or more beams (*multibeam spot system*). The multibeam satellite system, similar to the terrestrial cellular system, employs antenna directivity to achieve better frequency reuse, at the expense of system complexity.

Radio paging is a non-speech, one-way (from base station toward travelers), personal selective calling system with alert, without message, or with defined messages such as numeric or alphanumeric. A person wishing to send a message contacts a system operator by public switched telephone network (PSTN), and delivers his message. After an acceptable time (queuing delay), a system operator forwards the message to the traveler, by radio repeater (FM broadcasting transmitter, VHF or UHF dedicated transmitter, satellite, or cellular radio system). After receiving the message, a traveler's small (roughly the size of a cigarette pack) receiver (pager) stores the message in its memory, and on demand either emits alerting tones or displays the message.

Global Distress Safety System (*GDSS*) geostationary and inclined orbit satellites transfer emergency calls sent by vehicles to the central earth station. Examples are: *COSPAS*, Search And Rescue Satellite Aided Tracking system, *SARSAT*, Geostationary Operational Environmental Satellites *GOES*, and SEarch and REscue Satellite *SERES*). The recommended frequency for this transmission is 406.025 MHz.

Global Positioning System (*GPS*), [ION, 1980, 1984, 1986, 1993]. U.S. Department of Defense Navstar GPS 24–29 operating satellites in inclined orbits emit L band (L1 = 1575.42 MHz, L2 = 1227.6 MHz) spread spectrum signals from which an intelligent microprocessor–based receiver extracts extremely precise time and frequency information, and accurately determines its own three-dimensional position, velocity, and acceleration, worldwide. The coarse accuracy of < 100 m available to commercial users has been demonstrated by using a handheld receiver. An accuracy of meters or centimeters is possible by using the precise (military) codes and/or differential GPS (additional reference) principals and kinematic phase tracking.

Glonass is Russia's counterpart of the U.S.'s GPS. It operates in an FDM mode and uses frequencies between 1602.56 MHz and 1615.50 MHz to achieve goals similar to GPS.

Other systems have been studied by the European Space Agency (*Navsat*), and by West Germany (*Granas, Popsat,* and *Navcom*). In recent years many payloads carrying navigation transponders have been put on board of GSO satellites; corresponding navigational signals enable an increase in overall availability and improved determination of users positions. The comprehensive project, which may include existing and new radionavigation payloads, has also been known as the *Wide Area Augmentation System* (*WAAS*).

LORAN-C is the 100 kHz frequency navigation system that provides a positional accuracy between 10 and 150 m. A user's receiver measures the time difference between the master station transmitter and secondary station signals, and defines his hyperbolic line of position. North American LORAN-C coverage includes the Great Lakes, Atlantic, and Pacific Coast, with decreasing signal strength and accuracy as the user approaches the Rocky Mountains from the east. Recently, new LORAN stations have been augmented, worldwide. Similar radionavigation systems are the 100 kHz *Decca* and 10 kHz *Omega*.

Dispatch two-way radio land mobile or satellite systems, with or without connection to the PSTN, consist of an operating center controlling the operation of a fleet of vehicles such as aircraft, taxis, police cars, tracks, rail cars, etc.

TABLE 2.2 The Comparison of Dispatch WAN/LAN Systems

Parameter	US	Sweden	Japan	Australia	CDPD	IEEE 802.11
TX freq, MHz						
Base	935–941	76.0–77.5	850–860	865.00–870.00	869–894	2400–2483
	851–866			415.55–418.05		2470–2499
Mobile	896–902	81.0–82.5	905–915	820.00–825.00	824–849	2400–2483
	806–821			406.10–408.60		2470–2499
Duplexing method	sFFDD[a]	sFDD	sFDD	sfFDD	FDD	TDD
Channel spacing, kHz	12.5	25.0	12.5	25.0	30.0	1000
	25.00			12.5		
Channel rate, kb/s	≤9.6	1.2	1.2	≤	19.2	1000
# of Traffic channel	480	60	799	200	832	79
	600					
Modulation type:						
Voice	FM	FM	FM	FM		
Data	FSK	MSK-FM	MSK-FM	FSK	GMSK	DQPSK

[a] sfFDD stands for semi-duplex, full duplex, Frequency Division Duplex.
Similar systems are used in the Netherlands, U.K., USSR, and France.
ARDIS is a commercial system compatible with U.S. specs. 25 kHz spacing; 2FSK, 4FSK, ≤19.2 kb/s.
MOBITEX/RAM is a commercial system compatible with U.S. specs. 12.5 kHz spacing; GMSK, 8.0 kb/s.
Source: 4U Communications Research Inc., 2000.06.10~00:09, c:/tab/dispatch.sys

A summary of some of the existing and planned terrestrial mobile radio systems, including MOBITEX RAM and ARDIS, is given in Table 2.2.

OmniTRACS dispatch system employs a Ku-band geostationary satellite located at 103° W to provide two-way digital message and position reporting (derived from incorporated satellite-aided LORAN-C receiver), throughout the contiguous U.S. (CONUS).

Cellular radio or public land mobile telephone systems offer a full range of services to the traveler similar to those provided by PSTN. The technical characteristics of some of the existing and planned systems are summarized in Table 2.3.

Vehicle Information System and Intelligent Highway Vehicle System are synonyms for the variety of systems and services aimed toward traffic safety and location. This includes: traffic management, vehicle identification, digitized map information and navigation, radio navigation, speed sensing and adaptive cruise control, collision warning and prevention, etc. Some of the vehicle information systems can easily be incorporated in mobile radio communications transceivers to enhance the service quality and capacity of respective communications systems.

Airwaves Management

The airwaves (frequency spectrum and the space surrounding us) are a limited natural resource shared by several different radio users (military, government, commercial, public, amateur, etc.). Its sharing (among different users, services described in the previous section, TV and sound broadcasting, etc.), coordination, and administration is an ongoing process exercised on national, as well as on international levels. National administrations (Federal Communications Commission (FCC) in the U.S., Department of Communications (DOC), now Industry Canada, in Canada, etc.), in cooperation with users and industry, set the rules and procedures for planning and utilization of scarce frequency bands. These plans and utilizations have to be further coordinated internationally.

The International Telecommunications Union (ITU) is a specialized agency of the United Nations, stationed in Geneva, Switzerland, with more than 150 government and corporate members, responsible for all policies related to Radio, Telegraph, and Telephone. According to the ITU, the world is divided into three regions: Region 1 — Europe, including the Soviet Union, Outer Mongolia, Africa, and the Middle East west of Iran; Region 2 — the Americas, and Greenland; and Region 3 — Asia (excluding

TABLE 2.3 Comparison of Cellular Mobile Radio Systems in Bands Below 1 GHz

Parameter	AMPS NAMPS	MCS L1 MCS L2	NMT 900	NMT 450	R.com 2000	C450	TACS UK	GSM	IS-54 IS-136	IS-95 USA	PDC Japan
TX freq, MHz											
Base	869–894	870–885	935–960	463–468	424.8–428	461–466	935–960	890–915	869–894	869–894	810–826
Mobile	824–849	925–940	890–915	453–458	414.8–418	451–456	890–915	935–960	824–849	824–849	890–915
Max b/m eirp, dBW	22/5	19/7	22/7	19/12	20.10	22/12	22/8	27/9	27/9	/–7	/5
Multiple access	F	F	F	F	F	F	F	F/T	F/T	F/C	F/T
Duplex method	FDD	FDD	FDD	FDD	FDD	FDD	FDD	FDD	FDD	FDD	FDD
Channel bw, kHz	30.0 10.0	25.0 12.5	12.5	25.0	12.5	20.0 10.0	25.0 12.5	200.0	30.0	1250	25
Channels/RF	1	1	1	1	1	1	1	8	3	42	3
Channels/band	832 2496	600 1200	1999	200	160	222	1000	125 × 8	832 × 3	n × 42	640 × 3
Voice/Traffic:											
comp. or kb/s	analog	analog	analog	analog	analog	analog	analog	RELP	VSELP	CELP	VSELP
	2:1	2:1	2:1	2:1	2:1	2:1	2:1	13.0	8.0	≤9.6	6.7
modulation	PM	PM	PM	PM	PM	PM	PM	GMSK π/4	B/OQ	π/4	π/4
kHz and/or kb/s	±12	±5	±5	±5	±2.5	±4	±9.5	270.833	48.6	1228.8	42.0
Control:	digital	digital	digital	digital	digital	digital	digital	digital	digital	digital	digital
modulation	FSK	FSK	FFSK	FFSK	FFSK	FSK	FSK	GMSK	π/4	B/OQ	π/4
bb waveform	Manch. NRZ	Manch. NRZ	NRZ	NRZ	Manch.	NRZ	NRZ	NRZ	NRZ		
kHz and/or kb/s	±8.0/10	±4.5/0.3	±3.5/1.2	±3.5/1.2	±1.7/1.2	±2.5/5.3	±6.4/8.0	270.833	48.6	1228.8	42.0
Channel coding;	BCH	BCH	B1 Hag.	B1 Hag.	Hag.	BCH	BCH	RS	Conv.	Conv.	Conv.
base→mobile	(40,28)	(43,31)	burst	burst	(19,6)	(15,7)	(40,28)	(12,8)	1/2	6/11	9/17
mobile→base	(48,36)	a.(43,31) p.(11,07)	burst	burst	(19,6)	(15,7)	(48,36)	(12,8)	1/2	1/3	9/17

Note: Multiple Access: F = Frequency Division Multiple Access (FDMA), F/T = Hybrid Frequency/Time DMA, F/C = Hybrid Frequency/Code DMA. π/4 corresponds to the π/4 shifted differentially encoded QPSK with α = 0.35 square root raised–cosine filter for IS–136 and α = 0.5 for PDC. B/OQ corresponds to the BPSK outbound and OQPSK modulation scheme inbound.
comp. or kb/s stands for syllabic compandor or speech rate in kb/s; kHz and/or kb/s stands for peak deviation in kHz and/or channel rate kb/s.
IS–634 standard interface supports AMPS, NAMPS, TDMA and CDMA capabilities.
IS–651 standard interface supports **A** GSM capabilities and **A+** CDMA capabilities.

Source: 4U Communications Research Inc., 2000.06.10–00:09

parts west of Iran and Outer Mongolia), Australia, and Oceania. Historically, these three regions have developed, more or less independently, their own frequency plans, which best suit local purposes. With the advent of satellite services and globalization trends, the coordination between different regions becomes more urgent. Frequency spectrum planning and coordination is performed through ITU's bodies such as: Comité Consultatif de International Radio (CCIR), now ITU-R, International Frequency Registration Board (IFRB), now ITU-R, World Administrative Radio Conference (WARC), and Regional Administrative Radio Conference (RARC).

ITU-R, through it study groups, deals with technical and operational aspects of radio communications. Results of these activities have been summarized in the form of reports and recommendations published every four years, or more often, [ITU, 1990]. IFRB serves as a *custodian* of a common and scarce natural resource — the *airwaves*; in its capacity, the IFRB records radio frequencies, advises the members on technical issues, and contributes on other technical matters. Based on the work of ITU-R and the national administrations, ITU members convene at appropriate RARC and WARC meetings, where documents on frequency planning and utilization, the *Radio Regulations,* are updated. The actions on a national level follow, see [RadioRegs, 1986], [WARC, 1992], [WRC, 1997]. The far-reaching impact of mobile radio communications on economies and the well-being of the three main trading blocks, other developing and third world countries, and potential manufacturers and users, makes the airways (frequency spectrum) even more important.

The International Telecommunications Union (ITU) recommends the composite *bandwidth-space-time* domain concept as a measure of spectrum utilization. The *spectrum utilization factor* $U = B \cdot S \cdot T$ is defined as a product of the frequency bandwidth B, spatial component S, and time component T. Since mobile radio communications systems employ single omnidirectional antennas, their S factor will be rather low; since they operate in a single channel arrangement, their B factor will be low; since new digital schemes tend to operate in a packet/block switching modes which are inherently loaded with a significant amount of overhead and idle traffic, their T factor will be low as well. Consequently, mobile radio communications systems will have a poor spectrum utilization factors.

The model of a mobile radio environment, which may include different sharing scenarios with fixed service and other radio systems, can be described as follows. Objects of our concern are *events* (for example, conversation using a mobile radio, measurements of amplitude, phase and polarization at the receiver) occurring in *time* $\{u^0\}$, *space* $\{u^1, u^2, u^3\}$, *spacetime* $\{u^0, u^1, u^2, u^3\}$, *frequency* $\{u^4\}$, *polarization* $\{u^5, u^6\}$, and *airwaves* $\{u^0, u^1, u^2, u^3, u^4, u^5, u^6\}$, see Table 2.4. The coordinate $\{u^4\}$ represents frequency resource, i.e., bandwidth in the spacetime $\{u^0, u^1, u^2, u^3\}$. Our goal is to use a scarce natural resource — the airwaves in an environmentally friendly manner.

TABLE 2.4 The Multidimensional Spaces Including the Airwaves

u^0	time		
u^1			
u^2	space	spacetime	
u^3			airwaves
u^4	frequency/bandwidth		
u^5	polarization		
u^6			
u^7			
u^8	Doppler		
u^9			
u^A	users: government/military, commercial/public, fixed/mobile, terrestrial/satellite ...		
u^B			
$\vdots$			
u^n			

Source: 4U Communications Research Inc. 2000.06.10~00:09, c:/tab/airwaves.1

When users/events are divided (sorted, discriminated) along the time coordinate u^0, the term *time division* is employed for function $f(u^0)$. A division $f(u^4)$ along the frequency coordinate u^4 corresponds to the *frequency division*. A division $f(u^0, u^4)$ along the coordinates (u^0, u^4) is usually called a *code division* or *frequency hopping*. A division $f(u^1, u^2, x^3)$ along the coordinates (u^1, u^2, u^3) is called the *space division*. Terrestrial cellular and multibeam satellite radio systems are vivid examples of the space division concepts. Coordinates $\{u^5, u^6\}$ may represent two orthogonal polarization components, horizontal and vertical or right-handed and left-handed circular; a division of users/events according to their polarization components may be called the *polarization division*. Any division $f(u^0, u^1, u^2, u^3, u^4, u^5, u^6)$ along the coordinates $(u^0, u^1, u^2, u^3, u^4, u^5, u^6)$ may be called the *airwaves division*. Coordinates $\{u^7, u^8, u^9\}$ may represent velocity (or Doppler frequency) components; a division of users/events according to their Doppler frequencies similar to the moving target indication (MTI) radars may be called the *Doppler frequency division*. We may also introduce coordinate $\{u^A\}$ to represent users, divide the airways along the coordinate $\{u^A\}$ (military, government, commercial, public, fixed, mobile, terrestrial, satellite, and others) and call it the *users division*. Generally, the segmentations of frequency spectra to different users lead to uneven use and uneven spectral efficiency factors among different segments.

In analogy with division, we may have time, space, frequency, code, airwaves, polarization, Doppler, users, $\{u^\alpha, \dots, u^\omega\}$ *access* and *diversity*. Generally, the signal space may be described by m coordinates $\{u^0, \dots, u^{m-1}\}$. Let each signal component has k degrees of freedom. At the transmitter site, each signal can be radiated via n_T antennas, and received at n_R receiver antennas. There is a total of $n = n_T + n_R$ antennas, two polarization components, and L multipath components, i.e., paths between transmitter and receiver antennas. Thus, the total number of degrees of freedom $m = k \times n \times 2 \times L$. For example, in a typical land mobile environment 16 multipath components can exist; if one wants to study a system with four antennas on the transmitter side and four antennas on the receiver side, and each antenna may employ both polarizations, then the number of degrees of freedom equals $512 \times k$. By selecting a proper coordinate system and using inherent system symmetries, one might be able to reduce the number of degrees of freedom to a manageable quantity.

Operating Environment

A general configuration of terrestrial FS radio systems, sharing the same space and frequency bands with FSS and/or MSS systems, is illustrated in Fig. 2.6. The emphasis of this contribution is on mobile systems; however, it should be appreciated that mobile systems may be required to share the same frequency band with fixed systems. A *satellite system* usually consists of many earth stations, denoted Earth Station 0 ... Earth Station 2 in Fig. 2.6, one or many space segments, denoted Space Segment 0 ... Space Segment N, and in the case of a *mobile satellite system* different types of mobile segments denoted by ♣ and ♠ in the same figure. Links between different Space Segments and mobile users of MSS systems are called *service links*; links connecting Space Segments and corresponding Earth Stations are called *feeder links*. FSS systems employ Space Segments and fixed Earth Station segments only; corresponding connections are called *service links*. Thus, technically similar connections between Space Segments and fixed Earth Station segments perform different functions in MSS and FSS systems and are referred to by different names. Administratively, the feeder links of MSS systems are often referred to as FSS.

Let us briefly assess spectrum requirements of an MSS system. There exist many possibilities of how and where to communicate in the networks shown in Fig. 2.6. Each of these possibilities can use different spatial and frequency resources, which one needs to assess for sharing purposes. For example, a mobile user ♣ transmits at frequency f_0 using a small hemispherical antenna toward Space Segment 0. This space segment receives a signal at frequency f_0, transposes it to frequency F_{n+0}, amplifies it and transmits it toward Earth Station 0. This station processes the signal, makes decisions on the final destination, sends the signal back toward the same Space Segment 0, which receives the signal at frequency f_{m+0}. This signal is transposed further to frequency F_{k+0} and is emitted via *inter-satellite link* $_0\text{ISL}_1$ toward Space Segment 1, which receives this signal, processes it, transposes it, and emits toward the earth and mobile ♠ at frequency F_1. In this process a quintet of frequencies $(f_0, F_{n+0}, f_{m+0}, F_{k+0}, F_1)$ is used in one direction.

Once the mobile ♠ receives the signal, its set rings and sends back the signal in reverse directions at a different frequency quintet (or a different time slot, or a different code, or any combination of time, code, and frequency), thus establishing the two-way connection. Obviously, this type of MSS system uses significant parts of the frequency spectrum.

Mobile satellite systems consist of two directions with very distinct properties. A direction from an Earth Station, also called a *hub* or *base station,* which may include a *Network Management Center* (NMC), toward the satellite space segment and further toward a particular mobile user is known as the *forward* direction. In addition, we will call this direction the *dispatch* direction, *broadcast* direction, or *division* direction, since the NMC dispatches/broadcasts data to different users and data might be divided in frequency (F), time (T), code (C), or a hybrid (H) mode. The opposite direction from a mobile user toward the satellite space segment and further toward the NMC is known as the *return* direction. In addition, we will call this direction the *access* direction, since mobile users usually need to make attempts to access the mobile network before a connection with NMC can be established; in some networks the NMC may poll the mobile users, instead. A connection between NMC and a mobile user, or between two mobile users, may consist of two or more hops, including inter-satellite links, as shown in Fig. 2.6.

While traveling, a customer — a *user of a cellular mobile radio system* — may experience sudden changes in signal quality caused by his movements relative to the corresponding base station and surroundings, multipath propagation, and unintentional jamming such as man-made noise, adjacent channel interference, and co-channel interference inherent in cellular systems. Such an environment belongs to the class of nonstationary random fields, on which experimental data is difficult to obtain; their behavior hard to predict and model satisfactorily. When reflected signal components become comparable in level to the attenuated direct component, and their delays comparable to the inverse of the channel bandwidth, *frequency selective fading* occurs. The reception is further degraded due to movements of a user relative to reflection points and the relay station, causing Doppler frequency shifts. The simplified model of this environment is known as the *Doppler Multipath Rayleigh Channel.*

The existing and planned cellular mobile radio systems employ sophisticated narrowband and wideband filtering, interleaving, coding, modulation, equalization, decoding, carrier and timing recovery, and multiple access schemes. The cellular mobile radio channel involves a *dynamic interaction* of signals arriving via different paths, adjacent and co-channel interference, and noise. Most channels exhibit some degree of memory, the description of which requires higher order statistics of — *spatial and temporal* — multidimensional random vectors (amplitude, phase, multipath delay, Doppler frequency, etc.) to be employed.

A model of a multi-hop satellite system that incorporates interference and nonlinearities is illustrated and described in Fig. 2.7. The signal flow in the forward/broadcast direction, from Base to Mobile User, is shown on the left side of the figure; the right side of the figure corresponds to the reverse/access direction. For example, in the forward/broadcast direction, the transmitted signal at the Base, shown in the upper left of the figure, is distorted due to nonlinearities in the RF power amplifier. This signal distortion is expressed via differential phase and differential gain coefficients DP and DG, respectively. The same signal is emitted toward the satellite space segment receiver denoted as point 2; here, noise N, interference I, delay τ, and Doppler frequency $_Df$ symbolize the environment. The signals are further processed, amplified and distorted at stage 3, and radiated toward the receiver, 4. Here again, noise N, interference I, delay τ, and Doppler frequency $_Df$ symbolize the environment. The signals are translated and amplified at stage 5 and radiated toward the Mobile User at stage 6; here, additional noise N, interference I, delay τ, and Doppler frequency $_Df$ characterize the corresponding environment. This model is particularly suited for a detailed analysis of the link budget and for equipment design purposes. A system provider and cell designer may use a statistical description of a mobile channel environment, instead.

An FSS radio channel is described as the Gaussian; the mean value of the corresponding radio signal is practically constant and its value can be predicted with a standard deviation of a fraction of a dB. A terrestrial mobile radio channel could exhibit dynamics of about 80 dB and its mean signal could be predicted with a standard deviation of 5 to 10 dB. This may require the evaluation of usefulness of existing radio channel models and eventual development of more accurate ones.

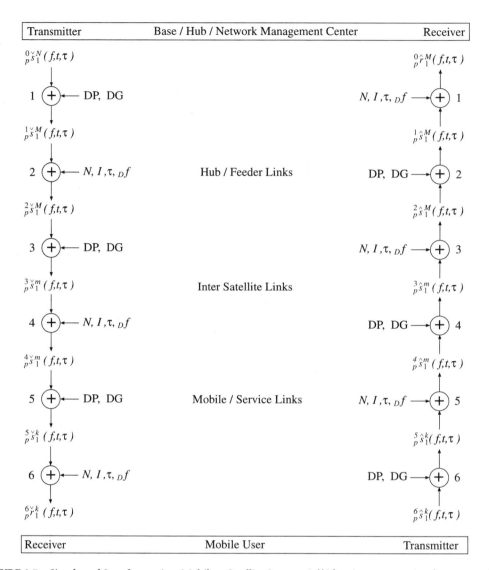

FIGURE 2.7   Signals and Interference in a Multihop Satellite System. ${}_{p}^{q}x_1^M(f,t,\tau)$ represents signals $x = r, s$, where r is the received and s is the sent/transmitted signal, $\check{x}$ represents the dispatch/forward direction and $\hat{x}$ represents access/return direction; p is the polarization of the signal at location q and the number of signal components ranges from 1 to M; f, t, τ are frequency, time and delay of a signal at the location q, respectively. DP, DG are Differential Phase and Differential Gain (include AM/AM and AM/PM); $N, I, \tau, {}_D f$ are the noise, interference, absolute delay and Doppler frequency, respectively.

Cell engineering and prediction of service area and service quality in an ever-changing mobile radio channel environment, is a very difficult task. The average path loss depends on terrain microstructure within a cell, with considerable variation between different types of cells (i.e., urban, suburban, and rural environments). A variety of models based on experimental and theoretic work have been developed to predict path radio propagation losses in a mobile channel. Unfortunately, none of them are universally applicable. In almost all cases, excessive transmitting power is necessary to provide adequate system performance.

The *first generation* mobile satellite systems employ geostationary satellites (or payload piggy-backed on a host satellite) with small 18 dBi antennas covering the whole globe. When the satellite is positioned directly above the traveler (at zenith), a near constant signal environment, known as *Gaussian channel*, is experienced. The traveler's movement relative to the satellite is negligible (i.e., Doppler frequency is practically zero). As the traveler moves — north or south, east or west — the satellite appears lower on the horizon. In addition to the direct path, many significant strength-reflected components are present, resulting in a degraded performance. Frequencies of these components fluctuate due to movement of traveler relative to the reflection points and the satellite. This environment is known as the *Doppler Ricean Channel.* An inclined orbit satellite located for a prolonged period of time above 45° latitude north and 106° longitude west, could provide travelers all over the U.S. and Canada, including the far North, a service quality unsurpassed by either geostationary satellite or terrestrial cellular radio. Similarly, a satellite located at 45° latitude north and 15° longitude east, could provide travelers in Europe with improved service quality.

Inclined orbit satellite systems can offer a low start-up cost, a near Gaussian channel environment, and improved service quality. Low orbit satellites, positioned closer to the service area, can provide high signal levels and short (a few milliseconds long) delays, and offer compatibility with the cellular terrestrial systems. These advantages need to be weighted against network complexity, inter-satellite links, tracking facilities, etc.

Terrestrial mobile radio communications systems provide signal dynamics of about 80 dB and are able to penetrate walls and similar obstacles, thus providing inside building coverage. Satellite mobile radio communications systems are power limited and provide signal dynamics of less than 15 dB; the signal coverage is, in most cases, limited to the outdoors.

Let us compare the efficiency of a Mobile Satellite Service (MSS) with the Fixed Satellite Service (FSS); both services are assumed to be using the GSO space segments. A user at the equator can see the GSO arc reaching ±81°; if satellites are spaces 2° apart along the GSO, then the same user can reach 81 satellites simultaneously. An MSS user employs a hemispherical antenna having gain of about 3 dBi; consequently, he can effectively use only one satellite, but prevent all other satellite users from employing the same frequency. An FSS user employs a 43 dBi gain antenna that points toward a desired satellite. By using the same transmit power as an MSS user, but employing larger and more expensive antenna, this FSS user can effectively transmit about 40 dB (ten thousand times) wider bandwidth, i.e., 40 dB more information. The FSS user can, by adding 3 dB more power into an additional orthogonal polarization channel, reuse the same frequency band and double the capacity. Furthermore, the same FSS user can use additional antennas to reach each of 81 available satellites, thus increasing the GSO are capacity by 80 times. Consequently, the FSS is power-wise 10,000 times more efficient and spatially about 160 times more efficient than corresponding MSS. Similar comparisons can be made for terrestrial systems. The convenience and smallness of today's mobile systems user terminals is traded for low spatial and power efficiency, which may carry a substantial economic price penalty. The real cost of a mobile system seems to have been subsidized by some means beyond the cost of cellular telephone and traffic charges (both often $0).

Service Quality

The primary and the most important measure of service quality should be *customer satisfaction.* The customer's needs, both current and future, should provide guidance to a service offerer and an equipment manufacturer for both the system concept and product design stages. In the early stages of the product life, mobile radio was perceived as a necessary tool for performing important tasks; recently, mobile/personal/handheld radio devices are becoming more like status symbols and fashion. Acknowledging the importance of every single step of the complex service process and architecture, attention is limited here to a few technical merits of quality:

1. *Guaranteed quality level* is usually related to a percentage of the service area coverage for an adequate percentage of time.
2. *Data service quality* can be described by the average bit error rate (e.g., BER < 10^{-5}), packet BER (PBER < 10^{-2}), signal processing delay (1 to 10 ms), multiple access collision probability (< 20%), the probability of a false call (false alarm), the probability of a missed call (miss), the probability of a lost call (synchronization loss), etc.

3. *Voice quality* is usually expressed in terms of the mean opinion score (MOS) of subjective evaluations by service users. MOS marks are: bad = 0, poor = 1, fair = 2, good = 3, and excellent = 4. MOS for PSTN voice service, pooled by leading service providers, relates the poor MOS mark to a single-to-noise ratio (S/N) in a voice channel of S/N ≈ 35 dB, while an excellent score corresponds to S/N > 45 dB. Currently, users of mobile radio services are giving poor marks to the voice quality associated with a S/N ≈ 15 dB and an excellent mark for S/N > 25 dB. It is evident that there is significant difference (20 dB) between the PSTN and mobile services. If digital speech is employed, both the speech and the speaker recognition have to be assessed. For more objective evaluation of speech quality under *real conditions* (with no impairments, in the presence of burst errors during fading, in the presence of random bit errors at BER = 10^{-2}, in the presence of Doppler frequency offsets, in the presence of truck acoustic background noise, in the presence of ignition noise, etc.), additional tests such as the diagnostic acceptability measure DAM, diagnostic rhyme test DRT, Youden square rank ordering, Sino-Graeco-Latin square tests, etc., can be performed.

Network Issues and Cell Size

To understand ideas and technical solutions offered in existing schemes, and in future systems one needs also to analyze the reasons for their introduction and success. Cellular mobile services are flourishing at an annual rate higher than 20%, worldwide. The first generation systems, (such as AMPS, NMT, TACS, MCS, etc.), use *frequency division multiple access* FDMA and digital modulation schemes for access, command and control purposes, and analog phase/frequency modulation schemes for the transmission of an analog voice. Most of the network intelligence is concentrated at fixed elements of the network including base stations, which seem to be well suited to the networks with a modest number of medium to large-sized cells. To satisfy the growing number of potential customers, more cells and base stations were created by the cell splitting and frequency reuse process. Technically, the shape and size of a particular cell is dictated by the base station antenna pattern and the topography of the service area. Current terrestrial cellular radio systems employ cells with 0.5 to 50 km radius. The maximum cell size is usually dictated by the link budget, in particular the grain of a mobile antenna and available output power. This situation arises in a rural environment, where the demand on capacity is very low and cell splitting is not economical. The minimum cell size is usually dictated by the need for an increase in capacity, in particular in downtown cores. Practical constraints such as real estate availability and price, and construction dynamics limit the minimum cell size to 0.5 to 2 km. However, in such types of networks, the complexity of the network and the cost of service grow exponentially with the number of base stations, while the efficiency of present handover procedures becomes inadequate. Consequently, the second generation of all-digital schemes, which handle this increasing idle traffic more efficiently, were introduced. However, handling of the information, predominantly voice, has not been improved significantly, if at all.

In the 1980s extensive studies of then existing AMPS- and NMT-based systems were performed, see Davis et al. (1984) and Mahmoud et al. (1989), and the references therein. Based on particular service quality requirements, particular radio systems and particular cell topologies, few empirical rules had been established. Antennas with an omnidirectional pattern in a horizontal direction, but with about 10 dBi gain in vertical direction provide the frequency reuse efficiency of N_{FDMA} = 1/12. It was anticipated that base station antennas with similar directivity in a vertical direction and 60° directivity in a horizontal direction (a cell is divided into six sectors) can provide the reuse efficiency N_{FDMA} = 1/4, which results in a threefold increase in the system capacity; if CDMA is employed instead of FDMA, an increase in reuse efficiency N_{FDMA} = 1/4 → N_{CDMA} = 2/3 may be expected. However, this does not necessarily mean that a CDMA system is more efficient than a FDMA system. The overall efficiency very much depends on spatiotemporal dynamics of a particular cell and the overall network.

Recognizing some of limitations of existing schemes and anticipating the market requirements, the research in *time division multiple access* (TDMA) schemes aimed at cellular mobile and DCT services, and in *code division multiple access* (CDMA) schemes aimed toward mobile satellite systems and cellular and personal mobile applications, followed with introduction of nearly ten different systems. Although employing different access schemes, TDMA (CDMA) network concepts rely on a smart mobile/portable unit that scans time slots (codes) to gain information on network behavior, free slots (codes), etc., improving frequency reuse and handover efficiency while hopefully keeping the complexity and cost of the overall network at reasonable levels. Some of the proposed system concepts depend on low gain (0 dBi) base station antennas deployed in a license-free, uncoordinated fashion; small size cells (10 to 1000 m in radius) and an emitted isotropic radiated power of about 10 mW (+10 dBm) per 100 kHz are anticipated. A frequency reuse efficiency of $N = 1/9$ to N to $1/36$ has been projected for DCT systems. $N = 1/9$ corresponds to the highest user capacity with the lowest transmission quality, while $N = 1/36$ has the lowest user capacity with the highest transmission quality. This significantly reduced frequency reuse capability of proposed system concepts, will result in significantly reduced system capacity, which needs to be compensated for by other means including new spectra.

In practical networks, the need for a capacity (and frequency spectrum) is distributed unevenly in space and time. In such an environment, the capacity and frequency reuse efficiency of the network may be improved by *dynamic channel allocation,* where an increase in the capacity at a particular hot spot may be traded for a decrease in the capacity in cells surrounding the hot spot, the quality of the transmission, and network instability. The first generation mobile radio communications systems used omnidirectional antennas at base stations. Today, three-sector 120°-wide cells are typical in a heavy traffic urban environment, while entry-level rural systems employ omnidirectional antennas; the most demanding environments with changing traffic patterns employ adaptive antenna solutions, instead.

To cover the same area (space) with smaller and smaller cells, one needs to employ more and more base stations. A linear increase in the number of base stations in a network usually requires an $(n(n-1)/2)$ increase in the number of connections between base stations, and increase in complexity of switches and network centers. These connections can be realized by fixed radio systems (providing more frequency spectra available for this purpose), or, more likely, by a cord (wire, cable, fiber, etc.).

An increase in overall capacity is attributed to

- increase in available bandwidth, particularly above 1 GHz, but to the detriment of other services,
- increased use of adaptive antenna solutions which, through spatial filtering, increase capacity and quality of the service, but at a significant increase in cost,
- trade-offs between service quality, vehicular vs. pedestrian environments, analog vs. digital voice, etc.

The *first generation* geostationary satellite system antenna beam covers the entire Earth (i.e., the cell radius equals ≈6500 km). The *second generation* geostationary satellites use large multibeam antennas providing 10 to 20 beams (cells) with 800 to 1600 km radius. Low orbit satellites such as Iridium use up to 37 beams (cells) with 670 km radius. The *third generation* geostationary satellite systems will be able to use very large reflector antennas (roughly the size of a baseball stadium), and provide 80 to 100 beams (cells) with a cell radius of ≈200 km. If such a satellite is tethered to a position 400 km above the Earth, the cell size will decrease to ≈2 km in radius, which is comparable in size with today's small size cell in terrestrial systems. Yet, such a satellite system may have the potential to offer an improved service quality due to its near optimal location with respect to the service area. Similar to the terrestrial concepts, an increase in the number of satellites in a network will require an increase in the number of connections between satellites and/or Earth network management and satellite tracking centers, etc. Additional factors that need to be taken into consideration include price, availability, reliability, and timeliness of the launch procedures, a few large vs. many small satellites, tracking stations, etc.

Coding and Modulation

The conceptual transmitter and receiver of a mobile system may be described as follows. The transmitter signal processor accepts analog voice and/or data and transforms (by analog and/or digital means) these signals into a form suitable for a double-sided suppressed carrier amplitude modulator [also called quadrature amplitude modulator (QAM)]. Both analog and digital input signals may be supported, and either analog or digital modulation may result at the transmitter output. Coding and interleaving can also be included. Very often, the processes of coding and modulation are performed jointly; we will call this joint process *codulation*. A list of typical modulation schemes suitable for transmission of voice and/or data over Doppler-affected Ricean channel, which can be generated by this transmitter is given in Table 2.5. Details on modulation, coding, and system issues can be found in Kucar (2000), Proakis, (1983), Sklar, (1988), and Van Trees, (1968–1971).

Existing cellular radio systems such as AMPS, TACS, MCS, and NMT employ hybrid (analog and digital) schemes. For example, in access mode, AMPS uses a digital codulation scheme (BCH coding and FSK modulation), while in information exchange mode, the frequency-modulated analog voice is merged with discrete *SAT* and/or *ST* signals and occasionally blanked to send a digital message. These hybrid codulation schemes exhibit a constant envelope and as such allow the use of power efficient radio frequency (RF) nonlinear amplifiers. On the receiver side, these schemes can be demodulated by an inexpensive, but efficient limiter/discriminator device. They require modest to high $C/N = 10 - 20$ dB, are very robust in adjacent (a spectrum is concentrated near the carrier) and co-channel interference (up to $C/I = 0$ dB, due to capture effect) cellular radio environment, and react quickly to the signal fade outages (no carrier, code, or frame synchronization). Frequency-selective and Doppler-affected mobile radio channels will cause modest to significant degradations known as the *random phase/frequency modulation*. By using modestly complex extended threshold devices C/N as low as 5 dB can provide satisfactory performance.

Tightly filtered codulation schemes, such as $\pi/4$ QPSK additionally filtered by a square root, raised-cosine filter, exhibit a nonconstant envelope that demands (quasi) linear, less D.C. power efficient amplifiers to be employed. On the receiver side, these schemes require complex demodulation receivers, a linear path for signal detection, and a nonlinear one for reference detection — differential detection or carrier recovery. When such a transceiver operates in a selective fading multipath channel environment, additional countermeasures (inherently sluggish equalizers, etc.) are necessary to improve the performance — reduce the *bit error rate floor*. These codulation schemes require modest $C/N = 8 - 16$ dB and perform modestly in adjacent and/or co-channel (up to $C/I = 8$ db) interference environment.

Codulation schemes employed in spread spectrum systems use low-rate coding schemes and mildly filtered modulation schemes. When equipped with sophisticated amplitude gain control on the transmit and receive side, and robust rake receiver, these schemes can provide superior $C/N = 4 - 10$ dB and $C/I < 0$ dB performance. Unfortunately, a single transceiver has not been able to operate satisfactorily in a mobile channel environment. Consequently, a few additional signals have been employed to achieve the required quality of the transmission. These pilot signals significantly reduce the spectrum efficiency in the forward direction and many times in the reverse direction. Furthermore, two combined QPSK-like signals have up to (4×4) different baseband levels and may look like a 16QAM signal, while three combined QPSK-like signals may look like a 64QAM signal. These combined signals, one information and two pilot signals, at user's transmitter output, for example, exhibit high peak factors and total power that is by 3 to 5 dB higher than the C/N value necessary for a single information signal. Additionally, inherently power inefficient linear RF power amplifiers are needed; these three signal components of a CDMA scheme may have been optimized for minimal cross-correlation and ease of detection. As such, the same three signals may not necessarily have states in the QAM constellation that optimize the peak-to-average ratio, and vice versa.

Speech Coding

Human vocal tract and voice receptors, in conjunction with language redundancy (coding), are well suited for face-to-face conversation. As the channel changes (e.g., from telephone channel to mobile radio channel), different coding strategies are necessary to protect against the loss of information.

TABLE 2.5   Modulation Schemes, Glossary of Terms

Abbreviation	Description	Remarks/Use
ACSSB	Amplitude Companded Single SideBand	Satellite transmission
AM	Amplitude Modulation	Broadcasting
APK	Amplitude Phase Keying modulation	
BLQAM	Blackman Quadrature Amplitude Modulation	
BPSK	Binary Phase Shift Keying	Spread spectrum systems
CPFSK	Continuous Phase Frequency Shift Keying	
CPM	Continuous Phase Modulation	
DEPSK	Differentially Encoded PSK (with carrier recovery)	
DPM	Digital Phase Modulation	
DPSK	Differential Phase Shift Keying (no carrier recovery)	
DSB-AM	Double SideBand Amplitude Modulation	
DSB-SC-AM	Double SideBand Suppressed Carrier AM	Includes digital schemes
FFSK	Fast Frequency Shift Keying ≡ MSK	NMT data and control
FM	Frequency Modulation	Broadcasting, AMPS, voice
FSK	Frequency Shift Keying	AMPS data and control
FSOQ	Frequency Shift Offset Quadrature modulation	
GMSK	Gaussian Minimum Shift Keying	GSM voice, data, and control
GTFM	Generalized Tamed Frequency Modulation	
HMQAM	Hamming Quadrature Amplitude Modulation	
IJF	Intersymbol Jitter Free ≡ SQORC	
LPAM	L-ary Pulse Amplitude Modulation	
LRC	LT symbols long Raised Cosine pulse shape	
LREC	LT symbols long Rectangularly EnCoded pulse shape	
LSRC	LT symbols long Spectrally Raised Cosine scheme	
MMSK	Modified Minimum Shift Keying ≡ FFSK	
MPSK	M-ary Phase Shift Keying	
MQAM	M-ary Quadrature Amplitude Modulation	A subclass of DSB-SC-AM
MQPR	M-ary Quadrature Partial Response	Radio-relay transmission
MQPRS	M-ary Quadrature Partial Response System ≡ MQPR	
MSK	Minimum Shift Keying	
m-h	multi-h CPM	
OQPSK	Offset (staggered) Quadrature Phase Shift Keying	
PM	Phase Modulation	Low capacity radio
PSK	Phase Shift Keying	4PSK ≡ QPSK
QAM	Quadrature Amplitude Modulation	
QAPSK	Quadrature Amplitude Phase Shift Keying	
QPSK	Quadrature Phase Shift Keying ≡ 4 QAM	Low capacity radio
QORC	Quadrature Overlapped Raised Cosine	
SQAM	Staggered Quadrature Amplitude Modulation	
SQPSK	Staggered Quadrature Phase Shift Keying	
SQORC	Staggered Quadrature Overlapped Raised Cosine	
SSB	Single SideBand	Low and High capacity radio
S3MQAM	Staggered class 3 Quadrature Amplitude Modulation	
TFM	Tamed Frequency Modulation	
TSI QPSK	Two-Symbol-Interval QPSK	
VSB	Vestigial SideBand	TV
WQAM	Weighted Quadrature Amplitude Modulation	Includes most digital schemes
XPSK	Crosscorrelated PSK	
π/4 DQPSK	π/4 shift DQPSK with $\alpha = 0.35$ raised cosine filtering	IS-54 TDMA voice and data
3MQAM	Class 3 Quadrature Amplitude Modulation	
4MQAM	Class 4 Quadrature Amplitude Modulation	
12PM3	12 state PM with 3 bit correlation	

Source: 4U Communications Research Inc., 2000.06.10~00:09, c:/tab/modulat.tab

In (analog) companded PM/FM mobile radio systems, speech is limited to 4 kHz, compressed in amplitude (2:1), pre-emphasized, and phase/frequency modulated. At a receiver, inverse operations are performed. Degradation caused by these conversions and channel impairments results in lower voice quality. Finally, the human ear and brain have to perform the estimation and decision processes on the received signal.

In digital schemes for sampling and digitizing of an analog speech (source) are performed first. Then, by using knowledge of properties of the human vocal tract and the language itself, a spectrally efficient source coding is performed. A high rate 64 kb/s, 56 kb/s, and AD-PCM 32 kb/s digitized voice complies with ITU-T recommendations for toll quality, but may be less practical for the mobile environment. One is primarily interested in 8 to 16 kb/s rate speech coders, which might offer satisfactory quality, spectral efficiency, robustness, and acceptable processing delays in a mobile radio environment. A glossary of the major speech coding schemes is provided in Table 2.6.

TABLE 2.6   Digitized Voice. Glossary of Terms

Abbreviation	Description	Remarks/Use
ADM	Adaptive Delta Modulation	
ADPCM	Adaptive Differential Pulse Code Modulation	Digital telephony, DECT
ACIT	Adaptive Code sub-band exclted Transform	GTE
APC	Adaptive Predictive Coding	
APC–AB	APC with Adaptive Bit allocation	
APC–HQ	APC with Hybrid Quantization	
APC–MQL	APC with Maximum Likelihood Quantization	
AQ	Adaptive Quantization	
ATC	Adaptive Transform Coding	
BAR	Backward Adaptive Reencoding	
CELP	Code Excited Linear Prediction	IS-95
CVSDM	Continuous Variable Slope Delta Modulation	
DAM	Diagnostic Acceptability Measure	
DM	Delta Modulation	A/D conversion
DPCM	Differential Pulse Code Modulation	
DRT	Diagnostic Rhyme Test	
DSI	Digital Speech Interpolation	TDMA FSS systems
DSP	Digital Signal Processing	
HCDM	Hybrid Companding Delta Modulation	
LDM	Linear Delta Modulation	
LPC	Linear Predictive Coding	
MPLPC	Multi Pulse LPC	
MSQ	Multipath Search Coding	
NIC	Nearly Instantaneous Companding	
PCM	Pulse Code Modulation	Digital Voice
PVXC	Pulse Vector eXcitation Coding	
PWA	Predicted Wordlength Assignment	
QMF	Quadrature Mirror Filter	
RELP	Residual Excited Linear Prediction	GSM
RPE	Regular Pulse Excitation	
SBC	Sub Band Coding	
TASI	Time Assigned Speech Interpolation	TDMA FSS systems
TDHS	Time Domain Harmonic Scalling	
VAPC	Vector Adaptive Predictive Coding	
VCELP	Vector Code Excited Linear Prediction	
VEPC	Voice Excited Predictive Coding	
VQ	Vector Quantization	
VQL	Variable Quantum Level coding	
VSELP	Vector–Sum Excited Linear Prediction	IS–136, PDC
VXC	Vector eXcitation Coding	

Source: 4U Communications Research Inc., 2000.06.10~00:09, c:/tab/voice.tab

TABLE 2.7 Comparison of Cordless Telephone (CT) Systems

Parameter	CT0	CT1/+	JCT	CT2/+	CT3	DECT	CDMA	PHT
				System Name				
TX freq, MHz								
Base	22,26,30,31,46,48,45	914/885	254	864–8, 994–8	944–948	1880–1900		1895–1907
Mobile	48,41,39,40,69,74,48	960/932	380	864–8, 944–8	944–948	1880–1990		1895–1907
Multiple access band	FDMA	FDMA	FDMA	F/TDMA	TDMA	TDMA	CDMA	F/TDMA
Duplexing method	FDD	FDD	FDD	TDD	TDD	TDD	FDD	TDD
Ch. spacing, kHz	1.7,20,25,40	25	12.5	100	1000	1728	1250	300
Channel rate, kb/s				72	640	1152	1228.80	384
Channels/RF	1	1	1	1	8	12	32	4
Channels/band	10,12,15,20,25	40,80	89	20	2	5		20
Burst/frame length, ms				1/2	1/16	1/10	n/a	1/5
Modulation type	FM	FM	FM	GFSK	GMSK	GMSK	B/QPSK	$\pi/4$
Coding				Cyclic, RS	CRC 16	CRC 16	Conv 1/2, 1/3	
Transmit power, mW				≤10	≤80	≤100	≤10	≤80
Transmit power steps				2	1	1	many	many
TX power range, dB				16	0	0	≥80	
Vocoder type	analog	analog	analog	ADPCM	ADPCM	ADPCM	CELP	ADPCM
Vocoder rate, kb/s				fixed 32	fixed 32	fixed 32	≤9.6	fixed 32
Max data rate, kb/s				32	ISDN 144	ISDN 144	9.6	384
Processing delay, ms	0	0	0	2	16	16	80	
[3] Minimum				1/25	1/15	1/15	1/4	
Average				1.15	1/07	1/07	2/3	
Maximum				[1] 1/02	[1] 1/02	[1] 1/02	3/4	
[4]				100 × 1	10 × 8	6 × 12	4 × 32	
[5] Minimum				4	5–6	5–6	[2] 32 (08)	
Average				7	11–12	11–12	85 (21)	
Maximum				[1] 50	[1] 40	[1] 40	96 (24)	

Note: [1] The capacity (in the number of voice channels) for a single isolated cell. [2] The capacity in parenthes may correspond to a 32 kbit/s vocoder. [3] Reuse efficiency. [4] Theoretical number of voice channels per cell and 10 MHz. [5] Practical number of voice channels per 10 MHz. Reuse efficiency and associate capacities reflect our own estimates.

Source: 4U Communications Research Inc., 2000.06.10~00:09 c:/tab/cordless.sys

At this point, a partial comparison between analog and digital voice should be made. The quality of 64 kb/s digital voice, transmitted over a telephone line, is essentially the same as the original analog voice (they receive nearly equal MOS). What does this *near equal MOS* mean in a radio environment? A mobile radio conversation consists of one (mobile to home) or a maximum of two (mobile to mobile) mobile radio paths, which dictate the quality of the overall connection. The results of a comparison between analog and digital voice schemes in different artificial mobile radio environments have been widely published. Generally, systems that employ digital voice and digital codulation schemes seem to perform well under modest conditions, while analog voice and analog codulation systems outperform their digital counterparts in fair and difficult (near threshold, in the presence of strong co-channel interference) conditions. Fortunately, present technology can offer a viable implementation of both analog and digital systems within the same mobile/portable radio telephone unit. This would give every individual a choice of either an analog or digital scheme, better service quality, and higher customer satisfaction. Trade-offs between the quality of digital speech, the complexity of speech and channel coding, as well as D.C. power consumption have to be assessed carefully, and compared with analog voice systems.

Macro and Micro Diversity

Macro diversity. Let us observe the typical evolution of a cellular system. In the beginning, the base station may be located in the barocenter of the service area (center of the cell). The base station antenna is omnidirectional in azimuth, but with about 6 to 10 dBi gain in elevation, and serves most of the cell area (e.g., > 95%). Some parts within the cell may experience a lower quality of service because the direct path signal may be attenuated due to obstruction losses caused by buildings, hills, trees, etc. The closest neighboring (the first tier) base stations serve corresponding neighboring area cells by using different sets of frequencies, eventually causing adjacent channel interference. The second closest neighboring (the second tier) base stations might use the same frequencies (frequency reuse) causing co-channel interference. When the need for additional capacity arises and/or a higher quality of service is required, the same nearly circular area may be divided into three 120°-wide sectors, six 60°-wide sectors, etc., all served from the same base station location; now, the same base station is located at the edge of respective sectors. Since the new sectorial antennas provide 5 dB and 8 dB larger gains than the old omnidirectional antenna, respectively, these systems with new antennas with higher gains have longer spatial reach and may cover areas belonging to neighboring cells of the old configuration. For example, if the same real estate (base stations) is used in conjunction with 120° directional (in azimuth) antennas, the new designated 120°-wide wedge area may be served by the previous base station and by two additional neighboring base stations now equipped with sectorial antennas with longer reach. Therefore, the same number of existing base stations equipped with new directional antennas and additional combining circuitry may be required to serve the same or different number of cells, yet in a different fashion. The mode of operation in which two or more base stations serve the same area is called the *macro diversity*. Statistically, three base stations are able to provide better coverage of an area similar in size to the system with a centrally located base station. The directivity of a base station antenna (120° or even 60°) provides additional discrimination against signals from neighboring cells, therefore, reducing adjacent and co-channel interference (i.e., improving reuse efficiency and capacity). Effective improvement depends on the terrain configuration, and the combining strategy and efficiency. However, it requires more complex antenna systems and combining devices.

Micro diversity is when two or more signals are received at one site (base or mobile):

1. *Space diversity* systems employ two or more antennas spaced a certain distance apart from one another. A separation of only $\lambda/2 = 15$ cm at $f = 1$ GHz, which is suitable for implementation on the mobile side, can provide a notable improvement in some mobile radio channel environments. Micro space diversity is routinely used on cellular base sites. Macro diversity, where in our example the base stations were located kilometers apart, is also a form of space diversity.

2. *Field-component diversity* systems employ different types of antennas receiving either the electric or the magnetic component of an electromagnetic signal.

3. *Frequency diversity* systems employ two or more different carrier frequencies to transmit the same information. Statistically, the same information signal may or may not fade at the same time at different carrier frequencies. Frequency hopping and very wide band signaling can be viewed as frequency diversity techniques.

4. *Time diversity* systems are primarily used for the transmission of data. The same data is sent through the channel as many times as necessary, until the required quality of transmission is achieved automatic repeat request (ARQ). *Would you please repeat your last sentence* is a form of time diversity used in a speech transmission.

The improvement of any diversity scheme is strongly dependent on the combining techniques employed, i.e., the selective (switched) combining, the maximal ratio combining, the equal gain combining, the feedforward combining, the feedback (Granlund) combining, majority vote, etc., see [Jakes, 1974].

Continuous improvements in DSP and MMIC technologies and broader availability of ever-improving CAD electromagnetics tools is making adaptive antenna solutions more viable than ever before. This is particularly true for systems above 1 GHz, where the same necessary base station antenna gain can be achieved with smaller antenna dimensions. An adaptive antenna could follow spatially shifting traffic patterns, adjust its gain and pattern, and consequently improve the signal quality and capacity.

Multiple Broadcasting and Multiple Access

Communications networks for travelers have two distinct directions: the *forward link* — from the base station (via satellite) to all travelers within the footprint coverage area, and the *return link* — from a traveler (via satellite) to the base station. In the forward direction a base station distributes information to travelers according to the previously established protocol, i.e., no multiple access is involved; this way of operation is also called *broadcasting*. In the reverse direction many travelers make attempts to access one of the base stations; this way of operation is also called *access*. This occurs in so-called *control channels*, in a particular time slot, at a particular frequency, or by using a particular code. If collisions occur, customers have to wait in a queue and try again until success is achieved. If successful (i.e., no collision occurred), a particular customer will exchange (automatically) the necessary information for call setup. The network management center (NMC) will verify the customer's status, his credit rating, etc. Then, the NMC may assign a channel frequency, time slot, or code, on which the customer will be able to exchange information with his correspondent.

The optimization of the forward and reverse links may require different coding and modulation schemes and different bandwidths in each direction.

In *forward link,* there are three basic distribution (multiplex broadcasting) schemes: one that uses discrimination in frequency between different users and is called *frequency division multiplex broadcasting* (FDMB); another that discriminates in time and is called *time division multiplex broadcasting* (TDMB); and the last having different codes based on spread spectrum signaling, which is known as *code division multiplex broadcasting* (CDMB). It should be noted that hybrid schemes using a combination of basic schemes can also be developed. All existing mobile radio communications systems employ an FDM component; consequently, only FDMB schemes are pure, while the other two schemes are hybrid, i.e., TDMB/FDM and CDMB/FDM solutions are used; the two hybrid solutions inherit complexities of both parents, i.e., the need for an RF frequency synthesizer and a linear amplifier for *single channel per carrier* (SCPC) FDM solution, and the need for TDM and CDM overhead, respectively.

In *reverse link,* there are three basic access schemes: one that uses discrimination in frequency between different users and is called *frequency division multiple access* (FDMA); another that discriminates in time and is called *time division multiple access* (TDMA); and the last having different codes based on spread spectrum signaling, which is known as *code division multiple access* (CDMA). It should be noted that hybrid schemes using combinations of basic schemes can also be developed.

A performance comparison of multiple access schemes is a very difficult task. The strengths of FDMA schemes seem to be fully exploited in narrowband channel environments. To avoid the use of equalizers, channel bandwidths as narrow as possible should be employed, yet in such narrowband channels, the quality of service is limited by the maximal expected Doppler frequency and practical stability of frequency sources. Current practical limits are about 5 kHz.

The strengths of both TDMA and CDMA schemes seem to be fully exploited in wideband channel environments. TDMA schemes need many slots (and bandwidth) to collect information on network behavior. Once the equalization is necessary (at bandwidths > 20 kHz), the data rate should be made as high as possible to increase frame efficiency and freeze the frame to ease equalization; yet, high data rates require high RF peak powers and a lot of signal processing power, which may be difficult to achieve in handheld units. Current practical bandwidths are about 0.1 to 1.0 MHz. All existing schemes that employ TDMA components are hybrid, i.e., the TDMA/FDM schemes in which the full strength of the TDMA scheme is not fully realized.

CDMA schemes need large spreading (processing) factors (and bandwidth) to realize spread spectrum potentials; yet, high data rates require a lot of signal processing power, which may be difficult to achieve in handheld units. Current practical bandwidths are up to about 5 MHz. As mentioned before, a single transceiver has not been able to operate satisfactorily in a mobile channel environment. Consequently, a few CDMA elementary signals, information and pilot ones, may be necessary for successful transmission. This multisignal environment is equivalent to a MQAM signaling scheme with a not necessarily optimal state constellation. Significant increase in the equipment complexity is accompanied with a significant increase in the average and peak transmitter power. In addition, an RF synthesizer is needed to accommodate the CDMA/FDM mode of operation.

Narrow frequency bands seem to favor FDMA schemes, since both TDMA and CDMA schemes require more spectra to fully develop their potentials. However, once the adequate power spectrum is available, the later two schemes may be better suited for a complex (micro) cellular network environment. Multiple access schemes are also message sensitive. The length and type of message, and the kind of service will influence the choice of multiple access, ARQ, frame and coding, among others.

System Capacity

The recent surge in the popularity of cellular radio and mobile service in general, has resulted in an overall increase in traffic and a shortage of available system capacity in large metropolitan areas. Current cellular systems exhibit a wide range of traffic densities, from low in rural areas to overloaded in downtown areas with large daily variations between peak hours and quiet night hours. It is a great system engineering challenge to design a system that will make optimal use of the available frequency spectrum, while offering a maximal traffic throughput (e.g., Erlangs/MHz/service area) at an acceptable service quality, constrained by the price and size of the mobile equipment. In a cellular environment, the overall system capacity in a given service area is a product of many (complexly interrelated) factors including the available frequency spectra, service quality, traffic statistics, type of traffic, type of protocol, shape and size of service area, selected antennas, diversity, frequency reuse capability, spectral efficiency of coding and modulation schemes, efficiency of multiple access, etc.

In the seventies, so-called analog cellular systems employed omnidirectional antennas and simple or no diversity schemes offering modest capacity, which satisfied a relatively low number of customers. Analog cellular systems of the nineties employ up to 60° sectorial antennas and improved diversity schemes. This combination results in a three- to fivefold increase in capacity. A further (twofold) increase in capacity can be expected from narrowband analog systems (25 kHz → 12.5 kHz) and nearly threefold increase in capacity from the 5 kHz-wide narrowband AMPS, however, slight degradation in service quality might be expected. These improvements spurred the current growth in capacity, the overall success, and the prolonged life of analog cellular radio.

Conclusion

In this contribution, a broad repertoire of terrestrial and satellite systems and services for travelers is briefly described. The technical characteristics of the dispatch, cellular, and cordless telephony systems are tabulated for ease of comparison. Issues such as operating environment, service quality, network complexity, cell size, channel coding and modulation (codulation), speech coding, macro and micro diversity, multiplex and multiple access, and the mobile radio communications system capacity are discussed.

Presented data reveals significant differences between existing and planned terrestrial cellular mobile radio communications systems, and between terrestrial and satellite systems. These systems use different frequency bands, different bandwidths, different codulation schemes, different protocols, etc. (i.e., they are not compatible).

What are the technical reasons for this incompatibility? In this contribution, performance dependence on multipath delay (related to the cell size and terrain configuration), Doppler frequency (related to the carrier frequency, data rate, and the speed of vehicles), and message length (may dictate the choice of multiple access) are briefly discussed. A system optimized to serve the travelers in the Great Plains may not perform very well in mountainous Switzerland; a system optimized for downtown cores may not be well suited to a rural radio; a system employing geostationary (above equator) satellites may not be able to serve travelers at high latitudes very well; a system appropriate for slow moving vehicles may fail to function properly in a high Doppler shift environment; a system optimized for voice transmission may not be very good for data transmission, etc. A system designed to provide a broad range of services to everyone, everywhere, may not be as good as a system designed to provide a particular service in a particular local environment — as a decathlete world champion may not be as successful in competitions with specialists in particular disciplines.

However, there are plenty of opportunities where compatibility between systems, their integration, and frequency sharing may offer improvements in service quality, efficiency, cost and capacity (and therefore availability). Terrestrial systems offer a low start-up cost and a modest per user in densely populated areas. Satellite systems may offer a high quality of service and may be the most viable solution to serve travelers in scarcely populated areas, on oceans, and in the air. Terrestrial systems are confined to two dimensions and radio propagation occurs in the near horizontal sectors. Barostationary satellite systems use the narrow sectors in the user's zenith nearly perpendicular to the Earth's surface having the potential for frequency reuse and an increase in the capacity in downtown areas during peak hours. A call setup in a forward direction (from the PSTN via base station to the traveler) may be a very cumbersome process in a terrestrial system when a traveler to whom a call is intended is roaming within an unknown cell. However, this may be realized earlier in a global beam satellite system.

References

Ariyavisitakul, S.L., Falconer, D.D., Adachi, F., and Sari, H., (Guest Editors), Special Issue on Broadband Wireless Techniques, *IEEE J. on Selected Areas in Commun.*, 17, 10, October 1999.

Chuang, J.C.-I., Anderson, J.B., Hattori, T., and Nettleton, R.W., (Guest Editors), Special Issue on Wireless Personal Communications: Part I, *IEEE J. on Selected Areas in Commun.*, 11, 6, August 1993, Part II, *IEEE J. on Selected Areas in Commun.*, 11, 7, September 1993.

Cimini, L.J. and Tranter W.H., (Guest Editors), Special Issues on Wireless Communication Series, *IEEE J. on Selected Areas in Commun.*, 17, 3, March 1999.

Cimini, L.J. and Tranter, W.H., (Guest Editors), Special Issue on Wireless Communications Series, *IEEE J. on Selected Areas in Commun.*, 17, 7, July 1999.

Cimini, L.J. and Tranter, W.H., (Guest Editors), Special Issue on Wireless Communications Series, *IEEE J. on Selected Areas in Commun.*, 17, 11, November 1999.

Cimini, L.J. and Tranter, W.H., (Guest Editors), Special Issue on Wireless Communications Series, *IEEE J. on Selected Areas in Commun.*, 18, 3, March 2000.

Cox, D.C., Hirade, K., and Mahmoud, S.A., (Guest Editors), Special Issue on Portable and Mobile Communications, *IEEE J. on Selected Areas in Commun.*, 5, 4, June 1987.

Cox, D.C. and Greenstein, L.J., (Guest Editors), Special Issue on Wireless Personal Communications, *IEEE Commun. Mag.*, 33, 1, January 1995.

Davis, J.H. (Guest Editor), Mikulski, J.J., and Porter, P.T. (Associated Guest Editors), King, B.L. (Guest Editorial Assistant), Special Issue on Mobile Radio Communications, *IEEE J. on Selected Areas in Commun.*, 2, 4, July 1984.

Del Re, E., Devieux Jr., C.L., Kato, S., Raghavan, S., Taylor, D., and Ziemer, R., (Guest Editors), Special Issue on Mobile Satellite Communications for Seamless PCS, *IEEE J. on Selected Areas in Commun.*, 13, 2, February 1995.

Graglia, R.D., Luebbers, R.J., and Wilton, D.R., (Guest Editors), Special Issue on Advanced Numerical Techniques in Electromagnetics. *IEEE Trans. on Antennas and Propagation*, 45, 3, March 1997.

Institute of Navigation (ION), *Global Positioning System*, Reprinted by The Institute of Navigation. Volume I. Washington, D.C., USA, 1980; Volume II. Alexandria, VA, USA, 1984; Volume III. Alexandria, VA, USA, 1986; Volume IV. Alexandria, VA, USA, 1993.

International Telecommunication Union (ITU), *Radio Regulations,* Edition of 1982, Revised in 1985 and 1986.
International Telecommunication Union (ITU), Recommendations of the CCIR, 1990 (also Resolutions and Opinions). *Mobile Radiodetermination, Amateur and Related Satellite Services,* Volume VIII, XVIIth Plenary Assembly, Düsseldorf, 1990. Reports of the CCIR, (also Decisions), *Land Mobile Service, Amateur Service, Amateur Satellite Service,* Annex 1 to Volume VIII, XVIIth Plenary Assembly, Düsseldorf, 1990. Reports of the CCIR, (also Decisions), *Maritime Mobile Service,* Annex 2 to Volume VIII, XVIIth Plenary Assembly, Düsseldorf, 1990.
Kucar, A.D. (Guest Editor), Special Issue on Satellite and Terrestrial Systems and Services for Travelers, *IEEE Commun. Mag.,* 29, 11, November 1991.
Kucar, A.D., Kato, S., Hirata, Y., and Lundberg, O., (Guest Editors), Special Issue on Satellite Systems and Services for Travelers, *IEEE J. on Selected Areas in Commun.,* 10, 8, October 1992.
Kucar, A.D. and Uddenfeldt, J., (Guest Editors), Special Issue on Mobile Radio Centennial, *Proceedings of the IEEE,* 86, 7, July 1998.
Mahmoud, S.A., Rappaport, S.S., and Öhrvik, S.O., (Guest Editors), Special Issue on Portable and Mobile Communications, *IEEE J. on Selected Areas in Commun.,* 7, 1, January 1989.
Mailloux, R.J., (Guest Editor), Special Issue on Phased Arrays, *IEEE Trans. on Antennas and Propagation,* 47, 3, March 1999.
Mitola, J. III, Bose, V., Leiner, B.M., Turletti, T., and Tennenhouse, D., (Guest Editors), Special Issue on Software Radios, *IEEE J. on Selected Areas in Commun.,* 17, 4, April 1999.
Oppermann, I., van Rooyen, P., and Kohno, R., (Guest Editors), Special Issue on Spread Spectrum for Global Communications I, *IEEE J. on Selected Areas in Commun.,* 17, 12, December 1999.
Oppermann, I., van Rooyen, P., and Kohno, R., (Guest Editors), Special Issue on Spread Spectrum for Global Communications II, *IEEE J. on Selected Areas in Commun.,* 18, 1, January 2000.
Rhee, S.B., Editor), Lee, W.C.Y., (Guest Editor), Special issue on Digital Cellular Technologies, *IEEE Trans. on Vehicular Technol.,* 40, 2, May 1991.
Steele, R., (Guest Editor), Special Issue on PCS: The Second Generation, *IEEE Commun. Mag.,* 30, 12, December 1992.
World Administrative Radio Conference (WARC), *FINAL ACTS of the World Administrative Radio Conference for Dealing with Frequency Allocations in Certain Parts of the Spectrum* (WARC-92), Málaga-Torremolinos, 1992). ITU, Geneva, 1992.
World Radiocommunications Conference (WRC), *FINAL ACTS of the World Radiocommunications Conference* (WRC-97. ITU, Geneva, 1997.

Further Information

This trilogy, written by participants in AT&T Bell Labs projects on research and development in mobile radio, is the Holy Scripture of diverse cellular mobile radio topics:

Jakes, W.C. Jr. (Editor), *Microwave Mobile Communications,* John Wiley & Sons, Inc., New York, 1974.
AT&T Bell Labs Technical Personnel, Advanced Mobile Phone Service (AMPS), *Bell System Technical Journal,* 58, 1, January 1979.
Lee, W.C.Y., *Mobile Communications Engineering,* McGraw-Hill Book Company, New York, 1982.

An in-depth understanding of design, engineering, and use of cellular mobile radio networks, including PCS and PCN, requires knowledge of diverse subjects such as three-dimensional cartography, electromagnetic propagation and scattering, computerized analysis and design of microwave circuits, fixed and adaptive antennas, analog and digital communications, project engineering, etc. The following is a list of books relating to these topics:

Balanis, C.A., *Antenna Theory Analysis and Design,* Harper & Row, Publishers, New York 1982; Second Edition, John Wiley & Sons, Inc., New York, 1997.
Bowman, J.J., Senior, T.B.A., and Uslenghi, P.L.E., *Electromagnetic and Acoustic Scattering by Simple Shapes,* Revised Printing. Hemisphere Publishing Corporation, 1987.

Hansen, R.C., *Phased Array Antennas,* John Wiley & Sons, Inc., New York, 1998.

James, J.R. and Hall, P.S., (Editors), *Handbook of Microstrip Antennas,* Volumes I and II. Peter Peregrinus Ltd., Great Britain, 1989.

Kucar, A.D., *Satellite and Terrestrial Wireless Radio Systems: Fixed, Mobile, PCS and PCN, Radio vs. Cable. A Practical Approach,* Stridon Press Inc., 2000.

Lo, Y.T. and Lee, S.W., (Editors), *The Antenna Handbook,* Volumes I–IV, Van Nostrand Reinhold, USA, 1993.

Mailloux, R.J., *Phased Array Antenna Handbook,* Artech House, Inc., Norwood, MA, 1994.

Proakis, John G., *Digital Communications,* McGraw-Hill Book Company, New York, 1983.

Silvester, P.P. and Ferrari, R.L., *Finite Elements for Electrical Engineers,* 3rd Edition, Cambridge University Press, Cambridge, 1996.

Sklar, B., *Digital Communications. Fundamentals and Applications,* Prentice-Hall Inc., Englewood Cliffs, NJ, 1988.

Snyder, J.P., *Map Projection — A Working Manual,* U.S. Geological Survey Professional Paper 1395, United States Government Printing Office, Washington: 1987. (Second Printing 1989).

Spilker, J.J., Jr., *Digital Communications by Satellite,* Prentice-Hall Inc., Englewood Cliffs, NJ, 1977.

Stutzman, W.L., Thiele, G.A., *Antenna Theory and Design,* John Wiley & Sons, Inc., New York, 1981. 2nd Edition, John Wiley & Sons, Inc., New York, 1998.

Van Trees, H.L., *Detection, Estimation, and Modulation Theory,* Part I, 1968, Part II, 1971, Part III, John Wiley & Sons, Inc., New York, 1971.

Walker, J., *Advances in Mobile Information Systems,* Artech House, Inc., Norwood, MA, 1999.

2.3 Broadband Wireless Access: High Rate, Point to Multipoint, Fixed Antenna Systems

Brian Petry

Broadband Wireless Access (BWA) broadly applies to systems providing radio communications access to a core network. Access is the key word because a BWA system by itself does not form a complete network, but only the access part. As the "last mile" between core networks and customers, BWA provides access services for a wide range of customers (also called subscribers), from homes to large enterprises. For enterprises such as small to large businesses, BWA supports such core networks as the public Internet, Asynchronous Transfer Mode (ATM) networks, and the Public Switched Telephone Network (PSTN). Residential subscribers and small offices may not require access to such a broad set of core networks — Internet access is likely BWA's primary access function. BWA is meant to provide reliable, high throughput data services as an alternative to wired access technologies.

This article presents an overview of the requirements, functions, and protocols of BWA systems and describes some of today's efforts to standardize BWA interfaces.

Fundamental BWA Properties

Currently, the industry and standards committees are converging on a set of properties that BWA systems have, or should have, in common. A minimal BWA system consists of a single base station and a single subscriber station. The base station contains an interface, or interworking function (IWF), to a core network, and a radio "air" interface to the subscriber station. The subscriber station contains an interface to a customer premises network and of course, an air interface to the base station. Such a minimal system represents the point-to-point wireless transmission systems that have been in use for many years. Interesting BWA systems have more complex properties, the most central of which is point-to-multipoint (P-MP) capability. A single base station can service multiple subscriber stations using the same radio channel. The P-MP property of BWA systems feature omnidirectional or shaped sector radio antennas at the base station that cover a geographic and spectral area that efficiently serves a set of customers given the allocation of radio spectrum. Multiple subscriber stations can receive the base station's downstream transmissions on the same radio channel. Depending on the density and data throughput requirements of subscribers in a given sector, multiple radio channels may be employed, thus overlaying sectors. The frequency bands used for BWA allow for conventional directional antennas. So, in the upstream transmission direction, a subscriber's radio antenna is usually highly directional, aimed at the base station. Such configuration of shaped sectors and directional antennas allow for flexible deployment of BWA systems and helps conserve radio spectrum by allowing frequency bands to be reused in nearby sectors.

With such P-MP functions and a sectorized approach, more BWA properties unfold and we find that BWA is similar to other other well-known access systems. A BWA deployment is cellular in nature, and like a cellular telephone deployment, requires complicated rules and guidelines that impact power transmission limits, frequency reuse, channel assignment, cell placement, etc. Also, since subscriber stations can share spectrum in both the upstream and downstream directions, yet do not communicate with each other using the air interface, BWA systems have properties very similar to hybrid fiber coaxial (HFC) access networks that coexist with cable television service. HFC networks also employ a base station (called a head end) and subscriber stations (called cable modems). And subscriber stations share channels in both downstream and upstream directions. Such HFC networks are now popularized by both proprietary systems and the Data-over-Cable System Interface Specifications (DOCSIS) industry standards [1]. In the downstream direction, digital video broadcast systems have properties similar to BWA. They employ base stations on the ground or onboard satellites: multiple subscribers tune their receivers to the same channels. With properties similar to cellular, cable modems, and digital video broadcast, BWA systems borrow many technical features from them.

BWA Fills Technology Gaps

Since BWA is access technology, it naturally competes with other broadband, high data rate access technologies, such as high data rate digital cellular service, digital subscriber line (DSL) on copper telephone wires, cable modems on coaxial TV cables, satellite-based access systems, and even optical access technologies on fiber or free space. To some, the application of BWA overlaps with these access technologies and also appears to fill in the gaps left by them. Following are some examples of technology overlaps where BWA fills in gaps.

High data rate digital cellular data service will be available by the time this book is published. This service is built "on top of" digital cellular telephone service. The maximum achievable data rate for these new "third generation" digital cellular systems is intended to be around 2.5 Mbps. At these maximum speeds, high data rate cellular competes with low-end BWA, but since BWA systems are not intended to be mobile, and utilize wider frequency bands, a BWA deployment should be able to offer higher data rates. Furthermore, a BWA service deployment does not require near ubiquitous service area coverage. Before service can be offered by mobile cellular services, service must be available throughout entire metropolitan areas. But for BWA, service can be offered where target customers are located before covering large areas. Thus, in addition to higher achievable data rates with BWA, the cost to reach the first subscribers should be much less.

Current DSL technology can reach out about 6 km from the telephone central office, but the achievable data rate degrades significantly as the maximum distance is reached. Currently, the maximum DSL data rate is around 8 Mbps. Asymmetric DSL (ADSL) provides higher data rates downstream than upstream, which is ideal for residential Internet access, but can be limiting for some business applications. BWA can fill in by providing much higher data rates further from telephone central offices. BWA protocols and deployment strategies enable the flexibility necessary to offer both asymmetric and symmetric services.

HFC cable modem technology, which is also asymmetric in nature, is ideal for residential subscribers. But many subscribers — potentially thousands — often share the same downstream channels and contend heavily for access to a limited number of available upstream channels. A key advantage of HFC is consistent channel characteristics throughout the network. With few exceptions, the fiber and coaxial cables deliver a consistent signal to subscribers over very long distances. BWA fills in, giving a service provider the flexibility to locate base stations and configure sectors to best service customers who need consistent, high data rates dedicated to them.

Satellite access systems are usually unidirectional, whereas less available bidirectional satellite-based service is more expensive. Either type of satellite access is asymmetric in nature: unidirectional service requires some sort of terrestrial "upstream," and many subscribers contend for the "uplink" in bidirectional access systems. Satellites in geostationary Earth orbits (GEO) impose a minimum transit delay of 240 ms on transmissions between ground stations. Before a satellite access system can be profitable, it must overcome the notable initial expense of launching satellites or leasing bandwidth on a limited number of existing satellites by registering many subscribers. Yet, satellite access services offer extremely wide geographic coverage with no infrastructure planning, which is especially attractive for rural or remote service areas that DSL and cable modems do not reach. Perhaps high data rate, global service coverage by low Earth orbiting (LEO) satellites will someday overcome some of GEO's limitations. BWA fills in by allowing service providers to locate base stations and infrastructure near subscribers that should be more cost effective and impose less delay than satellite services.

Optical access technologies offer unbeatable performance in data rate, reliability, and range, where access to fiber-optic cable is available. But in most areas, only large businesses have access to fiber. New technology to overcome this limitation, and avoid digging trenches and pulling fiber into the customer premises is free space optical, which employs lasers to extend between a business and a point where fiber is more readily accessible. Since BWA base stations could also be employed at fiber access points to reach non-fiber-capable subscribers, both BWA and free space optical require less infrastructure planning such as digging, tunneling, and pulling cables under streets. Although optical can offer an order of magnitude increase in data rate over the comparatively lower frequency/higher wavelength BWA radio communications, BWA can have an advantage in some instances because BWA typically has a longer range and its sector-based coverage allows multiple subscribers to be serviced by a single base station.

Given these gaps left by other broadband access technologies, even with directly overlapping competition in many areas, the long-term success of BWA technology is virtually ensured.

BWA Frequency Bands and Market Factors

Globally, a wide range of frequency bands are available for use by BWA systems. To date, systems that implement BWA fall into roughly two categories: those that operate at high frequencies (roughly 10 to 60 GHz) and those that operate at low frequencies (2 to 11 GHz). Systems in the low frequency category may be further subdivided into those that operate in licensed vs. unlicensed bands. Unlicensed low frequency bands are sometimes considered separately because of the variations of emitted power restrictions imposed by regulatory agencies and larger potential for interference by other "unlicensed" technologies. The high frequencies have significantly different characteristics than the low frequencies that impact the expense of equipment, base station locations, range of coverage, and other factors. The key differing characteristics in turn impact the type of subscriber and types of services offered as will be seen later in this article.

Even though available spectrum varies, most nationalities and regulatory bodies recognize the vicinity of 30 GHz, with wide bands typically available, for use by BWA. In the United States, for instance, the FCC has allocated Local Multipoint Distribution Service (LMDS) bands for BWA. That, coupled with the availability of radio experience, borrowed from military purposes and satellite communications, influenced the BWA industry to focus their efforts in this area. BWA in the vicinity of 30 GHz is thus also a target area for standardization of interoperable BWA systems. Available spectrum for lower frequencies, 2 to 11 GHz, varies widely by geography and regulatory body. In the United States, for instance, the FCC has allocated several bands called Multipoint/Multichannel Distribution Services (MDS) and licensed them for BWA use. The industry is also targeting the lower spectrum, both licensed and unlicensed, for standardization.

Radio communications around 30 GHz have some important implications for BWA. For subscriber stations, directional radio antennas are practical. For base stations, so are shaped sector antennas. But two key impairments limit how such BWA systems are deployed: line-of-site and rain. BWA at 30 GHz almost strictly requires a line-of-sight path to operate effectively. Even foliage can prohibitively absorb the radio energy. Some near line-of-sight scenarios, such as a radio beam that passes in close proximity to reflective surfaces like metal sheets or wet roofs, can also cause significant communications impairments. Rain can be penetrated, depending on the distance between subscriber and base station, the droplet size, and rate of precipitation. BWA service providers pay close attention to climate zones and historical weather data to plan deployments. In rainy areas where subscribers require high data rate services, small cell sizes can satisfy a particular guaranteed service availability. Also, to accommodate changing rain conditions, BWA systems offer adaptive transmit power control. As the rate of precipitation increases, transmit power is boosted as necessary. The base station and subscriber station coordinate with each other to boost or decrease transmit power.

Impairments aside, equipment cost is an important issue with 30 GHz BWA systems. As of today, of all the components in a BWA station, the radio power amplifier contributes most to system cost. Furthermore, since the subscriber antenna must be located outdoors (to overcome the aforementioned impairments), installation cost contributes to the equation. A subscriber installation consists of an indoor unit (IDU) that typically houses the digital equipment, modem, control functions, and interface to the subscriber network, and an outdoor unit (ODU), which typically houses the amplifier and antenna. Today these factors, combined with the aforementioned impairments, typically limit the use of 30 GHz BWA systems to businesses that both need the higher-end of achievable data rates and can afford the equipment. BWA technology achieves data rates delivered to a subscriber in a wide range, 2 to 155 Mbps. The cost of 30 GHz BWA technology may render the lower end of the range impractical. However, many people project the cost of 30 GHz BWA equipment to drop as the years go by, to the point where residential service will be practical.

In the lower spectrum for BWA systems, in the range of approximately 2 to 11 GHz, line-of-sight and rain are not as critical impairments. Here, a key issue to contend with is interference due to reflections, also called multipath. A receiver, either base station or subscriber, may have to cope with multiple copies of the signal, received at different delays, due to reflections off buildings or other large objects in a sector. So, different modulation techniques may be employed in these lower frequency BWA systems, as opposed to high frequency systems, to compensate for multipath. Furthermore, if the additional expense can be justified, subscribers and/or base stations, could employ spatial processing to combine the main signal with its reflections and thus find a stronger signal that has more data capacity than the main signal by itself. Such spatial processing requires at least two antennas and radio receivers. In some cases, it may even be beneficial for a base station to employ induced multipath, using multiple transmit antennas, perhaps aimed at reflective objects, to reach subscribers, even those hidden by obstructions, with a better combined signal than just one.

Unlike BWA near 30 GHz, BWA in the lower spectrum today has the advantage of less expensive equipment. Also, it may be feasible in some deployments for the subscriber antenna to be located indoors. Further, the achievable data rates are typically lower than at 30 GHz, with smaller channel bandwidths, in the range of about 2 to 15 Mbps. Although some promise 30 GHz equipment costs will drop, these factors make lower frequency BWA more attractive to residences and small businesses today.

Due to the differing requirements of businesses and residences and the different capabilities of higher frequency BWA vs. lower, the types of service offered is naturally divided as well. Businesses will typically subscribe to BWA at the higher frequencies, around 30 GHz, and employ services that carry guaranteed quality of service for both data and voice communications. In the business category, multi-tenant office buildings and dwellings are also lumped in. At multi-tenant sites, multiple paying subscribers share one BWA radio and each subscriber may require different data or voice services. For data, Internet Protocol (IP) service is of prime importance, but large businesses also rely on wide area network technologies like asynchronous transfer mode (ATM) and frame relay that BWA must efficiently transport. To date, ATM's capabilities offer practical methods for dedicating, partitioning, and prioritizing data flows, generally called quality of service (QoS). But as time goes on (perhaps by this reading) IP-based QoS capabilities will overtake ATM. So, for both residential and business purposes, IP service will be the data service of choice in the future. Besides data, businesses rely on traditional telephony links to local telephone service providers. Business telephony services, for medium-to-large enterprises, utilize time division multiplexed (TDM) telephone circuits on copper wires to aggregate voice calls. Some BWA systems have the means to efficiently transport such aggregated voice circuits. Due to the economic and performance differences between low frequency BWA and high frequency BWA, low frequency BWA generally carries residential- and small business-oriented services, whereas high frequency BWA carries small- to large-enterprise services.

Since BWA equipment for the lower frequencies may be less expensive and less sensitive to radio directionality, and therefore more practical to cover large areas such as residential environments, subscriber equipment can potentially be nomadic. Nomadic means that the equipment may be moved quickly and easily from one location to another, but is not expected to be usable while in transit. Whereas at the higher frequencies, with more expensive subscriber equipment, the decoupling of indoor and outdoor units, the highly directional nature of radio communications in that range, and subscriber-oriented services provisioned at the base station, subscriber stations are fixed. Once they are installed, they do not move unless the subscriber terminates service and re-subscribes somewhere else.

Standards Activities

Several standards activities are under way to enable interoperability between vendors of BWA equipment. The standardization efforts are primarily focused on an interoperable "air interface" that defines how compliant base stations interoperate with compliant subscriber stations. By this reading, some of the standards may have been completed — the reader is encouraged to check the status of BWA standardization. Some standards groups archive contributions by industry participants and those archives, along with the actual published standards, provide important insights into BWA technology. Currently, most activity is centered around the Institute for Electronics and Electrical Engineers (IEEE) Local Area Network/Metropolitan Area Network (LAN/MAN) Standards Committee (LMSC), which authors the IEEE 802 series of data network standards. Within LMSC, the 802.16 working group authors BWA standards. The other notable BWA standards effort, under the direction of the European Telecommunications Standards Institute (ETSI), is a project called Broadband Radio Access Networks/HyperAccess (BRAN/HA). The IEEE LMSC is an organization that has international membership and has the means to promote their standards to "international standard" status through the International Organization for Standardization (ISO) as does ETSI. But ETSI standards draw from a European base, whereas LMSC draws from a more international base of participation. Even so, the LMSC and BRAN/HA groups, although they strive to develop standards each with a different approach, have many common members who desire to promote a single, international standard. Hopefully, the reader will have discovered that the two groups have converged on one standard that enables internationally harmonized BWA interoperability.

To date, the IEEE 802.16 working group has segmented their activities into three main areas: BWA interoperability at bands around 30 GHz (802.16.1), a recommended practice for the coexistence of BWA systems (802.16.2) and BWA interoperability for licensed bands between 2 and 11 GHz (802.16.3). By the time this book is published, more standards activities may have been added, such as interoperability for some unlicensed bands. The ETSI BRAN/HA group is focused on interoperability in bands around 30 GHz.

Past standards activities were efforts to agree on how to adapt existing technologies for BWA: cable modems and digital video broadcast. A BWA air interface, as similar to DOCSIS cable modems as possible, was standardized by the radio sector of the International Telecommunications Union (ITU) under the ITU-R Joint Rappateur's Group (JRG) 9B committee [2]. The Digital Audio-Video Council (DAVIC) has standardized audio and video transport using techniques similar to BWA [3]. Similarly, the Digital Video Broadcasting (DVB) industry consortium, noted for having published important standards for satellite digital video broadcast, has also published standards, through ETSI, for terrestrial-based digital television broadcast over both cable television networks and wireless. DVB has defined the means to broadcast digital video in both the "low" (<10 Gbps) and "high" (>10 Gbps) BWA spectra [4, 5]. These standards enabled interoperability of early BWA deployment by utilizing existing subsystems and components. Technology from them provided a basis for both the IEEE LMSC and ETSI BRAN/HA standardization processes. However, the current IEEE and ETSI efforts strive to define protocols with features and nuances more particular to efficient BWA communications.

Technical Issues: Interfaces and Protocols

A BWA access network is perhaps best described by its air interface: what goes on between the base station and subscriber stations. Other important interfaces exist in BWA systems, such as:

- interfaces to core networks like ATM, Frame Relay, IP, and PSTN
- interfaces to subscriber networks like ATM, Ethernet, Token Ring, and private branch exchange (PBX) telephone systems
- the interface between indoor unit (IDU) and outdoor unit (ODU)
- interfaces to back-haul links, both wired and wireless, for remote base stations not co-located with core networks
- air interface repeaters and reflectors

These other interfaces are outside the scope of this article. However, understanding their requirements is important to consider how a BWA air interface can best support external interfaces, particularly how the air interface supports their unique throughput, delay, and QoS requirements.

Protocols and Layering

Network subsystems following the IEEE LMSC reference model [6] focus on the lower two layers of the ISO Basic Reference Model for Open Systems Interconnection [7]. The air interface of a BWA system is also best described by these two layers. In LMSC standards, layers one and two, the physical and data link layers, are typically further subdivided. For BWA, the important subdivision of layer 2 is the medium access control (MAC) sublayer. This layer defines the protocols and procedures by which network nodes contend for access to a shared channel, or physical layer. In a BWA system, since frequency channels are shared among subscriber stations in both the downstream and upstream directions, MAC layer services are critical for efficient operation. The physical layer (PHY) of a BWA system is responsible for providing a raw communications channel, employing modulation and error correction technology appropriate for BWA.

Other critical functions, some of which may reside outside the MAC and PHY layers, must also be defined for an interoperable air interface: security and management. Security is divided two areas: a subscriber's authorized use of a base station and associated radio channels and privacy of transported data. Since the communications channel is wireless, it is subject to abuse by intruders, observers, and those seeking to deny service. BWA security protocols must be well defined to provide wire-like security and allow for interoperability. Since to a great extent, HFC cable TV access networks are very similar to BWA regarding security requirements, BWA borrows heavily from the security technology of such cable systems. Similarly, interoperable management mechanisms and protocols include the means to provision, control and monitor subscribers stations and base stations.

The Physical Layer

The physical layer (PHY) is designed with several fundamental goals in mind: spectral efficiency, reliability, and performance. However, these are not independent goals. We can not have the best of all three because each of those goals affects the others: too much of one means too little of the others. But reliability and performance levels are likely to be specified. And once they are specified, spectral efficiency can be somewhat optimized. One measure of reliability is the bit error ratio (BER), the ratio of the number of bit errors to the number of non-errored bits, delivered by a PHY receiver to the MAC layer. The physical layer must provide for better than 10^{-6} BER, and hopefully closer to 10^{-9}. The larger error ratio may only be suitable for some voice services, whereas a ratio closer to the lower end of the range is required for reliable data services that could offer equivalent error performance as local area networks (LANs). Reliability is related to availability. Business subscribers often require contracts that guarantee a certain level of availability. For instance, a service provider may promise that the air interface be available to offer guaranteed reliability and performance 99.99% (also called "four nines") of the time.

Performance goals specify minimum data rates. Since, in BWA systems, the spectrum is shared by subscribers, and allocation of capacity among them is up to the MAC layer, the PHY is more concerned with the aggregate capacity of a single radio channel in one sector of a base station than for capacity to a given subscriber. But if one subscriber would offer to purchase all the available capacity, the service provider would undoubtedly comply. For instance, a capacity goal currently set by the BRAN/HA committee is 25 Mbps on a 28 MHz channel. Without considering deployment scenarios, however, PHY goals are meaningless. Obviously, higher capacity and reliability could be better achieved by shorter, narrower sectors (smaller cells) rather than wider, longer sectors (larger cells). And the same sized sector in a rainy, or obstructed, terrain offers less guaranteed capacity than one in the flattest part of the desert. In any case, the industry seems to be converging on a goal to provide at least 1 bps/Hz capacity in an approximately 25 MHz wide channel with a BER of 10^{-8}. Many deployments should be able to offer much greater capacity.

In addition to such fundamental goals, other factors affect the choice of PHY protocols and procedures. One is duplex mode. The duplex mode can affect the cost of equipment, and some regulatory bodies may limit the choice of duplex mode in certain bands. Three duplex modes are considered for BWA: frequency division duplex (FDD), time division duplex (TDD), and half-duplex FDD (H-FDD). In FDD, a radio channel is designated for either upstream- or downstream-only use. Some bands are regulated such that a channel could only be upstream or downstream, thus requiring FDD if such bands are to be utilized. In TDD mode, one channel is used for both upstream and downstream communications. TDD-capable BWA equipment thus ping-pongs between transmit and receive mode within a single channel; all equipment in a sector is synchronized to divisions between transmit and receive. TDD is useful for bands in which the number of available, or licensed, channels is limited. TDD also allows for asymmetric service without reconfiguring the bandwidth of FDD channels. For instance, a service provider may determine that a residential deployment is more apt to utilize more downstream bandwidth than upstream. Then, rather than reallocating or sub-channeling FDD channels, the service provider can designate more time in a channel for downstream communications than upstream. Additionally, TDD equipment could potentially be less expensive than FDD equipment since components may be shared between the upstream and downstream paths and the cost of a duplexor may be eliminated. However, the third option, H-FDD, is a reasonable compromise between TDD and FDD. In H-FDD mode, a subscriber station decides when it can transmit and when it can receive, but cannot receive while transmitting. But the base station is usually full duplex, or FDD. For subscriber stations, H-FDD equipment can achieve the same cost savings as TDD, and offers the flexibility of asymmetric service. But H-FDD does not require all subscribers in a sector to synchronize on the same allocated time between transmit and receive.

Another important factor affecting spectral efficiency, upgradability, and flexible deployment scenarios, is adaptive modulation. In BWA, the channel characteristics vary much more widely than wired access systems. Rain, interference and other factors can affect subscriber stations individually in a sector, whereas in wired networks, such as HFC cable TV, the channel characteristics are consistent. Thus, to make good use of available bandwidth in favorable channel conditions, subscribers that can take advantage of higher data rates should be allowed to do so. And when it rains in one portion of a sector, or other impairments such as interference occur, subscriber stations can adapt to the channel conditions by reducing the data rate (although transmit power level adjustment is usually the first adaptive tool BWA stations use when it rains). Besides adapting to channel conditions, adaptive modulation facilitates future deployment of newer modulation techniques while retaining compatibility with currently installed subscriber stations. When the service provider upgrades a base station and offers better modulation to new customers, not all subscriber stations become obsolete. To achieve the most flexibility in adaptive modulation, BWA employs "per-subscriber" adaptive modulation to both downstream and upstream communications. Per-subscriber means that each subscriber station can communicate with the base station using a different modulation technique, within the same channel. Some BWA equipment offers per-subscriber adaptive modulation in both the downstream and upstream directions. But other equipment implements a compromise that allows for equipment or components, similar to cable modems or digital video broadcast systems, to require all subscribers to use the same modulation in the downstream direction at any one point in time. Most BWA equipment implements adaptive modulation in the upstream direction. The overriding factor for the PHY layer, with regard to adaptive modulation, is burst mode. Adaptive modulation generally requires burst mode communications at the PHY layer. Time is divided into small units in which stations transmit independent bursts of data. If the downstream employs per-subscriber adaptive modulation, the base station transmits independent bursts to the subscribers. Each burst contains enough information for the receiver to perform synchronization and equalization. However, if per-subscriber adaptive modulation is not employed in the downstream direction, the base station can transmit in continuous mode, in very large, continuous chunks, each chunk potentially containing data destined for multiple subscribers. In burst mode downstream communications, the base station informs subscriber stations, in advance, which burst is theirs. In this way, a subscriber station is not required to demodulate each burst to discover which bursts are for the station, but only to demodulate the "map." The base station encodes the map using the least common denominator modulation type so all subscriber stations can decode it. Conversely, continuous mode downstream, in which per-subscriber adaptive modulation is not used, requires all subscriber stations to demodulate prior to discovering which portions of data are destined for the station. So, per-subscriber adaptive modulation in the downstream affords more flexibility, but a continuous mode downstream may also be used. The standards efforts currently are attempting to work out how both downstream modes may be allowed and yet still have an interoperable standard.

Burst size and the choice of continuous downstream mode in turn affect the choice of error correction coding. Some coding schemes are more efficient with large block sizes, whereas others are more efficient with smaller block sizes.

The fundamental choice of modulation type for BWA varies between the upper BWA bands (~30 GHz) and lower bands (~2 to 11 GHz). In the upper bands, the industry seems to be converging on Quadrature Phase Shift Keying (QPSK) and various levels of Quadrature Amplitude Modulation (QAM). These techniques may also be used in the lower bands, but given the multipath effects that are much more prevalent in the lower bands, BWA equipment is likely to employ Orthogonal Frequency Division Multiplexing (OFDM) or Code Division Multiple Access (CDMA) technology that have inherent properties to mitigate the effects of multipath and spread transmit energy evenly throughout the channel spectrum.

The Medium Access Control Layer

The primary responsibility of the Medium Access Control Layer (MAC) is to allocate capacity among subscriber stations in a way that preserves quality-of-service (QoS) requirements of the services it transports. For instance, traditional telephony and video services could require a constant, dedicated capacity with fixed delay properties. But other data transport services could tolerate more bursty capacity allocations and a higher degree of delay variation. ATM service is notable for its QoS definitions [8]. Although not mature as of this writing, the Internet Protocol (IP) QoS definitions are also notable [9,10]. Though QoS-based capacity allocation is a complex process, the BWA MAC protocol defines the mechanisms to preserve QoS as it transports data. Yet the MAC protocol does not fully define *how* MAC mechanisms are to be used. At first glance, this does not seem to make sense, but it allows the MAC protocol to be defined in as simple terms as possible and leave it up to implementations of base stations and subscriber stations how to best utilize the mechanism that the protocol defines. This approach also allows BWA vendors to differentiate their equipment and still retain interoperability. To simplify capacity allocation, the smarts of QoS implementation reside in the base station, since it is a central point in a BWA sector and is in constant communication with all of the subscriber stations in a sector. The base station is also administered by the service provider, and therefore can serve as the best point of control to keep subscribers from exceeding their contractual capacity limitations and priorities.

Capacity Allocation Mechanisms — An overview of the mechanisms employed by the MAC layer to allocate capacity follows. In the downstream direction, the MAC protocol informs subscriber stations what data belongs to what subscriber by means of per-subscriber addressing and within a subscriber, by per-data-flow addressing. All subscribers in a sector "listen" to the downstream data flow and pick off transmissions belonging to them. If the downstream channel employs per-subscriber adaptive modulation, some subscriber stations may not be able to decode the modulation destined to other subscribers. In this case, the base station informs subscribers what bursts it should observe, with a downstream "map." The downstream map indicates what offsets in a subsequent transmission may contain data for the specified subscriber. The MAC must communicate this information to the PHY layer to control its demodulation.

For upstream capacity allocation and reservation, the MAC employs slightly more complicated schemes. The upstream channel is the central point of contention: all subscriber stations in a channel are contending for access to transmit in the upstream channel. Some subscribers require constant periodic access, others require bursty access with minimum and maximum reservation limits. Still other data flows may not require any long-standing reservations but can request a chunk of capacity when needed and survive the inherent access delay until the base station satisfies the request. On top of these varying data flow requirements, which are specified by subscriber stations and granted by the base station, priorities increase complications. The base station administers both priorities and QoS parameters of each data flow in each subscriber station. How a base station keeps track of all the flows of subscribers and how it actually meets the reservation requirements is usually beyond the scope of the BWA air interface in standards documents. But base stations likely employ well-known queuing algorithms and reservation lists to ensure that it assigns capacity fairly and meets subscribers' contractual obligations. Yet, as mentioned earlier, room is left for BWA base station vendors to employ proprietary "tricks" to differentiate their equipment from others. To communicate capacity allocation to subscribers, the base station divides time into multi-access frames (e.g., on the order of 1 to 5 milliseconds) in which multiple subscribers are assigned capacity. To accomplish this, a fixed allocation unit, or time slot, is defined. So, the upstream channel is divided into small, fixed-length time slots (e.g., on the order of 10 microseconds) and the base station periodically transmits a "map" of slot assignments to all subscribers in a channel. The slot assignments inform the subscriber stations which slots are theirs for the upcoming multi-access frame.

Periodically, a set of upstream slots is reserved for "open contention." That is, any subscriber is authorized to transmit during an open contention period. A subscriber can utilize open contention for initial sign-on to the network (called "registration"), to transmit a request for upstream capacity, or even to transmit a small amount of data. Since a transmission may collide with that of another subscriber station, a collision avoidance scheme is used. A subscriber station initiates transmission in a randomly chosen open contention slot, but cannot immediately detect that its transmission collided with another. The only way a subscriber station can determine if its transmission collided is if it receives no acknowledgment from the base station. In this case, the subscriber backs off a random number of open contention slots before attempting another transmission. The process continues, with the random number range getting exponentially larger on each attempt, until the transmission succeeds. The random back-off interval is typically truncated at the sixteenth attempt, when the subscriber station starts over with its next attempt in the original random number range. This back-off scheme is called "truncated binary exponential back-off," and is employed by popular MAC protocols such as Ethernet [11].

To mitigate the effects of potentially excessive collisions during open contention, the MAC protocol defines a means to request bandwidth during assigned slots in which no collision would happen. For instance, active subscriber stations may receive from the base station a periodic slot for requesting capacity or requesting a change in a prior reservation. This form of allocation-for-a-reservation-request is called a "poll." Also, the MAC protocol provides a means to "piggy-back" a request for capacity with a normal upstream data transmission. Subscriber stations that have been inactive may receive less frequent polls from the base station so as to conserve bandwidth. So, with a means for contentionless bandwidth reservation, the only time subscriber stations need to use the open contention window is for initial registration.

Slot-based reservations require that the base stations and subscribers be synchronized. Of course, the base station provides a timing base for all subscriber stations. To achieve accurate timing, subscriber stations need to determine how far they are from the base station so their transmissions can be scheduled to reach the base station at the exact point in time, relative to each other. The procedure to determine this distance, which is not really a measured linear distance, but a measurement of time, is called "ranging." Each subscriber station, coordinating with the base station, performs ranging during its registration process.

To maintain efficient use of bandwidth and accommodate PHY requirements of transmit power control, and flexible duplex modes, the MAC protocol performs even more gyrations. If interested, the reader is encouraged to read BWA MAC protocol standards, or drafts in progress, to learn more.

Automatic Repeat Request (ARQ) Layer

Some BWA systems trade off the bandwidth normally consumed by the PHY's error correction coding for the potential delays of ARQ protocol. An ARQ protocol employs sequence numbering and retransmissions to provide a reliable air link between base station and subscriber. ARQ requires more buffering in both the base station and subscriber station than systems without ARQ. But even with a highly-coded PHY, some subscriber stations may be located in high interference or burst-noise environments in which error correction falls apart. In such situations, ARQ can maintain performance, or ensure the service meets contractual availability and reliability requirements. Standards groups seem to be converging on allowing the use of ARQ, but not requiring it. The MAC protocol is then specified so that when ARQ is not used, no additional overhead is allocated just to allow the ARQ option.

Conclusion

This article has provided an overview of how BWA fits in with other broadband access technologies. It was also a short primer on BWA protocols and standards. To learn more about BWA, the reader is encouraged to read currently available standards documents, various radio communications technical journals, and consult with vendors of BWA equipment.

References

1. SCTE SP-RFI-105-981010: *Data-Over-Cable Service Interface Specifications: Radio Frequency Interface Specification,* The Society of Cable Telecommunications Engineers, Exton, Pennsylvania, 1999.
2. Draft Recommendation F.9B/BWA.- Radio transmission systems for fixed broadband wireless access (BWA) based on cable modem standard, International Telecommunications Union, Geneva, 1999.
3. DAVIC 1.4.1 *Specification Part 8: Lower Layer Protocols and Physical Interfaces,* Digital Audio-Visual Council, Geneva, 1999.
4. ETS 300 748, *Digital Video Broadcasting (DVB): Multipoint Video Distribution Systems (MVDS) at 10 GHz and above,* European Telecommunications Standards Institute, Geneva, 1997.
5. ETS 300 749, *Digital Video Broadcasting (DVB): Digital Video Broadcasting (DVB); Microwave Multipoint Distribution Systems (MMDS) below 10 GHz,* European Telecommunications Standards Institute, Geneva, 1997.
6. IEEE Std 802-1990, *IEEE Standards for Local and Metropolitan Area Networks: Overview and Architecture,* Institute for Electrical and Electronics Engineers, Piscataway, NJ, 1990.
7. ISO/IEC 7498-1:1994, *Information Technology — Open Systems Interconnection — Basic Reference Model: The Basic Model,* International Organization for Standardization, Geneva, 1994.
8. Bermejo, L. P. et al., Service characteristics and traffic models in broadband ISDN, *Electrical Commun.,* 64-2/3, 132–138, 1990.

9. Blake, S. et al., *RFC-2475 An Architecture for Differentiated Service*, Internet Engineering Task Force, 1998.
10. Braden, R. et al., *RFC-2205 Resource ReSerVation Protocol (RSVP) — Version 1 Functional Specification*, Internet Engineering Task Force, 1997.
11. IEEE Std 802.3, *Information Technology — Telecommunications and information exchange between systems — Local and metropolitan area networks — Specific requirements — Part 3: Carrier sense multiple access with collision detection (CSMA/CD) access method and physical layer specifications*, Institute for Electronics and Electrical Engineers, Piscataway, NJ, 1998.

2.4 Digital European Cordless Telephone

Saf Asghar

Cordless technology, in contrast to cellular radio, primarily offers access technology rather than fully specified networks. The digital European cordless telecommunications (DECT) standard, however, offers a proposed network architecture in addition to the air interface physical specification and protocols but without specifying all of the necessary procedures and facilities. During the early 1980s a few proprietary digital cordless standards were designed in Europe purely as coexistence standards. The U.K. government in 1989 issued a few operator licenses to allow public-access cordless known as telepoint. Interoperability was a mandatory requirement leading to a common air interface (CAI) specification to allow roaming between systems. This particular standard (CT2/CAI), is described elsewhere in this book. The European Telecommunications Standards Institute (ETSI) in 1988 took over the responsibility for DECT. After formal approval of the specifications by the ETSI technical assembly in March 1992, DECT became a European telecommunications standard, ETS300-175 in August 1992. DECT has a guaranteed pan-European frequency allocation, supported and enforced by European Commission Directive 91/297. The CT2 specification has been adopted by ETSI alongside DECT as an interim standard I-ETSI 300 131 under review.

Application Areas

Initially, DECT was intended mainly to be a private system, to be connected to a private automatic branch exchange (PABX) to give users mobility, within PABX coverage, or to be used as a single cell at a small company or in a home. As the idea with telepoint was adopted and generalized to public access, DECT became part of the public network. DECT should not be regarded as a replacement of an existing network but as created to interface seamlessly to existing and future fixed networks such as public switched telephone network (PSTN), integrated services digital network (ISDN), global system for mobile communications (GSM), and PABX. Although telepoint is mainly associated with CT2, implying public access, the main drawback in CT2 is the ability to only make a call from a telepoint access point. Recently there have been modifications made to the CT2 specification to provide a structure that enables users to make and receive calls. The DECT standard makes it possible for users to receive and make calls at various places, such as airport/railroad terminals, and shopping malls. Public access extends beyond telepoint to at least two other applications: replacement of the wired local loop, often called cordless local loop (CLL), (Fig. 2.8) and neighborhood access, Fig. 2.9. The CLL is a tool for the operator of the public network. Essentially, the operator will install a multiuser base station in a suitable campus location for access to the public network at a subscriber's telephone hooked up to a unit coupled to a directional antenna. The advantages of CLL are high flexibility, fast installation, and possibly lower investments. CLL does not provide mobility. Neighborhoods access is quite different from CLL. Firstly, it offers mobility to the users and, secondly, the antennas are not generally directional, thus requiring higher field strength (higher output power or more densely packed base stations). It is not difficult to visualize that CLL systems could be merged with neighborhood access systems in the context of establishments, such as supermarkets, gas stations, shops, etc., where it might be desirable to set up a DECT system for their own use and at the

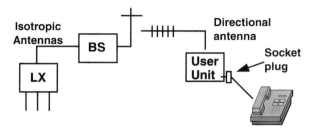

FIGURE 2.8

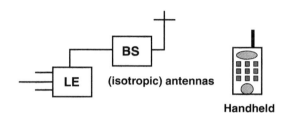

FIGURE 2.9

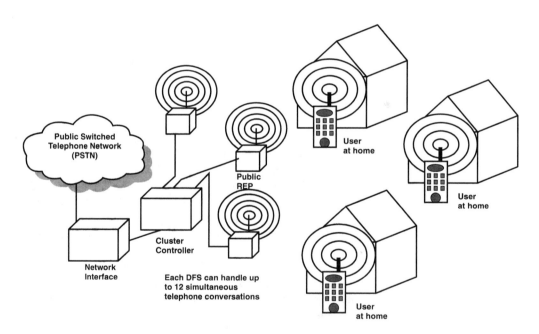

FIGURE 2.10

same time also provide access to customers. The DECT standard already includes signaling for authentication, billing, etc. DECT opens possibilities for a new operator structure, with many diversified architectures connected to a global network operator (Fig. 2.10). DECT is designed to have extremely high capacity. A small size is used, which may seem an expensive approach for covering large areas. Repeaters placed at strategic locations overcome this problem.

DECT/ISDN Interworking

From the outset, a major objective of the DECT specification was to ensure that ISDN services were provided through the DECT network. Within the interworking profile two configurations have been defined: DECT end system and DECT intermediate system. In the end system the ISDN is terminated in the DECT fixed system (DFS). The DFS and the DECT portable system (DPS) may be seen as ISDN terminal equipment (TE1). The DFS can be connected to an S, S/T, or a P interface. The intermediate system is fully transparent to the ISDN. The S interface is regenerated even in the DPS. Both configurations have the following services specified: 3.1-kHz telephony, i.e., standard telephony; 7-kHz telephony; i.e., high-quality audio; video telephony; group III fax, modems, X.25 over the ISDN; and telematic services, such as group IV fax, telex, and videotax.

DECT/GSM Interworking

Groupe Speciale Mobile (GSM) is a pan-European standard for digital cellular radio operation throughout the European community. ETSI has the charter to define an interworking profile for GSM and DECT. The profile describes how DECT can be connected to the fixed network of GSM and the necessary air interface functions. The users obviously benefit from the mobility functions of GSM giving DECT a wide area mobility. The operators will gain access to another class of customer. The two systems when linked together will form the bridge between cordless and cellular technologies. Through the generic access profile, ETSI will specify a well-defined level of interoperability between DECT and GSM. The voice coding aspect in both of these standards is different; therefore, this subject will be revisited to provide a sensible compromise.

DECT Data Access

The DECT standard is specified for both voice and data applications. It is not surprising that ETSI confirmed a role for DECT to support cordless local area network (LAN) applications. A new technical committee, ETSI RES10 has been established to specify the high performance European radio LAN similar to IEEE 802.11 standard in the U.S. (Table 2.8).

TABLE 2.8   DECT Characteristics

Parameters	DECT
Operating frequency, MHz	1880–1990 (Europe)
Radio carrier spacing, MHz	1.728
Transmitted data rate, Mb/s	1.152
Channel assignment method	DCA
Speech data rate, kb/s	32
Speech coding technique	ADPCM G.721
Control channels	In-call-embedded (various logical channels C, P, Q, N)
In-call control channel data rate, kb/s	4.8 (plus 1.6 CRC)
Total channel data rate, kb/s	41.6
Duplexing technique	TDD
Multiple access-TDMA	12 TDD timeslots
Carrier usage-FDMA/MC	10 carriers
Bits per TDMA timeslot, b	420 (424 including the 2 field)
Timeslot duration (including guard time), μs	417
TDMA frame period, ms	10
Modulation technique	Gaussian filtered FSK
Modulation index	0.45–0.55
Peak output power, mW	250
Mean output power, mW	10

How DECT Functions

DECT employs frequency division multiple access (FDMA), time division multiple access (TDMA), and time division duplex (TDD) technologies for transmission. Ten carrier frequencies in the 1.88- and 1.90-GHz band are employed in conjunction with 12 time slots per carrier TDMA and 10 carriers per 20 MHz of spectrum FDMA. Transmission is through TDD. Each channel has 24 times slots, 12 for transmission and 12 for receiving. A transmission channel is formed by the combination of a time slot and a frequency. DECT can, therefore, handle a maximum of 12 simultaneous conversations. TDMA allows the same frequency to use different time slots. Transmission takes place for ms, and during the rest of the time the telephone is free to perform other tasks, such as channel selection. By monitoring check bits in the signaling part of each burst, both ends of the link can tell if reception quality is satisfactory. The telephone is constantly searching for a channel for better signal quality, and this channel is accessed in parallel with the original channel to ensure a seamless changeover. Call handover is also seamless, each cell can handle up to 12 calls simultaneously, and users can roam around the infrastructure without the risk of losing a call. Dynamic channel assignment (DCA) allows the telephone and base station to automatically select a channel that will support a new traffic situation, particularly suited to a high-density office environment.

Architectural Overview

Baseband Architecture

A typical DECT portable or fixed unit consists of two sections: a baseband section and a radio frequency section. The baseband partitioning includes voice coding and protocol handling (Fig. 2.11).

Voice Coding and Telephony Requirements

This section addresses the audio aspects of the DECT specification. The CT2 system as described in the previous chapter requires adaptive differential pulse code modulation (ADPCM) for voice coding. The DECT standard also specifies 32-kb/s ADPCM as a requirement. In a mobile environment it is debatable whether the CCITT G.721 recommendation has to be mandatory. In the handset or the mobile it would be quite acceptable in most cases to implement a compatible or a less complex version of the recommendation. We are dealing with an air interface and communicating with a base station that in the residential situation terminates with the standard POTS line, hence compliance is not an issue. The situation changes in the PBX, however, where the termination is a digital line network. DECT is designed for this case, hence compliance with the voice coding recommendation becomes important. Adhering to this strategy for the base station and the handset has some marketing advantages.

G.721 32-kb/s ADPCM from its inception was adopted to coexist with G.711 64-kb/s pulse code modulation (PCM) or work in tandem, the primary reason being an increase in channel capacity. For

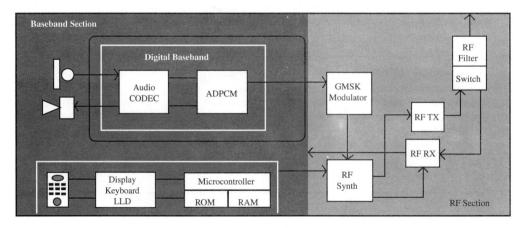

FIGURE 2.11

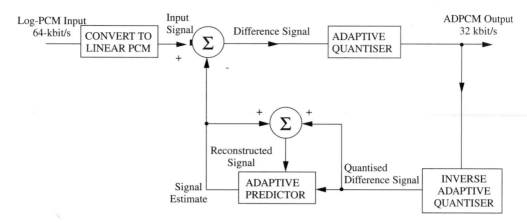

FIGURE 2.12   ADPCM encoder.

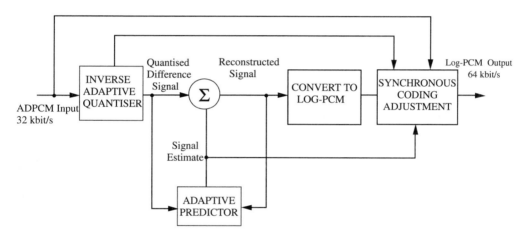

FIGURE 2.13   ADPCM decoder.

modem type signaling, the algorithm is suboptimal in handling medium-to-high data rates, which is probably one of the reasons why there really has not been a proliferation of this technology in the PSTN infrastructure. The theory of ADPCM transcoding is available in books on speech coding techniques, e.g., O'Shaughnessy, 1987.

The ADPCM transcoder consists of an encoder and a decoder. From Figs. 2.12 and 2.13 it is apparent that the decoder exists in the encoder structure. A benefit derived from this structure allows for efficient implementation of the transcoder.

The encoding process takes a linear speech input signal (the CCITT specification relates to a nonwireless medium such as a POTS infrastructure), and subtracts its estimate derived from earlier input signals to obtain a difference signal. This difference signal is 4-b code with a 16-level adaptive quantizer every 125 μs, resulting in a 32-kb/s bit stream. The signal estimate is constructed with the aid of the inverse adaptive quantizer that forms a quantized difference signal that added to the signal estimate is also used to update the adaptive predictor. The adaptive predictor is essentially a second-order recursive filter and a sixth-order nonrecursive filter,

$$S_0(k) = \sum_{i=1}^{2} a_i(k-1)\varepsilon_r(k-i) + \sum_{i=1}^{6} b_i(k-1)d_q(k-i) \tag{2.1}$$

where coefficients a and b are updated using gradient algorithms.

As suggested, the decoder is really a part of the encoder, that is, the inverse adaptive quantizer reconstructs the quantized difference signal, and the adaptive predictor forms a signal estimate based on the quantized difference signal and earlier samples of the reconstructed signal, which is also the sum of the current estimate and the quantized difference signal as shown in Fig. 2.13. Synchronous coding adjustment tries to correct for errors accumulating in ADPCM from tandem connections of ADPCM transcoders.

ADPCM is basically developed from PCM. It has good speech reproduction quality, comparable to PSTN quality, which therefore led to its adoption in CT2 and DECT.

Telephony Requirements

A general cordless telephone system would include an acoustic interface, i.e., microphone and speaker at the handset coupled to a digitizing compressor/decompressor analog to uniform PCM to ADPCM at 32 kb/s enabling a 2:1 increase in channel capacity as a bonus. This digital stream is processed to be transmitted over the air interface to the base station where the reverse happens, resulting in a linear or a digital stream to be transported over the land-based network. The transmission plans for specific systems are described in detail in Tuttlebee, 1995.

An important subject in telephony is the effect of network echoes [Weinstein, 1977]. Short delays are manageable even if an additional delay of, say, less than 15 μs is introduced by a cordless handset. Delays of a larger magnitude, in excess of 250 μs (such as satellite links [Madsen and Fague, 1993]), coupled to cordless systems can cause severe degradation in speech quality and transmission; a small delay introduced by the cordless link in the presence of strong network echoes is undesirable. The DECT standard actually specifies the requirement for network echo control. Additional material can be obtained from the relevant CCITT documents [CCITT, 1984–1985].

Modulation Method

The modulation method for DECT is Gaussian filtered frequency shift keying (GFSK) with a nominal deviation of 288 kHz [Madsen and Fague, 1993]. The *BT*, i.e., Gaussian filter bandwidth to bit ratio, is 0.5 and the bit rate is 1.152 Mb/s. Specification details can be obtained from the relevant ETSI documents listed in the reference section.

Digital transmission channels in the radio frequency bands, including the DECT systems, present serious problems of spectral schemes congestion and introduce severe adjacent/co-channel interference problems. There were several schemes employed to alleviate these problems: new allocations at high frequencies, use of frequency-reuse techniques, efficient source encoding, and spectrally efficient modulation techniques.

Any communication system is governed mainly by two criteria, transmitted power and channel bandwidth. These two variables have to be exploited in an optimum manner in order to achieve maximum bandwidth efficiency, defined as the ratio of data rate to channel bandwidth (units of bit/Hz/s) [Pasupathy, 1979]. GMSK/GFSK has the properties of constant envelope, relatively narrow bandwidth, and coherent detection capability. Minimum shift keying (MSK) can be generated directly from FM, i.e., the output power spectrum of MSK can be created by using a premodulation low-pass filter. To ensure that the output power spectrum is constant, the low-pass filter should have a narrow bandwidth and sharp cutoff, low overshoot, and the filter output should have a phase shift $\pi/2$, which is useful for coherent detection of MSK; see Fig. 2.14.

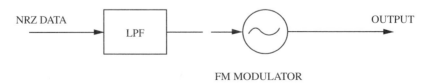

FIGURE 2.14   Premodulation baseband-filtered MSK.

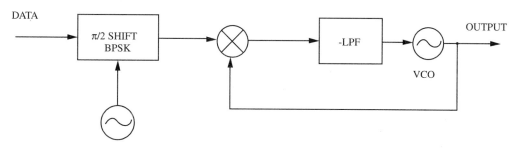

FIGURE 2.15   PLL-type GMSK modulator.

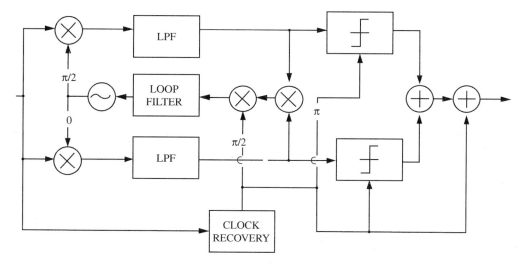

FIGURE 2.16   Costas Loop.

Properties of GMSK satisfy all of these characteristics. We replace the low-pass filter with a premodulation Gaussian low-pass filter [Murota and Hirade, 1981]. As shown in Fig. 2.15, it is relatively simple to modulate the frequency of the VCO directly by the baseband Gaussian pulse stream, however, the difficulty lies in keeping the center frequency within the allowable value. This becomes more apparent when analog techniques are employed for generating such signals. A possible solution to this problem in the analog domain would be to use a phase-lock loop (PLL) modulator with a precise transfer function. It is desirable these days to employ digital techniques, which are far more robust in meeting the requirements talked about earlier. This would suggest an orthogonal modulator with digital waveform generators [de Jager and Dekker, 1978].

The demodulator structure in a GMSK/GFSK system is centered around orthogonal coherent detection, the main issue being recovery of the reference carrier and timing. A typical method, is described in de Buda, 1972, where the reference carrier is recovered by dividing by four the sum of the two discrete frequencies contained in the frequency doubler output, and the timing is recovered directly from their difference. This method can also be considered to be equivalent to the Costas loop structure as shown in Fig. 2.16.

In the following are some theoretical and experimental representations of the modulation technique just described. Considerable literature is available on the subject of data and modulation schemes and the reader is advised to refer to Pasupathy (1979) and Murota and Hirade (1981) for further access to relevant study material.

Radio Frequency Architecture

We have discussed the need for low power consumption and low cost in designing cordless telephones. These days digital transmitter/single conversion receiver techniques are employed to provide highly

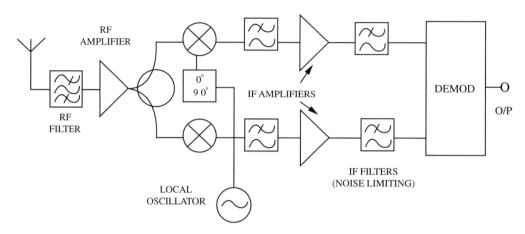

FIGURE 2.17   Direct conversion receiver architecture.

accurate quadrature modulation formats and quadrature downconversion schemes that allow a great deal of flexibility to the baseband section. Generally, one would have used digital signal processors to perform most of the demodulation functions at the cost of high current consumption. With the advent of application-specific signal processing, solutions with these techniques have become more attractive.

From a system perspective, range, multipath, and voice quality influence the design of a DECT phone. A high bit rate coupled with multipath reflections in an indoor environment makes DECT design a challenging task. The delay spread (multipath) can be anywhere in the 100 to 200 ns range, and a DECT bit time is 880 ns. Therefore, a potential delay spread due to multipath reflections is 1 to 20% of a bit time. Typically, antenna diversity is used to overcome such effects.

DECT employs a TDMA/TDD method for transmission, which simplifies the complexity of the radio frequency end. The transmitter is on for 380 ms or so. The receiver is also only on for a similar length of time.

A single conversion radio architecture requires fast synthesizer switching speed in order to transmit and receive on as many as 24 timeslots per frame. In this single conversion transmitter structure, the synthesizer has to make a large jump in frequency between transmitting and receiving, typically in the order of 110 MHz. For a DECT transceiver, the PLL synthesizer must have a wide tuning bandwidth at a high-frequency reference in addition to good noise performance and fast switching speed. The prescaler and PLL must consume as low a current as possible to preserve battery life.

In the receive mode the RF signal at the antenna is filtered with a low-loss antenna filter to reduce out-of-band interfering signals. This filter is also used on the transmit side to attenuate harmonics and reduce wideband noise. The signal is further filtered, shaped, and downconverted as shown in Fig. 2.17. The signal path really is no different from most receiver structures. The challenges lie in the implementation, and this area has become quite a competitive segment, especially in the semiconductor world.

The direct conversion receiver usually has an intermediate frequency nominally at zero frequency, hence the term zero IF. The effect of this is to fold the spectrum about zero frequency, which result in the signal occupying only one-half the bandwidth. The zero IF architecture possesses several advantages over the normal superheterodyne approach. First, selectivity requirements for the RF filter are greatly reduced due to the fact that the IF is at zero frequency and the image response is coincident with the wanted signal frequency. Second, the choice of zero frequency means that the bandwidth for the IF paths is only half the wanted signal bandwidth. Third, channel selectivity can be performed simply by a pair of low-bandwidth low-pass filter.

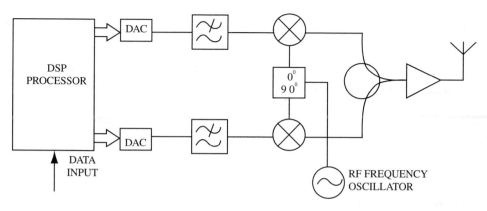

FIGURE 2.18   Transmit section.

For the twin IF chains of a direct conversion receiver, automatic gain control (AGC) is always required due the fact that each IF channel can vary between zero and the envelope peak at much lower rates than the highest signal bandwidth frequency. An additional requirement in newer systems is received signal strength indication (RSSI) to measure the signal or interference level on any given channel.

Moving on to the transmitter architecture (shown in Fig. 2.18 is a typical I-Q system), it is safe to say that the task of generating an RF signal is much simpler than receiving it. A transmitter consists of three main components: a final frequency generator, a modulator, and the power amplifier. These components can all be combined in common circuits, i.e., frequency synthesizer with inbuilt modulator. The problem of generating a carrier at a high frequency is largely one of frequency control. The main approach for accurately generating an output frequency from a crystal reference today is the PLL, and there is considerable literature available on the subject [Gardner, 1979]. In the modulation stage, depending upon the tightness of the phase accuracy specification of a cordless system, it may be necessary to apply tight control on the modulation index to ensure that the phase path of the signal jumps exactly in 90° increments.

Defining Terms

AGC: Automatic gain control.

ARQ: Automatic repeat request.

AWGN: Additive white Gaussian noise.

BABT: British approvals board for telecommunications.

Base Station: The fixed radio component of a cordless link. This may be single-channel (for domestic) or multichannel (for Telepoint and business).

BER: Bit error rate (or ratio).

CCITT: Comitè Consultatif International des Tèlègraphes et Tèlèphones, part of the ITU.

CEPT: Conference of European Posts and Telecommunications Administrations.

CPFSK: Continuous phase frequency shift keying.

CPP: Cordless portable part; the cordless telephone handset carried by he user.

CRC: Cyclic redundancy check.

CT2:   Second generation cordless telephone-digital.

D Channel: Control and information data channel (16 kb/s in ISDN).

DCT: Digital cordless telephone.

DECT: Digital European cordless telecommunications.

DLC: Data link control layer, protocol layer in DECT.

DSP: Digital signal processing.

DTMF: Dual tone multiple frequency (audio tone signalling system).

ETSI: European Telecommunications Standards Institute.

FDMA: Frequency division multiple access.

FSK: Frequency shift keying.

GMSK: Gaussian filtered minimum shift keying.

ISDN: Integrated services digital network.

ITU: International Telecommunications Union.

MPT 1375: U.K. standard for common air interface (CAI) digital cordless telephones.

MSK: Minimum shift keying.

PSK: Phase shift keying.

RES 3: Technical subcommittee, radio equipment and systems 3 of ETSI, responsible for the specification of DECT.

RSSI: Received signal strength indication.

SAW: Surface acoustic wave.

TDD: Time division duplex.

TDMA: Time division multiple access.

References

Cheer, A.P. .1985. Architectures for digitally implemented radios, IEE Colloquium on Digitally Implemented Radios, London.

Comité Consultatif International des Télégraphes et Téléphones. 1984, "32 kbits/sec Adaptive Differential Pulse Code Modulation (ADPCM)," CCITT Red Book, Fascicle III.3, Rec. G721.

Comité Consultatif International des Télégraphes et Téléphones, 1984–1985. *General Characteristics of International Telephone Connections and Circuits*, CCITT Red Book, Vol. 3, Fascicle III.1, Rec. G101–G181.

de Buda, R. 1972. Coherent demodulation of frequency shifting with low deviation ratio. *IEEE Trans.* COM-20 (June):466–470.

de Jager, F. and Dekker, C.B. 1978. Tamed frequency modulation. A novel method to achieve spectrum economy in digital transmission, *IEEE Trans. in Comm.* COM-20 (May):534–542.

Dijkstra, S. and Owen, F. 1994. The case for DECT, *Mobile Comms. Int.* 60–65.

European Telecommunications Standards Inst. 1992. RES-3 DECT Ref. Doc. ETS 300 175-1 (Overview). Oct. ETSI Secretariat, Sophia Antipolis Ceder, France.

European Telecommunications Standards Inst. 1992. RES-3 DECT Ref. Doc. ETS 300 175-2 (Physical Layer) Oct. ETSI Secretariat, Sophia Antipolis Ceder, France.

European Telecommunications Standards Inst. 1992. RES-3 DECT Ref. Doc. ETS 300 175-3 (MAC Layer) Oct. ETSI Secretariat, Sophia Antipolis Ceder, France.

European Telecommunications Standards Inst. 1992. RES-3 DECT Ref. Doc. ETS 300 175-4 (Data Link Control Layer) Oct. ETSI Secretariat, Sophia Antipolis Ceder, France.

European Telecommunications Standards Inst. 1992. RES-3 DECT Ref. Doc. ETS 300 175-5 (Network Layer) Oct. ETSI Secretariat, Sophia Antipolis Ceder, France.

European Telecommunications Standards Inst. 1992. RES-3 DECT Ref. Doc. ETS 300 175-6 (Identities and Addressing) Oct. ETSI Secretariat, Sophia Antipolis Ceder, France.

European Telecommunications Standards Inst. 1992. RES-3 DECT Ref. Doc. ETS 300 175-7 (Security Features) Oct. ETSI Secretariat, Sophia Antipolis Ceder, France.

European Telecommunications Standards Inst. 1992. RES-3 DECT Ref. Doc. ETS 300 175-8 (Speech Coding & Transmission) Oct. ETSI Secretariat, Sophia Antipolis Ceder, France.

European Telecommunications Standards Inst. 1992. RES-3 DECT Ref. Doc. ETS 300 175-9 (Public Access Profile) Oct. ETSI Secretariat, Sophia Antipolis Ceder, France.

European Telecommunications Standards Inst. 1992. RES-3 DECT Ref. Doc. ETS 300 176 (Approval Test Spec) Oct. ETSI Secretariat, Sophia Antipolis Ceder, France.

Gardner, F.M. 1979. *Phase Lock Techniques*, Wiley-Interscience, New York.

Madsen, B. and Fague, D. 1993. Radios for the future: designing for DECT, *RF Design*. (April):48–54.

Murota, K. and Hirade, K. 1981. GMSK modulation for digital mobile telephony, *IEEE Trans*. COM-29(7):1044–1050.

Olander, P. 1994. DECT a powerful standard for multiple applications, *Mobile Comms. Int*. 14–16.

O'Shaughnessy, D. 1987. *Speech Communication*, Addison-Wesley, Reading, MA.

Pasupathy, S. 1979. Minimum shift keying: spectrally efficient modulation, *IEEE Comm. Soc. Mag*. 17(4):14–22.

Tuttlebee, W.H.W., ed., 1995. *Cordless Telecommunications Worldwide*, Springer-Verlag, Berlin.

Weinstein, S.B. 1977. Echo cancellation in the telephone network, *IEEE Comm. Soc. Mag*. 15(1):9–15.

2.5   Wireless Local Area Networks (WLAN)

Jim Paviol, Carl Andren, and John Fakatselis

Wireless Local Area Networks (WLANs) use radio transmissions to substitute for the traditional coaxial cables used with wired LANs like Ethernet. The first generation WLAN products were targeted as wired-LAN extensions. They were originally intended to save money on relocation expenses and demonstrated the utility of wireless laptop operations. Wireless data technology and its application to local-area networks introduced mobile computing. Centrally controlled wireless networks are most often used as part of a larger wired network. A radio base station or Access Point (AP) arbitrates access to the remote wireless stations by means of packetized data. In contrast, in a Peer-to-Peer wireless network, an ad hoc network can be formed at will by a group of wireless stations. New forms of network access protocols, such as Carrier Sense Multiple Access/Collision Avoidance (CSMA/CA) are needed for low error rate operation of wireless networks. Roaming is one of the main advantages of wireless networks, allowing users to freely move about while maintaining connectivity.

The WLAN is used in four major market segments, "Vertical" with factory, warehouse, and retail uses; "Enterprise" with corporate infrastructure mobile Internet uses; "SOHO" (Small Office/Home Office) with small rented space businesses; and "Consumer" with emerging uses. General WLAN trends are shown in Fig. 2.19.

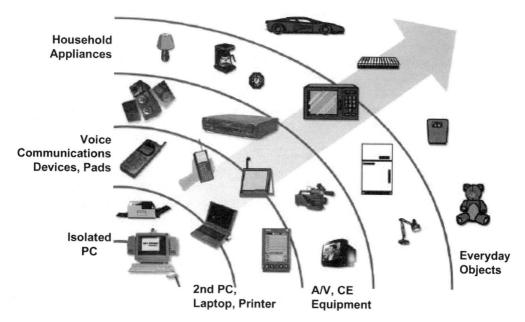

FIGURE 2.19   Commercial WLAN evolution.

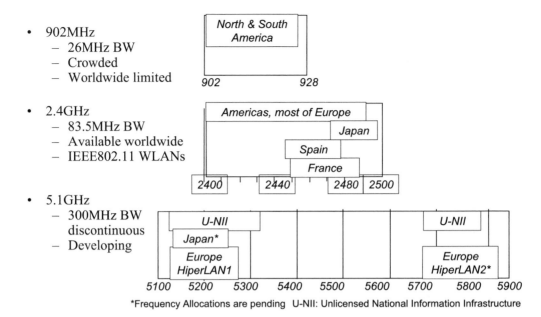

- 902MHz
  - 26MHz BW
  - Crowded
  - Worldwide limited

- 2.4GHz
  - 83.5MHz BW
  - Available worldwide
  - IEEE802.11 WLANs

- 5.1GHz
  - 300MHz BW discontinuous
  - Developing

*Frequency Allocations are pending   U-NII: Unlicensed National Information Infrastructure

FIGURE 2.20   Unlicensed ISM RF band details.

WLAN RF ISM Bands

In 1985 the Federal Communications Commission (FCC) in the United States defined the ISM (Industrial, Scientific, and Medical) frequency bands allowing unlicensed spread-spectrum communications. Three of the ISM bands are illustrated in Fig. 2.20 with frequencies at 900 MHz, 2.4 GHz, and 5 GHz.

Most important to ISM band WLAN users is that no license for operation is required when the signal transmission is per the guidelines specified by the FCC or other regulatory agencies. Spread-spectrum technology in the ISM band is used to minimize interference and offers a degree of interference immunity from other jamming signals or noise. Other non-spread commercial applications have existed in the ISM bands for many years such as microwave ovens at 2.4 GHz. This is a major potential interference to WLAN in the 2.4 GHz ISM band and has been accounted for in the system design.

The IEEE 802.11 committee selected the 2.4 GHz ISM band for the first WLAN global standard. Unlike the 900 MHz band, 2.4 GHz is available worldwide; 2.4 GHz also has more available bandwidth than the 900MHz band, and will support higher data rates and multiple adjacent channels in the band. In comparison with the 5.7 GHz band, it offers a good balance of equipment performance and cost. Increasing the transmit frequency impacts the power dissipation, availability of parts/processes and limits the indoor range. The 2.4 GHz band is ideal for a WLAN high-speed, unlicensed data link.

The 900 MHz band has been in use for some time, and component prices are very reasonable. Many cordless phones use this band. The 900 MHz band is quite crowded and it does not have global spectrum allocation. The 2.4 GHz band is less crowded, has global allocations, and the associated technology is very cost effective. This is the band in which IEEE 802.11b, Bluetooth, and HomeRF operate.

The 2.4 GHz band is most heavily used for WLAN. Operating channels are shown in Fig. 2.21.

The 5 GHz band has two 100 MHz segments for unlicensed use collectively known as the Unlicensed National Information Infrastructure (UNII) bands. There is a similar allocation in Europe, but it is currently reserved for devices that operate in compliance with the HIPERLAN standard. 5 GHz components are more expensive, and radio propagation has higher losses and more severe multipath at these frequencies. These impairments can be overcome, and systems offering data rates in excess of 50 Mbps in the 5 GHz band will become mainstream in the next several years.

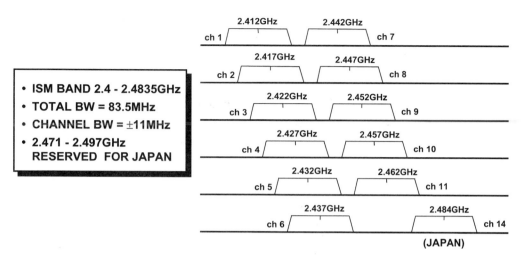

FIGURE 2.21 Operating channels for direct sequence.

WLAN Standardization at 2.4 GHz: IEEE 802.11b

Wired LAN standards were developed by the IEEE 802 committee such as IEEE 802.1 Systems Management/Networking, IEEE 802.3 Ethernet, IEEE 802.4 Token Ring, and IEEE 802.6 Metropolitan Area Networks. In 1990, the IEEE 802.11 Wireless LAN Working Group was formed and has in excess of 100 active voting members with global representation. It ratified the 802.11b high rate 2.4 GHz WLAN standard in 1999. The 802.11 standard fostered development of interoperable, inexpensive, and flexible equipment in the 2.4 GHz ISM band. Specified data rates for the IEEE 802.11 2.4 GHz WLAN are 1 Mbps, 2 Mbps, 5.5 Mbps, and 11 Mbps. Spread-spectrum technology is specified in the 802.11 standard transceiver to provide a robust solution in a multi-user environment. One advantage to spread-spectrum techniques in the ISM bands are seen in the allowable transmit power levels. System transmitter power for IEEE 802.11 WLANs must conform to a regulatory agency's specified levels for unlicensed operation. As an example, the FCC states that non-spread-spectrum applications in this band are limited to a 50 mV/m at 3 meters. This translates into 0.7 mW into a dipole antenna. Spread-spectrum applications in the U.S. are allowed up to 1 watt of transmit power, clearly giving it a higher signal strength advantage over non-spread systems. The low spectral power density of a spread-spectrum system also limits interference to other in-band users.

Segmentation of a data communication system into layers allows different approaches from various vendors as long as the responsibilities of the individual layer are met. The IEEE 802.11 specification focuses on the Media Access Control (MAC) and Physical (PHY) layers for WLANs.

The MAC layer controls the protocol and physical layer management. The protocol used for IEEE 802.11 WLANs is the CSMA/CA (Carrier Sense Multiple Access/Collision Avoidance).

The Physical Layer controls the wireless transmission and reception of digital data from the MAC. It is the transceiver or radio for the WLAN. IEEE 802.11 specifies three different physical layer options, Direct Sequence Spread Spectrum (DSSS), Frequency Hopping Spread Spectrum (FHSS), and Diffused Infrared (DFIR). The DFIR method has the shortest range of operation and is limited to indoor operation due to interference from sunlight. DSSS and FHSS are RF technologies that must conform to the standards set by the regulatory agencies of various countries such as the FCC. These impact items such as the allowable bandwidth and transmit power levels.

Another feature of IEEE 802.11 is that the data is packetized. Packetized data is a fixed number of data bytes sent in a single radio transmission of finite length. The data is grouped in frames up to 2304 bytes in length. A common data length is 1500 bytes. A header and preamble are attached in front of the data frame for control information. The preamble is the initial sequence at the start of the radio transmission that allows the demodulator to synchronize its timing and recognize key information concerning the data that follows. Short and long preambles exist. These are specified in greater detail within the standard. The packetization supports the CSMA/CA protocol.

Frequency Hopped (FH) vs. Direct Sequence Spread Spectrum (DSSS)

FH uses a form of FSK modulation called GFSK (Gaussian Frequency Shift Keying). The baseline 1 Mbps data rate for FH IEEE 802.11 has a 2-level GFSK modulation scheme. The symbol {1} is the center carrier frequency plus a peak deviation of (f+), whereas the symbol {o} is the center carrier frequency minus a peak deviation of (f–). The carrier frequency hops every 400 ms over the channel bandwidth per a prescribed periodic PN code. This channel is divided into 79 sub-bands. Each sub-band has a 1 MHz bandwidth. The minimum hop rate of 2.5 hops/s allows several complete data packets or frames to be sent at one carrier frequency before a hop.

DSSS in IEEE 802.11 specifies a DBPSK and DQPSK (D = Differential) modulation for 1 and 2 Mbps data rates. Differential techniques use the received signal itself to demodulate the signal by delaying one symbol period to obtain clock information. In DBPSK, a logic 1-bit input initiates a 180-degree phase change in the carrier and a ϕ-bit initiates no phase change in the carrier. DQPSK has a 0-, 90-, or 180-degree phase transition on each symbol.

The carrier frequency in an IEEE 802.11 DSSS transmitter is spread by an 11-bit Barker code. The chipping rate is 11 MHz for a 1 Mbit data rate. This yields a processing gain of 11. The main lobe spacing is twice the chip rate and each side lobe is the chip rate as shown in Fig. 2.22. The DSSS receiver will filter the side lobes, downconvert the main lobe spectral component to baseband, and use a copy of the PN code in a correlator circuit to recover the transmitted signal. The FH scheme has been limited to 1 Mbit and 2 Mbit by technical and regulatory issues. This is expected to limit the use of FH systems compared with the more versatile high-rate DSSS modulation in future applications. An illustration contrasting frequency hopping to spread spectrum is shown in Fig. 2.22.

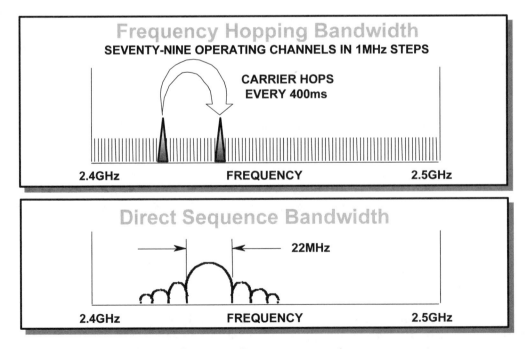

FIGURE 2.22   IEEE 802.11 frequency hopping vs. direct sequence spread spectrum.

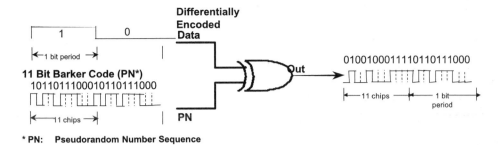

FIGURE 2.23   Direct sequence spectrum spreading. RF energy is spread by XOR of data with PN sequence.

Direct Sequence Spread Spectrum (DSSS) Energy Spreading

In Direct Sequence Spread Spectrum (DSSS) systems the spreading of the data is achieved by modulating the data with a Pseudo-random Number (PN) sequence of binary values called a PN code. If the PN code has a bandwidth much higher than the bandwidth of the data (approximately ×10 or greater), then the bandwidth of the modulated signal will have a spectrum that is nearly the same as a wideband PN signal. An 11 bit Barker Code is used with the IEEE 802.11 DSSS PN spreading. A Barker code was chosen for its unique short code properties.

By multiplying the information-bearing signal by the PN signal, each information bit is chopped up into a number of small time increments called "chips," as illustrated in the waveform diagram shown in Fig. 2.23. The rate at which the PN code is clocked to spread the data is called the chip rate. The PN sequence is a periodic binary sequence with a noise-like waveform. The acquisition of the data in the receiver is achieved by correlation of the received signal with the same PN code that was used to spread the signal at the transmitter. In DSSS systems the data is primarily PSK modulated before spreading. The spreading produces a "Processing Gain" dependent upon the PN spreading code to symbol rate ratio. This value is a minimum of 10 dB for the IEEE 802.11 DSSS WLAN waveform. The spectrum after this PN code is used is wider and lower in signal level as illustrated in Fig. 2.24.

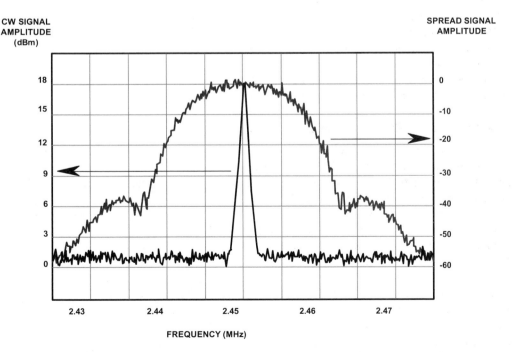

FIGURE 2.24   Direct sequence spread spectrum.

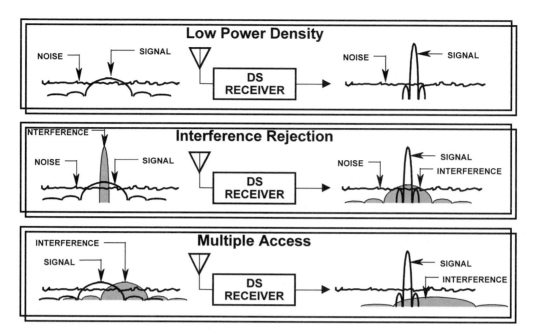

FIGURE 2.25   Direct sequence spread spectrum properties.

The primary advantage of a spread-spectrum system is its ability to reject interference whether it be the unintentional interference by another user transmitting on the same channel, or the intentional interference by a hostile transmitter attempting to jam the transmission. Due to its spread characteristic, a DSSS signal appears as noise to all receivers except the one meant to receive the signal. The intended receiver is able to recover the spread signal by means of correlation, which simultaneously recovers the signal of interest and suppresses the background noise by the amount of processing gain.

The reception of the DSSS signal in the presence of a narrow band interferer is accomplished by de-spreading the signal of interest while spreading the energy of the interfering signal by the amount equal to the processing gain. The result of the de-spreading process with an interference source is illustrated in Fig. 2.25 for narrow band interfering signals. Likewise, recovery of the signal of interest in presence of another spread signal having a different PN code is achieved by further spreading the interference while de-spreading the signal with the matching PN code. The worst case interference to DSSS systems is a narrow band interference signal in the middle of the spectrum of the spread signal.

Modulation Techniques and Data Rates

The 1 Mbit data rate is formed using Binary Phase Shift Keying (BPSK) and the 2 Mbit data rate uses Quadrature Phase Shift Keying (QPSK). QPSK doubles the data rate by increasing the number of bits per symbol from one (BPSK) to two (QPSK) within the same bandwidth. Both rates use BPSK for the acquisition header known as the preamble.

The high rate modulation for 5.5 and 11 mbps uses a form of M-Ary Orthogonal keying called Complementary Code Keying (CCK). CCK provides coding gain by the use of modified Walsh functions applied to the data stream. Two forms of this are used to provide multiple rates for stressed links. Altogether, four data rates are provided. To make 11 Mbit CCK modulation, the input is formed into bytes and then subgrouped into 2 bits and 6 bits. The 6 bits are used to pick one of 64 complex vectors of 8-chip length and the other 2-bits DQPSK modulate the whole symbol vector. For 5.5 Mbit CCK mode, the incoming data is grouped into 4 bit nibbles where 2 of those bits select the spreading function out of a set of 4 (the 4 having the greatest distance of the 11 Mbit set) while the remaining 2 bits set the QPSK polarity of the symbol. The spreading sequence modulates the carrier by driving the I and Q modulators. Figure 2.26 illustrates modulation at the four data rates of 1, 2, 5.5, and 11 Mbit.

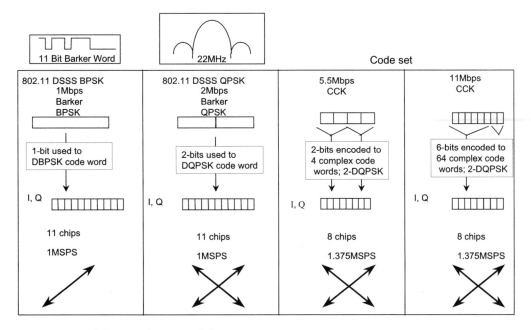

FIGURE 2.26 Modulation techniques and data rates.

Carrier Sense Multiple Access/Collision Avoidance (CSMA/CA)

The wireless environment offers a greater challenge to the WLAN designer when compared with the wired LAN environment. Wired LANs use a CSMA/CD (Carrier Sense Multiple Access/Collision Detect) protocol. Data collisions in a wired environment produce a unique voltage that can be monitored. This allows for random back-off time periods before the system initiates a data resend.

IEEE 802.11 WLANs use the CSMA/CA (Collision Avoidance) protocol. This protocol avoids data collisions by having the system listen to the channel and wait before sending a message. With CSMA/CA, only one node may talk at a time. This highlights the importance of performing a Clear Channel Assessment (CCA) to determine if the medium or channel is clear to transmit. Although the MAC controls the CSMA/CA protocol, the responsibility falls upon the physical layer to perform CCA.

IEEE 802.11 states that the physical layer must be able to provide at least one of three specified methods for CCA. CCA mode 1 simply detects energy above a programmable level. If no signal is present, the channel is clear to transmit. If the signal is present, the system will wait a set time period to check the channel again. CCA mode 2 provides the carrier sense function. Since this is a spread spectrum system, correlating the received signal with the 11-bit PN code performs carrier sense. No correlation indicates that the channel is clear to transmit. Correlation with a signal shows that the channel is busy and that the system will back off for a time period and try again. CCA mode 3 combines modes 1 and 2 by reporting a busy medium with both detection of energy and carrier sense. Figure 2.27 illustrates a four-station scenario where radio collisions are avoided using CSMA-CA.

Packet Data Frames in DSSS

The 802.11 WLAN standard specifies data packetization. This means that the data is segmented into frames with a preamble and header attached at the start of each frame. The preamble allows carrier and correlation lock as well as user identification. The header contains management and control information for the data transmission.

The sync field is made up of 128 one bits. Note that all bits are processed by a scrambler function, which is part of the IEEE 802.11 spread spectrum physical layer so this original pattern of ones will be altered. The purpose of the sync field is to allow the receiver to lock on to the signal and to correlate to the PN code.

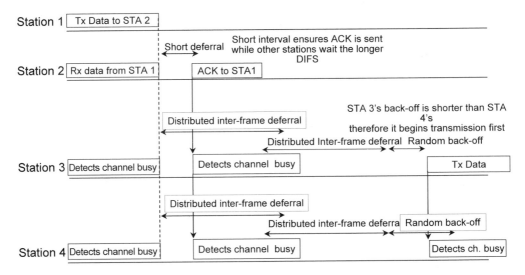

FIGURE 2.27   Carrier sense multiple access collision avoidance (CSMA/CA).

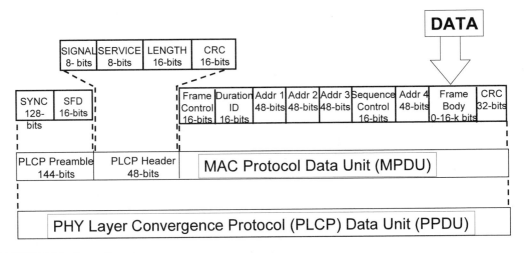

FIGURE 2.28   Frame format.

The start frame delimiter (SFD) field initiates the start of the data frame. The signal field indicates the data packet data wave. Note that the preamble and header are always transmitted as a DBPSK waveform. The length field defines the data packet size with a maximum length of 2304 bytes. Finally, one CRC (Cyclic Redundancy Check) protects the signal, service, and length fields with a frame check sequence and another protects the payload.

At the end of the data packet, the receiver will send an acknowledgment (ACK) indicating successfully transmitted data. If a data packet were lost either by multipath fading or interference, the sender would retransmit the packet. Figure 2.28 details frame format, preamble, header, and data.

IEEE 802.11 Network Modes

There are two network modes: Ad Hoc and Infrastructure. Ad Hoc mode permits users within range to set up a network among them without any infrastructure. Infrastructure mode uses an Access Point (AP) to coordinate the users and allow access to the wired network services.

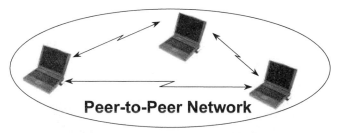

Peer-to-Peer Network

Basic Service Set (BSS)
Distributed Coordination Function (DCF)

FIGURE 2.29   IEEE 802.11 ad hoc mode.

Ad Hoc Mode

The Distributed Coordination Function (DCF) forms the basis to implement Ad Hoc networking. The provisions in the standard allow the creation and dissolution of an Ad Hoc network to be straightforward for users to set up. The CSMA/CA medium access method provides for fair access to the radio channel among all of the users. All data exchanges are directly between the individual stations.

When a single station is designated as the coordination function, the network is known as a Basic Service Set (BSS). This is illustrated in the peer-to-peer AdHoc network in Fig. 2.29.

Infrastructure Mode

An Access Point (AP) is a device that connects the wireless stations to the distribution system in a network. It typically will be configured to be the single Coordination Function in a BSS. The network planning for a large installation involves site surveys to do the cellular radio planning needed to determine the number and location of Access Points. The channel assignments for each AP can be optimized to reduce interference from adjacent cells depending on the physical layout. Many times, the installation will utilize a wired Ethernet network to connect a number of APs to the network server. Each AP will manage the traffic within its BSS. Stations in adjacent cells will recognize when the packet is not intended for its BSS. The same DCF mechanisms provided in the ad hoc mode help to manage radio interference from adjacent cells. Mechanisms such as Authentication, Association, and Wired Equivalency Privacy (WEP) provide security for the overall network. The authentication process is used to verify the identity of stations that are allowed access to the network. As users move between the various radios cells and they can no longer communicate with their AP, they can scan the channels to look for a new AP to associate with. With proper radio cell-planning, users can be assured of constant coverage as they roam through the facility. An infrastructure mode illustration is shown in Fig. 2.30.

"Hidden" Nodes

The planning for a WLAN ranges from none for an ad hoc IBSS, to careful radio surveys for AP-based infrastructure systems in large enterprise computing environments. It is impossible to plan for perfect radio coverage given the uncertain and time-varying conditions in the channel. Movement and location as well as indoor topology will affect radio coverage. In providing complete coverage of the facility there will be situations where there are two stations that can communicate with their common AP, but are out of range to hear each other. This is known as the Hidden Node problem. This could cause additional collisions to occur at the AP, since when one station is transmitting, the other will determine the channel is clear because it is out of range. After the normal deferral time it will transmit and will interfere with a packet from the other station. The standard provides an optional mechanism called Virtual Carrier Sense (VCS) to reduce the collision due to the hidden node problem. This mode is based on using a Request To Send (RTS) and Clear To Send (CTS) exchange between the station and the AP. When the station has determined it is clear to transmit, it will send a short request to send a message with a

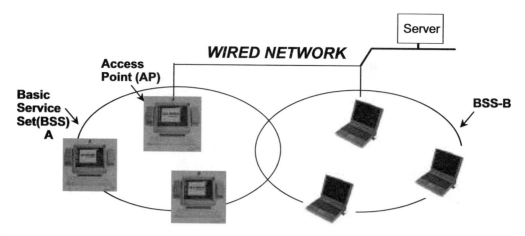

FIGURE 2.30   IEEE 802.11 infrastructure mode.

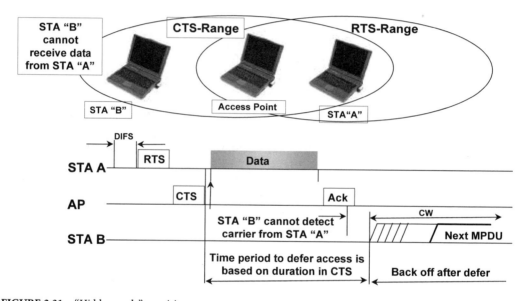

FIGURE 2.31   "Hidden node" provisions.

transaction duration included. The AP will respond with a Clear To Send (CTS) that also includes the transaction duration. The other station that is hidden cannot hear the RTS, but can hear the CTS. It will read the duration field and not begin looking for a clear channel until that time has elapsed. Hidden nodes are illustrated in Fig. 2.31 showing Stations A and B with an access point.

Point Coordination Function (PCF)

The Point Coordination Function (PCF) is an optional mode that can be selected in the installation of an 802.11 Network. This mode is provided to optimize the network throughput. Rather than having each station contend for the channel using the CCA and the random back-off periods, the AP defines contention-free periods using a beacon frame sent at a regular interval. The Network Allocation Vector (NAV) is a variable that is transmitted in the control frames to tell all of the stations the duration to defer from accessing the channel. It is used to define the length of the contention-free period. The AP will then poll each station during the contention-free period. The poll will send data if there is some waiting for transmission to that station and request data from the station. As each station is polled, it will acknowledge reception and will include data if there is some pending transmission to the AP. One of the benefits of

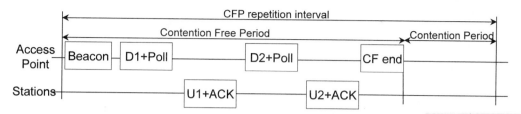

FIGURE 2.32 IEEE 802.11 point coordination function (PCF).

the PCF mode is that it allows the network planner to improve the probability of the delivery of data in a certain time bound. This is critical for real time data such as voice, audio, video, and multimedia. Minimizing the uncertainties of when a station can get the channel that exists in the DCF mode optimizes the system performance. If a packet of audio or video data has to be retransmitted due to lost packets and extended deferral times, the quality of the audio or video will suffer. The network can be optimized to permit a certain amount of retries within the contention-free period to improve the quality of service to the end application. The Point Coordination Function timing is illustrated in Fig. 2.32.

WLAN Security

Since they can be received outside of the controlled facility, wireless networks are more vulnerable to interference and theft than wired networks. The data security field known as cryptography is rapidly growing as more systems convert to wireless operation. Spread spectrum offers a little security by spreading the signal over a wide bandwidth.

There are a number of mechanisms provided in 802.11 to minimize the chances of someone either logging on the network without authorization or receiving and using data received off the air from authorized users. As an option the data can be encrypted to prevent someone who is receiving packets from the network to be able to interpret the data. The keys for the security are distributed to the users in the network by a secure key management procedure. Without the key the snooper will have to resort to complex code-breaking techniques to retrieve the original data. The level of encryption is defined as Wired Equivalency Privacy (WEP). It is strong enough to require effort equivalent to that required to get data from a wired LAN. This algorithm is licensed from RSA Data Security. The encryption mechanism is used in the authentication process as well.

Data Encryption

The WLAN Security: Authentication diagram illustrates a technique for authenticating the identity using the encryption features. In the "Challenge and Response Protocol" shown, the station transmits a "challenge" random message(r) to the AP. The AP receives the message, encrypts it by using a network algorithm (fK1), and transmits the encrypted information (y) "response" (fK1) back to A. System A has access to the network algorithm (Y) and compares it to the received response. If $y = y'$, the identification procedure criteria has been met and data transmission will follow. It does not, then A will issue a new challenge with a different random message to B. Note that A and B share a common (private) key, k1. After the authentication process is successfully completed the station will then associate itself with an AP. The association tells the overall network which AP services any station. Once identification has occurred, both systems communicate using network encryption algorithms. In this case, the station encrypts data (x) and transmits a cipher (y) over the channel. The AP or another station has access to the encryption method and decodes the cipher to obtain the data. Because the stations in a BSS share a common key (k1), this type of encryption is called private-key cryptography.

Also included in the standard is an Integrity Check Vector (ICV). This is a variable added to an encrypted data packet as an additional error-detection mechanism. After the decryption process the transmitted ICV is compared with the ICV calculated from the plain text as an error check. Figure 2.33 illustrates data encryption with a block diagram.

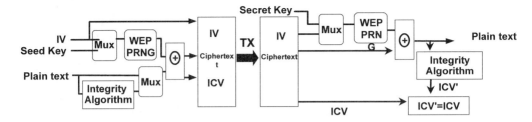

FIGURE 2.33   WLAN security: data encryption.

5 GHz WLAN

With the FCC rule and order establishing the Unlicensed National Information Infrastructure (U-NII), 300 more MHz of bandwidth were made available for WLAN users. The purpose was to expand the access of people to information without a lot of infrastructure build out. One of the frequently sited scenarios is the distribution of Internet access to schools without having to wire classrooms. The FCC chose to put in a minimum of specifications for the waveforms to be used in the band.

Spread Spectrum is not a requirement. Transmit power and power spectral density are the primary specifications. No channelization or spectrum-sharing rules were included in the 15-part regulation. It remains to be seen how the various devices and standards will coexist in this new band. The band is split into three 100 MHz bands for defining maximum output power and spurious emissions levels. There are users of licensed bands on all sides of these new bands that argued for protection against interference from unlicensed devices. This drives the channel band edges for carriers and defines radio requirements for suppression of spurious emissions. The power levels allowed are 50 mW, 250 mW, and 1 W in the lowest, middle, and upper bands, respectively. A change was made in 1998 to permit higher antenna gain with a reduction in output power. This extends the range for point-to-point links.

In Europe 150 MHz of bandwidth is set aside for HIPERLAN1 devices. These are WLAN devices with the same functional requirements as an 802.11a WLAN. As opposed to the FCC, the ETSI regulation requires all devices to meet the HIPERLAN standard for the PHY and MAC layers. The maximum data rate is 54 Mbit and the modulation is GMSK. Equalization is required for reliable operation. A small number of HIPERLAN products have been introduced since the standard was approved.

ETSI is defining a standard for the 5 GHz band that is oriented to wireless ATM traffic. This is suitable for Quality of Service applications and for wireless connectivity to an ATM system. The final spectral allocation has not been made for these devices.

Japan has started the development of standards for WLAN devices in the 5 GHz band. They have decided that devices will be required to use the same physical layer implementation that is defined in the IEEE 802.11a standard.

RF Link Considerations

The radio link performance can be characterized as consisting of radio design variables and link variables. When all the variables are understood the link performance can be determined. The most important WLAN link performance measure is the Packet Error Rate (PER), which is usually expressed as a percentage. Most radio link parameters are given at a packet error rate of 0.1 or 10 percent. This is the highest practical acceptable packet error rate and provides the design maximum. The IEEE 802.11 standard specifies an 8% PER for measurements.

All of the radio design variables combine to provide a given performance on a given link. The transmit power together with the receiver noise floor determines the ultimate range of the radio. Higher transmit power provides greater range but also higher battery drain. The antennas for wireless systems are generally omni-directional to allow mobility. The higher data rate requires a higher signal-to-noise ratio for the same error rate. The bit error rate and the length of the packet determine the packet error rate. Longer packets require lower bit error rates. Different modulation schemes require more or less power to achieve the same bit error rate. For instance FSK requires more power to achieve the same bit error rate as CCK. The link variables include the range, which is the distance from the transmitter to the receiver, the multipath environment, and the interference environment. Missed packets and corrupted (including recoverable) packet data are included in PER.

Radio signals radiated by an ideal isotropic antenna weaken with the square of the distance as they travel through free space (square power law). The attenuation also increases with frequency. At 2.4 GHz (Lambda = 0.125 m = 5") the path loss in free space is about 40dB for 1m. Propagation of RF signals in the presence of obstacles are governed by three phenomena: reflection, diffraction, and scattering. Reflection occurs when the dimension of obstacles are large compared to the wavelength of the radio wave. Diffraction occurs when obstacles are impenetrable by the radio wave. Based on Huygen's principle, secondary waves are formed behind the obstructing body even though there is no line of sight. Scattering occurs where the obstacles have dimensions that are on the order of the wavelength. The three propagation mechanisms all have impact on the instantaneous received signal in all different directions from the transmitting antenna.

Measurements have shown that propagation loss between floors does not increase linearly (in dB) with increasing separation of floors. Rather, the propagation loss between floors starts to diminish with increasing separation of floors. This phenomenon is thought to be caused by diffraction of radio waves along the side of a building as the radio waves penetrate the building's windows. Values for wall and door attenuation are shown in Table 2.9 and a plot of attenuation between building floors is shown in Fig. 2.34.

TABLE 2.9   Indoor Propagation Path Loss

2.4 GHz signal attenuation through:	
Window in brick wall	2 dB
Metal frame, glass wall into building	6 dB
Office wall	6 dB
Metal door in office wall	6 dB
Cinder wall	4 dB
Metal door in brick wall	12.4 dB
Brick wall next to metal door	3 dB

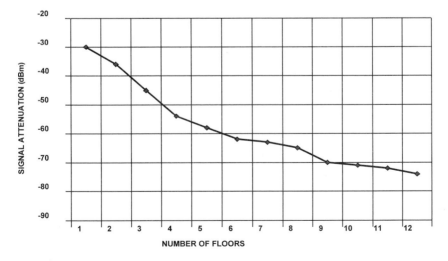

FIGURE 2.34   Path loss between building floors.

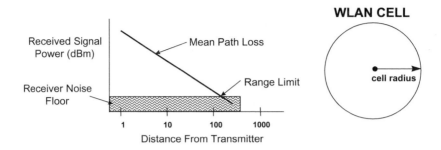

$$\text{Loss dB} = 40.2 + 10 \cdot \log (r^n)$$

- Where Loss is transmit power/received power in dB
  - r is the cell radius
  - n is 2 for free space

FIGURE 2.35 Path loss — free space. Loss dB = 40.2 + 10 • log (r^n) where Loss is transmit power/received power in dB; r is the cell radius; and n is 2 for free space.

WLAN Power, Sensitivity, and Range

The range of a WLAN radio is influenced by data rate, bit error rate requirements, modulating waveform S/N ratio, power amplification, receiver sensitivity, and antenna gain. At higher data rates more power is required. The log of the data increase is the amount of power increase in dB to achieve the same range. Reliability is also a function of power. Higher power produces a lower bit error at the same range, or more range at the same bit error rate. The waveform is a significant contributor to performance. Phase Shift Keying (PSK) type waveforms are the most power efficient. Frequency Shift Keying (FSK) requires almost twice as much transmitted power to achieve the same range.

IEEE 802.11 sensitivity for 11 Mbit CCK QPSK is specified as a minimum of –76 dBm. Typical radios attain 6 dB better than the minimum levels. Lower data rates such as 5.5 Mbit CCK and 2 Mbit QPSK have better sensitivities by approximately 5 dB than the 11 Mbit levels. The best sensitivity is obtained with the 1 Mbit BPSK. Values are typically 3 dB better than the 2 Mbit, QPSK thus 1 Mbit is used for the header/preamble information.

In an ideal propagation environment such as free space (i.e., no reflectors), the transmitted power reaches the receiver some distance away attenuated as a function of the distance, r. As shown in Fig. 2.35, the Constant, 40.2, is used for 2.4 GHz propagation and changes insignificantly across the ISM band. The exponent of 2 is for free space and increases with multipath. As the receiver moves away from the transmitter, the received signal power reduces until it dips into the receiver noise floor, at which time the error rate becomes unacceptable. This is the first order determination of the largest a cell can be. This model can be used for unobstructed line-of-sight propagation with highly directional antennas where the antenna gain allows a propagation path that is miles long.

Signal Fading and Multipath

As a transmitted radio wave undergoes reflection, diffraction, and scattering it reaches the receiving antenna via more than one path giving rise to a phenomenon called multipath. The multiple paths of the received signals cause them to have varying signal strengths as well as having different time delays (phase shifts), also known as delay spread. These signals are summed together (vector addition) by the receiving antenna according to their random instantaneous phase and strength giving rise to what is known as small-scale fading. Small-scale fading is a spatial phenomenon that manifests itself in the time domain having Rayleigh distribution; hence it is called Rayleigh fading. Small-scale fading produces

- **REFLECTION, DIFFRACTION AND SCATTERING CAUSE MULTIPATH**
- **MULTIPATH SMALL SCALE FADING**
- **MULTIPATH DELAY SPREAD**
- **RAYLEIGH AND RICIAN FADING**

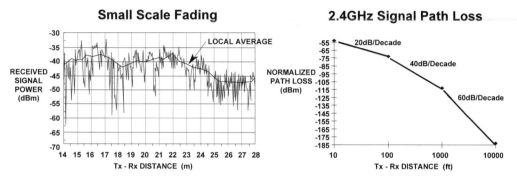

FIGURE 2.36 Signal fading and multipath. Reflection, diffraction, and scattering cause multipath; multipath small scale fading; multipath delay spread; Rayleigh and Rician fading.

instantaneous power levels that may vary as much as 30 or 40dB while the local average signal level changes much more slowly with distance.

Just as the power law relationship between distance and received power is applied to path loss in free space, it may be used in the presence of obstacles. A general propagation loss model for local average received power uses a parameter, n, to denote the power law relationship where n = 2 for free space, and is generally higher for indoor wireless channels.

The "2.4 GHz Signal Path Loss" curve in Fig. 2.36 represents various path losses with different values of n. The first segment, n = 2, of the curve, loss is primarily free space loss. The second and last segments of the curve have values of 4 and 6 for n, respectively, representing more lossy channels. The instantaneous drop of the signal power as it transitions from –40 dB/dec to –60 dB/dec is typical of a signal loss when a receiver loses line-of-sight to its respective transmitter.

In a multipath condition where the receiver also has a line-of-sight path to the transmitter, the statistical distribution of the local average signal level follows Rician distribution. Rician distribution is based on a factor, k, which specifies the ratio of direct path versus multipath power levels. Multipath is illustrated in Fig. 2.36.

Log Normal and Rayleigh Fading

The mechanism of the multipath fading can be viewed as being caused by two separate factors: the product of the reflection coefficients and the summation of the signal paths. These two mechanisms produce separate fading characteristics and can be described by their probability distribution functions. The first is characterized as having a Log Normal distribution and is called Log Normal fading. The second mechanism, the sum of the signal paths, produces a Rayleigh probability distribution function and is called Rayleigh fading. Figure 2.37 illustrates both multipath mechanisms.

Significant effort has gone into characterizing the multipath environment so that effective radio structures can be designed that operate in difficult high reflection environments.

Effects of Multipath Fading

The value of (n) is 2 for free space propagation, but in general, (n) can take non-integer values greater or less than 2 depending on the environment. k is a log normal random variable that is added to model the variability in the environment due to the different amounts and types of material through which the signal travels. The uncertainty is shown in Fig. 2.38.

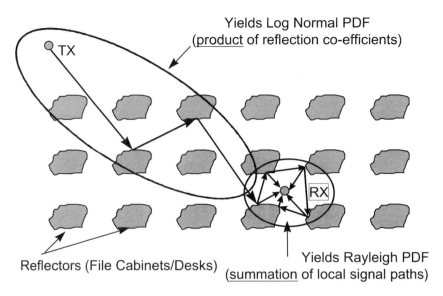

FIGURE 2.37   Log normal and Rayleigh fading.

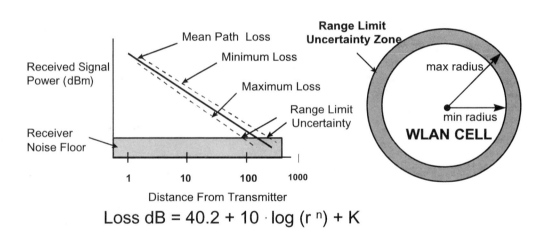

$$\text{Loss dB} = 40.2 + 10 \cdot \log (r^n) + K$$

FIGURE 2.38   Effects of multipath fading.

Residential: n = 1.4–4.0, with n = 2.8 typical
Residential: Standard deviation of the log normal distribution 7–12 dB with 8 dB typical
Office: n = 1.74–6.5, with n = 3.7 typical
Office: Standard deviation of the log normal distribution 6–16 dB with 10 dB typical
Light industrial: n = 1.4–4.0, with n = 2.2 typical (open plan),
Light Industrial: Standard deviation of the log normal distribution: 4–12 dB with 10 dB typical

Delay Spread Craters
Multipath fading has been the chief performance criteria for selecting a new high-data-rate 802.11 DSSS standard waveform. Many independent multipath surveys published in the IEEE literature are in agreement in showing that high delay-spread holes exist anywhere within a cell. These holes can be a real difficulty for cell planners because stations may fail to operate even a short distance from the cell center. In commercial environments it is common to see the multipath spread reach 100 nsec (rms). Multipath spread is one unit of measurement for the Rayleigh fading characteristic. A 100-nsec multipath spread is commonly observed in cafeterias, atriums, and open Wal-mart-like structures. Craters are illustrated in Fig. 2.39.

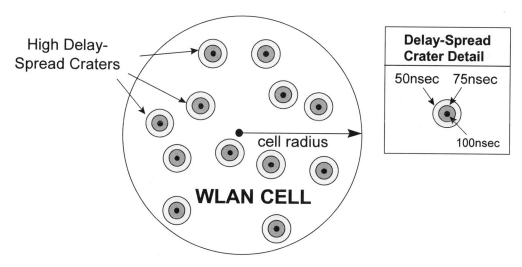

FIGURE 2.39   Delay spread craters.

Multipath Mitigation

By itself, antenna diversity provides minimal relief from multipath craters. If a station is located within a crater, two antennas tend to see the same degree of multipath.

In a multipath crater, the SNR is often high. The average signal power in the crater is the same as described by the mean path loss shown before. If the crater is not near the cell boundary, the SNR is good. However, the multipath components can highly distort the signal so the conventional receiver still fails.

A simple parallel is the distortion caused by audio echoes. A speaker becomes unintelligible when severe echoes exist, even in the absence of other noise. The equivalent effect occurs in a receiver. The multipath echoes cause the code words and DSSS spreading chips to overlap in time. This self-interference causes receiver paralysis even in the absence of noise, unless the receiver has been designed to correct the echoes.

Impulse Response Channel Models

In a room environment, two measurements of the same radio in the same place may not agree. This is due to the changing position of the people in the room and slight changes in the environment which, as we have seen, can produce significant changes in the signal power at the radio receiver. A consistent channel model is required to allow the comparison of different systems and to provide consistent results. Simulations may be run in software against models of the radio. But more valuable are hardware simulators that can be run against the radio itself. These simulators operate on the output of the radio and produce a simulated signal for the receiver from the transmitted signal. To be of value for comparison, a standard model should be used.

The IEEE 802.11 committee has been using quite a good simulation model that can be readily generalized to many different delay spreads.

Another model, which is more easily realized in hardware simulations, is the JTC '94 model. This model is a standard that provides three statistically based profiles for residential, office, and commercial environments. Two obscured line-of-sight profiles are provided for residential and commercial environments; these are illustrated in Fig. 2.40.

Interference Immunity and Processing Gain

Another significant parameter to be considered in the link is interference. In the ISM band the main source of interference is the microwave oven. Radios must be designed to operate in the presence of microwave ovens. The spread spectrum nature of the waveform allows narrowband interference to be

- IEEE802.11 Model
  - Ideal for software simulation -> continuously variable delay spread
  - Largest number of paths, min 4 per symbol
  - Purely exponential decay - no OLOS profiles

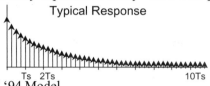

- JTC '94 Model
  - Ideal for hardware simulation (real time)
  - Up to 8 paths
  - 9 statistically based profiles, 3 each residential, office, commercial
  - 2 OLOS profiles (Residential C and Commercial B)

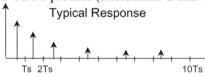

FIGURE 2.40   Impulse response channel models.

tolerated. Processing gain of 10 dB is available to provide protection from narrowband interferers of any type. Rate changes to lower data rates can be used to allow higher tolerance to microwave energy. The IEEE 802.11 protocol is designed to enable operation between microwave energy pulses. Within the 802.11 protocol is the ability to change frequencies to avoid a problem channel.

Other radios in the band can cause interference. Two types of interference must be considered. First is co-channel interference of our own system. This is the energy from a nearby cell of the same system that is on the same frequency. Frequency planning and good layout of access points can minimize this interference and keep it from being a system constraint. The second type of interference is from other systems. These might be Direct Sequence Spread Spectrum or Frequency Hop Spread Spectrum. Two mechanisms are available to mitigate this interference. The first is Clear Channel Assessment (CCA) provided in the IEEE 802.11 standard. MAC layer protocol provides collision avoidance using CSMA-CA. The second is processing gain, which provides some protection from Frequency Hop Spread Spectrum radios, which appear as narrowband interferers. Some frequency planning is always required and not all systems will coexist without some performance degradation.

The radiated energy from a microwave oven can interfere with the wireless transmission of data. The two plots shown in Fig. 2.41 illustrate that the radiated energy from a microwave oven is centered in the 2.4 GHz ISM band.

The plots show the results of leakage tests conducted at 9.5 inches and 10 feet from a microwave oven. The test shown on the left plot was conducted at a distance of 9.5 inches from the oven. The leakage test shown on the right plot was conducted at a distance of 10 feet.

The two cases are typical plots selected from more that 20 leakage tests performed on different brands of ovens, at different distances and different antenna angles. The measurements show that the energy drops off dramatically when the distance between the measurement antenna and the oven exceeds six feet. At a distance of ten feet, the energy level is typically below the –20 dBm level as illustrated in the plot on the right.

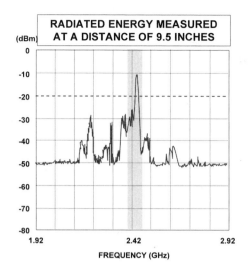

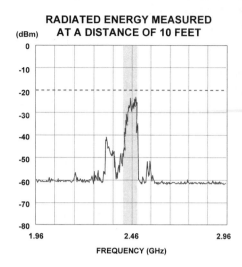

FIGURE 2.41 Microwave oven interference.

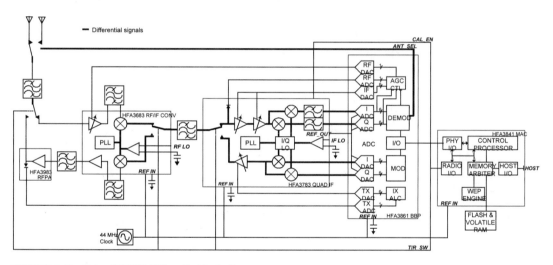

FIGURE 2.42 Intersil PRISM II® radio block diagram.

WLAN System Example: PRISM® II

The IEEE 802.11b high rate WLAN has been implemented with the Intersil PRISM® (Personal Radio using Industrial Scientific Medical bands) SiGe chipset. The block diagram is shown in Fig. 2.42. This implementation uses five chips and has a total component count of approximately 200. The major chips include an RF converter that operates using an RF frequency of 2.400 to 2.484 GHz and an IF frequency of 374 MHz. A synthesizer is included in the IC with ceramic filters and a VCO provided at the IC inputs. A PA boosts the signal to approximately +20 dBm prior to final T/R switching, filtering, and any diversity switching. An IF IC converts the signal to baseband or modulates the transmit signal following filtering with a single SAW filter at 374 MHz. The MODEM IC implements the IEEE 802.11 CCK modulation with special circuits for multipath corrections. Finally a digital IC implements the Media Access Controller (MAC) function for 11 Mbit data and interfaces with the computer/controller.

A brief description of the signal paths begins on the upper left that the dual antennas which may be selected for diversity with a command from the BaseBand Processor (BBP). The desired antenna is routed to a ceramic "roofing" filter to attenuate out-of–band signals and attenuate the 1.6 GHz image frequencies. Front-end losses with Diversity, T/R switches, and filters are typically 4 dB. The 3 dB NF LNA has a selectable high gain of +15dB or low gain of –15dB controlled by the BBP depending upon signal level. The range of signals from –90 dBm to –30 dBm uses the LNA high gain mode and –29 dBm to –4 dBm (IEEE 802.1 maximum) uses low gain on the LNA. The first active Gilbert Cell mixer has a gain of +8 dB and NF of 9 dB. A differential 374 MHz SAW filter with 8 dB loss follows the RF converter. An IF demodulator then processes the –80 dBm to –20 dBm signal with a linearized AGC stage prior to the final baseband mixer conversion. The signal is low pass filtered and passed to the Base Band Processor (BBP). Digital BBP processing includes I/Q A/D, interpolating buffers with digital NCO phase locked carrier recovery and Rake receiver/Equalizer processing. Finally the received data is processed with the MAC CPU in accordance with the IEEE 802.11 protocol.

The transmit function begins with data to the MAC being processed for BBP modulation. The data modulator includes a self-synchronizing scrambler function that removes the periodicity of the short 11 chip Barker sequence. A BBP transmit Data Formatter and Spreading Table provides the PN spreading function. A digital filter then reduces the DQPSK –13 dB Side Lobe Level (SLL) to approximately –45 dBc. This digital baseband waveform is then sent to the BBP I and Q D/A converters. The IF chip's modulator upconverts these signals to the 374 MHz IF using an active Gilbert Cell mixer and combiner. This modulated signal is then Automatic Level Controlled (ALC'd) using a feedback algorithm with the HFA3983 PA's detector to maintain the IEEE 802.11 30 dBc SLL. This IF signal is filtered by the same SAW as used in the receive path. A final RF upconversion from 374 MHz to 2.4 GHz is performed by the RF converter using the integrated on-chip synthesizer. The output level of approximately –5 dBm is filtered for image and harmonics and applied to the PA and finally the T/R and Diversity Switches. Schematics, application notes, and additional details for this example design may be obtained at www.Intersil.com.

2.6 Wireless Personal Area Network Communications: An Application Overview

Thomas M. Siep and Ian C. Gifford

Advancing technology is providing electronic aids that are increasingly becoming personal extensions of human beings. A prime example is the cellular telephone. It has replaced the pay phone as the means by which many of us maintain voice contact when away from our homes or businesses. The cell phone is a personal device and is customized to us. The pay phone is a generic, public device. The public device requires us to supply all necessary personal information, such as phone number and billing information, for its use. The personal device retains this slowly changing information for us. Each one of our personal information devices retains our stored information, ready for use. The number of these devices is increasing: Personal Digital Assistants (PDAs), lightweight laptop PCs, and pagers have joined the ranks of devices that keep our store of personal information.

If the stored personal information never changed, there would be no reason for anything but initial data entry. This is not the case. People move, appointments change, and access procedures are updated. This requires the modification of information in an information database retained in each of our devices.

The increasing number of electronic aids with their disparate and overlapping information databases fosters the need to interconnect them. There may be many devices within a relatively small area. This area is, however, more closely tied to an individual than any particular geographic location.

The obvious solution is to interconnect these increasingly intelligent devices and allow them to synchronize their information databases. Special-purpose wires and cables are the traditional method of interconnection. These are the wires that interconnect personal devices rather than connecting to public or local area networks. The problem is, nobody likes wires with personal, portable devices. They get forgotten, lost, or broken. Plus, they are inconvenient to use. The solution to that problem is to use a wireless communications technique.

Three different wireless communications schemes could solve the problem: satellite, metropolitan, and short distance. The first two technologies use service-provider infrastructure and cost money each time they are used. The short distance infrastructure may be owned by the user and is "free" once the equipment is paid for.

Several short distance wireless solutions exist in the marketplace. These include the IEEE Standards for Wireless Local Area Networks (P802.11[5]) and Wireless Personal Area Networks (P802.15). Project 802.15 defines the WPAN™ topology. Whereas P802.11 is concerned with features such as Ethernet-matching speed and handoff support for devices in a localized area, WPAN communications are even more localized in their purview. They are concerned only with the immediate area about the person using the device. This concept has been dubbed a Personal Area Network. The untethered version of this is, of course, called a Wireless Personal Area Network™ or WPAN.

WPAN communications provide a Personal Operating Space (POS). This is the space about a person or object that typically extends up to 10 meters in all directions and envelops the person whether stationary or in motion. WPAN communications are the wireless "last meter" for the so-called wearable computers or pervasive computing devices that began to emerge in the late 1990s.

The term WPAN was coined by the IEEE based on some of their members' work done in the IEEE Computer Society's IEEE 802 LAN/MAN Standards Committee. The IEEE has identified the Bluetooth™ technology as a solution for their applications for WPAN communications. The Bluetooth technology is a specification for small form factor, low-cost, wireless communication and networking between PCs, mobile phones, and other portable devices.[6]

The WPAN technology examined here is based on the work of the IEEE 802.15 Working Group. Other topologies and implementations exist for personal area networking, but generally they follow the characteristics of the protocol stack explained here.

Applications for WPAN Communications

WPAN communications are primarily for the user's personal convenience. Its purpose is to replace wires between objects that are more or less within easy reach. It may also hook to the larger network world when/if/while convenient. This wire replacement technology is intended as the primary method of connection between a large variety of devices, many with limited capabilities.

The Nature of WPAN Services

WPAN communications exist to tie together closely related objects, a function that is fundamentally different than the goal of the Wireless Local Area Network (WLAN). The purpose of WLANs is to provide connectivity to the Ethernet plug in the wall at the workplace and Ethernet-like connectivity in ad hoc situations, such as conferences. Devices that attach to the WLANs are usually high-capability devices. Devices such as laptops and desktop computers are relatively expensive. Wireless connectivity has been justifiable for business entities as an infrastructure cost. The essential difference between the two technologies is topology: WLAN communications are outward-looking and WPAN are inward-looking. WLANs are oriented to connecting to the world as a whole and WPAN communications seek to unify communications among personal devices.

In addition, WPAN technology is focused on supporting a certain kind of mobility. The backbone that WLANs attach to tends to stay in one place. WLAN devices may move about, but the infrastructure

---

[5]The name IEEE P802.11 derives from the naming scheme used by the Institute of Electrical and Electronic Engineers for their standards bodies and resulting standards documents. Project 802 (P802) is the standards group that is concerned with Local and Metropolitan Area Networks (LAN/MAN). There are many aspects of communications that are covered by the charter of this organization, each of which has a "dot" number. A standard that many readers may be familiar with is P802.3 or Ethernet.

[6]See http:/www.bluetooth.com for more information.

FIGURE 2.43 WPAN™ application hardware.

stays put (at least in the short term). People don't. They are their own infrastructure, and they expect the entire system to move with them wherever they go.

This kind of mobility has some interesting legal implications. It drives the need for a single standard that meets the worldwide regulatory requirements. WPAN technology is designed such that a single technology meets the spectrum power requirements of the world, since users don't want to break the law by crossing a border.

Application Example

Figure 2.43 shows the hardware involved in an example WPAN communications system. In this example there are three devices that are able to participate in the Wireless Personal Area Network:

- A headset consisting of an in-the-ear speaker, a microphone, and an embedded WPAN radio transceiver.
- A cellular telephone that has WPAN capabilities via an embedded radio transceiver as well as its normal macro-cellular radio.
- A laptop computer in the briefcase that has a WPAN embedded radio transceiver or interface card with radio transceiver.

Each of these devices is capable of communicating with the other two.

Figure 2.44 demonstrates the WPAN hardware in action. It shows an example of how this kind of system can allow the assemblage of currently existing technologies into a new and powerful capability. The scene is a traveler hurrying in an airport who receives an urgent e-mail that requires an immediate answer before embarking on a plane. The following numbered list corresponds to the circled numbers in Fig. 2.44.

1. The cell phone receives the urgent e-mail. Since the cell phone is set up to handle e-mails via a laptop (if available), it does not ring the phone.
2. The cell phone, acting as a gateway, sends a message via a WPAN link and wakes up the laptop computer in the traveler's briefcase, then relays the e-mail to it.

FIGURE 2.44 WPAN™ In action.

3. The laptop software inspects the incoming e-mail, determines that it is urgent, and translates the text to synthetic speech.
4. The laptop then uses the WPAN link with the headset that is normally used by the cell phone for voice communications and sends the translated e-mail to it.
5. The traveler then dictates a reply to the e-mail to the laptop via the WPAN link.
6. The laptop translates the voice to text and formats a reply e-mail.
7. The reply e-mail is then handed off to the cell phone for transmission.
8. The e-mail reply is sent within seconds of the arrival of the original e-mail, without the traveler having to stop running.

None of the actions described by Fig. 2.44 requires a WPAN connection. They all *could* be done with wires. The question is *would* they be done with wires? For the vast majority of mobile users the answer is "no": this functionality requires wireless links to be feasible.

WPAN Architecture

As shown in the example above, a WPAN communications system can provide connectivity among a wide variety of personal devices, including personal mobile devices such as headsets, phones, PDAs, notebook computers, and the like. A WPAN can further provide connectivity between those devices and stationary electronic devices such as data access points to wire line LAN installations, or wireless modems connected to the PSTN.

WPAN systems typically use a short-range radio link that has been optimized for power-cautious, battery-operated, small size, lightweight personal devices. WPAN communications generally support both circuit switched channels for telephony grade voice communication and packet switched channels for data communications. As demonstrated in the example above, a cellular phone may use the circuit switched channels to carrying audio to and from a headset, while at the same time, use a packet switched channel to exchange data with a notebook computer.

WPAN Hardware

General-use WPANs operate in the worldwide unlicensed Industrial, Scientific, Medical (ISM) band at 2.4 GHz. A frequency-hopping technique is utilized to satisfy regulatory requirements and combat interference and fading.

The WPAN system consists of a radio unit, a link control unit, and a support unit for link management and host terminal interface functions. The discussion here is limited to the specifications of the WPAN Physical (PHY) layer and Media Access Control (MAC) sublayer that carries out the baseband protocols and other low-level link routines. This corresponds to the protocol layers covered by the IEEE P802.15 WPAN Standard.

The IEEE P802.15 architecture consists of several components that interact to provide a WPAN that supports station mobility transparently to upper layers. The following definitions are taken from the Standard:

Physical Layer (PHY): Protocol layer that directly controls radio transmissions.
Medium Access Control (MAC): Protocol layer that determines when and how to use the PHY.
Master Station: Any *device* that contains an IEEE P802.15 conformant MAC and PHY interface. It provides identity and clocking to WPAN peer devices.
Slave Station: Any *device* that contains an IEEE P802.15 conformant MAC and PHY interface. It receives and transmits to the master station only. Transmissions from the slave occur only when the master permits.
Piconet: Two or more WPAN Stations sharing the same RF channel form a *piconet*, controlled by one (and only one) master.
Scatternet: Multiple independent and non-synchronized piconets that may have some overlapping membership.

The discussions below will use these terms to describe the topology and behavior of WPAN systems.

Ad Hoc Networks

The WPAN networks are not created *a priori* and have a limited life span. They are created on an as-needed basis when necessary, then subsequently abandoned. These *ad hoc networks* are established whenever applications need to exchange data with matching applications in other devices. A WPAN will likely cease to exist when the applications involved have completed their task.

WPAN units that are within range of each other can set up ad hoc connections. WPANs™ support both point-to-point and point-to-multipoint connections. In principle, each unit is a peer with the same hardware capabilities. Two or more WPAN units that share a channel form a piconet with one unit being the master.

Each piconet is defined by a different frequency-hopping channel. This channel is a narrow band of radio transmission that is continually moving from frequency to frequency ("hopping") in a pseudoran-dom fashion. All units participating in the same piconet are synchronized to this channel.

A WPAN vs. a WLAN

As mentioned above, the IEEE also defines another type of short-distance wireless technology. It is IEEE P802.11 Wireless Local Area Networks (WLAN). On the surface the WLAN and WPAN seem quite similar.

WLAN technologies are "instant infrastructure" specifically designed for interconnecting peer devices in and around the office or home. A WPAN device is a cable replacement "communications bubble" that is designed to travel from country to country, be used in cars, airplanes, and boats. As a result, the topologies are different.

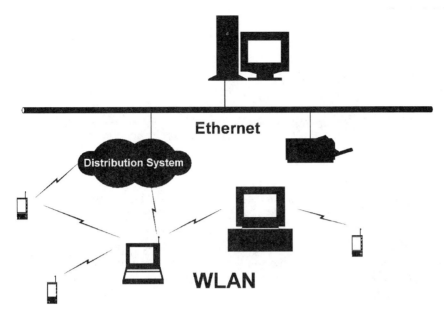

FIGURE 2.45 Wireless Local Area Network Topology.

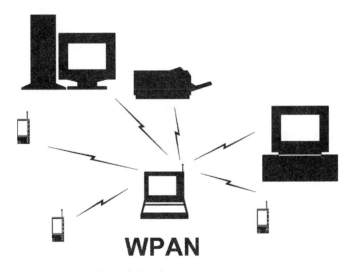

FIGURE 2.46 Wireless Personal Area Network Topology.

A typical WLAN topology is shown in Fig. 2.45. It is essentially an extension of the wired LAN. Each of the communicating devices is an equal partner in the communications network. As such, any device can communicate directly with any other device, provided the radio waves can reach between them. If they cannot reach, there are provisions in the standard for relaying messages. There is no requirement for a hard-wired backbone, but it is typical for most installations. The topology is range-limited, with no upper ceiling on the number of participating units.

The WPAN topology is simpler. Figure 2.46 illustrates the WPAN topology. The devices interconnected are identical. How they are interconnected is different. Instead of a free-for-all competition for the airwaves, the WPAN has a master device that directs the traffic of its slaves through it. The slaves are "good children"; they never speak unless spoken to. Any device that cannot be directly reached through the master cannot be communicated with: there is no provision for message relay in the protocol.

The master/slave topology has both disadvantages and advantages. It is a disadvantage for two slaves that wish to converse to have to go through the master to do it. There are many advantages, however. Among them are:

- Spectral efficiency
- Simplicity
- Cost

The orderly use of the airwaves by WPAN technology yields a higher utilization of the spectrum. The WLAN uses a technique of Carrier Sense/Collision Avoidance (CS/CA) to assure fair sharing of the airwaves. It can use several schemes, the simplest of which is to Listen Before Talk (LBT). The WLAN device never knows when the medium will be occupied, so it must listen to the current environment before it transmits. In the event that the airwaves are busy, the device is required to do an exponential back off and retry.

The WPAN technology has a more controlled approach to using the airwaves. It establishes the master of the piconet, which dictates when each device can use the airwaves. There is no "lost time" in waiting for clear air or collisions due to two partners in the same net transmitting at the same time to the same device because they are out of range of each other.

Implementations of WPAN communications are inherently simpler. Without the need to support LAN technology, the WPAN implementation can forgo the extra buffers and logic that LANs require. The WPAN world is smaller and therefore simpler.

This simplicity of topology yields a less expensive system cost. Having simpler logic and less capability allows designers of WPAN systems to create communication subsystems that are a smaller percentage of total system cost. This, in turn, allows the technology to be embedded in more low cost devices. The effect of that will be to increase volume, which brings down the cost again.

WPAN Scatternet

Sometimes more complex communications schemes are required of the devices that support WPANs. In that case, the WPAN devices can form a *scatternet*: multiple independent and non-synchronized piconets. These are high aggregate-capacity systems. The scatternet structure also makes it possible to extend the radio range by simply adding additional WPAN units acting as bridges at strategic places.

To integrate the IEEE P802.15 architecture with a traditional wired LAN (or even Wireless LAN), an application is utilized to create an "*access point*." An access point is the logical point at which data to/from an integrated non-802.15 wired LAN leave/enter the IEEE 802.15 WPAN piconet. For example, an access point is shown in Fig. 2.45 connecting a WLAN to a wired 802 LAN: the same construct can be implemented with a WPAN, with the restrictions inherent in it.

WPAN Protocol Stack

The IEEE Project 802 LAN/MAN Standards Committee develops wired and wireless standards. IEEE 802 standards deal with the physical and data link layers as defined by the International Organization for Standardization (ISO) Open Systems Inter-connection (OSI) Basic Reference Model (ISO/IEC 7498-1: 1994).

Currently, IEEE Project 802 standards define eight types of medium access technologies and associated physical media, each appropriate for particular applications or system objectives (see Fig. 2.47). The most popular of these standards is IEEE 802.3, the Ethernet standard.

WPAN Open Protocol Stack

All applications already developed by vendors according to the Bluetooth Foundation Specification can take immediate advantage of hardware and software systems that are compliant with the P802.15 Standard. The specification is open, which makes it possible for vendors to freely implement their own (proprietary) or commonly used application protocols on the top of the P802.15-specific protocols. Thus,

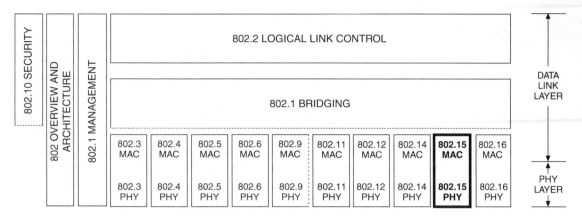

FIGURE 2.47 The IEEE P802 family of standards.

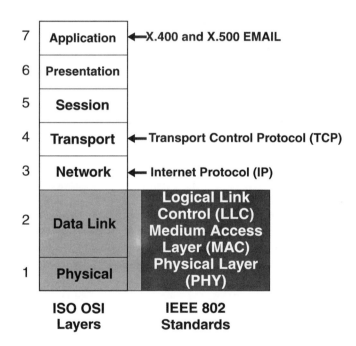

FIGURE 2.48 The relationship between ISO-OSI and IEEE P802.

the open specification permits the development of a large number of new applications that take full advantage of the capabilities of the WPAN technology.

The P802.15 Architecture

Project 802 LAN/MAN standards cover the two lowest layers of the ISO OSI protocol description.

Figure 2.48 shows the relationship of the ISO model to the purview of Project 802. Note that the lowest two levels of the ISO stack correspond to the three layers of the IEEE stack. The reason for that is that the interface that defines the Logical Link Control (LLC) is common for all 802 standards. P802.2 documents this interface and the other portions of the 802 family of standards just refer to it, rather than replicating it for each standard. Since the WPAN standard did not start out with the IEEE architecture in mind, the P802.15 group found it necessary to do a mapping from the Bluetooth structure to one that is compatible with the rest of the 802 family. Figure 2.49 shows this mapping. The radio definition of

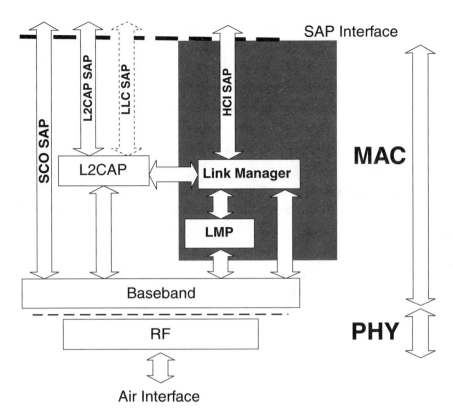

FIGURE 2.49   Bluetooth™ to P802.15 mapping.

the Bluetooth specification directly corresponds to the PHY of a P802.15 standard. The MAC is made up of three explicit entities and one implicit entity. The Logical Link Control and Adaptation Protocol (L2CAP), Link Manager Protocol (LMP), and Baseband (BB) are specifically defined in the specification. The Link Manager (LM) is mentioned in several sections of the specification, but never explicitly defined. The P802.15 group has created a set of System Definition Language (SDL[7]) graphics that explicates this and other implied structures in the Bluetooth specification.

PHY Layer

The *physical layer* (PHY) enables the physical radio link between WPAN units. The P802.15 RF system is a Frequency-Hopping Spread Spectrum system where packets are transmitted in defined time slots on defined frequencies. This hopping technique is used to achieve the highest possible robustness for noisy radio environments. The rate of hopping for this packet-based protocol is 1600 hops per second.

The P802.15 transceiver operates in the 2.4 GHz ISM band at a gross data rate of 1 Mbps. The required nominal range of a P802.15 radio is 10 meters with 0 dBm output power.

A single unit can support a maximum data transfer rate of 723 Kbps or a maximum of 3 voice channels. A mixture of voice and data is possible to support multimedia applications. The modulation is Gaussian Frequency Shift Keying (GFSK) with BT = 0.5.

The channel is represented by a pseudorandom hopping sequence. The channel is divided into time slots, where each slot corresponds to one RF hop. Time-division duplexing (TDD) is used for access to the channel.

The data is conveyed in packets. Each packet consists of an access code, header, and payload. Sixteen different packet types can be defined.

---

[7]ITU-T Recommendation Z.100.

MAC Sublayer

The *Media Access Control* (MAC) portion of a P802 protocol decides when transmissions will occur and what will be transmitted. It marries the raw transmitter (the PHY) to the layer that establishes logical connections between devices, the Logical Link Controller (LLC). It sets the topology of the network.

The P802.15 MAC uses the Bluetooth ad hoc piconet and scatternet concepts. Both point-to-point and point-to-multipoint connections are supported.

In the P802.15 network all units have essentially the same capabilities, with identical hardware and software interfaces. Each device is distinguished by a unique IEEE 48-bit address. At the start of a connection the initializing unit is temporarily assigned as master. This assignment is valid only during this connection. Every unit in the piconet uses the master identity and clock to track the hopping channel. Inter-piconet communication is achieved by selecting the proper master identity and clock offset to synchronize with the channel of the desired piconet. A corresponding set of identity and clock offsets is available for each piconet. A unit can act as a slave in several piconets. A master is also allowed to be a slave in another piconet.

Connection Types

The P802.15 standard provides two different kinds of logical connection types, Synchronous Connection-Oriented (SCO) and Asynchronous Connection-Less (ACL), which can be transmitted in a multiplexing manner on the same RF link. ACL packets are used for data only, while the SCO packet can contain audio only or a combination of audio and data.

The SCO link is a symmetric point-to-point link between the master and a specific slave. The SCO reserves two consecutive time slots (forward and return) at fixed intervals. It is considered a circuit-switched connection. Audio data can be transferred between one or more P802.15 devices, making various usage models possible. The typical model is relatively simple where two 802.15 devices send and receive audio data between each other by opening an audio link.

The ACL link supports symmetrical and asymmetrical packet-switched point-to-multipoint connections typically used for data. All data packets can be provided with differing levels of FEC or CRC error correction and can be encrypted. Furthermore, different and multiple link types may apply between different master-slave pairs of the same piconet and the link type may change arbitrarily during a session.

Packets

Compared to wired physical media channels, the data packets defined by the MAC sublayer are smaller in size. The MAC sublayer segments large packet transmissions into multiple smaller packets. Similarly, on the receiving end it reassembles the multiple smaller packets back into the larger packet. Segmentation and reassembly (SAR) operations are used to improve efficiency by supporting a maximum transmission unit (MTU) size larger than the largest packet that can be transmitted.

A packet consists of three fields: access code, header, and a variable length payload. The access code, described in the PHY layer, is used for synchronization, DC offset compensation, and identification. The packet header contains link-control information.

The channel is divided into time slots, each 625 microseconds in length. The packet must be completely transmitted in that time period. The slots are divided into two alternating groups, "even" and "odd." Even slots are master-to-slave, and odd slots are slave-to-master slots. Only the slave that was addressed in the preceding master-to-slave slot can transmit ACL data in the slave-to-master slot.

As stated above, the MAC sublayer provides connection-oriented and connectionless data services to upper layer protocols. Both of these services are supported with protocol multiplexing, segmentation and reassembly operation, and group abstraction capabilities. These features permit higher level protocols and applications to transmit and receive data packets up to 64 kilobytes in length.

In Case of Error

There are three error-correction schemes defined for 802.15: 1/3 rate FEC, 2/3 rate FEC, and ARQ scheme for data. The purpose of the FEC scheme on the data payload is to reduce the number of retransmissions. The FEC of the packet payload can be turned on or off, while the packet header is always protected by a 1/3 rate FEC. In the case of a multi-slave operation, the ARQ protocol is carried out for each slave independently.

Two speech coding schemes, continuously variable slope delta (CVSD) modulation and logarithmic Pulse Coded Modulation (logPCM), are supported, both operating at 64 Kbit/s. The default is CVSD. Voice is never retransmitted, but CVSD is very resistant to bit errors, as errors are perceived as background noise, which intensifies as bit errors increase.

Security and Privacy

To provide user protection and information secrecy, the WPAN communications system provides security measures at the physical layer and allows for additional security at the applications layer. P802.15 specifies a base-level encryption, which is well suited for silicon implementation, and an authentication algorithm, which takes into consideration devices that do not have the processing capabilities. In addition, future cryptographic algorithms can be supported in a backwards-compatible way using version negotiation.

The main security features are:

- Challenge-response routine for authentication
- Session key generation, where session keys can be exchanged at any time during a connection
- Stream-cipher

Four different entities are used for maintaining security at the link layer: a public address that is unique for each unit (the IEEE 48-bit address), two secret keys (authentication and encryption), and a random number that is different for each new connection. In a point-to-multipoint configuration the master may tell several slave units to use a common link key and broadcast the information encrypted. The packet payload can be encrypted for protection. The access code and packet header are never encrypted.

History of WPANs and P802.15

The chain of events leading to the formation of IEEE 802.15 began in June 1997 when the IEEE Ad Hoc "Wearables" Standards Committee was initiated during an IEEE Standards Board meeting. The purpose of the committee was to "encourage development of standards for wearable computing and solicit IEEE support to develop standards." The consensus recommendation was to encourage such standards development in the IEEE Portable Applications Standards Committee (PASC).

During the PASC Plenary Meeting in July 1997, an IEEE Ad Hoc Committee was assembled (17 attendees) to discuss "Wearables" Standards. The committee identified several areas that could be considered for standardization, including short range wireless networks or Personal Area Networks (PANs), peripherals, nomadicity, wearable computers, and power management. Of these, the committee determined that the best area of focus was the wireless PAN because of its broad range of application. The IEEE Ad Hoc "Wearables" Standards Committee met twice more — once in December 1997 and again in January 1998. During the January 1998 meeting it was agreed that the 802 LAN/MAN Standards Committee (LMSC) was probably a more suitable home for the group's initial activities, especially with a WPAN focus. Two delegates were sent to the IEEE 802.11 interim meeting that same month to get reactions and to gain support for the proposal. At that meeting it was agreed to propose the formation of a study group under 802.11 at the March plenary of 802. A WPAN tutorial was organized to socialize the idea within 802. The result was that in March 1998, the "Wearables" Standards Ad Hoc Committee under PASC became the IEEE 802.11 Wireless Personal Area Network (WPAN) Study Group within LMSC with the goal of developing a Project Authorization Request (PAR) for the development of a WPAN standard.

At the time the study group was formed, there had been no other publicized initiatives in the WPAN space other than the Infrared Data Association's IrDA specification. IrDA's line-of-sight restrictions do not allow it to adequately address the problem. By the time the work of the study group concluded a year later, both the Bluetooth Special Interest Group and HomeRF™ were active in developing WPAN specifications. By March 1999, when the study group and 802.11 submitted the PAR to the 802 executive committee for approval, Bluetooth had over 600 adopter companies and HomeRF had over 60. Because of the significance of these groups in the WPAN market space, it was felt that the standards development process would be better served if a new working group were formed to address the problem rather than pursue it as a task group under 802.11. The PAR was approved and the new group was designated as IEEE 802.15 Working Group for Wireless Personal Area Networks.

Conclusions

Wireless Personal Area Networks will proliferate early in the next millennium and the IEEE 802.15 Working Group for Wireless Personal Area Networks (WPANs) is providing the leadership in the IEEE 802 standards committees to establish open standards for WPANs. WPANs, which are synonymous with the Bluetooth Foundation Specification v1.0, are relatively new emerging wireless network technologies and as such the applications are evolving.

The first standard derived by 802.15 from the Bluetooth Version 1.0 Specification Foundation Core, and Bluetooth Version 1.0 Specification Foundation Profiles is addressing the requirements for Wireless Personal Area Networking (WPAN) for a new class of computing devices. This class, collectively referred to as pervasive computing devices, includes PCs, PDAs, peripherals, cell phones, pagers, and consumer electronic devices to communicate and interoperate with one another. The authors anticipate that the IEEE Standards Board on or before March 2001 will approve this standard. The 802.15 working group is paving the way for Wireless Personal Area Network standards that will be — Networking the World™.

2.7   Satellite Communications Systems

Ramesh K. Gupta

The launch of commercial communications satellite services in the early 1960s ushered in a new era in international telecommunications that has affected every facet of human endeavor. Although communications satellite systems were first conceived to provide reliable telephone and facsimile services among nations around the globe, today satellites provide worldwide TV channels (24 hours a day), global messaging services, positioning information, communications from ships and aircraft, communications to remote areas, disaster relief on land and sea, personal communications, and high-speed data services including Internet access. The percentage of voice traffic being carried over satellites — which stood at approximately 70 percent in the 1980s — is rapidly declining with the advent of undersea fiber-optic cables, and as new video and data services are added over existing satellite networks [1].

The demand for fixed satellite services and capacity continues to grow. Rapid deployment of private networks using very small aperture terminals (VSATs) — which interconnect widely dispersed corporate offices, manufacturing, supply, and distribution centers — has contributed to this growth. Approximately half a million VSATs [1] are in use around the world today employing a low-cost terminal with a relatively small aperture antenna (1 to 2 meters). These terminals are inexpensive and easy to install, and are linked together with a large central hub station, through which all the communications take place. The majority of these VSAT applications use data rates of 64 or 128 kb/s. Demand for higher rate VSAT services with data rates up to 2 Mb/s is now growing due to emerging multimedia services, fueled by unprecedented growth in the Internet.

In the last two decades, wireless communications has emerged as one of the fastest growing services. Cellular wireless services, which have been leading the growth over the last decade, achieved cumulative annual growth rates in excess of 30 percent [2], with more than 200 million subscribers worldwide. To complement this rapidly emerging wireless telecommunications infrastructure around the world, a number of satellite-based global personal communications systems (PCS) [3–5] have been deployed or are at advanced stages of implementation and deployment. Examples include low earth orbit (LEO) systems such as Iridium and Globalstar, medium earth orbit (MEO) systems such as ICO Global, and geostationary (GEO) systems such as Asia Cellular Satellite System (ACeS) and Thuraya. These systems are designed to provide narrowband voice and data services to small handheld terminals. Some of these systems have experienced problems with market penetration. These satellite systems will be discussed in detail in the next chapter.

In parallel with these developments, rapid growth in Internet traffic is creating an exponential increase in the demand for transmission bandwidth. For example, use of the Internet is projected to grow from its present estimate of 100 million households to more than 300 million households in the next few years. Telecommunications networks are currently being designed to support many new applications, including high-speed data, high-resolution imaging, and desktop videoconferencing, all of which require large transmission bandwidths. To serve these markets, the U.S. and other countries have been actively developing satellite-based broadband services for business and home users. One major advantage of satellite systems has been their ability to provide "instantaneous infrastructure," particularly in underserved areas. Currently, most satellite-based broadband services are being offered at Ku-band, using broadband VSAT-type terminals. Additional systems are being proposed for the recently available Ka- and V-Bands [6–7]. Deployment of these broadband systems could help overcome the "last mile" problem (also referred to as the "first mile" problem from the customer's perspective) encountered in the developed countries, in addition to offering a host of new services. In the U.S. alone since 1995, 14 Ka-band (30/20-GHz) and 16 V-band (50/40-GHz) satellite systems have been proposed to the Federal Communications Commission (FCC) in response to the Ka- and V-band frequency allotments. In addition, in 1996 Sky Station International proposed a V-band system employing a stabilized stratospheric platform at altitudes of 20 to 30 km. Deployment of the proposed systems will accelerate the implementation of broadband wireless infrastructure in regions of the world where terrestrial telecommunications infrastructure is inadequate for high-speed communications.

This chapter reviews the evolution of communications satellite systems, addresses various service offerings, and describes the characteristics of the newly proposed Ka- and V-band broadband satellite systems.

Evolution of Communications Satellites

Communications satellites have experienced an explosive growth in traffic over the past 25 years due to a rapid increase in global demand for voice, video, and data traffic. For example, the International Telecommunications Satellite Organization's INTELSAT I (Early Bird) satellite, launched in 1965, carried only one wideband transponder operating at C-band frequencies (6-GHz uplink and 4-GHz downlink) and could support 240 voice circuits. A large 30-m-diameter antenna was required. In contrast, the communications payload of the INTELSAT VI series of satellites, launched from 1989 onward, provides 50 distinct transponders operating over both C- and Ku-bands [8], providing the traffic-carrying capacity of approximately 33,000 telephone circuits. By using digital compression and multiplexing techniques, INTELSAT VI is capable of supporting 120,000 two-way telephone channels and three television channels [9]. The effective isotropically radiated power (EIRP) of these satellites is sufficient to allow use of much smaller earth terminals at Ku-band (E1: 3.5 m) and C-bands (F1: 5 m), together with other larger earth stations. Several technological innovations have contributed to this increase in number of active transponders required to satisfy growing traffic demand. A range of voice, data, and facsimile services is available today through modern communications satellites to fixed and mobile users. The largest traffic growth areas in recent years have been video and data traffic. The satellite services may be classified into three broad categories: Fixed Satellite Services, Direct Broadcast Satellite Services, and Mobile Satellite Services.

Fixed Satellite Services (FSS)

With FSS, signals are relayed between the satellite and relatively large fixed earth stations. Terrestrial landlines are used to connect long-distance telephone voice, television, and data communications to these earth stations. Figure 2.50 shows the FSS configuration. The FSS were the first services to be developed for global communications. The International Telecommunications Satellite Organization (INTELSAT), with headquarters in Washington, D.C., was formed to provide these services. INTELSAT has more than

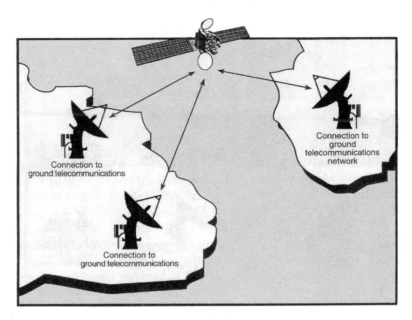

FIGURE 2.50   Fixed Satellite Communications Services (FSS).

130 signatories today, the largest being the U.S. signatory — COMSAT Corporation. With satellites operating over the Atlantic, Indian, and Pacific Ocean regions, INTELSAT provides a truly global communications service to the world.

Over the past several years, with privatization of the communications industry, new global satellite operators have emerged, including PanAmSat, Skynet, and New Skies. Several regional satellite systems have been developed around the world for communications among nations within a continent or nations sharing common interests. Such systems include Eutelsat, Arabsat, and Asiasat. For large countries like the United States, Canada, India, China, Australia, and Indonesia FSS are being used to establish communications links for domestic use. The potential for global coverage, and the growth in communications traffic as world economies become more global, has attracted many private and regional satellite operators, with several new systems being planned beyond the year 2000. According to a recent International Telecommunication Union (ITU) report (March 1998), 75 percent of all communications traffic is now provided under competitive market conditions, as compared to just 35 percent in 1990. The FSS segment is expected to grow significantly as new satellite operators introduce services at higher frequency bands.

Direct Broadcast Satellite (DBS) Services

Direct broadcast satellites use relatively high-power satellites to distribute television programs directly to subscriber homes (Fig. 2.51), or to community antennas from which the signal is distributed to individual homes by cable. The use of medium-power satellites such as Asiasat has stimulated growth in the number of TV programs available in countries like India and China. Launch of DBS services such as DIRECTV (Hughes Networks) and Dish Network (Echostar) provides multiple pay channels via an 18-in. antenna. The growth of this market has been driven by the availability of low-cost receive equipment consisting of an antenna and a low-noise block downconverter (LNB). This market segment has seen impressive growth, with direct-to-home TV subscribers exceeding 10 million today. The future availability of two-way connection through an 18-in. satellite dish is expected to place the consumer subscribers in direct competition with cable for home-based interactive multimedia access.

Mobile Satellite Services (MSS)

With MSS, communication takes place between a large fixed earth station and a number of smaller earth terminals fitted on vehicles, ships, boats, or aircraft (Fig. 2.52). The International Maritime Satellite

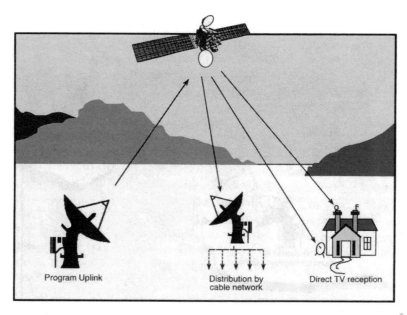

FIGURE 2.51   Direct Broadcast Satellite Communications Services (DBS). Direct-to-Home (DTH) services are also offered with similar high-power Ku-band satellite systems.

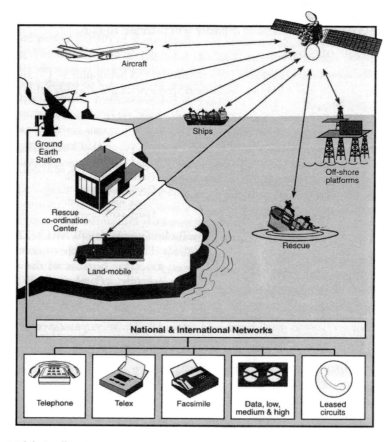

FIGURE 2.52   Mobile Satellite Communications Services (MSS).

TABLE 2.10 S-, C-, and Ku- Band Frequency Allocations

Frequency Band (GHz)	Band Designation	Service
2.500–2.655	S	Fixed regional systems, space-to-earth
2.655–2.690		Fixed regional systems, earth-to-space
3.625–4.200	C	Fixed, space-to-earth
5.925–6.245		Fixed, earth-to-space
11.700–12.500	Ku	Fixed, space-to-earth
14.000–14.500		Fixed, earth-to-space

Organization (Inmarsat), headquartered in London [10], was created in 1976 to provide a space segment to improve maritime communications, improve distress communications, and enhance the safety of life at sea. Subsequently, Inmarsat's mandate was extended to include aeronautical services and land-mobile services to trucks and trains. Launch of the Inmarsat 2 series of satellites began in 1989 and improved communications capacity by providing a minimum of 150 simultaneous maritime telephone calls. This number was further increased by using digital technology. With the launch of Inmarsat 3 spacecraft, which cover the earth with seven spot beams, communications capacity increased 10-fold.

The mobile services, which stimulated the growth of personal communications services (PCS) [4,5], have witnessed intense competition in the last decade. Failure of Iridium systems to provide low-cost services has created uncertainty in the market regarding the business viability of such systems. However, some of these systems are likely to modify their service offerings to data communications, rather than voice or low-data-rate messaging services. The mobile networks may evolve in the future to a universal mobile telecommunications system (UMTS) and IMT-2000 systems.

Frequency Allocations

The frequencies allocated for traditional commercial FSS in popular S-, C-, and Ku-bands [11] are listed in Table 2.10.

At the 1992 World Administrative Radio Conference (WARC) [11,12], the L-band frequencies were made available for mobile communications (Fig. 2.53). The L-band allocations have been extended due to a rapid increase in mobile communications traffic. Recently, a number of new satellite systems such as Iridium, Globalstar, ICO-Global [5], and Ellipso were proposed to provide personal communications services to small handheld terminals. These systems used low earth orbits (LEO) between 600 and 800 km (Iridium, Ellipso, Globalstar), or an intermediate (10,000 km) altitude circular orbit (ICO). In addition, a number of geostationary (GEO) systems such as Asia Cellular Satellite System (ACeS) and Thuraya were launched or are in the process of being implemented. A detailed discussion of these systems is presented in the next section.

The 1997 World Radio Conference (WRC-97) adopted Ka-band frequency allocations for geostationary orbit (GSO) and non-geostationary orbit (NGSO) satellite services. Some of these bands require coordination with local multipoint distribution services (LMDS) and/or NGSO MSS feeder links (Fig. 2.54). In the United States, the Federal Communications Commission (FCC) has allocated V-band frequencies for satellite services. In November 1997, the ITU favored the use of 47.2- to 48.2- GHz spectrum for stratospheric platforms.

The Ka- and V-band frequency allocations are given in Table 2.11.

Satellite Orbits

Commercial satellite systems were first implemented for operation in geostationary (GEO) orbit 35,700 m above the earth. With GEO satellite systems, a constellation of three satellites is sufficient to provide "approximately global" coverage. In the 1990s, satellite systems (lead by Iridium) were proposed using low earth orbit (LEO) satellites (at 700- to 1800-km altitude) and medium earth orbit (MEO) satellites (at 9,000- to 14,000-km altitude). Figure 2.55 shows the amount of earth's coverage obtained from LEO, MEO, and GEO orbits. The higher the satellite altitude, the fewer satellites are required to provide global

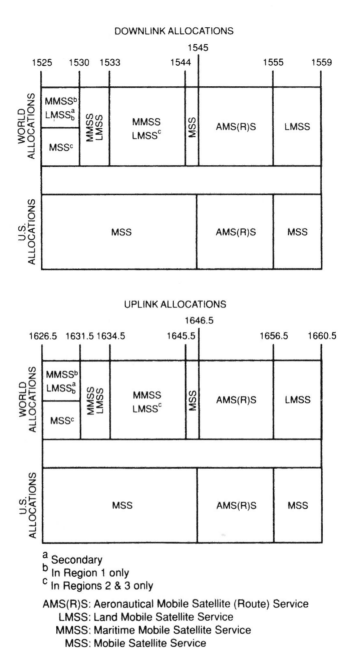

FIGURE 2.53   WARC-92 L-Band mobile satellite service allocations.

coverage. The minimum number of satellites required for a given orbital altitude and user elevation angle is shown in Fig. 2.56 and the number of orbital planes is shown in Fig. 2.57. LEO satellites tend to be smaller and less expensive than MEOs, which are in turn likely to be less expensive than GEO satellites. In terms of space segment implementation, this represents an important system/cost trade-off. For example, the benefit of smaller and less expensive satellites for LEO, as compared to GEO, is offset by the greater number of satellites required.

Another key consideration is the link margin required for the terminal for different satellite orbits. The LEO system offers the advantage of lower satellite power and smaller satellite antennas; however, the user terminals must track the satellites, and an effective handover from satellite to satellite must be

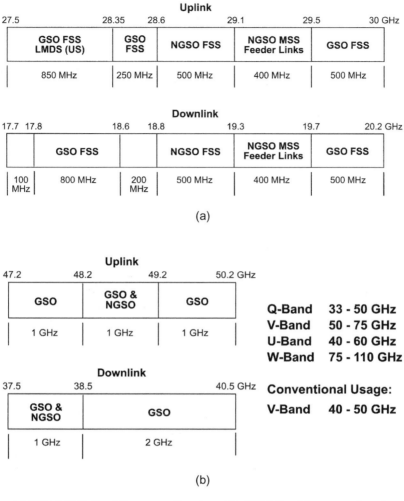

FIGURE 2.54 (a) WARC-97 Ka-Band frequency allocations; (b) FCC V-Band frequency allocations.

TABLE 2.11 Ka- and V-Band Frequency Allocations

Frequency Band (GHz)	Band Designation	Service
17.8–18.6	Ka	GSO FSS, space-to-earth
19.7–20.2	Ka	GSO FSS, space-to-earth
18.8–19.3	Ka	NGSO FSS, space-to-earth
19.3–19.7	Ka	NGSO MSS feeder links, space-to-earth
27.5–28.35	Ka	GSO FSS, earth-to-space (coordination required with LMDS)
28.35–28.6	Ka	GSO FSS, earth-to-space
29.5–30.0	Ka	GSO FSS, earth-to-space
28.6–29.1	Ka	NGSO FSS, earth-to-space
29.1–29.5	Ka	NGSO MSS feeder links, earth-to-space
38.5–40.5	V	GSO, space-to-earth
37.5–38.5	V	GSO and NGSO, space-to-earth
47.2–48.2	V	GSO, earth-to-space
49.2–50.2	V	GSO, earth-to-space
48.2–49.2	V	GSO and NGSO, earth-to-space

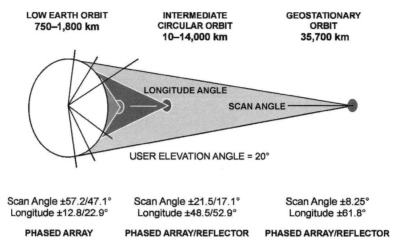

FIGURE 2.55   Relative Earth coverage by Satellites in LEO, MEO, and GEO orbits [4].

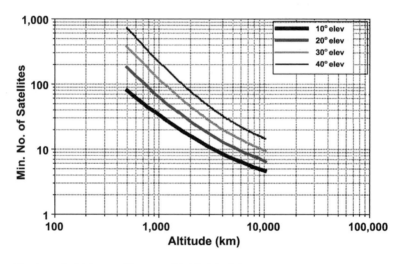

FIGURE 2.56   Number of orbiting satellites vs. orbital height [5].

designed into the system. To preserve the link margins, the spot size of the beams must be kept small. This requires larger satellite antennas, the further out the satellite is placed. For example, required satellite antenna diameters range from 1 m for LEO systems, to 3 to 4 m for MEO systems, and to 10 to 12 m for GEO systems.

The important characteristics of the three orbital systems are summarized in Table 2.12. System costs, satellite lifetime, and system growth play a significant role in the orbit selection process. On the other hand, round-trip delay, availability, user terminal antenna scanning, and handover are critical to system utility and market acceptance.

INTELSAT System Example

The evolutionary trends in INTELSAT communications satellites are shown in Fig. 2.58. This illustration depicts an evolution from single-beam global coverage to multibeam coverages with frequency reuse. The number of transponders has increased from 2 on INTELSAT I to 50 on INTELSAT VI, with a corresponding increase in satellite EIRP from 11.5 dBW to 30 dBW in the 4-GHz band, and in excess of

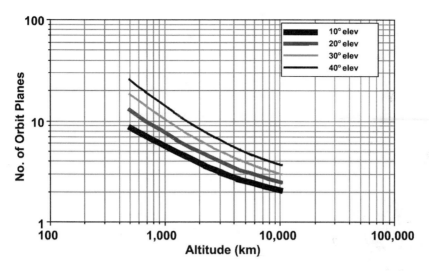

FIGURE 2.57   Number of orbit planes vs. oribital height [5].

TABLE 2.12   LEO, MEO, and GEO Satellite System Characteristics [5]

Characteristic	LEO	MEO	GEO
Space Segment Cost	Highest	Medium	Lowest
System Cost	Highest	Medium	Lowest
Satellite Lifetime, Years	5–7	10–12	10–15
Terrestrial Gateway Cost	Highest	Medium	Lowest
Overall System Capacity	Highest	Medium	Lowest
Round-Trip Time Delay	Medium	Medium	Longest
Availability/Elevation Angles	Poor	Best	Restricted
Operational Complexity	Complex	Medium	Simplest
Handover Rate	Frequent	Infrequent	None
Building Penetration	Limited	Limited	Very limited
Wide Area Connectivity	Intersatellite links	Good	Cable connectivity
Phased Start-up	No	Yes	Yes
Development Time	Longest	Medium	Shortest
Deployment Time	Longest	Medium	Short
Satellite Technology	Highest	Medium	Medium

50 dBW at Ku-band. During the same time frame, earth station size has decreased from 30 m (Standard A) to 1.2 m (Standard G) for VSAT data services (Fig. 2.59). Several technological innovations have contributed to the increase in the number of active transponders required to satisfy the growing traffic demand [13]. The development of lightweight elliptic function filters resulted in the channelization of allocated frequency spectrum into contiguous transponder channels of 40 and 80 MHz. This channelization provided useful bandwidth of 36 and 72 MHz, respectively, and reduced the number of carriers per transponder and the intermodulation interference generated by the nonlinearity of traveling wave tube amplifiers (TWTAs). For example, using filters and modifying the TWTA redundancy configuration resulted in the provision of twenty 40-MHz transponders in the 6/4-GHz frequency band for INTELSAT-IVA satellites, as compared to twelve for INTELSAT IV.

Traffic capacity was further increased with the introduction of frequency reuse through spatial and polarization isolation. In INTELSAT V, for example, 14/11 GHz (Ku) band was introduced and fourfold frequency reuse was achieved at C-band. The use of spatially separated multiple beams also increased antenna gain due to beam shaping, hence increasing the EIRP and gain-to-noise temperature ratio (G/T) for these satellites [14]. This was made possible by significant advances in beam-forming and reflector

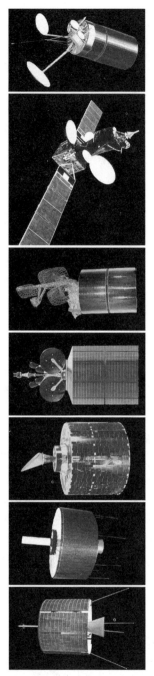

The improved design of INTELSAT satellites has yielded increased capacity and reduced costs for service.

INTELSAT DESIGNATION	I	II	III	IV	IV-A	V	V-A	VI
Year of First Launch	1965	1967	1986	1971	1975	1980	1985	1989
Prime Contractor	Hughes	Hughes	TRW	Hughes	Hughes	Ford Aerospace	Ford Aerospace	Hughes
Width Dimensions, m.(Undeployed)	0.7	1.4	1.4	2.4	2.4	2.0	2.0	3.6
Height Dimensions, m.(Undeployed)	0.6	0.7	1.0	5.3	6.8	6.4	6.4	6.4
Launch Vehicles	Thor Delta	Thor Delta	Thor Delta	Atlas Centaur	Atlas Centaur	Atlas Centaur Ariane 1, 2	Atlas Centaur Ariane 1, 2	Ariane 4 or NASA STS (Shuttle)
Design Lifetime, Years	1.5	3	5	7	7	7	7	14
Bandwidth, MHz	50	130	300	500	800	2,144	2,250	3,300
Capacity								
Voice Circuits	240	240	1,500	4,000	6,000	12,000	15,000	120,000
Television Channels	-	-	-	2	2	2	2	3

FIGURE 2.58 INTELSAT communications satellite trends (courtesy INTELSAT).

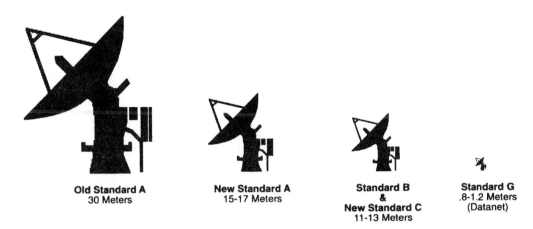

Old Standard A
30 Meters

New Standard A
15-17 Meters

Standard B
&
New Standard C
11-13 Meters

Standard G
.8-1.2 Meters
(Datanet)

FIGURE 2.59 INTELSAT Earth station size trend.

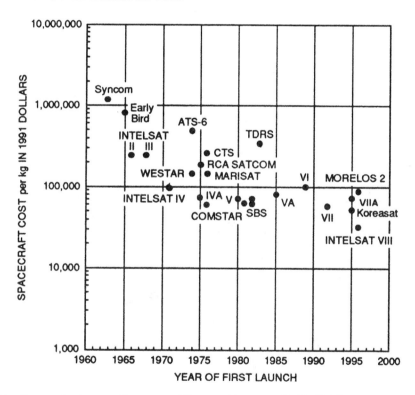

FIGURE 2.60 Communications satellite cost per kilogram. (Courtesy of COMSTAT.)

technologies. The increase in G/T on the uplink and the EIRP on the downlink, offer link budget advantages, enabling reductions in the earth terminal power amplifiers for a given antenna size. Consequently, earth terminal costs could be reduced and terminals could be located closer to customer premises. Transition also occurred from the analog frequency-division multiple access (FDMA) techniques to time-division multiple access (TDMA) transmission using digitally modulated quadrature phase shift keying (QPSK) signals. The multibeam satellite systems require onboard switch matrixes to provide connectivity among the isolated beams. In the INTELSAT V spacecraft, these interconnections were established by using electromechanical static switches that could be changed by ground command. One disadvantage of static interconnections is the inefficient use of satellite transponder capacity when there is not enough traffic to fill the capacity for the selected routing. In the INTELSAT VI spacecraft, satellite utilization efficiency was enhanced by providing cyclic and dynamic interconnection among six isolated beams using redundant microwave switch matrixes (MSMs) [15]. Satellite-switched TDMA (SS-TDMA) operation provided dynamic interconnections, allowing an earth station in one beam to access earth stations in all six beams in a cyclic manner in each TDMA frame [16]. Although INTELSAT VI MSMs were realized using hybrid MIC technology, use of GaAs monolithic microwave integrated circuit (MMIC) technology was demonstrated for such systems because it offered reproducibility, performance uniformity, and high reliability [17].

Improvements in quality of service and satellite operational flexibility can be achieved by using onboard regeneration and signal processing, which offer additional link budget advantages and improvements in bit error ratio performance through separation of additive uplink and downlink noise. Use of reconfigurable, narrow, high-gain pencil beams with phased-array antennas offers the additional flexibility of dynamic transfer of satellite resources (bandwidth and EIRP). Since these active phased-array antennas require several identical elements of high reliability, MMIC technology makes them feasible [18,19]. These technological developments in microwave and antenna systems, which have helped improve satellite capacity per kilogram of dry mass and reduced cost (Fig. 2.60), have positioned satellite systems to launch even higher capacity broadband satellites. Advances in digital device technologies have lowered the cost of TDMA terminals and made them relatively easy to maintain.

Broadband and Multimedia Satellite Systems

With the introduction of new services, the public switched telephone network (PSTN) has evolved toward Integrated Services Data Networks (ISDNs) and asynchronous transfer mode (ATM) [20]. ATM offers a suite of protocols that are well suited to handling a mix of voice, high-speed data, and video information, making it very attractive for multimedia applications. One of the unique virtues of satellite networks is that the satellite offers a shared bandwidth resource, which is available to many users spread over a large geographical area on earth. This forms the basis for the concept of bandwidth-on-demand, in which terminals communicate with all other terminals (full-mesh connectivity), but use satellite capacity on an as-needed basis. By using multifrequency TDMA to achieve high-efficiency management and flexibility, commercial multiservice networks are now implemented using multi-carrier, multi-rate, TDMA, bandwidth-on-demand products such as COMSAT Laboratories' LINKWAY™ 2000 or 2100 mesh networking platforms. Each of these platforms is capable of providing flexible data rates (up to 2 Mb/s) and can provide ATM, Frame Relay, ISDN, SS7, and IP interfaces [21].

NASA's advanced Communications Technology Satellite (ACTS), which was launched in September 1993, demonstrated new system concepts, including the use of Ka-band spectrum, narrow spot beams (beamwidths between 0.25° and 0.75°), independently steerable hopping beams for up- and downlinks, a wide dynamic range MSM, and an onboard baseband processor. In addition to these technology developments, several experiments and demonstrations have been performed using ACTS system. These include demonstration of ISDN, ATM, and IP using Ka- band VSATs, high-definition video broadcasts, health care and long-distance education, and several propagation experiments planned through September 2000. The success of the ACTS program, together with development of key onboard technologies, a better understanding of high-frequency atmospheric effects, and the availability of higher frequency spectrum has resulted in proposals for a number of Ka- and V-band satellite systems [5,6] for future multimedia services (Fig. 2.61).

Proposed Ka-Band Systems

The 14 proposed Ka-band systems in the U.S. are Astrolink, Cyberstar, Echostar, Galaxy/Spaceway, GE*Star, KaStar, Millennium, Morning Star, NetSat 28, Orion (F7, F8, and F9), PanAmSat (PAS-10 and PAS-11), Teledesic, VisionStar, and VoiceSpan. Table 2.13 gives an overview of these systems. Out of these original filings, some systems have been combined and others (for example, VoiceSpan), or have been withdrawn. The total number of satellites proposed for GEO systems is 69, in addition to 288 satellites proposed by Teledesic in a LEO constellation.

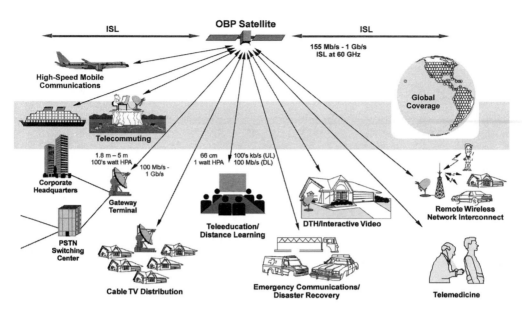

FIGURE 2.61 Multimedia system concept. (Courtesy of COMSTAT.)

TABLE 2.13 Proposed Ka-band Satellite Communications Systems [6]

System	Orbit	Coverage	No. of Satellites	Capacity (Gb/s)	Intersatellite Link	Onboard Switching	Capital Investment ($B)
Astrolink	GEO	Global	9	9.6	1 Gb/s	FPS	4.00
Cyberstar	GEO	Limited Global	3	4.9	1 Gb/s	BBS	1.05
Echostar	GEO	U.S.	2	5.8	120 MHz	BBS	0.34
Galaxy/Spaceway	GEO	Global	20	4.4	1 Gb/s	BBS	5.10
GE*Star	GEO	Limited Global	9	4.7	None	None	4.00
KaStar	GEO	U.S.	2	7.5	155 Mb/s	FPS	0.65
Millennium[a]	GEO	U.S./Americas	4	5.2	1 Gb/s	FPS	2.30
Morning Star	GEO	Limited Global	4	0.5	None	None	0.82
NetSat 28	GEO	CONUS	1	772.0	None	Optical SW	0.25
Orion	GEO	U.S./IOR	3	2.9	TBD	FPS	0.73
PanAmSat	GEO	AOR	2	1.2	None	None	0.41
Teledesic/Celestri	LEO	Global	288[a]	13.3	1 Gb/s	FPS	9.00
VisionStar	GEO	CONUS	1	1.9	None	None	0.21
VoiceSpan[b]	GEO	Limited Global	12	5.9	0.5 Gb/s	FPS	N/A

Note: FPS: fast packet switch, BBS: baseband switch.
[a] Subject to revision as a result of Teledesic/Celestri combination.
[b] Withdrawn in 1997.

It should be noted, however, that development of these systems is contingent on overcoming significant challenges in terms of frequency coordination, technology readiness, and financing. Considering the current status and activities of the proponents, only a few of the proposed systems are expected to be deployed in the 2002 to 2005 time frame.

Proposed V-Band Systems

In September 1996, Motorola Satellite Systems, Inc. filed an application with the FCC requesting authorization to deploy a system of 72 LEO satellites (M-Star) which would operate in V-band and provide multimedia and other services. Motorola's application was followed by 15 other applications for V-band satellite systems, filed by 13 U.S. companies in 1997. Hughes Communications, Inc. alone filed for three

TABLE 2.14   Proposed V-band Satellite Communications Systems [6]

System	Orbit	Coverage	No. of Satellites	Capacity (Gb/s)	ISL Capacity	Onboard Switching	Capital Investment ($B)
Aster	GEO	Global	25	6.2	Optical	SSTDMA & BBS	2.4
CAI	GEO	CONUS	1	1.86	None	Bent-Pipe	0.3
CyberPath	GEO	Global	10	17.9	447.5 Mb/s	FPS	1.2[a]
Expressway	GEO	Global	14	65.0	Optical	SSTDMA (BB)	3.8
GESN	GEO & MEO	Global	4 & 15	50.0 & 75.0	Optical 2.5 Gb/s	BBS	3.4
GE[a]StarPlus	GEO	Global	11	65.0	Optical	Bent Pipe	3.4
GS-40	LEO	Global	80	1.0	None	Bent Pipe	Not Avail.
Leo One USA	LEO	Global	48	0.007	None	BBS	0.03
M-Star	LEO	Global	72	3.7	830 Mb/s	Bent-Pipe	6.4
OrbLink	MEO	Global	7	75.0	1.244 Gb/s	Bent-Pipe	0.9
Pentriad	HEO	Limited global	9	30.2	None	Bent-Pipe	1.9
Q/V-Band	GEO	Global	9	31.3	Optical & Radio	FPS	4.8
SpaceCast	GEO	Limited global	6	60.0	Optical 3 Gb/s	SSTDMA (RF)	1.7
StarLynx	GEO & MEO	Global	4 & 20	5.9 & 6.3	Optical 3 Gb/s	BBS	2.9
VBS	LEO	Global	72	8.0	Optical 1 Gb/s	FPS	1.9
V-Stream	GEO	Global	12	32.0	1 GHz	Bent-Pipe	3.5

Note: HEO: highly elliptical orbit, FPS: fast packet switching, BBS: baseband switching.
[a] For only four satellites.

of these systems (four including a PanAmSat filing). The 1997 filings were in response to a deadline of September 30, 1997, set by the FCC for such applications. A majority of the V-band applications were filed by the same companies that had filed for multimedia Ka-band systems in 1995. Generally, the V-band systems, filed by those who had already planned to operate at Ka-band are intended to supplement the capacity of the Ka-band systems, and especially to provide service at higher data rates to regions of high capacity demand.

The 16 proposed V-band systems are Aster (Spectrum Astro, Inc.), CAI (CAI Satellite Communications, Inc.), CyberPath (Loral Space and Communications), Expressway, SpaceCast, and StarLynx (Hughes Communications), GESN (TRW), GE*StarPlus (GE Americom), GS-40 (Globalstar), Leo One USA (Leo One USA Corp.), M-Star (Motorola), OrbLink (Orbital Sciences), Pentriad (Denali Telecom), Q/V-Band (Lockheed Martin), V-Band Supplement (VBS) (Teledesic Corp.), and V-Stream (PanAmSat). Table 2.14 lists some of the characteristics of these systems [6].

Key Technologies

The majority of the proposed systems employ either FDMA/TDMA or TDMA transmission. Uplink bit rates vary from 32 kb/s to 10 Mb/s for a typical user. Most systems propose to employ advanced technologies — in particular digital signal processing and onboard switching — owing to the successful technology demonstration provided by the ACTS program. Demonstrated ACTS technologies include hopping beams, onboard demodulation/remodulation, onboard FEC decoding/coding, baseband switching, and adaptive fade control by FEC coding, and rate reduction. However, some of the proposed systems are vastly more complex than ACTS (a system capacity of 220 Mb/s in ACTS vs 10 Gb/s in the proposed systems, for example) and also employ new processing/switching technologies such as multicarrier demultiplexer/demodulators for several thousand carriers, fast packet switching (FPS), and intersatellite links (ISLs). In addition, network control functions traditionally performed by a ground control center will be partially implemented by an onboard network controller. Three of the key technology areas that influence payload and terminal design — onboard processing, multibeam antennas, and propagation effects — are discussed in the subsections that follow.

Onboard Processing

The majority of the proposed Ka- and V-band systems employ onboard baseband processing/switching [4], although some will use "bent-pipe" or SS-TDMA operation onboard the satellites. A majority of the processing systems will employ fast packet switching, which is also referred to as cell switching, packet switching, ATM switching, and packet-by-packet routing in the FCC filings. The remainder of the processing systems are currently undecided regarding their baseband switching (BBS) mechanisms and will probably use either FPS or circuit switching. Along with onboard baseband switching, most of the processing satellites will employ digital multicarrier demultiplexing and demodulation (MCDD). Onboard baseband switching allows optimized transmission link design based on user traffic volume, and flexible interconnection of all users in the network at the satellite. ISLs will provide user-to-user connection in many of these systems without assistance from ground stations.

Gallium arsenide (GaAs) monolithic microwave integrated circuit (MMIC) technology has been used successfully to develop microwave switch matrix (MSM) arrays for SS-TDMA operation and RF demodulator/remodulator hardware. Further development of low-power, application-specific integrated circuits (ASICs) with high integration densities and radiation tolerance is critical to the realization of relatively complex onboard processing and control functions.

Multibeam Antennas

The design of the satellite antennas depends on the beam definition, which in turn is a function of system capacity and projected traffic patterns. Several systems require a large number of fixed, narrow spot beams covering the whole service area and designed to deliver a high satellite EIRP of 50 to 60 dBW to user terminals. A single reflector with a large number of feeds may provide such coverage. However, if scanning loss is excessive due to the large number of beams scanned in one direction, multiple reflectors may be required. The coverage area is divided into a number of smaller areas, with each reflector boresight at the center of the corresponding area. Similarly, single or multiple phased arrays may be used. The phased array may have lower scan loss and, with the flexibility of digital beam formers, a single phased array can handle a large number of beams with no added complexity. If the system design calls for a small number of hopping beams instead of large number of fixed beams, the phased array solution becomes more attractive due to the flexibility and reliability of the beam-former vs. the switching arrangement in a focal-region-fed reflector antenna. For a small number of beams at a time, the microwave beam-former becomes a viable alternative and the choice between the microwave and the digital beam-formers becomes a payload system issue.

At Ka- and V-band, the manufacturing tolerance of the array feed elements and the surface tolerance of the reflectors play an important role in overall antenna performance. Waveguide-based elements are well developed at Ka-band, but lighter weight, lower profile printed circuit elements may need further development for space applications. For a large number of beams and a large number of frequency reuses, co- and cross-polarization interference becomes a major issue that imposes severe restrictions on the antenna sidelobe and cross-polarization isolations. Although the receive antenna may benefit from statistical averaging based on user distribution, the transmit antenna must satisfy a certain envelope in order to meet the required interference specifications.

Propagation Effects

Line-of-sight, rain attenuation, and atmospheric propagation effects are not significant at L-, S- and C-bands. At high elevation angles the communications between satellites and terminals at L- and S-bands is very reliable. In the mobile environment, however, multipath effects and signal blockages by buildings cause signal fades that require cooperative users willing to change their location. In addition, the links must be designed at worst-case user elevation and for operation at the beam edge. Because of these considerations, a 10- to 15-dB link margin is designed into the systems.

In comparison, the troposphere can produce significant signal impairments at the Ku-, Ka- and V-band frequencies, especially at lower elevation angles, thus limiting system availability and performance [7, 22]. Most systems are expected to operate at elevation angles above about 20°. Tropospheric radio wave propagation factors that influence satellite links include gaseous absorption, cloud attenuation, melting layer attenuation, rain attenuation, rain and ice depolarization, and tropospheric scintillation. Gaseous absorption and cloud attenuation determine the clear-sky performance of the system. Clouds are present for a large fraction of an average year, and gaseous absorption varies with the temperature and relative humidity. Rain attenuation — and to some extent melting layer attenuation — determine the availability of the system. Typical rain time is on the order of 5 to 10 percent of an average year. Depolarization produced by rain and ice particles must be factored into the co-channel interference budgets in frequency reuse systems. Signal amplitude fluctuations due to refractive index inhomogenieties (known as *tropospheric scintillation*) are most noticeable under hot, humid conditions at low elevation angles. Scintillation effects must be taken into account in establishing clear-sky margins and in designing uplink power control systems. Figure 2.62 shows the combined clear sky attenuation distribution at Ka- and V-band frequencies for a site in the Washington, D.C., area at an elevation angle of 20°. Figure 2.63 shows the rain attenuation distribution for the same site. It can be seen that the required clear-sky margin can be several dBs at V-band, especially at 50 GHz, due to the elevated oxygen absorption. Figure 2.63 indicates that, to achieve reasonable availabilities, rain fade mitigation must be an integral part of the system design. Rain fade mitigation can be accomplished through power control, diversity, adaptive coding, and data rate reduction.

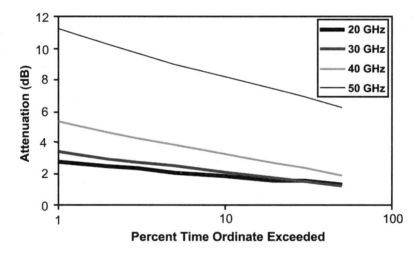

FIGURE 2.62   Probability distribution of clear-sky attenuation at 20° elevation, Washington, D.C. [22].

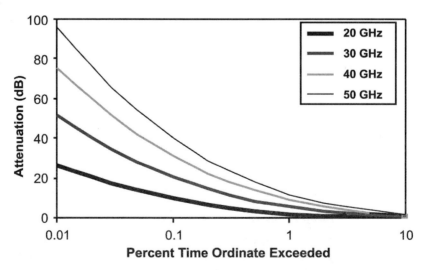

FIGURE 2.63   Probability distribution of rain attenuation at 20° elevation, Washington, D.C. [22].

User Terminal Characteristics

A variety of user terminals have been proposed for the Ka- and V-band systems. A typical user terminal operates at an uplink bit rate between 128 kb/s and 1 Mb/s for Ka-band systems, and up to 10 Mb/s for V-band systems. These terminals employ small-aperture antennas with diameters of 50 to 66 cm (Ka-band), and as small as 20 cm for V-band, as well as a solid-state power amplifier (SSPA) of 1 to 10 W. Uplink power control is required for all of these terminals. Terminal antennas for LEO and MEO systems must provide tracking capability to perform handovers every few minutes. All RF components (SSPA, LNA, and up-downconverters) are integrated into a small outdoor unit (ODU) and antenna system. Development of low-cost antennas and low-cost, low-power consuming RF integrated circuits (RFICs) at Ka- and V-bands is critical to substantially reducing the cost of the terminals to targets of $1,000 to $3,000. Gateway terminals employ a larger antenna with a diameter of 2.4 to 7 m and an HPA of 50 to 200 W.

Summary

The overall telecommunications market is growing very rapidly, largely because of increased demand for video traffic and exponential growth in the Internet. Multimedia services using satellites are now beginning to emerge. Growth in international trade, reduced prices due to privatization of telecommunications services worldwide, access to the World Wide Web, and significant drops in prices of desktop and portable computers are all contributing to a heavy demand for such services. Satellites also provide "instant infrastructure" and are therefore viewed as a cost-effective solution for providing wide area coverage for developing countries. In the developed world, satellites could provide effective "last mile" connection to businesses and homes for broadband data.

A number of Ka- and V-band systems have been proposed, and many of Ka-band systems are at advanced stages of implementation. Many of these systems require capital investment of several billions of dollars to implement. As these systems become operational in the 2003 to 2010 time frame, several technological and market challenges must be overcome. Perhaps the most important challenge is to develop a low-cost ($1,000 to $3,000) customer premises terminal. Low-cost RF integrated circuits, multichip packaging techniques, and low-cost Ka-band antennas must be developed to meet these cost goals. While significant progress continues to be reported in Ka-band component technology, the use of V-band presents additional challenges.

Acknowledgments

The author gratefully acknowledges the contributions of many of his colleagues at COMSAT Laboratories to the work described herein. In particular, the author thanks P. Chitre, A. Dissanayake, J. Evans, T. Inukai, and A. Zaghloul of COMSAT Laboratories for many useful discussions.

References

1. Evans, J., Network interoperability meets multimedia, *Satellite Commun.*, 30–36, February 2000.
2. Pontano, B., Satellite communications: services, systems, and technologies. *1998 IEEE MTT-S International Microwave Symposium Digest,* June 1998, 1–4.
3. Evans, J., Satellite and personal communications, 15th AIAA International Communications Satellite Systems Conference, San Diego, CA, Feb/Mar 1994, 1013–1024.
4. Evans, J., Satellites systems for personal communications, *IEEE Antenna and Propagation Mag.*, 39, 7–10, June 1997.
5. Williams, A., Gupta, R., and Zaghloul, A., Evolution of personal handheld satellite communications, *Applied Microwave and Wireless*, 72–83, Summer 1996.
6. Evans, J., The U.S. filings for multimedia satellites: a review, *Int. J. of Satellite Commun.*, 18, 121–160, 2000.

7. Evans, J. and Dissanayake, A., The prospectus for commercial satellite services at Q- and V-band, *Space Commun.*, 15, 1–19, 1998.

8. Bennett, S. and Braverman, D., INTELSAT VI — A continuing evolution, *Proc. IEEE*, 72, 11, 1457–1468, November 1984.

9. Wong, N., INTELSAT VI — A 50-channel communication satellite with SS-TDMA, *Satellite Communications Conference*, Ottawa, Canada, June 1983, 20.1.1–20.1.9.

10. Gallagher, B., ed., *Never Beyond Reach — The World of Mobile Satellite Communications*, International Maritime Satellite Organization, London, U.K., 1989.

11. Reinhart, E. and Taylor, R., Mobile communications and space communications, *IEEE Spectrum*, 27–29, February 1992.

12. Evans, J., Satellite and personal communications, *15th AIAA International Communications Satellite Systems Conference*, San Diego, CA, Feb.–Mar. 1994, 1013–1024.

13. Gupta, R. and Assal, F., Transponder RF Technologies Using MMICs for Communications Satellites, *14th AIAA International Communications Satellite Systems Conference*, Washington, D.C., March 1992, 446–454.

14. Sorbello, R. et al., A Ku-Band Multibeam Active Phased Array for Satellite Communications, *14th AIAA International Communications Satellite Systems Conference*, Washington, D.C., March 1992.

15. Assal, F., Gupta, R., Betaharon, K., Zaghloul, A., and Apple, J., A wideband satellite microwave switch matrix for SS-TDMA communications, *IEEE J. Selected Areas Comm.*, SAC-1(1), 223–231, Jan. 1983.

16. Gupta, R., Narayanan, J., Nakamura, A., Assal, F., and Gibson, B., INTELSAT VI On-board SS-TDMA subsystem design and performance, *COMSAT Technical Review*, 21, 1, 191–225, Spring 1991.

17. Gupta, R., Assal, F., and Hampsch, T., A microwave switch matrix using MMICs for satellite applications, IEEE MTT-S International Microwave Symposium, Dallas, TX, May 1990, *Digest*, 885–888.

18. Gupta, R. et al., Beam-forming matrix design using MMICs for a multibeam phased-array antenna, IEEE GaAs IC Symposium, October 1991.

19. Gupta, R., MMIC insertion in communications satellite payloads, Workshop J: GaAs MMIC System Insertion and Multifunction Chip Design, IEEE MTT-S International Microwave Symposium, Boston, MA, June 1991.

20. Chitre, D., Gokhale, D., Henderson, T., Lunsford, J., and Matthews, N., Asynchronous Transfer Mode (ATM) Operation via Satellites: Issues, Challenges, and Resolutions, *Int. J. of Satellite Commun.*, 12, 211–222, 1994.

21. Chitre, D., Interoperability of satellite and terrestrial networks, IEEE Sarnoff Symposium, April 2000.

22. Dissanayake, A., Allnutt, J., and Haidr, F., A prediction model that combines rain attenuation and other propagation impairments along the Earth-satellite path, *IEEE Trans. on Antennas and Propagation*, 45, 10, October 1997.

2.8 Satellite-Based Cellular Communications

Nils V. Jespersen

Communication satellite systems designed to serve the mobile user community have long held the promise of extending familiar handheld cellular communication to anywhere a traveler might find himself. One impetus to the fulfillment of this dream has been the success of the Inmarsat system of communication satellites. Founded in 1979 as an international consortium of signatories, Inmarsat provides worldwide communication services to portable and transportable terminals, thereby meeting one of its mandates by enhancing safety on the high seas and other remote areas. Although it does not support handheld, cellular-like operations (due to limitations in the satellite design), the current Inmarsat

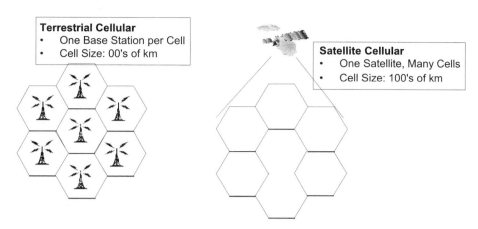

FIGURE 2.64   Comparison: terrestrial vs. satellite cellular.

system is a successful business. Clearly, the next logical step would be to enhance the capabilities of the space segment to provide cell-phone utility with even greater ubiquity than available terrestrially. What we would then accomplish is to, essentially, raise the familiar cellular base station several hundreds of kilometers high and, thereby, extend the coverage many times over (Fig. 2.64).

Many elements comprise the optimum solution to a satellite cellular system, not the least of which is the business aspect. The best technical solution does not necessarily result in a successful overall solution. A significant market consolidation is in process at the time of this writing.

In this discussion, we will focus on the systems that are intended to provide voice and/or data services similar to terrestrial cellular communications. As such, we will not be discussing systems intended to provide low-rate messaging and asset tracking services (the so-called "little-LEO" systems such as Orbcomm and LEO-One). Of the many mobile satellite systems proposed in 1995 to provide cellular-like service, only a handful remain as viable. We will examine the salient characteristics of satellite cellular system design and, then consider some specific system examples.

Driving Factors

How a particular cellular satellite system is configured will depend on a variety of factors along with the importance assigned to each characteristic. In this section we will explore some of these top-level design drivers and consider their impact on the overall system design. While this listing is far from exhaustive, the parameters discussed do tend to be common considerations in every cellular satellite system design.

The User Terminal (UT)

The baseline assumption in this discussion is that cellular satellite system users interface with the system using "disadvantaged" terminals. By this terminology we mean that the UT is "disadvantaged" from an electronic design standpoint, implying that the terminal is small, lightweight, and convenient to carry on the user's person. This convenience in the UT is achieved at the cost of both antenna gain and transmit power (meaning low gain-to-system-temperature [G/T] and low effective isotropic radiated power [EIRP]). The UT G/T must be high enough to ensure acceptable, low-noise reception of the signal from the satellite. Similarly, the UT EIRP must be high enough to ensure that the satellite will be able to reliably process and translate the signal from the UT. We refer to this basic design consideration as "closing the link." Far from being a "nice to have" characteristic, in the commercial world of handheld wireless communications, UT convenience plays a *significant* role in customer acceptance; a fact borne out within the Iridium cellular satellite system (discussed below). A small UT, in turn, makes the cellular satellite system engineer's job harder in that the performance burden now shifts to the space segment. The spacecraft, now, has to have both higher EIRP and G/T in order that the subscriber can have a small, sleek terminal that easily fits in his pocket. The impact on the system is, therefore, more complexity and cost.

Target Market

Arguably, the most important parameter to identify is the market that the cellular satellite system is targeted to serve. Military needs tend to be very different from those typically considered for a commercial system. For instance, guaranteed availability, reliable communication through multiple layers of jungle canopy, nuclear event hardening and probability-of-intercept, which are often found in the requirements levied by the Department of Defense (DoD), are rarely, if ever, discussed in reference to commercial systems. Another consideration is whether the system is intended to serve a worldwide user group (including, perhaps, coverage of polar regions and open oceans), or the users are located in a general geographic region (e.g., Europe, Asia, specific landmasses). The state of terrestrial communication infra-structures in the target market — whether wired or wireless — must also be identified and evaluated.

These market considerations tend to drive several high-profile system design decisions. A DoD system, with a need to penetrate jungle canopy, would drive the selection of a particular operating frequency band to lower frequencies (e.g., UHF). Satellite orbit type, for instance, would need to be of either the low earth or medium earth orbiting type (LEO or MEO) if worldwide coverage is required. Regional coverage, on the other hand, could be met with a satellite at geosynchronous earth orbit (GEO). The existence of adjacent terrestrial cellular coverage might influence the selection of a particular radio air interface structure (AIS). For instance, if the satellite cellular coverage area includes population centers operating on a particular AIS (Global System for Mobile communication, or GSM, for instance), then it might be prudent to base the satellite AIS on something similar in order to simplify UT design and, also facilitate roaming arrangements.

Expected Subscriber Population

An objective, well-researched estimate of the targeted subscriber population is an essential element in the viability of a cellular satellite system. Considerations include:

- Expected total number of subscribers.
- Expected peak number of users in view of a given satellite compared to the expected average user loading.
- Geographic distribution of the user population: whether they are evenly dispersed over the defined coverage area or concentrated in specific locations (population centers).

The total number of subscribers drives design elements such as the capacity of the overall system authentication registers: the so-called home location registers (HLR) and the visitor location registers (VLR). On the other hand, the number of active users in the satellite field-of-view has a direct impact on the satellite architecture. Generally, the satellite downlink tends to be the limiting constraint, i.e., the number of users that can be accommodated is based on the amount of available power in the downlink transmitter, since each downlink user signal receives a share of the total transmit power. In systems based on frequency division multiple access (FDMA), or a combination FDMA/TDMA (Time Division Multiple Access), it is important that the transmit system is operated in the linear region of the transmitter gain characteristic. As more carriers are added to the composite downlink signal (assuming equal power per carrier as necessary to close the link to the individual UT), the transmitter is driven closer to, and sometimes into, the nonlinear operating region. A transmitter driven with a multicarrier signal, at a nonlinear operating point, will generate intermodulation distortion (IMD) products due to the mixing of the individual carriers with one another (a classical heterodyning phenomenon). In a system trans-mitting many (e.g., hundreds) of carriers, particularly when the carriers have modulation, the IMD tends to take on a random characteristic and tends to appear as a uniform noise spectrum that is band limited by the output filter. The resulting effect is an increase in the transmitted noise of the system. This noise is characterized by the Noise Power Ratio (NPR) of the system and has a direct bearing on the downlink carrier-to-noise (C/N) ratio. System designs often include some type of automatic power control adjust-ment in order to maximize the number of downlink user signals that can be supported while maintaining the NPR at an acceptable level. Such a power control loop generally involves a measurement of received power level at the individual UT, which is then reported back to the Network Control Center (NCC) by way of a parallel control channel.

Assuming that the satellite transmitter can accommodate the expected peak traffic load, the next limitation becomes the switching capability within the satellite channelization equipment. The required switching capacity will generally dictate whether this channelization equipment is implemented in an analog or digital fashion. Lower capacity systems, such as the Inmarsat-3, are well served by conventional analog channelization and switching technology. By contrast, digital channelization and switching technology is required in modern cellular satellite systems that serve thousands of simultaneous users and have frequent associated call setups and teardowns.

Distribution of users in the satellite field of view drives the design of the antenna (cell coverage) pattern. In satellite cellular systems, the bulk of which operate in the crowded L-band or S-band spectra, efficiency of spectrum use is a critical concern. Operators of a proposed cellular satellite system must apply for an allocation of the spectrum pool, usually as a result of common meetings of the World Radio Conference (WRC) wherein many operators negotiate for spectrum rights. Since, inevitably, granted allocations are less than that requested, a major motivation exists to design the proposed system with a high level of frequency reuse. Frequency reuse implies that multiple cells (beams) in the satellite system can use and reuse the same frequency but with different user signals modulated onto them. Typical reuse values, for regional systems like ACeS and Thuraya, are on the order of 20 times. Thus, the same carrier frequency can be reused 20 or more times over the satellite field of view, thereby increasing the available capacity (the number of UTs that can be simultaneously served) by that same amount. By the same token, the link from the satellite to the Feeder, or Gateway, station must have an adequate spectrum allocation to accommodate the required terrestrial connectivity for the mobile UTs. Since Satellite-to-Feeder links are most often implemented in a higher frequency band (e.g., C, Ku or Ka band), wider operating bandwidths are generally more easily obtained and coordinated.

An operational scenario where many potentially active users are located in a concentrated area (population center), can put a strain on the amount of available spectrum. This concentration can lead to the need for smaller cell sizes (narrower beam patterns), which, in turn, drives the size of the satellite antenna (making it larger and more difficult to accommodate on the spacecraft). Given a specific coverage area, or satellite field of view, smaller cells mean more cells and, therefore, more switching hardware on the spacecraft. This increase in hardware means that the spacecraft becomes more complex and costly.

The users through busy signals, if reached, will negatively observe system capacity limits. How frequently a user encounters a busy signal is referred to as "call blocking rate" and is one measure of the system's quality of service (QoS). The Quality of Service Forum[1] defines QoS as "… consistent, predictable telecommunication service." Predictable service is what the subscriber demands and is, ultimately, the measure that will determine the success of the system. The elements that need to be assessed (estimated) for a planned cellular satellite system, in order to gain an estimate of required system capacity, include:

- The number of call attempts per second
- The average call holding time
- The expected distribution of call types
  - Percentage of Mobile Originated (MO) calls
  - Percentage of Mobile Terminated (MT) calls
  - Percentage of Mobile-to-Mobile (MM) calls

All of these factors impact system design decisions regarding:

- Satellite switching speed
- Satellite switching capacity
- Onboard buffering and memory capacity

Operating Environment

For a commercial system, putting aside political issues such as frequency coordination for the moment, we would want to select an operating frequency band that can penetrate at least *some* buildings. Additionally, the operating band should permit acceptable performance to be obtained with nondirectional antennas on the UT (i.e., a tolerably low space spreading loss with reasonably sized electrically small antennas). These constraints tend to drive toward lower frequency bands (e.g., L-band or S-band, which are roughly 1500 MHz and 2500 MHz, respectively).

Link Margin Considerations

During the system design phase, the radio link between the satellite and the UT has to be given a great deal of detailed consideration. Both the typical and disadvantaged user conditions need to be handled. For instance, the typical case might find the user with a clear line of sight to the satellite. On the other hand, a disadvantaged condition might be where the user is in the middle of an office building, in a city (a "concrete jungle" where multipath propagation could be an issue) and, perhaps, at the edge of the satellite's coverage area (low elevation angle, greatest path distance to the satellite and, therefore, greatest signal loss). Further, meteorological conditions such as rain and, to a lesser extent, snow will degrade the link performance in proportion to both the rate of precipitation and the frequency of operation. Operation in a tropical area would clearly be more affected by rain attenuation than if the system were intended to serve, say, Northern Africa. In all of this, link margin is "king." If we take the traditional approach to assessing the amount of available link margin in the system, we can define link margin (nonrigorously, and in decibel terms) as

$Link\ Margin = (C/N)_{received} - (C/N)_{required}$, where:

- $(C/N)_{received}$ is the ratio of the power of the transmitted signal (at the receiver) to the total noise power (e.g., thermal noise, interference, and other degradations) impinging on the receiver. In the example of the link going to the UT, the satellite is the transmitter and the UT is the receiver. Mobile service providers will often refer to this transmission direction as the "Forward" direction. Transmission *from* the UT *to* the satellite is, likewise, called the "Return" transmission.
- $(C/N)_{required}$ is the minimum ratio of received signal to total noise required at the receiver in order to accomplish acceptable detection with the modulation method selected.

We can gain additional insight into what this equation means by breaking $(C/N)_{received}$ into its constituent components. Thus,

$Link\ Margin = \{EIRP - L_{impairments} + (G/T)\} - (C/N)_{required}$ where:

- *EIRP* is the Effective Isotropic Radiated Power of the transmitter (e.g., the satellite in the case of the link from the satellite to the UT).
- (G/T) is the figure of merit often applied to satellite receiving stations, in this case the UT. (G/T) is the ratio of the passive antenna gain (G), at the receiving station in the direction of the incoming signal, to the total system noise of the station expressed as an equivalent temperature (T). The primary component of the system noise is, usually, the noise added by the passive components (following the antenna) and the noise figure of the first amplifier, or low-noise amplifier (LNA), in the receiver. Additionally, the noise contributed by the subsequent components in the UT receiving chain is referred back to the antenna terminal (suitably scaled by the gains of the components ahead in the signal flow).
- $L_{impairments}$ is the combination of losses and effects on the transmission channel that tend to degrade the overall received signal quality.

In other words, link margin can be thought of as the excess desired signal power available at the receiver once we have accounted for the signal strength required by the demodulator in the receiver (commensurate with the chosen modulation format), the inevitable thermal noise, and the impairments suffered along the way. Some of the impairments that have the greatest detrimental impact on the link margin are:

- Transmitter intermodulation, as characterized by the NPR.
- Spreading loss: path loss, which is proportional to the distance between the UT and the satellite; very much driven by the slant range, or elevation angle at the UT.
- Atmospheric loss: signal absorption in the propagation path; proportional to the carrier frequency and the amount of moisture in the air.
- Polarization mismatch: orientation of the UT antenna relative to the satellite antenna.
- Body losses: absorption and blockage of the communication signal by the user's body.
- Multipath interference: in CDMA systems, rake receivers can actually take advantage of multipath to *enhance* the received signal. In most other multiple access techniques, multipath is detrimental.
- Co-channel interference: leakage from the other channels on the same frequency but in a different beam (or, for a Code Division Multiple Access, or CDMA-based system, signals in the same beam but with a different code).
- Adjacent channel interference: leakage from a channel on an adjacent frequency.
- Digital implementation loss: effects such as spectrum truncation due to the finite filtering bandwidth in the receiver's demodulator.

Each of these factors must be quantified and accounted for in the overall impairment budget in order to ensure the link margin needed to provide acceptable service.

Implementation Loss — It is of particular interest to consider link margin in light of the modern trend toward digital satellite systems where a substantial amount of onboard signal processing takes place. Many of these digitally processed systems will demultiplex, demodulate, process, switch, remodulate, and remultiplex the communication signals. These digital transponders stand in contrast to traditional "bent pipe" transponder approaches where the uplinked signal is merely filtered and translated in frequency before being downlinked. In systems that are all digital, the primary performance measure of interest is the demodulated E_b/N_0, which is the signal energy per bit (E_b) divided by the noise density (N_0). We can relate the classical signal- (or carrier-) to-noise ratio to E_b/N_0 by way of the raw transmission bit rate and the pre-detection bandwidth, as:

$$(C/N)_{dB} = (E_b/N_0)_{dB} + (R/W)_{dB}, \text{ where:}$$

- R is the bit rate in bits per second, and
- W is the pre-detection bandwidth in Hz.

Typical performance curves plot bit error rate as a function of E_b/N_0, the so-called "waterfall curves," as shown in Fig. 2.65.

Each impairment that we discussed above could be considered as a contributor to the overall digital implementation loss. In effect, for digital modulation schemes (e.g., m-ary PSK or n-ary QAM, where m and n are integers) there are three contributors to implementation loss that can be quantified and modeled as a characteristic of every block in the overall system. These contributors are:

- Additive White Gaussian Noise (AWGN): Thermal noise, interference, and other random noise effects that degrade the received signal-to-noise ratio (and, therefore, directly affect the E_b/N_0.
- Phase Distortion: Nonlinearities in the transmission of phase information.
- Amplitude Distortion: Nonlinearities in the transmission of amplitude information.

In fact, all of the impairments we have discussed could be couched within the above three terms. The amount of implementation loss suffered, at any given stage, will be a strong function of the type of modulation chosen. For instance, m-ary Phase Shift Keyed (PSK), which has the intelligence coded in the phase of the signal only, will be much more affected by phase distortion than by amplitude distortion. Generally speaking, one can hard-limit a PSK signal and not lose its detectability. On the other hand, an n-ary Quadrature Amplitude Modulated (QAM) signal uses both amplitude and phase to code the intelligence. Such a signal would, clearly, be sensitive to both amplitude distortion as well as phase

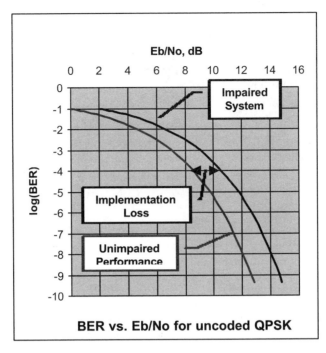

FIGURE 2.65   Typical "waterfall curve"; bit error rate vs. Eb/No for uncoded QPSK.

distortion. Each contributor to the overall implementation loss tends to shift the waterfall curve to the right (as shown in Fig. 2.65), implying a need for higher E_b/N_0 for a given (desired) bit error rate. As a consequence, for digital systems, there is justification for assessing link margin solely in terms of the various implementation losses encountered throughout the system. That is not to say that the conventional measures we previously discussed are obsolete. On the contrary, if we have limited control over the signals to be carried within the cellular satellite system we are designing, then the conventional measures offer the best common ground on which to specify the system performance. On the other hand, if we have complete control over both the ground and the space segments of the system, and it is an all-digital system (with a modulation format of our own choosing), we can streamline the entire analysis process, and drive more directly to the bottom line (i.e., E_b/N_0), if we model each segment in terms of its digital implementation loss to the end-to-end performance.[2]

   The success of a given system hinges on customer satisfaction, and a large part of that satisfaction is derived from the ability to make successful calls "most of the time." The larger the link margin (i.e., the more degradation the link can suffer before communication is no longer possible), the more frequently a user will be able to complete successful calls, and the more often he is likely to use the system. The link design has to contain enough margin to cope with these "most of the time" situations, which is why cellular satellite system links are, as a rule, designed on a statistical basis.[3] No hard and fast rule exists here as the trade-off is subject to interrelated and, often, subjective criteria. Current cellular satellite systems, regardless of whether LEO, MEO, or GEO, tend to present the average user with about 10 dB of link margin after all impairments have been considered. This somewhat counter intuitive result (given that GEO constellation orbits are some 40 to 50 times higher than the orbits of LEO systems) is due to the fact that the lower-complexity LEO satellites, as compared to GEO satellites, are smaller (lower overall transmit power capability) and must cover a broader angular field of view (lower antenna gain).

   If sound engineering conservatism prevails against the sometimes exuberant optimism of the marketing department, the average user will be defined as one who is located closer to the edge of a typical cell beam (rather than at the peak), does not have his UT antenna ideally oriented with respect to the spacecraft, and will probably be located inside of a building (but not too far away from a window). Such a design approach will tend to meet the "most of the time" criterion.

Other methods exist to address the five to ten percent of the conditions when the user is not ideally positioned. These methods generally include the assumption of a cooperative user. As an example, for a mobile terminated (MT) call (i.e., one in which the mobile user has an incoming call) the paging signal can be issued at a higher signal strength which, in turn, requests the user to move to a better line-of-sight position with respect to the satellite.

Latency

Latency refers to the amount of communication delay a cellular satellite system user experiences during a conversation or a data transaction. The effect of "substantial" latency can range from mild irritation (users at each end of the link talking over one another) to major transmission inefficiencies (multiple retransmissions under an IP environment), including dropped transactions, which could be disastrous at several levels ranging from economic to human safety considerations. The latency equation contains several contributing factors including:

- Propagation delay: the transit time (determined by the speed of light) for the signal to travel between the satellite and the UT. The total delay is, necessarily, twice the one-way delay since the signal must go from earth to satellite and back down again. The approximate round trip (earth-satellite-earth) delay for the three main orbital configurations are:
  - GEO (orbital altitude of approximately 35,900 km): 239 ms.
  - MEO (orbital altitude of approximately 10,300 km): 69 ms.
  - LEO (orbital altitude of approximately 700 to 1400 km): 6 ms.
- Coding delay: in digital systems the data to be communicated (be it digitized voice or computer files) is coded for various reasons (error correction and data compression are the most common reasons beside encryption and security concerns). The coding process implies that blocks of data must be stored before coding. Received blocks must, likewise, be buffered in blocks before decoding can take place. This buffering/coding process adds delay to the communication transmission. The amount of coding delay added to the transmission varies according to number of coding layers and the type of voice coder-decoder (codec or vocoder) employed. This coding delay can easily range between 50 to 150 ms.
- Relay delays: if a particular call needs to be routed through multiple nodes (for instance, satellite-to-satellite (ISLs), ground station-to-ground station, or relay station-to-relay station. Sometimes channels are demultiplexed and remultiplexed at these intermediate points (depending on the routing required and the multiplexing hierarchy required to effect the relay path). Again, the magnitude of the delay depends upon the path taken, but it will generally range between 10 and 40 ms.
- Other system delays: Aggregated digital communication channels, in the terrestrial infrastructure, are frequently buffered for purposes such as synchronization. Also, echo cancellers and the relative placement of the channel carrier (e.g., if placed, particularly, at the band edge of the channel filters) contribute varying amounts of delay. These processes also contribute to the overall latency that the user experiences and can range from negligible to 120 ms (for the worst combinations of echo cancellers and channel filter group delay).
  - Emerging terrestrial systems are being designed to accommodate network protocol infrastructures such as H.323, which includes support for "Voice over IP" (VoIP). Considerations, which include variable grades of service, are an integral part of H.323 and imply additional latency contributions. These considerations will apply directly to cellular satellite systems that are configured to operate in compatible packetized modes.
  - Concerns about the impact of latency on Asynchronous Transmission Mode (ATM) and TCP/IP communications over the satellite channel have interested many workers in the field. Many different approaches have been analyzed and experimentally evaluated to deal with the transmission

inefficiencies that could potentially arise.[4-7] It is evident that future cellular satellite systems will be required to have the capability of accommodating bursty packet modes such as these.

The relative impact of the various latency factors will, clearly, vary with the type of system. For a GEO system the latency is, generally, dominated by the propagation delay, which, in turn, is a direct function of the higher altitude of the satellite. LEO and MEO systems, with their lower orbital altitudes, have correspondingly lower values of propagation delay. This factor has been one of the main reasons that have caused several cellular satellite system designers to choose LEO or MEO constellations. On the basis of propagation delay alone, the decision would appear to be a "no-brainer" in favor of MEO or, especially LEO (particularly since the lower altitudes also provide a potential for lower spreading loss and, possibly, greater link margin). The subtleties of the other latency contributors, however, could easily conspire to erase the apparent advantage of the lower altitudes. For example, consider two users with LEO cellular satellite service located at opposite extremes of a GEO coverage area (separated by, say, 4000 km). If we assume that the two systems (LEO and GEO) have equal coding delays, the LEO path might have to traverse as many as four satellite planes and, perhaps, as many intermediate ground stations (if the system does not include ISLs). Consequently, with each relay making its own contribution to the overall latency, it is easy to see how a point-to-point LEO communication could enter the realm of the GEO propagation delay. This discussion is not intended to favor one type of system over another with respect to latency, but merely to point out that one must consider the full complement of effects in carrying out an impartial trade study.

Orbit Altitude and the Van Allen Radiation Belts

Though considerations of latency, individual satellite coverage, and link margin all tend to enter into the selection of system orbit altitude, another physical constraint exists as to the placement of the specific orbit. This constraint is placed by the location of the Van Allen radiation belts. These radiation belts (first discovered and characterized by Van Allen in 1959) consist of two annular rings of radiation that encircle the Earth and are centered on a plane defined by the equatorial latitude.[8-9] Sensitive spacecraft electronic equipment is readily damaged by high doses of radioactivity. Mitigation techniques usually involve component shielding of one form or another (e.g., thick metal boxes augmented with spot shielding with strips of tantalum), all of which imply additional spacecraft mass. Since mass is a premium item in spacecraft design (launch costs are proportional to launch mass), it is, clearly, desirable to make the satellite as light as possible. Flying a satellite within the Van Allen belts would entail massive amounts of shielding in order to achieve reasonable mission life. Consequently, orbits tend to be placed either to avoid the Van Allen belts altogether or to make only occasional (and very rapid) transitions through them. A schematic view of the radiation belt locations, from a polar perspective, is shown in Fig. 2.66. This figure also shows the relative locations of the three most common satellite orbits.

Other orbit types have been proposed for use in cellular satellite designs. Most notably, the elliptical orbit is planned for use in the Ellipso system.[10] In this system, a pair of elliptical orbits (perigee of 520 km and apogee of 7800 km) slips between the two Van Allen belts during the north-south transit.

Operating Environment Summary

Within the context of evaluating the operating environment of the system, we are primarily interested in assessing the amount of available link margin and determining the best way to maximize it. Higher link margin, therefore, primarily drives:

- The size of the satellite antenna (dramatically impacting cost, mass, and complexity).
- The optimum frequency band of operation (which, more often than not, has political ramifications that swamp the technical considerations).
- The power and linearity performance of the satellite transmitter (also a major cost element of the system).
- The noise performance (sensitivity) of the satellite LNA and receive system.

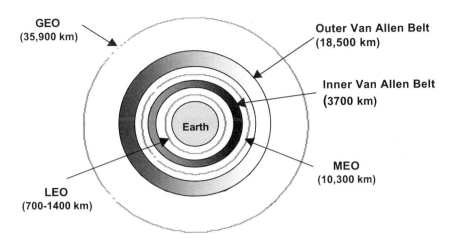

FIGURE 2.66 Locations of Van Allen radiation belts and typical cellular satellite orbits.

Where system latency is determined to be of high priority with respect to cellular satellite system performance, the main impacted design parameters tend to be:

- Satellite Orbit Altitude: impacts propagation delay, which tends to exhibit the greatest variability of all of the components of the latency equation.
- System coding approach: while important with respect to the latency assessment, may be more strongly driven by requirements to provide error detection and correction (EDAC) as well as communication security.

Service Offerings

Another major system design driver relates to the types of services that are to be offered to the user community. Many of the cellular satellite systems currently fielded were designed, primarily, to provide voice communication services with some facility to provide low-rate data (9.6 kb/s or less) and facsimile services. The fact that some systems were not "future-proofed" (i.e., physically incapable of supporting anything other than voice and low-rate data) has proven to be a considerable shortcoming and, in fact, has contributed to the demise of at least one early system. The explosive advent of the Internet in recent years has given rise to major advances in the wired communications infrastructure on a worldwide basis. The ubiquity of data communications at virtually every social stratum has fueled the tremendous growth of so-called e-commerce, a market that is forecasted to grow from $233 billion to over $1.44 trillion between the years 2000 and 2003.[11] Of this huge market, a very significant $200 billion slice (in 2004) is predicted to be transacted over some sort of wireless infrastructure.[12] Consequently, there exists a large incentive for a satellite cellular system operator to ensure that the system is capable of supporting wireless data services at attractive data rates (on the order of 100 kb/s or more).

If the system is to support data services, then there is a decision to be made as to whether the services will be sold as circuit switched or packet switched. Circuit switched services represent the traditional approach to providing communication services: a channel is allocated, and dedicated, to the customer for the duration of the call. The customer is subsequently billed for the amount of time that the call was active, whether or not any data was passed during that time. Packet switching, on the other hand, is metered on the basis of the amount of data that is transmitted. The customer does not use any system resources while idle. It is a "service-on-demand" paradigm in a mode of being "always connected" (similar to having a computer on a local area network in the wired world). Compatibility with data transmission frameworks, such as the Transmission Control Protocol/Internet Protocol (TCP/IP), is more streamlined and more efficient in a packet-based system. Infrastructures such as those developed for the General Packet Radio Service (GPRS) need to be incorporated into the design of the cellular satellite system if it is to support packet switching.

In summary, decisions on service offerings involve considerations as to whether the service backbone is to be voice-only or voice and data. If data is included, then the choice between circuit switching and packet switching needs to be considered. Besides the network equipment complement needed to execute these services, these decisions drive the basic system parameters of:

- Air interface: TDMA, FDMA, CDMA.
- Channel spacing: Higher data rates require wider spectral bandwidths which, in turn, mean that the channels need to be spaced further apart. CDMA approaches, likewise, require wider spread bandwidths for higher data rates.
- Modulation type: Bandwidth-efficient modulation methods should be employed in order to conserve system resources by transmitting the most amount of data for the lowest power and the narrowest bandwidth. Multilevel modulation types can be traded, such as various types of PSK or QAM.

Value Proposition

Will the typical user of the cellular satellite system be a business traveler, or will the business case assume a broader user population (penetrating lower economic strata)? Experience has shown that a target subscriber population based primarily on the affluent, worldwide business traveler places costly constraints on the system design and may not, necessarily, be attractive to the targeted consumer. For one thing, such a business case requires a worldwide coverage of satellites. This kind of coverage, in turn, dictates many satellites with narrow beams covering the entire surface of the earth. The most obvious approach to such worldwide coverage is to deploy a large fleet of satellites in a LEO configuration. The LEO constellation, if properly configured, provides the desired continuous global coverage, but it turns out to be extremely expensive to deploy such a system. Although the individual satellites, in a LEO system, tend to be small and inexpensive (in a relative sense considering the overall context of spacecraft technology), many satellites are required in order to achieve the desired coverage (tens to hundreds). The network infrastructure also demands a large number of gateways, with active features to track the rapidly moving satellites, in order to handle the connection of the mobile traffic to the public wired backbones. Clearly, the satellites could be designed to perform the call-by-call traffic switching, including relaying between satellites via intersatellite links (ISLs), which, in turn reduces the quantity of required gateways, but this approach adds complexity to the space segment and increases the overall system cost. Additionally, it is virtually required to have the entire constellation in place before a reasonable level of service can be offered for sale. A gradual ramp-up in service, based on a partially deployed constellation, is very difficult (if not impossible).

A system design based on the MEO configuration also has the potential of providing worldwide, or nearly worldwide, coverage with fewer satellites (on the order of tens or less) than a LEO version. MEO spacecraft tend to be more complex than LEO spacecraft, but less complex than their GEO counterparts. Fewer gateway stations are required to service a MEO constellation (relative to LEO) and, since the spacecraft are at a higher altitude, they move relatively slowly so that tracking is easier. Also because of the slow orbit, call handoffs between spacecraft tend to be less frequent. As a consequence, the total system cost for a MEO system tends to be somewhat lower than the cost of an equivalent LEO system. Also, depending on the actual orbits designed into the system and the quantity of spacecraft per orbital plane, MEO systems have the possibility of offering start-up service to selected areas on the Earth. Thus, some revenue can be returned to the enterprise prior to the complete deployment of the satellite constellation.

Total system cost is a major issue with regard to service pricing. The price point is generally set so as to provide a profitable return within a set time frame consistent with the business plan of the enterprise. It is evident that the base system cost is a large part of the initial investment against which a threshold to profitability is set. Obviously, the sooner the system finances cross this profitability threshold the sooner the system can be declared a success, making for a happy investor community. In this regard, GEO systems tend to be more cost effective, with a greater probability of being profitable at a lower price point. While GEO spacecraft tend to be larger, more complex and more costly than those destined for either LEO or MEO service, only one gateway is required to service the communication links. Connectivity is instantly available with the arrival of the spacecraft on station, and commercial service can generally be offered within a few months thereafter. Although GEO systems are, by nature, regional, careful selection of both service region and market mean that revenues can start flowing relatively quickly. Since the overall cost of deploying a regional GEO cellular satellite system is, by and large, lower than either a LEO or a MEO system, the threshold to profitability is potentially closer. The downside is, of course, that regional GEO systems, individually, cannot provide global coverage. Theoretically (except for the Polar Regions), three spacecraft placed at 120° longitude intervals can provide global coverage. The requisite narrow beams needed to cover the Earth field of view would, however, require multiple, very large and possibly impractical antennas in order to close the link to UTs. Alternatively, one could take the approach of concentrating on the major potential revenue producing areas on the Earth's surface (major landmasses), largely ignoring the open ocean areas. Under such a scenario "Earth coverage" could be accomplished with four to six GEO spacecraft, with two collocated spacecraft at longitudes over North-South America, and over Scandinavia-Europe-Africa.

Regardless of the approach taken, a successful cellular satellite system value proposition hinges on meeting the defined technical and service requirements with a system embodiment that minimizes system implementation cost and maximizes return on investment. The probability of obtaining a positive return on investment is inversely proportional to the risk taken in the design and deployment of the system. Several approaches can be taken to reduce the overall risk of the project. These risk reduction approaches include:[13]

- Minimizing development cost: reuse previously qualified designs to the maximum extent possible without compromising required system performance. Maximum reuse of qualified hardware also enhances (shortens) project schedules, and speeds system deployment and time to market.
- Maximizing system flexibility: anticipate future developments and plan to accommodate them. For instance, future data services will require wider channel bandwidths. A design that includes wider channel bandwidths will serve both current needs (e.g., voice transmission) in addition to allowing system migration to data services in the future.
- Well-researched business plan: diligently considered markets, along with a firm financing plan, helps to ensure steady progress in the deployment of the system.

Approaches

Having considered the fundamental driving factors of a cellular satellite system design, we now turn to specific implementations and the trade-off considerations that they imply. For convenience, a summary of the driving parameters already considered is shown in Table 2.15.

Regardless of the system embodiment selected, certain elements are common and will be found in any cellular satellite system. These main elements are depicted in Fig. 2.67.

Technology will, inevitably, continue to evolve. Unfortunately, once a satellite system is launched it is impractical, in most cases, to make modifications to the space segment. The typical 10 to 15 year on-orbit design lifetime of a spacecraft is close to an eternity when viewed against the backdrop of historic technology trends. These two incompatible facts make it incumbent upon system designers to anticipate the future and strive to make accommodations for upgrades to the extent practical. As we have previously discussed, system flexibility is key to future-proofing the design. Flexibility allows the system to support evolving services with minimum modification. For instance, a space segment able to support wideband channels will be able to grow with the relentless trend toward higher data rates. Beam-forming flexibility (allowing the coverage area to evolve with the market) is another approach to future proofing, either by way of digitally programmed onboard beamforming, ground-based beamforming, or by way of an overdesigned conventional analog approach. In the next section we will look at some specific system designs, the embodiments of which are the results of decisions made by previous cellular satellite system designers after having grappled with the concepts we have discussed.

TABLE 2.15 Summary of Cellular Satellite System Driving Parameters

Driving Parameter	Trade Issues
User terminal convenience	Satellite EIRP and G/T
Target market and its location	Operating frequency
	Orbit type
	Air interface structure
Subscriber population and geographic distribution	System capacity
	Authentication register size
	Satellite aggregate EIRP
	Multiple access method
	Satellite linearity requirements
	Satellite switching speed
	Satellite switching capacity
	Satellite memory capacity
	Channelization approach
	Degree of frequency reuse
	Cell size
Operating environment	Multiple access method
	Satellite EIRP and G/T
	Modulation method
	Phase and amplitude linearity
	Available link margin
	Orbit type
	Latency
	Band of operation
	Coding method (digital system)
	High penetration alerting method
Service offerings	Air interface structure
	Channel spacing
	Channel bandwidth
	Modulation type
Value proposition (profitability)	Overall system cost
	Risk element (amount of new technology)
	Incremental revenue possibilities
	System flexibility (future proofing)
	Business plan
	Funding base

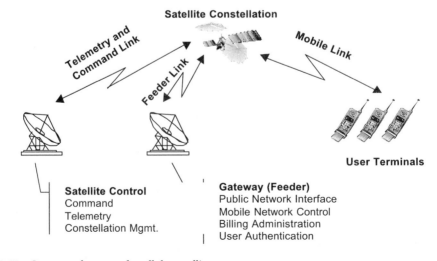

FIGURE 2.67 Common elements of a cellular satellite system.

Example Architectures

LEO

In this section we will examine two of the so-called "Big LEO" cellular satellite systems. These systems are "big" since they have relatively large, complex spacecraft and are designed to handle large quantities of information. These characteristics stand in contrast to the antithetical "Little LEOs" whose main mission is to provide short message service, paging and asset tracking, and are not considered cellular satellite systems for our purposes here. Common to all LEO cellular satellite systems is a large quantity of satellites in orbits that range from 700 km to about 1400 km (avoiding the inner Van Allen belt). LEO satellites also tend to be the simplest (or, at least, the lightest) of the major communication satellite types, and tend to have the shortest service life (5 to 7 years) due to limited onboard capacity fuel. Although somewhat counterintuitive, Big LEO systems tend to be the most costly to deploy (of the LEO, MEO, GEO varieties), mainly due to the large quantity of satellites required and the extensive, globally distributed ground infrastructure required to support the system. The main marketing point for LEOs is the low latency between the ground and the satellite due to the close proximity of the satellite to the user. The two best known of the Big LEOs are Iridium and Globalstar.

Iridium[14,15]

The Iridium cellular satellite system is owned by Iridium, LLC, of which Motorola is an 18% owner. The main contractors supplying system hardware are Raytheon (main mission antennas), Lockheed Martin (spacecraft bus), and Scientific Atlanta (Earth terminal equipment). Sixty-six active satellites, in 6 polar orbital planes of 11 satellites each, ring the Earth at an altitude of 780 km. The Iridium system is targeted at providing, primarily, voice service to the globe-trotting business professional. But the system is also designed to support low-rate data communications as well as facsimile and paging.

Iridium uses a protocol stack that is partially built on GSM and partially Iridium unique. Therefore, the system is compatible with the GSM infrastructure at the service level, even though the physical layer (the radio link) is not in accordance with GSM standards. The main elements of the Iridium system are shown in Fig. 2.68.

The main mission links (to the UTs) is accomplished by a combined FDMA/TDMA time division duplex (TDD) method in the L-band (specifically, 1610 to 1626.5 MHz) and with QPSK modulation. Voice communication is digitized in a vocoder, and the individual voice channels operate at a data rate of 2.4 kb/s. Traffic is passed between the individual satellites for the purposes of traffic routing and call hand-off, as one satellite transits out of view and another is needed to pick up the connection. These cross-links are also operated at K-band (23 GHz), but at the higher data rate of 25 Mb/s. One of the

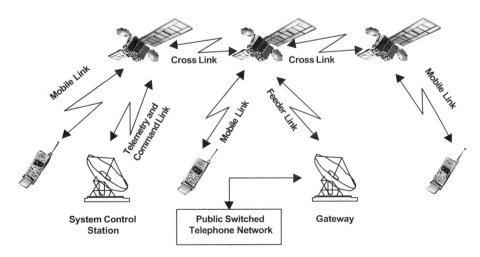

FIGURE 2.68 Elements of the Iridium communication system.

motivations for including the complexities of cross-links in the spacecraft design was to enable efficient call routing (less dependent upon the physical location of the gateways) which, in turn, also allowed full service coverage to the oceanic regions.

Globally distributed gateways, each with a 3.3 m antenna, provide the interface between the Iridium system and the public wired networks. Links between the satellites and these stations is done via the spacecraft feeder antennas on a K-band carrier (19 GHz down and 28 GHz up on QPSK modulation at a coded data rate of 6.25 Mb/s). Call setup and teardown is controlled at these local gateways, and this is also where billing records are generated. Each gateway also includes the necessary user authentication equipment (HLR, VLR) as well as the necessary infrastructure to enable UT position location. Position location is important for political reasons, among other things, so that local jurisdictions can maintain control over telecommunication traffic in their respective regions. At least two gateway antennas are required at each site in order to properly track and smoothly maintain contact with spacecraft in the field of view.

Management of the whole system is done at the system control station, which is physically located in Lansdowne, Virginia. Here the network infrastructure is monitored and controlled. This station also takes care of satellite maintenance and control, monitors status and system health, and serves as the center for any troubleshooting required.

The 700-kg Iridium spacecraft is very sophisticated. Its design includes a complex digital signal processor that demodulates incoming signals, switches them, and remodulates them as needed. One great advantage of this method is that uplink implementation loss can be significantly isolated from the downlink, thereby improving the overall bit error rate performance. The downside to this sophistication is, naturally, a major increase in the complexity and cost of the spacecraft. Three main mission antennas form the 48 cellular beams per satellite by way of direct radiating phased array technology. These beams cover a footprint on the Earth some 4700-km across. Each satellite has the switching capacity to handle 3840 simultaneous calls, but power considerations limit the practical number to around 1100.

The Iridium system has fallen on hard times as of this writing with the service having been discontinued and plans being made to actually deorbit the satellite constellation. This system is an object lesson in how even the most sophisticated, well-engineered system can become a failure if mishandled from a business perspective. Of the many problems that beset the Iridium system, some of the more significant ones include:

- Market share erosion: at the time of conception (1988–1990) the enormous build-out of inexpensive terrestrial cellular was not anticipated. This problem is the same one that contributed to the lackluster business performance of the American Mobile Satellite system.

- High system cost: estimates vary, but the cost of the Iridium system was some $5 to $7 billion. This high cost, in turn, implied the need for high service cost ($3 to $5 per minute) if returns were to be realized according to the business plan schedule. Additionally, the acquisition cost of the UT to the subscriber was high ($2000 to $3000). This combination of high service and terminal costs put the system out of reach to all but the most affluent consumers.

- Market overestimation: Iridium targeted the global business traveler who would often be outside of terrestrial cellular coverage. Most high-intensity business destinations today are well served by inexpensive terrestrial cellular coverage, thereby diminishing the need for Iridium's ubiquitous coverage.

- Ineffective marketing: Obtaining Iridium service was not a streamlined process. Also, advertising was underwhelming and tended to depict users in polar ice fields or deserts (a very limited revenue population).

- Poor UT form factor: The Iridium UT has been variously described as a "brick" or a "club." The bottom line is that the UT was awkward and heavy in a user environment used to small pocket-sized cellular telephones. The UT was designed by engineers with very limited regard to user appeal and aesthetics.

Globalstar[15,16]

Globalstar is one of the other Big LEO systems in process of being fielded. This system is owned by a partnership of several well-known companies led by Loral and Qualcomm as the general partners. System hardware suppliers include Alcatel (gateway equipment) and Alenia (satellite integration).

The Globalstar system consists of 48 satellites, in eight planes of six active satellites apiece, flying at an altitude of 1410 km. Thus, the Globalstar orbits are the highest of the Big LEOs fielded to date, which also means that the individual satellites move more slowly and are, consequently, easier to track at the gateway stations. Each orbit is inclined at 52°, thereby concentrating service resources in a band bordered by the 70° north and south latitudes. These latitude limits were carefully chosen in that the majority of the Earth's population is located in this region. The system uses a CDMA air interface that is based on the popular IS-95 terrestrial standard, which includes such features as soft handoffs, dynamic power control (essential to preserve capacity in a CDMA system, and preserves handset battery life in the bargain), and soft capacity limits (2000 to 3000 simultaneous calls per satellite). The intelligence signal is spread to 1.25 MHz, and these CDMA channels are spaced, in an FDMA fashion, on 1.23 MHz centers. Globalstar is targeting, essentially the same customer base as Iridium (i.e., the global business traveler), and offers an equivalent suite of services. For instance, the Globalstar offerings include variable-rate voice (2.4, 4.8, and 9.6 kb/s), low-rate data (on the order of 7.2 kb/s), facsimile, paging, and position location.

Mobile links are realized at S-band down (2483.5 to 2500.0 MHz) and L-band up (1610.0 to 1626.5 MHz) with QPSK modulation. The satellites are simple, bent-pipe transponders (amplification and translation only) with gateway links (satellite–gateway) at C-band (5019 to 5250 MHz up, and 6875 to 7055 MHz down). Rake receivers are included in both the UTs and the gateways in order to use multipath signals to advantage in strengthening the links. As a result, typical forward E_b/N_0 (the weakest link due to satellite power limitations) is reported to be on the order of 4 dB, while the return link weighs in at around 6 dB.

The main elements of the Globalstar system are depicted in Fig. 2.69. The Gateway Operations Control Center (GOCC) and the Satellite Operations Control Center (SOCC) are collocated in San Jose, CA. The individual, globally distributed gateways, with their 5.5 m tracking antennas, perform the local functions of the network administration. This administration includes user authentication (HLR, VLR functions), call control and local switching, local PSTN connections, and satellite TT&C processing.

Each satellite has a pair of direct radiating, fixed-beam phased array antennas to form the L-band cellular beam patterns. The transmit antenna consists of 91 elements, each having its own SSPA, phased with a fixed beam former into 16 transmit beams. The 16 receive beams (which are congruent with the transmit beams) are, similarly formed with a fixed, low power beam-forming system, but the array consists of only 61 elements, each with its own LNA. The graceful degradation properties inherent in a phased array mean that redundant SSPAs and LNAs are not required. The satellites have a design lifetime of

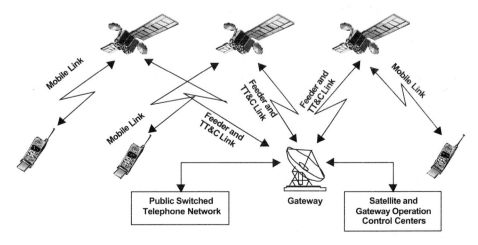

FIGURE 2.69   Elements of the Globalstar communication system.

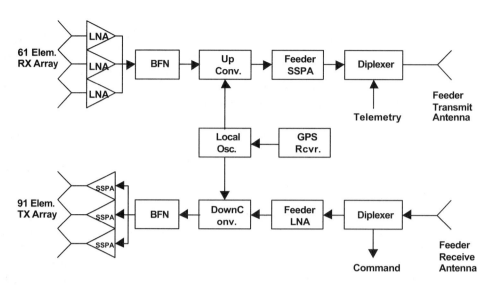

FIGURE 2.70   General architecture of the Globalstar satellite.

between 7 and 8 years. Figure 2.70 shows a generalized depiction of the Globalstar satellite architecture. Immediately obvious is the simplicity of the design. There is no onboard processing here; all of the technology is low risk. A hallmark of the Globalstar system is that the complexity, where needed, is relegated to the ground segment.

The Globalstar system is active at the time of this writing. Its commercial success has not been established at this stage. As a Big LEO system, one would think that Globalstar would approach service launch with great trepidation, given the negative example of the Iridium system. On the other hand, there are a number of factors about Globalstar that make its value proposition quite different, portending a brighter future. Some of these factors include:

- Simple architecture: The space segment is constructed with low risk technology. The complexity of the system is kept on the ground where upgrades, as necessary, are readily accomplished. This approach keeps the overall cost of the system down, and permits lower service pricing to be offered. In fact, the start of service has been offered at $1.79 per minute, and a handset price of around $1500,[17] a far cry from Iridium's $7 per minute and a $3000 handset.

- Proven air interface: The CDMA-based IS-95 air interface has been well proven in terrestrial cellular systems. Additionally, this approach makes the job of constructing multimode UTs (terrestrial-satellite) easier.

- Phased service rollout: Globalstar observed the techniques employed by its Iridium forerunner and decided to gradually roll out service in well-researched market areas. This technique allowed the company to gain experience and make corrections as needed with a fault-tolerant, methodical process.

- Revenue area focus: Globalstar is not attempting to service low population areas like the oceans or the Polar regions. The focus is on the parts of the globe where the people are. Precious resources are not expended in areas where return is marginal.

- More aesthetically pleasing UT: Although not quite the size of a modern-day cellular handset, and the antenna is still quite a bit larger than desirable, the Globalstar UT is much smaller than its Iridium counterpart. Broad user acceptance has still to be proven, but the prospects are good.

MEO

MEO systems, in general, tend to be orbitally located in the region between the inner and outer Van Allen radiation belts (around 10,300 km). Consequently, they tend to take on a blend of characteristics of which some are LEO-like and others are GEO-like. For instance, MEO propagation delay works out to be around 40 to 50 ms, not as good as LEO, but clearly better than GEO. The satellites also tend to be fairly complex and approach the GEOs in terms of design life (on the order of 10 to 12 years). Because of their relatively high orbit, MEO satellites move fairly slowly across the sky, greatly simplifying tracking requirements and reducing the number of handoffs during a typical call holding period. Two of the better-known MEO systems are ICO and Ellipso, and we will consider them in this section.

ICO[10,18]

The design of the ICO system began with a study program conducted by Inmarsat in the early part of the 1990s. Dubbed "Project 21," Inmarsat's objective was to move into the satellite cellular communications business. LEO, MEO, and GEO constellations were considered during the study, which eventually settled on a MEO configuration (an "intermediate circular orbit," or ICO), subsequent to which the project was spun off as a separate company. ICO was owned by ICO Global Communications Holdings, Ltd., with an additional 17 subsidiaries. That was until the company sought bankruptcy protection around the same time as Iridium faced a similar difficulty. Recently,[19] Craig McCaw and his Eagle River organization, along with Subash Chandra of ASC Enterprises (Ascel), both saw potential in the system and have, essentially, taken over the company. The new owner organization is known as New Satco Holdings, Inc. The major equipment contractors for the system include Hughes Space and Communications (satellites) and a team, lead by NEC, which includes Ericsson and Hughes Network Systems (ground infrastructure).

The original focus of the ICO system was to complement terrestrial cellular communication by servicing customers who reside outside of normal cellular coverage, providing cellular extension service to those who often travel outside of terrestrial cellular, maritime customers, and also government users. In other words, ICO was already targeting a fairly broad market. Craig McCaw, on the other hand, saw opportunities to enhance the system to provide data services. It has been reported that he is driving modifications into the design so that packet-based medium data rate services are supported (GPRS-like rates of around 384 kb/s), with early support for Wireless Application Protocol (WAP) services. WAP enables thin clients, like cell phones, to access enterprise services like e-mail and Internet information.

The ICO constellation consists of 10 active satellites in two planes of five satellites each. The orbital planes are both circular and inclined at 45°. The satellites communicate with 12 gateways (ground stations), called Satellite Access Nodes (SANs), spread around the world. Each SAN (Fig. 2.71) is equipped with five antennas to track as many satellites as might be in view at any given time. Due to the higher altitude of the constellation, each satellite is able to cover around 25% of the Earth's surface at any given time. The air interface is very similar to the terrestrial cellular IS-136 (D-AMPS) standard, in that it is

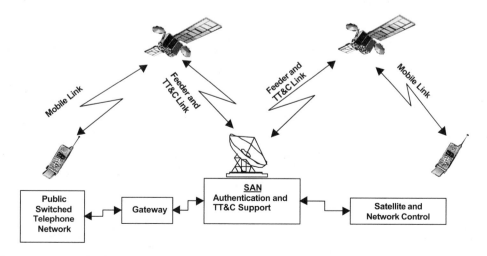

FIGURE 2.71   Generalized ICO system architecture.

an FDMA/TDMA scheme (QPSK modulation) that can support as many as 40 time slots (nominal-rate users) on five subcarriers within a 156 kHz channel bandwidth. The fact that the channel bandwidth is reasonably wide (approximately 156 kHz) is the reason that the system is readily modified to accept GPRS-like (and even EDGE-like) waveforms. There is, therefore, scope to provide packet-based medium data rate services in the ICO system, and this is one of the "future-proofing" characteristics that were designed in.

Mobile downlinks are at L-band (1980 to 2010 MHz) while the uplinks are at S-band (2170 to 2200 MHz). The aggregated signals, shipped between the satellites and the SANs, are transmitted at C-band (5187 to 5237 MHz up, and 7018 to 7068 MHz down).

The ICO Network Management Center (NMC) is located in Tokyo, Japan, while the Satellite Control Center (SCC) has been placed in Uxbridge, England. The SCC monitors the health and status of the spacecraft, and also takes care of any orbital adjustments that might become necessary in the course of the mission life.

The ICO spacecraft design is fairly sophisticated. Moderately large (2600 kg and consuming around 8700 W), the satellite is based on the HS 601 design, which has been extensively deployed for GEO missions. There are two mobile link antennas, one for transmit and the other for receive, each measuring around two meters in diameter. These antennas are direct radiating arrays driven by a sophisticated digital beamformer (DBF), which allows the antenna patterns (cells) to be dynamically shaped in order to respond to changing loading needs. This design also means that complexity (cost) has been added to the spacecraft, and that it must be supported by an intricate calibration infrastructure as well. Each antenna subsystem consists of 127 radiating elements and forms around 163 beams that, together, cover a ground surface diameter of around 12,900 km. The system frequency reuse factor is about four times with this design. Each satellite has switching capacity for around 6000 voice channels, although power constraints limit the actual capacity to around 4500 circuits. On the other hand, given the system modifications in process under McCaw's direction, the system is poised to enter the packet switched regime. Entirely different capacity calculations are possible under such an operation environment.

ICO's development has been a mixed bag, and apart from McCaw and Chandra's involvement, the value proposition is uncertain. Factors include:

- Higher orbit, slower movement, fewer handoffs: This characteristic makes for easier tracking and allows certain simplifications in the ground infrastructure.
- Smaller constellation (with respect to LEO): Fewer satellites to build and launch, but each satellite is heavier and more complex.
- Wider field of view: Higher orbit sees more of the Earth, and can cover oceanic regions (a mixed blessing).
- Fewer gateways required: Lower infrastructure cost than LEO, but higher cost than a GEO system.
- IS-136 (D-AMPS)-like air interface: Allows adaptation of standard terrestrial cellular hardware and easier manufacture of multimode handsets.

ICO has clearly had a tough time getting started, and for a time it appeared that the system would suffer the same fate as Iridium. The system is relatively expensive (included a significant amount of innovative technology). Again, estimates vary, but the overall system cost around $3 billion to $5 billion in its original state. McCaw's modifications will add additional cost, and he has invited other investors to participate, but his Eagle River organization is financially powerful enough to drive progress forward on its own if need be. With the further innovations to wireless packet data support, the future of the ICO system will be interesting to watch as it unfolds.

Ellipso[10,20,21]

Ellipso has also been placed in the class of the "Big LEOs." The system orbit design is unique, to the point of actually having been patented (U.S. patent #5,582,367). Due to the altitude (mean altitude, in the case of the elliptical planes) we class the Ellipso system here as a MEO system.

Mobile Communications Holdings, Inc. owns Ellipso with major contractors Lockheed Martin (ground) and Harris (space) supplying equipment to the system. Services are provided globally, though biased to favor populated areas, through satellites in three orbital planes; two elliptical ones that are called "Borealis" and one circular called "Concordia." The orbits are optimized to provide regional coverage proportional to the distribution of population on the surface of the earth. The Borealis orbits are elliptical, sun-synchronous, inclined at 116°, and each contain five satellites. These orbits each have a perigee at 520 km and an apogee at 7846 km. The Concordia orbit, on the other hand, is equatorial, circular, and has seven equally spaced satellites. Concordia's altitude is 8060 km. Both of these orbits are well within the band separating the two Van Allen radiation belts.

Ellipso's guiding philosophy is to perform all system trades with an eye toward the lowest end cost to the subscriber. Its services (including voice, messaging, positioning, and Internet access) are targeted toward "everyman." In other words, in contrast to the target customers of Iridium and ICO, Ellipso wants to reach deeper into the market and not just focus on the affluent, globetrotting businessman.

Like the ICO system, Ellipso provides a mobile user downlink at L-band (1610 to 1621.35 MHz) with the uplink placed at S-band (2483.5 to 2500 MHz). Feeder (ground station) communication, on the other hand, is done with the uplink at Ku-band (15450 to 15650 MHz) and the downlink at high C-band, nearly X-band (6875 to 7075 MHz). Each satellite forms a cell beam pattern consisting of 61 spot beams incorporating a high degree of frequency reuse (by way of cell isolation and orthogonal coding) across the coverage pattern. The system air interface is based on Third-Generation (3G) wideband CDMA (W-CDMA) with an occupied bandwidth of 5 MHz. This is a technology that is only just starting to be deployed in the terrestrial cellular world at the time of this writing. Consequently, the Ellipso design exhibits a tremendous degree of forethought and "future-proof" planning. Because of its W-CDMA infrastructure, Ellipso is poised to launch services with a 3G infrastructure in place and is, consequently, ready to provide high data rate (up to 2 Mb/s) packet-based services.

A general diagram of the overall Ellipso system is shown in Fig. 2.72. The central System Coordination Center takes care of the system network planning and monitoring. The Ground Control Stations (GCS) provide the gateway function as the interface to the communication signals going to, and coming from, the mobile stations via the satellites. Regional Network Control Stations provide local network control functions and collection of billing records. Associated with each GCS is an Ellipso Switching Office (ESO) that provides the interface between the PSTN and the Ellipso system. The ESO, additionally houses the HLR and VLR, and takes care of the user authentication function.

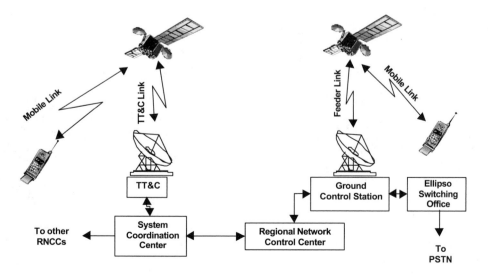

FIGURE 2.72   General diagram of the Ellipso MEO system.

All Ellipso satellites are of identical design, regardless of the orbit into which they are placed. Simplicity is the driver behind the satellite design as well. They are straightforward bent pipe translators with separate transmit and receive, direct radiating, fixed beamformed phased array antennas with 127 radiating elements in each planar array. Each satellite is of medium size with a mass of around 700-kg which, in turn, keeps launch costs down.

The success of Ellipso has yet to be seen at this juncture. The system does, however, contain many of the features important to an attractive value proposition. This is another system that will be interesting to watch as development unfolds. Some of Ellipso's more interesting value features include:

- Low system cost: The system designers are fanatical about keeping costs down as they are keenly aware of the connection between system deployment cost and the ultimate price point that can be offered to the subscriber.
- Wideband 3G-compatible air interface: A very forward-looking design feature. The system will be ready for enhanced 3G services, including always connected, high-speed packet-based Internet access. This feature has the potential of being very attractive to subscribers.
- Complexity on the ground: The space segment is designed as simply as practical in the overall system context. Complexity is kept on the ground where system upgrades are more readily accomplished.
- Low price point and large target market: The designers are targeting average consumers and not focusing on the lower quantity of affluent business travelers.
- Revenue area concentration: System design focuses on areas on the Earth where the bulk of the people are. Resources are not wasted over large ocean regions and areas of sparse population.
- Phased deployment: The characteristics of the Ellipso orbits are such that service can be initiated with a partially deployed constellation of satellites. In effect, early revenues can be generated that will help to pay for the remainder of the system deployment.

GEO

Two of the most advanced GEO-based cellular satellite systems are the ACeS and the Thuraya systems. Both of these systems are slated to provide commercial service offerings in the 2000 to 2001 time frame.

ACeS[10,22]

ACeS (Asia Cellular Satellite) is a GEO cellular satellite system conceived, designed, and backed by a partnership consisting of P.T. Pasifik Satelit Nusantara (of Indonesia), the Philippine Long Distance Telephone Company (PLDT), Lockheed Martin Global Telecommunications (LMGT), and Jasmine International Overseas Company, Ltd. (of Thailand).

The ACeS primary target market is the Southeast Asian area comprising the 5000 islands of Indonesia in the south, Northern China in the north, Pakistan in the west, and Japan in the east (see Fig. 2.73). The coverage area encompasses some three billion people, many of who have little or no access to a wired communication infrastructure. The Indonesian archipelago, for instance, is an expanse of islands that stretches some 4000 miles from east to west. It is not difficult to imagine the tremendous challenge of building a wired infrastructure to interconnect such a country. As such, the ACeS service is focused on a tightly defined market region, widely recognized as a rapidly expanding industrial world sector. While the typical ACeS user is perceived to be an active business traveler, the pricing of the planned services is expected to be at a level well within the reach of middle-class business people. Consequently, ACeS has a large addressable user population. This approach stands in stark contrast to some systems that target the high-end traveling businessman and seek to provide complete global coverage. In itself, a global coverage system, like some low earth orbiting (LEO) systems, requires a large quantity of satellites (along with sophisticated hand off and traffic management methods) and has an attendant high implementation price tag. For instance, the Iridium system is reported to have cost some $7 billion, about an order of magnitude more than the final cost of the ACeS system.

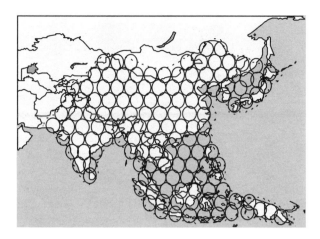

FIGURE 2.73   ACeS coverage and beam.

ACeS is designed to operate in a clearly defined geographical coverage area, therefore a regional GEO-based satellite system was chosen. Such a well-defined coverage area ensures that a maximum amount of precious satellite resources is concentrated on the desired revenue-producing areas. The satellite air interface standard was based on the ubiquitous Global System for Mobile Communication (GSM) terrestrial cellular standard in order to take advantage of its feature-rich suite of services, as well as the availability of a large quantity of standard supporting hardware. A GSM-based system also eases the integration of terrestrial hardware (e.g., dual-mode ACeS-GSM handsets) and facilitates intersystem roaming (based on GSM Subscriber Identity Module [SIM] cards). A mutually beneficial cooperative effort between LMGT and Ericsson (a world-class supplier of mobile communication equipment), assured the definition of an optimally tailored AIS. Further, the earth–satellite

FIGURE 2.74   AceS.

links were conservatively designed to ensure good service to disadvantaged users; frequently dropped calls are death to service acceptance and customer loyalty. An essential component of customer acceptance, facilitated by the strong links, is a handset form factor that is comparable to what is expected in modern cellular handsets. Figure 2.74 shows a picture of the ACeS handset in the form factor to be used at service launch. Other terminal types are also in development. Additionally, low cost of both service (airtime) and equipment is essential to customer uptake and heavy system use. Again, the modest deployment cost of a regional GEO system implementation aids in keeping the costs low.

The ACeS system has two main components, viz. the ground segment and the space segment, where the ground segment is further subdivided into the back-haul and control function (implemented at C-band) and the L-band user link function (see Fig. 2.75). There is one Satellite Control Facility (SCF) for each spacecraft. The Network Control Center (NCC) provides the overall control and management of the ACeS system, including such functions as resource management, call setup and teardown, call detail records, and billing support (customer management information system). Regional gateways, operated by the various National Service Providers (NSPs) manage the subset of system resources as allocated by the NCC. These gateways also provide the local interface and billing to the actual system users, and also provide connectivity between ACeS users and the wired infrastructure (Public Switched Telephone Network, Private Networks, and/or the Public Land Mobile Network). The user segment consists of handheld, mobile, or fixed terminals. These terminals can be configured to provide basic digital voice, data, and fax services. Further, since the system is based on GSM at the physical layer (in particular, 200 kHz Time Division Multiple Access, or TDMA, channels), it is future proofed in the sense

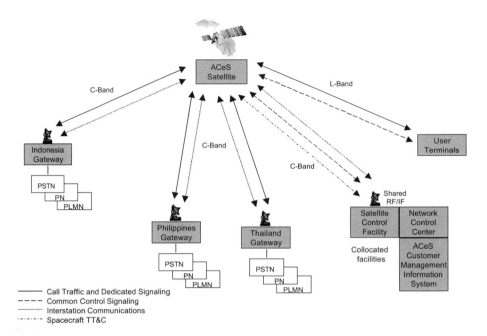

FIGURE 2.75   The ACeS system functions.

FIGURE 2.76   ACeS spacecraft: deployed.

that it will support General Packet Radio System (GPRS) and Enhanced Data for GSM Evolution (EDGE) upgrades with *no change required in the space segment.* This feature is a very important characteristic of the ACeS system.

The ACeS spacecraft, dubbed Garuda-1, is one of a family of Lockheed Martin modular spacecraft in the A2100-series (see Fig. 2.76). In particular, Garuda-1 is an A2100AXX, one of the largest models. This spacecraft is three-axis stabilized, with a bus subsystem that has been applied across numerous other spacecraft programs (e.g., Echostar, GE, LMI, and others) and, consequently, has a significant amount of application heritage. The payload is designed with two separate L-band antennas on the user link side. One antenna is dedicated to the transmit function (forward link) with the other, naturally, for receive (return link). This separation of antenna functions was done to minimize the probability of receive-side interference from passive intermodulation (PIM) in the transmit side. Large (12 meter) projected antenna apertures (two of them) provide the underpinnings for strong user links, crucial to reliable call completion under a variety of user circumstances. The result is high aggregate effective isotropic radiated power (EIRP), at 73 dBW, and high G/T, at +15.3 dB/K.

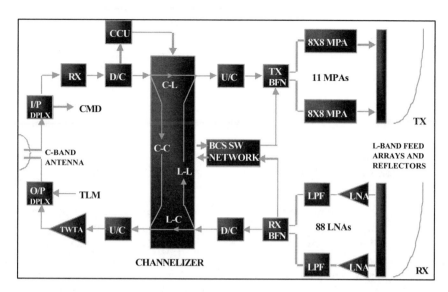

FIGURE 2.77 ACeS payload block diagram.

Figure 2.77 illustrates the major block functions of the Garuda-1 communication subsystem (CSS). As noted earlier, the user link is closed in the conventional mobile satellite system L-band. Narrow beams (140) are formed with a low risk, low power analog beamforming network (BFN) approach, in both forward and return directions. This beam design gives rise to a composite pattern that provides 20-times frequency reuse, thereby efficiently conserving valuable spectrum. A beam congruency system (BCS) operates in conjunction with ground beacons to ensure proper overlap of the corresponding transmit and receive beams. The L-band transmit amplification subsystem is implemented with a distributed set of Butler matrix based amplifier blocks called matrix power amplifiers (MPA). The MPA construct provides equal loading for all amplifiers in the block, which, in turn, minimizes phase and amplitude variation across the block. Small variations are key to good isolation between beams, giving good frequency reuse performance and, hence, maximum traffic capacity.

At the heart of the CSS sits the digital channelizer. This channelizer performs the important function of filtering the TDMA traffic channels and, also, of routing the individual traffic bursts as needed (particularly for the case of direct mobile-to-mobile communication).

The balance of the CSS (shown to the left of the channelizer) is the C-band back-haul transmit and receive equipment. This equipment is, largely, heritage being very similar to that used on previous direct broadcast and fixed service satellites.

The ACeS system has several important characteristics that hallmark a successful system. Among these characteristics are:

- Low infrastructure cost: Being a GEO system, once the satellite is on-station, instant connectivity is possible. The entire system, both space and ground segments, are reported to be under $1 billion (about an order of magnitude less than most Big LEO systems).

- Low targeted service cost: Low infrastructure cost allows lower price points to be charged while still allowing the business to turn a profit. This approach also means that a greater market can be attracted for the service.

- Well-researched target market: The Asia-Pacific region includes a multitude of islands that are not well connected by any kind of terrestrial infrastructure. This area is also one of high industrial growth which, in turn, needs good communication for support.

- Aesthetically-pleasing UT: The satellite bears the burden of providing the link margin, so that the UT can be sized similar to terrestrial cellular telephones. On the downside, complexity is added to the space segment thereby adding cost to the satellite. Since only one satellite is involved, this added complexity is easier to bear.

- Well-chosen air interface: The ACeS air interface is based around the GSM standard, implying the ability to directly use terrestrial hardware in the terrestrial equipment (thereby reducing development cost).
  - Wideband: 200 kHz channels (the GSM standard) will support future 2.5G services (GPRS and EDGE) without modification to the space segment. This is a design forethought that helps to future-proof the system.
  - Adjacent service compatibility: The adjacent terrestrial cellular services are largely based on GSM. Consequently, multimode UTs can be provided in a cost-effective manner.

Thuraya[10,23,24]

A partnership group led by Etisalat (Emirates Telecommunication Corporation) of the United Arab Emirates owns the Thuraya system. The objective of the system is to provide regional cellular satellite service to an area that includes Continental Europe, Northern Africa, the Middle East, and India. The Thuraya coverage area is adjacent to the ACeS coverage area, and also shares many of the same design features seen in that system. The target subscriber population includes national and regional roamers in an area (desert) that is not well served by terrestrial cellular service. Types of services to be offered include voice, facsimile, low-rate data, short messaging, and position determination. Major suppliers of the system components include Hughes Space and Communications (space segment), Hughes Network Systems (ground infrastructure), and Ericsson (network switching equipment).

The Thuraya system has one GEO satellite positioned at 44° E longitude (a second satellite is, reportedly, in process with the intended placement at 28.5° E. Mobile users connect with the Thuraya system through handheld UTs operating at L-band (1626.5 to 1660.5 MHz uplink and 1525 to 1559 MHz downlink). The aggregated signals destined for connection to the public wired infrastructure (PSTNs, etc.) are connected between the satellite and the Gateway stations at C-band (6425 to 6725 MHz up and 3400 to 3625 MHz down). The air interface is similar to that developed by Hughes Network Systems for the iCO system. That is, it is similar to the terrestrial cellular IS-136 (D-AMPS) system with Offset QPSK modulation, although the higher protocol elements are GSM compatible. Consequently, the Thuraya system is GSM compatible on a network and service level. Figure 2.78 shows the main elements that comprise the system.

The Gateway station serves as the main interface for communication signals into the public infrastructure. The system has a primary gateway that is located in Sharjah, UAE. Accommodation for several regional gateway stations is provided in the network infrastructure. The main functions performed by the gateway include user authentication (HLR and VLR), call control (setup and teardown), billing records, resource allocation, and roaming support. All interface conditioning required to connect to PSTNs, PLMNs, etc., are also handled in the gateway.

The Satellite Control Facility is actually classed as part of the space segment in that it handles the monitoring and control of the satellite. All telemetry is monitored in the SCF, and the required bus and payload commands are initiated at this facility. The Network Control Center, as the name implies, handles all of the network administration functions (routing, congestion control, and similar functions). Any payload commands required in this context are translated by the SCF and sent to the satellite.

The satellite is a very sophisticated element of the Thuraya system. It is a very large spacecraft (4500 kg) with a mission life of 12 years. The spacecraft is based on the well-known HS-601, but with several enhancements. For instance, solar concentrators are used on the solar arrays in order to enhance collection. This modification is done in order to help supply the large quantity of electrical power required (12.5 kW). The satellite is capable of switching about 13,750 simultaneous voice circuits. Mobile links interface to the ground through a single 12.25-meter projected aperture reflector. Passive intermodulation (PIM), i.e., unwanted mixing noise from the transmitter entering the receiver, is normally a central concern in systems such as these. Mitigation methods in other systems (e.g., Inmarsat-3 and ACeS) have included the use of two separate antenna apertures. The engineering of the Thuraya system has solved

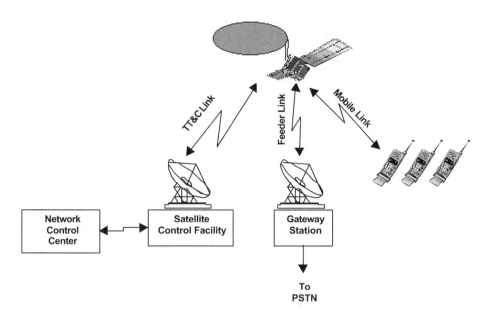

FIGURE 2.78   General diagram of the Thuraya system.

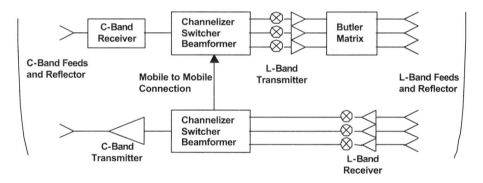

FIGURE 2.79   Major elements of the Thuraya communications payload.

that problem, and only one large reflector is required (serving both transmit and receive functions). The 256 cellular beams are generated by a state-or-the-art digital beamformer, which allows dynamic beam-shaping as required to meet variations in traffic loading, as required. This digital beamformer, though adding a great deal of flexibility to the system, comes at a price of added complexity (cost), power consumption, and a necessary dynamic calibration infrastructure. Some of the blocks of the communication subsystem of the Thuraya satellite are shown in Fig. 2.79.

As a GEO system, Thuraya has many of the same positive value characteristics previously described:

- Low infrastructure cost: Thuraya has the instant connectivity trait common to GEO systems where only one satellite is required. The system cost is reported to be around $1 billion.
- Low targeted service cost: Low price points are planned for system services (reportedly, on the order of $0.50 per minute). This approach also means that a greater market can be attracted for the service.
- Target market: The northern Africa area contains a lot of open territory (desert) where a significant industry is conducted (e.g., petroleum). Existing terrestrial infrastructure is, clearly, not adequate, so a good business opportunity exists. On the other hand the potential for generating significant revenue from European coverage area is questionable given the ubiquity of GSM in that region.

- Aesthetically-pleasing UT: Thuraya, with its 12.25-m antenna aperture, allows the UT to be sized similar to terrestrial cellular telephones. On the downside, complexity and cost accrues to the satellite as a result. As we noted in the ACeS case, only one satellite is involved so this added complexity is easier to bear.
- Air interface: Thuraya's air interface shares many of the characteristics of the D-AMPS IS-136 standard, implying the ability to directly use terrestrial hardware in the terrestrial equipment (thereby reducing development cost). It is also relatively wideband (around 156 kHz) and is, therefore, suited to support 2.5G services as they become available.
- GSM network infrastructure: The network services are based upon the GSM model and can support GSM services as a result.

Trends

Satellite cellular service developments have, by and large, mimicked the advancements in terrestrial cellular,[25-27] albeit with a predictable delay. Clearly, the satellite service infrastructure takes longer to develop and field than its terrestrial counterpart. Cellular satellite service providers have continued to look to terrestrial developments for the "next step." Activities are underway to adapt existing systems, and to incorporate future enhancements, to effectively support data transmission. The immense data communication infrastructure demands seen today are being fueled (akin to gasoline being fire-hosed onto a blaze!) by the explosive growth of the Internet. Cellular satellite system developers aim to have a part in the ongoing explosion. Internet traffic, from the user's perspective, is primarily a bursty form of communication based on packet switching. Circuit switched systems are not efficient in bursty packet mode, so cellular satellite systems are being adapted to communicate in packet mode and at higher data rates. Mobile data communications will require packet data access on the order of 100 kb/s or higher. GPRS, in GSM, accommodates dynamically adjustable rates and, in eight-slot full-rate mode will support a peak rate on the order of 115 kb/s. If EDGE is included (with adaptive coding and its 8PSK modulation approach), it will squeeze 384 kb/s into a 200 kHz GSM channel. Plainly, cellular satellite systems based on GSM will readily be able to support GPRS and EDGE.

3G W-CDMA terrestrial systems are being deployed in order to support mobile data rates as high as 2 Mb/s. One cellular satellite system (Ellipso) is already designed to provide such support. Others may follow. Teledesic, the wideband system owned by Bill Gates and Craig McCaw, is a LEO system specifically designed to provide high-speed data to the home fixed locations), but it is not a cellular satellite system under the criteria we have used here.

Other terrestrial enhancements are in the works, and it is likely that their incorporation into the cellular satellite world will occur at an accelerated pace. These enhancements include:

- WAP: The Wireless Application Protocol, which is a protocol stack specifically designed to allow "thin clients" (limited capability devices like cellular telephones) to take greater advantage of the Internet. WAP allows a more streamlined approach for cell phones to receive and transmit e-mail, browse Internet Websites, interact with corporate enterprise structures (Intranet services), and other well-established Internet-based activities.
- Bluetooth: A wireless picocell communication method that allows direct synchronization between Bluetooth enabled devices. Initially proposed as a method for interconnecting personal computers, printers, data devices, cell phones, and digital assistants over a spread-spectrum link in the unlicensed ISM band (Industrial, Scientific, and Medical band; the same band used by microwave ovens, for example). It is obvious to see how the use of Bluetooth can be extended to other applications (for example, as an intermediate link between a cellular satellite and a user who might be shopping in a mall.

These types of enhancements will only serve to heighten the need for, and the growth of cellular satellite systems.[28] It is clear that future applications are only limited by the imagination of entrepreneurs and system designers.

References

1. Quality of Service Forum web page, http://www.QoSforum.com
2. Butash, T. C., Lockheed Martin Manassas, private communication, 2000. I am grateful to Dr. Butash for his clear insight into all-digital satellite system assessment.
3. Matolak, D., Lockheed Martin Global Telecommunications, private communication, 1998.
4. Kadowaki, N., et al., ATM transmission performance over the trans-Pacific HDR satcom link, *Proc. Second International Workshop on Satellite Communications in the Global Information Infrastructure*, 63, 1997.
5. Fitch, M., ATM over satellite, *Proc. Second International Workshop on Satellite Communications in the Global Information Infrastructure*, 67, 1997.
6. Falk, A., TCP over satellite, *Proc. Second International Workshop on Satellite Communications in the Global Information Infrastructure*, 74, 1997.
7. Johnson, Lt. G.W., and Wiggins, ET1 M.D., Improved Coast Guard communications using commercial satellites and WWW technology, *Proc. Fifth International Mobile Satellite Conference*, 519, 1997.
8. Logsdon, T., *Mobile Communication Satellites*, McGraw-Hill, Inc., New York, New York, 1995, 131.
9. Stern, P. and Mauricio, P. The exploration of the Earth's magnetosphere, *NASA Web tutorial*, 1998, http://www-spof.gsfc.nasa.gov/Education
10. Miller, B., Satellites free the mobile phone, *IEEE Spectrum*, 26, March, 1998.
11. eGlobal Report, March 2000, http://www.emarketer.com
12. Lonergan, D., Strategy Analytics, Inc. Web-based report, January, 2000, http://www.strategyanalytics.com
13. Elizondo, E. et al., Success criteria for the next generation of space based multimedia systems, *Satellite Communications Symposium of the International Astronautical Federation*, Session 33, September 1998.
14. Maine, K., et al., Overview of Iridium satellite network, Motorola Satellite Communications Division, Chandler, AZ, 1995.
15. Lloyd's Satellite Constellations, *Big LEO Tables*, http://www.ee.surrey.AC.uk/Personal/L.Wood/constellations/tables/
16. Dietrich, F.J., The Globalstar satellite cellular communication system design and status, *Proc. Fifth International Mobile Satellite Conference*, 139, 1997.
17. Wilkinson, G., Satellite companies face murky fate after Iridium's demise, *Total Telecom*, 24, March 2000.
18. ICO Global Communications, http://www.icoglobal.com/
19. Foley, T., Mobile & satellite: McCaw unveils WAP strategy for new ICO, *Communications Week International*, 03, April 2000.
20. Draim, J.E., et al., Ellipso — An affordable global, mobile personal communications system, *Proc. Fifth International Mobile Satellite Conference*, 153, 1997.
21. Ellipso web site, http://www.ellipso.com
22. Nguyen, N.P., et al., The Asia Cellular Satellite System, *Proc. Fifth International Mobile Satellite Conference*, 145, 1997.
23. Alexovich, A., et al., The Hughes geo-mobile satellite system, *Proc. Fifth International Mobile Satellite Conference*, 159, 1997.
24. Thuraya web site, http://www.thuraya.com
25. The future: Down the road for PCS computing, *PCS Data Knowledge Site*, Intel, 1998, http://www.pcsdata.com/future.htm
26. Route to W-CDMA, *Ericsson Mobile Systems*, Ericsson, December, 1998, http://www.ericsson.com/wireless/products
27. Cadwalader, D., Toward global IMT-2000, *Ericsson Wireless NOW!*, January, 1998, http://www.ericsson.com/WN/wn1-98/imt2000.html
28. Tuffo, A.G., From the 'outernet' to the Internet, *18th AIAA International Communications Satellite Systems Conference*, April, 2000.

2.9 Electronic Navigation Systems

Benjamin B. Peterson

In an attempt to treat electronic navigation systems in a single chapter, one must either treat lightly or completely ignore many aspects of the subject. Before going into detail on some specific topics, I would like to point out what is and is not in this chapter and why, and where interested readers can go for missing topics or more details. The chapter begins with fairly brief descriptions of the U.S. Department of Defense satellite navigation system, NAVSTAR Global Positioning System (GPS), its Russian counterpart, GLObal NAvigation Satellite System, (GLONASS), and LORAN C. For civil use of GPS and GLONASS there are numerous existing and proposed augmentations to improve accuracy and integrity which will be mentioned briefly. These include maritime Differential GPS, the FAA Wide Area Augmentation System (WAAS), and Local Area Augmentation System (LAAS) for aviation. The second half of the chapter looks in detail at how position is determined from the measurements from these systems, the relationships between geometry, and the statistics of position and time errors. It also considers how redundant information may be optimally used and how information from multiple systems can be integrated.

GPS and its augmentations were chosen because they already are and will continue to be the dominant radionavigation technologies well into the next century. GLONASS, could potentially become significant, by itself, and potentially integrated with GPS. LORAN C is included while other systems were not for a variety of reasons. The combination of its long history and military significance dating back to World War II, and its large user base, has resulted in a wealth of research effort and literature. Because the basic principle of LORAN C, i.e., the measurement of relative times of arrival of signals from precisely synchronized transmitters, is the same as that of GPS, it is both interesting and instructive to analyze the solution for position of both systems in a unified way. While the future of LORAN C in the United States is uncertain (DoD/DoT, 1999), it is expanding in Europe and Asia and may remain significant on a worldwide basis for many years. Since LORAN C has repeatable accuracy adequate for many applications, because it is more resistant to jamming, has failure modes independent of those of GPS, and because its low frequency signals penetrate locations like urban canyons and forests better than those of GPS, analysis of its integration with GPS is felt useful and is included.

Of other major systems, VHF Omnidirectional Range (VOR), Instrument Landing System (ILS), Distance Measuring Equipment (DME), and Microwave Landing System (MLS) are relatively shorter range aviation systems that are tentatively planned to be phased out in favor of GPS/WAAS/LAAS beginning in 2008. (DoD/DoT, 1999). Due to space and because their principles of operation differ from GPS they are not included. The U.S. Institute of Navigation in its 50th Anniversary issue of Navigation published several excellent historical overviews. In particular, see Enge (1995) for sections on VOR, DME, ILS, and MLS, and Parkinson (1995) for the history of GPS.

Recently, there has been a significant and simultaneous reduction in the printing of U.S. Government documents and increased distribution of this information via the Internet. Details on Internet and other sources of additional information are included near the end.

The Global Positioning System (NAVSTAR GPS)

GPS is a U.S. Department of Defense (DoD) developed, worldwide, satellite-based radionavigation system that will be the DoD's primary radionavigation system well into the 21st century. GPS Full Operational Capability (FOC) was declared on July 17, 1995 by the Secretary of Defense and meant that 24 operational satellites (Block II/IIA) were functioning in their assigned orbits and the constellation had successfully completed testing for operational military functionality.

GPS provides two levels of service — a Standard Positioning Service (SPS) and a Precise Positioning Service (PPS). SPS is a positioning and timing service, which is available to all GPS users on a continuous, worldwide basis. SPS is provided on the GPS L1 frequency, which contains a coarse acquisition (C/A) code and a navigation data message. The current official specifications state that SPS provides, on a daily basis, the capability to obtain horizontal positioning accuracy within 100 meters (95% probability) and 300 meters (99.99% probability), vertical positioning accuracy within 140 meters (95% probability), and timing accuracy within 340 nanoseconds (95% probability). For most of the life of GPS, the civil accuracy was maintained at approximately these levels through the use of Selective Availability (SA) or the intentional degradation of accuracy via the dithering of satellite clocks. In his Presidential Decision Directive in 1996, President Clinton committed to terminating SA by 2006. On May 1, 2000, in a White House press release (White House, 2000) the termination of SA was announced and a few hours later it was turned off. Work is starting on a new civil GPS signal specification with revised accuracy levels. Very preliminary data, as this is being written in late May 2000, indicate 95% horizontal accuracy of approximately 7 meters will be possible under some conditions. When SA was on, it was the dominant error term; accuracy for all receivers with a clear view of the sky was easily predictable and independent of other factors. Now, it is expected that terms such as multipath and ionospheric delay, which vary greatly with location, time, and receiver and antenna technology will dominate, and exact prediction of error statistics will more difficult.

The GPS L1 frequency also contains a precision (P) code that is not a part of the SPS. PPS is a highly accurate military positioning, velocity, and timing service which is available on a continuous, worldwide basis to users authorized by the DoD. PPS is the data transmitted on GPS L1 and L2 frequencies. PPS is designed primarily for U.S. military use and is denied to unauthorized users by the use of cryptography. Officially, P-code-capable military user equipment provides a predictable positioning accuracy of at least 22 meters (2 drms) horizontally and 27.7 meters (2 sigma) vertically, and timing/time interval accuracy within 90 nanoseconds (95% probability).

The GPS satellites transmit on two L-band frequencies: L1 = 1575.42 MHz and L2 = 1227.6 MHz. Three pseudo-random noise (PRN) ranging codes are in use giving the transmitted signal direct sequence, spread spectrum attributes. The coarse/acquisition (C/A) code has a 1.023 MHz chip rate, a period of one millisecond (ms), and is used by civil users for ranging and by military users to acquire the P-code. Bipolar-Phase Shift Key (BPSK) modulation is utilized. The transmitted PRN code sequence is actually the Modulo-2 addition of a 50 Hz navigation message and the C/A code. The SPS receiver demodulates the received code from the L1 carrier, and detects the differences between the transmitted and the receiver-generated code. The SPS receiver uses an exclusive or truth table, to reconstruct the navigation data, based upon the detected differences in the two codes. Ward (1994–1995) contains an excellent description of how receivers acquire and track the transmitted PRN code sequence.

The precision (P) code has a 10.23 MHz rate, a period of seven days, and is the principle navigation ranging code for military users. The Y-code is used in place of the P-code whenever the anti-spoofing (A-S) mode of operation is activated. *Anti-spoofing* (A-S) guards against fake transmissions of satellite data by encrypting the P-code to form the Y-code. A-S was exercised intermittently through 1993 and implemented on January 31, 1994. The C/A code is available on the L1 frequency and the P-code is available on both L1 and L2. The various satellites all transmit on the same frequencies, L1 and L2, but with individual code assignments.

Each satellite transmits a navigation message containing its orbital elements, clock behavior, system time, and status messages. In addition, an almanac is provided that gives the approximate data for each active satellite. This allows the user set to find all satellites once the first has been acquired. Tables 2.16 and 2.17 include examples of ephemeris and almanac information for the same satellite at approximately the same time.

The nominal GPS constellation is composed of 24 satellites in six orbital planes, (four satellites in each plane). The satellites operate in circular 20,200-km altitude (26,570 km radius) orbits at an inclination angle of 55° and with a 12-hour period. The position is therefore the same at the same sidereal time each day, i.e., the satellites appear four minutes earlier each day.

The GPS Control segment consists of five Monitor Stations (Hawaii, Kwajalein, Ascension Island, Diego Garcia, Colorado Springs), three Ground Antennas, (Ascension Island, Diego Garcia, Kwajalein), and a Master Control Station (MCS) located at Falcon AFB in Colorado. The monitor stations passively

TABLE 2.16   Ephemeris Information

Satellite Ephemeris Status for PRN07
Ephemeris Reference Time = 14:00:00 02/10/1995, Week Number = 0821
All Navigation Data is Good, All Signals OK
Code on L2 Channel = Reserved, L2 P Code Data = On
Fit Interval = 4 hours, Group Delay = 1.396984e-09 s
Clock Correction Time = 14:00:00 02/10/1995, af0 = 7.093204e-04 s
af1 = 4.547474e-13 s/s, af2 = 0.000000e+00 s/s2
Semi-Major Axis = 26560.679 km, Eccentricity = 0.007590
Mean Anom = −23.119829, Delta n = 2.655361e-07, Perigee = −144.510162
Inclination = 55.258496, Inclination Dot = 1.100943e-08
Right Ascension = −10.880005, Right Ascension Dot = −4.796266e-07
Crs = 5.3750e+01 m, Cuc = 1.6691e-04, Crc = 2.7141e+02 m, Cic = 0.0000e+00

TABLE 2.17   Almanac Information

SATELLITE ALMANAC STATUS FOR PRN07
Almanac Validity = Valid Almanac at 09:06:07 01/10/1995
Semi-Major Axis = 26560.168 km
Eccentricity = 0.007590
Inclination = 55.258966
Right Ascension = -10.829880
Right Ascension Rate = -4.655885e-07/s
Mean Anomaly = -172.426664
Argument of Perigee = -144.590529
Clock Offset = 7.095337e-04 s, Clock Drift = 0.000000e+00 s/s
Navigation Health Status is: All Data OK
Signal Health Status is: All Signals OK

track all satellites in view, accumulating ranging data. This information is processed at the MCS to determine satellite orbits and to update each satellite's navigation message. Updated information is transmitted to each satellite via the Ground Antennas.

The GPS system uses time of arrival (TOA) measurements for the determination of user position. A precisely timed clock is not essential for the user because time is obtained in addition to position by the measurement of TOA of four satellites simultaneously in view. If altitude is known (e.g., for a surface user), then three satellites are sufficient. If a stable clock (say, since the last complete coverage) is keeping time, then two satellites in view are sufficient for a fix at known altitude. If the user is, in addition, stationary or has a known speed then, in principle, the position can be obtained by the observation of a complete pass of a single satellite. This could be called the "transit" mode, because the old Transit system used this method. In the case of GPS, however, the apparent motion of the satellite is much slower, requiring much more stability of the user clock.

The current (May 2000) GPS constellation consists of 28 satellites counting 27 operational and one recently launched and soon to become operational. Figures 2.80 and 2.81 illustrate the history of the constellation including both the limiting dates of the operational periods, individual satellites, and the total number available as a function of time.

GPS Augmentations

Many augmentations to GPS exist to improve accuracy and provide integrity. These are both government operated (for safety of life navigation applications) and privately operated. This section will only deal with government operated systems. The recent termination of Selective Availability has sparked some debate on the need for GPS augmentation, but the most prevalent views are that stand-alone GPS will not satisfy all accuracy and integrity requirements, and that some augmentation will continue to be necessary. Since the dominant post-SA errors vary much more slowly than SA-induced errors did, the information bandwidth requirements to meet specified accuracy levels have certainly been reduced. Since

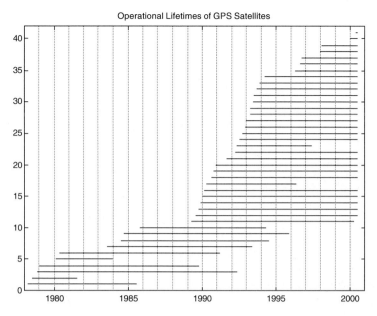

FIGURE 2.80 Operational lifetimes of GPS satellites.

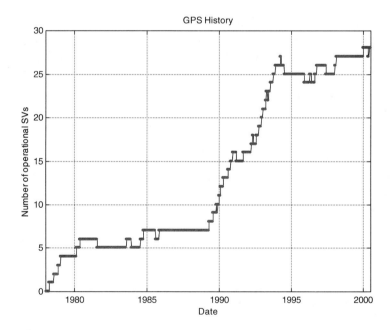

FIGURE 2.81 GPS history.

integrity requirements specify maximum times to warn users of system faults, these warning times and time to first fix specifications will now drive information bandwidth requirements.

The simplest and first to become operational were maritime Differential GPS stations. Existing radio-beacons in the 285 to 325 kHz band were converted to transmit differential corrections via a Minimum Shift Keyed (MSK) modulation scheme. A base station receiver in a fixed, known location measures pseudo-range errors relative to its own clock and known position. Messages containing the time of observation, these pseudo-range errors, and their rate of change are then transmitted at either 100 or

200 baud. Parity bits are added, but no forward error correction is used. If the user receiver notes a data error, it merely waits for the next message. The user receiver uses the corrections extrapolated ahead in time based on the age of the correction and its rate of change. For a description of the message format the reader is referred to (RTCM, 1992.)

With SA on, satellite clock dithering was the dominant error and was common to all users. It is also assumed that over the few hundred kilometers or less where the signal can be received, that errors such as ephemeris and ionospheric and tropospheric delay are correlated between base station and user receiver as well and can be considerably reduced. The algorithms to predict these delays are disabled in both the base station and user receivers.

Because the corrections are measured relative to the base station clock, they are relative and not absolute corrections. This means they must have corrections for all satellites used for a position and all of these corrections must be from the same base station. For fixed position, precise time users, DGPS provides no improvement in accuracy. For moving users, their solution for time will track the base station clock, and since this base station receiver is in a fixed known location, the time solution in the moving receiver will see improved accuracy as well.

The U. S. Coast Guard operates maritime DGPS along the coasts and rivers. In addition, the system is being expanded inland in support of the Federal Railroad Administration in the National DGPS project. Many foreign governments worldwide operate compatible systems primarily for maritime applications. RTCM type DGPS messages are also transmitted by modulating LORAN C transmitters in Europe. Additional details on this system known as EUROFIX are included in the later section on LORAN C.

Three Satellite Based Augmentation Systems (SBASs) for aviation users are in various stages of development. These include:

a. the Wide Area Augmentation System (WAAS) by the U.S. FAA,
b. the European Geostationary Navigation Overlay System (EGNOS) jointly by the European Union, the European Space Agency (ESA), and EUROCONTROL, and
c. MTSAT Satellite Based Augmentation System (MSAS) by the Japan Civil Aviation Bureau (JCAB).

These systems are intended for the enroute, terminal, non-precision approach, and Category I (or near Category I) precision approach phases of flight. All of these systems are designed to be compatible, and to the level of detail in this chapter, are equivalent. For additional details on all three systems the reader is referred to Walter (1999) and to RTCA (1999) for the WAAS signal specification. In SBASs corrections need to be applied over the very large geographic area in which the satellite signal can be received. In this case the errors due to satellite ephemeris and ionospheric delay are not the same for all users and cannot be combined into one overall correction. Separate messages are provided for satellite clock corrections, vector corrections of satellite position and velocity, integrity information, and vertical ionospheric delay estimates for selected grid points. These grid points are at 5 degree increments in latitude and longitude, except larger increments occur in extreme northern or southern latitudes. The user receiver calculates its estimates of ionospheric delay by first determining the ionospheric pierce point of the propagation path from satellite to user receiver, interpolating between grid points, and then correcting for slant angle.

These corrections are or will be transmitted from geostationary satellites at GPS L1 frequencies with signal characteristics similar to GPS. The SBASs will provide additional ranging signals and transmit their own ephemeris information to improve fix availability. Message symbols at 500 symbols per second will be modulo 2 added to a 1023 bit PRN code. The baseline data rate is 250 bits per second. This data is rate $\frac{1}{2}$ convolutional encoded with a Forward Error Correction code resulting in 500 symbols per second. The data will be sent in 250-bit blocks or one block per second. The data block contains an eight-bit preamble, a six-bit message type, a 212-bit message, and 24 bits of CRC parity.

For Category II and III precision approach, the U.S. FAA will implement the Local Area Augmentation System (LAAS). The transmitted data will include pseudo-range correction data, integrity parameters, approach data, and ground station performance category. The broadcast will be in the 108 to 117.95 MHz band presently used for VOR and ILS systems.

The modulation format is a differentially encoded, eight-phase-shift-keyed (D8PSK) scheme. The broadcast uses an eight-time-slot-per-half-second, fixed-frame, time-division-multiple-access (TDMA) structure. The total data rate of the system is 31,500 bits per second. After the header and cyclic redundancy check (CRC), the effective rate of the broadcast is 1776 bits (222 bytes) of application data per time slot, or a total of 16×1776 BPS equaling 28,416 application data bits per second. For additional information on LAAS the reader is referred to Braff (1997) and Skidmore (1999).

Global Navigation Satellite System (GLONASS)

GLONASS is the Russian parallel to GPS and has its origins in the mid-1970s in the former Soviet Union. Like GPS, the system has been primarily developed to support the military, but in recent years has been broadened to include civilian users. In September 1993, Russian President Boris Yeltsin officially proclaimed GLONASS to be an operational system and the basic unit of the Russian Radionavigation Plan. The well-publicized political, economic, and military uncertainties within Russia have undoubtedly hampered the implementation and maintenance of the system. These issues combined with the poor reliability record of the early satellites, raised questions relative to its eventual success. Figures 2.82 and 2.83 show data on the operational life of those satellites that became operational from the first one in 1982 through the present. It is also believed that attempts were made to place approximately 10 or more other satellites into operation resulting in failure for an number of reasons (Dale, 1989). However, in the mid-1990s, the reliability of the satellites considerably improved, and many more satellites were put in orbit such that in early 1996 there were 24 operational satellites. The triple launch on December 30, 1998 has been the only launch since late 1995, and the constellation has degraded to only 10 operational satellites, again raising concerns on the future of the system. Publicly, the Russian Federation remains fully committed to the system. To date, GLONASS receivers have been quite expensive and have only been produced in limited quantities.

While similar in many respects to GPS, GLONASS does have important differences. While all GPS satellites transmit at the same frequencies and are distinguished by their PRNs, all GLONASS satellites transmit the same PRN and are distinguished by their frequencies. The L1 transmitted frequencies in MHz are given by $1602 + 0.5625 \times$ Channel Number, which extends from 1602.5625 to 1615.5 MHz. L2 frequencies are $^{7}/_{9}$ L1 and in MHz are given by $1246 + 0.4375 \times$ Channel Number or from 1246.4375 to

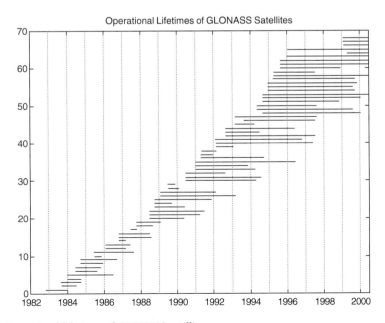

FIGURE 2.82 Operational lifetimes of GLONASS satellites.

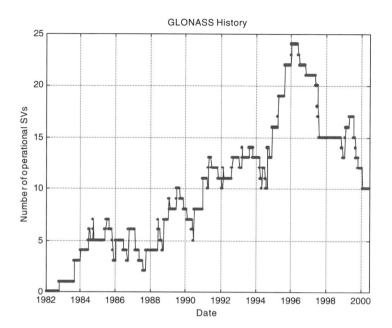

FIGURE 2.83   GLONASS history.

1256.5 MHz. GLONASS satellites opposite each other in the same orbit plane have been assigned the same channel number and frequency to limit spectrum use. Like GPS, the C/A code is transmitted on L1 only, and P code on both L1 and L2. The C/A code is 511 chips long at 511 KBPS for a length of 1 ms. The P code is at 5.11 MBPS. Like GPS, the actual modulation is the Modulo-2 addition of the PRN sequence and 50 BPS data and BPSK modulation is used.

The GLONASS planes have a nominal inclination of 64.8° compared to 55° for GPS, which gives slightly better polar fix geometry at the cost of fix geometry at lower latitudes. The 24 slots are in three planes of eight slots each. The orbital altitude is 25,510 km or 1050 km less than GPS. The orbit period is 11 hours, 15 minutes.

Rather than transmitting ephemeris parameters as in GPS, GLONASS satellites transmit actual position, velocity, and acceleration in Earth Centered, Earth Fixed (ECEF) coordinates that are updated on half-hour intervals. The user receiver integrates using Runge-Kutta techniques for other times. All monitor sites are within the former Soviet Union, which limits accuracy of both ephemeris and time and would delay user notification of satellite failures. The system produces both high accuracy signals for Russian military use only and lesser accuracy for civilian use. The civilian use signals are not degraded to the same extent as GPS was with SA, which resulted in significantly better accuracy than civil GPS during the period GLONASS had a complete constellation. Military accuracy is classified. For more details the reader is referred to the GLONASS Interface Control Document (CSIC, 1998) available at http://www.rssi.ru/sfcsic/sfcsic_main.html

LORAN C History and Future

Early in World War II, both the U.S. and Great Britain recognized the need for an accurate, long range, radionavigation system to support military operations in Europe. As a result the British developed Gee and the U.S. developed LORAN (Long Range Navigation). Both were pulsed, hyperbolic systems with time differences between master and secondary transmitters measured by an operator matching envelopes on a delayed sweep CRT. Gee operated at several carrier frequencies from 20 to 85 MHz and had a pulse width of 6 usec. Standard LORAN (or LORAN A) operated at 1.95 MHz and had a pulse width of 45 usec. The first LORAN A chain was completed in 1942 and by the end of the war 70 stations were operating. For details on the World War II LORAN effort see Pierce (1948).

It was recognized that lower frequencies would propagate longer distances, and near the end of the war testing began on a 180-kHz system. The pace of development slowed considerably after the war. The band from 90 to 110 kHz was established by international agreement for long-range navigation. In 1958, system tests started on the first LORAN C chain, which consisted of stations at Jupiter, FL, Carolina Beach, NC, and Martha's Vineyard, MA.

Over the next two decades LORAN C coverage was expanded by the U.S. Coast Guard in support of the Department of Defense to much of the U.S. (including Alaska and Hawaii), Canada, northwest Europe, the Mediterranean, Korea, Japan, and Southeast Asia. The Southeast Asia chain ceased operations with the fall of South Vietnam in 1975. In 1974 LORAN C was designated as the primary means of navigation in the U.S. Coastal Confluence Zone (CCZ), the cost of civilian receivers declined sharply, and the civil maritime use of the system became very large. In the late 1980s at the request of the Federal Aviation Administration, the Coast Guard built four new stations in the mid-continent and established LORAN C coverage for the entire continental U.S.

Other countries are developing and continuing LORAN C to meet their future navigational needs. Many of the recent initiatives have taken place as a result of the termination of the U.S. DoD requirement for overseas LORAN C. This need came to an end as of December 31, 1994. With the introduction of GPS, many countries have decided that it is in their own best interests not to have their navigational needs met entirely by a U.S. DoD-controlled navigation system. Many of these initiatives have resulted in multilateral agreements between countries, which have common navigational interests in those geographic areas where LORAN C previously existed to meet U.S. DoD requirements (e.g., northern Europe and the Far East). The countries of Norway, Denmark, Germany, Ireland, the Netherlands, and France established a common LORAN C system under the designation The Northwest European LORAN C System (NELS). Recently the governments of Italy and the United Kingdom have applied for membership. This system is presently comprised of eight stations forming four chains. In conjunction with this system, respective foreign governments took over operation of the former USCG stations of B0 and Jan Mayen, Norway; Sylt, Germany; and Ejde, Faroe Islands (Denmark) as of 31 December 1994. The two French stations formerly operated by the French Navy in the rho-rho mode were reconfigured and included and two new stations in Norway were constructed. A planned ninth station at Loop Head, Ireland has yet to be constructed. The former USCG stations at Angissoq, Greenland and Sandur, Iceland were closed.

An important feature of NELS is the transmission of DGPS corrections by pulse position modulation. This concept, called EUROFIX, was developed by Professor Durk Van Willigen and his students at Delft University of Technology in the Netherlands. The last six pulses in each group are modulated in a three-state pattern of prompt and 1 usec advance or retard. Of the possible 729 combinations, 142 of which are balanced, 128 balanced sequences are used to transmit a seven-bit word. Extensive Reed Solomon error correction is used resulting in a very robust data link, although at a somewhat lower data rate than the MSK beacons. With Selective Availability on, the horizontal accuracy was 2 to 3 meters, 2DRMS, or slightly worse than conventional DGPS due to the temporal decorrelation of SA. With SA off, the performance should be virtually comparable. EUROFIX has been operating at the Sylt, Germany transmitter since February, 1997, expansion to other NELS transmitters is planned for the near future. In a joint FAA/USCG effort, preliminary testing is underway of transmitting WAAS data via a much higher data rate LORAN communications channel. If LORAN C survives in the long term, it is likely to be both as a navigation system and as a communications channel to enhance the accuracy and integrity of GPS.

In the Far East, an organization was formed called the Far East Radionavigation Service (FERNS). This organization consists of the countries of Japan, The Peoples Republic of China, The Republic of Korea, and the Russian Federation. Japan took over operation of the former USCG stations in its territory and they are currently being operated by the Japanese Maritime Safety Agency (JMSA). In the Mediterranean Sea area, the four USCG stations were turned over to the host countries. The two stations in Italy, Sellia Marina and Lampedusa, are currently being operated. The stations at Kargaburun, Turkey and Estartit, Spain remain off air. Saudi Arabia and India are currently each operating two LORAN chains.

There is ongoing debate at high levels concerning the future of LORAN C in the United States. According to the 1999 Federal Radionavigation Plan (DoD/Dot, 1999), "While the administration continues to evaluate the long term need for continuation of the LORAN-C navigation system, the Government will operate the LORAN C system in the short term. The U. S. Government will give users reasonable notice if it concludes that LORAN C is not needed or is not cost effective, so that users will have the opportunity to transition to alternative navigational aids."

LORAN Principles of Operation

Each chain consists of three or more stations, including a master and at least two secondary transmitters. (Each master-secondary pair enables determination of one LOP, and two LOPs are required to determine a position. The algorithms to convert these LOPs into latitude and longitude are discussed in a later section.) Each LORAN C chain provides signals suitable for accurate navigation over a designated geographic area termed a *coverage area.*

The stations in the LORAN chain transmit in a fixed sequence that ensures that Time Differences (TDs) between receiving master and secondary can be measured throughout the coverage area. The length of time in usec over which this sequence of transmissions from the master and the secondaries takes place is termed the Group Repetition Interval (GRI) of the chain. All LORAN C chains operate on the same frequency (100 kHz), but are distinguished by the GRI of the pulsed transmissions.

The LORAN C system uses pulsed transmission, nine pulses for the master and eight pulses for the secondary transmissions. Figure 2.84 shows this overall pulse pattern for the master and three secondary transmitters (X, Y, and Z). Coding delay is the time between when a secondary receives the master signal and when it transmits. The time differences measured on the baseline extension beyond the secondary should be coding delay. Emission delay is the difference in time of transmission between master and secondary and is the sum of coding delay and baseline length (converted to time). Shown in Fig. 2.85 is an exploded view of the LORAN C pulse shape. It consists of sine waves within an envelope referred to as a t-squared pulse. (The equation for the envelope is also included in Fig. 2.85.) This pulse will rise from zero amplitude to maximum amplitude within the first 65 usec and then slowly trails off or decays over a 200 to 300 usec interval. The pulse shape is designed so that 99% of the radiated power is contained within the allocated frequency band for LORAN C of 90 kHz to 110 kHz. The rapid rise of the pulse allows a receiver to identify one particular cycle of the 100 kHz carrier. Cycles are spaced approximately 10 usec apart. The third cycle of this carrier within the envelope is used when the receiver matches the cycles. The third zero crossing (termed the positive 3rd zero crossing) occurs at 30 usec into the pulse. This time is both late enough in the pulse to ensure an appreciable signal strength and early enough in the pulse to avoid sky wave contamination from those skywaves arriving close after the corresponding ground wave.

Within each pulse group from the master and secondary stations, the phase of the radio frequency (RF) carrier is changed systematically from pulse-to-pulse in the pattern shown in Fig. 2.85 and Table 2.18. This procedure is known as phase coding. The patterns A and B alternate in sequence. The pattern of phase coding differs for the master and secondary transmitters. Thus, the exact sequence of pulses is actually matched every two GRIs, an interval known as a phase code interval (PCI).

Phase coding enables the identification of the pulses in one GRI from those in an earlier or subsequent GRI. Just as selection of the pulse shape and standard zero crossing enable rejection of certain early sky waves interfering with the same pulse, phase coding enables rejection of late sky waves interfering with

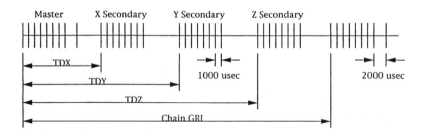

FIGURE 2.84   LORAN C pulse pattern.

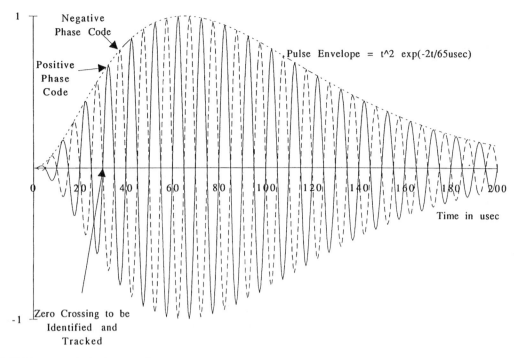

FIGURE 2.85 Ideal LORAN pulse.

TABLE 2.18 LORAN C Phase Codes.

GRI Interval	Master	Secondary
A	+ + − − + − + − +	+ + + + + − − +
B	+ − − + + + + + −	+ − + − + + − −

the next pulse. Since seven of the fourteen pulses that come immediately before another pulse have the same phase code and seven are different, late sky waves interfering with the next pulse will average to zero. Because the master and secondary signals have different phase codes, the LORAN receiver can distinguish between them.

Position Solutions from Radionavigation Data

In general, closed-form solutions that allow direct calculation of position from radionavigation observables do not exist. However, the inverse calculations can be done. Given one's position and clock offset, one can predict GPS or GLONASS pseudo-ranges or LORAN TDs or TOAs. These expressions are nonlinear but can be linearized about a point and the problem solved by iteration. The basic approach can be summarized by:

a. Assume a position,
b. Calculate the observables one should have seen at that point,
c. Calculate rate of change of observables (partial derivatives) as position is varied,
d. Compare calculated to measured observables,
e. Compute a new position based on differences between calculated and measured observables,
f. Depending on exit criteria, either exit algorithm or return to "b" above using the calculated position as the new assumed position.

We will look at the details of each of these steps and the expected statistics of the results for a number of cases. We will start with the example of processing three LORAN TOAs (or two TDs), but the basic principles apply to GPS as well. We will then consider overdetermined solutions and the techniques for assuring the solutions are optimal in some sense. We will also consider integrated solutions and Kalman filter based solutions.

For LORAN the calculated Times of Arrival (TOAs) are given by:

For Master $\qquad$ $\mathrm{TOA_m} = d_i,$
and for secondaries $\quad \mathrm{TOA_i} = d_i + ED_i,$ etc.

where d_m and d_i are the propagation times from the master and i^{th} secondary stations to the assumed position and ED_i is Emission Delay. (Note: Secondary stations are designated by the letters V, W, X, Y, and Z, but because we will also want to use x and y for position variables, in order to avoid confusion, we will use integer subscripts to denote secondaries.) The LORAN propagation times are the sums of three terms:

a. *Primary Factor* (PF) or the geodesic distance to the station divided by a nominal phase velocity. See WGA (1982, 57–61) for an algorithm that will work to distances up to 3000 nautical miles.
b. *Secondary Factor* (SF) which takes into account the additional phase lag due to the signal following a curved earth surface over an all-seawater path (see COMDT(G-NRN) 1992, II-14 for formula), and
c. *Additional Secondary Factor* (ASF) which takes into account the difference between the actual path traveled and an all-seawater path. This factor can be calculated based on conductivity maps, or can be obtained by tables (from the U. S Defense Mapping Agency) or from charts published by the Canadian government. The latter two are based both on calculations and observations.

A typical LORAN receiver measures time differences (TDs) vs. TOAs and the receiver need not necessarily solve for time. In GPS or GLONASS, since the transmitters move rapidly with time, in order to know their location it is essential that an explicit solution for time be made. In GPS each satellite transmits ephemeris parameters which, when substituted in standard equations (NATO, 1991, A-3-26 and 27), give the satellite location as a function of time expressed in Earth Centered, Earth Fixed (ECEF) coordinates. In this system, the positive x axis goes from the earth's center to 0° latitude and longitude, the positive y axis goes from the earth's center to 0° latitude and 90°E longitude, and the positive z axis goes from the earth's center to 90°N latitude. For GLONASS, the satellites transmit their location, velocity, and acceleration, in ECEF coordinates, and validate at half-hour intervals. The user is expected to numerically integrate the equations on motion to obtain position for a particular time.

The receiver measures the time of arrival of the satellite signal, and since the signal is tagged precisely in time, the difference in time of transmission and time of arrival (converted to distance) is a pseudo-range. "Pseudo" is used as the time of arrival is relative to the receiver clock, which cannot be assumed exact but must be solved for as part of the solution. These pseudo-ranges are then corrected for both ionospheric and tropospheric delay. In civil C/A code receivers, ionospheric delay parameters are trans-mitted and the delay calculated via an algorithm (NATO, 1991, p A-6-31). For dual frequency receivers, the ionospheric delay is assumed to be inversely proportional to frequency and the delay determined from the difference in pseudo-ranges at the two frequencies. For differential GPS, ionospheric delay is assumed to be the same at the reference station and the user's location and the algorithm is disabled at both locations.

Solution Using Two TDs (LORAN only)

TDs at the assumed position are calculated (TD_p) and subtracted from observations (TD_o) . i.e.;

$$\mathrm{TD}_i = \mathrm{TOA}_i - \mathrm{TOA}_m$$

$$\Delta\mathrm{TD} = \begin{bmatrix} \Delta\mathrm{TD}_1 \\ \Delta\mathrm{TD}_2 \end{bmatrix} = \mathrm{TD}_o - \mathrm{TD}_p$$

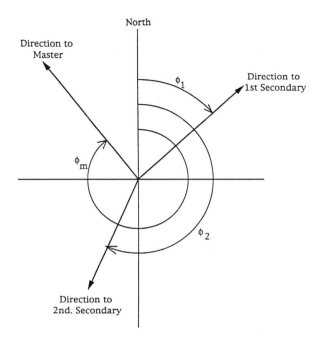

FIGURE 2.86 Definition of LORAN azimuths.

This difference in TDs can be linearly related to a difference in position $\Delta P = \begin{bmatrix} \Delta x \\ \Delta y \end{bmatrix}$ by

$$\begin{bmatrix} \sin(\phi_m) - \sin(\phi_1) & \cos(\phi_m) - \cos(\phi_1) \\ \sin(\phi_m) - \sin(\phi_2) & \cos(\phi_m) - \cos(\phi_2) \end{bmatrix} \begin{bmatrix} \Delta x \\ \Delta y \end{bmatrix} = c \begin{bmatrix} \Delta TD_1 \\ \Delta TD_2 \end{bmatrix}$$

where ϕ_m, ϕ_1, and ϕ_2 are the azimuths to the transmitters measured in a clockwise direction from the north or positive y direction as shown in Fig. 2.86. While exact calculations taking into account the ellipsoidal nature of the shape of the earth must be used in the calculation of predicted TDs, only approximate expressions based on a spherical earth are necessary to determine azimuths. The expression above can be expressed in matrix form:

$$A_{TD} \Delta P = c \Delta TD$$

and solved by:

$$\Delta P = c A_{TD}^{-1} \Delta TD$$

At this point a new assumed position would be calculated (using an appropriate conversion of meters to degrees), the predicted TDs at that location determined, and if they were within a selected tolerance of the measured, the algorithm would stop, otherwise a new solution would be determined. An alternative criteria based on the movement of the assumed position, $|\Delta P|$, can be used as well and will be more appropriate in an over-determined solution because exact match of TDs is not possible. Figures 2.87 and 2.88 show typical convergence of this type of algorithm. Since the algorithm is based on the assumption that the TD grid is linear in the vicinity of the assumed position, we need to limit the step size so that the position does not jump to radically different geometry in one step. As can be seen in Fig. 2.87 all

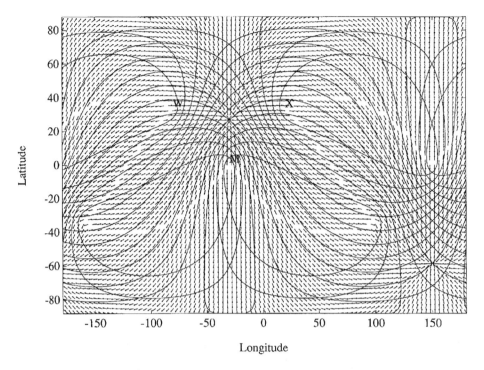

FIGURE 2.87 Convergence of position solution algorithn.

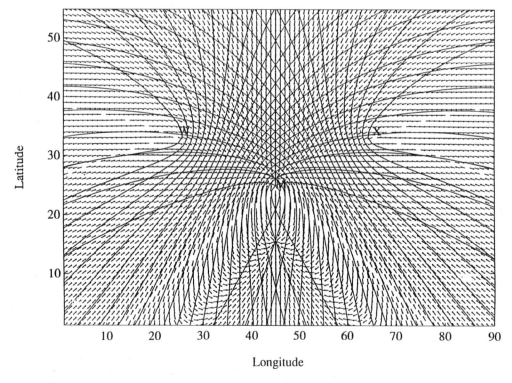

FIGURE 2.88 Example of position solution with close by ambiguous position.

lines of constant TD between two transmitters are closed curves on the earth's surface. Two of these curves intersect, either do not intersect, or intersect exactly twice. Since we assume we are starting with a set of TDs that exist, the latter applies. The initial assumed position is what determines which of these two ambiguous positions the algorithm will converge to. Essentially since the algorithm is based on the gradient of the TDs with respect to change in position and that gradient changes sign on the baseline extension, the position will move to the solution it can reach without crossing a baseline extension. The direction of the gradient, and hence the final solution also changes as the lines of constant TD become parallel as at 30°N and 10°E in Fig. 2.88.

Frequently, as shown in the example in Fig. 2.87, one of these ambiguous positions is well beyond the range the signals can be received and can easily be eliminated. This is not true when one of the solutions is near the baseline extension as shown in Fig. 2.88. In this example, it is necessary to have prior knowledge that the actual position is north or south of the Master station.

Solution Using TOAs

This method yields a solution for receiver clock bias (modulo one Phase Code Interval) and even though not generally implemented in present receivers, is presented for a number of reasons:

a. It is functionally equivalent to the GPS position solution but is presented first because it is a two- vs. three-dimensional and is somewhat easier to visualize.

b. For the exactly determined (two-TD or three-TOA) case it is exactly equivalent to the TD solution and therefore we can extend results from TOA or pseudo-range analysis to the TD case. Early LORAN TD analysis was typically done in scalar form before computer programs such as spread-sheets and MatLab™ made matrix calculations trivial. GPS fix analysis has typically been expressed in matrix form resulting is simpler expressions.

c. For over-determined solutions, since we can assume TOA errors are statistically independent, but cannot make the same assumption for TD errors, the solution is slightly easier to implement and analyze.

d. Solutions using Kalman filters, integrated GPS/LORAN, or a precise clock with only two TOAs, will typically be based on TOAs vs. TDs.

The vector difference between observed (TOA_o) and predicted (TOA_o) TOAs are calculated via:

$$\Delta\text{TOA} = \text{TOA}_o - \text{TOA}_p$$

Equation is replaced by:

$$-\begin{bmatrix} \sin(\phi_m) & \cos(\phi_m) & 1 \\ \sin(\phi_1) & \cos(\phi_1) & 1 \\ \sin(\phi_2) & \cos(\phi_2) & 1 \end{bmatrix} \begin{bmatrix} \Delta x \\ \Delta y \\ c\Delta t \end{bmatrix} = c \begin{bmatrix} \Delta\text{TOA}_m \\ \Delta\text{TOA}_1 \\ \Delta\text{TOA}_2 \end{bmatrix}$$

or:

$$a_{\text{TOA}}\Delta P = c\Delta\text{TOA}$$

and solved by:

$$\Delta P = c A_{\text{TOA}}^{-1} \Delta\text{TOA}.$$

It is useful to think of the sines and cosines in the matrix as the cosines of the angles between the positive direction of the coordinate axes and the directions to the transmitters. This concept can be easily extended to the three-dimensional case in GPS or GLONASS solutions. For a normal GPS-only solution it is usually more convenient to express both satellite and user positions in an Earth Centered, Earth Fixed (ECEF) coordinate system. In this case the direction cosines are merely the difference of the particular coordinate divided by the range to each satellite. After position solution, an algorithm is used to convert from (xyz) to latitude, longitude, and altitude. An alternative that will prove useful for meaningful Dilution of Precision (DOP) calculations or for integrated GPS/LORAN receivers is to use a east/north/altitude coordinate system and to solve for the azimuth (AZ) and elevation (EL) of the satellites. In this case the direction cosines for each satellite become:

East: $\sin(AZ)\cos(EL)$
North: $\cos(AZ)\cos(EL)$
Altitude: $\sin(EL)$

Error Analysis

With only three TOAs this solution will be exactly the same as the two-TD solution above. In both of these solutions after the algorithm has converged, we can consider ΔTOA or ΔTD as measurement errors and ΔP as the position error due to these errors. The covariance of the errors in position (and in time for TOA processing) can be expressed as functions of the direction cosine matrix and the covariance of the TDs or TOAs.

$$\Delta P\,\Delta P^T = c^2\,A_{TD}^{-1}\,\Delta TD\,\Delta TD^T\left(A_{TD}^{-1}\right)^T$$

$$\mathrm{cov}\left(\Delta P\right) = E\left\{\Delta P\,\Delta P^T\right\} = c^2\,A_{TD}^{-1}\,E\left\{\Delta TD\,\Delta TD^T\right\}\left(A_{TD}^{-1}\right)^T = c^2\,A_{TD}^{-1}\,\mathrm{cov}\left(\Delta TD\right)\left(A_{TD}^{-1}\right)^T$$

$$\mathrm{cov}\left(\Delta P\right) = c^2\,A_{TOA}^{-1}\,\mathrm{cov}\left(\Delta TOA\right)\left(A_{TOA}^{-1}\right)^T$$

Given that the covariances of the measurements are known it becomes a simple task to evaluate the covariance of the position errors. We will first assume each TOA measurement is uncorrelated, i.e.,

$$\mathrm{cov}\left(\Delta TOA\right) = \begin{bmatrix} \sigma_m^2 & 0 & 0 \\ 0 & \sigma_1^2 & 0 \\ 0 & 0 & \sigma_2^2 \end{bmatrix}$$

and

$$\mathrm{cov}\left(\Delta TD\right) = \begin{bmatrix} \sigma_m^2 + \sigma_1^2 & \sigma_m^2 \\ \sigma_m^2 & \sigma_m^2 + \sigma_2^2 \end{bmatrix}$$

For simplicity we will also assume each TOA measurement has the same variance (σ_{TOA}^2). While this assumption is not in general valid, particularly for LORAN C, it is useful in that it allows us to separate out purely fix geometry from signal-in-space issues. Under this assumption:

$$\mathrm{cov}\left(\Delta TOA\right) = \sigma_{TOA}^2\,I$$

and

$$\mathrm{cov}\left(\Delta TD\right) = \sigma_{TOA}^2 \begin{bmatrix} 2 & 1 \\ 1 & 2 \end{bmatrix}.$$

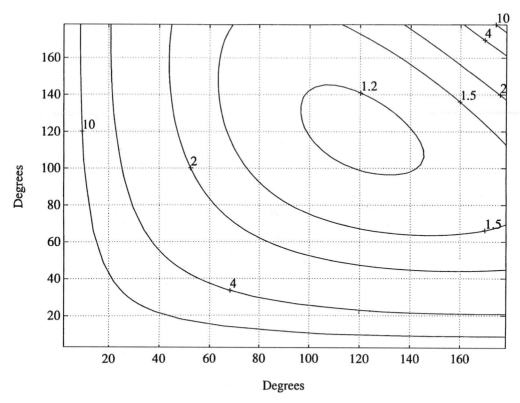

FIGURE 2.89 HDOP as function of angles from master to secondaries.

For the particular case of the same number of measurements and unknowns, these both yield the same results:

$$\text{cov}\left(\Delta P\right) = \begin{bmatrix} \sigma_x^2 & \sigma_{xy} & c\sigma_{xt} \\ \sigma_{xy} & \sigma_y^2 & c\sigma_{yt} \\ c\sigma_{xt} & c\sigma_{yt} & c^2\sigma_t^2 \end{bmatrix} = c^2\,\sigma_{\text{TOA}}^2\, A_{\text{TOA}}^{-1}\left(A_{\text{TOA}}^{-1}\right)^T = c^2\,\sigma_{\text{TOA}}^2\left[A_{\text{TOA}}^T\, A_{\text{TOA}}\right]^{-1}$$

What is significant is that the matrix $G = [A_{\text{TOA}}^T A_{\text{TOA}}]^{-1}$ is a dimensionless multiplier that relates our ability to measure the range to a transmitter (LORAN station or GPS satellite). In GPS, (NATO GPS 1991) $\sqrt{\text{trace}(G)}$ is defined as the Geometric Dilution of Precision (GDOP), which includes the time term. (NATO GPS 1991). In LORAN C, since convention has been not to explicitly solve for time, GDOP is normally considered as $\sqrt{G_{11} + G_{22}}$, which only contains horizontal position terms and is equivalent to Horizontal Dilution of Precision (HDOP) in GPS. Since we want to consider both LORAN and GPS in a consistent, integrated way, we will use HDOP and reserve GDOP for expressions including time term. The scalar expression for this quantity is

$$\text{HDOP} = \frac{1}{\sqrt{2}\sin\left(C\right)}\sqrt{\frac{1}{\sin^2\left(A/2\right)} + \frac{1}{\sin^2\left(B/2\right)} + \frac{\cos\left(C\right)}{\sin\left(A/2\right)\sin\left(B/2\right)}}$$

where A and B are the angles from the Master to each secondary and $C = \frac{A+B}{2}$. Figure 2.89 shows contours of constant HDOP as functions of these two angles. While the horizontal accuracy in one dimension is well defined ($\sigma_x^2 = G_{11} c^2 \sigma_{TOA}^2$, for example), the two-dimensional accuracy is slightly more complex. Generally the terms drms and 2 drms accuracy are used to refer to the one-sigma and two-sigma accuracies. For example:

$$2 \text{ drms horizontal accuracy} = 2 \text{ HDOP c } \sigma_{TOA}$$

A circle of this radius should contain between 95.4% and 98.2% of the fixes depending on the eccentricity of the error ellipse describing the distribution of the fixes. For example, if $\sigma_x = \sigma_y$ and $\sigma_{xy} = 0$, the error ellipse becomes a circle and one of radius 2 drms contains $[1 - \exp(-4)] = 98.2\%$. In the other extreme, when the ellipse collapses to a line ($\sigma_{xy} \cong \pm \sigma_x \sigma_y$) the 2 drms circle contains the probability within two standard deviations of the mean of a scalar normal variable or 95.4%.

The relationship between geometry and accuracy can best be explained via numerical example. Shown below are spreadsheet calculations for a Master and two secondaries at 0°, 80°, and 300° respectively.

	Bearing	East(Sine)	North(Cosine)	Time
Master	0	0	1	1
Sec. #1	80	0.9848	0.1736	1
Sec. #2	300	−0.8660	0.5000	1
		INVERSE(A^TA)		
TOA		0.7122	0.6748	−0.4047
		0.6748	3.5258	−1.9937
		−0.4047	−1.9937	1.4616
TD		sin(M) −sin(S)	cos(M)−cos(S)	
		−0.9848	0.8264	
		0.8660	0.5000	
		R(TD)		
		2	1	
		1	2	
		INV(A) R INV(A^T)		
		0.7122	0.6748	
		0.6748	3.5258	

Assuming $\sigma_{TOA} = 100$ ns or equivalently c $\sigma_{TOA} = 30$ meters, we can make specific calculations regarding repeatable accuracy. The standard deviations in the east and north directions and time are:

$$EDOP = \sqrt{0.7122} \quad \sigma_x = 30m \text{ EDOP} = 25.3m,$$

$$NDOP = \sqrt{3.5258} \quad \sigma_y = 30m \text{ NDOP} = 56.3m,$$

$$TDOP = \sqrt{1.4616} \quad \sigma_t = 100ns \text{ TDOP} = 120.9ns,$$

$$HDOP = \sqrt{0.7122 + 3.5258} = 2.058,$$

and the 2 drms accuracy = $2 \times 30m \times 2.058 = 123.5m$.

Error Ellipses (Pierce, 1948)

Frequently the scalar 2 drms accuracy does not adequately describe the position accuracy and the distribution of the error vector provides useful additional information. The probability density of two jointly normal random variables is given by:

$$p(x,y) = \frac{1}{2\pi\sqrt{\sigma_x^2\sigma_y^2 - \sigma_{xy}^2}} \exp\left[-\frac{\sigma_x^2\sigma_y^2}{2(\sigma_x^2\sigma_y^2 - \sigma_{xy}^2)}\left(\frac{x^2}{\sigma_x^2} - \frac{2xy\sigma_{xy}}{\sigma_x^2\sigma_y^2} + \frac{y^2}{\sigma_y^2}\right)\right]$$

and the ellipses

$$\frac{x^2}{\sigma_x^2} - \frac{2xy\sigma_{xy}}{\sigma_x^2\sigma_y^2} + \frac{y^2}{\sigma_y^2} = 2\left(1 - \frac{\sigma_{xy}^2}{\sigma_x^2\sigma_y^2}\right)c^2$$

are curves of constant probability density such that the probability a sample lies within the ellipse is $1 - \exp(-c^2)$. For algebraic simplicity this equation will be rewritten:

$$Ax^2 + 2Hxy + By^2 = C$$

which can be rotated by angle (Ω) into a (ξ,η) coordinate system aligned with the ellipse axes;

$$\frac{\xi^2}{\alpha^2} + \frac{\eta^2}{\gamma^2} = C$$

where

$$\tan(2\Omega) = \frac{-2H}{B-A},$$

$$\alpha^2 + \gamma^2 = \frac{A+B}{AB-H^2}, \text{ and}$$

$$\alpha^2 - \gamma^2 = \frac{(B-A)\sec(2\Omega)}{AB-H^2}.$$

Taking the sum and difference of the last two expressions;

$$2\alpha^2 = \frac{A\left[1 - \sec(2\Omega)\right] + B\left[1 + \sec(2\Omega)\right]}{AB-H^2}, \text{ and}$$

$$2\gamma^2 = \frac{A\left[1 + \sec(2\Omega)\right] + B\left[1 - \sec(2\Omega)\right]}{AB-H^2}.$$

In the example above, $\Omega = -12.8°$. To draw an ellipse that contains $[1 - \exp(-4)] = 98.17\%$ of the samples, $\alpha = 63.4$m and $\gamma = 162.8$m. (Note: α is the length of the ellipse axis rotated by angle Ω from the original positive x axis.) Figures 2.90 and 2.91 illustrate an example of 2000 fixes based on the above geometry and TOA statistics. In this example 1926 or 96.3% are contained in the 2 drms circle of radius 123.5 m.

Figure 2.92 is an illustration of error ellipses (not to scale) and lines of constant time difference for a typical triad. The main point is that when geometry becomes poor in the fringes of the coverage area, representing the accuracy by circles of increasing 2 drms radii does not truly reflect one's knowledge of accuracy. In many examples from both air and marine navigation, cross-track accuracy is far more important than along-track accuracy. Knowledge of the orientation of the error ellipse may be important.

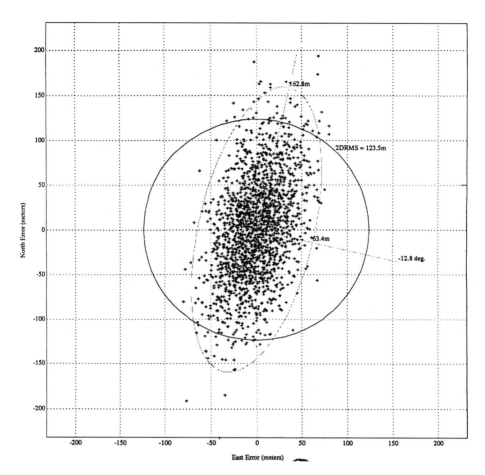

FIGURE 2.90 Error ellipse and distribution of fix errors.

In GPS or GLONASS, exactly the same error analysis and ellipses apply, the main difference being that the geometry at any particular point on the earth's surface has constantly changing geometry as satellites move and are added/subtracted from the available constellation. When we look at distributions of GPS or GLONASS fixes they tend to appear circularly symmetric, not because the instantaneous distributions are that way, but because the averaged distributions over long periods such as a day or longer become circularly symmetric.

Over-Determined Solutions

In the previous examples the issue of the distribution of measurement errors had no impact on the method used to find a solution, only on the error analysis of that solution. We assumed a normal distribution on these errors not necessarily because we could guarantee that distribution, but because it allowed us to make meaningful predictions. When the number of observables exceeds the number of unknowns, we have choices in how to process the observations and the optimum choice depends both on the distribution of our measurement errors and what parameter we are trying to optimize. Since the GPS space segment now has 28 satellites, commonly there are more than the minimum of three or four required.

If the errors in the measured pseudo-ranges or TOAs are statistically independent and have the same variance, then the solution that minimizes the sum of the squares of the pseudo-range residuals is the linear least squares solution.

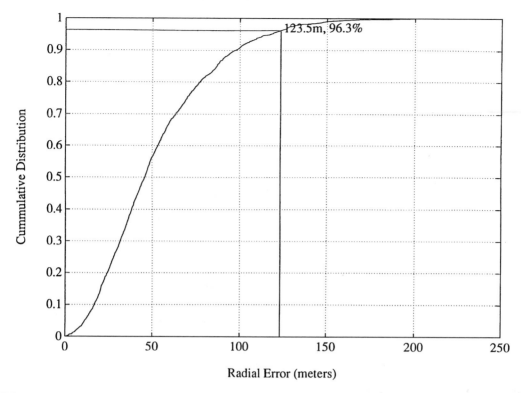

FIGURE 2.91

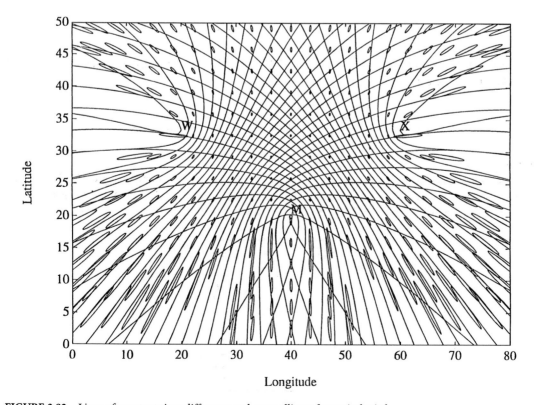

FIGURE 2.92   Lines of constant time difference and error ellipses for typical triad.

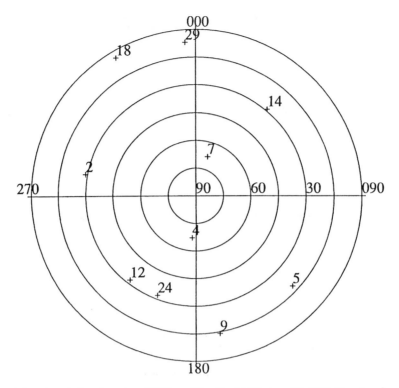

FIGURE 2.93   Azimuth and elevation plot of GPS satellites for 1400Z, Aug. 28, 1995, New London, CT.

$$\Delta P = \left(A^T A\right)^{-1} A^T \Delta PR$$

where A is the matrix of direction cosines and ΔPR is the difference between measured and calculated pseudo-ranges. The covariance of the position error is now:

$$\text{cov}\left(\Delta P\right) = \left(A^T A\right)^{-1} A^T \text{cov}\left(\Delta PR\right) A \left(A^T A\right)^{-1} = \sigma_{PR}\left(A^T A\right)^{-1}$$

where $\text{cov}(\Delta P) = \sigma_{PR} I$.

Figure 2.93 shows the azimuth and elevation plot of GPS satellites for 1400Z, August 28, 1995 in New London CT as a typical example. Eliminating the low elevation satellite (#18) due to potential multipath and using the nine remaining listed in the table below, we can calculate expected accuracies.

SVN	Elevation	Azimuth
5	18°	317°
12	32°	232°
7	68°	73°
24	32°	249°
14	29°	50°
29	7°	94°
2	29°	169°
9	14°	280°
4	67°	265°

East	North	Altitude	Time
cos(EL)sin(AZ)	cos(EL)cos(AZ)	sin(EL)	
−0.6486	0.6956	0.309	1
−0.6683	−0.5221	0.5299	1
0.3582	0.1095	0.9272	1
−0.7917	−0.3039	0.5299	1
0.6700	0.5622	0.4848	1
0.9901	−0.0692	0.1219	1
0.1669	−0.8586	0.4848	1
−0.9556	0.1685	0.2419	1
−0.3892	−0.0341	0.9205	1
$G = inv(A^T A)$			
0.2533	−0.0192	0.0223	0.0239
−0.0192	0.5248	0.1160	−0.0466
0.0223	0.1160	1.6749	−0.8404
0.0239	−0.0466	−0.8404	0.5380

$$\text{GDOP} = \sqrt{\text{trace}(G)} = 1.73$$

$$\text{PDOP} = \sqrt{G_{11} + G_{22} + G_{33}} = 1.57$$

$$\text{HDOP} = \sqrt{G_{11} + G_{22}} = 0.88$$

$$\text{VDOP} = \sqrt{G_{33}} = 1.29$$

Exactly what these DOPs imply in terms of accuracy depends on what version of GPS/DGPS service one is using.

For the Standard Positioning Service (SPS) with Selective Availability (SA) on, typically the value 32 meters was used for the one sigma error. Since this value depended almost entirely on policy, advances in technology did not change its value. In Conley (1999) we can get a glimpse of what accuracies will be expected in the short term, and how they may improve in the future. They report less than 1.4 meters rms for the on-orbit clocks and 59 centimeters for the ephemeris contribution to user range error (URE) in 1999. For SVN 43, the first operational Block IIR satellite, the combined clock and ephemeris errors were 70 centimeters RMS. This means that the errors in predicting ionospheric delay at about five meters, the errors in predicting tropospheric delay at about two meters, and multipath now dominate the statistics. DGPS will remove most of the residual errors due to satellite clock and ephemeris, and ionospheric and tropospheric delay leaving multipath as the most common dominant error source. The multipath errors are highly dependent on receiver and antenna design and antenna placement. Preliminary data after the termination of SA, from fixed sites, with high quality receivers, and where some care has been taken to minimize multipath, is indicating 7 meter 2 drms accuracy may be achievable. The performance on typical installations with low cost receivers on moving platforms will be considerably worse. In the GPS modernization program to be implemented over the next decade, the addition of C/A code on L2 and third civil frequency, will mean civil receivers will measure ionospheric delay via differential pseudo-range as in dual frequency, military receivers. This combined with other improvements in technology is expected to improve civil GPS accuracy to better than five meters 2 drms.

For Precise Positioning Service users, NATO GPS (1991) specifies 13.0 meters (95%) URE, which would imply a standard deviation of 6.5 meters assuming a normal distribution. The actual performance, while classified, is generally acknowledged to be considerably better than this 1991 value, and will continue improve with technology.

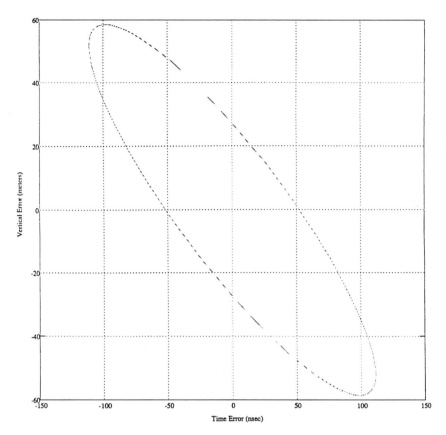

FIGURE 2.94 Typical joint distribution of GPS altitude and time errors.

The negative correlation ($\rho = -0.8404/\sqrt{1.6749 * 0.5380} = -0.885$) between the altitude and time errors is common because the satellites used (by receivers on the earth's surface) all must lie within a cone starting at some (elevation mask) angle above the horizon. Figure 2.94 shows the error ellipse for the joint time and altitude distribution. By using the first (east), second (north), and fourth (time) columns in the matrix A above we can calculate the expected DOPs for a GPS receiver in a fixed (known) altitude mode. HDOP is virtually unchanged but TDOP improves from 0.73 to 0.34. Similarly, if we assume we know time accurately and are only solving for three-dimensional position, we use the first three columns and VDOP improves from 1.29 to 0.60. Realistically, we will not know time exactly because we use the GPS to get time. We can improve vertical accuracy somewhat by propagating clock bias ahead via a Kalman filter as described in a later section, but can never get to zero clock error and this VDOP.

Weighted Least Squares

When the covariance of the measurement error is not a constant times an identity matrix, the optimum solution is a weighted vs. linear least squares solution.

$$\Delta P = \left(A^T W A \right)^{-1} A^T W \Delta P R$$

Now the covariance of the error is given by

$$E\left\{ \Delta P \Delta P^T \right\} = E\left\{ \left(A^T W A \right)^{-1} A^T W \Delta P R \, \Delta P R^T W^T A \left(A^T W A \right)^{-1} \right\}$$

This error is minimized if

$$W = R^{-1}$$

And this minimum error is given by

$$E\left\{\Delta P \Delta P^T\right\} = \left(A^T W A\right)^{-1}$$

For example, if we assume the standard deviation of each TOA measurement is the same for a three TD fix we should use a weighting matrix given by

$$W = R^{-1} = \begin{bmatrix} \sigma_m^2 + \sigma_1^2 & \sigma_m^2 & \sigma_m^2 \\ \sigma_m^2 & \sigma_m^2 + \sigma_2^2 & \sigma_m^2 \\ \sigma_m^2 & \sigma_m^2 & \sigma_m^2 + \sigma_3^2 \end{bmatrix}^{-1} = \frac{1}{\sigma_{TOA}} \begin{bmatrix} 0.75 & -0.25 & -0.25 \\ -0.25 & 0.75 & -0.25 \\ -0.25 & -0.25 & 0.75 \end{bmatrix}$$

When this matrix is multiplied times the TD observation vector, the correlation of the errors in the resulting vector is removed. It can be shown that weighted least squares processing of the TD measurements with the weighting matrix above is exactly equivalent to linear least squares processing of TOA measurements. For GPS, GLONASS, or DGPS there may be some advantages to weighted least squares processing. For example, low elevation satellites are subject to larger pseudo-range errors due to multipath and ionosphere delay modeling errors. The typical approach is to reject satellites below some elevation mask threshold of five to ten degrees. An alternative method using low elevation satellites with lower weights may help eliminate occasional spikes in GDOP. In integrated GPS/GLONASS, weighted least squares would have advantages in assigning different weights according to the relative accuracy of the two system's pseudo-ranges.

In LORAN, TOAs do not have same variance, TDs are not independent, and the position solution should be designed accordingly. Figures 2.95, 2.96, and 2.97 show the results of 1166 LORAN fixes over 6.5 hours using the 9960 chain at a stationary location in New York Harbor. The LOCUS™ receiver used was modified to use a Cesium time standard and provided TOA data. The observed TOA standard deviations were:

Seneca, NY	(175 nm)	7.6 ns
Caribou, ME	(452 nm)	100.6 ns
Nantucket, MA	(186 nm)	8.6 ns
Carolina Beach, NC	(438 nm)	21.1 ns
Dana, IN	(618 nm)	60.0 ns

(*Note:* Even though distances are comparable, the Carolina Beach TOAs are much better than those of Caribou because the path is seawater vs. land. In addition, Carolina Beach TDs are monitored and controlled using data from nearby Sandy Hook, NJ and Caribou is controlled using data from a monitor in Maine.)

The 2 drms repeatable accuracy is seen to be 21.8 meters for linear least squares and 7.3 meters for weighted least squares. Since for this particular data set, the Caribou and Dana signals are virtually ignored (Caribou is weighted $(7.6/100.6)^2 = 0.0057$ relative to Seneca), comparable accuracy to weighted least squares would have been obtained by totally ignoring Caribou and Dana data and doing three-TOA/two-TD fixes. In general using all data provides for a much more reliable solution. Figure 2.98 shows a histogram of time difference data of the 5930Y (Caribou/Cape Race) baseline as measured in

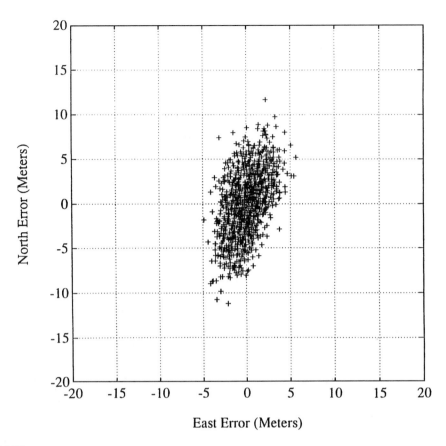

FIGURE 2.95

New London, CT. What is seen for marginal signals like Cape Race at 875 nm from New London, is frequent cycle errors resulting in 10 and 20 usec TD errors. Any kind of standard least squares algorithm will not work. Assuming an over-determined solution, there are two basic alternatives; first, a maximum likelihood algorithm that takes into account the potential for local maxima in the likelihood function due to cycle errors, and second, Receiver Autonomous Integrity Monitoring (RAIM). In RAIM, a fix using all measurements is calculated and the residuals compared to a threshold. For RAIM to be valid, assuming N measurements, each of the N fixes available from each subset of N-1 measurements must have acceptable geometry. For GPS, if we have two or more measurements than unknowns, the potential exists to identify and remove bad data via residual analysis on these N over-determined fixes. Since for LORAN, cycle errors have predictable behavior, we may be able to identify and remove them with only one measurement more than the number of unknowns.

Kalman Filters

This section points out the basic principles of Kalman Filters and specifically how they can be used in radionavigation systems. All of our analysis above has assumed we process each set of data independently. We know that parameters such as clock frequency and our velocity can only change at finite rates and therefore we ought to be able to do better using all past measurements. When we integrate LORAN TOAs or GLONASS pseudo-ranges with GPS pseudo-ranges we may not be able to assume the various systems are exactly synchronized in time. Therefore we cannot only solve for one time term, but we do know the drift between systems is extremely slow and need to somehow exploit that knowledge. Kalman filters allow us to use past measurements together with a priori knowledge of platform and clock dynamics to obtain optimum values of position and velocity. Typically, Kalman filters have found extensive use in the

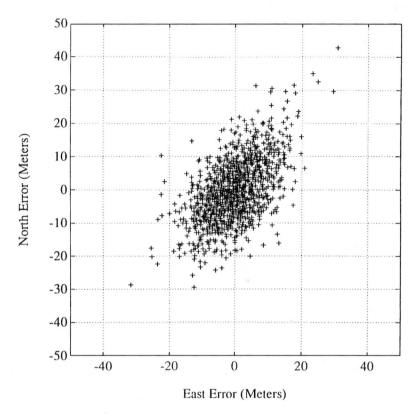

FIGURE 2.96

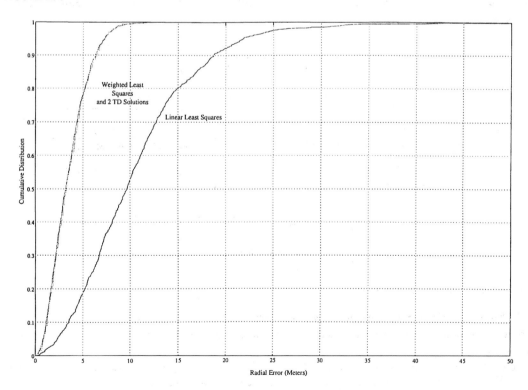

FIGURE 2.97

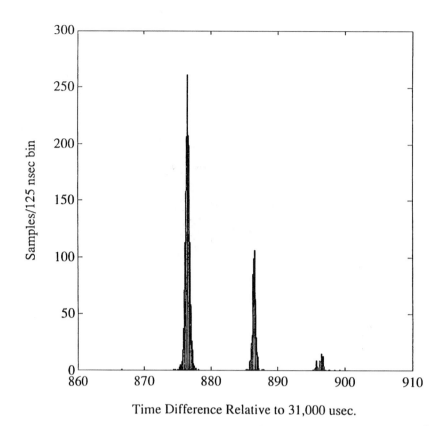

FIGURE 2.98 Histogram of time difference data of the 5930Y (Caribou/Cape Race) baseline measured in New London, CT.

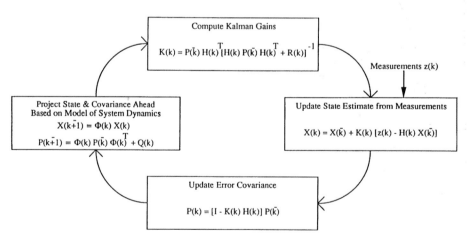

FIGURE 2.99 Kalman filter structure.

integration of inertial and electronic navigation systems. Here we will concentrate on stand-alone electronic navigation systems. The reader is referred to more extensive references such as Brown (1992) for more detail.

Figure 2.99 shows the basic structure of a discrete Kalman filter. The process starts with initial estimates of the state $[X(0)]$ and the error covariance $[P(0)]$. The other variables are:

R(k) is the covariance of the measurement errors. (They are assumed white, this is not valid for SA-dominated pseudo-range errors, which are correlated over minutes. Strictly speaking this correlation should be modeled via additional state variables in system model, but normally is not.)

H(k) Matrix of direction cosines and ones (as A above) that relate pseudo-range or TOA errors to positions and clock bias and Doppler's to velocity and frequency errors.

z(k) Pseudo-range (or TOA) and Doppler measurements

K(k) Kalman gains

Φ(k) State transition matrix

Q(k) Covariance of noise driving state changes. This matrix controls allowed accelerations and clock frequency and phase variations. Smaller Q(k) will result in smaller estimates of state error covariance and less weighting of measurements. Larger Q(k) will result in faster system response to measurements. Conventional wisdom suggests using larger Q when uncertain of exact model.

A typical state space model for stand-alone GPS would have eight states, the spatial coordinates and their velocities, and the clock offset and frequency. The individual pseudo-range measurements can be processed sequentially, which means that the Kalman gains can be calculated as scalars without the need for matrix inversions. There is no minimum number of measurements required to obtain an updated position estimate. The measurements are processed in an optimum fashion and if not enough for good geometry, the estimate of state error variance [P(k)] will grow. If two satellites are available, the clock bias terms are just propagated forward via the state transition matrix.

The structure is ideal for integrated systems where the offset between GPS and GLONASS time may only be known approximately initially, but is known to be virtually constant. This offset would then be estimated as a state variable with some initial uncertainty but with very small Q(k) driving changes. The same would be true for LORAN/GPS time offsets and the concept could be extended to ASFs for individual stations as well. The ASFs would be represented by very slowly varying states to be estimated. If GPS pseudo-ranges became unavailable such as in an urban canyon, mountain valley, or under heavy foliage cover, the LORAN would now be well calibrated.

Defining Terms

Accuracy: The degree of conformance between the estimated or measured position and/or velocity of a platform at a given time and its true position or velocity. Radionavigation system accuracy is usually presented as a statistical measure of system error and is specified as:

Predictable: The accuracy of a radionavigation system's position solution with respect to the charted solution. Both the position solution and the chart must be based upon the same geodetic datum.

Repeatable: The accuracy with which a user can return to a position whose coordinates have been measured at a previous time with the same navigation system.

Relative: The accuracy with which a user can measure position relative to that of another user of the same navigation system at the same time.

Availability: The availability of a navigation system is the percentage of time that the services of the system are usable. Availability is an indication of the ability of the system to provide usable service within the specified coverage area. Signal availability is the percentage of time that navigational signals transmitted from external sources are available for use. Availability is a function of both the physical characteristics of the environment and the technical capabilities of the transmitter facilities.

Block I and Block II Satellites: The Block I is a GPS concept validation satellite; it does not have all of the design features and capabilities of the production model GPS satellite, the Block II. The FOC 24 satellite constellation is defined to consist entirely of Block II/IIA satellites.

Coordinated Universal Time (UTC): UTC, an atomic time scale, is the basis for civil time. It is occasionally adjusted by one-second increments to ensure that the difference between the uniform time scale, defined by atomic clocks, does not differ from the earth's rotation by more than 0.9 seconds.

Differential: A technique used to improve radionavigation system accuracy by determining positioning error at a known location and subsequently transmitting the determined error, or corrective factors, to users of the same radionavigation system, operating in the same area.

Dilution of Precision (DOP):   The magnifying effect on radionavigation position error induced by mapping ranging errors into position and time through the position solution. The DOP may be represented in any user local coordinate desired. Examples are HDOP for local horizontal, VDOP for local vertical, PDOP for all three coordinates, TDOP for time, and GDOP for position and time.

Distance Root Mean Square (drms):   The root-mean-square value of the distances from the true location point of the position fixes in a collection of measurements. As used in this Chapter 2 drms is the radius of a circle that contains at least 95 percent of all possible fixes that can be obtained with a system at any one place. Actually, the percentage of fixes contained within 2 drms varies between approximately 95.5% and 98.2%, depending on the degree of ellipticity of the error distribution.

Hyperbolic Navigation System: A navigation system that produces hyperbolic lines of position (LOPs) through the measurement of the difference in times of reception (or phase difference) of radio signals from two or more synchronized transmitters.

Integrity: Integrity is the ability of a system to provide timely warnings to users when the system should not be used for navigation.

Multipath Transmission: The propagation phenomenon that results in signals reaching the receiving antenna by two or more paths. When two or more signals arrive simultaneously, wave interference results. The received signal fades if the wave interference is time varying or if one of the terminals is in motion.

Pseudo-range: The difference between the ranging signal time of reception (as defined by the receiver's clock) and the time of transmission contained within the satellite's navigation data (as defined by the satellite's clock) multiplied by the speed of light.

Receiver Autonomous Integrity Monitoring (RAIM):   A technique whereby a civil GPS receiver/processor determines the integrity of the GPS navigation signals without reference to sensors or non-DoD integrity systems other than the receiver itself. This determination is achieved by a consistency check among redundant pseudo-range measurements.

Selective Availability (SA):   The denial of full GPS accuracy to civil users by manipulating navigation message orbit data (epsilon) and/or satellite clock frequency (dither).

References

Braff, R. Description of the FAA's Local Area Augmentation System (LAAS), *Navigation*, 44, 4, 411–424, Winter 1997.

Brown, R. G. and Hwang, P. Y. C., *Introduction to Random Signals and Applied Kalman Filtering*, 2nd Ed., John Wiley & Sons, New York, 1992.

Conley, R. and Lavrakas, J. W., The World after Selective Availability, *Proceedings of ION GPS '99, 14–17*, 1353–1361, Nashville, TN, September 1999.

CSIC, *GLONASS Interface Control Document, Version 4.0*, The Ministry of Defence of the Russian Federation Coordination Scientific Information Center, 1998. Available in electronic form at http://www.rssi.ru/SFCSIC/SFCSIC_main.html

Dale, S, Daly, P., and Kitching, I., Understanding signals from GLONASS navigation satellites, *Int. J. Satellite Commun.*, 7, 11–22, 1989.

U.S. Departments of Defense and Transportation, (DoD/Dot, 1994) *1994 Federal Radionavigation Plan*, U.S. Departments of Defense and Transportation, NTIS Report DOT-VNTSC-RSPA-95-1/DoD-4650.5, 1995. Available in electronic form (Adobe Acrobat) from USCG NAVCEN (www.navcen,uscg.mil).

Enge, P., Swanson, E., Mullin, R., Ganther, K., Bommarito, A., and Kelly, R., Terrestrial radionavigation technologies, *Navigation*, 42, 1, 61–108, Spring, 1995.

Feairheller, S., The Russian GLONASS system, a U.S. Air Force study, *Proceedings of Institute of Navigation GPS-94*, Salt Lake City, September 1994.

NATO Navstar GPS Technical Support Group (NATO GPS 1991), *Technical Characteristics of the Navstar GPS*, June 1991.

Parkinson, B., Stansell, T., Beard, R., and Gromov, K., A history of satellite navigation, *Navigation*, 42, 1, 109–164, Spring 1995.

Pierce, J. A., Mckenzie, A. A., and Woodward, R. H., Eds., *LORAN*, MIT Radiation Laboratory Series, McGraw-Hill, New York, 1948.

RTCA, *Minimum Aviation Performance Standards for Local Area Augmentation System*, RTCA/DO-245, September 1998.

RTCA, *Minimum Operational Performance Standards for GPS/Wide Area Augmentation System Airborne Equipment*, RTCA/DO 229B, October 1999.

RTCM Special Committee 104, *RTCM Recommended Standard for Differential Navstar GPS Service*, Version 2.0, December 10, 1992.

Skidmore, T. A., Nyhus, O. K., and Wilson, A. A., An overview of the LAAS VHF data broadcast, *Proceedings of ION GPS '99, 14–17*, 671–680, Nashville, TN, September 1999.

U.S. Air Force, GPS Joint Program Office, (USAF JPO, 1995) *Global Positioning System, Standard Positioning Service, Signal Specification*, June, 1995. Available in electronic form (Adobe Acrobat) from USCG NAVCEN (www.navcen,uscg.mil).

U. S. Coast Guard, (USCG, 1992), *LORAN C User Handbook*, COMDTPUB P16562.6, COMDT(G-NRN), 1992. Available in electronic form from USCG NAVCEN (www.navcen,uscg.mil).

Walter, T. and Bakery El-Arini, M., Eds., *Selected Papers on Satellite based Augmentation Systems (SBASs)*, VI in the GPS Series, Institute of Navigation, Alexandria, VA, 1999.

Ward, Philip W., GPS receiver RF interference monitoring, mitigation and analysis techniques, *Navigation*, 41, 4, 367–391, Winter 1994–1995.

White House Press Secretary, Statement by the President Regarding the United States Decision to Stop Degrading Global Positioning System Accuracy, White House, May 1, 2000.

Wild Goose Association (WGA, 1982), On the calculation of geodesic arcs for use with LORAN, *Wild Goose Association Radionavigation Journal*, 57–61, 1982.

Further Information

Government Agencies

1. **U.S. Coast Guard Navigation Information Service (NIS)**

   Address: Commanding Officer
              USCG Navigation Center
              7323 Telegraph Road
              Alexandria, VA 22315
   Internet: http://www.navcen.uscg.mil/navcen.htm
   Phone: (703) 313-5900 (NIS Watchstander)

Formerly the GPS Information Center (GPSIC), NIS has been providing information since March, 1990. The NIS is a public information service. At the present time, there is no charge for the information provided. Hours are continuous: 24 hours a day, 7 days a week, including federal holidays.

The mission of the Navigation Information Service (NIS) is to gather, process, and disseminate timely GPS, DGPS, Omega, LORAN-C status, and other navigation related information to users of these navigation services. Specifically, the functions to be performed by the NIS include the following:

Provide the Operational Advisory Broadcast Service (OAB).

Answer questions by telephone or written correspondence.

Provide information to the public on the NIS services available.

Provide instruction on the access and use of the information services available.

Maintain tutorial, instructional, and other relevant handbooks and material for distribution to users.
Maintain records of GPS, DGPS, and LORAN-C broadcast information, and databases or relevant
   data for reference purposes.
Maintain bibliography of GPS, DGPS, and LORAN-C publications.

The NIS provides a watchstander to answer radionavigation user inquiries 24 hours a day, seven days a
week and disseminates general information on GPS, DGPS, Loran-C, and Radiobeacons through various
mediums.

   2. **NOAA, National Geodetic Survey**
      Address: 1315 East-West Highway, Station 09202
               Silver Spring, MD 20910
      Phone: (301) 713-3242;
      Fax: (301) 713-4172
      Monday through Friday, 7:00 a.m.–4:30 p.m., Eastern Time
      Internet: http://www.ngs.noaa.gov/

NOAA is modernizing the Nation's Spatial Reference System, providing horizontal and vertical positions
for navigation and engineering purposes. By implementing a new system called CORS (Continuously
Operating Reference Stations), NOAA will continuously record carrier phase and pseudo-range measurements for all GPS (Global Positioning System) satellites at each CORS site. A primary objective of
CORS is to monitor a particular site, which is determined to millimeter accuracy and provide local users
a tie to the National Spatial Reference System. NOAA will also use the CORS sites along with other
globally distributed tracking stations in computing precise GPS orbits.

   NOAA is also the federal agency responsible for providing accurate and timely Global Positioning
System (GPS) satellite ephemerides ("orbits") to the general public. The GPS precise orbits are derived
using 24-hour data segments from the global GPS network coordinated by the International Geodynamics
GPS Service (IGS). The reference frame used in the computation is the International Earth Rotation
Service Terrestrial Reference Frame (ITRF). In addition, an informational summary file is provided to
document the computation and to convey relevant information about the observed satellites, such as
maneuvers or maintenance. The orbits generally are available two to six days after the date of observation.
   Also available:

   • Software programs to compute, verify, or adjust field surveying observations; convert coordinates
     from one geodetic datum to another; or assist in other specialized tasks utilizing geodetic data.
   • Listings of publications available on geodesy, mapping, charting, photogrammetry, and related
     topics, such as plane coordinate systems and numerical analysis.
   • Information on NGS programs to assist users of geodetic data, including technical workshops and
     the geodetic advisor program.
   • Information on survey data recently incorporated into NSRS.
   • Information on the location of NGS field parties.

   3. **U.S. Naval Observatory (USNO)**
      Internet:http://www.usno.navy.mil/

USNO is the official source of time used in the United States. USNO timekeeping is based on an ensemble
of cesium beam and hydrogen maser atomic clocks. USNO disseminates information on time offsets of
the world's major time services and radionavigation systems.

   4. **USSPACECOM GPS Support Center**
      300 O'Malley
      Suite 41
      Schriever AFB, CO 80912-3041
      http://www.peterson.af.mil/usspace/gps_support

The **GPS Support Center** (GSC) is the DoD's focal point for operational issues and questions concerning military use of GPS. The GSC is responsible for:

a. Receiving reports and coordinating responses to radio frequency interference in the use of GPS in military operations.
b. Providing prompt responses to DoD user problems or questions concerning GPS.
c. Providing official USSPACECOM monitoring of GPS performance provided to DoD users on a global basis.
d. Providing tactical support for planning and assessing military missions involving the use of GPS.

As the DoD's focal point for GPS operational matters, the GSC serves as U.S. Space Command's interface to the civil community, through the U.S. Coast Guard's Navigation Center and Federal Aviation Administration's National Operations Command Center.

5. **The Ministry of Defence of the Russian Federation**
   Coordination Scientific Information Center
   Internet: http://www.rssi.ru/SFCSIC/SFCSIC_main.html

The mission of the Coordination Scientific Information Center is to plan, manage, and coordinate the activities on

- use of civil-military space systems (navigation, communications, meteorology etc.);
- realization of Russian and international scientific and economic space programs;
- realization of programs of international cooperation;
- conversional use of military space facilities, as well as to provide the scientific-informational, contractual and institutional support of these activities.

2.10 Avionics

James L. Bartlett

The term *avionics* was originally coined by contracting *avi*ation and elect*ronics*, and has gained widespread usage over the years. Avionics differs from many of the other wireless applications in several important areas. Avionics applications typically require functional integrity and reliability that is orders of magnitude more stringent than most commercial wireless applications, and rigor of these requirements is matched or exceeded only by the requirements for space and/or certain military applications. The need for this is readily understood when one compares the impact (pun intended), of the failure of a cellular phone call, and the failure of an aircraft Instrument Landing System on final approach during reduced visibility conditions. Avionics must function in environments that are more severe than most other wireless applications as well. Extended temperature ranges, high vibration levels, altitude effects including corona and high energy particle upset (known as single event upset), of electronics are all factors that must be considered in the design of products for this market. Quantities for this market are typically very low when compared to commercial wireless applications, e.g., the number of cell phones manufactured every single working day far exceeds the number of aircraft that are manufactured in the world in a year. Wireless systems for avionics applications cover an extremely wide range in a number of dimensions, including frequency, system function, modulation type, bandwidth, and power. Due to the number of systems aboard a typical aircraft, Electromagnetic Interference (EMI) and Electromagnetic Compatibility (EMC) between systems is a major concern, and EMI/EMC design and testing is a major factor in the flight certification testing of these systems.

Navigation, Communications, Voice, and Data

VLF (Very Low Frequency) is not used in civil aviation, but is used for low data rate transmission to and from strategic military platforms, due to its ability to transmit reliably worldwide, including water penetration adequate to communicate with submerged submarines. These are typically very high power systems, with 100 to 200 kilowatts of transmit power.

HF Communications (High Frequency) 2 to 30 MHz, has been used since the earliest days of aviation, and continues in use on modern aircraft, both civil and military. HF uses Single Sideband, Suppressed Carrier modulation, with a relatively narrow modulation bandwidth of about 2.5 kHz, typically several hundred-watt transmitters, and is capable of worldwide communications. Because propagation conditions vary with frequency, weather, ionosphere conditions, time of day, and sun spot activity, the whole HF band will generally not be available for communications between any two points on earth at any given time, but some portion will be suitable. Establishing a usable HF link frequency between two stations is an essential part of the communications protocol in this band, and until recently, required a fair amount of time. Modern HF systems have Automatic Link Establishment (ALE) software and hardware to remove that burden from the operator. Prior to the advent of communications satellites, HF was the only means of communicating with an aircraft for large parts of transoceanic flights.

VHF Communications (Very High Frequency) as used in the aviation world, comprises two different frequency bands. The military uses 30 to 88 MHz, with narrowband FM modulation, primarily as an air-to-ground link for close air support. RF power for this usage ranges from 10 to 40 watts. Both civil and military aircraft use the 116 to 136 MHz band with standard double sideband AM modulation for air traffic control purposes, typically with 10 to 30 watts of carrier power.

UHF Communications (Ultra High Frequency) as used in the aviation world actually bridges the VHF and UHF regions, operating from 225 to 400 MHz. Various radios in use here range in power output from 10 to 100 watts of carrier power for FM, and 10 to 25 watts of AM carrier power. This band is used for military communications, and many waveforms and modulation formats are in use here, including, AM, FM, and a variety of pulsed, frequency hopping, antijam waveforms using various protocols and encryption techniques.

SatCom, short for Satellite Communications, is used for aircraft data link purposes, as well as for communications by the crew and passengers. Passenger telephones use this service. Various satellites are used, including INMARSAT, and Aero-H and Aero-I, all operating at L band.

Data Links are a subset of the overall communications structure, set up to transmit digital data to and from aircraft. They cover an extremely wide range of operating frequencies, data rates, and security features. Data Link applications may be a shared use of an existing radio, or use a specialized transceiver dedicated to the application. Examples include the ARINC Communication Addressing and Reporting System (ACARS) which uses the existing VHF Comm radio on civil aircraft, and the Joint Tactical Information Distribution System (JTIDS) which is a secure, fast hopping L band system that uses a dedicated transceiver.

Military data link usage is an application area that has been dominated by military applications, but is now seeing an increasing number of commercial uses. Virtually all of the network management and spread-spectrum methods now coming into use for personal communications were originally pioneered by military secure data link applications. Code Division Multiple Access (CDMA), Time Division Multiple Access (TDMA), Bi-Phase Shift Keying (BPSK), Quadrature Phase Shift Keying (QPSK), Minimum Shift Keying (MSK), or Continuous Phase Shift Keying (CPSM), Gaussian Minimum Shift Keying (GMSK), various error detecting and correcting codes, along with interleaving and various forms of data redundancy have all found applications in military data link applications. A great deal of effort has been devoted to classes of orthogonal code generation and secure encryption methods for use in these systems. It is common to find systems using a variety of techniques together to enhance their redundancy and antijam nature.

One system, for example, uses the following techniques to increase the antijam capabilities of the system. Messages are encoded with an error detecting and correcting code, to provide a recovery capability for bit errors. The message is then further encoded using a set of orthogonal CCSK code symbols representing a smaller number of binary bits, i.e., 32 chips representing 5 binary bits, giving some level of CDMA spreading, which further increases the probability of successful decoding in the face of jamming. The result is then multiplied with a pseudo-random encryption code, further pseudo-randomizing the information in a CDMA manner. The resultant bit stream is then modulated on a carrier in a CPSM format, and transmitted symbol by symbol, with each symbol transmitted on a separate frequency in a pulsed, frequency hopped TDMA sequence determined by another encryption key. The message itself may also be parsed into short segments and interleaved with other message segments to time spread it still further, and repeated with different data interleaving during different time slots in the TDMA network arrangement to provide additional redundancy. Such systems have a high degree of complexity, and a large throughput overhead when operating in their most secure modes, but can deliver incredible robustness of message delivery in the face of jamming and/or spoofing attempts by an adversary. Additionally, these systems are incredibly difficult to successfully intercept and decode, which is frequently just as important to the sender as the successful delivery of the message to its intended user.

Navigation and identification functions comprise a large share of the avionics aboard a typical aircraft. There are systems used for enroute navigation, including Long Range Radio Navigation (LORAN), Automatic Direction Finder (ADF), Global Positioning System (GPS), Global Orbital Navigation Satellite System (GLONASS), Distance Measuring System (DME), and Tactical Air Navigation (TACAN), systems used for approach navigation, including Marker Beacon, VHF Omni Range (VOR), Instrument Landing System (ILS), Microwave Landing System (MLS), Radar Altimeter, Ground Collision Avoidance System (GCAS), and GPS, and systems used for identification of aircraft and hazards, including Air Traffic Control Radar Beacon System (ATCRBS), Mode S, Identification Friend or Foe (IFF), Traffic Collision Avoidance System (TCAS), and radar.

Many of the radionavigation systems rely on varying complexity schemes of spatial modulation for their function. In almost all cases, there are multiple navaids functional at a given airport ground station, including Marker Beacons, a DME or TACAN, a VOR and an ILS system.

In a typical example, the VHF Omni Range (VOR) system operates at VHF (108 to 118 MHz), to provide an aircraft with a bearing to the ground station location in the following manner. The VOR transmits continuously at its assigned frequency, with two antennas (or the electronically combined equivalent), providing voice transmission and code identification to ensure that the aircraft is tracking the proper station. The identification is in Morse code, and is a 2- or 3-letter word repeated with a modulation tone of 1020 Hz. The transmitted signal from the VOR station contains a carrier that is amplitude-modulated at a 30 Hz rate by a 30 Hz variable signal, and a 9960 Hz subcarrier signal that is frequency modulated between 10440 and 9480 Hz by a 30 Hz reference signal. The reference signal is radiated from a fixed omnidirectional antenna, and thus contains no time-varying spatial modulation signal. The variable signal is radiated from a rotating, (electrically or mechanically), semi-directional element driven at 1800 rpm or 30 Hz, producing a spatial AM modulation at 30 Hz. The phasing of the two signals is set to be in phase at magnetic north, 90 degrees out of phase at magnetic east, 180 degrees when south, and 270 degrees when west of the VOR station. VOR receivers function by receiving both signals, comparing their phase, and displaying the bearing to the station to the pilot. An area directly above a VOR station includes an area that has no AM component. This cone-shaped area is referred to as the "cone of confusion," and most VOR receivers contain delays to prevent "Invalid Data" flags from being presented during the brief flyover time.

The Instrument Landing System (ILS) contains three functions, Localizer, Glideslope, and Marker Beacon. Three receivers are used in this system: a VHF localizer, a UHF glideslope, and a VHF marker beacon. Their functions are described in the following paragraphs.

The marker beacon receiver functions to decode audio and provide signaling output to identify one of three marker beacons installed near the runway. The outer beacon is typically about 5 miles out, the middle approximately 3500 feet, and the inner just off the runway end. These beacons radiate a narrow beamwidth, 75 MHz signal in a vertical direction, and each has a different distinct modulation code so the receiver can identify which one it is flying over.

The localizer transmitter is located at the far end of the runway, and radiates two intersecting lobes in the 108 to 112 MHz frequency band along the axis of the runway, with the left lobe modulated at

90 Hz, and the right lobe modulated at 150 Hz. The installation is balanced so that the received signal will be equally modulated along the centerline of the runway. The total beamwidth is approximately 5 degrees wide. The localizer receiver uses this modulation to determine the correct approach path in azimuth. This pattern also extends beyond the departure end of the runway, and this is called the back course. In the case of an aircraft flying an inbound back course, the sensed modulations will be reversed, and there are no marker beacons or glideslope signals available on the back course. In the event of a missed approach and go around, the side lobes of the localizer antenna pattern could present erroneous or confusing bearing data and validity flags, so most installations utilize a localizer go-around transmitter on each side of the runway that radiates a signal directed outward from the side of the runway that is modulated with a go-around level modulation that identifies the signal and masks the sidelobe signals.

The glideslope transmitter is located near the near end of the runway, and similarly radiates two intersecting lobes, one on top of the other, in the 329 to 335 MHz frequency band along the axis of, and at the angle of the desired approach glideslope, usually approximately 3 degrees, with the upper lobe modulated at 90 Hz, and the lower lobe modulated at 150 Hz. The installation is balanced so that the received signal will be equally modulated along the centerline of the glideslope. The total beamwidth is approximately 1.4 degrees wide. The localizer receiver uses this modulation to determine the correct glideslope path. The combination of localizer and glideslope modulations for the approach is illustrated in Fig. 2.100 below. The back course, marker beacons, and go-around patterns are deleted for simplicity of visualization.

Another example of an avionics system in widespread use that has significant RF content is the Traffic Alert and Avoidance System, TCAS. TCAS functions in close harmony with the Air Traffic Control Radar Beacon System (ATCRBS) at 1030 and 1090 MHz, which has been in operation for several decades. The ATCRBS system comprises a system of ground-based interrogators and airborne responders, which serve to provide the Air Traffic Control (ATC) system with the necessary information to safely manage the air space.

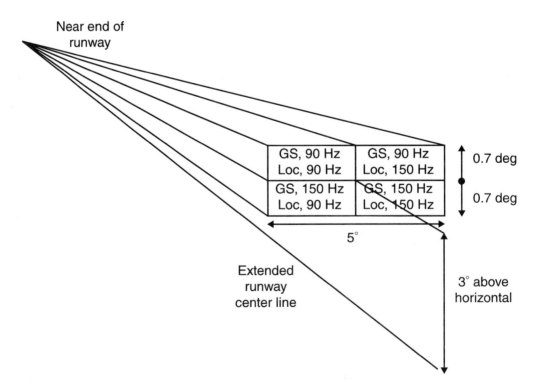

FIGURE 2.100   Glide slope and localizer modulation.

There are several interrogation/reply modes incorporated into the ATCRABS/TCAS system. Mode A and Mode C interrogations use a combination of pulse position and width coding, and replies are in a pulse position/timing format. A Mode A response identifies the aircraft responding. Aircraft equipped with Mode C transponders that are interrogated by the Air Traffic Control's Secondary Surveillance Radar system respond with a coded pulse train that includes altitude information, allowing automated altitude tracking by the ATC system. TCAS incorporates Mode C and Mode Select, or Mode S as it is more commonly known, which incorporates a data link capability, and interrogates these transponders in other aircraft. Mode S uses pulse position coding for interrogations and replies, and Differential Phase Shift Keying (DPSK) to send data blocks containing 56 or 112 data chips. All modes of transponder responses to an interrogation are designed to be tightly controlled in time so that slant range between the interrogator and responder are easily calculated. Altitude is directly reported in both Mode C and Mode S responses as well. Specific details of all of the allowable interrogation, reply, and data transmissions would take up far too much room to include here. As with most essential avionics equipment, passenger aircraft will have two systems installed for redundancy.

TCAS provides for a surveillance radius of 30 miles or greater and displays detected aircraft in four categories to the pilot: Other, Proximate, TA, and RA. Each has its own unique coding on the display. Intruders classified as "Other" are outside of 6 nautical miles, or more than ±1200 feet of vertical separation, and their projected flight path does not take them closer than these thresholds. A "Proximate" aircraft is within 6 nmi, but the flight path does not represent a potential conflict. TCAS provides the pilot with audible advisory messages for the remaining two categories; traffic advisories (TA) for "Proximate" aircraft when they close to within 20 to 48 seconds of Closest Point of Approach (CPA) announced as "Traffic, Traffic," and Resolution Advisories (RA) for aircraft whose flight path is calculated to represent a threat of collision. Resolution advisories are given 15 to 35 seconds prior to CPA, depending on altitude. RAs are limited to vertical advisories only; no turns are ever commanded. RAs may call for a climb, a descent, or to maintain or increase the rate of climb or descent. RAs are augmented by displaying a green arc on the vertical speed indicator, which corresponds to the range of safe vertical speeds.

TCAS antennas are multielement, multi-output port antennas, matched quite closely in phase and gain so that the receiver may determine the *angle of arrival* of a received signal. With altitude, slant range, and angle of arrival of a signal known, the present position, relative to own aircraft is computed. Based on changes between replies, the closest point of approach (CPA) is computed, and the appropriate classification of the threat is made. If both aircraft are equipped with Mode S capability, information is exchanged and RAs are coordinated to ensure that both aircraft do not make the same maneuver, requiring a second RA to resolve.

From an RF and microwave point of view, there are several challenging aspects to a TCAS design.

1. The antenna requires four elements located in close proximity within a single enclosure that radiate orthogonal cardiod patterns matched in gain and phase quite closely, i.e., 0.5 dB and 5 degrees, typically. With the high amount of mutual coupling present, this presents a challenging design problem.

2. Because of the high traffic density that exists within the U.S. and Europe, the potential for multiple responses and interference between aircraft with the same relative angle to the interrogating aircraft but different ranges is significant. In order to prevent this, a "Whisper Shout" algorithm has been implemented for TCAS interrogations. This involves multiple interrogations, starting at low power, with each subsequent interrogation being one dB higher in power. This requires control of a high power (typically in excess of 1 kW) transmitter output over a 20 dB range, with better than one dB linearity. Attaining this linearity with high-speed switches and attenuators capable of handling this power level is not an easy design task. The whisper shout pulse sequence is illustrated in Fig. 2.101.

Each interrogation sequence consists of four pulses, S1, P1, P3, and P4. Each pulse is 0.8 υSec in width, with a leading edge spacing of 2.0 υSec for pulses S1 to P1 and P3 to P4. There is a 21 υSec spacing from P1 to P3, (compressed in the figure). Each interrogation sequence will be increased in power level in

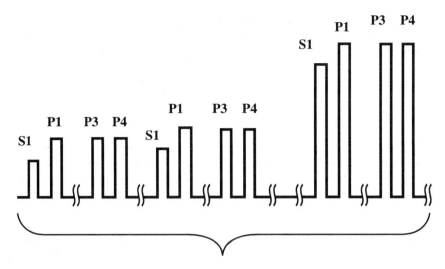

FIGURE 2.101   Mode C interrogation whisper shout sequence (all call).

one-dB steps over a 20-dB range. A transponder will only reply to pulse sequences that meet the timing requirements, and whose amplitude is such that S1 is below the receiver noise floor, and P1 is above a minimum trigger level. This effectively limits the number of responses to the whisper shout sequence from any aircraft to about two, thus limiting the inter-aircraft interference.

Entertainment is the latest application of microwave technology to the aircraft industry. While the traditional definition of avionics has referred to cockpit applications, in recent years there has been a move to receive Direct Broadcast TV services, and provide Internet access services to passengers via individual displays at each seat. These applications require steerable, aircraft mounted antennas at X or Ku band capable of tracking the service provider satellites, low noise downconverters, some sort of closed loop tracking system, typically utilizing the aircraft's inertial navigation system, and a receiver(s) with multiple tunable output channels, plus the requisite control at each seat to select the desired channel. Such a system is thus much more complex and expensive than a household Direct Broadcast System.

Military aircraft utilize all of the avionics systems listed above, (with the exception of entertainment), and depending on the mission of the individual airframe, may contain systems to spoof, jam, or just listen to other users of any or all of these systems. The radar systems on military aircraft may be much more powerful and sophisticated than those required for civil use, and serve a variety of weapons or intelligence gathering functions in addition to the basic navigation/weather sensing functions of civil aircraft. In fact, virtually all of the functions used by civil aircraft may be exploited or denied for a military function, and systems exist on various platforms to accomplish just that.

2.11   Radar

2.11.1   Continuous Wave Radar

James C. Wiltse

Continuous wave (CW) radar employs a transmitter that is on all or most of the time. Unmodulated CW radar is very simple and is able to detect the **Doppler-frequency shift** in the return signal from a target that has a component of motion toward or away from the transmitter. While such a radar cannot measure range, it is used widely in applications such as police radars, motion detectors, burglar alarms, proximity fuzes for projectiles or missiles, illuminators for semiactive missile guidance systems (such as the Hawk surface-to-air missile), and scatterometers (used to measure the scattering properties of targets or clutter such as terrain surfaces) [Nathanson, 1991; Saunders, 1990; Ulaby and Elachi, 1990].

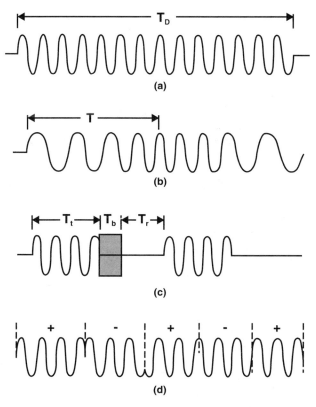

FIGURE 2.102   Waveforms for the general class of CW radar: (a) continuous sine wave CW; (b) frequency modulated CW; (c) interrupted CW; (d) binary phase-coded CW. (*Source:* F. E. Nathanson, *Radar Design Principles*, New York: McGraw-Hill, 1991, p. 450. With permission.)

Modulated versions include frequency-modulated (FM/CW), interrupted frequency-modulated (IFM/CW), and phase-modulated. Typical waveforms are indicated in Fig. 2.102. Such systems are used in altimeters, Doppler navigators, proximity fuzes, over-the-horizon radar, and active seekers for terminal guidance of air-to-surface missiles. The term *continuous* is often used to indicate a relatively long waveform (as contrasted to pulse radar using short pulses) or a radar with a high duty cycle (for instance, 50% or greater, as contrasted with the typical duty cycle of less than 1% for the usual pulse radar). As an example of a long waveform, planetary radars may transmit for up to 10 hours and are thus considered to be CW [Freiley et al., 1992]. Another example is interrupted CW (or **pulse-Doppler**) radar, where the transmitter is pulsed at a high rate for 10 to 60% of the total time [Nathanson, 1991]. All of these modulated CW radars are able to measure range.

The first portion of this section discusses concepts, principles of operation, and limitations. The latter portion describes various applications. In general, CW radars have several potential advantages over pulse radars. Advantages include simplicity and the facts that the transmitter leakage is used as the local oscillator, transmitter spectral spread is minimal (not true for wide-deviation FM/CW), and peak power is the same as (or only a little greater than) the average power. This latter situation means that the radar is less detectable by intercepting equipment.

The largest disadvantage for CW radars is the need to provide antenna isolation (reduce spillover) so that the transmitted signal does not interfere with the receiver. In a pulse radar, the transmitter is off before the receiver is enabled (by means of a duplexer and/or receiver-protector switch). Isolation is frequently obtained in the CW case by employing two antennas, one for transmit and one for reception. When this is done, there is also a reduction of close-in clutter return from rain or terrain. A second

TABLE 2.19 Doppler Frequencies for Several Transmitted Frequencies and Various Relative Speeds (1 m/s = 2.237 mph)

Microwave Frequency—f_T	Relative Speed			
	1 m/s	300 m/s	1 mph	600 mph
3 GHz	20 Hz	6 kHz	8.9 Hz	5.4 kHz
10 GHz	67 Hz	20 kHz	30 Hz	17.9 kHz
35 GHz	233 Hz	70 kHz	104 Hz	63 kHz
95 GHz	633 Hz	190 kHz	283 Hz	170 kHz

disadvantage is the existence of noise sidebands on the transmitter signal which reduce sensitivity because the Doppler frequencies are relatively close to the carrier. This is considered in more detail below.

CW Doppler Radar

If a sine wave signal were transmitted, the return from a moving target would be Doppler-shifted in frequency by an amount given by the following equation:

$$f_d = \frac{2v_r f_T}{c} = \text{Doppler frequency} \tag{2.2}$$

where f_T = transmitted frequency; c = velocity of propagation, 3×10^8 m/s; and v_r = radial component of velocity between radar and target.

Using Eq. (2.2) the Doppler frequencies have been calculated for several speeds and are given in Table 2.19.

As may be seen, the Doppler frequencies at 10 GHz (X-band) range from 30 Hz to about 18 kHz for a speed range between 1 and 600 mph. The spectral width of these Doppler frequencies will depend on target fluctuation and acceleration, antenna scanning effects, frequency variation in oscillators or components (for example, due to microphonism from vibrations), but most significantly, by the spectrum of the transmitter, which inevitably will have noise sidebands that extend much higher than these Doppler frequencies, probably by orders of magnitude. At higher microwave frequencies the Doppler frequencies are also higher and more widely spread. In addition, the spectra of higher frequency transmitters are also wider, and, in fact, the transmitter noise-sideband problem is usually worse at higher frequencies, particularly at millimeter wavelengths (i.e., above 30 GHz). These characteristics may necessitate frequency stabilization or phase locking of transmitters to improve the spectra.

Simplified block diagrams for CW Doppler radars are shown in Fig. 2.103. The transmitter is a single-frequency source, and leakage (or coupling) of a small amount of transmitter power serves as a local oscillator signal in the mixer. This is called homodyning. The transmitted signal will produce a Doppler-shifted return from a moving target. In the case of scatterometer measurements, where, for example, terrain reflectivity is to be measured, the relative motion may be produced by moving the radar (perhaps on a vehicle) with respect to the stationary target [Wiltse et al., 1957]. The return signal is collected by the antenna and then also fed to the mixer. After mixing with the transmitter leakage, a difference frequency will be produced which is the Doppler shift. As indicated in Table 2.19, this difference is apt to range from low audio to over 100 kHz, depending on relative speeds and choice of microwave frequency. The Doppler amplifier and filters are chosen based on the information to be obtained, and this determines the amplifier bandwidth and gain, as well as the filter bandwidth and spacing. The transmitter leakage may include reflections from the antenna and/or nearby clutter in front of the antenna, as well as mutual coupling between antennas in the two-antenna case.

The detection range for such a radar can be obtained from the following [Nathanson, 1991]:

$$R^4 = \frac{\bar{P}_T G_T L_T A_e L_R L_p L_a L_s \delta_T}{\left(4\pi^2\right) k T_s b \left(S/N\right)} \tag{2.3}$$

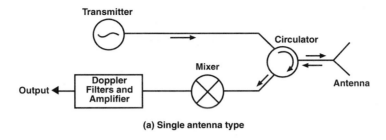

(a) Single antenna type

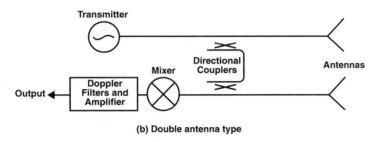

(b) Double antenna type

FIGURE 2.103 Block diagrams of CW-Doppler radar systems: (a) single antenna type; (b) double antenna type.

where R = the detection range of the desired target.

$\bar{P}_T$ = the average power during the pulse.

G_T = the transmit power gain of the antenna with respect to an omnidirectional radiator.

L_T = the losses between the transmitter output and free space including power dividers, waveguide or coax, radomes, and any other losses not included in A_e.

A_e = the effective aperture of the antenna, which is equal to the projected area in the direction of the target times the efficiency.

L_R = the receive antenna losses defined in a manner similar to the transmit losses.

L_p = the beam shape and scanning and pattern factor losses.

L_a = the two-way-pattern propagation losses of the medium; often expressed as $\exp(-2\propto R)$, where $\propto$ is the attenuation constant of the medium and the factor 2 is for a two-way path.

L_s = signal-processing losses that occur for virtually every waveform and implementation.

δ_T = the radar cross-sectional area of the object that is being detected.

k = Boltzmann's constant (1.38×10^{-23} W-s/K).

T_s = system noise temperature.

b = Doppler filter or *speedgate* bandwidth.

S/N = signal-to-noise ratio.

$S_{\min}$ = the minimum detectable target-signal power that, with a given probability of success, the radar can be said to *detect*, *acquire*, or *track* in the presence of its own thermal noise or some external interference. Since all these factors (including the target return itself) are generally noise-like, the criterion for a detection can be described only by some form of probability distribution with an associated probability of detection P_D and a probability that, in the absence of a target signal, one or more noise or interference samples will be mistaken for the target of interest.

While the Doppler filter should be a matched filter, it usually is wider because it must include the target spectral width. There is usually some compensation for the loss in detectability by the use of post-detection filtering or integration. The S/N ratio for a CW radar must be at least 6 dB, compared with the value of 13 dB required with pulse radars when detecting steady targets [Nathanson, 1991, p. 449].

The Doppler system discussed above has a maximum detection range based on signal strength and other factors, but it cannot measure range. The rate of change in signal strength as a function of range has sometimes been used in fuzes to estimate range closure and firing point, but this is a relative measure.

FM/CW Radar

The most common technique for determining target range is the use of frequency modulation. Typical modulation waveforms include sinusoidal, linear sawtooth, or triangular, as illustrated in Fig. 2.104. For a linear sawtooth, a frequency increasing with time may be transmitted. Upon being reflected from a stationary point target, the same linear frequency change is reflected back to the receiver, except it has a time delay that is related to the range to the target. The time is $T = (2R)/c$, where R is the range. The

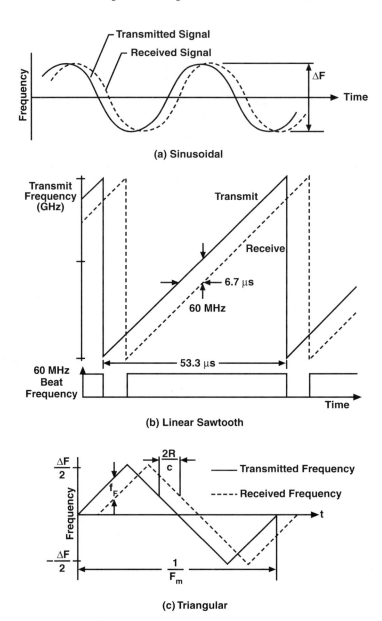

FIGURE 2.104 Frequency vs. time waveforms for FM/CW radar: (a) sinusoidal, (b) linear sawtooth, (c) triangular modulations.

received signal is mixed with the transmit signal, and the difference or beat frequency (F_b) is obtained. (The sum frequency is much higher and is rejected by filtering.) For a stationary target this is given by

$$F_b = \frac{4R}{c} \cdot \Delta F \cdot F_m \tag{2.4}$$

where ΔF = frequency deviation and F_m = modulation rate.

The beat frequency is constant except near the turnaround region of the sawtooth, but, of course, it is different for targets at different ranges. (If it is desired to have a constant intermediate frequency for different ranges, which is a convenience in receiver design, then the modulation rate or the frequency deviation must be adjusted.) Multiple targets at a variety of ranges will produce multiple-frequency outputs from the mixer and frequently are handled in the receiver by using multiple range-bin filters.

If the target is moving with a component of velocity toward (or away) from the radar, then there will be a Doppler frequency component added to (or subtracted from) the difference frequency (F_b), and the Doppler will be slightly higher at the upper end of the sweep range than at the lower end. This will introduce an uncertainty or ambiguity in the measurement of range, which may or may not be significant, depending on the parameters chosen and the application. For example, if the Doppler frequency is low (as in an altimeter) and/or the difference frequency is high, the error in range measurement may be tolerable. For the symmetrical triangular waveform, a Doppler less than F_b averages out, since it is higher on one-half of a cycle and lower on the other half. With a sawtooth modulation, only a decrease or increase is noted, since the frequencies produced in the transient during a rapid flyback are out of the receiver passband. Exact analyses of triangular, sawtooth, dual triangular, dual sawtooth, and combinations of these with noise have been carried out by Tozzi [1972]. Specific design parameters are given later in this chapter for an application utilizing sawtooth modulation in a **missile terminal guidance seeker**.

For the case of sinusoidal frequency modulation the spectrum consists of a series of lines spaced away from the carrier by the modulating frequency or its harmonics. The amplitudes of the carrier and these sidebands are proportional to the values of the Bessel functions of the first kind (J_n, $n = 0, \ldots 1, \ldots 2, \ldots 3, \ldots$), whose argument is a function of the modulating frequency and range. By choosing a particular modulating frequency, the values of the Bessel functions and thus the characteristics of the spectral components can be influenced. For instance, the signal variation with range at selected ranges can be optimized, which is important in fuzes. A short-range dependence that produces a rapid increase in signal, greater than that corresponding to the normal range variation, is beneficial in producing well-defined firing signals. This can be accomplished by proper choice of modulating frequency and filtering to obtain the signal spectral components corresponding to the appropriate order of the Bessel function. In a similar fashion, spillover and/or reflections from close-in objects can be reduced by filtering to pass only certain harmonics of the modulating frequency (F_m). Receiving only frequencies near $3F_m$ results in considerable spillover rejection, but at a penalty of 4 to 10 dB in signal-to-noise [Nathanson, 1991].

For the sinusoidal modulation case, Doppler frequency contributions complicate the analysis considerably. For details of this analysis the reader is referred to Saunders [1990] or Nathanson [1991].

Interrupted Frequency-Modulated CW (IFM/CW)

To improve isolation during reception, the IFM/CW format involves preventing transmission for a portion of the time during the frequency change. Thus, there are frequency gaps, or interruptions, as illustrated in Fig. 2.105. This shows a case where the transmit time equals the round-trip propagation time, followed by an equal time for reception. This duty factor of 0.5 for the waveform reduces the average transmitted power by 3 dB relative to using an uninterrupted transmitter. However, the improvement in the isolation should reduce the system noise by more than 3 dB, thus improving the signal-to-noise ratio [Piper, 1987]. For operation at short range, Piper points out that a high-speed switch is required [1987]. He also points out that the ratio of frequency deviation to beat frequency should be an even integer and that the minimum ratio is typically 6, which produces an out-of-band loss of 0.8 dB.

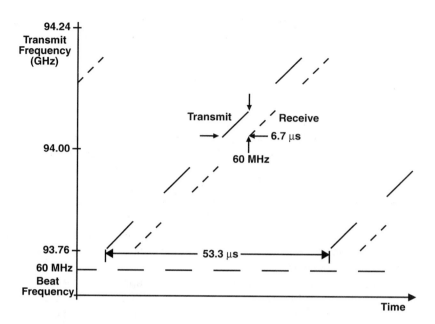

FIGURE 2.105 Interrupted FM/CW waveform. (*Source:* S.O. Piper, "MMW seekers," in *Principles and Applications of Millimeter Wave Radar*, N. Currie and C. E. Brown, Eds., Norwood, Mass.: Artech House, 1987, p. 683. With permission.)

IFM/CW may be compared with pulse compression radar if both use a wide bandwidth. Pulse compression employs a "long" pulse (i.e., relatively long for a pulse radar) with a large frequency deviation or "chirp." A long pulse is often used when a transmitter is peak-power limited, because the longer pulse produces more energy and gives more range to targets. The frequency deviation is controlled in a predetermined way (frequently a linear sweep) so that a matched filter can be used in the receiver. The large time-bandwidth product permits the received pulse to be compressed in time to a short pulse in order to make an accurate range measurement. A linear-sawtooth IFM/CW having similar pulse length, frequency deviation, and pulse repetition rate would thus appear similar, although arrived at from different points of view.

Applications

Space does not permit giving a full description of the many applications mentioned at the beginning of this chapter, but several will be discussed.

Radar Proximity Fuzes

Projectiles or missiles designed to be aimed at ships or surface land targets often need a height-of-burst (HOB) sensor (or target detection device) to fire or fuze the warhead at a height of a few meters. There are two primary generic methods of sensing or measuring height to generate the warhead fire signal. The most obvious, and potentially the most accurate, is to measure target round trip propagation delay employing conventional radar ranging techniques. The second method employs a simple CW Doppler radar or variation thereof, with loop gain calibrated in a manner that permits sensing the desired burst height by measurement of target return signal amplitude and/or rate of change. Often the mission requirements do not justify the complexity and cost of the radar ranging approach. Viable candidates are thus narrowed down to variations on the CW doppler fuze.

In its simplest form, the CW Doppler fuze consists of a fractional watt RF oscillator, homodyne detector, Doppler amplifier, Doppler envelope detector, and threshold circuit. When the Doppler envelope amplitude derived from the returned signal reaches the preset threshold, a fire signal is generated. The height at which the fire signal occurs depends on the radar loop gain, threshold level, and target reflectivity. Fuze gain is designed to produce the desired height of burst under nominal trajectory angle and target reflectivity conditions, which may have large fluctuations due to glint effects, and deviations from the desired height due to antenna gain variations with angle, target reflectivity, and fuze gain tolerances are accepted. A loop gain change of 6 dB (2 to 1 in voltage), whether due to a change in target reflection coefficient, antenna gain, or whatever, will result in a 2 to 1 HOB change.

HOB sensitivity to loop gain factors can be reduced by utilizing the slope of the increasing return signal, or so-called rate-of-rise. Deriving HOB solely from the rate-of-rise has the disadvantage of rendering the fuze sensitive to fluctuating signal levels such as might result from a scintillating target. The use of logarithmic amplifiers decreases the HOB sensitivity to the reflectivity range. An early (excessively high) fire signal can occur if the slope of the signal fluctuations equals the rate-of-rise threshold of the fuze. In practice a compromise is generally made in which Doppler envelope amplitude and rate-of-rise contribute in some proportion of HOB.

Another method sometimes employed to reduce HOB sensitivity to fuze loop gain factors and angle of fall is the use of FM sinusoidal modulation of suitable deviation to produce a range correlation function comprising the zero order of a Bessel function of the first kind. The subject of sinusoidal modulation is quite complex, but has been treated in detail by Saunders [1990, pp. 1422–1446 and 144.41]. The most important aspects of fuze design have to do with practical problems such as low cost, small size, ability to stand very high-g accelerations, long life in storage, and countermeasures susceptibility.

Police Radars

Down-the-road police radars, which are of the CW Doppler type, operate at 10.25 (X-Band), 24.150 (K-Band), or in the 33.4 to 36.0 GHz (Ka-Band) range, frequencies approved in the United States by the Federal Communications Commission. Half-power beamwidths are typically in the 0.21 to 0.31 radian (12 to 18°) range. The sensitivity is usually good enough to provide a range exceeding 800 meters. Target size has a dynamic range of 30 dB (from smallest cars or motorcycles to large trucks). This means that a large target can be seen well outside the antenna 3-dB point at a range exceeding the range of a smaller target near the center of the beam. Range is not measured. Thus there can be uncertainty about which vehicle is the target. Fisher [1992] has given a discussion of a number of the limitations of these systems, but in spite of these factors probably tens of thousands have been built.

The transmitter is typically a Gunn oscillator in the 30 to 100 mW power range, and antenna gain is usually around 20 to 24 dB, employing circular polarization. The designs typically have three amplifier gains for detection of short, medium, or maximum range targets, plus a squelch circuit so that sudden spurious signals will not be counted. Provision is made for calibration to assure the accuracy of the readings. Speed resolution is about 1 mph. The moving police radar system uses stationary (ground) clutter to derive the patrol car speed. Then closing speed, minus patrol car speed, yields target speed.

The limitations mentioned about deciding which vehicle is the correct target have led to the development of laser police radars, which utilize much narrower beamwidths, making target identification much more accurate. Of course, the use of microwave and laser radars has spawned the development of automotive radar detectors, which are also in wide use.

Altimeters

A very detailed discussion of FM/CW altimeters has been given by Saunders [1990, pp. 14.34–14.36], in which he has described modern commercial products built by Bendix and Collins. The parameters will be summarized below and if more information is needed, the reader may want to turn to other references [Saunders, 1990; Bendix Corp., 1982; Maoz et al., 1991; and Stratahos, 2000]. In his material, Saunders gives a general overview of modern altimeters, all of which use wide-deviation FM at a low modulation frequency. He discusses the limitations on narrowing the antenna pattern, which must be wide enough to accommodate attitude changes of the aircraft. Triangular modulation is used, since for this waveform the Doppler averages out, and dual antennas are employed. There may be a step error or quantization in height (which could be a problem at low altitudes), due to the limitation of counting zero crossings. A difference of one zero crossing (i.e., 1/2 Hz) corresponds to 3/4 meter for a frequency deviation of 100 MHz. Irregularities are not often seen, however, since meter response is slow. Also, if terrain is rough, there will be actual physical altitude fluctuations. Table 2.20 shows some of the altimeters' parameters.

TABLE 2.20   Parameters for Two Commercial Altimeters

Modulation Frequency	Frequency Deviation	Prime Power	Weight (pounds)	Radiated Power
Bendix ALA-52A	150 Hz	130 MHz	30 W	11*
Collins ALT-55	100 kHz	100 MHz	8	350 mW

*Not including antenna and indicator.

These altimeters are not acceptable for military aircraft, because their relatively wide-open front ends make them potentially vulnerable to electronic countermeasures. A French design has some advantages in this respect by using a variable frequency deviation, a difference frequency that is essentially constant with altitude, and a narrowband front-end amplifier [Saunders, 1990].

Doppler Navigators

These systems are mainly sinusoidally modulated FM/CW radars employing four separate downward looking beams aimed at about 15 degrees off the vertical. Because commercial airlines have shifted to nonradar forms of navigation, these units are designed principally for helicopters. Saunders [1990] cites a particular example of a commercial unit operating at 13.3 GHz, employing a Gunn oscillator as the transmitter, with an output power of 50 mW, and utilizing a 30-kHz modulation frequency. A single microstrip antenna is used. A low-altitude equipment (below 15,000 feet), the unit weighs less than 12 pounds. A second unit cited has an output power of 300 mW, dual antennas, dual modulating frequencies, and an altitude capability of 40,000 feet.

Millimeter-Wave Seeker for Terminal Guidance Missile

Terminal guidance for short-range (less than 2 km) air-to-surface missiles has seen extensive development in the last decade. Targets such as tanks are frequently immersed in a clutter background that may give a radar return that is comparable to that of the target. To reduce the clutter return in the antenna footprint, the antenna beamwidth is reduced by going to millimeter wavelengths. For a variety of reasons the choice is usually a frequency near 35 or 90 GHz. Antenna beamwidth is inversely proportional to frequency, so in order to get a reduced beamwidth we would normally choose 90 GHz; however, more deleterious effects at 90 GHz due to atmospheric absorption and scattering can modify that choice. In spite of small beamwidths, the clutter is a significant problem, and in most cases signal-to-clutter is a more limiting condition than signal-to-noise in determining range performance. Piper [1987] has done an excellent job of analyzing the situation for 35- and 90-GHz pulse radar seekers and comparing those with a 90-GHz FM/CW seeker. His FM/CW results will be summarized below.

In his approach to the problem, Piper gives a summary of the advantages and disadvantages of a pulse system compared to the FM/CW approach. One difficulty for the FM/CW can be emphasized here. That is the need for a highly linear sweep, and, because of the desire for the wide bandwidth, this requirement is accentuated. The wide bandwidth is desired in order to average the clutter return and to smooth the glint effects. In particular, glint occurs from a complex target because of the vector addition of coherent signals scattered back to the receiver from various reflecting surfaces. At some angles the vectors may add in phase (constructively) and at others they may cancel, and the effect is specifically dependent on wavelength. For a narrowband system, glint may provide a very large signal change over a small variation of angle, but, of course, at another wavelength it would be different. Thus, very wide bandwidth is desirable from this smoothing point of view, and typical numbers used in millimeter-wave radars are in the 450- to 650-MHz range. Piper chose 480 MHz.

Another trade-off involves the choice of FM waveform. Here the use of a triangular waveform is undesirable because the Doppler frequency averages out and Doppler compensation is then required. Thus the sawtooth version is chosen, but because of the large frequency deviation desired, the difficulty of linearizing the frequency sweep is made greater. In fact many components must be extremely wideband,

and this generally increases cost and may adversely affect performance. On the other hand, the difference frequency (F_b) and/or the intermediate frequency (F_{IF}) will be higher and thus further from the carrier, so the phase noise will be lower. After discussing the other trade-offs, Piper chose 60 MHz for the beat frequency.

With a linear FM/CW waveform, the inverse of the frequency deviation provides the theoretical time resolution, which is 2.1 ns for 480 MHz (or range resolution of 0.3 meter). For an RF sweep linearity of 300 kHz, the range resolution is actually 5 meters at the 1000-meter nominal search range. (The system has a mechanically scanned antenna.) An average transmitting power of 25 mW was chosen, which was equal to the average power of the 5-W peak IMPATT assumed for the pulse system. The antenna diameter was 15 cm. For a target radar cross-section of 20 m² and assumed weather conditions, the signal-to-clutter and signal-to-noise ratios were calculated and plotted for ranges out to 2 km and for clear weather or 4 mm per hour rainfall. The results show that for 1 km range the target-to-clutter ratios are higher for the FM/CW case than the pulse system in clear weather or in rain, and target-to-clutter is the determining factor.

Summary Comments

From this brief review it is clear that there are many uses for CW radars, and various types (such as fuzes) have been produced in large quantities. Because of their relative simplicity, today there are continuing trends toward the use of digital processing and integrated circuits. In fact, this is exemplified in articles describing FM/CW radars built on single microwave integrated circuit chips [Maoz et al., 1991; Chang et al., 1995; Haydl et al., 1999; Menzel et al., 1999].

Defining Terms

Doppler-frequency shift: The observed frequency change between the transmitted and received signal produced by motion along a line between the transmitter/receiver and the target. The frequency increases if the two are closing and decreases if they are receding.

Missile terminal guidance seeker: Located in the nose of a missile, a small radar with short-range capability which scans the area ahead of the missile and guides it during the terminal phase toward a target such as a tank.

Pulse Doppler: A coherent radar, usually having high pulse repetition rate and duty cycle and capable of measuring the Doppler frequency from a moving target. Has good clutter suppression and thus can see a moving target in spite of background reflections.

References

Bendix Corporation, Service Manual for ALA-52A Altimeter; Design Summary for the ALA-52A, Bendix Corporation, Ft. Lauderdale, FL, May 1982.

K.W. Chang, H. Wang, G. Shreve, J.G. Harrison, M. Core, A. Paxton, M. Yu, C.H. Chen, and G.S. Dow, Forward-looking automotive radar using a W-band single-chip transceiver, *IEEE Transactions on Microwave Theory and Techniques*, 43, 1659–1668, July, 1995.

Collins (Rockwell International), ALT-55 Radio Altimeter System; Instruction Book, Cedar Rapids, IA, October 1984.

P.D. Fisher, Improving on police radar, *IEEE Spectrum*, 29, 38–43, July 1992.

A.J. Freiley, B.L. Conroy, D.J. Hoppe, and A.M. Bhanji, Design concepts of a 1-MW CW X-band transmit/receive system for planetary radar, *IEEE Transactions on Microwave Theory and Techniques*, 40, 1047–1055, June 1992.

W.H. Haydl, et al., Single-chip coplanar 94 GHz FMCW radar sensors, *IEEE Microwave and Guided Wave Lett.*, 9, 73–75, February 1999.

B. Maoz, L.R. Reynolds, A. Oki, and M. Kumar, FM-CW radar on a single GaAs/AlGaAs HBT MMIC chip, *IEEE Microwave and Millimeter-Wave Monolithic Circuits Symposium Digest*, 3–6, June 1991.

W. Menzel, D. Pilz, and R. Leberer, A 77-GHz FM/CW radar front-end with a low-profile low-loss printed antenna, *IEEE Trans. Microwave Theory and Techniques,* 47, 2237–2241, December 1999.

F.E. Nathanson, *Radar Design Principles,* New York: McGraw-Hill, 1991, 448–467.

S.O. Piper, MMW seekers, in *Principles and Applications of Millimeter Wave Radar,* N. C. Currie and C. E. Brown, Eds., Norwood, MA: Artech House, 1987, chap. 14.

W.K. Saunders, CW and FM radar, in *Radar Handbook,* M.I. Skolnik, Ed., New York: McGraw-Hill, 1990, chap. 14.

G.E. Stratakos, P. Bougas, and K. Gotsis, A low cost, high accuracy radar altimeter, *Microwave J.,* 43, 120–128, February 2000.

L.M. Tozzi, Resolution in frequency-modulated radars, Ph.D. thesis, University of Maryland, College Park, 1972.

F.T. Ulaby and C. Elachi, *Radar Polarimetry for Geoscience Applications,* Norwood, MA: Artech House, 1990, 193–200.

J.C. Wiltse, S.P. Schlesinger, and C.M. Johnson, Back-scattering characteristics of the sea in the region from 10 to 50 GHz, *Proceedings of the IRE,* 45, 220–228, February 1957.

Further Information

For a general treatment, including analysis of clutter effects, Nathanson's [1991] book is very good and generally easy to read. For extensive detail and specific numbers in various actual cases, Saunders [1990] gives good coverage. The treatment of millimeter-wave seekers by Piper [1987] is excellent, both comprehensive and easy to read.

2.11.2 Pulse Radar

Melvin L. Belcher, Jr. and Josh T. Nessmith

Overview of Pulsed Radars

Basic Concept of Pulse Radar Operation

The basic operation of a pulse radar is depicted in Fig. 2.106. The radar transmits pulses superimposed on a radio frequency (RF) carrier and then receives returns (reflections) from desired and undesired scatterers. Scatterers corresponding to desired targets may include space, airborne, and sea- and/or surface-based vehicles. They can also include the earth's surface and the atmosphere in remote sensing applications. Undesired scatterers are termed *clutter.* Clutter sources include the earth's surface, natural and man-made discrete objects, and volumetric atmospheric phenomena such as rain and birds. Short-range/low-altitude radar operation is often constrained by clutter since the multitude of undesired returns

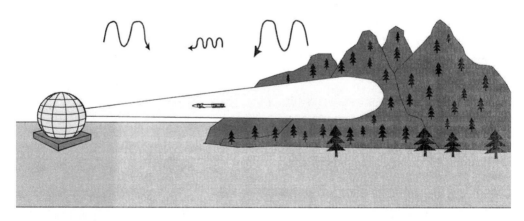

FIGURE 2.106   Pulse radar.

masks returns from targets of interest such as aircraft. Conversely, volumetric atmospheric phenomena may be considered as targets for weather radar systems.

The range, azimuth angle, elevation angle, and range rate can be directly measured from a return to estimate target metrics, position and velocity, and to support tracking. Signature data to support non-cooperative target identification or environmental remote sensing can be extracted by measuring the amplitude, phase, and polarization of the return.

Pulse radar affords a great deal of design and operational flexibility. Pulse duration, pulse rate, and pulse bandwidth can be tailored to specific applications to provide optimal performance. Modern computer-controlled multiple-function radars exploit this capability by choosing the best waveform from a repertoire for a given operational mode and interference environment automatically.

Radar Applications

The breadth of pulse radar applications is summarized in Table 2.21 in terms of operating frequencies. Radar applications can also be grouped into search, track, and signature measurement applications. Search radars are used for surveillance tracking but have relatively large range and angle errors. The search functions favor broad beamwidths and low bandwidths in order to efficiently search over a large spatial volume. As indicated in Table 2.21, search is preferably performed in the lower frequency bands. The antenna pattern is typically narrow in azimuth and has a cosecant pattern in elevation to provide acceptable coverage from the horizon to the zenith.

Tracking radars are typically characterized by a narrow beamwidth and moderate bandwidth in order to provide accurate range and angle measurements on a given target. The antenna pattern is a pencil beam with approximately the same dimensions in azimuth and elevation. Track is usually conducted at the higher frequency bands in order to minimize the beamwidth for a given antenna aperture area. After each return from a target is received, the range and angle are measured and input into a track filter. Track filtering smoothes the data to refine the estimate of target position and velocity. It also predicts the target's flight path to provide range gating and antenna pointing control to the radar system.

Signature measurement applications include remote sensing of the environment as well as the measurement of target characteristics. In some applications, synthetic aperture radar (SAR) imaging is conducted from aircraft or satellites to characterize land usage over broad areas. Moving targets that

TABLE 2.21   Radar Bands

Band	Frequency Range	Principal Applications
HF	3–30 MHz	Over-the-horizon radar
VHF	30–300 MHz	Long-range search
UHF	300–1000 MHz	Long-range surveillance
L	1000–2000 MHz	Long-range surveillance
S	2000– 4000 MHz	Surveillance
		Long-range weather characterization
		Terminal air traffic control
C	4000–8000 MHz	Fire control
		Instrumentation tracking
X	8–12 GHz	Fire control
		Air-to-air missile seeker
		Marine radar
		Airborne weather characterization
Ku	12–18 GHz	Short-range fire control
		Remote sensing
Ka	27– 40 GHz	Remote sensing
		Weapon guidance
V	40–75 GHz	Remote sensing
		Weapon guidance
W	75–110 GHz	Remote sensing
		Weapon guidance

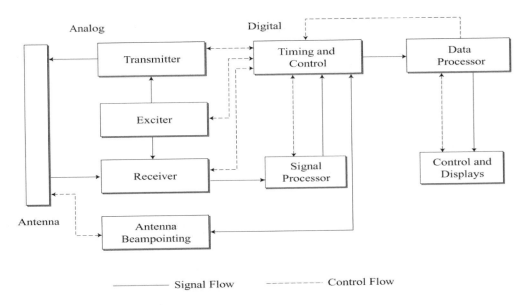

FIGURE 2.107   Radar system architecture.

present changing aspect to the radar can be imaged from airborne or ground-based radars via inverse synthetic aperture radar (ISAR) techniques. As defined in the subsection Resolution and Accuracy, cross-range resolution improves with increasing antenna extent. SAR/ISAR effectively substitutes an extended observation interval over which coherent returns are collected from different target aspect angles for a large antenna structure that would not be physically realizable in many instances.

In general, characterization performance improves with increasing frequency because of the associated improvement in range, range rate, and cross-range resolution. However, phenomenological characterization to support environmental remote sensing may require data collected across a broad swath of frequencies.

A multiple-function **phased array** radar generally integrates these functions to some degree. Its design is usually driven by the track function. Its operational frequency is generally a compromise between the lower frequency of the search radar and the higher frequency desired for the tracking radar. The degree of signature measurement implemented to support such functions as noncooperative target identification depends on the resolution capability of the radar as well as the operational user requirements. Multiple-function radar design represents a compromise among these different requirements. However, implementation constraints, multiple-target handling requirements, and reaction time requirements often dictate the use of phased array radar systems integrating search, track, and characterization functions.

Critical Subsystem Design and Technology

The major subsystems making up a pulse radar system are depicted in Fig. 2.107. The associated interaction between function and technology is summarized in this subsection.

Antenna

The radar antenna function is to first provide spatial directivity to the transmitted EM wave and then to intercept the scattering of that wave from a target. Most radar antennas may be categorized as mechanically scanning or electronically scanning. Mechanically scanned reflector antennas are used in applications where rapid beam scanning is not required. Electronic scanning antennas include phased arrays and frequency scanned antennas. Phased array beams can be steered to any point in their field of view, typically within 10 to 100 microseconds, depending on the latency of the beam-steering subsystem and the switching time of the phase shifters. Phased arrays are desirable in multiple function radars since they can interleave search operations with multiple target tracks.

There is a Fourier transform relationship between the antenna illumination function and the far-field antenna pattern analogous to spectral analysis. Hence, tapering the illumination to concentrate power near the center of the antenna suppresses sidelobes while reducing the effective antenna aperture area. The phase and amplitude control of the antenna illumination determines the achievable sidelobe suppression and angle measurement accuracy.

Perturbations in the illumination due to the mechanical and electrical sources distort the illumination function and constrain performance in these areas. Mechanical illumination error sources include antenna shape deformation due to sag and thermal effects as well as manufacturing defects. Electrical illumination error is of particular concern in phased arrays where sources include beam steering computational error and phase shifter quantization. Control of both the mechanical and electrical perturbation errors is the key to both low sidelobes and highly accurate angle measurements. Control denotes that either tolerances are closely held and maintained or that there must be some means for monitoring and correction. Phased arrays are attractive for low sidelobe applications since they can provide element-level phase and amplitude control.

Transmitter

The transmitter function is to amplify waveforms to a power level sufficient for target detection and estimation. There is a general trend away from tube-based transmitters toward solid-state transmitters. In particular, solid-state transmit/receive modules appear attractive for constructing phased array radar systems. In this case, each radiating element is driven by a module that contains a solid-state transmitter, phase shifter, low-noise amplifier, and associated control components. Active arrays built from such modules appear to offer significant reliability advantages over radar systems driven from a single transmitter. Microwave tube technology offers substantial advantages in power output over solid-state technology. However, there is a strong trend in developmental radars toward use of solid-state transmitters due to production base as well as performance considerations.

Receiver and Exciter

This subsystem contains the precision timing and frequency reference source or sources used to derive the master oscillator and local oscillator reference frequencies. These reference frequencies are used to downconvert received signals in a multiple-stage superheterodyne architecture to accommodate signal amplification and interference rejection. Filtering is conducted at the carrier and intermediate frequencies in processing to reject interference outside the operating band of the radar. The receiver front end is typically protected from overload during transmission through the combination of a circulator and a transmit/receive switch.

The exciter generates the waveforms for subsequent transmission. As in signal processing, the trend is toward programmable digital signal synthesis because of the associated flexibility and performance stability.

Signal and Data Processing

Digital processing is generally divided between two processing subsystems, i.e., signals and data, according to the algorithm structure and throughput demands. Signal processing includes pulse compression, Doppler filtering, and detection threshold estimation and testing. Data processing includes track filtering, user interface support, and such specialized functions as electronic counter-counter measures (ECCM) and built-in test (BIT), as well as the resource management process required to control the radar system.

The signal processor is often optimized to perform the repetitive complex multiply-and-add operations associated with the fast Fourier transform (FFT). FFT processing is used for implementing **pulse compression** via fast convolution and for Doppler filtering. Pulse compression consists of matched filtering on receive to an intrapulse modulation imposed on the transmitted pulse. As delineated subsequently, the imposed intrapulse bandwidth determines the range resolution of the pulse while the modulation format determines the suppression of the waveform matched-filter response outside the nominal range resolution extent. Fast convolution consists of taking the FFT of the digitized receiver output, multiplying it by the stored FFT of the desired filter function, and then taking the inverse FFT of the resulting product.

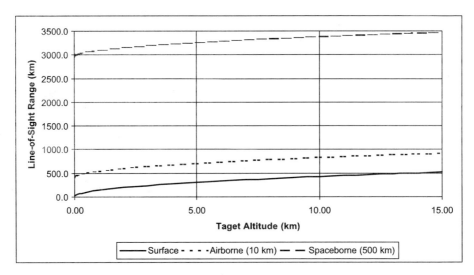

FIGURE 2.108 Maximum line-of-sight range for surface-based radar, an airborne surveillance radar, and a space-based radar.

Fast convolution results in significant computational saving over performing the time-domain convolution of returns with the filter function corresponding to the matched filter. The signal processor output can be characterized in terms of range gates and Doppler filters corresponding approximately to the range and Doppler resolution, respectively.

In contrast, the radar data processor typically consists of a general-purpose computer with a real-time operating system. Fielded radar data processors range from microcomputers to mainframe computers, depending on the requirements of the radar system. Data processor software and hardware requirements are significantly mitigated by off-loading timing and control functions to specialized hardware. This timing and control subsystem typically functions as the two-way interface between the data processor and the other radar subsystems. The increasing inclusion of BIT (built-in-test) and built-in calibration capability in timing and control subsystem designs promises to result in significant improvement in fielded system performance. The trend is toward increasing use of commercial off-the-shelf digital processing elements for radar applications and tighter integration of the signal and data processing functions.

Radar Performance Prediction

Radar Line-of-Sight

With the exception of over-the-horizon (OTH) radar systems, which exploit either sky-wave bounce or ground-wave propagation modes and sporadic ducting effects at higher frequencies, surface and airborne platform radar operation is limited to the refraction-constrained line of sight. Atmospheric refraction effects can be closely approximated by setting the earth's radius to 4/3 its nominal value in estimating horizon-limited range. The resulting line-of-sight range is depicted in Fig. 2.108 for a surface-based radar, an airborne surveillance radar, and a space-based radar.

As evident in the plot, airborne and space-based surveillance radar systems offer significant advantages in the detection of low-altitude targets that would otherwise be masked by earth curvature and terrain features from surface-based radars. However, efficient clutter rejection techniques must be used in order to detect targets since surface clutter returns will be present at almost all ranges of interest.

Radar Range Equation

The radar range equation is commonly used to estimate radar system performance, given that line-of-sight conditions are satisfied. This formulation essentially computes the signal-to-noise ratio (S/N) at the output of the radar signal processor. In turn, S/N is used to provide estimates of radar detection and position measurement performance as described in subsequent subsections. S/N can be calculated in terms of the number of pulses coherently integrated over a single coherent processing interval (CPI) using the radar range equation such that

$$S/N = \frac{P_D A T_p N_p \sigma}{\left(4\pi\right)^2 R^4 L_t L_{rn} L_{sp} k T_s} \qquad (2.5)$$

where P is peak transmitter power output, D is directivity of the transmit antenna, A is effective aperture area of the receive antenna in meters squared, T_p is pulse duration, σ is **radar cross-section** in square meters, N_p is the number of coherently integrated pulses within the coherent processing interval, R is range to target in meters, L^t is system ohmic and nonohmic transmit losses, L_{rn} is system nonohmic receive losses, L^{sp} is signal processing losses, k is Boltzmanns constant (1.38×10^{-23} degrees K), and T^s is system noise temperature, including receive ohmic losses (kelvin).

At X-band and above it may also be necessary to include propagation loss due to atmospheric absorption [Blake, 1986]. This form of the radar range equation is applicable to radar systems using pulse compression or pulse Doppler waveforms as well as the unmodulated single-pulse case. In many applications, average power is a better measure of system performance than peak power since it indicates the S/N improvement achievable with pulse integration over a given interval of time. Hence, the radar range equation can be modified such that

$$S/N = \frac{P_a DA T_c \sigma}{\left(4\pi\right)^2 R^4 L_t L_{rn} L_{sp} k T_s} \qquad (2.6)$$

where P_a is average transmitter power and Tc is coherent processing interval (CPI).

The portion of time over which the transmitter is in operation is referred to as the radar duty cycle. The average transmitter power is the product of duty cycle and peak transmitter power. Duty cycle ranges from less than 1% for typical **noncoherent** pulse radars to somewhat less than 50% for high pulse repetition frequency (PRF) pulse Doppler radar systems. The CPI is the period over which returns are collected for **coherent** processing functions such as pulse integration and Doppler filtering. The CPI can be estimated as the product of the number of coherently integrated pulses and the interval between pulses. Noncoherent pulse integration is less efficient and alters the statistical character of the signal and interference.

Antenna Directivity and Aperture Area
The directivity of the antenna is

$$D = \frac{4\pi A \eta}{\lambda^2} \qquad (2.7)$$

where η is aperture efficiency and λ is radar carrier wavelength. Aperture inefficiency is due to the antenna illumination factor.

The common form of the radar range equation uses power gain rather than directivity. Antenna gain is equal to the directivity divided by the antenna losses. In the design and analysis of modern radars, directivity is a more convenient measure of performance because it permits designs with distributed active elements, such as solid-state phased arrays, to be assessed to permit direct comparison with passive antenna systems. Beamwidth and directivity are inversely related; a highly directive antenna will have a narrow beamwidth. For typical design parameters,

$$D = \frac{10^7}{\theta_{az}\,\theta_{el}} \qquad\qquad (2.8)$$

where θ_{az} and θ_{el} are the radar azimuth and elevation beamwidths, respectively, in milliradians. The directivity then gives the power density relative to an isotropic radiator.

Radar Cross-Section

In practice, the *radar cross-section* (RCS) of a realistic target must be considered a random variable with an associated correlation interval. Targets are composed of multiple interacting scatters so that the composite return varies in magnitude with the constructive and destructive interference of the contributing returns. The target RCS is typically estimated as the mean or median of the target RCS distribution. The associated correlation interval indicates the rate at which the target RCS varies over time. RCS fluctuation degrades target detection performance at moderate to high probability of detection.

The median RCS of typical targets is given in Table 2.22. The composite RCS measured by a radar system may be composed of multiple individual targets in the case of closely spaced targets such as a bird flock.

Loss and System Temperature Estimation

Sources of S/N loss include ohmic and nonohmic (mismatch) loss in the antenna and other radio frequency components, propagation effects, signal processing deviations from matched filter operation, detection thresholding, and search losses. Scan loss in phased array radars is due to the combined effects of the decrease in projected antenna area and element mismatch with increasing scan angle.

Search operations impose additional losses due to target position uncertainty. Because the target position is unknown before detection, the beam, range gate, and Doppler filter will not be centered on the target return. Hence, straddling loss will occur as the target effectively straddles adjacent resolution cells in range and Doppler. Beamshape loss is a consequence of the radar beam not being pointed directly at the target so that there is a loss in both transmit and receive antenna gain. In addition, detection threshold loss associated with radar system adaptation to interference must be included [Nathanson, 1991].

TABLE 2.22 Median Target RCS (m²)

Carrier Frequency, GHz	1–2	3	5	10	17
Aircraft (nose/tail avg.)					
Small propeller	2	3	2.5		
Small jet (Lear)	1	1.5	1	1.2	
T38-twin jet, F5	2	2–3	2	1–2/6	
T39-Sabreliner	2.5		10/8	9	
F4, large fighter	5–8/5	4–20/10	4	4	
737, DC9, MD80	10	10	10	10	10
727, 707, DC8-type	22–40/15	40	30	30	
DC-10-type, 747	70	70	70	70	
Ryan drone				2/1	
Standing man (180 lb)	0.3	0.5	0.6	0.7	0.7
Automobiles	100	100	100	100	100
Ships-incoming ($\times 10^4$ m²)					
4K tons	1.6	2.3	3.0	4.0	5.4
16K tons	13	18	24	32	43
Birds					
Sea birds	0.002	0.001–0.004	0.004		
Sparrow, starling, etc.	0.001	0.001	0.001	0.001	0.001

Note: Slash marks indicate different set.

Source: F.E. Nathanson, *Radar Design Principles,* 2nd ed., New York: McGraw-Hill, 1991. With permission.

TABLE 2.23 Typical Microwave Loss and System Temperature Budgets

	Mechanically Scanned Reflector Antenna	Electronically Scanned Slotted Array	Solid-State Phased Array
Nominal losses			
Transmit loss, L_t (dB)	1	1.5	0.5
Nonohmic receiver loss, L_r (dB)	0.5	0.5	0.1
Signal processing loss, L_{sp} (dB)	1.4	1.4	1.4
Scan loss (dB)	N/A	N/A	30 log [cos (scan angle)]
Search losses, L_{DS}			
Beam shape (dB)	3	3	3
Range gate straddle (dB)	0.5	0.5	0.5
Doppler filter straddle (dB)	0.5	0.5	0.5
Detection thresholding (dB)	1	1	1
System noise temperature (kelvin)	500	600	400

System noise temperature estimation corresponds to assessing the system thermal noise floor referenced to the antenna output. Assuming the receiver hardware is at ambient temperature, the system noise temperature can be estimated as

$$T_s = T_a + 290 \left(L_{ro} F - 1 \right) \tag{2.9}$$

where T_a is the antenna noise temperature, L_{ro} is receive ohmic losses, and F is the receiver noise figure.

In phased array radars, the thermodynamic temperature of the antenna receive beamformer may be significantly higher than ambient, so a more complete analysis is required. The antenna noise temperature is determined by the external noise received by the antenna from solar, atmospheric, earth surface, and other sources.

Table 2.23 provides typical loss and noise temperature budgets for several major radar classes. In general, loss increases with the complexity of the radar hardware between the transmitter/receiver and the antenna radiator. Reflector antennas and active phased arrays impose relatively low loss, while passive array antennas impose relatively high loss.

Resolution and Accuracy

The fundamental resolution capabilities of a radar system are summarized in Table 2.24. In general, there is a trade-off between mainlobe resolution corresponding to the nominal range, Doppler, and angle resolution, and effective dynamic range corresponding to suppression of sidelobe components. This is evident in the use of weighting to suppress Doppler sidebands and angle sidelobes at the expense of broadening the mainlobe and S/N loss.

Cross range denotes either of the two dimensions orthogonal to the radar line of sight. Cross-range resolution in real-aperture antenna systems is closely approximated by the product of target range and radar beamwidth in radians. Attainment of the nominal ISAR/SAR cross-range resolution generally requires complex signal processing to generate a focused image, including correction for scatterer change in range over the CPI.

The best accuracy performance occurs for the case of thermal noise-limited error. The resulting accuracy is the resolution of the radar divided by the square root of the S/N and an appropriate monopulse or interpolation factor. In this formulation, the single-pulse S/N has been multiplied by the number of pulses integrated within the CPI as indicated in Eqs. (2.5) and (2.6).

In practice, accuracy is also constrained by environmental effects, target characteristics, and instrumentation error as well as the available S/N. Environmental effects include multipath and refraction. Target glint is characterized by an apparent wandering of the target position because of coherent interference effects associated with the composite return from the individual scattering centers on the target.

TABLE 2.24 Resolution and Accuracy

Dimension	Nominal Resolution	Noise-Limited Accuracy
Angle	$\dfrac{\alpha\lambda}{d}$	$\dfrac{\alpha\lambda}{dK_m\sqrt{2S/N}}$
Range	$\dfrac{\alpha C}{2B}$	$\dfrac{\alpha C}{2BK_i\sqrt{2S/N}}$
Doppler	$\dfrac{\alpha}{\text{CPI}}$	$\dfrac{\alpha}{\text{CPI}\,K_i\sqrt{2S/N}}$
SAR/ISAR	$\dfrac{\alpha\lambda}{2\Delta\theta}$	$\dfrac{\alpha\lambda}{2\Delta\theta K_i\sqrt{2S/N}}$

Note: α, taper broadening factor, typically ranging from 0.89 (unweighted) to 1.3 (Hamming); d, antenna extent in azimuth/elevation; B, waveform bandwidth; K_m, monopulse slope factor, typically on the order of 1.5; Ki, interpolation factor, typically on the order of 1.8; $\Delta\theta$, line-of-sight rotation of target relative to radar over CPI.

Instrumentation error is minimized with alignment and calibration but may significantly constrain track filter performance as a result of the relatively long correlation interval of some error sources.

Radar Range Equation for Search and Track

The radar range equation can be modified to directly address performance in the two primary radar missions: search and track.

Search performance is basically determined by the capability of the radar system to detect a target of specific RCS at a given maximum detection range while scanning a given solid angle extent within a specified period of time. S/N can be set equal to the minimum value required for a given detection performance, S/N^*r, while R can be set to the maximum required target detection range, R_{max}. Manipulation of the radar range equation results in the following expression:

$$\frac{P_a A}{L_t L_r L_{sp} L_{os} T_s} \geq \left(\frac{S}{N}\right)\frac{R_{max}^4 \Omega}{\sigma T_{fs}} \cdot 16k \tag{2.10}$$

where Ω is the solid angle over which search must be performed (steradians), T_{fs} is the time allowed to search Ω by operational requirements, and L_{os} is the composite incremental loss associated with search.

The left-hand side of the equation contains radar design parameters, while the right-hand side is determined by target characteristics and operational requirements. The right-hand side of the equation is evaluated to determine radar requirements. The left-hand side of the equation is evaluated to determine if the radar design meets the requirements.

The track radar range equation is conditioned on noise-limited angle accuracy as this measure stresses radar capabilities significantly more than range accuracy in almost all cases of interest. The operational requirement is to maintain a given data rate track providing a specified single-measurement angle accuracy for a given number of targets with specified RCS and range. Antenna beamwidth, which is proportional to the radar carrier wavelength divided by antenna extent, impacts track performance since the degree of S/N required for a given measurement accuracy decreases as the beamwidth decreases. Track performance requirements can be bounded as

$$\frac{P_a A^3}{\lambda^4 L_t L_r L_{sp} T_s} k_m^2 \eta^2 \geq 5k \frac{rN_t R^4}{\sigma\sigma_\theta^2} \tag{2.11}$$

where r is the single-target track rate, N_t is the number of targets under track in different beams, σ_θ is the required angle accuracy standard deviation (radians), and σ is the RCS. In general, a phased array radar antenna is required to support multiple target tracking when $Nt > 1$.

Incremental search losses are suppressed during single-target-per-beam tracking. The beam is pointed as closely as possible to the target to suppress beamshape loss. The tracking loop centers the range gate and Doppler filter on the return. Detection thresholding loss can be minimal since the track range window is small, though the presence of multiple targets generally mandates continual detection processing.

Radar Waveforms

Pulse Compression

Typical pulse radar waveforms are summarized in Table 2.6. In most cases, the signal processor is designed to closely approximate a matched filter. As indicated in Table 2.25, the range and Doppler resolution of any match-filtered waveform are inversely proportional to the waveform bandwidth and duration, respectively. Pulse compression, using modulated waveforms, is attractive since S/N is proportional to pulse duration rather than bandwidth in matched filter implementations. Ideally, the intrapulse modulation is chosen to attain adequate range resolution and range sidelobe suppression performance while the pulse duration is chosen to provide the required sensitivity. Pulse compression waveforms are characterized as having a time bandwidth product (TBP) significantly greater than unity, in contrast to an unmodulated pulse, which has a TBP of approximately unity.

Pulse Repetition Frequency

The radar system pulse repetition frequency (PRF) determines its ability to unambiguously measure target range and range rate in a single CPI as well as determining the inherent clutter rejection capabilities of the radar system. In order to obtain an unambiguous measurement of target range, the interval between radar pulses (1/PRF) must be greater than the time required for a single pulse to propagate to a target at a given range and back. The maximum unambiguous range is then given by $C/(2 \cdot PRF)$ where C is the velocity of electromagnetic propagation.

Returns from moving targets and clutter sources are offset from the radar carrier frequency by the associated Doppler frequency. As a function of range rate, $R\cdot$, the Doppler frequency, f_D, is given by $2R\cdot/\lambda$.

TABLE 2.25 Selected Waveform Characteristics

	Comments	Time Bandwidth Product	Range Sidelobes (dB)	S/N Loss (dB)	Range/Doppler Coupling	ECM/EMI Robustness
Unmodulated	No pulse compression	~1	Not applicable	0	No	Poor
Linear frequency modulation	Linearly swept over bandwidth	>10	Unweighted: −13.5 Weighted: >− 40[a]	0 0.7–1.4	Yes	Poor
Nonlinear FM	Multiple variants specific	Waveform specific	Waveform specific	0	Waveform	Fair
Barker	N-bit biphase	≤13 (N)	−20 log(N)	0	No	Fair
LRS	N-bit biphase	~N; > 64/pulse[a]	~ −10 log (N)	0	No	Good
Frank	N-bit polyphase (N = integer²)	~N	~ −10 log (π²N)	0	Limited	Good
Frequency coding	N subpulses noncoincidental in time and frequency	~N²	Waveform specific • Periodic • Pseudorandom	Waveform specific 0.7–1.40 0		

[a] Constraint due to typical technology limitations rather than fundamental waveform characteristics.

A coherent pulse train samples the return Doppler modulation at the PRF. Most radar systems employ parallel sampling in the in-phase and quadrature baseband channels so that the effective sampling rate is twice the PRF. The targets return is folded in frequency if the PRF is less than the target Doppler.

Clutter returns are primarily from stationary or near-stationary surfaces such as terrain. In contrast, targets of interest often have a significant range rate relative to the radar clutter. Doppler filtering can suppress returns from clutter. With the exception of frequency ambiguity, the Doppler filtering techniques used to implement pulse Doppler filtering are quite similar to those described for CW radar in Section 2.11.1. Ambiguous measurements can be resolved over multiple CPIs by using a sequence of slightly different PRFs and correlating detections among the CPIs.

Detection and Search

Detection processing consists of comparing the amplitude of each range gate/Doppler filter output with a threshold. A detection is reported if the amplitude exceeds that threshold. A false alarm occurs when noise or other interference produces an output of sufficient magnitude to exceed the detection threshold. As the detection threshold is decreased, both the detection probability and the false alarm probability increase. S/N must be increased to enhance detection probability while maintaining a constant false alarm probability.

As noted in the subsection on Radar Cross Section, RCS fluctuation effects must be considered in assessing detection performance. The Swerling models, which use chi-square probability density functions (PDFs) of 2 and 4 degrees of freedom (DOF), are commonly used for this purpose. The Swerling 1 and 2 models are based on the 2 DOF PDF and can be derived by modeling the target as an ensemble of independent scatterers of comparable magnitude. This model is considered representative of complex targets such as aircraft. The Swerling 3 and 4 models use the 4 DOF PDF and correspond to a target with a single dominant scatterer and an ensemble of lesser scatterers. Missiles are sometimes represented by Swerling 2 and 4 models. The Swerling 1 and 3 models presuppose slow fluctuation such that the target RCS is constant from pulse to pulse within a scan. In contrast, the RCS of Swerling 2 and 4 targets is modeled as independent on a pulse-to-pulse basis.

Single-pulse detection probabilities for nonfluctuating, Swerling 1/2, and Swerling 3/4 targets are depicted in Fig. 2.109. This curve is based on a typical false alarm number corresponding approximately to a false alarm probability of 10^{-6}. The difference in S/N required for a given detection probability for a fluctuating target relative to the nonfluctuating case is termed the fluctuation loss.

The detection curves presented here and in most other references presuppose noise-limited operation. In many cases, the composite interference present at the radar system output will be dominated by clutter returns or electromagnetic interference such as that imposed by hostile electronic countermeasures. The standard textbook detection curves cannot be applied in these situations unless the composite interference is statistically similar to thermal noise with a Gaussian PDF and a white power spectral density. The presence of non-Gaussian interference is generally characterized by an elevated false alarm probability. Adaptive detection threshold estimation techniques are often required to search for targets in environments characterized by such interference [Nathanson, 1991].

Estimation and Tracking

Measurement Error Sources

Radars measure target range and angle position and, potentially, Doppler frequency. Angle measurement performance is emphasized here since the corresponding cross-range error dominates range error for most practical applications. Target returns are generally smoothed in a tracking filter, but tracking performance is ultimately determined by the measurement accuracy and associated error characteristics of the subject radar system. Radar measurement error can be characterized as indicated in Table 2.26.

The radar design and the alignment and calibration process development must consider the characteristics and interaction of these error components. Integration of automated techniques to support alignment and calibration is an area of strong effort in modern radar design that can lead to significant performance improvement in fielded systems.

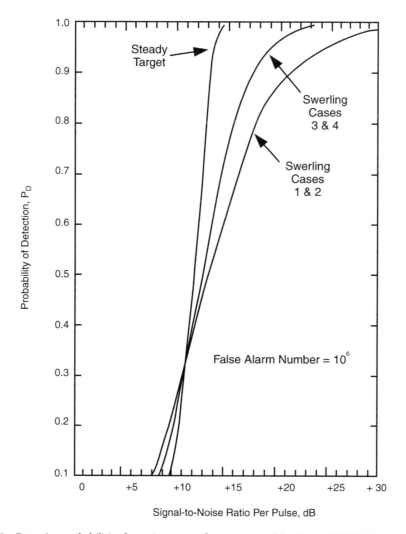

FIGURE 2.109 Detection probabilities for various target fluctuation models. (*Source:* F. E. Nathanson, *Radar Design Principles,* 2nd ed., New York: McGraw-Hill, 1991, p. 91. With permission.)

TABLE 2.26 Radar Measurement Error

Random errors	Those errors that cannot be predicted except on a statistical basis. The magnitude of the random error can be termed the *precision* and is an indication of the repeatability of a measurement.
Bias errors	A systematic error, whether due to instrumentation or propagation conditions. A nonzero mean value of a random error.
Systematic error	An error, whose quantity can be measured and reduced by calibration.
Residual systematic error	Those errors remaining after measurement and calibration. A function of the systematic and random errors in the calibration process.
Accuracy	The magnitude of the rms value of the residual systematic and random errors.

As indicated previously, angle measurement generally is the limiting factor in measurement accuracy. Target azimuth and elevation position is primarily measured by a monopulse technique in modern radars though early systems used sequential lobing and conical scanning. Specialized monopulse tracking radars utilizing reflectors have achieved instrumentation and *S/N* angle residual systematic error as low as 50 μrad. Phased array antennas have achieved a random error of less than 60 μrad, but the composite

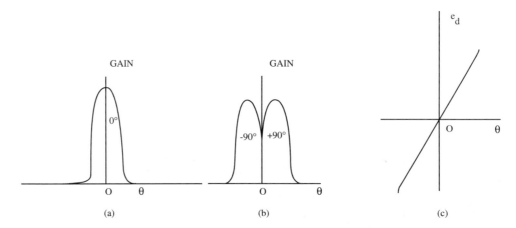

FIGURE 2.110   Monopulse beam patterns and difference voltage: (a) sum (S); (b) difference (D); (c) difference voltage.

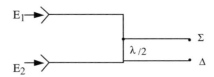

FIGURE 2.111   Monopulse comparator.

systematic residual errors remain to be measured. The limitations are primarily in the tolerance on the phase and amplitude of the antenna illumination function.

Figure 2.110 shows the monopulse beam patterns. The first is the received sum pattern that is generated by a feed that provides the energy from the reflector or phased array antenna through two ports in equal amounts and summed in phase in a monopulse comparator shown in Fig. 2.111. The second is the difference pattern generated by providing the energy through the same two ports in equal amounts but taken out with a phase difference of π radians, giving a null at the center. A target located at the center of the same beam would receive a strong signal from the sum pattern with which the target could be detected and ranged. The received difference pattern would produce a null return, indicating the target was at the center of the beam. If the target were off the null, the signal output or difference voltage would be almost linear and proportional to the distance off the center (off-axis), as shown in the figure. This output of the monopulse processor is the real part of the dot product of the complex sums and the difference signals divided by the absolute magnitude of the sum signal squared, i.e.,

$$ e_d = \mathrm{Re}\left[\frac{\Sigma \cdot \Delta}{\left| \Sigma \right|^2} \right] \tag{2.12} $$

The random instrumentation measurement errors in the angle estimator are caused by phase and amplitude errors of the antenna illumination function. In reflector systems, such errors occur because of the position of the feedhorn, differences in electrical length between the feed and the monopulse comparator, mechanical precision of the reflector, and its mechanical rotation. In phased array radars, these errors are a function of the phase shifters, time delay units, and combiners between the antenna elements and the monopulse comparator as well as the precision of the array. Although these errors are random, they may have correlation intervals considerably longer than the white noise considered in the thermal-noise random error and may depend upon the flight path of the target. For a target headed radially from or toward the radar, the correlation period of angle-measurement instrumental errors is essentially the tracking period. For crossing targets, the correlation interval may be pulse to pulse.

As in the estimate of range, the propagation effects of refraction and multipath also enter into the tracking error. The bias error in range and elevation angle by refraction can be estimated as

$$\Delta R = 0.007\, N_s\, cosecant\, E_o\, \big(meters\big)$$

$$\Delta E_o = N_s \cot E_o\, \big(\mu rad\big)$$

(2.13)

where N_s is the surface refractivity and E_o is the elevation angle [Barton and Ward, 1984].

One can calculate the average error in multipath. However, one cannot correct for it as in refraction since the direction of the error cannot be known in advance unless there are controlled conditions such as in a carefully controlled experiment. Hence, the general approach is to design the antenna sidelobes to be as low as feasible and accept the multipath error that occurs when tracking close to the horizon. There has been considerable research to find means to reduce the impact, including using very wide bandwidths to separate the direct path from the multipath return as well as specialized track filtering techniques that accommodate multipath effects.

Tracking Filter Performance

Target tracking based on processing returns from multiple CPIs generally provides a target position and velocity estimate of greater accuracy than the single-CPI measurement accuracy delineated in Table 2.24. In principle, the error variance of the estimated target position with the target moving at a constant velocity is approximately $4/n \cdot \sigma_m^2$ where n is the number of independent measurements processed by the track filter and sm is the single measurement accuracy. In practice, the variance reduction factor afforded by a track filter is often limited to about an order of magnitude because of the reasons summarized in the following paragraphs.

Track filtering generally provides smoothing and prediction of target position and velocity via a recursive prediction-correction process. The filter predicts the target's position at the time of the next measurement based on the current smoothed estimates of position, velocity, and possibly acceleration. The subsequent difference between the measured position at this time and the predicted position is used to update the smoothed estimates. The update process incorporates a weighting vector that determines the relative significance given the track filter prediction versus the new measurement in updating the smoothed estimate.

Target model fidelity and adaptivity are fundamental issues in track filter mechanization. Independent one-dimensional tracking loops may be implemented to control pulse-to-pulse range gate positioning and antenna pointing. The performance of one-dimensional polynomial algorithms, such as the alpha-beta filter, to track targets from one pulse to the next and provide modest smoothing is generally adequate. However, one-dimensional closed-loop tracking ignores knowledge of the equations of motion governing the target so that their smoothing and long-term prediction performance is relatively poor for targets with known equations of motion. In addition, simple one-dimensional tracking-loop filters do not incorporate any adaptivity or measure of estimation quality.

Kalman filtering addresses these shortcomings at the cost of significantly greater computational complexity. Target equations of motion are modeled explicitly such that the position, velocity, and potentially higher-order derivatives of each measurement dimension are estimated by the track filter as a state vector. The error associated with the estimated state vector is modeled via a covariance matrix that is also updated with each iteration of the track filter. The covariance matrix determines the weight vector used to update the smoothed state vector in order to incorporate such factors as measurement *S/N* and dynamic target maneuvering.

Smoothing performance is constrained by the degree of *a priori* knowledge of the targets kinematic motion characteristics. For example, Kalman filtering can achieve significantly better error reduction against ballistic or orbital targets than against maneuvering aircraft. In the former case the equations of motion are explicitly known, while the latter case imposes motion model error because of the presence of unpredictable pilot or guidance system commands. Similar considerations apply to the fidelity of the track filters model of radar measurement error. Failure to consider the impact of correlated measurement errors may result in underestimating track error when designing the system.

Many modern tracking problems are driven by the presence of multiple targets which impose a need for assigning measurements to specific tracks as well as accommodating unresolved returns from closely spaced targets. Existing radars generally employ some variant of the nearest-neighbor algorithm where a measurement is uniquely assigned to the track with a predicted position minimizing the normalized track filter update error. More sophisticated techniques assign measurements to multiple tracks if they cannot clearly be resolved or make the assignment on the basis on several contiguous update measurements.

Defining Terms

Coherent: Integration where magnitude and phase of received signals are preserved in summation.
Noncoherent: Integration where only the magnitude of received signals is summed.
Phased array: Antenna composed of an aperture of individual radiating elements. Beam scanning is implemented by imposing a phase taper across the aperture to collimate signals received from a given angle of arrival.
Pulse compression: The processing of a wideband, coded signal pulse, of initially long time duration and low-range resolution, to result in an output pulse of time duration corresponding to the reciprocal of the bandwidth.
Radar cross section (RCS):   Measure of the reflective strength of a radar target; usually represented by the symbol σ, measured in square meters, and defined as 4π times the ratio of the power per unit solid angle scattered in a specified direction of the power unit area in a plane wave incident on the scatterer from a specified direction.

References

Barton, D.K. and Ward, H.R., *Handbook of Radar Measurement,* Artech, Dedham, Mass., 1984.
Blake, L.V., *Radar Range-Performance Analysis,* Artech, Dedham, Mass., 1986.
Eaves, J.L. and Reedy, E.K., Eds., *Principles of Modern Radar,* Van Nostrand, New York, 1987.
Morris, G.V., *Airborne Pulsed Doppler Radar,* Artech, Dedham, Mass., 1988.
Nathanson, F.E., *Radar Design Principles,* 2nd ed., McGraw-Hill, New York, 1991.

Further Information

Skolnik, M.I., Ed., *Radar Handbook,* 2nd ed., McGraw-Hill, New York, 1990.
IEEE Standard Radar Definitions, IEEE Standard 686-1990, April 20, 1990.

2.12   Electronic Warfare and Countermeasures

Robert D. Hayes

Electronic warfare (EW) in general applies to military actions. Specifically, it involves the use of electro-magnetic energy to create an advantage for the friendly side of engaged forces while reducing the effectiveness of the opposing hostile armed forces.

Electronic warfare support measures (ESM) or electronic warfare support (ES) are the actions and efforts taken to listen, to search space and time, to intercept and locate the source of radiated energy, analyze, identify and characterize the electromagnetic energy. Tactical employment of forces is established and plans executed for electronic countermeasures.

Electronic countermeasures (ECM) or electronic attack (EA) are those action taken to reduce the enemy's effective use of the electromagnetic spectrum. These actions may include jamming, electronic deception, false targets, chaff, flares, transmission disruption, and changing tactics.

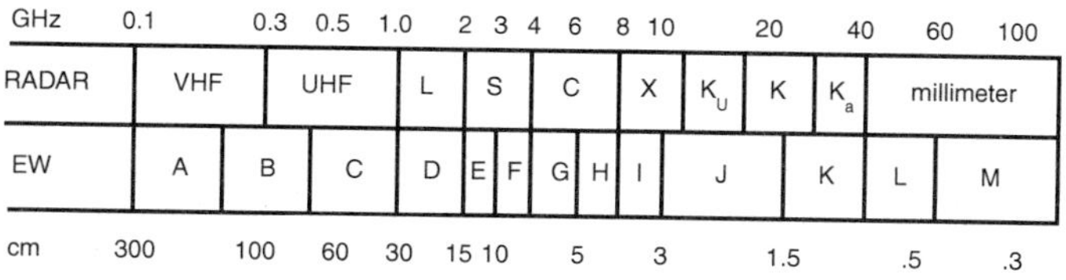

GHz	0.1		0.3	0.5	1.0	2	3	4	6	8	10		20		40	60	100
RADAR		VHF		UHF		L	S		C	X	K$_U$	K		K$_a$		millimeter	
EW		A	B	C		D	E	F	G	H	I		J		K	L	M
cm		300		100	60	30		15	10		5		3		1.5	.5	.3

FIGURE 2.112 Frequency designaton for radar and EW bands.

Electronic counter-countermeasures (ECCM) or electronic protection (EP) are actions taken to reduce the effectiveness of enemy used ECM to enhance the use of the electromagnetic spectrum for friendly forces. These actions are of a protective nature and are applied to tactical and strategic operations across the equipments used for sonar, radar, command, control, communications, and intelligence.

Radar and Radar Jamming Signal Equations

Electronic jamming is a technique where a false signal of sufficient magnitude and bandwidth is introduced into a receiving device so as to cause confusion with the expected received signal and create a loss of information. For a communication receiver, it simply overpowers the radio transmission; for radar, the jamming signal overrides the radar echo return from an object under surveillance or track. The jamming equation is the same for both communications and radar systems. In the case of a radar jammer, there are several operational scenarios to consider.

The radar jammer signal level must be determined for the various applications and jammer location. When the jammer is located on the target (self-protection), the jammer signal level must exceed the effective radar cross-section of the target to form an adequate screen; when the jammer is located away from the vehicle under protection (stand-off jammer), the the jammer should simulate the expected radar cross-section observed at a different range and different aspect.

The monostatic radar received signal power equation representing the signal received at an enemy detection radar, and the received jamming power transmitted toward the same enemy detection radar, are presented in the equations given in Table 2.27. Each quantity in equations is defined in the table. These quantities are dependent on electrical, polarization, material, shape, roughness, density, and atmospheric parameters.

Background interference is encountered in every branch of engineering. The received radar/communication signal power will compete with interfering signals which will impose limitations on performance of the detecting electronic system. There are many types of interference. A few are channel cross talk, nonlinear harmonic distortion, AM and FM interference, phase interference, polarization crosstalk, and noise. The common limiting interference in electronic systems is noise.

There are many types of noise. Every physical body not at absolute zero temperature emits electromagnetic energy at all frequencies, by virtue of the thermal energy the body possesses. This type of radiation is called thermal noise radiation. Johnson noise is a result of random motion of electrons within a resistor; semiconductor noise occurs in transistors; shot noise occurs in both transmitter and receiver vacuum tubes as a result of random fluctuations in electron flow. These fluctuations are controlled by a random mechanism and, thus, in time are random processes described by the Gaussian distribution. The signal is random in phase and polarization, and usually are broad in frequency bandwidth. The average noise power is given by

$$P_N = kT\,B$$

where

P_N = power, watts
k = Boltzman's constant
T = temperature, absolute degrees.

TABLE 2.27

Monostatic Radar Equation	Jamming Radar Equation
$$P_r = \dfrac{P_t G^2 \sigma^2 \lambda^2 F^4 E}{(4\pi)^3 R^4 L_t L_r} e^{-2\alpha R}$$	$$P_r = \dfrac{P_t G_t G_r \lambda^2 F^2 E}{(4\pi)^2 R^2 L_t L_r} e^{-\alpha R}$$

Quantity	Dependent Function
P_t = transmitter power, W	ΔF = frequency spectrum
	ψ = phase, coherent
L_t = loss, transmitter line	Equipment path
L_r = loss, receiver line	Equipment path
E = receiver gain	Processing techniques
G = antenna gain	p = polarization
	λ = wavelength, RF carrier
	l = length
	w = width
	c = curvature
	ϕ = elevation angle
	θ = azimuth angle
σ = point target RSC, m^2	p
	λ
	l
	w
	c
	ϕ
	θ
	r = rotation
	s = surface roughness
	m = motion
	ϵ = dielectric constant
	μ = permeability
σ_0 = area extended target RCS, m^2/m^2	p
	λ
	ϕ
	θ
	s
	m
	ϵ
	τ = pulse length
	T = temperature
	d = density
σ_u = volume extended target RCS, m^2/m^3	p
	λ
	ϕ
	θ
	s
	m
	ϵ

Quantity	Dependent Function
σ_u = volume extended target RCS, m^2/m^3	p
	μ
	τ
	T
	d
α_0 = clear air loss, Np	λ
	T
	d
	P = pressure
	ρ = gas constant
α = rain, dust, snow, hail, etc., loss, Np	p = polarization
	λ
	c
	m
	m
	ϵ
	T
	d
	P
	ρ
	n = number of particles, or rain rate
	R = range, path length
	β = differential polarization phase shift
F = electric field propagation factor—multipath	p
	λ
	c
	ϕ
	θ
	R
	s
	ϵ
	μ
	ψ
	β
R = range, path length, m	n = index of refraction spherical geometry

Antenna gain = G =

$$\frac{28,000}{\theta\phi}$$

where $\theta = 75\dfrac{\lambda}{\omega}$

and $\phi = 75\ \dfrac{\lambda}{l}$, in degrees

Noise jamming has been introduced into EW from several modes of operation. *Spot noise jamming* requires the entire output power of the jammer to be concentrated in a narrow bandwidth ideally identical to the bandwidth of the victim radar/communication receiver. It is used to deny range, conversation, and sometimes angle information. *Barrage noise jamming* is a technique applied to wide frequency spectrum to deny the use of multiple channel frequencies to effectively deny range information, or to cover a single wideband system using pulse compression or spread-spectrum processing. *Blinking* is a technique using two spot noise jammers located in the same cell of angular resolution of the victim system. The jamming transmission is alternated between the two sources causing the radar antenna to oscillate back and forth. Too high a blinking frequency will let the tracker average the data and not oscillate, while too low a frequency will allow the tracker to lock on one of the jammers.

In free space the radar echo power returned from the target varies inversely as the fourth power of the range, whereas the power received from a jammer varies as the square of the range from the jammer to the victim radar. There will be some range where the energy for the noise jammer is no longer great enough to hide the friendly vehicle. This range of lost protection is called the burn-through range.

In both EW technology and target identification technology, the received energy, as compared to the received power, better defines the process of signal separation from interference. The received radar energy will be given by

$$S = P\tau$$

where τ is the transmitted pulse length. When the radar receiver bandwidth is matched to the transmitter pulse length, then

$$S = \frac{P_r}{B_r}$$

where B_r is the bandwidth of radar receiver. The transmitted jamming energy will be given by

$$J = \frac{P_j}{B_j}$$

where B_j is the bandwidth of jamming signal.

If for a moment, the system losses are neglected, the system efficiency is 100%, and there are negligible atmospheric losses, then the ratio of jamming energy received to radar energy energy

$$\frac{J}{S} = \frac{4\pi P_j G_j G_r \lambda^2 F_j^2 R_r^4 B_r}{P_r G_r^2 \sigma \lambda^2 F_r^4 R_j^2 B_j}$$

for the stand off jammer, and

$$\frac{J}{S} = \frac{4\pi P_j G_j B_r R^2}{P_r G_r \sigma F^2 B_j}$$

for the self-protect jammer. For whatever camouflage factor is deemed sufficient (example $J/S = 10$), the the burn-through range can be determined from this self-protect jamming equation.

Radar Receiver Vulnerable Elements

Range gate pulloff (RGPO) is a deception technique used against pulse tracking radars using range gates for location. The jammer initially repeats the skin echo signal with minimum time delay at a high power

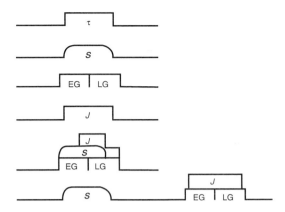

FIGURE 2.113   Range gate pulloff concept.

to capture the receiver automatic gain control circuitry. The delay is progressively changed, forcing the tracking gates to be pulled away from (walked off) the skin echo. Frequency memory loops (FMLs) or transponders provide the variable delay. The deceptive pulse can be transmitted before the radar pulse is received if the pulse repetition rate is stored and repeated at the proper time. In this technique the range gate can be either pulled off or pulled in.

One implementation is shown in Fig. 2.113. A split gate (early/late gates) range tracking scheme has two gates that are positioned by an automatic tracking loop so that equal energy appears in each gate. The track point is positioned on the target centroid. First, there is an initial dwell at the target range long enough for the tracker to capture the radar receiver. The received echo has S energy. Then, a jamming signal is introduced which contains J amount of energy. The energy in the early and late gates are compared and the difference is used to reposition the gates so as to keep equal energy in each gate. In this manner, the range gate is continually moved to the desired location. When RGPO is successful, the jamming signal does not compete with the target's skin return, resulting with an almost infinite J/S ratio.

Velocity gate pulloff (VGPO) is used against the radar's received echo modulation frequency tracking gates in a manner analogous to the RGPO used against the received echo signal amplitude tracking gates. Typically in a Doppler or moving target indicator (MTI) system there are a series of fixed filters or a group of frequency tracking filters used to determine the velocity of a detected moving target. By restricting the band width of the tracking filters, the radar can report very accurate speeds, distinguish between approaching and receding targets, and reduce echos from wideband clutter. Typically, the tracking filters are produced by linearly sweeping across the frequency domain of interest to the detecting radar. The ECM approach to confuse this linear sweep is to transmit a nonlinear frequency modulated sweep which could be parabolic or cubic. The jamming signal will cause the Doppler detector to be walked off of the true frequency as shown in Fig. 2.114.

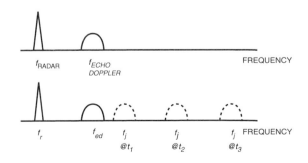

FIGURE 2.114   Frequency gate pulloff concept.

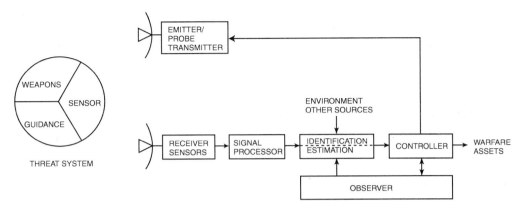

FIGURE 2.115 Surgical countermeasure block diagram.

Another technique used to generate a frequency shift is to modulate the phase of the transmitted jamming signal. The instantaneous frequency of any modulated signal is directly related to the change in phase.

$$2\pi f = \frac{d\theta}{dt}$$

When the phase is given by $\theta = 2\pi f_c t + kt$ where f_c is the carrier frequency and k the change of phase with time, then

$$\frac{d\theta}{dt} = 2\pi f_c + k$$

and

$$f = f_c + \frac{k}{2\pi}$$

This type of modulation has been referred to as serrodyne modulation and used with control voltage modulation of an oscillator or traveling wave tube (TWT).

An inverse-gain jammer is a repeater in which a signal is directed back toward the enemy radar to create false targets by varying the jammer transmitted signal level inversely with the magnitude of the received radar signal. To defeat conical scan trackers, the transponder signal have been modulated at the nutation of the scan rate to deceive angle tracking.

Most ECM techniques are open-loop control functions which add signals to the enemy radar return signal so as to confuse or deceive the weapon's ability to track. One program, named surgical counter-measures, employs closed-loop techniques to slice into the weapon tracking loop and control that loop from a stand off position. The basic elements of the EW system consist of the sensor, signal processor, controller, observer, and ECM output. In addition to being closed loop, the system is recursive and adaptable. The sensor observes the threat radar feeds the intercepted or reflected signals to the signal processor which will determine spectrum, carrier frequency, angle of arrival, pulse train, etc. and establish the electronic characteristic of the threat. The processed information is made available to the observer and the controller. The controller will generate ECM signals to probe, update, and fine tune the techniques generator in the ECM output. The observer notes the effects of the jogging introduced in the threat radar tracker and updates the controlled tracking information so as not to cause a track break, or lock pull off until the proper time in the engagement. The observer is the important feedback loop to determine the necessary refinements to previous observations and control equations.

The surgical countermeasure technique has been used successfully against both a conical-scan-on-receive only (COSRO) tracker and in conjunction with a two-transmitter blinker ECM suite to cause break lock and transfer tracking (see Timberlake, 1986).

Radar Antenna Vulnerable Elements

Radar backscatter from a variety of antennas was measured by Hayes at Georgia Institute of Technology for the U.S. Air Force Avionics Laboratory, Wright-Patterson Air Force Base, Ohio [Hayes and Eaves, 1966].

Typical measured data are presented in Figs. 2.116 and 2.117 for an AN/APS-3 antenna. This antenna is an 18-in. parabolic dish with a horizontally polarized double-dipole feed built upon a waveguide termination for use at a wavelength of 3.2 cm with an airborne radar. The instrument measuring radar operated at a wavelength of 3.2 cm, and transmitted either horizontally or vertically polarized signals, while receiving simultaneously both horizontally and vertically polarized signals.

First, the APS-3 antenna was terminated with a matched waveguide load and the backscatter patterns for horizontally polarized and vertically polarized echos obtained for both horizontal polarized transmissions and second, for vertical polarized transmissions. Then an adjustable short-circuit termination was placed at the end of the waveguide output port. The position of the waveguide termination greatly influenced the backscatter signal from the APS-3 antenna when the polarization of the interrogating measurement radar matched the operational horizontal polarization of the antenna. Tuning of the adjustable short circuit produced returns both larger and smaller then those returns recorded when the matched load was connected to the output port. The waveguide termination had very little effect on the backscatter when the interrogating transmission was vertical polarization (cross pol). In addition to the

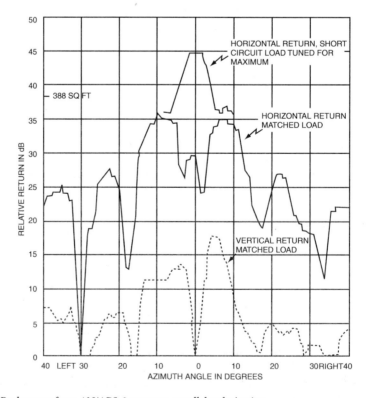

FIGURE 2.116   Backscatter from AN/APS-3 antenna, parallel polarization.

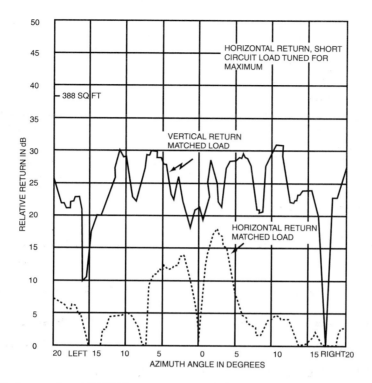

FIGURE 2.117 Backscatter from AN/APS-3 antenna, cross polarization.

APS-3, there were several other antennas and pyramidal horns (both in-band and out-of-band) investigated. The W shape in the center of the main beam is typical of backscatter patterns when the waveguide is terminated in a matched load, and when the interrogating measurement radar matched the operational frequency and polarization of the antenna under test.

Butler [1981] of Hughes Radar Systems has presented an analytical model of an antenna consisting of two scattering mechanisms. One component is the backscatter from the antenna surfaces (one-way scattering), and the second component is scattering from the feed system, which is coupled to the waveguide input/output port (round-trip scattering). It has been shown that the amplitude and phase relationships of the two scattering mechanisms will interact and can create this W shape in the equivalent main beam of an impedance-matched antenna when interrogated at the designed operating frequency and polarization.

A reflector antenna having a curved surface will generate a polarized signal, which is orthogonal to the polarization of the signal impinging upon the curved surface, when the impinging signal is not parallel to the surface. This new (cross-polarized) signal is a result of the boundary value requirement to satisfy a zero electrical field at the surface of a conductor. For a parabolic reflector antenna designed for operation with a horizontal polarization, there will be a vertically polarized component in the transmitted signal, the received signal, and the reflected signal. Curved surfaces produce cross-polarized components regardless of frequency and polarization. Thus, if horizontal polarized signals impinge upon a curved surface, vertical polarized signals as well as horizontal polarization are produced; vertical polarized impinging signals produced horizontal signals; right-circular polarized impinging signals produce a cross-polarized left-circular polarized signals, etc.

An example of parallel- and cross-polarized signals from a parabolic reflector antenna is shown in Fig. 2.118. These data are provided as a courtesy of Millitech Corp. Please note that the antenna is designed to operate at 95 GHz, showing that the effect of cross-polarization will occur even at millimeter waves. The orthogonally polarized signals have four peaks (one in each quadrant), which are on the 45 degree lines to the principle axis, and are within the main beam of the principle polarized signal. The magnitude

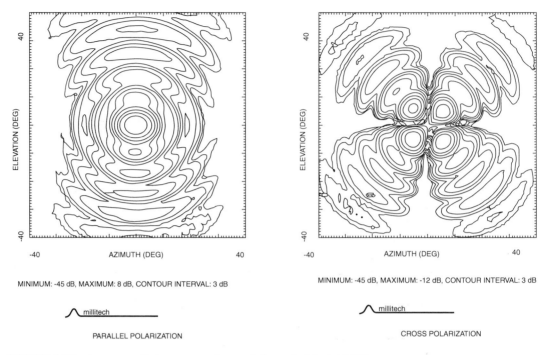

MINIMUM: -45 dB, MAXIMUM: 8 dB, CONTOUR INTERVAL: 3 dB

MINIMUM: -45 dB, MAXIMUM: -12 dB, CONTOUR INTERVAL: 3 dB

PARALLEL POLARIZATION

CROSS POLARIZATION

FIGURE 2.118   Antenna radiation patterns for parabola, conical feed, 94 GHz.

of these peaks is typically 13 to 20 dB below the peak of the principle polarized signal. The cross-polarized signal is theoretically zero on the two principle plane axis, so when perfect track lockup is made, cross-polarization will have no affect.

Injecting a cross-polarized signal into a system will have the same effect as entering a monopulse tracker. That is to say, the receiving system can not detect the difference between parallel and cross-polarized signals and yet the cross-polarized pattern has nulls on the major axis, and peaks off of boresight. When the cross-polarized signal dominates, the track should go to an angle about at the 10 dB level of the normal sum main beam.

Cross-eye is a countermeasure technique employed against conical scan and monopulse type angle-tracking radars. The basic idea is to intercept the enemy-propagated signal, phase shift the signal, then retransmit the signal to make it appear that the signal represents a reflected return emanating from an object located at an angular position different from the true location. In generating the retransmitted signal, there are two basic approaches to signal level configuration: saturated levels and linear signal levels. A typical implementation of a four antenna cross-eye system is shown in Fig. 2.119. Let us assume

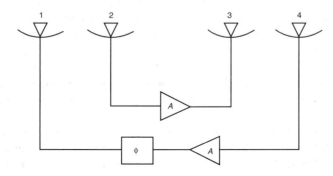

FIGURE 2.119   Cross-eye system block diagram.

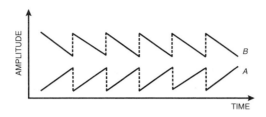

FIGURE 2.120   Saw-tooth modulation diagram for cross-eye.

that the two transmit antennas radiate about the same amount of power (slight difference could occur due to line loss, different amplifier gain, etc.). The signal received by antenna 2 is simply amplified with no phase shift and retransmitted out antenna 3, whereas the signal received by antenna 4 is amplified and phase shifted 180 degrees before retransmitted out of antenna 1.

The two wave fronts (generated by antennas 1 and 3) are of equal amplitude and 180 degrees out of phase, thus their sum will be zero on boresight, while their differences will be finite. Radars, such as a monopulse tracker, which generate sum and difference signals will interpret the results of the two wavefronts as a tracking error. The track point will drift, moving the target off boresight in such a manner as to cause the radar lock-up point to be off target, where the difference signal becomes zero.

The maximum tracking error for a saturated cross-eye is about $0.5\theta_{3\,dB}$ where $\theta_{3\,dB}$ is the antenna half-power beamwidth of the sum pattern of the radar. This is the angular position where the difference pattern has peaks and zero slope.

The addition of sawtooth modulation to the cross-eye signal can produce multiple beamwidth tracking errors in excess of the $0.5\theta_{3\,dB}$. Figure 2.120 shows two sawtooth wave forms. The RF power from one antenna is linearly increased as a function of time, while the RF power from the other antenna is linearly decreased with time. At the end of the ramp, the process is repeated, creating the sawtooth. When the antenna is given sufficient momentum during the first half of the modulation cycle to enable the antenna to traverse the singularity at $0.5\theta_{3\,dB}$, tracking pull-off will occur. Successive periods of modulation can cause the monopulse tracker to lock onto successive lobes of the sum pattern.

In linear cross-eye, the amplifiers are never allowed to go into saturation. Linear cross-eye differs from saturated cross-eye in that the sum pattern is zero for all pointing angles, and the difference pattern is never zero. No stable track points exist with linear cross-eye and large tracking errors can occur, resulting in large angular breaklocks.

The two most common types of angle tracking used today are sequential lobing and simultaneous lobing.

A common implementation of sequential lobing is to rotate a single antenna lobe about the boresight axis in a cone; the signal amplitude will be modulated at the conical scan frequency. The direction off of boresight is determined by comparing the phase of the scan modulation with an internal reference signal. As the source is brought on to axis, the modulation will go to zero. The source of the RF signal can be either an active source such as a beacon or a communication transmitter, or the target scattering, as in a radar application.

In a typical radar, the conical scan rate is chosen early in the system design. The rate should be slow relative to the radar PRF, so as to have ample samples to determine the error amplitude. However, the scan rate must be high enough to track source motions in the expected dynamic engagement. The list on trade-offs puts the scan rates typically in the range of 10 to 1000 Hz, where the weight and size of the antenna play an important role.

The most effective counter measure is to introduce an interference signal into the tracker that has the same frequency as the tracker's scan rate, but is 180 degrees out of phase. This is simple to accomplish when the victim's scan rate is known to the ECM system. When the exact scan frequency is not known, then the ECM system must vary the jamming modulation over the expected range of the victim's operation with enough power and phase difference to cause mistracking. A technique known as swept

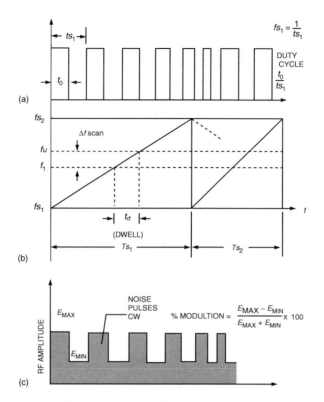

FIGURE 2.121 Swept square wave for conical scan pulloff.

square wave (SSW) has been successful in pulling radar conical scanners off track. The critical item is to keep within the audio bandwidth of the tracker control loop. There can be several variables in attempting to penetrate the tracking loop controls. The sweep frequency may be linear, nonlinear, or triangular, and the sweep time may have a varying time base and dwell times.

For radar trackers which do not scan the transmitted beam, but instead scan a receiving antenna, the scanning rate is not usually known. It can be assumed with reasonably accurate results that the transmitted beam will be close to boresight of the conical-scan-on-receive-only (COSRO) tracker, so that a repeated RF signal directed back toward the illuminating beam will intercept the COSRO beam. This energy transmitted toward the victim radar should be at the same RF carrier frequency and amplitude modulated with sufficient noise spectrum so as to cover the audio bandwidth of the tracking control system. The use of pulsed and CW TWTs as jamming noise power sources is common.

In a radar system employing track while scan (TWS), the tracking function is performed in a signal processor/computer by keep track of the reported position in range, azimuth, and elevation on a scan-to-scan basis. The antenna must move reasonable fast over the field-of-view in order to report target locations on a sufficiently timely basis. The antenna does not produce continues tracking, and thus does not remain fixed in azimuth or elevation for any long period of time. The countermeasure to a TWS radar is to move fast, turn fast, dispense chaff, noise jam, or some cover technique.

Simultaneous lobing or monopulse techniques, reduce the number of pulses required to measure the angular error of a target in two dimensions, elevation and azimuth. A minimum of three pulses are required in monopulse tracking, whereas more than four pulses are required for conical scan trackers. In the amplitude comparison monopulse, each pulse is measured in a sum and difference channel to determine the magnitude of the error off boresight, and then the signals are combined in a phase sensitive detector to determine direction. In a four horn antenna system, a sum pattern is generated for transmission and reception to determine range and direction. Two difference patterns, elevation and azimuth, are generated on reception for refining the error off of boresight. The receiving difference patterns produce a passive receiving tracking system, and there is no scanning. Difference patterns are generated by two patterns (for each plane of tracking) offset in angle by the quantity squint angle, $u_s = \frac{\pi d}{\lambda} \sin \theta s$, and 180° phase difference, such that on boresight ($u = 0$) the difference pattern has a null, and this null is in the center of the sum pattern. The use of the function u instead of the angle θ keeps the offset functions mirror images. There is no absolute optimum value of squint angle since the desire may be to have the best linearity of the track angle output, or to have the maximum sensitivity, with all the possible illumination functions on the reflecting surface [Rhodes, 1980].

The monopulse tracker is almost completely immune to amplitude modulations.

The two beams which form the difference pattern (one beam +, and one beam −), cross each other at less than 3 dB below their peaks for maximum error slope-sum pattern product and to give increased angle tracking accuracy. This can be seen in Fig. 5.11 in Skolnik [1962] and in Fig. 6.1 in Rhodes [1980], where a comparison is shown between squint angle and the two difference beam magnitude cross-over.

The difference pattern has a minimum on boresight which coincides with the peak of the sum pattern on boresight. The difference pattern is typically 35 to 40 dB below the sum pattern. These pattern values will typically improve the identity of the target location by 25:1 over a simple track-the-target in the main beam technique. Some ECM techniques are to fill the null of the difference channel so that the tracking accuracy is reduced from the 25:1 expected. If two transmitters are separated by the 3 dB width of the sum pattern (about the same as the peak of the two difference patterns), then a standard ECM technique of blinking of the two transmitters works very well.

Squint angle is a parameter within the control of the victim antenna designer; it may be chosen so as to optimize some desired function such as the best tracking linearity angle output, or maximun sensitivity on boresight. With the many options available to the antenna designer of the victim radar, the ECM designer will have to make some general assumptions. The assumption is that the squint angle off boresight is 0.4 of the half-power antenna beamwidth, given by setting the two difference patterns at a crossover of 2 dB, as shown in Fig. 2.122. (A squint of 0.5 the half-power angle is given by a 3 dB crossover). Two signals of equal amplitude (these may be radar backscatters or electronic sources, it makes no difference) and separated by 0.6 of the sum half-power beamwidth are introduced into the monopulse tracker. A typical monopulse tracker victim radar will seek the positive-going zero crossing of the error signal as a track point. In an example considering these system parameters, when the phase difference of these two signals is 90° or less, the track center position is between the two signals. When the phase difference between the two signals is between 90° and 180°, the positive going error signal is both to the left and the right outside the signal sources and the radar track will be outside of the sources.

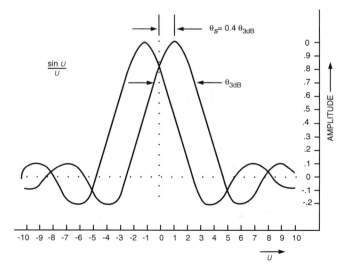

FIGURE 2.122   Squint angle monopulse (sin u)/u patterns.

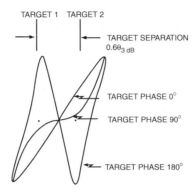

FIGURE 2.123 Two targets under track by monopulse antenna.

This can be seen in Fig. 2.123. (E. Rhodes, private communication, Georgia Institute of Technology.) Neither target is sought, and wandering of the boresight center is expected. Complex targets such as multiscattering facet aircraft, can produce 180° differences by slight changes in angle of attack. At X-Band, 180° is only 1.5 cm. Thus, it is shown that in a buddy system, a monopulse radar tracker can be made to track off of the two sources by slight changes in flight course.

Radar Counter-Countermeasures

The concept of counter-countermeasures is to prevent, or at least decrease, the detection, location, and tracking processes so that the mission can be accomplished before the victim group can defeat the attackers. Reducing the radar scattering cross-section has produced real-time advantages in combat. This technique, commonly referred to as stealth, has become a part of all advanced systems. The concept is to reduce the cross-section of the weapon's delivery system to such a low value that the detection range is so short that the defense system does not have time to react before the mission is completed.

The basic concepts of stealth are to make the object appear to have a characteristic impedance which is the same as free space, or to absorb all energy which impinges upon our platform, or to reflect the energy in some obscure direction. This is not a new concept. The Germans were using "paints" and shaping techniques on their U-boats during WW II to reduce 3-GHz radar reflections. A semiflexable material was developed at the MIT Radiation Labs in the U.S. during WW II called HARP. The material consisted of copper, aluminum, and ferromagnetic material in a nonconduction rubber binder and made into cloth, sometimes with conductive backing. HARP-X was used at the X-Band. Several techniques are appropriate in this process. First, the impinging signal can be reflected off axis so that no signals are returned in the direction of the transmitter/receiver, or multisignals are generated from multi-surfaces which will be out of phase and appear to cancel in the direction of the transmitter/receiver. These techniques are accomplished by physical configuration and shaping of our platform. The second approach is to bend, absorb, or alter the propagation of the impinging wave in the material covering our platform. Radar absorbing material, RAM, techniques are accomplished by material changes: typically, changing the complex dielectric constant, the dielectric-loss tangent, the magnetic permeability, and/or the magnetic-loss tangent. The Salisbury screen [Salisbury, 1952], a single layer of dielectric material backed by a reflecting surface, is often used as a model.

These devices create a resonant impedance to reduce reflections at multiple frequencies corresponding to multiple quarter-wavelenghts of the dielectric material. Multilayers, or Jaumann absorbers, are generally more frequency broadband absorbers and may even be graded multidielectric or multi-magnetic materials, and thus, increase the effectiveness at angles off of normal. Knott has shown that the thickness of the absorber can be reduced and the bandwidth considerable expanded if the sheets are allowed to have capacitive reactance in addition to pure resistance [Knott and Lunden, 1995].

The physics of RAM is well described by the Fresnel reflection and transmission equations. The material composite is changing daily, driven by application, and include concerns such as weight, weather durability, flexibility, stability, life cycle, heat resistance, and cost. Manufacturers of RAM material such as Emerson and Cuming, Emerson Electric Rantic, Plussy, G.E.C., and Siemens should be consulted.

The application of electronic coding of the transmitter signal and new techniques in processing the received signals have lead to a system concept referred to as low probability of intercept (LPI). There are many subtechnologies under this great umbrella, such as spread spectrum and cover pulses on transmission, pulse compression, frequency-coded and phase-coded CHIRP, frequency hopping, and variable pulse repetition frequencies. Barton [1975, 1977] has collected a number of papers describing the state of unclassified technology before 1977 in two ready references. A number of publications, primarily U.S. Naval Research Lab reports and IEEE papers, on coding techniques and codes themselves have been collected by Lewis et al. [1986].

In developing an LPI system concept, the first item will be to reduce the peak power on transmission so that the standoff detection range is greatly reduced. In order to have the same amount of energy on target, the transmitter pulse length must be increased. To have the same resolution for target detection, identification, and tracking, the new long pulse can be subdivided on transmission and coherently compressed to construct a much shorter pulse on reception. The compressed pulse produces an effective short pulse for greater range resolution and a greater peak signal which can exceed noise, clutter, or other interference for improved target detection.

One of the earliest pulse compression techniques was developed at the Bell Telephone Laboratories [Klauder et al., 1960; Rhodes, 1980]. The technique was called CHIRP and employs frequency modulation. In a stepped modulation, each separate frequency would be transmitted for a time that would correspond to a pulse length in a pulsed radar system. Upon reception, the separate frequencies are folded together to effectively form one pulse width containing the total energy transmitted. In a linear modulation, the frequency is continuous sweep over a bandwidth in a linear manner during the transmitted pulse length. The received signal can be passed through a matched filter, a delay line inversely related to frequency, an autocorrelater, or any technique to produce an equivalent pulse length equal to the reciprocal of the total bandwidth transmitted. The amplitude of this compressed signal is equal to the square root of the time-frequency-bandwidth product of the transmitted signal. The errors are limited to system component phase and frequency stabilities and linearities over the times to transmit and to receive. When using the Fourier transforms to transform from the frequency domain to the time and thus range domain, sidelobes are generated, and these sidelobes appear as false targets in range. Sidelobe suppression techniques are the same as employed in antenna sidelobe suppression design.

One of the first biphase coded wave forms proposed for radar pulse compression was the Barker code. See for example, Farnett et al. [1970]. This code changes the phase of the transmitted signal 180° at a time span corresponding to a pulse length of a standard pulse-transmitter radar. The Barker codes are optimum in the sense that the peak magnitude of the autocorrelation function is equal to the length of the coded pulse (or the number of subpulses), and the sidelobes are less than one. When the received signal is reconstructed from the phase domain, the range resolution is the same as the pulse length (or the length of a Barker subpulse) of the standard pulse-transmitter radar.

The maximum length of a Barker code is 13, and this code has sidelobes of −22.3 dB. The phase will look like +++++−−++−+−+, or −−−−−++−−+−+−.

There are many phase codes discussed by Lewis et al. (1986), which are unique and thus have unique autocorrelation functions. The compression ratios can be very large and the sidelobes very low. The price to be paid is equipment phase stability in polyphase code implementation, length of time to transmit, minimum range due to the long transmitted pulse, and equipment for reception signal processing. The advantages gained are high range resolution, low peak power, forcing the observer to use more power in a noise mode because of your radar uses wider bandwidth spread over segments of narrow bandwidths.

Another technique found helpful in reducing interference is to change the transmitted frequency on a pulse-by-pulse basis. This frequency hopping is effective, does require that the receiver and local

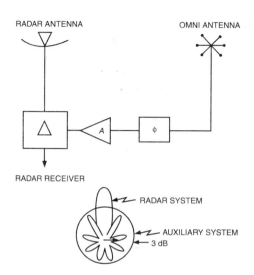

FIGURE 2.124 Sidelobe interference reduction concept.

oscillator be properly tuned so that the receiver is matched to the return signal for efficient operation. When Doppler or moving target indicators are employed, then the frequency hopping is done in batches. There must be enough pulses at any one frequency to permit enough resolution in the Fourier transform (FT) or the proper number of filter banks to resolve the Doppler components. If the transmitter can handle the total bandwidth and keep each frequency separate so there is no cross talk in the FTs or in the filters, then this is very effective jamming reduction. Today's digital systems are set up for frequency groups like 32, 64, or 128 and then repeat the frequency hopping in the same frequency direction, or the reverse direction. This reduces velocity gate pull off.

If the countermeasure system has identified your pulse repetition frequency (PRF) and thus can lock on your radar's range gate, then there are several techniques to reduce the EW range gate pull off. One technique is to place a noise pulse over the standard pulse. This noise pulse is longer than the system range resolution pulse and is not symmetrical around the range pulse. Only your system should know were your own target identification pulse is in the chain of PRF pulses. Another technique that shows promise, is to change the PRF after a batch of pulses have been transmitted. Typically the batch length is controlled in a digital manner. Again, as in the case of frequency hopping, PRF hopping still must permit processing in range for resolution, in frequency for Doppler, and in polarization (or other identifiers) for target identification.

So far as our system is concerned, there is no difference if a jamming or interference signal enters the main beam of our system antenna or a sidelobe of our system antenna. From the jammer's point of view, there needs to be 20 to 30 dB more power or 20 to 30 dB closer in range if the jammer is directed at a sidelobe instead of the main antenna lobe.

Suppression of sidelobe jamming can be accomplished by employing omnidirectional auxiliary antennas having gains greater than the highest sidelobe to be suppressed, any yet lower than the main beam of the antenna. The auxiliary antenna feeds a receiver identical to the receiver used by the radar. The detected output of the auxiliary receiver and the radar receiver are compared, and when the auxiliary receiver output is larger than the radar then the radar is blanked in all range cells where this occurs. The auxiliary antenna should be placed as close as possible to the phase center of the radar antenna to insure that the samples of the interference in the auxiliary antenna can be correlated with interference in the main radar antenna. This requires that the time of arrival of the signals to be compared must be less than the reciprocal of the smaller of the two bandwidths, in the radar receiver or the interference signal.

Chaff

Cylindrical metal rods scatter energy in the same manner as a dipole radiates energy. The same parameters affect the efficiency, beamwidth, loss, and spatial patterns. Thus, in this discussion of chaff as a scattering element, consideration will be given to rod length in terms of excitation frequency, length-to-diameter ratio, impedance or loss-scattering pattern as a function of polarization, and interaction (or shielding) between adjacent elements.

The aerodynamic characteristics are determined by material, density, shape, drag coefficient, and atmospheric parameters such as temperature, pressure, and relative humidity. In addition to these parameters, the movement of a rod is strongly influenced by local winds, both horizontal and vertical components, and by wind turbulence and shear.

When all of these variable parameters are defined, the radar scattering characteristic of a collection of rods forming a cloud of chaff can be calculated. The radar parameters of interest in this presentation are the scattering cross-section from the cloud, the power spectral density to define the frequency spread, Doppler frequency components, polarization preference, signal attenuation through the cloud, band-width, and length of time that these clouds can be expected to exist.

From antenna theory, the tuned half-wave dipole produces the maximum scattering coefficient. There are many books for your reference reading, examples are Terman [1943]; Kraus [1950]; Thourel [1960]; Jasik [1961]; Eustace [1975–1991] and Van Vleck et al. [1947].

When the dipole is observed broadside, and with a polarization parallel to the rod, then the maximum radar cross-section is observed and has a value of

$$\sigma = 0.86\lambda^2$$

As the dipole tumbles and is observed over all possible orientation then, the average radar cross-section has been calculated and found to be

$$\sigma = 0.18\lambda^2$$

Since the rod has a finite diameter, the maximum scattering cross-section occurs for lengths slightly less than $\lambda/2$. A reasonable value of 0.48λ for the length will be assumed to compensate for the end effects of the rod. The first resonance is plotted as a function of the length-to-diameter ratio in the June, 1976 edition of ICH [Eustace, 1975–1991]. The equations relating radar cross-section to the rod length and diameter in terms of wavelength are complicated. The radar cross-section has been shown to be directly related to the ratio of physical length-to-radar wavelength, and inversely related to the logarithm of the physical length-to-radius ratio in Van Vleck et al. [1947]. This article is a classic, and along with their references, is the place to begin.

The polarization of the observing radar and the alignment of the chaff material is important since horizontally aligned rods fall more slowly than vertically aligned rods. Under many applications, one end of the chaff filaments is weighted, causing these rods to be vertically aligned shortly after release thus enlarging the chaff cloud rapidly in the vertical elevation plane as well as making the radar cross-section of the lower elevation portion of the cloud more susceptible to vertically polarized waves and the upper elevation section of the chaff cloud having a higher cross-section to horizontally polarized waves.

Chaff elements have been constructed in all sizes and shapes. Early manufactured elements were cut from sheets of metal and were rectangular in cross-section and cut to half-wave dipoles in length; some were bent in V shape, some were long to form ribbons, some were made as ropes; some were cut from copper wire and some from aluminum. In the past decade or two, glass rods coated with aluminum have become very popular and this configuration will be assumed typical for this discussion. There has been a reasonable amount of experimental work and calculations made for dipoles constructed of both 1-mil- and 1/2-mil-diameter glass rods for use as chaff. The metal coating should be at least 3 skin depths at the designed frequency. One skin depth is defined as the thickness there the signal has decreased to $1/e$ of the original value.

There has been limited amounts of work done at 35 and 95 GHz as compared to much work done at lower frequencies. Consider that the wavelength at 35 GHz is only 0.857 cm and a half wavelength is only 0.1687 in. Thus, a cloud of dipoles will be a lot of very short elements of material. In order to have a reasonable length-to-diameter ratio and reduce end effects as well as to provide good electrical bandwidth, it will be assumed that 1/2-mil-diameter dipoles would be acceptable. From these dimensions, manufacturing of large bundles of these dipoles is not a trivial matter. Some areas of concern are: (1) the plating of the small diameter glass will not be simple and most likely will produce considerable variations in weight and L/D, with results in varying fall rates and changes in bandwidth of the scattering signal, (2) to consistently cut bundles of pieces of elements this short in length requires considerable ability; and (3) it is required that each dipole be independent and not mashed with several other dipoles or the result is reduction in radar cross-section and bird nesting. These effects result in lost efficiency. One note: do shop around because technical and manufacturing capability are advancing at a rapid rate.

There are two types of antennas that are discussed in the literature on antennas: the thin-wire antenna, and the cylindrical antenna. The thin-wire antenna model, where assumed length-to-diameter is greater than 5000, is used to predict general scattering patterns and characteristic impedance as a function of rode length and frequency. The cylindrical antenna will show the effect of wire diameter which in turn reveals changes in the reactive portion of the terminal impedance and thus changes in gain, propagation and scattering diagrams, filling in of pattern nulls, reduction of sidelobes, and wider frequency bandwidth. It has been shown that a thin-wire half-wave dipole, excited at the center, and at resonance, has an impedance in ohms of

$$Z = 73.2 + j42.5$$

When the length is not half-wave, and the diameter is not negligible, then to a first-order approximation

$$Z = 73.2 + j42.5 \pm j120 \left[\ln \frac{L}{D} - 0.65 \right] \cot \frac{2\pi L}{\lambda}$$

As is the response in all tuned circuits, the bandpass characteristics are primarily determined by the reactive elements as the frequency moves away from the resonant frequency. In low Q circuits, the real part of the impedance also has a notable effect on the total impedance, and thus, also on the bandpass characteristics. To a first-order effect, the real part of the half-wave dipole does not change drastically (less than 10 percent) from 73.2 Ω, however, the reactive part varies as ($\ln L/D$). As shown by VanVleck et al. [1947], the radar cross-section off of resonance is expressed in reciprocal terms of $\ln$(rod length/rod radius). The effective half-power bandwidth of dipole scatters is thus usually referenced to the ratio of dipole length to dipole diameter as shown in Fig. 2.125. As it is desirable to have some bandwidth in

PERCENT BANDWIDTH	LENGTH/DIAMETER
12	3000
13	1650
14	1050
15	740
16	520
17	400
18	320
19	260
20	220
21	180
22	150
23	140
24	120
25	110

FIGURE 2.125 Backscatter bandwidth response for chaff.

television antennas, it is also desirable to have some bandwidth in the radar scattering from chaff, which leads to a measurable diameter of the rods or width of the ribbon strips used for chaff.

Consider an application at X-Band in which it is desirable to screen against a wideband pulse compression radar having a bandwidth of 1.5 GHz. The rods will be cut to 0.48λ. From Fig. 2.125, the value of $L/D = 750$,

$$L/D = 750 = 0.48\lambda/D,$$

where D = 0.48 × 3 cm/750 = 0.00192 cm = 0.000756 inches = 0.756 mil.

Consider now the generation of a chaff cloud having a radar backscatter cross-section per unit volume of –60-dB m²/m³. This value of cross-section is comparable to the radar return from 10 mm/h of rain observed at X-Band [Currie, Hayes, and Trebits, 1992]. These small aluminum-coated glass cylinders fall very slowly in still air and have been known to stay aloft for hours. The movement is so strongly influenced by local wind conditions that meteorologists in several countries (including the U.S.S.R., Germany, and the U.S.) have used chaff clouds to determine wind profiles [Nathenson, 1969]. Local conditions give extreme wind variations even when the local wind structure is free from perturbations such as squalls, fronts, thunderstorms, etc. Some wind profile models include terrain surface material such as water, grass, dirt, and surface roughness, time of day, time of year, northern hemisphere, and other parameters shown to affect a particular area. At altitudes above 500 feet, the profile of wind has been shown to very mostly in a horizontal manner, with little vertical constantly defined variations [AFCRC, 1960]. The strong jet streams at around 35,000 to 40,000 feet define the top most layer typical of zero water vapor levels, zero oxygen levels, definable temperature profiles, wind stability, and pressure profiles. This being the case, it behooves the EW engineer to apply local meteorological records whenever possible.

One-mil diameter aluminum-coated, glass-filament chaff, for example will adjust to sharp wind gusts of up to 50 ft/s within 15 ms [Coleman, 1976–1977]. The vertical fall rate of chaff is proportional to the weight and air resistance, thus the important chaff parameters are diameter, length, and coating. A typical approach is to use a mixture of dipole diameter sizes so that a vertical cloud is generated from different fall rates. For the example above, the diameter mixture could be between 0.5- and 1.0-mil coated glass.

Consider a typical wind condition of a 5 knot horizontal wind, and a wind profile of 1/5 power relationship at a height below 500 feet. Coleman [1976–1977] has shown that a cloud mixture of 0.5 to 1.0 mil chaff would stay aloft for about 20 min when dispensed at 500 feet. The chaff cloud will smear in both elevation and in horizontal dimensions. The volume of the ellipsoid generated by the horizontal winds and the vertical fall rate is given by

$$\text{volume} = V = \frac{4}{3}\pi abc$$

In this example, α = 100 ft and $\beta = \chi$ = 300 ft. The resulting volume is 1,068,340 m³.

The radar cross-section of such a cloud is given by

$$\sigma_T = V\sigma_V$$
$$= \left(1.068 \times 10^6\right)\left(10^{-6}\right)$$
$$= 1.068\,\text{m}^2$$

This is not a strong target. But we know that 10 mm/hr. of rain is not a strong target at X-Band. The big problem with rain is a deep volume producing a large amount of attenuation. You notice here that the chaff cloud is only 600 feet across; this would not be a typical rain cloud. Perhaps the comparison should be for the chaff cloud to cover (hide) a 30 dB cross-section vehicle.

The number of dipoles in a cloud of chaff is given be

$$N = \frac{\sigma_T}{\sigma_d}$$

where σ_T is the total cross-section, and σ_d the cross-section of each dipole

$$N = \frac{1000}{0.18(3/100)^2}, \quad \frac{m^2}{m^2}$$

$$= 6.17 \times 10^6$$

The density of dipoles is given by

$$\frac{N}{V} = \frac{6.17 \times 10^6}{1.068 \times 10^6}$$

$$= 5.78 \text{ dipoles}/m^3$$

This is not a dense chaff cloud, and thus you can see the value of such a cloud to hide a good size target. Of course, there is a short fall; the cloud may last only 20 min and will drift several thousand feet in this time. This requires that there be proper area and time planning for chaff use. If the volume to be covered is larger than the $100 \times 300 \times 300$ ft, then, several clouds need to be dispersed.

In the International Countermeasure Handbook [EW Comm., 1975], the author has presented a graph to be used to determine the signal attenuation produced by a chaff cloud. The graph relates normalized two-way attenuation per wavelength squared to the number of half-way dipoles per cubic meter, as a function of the dipole orientation K relative to the radar polarization. When the dipoles are uniformly distributed and randomly oriented, K can be as low as 0.115; when the dipoles are aligned with the E-field, $K = 1.0$ and the loss is 10 dB higher; when the dipoles are almost cross-polarized to the radar, $K = 0.0115$ the attenuation will be 10 dB less than the random orientation. For $K = 0.115$, the attenuation relationship is given by

$$L/\lambda^2 = N/m^3$$

in units of dB/meter cube = number/cubic meter. Typical ranges of cloud density range from 10^{-7} to 10^{+2} dipoles per cubic meter. For the example given above, a cloud density of 5.78 dipoles/m³ cube will give a value of $L/\lambda^2 = 5.78$ For a two-way loss per meter of chaff cloud, $L = 0.0052$ dB/m at $\lambda = 0.03$ m; giving 0.5 dB of attenuation in a 100-m cloud.

Expendable Jammers

Expandable jammers are a class of electronic devices which are designed to be used only once. These devices are typically released from chaff/flare dispensers much as chaff is deployed. The electronic package can be deployed as a free-falling object, or released with a parachute for slow fall rates. Such packages have been used as screening devices and as simulated false targets. The RF energy can be continuous, pulsed, or modulated to respond as a repeater or a false radar. Expandable jammers have been configured as spot noise jammers and have been deployed as barrage noise jammers depending upon the mission and victim radars.

Defining Terms

P = power, watts
k = Boltzman constant = 1.38×10^{-23} joules/K
T = temperature = absolute temperature, K
S = energy = power x transmitted pulse length, joules
B = bandwith, Hz
f = frequency, Hz
Hz = 1 cycle per second
dB = 10 times logarithm of power ratio
Z = impedance, ohms

References

AFCRC. 1960. *Geophysics Handbook,* Chap. 5. Air Force Cambridge Research Center, MA.

Barton, D.K. 1975. Pulse compression. *RADAR,* Vol. 3. ARTECH House, Norwood, MA.

Barton, D.K. 1977. Frequency agility and diversity. *RADAR,* Vol. 6. ARTECH House, Norwood, MA.

Butler, W.F. 1981. Antenna backscatter with applications to surgical countermeasures. Contract N00014-81-C-2313, Hughes Aircraft, September, El Segundo, CA.

Coleman, E.R. 1976–1977. *International Countermeasures Handbook.* EW Communications, Inc., Palo Alto, CA.

Currie, Hayes, and Trebets. 1992. *Millimeter-Wave Radar Clutter.* ARTECH House, Norwood, MA.

Eustace, H.F., ed. 1975–1991. *The Countermeasures Handbook,* EW Communications Inc., Palo Alto, CA.

EW Comm. 1975. *International Countermeasures Handbook,* June, EW Communications, Inc., Palo Alto, CA, 227.

Farnett, Howard, and Stevens. 1970. Pulse-compression radar. In *Radar Handbook.* ed. Skolnik, Chap. 20. McGraw-Hill, New York.

Hayes, R.D. and Eaves, J.L. 1966. Study of polarization techniques for target exhancement. Georgia Institute of Technology Project A-871, March, AD 373522.

Jasik, H. 1961. *Antenna Engineering Handbook.* McGraw-Hill, New York.

Klauder, Prince, Darlington, and Albersheim, 1960. *Bell Sys. Tel. J.* 39(4).

Knott, E.F. and Lunden, C.D. 1995. The two-sheet capacitive Jaumann absorber. *IEEE Ant. and Prop. Trans* 43(11).

Kraus, J.D. 1950. *Antennas.* McGraw-Hill, New York.

Lewis, B.L., Kretschmer, F.F., Jr., and Shelton, W.W. 1986. *Aspects of Radar Signal Processing.* ARTECH House, Norwood, MA.

Nathenson. 1969. *Radar Design Principles.* McGraw-Hill, New York.

Rhodes, D.R. 1980. *Introduction to Monopulse.* ARTECH House, Norwood, MA.

Salisbury, W.W. 1952. Absorbent body for electromagnetic waves. U.S. Patent 2,599,944, June 10.

Skolnik, M.I. 1962. *Introduction to Radar Systems,* 2nd ed. McGraw-Hill, New York.

Terman, F.E. 1943. *Radio Engineers Handbook.* McGraw-Hill, New York.

Thourel, L. 1960. *The Antenna.* Chapman and Hall, New York.

Timberlake, T. 1986. *The International Countermeasure Handbook,* 11th ed. EW Communications Inc., Palo Alto, CA.

Van Vleck, Bloch, and Hamermesh. 1947. The theory of radar reflection from wires or thin metallic strips. *J. App. Phys.* 18(March).

Further Information

Applied ECM, Vols. 1 and 2, 1978, EW Engineering, Inc., Dunn Loring, VA.

International Countermeasures Handbook, 11th ed. 1986, EW Communications, Inc.

Introduction to Radar Systems, 2nd ed. 1962, McGraw-Hill.

Modern Radar Systems Analysis, 1988, ARTECH House.

2.13   Automotive Radar

Madhu S. Gupta

Scope: An automotive radar, as the name suggests, is any radar that has an application in automobiles and other autonomous ground vehicles. As a result, it represents a large and heterogeneous class of radars that are based on different technologies (e.g., laser, ultrasonic, microwave), perform different functions (e.g., obstacle and curb detection, collision anticipation, adaptive cruise control), and employ different operating principles (e.g., pulse radar, FMCW radar, microwave impulse radar). This article is limited to microwave radars that form a commercially significant subset of automotive radars. Microwave radars have an advantage over the laser and infrared radars (or "lidar") in that they are less affected by the presence of precipitation or smoke in the atmosphere. They also have the advantage of being unaffected by air temperature changes that would degrade an ultrasonic radar.

Need: The need for automotive radars can be understood at three different levels. At the national level, the statistics on traffic fatalities, injuries, and property loss due to vehicle accidents, and estimates of their fractions that are preventable with technological aids, has encouraged the development of automotive radar. The economic value of those losses, when compared with the dropping cost of automotive radar, leads to a cost-benefit analysis that favors their widespread deployment. At the level of the automotive manufacturer, radar is another "feature" for the consumer to purchase, that could be a possible source of revenue and competitive advantage. It is also a possible response to regulatory and public demands for safer vehicles. At the level of vehicle owners, automotive radar has an appeal as a safety device, and as a convenient, affordable gadget. Of greater practical importance is the potential for radar to lower the stress in driving and decrease the sensory workload of the driver by taking over some of the tasks requiring attentiveness, judgment, and skill.

Antecedents: Lower-frequency electronic systems have had a presence for decades in various automobile applications, such as entertainment, fuel injection, engine control, onboard diagnostics, antitheft systems, antiskid brakes, cruise control, and suspension control. The earliest microwave frequency products introduced in automobiles were speed radar detectors in the 1970s, followed by cellular telephones in the 1980s, and direct satellite receivers and GPS navigational aids in the 1990s. The use of microwave vehicular radars can also be traced back to the 1970s, when rear obstacle detection systems were first installed, mostly in trucks. Some of those radars evolved from simple motion detectors and security alarms that employed Gunn diodes and operated typically in the X-band around 10 GHz. The modern automotive radar owes its origin to three more recent technological advancements: low-cost monolithic microwave devices and hybrid or monolithic circuits; microprocessors, and digital signal processing.

Status: Automobile radars have been developed by a number of different companies in Europe, Japan, and the United States, since the 1970s. Automotive radars for speed measurement and obstacle detection have been available for more than a decade as an aftermarket product, and have been installed primarily on trucks and buses in quantities of thousands. Only now (in 2000) are automotive radars becoming available as original equipment for installation on trucks, buses, and automobiles. However, other types of radars are still in the development, prototyping, field-testing, and early introduction stages. Some of the most important challenges for the future in the development of automotive radar include (a) development of more advanced human interface, (b) reduction of cost, (c) meeting user expectations, and (d) understanding the legal ramifications of equipment failure.

Classification

An automotive radar can be designed to serve a number of different functions in an automobile, and can be classified on that basis as follows.

1. *Speed Measuring Radar.* Vehicle speed is measured with a variety of types of speedometers, many of which are based on measuring the rate of revolution of the wheels, and are therefore affected by tire size and wheel slippage. By contrast, a speed measuring radar can determine the true ground

speed of a vehicle, that may be required for the instrument panel, vehicle control (e.g., detection of skidding and antilock braking), and other speed-dependent functions.

2. *Obstacle Detection Radar.* Such a radar is essentially a vision aid for the driver, intended to prevent accidents by monitoring regions of poor or no visibility, and providing a warning to the driver. It can perform functions such as curb detection during parking, detection of obstacles in the rear of a vehicle while backing, and sensing the presence of other vehicles in blind zones (regions around the vehicle that are not visible to the driver via side- and rear-view mirrors) during lane changing.

3. *Adaptive Cruise Control Radar.* A conventional cruise control maintains a set vehicle speed regardless of the vehicular environment, and as a result, is ill suited in a heavy traffic environment where vehicles are constantly entering and leaving the driving lane. An adaptive cruise control, as the name suggests, adapts to the vehicular environment, and maintains a headway (clearance between a vehicle and the next vehicle ahead of it) that is safe for the given speeds of both vehicles.

4. *Collision Anticipation Radar.* This class of radars senses the presence of hazardous obstacles (other vehicles, pedestrians, animals) in the anticipated path of the vehicle that are likely to cause collision, given the current direction and speed of the vehicle. It is therefore potentially useful both under conditions of poor atmospheric visibility (e.g., in fog, sandstorm), as well as poor judgment on the part of the driver (unsafe headway, high speed). Its purpose may be to warn the driver, to deploy airbags or other passive restraints, or take over the control of vehicle speed.

5. *Other Vehicular Monitoring and Control Radars.* Many other vehicular control functions, such as vehicle identification, location, fleet monitoring, station keeping, guidance, and path selection, can be performed with the aid of radar. Such radars may be placed on board the vehicle, or on the ground with the vehicle carrying a radar beacon or reflector, that may be coded to allow vehicle identification. Still other variations of automotive radars can be envisaged for special purposes, such as control of vehicles constrained to move along tracks, or joyride vehicles in an amusement park.

The following is a discussion of each of the first four types of radars, including their purpose, principle of operation, requirements imposed on the radar system and its constituent parts, limitations, and expected future developments. The emphasis is on the RF front-end of the radar, consisting of the transmitter, the receiver, and the antenna. The adaptive cruise control and the collision anticipation radars share a number of characteristics and needs; they are therefore discussed together as forward-looking radars.

History of Automotive Radar Development

The idea of automotive radars is almost as old as the microwave radars themselves. The history of their development may be divided into four phases as follows.

1. *Conceptual Feasibility Phase.* The earliest known experiments, carried out in the late 1940s soon after the Second World War, involved a wartime radar mounted on top of an automobile, with the cost of the radar far exceeding that of the automobile. The idea that the radar could be used to control a vehicle was considered sufficiently novel that the U.S. Patent Office issued patents on that basis. An example of such early efforts is a patent (# 2,804,160), entitled "Automatic Vehicle Control System," issued to an inventor from Detroit, Michigan, that was filed in January 1954 and issued in August 1957.

2. *Solid-State Phase.* The presence of a microwave tube in the radar for generating the transmitted signal, and the high voltage supply they required, was known to be a principal bottleneck that impacted the size, cost, reliability, and safety of automotive radar. Therefore, the discovery of solid-state microwave sources in the form of two-terminal semiconductor devices (transferred-electron devices and avalanche transit-time diodes) in the mid-1960s led to a resurgence of interest in developing automotive radar. Since solid-state microwave devices were then one of the more

expensive components of radar, the possibility of using the device, employed simultaneously as the source of the microwave signal and as a mixer by taking advantage of its nonlinearity, was considered another desirable feature. Such radars were the subject of numerous experimental and theoretical studies, and even reached the stage of limited commercial production. Radars based on BARITT diodes were deployed on trucks in the 1970s to warn the driver of obstacles in the rear.

3. *MMIC Phase.* The production of monolithic microwave integrated circuits in the 1980s made both the microwave devices, and the associated circuitry, sufficiently small and inexpensive that the weight and cost of microwave components in radar was no longer an issue. The prototype radars in this period were based both on Gunn diodes and on GaAs MMICs that employed MESFETs as the active device and source of microwaves. At the same time, the technology of microstrip patch antennas, that are particularly convenient for automotive use, became sufficiently developed. A variety of volume production problems had to be solved in this phase, including those of maintaining frequency stability and packaging, while keeping the cost down. The focus of development thus shifted to other aspects of the radar design problem: reliability of the hardware and its output data; operability of the radar under unfavorable conditions; affordability of the entire radar unit; and extraction of useful information from radar return by signal processing. Prototype units of various types were field tested, and deployed on a limited scale both for evaluation and feedback as well as in special markets such as in bus fleets.

4. *Product Development Phase.* Automotive radar became economically viable as a potential consumer product due to the decreasing ratio of the cost of the radar to the cost of the automobile. Moreover, the economic risk of radar development decreased due to the availability of specific frequency bands for this application, and the confidence resulting from the results of some field tests. As a result, automotive radar is now (in year 2000) available as an "option" on some vehicles, and work continues on enhancing its capabilities for other applications, in human factors engineering, and for integration into the vehicle.

Speed-Measuring Radar

Operating Principle: The simplest method of measuring the true ground speed of a vehicle is by the use of a Doppler radar. In this radar, a microwave signal at frequency f is transmitted from on board the vehicle, aimed toward the ground; the signal scattered back by the ground is collected by a receiver also on board the vehicle. Given the large speed c of microwave signals, and the short distance between the antenna (typically mounted under the carriage) and the ground, the signal transit time is short. The returned signal is shifted in frequency due to Doppler effect, provided the vehicle and the scattering patch of the ground are moving with respect to each other, with a velocity component along the direction of propagation of the microwave signal. To ensure that, the microwave beam is transmitted obliquely toward the ground, making an angle q with respect to the horizontal, and in the plane formed by the ground velocity v of the vehicle and the perpendicular from the vehicle to the ground. Then the vehicle velocity component along the direction of propagation of the signal is v cos q, and the Doppler frequency shift is 2f(v/c)cosq, proportional both to the transmitted signal frequency f and to the vehicle velocity v. The Doppler shift frequency is extracted by mixing the returned signal with the transmitted signal, and carrying out filtering and signal processing. Typically (for a carrier frequency of 24 GHz and a tilt angle q = 30°), it lies in the range of 35 Hz to 3.5 kHz for vehicle speeds in the range of 1 to 100 miles per hour.

Error Sources: Several sources of error in speed estimation can be identified from the above discussion.

1. *Vehicle tilt.* Uneven loading of a vehicle, air pressure in the tires, and non-level ground can all contribute to a change in the tilt angle q of the beam, and hence in the estimated vehicle speed. A well-known technique for correcting this error is to employ a so-called Janus configuration in the forward and reverse directions (named after a Greek God with two heads). In this scheme, two microwave beams are transmitted from the transmitter, one in the forward and the other in the reverse direction, each making an angle q with the horizontal. The Doppler frequency shift

has the same magnitude for each signal, but a tilt of the vehicle makes q one beam larger while simultaneously making the other smaller. The correct ground velocity, as well as the tilt angle of the vehicle, can be deduced from the sum and difference of the two Doppler shift frequencies.

2. *Nonzero beam width.* Any reasonably sized antenna will produce a beam of finite width, so that the angle q between the transmitted signal and the horizontal is not a constant. A spread in the values of q produces a corresponding spread in the values of the Doppler shift frequency.

3. *Vertical vehicle velocity.* Vibrations of the vehicle and uneven ground will cause the vehicle of have a velocity with respect to the ground in the vertical direction. This velocity also has a component in the direction of propagation of the microwave signal, and thus modulates the Doppler shift frequency.

4. *Surface roughness.* Although surface roughness is essential for producing scattering in the direction of the receiver, it introduces errors because the surface variations appear as variations of q as well as of vehicle height, and therefore an apparent vertical vehicle velocity.

Extensive signal processing is required to extract an accurate estimate of vehicle velocity in the presence of these error sources. Current Doppler velocity sensors employ digital signal processing and are capable of determining the velocity with 99% certainty.

Obstacle-Detection Radar

Purpose: A driver is unable to view two principal areas around the vehicle that are a potential source of accidents: one is behind the vehicle, and the other on the two sides immediately behind the driver. The need to view these areas arises only under specific conditions: when driving in reverse and when changing lanes. The exact boundaries of these areas depend on the style of vehicle, the placement and setting of the viewing mirrors, and the height and posture of the driver.

Mission Requirements: Since obstacle detection radar needs to operate over a small range (typically less than 10 m), cover a wide area, and does not need to determine the exact location of the obstacle in that area, its operating frequencies can be lower, where the antenna beamwidth is wide.

Adaptive Cruise Control Radar

Purpose: The adaptive cruise control (to be abbreviated hereafter as ACC) radar is so called because it not only controls the speed of a vehicle but also adapts to the speed of a vehicle ahead. The ACC radar controls the vehicle speed, subject to driver override, so as to maintain a safe distance from the nearest in-path vehicle ahead (the "lead" vehicle). If there are no lead vehicles within the stopping distance of the vehicle, the ACC functions as a conventional cruise control that maintains a fixed speed set by the driver. With lead vehicles present within the stopping distance, the system governs the acceleration and braking so as to control both the speed and the headway. Such radar has also been referred to by several other names such as intelligent cruise control (ICC), autonomous intelligent cruise control (AICC), and others.

Mission Requirements: First and foremost, the ACC radar must be capable of distinguishing between the closest vehicle ahead in the same lane and all other vehicles and roadside objects. As a result, a high accuracy in range and angular resolution is necessary. Second, it must acquire sufficient information to establish the minimum safe distance from the target vehicle, S_{min}. At the simplest level, S_{min} equals the stopping distance of the vehicle carrying the radar (the "host" vehicle), minus the stopping distance of the lead vehicle, along with an allowance for the distance traveled within the reaction time of the driver initiating the stopping action. If v_r and a_r are the velocity and deceleration of the host vehicle, v_t and a_t those of the targeted lead vehicle, and T_r the reaction time of the driver, then the minimum safe distance to the target can be estimated approximately as

$$S_{min} = \left(v_r^2 / 2a_r \right) - \left(v_t^2 / 2a_t \right) + v_r T_r$$

The distance calculated by this equation is subject to large uncertainty and additional safety margin, since the deceleration of each vehicle depends on the brake quality, road conditions, vehicle loading, and tire condition, while the reaction time depends on the driver's age, health, state of mind, training, and fatigue. However, the equation does show that to determine whether the following distance is safe requires not only the distance to the target but also the ground speed of the vehicle as well as the relative velocity with respect to the target. In particular, it is the radial component of the velocity that is pertinent, and the sign of the velocity is also important because it determines whether the vehicles are approaching or receding. This defines the minimum information that the ACC system must be designed to acquire.

Collision Anticipation Radar

Purpose: The purpose of collision anticipation radar is to sense an imminent collision. Several different variations of collision anticipation radars have been considered, differing in the use of the information gathered by the radar, and hence the definition of "imminent." For example, if the purpose of the radar is to initiate braking, the possibility of a collision must be sensed as early as possible to allow time for corrective action; if its purpose is to serve as a crash sensor for deploying an inflatable restraint system (commonly called airbag), only a collision that is certain is of interest. The different functions have been given various names, such as collision warning (CW), collision avoidance (CA), collision reduction (CR) radar, and others, but the nomenclature is not consistent, and is sometimes based on marketing rather than on technical differences. The following discussion illustrates some applications.

Collision Warning Application: Traffic accident analyses show that a significant fraction of traffic accidents (30% of head-on collisions, 50% of intersection accidents, and 60% of rear-end collisions) can be averted if the drivers are provided an extra 0.5 s to react. The purpose of the collision warning radar is to provide such an advanced warning to the driver. In order to perform that function, the radar must resolve, classify, and track multiple targets present in the environment; collect range, speed, and acceleration information about individual targets; use the past and current information to predict the vehicular paths over a short interval; estimate the likelihood of an accident; and present the situational awareness information to the driver through some human interface. Moreover, these functions must be performed in real time, within milliseconds, and repeated several times per second for updating.

Crash Sensing Application: Passive restraints (such as airbags and seat belt tensioners) used to protect vehicle occupants from severe injuries, typically employ several mechanical accelerometers to sense the velocity change in the passenger compartment. A radar used as an electronic acceleration sensor can have a number of advantages, such as sophisticated signal processing, programmability to customize it for each vehicle structure, self-diagnosis and fault indication, and data recording for accident reconstruction. Since the passive restraints require only about 30 ms to deploy, the ranges and time intervals of interest are shorter than in collision warning applications.

Radar Requirements: In each case, the collision anticipation radar must detect objects in the forward direction (in the path of the vehicle), and acquire range and velocity data on multiple targets. In this respect, the radar function is similar to that of ACC radar, and there are many similarities in the design considerations for the RF front end of the two types of radars. The system requirements and radar architecture for the two are therefore discussed together in the following.

RF Front-End for Forward-Looking Radars

ACC and collision anticipation radars have a number of similarities in the areas they monitor, information about the vehicular environment they must acquire, and the constraints under which they must operate. The major differences between them lie in the signal processing carried out on the radar return; the range of parameter values of interest; and the manner in which the collected information is utilized. From a user perspective, the primary difference between them stems from their roles: whereas the ACC radar is a "convenience" feature, a collision radar is thought as a "safety" device; consequently, the legal liability in case of malfunction is vastly different.

Radar Requirements: Some of the requirements to be satisfied by a forward-looking radar follow directly from its expected mission. The expected radar range is the distance to the lead vehicles of interest, and therefore lies in an interval of 3 m to perhaps as much as 200 m. The uncertainty in the measured value of range should not exceed 0.5 m. Since the lead vehicle is expected to be traveling in the same direction as the host vehicle, the maximum relative velocity of the vehicles can be expected to lie in the interval of +160 Km/hr to −160 Km/hr. If the permissible uncertainty in the measurement of this relative speed is 1% at the maximum speed, a speed measurement error of up to 1.5 Km/hr is acceptable. If the information must be refreshed upon a change of 2 m in the range even at the highest speed, the radar needs to update the range and speed information once every 50 ms.

Environmental Complexity: The conditions under which the forward-looking radars operate is made complex by four features of the roadway environment. First, the radar must operate under harsh ambient conditions with respect to temperature, humidity, mechanical vibration and acceleration, and electromagnetic interference. Second, it must operate in inclement weather, caused by rain, sleet, snow, hail, fog, dust storm, and smoke. Third, the roadside scene is rapidly changing, and includes a large number of both potentially hazardous and harmless objects, some moving and others stationary, and the radar must identify and discriminate between them in order to maintain a low probability of false alarm. Fourth, due to road curvature and steering, and tangential components of vehicle velocities, the discrimination between in-path and off-path objects becomes more involved, and requires computationally intensive prediction algorithms. Therefore, features like real-time signal processing, robust algorithms, and built-in fault detection, are essential.

Frequency Selection: Although some of the earlier radars were designed for operation around 16 and 35 GHz, virtually all current developments employ V-band frequencies for forward-looking radars. Within the V-band, several different frequencies have been used in the past decade, including 77 GHz for U.S. and European systems, and 60 GHz in some Japanese systems. Three factors dictate the choice of millimeter wave range for this application. First, the range resolution of the radar is governed primarily by the bandwidth of the transmitted signal, and a resolution of the order of 1 meter requires a minimum bandwidth of around 150 MHz. Such a large bandwidth is not available below the millimeter wave frequency range. Second, for a given performance level (such as antenna directivity), a higher frequency permits the use of smaller antennas and circuits, thereby reducing size, weight, and cost of the radar. Third, the higher atmospheric absorption of the millimeter wave signals is not a concern for short-range applications like automotive radars. However, the spray from other vehicles can impact the visibility of vehicles on the roadway at frequencies that lie in the water absorption band.

Signal Modulation: Although several different types of radar signal modulations have been evaluated over the years as possible candidates for forward-looking applications, most developers believe the frequency-modulated continuous-wave (FMCW) radar is the best overall choice. Some of the reasons for this choice of transmitted signal modulation include ease of modulation, simpler Doppler information extraction, higher accuracy for the short-range targets, and lower power rating for the active device generating the microwave signal for a given average transmitted power. In an FMCW radar, the frequency of the transmitted signal is changed with time in a prescribed manner (such as a linear ramp or a triangular variation), and the difference between the transmitted and returned signal frequencies is a measure of the range of the reflecting target.

Antenna Performance Requirements: Both the ACC and the collision anticipation radars require an antenna that meets several demanding specifications:

1. The size of the antenna should be small for cost and vehicle styling reasons.
2. The antenna beam should be narrow enough in azimuth that it can resolve objects in the driving lane from those in nearby lanes or adjacent to the roadway, as well as in elevation so as to distinguished on-road objects from overhead signs and bridges.
3. The sidelobes of the antenna should be sufficient low that small objects (like motorcycles) in the same lane are not masked by large objects (like trucks) in neighboring lanes.
4. The antenna should preferably be planar for ruggedness and ease of integration.

Choice of Antennas: Given the antenna beamwidth requirements and the limited physical dimensions permissible, some form of antenna scanning is needed for monitoring the relevant space, while at the same time maintaining the spatial discrimination. At least three choices are available for scanning. The first, and possibly the most versatile and powerful option is electronic scanning, using a phased array; with the presently available technologies, the cost of this option is prohibitive. Second, mechanically steered antennas have been developed that use reflectors. Third, synthetic aperture antenna techniques have been used that allow for sophisticated signal processing and information acquisition.

Signal Processing Needs: One of the most challenging aspects of automotive radar design, on which work presently continues, is the processing of the returned radar signals. Sophisticated signal processing algorithms have been employed to achieve many of the characteristics desired in radar performance, including the following:

1. Resolving small and large objects at different ranges and velocities.
2. Rejecting reflections from stationary objects.
3. Rejecting reflections from vehicles traveling in opposite directions.
4. Extracting target range despite road curvature.
5. Obtaining high performance despite low-cost components with low-performance.

This software must be capable of simultaneously separating and tracking a dozen or more different targets within the field of view.

The Radar Assembly: The need for large volume production and cost considerations dictate the materials, technologies, and processes usable in the radar. The complete radar can be conceptually subdivided into two parts: the RF part including the antenna, transmitter, and receiver, and the baseband part, consisting of signal processing, power supply, microprocessors, displays, cables, and the packaging and housing. The baseband part, as is typical of other automotive electronics, employs silicon chips, automated assembly methods such as flip-chip, and lightweight plastic housing. The cost of the front-end RF module at millimeter wave frequencies is usually minimized by use of a single hybrid assembly for the entire module, low-power Gunn devices as millimeter wave sources, silicon Schottky barrier diodes as mixers, and metalized injection-molded plastic waveguide cavities. More recently, as the cost of monolithic millimeter wave integrated circuit (MMIC) chips has decreased, a monolithic RF front end becomes cost effective, and can be integrated in a hybrid microstrip circuit. Chipsets for this purpose have been developed by a number of campanies.

Other Possible Types of Automotive Radars

Several other types of radar architectures and types have been proposed for automotive applications, and have reached different stages of development. Some of the promising candidates are briefly summarized here.

1. *Noise Radar.* Given the large number of automotive radars that may be simultaneously present in a given environment, the need to minimize the likelihood of false alarms due to interference between them is important. Several different types of noise (or noise-modulated) radars have been advanced for this purpose. One proposed scheme employs a transmitted signal modulated by random noise, and a correlation receiver that determines the cross-correlation between the returned signal modulation and the transmitted signal modulation, to separate the returned signal from the noise due to the system, ambient, and other radars. In still other schemes, the transmitted signal can be broadband noise, or a CW signal modulated by a binary random signal.

2. *Micropower Impulse Radar (MIR).* Another class of radar that has been proposed for several automotive uses is the so-called micropower impulse radar (MIR). Its most distinguishing characteristic is that it transmits a very short pulse of electromagnetic energy, having a duration typically on the order of 0.1 nanoseconds, and a rise-time measured in picoseconds. As a result,

its spectrum occupies a bandwidth of several GHz, creating a so-called "ultra-wideband" (UWB) system. Consequently, to avoid interference problems, the transmitted power is kept very low, on the order of a microwatt. As a further consequence of that choice, it is necessary to integrate over a large number of pulses to improve the signal-to-noise ratio of the receiver; however, the pulse repetition rate can be random, so that multiple radars will not interfere with each other due to their distinctive pulse patterns. In addition, the receiver is gated in time, so that it receives radar signal echoes only over a narrow time window. The received signal is thus limited to echoes from targets lying at a pre-selected distance from the transmitter, allowing the echoes from smaller nearby targets to be distinguished in the presence of those from large faraway objects. Another consequence of the low power transmission is long battery life, and hence the small volume, weight, and low cost. Some consequences of ultra-wideband operation are high range resolution, substantial scattering cross-section of targets at any observation angle, and signal penetration through dielectric materials.

3. *Interferometric Radar.* A radar system in which signals reflected from a target are simultaneously received by two physically separated antennas and are processed to determine the phase difference between them, can be used to estimate the target range over short distances. Such systems can employ a CW signal and a two-channel digital correlator, making them very simple, and can distinguish between approaching and receding targets. Since the range cannot be determined unambiguously from a known phase difference, the system is limited in its capability without signal processing to extract additional information from the returned signals.

4. *Cooperative Radar Systems.* Targets that modify the returned signal in known ways, e.g., by modulation or frequency translation, to allow it to be easily detected and distinguished against other signals, are called cooperative targets. Radar systems with significantly higher performance capabilities can be developed if, as a result of common industry standards or regulatory directives, the vehicles incorporate appropriate means for enhancing returned radar signal, or carry a radar-interrogable distinguishing code, much like a license plate. In such a system, the ease of vehicle tracking can greatly improve the reliability and robustness of the radar system. The principal limitation of such sensors is their inability to handle targets with or without damaged reflectors, beacons, or other cooperative mechanisms.

Future Developments

Some of the major areas of emphasis in the future development of automotive radars are as follows.

1. *Radar Chipsets and MMICs.* With the frequency allocation and markets more definite, many manufacturers have developed chipsets for automotive radar that include GaAs MMIC chips for the front-end RF assembly. The availability of volume manufacturing capability, ease of incorporating additional functionality, and the resulting cost reductions make this a promising approach.

2. *Phased-array antennas.* With presently available technologies, phased-array antennas are not viable candidates for consideration in automotive radar applications due to their high cost. When low-cost phased-array antennas become commercially available, automotive radar can be expected to undergo a significant transformation and to attain a multifunction capability.

3. *Signal Processing Capability.* Digital processing of complex returned signals makes possible the characterization of more complex vehicular environments, in which large numbers of objects can be tracked. Improved algorithms, and the ability to carry out extensive real-time processing with inexpensive processor chips, will allow the radar to serve more sophisticated functions.

4. *Human Interface Enhancement.* The willingness of a driver to relinquish partial control of the vehicle to the radar depends only partly on attaining a high reliability, robustness, and low false-alarm probability. User acceptance will depend strongly on the quality of human interface through which the information gathered by an automotive radar is presented and utilized.

References

History

1. H. P. Groll and J. Detlefsen, History of automotive anticollision radars, *IEEE Aerospace and Electronic Systems Magazine*, 12, 8, 15–19, August 1997.
2. D. M. Grimes and T. O. Jones, Automobile radar: A brief review, *Proc. IEEE*, 62, 6, 804–822, June 1974.
3. M. S. Gupta, et al., Noise considerations in self-mixing IMPATT-diode oscillators for short-range Doppler radar applications, *IEEE Trans. Microwave Theory & Techniques*, MTT-22, 1, 37–43, January 1974.

Speed Measuring Radar

1. P. Heide, et al., A high performance multisensor system for precise vehicle ground speed measurement, *Microwave Journal*, 7, 22–34, July 1996.
2. P. Descamps, et al., Microwave Doppler sensors for terrestrial transportation applications, *IEEE Trans. Vehicular Technology*, 46, 1, 220–228, February 1997.
3. P. M. Schumacher, Signal processing enhances Doppler radar performance, *Microwaves & RF*, 30, 7, 79–86, June 1991.

Obstacle-Detection Radar

1. J. C. Reed, Side zone automotive radar, *IEEE Aerospace and Electronic Systems Magazine*, 13, 6, 3–7, June 1998.

Adaptive Cruise Control Radar

1. P. L. Lowbridge, Low cost millimeter-wave radar systems for intelligent vehicle cruise control applications, *Microwave Journal*, 38, 10, 20–33, October 1995.
2. L. H. Eriksson and S. Broden, High performance automotive radar, *Microwave Journal*, 39, 10, 24–38, October 1996.
3. M. E. Russel, et al., Millimeter wave radar sensor for automotive intelligent cruise control (ICC), *IEEE Trans. Microwave Theory & Techniques*, 45, 12, 2444–2453, December 1997.
4. A. G. Stove, Automobile radar, *Applied Microwaves*, 5, 2, 102–115, Spring 1993.

Collision Anticipation Radar

1. C. D. Wang and S. Halajian, Processing methods enhance collision-warning systems, *Microwaves & RF*, 36, 3, 72–82, March 1997.

RF Front-end for Forward-Looking Radars

1. W. H. Haydl, Single-chip coplanar 94-GHz FMCW radar sensors, *IEEE Microwave and Guided Wave Letters*, 9, 2, 73–75, February 1999.
2. D. D. Li, S. C. Luo, and R. M. Knox, Millimeter-wave FMCW radar transceiver/antenna for automotive applications, *Applied Microwaves and Wireless*, 11, 6, 58–68, June 1999.

Other Possible Types of Automotive Radars

1. Azevedo, S. and McEwan, T. E., Micropower impulse radar: A new pocket-sized radar that operates up to several years on AA batteries, *IEEE Potentials*, 16, 2, 15–20, April–May 1997.
2. G. Heftman, Macroapplications seen for micro radar, *Microwaves & RF*, 39, 5, 39–40, May 2000.
3. B. M. Horton, Noise modulated distance measuring systems, *Proc. IRE*, 47, 5, 821–828, May 1959.
4. I. P. Theron, et al., Ultrawide-band noise radar in the VHF/UHF band, *IEEE Transactions on Antennas and Propagation*, 47, 6, 1080–1084, 1999.
5. A. Benlarbi and Y. Leroy, A novel short-range anticollision radar, *Microwave and Optical Technology Letters*, 7, 11, 519–521, August 5, 1994.
6. F. Sterzer, Electronic license plate for motor vehicles, *RCA Review*, 35, 2, 167–175, June 1974.

Future Developments

1. Mueller, Jan-Erik GaAs HEMT MMIC chip set for automotive radar systems fabricated by optical stepper lithography, *IEEE Journal of Solid-State Circuits*, 32, 9, 1342–1349, September 1997.
2. Verweyen, L. Coplanar transceiver MMIC for 77 GHz automotive applications based on a nonlinear design approach, *1998 IEEE Radio Frequency Integrated Circuits Symposium Digest*, Baltimore, MD, June 1998, 33–36.
3. T. Shimura, et al., 76 GHz Flip-Chip MMICs for automotive radars, *1998 IEEE Radio Frequency Integrated Circuits (RFIC) Symposium Digest*, Baltimore, MD, June 1998, 25–28.

2.14   New Frontiers for RF/Microwaves in Therapeutic Medicine

Arye Rosen, Harel D. Rosen, and Stuart D. Edwards

The use of RF/microwaves in therapeutic medicine has increased dramatically in the last few years. RF and microwave therapies for cancer in humans are well documented, and are presently used in many cancer centers. RF treatments for supraventricular arrhythmias, and more recently for ventricular tachycardia (VT) are currently employed by major hospitals. RF/microwaves are also used in human subjects for the treatment of benign prostatic hyperplasia (BPH), and have gained international approval, including approval by the United States Food and Drug Administration (FDA). In the last two years, several otolaryngological centers in the United States have been utilizing RF to treat upper airway obstruction and alleviate sleep apnea. Despite these advances, considerable efforts are being expended on the improvement of such medical device technology. Furthermore, new modalities such as microwave-aided liposuction, tissue anastomoses in conjunction with microwave irradiation in future endoscopic surgery, RF/microwaves for the enhancement of drug absorption, and microwave septic wound treatment are continually being researched.

RF/Microwave Interaction with Biological Tissue

Definitions: In this chapter, we detail two types of thermal therapies: RF and microwaves. We define RF as frequencies in the range between hundreds of KHz to a few MHz, and microwaves as those in the range of hundreds of MHz to about 10 GHz.

RF Energy[1-7]

The history of the effect of RF current on tissue began in the 1920s, with the work of W.T. Bovie who studied the use of RF for cutting and coagulating. The first practical and commercially available RF lesion generators for neurosurgery were built in the early 1950s by S. Aranow and B.J. Cosman at around 1 MHz.[1,2] The controlled brain lesions made had smooth borders, an immediate improvement over those obtained with DC. As far as the choice of frequency, Alberts et al. have shown that frequencies of up to 250 KHz have stimulating effects on the brain. Thus, RF above 250 KHz was indicated.[4-7] The RF generator is a source of RF voltage between two electrodes. When the generator is connected to the tissue to be ablated, current will flow through the tissue between the active and dispersive electrodes. The active electrode is connected to the tissue volume where the ablation is to be made, and the dispersive electrode is a large-area electrode forcing a reduction in current density in order to prevent tissue heating. The total RF current, I_{RF} is a function of the applied voltage between the electrodes connected to the tissue and the tissue conductance. The heating distribution is a function of the current density. The greatest heating takes place in regions of the highest current density, J. The mechanism for tissue heating in the RF range of hundreds of KHz is primarily ionic. The electrical field produces a driving force on the ions in the tissue electrolytes, causing the ions to vibrate at the frequency of operation. The current density is $J = \sigma E$, where σ is the tissue conductivity. The ionic motion and friction heats the tissue, with a heating power per unit volume equal to J^2/σ. The equilibrium temperature distribution, as a function of distance from the electrode tip, is related to the power deposition, the thermal conductivity of the target tissue, and the heat sink, which is a function of blood circulation. The lesion size is, in turn, a function of the volume temperature. Many theoretical models to determine tissue ablation volume as a function of tissue type are available, but none is as good as actual data.

Microwave Energy

The need for accurate data on permittivity at microwaves and millimeter waves has long been recognized[8] and since 1980 many papers have appeared giving fairly extensive coverage of data up to 18 GHz.[9-13] The most recent tabulations of complex permittivity of various biological tissues are reported by Duck[14] and Gabriel et al.[15] Many of the applications and recent advances in the knowledge of the dielectric properties of tissues have been reviewed in the literature.[9,12,16,17] Since 1950, efforts have been directed toward characterization of a variety of tissues at microwave frequencies. Among many reported works in the literature Gabriel et al.[18] provide detailed measurements of a variety of tissues up to 20 GHz; they also fit the measured results to a Cole-Cole model with multiple relaxation time constants.[19] Knowledge of the dielectric properties of biological tissues and the basic physical properties of water in tissue at microwave frequencies is essential in order to predict the interaction between field and tissue, which in turn, provides the basis for some of the thermal applications of microwaves described in this chapter.

For sinusoidal fields of frequency f, the permittivity and conductivity are conveniently represented by a single parameter, the complex permittivity ε^*,

$$\varepsilon^* = \varepsilon' - j\varepsilon'' \tag{2.14}$$

where

$$\varepsilon'' = \frac{\sigma}{w\varepsilon_0} \tag{2.15}$$

and

$$w = 2\pi f \tag{2.16}$$

The real part ε' is referred to as the "relative permittivity" and is related to the energy storage. The imaginary part ε'' is called the "dielectric loss," and corresponds to the power absorption in terms of electromagnetic field interaction with matter; the former influences the phase of the transmitter wave, whereas the latter impacts its amplitude.

When a constant voltage is suddenly impressed on a system initially at equilibrium, for simple systems, the resulting response is usually found to be an exponential function of time. This response, for example, may be the charge buildup at an interface between two different dielectrics, or the alignment of dipoles with an applied electric field. When the voltage is removed, the system relaxes exponentially to its original state. In the general case of an alternating voltage of field, it can be shown[20,21,22] that the complex permittivity of such simple systems varies with frequency, and can be expressed in the form,

$$\varepsilon^*(w) = \varepsilon_\infty + \frac{\varepsilon_S - \varepsilon_\infty}{1 + j\omega\tau} \tag{2.17}$$

where τ is the time constant of the exponential relaxation process, ε_∞ is the permittivity at $\omega \ll 1/\tau$, and ε_s is the permittivity at $\omega \gg 1/\tau$. This is the Debye dispersion equation. The characteristic frequency f_c is defined as,

$$f_c = 1/2\pi\tau \tag{2.18}$$

By separating the Debye equation into real and imaginary parts,

$$\varepsilon' = \varepsilon_\infty + \frac{\varepsilon_s - \varepsilon_\infty}{1 + \left(\dfrac{f}{f_c}\right)^2} \tag{2.19}$$

$$\varepsilon'' = \frac{\left(\varepsilon_s - \varepsilon_\infty\right)f/f_c}{1 + \left(\dfrac{f}{f_c}\right)^2} \tag{2.20}$$

or

$$\sigma_d = \frac{\left(\varepsilon_s - \varepsilon_\infty\right)2\pi\varepsilon_0 f^2}{f_c\left(1 + \left(\dfrac{f}{f_c}\right)^2\right)} \tag{2.21}$$

where the subscript d denotes the conductivity due to a Debye relaxation process. The measured conductivity, σ_m, will be higher if there are other loss mechanisms in the material, i.e.,

$$\sigma_m = \sigma_d + \sigma_s \tag{2.22}$$

where σ_s may be the conductivity due to ions and any other contributions at frequencies well below f_c.

If for $f \gg f_c$ we denote σ_d by σ_∞, then

$$\left(\sigma_\infty - \sigma_s\right) = 2\pi f_c\left(\varepsilon_s - \varepsilon_\infty\right)\varepsilon_0 \tag{2.23}$$

which states that the total change in conductivity is proportional to the total change in permittivity and to the characteristic frequency.

The dielectric properties of most materials are not exactly described by the Debye equation. The dielectric properties can then often be approximated empirically by the Cole-Cole equation[19,22]

$$\varepsilon^* = \varepsilon_\infty + \frac{\varepsilon_s - \varepsilon_\infty}{\left(1 + jw\tau\right)^{1-\alpha}} \tag{2.24}$$

It can be shown that this equation is valid for a distribution of relaxation times about a mean value. The Cole-Cole parameter α ranges from 0 to 1 and is an indication of the spread of relaxation times (for $\alpha = 0$, the Cole-Cole equation reduces to the Debye equation).

Depth of penetration (δ) of energy into tissue is defined as the distance in which the power density of a plane wave is reduced by the factor e^{-2}, which numerically is 0.135. Since the power density diminishes as e^{-aL} as the microwave energy travels into the lossy material, the attenuation factor (a) is inversely related to the depth of penetration: (a = 1/δ). The generally accepted value for the depth of penetration of 2450 MHz energy into muscle tissue is 17 mm, and the corresponding attenuation constant is 0.118 for distances in millimeters. For a non-expanding plane wavefront, the reduction of power density as it penetrates muscle tissue is shown in Fig. 2.126[71] (Table 2.28[70,74]). Figure 2.126 shows, for example, that 11% of the total input power is absorbed in the first millimeter of penetration, and that a total of 21% of the input is converted to heat in the first two millimeters of the tissue. Microwave antennas that are

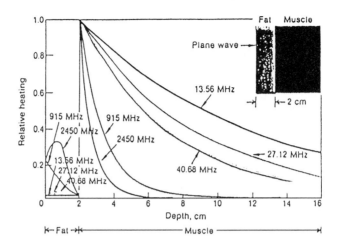

FIGURE 2.126 Calculated relative heating in fat and muscle as a function of distance for five frequencies. (After Paglione.[74])

TABLE 2.28 Relative Permittivity and Conductivity of Biological Media at Microwave Frequencies[70,74]

Frequency (MHz)	Wavelength (cm)	High Water Content Media		Low Water Content Media	
		c	σ (S/m)	c	σ (mS/m)
10	3,000	160	0.625	—	—
100	300	71.7	0.889	7.5	19.1–75.9
300	100	54	1.37	5.7	31.6–107
915	32.8	51	1.60	5.6	55.6–147
2,450	12.2	47	2.21	5.5	96.4–213
3,000	10	46	2.26	5.5	110–234
5,000	6	44	3.92	5.5	162–309
10,000	3	39.9	10.3	4.5	324–549

utilized in therapeutic medicine, however, are in the near field. The penetration depth is considerably lower depending on the antenna type).[23]

Previous studies[24] have shown that heat-induced damage to biological tissue is dependent on both the temperature and its duration. The temperature threshold for damage rises as the duration of exposure is shortened.

Test Fixture Structures for Biological Tissue Characterization[25]

A number of techniques have been established for microwave characterization of biological tissues. These techniques are subdivided in TEM transmission lines (i.e., coaxial lines) and non-TEM structures. The majority of published work deals with coaxial transmission lines either as a dielectric loaded[26,27] or an open-ended[28] coaxial line. In both approaches, the change in the terminating impedance causes change in the input reflection coefficient of the line. Different complex permittivity of the tissue under test causes change in the capacitance and conductance of termination, hence impacting amplitude and phase of the reflected wave. This technique is popular and is extensively developed by Burdette et al.[28] and Stuchly et al.[29]

Dielectric loaded waveguide structures such as circular or rectangular cross-section metallic waveguides are the second most popular structures for tissue characterization. Steel et al.[30,31] reported a technique for characterization of liquids and solids. For liquids measurements the sample is contained in a length of waveguide and a moving short circuit enables the liquid thickness to be varied. A microwave signal is applied to the sample, the modulus of the reflected signal is recorded as a function of sample length, and a least-squares curve-fitting analysis of data enables various parameters to be obtained. For measuring solid samples, an automated slotted line is used to record the standing wave ratio in front of the sample, the latter terminated by a short circuit.

The above techniques all use the reflecting wave from a terminating impedance and the transmission line theory to extract the electrical parameters of the tissues under test. On the other hand, methods exist that use tissue samples to change the resonance frequency and quality factor of a cavity resonator.[32] More accurate results can be achieved by using simpler setups. However, the resonance methods are only applicable to discrete frequency points corresponding to the resonance frequencies of the modes of interest. For instance, Land and Campbell[32] present a technique that uses simple formulas that relate the complex permittivity of a small piece of tissue to the change in resonance frequency and quality factor of a cylindrical cavity. The cavity is filled with PTFE and resonates for TM_{010} mode and has a diameter of 50.8 mm and a length of 7.6 mm. Three 1.5 mm diameter sample holes are provided in the cavity. The cavity resonates at 3.2 Ghz. The microwave setup is very simple, i.e., a signal generator, frequency meter, directional coupler, attenuator, diode detector, and a voltmeter are used in the experiment.

Tissue Characterization through Reflection Measurements[25]

In order to characterize various biological tissues, three distinctive methods have been reported. The first method is based on voltage standing wave ratio measurements using slotted line waveguides.[31] The second approach is based on an impedance analyzer and is suitable for microwave frequencies.

Finally, the most popular approach in the last 20 years has been the use of a network analyzer.[28,33-38] With the advent of the automatic network analyzer (ANA) (e.g., Hewlett-Packard's HP8510, HP8753, HP8720), accurate measurements of scattering parameters have been extended to frequencies beyond 10 GHz (since 1984). The advantage of this technique is fast and accurate measurements of coaxial-based structures. The majority of attempts to accurately characterize complex permittivity of biological tissues in the last 10 years are based on these families of ANA.[18,38,39]

Microwave Antenna in Therapeutic Medicine: Issues

Biomedical antenna designs have typically addressed the applications of microwave hyperthermia for the treatment of malignant tumors, microwave catheter ablation for the treatment of cardiac arrhythmia, microwave balloon angioplasty, or microwave-assisted liposuction. The analytical basis for much of this work has been based on the lossy transmission line analysis developed by King et al.,[40,41] which allowed the calculation of input impedance and near fields for simple antenna geometries. Several researchers have refined this work to provide improved accuracy or wider applicability. For example, Iskander and Tumeh developed an iterative approach to designing multi-sectional antennas based on an improved King method[42] and have used this approach to compare different antennas.[43] Debicki and Astrahan developed correction factors to allow accurate modeling of the input impedance for electrically small multi-section antennas,[44] and Su and Wu have refined the King approach to determine the input impedance characteristics of coaxial slot antennas.[45] Casey and Bansal used a different approach than King to compute near fields of an insulated antenna using direct numerical integration of a surface integral.[46]

A problem inherent in many biomedical antenna designs is the effect of heating along the transmission line due to current flow on the outer conductor of the coaxial transmission line. In most applications this effect is undesirable since thermal energy is being delivered to healthy tissue outside the intended treatment area. Moreover, magnitude of this effect is a strong function of the insertion depth of the antenna. Similarly, the antenna input impedance also varies with insertion depth. Hurter et al.[67] proposed the use of a sleeve Balun to present a high impedance to the current on the outer conductor, thus concentrating the microwave energy at the antenna tip. Temperature profile measurements made in a phantom using a fiber-optic temperature probe clearly show improved localization of thermal energy delivery.

RF/Microwaves in Therapeutic Medicine

RF/Microwave Ablation for the Treatment of Cardiac Arrhythmias[47,48]

Cardiac arrhythmias can result from a variety of clinical conditions, but at their root is an abnormal focus, or pathway, of electrical activity. Abnormal sources of electrical activity most commonly occur at

TABLE 2.29   Energy Sources for Catheter Ablation[47]

	Direct Current	Radiofrequency	Microwave
Waveform	Monophasic, damped sinusoidal	Continuous unmodulated sinusoidal	N/A[a]
Frequency	DC	550–750 kHz	915, 2450 MHz
Voltage V	2000–3000 V	<100 V	N/A
Mechanism of injury	Passive heating, baro trauma, electric field effects	Resistive heating	Radiant heating
Sparking, barotrauma	Yes	No	No
General anesthesia	Yes	No	No
Lesion size	Moderate	Small	Unknown
Control of injury	Low	High	High

[a] N/A = data not available.

or above the AV-node, and are thus deemed supraventricular tachy-arrhythmias. Alternatively, abnormal ventricular foci cause ventricular tachycardia. The presence of abnormal conduction pathways can also result in an uncontrolled cycling of electrical activity resulting from retrograde signal conduction through the myocardium (reentry tachyarrhythmias). Reentry can occur within the AV-node (AVNRT), or via accessory conduction pathways (AP). Regardless of the specific etiology, once the source of the arrhythmia has been identified, destruction of the abnormal cardiac tissue is curative. The goal of ablation is to modify the electrical system of the heart by converting electrically active cardiac tissue to electrically inactive scar tissue. The scar or lesion that forms then blocks the focus or accessory pathway and prevents the tachycardia. Various energy forms have been used to create such localized tissue injury, including direct current (DC), radiofrequency (RF), and microwave energy (Table 2.29).[47,48]

The clinical use of DC ablation dates back to 1982. An electrode catheter is placed at the desired location, and a DC shock is applied. Although complete ablation occurs in up to 65% of patients, DC ablation is fraught with complications. Hypotension, perforation, cardiac tamponade, embolization, pericarditis, and ventricular tachyarrhythmias have been reported in as many as 10% of patients. Mortality associated with DC ablation may be as high as 5% in some patient groups. RF ablation was developed with the hope of decreasing the risks associated with DC application. In RF ablation, lesion formation results from resistive tissue heating at the point of contact with the RF electrode (Fig. 2.127). This heating is thought to lead to coagulation necrosis and permanent tissue damage. If there is poor tissue contact, RF current can not be coupled to the underlying tissue, and the desired effect of tissue heating is lost. Overall success rates for RF ablation have been reported to be as high as 90% for AV junction ablation, and as high as 95% when applied to re-entry-mediated tachycardia. Furthermore, RF ablation has not been reported to result in serious side effects.

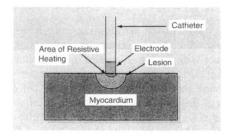

FIGURE 2.127   Mechanism of RF ablation. When RF current is delivered to the tip of a catheter electrode, resistive heating occurs along a small rim of tissue in direct contact with the electrode. A lesion is created as heat conducts passively away from this zone and the surrounding myocardium is heated to a temperature where cell death occurs (~50°C). Lesion size is therefore a function of the size of the electrode and the resulting temperature at the electrode-tissue interface.

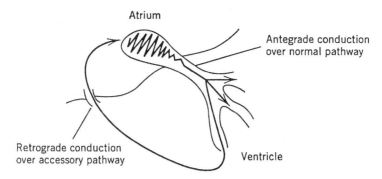

FIGURE 2.128 An arrhythmic circuit associated with the Wolff-Parkinson White syndrome. In this syndrome, there is a connection between the atrium and ventricle outside the normal V nodal pathway (accessory pathway). A tachycardia circuit can develop if an impulse conducts antegrade (forward) via the normal V node pathway and is able to conduct retrograde (reverse) from the ventricle to the atrium via the accessory pathway. Catheter ablation successfully treats these arrhythmias because it interrupts accessory pathway conduction without interfering with normal AV nodal conduction.

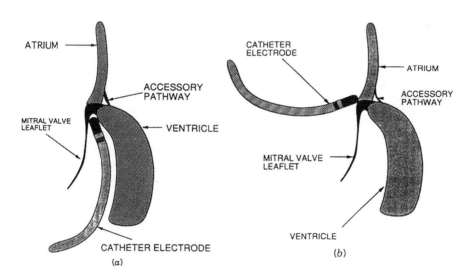

FIGURE 2.129 Diagrams of electrode positions used in RF catheter ablation of accessory pathways: (a) for the ventricular approach a catheter is passed retrograde across the aortic valve and positioned under the mitral leaflet; (b) for the atrial approach a catheter is passed across the interatrial septum (trans-septal catheterization) and positioned on top of the mitral valve leaflet. Electrical mapping confirms the site of the accessory pathway prior to the delivery of RF energy.

The first supraventricular tachycardias targeted for RF ablation were those associated with the Wolff-Parkinson-White syndrome. In this condition the anatomic basis for supraventricular tachycardias is an accessory connection or pathway that connects the atrium and ventricle outside the normal AV conduction pathway (Fig. 2.128). These accessory pathways cross between the atrium and the ventricle at the level of the mitral and tricuspid anulus. RF energy delivered to the mitral or tricuspid anulus either from the ventricular or atrial aspect can ablate these pathways (Fig. 2.129). Success rates of over 90% have been reported using either approach.

Recently, encouraging results in the treatment of ventricular arrhythmias occurring as a consequence of diffuse processes such as myocardial ischemia or infarction have been published. The search for ablation modalities capable of safely generating even larger lesions has spawned an interest in microwave ablation

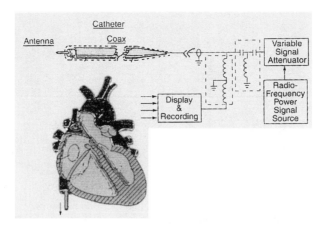

FIGURE 2.130   Microwave system used for myocardial tissue ablation.

(Fig. 2.130). Unlike DC and RF techniques, which generate lesions of relatively limited size and penetration, microwave energy might allow for greater tissue penetration, and thus a greater volume of heating. Microwave ablation systems are currently being developed.

Microwave hyperthermia has been useful in radiation oncology for the treatment of various solid tumors.[49,50,51] The cardiac applications of this modality have only recently been explored. Microwave energy using either 915 or 2450 MHz has been studied in an attempt to enlarge myocardial lesions in catheter ablation.[52,53,54] Microwave energy is delivered down the length of a coaxial cable that terminates in an antenna capable of radiating the energy into tissue. Radiant energy will cause the water molecules in myocardial tissue to oscillate, producing tissue heating and cell death. The higher frequency of microwave energy allows for greater tissue penetration and theoretically a greater volume of heating than that possible with RF, which produces direct ohmic or resistive heating.

Wonnell and coworkers studied the effects of microwave energy for cardiac ablation using a helical antenna mounted on a coaxial cable (2.44 mm o.d.).[55] High-frequency current at 2450 MHz was delivered via the helical antenna into a tissue-equivalent phantom model. The temperature distribution profile was measured around the antenna as well as into surrounding volume (the depth of penetration). The volume of heating for the microwave catheter system was 11 times greater than that of an RF electrode catheter at the same surface temperature. In addition, the microwave catheter penetrated an area that was twice as large as that penetrated by the RF catheter. These data suggest that microwave energy will produce larger lesions than RF because a greater volume of tissue is being heated. An additional theoretical advantage of the microwave system is that direct tissue contact is not crucial for tissue heating since heating occurs via radiation, and not via direct ohmic heating as seen with RF. Using this system, preliminary studies in six animals demonstrated that complete heart block could be achieved in all six animals by directing microwave energy (50 W at 2450 GHz for about 200 seconds) to the atrioventricular junction.

We evaluated helical and whip antenna designs in a tissue-equivalent phantom at 915 MHz and 2450 MHz utilizing a coaxial cable (0.06 in o.d.).[56] All catheters were measured utilizing a network analyzer prior to placing them in the phantom model. Such analysis demonstrated the great variability in tuning of these microwave catheters. Microwave ablation catheters have suffered from imperfect tuning leading to inefficient radiation of energy. Consequently, there is generation of heat along the length of the catheter rather than radiation of energy into tissue. Little heating into tissue was observed in poorly tuned catheters. Such analysis underscores the critical importance of proper tuning of microwave catheters prior to any further studies.

A perfusion chamber containing a muscle-equivalent phantom was constructed and placed in a saline bath held at 37°C. The muscle-equivalent phantom consisted of TX150, polyethylene powder, NaCl, and water. Ablation catheters were placed on the surface of the phantom material. Temperature measurements

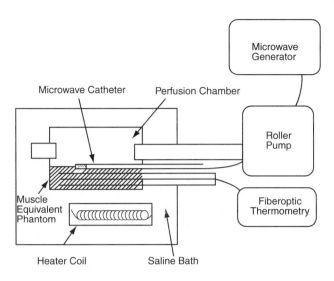

FIGURE 2.131 Flow-phantom model for cardiac ablation catheters.

were performed using a 12-channel Luxtron fiber-optic thermometry system. Probes were placed beneath the surface of the phantom. Saline at a constant temperature of 37°C was infused at a flow rate of 4 L/min across the surface of the phantom. This model simulates the heart where the phantom material has the dielectric properties of cardiac muscle and the saline properties of blood (Fig. 2.131).

Temperature curves were plotted from probes placed 1, 2.5, 5, and 7.5 mm from the point of maximal heating on the microwave catheter. Thermal profiling of these catheters demonstrated volume heating. Heating was proportional to power duration and to surface temperature. In addition to the volume heating, conductive heating was also present as a result of the increased temperature at the catheter-phantom interface.[56]

In vivo ablation using microwaves was performed on canine left ventricular myocardiums. A power of 80 W was delivered for a total of 5 min. Mean lesion size measured 435 ± 236 mm³, which was similar in size to lesions created with small-tipped RF catheters. The microwave ablation catheters, as presently designed, were not capable of producing lesions larger than those produced by RF catheters.[56]

Practical problems remain to be solved before microwaves become a useful clinical energy source. These problems include (1) power loss in the coaxial cable, (2) resultant heating of the coaxial cable during power delivery that has led to a breakdown in the dielectric and catheter material, (3) inefficiency of the radiating antenna, and (4) lack of a unidirectional antenna that can radiate energy into tissue and not the circulating blood pool. At the present time microwave catheter systems are poorly efficient radiators of energy into cardiac tissue. These obstacles will have to be overcome before microwaves supplant radio frequency as the preferred energy source for cardiac ablation.

RF/Microwave Treatment of BPH[57]

Benign prostatic hypertrophy (BPH) is an enlargement of the prostate gland which can lead to compression of the urethra, and thus cause urinary tract obstruction. The prostate gland is an organ at the base of the male bladder that surrounds the urethra and produces seminal fluid. Overgrowth of prostatic tissue leads to compression of the urethra. BPH is among the most common medical conditions affecting men over the age of 50. In fact, over 50% of men over 50 years of age have enlarged prostates. Symptoms of urinary tract obstruction (frequent urination, decreased urine flow, nocturia, dribbling, discomfort, pain) most commonly begin at 65 to 70 years of age.

Although drug therapy may be effective for patients with early stages of BPH, many men will need invasive intervention for relief of symptoms. Surgical excision of prostatic tissue has been the standard care for more advanced forms of BPH. Procedures such as prostatectomy and transurethral resection of the prostate, however, carry significant risks. To minimize hazards such as hemorrhage, coagulopathies,

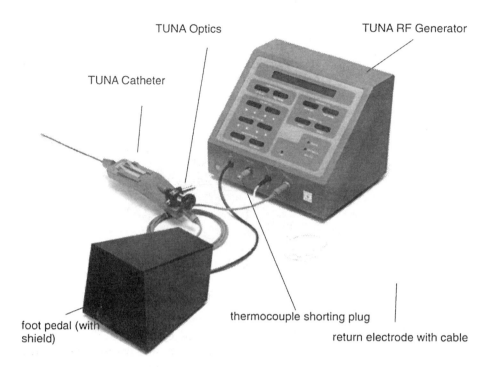

FIGURE 2.132   TUNA RF generator unit (with permission of VidaMed, Inc.).

pulmonary emboli, bladder perforation, incontinence, infection, urethral stricture, retention of prostatic chips, infertility, and retrograde ejaculation, minimally invasive alternatives have been developed and are being investigated. Transurethral RF and microwave procedures are becoming promising alternatives to surgical intervention. The goal of therapy is to decrease the volume of prostatic tissue. RF Transurethral Needle Ablation (TUNA) (Fig. 2.132) involves the introduction of interstitial needle electrodes directly into prostatic tissue. This technology uses RF (460 KHz) with excellent control of the RF thermal energy applied to the tissue. The TUNA catheter used is 24.1 cm long and 21 French. Through the tip of the catheter, two needles (electrodes) oriented 40° apart, can be deployed. The electrode-needles are shaped to facilitate passage through tissue. They are thin, and thus can be directed from the catheter through intervening tissue with a minimum of trauma to normal tissue. Each electrode-needle is enclosed within a longitudinally adjustable sleeve acting as a shield to prevent exposure of the tissue adjacent to the sleeve to the RF current, thus preserving the urethra by reducing the possibility of a rise in its temperature. The sleeve is also used to control the tissue interface, and therefore the ablation volume. Both the electrode-needle and the sleeve are locked into position. The TUNA catheter needle acts as the thermal electrode, and a grounding pad that is placed in back of the subject under treatment closes the RF circuit to the power supply (Fig. 2.133).

Thermocouples are located at the shield tip below each needle, and at the catheter bullet head (in order to record ablation temperature and prostatic urethral temperature), respectively. The RF unit (VidaMed, Inc.) includes an RF generator with the following readouts: RF power level, ablation time, impedance, and six thermocouple readouts (Fig. 2.132). The TUNA catheter (Fig. 2.133a) includes direct fiber-optic vision, as well as provisions for introducing electrode-needles at various angles (Fig. 2.133b).

Transurethral Microwave Thermotherapy (TUMT) (Fig. 2.134a, b, c) has also shown promise as a therapeutic modality for the treatment of BPH. This technique uses a microwave delivery system housed within a transurethral catheter. Its goal is to selectively destroy prostatic tissue without damaging the urethral mucosa or structures surrounding the treatment area. At microwave frequencies, temperatures in the target tissue can be raised to as high as 45 to 70°C without damaging periprostatic tissue. TUMT is used routinely worldwide.

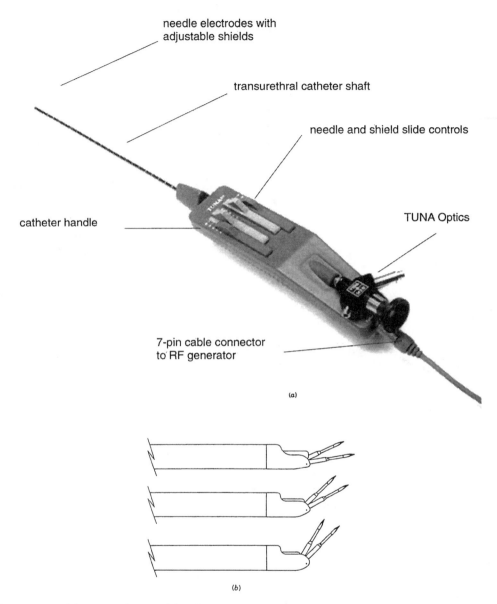

needle electrodes with
adjustable shields

transurethral catheter shaft

needle and shield slide controls

TUNA Optics

catheter handle

7-pin cable connector
to RF generator

(a)

(b)

FIGURE 2.133 (a) TUNA catheter with handle incorporating direct fiber-optic vision (with permission of VidaMed, Inc.). (b) Electrodes and needles at various angles (with permission of VidaMed, Inc.).

Microwave Balloon Catheter Techniques

Microwave Balloon Angioplasty[58,59]

Atherosclerosis, with its resultant occlusion of coronary blood flow, remains a leading cause of morbidity and mortality. For many patients with advanced disease, or in whom pharmacologic management has failed, percutaneous transluminal balloon angioplasty (PTCA) has offered an effective alternative to coronary bypass surgery. The efficacy of PTCA, however, has been limited by restenosis rates ranging from 17 to 47%, as well as by a risk of arterial dissection and/or thrombus formation. Furthermore, acute occlusion, resulting from elastic recoil at the angioplasty site, can occur in as many as 5% of patients undergoing PTCA. Such patients require emergency heart surgery. Microwave Balloon Angioplasty (MBA), the first microwave application in cardiology, was developed with the ultimate goal of decreasing both acute and long-term restenosis risks.

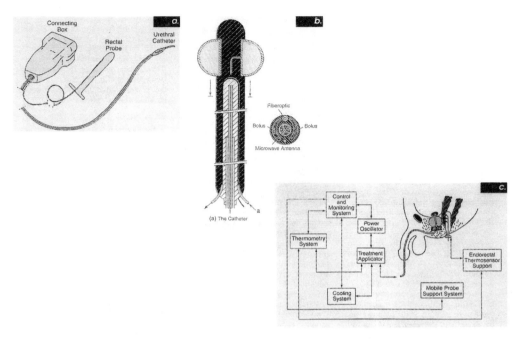

FIGURE 2.134   (a) Schematic representation of the treatment catheter; (b) cutaway view; (c) Prostatron treatment functional diagram. (With permission from Technemed Medical Systems.)

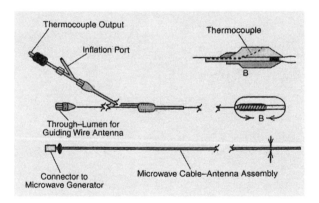

FIGURE 2.135   Microwave balloon angioplasty system.

MBA, like PTCA, employs a balloon catheter that is advanced to the site of arterial stenosis. While PTCA uses only the pressure generated by balloon inflation to dilate the affected artery, MBA takes advantage of the volume heating properties of microwave emitters. In MBA, a microwave cable-antenna assembly is threaded through the catheter, with the antenna centered in the balloon portion of the catheter (Fig. 2.135). By heating the tissue as the balloon is inflated, it was hoped that a patent vessel would be created that would be resistant to both acute and chronic reocclusion. Early *in vivo* studies, at 2.45 GHz, were conducted to assess the effects of various energy levels upon normal and atherosclerotic rabbit iliac arteries. Research on the therapeutic potential was subsequently conducted on atherosclerotic rabbit iliac arteries using microwave energy to raise the balloon surface temperature to 70 to 85°C. When compared to simultaneously performed conventional angioplasty, MBA at 85°C produced significantly wider luminal diameters, both immediately after angioplasty and 4 weeks after the procedure (Fig. 2.136). Further work, utilizing mongrel dogs with thrombin-induced coronary occlusion, has demonstrated the feasibility of MBA as a treatment modality for coronary thrombosis. MBA of such coronary thrombi in dogs resulted

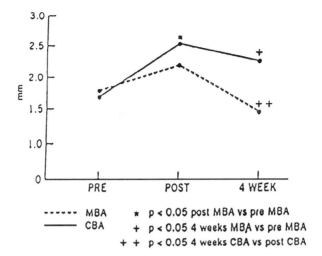

FIGURE 2.136   Microwave thermal angioplasty (85°C).

in patent vasculature with the added benefit of an organized and stabilized thrombus. Although the technique described was successful in animal studies, it has not yet found its way into clinical use. However, microwave balloon angioplasty has recently been suggested for applications in carotid stenosis and occlusions in peripheral circulation.

Microwave Balloon Catheters in the Treatment of Benign Prostatic Hypertrophy[60]
Localized microwave hyperthermia has been used for more than a decade to treat cancer of the prostate and since 1985 to treat BPH. The initial hyperthermia treatments used microwave applicators that heated the prostate via the rectum, but today transurethral applicators are favored. Transurethral applicators are usually placed inside liquid-cooled catheters and the temperatures produced inside the treated prostate can be measured noninvasively with a microwave radiometer.

With balloon catheters it is possible to produce both high therapeutic temperatures throughout the prostate gland without causing tissue burning, and biological stents in the urethra in a single treatment session. Compared to conventional microwave catheters, the distances microwaves have to travel through the prostate to reach the outer surface of the gland are reduced by the use of balloon catheters, as is the radial spreading of the microwave energy (Fig. 2.137). Furthermore, compression of the gland tissues reduces blood flow and its cooling effect within the gland. Also, since catheter balloons make excellent contact with the urethra, much better than do conventional catheters, the urethra is well cooled by the cooling liquid within the balloon, and is therefore well protected from thermal damage.

Microwave Balloon Catheters in the Treatment of Cancer[60]
Interstitial hyperthermia is usually combined with radiation therapy using radioactive seeds that are inserted into the tumor via the same tubing that is used for the hyperthermia (Fig. 2.138). A typical treatment sequence is brachytherapy (irradiation of the tumor with the seeds) followed by hyperthermia, followed again by brachytherapy. The interstitial hyperthermia enhances the efficacy of brachytherapy because (1) hyperthermia interferes with the repair of cells that have been sublethally damaged by the ionizing radiation, (2) cells in the S phase of the cell cycle and hypoxic tumor cells tend to be resistant to ionizing radiation but sensitive to heat, and (3) hyperthermia can be effective in oxygenating radiation-resistant hypoxic cells.

Intersitial arrays using conventional applicators are useful only for treating small tumors, because each interstitial applicator can heat and irradiate only a small volume of tissue, and the number of applicators that can be inserted into a tumor is limited because of their invasive nature. Interstitial applicators using balloons, on the other hand, can heat much larger volumes of tissues than can conventional interstitial applicators, making it possible to treat larger tumors. A catheter with a deflated balloon at its tip is

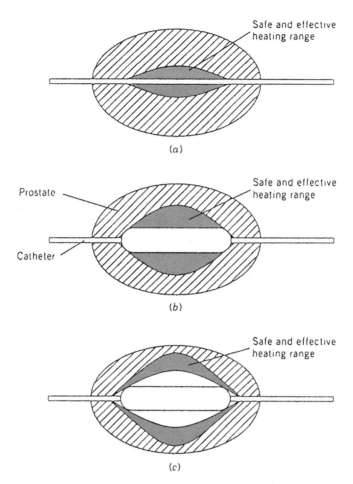

FIGURE 2.137 Safe and effective heating ranges in prostate glands with regular microwave catheters and with microwave balloon catheters. Microwave heating patterns: (a) with regular catheter; (b) with balloon catheter; (c) with balloon catheter and water cooling. (After Sterzer.[60])

inserted into the tumor volume to be heated, or where applicable, into a natural opening of the body such as the urethra, rectum, or vagina; the balloon is inflated, and radioactive seeds or a microwave antenna are inserted through the center lumen of the balloon catheter (Fig. 2.139).

RF in the Treatment of Obstructive Sleep Apnea[61,62,63]

Obstructive Sleep Apnea (OSA) is a disorder diagnosed when an individual's upper airway becomes intermittently blocked during sleep and breathing becomes interrupted. Approximately 20 million Americans are estimated to suffer from OSA, and over half of these are between the ages of 30 and 60 years. During sleep, there is a relaxation of the structures surrounding the pharynx/throat. Breathing becomes interrupted (apnea) when these anatomical structures relax in a position that occludes airflow. The most commonly involved structures include the soft palate, the base of the tongue, and the tonsils/adenoids. Enlarged turbinates within the nose can serve to further impede airflow.

OSA and its resultant interruption of normal sleep patterns have a wide range of clinical effects. Patients may experience daytime sleepiness, most hazardous while driving or during work. They may also exhibit personality changes, difficulty concentrating, memory difficulties, headaches, or sexual dysfunction. Sleep apnea is also associated with increased rates of systemic and pulmonary hypertension, stroke, heart failure, and myocardial infarction.

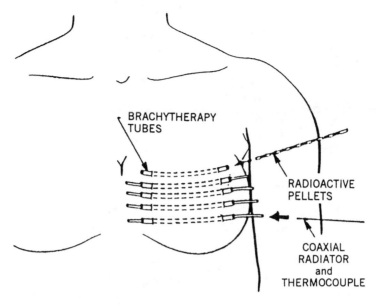

FIGURE 2.138 Interstitial hyperthermia combined with brachytherapy for treating breast cancer. (After Sterzer.[60])

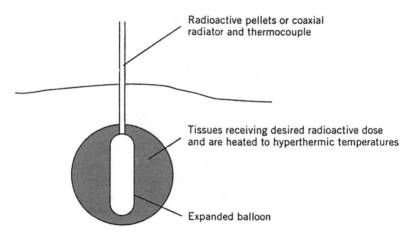

FIGURE 2.139 Interstitial microwave balloon catheter for combined localized hyperthermia and brachytherapy treatment of cancer. (After Sterzer.[60])

Treatment depends on the severity and frequency of symptoms. Some mild cases may be managed with weight loss alone. Often, however, further intervention is needed. Conventional management has relied upon dental appliances to maintain an open airway, ventilators to provide Continuous Positive Airway Pressure (CPAP), and attempts at surgical correction of the airway obstruction. Though effective, dental appliances and CPAP are both uncomfortable, and suffer from relatively low patient compliance rates (40 to 70%). Surgical correction may involve either excision of "excess tissue" (uvulopalatopharyngoplasty) or more involved maxillofacial surgery. Surgical cure rates have been reported to range between 30 and 75%.

Recently, Somnus Medical Technologies has developed an RF system (Somnoplasty™) that uses needle electrodes to create precise regions of submucosal tissue coagulation. Thus, both the tissue volume and its resulting airway obstruction are reduced. Applicator probes have been developed to target specific tissues including the base of the tongue (Fig. 2.140a), the uvula (Fig. 2.140b) and soft palate (Fig. 2.140c),

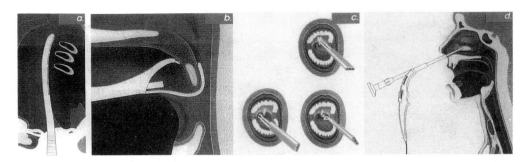

FIGURE 2.140 (a) Tongue Somnoplasty; (b) Uvula Somnoplasty; (c) Palatal Somnoplasty; (d) Turbinate Somnoplasty (with permission of Somnus, Inc.).

and nasal turbinates (Fig. 2.140d). Somnoplasty is designed to be performed on an outpatient basis, under local anesthesia, and is expected to boast such benefits as immediate results, little postoperative edema or discomfort, and no permanent scarring.

In the paper entitled "Radiofrequency Volumetric Reduction of the Tongue — A Porcine Pilot Study for the Treatment of Obstructive Sleep Apnea Syndrome," Powell et al.,[64] reported on the use of RF for the volumetric reduction of the tongue. Powell's three-stage pilot study investigated both the *in vitro* and *in vivo* effects of RF, delivered via a customized needle electrode. Volumetric measurements were performed using implanted ultrasonic crystals positioned around the treatment site. Changes in tissue volume could then be assessed both before and after the delivery of RF energy. To establish the feasibility of the technique, the initial stage of the project used two bovine tongues (*in vitro*). A single 0.05-in. diameter needle electrode delivered 30 kj over a 20 minute period at two sites per tongue. Volume reductions of between 12.8 and 26.7 % were noted immediately after the procedure, with an additional 4% reduction noted after 4 hours. The second stage was conducted using pigs, *in vivo*, and demonstrated that volume reduction increases as the amount of energy delivered is increased from 6.8 to 40 kj. Finally, in the third stage, an *in vivo* porcine model was again used, this time assessing clinical efficacy of the procedure by measuring both tissue volume changes, and histological changes. RF tissue reduction was performed on 9 pigs, with 3 additional pigs serving as controls. An 0.035-in. diameter needle electrode was used to deliver 2.4 kj over $6^{\pm}1.20$ minutes. Immediately after the procedure, a mean volume shrinkage of 7.02% was described. By 24 hours after the procedure, edema resulted in a 4 to 6% increase in tissue volume, thus returning nearly to baseline volumes. Subsequently, however, a progressive volume reduction of up to 26.3% was identified over the following 10 days. Animals were sacrificed at between 1 hour and 5 weeks after the procedure. Lesions were described as spherical, well-defined regions of tissue destruction, initially demonstrating edema and hemorrhage. As the lesion healed, scar formation occurred along with neovascularization. Tissue and vessels surrounding the lesion remained intact and viable. Given the seeming success of this technique in the animal model, RF tissue reduction may offer a promising alternative to the conventional management of obstructive sleep apnea.

Microwave-Aided Liposuction (MAL)[59,65]

Liposuction is used for aesthetic and reconstructive surgery. Its uses include the undermining of large flaps while preserving vascular attachments, removing lipomas, treating gynecomastia, and improving axillary hyperhidrosis. The application of microwave for aided liposuction may reduce some problems associated with standard mechanical liposuction, including blood loss, fluid shifts, and systemic effects.

Dry-technique liposuction vs. microwave-aided dry-technique liposuction — Preliminary work has been conducted, in swine, to compare the effects of the dry-technique liposuction vs. microwave-aided dry-technique liposuction. The "non-microwave" dry-technique liposuction performed at the two cephalad sites yielded typical fat debris that grossly appeared to be mixed with a noticeable amount of blood. The "microwave" liposuction performed at the two caudal sites yielded fat that differed considerably in quality and texture from tissue extracted using the dry-technique. The duration of microwave-aided suctioning appeared to be related to the histologic changes observed in the subcutaneous fat derived from the caudal sites. The fat initially removed during the first 30 seconds grossly appeared similar to conventionally suctioned fat. However, the fat removed as the duration of microwave-aided liposuction increased from 30 seconds to 2 minutes appeared increasingly softened. The longest duration of microwave suctioning, from 2 to 4 minutes, yielded fat that grossly appeared to be fused into an opaque, amorphous melted state.

The tumescent technique for liposuction surgery — The "tumescent technique" of liposuction was introduced in 1986. Use of the Klein needle has allowed the anesthetic solution to be rapidly injected through the same incision used for liposuction, efficiently anesthetizing large subcutaneous areas, thus eliminating the need and risks of general anesthesia. Injection of a large volume of dilute lidocaine produces a swelling and firmness of the site to be aspirated, which greatly facilitates fat removal. The small (3 to 4 mm) cannulas produce less trauma and therefore result in less blood loss, bruising, and discomfort. The basic technique was later expanded, and much larger volumes of lidocaine were administered, resulting in the capability of aspirating significantly greater volumes of tissue with a minimum increase in blood loss. This was achieved with serum lidocaine levels well below the toxicity range. We believe that using microwave volume heating will further enhance and benefit the tumescent technique.

Tumescent-technique liposuction vs. microwave-aided tumescent-liposuction — A similar protocol was followed at corresponding sites on the left side of the swine. The only modification was to employ tumescent liposuction instead of the dry technique that was used for sites on the right side. The solution used for tumescence consisted of 1000 cc of normal saline, combined with 60 cc of 1% lidocaine with epinephrine. Approximately 250 cc of this solution was infiltrated into each of the four sites prior to liposuction. The conventional "non-microwave" tumescent liposuction performed at the two cephalad sites yielded fat typically seen in such procedures; there was also less bleeding than seen with the dry technique.

Tumescent liposuction combined with microwaves between 30 watts and 40 watts yielded a transformation in the fat suctioned, enabling easier fat removal with less bleeding in comparison to both conventional dry and tumescent liposuction without microwaves.

Cannula design — The cannula utilized was a Byron Accelerator III type cannula, which was modified to hold a microwave semirigid coaxial cable having a whip antenna at the distal end (Fig. 2.141). The tip of each cannula distal of the suction port was modified by removal of its metal tip, which was replaced

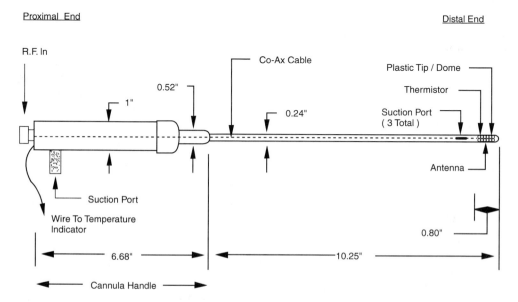

FIGURE 2.141 Microwave-aided liposuction cannula.

with a dome made of plastic in order to facilitate microwave radiation. The suction port in the proximal end of the cannula handle was converted to accept the semirigid coaxial cable/antenna structure. Suction was effectuated through a new port installed in the cannula handle.

The system used in our preliminary experiments was designed for use at 2.45 GHz while immersed in a tissue phantom. With the modified liposuction cannula and antenna, we have measured return losses as low as −37 db.

Antenna considerations[66] — The antenna design for Microwave Aided Liposuction (MAL) presents some interesting challenges. The antenna must deliver microwave energy to heat the treated volume of fat. Unlike traditional biological antenna designs, the MAL antenna radiates in close proximity to a metallic cannula. The cannula serves two primary functions. First, the openings in the end of the cannula are sharp to facilitate fat removal via mechanical cutting. Second, removed fat is suctioned through the cannula. Clearly, the cannula imposes a more complex antenna geometry than exists for traditional biological antennas. Furthermore, as opposed to microwave hyperthermia application, for example, it may actually be desirable to allow heating to occur along the transmission line since that heating will prevent the coagulation of fat being suctioned through the cannula. Analytical antenna design approaches become increasingly difficult under these design constraints. Therefore, it is appealing to consider the use of accurate simulation tools to design the antenna and to analyze the antenna performance. This approach was used by Labonte et al. who implemented a finite-element method (FEM) in the frequency domain to compare the near field radiation patterns of several types of antennas including the dielectric tip monopole, open tip monopole, and metal tip monopole.[67,68] In our future work, we propose to use a finite-element time domain (FDTD) code by REMCOM, Inc. to model the MAL antenna. This code will allow the computation of input impedance, near field values, and specific absorption rate (SAR).

Tissue Anastomoses Utilizing Biological Solder in Conjunction with Microwave Irradiation in Future Endoscopic Surgery[69]

Endoscopic surgery is revolutionizing many surgical procedures. For example, laparoscopic surgical procedures, particularly laparoscopic cholecystectomy, have gained widespread acceptance. Further expansion of the endoscopic approach is inevitable. Although minimal access surgery is advantageous to patients, the technical problems imposed by the limited access are pushing existing tissue closure technologies (mechanical stapling devices and hand-sewn sutures) to their limits. The laparoscopic closure of an incision made in the bile duct for removal of stones is an example of the shortcomings of current technologies. Closure of this incision with laparoscopically placed sutures is difficult and post-operative bile leakage may result. Mechanical stapling devices for this purpose are beyond currently available technology.

To enhance a tissue anastomosis with microwaves, the tissue temperature must be kept below the threshold for damage, while the biological solder is heated above 60°C. Microwave anastomosis may also prove useful for vascular repairs, for example. A microwave antenna can be positioned inside an artery, and solder (albumin) is then placed on the outside of the vessel and in any small gaps between the arterial segments undergoing repair. The successful results *in vitro* have encouraged the preliminary investigation in a rabbit model. Early results *in vivo*, however, have indicated the need for a dry environment. More research is needed to evaluate the full potential of the microwave anastomosis technique.

Nerve Ablation for the Treatment of Gastroesophageal Reflux Disease

Gastroesophageal reflux disease (GERD) results from the chronic backward flow of stomach contents into the esophagus. The acid, bile, and digestive enzymes cause irritation of the esophagus and symptoms of heartburn, regurgitation, chest pain, voice disorders, and swallowing problems.

Normally, the muscular valve (lower esophageal sphincter or LES) at the junction of the esophagus and stomach prevents reflux from occurring. Reflux of stomach contents occurs when the LES and diaphragm are unable to provide enough tone or force to squeeze adequately on the esophagus. This may happen in some patients in whom the muscles have weakened over time or in those patients with hiatal hernias. The barrier function in these patients is completely lost, and reflux is present throughout the day.

The majority of patients with GERD, however, have normal LES and diaphragm pressures, yet the sphincter muscles relax frequently throughout the daytime to cause reflux. The relaxation events permit excessive reflux of stomach contents and the patient develops significant symptoms of GERD.

This abnormal event is a neurological reflex, termed transient lower esophageal relaxation (tLESR),[75–79] and is the cause of GERD in over 80% of patients. A tLESR is prompted when there is stretching of the stomach wall, as after a meal. The stretch receptors generate a nerve impulse, which travels upward within the myenteric plexus of the gastroesophageal junction. The myenteric plexus is a network of very small nerves lying between the layers of the stomach and esophagus musculature. The impulses travel through the LES, into the esophagus, and then join the vagus nerve on their way to the brain. When the brain receives these signals, a motor signal is sent to the LES causing prolonged relaxation.

There are hundreds of peer-reviewed scientific publications addressing the importance of tLESR in the development of GERD. Many investigators have collaborated to study the delivery of radio frequency energy for the treatment of GERD.

Investigators at Stanford[80] have recently performed radio frequency ablation of the stomach cardia (Fig. 2.142a,b) in Yucatan mini-pigs to establish the effect on these nerve pathways. These nerve fibers course between the muscle layers of the LES and cardia. These investigators have demonstrated a statistically significant effect of delivering radio frequency energy to the cardia on the parameter of gastric yield pressure. This test is directly related to tLESRs. The stomach is stretched with carbon dioxide gas until the LES yields or relaxes in response to pressure. Yield pressures were higher in all animals after treatment, indicating that the nerve reflex arc was modulated to have a higher threshold for stimulation, or a lower frequency of transmission to the brain.

Step-by-Step Treatment for Reflux — Research Carried Out at Conway Stuart Medical, Inc.

1. **Position, Inflate, Deploy, Irrigate, Treat.** The physician positions the catheter, inflates the balloon, deploys the needles and begins irrigation. During treatment, radio frequency energy is delivered in a controlled manner to the tissue surrounding the needle electrodes (Fig. 2.143a);
2. **Treatment at multiple levels.** The treatment sequence is repeated to create well-defined coagulative lesions along the length of the lower esophageal sphincter and cardia (Fig. 2.143b);
3. **Resorption and shrinking.** Over the next few weeks, the coagulated tissue resorbs and shrinks, increasing resistance to reflux (Fig. 2.143c).

In an abstract entitled "Augmentation of lower esophageal sphincter pressure and gastric yield pressure after radiofrequency energy delivery to the lower esophageal sphincter muscle: a porcine model," Drs. D.S. Utley, M.A. Vierra, M.S. Kim, and G. Triadafilopoulos of VA Palo Alto Health Care System and Stanford University in Palo Alto, CA[80] report on their investigation of the technique of endoscopic, submucosal radio frequency energy (Rfe) delivery to the lower esophageal sphincter (LES) as a possible alternative treatment of GERD. Utilizing a porcine reflux model, they determined the effects of Rfe on LES pressure (LESP) and gastric yield pressure (GYP). In summary, Rfe is a promising new modality in the endoscopic treatment of GERD (Fig. 2.143).

RF in the Treatment of Solid Organ Tumors

RITA Medical Systems, Inc. has developed a controlled tissue ablation system to treat solid organ tumors using minimally invasive RF ablation technology. This system includes an RF generator and a family of electrodes for the treatment of solid tumors. The controlled application of RF energy through an electrode placed directly in the tumor heats tissue to the required target temperature.

Each electrode consists of a thin hollow stainless steel shaft that acts as a primary electrode and also allows the introduction into the tumor of a curved array of secondary electrodes. The secondary electrodes have temperature sensors mounted on the tips to provide temperature feedback.

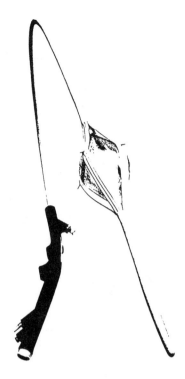

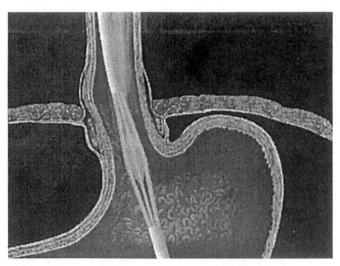

FIGURE 2.142   (a) Catheter used for nerve ablation in the treatment of gastroesophageal reflux disease; (b) Catheter positioning within the gastric cardia to deliver radiofrequency energy for nerve ablation.

The RF generator delivers up to 50 watts to destroy or ablate the tumor. The operator sets the desired temperature; the generator automatically adjusts the power to attain the proper temperature and displays delivered power, impedance, and temperature.

The unique features of RF in treating solid tumors are as follows (Fig. 2.144):

- *Minimally invasive* — many procedures can be performed through a laparascopic or even a percutaneous approach, frequently on an outpatient basis.
- *Creates large volume of ablated tissue* — the current device can ablate a spherical area of around 3 cm, approximating the size and shape of many cancerous lesions.

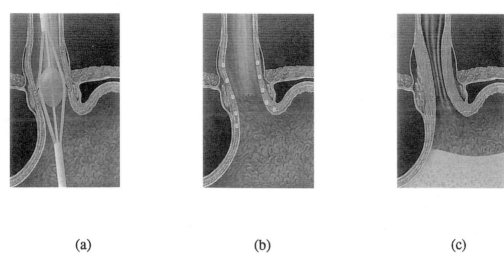

(a)   (b)   (c)

FIGURE 2.143   Step-by-step treatment for reflux.

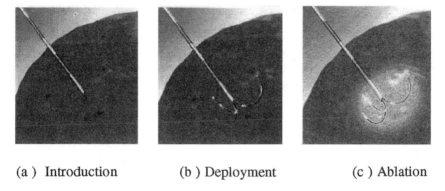

(a) Introduction   (b) Deployment   (c) Ablation

FIGURE 2.144   RF in the treatment of solid organ tumors. (With permission from RITA Medical Systems Inc.)

- *Temperature feedback leads to predictability and controllability* — the system provides temperature feedback at the periphery of the ablation volume to confirm tissue destruction. It also provides impedance feedback, which can be used to guide the application of RF power in order to ablate the tumor.

Application of RF Thermal Arthroscopy[72]

The combined use of RF energy and arthroscopy has been, in recent years, successfully advanced by ORATEC Interventions, Inc. They have accumulated an extensive collection of case reports, and from among these we will discuss one important clinical application. This method will serve as an example of the significance of RF thermal arthroscopy techniques.

In a report entitled, "Arthroscopic shoulder stabilization using suture anchors and capsular shrinkage," Jeffrey S. Abrams, M.D. of Princeton Orthopedic and Rehabilitation Associates, describes the treatment of recurrent shoulder joint instability after traumatic injury. The ligaments that normally anchor the upper arm to the shoulder joint can tear away from their attachments on the joint, (glenoid cavity) as a result of trauma. Without these ligamentous attachments (the labrum) the shoulder joint can repeatedly dislocate. Such recurrent instability not only eliminates recreational sports for the patients but can significantly affect their daily life.[73]

The traditional approach to such injury involves orthopedic surgery to reattach the torn labrum and restore tension to the damaged ligaments, thus holding the joint together firmly. Laporoscopic techniques, in general, reduce surgical time, decrease morbidity, and can increase success rates. The arthroscopic technique for shoulder stabilization still involves the use of sutures to anchor the labrum back to the edge of the glenoid. RF electrothermal technology is then used to shrink the tissues of the ligamentous joint capsule and thus increase the tension on these ligaments. Increased tension on the ligaments, in turn, stabilizes the shoulder joint. Such procedures can be performed on an outpatient basis and require minimal postoperative medication.

Conclusions

In this chapter, we have reviewed a few of the existing applications of RF/microwaves in medicine. We have indicated with some detail the new applications currently under investigation. A more detailed discussion of some of the topics can be found in the book entitled *New Frontiers in Medical Device Technology* edited by Arye Rosen and Harel D. Rosen, published by John Wiley and Sons, 1995 as part of the Wiley Series in Microwave and Optical Engineering/Kai Chang, Series Editor.

Acknowledgments

We wish to recognize the assistance and advice of many who, since the early 1980s, have participated in research in the areas of RF/microwaves in medicine, some of whose research was covered in this chapter: from Jefferson Medical College, Drs. Paul Walinsky and Arnold J. Greenspon; from MMTC, Dr. Fred Sterzer and Mr. Dan Mawhinney; from Temple University, Dr. William Santamore; from the University of Pennsylvania, Dr. Louis Bucky. Gratitude is also due to Mr. John Hendrick of VidaMed, Inc., Mr. Hugh Sharkey of ORATEC, Interventions, Inc., and Mr. Barry Cheskin of RITA Medical Systems, Inc., who were so kind to furnish some of the material; and to Mr. Walter Janton for his technical skills and invaluable support. Finally, recognition is due to Mrs. Daniella Rosen for her contribution to the research on microwave-assisted liposuction, and for revising this manuscript again and again.

References

1. Cosman BJ, Cosman ER: *Guide to Radio Frequency Lesion Generation in Neurosurgery.* Radionics, Inc., 1974.
2. Cosman ER, Cosman BJ: *Methods of making nervous system lesions, Medical Therapy of Movement Disorders Guide to Radio Frequency Lesion Generation in Neurosurgery.* Radionics, Inc., 1974.
3. Alberts WW, Wright EW Jr, Feinstein B, Gleason CA: Sensory responses elicited by subcortical high frequency electrical stimulation in man. *J. Neurosurg.,* 36, 80–82, 1972.
4. Dieckmann G, Gabriel E, Hassler R: Size, form, and structural peculiarities of experimental brain lesions obtained by thermocontrolled radiofrequency. *Confin. Neurol.,* 26, 134–142, 1965.
5. Brodkey JS, Miyazaki Y, Ervin FR, Mark VH: Reversible heat lesions with radiofrequency current: A method of stereotactic localization. *J. Neurosurg.,* 21, 49–53, 1964.
6. Fager CA: Surgical treatment of involuntary movement disorders, *Lahey Clin. Found. Bull.,* 22, 79–83, 1973.
7. Hurt RW, Ballantine HT Jr: Stereotactic anterior cingulate lesions for persistent pain: A report on 68 cases. *Clin. Neurosurg.,* 21, 334–351, 1974.
8. Schwan HP, Electrical properties of tissues and cell suspensions, *Advanced Phys. Med. Biol.,* 5, 147–209, 1957.
9. Stuchly MA, Stuchly SS: Dielectric properties of biological substances-tabulated, *J. Microwave Power,* 15, 19–26, 1980.
10. Foster KR, Schepps JL, Schwan HP: Microwave dielectric relaxation in tissue: a second look, *Biophys. J.,* 29, 271–281, 1980.
11. Thuari M, Steel MC, Sheppard RJ, Grant HE: Dielectric properties of developing rabbit brain at 37 degrees C, *Bioelectromagnetics,* 6, 235–242, 1985.

12. Schwan HP, Foster KR: RF-field interactions with biological systems: electrical properties and biophysical mechanism, *Proc. IEEE,* 68, 104–113, 1980.

13. Pethig R, Kell DB: The passive electrical properties of biological systems: their significance in physiology, biophysics, and biotechnology, *Phys. Med. Biol.,* 32, 933–970, 1987.

14. Duck, FA, *Physical Properties of Tissue: A Comprehensive Reference Book,* Academic Press, New York, 1990.

15. Gabriel C, Gabriel S, Corthout E: The dielectric properties of biological tissues: I. Literature review, *Phys. Med. Biol.,* 41, 2231–2249, 1996.

16. Grant HE, Sheppard RJ, South FP, *Dielectric Behaviour of Biological Molecules in Solution,* Oxford University Press, Oxford, 1978.

17. Foster KR, Schepps JL: Dielectric properties of tumor and normal tissues at radio through microwave frequencies, *J. Microwave Power,* 16, 107–119, 1981.

18. Gabriel S, Lau RW, Gabriel C: The dielectric properties of biological tissues: II. measurements in the frequency range 10 Hz to 20 Ghz, *Phys. Med. Biol.,* 41, 2251–2269, 1996.

19. Cole KS, Cole RH, Dispersion and absorption in dielectrics: I. alternating current characteristics, *J. Chem. Phys.,* 9, 341–351, 1941.

20. Debye P, Huckel E: *Phys. Z.,* 24, 185, 1923.

21. Frohlich H: *Theory of Dielectrics,* Oxford University Press, Oxford, England, 1949.

22. Daniel VV: *Dielectric Relaxation,* Academic Press, London, 1967.

23. Sterzer F et al.: RF therapy for malignancy, *IEEE Spectrum,* 17, 12, 32–37, Dec. 1980.

24. Lele, PP: Induction of deep, local hyperthermia by ultrasound and electromagnetic fields: problems and choices. *Radiat. Environ. Biophy.,* 17, 205, 1980.

25. Tofighi M-R: A two-port microstrip test fixture for measurement of complex permittivity of biological tissues at microwave and millimeter wave frequencies, PhD. Dissertation, Department of ECE, Drexel University, Philadelphia, PA.

26. Roberts S, Von Hippel A: A new method for measuring dielectric constant and loss in the range of centimeter waves, *J. Appl. Phys.,* 17, 610–616, 1946.

27. Westphal: Dielectric measuring techniques, in *Dielectric Materials and Applications,* A.R. von Hippel, Ed., Wiley, New York, 63–122, 1954.

28. Burdette EC, Cain FL, Seals J: *In vivo* probe measurement technique for determining dielectric properties at VHF through microwave frequencies, *IEEE Trans. Microwave Theory and Tech.,* MTT-28, 414–424, 1980.

29. Stuchly SS, Sibbald CL, Anderson JM: A new aperture admittance model for open-ended waveguides, *IEEE Trans. Microwave Theory and Tech.,* MTT-42, 192–198, 1994.

30. Steel MC, Sheppard RJ, The dielectric properties of rbbit tissue, pure water and various liquids suitable for tissue phantoms at 35 Ghz, *Phys. Med. Biol.,* 33, 467–472, 1988.

31. Steel MC, Sheppard RJ, Collin R: Precision waveguide cells for the measurement of complex permittivity of lossy liquids and biological tissue at 35 Ghz, *Journal of Physics. E. Scientific Instruments,* 20, 872–877, 1987.

32. Land DV, Campbell AM: A quick accurate method for measuring the microwave dielectric properties of small tissue samples, *Phys. Med. Biol.,* 37, 183–192, 1992.

33. Athey TW, Stuchly MA, Stuchly SS: Measurment of radio frequency permittivity of biological tissues with an open-ended coaxial line: part I, *IEEE Trans. Microwave Theory and Tech.,* MTT-30, 82–86, 1982.

34. Kraszewski A, Stuchly MA, Stuchly SS, Smith M: *In vivo* and *in vitro* dielectric properties of animal tissues at radio frequencies, *Bioelectromagnetics,* 3, 421–432, 1982.

35. Nyshadham A, Sibbald CL, Stuchly SS, Permittivity measurements using open-ended sensors and reference liquid calibration — an uncertainty analysis, *IEEE Trans. Microwave Theory and Tech.,* MTT-40, 305–314, 1992.

36. Misra DM, Chabbra M, Epstein BR, Mirotznik M, Foster KR: Noninvasive electrical characteriation of materials at microwave frequencies using an open-ended coaxial line: test of an improved calibration technique, *IEEE Trans. Microwave Theory and Tech.,* MTT-38, 8–13, 1990.

37. Gabriel C, Chan TYA, Grant EH: Admittance models for open ended coaxial probes and their place in dielectric spectroscopy, *Phys. Med. Biol.,* 39, 2183–2199, 1994.

38. Bao J, Lu S, Hurt DW: Complex dielectric measurements and analysis of brain tissues in the radio and microwave frequencies, *IEEE Trans. Microwave Theory and Tech.,* MTT-45, 1730–1740, 1997.

39. Daryoush A, private communications.

40. King RWP, Smith GS: *Antennas in Matter,* The MIT Press, Cambridge, MA, 1981.

41. King RWP, Trembly BS, Strohbehn JW: The electromagnetic field of an insulated antenna in a conducting or dielectric medium, *IEEE Trans. on Microwave Theory and Techniques,* MTT-31, 7, 574–583, July 1983.

42. Iskander MF, Umeh AM: Design optimization of interstitial antennas, *IEEE Trans. Biomed. Eng.,* 36, 238–246, Feb. 1989.

43. Rumeh AM, Iskander MF: Performance comparison of available interstitial antennas for microwave hyperthermia, *IEEE Trans. on Microwave Theory and Techniques,* 37, 7, 1126–1133, July 1989.

44. Debicki PS, Astrahan MA: Calculating input impedance of electrically small insulated antennas for microwave hyperthermia, *IEEE Trans. on Microwave Theory and Techniques,* 41, 2, 357–360, February 1993.

45. Su DW-F, Wu L-K, Input impedance characteristics of coaxial slot antennas for interstitial microwave hyperthermia, *IEEE Trans. on Microwave Theory and Techniques,* 47, 3, 302–307, March 1999.

46. Casey JP, Bansal R: The near field of an insulated dipole in a dissipative dielectric medium, *IEEE Trans. on Microwave Theory and Techniques,* 34, 4, 459–463, April 1986.

47. Greenspon AJ, Walinsky P, Rosen A: Catheter ablation for the treatment of cardiac arrhythmias, in *New Frontiers in Medical Technology,* John Wiley & Sons, Inc., New York, 1995.

48. Rosenbaum RM, Greenspon AJ, Hsu S, Walinsky P, Rosen A: RF and microwave ablation for the treatment of ventricular tachycardia, *IEEE MTT-S Digest,* 1993.

49. Sterzer F: Localized hyperthermia treatment of cancer, *RCA Rev.,* 42, 727, 1981.

50. Hahn GM: *Hyperthermia and Cancer,* Plenum Press, New York, 1982.

51. Storm FK: *Hyperthermia in Cancer Therapy.* GK Hall, Boston, 1983.

52. Walinsky P, Rosen A, Greenspon AJ: Method and apparatus for high frequency catheter ablation. U.S. Pat. 4,641,649.

53. Walinsky P, Rosen A, Martinez-Hernandez A, Smith D, Nardone DO, Brevette B: Microwave balloon angioplasty, *The J. of Invasive Cardiology,* 3, 3, May/June 1991.

54. Langberg JJ, Wonnell TL, Chin M, et al.: Catheter ablation of the atrioventricular junction using a helical microwave antenna: A novel means of coupling energy to the endocardium. *PACE,* 14, 2105, 1991.

55. Wonnell TL, Stauffer PR, Langberg JJ: Evaluation of microwave and RF catheter ablation in a myocardial equivalent phantom model. *IEEE Trans. Biomed. Eng.,* 39, 1086, 1992.

56. Rosen A, Walinsky P, Smith D, Kosman Z, Martinez A, Sterzer F, Presser A, Mawhinney D, Chou J-S, Goth P: Studies of microwave thermal balloon angioplasty in rabbits, *IEEE MTT-S Digest,* 1993.

57. Rosen A, Rosen HD: The efficacy of transurethral thermal ablation in the management of benign prostatic hyperplasia, in *New Frontiers in Medical Technology,* John Wiley & Sons, Inc., New York, NY, 1995.

58. Rosen A, Walinsky P: Microwave balloon angioplasty, in *New Frontiers in Medical Technology,* John Wiley & Sons, Inc., New York, NY, 1995.

59. Rosen HD, Rosen A: RF/Microwaves, a hot topic in medicine, *IEEE Potentials,* August/September 1999.

60. Sterzer F: Localized heating of deep-seated tissues using microwave balloon catheters, in *New Frontiers in Medical Technology,* John Wiley & Sons, Inc., New York, 1995.

61. Schmidt-Nowara W et al.: Oral appliances for the treatment of snoring and Obstructive Sleep Apnea, *Sleep,* 18, 6, 501–510, 1995.

62. Simmons FB, Guilleminault C, Miles LE: The platopharyngoplasty operation for snoring and sleep apnea: An interim report. *Otolaryngol. Head Neck Surg.*, 92, 375–380, 1984 and Conway W, Fujita S, Zorick F, et al.: Uvulopalatopharyngoplasty: One year follow-up. *Chest*, 88, 385–387, 1985.

63. Riley et al. 1995, op. cit. The description of this approach and the outcomes data are taken from this paper, though the team has published several papers on the method since 1988.

64. Powell NB, Riley RW, Troell RJ, Blumen MB, and Guilleminault C: Radiofrequency volumetric reduction of the tongue — A porcine pilot study for the treatment of obstructive sleep apnea syndrome, laboratory and animal investigations, *Chest*, 111, 1348–55, 1997.

65. Rosen A, Rosen D, Tuma G, Bucky L,: RF/Microwave aided tumescent liposuction, *IEEE MTT Transactions*, (Nov. 2000), in press.

66. Jemison W et al.: New antenna design for microwave assisted liposuction, to be published; Private communications.

67. Hurter W, Reinbold F, Lorenz WJ: A dipole antenna for interstitial microwave hyperthermia, *IEEE Trans. on Microwave Theory and Techniques*, 34, 4, 459–463, April 1986.

68. Labonte S, Blais A, Legault SR, Ali HO, Roy L: Monopole antennas for microwave catheter ablation, *IEEE Trans. on Microwave Theory and Techniques*, 44, 10, 1832–1840, Oct. 1996.

69. Santamore W: Private communication.

70. *Health aspects of radio frequency and microwave radiation exposure, Part 1*, 77-EHD-13, National Health and Welfare, Canada, November 1977.

71. Paglione R: Medical applications of microwave energy, *RCA Engineer*, 27, 5, 17–21, Sept./Oct. 1982.

72. Sharkey H: Private communication.

73. Abrams JS: Arthroscopic shoulder stabilization using suture anchors and capsular shrinkage, ORATEC Interventions, Inc., Applications in Electrothermal Arthroscopy, Case Report.

74. Tell RA: Microwave Energy Absorption in Tissue, *Techn. Rep. PB*, Environmental Protection Agency, Feb. 1972.

75. Mittal RK, Holloway RH, Penagini R, Blackshaw LA, Dent D: Transient lower esophageal sphincter relaxation. *Gastroenterology* 1995; 109:601–610.

76. Rawahara H, Dent J, Davidson G: Mechanisms responsible for gastroesophageal reflux in children. *Gastroenterology* 1997; 113:399–408.\

77. Panagini R, Bianchi PA: Effect of morphine on gastroesophageal reflux and transient lower esophageal sphincter relaxation. *Gastroenterology* 1997; 113:409–414.

78. Blackshaw LA, Haupt JA, Omari T, Dent J: Vagal and sympathetic influences on the ferret lower esophageal sphincter. *J Autonomic Nerv System* 1997; 66:179–188.

79. Stakeberg J, Lehman A: Influence of different intragastric stimuli on triggering of transient lower esophageal sphincter relaxation in the dog. *Neurogastroenterol* 1999; 11:125–132.

80. Private communications with Edwars SD.

3

Systems Considerations

Avram Bar-Cohen
University of Minnesota

Karl J. Geisler
University of Minnesota

Allan D. Kraus
Allan D. Kraus Associates

John M. Osepchuk
Full Spectrum Consulting

Ronald C. Petersen
Lucent Technologies Inc./Bell Labs

John F. Sevic
UltraRF, Inc.

Leland M. Farrer
Cadence Design Systems, Inc.

Brent A. McDonald
Dell Computer Corp.

George K. Schoneman
Rockwell Collins

Daniel E. Jenkins
Dell Computer Corp.

Mike Golio
Rockwell Collins

Arun Ramakrishnan
University of Maryland

Toby Syrus
University of Maryland

Michael Pecht
University of Maryland

3.1 Thermal Analysis and Design of Electronic Systems

Avram Bar-Cohen, Karl J. Geisler, and Allan D. Kraus

Motivation

In the thermal control of RF devices, it is necessary to provide an acceptable *microclimate* for a diversity of devices and packages that vary widely in size, power dissipation, and sensitivity to temperature.

0-8493-8592-X/01/$0.00+$.50
© 2001 by CRC Press LLC

Although the thermal management of all electronic components is motivated by a common set of concerns, this diversity often leads to the design and development of distinct thermal control systems for different types of electronic equipment. Moreover, due to substantial variations in the performance, cost, and environmental specifications across product categories, the thermal control of similar components may require widely differing thermal management strategies.

The prevention of catastrophic thermal failure (defined as an immediate, thermally induced, total loss of electronic function) must be viewed as the primary and foremost aim of electronics thermal control. Catastrophic failure may result from a significant deterioration in the performance of the component/system or from a loss of structural integrity at the relevant packaging levels. In early microelectronic systems, catastrophic failure was primarily *functional* and thought to result from changes in the bias voltage, *thermal runaway* produced by regenerative heating, and dopant migration, all occurring at elevated transistor junction temperatures. While these failure modes may still occur during the device development process, improved semiconductor simulation tools and thermally compensated devices have largely quieted these concerns and substantially broadened the operating temperature range of today's RF devices.

In microelectronic, microwave, and RF components, the levels of integration and device density on the chips, as well as frequencies of operation, continue to increase. The most critical heat-producing component for most RF systems is the power amplifier (PA) stage. Output power required from these stages ranges from less than 1 watt for some handheld commercial applications to greater than 1 kW (multiple parallel stages) for certain military, avionics, and data link applications. Single transistor output power levels are as high as 100 W to 200 W for applications ranging from commercial base stations to avionics, satellite communications, and military. Amplified efficiencies for the highest output power requirements are typically in the 15 to 35% range. To facilitate effective thermal management for such high power levels, PA operation must be pulsed with low duty cycles (reducing thermal power dissipation requirements). Improved thermal performance can be translated into higher duty cycles and therefore into greater data transfer or more efficient use of bandwidth. For these kinds of applications, performance is already limited primarily by the maximum achievable heat flux. Improvements in that figure of merit automatically and immediately translate into improved system performance.

More generally, however, thermal design is aimed at preventing thermally induced physical failures through reduction of the temperature rise above ambient and minimization of temperature variations within the packaging structure(s). With RF integrated circuits or discrete RF high-performance devices, maximum frequency of operation, noise figure, power saturation levels, and nonlinear behavior are all affected by temperature. The use of many low-temperature materials and the structural complexity of chip packages and printed circuit boards has increased the risk of catastrophic failures associated with the vaporization of organic materials, the melting of solders, and thermal-stress fractures of leads, joints, and seals as well as the fatigue-induced delamination and fracture or creep-induced deformation of encapsulants and laminates. To prevent catastrophic thermal failure, the designer must know the maximum allowable temperatures, acceptable internal temperature differences, and the power consumption/dissipation of the various components. This information can be used to select the appropriate fluid, heat transfer mode, and inlet temperature for the coolant and to thus establish the thermal control strategy early in the design process.

After the selection of an appropriate thermal control strategy, attention can be turned to meeting the desired system-level reliability and the target failure rates of each component and subassembly. Individual solid-state electronic devices are inherently reliable and can typically be expected to operate, at room temperature, for some 100,000 years, i.e., with a base failure rate of 1 FIT (failures in 10^9 h). However, since the number of devices in a typical radio component is rapidly increasing and since an RF system may consist of many tens to several hundreds of such components, achieving a system Mean Time Between Failures of several thousand hours in military equipment and 40,000 to 60,000 hours in commercial systems is a most formidable task.

Many of the failure mechanisms, which are activated by prolonged operation of electronic components, are related to the local temperature and/or temperature gradients, as well as the thermal history of the package.[1] Device-related functional failures often exhibit a strong relationship between failure rate and operating temperature. This dependence can be represented in the form of an exponential Arrhenius relation, with unique, empirically determined coefficients and activation energy for each component type and failure mechanism. In the normal operating range of microelectronic components, a 10 to 20°C increase in chip temperature may double the component failure rate, and even a 1°C decrease may then lower the predicted failure rate associated with such mechanisms by 2 to 4%.[2]

Unfortunately, it is not generally possible to characterize thermally induced structural failures, which develop as a result of differential thermal expansion among the materials constituting a microwave package, in the form of an Arrhenius relation. Although these mechanical stresses may well increase as the temperature of the component is elevated, thermal stress failures are, by their nature, dependent on the details of the local temperature fields, as well as the assembly, attachment, and local operating history of the component. Furthermore, thermal stress generation in packaging materials and structures is exacerbated by power transients, as well as by the periodically varying environmental temperatures, experienced by most electronic systems, during both qualification tests and actual operation. However, stress variations in the elastic domain or in the range below the fatigue limit may have little effect on the component failure rate. Consequently, the minimization of elimination of thermally induced failures often requires careful attention to both the temperature and stress fields in the electronic components and necessitates the empirical validation of any proposed thermostructural design criteria.

Thermal Packaging Options

When the heat flux dissipated by the electronic component, device, or assembly is known and the allowable temperature rise above the local ambient condition is specified, the equations of the following sections can be used to determine which heat transfer process or combination of processes (if any) can be employed to meet the desired performance goals. Figure 3.1 shows the variation of attainable temperature differences with surface heat flux for a variety of heat transfer modes and coolant fluids.

Examination of Fig. 3.1 reveals that for a typical allowable temperature difference of 60°C between the component surface and the ambient, "natural" cooling in air — relying on both free convection and radiation — is effective only for heat fluxes below approximately 0.05 W/cm². Although forced convection cooling in air offers approximately an order-of-magnitude improvement in heat transfer coefficient, this thermal configuration is unlikely to provide heat removal capability in excess of 1 W/cm² even at an allowable temperature difference of 100°C.

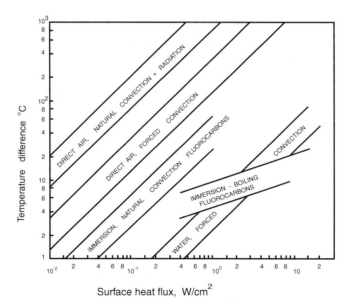

FIGURE 3.1 Temperature differences attainable as a function of heat flux for various heat transfer modes and various coolant fluids. (From Ref. 3, with permission.)

To facilitate the transfer of moderate and high heat fluxes from component surfaces, the thermal designer must choose between the use of finned, air-cooled heat sinks and direct or indirect liquid cooling. Finned arrays and sophisticated techniques for improving convective heat transfer coefficients can extend the effectiveness of air cooling to progressively higher component heat fluxes, but often at ever-increasing weight, cost, and volume penalties. Alternately, reliance on heat transfer to liquids flowing at high velocity through so-called "cold plates" can offer a dramatic improvement in the transferable heat flux even at temperature differences as low as 10°C, when the conduction resistance in the cold plate wall is negligible.

A similar high heat flux capability is offered by boiling heat transfer to perfluorinated (FCs) and hydrofluoroether (HFEs) liquids. The high dielectric strength and low dielectric constant of these liquids make it possible to implement this approach for a wide range of components. Direct liquid contact allows the removal of component heat fluxes in excess of 10 W/cm² with saturated pool boiling at temperature differences typically less than 20°C. Natural convection (i.e., non-boiling) immersion cooling can also offer significant advantages and, as seen in Fig. 3.1, serves to bridge the gap between direct air cooling and cold plate technology.

Unfortunately, when addressed within stringent cost targets, the cooling requirements of 21st-century microelectronic, microwave, and RF components cannot be met by today's thermal packaging technology. Rather, ways must be sought to improve on currently available technology, to leverage and combine the best features of existing thermal packaging hardware, and to introduce unconventional, perhaps even radical, thermal solutions into the electronic product family. In so doing, attention must be devoted to three primary issues:

- Highly effective air cooling — removing dissipated power from one or several high-performance components within minimal volumes and with low air-side pressure drops.
- Heat spreading — transporting heat from the relatively small area of the device to the substrate, card, or board, or to a relatively large heat sink or cold plate base.
- Interfacial heat transfer — transferring heat across the thermal resistances between the device and the next level of thermal packaging.

Attention now turns to a detailed discussion of basic heat transfer and the determination of the various types of thermal resistances often encountered in electronic equipment.

Thermal Modeling

To determine the temperature differences encountered in the flow of heat within electronic systems, it is necessary to recognize the relevant heat transfer mechanisms and their governing relations. In a typical system, heat removal from the active regions of the device(s) may require the use of several mechanisms, some operating in series and others in parallel, to transport the generated heat to the coolant or ultimate heat sink. Practitioners of the thermal arts and sciences generally deal with four basic thermal transport modes: conduction, convection, phase change, and radiation.

Conduction Heat Transfer

One-Dimensional Conduction

Steady thermal transport through solids is governed by the Fourier equation, which in one-dimensional form, is expressible as

$$q = -kA\frac{dT}{dx} \qquad (3.1)$$

where q is the heat flow, k is the thermal conductivity of the medium, A is the cross-sectional area for the heat flow, and dT/dx is the temperature gradient. As depicted in Fig. 3.2, heat flow produced by a negative temperature gradient is considered positive. This convention requires the insertion of the minus sign in Eq. (3.1) to assure a positive heat flow, q. The temperature difference resulting from the steady

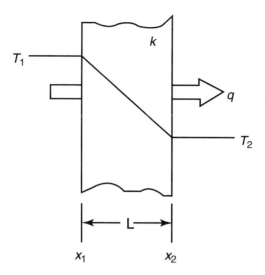

FIGURE 3.2   One-dimensional conduction through a slab. (From Ref. 4, with permission.)

state diffusion of heat is thus related to the thermal conductivity of the material, the cross-sectional area, and the path length, L, according to

$$\left(T_1 - T_2\right)_{cd} = q\,\frac{L}{kA} \tag{3.2}$$

The form of Eq. (3.2) suggests that, by analogy to Ohm's Law governing electrical current flow through a resistance, it is possible to define a thermal resistance for conduction, R_{cd}, as

$$R_{cd} \equiv \frac{\left(T_1 - T_2\right)}{q} = \frac{L}{kA} \tag{3.3}$$

One-Dimensional Conduction with Internal Heat Generation

Situations in which a solid experiences internal heat generation, such as that produced by the flow of an electric current, give rise to more complex governing equations and require greater care in obtaining the appropriate temperature differences. The axial temperature variation in a slim, internally-heated conductor whose edges (ends) are held at a temperature T_o, is found to equal

$$T = T_o + q_g\,\frac{L^2}{2k}\left[\left(\frac{x}{L}\right) - \left(\frac{x}{L}\right)^2\right] \tag{3.4}$$

When the volumetric heat generation rate, q_g, in W/m³, is uniform throughout, the peak temperature is developed at the center of the solid and is given by

$$T_{max} = T_o + q_g\,\frac{L^2}{8k} \tag{3.5}$$

Alternatively, since q_g is the volumetric heat generation, $q_g = q/LW\delta$, the center-edge temperature difference can be expressed as

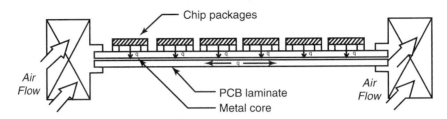

FIGURE 3.3   Edge-cooled printed circuit board populated with components. (From Ref. 4, with permission.)

$$T_{max} - T_o = q \frac{L^2}{8kLW\delta} = q \frac{L}{8kA} \tag{3.6}$$

where the cross-sectional area, A, is the product of the width, W, and the thickness, δ. An examination of Eq. (3.6) reveals that the thermal resistance of a conductor with a distributed heat input is only one quarter that of a structure in which all of the heat is generated at the center.

In the design of airborne electronic system and equipment to be operated in a corrosive or damaging environment, it is often necessary to conduct the heat dissipated by the components down into the substrate or printed circuit board and, as shown in Fig. 3.3, across the substrate/PCB to a cold plate or sealed heat exchanger. For a symmetrically cooled substrate/PCB with approximately uniform heat dissipation on the surface, a first estimate of the peak temperature at the center of the board, can be obtained using Eq. (3.6).

This relation can be used effectively in the determination of the temperatures experienced by conductively cooled substrates and conventional printed circuit boards, as well as PCBs with copper lattice on the surface, metal cores, or heat sink plates in the center. In each case it is necessary to evaluate or obtain the effective thermal conductivity of the conducting layer. As an example, consider an alumina substrate, 0.20 m long, 0.15 m wide, and 0.005 m thick with a thermal conductivity of 20 W/mK, whose edges are cooled to 35°C by a cold plate. Assuming that the substrate is populated by 15 compounds, each dissipating 2 W, the substrate center temperature will be found to be equal to 85°C when calculated using Eq. (3.6).

Spreading Resistance

In component packages that provide for lateral spreading of the heat generated in the device(s), the increasing cross-sectional area for heat flow at successive "layers" adjacent to the device reduces the internal thermal resistance. Unfortunately, however, there is an additional resistance associated with this lateral flow of heat. This, of course, must be taken into account in the determination of the overall component package temperature difference.

For the circular and square geometries common in many applications, Negus et al.[5] provided an engineering approximation for the spreading resistance of a small heat source on a thick substrate or heat spreader (required to be 3 to 5 times thicker than the square root of the heat source area), which can be expressed as

$$R_{sp} = \frac{0.475 - 0.62\epsilon + 0.13\epsilon^3}{k\sqrt{A_c}} \tag{3.7}$$

where ϵ is the ratio of the heat source area to the substrate area, k is the thermal conductivity of the substrate, and A_c is the area of the heat source.

For relatively thin layers on thicker substrates, such as encountered in the use of thin lead frames, or heat spreaders interposed between the device and substrate, Eq. (3.7) cannot provide an acceptable prediction of R_{sp}. Instead, use can be made of the numerical results plotted in Fig. 3.4 to obtain the requisite value of the spreading resistance.

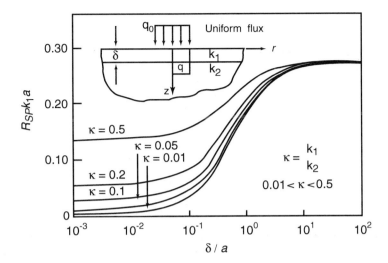

FIGURE 3.4 The thermal resistance for a circular heat source on a two-layer substrate. (From Ref. 6, with permission.)

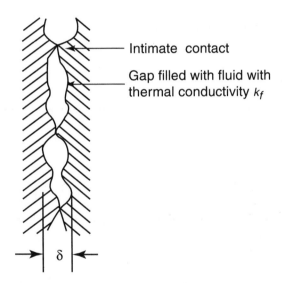

FIGURE 3.5 Physical contact between two nonideal surfaces. (From Ref. 4, with permission.)

Interface/Contact Resistance

Heat transfer across the interface between two solids is generally accompanied by a measurable temperature difference, which can be ascribed to a contact or interface thermal resistance. For perfectly adhering solids, geometrical differences in the crystal structure (lattice mismatch) can impede the flow of photons and electrons across the interface, but this resistance is generally negligible in engineering design. When dealing with real interfaces, the asperities present on each of the surfaces, as shown in an artist's conception in Fig. 3.5, limit actual contact between the two solids to a very small fraction of the apparent interface area. The flow of heat across the gap between two solids in nominal contact is, thus, seen to involve solid conduction in the areas of actual contact and fluid conduction across the "open" spaces. Radiation across the gap can be important in a vacuum environment or when the surface temperatures are high.

The total contact conductance, h_{co}, is taken as the sum of the solid-to-solid conductance, h_c, and the gap conductance, h_g.

$$h_{co} = h_c + h_g \qquad (3.8)$$

The contact resistance based on the apparent contact area, A_a may be defined as

$$R_{co} \equiv \frac{1}{h_{co} A_a} \qquad (3.9)$$

In Eq. (3.8), h_c is given by Yovanovich and Antonetti[6] as

$$h_c = 1.25 k_s \left(\frac{m}{\sigma}\right)\left(\frac{P}{H}\right)^{0.95} \qquad (3.10)$$

where P is the contact pressure and H is the micro-hardness of the softer material (both in Pa), k_s is the harmonic mean thermal conductivity for the two solids with thermal conductivities, k_1 and k_2,

$$k_s = \frac{2 k_1 k_2}{k_1 + k_2}$$

σ is the effective rms surface roughness developed from the surface roughnesses of the two materials, σ_1 and σ_2,

$$\sigma = \sqrt{\sigma_1^2 + \sigma_2^2}$$

and m is the effective absolute surface slope composed of the individual slopes of the two materials, m_1 and m_2,

$$m = \sqrt{m_1^2 + m_2^2}$$

In the absence of detailed information, the σ/m ratio can be assumed to fall into the range of 5 to 9 microns for relatively smooth surfaces.[7]

For normal interstitial gases around atmospheric pressure, h_g in Eq. (3.8) is given by

$$h_g = \frac{k_g}{Y} \qquad (3.11)$$

where k_g is the thermal conductivity of the gap fluid, and Y is the distance between the mean planes given by

$$Y = 1.185 \left[-\ln\left(3.132 \frac{P}{H} \right) \right]^{0.547} \sigma$$

Equations (3.10) and (3.11) can be added, and then in accordance with (3.9), the total contact resistance becomes

$$R_{co} \equiv \left\{ \left[1.25 k_s \left(\frac{m}{\sigma}\right)\left(\frac{P}{H}\right)^{0.95} + \frac{k_g}{Y} \right] A_a \right\}^{-1} \qquad (3.12)$$

Transient Heating or Cooling

An externally heated solid of relatively high thermal conductivity, that is experiencing no external cooling, will undergo a constant rise in temperature according to

$$\frac{dT}{dt} = \frac{q}{mc} \tag{3.13}$$

where q is the rate of internal heat generation, m is the mass of the solid, and c is the specific heat of the solid. Equation (3.13) assumes that all of the mass can be represented by a single temperature and this relation is frequently termed the "lumped capacity" solution for transient heating.

Expanding on the analogy between thermal and electrical resistances, the product of mass and specific heat can be viewed as analogous to electrical capacitance and thus to constitute the "thermal capacitance."

When the same solid is externally cooled, the temperature rises asymptotically toward the steady-state temperature, which is itself determined by the external resistance to the heat flow, R_{ex}. Consequently, the time variation of the temperature of the solid is expressible as

$$T(t) = T(t=0) + qR_{ex}\left[1 - e^{-t/mcR_{ex}}\right] \tag{3.14}$$

The lump capacitance model is accurate when the ratio of the internal conduction resistance of a solid to the external thermal resistance is small. This ratio is represented by the Biot number, and the criterion for applicability of the lumped capacitance model is given as

$$Bi = \frac{hL_c}{k} < 0.1 \tag{3.15}$$

where the characteristic length, L_c, is typically defined as the ratio of the solid's volume to its surface area. More generally, L_c should be taken as the distance over which the solid experiences its maximum temperature difference.[8]

Convective Heat Transfer

The Heat Transfer Coefficient

Convective thermal transport from a surface to a fluid in motion can be related to the heat transfer coefficient, h, the surface-to-fluid temperature difference and the "wetted" surface area, A, in the form

$$q = hA\left(T_s - T_{fl}\right) \tag{3.16}$$

The differences between convection to a rapidly moving fluid, a slowly flowing or stagnant fluid, as well as variations in the convective heat transfer rate among various fluids, are reflected in the values of h. For a particular geometry and flow regime, h may be found from available empirical correlations and/or theoretical relations. Use of Eq. (3.16) makes it possible to define the convective thermal resistance, as

$$R_{cv} \equiv \frac{1}{hA} \tag{3.17}$$

Dimensionless Parameters

Common dimensionless quantities that are used in the correlation of heat transfer data are the *Nusselt number*, Nu, which relates the convective heat transfer coefficient to the conduction in the fluid where the subscript, fl, pertains to a fluid property,

$$\text{Nu} \equiv \frac{h}{k_{fl}/L} = \frac{hL}{k_{fl}} \tag{3.18}$$

the *Prandtl number*, Pr, which is a fluid property parameter relating the diffusion of momentum to the conduction of heat,

$$\text{Pr} \equiv \frac{c_p \mu}{k_{fl}} \tag{3.19}$$

the *Grashof number*, Gr, which accounts for the buoyancy effect produced by the volumetric expansion of the fluid,

$$\text{Gr} \equiv \frac{\rho^2 \beta g L^3 \Delta T}{\mu^2} \tag{3.20}$$

and the *Reynolds number*, Re, which relates the momentum in the flow to the viscous dissipation,

$$\text{Re} \equiv \frac{\rho V L}{\mu} \tag{3.21}$$

Natural Convection

Despite increasing performance demands and advances in thermal management technology, direct air-cooling of electronic equipment continues to command substantial attention. Natural convection is the quietest, least expensive, and most reliable implementation of direct fluid cooling. In more demanding systems, natural convection cooling with air is often investigated as a baseline design to justify the application of more sophisticated techniques.

In natural convection, fluid motion is induced by density differences resulting from temperature gradients in the fluid. The heat transfer coefficient for this regime can be related to the buoyancy and the thermal properties of the fluid through the *Rayleigh number*, Ra, which is the product of the Grashof and Prandtl numbers,

$$\text{Ra} = \frac{\rho^2 \beta g c_p}{\mu k_{fl}} L^3 \Delta T \tag{3.22}$$

where the fluid properties, ρ, β, c_p, μ, and k are evaluated at the fluid bulk temperature, and ΔT is the temperature difference between the surface and the fluid.

Empirical correlations for the natural convection heat transfer coefficient generally take the form

$$h = C\left(\frac{k_{fl}}{L}\right)(\text{Ra})^n \tag{3.23}$$

where n is found to be approximately 0.25 for $10^3 < \text{Ra} < 10^9$, representing laminar flow, 0.33 for $10^9 < \text{Ra} < 10^{12}$, the region associated with the transition to turbulent flow, and 0.4 for $\text{Ra} > 10^{12}$ when strong turbulent flow prevails. The precise value of the correlating coefficient, C, depends on fluid, the geometry of the surface, and the Rayleigh number range. Nevertheless, for natural convection in air from common plate, cylinder, and sphere configurations, it has been found to vary in the relatively narrow range of 0.45 to 0.65 for laminar flow and 0.11 to 0.15 for turbulent flow past the heated surface.[3]

Vertical Channels — Vertical channels formed by parallel printed circuit boards (PCBs) or longitudinal fins are a frequently encountered configuration in natural convection cooling of electronic equipment. The historical work of Elenbaas,[9] a milestone of experimental results and empirical correlations, was the first to document a detailed study of natural convection in smooth, isothermal parallel plate channels. In subsequent years, this work was confirmed and expanded both experimentally and numerically by a number of researchers, including Bodoia[10], Sobel et al.,[11] Aung,[12] Aung et al.,[13] Miyatake and Fujii,[14] and Miyatake et al.[15]

These studies revealed that the value of the Nusselt number lies between two extremes associated with the separation between the plates or the channel width. For wide spacing, the plates appear to have little influence upon one another and the Nusselt number in this case achieves its *isolated plate limit*. On the other hand, for closely spaced plates or for relatively long channels, the fluid attains its *fully developed* velocity profile and the Nusselt number reaches its *fully developed limit*. Intermediate values of the Nusselt number can be obtained from a family of composite expressions developed by Bar-Cohen and Rohsenow[16] and verified by comparison to numerous experimental and numerical studies.

For an isothermal channel, at the fully developed limit, the Nusselt number takes the form

$$Nu = \frac{El}{C_1} \tag{3.24}$$

where *El* is the Elenbaas number, defined as

$$El \equiv \frac{c_p \rho^2 g \beta \left(T_w - T_{amb}\right) b^4}{\mu k L} \tag{3.25}$$

where *b* is the channel spacing, *L* is the channel length, and $(T_w - T_{amb})$ is the temperature difference between the channel wall and the ambient, or channel inlet. For an asymmetric channel, or one in which one wall is heated and the other is insulated, the appropriate values of C_1 is 12, while for symmetrically heated channels, $C_1 = 24$.

For an isoflux channel, at the fully developed limit, the Nusselt number has been shown to take the form

$$Nu = \sqrt{\frac{El'}{C_1}} \tag{3.26}$$

where the modified *Elenbaas number*, *El'*, is defined as

$$El' \equiv \frac{c_p \rho^2 g \beta q'' b^5}{\mu k^2 L} \tag{3.27}$$

where q'' is the heat flux leaving the channel wall(s). When this Nusselt number is based on the maximum wall temperature $(x = L)$, the appropriate values of C_1 are 24 and 48 for the asymmetric and symmetric cases, respectively. When based on the mid-height $(x = l/2)$ wall temperature, the asymmetric and symmetric C_1 values are 6 and 12, respectively.

In the limit where the channel spacing is very large, the opposing channel walls do not influence each other either hydrodynamically or thermally. This situation may be accurately modeled as heat transfer from an isolated vertical surface in an infinite medium. Natural convection from an isothermal plate can be expressed as

$$Nu = C_2 El^{1/4} \tag{3.28}$$

where McAdams[17] suggests a C_2 value of 0.59. However, more recent research suggests that the available data could be better correlated with $C_2 = 0.515$.[8] Natural convection from an isoflux plate is typically expressed as

$$Nu = C_2 El'^{1/5} \tag{3.29}$$

with a leading coefficient of 0.631 when the Nusselt number is based on the maximum $(x = L)$ wall temperature and 0.73 when the Nusselt number is based on the mid-height $(x = l/2)$ wall temperature.

Composite Equations — When a function is expected to vary smoothly between two limiting expressions, which are themselves well defined, and when intermediate values are difficult to obtain, an approximate composite relation can be obtained by appropriately summing the two limiting expressions. Using the Churchill and Usagi[18] method, Bar-Cohen and Rohsenow[19] developed composite Nusselt number relations for natural convection in parallel plate channels of the form

$$Nu = \left[\left(Nu_{fd} \right)^{-n} + \left(Nu_{ip} \right)^{-n} \right]^{-1/n} \tag{3.30}$$

where Nu_{fd} and Nu_{ip} are Nusselt numbers for the fully developed and isolated plate limits, respectively. The correlating exponent n was given a value of 2 to offer good agreement with Elenbaas'[9] experimental results.

For an isothermal channel, combining Eqs. (3.24) and (3.28) yields a composite relation of the form

$$Nu = \left[\frac{C_3}{El^2} + \frac{C_4}{\sqrt{El}} \right]^{-1/2} \tag{3.31}$$

while for an isoflux channel, Eqs. (3.26) and (3.29) yield a result of the form

$$Nu = \left[\frac{C_3}{El'} + \frac{C_4}{El'^{2/5}} \right]^{-1/2} \tag{3.32}$$

Values of the coefficients C_3 and C_4 appropriate to various cases of interest appear in Table 3.1. It is to be noted that the tabulated values reflect a change in the value of the coefficient from 0.59 originally used by Bar-Cohen and Rohsenow[19] in the isolated, isothermal plate limit to the more appropriate 0.515 value.

In electronic cooling applications where volumetric concerns are not an issue, it is desirable to space PCBs far enough apart that the isolated plate Nusselt number prevails along the surface. In lieu of choosing an infinite plate spacing, the composite Nusselt number may be set equal to 99%, or some other high fraction of its associated isolated plate value. The composite Nusselt number relation may then be solved for the appropriate channel spacing.

For an isothermal channel, the channel spacing that maximizes the rate of heat transfer from individual PCBs takes the form

TABLE 3.1 Appropriate Values for the C_i coefficients appearing in Eqs. (3.24)–(3.39).

Case	C_1	C_2	C_3	C_4	C_5	C_6	C_7
Isothermal							
Symmetric heating	24	0.515	576	3.77	4.43	0.00655	2.60
Asymmetric heating	12	0.515	144	3.77	3.51	0.0262	2.06
Isoflux							
Symmetric heating							
Maximum temp.	48	0.63	48	2.52	9.79	0.105	2.12
Midheight temp.	12	0.73	12	1.88	6.80	0.313	1.47
Asymmetric heating							
Maximum temp.	24	0.63	24	2.52	7.77	0.210	1.68
Midheight temp.	6	0.73	6	1.88	5.40	0.626	1.17

$$b_{max} = \frac{C_5}{P^{1/4}} \tag{3.33}$$

where

$$P = \frac{c_p \rho^2 g \beta (T_w - T_{amb})}{\mu k L} = \frac{El}{b^4}$$

while for an isoflux channel, the channel spacing that minimizes the PCB temperature for a given heat flux takes the form

$$b_{max} = \frac{C_5}{R^{1/5}} \tag{3.34}$$

where

$$R = \frac{c_p \rho^2 g \beta q''}{\mu k^2 L} = \frac{El'}{b^5}$$

Values of the coefficient C_5 appropriate to various cases of interest appear in Table 3.1.

Optimum Spacing — In addition to being used to predict heat transfer coefficients, the composite relations presented may be used to optimize the spacing between plates. For isothermal arrays, the optimum spacing maximizes the total heat transfer from a given base area or the volume assigned to an array of plates or printed circuit boards. In the case of isoflux parallel plate arrays, the total array heat transfer for a given base area may be maximized by increasing the number of plates indefinitely, though the plate will experience a commensurate increase in temperature. Thus, it is more appropriate to define the optimum channel spacing for an array of isoflux plates as the spacing that will yield the maximum volumetric heat dissipation rate per unit temperature difference. Despite this distinction, the optimum spacing is found in the same manner.

The total heat transfer rate from an array of vertical, single-sided plates can be written as

$$\frac{Q_T}{LsWk\Delta t} = \left(\frac{Nu}{b(b+d)} \right) \tag{3.35}$$

where the number of plates, $m = W/(b + d)$, d is the plate thickness, W is the width of the entire array, and s is the depth of the channel. The optimum spacing may be found by substituting the appropriate composite Nusselt number equation into Eq. (3.35), taking the derivative of the resulting expression with respect to b, and setting the result equal to zero. Use of an isothermal composite Nusselt number in Eq. (3.35) yields a relation of the form

$$\left(2b + 3d - C_6 P^{3/2} b^7\right)_{opt} = 0 \tag{3.36}$$

or

$$b_{opt} = \frac{C_7}{P^{1/4}} \ (d = 0) \tag{3.37}$$

where d, the plate thickness, is negligible. Use of an isoflux composite Nusselt number yields

$$\left(b + 3d - C_6 R^{3/5} b^4\right)_{opt} = 0 \tag{3.38}$$

or

$$b_{opt} = \frac{C_7}{R^{1/5}} \ \left(d = 0\right) \tag{3.39}$$

Values of the coefficients C_6 and C_7 appropriate to various cases of interest appear in Table 3.1.

Limitations — These smooth-plate relations have proven useful in a wide variety of applications and have been shown to yield very good agreement with measured empirical results for heat transfer from arrays of PCBs. However, when applied to closely spaced printed circuit boards, where the spacing is of the order of the component height, these equations tend to under-predict heat transfer in the channel due to the presence of between-package "wall flow" and the non-smooth nature of the channel surfaces.[20]

Forced Convection

For forced flow in long or very narrow parallel-plate channels, the heat transfer coefficient attains an asymptotic value (a fully developed limit), which for symmetrically heated channel surfaces is approximately equal to

$$h = \frac{4k_{fl}}{d_e} \tag{3.40}$$

where d_e is the *hydraulic diameter* defined in terms of the flow area, A, and the wetted perimeter of the channel, P_w

$$d_e \equiv \frac{4A}{P_w}$$

In the inlet zones of such parallel-plate channels and along isolated plates, the heat transfer coefficient varies with the distance from the leading edge. The low-velocity, or laminar flow, average convective heat transfer coefficient for $Re < 2 \times 10^5$ is given by[3]

$$h = 0.664 \left(\frac{k}{L}\right) Re^{1/2} Pr^{1/3} \tag{3.41}$$

where k is the fluid thermal conductivity and L is the characteristic dimension of the surface. This heat transfer coefficient decreases asymptotically toward the fully developed value given by Eq. (3.40).

A similar relation applies to flow in tubes, pipes, ducts, channels and/or annuli with the equivalent diameter, d_e, serving as the characteristic dimension in both the Nusselt and Reynolds numbers. For laminar flow, Re ≤ 2100

$$\frac{\bar{h}d_e}{k} = 1.86\left[\text{RePr}\left(\frac{d_e}{L}\right)\right]^{1/3}\left(\frac{\mu}{\mu_w}\right)^{0.14} \tag{3.42}$$

which is attributed to Sieder and Tate[21] and where μ_w is the viscosity of the convective medium at the wall temperature. Observe that Eqs. (3.41) and (3.42) show that the heat transfer coefficient from the surface to the fluid is highest for short channels and decreases as L increases.

In higher velocity turbulent flow, the dependence of the convective heat transfer coefficient on the Reynolds number increases and, in the range Re $\geq 3 \times 10^5$, is typically given by[3]

$$h = 0.036\left(\frac{k}{L}\right)(\text{Re})^{0.80}(\text{Pr})^{1/3} \tag{3.43}$$

In pipes, tubes, channels, ducts and/or annuli, transition to turbulent flow occurs at an equivalent diameter-based Reynolds number of approximately 10,000. Thus, the flow regime bracketed by

$$2100 \leq \text{Re} \leq 10,000$$

is usually referred to as the transition region. Hausen[22] has provided the correlation

$$\frac{hd_e}{k} = 0.116\left[\text{Re}-125\right](\text{Pr})^{1/3}\left(1+\frac{d_e}{L}\right)^{2/3}\left(\frac{\mu}{\mu_w}\right) \tag{3.44}$$

and Sieder and Tate[21] give for turbulent flow

$$\frac{hd_e}{k} = 0.23(\text{Re})^{0.80}(\text{Pr})^{1/3}\left(\frac{\mu}{\mu_w}\right) \tag{3.45}$$

Additional correlations for the coefficient of heat transfer in forced convection for various configurations may be found in the heat transfer textbooks.[8,23–25]

Phase Change Heat Transfer

When heat exchange is accompanied by evaporation of a liquid or condensation of a vapor, the resulting flow of vapor toward or away from the heat transfer surface and the high rates of thermal transport associated with the latent heat of the fluid can provide significantly higher heat transfer rates than single-phase heat transfer alone.

Boiling

Boiling heat transfer displays a complex dependence on the temperature difference between the heated surface and the saturation temperature (boiling point) of the liquid. In nucleate boiling, the primary region of interest, the ebullient heat transfer rate is typically expressed in the form of the Rohsenow[26] equation

$$q = \mu_f h_{fg} \sqrt{\frac{g(\rho_f - \rho_g)}{\sigma}} \left[\frac{c_{pf}}{C_{sf} Pr_f^{1.7} h_{fg}} \right]^{1/r} (T_S - T_{sat})^{1/r} \tag{3.46}$$

where $1/r$ is typically correlated with a value of 3, and C_{sf} is a function of characteristics of the surface/fluid combination. Rohsenow recommended that the fluid properties in Eq. (3.46) be evaluated at the liquid saturation temperature.

For pool boiling of the dielectric liquid FC-72 ($T_{sat} = 56°C$ at 101.3 kPa) on a plastic-pin-grid-array (PPGA) chip package, Watwe et al.[27] obtained values of 7.47 for $1/r$ and 0.0075 for C_{sf}. At a surface heat flux of 10 W/cm², the wall superheat at 101.3 kPa is nearly 30°C, corresponding to a average surface temperature of approximately 86°C.

The departure from nucleate boiling, or "Critical Heat Flux" (CHF), places an upper limit on the use of the highly efficient boiling heat transfer mechanism. CHF can be significantly influenced by system parameters such as pressure, subcooling, heater thickness and properties, and dissolved gas content. Watwe et al.[27] presented the following equation to predict the pool boiling critical heat flux of dielectric coolants from microelectronic components and under a variety of parametric conditions.

$$CHF = \left\{ \frac{\pi}{24} h_{fg} \sqrt{\rho_g} \left[\sigma_f g (\rho_f - \rho_g) \right]^{1/4} \right\} \left(\frac{\delta \sqrt{\rho_h c_{ph} k_h}}{\delta \sqrt{\rho_h c_{ph} k_h} + 0.1} \right)$$
$$\times \left\{ 1 + \left[0.3014 - 0.01507 L'(P) \right] \right\} \left\{ 1 + 0.03 \left[\left(\frac{\rho_f}{\rho_g} \right)^{0.75} \frac{c_{pf}}{h_{fg}} \right] \Delta T_{sub} \right\} \tag{3.47}$$

The first term on the right-hand side of Eq. (3.47) is the classical Kutateladze-Zuber prediction, which is the upper limit on the saturation value of CHF on very large horizontal heaters. The second term represents the effects of heater thickness and thermal properties on the critical heat flux. The third term in Eq. (3.47) accounts for the influence of the heater size, where

$$L' = L \sqrt{\frac{g(\rho_f - \rho_g)}{\sigma_f}} \tag{3.48}$$

This third term is only to be included when its value is larger than unity (i.e., $0.3014 - 0.01507 L' > 0$) as small heaters show an increase in CHF over larger heaters. The last term is an equation representing the best-fit line through the experimental data of Watwe et al.[27] and represents the influence of subcooling on CHF. The pressure effect on CHF is embodied in the Kutateladze-Zuber and the subcooling model predictions, which make up Eq. (3.47), via the thermophysical properties. Thus, Eq. (3.47) can be used to estimate the combined influences of various system and heater parameters on CHF. The critical heat flux, under saturation conditions at atmospheric pressure, for a typical dielectric coolant like FC-72 and for a 1-cm component is approximately 15 W/cm². Alternately, at 2 atm and 30°C of subcooling CHF for FC-72 could be expected to reach 22 W/cm².

Condensation
Closed systems involving an evaporative process must also include some capability for vapor condensation. Gerstmann and Griffith[28] correlated film condensation on a downward-facing flat plate as

$$Nu = 0.81 Ra^{0.193} \qquad 10^{10} > Ra > 10^8 \tag{3.49}$$

$$\text{Nu} = 0.69 Ra^{0.20} \qquad 10^8 > Ra > 10^6 \tag{3.50}$$

where,

$$Nu \equiv \frac{h}{k}\left(\frac{\sigma}{g\left(\rho_f - \rho_g\right)}\right)^{1/2} \tag{3.51}$$

$$Ra \equiv \frac{g\rho_f\left(\rho_f - \rho_g\right)h_{fg}}{k\mu\Delta T}\left(\frac{\sigma}{g\left(\rho_f - \rho_g\right)}\right)^{3/2} \tag{3.52}$$

The Nusselt number for laminar film condensation on vertical surfaces was correlated by Nusselt[29] and later modified by Sadasivan and Lienhard[30] as:

$$\text{Nu} = \frac{hL}{k_f} = 0.943\left[\frac{g\Delta\rho_{fg}L^3 h'_{fg}}{k_f v_f\left(T_{sat} - T_c\right)}\right]^{1/4} \tag{3.53}$$

where

$$h'_{fg} = h_{fg}\left(1 + C_c Ja\right)$$

$$C_c = 0.683 - \frac{0.228}{Pr_l}$$

$$Ja = \frac{c_{pf}\left(T_{sat} - T_c\right)}{h_{fg}}$$

Phase Change Materials

In recent years there has been growing use of solid-liquid phase change materials (PCM) to help mitigate the deleterious effects of transient "spikes" in the power dissipation and/or environmental load imposed on RF modules. This is of particular importance for outdoor modules, where PCMs can serve to smooth diurnal variations in the air temperature and solar radiations. To determine the mass of PCM needed to absorb a specified thermal load at a constant (melting) temperature, it is necessary to obtain the latent heat of fusion of that material and insert it in the following relation

$$m = \frac{Q}{h_{fs}} \tag{3.54}$$

Flow Resistance

The transfer of heat to a flowing gas or liquid that is not undergoing a phase change results in an increase in the coolant temperature from an inlet temperature of T_{in} to an outlet temperature of T_{out}, according to

$$T_{out} - T_{in} = \frac{q}{\dot{m}c_p} \tag{3.55}$$

Based on this relation, it is possible to define an effective flow resistance, R_{fl}, as

$$R_{fl} \equiv \frac{1}{\dot{m}c_p} \qquad (3.56)$$

where $\dot{m}$, the mass flow rate, is given in kg/s.

In multicomponent systems, determination of individual component temperatures requires knowledge of the fluid temperature adjacent to the component. The rise in fluid temperature relative to the inlet value can be expressed in a flow thermal resistance, as done in Eq. (3.56). When the coolant flow path traverses many individual components, care must be taken to use R_{fl} with the total heat absorbed by the coolant along its path, rather than the heat dissipated by an individual component. For system-level calculations aimed at determining the average component temperature, it is common to base the flow resistance on the average rise in fluid temperature, that is, one-half the value indicated by Eq. (3.56).

Radiative Heat Transfer

Unlike conduction and convection, radiative heat transfer between two surfaces or between a surface and its surroundings is not linearly dependent on the temperature difference and is expressed instead as

$$q = \sigma A F \left(T_1^4 - T_2^4 \right) \qquad (3.57)$$

where F includes the effects of surface properties and geometry and σ is the Stefan-Boltzman constant, $\sigma = 5.67 \times 10^{-8}$ W/m^{2}K^4. For modest temperature differences, this equation can be linearized to the form

$$q = h_r A \left(T_1 - T_2 \right) \qquad (3.58)$$

where h_r is the effective "radiation" heat transfer coefficient

$$h_r = \sigma F \left(T_1^2 + T_2^2 \right) \left(T_1 + T_2 \right) \qquad (3.59)$$

and, for small $\Delta T = T_1 - T_2$, h_r is approximately equal to

$$h_r = 4\sigma F \left(T_1 T_2 \right)^{3/2} \qquad (3.60)$$

where T_1 and T_2 must be expressed in absolute degrees Kelvin. It is of interest to note that for temperature differences of the order of 10 K with absolute temperatures around room temperature, the radiative heat transfer coefficient, h_r, for an ideal (or "black") surface in an absorbing environment is approximately equal to the heat transfer coefficient in natural convection of air.

Noting the form of Eq. (3.58), the radiation thermal resistance, analogous to the convective resistance, is seen to equal

$$R_r \equiv \frac{1}{h_r A} \qquad (3.61)$$

Environmental Heat Transfer

In applying the foregoing thermal transport relations to microwave equipment located outdoors, attention must be devoted to properly characterizing the atmospheric conditions and including both incoming solar radiation and outgoing night-sky radiation in the heat balance relations. While best results will be obtained by using precise environmental specifications for the "microclimate" at the relevant location, more general specifications may be of use in early stages of product design. The external environment can vary in temperature from −50°C to +50°C, representing the polar regions at one extreme and the subtropical deserts at the other, and experience a range in air pressure from 76 kPa (11 psi), at high plateaus, to 107 kPa (15.5 psi), in deep rift valleys. Incident solar fluxes at noon can reach 1 kW/m² on a horizontal surface, but more typically may average 0.5 kW/m², of both direct and diffuse radiation, during the peak solar hours. The outgoing long-wave radiation from an outdoor module exposed to the clear nighttime sky falls in the range of 0.01 to 0.1 kW/m².[31] It may be anticipated that convective heat transfer coefficients on large exposed surfaces at sea level will attain values of 6 W/m²K for still air to 75 W/m²K, at wind velocities approaching 100 km/h. To determine the surface temperature of an outdoor module, use can be made of the heat balance relation equating the incoming heat — from the microwave components and solar load — with the outgoing heat — by radiation and convection, as

$$q_{rf} + q_{solar} = q_{rad} + q_{conv} \tag{3.62}$$

or

$$
\begin{aligned}
q_{rf} &= q_{rad} + q_{conv} - q_{solar} \\
&= \sigma A_{surf} F\left(T_{surf}^4 - T_{sky}^4\right) + h_{conv} A_{surf}\left(T_{surf} - T_{amb}\right) - \alpha A_{surf} S
\end{aligned}
\tag{3.63}
$$

or

$$T_{surf} = \frac{\dfrac{q_{rf}}{A_{surf}} + \alpha S}{\sigma F\left(T_{surf}^2 + T_{sky}^2\right)\left(T_{surf} + T_{sky}\right) + h_{conv}} + T_{sky} \tag{3.64}$$

where S is the solar incidence (W/m²) and T_{sky} is the effective sky temperature (K) (typically equal to ambient temperature during the day and up to 20 K below the air temperature on a dry, clear night).

Thermal Resistance Networks

The expression of the governing heat transfer relations in the form of thermal resistances greatly simplifies the first-order thermal analysis of electronic systems. Following the established rules for resistance networks, thermal resistances that occur sequentially along a thermal path can be simply summed to establish the overall thermal resistance for that path. In similar fashion, the reciprocal of the effective overall resistance of several parallel heat transfer paths can be found by summing the reciprocals of the individual resistances. In refining the thermal design of an electronic system, prime attention should be devoted to reducing the largest resistances along a specified thermal path and/or providing parallel paths for heat removal from a critical area.

While the thermal resistances associated with various paths and thermal transport mechanisms constitute the "building blocks" in performing a detailed thermal analysis, they have also found widespread application as "figures-of-merit" in evaluating and comparing the thermal efficacy of various packaging techniques and thermal management strategies.

Chip Module Thermal Resistance

Definition

The thermal performance of alternative chip packaging techniques is commonly compared on the basis of the overall (junction-to-coolant or junction-to-ambient) thermal resistance, R_{ja}. This packaging figure-of-merit is generally defined in a purely empirical fashion,

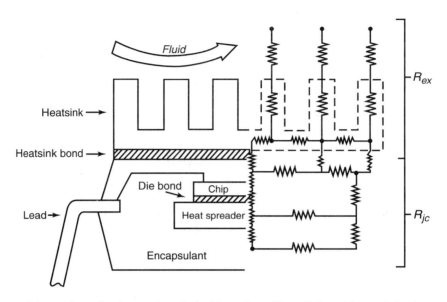

FIGURE 3.6 Primary thermal resistances in a single chip package. (From Ref. 4, with permission.)

$$R_{ja} \equiv \frac{T_j - T_{fl}}{q_c} \qquad (3.65)$$

where T_j and T_{fl} are the junction and coolant (fluid) temperatures, respectively, and q_c is the chip heat dissipation.

Unfortunately, however, most measurement techniques are incapable of detecting the actual junction temperature, that is, the temperature of the small volume at the interface of p-type and n-type semiconductors. Hence, this term generally refers to the average temperature or a representative temperature on the chip.

Examination of various packaging techniques reveals that the junction-to-coolant thermal resistance is, in fact, composed of an internal, largely conductive resistance and an external, primarily convective resistance. As shown in Fig. 3.6, the internal resistance, R_{jc}, is encountered in the flow of dissipated heat from the active chip surface through the materials used to support and bond the chip and on to the case of the integrated circuit package. The flow of heat from the case directly to the coolant, or indirectly through a fin structure and then to the coolant, must overcome the external resistance, R_{ex}.

Internal Thermal Resistance

As previously discussed, conductive thermal transport is governed by the Fourier Equation, which can be used to define a conduction thermal resistance, as in Eq. (3.3). In flowing from the chip to the package surface or case, the heat encounters a series of resistances associated with individual layers of materials, starting with the chip (silicon, galium arsenide, indium phosphide, etc.) and continuing thru solder, copper, alumina, and epoxy, as well as the contact resistances that occur at the interfaces between pairs of materials. Although the actual heat flow paths within a chip package are rather complex and may shift to accommodate varying external cooling situations, it is possible to obtain a first-order estimate of the internal resistance by assuming that power is dissipated uniformly across the chip surface and that heat flow is largely one-dimensional. To the accuracy of these assumptions, Eq. (3.67)

$$R_{jc} = \frac{T_j - T_c}{q_c} = \sum \frac{\Delta x}{kA} \qquad (3.66)$$

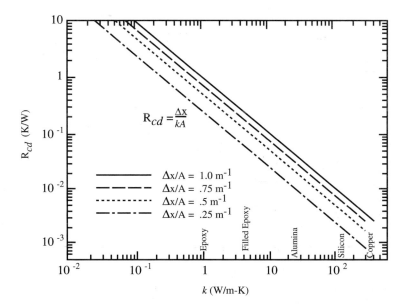

FIGURE 3.7   Conductive thermal resistances for packaging materials. (From Ref. 4, with permission.)

can be used to determine the internal chip module resistance, where the summed terms represent the conduction thermal resistances posed by the individual layers, each with thickness Δx. As the thickness of each layer decreases and/or the thermal conductivity and cross-sectional area increase, the resistance of the individual layers decreases. Values of R_{cd} for packaging materials with typical dimensions can be found via Eq. (3.67) or Fig. 3.7, to range from 2 K/W for a 1000 mm^2 by 1-mm thick layer of epoxy encapsulant to 0.0006 K/W for a 100 mm^2 by 25-micron (1-mil) thick layer of copper. Similarly, the values of conduction resistance for typical "soft" bonding materials are found to lie in the range of approximately 0.1 K/W for solders and 1 to 3 K/W for epoxies and thermal pastes for typical $\Delta x/A$ ratios of 0.25 to 1.0.

Comparison of theoretical and experimental values of R_{jc} reveals that the resistances associated with compliant, low thermal conductivity bonding materials and the spreading resistances, as well as the contact resistances at the lightly loaded interfaces within the package, often dominate the internal thermal resistance of the chip package. It is, thus, not only necessary to correctly determine the bond resistance but to also add the values of R_{sp}, obtained from Eq. (3.7) and/or Fig. 3.4, and R_{co} from Eq. (3.9) or (3.12) to the junction-to-case resistance calculated from Eq. (3.67). Unfortunately, the absence of detailed information on the voidage in the die-bonding and heat-sink attach layers and the present inability to determine, with precision, the contact pressure at the relevant interfaces, conspire to limit the accuracy of this calculation.

External Resistance

An application of Eq. (3.41) or (3.43) to the transfer of heat from the case of a chip module to the coolant shows that the external resistance, $R_{ex} = 1/hA$, is inversely proportional to the wetted surface area and to the coolant velocity to the 0.5 to 0.8 power and directly proportional to the length scale in the flow direction to the 0.5 to 0.2 power. It may, thus, be observed that the external resistance can be strongly influenced by the fluid velocity and package dimensions and that these factors must be addressed in any meaningful evaluation of the external thermal resistances offered by various packaging technologies.

Values of the external resistance for a variety of coolants and heat transfer mechanisms are shown in Fig. 3.8 for a typical component wetted area of 10 cm^2 and a velocity range of 2 to 8 m/s. They are seen to vary from a nominal 100 K/W for natural convection in air, to 33 K/W for forced convection in air, to 1 K/W in fluorocarbon liquid forced convection and to less than 0.5 K/W for boiling in fluorocarbon

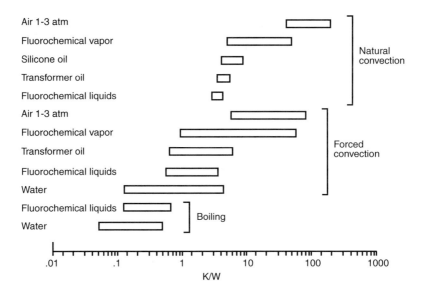

Note: For wetted area = 10 cm^2

FIGURE 3.8 Typical external (convective) thermal resistances for various coolants and cooling modes. (From Ref. 4, with permission.)

liquids. Clearly, larger chip packages will experience proportionately lower external resistances than the displayed values. Moreover, conduction of heat through the leads and package base into the printed circuit board or substrate will serve to further reduce the effective thermal resistance.

In the event that the direct cooling of the package surface is inadequate to maintain the desired chip temperature, it is common to attach finned heat sinks, or compact heat exchangers, to the chip package. These heat sinks can considerably increase the wetted surface area, but may act to reduce the convective heat transfer coefficient by obstructing the flow channel. Similarly, the attachment of a heat sink to the package can be expected to introduce additional conductive resistances, in the adhesive used to bond the heat sink and in the body of the heat sink. Typical air-cooled heat sinks can reduce the external resistance to approximately 10 to 15 K/W in natural convection and to as low as 3 to 5 K/W for moderate forced convection velocities.

When a heat sink or compact heat exchanger is attached to the package, the external resistance accounting for the bond-layer conduction and the total resistance of the heat sink, R_{hs}, can be expressed as

$$R_{ex} = \frac{T_c - T_{fl}}{q_c} = \sum \left(\frac{\Delta x}{kA} \right)_b + R_{hs} \tag{3.67}$$

where R_{hs}

$$R_{hs} = \left[\frac{1}{nhA_f\eta} + \frac{1}{h_b A_b} \right]^{-1} \tag{3.68}$$

is the parallel combination of the resistance of the *n* fins

$$R_f = \frac{1}{nhA_f\eta} \tag{3.69}$$

and the *bare* or base surface not occupied by the fins

$$R_b = \frac{1}{h_b A_b}$$
(3.70)

Here, the base surface is $A_b = A - A_f$ and use of the heat transfer coefficient, h_b, is meant to recognize that the heat transfer coefficient that is applied to the base surfaces is not necessarily equal to that applied to the fins.

An alternative expression for R_{hs} involves an *overall surface efficiency, η_o,* defined by

$$\eta_o = 1 - \frac{nA_f}{A}\left(1-\eta\right)$$
(3.71)

where A is the total surface composed of the base surface and the finned surfaces of n fins

$$A = A_b + nA_f$$
(3.72)

In this case, it is presumed that $h_b = h$ so that

$$R_{hs} = \frac{1}{h\eta_o A}$$
(3.73)

In an optimally designed fin structure, η can be expected to fall in the range of 0.50 to 0.70.[4] Relatively thick fins in a low velocity flow of gas are likely to yield fin efficiencies approaching unity. This same unity value would be appropriate, as well, for an unfinned surface and, thus, serve to generalize the use of Eq. (3.68) to all package configurations.

Total Resistance — Single Chip Packages

To the accuracy of the assumptions employed in the preceding development, the overall single chip package resistance, relating the chip temperature to the inlet temperature of the coolant, can be found by summing the internal, external, and flow resistances to yield

$$R_{jaj} = R_{jc} + R_{ex} + R_{fl}$$

$$= \sum \frac{\Delta x}{kA} + R_{int} + R_{sp}$$
(3.74)

$$= \frac{1}{\eta h A} + \left(\frac{Q}{q}\right)\left(\frac{1}{2\rho Q c_p}\right)$$

In evaluating the thermal resistance by this relationship, care must be taken to determine the effective cross-sectional area for heat flow at each layer in the module and to consider possible voidage in any solder and adhesive layers.

As previously noted in the development of the relationships for the external and internal resistances, Eq. (3.75) shows R_{ja} to be a strong function of the convective heat transfer coefficient, the flowing heat capacity of the coolant, and geometric parameters (thickness and cross-sectional area of each layer). Thus, the introduction of a superior coolant, use of thermal enhancement techniques that increase the local heat transfer coefficient, or selection of a heat transfer mode with inherently high heat transfer coefficients (boiling, for example) will all be reflected in appropriately lower external and total thermal resistances. Similarly, improvements in the thermal conductivity and reduction in the thickness of the relatively low conductivity bonding materials (such as soft solder, epoxy, or silicone) would act to reduce the internal and total thermal resistances.

Weighted-Average Modification of R_{jc}

The commonly used junction-to-case thermal resistance, relying on just a single case temperature, car be used with confidence only when the package case is nearly isothermal. In a more typical packaging configuration, when substantial temperature variations are encountered among and along the external surfaces of the package,[32–34] the use of the reported R_{jc} can lead to erroneous chip temperature predictions. This is especially of concern in the analysis and design of plastic chip packages, due to the inherently high thermal resistance of the plastic encapsulant and the package anisotropies introduced by the large differences in the conductivity between the lead frame and/or heat spreader and the plastic encapsulant. Since R_{jc} is strictly valid only for an isothermal package surface, a method must be found to address the individual contributions of the various surface segments according to their influence on the junction temperature.

Following Krueger and Bar-Cohen,[35] it is convenient to introduce the expanded R_{jc} methodology with a thermal model of a chip package that can be approximated by a network of three thermal resistances connected in parallel, from the chip to the top, sides, and bottom of the package, respectively. This type of compact model is commonly referred to as a "star network" and, in this model, the heat flow from the chip is

$$q = q_1 + q_2 + q_3 \tag{3.75}$$

or

$$q = \frac{T_j - T_1}{R_1} + \frac{T_j - T_2}{R_2} + \frac{T_j - T_3}{R_3} \tag{3.76}$$

This compact model of an electronic device is shown schematically in Fig. 3.9.

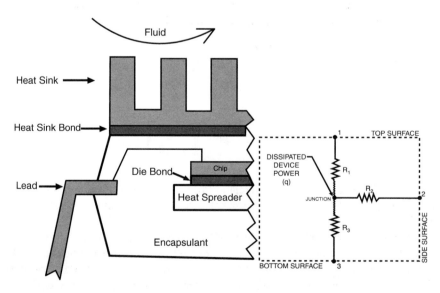

FIGURE 3.9 Geometry of a 28-lead PLCC device. (a) The compact model schematic, and (b) the actual device cross-section.[35]

Equation (3.77) can be rearranged to yield the dependence of the chip (or junction) temperature on the temperature of the three surface segments as

$$T_j = \left(\frac{R_2 R_3}{R_s}\right) T_1 + \left(\frac{R_3 R_1}{R_s}\right) T_2 + \left(\frac{R_1 R_2}{R_s}\right) T_3 + \left(\frac{R_1 R_2 R_3}{R_s}\right) q \tag{3.77}$$

where $R_s = R_1 R_2 + R_1 R_3 + R_2 R_3$

Equation (3.78) may be generalized to admit n-distinct elements along the package surface, or

$$T_j = \sum_{k=1}^{n} I_k T_k + I_{n+1} q \tag{3.78}$$

A comparison of Eqs. (3.78) and (3.79) shows that the coefficients of the specified surface temperatures, the I_k's are totally determined by the internal resistances of the chip package

$$I_1 = \frac{R_2 R_3}{R_s} \qquad I_2 = \frac{R_3 R_1}{R_s}$$
$$I_3 = \frac{R_1 R_2}{R_s} \qquad I_4 = \frac{R_1 R_2 R_3}{R_s} \tag{3.79}$$

The temperature coefficients needed to generate a junction temperature relation of the form shown in Eq. (3.79) can thus be determined from previously calculated internal resistance or, in the absence of such values, by extraction from empirical data or numerical results for the junction temperature. Furthermore, it is to be noted that the sum of the coefficients of the various surface temperature is equal to unity and that the power dissipation coefficient, $I_{n+1} q$, is, in fact, the familiar R_{jc} isothermal, junction-to-case thermal resistance. Consequently, Eq. (3.79) may be rewritten as

$$T_j = \sum_{k=1}^{n} I_k T_k + R_{jc} q \tag{3.80}$$

or, returning to R_{jc}

$$R_{jc} = \frac{T_j - \sum_{k=1}^{n} I_k T_k}{q} = \frac{T_j - \overline{T}_c}{q} \tag{3.81}$$

where $\overline{T}_c$ is the average case temperature

$$\overline{T}_c = \frac{\sum_{k=1}^{n} A_k T_k}{A_T} \tag{3.82}$$

where A_k is the surface area of the kth surface and A_T is the surface area of the entire package.

In many applications, chip packages are cooled selectively along particular exposed surfaces. One such example is a package cooled from the top and side surfaces while the bottom surface is insulated. The thermally active surfaces may vary from application, to application and the thermal analyst needs to quantify the effect of thermally insulating one or more areas on a package of known thermal resistance.

For the assumptions used in the development of the expanded R_{jc} model, insulation of surface-m results in zero heat flow through resistance, R_m. This causes the temperature of surface-m to equal the chip temperature. With this in mind, the junction temperature for a package with a single insulated surface given by Eq. (3.81) is found to equal

$$T_j = \sum_{k \neq m}\left(\frac{I_k}{1-I_m}\right)T_k + \left(\frac{I_{n+1}}{1-I_m}\right)q \tag{3.83}$$

The weighted average case temperature for this thermal configuration is found to equal

$$\bar{T}_c = \sum_{k \neq m}\left(\frac{I_k}{1-I_m}\right)T_k \tag{3.84}$$

and the modified junction to case resistance, R_{jc}^* is

$$R_{jc}^* = \frac{R_{jc}}{1-I_m} \tag{3.85}$$

Multichip Modules

The thermostructural complexity of multichip modules in current use hampers effective thermal characterization and introduces significant uncertainty in any attempt to compare the thermal performance of these packaging configurations. Variations in heat generation patterns across the active chips (reflecting differences in functional requirements and the duty cycle among the macrocells constituting a particular chip), as well as nonuniformities in heat dissipation among the chips assembled in a single module, further complicate this task. While nonthermal issues (e.g., electromagnetic crosstalk) may often be the dominant limiting factors in RF multichip modules, the high power associated with microwave devices makes it essential that the thermal performance of this packaging configuration be analyzed and reported in a consistent manner. To establish a common basis for comparison of multichip modules, it is possible to neglect the on-chip and chip-to-chip variations and consider that the heat generated by each chip flows through a unit cell of the module structure to the external coolant.[36,37] For a given structure, increasing the area of the unit cell allows heat to spread from the chip to a larger cross-section, reducing the heat flux at some of the thermally critical interfaces and at the convectively cooled surfaces. Consequently, the thermal performance of a multichip module can be best represented by the area-specific thermal resistance, i.e., the temperature difference between the chip and the coolant divided by the substrate heat flux, expressed in units of K/(W/cm²). This figure of merit is equivalent to the inverse of the overall heat transfer coefficient, U, commonly used in the compact heat exchanger literature. Despite significant variation in design and fabrication, the leading edge water-cooled and air-cooled modules of the late 1980s provided a specific thermal resistance of approximately 20°C for every watt per square centimeter at the substrate. A decade later, the thermally best multichip modules of the 1990s offered specific thermal resistance values between 5 and 10 K/(W/cm²).

Radar System Applications

The frequent demand for high radiated power in civilian and military radar systems, for the ranging and detection of remote objects, has led to the development of many high-energy microwave systems. Due to the inefficiencies inherent in electrical-to-microwave energy conversion and the management of radio-frequency (RF) energy, the operation of such radar equipment often results in significant heat dissipation. To avoid catastrophic failures, to achieve the reliability targets of the system, and to satisfy the frequency stability requirements of the RF tubes, system requirements commonly specify temperature control to within several degrees Celsius around 150°C. These thermal requirements necessitate the use of aggressive thermal management techniques, including the use of high flow rate liquid forced convection, pool and flow boiling, and high pressure-drop, air-cooled compact heat exchangers.[3]

In traditional, mechanically steered radar systems, much of the total power dissipation occurs in the power tubes (e.g., klystron, gyrotrons, amplitrons), with secondary dissipation in the rotary joints and wave guides. Heat release along the collector surfaces of the tubes is proportional to the RF power of the system and can range from 1 kW to as much as 20 kW, with a peak local flux of several watts/cm² to, perhaps, several thousand watts/cm², at operating temperatures of 150°C.[3] Similar, though less severe thermal problems are encountered in the RF rotary joints and waveguides, where heat fluxes of 5 W/cm² to 10 W/cm², with allowable temperatures of 150°C, may be encountered.

Growing applications of active antenna array elements, utilizing amplifier diodes and solid-state phase shifters to provide electronic steering of the radiated RF beam, have introduced new dimensions into the thermal control of microwave equipment. In such "phased array" radar systems, high power tubes and waveguides can be eliminated with low power RF delivered to each antenna element. The low conversion efficiency of the amplifier diodes results in diode heat dissipation between 1 W and 10 W and local heat fluxes comparable to power transistors. In "phased array" antenna, precise shaping of the radiated wave often requires relatively cool (<90°C) ferrite phase shifters and no significant temperature variations (<10°C) across the elements used in the array.

References

1. M. Pecht, P. Lall, and E. Hakim, The influence of temperature on integrated circuit failure mechanisms, *Advances in Thermal Modeling of Electronic Components and Systems,* vol. 3, A. Bar-Chen and A.D. Kraus, Eds., ASME Press, New York, 1992.
2. R.A. Morrison, Improved avionics reliability through phase change conductive cooling, *Proceedings, IEEE National Telesystems Conference,* 1982, B5.6.1–B5.6.6.
3. A.D. Kraus and A. Bar-Cohen, *Thermal Analysis and Control of Electronic Equipment,* McGraw-Hill, New York, 1983.
4. A.D. Kraus and A. Bar-Cohen, *Design and Analysis of Heat Sinks,* John Wiley & Sons, New York, 1995.
5. K.J. Negus, M.M. Yovanovich, and J.V. Beck, On the non-dimensionalization of constriction resistance for semi-infinite heat flux tubes, *J. Heat Trans.,* 1111, 804–807, 1989.
6. M.M. Yovanovich and V.W. Antonetti, Application of thermal contact resistance theory to electronic packages, in *Advances in Thermal Modeling of Electronic Components and Systems,* A. Bar-Cohen and A.D. Kraus (eds.), Hemisphere Publishing Co, New York, 1988, 79, 28.
7. M.M. Yovanovich, personal communication, 1990.
8. F.P. Incropera and D.P. Dewitt, *Introduction to Heat Transfer,* John Wiley and Sons, New York, 1996.
9. W. Elenbaas, Heat dissipation of parallel plates by free convection, *Physica,* 9, 1, 665–671, 1942.
10. J.R. Bodoia and J.F. Osterle, The development of free convection between heated vertical plates, *J. Heat Transfer,* 84, 40–44, 1964.
11. N. Sobel, F. Landis, and W.K. Mueller, Natural convection heat transfer in short vertical channels including the effect of stagger, *Proc. 3d Int. Heat Transfer Conf.,* Chicago, 2, 1966, 121–125.
12. W. Aung, Fully developed laminar free convection between vertical plates heated asymmetrically, *Int. J. Heat Mass Transfer,* 15, 40–44, 1972.
13. W. Aung, L.S. Fletcher, and V. Sernas, Developing laminar free convection between vertical flat plates with asymmetric heating, *Int. J. Heat Mass Transfer,* 15, 2293–2308, 1972.
14. O. Miyatake and T. Fujii, Free convection heat transfer between vertical parallel plates — one plate isothermally heated and the other thermally insulated, *Heat Transfer Jpn. Res.,* 3, 30–38, 1972.
15. O. Miyatake, T. Fujii, M. Fujii, and H. Tanaka, Natural convection heat transfer between vertical parallel plates — one plate with a uniform heat flux and the other thermally insulated, *Heat Transfer Jpn. Res.,* 4, 25–33, 1973.

16. A. Bar-Cohen and W.M. Rohsenow, Thermal optimum spacing of vertical, natural convection cooled, vertical plates, *J. Heat Trans.,* 106, 116–122, 1984.

17. W.H. McAdams, *Heat Transmission,* McGraw-Hill, New York, 1954.

18. S.W. Churchill and R. Usagi, A. general expression for the correlation of rates of transfer and other phenomena, *AIChE J.,* 18, 6, 1121–1138, 1972.

19. A. Bar-Cohen and W.M. Rohsenow, Thermally optimum spacing of vertical, natural convection cooled, parallel plates, *Heat Transfer in Electronic Equipment,* ASME HTD 20, ASME WAM, Washington, D.C., 11–18, 1981.

20. A. Bar-Cohen, Bounding relations for natural convection heat transfer from vertical printed circuit boards, *Proceedings of the IEEE,* September 1985.

21. E.N. Sieder and G.E. Tate, Heat transfer and pressure drops of liquids in tubes, *Ind. Eng. Chem.,* 28, 1429–1435, 1936.

22. H. Hausen, Darstelling des wärmeuberganges in rohren durch verallgemeinerte potenzbeziehungen, *VDI Z,* 4, 91–98, 1943.

23. A. Bejan, *Heat Transfer,* John Wiley & Sons, New York, 1993.

24. J.P. Holman, *Heat Transfer,* McGraw-Hill, New York, 1990.

25. J.H. Lienhard, *A Heat Transfer Textbook,* Prentice-Hall, Englewood Cliffs, NJ, 1987.

26. W.M. Rohsenow, A method of correlating heat transfer data for surface boiling of liquids, *Transactions of ASME* (reprinted in *3rd ASME/JSME Thermal Engineering Joint Conference,* 1, 503–512), 74, 969–976, 1951.

27. A.A. Watwe, A. Bar-Cohen, and A. McNeil, Combined pressure and subcooling effects on pool boiling from a PPGA chip package, *ASME J. Electronics Packaging,* March 1997.

28. F. Gertsmann and P. Griffith, Laminar film condensation on the underside of horizontal and inclined surfaces, *Int. J. Heat Mass Trans.,* 10, 567–580, 1966.

29. W.Z. Nusselt, Die oberflächencondensation der wasserdamfes, *Z. Ver. Deut. Ing.,* 60, 541–569, 1916.

30. P. Sadasivan and J.H. Lienhard, Sensible heat correction in laminar film boiling and condensation, *J. Heat Trans.,* 109, 545–546, 1987.

31. C. Rambach and A. Bar-Cohen, Nocturnal heat rejection by skyward radiation, in *Future Energy Production Systems,* J.C. Denton and N. Afgan, eds., Hemisphere Publishing, New York, 1976, 713–726.

32. J.A. Andrews, Package thermal resistance model dependency on equipment design, *IEEE CHMT Trans.,* 11, 4, 528–537, 1988.

33. S.S. Furkay, Thermal characterization of plastic and surface mount components, *IEEE CHMT Trans.,* 11, 4, 521–527, 1988.

34. E.A. Wilson, Factors influencing the interdependence of R_{jc} and R_{ca}, *Proceedings of the Second International Electronic Packaging Society Meeting,* 247–255, 1981.

35. W.B. Krueger and A. Bar-Cohen, Thermal characterization of a PLCC — expanded R_{jc} methodology, *IEEE CHMT Trans.,* 15, 5, 691–698, 1992.

36. A. Bar-Cohen, Thermal management of air- and liquid-cooled multichip modules, *IEEE CHMT Trans.,* CHMT-10, 2, 159–175, 1987.

37. A. Bar-Cohen, Addendum and correction to thermal management of air- and liquid-cooled multichip modules, *IEEE CHMT Trans.,* CHMT-11, 3, 333–334, 1988.

3.2 Safety and Environmental Issues

John M. Osepchuk and Ronald C. Petersen

This section is aimed at providing the modern microwave engineer with fundamental facts about the bioeffects and potential hazards of exposure to microwave energy. This subject had a tortured history in the last half of the 20th century and is often characterized by misinformation, especially in the modern media. The research in this field, though very extensive, is often of poor quality and often marred by

TABLE 3.2 Approximate Dielectric Parameters for Muscle Tissue at Various Frequencies[a]

Frequency (MHz)	Relative Dielectric Constant (ε_r)	Conductivity (σ) (S/m)	Penetration Depth (δ) (cm)
0.1	1850	0.56	213
1.0	411	0.59	70
10	131	0.68	13.2
100	79	0.81	7.7
1000	60	1.33	3.4
10,000	42	13.3	0.27
100,000	8	60	0.03

[a] Muscle tissue, field parallel to tissue fibers.[3]

microwave artifacts. Scientists and engineers of all disciplines critically review this literature and derive, by broad consensus and due process of the IEEE (the Institute of Electrical and Electronics Engineers), sound safety standards with reasonable safety factors. The standards literature deals not only with potential effects on the human body but also allied effects on fuel, electro-explosive devices (EEDs), and medical devices. Related subjects are the expanding field of medical applications of microwave energy, as in hyperthermia, and the growing threat of interference between microwave systems — especially between wireless LANs operating in the ISM (Industrial, Scientific, and Medical) band of 2.40 to 2.5 GHz, where many microwave-power systems operate, including roughly 200 million microwave ovens in the world. Although science-based standards are the rational basis for developing safety practices in the presence of microwave energy, there is some confusion presented by recent ideas such as the Precautionary Principle. Thus, there is a need for continuing education on this subject from responsible professional organizations such as the IEEE.

Characteristics of Biological Tissue and RF Absorption Properties

Modern IEEE standards for safe exposure to RF electromagnetic energy cover frequencies up to 300 GHz and down to at least 3 kHz. Although the term "microwaves" usually means frequencies well above 100 or 300 MHz, related bioeffects/hazards exist down to roughly 100 kHz. These are thermal in nature. Below 100 kHz the dominant effects are electrostimulatory in nature. We will confine our attention to thermal effects and refer the reader to an authoritative treatment on the subject of electrostimulation by Reilly.[1]

In order to understand any bioeffect caused by exposure of a biological body to microwave/RF energy one needs to have an idea of the distribution of the internal E and B fields generated by the exposure. In turn one must have some information about the dielectric properties of the tissues in the biological body. Later we will refer to modern computer modeling of the absorption using elaborate anatomically correct models of the body that are heterogeneous in general with muscle, bone, skin, etc. Some simple but important properties of absorption can be gained from simple models.

The dielectric properties of various tissues have been tabulated in a popular reference called the *Radiation Dosimetry Handbook,* edited by Durney et al.[2] In recent years these data, particularly for bone, have been improved by Gabriel et al.[3] From these sources we can tabulate the approximate values for muscle-like tissue as in Table 3.2.

In Table 3.2, the relative dielectric constant is shown as well as the conductivity. The complex permittivity is given by:

$$\varepsilon = \varepsilon_r \varepsilon_o + j\frac{\sigma}{\omega} \qquad (3.86)$$

where $\varepsilon_o = 8.86 \times 10^{-12}$ f/m. The penetration depth is that at which a plane wave is attenuated by a factor of e in E-field or 8.69 dB. Its classical derivation is presented in Durney et al.[2] We see that penetration at low RF frequencies is considerably more than 10 cm, but above 6 GHz the penetration depth rapidly decreases to a millimeter or less in the millimeter-wave range of the microwave spectrum. The penetration depth is one basic factor in determining how much energy reaches deep into the body. The other factor is the reflection at the external surface, or at lower frequencies the shunting of the electric field by a conducting body. For a small spherical object, Schwan[4] has shown that at 60 Hz, the internal E field is nearly six orders of magnitude less than the external E field, even though the theoretical penetration depth is quite large at low frequencies. It has been estimated by Osepchuk[5] that only around the "resonance" frequency of man, i.e., around 100 MHz, is the internal E field deep in the body within one order of magnitude of the external field. At very high frequencies, e.g., in the millimeter-wave range, the E field deep in the body is many orders of magnitude below the external field because the penetration depth is only one millimeter or less.

The principles of modern dosimetry have recently been reviewed by Chou et al.[6] The specific absorption rate (SAR), i.e., the mass averaged rate of energy absorption in tissue, is related to E by

$$SAR = \frac{\sigma |E|^2}{\rho} \quad W/kg \tag{3.87}$$

where σ is the conductivity of the tissue in Siemens per meter, ρ is the density in kg per cubic meter, and E is the rms electric field strength in volts per meter. SAR is thus a measure of the electric field at the point under study and it is also a measure of the local heating rate dT/dt, viz.:

$$\frac{dT}{dt} = \frac{SAR}{c} \quad °C/s \tag{3.88}$$

where c is the specific heat capacity of the tissue in J/kg °C. Thus, a SAR of 1 W/kg is associated with a heating rate less than 0.0003°C per second in muscle tissue ($c \cong 3.5$ kJ/kg°C). Clearly this is a very small heating rate since even without blood or other cooling it would take more than one hour to increase the temperature 1 degree Celsius.

The SAR concept is a key concept in planning and analysis of experiments, both *in vivo* and *in vitro*, as well as the formulation of safe exposure limits for humans. Both the local SAR value and the whole-body average are important in these endeavors. There is an extensive literature on the calculation of whole-body average SAR for various models of animals, including man, especially those based on ellipsoids, which are summarized in the *Dosimetry Handbook*.[2] In Fig. 3.10 we show the calculated SAR for an average man based on such a model, when exposed to a plane wave of power density of 1 mW/cm². Shown is the whole-body-averaged SAR vs. frequency for three polarizations; E — in which the E-field is parallel to the main axis, H — in which the H-field is parallel to the main axis, and k — in which the direction of propagation is parallel to the main axis of the body. We see that there is a low-Q resonance at about 70 to 80 MHz for a standard man (and at about half that frequency when standing on a conducting ground plane). The peak absorption is highest for E polarization and is equal to about 0.2 W/kg per mW/cm² incident power density. At high frequencies the SAR decreases to an asymptotic "quasi-optical" value 5 to 6 times lower than the SAR peak. At very low frequencies the SAR varies as f^2, as expected. The peak SAR values for animals are higher, e.g., for a mouse at its resonance frequency of about 2 GHz, the peak SAR is somewhat over 1.0 W/kg per mW/cm².

The SAR distributions within the body are quite complicated even when resulting from exposure to the simplest plane wave in the far field. Depending upon the body position and frequency, it is possible that there are one or more "hot spots" of SAR peaks within the animal body. Based upon the work of

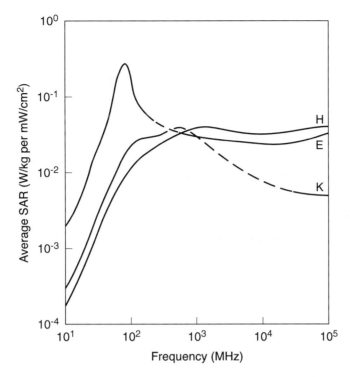

FIGURE 3.10 Calculated whole-body average SAR versus frequency for models of the average man for three standard polarizations. The incident power density is 1 mW/cm^2. (From Durney, et al.[2])

Kritikos and Schwan,[7] however, such internal SAR peaks are very unlikely for man, but are more probable for small animals.

Bioeffects and Hazards of Microwave/RF Energy

More than 20,000 papers have been published in world literature about the bioeffects of microwave/RF energy. Many deal with experiments with animals, i.e., *in vivo*, especially at frequencies between 10 MHz and 60 GHz. The early literature before 1980 showed a major share of these experiments with small animals were done at 2.45 GHz, at which, as the microwave-oven frequency, there is an inexpensive source of power. Substantial literature is also available about *in vitro* experiments in which small samples of tissue or cells are exposed in a variety of exposure systems. These occur more after 1980. Epidemiological studies are few and tend to have been done in recent years. For most, exposure assessment is at best questionable and at worst nonexistent. One should not forget that a large amount of human exposure data exist from the history[8] of diathermy at 27 MHz and at 2.45 GHz. Millions of patients were treated with up to 125 watts of power applied to various parts of the body for 15 to 30 minutes. In recent years, diathermy has been less popular, perhaps because of the growth of electrophobia directed toward the "radiation" aspect of "microwave" energy. In the last few decades, however, this human exposure data has been tremendously expanded in the medical area by the widespread use of MRI (magnetic resonance imaging) in which RF energy at VHF frequencies and of the order of 100 watts is applied to the body. Lastly there are a variety of other medical applications[9] of microwave/RF energy such as hyperthermia for the treatment of cancer.

To better understand experiments in microwave exposure as well as their relation to safety standards, it is useful to refer to the "exposure diagram" of Fig. 3.11. In this diagram, with log-log coordinates of power (or power density or SAR) on the ordinate and time on the abscissa, we can draw the threshold

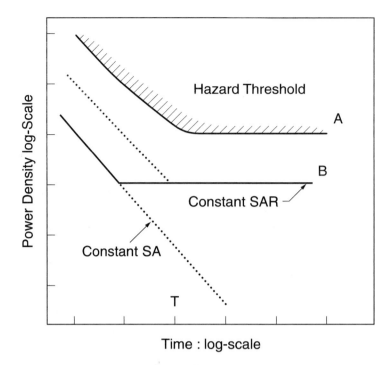

FIGURE 3.11   Thresholds for various effects and hazards expressed as a function of time.

for various effects and hazards. For example, to heat a finite sample to a given temperature, the threshold is a constant SAR for long periods of time while for small periods of time, during which no heat is lost from the sample, the threshold curve is a line of constant SA (specific absorption = SAR × time) which is at 45 degrees from the horizontal in Fig. 3.11. The intersection of the two lines, constant SAR and constant SA, determines the applicable thermal time constant or associated "averaging time" in exposure standards. Similar curves would result for the threshold for burns using the classic data of Henriques and Moritz[10] for threshold temperature for burns, which is around 60°C for 5 seconds but approaching 45°C for long exposure times, where 45°C is also the threshold temperature for pain sensation in man.

Thus, much early data described by such threshold curves described the lethal thresholds of exposure of animals in terms of power density and exposure duration, usually at room temperature. For example, Michaelson[11] found that for a dog the lethal threshold at 2.8 GHz was about 2 to 4 hours at 165 mW/cm². On the other hand Addington[12] found that at 200 MHz, for a dog the lethal threshold was only 20 minutes at 220 mW/cm². In the Soviet literature[13] the lethal threshold reported for the rat was 40 mW/cm² for 90 minutes at 3 GHz but at 70 MHz, the threshold was about 1000 mW/cm² for 100 minutes. Although not appreciated at those early times, these results appear quite reasonable in terms of the expected SAR absorption curves and respective resonance frequencies for the various size animals. This physical understanding based on heating was further strengthened when experiments[14] with fruit flies (Drosophilae) showed no effect when exposed at 2.45 GHz with over 6500 mW/cm² and 45 minutes exposure duration. This result is eminently reasonable to the engineer well acquainted with the absorption cross-section theory that shows absorption decreasing rapidly as the square of the animal dimension. (It also explains the mystifying — to the layman — observation that small ants are not perturbed in an operating microwave oven). In 1971 Samaras et al.[15] demonstrated the expected, but still dramatic, dependence on environmental temperature. At room temperature the lethal threshold for a rat at 2.45 GHz for a 17-minute exposure was 100 mW/cm², but at freezing temperatures below 0°C that same power density was life preserving for the rat.

In 1979 Tell and Harlen[16] analyzed a host of data in the literature demonstrating thermal effects in animals. They showed how all the animal data seemed to form a coherent picture and extrapolated to man so as to suggest that, at least for frequencies above 1 GHz, 100 mW/cm^2 was a conservative estimate of the threshold for a 1°C core temperature rise for exposure durations of more than an hour. Their analysis suggested also that the thermal time constant for whole-body heating of man is an hour or more. In the last decade, however, actual experiments with humans by Adair[17] have shown that exposures at 450 MHz and 2.45 GHz for 45 minutes showed no core temperature rise at power densities corresponding to at least ten times the basis of the current safety limits (expressed as 4 W/kg). Adair is proceeding with further experiments with humans in the resonance frequency range below 100 MHz.

Many endpoints of the health of animals have been studied in the thousands of experiments reported in the literature. We will not specifically review all categories but special mention of "cataracts" is justified in view of the myths that were attached to this subject in the 1970s. Cataracts, i.e., opacities of the lens that interfere with vision, are probably the most studied effect associated with exposure to microwave energy. The hard science on the subject shows that the threshold[18] for cataracts in the rabbit is of the order of 180 mW/cm^2 for an exposure duration of one-half hour or more at 2.45 GHz. This result was obtained when the rabbit was restrained or under anesthesia, and only when the energy was applied locally to the eyes with near-field applicators such as diathermy antennas. At X-band, attempts to produce cataracts resulted in skin burns first around the eyes. Attempts to produce cataracts at UHF resulted in the death of the animal before the cataract could be produced. Long-term exposures of the rabbit by Guy et al.[19] at 2.45 GHz and 10 mW/cm^2 showed no ocular damage. Microwave-induced cataracts have not been demonstrated in primates, but in the last two decades there have been some reports[20] of corneal damage from high peak power pulsed fields at moderate average power densities around 10 mW/cm^2. Attempts to replicate these findings have failed.[21]

The extensive literature on microwave bioeffects has been surveyed often. Some of the classic papers are reproduced, along with extensive bibliographies and commentaries in a reprint volume[22] produced by the IEEE Committee on Man and Radiation (COMAR). Other good reviews of the literature before 1990 include a special issue[23] of the *Proceedings of the IEEE* and an extensive review[24] by the EPA.

Over the years Polson and Heynick[25] have produced critical reviews of the literature, under the sponsorship of the U. S. Air Force. An excellent text was authored by Michaelson and Lin[26] and multi-author books have been edited by Gandhi[27] and Polk and Postow.[28] Although most confirmed bioeffects are associated with significant temperature rise in experimental animals there is one exception, the microwave auditory effect.[29] It has been shown that exposure of the head to microwave pulses results in audible clicks above a threshold of roughly 40 µJ/cm^2 energy density incident on the head at 2.45 GHz. This effect is not believed hazardous but it has been used in some safety guidelines to set limits for exposures to pulsed fields. Special mention should be made of DeLorge's[30] studies of the disruption of food-motivated learned behavior of animals. Thresholds for this effect, which is believed to be the most sensitive and reproducible known effect, have been the principal basis for most modern safety standards beginning with the series of C95 standards produced under IEEE sponsorship. Disruption occurs reliably at whole-body-averaged SARs between 2 to 3 and 9 W/kg across frequency and animal species from mice to baboons (see Table 3.3).[43,48]

TABLE 3.3 Comparison of Power Density and SAR Thresholds for Behavioral Disruption in Trained Laboratory Animals

Species and Conditions	225 MHz (CW)	1.3 GHz (Pulsed)	2.45 GHz (CW)	5.8 GHz (Pulsed)
Norwegian Rat				
Power Density	—	10 mW/cm^2	28 mW/cm^2	20 mW/cm^2
SAR	—	2.5 W/kg	5.0 W/kg	4.9 W/kg
Squirrel Monkey				
Power Density	—	—	45 mW/cm^2	40 mW/cm^2
SAR	—	—	4.5 W/kg	7.2 W/kg
Rhesus Monkey				
Power Density	8 mW/cm^2	57 mW/cm^2	67 mW/cm^2	140 mW/cm^2
SAR	3.2 W/kg	4.5 W/kg	4.7 W/kg	8.4 W/kg

It needs to be said that this field has been fraught with much confusion and much poor quality literature. In 1977 Senator Stevenson[31] stated: "I have never gotten into a subject on which there has been so much disagreement and so much confessed lack of knowledge." He was speaking of the field of microwave bioeffects. Later Foster and Guy[32] and Foster and Pickard[33] wrote critical papers pointing out the prevalence of many papers in the literature that could not be replicated or confirmed. They wondered when such research, which never found robust confirmed effects, would cease. An example of such literature is that of the former Soviet Union, and to some extent Germany, which reports frequency-sensitive effects of millimeter-wave radiation at low levels of around 1 mW/cm². These reports led to an extensive practice of mm-wave therapy for medical purposes in Russia and the Ukraine, but neither the research nor the medical practice has been found valid in the West. (See the discussion in Reference 27.) More recently there was a report[34] from Russia of bioeffects at extraordinarily low levels of power density $\sim 10^{-19}$ W/cm² at a millimeter-wave frequency. We[35] have shown, however, that this extraordinary claim is most probably invalid because of the lack of control of significant energy at the harmonic frequencies. This is only one example of the presence of microwave artifacts that mar many of the papers in the literature. Other artifacts include the great nonuniformity of microwave heating of objects, which is often neglected. Thus, it often is reported that the object temperature is of some value when in actuality the object has a wide spatial variation in temperature. Unfortunately these artifacts not only appear in *in vitro* studies but also in *in vivo* studies that have become increasingly prevalent in this field. One recent study[36] of a large number of transgenic mice has implied a connection of low-level microwave exposure with a cancer (lymphoma). Unfortunately the experiment was done in a metal enclosure and it is known that exposures in metal cavities, lightly-loaded, most probably are chaotic and unpredictable.

Some recent reviews of the field have tended to ignore the past bulk of literature on confirmed effects and instead focus on more recent controversial claims of low-level or "athermal" effects, particularly for ELF amplitude-modulated RF/microwave exposures where it is claimed that the modulation frequency is important. These reviews include that by the Royal Scientific Society in Canada[37] and the recent report of a panel in the U.K. chaired by Sir William Stewart. [38] These claims of "athermal" effects in general are characterized by lack of replication and by the presence of artifacts. There are valid scientific considerations that make such claims implausible. The extensive paper by Valberg et al.[39] has shown that claims of low-level mechanisms are implausible at low frequencies. It is worthwhile to recall that similar claims of "specific," rather than "thermal," effects were prevalent during the first half of the 20th century. The challenge presented in the review by Bierman[40] in the 1940s remains as useful today as it was then, viz.: the burden of proof remains on those who claim other than heating as a mechanism for observed microwave bioeffects.

Standards for the Safe Use of Microwave Energy

It is useful, at the outset, to define some terms prevalent in the standards world, especially that of the IEEE.[41] There are three levels of authority in standards documents. First there are *standards* that imply mandatory actions and rules. The word *shall* is used often. Secondly there are *recommended practices*. These describe preferred actions and rules. The word *should* is used often. Lastly there are *guides*. In these documents alternative actions, procedures, policies, and rules are discussed. The word *may* is used and the choice of various options is at the discretion of the user.

In addition it is useful to classify standards by their differing purpose or target for control. The most fundamental type of standard is the *exposure* standard. This sets limits on safe exposure of people in terms of easily measurable quantities such as field strength, power density, and induced current. These limits, usually called maximum permissible exposure (MPE) levels or investigation levels, are expressed as a function of time or exposure duration. The MPEs are derived from the limiting dosimetric quantities such as SAR, energy density or SA, and current density. The key feature of *exposure* standards is that they are rules that apply to people and it is implied, to some extent, that exposure is voluntary or at least the subject is aware of the exposure or acknowledges the exposure. In the IEEE system, exposure limits are presented in standards documents that also contain the definitions and rules necessary for their implementation. Other organizations may apply different names to their documents, such as *guidelines,* and they may not have the strict definitions, background, and guidance for their implementation that an IEEE standard has.

A different type of standard is the *product performance* standard. This applies to a product and it is designed to ensure that potential exposures of people who use the product will be well below the MPEs found in exposure standards. Examples include the laser standards and the microwave oven standard. The laser standards define accessible emission limits per stated class of laser where the classes are based on potential hazard. The microwave oven standard limits the emission of microwave energy as measured at any point 5 cm from the external surface of the oven.

Although not commonly used today, there is another type of standard that may find greater use in the future. This is the *environmental* standard in which limits are set for EM fields in the environment without a specified dependence on an exposure time, except for possible exemptions at sites where passage is transient, e.g., on bridges, etc. The basis for an *environmental* standard would be considerations of possible interference effects, side effects such as arcing at ends of long cables near a transmitting tower, and possible effects on local flora and fauna, e.g., endangered bird species. In addition, consideration would be made of the psychosocial factors deriving from the concerns of the people who live or work in the environment in question.

Because interference with electronic devices and equipment occurs at far lower EM field levels than biological exposure phenomena, there is also the possibility that exposures to EM fields could lead to *hazardous RFI*, e.g., with medical devices like the implanted pacemaker. Although limiting exposure could control this, it is more logical to control such hazards by limiting the *susceptibility* of medical devices. In practice, often there are applied control measures on both the "radiator" and the "victim" even though the preferred (e.g., by the FCC) emphasis should be on limiting susceptibility of the "victim" device or system.

Another class of standard document may be that for prevention of accidental ignition of fuels or electro-explosive devices. This may take the form of limiting the approach of users of EEDs to known high-power transmitters, i.e., safe distances for use of EEDs.

The history of development of microwave exposure standards has been recently presented[42] by the present authors, especially that of the development of the ANSI/IEEE series of C95 standards. In the early 1990s a survey by Petersen[43] showed that exposure limits in the microwave frequency range of various organizations around the world were not greatly different (i.e., within a few dB). In the United States, in addition to the IEEE, safety documents for microwave exposure are also produced[44] by the NCRP (National Council on Radiation Protection and Measurement) as well as the ACGIH (American Conference of Governmental and Industrial Hygienists). In Europe safety documents have been produced by various organizations including CENELEC. In recent years European countries, besides developing national safety documents, have made increasing reference to guidelines published by ICNIRP (International Commission on Non-Ionizing Radiation Protection). The most recent ICNIRP guidelines[45] are in fair agreement with IEEE standards in the microwave range (both are based on the same SAR thresholds) but are far more conservative at low frequencies. Extensive criticism[46] has been leveled at ICNIRP for the excessive safety factor at low frequencies. Finally, the IEC (International Electrotechnical Commission) develops product and measurement standards. Recently a new umbrella committee, TC-106, has been created[47] to oversee activity within the IEC in the area of EM energy safety.

In this review, we discuss in some detail only the requirements in the latest C95.1 standard[48] issued by the IEEE Standards Coordinating Committee 28 (SCC-28). This standard was developed under the due process rules of the IEEE with extensive documentation of all deliberations and with procedures assuring a broad consensus. At present several hundred people of all disciplines are involved in the various subgroups of SCC-28 with about 15% non-U.S. participation, a figure that is growing rapidly. As suggested by the exposure diagram in Fig. 3.12, exposure limits involve an *averaging time.* Thus, the maximum permissible exposures (MPEs) are in terms of field-strength (squared) or power density (or internal SAR), which are time averaged over any contiguous period of time equal to the averaging time. Thus, in the microwave range, and in the controlled environment, all exposures are averaged over six minutes. This

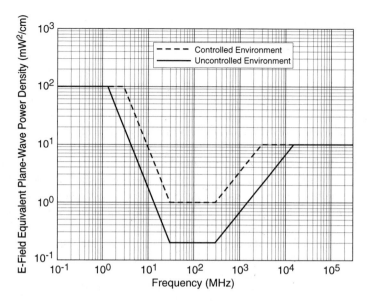

FIGURE 3.12   The IEEE C95.1-1991 MPEs (in terms of the E-field plane-wave equivalent power density). The lower curve is for the uncontrolled environment.

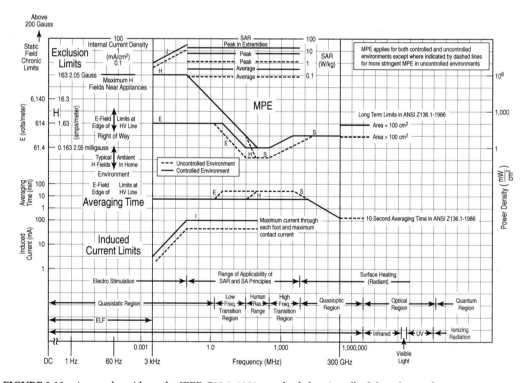

FIGURE 3.13   A capsule guide to the IEEE C95.1-1991 standard showing all of the relevant features.

means that higher exposures are permitted for short periods of time as long as the time average is below the MPE.

The basic rules of the latest C95.1 standard are presented in Figs. 3.12 and 3.13. In Fig. 3.12 the MPEs for the controlled and for the uncontrolled environment are shown in terms of the E-field plain-wave equivalent power as a function of frequency. In Fig. 3.13, a "capsule" presentation of most of the features of the C95.1 standard is shown. Some key observations are:

1. The use of two "tiers," for controlled and uncontrolled environments applies only around the resonance range for man, where there may be maximum concern about exposures. The lower tier applies when the exposed people are unaware of their exposure and are not competent to address safety measures. The use of two tiers originated[49] in an attempt to apply an exposure standard conservatively to the environment. In the future the lower tier may disappear from standards like C95.1.

2. The use of the SAR and SA concepts as basic restrictions applies only between 100 kHz and 6 GHz, in which range there are significant internal fields of RF exposure. Below 100 kHz the limits derive from electrostimulation phenomena. Above 6 GHz there is little penetration (e.g., less than two millimeters) into the body, and surface absorption determines bioeffects and hazards.

3. Averaging times are ramped with frequency in accordance with the thermal time constants of the body. In the last few years, two intensive conferences[50] between the laser and microwave communities have resulted in plans to improve the agreement between laser and microwave standards in the transition range around 300 GHz. This involves tailoring the ramp in averaging time to better fit the dependence of thermal time constants versus frequency.

4. In response to a request from the IEEE MTT Society, a limit on peak exposure levels for very short pulses was introduced. It consists of a value of 100 kV/m maximum as well as an energy limit for short pulses. These are based on studies by the U. S. Air Force that showed that the potential hazards of short pulses were of the "stun" variety and associated with thermal phenomena. The peak limits are well above theoretical thresholds for the auditory effect.

5. Limits on induced or contact currents are included, but these are continually being expanded and refined to include spark-discharge phenomena as well as point contact currents.

6. Relaxation of limits for partial-body exposure is included in an incomplete form and on a tentative basis. Exposure of the testes and eyes poses a risk, but this will likely be eliminated if improved ramps for averaging time at high frequencies eliminate undesirable high energy content in short duration exposures.

7. The basic SAR limits are 0.4 W/kg average and 8 W/kg spatial peak. These were derived from far-field animal exposure studies associated with behavioral disruption. The 8 W/kg (in any 1 g of tissue in the shape of a cube) peak spatial average derives from the 20:1 peak-to-average ratio of the SAR distributions observed in experimental animals under far-field exposure conditions. In C95.1 these have been extended in a doubly conservative manner to the control of localized exposures from sources such as cellular phones, first by invoking the lower tier reduction of 5 and then by assuming that the spatial peak from whole-body experiments should apply to partial-body exposure. In the C95.1 standard, the result is the so-called "low-power" exclusion clauses.

Besides C95.1, important C95 standards include a recently issued revision[52] of the standard on warning symbols and the C95.3 measurement standard,[53] which is presently being revised and balloted.

Product performance standards for microwave technologies arose after the passage of the Radiation Control for Health and Safety Act of 1968 (P. L. 90-602).[53] This led to the emission standard[54] for microwave ovens and to some control of industrial and other microwave sources. All the currently applicable regulations are accessible at Internet site http://www.fda.gov/cdrh/radhlth/index.html. It is unlikely that oven limits will be changed for health reasons. Because of an impending RFI scenario[56] involving microwave ovens and wireless (e.g., Bluetooth) applications operating in the oven (ISM) band there are some pressures to voluntarily limit emissions from ISM equipment.

In the U. S., most civilian applications of microwave/RF energy are now regulated by FCC rules,[57] which are based in large part on the C95.1 standards but also on the NCRP guidelines at higher microwave frequencies. The FCC is concerned not only with regulation of the environment around transmitting towers and antennas, but also the microwave/RF exposure to users of wireless handsets. In this sense the

FCC has taken on the responsibility that normally, under P.L. 90-602, would fall to the FDA. In any case the FCC as well as the FDA support the derivation of product standards through the IEEE system. Although the FDA has the authority to develop performance standards for all microwave equipment including wireless phones, it has instead supported the creation of a new committee to develop such standards, IEEE Standards Coordinating Committee 34.

SCC-34 is a relatively new committee having been established in 1995 for the purpose of developing product performance standards relative to the safe use of electromagnetic energy. The committee uses the exposure criteria and basic restrictions developed by SCC-28, and in some cases, by other committees, to develop standardized assessment procedures, emission limits, etc., to allow manufacturers to readily ensure that their products comply with these criteria. The goal is to develop unambiguous procedures that yield repeatable results, e.g., similar to the procedure for certifying compliance of microwave ovens. So far three subcommittees have been established. The subjects being addressed are pleasure boat radar, protocols for assessing the efficacy of RF protective garments, and protocols for certifying compliance of wireless handsets.

Protocols for assessing the peak spatial average SAR associated with the use of handheld radio transceivers used for personal communications, e.g., cell phones, are especially important. Since 1993, when a guest on a TV talk show alleged that his wife's brain tumor was exacerbated by the use of a cell phone, the media has focused on this issue. Inordinate attention continues to be given to preliminary results of every study reported to even suggest an association between untoward medical effects and the use of cell phones, even when the exposure paradigm is completely different. Recent attention has focused on differences between SAR measurement results reported by different laboratories for the same phone. By a major leap of logic small differences, which are not unexpected in light of the different protocols being used to test cell phones, are translated to a theme of uncertainty about cell phone safety. This seems to occur more in this field than in many others, i.e., a focus on uncertainty related to small differences in analytical or measurement results while completely ignoring the issue of how far below established safety criteria (exposure) the results may be.

Following the 1993 program that first raised the safety issue, a $25 to 27 million research program was established. This program, funded by industry through a blind trust, was to review the literature, develop a research agenda, and fund the studies necessary to address criticisms levied against judgments based on interpretations of the extant literature. One such criticism, unwarranted as it may be, was that there was a lack of studies at exactly the same frequencies, modulations, and exposures as those associated with cell phones. Part of the research program was the establishment of a dosimetry working group to develop uniform protocols for assessing exposure from wireless handsets. When funding was withdrawn for this particular project, the working group, which by then included representatives from most handset manufacturers, a number of test houses, and academia, evolved into Subcommittee 2 of SCC-34.

Two separate recommended practices are being developed by SCC-34. One is based on experimental techniques, the other on numerical techniques. The experimental technique utilizes robot-controlled miniature electric field probes to scan and measure the SAR in a homogeneous tissue-simulating liquid-filled anthropomorphic model of the human head. The numerical technique applies the FDTD method of solving Maxwell's equations in a heterogeneous representation of the human head developed from CT and MRI scans of humans. Models with resolutions of $2 \times 2 \times 2$ mm,[58] $1.1 \times 1.1 \times 1.4$ mm using sub-gridding in some regions,[59] and $0.9 \times 0.9 \times 1.5$ mm[60] have been reported. Both techniques have advantages and disadvantages. For example, the experimental technique uses the actual cell phone as the source, but the homogeneous head model is less than an ideal representation. Head size, the dielectric properties of the tissue simulant, and the thickness of the spacer representing the pinna are standardized to represent a worst case situation that overestimates the SAR induced in the brain. With the numerical technique, the head model is an accurate representation of a human head, but the handset has to be modeled, usually as a simple metal box with an appropriate antenna. CAD files of actual phones complete with some internal structures have been used and differences between these and the results from the simple model are being investigated. An advantage of the computational technique is that it can be applied at the design stage to optimize antenna performance while ensuring that the peak SAR is below the specified limit.

Because most manufacturers are now using the experimental technique to certify their products, mainly because measurement systems are available commercially, the initial effort has been directed to completing this practice. Much of the information needed to complete the document was not available in the literature, e.g., assessment of the uncertainty associated with each component of the system and the overall assessment uncertainty, but was developed in the laboratories of the committee members as the practice developed. This included series of inter-laboratory comparisons of canonical models such as standard half-wave dipoles above a flat phantom or sphere and the development of a generic phone by three of the manufacturers for further inter-laboratory comparisons. The committee has also been working with similar groups, e.g., CENELEC, to ensure international harmonization. Both the FDA and the FCC play an active role on the committee. In fact, the FCC uses the drafts to develop their own requirements with which manufacturers must comply. Future projects will be the extension of the handset protocols to other wireless devices, e.g., wireless modems and body-mounted radio transceivers. Having standard protocols in place should mitigate some of the media-driven concern about wireless devices exemplified by a recent series of TV "specials" calling attention to the uncertainty of cell phone safety — in this case the uncertainty of the peak SAR.

In recent years there has been a trend toward international harmonization of standards in the microwave/RF area. IEEE SCC-28 and SCC-34 are very involved in this activity both by expansion of non-U.S. membership and by closer liaison with other international groups such as the IEC and ICNIRP. In this global activity it is the intent of the IEEE to ensure that the benefits of the IEEE standards process, with due process and broad consensus of all stakeholders, are recognized. The importance of rational science-based standards cannot be overstated. Excessive safety factor, as in the ICNIRP '98 guidelines, can unduly suppress technology, e.g., antitheft and weapons detector systems. Uneven safety factors across the spectrum can give unfair advantage to one technology versus another that uses a different part of the spectrum and, almost as important, convey a sense of uncertainty about the science. In addition, ultra-conservative safety limits can lead to inferior performance and loss of benefit and reliability. This could be important if it involves emergency communications, for example.

Risk Assessment and Public Education

The fears of microwaves along with a generalized electrophobia were promoted, in large measure, by the writings of Paul Brodeur.[61] In succession there have been waves of fear targeted at microwave ovens, radars, microwave relay towers, visual-display terminals, power lines, electric blankets, police radar, cellular phones, and wireless base stations. In the 1980s historian Nicholas Steneck[62] attacked the C95 standards as being too science based. He felt exposure limits should be as low as possible and not based on actual hazard thresholds. When the power line (50 Hz) controversy erupted in the nineties, Nair and Morgan[63] proposed the practice of "prudent avoidance," just in case the weak allegations of a cancer link were true. As documented by Park,[64] however, the power-line scare was baseless and derived from "voodoo science."

Now, in the context of a mushrooming global wireless technology, a plethora of proposals to apply the same caution with a new name has arisen, viz. the Precautionary Principle.[65] It has been endorsed by the Stewart panel[38] in the U.K. for use in processing applications for wireless base stations. Despite the new name this proposal is simply another tool for attacking reliance on science-based standards. The Electromagnetic Energy Association is preparing[66] a formal rebuttal of this concept for application to EM energy. In response to a request from members of a federal interagency committee charged with writing a report at the conclusion of the RAPID program on EMF research, IEEE SCC-28 submitted advice[67] that discussions of cautionary polices and procedures should be done at the lowest level of authority — i.e., in preparing *guides* that present optional alternatives that individuals or organizations can adopt on an ad hoc discretionary basis. In this way there is no acknowledgment of such a cautionary principle as voiding or superseding reliance on standards.

This new principle simply adds to the forces of electrophobia. It raises the level of fear of microwaves and stimulates the sale of a wide variety of allegedly protective devices and services. This amounts to an exploitation of electrophobia and in general is to be discouraged. These devices include measurement devices, shields, books, newsletters, etc. Besides selling oven leakage detectors, discouraged by both the FDA and leading consumer organizations, the exploiting parties give advice on how to minimize exposure including "running" out of the kitchen after turning on the microwave oven. The IEEE COMAR committee[68] continues to fight this phobia and allied misinformation through a series of technical information statements. Other organizations also provide useful information.

A partial list of useful Websites includes:

CTIA:	http://www.WOW-COM.com
EEA:	http://www.elecenergy.com
FCC:	http://www.fcc.gov/oet.rfsafety/cellpcs.html
FDA:	http://www.fda.gov/cdrh/radhlth/index.html
ICNIRP:	http://www.icnirp.de
IEEE/COMAR:	http://homepage.seas.upenn.edu/~kfoster/comar.html
IMPI:	http://www.impi.org
IEEE SCC-28:	http://grouper.IEEE.org/groups/scc28
IEEE SCC-34:	http://grouper.IEEE.org/groups/scc34
Dr. Moulder/FAQ:	http://www.mcw.edu/grcr/cop.html
WHO:	http://www.who.ch/
IEC:	http://www.iec.ch

These are some of the Websites where information useful to microwave engineers can be obtained. The FDA Website also links to an educational site maintained by a professor at the University of Virginia. It contains much useful information, but errs in recommending the use of inexpensive oven leakage detectors and presenting a less than complete explanation of superheating of liquids in microwave ovens. On the Website for the FCC, much detailed information is available including a series of reports, fact sheets, and an interactive database containing the dielectric properties for a number of biological tissues.

The site for IMPI (International Microwave Power Institute) can be useful in exploring power (non-communications) applications of microwave energy. These include medical applications, food heating and processing, microwave discharge lamps for UV curing and lighting, and microwave power transmission, some of it for solar power satellites. Other future applications include the comfort heating of animals and man through microwaves that can contribute toward solving energy crises. The other sites are self-explanatory in their functions.

Conclusions

The modern microwave engineer needs to be knowledgeable about safety standards in general, and in particular, the IEEE C95 series of standards. Microwave engineers, through their professional societies as well as employers, should support and participate in the increasingly global activities of the IEEE in developing standards for the safe use of electromagnetic energy. The financial support of all IEEE activities in global standards and public education will be critical in the future development of microwave technology. With rational standards and an educated public all microwave technologies, both for communications and power applications, can be realized for optimum benefit to mankind.

References

1. Reilly, J. P., *Applied Bioelectricity: From Electrical Stimulation to Electropathology,* Springer Verlag, New York, 1998.
2. Durney, C. H., Massoudi, H., and Iskander, M. F., *Radiofrequency Radiation Dosimetry Handbook,* Fourth Edition, Report USAFSAM-TR-85-73, USAF School of Aerospace Medicine, Brooks AFB, TX, 1986.

3. Gabriel, S., Lau, R. W., and Gabriel, C., The dielectric properties of biological tissues: III Parametric models for the dielectric spectrum of tissues, *Phys. Med. Biol.*, 41, 2271–2293, 1996. These properties can also be downloaded from Internet site: http://www.brooks.af.mil/AFRL/HED/hedr/reports/dielectric/home.html.

4. Schwan, H. P., History of the genesis and development of the study of low energy electromagnetic fields, in Grandolof, M., Michaelson, S., and Rindi, A., (eds.), *Biological Effects and Dosimetry of Non-Ionizing Radiation*, Plenum Publishing Company, New York, 1981, 1–17.

5. Osepchuk, J. M., Basic characteristics of microwaves, *Bull. N. Y. Acad. Med.*, 55, 976–998, 1979.

6. Chou, C. K., Bassen, H., Osepchuk, J., Balzano, Q., Petersen R., Meltz, M., Cleveland, R., Lin, J. C., and Heynick, L., Radio frequency electromagnetic exposure: tutorial review on experimental dosimetry, *Bioelectromagnetics*, 17, 195–206, 1996.

7. Kritikos, H. N. and Schwan, H. P., The distribution of heating potential inside lossy spheres, *IEEE Trans. Bio, Med. Eng.*, BME-22, 457–463, 1975.

8. Guy, A. W., Lehmann, J. F., and Stonebridge, J. B., Therapeutic applications of microwave power, *Proc. IEEE*, 62, 55–75, 1975.

9. Durney, C. H. and Christensen, D. A., Hyperthermia for Cancer Therapy, in Gandhi, O. P., Ed., *Biological Effects and Medical Applications of Electromagnetic Energy*, Prentice-Hall, Englewood Cliffs, NJ, 1990, 436–477.

10. Moritz, A. R. and Henriques, F. C., Jr., Studies of thermal injury II, The relative importance of time and surface temperature in the causation of cutaneous burns, *Am. J. Pathol.*, 23, 695–720, 1947.

11. Michaelson, S.M., Thomson, R. A. E., and Howland, J. W., Physiologic aspects of microwave irradiation of mammals, *Am. J. Physiol.*, 201, 351, 1961.

12. Addington, C. H., Osborn, C., Swartz, G., Fischer, F., Neubauer, R. A., and Sarkees, Y. T., Biological effects of microwave energy at 200 megacycles, in Peyton, M. F., Ed., *Proc. 4th Annual Tri-Service Conf. Biol. Effects of Microwave Radiation*, Plenum Press, New York, 1961, 177.

13. Minin, B. A., *Microwaves and Human Safety*, I. Translation from U. S. Joint Publications Research Service, JPRS 65506-1, U.S. Department of Commerce, 1965.

14. Pay, T.L., Microwave effects on reproductive capacity and genetic transmission in Drosophila melanogaster, *J. Microwave Power*, 7, 175, 1972.

15. Samaras, G. M., Muroff, L. R., and Anderson, G. E., Prolongation of life during high-intensity microwave exposures, *IEEE Trans. on Microwave Theory and Techniques*, MTT-19, 2, 245–247, 1971.

16. Tell, R. A. and Harlen, F., A review of selected biological effects and dosimetric data useful for development of radiofrequency safety standards for human exposure, *J. Microwave Power*, 14, 4, 405–424, 1979.

17. Adair, E.R., Mylacraine, K. S., and Kelleher, S. A., Human exposure at two radiofrequencies (450 and 2450 MHz): similarities and differences in physiological response, *Bioelectromagnetics*, 20, 12, 1999.

18. Carpenter, R. L., Ocular effects of microwave radiation, *Bull. NY Acad Med.*, 55, 1048–1057, 1979.

19. Guy, A. W., Kramar, P. O., Harris, C. A., and Chou, C. K., Long-term 2450 MHz CW microwave irradiation of rabbits: methodology and evaluation of ocular and physiologic effects, *J. Microwave Power*, 15, 37–44, 1980.

20. Kues, H. A., Hirst, L. W., Lutty, G. A., D'Anna, S. A., and Dunkelberger, G. R., Effects of 2.34 GHz microwaves on primate corneal endothelium, *Bioelectromagnetics*, 6, 177–188, 1985.

21. Kamimura, Y., Saito, K-i, Saiga, T., and Amenmiya, Y., Effect of 2.45 GHz microwave irradiation on monkey eyes, *IEICE Trans. Commun.*, E77-B, 762–764, 1994.

22. Osepchuk, J. M. Ed., *Biological Effects of Electromagnetic Radiation*, John Wiley & Sons, New York, 1983.

23. Gandhi, O. P. Ed., Special issue on biological effects and medical applications of electromagnetic energy, *Proc. IEEE*, 68, 1980.

24. Elder, J. A. and Cahill, D. F., *Biological Effects of Radiofrequency Radiation*, Report EPA-600/8-83-026F, 1984.

25. Polson, P. and Heynick, L. N., Overview of the radiofrequency radiation (RFR) bioeffects database, in Klauenberg, B. J., Grandolfo, M., and Erwin, D. N., *Radiofrequency Radiation Standards*, Plenum Press, New York, 337–390, 1995.

26. Michaelson, S. M., and Lin, J. C., *Biological Effects and Health Implications of Radiofrequency Radiation*, Plenum Press, New York, 1987.

27. Gandhi, O. P. Ed., *Biological Effects and Medical Applications of Electromagnetic Energy*, Prentice-Hall, Englewood Cliffs, NJ, 1990.

28. Polk, C. and Postow, E., *CRC Handbook of Biological Effects of Electromagnetic Fields*, CRC Press, Boca Raton, FL, 1995.

29. Lin, J. C., *Microwave Auditory Effects and Applications*, Charles C. Thomas, Springfield, IL, 1978.

30. DeLorge, J. O. and Ezell, C. S., Observing responses of rats exposed to 1.28 GHz and 5.4 GHz microwaves, *Bioelectromagnetics*, 1, 183–198, 1980.

31. Stevenson, Sen. A., U.S. Senate Hearing Record on Oversight of Radiation Health and Safety, June 16–29, 1977.

32. Guy, A. W. and Foster, K. R., The microwave problem, *Scientific American*, 255, 3, 32–39, 1986.

33. Foster, K. R. and Pickard, W. F., The risks of risk research, *Nature* 330, 531–532, 1987.

34. Belyaev, I. Y., Sheheglov, V. S., Alipov, Y. D., and Polunin, V. A., Resonance effect of millimeter waves in the power range of 10^{-19} to 3×10^{-3} W/cm² on *Escherichia coli* cells at different concentrations, *Bioelectromagnetics* 17, 4, 312–321, 1996.

35. Osepchuk, J. M. and Petersen, R. C., Comments on Resonance effect of millimeter waves in the power range from 10^{-19} to 3×10^{-3} W/cm² on escheria coli cells at different concentrations, Belyaev et al., *Bioelectromagnetics*, 17, 312–321 (1996), *Bioelectromagnetics*, 18, 7, 527–528, 1997.

36. Repacholi, M. H., Basten, A., Gebski, V., Noonan, D., Finnie, J., and Harris, A. W., Lymphomas in Eμ-Pim 1 transgenic mice exposed to pulsed 900 MHz electromagnetic fields, *Radiat. Res.*, 147, 631–647, 1997.

37. Royal Society of Canada Expert Panel Report, *A Review of the Potential Health Risks of Radiofrequency Fields from Wireless Telecommunication Devices*, Royal Society of Canada, RSC.EPR 99-1, 1999.

38. Independent Expert group on Mobile Phones, Stewart, Sir William, Chairman, *Mobile Phones and Health*, NRPB, Chilton, Didcot, Oxon OX11 0RQ, U.K., 2000.

39. Valberg, P., Kavet, R., and Rafferty, C. N., Can low-level 50/60 Hz electric and magnetic fields cause biological effects?, *Radiation Research*, 148, 2–21, 1997.

40. Mortimer, B. and Osborne, S. L., *J. Am. Med. Assoc.* 104, 1413, 1935.

41. IEEE Standards Association, *IEEE Standards Manual*, IEEE, Piscataway, NJ, 2000.

42. Osepchuk, J. M. and Petersen, R. C., Past may help solve electromagnetic issue, *Forum* 9, 2, 99–103, 1994.

43. Petersen, R. C., Radiofrequency/microwave protection guides, *Health Physics* 61, 59–67, 1991.

44. Petersen, R. C., Radiofrequency safety standards-setting in the United States, in *Electricity and Magnetism in Biology and Medicine*, Bersani, F. Ed., Plenum Press, New York, 1999, 761–764.

45. International Commission on Non-Ionizing Radiation Protection (ICNIRP), Guidelines for limiting exposure to time-varying electric, magnetic, and electromagnetic fields (up to 300 GHz), *Health Physics* 74, 4, 494–522, 1998.

46. Various authors, Series of letters commenting on ICNIRP '98 guidelines, *Health Physics* 76–77, 1998–1999.

47. ACEC Task Force on Human Exposure to EM Fields, *ACEC TF EMF (Convenor) 21*, International Electrotechnical Commission (IEC), 1999-05.

48. IEEE, *IEEE Standard for Safety Levels with Respect to Human Exposure to Radio Frequency Electromagnetic Fields, 3 kHz to 300 GHz*, IEEE Std. C95.1, 1999 Edition, Piscataway, NJ, 1999.

49. Osepchuk, J. M., Impact of public concerns about low-level EMF-RFR database, in *Radiofrequency Radiation Standards*, Klauenberg, B. J., Grandolfo, M., and Erwin, D., Eds., Plenum Press, New York, 1995, 425–426.

50. *Digest, Infrared Lasers & Millimeter Waves Workshop: The Links Between Microwaves & Laser Optics,* Air Force research Laboratories, Brooks AFB, TX, January 1997.

51. IEEE, *IEEE C95.2-1999, IEEE Standard for Radio Frequency Energy and Current Flow Symbols,* IEEE-SA, Piscataway, NJ, 1999.

52. IEEE, *IEEE C95.3-1991 (Reaff 1997), IEEE Recommended Practice for the Measurement of Potentially Hazardous Electromagnetic Fields-RF and Microwave,* IEEE-SA, Piscataway, NJ, 1991.

53. Public Law 90-602, *Radiation Control for Health and Safety Act of 1968,* Approved by Congress, October 18, 1968.

54. Osepchuk, J. M., Review of microwave oven safety, *J. Microwave Power* 13, 13–26, 1978.

55. The Internet site for FDA/CDRH is: http://www.fda.gov/cdrh/radhlth/index.html.

56. Osepchuk, J. M., The Bluetooth threat to microwave equipment, *Microwave World,* 20, 1, 4–5, 1999.

57. FCC, *Report and Order FCC 96-326,* In the matter of guidelines for evaluating the environmental effects of radiofrequency radiation, ET Docket No. 93–62, August 1, 1996.

58. Dimbylow, P. J. and Mann, S. M., SAR calculations in an anatomically based realistic model of the head for mobile communication transceivers at 900 MHz and 1.8 GHz, *Physics in Medicine and Biology,* 39, 1537-1553, 1994

59. Okoniewski, M. and Stuchly, M.A., A study of the handset antenna and human body interaction, *IEEE Trans. MTT,* 44, 10, 1855-1864, Oct. 1996.

60. Olley, P. and Excell, P. S., Classification of a high resolution voxel image of a human head, *Voxel Phantom Development,* Proceedings of an International Workshop held at the National Radiological Protection Board, Chilton, U.K., P. J. Dimbylow, Ed., 1995.

61. Brodeur, P., *The Zapping of America,* William Morrow, New York, 1978.

62. Steneck, N. H., *The Microwave Debate,* MIT Press, Cambridge, MA, 1985.

63. Morgan, M. G., Prudent avoidance, *Public Utilities Fortnightly,* 26–29, 1992.

64. Park, R., *Voodoo Science,* Oxford University Press, New York, 2000.

65. Foster, K. R., Vecchia, P., and Repacholi, M. H., Science and the Precautionary Principle, *Science,* 979–981, May 12, 2000.

66. Electromagnetic Energy Association, *Fact Sheet on Cautionary Policies,* Washington, D.C., in press.

67. IEEE SCC-28, *Response of: Institute of Electrical and Electronic Engineers Standards Coordinating Committee 28 to: Interagency Committee re: Advice on Mitigation of Risks Related to EMF Exposure,* December 1999.

68. Osepchuk, J. M., COMAR after 25 years: still a challenge!, *IEEE Eng. in Medicine and Biology,* 15, 3, 120–125, 1996.

3.3 Signal Characterization and Modulation Theory

John F. Sevic

Spectral efficiency is of paramount importance when considering the design of virtually all commercial wireless communication systems, whether for voice, video, or data. Spectral efficiency can be measured by the number of users per unit of spectrum or by the number of bits that can be represented per unit of spectrum. In general, wireless service providers are interested in maximizing both the number of users and the number of bits per unit spectrum, which in both cases results in the maximum revenue.

Spectral efficiency can be increased by using several methods including signal polarization, access method, modulation method, and signal coding technique. The first method is commonly adopted in satellite communication systems, where, for example, the uplink and downlink may be right-hand and left-hand circularly polarized, respectively.

Access method refers to how a common resource, such as frequency, is shared among each of the users of the system. Frequency domain multiple access (FDMA) is the basis for all commercial broadcast and most wireless communication systems. With FDMA, each user is allocated a particular section of spectrum that is devoted to that user only. The first-generation cellular phone system in the United States, called advanced mobile phone system (AMPS) has approximately 600 30-kHz channels, each one of which can be utilized for voice and low data-rate communication [1].

Users can also be allocated a certain segment of time, called a slot, leading to time-domain multiple access (TDMA). With TDMA, users share a common frequency and are assigned one, or in some cases more than one slot out of several available slots. In this fashion spectral efficiency is increased by segregating users in time. Second-generation wireless communication systems based on TDMA include the North American Digital Cellular system (NADC) and the Global Standard for Mobile Communications (GSM), which have three slots and eight slots, respectively [2,3].

Code-division multiple access (CDMA) results when each user is assigned a unique code, which is orthogonal to all the other available codes. The Walsh function has been adopted as the orthogonal code set for first-generation CDMA systems. At the receiver, the signal is correlated with a known Walsh function, and the transmitted information thus extracted. The ability of these systems to improve spectral efficiency relies on the development of robust orthogonal functions, since any cross-correlation results in performance degradation. Many wireless systems are hybrids of two access methods. For example, the first CDMA-based wireless system, developed by Qualcomm, is based on FDMA and CDMA [4].

Modulation is the process of impressing an information source on a carrier signal. Three characteristics of a signal can be modulated: amplitude, frequency, and phase. In many types of modulation, two characteristics are modulated simultaneously. Analog modulation results when the relationship between the information source and the modulated signal is continuous. Digital modulation results when the modulated characteristic assumes certain prescribed discrete states.

Signal coding techniques are many and varied, and require detail beyond what can be presented in this introductory chapter. Coding can impact the characteristics of the signal, as will be examined in more detail with CDMA.

The purpose of this chapter is to provide an introduction to the representation and characterization of signals used in contemporary wireless communication systems. The time domain representation of an information-bearing signal determines uniquely what the impact of nonlinear amplification will be, so the study begins with a review of time domain signal analysis techniques. Since the effects of nonlinear amplification are of most interest in the frequency domain, the signal analysis review will also include frequency domain methods. Random process theory is an integral element of digital modulation theory, and will also be covered. Following this, a review of several types of modulations will be given, with both a time domain and frequency domain complex envelope description. The impact of filtering, for spectral efficiency improvement, will be assessed. A probabilistic time domain method of characterizing the envelope of a signal, called the envelope distribution function (EDF), will be introduced. This function is more useful than the peak-to-average ratio in estimating the impact of a PA on a signal. A complete reference section is given at the end of the chapter.

Complex Envelope Representation of Signals

Using Fourier analysis, any periodic signal can be exactly represented as an infinite summation of harmonic phasors [5]. Most often a signal $x(t)$ is approximated as a finite summation of harmonic phasors, in which case

$$x(t) = \sum_{k=-\infty}^{\infty} \tilde{a}_k e^{j\omega kt} \approx \sum_{k=-Q}^{Q} \tilde{a}_k d^{j\omega kt} \qquad (3.89a)$$

where

$$\tilde{a}_k = \frac{1}{T} \int_{t}^{t+T} x(\tau) e^{-j2\pi f\tau} d\tau \qquad (3.89b)$$

and Q is chosen sufficiently large to accurately represent the signal under consideration. A physical basis for this approximation is that all systems have an essentially low-pass response.

Parsavel's theorem states that average power is invariant with respect to which domain it is calculated in, and is expressed as

$$\overline{P} = \frac{1}{T} \int_{t}^{t+T} \left| x(\tau) \right|^2 d\tau = \sum_{k=-\varphi}^{\infty} \left| \tilde{a}_k \right|^2 \approx \sum_{k=-Q}^{Q} \left| \tilde{a}_k \right|^2 \tag{3.90}$$

In many instances, calculation of average power may be easier to evaluate in either the time domain or the frequency domain representation.

The modulation property illustrates the frequency-shifting nature of time domain multiplication. This is expressed as

$$m(t)\cos(2\pi f_o t) \Rightarrow \frac{1}{2}\left[M(f - f_o) + M(f + f_o) \right] \tag{3.91}$$

where $M(f)$ is the Fourier transform of $m(t)$. Note that both upper and lower sidebands are generated, indicating the presence of a negative frequency component.

In many instances, knowledge of the envelope of a signal is sufficient for its characterization and for assessing the associated impact of a nonlinear PA. An arbitrarily modulated signal can be represented in the time-domain as

$$x(t) = real\left\{ m(t)\exp\left[2\pi f_o t + \phi(t) \right] \right\} \tag{3.92}$$

where $m(t)$ and $\phi(t)$ describe the time-varying amplitude and phase of the information signal, respectively, and f_o is the carrier frequency [6]. Note that frequency modulation results by differentiation of the phase modulation. The complex envelope of Eq. (3.92) is

$$\tilde{x}(t) = i(t) + jq(t) = m(t)e^{j\phi(t)} \tag{3.93}$$

where $i(t)$ and $q(t)$ are defined as the in-phase and quadrature components of the complex envelope. Although there are other methods available of representing modulated signals, Eq. (3.93) is adopted here due to the elegant geometric interpretation afforded by the complex plane representation. The components of Eq. (3.93) are calculated from

$$i(t) = real\left\{ x(t)e^{-j2\pi f_0 t} \right\} \tag{3.94a}$$

$$q(t) = imag\left\{ \hat{x}(t)e^{-j2\pi f_0 t} \right\} \tag{3.94b}$$

where $\hat{x}(t)$ is the Hilbert transform of $x(t)$ [7]. Since statistically independent signals are orthogonal, calculation of the Hilbert transform is often unnecessary, and instead two statistically independent data sources for i, (t) and $q(t)$ can be used. Figure 3.14 shows the spectrum of an arbitrary complex envelope equivalent power density spectrum. Note that the spectrum is not even-symmetric about the y-axis, with the resultant requirement that the time domain signal Eq. (3.93) is in general complex.

Linear network analysis using the complex envelope is similar to conventional network analysis. Taking the Fourier transform of Eq. (3.93), and denoting the complex transfer function as $\tilde{H}(f)$, we have

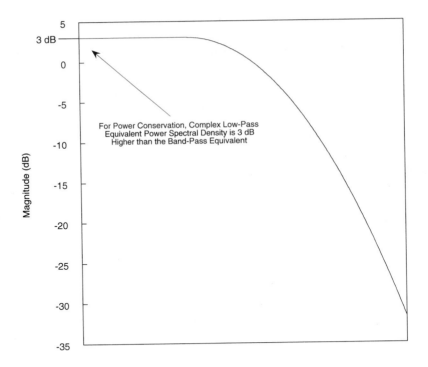

FIGURE 3.14   Power spectral density of an arbitrary complex envelope equivalent signal.

$$\tilde{Y}(f) = \tilde{H}(f)\tilde{X}(f) \tag{3.95a}$$

This response can also be calculated directly in the time domain using the convolution integral

$$\tilde{y}(t) = \int_{-\infty}^{\infty} \tilde{h}(\tau)\tilde{x}(t-\tau)d\tau \tag{3.95b}$$

The equivalent band-pass time domain response is found by

$$y(t) = real\left\{\tilde{v}(t)e^{j2\pi f_0 t}\right\} \tag{3.96}$$

Representation and Characterization of Random Signals

Consider a random signal $x(t)$ that could describe either a voltage or current. In general the associated *n-dimensional* joint probability density function is required to describe $x(t)$ over n time instants [8]. Since the expected amplitude and power of a signal are most often of interest, the first- and second-moments are sufficient for representation and characterization of $x(t)$. The first- and second-moments of $x(t)$ are

$$\bar{x}(t) = \int_{-\infty}^{\infty} \tau f(\tau)d\tau = \sqrt{DC\ Power} \tag{3.97a}$$

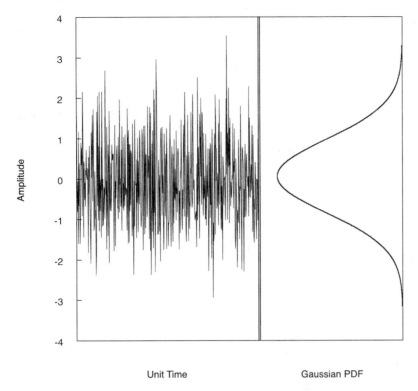

FIGURE 3.15 Illustration of the meaning of random signal having a Gaussian amplitude distribution. In this example the Gaussian pdf represents the probability of the signal amplitude being less than a given value. For example, since the pdf is symmetric about the time axis, this signal has zero mean. Since the noise in this example is uncorrelated, it is defined as white. Similarly, a colored noise process will exhibit correlation.

$$\overline{x^2} = \int\limits_{-\infty}^{\infty} \tau^2 f(\tau)d\tau = Total\ Average\ Power \tag{3.97b}$$

where $f(\tau)$ is the associated probability distribution function (pdf) of $x(t)$. The pdf describes the probability of the instantaneous amplitude of $x(t)$ being less than a specified value. This idea is illustrated with a Gaussian pdf in Fig. 3.15. Note that the amplitude of this signal is concentrated around its mean value, which is zero.

Since the average value and average power of $x(t)$, as determined by Eq. (3.97), are based on ensembles, they are defined as ensemble averages. The average value and average power of a signal can also be calculated with respect to time, which is the method most engineers are familiar with. Signals are said to be ergodic when their ensemble and time averages are the same. All of the signals described in this chapter are ergodic.

Probabilistic characterization of how rapidly a signal changes over time, and hence its spectral distribution, is described by its autocorrelation function

$$R_x(\tau) = \overline{x(t)x(t+\tau)} = \iint\limits_{\infty} \tau_1\tau_2 f(\tau_1,\tau_2)d\tau_1 d\tau_2 \tag{3.98}$$

The autocorrelation function evaluated at $\tau = 0$ is the total average power in the signal, as Eq. (3.97b) shows. Using Eq. (3.97a) and Eq. (3.97b), which represent average amplitude and total average power, respectively, gives the AC power of the signal

$$P_{AC} = R_x(0) - (\bar{x})^2 = \overline{x^2} - (\bar{x})^2 \tag{3.99}$$

Using the Wiener-Khintchine theorem, the spectral distribution of power of $x(t)$ is evaluated by taking the Fourier transform of the autocorrelation function

$$S_x(f) = \int_{-\infty}^{\infty} R_x(\tau) e^{j2\pi f\tau} d\tau \tag{3.100}$$

where $S_x(f)$ is the power spectral density (PSD) of $x(t)$.

In the case of an arbitrary autocorrelation function, $R_x(\tau)$, the PSD is expressed as

$$S_x(f) = S_{pulse}(f) S_{corr}(f) \tag{3.101}$$

where $S_{pulse}(f)$ is the PSD of the pulse function representing $x(t)$ and $S_{corr}(f)$ is the PSD of the autocorrelation function of the data [9]. From this expression, it is clear that the spectral characteristics of a signal are influenced not only by the pulse characteristics, but also any correlation between adjacent pulses.

Let $x(t)$ be the input to a linear system and let $y(t)$ be the associated output. The impulse response, $h(t)$, is described in the frequency domain as $H(f)$. The average amplitude of $y(t)$ is

$$\bar{y}(t) = \tilde{H}(0) \bar{x}(t) \tag{3.102}$$

and the average power of $y(t)$ is

$$\overline{y^2}(t) = R_y(0) = \int_{-\infty}^{\infty} |H(f)|^2 df = \int_{-\infty}^{\infty} |H(f)|^2 S_x(f) df \tag{3.103}$$

The output autocorrelation function is

$$R_y(\tau) = h(\tau) * h(-\tau) * R_x(\tau) \tag{3.104}$$

where * denotes convolution. Taking the Fourier transform gives the output PSD in terms of the input PSD

$$S_y(f) = |H(f)|^2 S_x(f) \tag{3.105}$$

Complex envelope equivalent analysis is done by replacing all variables with the associated complex envelope representation and restricting the integration of Eq. (3.103) to positive frequency [6]. Note that this analysis gives an example of how choosing the domain in which to carry out an analysis can greatly simplify the effort involved.

Modulation Theory

Modulation theory provides a framework for representing and characterizing the time domain and frequency domain characteristics of an information-bearing signal, and the subsequent impact of non-linear amplification of the signal. The general analysis method to be followed is to describe the signal in the time domain, using the geometric interpretation of Eq. (3.93), and then determine resultant signal degradation by characterization in the frequency domain.

Access method, modulation, and coding each directly impact the spectral efficiency of a signal. From Fourier analysis it is also clear, therefore, that the time domain characteristics are also impacted, due to the inverse relationship between the time domain and frequency domain. In other words, a signal that varies rapidly in time, due to access method, modulation, or coding, will be wider in extent in the frequency domain than a slowly varying signal. Note here that signal refers to the envelope of the carrier, and the extent of the signal in the frequency domain refers to the associated spectral description of the envelope.

As Eq. (3.93) indicates, modulation can be interpreted geometrically by associating a position in the complex plane with the instantaneous value of the information source. For analog modulation, a continuous range of values is possible; digital modulation allows only certain locations to be occupied. This mapping operation will directly establish the resultant time domain characteristics of the modulation. Many modulations are based on phase since they are relatively impervious to amplitude-related noise disturbances, FM being an example of this. When phase modulation is digital, the characteristics of the digital pulse will influence the signal as well, as Eq. (3.101) shows. Since pulses may be rapidly varying, it is expected that some type of filtering will be necessary to give acceptable spectral efficiency. Thus, to describe a modulation, the mapping method and any associated band-limiting filtering must be considered to provide a complete time domain description of the complex envelope.

Analog Modulation

The simplest modulation is analog double sideband suppressed carrier (DSB-SC). When the information source is a sinusoid, DSB-SC is the classical two-tone intermodulation test signal. The complex envelope representation of DSB-SC is

$$\tilde{x}(t) = m(t) \tag{3.106}$$

where $m(t)$ is the information signal. Since the envelope of DSB-SC varies in direct proportion to $m(t)$, it is not a constant envelope modulation. Note also that, from a geometric interpretation using Eq. (3.93), DSB-SC requires only one dimension, meaning there is no phase modulation.

All of the first-generation cellular systems, such as AMPS and ETACS, are based on FM, with a complex envelope representation of

$$\tilde{x}(t) = \exp\left[jk_f \int_{-\infty}^{t} m(\tau) d\tau \right] \tag{3.107}$$

where k_f is the frequency-deviation constant and $m(t)$ is the information signal. Since the magnitude of the complex exponential is unity, FM is a constant envelope modulation. A geometric interpretation of FM shows that it would be a unit circle, with the speed in movement about this circle proportional to the instantaneous value of the information signal. Carson's rule shows that the spectral efficiency of FM is less than DSB-SC [6]. In general, constant envelope modulation is not as spectrally efficient as modulation that exhibits a time-varying envelope.

Discontinuous Phase-Shift Keying

Virtually all second- and third-generation wireless communication systems are based on digital modulation, with digital phase modulation being the most common. Digital phase modulation is commonly referred

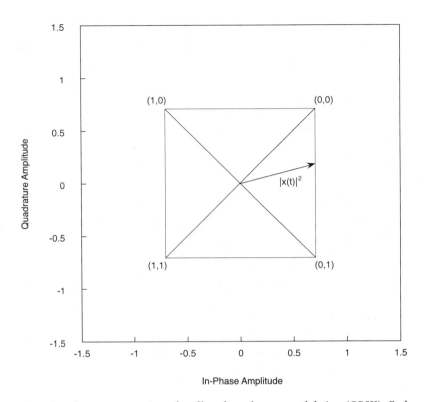

FIGURE 3.16 Complex plane representation of unfiltered quadrature modulation (QPSK). Each corner of the constellation diagram represents two data bits. Depending on the data sequence, each transition goes through a phase change of either ± 90° or ± 180°, which always causes a step change in the signal envelope.

to as phase-shift keying (PSK). PSK consists of two DSB-SC signals in quadrature, and can be represented using Eq. (3.93) with in-phase and quadrature components each generated from a digital data source. Each data source is usually generated by multiplexing a serial data stream, which, if necessary, is already coded, such as with CDMA. Quadrature digital modulation is represented in the complex plane as shown in Fig. 3.16, where the horizontal axis represents the real part of Eq. (3.93) and the vertical axis represents the imaginary part of Eq. (3.93). The trajectory is the instantaneous envelope of the signal and for the constellation given, there are four unique phases, with each phase representing a unique combination of two bits. Each combination of bits is defined as a symbol. Some third-generation wireless systems have adopted PSK with eight unique locations, giving 8-PSK, with the result that each location now represents a unique combination of three bits. Note also from Fig. 3.16 that PSK modulation is constant envelope.

The complex envelope representation of quadrature digital modulation is

$$\tilde{x}\left(kT_b\right) = i\left(kT_b\right) + jq\left(kT_b\right) \qquad (3.108)$$

where

$$i(kT_b) = \sum_{k=-\infty}^{\infty} a(kT_b) f(t - kT_b - \tau_f) \qquad (3.109a)$$

$$q(kT_b) = \sum_{k=-\infty}^{\infty} b(kT_b) g(t - kT_b - \tau_g) \qquad (3.109b)$$

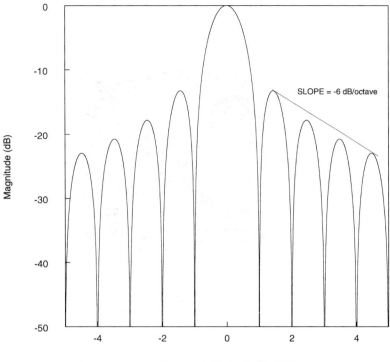

FIGURE 3.17

and $a(kT_b)$ and $a(kT_b)$ represent unit-amplitude data sources, f and g are pulse functions, T_b is the bit rate, and τ_f and τ_g are arbitrary phase offsets. The functions f and g are usually rectangular pulse streams with zero mean and zero correlation between pulses. In this case, the PSD of Eqs. (3.109a) and (3.109b) is given as

$$S_i\left(f\right) = T_b\mathrm{sinc}^2\left(\pi T_b f\right) \tag{3.109c}$$

$$S_q\left(f\right) = T_b\mathrm{sinc}^2\left(\pi T_b f\right) \tag{3.109d}$$

In Fig. 3.17 it is shown that Eqs. (3.109c) and (3.109d) exhibit a −6 dB/octave roll off. A communication system based on rectangular pulses would therefore exhibit very poor spectral efficiency due to the relatively large energy present in the spectrum in the neighboring sidelobes. To resolve this problem, band-limiting filtering is used. Although it is possible to use arbitrary band-limiting filters, such as the Chebyshev low-pass response, the Nyquist response is often adopted since it exhibits certain characteristics amenable to demodulation of the signal at the receiver [10-12].

A significant consequence of band-limiting a discontinuous signal, such as that exhibited by PSK with rectangular pulses, is that envelope variations will be introduced. Thus, although PSK modulation is constant envelope, the necessity of band-limiting induces envelope variations, which gives a band-limited PSK signal properties of both phase modulation and amplitude modulation. The degree of amplitude variation is directly proportional to the degree of band limiting. Figure 3.18 illustrates the impact of band limiting on the QPSK constellation of Fig. 3.16, where the amplitude variations are apparent. Band limiting will necessarily require a reduction in the efficiency of a PA, in order to support the peak excursions of the envelope.

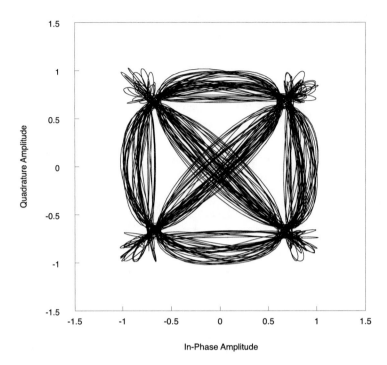

FIGURE 3.18 Constellation diagram of band-limited QPSK. The transient response of the band-limiting filter increases the peak-average ratio of the signal. The more severe the band limiting, yielding increased spectral efficiency, the higher the peak-to-average ratio becomes.

The Nyquist filter response exhibits simultaneous band limiting and zero inter-symbol interference (ISI). Zero ISI is desirable, though not necessary, for a digital communication system to give acceptable performance. To understand the impact of inter-symbol interference, consider the convolution integral Eq. (3.95b), where it is seen that convolution is essentially a summation operation that accounts for the past state of a signal or system at the present time. Use of an arbitrary filter response will result in interference of the present state of a signal due to its past values, leading to the potential for an error in the actual value of the signal at the present time. The impulse response of a Nyquist filter has zero crossings at multiples of the symbol rate, and thus does not cause inter-symbol interference [11,12].

The most common form of Nyquist filter currently adopted is the raised cosine response. The raised-cosine response is expressed in the frequency domain as

$$
H(f) = \begin{cases} T_s, & |f| \le \dfrac{1}{2T_s} - \alpha \\[2mm] T_s \cos^2\left[\dfrac{\pi}{4\alpha}\left(|f| - \dfrac{1}{2T_s} + \alpha\right)\right], & \dfrac{1}{2T_s} - \alpha \le |f| \le \dfrac{1}{2T_s} + \alpha \\[2mm] 0, & |f| > \dfrac{1}{2T_s} + \alpha \end{cases}
\tag{3.110}
$$

where T_s is the symbol rate in seconds and α is the excess-bandwidth factor [6,13]. By adjusting α, the spectral efficiency of a digitally modulated signal can be controlled, while maintaining zero ISI. Figure 3.19 shows the frequency response of Eq. (3.110) for various values of α. At the Nyquist limit of $\alpha = 0$, all energy above half of the main lobe is removed. This results in maximum spectral efficiency but the largest time domain peak-to-average ratio.

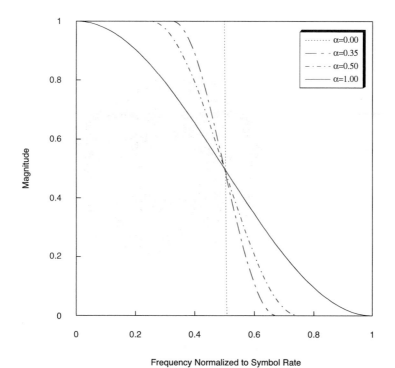

FIGURE 3.19 Raised-cosine filter response for several excess-bandwidth factors. The frequency axis is normalized to the symbol rate. NADC uses an $\alpha = 0.35$.

To maximize receiver signal-to-noise ratio (SNR), many wireless communication systems split the filter response equally between the transmitter and receiver [9,10]. The resultant response is called the square-root raised-cosine response, which in the time domain is expressed as

$$h(t) = 4\alpha \frac{\cos\left[(1+\alpha)\dfrac{\pi}{T_s}t\right] + \dfrac{\sin\left[(1-\alpha)\dfrac{\pi}{T_s}t\right]}{\dfrac{4\alpha}{T_s}t}}{\left(\pi T_s^{0.5}\right)\left[\left(\dfrac{4\alpha}{T_s}t\right)^2 - 1\right]} \tag{3.111}$$

where the variables are defined as above in Eq. (3.110) [11].

Many digital wireless standards specify the symbol rate and excess-bandwidth factor to specify the root-raised cosine filter. This can lead to ambiguity in specifying the filter response insofar as the length of the impulse response is not specified. An impulse response that is too short will exhibit unacceptable band-limiting performance and can lead to incorrect characterization of spectral regrowth. Alternatively, some standards specify a band-limiting impulse response in the form of a digital filter to ensure that the appropriate frequency domain response is used [4,14].

Several wireless communication systems are based on discontinuous PSK modulation. These include the North American Digital Cellular standard (NADC), the Personal Digital Cellular System (PDC), the Personal Handyphone System (PHS), and cdmaOne [2,15,4]. The first three of these systems have adopted $\pi/4$-DQPSK modulation with a root-raised cosine filter response. The cdmaOne system has adopted

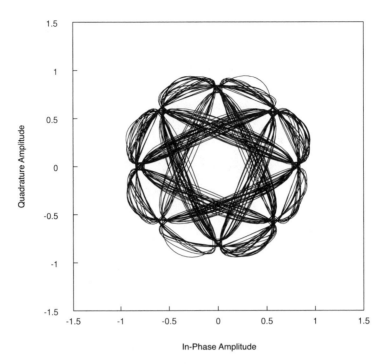

FIGURE 3.20   Constellation diagram for π/4-DQPSK with α = 0.35. This modulation is used for NADC, PDC, and PHS.

QPSK for the forward-link and offset QPSK (O-QPSK) for the reverse link, in both cases using a Chebyshev filter response.

π/4-DQPSK, a derivative of QPSK, differentially encodes the symbols, and adds a π/4 rotation at each symbol instant. This results in an envelope that avoids the origin, thus decreasing the peak-average of the signal. This is done presumably to reduce the PA back-off requirement, but as recent work has illustrated, this is not achieved in practice due to the original assumptions made [16]. Figure 3.20 shows the constellation diagram for π/4-DQPSK with α = 0.35, corresponding to the NADC standard. Note that at each of the constellation points, the trajectories do not overlap; this is inter-symbol interference. At the receiver, this signal will be convolved with the remaining half of the raised-cosine filter, leading to the desired response given by Eq. (3.110). PDC and PHS are also based on π/4-DQPSK, with an excess bandwidth factor of α = 0.50. Their constellation diagrams are similar to the NADC diagram shown in Fig. 3.20.

Figure 3.21 shows the constellation diagram for reverse-link cdmaOne, which is based on O-QPSK. Like π/4-DQPSK, O-QPSK was developed to reduce the peak-to-average ratio of the signal. It has been shown that this approach, while reducing the peak-to-average ratio, does not necessarily reduce the PA linearity requirements [16]. Note also in Fig. 3.21 the significant ISI, due to the fact that a Nyquist filter was not used. The forward link for cdmaOne uses QPSK modulation, which is shown in Fig. 3.18. As with the reverse link, the forward link uses a Chebyshev response specified by the standard.

To increase spectral efficiency, it is possible to add four additional phase locations to QPSK, called 8-PSK, with each location corresponding to a unique combination of three bits. Figure 3.22 shows the constellation diagram for the EDGE standard, which is based on 8-PSK and a specially developed filter that exhibits limited response in both the time domain and frequency domain [17]. EDGE was developed to increase the system capacity of the current GSM system while minimizing the requirements for linearity. In contrast to GSM, which is constant envelope, the EDGE systems provides increased spectral efficiency at the expense of envelope variations. However, EDGE intentionally minimizes the resultant

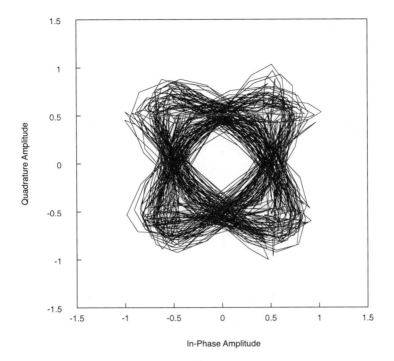

FIGURE 3.21 Constellation diagram for O-QPSK using the filter response specified by the IS-95 standard for cdmaOne.

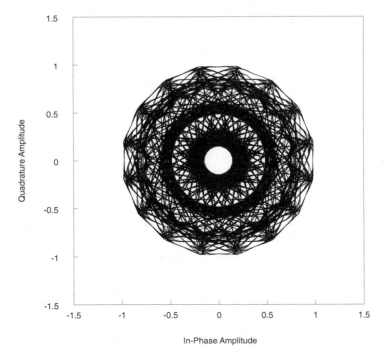

FIGURE 3.22 Constellation diagram for EDGE.

amplitude modulation to ensure that existing GSM amplification systems, optimized for constant envelope modulation, could still be used.

Continuous Phase-Shift Keying

With the exception of EDGE, each of the PSK modulations considered in the previous section exhibited
significant out-of-band energy due to the discontinuous transitions. To reduce the out-of-band energy,
band limiting was required, which resulted in envelope variations. An alternative modulation process is
continuous-phase modulation (CPM), which requires that the phase transitions from one symbol to the
next be continuous [9,13]. This results in a signal that is intrinsically band limited while maintaining a
constant envelope. The price paid is a potential reduction in spectral efficiency. Two types of CPM will
be examined in this section: minimum-shift keying (MSK) and Gaussian MSK (GMSK).

The complex envelope representation of MSK is

$$x(t) = \cos\left[a\left(kT_b \frac{\pi t}{2T_b} \right) \right] + j\sin\left[b(kT_b) \frac{\pi t}{2T_b} \right] \tag{3.112}$$

$a(kT_b)$ and $b(kT_b)$ represent unit-amplitude data sources as defined in Eq. (3.109). This modulation is
identical to O-QSPK using a half-wave sinusoidal pulse function instead of a rectangular pulse function.
Taking the magnitude squared of each term of Eq. (3.112) shows that it is indeed a constant envelope.

GMSK uses Gaussian pulse shaping, and has a complex envelope representation of

$$x(t) = \exp\left[jk_f \int_{-\infty}^{t} i(\tau)d\tau \right] = \exp\left[jk_f \int_{-\infty}^{t} \sum_{k=-\infty}^{\infty} a(kT_b)f(\tau - kT_b)d\tau \right] \tag{3.113}$$

where $f(\tau - kT_b)$ is a Gaussian pulse function and $a(kT_b)$ is a unit-amplitude data source as defined in
Eq. (3.109). Since the Fourier transform of a Gaussian pulse in the time domain is a Gaussian pulse, it
is seen that this modulation will exhibit intrinsic band limiting, in contrast to PSK. In Fig. 3.23 the GMSK
constellation diagram is illustrated, where it is seen that the envelope is constant. The information is
contained in how rapidly the phase function moves from one location on the circle to another, in a
fashion similar to FM. GMSK is used in the Global Standard for Mobile Communications (GSM) wireless
system [3].

Probabilistic Envelope Characterization

The complex trajectory of a signal, determined by the modulation technique, band limiting, and signal
coding used, is a time parametric representation of the instantaneous envelope of the signal. As such,
the duration of a particular envelope level in combination with the transfer characteristics of the power
amplifier (PA) will establish the resultant instantaneous saturation level. If the average spectral regrowth
exhibited by a PA is considered a summation of many instantaneous saturation events, it follows that
the more often an envelope induces saturation, the higher the average spectral regrowth will be. It is for
this reason that the peak-to-average ratio, though widely used, is ill suited for estimating and comparing
the linearity requirements of a PA [16].

This section introduces a method for probabilistically evaluating the time domain characteristics of a
an arbitrary signal to establish its associated peak-to-average ratio and instantaneous envelope power
distribution. The envelope distribution function (EDF) is introduced to estimate the peak power capability required of the PA and compare different signals. The EDF for many of the wireless systems presently
in use are examined.

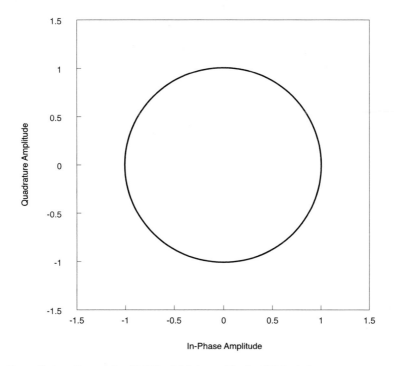

FIGURE 3.23   Constellation diagram for GMSK, which is used in the GSM wireless system.

The Envelope Distribution Function

Let $\tilde{p}(t)$ be the instantaneous power of a signal $\tilde{x}(t)$, with an associated instantaneous power probability distribution function $\varphi(\tilde{p})$. The probability of the instantaneous power exceeding the mean signal power is

$$\Pr\left[\text{instantaneous power} > \text{mean power}\right] = \Psi\left(\tilde{p}\right) = 1 - \int_{E[\tilde{p}]}^{\infty} \varphi\left(\tilde{p}\right) d\tilde{p} \qquad (3.114)$$

where $E[\tilde{p}]$ is the average power of $\tilde{p}(t)$. This function is defined as the envelope distribution function (EDF). In practice, the EDF is evaluated numerically, although it is possible to generate closed-form expressions. A specified probability of the EDF, typically 10^{-6}, is referred to as the peak-to-average ratio, σ. This is defined as

$$\sigma = \frac{\text{EDF @ } 10^{-6}}{E\left[\tilde{p}(t)\right]} \qquad (3.115)$$

A gradual roll off of the EDF indicates the instantaneous power is less likely to be near the mean power of the signal. This characteristic implies enhanced linearity performance to minimize increased distortion associated with the relative increase in the instantaneous clipping of the signal. Alternatively, for a given linearity requirement, the average power of the signal must necessarily decrease, resulting in lower efficiency. The amount that the average power must be reduced is referred to as output back off, and is usually specified in dB normalized to the 1 dB compression point of the PA under single-tone excitation. Finally, note that the EDF only has meaning for signals with a time-varying envelope.

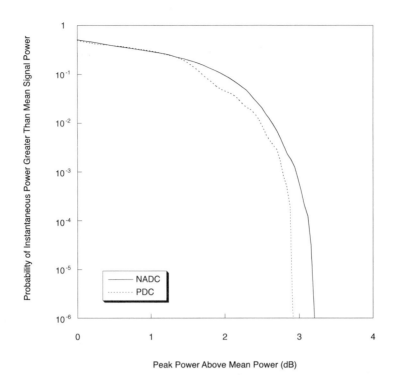

FIGURE 3.24 The EDF for NADC and PDC. NADC uses π/4-DQPSK with α = 0.35 and PDC uses π/4-DQPSK with α = 0.50. Note PDC has a slightly lower peak-to-average ratio due to the more gradual roll off of the filter response.

The EDF for Various Wireless Systems

In this section, Eq. (3.114) is used to examine the EDF for several of the wireless systems described earlier. NADC, PDC, and EDGE are considered first. The EDF for CDMA-FL and CDMA-RL using only a pilot tone is considered next. The impact of traffic channel, synchronization channel, and paging channels on the CDMA-FL EDF is then illustrated, where it will be seen that a significant increase in the peak-to-average ratio results, making this signal difficult to amplify. For comparison purposes, the EDF for a two-tone signal and complex Gaussian noise will also be shown. The EDF for each of these signals is shown in Figs. 3.24 through 3.29.

Summary

Contemporary microwave circuit design requires a basic understanding of digital modulation theory in order to meet the needs of a customer who ultimately speaks in terms of communication theory. This chapter was intended to provide a brief overview of the signal analysis tools necessary for the microwave engineer to understand digital modulation and how it impacts the design and characterization of microwave circuits used in contemporary wireless communication systems.

Complex envelope analysis was introduced as a means to describe arbitrarily modulated signals, leading to a geometric interpretation of modulation. The necessity and subsequent implications of band-limiting PSK signals were discussed. As an alternative to PSK, CPM was also introduced.

Signal analysis methods are often used to simplify the design process. Although the peak-to-average ratio of a signal is widely used to estimate of the linearity requirements of a PA, it was shown that this metric is ill suited in general for this purpose due to the random distribution of instantaneous power of a signal. The envelope distribution function (EDF) was introduced as means to compare various signals and to provide a more accurate estimate of the required linearity performance of a PA.

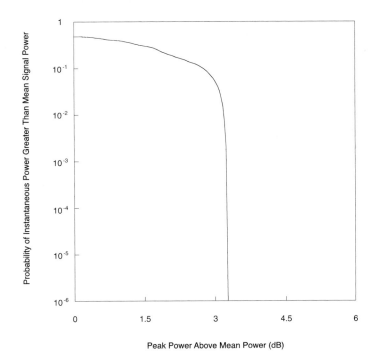

FIGURE 3.25 The EDF for EDGE (with all time slots active). Compare to Fig. 3.28, where the EDF for a two-tone signal is shown.

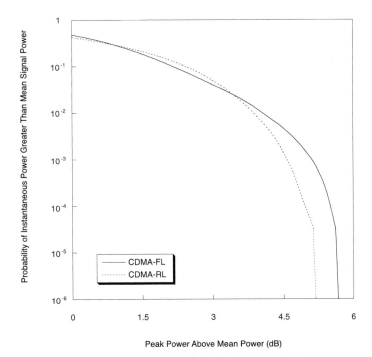

FIGURE 3.26 The EDF for forward-link CDMA and reverse-link CDMA with pilot tone only. Note that although CDMA-RL has a lower peak-to-average ratio, in areas of high probability of occurrence it has a higher peak power and results in higher spectral regrowth than CDMA-FL. This comparison clearly illustrates the advantages of using the EDF over the peak-to-average ratio.

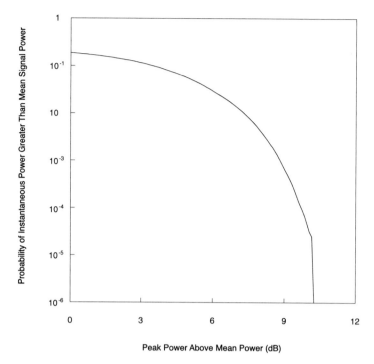

FIGURE 3.27   The EDF for forward-link CDMA with six traffic channels active and synchronization, paging, and pilot also active. Compare to Fig. 3.26 and observe the significant increase in peak-to-average ratio.

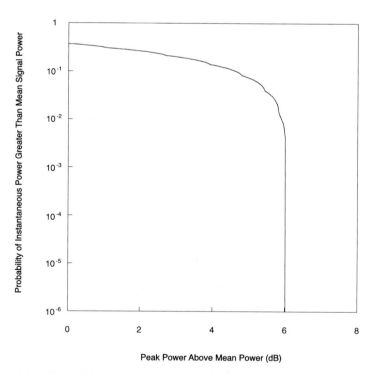

FIGURE 3.28   The EDF for a two-tone signal. Note that like the EDGE signal, the two-tone signal exhibits a gradual roll off, leading to increased distortion with respect to a similar signal with the same peak-to-average ratio but a faster roll off, such as CDMA-FL.

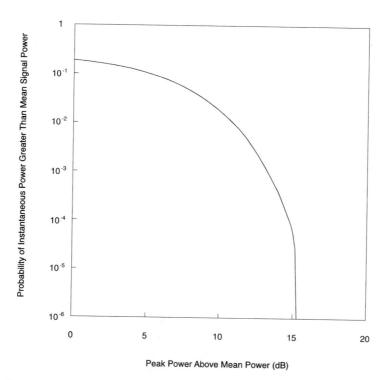

FIGURE 3.29 The EDF for a complex white Gaussian noise signal (commonly used for noise-to-power ratio characterization). The theoretical peak-to-average of this signal is infinite, but in practice is approximately 10 dB to 15 dB, depending on the length the sequence used.

References

1. W. C. Y. Lee, *Mobile Cellular Telecommunications Systems*, New York: McGraw-Hill, 1989.
2. EIA/TIA Interim Standard, Recommended Minimum Performance Standards for 800 MHz Dual-Mode Mobile Stations, EIA/TIA/IS-55, 1991.
3. M. R. L. Hodges, The GSM Radio Interface, *British Telecom. Tech. Journal*, 8, 31–43, January 1990.
4. EIA/TIA Interim Standard, *Mobile Station — Base Station Compatibility Standard for Dual-Mode Wideband Spread Spectrum Cellular System*, TIA/EIA/IS-95, 1993
5. R. N. Bracewell, *The Fourier Transform and Its Application*, New York: McGraw-Hill, 1978.
6. L. W. Couch, *Digital and Analog Communications*, 4th ed. New York: Macmillan, 1994.
7. M. Jeruchim, P. Balaban, and K. Shanmugan, *Simulation of Communication Systems*, New York: Plenum Press, 1992.
8. A. Papoulis, *Probability, Random Variables, and Stochastic Processes*, 3rd ed., New York: McGraw-Hill, 1991.
9. M. K. Simon, S. M. Hinedi, and W. C. Lindsey, *Digital Communication Techniques: Signal Design and Detection*, New York: Prentice-Hall, 1994.
10. M. Schwartz, *Information Transmission, Modulation, and Noise*, 3rd ed., New York: McGraw-Hill, 1980.
11. W. R. Bennett and J. R. Davey, *Data Transmission*, New York: McGraw-Hill, 1965.
12. H. Nyquist, Certain topics in telegraph transmission theory, *Transactions of the AIEE*, 47, 117-141, February 1928.
13. J. G. Proakis, *Digital Communications*, 2nd ed., New York: McGraw-Hill, 1992.
14. R. Schaefer and A. Openheim, *Discrete Time Signal Processing*, New York: McGraw-Hill, 1992.

15. Research and Development Center for Radio Systems (RCR), *Digital Cellular Communication Systems*, RCR STD-27, April 1991.

16. J. F. Sevic and M. B. Steer, On the Significance of Envelope Peak-to-Average Ratio for Estimating the Spectral Regrowth Characteristics of an RF/Microwave Power Amplifier, *IEEE Transactions on Microwave Theory and Techniques*, in press.

17. ETSI Standard GSM 11.21: Digital Cellular Telecommunications System BSS Radio Aspects, ETSI, 1999.

3.4 Cost Modeling

Leland M. Farrer

In the commercial industry, the ability to accurately estimate what a product can be built and sold for at a competitive profit, is a challenging task. Accurate cost modeling will simplify and organize the process into a powerful tool for the person entrusted with the task of quoting a project.

Each time a project manager is asked to quote on a future project, he is tasked with producing a cost for the development and cost for the end product. It does not seem to matter what industry you are in, the process is the same, and in the end, most managers succumb to estimating by experience rather than using a detailed costing model. The costing model is used to minimize the potential error by homing in on a model design that matches the market and, with feedback, can produce very good correlation with the end result.

The three major divisions of models include *product* focused, *service* focused, and *idea* focused. The classification of the three shows that the major emphasis is placed on *material, labor,* or *technology.*

1. Product-focused models are those where material costs are highest; where the added labor component has been minimized to the highest yield per hour spent. Examples include any production-line product that is automated.
2. Service-focused models are those where labor cost is highest. An example is a firmware product.
3. Idea- or technology-focused models involve an emerging technology with little track record. Examples are early life cycle inventions.

The cost model for all estimating has the following elements:

Bill of materials + Process time + Yield margin + Overhead + Profit = COST

The elements can be interconnected and developed into a spreadsheet application. Many organizations will have the company database connected to the cost model so that all elements are current and real time. A listing of the elements would be like this:

1. Bill of material (BOM) = Raw material cost (no labor for assembly or test) in $$
2. Process time = Operating cost per hour (labor) x Production cycle time = $$
3. Yield margin = (BOM + Process time)/Yield factor = $$
4. Overhead = (Process time + Yield margin) x Burden = $$. Overhead should also contain marketing costs as well as warranty or field returns as part of the burden element.
5. Profit = (BOM + Process time + Yield margin + Overhead) x Percent of markup = $$
6. Cost = the $$ value is the total of all elements.

Complimentary to the cost model is the risk assessment model. Where the cost model covers everything you know about, risk modeling evaluates and puts a dollar value on the things you don't know about that can surprise you. If risk modeling is done, it can be included with the other elements as a cost consideration.

BOM (Bill of Materials)

The bill of material is usually the total cost of all purchased parts such as resistors, semiconductors, etc., and fabricated cost such as PCB, shields, and outside fabricated parts. Depending on the internal requirement, all of these costs may be listed as raw materials with no handling markup. Some cost models will use this as the cost of goods for a product estimation and have a 20% handling charge on the dollar value as a burden element, independent of the G&A burden burden. This allows a softer reliance on hard yield and overhead elements later, but does mask the true cost. As with any production, the BOM costs will vary greatly with volume numbers. BOM costs at a quantity of 1000 will be much higher than at 1,000,000. Similarly, domestic BOM costs will be different than the same costs offshore.

Process Time

Depending on the item being estimated, the process time might be a resource output from a Gantt chart of all of the labor needed to design, test, and commission a prototype. This gets more difficult for development projects involving design review meetings and where many of the tasks are interrelated with critical path links. In complicated multiphase projects, the estimation of critical paths and margins to accommodate some slippage will become a requirement of the overall labor element of cost modeling. The cost element will be a total of all the individual worker hours at their unburdened individual rates.

Many models are available for setting up the production process. The decision regarding the method used will be based on industry needs and the maturity of the product.

In a manufacturing model, it might be the average cost per hour to operate a machine. The process time would then be the cost per hour times the cycle time to produce the item. Two types of process times must be considered each with a different cost impact. Variable (recurring) costs allow an incremental cost based on quantity changes. Fixed (nonrecurring) process cost is the base cost of holding the doors open (rent, etc.), and must be added. The final process time cost is the result of adding the percentage of use of the fixed cost and the increase due to volume increases.

Yield Margin

Yield margin is the amount of acceptable product vs. the number that failed in the manufacturing process. In the semiconductor industry, it is the percentage or dollar value of good devices vs. those units lost in wafer test, packaging, operational test, etc. In assembly, it is the number of units needed to set up the line, or parts lost in transport, QC, or customer rejection. If you need to deliver 100 units, and experience for like products have a total reject rate of 3%, then you will need a 3% margin (103) for proper yield. The cost, then, is the value of the rejected units spread across the 100 good units delivered.

Overhead

Overhead = Burden = $$(%). There is no easy number here. It must be calculated for each industry, geographical location, and will change as a company grows. For fledgling companies, this cost is shared by just a few individuals and can be quite high. If you are renting a location for $2 to $3 per square foot triple net, and have 5000 ft^2, this can be a significant number for 10 people. "Triple net" is a lease requiring the tenant to pay, in addition to a fixed rental, the expenses of the property leases such as taxes, insurance, maintenance, utilities, cleaning, etc. In this example, that is $10,000 to $15,000 plus capital equipment amortization, insurance, utilities, facilities maintenance, etc., plus the administrative staff, secretaries, sales staffs, etc. These are items that are not directly adding labor to produce the product. If you are quoting on the development of a product and expect it to take 3 months of time and all of your direct labor people, then the overhead cost for the three months will be the number to use. The number can be proportionally less if only some of the resources are used. When commissions are used, it can be added as part of the burden factor, although it is more commonly added after the cost is computed and becomes part of the cost given to the customer. Usually when a company is stable in resources and size, the burden number can be expressed as a percentage of man-hours. If not, it must be computed on an individual basis of labor cost for each direct labor resource used.

Profit

Profit = (BOM + Process time + Yield margin + Overhead) × Percent of markup = $$(%). How much do you want to make? There are many ways to look at this number. If your market is highly variable, you may need a higher number to cover the low-income periods. If you are in the semiconductor industry, the number is part of a strategy for growth. Competition will also force some consideration of how much to charge. Generally the final number is a percentage of the cost of doing business and marketing opportunity based upon the customers' needs and what the market will tolerate. A point to remember for costing is the constant fluctuation of the exchange rates for foreign currency. Where the estimate covers a significant period of time, care must be taken to get professional assistance in this area. This could involve tying the value of the currency to the dollar or buying foreign funds to cover the exposure.

Cost

Now that we have put all of the elements together, the other side of the equation equals *cost*. It is now time to evaluate this number. Does the dollar amount meet the needs of the customer? Will the company be underbid by the competition? Are we leaving money on the table? If the answer to any of these is no, go back and see where the critical element is that will turn them into yeses. Remember, not all quotes are good ones. Some will not be good for you or your business. Most are good ones and it is up to you to find the right balance of value elements for your cost model equation.

Although this is not an exhaustive list of elements, it serves as a basis. Cost modeling is an extremely valuable tool in the quoting process, as well as a monitor of ongoing business health. Let's look at the major differences in the marketplace.

Product-Focused Models

When a technology has matured to the point where a high level of automation is implemented or where the percentage of material cost to labor contribution is high, the model can be very accurate and margins can be predicted to a high level of certainty. The BOM cost can be easily obtained from past purchases or the company's materials resource planning (MRP) database. The process time of the new design can be compared to the existing product and weighted accordingly. The yield and margins applied and the cost to produce can be generated from existing production runs. The chip manufacturing industry is an excellent example of this type of market, as are low-priced consumer market products. The material costs can be very accurately determined and process times are minimized as much as possible. In these markets, the volumes are high and per-unit costs are highly refined and generally low. This market is highly mature and cost models are finely tuned so that extracting the final profitability dollar amount has low risk. One key control for this model is feedback from the production line to confirm yields. Statistical Process Control (SPC) can be employed to measure performance and flag when corrective action is necessary. The run numbers from yield and process time allow confirmation of the model accuracy for future cost model estimating. With the high commitment of capital equipment and process time, cost modeling is necessary to minimize the loading risk. This model is well suited for fixed bid as well as projects with a time and material (T & M) cap.

Service-Focused Models

When the major contribution of the product is labor, the estimation is only as good as the reliability of the workers and their task execution. The elements of the model are the same, but the focus on past experience and more margin is required to compensate for this uncertainty. Examples of this are electronic prototype construction, software development, and warranty service. This model will display a wider swing in uncertainty relative to the product-focused model. Many service industries use this weighted model, and it is also taxed with individual performance estimating as a requirement. It is not helpful when worker substitution and different skill levels are factors, unless added margin is allowed.

Idea-/Technology-Focused Models

These can be the highest risk models. This is where invention is needed or relied upon for success. This can be seen in Gantt charts where the task "something magic happens here" is implied. It is also the area where the Intellectual Property (IP) (the idea) is not fully developed or needs refinement to make it ready for production. An example would be taking a reference design circuit and making an application-specific product. Due to the uncertainty of the elements of this model, the estimation is likely to be on the conservative side unless an investment in market penetration or a strategic alliance exists or is desired, in which case a more aggressive costing might be undertaken to secure the contract against competition.

Here again is the reliance on past experience. Regular reviews are needed of completed project performance to ensure that the model elements are operating properly. Failure to continually review and update the model elements on a daily, weekly, and monthly basis will result in an out-of-control erosion of the margins and profit.

Feedback

Probably the most important and often overlooked feature for accurate cost modeling is the use of feedback in the maintenance of the cost model. This can take several forms and is a featured process in all ISO 9000 documentation. Production feedback is used regularly on the factory floor at QA stations to ensure the quality of the product. No less important is the feedback from a development project to indicate the fiscal health of the team in meeting the milestones with calendar, material, and labor costs on track. This information not only needs to get back to the project managers for online project correction, but, also needs to get back to the contract administrators who generate the quotes to evaluate and adjust the cost model elements as necessary. Examples of corrections are the need for margin in the labor element if a unique skill set is required, or when vendor performance is in the critical path to meet a scheduled milestone and a calendar slip is likely. All feedback that will affect the performance accuracy must be routinely fed back into the cost model for correction.

Refinement of the Cost Model

In conclusion, the cost model is a living tool. Cost model effectiveness is only as good as the accuracy of the information going into each of the elements and the constant feedback to fine-tune it. A model that is good in a bull market may not be the one you need for the downturn era. As your company grows, the model will need to constantly evolve to keep pace with your needs. Putting the cost model into a spreadsheet will allow effective use and permit a lot of "what if" investigation as well as cost model evolution.

References

ECOM, European Space Agency, (ESA) Cost Modeling, A general Description http://www.estec.esa.nl/eawww/ecom/article/ecom.htm.

Barringer and Associates, Life Cycle Cost Modelling, Freeware, http://www.barringer1.com/MC.htm.

Skitmore, M., ed., *Cost Modelling,*

3.5 Power Supply Management

Brent A. McDonald, George K. Schoneman, and Daniel E. Jenkins

Background

A Short History (1970–2000)

1970 — Switching frequencies were normally 50 KHz or less. Bipolar transistors were predominately used as the power switching elements. Royer oscillators were popular during this period.[1]

1980 — Switching frequencies were normally 100 KHz or less. MOSFET transistors were predominately used as the power switching elements. Control ICs were introduced. Constant frequency pulse width modulation became popular. Schottky diodes were introduced.

1990 — Switching frequencies were normally 500 KHz or less. Current mode control became popular (Peak Current Mode Control and Average Current Mode Control). Synchronous rectifiers were introduced. Commercial Off The Shelf (COTS) became popular. Zero-voltage and zero-current switching techniques were introduced.

2000 — Technology trends have not changed much since 1990. Switching frequencies are still 500 KHz or less. Current mode control remains on the forefront of many designs. Synchronous rectifiers are gaining popularity.

Future — Consumer PC voltages are projected to be as low as 0.5 V by the year 2010. These trends are suggesting dense hybridized power supplies co-located with the microprocessor. The automotive industry is going toward the more electric and all electric car.

Power Supply Types

Most power supplies are configured to supply a well-regulated, constant voltage output, however, this does not have to be the case. Theoretically, a power supply can be configured to deliver constant current or power. Since most applications require some type of regulated voltage source, this topic will be addressed from that point of view.

Obviously, a power supply does not produce power for its load. It takes power from a source and reconditions the power so that the load can operate without damage or interference. If a power supply is required, it must be assumed that the power source is in some way not suitable to the device(s) it is powering.

This section will discuss three basic types of power supplies: linear regulators, charge pumps, and switch mode power supplies (SMPS). Each of these has an appropriate range of application and its own unique operational characteristics. A linear regulator takes a DC voltage with a higher magnitude than required by the load and steps it down to the intended level. The linear regulator typically accomplishes this by dropping the excess voltage across some kind of series pass element, commonly known as a series pass regulator. The second most common form of linear regulator is the shunt regulator. This approach often utilizes a fixed series impedance with a variable shunt impedance. Since the source voltage and load may vary, a feedback loop is usually used to minimize these affects on the output voltage. Figure 3.30 illustrates these two types of regulators.

A charge pump is capable of boosting a voltage to a higher level. A simplified charge pump is shown in Fig. 3.31. These voltages may not be as well regulated as the SMPS or linear regulator approaches since their output regulation is dependant on items such as the forward voltage drop of the diodes and the initial tolerance of the power source being used. When turned on, the inrush current may be very high. Essentially, the voltage sources are looking into a capacitor. The currents will be limited by the rate of rise of the source, the impedance of the diodes, and any parasitic resistance and inductance in series with the capacitor.

An SMPS operates by taking a DC voltage and converting it into a square wave and then passing that square wave through a low pass filter. In many applications, this allows for a more efficient transfer of power than can be achieved with a linear.

Figure 3.32 lists a general comparison between these three types of power supplies.

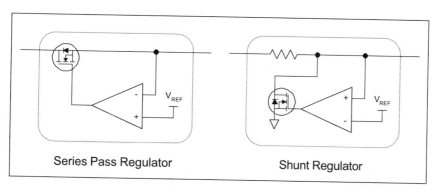

FIGURE 3.30 Two basic types of linear regulators.

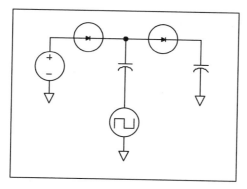

FIGURE 3.31 Typical charge pump.

Linear	SMPS	Charge Pump
(Based on a 1V low drop out device)		
No output voltage inversions possible	Output voltage inversions possible	Output voltage inversions possible
Low parts count	High parts count	Low parts count
Low design complexity	High design complexity	Low design complexity
Efficiency: 67% to 91%	Efficiency: 80% to 96%	Efficiency: 80%
Limited input voltage range	Wide input voltage range	Limited input voltage range
EMI Filter: Small	EMI Filter: Large	EMI Filter: Small to medium

FIGURE 3.32 Power supply comparison (2V < Vout < 10V, 15W).

Why Use a Power Supply?

Some power sources have voltages that vary 2:1 in normal operation and more than 10:1 during some transient conditions. Very few loads can function properly over these ranges and most will be damaged. A power supply can provide an interface to make these line voltage variations transparent, or at least tolerable to the load's operation.

Some applications such as transmitters, stepper motors, processors, etc., have large step loads. This may require lower output impedance than the source alone can provide. In other words, if the load requires a sudden burst of current, it is desirable that the output voltage remain in regulation. A low impedance capacitor bank can provide this step current. Alternatively, the control loop can sometimes be modified to supply such step loads.

System Issues: Specifications and Requirements

Input Voltage

This is an area where many mistakes are made in the initial design. The input voltage, or source, is rarely as well behaved as the title indicates. As an example, let's examine the average automobile electrical system as it exists in year 2000. The owner's manual will call it a "12 Volt" system, which sounds good, but when the alternator is charging, the voltage is typically ~13.8 Volts (~15% increase). That is "typical," but actual numbers could range as high as 15 Volts (~25% increase). And it gets worse! Those voltages are defined as "steady state" or "normal operation." When heavy loads on the system are removed (headlights, starter motor, cooling/heating fans, air-conditioning, horn, etc.) the voltage on the bus can exceed 60 Volts (500% increase!) for durations > 100 ms.

Standard household power is equally inhospitable. Lightning, power surges, and brownouts can result in a very wide input-voltage range that your design may have to tolerate.

Battery-operated units are somewhat better, but attaching and removing battery chargers can cause large voltage spikes. Due to the internal impedance of the battery, transient step loads will cause degradation in the battery bus voltage.

Output

Power supplies are designed to power a specified load. In other words, there is no such thing as "one size fits all." For example, a supply that powers a 50 W load with 85% efficiency would be totally inappropriate for a 1 kW load.

A reactive load will introduce additional phase shift that a resistive load will not. This additional phase shift can create unstable or marginally stable systems. The module designer will often add significant capacitance to decouple one circuit from another. If this capacitance is large enough and not accounted for in the selection of the power supply, reduced stability margins and/or a compromised transient performance can result. Likewise, if inductors are used to de-couple the circuits, an inductive load can result and a similar detrimental effect on stability and transient performance is possible.

If the load exhibits large/fast step changes, the equivalent series resistance (ESR) of the output capacitors may cause the output voltage to drop below the valid operating range. The control loop will then try to correct for the error and overshoot can result.

A common misconception is that the DC accuracy of the output voltage is primarily a function of the reference. Usually, the output voltage is resistively divided down before being compared to the reference. If this is the case, the tolerance of the divider resistors needs to be accounted for. Offset voltages, offset currents, bias currents, and the location of the gains in the loop relative to the reference voltage can also significantly affect the DC settling point. This is all further complicated by the fact that these are all temperature-dependent terms.

The output ripple voltage waveforms of SMPS are harmonically rich. Generally, they will take the shape of a triangle or trapezoid. The fundamental frequency of these waveforms is at the switching frequency and will usually have harmonic content both above and below the fundamental. The content below the fundamental can be the result of input voltage variations, load variations, or injected noise. The ripple voltage specification listed on a data sheet or in a requirements document will probably give the maximum peak-to-peak voltage at the switching frequency or the RMS voltage. Often, larger spikes can be present for which the ripple voltage specification does not account. Figure 3.33 shows a typical output ripple voltage waveform and the frequency spectrum of that waveform. This particular case is a 20 V input, 3.3 V output, operating at 18 W.

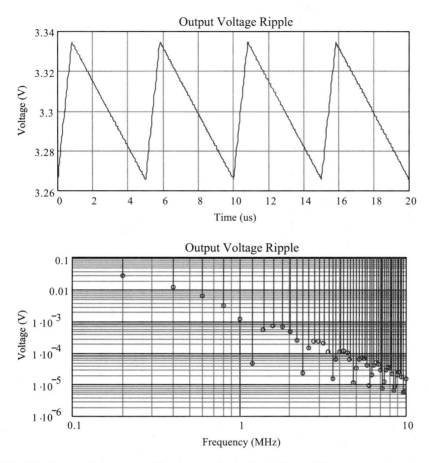

FIGURE 3.33 Waveform and spectrum of the output voltage of a 20 V to 3.3 V power supply at a load of 18 W.

Environment: Thermal, EMI/EMC

Looking at efficiency numbers can be misleading. Attention needs to be given to what the actual systems requirements are. In the case of a battery-operated system, efficiency may be of utmost importance. In this case, maximizing the time between battery replacements (or recharges) is a must. It may be particularly important to have some kind of ultra-low-power, stand-by mode to prevent the needless waste of power when the supply is not being used. A system running off a generator or a low impedance power source probably is not as driven by the actual efficiency, but by the overall power dissipation. For example, it would not be uncommon to see a supply operate at 90% efficiency at the full-specified load. However, if that load drops by a factor of 10, the efficiency may drop into the 60% range. The reason for this is that the power supply usually requires some fixed level of overhead power. The larger the percentage of the actual load that the overhead represents, the lower the efficiency. However, this is usually inconsequential. After all, if the source can provide the full load, it should have no problem powering a lighter load. Therefore, the efficiency at a light load may be irrelevant. The thermal loading of the box is what matters.

EMI/EMC can be divided into several categories: conducted emissions, radiated emissions, conducted susceptibility, and radiated susceptibility. A conducted emissions specification will define how much current can be put on the power lines for a given frequency. Figure 3.34 shows a typical input current waveform for a Buck derived SMPS. This particular case is a 20 V input, 3.3 V output, operating at 18 W.

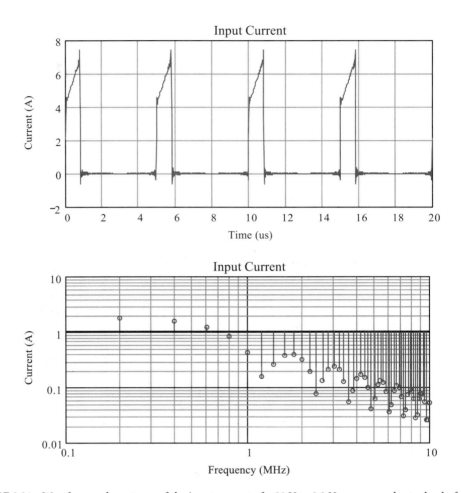

FIGURE 3.34   Waveform and spectrum of the input current of a 20 V to 3.3 V power supply at a load of 18 W.

The EMI filter must attenuate this current spectrum per the system specifications. A two-pole LC filter is often used as a method for controlling emissions, such as those shown in Fig. 3.34.

A radiated emissions requirement will define the level of E and H field that can be radiated from the unit under test. A SMPS can be a significant source of these types of emissions, because of the high $\frac{dV}{dt}$'s and $\frac{dI}{dt}$'s present. Typically speaking, if the conducted emissions are controlled, the radiated emission requirement is usually met. If necessary, shielded magnetics may be used to limit the H fields.

Conducted susceptibility defines the disturbance levels the power lines can be subjected to while the unit under test must continue to operate properly. Figure 3.35 shows an example of this kind of requirement.

In this case, the chart shows the RMS voltage that can be put on the power line for a range of frequencies. For example, the graph may start at 20 Hz, start to roll off at 2 kHz, and end at 50 kHz. (For a specific example see MIL-STD-461C.[2]) Many times, conducted susceptibility can be addressed by using an input-voltage-feed-forward term in the power supply control loop (this will be discussed in greater detail later).

Radiated susceptibility defines how much radiated E and H fields the unit under test can be exposed to and expected to operate properly. Ensuring EMC here is generally a mechanical housing issue.

Cost

Cost considerations are one of the most difficult areas to predict. So much of cost is dependent on the technology of the day. However, it can be safely stated that reducing size will increase both recurring and nonrecurring cost. Increasing efficiency will increase nonrecurring cost and may or may not increase

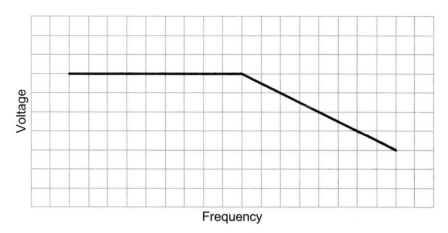

FIGURE 3.35   General conducted susceptibility curve.

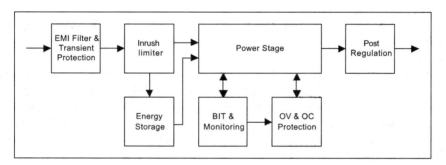

FIGURE 3.36   Power supply block diagram.

recurring cost. Typically speaking, the power inductor, bulk input and output capacitors, integrated circuits, and printed wiring boards will represent the majority of the cost.

Block Diagram

Figure 3.36 shows the basic blocks that make up a power supply. In reality, the blocks may be more integrated or arranged differently to accommodate various system requirements. For example, some systems have no requirement for energy storage or it is done in series with the power stage instead of in parallel.

The EMI (Electro Magnetic Interference) filter and transient protection serve to protect the power supply from excessive line voltages and to ensure EMC (Electro Magnetic Compatibility). Many power lines have short duration, high-voltage spikes (e.g., lightning). Designing a power supply so that all of the relevant components can withstand these elevated voltages is usually unnecessary and will likely result in reduced efficiency and increased size. Normally, the energy contained in such a burst is low enough that a transient suppressor can be used to absorb it.

The inrush limiter serves to limit peak currents that occur due to the rapid charging of the input capacitance during start up or "hot swap." Hot swap is a special start-up condition that occurs when a power supply is removed and reinserted without the removal of power. Once a power source is turned on, it can take several milliseconds before its voltage is up and good. This time reduces the peak current stresses due to the reduced $\frac{dV}{dt}$ that the capacitors see. If a power supply is plugged into a source that is already powered up, significantly higher inrush currents can result.

Some power systems require the output voltages to be unaffected by short duration power outages (less than 200 ms for the purposes of this paper). An energy storage module accomplishes this. If the

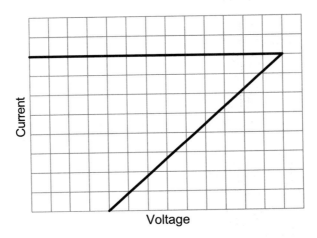

FIGURE 3.37   General fold-back current limit curve.

duration is longer, a UPS (Uninterruptible Power Supply) may be used. In either case, some type of energy storage element (e.g., battery, capacitor bank) is charged and in the event of a loss of primary power, the supply runs off of the stored energy. The power stage will be discussed in detail later.

The BIT and monitoring typically monitors for output overvoltage, under-voltage, and over-current. Overvoltage protection is necessary to limit the propagation of damage in a system. In order to do this, something needs to be done to clear the fault. Sometimes a series switch is used to remove power from the supply. If this is the case, the switch normally latches open and can only be reset by cycling the primary power on and off. Another common method is a crow bar. A fuse is placed in series with the power line. If an overvoltage is detected, a short is placed on the output of the supply that will force the fuse to open. The downside of doing this is that it becomes a maintenance item, however, it does a good job of guaranteeing that the overvoltage will be cleared.

Over-current is also used to prevent the propagation of damage. Two common ways of addressing this are a brick wall current limit or fold back. Limiting the maximum current that the supply can put out makes a brick wall limit. For example, if the current limit is set at 6 A, the supply will never put out more than 6 A, even if a dead short is placed on the output. This works well, but sometimes the thermal loading during current limit can be more than the system can handle. In such a case, a fold-back limit may be more appropriate. A typical characteristic is shown in Fig. 3.37.

Notice that when the current hits the maximum level and the output voltage begins to drop, the current limit point also falls. This works to reduce the power dissipation in the supply. Care needs to be used when working with fold-back limiting. It is possible to get the supply latched in a low voltage state when there is no fault.

Under-voltage is used to prevent erroneous operation. Often a signal is sent out to report an under-voltage condition. This signal can be used to hold a processor in reset or to inhibit another operation. Latching a supply off on an under-voltage is difficult, due to the danger that the supply may never start up!

In an effort to reduce size, many SMPS converters rely on a transformer as an easy way to produce multiple outputs. The basic idea is for one of the outputs to be tightly regulated through a control loop and the auxiliary outputs to track by transformer turns ratio. The problem is variations in the inductor DC Resistance (DCR), diode forward voltage drop, transformer coupling, load, temperature, etc., can all cause the auxiliary voltages to vary. This additional variation appears on top of the initial tolerance established by the control loop. The bottom line is the auxiliary windings may not have sufficient accuracy. If this is the case, there are post-regulation methods that can tighten the output tolerance with a minimal efficiency hit.

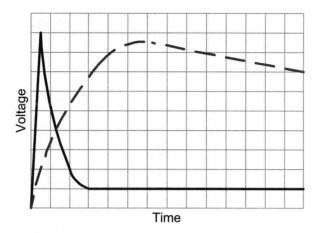

FIGURE 3.38 Typical start-up characteristics without soft start.

Power Supplies

Soft start is an area that can often be a problem. Figure 3.38 illustrates this problem. The solid trace is the error amplifier output voltage. This signal controls the output voltage level. The dashed trace is the output voltage.

When a power supply is commanded to turn on, the error amplifier will attempt to raise the output voltage as quickly as possible. Once the output voltage is in regulation, the error amplifier must slew quickly to avoid overshoot. The problem is compounded by the fact that stability considerations demand that the error amplifier be slowed down. This usually results in an overvoltage.

Active circuitry can be added to force the error amplifier's output to rise slowly to the level required for proper regulation. There are a variety of ways to accomplish this, however, it is essential that it be addressed in most systems.

When talking about SMPS, it is customary to talk about duty cycle (herein referred to as D). This is the percentage of time that the main switch is on per cycle. The remainder of the cycle is often referred to as 1-D.

SMPS are often categorized according to the state of the inductor current. Figure 3.40 shows the inductor current for a simple buck converter in steady state operation. The first picture shows discontinuous conduction mode (DCM). This condition occurs when the inductor current goes to zero during the 1-D time. A

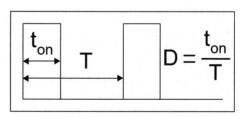

FIGURE 3.39 General duty cycle definition.

special case exists when the inductor current just reaches zero. This case is shown in the second picture and is called critical conduction. The third picture shows continuous conduction mode (CCM). This condition occurs when the inductor current is greater than zero throughout the cycle.

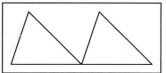

FIGURE 3.40 Basic inductor current waveforms (DCM, Critical, CCM).

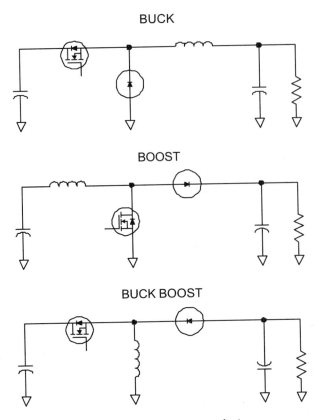

FIGURE 3.41   Basic non-isolated DC-DC converter power stage topologies.

Figure 3.41 shows the power stages for the basic non-isolated DC-to-DC switching power supply topologies. In each case, there are two switches (MOSFET, diode), an inductor, and an output capacitor. The input capacitor is shown to represent the low impedance input required for operation. The resistor on the output models the load, however, the load may not necessarily be resistive.

The CCM buck converter is the simplest to understand of the three approaches. It takes an input voltage and steps it down. Essentially, the MOSFET and diode create a square wave at the input to an LC filter. The high frequency content of the square wave is filtered off by the LC, resulting in a DC voltage across the output capacitor. Since the AC portion of the inductor current is the same as the output capacitor current, it is relatively easy to reduce the output ripple voltage by increasing the inductance. Figures 3.42 and 3.43 show the MOSFET and diode current, respectively. Please note that these waveforms have an identical shape for each of the three topologies.

The CCM boost converter operation is a little more complex. This time the MOSFET shorts the inductor to ground, allowing the inductor current to ramp up (Fig. 3.42 is the MOSFET current). When the MOSFET turns off, the inductor current continues to flow by turning the diode on (Fig. 3.43 is the diode current). In order for the inductor current to continue to flow, the output voltage must rise higher than the input.

The CCM buck-boost (or flyback) operates in an energy transfer mode. When the main MOSFET closes, the inductor charges. This charge action is shown in Fig. 3.42. Once the MOSFET opens, the current in the inductor continues to flow and turns the diode on, transferring charge to the output capacitor. In the non-coupled approach, this always results in a voltage inversion on the output. Figure 3.43 shows the diode current.

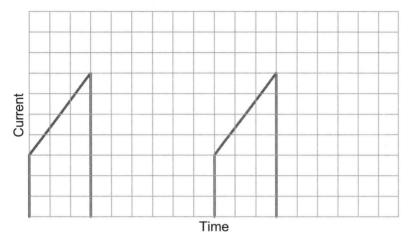

FIGURE 3.42 MOSFET current waveform for a SMPS.

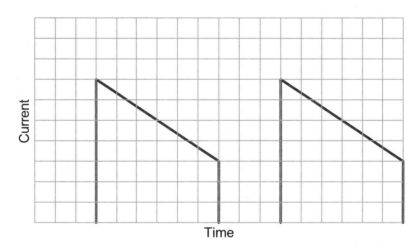

FIGURE 3.43 Diode current waveform for a SMPS.

The duty cycle required to generate a given output voltage is different in each topology and is dependent on the mode of operation. When in CCM, D is defined only in terms of input and output voltage. If in DCM, the relationship is more complex. The main point here is in CCM, the output voltage is independent of load. In DCM, it has the additional dependence on inductance, switch frequency, and load.

Duty Cycle	Buck	Boost	Buck-Boost
CCM	$\dfrac{V_{out}}{V_{in}}$	$\dfrac{V_{out} - V_{in}}{V_{out}}$	$\dfrac{V_{out}}{V_{out} - V_{in}}$
DCM	$V_{out} \cdot \sqrt{\dfrac{2 \cdot L \cdot f_s}{R \cdot V_{in} \cdot (V_{in} - V_{out})}}$	$\dfrac{1}{V_{in}} \cdot \sqrt{\dfrac{2 \cdot L \cdot f_s \cdot V_{out} \cdot (V_{out} - V_{in})}{R}}$	$\dfrac{V_{out}}{V_{in}} \sqrt{\dfrac{2 \cdot L \cdot f_s}{R}}$

FIGURE 3.44 Duty cycle expressions for Buck, Boost, and Buck-Boost converters.

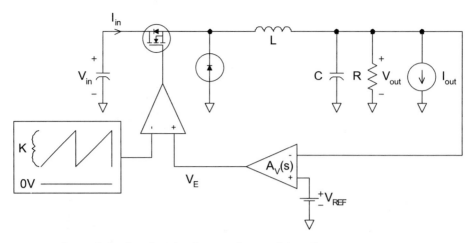

FIGURE 3.45   Basic non-isolated Buck with voltage mode control (VMC).

$$
\begin{bmatrix} \dfrac{V_{in}}{I_{in}} & \dfrac{I_{in}}{I_{out}} \\[2ex] \dfrac{V_{out}}{V_{in}} & \dfrac{V_{out}}{I_{out}} \end{bmatrix} = \begin{bmatrix} \dfrac{s^2 + \dfrac{s}{R \cdot C} + \dfrac{1}{L \cdot C} \cdot \left(1 + \dfrac{v_{in}}{V_{RP}} \cdot A_v(s)\right)}{\dfrac{d^2}{L} \cdot \left[s + \dfrac{1}{R \cdot C} \cdot \left(1 - \dfrac{v_{in}}{V_{RP}} \cdot A_v(s)\right)\right]} & \dfrac{\dfrac{d}{L \cdot C \cdot R} \cdot \left[s \cdot L \cdot \dfrac{v_{in}}{V_{RP}} \cdot A_v(s) + R \cdot \left(1 + \dfrac{v_{in}}{V_{RP}} \cdot A_v(s)\right)\right]}{s^2 + \dfrac{s}{R \cdot C} + \dfrac{1}{L \cdot C} \cdot \left(1 + \dfrac{v_{in}}{V_{RP}} \cdot A_v(s)\right)} \\[6ex] \dfrac{\dfrac{d}{L \cdot C}}{s^2 + \dfrac{s}{R \cdot C} + \dfrac{1}{L \cdot C} \cdot \left(1 + \dfrac{v_{in}}{V_{RP}} \cdot A_v(s)\right)} & \dfrac{\dfrac{-s}{C}}{s^2 + \dfrac{s}{R \cdot C} + \dfrac{1}{L \cdot C} \cdot \left(1 + \dfrac{v_{in}}{V_{RP}} \cdot A_v(s)\right)} \end{bmatrix}
$$

FIGURE 3.46   Voltage Mode Control (VMC) transfer functions.

Figure 3.45 shows one of the simplest standard methods for regulating the output voltage of a SMPS, voltage mode control (VCM). For simplicity, the Buck typology in CCM has been chosen.

A constant duty cycle applied to the MOSFET switch will produce a DC output voltage equal to the input voltage multiplied by duty cycle. If the duty cycle goes up, the output voltage will go up. If it goes down, the output voltage will also go down. The device driving the MOSFET is a comparator. The inputs to the comparator are a signal proportional to the output voltage (V_E) and a voltage ramp. If V_E goes up, D goes up and raises the output. Likewise, if V_E goes down, D goes down and the output falls. Some type of compensation is placed between the output and V_E to close the control loop (labeled $A_v(s)$ in the diagram). This compensation will control the stability and transient performance of the supply. Figure 3.46 contains the mathematical definition of several transfer functions for the above system.

These equations are based on a modeling technique called state space averaging.[3] Practically, they only apply to 1/10 to 1/3 of the switch frequency. They are based on an averaged linearized model of the power supply. They do not contain valid information about the large signal response of the system, only the small signal. When this method is applied to a DCM topology, the standard approach may need to be modified to obtain a more accurate result.[4]

If the voltage ramp that is fed into the comparator is proportional to the input voltage (generally referred to as input voltage feedforward), these equations simplify to the following:

$$\begin{bmatrix} \dfrac{V_{in}}{I_{in}} & \dfrac{I_{in}}{I_{out}} \\[3mm] \dfrac{V_{out}}{V_{in}} & \dfrac{V_{out}}{I_{out}} \end{bmatrix} = \begin{bmatrix} -\dfrac{R}{d^2} & \dfrac{\dfrac{d}{L\cdot C\cdot R}\left[s\cdot L\cdot A_v(s)\cdot K + R\cdot\left(1 + A_v(s)\cdot K\right)\right]}{s^2 + \dfrac{s}{R\cdot C} + \dfrac{1}{L\cdot C\cdot R}\cdot\left(1 + A_v(s)\cdot K\right)} \\[8mm] 0 & \dfrac{\dfrac{-s}{C}}{s^2 + \dfrac{s}{R\cdot C} + \dfrac{1}{L\cdot C\cdot R}\cdot\left(1 + A_v(s)\cdot K\right)} \end{bmatrix}$$

FIGURE 3.47 Voltage Mode Control (VMC) transfer functions with input voltage feedforward.

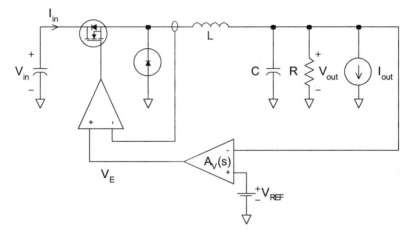

FIGURE 3.48 Basic non-isolated Buck with peak current mode control.

The most practical result of this simplification is that perfect input transient rejection is achieved. In other words, the output will not be perturbed by variations on the input. Please keep in mind that these equations are simplified and idealized. Perfect input voltage transient rejection is not possible. Some input variations will reach the output due to delay times, parasitic effects, etc.

Figure 3.48 shows one of the most popular ways to control a power supply, peak current mode control (CMC). The fundamental difference between CMC and VMC is that the ramp used for comparison to V_E is now directly proportional to the inductor current. This action creates an inherent feedforward term that serves to reject input voltage transients. However, it does not provide the ideal rejection that the VMC with input voltage feedforward does. This can be achieved by adding slope compensation. If a constant slope (exactly equal to $\frac{1}{2}V_{out}/L$, or $\frac{1}{2}$ the down slope) is added to the ramp, derived from the inductor current, the ideal input transient rejection is achieved.[5] Again, this is an ideal approximation. In reality, delay times and circuit parasitics will allow some of the input voltage variations to reach the output.

The following are the linearized state space averaged equations with the slope compensation equal to $\frac{1}{2}$ the downslope of the inductor current. The technique used for modeling the current loop does not model the sub-harmonic oscillation, which is described widely in the literature. Models that are more complex are needed to predict this effect.[6] (Please note: R_s/N is the gain of the inductor current to the input of the comparator, f_s is the switching frequency of the converter.)

$$
\begin{bmatrix}
\dfrac{V_{in}}{I_{in}} & \dfrac{I_{in}}{I_{out}} \\[2ex]
\dfrac{V_{out}}{V_{in}} & \dfrac{V_{out}}{I_{out}}
\end{bmatrix}
=
\begin{bmatrix}
-\dfrac{R}{d^2}\;\dfrac{\dfrac{d}{R\cdot C}\cdot\dfrac{2\cdot L\cdot A_v(s)\cdot f_s\cdot N}{R_s}+d\;\cdot s+\dfrac{d}{L\cdot C}\cdot\left[1+d+2\cdot f_s\cdot L\cdot\dfrac{A_v(s)\cdot N}{R_s}-\dfrac{1}{R}\right]}{s^2+\dfrac{1}{R\cdot C}+2\cdot f_s\;\cdot s+\dfrac{1}{L\cdot C}\cdot\left[1+d+2\cdot f_s\cdot L\cdot\dfrac{A_v(s)\cdot N}{R_s}+\dfrac{1}{R}\right]} \\[6ex]
0 \qquad\qquad\qquad \dfrac{\dfrac{-1}{C}\cdot s+2\cdot f_s}{s^2+\dfrac{1}{R\cdot C}+2\cdot f_s\;\cdot s+\dfrac{1}{L\cdot C}\cdot\left[1+d+2\cdot f_s\cdot L\cdot\dfrac{A_v(s)\cdot N}{R_s}+\dfrac{1}{R}\right]}
\end{bmatrix}
$$

FIGURE 3.49 Transfer functions of a basic non-isolated Buck with peak CMC and optimal slope compensation.

Conclusion

Simply put, a DC-to-DC power supply takes one DC voltage and puts out a different one. However, the process of getting there can be much more complex. Consideration needs to be given to the system and environment in which the supply is expected to operate (e.g., input voltage, EMI/EMC, energy hold up, output ripple, efficiency, thermal management, battery life, temperature, health monitoring, stability, transient response, etc.). These types of issues all impact the type of supply needed (e.g., linear, charge pump, SMPS) and how that supply needs to be configured for a given application.

As stated earlier, the trend for power seems to be toward lower output voltages at higher load currents. This trend is going to demand that power supply efficiency be maximized. This is always important, however, when the required output voltage drops, the detrimental effect of circuit parasitics (e.g., resistance, diode forward voltage drops) on efficiency becomes larger. This coupled with the fact that the load currents are not dropping, but rising, creates a greater thermal load on the system. In the case of microprocessors, this will likely result in dense hybridized power supplies that will be colocated with the processor.

References

1. Pressman, A. I., *Switching Power Supply Design,* McGraw-Hill, New York, 1991, 248–264.
2. MIL-STD-461C, Electromagnetic Emission and Susceptibility Requirements for the Control of Electromagnetic Interference, Department of Defense, The United States of America, August 4, 1986, Part 2, 13.
3. Mitchell, D. M., *Switching Regulator Analysis,* McGraw-Hill, New York, 1988, 51–73.
4. Sun, J., Mitchell, D. M., and Jenkins, D. E., Delay effects in averaged modeling of PWM converters, Power Electronic Specialists Conference, 1999. PESC99. 30th Annual IEEE, 1999.
5. Modeling, Analysis and Compensation of the Current-Mode Converter, Unitrode Application Note, U97, 1997, Section 3, 43–48.
6. Ridley, R. B., A New, Continuous-Time Model For Current-Mode Control, *IEEE Trans. on Power Electr.,* 6, 2, 271-280, April 1991.

3.6 Low Voltage/Low Power Microwave Electronics

Mike Golio

The development and manufacture of low voltage/low power RF and microwave electronics has been the subject of intense focus over the past few decades. The development of active circuits that consume lower power at lower voltage levels is consistent with the general trends of semiconductor scaling. From the simple relationship

$$E = V/d \qquad (3.116)$$

where E is the electric field, V is the voltage, and d is the separation distance, it can be seen that smaller devices (reduced d) require less voltage to achieve the same electric field levels. Since to first order it is electric field in a material that dominates electron transport properties, electric field levels must be preserved when devices are scaled. Therefore, for each reduction in critical dimension (gate length or base width) reductions in operating voltage have been realized.

Because semiconductor physics implies an inverse relationship between device size and required voltage level, and because transistor scaling has been a major driver for integrated circuit (IC) technology since the earliest transistors were introduced, the trend toward reduced power supply voltage has been ongoing since the beginning of the semiconductor industry. The explosive growth of portable wireless products, however, has brought a sense of urgency and technical focus to this issue. Portable product requirements have forced technologists to consider battery drain and methods to extend battery lifetime.

Since the total global volume requirements for digital ICs far exceeds the requirements for RF and microwave ICs, digital applications have been the primary drivers for most IC device technologies. The continuing advances of CMOS technology in particular have supported the low voltage/low power trend. Bipolar, BiCMOS, and more recently SiGe HBT developments have similarly supported this trend. For digital circuitry, the low voltage/low power trend is essentially without negative consequences. Each reduction in device size and power supply requirement has led to significant performance improvements. The reduced device size leads to smaller, lighter, cheaper, and faster circuits while the reduced voltage leads to reduced power consumption and improved reliability. The combination of these advantages provides the potential for higher levels of integration, which leads to the use of more complex and sophisticated architectures and systems.

For RF and microwave circuitry, however, the push toward lower voltage and lower power consumption is counter to many other radio requirements and presents challenges not faced by digital circuit designers. Although smaller, lighter, cheaper, and faster are generally desirable qualities for RF as well as digital applications, reduced voltage can produce negative consequences for RF circuit performance. In particular, reduced voltage leads to reduced dynamic range, which can be especially critical to the development of linear power amplifiers.

An imperative to develop and produce low voltage/low power RF products requires that every aspect of radio component development be considered. The ultimate success of a low voltage/low power RF product strategy is affected by all of the following issues:

- Materials technology (GaAs, Si, epitaxiy, implant, heterojunctions).
- Device technology (BJT, HBT, MESFET, MOSFET, HEMT).
- Circuit technology (matching and topology challenges for low voltage).
- Radio system design issues (impact of system and radio architecture on performance).

Motivations for Reduced Voltage

Commercial handheld products are clearly a significant driver for the low voltage/low power imperative. Reduced power consumption translates into longer battery lifetime and/or smaller, lighter products through reduction of battery size. Consumers value both of these attributes highly, which has led to the widespread acceptance of portable pagers, cell phones, and wireless personal area network products. An additional collateral benefit to reduced voltage operation is improved product reliability.

For vehicle-based products, the low power imperative is less compelling since battery size and lifetime issues are not relevant to these applications. Smaller and lighter electronics, however, are still valued in these applications. Power supplies represent a fixed payload that must be transported throughout the lifetime of the vehicle. In the case of jet aircraft, a single additional pound of payload represents tens of thousands of dollars of jet fuel over the life of the aircraft. A reduction in the size and weight of power supplies also represents payload that can be displaced with additional safety, communication, or navigation equipment. As in the case of handheld equipment, reduced power consumption will contribute to improved product reliability.

Even for many fixed site applications, reduced power consumption can provide benefits. Particularly in urban environments, base stations must be located in high-cost real estate areas (at the top of skyscrapers, for example). Rental of this space is often billed on a per square foot basis, so reduction in size carries some cost advantage. Increased reliability also leads to reduced maintenance costs.

Advantages of low voltage/low power RF circuitry are also significant for military applications. Supply line requirements during a military operation are often a determining factor in the outcome of battle. Transport vehicles are limited in number and capacity, so extended battery lifetime translates directly into decreased battlefield transport capacity requirements for battery resupply. This reduction in battery resupply transport requirements frees up valuable resources to transport additional lifesaving materials. The advantage is also directly felt by the foot soldier who is able to achieve greater mobility and/or additional electronic functionality (communication or navigation equipment).

As mentioned in the introduction, low voltage/low power trends have been almost without negative impact for most digital applications. Since most portable radio products are comprised of both digital ICs and RF parts, manufacturers would prefer to run all of the electronics from a single power supply. RF and microwave circuitry has had to follow the trend to low voltage, even when that trend brings on increasing challenges. This requirement to conform is exacerbated by the fact that the volume of digital circuitry required from semiconductor factories is far higher than the volume of RF circuitry. Applications that would benefit from high voltage operation comprise a nearly insignificant portion of the total semiconductor market so get little attention from most semiconductor manufacturers. The RF designer often has little choice but to use a low voltage IC process.

Semiconductor Materials Technology

Low voltage/low power operation requires improved efficiency, low parasitic resistance, low parasitic capacitance, and precise control of on-voltage. The selection of the semiconductor material impacts all of these characteristics.

Improvements in efficiency and reductions in parasitic resistance can be achieved by using materials that exhibit increased carrier mobility and velocity. Figure 3.50 plots the carrier velocity in bulk

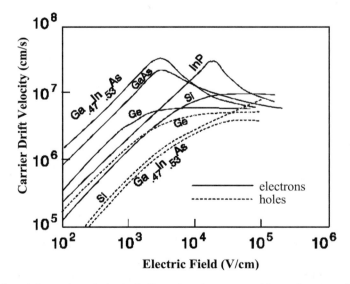

FIGURE 3.50 A plot of the carrier velocity in bulk semiconductor material as a function of the applied electric field for several commonly used semiconductors.

semiconductor material as a function of the applied electric field for several commonly used semiconductors. The carrier (electron or hole) mobility is given by

$$\mu_0 = \frac{v}{E} \qquad \text{for small E} \qquad (3.117)$$

where v is the carrier velocity in the material and E is the electric field. The figure illustrates that III-V semiconductor materials such as GaAs, GaInAs, and InP exhibit superior electron velocity and mobility characteristics relative to silicon. For this reason, devices fabricated in III-V materials are expected to exhibit some advantages in achieving high efficiency and low parasitic resistance.

Low parasitic capacitance is also a desirable characteristic to insure low loss and minimal battery drain. Because GaAs and InP offer semi-insulating substrates, they exhibit performance advantages over the much more conductive, and hence poorer performing silicon substrates. Advanced materials strategies such as silicon-on-insulator offer potential advantages even over III-V materials, but at increased cost and complexity.

Precise control of on-voltage becomes increasingly important as supply voltages diminish. Low voltage operation means that the total voltage swing that can be realized by the RF signal is reduced. Bipolar transistors offer clear, significant advantages over FETs with respect to this characteristic since bipolar on-voltage characteristics are determined primarily by material band gaps while FET on-voltage characteristics are determined by material process parameters such as thickness and charge densities. When other considerations indicate that a FET structure is appropriate, however, heterostructure buffer layers improve the on-voltage control of FETs. Within either the FET or bipolar class of semiconductor devices, Silicon process control tends to be superior to Gallium Arsenide or other III-V process control.

Heterostructure semiconductor materials offer other potential advantages for low voltage design. In FET structures, heterostructure buffer layers provide greater carrier confinement. This can reduce substrate parasitics. Similarly, in bipolar transistors, heterostructures provide the potential to reduce unwanted minority carrier concentrations in critical portions of the device. In both the FET and bipolar case improved output conductance and greater linearity results from the use of properly designed heterostructures. Heterostructure technology also offers the potential of realizing high current and high breakdown simultaneously in a device by placing the right material in the right place.

From a purely technical performance perspective, the above discussion argues primarily for the use of III-V heterostructure devices. These arguments are not complete, however. Most commercial wireless products also have requirements for high yield, high volume, low cost, and rapid product development cycles. These requirements can overwhelm the material selection process and they favor mature processes and high volume experience. The silicon high volume manufacturing experience base is far greater than that of any III-V semiconductor facility. Silicon digital parts are routinely produced in volumes that dwarf total global III-V semiconductor device requirements. This digital experience is easily leveraged into the production of silicon RF and microwave parts, giving silicon materials a clear advantage in these areas.

No short cut for the development of experience and maturity with advanced material structures is available. As demand for diminishing power consumption RF products continues to grow, however, experience will be gained and the fundamental physical advantages of III-V heterostructure devices will lead to increasing presence in commercial wireless products.

Semiconductor Device Technology

Device technology decisions influence many of the same performance metrics that are affected by material decisions. Efficiency, parasitic resistance, and on-voltage control are all affected by the choice of device. Additional device considerations may depend on the particular circuit application being considered. For an oscillator application, for example, the 1/f noise characteristics of the transistor are likely to be critical to achieving the required circuit level performance. These same transistor characteristics may be of little interest for many other circuit applications.

TABLE 3.4   Semiconductor Devices are Evaluated against Several Important Device Characteristics for Portable Power Amplifer Applications

	MESFET/HEMT	III-V HBT	Si BJT	Si:Ge HBT	Power MOSFET (LDMOS)
		Low Voltage Related Characteristics			
Parasitic loss	Very good	Very good	Moderate	Moderate	Moderate
Single-polarity supply	No[a]	Yes	Yes	Yes	Yes
Power added efficiency	Excellent	Very good	Moderater	Moderate	Good for f<~2 GHz
Linearity	Excellent[b]	Very good	Moderate	Very good	Moderate
Power density	Moderate	Excellent	Very good	Excellent	Moderate
		General Characteristics			
Cost	Moderate to high	Moderate to high	Low to moderate	Moderate to high	Low to moderate
Maturity	Good	Good	Excellent	Moderate	Very good

[a] e-mode MESFETs and HEMTs do make use of single-polarity supply voltage, but at a cost of more difficult manufacturing and with a more limited dynamic range.

[b] MESFET devices manufactured using epi-taxial material (as opposed to ion-implanted) as well as HEMTs exhibit excellent linearity.

Table 3.4 evaluates most modern transistor technologies against a list of performance characteristics that are of interest for portable power amplifier applications. The performance metrics considered in the table include: parasitic loss, ability to easily use a single polarity power supply, power added efficiency, linearity, cost, and maturity. As is readily seen from the table, no particular device type excels in all areas and all devices exhibit at least some significant disadvantages. The relative values of each of the metrics are not equal and are not identical for all applications. Meeting linearity requirements, for example, is central to the power amplifier (PA) design process for many digitally modulated portable radios, while the linearity requirements for many analog radios is easily achieved and therefore less valued.

Parasitic loss is clearly important to reduce undesirable current drain. The III-V device technologies offer advantages over silicon technologies in this area because of their material advantages.

Single polarity power supplies are desirable since other radio components do not require a negative supply. The generation of a negative voltage required to bias a transistor adds to the parts count, cost, and current drain of the radio. HBTs, BJTs, and LDMOS parts have a clear advantage over conventional MESFETs and HEMTs in this area. Enhancement mode HEMTs offer a potential solution to this problem, but at a cost to manufacturing ability and dynamic range.

Linearity requirements are typically driven by government regulations to keep wireless products from interfering with each other. Linearity specifications are not consistent for all radio architectures and are sometimes the subject of erroneous discussions in the technical literature. The linearity of the Gummel plot [log(I_c) vs. log(I_b)] of a BJT or HBT, for example, is not equivalent or translated into RF linearity. It is also difficult to measure linearity and determine an absolute performance limit for a particular device. RF circuit tuning performed to optimize power, gain, or efficiency is not necessarily optimum for achieving high linearity. MESFETs fabricated using epitaxial material (as opposed to ion-implanted material) and HEMTs exhibit slight advantages in linearity over other devices. HBTs and BJTs have also demonstrated competitive linearity characteristics.

Once power, gain, and linearity specifications are met, efficiency often becomes the key distinguishing figure of merit for modern wireless power amplifiers. Improvements in PA efficiency are translated directly into reduced battery drain and longer battery lifetime. HEMTs hold certain advantages in this area while silicon BJTs have not produced efficiency values that are as competitive. Both LDMOS and HBT parts have also produced competitive efficiencies for certain applications.

High power density is desirable since it leads to reduced part size. Smaller devices not only reduce the size of the final product, but more importantly, reduce the cost of the device since more devices can be processed on a single wafer. HBTs demonstrate the highest power density with adequate breakdown characteristics of any wireless PA devices today. Ion-implanted MESFETs and LDMOS parts exhibit very low power densities compared to HBTs.

As in the case of material considerations, the technical characteristics of devices must be weighed against the cost and maturity requirements for wireless product development. Although cost is a primary driver for many radio applications today, it may be difficult to determine fabrication costs of emerging device technologies. For most parts utilized in high volume commercial applications, the cost of fabrication is not closely related to the purchase price for the part. The market supply and demand forces determine purchase price — the price is set as high as the market will bear. If the fabrication costs can be brought below the market price with a comfortable margin, the part will continue to be manufactured. If not, the part will be discontinued. Large semiconductor manufacturing companies may choose to subsidize new technologies for several years before deciding to discontinue them. Although difficult to quantify, cost is critical to the long-term success of a device. Silicon transistors have historically come with a cost advantage over III-V devices. The high volume experience of silicon fabrication facilities and greater maturity of the processes contribute significantly to the low cost advantage.

Examination of Table 3.4 and the discussion of the preceding paragraphs illustrate the difficulty in choosing the optimum device for low voltage/low power applications. Although important, the low voltage characteristics of the part are not the only characteristics that must be considered. No transistor is optimal for all requirements and each device choice carries implied compromises for the final product.

Although not included Table 3.4, the frequency of the application can also be a critical performance characteristic in the selection of device technology. Figure 3.51 illustrates the relationship between frequency of the application and device technology choice. Because of the fundamental material characteristics illustrated in Fig. 3.50, silicon technologies will always have lower theoretical maximum operation frequencies than III-V technologies. The higher the frequency of the application, the more likely the optimum device choice will be a III-V transistor over a silicon transistor. Above a certain frequency (labeled f_{III-V} in Fig. 3.51), III-V devices dominate the transistors of choice, with silicon playing no significant role in the RF portion of the product. In contrast, below a certain frequency (labeled f_{Si} in Fig. 3.51) the cost and maturity advantages of silicon provide little opportunity for III-V devices to compete. In the transition spectrum between f_{Si} and f_{III-V}, the device technology is not an either/or choice, but rather silicon and III-V devices coexist. Although silicon devices are capable of operating above frequency f_{Si}, this operation is often gained at the expense of current drain. As frequency is increased above f_{Si} in the transition spectrum, efficiency advantages of gallium arsenide and other III-V devices provide competitive opportunities for these parts. The critical frequencies, f_{Si} and f_{III-V} are not static frequency values. Rather, they are continually being moved upward by the advances of silicon technologies — primarily by decreasing critical device dimensions.

The process of choosing device technology is compounded further if higher levels of integration are implemented. Table 3.5 is a companion to Table 3.4. Again, several semiconductor device technologies are ranked in terms of desirable device characteristics. In contrast to Table 3.4, Table 3.5 considers characteristics of interest for integrated radio front-end applications. Although some characteristics are common to both tables, many are not. The evaluation of which technology is superior is likely to change when Table 3.5 characteristics are used to make a device technology choice instead of those listed in Table 3.4. The final column of Table 3.5 replaces LDMOS characteristics with those of CMOS. This change in device technologies that are considered for the application is appropriate and indicative of the problem of choosing an optimum device technology. LDMOS is not a likely device technology for most integrated radio applications while CMOS is not a viable power amplifier technology for most portable wireless products.

Circuit Design

RF circuit performance is not generally improved by the reduction of voltage or current. Virtually all types of circuits exhibit degradation in performance with reduced DC power. As an example, when either

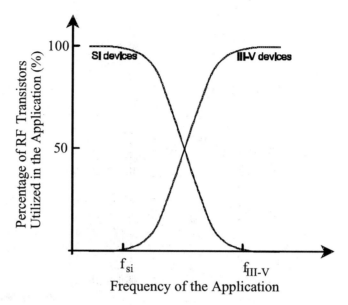

FIGURE 3.51   An illustration of the relationship between frequency of the application and device technology choice.

TABLE 3.5   Semiconductor Devices are Evaluated against Several Important Device Characteristics for Radio Integration Applications

	MESFET/HEMT	III-V HBT	Si BJT	Si:Ge HBT	CMOS
Low Voltage Related Characteristics					
Single-polarity supply	No[a]	Yes	Yes	Yes	Yes
Turn-on voltage control	Moderate	Very good	Excellent	Excellent	Good
General Characteristics					
Cost	Moderate to high	Moderate to high	Low to moderate	Moderate to high	Low
Maturity	Good	Good	Excellent	Moderate	Excellent
RF integration capability[b]	Very good	Very good	Excellent for f<3 GHz	Excellent for f<3 GHz	Excellent for f<2 GHz
Noise figure	Excellent	Very good	Good	Very good	Moderate
Phase noise	Poor	Good	Excellent	Excellent	Good

  [a] e-mode MESFETs and HEMTs do make use of single-polarity supply voltage, but at a cost of more difficult manufacturing and with a more limited dynamic range.

  [b] Comparing Silicon RF integration capability to III-V RF integration capability is difficult. Silicon process maturity allows for far higher levels of integration (in terms of numbers of transistors), but becomes of limited value above a few GHz because of the parasitics of the Si substrate.

current or voltage is reduced, a typical RF amplifier circuit will exhibit degradation of gain, linearity, and dynamic range.

If RF power levels are to be maintained, power amplifier load lines must be altered as supply voltage is reduced. Figure 3.52 illustrates the issue of load line and reduced voltage. The load line labeled "a" represents an optimum, idealized, high-efficiency load line for 6 volt power supply operation. The slope of the load line is determined primarily by the matching characteristics between the device and the load. As voltage is reduced from 6 volts to 3 volts, preservation of the slope of the load line would imply a reduction in power by ~3 dB (load line labeled "b"). Power is estimated to first order as the area underneath the load line and contained by the current-voltage plot. If power is to be maintained, the load line for the 3 volt supply case must be made steeper. This is indicated by load line "c." With a steeper

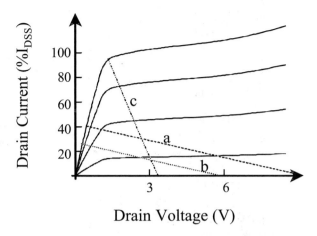

FIGURE 3.52 An illustration of the effect that lowering supply voltage has on power amplifier load line. Load line *a* represents an optimum high efficiency load line for 6-volt supply voltage. Load line *b* results from reducing the supply voltage to 3 volts without modifying the load line slope. The result of using load line *b* is a reduction in output power by more than 3 dB. Load line *c* represents a 3-volt load line where output power level is maintained. The result of using load line *c* is reduced efficiency from the load line *a* case.

load line, the average current along the load line (which represents DC current drain) is increased, while the intersection of the load line and the I-V curve is raised in voltage. When RF power levels are held constant, moving from load line "a" to load line "c" will tend to have the undesirable effect of reducing the amplifier efficiency.

To compensate for this problem, low voltage devices are designed to be larger than their higher voltage counterparts. Increased periphery devices provide higher peak currents required to maintain power level with decreased voltage. Larger devices, however, exhibit lower impedance that must be matched. Thus, low voltage power amplifier parts have high transformation ratios. Typical 1-4 Watt power amplifier parts for commercial portable wireless products exhibit output impedance of less than a few ohms. High Q matching elements provide some advantage in achieving required transformations but come at high cost and reduced integration.

As designers struggle to maintain power, gain, and efficiency with diminishing supply voltage, linearity must also be maintained. This requirement brings increasing importance to the use of linearization circuitry, feedforward, and pre-distortion schemes. Active bias control is also a valuable circuit design technique for many applications. Bias schemes that cause the transistor bias to rise with increased incident RF power help to provide needed maximum power while maintaining efficiency when lower power levels are needed.

Even for low power RF applications, bias control is critical when limited power supply voltage is available. The problem is compounded by device-to-device variability in current-voltage characteristics. Device stacking opportunities are limited and current control bias schemes, as opposed to voltage control schemes, are often preferred.

Radio and System Architecture

The ongoing communication technology revolution continues to demand greater bit transfer rates and higher bandwidths from wireless products. These demands lead to greater required linearity and higher frequencies of operation from the RF portion of these products. Broader bandwidth, higher frequency operation contributes to increases in battery drain creating a greater challenge to reduce operating voltage and power consumption.

Providing access to significantly increased information content also forces requirements for greater functionality causing a need for greater circuit complexity. Increasing functionality of the product is viewed as a potential method for establishing competitive advantage in an otherwise regulated environment. Built-in address books and more sophisticated displays are just two of the features appearing on new products. Since each new function requires power, this trend also works against the low power consumption imperative.

Other radio architecture trends, such as the movement toward digital modulation, can ease the low voltage/low power design issues. The improved signal-to-noise advantages offered by digital modulation results in reduced output power requirements. Digital modulation schemes can also provide the potential for pulsed (versus CW) operation, reducing power consumption even more.

Some proposed and emerging wireless systems will have a dramatic effect on the war on power consumption, which will change the requirements (and the technologies of choice) entirely.

Micro-cell telephone systems and small area Wireless Personal Area Networks (see Chapter 2) will dramatically reduce the battery requirements for these wireless systems. A smaller geographic cell size reduces the required RF output power of both subscriber and base station amplifiers. The required subscriber transmission power is reduced since the total distance that information must be transmitted is reduced. The base station power is also reduced for the same reason and because the total number of users that will be within the smaller cell is reduced. These reductions in required transmission power translate into dramatically reduced power consumption requirements. The power consumption of the RF portion of the radio becomes insignificant compared to the rest of the electronics. The device and material selection process can also become inconsequential. Although from a consumer point of view, this solution offers a significant advantage with no clear downside, the micro-cell infrastructure installation represents a significant investment for the service provider.

In contrast, satellite cellular systems (see Chapter 2) present significant challenges due to the high instantaneous output power requirements of the PA and low minimum noise figure of the front end LNA.

Although the choice of semiconductor material, type of transistor, and circuit topology all affect low voltage design, the radio architecture and system concept can dominate important low voltage performance specifications such as required battery size and battery lifetime.

Limits to Reductions in Voltage

There are some implications to the reduction of battery size and weight and that will impose serious constraints on the minimum useful operating voltage for portable RF products.

As discussed previously in this section, a primary driver for reduced voltage is the successful evolution of low voltage digital IC technologies. The advantage of running digital and RF circuitry from a single power supply is clear. Issues that limit DC power requirements for RF circuits, however, are fundamentally different than those that limit DC power requirements for associated digital circuits. Digital circuitry is required to store and analyze information that is encoded in a binary manner. This can be accomplished theoretically by the presence or absence of a small charge (even a single electron). Although practical considerations make a single electron memory improbable, and movement of even one electron into and out of storage still requires energy, it is clear that binary data can be manipulated with extremely small amounts of energy. The digital system requirements do not impose arbitrary power requirements on the strength of the digital signal.

In contrast to the digital case, minimum RF power requirements are implied by the radio system design. The RF signal must be transmitted and received across a specified distance or range of distances. Because power is lost in the radiation process, RF circuits must be able to handle power levels that are determined by the propagation media and transmitter-to-receiver separation. If 1 W of transmission power is required for the remote receiver to successfully detect the signal, then reductions in RF power below this level will cause the radio to fail. Unlike the digital situation where any discernable bit is as good as any other bit, the minimum RF transmission power must be maintained. For RF circuitry, increased efficiency and/or increased current must accompany a reduction in voltage.

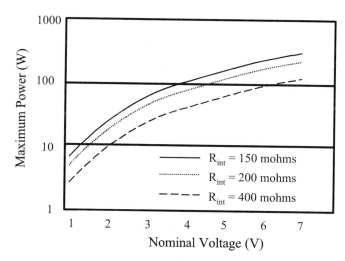

FIGURE 3.53 The maximum power capability for a battery as a function of nominal voltage and internal resistance. Maximum power for this figure is defined as the nominal voltage times the short circuit current.

Most portable units operate at efficiency levels within 2 to 10 percent of the theoretical limits. The implication of near theoretical efficiency is that voltage reductions necessarily involve increased current requirements. If voltage is halved, current must be doubled.

As battery current requirements are increased, the internal resistance becomes a limiting factor in the total power the battery is capable of delivering. For most applications, batteries have been considered ideal voltage sources, but as low voltage electronics trends continue, this assumption will begin to break down.

Figure 3.53 presents the maximum power capability for a battery as a function of nominal voltage and internal resistance. Maximum power for Fig. 3.53 is defined quite optimistically as the nominal voltage times the short circuit current. Although internal resistance is a function of the chemistry as well as the number and size of the battery cells, the values used in Fig. 3.53 are typical for nickel-metal-hydride and Li-ion batteries in common use for portable wireless products today. Further, the trends plotted in the figure will hold for all batteries, regardless of battery chemistry or size. It is clear from the figure that as battery voltages are reduced below approximately 2 volts, the maximum power available from them is reduced to levels on the order of that required from a portable cellular PA alone. When efficiency and other circuit requirements are accounted for, battery power levels are inadequate to support the product.

Summary

Reductions in voltage and power consumption are important to the development of portable wireless products that are smaller, lighter, and have long battery lifetimes. Low voltage supply requirements are also being driven by a need to make RF ICs compatible with digital ICs. Careful consideration and improvement in the choice of semiconductor material, device type, and circuit topology can help to achieve reductions in voltage. System design and choice of radio architecture also has impact on low voltage/low power electronics. These latter issues have potentially the largest impact on low voltage requirements. The ultimate limit to decreasing battery voltage, however, will be determined by transmitter power requirements and achievable internal resistance of batteries.

The discussion in this section has focused on commercial wireless products. These same trends, however, also describe military, satellite, or avionics communications products — both today and over the past several decades. All of these systems are being driven by the demand to move increasing amounts of information between points via a wireless system.

References

M. Golio, Low Voltage Electronics for Portable Wireless Applications: An Industrial Perspective, 1998 IEEE MTT-S International Microwave Symposium Digest, Baltimore, MD, IEEE Press, June 7–12, 1998.

M. Golio, Low Voltage Microwave Electronics, Santa Clara Valley IEEE MTT Short Course, Santa Clara, CA, Santa Clara Chapter IEEE, MTT, April 4, 1998.

M. Golio, Active Devices for Commercial Subscriber Power Amplifier Products, 1996 URSI Conference, Lille, France, 28 August–5 September 1996.

D. Halchin, M. Golio, C. Farley, Issues with Measurements, Design, and Modeling of Devices for Wireless Applications, Wireless Conference, Sedona, AZ, September 30, 1996.

E. Pettenpaul, New Horizons for the Microwave Business in Europe, *Proceedings of the European Microwave Conference*, Madrid, Spain, 1–5, 1993.

Y. Kitahara, Personal Mobile Communications in Microwave Band, *Proceedings of the European Microwave Conference*, Madrid, Spain, 94–95, 1993.

R. C. Dixon, Portable Communications Needs and No-Nos, *GaAs IC Symposium Digest*, 29–32, 1993.

P. Briere, E. Perea, O. de Saint Leger, System and Technology Aspects of Future Digital Cellular Phones in Europe, *GaAs IC Symposium Digest*, Monterey, CA, 33–36, 1991.

E. Pettenpaul, L. Scharf, K.J. Schopf, Enhanced GaAs Device Concepts for the New Digital Mobile Communication Systems, *Proceedings of the European Microwave Conference*, Madrid, Spain, 132–134, 1993.

G. P. Donzelli, GaAs versus Silicon Technology, *Proceedings of the European Microwave Conference*, Madrid, Spain, 110–112, 1993.

C. Kermarrec, G. Dawe, T. Tewksbury, B. Meyerson, D. Harame, M. Gilbert, SiGe Technology: Applications to Wireless Digital Communications, *IEEE Microwave and Millimeter-Wave Monolithic Circuits Symposium Digest*, San Diego, CA, 1–4, 1994.

M. Hirayama, K. Honjo, M. Obara, N. Yokoyama, HBT IC Technologies and Applications in Japan, *GaAs IC Symposium Digest*, 3–6, 1991.

D. Whipple, The CDMA Standard, *Applied Microwave & Wireless*, 6, 24–37, Spring 1994.

M. Sullivan, Wireless LANs, Status and Future, *Applied Microwave & Wireless*, 6, 54–61, Spring 1994.

Low Voltage Silicon Bipolar Transistors, *Microwave Journal*, 37, 126–128, June 1994.

T. Yamaguchi, T. Archer, R. Johnston, J. Lee, Process and Device Optimization of an Analog Complementary Bipolar IC Technology with 5.5-Ghz fT PNP Transistors, *IEEE Trans. Electron. Devices*, 41, 1019–1026, June 1994.

D. Vook, T. Kamins, G. Burton, P. Vande Voorde, H. Wang, R. Coen, J. Lin, D. Pettengill, P. Yu, S. Rosner, J. Turner, S. Laderman, H-S. Fu, A. Wang, Double-Diffused Graded SiGe-Base Bipolar Transistors, *IEEE Trans. Electron. Devices*, 41, 1013–1018, June 1994.

V. Nair, R. Vaitkus, D. Scheitlin, J. Kline, H. Swanson, Low Current GaAs Integrated Down Converter for Portable Communication Applications, *GaAs IC Symposium Digest*, IEEE Press, 41–44, 1993.

V. Nair, Low Current Enhancement Mode MMICs for Portable Communication Applications, *GaAs IC Symposium Digest*, IEEE Press, 67–70, 1989.

K. Kanazawa, M. Kazumura, S. Nambu, G. Kano, I. Teramoto, A GaAs Double-Balanced Dual-Gate FET Mixer IC for UHF Receiver Front-End Applications, *IEEE Trans. Microwave Theory & Techn.*, MTT-33, 1548–1553, December 1985.

O. Ishikawa, Y. Ota, M. Maeda, A. Tezuka, H. Sakai, T. Katoh, J. Itoh, Y. Mori, M. Sagawa, M. Inada, Advanced Technologies of Low-Power GaAs ICs and Power Modules for Cellular Telephones, *GaAs IC Symposium Digest*, IEEE Press, 131–134, 1992.

P. Philippe, M. Pertus, A 2 Ghz Enhancement Mode GaAs Down Converter IC for Satellite TV Tuner, *Proceedings of the European Microwave Conference*, Madrid, Spain, 61–64, 1993.

H. Heaney, F. McGrath, P. O'Sullivan, C. Kermarrec, Ultra Low Power Low Noise Amplifiers for Wireless Communications, *GaAs IC Symposium Digest*, IEEE Press, 49–51, 1993.

J.O. Plouchart, H. Wang, M. Riet, HBT Monolithic Integrated Phase Locked Oscillator for DCS 1800 Mobile Communications, *Proceedings of the European Microwave Conference*, Madrid, Spain, 834–836, 1993.

M. Nakatsugawa, Y. Yamaguchi, M. Muraguchi, An L-Band Ultra Low Power Consumption Monolithic Low Noise Amplifier, *GaAs IC Symposium Digest*, IEEE Press, 45–48, 1993.

K. R. Cioffi, Monolithic L-Band Amplifiers Operating at Milliwatt and Sub-Milliwatt DC Power Consumption, *IEEE Microwave and Millimeter-Wave Monolithic Circuits Symposium Digest*, Albuquerque, NM, 9–12, 1992.

H. Takeuchi, M. Muraoka, T. Hatakeyama, A. Matsuoka, M. Honjo, S. Miyazaki, K. Tanaka, T. Nakata, A Si Wide-Band MMIC Amplifier Family for L-S Band Consumer Product Applications, *IEEE Microwave Theory and Techniques Symposium Digest*, 1283–1284, 1991.

Y. Imai, M. Tokumitsu, A. Minakawa, Design and Performance of Low-Current GaAs MMICs for L-Band Front-End Applications, *IEEE Trans. Microwave Theory & Techn.*, 39, 209–214, February 1991.

Y. Imai, M. Tokumitsu, A. Minakawa, T. Sugeta, M. Aikawa, Very Low-Current and Small-Size GaAs MMICs for L-Band Front-End Applications, *GaAs IC Symposium Digest*, IEEE Press, 71–74, 1989.

W. Baumberger, A Single Chip Rejecting Downconverter for the 2.44 Ghz Band, *GaAs IC Symposium Digest*, 37–40, 1993.

B. Khabbaz, A. Douglas, J. DeAngelis, L. Hongsmatip, V. Pelliccia, W. Fahey, G. Dawe, A High Performance 2.4 Ghz Transceiver Chip-Set for High Volume Commercial Applications, *IEEE Microwave and Millimeter-Wave Monolithic Circuits Symposium Digest*, San Diego, CA, 11–14, 1994.

M. Wang, M. Carriere, P. O'Sullivan, B. Maoz, A Single-Chip MMIC Transceiver for 2.4 Ghz Spread Spectrum Communication, *IEEE Microwave and Millimeter-Wave Monolithic Circuits Symposium Digest*, San Diego, CA, 19–22, 1994.

J. Moniz, B. Maoz, Improving the Dynamic Range of Si MMIC Gilbert Cell Mixers for Homodyne Receivers, *IEEE Microwave and Millimeter-Wave Monolithic Circuits Symposium Digest*, San Diego, CA, 103–106, 1994.

R. Herman, A. Chao, J. Deichsel, C. Mason, A. McKay, Highly Integrated Mixed-Signal L-Band Configurable Receiver Array, *IEEE Microwave and Millimeter-Wave Monolithic Circuits Symposium Digest*, San Diego, CA, 23–26, 1994.

T. Apel, E. Creviston, S. Ludvik, L. Quist, B. Tuch, A GaAs MMIC Transceiver for 2.45 Ghz Wireless Commercial Products, *IEEE Microwave and Millimeter-Wave Monolithic Circuits Symposium Digest*, San Diego, CA, 15–18, 1994.

R. Meyer, W. Mack, A 1-Ghz BiCMOS RF Front-End IC, *IEEE J. of Solid-State Circuits*, 29, 350–355, March 1994.

K. Sakuno, M. Akagi, H. Sato, M. Miyauchi, M. Hasegawa, T. Yoshimasu, S. Hara, A 3.5W HBT MMIC Power Amplifier Module for Mobile Communications, *IEEE Microwave and Millimeter-Wave Monolithic Circuits Symposium Digest*, San Diego, CA, 63–66, 1994.

Y. Ota, M. Yanagihara, T. Yokoyama, C. Azuma, M. Maeda, O. Ishikawa, Highly Efficient, Very Compact GaAs Power Module for Cellular Telephone, *IEEE Microwave Theory and Techniques Symposium Digest*, Albuquerque, NM, 1571–1520, 1992.

M. Maeda, M. Nishijima, H. Takegara, C. Adachi, H. Fujimoto, Y. Ota, O. Ishikawa, A 3.5 V, 1.3 W GaAs Power Multi Chip IC for Cellular Phone, *GaAs IC Symposium Digest*, Atlanta, GA, 53–56, 1993.

A. Herrera, E. Artal, E. Puechberty, D. Masliah, High Efficiency, Highly Compact L-Band Power Amplifier for DECT Applications, *Proceedings of the European Microwave Conference*, Madrid, Spain, 155–157, 1993.

S. Makioka, K. Tateoka, M. Yuri, N. Yoshikawa, K. Kanazawa, A High Efficiency GaAs MCM Power Amplifier for 1.9 Ghz Digital Cordless Telephones, *IEEE Microwave and Millimeter-Wave Monolithic Circuits Symposium Digest*, San Diego, CA, 51–54, 1994.

T. Kunihisa, T. Yooyama, H. Fujimoto, K. Ishida, H. Takehara, O. Ishikawa, High Efficiency, Low Adjacent Channel Leakage GaAs Power MMIC for Digital Cordless Telephone, *IEEE Microwave and Millimeter-Wave Monolithic Circuits Symposium Digest*, San Diego, CA, 55–58, 1994.

T. Yoshimasu, N. Tanaba, S. Hara, An HBT MMIC Linear Power Amplifier for 1.9 Ghz Personal Communications, *IEEE Microwave and Millimeter-Wave Monolithic Circuits Symposium Digest*, San Diego, CA, 59–62, 1994.

P. O'Sullivan, G. St. Onge, E. Heaney, F. McGrath, C. Kermarrec, High Performance Integrated PA, T/R Switch for 1.9 Ghz Personal Communications Handsets, *GaAs IC Symposium Digest*, Albuquerque, NM, 33–35, 1993.

S. Dietsche, C. Duvanaud, G. Pataut, J. Obregon, Design of High Power-Added Efficiency FET Amplifiers Operating with Very Low Drain Bias Voltages for Use in Mobile Telephones at 1.7 Ghz, *Proceedings of the European Microwave Conference*, Madrid, Spain, 252–254, 1993.

J. Griffiths, 3 Volt MMIC Power Amplifier for Wireless Phones, *Applied Microwave & Wireless*, 6, 38–53, Spring 1994.

K. Inosako, K Matsunaga, Y. Okamoto, M. Kuzuhara, Highly Efficient Double-Doped Heterojunction FET's for Battery-Operated Portable Power Applications, *IEEE Electron. Device Letters*, 15, 248–250, July 1994.

M. Golio, Large Signal Microwave Device Modeling Challenges for High Volume Commercial Communications Applications, *Workshop Proceedings of the European Microwave Conference: Advanced Microwave Devices, Characterisation and Modelling*, Madrid, Spain, 30–33, 1993.

R. Darling, Subthreshold Conduction in Uniformly Doped Epitaxial GaAs MESFETs, *IEEE Trans. Electron Devices*, 36, 1264–1273, July 1989.

M. Golio, Characterization, Parameter Extraction and Modeling for High Frequency Applications, *Proceedings of the European Microwave Conference*, Madrid, Spain, pp. 69-72, 1993.

J. Pedro, J. Perez, A Novel Non-Linear GaAs FET Model for Intermodulation Analysis in General Purpose Harmonic-Balance Simulators, *Proceedings of the European Microwave Conference*, 714–716, 1993.

T. Yoshimasu, K. Sakuno, N. Matsumoto, E. Suematsu, T. Tsukao, T. Tomita, Low Current GaAs MMIC Family with a Miniaturized Band-Stop Filter for Ku-Band Broadcast Satellite Applications, *GaAs IC Symposium Digest*, 147–150, 1991.

N. Shiga, T. Sekiguchi, S. Nakajima, K. Otobe, N. Kuwata, K. Matsuzaki, H. Hayashi, MMIC Family for DBS Downconverter with Pulse-Doped GaAs MESFETs, *GaAs IC Symposium Digest*, 139–142, 1991.

C. Caux, P. Gamand, M. Pertus, A Cost Effective True European DBS Low Noise Converter, *GaAs IC Symposium Digest*, 143–145, 1991.

K. W. Kobayashi, R. Esfandiari, M. Hafizi, D. Streit, A. Oki, M. Kim, GaAs HBT Wideband and Low Power Consumption Amplifiers to 24 Ghz, *IEEE Microwave and Millimeter-Wave Monolithic Circuits Symposium Digest*, 85–88, 1991.

J. M. Golio, M. Miller, J. Staudinger, W.B. Beckwith, E.N. Arnold, *Microwave MESFETs and HEMTs*, Norwood, MA: Artech House, 1991.

R. A. Pucel, D. Masse, R. Bera, Performance of GaAs FET Mixers at X-band, *IEEE Trans. Microwave Theory & Techn.*, MTT-24, 351–360, June 1976.

S. Maas, B. Nelson, D. Tait, Intermodulation in Heterojunction Bipolar Transistors, *IEEE Microwave Theory and Techniques Symposium Digest*, IEEE Press, 91–93, 1991.

M. Kim, A. Oki, J. Camou, P. Chow, B. Nelson, D. Smith, J. Canyon, C. Yang, R. Dixit, B. Allen, 12–40 Ghz Low Harmonic Distortion and Phase Noise Performance of GaAs Heterojunction Bipolar Transistors, *GaAs IC Symposium Digest*, 117–120, 1988.

W. Liu, T. Kim, P. Ikalainen, A. Khatibzadeh, High Linearity Power X-Band GaInP/GaAs Heterojunction Bipolar Transistor, *IEEE Electron. Device Letters*, 15, 190–192, June 1994.

3.7 Productivity Initiatives

Mike Golio

Productivity initiatives have become a significant part of the engineering design system utilized by the RF/microwave industry. There are an abundance of these initiatives that have modified and molded the microwave engineer's methodologies and work habits in the past 10 to 15 years. The popularity of these initiatives is so pervasive among companies that an entire industry of consultant trainers has emerged to help companies launch and implement successful programs to improve corporate efficiency and performance. Popular initiatives include:

- Design for Manufacturability
- Design to Cost
- Total Customer Satisfaction
- Six Sigma
- Just-In-Time Delivery
- Total Cycle Time Reduction
- Total Quality Management
- Lean Electronics
- Best Place to Work

All of the initiatives are designed to improve productivity — and therefore profit. Some initiatives focus primarily on the use of new tools and metrics (such as Six Sigma) to improve manufacturing productivity. Just-In-Time initiatives focus on inventory reduction and procurement efficiency. Initiatives like Lean Electronics represent a collection of productivity techniques coupled together by an underlying theme of efficiency. Finally, some initiatives focus primarily on direct alteration of the attitudes and perceptions of the workforce toward their customers and coworkers (such as Total Customer Satisfaction and Best Place to Work). To some extent, all major productivity initiatives are culture-changing exercises.

Although it is not possible within a single article to discuss (or even list) all of the specific productivity initiatives that have invaded the workplace, it is possible to discuss some of the fundamental engineering principles and tools that lead to efficient engineering practice.

Customizing Initiatives for the Organization

Improved productivity and profit are undeniably compelling reasons for a company to undertake the task of changing its culture. But the successful implementation of a productivity initiative must be tailored to the specific organization. Implementation of any successful initiative must account for: 1) the market and market drivers for the organization undergoing change, and 2) the existing culture of that organization.

Figure 3.54 illustrates how particular microwave/RF markets affect the primary focus and development time constants of the technology development process. The shift of technical focus is dramatic as technologies move from a proof-of-concept to a high volume manufacturing stage. A similarly dramatic shift in the acceptable development time frames is associated with this maturity process.

RF and microwave technology typically travels a maturation path that starts in a university, industry, or government R&D lab with a feasibility demonstration. These proof-of-concept efforts sometimes take decades or even lifetimes to achieve. This type of work is focused primarily on the development of a one-time proof-of-concept demonstration. Although the advocates of the new technology may have visions of high volume manufacturing at a future date, the initial goal is simply to prove that the idea has potential merit.

Promising technologies with demonstrated proof-of-concept merit typically migrate into low volume, high performance applications. Often government/military needs drive these early applications. In this stage of maturity, the research and development is likely to be focused on optimization of performance and on reliability issues. Development time frames often run for several years.

As success is reached at each stage of technology development, new applications emerge with different technical foci and shorter acceptable development times. Communications Satellites and Avionics are example applications that form the center of the technology maturation chart of Fig. 3.54. Manufacturing volumes for these applications are still low enough that nonrecurring engineering costs (NRE) usually dominate the technical focus of the development process, which can still run in time frames of years.

Applications such as automotive, industrial, and wireless base station electronics represent the next stages of typical technology maturation. These applications often support high enough volumes to bring increased focus onto development schedule duration and recurring costs (RE).

Portable wireless products represent an application that is dominated by manufacturability issues and reduced cycle times. Product development cycles are typically measured in months while cost, yield, and factory throughput dominate the focus of the technical workforce.

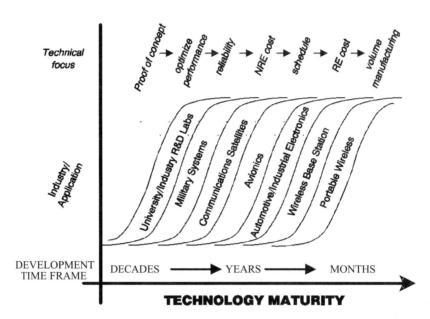

FIGURE 3.54 An illustration of a typical maturation path for emerging RF technologies. At each phase of the maturity process, technical focus and development time frames are radically different. Productivity initiatives should account for these differences if they are to be implemented successfully.

A successful *productivity initiative* needs to address the primary technical focus and development time constants of the industry where it is being applied. From the discussion above, it is clear that any profit-driven productivity initiative needs to account for the unique constraints of the organization's target market. Initiatives with primary focus on statistical process control and high manufacturing yields may be critical to an organization working in the portable wireless industry, but have very little applicability to a research and development lab with a charter to perform long-term, high-risk research of new technologies.

Even within a given market segment, application of productivity initiatives must be customized to specific organizations. Different countries, regions of a country, companies, and even organizations within a large company all have different roles, different cultures, and different levels of productivity awareness and maturity. Since these initiatives are culture-changing processes, it is important that the existing culture is understood so that appropriate steps are taken to mature the culture toward the ultimate goal. An organization that has no quality and productivity metrics in place may not be ready to take the same steps toward a fully integrated, customer-focused, quality manufacturing system that an organization that has already spent several years moving down that path should take.

Productivity and Marketing

For most product development applications, the engineering process begins with marketing. Productivity initiatives that focus on customer awareness and customer relationships (Total Customer Satisfaction, for example) should address some of the marketing issues engineers will be concerned with. Marketing establishes the groundwork for design for manufacturability efforts.

Prior to any significant engineering planning it is important to identify the required sales volumes, cost targets, and acceptable development schedule (i.e., required time-to-market).

As an example, assume that an 18 GHz radio product is scheduled to go into production in 18 months with a cost target of $100/unit at a quantity of 100,000 units per month. The targeted time-to-market dictates that little new technology will be developed. In only 18 months, there will barely be time to identify a design, find components, assemble and align a working breadboard, and develop a production manufacturing and test plan consistent with an existing production facility. The limit placed on the amount of technology development that is feasible in such a project extends not only to the radio components, but also to the manufacturing and test technologies needed in production. An analysis of the quantity requirements (assuming two shifts working 5 days a week) leads to the result that the radio product must be manufactured at the rate of approximately one unit every 12 seconds. Clearly, little tuning or manual alignment will be acceptable. The quantity and cost targets dictate an automated manufacturing and test environment. Locating off-the-shelf parts available in the quantities required will be critical. Assuming a reasonable product life cycle, the recurring cost of manufacturing dominates this engineering effort. Nonrecurring cost of development is likely to be insignificant as compared to the goal of achieving the required development cycle time.

In contrast, if a similar radio product is to be produced in quantity 4 for an R&D demonstration as part of a three-year research project, the entire engineering approach changes. Optimum technical performance may be the primary focus of the effort. A premium is likely to be paid (or even required) for utilizing new technologies, new architectures, or new methodologies. Recurring manufacturing cost has little meaning while nonrecurring cost is the determining factor in the profitability of the program. Delays in schedule of months or years may be acceptable provided the innovation is great enough.

When optimizing for profit, nonrecurring costs, recurring costs, performance, and schedule are all part of the formula. The engineering process must compare the options to the marketing requirements. In consumer applications where time-to-market is critical, a rapidly developed, nonoptimized design may provide enough performance at a low enough cost to succeed. The critical questions for these cases may be, "How long will it take to achieve the minimum performance?" and "How long will it take to achieve the lowest cost?" In general, speed and technology insertions are inversely related.

Planning and Scheduling

Although planning and scheduling are not explicitly addressed in many productivity initiatives, they are the foundation for all engineering efforts. Without good planning, other productivity efforts can produce only modest success at best. Planning and scheduling are also required for the development of important productivity metrics.

Planning and scheduling, are the processes of deciding in advance what work is required and when and where the work will be performed. The scheduling process seeks estimates of the time to do the work, identification of the organizations or people who will perform the work, and identification of the required resources.

Many scheduling tools, methodologies, and representations can be used to perform the planning and scheduling functions. Gantt systems are perhaps the most common scheduling tools used in engineering. Regardless of the methodology, the baseline schedule represents the yardstick against which actual work is measured. If improvements in engineering productivity are to be quantified, it is critical that planning and scheduling be done as accurately as possible.

Table 3.6 lists several desirable characteristics for a good engineering planning and scheduling process.

TABLE 3.6 Desirable Characteristics of a Good Engineering Planning and Scheduling Process

Characteristic of Schedule and Planning Process	Comments
Formal process	Involve representation from all organizations that will be required to release the product
Simple process	Should be simple enough to encourage use by many contributors
Work tasks	Work should be broken down into tasks small enough to establish estimates of the level of effort
Meaningful objectives	Schedule must identify meaningful, measurable milestones that mark the end of each work task

The planning process should be a formal, concise process that involves representation from all organizations that will contribute directly to the final release of the product. This will include not only project managers and engineers, but may also require representation from manufacturing, systems engineering, quality assurance, reliability, marketing, contracts, and sales.

Simplicity is another valued characteristic of a good planning and scheduling process. The scheduling system needs to be used and updated by a large number of people. Complexity or confusion should not discourage use of the system. Unneeded complexity will also add cost to the process. The scheduling system should provide the required information at the lowest cost.

All schedules start from the documentation of clearly identified objectives. A well-defined project plan expressed in a work breakdown structure (WBS) can accomplish this. The WBS should be comprised of work tasks that end with the completion of a meaningful milestone. The work tasks that comprise the WBS must be developed with enough understanding to establish accurate estimates of the time and effort required to complete the task.

Establishing meaningful milestones is a critical task for effective scheduling. A milestone that does not represent a meaningful product, accomplishment, or event to the performer is of very limited value.

Design

The goal of design is to develop a blueprint for successful production. From this perspective, design teams can be viewed as a factory whose product output is a production blueprint. The design factory must develop a road map to convert intellectual property, labor, and/or materials into a competitive product. This description of design is far broader than the conventional view of design as a process that produces an electrical circuit description or demonstration breadboard. Whether the product is a research report or a commercial wireless telephone, this design factory concept can be used to some advantage.

As in the case of other manufacturing facilities, economy and efficiency can be obtained by loading the factory, but this loading will adversely affect the throughput time for a product and the ability to effect change. Since the goal of all productivity initiatives is to bring about cultural change, close examination of the loading of the design factory is required. If every engineer in the design factory is loaded with work accounting for a 110% workload factor, there will be little capacity to learn new concepts and methods. Yet such learning is required to realize successful productivity initiatives. Change will come very slowly for a loaded design factory. If the design factory loading is reduced, new initiative efforts can fill the remaining capacity and change may occur more quickly.

Important parameters to monitor for design factories include speed, cost, and value of the designs produced by the factory. Clearly, as in any other factory, access to required resources and raw materials is required to fuel the factory. For the design factory, these resources include appropriate office environments, office support, computers, design tools, literature access, etc. The efficient design factory must also be maintained with appropriate support organizations. For many engineering applications these organizations include marketing, quality assurance, reliability, and sales. An important concept unique to the design factory, which should be considered when launching productivity initiatives, is that the design factory improves its speed, cost, and value through learning. The design factory machinery (engineers) thrive on learning and teaching. This fact leads to some important concepts related to productivity initiatives.

We learn in many ways, especially by failure. Finding a failure quickly is often critical to ultimate timely success. For the entire design factory to learn from failure, however, requires that failures are identified and made public. There is a natural resistance to this in most engineering cultures, since failures have historically not been rewarded. It is important that the cost and value of failures be understood and that high value/low cost failures be rewarded. Early fast failures often stimulate the design machine to identify and apply appropriate resources to the most critical problems, thus developing successful designs more quickly. Rewarding such failure will create an environment that will favor fast design.

The need for learning also implies a need for intellectual stimulation and creativity. Diversity is a key ingredient to the development of a stimulating engineering culture. Diversity in both training and experience helps to fuel the creative machinery.

Productivity Metrics for Design — Earned Value

A design is only good to the extent that it helps accomplish the cost and schedule targets as well as the technical targets. Meeting performance and missing the cost or schedule can be a failure while relaxed performance at an earlier date and lower cost may provide success. Design metrics should provide a yardstick to measure cost, schedule, and performance variances during the design process. Earned Value Analysis (**EVA**) provides a simple method to develop design performance metrics that indicate how the design factory is performing against its original plan.

To perform EVA, a baseline plan that includes the following is required:

1. Planned budget to execute
2. Planned schedule
   - Detailed work breakdown structure (WBS)
   - Measurable milestones
3. Actual budget to execute
4. Actual schedule required to perform

Figure 3.55 illustrates an initial baseline plan for a design project prior to initiating work on any of the tasks. The example plan is comprised of four separate tasks with estimated levels of effort of 80, 60, 100, and 50 hours. The plan could be quantified in terms of monetary values or other measurable units instead of hours. The example plan is also divided into two distinct work periods. The x-axis on the chart is calendar time. The task hours may not correspond directly with the calendar time since task hours may represent part-time effort of an individual engineer or may represent several engineers' efforts on a full-time basis.

The plan defines the Budgeted Cost of Work Scheduled (**BCWS**). BCWS is simply the budget or plan. The BCWS for an interval of time is the sum of the work of all tasks scheduled to be performed in that period. For the example in Fig. 3.55, the BCWS for period 1 is the sum of 80 hours for task A, plus 60 hours for task B, plus 30 hours for task C. For period 1, BCWS = 170 hours.

The Budget at Completion (**BAC**) is also defined in terms of the plan and is simply the sum of all tasks for the entire project. For the example in Fig. 3.55, BAC = 290 hours.

As work progresses on the project, actual accomplishments and levels of effort can be compared to the baseline schedule. Figure 3.56 illustrates a possible representation of the progress through the end of period 1. In the figure, black bars are used to fill the work tasks to the extent that they have been completed to date.

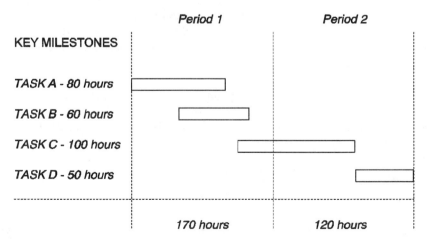

FIGURE 3.55 An illustration of an initial baseline plan for a design project prior to initiating work on any of the tasks.

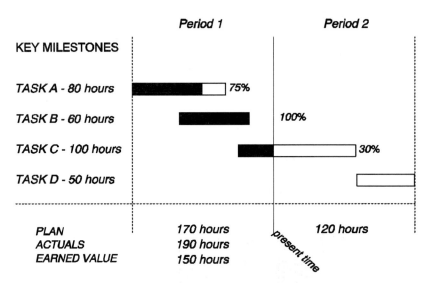

FIGURE 3.56 A representation of the progress completed on the initial baseline plan of Fig. 3.55 through the end of period 1. In the figure, black bars are used to fill the work tasks to the extent that they have been completed to date.

The Actual Cost of Work Performed (**ACWP**) is the actual costs incurred for a task or period. From Fig. 3.56, for period 1, ACWP = 190 hours. This value is noted on the figure, but is not represented in the graphics. Actual hours worked to achieve the progress reported is determined from other charging reports and simply recorded on the graph of Fig. 3.56. The 190 hours ACWP represents a variance from the baseline plan of 20 additional hours.

Figure 3.56 also indicates that the tasks originally planned for period 1 were not completed. Although tasks B and C were completed to the level indicated in the original plan, task A was scheduled for completion before the end of period 1 and is actually only 75% complete. The Budgeted Cost of Work Performed (**BCWP**) captures the earned value of the effort. The BCWP takes actual accomplishments and assigns the value that the original plan indicated it would cost to do this task. In Fig. 3.56, BCWP for period 1 is obtained by summing up the costs identified by the black bars of each task. The earned value for period 1 is 150 hours. This represents a deficiency from the baseline plan of 20 hours.

From Figs. 3.55 and 3.56 and the definitions given above, several important performance metrics can be defined. Figure 3.57 illustrates some of these metrics graphically.

Schedule Variance (**SV**) indicates whether the project is ahead or behind schedule. A schedule variance can exist for several reasons and does not provide any indication about the cost performance. The Schedule Variance is defined as

$$SV = BCWP - BCWS \qquad (3.118)$$

The Schedule Performance Index (**SPI**) indicates the schedule efficiency. Schedule Performance Index is the ratio of earned value to planned effort expressed as

$$SPI = \frac{BCWP}{BCWS} \qquad (3.119)$$

Cost Variance (**CV**) indicates whether the project is over or below budget. A cost variance can exist for several different reasons and does not provide any indication about the schedule performance. The Cost Variance is defined as

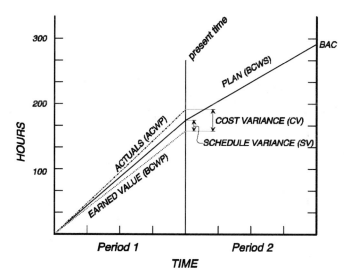

FIGURE 3.57 A graphical illustration of key metrics from an earned value program evaluation.

$$CV = BCWP - ACWP \tag{3.120}$$

Cost Efficiency is quantified with the Cost Performance Index (**CPI**). Cost Performance Index is the ratio of earned value to actual effort expressed as

$$CPI = \frac{BCWP}{ACWP} \tag{3.121}$$

The earned value quantities defined above provide a method to evaluate design performance against the original plan. Evaluation of the cost and schedule variance and efficiency is a first step to improving design performance.

A project that is on schedule and at budget will exhibit CPI and SPI values of unity. A low value for SPI indicates the project is behind schedule while a low value of CPI indicates the project is over budget. Schedule and budget performance variance, however, can have many different root causes, including the use of a poor plan. It is important that the EVA is followed by an effort to determine the root cause of any poor performance metrics before action to remedy the problem is adopted. For example, if schedule metrics are in deficit with the baseline plan while budget metrics are consistent with plan, this is an indication that the level of effort applied to the program is less than planned. Acquisition or identification of additional resources for the remainder of the program could lead to successful completion of the program (both budget and schedule goals met) for this case. If both schedule and budget metrics show a deficit from plan, there is an indication that the work is not on target for successful completion. Careful review of the original plan as well as the execution issues related to each task is required.

Manufacturing

The most widely known productivity initiatives focus on manufacturing metrics and yield analysis tools. These initiatives make extensive use of statistical process control. Statistical process control uses sampling and statistical methods to monitor the quality of an ongoing process such as a production operation.

A graphical display of statistical process variation known as a control chart provides the basis for determining when process variation is due to common causes (random variations) or to assignable,

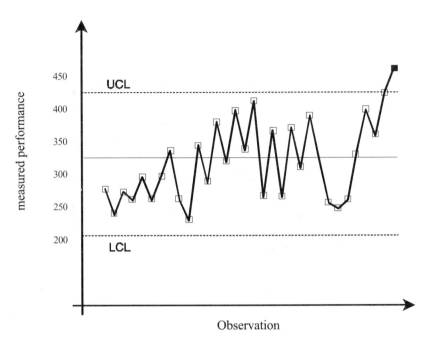

FIGURE 3.58 An example control chart.

preventable causes. If our measurement equipment has enough accuracy, all processes will exhibit variations. The goal of reviewing control charts is to distinguish the variations intrinsic in the process from those that can be assigned a caus and, therefore, can be controlled. An example control chart is presented in Fig. 3.58. Control charts plot sequential evolution of time along the x-axis and the statistic that is being measured along the y-axis. Control limits (random variations inherent in the process) are statistically determined by careful observation of the process. These limits are plotted as bounds of expected process behavior. The mean value of the process is also plotted on the control chart. When variations exceed the control limits, the process is assumed to have shifted and adjustments in the process must be made. Variations within the control limits are assumed to be inherent to the process and do not require process adjustments.

Control limits must represent actual process variability inherent in the process. Customer requirements or goals should never be used to establish control limits. To define control limits requires that an appropriate history of the process be collected.

Six Sigma

Six Sigma is the name applied to the manufacturing initiative that has become by far the most pervasive in the industry. This productivity initiative is a quality improvement and a business strategy that was pioneered at Motorola in 1987. The Greek letter sigma designates the standard deviation from the average in a Gaussian distribution of any process or procedure. The common measurement unit of Six Sigma is "defects per million units." The sigma value of a process indicates how often defects are likely to occur. Three sigma means 66,807 defects per million. Four sigma translates into 6,210 defects. Six Sigma quality implies only 3.4 defects per million. Ideally, product quality increases and cost decreases as sigma rises.

Although Six Sigma programs can be described in elaborate mathematical detail, the attainment of "six sigma quality" is not precisely defined. There are many ways to calculate defects per million units of a process. A unit, for example, can be almost anything: a system, a module, a component within a module, a part within a component, etc. Similarly, a process or procedure can be defined in many ways. The action of soldering a component to a board can be considered to be one single process, or can be

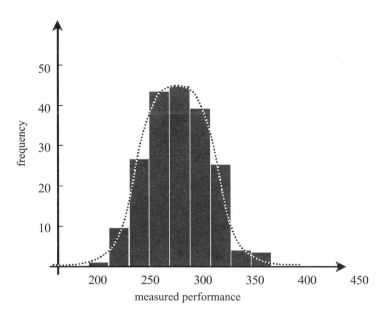

FIGURE 3.59 Example distribution statistics for a process. The observed statistics appear to approximate a normal distribution function.

broken down into several — picking the component, placing it, applying heat, etc. The choice of definition for a unit and a process can have a dramatic effect on the "defects per million" calculation. When an auto manufacturer claims to have achieved six sigma production standards, it does not mean that they produce only 3.4 defective cars per million. The 3.4 figure applies to the number of opportunities for errors. If a car consists of 10,000 different components and each one is assembled using 10 defined procedures, the opportunities for error totals one hundred thousand per auto. Six sigma production quality would imply 3.4 defects for every 10 cars.

Despite these ambiguities, Six Sigma programs have led to dramatic cost savings and competitive advantages to many companies that have embraced the philosophy. The list of companies that credit Six Sigma programs with significant cost and cycle time improvements is nearly endless.

Implementation of a Six Sigma program involves training employees to use statistical process control tools to achieve their defect reduction goals. Figures of merit are defined in terms of the statistical distribution of processed parts and these figures are monitored in an attempt to gain better control of the process.

Figure 3.59 illustrates the distribution statistics for a process. Plotted along the x-axis is the range of measured performance for the process of interest. The performance metric used for the process should be easily and quickly measured. This should be done in order to keep the measurement process from having a significant negative impact on the production cycle. Automation of the measurement and implementation of the measurement as a required part of the manufacturing process flow is important. The metric may not be a direct measurement of the actual desirable result, but it should reflect the quality and variability of the process (i.e., if the metric is controlled, the process should be controlled).

Plotted along the y-axis of Fig. 3.59 is the frequency that a measurement was observed within the sample of process measurements. For a process that is dominated by random variations, the frequency of observation versus measurement value plot is expected to describe a normal distribution. As seen from the figure, the normal distribution peaks at a specific (mean) value and drops off toward zero symmetrically in either direction away from the mean. The mean value of a distribution provides a good description of the central tendency of the process. A second important parameter describing a distribution is the standard deviation. Standard deviation provides a measure of the magnitude of the random variation inherent in a process.

Six Sigma Mathematics

The normal distribution function (also called the Gaussian distribution function after the German mathematician, Carl Friedrich Gauss) is the fundamental mathematical equation for Six Sigma metrics. A normal distribution function is the indefinite integral of the normal density function, the graph of which exhibits a typical bell shape. The normal distribution function can be described by

$$F_N(x') = k \int_{-\infty}^{x'} \exp\left[\frac{-(x-\mu)^2}{2\omega^2}\right] dx \tag{3.122}$$

where the constant k is called the normalizing coefficient and adjusts the amplitude of the function so that the integral of the distribution over all possible values of x will give a probability value of unity. The constant k can be expressed as

$$k = \frac{1}{\sigma\sqrt{2\pi}} \tag{3.123}$$

In Eqs. (3.122) and (3.123), μ is the mean, or average, of the distribution. For normal distributions, the mean value is also the point where the function reaches its maximum. The quantity σ is the standard deviation and characterizes the width of the distribution. For engineering processes, the mean value is typically the manufacturing target value and the sigma corresponds to the variability in the process. An ideal probability distribution would have a σ value that approached zero.

Two important metrics to analyze relative to Six Sigma performance are the C_p and C_{pk}. The C_p metric is referred to as the Process Capability Index and is given by

$$C_p = \frac{CPU + CPL}{2} = \frac{USL - LSL}{6\sigma} \tag{3.124}$$

where CPU is defined as

$$CPU = \frac{USL - \mu}{3\sigma} = \frac{z_U}{3} \tag{3.125}$$

CPL is defined as

$$CPL = \frac{\mu - LSL}{3\sigma} = \frac{z_L}{3} \tag{3.126}$$

and the quantities LSL and USL represent the lower and upper spec limits. The quantities z_U and z_L that appear in Eqs. (3.125) and (3.126) are derived from normalizing the x values of the observed measurement space to the process standard deviation using the transformation

$$z = \frac{x - \mu}{\sigma} \tag{3.127}$$

The C_{pk} metric is called the Process Bias and is similarly given by

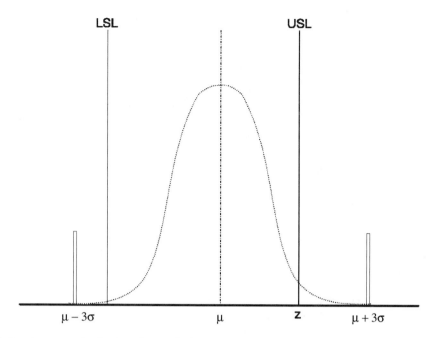

FIGURE 3.60 A process distribution chart with the quantities USL, LSL, μ, and σ illustrated on the plot. In this example, USL and LSL are not symmetric with respect to the mean while both are within the 6σ limits.

$$C_{pk} = \min\left(CPU, CPL\right) \tag{3.128}$$

For a process to be of Six Sigma quality, the following conditions must hold:

$$C_p \geq 2 \tag{3.129}$$

and

$$C_{pk} \geq 1.5 \tag{3.130}$$

A process distribution chart is illustrated in Fig. 3.60 with the quantities USL, LSL, μ, and σ illustrated on the plot. In this example, USL and LSL are not symmetric with respect to the mean while both are within the 6σ limits.

The Six Sigma goal may not be appropriate for every process as applied to every manufacturing task. The initiative to train all of the workers in a factory to monitor process variations and to understand the statistical metrics listed above, however, almost certainly has merit. This training procedure and heightened awareness of yield and process variation issues may be the real value of Six Sigma initiatives.

References

Reinertsen, D., *Managing the Design Factory,* The Free Press, New York, 1997.

Montgomery, D. C., *Introduction to Statistical Quality Control,* John Wiley & Sons, New York, 1991.

Hahn, G. J. and Shapiro, S S., *Statistical Models in Engineering,* John Wiley & Sons, New York, 1967.

Womack, J. P. and Jones, D. T., Beyond Toyota: How to root out waste and pursue perfection, *Harvard Business Review,* Sept.–Oct. 1996.

3.8 Electronic Hardware Reliability

Arun Ramakrishnan, Toby Syrus, and Michael Pecht

Reliability is the ability of a product to perform as intended (i.e., without failure and within specified performance limits) for a specified time, in its life cycle application environment. To achieve product reliability over time demands an approach that consists of a set of tasks, each requiring total engineering and management commitment and enforcement. These tasks impact electronic hardware reliability through the selection of materials, structural geometries and design tolerances, manufacturing processes and tolerances, assembly techniques, shipping and handling methods, operational conditions, and maintenance and maintainability guidelines.[1] The tasks are as follows:

1. Define realistic product requirements and constraints determined by the life cycle application profile, required operating and storage life, performance expectations, size, weight, and cost. The manufacturer and the customer must jointly define the product requirements in the light of both the customer's needs and the manufacturer's capability to meet those needs.

2. Define the product life cycle environment by specifying all relevant assembly, storage, handling, shipping, and operating conditions for the fielded product. This includes all stress and loading conditions.

3. Characterize the materials and the manufacturing and assembly processes. Variabilities in material properties and manufacturing processes can induce failures. A knowledge of the variability is required to assess design margins and possible trade-offs with weight, size, and cost.

4. Select the parts required for the product, using a well-defined assessment procedure that ensures that the parts selected have sufficient quality and integrity, are capable of delivering the expected performance and reliability in the application, and will be available to sustain the product throughout its life cycle.

5. Identify the potential failure sites and failure mechanisms by which the product can be expected to fail. Critical parts, part details, and potential failure modes and mechanisms must be identified early in the design, and appropriate measures must be implemented to assure design control. Potential architectural and stress interactions must also be defined and assessed.

6. Design to the usage and process capability of the product (i.e., the quality level that can be controlled in manufacturing and assembly), considering the potential failure sites and failure mechanisms. The design stress spectra, the part test spectra, and the full-scale test spectra must be based on the anticipated life cycle usage conditions. The proposed product must survive the life cycle environment, be optimized for manufacturability, quality, reliability, and cost-effectiveness, and be available to the market in a timely manner.

7. Qualify the product manufacturing and assembly processes. Key process characteristics in all the manufacturing and assembly processes required to make the part must be identified, measured, and optimized. Tests should be conducted to verify the results for complex products. The goal of this step is to provide a physics-of-failure basis for design decisions, with an assessment of all possible failure mechanisms for the anticipated product. If all the processes are in control and the design is valid, then product testing is not warranted and is therefore not cost effective. This represents a transition from product test, analysis, and screening to process test, analysis, and screening.

8. Monitor and control the manufacturing and assembly processes addressed in the design, so that process shifts do not arise. Each process may involve screens and tests to assess statistical process control.

9. Manage the life cycle usage of the product using closed loop management procedures. This includes realistic inspection and maintenance procedures.

Product Requirements and Constraints

A product's requirements and constraints are defined in terms of customer demands and the company's core competencies, culture, and goals. If the product is for direct sale to end users, marketing usually takes the lead in defining the product's requirements and constraints, through interaction with the customer's marketplace, examination of the current product sales figures, and analysis of the competition. Alternatively, if the product is a subsystem that fits within a larger product, the requirements and constraints are determined by the product into which the subsystem fits. The results of capturing product requirements and constraints allow the design team to choose product parts that conform to product-specific and company objectives.

The definition process begins with the identification of an initial set of requirements and constraints defined by either the marketing activity (or in some cases by a specific customer), or by the product into which the subsystem fits. The initial requirements are formulated into a requirements document, where they are prioritized. The requirements document needs to be approved by several groups of people, ranging from engineers to corporate management to customers (the specific people involved in the approval will vary with the company and the product). Once the requirements are approved, the engineering team prepares a preliminary specification indicating the exact set of requirements that are practical to implement. Disconnects between the requirements document and the preliminary specification become the topic of trade-off analyses (usually cost/performance trade-offs), and if, after analyses and negotiation, all the requirements cannot be implemented, the requirements document may be modified. When the requirements document and the preliminary specifications are agreed upon, a final specification is prepared and the design begins.

The Product Life Cycle Environment

The product life cycle environment goes hand-in-hand with the product requirements. The life cycle environment affects product design and development decisions, qualification and specification processes, parts selection and management, quality assurance, product safety, warranty and support commitments, and regulatory conformance.

The product life cycle environment describes the assembly, storage, handling, and scenario for the use of the product, as well as the expected severity and duration of these environments, and thus contains the necessary load input information for failure assessment and the development of design guidelines, assembly guidelines, screens, and tests. Specific load conditions include absolute temperatures, temperature ranges, temperature cycles, temperature gradients, humidity levels, pressure levels, pressure gradients, vibrational or shock loads and transfer functions, chemically aggressive or inert environments, acoustic levels, sand, dust, and electromagnetic radiation levels. In electrical systems, stresses caused by power, current, and voltage should also be considered. These conditions may influence the reliability of the product either individually or in combination with each other. Since the performance of a product over time is often highly dependent on the magnitude of the stress cycle, the rate of change of the stress, and the variation of the stress with time and space, the interaction between the application profile, and the internal conditions must be specified in the design.

The product life cycle environment can be divided into three parts: the application and life profile conditions, the external conditions under which the product must operate, and the internal product-generated stress conditions. The application and life profile conditions include the application length, the number of applications in the expected life of the product, the product utilization or non-utilization profile (storage, testing, transportation), the deployment operations, and the maintenance concept or plan. This information is used to group usage platforms (whether the product will be installed in a car, boat, airplane, satellite, or underground), to develop duty cycles (on-off cycles, storage cycles, transportation cycles, modes of operation, and repair cycles), to determine design criteria, to develop screens and test guidelines, and to develop support requirements to sustain attainment of reliability and maintainability objectives.

The external operational conditions include the anticipated environment(s) and the associated stresses that the product will be required to survive. These conditions are determined through experimentation and through the use of numerical simulation techniques. Experiments are performed by creating environmental parameter monitoring systems, consisting of sensors placed near and within the product that are capable of monitoring the loads that the product experiences. A sensor's function is to convert a physical variable input into, in most cases, an electrical output that is directly related to the physical variable. Signals can be transmitted to either local or remote output devices, enabling data to be collected in a safe and secure manner. Numerical simulation techniques combine material properties, geometry, and product architecture information with environmental data to determine the life cycle environment based on external stresses. Whenever credible data is not available, the worst-case design load must be estimated. A common cause of failure is the use of design factors related to average loads, without adequate consideration being given to the extreme conditions that may occur during the product's life cycle.[2]

The internal operational conditions are associated with product-generated stresses, such as power consumption and dissipation, internal radiation, and release or outgassing of potential contaminants. If the product is connected to other products or subsystems in a system, the stresses associated with the interfaces (i.e., external power consumption, voltage transients, voltage spikes, electronic noise, and heat dissipation) must also be included.

Life cycle stresses can cause strength degradation in materials, for example, combined stresses can accelerate damage and reduce the fatigue limit. In such cases, protective measures must be taken to mitigate the life cycle environment by the use of packaging, provision of warning labels and instructions, and protective treatment of surfaces. The measures to be taken must be identified as appropriate to assembly, storage, transportation, handling, operation, and maintenance. Protection against extreme loads may not always be possible, but should be considered whenever practicable. When overload protection is provided, a reliability analysis should be performed on the basis of the maximum anticipated load, keeping the tolerances of the protection system in mind.[2] If complete protection is not possible, the design team must specify appropriate maintenance procedures for inspection, cleaning, and replacement.

An example of the scenario for use of a product is a flight application, which can involve engine warm-up, taxi, climb, cruising, maneuvers, rapid descent, and emergency landing. Each part of the application will be associated with a set of load conditions, such as time, cycles, acceleration, velocity, vibration, shocks, temperature, humidity, and electrical power cycles. Together, these loads comprise a load history of the product.

Characterization of Materials, Parts, and Manufacturing Processes

Design is intrinsically linked to the materials, parts, interfaces, and manufacturing processes used to establish and maintain the functional and structural integrity of the product. It is unrealistic and potentially dangerous to assume defect-free and perfect-tolerance materials, parts, and structures. Materials often have naturally occurring defects, and manufacturing processes can introduce additional defects in the materials, parts, and structures. The design team must also recognize that the production lots or vendor sources for parts that comprise the design are subject to change, and variability in parts characteristics is likely to occur during the fielded life of a product.

Design decisions involve the selection of parts, materials, and controllable process techniques using processes appropriate to the scheduled production quantity. Any new parts, materials, and processes must be assessed and tested before being put into practice, so that training for production personnel can be planned, quality control safeguards can be set up, and alternative second sources can be located. Often, the goal is to maximize part and configuration standardization, to increase package modularity for ease in fabrication, assembly, and modification, to increase flexibility of design adaptation to alternate uses, and to utilize common fabrication processes. Design decisions also involve choosing the best material interfaces and the best geometric configurations, given the product requirements and constraints.

Parts Selection and Management

Product differentiation, which determines market share gain and loss, often motivates a company to adopt new technologies and insert them into their mainstream products. However, while technological advances continue to fuel product development, two factors, management decisions regarding when and

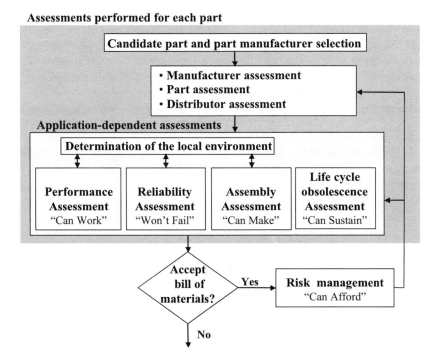

FIGURE 3.61 Parts selection and management methodology.

how a new technology will be used, and accurately assessing risks associated with a technology, differentiate the winners from the losers. Few companies have failed because the right technology was not available; far more have failed when a technology was not effectively managed.

The methodology, shown in Fig. 3.61, provides an "eyes-on, hands-off" approach to parts selection and management, which enables companies to:

- employ risk assessment and mitigation techniques to address technology insertion;
- organize and conduct fact-finding processes to select parts with improved quality, integrity, application-specific reliability, and cost effectiveness;
- make an informed company-wide decision about parts selection and management, based upon company resources, policies, culture, goals, and customer demands;
- understand and evaluate the local environment the part sees within a product's life cycle, and thereby choose the most appropriate technique to fit the part to its intended environmental requirements;
- maximize product supportability by preparing for and meeting the challenge of parts becoming obsolete during product life; and
- improve supply-chain interactions and communications with regulatory agencies in order to minimize time to profit.

Candidate Part and Part Manufacturer Selection

A candidate part is one that conforms to the functional, electrical, and mechanical requirements of the product, considering a company's product requirements, technology direction, and development. In addition, a candidate part must conform to availability and cost constraints. Availability of an electronic part is a measure of the ease with which the part can be procured. Availability is assessed by determining the amount of inventory at hand, the number of parts required for units in production and forecasted, the economic order quantity for the part(s), the lead time(s) between placing an order for the part(s) and receiving the part(s), production schedules and deadlines, and part discontinuation plans. The cost of the part is assessed relative to the product's budget during candidate part selection. In many cases, a part similar to the required one will have already been designed and tested. This "preferred part" is typically mature, in the sense that the variabilities in manufacturing, assembly, and field operation that could cause problems will have already been identified and corrected. Many design groups maintain a list of preferred parts of proven performance, cost, availability, and reliability.

Manufacturer, Part, and Distributor Assessment

In the manufacturer assessment, the part manufacturer's ability to produce parts with consistent quality is evaluated, and in the part assessment, the candidate part's quality and integrity is gauged. The distributor assessment evaluates the distributor's ability to provide parts without affecting the initial quality and integrity, and to provide certain specific services, such as part problem and change notifications. The equipment supplier's parts selection and management team defines the minimum acceptability criteria for this assessment, based on the equipment supplier's requirements. If the part satisfies the minimum acceptability criteria, the candidate part then moves to "application-dependent assessments."

If the part is found unacceptable due to nonconformance with the minimum acceptability criteria, some form of equipment supplier intervention may be considered.[3,4] If equipment supplier intervention is not feasible due to economic or schedule considerations, the candidate part may be rejected. If, however, equipment supplier intervention is considered necessary, then the intervention action items should be identified, and their cost and schedule implications should be analyzed through the "risk management" process step.

Performance Assessment

The goal of performance assessment is to evaluate the part's ability to meet the functional, mechanical, and electrical performance requirements of the product. In order to increase performance, products often incorporate features that tend to make them less reliable than proven, lower-performance products. Increasing the number of parts, although improving performance, also increases product complexity, and may lead to lower reliability unless compensating measures are taken.[5] In such situations, product reliability can be maintained only if part reliability is increased or part redundancy is built into the product. Each of these alternatives, in turn, must be assessed against the incurred cost. The trade-off between performance, reliability, and cost is a subtle issue, involving loads, functionality, system complexity, and the use of new materials and concepts.

In general, there are no distinct stress boundaries for parameters such as voltage, current, temperature, and power dissipation, above which immediate failure will occur and below which a part will operate indefinitely.[6] However, there is often a minimum and a maximum stress limit beyond which the part will not function properly, or at which the increased complexity required will not offer an advantage in cost effectiveness. Part manufacturers' ratings or users' procurement ratings are generally used to determine these limiting values. Equipment manufacturers who integrate such parts into their products need to adapt their design so that the parts do not experience conditions beyond their absolute maximum ratings, even under the worst possible operating conditions (e.g., supply voltage variations, load variations, and signal variations).[7] It is the responsibility of the parts selection and management team to establish that the electrical, mechanical, and functional performance of the part is suitable for the operating conditions of the particular product. If a product must be operated outside the manufacturer-specified operating conditions, then uprating* may have to be considered.

Part manufacturers need to assess the capability of a part over its entire intended life cycle environment, based on the local environment that is determined. If the parametric and functional requirements of the

*The term uprating was coined by Michael Pecht to distinguish it from *upscreening*, which is a term used to describe the practice of attempting to create a part equivalent to a higher quality by additional screening of a part (e.g., screening a JANTXV part to JANS requirements).

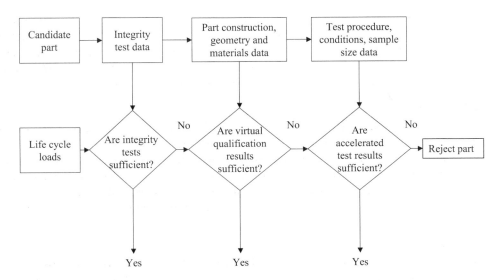

FIGURE 3.62 Reliability assessment process.

system cannot be met within the required local environment, then the local environment may have to be modified, or a different part may have to be used.

Reliability Assessment

Reliability assessment results provide information about the ability of a part to meet the required performance specifications in its life cycle application environment for a specified period of time. Reliability assessment is conducted through the use of integrity test data, virtual qualification results, or accelerated test results. The reliability assessment process is shown in Fig. 3.62.

Integrity is a measure of the appropriateness of the tests conducted by the manufacturer and of the part's ability to survive those tests. Integrity monitoring tests are conducted by the part manufacturer to monitor part/process changes and the ongoing material or process changes specific to the part. Integrity test data (often available from the part manufacturer) is examined in light of the application life cycle stresses and the applicable failure modes and mechanisms. If the magnitude and duration of the application life cycle loads are less severe than those of the integrity tests, and if the test sample size and results are acceptable, then the part reliability is acceptable. However, if the magnitude and duration of the application life cycle loads are not as severe as those of the integrity tests, then integrity test data cannot be used to validate part reliability in the application, and virtual qualification should be considered.

Virtual qualification is a simulation-based methodology used to identify the dominant failure mechanisms associated with the part under the life cycle loads, to determine the acceleration factor for a given set of accelerated test parameters, and to determine the time-to-failures corresponding to the identified failure mechanisms. Virtual qualification allows the operator to optimize the part parameters (e.g., dimensions, materials) so that the minimum time-to-failure of any part is greater than the expected product life.

If virtual qualification proves insufficient to validate part reliability, accelerated testing should be performed. Once the appropriate test procedures, conditions, and sample sizes are determined, accelerated testing can be conducted by either the part manufacturer, the equipment supplier, or third-party test facilities. Accelerated testing results are used to predict the life of a product in its field application by computing an acceleration factor that correlates the accelerated test conditions and the actual field conditions. Whether integrity test data, virtual qualification results, accelerated test results, or a combination thereof are used, each applicable failure mechanism to which the part is susceptible must be addressed.

If part reliability is not ensured through the reliability assessment process, the equipment supplier must consider an alternate part or product redesign. If redesign is not considered a viable option, the part should be rejected, and an alternate part must be selected. If the part must be used in the application, redesign options may include thermal management techniques, vibration damping, and modifying assembly parameters. If product design changes are made, part reliability must be reassessed.

Assembly Issues

A part may be unacceptable from an assembly viewpoint if (1) it is incompatible with the assembly equipment or process; (2) it is impossible or impractical to wire the part into the product (routing compatibility), or (3) it cannot be acceptably tested or reworked. Assembly compatibility addresses whether a product that contains the part can be manufactured (assembled). Routing compatibility assesses if the candidate part can be routed within a specific application on the selected board. Test and rework acceptability assess whether the candidate part can be adequately and economically tested and reworked during assembly.

Assembly Compatibility

Parts must conform to a range of manufacturability constraints associated with their assembly into products. Assembly manufacturability constraints that must be considered when designing a product fall into three categories:

- Assembly process compatibility: involves comparing the part's size, shape, and mounting method to the process that will be used to assemble the boards containing the part.
- Proximity to other structures: involves checking the location of the component relative to other parts assembled on the board and the edge of the board. Proximity checking includes evaluating the orientation (rotation) of the part.
- Artwork verification: involves checking the board layout for the correct orientation and location of fiducials (alignment marks), alignment holes, and other structures necessary to facilitate assembly.

Assembly compatibility and proximity checking can produce three possible outcomes: cannot be assembled, can be assembled with a corresponding cost and yield penalty, and can be assembled with no cost or yield penalties. Artwork verification is decoupled from part selection.

Routing Compatibility

Routing compatibility pertains to the layout and routing of an application. If the selection of a particular part causes significant layout or routing problems within the board, the part may be rejected. Rejection of a part is usually based on its use of routing resources within the board. Two routing issues must be considered:

- How much board area is required to wire the part to the rest of the product?
- How many layers of the board are required to "escape route" the part?

Escape routing is only applicable if the part has an area array format connection to the board, for example, a flip chip or ball grid array package. A component is virtually always "routable," given a sufficient number of board layers. If the rest of the parts on the board are known, routing estimation techniques can be used to determine the effective routing limited footprint of a part under the constraints posed by the board design rules (lines, spaces, via/hole capture pad diameter) and layer count. If a candidate part exceeds the fraction of board wiring resources budgeted to it based on board growth and cost constraints, it may be rejected.

A limiting requirement for some parts is escape routing. If a part's I/Os are in an area array format (as opposed to a peripheral format), the part cannot be wired into the product until all of its I/Os are routed out from under the part. The process of liberating I/Os from an array is called escape routing.

Test and Rework Acceptability

Test and rework costs are important criteria in determining whether a part is acceptable or not. The cost of testing the part (to a specified quality level) prior to assembly and the cost of replacing the part if it needs to be repaired after it is assembled must be considered.

The cost of testing a part is related to what level of testing is performed by the part manufacturer, whether the part is in a package or bare, the function that the part performs, the number of gates or bits contained in the part, and the test equipment. If the part does not come from the manufacturer fully tested (e.g., a bare die), then test costs may need to be assessed. Test costs include the cost of creating the test patterns (or obtaining them from the manufacturer) and the cost of applying the test to the part. Predicting testing costs is of little value unless the corresponding test coverage (fraction of defects detected by the test) is also predicted.

Another key assembly-related cost is the cost of replacing a part that has been identified as defective during the assembly process. The cost of removing a defective part is a function of how the part is mounted to the board, the size of the part, and its proximity to other parts.

Life Cycle Mismatch Assessment

Lengthy design, qualification, and production processes inherent in electronic industries often cause parts to become obsolete before the first product is produced.[8] Furthermore, in order to cater to market demands and remain competitive, part manufacturers often introduce new parts and discontinue older parts. In general, electronic products go through six phases during their life cycles: design, manufacturing, growth, maturity, decline, and discontinuation. A life cycle mismatch occurs between a product and its constituent parts if the parts are not available to support the product throughout its life cycle. When factors such as lead time, risk of part obsolescence, or estimation of the product market are ignored or improperly judged during the design phase, the consequences can be costly. The obsolete part can inhibit the functioning of the product, idle the assembly line, lead to dissatisfied customers, and cause a loss of reputation for the company. The net outcome can be a financial loss for the company.

A successful life cycle mismatch assessment process is one that prevents, if possible, the selection of parts that are already obsolete or soon to be discontinued. This strategy reduces the risk associated with a life cycle mismatch between a product and its parts. The part selection depends on the degree of mismatch and the flexibility to adopt an obsolescence management strategy (e.g., redesign, lifetime buy, buy from aftermarket sources, part substitution). The strategy is intended to mitigate obsolescence risks associated with using the part at some future point in the life cycle of the product. If the equipment supplier finds the life cycle mismatch between part and product unacceptable, the part is unsuitable and should be rejected.

Risk Management

After a part is accepted, resources must be applied to managing the life cycle of the part, including supply chain management, obsolescence assessment, manufacturing and assembly feedback, manufacturer warranties management, and field failure and root-cause analysis. It is important to consider the process of managing the part and all the risks associated with the long-term use of the part throughout its life cycle during the part selection process. The risk management process is characterized using the risks identified in the parts selection process to determine the resources needed to support a part throughout its application life cycle, thus minimizing the probability of a failure. The key metric used to determine whether risks should be managed or not is resources, which include time, data, opportunity, and money.

The risks associated with including a part in the product fall into two categories:

- Managed risks: risks that the product development team chooses to proactively manage by creating a management plan and performing a prescribed regimen of monitoring the part's field performance, manufacturer, and manufacturability.
- Unmanaged risks: risks that the product development team chooses not to proactively manage.

If risk management is considered necessary, a plan should be prepared. The plan should contain details about how the part is monitored (data collection), and how the results of the monitoring feed back into various parts selection and management processes. The feasibility, effort, and cost involved in management processes prior to the final decision to select the part must be considered.

Feedback regarding the part's assembly performance, field performance, and sales history may be essential to ascertain the validity of the predictions made during the part selection process. If the feedback calls for changes in selection criteria, they should be incorporated into the part selection process. Prospective parts should be judged based on the altered part selection criteria. Part monitoring data may also be needed to make changes in parts that are already in use. For example, part monitoring field data might indicate that a change in operating conditions is required for the part to perform satisfactorily.

Failure Modes and Mechanisms

Failure mechanisms are the physical processes by which stresses can damage the materials used to build the product. Investigation of the possible failure modes and mechanisms of the product aids in developing failure-free and reliable designs. The design team must be aware of all possible failure mechanisms in order to design hardware capable of withstanding loads without failing. Failure mechanisms and their related models are also important for planning tests and screens to audit the nominal design and manufacturing specifications, as well as the level of defects introduced by excessive variability in manufacturing and material parameters. Numerous studies focusing on material failure mechanisms and physics-of-failure based damage models and their role in obtaining reliable electronic products have been illustrated in a series of tutorials comprising all relevant wearout and overstress failures.[9-23]

Catastrophic failures occurring due to a single occurrence of a stress event when the intrinsic strength of the material is exceeded are termed overstress failures. Failure mechanisms due to monotonic accumulation of incremental damage beyond the endurance of the material are termed "wearout mechanisms."[24] When the damage exceeds the endurance limit of the component, failure will occur. Unanticipated large stress events can either cause an overstress (catastrophic) failure, or shorten life by causing the accumulation of wearout damage. Examples of such stresses are accidental abuse and acts of God. On the other hand, in well-designed and high-quality hardware, stresses should cause only uniform accumulation of wearout damage; the threshold of damage required to cause eventual failure should not occur within the usage life of the product.

Electrical performance failures can be caused by individual components with improper electrical parameters, such as resistance, impedance, capacitance, or dielectric properties, or by inadequate shielding from electromagnetic interference (EMI) or particle radiation. Failure modes can manifest as reversible drifts in transient and steady-state responses, such as delay time, rise time, attenuation, signal-to-noise ratio, and crosstalk. Electrical failures, common in electronic hardware, include overstress mechanisms due to electrical overstress (EOS) and electrostatic discharge (ESD) such as dielectric breakdown, junction breakdown, hot electron injection, surface and bulk trapping, and surface breakdown and wearout mechanisms such as electromigration and stress-driven diffusive voiding.

Thermal performance failures can arise due to incorrect design of thermal paths in an electronic assembly. This includes incorrect conductivity and surface emissivity of individual components, as well as incorrect convective and conductive paths for heat transfer. Thermal overstress failures are a result of heating a component beyond critical temperatures such as the glass-transition temperature, melting point, fictive point, or flash point. Some examples of thermal wearout failures are aging due to depolymerization, intermetallic growth, and interdiffusion. Failures due to inadequate thermal design may be manifested as components running too hot or too cold and causing operational parameters to drift beyond specifications, although the degradation is often reversible upon cooling. Such failures can be caused either by direct thermal loads or by electrical resistive loads, which in turn generate excessive localized thermal stresses. Adequate design checks require proper analysis for thermal stress, and should include conductive, convective, and radiative heat paths.

Mechanical performance failures include those that may compromise the product performance without necessarily causing any irreversible material damage, such as abnormal elastic deformation in response to mechanical static loads, abnormal transient response (such as natural frequency or damping) to dynamic loads, and abnormal time-dependent reversible (anelastic) response, as well as failures that cause material damage, such as buckling, brittle and/or ductile fracture, interfacial separation, fatigue crack initiation, and propagation, creep, and creep rupture. For example, excessive elastic deformations in slender structures in electronic packages due to overstress loads can sometimes constitute functional failure, such as excessive flexing of interconnection wires, package lids, or flex circuits in electronic devices, causing shorting and/or excessive crosstalk. However, when the load is removed, the deformations (and consequent functional abnormalities) disappear completely without any permanent damage.

Radiation failures are principally caused by uranium and thorium contaminants, and secondary cosmic rays. Radiation can cause wearout, aging, embrittlement of materials, and overstress soft errors in electronic hardware, such as logic chips. Chemical failures occur in adverse chemical environments that result in corrosion, oxidation, or ionic surface dendritic growth. There may also be interactions between different types of stresses. For example, metal migration may be accelerated in the presence of chemical contaminants and composition gradients, and thermal loads can accelerate a failure mechanism due to a thermal expansion mismatch.

Failure modes and effects analysis (FMEA) is an evaluation process for analyzing and assessing the potential failures in a product. Its objectives are to:

1. identify the causes and effects of each failure mode in every part in the product;
2. ascertain the effects of each failure mode on product operation and personnel safety;
3. assess each potential failure according to the effects on other portions of the systems; and,
4. provide a recommendation to eliminate the causes of the failure modes or compensate for their effects.

Failure effects may be considered at subsystem and at overall system levels.

The two approaches to FMEA are functional and hardware. The functional approach, which should be used when the product definition has been identified, begins with the initial product indenture level, and proceeds downward through lower levels. The top level shows the gross operational requirements of the product, while the lower levels represent progressive expansions of the individual functions of the preceding level. This documentation is prepared down to the level necessary to establish the hardware, software, facilities, and personnel and data requirements of the system.

The hardware approach to FMEA should be used when the design team has access to schematics, drawings, and other engineering and design data normally available once the system has matured beyond the functional design stage. This approach begins with obtaining all the information available on the design, including specifications, requirements, constraints, intended applications, drawings, stress data, test results, and so on, to the extent they are available at that time. The approach then proceeds in a part level-up fashion.

Once the approach for the analysis is selected, the product is defined in terms of a functional block diagram and a reliability block diagram. If the product operates in more than one mode in which different functional relationships or part operating modes exist, then these must be considered in the design. FMEA should involve an analysis of possible sneak circuits in the product, that is, an unexpected path or logic flow that can initiate an undesired function or inhibit a desired function. Effects of redundancy must also be considered by evaluating the effects of the failure modes assuming that the redundant system or subsystem is or is not available. The FMEA is then performed using a worksheet, and working to the part or subsystem level considered appropriate, keeping the design data available in mind. A fish-bone diagram of the product, showing all the possible ways in which the product can be expected to fail, is often used in the process. The analysis should take all the failure modes of every part into account, especially when the effects of a failure are serious (e.g., high warranty costs, reliability reputation, safety). FMEA should be started as soon as initial design information is available, and should be performed iteratively as the design evolves, so that the analysis can be used to improve the design and to provide documentation of the eventually completed design.

Design Guidelines and Techniques

Generally, products are replaced with other products, and the replaced product can be used as a baseline for comparisons with products to be introduced. Lessons learned from the baseline comparison product can be used to establish new product parameters, to identify areas of focus in new product designs, and to avoid the mistakes of the past.

Once the parts, materials, processes, and stress conditions are identified, the objective is to design a product using parts and materials that have been sufficiently characterized in terms of how they perform over time when subjected to the manufacturing and application profile conditions. Only through a methodical design approach using physics-of-failure and root-cause analysis can a reliable and cost-effective product be designed. A physics-of-failure based reliability assessment tool must exhibit a diverse array of capabilities:

1. It should be able to predict the reliability of components under a wide range of environmental conditions.
2. It should be able to predict the time-to-failure for fundamental failure mechanisms.
3. It should consider the effect of different manufacturing processes on reliability.

All of these can be accomplished by the use of tools such as virtual qualification and accelerated testing. Design guidelines that are based on physics-of-failure models can also be used to develop tests, screens, and derating factors. Tests based on physics-of-failure models can be designed to measure specific quantities, to detect the presence of unexpected flaws, and to detect manufacturing or maintenance problems. Screens can be designed to precipitate failures in the weak population while not cutting into the design life of the normal population. Derating or safety factors can be determined to lower the stresses for the dominant failure mechanisms.

In using design guidelines, there may not be a unique path to follow. Instead, there is a general flow in the design process. Multiple branches may exist, depending on the input design constraints. The design team should explore an adequate number of these branches to gain confidence that the final design is the best for the prescribed input information. The design team should also assess the use of guidelines for the complete design, and not those limited to specific aspects of an existing design. This does not imply that guidelines cannot be used to address only a specific aspect of an existing design, but the design team may have to trace through the implications that a given guideline suggests.

Protective Architectures

In designs where safety is an issue, it is generally desirable to design in some means for preventing a part, structure, or interconnection from failing, or from causing further damage when it fails. Fuses and circuit breakers are examples of elements used in electronic products to sense excessive current drain and to disconnect power from the concerned part. Fuses within circuits safeguard parts against voltage transients or excessive power dissipation, and protect power supplies from shorted parts. As another example, thermostats can be used to sense critical temperature limiting conditions, and to shut down the product or a part of the system until the temperature returns to normal. In some products, self-checking circuitry can also be incorporated to sense abnormal conditions and operate adjusting means to restore normal conditions, or to activate switching means to compensate for the malfunction.[6]

In some instances, it may be desirable to permit partial operation of the product after a part failure in preference to total product failure. By the same reasoning, degraded performance of a product after failure of a part is often preferable to complete stoppage. An example is the shutting down of a failed circuit whose function is to provide precise trimming adjustment within a deadband* of another control

---

*When the input in a control system changes direction, an initial change in the input has no effect on the output. This amount of side-to-side play in the system for which there is no change in the output is referred to as the deadband. The deadband is centered about the output.

product; acceptable performance may thus be achieved, perhaps under emergency conditions, with the deadband control product alone.[6]

Sometimes, the physical removal of a part from a product can harm or cause failure in another part by removing either load, drive, bias, or control. In such cases, the first part should be equipped with some form of interlock mechanism to shut down or otherwise protect the second part. The ultimate design, in addition to its ability to act after a failure, should be capable of sensing and adjusting for parametric drifts to avert failures.

In the use of protective techniques, the basic procedure is to take some form of action, after an initial failure or malfunction, to prevent additional or secondary failures. By reducing the number of failures, techniques such as enhancing product reliability can be considered, although they also affect availability and product effectiveness. Equally important considerations are the impacts of maintenance, repair, and part replacement. For example, if a fuse protecting a circuit is replaced, the following questions need to be answered: What is the impact when the product is reenergized? What protective architectures are appropriate for post-repair operations? What maintenance guidance must be documented and followed when fail-safe protective architectures have or have not been included?

Stress Margins

A properly designed product should be capable of operating satisfactorily with parts that drift or change with variables such as time, temperature, humidity, pressure, altitude, etc., as long as the interconnects and the other parameters of the parts are within their rated tolerances. To guard against out-of-tolerance failures, the design team must consider the combined effects of tolerances on parts to be used in manufacture, of subsequent changes due to the range of expected environmental conditions, of drifts due to aging over the period of time specified in the reliability requirement, and of tolerances in parts used in future repair or maintenance functions. Parts and structures should be designed to operate satisfactorily at the extremes of the parameter ranges, and allowable ranges must be included in the procurement or reprocurement specifications.

Statistical analysis and worst-case analysis are methods of dealing with part and structural parameter variations. In statistical analysis, a functional relationship is established between the output characteristics of the structure and the parameters of one or more of its parts. In worst-case analysis, the effect that a part has on product output is evaluated on the basis of end-of-life performance values or out-of-specification replacement parts.

Derating

Derating is a technique by which either the operational stresses acting on a device or structure are reduced relative to the rated strength, or the strength is increased relative to the allocated operating stress levels. Reducing the stress is achieved by specifying upper limits on the operating loads below the rated capacity of the hardware. For example, manufacturers of electronic hardware often specify limits for supply voltage, output current, power dissipation, junction temperature, and frequency. The equipment design team may decide to select an alternative component or make a design change that ensures that the operational condition for a particular parameter, such as temperature, is always below the rated level. The component is then said to have been derated for thermal stress.

The derating factor, typically defined as the ratio of the rated level of a given stress parameter to its actual operating level, is actually a margin of safety or margin of ignorance, determined by the criticality of any possible failures and by the amount of uncertainty inherent in the reliability model and its inputs. Ideally, this margin should be kept to a minimum to maintain the cost effectiveness of the design. This puts the responsibility on the reliability engineer to identify the rated strength, the relevant operating stresses, and the reliability as unambiguously as possible.

To be effective, derating criteria must target the right stress parameter to address modeling of the relevant failure mechanisms. Field measurements may also be necessary, in conjunction with modeling simulations, to identify the actual operating stresses at the failure site. Once the failure models have been quantified, the impact of derating on the effective reliability of the component for a given load can be determined. Quantitative correlations between derating and reliability enable design teams and users to effectively tailor the margin of safety to the level of criticality of the component, leading to better and more cost-effective use of the functional capacity of the component.

Redundancy

Redundancy permits a product to operate even though certain parts and interconnections have failed, thus increasing its reliability and availability. Redundant configurations can be classified as either active or standby. Elements in active redundancy operate simultaneously in performing the same function. Elements in standby redundancy are designed so that an inactive one will, or can, be switched into service when an active element fails. The reliability of the associated function increases with the number of standby elements (optimistically assuming that the sensing and switching devices of the redundant configuration are working perfectly, and failed redundant components are replaced before their companion components fail). One preferred design alternative is that a failed redundant component can be repaired without adversely impacting the product operation, and without placing the maintenance person or the product at risk.

A design team may often find that redundancy is:

- the quickest way to improve product reliability if there is insufficient time to explore alternatives, or if the part is already designed;
- the cheapest solution, if the cost of redundancy is economical in comparison with the cost of redesign; and/or
- the only solution, if the reliability requirement is beyond the state of the art.

On the other hand, in weighing its disadvantages, the design team may find that redundancy will:

- prove too expensive, if the parts, redundant sensors, and switching devices are costly;
- exceed the limitations on size and weight, (for example, in avionics, missiles, and satellites);
- exceed the power limitations, particularly in active redundancy;
- attenuate the input signal, requiring additional amplifiers (which increase complexity); and/or
- require sensing and switching circuitry so complex as to offset the reliability advantage of redundancy.

Qualification and Accelerated Testing

Qualification includes all activities that ensure that the nominal design and manufacturing specifications will meet or exceed the desired reliability targets. Qualification validates the ability of the nominal design and manufacturing specifications of the product to meet the customer's expectations, and assesses the probability of survival of the product over its complete life cycle. The purpose of qualification is to define the acceptable range of variabilities for all critical product parameters affected by design and manufacturing, such as geometric dimensions, material properties, and operating environmental limits. Product attributes that are outside the acceptable ranges are termed defects, since they have the potential to compromise product reliability.[25]

Qualification tests should be performed only during initial product development, and immediately after any design or manufacturing changes in an existing product. Once the product is qualified, routine lot-to-lot requalification is redundant and an unnecessary cost item. A well-designed qualification procedure provides economic savings and quick turnaround during development of new products or mature products subject to manufacturing and process changes.

Investigating failure mechanisms and assessing the reliability of products where long lives are required may be a challenge, since a very long test period under the actual operating conditions is necessary to obtain sufficient data to determine actual failure characteristics. One approach to the problem of obtaining meaningful qualification data for high-reliability devices in shorter time periods is to use methods such as virtual qualification and accelerated testing to achieve test-time compression. However, when qualifying the reliability of a product for overstress mechanisms, a single cycle of the expected overstress load may be adequate, and acceleration of test parameters may not be necessary. This is sometimes called proof-stress testing.

Virtual Qualification

Virtual qualification is a process that requires significantly less time and money than accelerated testing to qualify a part for its life cycle environment. This simulation-based methodology is used to identify the dominant failure mechanisms associated with the part under life cycle loads, to determine the acceleration factor for a given set of accelerated test parameters, and to determine the time-to-failure corresponding to the identified failure mechanisms. Each failure model comprises a stress analysis model and a damage assessment model. The output is a ranking of different failure mechanisms, based on the time to failure. The stress model captures the product architecture, while the damage model depends on a material's response to the applied stress. This process is therefore applicable to existing as well as new products. The objective of virtual qualification is to optimize the product design in such a way that the minimum time-to-failure of any part of the product is greater than its desired life. Although the data obtained from virtual qualification cannot fully replace that obtained from physical tests, it can increase the efficiency of physical tests by indicating the potential failure modes and mechanisms that the operator can expect to encounter.

Ideally, a virtual qualification process will involve identification of quality suppliers, computer-aided physics-of-failure qualification, and a risk assessment and mitigation program. The process allows qualification to be readily incorporated into the design phase of product development since it allows design, test, and redesign to be conducted promptly and cost effectively. It also allows consumers to qualify off-the-shelf components for use in specific environments without extensive physical tests. Since virtual qualification reduces emphasis on examining a physical sample, it is imperative that the manufacturing technology and quality assurance capability of the manufacturer be taken into account. The manufacturer's design, production, test, and measurement procedures must be evaluated and certified. If the data on which the virtual qualification is performed is inaccurate or unreliable, all results are suspect. In addition, if a reduced quantity of physical tests is performed in the interest of simply verifying virtual results, the operator needs to be confident that the group of parts selected is sufficient to represent the product. Further, it should be remembered that the accuracy of the results using virtual qualification depends on the accuracy of the inputs to the process, i.e., the accuracy of the life cycle loads, the choice of the failure models used, the choice of the analysis domain (e.g., 2D, pseudo-3D, full 3D), the constants in the failure model, the material properties, and so on. Hence, in order to obtain a reliable prediction, the variabilities in the inputs should be specified using distribution functions, and the validity of the failure models should be tested by conducting accelerated tests.

Accelerated Testing

Accelerated testing involves measuring the performance of the test product at loads or stresses that are more severe than would normally be encountered, in order to enhance the damage accumulation rate within a reduced time period. The goal of such testing is to accelerate time-dependent failure mechanisms and the damage accumulation rate to reduce the time to failure. The failure mechanisms and modes in the accelerated environment must be the same as (or quantitatively correlated with) those observed under actual usage conditions, and it must be possible to quantitatively extrapolate from the accelerated environment to the usage environment with some reasonable degree of assurance.

Accelerated testing begins by identifying all the possible overstress and wearout failure mechanisms. The load parameter that directly causes the time-dependent failure is selected as the acceleration parameter, and is commonly called the accelerated load. Common accelerated loads include thermal loads, such as temperature, temperature cycling, and rates of temperature change; chemical loads, such as humidity, corrosives, acid, and salt; electrical loads, such as voltage, or power; and mechanical loads, such as vibration, mechanical load cycles, strain cycles, and shock/impulses. The accelerated environment may include a combination of these loads. Interpretation of results for combined loads requires a quantitative understanding of their relative interactions and the contribution of each load to the overall damage.

Failure due to a particular mechanism can be induced by several acceleration parameters. For example, corrosion can be accelerated by both temperature and humidity; and creep can be accelerated by both mechanical stress and temperature. Furthermore, a single accelerated stress can induce failure by several wearout mechanisms simultaneously. For example, temperature can accelerate wearout damage accumulation not only by electromigration, but also by corrosion, creep, and so on. Failure mechanisms that dominate under usual operating conditions may lose their dominance as the stress is elevated. Conversely, failure mechanisms that are dormant under normal use conditions may contribute to device failure under accelerated conditions. Thus, accelerated tests require careful planning in order to represent the actual usage environments and operating conditions without introducing extraneous failure mechanisms or nonrepresentative physical or material behavior. The degree of stress acceleration is usually controlled by an acceleration factor, defined as the ratio of the life under normal use conditions as compared to that under the accelerated condition. The acceleration factor should be tailored to the hardware in question, and can be estimated from an acceleration transform (i.e., a functional relationship between the accelerated stress and the life cycle stress) in terms of all the hardware parameters.

Once the failure mechanisms are identified, it is necessary to select the appropriate acceleration load; to determine the test procedures and the stress levels; to determine the test method, such as constant stress acceleration or step-stress acceleration; to perform the tests; and to interpret the test data, which includes extrapolating the accelerated test results to normal operating conditions. The test results provide qualitative failure information for improving the hardware through design and/or process changes. Accelerated testing includes:

- Accelerated test planning and development: used to develop a test program that focuses on the potential failure mechanisms and modes that were identified in the first phase as the weak links under life cycle loads. The various issues addressed in this phase include designing the test matrix and test loads, analysis, design and preparation of the test vehicle, setting up the test facilities (e.g., test platforms, stress monitoring schemes, failure monitoring and data acquisition schemes), fixture design, effective sensor placement, and data collection and post-processing schemes.

- Test vehicle characterization: used to identify the contribution of the environment on the test vehicle in the accelerated life tests.

- Accelerated life testing: evaluates the vulnerability of the product to the applied life cycle due to wearout failure mechanisms. This step yields a meaningful assessment of life cycle durability only if it is preceded by the steps discussed above. Without these steps, accelerated life testing can only provide comparisons between alternate designs if the same failure mechanism is precipitated.

- Life assessment: used to provide a scientific and rational method to understand and extrapolate accelerated life testing failure data to estimate the life of the product in the field environment.

Detailed failure analysis of failed samples is a crucial step in the qualification and validation program. Without such analyses and feedback to the design team for corrective action, the purpose of the qualification program is defeated. In other words, it is not adequate simply to collect failure data. The key is to use the test results to provide insights into, and consequent control over, relevant failure mechanisms and to prevent them cost effectively.

Manufacturing Issues

Manufacturing and assembly processes can significantly impact the quality and reliability of hardware. Improper assembly and manufacturing techniques can introduce defects, flaws, and residual stresses that act as potential failure sites or stress raisers later in the life of the product. If these defects and stresses can be identified, the design analyst can proactively account for them during the design and development phase.

Auditing the merits of the manufacturing process involves two crucial steps. First, qualification procedures are required, as in design qualification, to ensure that manufacturing specifications do not compromise the long-term reliability of the hardware. Second, lot-to-lot screening is required to ensure that the variabilities of all manufacturing-related parameters are within specified tolerances.[25,26] In other words, screening ensures the quality of the product by precipitating latent defects before they reach the field.

Process Qualification

Like design qualification, process qualification should be conducted at the prototype development phase. The intent is to ensure that the nominal manufacturing specifications and tolerances produce acceptable reliability in the product. The process needs requalification when process parameters, materials, manufacturing specifications, or human factors change.

Process qualification tests can be the same set of accelerated wearout tests used in design qualification. As in design qualification, overstress tests may be used to qualify a product for anticipated field overstress loads. Overstress tests may also be exploited to ensure that manufacturing processes do not degrade the intrinsic material strength of hardware beyond a specified limit. However, such tests should supplement, not replace, the accelerated wearout test program, unless explicit physics-based correlations are available between overstress test results and wearout field-failure data.

Manufacturability

The control and rectification of manufacturing defects has typically been the concern of production and process-control engineers, but not of the design team. In the spirit and context of concurrent product development, however, hardware design teams must understand material limits, available processes, and manufacturing process capabilities in order to select materials and construct architectures that promote producibility and reduce the occurrence of defects, increasing yield and quality. Therefore, no specification is complete without a clear discussion of manufacturing defects and acceptability limits. The reliability engineer must have clear definitions of the threshold for acceptable quality, and of what constitutes nonconformance. Nonconformance that compromises hardware performance and reliability is considered a defect. Failure mechanism models provide a convenient vehicle for developing such criteria. It is important for the reliability analyst to understand what deviations from specifications can compromise performance or reliability, and what deviations are benign and can be accepted.

A defect is any outcome of a process (manufacturing or assembly) that impairs or has the potential to impair the functionality of the product at any time. The defect may arise during a single process or may be the result of a sequence of processes. The yield of a process is the fraction of products that are acceptable for use in a subsequent process in the manufacturing sequence or product life cycle. The cumulative yield of the process is approximately determined by multiplying the individual yields of each of the individual process steps. The source of defects is not always apparent, because defects resulting from a process can go undetected until the product reaches some downstream point in the process sequence, especially if screening is not employed.

It is often possible to simplify the manufacturing and assembly processes in order to reduce the probability of workmanship defects. As processes become more sophisticated, however, process monitoring and control are necessary to ensure a defect-free product. The bounds that specify whether the process is within tolerance limits, often referred to as the process window, are defined in terms of the independent variables to be controlled within the process and the effects of the process on the product or the dependent product variables. The goal is to understand the effect of each process variable on each product parameter in order to formulate control limits for the process, that is, the points on the variable scale where the defect rate begins to possess a potential for causing failure. In defining the process window, the upper and lower limits of each process variable beyond which it will produce defects must be determined. Manufacturing processes must be contained in the process window by defect testing, analysis of the causes of defects, and elimination of defects by process control, such as closed-loop corrective action systems. The establishment of an effective feedback path to report process-related defect data is critical. Once this is done and the process window is determined, the process window itself becomes a feedback system for the process operator.

Several process parameters may interact to produce a different defect than would have resulted from the individual effects of these parameters acting independently. This complex case may require that the interaction of various process parameters be evaluated in a matrix of experiments. In some cases, a defect cannot be detected until late in the process sequence. Thus, a defect can cause rejection, rework, or failure of the product after considerable value has been added to it. These cost items due to defects can affect return on investments by adding to hidden factory costs. All critical processes require special attention for defect elimination by process control.

Process Verification Testing

Process verification testing is often called screening. Screening involves 100% auditing of all manufactured products to detect or precipitate defects. The aim is to preempt potential quality problems before they reach the field. In principle, this should not be required for a well-controlled process. When uncertainties are likely in process controls, however, screening is often used as a safety net.

Some products exhibit a multimodal probability density function for failures, with a secondary peak during the early period of their service life due to the use of faulty materials, poorly controlled manufacturing and assembly technologies, or mishandling. This type of early-life failure is often called infant mortality. Properly applied screening techniques can successfully detect or precipitate these failures, eliminating or reducing their occurrence in field use. Screening should only be considered for use during the early stages of production, if at all, and only when products are expected to exhibit infant mortality field failures. Screening will be ineffective and costly if there is only one main peak in the failure probability density function. Further, failures arising due to unanticipated events such as acts of God (lightning, earthquakes) may be impossible to screen cost effectively.

Since screening is done on 100% basis, it is important to develop screens that do not harm good components. The best screens, therefore, are nondestructive evaluation techniques, such as microscopic visual exams, X-rays, acoustic scans, nuclear magnetic resonance (NMR), electronic paramagnetic resonance (EPR), and so on. Stress screening involves the application of stresses, possibly above the rated operational limits. If stress screens are unavoidable, overstress tests are preferred to accelerated wearout tests, since the latter are more likely to consume some useful life of good components. If damage to good components is unavoidable during stress screening, then quantitative estimates of the screening damage, based on failure mechanism models must be developed to allow the design team to account for this loss of usable life. The appropriate stress levels for screening must be tailored to the specific hardware. As in qualification testing, quantitative models of failure mechanisms can aid in determining screen parameters.

A stress screen need not necessarily simulate the field environment, or even utilize the same failure mechanism as the one likely to be triggered by this defect in field conditions. Instead, a screen should exploit the most convenient and effective failure mechanism to stimulate the defects that can show up in the field as infant mortality. Obviously, this requires an awareness of the possible defects that may occur in the hardware and extensive familiarity with the associated failure mechanisms.

Unlike qualification testing, the effectiveness of screens is maximized when screens are conducted immediately after the operation believed to be responsible for introducing the defect. Qualification testing is preferably conducted on the finished product or as close to the final operation as possible; on the other hand, screening only at the final stage, when all operations have been completed, is less effective, since failure analysis, defect diagnostics, and troubleshooting are difficult and impair corrective actions. Further, if a defect is introduced early in the manufacturing process, subsequent value added through new materials and processes is wasted, which additionally burdens operating costs and reduces productivity. Admittedly, there are also several disadvantages to such an approach. The cost of screening at every manufacturing station may be prohibitive, especially for small batch jobs. Further, components will experience repeated screening loads as they pass through several manufacturing steps, which increases the risk of accumulating wearout damage in good components due to screening stresses. To arrive at a screening matrix that addresses as many defects and failure mechanisms as feasible with each screen test, an optimum situation must be sought through analysis of cost effectiveness, risk, and the criticality of the defects. All defects must be traced back to the root cause of the variability.

Any commitment to stress screening must include the necessary funding and staff to determine the root cause and appropriate corrective actions for all failed units. The type of stress screening chosen should be derived from the design, manufacturing, and quality teams. Although a stress screen may be necessary during the early stages of production, stress screening carries substantial penalties in capital, operating expense, and cycle time, and its benefits diminish as a product approaches maturity. If almost all of the products fail in a properly designed screen test, the design is probably incorrect. If many products fail, a revision of the manufacturing process is required. If the number of failures in a screen is small, the processes are likely to be within tolerances and the observed faults may be beyond the resources of the design and production process.

Summary

Reliability is not a matter of chance or good fortune; rather, it is a rational consequence of conscious, systematic, rigorous efforts at every stage of design, development, and manufacture. High product reliability can only be assured through robust product designs, capable processes that are known to be within tolerances, and qualified components and materials from vendors whose processes are also capable and within tolerances. Quantitative understanding and modeling of all relevant failure mechanisms provide a convenient vehicle for formulating effective design, process, and test specifications and tolerances.

The physics-of-failure approach is not only a tool to provide better and more effective designs, but it also helps develop cost-effective approaches for improving the entire approach to building electronic products. Proactive improvements can be implemented for defining more realistic performance requirements and environmental conditions, identifying and characterizing key material properties, developing new product architectures and technologies, developing more realistic and effective accelerated stress tests to audit reliability and quality, enhancing manufacturing-for-reliability through mechanistic process modeling and characterization to allow proactive process optimization, increasing first-pass yields, and reducing hidden factory costs associated with inspection, rework, and scrap.

When utilized early in the concept development stage of a product's development, reliability serves as an aid to determine feasibility and risk. In the design stage of product development, reliability analysis involves methods to enhance performance over time through the selection of materials, design of structures, choice of design tolerance, manufacturing processes and tolerances, assembly techniques, shipping and handling methods, and maintenance and maintainability guidelines. Engineering concepts such as strength, fatigue, fracture, creep, tolerances, corrosion, and aging play a role in these design analyses. The use of physics-of-failure concepts coupled with mechanistic and probabilistic techniques is often required to understand the potential problems and trade-offs, and to take corrective actions. The use of factors of safety and worst-case studies as part of the analysis is useful in determining stress screening and burn-in procedures, reliability growth, maintenance modifications, field testing procedures, and various logistics requirements.

Defining Terms

Accelerated testing: Tests conducted at stress levels that are more severe than the normal operating levels, in order to enhance the damage accumulation rate within a reduced time period.

Damage: The failure pattern of an electronic or mechanical product.

Derating: Practice of subjecting parts to lower electrical or mechanical stresses than they can withstand in order to increase the life expectancy of the part.

Failure mode: Any physically observable change caused by a failure mechanism.

Failure mechanism: A process through which a defect nucleates and grows as a function of stresses such as thermal, mechanical, electromagnetic, or chemical loadings, i.e., creep, fatigue, wear, and so on, which results in degradation or failure of a product.

Integrity: A measure of the appropriateness of the tests conducted by the manufacturer and the part's ability to survive those tests.

Overstress failures: Catastrophic sudden failures due to a single occurrence of a stress event that exceeds the intrinsic strength of a material.

Product performance: The ability of a product to perform as required according to specifications.

Qualification: All activities that ensure that the nominal design and manufacturing specifications will meet or exceed the reliability goals.

Quality: A measure of a part's ability to meet the workmanship criteria of the manufacturer.

Reliability: The ability of a product to perform as intended (i.e., without failure and within specified performance limits) for a specified time, in its life cycle application environment.

Wearout failures: Failures due to accumulation of incremental damage, occurring when the accumulated damage exceeds the material endurance limit.

References

1. Pecht, M., *Integrated Circuit, Hybrid, and Mutichip Module Package Design Guidelines — A Focus on Reliability*, John Wiley & Sons, New York, 1994.
2. O'Connor, P., *Practical Reliability Engineering*, John Wiley & Sons, New York, 1991.
3. Jackson, M., Mathur, A., Pecht, M., and Kendall, R., Part Manufacturer Assessment Process, *Quality and Reliability Engineering International*, 15, 457, 1999.
4. Jackson, M., Sandborn, P., Pecht, M., Hemens-Davis, C., and Audette, P., A risk-informed methodology for parts selection and management, *Quality and Reliability Engineering International*, 15, 261, 1999.
5. Lewis, E.E., *Introduction to Reliability Engineering*, John Wiley & Sons, New York, 1996.
6. Sage, A.P. and Rouse, W.B., *Handbook of Systems Engineering and Management*, John Wiley & Sons, New York, 1999.
7. IEC Standard 60134, *Rating systems for electronic tubes and valves and analogous semiconductor devices*, (Last reviewed in July 1994 by the IEC Technical Committee 39 on Semiconductors), 1961.
8. Stogdill, R. C., Dealing with obsolete parts, *IEEE Design & Test of Computers*, 16, 2, 17, 1999.
9. Dasgupta, A. and Pecht, M., Failure mechanisms and damage models, *IEEE Transactions on Reliability*, 40, 5, 531, 1991.
10. Dasgupta, A. and Hu, J.M., Failure mechanism models for brittle fracture, *IEEE Transactions on Reliability*, 41, 3, 328, 1992.
11. Dasgupta, A. and Hu, J.M., Failure mechanism models for ductile fracture, *IEEE Transactions on Reliability*, 41, 4, 489, 1992.
12. Dasgupta, A. and Hu, J.M., Failure mechanism models for excessive elastic deformation, *IEEE Transactions on Reliability*, 41, 1, 149, 1992.
13. Dasgupta, A. and Hu, J.M., Failure mechanism models for plastic deformation, *IEEE Transactions on Reliability*, 41, 2, 168, 1992.
14. Dasgupta, A. and Haslach, H. W., Jr., Mechanical design failure models for buckling, *IEEE Transactions on Reliability*, 42, 1, 9, 1993.
15. Engel, P.A., Failure models for mechanical wear modes and mechanisms, *IEEE Transactions on Reliability*, 42, 2, 262, 1993.
16. Li, J. and Dasgupta, A., Failure mechanism models for material aging due to interdiffusion, *IEEE Transactions on Reliability*, 43, 1, 2, 1994.
17. Li, J. and Dasgupta, A., Failure-mechanism models for creep and creep rupture, *IEEE Transactions on Reliability*, 42, 3, 339, 1994.
18. Dasgupta, A., Failure mechanism models for cyclic fatigue, *IEEE Transactions on Reliability*, 42, 4, 548, 1993.
19. Young, D. and Christou, A., Failure mechanism models for electromigration, *IEEE Transactions on Reliability*, 43, 2, 186, 1994.

20. Rudra, B. and Jennings, D., Failure mechanism models for conductive-filament formation, *IEEE Transactions on Reliability*, 43, 3, 354, 1994.

21. Al-Sheikhly, M. and Christou, A., How radiation affects polymeric materials, *IEEE Transactions on Reliability*, 43, 4, 551, 1994.

22. Diaz, C., Kang, S.M., and Duvvury, C., Electrical overstress and electrostatic discharge, *IEEE Transactions on Reliability*, 44, 1, 2, 1995.

23. Tullmin, M. and Roberge, P.R., Corrosion of metallic materials, *IEEE Transactions on Reliability*, 44, 2, 271, 1995.

24. Upadhyayula, K. and Dasgupta, A., Guidelines for physics-of-failure based accelerated stress testing, *Annual Reliability and Maintainability Symposium 1998 Proceedings, International Symposium on Product Quality and Integrity*, 345, 1998.

25. Pecht, M., Dasgupta, A., Evans, J. W., and Evans, J. Y, *Quality Conformance and Qualification of Microelectronic Packages and Interconnects*, John Wiley & Sons, New York, 1994.

26. Kraus, A., Hannemann, R., Pecht, M., *Semiconductor Packaging: A Multidisciplinary Approach*, John Wiley & Sons, New York, 1994.

Further Information

Microelectronics Reliability: http://www.elsevier.com/locate/microrel
IEEE Transactions on Reliability: http://www.ewh.ieee.org/soc/rs/transactions.htm

4

Microwave Measurements

R.E. Ham
Consultant

Joseph Staudinger
Motorola

Alfy Riddle
Macallan Consulting

J. Stevenson Kenney
Georgia Institute of Technology

John F. Sevic
UltraRF, Incorporated

Anthony E. Parker
Macquarie University

James G. Rathmell
University of Sydney

Jonathan B. Scott
Agilent Technologies

Jean-Pierre Lanteri
TycoElectronics

Christopher Jones
TycoElectronics

John R. Mahon
TycoElectronics

4.1   Linear Measurements

R.E. Ham

Microwave and RF measurements can be classified in two distinct but often overlapping categories: signal measurements and network measurements. Signal measurements include observation and determination

0-8493-8592-X/01/$0.00+$.50
© 2001 by CRC Press LLC

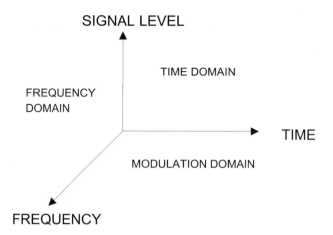

FIGURE 4.1 Signals are characterized by three different types of measurements.

of the characteristics of waves and waveforms. These parameters can be obtained in the time, frequency, or modulation domain. Network measurement determines the terminal and signal transfer characteristics of devices and systems with any number of ports.

Signal Measurements

Signal measurements are taken in any one or more of three measurement planes as illustrated in Fig. 4.1. The most common measurement at low frequencies is in the time domain where the amplitude of a signal waveform is observed with respect to time. The instrument used for this is an oscilloscope. By continuing to observe the amplitude of the signal over a small frequency range, the spectral components of the signal are obtained. This measurement is normally made with a spectrum analyzer. Determining the instantaneous frequency of a signal versus time is a modulation domain measurement.

Time Domain

Observation of RF and microwave signals with an analog oscilloscope is limited by the speed of response of the instrument circuits and of the display. Building such an instrument for operation beyond a few hundred megahertz is very difficult and expensive. For observing very high-speed waveforms signal sampling techniques are incorporated.

A sampling oscilloscope measures the value of a waveform at a particular time and digitally stores the sample data for display. If the sampling can be performed fast enough, the entire waveform shape can be recreated from the sample data. This is done at relatively low frequencies; however, as the frequency increases it is not possible to capture enough points during one waveform occurrence. By delaying subsequent samples to be taken a very short time later during the occurrence of another cycle of the waveform, the recurring waveform shape can ultimately be reconstructed from the stored digital data.

Note the qualification that the waveform shape that can be measured by the high-speed sampling technique must be recurring. This makes capturing a onetime occurrence very difficult and, even if points in a gigahertz waveform can be captured, this does not mean that one cycle of the waveform can be captured.

Frequency Domain

The number of measurements that must be made on a signal over a specified period of time is a function of the stability and modulation placed on the signal. The exact measurement of the frequency of a stable and spectrally pure signal is performed with a frequency counter and measurements are normally made a few times per second. Direct counting circuits are available well into the lower microwave frequency range. At high microwave frequencies counters use conversion oscillators and mixers to heterodyne the signal down in frequency to where it can be directly counted. Microprocessor controllers and knowledge of the exact frequency of the conversion oscillators enables an exact signal frequency to be calculated.

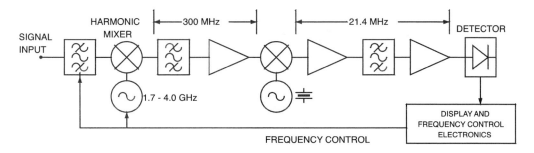

OSCILLATOR HARMONIC NUMBER (n)	EFFECTIVE OSCILLATOR FREQUENCY (GHz)	TUNED RF FREQUENCY (GHz)
1	1.7 - 4.0	2.0 - 4.3
2	3.4 - 8.0	3.7 - 8.3
3	5.1 - 12.0	5.4 - 12.3
4	6.8 - 14.0	7.1 - 14.3
5	8.5 - 20.0	8.8 - 20.3

FIGURE 4.2 Simplified block diagram of a microwave spectrum analyzer.

A spectrum analyzer [1] is used to make frequency domain measurements of complex signals and signals with characteristics that vary with time. This is basically a swept frequency filter with a detector to determine the signal amplitude within the bandwidth of the filter and some means of displaying or storing the measured information. To increase the selectivity and dynamic range of such a basic instrument, heterodyne conversions are used. Figure 4.2 is the block diagram of a typical microwave spectrum analyzer.

The first intermediate frequency is chosen to permit a front-end filter to eliminate the image from the first mixer. In this case, 300 MHz is chosen because the tunable filter, usually a YIG device, will have considerable attenuation at the image frequency 600 MHz away from the desired signal. The second intermediate frequency is chosen because reasonably selective filters can be constructed to enable resolving signal components that are close to each other. Additionally, detector and signal processing components, such as digital signal processors, can be readily constructed at the lower frequency.

Because the normal frequency range required from a microwave spectrum analyzer is many octaves wide, multiple first conversion oscillators are required; however, this is an extremely expensive approach. Spectrum analyzers use a harmonic mixer for the first conversion and the first filter is tuned to eliminate the products that would be received due to the undesired harmonics of the conversion oscillator. Note the list of harmonic numbers (n) and the resulting tuned frequency of the example analyzer. As the harmonic number increases the sensitivity of the analyzer decreases because the harmonic mixer efficiency decreases with increasing n.

The most important spectrum analyzer specifications are:

1. Frequency tuning range — to include all of the frequency components of the signal to be measured.
2. Frequency accuracy and stability — to be more stable and accurate than the signal to be measured.
3. Sweep width — the band of frequencies over which the unit can sweep without readjustment.
4. Resolution bandwidth — narrow enough to resolve different spectral components of the signal.
5. Sensitivity and/or noise figure — to observe very small signals or small parts of large signals.
6. Sweep rate — maximum sweep rate is established by the settling time of the filter that sets the resolution bandwidth.
7. Dynamic range — the difference between the largest and smallest signal the analyzer can measure without readjustment.
8. Phase noise — a signal with spectral purity greater than that of the analyzer conversion oscillators cannot be characterized.

Spectrum analyzers using other than swept frequency techniques can be made. For example, high speed sampling methods used with digital signal processors (DSP) calculating the Fast Fourier Transform (FFT) are readily implemented; however, the speed of operation of the logic circuits limits the upper frequency of operation. This is a common method of intermediate frequency demodulation and the useable frequency will move upward with semiconductor development.

Modulation Domain

Modulation domain measurements [2] yield the instantaneous frequency of a signal as a function of time. Two examples of useful modulation domain data are the instantaneous frequency of a phase-locked oscillator as the loop settles and the pulse repetition rate of a fire control radar as it goes from search mode (low pulse repetition frequency or PRF) to lock and fire mode (high PRF).

A modulation domain analyzer establishes the exact time at which a desired event occurs and catalogs the time. The event captured in a phase-locked oscillator is the zero crossing of the oscillator output voltage. For a radar it is the leading edge of each pulse. From this information the event frequency is calculated. Various other modulation domain analyzers can be made with instantaneous frequency correlators and frequency discriminators.

Network Measurements

Low frequency circuit design and performance evaluation is based upon the measurement of voltages and currents. Knowing the impedance level at a point in a circuit to be the ratio of voltage to current, a voltage or current measurement can be used to calculate power. By measuring voltage and current as a complex quantity, yielding complex impedances, this method of circuit characterization can be used at relatively high frequencies even with the limitations of nontrivial values of circuit capacitive and inductive parasitics. When the parasitics can no longer be treated as lumped elements, distributed circuit concepts must be used.

A simple transmission line such as the coaxial line in Fig. 4.3a can, if physically very small in all dimensions with respect to a wavelength, be modeled as a lumped element circuit as shown in Fig. 4.3b; however, as the size of the line increases relative to the wavelength, it becomes necessary to use an extremely complex lumped element model or to use the transmission line equations for the distributed line. The concept of a transmission line accounts for the transformation of impedances between circuit points and for the time delay between points that must be considered when the circuit size approaches

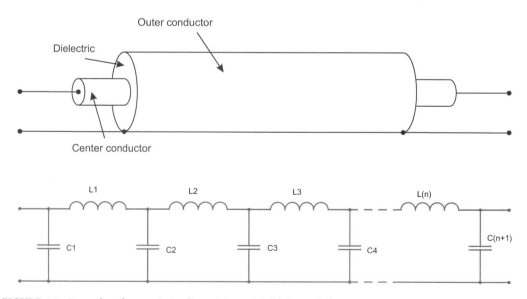

FIGURE 4.3   Examples of transmission lines: (a) coaxial; (b) lumped element.

a significant fraction of a wavelength of the frequency being measured; hence, RF and microwave measurements are primarily based upon transmission line concepts and measurements.

The basic quantities measured in high frequency circuits are power, impedance, port-to-port transfer functions of n-port devices, frequency, and noise [3, 4].

Power

Microwave power cannot be readily detected with equipment used at lower frequencies such as voltmeters and oscilloscopes [5]. The RF and microwave utility of these instruments are limited by circuit parasitics and the resultant limited frequency response. Central to all microwave measurements is the determination of the microwave power available at ports in the measurement circuit. To facilitate measurements, a characteristic impedance or reference resistance is assumed. The instruments used to measure microwave and RF power typically have a 50-ohm input and output impedance at the frequency being measured.

Diode detectors sense the amplitude of a signal. By establishing the input impedance of a diode detector, the power of a signal at a test port can be measured. The diode detector shown in Fig. 4.4 allows current to pass through the diode when the diode is forward biased and prevents current from flowing when the diode is reverse biased. The average of the current flow when forward biased results in a DC output from the lowpass RC filter that is proportional to the amplitude of the input voltage. Note that as the diode junction area must be small to minimize the parasitic junction capacitance that would short the signal

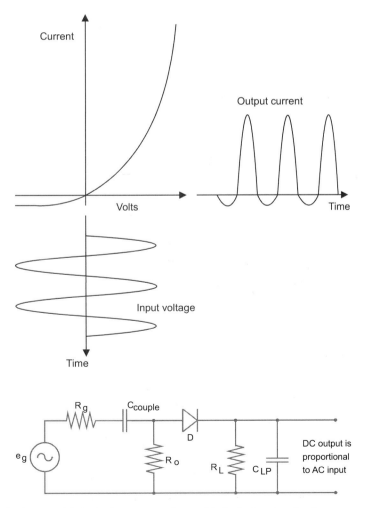

FIGURE 4.4 Diode detector: (a) diode detector waveforms; (b) diode detector circuit.

across the diode, the load resistor must be a relatively large value to minimize the diode current; therefore, the impedance seen looking into the diode detector is established primarily by the resistor placed across the detector input. If the input voltage is less than that where the diode current becomes linearly proportional to the input voltage, the diode is in a predominantly square law region and the voltage out of the detector is proportional to the input power in decibels. This square law range typically extends over a 50-dB range from –60 dBm to –10 dBm in a 50-ohm system. Diodes are used in the linear range up to about 10 dBm. The one significant disadvantage of the diode detector is the temperature sensitivity of the diode. The diode detector response can be very fast, but it cannot easily be used for accurate power measurement.

The most accurate RF and microwave power measuring devices are thermally dependent detectors. These detectors absorb the power and by either measuring the change in the detector temperature or the change in the resistance of the detecting device with a change in temperature, the power absorbed by the detector can be accurately determined.

The primary thermally dependent detectors are the bolometer and the thermistor. They are placed across the transmission media as a matched impedance termination. A bridge as shown in Fig. 4.5(a) can be used to detect a change in the resistance of the bolometer. To increase the detector sensitivity, two units can be placed in parallel for the RF/microwave signal and in series for the change in DC resistance as shown in Fig. 4.5(b). Unfortunately, this circuit can also be used as a thermometer; therefore, an

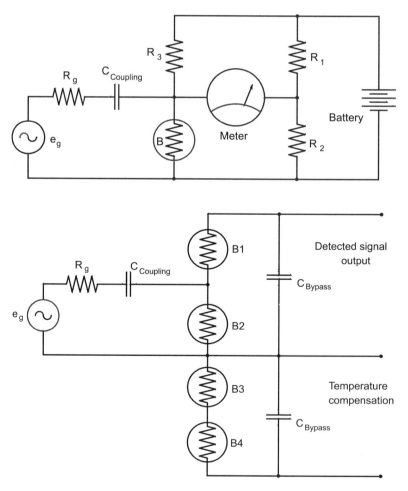

FIGURE 4.5 Thermally dependent detector circuits: (a) bolometer in a bridge circuit; (b) temperature compensated bolometer head; (c) self-balancing bridge circuit.

identical pair of bolometer detectors are normally placed in close thermal proximity but only one of the detectors is used to detect signal power. The other detector is used to detect environmental temperature changes so that the difference in temperature change is due to the signal power absorbed in the upper detector.

To maintain a constant impedance looking into the bolometer elements, a bias current is passed through the elements to increase their temperature above operational ambient. The resistance of the detectors is compared to a fixed resistance in a bridge. The bridge error is used to adjust the bias current in the bolometers. The bias energy that must be removed from the detector to maintain a constant resistance is equal to the amount of signal energy absorbed by the detector; therefore, the meter can be calibrated in power by knowing the amount of bias power applied to the detector. Figure 4.5(c) is a simplified example of a self-balancing bridge circuit.

Impedance

Consider a very simple transmission line, two parallel pieces of wire spaced a uniform distance and in free space, as shown in Fig. 4.6. A DC voltage with a source resistance R_g and series switch is connected to terminal 1 and a resistor R_L is placed across terminal two.

First, let the length of the wires be zero. Close the switch. If the load resistor R_L is equal to the source resistance R_g, the condition necessary for maximum power transfer from a source to a load, then the voltage across the load R_L is $e_g/2$. This is the voltage that will be measured from a signal generator when the output is terminated in its characteristic impedance, commonly called R_o. The signal power from the signal generator, and also the maximum available power from the generator, is then e_g^2/R_g. If R_L is a short circuit the output voltage is zero. If R_L is an open circuit the output voltage is 2 times $e_g/2$ or e_g.

Now let the line have a length, L. When the switch is closed, a traveling wave of voltage moves toward the load resistor at the speed of light, c. At time t, the wave has moved down the line a distance ct. A wave of current travels with the wave of voltage. If the characteristic impedance of this parallel transmission line is Z_o and the load resistance is equal to Z_o, then the current traveling with the voltage wave has a value at any point along the line of the value of the voltage at that point divided by Z_o. For this special case, when the wave reaches the load resistor, all of the energy in the wave is dissipated in the resistor; however, if the resistor is not equal to Z_o there is energy in the wave that must go someplace as it is not dissipated in the load resistor.

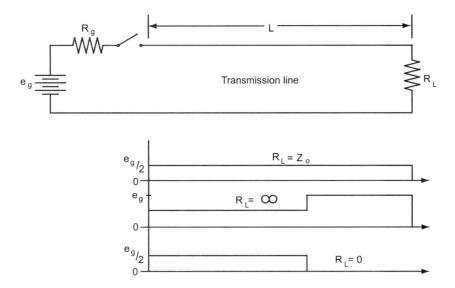

FIGURE 4.6 Switched DC line voltage at time > length/velocity for various impedances at the end of the line.

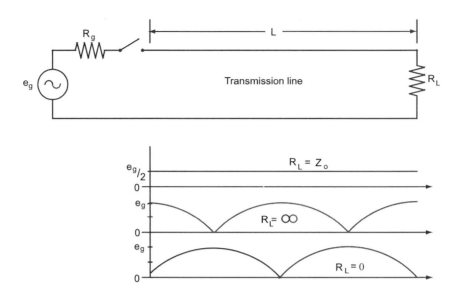

FIGURE 4.7   Waveforms on the line for a sinusoidal source and various impedances at the end of the line.

This mismatch between the characteristic impedance of the line and the terminating load resistor results in a reflected wave that travels back toward the voltage source. If the load resistor is a short circuit, the voltage at the end of the line must equal zero at all times. The only way for this to occur is for the reflected voltage at the end of the wire to be equal to –1 times the incident voltage at that same point. If the load is an open circuit the reflected voltage will be exactly equal to the incident voltage; hence the sum of the incident and reflected voltages will be twice the value of the incident voltage at the end of the line. Note the similarity of these three cases to those of the zero length line.

Now replace the DC voltage source and switch with a sinusoidal voltage source as in Fig. 4.7. The voltages shown are the RMS values of the vector sum of the incident and reflected waves. As the source voltage varies, the instantaneous value of the sinusoidal voltage between the wires travels down the wires. The ratio of the traveling voltage wave to the traveling current wave is the characteristic impedance of the transmission line. If the terminating impedance is equal to the line characteristic impedance, there is no wave reflected back toward the generator; however, if the termination resistance is any value other than Z_o there is a reflected wave. If R_L is a real impedance and greater than Z_o, the reflected wave is 180° out of phase with the incident wave. If R_L is a real impedance and is less than Z_o, the reflected wave is in phase with the incident wave. The amount of variation of R_L from Z_o determines the magnitude of the reflected wave. If the termination is complex, the phase of the reflected wave is neither zero nor 180°.

Assuming the generator impedance R_s is equal to the line characteristic impedance, so that a reflected wave incident on the generator does not cause another reflected wave, sampling the voltage at any point along the transmission line will yield the vector sum of the incident and reflected waves. With a matched impedance $(R_L = Z_o)$ termination the magnitude of the AC voltage along the line is a constant. With a short circuit termination, the voltage magnitude at the load will be zero and, moving back toward the generator, the voltage one-half wavelength from the end of the line will also be zero. With an open circuit there is a voltage maxima at the end of the line and a minima on the line one-quarter wavelength back toward the generator.

The complex reflection coefficient Γ is the ratio of the reflected wave to the incident wave; hence it has a magnitude ρ between 0 and 1 and an angle θ between +180° and –180°. The reflection coefficient as a function of the measured impedance Z_L with respect to the measurement system characteristic impedance Z_o is

$$\Gamma = \frac{Z_L - Z_o}{Z_L + Z_o} = \rho\left(\sin\theta + j\cos\theta\right)$$

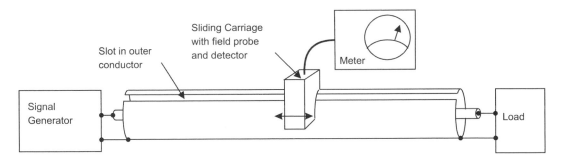

FIGURE 4.8 A slotted line is used to measure the impedance of an unknown load.

Slotted Line

Determination of the relative locations of the minima and maxima along the line, or similarly the determination of the magnitude of waves traveling toward and away from the load resistor, is the basis for the measurement of RF and microwave impedance and the most basic instrument used for making this measurement is the slotted line. The slotted line is a transmission line with a slit in the side that enables a probe to be inserted into the transmission mode electromagnetic field as shown in Fig. 4.8. A diode detector placed within the sliding probe provides a DC voltage that is proportional to the magnitude of the field in the slotted line. As the probe is moved along the line, the minimum and maximum field positions and magnitudes can be determined. The ratio of the maximum field magnitude to the minimum field magnitude is the standing wave ratio (SWR). SWR is normally stated as a scalar quantity and is

$$SWR = \frac{1+\rho}{1-\rho}$$

Before placing an unknown impedance at the measurement terminal of the slotted line, the line is calibrated with a short circuit. This establishes a measurement plane at the short circuit. Any measurement made after calibrating with this reference short is made at the plane of the short circuit. A phase reference is located at the position on the slotted line of a minimum voltage measurement. The distance between two minimum voltage measurement locations is one-half wavelength at the measurement frequency.

If the short circuit is replaced with an open circuit, the minimum voltage locations along the line are shifted by one-quarter wavelength. The difference between the phase of a reflected wave of an open and a short circuit is 180°; hence, the distance between two minimum measurements represents 360° of phase shift in the reflected wave. Note that it is very difficult to use an open circuit for a reference at high frequencies because fringing and radiated fields at the end of the transmission line result in phase and amplitude errors in the reflected wave.

The impedance to be measured now replaces the calibrating short circuit. The new minimum voltage location is found by moving the detector carriage along the slotted line. The distance the minimum voltage measurement moves from the short circuit reference location is ratioed to 180° at a quarter of a wavelength shift (For example, a minimum shift of one-eighth wavelength results from a reflection coefficient phase shift of 90°). This is the phase difference between the forward and reflected waves on the transmission line. Either way the minimum moves from the short circuit calibrated reference point is a shift from 180° back toward 0°. If the shift is toward the load, then the actual phase of the reflection coefficient is −180° plus the shift. If the shift is toward the generator from the reference point, the actual phase of the reflection coefficient is 180° minus the shift.

The best method of visualizing complex impedances as a function of the complex reflection coefficient is the Smith Chart [6, 7, 8]. A simplified Smith Chart is shown in Fig. 4.9. The distance from the center of the chart to the outside of the circle is the reflection coefficient ρ. The minimum value of ρ is 0 and the maximum value is 1. If there is no reflection, the impedance is resistive and equal to the characteristic

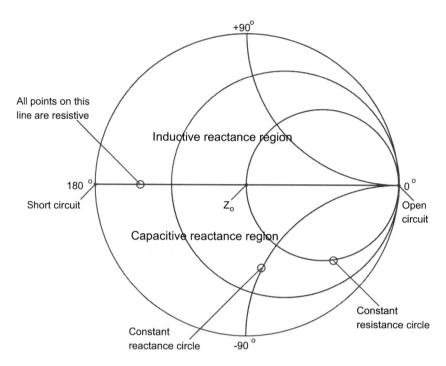

FIGURE 4.9 The Smith Chart is a plot of all nonnegative real impedances.

impedance of the transmission line or slotted line. If the reflected wave is equal to the incident wave, the reflection coefficient is one and the impedance lies on the circumference of the circle. If the angle of the reflection coefficient is zero or 180°, the impedance is real and lies along the central axis. Reflection coefficients with negative angles have capacitive components in the impedance and those with positive angles have inductive components.

Directional Coupler

Slotted lines must be on the order of a wavelength long. Additionally, they do not lend themselves to computer-controlled or automatic measurements. Another device for measuring the forward and reflected waves on a transmission line is the directional coupler [9]. Physically this is a pair of open transmission lines that are placed close enough for the fields generated by a propagating wave in one line to couple to the other line, hence inducing a proportional wave in the second line. The coupler is a four-port device. Referencing Fig. 4.10, a wave propagating to the right in line one couples to line 2 and propagates to the left. A wave propagating in line 1 to the left couples to line 2 and propagates to the right; therefore, the outputs from ports 3 and 4 are proportional to the forward and reverse wave propagating in line 1.

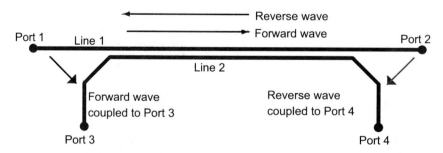

FIGURE 4.10 A directional coupler separates forward and reverse waves on a transmission line.

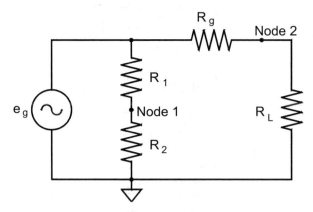

FIGURE 4.11 A resistive bridge can be used to measure the reverse wave on a transmission line.

The primary specifications for a coupler are its useful frequency range, the attenuation of the coupled wave to the coupled ports (coupling), and the attenuation of a signal traveling in the opposite direction to the desired signal at the desired signal's coupled port (directivity). For example, a 10-dB coupler with a 10-dBm signal propagating in the forward direction in line 1 will output a 10-dBm signal at port 3. If the directivity of the coupler is 30 dB there will also be a −40-dBm signal resulting from the forward wave at the reverse wave port, port 4. If the forward wave is properly terminated with the system impedance, there will be no reverse wave on line 1; hence, there will not be an output at port 4 due to a reverse wave.

Note that power must be conserved through the coupler. Therefore, if in the example above, 1.54 dBm is coupled from the forward signal in line 1 to port 3, there will be only a 90-dBm output from port 2. This power must be taken into account in the measurement. The greater the attenuation to the coupled ports, the less the correction will be. Normally 20- or 30-dB couplers are used so the correction is minimal and, in many cases, small enough to be ignored.

By measuring the power from the forward and reverse coupled ports, the magnitude of the reflection coefficient and the SWR can be calculated. Typically the most common indication of the quality of the power match of a device being measured is the attenuation of the reflected wave. This is

$$RL = 10 * \log_{10}\left(\frac{P_{Forward}}{P_{Reverse}}\right)$$

As power is proportional to voltage squared, when the termination resistance is equal on all ports, the return loss can also be expressed as a voltage ratio

$$RL(dB) = 10 * \log_{10}\left(\frac{V_{Forward}^2/R_o}{V_{Reverse}^2/R_o}\right) = -20 * \log_{10}\left(\frac{V_{Reverse}}{V_{Forward}}\right) = -20 * \log_{10}(\rho)$$

Hence the return loss is the magnitude of the reflection coefficient ρ in decibels.

Resistive Bridge

The directional coupler is functionally equivalent to a bridge circuit, the primary difference being that the only losses in the transmission line coupler are from parasitics and can be designed to be very small. Referencing Fig. 4.11, the voltage drop across R_g when R_L equals R_g is $e_g/2$. For this case, the equivalent reflected wave amplitude is zero. By summing circuit voltages it is found that

$$\Gamma = \frac{e_g - e_g/2}{e_g/2} = \frac{V_{Reverse}}{V_{Forward}}$$

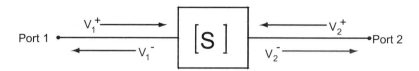

FIGURE 4.12   S-parameters are defined by forward and reverse voltage waves.

By placing a series circuit of two equal resistors across e_g, node 1 has a voltage of $e_g/2$. The voltage between node 1 and node 2 is equal to the reflected wave. Note that this is the standard resistive bridge circuit.

Network Analyzers

General RF and microwave network analyzers (NWA) measure scattering parameters (s-parameters). These measurements use a source with a well-defined impedance equal to the system impedance and all ports of the device under test (DUT) are terminated with the same impedance. The output port being measured is terminated in the test channel of the network analyzer that has an input impedance equal to the system characteristic impedance. Measurement of system parameters with all ports terminated minimizes the problems caused by short-circuit, open-circuit, and test-circuit parasitics that cause considerable difficulty in the measurement of Y- and h-parameters at very high frequencies. S-parameters can be converted to Y- and h-parameters.

Figure 4.12 illustrates a two-port device under test. If the generator is connected to port 1 and a matched load to port 2, the incident wave to the DUT is V_1^+. A wave reflected from the device back to port one is V_1^-. A signal traveling through the DUT and toward port 2 is V_2^-. Any reflection from the load (zero if it is truly a matched load) is V_2^+. The s-parameters are defined in terms of these voltage waves:

$s_{11} = V_1^-/V_1^+ =$ Input terminal reflection coefficient, Γ_1
$s_{21} = V_2^-/V_1^+ =$ Forward gain or loss

By moving the signal generator to port 2 and terminating port 1, the other two port s-parameters are measured:

$s_{12} = V_1^-/V_2^+ =$ Reverse gain or loss
$s_{22} = V_2^-/V_2^+ =$ Output terminal reflection coefficient, Γ_2

The s-matrix is then

$$[S] = \begin{bmatrix} s_{11} & s_{12} \\ s_{21} & s_{22} \end{bmatrix}$$

where

$$\begin{bmatrix} V_1^- \\ V_2^- \end{bmatrix} = [S] \begin{bmatrix} V_1^+ \\ V_2^+ \end{bmatrix}$$

Scalar Analyzer

A scalar network analyzer, Fig. 4.13, with resistor-loaded diode probes or power meters is used to measure scalar return loss and gain. Diode detectors are either used in the square law range as power detectors or logarithmic amplifiers are used in the analyzer to produce nominally a 50 dB dynamic range of measurement. A spectrum analyzer with a tracking test generator can be used as a scalar analyzer with up to 90 dB of dynamic range.

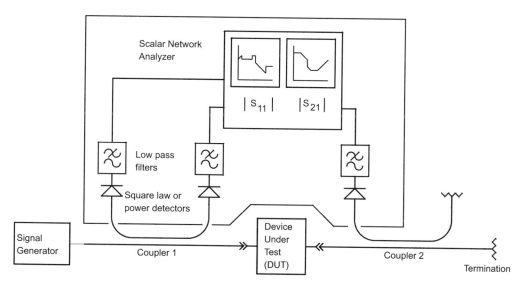

FIGURE 4.13   A scalar network analyzer can measure the magnitude of gain and return loss.

Gains and losses are calculated in scalar analyzers by adding and subtracting relative power levels in decibels. Note that this can only establish the magnitude of the reflection coefficient so that an absolute impedance cannot be measured. To establish the impedance of a device, the phase angle of the reflected wave relative to the incident wave must be known. To measure the phase difference between the forward and reflected wave, a phase meter or vector network analyzer is used.

Vector Heterodyne Analyzer

Accurate direct measurement of the phase angle between two signals at RF and microwave frequencies is difficult; therefore, most vector impedance analyzers downconvert the signals using a common local oscillator. By using a common oscillator the relative phase of the two signals is maintained. The signal is ultimately converted to a frequency where rapid and accurate comparison of the two signals yields their phase difference. In these analyzers the relative amplitude information is maintained so that the amplitude measurements are also made at the low intermediate frequency (IF).

The vector network analyzer (VNA) is a multichannel phase-coherent receiver with a tracking signal source. When interfaced with various power splitters and couplers, the channels can measure forward, reverse, and transmitted waves. As the phase and amplitude information is available on each channel, parameters of the device being measured can be computed. The most common VNA configuration measures the forward and reflected waves to and from a two-port device. From these measurements, the two-port scattering matrix can be computed.

The automatic vector network analyzer performs these operations under the supervision of a computer requiring the operator to input instructions relating to the desired data. The computer performs the routine "housekeeping."

The use of computers also facilitates extensive improvement in measurement accuracy by measuring known high-quality components, calculating nonideal characteristics of the measurement system, and applying corrections derived from these measurements to data from other devices. In other words, the accuracy of a known component can be transferred to the measurement accuracy of an unknown component.

With the measurement frequency accurately known and the phase and amplitude response measured and corrected, the Fourier transform of the frequency domain yields the time domain response. A very useful measurement of this type transforms the s_{11} frequency domain data to a time domain response

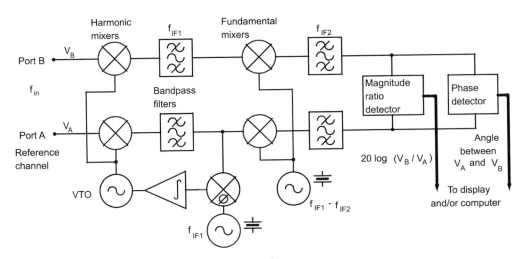

FIGURE 4.14   A vector network analyzer measures complex ratios.

with the same information as time domain reflectometry; that is, deviations from the characteristic impedance can be seen over the length of the measured transmission media.

The simplified block diagram of a typical multichannel VNA is shown in Fig. 4.14. There are two channels fed from the test set. The inputs are converted first to a low intermediate frequency such as 20 MHz and then to 100 kHz before being routed to phase detectors. The first conversion oscillator is followed by a comb generator and the oscillator is phase locked to the mixer output so the unit will frequency track the test source.

Multiple methods of generating the conversion oscillator voltages are used. Low RF frequency analyzer signal generators commonly generate a test signal plus another output that is frequency offset by the desired IF frequency. This can be done with offset synthesizers or by mixing a common oscillator with a stable oscillator at the IF frequency and selecting the desired mixing product using phasing or filtering techniques.

For microwave analyzers, because of the high cost of oscillators and the wide frequency coverage required, a more common method of generating conversion oscillators is to use a low frequency oscillator and a very broadband frequency multiplier. A harmonic of the low frequency conversion oscillator is offset by an oscillator equal to the IF frequency and the conversion oscillator is then phase locked to the reference channel of the NWA. The reference channel signal is normally the forward wave voltage derived from a directional coupler in an impedance measurement.

The outputs of the synchronous detectors supply the raw data to be converted to a format compatible with the computer. Corrections and manipulation of the data to the required output form is then done by the processors.

The test set supplies the first mixer inputs with the sampled signals necessary to make the desired measurement and there are many possible configurations. The most versatile is the two-port scattering matrix test set. This unit enables full two-port measurements to be made without the necessity of changing cable connections to the device. The simplified block diagram of a two-port s-parameter test set is shown in Fig. 4.15. The RF/microwave input is switched between port 1 and port 2 measurements. In each case the RF is split into a reference and test channel. The reference channel is fed directly to a reference channel converter. The test channel feeds the device under test by way of a directional coupler. The coupler output sampling the reflected power is routed to the test channel converter. Sampled components of incident and reflected power to both the input and output of the test device are available for processing.

In a full two-port measurement, multiple error terms can be identified, measured, and then used to translate the accuracy of calibration references to the measured data from the device under test. For

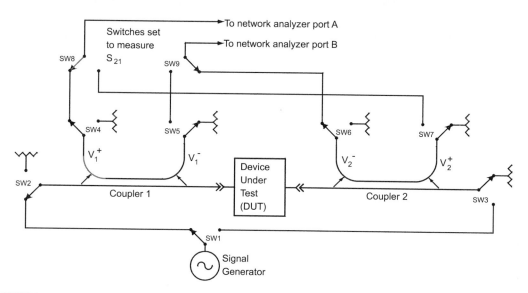

FIGURE 4.15 A two-port s-parameter test set can measure all four s-parameters without moving the DUT.

example, if the load used is not ideal, there will be some reflection back into the DUT. If the source generator impedance is not ideal, any reflections from the input of the DUT back to the generator will result in a further contribution to the incident DUT voltage. The couplers are also nonideal and have phase and amplitude errors.

By measuring the full two-port s-parameters of a set of known references such as opens, shorts, matched loads, known lengths of transmission lines, and through and open circuited paths, a system of equations can be derived that includes the error terms. If 8 error terms are identified, then 8 equations with 8 unknowns can be derived. The error terms can then be solved for and applied to the results of the measurement of an unknown two-port device to correct for measurement system deviations from the ideal.

Vector Six-Port Analyzer

A combination of couplers and power dividers, having 0°, 90°, and 180° differences in their output signals can be used to construct a circuit with multiple outputs where the power from the outputs can be used in a system of n equations with n unknowns. An example of this circuit is shown in Fig. 4.16. In a properly designed circuit, among the solutions to the system of equations will be the magnitudes and relative phase of the forward and reflected wave. The optimum number of ports for such a device is six; hence, a passive six-port device with diode or power detectors on four of the ports can be used as a vector impedance analyzer [10].

The six-port analyzer has limited bandwidth, usually no more than an octave, because the couplers and power dividers [11] have the same limitation in frequency range to maintain the required amplitude and phase characteristics; however, the low cost of the six-port analyzer makes it attractive for narrowband and built-in test applications.

Typically, measurement test set deviations from the ideal are even more prevalent with the six-port analyzer than for the frequency converting VNA; therefore, use of known calibration elements and the application of the resultant error correction terms is very important for the six-port VNA. The derivation of the error terms and their application to measurement correction is virtually the same for the two analyzers.

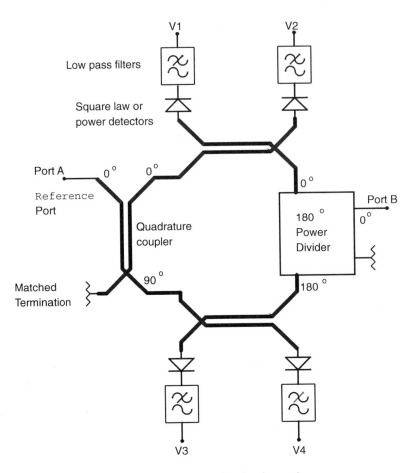

FIGURE 4.16 A six-port network can be used as a narrow band vector analyzer.

References

1. M. Engelson and F. Telewski, *Spectrum Analyzer Theory and Applications*, Artech House, Dedham, MA, 1974.
2. Agilent Technologies, Inc., *Operating Reference Manual for HP 53310A Modulation Domain Analyzer*, Agilent, Santa Clara, CA.
3. S. F. Adam, *Microwave Theory and Applications*, Prentice-Hall, Englewood Cliffs, NJ, 1969.
4. T. S. Laverghetta, *Modern Microwave Measurements and Techniques*, Artech House, Dedham, MA, 1989.
5. J. G. Webster (ed.), *Wiley Encyclopedia of Electrical and Electronics Engineering*, Vol. 13, John Wiley & Sons, New York, 1999, 84–90.
6. P. H. Smith, Transmission line calculator, *Electronics*, 12, 29, January 1939.
7. F. E. Terman, *Electronic and Radio Engineering*, McGraw-Hill, New York, 1955, 100.
8. S. Ramo, J. R. Winnery, and T. Van Duzer, *Fields and Waves in Communications Electronics*, 2nd ed., John Wiley & Sons, New York, 1988, 229–238.
9. G. L. Matthaei, L. Young, and E. M. T. Jones, *Microwave Filters, Impedance-Matching Networks, and Coupling Structures*, Artech House, Dedham, MA, 1980, 775–842.
10. G. F. Engen, A (Historical) Review of the Six-Port Measurement Technique, *IEEE Transactions on Microwave Theory and Technique*, December 1997, 2414–2417.
11. P. A. Rizzi, *Microwave Engineering: Passive Circuits*, Prentice Hall, Englewood Cliffs, NJ, 1988, 367–404.

4.2 Network Analyzer Calibration

Joseph Staudinger

Vector network analyzers (VNA) find very wide application as a primary tool in measuring and characterizing circuits, devices, and components. They are typically applied to measure small signal or linear characteristics of multi-port networks at frequencies ranging from RF to beyond 100 GHz (submillimeter in wavelength). Although current commercial VNA systems can support such measurements at much lower frequencies (a few Hz), higher frequency measurements pose significantly more difficulties in calibrating the instrumentation to yield accurate results with respect to a known or desired electrical reference plane. For example, characterization of many microwave components is difficult since the devices cannot easily be connected directly to VNA-supporting coaxial or waveguide media. Often, the device under test (DUT) is fabricated in a noncoaxial or waveguide medium and thus requires fixturing and additional cabling to enable an electrical connection to the VNA (Fig. 4.17). The point at which the DUT connects with the measurement system is defined as the DUT reference plane. It is generally the point where it is desired that measurements be referenced. However, any measurement includes not only that of the DUT, but contributions from the fixture and cables as well. Note that with increasing frequency, the electrical contribution of the fixture and cables becomes increasingly significant. In addition, practical limitations of the VNA in the form of limited dynamic range, isolation, imperfect source/load match, and other imperfections contribute systematic error to the measurement. To lessen the contribution of systematic error, remove contributions of cabling and fixturing, and therefore enhance measurement accuracy, the VNA must first be calibrated though a process of applying and measuring standards in lieu of the DUT.

Basic measurements consist of applying a stimulus and then determining incident, reflected, and transmitted waves. Ratios of these vector quantities are then computed via post processing yielding network scattering parameters (S-parameters). Most VNAs support measurements on one- and two-port networks, although equipment is commercially available that supports measurements on circuits with more than two ports as well as on differential networks.

VNA Functionality

A highly simplified block diagram illustrating the functionality of a vector network is provided in Fig. 4.18. Generally, a VNA includes an RF switch such that the RF stimulus can be applied to either port 1 or 2, thereby allowing full two-port measurements without necessitating manual disconnection of the DUT and reversing connections. RF couplers attached at the input and output ports allow

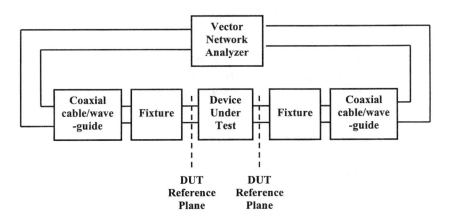

FIGURE 4.17 Typical measurement setup consisting of a device under test embedded in a fixture connected to the vector network analyzer with appropriate cables.

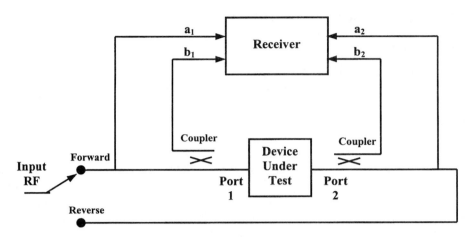

FIGURE 4.18   High simplified VNA block diagram.

measuring reflected voltages. With the RF signal applied in the forward direction (i.e., to port 1), samples of the incident (a_1) and reflected signals at port 1 (b_1) are routed to the receiver. The transmitted signal b_2 reaching port 2 is also directed to the receiver. The receiver functions to downconvert these signals to a lower frequency, which enables digitization and post-processing. Assuming ideal source and load terminations such that a_2 is equal to zero, two scattering parameters can be defined:

$$S_{11} = \frac{b_1}{a_1} \text{ and } S_{21} = \frac{b_2}{a_1}.$$

In reverse operation, the RF signal is directed to port 2 and samples of signals a_2, b_2, and b_1 are directed to the receiver. Assuming ideal source and load terminations such that a_1 is equal to zero, the remaining two scattering parameters are defined:

$$S_{22} = \frac{b_2}{a_2} \text{ and } S_{12} = \frac{b_1}{a_2}.$$

Sources of Measurement Uncertainties

Sources of uncertainty or error in VNA measurements are primarily the result of systematic, random, and drift errors. The latter two effects tend to be unpredictable and therefore cannot be removed from the measurement. They are the results of factors such as system noise, connector repeatability, temperature variations, and physical changes within the VNA. Systematic errors, however, arise from imperfections within the VNA, are repeatable, and can be largely removed through a process of calibration. Of the three, systematic errors are generally the most significant, particularly at RF and microwave frequencies. In calibration, such errors are quantified by measuring characteristics of known devices (standards). Hence, once quantified, systematic errors can be removed from the resulting measurement. The choice of calibration standards is not necessarily unique. Selection of a suitable set of standards is often based on such factors as ease of fabrication in a particular medium, repeatability, and the accuracy to which the characteristics of the standard can be determined.

Modeling VNA Systematic Errors

A mathematical description of systematic errors is accomplished using the concept of error models. The error models are intended to represent the most significant systematic errors of the VNA system up to

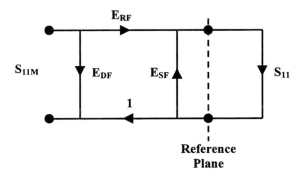

FIGURE 4.19 Typical one-port VNA error model for reflection coefficient measurements.

the reference plane — the electrical plane where standards are connected (Fig. 4.17). Hence, contributions from cables and fixturing in the measurement, up to the reference plane, are accounted for as well.

A flow graph illustrating a typical error model for one-port reflection measurements is depicted in Fig. 4.19. The model consists of three terms, E_{DF}, E_{RF}, and E_{SF}. The term S_{11M} represents the reflection coefficient measured by the receiver within the VNA. The term S_{11} represents the reflection coefficient of the DUT with respect to the reference plane (i.e., the desired quantity).

The three error terms represent various imperfections. Term E_{DF} accounts for directivity in that the measured reflected signal does not consist entirely of reflections caused by the DUT. Limited directivity of the coupler and other signal leakage paths result in other signal components vectorally combining with the DUT reflected signal. Term E_{SF} accounts for source match in that the impedance at the reference plane is not exactly the characteristic impedance (generally 50 ohms). Term E_{RF} describes frequency tracking imperfections between reference and test channels.

A flow graph illustrating a typical error model for two-port measurement, accounting for both reflection and transmission coefficients is depicted in Fig. 4.20. The flow graph consists of both forward (RF signal applied to port 1) and reverse (RF signal applied to port 2) error models. The model consists of twelve terms, six each for forward and reverse paths. Three more error terms are included in addition to those shown in the one-port model, (E_{LF}, E_{TF}, and E_{XF} for the forward path, and similarly E_{LR}, E_{TR}, and E_{XR} for the reverse path). As before, reflection as well as transmission coefficients measured by the receiver within the VNA are denoted with an M subscript (e.g., S_{21M}). The desired two-port S-parameters referenced with respect to port 1/2 reference planes are denoted as S_{11}, S_{21}, S_{12}, and S_{22}. The transmission coefficients are ratios of transmitted and incident signals. Error term E_{LF} accounts for measurement errors resulting from an imperfect load termination. Term E_{TF} describes transmission frequency tracking errors. The term E_{XF} accounts for isolation in that a small component of the transmitted signal reaching the receiver is due to finite isolation where it reaches the receiver without passing through the DUT. The error coefficients for the reverse path are similarly defined.

Calibration

From the above discussion, it is possible to mathematically relate uncorrected scattering parameters measured by the VNA (S_M) to the above-mentioned error terms and the S-Parameters exhibited by the DUT (S). For example, with the VNA modeled for one-port measurements as illustrated in Fig. 4.19, the reflection coefficient of the DUT (S_{11}) is given by:

$$S_{11} = \frac{S_{11M} - E_{DF}}{E_{SF}\left(S_{11M} - E_{DF}\right) + E_{RF}}$$

Similarly, for two-port networks, DUT S-parameters can be mathematically related to the error terms and uncorrected measured S-parameters. DUT parameters S_{11} and S_{21} can be described as functions of

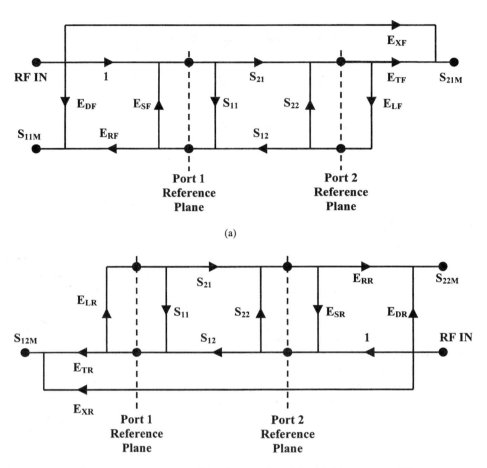

FIGURE 4.20   Typical two-port VNA error model; (a) forward model, and (b) reverse model.

S_{11M}, S_{21M}, S_{12M}, S_{22M} and the six forward error terms. Likewise, S_{12} and S_{22} are functions of the four measured S-parameters and the six reverse error terms. Hence, when each error coefficient is known, the DUT S-parameters can be determined from uncorrected measurement.

Therefore, calibration is essentially the process of determining these error coefficients. This is accomplished by replacing the DUT with a number of standards whose electrical properties are known with respect to the desired reference plane (the reader is referred to [1-5] for additional information). Additionally, since the system is frequency dependent, the process is repeated at each frequency of interest.

Calibration Standards

Determination of the error coefficients requires the use of several standards, although the choice of which standards to use is not necessarily unique. Traditionally, short, open, load, and through (SOLT) standards have been applied, especially in a coaxial medium that facilitates their accurate and repeatable fabrication. Electrical definitions for ideal and lossless SOLT standards (with respect to port 1 and 2 reference planes) are depicted in Fig. 4.21. Obviously, and especially with increasing frequency, it is impossible to fabricate standards such that they are (1) lossless and (2) exhibit the defined reflection and transmission coefficients at these reference planes. Fabrication and physical constants dictate some nonzero length of transmission line must be associated with each (Fig. 4.22). Hence, for completeness, the characteristics of the transmission line must be (1) known, and (2) included in defining the parameters of each standard. Wave propagation is described as

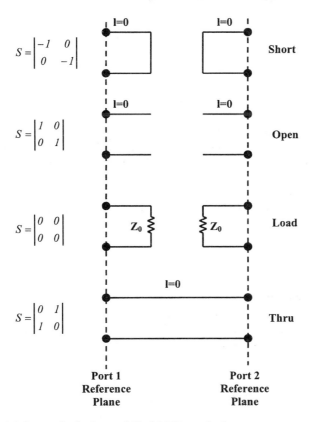

FIGURE 4.21   Electrical definition for lossless and ideal SOLT standards.

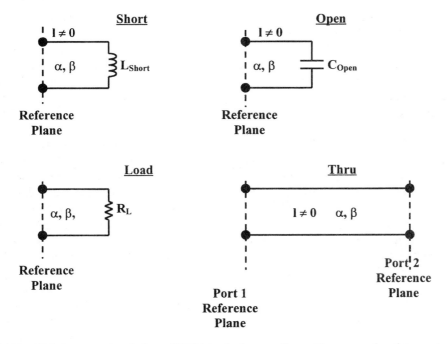

FIGURE 4.22   High frequency descriptions of SOLT standards generally consider nonzero length transmission lines, loss mechanisms, and fringing field effects associated with the open standard.

$$V\left(z\right) = Ae^{-\gamma z} + Be^{\gamma z}$$

where γ is the propagation constant defined as

$$\gamma = \alpha + j\beta$$

Assuming the electrical length of the transmission line associated with the standards is short, losses become small and perhaps α can be neglected without significant degradation in accuracy. Alternatively, commercial VNA manufacturers often describe the transmission line in terms of a delay coefficient with a small resistive loss component. The open standard exhibits further imperfections since the electric field pattern at the open end tends to vary with frequency. The open-end effect is often described in terms of a frequency-dependent fringing capacitance (C_{Open}) expressed in terms of a polynomial expansion taking the form:

$$C_{Open} = C_0 + C_1 F + C_2 F^2 + + C_3 F^3 + \ldots$$

where C_0, C_1, ... are coefficients and F is frequency.

The load termination largely determines forward and reverse directivity error terms (E_{DF} and E_{DR}). Considering the error models in Figs. 4.19 and 4.20, with the load standard applied on port 1, forward directivity error takes the following form:

$$E_{DF} = S_{11M} - \frac{S_{11Load} E_{RF}}{1 - E_{SF} S_{11Load}}$$

where $S_{11\ Load}$ is the actual reflection coefficient of the load standard. Ideally, the load standard should exhibit an impedance of Z_0 (characteristic impedance) and thus a reflection coefficient of zero (i.e., $S_{11\ Load} = 0$) in which case E_{DF} becomes the measured value of S_{11} with the load standard connected to port 1. High quality coaxial-based fixed load standards exhibiting high return loss over broad bandwidths are generally commercially available, especially at RF and microwave frequencies. At higher frequencies and/or where the electrical performance of the fixed load terminations is inadequate, sliding terminations are employed. Sliding terminations use mechanical methods to adjust the electrical length of a transmission line associated with the load standard. Neglecting losses in the transmission line, the above expression forms a circle in the S_{11} measurement plane as the length of the transmission line is varied. The center of the circle defines error term E_{DF} (Fig. 4.23).

Often it is desirable to characterize devices in noncoaxial media. For example, measuring the characteristics of devices and circuits at the wafer level by connecting microwave probes directly to the wafer. Other situations arise where components cannot be directly probed but must be placed in packages with coaxial connectors and it is desirable to calibrate the fixture/VNA at the package/fixture interface. Although fabrication techniques favor SOLT standards in coax, it is difficult to realize them precisely in other media such as microstrip and hence non-SOLT standards are more appropriate. Presently, standards based on one or more transmission lines and reflection elements have become popular for RFICs and MMICs. Fundamentally, they are more suitable for MMICs and RFICs since they rely on fabricating transmission lines (in microstrip, for example), where the impedance of the lines can be precisely determined based on physical dimensions, metalization, and substrate properties. The TRL (thru, reflect, line) series of standards have become popular as well as variations of it such as LRM (line, reflect, match), and LRL (line, reflect, line) to name but a few. In general, TRL utilizes a short length thru (sometime assumed zero length), a highly reflective element, and a nonzero length transmission line. One advantage

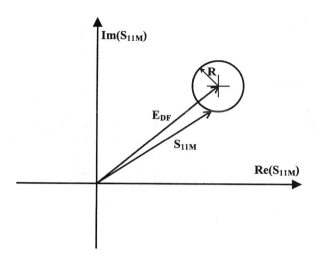

FIGURE 4.23 Characterizing directivity error terms using a sliding load termination.

of this technique is that a complete electrical description of each standard is not necessary. However, each standard is assumed to exhibit certain electrical criteria. For example, the length of the thru generally must be known, or alternatively, the thru may in many cases be fabricated such that its physical dimensions approach zero length at the frequencies of interest and are therefore insignificant. The characteristic impedance of the line standard is particularly important in that it is the major contributor in defining the reference impedance of the measurement. Its length is also important. Lengths approaching either 0° or multiples of 180° (relative to the length of the thru) are problematic and lead to poor calibrations. The phase of the reflection standard is not critical, although its phase generally must be known to within one-quarter of a wavelength.

In the interest of reducing hardware cost, a series of VNAs are commercially available based on a receiver architecture containing three rather than four sampling elements. In four-sampling receiver architecture, independent measurements are made of a_1, b_1, a_2, and b_2. Impedance contributions of the internal switch that routes the RF stimulus to port 1 for forward measurements and to port 2 for reverse measurements can be accounted for during the calibration process. In a three-sampling receiver architecture, independent measurements are made on b_1, b_2 and on a combined a_1 and a_2. This architecture is inherently less accurate than the former in that systematic errors introduced by the internal RF switch are not fully removed via TRL calibration, although mitigating this effect to some extent is possible [5]. However, it should be noted that this architecture provides measurement accuracy that is quite adequate for many applications.

References

1. Staudinger, J., A two-tier method of de-embedding device scattering parameters using novel techniques, Master Thesis, Arizona State University, May 1987.
2. Lane, R., De-Embedding Device Scattering Parameters, *Microwave J.*, Aug. 1984.
3. Fitzpatrick, J., Error Models For Systems Measurement, *Microwave J.*, May 1978.
4. Operating and Programming Manual For the HP8510 Network Analyzer, Hewlett Packard, Inc., Santa Rosa, CA.
5. Metzger, D., Improving TRL* Calibrations of Vector Network Analyzers, *Microwave J.*, May 1995.

4.3   Noise Measurements

Alfy Riddle

Fundamentals of Noise

Statistics

Noise is a random process. There may be nonrandom system disturbances we call noise, but this section will consider noise as a random process. Noise can have many different sources such as thermally generated resistive noise, charge crossing a potential barrier, and generation-recombination (G-R) noise [1]. The different noise sources are described by different statistics, the thermal noise in a resistor is a Gaussian process while the shot noise in a diode is a Poisson process. In the cases considered here, the number of noise "events" will be so large that all noise processes will have essentially Gaussian statistics and so be represented by the probability distribution in Eq. (4.1).

$$p(x) = 1 / (2\pi\sigma^2) \, e^{-x^2/2\sigma^2} \tag{4.1}$$

The statistics of noise are essential for determining the results of passing noise through nonlinearities because the nonlinearity will change the noise distribution [2]. Noise statistics are useful even in linear networks because multiple noise sources will require correlation between the noise sources to find the total noise power. Linear networks will not change the statistics of a noise signal even if the noise spectrum is changed.

Bandwidth

The noise energy available from a hot resistor is given in Eq. (4.2), where $h = 6.62 \times 10^{-34}$ J s, T is in degrees Kelvin, and $k = 1.38 \times 10^{-23}$ J/degree K [1]. N is in joules, or watt-seconds, or W/Hz, which is noise power spectral density. For most of the microwave spectrum $hf \ll kT$ so Eq. (4.2) reduces to Eq. (4.3).

$$N = hf / \left(e^{hf/kT} - 1 \right) \tag{4.2}$$

$$N \cong kT \qquad\qquad f \ll kT/h \tag{4.3}$$

The noise power available from the hot resistor will be the integration of this energy, or spectral density, over the measuring bandwidth as given in Eq. (4.4).

$$P = {}_{f1}\!\int^{f2} N \, df \tag{4.4}$$

As the frequency increases, N reduces so the integration in Eq. (4.4) will be finite even if the frequency range is infinite. Note that for microwave networks using cooled circuits, quantum effects can become important at relatively low frequencies because of the temperature-dependent condition in Eq. (4.3). For a resistor at microwave frequencies and room temperature, N is independent of frequency so the total power available is simply $P = kT(f_2 - f_1)$, or $P = kTB$, as shown in Eq. (4.5), where B is the bandwidth. Figure 4.24 shows a resistor with an available thermal power of kTB, which can be represented either as a series voltage source with $e_n^2 = 4\,kTRB$ or a shunt current source with $i_n^2 = 4\,kTB/R$, where the squared value is taken to be the mean-square value. At times it is tempting to represent e_n as $\sqrt{(4\,kTB)}$, but this is a mistake because e_n is a random variable, not a sinusoid. The process of computing the mean-square value of a noise source is important for establishing any possible correlation with any other noise source in the system [1,16]. Representing a noise source as an equivalent sinusoidal voltage can result in an error due to incorrect accounting of correlation.

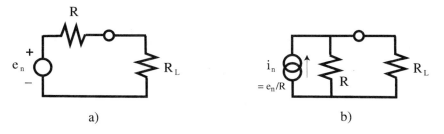

FIGURE 4.24 Equivalent thermal noise sources: (a) voltage and (b) current.

$$P = kT\,B \qquad\qquad f \ll kT/h \qquad\qquad (4.5)$$

When noise passes through a filter we must repeat the integration of Eq. (4.4). Two useful concepts in noise measurement are noise power per hertz and equivalent noise bandwidth. Noise power per hertz is simply the spectral density of the noise, or N in the above equations because it has units of watt-seconds or joules. Spectral densities are also given in V^2/Hz and A^2/Hz. The equivalent noise bandwidth of a noise source can be found by dividing the total power detected by the maximum power detected per hertz, as shown in Eq. (4.6).

$$B_e = P/\mathrm{Max}\{N\} \qquad\qquad (4.6)$$

The noise equivalent bandwidth of a filter is especially useful when measuring noise sources with a spectrum analyzer. The noise equivalent bandwidth of a filter is defined by integrating its power transfer function, $|H(f)|^2$, over all frequency and dividing by the peak of the power transfer function, as shown in Eq. (4.7).

$$B_e = {}_0\!\int^{\infty} |H(f)|^2 \, df \Big/ \mathrm{Max}\{|H(f)|^2\} \qquad\qquad (4.7)$$

Power meters are often used with bandpass filters in noise measurements so that the noise power has a well-defined range. The noise equivalent bandwidth of the filter can be used to convert the noise power back to a power/Hz spectral density that is easier to use in computations and comparisons. As an example, a first-order bandpass filter has a $B_e = \pi/2\ B_{-3}$, where B_{-3} is the −3 dB bandwidth. The noise equivalent bandwidth is greater than the 3 dB bandwidth because of the finite power in the filter skirts.

Detection

The most accurate and traceable measurement of noise power is by comparison with thermal standards [3]. In the everyday lab the second best method for noise measurement is a calibrated power meter preceded by a filter of known noise bandwidth. Because of its convenience, the most common method of noise power measurement is a spectrum analyzer. This most common method is also the most inaccurate because of the inherent inaccuracy of a spectrum analyzer and because of the nonlinear processes used in a spectrum analyzer for power estimation. As mentioned in the section on statistics, nonlinearities change the statistics of a noise source. For example, Gaussian noise run through a linear envelope detector acquires a Rayleigh distribution as shown in Fig. 4.25.

The average of the Rayleigh distribution is not the standard deviation of the Gaussian, so a detector calibrated for sine waves will read about 1 dB high for noise. Spectrum analyzers also use a logarithmic amplifier that further distorts the noise statistics and accounts for another 1.5 dB of error. Many modern spectrum analyzers automatically correct for these nonlinear errors as well as equivalent bandwidth when put in a "Marker Noise" mode [4].

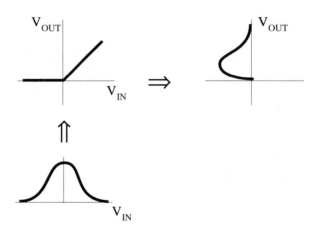

FIGURE 4.25   Nonlinear transformation of Gaussian noise.

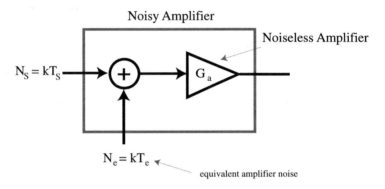

FIGURE 4.26   System view of amplifier noise.

Noise Figure and Y-Factor Method

At high frequencies it is far easier to measure power flow than it is to measure individual voltage and current noise sources. All of a linear device's noise power can be considered as concentrated at its input as shown in Fig. 4.26 [1].

We can lump all of the amplifier noise generators into an equivalent noise temperature with an equivalent input noise power per hertz of $N_a = kT_e$. As shown in Fig. 4.26, the noise power per hertz available from the noise source is $N_S = kT_S$. In system applications the degradation of signal-to-noise ratio (SNR) is a primary concern. We can define a figure of merit for an amplifier, called the noise factor (F), which describes the reduction in SNR of a signal passed through the amplifier shown in Fig. 4.26 [5]. The noise factor for an amplifier is derived in Eq. (4.8).

$$F = SNR_{IN}/SNR_{OUT} = S_{IN}/kT_S \Big/ \Big(G_a S_{IN} \Big/ \big(G_a k(T_S + T_e)\big)\Big) = (T_S + T_e)/T_S = 1 + T_e/T \quad (4.8)$$

Eq. (4.8) is very simple and only contains the amplifier equivalent temperature and the source temperature. F does vary with frequency and so is measured in a narrow bandwidth, or spot. Note that F is not a function of measurement bandwidth. Eq. (4.8) also implies that the network is tuned for maximum available gain, which happens by default if all the components are perfectly matched to 50 ohms and used in a 50-ohm system.

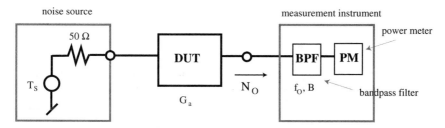

FIGURE 4.27 Test setup for Y-factor method.

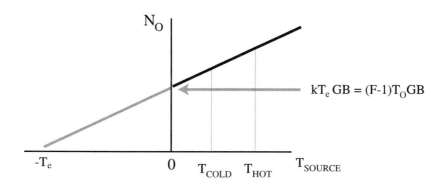

FIGURE 4.28 Output noise power vs. source temperature.

Device noise factor can be measured with the setup shown in Fig. 4.27 [1]. The Y-Factor method takes advantage of the fact that as the source temperature is varied, the device noise output, N_O, varies yet the device noise contribution remains a constant. Figure 4.28 shows that as T_S changes the noise power measured as the power meter changes according to Eq. (4.9).

$$N_O\left(T_S\right)=\left(k\,T_S\,G_a+k\,T_e\,G_a\right)B \qquad (4.9)$$

The value of $N_O(T_S = 0)$ gives the noise power of the device alone. By using two known values of T_S, a cold measurement at $T_S = T_{COLD}$, and a hot measurement at $T_S = T_{HOT}$, the slope of the line in Fig. 4.28 can be derived. Once the slope is known, the intercept at $T_S = 0$ can be found by measuring $N_O(T_{COLD})$ and $N_O(T_{HOT})$. The room temperature, T_O, is also needed to serve as a reference temperature for the device noise factor, F. For a room temperature $F = 1 + T_e/T_O$, we can define $T_e = (F - 1)\,T_O$. The following equations derive the Y-Factor method. Equation (4.10) is the basis for the Y-Factor method.

$$Y=\frac{N_O\left(T_{HOT}\right)}{N_O\left(T_{COLD}\right)}=\frac{\left(k\,T_{HOT}\,G_a+\left(F-1\right)k\,T_O\,G_a\right)B}{\left(k\,T_{COLD}\,G_a+\left(F-1\right)k\,T_O\,G_a\right)B}=\frac{T_{HOT}+\left(F-1\right)T_O}{T_{COLD}+\left(F-1\right)T_O} \qquad (4.10)$$

Solving for F we get Eq. (4.11) which can be solved for T_e as shown in Eq. (4.12).

$$F=1+\frac{T_{HOT}-Y\,T_{COLD}}{\left(Y-1\right)T_O}=1+T_e/T_O \qquad (4.11)$$

$$T_e = \frac{T_{HOT} - Y\, T_{COLD}}{(Y-1)} \tag{4.12}$$

Equation (4.12) can be rearranged to define another useful parameter known as the equivalent noise ratio, or ENR, of a noise source as shown in Eq. (4.13) [1].

$$F = \frac{(T_{HOT}/T_O - 1) + Y(1 - T_{COLD}/T_O)}{(Y-1)} = \frac{ENR + Y(1 - T_{COLD}/T_O)}{(Y-1)} = \left.\frac{ENR}{(Y-1)}\right|_{T_{COLD} = T_O} \tag{4.13}$$

Note that when T_{COLD} is set to the reference temperature for F, which the IEEE gives as $T_O = 290°$ Kelvin, then the device noise factor has a simple relationship to both ENR and Y [1,3].

Practically speaking, the noise factor is usually given in decibels and called the noise figure, $NF = 10 \log F$. While the most accurate noise sources use variable temperature loads, the most convenient variable noise sources use avalanche diodes with calibrated noise power versus bias current [6]. The diode noise sources usually contain an internal pad to reduce the impedance variation between on (hot) and off (cold) states. Also, the diodes come with an ENR versus frequency calibration curve.

Phase Noise and Jitter

Introduction

The noise we have been discussing was broadband noise. When noise is referenced to a carrier frequency it appears to modulate the carrier and so causes amplitude and phase variations in the carrier [7]. Because of the amplitude-limiting mechanism in an oscillator, oscillator phase modulation noise (PM) is much larger than amplitude modulation noise (AM) close enough to the carrier to be within the oscillator loop bandwidth. The phase variations, caused by the noise at different offset, or modulation frequencies create a variance in the zero crossing time of the oscillator. This zero crossing variance in the time domain is called jitter and is critical in digital communication systems. Paradoxically, even though jitter is easily measured in the time domain and often defined in picoseconds it turns out to be better to specify jitter in the frequency domain as demonstrated by the jitter tolerance mask for an OC-48 SONET signal [8]. The jitter plot shown in Fig. 4.29 can be translated into script L versus frequency using the equations in the following section [9].

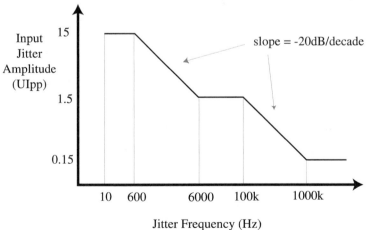

FIGURE 4.29　SONET Category II jitter tolerance mask for OC-48.

Jitter is best specified in the frequency domain because systems are more sensitive to some jitter frequencies than others. Also, jitter attenuators, which are simply narrowband phase-locked loops (PLLs), have well-defined frequency domain transfer functions that can be cascaded with the measured input jitter to derive the output jitter.

$$V_O(t) = V_C \left\{ 1 + m(t) \right\} \cos\left[\omega_C t + \beta(t) \right] \tag{4.14}$$

Mathematical Basics

Consider the time domain voltage given in Eq. (4.14). This signal contains both amplitude modulation, $m(t)$, and phase modulation, $\beta(t)$ [10]. If we let $m(t) = m_1(t) \cos(\omega_m t)$, $\beta(t) = \beta_1(t) \sin(\omega_m t)$, and we define $|\beta_1(t)| \ll 1$, then Eq. (4.14) can be expanded into AM and PM sidebands as shown in Eq. (4.15).

$$
\begin{aligned}
V_O(t) &\cong V_C \cos\left[\omega_C t \right] \\
&+ V_C m_1(t)/2 \left\{ \cos\left[\omega_C t + \omega_m t \right] + \cos\left[\omega_C t - \omega_m t \right] \right\} \\
&+ V_C \beta_1(t)/2 \left\{ \cos\left[\omega_C t + \omega_m t \right] - \cos\left[\omega_C t - \omega_m t \right] \right\}
\end{aligned}
\tag{4.15}
$$

We can let $m_1(t)$ and $\beta_1(t)$ be fixed amplitudes as when sinusoidal test signals are used to characterize an oscillator, or we can let $m_1(t)$ and $\beta_1(t)$ be slowly varying, with respect to ω_m, noise signals. The latter case gives us the narrowband Gaussian noise approximation, which can represent an oscillator spectrum when the noise signals are summed over all modulation frequencies, ω_m [7].

Several notes should be made here. First, as $\beta_1(t)$ becomes large the single sidebands of Eq. (4.15) expand into a Bessel series that ultimately generates a flat-topped spectrum close into the average carrier frequency. This flat-topped spectrum is essentially the FM spectrum created by the large phase excursions that result from the $1/f^3$ increase in phase noise at low modulating frequencies. Second, if $|m_1(t)| = |\beta_1(t)|$, and they are fully correlated, then by altering the phases between $m_1(t)$ and $\beta_1(t)$ we can cancel the upper or lower sideband at will. This second point also shows that a single sideband contains equal amounts of AM and PM, which is useful for testing and calibration purposes [11].

Oscillator noise analysis uses several standard terms such as AM spectral density, PM spectral density, FM spectral density, script L, and jitter [10,12]. These terms are defined in Eqs. (4.16) through (4.20). The AM spectral density, or $S_{AM}(f_m)$, shown in Eq. (4.16) is derived by computing the power spectrum of $m(t)$, given in Eq. (4.14), with a 1-hertz-wide filter. S_{AM} is called a spectral density because it is on a 1-hertz basis. Similarly, the PM spectral density, or $S_\phi(f_m)$ in radians²/Hz, is shown in Eq. (4.17) and is derived by computing the power spectrum of $\beta(t)$ in Eq. (4.14). The FM spectral density, or $S_{FM}(f_m)$ in Hz²/Hz, is typically derived by using a frequency discriminator to measure the frequency deviations in a signal. Because frequency is simply the rate of change of phase, FM spectral density can be derived from PM spectral density as shown in Eq. (4.18). Script L is a measured quantity usually given in dBc/Hz and best described by Fig. 4.30. It is important to remember that the definition of script L requires the sidebands to be due to phase noise. Because script L is defined as a measure of phase noise it can be related to the PM spectral density as shown in Eq. (4.19). Two complications arise in using script L. First, most spectrum analyzers do not determine if the sidebands are only due to phase noise. Second, the constant relating script L to S_ϕ is 2 if the sidebands are correlated and $\sqrt{2}$ if the sidebands are uncorrelated and the spectrum analyzer does not help in telling these two cases apart. Jitter is simply the rms value of the variation in zero crossing times of a signal compared with a reference of the same average frequency. Of course, the jitter of a signal can be derived by accumulating the phase noise as shown in Eq. (4.20). In Eqs. (4.16) through (4.20) $\Im\{x\}$ denotes the Fourier transform of x [7]. Most of these terms can also be defined from the Fourier transform of the autocorrelation of $m(t)$ or $\beta(t)$. The spectral densities are typically given in dB using a 1-Hz measurement bandwidth, abbreviated as dB/Hz.

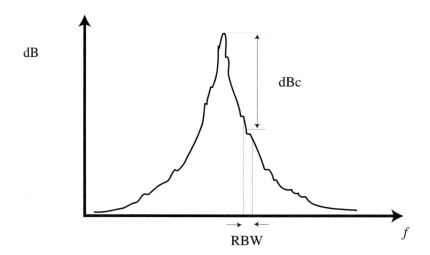

FIGURE 4.30　Typical measured spectrum on a spectrum analyzer.

$$S_{AM}(f_m) = \Im\{m(t)\}\,\Im\{m*(t)\} = m_1^2/2\,\delta(f_m - f_a)\Big|_{m(t)=m_1\cos(\omega_a t)} \tag{4.16}$$

$$S_\phi(f_m) = \Im\{\beta(t)\}\,\Im\{\beta*(t)\} = \beta_1^2/2\,\delta(f_m - f_a)\Big|_{\beta(t)=\beta_1\cos(\omega_a t)} \tag{4.17}$$

$$S_{FM}(f_m) = f_m^2\,S_\phi(f_m) \qquad\qquad \text{because } \omega_m = d\phi/dt \tag{4.18}$$

$$\text{script } L(f_m) = P_{SSB}(f_m)/\text{Hz}/P_C = S_\phi(f_m)/2\Big|_{\text{when phase noise has correlated sidebands}} \tag{4.19}$$

$$\text{jitter} = \sqrt{\left({}_0\!\int^\infty S_\phi(f_m)\,df_m\right)} = \beta_1/\sqrt{2}\Big|_{\beta(t)=\beta_1\cos(\omega_m t)} \tag{4.20}$$

In the above equations f_m indicates the offset frequency from the carrier. In Eq. (4.19) P_{SSB} is defined as phase noise, but often is just the noise measured by a spectrum analyzer close to the carrier frequency, and P_C is the total oscillator power. The jitter given in Eq. (4.20) is the total jitter that results from a time domain measurement. Jitter as a function of frequency, f_m, is just the square root of $S_\phi(f_m)$. Jitter as a function of frequency can be translated to various other formats, such as degrees, radians, seconds, and unit intervals (UIs), using Eq. (4.21) [9].

$$\text{jitter} = \begin{cases} \beta_{RMS} \text{ in radians} \\ UI = \beta_{RMS}/2\pi \text{ in unit intervals} \\ UI/f_C = \beta_{RMS}/\omega_C \text{ in seconds} \\ 360\,UI = 360\beta_{RMS}/2\pi \text{ in degrees} \end{cases} \tag{4.21}$$

For most free-running oscillators the $1/f^3$ region of the phase noise dominates the jitter so integrating the $1/f^3$ slope gives $UI \approx f_a/2\; 10^{\text{scriptL}(f_a)/10}$ where f_a is any frequency on the $1/f^3$ slope and script L is in dBc/Hz. For PLL-based sources with large noise pedestals, a complete integration should be done.

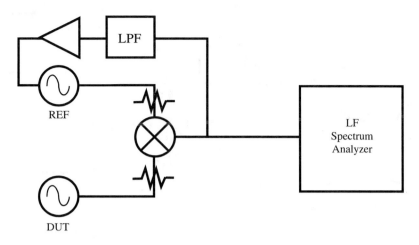

FIGURE 4.31   PLL phase noise measurement.

Phase Noise Measurements

Phase noise is typically measured in one of three ways: spectrum analyzer, PLL, or transmission line discriminator [1,10,13-15]. The spectrum analyzer is the easiest method of measuring script L(f_m) for any oscillator noisier than the spectrum analyzer reference source. Figure 4.30 shows a typical source spectrum. Care must be taken to make the resolution bandwidth, RBW, narrow enough to not cover a significant slope of the measured noise [1]. The spectrum analyzer cannot distinguish between phase and amplitude noise, so reporting the results as script L only holds where $S_\phi > S_{AM}$, which usually means within the $1/f^3$ region of the source. Spectrum analyzer measurements can be very tedious when the oscillator is noisy enough to wander significantly in frequency.

PLL-based phase noise measurement is used in most commercial systems [14]. Figure 4.31 shows a PLL-based phase noise test set. The reference oscillator in Fig. 4.31 is phase locked to the device under test (DUT) through a low pass filter (LPF) with a cutoff frequency well below the lowest desired measurement frequency. This allows the reference oscillator to track the DUT and downconvert the phase noise sidebands without tracking the noise as well. The low frequency spectrum analyzer measures the noise sidebands and arrives at a phase noise spectral density by factoring in the mixer loss or using a calibration tone [11]. A PLL system requires the reference source to be at least as quiet as the DUT. Another DUT can be used as a reference with the resulting noise sidebands increasing by 3 dB, but usually the reference is much quieter than the DUT so fewer corrections have to be made.

A transmission line frequency discriminator can provide accurate and high-resolution phase noise measurements without the need for a reference oscillator [13]. The discriminator resolution is proportional to the delay line delay, τ. The phase shifter is adjusted so that the mixer signals are in quadrature, which means the mixer DC output voltage is set to the internal offset voltage (approximately zero). Transmission line discriminators can be calibrated with an offset source of known amplitude, as discussed previously, or with a source of known modulation sensitivity [11] (see Fig. 4.32). The disadvantages of a transmission line discriminator are that high source output levels are required to drive the system (typically greater than 13 dBm), and the system must be retuned as the DUT drifts. Also, it is important to remember that the discriminator detects FM noise which is related to phase noise as given in Eq. (4.18) and shown in Fig. 4.33.

Summary

Accurate noise measurement and analysis must recognize that noise is a random process. While nonlinear devices will affect the noise statistics, linear networks will not change the noise statistics. Noise statistics are also important for analyzing multiple noise sources because the correlation between the noise sources

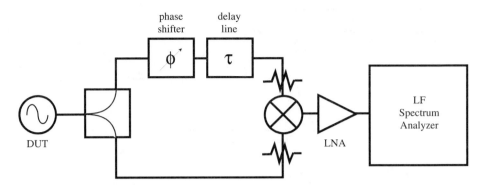

FIGURE 4.32   Transmission line discriminator.

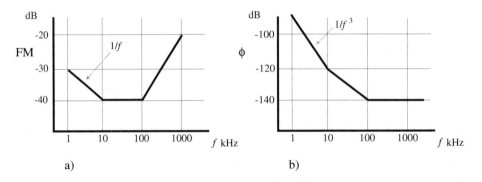

FIGURE 4.33   (a) FM and (b) phase noise spectral densities for the same device.

must be considered. At very high frequencies it is easier to work with noise power flow than individual noise voltage and current sources, so methods such as the Y-Factor technique have been developed for amplifier noise figure measurement. Measuring oscillator noise mostly involves the phase variations of a source. These phase variations can be represented in the frequency domain as script L, or in the time domain as jitter. Several techniques of measuring source phase noise have been developed which trade off accuracy for cost and simplicity.

References

1. Ambrozy, A., *Electronic Noise*, McGraw-Hill, New York, 1982.
2. Papoulis, A., *Probability, Random Variables and Stochastic Processes*, McGraw-Hill, New York, 1965.
3. Haus, H.A., IRE Standards on Methods of Measuring Noise in Linear Twoports, *Proc. IRE*, 60–68, Jan. 1960.
4. Staff, *HP 8560 E-Series Spectrum Analyzer User's Guide*, Hewlett-Packard, Dec. 1997.
5. Friis, H.T., Noise Figures of Radio Receivers, *Proc. IRE*, 419–422, July 1944.
6. Pastori, W.E., A Review of Noise Figure Instrumentation, *Microwave J.*, 50–60, Apr. 1983.
7. Carlson, B.A., *Communication Systems*, McGraw-Hill, New York, 1975.
8. ANSI, Telecommunications-Synchronous Optical Network (SONET)- Jitter at Network Interfaces, *ANSI T1.105.03-1994*, 1994.
9. Adler, J.V., Clock Source Jitter: A Clear Understanding Aids Clock Source Selection, *EDN*, 79–86, Feb. 18, 1999.
10. Ondria, J.G., A Microwave System for Measurement of AM and FM Noise Spectra, *IEEE Trans. MTT*, 767–781, Sept. 1968.

11. Buck, J.R., and Healey, D.J. III, Calibration of Short-Term Frequency Stability Measuring Apparatus, *Proc. IEEE*, 305–306, Feb. 1966.
12. Blair, B.E., *Time and Frequency: Theory and Fundamentals*, U.S. Dept. Commerce, NBS Monograph 140, 1974.
13. Schielbold, C., Theory and Design of the Delay Line Discriminator for Phase Noise Measurement, *Microwave J.,* 103–120, Dec. 1983.
14. Harrison, D.M., Howes, M.J., and Pollard, R.D., The Evaluation of Phase Noise in Low Noise Oscillators, *IEEE MTT-S Digest*, 521–524, 1987.
15. Staff, *Noise Measurements Using the Spectrum Analyzer, Part One: Random Noise*, Tektronix, Beaverton, 1975.
16. Haus, H.A. and Adler, R.B., *Circuit Theory of Linear Noisy Networks*, MIT Press, Cambridge, MA, 1959.

4.4 Nonlinear Microwave Measurement and Characterization

J. Stevenson Kenney

While powerful methods have been developed to analyze complex linear circuits, it is unfortunate that almost all physical systems exhibit some form of nonlinear behavior. Often the nonlinear behavior of a microwave circuit is detrimental to the signals that pass through it. Such is the case with distortion within a microwave power amplifier. In some cases nonlinearities may be exploited to realize useful circuit functions, such as frequency translation or detection. In either case, methods have been devised to characterize and measure nonlinear effects on various signals. These effects are treated in this chapter and include:

- Harmonic Distortion
- Gain Compression
- Intermodulation Distortion
- Phase Distortion
- Adjacent Channel Interference
- Error Vector Magnitude

Many of the above characterizations are different manifestations of nonlinear behavior for different types of signals. For instance, both analog and digital communication systems are affected by *intermodulation distortion*. However, these effects are usually measured in different ways. Nevertheless, some standard measurements are used as figures of merit for comparing the performance to different circuits. These include:

- Output Power at 1 *dB* Gain Compression
- Third Order Intercept Point
- Spurious Free Dynamic Range
- Noise Power Ratio
- Spectral Mask Measurements

This chapter treats the characterization and measurement of nonlinearities in microwave circuits. The concentration will be on standard techniques for analog and digital communication circuits. For more advanced techniques, the reader is advised to consult the references at the end of this section.

Mathematical Characterization of Nonlinear Circuits

To analyze the effects of nonlinearities in microwave circuits, one must be able to describe the input-output relationships of signals that pass through them. Nonlinear circuits are generally characterized by input-output relationships called *transfer characteristics*. In general, any memoryless circuit described by transfer characteristics that does not satisfy the following definition of a *linear* memoryless circuit is said to be *nonlinear*.

$$v_{out}(t) = A v_{in}(t),$$

(4.22)

where v_{in} and v_{out} are the input and output time-domain waveforms and A is a constant independent of time. Thus, one form of a nonlinear circuit has a transfer characteristic of the form

$$v_{out}(t) = g\left[v_{in}(t)\right].$$

(4.23)

The form of $g(v)$ will determine all measurable distortion characteristics of a nonlinear circuit. Special cases of nonlinear transfer characteristics include:

- Time Invariant: g does not depend on t
- Memoryless: g is evaluated at time t using only values of v_{in} at time t

Nonlinear Memoryless Circuits

If a transfer characteristic includes no integrals, differentials, or finite time differences, then the instantaneous value at a time t depends only on the input values at time t. Such a transfer characteristic is said to be *memoryless*, and may be expressed in the form of a power series

$$g(v) = g_0 + g_1 v + g_2 v^2 + g_3 v^3 + \dots$$

(4.24a)

where g_n are real-valued, time-invariant coefficients. Frequency domain analysis of the output signal $v_{out}(t)$ where $g(v)$ is expressed by Eq. (4.24a) yields a Fourier series, whereby the harmonic components are governed by the coefficients G_n. If $v_{in}(t)$ is a sinusoidal function at frequency f_c with amplitude V_{in}, then the output signal is a harmonic series of the form

$$v_{out}(t) = G_0 + G_1 V_{in} \cos(2\pi f_c t) + G_2 V_{in}^2 \cos(4\pi f_c t) + G_3 V_{in}^3 \cos(6\pi f_c t) + \dots$$

(4.24b)

The coefficients, G_n are functions of the coefficients g_n, and are all real. The extent that the coefficients g_n are nonzero is called the *order* of the nonlinearity. Thus, from Eq. (4.24b), it is seen that an nth order system will produce harmonics of nth order of amplitude $G_n V_{in}^n$.

Nonlinear Circuits with Memory

As described in Eq. (4.24a), $g(v)$ is said to be *memoryless* because the output signal at a time t depends only on the input signal at time t. If the output depends on the input at times different from time t, the nonlinearity is said to have *memory*. A nonlinear function with a finite memory (i.e., a *finite impulse response*) may be described as

$$v_{out}(t) = g\left[v_{in}(t), v_{in}(t - \tau_1), v_{in}(t - \tau_2), \dots v_{in}(t - \tau_n)\right].$$

(4.25)

The largest time delay, τ_n, determines the length of the memory of the circuit. *Infinite impulse response* nonlinear systems may be represented as functions of integrals and differentials of the input signal

$$v_{out}(t) = g\left[v_{in}(t), \int_{-\infty}^{t} v_{in}(\tau) d\tau, \frac{\partial^n v_{in}}{\partial t^n}\right].$$

(4.26)

The most general characterization of a nonlinear system is the *Volterra Series*.[1] Consider a linear circuit that is stimulated by an input signal $v_{in}(t)$. The output signal $v_{out}(t)$ is then given by the convolution with the input signal $v_{in}(t)$ and the *impulse response* $h(t)$. Unless the impulse response takes the form of the *delta function* $\delta(t)$, the output $v_{out}(t)$ depends on values of the input $v_{in}(t)$ at times other than t, i.e., the circuit is said to have memory.

$$v_{out}(t) = \int_{-\infty}^{\infty} v_{in}(\tau) h(t-\tau) d\tau. \tag{4.27a}$$

Equivalently, in the frequency domain,

$$V_{out}(f) = V_{in}(f) H(f). \tag{4.27b}$$

In the most general case, a nonlinear circuit with reactive elements can be described using a Volterra series, which is said to be a power series with *memory*.

$$v_{out}(t) = g_0 + \int_{-\infty}^{\infty} v_{in}(\tau) g_1(t-\tau) d\tau + \int_{-\infty}^{\infty}\int_{-\infty}^{\infty} v_{in}(t-\tau_1) v_{in}(t-\tau_2) g_2(\tau_1,\tau_2) d\tau_1 d\tau_2 + \ldots \tag{4.28a}$$

An equivalent representation is obtained by taking the n-fold Fourier transform of Eq. (4.28a)

$$V_{out}(f_1,f_2,\ldots) = G_0 \delta(f_1) + G_1(f_1) V_{in}(f_1) + G_2(f_1,f_2) + \ldots \tag{4.28b}$$

Notice that the Volterra series is applicable to nonlinear effects on signals with discrete spectra (i.e., a signal consisting of a sum of sinusoids). For instance, the DC component of the output signal is given by $g_0 = G_0$, while the fundamental component is given by $G_1(f_1) V_{in}(f_1)$, where G_1 and V_{in} are the Fourier transforms of the impulse response g_1 and v_{in}, respectively, evaluated at frequency f_1. The higher order terms in the Volterra series represent the harmonic responses and intermodulation response of the circuit.

Fortunately, extraction of high order Volterra series representations of nonlinear microwave circuits is rarely required to gain useful information on the deleterious and/or useful effects of distortion on common signals. Such simplifications often involve considering the circuit to be memoryless, as in Eq. (4.24a,b), or having finite order, or having integral representations, as in Eq. (4.26).

Harmonic Distortion

A fundamental result of the distortion of nonlinear circuits is that they generate frequency components in the output signal that are not present in the input signal. For sinusoidal inputs, the salient characteristic is *harmonic distortion*, whereby signal outputs consist of integer multiples of the input frequency.

Harmonic Generation in Nonlinear Circuits

As far as microwave circuits are concerned, the major characteristic of a nonlinear circuit is that the frequency components of the output signal differ from those of the input signal. This is readily seen by examining the output of a sinusoidal input from Eq. (4.29).

$$v_{out}(t) = g_0 + g_1 A \cos(2\pi f_c t) + g_2 A^2 \cos^2(2\pi f_c t) + g_3 A^3 \cos^3(2\pi f_c t) + \cdots$$

$$= g_0 + \frac{g_2 A^2}{2} + \left(g_1 A + \frac{3 g_3 A^3}{4} \right) \cos(2\pi f_c t) + \frac{a g_2 A^2}{2} \cos(4\pi f_c t) + \frac{g_3 A^3}{4} \cos(6\pi f_c t) + \cdots \tag{4.29}$$

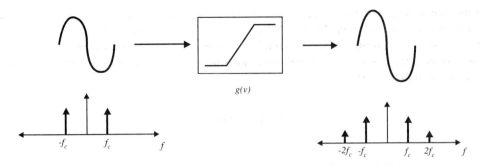

FIGURE 4.34 Effects of a nonlinear transfer characteristic on a sinusoidal input: harmonic distortion.

TABLE 4.1 Effect of Nonlinearities on Carrier Term by Term

Term	Amplitude	Qualitative Effect
DC	$g_0 + g_2A^2/2$	Small offset added due to RF detection
Fundamental	$20\log(g_1A + 3\ g_3A^3/4)$	Amplitude changed due to compression
2nd Harmonic	$40\log(g_2A^2/2)$	2:1 slope on P_{in}/P_{out} curve
3rd Harmonic	$60\log(g_3A^3/4)$	3:1 slope on P_{in}/P_{out} curve

It is readily seen that, along with the *fundamental* component at a frequency of f_c, there exists a DC component, and *harmonic* components at integer multiples of f_c. The output signal is said to have acquired *harmonic distortion* as a result of the nonlinear transfer characteristic. This is illustrated in Fig. 4.34. The function represented by $g(v)$ is that of an ideal limiting amplifier. The net effect of the terms are summarized in Table 4.1.

Measurement of Harmonic Distortion

While instruments are available at low frequencies to measure the *total harmonic distortion* (*THD*), the level of each harmonic is generally measured individually using a spectrum analyzer. Such a setup is shown in Fig. 4.35.

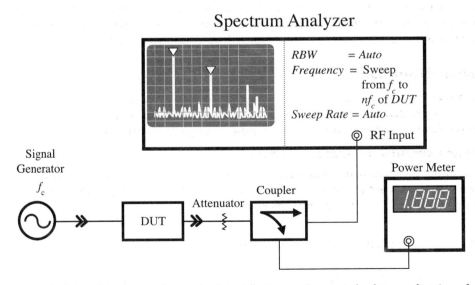

FIGURE 4.35 Setup used to measure harmonic distortion. Because harmonic levels are a function of output amplitude, a power meter is needed to accurately characterize the harmonic distortion properties.

Harmonic levels are usually measured in a relative manner by placing a marker on the fundamental signal and a delta marker at the nth harmonic frequency. When measured in this mode, the harmonic level is expressed in *dBc*, which designates *dB* relative to carrier (i.e., the fundamental frequency) level. While it is convenient to set the spectrum analyzer sweep to include all harmonics of interest, it may be necessary to center a narrow span at the harmonic frequency in order to reduce the *noise floor* on the spectrum analyzer. An attenuator may be needed to protect the spectrum analyzer from overload. Note that the power level present at the spectrum analyzer input includes all harmonics, not just the ones displayed on the screen. Finally, it is important to note that spectrum analyzers have their own nonlinear characteristics that depend on the level input to the instrument. It is sometimes difficult to ascertain whether measured harmonic distortion is being generated within the device or with the test instrument. One method to do this is to use a step attenuator at the output of the device and step up and down. If distortion is being generated with the spectrum analyzer, the harmonic levels will change with different attenuator settings.

Gain Compression and Phase Distortion

A major result of changing impedances in microwave circuits is signal gain and phase shift that depend on input amplitude level. A change in signal gain between input and output may result from signal clipping due to device current saturation or cutoff. Insertion phase may change because of nonlinear resistances in combination with a reactance. Though there are exceptions, signal gain generally decreases with increasing amplitude or power level. For this reason, the *gain compression* characteristics of microwave components are often characterized. Phase distortion may change either way, so it is often described as *phase deviation* as a function of amplitude or power level.

Gain Compression

Referring back to Eq. (4.29), it is seen that, in addition to harmonic distortion, the level fundamental signal has been modified beyond that dictated by the linear term, g_1. This effect is described as *gain compression* in that the gain of the circuit becomes a function of the input amplitude A. Figure 4.36 illustrates this result. For small values of A, the g_1 term will dominate, giving a 1:1 slope when the output power is plotted against the input power on a *log* (i.e., *dB*) scale. Note that the power level of the nth harmonic plotted in like fashion will have an n:1 slope.

Gain compression is normally measured on a *bandpass* nonlinear circuit.[2] Such a circuit is illustrated in Fig. 4.37. It is interesting to note that an ideal limiting amplifier described by Eq. (4.30) when heavily

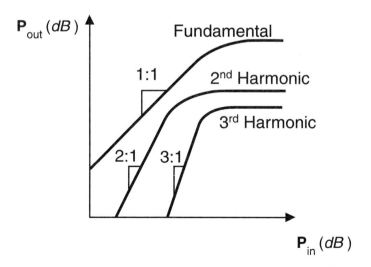

FIGURE 4.36   Output power vs. input power for a nonlinear circuit.

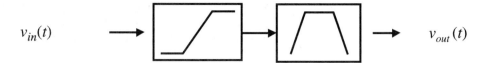

FIGURE 4.37   Bandpass nonlinear circuit.

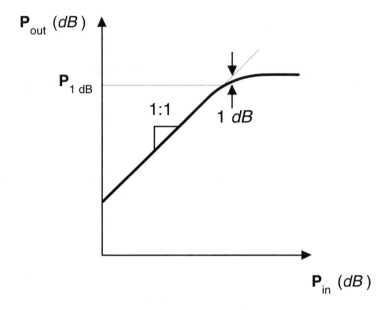

FIGURE 4.38   Gain compression of a bandpass nonlinear circuit. A figure of merit P_{1dB} is the output power at which the gain has been reduced by 1 dB.

overdriven at the input will eventually produce a square wave at the output, which is filtered by the bandpass filter. Note that the amplitude of the fundamental component of a square wave is at a level of $4/\pi$ times, or 2.1 dB greater than the amplitude of the square wave set by the clipping level.

$$v_{out}(t) = \begin{cases} g_1 v_{in}(t) & v_{out} < v_{\lim} \\ v_{\lim} & otherwise \end{cases} \tag{4.30}$$

For a general third-order nonlinear transfer characteristic driven by a sinusoidal input, the bandpass output is given by

$$v_{out}(t) = \left(g_1 A + \frac{3 g_3 A^3}{4} \right) \cos\left(2\pi f_c t \right) \tag{4.31}$$

A bandpass nonlinear circuit may be characterized by the power output at 1 dB gain compression, P_{1dB} as illustrated in Fig. 4.38.

Phase Distortion

Nonlinear circuits may also contain reactive elements that give rise to *memory* effects. It is usually unnecessary to extract the entire Volterra representation of a nonlinear circuit with reactive elements if a few assumptions can be made. For bandpass nonlinear circuits with memory effects of time duration of the order of the period of the carrier waveform, a simple model is often used to describe the phase deviation versus amplitude:

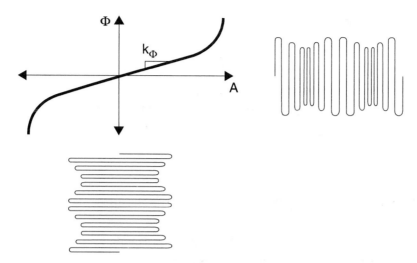

FIGURE 4.39 Effect of *AM-PM* distortion on a modulated signal. Input signal has *AM* component only. Output signal has interrelated *AM* and *FM* components due to the *AM-PM* distortion of the circuit.

$$v_{out}(t) = A(t)\cos\left\{2\pi f_c t + \Phi\left[A(t)\right]\right\}. \tag{4.32}$$

Equation (4.32) represents the *AM-PM* distortion caused by short-term memory effects (i.e., small capacitances and inductances in microwave circuits). The *effects* of *AM-PM* on an amplitude-modulated signal is illustrated in Fig. 4.39.

For the case of input signals with small deviations of amplitude ΔA, the phase deviation may be considered linear, with a proportionality constant k_ϕ as seen in Fig. 4.39. For a sinusoidally modulated input signal, an approximation for small modulation index FM signals may be utilized. One obtains the following expression for the output signal:

$$v_{out}(t) = \left[A + \Delta A \sin\left(2\pi f_m t\right)\right]\cos\left\{2\pi f_c t + k_\phi A + k_\phi \Delta A \sin\left(2\pi f_m t\right)\right\}$$

$$\approx A\cos\left(2\pi f_c t + k_\phi A\right)\sum_{n=0}^{\infty} J_n\left(k_\phi \Delta A\right)\cos\left(2n\pi f_m t\right) \tag{4.33}$$

where J_n is the nth order Bessel function of the first kind.[3] Thus, like amplitude distortion, *AM-PM* distortion creates sidebands at the harmonics of the modulating signal. Unlike amplitude distortion, these sidebands are not limited to the first sideband. Thus, *AM-PM* distortion effects often dominate the out-of-band interference beyond $f_c \pm f_m$ as seen in Fig. 4.40.

The *FM* modulation index k_ϕ may be used as a figure of merit to assess the impact of *AM-PM* on signal with small amplitude deviations. The relative level of the sidebands may be calculated from Eq. (4.33). It must be noted that two sidebands nearest to the carrier may be masked from the *AM* components of the signal, but the out-of-band components are readily identified.

Measurement of Gain Compression and Phase Deviation

For bandpass components where the input frequency is equal to the output frequency, such as amplifiers, gain compression and phase deviation of a nonlinear circuit are readily measured with a *network analyzer* in power sweep mode. Such a setup is shown in Fig. 4.41. P_{1dB} is easily measured using delta markers by placing the reference marker at the beginning of the sweep (i.e., where the DUT is not compressed), and moving the measurement marker where $\Delta Mag(S_{21}) = -1\ dB$. Sweeping at too high a rate may affect the

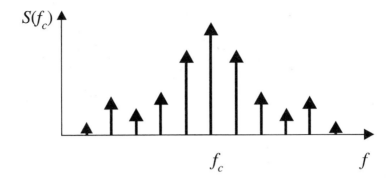

FIGURE 4.40 Output components of an amplitude-modulated signal distorted by *AM-PM* effects.

Vector Network Analyzer

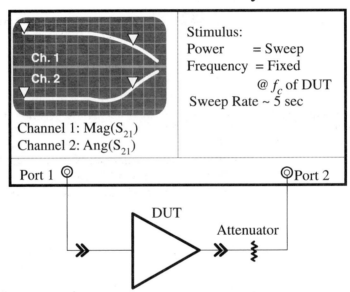

FIGURE 4.41 Setup used to measure gain compression and *AM-PM*.

readings. The sweep must be slow enough so that steady-state conditions exist in both the thermal case and the DC bias network within the circuit. Sweeper retrace may also affect the first few points on the trace. These points must be neglected when setting the reference marker.

The *FM* modulation index is often estimated by measuring the phase deviation at 1 *dB* gain compression $\Delta\Phi(P_{1dB})$

$$k_\phi \approx \frac{\Delta\Phi(P_{1dB})}{2Z_0\sqrt{P_{1dB}}}. \qquad (4.34)$$

For circuits such as mixers, where the input frequency is not equal to the output frequency, gain compression may be measured using the network analyzer with the measurement mode setup for frequency translation. The operation in this mode is essentially that of a *scalar network analyzer*, and all phase information is lost. *AM-PM* effects may be measured using a *spectrum analyzer* and fitting the sideband levels to Eq. (4.33).

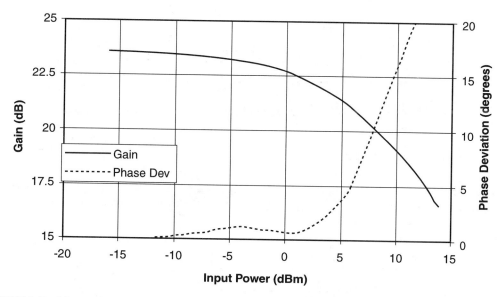

FIGURE 4.42 Measured gain compression and *AM-PM* of a 0.5 W 1960 MHz GaAs MESFET power amplifier IC using an HP8753C Vector Network Analyzer in power sweep mode.

The gain compression and phase deviation of a GaAs power amplifier is shown in Fig. 4.42. P_{1dB} for this amplifier is approximately 23 *dBm* or 0.5 W. The phase deviation $\Delta\Phi$ is not constant from low power to P_{1dB}. Nevertheless, as a figure of merit, the modulation index k_ϕ may be calculated from Eq. (4.34) to be 0.14°/V. Notice that for higher power levels, the amplifier is well into compression, and the phase deviation occurs at a much higher slope than k_ϕ would indicate.

Intermodulation Distortion

When more than one frequency component is present in a signal, the distortion from a nonlinear circuit is manifested as *intermodulation distortion (IMD)*.[4] The *IMD* performance of microwave circuits is important because it can create unwanted interference in adjacent channels. While *bandpass* filtering can eliminate much of the effects of harmonic distortion, intermodulation distortion is difficult to filter out because the IMD components may be very close to the carrier frequency. A common figure of merit is *two-tone* intermodulation distortion.

Two-Tone Intermodulation Distortion

Consider a signal consisting of two sinusoids

$$v_{in}(t) = A\cos(2\pi f_1) + A\cos(2\pi f_2). \tag{4.35}$$

Such a signal may be represented in a different fashion by invoking well-known trigonometric identities.

$$v_{in}(t) = A\cos(2\pi f_c)\cos(2\pi f_m), \tag{4.36a}$$

where

$$f_c = \frac{f_1 + f_2}{2} \quad \text{and} \quad f_m = \left|\frac{f_1 - f_2}{2}\right| \tag{4.36b}$$

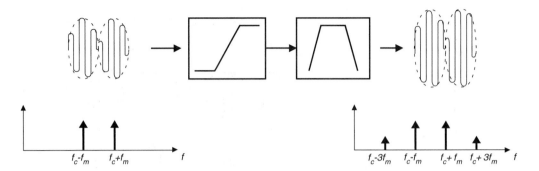

FIGURE 4.43 Intermodulation distortion of a two-tone signal. The output bandpass signal contains the original input signal as well as the harmonics of the envelope at the sum and difference frequencies.

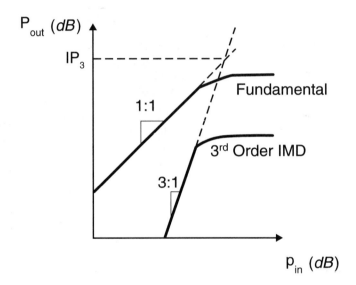

FIGURE 4.44 Relationship between signal output power and intermodulation distortion product power levels. Extrapolating the trends, a figure of merit called the *third order intercept point* (IP_3) is obtained.

Applying such a signal to a memoryless nonlinearity as defined in Eq. (4.24), one obtains the following result:

$$v_{out} = \left[\left(g_1 A + \frac{3g_3 A^3}{4}\right)\cos\left(2\pi f_m t\right) + \frac{3g_3 A^3}{4}\cos\left(6\pi f_m t\right)\right]\cos\left(2\pi f_c t\right). \tag{4.37}$$

Thus, it is seen that the IMD products near the input carrier frequency are simply the odd-order harmonic distortion products of the modulating *envelope*. This is illustrated in Fig. 4.43.

Third Order Intercept Point

Referring to Fig. 4.44, note that the output signal varies at a 1:1 slope on a *log-log* scale with the input signal, while the IMD products vary at a 3:1 slope. Though both the fundamental and the IMD products saturate at some output power level, if one were to extrapolate the level of each and find the intercept point, the corresponding output power level is called the *third order intercept point* (IP_3). Thus, if the IP_3 of a nonlinear circuit is known, the IMD level relative to the output signal level may be found from

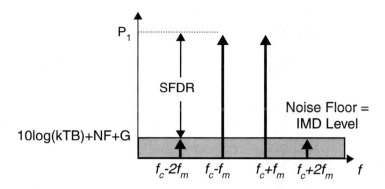

FIGURE 4.45 An illustration of spurious free dynamic range, which defines the range of signal levels where the worst case signal-to-noise ratio is defined by the noise floor of the system, rather than the *IMD* level.

$$IMD_{dBc} = 2\left(P_{out,dBm} - IP_{3,dBm}\right).$$ (4.38)

It must be noted that 3rd order *IMD* is only dominant for low levels of distortion (<10 dB below P_{1dB}). At higher levels, 5th and higher order *IMD* effects can also produce sidebands at the 3rd order frequency. The net result is that the relative *IMD* level will change at a rate greater than 2:1 compared to carrier level. Care should be taken to avoid extrapolating IP_3 from points where this may be occurring. Another point of caution is *AM-PM* effects. In theory, the sidebands produced by phase modulation are in quadrature with those produced by *AM* distortion, and thus should add directly to the *IMD* power. However, the author's experience has shown that these *AM-PM* products can be rotated in phase and thus vector added to the *AM* sidebands. Since the *FM* sidebands are antiphase, one *FM* sideband adds constructively to the *AM* sidebands, while the other adds destructively. The net effect is an imbalance in the *IMD* levels from lower to higher sideband frequencies. Most specifications of *IMD* level will measure the worst case of the two.

For a limiting amplifier, an often used rule of thumb may be derived that predicts a relationship between P_{1dB} and IP_3.[4]

$$IP_3 = P_{1dB} + 9.6 \ dB.$$ (4.39)

While this may not be rigorously relied upon for every situation, it is often accurate within ± 2 dB for small-signal amplifiers and class-A power amplifiers.

Dynamic Range

Because intermodulation distortion generally increases with increasing signal levels, IP_3 may be used to establish the *dynamic range* of a system. The signal level at which the *IMD* level meets the noise floor is defined at the *spurious free dynamic range* (*SFDR*).[5] This is illustrated in Fig. 4.45.

The *SFDR* of a system with gain G may be derived from IP_3 and the *noise figure NF*

$$SFDR = \frac{2IP_3 - 2\left[10\log\left(kT_{eq}B\right) + NF + G\right]}{3},$$ (4.40)

where k is Boltzman's constant, T_{eq} is the equivalent input noise temperature, and B is the bandwidth of the system.

Intermodulation Distortion of Cascaded Components

The question often arises when two components are cascaded of what effect the driving stage *IMD* has on the total *IMD*. This is shown in Fig. 4.46. To the degree that the *IMD* products produced by the nth

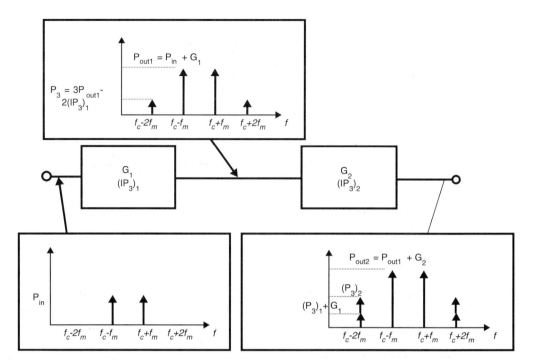

FIGURE 4.46 Effect of cascaded *IMD* levels. The *IMD* from the first stage may be power added to those of the second stage with levels adjusted for the gain of the stage.

stage are uncorrelated with those of the $n+1$ stage, the output *IMD* may be calculated as the power addition of 3rd order *IMD* levels (P_3) with levels adjusted accordingly for gain.

$$\left(P_3\right)_{n+1} = 10 \log\left[10^{\left[\left(P_3\right)_n + G_1\right]/10} + 10^{\left[3\left(P_1\right)_{n+1} - 2\left(IP_3\right)_{n+1}/10\right]}\right]. \tag{4.41}$$

Measurement of Intermodulation Distortion

IMD is normally measured with two-signal generators and a *spectrum analyzer*. Such a setup is shown in Fig. 4.47. Care must be taken to isolate the signal generators, as *IMD* may result from one output mixing with signal from the opposing generator. The carrier levels should be within 0.5 *dB* of each other for accurate *IMD* measurements. Also, it is usually recommended that a power meter be used to get an accurate reading of output power level from the DUT. Relative *IMD* level is measured by placing a reference marker on one of the two carrier signals, and placing a delta marker at either sideband. Finally, the input level must be maintained well below the input IP_3 of the spectrum analyzer to insure error-free reading of the DUT.

Multicarrier Intermodulation Distortion and Noise Power Ratio

While two-tone intermodulation distortion serves to compare the linearity of one component to another, in many applications, a component will see more than two carriers in the normal operation of a microwave system. Thus direct measurement of multitone *IMD* is often necessary to insure adequate carrier-to-interference level within a communication system.

Peak-to-Average Ratio of Multicarrier Signals

The major difference between two-tone signals and multitone signals is the *peak-to-average (pk/avg)* power ratio.[6] From Eq. (4.35), it is clear that the average power of a two-tone signal is equal to the sum

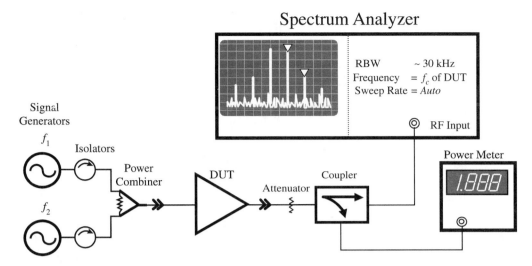

FIGURE 4.47   Setup used to measure two-tone intermodulation distortion (*IMD*).

of powers from the individual carriers. However, from Eq. (4.36), one may derive that the *peak envelope power* (*PEP*) is four times the level of the individual carriers. Thus, it is said that the *pk/avg* ratio of a two-tone signal is a factor of 2, or 3 *dB*. From inductive reasoning, it is then clear that the *pk/avg* ratio of an *n*-tone signal is

$$pk/avg = 10\log(n).$$ (4.42)

While the absolute peak of a multicarrier is dependent only on the number of carriers, the probability distribution of the *pk/avg* ratio depends on the modulation. Figure 4.48 shows the difference between 16 phase-aligned tones, and 16 carriers with randomly modulated phases. In general, multiple modulated signals encountered in communication systems will mimic the behavior of random phase modulated sinusoids. Phase aligned sinusoids may be considered a worst case condition. As the number of carriers increases, and if their phases are uncorrelated, the Central Limit Theorem predicts that the distribution of *pk/avg* approaches that of white Gaussian random noise.[7] The latter signal is treated in the next section.

Noise Power Ratio

For many systems, including those that process multicarrier signals, white Gaussian noise is a close approximation for the real-world signals. This is a result of the *Central Limit Theorem,* which states that the probability distribution of a sum of a large number of random variables will approach the Gaussian distribution, regardless of the distributions of the individual signals.[7] One metric that has been employed to describe the *IMD* level one would expect in a dense multicarrier environment is the *noise power ratio*. This concept is illustrated in Fig. 4.49.

Measurement of Multitone IMD and Noise Power Ratio

Thus it is clear that power ratings for components must be increased for peak power levels given by Eq. (4.42). Furthermore, two-tone intermodulation distortion may not be indicative of *IMD* of multitone signals. Measurement over various power levels is the only way to accurately predict multitone IMD. Figure 4.50 illustrates a setup that may be used to measure multitone intermodulation.

The challenge in measuring *NPR* is creating the signal. It is clear that, to get an accurate indication of *IMD* performance, the signal bandwidth must not exceed the bandwidth of the device under test. Furthermore, to measure *NPR*, one must notch out the noise power over a bandwidth approximating one channel *BW*. As an example, an *NPR* measurement on a component designed for North American Digital Cellular System (IS-136) ideally would produce a 25-MHz wide noise source with one channel

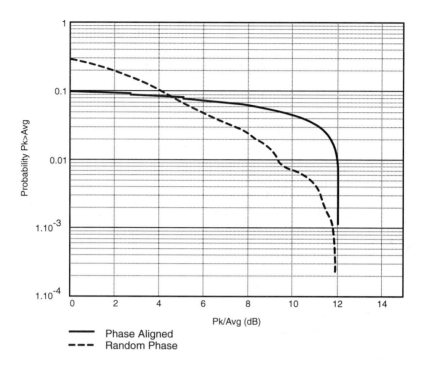

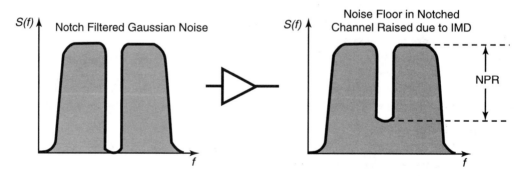

FIGURE 4.48 Distribution of *peak-to-average* ratio of a phase-aligned 16 carrier signal and a random-phase 16 carrier signal. The *y*-axis shows the probability that the signal exceeds a power level above average on the *x*-axis. While both signals ultimately have the same *pk/avg* ratio, their distributions are much different.

FIGURE 4.49 An illustration of *noise power ratio*. NPR is essentially a measure of the carrier-to-interference level experienced by multiple carriers passing through a nonlinear component.

of bandwidth equal to 30 *kHz*. The *Q* of a notch filter to produce such a signal would be in excess of 25,000. Practical measurements employ filters with *Q*s around 1000, and are able to achieve more than 50 *dB* of measurement range. Such a setup is shown in Fig. 4.51.

Distortion of Digitally Modulated Signals

While standard test signals such as a two-tone or band-limited Gaussian noise provide relative figures of merit of the linearity of a nonlinear component, they cannot generally insure compliance with government or industry system-compatibility standards. For this reason, methods have been developed to measure and characterize the intermodulation distortion of the specific digitally modulated signals used in various systems. Table 4.2 summarizes the modulation formats for North American digital cellular telephone systems.[8,9]

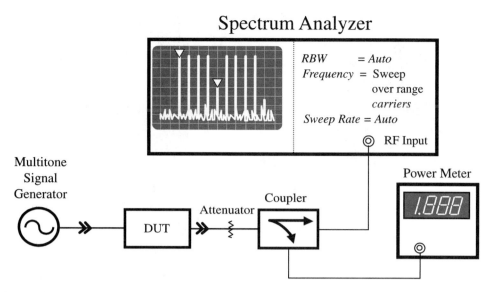

FIGURE 4.50 Measurement setup for multitone *IMD*. Tones are usually spaced equally, with the middle tone deleted to allow measurement of the worst-case *IMD*.

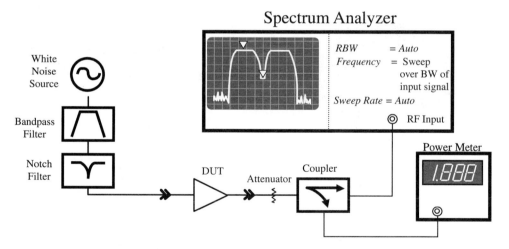

FIGURE 4.51 Noise Power Ratio measurement setup. The rejection of the notch filter should be at least 10 *dB* below the *NPR* level to avoid erroneous measurement.

TABLE 4.2 Modulation Formats for North American Digital Cellular Telephone Systems

Standard	Multiple Access Mode	Channel Power Output	Modulation	Channel Bandwidth
IS-136[8]	TDMA	+28 dBm	π/4-DQPSK	30 kHz
IS-95[9]	CDMA	+28 dBm	OQPSK	1.23 MHz

Intermodulation Distortion of Digitally Modulated Signals

Amplitude and phase distortion affect digitally modulated signals the same way they affect analog modulated signals: gain compression and phase deviation. This is readily seen in Fig. 4.52. Because both amplitude and phase modulation are used to generate digitally modulated signals, they are often expressed

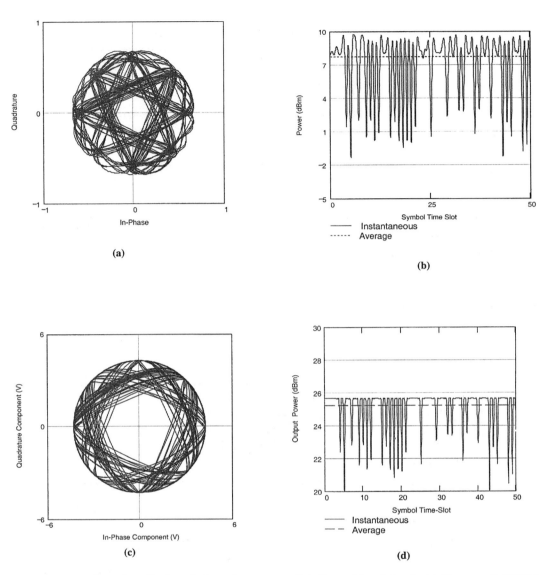

FIGURE 4.52 Effect of amplitude and phase distortion on digitally modulated signals. (a) Shows a π/4 DQPSK signal constellation, and its associated power envelope in (b). When such a signal is passed through a nonlinear amplifier, the resulting envelope is clipped (d), and portions of the constellation are rotated (c).

as a *constellation* plot, with the in-phase component $I = A\cos\phi$ envelope plotted against the quadrature component $Q = A\sin\phi$. The instantaneous power envelope is given by

$$P(t) = I(t)^2 + Q(t)^2 = A(t)^2. \tag{4.43}$$

When the envelope is clipped and/or phase rotated, the resulting *IMD* is referred to as *spectral regrowth*. Figure 4.53 shows the effect of nonlinear distortion on a digitally modulated signal. The out-of-band products may lie in adjacent channels, thus causing interference to other users of the system. For this reason, the *IMD* of digitally modulated signals are often specified as *adjacent channel power ratio* (*ACPR*).[10]

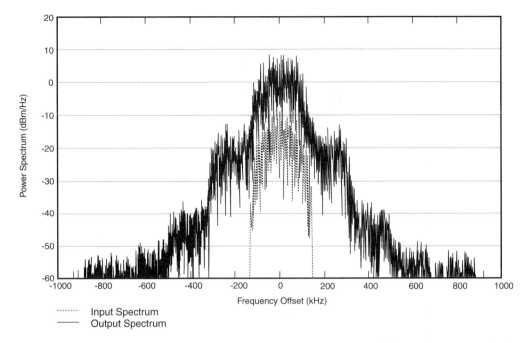

FIGURE 4.53 Effect of nonlinear distortion on a digitally modulated signal. The lower power input signal to a power amplifier has a frequency spectrum that is well contained within a specified channel bandwidth. IMD due to nonlinear distortion creates out-of-band products that may fall within the adjacent channels, causing interference to other users of the system.

ACPR may be specified in a number of ways, depending on the system architecture. In general, *ACPR* is given by

$$ACPR = \frac{I_{adj}}{C_{ch}} = \frac{\int_{f_o - B_{adj}}^{f_o - B_{adj}} S(f) df}{\int_{-B_{ch}/2}^{B_{ch}/2} S(f) df}, \tag{4.44}$$

where I_{adj} is the total interference power in a specified adjacent channel bandwidth, B_{adj} at a given frequency offset f_o from the carrier frequency, and C_{ch} is the channel carrier power in the specified channel bandwidth B_{ch}. Note that the carrier channel bandwidth may be different from the interference channel bandwidth because of regulations enforcing interference limits between different types of systems. Furthermore, the interference level may be specified in more than one adjacent channel. In this case, the specification is referred to as the *alternate channel power ratio*. Table 4.3 shows ACPR specifications for various digital cellular standards.

In addition to the out-of-band interference due to the intermodulation distortion in-band interference will also result from nonlinear distortion. The level of the in-band interference is difficult to measure directly because it is superimposed on the channel spectrum. However, when the signal is demodulated,

TABLE 4.3 ACPR and EVM Specifications for Digital Cellular Subscriber Equipment

Standard	ADJ. CH. PWR	ALT. CH. PWR	EVM
IS-136	−26 dBc/30 kHz @ ± 30 kHz	−45 dBc/30 kHz @ ± 60 kHz	12.5%
IS-95	−42 dBc/30 kHz @ > ± 885 kHz	−54 dBc/30 kHz @ > ± 1.98 MHz	23.7%

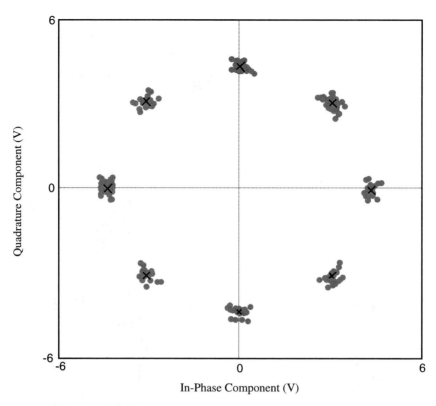

FIGURE 4.54 Errors in the demodulated *I-Q* constellation may result from the in-band *IMD* products. The *rms* summation of errors from the desired location (given by the × markers) give the *error vector magnitude* of the signal distortion.

errors in the output *I-Q* constellation occur at the sample points. This is shown in Fig. 4.54. Because the demodulator must make a decision as to which symbol (i.e., which constellation point) was sent, the resulting errors in the *I-Q* vectors may produce a false decision, and hence cause *bit errors*.

There are two methods to characterize the level *I-Q* vector error: *error vector magnitude* (*EVM*), and a quality factor called the ρ-*factor*.[11] Both *EVM* and ρ-*factor* provide an indication of signal distortion, but they are calculated differently. *EVM* is the *rms* sum of vector errors divided by the number of samples.

$$EVM = \frac{1}{n}\sqrt{\sum_n \left[\left|I(t_n) - S_{In}\right|^2 + \left|Q(t_n) - S_{Qn}\right|^2\right]}, \qquad (4.45)$$

where the *I-Q* sample points at the nth sample windows are given by $I(t_n)$ and $Q(t_n)$, and the nth symbol location point in-phase and quadrature components are given by S_{In} and S_{Qn} respectively.

Whereas *EVM* provides an indication of *rms* % error of the signal envelope at the sample points, ρ-*factor* is related to the waveform quality of a sig al. It is related to *EVM* by

$$\rho = \frac{1}{1 - EVM^2} \qquad (4.46)$$

Measurement of ACPR, EVM, and Rho-Factor

ACPR may be measured using a setup similar to those for measuring *IMD*. The major difference involves generating the test signal. Test signals for digitally modulated signals must be synthesized according to system standards using an *arbitrary waveform generator* (*AWG*), which generates *I*- and *Q*-baseband envelopes. In the most basic form, these are high speed digital-to-analog converters (*DACs*). The files used to generate the envelope waveforms may be created using commonly available mathematics software, and are built in many commercially available *AWGs*. The *I*- and *Q*-baseband envelopes are fed to an *RF* modulator to produce a modulated carrier at the proper center frequency.

In the case of *CDMA* standards, deviations between test setups can arise from different selections of Walsh codes for the traffic channels. While a typical *CDMA* downlink (base station to mobile) signal has a *pk/avg* of approximately 9.5 *dB*, it has been shown that some selections of Walsh codes can result in peak-to-average ratios in excess of 13 *dB*.[12] Measurement of *EVM* is usually done with a *vector signal analyzer* (*VSA*). This instrument is essentially a receiver that is flexible enough to handle a variety of frequencies and modulation formats. Specialized software is often included to directly measure *EVM* or rho-factor for well-known standards used in microwave radio systems.

Summary

This section has treated characterization and measurement techniques for nonlinear microwave components. Figures of merit were developed for such nonlinear effects as harmonic level, gain compression, and intermodulation distortion. While these offer a basis for comparison of the linearity performance between like components, direct measurement of adjacent channel power and error vector magnitude are preferred for newer wireless systems. Measurement setups for the above parameters were suggested in each section. For more advanced treatment, the reader is referred to the references at the end of this section.

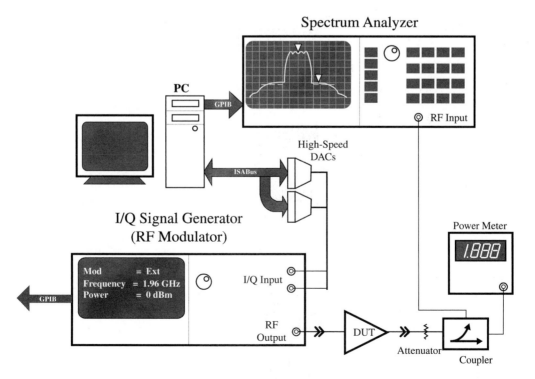

FIGURE 4.55 Measurement setup for ACPR. Waveforms are created using PCs or specialized arbitrary waveform generators. In either case, the baseband waveform must be upconverted to the center frequency of the DUT.

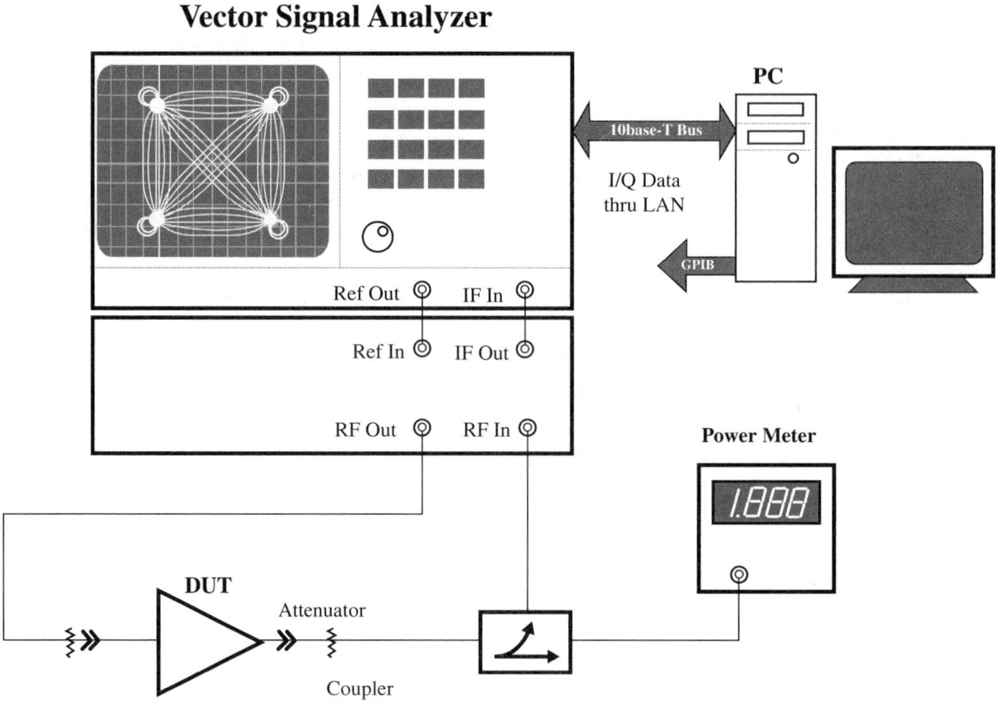

Vector Signal Analyzer

FIGURE 4.56 Setup for measuring *EVM*. The *VSA* demodulates the *I-Q* waveform and calculates the deviation from ideal to calculate *EVM* and ρ-*factor* as given in Eq. (4.45) and Eq. (4.46), respectively.

References

1. Maas, S.A., *Nonlinear Microwave Circuits*, Artech House, Boston, 1988.
2. Blachman, N.M., Band-pass nonlinearities, *IEEE Trans. Information Theory*, IT-10, 162–64, April, 1964.
3. Andrews, L.C., *Special Functions of Mathematics for Engineers*, 2nd ed., McGraw-Hill, New York, 1992, chap. 6.
4. Cripps, S.C., *RF Power Amplifiers for Wireless Communications*, Artech House, Boston, 1999, chap. 7.
5. Carson, R.S., *Radio Concepts: Analog*, John Wiley & Sons, New York, 1990, chap. 10.
6. Kenney, J.S., and Leke, A., Design considerations for multicarrier CDMA base station power amplifiers, *Microwave J.*, 42, 2, 76–86, February, 1999.
7. Papoulis, A., *Probability, Random Variables, and Stochastic Processes*, 3rd ed., McGraw-Hill, New York, 1991, chap. 8.
8. IS-136 Interim Standard, Cellular System Dual-Mode Mobile Station — Base Station Compatibility Standards, Telecommunications Industry Assoc.
9. IS-95 Interim Standard, Mobile Station — Base Station Compatibility Standard for Dual-Mode Wideband Spread Spectrum Cellular Systems, Telecommunications Industry Assoc.
10. Kenney, J.S. and Leke, A., Power amplifier spectral regrowth for digital cellular and PCS applications, *Microwave J.*, 38, 10, 74–92, October 1995.
11. Lindsay, S.A., Equations derive error-vector magnitude, *Microwaves & RF*, April, 1995, 158–67.
12. Braithwaite, R.N., Nonlinear amplification of CDMA waveforms: an analysis of power amplifier gain errors and spectral regrowth, *Proc. 48th Annual IEEE Vehicular Techn. Conf.*, 2160–66, 1998.

4.5 Theory of High-Power Load-Pull Characterization for RF and Microwave Transistors

John F. Sevic

In both portable and infrastructure wireless systems the power amplifier often represents the largest single source of power consumption in the radio. While the implications of this are obvious for portable applications, manifested as talk-time, it is also important for infrastructure applications due to thermal management, locatability limitations, and main power limitations. Significant effort is devoted toward developing high-performance RF and microwave transistors and circuits to improve power amplifier efficiency. In the former case, an accurate and repeatable characterization tool is necessary to evaluate the performance of the transistor. In the latter case, it is necessary to determine the source and load impedance for the best trade-off in overall performance. Load-pull is presently the most common technique, and arguably the most useful for carrying out these tasks. In addition, load-pull is also necessary for large-signal model development and verification.

Load-pull as a design tool is based on measuring the performance of a transistor at various source and/or load impedances and fitting contours, in the gamma-domain, to the resultant data; measurements at various bias and frequency conditions may also be done. Several parameters can be superimposed over each other on a Smith chart and trade-offs in performance established. From this analysis, optimal source and load impedances are determined.

Load-pull can be classified by the method in which source and load impedances are synthesized. Since the complex ratio of the reflected to incident wave on an arbitrary impedance completely characterizes the impedance, along with a known reference impedance, it is convenient to classify load-pull by how the reflected wave is generated.

The simplest method to synthesize an arbitrary impedance is to use a stub tuner. In contrast to early load-pull based on this method, contemporary systems fully characterize the stub tuner *a priori*, precluding the need for determining the impedance at each load-pull state [1]. This results in a significant reduction in time and increases the reliability of the system. This method of load-pull is defined as passive-mechanical. Passive-mechanical systems are capable of presenting approximately 50:1 VSWR, with respect to 50 Ω, and are capable of working in very high power environments. Repeatability is better than −60 dB. Maury Microwave and Focus Microwave each develop passive-mechanical load-pull systems [2,3]. For high-power applications, e.g., > 100 W, the primary limitation of passive-mechanical systems is self-heating of the transmission line within the tuner, with the resultant thermally induced expansion perturbing the line impedance.

Solid-state phase-shifting and attenuator networks can also be used to control the magnitude and phase of a reflected wave, thereby effecting an arbitrary impedance. This approach has been pioneered by ATN Microwave [4]. These systems can be based on a lookup table approach, similar to the passive-mechanical systems, or can use a vector network analyzer for real-time measurement of tuner impedance. Like all passive systems, the maximum VSWR is limited by intrinsic losses of the tuner network. Passive-solid-state systems, such as the ATN, typically exhibit a maximum VSWR of 20:1 with respect to 50 Ω. These systems are ideally suited for medium power applications and noise characterization (due to the considerable speed advantage over other types of architectures).

Tuner and fixture losses are the limiting factor in achieving a VSWR in excess of 50:1 with respect to 50 Ω. This would be necessary not only for characterization of high-power transistors, but also low-power transistors at millimeter-wave frequencies, where system losses can be significant. In these instances, it is possible to synthesize a reflected wave by sampling the wave generated by the transistor traveling toward the load, amplifying it, controlling its magnitude and phase, and reinjecting it toward the transistor. Systems based on this method are defined as active load-pull. Although in principle active load-pull can be used to create very low impedance, the power necessary usually limits the application of this method to millimeter-wave applications [5,6]. Because active load-pull systems are capable of placing any reflection coefficient on the port being pulled (including reflections greater than unity) these systems can be very unstable and difficult to control. Instability in a high-power load-pull system can lead to catastrophic failure of the part being tested.

The present chapter is devoted to discussing the operation, setup, and verification of load-pull systems used for characterization of high-power transistors used in wireless applications. While the presentation is general in that much of the discussion can be applied to any of the architectures described previously, the emphasis is on passive-mechanical systems. There are two reasons for limiting the scope. The first reason is that passive-solid-state systems are usually limited in the maximum power incident on the tuners, and to a lesser extent, the maximum VSWR the tuners are capable of presenting. The second reason is that currently there are no active load-pull systems commercially available. Further, it is unlikely that an active load-pull system would be capable of practically generating the sub 1 Ω impedances necessary for characterization of high-power transistors.

The architecture of the passive-mechanical system is discussed first, with a detailed description of the necessary components for advanced characterization of transistors, such as measuring input impedance and ACPR [7]. Vector network analyzer calibration, often overlooked, and the most important element of tuner characterization, is presented next. Following this, tuner, source, and load characterization methods are discussed. Fixture characterization methods are also presented, with emphasis on use of pre-matching fixtures to increase tuner VSWR. Finally, system performance verification is considered.

System Architecture for High-Power Load-Pull

Figure 4.57 shows a block diagram of a generalized high-power automated load-pull system, although the architecture can describe any of the systems discussed in the previous section. Sub-harmonic and harmonic tuners are also included for characterization of out-of-band impedances [8]. The signal sample ports are used to measure the incident and reflected voltage waves at the source-tuner interface and the incident voltage wave at the load. The signals at each of these ports are applied to the equipment necessary to make the measurements the user desires. Each of these blocks is described subsequently.

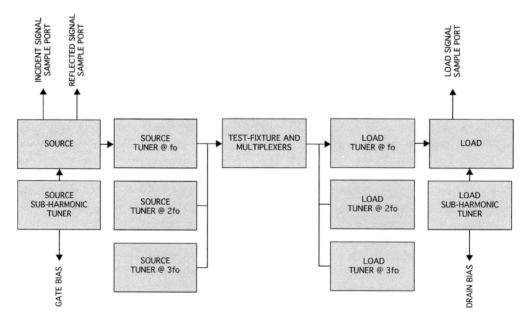

FIGURE 4.57 Block diagram of a generalized high-power load-pull system, illustrating the source, tuners, test-fixture, and load. The incident, reflected, and load signals are sampled at the three sampling points shown. Also shown, though not necessary, are harmonic and sub-harmonic tuners.

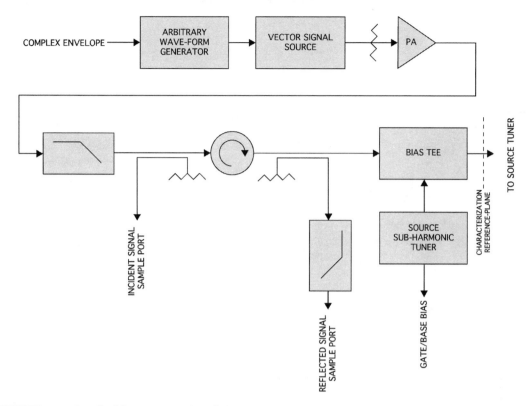

FIGURE 4.58 Detail of the source portion of Fig. 4.57.

The source block of Fig. 4.57 usually includes all of the components necessary for generating the signal, leveling its power, providing gate/base bias for the device under test, and providing robust sampling points for the measurement equipment. Figure 4.58 shows the details of a typical source block. For flexibility and expediency in applying arbitrarily modulated signals, an arbitrary waveform generator and vector signal source are shown. The signal is typically created using MATLAB, and can represent not only digitally modulated signals, but also the more conventional two-tone signal. The signal is applied to a reference PA, which must be characterized to ensure that it remains transparent to the DUT; for high-power applications this is often a 50 W to 100 W PA.

Following the reference PA is a low-pass filter to remove harmonics generated from the source and/or reference PA. Next are the sampling points for the incident and reflected waves, which is done with two distinct directional couplers. Since the source tuner may present a high reflection, a circulator to improve directivity separates each directional coupler; the circulator also protects the reference PA from reflected power. The circulator serves to present a power-invariant termination for the source tuner, the impedance of which is critical for sub 1 Ω load-pull. The bias-tee is the last element in the source block, which is connected to the gate/base bias source via a low-frequency tuner network for sub-harmonic impedance control. Since the current draw of the gate/base is typically small, remote sensing of the power supply can be done directly at the bias-tee.

Although components within the source block may have type-N or 3.5 mm connectors, interface to the source tuner is done with an adapter to an APC 7 mm connector. This is done to provide a robust connection and to aid in the VNA characterization of the source block. Depending on the measurements that are to be made during load-pull, a variety of instruments may be connected to the incident and reflected sample ports, including a power meter and VNA. The former is required for real-time leveling and the latter for measuring the input impedance to the DUT [9].

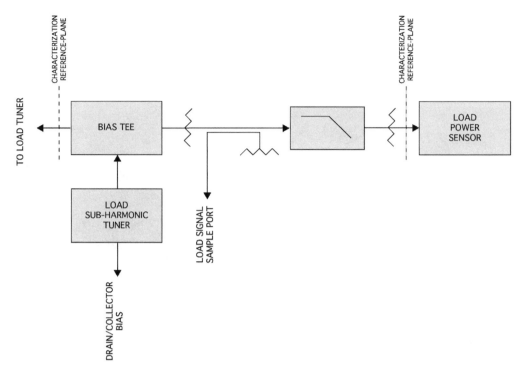

FIGURE 4.59   Detail of the load portion of Fig. 4.57.

The load block of Fig. 4.57 usually includes a port for sampling the load signal of the DUT and the padding and filtering necessary to interface the load signal to a power sensor. Figure 4.59 shows the details of a typical load block. The bias-tee comes first. Although remote-sense can be sampled here, in situations where significant current is required, the remote-sense should be sampled directly on the DUT test fixture. For a load-pull system capable of 100 W average power, the attenuator following the bias-tee should be appropriately rated and exhibit at least 30 dB attenuation.

The load signal is sampled at a directional coupler after the high-power pad. A spectrum analyzer is often connected at this port, and it may be useful to use a low coupling factor, e.g., −30 dB, to minimize the padding necessary in front of the spectrum analyzer. This results in an optimal dynamic range of the system for measuring ACPR. Following the directional coupler is a low-pass filter, to remove harmonics,[1] which is followed by another attenuator. This attenuator is used to improve the return loss of the filter with respect to the power sensor. As with the source block, interface to the load tuner and power sensor are done with APC 7 mm connectors to improve robustness and power-handling capability.

The DUT test-fixture is used to interface the source and load tuners to a package. For cost and package de-embedding reasons, it is useful to standardize on two or three laboratory evaluation packages. For hybrid circuit design, it is useful to design a test fixture with feeds and manifolds identical to those used in hybrid to mitigate de-embedding difficulties. The collector/drain side of the test fixture should also have a sampling port for remote sensing of the power supply.

After the load-pull system has been assembled, it is recommended that the maximum expected power be applied to the system and changes in impedance be measured due to tuner self-heating. This may be significant where average powers exceed 100 W or peak powers exceed several hundred watts. Any impedance change will establish the upper power limit of the system with respect to impedance accuracy.

---

[1]Although a filter is not necessary, characterization of a DUT in significant compression will result in the average power detected by the power sensor including fundamental and harmonic power terms. When the DUT is embedded into a matching network, the matching network will usually attenuate the harmonics; thus, inclusion of the low-pass filter more closely approximates the performance that will be observed in practice.

Characterization of System Components

Each of the blocks described in the previous section must be characterized using s-parameters in order for a load-pull system to function properly. In this section, the characterization procedure for each of the sections of Fig. 4.57 is described, with emphasis on calibration of the vector network analyzer and the characterization of the transistor test fixture. Two-tier calibration and impedance re-normalization are considered for characterizing quarter-wave pre-matching test fixtures.

Vector Network Analyzer Calibration Theory

Due to the extremely low impedances synthesized in high-power load-pull, the vector network analyzer (VNA) calibration is the single most important element of the characterization process. Any errors in the measurement or calibration, use of low quality connectors, e.g., SMA or type-N, or adoption of low-performance calibration methods, e.g., SOLT, will result in a significant reduction in accuracy and repeatability. Only TRL calibration should be used, particularly for tuner and fixture characterization. Use of high-performance connectors is preferred, particularly APC 7 mm, due to its repeatability, power handling capability, and the fact that it has a hermaphroditic interface, simplifying the calibration process.

Vector network analysis derives its usefulness from its ability to characterize impedance based on ratio measurements, instead of absolute power and phase measurements, and from its ability to characterize and remove systematic errors due to nonidealities of the hardware. For a complete review of VNA architecture and calibration theory, the reader is encouraged to review notes from the annual ARFTG Short-Course given in November of each year [10,11].

Figure 4.60 shows a signal-flow graph of the forward direction of a common VNA architecture, where six systematic error terms are identified. An identical flow-graph exists for the reverse direction, with six additional error terms. Consider the situation where it is required to measure an impedance that exhibits a near total reflection, such as a load tuner set for 1 Ω. Assuming a 50 Ω reference impedance, nearly all of the incident power is reflected back toward the VNA, along with a phase shift of 180°. Consider what happens when the reflected wave is sampled at the VNA, denoted as b_{1M} in Fig. 4.60. If there is any re-reflection of the reflected wave incident at the VNA, an error will occur in measuring the actual impedance of the load. The ability of a VNA to minimize this reflected power is characterized by its residual source match, which is the corrected source impedance looking into the VNA. The uncorrected source impedance looking into the VNA is characterized by the E_{sf} term in the flow graph of Fig. 4.60.

Continuing with this example, Fig. 4.61 shows a plot of the upper bound on apparent load impedance versus the residual source match (with respect to a reference impedance of 50 Ω and an actual impedance of 1 Ω). For simplicity, it is assumed that the residual source match is in phase with the reflected signal.

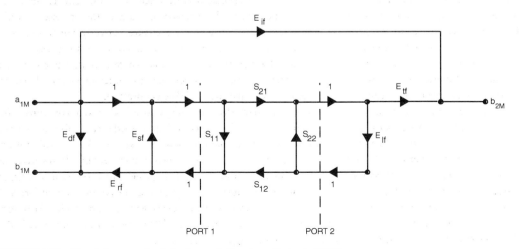

FIGURE 4.60 Signal-flow graph of the forward direction of a typical VNA.

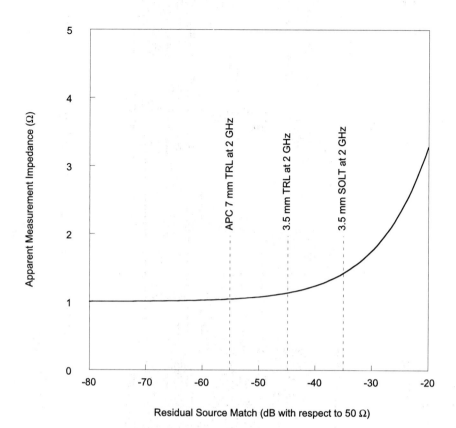

FIGURE 4.61   The influence of residual source match on the ability of a VNA to resolve a 1 Ω impedance with a 50 Ω reference impedance. The calibration performance numbers are typical for an HP 8510C with an 8514B test-set operating a 2 GHz.

Also shown are typical residual source match performance numbers for an HP 8510C using an HP 8514B test set. From this graph it is clear that use of low-performance calibration techniques will result in latent errors in any characterization performed using a DUT with reflection VSWR near 50:1. Using a 3.5 mm SOLT calibration can result in nearly 20% uncertainty in measuring impedance. Note that TRL*, the calibration method available on low-cost VNAs, offers similar performance to 3.5 mm SOLT, due to its inability to uniquely resolve the test-set port impedances. This limitation is due to the presence of only three samplers instead of four, and does not allow switch terms to be measured directly. For this reason, it is recommended that three-sampler architectures not be used for the characterization process.

Similar arguments can be made for the load reflection term of Fig. 4.60, which is characterized by the residual load match error term. Identical error terms exist for the reverse direction too, so that there are a total of four error terms that are significant for low impedance VNA calibration.

TRL calibration requires a thru line, a reflect standard (known only within λ/4), and a delay-line. The system reference impedances will assume the value of the characteristic impedance of the delay-line, which if different from 50 Ω, must be appropriately re-normalized back to 50 Ω [12–15]. TRL calibration can be done in a variety of media, including APC 7 mm coaxial waveguide, rectangular/cylindrical waveguide, microstrip, and stripline. Calibration verification standards, which must be used to extract the residual error terms described above, are also easily fabricated. Figure 4.62 shows the residual forward source and load match response of an APC 7 mm calibration using an HP 8510C with an HP 8514B test set. These were obtained with a 30 cm offset-short airline and 30 cm delay-line, respectively [16,17,18]. The effective source match is computed from the peak-peak ripple using

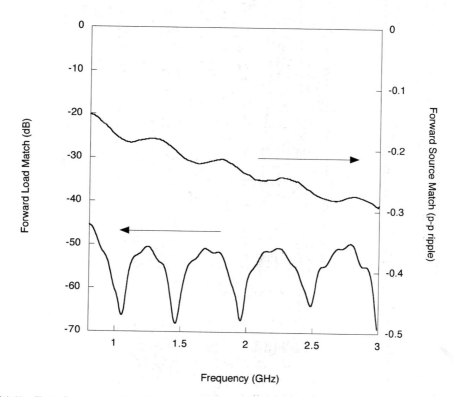

FIGURE 4.62 Typical response of an APC 7 mm TRL calibration using an offset-short and delay-line to extract source match and load match, respectively. This data was taken from an HP 8510C with an HP 8514B test set.

$$E_{sf} = 10 * \log_{10}\left[\frac{1 - 10^{-\frac{p-p\ ripple}{20}}}{1 + 10^{-\frac{p-p\ ripple}{20}}} \right] \tag{4.47}$$

where it is seen that better than −53 dB source match is obtained across the band. Due to finite directivity, 6 dB must be subtracted from the plot showing the delay-line response, indicating that better than −56 dB load match is obtained except near the low end of the band. Calibration performance such as that obtained in Fig. 4.62 is necessary for accurate tuner and fixture characterization, and is easily achievable using standard TRL calibration.

For comparison purposes, Figs. 4.63 and 4.64 show forward source and load match for 3.5 mm TRL and SOLT calibration, respectively. Here it is observed that the source match of the 3.5 mm TRL calibration has significantly degraded with respect to the APC 7 mm TRL calibration and the 3.5 mm SOLT calibration has significantly degraded with respect to the 3.5 mm TRL calibration.

Proper VNA calibration is an essential first step in characterization of any component used for high-power load-pull characterization, and is particularly important for tuner and fixture characterization. All VNA calibrations should be based on TRL and must be followed by calibration verification to ensure that the calibration has been performed properly and is exhibiting acceptable performance, using the results of Fig. 4.62 as a benchmark. Averaging should be set to at least 64. Smoothing should in general be turned off in order to observe any resonances that might otherwise be obscured. Although APC 7 mm is recommended, 3.5 mm is acceptable when used with a TRL calibration kit. Under no circumstances should type-N or SMA connectors be used, due to phase repeatability limitations and connector reliability limitations.

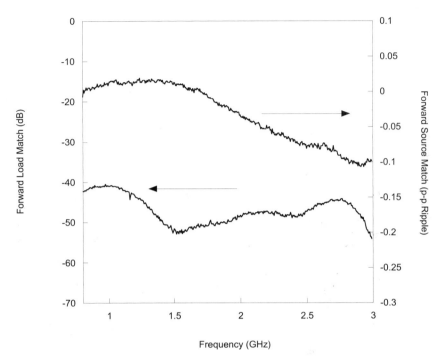

FIGURE 4.63 Typical response of a 3.5 mm TRL calibration using an offset-short and delay-line to extract source match and load match, respectively. This data was taken from an HP 8510C with an HP 8514B test set.

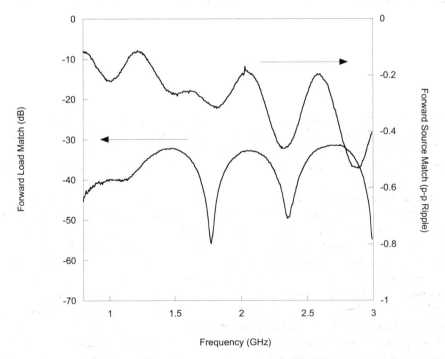

FIGURE 4.64 Typical response of a 3.5 mm SOLT calibration using an offset-short and delay-line to extract source match and load match, respectively. This data was taken from an HP 8510C with an HP 8514B test set.

S-Parameter Characterization of Tuners

Tuner characterization begins with proper calibration of the VNA, as described in the previous section. It is suggested at this point that any adapters on the tuner be serialized and alignment marks made to ensure that in the event of removal, they can be replaced in their original positions. Replacement of an adapter, for any reason, will require a new tuner characterization. Tuners should be leveled using a bubble-level and should be positioned such that the VNA test-port cables are not flexed. Proper torquing of all connector interfaces is essential. Since the tuner files usually consist of a small number of frequencies with respect to the number of frequencies present in a typical VNA calibration, it is appropriate to increase the number of averages to 128 or 256.

It is generally most useful to characterize a tuner without any additional components attached, such as a bias-tee, in order to maintain maximum flexibility in the use of the tuner subsequent to the characterization. For tuners that are being characterized for the first time, it is recommended that they be fully evaluated for insertion loss, minimum and maximum VSWR, and frequency response to ensure they are compliant with the manufacturer's specifications.

After characterization the tuner file should be verified by setting the tuner for arbitrary impedances near the center and edge of the Smith Chart over 2π radians. The error should be less than 0.2% for magnitude and $0.1°$ for phase. Anything worse than this may indicate a problem with either the calibration (verify it again) or the tuner.

S-Parameter Characterization of System Components

Characterization of system components consists of creating one-port and two-port s-parameter files of the source block and load block, as shown in Figs. 4.57 and 4.58, respectively. Each of these figures show suggested reference-planes for characterization of the network. Since the reflection coefficient of each port of the source and load blocks is in general small with respect to that exhibited by tuners, the VNA calibration is not as critical[2] as it is for tuner characterization. Nevertheless, it is recommended to use the same calibration as used for the tuner characterization and to sweep a broad range of frequencies to eliminate the possibility of characterization in the future at new frequencies.

If possible, each component of the source and load blocks should be individually characterized prior to integration into their respective block. This is particularly so for circulators and high-current bias-tees, which tend to have limited bandwidth. The response of the source and load block should be stored for future reference and/or troubleshooting.

Fixture Characterization to Increase System VSWR

In the beginning of this section it was indicated that high-power load-pull may require source and load impedances in the neighborhood of 0.1 Ω. This does not mean that the DUT may require such an impedance as much as it is necessary for generating closed contours, which are useful for evaluation of performance gradients in the gamma domain. A very robust and simple method of synthesizing sub 1 Ω impedances is to use a quarter-wave pre-matching network characterized using numerically well-defined two-tier calibration methods. To date, use of quarter-wave pre-matching offers the lowest impedance, though it is limited in flexibility due to bandwidth restrictions. Recently, commercially available passive mechanical systems cascading two tuners together have been made available offering octave bandwidths, though they are not able to generate impedances as low as narrowband quarter-wave pre-matching. In this section, a robust methodology for designing and characterizing a quarter-wave pre-matching network capable of presenting 0.1 Ω at 2 GHz is described [16,18]. It is based on a two-tier calibration with thin-film gold on alumina substrates (quarter-wave pre-matching networks on soft substrates are not recommended due to substrate variations and repeatability issues over time).

---

[2]If the magnitude of the reflection coefficient approaches the residual directivity of the VNA calibration, then errors may occur.

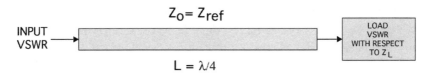

FIGURE 4.65 Network to describe the mismatch invariance property of lossless networks.

The theory of quarter-wave pre-matching begins with the mismatch invariance property of lossless networks [19]. Consider the quarter-wave line of characteristic impedance Z_{ref} shown in Fig. 4.65. This line is terminated in a mismatch of $VSWR_{load}$ with an arbitrary phase. The reference impedance of $VSWR_{load}$ is Z_L. The mismatch invariance property of lossless networks shows that the input VSWR is identical to the load VSWR, but it is with respect to the quarter-wave transformed impedance of Z_L. Thus, the minimum achievable impedance, which is real valued, is the impedance looking into the quarter-wave line when it is terminated in Z_L divided by $VSWR_{load}$. This is expressed as

$$
R_{in,min} = \frac{\dfrac{Z_{ref}^2}{Z_L}}{VSWR_{load}} \tag{4.48}
$$

Suppose it is desired to synthesize a minimum impedance of 0.1 Ω, which might be required for characterizing high power PCS and UMTS LDMOS transistors. If a typical passive-mechanical tuner is capable of conservatively generating a 40:1 VSWR, then the input impedance of the quarter-wave line must be approximately 4 Ω, requiring the characteristic impedance of the quarter-wave line to be approximately 14 Ω, assuming a Z_L of 50 Ω. To the extent that the minimum impedance deviates from the ideal is directly related to fixture losses. Thus, the importance of using a low-loss substrate and metal system is apparent.

Full two-port characterization of each fixture side is necessary to reset the reference plane of each associated tuner. Several methods are available to do this, including analytical methods based on approximate closed-form expressions, full-wave analysis using numerical techniques, and employment of VNA error correction techniques [20,21,22]. The first method is based on approximations that have built-in uncertainty, as does the second method, in the form of material parameter uncertainty. The third method is entirely measurement based, and relies on well-behaved TRL error correction mathematics to extract a two-port characterization of each fixture half from a two-tier calibration. More importantly, using verification standards, it is possible to quantify the accuracy of the de-embedding, as described in the section on VNA calibration.

Using the error-box formulation of the TRL calibration it is possible to extract the two-port characteristics of an arbitrary element inserted between two reference planes of two different calibrations [11]. The first tier of the calibration is usually done at the test-port cables of the VNA. The second tier of the calibration is done in the media that matches the implementation of the test fixture, which is usually microstrip. Figure 4.66 illustrates the reference-plane definitions thus described. The second tier of the calibration will have its reference impedance set to the impedance of the delay standard, which is the impedance of the quarter-wave line. Although there are many methods of determining the characteristic impedance of a transmission line, methods based on estimating the capacitance per unit length and phase velocity are well suited for microstrip lines [12,15]. The capacitance per unit length and phase velocity uniquely describe the quasi-TEM characteristic impedance as

$$
Z_O = \frac{1}{v_p C} \tag{4.49}
$$

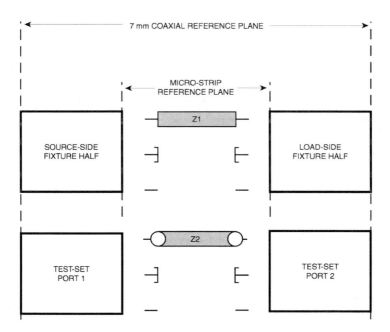

FIGURE 4.66 Reference-plane definitions for a two-tier calibration used for fixture characterization. The first tier is based on a TRL APC 7 mm calibration and the second tier is based on a microstrip TRL calibration.

Once the characteristic impedance of the delay-line is known, the s-parameters can be re-normalized to 50 Ω to make them compatible with the 50 Ω reference impedance that most automated load-pull systems use [2,3,15].

Figure 4.67 shows the forward source and load match of the second tier microstrip calibration used in the pre-matching fixture described in References 16 and 18. This fixture was intended to present 0.1 Ω at 2 GHz with extremely high accuracy. From the verification data, the resultant source match is better than –45 dB across the band and the resultant load match is better than –52 dB across the band. Comparing these results with Fig. 4.61 shows that the uncertainty is very low.

A significant advantage of using a transforming network to increase system VSWR, whether it be a quarter-wave line or an additional cascaded tuner, is that the two-port characterization of each element is done at manageable impedance levels. Characterization of a tuner presenting a 50:1 VSWR in direct cascade of a quarter-wave pre-match network would result in a significant increase in measurement uncertainty since the VNA must resolve impedances near 0.1 Ω. Segregating the characterization process moves the impedances that must be resolved to the 1 Ω to 2 Ω range, where the calibration uncertainty is considerably smaller.

The final step of the fixture verification process is to verify that the two-tier calibration has provided the correct two-port s-parameter description of each fixture half. Figure 4.68 shows each fixture half cascaded using the port definitions adopted by NIST Multical™ [15]. With microstrip, an ideal thru can be approximated by butting each fixture half together and making top-metal contact with a thin conductive film. When this is not possible, it is necessary to extract a two-port characterization of the thru. The cascaded transmission matrix is expressed as

$$\begin{bmatrix} A_{11} & B_{12} \\ C_{21} & D_{22} \end{bmatrix}_{cascade} = \begin{bmatrix} A_{11} & B_{12} \\ C_{21} & D_{22} \end{bmatrix}_{source} \begin{bmatrix} 1 & 0 \\ 0 & 1 \end{bmatrix}_{thru} \begin{bmatrix} A_{11} & B_{12} \\ C_{21} & D_{22} \end{bmatrix}_{load} \qquad (4.50)$$

where the middle matrix of the right-hand side is the transmission matrix of a lossless zero phase-shift thru network. Converting the cascade transmission matrix back to s-parameter form yields the predicted

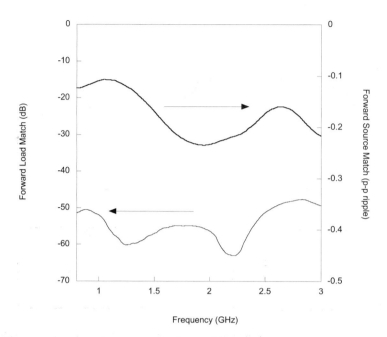

FIGURE 4.67   Microstrip TRL calibration using an offset-short and delay-line to extract source match and load match, respectively. This data was taken from an HP 8510C with an HP 8514B test set.

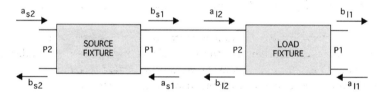

FIGURE 4.68   Port and traveling-wave definitions for cascading the source-fixture and load-fixture to examine the accuracy of the two-tier calibration fixture characterization.

response of the cascaded test-fixture, which can then be compared to the measurements of the cascade provided by the VNA.

Figure 4.69 shows the measured and predicted cascade magnitude response of a typical PCS quarter-wave pre-matching fixture based on an 11 Ω quarter-wave line; the phase is shown in Fig. 4.70 [16,18]. The relative error across the band is less than 0.1%. This type of fixture characterization performance is necessary to minimize error for synthesizing sub 1 Ω impedances.

System Performance Verification

Just as verification of VNA calibration is essential, so too is verification of overall load-pull system performance essential. Performance verification can be done with respect to absolute power or with respect to power gain. The former is recommended only occasionally, for example when the system is assembled or when a major change is made. The latter is recommended subsequent to each power calibration. Each of the methods will be described in this section.

Absolute power calibration is done by applying a signal to the source tuner via the source block of Fig. 4.58. After appropriately padding a power sensor, it is then connected to DUT side of the source tuner and, with the tuners set for 1:1 transformation, the resultant power is compared to what the overall cascaded response is expected to be.

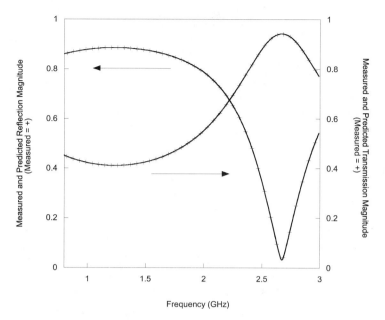

FIGURE 4.69 Forward reflection and transmission magnitude comparison of measured and cascaded fixture response. The error is so small the curves sit on top of each other.

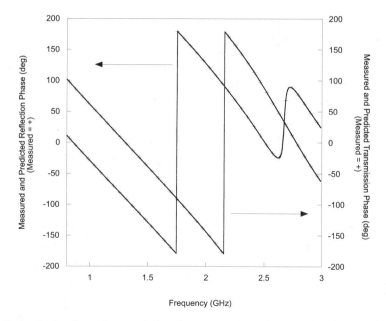

FIGURE 4.70 Forward reflection and transmission phase comparison of measured and cascaded fixture response. The error is so small the curves sit on top of each other.

This procedure is repeated for the load tuner except that the signal is injected at the DUT side of the load tuner and the power sensor is located as shown in Fig. 4.59. Splitting this verification in two steps assists in isolating any issues with either the source or load side. It is also possible to vary the impedance of each tuner and calculate what the associated available gain or power gain is, although this step is more easily implemented in the power gain verification.

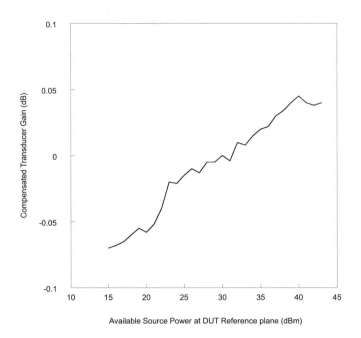

FIGURE 4.71 Measured transducer gain under the condition of conjugate match with mismatch loss compensation included.

Power gain verification starts with a two-port characterization of a known mismatch standard. The simplest way to implement this standard is to use one of the tuners, and then set the other tuner for the conjugate of this mismatch. In this case, the mismatch standard is an ideal thru, similar to the one used in fixture verification described in the previous section. Since it is unlikely that both the source and load tuners would have identical impedance domains, the measured loss must be compensated to arrive at actual loss. To compensate for this, the mismatch loss is computed as

$$G_{mm} = 10 \log_{10}\left[\frac{\left(1-\left|\Gamma_s\right|^2\right)\left(1-\left|\Gamma_l\right|^2\right)}{\left|1-\Gamma_s\Gamma_l\right|^2} \right] \tag{4.51}$$

where Γ_s and Γ_l are the source and load reflection coefficients, respectively, looking back into each tuner. Figure 4.71 shows a typical response of an entire cascade, including the quarter-wave pre-matching network. A transducer gain response boundary of ±0.1 dB is typical, and ±0.2 should be considered the maximum.

Summary

Load-pull is a valuable tool for evaluating high-power RF and microwave transistors, designing power amplifiers, and verifying large-signal model performance and validity domains. To enhance the reliability of the data that a load-pull system provides, it is essential that high performance VNA calibration techniques be adopted. Further, as emphasized in the present section, treating each section of the load-pull separately is useful from a measurement perspective and from a problem resolution perspective. In the former case, it was shown that measuring quarter-wave pre-matching networks and tuners separately reduces the uncertainty of the calibration. In the latter case, it was shown that characterization of each section individually allows its performance to be verified prior to integrating it within the entire system.

The central theme of this section has been the VNA and its associated calibration. Due to the extremely low impedances synthesized in high-power load-pull, the VNA calibration is the single most important element of the characterization process. Any errors or uncertainty encountered in the VNA calibration will be propagated directly into the load-pull characterization files and may result in erroneous data, particularly if system performance verification is not performed.

To present the sub 1 Ω impedances necessary for evaluation of high-power transistors, transforming networks are required. These can be implemented using an impedance transforming network, such as a quarter-wave line, or by cascading two tuners together. The former offers the highest VSWR at the expense of narrow bandwidth, while the latter is in general more flexible. In either case, high performance and reliable characterization methods are necessary to attain the best possible results for using load-pull as a verification and design tool.

Acknowledgments

Kerry Burger (Philips), Mike Majerus (Motorola), and Gary Simpson and John King (Maury Microwave) have, in many ways, influenced the content of this section. Their support and friendship is happily acknowledged.

References

1. J. M. Cusak et al., Automatic load-pull contour mapping for microwave power transistors, *IEEE Transactions on Microwave Theory and Techniques*, 1146–1152, December 1974.
2. *Automated Tuner System User's Manual*, v.1.9, Maury Microwave Corporation, 1998.
3. *Computer Controlled Tuner System User's Manual*, v. 6.0, Focus Microwave Corporation, 1998.
4. *LP2 Automated Load-Pull System User's Manual*, ATN Microwave Corporation, 1997.
5. F. Larose, F. Ghannouchi, and R. Bosisio, A new multi-harmonic load-pull method for non-linear device characterization and modeling, *Digest of the IEEE International Microwave Symposium Digest*, 443–446, June 1990.
6. F. Blache, J. Nebus, P. Bouysse, and J. Villotte, A novel computerized multi-harmonic load-pull system for the optimization of high-efficiency operating classes in power transistors, *IEEE International Microwave Symposium Digest*, 1037–1040, June 1995.
7. J. Sevic, R. Baeten, G. Simpson, and M. Steer, Automated large-signal load-pull characterization of adjacent-channel power ratio for digital wireless communication system, *Proceedings of the 45th ARFTG Conference*, 64–70, November 1995.
8. J. Sevic, K. Burger, and M. Steer, A novel envelope-termination load-pull method for the ACPR optimization of RF/microwave power amplifiers, *Digest of the IEEE International Microwave Symposium Digest*, 723–726, June 1998.
9. G. Simpson and M. Majerus, Measurement of large-signal input impedance during load-pull, *Proceedings of the 50th ARFTG Conference*, 101–106, December 1997.
10. D. Rytting, ARFTG Short-Course: Network Analyzer Calibration Theory, 1997.
11. R. Marks, Formulation of the basic vector network analyzer error model including switch terms, *Proceedings of the 50th ARFTG Conference*, 115–126, December 1997.
12. R. Marks and D. Williams, Characteristic impedance measurement determination using propagation measurement, *IEEE Microwave and Guided Wave Letters*, 141–143, June 1991.
13. G. Engen and C. Hoer, Thru-reflect-line: an improved technique for calibrating the dual six-port automatic network analyzer, *IEEE Transactions on Microwave Theory and Techniques*, 987–993, December 1979.
14. R. Marks, A multi-line method of network analyzer calibration, *IEEE Transactions on Microwave Theory and Techniques*, 1205–1215, July 1990.
15. *MultiCal™ User's Manual*, v. 1.0, National Institute of Standards and Technology, 1997.
16. J. Sevic, A sub 1 Ω load-pull quarter-wave pre-matching network based on a two-tier TRL calibration, *Proceedings of the 52nd ARFTG Conference*, 73–81, December 1998.

17. D. Balo, Designing and calibrating RF fixtures for SMT devices, *Hewlett-Packard 1996 Device Test Seminar*, 1996.

18. J. Sevic, A sub 1 Ω load-pull quarter-wave pre-matching network based on a two-tier TRL calibration, *Microwave Journal*, 122–132, March 1999.

19. R. Collin, *Foundations for Microwave Engineering*, McGraw-Hill: New York, 1966.

20. B. Wadell, *Transmission Line Design Handbook*, Artech House: Boston, 1991.

21. *EM User's Manual*, v. 6.0, Sonnet Software, Inc., Liverpool, NY, 1999.

22. *HP 8510C User's Manual*, Hewlett-Packard Company, 1992.

4.6 Pulsed Measurements

Anthony E. Parker, James G. Rathmell, and Jonathan B. Scott

Pulsed measurements ascertain the radio-frequency (RF) behavior of transistors or other devices at an unchanging bias condition. A pulsed measurement of a transistor begins with the application of a bias to its terminals. After the bias has settled to establish a **quiescent condition**, it is perturbed with pulsed stimuli during which the change in terminal conditions, voltage and current, is recorded. Sometimes a RF measurement occurs during the pulse. The responses to the pulse stimuli quantify the behavior of the device at the established quiescent point. **Characteristic curves**, which show the relationship between terminal currents or RF parameters and the instantaneous terminal potentials, portray the behavior of the device.

Pulsed measurement of the characteristic curves is done using short pulses with a relatively long time between pulses to maintain a constant quiescent condition. The characteristic curves are then specific to the quiescent point used during the measurement. This is of increasing importance with the progression of microwave-transistor technology because there is much better consistency between characteristic curves measured with pulses and responses measured at high frequencies. When the behavior of the device is bias- or rate-dependent, pulsed measurements yield the correct high-frequency behavior because the bias point remains constant during the measurement. Pulse techniques are essential for characterizing devices used in large-signal applications or for testing equipment used in pulse-mode applications. When measurements at high potentials would otherwise be destructive, a pulsed measurement can safely explore breakdown or high-power points while maintaining a bias condition in the safe-operating area (SOA) of the device. When the device normally operates in pulse mode, a pulsed measurement ascertains its true operation.

The response of most microwave transistors to high-frequency stimuli depends on their operating conditions. If these conditions change, the characteristic curves vary accordingly. This causes **dispersion** in the characteristic curves when measured with traditional curve-tracers. The operating condition when sweeping up to a measurement point is different than that when sweeping down to the same point. The implication is that any change in the operating conditions during the measurement will produce ambiguous characteristic curves.

Mechanisms collectively called **dispersion effects** contribute to dispersion in characteristic curves. These mechanisms involve thermal, rate-dependent, and electron trapping phenomena. Usually they are slow acting, so while the operating conditions of the device affect them, RF stimuli do not. Even if the sequence of measurement precludes observation of dispersion, dispersion effects may still influence the resulting characteristic curves.

Pulsed measurements are used to acquire characteristic curves that are free of dispersion effects. The strategy is to maintain a constant operating condition while measuring the characteristic curves. The pulses are normally short enough to be a signal excursion rather than a change in bias, so dispersion effects are negligible. The period between pulses is normally long enough for the quiescent condition of the device to recover from any perturbation that may occur during each pulse.

Pulse techniques cause less strain, so are suitable for extending the range of measurement into regions of high power dissipation and electrical breakdown. Pulse techniques are also valuable for experiments in device physics and exploration of new devices and material systems at a fragile stage of development.

Stresses that occur when operating in regions of breakdown or overheating can alter the characteristic curves permanently. In many large-signal applications, there can be excursions into both of these regions for periods brief enough to avoid undue stress on the device. To analyze these applications, it is desirable to extend characteristic curves into the stress regions. That is, the measurements must extend as far as possible into regions that are outside the SOA of the device. This leads to another form of dispersion, where the characteristic curves change after a measurement at a point that stresses the device.

Pulsed measurements can extend to regions outside the SOA without stressing or damaging the device. If the pulses are sufficiently short, there is no permanent change in the characteristic curves. With pulses, the range of the measured characteristic curves can often extend to completely encompass the signal excursions experienced during the large-signal operation of devices.

In summary, pulsed measurements yield an extended range of characteristic curves for a device that, at specific operating conditions, corresponds to the high-frequency behavior of the device. The following sections present the main principles of the pulse technique and the pulse-domain paradigm, which is central to the technique. The pulse-domain paradigm considers the characteristic curves to be a function of quiescent operating conditions. Therefore, the basis for pulse techniques is the concept of measurements made in **isodynamic** conditions, which is effectively an invariable operating condition. A discussion is included of the requirements for an isodynamic condition, which vary with the transistor type and technology. There is also a review of pulsed measurement equipment and specifications in terms of cost and complexity, which vary with application. Finally, there is an examination of various pulsed measurement techniques.

Isothermal and Isodynamic Characteristics

For the analysis of circuit operation and the design of circuits, designers use transistor **characteristics**. The characteristics consist of **characteristic curves** derived from measurements or theoretical analysis. These give the relationship between the variable, but interdependent terminal conditions and other information that describes the behavior of the device. To be useful, the characteristics need to be applicable to the operating condition of the device in the circuit.

In all circuits, when there is no signal, the device operates in a **quiescent condition** established by bias networks and power supplies. The **DC characteristics** are characteristic curves obtained with slow curve tracers, conventional semiconductor analyzers, or variable power supplies and meters. They are essentially data from a set of measurements at different bias conditions. Consequently, the quiescent operating point of a device is predictable with DC characteristics derived from DC measurements. Figure 4.72 shows a set of DC characteristics for a typical microwave MESFET. This figure also shows the very different set of **pulsed characteristics** for the same device made at the indicated quiescent point. The pulsed characteristics give the high-frequency behavior of the MESFET when biased at that quiescent point.

A clear example of a dispersion effect that causes the observed difference between the DC and pulsed characteristics is heating due to power dissipation. When the characteristics are measured at a slow rate (≈ 10 ms per point), the temperature of the device at each data point changes to the extent that it is heated by the power being dissipated at that point. Pulsed characteristics are determined at the constant temperature corresponding to the power dissipation of a single bias point. This measurement at constant temperature is one made in **isothermal** conditions.

In general, device RF characteristics should be measured in a constant bias condition that avoids the dynamics of thermal effects and any other dispersion effects that are not invoked by a RF signal. Such a measurement is one made in **isodynamic** conditions.

Small-Signal Conditions

Devices operating in **small-signal** conditions give a nearly linear response, which can be determined by **steady-state RF** measurements made at the quiescent point. A network analyzer, operated in conjunction with a bias network, performs such a measurement in isodynamic conditions. Once the quiescent condition is established, RF measurements characterize the terminal response in terms of small-signal

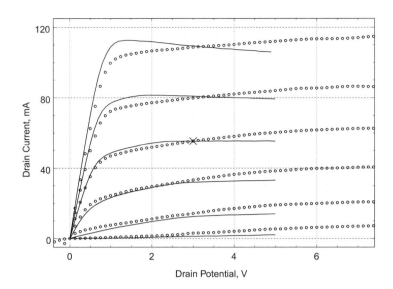

FIGURE 4.72 Characteristic curves for a MESFET. Shown are the DC characteristics (−) and the pulsed characteristics (O), with 300 ns pulses separated by 200 ms quiescent periods, for the quiescent point $V_{DS} = 3.0$ V, $I_D = 55.4$ mA (×). Gate-source potential from −2.0 to +0.5 V in 0.5 V steps is the parameter.

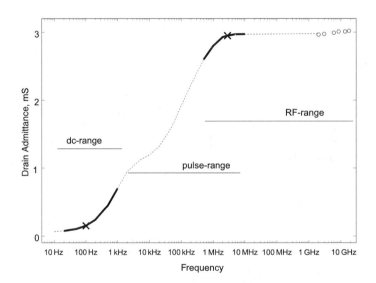

FIGURE 4.73 Frequency variation of drain-source admittance for the typical MESFET of Fig. 4.72 at the quiescent point $V_{DS} = 3.0$ V, $I_D = 55.4$ mA. An indicative response (- -) connects measured $\Re(Y_{22})$ from a RF network analyzer (O) and calculation from the pulsed and DC data in Fig. 4.72 (×). Also indicated are the typical frequency ranges applicable to DC, pulsed, and RF measurements.

parameters, such as Y-parameters. A different set of small-signal parameters is required for each quiescent condition.

It is not possible to correlate the small-signal parameters with the DC characteristics when there are dispersion effects. For example, the output conductance (drain-source admittance) of a typical MESFET varies with frequency as shown in Fig. 4.73. For this device, the small-signal conductance varies little with frequency above about 1 MHz. The conductance is easily determined from the real part of Y_{22} measured with a network analyzer. The conductance can also be determined from the slope of the pulsed characteristics at the quiescent point. The data from short pulses, shown in Fig. 4.72 in the regime of 1 to 10 MHz, give an **isodynamic characteristic** for this typical device because the calculated conductance is the same as that measured at higher frequencies. With longer pulses, corresponding to lower frequencies, dispersion effects influence the conductance significantly. The characteristics measured at rates below 1 MHz can vary with the type of measurement because each point affects the subsequent point. The dispersion effects are prominent at the slow 10 to 1000 Hz rate of curve-tracer operation, which is why dispersion is observed in curve-tracer measurements. True DC measurements usually require slower rates.

Thermal Model

Thermal dispersion has a significant influence on the output conductance. To quantify this, consider the relationship between the terminal current i_T [A] and voltage v_T [V]. The small-signal terminal conductance is $g = d\, i_T/d\, v_T$ [S]. To explore the influence of thermal dispersion on this parameter, assume that the terminal current is a linear function of temperature rise ΔT [K] with thermal-coefficient λ [1/K], so that

$$i_T = i_O\left(1 - \lambda\Delta T\right). \tag{4.52}$$

The thermodynamic rate equation relates the temperature rise of the device to time t and heat flow due to power dissipation $Q = i_T\, v_T$ [W]:

$$mC\, R_T\, d\Delta T\big/dt + \Delta T = R_T\, Q. \tag{4.53}$$

The term mC [J/K] is the product of mass and heat capacity of the thermal path to ambient temperature, and R_T [K/W] is the thermal resistance of the path. There is a time constant $\tau = mC\, R_T$ [s] associated with this rate equation.

With **isothermal** conditions, the temperature rise remains constant during operation of the device. This occurs when the rate of change of power dissipation, due to signal components, is either much faster or much slower than the thermal time constant. With high-frequency signals, it is the quiescent power dissipation at the quiescent terminal current, I_T, and voltage, V_T, that sets the temperature rise. The rate Eq. (4.53) reduces to $\Delta T = R_T\, Q$ where $Q = I_T\, V_T$ is constant. The isothermal terminal current [Eq. (4.52)] is then:

$$i_T = i_O\left(1 - \lambda R_T I_T V_T\right). \tag{4.54}$$

The terminal conductance determined by small-signal RF measurement is then:

$$g = di_T\big/dv_T = di_O\big/dv_T\left(1 - \lambda R_T I_T V_T\right). \tag{4.55}$$

During measurement of DC characteristics, which are made at rates slower than the thermal time constant, the rate Eq. (4.53) reduces to $\Delta T = R_T\, i_T\, v_T$. This is different at each measurement point, so the DC terminal current [Eq. (4.52)] becomes:

$$i_T = i_O\left(1 - \lambda R_T\, i_T v_T\right). \tag{4.56}$$

An estimate of the terminal conductance from DC characteristics would be

$$G = di_O\big/dv_T\left(1 - \lambda R_T\, i_T\, v_T\right) - \lambda R_T\, i_O\left(i_T + G v_T\right). \tag{4.57}$$

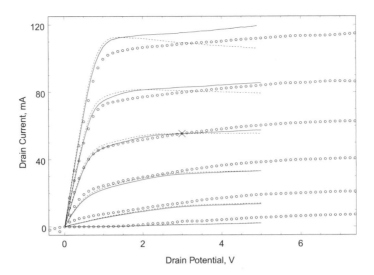

FIGURE 4.74 The MESFET characteristic curves shown in Fig. 4.72 with a set of DC characteristics (−) normalized to the temperature due to power dissipation at the quiescent point, $V_{DS} = 3.0$ V, $I_D = 55.4$ mA (×). Also shown are the raw DC characteristics (- -) and the pulsed characteristics (O) for the quiescent point V_{DS}, I_D. Gate-source potential from −2.0 to +0.5 V in 0.5 V steps is the parameter.

The difference between the small-signal conductance g in Eq. (4.55) and the DC conductance G in Eq. (4.57) is due to thermal dependence. If $\lambda = 0$, then $g = G$. Without knowing λR_T it is not possible to determine the small-signal conductance from the DC characteristics.

Figure 4.74 shows an attempt to determine the pulsed characteristics from the DC characteristics. The thermal effect is removed from the DC characteristics with a value of $\lambda R_T = 0.3$ W^{-1} determined from a complete set of pulsed characteristics made over many quiescent points. Multiplying the drain current of each point (v_{DS}, i_D) in the DC characteristics by $(1 - \lambda R_T I_D V_{DS})/(1 - \lambda R_T i_D v_{DS})$ normalizes it to the temperature of the quiescent point (V_{DS}, I_D) used in the pulsed measurement. Figure 4.74 demonstrates that although temperature, explained by the simple model above, is a dominant effect, other dispersion effects also affect the characteristics. The ambient-temperature DC characteristics exhibit changes in threshold potential, transconductance, and other anomalous characteristics, which occur because electron trapping, breakdown potentials, and leakage currents also vary with bias. The pulsed measurements made in isodynamic conditions are more successful at obtaining characteristic curves that are free of these dispersion effects.

Large-Signal Conditions

Transistors operating in **large-signal** conditions operate with large signal excursions that can extend to limiting regions of the device. Large-signal limits, such as clipping due to breakdown or due to excessive input currents, can be determined from an extended range of characteristic curves. Steady-state DC measurements are confined to operating regions in the SOA of the device. It is necessary to extrapolate to breakdown and high-power conditions, which may prompt pushing the limits of measurements to regions that cause permanent, even if non-destructive, damage. The stress of these measurements can alter the characteristics and occurs early in the cycle of step-and-sweep curve tracers, which leads to incorrect characteristic curves in the normal operating region. The observed dispersion occurs in the comparison of the characteristics measured over moderate potentials, measured before and after a stress.

Pulsed measurements extend the range of measurement without undue stress. Figure 4.75 shows characteristic curves of a HEMT that encompasses regions of breakdown and substantial forward gate potential. The diagram highlights points in these regions, which are those with significant gate current.

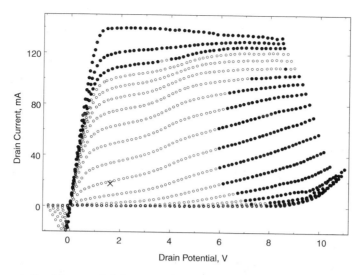

FIGURE 4.75 An example of the extended characteristic curves for an HEMT obtained with 500 ns pulses at 10 ms intervals for the quiescent point $V_{DS} = 1.6$ V, $I_D = 17.7$ mA ($\times$). The solid points • are those for which the magnitude of gate current is greater than 1 mA. Gate-source potential from -3.0 to $+1.5$ V in 250 mV steps is the parameter.

The extended characteristics are essential for large-signal applications to identify the limits of signal excursion. The pulsed characteristics in the stress regions are those that would be experienced during a large-signal excursion because the measurement is made in isodynamic conditions set by the operating point. There is little correlation between these and an extrapolation from DC characteristics because the stress regions are significantly affected by bias conditions and temperature.

Pulsed Measurements

Dispersion effects in microwave devices generate a rich dynamic response to large signals and changing operating conditions. The dynamic behavior affects the DC and high-frequency characteristics but is not observable in either. Thus, pulsed measurement techniques are required to quantify various aspects of the dynamic behavior.

The pulsed current/voltage (**pulsed-I/V**) characteristics are characteristic curves determined from an isodynamic measurement with short pulses separated by long relaxation periods at a specific quiescent point. Each quiescent point has its own pulsed-I/V characteristics, so a complete characterization of a device requires pulsed-I/V measurements over various quiescent points. Dispersion effects do not affect each pulsed-I/V characteristic but do affect the variation between characteristics measured in different quiescent conditions.

The pulsed characteristics vary with pulse length. Short pulses produce isodynamic pulsed-I/V characteristics, and very long pulses produce DC characteristics. A **time domain** pulsed measurement, performed by recording the variation of terminal conditions during a measurement pulse, can trace the transition from isodynamic to DC behavior. The time constants of the dispersion effects are present in the time domain characteristic. Note that the range of time domain measurements is limited to the SOA for the long pulses used.

Isodynamic small-signal parameters are determined from **pulsed-RF** measurements. During the measurement pulse, a RF signal is applied and a pulsed vector network analyzer determines the scattering parameters. The terminal potentials during each pulse are the **pulsed bias** for the RF measurement. Each operating point, at which the device relaxes between pulses, has its own set of pulsed-bias points and corresponding RF parameters. Pulsed-RF characteristics give small-signal parameters, such as reactance and conductance, as a surface function of terminal potentials. There is a small-signal parameter surface for each quiescent operating point and the dispersion effects only affect the variation of each surface with quiescent condition. Pulsed-RF measurements are also required for pulse-operated equipment, such as pulsed-radar transmitters, that have off-state quiescent conditions and pulse to an on-state condition that may be outside the SOA of the device.

Pulse timing and potentials vary with the measurement type. The periods required for isodynamic conditions and safe-operating potentials for various types of devices are discussed in the next section. The complexity and cost of pulse equipment, which also varies with application, is discussed in the subsequent section.

Relevant Properties of Devices

Three phenomena present in active devices that cause measurement problems best addressed with pulsed measurements. These are the SOA constraint, thermal dependency of device characteristics, and dependency of device characteristics upon charge trapped in and around the device. The following discusses these phenomena and identifies devices in which they can be significant.

Safe-Operating Area

The idea of a safe operating area is simply that operating limits exist beyond which the device may be damaged. The SOA limits are generally bounds set by the following four mechanisms:

- A maximum voltage, above which a mechanism such as avalanche breakdown can lead to loss of electrical control or direct physical alteration of the device structure.
- A maximum power dissipation, above which the active part of the device becomes so hot that it is altered physically or chemically.
- A maximum current, above which some part of the device like a bond wire or contact region can be locally heated to destruction.
- A maximum current-time product, operation beyond which can cause physical destruction at local regions where adiabatic power dissipation is not homogeneous.

It is important to realize that damage to a device need not be catastrophic. The above mechanisms may change the chemical or physical layout of the device enough to alter the characteristics of the device without disabling it.

Pulsed-I/V measurements offer a way to investigate the characteristics of a device in areas where damage or deterioration can occur, because it is possible to extend the range of measurements under pulsed conditions, without harm. This is not a new idea — pulsed capability has been available in curve tracers for decades. These pulsed systems typically have pulses no shorter than a few milliseconds or a few hundred microseconds. However, shorter pulses allow further extension, and for modern microwave devices, true signal response may require sub-microsecond stimuli.

There are time constants associated with SOA limitations. For example, the time constant for temperature rise can allow very high power levels to be achieved for short periods. After that time, the device must be returned to a low-power condition to cool down. The SOA is therefore much larger for short periods than it is for steady DC conditions. Figure 4.76 shows successful measurement of a 140 μm^2 HBT well beyond the device SOA. The example shows a sequence of measurement sweeps with successively increasing maximum collector potential. There is no deterioration up to 7.5 V, which is an order of magnitude above that which would rapidly destroy the device under static conditions. The sweeps to a collector potential greater than 7.5 V alter the device so its characteristics have a lower collector current in subsequent sweeps. Shorter pulses may allow extension of this limit.

Different active devices are constrained by different SOA limits. For instance, GaN FETs are not usually limited by breakdown, whereas certain III-V HBTs are primarily limited by breakdown; silicon devices suffer more from a current-time product limit than do devices in the GaAs system. Pulsed I/V measurements provide a way for device designers to identify failure mechanisms, and for circuit designers to obtain information about device characteristics in regions where signal excursions occur, which are outside the SOA.

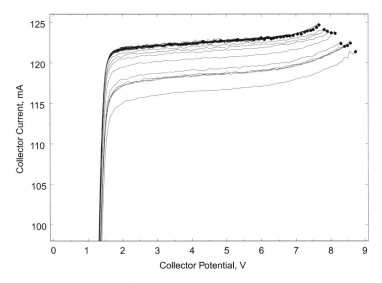

FIGURE 4.76 A single collector characteristic measured on a 140 μm² III-V HBT with sequentially increasing maximum voltage (shown by •) applied in 1 μs pulses. Note the progressive deterioration above a certain instantaneous dissipation level.

Thermal Dispersion

GaAs devices, both FETs and HBTs, have greater thermal resistance than do their silicon counterparts. They tend to suffer larger changes in characteristics per unit change in junction temperature. Perhaps the first need for pulsed-I/V measurements arose with GaAs MESFETs because of the heating that occurs in simple DC measurement of these devices. Such a measurement turns out to be useless in high-frequency terms because each part of the measurement is at a vastly different temperature. This does not represent device characteristics in a RF situation where the temperature does not perceptibly change in each signal period. The sole utility of DC characteristics is to help predict quiescent circuit conditions.

A pulsed-I/V measurement can approach isothermal conditions, and can circumvent this problem. Figure 4.72, showing the DC and pulsed characteristics of a simple MESFET, exemplifies the difference. It is remarkable that the characteristics are for the same device.

Silicon devices, both FET and BJT, are relatively free of thermal dispersion effects, as are GaN FETs. The susceptibility of any given device, and the pulse duration and duty cycle required to obtain isothermal data, must be assessed on a case-by-case basis. Methods for achieving this are explored in the later discussion of measurement techniques.

Charge Trapping

Temperature is not the only property of device operation that can give rise to dispersion. Charge trapped in substrate or defects is particularly troublesome in FETs. Rather than power dissipation, currents or junction potentials can control slow-moving changes in the device structure. These phenomena are not as well understood as their thermal counterparts.

Exposing charge-trapping phenomena that may be influencing device performance is more difficult, but is still possible with an advanced pulsed-I/V system. One method is to vary the quiescent conditions between fast pulses, observing changes in the pulsed characteristic as quiescent fields and currents are varied independently, while holding power dissipation constant. Figure 4.77 shows two pulsed characteristics measured with identical pulse-stimulus regimes, but with different quiescent conditions. Since the power dissipation in the quiescent interval is unchanged, temperature does not vary between the two experiments, yet the characteristics do. The difference is attributed to trapped charge exhibiting a relatively long time constant.

Charge-trapping dispersion is most prevalent in HEMTs, less so in HFETs and MESFETs, and has yet to be reported in bipolar devices such as HBTs.

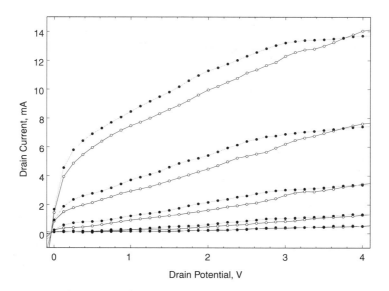

FIGURE 4.77 Two pulsed-I/V characteristics for the same GaAs FET measured at different quiescent conditions, $V_{DS} = 1.1$ V, $I_D = 0.4$ mA (O) and $V_{DS} = 2.2$ V, $I_D = 0.2$ mA (•). They have identical power dissipation. The measurements used 300 ns pulses separated by 200 ms quiescent periods. Gate-source potential from 0.0 to +0.5 V in 125 mV steps is the parameter.

Time Constants

Avalanche effects can operate extremely quickly — much faster than any pulse system — so SOA measurements exhibit the thermal and charge trapping time constants. Thermal effects typically have several time constants associated with them, each associated with the thermal capacity of some part of the system, from the active region to the external heat sink. Small devices typically have time constants of the order of one microsecond; larger devices may have their smallest significant time constant ten times larger than this. Package time constants tend to be of the order of milliseconds to tens or hundreds of milliseconds. External heat sinks add long time constants, though anything above a few seconds is disregarded or treated as environmental drift, since measurement or control of such external temperature is straightforward.

Charge trapping phenomena are more variable. Indeed, there are reports of devices susceptible to disruption from charge stored apparently permanently, after the fashion of flash memory. Values of the order of hundreds of microseconds are common, ranging up to milliseconds and longer.

Because of the wide variation of time constants, it is hard to know *a priori* what settings are appropriate for any measurement, let alone what capability ought to be specified in an instrument to make measurements. Values of less than 10 µs for pulse width and 1 ms for quiescent time might be marginally satisfactory, while 500 ns pulses with 10 ms quiescent periods would be recommended.

Pulsed-I/V and Pulsed-RF Characteristics

Pulsed-I/V measurement is sometimes accompanied by pulsed-RF measurements. The RF equipment acquires the raw data during the pulse stimulus part of the measurement. Given that pulsed-I/V systems characterize devices in isodynamic conditions, the need for measurement at microwave frequencies, simultaneously with pulse stimuli, might be questioned. The problem is that it may not be possible to infer the reactive parameters for a given quiescent point from static S-parameters that are measured over a range of DC bias conditions. This is because of significant changes in RF behavior linked to charge trapping or thermal dispersion effects.

Figure 4.78 compares S-parameters of an HBT measured at a typical operating point (well within the SOA) using a DC bias and using a 1µs pulsed bias at the same point with the device turned off between pulses. The differences, attributed to temperature, indicate the impact of dispersion effects on RF characteristics.

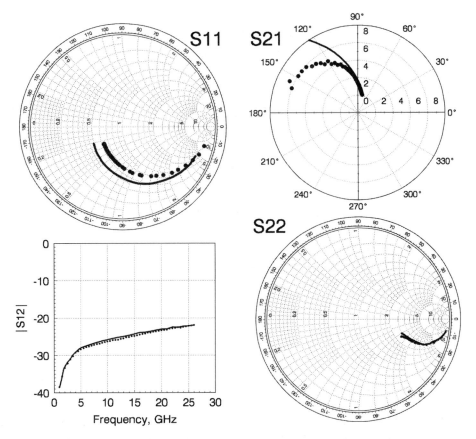

FIGURE 4.78 S-parameters measured at the same bias point with off-state and on-state quiescent conditions. The on-state parameters are from static, or DC measurements (–) and the off-state parameters are from measurements in a pulsed bias at the same point with off-state quiescent periods (•).

In addition, S-parameters cannot be gathered at bias points outside the SOA without pulse equipment. Pulse amplifiers often operate well beyond the SOA, so that a smaller, less expensive device can be used. This is possible when the duration of operation beyond SOA is brief, but again, it is not possible to characterize the device with DC techniques. For many of these applications, pulsed-RF network analyzers have been developed. These can measure the performance of the transistor during its pulsed operating condition.

Pulsed Measurement Equipment

Pulsed measurement systems comprise subsystems for applying bias, pulse, and RF stimuli, and for sampling current, voltage, and RF parameters. Ancillary subsystems are included to synchronize system operation, provide terminations for the device under test (DUT), and store and process data. A simple system can be assembled from individual pulse generators and data acquisition instruments. More sophisticated systems generate arbitrary pulse patterns and are capable of measurements over varying quiescent and pulse timing conditions. Pulsed-I/V systems can operate as stand-alone instruments or can operate in a pulsed-RF system to provide the pulsed bias.

System Architecture

The functional diagram of a pulsed measurement system, shown in Fig. 4.79, includes both pulsed-I/V and pulsed-RF subsystems. Pulse and bias sources, voltage and current sampling blocks, and associated timing generators form the pulsed-I/V subsystem. A pulsed-RF source and mixer-based vector network

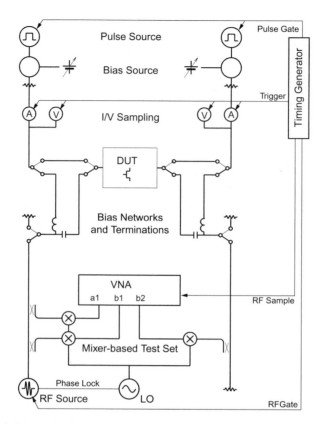

FIGURE 4.79 Simplified diagram of a generic pulsed measurement system. Alternative connections provide load terminations when there is no pulsed-RF test set or directly connect the pulsed-I/V subsystem to the DUT.

analyzer form the pulsed-RF subsystem. The DUT is connected directly to the pulsed-I/V subsystem, or to bias networks that connect the pulsed-RF subsystem or RF terminations.

Pulsed-I/V System

Steady-state DC semiconductor parameter analyzers provide a source-monitor unit for each terminal of the DUT. The unit sources one of voltage or current while monitoring the other. In a pulsed measurement system, a pulsed voltage is added to a bias voltage and applied to the device. It is not practical to control the source potential within short pulse periods, so in order to ascertain the actual terminal conditions, both voltage and current are monitored. If a precise potential is required, then it is necessary to iterate over successive pulses, or to interpolate data from a range of pulsed measurements, or use longer pulse periods.

Simple systems use a pulse generator as the pulse source. Stand-alone pulse generators usually provide control of pulse and quiescent levels, so a single pulse point is measured during each test run. Such a system is easily assembled with pulse generators and is operated from their front panels. A single-point measurement mode is also employed by high-power pulsers that deliver high current pulses by dumping charge from capacitors, which are precharged during the quiescent period.

Systems that measure several pulse points in sequence use computer controlled arbitrary function generators to provide pulse and quiescent potentials. The function generators are essentially digital memory delivering values to a digital-to-analog converter. Pulse values are stored in every second memory location and the quiescent value is stored in every other location. A timing generator then clocks through successive potentials at the desired pulse and quiescent time intervals. The quiescent potential is either simply delivered from the pulse generators or it is delivered from bench power supplies or other computer controlled digital-to-analog converters. In the latter cases, a summing amplifier adds the pulse and quiescent potentials and drives the DUT. This architecture extends the pulse power capability of the system. Whereas the continuous rating of the amplifier dictates the maximum quiescent current delivered to the device, the pulse range extends to the higher transient current rating of the amplifier.

In most systems, either data acquisition digitizers or digital oscilloscope channels sample current and voltage values. In a simple setup, an oscilloscope will display the terminal conditions throughout the pulse and the required data can be read on screen or downloaded for processing. Oscilloscope digitizers tend to have resolutions sufficient for displaying waveforms, but insufficient for linearity or wide dynamic range measurements. Data acquisition digitizers provide wider dynamic range and ability to sample at specific time points on each pulse or throughout a measurement sequence. When several pulse points are measured in sequence, the digitizers record pulse data from each pulse separately or time domain data from several points across each pulse. Either mode is synchronized by appropriate sampling triggers provided by a timing generator.

The position of the voltage and current sensors between the pulse source and the DUT is significant. There are transmission line effects associated with the cabling between the sensing points and the digitizers. The cable lengths and types of terminations will affect the transient response of, and hence the performance of, the pulse system. An additional complication is introduced when the DUT must be terminated for RF stability. A bias network is used but this introduces its own transient response to the measured pulses. For example, the initial 100 ns transient in Fig. 4.84 is generated by the bias network and is present when the DUT is replaced by a 50 Ω load.

Current is sensed by various methods that trade between convenience and pulse performance. With a floating pulse source, a sense resistor in the ground return will give the total current delivered by the source. There is no common-mode component in this current sensor, so a single-ended digitizer input is usable. The current reading will include, however, transient components from the charging of capacitances associated with cables between the pulser and the DUT. Low impedance cables can ameliorate this problem. Alternatively, hall-effect/induction probes placed near the DUT can sense terminal current. These probes have excellent common-mode immunity but tend to drift and add their own transient response to the data. A stable measurement of current is possible with a series sense resistor placed in line near the DUT. This eliminates the effect of cable capacitance currents, but requires a differential input with very good common-mode rejection. The latter presents a severe limitation for short pulses because common-mode rejection degrades at high frequency.

Data collection and processing in pulse systems is different than that of slow curve tracers or semiconductor parameter analyzers. The latter usually measure over a predefined grid of step-and-sweep values. If the voltage grid is defined, then only the current is recorded. The user relies on the instrument to deliver the specified grid value. In pulse systems, a precise grid point is rarely reached during the pulse period. The pulse data therefore includes measured voltage and current for each terminal. An important component in any pulse system is the interpretation process that recognizes that the pulse data do not lie on a regular grid of values. One consequence of this is that an interpolation process is required to obtain traditional characteristic curves.

Pulsed-RF System

Pulsed-RF test sets employ vector network analyzers with a wideband intermediate frequency (IF) receiver and an external sample trigger.[1] The system includes two RF sources and a mixer-based S-parameter test set. One source provides a continuous local oscillator signal for the mixers, while the other provides a gated RF output to the DUT. The local oscillator also provides a phase reference, so that a fast sample response is possible.

The pulsed bias must be delivered through bias networks, which are essential for the pulsed-RF measurement. During a pulsed-I/V measurement, the RF source is disabled and the RF test set provides terminations for the DUT. Pulsed-RF measurements are made one pulse point at a time. With the pulsed bias applied, the RF source is gated for a specified period during the pulse and the network analyzer is

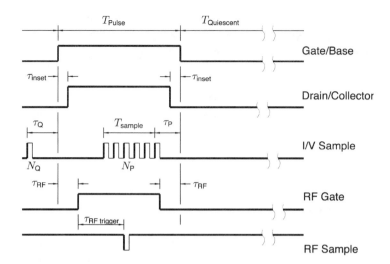

FIGURE 4.80 Generic timing diagram for each pulsed measurement pulse event.

triggered to sample the RF signals. The same pulse point is measured often enough for the analyzer to work through its frequency list and averaging requirements.

Technical Considerations

A trade between cost, complexity, and technical performance arises in the specification and assembly of pulsed measurement systems. Important considerations are pulse timing capability, measurement resolution and range, total time required for a measurement task, and the flexibility of the pulse sequencing.

Pulse Events

Pulsed measurement systems produce a continuous, periodic sequence of **pulse events**. The generic timing of each part of a pulse event is shown in Fig. 4.80. Each pulse event provides a pulse stimulus and a quiescent period. The period of the pulse, T_{Pulse}, ranges from 10 ns to 1 s. Typically, pulsed-I/V measurements require 200 to 500 ns pulses, and true DC measurements require periods of 100 ms or more. To achieve sub-100 ns pulses, usually the DUT is directly connected to a pulse generator to avoid transmission-line effects. Quiescent periods, $T_{Quiescent}$, range from 10 μs to 1 s and often must be longer than 1 ms for isodynamic pulsed-I/V measurements.

One or both terminals of the DUT may be pulsed. In some systems, the pulse width on the second terminal is inset relative to the first, by τ_{inset}, which gives some control over the trajectory of the initial pulse transient to avoid possible damage to the DUT.

Samples of current and voltage occur some time, τ_P, before the end of the pulse. Some systems gather a number, N_P, of samples over a period, T_{sample}, which may extend over the entire pulse if a time domain transient response is measured. The number of samples is limited by the sampling rate and memory of the digitizers. A measurement of the quiescent conditions some time, τ_Q, before the start of each pulse may also be made.

For pulsed-RF measurements, the RF source is applied for a period that is inset, by τ_{RF}, within the pulsed bias. A RF trigger sequences sampling by the network analyzer. The RF source is disabled during pulsed-I/V measurements.

Measurement Cycles

A pulsed **measurement cycle** is a periodic repetition of a sequence of pulse events. A set of pulse points, required to gather device characteristics, is measured in one or more measurement cycles. With single pulse-point measurements, there is only one pulse event in the sequence and a separate measurement cycle is required for each pulse point. This is the case with pulsed-RF measurements, with high-power pulsers, or with very-high-speed pulse generators. With arbitrary function generators, the measurement cycle is a sequence of pulse events at different pulse points; so one cycle can measure several pulse points.

Measurement cycles should be repeated for a **stabilizing period** to establish the bias condition of the measurement cycle, which is a steady-state repetition of pulse events. Then the cycle is continued while data are sampled. Typical stabilization periods can range from a few seconds to tens of seconds. These long times are required for initial establishment of stable operating conditions, whereas shorter quiescent periods are sufficient for recovery from pulse perturbations.

When several pulse points are measured in each cycle, the pulse stimulus is a steady-state repetition, so each pulse point has a well-known initial condition. Flexible pulse systems can provide an arbitrary initial condition within the cycle or use a pseudo-random sequencing of the pulse points. These can be used to assess the history dependence or isodynamic nature of the measurements. For example, it may be possible to precede a pulse point with an excursion into the breakdown region to assess short-term effects of stress on the characteristic.

Bias Networks

The most significant technical limitation to pulsed measurement timing is the bias network that connects the DUT to the pulse system. The network must perform the following:

- Provide RF termination for the DUT to prevent oscillations
- Pass pulsed-bias stimuli to the DUT
- Provide current and voltage sample points
- Control transients and overshoots that may damage the DUT

These are contradictory requirements that must be traded to suit the specific application. In general, the minimum pulse period is dictated by the bias network.

For very-fast pulsed measurements, less than 100 ns, the pulse generator is usually connected directly to the DUT.[2] The generator provides the RF termination required for stability, and current and voltage are sensed with a ground-return sense resistor and a high impedance probe, respectively. Pulsed-RF measurements are not contemplated with this arrangement.

Systems that are more flexible use a modified bias network similar to that shown in Fig. 4.81. The DC-blocking capacitor must be small, so that it does not draw current for a significant portion of the pulsed bias, but must be large enough to provide adequate termination at RF frequencies. The isolating inductor must be small, so that it passes the pulsed bias, but must also be large enough to provide adequate RF isolation. In this example, the DUT is connected to a RF termination provided by a load or network analyzer. The DC-blocking capacitor, 30 pF, and isolating inductor, 70 nH, values are an order of magnitude smaller than are those in conventional bias networks. The network provides a good RF path for frequencies above 500 MHz and does not significantly disturb pulses longer than 100 ns. Modifying the network to providing a RF path at lower frequencies will disturb longer pulses.

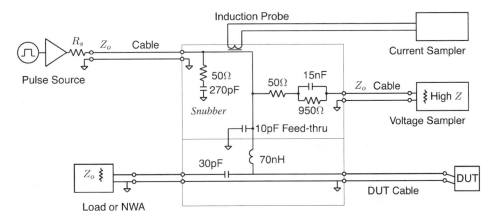

FIGURE 4.81 Schematic of a bias network that provides RF termination and pulsed bias feed with voltage and current measuring points.

The pulsed bias is fed to the bias network in Fig. 4.81 through a cable that will introduce transmission line transients. To control these, the source output impedance can provide line termination. Although this can provide significant protection from transients when fragile devices are being measured, it will limit the voltage and current range of the pulses. An alternative is to provide a termination at the bias network end of the cable with a series resistor-capacitor snubber. The values shown in this example are suitable for suppressing the 10 ns transients associated with a 1 m cable.

Voltage sampling in Fig. 4.81 is through a frequency-compensated network that provides isolation between the RF path and the cable connected to the voltage sampling digitizer. Without this isolation, the capacitance of the cable would load the pulsed bias waveform, significantly increasing its rise time. The voltage sample point should be as close as possible to the DUT to reduce the effect of the return pulse reflected from the DUT. The network in this example sets a practical limit of about 15 cm on the length of the cable connecting the DUT to the bias network.

In general, bias networks that provide RF terminations or pulsed-RF capability will limit the accuracy of measurements in the first 100 to 200 ns of a pulse. With such an arrangement, the pulse source need not produce rise times less than 50 ns. Rather, shaped rising edges would be beneficial in controlling transients at the DUT.

Current measurement with series sense resistors will add to the output impedance of the pulse source. Usually a capacitance of a few picofarads is associated with the sense or bias network that will limit resistance value for a specified rise time.

Measurement Resolution

Voltage and current ranges are determined by the pulse sources. Summing amplifiers provide a few hundred milliamps at 10 to 20 V. High-power, charge-dumping pulsers provide several amps and 50 V. Current pulses are achieved with series resistors and voltage sources. These limit the minimum pulse time. For example, a 1 kΩ resistor may be used to set a base current for testing bipolar transistors. With 10 pF of capacitance associated with the bias network, the minimum rise time would be of the order of 10 μs. An isodynamic measurement would need to use short collector-terminal pulses that are inset within long base-terminal pulses.

There is no practical method for implementing current limiting within the short time frame of pulses other than the degree of safety afforded by the output impedance of the pulse source.

Measurement resolution is determined by the sampling digitizers and current sensors. Oscilloscopes provide 8-bit resolution with up to 11-bit linearity, which provides only 100 μA resolution in a 100 mA range. The 12-bit resolution, with 14-bit linearity, of high-speed digitizers may therefore be desirable. To achieve the high resolutions, averaging is often required. Either the pulse system can repeat the measurement cycle to accumulate averages, or several samples in each pulse can be averaged.

Measurement Time

Measurement speed, in the context of production-line applications, is optimized with integrated systems that sequence several pulse points in each measurement cycle. As an example, acquiring 1000 pulse points with 1 ms quiescent periods, 500 ns pulse periods, and an averaging factor of 32 will necessarily require 32 s of pulsing. With a suitable stabilization period, and overhead in instrument setup and data downloading, this typical pulsed-I/V measurement can be completed in just less than one minute per quiescent point.

Single-point measurement systems have instrument setup and data downloading overhead at each pulse point. A typical 1000-point measurement usually requires substantially more than ten minutes to complete; especially when data communication is through GPIB controllers.

A pulsed-RF measurement is also slow because the network analyzer must step through its frequency list, and requires a hold-off time between RF sampling events. A typical pulsed-RF measurement with a 50-point frequency list, an averaging factor of 32, and only 100 pulse points, would take about half a minute to complete.

Commercial Measurement Systems

Figure 4.82 graphically portrays the areas covered in a frequency/signal level plane by various types of instruments used to characterize devices. The curve tracer, epitomized perhaps by the HP4145 and

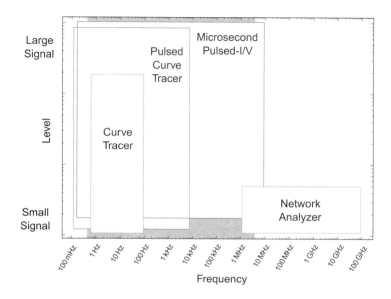

FIGURE 4.82 Relative position of various types of measurement equipment, including pulsed-I/V systems, in terms of measurement frequency and signal level. The shaded area indicates the frequency range of dispersion effects.

numerous analog predecessors made by companies such as Tektronix, cover the most basic measurement range. Beyond this range, instruments with some pulse capability, such as the HP4142 or HP4155/56, offer very wide capability, but this is still at speeds below that required for isodynamic characterization. Network analyzers reach millimeter-wave frequencies but perform small-signal measurements by definition. Between these, pulsed-I/V systems such as those described below have the advantage of large-signal capability and speeds sufficient to give isodynamic characteristics.

The majority of pulsed measurements reported in the literature to date have been made with experimental equipment, or with systems under development. Three sub-microsecond systems are commercially available. These come with a range of options that require some assessment before purchase. This is partly a consequence of the immature nature of pulsed-I/V instrumentation (in comparison to conventional curve tracers), and partly a result of pulsed-I/V measurement being a more complicated problem.

Before reviewing the available systems, it is useful to identify an intrinsic problem for pulsed measurements. The performance limit on pulsed-I/V systems is frequently the DUT connection network and the form of the stimulus, not the measurement system itself.

Network analyzers achieve very high-frequency resolution with a narrowband stimulus and receiver, which allows them to minimize noise and apply vector calibration techniques to eliminate parasitic disturbances. They define a measurement plane, behind which any fixed error is identified and eliminated by postprocessing of the data. They can also allow the DUT to come to a steady state during the measurement. Pulse systems conversely use a stimulus that contains many frequency components from the slow pulse repetition rate up to many times the fundamental component in the fast pulse. The measurement is both of wide bandwidth, and therefore noisy, and at high frequencies. Viewed in the time domain, the pulse width is limited by the charging of the unknown capacitance in the bias network, which can be minimized but not eliminated. For example, bias networks may contribute sufficient parasitic capacitance to limit pulsed measurements to 500 ns, or slower, with a pulse source impedance of 50 Ω. The situation is worse for current drive, and may be worse still, because of transients, for a voltage drive that does not match transmission line impedance. Thus, the system is infrequently limited by the minimum width of the pulse from the instrument, and some judgment needs to be exercised in each measurement setup.

GaAs Code

GaAs Code Ltd., based in Cambridge, England, offers a low-end pulsed-I/V measurement system.[3] It is controlled by a PC via a serial interface. Hardware cost is on the order of US$20,000. Specifications range from ±10 V, 0.5 A, 2.5 W up to +25 V, 1 A, 6 W, with output impedance at or above 10 Ω. Pulse width is from 100 ns to 1 ms. A higher power model is under development. Software supplied by GaAs Code allows control, plus generation of graphs that can be printed or incorporated into documents under the Windows operating system. The instrument works with various modeling software programs supplied by GaAs Code. No provision is made for synchronization with a network analyzer for pulsed-RF measurements.

Macquarie Research

Macquarie Research Ltd. offers an Arbitrary Pulsed-I/V Semiconductor Parameter Analyzer (APSPA).[4] The hardware is largely commercial VXI modules. Control is via proprietary software running on an embedded controller. Each measurement cycle can cover up to 2048 pulse points, which, together with the integrated bus architecture, gives fast measurement turnaround. System cost (hardware and software) is on the order of US$100,000. Specifications start at ±20 V, ±0.5 A and rise to 3 A in the VXI rack or to 50 V and 10 A with an external Agilent K-series pulse source. Output impedance ranges from less than 1 Ω to 50 Ω in discrete steps, depending upon options. Pulse timing is from 100 ns to greater than 1 s in 25 ns steps, with pseudo-random, arbitrary sequencing, and scripting capability. A 50 V, 5 A high-speed module, and support for low-cost digitizers, are under development. The proprietary software produces data files but does not support data presentation. Synchronization with an Agilent HP85108A pulsed network analyzer is included for routine pulsed-RF measurements.

Agilent Technologies

Agilent Technologies offers a pulsed-I/V system as a subsection of their pulsed modeling system.[5] The pulsed-I/V subsystem is composed of rack-mounted instruments controlled by a workstation running IC-CAP software. System cost is on the order of US$500,000 inclusive of the RF and pulsed subsystems, and software. The DC and pulsed-I/V system is approximately half of that cost, the pulsed-I/V subsystem constituting about US$200,000. Specifications are ±100 V at 10 A with an output resistance of about 1 Ω, based exclusively on K49 Pulse Sources. Pulse width is effectively limited by a lower bound of 800 ns. Data presentation and S-parameter synchronization are inherent in the system. A difficulty of the use of GPIB and K49s driven by conventional pulse generators is the overall measurement time, which at best is about 2 orders of magnitude slower than integrated multipoint systems. Only one pulse point is possible in each measurement cycle.

Measurement Techniques

With flexible pulsed measurement systems, a wide range of measurements and techniques is possible. Consideration needs to be given to what is measured and the measurement procedures, in order to determine what the data gathered represents. The following sections discuss different aspects of the measurement process.

The Pulse-Domain Paradigm and Timing

A general pulsed-I/V plane can be defined as the grid of terminal voltages pulsed to and from a particular quiescent condition. For isodynamic pulsing, a separate pulsed characteristic would be measured for each quiescent condition.

At each pulse point on an I/V-plane, measurements can be characterized in terms of the following:

- The quiescent point pulsed from, defined by the established bias condition and the time this had been allowed to stabilize.
- The actual pulse voltages, relative to the quiescent voltage, the sequence of application of the terminal pulses, and possibly the voltage rise times, overshoot, and other transients.

- The position in time of sampling relative to the pulses.
- The type of measurements made; voltage and current at the terminals of the DUT, together with RF parameters at a range of frequencies.

Thus, if a number of quiescent conditions are to be considered, with a wide range of pulsed terminal voltages, a large amount of data will be generated. The time taken to gather this data can then be an important consideration. Techniques of overnight batch measurements may need to be considered, together with issues such as the stability of the measurement equipment. Equipment architecture can be categorized in terms of the applications to measurement over a generalized I/V-plane. Those that allow arbitrary pulse sequences within each measurement cycle enable an entire I/V-plane to be rapidly sampled. Systems intended for single pulses from limited quiescent conditions may facilitate precise measurement of a small region in the I/V-plane, but this is at the expense of speed and flexibility.

In the context of isodynamic pulsing, the most important consideration in interpreting the measured data is the sample timing. This is the time of current and voltage sampling relative to the application of the voltage pulses. As it is often information on time-dependent dispersion effects that is gathered, it is important to understand the time placement of sampling relative to the time constants of these rate-dependent effects.

For an investigation of dispersion effects, time domain pulse-profile measurements are used. Terminal currents and voltages are repeatedly sampled, from before the onset of an extended pulse, until after dispersion effects have stabilized. This can involve sampling over six decades of time and hence produces large amounts of data. From such data, the time constants of dispersion effects can be extracted. From pulse-profile measurements of a range of pulse points, and from a range of initial conditions, the dependence of the dispersion effects upon initial and final conditions can be determined.

For isodynamic measurements unaffected by dispersion, sampling must be done quickly after the application of the pulse, so that dispersion effects do not become significant. Additionally, the relaxation time at the quiescent condition, since the application of the previous pulse, must be long enough that there are no residual effects from this previous pulse. The device can then be considered to have returned to the same quiescent state. Generally, sampling must be done at a time, relative to pulse application, at least two orders of magnitude less than the time constants of the dispersion effects (for a less than 1% effect). Similarly, the quiescent time should be at least an order of magnitude greater than these time constants.

Note that for hardware of specific pulse and sampling speed limitations, there may be some dispersion effects too fast for observation. Thus, this discussion refers to those dispersion time constants greater than the time resolution of the pulse equipment.

Quantification of suitable pulse width, sample time, and quiescent time can be achieved with reference to the time constants observed in a time domain pulse profile. For example, for dispersion time constants in the 10 to 100 μs range, a pulse width of 1 μs with a quiescent time of 10 ms might be used. Sampling might be done 250 ns after pulse application, to allow time for bias network and cable transients to settle.

In the absence of knowledge of the applicable dispersion time constants, suitable pulse and quiescent periods can be obtained from a series of pulsed measurements having a range of pulse, sample, and quiescent periods. Observation of sampled current as a function of these times will reflect the dispersion effects present in a manner similar to that achievable with a time domain pulse-profile measurement.[6]

A powerful technique for verifying isodynamic timing is possible with measurement equipment capable of pulsing to points on the I/V-plane in a random sequence. If the quiescent time of pulse relaxation is insufficient, then the current measurement of a particular pulsed voltage will be dependent upon the particular history of previous pulse points. In conventional measurement systems, employing step-and-sweep sequencing whereby pulse points are swept monotonically at one terminal for a stepping of the other terminal, dispersion effects vary smoothly and are not obvious immediately. This is because adjacent points in the I/V-plane are measured in succession and therefore have similar pulse histories.

If, however, points are pulsed in a random sequence, adjacent points in the I/V-plane each have a different history of previous pulses. If pulse timing does not give isodynamic conditions, then the

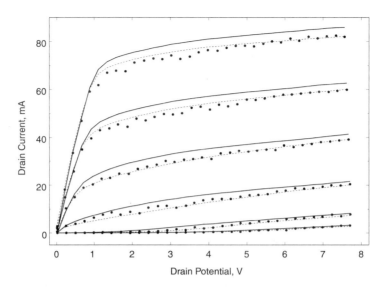

FIGURE 4.83 Characteristic curves for a MESFET measured with three different pulse sequences: a step-and-sweep with 1 µs pulses and 1 µs quiescent periods (- -), the same pulses sequenced in pseudo-random order (•), and an isodynamic measurement. The latter used 800 ns pulses with 1 ms quiescent periods.

dispersion effects resulting from the pulse history will be evident in the characteristic curves. Adjacent points, having different pulse histories, will have dispersion effects of differing magnitude and hence markedly different values of current. This is observed in Fig. 4.83, showing isodynamic and non-isodynamic measurement of the characteristics of a particular device. The non-isodynamic sets of characteristics were measured with the same pulse timing. One characteristic was measured by sweeping the drain-terminal pulse monotonically for different gate-terminal pulse settings. The other characteristic was measured as a random sequence of the same pulses. The smooth shape of the former does not suggest dispersion effects. The apparently noisy variation between adjacent points in the latter indicates history-dependent dispersion is in effect.

Thus, by random sequencing of the pulse points, isodynamic timing can be verified. To obtain isodynamic characteristics, shown in Fig. 4.83, the quiescent relaxation time was increased and the pulse time reduced, until both curves became smooth and identical. That is, until there is no observable history-dependent dispersion.

General Techniques

Within the context of the pulse-domain paradigm discussed in the previous section, and the available equipment, a number of specific measurement techniques and issues arise. These are affected by the equipment limitations and influence the data gathered. A number of these techniques and issues are discussed here.

Interpolation and Iteration

Often measurements are desired at a particular pulse point or specific grid of points. For a target pulse voltage, the actual voltage at the DUT at a certain time will usually be less. This results from various hardware effects such as amplifier output impedance and amplifier time constants, as well as cabling and bias network transients. Voltage drop across amplifier output impedance could be compensated for in advance with known current, but this current is being measured. This is why pulsed voltages need to be measured at the same time as the device currents.

If measurements are desired at specific voltage values, then one of two approaches can be used. Firstly, over successive pulses, the target voltage values can be adjusted to iterate to the desired value. This necessarily involves a measurement control overhead and can require considerable time for many points.

If the thermal noise implicit in using wide-bandwidth digitizers is considered, it is of dubious value to iterate beyond a certain point.

Alternatively, if a grid of pulse points is sampled, covering the range of points of interest, then the device characteristics at these particular points can be interpolated from the measured points. Without iteration, these measured points can be obtained quickly. A least-squares fit to a suitable function can then be used to generate characteristics at as many points as desired. Thus, provided the sampled grid is dense enough to capture the regional variation in characteristics, the data gathering is faster. The main concept is that it is more efficient to rapidly gather an entire I/V-plane of data and then post-process the data to obtain specific intermediate points.

Averaging

The fast pulses generally required for isodynamic measurement necessitates the use of wide-bandwidth digitizers. Voltage and current samples will then contain significant thermal noise. A least-squares fit to an assumed Gaussian distribution to an I/V-grid can be employed to smooth data. Alternatively, or additionally, averaging can be used.

Two types of averaging processes present themselves. The first process is to average multiple samples within each pulse. This assumes a fast digitizer and that there is sufficient time within the pulse before dispersion becomes significant. If dispersion becomes significant over the intra-pulse period of sampling, then averaging cannot be employed unless some assumed model of dispersion is applied (a simple fitted time constant may suffice). An additional consideration with intra-pulse averaging is that voltage value within a pulse cannot be considered constant. The measurement equipment providing the voltage pulse has nonzero output impedance and time constants. Thus, the actual voltage applied to the DUT will vary (slightly) during the voltage pulse. Consecutive samples within this pulse will then represent the characteristics for different voltage values. These are valid isodynamic samples if the sample timing is still below the time constants of dispersion effects. However, they could not be averaged unless the device current could be modeled as a linear function of pulsed voltages (over the range of voltage variation).

The second averaging process is to repeat each pulse point for as many identical measurements as required and average the results. Unlike intra-pulse averaging, this inter-pulse averaging will result in a linear increase in measurement time, in that a measurement cycle is repeated for each averaging. Issues of equipment stability also need to be considered. Typically, both intra- and inter-pulse averaging might be employed. With careful application, averaging can provide considerable improvement in the resolution of the digitizers used, up to their limit of linearity.

Pseudo-Random Sequencing

As previously discussed, randomizing the order of the sequence of pulse points can provide a means of verifying that the quiescent relaxation time is sufficient. It can also provide information on the dispersion effects present. In this, a sequence of voltage pulse values is determined for the specified grid of terminal values. These are first considered a sweeping of one terminal for a stepping of the other. To this sequence, a standard pseudo-randomizing process is used to re-sequence the order of application of pulses. As this is deterministic for a known randomizing process, it is repeatable. This sequence is then applied to the DUT. Upon application of pulses, this random pulse sequence can help identify non-isodynamic measurement timing.

Additionally, if dispersion is present in the measured data, the known sequence of pulse points can provide information on history-dependent dispersion. With step-and-sweep sequencing of pulses, the prior history of each pulse is merely the similar adjacent pulse points. This represents an under-sampling of the dispersion effects. With random sequencing, consecutive pulse points have a wide range of earlier points, providing greater information on the dispersion effects.

Thus, for the known sequence of voltage pulses and the non-isodynamic pulse timing, a model of the dispersion effects can be fitted. These can then be subtracted to yield isodynamic device characteristics. This, however, only applies to the longer time-constant effects and requires that the timing be close to that of the time constants of the dispersion effects.

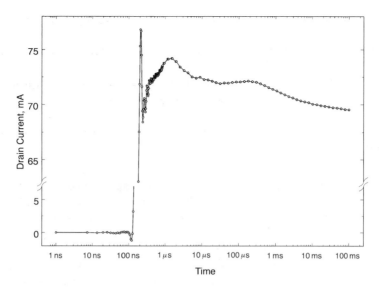

FIGURE 4.84 Transient response measured with eight repeated measurements at 50 ns intervals. Each repetition is shifted by 6.25 ns to give the composite response shown.

Pulse Profile

In normal isodynamic pulsing, pulse widths are kept shorter than the time constants of applicable dispersion effects. Relaxation periods between pulses, at the quiescent condition, are longer than these times. Typically, pulse widths of 1 μs and quiescent periods of 100 ms might be used.

In a pulse profile measurement, an extended pulse width of 0.1 to 1 s might be used, so that the dispersion effects can be observed. All dispersion time constants greater than the pulse rise and settling time are then observable. Quiescent periods between these extended pulses still need to be long, so that subsequent pulses can be considered as being from the same bias condition.

Plotted on a logarithmic time axis, the dispersion effects can be seen as a variation of device output current with time (see Fig. 4.84). Typically, output current might rise in the first 1 to 10 μs period due to junction heating and trapping effects, then fall due to channel heating. Time constants of the amplifier driving the pulses might need to be deconvolved before identifying those of the DUT alone. From such a plot, it can first be identified where isodynamic conditions apply. That is, how soon after pulse application sampling needs to be done before dispersion effects become significant. How long these dispersion effects take to stabilize will indicate how long the quiescent periods of isodynamic pulsing need to be. Secondly, values for dispersion time constants can be extracted from the data, together with other parameters applicable to a particular dispersion model.

Note that because the extended pulse widths of pulse profile measurements are intended to bring into effect heating and dispersion, the range of pulse points on the I/V-plane must be restricted. With isodynamic pulsing, it is possible to pulse to voltages well outside the SOA of the DUT. This is because the short pulses do not invoke the time-dependent thermal and current damage of static conditions. With pulse profile measurements, pulse widths extend to essentially static periods and so voltages must be restricted to the SOA for static measurements (although pulse profile techniques could be used to observe destruction outside the SOA).

Equipment issues influence pulse profile measurements in several ways. The first is pulse duration. Systems employing capacitor charge dumping for pulsing will be limited in the length of time that they can hold a particular output voltage. The second is output rise and settling times. Bias network and cable transients and the response time of data measurement will limit the earliest time after pulse application for which valid samples can be taken. This, typically, might be of the order of 100 ns, although with restrictions on application might extend down to 1 ns. This necessarily affects the range of dispersion effects observable to those having time constants greater than perhaps an order of magnitude more than this minimum time resolution.

Digitizer speed and bandwidth are another major issue in pulse profile measurements. A wide bandwidth is necessary so that sample values accurately reflect DUT conditions. In isodynamic pulsing, only one time point need be sampled, with a long time before the next pulse. With a pulse profile, it is desirable to repeatedly sample the pulse as fast as possible to observe variation with time. Sampling speed needs to be perhaps an order of magnitude faster than the time constant to be observed. Additionally, if bandwidth, jitter, and stability permit, an *equivalent time* sampling may be used. In this, repeated pulse profile measurements are performed, with sample times relative to pulse onset shifted slightly with each successive pulse. As an example, a 20 MHz digitizer, sampling at 50ns intervals, might be applied to eight successive, identical pulses. Sampling is commenced 50 ns before the start of the first pulse, but offset an accumulating 6.25 ns on successive pulses. The sum of these then represents sampling at a rate of 160 MHz. This assumes the bandwidth of the digitizer input track-and-hold circuit is sufficient.

Sampling at a rate of 160 MHz generates a large amount of data when applied to a 1s long pulse. However, as the dispersion processes to be observed tend to be exponential in effect over time, then it is not necessary to continue sampling at this rate for the entire pulse profile. The sampling period needs to be less than 70% of the time constant to be observed, but typically sampling would be an order of magnitude faster for better amplitude resolution in noisy conditions. Thus, sampling may begin at 10 ns intervals for the first 100 ns, but then continue at every 100 ms toward the end of the 1 s pulse. Such logarithmic placement of sampling over the pulse is possible with digitizers that allow arbitrary triggering and systems that can generate arbitrary trigger signals. With such a system, sampling would be performed at a linear rate initially while requiring samples as fast as possible, reducing to a logarithmic spacing over time. For example, with a 20 MHz digitizer, sampling might be done every 50 ns for the first 1 µs, but then only ten samples per decade thereafter. This would give only 80 samples over a 1 s pulse, rather than the excessive 20 M samples from simple linear sampling. In this way, data can be kept to a manageable but adequate amount.

Output Impedance

In testing a device, whether the terminal current or voltage is the dependent variable or the independent variable is subjective and conditional upon the type of device (BJT or FET). However, pulsed measurement systems are usually implemented with sources of voltage pulses, for practical reasons. Thus, it is desirable to have negligible output impedance in the pulse generator or driving amplifier.

There exist, however, some situations where it is desirable to have significant output impedance in the pulse driver. For example, in testing FETs with very fast pulses, it is usually necessary to use a 50 Ω output impedance with the gate-terminal pulser to prevent RF oscillations.

When current is the more convenient independent variable, a large driver output impedance can simulate a current source. With bipolar devices (BJTs and HBTs), it is desirable to perform measurements at particular values of base current. This is a very strong function of base emitter voltage and hence difficult to control with a voltage source. With a large source resistance (e.g., 10 kΩ) in the base voltage driver, a reasonable current source can be approximated and base current controlled. This will necessarily severely limit the rise time of a base terminal pulse, so that typically this pulse would be first applied and allowed to stabilize before a fast pulse is applied to the collector terminal. This is fine for investigating isodynamic collector current in relation to dispersion effects due to collector voltage and power dissipation. However, the long base current pulse implies that base voltage and current-related dispersion effects are not isodynamic.

Output impedance is also used for current limiting and for safe exploration of the I/V-plane. The diode characteristic of the FET gate junction during forward conduction and breakdown means that gate current can become very large. Having 50 Ω in the gate-terminal pulser will limit this current to 20 mA typically. Similarly, 50 Ω in the drain-terminal pulser will limit drain current for a particular voltage pulse and constrain DUT output behavior to follow the load line determined by this 50 Ω load impedance and the applied voltage pulse. In this way, pulse voltage can be slowly increased to explore expanded regions of device operation safely. It will also curb transients.

Extending the Data Range

An important aspect of pulsed testing is that a wider range of data points can be tested. Beyond a certain range of terminal potentials or power, damage can be done to a device because of excessive temperature or current density. As the DUT temperature is a function of the time for which a given power level is applied, the shorter a pulse, the greater the voltage and/or instantaneous power that can be applied.

The conventional SOA of a device is that part of the I/V-plane for which the device can withstand static or continuous application of those voltage levels. Pulsed testing then extends this region, in particular to regions that are outside the static SOA, but are still encountered during normal RF operation of the device. This gives an extended range of data for use in modeling device operation, not only for isodynamic I/V characteristics, but also for RF parameters for extraction of parasitic resistances and capacitances. With a pulsed S-parameter system coupled with a pulsed-I/V system, the voltage pulses can take the DUT to an isothermal point outside the static SOA, where S-parameters can then be measured during this pulse.

Repetition

The characteristics of a device can change due to the manner in which it is used. For example, an excursion into a breakdown region can alter, although not damage, a device, permanently modifying its characteristics. To investigate such phenomena, an I/V-grid can be measured before and after such an excursion. Changes in the device characteristics can then be observed in the difference between the measurements.[7]

Of use in such investigations is the ability to specify an arbitrary list of pulse points. In this case, the list of points in the I/V-plane to be pulsed to would first list the regular grid, then the points of breakdown excursion, and then repeat the same regular grid points. Additionally, scripting capabilities might be used to create a series of such measurements.

Onion-Ring Destructive Testing

Often it is desired to test a device until destruction. An example of this might be breakdown measurements. Sometimes it is difficult not to destroy a *fragile* device during testing — especially devices fabricated with an immature technology. In either case, it is desirable to structure the sequence of pulse points from safe voltage and power levels to increasing levels up to destruction. It is essential in this that all data up to the point of device destruction is preserved.

Here again, scripting capabilities and the use of a list of pulse points allow measurements to be structured as a series of layers of pulse points, increasing in power and/or voltage level. In this way, the characteristics of a device can be explored as an extension, in layers, of the safe device operation or constant power level. Inter-pulse averaging and a waiting period for device stabilization would not normally be used in this form of measurement.

Quiescent Measurement

It is important to measure the bias point representing the isodynamic conditions of the DUT. This is the terminal voltage and current before each pulse and as such gives the quiescent thermal and trapping state of the device. This needs to be measured as part of the pulse exercise if the pulse sequence used is such that the average device temperature is raised.

The time spent at the quiescent point is usually quite long, affording opportunity for considerable averaging. Additionally, when pulsing too many points of the I/V-plane, the quiescent point can be measured many times. Thus, a comparatively noise-free measurement can be obtained.

Sample points for quiescent data would usually be placed immediately before a pulse. Several samples would be taken and averaged. It is assumed that the relaxation time at the quiescent condition, since the previous pulse, is very much greater than all relevant dispersion-effect time constants (unless these time constants are themselves being investigated). This is necessary if the samples are to be considered as representing a bias condition, rather than a transient condition.

Alternatively, or additionally, some samples might be taken immediately after a pulse. For these post-pulse samples to be considered to represent the bias condition, the pulse must be short enough for no significant dispersion effects to have occurred. Notwithstanding this, there may be useful information in observing relaxation after a pulse and in the change in device current immediately before and after a return from a pulse.

Timing

A number of different timing parameters can be defined within the paradigm of pulse testing. Referring to Fig. 4.80, a basic pulse cycle consists of an extended time at the quiescent bias point ($T_{Quiescent}$) and a (usually) short time at particular pulsed voltage levels (T_{Pulse}). In this diagram, T_{Pulse} refers to the time for which the gate or base voltage pulse is applied. The sum of these two times is then the **pulse event time** and the inverse of this sum would be the **pulse repetition frequency** for continuous pulsing.

A third timing parameter, τ_{inset}, reflects the relationship of the drain/collector pulse to the gate/base pulse. These voltage pulses need not be coincident, but will normally overlap. Often the gate pulse will be applied before the drain pulse is applied — an inset of 100 ns is typical. Sometimes it might be necessary for the drain pulse to lead the gate pulse in order to control the transition path over the I/V-plane. Thus, the parameter τ_{inset} might be positive or negative and might be different for leading and trailing pulse edges. In a simple system, it is most easily set to zero so that the terminal pulses are coincident.

These three parameters define pulse event timing — the times for which terminal voltage pulses are applied and the quiescent relaxation time. Note that actual voltage pulses will not be square shaped. For single-point pulsing, there might only be one pulse event, or a sequence of identical pulse events. For generalized pulsing over the I/V-plane, a measurement cycle may be an arbitrary sequence of different pulse points, all with the same cycle timing.

The number of sample points within a basic pulse event could be specified as both a number of samples within the pulse (N_P) and as a number of samples of the quiescent condition (N_Q). Typically these would be averaged, except in the case of a pulse profile measurement. The placement of these sample points within the pulse cycle need also be specified.

If the pulsed-I/V system is to be coupled with a pulsed-RF system, such as the Agilent Technologies HP85108, then relative timing for this needs to be specified. Figure 4.80 defines a time for application of the RF signal relative to the gate voltage pulse and a trigger point within this for RF sampling. These two signals can be supplied to the HP85108 for synchronization.

The above times would refer to the pulse event timing at the terminals of the DUT. Various instrument and cabling delays might require that these times be individually adjusted when referred to the pulse amplifiers and sample digitizers. Different signal paths for current and voltage digitizers might require separate triggers for these.

General Techniques

As well as the various measurement techniques just discussed, there exists a range of practical issues. For example, with combined pulsed-I/V and pulsed-RF systems, the RF must be turned off while measuring DUT current. This means that experiment times are longer than might be expected, as the pulsed-I/V and pulsed-RF data are gathered separately.

Another consideration is that the applied voltage pulses are not square shaped. Instrumentation and cable termination issues result in pulses having significant rise and fall times and in particular overshoot and settling. The devices being tested are generally fast enough to respond to the actual instantaneous voltages, rather than an averaged rectangular pulse. First, this means that sampling of both voltage and current must be performed, and that this must be at the same time. Second, as any pulse overshoot will be responded to, if this voltage represents a destructive level then damage may be done even when the target voltage settled to is safe. This particularly applies to gate voltage pulses approaching forward conduction or breakdown.

Also arising from the fact that the DUT is far faster in response than the pulse instrumentation, is the issue of pulsing trajectory. In pulsing from a bias point to the desired pulse point, the DUT will follow a path of voltage and current values across the I/V-plane, between the two points. Similarly, a path is followed in returning from the pulse point to the bias point. The actual trajectory followed between these two points will be determined by the pulse rise and fall times, overshoot and other transients, and by the relative inset of gate and drain pulses (Fig. 4.80).

A problem can arise if, in moving between two safe points on the I/V-plane, the trajectory passes through a destructive point. An example is pulsing to a point of low drain voltage and high current from a bias point of high drain voltage and low current. Here drain voltage is pulsing to a lower voltage while gate voltage is pulsing to a higher value. If the gate pulse is applied first, then the DUT will move through a path of high voltage and high current. This is a problem if it represents destructive levels and is dependent upon trajectory time. A similar problem exists in returning from the pulse point to the bias point. In general, because gate/drain coincidence cannot be sufficiently well controlled, consideration need be given to the trajectories that may be taken between two points on the I/V-plane and the suitability of these. With appropriate choice of leading and trailing overlaps between the gate and drain pulses, this trajectory can be controlled.

Data Processing

Having gathered data through pulsed measurements, various processing steps can follow. In this, reference need again be made to the pulse domain paradigm. In the simplest case, the data consists of a grid of pulse points for a fixed bias point, sampled free of dispersion effects. To this could be added further data of grids for multiple bias points. Rate dependence can be included with data from pulse profile measurements and grids with delayed sample times. In this way, the data can be considered as a sampling of a multidimensional space. The dimensions of this space are the terminal currents and voltages, both instantaneous and average, together with sample timing and temperature. RF parameters at a range of frequencies can also be added to this.

Processing of this data can be done in two ways. First, the data can be considered as raw and processed to clean and improve it. Examples of this form of processing are interpolation and gridding. Second, data can be interpreted against specific models. Model parameter extraction is the usual objective here. However, to fully use the information available in the pulsed data, such models need to incorporate the dispersion effects within the pulse domain paradigm.

Interpolation and Gridding

Data over the I/V-plane can be gathered rapidly about a grid of target pulse points. The grid of voltage values represents raw data points. Instrument output impedance and noise usually differentiate these from desired grid points. Interpolation and gridding can translate this data to the desired grid.

Data can be gathered rapidly if the precision of the target pulse-voltage values is relaxed. The data still represents accurate samples, however the actual voltage values will vary considerably. This variation is not a problem in model extraction, but can be a problem in the comparison of different characteristic curves (for different quiescent conditions) and the display of a single characteristic curve for a specified terminal voltage.

Gridding is performed as the simple two-dimensional interpolation of current values as a function of input and output pulse-voltage values. A second- or third-order function is usually used. The interpolated voltage values represent a regular grid of desired values, whereas the raw data values are scattered. A least-squares fit can be used if a noise model is assumed, such as thermal noise. Nothing is assumed about the underlying data, except for the noise model and the assumption that the data local variation can be adequately covered by the interpolation function used.

Intrinsic Characteristics

The simplest of models for data interpretation all assume series access resistances at each terminal. Fixed resistances can be used to model probe and contact resistances, as connecting external terminals to an idealized internal nonlinear device. For measured terminal current and assumed values of resistances, the voltage across the terminal access resistances is calculated and subtracted to give intrinsic voltages. These voltages can then be used in model interpretation.

For example, consider a FET with gate, drain, and source access resistances of R_G, R_D, and R_S respectively. If the measured terminal voltages and currents are v_{GS}, i_G, v_{DS}, and i_D respectively, then the intrinsic voltages can be obtained as:

$$v_{DS'} = v_{DS} - i_D R_D - \left(i_D + i_G\right) R_S,$$
$$v_{GS'} = v_{GS} - i_G R_G - \left(i_D + i_G\right) R_S. \tag{4.58}$$

If v_{GS}, i_G, v_{DS} and i_D are raw data, then a set of $v_{DS'}$, $v_{GS'}$ values can be used to obtain a grid of intrinsic data. This is easy to do with copious amounts of data gathered over the I/V-plane.

Interpretation

The data, raw or gridded, can be used to extract information on specific effects under investigation. In the simplest case, small-signal transconductance and conductance can be obtained as gradients, such as di_D/dv_{GS} and di_D/dv_{DS} in the case of a FET. These could then be used in circuit design where the device is being operated at a specific bias point. A second example is in the extrapolation of plots of voltage and current ratios to give estimates of terminal resistances for use in determining intrinsic values. The advantage of pulsed testing here is that an extended range of data can be obtained, extending outside the static SOA.

Another example of data interpretation is the use of measured history dependence to give information on dispersion effects. If, in pulsed testing, the quiescent relaxation time is insufficient, then pulse samples will be affected by dispersion. The use of shuffling of the pulse sequence enhances sampling of dispersion. Models of dispersion can then be fitted to this data to extract parameters for dispersion, as a function of terminal voltages and of pulse timing.

Modeling

The paradigm of pulsed testing assumes that DUT terminal currents are functions of both instantaneous and of average terminal voltages. This means that device response to RF stimuli will be different for different average or bias conditions. Pulsed testing allows separation and measurement of these effects.

A model of device behavior, for use in simulation and design, must then either incorporate this bias dependence or be limited to use at one particular bias condition. The latter is the usual case, where behavior is measured for a particular bias condition, for modeling and use at that bias condition.

If a model incorporates the bias-dependent components of device behavior, the wider sample space of pulsed testing can be utilized in model parameter extraction. From I/V-grids sampled for multiple bias conditions, the bias dependency of terminal current can be extracted as a function of both instantaneous and bias terminal voltages. From pulse profile measurements, dispersion effects can be modeled in terms of average terminal voltages, where this average moves from quiescent to pulse target voltage, over the pulse period, according to a difference equation and exponential time constants. The actual parameter extraction consists of a least-squares fit of model equations to the range of data available, starting from an initial guess and iterating to final parameter values. The data used would be I/V-grids, pulse profiles, and RF measurements over a range of frequencies, at a range of bias points, depending on the scope of the model being used. Important in all this is a proper understanding of what the sampled DUT data represents, in the context of the pulse domain paradigm, and of how the data is being utilized in modeling.

Empirical models that account for dispersion effects must calculate terminal currents in terms of the instantaneous and time-averaged potentials. In the case of a FET, the modeled drain current is a function of the instantaneous potentials v_{GS} and v_{DS}, the averaged potentials $\langle v_{GS} \rangle$, $\langle v_{DS} \rangle$ and average power $\langle i_{DS} v_{DS} \rangle$. The time averages are calculated over the time constants of the relevant dispersion effects. A model of thermal dispersion is:

$$i_{DS} = i_O \left(1 - \lambda R_T \left\langle i_{DS} v_{DS} \right\rangle \right), \tag{4.59}$$

where i_O includes other dispersion effects in a general form

$$i_O = I \left(v_{GS}, v_{DS}, \left\langle v_{GS} \right\rangle, \left\langle v_{DS} \right\rangle \right). \tag{4.60}$$

With a suitable value of λR_T, the thermal effects present in the characteristics of Fig. 4.72 can be modeled and the other dispersion effects can be modeled with the correct function for i_O in Eq. (4.60). The DC characteristics are given by the model when the instantaneous and time-averaged potentials track each other such that $\left\langle v_{GS} \right\rangle = v_{GS}$, $\left\langle v_{DS} \right\rangle = v_{DS}$, and $\left\langle i_{DS} v_{DS} \right\rangle = i_{DS} v_{DS}$. In this case, the model parameters can be fitted to the measured DC characteristics and would be able to predict the apparently negative drain conductance that they exhibit. In other words, the DC characteristics are implicitly described by

$$I_{DS} = I \left(V_{GS}, V_{DS}, V_{GS}, V_{DS}\right)\left(1 - \lambda R_T I_{DS} V_{DS}\right). \tag{4.61}$$

Of course, this would be grossly inadequate for modeling RF behavior, unless the model correctly treats the time-averaged quantities as constants with respect to high-frequency signals.

For each quiescent point ($\left\langle v_{GS} \right\rangle, \left\langle v_{DS} \right\rangle$), there is a unique set of isodynamic characteristics, which relate the drain current i_{DS} to the instantaneous terminal-potentials v_{GS} and v_{DS}. Models that do not provide time-averaged bias dependence must be fitted to the isodynamic characteristics of each quiescent condition individually. Models in the form of Eqs. (4.59) and (4.60) simultaneously determine the quiescent conditions and the appropriate isodynamic characteristics.[8,9] Pulsed measurements facilitate this characterization and modeling of device RF behavior with bias dependency.

Defining Terms

Characteristic curves: For FETs/HBTs, a graph showing the relationship between drain/collector current (or RF parameters) as a function of drain/collector potential for step values of gate/base potential.

Bias condition: For a device, the average values of terminal potential and currents when the device is operating with signals applied.

Dispersion effects: Collective term for thermal, rate-dependent, electron trapping and other anomalous effects that alter the characteristic curves with the bias condition changes.

DC characteristics: Characteristic curves relating quiescent currents to quiescent terminal potentials.

Isodynamic characteristic: Characteristic curves relating instantaneous terminal currents and voltages for constant, and equal, bias and quiescent conditions.

Isothermal characteristic: Characteristic curves relating instantaneous terminal currents and voltages for constant operating temperature.

Pulsed bias: Pulsed stimulus that briefly biases a device during a pulsed-RF measurement.

Pulsed characteristics: Characteristic curves measured with pulsed-I/V or pulsed-RF measurements.

Pulsed-I/V measurement: Device terminal currents and voltages measured with pulse techniques.

Pulsed-RF measurement: Device RF parameters measured with pulse techniques.

Quiescent condition: For a device, the value of terminal potential and currents when the device is operating without any signals applied.

References

1. Teyssier, J.-P., et al., 40-GHz/150-ns Versatile pulsed measurement system for microwave transistor isothermal characterization, *IEEE Trans. MTT*, 46, 12, 2043–2052, Dec. 1998.
2. Ernst, A.N., Somerville, M.H., and del Alamo, J.A., Dynamics of the kink effect in InAlAs/InGAs HEMT's, *IEEE Electron Device Letters*, 18, 12, 613–615, Dec. 1997.

3. GaAs Code Ltd, Home page, 2000. [Online]. Available: URL: http://www.gaascode.com/.

4. Macquarie Research Ltd, Pulsed-bias semiconductor parameter analyzer, 2000. [Online]. Available: URL: http://www.elec.mq.edu.au/cnerf/apspa.

5. Agilent Technologies, HP85124 pulsed modeling system and HP85108 product information, 2000. [Online]. Available: URL: http://www.agilent.com.

6. Parker, A.E. and Scott, J.B., Method for determining correct timing for pulsed-I/V measurement of GaAs FETs, *IEE Electronics Letters*, 31, 19, 1697–1698, 14 Sept. 1995.

7. Scott, J.B., et al., Pulsed device measurements and applications, *IEEE Trans. MTT*, 44, 12, 2718–2723, Dec. 1996.

8. Parker, A.E. and Skellern, D.J., A realistic large-signal MESFET model for SPICE, *IEEE Trans. MTT*, 45, 9, 1563–1571, Sept. 1997.

9. Filicori, F., et al., Empirical modeling of low frequency dispersive effects due to traps and thermal phenomena in III-V FET's, *IEEE Trans. MTT*, 43, 12, 2972–2981, Dec. 1995.

4.7 Microwave On-Wafer Test

Jean-Pierre Lanteri, Christopher Jones, and John R. Mahon

On-Wafer Test Capabilities and Applications

Fixtured Test Limitations

Until 1985 the standard approach to characterize at microwave frequencies and qualify a semiconductor wafer before shipping was to dice it up, select a few devices, typically one in each quadrant, assemble them, and then test them in a fixture, recording s-parameters or power levels. Often, the parts were power transistors, the most common RF/microwave product then, and a part was used as a sample. For Gallium Arsenide (GaAs) Monolithic Microwave Integrated Circuits (MMICs), a transistor was similarly used for test coupon, or the MMIC itself. Typically, the parts were assembled in a leaded metal ceramic package, with epoxy or eutectic attach, and manually wedge bonded with gold wires for RF and bias connections. The package was then manually placed in a test fixture and held down by closing a clamp on the leads and body. The fixture was connected to the test equipment, typically a Vector Network Analyzer (VNA) or a scalar power meter, by Radio Frequency (RF) coaxial cables to present a 50 Ohms environment at the end of the coaxial cables. The sources of test uncertainty were numerous:

- Part placement in the package and bond wire loop profile, manually executed by an operator, lead to bond wire length differences and therefore matching variations for the Device Under Test (DUT).
- Package model inaccuracy and variability from package to package.
- RF and ground contacts through physical pressure of the clamp, applying force to the body of the package and the leads, with variable results for effective lead inductance and resistance, and potential oscillations especially at microwave frequencies.
- Fixture de-embedding empirical model for the connectors and transmission lines used on the RF ports.
- Calibration of the test equipment at the connectorized interface between the RF cables and the test fixture, not at the part or package test planes.

Most of these technical uncertainties arise because the calibration plane is removed from the product plane and the intermediate connection is not well characterized or not reproducible.

The main drawbacks of fixtured tests from a customer and business perspective were:

- Inability to test the very product shipped, only a "representative" sample is used due to the destructive nature of the approach. Especially for MMICs where the yield loss can be significant, this can lead to the rejection of many defective modules and products after assembly, at a large loss to the user.
- Cost of fixtured test; sacrificing parts and packages used for the test.

- Long cycle time; typically a day or two are needed for the parts to make it through assembly.
- Low rate production test; part insertion in a fixture is practically limited to a part per minute.

A first step was to develop test fixtures for bare die that could be precisely characterized. One solution was a modular fixture, where the die is mounted on an insert of identical length, which is sandwiched between two end pieces with transmission line and connector. The two end pieces can be fully characterized with a VNA to the end point of the transmission lines by Short-Open-Load-Thru (SOLT) or Thru-Reflect-Line (TRL) calibrations; wire bonding to preset inserts or between the two end-pieces butted together. Then the die is attached to the insert, assembled in between the end pieces, and wire bonded to the transmission lines. This approach became the dominant one for precise characterization and model extraction. The main advances were removal of die placement, package, lead contact and fixture as sources of variability, at the expense of a complex assembly and calibration process. The remaining limitations are bond loop variation, and destructiveness, and the length and cost of the approach, preventing its use in volume applications such as statistical model extraction or die acceptance tests.

On-Wafer Test Enabler: Coplanar Probes

The solution to accurate, high volume microwave testing of MMICs came from Cascade Microtech, the first company to make RF and microwave probes commercially available, along with extensive application support; their history and many useful application notes are provided on their Website (*www.cascademicrotech.com*). On-wafer test was common place for DC and digital applications, with high pin count probe cards available, based upon needles mounted on metal or ceramic blades. Although a few companies had developed RF frequency probes for their internal use, they relied on shortened standard DC probes, not the coplanar Ground-Signal-Ground (G-S-G) structure of Cascade Microtech's probes, and were difficult to characterize and use at microwave frequencies. The breakthrough idea to use a stable GSG configuration up to the probe tip enabled a reproducible 50 Ohms match to the DUT, leading to highly reproducible, nondestructive microwave measurements at the wafer level.[1,2] All intermediate interconnects were eliminated, along with their cost, delay, and uncertainty, provided that the DUT was laid out with the proper GSG inputs and outputs. Calibration patterns (Short, Open, Load, Thru, Line Stub) available on ceramic substrates or fabricated on the actual wafers provided standard calibration to the probe tips.[3,4] A few years later, PicoProbe (*www.picoprobe.com*) introduced a different mechanical embodiment of the same GSG concept.

About the same time, automatic probers with top plates fitted with probe manipulators for Cascade Microtech's probes became available. Agilent (then Hewlett Packard) introduced the 8510 Vector Network Analyzer, a much faster and easier way to calibrate microwave test equipment, and 50 Ohms matched MMICs dominated microwave applications. These events combined to completely change the characterization and die selection process in the industry. By the late 1980s, many MMIC suppliers were offering wafer qualification based upon RF test results on standard transistor cells in a Process Control Monitor (PCM) and providing RF tested Known Good Dies (KGD) to their customers.

On-Wafer Test Capabilities

At first, RF on-wafer testing was used only for the s-parameter test, for two port devices up to 18 GHz. Parameters of interest were gain, reflection coefficients, and isolation. Soon RF switching was introduced to test complex MMICs in one pass, switching the two ANA ports between multiple DUT ports. Next came noise figure test on-wafer, using noise source and figure meter combined with ANA. Power test on-wafer required a new generation of equipment, pulsed vector analyzers, to become reliable, and provided pulsed power, power droop, and phase droop.[5] Soon many traditional forms of microwave test equipment were connected to the DUT through complex switching matrixes for stimuli and responses, such as multiple sources, amplifiers, spectrum analyzers, yielding intermodulation distortion products. Next came active source pull equipment, and later on active load pull,[6] from companies such as ATN Microwave (*www.atnmicrowave.com*) and Cascade Microtech. The maximum s-Parameter test frequency kept increasing, to 26 GHz, then 40 GHz, 50 GHz, and 75 GHz. In the late 1990s new parameters such

TABLE 4.4 On-Wafer RF Test Capabilities Evolution

Year	Product	Configuration	Test Capability	Equipment
1985	Amplifier	2-Port	18 GHz s-Parameters	ANA
1987	Amplifier	Switched Multi-Port	26 GHz s-Parameters	ANA + Switch Matrix
1989	LNA	2-Port	Noise Figure	ANA + Noise System
1990	HPA	2-Port	Pulsed Power	Pulsed Power ANA
1991	Amplifier	2-Port	Intermodulation	Spectrum Analyzer
1991	LNA	2-Port, Zin Variable	Noise Parameters	Active Source Pull, ANA
1992	Mixer	3-Port	Conversion Parameters	ANA, Spectrum Analyzer
1993	HPA	2-Port, Zout Variable	Load Power Contours	Active Load Pull, ANA
1995	T/R Module	Switched Multi-Port	40 GHz s-Par, NF, Power	ANA, Noise, Spectrum
1998	Transceiver	Multi-Port	Modulation Parameters	Vector Signal Analyzer
1999	Amplifier	2-Port	110 GHz s-Parameters	ANA

TABLE 4.5 On-Wafer RF Test Applications

Application	DUT	Technique	Test	Test time/DUT	Volume/year
FET Model Development	Standard transistor	Transistor library	PCM transistor	MMIC or transistor	Assembly or package
Statistical Model Extraction	Source or load pull	s-par, NF, PP, set load	s-parameters, 50 Ohms	s-parameters, NF, PP	s-parameters, NF, PP
Process Monitoring	Noise parameters, load contours	Small and large signal models	Small signal model	Test specification	Test specification
Know Good Die Test	10 min	1 min	10 s	10–30 s	10–60 s
Module or Carrier Test	100s	1000s	10,000s	100,000s	100,000s

as Noise Power Ratio (NPR) and Adjacent Channel Power Ratio (ACPR) were required and could be accommodated by digitally modulated synthesizers and vector signal analyzers (Table 4.4). Today, virtually any microwave parameter can be measured on-wafer, including s-parameters up to 110 GHz.

On-Wafer RF Test Applications:

On-wafer test ease of use, reasonable cost, and extensive parameter coverage has led to many applications in MMIC development and production, from device design and process development to high volume test for Known Good Die (KGD). The main applications are summarized in Table 4.5. Of course, all of the devices to test need to have been designed with one of the standard probe pad layouts (S-G-S, G-S, or S-G) to allow for RF probing.

1. Model development and statistical model extraction is often performed on design libraries containing one type of element, generally Field Effect Transistors (FET), but sometimes inductors or capacitors, implemented in many variations that are characterized to derive a parametric model of the element.[7] The parts must be laid out with G-S-G (or G-S only for low microwave frequencies) in a coplanar and/or microstrip configuration. This test task would have taken months ten years ago, and is now accomplished in a few days. The ability to automatically perform all these measurements on significant sample sizes has considerably increased the statistical relevance of the device models. They are stored in a statistical database automatically used by the design and yield simulation tools. This allows first pass design success for complex MMICs.

2. Process monitoring is systematically performed on production wafers, sample testing a standard transistor in a Process Control Monitor (PCM) realized at a few places on each wafer. The layout is in a coplanar configuration that does not require back-side ground vias and therefore can be tested in process. Each time, a small signal model is extracted. Very good agreement between the tested s-parameters and the calculated ones from the extracted model can be seen in Fig. 4.85.

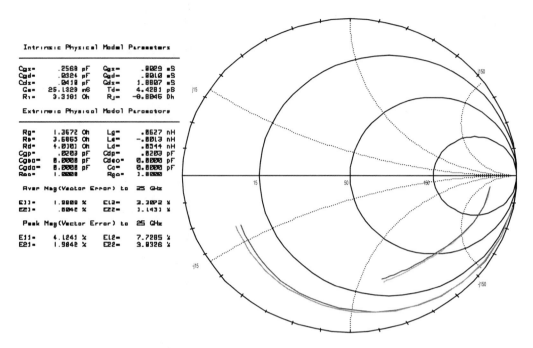

FIGURE 4.85 Equivalent circuit FET model extraction and fit with measurement.

The results are used during fabrication for pass/fail screening of wafers on RF parameters, and supplement the statistical model data.

3. On-wafer test is a production tool for dies, typically 100% RF tested when sold as is — as KGD — or used in expensive packages or modules. This is the norm for high power amplifiers in expensive metal ceramic packages, MMICs for Transmit/Receive (T/R) modules, bumped parts for flip-chip assembly, and military applications. The RF parameters of interest are measured at a few points across the DUT bandwidth, as seen in Fig. 4.86, and used to make the pass/fail decision. The rejected dies on the wafer are either marked with an ink dot, or saved in an electronic wafer map, as seen in Fig. 4.87, which is used by the pick-and-place equipment to pick the passing devices. Final RF test on-wafer is usually not performed on high volume products. These achieve high yields and are all assembled in inexpensive packages, therefore it is easier and cheaper to plastic package all parts on the wafer to test them on automatic handlers and take the yield at this point.

4. The same "on-wafer" test application is used when testing packages, carriers, or modules manufactured in array form on ceramic or laminate substrates, or leadless packages held in an array format by a test fixture.

Test Accuracy Considerations

In any test environment, three important variables to consider are accuracy, speed, and repeatability. The order of importance of these variables is based on price of the device, volume, and specification limits. High test speed is beneficial when it reduces the test cost-per-part and provides greater throughput without reaching an unacceptable level of accuracy and repeatability. Perfect accuracy would seem ideal, although in a high volume manufacturing environment "accuracy" is usually based on agreement between test results of two or more parties, primarily the vendor and end customer, for a specific product. The end customer, utilizing their available methods of measurement, usually defines most initial device specifications and sets the reference "accuracy," defining what parts work in the specific customer application. If due to methodology differences, a vendor's measurement is incompatible with that of a customer, yield and output can be affected without any benefit to the customer. It is not always beneficial,

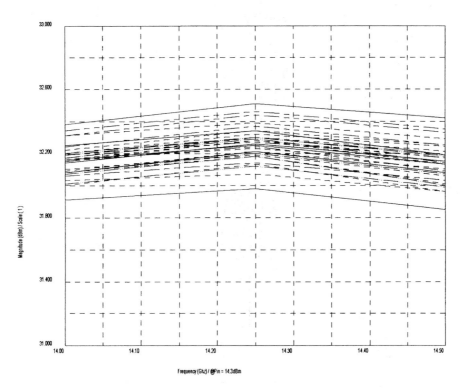

FIGURE 4.86 Pout response of Ku band PAs across a wafer.

in this environment, to provide a more "accurate" method of measuring a product if the end customer is not testing it in the same fashion. Repeatability of the supplier measurement and correlation with the customer result are the more important criteria in that case.

Accuracy and repeatability considerations of any measurement system can be broken down into four primary parts, discussed in detail in the next sections.

Test Equipment Manufacturer

The manufacturer tolerances and supplied instrument error models are the first places to research when selecting the appropriate system. Most models will provide detail information on performance, dynamic range, and accuracy ratings of the individual instruments. Vendors like Agilent, Anritsu, Tektronix, and Boonton, to name a few, provide most hardware resources needed for automatic testing. There are many varieties of measurement instruments available on the market today. The largest single selection criterion of these is the frequency range. The options available diminish and the price increases dramatically as the upper frequency requirements increase. In the last decade many newer models with faster processors, countless menu levels, and more compact enclosures have come on the market making selections almost as difficult as buying a car. Most vendors will be competitive with each other in these matters. More important is support availability, access to resources when questions and problems arise, and software compatibility. Within the last decade many vendors have adopted a standard language structure for command programming of instruments known as SCPI (pronounced Skippy). This reduces software modification requirements when swapping instrumentation of one vendor with another. Some vendors have gone so far as to option the emulation of a more established competitor's model's instrument language to help inject their products into the market.

System Integration

Any system requiring full parametric measurement necessitates a complex RF matrix scheme to integrate all capabilities into a single function platform. Criteria such as frequency range, power levels, and device

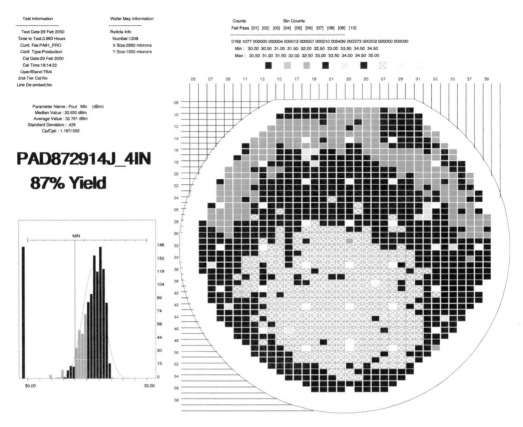

FIGURE 4.87   Wafter map of known good dies from on-wafer test.

interface functionality drive the requirements of a RF matrix. Highly integrated matrices can easily exhibit high loss and poor matches that increase with frequency if care is not taken in the construction. These losses and mismatches can significantly degrade the accuracy of a system regardless of the calibration technique used. Assuming moderate power levels are to be used, frequency range is by far the most critical design consideration.

A system matrix must outperform the parts being tested on it. For complex systems requiring measurements such as intermodulation, harmonics, noise figure, or high port-to-port isolation, mechanical switches are the better alternative over solid state. Solid state switches would likely add their own performance limitations to the critical measurements being performed and cause erroneous results. Mechanical switches also have limitations to be considered. Although most mechanical switches have excellent transfer, isolation, and return loss characteristics, there is one issue that is sometimes overlooked. The return loss contact repeatability can easily vary by ± 5 milliunits and is additive based on the number of switches in series. To remove this error, directional couplers could be placed last in the matrix closest to the DUT and multiplexed to a common measurement channel within the network analyzer. This deviates from a conventional 2-port ANA configuration, but is worth consideration when measuring low VSWR devices.

Calibration Technique

Regardless of the environment, the level of system complexity and hardware resources can be minimized depending on the accuracy and speed requirements. Although the same criteria applies to both fixture and wafer environments, for optimum accuracy, errors can be minimized by focusing efforts on the physical limitations of the system integration, the most important being source and load matches presented to the DUT. By minimizing these parameter interactions, the accuracy of a scalar system can approach that of a full vector corrected measurement system.

The level of integration and hardware availability dictates the calibration requirements and capabilities of any test system. Simple systems designed for only one or two functions may necessitate assumptions in calibration and measurement errors. As an example, performing noise figure measurements on wafer using only a scalar noise figure system required scalar offsets be applied to attribute the loss of the probe environment, which cannot be dynamically ascertained through an automated calibration sequence. The same can also apply to a simple power measurement system consisting of only a RF source and a conventional power meter and assuming symmetry of input and output probes. These methods can and are used in many facilities, but can create large errors if care is not taken to minimize mismatch error terms that often come with contact degradation from repeated connections.

To obtain high accuracy up to the probe interface in a wafer environment requires a two-tier calibration method for certain measurements since it is usually difficult to provide a noise source or power sensor connection at the wafer plane. The most effective measurement tool for this second-tier calibration is a vector network analyzer. It not only provides full vector correction to the tips of the RF probes, but when the resulting vector measurements are used in conjunction with other measurement, such as noise figure and power, it can compensate for dynamic vector interactions between the measurement system and the device being tested. Equation (4.62), the vector relationship to the corrected input power (P_{A1}), and Eq. (4.63), the scalar offset normally applied in a simpler system, illustrate the relationship that would not be taken into account during a scalar power measurement when trying to set a specific input power level to the DUT. Usually a simple offset, P_{offset}, is added to the raw power measured at port A_0, (P_{A0}) to correct for the incident power at the device input A_1 (P_{A1}) . This can create a large error when poor or moderate matches are present.

As an example, a device with a 10 dB return loss in a system with a 15 dB source match, not uncommon in a wafer environment, can create an error of close to ±0.5 dB in the input power setting when system interactions are ignored.

$$P_{A1} = \left| \frac{P_{A0}}{\left(1 - E_{sf} S_{11a}\right)^2} \right| \left(P_{offset}\right) \tag{4.62}$$

$$P_{A1} = P_{A0} \left(P_{offset}\right) \tag{4.63}$$

A similar comparison can be shown for the noise figure. Equations (4.64) and (4.65) illustrate the difference between the vector and scalar correction of the raw noise figure (R_{NF}) as measured by a standard noise figure meter. Depending on the system matches and the noise source gamma, the final corrected noise figure (C_{NF}) could vary considerably.

$$C_{NF} = R_{NF} + 10 LOG \left(\frac{\left(\left|E_{10}^2\right|\right)\left(1 - \left|G_{ns}^2\right|\right)}{\left(\left(1 - \left|E_{sf} + \left(E_{10}^2 G_{ns} / \left(1 - G_{ns} E_{df}\right)\right)\right|^2\right)\right)\left(\left|\left(1 - E_{df} G_{ns}\right)^2\right|\right)} \right) \tag{4.64}$$

$$C_{NF} = R_{NF} + 10 LOG \left(\left|E_{10}^2\right|\right) \tag{4.65}$$

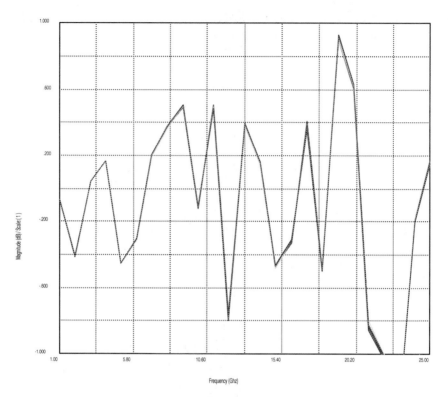

FIGURE 4.88 S21 vector to scalar measurement comparison of discrete FET (mismatched).

For small signal correction, the forward path of the standard 12 Term, Full 2-Port Error model as given in Fig. 4.20 of section 4.2 (Network Analyzer Calibration),[8] is applied. Equation (4.66) gives the derivation of the actual forward transmission (S_{21a}) from these error terms combined with raw measured data. By minimizing the mismatched terms E_{sf}, E_{lf}, E_{sr}, E_{lr}, E_{xf}, and E_{df}, detailed in section 4.2, Eq. (4.66) simplifies to Eq. (4.67). This simplified term is essentially the calculation used in standard scalar measurement systems and reflects an ideal environment. A further level of accuracy can be obtained when dealing with scalar systems that is very dependent on the type of device being tested. Looking at Eq. (4.66) it can be seen that in deriving S_{21a} many relationships between the error terms and measured values provide products that can further minimize errors based on the return loss components of the DUT as well as isolation in the reverse path. This makes an active device with good return losses and high reverse isolation a good candidate for a scalar measurement system when only concerned with gain as the functional pass/fail criteria. On the other hand, a switch or other control product has a potential for being a problem due to the symmetrical nature of the device if care is not taken to minimize the match terms. An even poorer candidate for a scalar system would be discrete transistors, which normally have not been tuned for optimum matching in the measurement environment. Figure 4.88 is an on-wafer measurement comparison of a discrete FET measurement using both full 2-port error correction as in Eq. (4.66) and the simplified scalar response Eq. (4.67) from 1 GHz to 25 GHz. The noticeable difference between these data sets is the "ripple" effect that is induced in the scalar corrected data, which stems from the vector sums of the error terms rotational relationship to the phase rotation of the measurement. Figure 4.89 shows the error terms E_{lf} and E_{lr} generated by multiple calibrations on the same vector test system used to measure the data in Fig. 4.88. Although the values seem reasonable, the error induced in the final measurement is significant.

This error is largely based on the poor input and output match of the discrete FET, as shown in Fig. 4.90, and their interaction with the system matches.

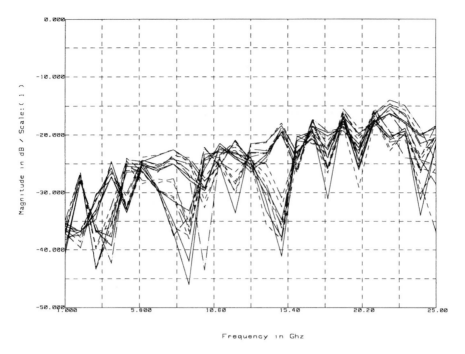

FIGURE 4.89 Elf and Elr error terms over a 5-month period.

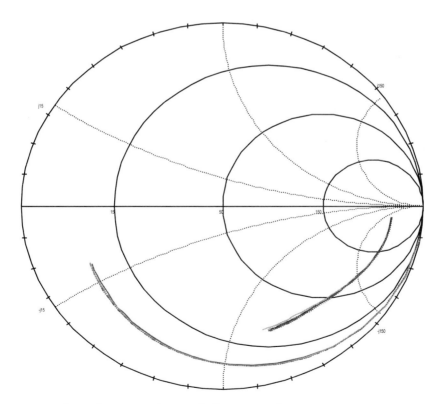

FIGURE 4.90 S11 and S22 of PCM FETs (mismatched) across a wafer.

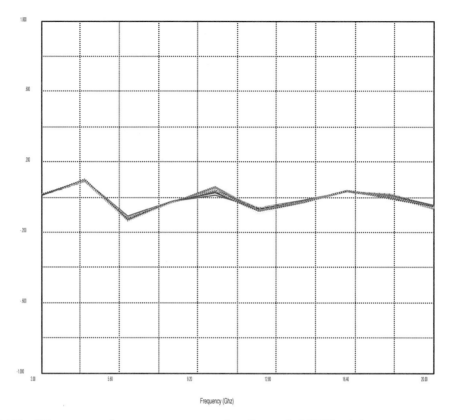

Frequency (Ghz)

FIGURE 4.91   S21 vector to scalar measurement comparison for matched SPDT switch.

Figure 4.91, an example of better scalar-to-vector correlation, is an on-wafer measurement of a single pole double throw switch comparison using both full 2-port error correction as in Eq. (4.66) and the simplified scalar response Eq. (4.67) from 2 GHz to 20 GHz. Although the system matches are comparable to the discrete FET measurement, the device input and output return losses are both below 15 dB (Fig. 4.92). This product minimizes the errors induced by system to DUT interactions thus giving errors much smaller than that of the discrete FET measurement of Fig. 4.88.

$$S_{21a} = \frac{\left(\left(S_{21m} - E_{xf}\right)/E_{tf}\right)\left(1 + \left(S_{22m} - E_{dr}\right)\left(E_{sr} - E_{lf}\right)/E_{rr}\right)}{\left(1 + \left(\left(S_{11m} - E_{df}\right)E_{sf}/E_{rf}\right)\right)\left(1 + \left(\left(S_{22m} - E_{dr}\right)E_{sr}/E_{rr}\right)\right) - \left(\left(S_{21m} - E_{xf}\right)\left(S_{12m} - E_{xr}\right)E_{lf}E_{lr}/E_{tf}E_{tr}\right)} \tag{4.66}$$

$$S_{21a} = \frac{S_{21m}}{E_{tf}}\bigg|_{E_{lf},E_{sf},E_{sr},E_{lr},E_{df} \to 0} \tag{4.67}$$

Dynamic Range

Dynamic range is the final major consideration for accuracy of a measurement system. Dynamic range of any measurement instrument can be enhanced with changes in bandwidth or averaging. This usually degrades the speed of the test. A perfect example of this is a standard noise figure measurement of a medium gain LNA using an HP 8970 noise figure meter. Noise figure was measured on a single device one hundred times using 8 averages. The standard deviation is .02 dB, the cost for this is a 1.1-second

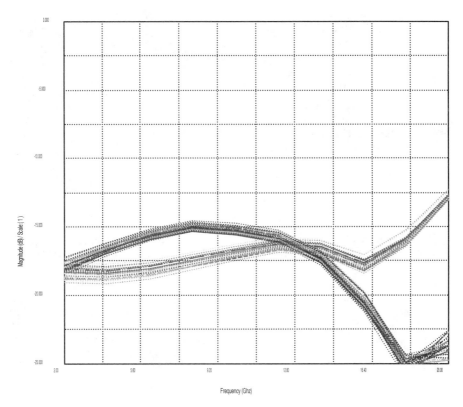

FIGURE 4.92 S11 and S22 of matched SPDT switch.

measurement rate. By comparison, the same device measured with no averaging resulted in a standard deviation of .07 dB, but the measurement rate was less than 500 milliseconds.

Other methods can be applied to enhance the accuracy of the measurement without losing the speed. Placing a high gain 2nd stage LNA between the DUT and noise receiver will increase the dynamic range of the system and minimize the standard deviation obtained without losing the speed enhancement. These types of decisions should be made based on the parts performance and some experimentation.

Another obvious example is bandwidth and span setting on a spectrum analyzer. Sweep rates can vary from 50 milliseconds to seconds if optimization is not performed based on the requirements of the measurement. As in the noise measurement, this also should be evaluated based on the parts performance and some experimentation.

Highly customized systems that are optimized for one device type can overcome many dynamic range and mismatch error issues with additional components such as amplifiers, filters, and isolators. This can restrict or limit the capabilities of the system, but will provide speed enhancements and higher device output rates with minimal impact on accuracy.

On-Wafer Test Interface

On-wafer test of RF devices is almost an ideal measurement environment. Test interface technologies exist to support vector or scalar measurements. Common RF circuits requiring wafer test are: amplifiers, mixers, switches, attenuators, phase shifters, and coupling structures. The challenge is to select the interface technology or technologies that deliver the appropriate performance/cost relationship to support your product portfolio. Selection of test interface of wafer probes will be based on the measurements made and the desired product environment. It is common for high gain amplifiers to oscillate or for narrowband devices to shift frequency due to lack of bypass capacitors or other external components. It

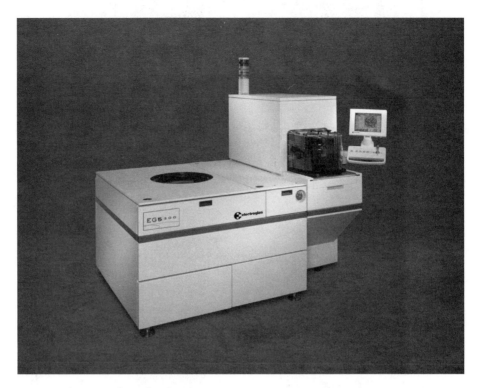

FIGURE 4.93   Production wafer prober for RF test.

is recommended to consider wafer test during the circuit design stage to assure the circuit layout satisfies wafer test requirements.

A typical wafer probe system incorporates a test system, wafer prober, RF probes, and DC probes. Figure 4.93 shows a photograph of a typical production wafer prober. This prober has cassette feed, auto alignment, and is configured for a test system "test head." The test head connects to the test interface, which mounts in the hole on the left side of the machine. This prober uses a ring-type probe card as shown in Fig. 4.94. Conventional RF probes are mounted to the prober top plate using micro-manipulators arranged in quadrants. This allows access to each of the four sides of the integrated circuit. Figure 4.95 shows a two-port high frequency setup capable of vector measurements. Wafer prober manufacturers offer different top plates for different probe applications. Specification of top plate configuration is necessary for new equipment purchases.

Probe calibration standards are necessary to de-imbed the probe from the measurement. Calibrated open, short, and load standards are required for vector measurements. Probe suppliers offer calibrated standards designed specifically for their probes. For scalar measurements or when using complex probe assemblies, alternative calibration standards can be used, but with reduced measurement accuracy. Alternative calibration standards may be a custom test structure printed on a ceramic substrate or on a wafer test structure. Scalar offsets can be applied for probe loss if you have a method of probe qualification prior to use. In general you have to decide if you are performing characterization or just a functionality screen of the device. This is important to consider early since measurement accuracy defines the appropriate probe technology, which places physical restrictions on the circuit layout.

When selecting the probe technology for any application you should consider the calibration approach, the maximum-usable frequency, the number of RF and DC connections required, the ability to support off-chip matching components, the cost of probes, and the cost of the calibration circuits. By understanding the advantages and limitations of each probe approach, an optimum technology/cost decision can be made. Remember that the prober top plate can be specified for ring frames or micro-manipulator type probes. Machine definition often dictates the types of probes to be used.

FIGURE 4.94 Ring-type RF probe card.

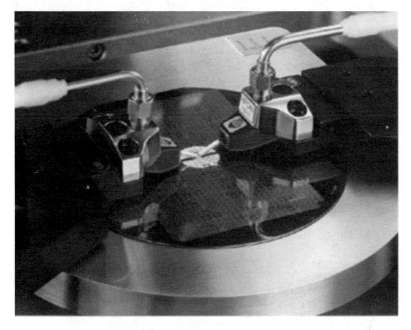

FIGURE 4.95 RF probes mounted on manipulators.

Traditional RF probes convert a coax transmission line into coplanar signal and ground probe points. This allows a coplanar or microstrip circuit with ground vias to be measured. These probes are offered as ground-signal and ground-signal-ground. They have been widely used for accurate high frequency measurements for many years. The ground-signal-ground probe offers improved performance above 12 GHz and can be used up to 100 GHz with proper construction. Probe spacing from signal to ground is referred to as the pitch. A common probe pitch is 0.006 in. Due to the small size, material selection significantly impacts RF performance and physical robustness. Many companies including Cascade Microtech and PicoProbe specialize in RF probes.

Cost considerations of probes are important. RF probes or membranes can cost anywhere from $300 to $3,000 each. This adds up quickly when you need multiple probes per circuit, plus spares, plus calibration circuits. When possible it is recommended to standardize the RF probe pitch. This will minimize setup time and the amount of hardware that has to be purchased and maintained. When custom probes are to be used, be prepared to incur the cost of probe and the calibration circuit development.

Wafer level RF testing using coplanar probing techniques can easily be accomplished provided the constraints of the RF probe design are incorporated into the circuit layout. This usually requires more wafer area be used for the required probe patterns and ground vias. These are standard and preferred design criteria for high frequency devices requiring on-wafer test. Devices without ground vias may require alternative interface techniques such as custom probes or membrane probes.

Although typical RF circuits have two or three RF ports and several DC, there are many that require increased port counts. Advanced probing techniques have been developed to support the need for increased RF and DC ports as well as the need for near chip matching and bypass elements. Probe manufacturers have responded by producing custom RF/DC probe cards allowing multiple functions per circuit edge. Figure 4.96 is an example of a single side four-port RF probe connected to a calibration substrate. Probe manufacturers have also secured the ability to mount surface mount capacitors on the end of probe tips to provide close bypass elements.

Another approach is Cascade Microtech's Pyramid Probe. It is a patented membrane probe technology that offers impedance lines, high RF and DC port count, and close location of external components.

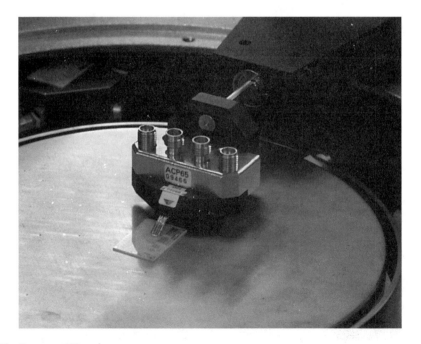

FIGURE 4.96   Four-part RF probe.

FIGURE 4.97   Cascade Microtech pyramid probe.

Figure 4.97 shows the Pyramid Probe with an off-chip bypass capacitor. One important aspect of the construction is that it incorporates an uninterrupted RF ground path throughout the membrane. This differs from the traditional coplanar probes that require the circuit to conduct the ground from one RF port to another. This allows for RF probing of lumped element circuits that do not utilize via holes and back side ground planes. This is becoming especially important to support developments such as chip scale packaging and multi-chip modules where the use of known good die is required for manufacturing.

For high volume devices where the circuit layout is optimized for the final package environment, considerations for on-wafer testing are secondary if not ignored. Products targeting the wireless market undergo aggressive die size reductions. Passive components such as capacitors, inductors, and resistors are often realized external to the integrated circuit. In this case the probes must be designed to simulate the packaged environment including the use of off chip components. Membrane technology is a good consideration for this. The membrane probe has the potential to emulate the package environment and external components that may be required at the final device level.

On-Wafer RF Test Benefits

The benefits of on-wafer RF testing are multiple and explain its success in the RF and microwave industry:

- Accuracy of RF test results with calibration performed at the probe tip, contact point to the DUT. The calibration techniques are now well established, supported by elaborate calibration standards, and easily implemented with software internally developed or purchased from the test equipment or probe vendors. This leads to accurate device models and higher first-pass design yields.
- Reproducibility of test results with stable impedance of the probe — be it 50 Ohms or a custom impedance — and automatic probe-to-pad alignment performed by modern wafer probers. Set probe placement on the pads during test and calibration is critical, especially above 10 GHz and for DUTs presenting a narrowband match.
- Nondestructive test of the DUT, allowing shipment of RF Known Good Dies to the user. This ability is key for multi-chip module or flip-chip onboard applications. The correlation between on-wafer and assembled device test results is excellent if the MMIC grounding is properly realized

and the DC biasing networks are similar. For example, our experience producing 6 GHz power devices shows a maximum 0.2 dB difference in output power between wafer and module levels.

- Short cycle time for product test or statistical characterization and model extraction of library components, allowing for successful yield modeling and prediction.
- High throughput with complete automation of test and probing activities, and low cost, decreased by a factor of 10 in 10 years, to well below one dollar for a complex DUT today.

Wafer probing techniques are in fact gaining in importance today and are used for higher volume applications as Chip Scale Packages, Chip Size Packages (CSP), and flip chip formats become more common, bypassing the traditional plastic packaging step and test handler. Another increasing usage of on-wafer test is for parts built in array formats such as multi-chip modules or ball grid arrays. For these applications, robust probes are needed to overcome the low planarity of laminate boards. Higher speed test equipment such as that used with automatic handlers is likely to become more prevalent in wafer level test to meet volume needs. The probing process must now be designed to form a continuous flow, including assembly, test, separation, sorting, and packaging.

References

1. Strid, E.W., 26 GHz Wafer Probing for MMIC Development and Manufacture, *Microwave Journal*, August 1986.
2. Strid, E.W., On-Wafer Measurements with the HP 8510 Network Analyzer and Cascade Microtech Wafer Probes, RF & Microwave Measurement Symposium and Exhibition, 1987.
3. Cascade Microtech Application Note, On-Wafer Vector Network Analyzer Calibration and Measurements. (www.cascademicrotech.com)
4. Cascade Microtech Technical Brief TECHBRIEF4-0694, A Guide to Better Network Analyzer Calibrations for Probe-Tip Measurements. (www.cascademicrotech.com)
5. Mahon, J.R. et al., On-Wafer Pulse Power Testing, ARFTG Conference, May 1990.
6. Poulin, D.D. et al., A High Power On-Wafer Pulsed Active Load Pull System, *IEEE Trans. Microwave Theory and Tech.*, MTT-40, 2412–2417, Dec. 1992.
7. Dambrine, G. et al., A New Method for Determining the FET Small Signal Equivalent Circuit, *IEEE Trans. Microwave Theory and Tech.*, MTT-36, 1151–1159, July 1988.
8. Staudinger, J., Network Analyzer Calibration, *CRC Modern Microwave and RF Handbook*, CRC Press, Boca Raton, FL, chap. 4.2, 2000.

4.8   High Volume Microwave Test

Jean-Pierre Lanteri, Christopher Jones, and John R. Mahon

High Volume Microwave Component Needs

Cellular Phone Market Impact

High volume microwave test has emerged in the early 1990s to support the growing demand for GaAs RFICs used in cellular phones. Prior to that date, most microwave and RF applications were military and only required 10,000s of pieces a year of a certain MMIC type, easily probed or tested by hand in mechanical fixtures. For most companies in this industry, the turning point for high volume was around 1995 when some RFIC parts for wireless telephony passed the million per year mark. Cellular phones have grown to over 300 million units shipped in 1999 and represent 80% of the volume of microwave and RF ICs manufactured, driving the industry and its technology.

The cellular phone needs in terms of volume, test cost, and acceptable defect rate demanded new test solutions (Table 4.6) be developed that relied on the following key elements:

1. "Low" frequency ICs, first around 900 MHz and later on around 1.8 and 2.4 GHz, with limited bandwidth, allowing simpler device interfaces and fewer test points over frequency. Previously, MMICs were mostly military T/R module functions with frequencies ranging from 2 to 18 Ghz, with 30% or more bandwidths. They were tested at hundreds of frequencies, requiring specialized fast ramping Automatic Network Analyzers (ANA) such as Agilent's HP8510 or HP8530.
2. Standard plastic packages, based upon injection molding around a copper lead frame, to reach the low cost required in product assembly and test. Most early RFICs used large gull wing Dual In-line Packages (DIP), then Small Outline IC packages (SOIC), later Small Outline Transistor packages (SOT), and today's Micro Leadframe Flatpack (MLF).
3. Automatic handlers from the digital world, typically gravity fed, leveraging the plastic packages for full automation and avoiding human errors in bin selection. Previous metal or ceramic packages were mostly custom, bulky, and could only be handled automatically by pick-and-place type handlers, such as the one made by Intercontinental Devices in the early 1990s, barely reaching throughputs of a few hundred parts per hour.
4. Highly repeatable, accurate, and durable device contact interface and test board, creating the proper impedance environment for the device while allowing mechanized handling of the part. Most products before that were designed as matched to 50 Ohm impedance in and out, where cellular phone products will most often need to be matched in the user's system, and therefore on the test board. Adding to the difficulty, many handlers converted from digital applications hold the part in the test socket with a bulky mechanical clamp that creates ground discontinuities in the test board and spread the matching components further apart than designed in the part application.
5. Faster Automatic Network Analyzer (ANA) test equipment through hardware and software advances, later supplanted by specialized RFIC testers. The very high volumes reached by some parts, over a million pieces a week, allow dedication of a customized system to their testing to reduce measurement time and cost. Therefore the optimum test equipment first evolved from a powerful ANA-based system (HP8510, for example) with noise figure meter, spectrum analyzer, and multiport RF switch matrix, to an ad hoc set of bench-top equipment around an integrated ANA or ANA/spectrum analyzer. Next appeared products inspired from the digital world concept of the "electronic pin" tester, with RF functionality at multiple ports, such as the HP84000, widely used today.
6. Large databases on networked workstations and PCs for test results collection and analysis. The value of the information does not reside in the pass or fail outcome of a specific part, but in the statistical trends and operational performance measures available to company management. They provide feedback on employee training, equipment and calibration reproducibility, equipment maintenance schedules, handler supplier selection, and packaging supplier tolerances to name a few.

Although the high volume techniques described in this chapter would apply to most microwave and RF components, they are best fitted for products that do not require a broadband matched environment and that are packaged in a form that can be automatically tested in high-speed handlers.

TABLE 4.6 Microwave and RF IC Test Needs Evolution

Year	Product	Application	Package	Price	Volume	Test Time	Test Cost	Escape Rate
1991	T/R Module	Radar	Carrier	$200	10K/Y	1 min	$30	1%
1993	T/R Switch	Radar/Com	Ceramic	$40	100K/Y	30 sec	$4	0.5%
1995	RF Switch	Com	Plastic	$10	Mil/Y	10 sec	$1	0.1%
1997	RF MMIC	Com	SOIC	$3	Mil/M	3 sec	$0.30	0.05%
1999	RF MMIC	Com	SOT	$1	Mil/W	1 sec	$0.10	0.01%

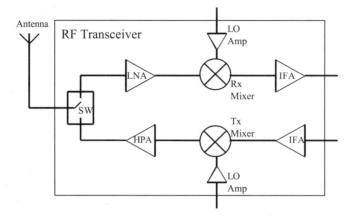

FIGURE 4.98 Typical RF transceiver building blocks.

High Volume RF Component Functions and Test Specifications

We will focus in this section on the different functions in the RF front end of a wireless phone to illustrate the typical products tested, their function, specification, and performance. The generic building blocks of a RF front end (Fig. 4.98) are switches (for antenna, Transmit/Receive (T/R), or band selection), input Low Noise Amplifiers (LNA), output Power Amplifiers (PA), up- and downconverters (typically comprising a mixer), Local Oscillator Amplifier (LOA), and Intermediate Frequency Amplifier (IFA). In most cases, these products are single band, either cellular or PCS, although new dual band components are appearing, requiring two similar tests in sequence, one for each band.

The test equipment should therefore be capable of measuring DC parameters, network parameters such as gain or isolation, and spectral parameters such as IMD for most high volume products. Noise figure is required for LNAs and downconverters, and output power for HPAs. Typically, two types of RFIC testers will handle most parts, a general purpose RFIC for converters and eventually switches, and a specialized one for HPAs.

Typical specifications for the various parts are provided below. No specification is very demanding on the test instrument in absolute terms, but the narrow range of acceptance for each one requires outstanding reproducibility of the measurements, part after part. This will be the limiting factor in escape rate in our experience.

TABLE 4.7 Typical Product Specifications for High Volume Test

Switch Parameters	Min	Max	LNA Parameters	Min	Max
Frequency Range	800 MHz	1000 MHz	Frequency Range	800 MHz	1000 MHz
Control Leakage	−10 uA	10 uA	Current Consumption	8 mA	12 mA
Insertion Loss		0.5 dB	Linear Gain	15 dB	18 dB
Isolation	25 dB		Noise Figure		2 dB
Input IP3	60 dBm		Input IP3	−4 dBm	
PA Parameters	Min	Max	Mixer Parameters	Min	Max
Frequency Range	800 MHz	1000 MHz	Frequency Range	800 MHz	1000 MHz
Linear Current	160 mA	200 mA	IF Frequency Range	DC	100 MHz
Linear Gain	27 dB	35 dB	Conversion Loss		7.5 dB
Pout @ Pin = −1 dBm	25 dBm	30 dBm	LO to RF Leakage	38 dB	
Current @ Pin = −1 dBm		300 mA	1 dB Compression	21 dBm	
1 dB Compression	22.5 dBm		IMD @ Pin = −10 dBm	65 dBc	

These specifications are dictated by the application and therefore largely independent of the technology used for fabrication of the RFIC. RFIC technology was predominantly GaAs Metal Semiconductor Field Effect Transistor (MESFET) until 1997, when GaAs Heterojunction Bipolar Transistor (HBT) appeared, soon followed by silicon products, in BiCMOS, SiGe BiCMOS, and CMOS technologies. The RF test is performed in a similar fashion for all implementation technologies of a given functionality.

High Volume Test Success Factors

The next sections will review in detail aspects of a successful back-end production of typical RF high volume parts; inexpensive, not too complex, packaged in plastic, produced at the rate of a million per week. The basic requirements addressed are:

- test equipment selection, balancing highest test speed with lowest test cost for the product mix
- automatic package handler keeping pace with the tester through parallel handling, and highly reliable
- part contactor performing at the required frequency, lasting for many contacts
- test software for fast set up of a new part with automatic revision control

Less obvious but key points for cost-effective high volume production are also discussed:

- tester, contactor, and test board calibration approach for reproducible measurements
- cost factors in a high volume test operation
- data analysis capabilities for relating yield to design or process
- test process monitoring tools, to ascertain the performance of the test operation itself

Test System Overview

Hardware: Rack and Stack vs. High Speed IC Testers

Hardware considerations are based on the measurement requirements of your product set. To evaluate this, the necessary test dimensions should be determined. These dimensions can include but are not limited to swept frequency, swept spectrum, modulation schemes, swept power, and DC stimulus.

Commercially available hardware instruments can be combined to perform most RF/DC measurement requirements for manufacturing applications. These systems better known as "Rack and Stack" along with widely available third party instrument control software can provide a quick, coarse start-up for measurement and data collection, ideally suited for engineering evaluation. As the measurements become more integrated, the complexity required may exceed the generic capabilities of the third party software and may have to be supplemented with external software that can turn the original software into nothing more than a cosmetic interface.

To take the "Rack and Stack" system to a higher level requires a software expertise in test hardware communication and knowledge of the optimum sequencing of measurement events. Most hardware in a rack and stack system provides one dimension of competence, for example a network analyzer's optimum performance is achieved during a swept frequency measurement, a spectrum analyzer is optimized for frequency spectrum sweeps with fixed stimulus. Taking these instruments to a different dimension or repeating numerous cycles within their optimum dimension may not provide the speed required. Some instruments do provide multiple dimensions of measurement, but usually there is a setup or state change required that can add to the individual die test time. Another often-ignored aspect of these types of instruments is the overhead of the display processor, which is important in an engineering environment but an unnecessary time consumer in a manufacturing environment.

Commercially available high volume test systems usually provide equivalent speed in all dimensions by integrating one or two receivers with independently controlled stimulus hardware, unlike a network analyzer where the stimulus is usually linked to the receiver. These high-speed receivers combined with independently controlled downconverters, for IF processing, perform all the RF measurements that normally would take multiple instruments in a rack and stack system. Since these receivers are plug-in

TABLE 4.8 Speed Comparison of Rack and Stack and High Speed IC Tester

Repeat Count	Measurement/Stimulus	Rack and Stack		High Speed IC	
		Each	Total	Each	Total
3 Times	Set RF Source #1 Stimulus	100 mS	300 mS	50 mS	150 mS
3 Times	Set RF Source #2 Stimulus	100 mS	300 mS	50 mS	150 mS
12 Times	Set Analyzer to Span	250 mS	3000 mS	50 mS	600 mS
12 Times	Acquire Output Signal	50 mS	600 mS	40 mS	480 mS
	Total Time		4200 mS		1380 mS

modules, whether for a PC back plane or a controlling chassis like a VXI card cage, they are also optimized for fast I/O performance and do not require a display processor, which can significantly impact the measurement speed. And since these receivers are usually based on DSP technology, complex modulation measurements such as ACPR can easily be made without additional hardware as would be required in most rack and stack systems.

In a normal measurement sequence of any complex device, the setting of individual stimulus far exceeds the time required to acquire the resulting output. A simple example of this would be a spectrum analyzer combined in a system with two synthesized sources to perform an intermodulation measurement at three RF frequencies. Accomplishing this requires extensive setting before any measurements can be made. Table 4.8 shows the measurement sequence and the corresponding times derived from a rack and stack system and a commercially available high speed IC measurement system for comparison. The measurement repeatability of these systems is equivalent for this example, therefore the bandwidth of the instrument setting is comparable.

As shown in the table, the acquisition of the output signal shows relatively no speed improvement with a difference of only 120 mS total. The most significant improvement is the setting of the acquisition span on the High Speed IC tester. This speed is the same as the setting of a RF stimulus since the only overhead is the setting of the LO source required for the measurement downconversion. The only optimization that could be performed with the rack and stack system would be higher speed RF sources having internal frequency list and power leveling capability. The change in span setting on a standard spectrum analyzer will always be a speed inhibitor since it is not its optimum dimension of performance.

From this type of table a point can be determined where the cost of a high-speed IC tester outweighs the speed increase it will yield. This criteria is based on complex multifunction devices that require frequent dimension hopping as described above. Other component types, such as filters requiring only broadband frequency sweeps in a single dimension, would show less speed improvement with an increase in frequency points since network analyzers are optimized for this measurement type.

Various vendors for high speed systems exist. Agilent Technologies (formerly Hewlett Packard), Roos Instruments, LTX, and Teradyne are just a few of the more well-known suppliers. The full system prices can range from a few hundred thousand dollars to well into the millions depending on the complexity/customization required.

A note of caution when purchasing a high speed IC tester: careful homework is warranted. Most IC testers are a three- to five-year commitment of capital dollars, and the one purchased should meet current and future product requirements. Close attention to measurement capabilities, hardware resources, available RF ports, DC pin count, and compatibility to existing test boards will avoid future upgrades, which are usually costly and delay time to market for new products if the required measurement capability is not immediately available.

System Software Integration

Software capabilities of third party systems require close examination, especially if it is necessary to integrate the outputs with existing resources on the manufacturing floor. Most high-speed IC testers focus on providing a test solution not a manufacturing solution. Network integration, software or test

TABLE 4.9 Test Handler Manufacturers and Type

Manufacturer	Pick and Place	Gravity	Turret
Aetrium		X	X
Asseco	X	X	
Delta Daymark	X	X	
Exatron	X	X	
Intercontinental Microwave	X		
Ismeca	X		X
MultiTest	X	X	
Roos		X	

plan control, and data file organization is usually taken care of by the end customer. This software usually provides little operator input error checking or file name redundancy checking when dealing with multiple systems. The output file structure should have all the information required available in the file. Most third party systems provide an ASCII file output, which supports STDF (Standard Test Data Format), an industry standard data format invented by Teradyne. As with the hardware, the software is fixed at a revision level. It is important to suggest improvements to the vendors to make the system more effective. Software revisions introduced by the vendor may not be available as fast as expected to correct observed deficiencies. It is still valuable to use the current revision level of the software to avoid known bugs and receive the best technical support.

RFIC Test Handlers

The primary function of the test handler is to move parts to the test site and then to sort them based on the test result. Package style and interface requirements will define what machines are available for consideration. The product will define the package and the handler is typically defined by the package. Common approaches include tube input — gravity handling, tray input — pick and place handling, and bulk input — turret handling. During the product design phase, selection of a package that works well with automation is highly recommended. The interface requirements are extremely critical for RF devices. Contact inductance, off chip matching components, and high frequency challenge our ability to realize true performance. The best approach is a vacuum pick up and plunge. This allows optimal RF circuit layout and continuous RF ground beneath the part.

Common test handler types and suppliers are listed in Table 4.9. Various options can be added to support production needs such as laser marking, vision inspection, and tape and reel. For specialized high volume applications, handlers are configured to accept lead frame input and tape and reel output providing complete reel-to-reel processing. When evaluating handlers for purchase, some extra time to identify process needs is very valuable. The machine should be configured for today's needs with the flexibility to address tomorrow's requirements. Base price, index time, jam rate, hard vs. soft tooling, conversion cost, tolerance to multiple package vendors, and vendor service should be considered. One additional quantitative rating is design elegance. An elegant design typically has the fewest transitions and fewest moving parts. Be cautious of machines that have afterthought solutions to hide their inherent limitations.

Contact Interface and Test Board

The test interface is comprised of a contactor and test board. The contactor provides compliance and surface penetration ensuring a low resistance connection is made to all device ports. Figure 4.99 shows a sectioned view of a pogo pin contactor. For RF applications the ideal contactor has zero electrical length and coupling capacitance. In the real-world contactors typically have 1 to 2 nH of series inductance and 0.2 to 0.4 pF of coupling capacitance. This can have significant impacts on electrical performance. Refer to Table 4.10 for a review of contactor manufacturers and parasitics. A more in-depth review of some available contactor approaches and suppliers is given in an article by Robert Crowley.[1] Parasitics of

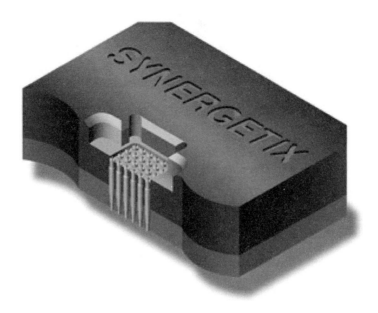

FIGURE 4.99   Pogo pin contactor.

TABLE 4.9   Test Contactor Manufacturers and Type

Manufacturer	Approach	Self Inductance	Mutual Inductance	Capacitance
Agilent	"YieldPro"	0.3 nH		0.17 pF
Aries	Microstrip Contact	0.01 pF	0.05 nH	0.04 pF
Exatron	Particle Interconnect	0.26 nH		0.024 pF
Johnstech International	"S" Contact	1.0 nH	0.2 nH	0.07 pF
Oz Tek	Pogo Pin	2.4 nH	0.4 nH	0.09 pF
Prime Yield	"Surface Mount Matrix"			
Synergetix	Pogo Pin	1.3 nH	0.1 nH	0.1 pF
Tecknit	"Fuzz Button"	2.7 nH	0.3 pF	0.3 pF

Note: Values supplied are typical values from manufacturer's catalog. Refer to manufacturer for specific information to support your specific needs.

contactors can typically be compensated for in series ports using filter networks. Shunt ports however, such as an amplifier ground reference, challenge the use of contactors because the electrical length cannot be removed. The additional electrical length often shifts performance in magnitude or frequency beyond the range where scalar offsets can be used.

Fine pitch packaging has increased the challenges associated with contactor manufacturing and lifetime. Packages such as TSSOP, SOT, SC70, and the new Micro Leadframe Flatpack (MLF) have pitches as small as 0.020 in. and may require a back-side ground connection. As contactor element size is reduced to support fine pitch packages, sacrifices are made in compliance and lifetime.

High frequency contactors are typically custom machined and assembled making them expensive. Suppliers are quoting $1000 to $4000 for a single contactor. If this expense is amortized over 500,000 parts, the cost per insertion is about one-half cent. This may be acceptable for some high value added part, but certainly not for all RF parts in general. Add to this the need to support your product mix and the need for spares and you will find that contactors can be more expensive than your capital test equipment. There is a true need for an industry solution to provide an affordable contactor with low parasitics, adequate compliance, tolerance to tin lead buildup.

The second half of the test interface is the test board, which interfaces the contactor to the test system. The test board can provide a simple circuit routing function or a matching circuit. It is common for RF circuits to utilize off-chip components for any non-active function. The production test board often requires tuning to compensate for contactor parasitics. This can result in a high Q matching circuit that increases measurement variability due to the interaction between the part, the contactor, and the test board. It is recommended to consider the contactor and test board during the product design cycle allowing the configuration to be optimized for robust performance.

High Volume Test Challenges

Required Infrastructure

The recommended facility for test of RF semiconductor components is a class 100,000 clean room with full ESD protection. RF circuits, especially Gallium Arsenide, are ESD sensitive to as little as 100 volts. Although silicon tends to be more robust than Gallium Arsenide, the same precautions should be taken. The temperature and humidity control aids test equipment stable operation and helps prolong the life of other automated equipment. Facility requirements include HVAC, lights, pressurized air and nitrogen, vacuum, various electrical resources, and network lines.

As volume increases the information system becomes a critical part of running the operation. The ideal system aids the decision process, communicates instructions, monitors inventory, tracks work in process, and measures equipment and product performance. The importance of information automation and integration cannot be overemphasized. It takes vision, skill, and corporate support to integrate all technical, manufacturing, and business systems.

The human resources are the backbone of any high volume operation. Almost any piece of equipment or software solution can be purchased, but it takes a talented core team to assemble a competitive operation and keep it running. Strengths are required in operations, software, and test systems, products, data analysis, and automation.

Accuracy and Repeatability Challenges

Measurement accuracy and repeatability are significant challenges for most high volume RF measurements. All elements of the setup may contribute to measurement inaccuracies and variability. The primary considerations are test system, the test board, the contactor, and the test environment.

For this discussion we will assume that all production setups are qualified for accuracy. This allows us to focus this discussion on variability.

Measuring Variability

Gauge Repeatability and Reproducibility (Gauge R&R) measurements can be used to measure variability. In this context the measurement system is referred to as the gauge. The gauge measurement is a structured approach that measures "x" products, "y" times, on "z" machines allowing the calculation of "machine" variability. Variability is reported in terms of repeatability and reproducibility. Repeatability describes variability within a given setup such as variability of contact resistance in one test lot. Reproducibility describes the variability between setups such as between different test systems or on different days. An overview of gauge measurement theory and calculations can be found in any statistical textbook.[2]

Figure 4.100 summarizes the sources of measurement variability within an automated test setup. The three locations are identified to allow easy gauge measurements.

Table 4.11 qualitatively rates the sources of measurement variability for repeatability and reproducibility. We can see that the system calibration and test board variations are large between setups while the contactor variations are large within a given setup. We will use these relationships in the case study to follow.

Variability is expressed in terms of standard deviation. This allows normalized calculations to be made. For example, the variability of any measurement is a combination of the variability of the product and the gauge. This can be expressed as:

$$\sigma^2\text{measured} = \sigma^2\text{product} + \sigma^2\text{gauge}$$

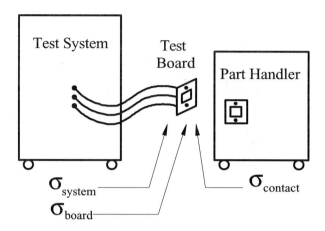

FIGURE 4.100   Sources of variability in an automated test setup.

TABLE 4.11   Repeatability and Reproducibility Comparison
for the Complete Test Environment

Source	Description	Repeatability (within a setup)	Reproducibility (between setups)
Test System	Calibration	Low	High
Test Board	Matching Circuit	Low	High
Contactor	Contact Resistance	High	Low

TABLE 4.12   Gauge Test Design

"Machine"	# Machines	"Product"	# Products	# "Measurements"
Test System	4	Part soldered to test board	3	3
Test Board	4	Loose parts	3	3
Handler Contact	3	Loose parts	10	3

Based on Fig. 4.100 the total variability of an automated test can be described as:

$$\sigma^2 total = \sigma^2 product + \sigma^2 system + \sigma^2 board + \sigma^2 contact$$

And for any expression of variability we can distinguish between repeatability and reproducibility as:

$$\sigma^2 gauge = \sigma^2 repeatability + \sigma^2 reproducibility$$

Table 4.12 recommends a gauge test design to characterize the components shown in Fig. 4.100. In this design we are measuring the "Machine" variation using "Products" and repetitive "Measurements." In all cases, stable product fixturing techniques are required for measurement accuracy. For the handler contact measurement, a single test setup is recommended.

Case Study

A low yielding product has been identified. Feedback from the test floor suggests the yield depends on the machine used and the day it was tested. These are the signs that yield is largely affected by variability. The following presents an analytical process that identifies the variability that must be addressed to improve yields.

TABLE 4.13 Measurement and Product Variability

Data Source	Variability	Repeatability (one setup)	Reproducibility (across setups)	Total
Production Data	Total	1.6 dB	1.0 dB	1.89 dB
Gauge R&R	System	0.09 dB	0.23 dB	0.25 dB
	Board	0.13 dB	0.75 dB	0.76 dB
	Contact	0.54 dB	0.00 dB	0.54 dB
Calculation	Product	1.50 dB	0.62 dB	1.62 dB

Step 1: Identify the Failure Mode— For this product we found one gain measurement to be more sensitive than others. In fact this single parameter was driving the final yield result. This simplifies the analysis allowing us to focus on one parameter.

Step 2: Quantify Measurement and Product Variability— A query of the test database showed 1086 production lots tested over a four-month span. For each production lot the average gain and standard deviation was reported. We define a typical gain value by taking the average of all production lot averages. Repeatability, or variability within any given test, was defined by finding the average of all production lot standard deviations. Reproducibility, or variability between tests, was found by taking the standard deviation of the average gain values for all production lots. Gauge R&R testing was conducted to determine the repeatability and reproducibility of the "system," "board," and "contact" as described previously. This allows calculation of product variability as shown in Table 4.13.

Step 3: Relate Variability to Yield— Relating variability to yield will define the product's sensitivity to the measurement. This will allow us to focus our efforts efficiently to maximize yield while supporting the customers' requirements. We can calculate yield to each spec limit using Microsoft Excel's NORMDIST function as follows:

Percent below upper spec limit = $Y(USL)$ = NORMDIST(USL, μ, σ, 1)
Percent above lower spec limit = $Y(LSL)$ = 1- NORMDIST(LSL, μ, σ, 1)

And we can calculate the final yield as follows:

Yield = $Y(USL) - (1 - Y(LSL))$

Prior to calculating yield we need to make some assumptions of how repeatability and reproducibility should be treated. For this analysis it is assumed that repeatability values will be applied to the standard deviation and reproducibility values will be used to shift the mean. Yield will be calculated assuming a worst case shift of the mean by one, two, and three standard deviations. The result will be plotted as Yield vs. Standard Deviation. The plot can be interpreted as the sensitivity of the parameter yield versus the measured variability of the test setup. This result is shown in Fig. 4.101 using the data in Table 4.13, the USL = 26.5 dB, the LSL = 21.5 dB, and the Average Gain = 23.1 dB.

Figure 4.101 quickly communicates the severity of the situation and identifies the test board as the most significant contributor. Looking at the product by itself we see that its yield can vary between 90% and 43%. Adding the test system variability makes matters worse. Adding the test board shows the entire process is not capable of supporting the specification. There are three solutions to this problem. Change the specifications, reduce the variability, or control the variability. Changing the specification requires significant customer involvement and communication. From the customer's point of view, if a product is released to production, specification changes are risky and avoided unless threat of line shutdown is evident. Reducing variability is where your effort needs to be focused. This may require new techniques and technology to achieve. In the process of reducing variability lessons learned can be applied across all products resulting in increased general expertise that can support existing and future products. The last method that can be applied immediately is to control variability. This is a process of tightly measuring and approving your measurement hardware from test systems to surface mount components. Everything

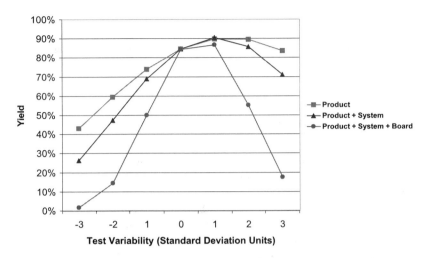

FIGURE 4.101   Yield vs. variability for test system elements.

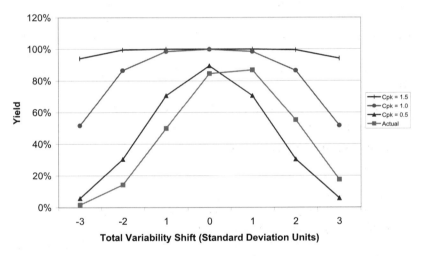

FIGURE 4.102   Yield vs. variability as function of Cpk.

gets qualified prior to use. This may take significant logistics efforts to put in place, but the yield improvements can be substantial.

This study is of an extreme case. To communicate this issue in a generic sense we can compare the same product case for various Cpk values. Figure 4.102 displays total variability vs. Cpk values of 0.5, 1.0, and 1.5. We see that the case study shape is similar to the Cpk = 0.5 curve with a mean offset. It also shows that the process can be supported by a Cpk = 1.5 or greater. Anything less requires control of variability.

Volume and Cost Relationship

In general, cost of test reduces with increasing volume. Your ability to model available capacity will allow accurate estimation of cost. A generic capacity equation is:

$$Capacity = \frac{(Time\ Available)(Efficiency)}{Test\ Time + Handling\ Time} \tag{4.68}$$

Time available can be a day, month, or year as long as all time units are consistent. Efficiency is a measure of productive machine time. Efficiency accounts for all downtime within the time available due to equipment calibration, handler jams, material tracking operations, or anything else. For time intervals greater than a week you will find that efficiency converges. A typical range for initial estimates is 60% to 70%. Focus or lack of focus can swing the initial range by ±20%.

Cost of testing can be calculated using the estimated capacity and costs or with the actual cost and volume. The baseline result is shown in Eq. (4.69).

$$Unit\ Cost = \frac{Cost}{Volume} = \frac{Facility + Equipment + Labor + Materials}{\left(Capacity\right)\left(Yield\right)} \tag{4.69}$$

Example Cost of Test: A complex part enters production. A $650,000 test system and a $350,000 handler are required and have been purchased. The estimated test and handling times are both one-half second. Based on Eq. (4.68) we can solve for the monthly capacity for varying efficiencies. This is shown in Table 4.14 for an average of 600 hours available per month.

We can see from Table 4.14 that there is a wide range of possible outcomes for capacity. In fact this is a very realistic result. If the objective was to install a monthly capacity of 1,600,000 parts, then the efficiency of operation defines if one or two systems are required. For this case an average of 74% efficiency will be required to support the job. Successful implementation requires consideration of machine design, vendor support, and operation skill sets to support 74% efficiency. If the efficiency cannot be met, then two systems need to be purchased.

Efficiency has little impact on the cost of test unless the volume is increased. This can be shown by expanding our example to calculate cost. We will assume fixed facilities and capital costs; variable labor and material costs; and 100% yield to calculate the cost per test insertion. The assumptions are summarized in Table 4.15.

Cost per insertion calculations are shown in Table 4.16 for varying volume and efficiencies.

Columns compare the cost per insertion to the volume of test. The improvements in cost are due to amortizing facility and capital costs across more parts. The impact is significant due to the high capital cost of the test system and handler. Rows compare the cost per insertion as compared to efficiency. The difference in cost is relatively low since the only savings are labor. For this dedicated equipment example,

TABLE 4.14 Efficiency vs. Capacity

Efficiency	Capacity
40%	864,000
60%	1,296,000
80%	1,728,000
100%	2,160,000

TABLE 4.15 Cost Assumptions

Cost	Assumption	Fixed or Volume Dependent
Facility	$ per square foot of floor space	Fixed
Capital	3 year linear depreciation	Fixed
Labor	Labor and fringe	Volume Dependent
Materials	General Consumables	Volume Dependent
Yield	Not used	

TABLE 4.16 Cost vs. Volume vs. Efficiency

Efficiency/Volume	100%	90%	80%	70%	60%
400,000	$0.096	$0.097	$0.098	$0.100	$0.101
800,000	$0.053	$0.054	$0.055	$0.056	$0.058
1,200,000	$0.038	$0.039	$0.040	$0.042	$0.043
1,600,000	$0.031	$0.032	$0.033	N/A	N/A
2,000,000	$0.027	N/A	N/A	N/A	N/A

TABLE 4.17　Monthly Capacity of Four Products with Varying Setup Time and Delivery Intervals

Setup/Delivery	10 min.	30 min.	1 hour	2 hours	4 hours
Monthly	1,294,531	1,291,680	1,287,360	1,278,720	1,261,440
Weekly	1,290,125	1,278,720	1,261,440	1,226,880	1,157,760
Daily	1,251,936	1,166,400	1,036,800	777,600	259,200

improving efficiency only has value if the capacity is needed. Given efficiency or capacity, the cost of test can be reduced by increasing volume through product mix.

Product Mix Impact

Product mix adds several challenges such as tooling costs and manufacturing setup time. Tooling costs include test boards, mounting hardware, product standards, documentation, and training. These costs can run as high as $10,000 or as low as the documentation depending on product similarity and your approach to standardization. Tooling complexity will ultimately govern your product mix through resource limitations. Production output, on the other hand, will be governed by setup time. Setup time is the time to break down a setup and configure for another part number. This can involve test system calibration, test board change and/or handler change. Typical setup time can take from ten minutes to four hours. The following example explores product mix, setup time, and volume.

Example: Setup Time — Assume that setup can vary between ten minutes and four hours, equal volumes of four products are needed, test plus handing time is 1.0 second, and the efficiency is 60%. Calculate the optimum output assuring deliveries are required at monthly, weekly, or daily intervals. To do this we subtract four setup periods from the delivery interval, calculate the test capacity of the remainder of the interval, and then normalize to one-month output. Table 4.17 summarizes the results.

As you may have expected, long setup times and regular delivery schedules can significantly reduce capacity. When faced with a high-mix environment everything needs to be standardized from fixturing to calibration files to equipment types and operating procedures.

Data Analysis Overview

Product Data Requirements and Database

Tested parameters for average RF devices can range from as little as 3 to as many as 30 depending on the functional complexity. In a high volume environment, where output can reach over 500,000 devices daily with a moderate product mix, methods to monitor and evaluate performance criteria have to provide efficient access to data sets with minimal user interaction. Questions such as "How high is the yield?" and "What RF parameters are failing most?" are important in any test facility, but can be very difficult to monitor and answer as volumes grow.

Many arguments have been made concerning the necessity of collecting parameter information on high yielding devices. To answer the two questions asked above, only limited information need be gathered. Most testers are capable of creating bin summary reports that can assign a bin number to a failure mechanism and output final counts to summarize the results.

The "binning" method may yield enough information for many circuits, but will not give insight into questions about tightness of parameter distributions, insufficient (or over-sufficient) amount of testing, test limits to change to optimize the yield, or possible change in part performance. These can only be answered with full data analysis packages either supplied by third parties or developed in-house. Standard histogram (Fig. 4.103) or wafer maps (Fig. 4.104) can answer the first question by providing distributions, standard deviation, average values, and when supplied with limit specifications, CP and CPK values. XY or correlation plots (Fig. 4.105) can answer the second question, but when dealing with 20 or so parameters, this can be very time consuming to monitor.

The last questions require tools focusing on multivariable correlation and historical analysis. Changing of limit specifications to optimize yield is a tricky process and should not be performed on a small sample

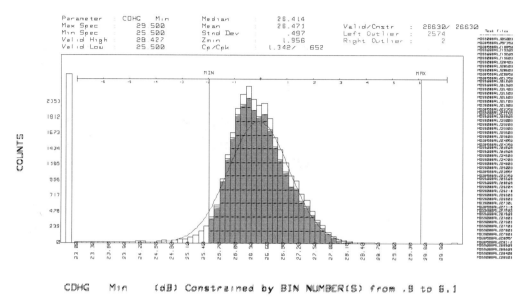

FIGURE 4.103 Histogram for distribution analysis.

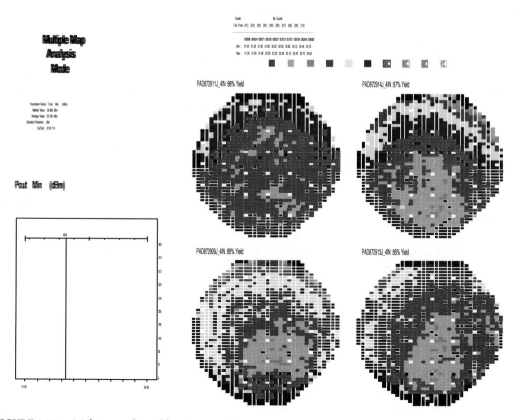

FIGURE 4.104 Wafer maps for yield pattern analysis.

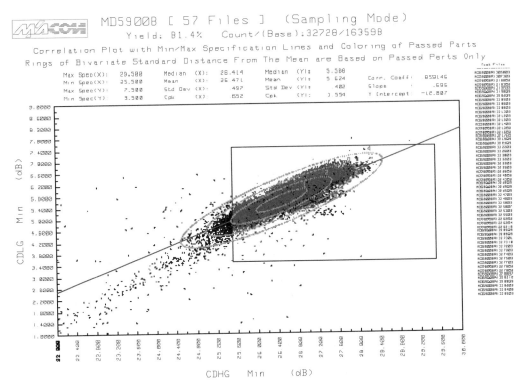

FIGURE 4.105 Scatter plot for parameter correlation analysis.

base. Nor should the interdependency of multiple parameters be ignored. Control charts such as Box Plots (Fig. 4.106) are ideal tools for monitoring performance variations over time.

These same tools when applied in real time can usually highlight problem parameters to help drill down to the problem at hand. Yield analysis tools displaying low yielding test parameters or single failure mechanisms are critical for efficient feedback analysis to the test floor as well as the product lines.

Database Tools

Analysis tools to quickly identify failure mechanisms are among the most important in high volume for quick feedback to the manufacturing floor. This requires that the database have full knowledge of not only the resulting data but also the high and low specifications placed on each individual parameter.

All databases, whether third party or custom, are depots for immense amounts of data with standard input and output utilities for organizing, feeding, and extracting information. The tools to display and report that information are usually independent of the database software.

Most third party database software packages can accommodate links to an exhaustive set of tools for extensive data analysis requirements. These external tools, again whether third party or custom, can be designed to provide fixed output reports for each device in question. But these databases usually require rigid data structures with fixed field assignments. Because of this, a high level of management for porting data, defining record structures, and organizing outputs is necessary when dealing with a continually changing product mix. Of course, if the application is needed for a few devices with compatible parameter tables, the management level will be minimal.

The alternative is creating a custom database structure to handle the dynamics of a high product mix for your specific needs. This is neither easy or recommended when starting fresh in today's market since it requires in-house expertise in selecting the appropriate platforms and data structures. But if the capability already exists and can handle the increased demand, it may be a more cost-effective path considering the available resources.

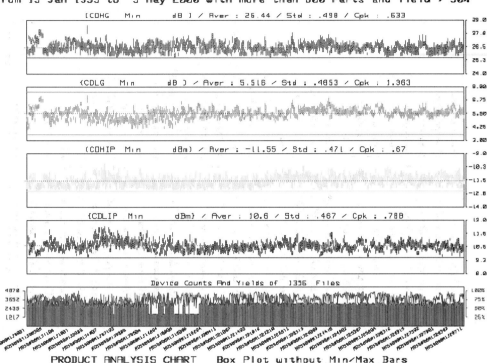

FIGURE 4.106   Multiple parameter control charts for product performance analysis.

An important note on the consideration of third party vs. in-house is the ability to implement software changes as the need arises. With third party platforms these changes may not be instituted until the next available revision or never if deemed highly custom. So be sure to select the appropriate mix to ensure this does not happen.

Regardless of the database option selected, data backups, network issues, and system integrity will still have to be maintained. Most systems today can use compression tools to maintain access to large amounts of data without the need to reload from externally archived tapes. Disc space is extremely cheap today. Even with high volume data collection requirements, information can be kept online for well over a year if necessary. More mature products can actually stop processing dense detailed information and only provide more condensed summary statistics used for tracking process uniformity.

Test Operation Data

To reduce the cost of testing and remain competitive in today's market, a constant monitoring of resource utilization is advantageous. A simple system utilization analysis can consist of a count test system, average cycle time of a device, and the quantity of parts in and parts out. This information is enough to get a rough idea of the average system utilization, but cannot give a complete picture when dealing with a large product base and package style mix. With detailed information of system throughput, pinpointing specific problem systems and focusing available resources to resolve the issues can be performed more efficiently. Output similar to the operational chart of Fig. 4.107 can show information such as efficiency and system utilization within seconds to evaluate performance issues.

Another important aspect of monitoring is the availability of resources to floor personnel to help them react to issues as fast as possible. During the course of a measurement sequence, potential problems could arise that require immediate response. A continuous yield display will react slowly to a degradation in

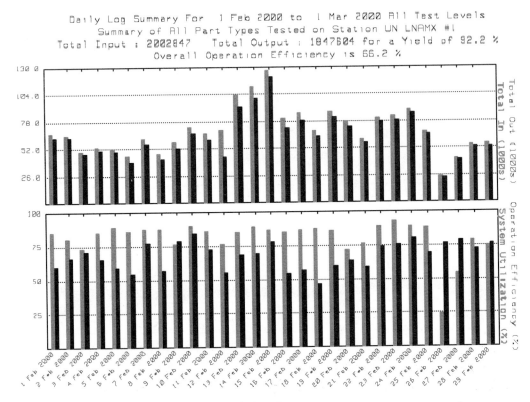

FIGURE 4.107 Yield and operation efficiency analysis tool.

contact or measurement performance, especially after thousands of devices have been tested. For this reason it is beneficial to have a sample or instantaneous yield reported during the test cycle to alert operators for quick reaction.

Conclusion

High volume microwave testing has become an everyday activity for all RFIC suppliers. Microwave test equipment vendors have developed equipment with acceptable accuracy and reproducibility, and satisfactory speed. Actual test software is robust and allows automatic revision tracking. Package handlers are improving although they are the throughput bottleneck for most standard RFICs, and do not accept module packages easily. Test contactors remain a technical difficulty, especially for high frequency or high power applications. In general, "hardware" solutions for microwave high volume testing exist today.

The remaining challenge is to reduce the customer's cost of quality and the supplier test cost with existing equipment. The ability to understand the customer specifications, the test system limitations, the test information available, and their interaction is key to test effectiveness improvement today. Analysis tools and methods to exploit the vast amount of data generated are essential to pinpoint the areas of possible improvement. These tools can highlight the fabrication process, the calibration process, the specification versus process limits, the package supplier, or the handler as the first area to focus upon for cost and quality improvement. This "software" side of people with the appropriate knowledge and tools to translate data into actionable information is where we expect the most effort and the most progress.

References

1. Crowley, R., Socket Developments for CSP and FBGA Packages, *Chip Scale Review*, May 1998.
2. Montgomery, D.C., Introduction to Statistical Quality Control, chap. 9.6, 455–460.

5

Circuits

Warren L. Seely
Motorola

Jakub Kucera
Infineon Technologies

Urs Lott
Acter AG

Anthony M. Pavio
Motorola

Charles Nelson
California State University

Mark Bloom
Motorola SPS

Alfy Riddle
Macallan Consulting

Robert Newgard
Rockwell Collins

Richard V. Snyder
RS Microwave

Robert J. Trew
U.S. Department of Defense

0-8493-8592-X/01/$0.00+$.50
© 2001 by CRC Press LLC

5.1   Receivers

Warren L. Seely

An electromagnetic signal picked up by an antenna is fed into a receiver. The ideal receiver rejects all unwanted noise including other signals. It does not add any noise or interference to the desired signal. The signal is converted, regardless of form or format, to fit the characteristics required by the detection scheme in the signal processor, which in turn feeds an intelligible user interface (Fig. 5.1). The unit must require no new processes, materials, or devices not readily available. This ideal receiver adds no weight, size, or cost to the overall system. In addition, it requires no power source and generates no heat. It has an infinite operating lifetime in any environment, and will never be obsolete. It will be flexible, fitting all past, present, and future requirements. It will not require any maintenance, and will be transparent to the user, who will not need to know anything about it in order to use it. It will be fabricated in an "environmentally friendly" manner, visually pleasing to all who see it, and when the user is finally finished with this ideal receiver, he will be able to recycle it in such a way that the environment is improved rather than harmed. Above all else, this ideal receiver must be wanted by consumers in very large quantities, and it must be extremely profitable to produce. Fortunately, nobody really expects to achieve all of these "ideal" characteristics, at least not yet! However, each of these characteristics must be addressed by the engineering design team in order to produce the best product for the application at hand.

Frequency

Receivers represent a technology with tremendous variety. They include AM, FM, analog, digital, direct conversion, single and multiple conversions, channelized, frequency agile, spread spectrum, chirp, frequency hopping, and others. The applications are left to the imaginations of the people who create them. Radio, telephones, data links, radar, sonar, TV, astronomical, telemetry, and remote control, are just a few of those applications. Regardless of the application, the selection of the operating frequencies is fundamental to obtaining the desired performance.

The actual receive frequencies are generally beyond the control of the design team, being dictated, controlled, and even licensed by various domestic or foreign government agencies, or by the customer. When a product is targeted for international markets, the allocated frequencies can take on nightmare qualities due to differing allocations, adjacent interfering bands, and neighboring country restrictions or allocations. It will usually prove impossible to get the ideal frequency for any given application, and often the allocated spectrum will be shared with other users and multiple applications. Often the spectrum is

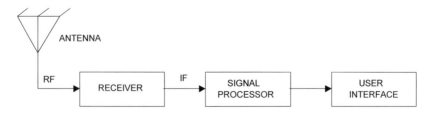

FIGURE 5.1   The receiver.

available for a price, usually to the highest bidder. Failure to utilize the purchased spectrum within a specified time frame may result in forfeiture of what is now an asset; an expensive mistake. This has opened up the opportunity to speculate and make (or lose) large sums of money by purchasing spectrum to either control a market or resell to other users. For some applications where frequency allocation is up to the user, atmospheric or media absorption, multipathing, and background noise are important factors that must be considered. These effects can be detrimental or used to advantage. An example includes cross links for use with communications satellites, where the cross link is unaffected by absorption since it is above the atmosphere. However, the frequency can be selected to use atmospheric absorption to provide isolation between ground signals and the satellite cross links. Sorting out these problems is time consuming and expensive, but represents a fundamental first step in receiver design.

Dynamic Range

The receiver should match the dynamic range of the desired signal at the receiver input to the dynamic range of the signal processor. Dynamic range is defined as the range of desirable signal power levels over which the hardware will operate successfully. It is limited by noise, signal compression, and interfering signals and their power levels.

Power and Gain

The power in any signal(s), whether noise, interference or the desired signal, can be measured and expressed in Watts (W), decibels referenced to 1 Watt (dBW), milliwatts (mW) or decibels referenced to one milliwatt (dBm). The power decibel is 10 times the LOG of the dimensionless power ratio. The power gain of a system is the ratio output signal power to the input signal power expressed in decibels (dB). The gain is positive for components in which the output signal is larger than the input, negative if the output signal is smaller. Negative gain is loss, expressed as attenuation (dB). The power gain of a series component chain is found by simple multiplication of the gain ratios, or by summing the decibel gains of the individual components in the chain. All of these relationships are summarized in Fig. 5.2.

Noise

Thermal noise arises from the random movement of charge carriers. The thermal noise power (n_T) is usually expressed in dBm (N_T), and is the product of Boltzman's constant (k), system temperature in degrees Kelvin (T), and a system noise bandwidth in Hertz (b_n). The system noise bandwidth (b_n) is defined slightly different from system bandwidth. It is determined by measuring or calculating the total system thermal average noise power ($n_{tot-ave}$) over the entire spectrum and dividing it by the system peak average noise power (n_{pk-ave}) in a 1 Hz bandwidth. This has the effect of creating a system noise bandwidth

$$Decibel = 10 LOG\left[\frac{p}{p_{ref}}\right]$$

$$g(-) = \frac{p_{out}}{p_{in}}$$

$$P(dBW) = 10 LOG\left[\frac{p(W)}{1W}\right]$$

$$G(dB) = 10 LOG(g)$$

$$P(dBm) = 10 LOG\left[\frac{p(mW)}{1mW}\right]$$

$$g_{total}(-) = g_1 * g_2 * .... * g_N = \frac{p_{Nout}}{p_{1in}}$$

$$0 dBW = 1 Watt$$

$$G_{total}(dB) = G_1(dB) + G_2(dB) + ... + G_N(dB) = \Sigma G_i$$

$$0 dBm = 1mW$$

$$Loss(dB) = -G(dB)$$

$$1000 mW = 1W$$

$$Loss(dB) = Attenuation(dB)$$

$$30 dBm = 0 dBW$$

FIGURE 5.2 Power and gain relationships.

$$n_T = kTb_n$$

$$b_n(Hz) = \frac{n_{tot-ave}(W)}{\underline{n}_{pk-ave}(W/Hz)}$$

$$k = 1.38*10^{-23}\frac{W\sec}{K}$$

$$T(K) = T({}^0C) + 273.15$$

$$n_T = 1.38*10^{-23}\frac{W\sec}{K}*(25^0C + 273.15)K*1Hz = 4.46*10^{-21}W = 4.46*10^{-18}mW$$

$$N_T = 10LOG(n_T) = -204dBW = -174dBm$$

FIGURE 5.3 Noise power relationships.

in which the noise is all at one level, that of the peak average noise power. For a 1 Hz system noise bandwidth at the input to a system at room temperature (25°C), the thermal noise power is about −174 dBm. These relationships are summarized in Fig. 5.3.

Receiver Noise

The bottom end of the dynamic range is set by the lowest signal level that can reasonably be expected at the receiver input and by the power level of the smallest acceptably discernable signal as determined at the input to the signal processor. This bottom end is limited by thermal noise at the input, and by the gain distribution and addition of noise as the signal progresses through the receiver. Once a signal is below the minimum discernable signal (MDS) level, it will be lost entirely (except for specialized spread spectrum receivers). The driving requirement is determined by the signal clarity needed at the signal processor. For analog systems, the signal starts to get fuzzy or objectionably noisy at about 10 dB above the noise floor. For digital systems, the allowable bit error rate determines the acceptable margin above the noise floor. Thus the signal with margin sets the threshold minimum desirable signal level.

Noise power at the input to the receiver will be amplified and attenuated like any other signal. Each component in the receiver chain will also add noise. Passive devices such as filters, cables, and attenuators will cause a drop in both signal and noise power alike. These passive devices also contribute a small amount of internally generated thermal noise. Thus the actual noise figure of a passive device is slightly higher than the attenuation of that component. This slight difference is ignored in receiver design since the actual noise figures and losses vary by significantly larger amounts. Passive mixers will generally have a noise figure about 1 dB greater than the conversion loss. Active devices can exhibit loss or gain, and signal and noise power at the input will experience the same effect when transferred to the output. However, the internally generated noise of an active device will be substantial and must be accounted for, requiring reasonably accurate noise figures and gain data on each active component.

The bottom end dynamic range of a receiver component cascade is easily described by the noise equations shown in Fig. 5.4. The first three equations for noise factor (f_n), noise figure (NF), and noise temperature (T_n) are equivalent expressions to quantify noise. The noise factor is a dimensionless ratio of the input signal-to-noise ratio and the output signal-to-noise ratio. Replacing the signal ratio with gain results in the final form shown. Noise figure is the decibel form of noise factor, in units of dB. Noise temperature is the conversion of noise factor to an equivalent input temperature that will produce the output noise power, expressed in Kelvin. Convention dictates using noise temperature when discussing antennas and noise figure for receivers and associated electronics. By taking the decibel equivalent of the noise factor, the expression for noise out (N_o) is obtained, where noise in (N_i) is in dBm and noise figure (NF) and gain (G) are in dB. The cascaded noise factor (f_t) is found from the sum of the added noise due to each cascaded component divided by the total gain preceding that element. Use the cascaded noise factor (f_t) followed by the noise out (N_o) equation to determine the noise level at each point in the receiver.

$$f_n = \frac{S_i/n_i}{S_o/n_o} = \frac{n_o}{g n_i}$$

$$NF = 10 LOG(f_n)$$

$$T_n = T(f_n - 1) \quad where\, T\ is\ in\ Kelvin$$

$$N_o = NF + G + N_i$$

$$f_t = f_1 + \frac{f_2 - 1}{g_1} + \frac{f_3 - 1}{g_1 * g_2} + \cdots + \frac{f_n - 1}{\Pi g_n}$$

$$\Delta f_{n-bandwidth} = \frac{g_1 f_1 + (f_2 - 1)^{b_{n2}}/b_{n1}}{g_1 f_1 + f_2 - 1} \qquad b_{n2} > b_{n1}$$

$$\Delta f_{n-image} = 1 + \frac{l_{ar}}{f_x}$$

$$f_{total} = f_{cascade} * \Delta f_{n-bandwidth} * \Delta f_{n-image}$$

FIGURE 5.4 Receiver noise relationships.

Noise factor is generally computed for a 1 Hz bandwidth and then adjusted for the narrowest filter in the system, which is usually downstream in the signal processor. Occasionally, it will be necessary to account for noise power added to a cascade when components following the narrowest filter have a relatively broad noise bandwidth. The filter will eliminate noise outside its band up to that filter. Broader band components after the filter will add noise back into the system depending on their noise bandwidth. This additional noise can be accounted for using the equation for $\Delta f_{n-bandwidth}$, where subscript 1 indicates the narrowband component followed by the wideband component (subscript 2). Repeated application of this equation may be necessary if several wideband components are present following the filter. Image noise can be accounted for using the relationship for $\Delta f_{n-image}$ where l_{ar} is the dimensionless attenuation ratio between the image band and desired signal band and f_x is the noise factor of the system up to the image generator (usually a mixer). Not using an image filter in the system will result in a $\Delta f_{n-image} = 2$ resulting in a 3 dB increase in noise power. If a filter is used to reject the image by 20 dB, then a substantial reduction in image noise will be achieved. Finally, the corrections for bandwidth and image are easily incorporated using the relationship for the cascaded total noise factor, f_{total}.

A simple single sideband (SSB) receiver example, normalized to a 1 Hz noise bandwidth, is shown in Fig. 5.5. It demonstrates the importance of minimizing the use of lossy components near the receiver front end, as well as the importance of a good LNA. A 10 dB output signal-to-noise margin has been established as part of the design. Using the −174 dBm input thermal noise level and the individual component gains and noise figures, the normalized noise level can be traced through the receiver, resulting in an output noise power of −136.9 dBm. Utilizing each component gain and working backwards from this point with a signal results in the MDS power level in the receiver. Adding the 10 dB signal-to-noise margin to the MDS level results in the signal with margin power level as it progresses through the receiver. The signal and noise levels at the receiver input and output are indicated. The design should minimize the gap between the noise floor and the MDS level. Progressing from the input toward the output, it is readily apparent that the noise floor gets closer to and rapidly converges with the MDS level due to the addition of noise from each component, and that lossy elements near the input hurt system performance. The use of the low noise amplifier as close to the front end of the cascade as possible is critical in order to mask the noise of following components in the cascade and achieve minimum noise figure. The overall cascaded receiver gain is easily determined by the difference in the signal levels from input to output.

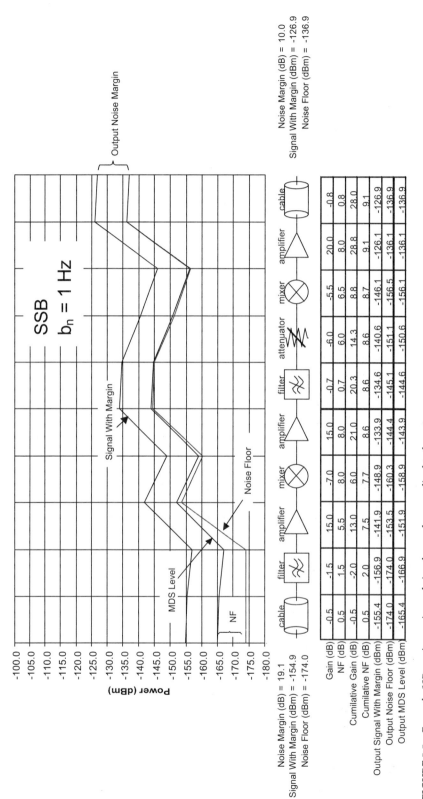

FIGURE 5.5   Example SSB receiver noise and signal cascade normalized to $b_n = 1$.

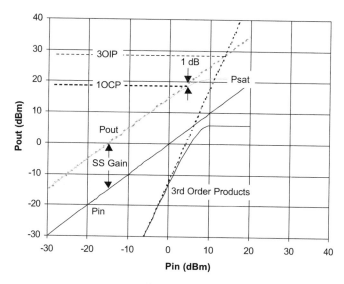

FIGURE 5.6 3OIP, P_{sat}, and 1 dB OCP.

The noise floor margins at both the input and output to the receiver are also easily observed, along with the receiver noise figure. Note also that the actual noise power does not drop below the thermal noise floor, which is always the bottom limit. Finally, the actual normalized signal with margin level of −154.9 dBm at the input to the receiver is easily determined.

Intermodulation

Referring to Fig. 5.6, the upper end of the dynamic range is limited by nonlinearity, compression, and the resulting intermodulation in the presence of interfering signals. The in-band two-tone output 3rd order intercept point (3OIP) is a measure of the nonlinearity of a component. This particular product is important because it can easily result in an undesired signal that is close to the desired signal, which would be impossible to eliminate by filtering. By definition, the 3OIP is found by injecting 2 equal amplitude signals (F_1 and F_2) that are not only close to each other in frequency, but are also both within the passband of the component or system. The 3rd order intermodulation products are then given by $\pm nF_1 \pm mF_2$ where $n + m = 3$. For 3rd order products, n and m must be 1 or 2. Since negative frequencies are not real, this will result in two different 3rd order products which are near each other and within the passband. The power in the inter-modulation products is then plotted, and both it and Pout are projected until they intersect, establishing the 3OIP. The desired signal is projected using a 1:1 slope, while the 3rd order products are projected using a 3:1 slope. The output saturation power (P_{sat}) is the maximum power a device will produce. The output 1 dB compression point (1 dB OCP) is the point at which the gain is lowered by 1 dB from small signal conditions due to signal compression on its way to complete saturation. In general, higher values mean better linearity and thus better performance. However, component costs rapidly increase along with these numbers, especially above saturated levels of about +15 dBm, limiting what can be achieved within project constraints. Thus, one generally wants to minimize these parameters in order to produce an affordable receiver. For most components, a beginning assumption of square law operation is reasonable. Under these conditions, the 1 dB OCP is about 3 dB below P_{sat}, and the 3OIP is about 10 dB above the 1 dB OCP. When the input signal is very small (i.e., small signal conditions), Pout increases on a 1:1 slope, and 3rd order products increase on a 3:1 slope. These numbers can vary significantly based on actual component performance and specific loading conditions. This whole process can be reversed, which is where the value of the concept lies. By knowing the small signal gain, Pin or Pout, and the 3OIP, all the remaining parameters, including 1OCP, P_{sat}, and 3rd order IM levels can be estimated. As compo-nents are chosen for specific applications, real data should be utilized where possible. Higher order

$$\frac{1}{P_{3iip,tot}} = \frac{1}{P_{3iip,1}} + \frac{g_1}{P_{3iip,2}} + \cdots + \frac{\Pi g_n}{P_{3iip,n}}$$

$$P_{3OIP} = P_{3IIP} + G_{ss}$$

$$SFDP = \frac{2}{3}\left(3IIP - MDS_{input}\right) = \frac{2}{3}\left(3OIP - MDS_{output}\right)$$

$$MSI_{out} = NOISEFLOOR_{out} + SPDR$$

FIGURE 5.7   Receiver 3OIP, P_{sat}, and 1 dB OCP cascade.

products may also cause problems, and should be considered also. Finally, any signal can be jammed if the interfering signal is large enough and within the receiver band. The object is to limit the receiver's susceptibility to interference under reasonable conditions.

Receiver Intermodulation

Analog receiver performance will start to suffer when in-band 3rd order products are within 15 dB of the desired signal at the detector. This level determines the maximum signal of interest (MSI). The margin for digital systems will be determined by acceptable bit error rates. The largest signal that the receiver will pass is determined by the saturated power level of the receiver. Saturating the receiver will result in severe performance problems, and will require a finite time period to recover and return to normal performance. Limiting compression to 1 dB will alleviate recovery.

Analyzing the receiver component cascade for 3OIP, P_{sat}, 1 dB OCP, and MSI will provide insight into the upper limits of the receiver dynamic range, allowing the designer to select components that will perform together at minimum cost and meet the desired performance (Fig. 5.7). The first equation handles the cascading of known components to determine the cumulated input 3rd order input intermod point. After utilizing this equation to determine the cascaded 3OIP up to the component being considered, the second equation can be utilized to determine the associated P_{3OIP}. Successive application will result in completely determining the cascaded performance. The last two equations determine the 3rd order IM spur free dynamic range (SPDR) and the maximum spur-free level or maximum signal of interest (MSI) at the output of the receiver.

An example receiver 3OIP and signal cascade is shown in Fig. 5.8. The results of the cumulative 3OIP are plotted. A gain reference is established by setting and determining the cumulative gain and matching it to the cumulative 3OIP at the output. A design margin of 20 dB is added to set the MSI power level for the cascade.

Receiver Dynamic Range

Combining the results for the noise and intermodulation from the above discussion into one graph results in a graphical representation of the dynamic range of the receiver (Fig. 5.9). Adjusting for the 6 MHz bandwidth moves the noise plots up by 10LOG(6e6) = 67.8 dB. The cumulative 3OIP and gain reference plot remain at the previously determined levels. The SPDR = 2(–5.5 dBm – (–97.1 dBm))/3 = 61.1 dB is calculated and then used to determine the MSI level = –69.1 dBm + 61.1 dB = –8 dBm at the output. The MSI level on the graph is set to this value, backing off to the input by the gain of each component.

The receiver gain, input, and output dynamic ranges, signal levels which can be easily handled, and the appropriate matching signal processing operating range are readily apparent, being between the MSI level and the signal with noise margin. The receiver NF, 3OIP, gain, and SPDR are easily determined from the plot. Weaknesses and choke points, as well as expensive parts are also apparent, and can now be attacked and fixed or improved. In general, components at or near the input to the receiver dominate the noise performance, and thus the lower bounds on dynamic range. Components at or near the output dominate the nonlinear performance, and thus the upper bounds on dynamic range.

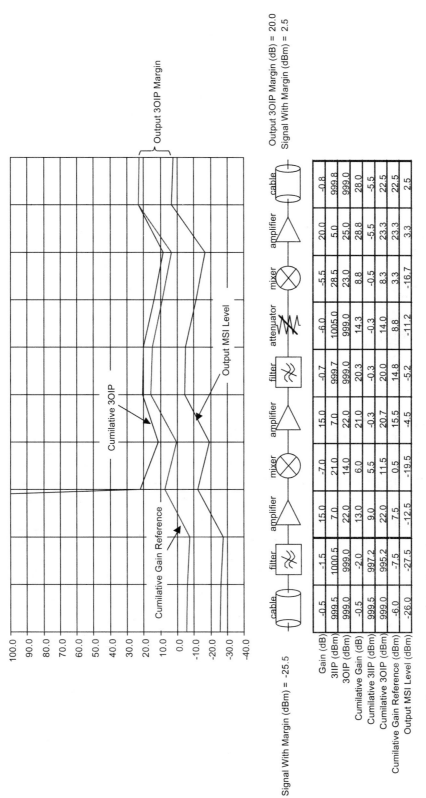

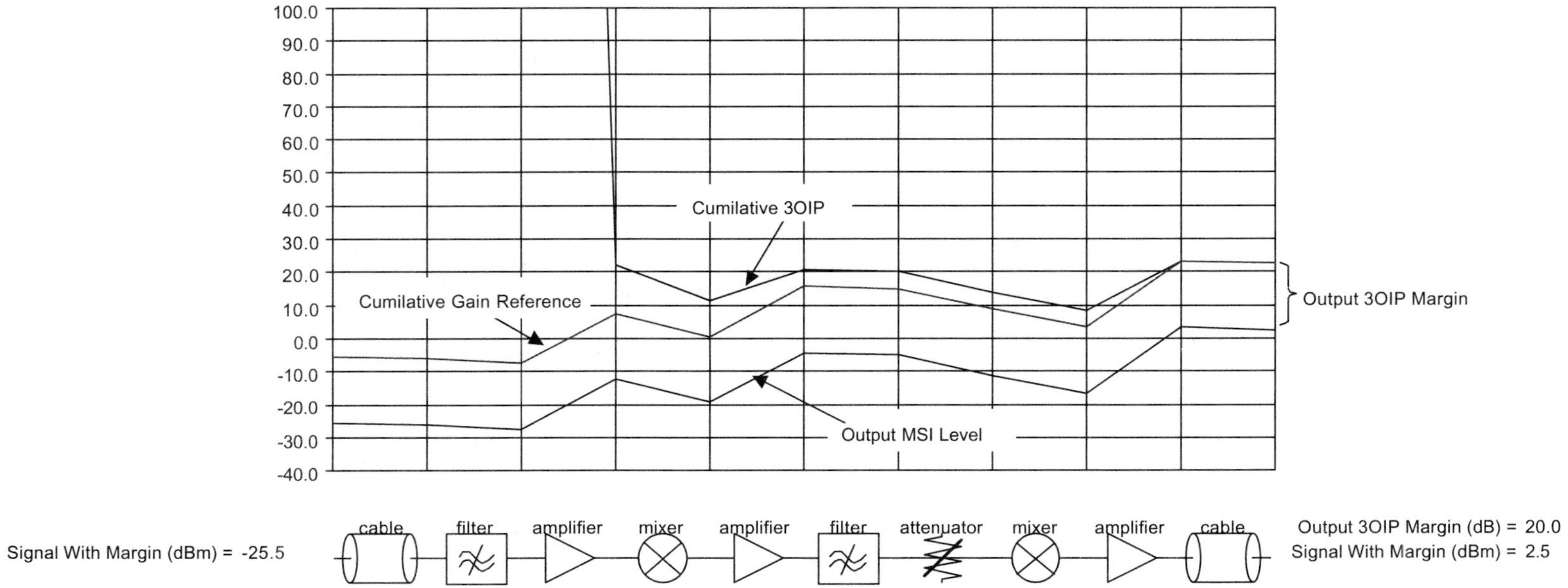
100.0
90.0
80.0
70.0
60.0
50.0
40.0
30.0
20.0
10.0
0.0
-10.0
-20.0
-30.0
-40.0
Cumilative 3OIP
Cumilative Gain Reference
Output MSI Level
Output 3OIP Margin
cable
filter
amplifier
mixer
amplifier
filter
attenuator
mixer
amplifier
cable
Signal With Margin (dBm) = -25.5
Output 3OIP Margin (dB) = 20.0
Signal With Margin (dBm) = 2.5

Gain (dB)	-0.5	-1.5	15.0	-7.0	15.0	-0.7	-6.0	-5.5	20.0	-0.8
3IIP (dBm)	999.5	1000.5	7.0	21.0	7.0	999.7	1005.0	28.5	5.0	999.8
3OIP (dBm)	999.0	999.0	22.0	14.0	22.0	999.0	999.0	23.0	25.0	999.0
Cumilative Gain (dB)	-0.5	-2.0	13.0	6.0	21.0	20.3	14.3	8.8	28.8	28.0
Cumilative 3IIP (dBm)	999.5	997.2	9.0	5.5	-0.3	-0.3	-0.3	-0.5	-5.5	-5.5
Cumilative 3OIP (dBm)	999.0	995.2	22.0	11.5	20.7	20.0	14.0	8.3	23.3	22.5
Cumilative Gain Reference (dBm)	-6.0	-7.5	7.5	0.5	15.5	14.8	8.8	3.3	23.3	22.5
Output MSI Level (dBm)	-26.0	-27.5	-12.5	-19.5	-4.5	-5.2	-11.2	-16.7	3.3	2.5

FIGURE 5.8 Example receiver 3OIP and signal cascade.

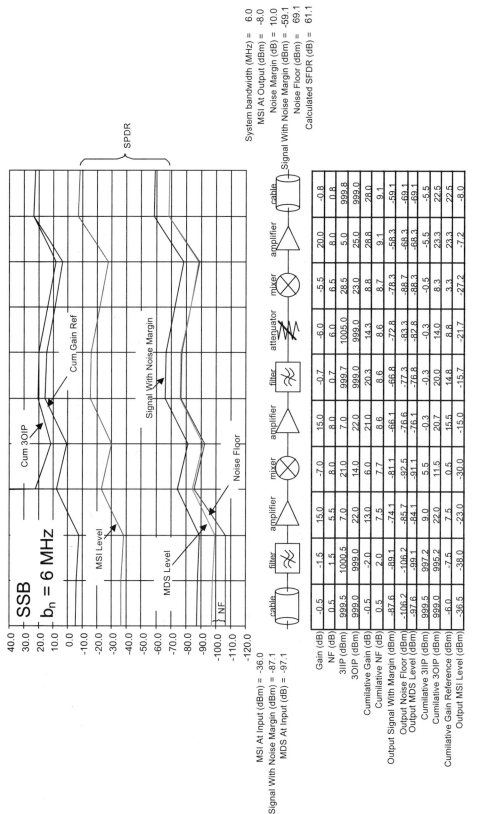

FIGURE 5.9 Example SSB receiver spur free dynamic range normalized to b_n = 6 MHz.

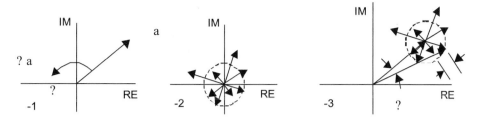

FIGURE 5.10 Phase noise, AM, and PM noise.

The use of 3OIP and noise floors is just one way commonly used to characterize the dynamic range of a receiver. Other methods include determining compression and saturation curves, desensitization, noise power ratios, and intercept analysis for other IM products such as 2OIP up to as high as possibly 15OIP. Specific applications will determine the appropriate analysis required in addition to the main SFDR analysis described above.

The LO Chain

A reference signal or local oscillator (LO) is generally required in order to up- or downconvert the desired signal for further processing. The design of the LO chain is tied to the receiver components by frequency, power level, and phase noise. The LO signal will have both amplitude and phase noise components, both of which will degrade the desired signal. Often, a frequency agile LO is required. The LO can be generated directly by an oscillator, multiplied or divided from another frequency source, created by mixing several signals, injection or phase locked to a reference source, digitally synthesized, or any combination thereof.

Amplitude and Phase Noise

A pure tone can be represented as a vector of a given amplitude (α) rotating with a fixed angular velocity (ω) as shown in Fig. 5.10. A random noise source can be viewed similarly, but has random phase equally distributed over time about 360°, and random amplitude based on a probability distribution. At any given instant in time the vector will change amplitude and angle. A plot of the noise vector positions for a relatively long period of time would appear solid near the origin and slowly fade away at larger radii from the origin (Fig. 5.10). A plot of the number of hits vs. distance from the origin would result in the probability distribution. This random noise is the same noise present in all electronic systems. Combining the pure tone with the random noise results in the vector addition of both signals (Fig. 5.10). At any given instance in time the combined angular velocity will change by $\Delta\omega$, and the amplitude will change by $\Delta\alpha$. The frequency jittering about ω and the amplitude wavering about α result in AM and PM noise components, which distort the signal of interest. Thus, phase noise is a measure of signal stability.

The design of the LO chain usually includes at least one final amplifier stage that is completely saturated. This sets the LO amplitude to a fixed level, minimizes temperature variations, and minimizes or eliminates the AM noise component. Saturation results in a gain reduction of several dB and simultaneously limits the maximum amplitude that can be achieved. Thus as the random noise vector changes the LO amplitude, the saturated amplifier acts to eliminate the output amplitude change. The AM contribution to noise is cleaned up.

The phase noise in the LO chain must be attacked directly at the source. Clean, low phase noise signal generation in oscillators is achieved by the use of very high Q filter components, incorporating bipolar devices as the active oscillator element, maximizing the source power generation, and careful design of the conduction cycle within the oscillator itself. Once a clean signal is created, it must be kept clean.

Frequency multiplying or dividing the signal will also multiply or divide the phase noise by a factor of 20*LOG(N) at any given offset from the base signal. Conversely, if the signal is multiplied or divided, then the spectrum is respectively stretched or contracted by the factor N. The mixing process also mixes the phase noise, but the net result will depend on the mixer types utilized. Injection locking will replicate the injection source modified by the multiplication or division factor N. A phase lock loop exhibits close in noise dependent on the reference source and loop circuitry, but the far out phase noise is set by the source used in the loop itself. Finally, the LO is utilized in the receiver chain to perform frequency conversion. The resulting converted signal will have components of phase noise from the original signal, the LO signal, and from noise in the mixing component.

The Potential for Trouble

Receivers are designed to work with very small signals, transforming these to much larger signals for handling in the signal processor. This inherent process is open to many pitfalls resulting in the vast majority of problems encountered in creating a viable product. It will only take a very small interfering or spurious signal to wreak havoc. Interfering signals generally can be categorized as externally generated, internally generated, conducted, electromagnetically coupled, piezoelectrically induced, electromechanically induced, and optically coupled or injected. Some will be fixed, others may be intermittent, even environmentally dependent. Most of these problem areas can be directly addressed by simple techniques, precluding their appearance altogether. However, ignoring these potential problem areas usually results in disaster, primarily because they are difficult to pinpoint as to cause and effect, and because eliminating them may be difficult or impossible without making major design changes and fabricating new hardware in order to verify the solution. This can easily turn into a long-term iterative nightmare. Additionally, if multiple problems are present, whether or not they are perceived as multiple problems or as a single problem, the amount of actual time involved in solving them will go up exponentially! Oscillator circuits are generally very susceptible to any and all problems, so special consideration should be given in their design and use. Finally, although the various cause, effect, and insight into curing problems are broken down into component parts in the following discussion, it is often the case that several concepts must be combined to correctly interpret and solve any particular problem at hand.

Electromechanical

Vibrations and mechanical shocks will result in physical relative movement of hardware. Printed circuit boards (PCBs), walls, and lids may bow or flutter. Cables and wire can vibrate. Connectors can move. Solder joints can fracture. PCBs, walls, and lids capacitively load the receiver circuitry, interconnects, and cabling. Movement, even very small deflections, will change this parasitic loading, resulting in small changes in circuit performance. In sensitive areas, such as near oscillators and filters, this movement will induce modulation onto the signals present. In phase-dependent systems, the minute changes in physical makeup and hence phase length of coaxial cable will appear as phase modulation. Connector pins sliding around during vibration can introduce both phase and amplitude noise. These problems are generally addressed by proper mechanical design methods, investigating and eliminating mechanical resonance, and minimizing shock susceptibility. Don't forget that temperature changes will cause expansion and contraction, with similar but slower effects.

Optical Injection

Semiconductor devices are easily affected by electromagnetic energy in the optical region. Photons impinging on the surface of an active semiconductor create extra carriers, which appear as noise. A common occurrence of this happens under fluorescent lighting common in many offices and houses. The 60 Hz "hum" is present in the light given off by these fixtures. The light impinges on the surface of a semiconductor in the receiver, and 60 Hz modulation is introduced into the system. This is easily countered by proper packaging to prevent light from hitting optically sensitive components.

Piezoelectric Effects

Piezoelectric materials are reciprocal, meaning that the application of electric fields or mechanical force changes the electromechanical properties, making devices incorporating these materials highly susceptible to introducing interference. Even properly mounted crystals or SAW devices, such as those utilized in oscillators, will move in frequency or generate modulation sidebands when subjected to mechanical vibration and shock. Special care should therefore be given to any application of these materials in order to minimize these effects. This usually includes working closely with the original equipment manufacturer (OEM) vendors to ensure proper mounting and packaging, followed by extensive testing and evaluation before final part selection and qualification.

Electromagnetic Coupling

Proper design, spacing, shielding, and grounding is essential to eliminate coupled energy between circuits. Improper handling of each can actually be detrimental to achieving performance, adding cost without benefit, or delaying introduction of a product while problems are solved. Proper design techniques will prevent inadvertent detrimental E-M coupling. A simple example is a reject filter intended to minimize LO signal leakage into the receiver, where the filter is capable of the required performance, but the packaging and placement of the filter allow the unwanted LO to bypass the filter and get into the receiver anyway.

It is physically impossible to eliminate all E-M resonant or coupled structures in hardware. A transmission line is created by two or more conductors separated by a dielectric material. A waveguide is created by one or more conductive materials in which a dielectric channel is present, or by two or more nonconductive materials with a large difference in relative dielectric constant. Waveguides do not have to be fully enclosed in order to propagate E-M waves. In order to affect the hardware, the transmission line or waveguide coupling must occur at frequencies that will interfere with operation of the circuits, and a launch into the structure must be provided. Properly sizing the package (a resonant cavity) is only one consideration. Breaking up long, straight edges and introducing interconnecting ground vias on multilayer PCBs can be very effective. Eliminating loops, opens and shorts, sharp bends, and any other "antenna like" structures will help.

E-field coupling usually is associated with high impedance circuits, which allow relatively high E-fields to exist. E-field or capacitive coupling can be eliminated or minimized by any grounded metal shielding. M-field coupling is associated with low impedance circuits in which relatively high currents and the associated magnetic fields are present. M-field or magnetic coupling requires a magnetic shielding material. In either case, the objective is to provide a completely shielded enclosure. Shielding metals must be thick enough to attenuate the interfering signals. This can be determined by E or M skin effect calculations. Alternatively, absorbing materials can also be used. These materials do not eliminate the basic problem, but attempt to mask it, often being very effective, but usually relatively expensive for production environments. Increased spacing of affected circuitry, traces, and wires will reduce coupling. Keeping the E-M fields of necessary but interfering signals orthogonal to each other will add about 20 dB or more to the achieved isolation.

Grounding is a problem that could be considered the "plague" of electronic circuits. Grounding and signal return paths are not always the same, and must be treated accordingly. The subject rates detailed instruction, and indeed entire college level courses are available and recommended for the serious designer. Basically, grounding provides a reference potential, and also prevents an unwanted differential voltage from occurring across either equipment or personnel. In order to accomplish this objective, little or no current must be present. Returns, on the other hand, carry the same current as the circuit, and experience voltage drops accordingly. A return, in order to be effective, must provide the lowest impedance path possible. One way to view this is by considering the area of the circuit loop, and making sure that it is minimized. In addition, the return conductor size should be maximized.

Summary

A good receiver design will match the maximum dynamic range possible to the signal processor. In order to accomplish this goal, careful attention must be given to the front end noise performance of the receiver and the selection of the low noise amplifier. Equally important in achieving this goal is the linearity of the back end receiver components, which will maximize the SFDR. The basic receiver calculations discussed above can be utilized to estimate the attainable performance. Other methods and parameters may be equally important and should be considered in receiver design. These include phase noise, noise power ratio, higher order intercepts, internal spurious, and desensitization.

Further Reading

Sklar, Bernard, Digital Communications Fundamentals and Applications, Prentice-Hall, Englewood Cliffs, NJ, 1988.

Tsui, Dr. James B., *Microwave Receivers and Related Components*, Avionics Laboratory, Air Force Write Aeronautical Laboratories, 1983.

Watkins-Johnson Company Tech-notes, Receiver Dynamic Range: Part 1, Vol. 14, No. 1.

Watkins-Johnson Company Tech-notes, Receiver Dynamic Range: Part 2, Vol. 14, No. 2.

Steinbrecher, D., Achieving Maximum Dynamic Range in a Modern Receiver, *Microwave Journal*, Sept 1985.

5.2 Transmitters

Warren L. Seely

A signal is generated by the frequency synthesizer and amplified by the transmitter (Fig. 5.11), after which it is fed to the antenna for transmission. Modulation and linearization may be included as part of the synthesized signal, or may be added at some point in the transmitter. The transmitter may include frequency conversion or multiplication to the actual transmit band. An ideal transmitter exhibits many of the traits of the ideal receiver described in the previous section of this handbook. Just as with the receiver, the task of creating a radio transmitter begins with defining the critical requirements, including frequencies, modulations, average and peak powers, efficiencies, adjacent channel power or spillover, and phase noise. Additional transmitter parameters that should be considered include harmonic levels, noise powers, spurious levels, linearity, DC power allocations, and thermal dissipations. Nonelectrical, but equally important considerations include reliability, environmentals such as temperature, humidity, and vibration, mechanicals such as size, weight, and packaging or mounting, interfaces, and even appearance, surface textures, and colors. As with most applications today, cost is becoming a primary driver in the design and production of the finished product. For most RF/microwave transmitters, power amplifier considerations dominate the cost and design concerns.

Safety must also be considered, especially when high voltages or high power levels are involved. To paraphrase a popular educational TV show; "be sure to read and understand all safety related materials that are applicable to your design before you begin. And remember, there is nothing more important than shielding yourself and your coworkers (assuming you like them) from high voltages and high levels of RF/microwave power." Another safety issue for portable products concerns the use of multiple lithium batteries connected in parallel. Special care must be taken to insure that the batteries charge and discharge independent of each other in order to prevent excessive I-R heating, which can cause the batteries to explode. It is your life — spend a little time to become familiar with these important issues.

ACP, Modulation, Linearity, and Power

The transmitter average and peak output powers are usually determined from link/margin analysis for the overall system. The transmitter linearity requirements are determined from the transmit power levels, phase noise, modulation type, filtering, and allowed adjacent channel power (ACP) spillover. Linearity

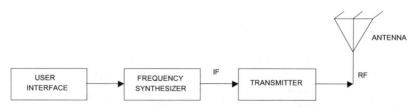

FIGURE 5.11   The transmitter.

is intimately tied to the transmitter saturated power, which in turn is tied to the 1 dB compression point. The need for high linearity is dependent on the maximum acceptable ACP in adjacent channels or bands. This spillover of power will cause interference in adjacent channels or bands, making it difficult or impossible to use that band. The maximum ACP may be regulated by government agencies, or may be left up to the user. Often the actual transmitted power requirement is less stringent than the linearity requirement in determining the necessary power handling capability or saturated power of the transmitter. In order to achieve required linearity, the transmitter may operate significantly backed off from the saturated power capability, even under peak power operation. Since the cost of a transmitter rapidly increases with its power handling capability, and the linearity requirements are translated into additional power requirements, a great deal of the cost of a transmitter may actually be associated with linearity rather than transmit power level. If the added cost to achieve necessary linearity through additional power capability is significant, linearizing RF transmitter can be cost effective.

Power

The single most important specification affecting the final system cost is often the transmitter saturated power, which is intimately linked to the transmitter linearity requirements. This parameter drives the power amplifier (PA) device size, packaging, thermal paths and related cooling methods, power supply, and DC interconnect cable sizes, weight, and safety, each of which can rapidly drive costs upward. The power level analysis may include losses in the cables and antenna, transmitter and receiver antenna gains, link conditions such as distance, rain, ice, snow, trees, buildings, walls, windows, atmospherics, mountains, waves, water towers, and other issues that might be pertinent to the specific application. The receiver capabilities are crucial in determining the transmitter power requirements. Once a system analysis has been completed indicating satisfactory performance, then the actual power amplifier (PA) requirements are known. The key parameters to the PA design are frequency, bandwidth, peak and average output power, duty cycle, linearity, gain, bias voltage and current, dissipated power, and reliability mean time to failure (MTBF, usually given as maximum junction temperature). Other factors may also be important, such as power added efficiency (PAE), return losses, isolations, stability, load variations, cost, size, weight, serviceability, manufacturability, etc.

Linearization

Linearity, as previously indicated, is intimately tied to the transmitter power. The need for high linearity is dependent on the maximum acceptable ACP in adjacent channels or bands. This spillover of power will cause interference in those bands, making it difficult or impossible to use that band. The ACP spillover is due to several factors, such as phase noise, modulation type, filtering, and transmit linearity. The basic methods used for linearization include the class A amplifier in back-off, feedforward, Cartesian and polar loops, adaptive predistortion, envelope elimination and recovery (EER), linear amplification using nonlinear components (LINC), combined analog locked-loop universal modulation (CALLUM), I-V trajectory modification, device tailoring, and Dougherty amplification. Each of these methods strives to improve the system linearity while minimizing the overall cost. The methods may be combined for further improvements. Economical use of the methods may require the development of application-specific integrated circuits (ASICs). As demand increases these specialized ICs should become available as building blocks, greatly reducing learning curves, design time, and cost.

Efficiency

Power added efficiency (η_a or *PAE*) is the dimensionless ratio of RF power delivered from a device to the load (p_{out}) minus the input incident RF power ($p_{incident}$) versus the total DC power dissipated in the device (p_{DC}). It is the most commonly used efficiency rating for amplifiers and accounts for both the switching and power gain capabilities of the overall amplifier being considered. High *PAE* is essential to reducing the overall cost of high power transmitter systems, as previously discussed in the power section

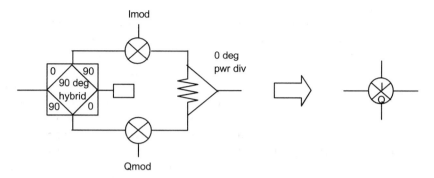

FIGURE 5.12 I-Q modulator block diagram.

above. As with power, *PAE* affects the PA device size, packaging, thermal paths and related cooling methods, power supply, and DC interconnect cable sizes, weight, and safety, each of which can rapidly drive up cost.

$$PAE = \eta_a = \frac{P_{load} - P_{incident}}{P_{DC}}$$

The I-Q Modulator

The I-Q modulator is a basic building block used in numerous applications, and is an essential element of many linearization methods. The basic block diagram is shown in Fig. 5.12, along with the associated symbol that will be used in the following discussions. The modulator consists of two separate mixers that are driven 90° out of phase with each other to generate quadrature (I and Q) signals. The 90° port is usually driven with the high level bias signal, allowing the mixer compression characteristic to minimize amplitude variation from the 90° hybrid. The configuration is reciprocal, allowing either up- or down-conversion.

Class A Amplifier in Back Off

An amplifier is usually required near the output of any transmitter. The linearity of the amplifier is dependent on the saturated power that the amplifier can produce, the amplifier bias and design, and the characteristics of the active device itself. An estimate of the DC power requirements and dissipation for each stage in the PA chain can be made based on the peak or saturated power, duty cycle, and linearity requirements. The maximum or saturated power (P_{sat} in dBW or dBm, depending on whether power is in W or mW) can be estimated (Fig. 5.13) from the product of the RMS voltage and current swings across the RF load, ($V_{sup} - V_{on}$)/2 and I_{on}/2, and from the loss in the output matching circuits (L_{out} in dB). As previously discussed in the receiver section, the 3OIP is about 6 dB above the saturated power for a square law device, but can vary by as much as 4 dB lower to as much as 10 dB higher for a given actual device. Thus it is very important to determine the actual 3OIP for a given device, using vendor data, simulation, or measurement. One must take into account the effects of transmitter components both prior to and after the PA, utilizing the same analysis technique used for receiver intermodulation. The ACP output intercept point (AOIP) will be closely correlated to the 3OIP, and will act in much the same way, except for the actual value of the AOIP. The delta between the AOIP and 3OIP will be modulation dependent, and must be determined through simulation or measurement at this time. Once this has been determined, it is a relatively easy matter to determine the required back off from P_{sat} for the output amplifier, often as much as 10 to 15 dB. Under these conditions, amplifiers that require a high intercept but in which the required peak power is much lower that the peak power that is available, linearization can be employed to lower the cost.

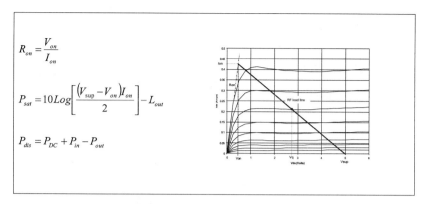

FIGURE 5.13   Class A amplifier saturated power estimation.

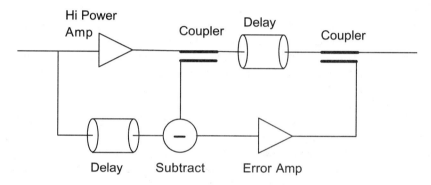

FIGURE 5.14   Feedforward amplifier.

Feed Forward

Although the feedforward amplifier (Fig. 5.14) is a simple concept, it is relatively difficult to implement, especially over temperature and time. The applied signal is amplified to the desired power level by the power amplifier, whose output is sampled. The PA introduces distortion to the system. A portion of the input signal is delayed and then subtracted from the sampled PA signal, nulling out the original signal, leaving only the unwanted distortion created by the PA. This error signal is then adjusted for amplitude and recombined with the distorted signal in the output coupler, canceling out the distortion created by the PA. The resulting linearity improvement is a function of the phase and amplitude balances maintained, especially over temperature and time. The process of generating an error signal will also create nonlinearities, which will limit the ultimate improvements that are attainable, and thus are a critical part of the design.

Cartesian and Polar Loops

The Cartesian loop (CL) (Fig. 5.15) is capable of both good efficiency and high linearity. The efficiency is primarily determined by the amplifier efficiency. The loop action determines the linearity achieved. A carrier signal is generated and applied to the input of the CL, where it is power divided and applied to two separate I-Q mixers. The high power carrier path is quadrature modulated and then amplified by the output PA, after which the distorted modulated signal is sampled by a coupler. The sample distorted signal is then demodulated by mixing it with the original unmodulated carrier, resulting in distorted modulation in quadrature or I-Q form. These distorted I and Q modulations are then subtracted from the original I and Q modulation to generate error I and Q modulation (hence the name Cartesian),

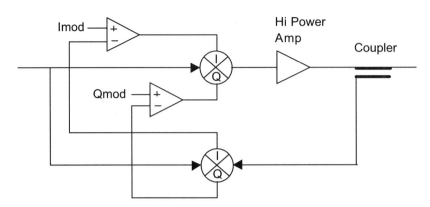

FIGURE 5.15 Cartesian loop.

which will continuously correct the nonlinearity of both the power amplifier and the I-Q modulator. Loop gain and phase relationships are critical to the design, and as with any feedback scheme, care must be taken to prevent loop oscillation. The I-Q modulators and the sampling coupler utilize 90-degree power dividers with limited bandwidth over which satisfactory performance can be attained. The loop delay ultimately will limit the attainable bandwidth to about 10%. Even with these difficulties, the CL is a popular choice. Much of the circuitry is required anyway, and can be easily integrated into ASICs, resulting in low production costs.

Whereas the Cartesian loop depends on quadrature I and Q signals, the related polar loop uses amplitude and phase to achieve higher linearity. The method is much more complex since the modulation correction depends on both frequency modulating the carrier as well as amplitude modulating it. Ultimately the performance will be worse than that of the Cartesian loop, as well as being more costly by a considerable margin. For these reasons, it is not used.

Fixed Predistortion

Fixed predistortion methods are conceptually the simplest form of linearization. A power amplifier will have nonlinearities that distort the original signal. By providing complimentary distortion prior to the PA, the predistorted signal is linearized by the PA. The basic concept can be divided into digital and transfer characteristic methods, both with the same objective. In the digital method (Fig. 5.16), digital signal processing (DSP) is used to provide the required predistortion to the signal. This can be applied at any point in the system, but is usually provided at baseband where it can be cheaply accomplished. The information required for predistortion must be determined and then stored in memory. The DSP then utilizes this information and associated algorithms to predistort the signal, allowing the PA to correct

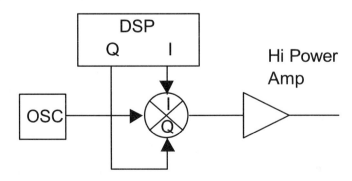

FIGURE 5.16 Fixed digital predistortion.

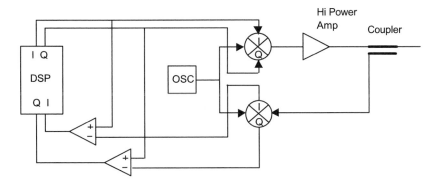

FIGURE 5.17 Adaptive predistortion.

the predistortion, resulting in high linearity. When hardware is used to generate the predistortion, the predistorting transfer characteristic must be determined, and appropriate hardware must be developed. There are no algorithms or methods to accomplish this, so it can be a formidable task. In either case, the improvements in linearity are limited by the lack of any feedback to allow for deviations from the intended operation, and by the ability to actually determine and create the required predistortion. In short, it is cheap, but don't expect dramatic results!

Adaptive Predistortion

Linearization by adaptive predistortion (Fig. 5.17) is very similar to fixed methods, with the introduction of feedback in the form of an error function that can be actively minimized on a continuous basis. The ability to change under operational conditions requires some form of DSP. The error signal is generated in the same way used for Cartesian loop systems, but is then processed by the DSP, allowing the DSP to minimize the error by modifying or adapting the applied predistortion. The disadvantages in this method center on the speed of the DSP and the inability of the system to react due to loop delay. It must see the error before it can correct for it.

Envelope Elimination and Recovery (EER)

The highly efficient envelope elimination and recovery amplifier (Fig. 5.18) accepts a fully modulated signal at its input and power divides the signal. One portion of the signal is amplitude detected and filtered to create the low frequency AM component of the original signal. The other portion of the signal is amplitude limited to strip off or eliminate all of the AM envelope, leaving only the FM component or carrier. Each of these components is then separately amplified using high-efficiency techniques. The amplified AM component is then utilized to control the FM amplifier bias, modulating the amplified FM carrier. Thus the original signal is recovered, only amplified. While this process works very well, an alternative is available that utilizes the DSP capabilities to simplify the whole process, cut costs, and improve performance. In a system, the input half of the EER amplifier can be eliminated and the carrier FM

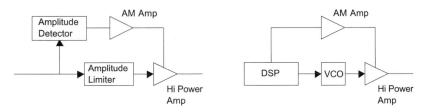

FIGURE 5.18 Envelope elimination and recovery.

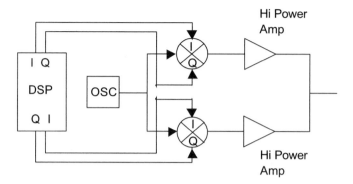

FIGURE 5.19   LINC transmitter.

modulated directly by DSP-generated tuning of a voltage-controlled oscillator (VCO). The DSP-generated AM is amplified and used to control the FM amplifier bias. The result is the desired modulated carrier.

Linear Amplification Using Nonlinear Components (LINC)

The LINC transmitter (Fig. 5.19) concept is quite simple. The DSP creates two separate amplitude and phase-modulated signals, each in quadrature (I-Q) format. These signals are upconverted by I-Q modulators to create two separate phase-modulated signals that are separately applied to high-efficiency output power amplifiers. The amplified FM signals are then combined at the output, the signals being such that all of the unwanted distortion is cancelled by combining 180° out of phase, and all of the desired signal components are added by combining in phase. The challenge in this method is in the DSP generation of the original pair of quadrature signals required for the desired cancellation and combination at the output of the transmitter. Another area of concern with this linearization method is the requirement for amplitude and phase matching of the two channels, which must be tightly controlled in order to achieve optimum performance.

Combined Analog Locked-Loop Universal Modulation (CALLUM)

The CALLUM linearization method is much simpler than it looks at first glance (Fig. 5.20). Basically, the top portion of the transmitter is the LINC transmitter discussed above. An output coupler has been added to allow sampling of the output signal, and the bottom half of the diagram delineates the feedback method that generates the two quadrature pairs of error signals in the same way as used in the Cartesian loop or EER methods. This feedback corrects for channel differences in the basic LINC transmitter, substantially improving performance. Since most of the signal processing is performed at the modulation frequencies, the majority of the circuit is available for ASIC implementation.

I-V Trajectory Modification

In I-V trajectory or cyclic modification, the idea is to create an active I-V characteristic that changes with applied signal level throughout each signal cycle, resulting in improved linear operation (see device tailoring below). A small portion of the signal is tapped off or sampled at the input or output of the amplifier and, based on the continuously sampled signal amplitude, the device bias is continuously modified at each point in the signal cycle. The power range over which high PAE is achieved will be compressed. This method requires a good understanding of the PA device, and excellent modeling. Also, the sampling and bias modification circuitry must be able to react at the same rate or frequencies as the PA itself while providing the required voltage or current to control the PA device. Delay of the sampled signal to the time the bias is modified is critical to obtaining performance. This method is relatively cheap to implement, and can be very effective in improving linearity.

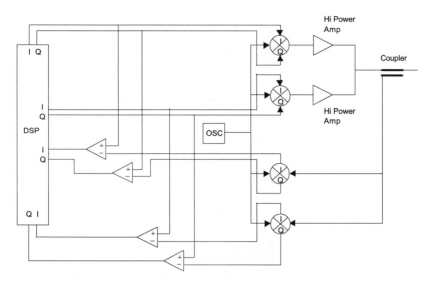

FIGURE 5.20   CALLUM.

Dougherty Amplification

The simplified form of the Dougherty amplifier, which maintains high PAE over a much wider power range than a single amplifier, is shown in Fig. 5.21. In the low power path, a 90° phase shifter is used to compensate for the 90° phase shifter/impedance inverter required in the high power path. The low power amplifier is designed to operate efficiently at a given signal level. The class C high power under low power conditions does not turn on, and thus represents high impedance at the input. The high power 90° phase shifter/impedance inverter provides partial matching for the low power amplifier under these conditions. As the signal level increases, the low power amplifier saturates, the class C high power amplifier turns on, and the power of both amplifiers sum at the output. Under these conditions the high power 90° phase shifter/impedance inverter matches the high power amplifier to the load impedance. Although the modulation bandwidths are not a factor in this technique, the bandwidth is limited by the phase and amplitude transfer characteristics of the 90° elements. This concept can be extended by adding more branches, or by replacing the low power amplifier with a complete Dougherty amplifier in itself.

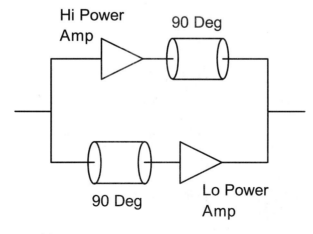

FIGURE 5.21   Dougherty amplifier.

Device Tailoring

For designers with access to a flexible semiconductor foundry service, linearity can be improved directly at the device level. The most obvious way of accomplishing this is by modifying the semiconductor doping to achieve the desired linearity while maintaining other performance parameters. In the ideal device, both the real and reactive device impedance would remain constant and linear (constant derivatives) as the I-V load line or trajectory is traversed for increasing amplitude signals. This linear operation would continue up to the signal amplitude at which both clipping and pinch-off simultaneously occur (ideal biasing). Thus the ideal device would be perfectly linear for any signal below P_{sat}. The objective should be to come as close to this ideal as possible in order to maximize linearity. At best this is a difficult task involving a great deal of device and process engineering. Another strategy might involve trying to minimize the amplitude-dependent parasitic effects such as leakage currents and capacitive or charge-related problems. A third strategy would be to modify the linearity by paralleling two or more devices of varying design together resulting in the desired performance. This is relatively easy to implement through device layout, with best results achieved when this is accomplished at the lowest level of integration (i.e., multiple or tapered gate lengths within a single gate finger, or a stepped or tapered base layer). The results can be quite dramatic with respect to linearity, and best of all the added recurring cost is minimal. As with trajectory modification, the power range for high efficiency operation is compressed.

Summary

The key transmitter parameters of ACP, modulation, linearity, and power are all tightly correlated. These parameters must be determined early in transmitter design so that individual component parameters can be determined and flow-down specifications can be made available to component designers. Linear operation is essential to controlling the power spill-over into adjacent channels (ACP). The basic linearization methods commonly used have been described. These include the class A amplifier in back-off, feedforward, Cartesian and polar loops, adaptive predistortion, envelope elimination and recovery (EER), linear amplification using nonlinear components (LINC), combined analog locked-loop universal modulation (CALLUM), I-V trajectory modification, device tailoring, and Dougherty amplification. Combining methods in such a way as to take advantage of multiple aspects of the nonlinear problem can result in very good performance. An example might be the combination of a Cartesian loop with device tailoring. Unfortunately, it is not yet possible to use simple relationships to calculate ACP directly from linearity requirements, or conversely, required linearity given the ACP. The determination of these requirements is highly dependent on the modulation being used. However, simulators are available that have the capability to design and determine the performance that can be expected.

Further Reading

Casadevall, F., The LINC Transmitter, *RF Design*, Feb. 1990.

Boloorian, M. and McGeeham, J., The Frequency-Hopped Cartesian Feedback Linear Transmitter, *IEEE Transactions on Vehicular Technology*, 45, 4, 1996.

Zavosh, F., Runton, D., and Thron, C., Digital Predistortion Linearizes CDMA LDMOS Amps, *Microwaves & RF*, Mar. 2000.

Kenington, P., Methods Linearize RF Transmitters and Power Amps, Part 1, *Microwaves & RF*, Dec. 1998.

Kenington, P., Methods Linearize RF Transmitters and Power Amps, Part 2, *Microwaves & RF*, Jan. 1999.

Correlation Between P1db and ACP in TDMA Power Amplifiers, *Applied Microwave & Wireless*, Mar. 1999.

Bateman, A., Haines, D., and Wilkinson, R., Linear Transceiver Architectures, Communications Research Group, University of Bristol, England.

Sundstrom, L. and Faulkner, M., Quantization Analysis and Design of a Digital Predistortion Linearizer for RF Power Amplifiers, *IEEE Transactions on Vehicular Technology*, 45, 4, 1996.

5.3 Low Noise Amplifier Design

Jakub Kucera and Urs Lott

Signal amplification is a fundamental function in all wireless communication systems. Amplifiers in the receiving chain that are closest to the antenna receive a weak electric signal. Simultaneously, strong interfering signals may be present. Hence, these low noise amplifiers mainly determine the system noise figure and intermodulation behavior of the overall receiver. The common goals are therefore to minimize the system noise figure, provide enough gain with sufficient linearity, and assure a stable 50 Ω input impedance at a low power consumption.

Definitions

This section introduces some important definitions used in the design theory of linear RF and microwave amplifiers. Further, it develops some basic principles used in the analysis and design of such amplifiers.

Gain Definitions

Several gain definitions are used in the literature for high-frequency amplifier designs.

The transducer gain G_T is defined as the ratio between the effectively delivered power to the load and the power available from the source. The reflection coefficients are shown in Fig. 5.22.

$$G_T = \frac{1-|\Gamma_S|^2}{|1-\Gamma_S \cdot S_{11}|^2} \cdot |S_{21}|^2 \cdot \frac{1-|\Gamma_L|^2}{|1-\Gamma_L \cdot \Gamma_{OUT}|^2}$$

The available gain G_{AV} of a two-port is defined as the ratio of the power available from the output of the two-port and the power available from the source.

$$G_{AV} = \frac{1-|\Gamma_S|^2}{|1-\Gamma_S \cdot S_{11}|^2} \cdot |S_{21}|^2 \cdot \frac{1}{|1-\Gamma_{OUT}|^2} \quad \text{with } \Gamma_{OUT} = S_{22} + \frac{S_{12} \cdot S_{21} \cdot \Gamma_S}{1-\Gamma_S \cdot S_{11}}$$

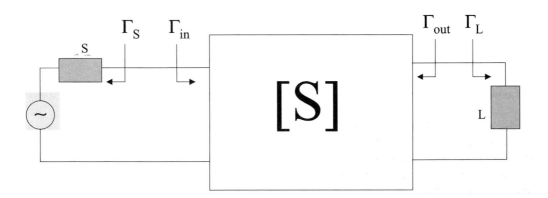

FIGURE 5.22   Amplifier block diagram. Z_S: source impedance, Z_L: load impedance, Γ_S: source reflection coefficient, Γ_{in}: input reflection coefficient, Γ_{out}: output reflection coefficient, Γ_L: load reflection coefficient.

The entire available power at one port can be transferred to the load, if the output is terminated with the complex conjugate load.

The available gain G_{AV} is a function of the two-port scattering parameters and of the source reflection coefficient, but independent of the load reflection coefficient Γ_L. The available gain gives a measure for the maximum gain into a conjugately matched load at a given source admittance.

The associated gain G_{ASS} is defined as the available gain under noise matching conditions.

$$G_{ASS} = \frac{1 - \left|\Gamma_{opt}\right|^2}{\left|1 - \Gamma_{opt} \cdot S_{11}\right|^2} \cdot \left|S_{21}\right|^2 \cdot \frac{1 - \left|\Gamma_L\right|^2}{\left|1 - \Gamma_L \cdot \Gamma_{OUT}\right|^2}$$

Stability and Stability Circles

The stability of an amplifier is a very important consideration in the amplifier design and can be determined from the scattering parameters of the active device, the matching circuits, and the load terminations (see Fig. 5.22). Two stability conditions can be distinguished: unconditional and conditional stability.

Unconditional stability of a two-port means that the two-port remains stable (i.e., does not start to oscillate) for any passive load at the ports. In terms of the reflection coefficients, the conditions for unconditional stability at a given frequency are given by the following equations

$$\left|\Gamma_{IN}\right| = \left|S_{11} + \frac{S_{12} \cdot S_{21} \cdot \Gamma_L}{1 - \Gamma_L \cdot S_{22}}\right| < 1$$

$$\left|\Gamma_{OUT}\right| = \left|S_{22} + \frac{S_{12} \cdot S_{21} \cdot \Gamma_S}{1 - \Gamma_S \cdot S_{11}}\right| < 1$$

$$\left|\Gamma_S\right| < 1 \quad \text{and} \quad \left|\Gamma_L\right| < 1$$

In terms of the scattering parameters of the two-port, unconditional stability is given, when

$$K = \frac{1 - \left|S_{11}\right|^2 - \left|S_{22}\right|^2 + \left|\Delta\right|^2}{2\left|S_{12} \cdot S_{21}\right|} > 1$$

and

$$\left|\Delta\right| < 1$$

with $\Delta = S_{11} \cdot S_{22} - S_{12} \cdot S_{21}$. K is called the stability factor.[1]

If either $\left|S_{11}\right| > 1$ or $\left|S_{22}\right| > 1$, the network cannot be unconditionally stable because the termination $\Gamma_L = 0$ or $\Gamma_S = 0$ will produce or $\left|\Gamma_{IN}\right| > 1$ or $\left|\Gamma_{OUT}\right| > 1$.

The maximum transducer gain is obtained under simultaneous conjugate match conditions $\Gamma_{IN} = \Gamma_S^*$ and $\Gamma_{OUT} = \Gamma_L^*$. Using

$$\Gamma_{IN} = S_{11} + \frac{S_{12} \cdot S_{21} \cdot \Gamma_L}{1 - \Gamma_L \cdot S_{22}} \quad \text{and} \quad \Gamma_{OUT} = S_{22} + \frac{S_{12} \cdot S_{21} \cdot \Gamma_S}{1 - \Gamma_S \cdot S_{11}}$$

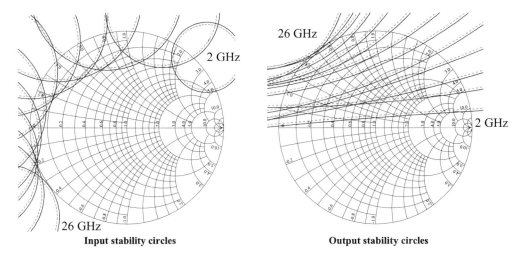

FIGURE 5.23 Source (input) and load (output) stability circles in the Smith chart for the MESFET NE710 over the frequency range from 2 to 26 GHz. Unstable region is indicated by the dotted line.

a closed-form solution for the source and load reflection coefficients Γ_S and Γ_L can be found. However, a simultaneous conjugate match having unconditional stability is not always possible if $K < 1^2$.

Conditional stability of a two-port means that for certain passive loads (represented as $\Gamma_L < 1$ or $\Gamma_S < 1$) oscillation may occur. These values of Γ_L and Γ_S can be determined by drawing the stability circles in a Smith chart. The source and load stability circles are defined as

$$\left|\Gamma_{IN}\right| = 1 \text{ and } \left|\Gamma_{OUT}\right| = 1$$

On one side of the stability circle boundary, in the Γ_L plane, $\left|\Gamma_{IN}\right| > 1$ and on the other side $\left|\Gamma_{IN}\right| < 1$. Similarly, in the Γ_S plane, $\left|\Gamma_{OUT}\right| > 1$ and on the other side $\left|\Gamma_{OUT}\right| < 1$. The center of the Smith chart ($\Gamma_L = 0$) represents a stable operating point, if $\left|S_{11}\right| < 1$, and an unstable operating point, if $\left|S_{11}\right| > 1$ (see Fig. 5.23). Based on these observations, the source and load reflection coefficient region for stable operation can be determined.

With unconditional stability, a complex conjugate match of the two-port is possible. The resulting gain then is called the maximum available gain (MAG) and is expressed as

$$MAG = \left|\frac{S_{21}}{S_{12}}\right| \cdot \left(K - \sqrt{K^2 - 1}\right)$$

The maximum stable gain MSG is defined as the maximum transducer gain for which $K = 1$ holds, namely

$$MSG = \left|\frac{S_{21}}{S_{12}}\right|$$

MSG is often used as a figure of merit for potentially unstable devices (see Fig. 5.24).

It must be mentioned, however, that the stability analysis as presented here in its classical form, is applicable only to a single-stage amplifier. In a multistage environment, the above stability conditions are insufficient, because the input or output planes of an intermediate stage may be terminated with active networks. Thus, taking a multistage amplifier as a single two-port and analyzing its K-factor is helpful, but does not guarantee overall stability. Literature on multistage stability analysis is available.[3]

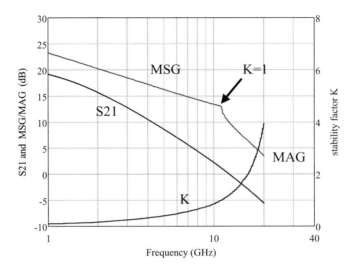

FIGURE 5.24   Maximum stable gain (MSG), maximum available gain (MAG), S_{21}, and stability factor K for a typical MESFET device with 0.6 μm gate length.

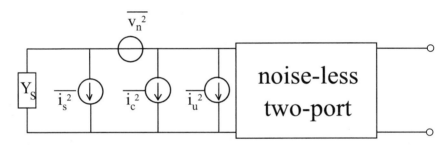

FIGURE 5.25   Representation of noisy two-port as noiseless two-port and two partly correlated noise sources (v_n and $i_c + i_u$) at the input.

Representation of Noise in Two-Ports

The LNA can be represented as a noise-free two-port and two partly correlated input noise sources i_n and v_n as shown in Fig. 5.25. The partial correlation between the noise sources i_n and v_n can be described by splitting i_n into a fully correlated part i_c and a noncorrelated part i_u as

$$i_n = i_c + i_u$$

The fully correlated part i_c is defined by the correlation admittance[4] Y_{cor}

$$i_c = Y_{cor} v_n$$

The source impedance $Z_s = R_s + jB_s$ shows thermal noise i_s which depends on the bandwidth as

$$\overline{i_s^2} = 4kTG_s \Delta f$$

Finally, the noise factor F can be expressed in terms of these equivalent input noise generators as

$$F = 1 + \left| Y_s + Y_{cor} \right|^2 \frac{\overline{v_n^2}}{\overline{i_s^2}} + \frac{\overline{i_u^2}}{\overline{i_s^2}}$$

Details on the use of the correlation matrix and the derivation of the noise factor from the noise sources of the two-port can be found in References 4 and 5.

Noise Parameters

The noise factor F of a noisy two-port is defined as the ratio between the available signal-to-noise power ratio at the input to the available signal-to-noise ratio at the output.

$$F = \frac{S_{in}}{N_{in}} \bigg/ \frac{S_{out}}{N_{out}}$$

The noise factor of the two-port can also be expressed in terms of the source admittance $Y_s = G_s + jB_s$ as

$$F = F_{min} + \frac{R_n}{G_s} \left| Y_s - Y_{opt} \right|^2$$

where F_{min} is the minimum achievable noise factor when the optimum source admittance $Y_{opt} = G_{opt} + jB_{opt}$ is presented to the input of the two-port, and R_n is the equivalent noise resistance of the two-port. Sometimes the values Y_s, Y_{opt}, and R_n are given relative to the reference admittance Y_0.

The noise performance of a two-port is fully characterized at a given frequency by the four noise parameters F_{min}, R_n, and real and imaginary parts of Y_{opt}.

Several other equivalent forms of the above equation exist, one of them describing F as a function of the source reflection coefficient Γ_s.

$$F = F_{min} + \frac{4R_n}{Z_0} \frac{\left| \Gamma_s - \Gamma_{opt} \right|^2}{\left| 1 + \Gamma_{opt} \right|^2 \cdot \left(1 - \left| \Gamma_s \right|^2 \right)}$$

When measuring noise, the noise factor is often represented in its logarithmic form as the noise figure NF

$$NF = 10 \log F$$

Care must be taken not to mix up the linear noise factor and the logarithmic noise figure in noise calculations.

Noise Circles

Noise circles refer to the contours of constant noise figure for a two-port when plotted in the complex plane of the input admittance of the two-port. The minimum noise figure is presented by a dot, while for any given noise figure higher than the minimum, a circle can be drawn. This procedure is adaptable in the source admittance notation as well as in the source reflection coefficient notation. Fig. 5.26 shows the noise circles in the source reflection plane.

Noise circles in combination with gain circles are efficient aids for circuit designers when optimizing the overall amplifier circuit network for low noise with high associated gain.

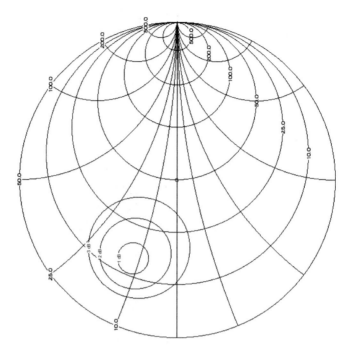

FIGURE 5.26   Noise circles in the input reflection coefficient plane.

Friis Formula: Cascading Noisy Two-Ports

When several noisy two-ports are connected in cascade, the overall noise characteristics are described by[6]

$$F_{tot} = F_1 + \frac{F_2 - 1}{G_1} + \frac{F_3 - 1}{G_1 \cdot G_2} + \ldots + \frac{F_i - 1}{G_1 \cdot G_2 \ldots G_{i-1}}$$

where F_i and G_i are noise factor and available gain of the ith two-port. The available gain depends on the output admittance of the previous stage.

Noise Measure M

The overall noise factor of an infinite number of identical cascaded amplifiers is $F = 1 + M$ with

$$M = \frac{F - 1}{1 - \dfrac{1}{G}}$$

M is here called the noise measure. The noise measure is useful for comparing the noise performance of devices or LNAs with different power gains.

Design Theory

The apparent structural simplicity of an LNA with its relatively few components is misleading. The design should be easy, but the trade-offs complicate the design. A simultaneous noise and power matching involves a more complicated matching network and the achievable dynamic range is often limited by the given low supply voltage and the maximum allowed current consumption. The LNA must provide enough gain so that the noise contributions from the following components become small. But the maximum tolerable gain is limited by the linearity requirements of the following receiver chain.

The most important design considerations in a high-frequency amplifier are stability, power gain, bandwidth, noise, and DC power consumption requirements.

A systematic mathematical solution, aided by graphical methods, is developed to determine the input and output matching network for a particular noise, gain, stability, and gain criteria. Unconditionally stable designs will not oscillate with any passive termination, while designs with conditional stability require careful analysis of the loading to assure stable operation.

Linear Design Procedure for Single-Stage Amplifiers

1. *Selection of device and circuit topology:* Select the appropriate device based on required gain and noise figure. Also decide on the circuit topology (common-base/gate or common-emitter/source). The most popular circuit topology in the first stage of the LNA is the common-emitter (source) configuration. It is preferred over a common-base (-gate) stage because of its higher power gain and lower noise figure. The common-base (-gate) configuration is a wideband unity-current amplifier with low input impedance ($\approx 1/g_m$) and high predominantly capacitive output impedance. Wideband LNAs requiring good input matching use common-base input stages. At high frequencies the input impedance becomes inductive and can be easily matched.

2. *Sizing and operating point of the active device:* Select a low noise DC operating point and determine scattering and noise parameters of the device. Typically, larger input transistors biased at low current densities are used in low noise designs. At RF frequencies and at a given bias current, unipolar devices such as MOSFET, MESFET, and HEMT are easier to match to 50 Ω when the device width is larger. Both, (hetero-) BJTs and FETs show their lowest intrinsic noise figure when biased at approximately one tenth of the specified maximum current density. Further decreasing the current density will increase the noise figure and reduce the available gain.

3. *Stability and RF feedback:* Evaluate stability of the transistor. If only conditionally stable, either introduce negative feedback (high-resistive DC parallel or inductive series feedback) or draw stability circles to determine loads with stable operation.

4. *Select the source and load impedance:* Based on the available power gain and noise figure circles in the Smith chart, select the load reflection coefficient Γ_L that provides maximum gain, the lowest noise figure (with $\Gamma_S = \Gamma_{opt}$), and good VSWR. In unconditionally stable designs, Γ_L is

$$\Gamma_L = \left(S_{22} + \frac{S_{12} \cdot S_{21} \cdot \Gamma_{opt}}{1 - \Gamma_{opt} \cdot S_{11}} \right)^*$$

In conditionally stable designs, the optimum reflection coefficient Γ_S may fall into an unstable region in the source reflection coefficient plane. Once Γ_S is selected, Γ_L is selected for the maximum gain $\Gamma_L = \Gamma_{OUT}$, and Γ_L must again be checked to be in the stable region of the load reflection coefficient plane.

5. *Determine the matching circuit:* Based on the required source and load reflection coefficients, the required ideal matching network can be determined. Depending on the center frequency, lumped elements or transmission lines will be applied. In general, there are several different matching circuits available. Based on considerations about insertion loss, simplicity, and reproducibility of each matching circuit, the best selection can be made.

6. *Design the DC bias network:* A suitable DC bias network is crucial for an LNA, which should operate over a wide temperature and supply voltage range and compensate parameter variations of the active device. Further, care must be given that no excessive additional high-frequency noise is injected from the bias network into the signal path, which would degrade the noise figure of the amplifier. High-frequency characteristics including gain, noise figure, and impedance matching are correlated to the device's quiescent current. A resistor bias network is generally avoided because of its poor supply rejection. Active bias networks are capable of compensating temperature effects and rejecting supply voltage variations and are therefore preferred.

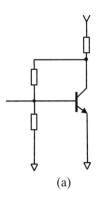
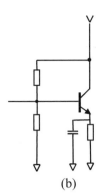

(a)   (b)

FIGURE 5.27   Passive bias network for bipolar amplifiers.

For bipolar circuits, a simple grounded emitter DC bias network is shown in Fig. 5.27a. The high-resistive bias network uses series feedback to stabilize the current of the active device against device parameter variations. However, the supply rejection of this network is very poor, which limits its applicability. A bypassed emitter resistor is often used at low frequencies to stabilize the DC bias point (Fig. 5.27b). At RF and microwave frequencies, the bypass capacitor can cause unwanted high-frequency instability and must be applied with care. Furthermore, an emitter resistor will degrade the noise figure performance of the amplifier, if the resistor is not fully bypassed at the signal frequency.

More advanced active bias circuits use (temperature compensated) bandgap references and generate a reference current, which is mirrored to the amplifying device through a high value resistor or an RF choke to minimize noise injection. Another popular method is to generate a proportional to absolute temperature (PTAT) current source. The amplifier gain is proportional to the transconductance which itself is proportional to the collector current and inversely proportional to the temperature ($g_m = qI_c/kT$). With the transistor biased with a current proportional to temperature, the gain remains roughly constant over temperature. Combining bandgap circuits with PTAT sources leads to excellent supply and temperature variation suppression.[7]

The implementation of appropriate bias methods for FET amplifiers is generally more involved. The most critical parameter affecting the bias point is the threshold voltage. Stable voltage reference and PTAT current sources are typically based on the Schottky diode barrier height and involve rather sophisticated circuitry.[8]

7. *Optimize entire circuit with lossy matching elements:* The final design optimization of the LNA circuit must include the nonidealities of the matching elements, parasitic components such as bond wire inductance, as well as fabrication tolerance aspects. This last design phase today is usually performed on a computer-aided design (CAD) system.

The dominant features of an LNA (gain, noise, matching properties) can be simulated with excellent accuracy on a linear CAD tool. The active device is characterized by its scattering parameters in the selected bias point, and the four noise parameters. The passive components are described by empirical or equivalent circuit models built into the linear simulation tool. If good models for elements like millimeter-wave transmission lines are not available, these elements must be described by their measured scattering parameters, too.

Alternatively, a nonlinear simulator with a full nonlinear device model allows direct performance analysis over varying bias points. The use of nonlinear CAD is mandatory for compression and intermodulation analysis.

Advanced CAD tools allow for direct numerical optimization of the circuit elements toward user-specified performance goals. However, these optimizers should be used carefully, because it can be very difficult to transform the conflicting design specifications into optimization goals. In many cases, an experienced designer can optimize an LNA faster by using the "tune" tools of a CAD package.

Practical Design of a Low Noise Amplifier

The last section presented a design procedure to design a stable low noise amplifier based on linear design techniques. In practice, there are nonidealities and constraints on component sizing that typically degrade the amplifier performance and complicate the design. In fact, the presented linear design method does not take power consumption versus linearity explicitly into account. Some guidelines are provided in this section that may facilitate the design.

Hybrid vs. Monolithic Integrated LNA

With the current trend to miniaturized wireless devices, LNAs are often fabricated as monolithic integrated circuits, usually referred to as MMIC (monolithic microwave integrated circuit). High volume applications such as cell phones call for even higher integration in the RF front end. Thus the LNA is integrated together with the mixer, local oscillator, and sometimes even parts of the transmitter or the antenna.[9]

Depending on the IC technology, monolithic integration places several additional constraints on the LNA design. The available range of component values may be limited, in particular the maximum inductance and capacitance values are often smaller than required. Integrated passive components in general have lower quality factors Q because of their small size. In some cases, the first inductor of the matching circuit must be realized as an external component.

The electromagnetic and galvanic coupling between adjacent stages is often high due to the close proximity of the components. Furthermore, the lossy and conducting substrate used in many silicon-based technologies increases coupling. At frequencies below about 10 GHz, transmission lines cannot be used for matching because the required chip area would make the IC too expensive, at least for commercial applications.

Finally, monolithic circuits cannot be tuned in production. On the other hand, monolithic integration also has its advantages. The placement of the components is well controlled and repeatable, and the wiring length between components is short. The number of active devices is almost unlimited and adds very little to the cost of the LNA. Each active device can be sized individually.

For applications with low volume where monolithic integration is not cost effective, LNAs can be built as hybrid circuits, sometimes called MIC (microwave integrated circuit). A packaged transistor is mounted on a ceramic or organic substrates. The matching circuit is realized with transmission lines or lumped elements. Substrates such as alumina allow very high quality transmission line structures to be fabricated. Therefore, in LNAs requiring ultimate performance, e.g., for satellite ground stations, hybrid circuit technology is sometimes used even if monolithic circuits are available.

Multistage Designs

Sometimes a single amplifier stage cannot provide the required gain and multiple gain stages must be provided. Multiple gain stages complicate the design considerably. In particular, the interstage matching must be designed carefully (in particular in narrowband designs) to minimize frequency shifts and ensure stability. The ground lines of the different gain stages must often be isolated from each other to avoid positive feedback, which may cause parasitic oscillations. Moreover, some gain stages may need some resistive feedback to enhance stability.

Probably the most widely used multistage topology is the cascode configuration. A low noise amplifier design that uses a bipolar cascode arrangement as shown in Fig. 5.28 offers performance advantages in wireless applications over other configurations. It consists of a common-emitter stage driving a common-base stage. The cascode derives its excellent high-frequency properties from the fact that the collector load of the common-emitter stage is the very low input impedance of the common-base stage. Consequently, the Miller effect is minimal even for higher load impedances and an excellent reverse isolation is achieved. The cascode has high output impedance, which may become difficult to match to 50 Ω. A careful layout of the cascode amplifier is required to avoid instabilities. They mainly arise from parasitic inductive feedback between the emitter of the lower and the base of the upper transistor. Separating the two ground lines will enhance the high-frequency stability considerably.

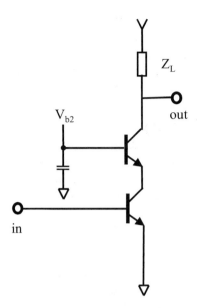

FIGURE 5.28   Cascode amplifier.

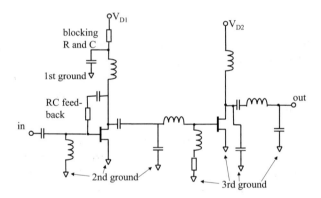

FIGURE 5.29   Design for stability.

Stability Considerations

Figure 5.29 shows a possible strategy for a stable design of a two-stage amplifier. Separated ground and supply lines of the two gain stages minimize positive feedback. RC parallel feedback further enhances in-band stability. Low frequency oscillations caused through unstable bias lines can be attenuated by adding small resistors and blocking capacitors into the supply line.

Feedback

Negative feedback is widely used in amplifier design to stabilize gain against parameter changes in the active device due to supply voltage variations and temperature changes. RF feedback is used in many LNAs to ensure high-frequency stability and make noise and power match coincident. A well-known technique is adding inductance at the emitter (source) of the active device. The inductance L interacts with the base-emitter (gate-source) capacitance C_{IN} and device transconductance g_m to produce a resistive component to the input impedance $g_m \frac{L}{C_{IN}}$, while no additional noise source is introduced (except for the parasitic series resistance of the inductor). Neglecting the Miller capacitance, the input impedance of an inductively degenerated FET stage is

$$Z_{IN} = \frac{1}{j\omega C_{IN}} + j\omega L + g_m \frac{L}{C_{IN}}$$

This method of generating a real term to the input impedance is preferable to resistive methods as only negligible additional noise is introduced. Moreover, the inductance has the helpful side effect of shifting the optimum noise match closer to the complex conjugate power match and reducing the signal distortion. However, the benefits are accompanied by a gain reduction.

Impedance Matching

Following the design procedure in the last section, the conditions for a conjugate match at the input and output ports are satisfied at one frequency. Hence, reactive matching inherently leads to a narrowband design. The input bandwidth is given by

$$BW = \frac{f_0}{Q_{IN}}$$

where f_0 is the center frequency and Q_{IN} is the quality factor of the input matching network. The bandwidth can be increased by increasing the capacitance or decreasing the inductance of the matching network.

Using multistage impedance transformators (lumped element filters or tapers) can broaden the bandwidth, but there is a given limit for the reflection coefficient-bandwidth product using reactive elements.[10] In reality, each matching element will contribute some losses, which directly add to the noise figure.

Select an appropriate matching network based on physical size and quality factor (transmission line length, inductance value): long and high-impedance transmission lines show higher insertion loss. Thus, simple matching typically leads to a lower noise figure.

At higher microwave and millimeter-wave frequencies, balanced amplifiers are sometimes used to provide an appropriate noise and power match over a large bandwidth.

Temperature Effects

Typically, LNAs must operate over a wide temperature range. As transistor transconductance is inversely proportional to the absolute temperature, the gain and amplifier stability may change considerably. When designing LNAs with S-parameters at room temperature, a stability margin should be included to avoid unwanted oscillations at low temperatures, as the stability tends to decrease.

Parasitics

Parasitic capacitance, resistance, or inductance can lead to unwanted frequency shifts, instabilities, or degradation in noise figure and gain and rarely can be neglected. Hence, accurate worst-case simulations with the determined parasitics must be made. Depending on the frequency, the parasitics can be estimated based on simple analytical formulas, or must be determined using suitable electromagnetic field-simulators.

Design Examples

In this section a few design examples of recently implemented low noise amplifiers for frequencies up to 5.8 GHz are presented. They all were manufactured in commercial IC processes.

A Fully Integrated Low Voltage, Low Power LNA at 1.9 GHz[11]

Lowest noise figure can only be achieved when minimizing the number of components contributing to the noise while simultaneously maximizing the gain of the first amplifier stage. Any resistive matching and loading will degrade the noise figure and dynamic behavior and increase power consumption.

In GaAs MESFET processes, the semi-insulating substrate and thick metallization layers allow passive matching components such as spiral inductors and metal-insulator-metal (MIM) capacitors with high

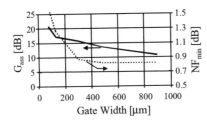
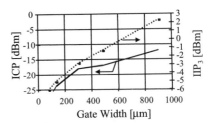

FIGURE 5.30 Simulated performance of an enhancement FET versus device width at a constant power dissipation.

quality factors. These lumped passive components are ideally suited for integrated impedance matching at low GHz frequencies. A fully integrated matching network improves the reproducibility and saves board space while it increases expensive chip area.

It is generally known that GaAs MESFETs have excellent minimum noise figures in the lower GHz frequency range. Still few designs achieve noise figures close to the transistor F_{min}. In fact, several factors prevent F_{min} being attained in practice. If a small input device is employed, a large input impedance transformation involving large inductance values is required. Larger MMIC inductors have higher series resistance and consequently introduce more noise. Further a simultaneous noise and power match often needs additional inductive source degeneration, again introducing noise and reducing gain. For a given maximum power dissipation, very large devices, in contrast, must be biased at very low current densities at which the F_{min} and the gain are degraded. Consequently a trade-off must be made for an optimum design.

The employed GaAs technology features three types of active devices: an enhancement and two depletion MESFETs with different threshold voltages. The enhancement device has a higher maximum available gain, a slightly lower minimum noise figure, but somewhat higher distortion compared to the depletion type. Another advantage of the enhancement FET is that a positive gate bias voltage can be used, which greatly simplifies single-supply operation.

Preliminary simulations are performed using a linear simulator based on measured S- and noise parameters of measured active devices at various bias points and using a scalable large signal model within a harmonic balance simulator in order to investigate the influence of the transistor gate width on the RF performance. The current consumption of the transistor is set at 5.5 mA independent of the gate width. The simulations indicate a good compromise between gain, NF, and intermodulation performance at a gate width of 300 µm (Fig. 5.30).

The LNA schematic is shown in Fig. 5.31. The amplifier consists of a single common-source stage, which uses a weak inductive degeneration at the source (the approximately 0.3 nH are realized with several parallel bondwires to ground). The designed amplifier IC is fabricated in a standard 0.6 µm E/D MESFET foundry process. The matching is done on chip using spiral inductors and MIM capacitors. The complete LNA achieves a measured 50 Ω noise figure of 1.1 dB at 1.9 GHz with an associated gain of 16 dB at a very low supply voltage of $V_{dd} = 1$ V and a total current drain of $I_{dd} = 6$ mA. Fig. 5.32 depicts the measured gain and 50 Ω noise figure versus frequency.

The low voltage design with acceptable distortion performance and reasonable power gain can only be achieved using a reactive load with almost no voltage drop.

Figure 5.33 shows, respectively, the supply voltage and supply current dependence of the gain and noise figure. As can be seen, the amplifier still achieves 10 dB gain and a 1.35 dB noise figure at a supply voltage of only 0.3 V and a total current consumption of 2.3 mA. Sweeping the supply voltage from 1 to 5 volts, the gain varies less than 0.5 dB and the noise figure less than 0.15 dB, respectively. IIP_3 and −1 dB compression point are also insensitive to supply voltage variations as shown in Fig. 5.34.

Below 1 V, however, the active device enters the linear region resulting in a much higher distortion.

Finally, the input and output matchings are measured for the LNA. At the nominal 1 V supply, the input and output return loss are −8 dB and −7 dB, respectively.

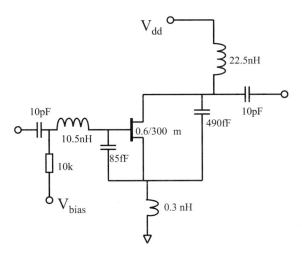

FIGURE 5.31   Schematic diagram of the low voltage GaAs MESFET LNA.

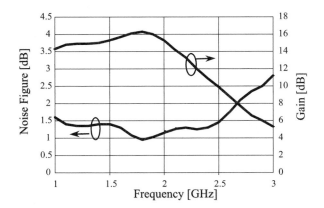

FIGURE 5.32   Measured gain and 50 Ω noise figure vs. frequency.

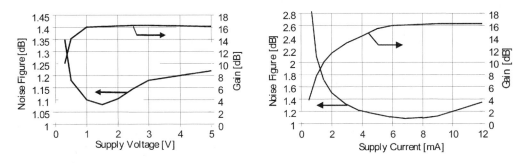

FIGURE 5.33   Measured gain and noise figure vs. supply voltage (I_{dd} = 6 mA) and vs. supply current (V_{dd} = 1 V).

A Fully Matched 800 MHz to 5.2 GHz LNA in SiGe HBT Technology

Bipolar technology is particularly well suited for broadband amplifiers because BJTs typically show low input impedances in the vicinity of 50 Ω and hence can be easily matched. A simplified schematic diagram of the monolithic amplifier is shown in Fig. 5.35. For the active devices of the cascode LNA large emitter

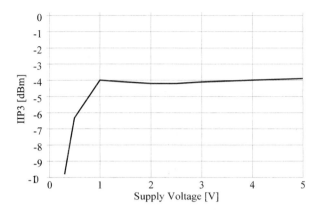

FIGURE 5.34   Measured input IP3 vs. supply voltage.

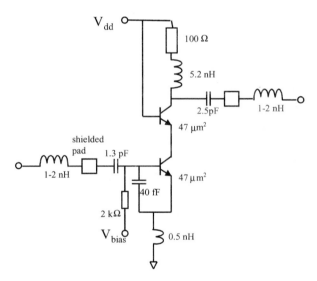

FIGURE 5.35   Schematic diagram of the SiGe HBT LNA.

areas (47 μm²), biased at low current densities are employed to simplify the simultaneous noise and power match. Input and output matching is consequently obtained simply by the aid of the bondwire inductance at the input and output ports and the chip ground.

The LNA was fabricated with MAXIM's GST-3 SiGe process and subsequently was mounted on a ceramic test package for testing. No additional external components are required for this single-supply LNA.

Figure 5.36 shows the 50 Ω noise figure and associated gain over the frequency range of interest. A relatively flat gain curve is measured from 500 MHz up to 3 GHz. Beyond 3 GHz the gain starts to roll off. The circuit features 14.5 dB of gain along with a 2 dB noise figure at 2 GHz. At 5.2 GHz, the gain is still 10 dB and the noise figure is below 4 dB. The input return loss is less than −10 dB between 2.5 and 6.5 GHz. At 1 GHz it increases to −6 dB.

The distortion performance of the amplifier was measured at the nominal 3 V supply for two frequencies, 2.0 and 5.2 GHz, respectively. At 2 GHz, the −1 dB compression point is +2 dBm at the output. At 5.2 GHz the value degrades to 0 dBm.

The LNA is comprised of two sections: the amplifier core and a PTAT reference. The core is biased with the PTAT to compensate for the gain reduction with increasing temperature. The gain is proportional

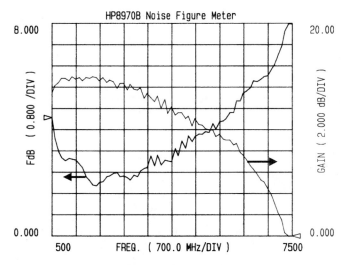

FIGURE 5.36 Measured LNA gain and noise figure vs. frequency ($V_{dd} = 3$ V, $I_{dd} = 8.8$ mA).

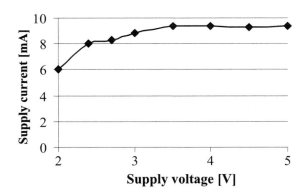

FIGURE 5.37 Supply current vs. supply voltage.

to the transconductance of the transistor, which itself is proportional to collector current and inversely proportional to temperature. The PTAT biasing increases the collector current with temperature to keep the gain roughly constant over temperature. Simultaneously, the biasing shows a good supply rejection as shown in Fig. 5.37.

A chip photograph of the 0.5×0.6 mm² large LNA is shown in Fig. 5.38.

A Fully Matched Two-Stage Low Power 5.8 GHz LNA[12]

A fully monolithic LNA achieves a noise figure below 2 dB between 4.3 GHz and 5.8 GHz with a gain larger than 15 dB at a DC power consumption of only 6 mW using the enhancement device of a standard 17 GHz f_T 0.6 μm E/D-MESFET process.

A schematic diagram of the integrated LNA core is shown in Fig. 5.39. The circuit consists of two common-source gain stages to provide enough power gain. The first stage uses an on-chip inductive degeneration of the source to achieve a simultaneous noise and power match, and to improve RF stability. Both amplifier stages are biased at the same current. The noise contributions of the biasing resistors are negligible.

The output of each stage is loaded with a band pass LC section to increase the gain at the desired frequency. The load of the first stage, together with the DC block between the stages, is also used for inter-stage matching.

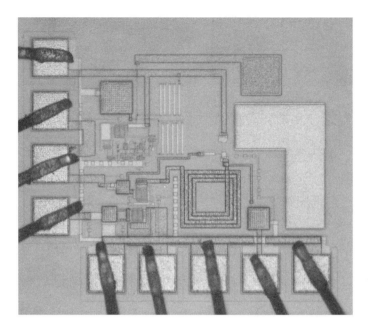

FIGURE 5.38   Chip photograph of the SiGe HBT LNA (0.5×0.6 mm²).

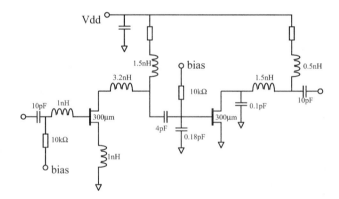

FIGURE 5.39   Schematic diagram of the low noise amplifier.

The DC biasing is done on-chip with a combination of E/D MESFETs (Fig. 5.40). The bias circuit is able to effectively stabilize the bias point for voltages from 1 V to beyond 4 V without any feedback network within the amplifier. It also can accurately compensate for threshold voltage variations.

The correlation of the threshold voltages of enhancement and depletion devices due to simultaneous gate recess etch of both types is used in the bias circuit to reduce the bias current variations over process parameter changes. Figure 5.41 shows the simulated deviation from the nominal current as a function of threshold voltage variations. The device current remains very constant even for extreme threshold voltage shifts.

If the RF input device is small, a large input impedance transformation is required. The third-order intercept point can be degraded and larger inductor values are needed sacrificing chip area and noise figures, due to the additional series resistance of the inductor. If instead a very large device is used, the current consumption is increased, unless the current density is lowered. Below a certain current density the device gain will decrease, the minimum noise figure will increase, and a reliable and reproducible biasing of the device becomes difficult as the device is biased close to the pinch-off voltage. To achieve high quality factors, all inductors are implemented using the two top wiring levels with a total metal

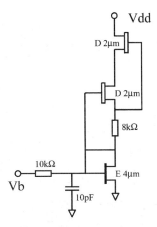

FIGURE 5.40 Schematic diagram of the employed bias circuit.

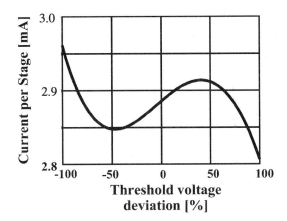

FIGURE 5.41 Simulated current dependence on threshold voltage variations.

thickness of 6 μm. The spiral inductors were analyzed using a 2.5D field simulator in order to accurately determine their equivalent circuit.

Sample test chips were mounted in a ceramic test package (Fig. 5.42) to investigate the influence of the bonding wires and the package parasitics.

In Figs. 5.43 and 5.44 the influence of the bond wires on the input and return loss, gain, and noise figure, respectively, is shown. The optimum input matching is shifted from 5.2 GHz to 5.8 GHz with the bond wire included. In an amplifier stage with moderate feedback one would expect the bond wire to shift the match toward lower frequencies. However, due to the source inductor the inter-stage matching circuit strongly interacts with the input port, causing a frequency shift in the opposite direction.

As expected, the gain curve of the packaged LNA (Fig. 5.44) is flatter and the gain is slightly reduced because of the additional ground inductance arising from the ground bond wires (approx. 40 pH).

At the nominal supply current of 6 mA the measured 50 Ω noise figure is 1.8 dB along with more than 15 dB gain from 5.2 GHz to 5.8 GHz as given in Fig. 5.44. For the packaged LNA the noise figure is slightly degraded due to losses associated with the package and connectors.

At 5.5 GHz the minimum noise figure of the device including the source inductor at the operating bias point is 1.0 dB and the associated gain is 8.5 dB. The minimum noise figure of an amplifier with two identical stages is therefore 1.6 dB. Thus, only a small degradation of the noise figure by the on-chip matching inductor is introduced at the input.

FIGURE 5.42 Photograph of the chip mounted in the test package.

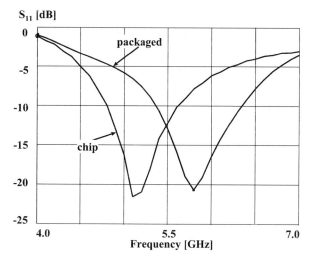

FIGURE 5.43 Input return loss vs. frequency of chip and packaged LNA.

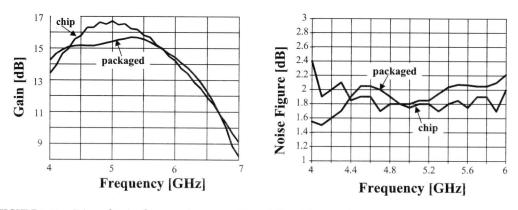

FIGURE 5.44 Gain and noise figure vs. frequency ($V_{dd} = 1$ V and $I_{dd} = 6$mA).

At 5.2 GHz a measured −1 dB compression point of 0 dBm at the output confirms the excellent distortion characteristics of GaAs MESFET devices at very low power consumption. The measured input referenced third order intercept point (IIP3) is -6 dBm.

0.25 μm CMOS LNAs for 900 MHz and 1.9 GHz[13,14]

CMOS technology starts to play a significant role in integrated RF transceivers for the low GHz range with effective gate lengths reaching the deep submicron regions. Competitive circuit performance at low power dissipation is becoming possible even for critical building blocks such as the LNA. In fact, quarter-micron CMOS seems to be the threshold where robust designs can be realized with current consumption competitive to BJT implementations. Further downscaling calls for a reduction in supply voltage which will ultimately limit the distortion performance of CMOS-based designs.

Designing a low noise amplifier in CMOS is complicated by the lossy substrate, which requires a careful layout to avoid noise injection from the substrate. The schematic diagrams of two demonstrated 0.25 μm CMOS LNAs for 900 MHz and 1.9 GHz are shown in Figs. 5.45a and 5.45b, respectively. Both circuits use two stages to realize the desired gain.

The first amplifier consisting of an externally matched cascode input stage and a transimpedance output stage consumes 10.8 mA from a 2.5 V supply. The cascode is formed using two 600-μm wide NMOS devices loaded by a 400 Ω resistor. The inductance of approximately 1.2 nH formed by the bondwire at the source of the first stage is used to simplify the matching of an otherwise purely capacitive input impedance. The directly coupled transimpedance output stage isolates the high-gain cascode and provides a good 50 Ω output matching. A simple biasing is included on the chip. At the nominal power dissipation and 900 MHz, the LNA achieves 16 dB gain and a noise figure of below 2 dB. The input and output return losses are −8 dB and −12 dB, respectively. The distortion performance of the LNA can well be estimated by measuring the input referred third order intercept point and the −1 dB compression point. They are −7 dBm and −20 dBm, respectively.

The 1.9 GHz LNA shown in Fig. 5.45b employs a resistively loaded common-source stage followed by a reactively loaded cascode stage. To use inductors to tune out the output capacitance and to realize the 50 Ω output impedance is a viable alternative to using the transimpedance output stage. The circuit employs a self-biasing method by feeding the DC drain voltage of the first stage to the gates. The supply rejection is consequently poor.

The LNA achieves 21 dB gain and a 3 dB noise figure while drawing 10.8 mA from a 2.7 V supply. In Fig. 5.46 the measured gain and noise figure versus frequency are plotted. At the nominal bias the input referred −1 dB compression point is −25 dBm, which corresponds to −4 dBm at the output. The input and output return loss are −5 dB and −13 dB, respectively.

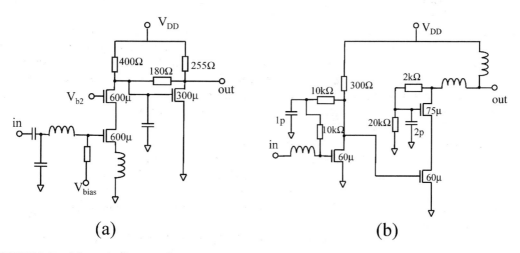

FIGURE 5.45   Schematic diagram of two 0.25 μm CMOS LNAs for 900 MHz (a) and 1900 MHz (b).

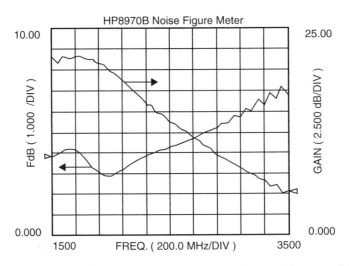

FIGURE 5.46   Measured gain and noise figure of the 1900 MHz CMOS LNA of Figure 5.24(b).

A comparison between the two amplifiers presented reveals some interesting points:

- The 900 MHz LNA explicitly makes use of the bondwire inductance to reduce the (otherwise purely capacitive) input impedance while the fist stage of the 1.9 GHz amplifier is connected to the chip ground. Both amplifiers use an external inductor for the input matching and both achieve a relatively poor input match.

- No explicit inter-stage matching is employed in either of the amplifiers. The 900 MHz amplifier uses the second stage as an impedance transformer.

- The 900 MHz amplifier employs ten times wider devices biased at lower current densities compared to its 1.9 GHz counterpart. As a consequence, the bias current becomes more sensitive to threshold voltage variations due to fabrication.

- At comparable power consumption the two amplifiers show roughly same distortion performance.

A Highly Selective LNA with Electrically Tunable Image Reject Filter for 2 GHz[15]

LNA designs with purely reactive passive components are inherently narrowband. IC technologies on high resistivity substrates allow reproducible passive components (inductors, capacitors, varactors, transmission lines) with excellent quality factors. They are well suited for designs to include a frequency selectivity which goes beyond a simple matching. In particular, amplifiers with adjustable image rejection can be realized. To show the potential of highly frequency selective LNAs as viable alternative to image reject mixers, an LNA for 1.9 GHz is demonstrated, which allows a tunable suppression of the image frequency. The schematic diagram of the circuit is shown in Fig. 5.47. The amplifier consists of two cascaded common-source stages loaded with LC resonant circuits. Undesired frequencies are suppressed using series notch filters as additional loads. Each of the two notch filters is formed by a series connection of a spiral inductor and a varactor diode. The two notches resonate at the same frequencies and must be isolated by the amplifier stages.

A careful design must be done to avoid unwanted resonances and oscillations. In particular, immunity against variations in the ground inductance and appropriate isolation between the supply lines of the two stages must be included. Only the availability of IC technologies with reproducible high-Q, low-tolerance passive components enables the realization of such highly frequency-selective amplifiers.

The LNA draws 9.5 mA from a 3 V supply. At this power dissipation, the input referred −1 dB compression point is measured at −24 dBm.

The measured input and output reflection is plotted in Fig. 5.48. The tuning voltage is set to 0 V. The excellent input match changes only negligibly with varying tuning voltage. The input matching shows a

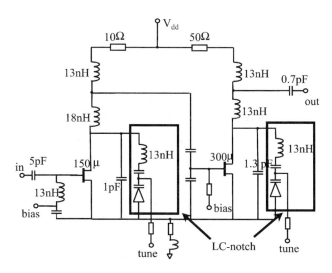

FIGURE 5.47 Schematic diagram of the selective frequency LNA at 2 GHz.

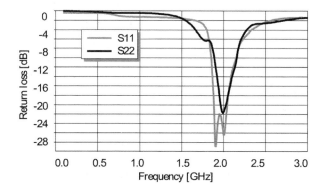

FIGURE 5.48 Measured input and output return loss of the 2 GHz selective LNA.

high-pass characteristic formed by the series C-L combination instead to the commonly used low-pass. So, the inductor can also act as a bias choke and the input matching can contribute to the suppression of lower frequency interferer. Moreover, the employed matching achieves better noise performance than the high-pass matching network.

The power gain vs. frequency for different notch tuning voltages is shown in Fig. 5.49. By varying the tuning voltage from 0.5 V to 1.5 V, the filter center frequency can be adjusted from 1.44 to 1.6 GHz. At all tuning voltages the unwanted signal is suppressed by at least 35 dB.

The temperature dependence of gain and noise figure was measured. The temperature coefficients of the gain and noise figure are −0.03 dB/°C and +0.008 dB/°C, respectively. The noise figure of the LNA at different temperatures is plotted in Fig. 5.50.

A chip photograph of the fabricated 1.6×1.0 mm^2 LNA is depicted in Fig. 5.51. More than 50% of the chip area is occupied by the numerous spiral inductors.

Future Trends

RF and microwave functions are increasingly often realized as integrated circuits (ICs) to reduce size and power consumption, enhance reproducibility, minimize costs, and enable mass production.

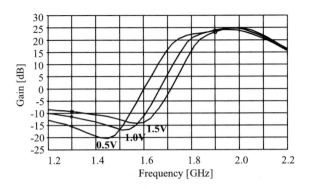

FIGURE 5.49   Selective amplifier gain vs. frequency for different notch filter control voltages.

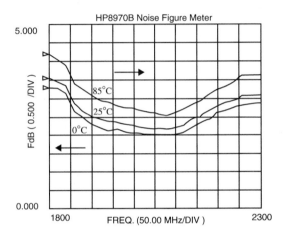

FIGURE 5.50   Amplifier noise figure at various temperatures.

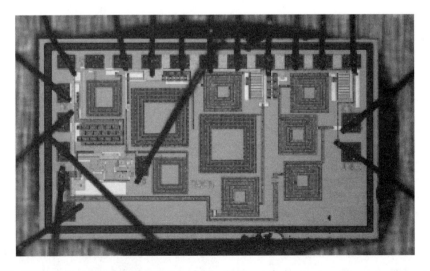

FIGURE 5.51   Chip photograph of the frequency selective LNA.

Design Approach

The classical noise optimization is based on linear methods and does not take power consumption and linearity requirements explicitly into account. Further, these methods offer only little guidance about how to select the active device dimensions. However, LNA circuit design practices are increasingly influenced by the improvements in the device models in terms of accuracy. Powerful optimization tools become available and eases the design procedure. However, a detailed understanding of the basic material will remain necessary for an efficient and robust LNA circuit design.

Device Models

The plurality of bias conditions applied to integrated circuits requires the flexibility of bias-dependent device models. State-of-the-art BJT models (such as Gummel-Poon) already work very well in RF simulations. More recently, sophisticated, semiempirical MOSFET models (such as BSIM3, MM9, or EKV) became suitable for RF simulations. Using accurate models, designs do not need to rely on sample scattering parameters of test devices and tolerance simulations can be implemented.

Circuit Environment

New RF design practices away from the 50 Ω impedance culture will affect the selection of the device size and operation point, but will leave the design procedure basically unchanged. The obstacles in the quest for higher integrated RF radios are the requirements on system noise figure, substrate crosstalk, and parasitic coupling. Trends to alleviate the unwanted coupling involve using fully differential circuit design, which in turn increases the power consumption.

IC Technologies

In recent years, the advances in device shrinking have made silicon devices (BJTs and more recently MOSFETs) become competitive with III-V semiconductors in terms of gain and minimum noise figure at a given power dissipation in the low GHz range.

The introduction of SiGe and SiC layers further enhance the cutoff frequencies and reduce power dissipation of silicon-based transistors. Furthermore, the use of thick (copper) metallization layers allow relatively low-loss passive components such as MIM capacitors and spiral inductors. Silicon-on-insulator (SOI) technologies will further cut substrate losses and parasitic capacitance and reduce bulk crosstalk.

With the scaling toward minimum gate length of below 0.25 μm, the use of CMOS has become a serious option in low-noise amplifier design. In fact, minimum noise figures of 0.5 dB at 2 GHz and cutoff frequencies of above 100 GHz for 0.12 μm devices[16] can easily compete with any other circuit technology. While intrinsic CMOS device F_{min} is becoming excellent for very short gate lengths, there remains the question of how closely amplifier noise figures can approach F_{min} in practice, particularly if there is a constraint on the allowable power consumption.

References

1. J. M. Rollett, Stability and power-gain invariance of linear two ports, *IEEE Trans. on Circuit Theory,* CT-9, 1, 29–32, March 1962, with corrections, CT-10, 1, 107, March 1963.
2. G. Gonzales, *Microwave Transistor Amplifiers Analysis and Design,* 2nd Edition, Prentice Hall, Englewood Cliffs, NJ, 1997.
3. G. Macciarella, et al., Design criteria for multistage microwave amplifiers with match requirements at input and output, *IEEE Trans. Microwave Theory and Techniques,* MTT-41, 1294–98, Aug. 1993.
4. G. D. Vendelin, A. M. Pavio, U. L. Rohde, *Microwave Circuit Design using Linear and Nonlinear Techniques,* 1st Edition, John Wiley & Sons, New York, 1990.
5. H. Hillbrand and P. H. Russer, An efficient method for computer aided noise analysis of linear amplifier networks, *IEEE Trans. on Circuit and Systems,* CAS-23, 4, 235–38, April 1976.
6. H. T. Friis, Noise figure for radio receivers, *Proc. of the IRE,* 419–422, July 1944.
7. H. A. Ainspan, et al., A 5.5 GHz low noise amplifier in SIGe BiCMOS, in *ESSCIRC98 Digest,* 80–83.

8. S. S. Taylor, A GaAs MESFET Schottky diode barrier height reference circuit, *IEEE Journal of Solid-State Circuits*, 32, 12, 2023–29, Dec. 1997.

9. J. J. Kucera, U. Lott, and W. Bächtold, A new antenna switching architecture for mobile handsets, *2000 IEEE Int'l Microwave Symposium Digest*, in press.

10. R. M. Fano, Theoretical limitations on the broad-band matching of arbitrary impedances, *Journal of the Franklin Institute*, 249, 57–83, Jan. 1960, and 139–155, Feb. 1960.

11. J. J. Kucera and W. Bächtold, A 1.9 GHz monolithic 1.1 dB noise figure low power LNA using a standard GaAs MESFET foundry process, *1998 Asia-Pacific Microwave Conference Digest*, 383–386.

12. J. J. Kucera and U. Lott, A 1.8 dB noise figure low DC power MMIC LNA for C-band, *1998 IEEE GaAs IC Symposium Digest*, 221–224.

13. Q. Huang, P. Orsatti and F. Piazza, Broadband, 0.25 μm CMOS LNAs with sub-2dB NF for GSM applications, *IEEE Custom Integrated Circuits Conference*, 67–70, 1998.

14. Ch. Biber, Microwave modeling and circuit design with sub-micron CMOS technologies, PhD thesis, Diss. ETH No. 12505, Zurich, 1998.

15. J. J. Kucera, Highly integrated RF transceivers, PhD thesis, Diss. ETH No. 13361, Zurich, 1999.

16. R. R. J. Vanoppen, et al., RF noise modeling of 0.25μm CMOS and low power LNAs, *1997 IEDM Technical Digest*, 317–320.

5.4 Microwave Mixer Design

Anthony M. Pavio

At the beginning of the 20th century, RF detectors were crude, consisting of a semiconductor crystal contacted by a fine wire ("whisker"), which had to be adjusted periodically so that the detector would keep functioning. With the advent of the triode, a significant improvement in receiver sensitivity was obtained by adding amplification in front of and after the detector. A real advance in performance came with the invention by Edwin Armstrong of the super regenerative receiver. Armstrong was also the first to use a vacuum tube as a frequency converter (mixer) to shift the frequency of an incoming signal to an intermediate frequency (IF), where it could be amplified and detected with good selectivity. The superheterodyne receiver, which is the major advance in receiver architecture to date, is still employed in virtually every receiving system.

The mixer, which can consist of any device capable of exhibiting nonlinear performance, is essentially a multiplier or a chopper. That is, if at least two signals are present, their product will be produced at the output of the mixer. This concept is illustrated in Fig. 5.52. The RF signal applied has a carrier frequency of w_s with modulation $M(t)$, and the local oscillator signal (LO or pump) applied has a pure sinusoidal frequency of w_p. From basic trigonometry we know that the product of two sinusoids produces a sum and difference frequency.

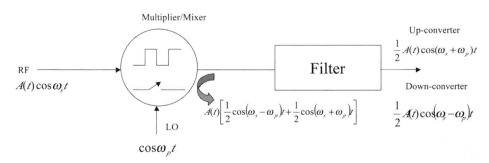

FIGURE 5.52   Ideal mixer model.

The voltage-current relationship for a diode can be described as an infinite power series, where V is the sum of both input signals and I is the total signal current. If the RF signal is substantially smaller than the LO signal and modulation is ignored, the frequency components of the signal are:

$$w_d = n w_p \pm w_s \tag{5.1}$$

As mentioned above, the desired component is usually the difference frequency ($|w_{p} + w_{s}|$ or $|f_{p} - f_{s}|$), but sometimes the sum frequency ($f_s + f_p$) is desired when building an up-converter, or a product related to a harmonic of the LO can be selected.

A mixer can also be analyzed as a switch that is commutated at a frequency equal to the pump frequency w_p. This is a good first-order approximation of the mixing process for a diode since it is driven from the low-resistance state (forward bias) to the high-resistance state (reverse bias) by a high-level LO signal.

The concept of the switching mixer model can also be applied to field-effect transistors used as voltage-controlled resistors. In this mode, the drain-to-source resistance can be changed from a few ohms to many thousands of ohms simply by changing the gate-to-source potential. At frequencies below 1 GHz, virtually no pump power is required to switch the FET, and since no DC drain bias is required, the resulting FET mixer is passive. However, as the operating frequency is raised above 1 GHz, passive FET mixers require LO drive powers comparable to diode or active FET designs.

Mixers can be divided into several classes: (1) single ended, (2) single balanced, or (3) double balanced. Depending on the application and fabrication constraints, one topology can exhibit advantages over the other types. The simplest topology (Fig. 5.53a) consists of a single diode and filter networks. Although there is no isolation inherent in the structure (balance), if the RF, LO, and IF frequencies are sufficiently separated, the filter (or diplexer) networks can provide the necessary isolation. In addition to simplicity, single diode mixers have several advantages over other configurations. Typically, the best conversion loss is possible with a single device, especially at frequencies where balun or transformer construction is difficult or impractical. Local oscillation requirements are also minimal since only a single diode is employed and DC biasing can easily be accomplished to reduce drive requirements. The disadvantages of the topology are: (1) sensitivity to terminations; (2) no spurious response suppression; (3) minimal tolerance to large signals; and (4) narrow bandwidth due to spacing between the RF filter and mixer diode.

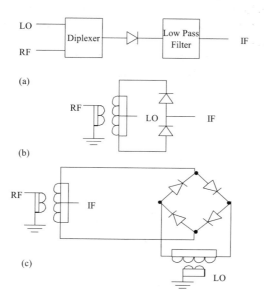

FIGURE 5.53 Typical mixer configurations. (a) Single ended; (b) single balanced; (c) double balanced.

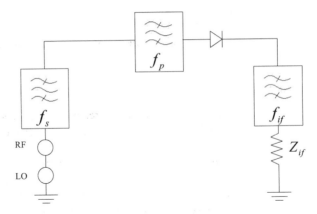

FIGURE 5.54   Filtering requirements for single-diode mixer.

The next topology commonly used is the single balanced structure shown in Fig. 5.53b. These structures tend to exhibit slightly higher conversion loss than that of a single-ended design, but since the RF signal is divided between two diodes, the signal power-handling ability is better. More LO power is required, but the structure does provide balance. The double-balanced mixer (Fig. 5.53c) exhibits the best large signal-handling capability, port-to-port isolation, and spurious rejection. Some high-level mixer designs can employ multiple-diode rings with several diodes per leg in order to achieve the ultimate in large-signal performance. Such designs can easily require hundreds of milliwatts of pump power.

Single-Diode Mixers

The single-diode mixer, although fondly remembered for its use as an AM "crystal" radio or radar detector during World War II, has become less popular due to demanding broadband and high dynamic range requirements encountered at frequencies below 30 GHz. However, there are still many applications at millimeter wave frequencies, as well as consumer applications in the microwave portion of the spectrum, which are adequately served by single-ended designs. The design of single-diode mixers can be approached in the same manner as multi-port network design. The multi-port network contains all mixing product frequencies regardless of whether they are ported to external terminations or terminated internally. With simple mixers, the network's main function is frequency component separation; impedance matching requirements are secondary (Fig. 5.54). Hence, in the simplest approach, the network must be capable of selecting the LO, RF, and IF frequencies (Fig. 5.55).

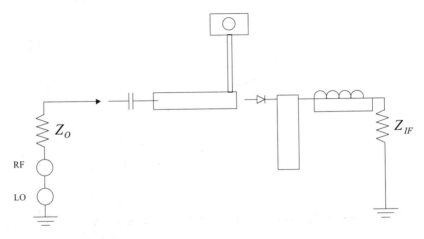

FIGURE 5.55   Typical single-ended mixer.

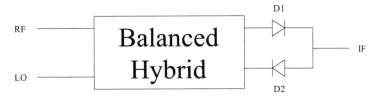

FIGURE 5.56 Single-balanced mixer topology.

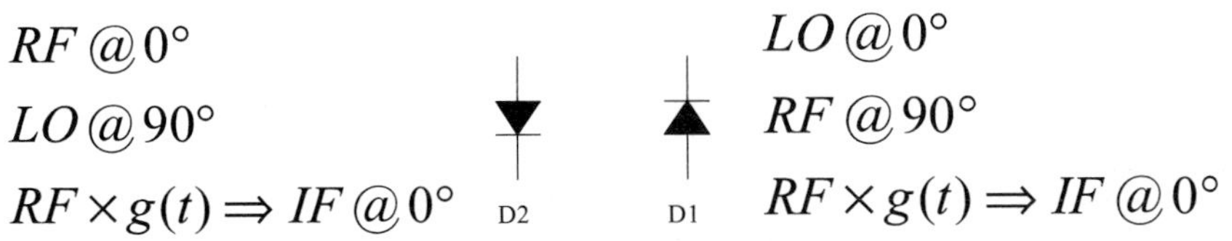

$$RF @ 0° \qquad\qquad LO @ 0°$$
$$LO @ 90° \qquad\qquad RF @ 90°$$
$$RF \times g(t) \Rightarrow IF @ 0° \quad_{D2} \qquad _{D1}\quad RF \times g(t) \Rightarrow IF @ 0°$$

FIGURE 5.57 Signal phase relationships in quadrature coupled hybrid single-balanced mixer.

However, before a network can be designed, the impedance presented to the network by the diode at various frequencies must be determined. Unfortunately, the diode is a nonlinear device; hence, determining its characteristics is more involved than determining an "unknown" impedance with a network analyzer. Since the diode impedance is time varying, it is not readily apparent that a stationary impedance can be found. Stationary impedance values for the RF, LO, and IF frequencies can be measured or determined if sufficient care in analysis or evaluation is taken.

Single-Balanced Mixers

Balanced mixers offer some unique advantages over single-ended designs such as LO noise suppression and rejection of some spurious products. The dynamic range can also be greater because the input RF signal is divided between several diodes, but this advantage is at the expense of increased pump power. Both the increase in complexity and conversion loss can be attributed to the hybrid or balun, and to the fact that perfect balance and lossless operation cannot be achieved.

There are essentially only two design approaches for single-balanced mixers; one employs a 180° hybrid, while the other employs some form of quadrature structure (Fig. 5.56). The numerous variations found in the industry are related to the transmission-line media employed and the ingenuity involved in the design of the hybrid structure. The most common designs for the microwave frequency range employ either a branch-line, Lange, or "rat-race" hybrid structure (Fig. 5.57). At frequencies below about 5 GHz, broadband transformers are very common, while at frequencies above 40 GHz, waveguide and MMIC structures become prevalent.

Double-Balanced Mixers

The most commonly used mixer today is the double-balanced mixer. It usually consists of four diodes and two baluns or hybrids, although a double-ring or double-star design requires eight diodes and three hybrids. The double-balanced mixer has better isolation and spurious performance than the single-balanced designs described previously, but usually requires greater amounts of LO drive power, are more difficult to assemble, and exhibit somewhat higher conversion loss. However, they are usually the mixer of choice because of their spurious performance and isolation characteristics.

A typical single-ring mixer with transformer hybrids is shown in Fig. 5.58. With this configuration the LO voltage is applied across the ring at terminals LO⁻ and LO⁺, and the RF voltage is applied across terminals RF⁻ and RF⁺. As can be seen, if the diodes are identical (matched), nodes RF⁻ and RF⁺ are virtual grounds; thus no LO voltage appears across the secondary of the RF transformer. Similarly, no

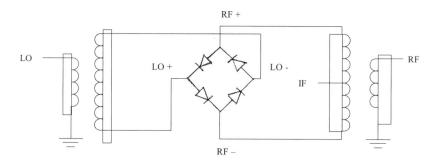

FIGURE 5.58 Transformer coupled double-balanced mixer.

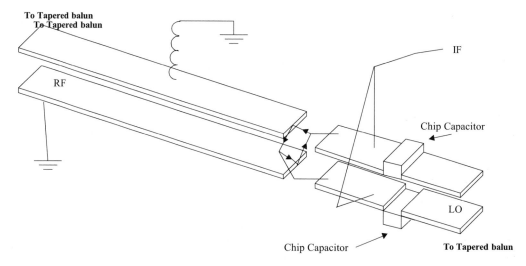

FIGURE 5.59 Double-balanced mixer center section.

RF voltage appears across the secondary of the LO balun. Because of the excellent diode matching that can be obtained with diode rings fabricated on a single chip, the L-to-R isolation of microwave mixers can be quite good, typically 30 to 40 dB.

Transmission-line structures which are naturally balanced, such as slotline and finline, can also be used as balanced feed in mixer design. However, all of the structures above, and the more complex transmission-line structures to follow, exhibit one major drawback compared to a transformer hybrid: There is no true RF center tap. As will be seen, this deficiency in transmission-line structures, extensively complicates the design of microwave-balanced mixers.

The lack of a balun center tap does indeed complicate the extraction of IF energy from the structure, but if the IF frequency is low, diplexing can be employed to ease performance degradation. This concept is illustrated in the following example of the center section of a double-balanced 2 to 12 GHz mixer (Fig. 5.59). It will be assumed that because of the soft-substrate transmission-line media and frequency range, a packaged diode ring with known impedances can be used. For Si diodes in this frequency range, the typical LO impedance range (magnitude) is on the order of 75, while the RF impedance is approximately 50. With these values in mind, microstrip-to-parallel plate transmission-line baluns can be fabricated on soft-substrate material.

As can be seen, both the RF and LO baluns terminate at the diode ring and provide the proper phase excitation. But since there is no center tap, the IF must be summed from the top and bottom of either balun. This summing is accomplished with bond wires that have high reactances at microwave frequencies but negligible inductances in the IF passband. Blocking capacitors form the second element in a high-pass filter, preventing the IF energy to be dissipated externally. An IF return path must also be provided at the terminals of the opposite balun. The top conductor side of the balun is grounded with a bond wire, providing a low-impedance path for the IF return and a sufficiently large impedance in shunt with the RF path. The ground-plane side of the balun provides a sufficiently low impedance for the IF return from the bottom side of the diode ring. The balun inductance and blocking capacitor also form a series resonant circuit shunting the IF output; therefore, this resonant frequency must be kept out of the IF passband.

The upper-frequency limit of mixers fabricated using tapered baluns and low parasitic diode packages, along with a lot of care during assembly, can be extended to 40 GHz. Improved "high-end" performance can be obtained by using beam-lead diodes. Although this design technique is very simple, there is little flexibility in obtaining an optimum port VSWR since the baluns are designed to match the magnitude of the diode impedance. The IF frequency response of using this approach is also limited, due to the lack of a balun center tap, to a frequency range below the RF and IF ports.

FET Mixer Theory

Interest in FET mixers has been very strong due to their excellent conversion gain and intermodulation characteristics. Numerous commercial products employ JFET mixers, but as the frequency of operation approaches 1 GHz, they begin to disappear. At these frequencies and above, the MESFET can easily accomplish the conversion functions that the JFET performs at low frequencies. However, the performance of active FET mixers reported to date by numerous authors has been somewhat disappointing. In short, they have not lived up to expectations, especially concerning noise-figure performance, conversion gain, and circuit-to-circuit repeatability. However, they are simple and low cost, so these sins can be forgiven.

Recently, growing interest is GaAs monolithic circuits is again beginning to heighten interest in active MESFET mixers. This is indeed fortunate, since properly designed FET mixers offer distinct advantages over their passive counterparts. This is especially true in the case of the dual-gate FET mixer; since the additional port allows for some inherent LO-to-RF isolation, it can at times replace single balanced passive approaches. The possibility of conversion gain rather than loss is also an advantage, since the added gain may eliminate the need for excess amplification, thus reducing system complexity.

Unfortunately, there are some drawbacks when designing active mixers. With diode mixers, the design engineer can make excellent first-order performance approximations with linear analysis; also, there is the practical reality that a diode always mixes reasonably well almost independent of the circuit. In active mixer design, these two conditions do not hold. Simulating performance, especially with a dual-gate device, requires some form of nonlinear analysis tool if any circuit information other than small-signal impedance is desired. An analysis of the noise performance is even more difficult.

As we have learned, the dominant nonlinearity of the FET is its transconductance, which is typically (especially with JFETs) a squarelaw function. Hence it makes a very efficient multiplier.

The small-signal circuit [1] shown in Fig. 5.60 denotes the principal elements of the FET that must be considered in the model. The parasitic resistances R_g, R_d, and R_s are small compared to R_{ds} and can be considered constant, but they are important in determining the noise performance of the circuit. The mixing products produced by parametric pumping of the capacitances C_{gs}, C_{dg}, and C_{ds} are typically small and add only second-order effects to the total circuit performance. Time-averaged values of these capacitances can be used in circuit simulation with good results.

This leaves the FET transconductance g_m, which exhibits an extremely strong nonlinear dependence as a function of gate bias. The greatest change is transconductance occurs near pinch off, with the most linear change with respect to gate voltage occurring in the center of the bias range. As the FET is biased toward I_{dss}, the transconductance function again becomes nonlinear. It is in these most nonlinear regions that the FET is most efficient as a mixer.

If we now introduce a second signal, V_c, such that it is substantially smaller than the pump voltage, across the gate-to-source capacitance C_{gs}, the nonlinear action of the transconductance will cause mixing action within the FET producing frequencies $|nw_p \pm w_1|$, where n can be any positive or negative integer. Any practical analysis must include mixing products at both the gate and drain terminal, and at a minimum, allow frequency components in the signal, image, LO, and IF to exist.

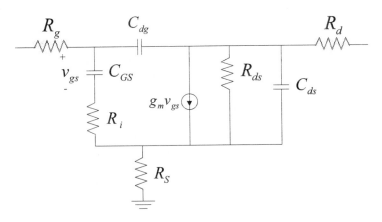

FIGURE 5.60   Typical MESFET model.

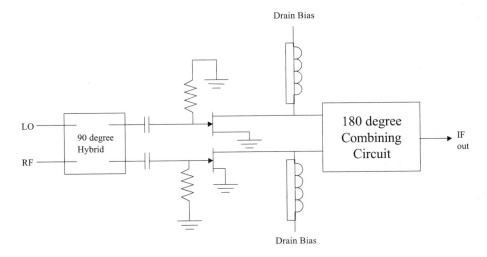

FIGURE 5.61   Typical FET single-balanced mixer.

Double-balanced FET mixers can also be designed using transformer hybrids [1]. Fig. 5.61 shows a typical balanced FET mixer, which can be designed to operate from VHF to SHF. An additional balun is again required because of the phase relationships of the IF signal. This structure is completely balanced and exhibits spurious rejection performance, similar to diode mixers constructed for the same frequency range. However, the intermodulation and noise-figure performance of such structures is superior to those of simple four-diode designs. For example, third-order intercept points in excess of 33 dBm, with associated gains of 6 dB, are common in such structures. High-level multiple-diode ring mixers, which would require substantially more LO power, would exhibit comparable intermodualtion characteristics, but would never exhibit any gain.

There are a variety of interesting mixer topologies in widespread use that perform vital system functions that cannot be simply classified as balanced mixers. Probably the most popular configuration is the image rejection or single-sideband mixer. However, a variety of subharmonically pumped and self-oscillating mixers are in limited use [1].

Reference

1. G. D. Vendelin, A. M. Pavio, and U. L. Rohde, The Design of Amplifiers Mixers and Oscillators Using the S-Parameter Method, John Wiley and Son, New York, 1990.

5.5 Modulation and Demodulation Circuitry

Charles Nelson

Some Fundamentals: Why Modulate?

Because this chapter uses a building block approach, it may seem to be a long succession of setting up straw men and demolishing them. To some extent, this imitates the development of radio and TV, which has been going on for most of the century just ended. A large number of concepts were developed as the technology advanced; each advance made new demands upon the hardware. At first, many of these advances were made by enthusiastic amateurs who had no fear of failure and viewed radio communication the way Hillary viewed Everest — something to be surmounted "because it was there." Since about World War II, there have been increasing numbers of engineers who understood these principles and could propose problem solutions that might have worked the first or second time they were tried. The author fondly hopes this book will help to grow a new cadre of problem solvers for the 21st century.

What probably first motivated the inventors of radio was the need for ships at sea to make distress calls. It may be interesting to note that the signal to be transmitted was a digital kind of thing called Morse Code. Later, the medium became able to transmit equally crucial analog signals, such as a soldier warning, "Watch out!! The woods to your left are full of the abominable enemy!" Eventually, during a period without widespread military conflict, radio became an entertainment medium, with music, comedy, and news, all made possible by businessmen who were convinced you could be persuaded, by a live voice, to buy soap, and later, detergents, cars, cereals not needing cooking, and so on. The essential low and high frequency content of the signal to be transmitted has been very productive for problems to be solved by radio engineers.

The man on radio, urging you to buy a "pre-owned" Cadillac, puts out most of his sound energy below 1000 Hz. A microphone observes pressure fluctuations corresponding to the sound and generates a corresponding voltage. Knowing that all radio broadcasting is done by feeding a voltage to an antenna, the beginning engineer might be tempted to try sending out the microphone signal directly. A big problem with directly broadcasting such a signal is that an antenna miles long would be required to transmit it efficiently. However, if the frequency of the signal is shifted a good deal higher, effective antennas become much shorter and more feasible to fabricate. This upward translation of the original message spectrum is perhaps the most crucial part of what we have come to call "modulation." However, the necessities of retrieving the original message from the modulated signal may dictate other inclusions in the broadcast signal, such as a small or large voltage at the center, or "carrier" frequency of the modulated signal. The need for a carrier signal is dictated by what scheme is used to transmit the modulated signal, which determines important facts of how the signal can be demodulated.

More perspective on the general problem of modulation is often available by looking at the general form of a modulated signal,

$$f(t) = A(t)\cos\theta(t).$$

If the process of modulation causes the multiplier A(t) out front to vary, it is considered to be some type of "amplitude" modulation. If one is causing the angle to vary, it is said to be "angle" modulation, but there are two basic types of angle modulation. We may write

$$\theta(t) = \omega_c t + \phi(t).$$

If then our modulation process works directly upon $\omega_c = 2\pi f_c$, we say we have performed "frequency" modulation. If, instead, we directly vary the phase factor $\phi(t)$, we say we have performed "phase" modulation. The two kinds of angle modulation are closely related, so that we may do one kind of operation to get the other result, by proper preprocessing of the modulation signal. Specifically, if we put the modulating signal through an integrating circuit before we feed it to a phase modulator, we come out with frequency modulation. This is, in fact, often done. The dual of this operation is possible but is seldom done in practice. Thus, if the modulating signal is fed through a differentiating circuit before it is fed to a frequency modulator, the result will be phase modulation. However, this process offers no advantages to motivate such efforts.

How to Shift Frequency

Our technique, especially in this chapter, will be to make our proofs as simple as possible; specifically, if trigonometry proves our point, it will be used instead of the convolution theorem of circuit theory. Yet, use of some of the aspects of convolution theory can be enormously enlightening to those who understand. Sometimes, as it will in this first proof, it may also indicate the kind of circuit that will accomplish the task. We will also take liberties with the form of our modulating signal. Sometimes we can be very general, in which case it may be identified as a function m(t). At other times, it may greatly simplify things if we write it very explicitly as a sinusoidal function of time

$$m(t) = \cos \omega_m t.$$

Sometimes, in the theory, this latter option is called "tone modulation," because, if one listened to the modulating signal through a loudspeaker, it could certainly be heard to have a very well-defined "tone" or pitch. We might justify ourselves by saying that theory certainly allows this, because any particular signal we must deal with could, according the theories of Fourier, be represented as a collection, perhaps infinite, of cosine waves of various phases. We might then assess the maximum capabilities of a communication system by choosing the highest value that the modulating signal might have. In AM radio, the highest modulating frequency is typically about $f_m = 5000$ Hz. For FM radio, the highest modulation frequency might be $f_m = 19$ kHz, the frequency of the so-called FM stereo "pilot tone."

In principle, the shifting of a frequency is very simple. This is fairly obvious to those understanding convolution. One theorem of system theory says that multiplication of time functions leads to convolution of the spectra. Let us just multiply the modulating signal by a so-called "carrier" signal. One is allowed to have the mental picture of the carrier signal "carrying" the modulating signal, in the same way that homing pigeons have been used in past wars to carry a light packet containing a message from behind enemy lines to the pigeon's home in friendly territory. So, electronically, for "tone modulation," we need only to accomplish the product

$$\phi(t) = A \cos \omega_m t \cos \omega_c t.$$

Now, we may enjoy the consequences of our assumption of tone modulation by employing trigonometric identities for the sum or difference of two angles:

$$\cos(A+B) = \cos A \cos B - \sin A \sin B \quad \text{and} \quad \cos(A-B) = \cos A \cos B + \sin A \sin B$$

If we add these two expressions and divide by two, we get the identity we need:

$$\cos A \cos B = 0.5 \left[\cos(A+B) + \cos(A-B) \right].$$

Stated in words, we might say we got "sum and difference frequencies," but neither of the original frequencies. Let's be just a little more specific and say we started with $f_m = 5000$ Hz and $f_c = 1$ MHz, as

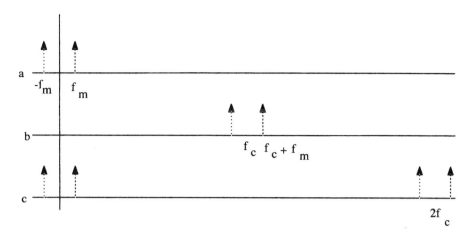

FIGURE 5.62 Unmodulated, modulated, and synchronously demodulated signal spectra. a. Spectrum of tone-modulating signal. b. Spectrum (positive part only) of double sideband suppressed carrier signal. c. Spectrum of synchronously detected DSB–SC signal (except for part near $-2f_c$).

would happen if a radio station whose assigned carrier frequency was 1 MHz were simply transmitting a single tone at 5000 Hz. In "real life," this would not be done very often, but the example serves well to illustrate some definitions and principles. The consequence of the mathematical multiplication is that the new signal has two new frequencies at 995 kHz and 1005 kHz. Let's now just add one modulating tone at 3333 Hz. We would have added two frequencies at 9666.667 kHz and 1003.333 kHz. However, if this multiplication was done purely, *there is no carrier frequency term present.* For this reason, we say we have done a type of "suppressed carrier" modulation. Also, furthermore, we have two *new* frequencies for *each* modulating frequency. We define all of those frequencies above the carrier as the "upper sideband" and all the frequencies below the carrier as the "lower sideband." The whole process we have done here is named "double sideband suppressed carrier" modulation, often known by its initials DSB–SC. Communication theory would tell us that the signal spectrum, before and after modulation with a single tone at a frequency f_m, would appear as in Fig. 5.62. Please note that the theory predicts equal positive and negative frequency components. There is no deep philosophical significance to negative frequencies. They simply make the theory symmetrical and a bit more intuitive.

Analog Multipliers, or "Mixers"

First, there is an unfortunate quirk of terminology; the circuit that multiplies signals together is in communication theory usually called a "mixer." What is unfortunate is that the engineer or technician who produces sound recordings is very apt to feed the outputs of many microphones into potentiometers, the outputs of which are sent in varying amounts to the output of a piece of gear, and *that* component is called a "mixer." Thus, the communication engineer's mixer multiplies and the other adds. Luckily, it will usually be obvious which device one is speaking of.

There are available a number of chips (integrated circuits) designed to serve as analog multipliers. The principle is surprisingly simple, although the chip designers have added circuitry which no doubt optimizes the operation and perhaps makes external factors less influential. The reader might remember that the transconductance g_m for a bipolar transistor is proportional to the collector current; its output is proportional to the g_m *and* the input voltage, so in principle one can replace an emitter resistor with the first transistor, which then controls the collector current of the second transistor. If one seeks to fabricate such a circuit out of discrete transistors, one would do well to expect a need to tweak operating conditions considerably before some approximation of analog multiplication occurs. Recommendation: buy the chip. Best satisfaction will probably occur with a "four-quadrant multiplier." The alternative is a "two-quadrant multiplier," which might embarrass one by being easily driven into cut-off.

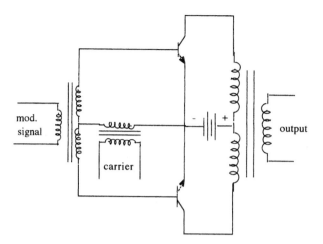

FIGURE 5.63   Balanced modulator.

Another effective analog multiplier is alleged to be the dual-gate FET. The width of the channel in which current flows depends upon the voltage on each of two gates which are insulated from each other. Hence, if different voltages are connected to the two gates, the current that flows is the product of the two voltages. Both devices we have discussed so far have the advantage of having some amplification, so the desired resulting signal has a healthy amplitude. A possible disadvantage may be that spurious signals one does *not* need may also have strong amplitudes.

Actually, the process of multiplication may be the byproduct of any distorting amplifier. One can show this by expressing the output of a distorting amplifier as a Taylor series representing output in terms of input. In principle, such an output would be written

$$V_o = a_0 + a_1\left(v_1 + v_2\right) + a_3\left(v_1 + v_2\right)^2 + \text{ smaller terms.}$$

One can expand $(v_1 + v_2)^2$ as $v_1^2 + 2v_1v_2 + v_2^2$, so this term yields second harmonic terms of each input plus the product of inputs one was seeking. However, the term $a_1(v_1 + v_2)$ also yielded each input, so the carrier here would not be suppressed. If it is fondly desired to suppress the carrier, one must resort to some sort of "balanced modulator." An "active" (meaning there is amplification provided) form of a balanced modulator may be seen in Fig. 5.63; failure to bias the bases of the transistors should assure that the voltage squared term is large.

One will also find purely passive mixers with diodes connected in the shape of a baseball diamond with one signal fed between first and third base, the other from second to home plate. Such an arrangement has the great advantage of not requiring a power supply; the disadvantage is that the amplitude of the sum or difference frequency may be small.

Synchronous Detection of Suppressed Carrier Signals

At this point, the reader without experience in radio may be appreciating the mathematical tricks but wondering, if one can accomplish this multiplication, can it be broadcast and the original signal retrieved by a receiver? A straightforward answer might be that multiplying the received signal by another carrier frequency signal such as cos $\omega_c t$ will shift the signal back exactly to where it started and also up to a center frequency of twice the original carrier. This is depicted in part c of Fig. 5.62. The name of this process is "synchronous detection." (In the days when it was apparently felt that communications enjoyed a touch of class if one used words having Greek roots, they called it "homodyne detection." If the reader reads a wide variety of journals, he/she may still encounter the word.) The good/bad news about synchronous detection is that the signal being used in the detector multiplication must have the *exact frequency and phase* of the original carrier, and such a signal is not easy to supply. One method is to send a "pilot" carrier, which is a small amount of the correct signal. The pilot tone is amplified until it is strong enough to accomplish the detection.

Suppose the pilot signal reaches high enough amplitude but is phase-shifted an amount θ with respect to the original carrier. We would then in our synchronous detector be performing the multiplication:

$$m(t)\cos\omega_c t\cos(\omega_c t+\theta).$$

To understand what we get, let us expand the second cosine using the identity for the sum of two angles,

$$\cos(\omega_c t+\theta)=\cos\omega_c t\cos\theta-\sin\omega_c t\sin\theta.$$

Hence, the output of the synchronous detector may be written as

$$m(t)\cos^2\omega_c t\cos\theta-m(t)\cos\omega_c t\sin\omega_c t\sin\theta=$$
$$(0.5)\big[m(t)\cos\theta(1-\cos2\omega_c t)-m(t)\sin\theta\sin2\omega_c t\big].$$

The latter two terms can be eliminated using a low-pass filter, and one is left with the original modulating signal, m(t), attenuated proportionally to the factor cos θ, so major attenuation does not appear until the phase shift approaches 90°, when the signal would vanish completely. Even this is not totally bad news, as it opens up a new technique called "quadrature amplitude modulation."

The principle of QAM, as it is abbreviated, is that entirely different modulating signals are fed to carrier signals that are 90° out of phase; we could call the carrier signals cos ω_ct and sin ω_ct. The two modulating signals stay perfectly separated if there is no phase shift to the carrier signals fed to the synchronous detectors. The color signals in a color TV system are QAM'ed onto a 3.58 MHz subcarrier to be combined with the black-and-white signals, after they have been demodulated using a carrier generated in synchronism with the "color burst" (several periods of a 3.58 MHz signal), which is cleverly "piggy-backed" onto all the other signals required for driving and synchronizing a color TV receiver.

Single Sideband Suppressed Carrier

The alert engineering student may have heard the words "single sideband" and be led to wonder if we are proposing sending one more sideband than necessary. Of course it is true, and SSB–SC, as it is abbreviated, is the method of choice for "hams," the amateur radio enthusiasts who love to see night fall, when their low wattage signals can bounce between the earth and a layer of ionized atmospheric gasses 100 or so miles up until they have reached halfway around the world. It turns out that a little phase shift is not a really drastic flaw for voice communications, so the "ham" just adjusts the variable frequency oscillator being used to synchronously demodulate incoming signals until the whistles and squeals become coherent, and then he/she listens.

How can one produce single sideband? For many years it was pretty naïve to say, "Well, let's just filter one sideband out!" This would have been very naïve because, of course, one does not have textbook filters with perfectly sharp cut-offs. Recently, however, technology has apparently provided rather good "crystal lattice filters" which are able fairly cleanly to filter the extra sideband. In general,

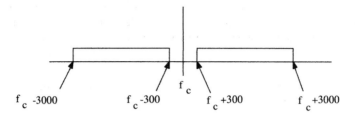

FIGURE 5.64 Double sideband spectrum for modulating signal 300–3000 Hz.

though, the single sideband problem is simplified if the modulating signal does not go to really deep low frequencies; a microphone that does not put out much below 300 Hz might have advantages, as it would leave a transition region of 600 Hz between upper and lower sidebands in which the sideband filter could have its amplitude response "roll off" without letting through much of the sideband to be discarded. Observe Fig. 5.64, showing both sidebands for a baseband signal extending only from 300 Hz to 3.0 kHz.

Another method of producing single sideband, called the "phase-shift method," is suggested if one looks at the mathematical form of just one of the sidebands resulting from tone modulation. Let us just look at a lower sideband. The mathematical form would be

$$v(t) = A\cos(\omega_c - \omega_m)t = A\cos\omega_c t\cos\omega_m t + A\sin\omega_c t\sin\omega_m t$$

Mathematically, one needs to perform DSB–SC with the original carrier and modulating signals (the cosine terms) and also with the two signals each phase shifted 90°; the resulting two signals are then added to obtain the lower sideband. Obtaining a 90° phase shift is not difficult with the carrier, of which there is only one, but we must be prepared to handle a band of modulating signals, and it is not an elementary task to build a circuit that will produce 90° phase shifts over a range of frequency. However, a reasonable job will be done by the circuit of Fig. 5.65 when the frequency range is limited (e.g., from 300 to 3000 Hz). Note that one does *not* modulate directly with the original modulation signal, but that the network uses each input frequency to generate two signals which are attenuated equal amounts and 90° away from each other. These voltages would be designated in the drawing as V_{xz} and V_{yz}. In calculating such voltages, the reader should note that there are two voltage dividers

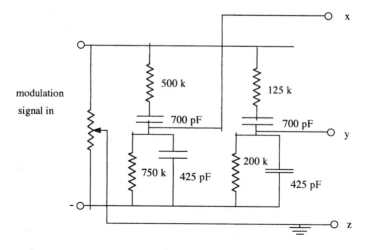

FIGURE 5.65 Audio network for single sideband modulator.

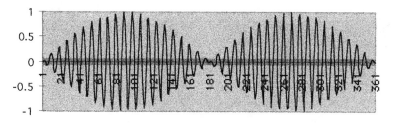

FIGURE 5.66   Double sideband suppressed carrier signal.

connected across the modulating voltage, determining V_x and V_y, and that from both of these voltages is subtracted the voltage from the center-tap to the bottom of the potentiometer. Note also that the resistance of the potentiometer is not relevant as long as it does not load down the source of modulating voltage, and that a good result has been found if the setting of the potentiometer is for 0.224 of the input voltage.

Amplitude Modulation as Double Sideband with Carrier

The budding engineer must understand that synchronous detectors are more expensive than many people can afford, and that a less expensive detection method is needed. What fills this bill much of the time is called the "envelope detector." Let us examine some waveforms, first for DSB–SC and then for a signal having a large carrier component. Figure 5.66 shows a waveform in which not very different carrier and modulating frequencies were chosen so that a spreadsheet plot would show a few details.

An ideal circuit we call an envelope detector would follow the topmost excursion of the waveform sketched here. Now, the original modulating signal was a sine wave, but the topmost excursion would be a *rectified* sinusoid, thus containing large amounts of harmonic distortion. How can one get a waveform that will be detected without distortion by an envelope detector? What was plotted was 1.0 cos ω_{ct} cos $\omega_m t$. We suspect we must add some amount of carrier B cos $\omega_c t$. The sum will be

$$\phi_{AM}\left(t\right) = B\cos\omega_c t + 1.0\cos\omega_{ct}\,t\cos\omega_m t = \cos\omega_c t\left[B + 1.0\cos\omega_m t\right].$$

This result is what is commonly called "amplitude modulation." Perhaps the most useful way of writing the time function for an amplitude modulation signal having tone modulation at a frequency f_m follows:

$$\phi_{AM}\left(t\right) = A\cos\omega_c t\left[1 + a\cos\omega_m t\right].$$

In this expression, we can say that A is the peak amplitude of the carrier signal that would be present if there were no modulation. The total expression inside the [ ] brackets can be called the "envelope" and the factor "a" can be called the "index of modulation." As we have written it, if the index of modulation were >1, the envelope would attempt to go negative; this would make it necessary, for distortion-free detection, to use synchronous detection. "a" is often expressed as a percentage, and when the index of modulation is less than 100%, it is possible to use the simplest of detectors, the envelope detector. We will look at the envelope detector in more detail a bit later.

Modulation Efficiency

It is good news that sending a carrier along with two sidebands makes inexpensive detection using an envelope detector possible. The accompanying bad news is that the presence of carrier does not contribute *at all* to useful signal output; the presence of a carrier only leads after detection to DC, which may be filtered out at the earliest opportunity. Sometimes, as in video, the DC is needed to set the brightness level, in which case DC may need to be added back in at an appropriate level.

To express the effectiveness of a communication system in establishing an output signal-to-noise ratio, it is necessary to define a "modulation efficiency," which, in words, is simply the fraction of output power that is put into sidebands. It is easily figured if the modulation is simply one or two purely sinusoidal tones; for real-life modulation signals, one may have to express it in quantities that are less easy to visualize.

For tone modulation, we can calculate modulation efficiency by simply evaluating the carrier power and the power of all sidebands. For tone modulation, we can write:

$$\phi_{AM}(t) = A\cos\omega_c t\left[1 + a\cos\omega_m t\right] =$$
$$A\cos\omega_c t + (aA)/2\left[\cos(\omega_c + \omega_m)t + \cos(\omega_c - \omega_m t)\right].$$

Now, we have all sinusoids, the carrier, and two sidebands of equal amplitudes, so we can write the average power in terms of peak amplitudes as:

$$P = 0.5\left[A^2 + 2 \times (aA/2)^2\right] = 0.5A^2\left[1 + a^2/2\right].$$

Then modulation efficiency is the ratio of sideband power to total power, for modulation by a single tone with modulation index "a," is:

$$\eta = \frac{(aA/2)^2}{0.5A^2(1 + a^2/2)} = \frac{a^2}{2 + a^2}.$$

Of course, most practical modulation signals are not so simple as sinusoids. It may be necessary to state how close one is to overmodulating, which is to say, how close to negative modulating signals come to driving the envelope negative. Besides this, what is valuable is a quantity we shall just call "m," which is the ratio of average power to peak power for the modulation function. For some familiar waveforms, if the modulation is sinusoidal, m = 1/2. If modulation were a symmetrical square wave, m = 1.0; any kind of symmetrical triangle wave has m = 1/3. In terms of m, the modulation efficiency is

$$\eta = \frac{ma^2}{1 + ma^2}$$

The Envelope Detector

Much of the detection of modulated signals, whether the signals began life as AM or FM broadcast signals or the sound or the video of TV, is done using envelope detectors. Figure 5.68 shows the basic circuit configuration.

The input signal is of course as shown in Fig. 5.67. It is assumed that the forward resistance of the diode is 100 ohms or less. Thus, the capacitor is small enough that it gets charged up to the peak values of the high frequency signal, but then when input drops from the peak, the diode is reverse-biased so the capacitor can only discharge through R. This discharge voltage is of course given by

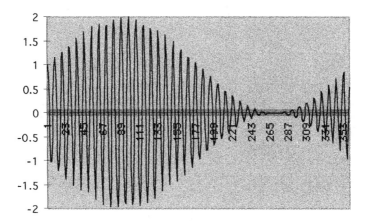

FIGURE 5.67 Amplitude-modulated signal.

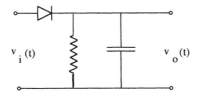

FIGURE 5.68 Simple envelope detector schematic.

$$V(0)\exp(-t/RC).$$

Now the problem in AM detection is that we must have the minimum rate of decay of the voltage be at least the maximum decay of the envelope of the modulated wave. We might write the envelope as a function of time:

$$E(t) = A(1 + a\cos\omega_{mt}t),$$

where A is the amplitude of the carrier before modulation and "a" is the index of modulation, which must be less than one for accurate results with the envelope detector. Then, when we differentiate, we get

$$\frac{dE}{dt} = -\omega_m A a \sin(\omega_m t).$$

We want this *magnitude* to be less than or equal to the maximum magnitude of the rate of decay of a discharging capacitor, which is E(0)/RC. For what is written as E(0), we will write the instantaneous value of the envelope, and the expression becomes

$$A(1 + a\cos\omega_{mt}t) \geq RC(\omega_m a A \sin(\omega_m t)).$$

The As cancel, and we have

$$RC \le \frac{1 + a\cos(\omega_m t)}{\omega_m a \sin(\omega_m t)};$$

our major difficulty occurs when the right-hand side has its minimum value.

If we differentiate with respect to $\omega_m t$, we get

$$\frac{\omega_m a \sin\left(\omega_m t \times \left(-\omega_m a \sin(\omega_m t) - \left(1 + a\cos(\omega_m t)\right)(\omega_m)^2 a \cos\omega_m t\right)\right)}{\left(\omega_m a \sin(\omega_m t)\right)^2}.$$

We set the numerator equal to zero to find its maximum. We find we have

$$-(\omega_m a)^2 \left[\sin^2(\omega_m t) + \cos^2(\omega_m t)\right] - a(\omega_m)^2 \cos\omega_m t$$

$$= -(\omega_m a)^2 - a(\omega_m)^2 \cos\omega_m t = 0.$$

Hence, the maximum occurs when $\cos\omega_m t = -a$, and of course by identity, at that time, $\sin\omega_m t = \sqrt{1-a^2}$. Inserting these results into our inequality for the time constant RC, we have

$$RC \le \frac{1-a^2}{\omega_m a\sqrt{1-a^2}} = \frac{\sqrt{1-a^2}}{\omega_m a}.$$

Example 5.1

Suppose we say 2000 Hz is the main problem in our modulation scheme, our modulation index is 0.5, and we choose R = 10k to make it large compared to the diode forward resistance, but not *too* large. What should be the capacitor C?

Solution   We use the equality now and get

$$C = \frac{\sqrt{1 - 0.5^2}}{0.5 \times 4000\pi \times 10,000} = 13.8 \text{ nF}.$$

Envelope Detection of SSB Using Injected Carrier

Single sideband, it might be said, is a very forgiving medium. Suppose that one were attempting synchronous detection using a carrier that was off by a Hertz or so, compared to the original carrier. Because synchronous detection works by producing sum and difference frequencies, 1 Hz error in carrier frequency would produce 1 Hz error in the detected frequency. Because SSB is mainly used for speech, it would be challenging indeed to find anything wrong with the reception of a voice one has only ever heard over a static-ridden channel. Similar results would also be felt in the following, where we add a carrier to the sideband and find that we have AM, albeit with a small amount of harmonic distortion.

Example 5.2

Starting with just an upper sideband B cos(ω_c + ω_m)t, let us add a carrier term A cos ω_c t, manipulate the total, and prove that we have an envelope to detect. First we expand the sideband term as

$$\Phi_{SSB}\left(t\right) = B\left[\cos\omega_c t\cos\omega_m t - \sin\omega_c t\sin\omega_m t\right].$$

Adding the carrier term A cos ω_c t and combining like terms, we have

$$\phi\left(t\right) = \cos\omega_c t\left(A + B\cos\omega_m t\right) - B\sin\omega_c t\sin\omega_m t.$$

In the first circuits class, we see that if we want to write a function of one frequency in the form E(t) cos (ω_c t + phase angle), the amplitude of the multiplier E is the square root of the squares of the coefficients of cos ω_ct and sin ω_ct. Thus,

$$E\left(t\right) = \sqrt{\left(A + B\cos\omega_m t\right)^2 + \left(B\sin\omega_m t\right)^2}$$
$$= \sqrt{A^2 + 2AB\cos\omega_m t + B^2\left(\cos^2\omega_m t + \sin^2\omega_m t\right)}.$$

Now, of course, the coefficient of B^2 is unity for all values of ω_mt. We find that best performance occurs if B ≪ A. Then we would have our expression for the envelope (and thus it is detectable using an envelope detector):

$$E\left(t\right) = \sqrt{A^2 + B^2 + 2AB\cos\omega_m t} = \sqrt{A^2 + B^2}\sqrt{1 + \frac{2AB}{A^2 + B^2}\cos\omega_m t}.$$

Our condition that B ≪ A allows us to say the coefficient of cos ω_mt is really small compared to unity. We use the binomial theorem to approximate the second square root: $(1 + x)^n \approx 1 + x/2 + (1/2)^2 (-1/2)x^2$ when x ≪ 1. Using our approximation, x ≈ (2B/A) cos ω_mt. In our expansion, the x term is the modulation term we were seeking, the x^2 term contributes second harmonic distortion. Using the various approximations, and stopping after we find the second harmonic (other harmonics *will* be present, of course, but in decreasing amplitudes), we have

$$\text{Detected } f\left(t\right) = B\cos\omega_m t - \left(1/2\right)\left(B^2/A\right)\cos^2\left(\omega_m t\right).$$

When we use trig identities to get the second harmonic, we get another factor of one half; the ratio of detected second harmonic to fundamental is thus (1/4)(B/A). Thus, for example, if B is just 10% of A, second harmonic is only 2.5% of fundamental.

Direct vs. Indirect Means of Generating FM

Let us first remind ourselves of basics regarding FM. We can write the time function in its simplest form as

$$\phi_{FM}\left(t\right) = A\cos\left(\omega_c t + \beta\sin\omega_m t\right).$$

Now, the alert reader might be saying, "Hold on! That looks a lot like phase modulation. If $\beta = 0$, the phase would increase linearly in time, as an unmodulated signal, but gets advanced or retarded a maximum of β." One needs to remember the definition of instantaneous frequency, which is

$$f_i = \frac{1}{2\pi}\frac{d}{dt}\left(\omega_c t + \beta\sin\omega_m t\right) = \frac{1}{2\pi}\left(2\pi f_c + \beta 2\pi f_m \cos\omega_m t\right) = f_c + \beta f_m \cos\omega_m t.$$

Thus, we can say that instantaneous frequency departs from the carrier frequency by a maximum amount βf_m, which is the so-called "frequency deviation." This has been specified as a maximum of 75 kHz for commercial FM radio but 25 kHz for the sound of TV signals.

Now, certainly, the concept of directly generating FM has an intellectual appeal to it. The problems of direct FM are mainly practical; if the very means of putting information onto a high frequency "carrier" is in varying the frequency, it perhaps stands to reason the center value of the operating frequency will not be well nailed down. Direct FM could be accomplished as in Fig. 5.69(a), but Murphy's law would be very dominant and one might expect center frequency to drift continually in one direction all morning and the other way all afternoon, or the like. This system is sometimes stabilized by having an FM detector called a discriminator tuned to the desired center frequency, so that output would be positive if frequency got high and negative for frequency low. Thus, instantaneous output could be used as an error voltage with a long time constant to push the intended center frequency toward the center, whether the average value is above or below.

The best known method of indirect FM gives credit to the man who, more than any other, saw the possibilities of FM and that its apparent defects could be exploited for superior performance, Edwin Armstrong. He started with a crystal-stabilized oscillator around 100 kHz, from which he obtained also a 90° phase-shifted version. A block diagram of just this early part of the Armstrong modulator is shown in Fig. 5.69(a).

The modulating signal is passed through an integrator before it goes into an analog multiplier, to which is also fed the *phase-shifted* version of the crystal-stabilized signal. Thus, we feed $\cos\omega_c t$ and $\sin\omega_c t \sin\omega_m t$ into a summing amplifier. The phasor diagram shows the two signals with $\cos\omega_c t$ as the reference. There is a small phase shift given by $\tan^{-1}(\beta\sin\omega_m t)$ where β here gives the maximum amount of phase shift as a function of time. To see how good a job we have done, we need to expand $\tan^{-1}(x)$ in a Taylor series. We find that

$$\tan^{-1}(x) \approx x - (x)^3\big/3 + (x)^5\big/5.$$

We see that we have a term proportional to the modulating signal (x) and others that must represent odd-order harmonic distortion, if one accounts for the fact that we have resorted to a subterfuge, using

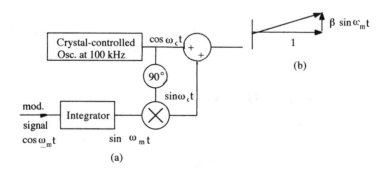

FIGURE 5.69   (a) Crystal-stabilized phase modulator; (b) phasor diagram.

a phase modulator to produce frequency modulation. Assuming that our signal finally goes through a frequency detector, we find that the amount of third harmonic as a fraction of the signal output is $\beta^2/4$. Now, in frequency modulation, the maximum amount of modulation which is permitted is in terms of frequency deviation, an amount of 75 kHz. The relation between frequency deviation and maximum phase shift is

$$\Delta f = \beta f_m,$$

where Δf is the frequency deviation, β is maximum phase shift, and f_m is modulation frequency. Since maximum modulation is defined in terms of Δf, the maximum value of β permitted will correspond to *minimum* modulation frequency. Let us do some numbers to illustrate this problem.

Example 5.3

Suppose we have a high fidelity broadcaster wishing to transmit bass down to 50 Hz with maximum third harmonic distortion of 1%. Find the maximum values of β and Δf.

Solution   We have $\beta^2/4 = 0.01$. Solving for β, we get $\beta = 0.2$.

Then, $\Delta f = 0.2 \times 50$ Hz $= 10$ Hz.

One can recall that the maximum value of frequency deviation allowed in the FM broadcast band is 75 kHz. Thus, use of the indirect modulator has given us much lower frequency deviation than is allowed, and clearly some kind of desperate measures are required. Such are available, but do complicate the process greatly. Suppose we feed the modulated signal into an amplifier which is not biased for low distortion, that is, its Taylor series looks like

$$a_1 x + a_2 x^2 + a_3 x^3, \text{ etc.}$$

Now the squared term leads to second harmonic, the cubed one gives third harmonic, and so on. The phase-modulated signal looks like $A \cos(\omega_c t + \beta \sin \omega_m t)$ and the term $a^2 x^2$ *not only doubles the carrier frequency, but also the maximum phase shift* β. Thus, starting with the rather low frequency of 100 kHz, we have a fair amount of multiplying room before we arrive in the FM broadcast band 88 to 108 MHz. Unfortunately, we may need different amounts of multiplication for the carrier frequency than we need for the depth of modulation. Let's carry on our example and see the problems that arise. First, if we wish to go from

$$\Delta f = 10 \text{ Hz to } 75{,}000 \text{ Hz},$$

that leads to a total multiplication of $75{,}000/10 = 7500$.

The author likes to say we are limited to frequency doublers and triplers. Let's use as many triplers as possible; we divide the 7500 by 3 until we get close to an even power of 2:

$$7500/3 = 2500; \ 2500/3 = 833, \ 833/3 = 278; \ 278/3 \approx 93, \ 93/3 = 31,$$

which is very close to $32 = (2)^5$.

So, to get our maximum modulation index, we need five each triplers and doublers. However, 7500×0.1 MHz $= 750$ MHz, and we have missed the broadcast band by about 7 times. One more thing we need is a mixer, after a certain amount of multiplication. Let's use all the doublers and one tripler to get a multiplication of $32 \times 3 = 96$, so the carrier arrives at 9.6 MHz. Suppose our final carrier frequency is 90.9 MHz, and because we have remaining to be used a multiplication of $3^4 = 81$, what comes out of the mixer must be

$$90.9 \big/ 81 - 1.122 \, \text{MHz.}$$

To obtain an output of 1.122 MHz from the mixer, with 9.6 MHz going in, we need a local oscillator of either 10.722 or 8.478 MHz. Note that this local oscillator needs a crystal control also, or the eventual carrier frequency will wander about more than is allowed.

Quick-and-Dirty FM Slope Detection

A method of FM detection that is barely respectable, but surprisingly effective, is called "slope detection." The principle is to feed an FM signal into a tuned circuit, not right at the resonant frequency but rather somewhat off the peak. Therefore, the frequency variations due to the modulation will drive the signal up and down the resonant curve, producing simultaneous amplitude variations, which then can be detected using an envelope detector. Let us just take a case of FM and a specific tuned circuit and find the degree of AM.

Example 5.4

We have an FM signal centered at 10.7 MHz, with frequency deviation of 75 kHz. We have a purely parallel resonant circuit with a Q = 30, with resonant frequency such that 10.7 MHz is at the lower half-power frequency. Find the output voltage for $\Delta f = +75$ kHz and for -75 kHz.

Solution   When we operate close to resonance, adequate accuracy is given by

$$V_o = \frac{V_i}{1 + j2Q\delta'}$$

where δ is the fractional shift of frequency from resonance. If now, 10.7 MHz is the lower half-power point, we can say that $2Q\delta = 1$.

$$\text{Hence, } \delta = 1 \big/ \left(2 \times 30 \right) = \left(f_o - 10.7 \, \text{MHz} \right) \big/ f_o \, ; \ f_o = 10.881 \, \text{MHz.}$$

Now, we evaluate the transfer function at 10.7 MHz $\pm$ 75.kHz.
We defined it as 0.7071 at 10.7 MHz. For 10.7 + 0.075 MHz, $\delta = (10.881 - 10.775)/ 10.881 = 9.774 \times 10^{-3}$, and the magnitude of the transfer function is $|1/(1 + j60\delta)| = 0.8626$.

Because the value was 0.7071 for the unmodulated wave, the modulation index in the positive direction would be

$$\left(0.8624 - 0.7071 \right) \big/ 0.7071 = 0.2196 \text{ or } 21.96\%.$$

For (10.7 − 0.075) MHz, $\delta = (10.881 - 10.625)/10.881 = 0.02356$, and the magnitude of the transfer function is $|1/(1 + j60)| = 0.5775$. The modulation index in the negative direction is $(0.7071 - 0.5775)/0.7071 = 18.32\%$. So, modulation index is not the same for positive as for negative indices. The consequence of such asymmetry is that this process will be subject to harmonic distortion, which is why this process is not quite respectable.

Lower Distortion FM Detection

We will assume that the reader has been left wanting an FM detector that has much better performance than the slope detector. A number of more complex circuits have a much lower distortion level than the slope detector. One, called the Balanced FM Discriminator, is shown in Fig. 5.70.

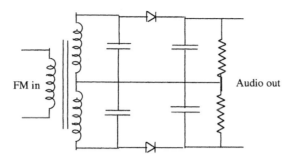

FIGURE 5.70 Balanced FM discriminator.

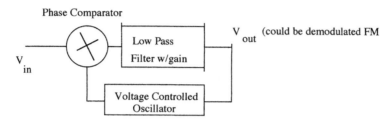

FIGURE 5.71 Basic phase-locked loop.

Basically, we may consider that the circuit contains two "stagger-tuned" resonant circuits, i.e., they are tuned equidistant on opposite sides of the center frequency, connected back to back. The result is that the nonlinearity of the resonant circuits balance each other out, and the FM detection can be very linear. The engineer designing an FM receiving system has a relatively easy job to access such performance; all that he/she must do is to spend the money to obtain high-quality components.

Phase-Locked Loop

The phase-locked loop is an assembly of circuits or systems that perform a number of functions to accomplish several operations, any one or more of the latter, perhaps being useful and to be capitalized upon. If one looks at a simple block diagram, one will see something like Fig. 5.71.

Thus, one function that will always be found is called a "voltage-controlled oscillator;" the linking of these words means that there is an oscillator which would run freely at some frequency, but that if a non-zero DC voltage is fed into a certain input, the frequency of oscillation will shift to one determined by that input voltage. Another function one will always find (although the nomenclature might vary somewhat) is "phase-comparison." The phase "comparator" will usually be followed by some kind of low-pass filter. Of course, if a comparator is to fulfill its function, it requires two inputs — the phases of which to compare. This operation might be accomplished in various ways; however, one method which might be understood from previous discussions is the analog multiplier. Suppose an analog multiplier receives the inputs $\cos \omega t$ and $\sin (\omega t + \phi)$; their product has a sine and a cosine. Now, a trigonometric identity involving these terms is

$$\sin A \cos B = 0.5 \left[\sin(A + B) + \sin(A - B) \right].$$

Thus, the output of a perfect analog multiplier will be $0.5[\sin (2\omega + \phi) + \sin \phi]$. A low-pass filter following the phase comparator is easily arranged; therefore, one is left with a DC term, which, if it is fed to the VCO in such a polarity as to provide negative feedback, will "lock" the VCO to the frequency of the input signal with a fixed phase shift of 90°.

Phase-locked loops (abbreviated PLL) are used in a wide variety of applications. Many of the applications are demodulators of one sort or another, such as synchronous detectors for AM, basic FM, FM–stereo detectors, and in very precise oscillators known as "frequency synthesizers." One of the early uses seemed to be the detection of weak FM signals, where it can be shown that they extend the threshold of usable weak signals a bit.[1] This latter facet of their usefulness seems not to have made a large impact, but the other aspects of PLL usefulness are very commonly seen.

Digital Means of Modulation

The sections immediately preceding have been concerned with rather traditional analog methods of modulating a carrier. While the beginning engineer can expect to do little or no design in analog communication systems, they serve as an introduction to the digital methods which most certainly will dominate the design work early in the 21st century. Certainly, analog signals will continue to be generated, such as speech, music, and video; however, engineers are finding it so convenient to do digital signal processing that many analog signals are digitized, processed in various performance-enhancing ways, and only restored to analog format shortly before they are fed to a speaker or picture tube. Digital signals can be transmitted in such a way as to use extremely noisy channels. Not long ago, the nightly news brought us video of the Martian landscape. The analog engineer would be appalled to know the number representing traditional signal-to-noise ratio for the Martian signal. The detection problem is greatly simplified because the digital receiver does not need at each instant to try to represent which of an infinite number of possible analog levels is correct; it simply asks, was the signal sent a one or a zero? *That* is simplicity.

Several methods of digital modulation might be considered extreme examples of some kind of analog modulation. Recall amplitude modulation. The digital rendering of AM is called "amplitude shift keying," abbreviated ASK.

What this might look like on an oscilloscope screen is shown in Fig. 5.72. For example, we might say that the larger amplitude signals represent the logic ones and smaller amplitudes represent logic zeroes. Thus, we have illustrated the modulation of the data stream 10101. If the intensity of modulation were carried to the 100% level, the signal would disappear completely during the intervals corresponding to zeroes. The 100% modulation case is sometimes called on–off keying and abbreviated OOK. The latter case has one advantage if this signal were nearly obscured by large amounts of noise; it is easiest for the digital receiver to distinguish between ones and zeroes if the difference between them is maximized. That is, however, only one aspect of the detection problem. It is also often necessary to know the timing of the bits, and for this one may use the signal to synchronize the oscillator in a phase-locked loop; if, for

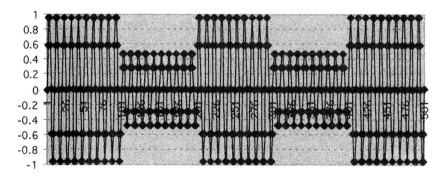

FIGURE 5.72   ASK (amplitude shift keying).

[1] Taub, H. and Schilling, D.L. *Principles of Communication Circuits,* 2nd Edition, McGraw-Hill, New York, 1986, 426–427.

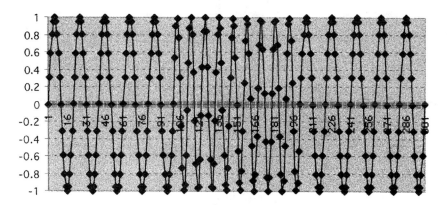

FIGURE 5.73   Frequency shift keying.

50% of the time, there is zero signal by which to be synchronized, the oscillator may drift significantly. In general, a format for digital modulation in which the signal may vanish utterly at intervals is to be adopted with caution and with full cognizance of one's sync problem. Actually, amplitude shift keying is not considered a very high performance means of digital signaling, in much the same was that AM is not greatly valued as a quality means of analog communication. What is mainly used is one or the other of the following methods.

Frequency Shift Keying

Frequency shift keying (abbreviated FSK) can be used in systems having very little to do with high data rate communications; for years it has been the method used in the simple modems one first used to communicate with remote computers. For binary systems, one just sent a pulse of one frequency for a logic one and a second frequency for a logic zero. If one was communicating in a noisy environment, the two signals would be orthogonal, which meant that the two frequencies used were separated by at least the data rate. Now, at first the modem signals were sent over telephone lines which were optimized for voice communications, and were rather limited for data communication. Suppose we consider that for ones we send a 1250 Hz pulse and for zeroes, we send 2250 Hz. In a noisy environment one ought not to attempt sending more than 1000 bits per second (note that 1000 Hz is the exact difference between the two frequencies being used for FSK signaling). Let us instead send at 250 bps. Twelve milliseconds of a 101 bit stream would look as in Figure 5.73.

It is not too difficult to imagine a way to obtain FSK. Assuming one does have access to a VCO, one simply feeds it two different voltage levels for ones and for zeroes. The VCO output is the required output.

Phase Shift Keying

Probably the most commonly used type of digital modulation is some form of phase shift keying. One might simply say there is a carrier frequency f_c and that logic zeroes will be represented by $-\sin 2\pi f_c t$, logic ones by $+\sin 2\pi f_c t$. If the bit rate is 40% of the carrier frequency, the data stream 1010101010 might look as in Fig. 5.74.

In principle, producing binary phase shift keying ought to be fairly straightforward, if one has the polar NRZ (nonreturn to zero, meaning a logic one could be a constant positive voltage for the duration of the bit, zero being an equal negative voltage) bit stream. If then, the bit stream and a carrier signal are fed into an analog multiplier, the output of the multiplier could indeed be considered $\pm\cos \omega_c t$, and the modulation is achieved.

Correlation Detection

Many years ago, the communications theorists came up with the idea that if one could build a "matched filter," that is, a special filter designed with the bit waveform in mind, one would startlingly increase the

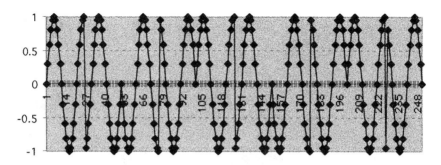

FIGURE 5.74 Phase shift keying.

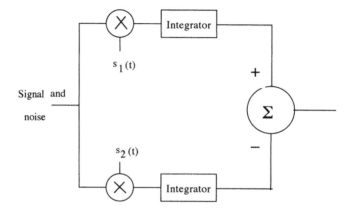

FIGURE 5.75 Correlation detector.

signal-to-noise ratio of the detected signal. Before long, a practically minded communications person had the bright idea that a correlation detector would do the job, at least for rectangular bits. For some reason, as one explains this circuit, one postulates two signals, $s_1(t)$ and $s_2(t)$, which represent, respectively, the signals sent for logic ones and zeroes. The basics of the correlation detector are shown in Fig. 5.75.

Now, a key consideration in the operation of the correlation detector is bit synchronization. It is crucial that the signal $s_1(t)$ be lined up perfectly with the bits being received. Then, the top multiplier "sees" sin $\omega_c t$ coming in one input, and $\pm$sin $\omega_c t$ + noise coming in the other, depending upon whether a one or a zero is being received. If it happens that a one is being received, the multiplier is asked to multiply sin $\omega_c t$(sin $\omega_c t$ + noise). Of course,

$$\sin^2 \omega_c\left(t\right) = \left(1/2\right)\left(1 + \cos 2\omega_c t\right).$$

In the integrator, this is integrated over one bit duration, giving a quantity said to be the energy of one bit. The integrator might also be considered to have been asked to integrate n(t) sin $\omega_c t$, where n is the noise signal. However, the nature of noise is that there is no net area under the curve of its waveform, so considering integration to be a summation, the noise output out of the integrator would simply be the last instantaneous value of the noise voltage at the end of a bit duration, whereas the signal output was bit energy, if the bit synchronization is guaranteed. Meanwhile, the output of the bottom multiplier was the *negative* of the bit energy, so with the signs shown, the output of the summing amplifier is twice the bit energy. Similar reasoning leads to the conclusion that if the instantaneous signal being received were a zero, the summed output would be *minus* twice the bit energy. It takes a rather substantial bit of theory to show that the noise output from the summer is *noise spectral density*. The result may be summarized that the correlation detector can "pull a very noisy signal out of the mud." And, we should

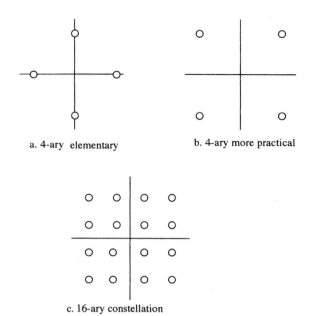

a. 4-ary elementary

b. 4-ary more practical

c. 16-ary constellation

FIGURE 5.76 Constellation showing carrier amplitudes and phase for M'ary signals.

assert at this point that the correlation detector can perform wonders for any one of the methods of digital modulation mentioned up to this point.

Digital QAM

Once the engineer has produced carrier signals that are 90° out of phase with each other, there is no intrinsic specification that the modulation must be analog, as is done for color TV. As a start toward extending the capabilities of PSK, one might consider that one sends bursts of several periods of $\pm\cos \omega_c t$ or $\pm\sin \omega_c t$. This is sometimes called "4-ary" transmission, meaning that there are four different possibilities of what might be sent. Thus, whichever of the possibilities is sent, it may be considered to contain two bits of information. It is a method by which more information may be sent without demanding any more bandwidth, because the duration of the symbol being sent may be no longer or shorter than it was when one was doing binary signaling, sending, for example, simply $\pm\cos \omega_c t$. This idea is sometimes represented in a "constellation," which, for the case we just introduced, would look like part a of Fig. 5.76. However, what is more often done is as shown in Fig. 5.76b, where it could be said that one is sending $\pm\cos (\omega_c t + 45°)$ or $\pm\cos (\omega_c t + 135°)$. It seems as though this may be easier to implement than the case of part a; however, the latter scheme lends itself well to sending 4 bits in a single symbol, as in Fig. 5.76c.

Strictly speaking, one might consider "a" to be the constellation for 4-ary PSK. This leads also to the implication that one could draw a circle with "stars" spaced 45° apart on it and one would have the constellation for 8-ary PSK. The perceptive or well-informed reader might have the strong suspicion that crowding more points on the circle makes it possible to have more errors in distinguishing one symbol from adjacent ones, and would be correct in this suspicion. Hence, M'ary communication probably more commonly uses "b" or "c," which should be considered forms of QAM.

5.6 Power Amplifier Circuits

Mark Bloom

The Power Amplifier (PA) is typically the last stage in a transmitter system. Its role is to provide the final amplification of signal power to a level that is large enough for microwave propagation through an appropriate antenna. In some systems, the PA is connected directly to an antenna, while in other systems isolators, filters, and switches may follow before the antenna is reached. Often the PA draws 50% or more of the total transmit-current required by a system. If current is an issue (and in most systems it is a key parameter) then the PA design is a critical one to ensure the system current budget is met. Also, in most modern commercial wireless systems, the PA and the associated driver amplifier determine the overall linearity of the transmit chain.

Design Analysis

As the first step in a PA design, the design must be analyzed and the specification determined in sufficient detail to allow accurate synthesis of the design.

Applications

As the starting point in a design analysis, the application must be carefully considered before the PA specification is defined. Typical considerations are as follows:

- Consumer, high-volume applications: These require *cheap* PAs. To achieve this, customer specifi-cations are barely met. Voltages available to drive the PA are typically low (2.7 to 4.7 V) or limited (i.e., no negative supply to bias a GaAs depletion-MESFET gate), and size is critically important — market pressure is generally forcing the cost and size down, and performance ever upward. Consumer PAs tend to fall into two broad categories of distinction: linear or saturated PAs. Certain process technologies and circuit topologies favor either linear or saturated applications.

- Non-consumer, low-volume applications: These require high performance, and typically require high reliability. The performance often cannot be compromised, which leads to a high cost. High reliability will impact PA design, often with much derating required, which will reduce efficiency.

Modulation Effects

One of the key considerations for a PA is the modulation scheme used. Many PA designs have been single-tone CW, both in simulation and in physical measurement. However, with the widespread implementation of nonconstant envelope modulation schemes for mass wireless markets (i.e., CDMA IS95/98 and NADC IS136), single-tone CW measurements are being used less. More wireless systems are being developed based on spectrally efficient nonconstant envelope modulation, which have a profound impact on PA design.

If a PA is to be used with constant-envelope modulation, then the PA can be operated close to saturation — typically the harmonic content limits the degree of compression permitted in a design. For instance, the widespread GSM wireless standard allows PAs to operate 4 to 5 dB into compression, leading to power-added efficiencies [PAE, see Eq. (5.3)] greater than 60% from a 3 V supply, with 35 dBm of RF output power.

However, if a nonconstant envelope modulation scheme is used, then spectral regrowth, or Adjacent Channel Power (ACP) typically manifests. This distortion can degrade BER for wireless users allocated adjacent (or alternate) frequency bands. Hence most nonconstant envelope schemes have stringent specifications on Adjacent Channel Power Ratio (ACPR). ACPR is defined as the relative difference between the users in-band output power and the users adjacent (or alternate) band output power. For CDMA (IS95/98) a PA can typically operate no more than 1 dB into compression, which limits PAE to around 55% [13]. Methods exist to increase PAE, but these rely on some form of linearization scheme and have proven slow to develop for consumer wireless products.

Typical PA Specification Parameters

Understanding the specification is critical in choosing the overall amplifier topology and methodology. The key parameters and their impact in design are listed below:

- Small signal gain (S_{21}). Under small-signal operation, a network analyzer can accurately determine small signal gain. Small signal gain is a vector quantity, with magnitude typically expressed in dB, and with phase expressed in degrees. Small signal gain is often the first point in considering how to budget the gain specification for a PA. As a rule of thumb, a power transistor with 20 GHz <

Ft < 40 GHz, sized to generate 1 W of RF power will have a small signal gain around 20 dB at 1 GHz when matched for gain using Surface-Mount-Technology (SMT) low-pass matching transformations. Thus gain at 2 GHz (an octave higher) would be 20 dB – 6dB = 14 dB (gain tends to fall off at the rate of –6 dB per octave of frequency).

- Small Signal Return Loss (S_{11} and S_{22}). Again, under small signal conditions, return loss can be calculated with a network analyzer. As will be shown later, output return loss (S_{22}) is often poor in a PA, as the match for good return loss is different for the match for maximum power. Output return loss for a PA can typically be between –5 dB and –10 dB. Input return loss (S_{11}), when matched for maximum gain, can typically be at least –10 dB, and often around –20 dB.

- Output power (P_{out}). The power delivered to a load (typically a 50 Ω termination) can be measured using microwave power meters. For high frequency measurements, units are typically expressed in "dBm," i.e., referred to 1 mW. For instance, 1 W = 30 dBm. Depending on bias and load, power increases linearly with the input power, until the amplifier begins to suffer gain compression — at this point, the gain starts to fall off as a function of input drive. Eventually no more power can be gained out of the PA, leading to the term "saturated output power."

- Efficiency (η). Measured as a percentage between 0% and 100%, two definitions of interest exist. DC-RF efficiency (also referred to a drain or collector efficiency) is simply the ratio of power delivered to a load and the DC power consumed by the PA to deliver that power:

$$\eta_{dc-rf} = \frac{P_{out}}{P_{dc}} \tag{5.2}$$

Power-added efficiency is a more interesting expression, as the input-power to the device is considered, i.e., it is the ratio of the amount of power added by the PA to the DC consumption:

$$\eta_{PowerAdded} = \frac{\left(P_{out} - P_{in}\right)}{P_{dc}} \tag{5.3}$$

By inspection of Eq. (5.3), power-gain effects the value of power-added efficiency calculated.

- Harmonic distortion. Depending on the bias point and class of operation, harmonic content from the PA will generally increase as drive level increases. Generated by clipping of the input signal, odd/even harmonics of the fundamental can become an issue in some systems. Typically harmonics must be at least –30 dBc (referenced to the fundamental). IP_3 and ACPR are phenomena related very closely to harmonic content — the same mechanisms explain all three of these forms of distortion.

Basic Power Amplifier Concept

The basic problem in designing a power amplifier is in regard to the output match. An inherent trade-off must be made between gain and output power. The problem stems from the fact that the output load a transistor needs for maximum power is *different* from maximum small signal gain.

Transducer power gain, G_T, is defined as the ratio of power delivered to the load to the power available from the source [1–3]. When the small-signal S-parameters and reflection coefficients are normalized to the same reference impedance, then transducer power gain is given as

$$G_T = \frac{P_L}{P_A} = \frac{\left(1-|\Gamma_g|^2\right)|S_{21}|^2\left(1-|\Gamma_L|^2\right)}{\left|\left(1-\Gamma_g S_{11}\right)\left(1-S'_{22}\Gamma_L\right)\right|^2} \tag{5.4}$$

where

$$S_{22}' = S_{22} + \frac{S_{12}S_{21}\Gamma_g}{1 - S_{11}\Gamma_g} \tag{5.5}$$

So as to simplify Eq. (5.4), assume the device is unilateral i.e., $S_{12} = 0$. Now unilateral transducer gain, G_{Tu}, is

$$G_{Tu} = \frac{\left(1 - \left|\Gamma_g\right|^2\right)\left|S_{21}\right|^2\left(1 - \left|\Gamma_L\right|^2\right)}{\left|1 - \Gamma_g S_{11}\right|^2 \left|1 - S_{22}\Gamma_L\right|^2} \tag{5.6}$$

By inspection of Eq. (5.6), G_{Tu} will be a maximum when $\Gamma_g = S_{11}^*$ and $\Gamma_L = S_{22}^*$. Then, maximum unilateral gain is

$$G_{Tu-max} = \frac{\left|S_{21}\right|^2}{\left(1 - \left|S_{11}\right|^2\right)\left(1 - \left|S_{22}\right|^2\right)} \tag{5.7}$$

Hence from inspection of Eqs. (5.6) and (5.7), unilateral transducer gain will be maximized when the input and output terminals are conjugately matched into the load and source. Remember that this derivation is strictly only true for a unilateral device. However, it is a very close approximation with modern device technology.

Now, the actual load required by the output device for maximum gain will be

$$\Gamma_{out} = \left(S_{22} + \frac{S_{21}S_{12}\Gamma_g}{1 - S_{11}\Gamma_g}\right)^* \tag{5.8}$$

Now consider a power amplifier. The load line determines the available output power, as in Fig. 5.77. The output power will be

$$P_{out} = \frac{1}{2}I_{dd} \cdot \left(V_{dd} - V_{knee}\right) \tag{5.9}$$

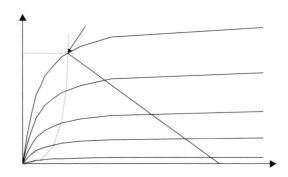

FIGURE 5.77 Load-line for a transistor (this example assumes a FET device).

So, since $P = IV$ and $V = IR$,

$$R_L = \frac{\left(V_{dd} - V_{knee}\right)^2}{2P_{out}} \tag{5.10}$$

Limits typically bound the maximum power from a device. For a depletion GaAs FET, I_{dd} is a maximum at a slightly positive gate voltage — too far forward, and the gate will conduct leading to device failure. V_{dd} is either fixed by the available voltage supply, or by the drain-source breakdown characteristics. The saturation voltage determines V_{knee}, which is a function of the size of the FET device and the technology.

In practical devices, Eqs. (5.8) and (5.10) yield very different results, i.e.,

$$\left| \left(S_{22} + \frac{S_{21}S_{12}\Gamma_g}{1 - S_{11}\Gamma_g} \right)^* \right| \neq \frac{\left(V_{dd} - V_{knee}\right)^2}{2P_{out}} \tag{5.11}$$

Hence the fundamental problem in trying to simultaneously match for gain and power.

Analysis of the Specification

The first step in a PA design is to understand the specification, and the customer requirements (which may be slightly different).

Basic Considerations

Initially, available voltages and currents must be considered. Many high-volume commercial PA products require low operating voltages around 3.0 V. This will greatly effect the design — a 3.0 V-supplied PA will have to draw current approximately four times higher than a 12 V-supplied PA. For a typical specification,

$$I_{DD} \approx \frac{P_{OUT}}{\left(V_{DD} - V_{knee}\right) \cdot \left(\dfrac{\eta}{100}\right)} \tag{5.12}$$

where,

> I_{DD} is defined in mA
> V_{dd} is defined in Volts
> V_{knee} is defined in Volts and can be assumed to be zero if V_{dd} is much larger
> P_{OUT} is specified in mW
> η is between 0 and 100%

Leading on from this, once the DC current requirements have been estimated, then the packaging requirements can be considered.

The key consideration here is that as I_{DD} increases, so does the effect of many circuit parasitics. Series-resistance will play a bigger role in external matching components.

Figure 5.78 shows this effect clearly. As the DC voltage available for the PA falls, the current increases. So, as the Q of the elements (as an example used in a low-pass output-match) lowers, the losses present dissipate more current, leading to a reduction in efficiency.

Thermal design, especially in regard to the active-transistor region and how it is grounded will need to be considered. Thermal design consists of ensuring that the transistor junction temperature, T_j, does

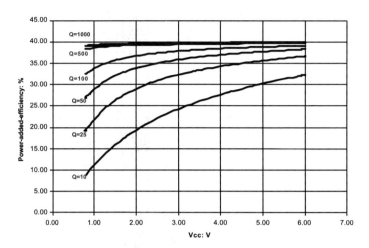

FIGURE 5.78   Effect of Q and Voltage on PAE (for a 1 W PA, with a single L-C low-pass output match with a certain Q).

not exceed the limit for reliable operation. Each foundry-specific device technology has a maximum junction temperature. Junction temperature can be found from

$$T_j = P_d \cdot \left(\theta_{jc} + \theta_{cs} + \theta_{sa} \right) + T_a \qquad (5.13)$$

where,

P_d = power dissipated in junction, which includes DC power and RF power (W)
θ_{jc} = junction-to-case thermal impedance (°C/W)
θ_{cs} = case-to-heat-sink thermal impedance (°C/W)
θ_{sa} = heat-sink-to-ambient thermal impedance (°C/W)
T_a = ambient temperature (°C).

From these simple considerations, an appropriate package can be determined. In the case of a package already specified, then the designer will simply have one less variable to optimize during the design synthesis.

Do not forget that the PA, whether packaged as a ceramic hybrid or a plastic-encapsulated MMIC, is part of an overall system. Consideration must be given to the particular system interface — is the PA package to be soldered to a board, or will epoxy be used? How will grounding of the PA be applied? Will the board ground-plane be a solid shunt of metal, or will board vias be employed to connect with a spatially separated ground-plane? How thick is the board, and of what material? How will it react thermally? Many of these seemingly basic questions are often not answered until the design has been fabricated — of course then it may be too late.

Now the designer has a basic understanding of the environment the PA will operate within. From this, the electrical specification can be analyzed in more detail.

Budgeting

Gain partitioning is the next step in the design. A specification may be for the complete PA to have more small signal gain than a single stage can provide. It is most likely that more than one stage will be required for this to be realized. Hence a gain budget is required to determine the rough gain (and power) levels in each stage.

For instance, consider the simple specification in Table 5.1. The first thing to consider is the maximum output power. Since it is unlikely to find a single transistor with 40 dB power gain, a multistage design is required.

TABLE 5.1   A Simple Specification

Parameter	Typical Value	Unit
Frequency	2.0	GHz
Maximum output power	36.0	dBm
Gain at maximum output power	40.0	dB
Amount of gain compression at maximum output power	3.0	dB
Efficiency	65.0	%
Technology of implementation	Bipolar, and V_{cc} = 3.6 V	

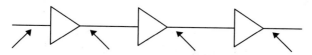

FIGURE 5.79   Gain/power budget.

As a rule-of-thumb example, assume a device has 20 dB of small signal gain, but is only conditionally stable. A designer will have to lose some gain (typically 5 dB) to achieve stability. Hence the output stage will have a gain of around 15 dB. At 3 dB gain compression, this implies the output stage will operate with a power gain of 15 dB − 3 dB = 12 dB. However, the previous stages will need to be slightly compressed to obtain good efficiency, so assume the output stage operates at 2 dB compression — now the power gain is 13 dB.

Consider Fig. 5.79. A hypothetical lineup is shown. Generally, gain of a transistor decreases as transistor active-area increases (since the device parasitics increase). Thus stage 1 and stage 2 have more gain than the output stage. Also note that the power gain in is 45 dB. To reduce this (and to help stability) gain in stages 1 and 2 will be lowered to bring the whole design to the 40 dB gain target.

Assume the output stage is running at η = 70%, stage 1 at η = 55%, and stage 2 at η = 40%. Now apply Eq. (5.12). The following collector currents can calculated:

OP Stage: 1580 mA

Stage 1: 101 mA

Stage 2: 4 mA

Thus total efficiency is 65.6%, which meets the specification. A spreadsheet is useful to perform these basic calculations — the degree of efficiency for each stage can be found very quickly.

Choice of Device

The next step is to estimate the transistor device areas. In this example it is a BiPolar technology, so emitter area is the active parameter. Assume the technology offers reliable operation up to 0.2 mA/μm^2. Then, the following device areas can be calculated from the DC currents above:

OP Stage: 7900 μm^2

Stage 1: 505 μm^2

Stage 2: 20 μm^2

Bias Point and Class of Operation

A PA must be designed to amplify in a certain class of operation. The class is determined by three key factors: Quiescent bias point, matching topology, and transistor configuration. The class determines the maximum potential efficiency η_{max} as well as the relative maximum potential output power P_{relmax}. The classes can also be grouped by their linearity — some classes are highly linear, while others generate a lot of harmonic distortion leading to degraded IP_3 or ACPR. Three main classes exist: A, B, and C. A fourth, class-AB, is a compromise between class-A and class-B. Some other classes exist (D,E,F,S) but these are specialized and generally are not in commercial use (due to their high distortion and bias/matching problems coupled with the complexities of their design). See Reference 1 for a more complete description.

The quiescent bias point of a transistor (V_{dq}, I_{dq} for a FET and V_{cq}, I_{cq} for a bipolar) determines the conduction angle, θ. An amplifier under class-A operation has $\theta = 360°$, a class-AB amplifier obeys $360° > \theta > 180°$ and a class-B amplifier has $\theta = 180°$. A class-C amplifier has $180° > \theta > 0°$.

The first step in determining the class to use is to consider η_{max}. From Reference 1, it can be shown that efficiency is given by

$$\eta = \frac{1}{4} \frac{\theta - \sin\theta}{4\sin\left(\theta/2\right) - \theta/2\cos\left(\theta/2\right)}. \qquad (5.14)$$

From Eq. (5.14), the classical values of η_{max} can be determined: Class-A has $\eta_{max} = 50\%$, class-B has $\eta_{max} = 78.5\%$, and class-C has $\eta_{max} = 100\%$. Note that to obtain η_{max} for a class-C, $\theta = 0°$.

Also, P_{relmax} can be calculated from by Reference 1 by

$$P_{out} \propto \frac{\theta - \sin\theta}{1 - \cos\left(\theta/2\right)}. \qquad (5.15)$$

By inspection of Eq. (5.15), P_{out} decreases as conduction angle decreases. Thus to obtain η_{max} for a class-C PA ($\theta = 0°$), $P_{out} = 0$ W. So, practically η_{max} cannot be achieved.

Conduction angle for a PA can be visualized by considering the DC-IV curves. Since V_{dq}, I_{dq} determine quiescent point, V_{dq}, I_{dq} can be superimposed on the DC-IV data.

As can be seen in Fig. 5.80, class-A has the lowest "clipping" of the waveform, and hence the least distortion. As the bias point I_{dq} deepens toward class-C, the harmonic distortion caused by clipping below threshold (on large-signal negative cycles), or from forward gate-conduction (on large-signal positive cycles) increases.

Topology

Several topologies exist for PAs. Topology refers to the matching techniques to be used within the overall PA lineup.

Reactive Matching

The terminal impedances of the transistor can be matched using reactive elements, such as capacitors and inductors. Reactive elements allow both narrow- and wide-band transformations. The frequency response is critical for an amplifier, especially the impedance presented at the harmonics of the fundamental frequency. The frequency response characteristics of reactive matching may be low pass, high pass, or band pass. These general considerations can be applied when choosing which to use:

- Inductors tend to have a lower Q than capacitors. So, a low pass structure will tend to be more lossy than a high pass. The loss of the matching structure can affect stability, noise figure, and PAE of the amplifier to a great extent, and so this must be considered during the design process. So, input match of a transistor may benefit from being lossy, especially with bipolar designs, which often have base ballasting anyway — the ballasting can be redistributed to be within the input match. However, in a PA, output match generally needs the lowest loss possible to maximize PAE, suggesting a high pass transformation.

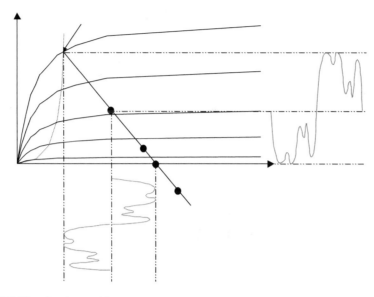

FIGURE 5.80   DC-IV and quiescent bias.

- Lossy structures at the input to an amplifier increase noise figure. Some modulation systems (such as IS-95) require very low noise figures, and so the topology may be affected by the noise specifications.
- PAs are often required to have low distortion, which equates to low harmonic content in the output signal. A low pass output match will tend to attenuate the harmonics considerably, leading to improved linearity.
- Inter-stage matches tend to be a compromise of factors — as power levels tend to be relatively high, low pass structures may be avoided due to loss. Also, a high pass structure may allow implicit DC blocking, which most amplifiers require. This leads to a lower passive component count, and reduced cost. To obtain high linearity though, a low pass structure may be beneficial.

As the points above show, choice of matching topology is often a compromise, with each transformer being carefully considered in regard to the specific specifications. It is not possible to generalize, so the whole design must be considered carefully.

Feedback

The use of series or shunt feedback can help ensure stability of a PA. Resistive feedback between drain and gate (typically a few hundred ohms) will increase stability margin, and increase bandwidth [6,7]. See Fig. 5.81 for an example. However, the gain of the transistor will be lower, leading to reduced power-added efficiency or the requirement to add another stage of amplification. Generally gate drain series feedback will not be used on the final amplifier stage due to the degradation in gain.

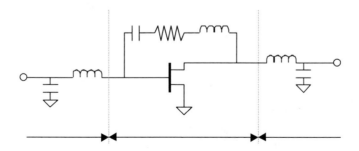

FIGURE 5.81   Series feedback around a simple FET amplifier.

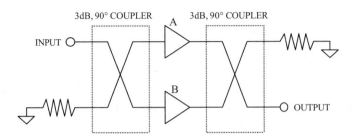

FIGURE 5.82　Diagram of the balanced amplifier.

Often a capacitor is required in series with the resistor. This is essential for blocking DC-bias voltages. The *R-C* time constant of the feedback will have a slight affect on the frequency response of the circuit. A series inductor can also be used with the resistor to adjust the frequency response significantly. Inductance is required in a broadband design where the frequency shaping of the feedback is critical to the gain response.

Balanced Power Amplifier

A power amplifier may be designed to operate in a balanced topology. A balanced amplifier consists of two identical amplifiers (A and B), with two couplers connecting the inputs and the outputs together. Figure 5.82 shows a balanced arrangement.

The advantage of a balanced amplifier is that unwanted reflections are terminated in a load, typically 50 Ω. One amplifier (B) is driven with the signal, while the other (A) is driven by a signal that is 90° out of phase. Any reflected signals are thus 180° out of phase after passing back through the coupler. Hence the unwanted reflections cancel out. The outputs are arranged opposite to the input, so the other output passes through the coupled port, thus recombining the signal in phase.

As shown in [8], transducer gain of a balanced amplifier can be shown to be

$$G_T = \frac{G_1 + G_2 + 2\left(G_1 G_2\right)^{1/2} \cos\left(\varphi_1 - \varphi_2\right)}{4} \tag{5.16}$$

where

$$S_{21A} = G_1^{1/2} \exp\left(j\varphi_1\right) \tag{5.17}$$

$$S_{21B} = G_2^{1/2} \exp\left(j\varphi_2\right) \tag{5.18}$$

So, if the amplifiers A and B are identical, then $G_1 = G_2$ and $\varphi_1 = \varphi_2$. Then Eq. (5.16) resolves to $G_T = G_1$. However, if one amplifier should be turned off or shut down, then G_T falls by a factor of 4 (–6 dB).

The benefit of the balanced power amplifier is that the output ports are very insensitive to mismatch as the return losses are much better than a single amplifier on its own. This results in a very stable PA that is insensitive to mismatch. Note that the output power is the sum of the two amplifiers, but the gain is reduced by twice the coupler loss (assuming identical couplers).

Modern couplers can be purchased in a variety of forms, including lumped ceramic surface mount, which fit the footprint of an 0805 SMT component.

Distributed Power Amplifier

Traveling wave, or distributed amplifiers have been around since the 1940s when the first patent was filed [9]. Since then, much work has been performed to optimize and improve the concept, especially in regard to power amplification [10].

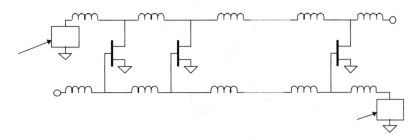

FIGURE 5.83 Distributed power amplifier topology.

The distributed amplifier concept uses the parasitic capacitance/inductance of a transistor as part of two artificial transmission lines. The transmission lines are a lumped equivalent of a distributed line, with some of the lumped elements formed in the transistor. Hence, as the amplifier appears to be a transmission line, it is matched into the required terminations over a very wide bandwidth. That is the benefit of this topology — bandwidths up to many octaves are possible.

Figure 5.83 shows a typical distributed PA topology. A number of FET devices, n, are connected in parallel, with inductances L_g and L_d between the gate and drains.

It has been shown [11] that gain, G, is

$$G = \frac{g_m^2 n^2 Z_o^2}{4}\left(1 - \frac{\alpha_g l_g n}{2}\right)^2 \tag{5.19}$$

where α_g is the effective gate-line attenuation per unit length, and l_g is the length of the gate transmission line per unit cell. From this, is it clear that the gain will increase as more FET stages are added. However, stages cannot be added indefinitely as each FET has parasitic resistance (typically R_i and R_{ds}) which will attenuate the signal — eventually the loss added will exceed the gain benefit of another stage. Note that as the bandwidth generally is very large, the overall gain (even with many FET segments) is fairly low.

Two unusual constraints exist that determine the maximum output power from a distributed power amplifier, in addition to the voltage-swing constraints with all power amplifiers. Firstly, gate periphery cannot be increased without adding more attenuation from the parasitic elements in the FET. Thus gate periphery is a trade-off between gain, bandwidth, and power. Secondly, the drain of each FET must see an impedance that is determined by the characteristic impedance of the drain line. Hence, each FET will not be able to see the impedance that it needs to satisfy maximum power output as in Eq. (5.10). This power mismatch can result in significant reductions in output power.

One final consideration is that as the frequency response is greater than an octave, harmonic terminations will not be correct to maximize efficiency. Thus a traveling wave power amplifier will have lower efficiency than a narrowband reactively matched design.

PA Architecture

The PA architecture may be one of several — Fig. 5.84 shows a tree of the typical types of PA. A MMIC is deemed to have active and passive components on a single substrate. A discrete contains minimal passive components. A plastic package typically only contains an active die, with passive components being external. A module contains an active die and passive discrete components. Table 5.2 details some of the key considerations with each type of PA.

Choice of Active Device Technology

The main active device technologies suitable for PAs are: GaAs MESFET, GaAs HEMT, GaAs HBT, Si MOSFET, Si Bipolar, and SiGe HBT.

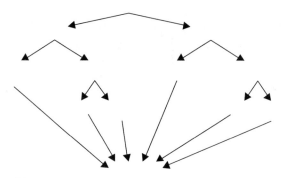

FIGURE 5.84   PA architectures.

TABLE 5.2   PA Architectures

Architecture	Benefit	Disadvantage	Applications
MMIC, Plastic	Low cost	Often will require many external components to get optimal performance Overall size may be large after all the externally required components are added Performance is very dependent on the board it is attached to High volume testing can be difficult	Up to 2–3 GHz Very high volume Requires lots of customer interaction to obtain optimal performance
MMIC, module, ceramic	Can be very small Can operate in many board environments with minimal modifications	High cost Multilayer ceramic technology expensive, which means very complex designs may be large	Up to millimeter wave High volume Often good for a "quick and easy" solution for the customer
MMIC, module, resin laminate	Can be very small Can operate in many board environments with minimal modifications Multilayer easy to implement	Medium cost Performance not as good as ceramic and generally lower E_r means distributed elements are large	Up to 2–3 GHz Very high volume Often good for a "quick and easy" solution for the customer
Discrete, plastic	Very low cost	Will require many external components to get optimal performance Overall size may be large Performance is very dependent on the board it is attached to High volume testing can be difficult	Up to 2–3 GHz High volume Requires huge customer interaction to obtain optimal performance
Discrete, module, ceramic	Can be small Can operate in many board environments with minimal modifications	High cost Multilayer ceramic technology expensive, which means very complex designs may be large Will require many external components to get optimal performance	Up to 20 GHz — beyond this requires very careful passive component selection High volume Often good for a "quick and easy" solution for the customer
Discrete, module, resin laminate	Can be small Can operate in many board environments with minimal modifications Multilayer easy to implement	Medium cost Performance not as good as ceramic and generally lower E_r means distributed elements are large Will require many external components to get optimal performance	Up to 2–3 GHz High volume Often good for a "quick and easy" solution for the customer Good if customer can tolerate a larger module

It is very difficult to subjectively determine which type of device is most suitable for any particular application. Up to 2 GHz, all can compete. Between 2 GHz and 20 GHz, most silicon technologies are unsuitable. Beyond 20 GHz only GaAs HEMT and HBT really perform well. Most millimeter-wave devices are HEMT based.

To cloud the issues, research papers may show a particular device technology that is performing beyond what is thought as "normal." This can often be facilitated by hand choosing the best device from a lot of material. Consideration must be made that in production, especially high-volume/low-cost things are much more limited. But remember that what is research today could be production in one or two years.

Gallium Arsenide Solutions

As a rule, GaAs performs better than silicon, but will cost more. High performance GaAs generally require epitaxially grown materials, which can cost around five times the price of a comparable silicon wafer. The majority of the worlds GaAs output is on four-inch material wafers, but GaAs six-inch wafers have been adopted by several foundries. As the demand for GaAs increases (driven primarily by the wireless consumer market) six-inch material will be adopted by more foundries.

Heterojunction Bipolar Transistors (HBT)
Generally, HBT technologies have high current density due to their vertical structure — they therefore consume less die area than a FET based technology. However, material structure tends to be more complex than a FET (but more forgiving of certain variations). A typical HBT will consist of seven or eight epitaxially grown layers, whereas an epitaxial FET may only consist of four (plus a super-lattice buffer region). Epitaxial material for HBT devices costs virtually the same as an epitaxial FET. This is because cost of material is dominated by the total thickness of the grown layers. This is similar for both HBT and FET. One consideration is that more care with alignment and registration is required for an HBT (compared to a long gate FET with $L_g = 1$ um).

Epitaxial High-Electron Mobility Transistors (HEMT)
HEMT technologies (based on short gate length and epitaxial material) offer best in absolute performance (i.e., noise figure, and PAE). Submicron gate lengths mean cost may be high. Note that most HEMT structures are very sensitive to key layer thickness, which is not such an issue for HBT devices. The sensitivity of layer variations can result in poor yield and hence a higher cost for the customer. Controlling this variation is one of the key issues in choosing a HEMT technology.

One clear benefit that a HEMT (or a MESFET) technology has over an HBT is that its on-resistance is a function of gate width (among other parameters). As a consequence, if very low voltage operation is required (sub-3 V) then the on-resistance plays a large part in determining efficiency. As Eq. (5.12) shows, V_{knee} (or saturation voltage) will have an effect determining overall efficiency. V_{knee} can be lowered if the on-resistance is reduced.

One drawback is that most HEMT (or MESFET) technologies require a negative gate voltage. This can prove complex and difficult to generate in a mobile subscriber unit, and most manufacturers would rather avoid the complexity of adding this control. Enhancement-Mode GaAs HEMT and MESFET devices are available, but these can be even more sensitive to material/process variations. Unless a device offers true enhancement-mode operation, a drain-switch is often still required to shut the device down. HBT devices do not have this concern — if the base is held at zero potential, there is no significant collector leakage.

Epitaxial Metal Semiconductor Field-Effect Transistors (MESFET)
A MESFET grown with epitaxial layers can perform almost as well as a HEMT for L-band applications. It does not have the same concerns over layer thickness as a HEMT, but it suffers some of the same drawbacks. Yield on this kind of device can be very high, and process variations can be minimized by the use of etch-stop layers.

Ion-Implanted MESFET
Ion implantation was used in the first MESFET devices to create the required channel and Ohmic regions. Once the mainstream GaAs device, it has fallen behind in terms of performance compared to epitaxial

devices. Ion implantation offers an extremely low-cost device, but suffers in a significant variation in performance from lot to lot. This is due to the intrinsic variation within the ion implantation depth profile. The use of epitaxial material eliminates this issue, but with the added cost that MBE/MOCVD implies.

Silicon Solutions

Silicon offers a low-cost solution for power amplifiers. Even though die are often larger than GaAs, the processed wafer cost of silicon is significantly lower than GaAs. Silicon offers a higher range of integration, coupled with a very well controlled and understood process. On-chip passive elements (predominantly inductors) tend to have more loss than GaAs, but the use of these elements in PA circuits tends to be minimal.

CMOS, Bipolar, and BiCMOS

Silicon CMOS has not had much success in power amplifiers due to its comparatively low breakdown and low current density [12]. Bipolar processes can have high figures of merit which allows them to perform comparatively well. However, current densities are lower than GaAs, leading to larger die. Larger die can also lead to stability problems, which a compact design would not suffer from. The use of aluminum interconnect (as opposed to gold used in GaAs) limits the current density significantly. However, with the aluminum/copper interconnect now becoming more common, this issue is less prevalent.

One benefit that silicon bipolar devices have over GaAs Bipolar devices is that the based emitter turn-on voltage, V_{be}, for silicon is around half of a GaAs HBT. Silicon has a V_{be} of around 0.7 V while GaAs V_{be} is around 1.4 V. This allows more flexibility in the circuit design as more devices can be "stacked" between the supply rail and ground. This is a major benefit in wireless systems operated from a battery, which may supply down to 2.8 V (or lower) DC.

Laterally Diffused Metal Oxide Semiconductor (LDMOS)

LDMOS is a very cheap silicon process that allows MOS devices to be integrated with passives onto a die. LDMOS uses a slow, well-controlled, and repeatable lateral diffusion process to effectively make the device gate much smaller than its drawn dimension. Since (to a first order) performance is inversely proportional to gate length, LDMOS offers very good RF and microwave performance. However, performance is currently limited to around 3 GHz as an upper maximum.

Silicon Germanium (SiGe)

By adding small amounts of germanium into a silicon bipolar process, SiGe devices can be made to perform well at RF/microwave frequencies [14]. The main drawback is that the lattice mismatch between Si and Ge is considerable. Thus devices can only be lightly doped with Ge, which leads to a "weak" heterojunction effect. Since PAs benefit from this heterojunction to achieve higher efficiency, SiGe falls behind GaAs in terms of performance. However, being silicon-based, it is relatively cheap, despite having a fairly complex process. Breakdown voltages in a SiGe process tend to be low, as the devices have often been designed for high-speed digital applications (where SiGe can have major benefits over GaAs). Low breakdown tends to be undesirable for a PA (leading to either device-failure or increased instability, especially under mismatched RF conditions). SiGe devices can be engineered to have higher breakdown, but only at the expense of RF performance. However, some markets such as wireless hand portables are evolving to operate the PA at lower power levels and from lower supply voltages. This could work to favor a SiGe solution over a GaAs solution.

References

1. Ha, T.T., *Solid State Microwave Amplifier Design*, Wiley, New York, 1981.
2. Vendelin, G.D., *Design of Amplifiers and Oscillators by the S-parameter Method*, Wiley, New York, 1982.
3. Bodway, G.E., Two-power power flow analysis using generalized scattering parameters, *Microwave Journal*, 10, 61, 1967.
4. Vendelin, G.D., Pavio, A.M., Rohde, U.L., *Microwave Circuit Design Using Linear and Nonlinear Techniques*, Wiley, New York, 1990.

5. Kraus, H.L., Bostian, C.W., and Raab, F.H., *Solid State Radio Engineering*, McGraw-Hill, New York, 1980.

6. Rigby, P.H., Suffolk, J.R., Peneglly, R.S., Broadband monolithic low-noise feedback amplifiers, *IEEE Microwave and Millimeter Circuits Symposium*, 71, 1983.

7. Jastrzebski, A.J., Bloom, M., Davies, A., Buck, J., and Pennington, D., Design of broad-band MMIC power amplifiers for 6–18 GHz, *IEE Colloquium*, IEE Digest No. 1991/191, 1991.

8. Soares, R. (ed), *GaAs MESFET Circuit Design*, Artech House, Boston, 1988.

9. Percival, W.S., Thermionic valve circuits, British Patent 460562, July 1936.

10. Ayasli, Y., Reynolds, L.D., Mozzi, R.L., Hanes, L.K., 2–20 GHz GaAs traveling-wave power amplifier, *IEEE Trans. MTT*, 32, 290, 1984.

11. Ayasli, Y., Mozzi, R.L., Vorhaus, L.D., Reynolds, L.D., Pucel, R.A., A monolithic GaAs 1–13 GHz GaAs traveling-wave amplifier, *IEEE Trans. MTT*, 30, 976, 1982.

12. Tsai, K-C., Gray, P.R., Techniques in designing CMOS power amplifiers for wireless communications, *JSSC*, July 1999.

13. RF Micro Devices Inc., A Linear, High Efficiency, HBT CDMA Power Amplifier, *Microwave Journal*, January 1997.

14. Crabbe, E.F., Comfort, J.H., Lee, W., Cressler, J.D., Meyerson, B.S., Megdanis, A.C., Sun, J.Y.-C., and Stork, J.M., 73-GHz Self-Aligned SiGe-base bipolar transistor with phosphorus-doped polysilicon emitters, *IEEE Elec. Dev. Lett.*, 16, 1980.

5.7 Oscillator Circuits

Alfy Riddle

Figure 5.85 shows a variety of styles and packaging options for RF and microwave oscillators. Oscillators serve two purposes: 1) to deliver power within a narrow bandwidth, and 2) to deliver power over a frequency range (i.e., they are tunable). Each purpose has many subcategories and a large range of specifications to define the oscillator. Table 5.3 gives a summary of oscillator specifications.

Fixed oscillators can be used for everything from narrowband power sources to precision clocks. Tunable oscillators are used as swept sources for testing, FM sources in communication systems, and the controlled oscillator in a PLL. Fixed tuned oscillators will have a power supply input and the oscillator

FIGURE 5.85 A picture of various RF and microwave oscillators. Top to bottom and across from the left there is a crystal oscillator and two YIG oscillators, two chip and wire oscillators in TO-8 cans, a microwave IC oscillator, and three discrete PCB VCOs in packages of decreasing size.

TABLE 5.3 Oscillator Specifications

Specification	Characteristic
Power	Minimum output power (over temperature)
	(Flatness over tuning band if tunable)
Frequency	Accuracy (in Hz or ppm)
	Drift over temperature in MHz/degree C
	Aging in ppm/time
	Phase noise in dbc/Hz (or jitter in picoseconds)
	Pulling in Hz (due to load variation)
	Pushing in Hz/V (due to power supply variation)
	Vibration sensitivity in Hz/g acceleration
Tunable	Bandwidth
	Modulation sensitivity in MHz/V
	Modulation sensitivity ratio (max/min sensitivity)
	Tuning range voltage
	Tuning speed in MHz/microsecond
Power Consumption	V, I DC
Package Style	

output, while tunable sources will have one or more additional inputs to change the oscillator frequency. Some tunable oscillators, particularly those using YIG resonators, will have a second tuning port for small deviations. The theory section will provide the background for understanding all the oscillator specifications.

Specifications

Power Output

Power output and frequency of oscillation are the most basic oscillator specifications [1]. Oscillators with maximal output power are used in industrial applications and usually have more noise due to their extracting as much power as possible from the resonator and thereby lowering the loaded resonator Q. Power output will vary over temperature, so some designs use a more saturated transistor drive or pass the oscillator signal through a limiter to achieve greater amplitude stability. Both of these actions also increase the oscillator noise and cost. Oscillators optimized for low noise, or jitter, usually have low output power to minimize resonator loading and so these designs rely on post-amplification stages to bring the oscillator power up to useful levels for transmitters and radars. As discussed in the theory section, oscillators create more near-carrier noise than amplifiers, so post-amplifiers usually have a minor impact on total oscillator noise. When an oscillator is tunable, the power flatness over the tuning range must also be specified.

Frequency Accuracy and Precision

The frequency accuracy of an oscillator encompasses a large number of sub-specifications because so many things affect an oscillator's frequency. Temperature, internal circuit noise, external vibration, load variations, power supply variations, as well as absolute component tolerance all affect frequency accuracy. We can consider only component tolerances for frequency accuracy and lump all the variations into oscillator precision.

The accuracy of the fundamental frequency of an oscillator is usually specified in ppm or parts per million. So a 2.488 GHz oscillator which is accurate to ±10 ppm will have an output frequency within ±24.88 kHz of 2.488 GHz at the stated temperature, supply voltage, and load impedance. Ambient temperature changes also change the oscillator frequency. The perturbation in a oscillator frequency from temperature is often given in MHz/degree C or ppm/degree C. Manufacturers use several techniques to compensate for temperature changes, such as using an oven to keep the oscillator at a constant temperature such as 70° C, building in a small amount of tuning that is either adjusted digitally or directly from a temperature sensor, and finally resonators can be built with temperature compensating capacitors or cavities [2]. Oscillator components also change with time, which causes a frequency drift due to aging. Aging is usually specified in ppm/year or some other time frame.

Power supply variation affects both the absolute accuracy of an oscillator frequency and the precision with which it maintains that frequency. The sensitivity of an oscillator to power supply variations is called "pushing" and is usually given in MHz/V. Drift in the supply voltage over temperature or with changes in the instrument state affect the accuracy of the frequency while noise on the power supply due to switching circuits will modulate the oscillation frequency through the same pushing mechanism. Time constants in the oscillator bias circuitry will cause the pushing factor to change as the modulation frequency increases, but this is rarely specified. Communication receivers and spectrum analyzers go to great lengths to filter oscillator power supplies.

Load variations also cause changes in oscillation frequency. Because the load on the oscillator output port has some finite coupling to the resonator, changes in the load reactance will change the resonator reactance and so change the oscillation frequency. Typically, a variable length of line is terminated in a standard return loss, such as 12 dB, and the oscillation frequency is measured as the line length is changed. Changing the line length creates a variable load that traces a circle on the Smith Chart. The maximum frequency change is quoted as the "pulling" for the oscillator at the given return loss. Precision oscillators will go through an isolator or a buffer stage to minimize pulling.

Oscillators with cavities and even suspended crystals can be affected by vibration. The vibration sensitivity specification depends on oscillator construction and mounting. Communication systems have been taken down by raindrops hitting the enclosure of an outdoor cavity oscillator. A vibration sensitivity in MHz/g can show sensitivity to vibration, but usually the frequency of the vibration is important as well.

Probably the most common specification of oscillator precision is phase noise or jitter [3]. Phase noise is the frequency domain equivalent of jitter in the time domain. Phase noise will be described in the theory section. Phase noise, FM noise, and jitter are all the same problem with different names. Because an oscillator contains a saturated gain element and a positive feedback loop, it will have very little gain for amplitude noise and a near infinite gain for phase noise. The amplitude and phase variations are with respect to the average oscillation frequency. If an oscillator is measured with a spectrum analyzer with sufficient resolution, the narrow line of the oscillator will appear broadened by noise which falls off at $1/f^3$ or $1/f^2$. The loop gain in an oscillator feedback loop reduces as $1/f^2$ for frequencies other than resonance. The additional $1/f$ factor comes from low frequency modulation within the device or resonator. The phase noise specification is usually given as script $L(f_m) = P_{SSB}(f_m)/Hz/P_C$, which is easily measured with a spectrum analyzer. The original script L definition noted that $P_{SSB}(f_m)/Hz$ was to be the phase noise power in one Hertz of bandwidth, but often people take the spectrum analyzer measurement and call it phase noise because phase noise dominates oscillator noise close to the carrier [4]. In reality, the spectrum analyzer cannot tell the difference between amplitude and phase noise. Whenever the phase noise approaches the noise floor or even a flat noise pedestal, it is likely that a significant amount of amplitude noise is present. In all of the above, f_m denotes the offset frequency from the carrier and corresponds to the frequency modulating the carrier. These same variations are called FM noise when measured with a frequency discriminator.

When digital systems are characterized, the time domain specification of jitter is more common than phase noise. Deviations in expected zero crossing times are measured and accumulated to give peak-to-peak and rms values. These values are given in picoseconds or in UI (unit intervals). UIs are just a fraction of the clock period, so UI = (jitter in picoseconds)/(clock period). Because phase noise shows the phase deviation at each frequency modulating the carrier, we can sum up all of the phase deviations and reach a total phase deviation which, when divided by 360°, also gives the jitter in UIs. This is the same as being able to compute the total power of a signal in frequency or in time. Often communication systems are more sensitive to jitter at certain modulation frequencies than others, so a tolerance plot of phase noise vs. offset frequency is given as a jitter specification [5]. The frequency domain view of jitter also makes it clear why PLLs work as "jitter attenuators." The narrow bandwidth of the PLL feedback loop filters

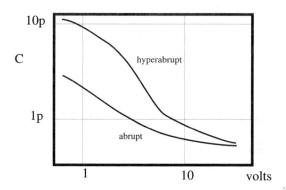

FIGURE 5.86 Varactor C-V plots.

higher modulation frequencies and so reduces the total phase deviations, but only if a significant amount of the jitter is due to frequencies above the loop bandwidth.

Tuning Bandwidth Specifications

For tunable oscillators there is an additional set of specifications. Typically broadband tunable oscillators have their bandwidth specified in terms of minimum and maximum frequency (e.g., f_{Max} and f_{Min}) with the center frequency not mentioned. Narrowband tunable oscillators, those with tuning bandwidths of 10% or less, have their center frequency and bandwidth specified. The tuning range is the voltage range of the tuning port for varactor-tuned oscillators (e.g., $V_{Max} - V_{Min}$) and the current range of the tuning ports for YIG tuned oscillators. Usually the minimum varactor voltage is greater than zero because the varactor diode needs reverse bias to maintain a high Q under the swing of the oscillator signal. The modulation sensitivity is the MHz change per volt at the tuning port. Often the modulation sensitivity is not equal to $(f_{Max} - f_{Min})/(V_{Max} - V_{Min})$ because the tuning sensitivity changes over the tuning range. The modulation sensitivity is usually measured at the center of the tuning range with a small voltage deviation. The modulation sensitivity ratio gives the ratio of maximum to minimum modulation sensitivity over the tuning range. This is especially important for varactor-tuned oscillators used in PLLs because the loop gain will vary by the modulation sensitivity ratio. At low voltages varactors have their maximum capacitance, as shown in Fig. 5.86 [6]. The capacitance rapidly decreases as the tuning voltage increases until a minimum capacitance plateau is reached. The large capacitance change at low voltages means that the oscillator will have a more rapid frequency change at lower tuning voltages than at higher tuning voltages. This also means that the modulation sensitivity is higher at the minimum output frequency than it is at the maximum output frequency. The doping profile of a varactor affects its capacitance vs. voltage curve. The simplest doping profile is an abrupt junction that gives the curve shown in Fig. 5.86. A hyper-abrupt junction C-V curve is also shown in Fig. 5.86. The hyper-abrupt curve will give the oscillator a more linear tuning characteristic, or modulation sensitivity ratio closer to unity. Another approach to linearizing a varactor oscillator is to shape the tuning voltage with an analog diode shaping network or with a digital lookup table [7-8]. YIG oscillators have an inherently linear tuning characteristic as given in Eq. (5.20) [9,56]. In Eq. (5.20), H_O is the magnetic bias field strength in Oersteds (Oe), H_a is the internal anisotropy field, and γ is 2.8 MHz/Oe.

$$f_{YIG} = \gamma \left(H_O \pm H_a \right) \tag{5.20}$$

The last specification is the tuning speed in MHz/second. This parameter determines the maximum modulation rate of an oscillator, or its agility in a frequency hopping application. Varactor-based oscillators can tune much faster than YIG based oscillators. For example, for 10 GHz oscillators, the tuning

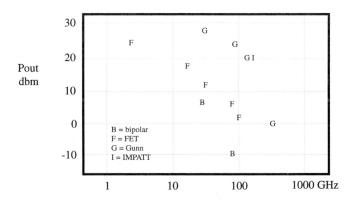

FIGURE 5.87 Device technology and power capability [10–12,57–63].

port of a VCO will typically have a bandwidth of 100 MHz, while the FM coil of a YIG oscillator will have a bandwidth of only 500 kHz.

Technologies and Capabilities

All of the characteristics discussed above depend on three aspects of oscillator technology: 1) active device; 2) resonator; and 3) packaging. Packaging mainly affects the oscillator's cost, size, temperature stability, susceptability to mechanical vibration, and susceptability to interference. Resonator technology mainly affects the oscillators cost, phase noise (jitter), vibration sensitivity, temperature sensitivity, and tuning speed. Device technology mainly affects the oscillator maximum operating frequency, output power, and phase noise (jitter).

Figure 5.87 shows various device technologies and their power vs. frequency capability [10–12,57–63]. While Fig. 5.87 shows fundamental frequency power, in many cases it is most cost effective to use a frequency multiplier to move a lower frequency oscillator up to a higher frequency [13]. While there is always a power loss from frequency multiplication, there is usually very little noise penalty because for equal resonator Qs, oscillator phase noise is proportional to the operating frequency [14]. Frequency multipliers can be simple resistive diode nonlinearities, tuned varactor diode circuits, or PLLs. Resistive nonlinearities are the simplest and broadest band, but have the most loss. Varactor multipliers can be extremely efficient and are used in the highest frequency multipliers [15]. At very high frequencies PLLs are implemented via subharmonic injection locking so that high frequency external frequency dividers and phase comparators are not needed [16,63]. The practical problem with injection locking is that without an external phase detector it is difficult to verify that the oscillator is in lock.

Most small signal oscillators are designed to source 0 to 20 dbm of power. Power oscillators are made to deliver watts of power, but frequency stability suffers due to extracting more power from the resonator and lowering the loaded Q. When stability is a concern, small signal oscillators are built and followed by a carefully designed chain of amplifiers.

Device Technologies

Bipolar Transistors
Silicon bipolar transistors are used in most low noise oscillators below 5 GHz. Hetero-junction bipolar transistors (HBTs) are common today and extend the bipolar range to 100 GHz as shown in Fig. 5.87. Bipolar transistors have high gains of over 20 dB at frequencies below 1 GHz and typically have $1/f$ corners in the kHz region. The $1/f$ corner of the oscillator phase noise is less than or equal to the $1/f$ corner of the active device because although the device low frequency noise modulates the device bias and causes phase modulation, it may not be enough to overcome the high frequency noise modulations. In crystal oscillators typically the resonator $1/f$ noise dominates [17,46]. The device $1/f$ noise corner scales with

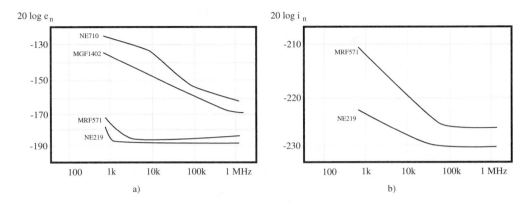

FIGURE 5.88   Device $1/f$ noise comparison for equivalent voltage input noise (a), and equivalent current input noise, (b) [31]. FET input current noise is not shown because it is so small.

f_{MAX} within a given technology, so smaller devices with higher f_{MAX} will have higher $1/f$ noise corners. A 1 kHz $1/f$ corner is typical for a 10 GHz transistor while a 100 MHz transistor will have a $1/f$ corner in the tens of Hertz. Device technology such as ion implantation will raise the $1/f$ noise corner to 100 kHz or higher by introducing traps in to the device. $1/f$ noise is very sensitive to device construction [18]. The $1/f$ noise in a bipolar transistor is concentrated in the base current, as shown in Fig. 5.88. The $1/f$ noise is due to traps at the base-emitter edge.

MOSFETs
CMOS integrated oscillators are becoming more common, although they are limited to the low GHz region [19,20]. Typically two transistors are used in a free running flip-flop configuration. CMOS transistors have lower gain, higher $1/f$ noise corners, and less output power than bipolar transistors, but they offer high integration density and low cost. The higher $1/f$ noise corner of CMOS is mitigated by the reduced modulation sensitivity of the device and by using balanced configurations to reduce noise modulation by symmetry [21].

JFETs
JFETs have excellent low frequency noise and limited gain relative to bipolar transistors. While JFETs are found in some extremely low noise discrete oscillators, they are not as common as other devices, especially above 200 MHz [22,54].

MESFETs and HEMTs
Above 5 GHz MESFETs and HEMTs are the most common 3-terminal oscillator engine. MESFETs and HEMTs have less gain than bipolar transistors at low frequencies, but have a much higher maximum frequency of operation, or f_{MAX}, as shown in Fig. 5.87. HEMT f_{MAX} is higher than that of MESFETs due to transistor construction that maximizes mobility and provides better channel confinement and control [6]. Both MESFETs and HEMTs have much higher low frequency $1/f$ noise than bipolar transistors, with corner frequencies in the 10 to 100 MHz range being typical for a 600 um device. As with bipolar transistors, the $1/f$ corner scales with f_{MAX} of the device within a given process. Specifically, the $1/f$ level scales with the channel volume beneath the gate structure. GaAs MESFETs and HEMTs have problems with surface traps because of the lack of a native oxide, and there are problems with substrate and channel traps due to the material layering inside the FET. These traps are thermally spread into an approximately continuous $1/f$ distribution at room temperature [23,24]. As shown in Fig. 5.88, the $1/f$ noise for a MESFET or HEMT is mostly in the equivalent gate voltage noise source. The reduced low frequency gain of FETs gives them less modulation sensitivity, so they typically have phase noise levels only 10 dB worse than bipolar oscillators even though the low frequency noise is often 30 dB worse.

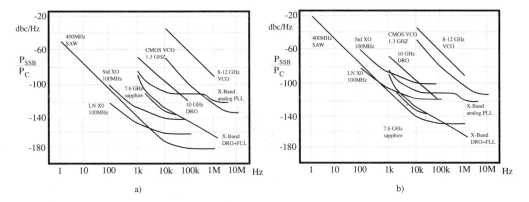

FIGURE 5.89 Oscillator noise performance of some microwave sources: (a) actual; and (b) referred to 10 GHz for comparison (scaled by 20 log[10 GHz/f$_{OSC}$]) [64–67].

Diodes

Diodes have the highest maximum useable frequencies for solid state devices, as shown in Fig. 5.87. There are many different types of diodes for generating negative resistances and negative conductances. The diode negative immittance cancels the positive loss of the resonator and allows an oscillation to build up from the noise within the device. Traditionally microwave oscillators have been designed as negative immittance devices because it is much easier to measure reflections than to set up feedback loops at microwave frequencies. The distinction between negative resistance devices, such as IMPATT diodes, and negative conductance devices, such as Gunn diodes, is important because the device IV characteristic determines how the device saturates and whether it is stable with a series resonator or shunt resonator.

IMPATT diodes generate negative resistances and so are used at series resonant points in waveguides and planar circuits [25]. The avalanche mode of IMPATT operation creates high power but at the cost of high noise levels. Gunn diodes have an inherently quiet Gunn domain negative conductance that requires a parallel resonant circuit [26]. Gunn diodes are among the quietest high frequency oscillators and exhibit excellent power into the 100s of GHz, as shown in Fig. 5.87. The bulk nature of the Gunn device means that no third terminal metallization is required. This lack of a third terminal and bulk mode of operation reduces the device 1/f noise. As with FETs, more advanced material structures, such as using InP rather than GaAs, maximize the high frequency performance of Gunn diodes [12].

Multipliers

Frequency multipliers will always be a way of generating the highest frequencies. Reactive multipliers using varactor diodes offer low noise and efficient power generation almost to 1 THz [27]. Resistive multipliers are simple, broadband alternatives to tuned reactive multipliers. Resistive multipliers also suffer significant conversion losses, but are commonly used in broadband instrumentation. All frequency multipliers will increase the phase noise by the same factor that they multiply frequency because frequency and phase are both multiplied, as shown in Eq. (5.21). In dB this would be 20 log N. For example, if the oscillator signal is $V_O(t) = A \, Cos(\omega_O t + \phi(t))$ then a times two multiplier would generate:

$$V_O(t)^2 = A^2 \left(\frac{1}{2} + \frac{1}{2} Cos\left(2\omega_O t + 2\phi(t) \right) \right) \tag{5.21}$$

Resonators

Table 5.3 shows an overview of resonator technologies for oscillators. Various abbreviations are used in the above table, with YIG being Yttrium-Iron-Garnet, TL being transmission line, DR being dielectric resonator, and SAW being surface acoustic wave. Resonator choice is a compromise of stability, cost, and size. Generally, Q is proportional to volume, so cost and size tend to increase with Q. Technologies such

TABLE 5.3   Resonators

Type	Q Range	Range (GHz)	Limitation	Benefit
LC	0.5–200	Hz–100 GHz	Q, lithography	cost
Varactor	0.5–100	Hz–100 GHz	Q, nonlinear, noise	tunable
Stripline	100–1000	MHz–100 GHz	size, lithography	cost, Q
Waveguide	1000–10,000	1–600 GHz	size, cost	Q
YIG	1000	1–50 GHz	cost, magnet, tuning speed	Q, tunable, linear
TL	200–1500	500 MHz–3 GHz	cost	Q, temperature stable
DR	5000–30,000	1–30 GHz	cost, size	Q, temperature stable
Sapphire	50 k	1–10 GHz	cost, size	Q
Quartz	100 k–2.5 M	kHz–500 MHz	frequency	Q, temperature stable
SAW	500 k	1 MHz–2 GHz	frequency, cost	Q

FIGURE 5.90   A picture of various resonators. From the left are three transmission line resonators for 500 MHz to 2 GHz operation, two dielectric resonators for 7 and 20 GHz operation, a 10 MHz crystal resonator, and a 300 MHz SAW resonator. The resonators are sitting on top of a 2.5 inch diameter dielectric cylindrical resonator for 850 MHz.

as quartz, SAW, YIG, and DR allow great reductions in size while achieving high Q by using acoustic, magnetic, and dielectric materials, respectively. Most materials change size with temperature, so temperature-stable cavities have to be made of special materials such as Invar or carbon fiber. Transmission line, dielectric resonator, and quartz resonators can easily have temperature coefficients below 10 ppm. Q changes with frequency for most resonators. Capacitors and dielectric resonators have Qs that decrease with frequency, while inductors and transmission line resonators have Qs that increase with frequency. Quartz resonators are an extremely mature technology with excellent Q, temperature stability, and low cost. Most precision microwave sources use a quartz crystal to control a high frequency tunable oscillator via a PLL. Oscillator noise power, and jitter, is inversely proportional to Q^2, making high resonator Q the most direct way to achieve a low noise oscillator.

Tunable resonators are very important because they offer the ability to transfer a reference frequency, with or without modulation, through a PLL. Tunable resonators also offer direct modulation and frequency agility for communication and test purposes. Varactor diodes are the most common device for tuning an oscillator. These devices are inexpensive, available in a variety of packages, and can be used at almost any frequency of interest. Varactors also offer rapid tuning for frequency hopping and high speed direct modulation. The only disadvantages of a varactor diode are low Q at high frequencies, low frequency noise, and a nonlinear tuning characteristic [6].

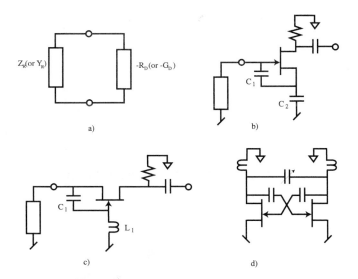

FIGURE 5.91   Oscillator configurations: a) diode; b) source feedback; c) gate feedback; and d) cross-coupled.

YIG resonators offer the advantages of tuning linearity and high Q. These resonators are excellent for instrumentation and special applications, but suffer from the needing a magnetic bias circuit, which increases the size and cost of the oscillator. Typically YIG resonators have both a broadband and a narrowband tuning port [9, 56]. The narrowband tuning port requires much less inductance and so can be tuned faster than the broadband port, making it more useful for modulation.

Theory

Introduction

A brief review of oscillator theory will aid in understanding the oscillator specifications mentioned in the first section. First, an overview of oscillator topologies is shown in Fig. 5.91. Traditionally, microwave oscillators have been viewed as one-port circuits with the active device presenting a negative immittance to the resonator, as shown in Fig. 5.91a [28]. The one-port philosophy is easy to measure with a slotted line or a network analyzer. Two-port oscillators are much more common at low frequencies, but probing voltages and currents in a feedback loop will always be difficult at the highest frequencies. Dielectric resonators have made feedback oscillators more common at microwave frequencies [29]. Integrated circuits have made the cross-coupled oscillator configuration, as shown in Fig. 5.91d, popular [30].

One-port analysis does allow confusion over the device acting as a negative impedance or admittance. Knowing the device type is essential for establishing a stable oscillation. For example, a negative conductance Gunn diode has the IV characteristic shown in Fig. 5.92. As the oscillation signal grows about the bias point it eventually extends into the positive resistance region. During saturation the load line becomes more horizontal, reducing the negative conductance. The resonator needs to have its minimum conductance at the resonance frequency, so that moving off the resonant frequency would require an increase in the device negative conductance. If a Gunn diode is loaded with a series resonant circuit, which has maximum conductance at resonance, noise will move the oscillation off the series resonance and onto a nearby parasitic parallel resonance where it will stabilize.

$$Z_D + Z_R = 0 \tag{5.22}$$

$$Y_D + Y_R = 0 \tag{5.23}$$

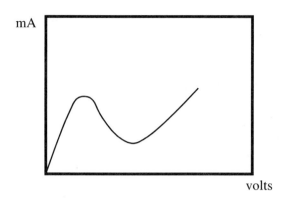

FIGURE 5.92 Gunn diode IV characteristic.

$$G\,H = 1 \qquad\qquad (5.24)$$

The various oscillation conditions can be defined by the preceding three equations. Eq. (5.22) describes the active device impedance, Z_D, canceling the resonator impedance, Z_R, to support an oscillation. Eq. (5.23) describes the active device admittance, Y_D, canceling the resonator admittance, Y_R, to support oscillation. Eq. (5.22) can be split into real and imaginary parts to give $R_D + R_R = 0$, and $X_D + X_R = 0$. Often X_D is small, so the equations reduce to the active device negative resistance canceling the resonator (and load) resistance while the resonator is just off center frequency enough to cancel the device reactance. From a one-port point of view, when the resonator reactance is zero, no net phase shifts occur from the oscillation signal as it reflects back and forth from the active device to the resonator. Eq. (5.24) is the oscillation equation for a feedback circuit with gain element G and feedback resonator H. Eq. (5.24) describes a positive feedback situation where the gain cancels the loss in the feedback while the net phase shift around the loop is zero. With zero phase shift around a loop and no loss, a signal will be sustained at the frequency of zero phase. The above equations describe a linear approximation to a stable oscillation. In reality each circuit is nonlinear. Oscillations start from noise in a circuit with positive feedback and grow until the circuit gain element saturates and the above equations are satisfied. Typically oscillators are set up so that the negative immittance, or gain, is 1.5 to 2 times greater than the circuit loss so that the device saturates into a stable oscillation without being driven so hard that its operating point changes excessively [2]. All of the above equations can be brought into a single oscillator theory [31].

At lower frequencies the oscillator output power can be predicted analytically [32]. At microwave frequencies the accuracy of the oscillator frequency and output power is very dependent on the CAD model used. Both harmonic balance and SPICE simulators can be used to predict oscillator output power and frequency, but component and circuit parasitics can make exact frequency predictions difficult. Linear simulators have a long history in oscillator design, as might be predicted by looking at Eqs. (5.22)–(5.24). The operating frequency, tuning range, as well as sensitivities to load, bias, and power supply variations can all be obtained from a linear simulator with bias dependent S-parameters for the active device.

Although many oscillator circuits are used, Figs. 5.91b and 5.91c show two of the most common discrete configurations. Figure 5.91b is used mostly with varactor-tuned inductors and dielectric resonators. Figure 5.91b is also used at lower frequencies where it is known as the Seiler oscillator [33]. Figure 5.91c is used mostly with YIGs, transmission line resonators, and dielectric resonators [34]. Simple analysis using an ideal transconductance for the device shows a negative resistance with capacitance at the gate of Fig. 5.91b and a negative conductance with shunt inductance at the source for Fig. 5.91c. Therefore, Fig. 5.91b operates in the inductive region of a series resonator, and Fig. 5.91c operates in the capacitive region of a parallel resonator. At microwave frequencies C_1 is simply the device capacitance. Close examination of Figs. 5.91b and 5.91c shows that they are both the same circuit. The only real difference is where the tuning takes place. Each circuit uses the device drain to provide load isolation and expects to have a relatively low impedance in the device drain to take power out. Eq. (5.25) shows how the negative input resistance of the source feedback oscillator changes with frequency, and that the residual reactance is capacitive. The input impedance in Fig. 5.91b is that from the resonator looking into the device node. Eq. (5.26) shows how the negative conductance of the gate feedback oscillator varies with frequency and that its residual susceptance is inductive. Note that the conductance is negative above the L_1C_1 resonance frequency and rapidly decreases with frequency. Both of these equations can be used for estimating the tuning bandwidth and oscillation frequency of an oscillator. The resonator used in either case will have its own immittance which can be compared to the device immittance using Eqs. (5.22) or (5.23) to determine if the oscillation condition is satisfied. CAD programs with S-parameter models will provide more accurate device characterization and tuning bandwidth analysis. For example, the configurations of Fig. 5.91 can be analyzed as one-ports for S11, $S11_D$. The resonator S11, $S11_R$, forms a reflection loop so that Eq. (5.24) is valid if $G = S11_D$ and $H = S11_R$.

$$Z_{IN} = -g_m / C_1 C_2 \omega^2 - j(C_1 + C_2)/C_1 C_2 \omega^2 \qquad (5.25)$$

$$Y_{IN} = g_m / (1 - \omega^2 L_1 C_1) + j\omega C_1 / (1 - \omega^2 L_1 C_1) \qquad (5.26)$$

Modulation, Noise, and Temperature

Many things perturb the oscillator frequency. To better study these perturbations consider the idealized oscillator shown in Fig. 5.93. Note that this oscillator is connected in a positive feedback configuration. The active device could be two FETs connected to create positive feedback, or it could be a single negative conductance shunting the resonator. The feedback configuration is chosen because it clarifies that the loaded Q of the oscillator is:

$$Q_L = 1/(\omega_o L_O G_O). \qquad (5.27)$$

The loaded Q is only determined by circuit losses and external loading. The loaded Q is unaffected by the device g_m or negative conductance. The total losses, G_O, will be made up of internal losses, G_I, and external loading, G_{EXT}, so $G_O = G_I + G_{EXT}$. A little math will show that the total oscillator Q is a parallel combination of the internal Q_I and the external Q_{EXT}, as shown in Eq. (5.28).

$$Q_L = (1/Q_I + 1/Q_{EXT})^{-1} \qquad (5.28)$$

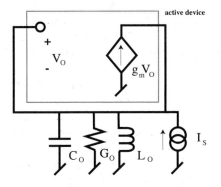

FIGURE 5.93 Idealized oscillator.

The total oscillator Q is important because it determines the sensitivity to perturbations in oscillator frequency. Unfortunately, all we can measure directly is the external Q of an oscillator. Load pulling, as given in the pulling sensitivity specification, can be used to compute the external Q of an oscillator [22]. Load variations affect the oscillation frequency by changing the values of C_O or L_O in the same way G_{EXT} changes the value of G_O. Typically load pulling is measured by placing a 12 dB return loss load on a line stretcher and rotating the load through all phases while recording the maximum frequency deviation. The C_O and L_O of Fig. 5.93 are made up of device capacitance, resonator elements, and load reactance. In low noise oscillators the resonator reactances dominate, but all oscillators have some influence from the other sources of reactance.

Power supply voltage variations change the oscillation frequency by changing the active device reactances and bias. The pushing specification in MHz/V defines the sensitivity to low frequency supply variations. Knowing the spectrum of the power supply variations allows the direct calculation of oscillator FM noise due to power supply variations through the pushing sensitivity. As the frequency of the supply variation is increased, the supply and bias filtering circuitry will reduce its influence on the oscillation frequency, so some measurements may have to be made if the power supply has large variations at high frequencies. Pushing can also be used as a technique to evaluate oscillator Q and minimize phase noise [69].

As discussed in the section on noise, the oscillator phase noise spectrum is the FM noise spectrum divided by f_m^2, where f_m is the modulating, or carrier offset frequency. Oscillator frequency variations are easily measured with a frequency discriminator or other FM detector such as a transmission line discriminator [35]. Spectrum analyzers are also very useful for measuring oscillator phase noise as long as the analyzer noise is much less than that of the measured device. Spectrum analyzers can measure script $L(f_m) = P_{SSB}(f_m)/Hz/P_C$ where the $P_{SSB}(f_m)/Hz$ is the phase noise power per Hertz within the $1/f^3$ region close to the carrier. Typically if the noise measurement is very near the carrier frequency, i.e., within the f^{-3} region, the noise is dominated by phase noise. For modulating noise such as power supply pushing the actual phase noise spectrum, $S_\phi(f_m)$, is twice script $L(f_m)$ because the sidebands are correlated [31].

All active devices contain internal noise sources due to resistance (thermal noise), charge crossing an energy barrier (shot noise), and traps (G-R noise) [18]. All of these noise sources perturb the device bias point no matter how well filtered the bias circuit. The most significant source of bias noise is due to traps [31]. These traps are spread in energy distribution by thermal and mechanical processes [23,36,37]. The spreading of the traps in bipolar transistors, FETs, and MOSFETs results in a $1/f$ low frequency noise spectrum in each device. FET based devices have $1/f$ variations in the drain current which converts to an equivalent input noise via the device transconductance. Bipolar devices have $1/f$ variations in the base current due to traps at the edge of the base-emitter junction [38]. The low frequency bias variations change the device reactances and so change the oscillation frequency [31,39–41]. The oscillator sensitivity to bias current or voltage variation is easily simulated and combined with measured device noise to provide a prediction of oscillator noise. Although simulators have estimates of device $1/f$ sources, low frequency noise sources are so dependent on device processing that measuring low frequency noise is the best way to verify noise performance. Once the bias sensitivity and device noise are determined, the oscillator internal phase noise can be computed in a similar manner to power supply pushing. Various schemes using symmetry, low frequency device loading, and even low frequency feedback exist for reducing total oscillator phase noise [42–45]. Even if device noise can be eliminated, resonators such as quartz crystals have their own $1/f$ noise sources that modulate the oscillator spectrum [46].

For voltage- or current-controlled oscillators the noise spectrum contains an additional term due to noise at the modulation port. Even modulating devices, such as varactor diodes, have internal noise that affects the oscillator spectrum. These noise sources can be accounted for just like the pushing and internal contributions. While there are many contributors to oscillator noise, each term is easily accounted for once the modulation sensitivity is known. Some noise contributions, such as power supply switching noise, can come through several paths so the total noise cannot just be a power summation of the individual contributions, but must include any correlation between contributions. In most oscillators one noise source will dominate [31]. For example, in broadband varactor-controlled oscillators, VCOs, the modulation noise usually dominates.

Figure 5.93 also shows a current source, I_S, which adds to the oscillation. This current source is useful for analyzing the oscillator response to high frequency noise and injection locking signals. The modulative low frequency noise sources discussed earlier operate in a multiplicative or nonlinear way, whereas noise sources at the oscillator frequency simply add to the oscillator noise in a linear manner [47]. For additive noise sources, increasing the oscillator power decreases the oscillator phase noise because the additive noise sources are fixed. However, for modulative noise sources increasing the oscillator power does not change the relative noise power because the modulation affects a fraction of the total oscillator power. This seemingly "linear" behavior is the result of any second order nonlinearity, and has caused confusion for many people analyzing oscillator noise. This same fact is why reducing even order nonlinearities through symmetry is effective in reducing oscillator noise.

If I_S only consists of high frequency noise due to thermal and shot noise in the device and circuit, we can produce a simple noise analysis of the oscillator. A more rigorous analysis is given in References 31 and 47. The positive feedback of the oscillator loop will cause the noise at the center of the resonator frequency to be amplified until the amplifier nonlinearity causes device saturation. The amplitude saturation does two things: 1) it reduces the device gain until Eq. (5.24) is satisfied; and 2) it reduces the loop gain to amplitude perturbations. I_S will cause both amplitude and phase variations in the oscillator carrier buildup. Amplitude saturation in the device will effectively strip the amplitude noise off the carrier as it passes through the active device, which means the amplitude noise of V_O will be only I_S passed through the resonator, or

$$S_{V\_AM}(f_m) = \left(S_{Is}(f_O + f_m) + S_{Is}(f_O - f_m)\right) \Big/ \left(2V_O^2/2\right) 1 \Big/ \left|G_O + j\left(\omega C_O - 1/(\omega L_O)\right)\right|^2, \quad (5.29)$$

which, near to the carrier, simplifies to

$$S_{V\_AM}(f_m) \approx S_{Is}(f_O + f_m) \Big/ \left(G_O^2 V_O^2/2 \left(1 + 4Q^2 f_m^2/f_O^2\right)\right), \quad (5.30)$$

where $V_O^2/2$ represents the carrier mean square level and $S_{Is}(f_O + f_m)$ is approximately equal to $S_{Is}(f_O - f_m)$.

The noise source, I_S, will perturb the carrier with an equal amount of amplitude and phase noise, which can be seen by envisioning each noise sideband as a phasor rotating about the carrier phasor. The limiting action of the active device does not attenuate the phase modulation of the carrier, so the full loop gain acts on the phase variations causing phase noise to be greatly amplified [48]. The phase noise amplification causes the phase noise to dominate amplitude noise near the carrier. Just as in the amplitude noise case, a low pass equivalent analysis can be performed on the circuit [49]. Solving the low pass equivalent form for phase noise variations is Leeson's model for phase noise [50]. Equation (5.31) gives the phase noise spectral density for the circuit of Fig. 5.93.

$$S_{V\_\phi}(f_m) \approx S_{Is}(f_O + f_m) \Big/ \left(G_O^2 V_O^2/2 \; 4Q^2 f_m^2/f_O^2\right), \quad (5.31)$$

The term $S_{Is}(f_O + f_m)/(G_O^2 V_O^2)$ is equal to $kTF/P_C (1 + f_k/f_m)$ in Leeson's analysis, where P_C is the carrier power and F is approximately the device noise figure. The $1 + f_k/f_m$ term accounts for $1/f$ noise modulation with f_k being less than or equal to the device $1/f$ noise corner. Eq. (5.14) is Leeson's equation. Although the preceding analysis assumed uncorrelated high frequency noise sidebands, the inclusion of a $1/f$ term in Leeson's result implies correlated sidebands and a factor of 2 between script L and phase noise spectral density, S_ϕ. Leeson's model for phase noise contains most of the important aspects of oscillator noise in a simple equation. More rigorous oscillator analysis appears with regularity, but Leeson's model continues to be useful and relevant [28,31,39]. Naturally, it is also possible to analyze oscillator noise in the time domain [51,52]. Several important aspects of oscillator noise predicted by Leeson's model are that the noise power decreases as Q^2, that the noise falls off as f_m^{-3} near the carrier, and that the noise power increases as f_O^2. The problems with Leeson's model are the approximations involved. Because oscillators are not linear, noise-matched amplifiers F cannot be used. Because the low frequency noise modulates the carrier, script L does not decrease in proportion to increases in P_C in well-designed, low-noise oscillators. And finally, f_k in the final oscillator depends on all the low frequency noise sources and modulation sensitivities and is usually less than or equal to the device $1/f$ noise corner for the dominant $1/f$ noise source.

$$\text{script } L(f_m) \approx kTF \big/ P_C \left(1 + f_k / f_m\right) \big/ \left(4Q^2 f_m^2 / f_O^2\right) \tag{5.32}$$

Injection Locking

Very high frequency oscillators are often phase locked by coupling a reference signal directly into the oscillator [28,53,55]. Understanding injection locking can help in understanding oscillator operation in general. In Fig. 5.93 we can let I_S be an injection locking signal, $I_S e^{j\omega t}$. Then $V_O = V_O e^{j\omega t + \phi}$, where the phase shift ϕ occurs if ω is different from the oscillator loop center frequency of ω_O. From linear circuit analysis we can derive Eq. (5.33), and Eq. (5.34) follows from Euler's identity.

$$V_O e^{j\omega t + \phi}\left[1 - \omega^2 L_O C_O + j\omega L_O \left(G_O - g_m\right)\right] = j\omega L_O I_S e^{j\omega t} \tag{5.33}$$

$$V_O\left[1 - \omega^2 L_O C_O + j\omega L_O \left(G_O - g_m\right)\right] = j\omega L_O I_S \left(\cos(\phi) - j\sin(\phi)\right) \tag{5.34}$$

By grouping real and imaginary parts, Eq. (5.34) can be expanded into Eqs. (5.35) and (5.36) which define the locking gain and the bandwidth of locking.

$$I_S = V_O \left(G_O - g_m\right) \big/ \cos(\phi) \tag{5.35}$$

$$\sin(\phi) = \left(1 / \omega L_O - \omega C_O\right) V_O / I_S \tag{5.36}$$

Equation (5.35) shows that as the device saturates and g_m approaches G_O, very little current, I_S, is required to maintain lock. However, as the injection frequency shifts away from the resonator frequency, ω_O, ϕ moves away from zero as shown in Eq. (5.36). As ϕ increases from zero to $\pm 90°$, the $\cos(\phi)$ term in Eq. (5.35) decreases to zero and an infinite injection current is required to maintain lock, so ϕ equal to $\pm 90°$ defines the locking bandwidth. Equation (5.36) can be used to translate the $\pm 90°$ limits into frequency limits that define the locking bandwidth. Equation (5.35) shows that very little injection locking signal is required at the band center while more signal is required as the frequency approaches the band edges. Another way of interpreting this is to say there is near infinite injection locking gain at the band center and zero locking gain at the band edges. When I_S is broadband noise this injection gain works to provide the commonly observed high levels of noise that decrease away from the carrier [48]. The relationship between the gain and offset frequency is linked through the loop phase shift. An injection locked oscillator is in fact a first order PLL.

Summary

Oscillators consist of an active device, a resonator, and a package. These three things determine the frequency, accuracy, available power, and cost of the source. For a simple task such as providing a sinusoid, oscillators require an inordinate number of specifications. The sections on specifications and theory try to provide a background for understanding oscillator requirements, while the technologies and capabilities section tries to show how modern devices and resonators are combined to meet oscillator specifications.

Acknowledgment

Thanks to Mark Shiman of Disman Bakner and Ron Korber of Stellex for providing various samples for the photographs.

References

1. Leier, R.M., and Patston, R.W., Voltage-Controlled Oscillator Evaluation for System Design, *MSN*, 102–125, Nov. 1985.
2. Rogers, R.G., *Low Phase Noise Microwave Oscillator Design*, Artech House, Boston, 1991.
3. Robins, W.P., *Phase Noise in Signal Sources*, Peter Peregrinus Ltd., London, U.K., 1982.
4. Blair, B.E., *Time and Frequency: Theory and Fundamentals*, U.S. Department of Commerce, NBS Monograph 140, 1974.
5. ANSI, *Telecommunications — Synchronous Optical Network (SONET) — Jitter at Network Interfaces*, T1.105.03-1994, 1994.
6. Bahl, I.J., and Bhartia, P.B., *Microwave Solid State Circuit Design*, John Wiley & Sons, New York, 1988.
7. Engineering Staff, *Nonlinear Circuits Handbook*, Analog Devices, Norwood, MA, 1976.
8. Huckleberry, B.E., Design Considerations for a Modern DTO, *Microwave Journal*, 291–295, May 1986.
9. Osbrink, N.K., YIG-Tuned Oscillator Fundamentals, *Microwave System Designer's Handbook*, 207–225, 1983.
10. Heins, M.S., Juneja, T., Fendrich, J.A., Mu, J., Scott, D., Yang, Q., Hattendorf, M., Stillman, G.E., and Feng, M., W-band InGaP/GaAs HBT MMIC Frequency Sources, *IEEE MTT-S Digest*, 239–242, 1999.
11. Dieudonne, J.-M., Adelseck, B., Narozny, P., and Dambkes, H., Advanced MMIC Components for Ka-Band Communication Systems: A Survey, *IEEE MTT-S Digest*, 409–415, 1995.
12. Eisele, H., and Haddad, G.I., Potential and Capabilities of Two-Terminal Devices as Millimeter- and Submillimeter-Wave Fundamental Sources, *IEEE MTT-S Digest*, 933–936, 1999.
13. Faber, M.T., Chramiec, J., and Adamski, M.E., *Microwave and Millimeterwave Diode Frequency Multipliers*, Artech House, Boston, 1995.
14. Scherer, D., Generation of Low Phase Noise Microwave Signals, *RF & Microwave Measurement Symposium*, Hewlett-Packard, Palo Alto, CA, 1983.
15. Bruston, J., Smith, R.P., Martin, S.C., Humphrey, D., Pease, A., and Siegel, P.H., Progress Towards the Realization of MMIC Technology at Submillimeter Wavelengths: A Frequency Multiplier to 320 GHz, *IEEE MTT-S Digest*, 399–402, 1998.
16. Roberts, M.J., Iezekiel, S., and Snowden, C.M., A Compact Subharmonically Pumped MMIC Self-Oscillating Mixer for 77 GHz Applications, *IEEE MTT-S Digest*, 1435–1438, 1998.
17. Bates, P.C., Measure Residual Noise in Quartz Crystals, *Microwaves & RF*, 95–106, Nov. 1999.
18. Ambrozy, A., Electronic Noise, McGraw-Hill, New York, 1982.
19. Banu, M., MOS Oscillators with Multi-Decade Tuning Range and Gigahertz Maximum Speed, *IEEE JSSC*, 1386–1393, Dec. 1988.
20. Svelto, F., Deantoni, S., and Castello, R., A 1.3 GHz Low-Phase Noise Fully Tunable CMOS LC VCO, *IEEE JSSC*, 356–361, Mar. 2000.
21. Aoki, H., and Shimasue, M., Noise Characterization of MOSFET's for RF Oscillator Design, *IEEE MTT-S Digest*, 423–426, 1999.
22. Vendelin, G., Pavio, A.M., and Rohde, U.L., *Microwave Circuit Design*, John Wiley & Sons, 1990.
23. Sodini, D., Touboul, A., Lecoy, G., and Savelli, M., Generation-Recombination Noise in the Channel of GaAs Schottky gate FET, *Electronics Letters*, 42–43, Jan. 22, 1976.
24. Hughes, B., Fernandez, N.G., and Gladstone, J.M., GaAs FETs with a Flicker Noise Corner Below 1 MHz, *WOCSEMMAD*, 20–23, 1986.

25. Goedbloed, J.J., Noise in IMPATT Diode Oscillators, *Philips Research Reports Supplement*, 1–115, 1973.
26. Sze, S.M., *Physics of Semiconductor Devices*, John Wiley & Sons, New York, 1981.
27. Crowe, T.W., Weikle, R.M., and Hesler, J.L., GaAs Devices and Circuits for Terahertz Applications, *IEEE MTT-S Digest*, 929–932, 1999.
28. Kurokawa, K., Noise in Synchronized Oscillators, *IEEE Trans MTT*, 234–240, Apr. 1968.
29. Popovic, N., Review of Some Types of Varactor Tuned DROs, *Applied Microwave and Wireless*, 62–70, Aug. 1999.
30. Abidi, A.A., Radiofrequency CMOS Circuits, *IEEE SCV-MTT Short Course*, Apr. 1997.
31. Riddle, A.N., Oscillator Noise: Theory and Characterization, N.C. State University, Raleigh, NC, PhD Dissertation, 1986.
32. Clarke, K.K., and Hess, D.T., *Communication Circuits: Analysis and Design*, Addison-Wesley, Reading, 1978.
33. Clapp, J.K., Frequency Stable LC Oscillators, *Proc. IRE*, 1295–1300, Aug. 1954.
34. Schiebold, C.F., An Approach to Realizing Multi-Octave Performance in GaAs-FET YIG-Tuned Oscillators, *IEEE MTT-S Digest*, 261–263, 1985.
35. Ondria, J.G., A Microwave System for Measurement of AM and FM Noise Spectra, *IEEE Trans. MTT*, 767–781, Sept. 1968.
36. Rohdin, H., Su, C.-Y., and Stolte, C., A Study of the Relationship Between Low Frequency Noise and Oscillator Phase Noise for GaAs MESFETs, *IEEE MTT-S Digest*, 267–269, 1984.
37. Christenssen, S., Lundstrum, I., and Svensson, C., Low Frequency Noise in MOS Transistors, *Solid-State Electronics*, 797–812, 1968.
38. van der Ziel, A., Noise in Solid State Devices and Lasers, *Proc IEEE*, 1178–1206, Aug. 1970.
39. Siweris, H.V., and Schiek, B., Analysis of Nosie Upconversion in Microwave FET Oscillators, *IEEE Trans. MTT*, 233–242, Mar. 1985.
40. Pucel, R.A, and Curtis, J., Near-Carrier Noise in FET Oscillators, *IEEE MTT-S Digest*, 282–284, 1983.
41. Dallas, P.A., and Everard, J.K.A., Characterization of Flicker Noise in GaAs MESFETs for Oscillator Applications, *IEEE Trans. MTT*, 245–257, Feb. 2000.
42. Chen, H.B., van der Ziel, A., and Amberiadis, K., Oscillators with Odd-Symmetry Characteristics Eliminate Low-Frequency Noise Sidebands, *IEEE Trans. CAS*, 807–809, Sept. 1984.
43. Riddle, A.N., and Trew, R.J., A Novel GaAs FET Oscillator with Low Phase Noise, *IEEE MTT-S Digest*, 257–260, 1985.
44. Tutt, M.N., Pavlidis, D., Khatibzadeh, A., and Bayraktaroglu, B., The Role of Baseband Noise and its Unconversion in HBT Oscillator Phase Noise, *IEEE Trans. MTT*, 1461–1471, July 1995.
45. Prigent, M., and Obregon, J., Phase Noise Reduction in FET Oscillators by Low-Frequency Loading and Feedback Circuitry Optimization, *IEEE Trans. MTT*, 349–352, Mar. 1987.
46. Gagnepain, J.J., Olivier, M., and Walls, F.L., Excess Noise in Quartz Crystal Resonators, *Proc. 36th Symp. Frequency Control*, 218–225, 1983.
47. Edson, W.A., Noise in Oscillators, *Proc. IRE*, 1454–1466, Aug. 1960.
48. Spaelti, A., Der Einfluß des Thermischen Widerstandrauschens und des Schroteffektes auf die Stoermodulation von Oscillatoren, *Bull. Schweiz Elektrotech. Verein*, 419–427, June 1948.
49. Egan, W.F., The Effects of Small Contaminating Signals in Nonlinear Elements used in Frequency Synthesis and Conversion, *Proc. IEEE*, 797–811, July 1981.
50. Leeson, D.B., A Simple Model of Feedback Oscillator Noise Spectrum, *Proc. IEEE*, 329–330, Feb. 1966.
51. Abidi, A.A., and Meyer, R.G., Noise in Relaxation Oscillators, *IEEE JSSC*, 794–802, Dec. 1983.
52. Lee, T.H., and Hajimir, A., Oscillator Phase Noise: A Tutorial, *IEEE JSSC*, 326–336, Mar. 2000.
53. Paciorek, L.J., Injection Locking of Oscillators, *Proc. IEEE*, 1723–1727, Nov. 1965.
54. Rohde, U.L., *Digital PLL Frequency Synthesizers*, Prentice-Hall, Englewood Cliffs, NJ, 1983.

55. Khanna, A.P.S., and Gane, E., A Fast-Locking X-Band Transmission Injection-Locked DRO, *IEEE MTT-S Digest*, 601–604, 1988.

56. Trew, R.J., Design Theory for Broad-Band YIG-Tuned FET Oscillators, *IEEE Trans MTT*, 8–14, Jan. 1979.

57. Prigent, M., Camiiade, M., Dataut, G., Raffet, D., Nebus, J.M., and Obregon, J., High Efficiency Free Running Class F Oscillator, *IEEE MTT-S Digest*, 1317–1324, 1995.

58. Maruhashi, K., Madihian, M., Desclos, L., Onda, K., and Kuzuhama, M., A K-Band monolithic CPW Oscillator Co-Integrated with a Buffer Amplifier, *IEEE MTT-S Digest*, 1321–1324, 1995.

59. Heins, M.S., Barbage, D.W., Fresina, M.T., Ahmari, D.A., Hartmann, Q.J., Stillman, G.E., and Feng, M., Low Phase Noise Ka-Band VCOs Using InGaP/GaAs HBTs and Coplanar Waveguide, *IEEE MTT-S Digest*, 255–258, 1997.

60. Eisele, H., Manns, G.O., and Haddad, G.I., RF Performance Characteristics of InP Millimeter-Wave N^+-N^--N^+ Gunn Devices, *IEEE MTT-S Digest*, 451–454, 1997.

61. Wollitzer, M., Buecher, J., and Luy, J.-F., High Efficiency Planar Oscillator with RF Power of 100 mW Near 40 GHz, *IEEE MTT-S Digest*, 1205–1208, 1997.

62. Siweris, H.J., Werthoff, A., Tischer, H., Schaper, U., Schaefer, A., Verweyen, L., Grave, T., Boeck, G., Schleichtweg, M., and Kellner, W., Low Cost GaAs PHEMT MMICs for Millimeter-Wave Sensor Applications, *IEEE MTT-S Digest*, 227–230, 1998.

63. Kadszus, S., Haydl, W.H., Neumann, M., Bangert, A., and Huelsmann, A., Subharmonically Injection Locked 94 GHz MMIC HEMT Oscillator using Coplanar Technology, *IEEE MTT-S Digest*, 1585–1588, 1998.

64. Galani, Z., Low Noise Microwave Sources for Radar and Missile Systems, Ultra Low Noise Microwave Sources Workshop, 1994.

65. Everard, J.K.A., and Page-Jones, M., Ultra Low Noise Microwave Oscillator with Low Residual Flicker Noise, *IEEE MTT-S Digest*, 693–696, 1995.

66. Vectron, *Frequency Control Products*, Vectron International, 1997.

67. Avantek, *Modular and Oscillator Components*, Avantek, 1989.

68. Feng, Z., Zhang, W., Su, B., Harch, K.F., Gupta, K.C., Bright, V., and Lee, Y.C., Design and Modeling of RF MEMS Tunable Capacitors Using Electro-Thermal Actuators, *IEEE MTT-S Digest*, 1507–1510, 1999.

69. Trans-Tech, Optimize DROs for Low Phase Noise, Application Note No. 1030, *Trans-Tech Temperature Stable Microwave Ceramics*, 1998.

5.8 Phase Locked Loop Design

Robert Newgard

The objective of this chapter is to present the fundamental considerations that go into phase locked loop (PLL) design. The PLL has been utilized in various systems for many years, but it wasn't until the development of integrated circuits in the 1970s that widespread use as a frequency synthesizer came about. With the expansion in the wireless industry and the ever-increasing demand for higher frequency systems, PLL design has now moved into the microwave realm. This chapter is by no means a complete treatise on PLL design, but rather a synopsis of PLL characteristics and design considerations.

The architecture of the frequency synthesizer is often dependent upon the design of the receiver and exciter, and the PLL design engineer must take this fact and system requirements into account. In addition, many trade-offs must be performed in order to implement a successful frequency synthesizer design; the appropriate injection frequencies must be provided, but these should be generated with consideration given to tuning speed (i.e., settling time), phase noise performance, spurious requirements, and channel spacing to name a few. The following sections provide guidelines related to these trade-offs, and while many details are left to the reader, it should be clear that successful PLL design is no accident.

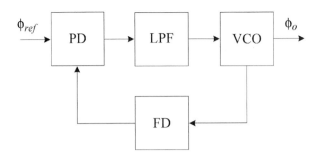

FIGURE 5.94   Phase locked loop.

Roles and Attributes of Phase Locked Loops

In modern wireless communications systems the phase locked loop plays a key role in the performance of the system. The primary function of the PLL is to generate a band of transmit and receive injection frequencies that allow the receiver or transmitter to resolve the required channel spacing. The injections may be synthesized by a combination of PLLs depending upon the system requirements. The spectral purity of the injection signal is a determining factor in the communication systems performance. The PLL generates the injection frequencies from a reference source that is typically a crystal controlled temperature compensated oscillator, but may be any stable reference oscillator (e.g., rubidium, cesium, etc.). The frequency accuracy and temperature stability of the output signal of the PLL is proportional to the frequency accuracy and temperature stability of the crystal oscillator.

The PLL is a negative feedback control system utilizing a phase detector (PD), lowpass filter (LPF), voltage controlled oscillator (VCO), and frequency divider (FD). The basic block diagram of a PLL is shown in Fig. 5.94. The VCO generates an output signal that is dependent upon a DC control voltage at its input. The PD compares the phase of the reference signal to the phase of the divided down VCO signal and generates a correction signal, which is proportional to the phase difference. The LPF's function is to: (1) attenuate the reference sidebands, (2) shape the phase noise, and (3) tailor the PLL's dynamics. Frequency selection of the output of the PLL is accomplished by varying the divisor of the FD. The frequency divider is typically programmable and has enough range to cover the desired amount of frequency tuning bandwidth.

The phase of the PLL's output signal is given by

$$\phi_o = \phi_{ref} N \tag{5.37}$$

where N is the division ratio of the frequency divider and is stepped in integer values. Because frequency is the time derivative of phase, the output frequency of the PLL is given by

$$F_O = \frac{d\phi_o}{dt} = \frac{d\phi_{ref}}{dt} N = F_{ref} N \tag{5.38}$$

which shows that the output frequency is an integer multiple of the reference frequency. The reference frequency is chosen to attain the desired channel spacing, since incrementing N increases the output frequency in multiples of the reference frequency.

The PLL makes a relatively unstable VCO track the phase of the reference signal, which is derived from a stable crystal oscillator. Free running voltage controlled oscillators drift with variation in temperature and power supply noise, as well as noise on the control voltage (a.k.a., tune voltage) line. The action of the feedback loop is to keep the VCO phase locked to the reference oscillator signal.

There are three primary problems that challenge the PLL designer. One is improving the frequency acquisition time or settling time (i.e., when a command occurs to change channels, the PLL takes time to move from the old frequency to the new one and acquire lock). The second is reducing sidebands and spurious signals from appearing on the PLL's output. Any discrete frequency components appearing on the VCO control line will modulate the VCO and appear as spurious sidebands on the output of the PLL. The primary discrete spurious frequency source is modulation of the VCO by the error signal, at the comparison frequency, coming from the output of the phase detector. These spurs are referred to as reference sidebands. Other sources of spurious signals are conducted signals on power supplies (e.g., VCO power supply, phase detector power supply, etc.), radiated signals (e.g., induced on the VCO tank coil and loop filter coils), and isolation from other signal sources (e.g., reverse isolation from the programmable divider to the PLL output). In addition to these spurious signal sources, mechanical vibration of the synthesizer assembly may induce unwanted sidebands on the VCO by physically modulating the printed wiring board and/or the VCO's tank coil. The spurious signals are reduced through good design of the PLL's dynamics, by using good RF shielding techniques, and by providing mechanical support to the assemblies. The third problem that challenges the PLL designer is phase noise performance. As mentioned before, the long-term frequency stability of the output signal of the PLL is determined by the frequency standard used, but the phase noise performance, and thus the short-term stability, is dependent upon the design of the PLL. In the receive path, in order to downconvert the modulated radio frequency (RF) signal, the output signal of the PLL (a.k.a., local oscillator or LO) is mixed with the RF signal. The phase noise of the LO is superimposed onto the intermediate frequency (IF) or baseband signal and thereby affects the receiver's selectivity. In the transmit path, because the LO is mixed with the IF signal or baseband signal to generate the modulated RF signal and again the phase noise of the LO is superimposed, the transmit noise floor or signal-to-noise ratio (SNR) is a function of the LO's phase noise. Also, the performance of the receiver in the presence of a strong adjacent channel signal is affected by the phase noise performance of the LO. The adjacent channel signals mix with the LO's phase noise and produce noise signals at the IF, thereby decreasing the receiver's selectivity. In order to better understand these design considerations, the designer needs to develop a clear understanding of the mathematical models that are used to characterize PLL behavior.

Transfer Function of the Basic PLL

By making the assumption that the PLL is continuous in time, basic feedback control theory utilizing Laplace Transforms can be utilized to determine the loop's behavior, provided that the loop bandwidth is much, much less than the reference frequency. While in practice it is true that the phase detector and frequency dividers are not continuous in time, it is necessary to make this assumption in order to model the stability of the PLL using the Laplace transform. When wide loop bandwidth synthesizers are designed, the sampling nature of the frequency divider and phase detector cannot be ignored. The time delay of these devices will introduce phase shift (i.e., reduction in phase margin), thereby affecting the dynamic performance of the PLL. Another assumption is that the PLL has reached steady state (i.e., it has reached a phase locked condition).

The characteristics described in this section do not address the acquisition of phase lock. The block diagram of a PLL and the gain of each of the functional blocks is shown in Fig. 5.95. The phase detector is shown as an adder and gain block in order to clarify the understanding of the functionality of the phase detector. The forward gain, G(s), is used represent the product of the transfer function of each individual block within the forward path of the PLL. Likewise, the feedback gain, H(s), represents the product of each individual transfer function within the feedback path of the PLL. The equations describing the PLL shown in Fig. 5.95, in terms of the transform variables, are

$$\phi_O(s) = \phi_e(s)G(s) \tag{5.39}$$

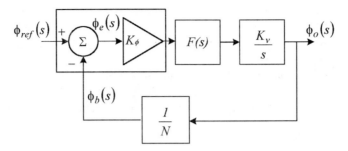

FIGURE 5.95 PLL gain block diagram.

$$\phi_b(s) = H(s)\phi_o(s) \tag{5.40}$$

$$\phi_e(s) = \phi_{ref}(s) - \phi_b(s) \tag{5.41}$$

where classical control theory notation is used. The overall closed-loop transfer function is found by solving the above equations.[1]

$$A_{CL}(s) = \frac{\phi_o(s)}{\phi_{ref}(s)} = \frac{G(s)}{1 + G(s)H(s)} \tag{5.42}$$

The denominator of the closed-loop response is defined as the characteristic equation. The forward and feedback gain of the PLL shown in Fig. 5.95 are

$$G(s) = K_\phi F(s)\frac{K_v}{s} \tag{5.43}$$

$$H(s) = \frac{1}{N} \tag{5.44}$$

Therefore, the transfer function of the PLL shown in Fig. 5.95 is

$$A_{CL}(s) = \frac{K_\phi F(s)\dfrac{K_v}{s}}{1 + K_\phi F(s)\dfrac{K_v}{s}\dfrac{1}{N}} \tag{5.45}$$

The transfer function given in Eq. (5.45) is referred to as the closed-loop response. The open-loop transfer function is defined as the ratio of the output of the feedback path $\phi_b(s)$ to the system error signal $\phi_e(s)$. The open-loop transfer function is used in the analysis of the PLL's stability.

$$A_{OL}(s) = \frac{\phi_b}{\phi_e} = G(s)H(s) = K_\phi F(s)\frac{K_v}{s}\frac{1}{N} = M\angle\alpha \tag{5.46}$$

The closed-loop and open-loop response, Eqs. (5.42) and (5.46) respectively, yield a phasor quantity for each unique complex parameter s. For the open-loop response, the magnitude is M and the phase angle is α. As can be seen from Eq. (5.46), the open-loop response appears in the denominator of the closed-loop response. The frequency at which the magnitude of the open-loop response equals one is used to determine the stability of the PLL. As described in the following section, the phase of the open-loop response at this point is critical in determining the loop stability.

Stability

There are many ways to evaluate the stability of a PLL, but a very popular method is to analyze the stability by plotting the open-loop gain and phase margin as a function of frequency. A feedback control system will become unstable if the magnitude of the open-loop response of the system exceeds unity at the frequency for which the open-loop phase shift is equal to ±180°. The magnitude of the open-loop response at this point is referred to as the gain margin. For a stable PLL the gain margin should be greater than 10 dB. Also, as a measure of relative stability, the phase margin of the PLL is 180° plus the phase angle where the magnitude of the open-loop response is equal to unity (i.e., 0 dB). The frequency at which this occurs is referred to as the open-loop bandwidth. In other words, the phase margin is the amount of phase shift at the loop bandwidth that would produce instability. For a stable PLL the phase margin should be greater than 30°. It is common practice to plot the log magnitude and phase margin of the open-loop transfer function to analyze stability. In designing the PLL, it is imperative that enough phase margin is allowed such that the loop's closed-loop gain response will not have peaking. The plot shown in Fig. 5.96 shows the closed-loop gain of a PLL with three values of phase margin (10°, 45°, and 60°). When adequate phase margin is not provided, the loop will be unstable. The PLL output signal can be observed on a spectrum analyzer. Any discrete spurs separated from the desired output signal by the loop bandwidth and harmonics thereof, indicates inadequate phase margin. The designer should make sure that variations in loop bandwidth, that occur as the PLL output signal is tuned, do not cause a loss in phase margin and thereby have an adverse affect on loop stability. Loop bandwidth variations are caused by changes in VCO gain, phase detector gain, component temperature coefficients, and loop division ratio.

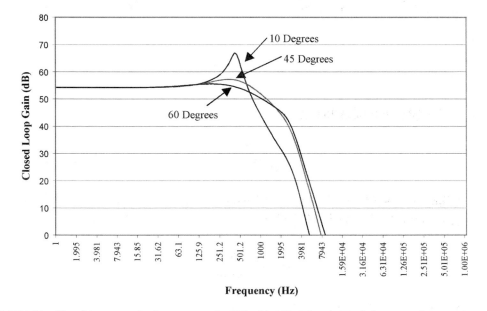

FIGURE 5.96 Closed-loop magnitude response of a PLL with 10°, 45°, and 60° of phase margin.

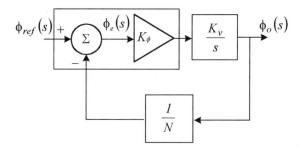

FIGURE 5.97   Type I first-order PLL.

Type and Order

It is imperative in modeling the transient and steady-state response of PLLs to develop an understanding of how the PLL will respond to various inputs. Most common PLL designs fall into two categories, Type I and Type II, albeit the PLL type is not limited. The type of system refers to the number of poles in the open-loop gain located at the origin (i.e., the number of perfect integrators in the PLL). The order of the system refers to the degree of the characteristic equation or the denominator of the closed-loop transfer function. As shown in Fig. 5.95, there are two blocks that are a function of frequency, the loop filter and VCO. Therefore, the filter block, F(s), is the factor that determines the type and order of the PLL. The control system examples that follow will further the reader's understanding of PLL design.

Type I First-Order Loop

The first PLL introduced is a type I, first-order; although it is not practical due to the fact that the sidebands caused by the error signal, $\phi_e(s)$, are in most cases too high without a loop filter. This is dependent upon many of the system parameters (e.g., reference signal frequency, VCO gain, division ratio, etc.). It is presented here as a basis for furthering the reader's understanding of the analysis of phase locked loops. A simplified block diagram is shown in Fig. 5.97. As can be seen from the block diagram, the only integrator is the VCO. The assumption is made at this point that the phase detector doesn't have a pole at the origin, but such is not always the case. The closed-loop transfer function is given by

$$A_{CL}(s) = \frac{K_\phi K_v}{s + \dfrac{K_\phi K_v}{N}} \tag{5.47}$$

The uncompensated loop bandwidth is commonly defined as

$$\omega_n = \frac{K_\phi K_v}{N} \tag{5.48}$$

which is, in this case type I first-order, the loop's bandwidth, since there is no compensation by a loop filter. Substituting eq. (5.48) into Eq. (5.49), the closed-loop transfer function becomes

$$A_{CL}(s) = \frac{N\omega_n}{s + \omega_n} \tag{5.49}$$

The open-loop transfer function is given by

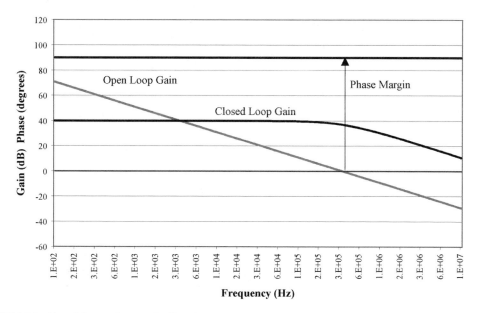

FIGURE 5.98 Type I first-order transfer functions.

$$A_{OL}(s) = \frac{K_{\phi}K_{v}}{sN} \tag{5.50}$$

The plot in Fig. 5.98 shows the open-loop gain and margin phase along with the closed-loop gain. It can be seen from the closed-loop gain that this example has 40 dB of gain inside the loop bandwidth. The loop bandwidth is defined, with respect to frequency, as the point where the open-loop gain equals one. The phase margin is equal to 90°, which is more than adequate for stability. The problem here is the reference sideband spurs will not be attenuated by the loop. Hence, it is imperative that the PLL designer adds a loop filter to the design and thereby the order of the PLL is increased. By examining the closed-loop transfer function, Eq. (5.49), it can be seen that it is a lowpass filter response with gain inside of the loop bandwidth. If the bandwidth of loop is narrow enough and the reference frequency is high enough the loop will provide attenuation, albeit the slope of the attenuation is 20 dB/decade. The limitations on the design due to system requirements, of vibration and settling time, typically force the designer to add additional filtering to the forward path of the PLL.

Type I Second-Order Loop

A lowpass filter is added in the forward path of the PLL in order to attenuate the reference sideband spurs. In this example, a single pole filter has been added and hence the PLL becomes a type I, second order loop. A simplified block diagram is shown in Fig. 5.99. The transfer function of the loop filter, F(s), is given by

$$F(s) = \frac{1}{RCs+1} \tag{5.51}$$

The closed-loop transfer function of the PLL is given by

$$A_{CL}(s) = \frac{K_{\phi}K_{v}F(s)}{s + \frac{K_{\phi}K_{v}F(s)}{N}} \tag{5.52}$$

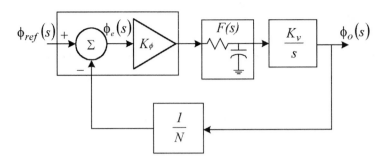

FIGURE 5.99   Type I second-order PLL.

By substituting Eq. (5.51) into Eq. (5.52) and simplifying, the order of the PLL is plainly seen

$$A_{CL}(s) = \frac{K_\phi K_v \dfrac{1}{RCs+1}}{s + \dfrac{K_\phi K_v \dfrac{1}{RCs+1}}{N}} = \frac{K_\phi K_v}{RCs^2 + s + \dfrac{K_\phi K_v}{N}} \tag{5.53}$$

The open-loop transfer function is given by

$$A_{OL}(s) = K_\phi \frac{K_v}{s} \frac{1}{RCs+1} \frac{1}{N} = \frac{K_\phi K_v}{s(NRCs+N)} \tag{5.54}$$

Figure 5.100 shows an example of the closed-loop gain, the open-loop gain and phase margin of a type I, second-order loop. The addition of the filter has added phase shift to the open-loop response, but at 75°, the phase margin is adequate for stability. The loop filter, however, still doesn't offer much filtering for reference signal spurs. Higher order filters are typically added to the PLL in order to provide the appropriate attenuation, but as can be seen in Fig. 5.100, the additional filtering adds phase shift. The

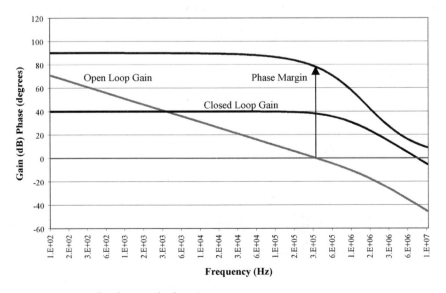

FIGURE 5.100   Type I second-order transfer functions.

goal here is to maximize the filter's attenuation while realizing minimum phase shift. An elliptic filter is often used due to the fact that they have higher selectivity (i.e., the passband is closest to the stopband) compared to other filters. The higher selectivity results in a minimization of phase shift.

Phase Errors for Type I and Type II PLL

Dependent upon the system requirements, the PLL will have to respond to various kinds of inputs (i.e., phase of reference signal, change in division ratio, etc.). The designer is required to know how the PLL will respond to these inputs when the loop has reached steady state. In classical control theory a system is characterized by its response to step changes in position, velocity, and acceleration. In PLL design these changes correspond to step changes in phase, frequency, and time-varying frequency. The steady state response is determined by using the Laplace final value theorem, which is

$$\text{Lim}_{t \to \infty}\left[\phi_e(t)\right] = \text{Lim}_{s \to 0}\left[s\phi_e(s)\right] \tag{5.55}$$

For the Type I PLL, shown in Fig. 5.97, the function $\phi_e(s)$ is the phase error signal generated within the phase detector and is referred to as the system error.

$$\phi_e(s) = \frac{1}{1 + \dfrac{K_\phi K_v}{sN}} \phi_{ref}(s) \tag{5.56}$$

A phase unit step function, u(t), is applied to the input and the Laplace transform gives

$$\phi_{ref}(s) = \frac{A}{s} \tag{5.57}$$

where A is the magnitude of the phase step in radians. This would represent the input signal shifting phase of A radians. By substituting Eqs. (5.56) and (5.57) into the Laplace final value theorem

$$\text{Lim}_{t \to \infty}\left[\phi_e(t)\right] = \text{Lim}_{s \to 0}\left[s\frac{1}{1 + \dfrac{K_\phi K_v}{sN}}\frac{A}{s}\right] = 0 \tag{5.58}$$

Thus, when a step phase change is applied to the Type I PLL, the final value of the system error is zero, which better be the case or we don't have a phase locked loop. Next a unit step function of frequency is applied to the PLL. Phase is the integral of frequency, therefore the reference signal becomes

$$\phi_{ref}(s) = \frac{A}{s^2} \tag{5.59}$$

Once again by substituting Eq. (5.56) and (5.59) into the Laplace final value theorem

$$\text{Lim}_{t \to \infty}\left[\phi_e(t)\right] = \text{Lim}_{s \to 0}\left[s\frac{1}{1 + \dfrac{K_\phi K_v}{sN}}\frac{A}{s^2}\right] = \frac{AN}{K_\phi K_v} \tag{5.60}$$

TABLE 5.5 System Phase Error

Input Signal ϕ_{ref}	Type I	Type II
Phase	0	0
Frequency	Constant	0
Time varying frequency	Continually increasing	Constant

Thus, when a step frequency change is applied to the Type I PLL, the final value of the system error is a constant, but as can be seen, this constant is dependent upon the magnitude of the change, along with the division ratio of the loop. What this means to the PLL designer is that for any given N or output frequency, there will be a phase error between the reference signal and the PLL output. If the system cannot tolerate this error and needs the PLL output to be phase coherent with the reference signal, the designer will have to use a Type II loop. Next we will examine the case of a Type I loop with a time varying frequency input. The reference signal is given by

$$\phi_{ref}(s) = \frac{A}{s^3} \tag{5.61}$$

Substituting into the Laplace Final Value Theorem

$$\text{Lim}_{t \to \infty}\left[\phi_e(t)\right] = \text{Lim}_{s \to 0}\left[s\frac{1}{1+\dfrac{K_\phi K_v}{sN}}\frac{A}{s^3}\right] = \infty \tag{5.62}$$

What this indicates is that the phase error signal is continually increasing. The Laplace Final Value Theorem can be applied to any Type of PLL and Table 5.5 is a quick reference to the system phase error within Type I and Type II loops.

Type II Third-Order Loop

With the introduction of an integrator and a single pole RC as the lowpass filter, F(s), the PLL becomes a Type II third order system. A popular configuration of the loop filter is shown in Fig. 5.101, which may be used with a proportional or pseudo-differential phase detector. These types of phase detectors will be discussed later in this section. In order to prevent slew rate limiting in the amplifier, capacitor C1 is commonly added to realize a single pole in front of the amplifier.

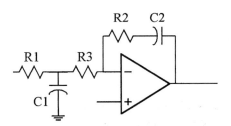

FIGURE 5.101   Integrator loop filter design.

The transfer function for the loop filter given in Fig. 5.101 is given by

$$F(s) = \left[\frac{sR_2C_2+1}{s(R_1+R_3)C_2}\right]\left[\frac{1}{s\left[\dfrac{R_1R_3}{R_1+R_3}\right]C_1+1}\right] \tag{5.63}$$

The zero is added to the filter transfer function to pull the phase margin up toward 90°. By substituting Eq. (5.63) into Eq. (5.52), the closed-loop transfer function becomes

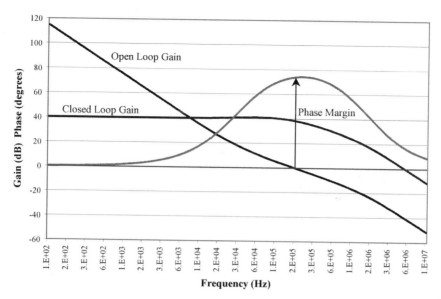

FIGURE 5.102 Type II second-order transfer functions.

$$A_{CL}(s) = \frac{K_\phi K_v \left(R_2 C_2 s + 1\right)}{\left(R_1 + R_3\right)C_1 C_2 s^3 + \left(R_1 + R_3\right)C_2 s^2 + \dfrac{K_\phi K_v R_2 C_2}{N}s + \dfrac{K_\phi K_v}{N}} \qquad (5.64)$$

Figure 5.102 shows the closed-loop gain, the open-loop gain and phase margin of a type II, third-order loop. The additional integrator causes the phase response to start from 0°. As mentioned earlier, the zero was added to the filter to move the phase response toward 90°. Due to the additional integrator, the useable loop bandwidth is narrower than the Type I PLL. The benefit of using the integrator cannot be seen in the closed-loop response but will be shown later in the section on filter design. Typically to achieve the necessary attenuation of the reference sidebands, the PLL needs to be designed with a higher order.

Higher Order Loops

While the above examples serve well to further the understanding of PLLs, practical requirements often drive the designer to higher order loops. To make the proper trade-off between settling time and spurious signals at the PLL output, higher order filters are often necessary to minimize the amount of phase shift and maximize the amount of reference spur attenuation. Higher order filters have a steeper attenuation characteristic thereby achieving less phase shift. Figure 5.103 illustrates this by plotting filter attenuation vs. the phase bandwidth (i.e., filter's frequency at which the phase response is equal to 45° divided by the frequency of the stop-band attenuation).

The system specifications, in some applications, require that the noise from the PLL meet a certain shape factor. The noise sources from within the loop can be tailored by the design of the loop's lowpass filter (e.g., dual stop-band filter). The noise shaping requirements typically forces the design to use high order filters.

Phase Noise

The phase noise performance of a PLL is a critical parameter in the design of any system. The phase noise model developed in this chapter is directed toward the identification of the major contributors to the overall phase noise in a PLL, and evaluating the relative contributions of the significant sources that

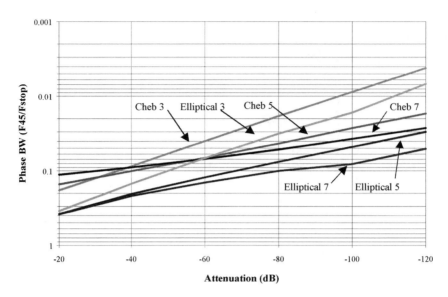

FIGURE 5.103   Lowpass filter phase shift comparisons.

contribute to the output power spectral density. In certain cases, the source of noise in the loop can be pinpointed, but often it is difficult to characterize the noise precisely enough to make the necessary trade-offs. Most PLL designers are familiar with the different sources of noise that exist; in particular these include the frequency standard, VCO, frequency divider, phase detector, and integrator/low pass filter (i.e., active filter).

The spectral characteristics of the oscillators (i.e., VCO and frequency standard) have been modeled in the past and are relatively well understood. Mathematically, an ideal sinewave can be described by the following equation

$$V(t) = V_o \sin(\omega t) \tag{5.65}$$

where V_o is the nominal amplitude and ω is the carrier frequency expressed in radians/second. In the real world, the sinewaves have error components related to both the phase and amplitude. A real sinewave signal is better modeled by

$$V(t) = \left[V_o + \varepsilon(t)\right] \sin\left[\omega t + \Delta\phi(t)\right] \tag{5.66}$$

where $\varepsilon(t)$ is the amplitude fluctuation and $\Delta\phi(t)$ is the randomly fluctuating phase noise term. Both of these terms, $\varepsilon(t)$ and $\Delta\phi(t)$, are stationary random processes and are narrowband with respect to ω. For the purpose of this discussion, the amplitude spectral density will be ignored since it is of negligible significance compared to the phase perturbations.

There are two types of fluctuating phase terms. The first is the discrete signal components, which appear in the spectral density plot. These are commonly referred to as spurious signals. The second type of phase instability is random in nature and is commonly called phase noise. There are many sources of random phase perturbations in any electronic system, such as thermal, shot, and flicker noise. One description of phase noise is the spectral density of phase fluctuations on a per-Hertz basis. The term spectral density describes the energy distribution as a continuous function, expressed in units of phase variance per unit bandwidth. The spectral density is described by the following equation

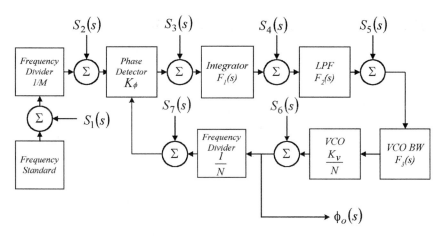

FIGURE 5.104 Phase noise sources in a PLL.

$$S_\phi\left(f_m\right) = \frac{\Delta\phi^2_{rms}\left(f_m\right)}{measurementBW} \tag{5.67}$$

The units of spectral density are rad²/Hz. The U.S. National Bureau of Standards has defined the single sideband spectral density as

$$L\left(f_m\right) = \frac{P_{ssb}}{P_s} \tag{5.68}$$

where P_{ssb} is the power in one hertz of bandwidth at one phase modulation sideband and P_s is the total signal power. The single sideband spectral density, $L(f_m)$, is directly related to the spectral density, $S_\phi(f_m)$, by

$$L\left(f_m\right) \cong \frac{1}{2}S_\phi\left(f_m\right) \tag{5.69}$$

This holds true only if the modulation sideband, P_{ssb}, is such that the total phase deviation is much less than 1 radian. $L(f_m)$ is expressed in dBc/Hz or dB relative to the carrier on a per hertz basis.

For the purpose of evaluating the noise performance of the PLL, each of the functional blocks is considered noiseless and a noise signal is added into the PLL at a summing junction in from of each of the functional blocks. In Fig. 5.104, the noise sources within the PLL are shown along with the gains of the various blocks. In evaluating the contribution of each of the noise sources to the overall noise at the output of the PLL, each one will be considered alone. Since these noise sources are independent, super-position may be used to determine the phase noise at the output of the PLL. The transfer function for each of the noise sources is easily written. Once again, the transfer functions are derived using classic control theory. Two additional gain blocks have been added to the basic block diagram. The first one is the additional filter after the loop integrator as discussed in the section on higher order PLLs. The second is the inclusion of the VCO's modulation bandwidth. Dependent upon the design of the VCO's input circuitry, the VCO's tune voltage input will have a finite bandwidth within which it will respond to an input signal. The 3 dB point of this response is defined as the modulation bandwidth. If the forward gain is defined as G(s) and the feedback is defined as H(s), then the closed-loop gain is given by

$$A_{CL}\left(s\right) = \frac{G\left(s\right)}{1 + H\left(s\right)G\left(s\right)} \tag{5.70}$$

Table 5.6 Phase Noise Sources Transfer Functions

Source	Transfer Function	Simplification $K(s) = \dfrac{K_\phi F_1(s)F_2(s)F_3(s)K_v}{N}$
Frequency Standard	$A_1(s) = \left[\dfrac{\dfrac{K_\phi F_1(s)F_2(s)F_3(s)K_v}{s + \dfrac{K_\phi F_1(s)F_2(s)F_3(s)K_v}{N}}}{}\right]\dfrac{1}{M}$	$A_1(s) = \left[\dfrac{K(s)}{s+K(s)}\right]\dfrac{N}{M}$
Reference Divider	$A_2(s) = \dfrac{K_\phi F_1(s)F_2(s)F_3(s)K_v}{s + \dfrac{K_\phi F_1(s)F_2(s)F_3(s)K_v}{N}}$	$A_2(s) = \left[\dfrac{K(s)}{s+K(s)}\right]N$
Phase Detector	$A_3(s) = \dfrac{F_1(s)F_2(s)F_3(s)K_v}{s + \dfrac{K_\phi F_1(s)F_2(s)F_3(s)K_v}{N}}$	$A_3(s) = \left[\dfrac{K(s)}{s+K(s)}\right]\dfrac{N}{K_\phi}$
Integrator	$A_4(s) = \dfrac{F_2(s)F_3(s)K_v}{s + \dfrac{K_\phi F_1(s)F_2(s)F_3(s)K_v}{N}}$	$A_4(s) = \left[\dfrac{K(s)}{s+K(s)}\right]\dfrac{N}{K_\phi F_1(s)}$
Lowpass Filter	$A_5(s) = \dfrac{F_3(s)K_v}{s + \dfrac{K_\phi F_1(s)F_2(s)F_3(s)K_v}{N}}$	$A_5(s) = \left[\dfrac{K(s)}{s+K(s)}\right]\dfrac{N}{K_\phi F_1(s)F_2(s)}$
VCO	$A_6(s) = \dfrac{s}{s + \dfrac{K_\phi F_1(s)F_2(s)F_3(s)K_v}{N}}$	$A_6(s) = \left[\dfrac{s}{s+K(s)}\right]$
Feedback Divider	$A_7(s) = \dfrac{K_\phi F_1(s)F_2(s)F_3(s)K_v}{s + \dfrac{K_\phi F_1(s)F_2(s)F_3(s)K_v}{N}}$	$A_7(s) = \left[\dfrac{K(s)}{s+K(s)}\right]N$

By applying Eq. (5.70) to the PLL, the closed-loop gain for each of the noise sources is determined. This is done to characterize the loop's overall phase noise performance. By plotting the PLL's response to the individual noise sources, the proper trade-off for the optimization of the loop's performance (i.e., phase noise and settling time) can be made. The transfer function for each of the noise sources is given in Table 5.6.

The complete equation for the output phase noise of the PLL as a function of frequency is given by

$$S_\phi(f_m) = |A_1|^2 S_1 + |A_2|^2 S_2 + |A_3|^2 S_3 + |A_4|^2 S_4 + |A_5|^2 S_5 + |A_6|^2 S_6 + |A_7|^2 S_7 \qquad (5.71)$$

where the S_x's are power spectral densities that are a function of the offset frequency, f_m, from the carrier and the A_x's are a function of the complex variable s. By substituting simplified equations from Table 5.6 into Eq. (5.71), some interesting conclusions can be drawn.

$$\Phi(s) = \frac{K(s)}{s+K(s)}\left[\frac{N}{M}S_1 + N(S_2 + S_7) + \frac{N}{K_\phi}S_3 + \frac{N}{K_\phi F_1}S_4 + \frac{N}{K_\phi F_1 F_2}S_5\right] + \frac{s}{s+K(s)}S_6 \qquad (5.72)$$

First, the frequency standard noise S_1, reference frequency divider S_2, phase detector noise S_3, integrator noise S_4, loop low pass filter noise S_5, and feedback divider noise S_7 are all acted upon by a common PLL function. All these noise sources are passed through a low pass filter. The actual cutoff frequency is determined by the designer in the choosing of the various parameters that establish the PLL compensated loop bandwidth, $K(s)$. The next noise component in Eq. (5.72) to be considered is that of the VCO. The loop acts upon the VCO phase noise as if it was passed through a high pass filter. At offset frequencies that are much less than the compensated loop bandwidth, the dominant noise sources are the digital noise and frequency standard noise. At offset frequencies that are much greater than the compensated loop bandwidth, the dominant noise source is the VCO noise. As the loop's compensated bandwidth is approached, the PLL's output noise is a summation of all the noise sources. Care needs to be taken in the designing of the loop parameters, such that peaking of the noise at the loop's bandwidth doesn't occur. A general rule of thumb is that the designer would like to set the loop bandwidth at the point where the VCO phase noise crosses the digital noise. In doing so, the optimum noise performance of the overall PLL can be achieved, but this is not always possible due to settling time requirements. Next, it looks very advantageous to increase M and thereby decrease the contribution of the frequency standard phase noise. But if increasing M lowers the reference frequency, then N must increase if the overall multiplication factor to the output of the PLL is to remain the same. This actually decreases the frequency standard noise, but increases the multiplication of the phase noise contribution of the integrator, phase detector, loop filter, and loop divider. Therefore, from Eq. (5.72) the designer would want to keep N as low as possible to keep the noise inside the loop bandwidth as low as possible.

The evaluation of the phase noise performance of a PLL can be an arduous task, but a simplified approach follows. The phase noise inside the loop bandwidth is multiplied by reference noise and digital noise. The noise outside of the loop bandwidth is primarily that of the VCO. While this certainly is an oversimplification, it is helpful in understanding the noise performance of a PLL. Writing the transfer function for each of the noise sources within the loop will help to clarify how the loop acts upon each noise source as well as what trade-off can be made in filter design for phase noise performance and settling time. Also, by looking at the individual contributions of each of the phase noise sources, the designer can determine where to focus their energy in reducing the overall phase noise performance of the PLL.

Phase Detector Design

The phase detector produces an output signal that is proportional to the phase difference between the reference input, ϕ_{ref}, and the phase of the divided down VCO signal, ϕ_o/N. The most commonly used phase detector is a phase-frequency detector. In an out-of-lock condition, the output of the phase-frequency detector latches (i.e., the AC component is removed), thereby the error signal goes to the low or high rail, depending upon the direction of phase error. The phase detector having the ability to perform frequency discrimination has greatly simplified the complexity of the PLL circuitry. In much of the literature written on PLL design, the analog phase detector is addressed. With the advent of VLSI design, the digital phase detector is most favored. There are three basic types of digital phase-frequency detectors: (1) the charge pump, (2) the proportional or pulse width modulated, and (3) the pseudo-differential.

There are many factors to consider when determining which kind of phase detector the designer should use (e.g., PLL's tuning bandwidth, settling time, power consumption, phase error, etc.), but the most prevalent phase-frequency detector has a charge pump output. The three kinds of digital phase-frequency detectors mentioned above will be discussed.

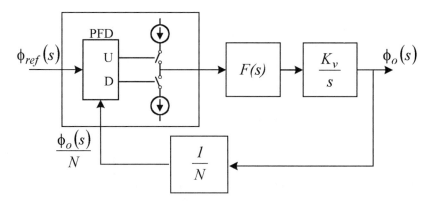

FIGURE 5.105   Charge pump phase detector PLL.

Charge Pump Phase-Frequency Detector

The charge pump phase-frequency detector is used predominately within the commercial PLL ASIC industry. The charge pump phase detector output is a current that has an average value equal to the system phase error.[2] In Fig. 5.105 the configuration of the PLL utilizing a charge pump phase-frequency detector is shown. One advantage of the charge pump phase-frequency detector is the reduction in complexity of the PLL. A second advantage is the ability to program the current, thereby being able to adjust the gain of the phase detector for optimum loop performance. A third advantage is that for narrow tuning bandwidth or fixed injection PLLs, only a passive filter on the charge pump output is required, thereby reducing the cost and size of the PLL. If the tuning bandwidth of the PLL is wide (e.g., octave) an op amp will usually have to be added to increase the dynamic range of the tuning voltage supplied to the VCO. A disadvantage of the charge pump phase frequency detector is the leakage current. All attempts must be made to reduce the leakage current on the charge pump's output. As the level of the leakage current increases, so will the amount of loop filtering needed to suppress the phase error spurious signals on the VCO's output.

The charge pump phase-frequency detector has three states: (1) sourcing current, (2) sinking current, and (3) high impedance or tri-state. The amount of current being sourced or sunk is defined to be I_ϕ. Since the phase detector operates over a 2π range, the gain of the phase detector is therefore $I_\phi/2\pi$. If ϕ_o/N is leading ϕ_{ref}, then the charge pump phase detector is sinking I_ϕ current, which is defined as the pull down current. If ϕ_o/N is lagging ϕ_{ref}, then the charge pump phase detector is sourcing I_ϕ current, which is defined as the pull up current. When the two input waveforms have nearly identical phase, the charge pump phase detector is tri-stated. This is illustrated in Fig. 5.106. The pull up and pull down

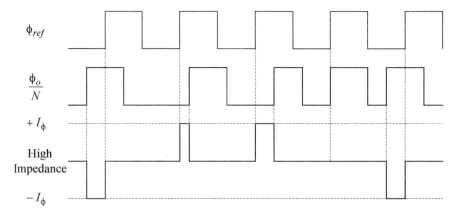

FIGURE 5.106   Charge pump phase detector output waveform.

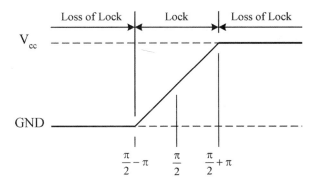

FIGURE 5.107 Proportional phase detector average voltage output.

currents must be equal for the gain of the phase detector to be linear. The charge pump current needs to be constant over the operating temperature of the system and operating voltage of the phase detector, due to the fact that the phase detector gain, and thereby the loop gain changes, are proportional to the charge pump current.

Proportional Phase-Frequency Detector

The proportional phase-frequency detector output is a variable pulse width rectangular wave, which has a duty cycle proportional to the phase difference between the two inputs. The phase range is 2π radians, with positive latching at either end. The gain curve of the proportional phase-frequency detector is shown in Fig. 5.107, which shows the average value of the phase-frequency detector output. The slope of the gain curve is approximately equal to V_{CC} divided by 2π, and is defined as K_ϕ. If the phase difference exceeds the $\frac{\pi}{2} \pm \pi$ range, thereby causing a loss-of-lock condition, the frequency discrimination capability of the phase-frequency detector causes the output to rail at VCC or GND, depending upon the sense of the phase error. An advantage of this type of phase-frequency detector is the gain linearity. A disadvantage of using the proportional phase-frequency detector is the amount of filtering needed to attenuate the reference sidebands, which will be discussed in further detail in the section on loop filter design. Another disadvantage is the phase error associated with a Type I loop. Of course an integrator may be added as the loop filter, thereby making the loop a Type II and removing the phase error associated with a change in frequency.

There are two common methods to extend the frequency tuning bandwidth of the PLL using a proportional phase detector, which are shown in Fig. 5.108. The first method is the use of an operational amplifier as an integrator. The second is to add a discrete amplifier, which extends the tuning range with less added noise than an operational amplifier. The amplifiers effectively step up the voltage that is input to the VCO, thereby reducing the amount of VCO gain necessary to cover a given tuning bandwidth.

Pseudo-Differential

The pseudo-differential phase detector is so called because the phase error information is contained in two signals, which must be combined in the loop filter, or an integrator. A very popular method is to have pulses that are normally at V_{cc} and pulsing low, but it is also possible to design the phase detector such that the signals are normally at ground and pulsing high. If ϕ_o/N is leading ϕ_{ref}, then the pseudo-differential phase detector output ϕ_v is pulsing low and output ϕ_r is predominately high with very narrow pulses. If ϕ_o/N is lagging ϕ_{ref}, then the pseudo-differential phase detector output ϕ_r is pulsing low and the output ϕ_v is predominately high with very narrow pulses. When the two input waveforms have identical phase, the pseudo-differential phase detector's outputs are both high with very narrow pulses. The pseudo-differential phase detector's waveforms are shown in Fig. 5.109. An advantage to using this type of phase-frequency detector is the reduced amount of filtering needed for attenuation of the reference sidebands. Another advantage is that with the introduction of the second integrator into the loop, the phase error between the VCO output and the reference signal approaches zero for a phase or frequency change, as was shown in the previous sections of this chapter. If the PLL being designed has a phase

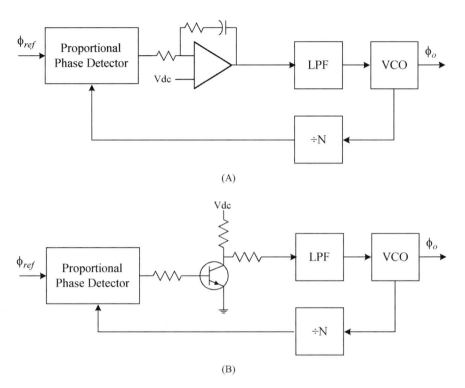

FIGURE 5.108   Proportional phase detector (A) Type II PLL and (B) Type I PLL.

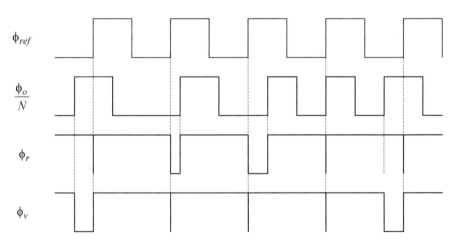

FIGURE 5.109   Pseudo-differential phase detector waveform.

inversion in the feedback, as is the case when a high-side injection mixer, is used in the feedback path of the loop to reduce the loop's division ratio, the outputs of the pseudo-differential phase detector will have to be flipped in order for the loop to lock. The output spurious performance of the PLL is determined by the loop filter design.

Loop Filter Design

Once the designer has determined the type of PLL needed from the system requirements, the next step is to determine the configuration of the forward path of the PLL. In the design of a PLL, the level of the

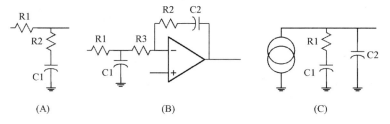

FIGURE 5.110 Loop filters: (A) lead-lag, (B) active filter integrator with zero, and (C) charge pump integrator.

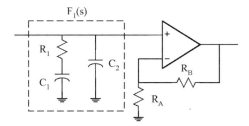

FIGURE 5.111 Basic forward path of a charge pump PLL.

reference sideband spur on the VCO output is determined by a number of factors within the PLL. The amount of filtering needed for a desired reference sideband spur level can be calculated by utilizing the formulas presented in this section. Though these equations are empirical and by no means exact, they are a good starting point for determining the performance (i.e., filter attenuation) that is needed. In the basic form, the loop filter can take on three forms: (1) a lead-lag filter, (2) an active filter integrator, and (3) a charge pump passive filter. These three filters are shown in Fig. 5.110. It is common practice to add an elliptic filter to the output of these basic filters, in order to achieve the needed attenuation of the reference sidebands. The configuration of the loop filter is dependent upon the type of phase-frequency detector circuitry used.

Charge Pump Phase Detector

As mentioned before the charge pump phase detector has become very popular in commercial PLL ASIC applications and is covered in the following section. A typical filter used with a charge pump phase detector is shown in Fig. 5.111. In any application where wide tuning is needed, a higher tuning voltage must be supplied to the VCO. The higher tune voltage is supplied by adding an amplifier stage. A closer examination of the transfer function, $F_1(s)$, results in the determination of the component values. The transfer function of the filter, $F_1(s)$, is

$$F_1(s) = \frac{R_1 C_1 s + 1}{s(C_1 + C_2)\left(sR_1 \frac{C_1 C_2}{C_1 + C_2} + 1\right)} = K \frac{\tau_z s + 1}{s(\tau_p s + 1)} \tag{5.73}$$

where

$$K = \frac{1}{C_1 + C_2} \tag{5.74}$$

$$\tau_z = R_1 C_1 \tag{5.75}$$

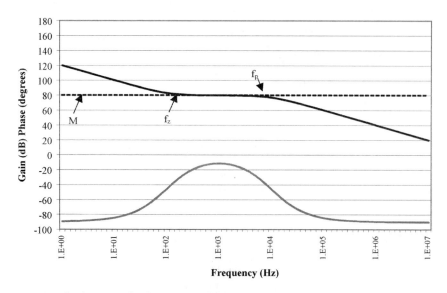

FIGURE 5.112   Transfer function of a charge pump filter.

$$\tau_p = R_1 \frac{C_1 C_2}{C_1 + C_2} \tag{5.76}$$

By plotting the transfer function, $F_1(s)$, as a function of frequency, the pole and zero break frequencies can easily be seen, as shown in Fig. 5.112.

In designing the loop filter for a specified loop bandwidth, the PLL's natural frequency will have to be adjusted by the filter gain. This will have to been done with the appropriate amount of phase margin. The pole and zero location and the filter gain determine the component values for the filter.[3] The magnitude of $F_1(s)$ at the geometric mean of the pole and zero frequencies is defined as

$$M \equiv \left| F_1 \left(j2\pi \sqrt{f_z f_p} \right) \right| \tag{5.77}$$

The component values for R_1, C_1, and C_2 may be calculated by the following formulas

$$R_1 = M \frac{f_p}{f_p - f_z} \tag{5.78}$$

$$C_1 = \frac{1}{2\pi f_p M} \tag{5.79}$$

$$C_2 = \frac{f_p - f_z}{2\pi f_p f_z M} \tag{5.80}$$

Once the component values of the filter, $F_1(s)$, are determined, the amount of additional filtering for attenuation of the reference sidebands will need to be calculated. It is important to note that the closed-loop transfer function will not predict the amount of filtering needed for attenuation of the reference side bands. The closed-loop response doesn't take into account all of the parameters that affect the level of the reference sideband signal on the output of the VCO. An approximate calculation of the needed filtering for a given reference sideband level may be calculated from the following formula

$$F(dB) = 20\log\left[\frac{\sin\left(n\pi\dfrac{V_{dcmax}}{R_{leak}I_{max}}\right)}{n\pi F_{ref}}mR_1 K_v I_{max}\right] + RSB(dB) \tag{5.81}$$

where n is the harmonic of the reference frequency to be filtered, V_{dcmax} is the maximum voltage of the charge pump, R_{leak} is the leakage resistance across the charge pump, I_{max} is the maximum charge pump current, F_{ref} is the reference frequency, m is either the gain or loss in the forward path (i.e., op amp/lead-lag network, etc.), and R_1 is the value of the resistor in the filter, $F_1(s)$.[4] An example of a charge pump phase detector utilizing the filter, $F_1(s)$, is given below.

Example 5.5
A charge pump phase detector utilizing the filter shown in Fig. 5.111 with the following parameters: maximum voltage of the charge pump of 5 Volts, maximum VCO gain of 40 MHz/Volt, a reference frequency of 1 MHz, a filter gain of 1, a leakage resistance of 100 kohm, a value of 200 ohms for R1, and the fundamental harmonic at a level of −70 dBc.

$$F(dB) = 20\log\left[\frac{\sin\left(\pi\dfrac{5}{1E5 \times 5E-3}\right)}{\pi \times 1E6}200 \times 40E6 \times 5E-3\right] + 70 = 62.04\ dB \tag{5.82}$$

Approximately 62 dB of additional attenuation will be needed at the reference frequency of 1 MHz. If a simple single pole filter is used, the designer will have to take care that it doesn't introduce excessive phase shift at the loop bandwidth is not introduced. It is also worth noting that if the amplifier shown in Fig. 5.113 is replaced with any filter design, the loading affect of that filter will have an effect on the

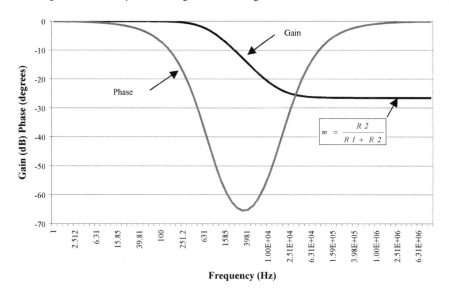

FIGURE 5.113 Lead-lag filter transfer function.

response of the integrator and lag network proceeding it. The designer may have to go to a higher order elliptic or Chebychev filter to minimize the amount of phase shift.

Proportional Phase Detector

The lead-lag filter shown in Fig. 5.110 (A) is commonly used with the proportional phase detector when a Type I loop is being designed. The transfer function for this filter is

$$F_1(s) = \frac{C_1 R_2 s + 1}{(R_1 + R_2) C_1 s + 1}$$
(5.83)

Using some typical values, the transfer function is plotted in Fig. 5.113. There is a zero located at

$$f_z = \frac{1}{2\pi C_1 R_2}$$
(5.84)

and a pole located at

$$f_p = \frac{1}{2\pi C_1 (R_2 + R_1)}$$
(5.85)

with the attenuation of the filter when $s \gg 1$ given as

$$m = \frac{R2}{R1 + R2}$$
(5.86)

The use of a proportional phase detector often requires that a gain stage be added to the phase detector output in order to meet the system's frequency tuning bandwidth. This is dependent upon the gain of the VCO. The amplifier can be a discrete amplifier or an operational amplifier, dependent upon the phase noise performance needed. The level of filter attenuation needed at the reference frequency may be approximated by the following equation

$$F(dB) = 20 \log \left[\frac{n^2 \pi F_{ref} 10^{\left(\frac{RSB}{20}\right)}}{m K_v V_{dd}} \right]$$
(5.87)

where V_{dd} is the maximum swing of the voltage into the LPF; V_{dd} is either directly out of the phase detector or after the added gain stage; K_v is the maximum gain of the VCO in Hz/Volt; n is the harmonic of interest; F_{ref} is the reference frequency in Hz. If a lead-lag network is used, m is the loss of the lead-lag network. RSB is the desired level of the reference side band at the VCO output. An example of a Type I loop utilizing an emitter follower amplifier with a lead-lag filter follows.

Example 5.6
A proportional phase detector in a Type I loop utilizing the lead-lag filter shown in Fig. 5.110 (A), and a discrete amplifier as shown in Fig. 5.108 (A) with a maximum voltage swing of 15 Volts, maximum VCO gain of 10 MHz/Volt, a reference frequency of 250 kHz, and a lead-lag loss of 0.047. The fundamental harmonic is desired to be at a level of –60 dBc, resulting in a filter needing a rejection of

$$F(dB) = 20\log\left[\frac{\pi \times 2.5E5 \times 10^{\left(\frac{-60}{20}\right)}}{0.047 \times 10E6 \times 15}\right] = -79.1 \tag{5.88}$$

In order to achieve −60 dBc sidebands, the filter following the discrete amplifier must have −79.1 dB of attenuation at 250 kHz. The next step is to determine the kind of low pass filter to be used following the lead-lag filter. The additional filter must meet the attenuation and minimize the amount of phase shift at the open-loop bandwidth. A chart, to aid in this decision is shown in Fig. 5.103, where the ratio of the filter's 45-degree point to frequency stop vs. the filter attenuation is plotted. As can be seen in Fig. 5.103, the minimum amount of phase shift for a −80 dB filter is a 7-pole elliptic.

Pseudo-Differential Phase Detector

A typical topology of the integrator used with the pseudo-differential phase detector (PDPD) is shown in Fig. 5.114. An advantage of the PDPD is the amount of attenuation needed following the integrator. A disadvantage is the effect the integrator has on the settling time. Because, the charge on the integrator capacitor, C_2, needs to change for every new output frequency of the PLL, dielectric absorption can increase the settling time. The use of this type of integrator is a common approach when the VCO has to have a wide frequency tuning bandwidth. The transfer function of this filter was given earlier in Eq. (5.63). The amount of filtering needed following the integrator can be calculated approximately by the following equation

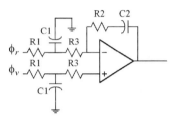

FIGURE 5.114 Pseudo-differential phase detector integrator.

$$F(dB) = 20\log\left[\frac{nF_{ref}10^{\left(\frac{RSB}{20}\right)}}{mK_v V_{off}}\right] \tag{5.89}$$

where V_{off} is the offset voltage of the op amp, K_v is the gain of the VCO in Hz/Volt, n is the harmonic of interest, F_{ref} is the reference frequency, m is the gain of the op amp, and RSB is the desired level of the reference side band at the VCO output.

Example 5.7

A PDPD in a Type II loop utilizing the circuit shown in Fig. 5.114 with a maximum op amp input offset voltage of 5 mVolts, maximum VCO gain of 10 MHz/Volt, a reference frequency of 250 kHz, an op amp gain of 1, and a desired fundamental harmonic at a level of −60 dBc. This results in a filter needing

$$F(dB) = 20\text{Log}\left[\frac{2.5E5 \times 10^{\left(\frac{-60}{20}\right)}}{20E6 \times 5E-3 \times 1}\right] = -46.0 \tag{5.90}$$

approximately 46 dB, of additional attenuation at the reference frequency of 250 kHz. In comparison to the example utilizing the proportional phase detector, the use of the integrator has reduced the amount of filtering needed by 33 dB.

Transient Response

So far the assumption has been made that the PLL is operating in a steady state. What is the PLL's response when a disturbance is introduced or what is the transient response? In most of the articles written on PLL design, a second order approximation is used to model the PLL's transient response to a change in phase or frequency. The most common is a change in the loop's division ratio to bring the PLL to a new frequency output. This change has two effects on the loop. One is, because the closed-loop response is dependent upon the value of the feedback divider, N, the loop bandwidth changes. The PLL has to acquire phase lock to the new VCO output frequency, which is known as the transient response. Of course, depending upon how large of change in frequency, the gain of the VCO will also affect the loop bandwidth. With the mathematical modeling software tools available to the designer today, the second order approximation is unnecessary, but is still useful in understanding the basic loop transient response.

In most systems, the PLL's transient phase error is the response of interest. Albeit many system specifications define the frequency error, Phillips[5] has shown the phase settling characteristic corresponds to the system performance better than the frequency settling characteristics. The transient phase response is the phase difference between the final value of the VCO and a steady-state signal, which has the same phase as the final value of the VCO. In the laboratory, the transient phase response is measured by mixing a signal generator and the VCO's output signal, both of which need to be phase-locked to the same frequency standard. The output of the mixer shows the phase difference and needs to be lowpass filtered in order to remove the summed output. As an example, Fig. 5.115 shows the result of a simulation of a PLL hopping between two frequencies. During the first hop, the loop has too little phase margin and there is excessive ringing. The second hop shows the phase response with a phase margin of 70°.

There are many factors within the loop that can affect the transient response. The nonlinearity of the VCO gain, which typically is less at the high end of its tuning range, will affect the loop bandwidth and thereby the phase margin. The frequency range and modulation bandwidth of the VCO will impact the transient response. Prepositioning of the VCO's control voltage, either with a digital control word into a DAC or by summing another PLL's tune voltage, can be done to reduce the amount of overshoot and thereby reduce the settling time. To prevent additional noise from being introduced into the PLL, careful design of the prepositioning circuitry must be done. The discrete or sampling nature of the phase detector and divider need to be considered in wide loop bandwidth designs. The continuous time approximation

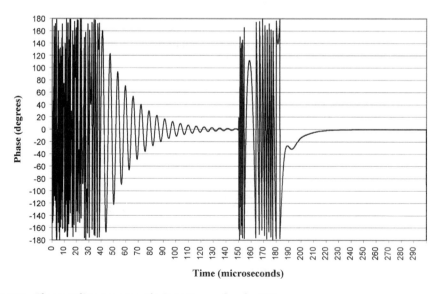

FIGURE 5.115 Phase settling response of a Type I second-order PLL.

is typically used in transient modeling, but a number of papers have been written that use z-transforms to model the loop's response.[6] Once again, if the loop bandwidth is kept low relative to the reference signal frequency, the discrete nature of the phase detector can be ignored. When a charge pump phase detector is used, the mismatch between current sources, I_ϕ and $-I_\phi$, the leakage current of the charge pump as well as the leakage current of the components used in the loop lowpass filter and board parasites, all have an effect on the transient response. If an operational amplifier is used in the loop filter design, the slew rate and voltage limits will impact the transient response.

Conclusion

In this chapter we have considered some of the fundamental design considerations that go into the design of a PLL. The design process is made up of a series of trade-offs (e.g., wide loop bandwidth for improved settling time, but narrow loop bandwidth for improved noise performance). There cannot be enough emphasis placed on the robustness of the PLL design. The designer needs to ensure that there is enough margin in the design parameters such that the loop works well over its operating temperature and other environmental conditions, as well as component tolerance. Modeling plays a key role in the development of the PLL and it is imperative that the designer have an understanding of those models and the limitations inherent in any mathematical model. It is a common practice to model higher order PLLs as second order systems to simplify the design process, but with the computer-aided design tools available to the designer today, this is an unnecessary simplification. When the designer is challenged by the system requirements, it is necessary to have the most comprehensive model possible.

References

1. Egan, W.F., *Frequency Synthesis by Phase Lock,* Robert E. Krieger Publishing, Malabar, FL, 1990.
2. Gardner, F.M., Charge-Pump Phase-Lock Loops, *IEEE Trans. Comm.,* COM-28, 1849–1858, Nov. 1980.
3. Opsahl, P.L., *Charge-Pump Filter Design,* Rockwell Collins, Inc., Frequency Control Team Internal Paper, 1999.
4. Mroch, A.B., *Charge-Pump Filter Attenuation,* Rockwell Collins, Inc., Frequency Control Team Internal Paper, 1999.
5. Phillips, D.E., Settling Time Specifications: Phase or Frequency?, *IEEE Military Communications Conference,* 3, 806–810, Oct. 1994.
6. Crawford, J.A., *Frequency Synthesizer Design Handbook,* Artech House, Inc., Boston, 1994.

5.9 Filters and Multiplexers

Richard V. Snyder

"Filter: a device or material for suppressing or minimizing waves or oscillations of certain frequencies"… per Webster. This definition, while accurate, is insufficient for microwave engineers. Microwave systems and components enhance and direct, as well as suppress waves and oscillations. Components such as circulators, mixers, amplifiers, oscillators, switches (in common with most complex systems) are in fact filters, in which inherent physical properties are represented as smaller, constituent networks embedded within larger "filtering" (response-determining) structures or systems. Inclusion of the concept of embedding is thus central to understanding microwave filters. To clarify "embedding": the terminals of a well-defined subnetwork are provided with a known interface to the rest of the system, thus providing selective suppression or enhancement of some oscillatory effect within the device or system. The subnetworks can be linear or nonlinear, passive or active, lumped or distributed, time dependent or not, reciprocal or nonreciprocal, chiral (handed) or nonchiral, or any combination of these or other properties of the basic constituent elements. The subnetworks are carefully defined (or charac-

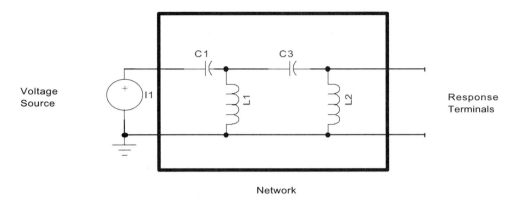

FIGURE 5.116

terized) so that the cascade response of a series of such subnetworks can be predicted (or analyzed) using software simulation tools employing a variety of methods, such as linear, harmonic balance, Volterra series, finite-element, method of moments, finite-difference time domain, etc. The careful definition normally involves the process called "synthesis," in which the desired response to a particular stimulus suggests a topological form for the subnetwork, followed by extraction of the specific elements of the subnetwork. The computed response of the synthesized network is compared to the desired response, with iteration as necessary using repeated synthesis or perhaps an optimization loop within the simulation tool. Hybrid combinations of these two iterative approaches are possible.

Analysis and Synthesis

The difference between prediction or "analysis" and definition or "synthesis" can be summarized as follows. The word analysis comes from the Greek *lysis*, a loosening, and *ana*, up; hence a loosening up of a complex. Synthesis, on the other hand, means the building up of a complex from parts or elements to meet prescribed excitation-response characteristics. Fig. 5.116 illustrates an example of the excitation network and response.

Another difference between analysis and synthesis must be considered. There is always a unique solution for an analysis, although it might be hard to find. Synthesis, on the other hand, might result in several networks with the specified response, or possibly no solution whatsoever. In general, solutions are not unique but some might be more realizable than others. Fig. 5.117 presents the general problem of synthesis. What combination of elements in Fig. 5.117 will give the prescribed response? It is important to realize that with a *finite* number of elements, in general the required response *cannot* be realized at all. Functions having a required variation over some band of frequencies and zero value for all other frequencies cannot be represented by a rational function of the form of a quotient of polynomials. Thus, it is necessary to modify the response requirements to include some *tolerance*.

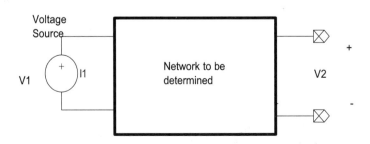

FIGURE 5.117

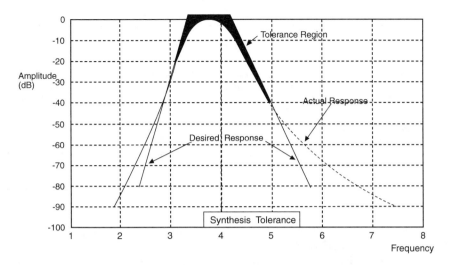

FIGURE 5.118

Figure 5.118 illustrates the imposition of a tolerance, or acceptable difference, between the desired response and the resultant response, for a particular synthesized characteristic. The approximation can take many possible forms. The approximations might require the magnitude squared of the voltage ratio to be the quotient of rational and even polynomials in ω. A typical quotient of polynomials might be as given in Fig. 5.118.

$$\left|G(\omega)\right|^2 = \frac{a_0\omega^6 + a_2\omega^4 + a_4\omega^2 + a_0}{b_0\omega^6 + b_2\omega^4 + b_4\omega^2 + b_0} \tag{5.91}$$

This is not a unique polynomial for realization. The are any number of other polynomial ratios of lower or higher degrees that may be used. The higher the degree of assumed polynomials, the better the approximation to the desired response, the smaller the tolerance region. The coefficients of Eq. (5.91) are determined by the solution of a set of simultaneous linear equations, as many simultaneous equations as there are unknown coefficients. In general, we cannot match the desired response characteristic at all points in the spectrum; rather, we must choose to match exactly at certain points and approximately over the remainder of the tolerance region. We can choose to match the points, the derivatives, or use other criteria that will be discussed herein. The response shown in Fig. 5.118 is amplitude vs. frequency. Typically, a network also has a desired time vs. frequency response. Generally these two requirements are interrelated and may not be specified independently. In very important classes of approximation, the amplitude and time responses are connected by the Hilbert transform, and thus to know either the amplitude or the time response is sufficient to enable determination of the other. This will be discussed in the section on Approximations. The polynomial in Eq. (5.91) has a simple dependence upon frequency ω (representing lumped elements in the network). Generally, microwave filters include distributed elements such as quarter-wave resonators, which display response characteristics dependent upon transcendental functions, such as tan (ω). The resultant synthesis polynomials require more specialized techniques for element extraction. It is possible that real networks will include both lumped and distributed elements, with very complex transfer function polynomials. The synthesis process can thus be quite complex, and a comprehensive coverage is beyond the scope of this article.

Types of Transfer Function

Before approximating a particular transfer function, one must determine the type of transfer function desired. These functions can be defined in terms of amplitude or time, expressing either as functions of

TABLE 5.7   Transfer Function Types

Transfer Function Type	Characteristics
Lowpass	Low loss region approximated over some bandwidth, prescribed rejection achieved at frequency some distance from the *highest* frequency in the low loss approximation region (Fig. 5.119a)
Highpass	Low loss region approximated over some bandwidth, prescribed rejection achieved at frequency some distance from the *lowest* frequency in the low loss approximation region (Fig. 5.119b)
Bandpass	Low loss region approximated over some bandwidth, prescribed rejection achieved at frequencies some distance from *both* the *highest and lowest* frequencies in the low loss approximation region (Fig. 5.119c)
Bandstop	Low loss approximated over two regions, extending downward toward DC and upward toward infinity. Prescribed rejection achieved at frequencies some distance *above* the lower low loss region and *below* the upper low loss region (Fig. 5.119d)

frequency. It is convenient to initially concentrate on amplitude vs. frequency transfer characteristics. Table 5.7 presents the four available types.

It should be understood that the above transfer functions in Table 5.7 are defined between any pair of input-output ports of a potentially multiport network. Transfer functions so defined are known as "two-port" transfer functions. We will initially restrict our efforts to such two-port circuits.

Approximations to Transfer Functions

It is not possible to achieve the flat passbands and abrupt transitions illustrated in Fig. 5.119 without using an infinite number of elements, each with zero resistance. We will discuss the properties of elements used to realize filters in a later section, but certainly the "Q" of available elements is less than infinity. Thus, some approximation to the idealized transfer functions must be made in order to implement a filter network falling within the allowable tolerance shown in Fig. 5.118. Essentially, the approximation procedure is directed toward writing mathematical expressions that approximate the ideal forms shown in Fig. 5.119. These expressions include polynomial functions that are substituted into the left side of Eq. (5.91) prior to element extraction. Some of the most common approximations will now be discussed. We will treat approximations to the amplitude response in some detail, and will briefly touch on approximations to phase or time delay.

Butterworth

The response function given by Eq. (5.92) is known as the nth order Butterworth or maximally flat form.

$$\left| G_{12}(j\omega) \right| = \frac{1}{\sqrt{1 + \omega^{2n}}} \tag{5.92}$$

From binomial series expansion

$$\left(1 \pm x\right)^{-n} = nx + \frac{(n+1)x^2}{2!} \mp \frac{n(n+1)(n+2)x^3}{3!} + \ldots, x^2 \leq 1 \tag{5.93}$$

We see that near $\omega = 0$

$$\left(1 + \omega^{2n}\right)^{-1/2} = 1 - 0.5\omega^{2n} + 0.375\omega^{4n} - 0.313\ \omega^{6n+\cdots} \tag{5.94}$$

and from this expression, the first $2n - 1$ derivatives are zero at $\omega = 0$. Thus, the magnitude

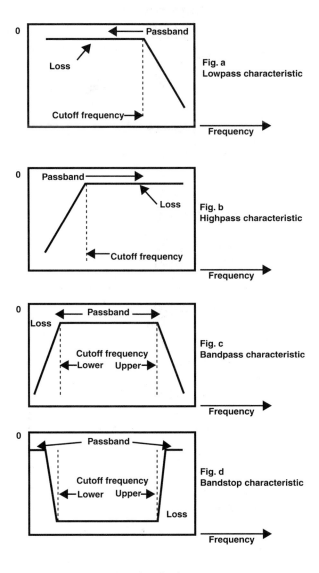

FIGURE 5.119

$$|G_{12}(j\,1)| = 0.707 \text{ for all values of n.} \tag{5.95}$$

The pole locations corresponding to the Butterworth response may be determined using analytic continuation of the binomial series expansion above. The poles of this function are defined by the equation

$$1 + \left(-s^2\right)^n = 0. \tag{5.96}$$

The poles so defined are located on a unit circle in the s plane and have symmetry with respect to both the real and the imaginary axes. Only the left half plane poles are used to form what is known as the all-pole response function that will yield the response required in Eq. (5.91).

The form of the Butterworth response is shown in Fig. 5.120a for several values of n. The 2n − 1 zero derivatives ensure the "maximally flat" passband characteristic.

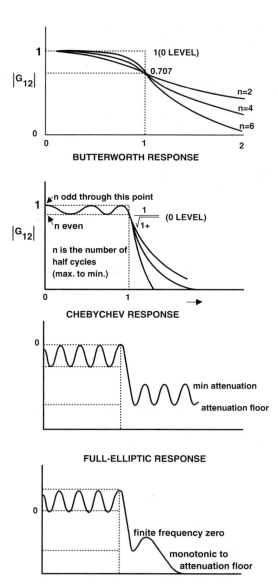

FIGURE 5.120

Chebychev

If a rippled approximation to the passband region of the ideal transfer function is acceptable, one can use the expression

$$\left|G_{12}\right|^2 = \frac{1}{1+\varepsilon^2 C_n^{2\omega} 2} \tag{5.97}$$

where $C_n(\omega)$ is the nth order Chebychev polynomial and $\varepsilon < 1$ is a real constant. These polynomials are defined in terms of a real variable ω as follows:

$$C_n(\omega) = \cos\left(n\cos^{-1}\omega\right) \tag{5.98}$$

The response of a Chebychev-approximated transfer function is shown is Fig. 5.120b. Analytic continuation can again be used to locate the poles, which will be found to be distributed on an ellipse, major and minor axes, respectively, the imaginary and real axes of the s-plane, s the normal Laplace transform variable $\sigma + j\omega$. (Remember that the Butterworth poles were distributed on a circle, same axes.) The reader is referred to many standard reference works for the details of element extraction, but the response of the Chebychev approximation will generally provide less passband performance but will achieve specified levels of stopband attenuation more quickly than the Butterworth approximation discussed previously. Butterworth and Chebychev transfer functions can be realized using single or resonant elements, with coupling only between adjacent elements. As such, the physical form for the network has the appearance of a ladder (see Fig. 5.116) and these circuits are known as "ladder" networks.

Elliptic Approximation

If one can utilize rippled approximations to both the amplitude of both passband and stopband regions, the resultant filter characteristics will display somewhat better passband performance coupled with steeper attenuation slopes, as compared to the Chebychev, but with attenuation slope characteristics with a level set on the minimum value of stopband. The response is shown in Fig. 5.120c. The stopband region contains finite-frequency transmission zeros. Filters designed with elliptic responses are derived from expressions containing elliptic functions. Such filters are sometimes termed "full-elliptic" or "Cauer parameter." They can be derived from many starting points, but it is important to note that for bandpass cases, narrow passbands are hard to achieve and for bandstop, narrow stopbands are difficult. Designs of this type require extra resonant elements and sometimes coupling between nonadjacent resonators. Typically, such designs are used to achieve specified minimum stopband levels in close proximity to the passband (5 to 10% away in frequency).

Quasi-Elliptic Approximation

This class of function is achieved with rippled approximation to the passband amplitude and a limited number of ripples (finite frequency transmission zeros) in the stopband. Typically, the filter stopband slope displays monotonicity beyond the few ripples. Filters in this category can achieve the improved passband response of elliptic designs, with stopband performance almost as steep as an elliptic, and also with maximum stopband levels considerably improved as compared to the full-elliptic approach. If the extra zeros are real axis, rather than real frequency, the filter will display improved passband flatness and more constant group delay. Filters in this category can be what is known as "non-minimum-phase," as the extra zeros can be located in the right half plane. The general response type is illustrated in Fig. 5.120d. These filters usually require coupling between nonadjacent resonators (sometimes this is called "cross-coupling"), but do not need the extra resonant elements nor do they display the realization difficulties of full-elliptic designs. Although this design approach has been known since the 1970s, recent advances in simulation tools and synthesis techniques have resulted in the emergence of this category as the "cutting edge" in filter design. Such filters are also known as "pseudo-elliptic."

Other Approximations

The aforementioned approaches have all started with approximating the amplitude portion of the transfer function. In some cases, it is desirable to approach the delay or phase as a function to be approximated. In the ladder-derived filters above, to know the amplitude is to have determined the phase/delay (and contrawise). In more complex structures (cross-coupled) it is possible to have some degree of control over both amplitude and phase/delay. It is also possible to adjust the amplitude or phase/delay properties with the cascade of an additional circuit, known as an "equalizer." When the adjustments are internal, as in the cross-coupled cases, the equalization is known as "self-equalized." In this case, the extra transmission zeros are located on the s-plane real axis (imaginary frequency). If the additional circuit is used for equalization of either amplitude or phase/delay, the descriptive term is "externally equalized." Some of the more common approximations are summarized in Table 5.8.

TABLE 5.8

Approximation Type	Salient Characteristics
Butterworth	Lowest loss at center frequency, bandwidth is defined as −3 dB relative to center, maximally flat (no ripples) in passband, stopband slope depends on order, good group delay performance, generally easy to realize as ladder structure. Poles distributed on a unit of the s-plane (Laplace variable space). This is an all-pole filter.
Chebychev	Rippled approximation to passband with reflections proportional to ripple level and higher than Butterworth, for the same order. Stopband slope depends on order, but steeper than Butterworth for the same order, poor group delay in passband (not constant), generally easy to realize as ladder structure. Poles distributed on an ellipse, major axis being the imaginary axis of the s-plane, with all poles within the unit circle of the s-plane. The ellipse intersects the Butterworth pole circle at two points on the imaginary axis. This is an all-pole filter.
Elliptic	Rippled approximation to both passband and stopband. Passband reflection proportional to ripple level, stopband slope proportional to stopband ripple level. Finite-frequency transmission zeros supplement the amplitude response attributable to the poles and affect the phase response. Ratio between stopband and passband widths can be smaller than for all-pole designs.
Quasi-Elliptic	Rippled approximation to passband, small number of ripples in stopband. Finite frequency or real-axis (imaginary frequency) zeros can be achieved, to emphasize either attenuation slopes, passband flatness and delay, or both to some extent. Easier to realize for narrow bandwidths than full-elliptic designs. Physical structure requires coupling between nonadjacent resonators and thus folding of the structure, for microwave implementations.
Max flat time delay (Bessel)	Analogous to Butterworth, but with the time-delay (group delay) function vs. frequency containing $2n - 1$ zero derivatives. Poles lie on an ellipse-like path outside the s-plane unit circle (as contrasted to the Chebychev locations inside the unit circle). Provides flat, constant group delay within the passband but poor attenuation slopes.
Gaussian	A Taylor-series approximation to a Gaussian magnitude function $|G_{12}(j\omega)| = e^{-\omega^2/2}$ is used to extract filters that have optimum transient overshoot characteristics and thus display minimum ringing when excited by a pulsed input signal. The group delay is not as flat as that of the Bessel design nor are the stopband slopes as steep. In common with Bessel designs, the filter will display high reflections at frequencies away from center frequency.
Transitional	These are filters with transfer functions between those defined by the classical polynomials such as Chebychev, Butterworth, etc. An example is Gaussian-Chebychev.

Element Types and Properties

There are many ways to classify the available elements. Perhaps the most basic is to characterize the element as "passive" (no D.C. required) or "active (D.C. required). Within the passive regime, classification includes "lumped," "distributed," "non-reciprocal" (and "reciprocal"), and combinations in which a particular element can display more than one of these characteristics.

"Lumped" elements are those that present capacitive, inductive, resistive, or gyrator responses. The element impedance is essentially a function of ω. Typically, the enumerated elements are predominantly capacitive, inductive, etc. but will also display bits of the other possibilities. For example, at low frequencies, the lead inductance of a capacitor is not important, but as frequency increases, the inductive reactance becomes a significant fraction of the element impedance, until at some frequency the capacitor behaves as a resonant circuit. It is possible to design networks with lumped element concepts all the way up to 100 Ghz or so, but the usual limitation is below 10 GHz. Lumped circuitry usually displays an intrinsically lowpass behavior *above* some frequency.

"Distributed" elements have impedance properties which are functions of tan (ω) or tanh (ω). These include quarter wavelength (or non-quarter wavelength) TEM mode resonators, waveguide resonators, cavities, dielectric resonators, and any structure built using essentially length-dependent techniques (as contrasted to length-independent but position-dependent lumped element circuits). The frequency range for "distributed" elements ranges from a few MHz to the terahertz range. Waveguide elements have the property that internal wavelength is not linearly related to actual free-space wavelength, and are thus

TABLE 5.9

Element Type	Frequency Range	Unloaded Q	Implementation
Inductor, lumped	Almost DC to 100 GHz	50–300 at room temperature, 1000s if superconductive	Coils (air and ferrite-loaded), helices, printed, shorted stubs, evanescent waveguide
Capacitor, lumped	To 100 GHz	50–1000 at room temperature, 1000 if superconductive	Multilayer, single layer, open stubs, coaxial
Resistor, lumped	DC–5 GHz	N/A (parasitic capacitance and inductance can be problems)	Metal, composition, chips
Stub or line, printed, TEM, suspended substrate, coplanar stripline, coplanar waveguide, finline, coaxial, other TEM or almost TEM lines	DC–100 GHz	100–500 at room, 1000 if superconductive	Microstrip, stripline, finline, CPS, CSS, SSS.
Evanescent	200 MHz–90 GHz	300–10,000 at room (no data on superconductive application	Below cutoff waveguide of various aspect ratios, machined sections resonated using various capacitive schemes
Dispersive (guided but non-TEM modes)	100 MHz to terahertz	1000–20,000 at room, 100,000 or more if superconductive	Waveguide, air or dielectric filled cavities with metal walls, dielectric resonators, multimode

termed "dispersive." Such elements display an intrinsically highpass response *below* a frequency known as the cutoff frequency (energy cannot freely propagate through the section below this frequency).

"Non-reciprocal" elements contain ferrimagnetic structures (circulators, isolators, various gyrators) and are used in conjunction with other elements. An additional hybrid element of interest is formed by a resonated short length of below-cutoff waveguide or other dispersive structure, and is termed an "evanescent" section. Such a below-cutoff section presents an essentially inductive equivalent circuit and can be resonated with a capacitor. The result is formation of a high-Q resonant circuit that can be embedded into a variety of filter circuits. These elements have impedance characteristics similar to lumped elements at frequencies well below cutoff, and similar to distributed, near cutoff.

Unloaded Q is an important property of any circuit element, and is a measure of the ability of the element to store energy without dissipation. High Q means low loss and is a desirable property. The above elements can be fabricated using superconductive material to obtain remarkably high unloaded Q values. Table 5.9 summarizes the properties of many common circuit elements. Application depends on various factors, including basic electrical specification, and ambient environment with concomitant difficulties associated with temperature, humidity, vibration, and shock. In general, lumped elements must be potted in place. Distributed and lumped elements have natural changes in impedance or resonant frequency as functions of temperature, which must be compensated using elements with opposite drift properties. Filters can be built that will be stable to no worse than 1 ppm per degree Centigrade, without the need for external stabilization. Vibration and shock must be damped or isolated from the circuitry. Humidity will affect resonant frequency as well as degrade performance over time. Typically, filter circuits are sealed to eliminate the presence of moisture and to prevent the intrusion of moisture as temperature changes, in a humid environment. Salt will degrade performance and must be eliminated through sealing and special plating systems. In general, filters can be built that combine the properties of the various lumped and distributed element types. This is a difficult, if not impossible synthesis problem but with the modern simulation and optimization tools, such globally-designed networks are practical and offer the optimum in electrical and environmental performance with associated production cost reduction. The cost reduction stems from the fact that performance over the full range of ambient environment

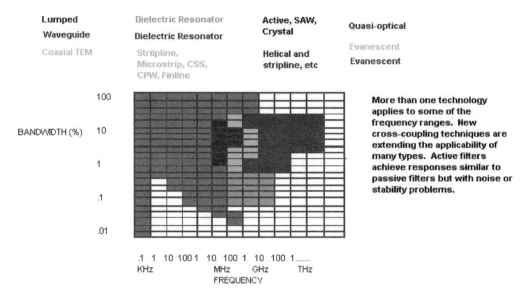

FIGURE 5.121

can be predicted, with sensitivity to production tolerances easily taken into account prior to "cutting metal." Tolerances are thus fit to the problem at hand, with proper care taken and waste minimized.

Filter Implementations

Figure 5.121 presents a Filter Selection Guide applicable to current technology.

Multiplexers

The interconnection of more than one filter at a common junction results in a network termed a "multiplexer." With one common port and two individual ports, we have a "diplexer." With three individual ports, a "triplexer," and so forth through quadruplexer, quintaplexer, sextaplexer, etc. The individual networks can be lowpass, highpass, bandpass, or bandstop. The common connection presents significant difficulty, as without proper precaution, the interaction between the individual filters causes severe degradation of the desired path transfer function.

Many techniques have evolved for performing the interconnection. A multiplexer is normally used if a wide spectrum must be accessed equally and instantaneously. Conventionally, multiplexers have had the disadvantage of requiring at least 3 dB excess loss ("crossover" loss) at frequencies common to two channels. Thus, the passband characteristics for contiguous structures always showed an insertion loss variation over the passband of at least 3 dB. To construct any multiplexer, it is necessary to connect networks to the constituent filters such that each filter appears as an open circuit to each other filter (see Fig. 5.122). While this is simple for narrowband channels, it is difficult for broadband or contiguous

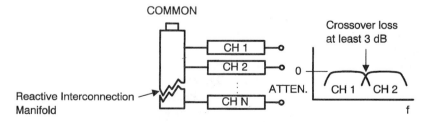

FIGURE 5.122

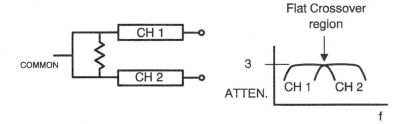

FIGURE 5.123

filters. Normally, the filters and the multiplexing network are synthesized as a set, with computer optimization being used to simulate the results before construction begins. Some of the more common multiplexing techniques include line lengths, circulators, hybrids, and transformers.

More recently, the multiplexer filter channels have been combined using power dividers (Fig. 5.123). This recent adaptation of always-available technology is due to newly available cheap and compact amplifier stages. Such gain blocks provide flat gain and low noise over wide bandwidths. In the case of two-way combining, conservation of energy means that the 3 dB insertion loss is still experienced, but on a flat-loss basis. Although each channel is subject to the additional 3 dB loss, it is essentially constant loss over each channel and thus the excess passband loss variation is less than 1 dB. Excess loss is defined as that loss not attributable to the individual channel filter roll off. This power divider based combining can be extended to triplexers (4.7 dB flat loss), quadruplexers (6 dB flat loss), etc. Because the loss variation is minimized, the overall insertion loss can frequently be made up using amplifiers, which display flat gain vs. frequency.

Filters can be multiplexed by parallel combination at both ends. For example, if two bandpass filters are multiplexed at both input and output, a network results that provides one input and one output, with two passbands essentially attenuating everything else. Such assemblies are useful in systems such as GPS that have two or more operating frequencies, with the requirement for isolation between the operating channels and adjacent, cluttered regions of the spectrum (Fig. 5.124). Another approach employs switched selection of filters. Hybrid combinations using multiplexers with power dividers, switches, and amplifiers are now possible (see Fig. 5.125). The interactions of these essentially reactive components can cause undesirable degradation of stopbands or passbands, if precautions are not taken. Available computer simulation techniques are sufficiently sophisticated that accurate prediction of performance and dimensions minimizes the time required to develop and deliver such complex assemblies.

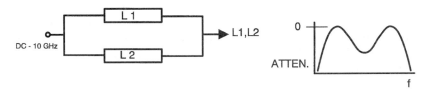

FIGURE 5.124

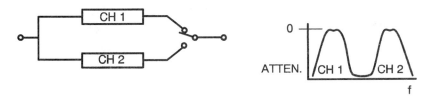

FIGURE 5.125

Interconnection of subcomponents or submodules within multiplexers is sometimes difficult, with parasitic lengths causing degradation of performance. Although the computer can predict these problems, sometimes the parasitics reach levels for which compensation cannot be effected. It is possible to use blind-mate interconnection of submodules to minimize both parasitic interconnections and spurious crosstalk. Thus, the physical structure, including all interactions, can be predicted accurately and the unacceptable interactions and crosstalk eliminated using the mechanical elegance and electrical isolation of blind-mate internal connections.

Multiplexer development is impacted heavily by network synthesis and computer simulation techniques. As it becomes possible to synthesize combinations of lumped, distributed, and evanescent elements as well as predict and compensate their interactions, multiplexers will shrink in size, increase in order (number of channels), and display improved performance in insertion loss, isolation, and bandwidth.

Simulation and Synthesis Software

The process of simulation involves four separate, but related steps:

1. Synthesis and analysis of a theoretical network compliant to specification, under idealized terminating conditions and with idealized construction.
2. Representation of the synthesized network by an appropriate set of very accurate lumped elements. For any circuit, this involves modeling the physical structure and computing the lumped elements that best represent the actual, electromagnetic structure (i.e., solving Maxwell's equations inside the proposed filter structure).
3. Optimizing the filter response with the stipulated terminating impedances (i.e., the complex source and load impedance), using the representation of the circuit as computed in step 2.
4. Revising the physical structure, if required, by iterating the analysis portion of step 1.

The solutions to Maxwell's equations that allow for derivation of the lumped equivalents requires the comparison of a set of scattering parameters describing the physical structure (computed using E-M) to a set describing the characteristics of an assumed lumped element topology (computed using linear simulation). The difference between the two sets is reduced using optimization [7]. The data set is stored, and is used in an iterative manner as described in step 4. All physical structures can be described by a set of lumped elements of arbitrary complexity. Unfortunately, not every set of lumped parameters describes a physically realizable structure, so care must be taken to assume a "realizable" lumped circuit topology.

Traditional filter designs proceed from the basis of network synthesis. Over the last 90 years or so, the application of matrix, transform, complex variable theory, and advanced algebra has led to many clever network topologies. Numerical methods have also advanced the design process, not only simplifying the calculation process but enabling determination of the design suitability through the use of linear simulators that essentially compute the response of the synthesized structure so that the computed response may be compared to the desired response. If it is found that the synthesis is inadequate, the design can be iterated without the necessity for actual laboratory experimentation. Synthesis techniques have been developed to a very high degree for networks consisting of linear lumped elements or linear distributed elements, but to a much lesser extent for combinations of lumped and distributed elements. This is because the natural frequency variation for a lumped element is in terms of $j\omega$, while the variation of a distributed element is in terms of $\tan j\omega$. Thus, it is difficult to perform a synthesis that requires extraction of elements based upon the location of poles and zeros in a complex plane, when the coordinates of the complex plane are different for lumped and distributed structures.

Linear Simulators

The availability of linear simulators, combined with mathematical optimization, has reduced the need for advanced synthesis development (probably to the detriment of our profession and certainly to the dismay of many). The various elements can be readily combined and calculated in the simulator, as long as the elements can be described in transfer matrix (S-parameter) format. However, most physical elements have complex matrix descriptions because the elements are embedded into the surrounding structure in such a way as to respond to more than one mode of excitation. For example, a simple waveguide resonant cavity is analogous to an L-C tank circuit, but the waveguide cavity will resonate at more than one frequency based on field distribution. Thus, computation of the analogous (or equivalent) L-C values for the waveguide cavity requires knowledge of the excitation field. Combining microwave elements, such as cavities, probes, irises, etc. with each other (or for that matter with R-L-C elements), thus requires inclusion *of* the effects of the excitation field and the effects *upon* the field of each of the microwave elements encountered within the composite structure. Accomplishing this requires solutions to Maxwell's equations within the structure.

Electromagnetic (E-M) Simulators

Fortuitously, numerical methods have been applied to the solution of Maxwell's equations resulting in the development of what have come to be known as E-M simulators. These programs employ techniques such as finite elements in frequency or time domains, method of moments, spectral domain, etc., combined with advanced gridding methods and various structure generation software. Although quite advanced, most of these simulators are far too slow to use in conjunction with the mathematical optimization techniques that originally reduced the need for developing new and elegant synthesis techniques. It is well known [7–10] that frequency-dependent equivalent circuits can be derived that are adequate lumped representations of distributed structures to some degree of accuracy. When these structures are so represented, the equivalent circuits depend on the aforementioned mutual interaction of excitation and element. When the response modes are widely separated in the frequency or space domains, a single-mode computation provides a sufficiently accurate representation to enable the resultant lumped circuit to be used for computation of the approximate response of the distributed element or some combination of elements.

Synthesis Software

There have been a few software packages created that automate the design process to a large extent. However, most practitioners elect to create custom software to facilitate the transition from theory to practical filter networks. Some of the currently available most notable packages include Filter (Eagleware) and Filpro (Middle Eastern Technical University in Turkey) [11]. Packages that integrate linear and electromagnetic simulation are available from several sources, but inclusion of filter synthesis as an integrated package is rarely available (Eagleware has such an integrated package).

Active Filters

Since about 1970, it has been possible to simulate a high-Q inductance using a bipolar or FET transistor to convert the output capacitance into equivalent input inductance. The introduction of DC as an external power source acts to compensate for the loss properties of the inductor and make available the inductive element for inclusion into filter circuits. Over the years, other techniques have been developed for using active elements to realize high-Q filter circuits. These filters differ from the better known low frequency op-amp-based filters in that the synthesis generally is identical to that used for conventional passive RF filters, in which there is no requirement for constant voltage or constant current sources (typical impedances are 20 to 150 ohms). Such active filters and multiplexers have been built from 100 MHz to over 10 GHz. Stability, noise figure, thermal stability, and power consumption are problems yet to be fully overcome. Miniature passive filters suffer from poor performance due to the low Q of the available elements. In principle, miniature active filters can be constructed that will provide great selectivity, low loss, etc. with the price being the need for DC power. Applications to handheld cell phones are now to be found, with considerable progress reported. Combinations of active and passive devices are also possible, with the passive elements being used in such a way as to provide stability for the active, high-Q components.

References

1. H. Blinchikoff and A. Zverev, *Filtering in the Time and Frequency Domains*, John Wiley and Sons, New York, 1976.
2. C. Matthaei, L. Young and E.M.T. Jones, *Microwave Filters, Impedance-Matching Networks and Coupling Structures*, McGraw-Hill, New York, 1964. This is the so-called black-book of filters and should be purchased by any serious student.
3. S. Frankel, *Multiconductor Transmission Line Analysis*, Artech House, Boston, 1977.
4. Craven and Skedd, *Evanescent Mode Microwave Components*, Artech House, Boston, 1987.
5. M.E. Van Valkenburg, *Introduction to Modern Network Synthesis*, John Wiley and Sons, New York, 1960.
6. J. Malherbe, *Microwave Transmission Line Filters*, Artech House, Boston, 1979.
7. R. V. Snyder, Embedded-Resonator Filters, Proceedings of the ESA-ESTEC Conference on Filter CAD, ESA, The Netherlands, Nov. 6–8, 1995.
8. R. V. Snyder, Inverted Resonator Evanescent Mode Filters, IEEE-MTT-S Symposium Proceedings, San Francisco MTT IMS, 1996.
9. N. Marcuvitz, *Waveguide Handbook*, Vol. 10, MIT RadLab Series, 1948.
10. R. V. Snyder, Filter Design Using Multimode Lumped Equivalents Extracted from E-M Simulations, MTT/ED Workshop on Global Simulators, La Rochelle, France, May 27, 1998.
11. N. Yildirim, FILPRO Manual, November, 1996, METU, Ankara, Turkey.

5.10   RF Switches

Robert J. Trew

Microwave switches are control elements required in a variety of systems applications. They are used to control and direct, under stimulus from externally applied signals, the flow of RF energy from one part of a circuit to another. For example, all radars that use a common send and receive antenna require an RF switch to separate the send and receive signals, which often differ in amplitude by orders of magnitude. The large difference between the send and receive signals places severe demands upon the switching device, which must be able to sustain the high power of the transmitted signal, as well as have low loss to the returning signal. Isolation is very important in this application since the switch must be able to protect the sensitive receive circuit from the large RF transmitted power. The isolation requirement places severe restrictions upon the switch, and high power radars generally use gas discharge tubes to implement the switch function. Phased-array radars generally use semiconductor transmit/receive modules and use large numbers of switches. A phased-array radar, for example, may require thousands or tens of thousands of switches to permit precise electronic control of the radiated beam. The distributed nature of a phased-array permits the switches to operate at lower power, but the devices still need to operate at power levels on the order of 1 to 10 watts. In general, switches can be manually or electronically switched from one position to the next. However, most microwave integrated circuit applications require switching times that cannot be achieved manually, and electronic control is desirable. Integrated circuit implementation is ideal for switching applications since a large number of components can easily be accommodated in a relatively small area.

Electronically controlled switches can be fabricated using pin diodes [1,2] or transistors, generally GaAs MESFETs [3]. Both types of switches are commonly employed. Switches fabricated using pin diodes have often been used in radar applications [4], achieving insertion slightly over 1 db in L-band, with isolation greater than 35 db. Broadband operation can also be obtained [5] and 6 to 18 GHz bandwidth with insertion loss less than 2 db, isolation greater than 32 db, and CW power handling in excess of 6 watts has been reported using pin diodes connected in a shunt circuit configuration. Such switches have also demonstrated the ability to be optically controlled [6]. GaAs MESFETs are commonly used to fabricate RF switches suitable for use in integrated circuit applications [3,7]. High performance is achieved and a

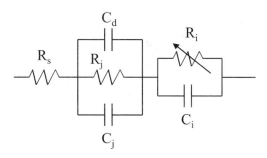

FIGURE 5.126   Schematic diagram for a PIN diode.

1 watt SPDT switch with insertion loss of 0.6 db and isolation greater than 20 db has been reported [8]. These switches often use multi-gate GaAs MESFETs specifically designed for switching applications [9] that permit switching control of high RF power with low gate voltage. Such IC switches have demonstrated the ability to handle large power levels [10] and RF power on the order of 38 dbm can be effectively controlled. Switching at extremely high frequency is also possible by replacing the GaAs MESFET with a HEMT, and high performance Q-band [11] and W-band [12] operation has been achieved. A comparison of the RF performance of MESFET and HEMT switches [13] indicates that HEMT devices generate more distortion than MESFET devices, but are useful at high millimeter-wave frequency. All semiconductor switches, whether fabricated using pin diodes or transistors, can be considered as two-state, one-port devices. Recently, a unified method for characterizing these networks has been presented [14].

PIN Diode Switches

A pin diode is a nonlinear device fabricated from a p^+nn^+ structure, as shown in Fig. 5.126. These devices are widely used in switch applications such as phase shifters [2] and have properties that result in low loss and high frequency performance. A pin diode can also be optically controlled [6], which is desirable for certain applications. The diode is a pn junction device with a lightly doped or undoped (intrinsic) region located between two highly doped contact regions. The presence of the intrinsic region yields operational characteristics very desirable for switching applications. That is, under reverse bias the intrinsic region produces very high values for breakdown voltage and resistance, thereby providing a good approximation to an "open" switching state. Both the breakdown voltage and off-state resistance are dependent upon the length of the intrinsic region, which is limited in design length only by transit-time considerations associated with the frequency of operation. Under forward bias, the conductivity of the intrinsic region is controlled or modulated by the injection of charge from the end regions and the diode conducts current, thereby providing the "on" switching state. The "on" resistance of the diode is controlled by the bias current and in forward bias, the diode has excellent linearity and low distortion.

An equivalent circuit for the PIN diode is shown in Fig. 5.127, and in operation the diode functions as a single-pole, double-throw (SPDT) switch, depending upon the bias state. Under reverse bias, the equivalent circuit reduces to that shown in Fig. 5.128, and under forward bias it reduces to the forward resistance R_f. The reverse bias resistance can be expressed as [3]

$$R_r = R_c + R_i + R_m \tag{5.99}$$

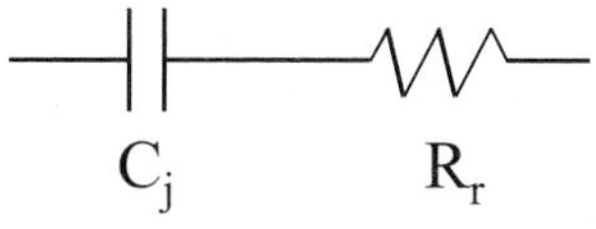

FIGURE 5.127   PIN diode equivalent circuit.

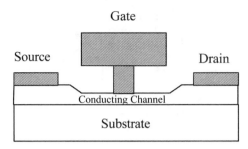

FIGURE 5.128   Reverse biased PIN diode equivalent circuit.

where R_c is the contact resistance of the metal semiconductor interfaces, R_i is the channel resistance of the intrinsic region, and R_m is the resistance of the contact metals. The resistance of the intrinsic region dominates and the reverse resistance becomes essentially that of the intrinsic region, which in terms of physical parameters can be expressed as

$$R_i \cong \frac{3\left(kT\right)L^2}{8qI_0 L_a^2} \qquad (5.100)$$

where L is the length of the intrinsic region, typically in the range of 1 to 100 μm. Depending upon design frequency, I_0 is the bias current, and La is the ambipolar diffusion length, which is a constant of the material [3]. The other parameters have their usual meanings. Note that the reverse resistance, which can be in the kΩ range, is inversely proportional to bias current, and decreases with the magnitude of the applied bias current. The greatest off-state resistance, therefore, occurs under low reverse bias voltage. Under reverse bias the intrinsic region is essentially depleted of free charge, so the series capacitance is simply the capacitance of the intrinsic region, and can be expressed as

$$C_i = \frac{\varepsilon A}{L} \qquad (5.101)$$

where A is the cross-sectional area of the diode. Note that the capacitance is constant under reverse bias.

Under forward bias the diode is dominated by the forward charge injection characteristics of the pn junction, and the diode can be represented as a resistance, with magnitude determined by the forward current. The on-state resistance can be expressed as

$$R_i = \frac{nkTA}{qI_0} \qquad (5.102)$$

where n is the ideality factor for the diode (given in the diode specifications). The resistance of the diode in forward bias is inversely proportional to bias current, and the lowest resistance is obtained at high currents. The impedance of the diode can be tuned for RF circuit matching by adjustment of the bias current.

The rate at which the pin diode can be switched from a low-impedance, forward biased condition to a high-impedance, reverse biased condition, is determined by the speed at which the free charge can be extracted from the diode. Diodes with longer intrinsic regions and larger cross-sectional areas will store more charge, and require, therefore, longer times to switch. The actual switching time has two components: the time required to remove most of the charge (called the delay time) from the intrinsic region, and the time during which the diode is changing from a low- to a high-impedance state (called the

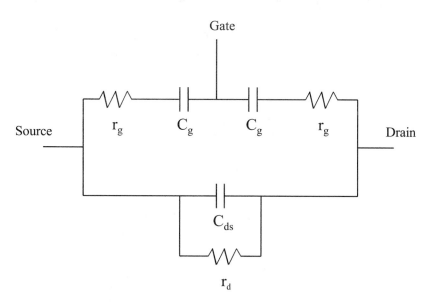

FIGURE 5.129 Schematic diagram for a MESFET switching element.

transition time). The transition time depends upon diode geometry and details of the diode doping profile, but is not sensitive to the magnitude of the forward or reverse current. The delay time is inversely proportional to the charge carrier lifetime. Diodes with short carrier lifetime have short delay times, but suffer from high values of forward bias resistance. High forward bias resistance increases the insertion loss for the diode, and this will produce attenuation of the signal through the device in the on-state.

MESFET Switches

A schematic diagram for a GaAs MESFET is shown in Fig. 5.129, and these devices are often used in switching applications. In general, a MESFET can be used in two different modes as passive or active elements. In the active mode the transistor is used as a three-terminal switch where the transistor is configured similar to an amplifier circuit. Either single-gate or dual-gate FETs can be used. The transistor is biased with a positive drain and a negative gate voltage, which are set so that the transistor is active. Switching action is accomplished by control of the transistor gain, which can be varied over several orders of magnitude. Dual-gate devices are particularly attractive for this application since the second gate can be used as a control port for efficient control of the gain.

In the passive mode of operation, the MESFET is configured to function as a passive two-terminal device, with the gate terminal acting as a port for only the control signal. That is, the RF signal is not applied to the gate and only travels between the drain and source terminals. The magnitude of the RF impedance between the drain and source terminals is controlled by a DC signal applied to the gate terminal. The drain-to-source impedance can be varied from a low value, obtained under open channel conditions when a zero potential is applied to the gate, to a high value, obtained when the gate is biased with a negative potential of sufficient amplitude to prevent current from flowing through the transistor. This occurs when the gate voltage achieves the transistor pinch-off voltage, which has a magnitude that is a function of the particular MESFET used.

In the passive mode the low-impedance state of the MESFET switch is dominated by the fully open conducting channel, and the open-channel resistance for the device is low. The equivalent circuit is essentially the "on" resistance for the transistor. In the high-impedance state the MESFET is dominated by the depleted channel or "off" resistance, which is large, and the switch has an equivalent circuit as shown in Fig. 5.130. The high-impedance state for the MESFET switch can be approximated with the simplified equivalent circuit shown in Fig. 5.131, where the "off" state resistance and capacitance are

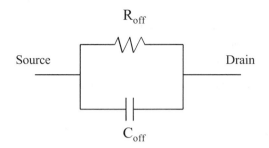

FIGURE 5.130   High-impedance, off-state equivalent circuit for a MESFET switch.

High Impedance State

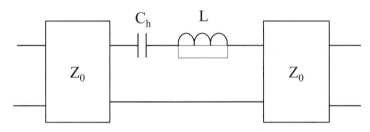

Low Impedance State

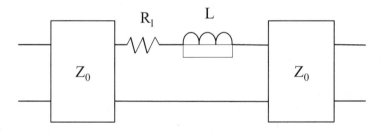

FIGURE 5.131   Simplified off-state equivalent circuit for a MESFET switch.

$$R_{off} = \frac{2r_d}{2+r_d\omega^2C_g^2r_g} \tag{5.103}$$

where ω is the radian frequency, and

$$C_{off} = C_{ds} + \frac{C_g}{2} \tag{5.104}$$

Note that the "off" state resistance is an inverse function of frequency and, therefore, the magnitude of the blocking resistance decreases as frequency increases. The performance of the switch will degrade at high frequency and switch design becomes more difficult.

Switching Circuits

There are two basic configurations used for single-pole, double-throw (SPDT) switches that are commonly used to control the flow of microwave signals along a transmission line. The basic configurations

High Impedance State

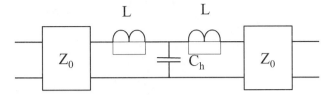

Low Impedance State

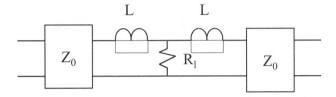

FIGURE 5.132 Simplified series connected switch circuit.

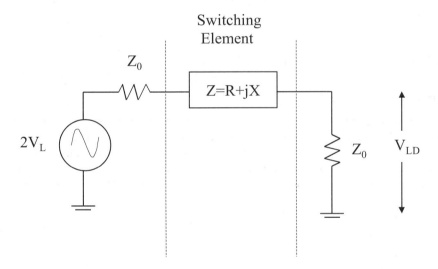

FIGURE 5.133 Simplified shunt connected switch circuit.

can be fabricated using either pin diodes or MESFET switching elements, and are realized by utilization of the diode or transistor in a series or shunt connection to the transmission line. A simplified equivalent circuit for a series connected switch is shown in Fig. 5.132, and a shunt connected switch is shown in Fig. 5.133. The two configurations are complimentary in that the low-impedance state of the series switch permits signal flow, while the high-impedance state of the shunt switch permits signal flow. In the "off" state for both configurations, the microwave power incident upon the switching device is primarily reflected back toward the source. A small fraction of the incident power is dissipated in the switching element and transmitted through the device toward the load. It is this fraction of the incident power that accounts for the insertion loss and the finite and nonideal isolation of the device. The fraction of microwave power that is transmitted through the device increases with frequency due to parasitic paths due to mounting, bonding, packaging, etc. elements, and switch isolation tends to degrade as operating frequency increases. It is possible, however, to minimize the parasitic signal flow by RF tuning and impedance compensation techniques.

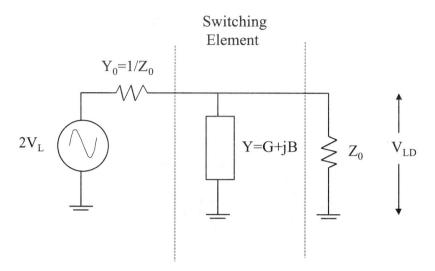

FIGURE 5.134   Equivalent circuit for a series connected switch.

FIGURE 5.135   Equivalent circuit for a shunt connected switch.

Insertion Loss and Isolation

Insertion Loss (IL) and isolation are important parameters that are used to characterize the performance of microwave switches. Insertion loss is defined as the ratio, generally in decibels, of the power delivered to the load in the "on" state of an ideal switch to the actual power delivered by the switch. The insertion loss can be calculated from consideration of the series and shunt equivalent circuits shown in Figs. 5.134 and 5.135. If V_L represents the voltage developed at the load for an ideal switch, the insertion loss can be written as,

$$IL = \left| \frac{V_L}{V_{LD}} \right|^2 \tag{5.105}$$

where, for the series configuration

$$V_{LD} = \frac{2V_L}{2 + Z/Z_0} \tag{5.106}$$

and

$$Z = R + jX \tag{5.107}$$

is the impedance of the switching device. The insertion loss is expressed as

$$IL = \left| \frac{2 + Z/Z_0^2}{2Z_0} \right| = \left| 1 + \frac{R + jX}{2Z_0} \right|^2 = 1 + \frac{R}{Z_0} + \frac{1}{4}\left(\frac{R}{Z_0} \right)^2 + \frac{1}{4}\left(\frac{X}{Z_0} \right)^2 \tag{5.108}$$

where R and X are the resistance and reactance of the switching device in the low-impedance state.

For the shunt configuration, the voltage across the load is

$$V_{LD} = \frac{2V_L Y_0}{2Y_0 + Y}$$

(5.109)

and the insertion loss is written

$$IL = \left| \frac{2Y_0 + Y}{2Y_0} \right|^2 = \left| 1 + \frac{G + jB}{2Y_0} \right|^2 = 1 + \frac{G}{Y_0} = \frac{1}{4}\left(\frac{G}{Y_0}\right)^2 + \frac{1}{4}\left(\frac{B}{Y_0}\right)^2$$

(5.110)

where $Y_0 = 1/Z_0$, and $Y = G + jB$.

Isolation is a measure of the off-state performance of the switch. It is defined as the ratio of microwave power delivered to the load for an ideal switch in the "on" state, to the actual power delivered to the load when the switch is in the "off" state. In order to calculate isolation, the insertion loss expressions given above are used with the real and reactive terms for the device in the low-impedance state interchanged with the appropriate device parameters for the high-impedance state.

Switch Design

Switch design procedures are based upon the principle that the switching element in the "on" and "off" states can be considered as a reactance or susceptance that can be included in a filter configuration. Switch design, therefore, makes use of filter design procedures and all approaches to filter design can be used. The "on" and "off" state equivalent circuits are used to embed the switch element in the filter design. Generally, the "on" or low-insertion loss state is considered first, and the network is designed to yield the desired pass-band performance. The "off" state can be considered as a detuned network, and the impedances are adjusted to achieve the desired isolation. This approach to switch design may require several iterations until satisfactory performance in both the "on" and "off" states are achieved. Mounting and lead reactances are considered in the design and are absorbed and incorporated into the filter network. The actual filter element values may, therefore, differ in value from the design values. The performance of the insertion loss and isolation will vary with tuning and the lowest insertion loss and greatest isolation generally are obtained over narrow bandwidth. Increased bandwidth produces degradation in switch performance. Bias control circuits and thermal handling are accomplished in the same manner as for amplifier circuits.

References

1. H.A. Watson, *Microwave Semiconductor Devices and Their Circuit Applications*, McGraw-Hill, New York, 1969.
2. S.K. Koul and B. Bhat, Microwave and Millimeter Wave Phase Shifters, in *Semiconductor and Delay Line Phase Shifters*, Norwood, MA, Artech House, 1991.
3. I. Bahl and P. Bhartia, *Microwave Solid State Circuit Design*, Wiley Interscience, New York, 1988.
4. M.E. Knox, P.J. Sbuttoni, J.J. Stangel, M. Kumar, and P. Valentino, Solid State 6x6 Transfer Switch for Cylindrical Array Radar, *1993 IEEE International Microwave Symposium Digest*, 1225–1228.
5. P. Omno, N. Jain, C. Souchuns, and J. Goodrich, High Power 6-18 Transfer Switch Using HMIC, *1994 International Microwave Symposium Digest*, 79–82.
6. C.K. Sun, C.T. Chang, R. Nguyen, and D.J. Albares, Photovoltaic PIN Diodes for RF Control — Switching Applications, *IEEE Trans. Microwave Theory Tech.*, 47, 2034–2036, Oct. 1999.
7. M. Shifrin, P. Katzin, and Y. Ayasli, High Power Control Components Using a New Monolithic FET Structure, *1989 IEEE Monolithic and Millimeter-Wave Integrated Circuits Symposium Digest*, 51–56.

8. T. Yamaguchi, T. Sawai, M. Nishida, and M. Sawada, Ultra-Compact 1 W GaAs SPDT Switch IC, *1999 IEEE International Microwave Symposium Digest*, 315–318.

9. H. Uda, T. Yamada, T. Sawai, K. Nogawa, and Yu. Harada, A High-Performance GaAs Switch IC Fabricated Using MESFET's with Two Kinds of Pinch-Off Voltages, *GaAs IC Symp. Digest,* 139–142, 1993.

10. M. Masuda, N. Ohbata, H. Ishiuchi, K. Onda, and R. Yamamoto, High Power Heterojunction GaAs Switch IC with P-1db of More Than 38 dbm for GSM Application, *GaAs IC Symp. Digest*, 229–232, 1998.

11. D.L. Ingram, K. Cha, K. Hubbard, and R. Lai, Q-Band High Isolation GaAs HEMT Switches, *IEEE GaAs IC Symposium Digest*, 289–292, 1996.

12. H. Takasu, F. Sasaki, H. Kawasaki, H. Tokuda, and S. Kamihashi, W-Band SPST Transistor Switches, *IEEE Microwave Guided Wave Letters*, 315–316, Sept. 1996.

13. R.H. Caverly, and K.J. Heissler, On-State Distortion in High Electron Mobility Transistor Microwave and RF Switch Control Circuits, *IEEE Trans. Microwave Theory Tech.*, 98–103, Jan. 2000.

14. I.B Vendik, O.G. Vendik, and E.L. Kollberg, Commutation Quality Factor of Two-State Switchable Devices, *IEEE Trans. Microwave Theory Tech.*, 802–808, May 2000.

6

Passive Technologies

Alfy Riddle
Macallan Consulting

Michael B. Steer
North Carolina State University

S. Jerry Fiedziuszko
Space Systems/LORAL

Karl R. Varian
Raytheon

Donald C. Malocha
University of Central Florida

Michael E. Majerus
Motorola

David Anderson
Maury Microwave Corporation

James B. West
Rockwell Collins

Jeanne S. Pavio
Motorola

0-8493-8592-X/01/$0.00+$.50
© 2001 by CRC Press LLC

6.1   Passive Lumped Components

Alfy Riddle

Lumped components such as resistors, capacitors, and inductors make up most of the glue that allows microwave discrete transistors and integrated circuits to work. Lumped components provide impedance matching, attenuation, filtering, DC bypassing, and DC blocking. More advanced lumped components such as chokes, baluns, directional couplers, resonators, and EMI filters are a regular part of RF and microwave circuitry. Figure 6.1 shows examples of lumped resistors, capacitors, inductors, baluns, and directional couplers. Surface mount techniques and ever-shrinking package sizes now allow solderable lumped components useful to 10 GHz.[1]

As lumped components are used at higher and higher frequencies, the intrinsic internal parasitics, as well as the package and mounting parasitics play a key role in determining overall circuit behavior both in and out of the desired frequency range of use. Mounting parasitics come from excess inductance caused by traces between the component soldering pad and a transmission line, and excess capacitance from relatively large mounting pads and the component body.

Resistors

Figure 6.2 shows a typical surface mount resistor. These resistors use a film of resistive material deposited on a ceramic substrate with solderable terminations on the ends of the component. Individual surface mount resistors are available in industry standard sizes from over 2512 to as small as 0201. The package size states the length and width of the package in hundredths of an inch. For example an 0805 package is 8 hundredths long by 5 hundredths wide or 80 mils by 50 mils, where a mil is 1/1000th of an inch. Some manufacturers use metric designations rather than English units, so an 0805 package would be 2 mm by 1.2 mm or a 2012 package. The package size determines the intrinsic component parasitics to the first order, and in the case of resistors determines the allowable dissipation. For resistors the most important specifications are power dissipation and tolerance of value.

FIGURE 6.1   Assorted surface mount lumped components.

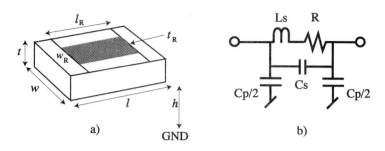

FIGURE 6.2 Surface mount resistor: a) component, b) schematic.

Resistance

Chip resistors have plated, wrap-around end contacts that overlap the resistive material deposited on the top surface of the ceramic carrier. Circuits built from film on alumina can have multiple thin or thick film resistors on one substrate. The same principles for determining resistance, heat dissipation limits, parasitic inductance, and parasitic capacitance apply to both chips and thin film circuits.[2] Standard chip resistors come in 10% and 1% tolerances, with tighter tolerances available.

The resistive material is deposited in a uniform thickness, t_R, and has a finite conductivity, σ_R. The material is almost always in a rectangle that allows the resistance to be calculated by Eq. (6.1).

$$R = l_R \big/ \left(\sigma_R \, w_R \, t_R \right) = R_p \, l_R \big/ w_R \tag{6.1}$$

Increasing the conductivity, material thickness, or material width lowers the resistance, and the resistance is increased by increasing the resistor length, l_R. Often it is simplest to combine the conductivity and thickness into $R_p = 1/(\sigma_R \, t_R)$, where any square resistor will have resistance R_p. Resistor physical size determines the heat dissipation, parasitic inductance and capacitance, cost, packing density, and mounting difficulty.

Heat Dissipation

Heat dissipation is determined mostly by resistor area, although a large amount of heat can be conducted through the resistor terminations.[3] Generally, the goal is to keep the resistor film below 150 degrees C and the PCB temperature below 100 degrees C. Derating curves are available from individual manufacturers with a typical curve shown in Fig. 6.3. Table 6.1 shows maximum power dissipation versus size for common surface mount resistors.

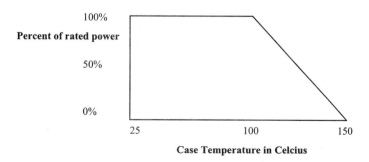

FIGURE 6.3 Surface mount resistor derating curve.

TABLE 6.1   Maximum Resistor
Dissipation vs. Size

Resistor Size	P_{Max}(Watts)
2512	1
2510	0.5
1210	0.25
1206	0.125
805	0.1
603	0.0625
402	0.0625

Intrinsic Inductive Parasitics

The resistor length and width determine its effective series inductance. This inductance can be computed from a transmission line model or as a ribbon inductor if the resistor is far enough above the ground plane. The ribbon inductor equation is given in Eq. (6.2), with all units MKS.[20]

$$L_S = 0.125 \, l_R \left(\ln\left[2 \, l_R / w_R \right] + 0.5 + 2 \, w_R / 9 \, l_R \right) \mu H \qquad (6.2)$$

For example, an 0805 package has an inductance of about 0.7 nH. If the resistor width equals the transmission line width, then no parasitic inductance will be seen because the resistor appears to be part of the transmission line.

Intrinsic Capacitive Parasitics

There are two types of capacitive parasitics. First, there is the shunt distributed capacitance to the ground plane, C_P. The film resistor essentially forms an RC transmission line and is often modeled as such. In this case, a first order approximation can be based on the parallel plate capacitance plus fringing of the resistor area above the ground plane, as shown in Eq. (6.3). The total permittivity, ε, is a product of the permittivity of free space, ε_O, and the relative permittivity, ε_R. An estimate for effective width, w_E, is $1.4 * w_R$, but this depends on the ground plane distance, h, and the line width, w. A more accurate method would be to just use the equivalent microstrip line capacitance.

$$C_P = \varepsilon \, w_E \, l_R / h \qquad (6.3)$$

For an 0805 resistor on 0.032" FR4 the capacitance will be on the order of 0.2 pF. As with the inductance, when the resistor has the same width as the transmission line, the line C per unit length absorbs the parasitic capacitance. When the resistor is in shunt with the line, the parasitic capacitance will be seen, often limiting the return loss of a discrete pad.

An additional capacitance is the contact-to-contact capacitance, C_S. This capacitance will typically only be noticed with very high resistances. Also, this capacitance can be dominated by mounting parasitics such as microstrip gap capacitances. C_S can be approximated by the contact area, the body length, and the body dielectric constant as in a parallel plate capacitor.

Capacitors

Multilayer chip capacitors are available in the same package styles as chip resistors. Standard values range from 0.5 pF up to microfarads, with some special products available in the tenths of picofarads. Parallel plate capacitors are available with typically lower maximum capacitance for a given size. Many different types of dielectric materials are available, such as NPO, X7R, and Z5U. Low dielectric constant materials, such as NPO, usually have low loss and either very small temperature sensitivity, or well-defined temperature variation for compensation, as shown in Table 6.2. Higher dielectric constant materials, such

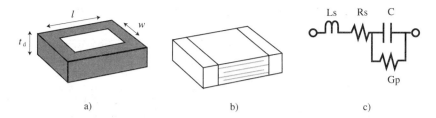

FIGURE 6.4   Capacitor styles and schematic: (a) parallel plate, (b) multilayer, and (c) schematic.

as X7R and Z5U, vary more with temperature than NPO. Z5U will lose almost half its capacitance at very low and very high temperatures. Higher dielectric constant materials, such as X7R and Z5U, also have a reduction in capacitance as voltage is applied.[5,23] The critical specification for these capacitors is the voltage rating. Secondary specifications include temperature stability, Q, tolerance, and equivalent series resistance (ESR).

Parallel Plate

Parallel plate capacitors, as shown in Fig. 6.4a, can use a thin dielectric layer mounted on a low resistance substrate such as silicon, or they can be a thick ceramic with plated terminations on top and bottom. These capacitors can be attached by soldering or bonding with gold wire. Some capacitors come with several pads, each pad typically twice the area of the next smaller, which allows tuning. These capacitors obey the parallel plate capacitance equation below.

$$C = \varepsilon w l / t_d \tag{6.4}$$

Parasitic resistances, R_S, for these capacitors are typically small and well controlled by the contact resistance and the substrate resistance. Parasitic conductances, G_P, are due to dielectric loss. These capacitors have limited maximum values because of using a single plate pair. The voltage ratings are determined by the dielectric thickness, t_d, and the material type. Once the voltage rating and material are chosen, the capacitor area determines the maximum capacitance. The parasitic inductance of these capacitors, which determines their self-resonance frequency, is dominated by the wire connection to the top plate. In some cases these capacitors are mounted with tabs from the top and bottom plate. When this occurs, the parasitic inductance will be the length of the tab from the top plate to the transmission line, as well as the length of the capacitor acting as a coupled transmission line due to the end launch from the tab.

Multilayer Capacitors

Multilayer chip capacitors are a sandwich of many thin electrodes between dielectric layers. The end terminations connect to alternating electrodes, as shown in Fig. 6.4b. A wide variety of dielectric materials are available for these capacitors and a few typical dielectric characteristics are given in Table 6.2.[23]

TABLE 6.2   Chip Capacitor Dielectric Material Comparison

Type	ε_R	Temp. Co. (ppm/degC)	Tol (%)	Range (pF in 805)	Voltage Coeff. (%)
NPO	37	0+/−30	1–20	0.5 p–2200 p	0
4	205	−1500+/−250	1–20	1 p–2200 p	0
7	370	−3300+/−1000	1–20	1 p–2200 p	0
Y	650	−4700+/−1000	1–20	1 p–2200 p	0
X7R	2200	+/−15%	5–20	100 p–1 μ	+0/−25
Z5U	9000	+22/−56%	+80/−20	0.01 μ–0.12 μ	+0/−80

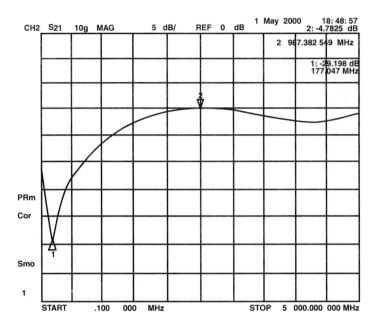

FIGURE 6.5   S21 of a shunt 0805 1000 pF multilayer capacitor, as if used for bypassing.

Table 6.2 shows that critical capacitors will have to use NPO dielectric (or alumina) because of tolerance and stability. DC blocking and bypass capacitors will often use X7R or Z5U dielectrics, but these dielectrics must be used with the knowledge that their capacitance varies significantly with temperature and has a voltage dependent nonlinearity.

$$C_p = (n-1)\varepsilon w_p l / t_L \tag{6.5}$$

Multilayer capacitors have a more complicated structure than parallel plate capacitors. Their capacitance is given by Eq. (6.5), where t_L is the dielectric layer thickness and w_p is the plate width. The series resistance of the capacitor, Rs, is determined by the parallel combination of all the plate resistance and the conductive loss, Gp, is due to the dielectric loss. Often the series resistance of these capacitors dominates the loss due to the very thin plate electrodes.

By using the package length inductance, as was given in Eq. (6.2), the first series resonance of a capacitor can be estimated. The schematic of Fig. 6.4c only describes the first resonance of a multilayer capacitor. In reality many resonances will be observed as the multilayer transmission line cycles through quarter- and half-wavelength resonances due to the parallel coupled line structure, as shown in Fig. 6.5.[4] The 1000 pF capacitor in Fig. 6.5 has a series inductance of 0.8 nH and a series resistance of 0.8 ohms at the series resonance at 177 MHz. The first parallel resonance for the measured capacitor is at 3 GHz. The equivalent circuit in Fig. 6.4c does not show the shunt capacitance to ground, C_p, caused by mounting the chip capacitor on a PCB. C_p may be calculated just as it was for a chip resistor in the section on Intrinsic Capacitive Parasitics. The Qs for multilayer capacitors can reach 1000 at 1 GHz.

Printed Capacitors

Printed capacitors form very convenient and inexpensive small capacitance values because they are printed directly on the printed circuit board (PCB). Figure 6.6 shows the layout for gap capacitors and interdigital capacitors. The capacitance values for a gap capacitor are very low, typically much less than 1 pF, and estimated by Eq. (6.6).[6] Gap capacitors are best used for very weak coupling and signal sampling because they are not particularly high Q. Equation (6.6) can also be used to estimate coupling between two circuit points to make sure a minimum of coupling is obtained.

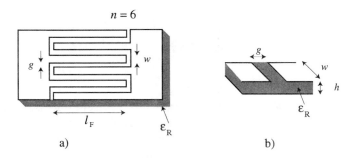

FIGURE 6.6 Printed capacitors: (a) Gap capacitor; (b) interdigital capacitor.

$$C_G = \varepsilon_O \, \varepsilon_R \, w \left(\ln\left[0.25 + \left(h/g \right)^2 \right] + g/h \, \mathrm{Tan}^{-1}\left[2h/g \right] \right) \Big/ \left(2\pi \right) \tag{6.6}$$

Interdigital capacitors, shown in Fig. 6.6a are a planar version of the multilayer capacitor. These capacitors have medium Q, are accurate, and are typically less than 1 pF. These capacitors can also be tuned by cutting off fingers. Equation (6.7) gives an estimate of the interdigital capacitance using elliptic functions, K.[7] Because interdigital capacitors have a distributed transmission line structure, they will show multiple resonances as frequency increases. The first resonance occurs when the structure is a quarter wavelength. The Q of this structure is limited by the current crowding at the thin edges of the fingers.

$$C_I = \varepsilon_O \left(1 + \varepsilon_R \right)\left(n - 1 \right) l_F \, K\left[\mathrm{Tan}\left(w\pi/\left(4\left(w + g \right) \right) \right)^4 \right] \Big/ K\left[1 - \mathrm{Tan}\left(w\pi/\left(4\left(w + g \right) \right) \right)^4 \right] \tag{6.7}$$

Inductors

Inductors are typically printed on the PCB or surface mount chips. Important specifications for these inductors are their Q, their self-resonance frequency, and their maximum current. While wire inductors have their maximum current determined by the ampacity of the wire or trace, inductors made on ferrite or iron cores will saturate the core if too much current is applied. Just as with capacitors, using the largest inductance in a small area means dealing with the parasitics of a nonlinear core material.[8,22]

Chip Inductors

Surface mount inductors come in the same sizes as chip resistors and capacitors, as well as in air-core "springs." "Spring" inductors have the highest Q because they are wound from relatively heavy gauge wire, and they have the highest self-resonance because of their air core. Wound chip inductors, shown in Fig. 6.7a, use a fine gauge wire wrapped on a ceramic or ferrite core. These inductors have a mediocre

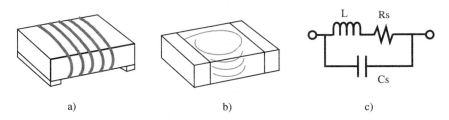

FIGURE 6.7 Chip inductors: (a) wound; (b) multilayer; and (c) schematic.

Q of 10 to 100 and a lowered self-resonance frequency because of the dielectric loading of the ceramic or ferrite core. However, these inductors are available from 1 nH to 1 mH in packages from 402 to 1812. The chip inductors shown in Fig. 6.7b use a multilayer ceramic technology, although some planar spiral inductors are found in chip packages. Either way, these inductors typically have lower Q than wound inductors, but Qs can still reach 100. Multilayer chip inductors use 805 and smaller packages with a maximum inductance of 470 nH. Although the self-resonance frequency of these inductors is high because of the few turns involved, the dielectric loading of the sandwich makes the resonance lower than that of an equivalent "spring" or even a wound inductor, as shown in Table 6.3 and Fig. 6.8. The solenoid formula given in Eq. (6.8) uses MKS units and describes both wound and "spring" inductors.[10] In Eq. (6.8), n is the number of turns, d is the coil diameter, and l is the coil length.

$$L = 9.825\, n^2\, d^2 \big/ \left(4.5\, d + 10\, l\right) \mu H \qquad (6.8)$$

Figure 6.8 shows an S21 measurement of several series inductors. The inductors are mounted in series to emphasize the various resonances. The 470 nH wound inductor used a 1008 package and had a first parallel resonance at 600 MHz. The 470 nH chip inductor used an 805 package and had a first parallel resonance of 315 MHz. The chip inductor shows a slightly shallower first null, indicating a higher series resistance and lower Q. Both inductors had multiple higher order resonances. The chip inductor also showed less rejection at higher frequencies. Figure 6.8b shows the S21 of a series "spring" inductor with a first parallel resonance of 3 GHz. No higher modes of the "spring" inductor can be seen because of the limited bandwidth of the network analyzer, only a fixture resonance is shown around 4 GHz. Table 6.3 shows a matrix of typical "spring," wound, and chip inductor capability.

Current Capability
Chip inductors have limited current-carrying capability. At the very least, the internal conductor size limits the allowable current. When the inductor core contains ferrite or iron, magnetic core saturation will limit the useful current capability of the inductor, causing the inductance to decrease well before conductor fusing takes place.

$$Q = \omega\, L \big/ Rs \qquad (6.9)$$

Parasitic Resistance
The inductor Q is determined by the frequency, inductance, and effective series resistance as shown in Eq. (6.9). The effective series resistance, Rs, comes from the conductor resistance and the core loss when a magnetic core is used. The conductor resistance is due to both DC and skin effect resistance as given by Eqs. (6.1) and (6.10).[9]

$$R_{SKIN} = l_R \big/ \left(\sigma_R\, \rho_R\, \delta\right), \qquad (6.10)$$

where ρ_R is the perimeter of the wire and δ is the skin depth,

$$\delta = Sqrt\left[\left(2.5\ E\ 6\right) \big/ \left(\sigma_R\, f\, \pi^2\right)\right]. \qquad (6.11)$$

When the skin depth is less than half the wire thickness, Eq. (6.10) should be used. For very thin wires the DC resistance is valid up to very high frequencies. For example, 1.4-mil thick copper PCB traces will show skin effect at 14 MHz, while 5 microns of gold on a thin film circuit will not show skin effect until almost 1 GHz.

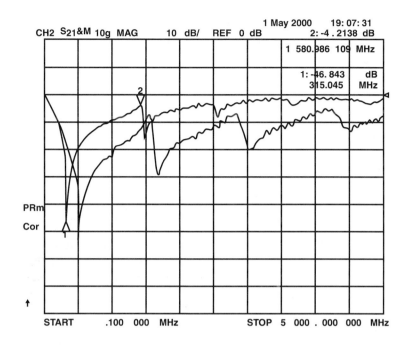

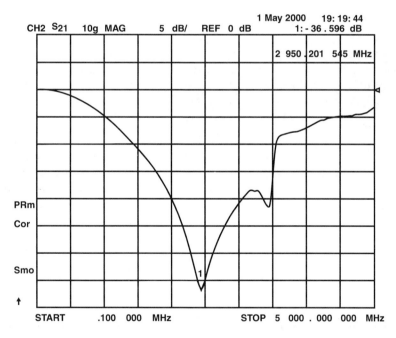

FIGURE 6.8 S21 of series inductors: (a) 470 nH wound and chip inductors; and (b) 19 nH spring inductor.

Parasitic Capacitance

The broadband inductor sweep of Fig. 6.8 shows resonances in wound, chip, and spring inductors. Manufacturers, and the schematic of Fig. 6.7, only show the first parallel resonance. Wound inductors can be modeled as a helical transmission line.[10] The inductance per unit length is the total solenoid inductance divided by the length of the inductor. The capacitance to ground can be modeled as a cylinder

TABLE 6.3   Inductor Matrix

Type	Size	Inductance Range	Q Range	Current Range	Resonance Range
Spring					
	Micro	1.6–13 nH	50–225	NA	4.5–6 GHz
	Mini	2.5–43 nH	90–210	NA	1.5–3 GHz
Wound					
	603	1.8–120 nH	10–90	0.3–0.7 A	1.3–6 GHz
	805	3.3–220 nH	10–90	0.3–0.8 A	0.8–6 GHz
	1008	3.3–10,000 nH	10–100	0.24–1 A	0.06–6 GHz
	1812	1.2–1000 µH	10–60	0.05–0.48 A	1.5–230 MHz
Multilayer					
	402	1–27 nH	10–200	0.3 A	1.6–6 GHz
	603	1.2–100 nH	5–45	0.3 A	0.83–6 GHz
	805	1.5–470 nH	10–70	0.1–0.3 A	0.3–6 GHz

of diameter equal to the coil diameter. The first quarter wave resonance of this helical transmission line is the parallel resonance of the inductor, while the higher resonances follow from transmission line theory. When the inductor is tightly wound, or put on a high dielectric core, the interwinding capacitance increases and lowers the fundamental resonance frequency.

$$L_{WIRE} = 200 \, l_W \left(\ln\left[4 \, l_W / d\right] + 0.5 \, d / l_W - 0.75 \right) nH \qquad (6.12)$$

Bond Wires and Vias
The inductance of a length of wire is given by Eq. (6.12).[1] This equation is a useful first order approximation, but rarely accurate because wires usually have other conductors nearby. Parallel conductors such as ground planes or other wires reduce the net inductance because of mutual inductance canceling some of the flux. Perpendicular conductors, such as a ground plane terminating a wire varies the inductance up to a factor of 2 as shown by the Biot-Savart law.[11]

Spiral Inductors
Planar spiral inductors are convenient realizations for PCBs and ICs. They are even implemented in some low-value chip inductors. These inductors tend to have low Q because the spiral blocks the magnetic flux; ground planes, which reduce the inductance tend to be close by; and the current crowds to the wire edges, which increases the resistance. With considerable effort Qs of 20 can be approached in planar spiral inductors.[12] Excellent papers have been written on computing spiral inductance values.[13,14] In some cases, simple formulas can produce reasonable approximations; however, the real challenge is computing the parallel resonance for the spiral. As with the solenoid, the best formulas use the distributed nature of the coil.[14] A first order approximation is to treat the spiral as a microstrip line, compute the total capacitance, and compute the resonance with the coupled wire model of the spiral.

Chokes

Chokes are an important element in almost all broadband RF and microwave circuitry. A choke is essentially a resonance-free high impedance. Generally, chokes are used below 2 GHz, although bias tees are a basic choke application that extends to tens of GHz. Schematically, a choke is an inductor. In the circuit a choke provides a high RF impedance with very little loss to direct current so that supply voltages and resistive losses can be minimized. Most often the choke is an inductor with a ferrite core used in the frequency range where the ferrite is lossy. A ferrite bead on a wire is the most basic form of choke. It is important to remember that direct current will saturate ferrite chokes just as it does ferrite inductors. At higher frequencies, clever winding tricks can yield resonance-free inductance over decades of bandwidth.[15] Ferrite permeability can be approximated by Eq. (6.13).[16]

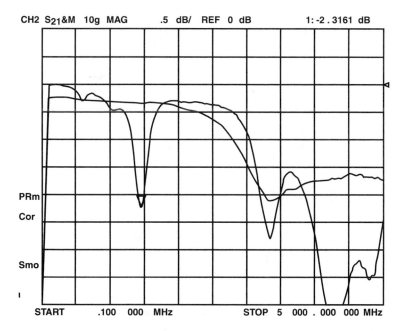

FIGURE 6.9 S21 of shunt 4.7 μH choke and 4.7 μH wound chip inductor, as if used in a DC bias application.

$$\mu\left(f_{MHz}\right)=1\bigg/\left(\,\mathrm{j}\,f_{MHz}\big/2250+1\bigg/\left(\mu_i\left(1+\mathrm{j}\,0.8\,\mu_i\,\,f_{MHz}\big/2250\right)\right)\right) \qquad (6.13)$$

For frequencies below 2250 MHz/μ_i, the ferrite makes a low loss inductor, but above this frequency the ferrite becomes lossy and appears resistive.[17] Figure 6.9 shows a comparison of a 4.7 μH choke and a 4.7 μH wound chip inductor. The high frequency resonances present in the wound inductor are absorbed by the ferrite loss. The best model for a ferrite choke is an inductor in parallel with a resistor. The resistance comes from the high frequency core loss, and the inductor comes from the low frequency ferrite permeability, core shape, and wire turns. As can be seen in Eq. (6.13), the real part of the permeability decreases as frequency increases, causing the loss term to dominate.

Ferrite Baluns

Ferrite baluns are a cross between transmission lines and chokes. While transmission line baluns can be used at any frequency, ferrite baluns are most useful between 10 kHz and 2 GHz. Basically, the balun is made from twisted or bifilar wire with the desired differential mode impedance, such as 50 or 75 ohms. The wire is wound on a ferrite core so any common mode currents see the full choke impedance, as shown in Fig. 6.10a. Two-hole balun cores are used because they provide the most efficient inductance per turn in a small size. The ferrite core allows the balun to work over decades of frequency, for example 10 MHz to 1000 MHz is a typical balun bandwidth. Also, the ferrite allows the baluns to be small enough to fit into T0-8 cans on a 0.25-in. square SMT substrate as shown in Fig. 6.1. Extremely high performance baluns are made from miniature coax surrounded by ferrite beads.[18]

Manufacturers design baluns for specific impedance levels and bandwidths. In order to understand the performance trade-offs and limitations, some discussion will be given. Figure 6.10a shows a typical ferrite and wire balun. Figure 6.10b shows a schematic representation of the balun in Fig. 6.10a. The balun is easiest to analyze using the equivalent circuit of Fig. 6.10c. The equivalent circuit shows that the inverting output is loaded with the even-mode choke impedance, Zoe. For low frequency symmetry in output impedance and gain, a choke is often added to the non-inverting output as shown in Fig. 6.10d.

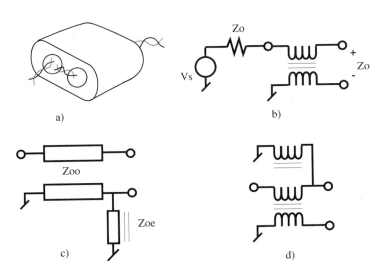

FIGURE 6.10   Ferrite balun (a), with schematic (b), equivalent circuit (c), and symmetric version *(d).

The configuration of Fig. 6.10d is especially useful when coaxial cable is used instead of twisted pair wire. This is because the coax shields the center conductor from the ferrite so the center conductor has infinite even-mode impedance to ground. For twisted pair baluns, the even-mode impedance to ground is finite and creates a lossy transmission line on the ferrite that attenuates the high frequencies. For broadest band operation, ferrite and wire baluns use the configuration of Fig. 6.10b because Fig. 6.10d causes excessive high frequency attenuation. Finally, transitions into and out of the balun, as well as imperfect wire impedance contribute to excess series inductance for the differential-mode impedance. The inductive S11 increases with frequency and is often tuned out with small capacitors across the twisted pair.[19]

References

1. Bahl, I, and Bhartia, P., *Microwave Solid State Circuit Design*, John Wiley & Sons, New York, 1988.
2. Chin, S., RLC Components, *Electronic Products*, 31–33, July 1989.
3. Florida RF Labs, *Component Reference Catalog*, Florida RF Labs, Stuart, FL, 2000.
4. ATC, *The RF Capacitor Handbook*, ATC, Huntington Station, NY, 1979.
5. Ingalls, M.W., Perspectives on Ceramic Chip Capacitors, *RF Design*, 45–53, Nov. 1989.
6. Pavlov, V.I., Gap Capacity in Microstrip, *Radioelectronica and Communication Systems*, 85–87, 1981.
7. Wolff, I., and Kibuuka, G., Computer Models for MMIC Capacitors and Inductors, *14th Euorpean Microwave Conference*, 853–859, 1984.
8. Grover, F.W., *Inductance Calculations*, Dover, New York, 1962.
9. Ramo, S., Whinnery, J.R., and van Duzer, T., *Fields and Waves in Communication Electronics*, John Wiley & Sons, New York, 1984.
10. Rhea, R.W., A Multimode High-Frequency Inductor Model, *Applied Microwaves and Wireless*, 70–80, Nov/Dec 1997.
11. Goldfard, M.E., and Pucel, R.A., Modeling Via Hole Grounds in Microstrip, *IEEE Microwave Guided Letters*, 135–137, June 1991.
12. Gecorgyan, S., Aval, O., Hansson, B., Jaconsson, H., and Lewin, T., Loss Considerations for Lumped Inductors in Silicon MMICs, *IEEE MTT-S Digest*, 859–862, 1999.
13. Remke, R.L., and Burdick, G.A., Spiral Inductors for Hybrid and Microwave Applications, *24th Electronic Components Conference*, 152–161, May 1974.
14. Lang, D., Broadband Model Predicts S-Parameters of Spiral Inductors, *Microwaves & RF*, 107–110, Jan. 1988.
15. Piconics, *Inductive Components for Microelectronic Circuits*, Piconics, Tyngsboro, 1998.

16. Riddle, A., Ferrite and Wire Baluns with under 1 dB Loss to 2.5 GHz, *IEEE MTT-S Digest*, 617–620, 1998.
17. Trans-Tech, *Microwave Magnetic and Dielectric Materials*, Trans-Tech, Adamstown, MD, 1998.
18. Barabas, U., On an Ultrabroadband Hybrid Tee, *IEEE Trans MTT*, 58-64, Jan. 1979.
19. Hilbers, A.H., High-Frequency Wideband Power Transformers, *Electronic Applications*, Philips, 30, 2, 65–73, 1970.
20. Boser, O., and Newsome, V., High Frequency Behavior of Ceramic Multilayer Capacitors, *IEEE Trans CHMT*, 437–439, Sept. 1987.
21. Ingalls, M., and Kent, G., Monolithic Capacitors as Transmission Lines, *IEEE Trans MTT*, 964–970, Nov. 1989.
22. Grossner, N.R., *Transformers for Electronic Circuits*, McGraw Hill, New York, 1983.
23. AVX, RF Microwave/Thin-Film Products, AVX, Kyocera, 2000.

6.2 Passive Microwave Devices

Michael B. Steer

Wavelengths in air at microwave and millimeter-wave frequencies range from 1 m at 300 MHz to 1 mm at 300 GHz. These dimensions are comparable to those of fabricated electrical components. For this reason circuit components commonly used at lower frequencies, such as resistors, capacitors, and inductors, are not readily available. The relationship between the wavelength and physical dimensions enables new classes of distributed components to be constructed that have no analogy at lower frequencies. Components are realized by disturbing the field structure on a transmission line, which results in energy storage and thus reactive effects. Electric (E) field disturbances have a capacitive effect and the magnetic (H) field disturbances appear inductive. Microwave components are fabricated in waveguide, coaxial lines, and strip lines. The majority of circuits are constructed using strip lines as the cost is relatively low since they are produced using photolithography techniques. Fabrication of waveguide components requires precision machining, but they can tolerate higher power levels and are more easily realized at millimeter-wave frequencies (30 to 300 GHz) than can either coaxial or microstrip components.

Characterization of Passive Elements

Passive microwave elements are defined in terms of their reflection and transmission properties for an incident wave of electric field or voltage. In Fig. 6.11(a) a traveling voltage wave with phasor $\mathbf{V}_1^+$ is incident at port 1 of a two-port passive element. A voltage $\mathbf{V}_1^-$ is reflected and $\mathbf{V}_2^-$ is transmitted. In the absence of an incident voltage wave at port 2 (the voltage wave $\mathbf{V}_2^-$ is totally absorbed by Z_0), at port 1 the element has a voltage reflection coefficient

$$\Gamma_1 = \mathbf{V}_1^- / \mathbf{V}_1^+$$

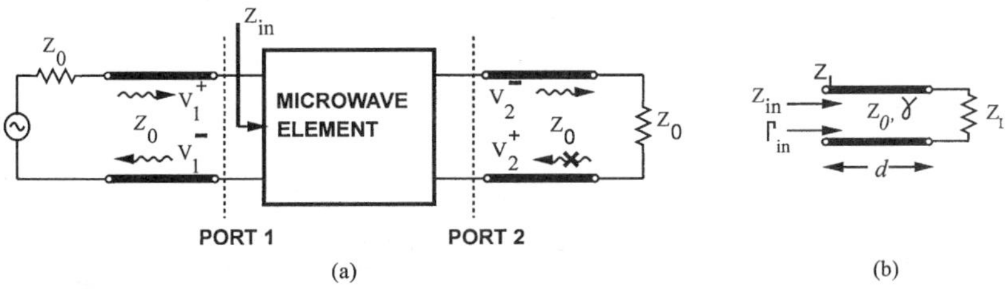

FIGURE 6.11 Incident, reflected and transmitted traveling voltage waves at (a) a passive microwave element, and (b) a transmission line.

and transmission coefficient

$$T = \mathbf{V}_2^- / \mathbf{V}_1^+ .$$

More convenient measures of reflection and transmission performance are the return loss and insertion loss as they are relative measures of power in transmitted and reflected signals. In decibels

$$\text{RETURN LOSS} = -20 \log \Gamma \ (\text{dB}) \quad \text{and} \quad \text{RETURN LOSS} = -20 \log T \ (\text{dB})$$

The input impedance at port 1, Z_{in}, is related to Γ by

$$Z_{in} = Z_0 \frac{1 + \Gamma_1}{1 - \Gamma_1} \quad \text{or by} \dots \ \Gamma = \frac{Z_L - Z_0}{Z_L + Z_0} \tag{6.14}$$

The reflection characteristics are also described by the voltage standing wave ratio (VSWR), a quantity that is more easily measured. The VSWR is the ratio of the maximum voltage amplitude $e|V_1^+| + |V_1^-|$ on the input transmission line to the minimum voltage amplitude $e|V_1^+| - |V_1^-|$. Thus

$$\text{VSWR} = \frac{1 + |\Gamma|}{1 - |\Gamma|}$$

These quantities will change if the loading conditions are changed. For this reason scattering (S) parameters are used that are defined as the reflection and transmission coefficients with a specific load referred to as the reference impedance. Thus

$$S_{11} = \gamma_1 \quad \text{and} \quad S_{21} = T$$

S_{22} and S_{12} are similarly defined when a voltage wave is incident at port 2. For a multiport $S_{pq} = \mathbf{V}_q^- / \mathbf{V}_p^+$ with all of the ports terminated in the reference impedance. Simple formulas relate the S parameters to other network parameters [1] (pp. 16–17), [2]. S parameters are the most convenient network parameters to use with distributed circuits as a change in line length results in a phase change. As well they are the only network parameters that can be measured directly at microwave and millimeter-wave frequencies. Most passive devices, with the notable exception of ferrite devices, are reciprocal and so $S_{pq} = S_{qp}$. A lossless passive device also satisfies the unitary condition: $\sum_p |S_{pq}|^2$ which is a statement of power conservation indicating that all power is either reflected or transmitted. A passive element is fully defined by its S parameters together with its reference impedance, here Z_0. In general the reference impedance at each port can be different.

Circuits are designed to minimize the reflected energy and maximize transmission at least over the frequency range of operation. Thus the return loss is high and the VSWR ≈ 1 for well-designed circuits. However, individual elements may have high reflections.

A terminated transmission line such as that in Fig. 6.11b has an input impedance

$$Z_{in} = Z_0 \frac{Z_L + jZ_0 \tanh \gamma d}{Z_0 + jZ_L \tanh \gamma d}$$

Thus a short section ($\gamma d \ll 1$) of short circuited ($Z_L = 0$) transmission line looks like an inductor, and looks like a capacitor if it is open circuited ($Z_L = \infty$). When the line is a quarter wavelength long, an open circuit is presented at the input to the line if the other end is short circuited.

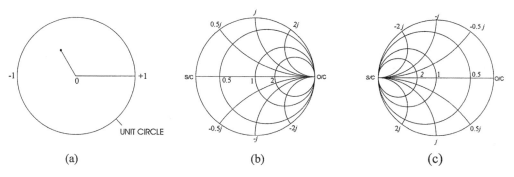

FIGURE 6.12   Smith charts: (a) polar plot of reflection coefficient; (b) normalized impedance Smith chart; and (c) normalized admittance Smith chart.

The Smith Chart

Scattering parameters can be conveniently displayed on a polar plot as shown in Fig. 6.12a. For passive devices $|S_{pq}| \leq 1$ and only for active devices can a scattering parameter be greater than 1. Thus the location of a scattering parameter, plotted as a complex number on the polar plot with respect to the unit circle indicates a basic physical property. As shown earlier in Eq. (6.14), there is a simple relationship between load impedance and reflection coefficient. With the reactance of a load held constant and the load resistance varied, the locus of the reflection coefficent is a circle as shown. Similarly, arcs of circles result when the resistance is held constant and the reactance varied. These lines result in the normalized impedance Smith chart of Fig. 6.12b.

Transmission Line Sections

The simplest microwave circuit element is a uniform section of transmission line that can be used to introduce a time delay or a frequency-dependent phase shift. More commonly it is used to interconnect other components. Other line segments used for interconnections include bends, corners, twists, and transitions between lines of different dimensions (see Fig. 6.13).

The dimensions and shapes are designed to minimize reflections and so maximize return loss and minimize insertion loss.

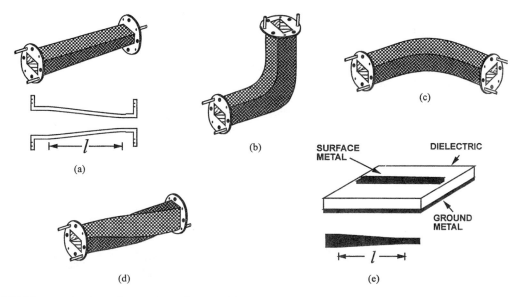

FIGURE 6.13   Sections of transmission lines used for interconnecting components: (a) waveguide tapered section, (b) waveguide E-plane bend, (c) waveguide H-plane bend, (d) waveguide twist, and (e) microstrip tapered line.

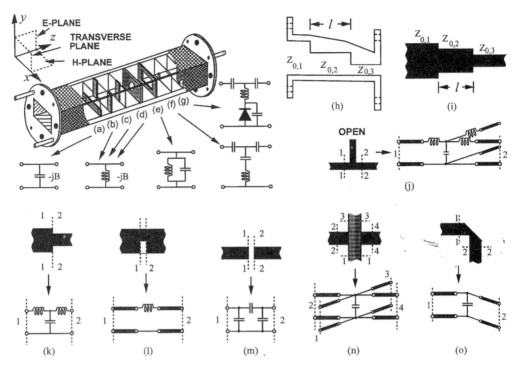

FIGURE 6.14 Discontinuities. Waveguide discontinuities:(a) capacitive E-plane discontinuity, (b) inductive H-plane discontinuity, (c) symmetrical inductive H-plane discontinuity, (d) inductive post discontinuity, (e) resonant window discontinuity, (f) capacitive post discontinuity, (g) diode post mount, and (h) quarter-wave impedance transformer; microstrip discontinuities: (i) quarter-wave impedance transformer, (j) open microstrip stub, (k) step, (l) notch, (m) gap, (n) crossover, and (o) bend.

Discontinuities

The waveguide discontinuities shown in Fig. 6.14a–f illustrate most clearly the use of E and H field disturbances to realize capacitive and inductive components. An E plane discontinuity, Fig. 6.14a, is modeled approximately by a frequency-dependent capacitor. H plane discontinuities (Figs. 6.14b and 6.14c) resemble inductors as does the circular iris of Fig. 6.14d. The resonant waveguide iris of Fig. 6.14e disturbs both the E and H fields and can be modeled by a parallel LC resonant circuit near the frequency of resonance. Posts in waveguide are used both as reactive elements (Fig. 6.14f), and to mount active devices (Fig. 6.14g). The equivalent circuits of microstrip discontinuities (Figs. 6.14j–o), are again modeled by capacitive elements if the E-field is interrupted and by inductive elements if the H field (or current) is disturbed. The stub shown in Fig. 6.14j presents a short circuit to the through transmission line when the length of the stubs is $\lambda_g/4$. When the stub is electrically short $\ll \lambda_g/4$ it introduces a shunt capacitance in the through transmission line.

Impedance Transformers

Impedance transformers interface two sections of line of different characteristic impedance. The smoothest transistion and the one with the broadest bandwidth is a tapered line as shown in Figs. 6.13a and 6.13e. This element tends to be very long as $l > \lambda_g$ and so step terminations called quarter-wave impedance transformers (see Fig. 6.14h) are sometimes used, although their bandwidth is relatively small centered on the frequency at which $l = \lambda_g$. Ideally $Z_{0,2} = \sqrt{Z_{0,1} Z_{0,3}}$.

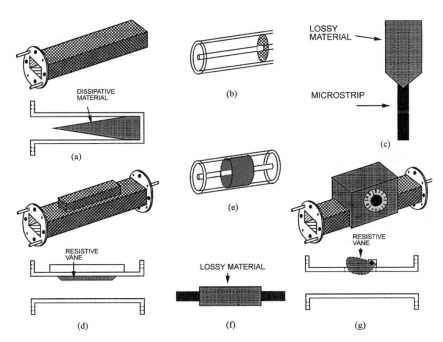

FIGURE 6.15 Terminations and attenuators: (a) waveguide matched load, (b) coaxial line resistive termination, (c) microstrip matched load, (d) waveguide fixed attenuator, (e) coaxial fixed attenuator, (f) microstrip attenuator, and (g) waveguide variable attenuator.

Terminations

In a termination, power is absorbed by a length of lossy material at the end of a shorted piece of transmission line (Figs. 6.15a and 6.15c). This type of termination is called a matched load as power is absorbed and reflections are small irrespective of the characteristic impedance of the transmission line. This is generally preferred as the characteristic impedance of transmission lines varies with frequency — particularly so for waveguides. When the characteristic impedance of a line does not vary much with frequency, as is the case with a coaxial line or microstrip, a simpler and smaller termination can be realized by placing a resistor to ground (Fig. 6.15b).

Attenuators

Attenuators reduce the signal level traveling along a transmission line. The basic design is to make the line lossy but with characteristic impedance approximating that of the connecting lines so as to reduce reflections. The line is made lossy by introducing a resistive vane in the case of a waveguide, Fig. 6.15d, replacing part of the outer conductor of a coaxial line by resistive material, Fig. 6.5e, or covering the line by resistive material in the case of a microstrip line Fig. 6.15f. If the amount of lossy material introduced into the transmission line is controlled, a variable attenuator is obtained, e.g., Fig. 6.15g.

Microwave Resonators

In a lumped element resonant circuit, stored energy is transfered between an inductor, which stores magnetic energy, and a capacitor, which stores electric energy, and back again every period. Microwave resonators function the same way, exchanging energy stored in electric and magnetic forms but with the energy stored spatially. Resonators are described in terms of their quality factor

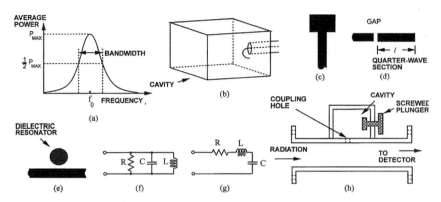

FIGURE 6.16   Microwave resonators: (a) resonator response, (b) rectangular cavity resonator, (c) microstrip patch resonator (d) microstrip gap-coupled reflection resonator, (e) transmission dielectric transmission resonator in microstrip, (f) parallel equivalent circuits, (g) series equivalent circuits, and (h) waveguide wavemeter.

$$Q = 2\pi f_0 \left| \frac{\text{Maximum energy stored in the cavity at } f_0}{\text{Power lost in the cavity}} \right|$$

where f_0 is the resonant frequency. The Q is reduced and thus the resonator bandwidth is increased by the power lost to the external circuit so that the loaded Q

$$Q_L = 2\pi f_0 \left| \frac{\text{Maximum energy stored in the resonator at } f_0}{\text{Power lost in the cavity and to the external circuit}} \right| = \frac{1}{1/Q + 1/Q_{\text{ext}}}$$

where Q_{ext} is called the external Q, Q_L accounts for the power extracted from the resonant circuit and is typically large. For the simple response shown in Fig. 6.16a the half power (3 dB) bandwidth is f_0/Q_L.

Near resonance, the response of a microwave resonator is very similar to the resonance response of a parallel or series *RLC* resonant circuit, Figs. 6.16f and 6.16g. These equivalent circuits can be used over a narrow frequency range.

Several types of resonators are shown in Fig. 6.16. Figure 6.16b is a rectangular cavity resonator coupled to an external coaxial line by a small coupling loop. Figure 6.16c is a microstrip patch reflection resonator. This resonator has large coupling to the external circuit. The coupling can be reduced and photolithographically controlled by introducing a gap as shown in Fig. 6.16d for a microstrip gap-coupled transmission line reflection resonator. The Q of a resonator can be dramatically increased by using a high dielectric constant material as shown in Fig. 6.16e for a dielectric transmission resonator in microstrip.

One simple application of cavity resonator is the waveguide wavemeter, Fig. 6.16h. Here the resonant frequency of a rectangular cavity is varied by changing the the physical dimensions of the cavity with a null of the detector indicating that the frequency corresponds to the cavity resonant frequency.

Tuning Elements

In rectangular waveguide the basic adjustable tuning element is the sliding short shown in Fig. 6.17a. Varying the position of the short will change resonance frequencies of cavities. It can be combined with hybrid tees to achieve a variety of tuning functions. The post in Fig. 6.19f can be replaced by a screw to obtain a screw tuner, which is commonly used in waveguide filters.

Sliding short circuits can be used in coaxial lines and in conjunction with branching elements to obtain stub tuners. Coaxial slug tuners are also used to provide matching at the input and output of active circuits. The slug is movable and changes the characteristic impedance of the transmission line. It is more

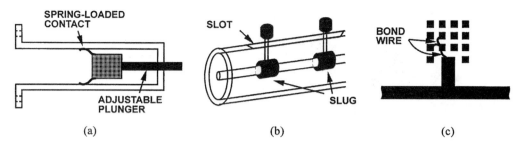

FIGURE 6.17 Tuning elements: (a) waveguide sliding short circuit, (b) coaxial line slug tuner, (c) microstrip stub with tuning pads.

difficult to achieve variable tuning in passive microstrip circuits. One solution is to provide a number of pads as shown in Fig. 6.15c which, in this case, can be bonded to the stub to obtain an adjustable stub length.

Variable amounts of phase shift can be inserted by using a variable length of line called a line stretcher, or by a line with a variable propagation constant. One type of waveguide variable phase shifter is similar to the variable attenuator of Fig. 6.15g with the resistive material replaced by a low-loss dielectric.

Hybrid Circuits and Directional Couplers

Hybrid circuits are multiport components that preferentially route a signal incident at one port to the other ports. This property is called directivity. One type of hybrid is called a directional coupler the schematic of which is shown in Fig. 6.18a. Here the signal incident at port 1 is coupled to ports 2 and 4 while very little is coupled to port 3. Similarly a signal incident at port 2 is coupled to ports 1 and 3, but very little power appears at port 4. The feature that distinguishes a directional coupler from other types of hybrids is that the power at the output ports are different. The performance of a directional coupler is specified by two parameters:

Coupling Factor $= P_1/P_4$
Directivity $\quad = P_4/P_3$

Microstrip and waveguide realizations of directional couplers are shown in Figs. 6.18b and 6.18c.

The power at the output ports of the hybrids shown in Figs. 6.19 and 6.20 are equal and so the hybrids serve to split a signal in half as well having directional sensitivity.

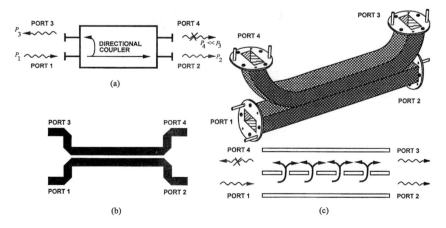

FIGURE 6.18 Directional couplers: (a) schematic, (b) microstrip directional coupler, (c) waveguide directional coupler.

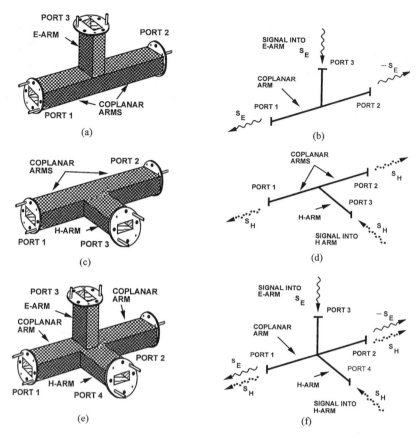

FIGURE 6.19 Waveguide hybrids:(a) E-plane tee and (b) its signal flow; (c) H-plane tee and (d) its signal flow; and (e) magic tee and (f) its signal flow. The negative sign indicates 180° phase reversal.

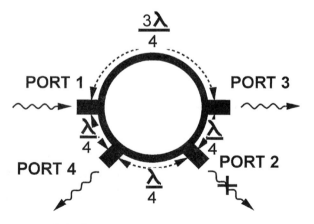

FIGURE 6.20 Microstrip rat race hybrid.

Ferrite Components

Ferrite components are nonreciprocal in that the insertion loss for a wave traveling from port A to port B is not the same as that from port B to port A.

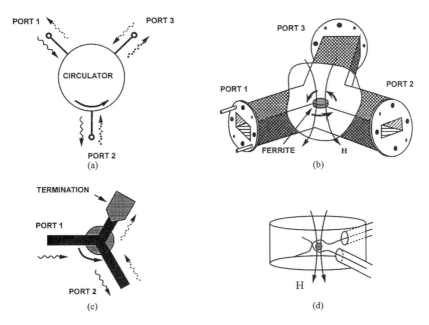

FIGURE 6.21 Ferrite Components: (a) Schematic of a circulator, (b) a waveguide circulator, (c) a microstrip isolator, and (d) a YIG tuned band-pass filter.

Circulators and Isolators

The most important ferrite component is a circulator, Figs. 6.21a and 6.21b. The essential element of a circulator is a piece of ferrite which, when magnetized, becomes nonreciprocal preferring progression of electromagnetic fields in one circular direction. An ideal circulator has the scattering matrix

$$S = \begin{vmatrix} 0 & 0 & S_{13} \\ S_{21} & 0 & 0 \\ & S_{32} & 0 \end{vmatrix}$$

In addition to the insertion and return losses, the performance of a circulator is described by its isolation, which is its insertion loss in the undesired direction. An isolator is just a three-port circulator with one of the ports terminated in a matched load as shown in the microstip realization of Fig. 6.21c. It is used in a transmission line to pass power in one direction but not in the reverse direction. It is commonly used to protect the output of equipment from high reflected signals. A four-port version is called a duplexer and is used in radar systems and to separate the received and transmitted signals in a transciever.

YIG Tuned Resonator

A magnetized YIG (Yttrium Iron Garnet) sphere shown in Fig. 6.21d, provides coupling between two lines over a very narrow band. The center frequency of this bandpass filter can be adjusted by varying the magnetizing field.

Filters and Matching Networks

Filters are combinations of microwave passive elements designed to have a specified frequency response. Typically a topology of a filter is chosen based on established lumped element filter design theory. Then computer-aided design techniques are used to optimize the response of the circuit to the desired response. Matching networks contain reactive elements and have the essential purpose of realizing maximum power transfer. They often interface a complex impedance termination to a resistance source. On a chip, and

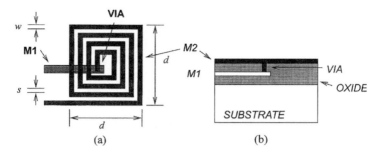

FIGURE 6.22 A spiral inductor fabricated ona monolithic integrated circuit: (a) top view; and (b) cross-section.

at RF frequencies, the matching network consists of parallel plate capacitors and spiral inductors such as that shown in Fig. 6.22.

Passive Semiconductor Devices

A semiconductor diode is modeled by a voltage dependent resistor and capacitor in shunt. Thus an applied DC voltage can be used to change the value of a passive circuit element. Diodes optimized to produce a voltage variable capacitor are called varactors. In detector circuits, a diode's voltage variable resistance is used to achieve rectification and, through design, produce a DC voltage proportional to the power of an incident microwave signal. The controllable variable resistance is used in a PIN diode to realize an electronically controllable switch.

Defining Terms

Characteristic impedance: Ratio of the voltage and current on a transmission line when there are no reflections.
Insertion loss: Power lost when a signal passes through a device.
Reference impedance: Impedance to which scattering parameters are referenced.
Return loss: Power lost upon reflection from a device.

References

1. G.D. Vendelin, A.M. Pavio, and U.L. Rohde, *Microwave Circuit Design Using Linear and Nonlinear Techniques*, Wiley: New York, 1990.
2. T.C. Edwards and M.B. Steer, *Foundations of Interconnect and Microstrip Design*, 3rd edition, Wiley: Chichester, 2000.

Further Information

The following books provide good overviews of passive microwave components: *Foundations of Interconnect and Microstrip Design* by T.C. Edwards and M.B. Steer, 3rd edition, Wiley, Chichester, 2000, *Microwave Engineering Passive Circuits* by P.A. Rizzi, Prentice Hall, Englewood Cliffs, N.J., 1988; *Microwave Devices and Circuits* by S.Y. Liao, 3rd edition, Prentice Hall, Englewood Cliffs, N.J., 1990; *Microwave Theory, Components and Devices* by J.A. Seeger, Prentice Hall, Englewood Cliffs, N.J., 1986; *Microwave Technology* by E. Pehl, Artech House, Dedham, MA, 1985; *Microwave Engineering and Systems Applications* by E.A. Wolff and R. Kaul, Wiley, New York, 1988; and Microwave Engineering by T.K. Ishii, 2 nd edition, Harcourt Brace Jovanovich, Orlando, Florida, 1989. *Microwave Circuit Design Using Linear and Nonlinear Techniques* by G.D. Vendelin, A.M. Pavio and U.L. Rohde, Wiley, New York, 1990, also provides a comphrensive treatment of computer-aided design techniques for both passive and active microwave circuits.

The monthly journals, *IEEE Transactions on Microwave Theory and Techniques*, *IEEE Microwave and Guided Wave Letters*, and *IEEE Transactions on Antennas and Propagation* publish articles on modeling and design of microwave passive circuit components. Articles in the first two journals are more circuit and component oriented while the third focuses on field theoretic analysis. These are published by The Institute of Electrical and Electronics Engineers, Inc. For subscription or ordering contact: IEEE Service Center, 445 Hoes Lane, PO Box 1331, Piscataway, New Jersey 08855-1331, U.S.A.

Articles can also be found in the biweekly magazine, *Electronics Letters*, and the bimonthly magazine, *IEE Proceedings Part H — Microwave, Optics and Antennas*. Both are published by the Institute of Electrical Engineers and subscription enquiries should be sent to IEE Publication Sales, PO Box 96, Stenage, Herts. SG1 2SD, United Kingdom. Telephone number (0438) 313311.

The *International Journal of Microwave and Millimeter-Wave Computer-Aided Engineering* is a quarterly journal devoted to the computer-aided design aspects of microwave circuits and has articles on component modeling and computer-aided design techniques. It has a large number of review-type articles. For subscription information contact John Wiley & Sons, Inc., Periodicals Division, PO Box 7247-8491, Philadelphia, Pennsylvania 19170-8491, U.S.A.

6.3 Dielectric Resonators

S. Jerry Fiedziuszko

Resonating elements are key to the function of most microwave circuits and systems. They are fundamental to the operation of filters and oscillators, and the quality of these circuits is basically limited by the resonator quality factor. Traditionally, microwave circuits have been encumbered by large, heavy, and mechanically complex waveguide structures that are expensive and difficult to adjust and maintain. Dielectric resonators, which can be made to perform the same functions as waveguide filters and resonant cavities, are, in contrast very small, stable, and lightweight. The popularization of advanced dielectric resonators roughly coincides with the miniaturization of many of the other associated elements of most microwave circuits. When taken together, these technologies permit the realization of small, reliable, lightweight, and stable microwave circuits.

Historically, guided electromagnetic wave propagation in dielectric media received widespread attention in the early days of microwaves. Surprisingly, substantial effort in this area predates 1920 and includes such famous scientists as Rayleigh, Sommerfeld, J.C. Bose, and Debye.[1] The term "dielectric resonator" first appeared in 1939 when R.D. Richtmyer of Stanford University showed that unmetalized dielectric objects (sphere and toroid) can function as microwave resonators.[2] However, his theoretical work failed to generate significant interest, and practically nothing happened in this area for more than 25 years. In 1953, a paper by Schlicke[3] reported on super high dielectric constant materials (~1,000 or more) and their applications at relatively low RF frequencies. In the early 1960s, researchers from Columbia University, Okaya and Barash, rediscovered dielectric resonators during their work on high dielectric materials (rutile), paramagnetic resonance and masers. Their papers[4,5] provided the first analysis of modes and resonator design. Nevertheless, the dielectric resonator was still far from practical applications. High dielectric constant materials such as rutile exhibited poor temperature stability causing correspondingly large resonant frequency changes. For this reason, in spite of high Q factor and small size, dielectric resonators were not considered for use in microwave devices.

In the mid-1960s, S. Cohn and his co-workers at Rantec Corporation performed the first extensive theoretical and experimental evaluation of the dielectric resonator.[6] Rutile ceramics were used for experiments that had an isotropic dielectric constant in the order of 100. Again, poor temperature stability prevented development of practical components.

A real breakthrough in ceramic technology occurred in the early 1970s when the first temperature stable, low-loss, Barium Tetratitanate ceramics were developed by Raytheon.[7] Later, a modified Barium Tetratitanate with improved performance was reported by Bell Labs.[8] These positive results led to the actual implementations of dielectric resonators as microwave components. The materials, however, were in scarce supply and not commercially available.

The next major breakthrough came from Japan when Murata Mfg. Co. produced $(Zr-Sn)TiO_4$ ceramics.[9] They offered adjustable compositions so that the temperature coefficient could be varied between +10 and −12 ppm/degree C. These devices became commercially available at reasonable prices. Afterward, the theoretical work and use of dielectric resonators expanded rapidly.

FIGURE 6.23   Dielectric resonators.

Microwave Dielectric Resonator

What is it? A dielectric resonator is a piece of high dielectric constant material usually in the shape of a disc that functions as a miniature microwave resonator (Fig. 6.23).

How does it work? The dielectric element functions as a resonator because of the internal reflections of electromagnetic waves at the high dielectric constant material/air boundary. This results in confinement of energy within, and in the vicinity of, the dielectric material which, therefore forms a resonant structure.

Why use it? Dielectric resonators can replace traditional waveguide cavity resonators in most applications, especially in MIC structures. The resonator is small, lightweight, high Q, temperature stable, low cost, and easy to use. A typical Q exceeds 10,000 at 4 GHz.

Theory of Operation

A conventional metal wall microwave cavity resonates at certain frequencies due to the internal reflections of electromagnetic waves at the air (vacuum)/metal boundary. These multiple reflections from this highly conductive boundary (electrical short) form a standing wave in a cavity with a specific electromagnetic field distribution at a unique frequency. This is called a "mode." A standard nomenclature for cavity modes is based on this specific electromagnetic field distribution of each mode. Since a metal wall cavity has a very well-defined boundary (short) and there is no field leaking through the wall, the associated electromagnetic field problem can be easily solved through exact mathematical analysis and modes for various cavity shapes (e.g., rectangular cavity or circular cavity) are precisely defined.

The TE (transverse electric) and TM (transverse magnetic) mode definitions are widely used. Mode indices e.g., TE_{113} (rectangular cavity analyzed in Cartesian coordinates) indicate how many of the electromagnetic field variations we have along each coordinate (in this case 1 along x and y, and 3 along z). The case of a dielectric resonator situation is more complicated. An electromagnetic wave propagating in a high dielectric medium and impinging on a high dielectric constant medium/air boundary will be reflected. However, contrary to a perfectly conducting boundary (e.g., highly conductive metal) this is a partial reflection and some of the wave will leak through the boundary to the other, low dielectric constant medium (e.g., air or vacuum). The higher the dielectric constant is of the dielectric medium, more of

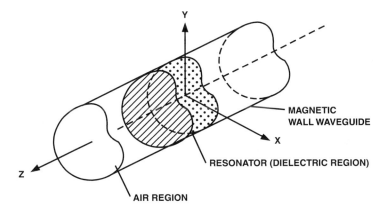

FIGURE 6.24   Magnetic wall waveguide below cut-off model of a dielectric resonator.

the electromagnetic wave is reflected and this boundary can be modeled not as a short (metal) but as an "open." As in a metal wall cavity, these internal reflections form a resonant structure called a dielectric resonator.

As in a conventional metal wall cavity, an infinite number of modes can exist in a dielectric resonator. To a first approximation, a dielectric resonator can be explained as a hypothetical magnetic wall cavity, which is the dual case of a metal (electric) wall cavity. The magnetic wall concept (on which the normal component of the electric field and tangential component of a magnetic field vanish at the boundary) is well known and widely used as a theoretical tool in electromagnetic field theory. In a very crude approximation, the air/high dielectric constant material interface can be modeled as such a magnetic wall (open circuit). Hence, the field distribution and resonant frequencies for such a resonator can be calculated analytically.

To modify this model we have to take into consideration that in actuality some of the electromagnetic field leaks out of the resonator and eventually decays exponentially in its vicinity. This leaking field portion is described by a mode subscript δ. Mode subscript δ is always smaller than one and varies with the field confinement in a resonator. If the dielectric constant of the resonator increases, more of the electromagnetic field is confined in the resonator and the mode subscript δ starts approaching one. The first modification of the magnetic wall model to improve accuracy was to remove two xy plane magnetic walls (Fig. 6.24), and to create a magnetic wall waveguide below cut off filled with the dielectric.[5,6,10]

This gave a calculated frequency accuracy for the $TE_{01\delta}$ mode of about 6%. Figure 6.24 shows the magnetic wall waveguide below cutoff with a dielectric resonator inside. Later, the circular wall was also removed (dielectric waveguide model) and the accuracy of calculations of resonant frequency was improved to 1 to 2%.[11,12] In an actual resonator configuration, usually some sort of metal wall cavity or housing is necessary to prevent radiation of the electromagnetic field and resulting degradation of resonator Q. This is illustrated in Fig. 6.25. Taking this into consideration, the model of the dielectric resonator assembly was modified, and accurate formulas for resonant frequency and electromagnetic field distribution in the structure were obtained through the mode matching method.[13]

In advanced models, additional factors such as dielectric supports, tuning plate, and microstrip substrate, can also be taken into account. The resonant frequency of the dielectric resonator in these configurations can be calculated using mode matching methods with accuracy much better than 1%.

The most commonly used mode in a dielectric resonator is the $TE_{01\delta}$ (in cylindrical resonator) or the $TE_{11\delta}$ (in rectangular resonator). The $TE_{01\delta}$ mode for certain Diameter/Length (D/L) ratios has the lowest resonant frequency, and therefore is classified as the fundamental mode. In general, mode nomenclature in a dielectric resonator is not as well defined as for a metal cavity (TE and TM modes). Many mode designations exist[14-16] and this matter is quite confusing as is true for the dielectric waveguide. In the authors opinion, the mode designation proposed by Y. Kobayashi[14] is the most promising and should be

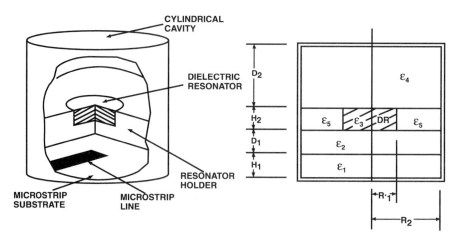

FIGURE 6.25   Practical configurations of a dielectric resonator and corresponding theoretical model.

adopted as a standard (this was addressed by MTT Standards Committee and presently two designations including Kobayashi's are recommended). Some of the modes and their field distributions are presented in Fig. 6.26.

The $TE_{01\delta}$ mode is the most popular and is used in single mode filters and oscillators. $HE_{11\delta}$ (HE indicates a hybrid mode) is used in high performance, dual mode filters,[17] directional filters and oscillators. TM mode is being used in cavity combiners and filters. Hybrid modes have all six components of the electromagnetic field. Also, low frequency filters (~800 MHz) for cellular telephone base stations were designed using the triple mode TM_{010}.[18]

Very high order modes called "whispering gallery modes" are finding applications at millimeter wave frequencies. These modes were first observed by Rayleigh in a study of acoustic waves.

Analogous propagation is possible in dielectric cylinders. Since these modes are well confined near the surface of the dielectric resonator, very high Q values at millimeter frequencies are posssible.

Coupling to Microwave Structures

An advantage of dielectric resonators is the ease with which these devices couple to common transmission lines such as waveguides and microstrip. A typical dielectric resonator in the $TE_{01\delta}$ mode can be transversely inserted into a rectangular waveguide. It couples strongly to the magnetic field, and acts as a simple bandstop filter. Figure 6.27 illustrates the magnetic field coupling into the microstrip. Coupling to a magnetic field in the waveguide can be adjusted by either rotating/tilting the resonator or moving a resonator toward the side of the waveguide. In microstrip line applications, a dielectric $TE_{11\delta}$ resonator couples magnetically and forms a bandstop filter. This is shown in Fig. 6.27. The coupling can be easily adjusted by either moving the resonator away (or toward center) from the microstrip or by lifting the resonator on a special support above the microstrip.

The resonant frequency of a dielectric resonator in this very practical case can be calculated using an equation derived in Reference 13. The resonant frequency in this topology can be adjusted to higher frequency with a metal screw or plate located above the resonator and perturbing the magnetic field, or down in frequency by lifting the resonator (moving it away from the ground plane). A typical range is in the order of 10%. Extra care must be taken, however, not to degrade the Q factor or temperature performance of the resonator by the closely positioned metal plate.

An interesting modification of the dielectric resonator is the so-called double resonator. This configuration is shown in Fig. 6.28. In this configuration, two halves of the ceramic disc or plate act as one resonator. Adjustment of the separation between the two halves of the resonator results in changes of the resonant frequency of the structure (Fig. 6.28). A much wider linear tuning range can be obtained in this configuration without degradation of the Q factor.[19-21]

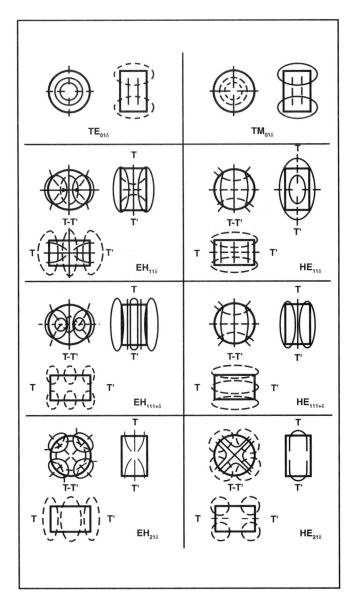

FIGURE 6.26 Modes in a dielectric resonator.

Ceramic Materials

The major problem with previously available high Q materials, such as rutile or rutile ceramics, was the poor temperature stability of the dielectric constant and the resulting instability of the resonant frequency of the dielectric resonators. Newly developed high Q Ceramics, however, have excellent temperature stability and an almost zero temperature coefficient is possible. The most popular materials are composed of a few basic, high Q compounds capable of providing negative and positive temperature coefficients. By adjusting proportions of the compounds and allowing for the linear expansion of the ceramic, perfect temperature compensation is possible. Basic properties of high quality ceramics developed for dielectric resonator applications are presented below in Table 6.4.[22-23]

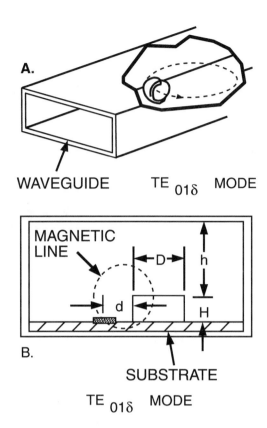

FIGURE 6.27 Magnetic field coupling of a dielectric to rectangular waveguide (top) and Microstrip line (bottom).

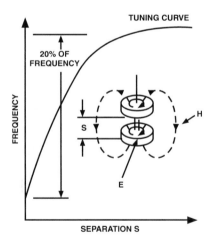

FIGURE 6.28 Double dielectric resonator configuration and tuning curve showing 20% tuning range.

Applications

The miniaturization of microwave components began with the introduction of microwave integrated circuits (MIC) and general advances in semiconductor technology, especially gallium arsenide. MIC components have become very common and presently more advanced, monolithic circuits (MMICs) are

TABLE 6.4 Basic Properties of High Quality Ceramics

Material Composition	Manufacturer	ε	Loss Tangent @ 4 GHz	Temp. Coeff. ppm/degree C
Ba Ti$_4$O$_9$	Raytheon, Transtech	38	0.0001	+4
Ba$_2$Ti$_9$O$_{20}$	Bell Labs	40	0.0001	+2
(Zr-Sn) TiO$_4$	Murata Tekelec Siemens Transtech	38	0.0001	−4 to +10 adj.
Ba(Zn$_{1/3}$ Nb$_{2/3}$)O$_2$– Ba(Zn$_{1/3}$ Ta$_{2/3}$)O$_2$	Murata	30	0.00004	0 to +10 adj.

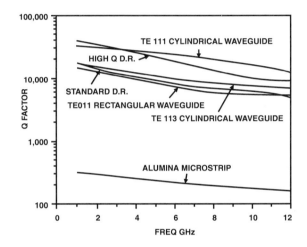

FIGURE 6.29 Achievable quality factor (Q) of various microwave resonator types (copper waveguide — 60 to 70% of theoretical Q is assumed).

being used in many applications. MIC/MMIC structures have suffered, however, from a lack of high Q miniature elements that are required to construct high performance, narrowband filters, and highly stable, fundamental frequency oscillators. Expensive and bulky coaxial and waveguide resonators made out of temperature-stable materials such as INVAR or graphite composites were the only solution in the past (Fig. 6.29). With the dielectric resonator described above, a very economical alternative, which also satisfies very stringent performance requirements, was introduced. Dielectric resonators find use as probing devices to measure dielectric properties of materials as well as the surface resistance of metals, and more recently high temperature superconductors (HTS). Additional applications include miniature antennas, where strongly radiating lower order resonator modes are successfully used.[24]

Filters

Simultaneously with advances in dielectric resonator technology, significant advances were made in microwave filter technology. More sophisticated, high performance designs (such as elliptic function filters) are now fairly common. Application of dielectric resonators in high quality filters is most evident in bandpass and bandstop filters. There are some applications in directional filters and group delay equalizers, but bandpass and bandstop applications using dielectric resonators dominate the filter field.

Bandpass and bandstop filter fields can be subdivided according to the dielectric resonator mode being used. The most commonly used mode is the TE$_{01\delta}$. The HE$_{11\delta}$ (degenerate) hybrid mode finds applications in sophisticated elliptic function filters and high frequency oscillators. This particular mode offers the

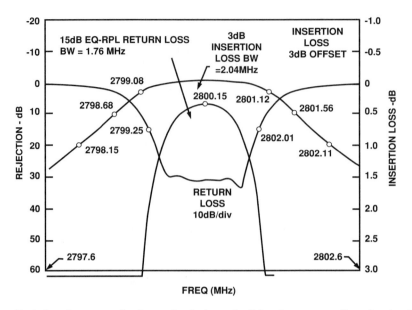

FIGURE 6.30 Typical performance of a four-pole, single made dielectric resonator filter showing insertion loss corresponding to Q factor of 9000.

advantage of smaller volume and weight (approximately $1/2$) when compared to a single mode device. This is possible since each resonator resonates in two orthogonal, independent modes.

Single Mode Bandpass Filters

The basic bandpass filter topology element generally can be described as a section of an evanescent-mode waveguide (waveguide below cutoff) in which the dielectric resonators are housed. This particular configuration was originally proposed by Okaya and Barash[5] and later expanded by Cohn[6] and Harrison.[25] The orientation of dielectric resonators can be either transverse or coaxial.

The transverse configuration yields a larger filter, but is presently preferred in practical designs because these can be more conveniently tuned with screws concentric with the resonators. Typical configurations of such filters are presented in Fig. 6.31. Actual performance of one of the filters is shown in Fig. 6.30. This particular design suffers from spurious responses on the high frequency side of the filter and suppressing techniques for these frequencies are necessary. The situation is worse when a microstrip transmission line is used to couple between the resonators.

The equivalent Q-factor of the filter is degraded and mounting of the dielectric resonator on special supports is usually necessary. Also, extra care must be taken to select proper dimensions of the dielectric resonator (e.g., D/L ratio) to place spurious modes as far as possible from the operating frequency. Sufficient spacing from metal walls of the housing is also important since close proximity of conductive walls degrades the high intrinsic Q of the dielectric resonator.

Dual Mode Filters

After reviewing literature in the dielectric resonator area, it is obvious that most attention has been directed toward analysis and applications of the fundamental $TE_{01\delta}$ mode. Higher order modes and the $HE_{11\delta}$ mode, which for certain ratios of diameter/length has a lower resonant frequency than that of the $TE_{01\delta}$ mode, are considered spurious and hard to eliminate. Even for a radially symmetrical mode like $TE_{01\delta}$, which has only 3 components of the electromagnetic field, rigorous analysis is still a problem, and various simplifying assumptions are required. The situation is much more complex for higher modes that are generally hybrid, are usually degenerate, and have all six components of the electromagnetic field.

FIGURE 6.31 Single mode dielectric resonator filters (2.8 GHz and 5.6 GHz).

A typical filter configuration using the $HE_{11\delta}$ mode (in line) is presented in Fig. 6.32.[26] Coupling between modes within a single cavity is achieved via a mode-coupling screw with an angular location of 45 degrees with respect to orthogonal tuning screws.

Intercavity coupling is provided by polarization-sensitive coupling slots. This arrangement is similar to that presently used in metal-cavity filters. The design is identical and the standard filter synthesis method can be used. Dielectric resonators are mounted axially in the center of each evanescent, circular cavity. A low-loss stable mounting is required to ensure good electrical and temperature performance.

Size comparison between traditional cavity filters and a dielectric resonator filter is shown in Fig. 6.33. Weight reduction by a factor of five and volume reduction by a factor of > 20 can be achieved. Spurious response performance of the 8-pole filters is similar to the TE_{111} mode cavity filter.

It was found that selection of diameter/length ratios greater than two yields optimum spacing of spurious responses. The $TE_{01\delta}$ mode is not excited because of the axial orientation of the resonator in the center of a circular waveguide.

One of the factors in evaluating a filter design, that is equal in importance to its bandpass characteristics, is its temperature stability. Since most of the electromagnetic field of a dielectric resonator is contained in the material forming the resonator, temperature properties of the filter are basically determined by properties of the ceramics. Typical temperature performance of the filters is in the order of $\pm 1 ppm/^\circ C$ with almost perfect temperature compensation possible.[28]

Dielectric Resonator Probe

The dielectric resonator probe configuration illustrated in Fig. 6.34 employs a dielectric resonator sandwiched between two conductive metal plates in a "post resonator" configuration. The "post resonator" is a special configuration of the dielectric resonator. In a "post resonator," the xy surfaces of the resonator are conductive (e.g., metalized).

The measured Q factor from this configuration is dominated by the losses from the conductive plates directly above and below the dielectric resonator and the dielectric loss (loss tangent) of the dielectric resonator materials. Either one of these two loss contributors can be calibrated out and the other one can be determined with a great accuracy. If we calibrate out the conductive loss (in metal plates) the dielectric loss can be determined.[27,28] In the other case, the loss tangent of the dielectric resonator is known and the conductivity of the nearby metal or superconductor is unknown.[29] As in the case of the other measurement, surface resistance is calculated from a measured Q value.

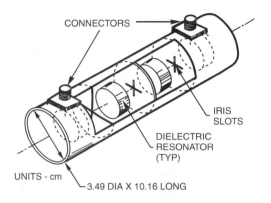

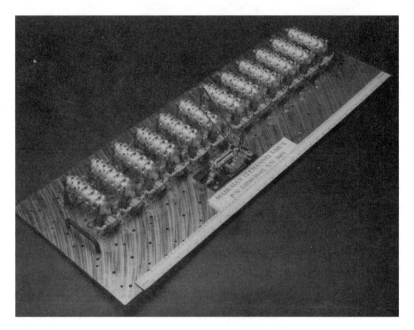

FIGURE 6.32 Dual mode bandpass filter configuration and multiplexer assembly using dual mode dielectric resonator filters.

The TE_{011} mode (post resonator) is used for these measurements since it is easily identified, relatively insensitive to small gaps between the dielectric and the test sample, and has no axial currents across any possible discontinuities in the probe fixture.

Diode Oscillators

Dielectric resonators have been used with Gunn diode oscillators to reduce frequency variations resulting from temperature or bias changes and to lower the oscillator phase noise. A typical configuration of such an oscillator is shown in Fig. 6.35. Stability in order of 0.5 ppm/degree C is achievable at 10 GHz.[30]

Gunn diode oscillators exhibit lower phase noise at higher frequencies when compared to FET oscillators. Their performance is inferior, however, when factors such as efficiency and required bias levels are taken into consideration. An interesting application of dielectric resonators, which were used for stabilization of a 4 GHz high power IMPATT oscillator, is shown in Fig. 6.36.[31] The novel oscillator configuration uses two dielectric resonators. This technique allows independent control of fundamental and harmonic frequencies of the oscillator.

FIGURE 6.33 Size comparisons between traditional cavity filters and dielectric resonator filter (single mode rectangular waveguide filter, dual mode circular cavity filter, and dual mode dielectric resonator filter are shown).

At very high millimeter wave frequencies, dielectric resonators are too small to be effectively controlled. Therefore, much larger resonators utilizing whispering gallery dielectric resonator modes are preferred.[32-38] An additional advantage of these resonators (higher order modes) is better confinement of the electromagnetic field inside the dielectric resonator and consequently, the higher Q factor.

Field Effect Transistor and Bipolar Transistor Oscillators

FET (or bipolar) oscillators using dielectric resonators are classified as reflection or feedback oscillators (Fig. 6.37).

For a reflection oscillator, initial design starts with either an unstable device or external feedback (low Q) to obtain negative resistance and reflection gain at the desired frequency. Next, a properly designed dielectric resonator is placed approximately one-half wavelength away from the FET or bipolar device. In this configuration, the dielectric resonator acts as a weakly coupled bandstop filter with a high external Q. Part of the output energy is reflected toward the device and such a self-injected oscillator will generate a signal at the resonant frequency of the dielectric resonator.

Typical reflection oscillators exhibit very good phase noise characteristics and frequency stability of approximately 1.5 ppm/degree C. Because of the reflective mode of operation, however, these designs are sensitive to load changes and require an output isolator or buffer amplifier.

Feedback oscillators can be divided into two classes: shunt feedback and series feedback (see Fig. 6.37). In these examples, a dielectric resonator actually forms the feedback circuit of the amplifying element,

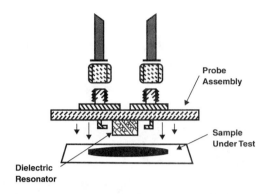

FIGURE 6.34   Illustration and photograph of the dielectric resonator probe.

usually a FET or bipolar transistor. In the shunt feedback arrangement, the resonator is placed between the output and input of the device (e.g., between gate and source, or gate and drain circuits). The conditions for oscillations are met at the resonant frequency of the dielectric resonator. In the shunt feedback scheme, however, the resonator is strongly coupled to the drain and gate transmission lines. Therefore, the loaded Q of the circuit is quite low and phase noise performance is degraded.

Another circuit that yields high stability and low phase noise is the series feedback oscillator.[39] This circuit consists of a high-gain, low-noise FET or bipolar transistor, a 50 ohm transmission line connected to the FET gate (bipolar-base) terminated with a 50 ohm resistor for out-of-band stability, a dielectric resonator coupled to the line and located at the specific distance from the gate (base), a shunt reactance connected to the FET source or drain (collector), and matching output impedance. Critical to the performance of this circuit is the placement of the dielectric resonator on the gate port, where it is isolated from the output circuits by the very low drain to gate capacitance inherent in the device. This isolation minimizes interaction between the output and input ports, which allows the resonator to be lightly coupled to the gate, resulting in a very high loaded Q, and therefore, minimum phase noise. A photograph of the typical dielectric resonator oscillator is shown in Fig. 6.37. The phase noise performance, which demonstrates suitability of such oscillators for stringent communication systems is presented in Fig. 6.38.

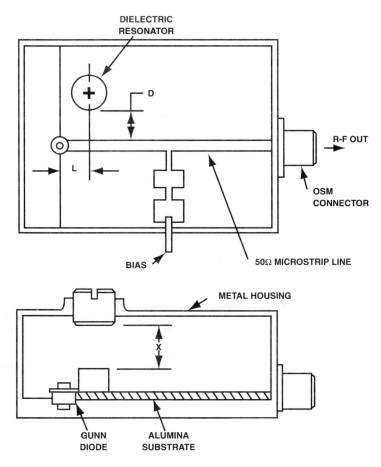

FIGURE 6.35 Gunn diode oscillator stabilized by a dielectric resonator.

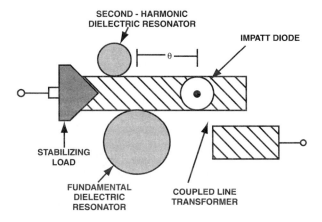

FIGURE 6.36 IMPATT diode oscillator stabilized by dielectric resonators.

To further increase stability of the oscillator and improve close-in phase noise, phase-locked systems are getting increasingly popular. In such systems, a tunable (e.g., varactor) dielectric resonator oscillator is phase locked to a low frequency crystal source or digital synthesizer (typically in ASIC form). This

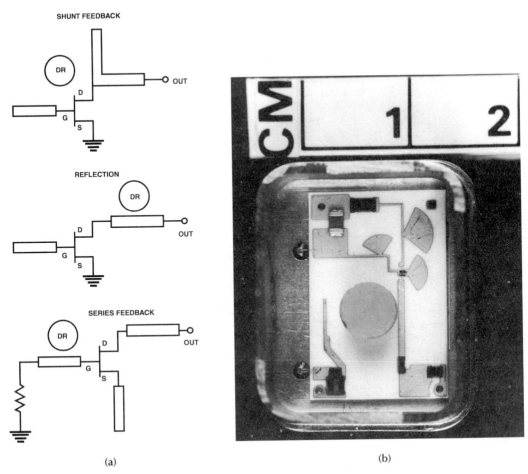

(a)   (b)

FIGURE 6.37 Basic configurations and photograph of dielectric resonator oscillator.

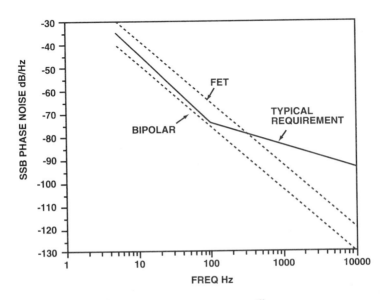

FIGURE 6.38 Phase noise characteristics of a dielectric resonator oscillator.

approach significantly enhances the overall performance of DROs allowing their use in all (even most stringent and demanding) communication systems (including satellite receivers).

Conclusions and Future Outlook

Applications of dielectric resonators in various microwave components are very cost effective and lead to significant miniaturization, particularly when MIC or MMIC structures are used. Excellent performance in filters and oscillators is currently being achieved. Dielectric resonators are widely used in wireless communication systems. Additional applications include dielectric or superconductor testing and antenna applications and radiating dielectric resonators. Miniature dielectric filled coaxial resonators are commonly used in wireless headsets (cellular and PCS phones). Recently available very high Q materials will extend commercial applications of dielectric resonators to much higher frequencies. Applications as high as 100 GHz are being reported.

Development of higher dielectric materials (80 to 100) has had a significant impact on lower frequency microwave devices (1 GHz region). Such dielectric resonators are used in practically all cellular and PCS base stations.

However, further material development is needed, mostly in dielectric materials with lower dielectric constants, which are used to mount dielectric resonators. New, low-loss plastics and adhesives should be developed to ensure that the excellent properties of dielectric resonator ceramics are not degraded.

Dielectric resonators are here to stay, and a wide variety of commercial wireless components using these elements is readily available. With the advent of new materials and improved circuit techniques, the field of dielectric resonators will continue to develop and will certainly be exciting in the future.

References

1. K.S. Packard, The Origin of Waveguides: A Case of Multiple Rediscovery, *IEEE Trans. Microwave Theory & Tech.*, MTT-32, 961–969, September 1984.
2. R.D. Richtmyer, Dielectric Resonator, *J. Appl. Phys.*, 10, 391–398, June 1939.
3. H.M. Schlicke, Quasi-degenerate Modes in High –ε Dielectric Cavities, *J. Appl. Phys.*, 24, 187–191, February 1953.
4. A. Okaya, The Rutile Microwave Resonator, *Proc. IRE*, 48, 1921, November 1960.
5. A. Okaya, L.F. Barash, The Dielectric Microwave Resonator, *Proc. IRE*, 50, 2081–2092, October 1962.
6. S.B.Cohn, Microwave Bandpass Filters Containing High Q Dielectric Resonators, *IEEE Trans. Microwave Theory & Tech.*, MTT-16, 218–227, April 1968.
7. D.J. Masse et al., A New Low Loss High-k Temperature Compensated Dielectric for Microwave Applications, *Proc. IEEE*, 59, 1628–1629, November 1971.
8. J. K. Plourde, D.F. Linn, H.M. O'Bryan Jr., and J. Thompson Jr, $Ba_2Ti_9O_{20}$ as a Microwave Dielectric Resonator, *J. Amer. Ceram. Soc.*, 58, 418–420, October-November 1975.
9. T. Nishikawa, Y. Ishikawa, and H. Tamura, Ceramic Materials for Microwave Applications, *Electronic Ceramics*, Spring Issue, Special Issue on Ceramic Materials for Microwave Applications, Japan, 1979.
10. S. J. Fiedziuszko and A. Jelenski, The Influence of Conducting Walls on Resonant Frequencies of the Dielectric Resonator, *IEEE Trans. Microwave Theory & Tech.*, MTT-19, 778, September 1971.
11. T. Itoh and R. Rudokas, New Method for Computing the Resonant Frequencies for Dielectric Resonators, *IEEE Trans. Microwave Theory & Tech.*, MTT-25, 52–54, January 1977.
12. M. W. Pospieszalski Cylindrical Dielectric Resonators and Their Applications in TEM Line Microwave Circuits, *IEEE Trans. Microwave Theory & Tech.*, MTT-27, 233–238, March 1979.
13. R.Bonetti and A. Atia, Resonant Frequency of Dielectric Resonators in Inhomogeneous Media, *1980 IEEE MTT-S Int. Symposium Digest*, 376–378, May 1980.

14. Y. Kobayashi, N. Fukuoka, and S. Yoshida, Resonant Modes in a Shielded Dielectric Rod Resonator, *Electronics & Communications in Japan*, 64-B, 11, 44–51, 1981.

15. K.A. Zaki, C.Chen, Field Distribution of Hybrid Modes in Dielectric Loaded Waveguides, *1985 IEEE MTT-S, Int. Symposium Digest*, 461–464, June 1985.

16. P. Wheless Jr. and D. Kajfez, The Use of Higher Resonant Modes in Measuring the Dielectric Constant of Dielectric Resonators, *IEEE MTT-S Int. Microwave Symposium Digest*, 473–476, June 1985.

17. S. J. Fiedziuszko, Dual Mode Dielectric Resonator Loaded Cavity Filters, *IEEE Trans. Microwave Theory & Tech.*, MTT-30, 1311–1316, September 1982.

18. T. Nishikawa, K. Wakino, H. Wada, and Y. Ishikawa, 800 MHz Band Dielectric Channel Dropping Filter Using TM_{010} Triplet Mode Resonance, *IEEE MTT-S Int. Microwave Symposium Digest*, 289–292, June 1985.

19. M. Stiglitz, Frequency Tuning of Rutile Resonators, *Proc. IEEE*, 54, 413–414, March 1966.

20. S. J. Fiedziuszko and A. Jelenski, Double Dielectric Resonator, *IEEE Trans. Microwave Theory and Tech.*, MTT-19, 79–780, September 1971.

21. S. Maj and M. Pospieszalski, A Composite Multilayered Cylindrical Dielectric Resonator, *IEEE MTT-S Int. Microwave Symposium Digest*, 190–192, June 1984.

22. Murata Mfg.Co — catalog

23. Transtech- catalog

24. D. Kajfez and P. Guillon, Editors, Dielectric Resonators, Artech House, 1986, available from Vector Fields, PO Box 757, University, MS 38677.

25. W.H. Harrison, A Miniature High Q Bandpass Filter Employing Dielectric Resonators, *IEEE Trans. Microwave Theory & Tech.*, MTT-16, 210–218, April 1968.

26. S. J. Fiedziuszko, Dielectric Resonator Design Shrinks Satellite Filters and Resonators, *Microwave Systems News*, August 1985.

27. B.W. Hakki, P.D. Coleman, Dielectric Resonator Method of Measuring Inductive Capacities in the Millimeter Range, *IRE Trans. Microwave Theory and Tech.*, MTT-8, 402–410, July 1960.

28. Y. Kobayashi and M. Katoh, Microwave Measurement of Dielectric Properties of Low-Loss Materials by the Dielectric Rod Resonator Method, *IEEE Trans. Microwave Theory and Tech.*, MTT-33, 586–592, July 1987.

29. S.J. Fiedziuszko and P.D. Heidemann, Dielectric Resonator Used as a Probe for High Tc Superconductor Measurements, *IEEE MTT-S International Microwave Symposium Digest*, Long Beach, CA, 1989.

30. T. Makino, Temperature Dependence and Stabilization Conditions of an MIC Gunn Oscillator Using Dielectric Resonator, *Trans. IECE Japan*, E62, 262–263, April 1979.

31. M. Dydyk, H. Iwer, Planar IMPATT Diode Oscillator Using Dielectric Resonator, *Microwaves & RF*, October 1984.

32. C. Chen and K.A. Zaki, Resonant Frequencies of Dielectric Resonators Containing Guided Complex Modes, *IEEE Trans. on Microwave Theory and Tech.*, MTT-36, 1455–1457, October 1988.

33. D. Cros and P. Guillon, Whispering Gallery Dielectric Resonator Modes for W-Band Devices, *IEEE Trans. on Microwave Theory and Tech.*, MTT-38, 1657–1674, November 1990.

34. D.G. Santiago, G. J. Dick, and A. Prata, Jr., Mode Control of Cryogenic Whispering Gallery Mode Sapphire Dielectric Ring Resonators, *IEEE Trans. on Microwave Theory and Tech.*, MTT-42, 52–55, January 1994.

35. E.N. Ivanov, D.G. Blair, and V.I. Kalinichev, Approximate Approach to the Design of Shielded Dielectric Disk Resonators with Whispering-Gallery Modes, *IEEE Trans. on Microwave Theory and Tech.*, MTT-41, 632–638, April 1993.

36. J. Krupka, D. Cros, M. Aubourg, and P. Guillon, Study of Whispering Gallery Modes in Anisotropic Single-Crystal Dielectric Resonators, *IEEE Trans. on Microwave Theory and Tech.*, MTT-42, 56–61, January 1994.

37. D. Cros, C. Tronche, P. Guillon, and B. Theron, W Band Whispering Gallery Dielectric Resonator Mode Oscillator, *1991 MTT-S Int. Microwave Symposium Digest*, 929–932, June 1991.

38. J. Krupka, K. Derzakowski, A. Abramowicz, M. Tobar, and R.G. Geyer, Complex Permittivity Measurements of Extremely Low Loss Dielectric Materials using Whispering Gallery Modes, *1997 MTT-S Int. Microwave Symposium Digest*, 1347–1350, June 1997.
39. K.J. Anderson, A.M. Pavio, FET Oscillators Still Require Modeling But Computer Techniques Simplify the Task, *Microwave Systems News*, September 1983.

6.4 RF MEMS

Karl R. Varian

Micro-Electro-Mechanical Systems (MEMS) are integrated circuit (IC) devices or systems that combine both electrical and mechanical components. MEMS are fabricated using typical IC batch-processing techniques with characteristic sizes ranging from nanometers to millimeters. RF MEMS are micro-electro-mechanical systems that interact with a radio frequency (RF) signal. The integration/implementation of RF MEMS provides engineers with an additional integration option for better performance, smaller size, and lower cost in their designs.

Figure 6.39 includes a selection of RF MEMS devices and application of those devices in circuits that will be discussed later in this chapter. Figure 6.39 includes static devices, such as transmission lines and resonators; active devices, such as switches and variable capacitors; and circuits, such as oscillators (fixed frequency and voltage controlled), and tunable filters. The range of frequencies covered by these circuits is from a few MHz into the millimeter wave region. Both static and active RF MEMS devices will be discussed, but the chapter emphasis will be on the active devices.

RF MEMS provide microwave and RF engineers with low insertion loss, high Q, small size, very low current consumption, and potentially low cost options to solving their design problems. The low insertion loss is obtained by replacing the moderate losses associated with semiconductors with lower metallic losses. Cost and size reduction is the result of utilizing semiconductor batch-processing techniques in RF MEMS manufacturing. Like existing semiconductor devices, the RF MEMS circuitry must be protected from the environment. But unlike semiconductors, the environmental protection is required due to either the mechanical movement and/or the mechanical fragility of the parts.

This section is an introduction to the terminology and technologies involved in the batch-processing, operation, and implementation of RF MEMS. The first section provides an RF MEMS technical overview,

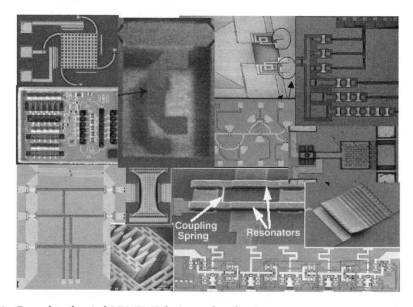

FIGURE 6.39 Examples of typical RF MEMS devices and applications.

based on a switch. The operational theory of a switch is conceptually similar to most RF MEMS devices. This section also includes a switch equivalent circuit model, actuation methods, and various performance issues (e.g., power handling, switching time, reliability, and packaging).

After the RF MEMS overview, the following sections cover fabrication, circuit elements, and typical circuit implementations. The fabrication section introduces the IC batch-processing technology, micro-machining and wafer bonding, utilized in the fabrication of RF MEMS. The section provides the background required to understand some of the manufacturing constraints and how RF MEMS devices are constructed. The next section covers RF MEMS circuit elements, describing various RF MEMS switches, variable capacitors, resonators, and transmission lines. The final section provides several typical RF MEMS implementation examples, such as: phase shifters, tunable filters, oscillators, and reconfigurable elements.

RF MEMS Technical Overview

The RF MEMS switch is a conceptually simple device, with actuation mechanisms common to RF MEMS technology. The most common actuation mechanism is electrostatic. A simple electromechanical model will be used to introduce basic switch concepts of "on" and "off" states. An electrostatic model for the switch will then be described introducing the concept of pull down voltage. An alternative switch actuation mechanism, electrothermal, will be discussed. The section will conclude with a discussion of several RF MEMS performance issues; power handling, switching time, reliability, and packaging.

One-Dimensional Electromechanical Model

Switch actuation can be conceptualized with the following one-dimensional electromechanical example (Fig. 6.40) two parallel plates (with zero mass); one supported above the other with a gap between them. The upper plate is held above the lower plate by some force, such as an ideal linear spring. This initial condition can be thought of as the switch "off" condition. When an electrostatic potential is applied between the two parallel plates, there is a resulting attractive electrostatic force. The spring force will balance out this attractive electrostatic force as the gap between the plates is reduced. The upper plate movement continues until the attractive electrostatic force overcomes the spring force, bringing the plates together. When the plates are together, the switch is in the "on" condition or activated. When the electrostatic force is removed, the spring restores the upper plate to the original noncontacting "off" position.

Electromechanical Switch Activation Characteristics[1]

The one-dimensional electromechanical model just described approximates the switch electromechanical motion. The model approximates the switch as a single rigid, but moveable, parallel-plate (switch body) suspended above a fixed ground plate by an ideal linear spring (a classical representation of a capacitor is as two parallel plates separated by a dielectric gap). This model has a single degree of freedom, which is the gap between the movable top plate and the fixed bottom plate. An important feature of this model

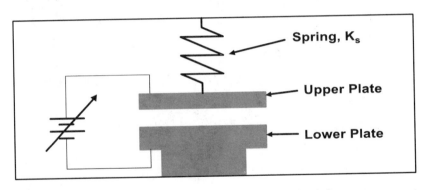

FIGURE 6.40   One-dimensional electromechanical model of an RF MEMS switch.

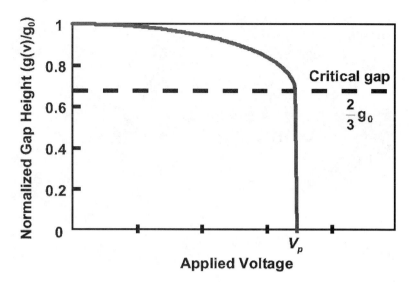

FIGURE 6.41 Gap height as a function of applied voltage. (*Source:* Goldsmith, C., Randall, J., Eshelman, S., Lin, T.H., Denniston, D., Chen, S., and Norvell, B., Characteristics of micromachined switches at microwave frequencies, in *1996 IEEE MTT-S International Microwave Digest*, 1141–1144, 1996.)

is its ability to correctly predict the pull-in of the membrane as a function of applied voltage. This switch motion can be described by the pressure balance equation:

$$P(g) = K_s(g_o - g) - (\varepsilon V^2 / 2g^2) \tag{6.15}$$

where P is the total pressure on the mechanical body of the switch, g is the height of the switch body above the bottom plate, g_o is the initial height of g with no applied field, and V is the applied electrostatic potential. The spring constant of the switch body, K_s, is determined by the Young's modulus and Poisson ratio of the membrane metal and the residual stress within the switch body. The gap permittivity, ε, is the permittivity of the dielectric material located between the upper and lower plates.

As the electrostatic field is applied to the switch, the switch membrane (switch body) starts to deflect downward, decreasing the gap g and increasing the electrostatic pressure on the membrane. At a critical height of 2/3 g_o, this mechanical system becomes unstable, causing the membrane to suddenly snap down onto the bottom plate. A graph, Fig. 6.41, of the gap height as a function of applied voltage is obtained by solving the above equation.

A key parameter associated with RF MEMS switches is pull down voltage. The pull down voltage is the voltage at which the membrane suddenly snaps down onto the bottom plate. The pull down voltage for the simple one-dimensional system can be solved as

$$V_p = \text{Sqrt}\left((8K_s g_o^3)/27\varepsilon\right) \tag{6.16}$$

From this equation, one can see that pull down voltage for the simple one-dimensional system, is related to the switch body spring constant, permittivity of the material in the gap, and to the initial membrane gap.

When the electrostatic force is removed from the switch, the tension in the metal membrane, K_s, pulls it back into the unactuated state.

Electrothermal Switch Actuation Characteristics

Besides the electrostatic mechanism just described, there also exists electrothermal actuators,[2-4] (Fig. 6.42). Electrothermal actuators use the thermal expansion mismatch of different beam materials to

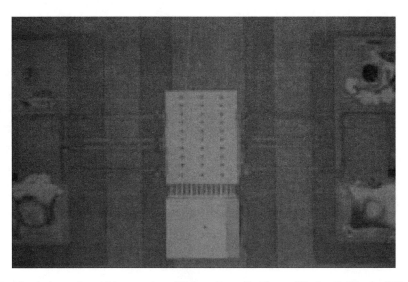

FIGURE 6.42   Electrothermal variable capacitor. (*Source:* Feng, Z., Zhang, W., Su, B., Harsh, K.F., Gupta, K.C., Bright, V., and Lee, Y.C., Design and modeling of RF MEMS tunable capacitors using electrothermal actuators, in *1999 IEEE MTT-S International Microwave Digest,* 1507–1510, 1999.)

produce a force that moves the plate up and down. Electrothermal actuator advantages are avoiding static charges from collecting on the plates, approximately linear capacitance tuning, and activation with lower driving voltage (below 5 volts). Electrothermal actuator disadvantages are a slower tuning speed, requirement for current draw, small displacements, and additional space requirements. This actuation method can be used when precise physical displacement is required. Variable capacitors utilizing the electrothermal actuation have demonstrated capacitive tuning ratios of around 5:1 with a 1.8 micron gap change.

Switching Time

RF MEMS (mechanical) switches are inherently slower than electronic switches, with switching speeds in the microsecond to millisecond range, depending on the material and switch construction. Membrane switches, with relatively smaller mass and spring constant, have the fastest demonstrated switching times, with rise times of 6 microseconds and fall times of 8 microseconds.[1] The best switching times reported for cantilever beams switches on the order of 20 microseconds.[5]

Power Handling

The RF power handling capability of RF MEMS switches is constrained by two major factors. One constraint is due to the classical switch current and/or voltage handling capability of the contact and contact area. The classical solution to this constraint has been to increase the contact area, use of special contact metals, sequential DC/RF switching, etc.

   The other major constraining factor is inadvertent switch actuation due to the RF power at the switch. The inadvertent switching occurs since RF MEMS switches are activated by an electrostatic potential. The average RF power carried on the line can generate enough electrostatic potential to cause the switch to be activated. Based on Eq. (6.16), the pull down voltage (and hence the switch power handling capability) can be increased by increasing the spring constant associated with either the beam or membrane, increasing the gap between the contact and RF line, or changing the permittivity in the gap. Since RF MEMS switches are mechanical switches, increasing the either the spring constant or the gap would probably decrease the switching speed, while the impact on size would be nominal.

   Potentially, the cantilever beam switch has a power handling advantage over the capacitive switch, since the bias electrode is separate from the contact region. This allows the cantilever beam switch designer to design the switch contact area separate from the bias electrode area. Typical power handling capability of cantilever beam switches have not been reported, while the power handling capability of the membrane switch is in the 2 to 9 watt range.

Reliability

Outside of physical damage to the mechanical structure, the major reliability concern of RF MEMS devices occurs when the desired device movement is restricted. This lack of movement is usually referred to as stiction. Stiction occurs when the restoring force associated with an activated microstructure is insufficient to overcome the force that is holding the microstructure in the activated position. Possible causes of stiction are:

1. Metal-to-metal — caused by van der Waal's forces between the two clean, smooth metal surfaces.
2. Micro-welding — caused by high current density within the device during hot switching of signals.
3. Dielectric charging — created by tunneling of electric charges into the dielectric which subsequently become trapped in the dielectric and screen the applied electric field.
4. Humidity — surface tension of water can exert enough force between films to cause unwanted sticking.
5. ESD — static electricity can cause unwanted MEMS actuation.
6. Inadequate micromachining — contaminants and residues left from an incomplete undercut can cause sticking of the MEMS.

Some, but not all of these stiction sources may be catastrophic and irreversible. Sometimes "stuck" RF MEMS devices can be freed by temperature cycling the device.

Packaging

Due to the sensitivity of RF MEMS devices to environmental damage, handling, and humidity, packaging is a key technology. State-of-the-art RF MEMS utilize conventional RF packaging techniques. This packaging consists of either solder or epoxy mounting the MEMS devices into hermetic RF packages. The difficulty associated with this packaging approach is that the devices are subjected to environmental influences until the package is sealed. To minimize exposing devices to environmental damage, alternative packaging approaches are a current research topic.

Fabrication Technology[6-8]

The major fabrication technologies used in RF MEMS batch-processing are micromachining and wafer bonding. Micromachining is the process of building three-dimensional structures either on or into a supporting wafer material (substrate). Wafer bonding is the building of three–dimensional structures by the stacking and attachment of wafers. Both micromachining and wafer bonding utilize typical batch-processing techniques used in traditional semiconductor wafer processing. Lesser-used technologies in RF MEMS fabrication such as laser micromachining, three-dimensional lithography, etc. will not be discussed. RF MEMS devices have been fabricated on a number of different substrate materials: silicon, gallium arsenide, quartz, etc., with the majority of the work utilizing silicon. Several different metallizations (gold, aluminum, copper, etc.) have been used. By leveraging the silicon batch-process technology, rapid progress has been made in RF MEMS advancement. The term silicon and substrate will be used interchangeably in the rest of this section, but not restricted to silicon exclusively.

Micromachining Processes

The micromachining processes used in the fabrication of RF MEMS structures are bulk micromachining, surface micromachining, and LIGA (a refinement of surface micromachining). To build an RF MEMS structure with micromachining, the wafer could be processed using conventional processes to create transmission lines, capacitors, resistors, inductors, transistors, etc. that are required before the RF MEMS processing starts. A resist layer is then deposited, patterned, and cured in the areas not requiring RF MEMS process (to protect this part of the circuit from the RF MEMS processing steps). This layer can be removed at the completion of RF MEMS fabrication.

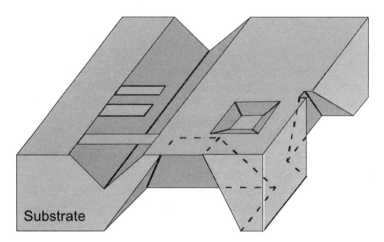

FIGURE 6.43 Example of bulk micromachining; note removal of substrate material and different etch depths. Courtesy of Raytheon.

Bulk micromachining involves the selective removal of the bulk silicon substrate material with either an anisotropic wet etch or by deep reactive ion etching (DRIE), as shown in Fig. 6.43. Wet etching normally produces features with sloped sides at an angle of 54.7° while DRIE produces nearly vertical walls. For bulk micromachining the wafer is coated with a resist layer, patterned, and cured. The areas that are not protected by a resist layer are then etched, either wet or DRIE, to the required depth. If the circuit requires features of different depths, then the process of applying a resist, patterning, and etching are repeated until all features are defined. The wafer surface is metalized, as required, after wafer etching using a resist, patterning, and depositioning process. Upon RF MEMS process completion on this surface, the remaining resist is removed. If "back-side" processing is required, the wafer can be turned over and the processing steps repeated. This technique has been used to generate low-loss transmission lines and resonant cavities (with wafer bonding techniques to be described later).

Surface micromachining involves the selective adding and removing of metal, dielectric, and sacrificial layers on the substrate surface. A resist layer is deposited, patterned, and cured on the wafer. Depending on the step, either a metal, dielectric, or sacrificial layer is then deposited, patterned, and etched. This sequence of steps is repeated until the required RF MEMS three-dimensional structure is completed. The structures that are to be suspended are then "released" by the removal of the sacrificial material under the "to be suspended structure." Surface micromachining has been used to generate three-dimensional suspended structures such as cantilever beams and membranes (see Fig. 6.44).

A refined form of surface micromachining is the process referred to as LIGA.[9,10] LIGA is an acronym that comes from the German name for the process, LIthographie Galvanoformung Abformung (lithography,

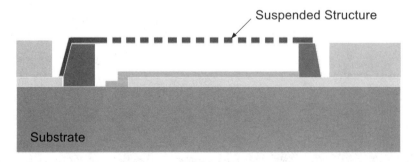

FIGURE 6.44 Example of surface micromachining. The structure is built up on the substrate surface. Courtesy of Raytheon.

FIGURE 6.45 These three gears are examples of structures that were built with the LIGA process. See http://mems.engr.wisc.edu/images/gears/

TABLE 6.5 RF MEMS Process Technology Comparison

Capability	Bulk (Wet and DRIE)	Surface	LIGA
Maximum structural thickness	Wafer(s) thickness	<50 µm	1000 µm
Planar geometry	Wet — rectangular DRIE — unrestricted	Unrestricted	Unrestricted
Minimum planar feature size	$\sqrt{2} \times$ depth	1 µm	3 µm
Side wall features	Wet — 54.74° slope DRIE — Limited by the dry etch	Limited by dry etch	0.2 µm runout over 400 µm
Surface and edge definitions	Excellent	Mostly adequate	Very good
Material properties	Very well controlled	Mostly adequate	Very good
Integration with electronics	Demonstrated	Demonstrated	Difficult
Capital investment and costs	Low	Moderate	High
Published knowledge	Very high	High	Moderate

electroplating, moulding). One of the first steps in the LIGA process is locating sacrificial metallic pads where the LIGA microstructures will occur. A resist layer is then deposited to the required depth, patterned with x-rays, and cured. Metal is then electroplated into the patterned areas of the resist layer making contact with the sacrificial metal layer. The microstructure metal is released by removing the resist layer and original metallic pad. This process has demonstrated finely defined microstructures of up to 1000 microns high and the major structures form with the LIGA process are gears (see Fig. 6.45).[11]

Comparisons of the three RF MEMS micromachining processes are shown in Table 6.5. The majority of the RF MEMS devices are fabricated with surface and bulk micromachining processes. The LIGA process is used presently primarily for micro-machine structures.

Wafer Bonding

Wafer bonding is used to generate three-dimensional RF MEMS structures that are buried in a substrate. Wafer bonding is the permanent bonding of two or more wafers. Wafer bonding techniques that will be discussed are fusion, anodic, and eutectic. Typical applications are resonant cavities and packaging.

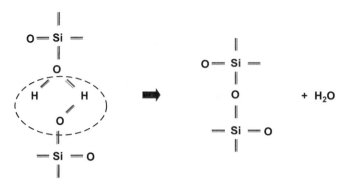

FIGURE 6.46   Chemical reaction during silicon-to-silicon fusion bonding.

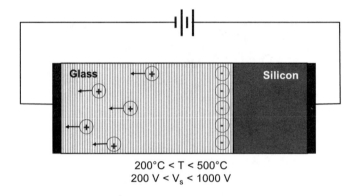

200°C < T < 500°C
200 V < V_s < 1000 V

FIGURE 6.47   Schematic representation of anodic bonding.

Silicon fusion bonding[12] occurs when pressure is applied to smooth, flat, clean, and hydrated surfaces. The resulting bond is a silicon-to-silicon bond with a water by-product (Fig. 6.46). The completed assemblies can then be annealed at temperatures in the range of 800 to 1200 degrees centigrade to increase the bond strength by an order of magnitude. The resulting bond strength reaches the bond strength of crystalline silicon: 10 to 20 MPa.

Anodic bonding occurs when an electrostatic potential of between 200 to 1000 volts is applied across a smooth glass-to-silicon interface at elevated temperatures of 200 to 500 degrees centigrade (Fig. 6.47). The resulting chemical reaction results in a trapped electrical charge in the glass at the glass-silicon interface. This trapped charge electrically pulls the silicon wafer into intimate contact with the glass substrate. The resulting bond strength is 2 to 3 MPa.

In eutectic bonding, a eutectic film is deposited on either of the wafers that are to be bonded together. The assembly is then placed in a vacuum chamber and heated to the eutectic temperature (Au-Si is 363°C) with pressure applied (Fig. 6.48). Under the appropriate conditions a eutectic bond is form between the two substrates. The resulting bond strength (with a Au-Si bond) has measured 148 MPa.

Devices, Components, and Circuits

This section will cover various RF MEMS devices. Emphasis will be on switches, followed by examples of various types of variable capacitors, transmission lines, and resonators. The section begins with a description of the two different types of switches, cantilever beam and membrane, followed by a discussion of both ohmic and capacitive switch contacts. Equivalent circuits and a comparison between the different contact approaches will then be presented. Based on the equivalent circuits, several figures of

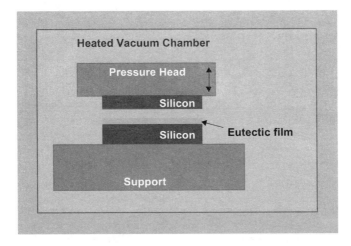

FIGURE 6.48 Schematic representation of eutectic bonding.

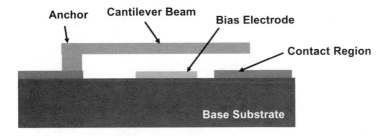

FIGURE 6.49 Cross-sectional view, cantilever beam switch.

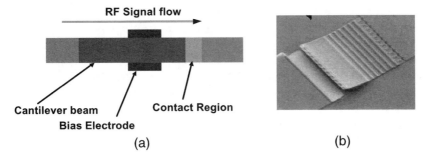

(a)   (b)

FIGURE 6.50 Cantilever beam switch, (a) overhead illustration and (b) perspective picture. (*Source:* Bozler, C. et al., MEMS Microswitch arrays for reconfigurable distributed microwave components, in *2000 IEEE MTT-S International Microwave Symposium Digest,* 153–156, 2000.)

merit will be introduced, which are used to compare various switches and technologies. The section concludes with a discussion of some typical physical switch dimensions.

Cantilever Beam and Membrane Switches

A cantilever beam switch[4,13-15] has one end of a "beam" anchored to a fixed point(s) while the other end of the beam is free to move (Fig. 6.49 through Fig. 6.51). With no electrostatic force applied, the cantilever beam is up, or in a minimum force state. The beam is actuated when a sufficient electrostatic potential is applied to a bias electrode (bias pad or control contact) located under the beam. With a sufficient electrostatic potential the cantilever beam is pulled into contact with the bias pad, and consequently the

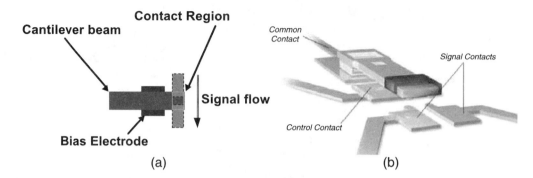

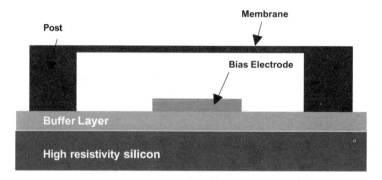

FIGURE 6.51 Cantilever beam relay, (a) overhead and (b) perspective view. (*Source:* K.R. Varian and Micromachined-relay illustration, Analog Devices, http://www.analog.com/industry/umic/relay.html.) Note: Control Contact and Bias electrode perform the same funciton.

FIGURE 6.52 Cross-sectional view, membrane switch.

end of the cantilever beam is pulled into contact in the contact region. For a solid metal cantilever beam, the beam will be in electrically connected to the contact region. A variation on this approach is for the end of the cantilever beam to be electrically isolated from the rest of the cantilever beam and the contact region to actually be two contact regions (Fig. 6.51) similar to a relay.[16] Therefore, when the switch is actuated, the two contact regions are connected together. The cantilever beam is restored to its original position when the electrostatic potential is removed. In most cases, the bias electrode and contact region are physically different as shown in Figs. 6.49 through 6.51.

A membrane switch consists of a thin membrane anchored on more than one side and allowed to flex in the middle as shown in Fig. 6.52. In the unactuated state, the membrane switch exhibits a high impedance due to the air gap between the membrane and bias electrode. Application of an electrostatic potential between the membrane and bias electrode causes the thin upper membrane to deflect downward due to the electrostatic attraction. When the electrostatic force is greater than the restoring tensile force, the top membrane deflects into the actuated position. In this state, the membrane is in contact with the bias electrode. The membrane is restored to its original position when the electrostatic potential is removed. In most cases, the bias electrode and contact area are the same.

Although not shown in Figs. 6.49 through 6.52, a feature that is usually added to the switches is a structure preventing the cantilever beam or membrane from shorting out to the bias pad. These structures, such as "bumps," can occur on either the cantilever beam, membrane, or substrate.

Besides the two different types of switches, there are two different contact mechanisms: ohmic and capacitive contacts. An ohmic contact occurs when there is metal-to-metal contact. A capacitive contact occurs when there is a dielectric layer located between the potentially contacting areas, preventing metal-to-metal contact.

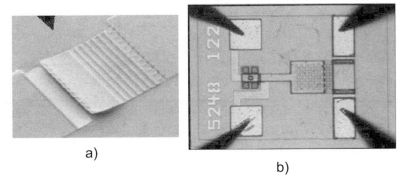

a)

b)

FIGURE 6.53 Photograph of ohmic contact (a) cantilever beam switch (*Source:* Bozler, C. et al., MEMS Microswitch arrays for reconfigurable distributed microwave components, in *2000 IEEE MTT-S International Microwave Symposium Digest,* 153–156, 2000), and (b) cantilever beam relay switch (*Source:* Hyman, D., Schmitz, A., Warneke, B., Hsu, T.Y., Lam, J., Brown, J., Schaffner, J., Walston, A., Loo, R.Y., Tangonan, G.L., Mehregany, M., and Lee, J., GaAs-compatible surface-micromachined RF MEMS switches, *Electronics Letters,* 35, 3, 224–226, February 4, 1999).

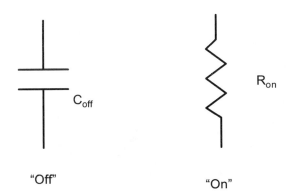

"Off" "On"

FIGURE 6.54 Simplified equivalent circuit of an ohmic contact switch.

When in the off condition, both contact types are characterized by an "off" capacitance, C_{off}. The "off" capacitance is proportional to the distance and dielectric properties of the material between the contact pad and the contact, the contact shape, and the contact area.

The contact illustrated in Figs. 6.49 through 6.52 were ohmic contacts. An example of an ohmic cantilever beam switch and an ohmic cantilever beam relay are shown in Fig. 6.53.[5,17] When actuated, an ohmic contact switch makes direct metal-to-metal contact. The "on" characteristic of an ohmic contact switch is characterized by the metal-to-metal contact resistance and is referred to as an "on" resistance, R_{on}. The combined simplified equivalent circuit for an ohmic contact switch is shown in Fig. 6.54. For metal-to-metal contacts, switching with RF present at the contacts (hot switching) may lead to degradation of the metal in the contact region. This issue is addressed either in the design of the switch, requiring the RF signal to remain below a critical level, or by removing the RF signal before switching (cold switching).

The frequency range of ohmic switch is from DC to the upper frequency limit. The upper frequency limit is set by the reactance of C_{off}.

To explain a capacitive switch, a capacitive membrane switch, Fig. 6.55, is used.[18-21] When a capacitive switch is actuated, the upper contact surface (membrane) is pulled into intimate contact with a dielectric layer that is located on the lower contact bias electrode, as shown in Fig. 6.55. The large area formed by

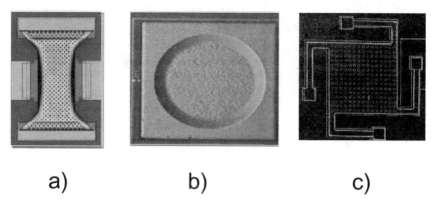

a)   b)   c)

FIGURE 6.56 Examples of membrane switches (or variable capacitors), to different scales. Figures a) and b) courtesy of Raytheon. (*Source:* Young, D.J. and Boser, B.E., A micromachined variable capacitor for Monolithic low-noise VCOs, in *Proceedings of 1996 Solid-State Sensor and Actuator Workshop*, Hilton Head Island, 86–89, 1996.)

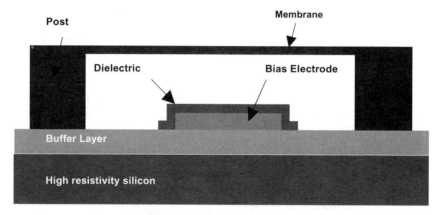

FIGURE 6.55 Cross-sectional view, membrane capacitive switch. Courtesy of Raytheon.

the metallic membrane-dielectric interface acts as a capacitor, preventing the metal-to-metal contact. Examples of capacitive membrane switches are shown in Fig. 6.56.[18,22]

In the design of a capacitive switch, the electrostatic field strength required to hold the membrane in "on" condition must not exceed the dielectric breakdown voltage.

An "on" capacitance, C_{on}, and an effective series "on" resistance, R_{on}, characterize the actuated capacitive switch. C_{on} is related to the membrane-dielectric-electrode area while R_{on} is related to the electrode and membrane ohmic losses. A simplified equivalent circuit of a capacitive switch is shown in Fig. 6.57. Capacitive switches are inherently band-limited switches. The reactance of C_{on} sets the lower frequency range of the capacitive switch while the upper frequency range is set by the reactance of C_{off}. A capacitive switch figure of merit is the ratio of C_{on}/C_{off} with acceptable ratios in the range of 40 to 100 presently being reported for RF MEMS capacitive switches. The larger the C_{on}/C_{off} ratio, the broader the frequency range over which the switch can work.

A general switch figure of merit is cutoff frequency, f_c, and is defined as

$$f_c = 1/\left(2\pi R_{on} C_{off}\right) \qquad (6.17)$$

This figure of merit is a measure of two critical switch characteristics, R_{on} and C_{off}, both of which should be as small as possible, consequently, the larger the cutoff frequency, the better the switch. As shown in

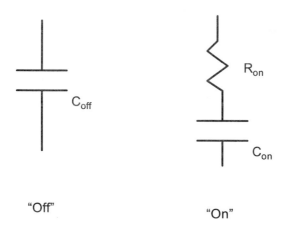

"Off"   "On"

FIGURE 6.57   Simplified equivalent circuit of a capacitive switch.

TABLE 6.6   Comparison of Switch Technologies

Device Type	R_{ON} (Ω)	C_{OFF} (fF)	F_C (GHz)
GaAs MESFET	2.3	249	280
GaAs pHEMT	4.7	80	420
GaAs P-I-N Diode	5.6	39	730
Capacitive Membrane MEMS	0.25	35	18000

Table 6.6, a table comparing the f_c of various switch technologies, the RF MEMS switch is more than a factor of 20 better than the next best switch technology, P-I-N diodes. An additional benefit over the P-I-N diode is negligible power consumption.

Typical Switch Construction Details

The ohmic cantilever beam switch has four basic components: a cantilever beam, an anchor point, a bias electrode region, and the RF contact, Figs. 6.49 through 6.51. The cantilever beam is anchored at one point, the anchor point, while the contact point is normally at the opposite end with the pull down electrode located in between. The distance between the pull down electrode and the anchor point and the cantilever beam spring constant determines the restoring force. Located between the pull down electrode and the bias electrode there is normally either a dielectric or mechanical stop. This prevents the bias electrode (the cantilever beam) from shorting out to the pull down electrode, maintaining an electrical separation between the RF and control circuitry. A DC ground return is required.

The capacitive membrane switch consists of an anchored, RF grounded membrane suspended above a dielectrically protected pull down electrode, Fig. 6.55. The membrane thickness is in the range of 0.2 to 2.0 microns. In the actuated state, the membrane is pulled down onto the dielectrically protected pull down electrode. As in the cantilever beam case, the restoring force is related to the distance between the anchor points and the pull down electrode. In RF capacitive switch design, the pull down electrode is usually an RF line and a DC ground return is required.

The membrane height for capacitive switches and the contact height for cantilever beam switches are in the range of 1.5 to 5 microns. Due to the normally larger contact area associated with capacitive switches, the membrane height is generally greater than cantilever beam contact height in order to minimize the off capacitance.

Figure 6.58 has typical response curves of insertion loss and isolation illustrating the RF electrical performance differences between an ohmic switch and a capacitive switch.[23] Recent measurements of the capacitive membrane switch indicate that its loss is less than the ohmic contact switch up to about 30 GHz. The loss of the ohmic contact switch is due to contact and "beam" losses. The major loss

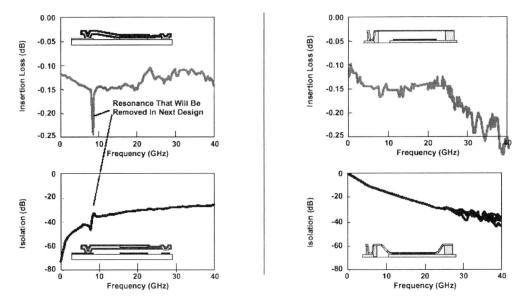

FIGURE 6.58 Insertion loss and isolation comparison between ohmic and capacitive switches verses frequency. (From B. Pierce, personal communication.)

mechanism for the capacitive membrane switch is related to the electrode ohmic losses. Above 25 GHz the loss of the capacitive switch increases due to either modeing of the CPW structure, mismatch losses, and dielectric losses. Recent measurements of a capacitive membrane switch have indicated combined ohmic and dielectric losses of less than 0.1 dB up to 30 GHz.[24]

The fundamental difference between these two switch losses is shown in the isolation measurements. For the ohmic switch, isolation is infinite at DC increasing to an isolation corresponding to the off capacitance at high frequencies. Meanwhile, the capacitive switch has a near zero isolation at low frequencies, limited by electrode ohmic losses, and the isolation increases, corresponding to the voltage ratio between the C_{off} reactance and the load.

Variable Capacitors

Many different types of RF MEMS variable capacitors have been developed. They cover both the digital and analog spectrum. A few conceptually different ones will be described, one digital and three analog variable capacitors.

The digital approach utilizes switches to select a discrete capacitor that is in series with the switch. If ideal ohmic switches were used, then activating a switch(s) would add the discrete capacitor(s) to the circuit. The capacitive tuning range over which this type of variable capacitor can work is limited by the series R_{on} and C_{off}.

If an ideal capacitive switch is used in place of the ohmic switch, then activating the switch(s) would add the series combination of the discrete capacitor(s) and C_{on} of the switch(s), as shown in Fig. 6.59.[25] The capacitive tuning range over which this type of variable capacitor can work is limited by the series combination of $C_{on} - R_{on}$ and C_{off}. This later approach to building a digital variable capacitor with capacitive switches was demonstrated with a tunable filter. See the tunable filter section, where the tunable elements of the filter are the digital variable capacitor.

Remembering the one-dimensional description of how a switch works, one recalls that as a potential is applied to the plates, the plates move closer together. This moving of the top plate continues until the height is reduced to 2/3 of the static state. When 2/3 of the static state height is reached, the switch becomes unstable and is pulled (collapses) into the fully activated state. In the region before the collapse, the movable plate (membrane) is a variable capacitor.[22] This type of variable capacitor has a limited

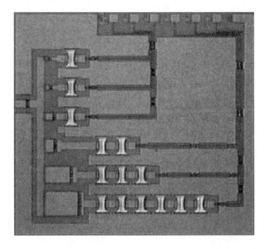

FIGURE 6.59 Six-bit variable capacitor, MEMS switches are in series with a fixed capacitor. (*Source:* Goldsmith, C.L., Malczewski, A. Yao, Z.J., Chen, S., Ehmke, J., and Hinzel, D.H., RF MEMS variable capacitors for tunable filters, *Internatinoal Journal of RF & Microwave Computer-Aided Engineering,* 9, 4, 362–374, 1999.)

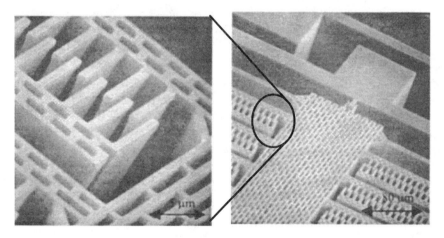

FIGURE 6.60 Variable capacitor using interdigitated metallic fingers. (*Source:* Yao, J.J., Park, S., and DeNatale, J., High tunning-ratio MEMS-based tunable capacitors for RF communications applications, in *Proceedings of 1998 Solid-State Sensor and Actuator Workshop,* Hilton Head Island, 124–127, 1998.)

tuning range and has been demonstrated with both electrostatic potential as described above and with electrothermal switches. The theoritical maximum capacitive tuning ratio is 1.5:1.

Another analog variable capacitor approach is to use interdigitated metallic fingers (see Fig. 6.60), to provide the capacitance.[26,27] One set of fingers, similar to a comb, is fixed while the other set is constrained to movement in only one direction. The movable comb structure is moved by the electrostatic force generated by fringing fields at the ends (both fixed and stationary) of the comb structure, and thus is independent of and not limited by the gap spacing between the comb structures. As the overlap of the variable capacitor interdigital fingers increases, this relatively uniform force results in a relatively linear voltage–capacitance curve. This mechanical configuration for an interdigitated variable capacitor allows a motion on the order of 10s of microns, thus a large tuning range for the variable capacitor, with relative small electrostatic fields. The interdigital variable capacitor has demonstrated a maximum capacitance of approximately 6 pF, a capacitance tuning ratio of greater than 4.55:1 over a voltage tuning range of 0 to 5.2 volts, and a series resonance of over 5 GHz.

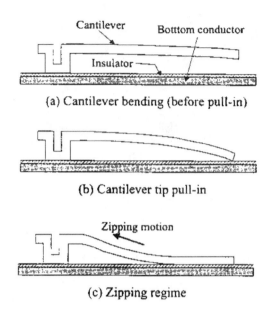

FIGURE 6.61 Capacitive cantilever beam "zipper" variable capacitor. (*Source:* Hung, E. and Senturia, S., Tunable capacitors with programmable capacitance-voltage characteristics, in *Proceedings of 1998 Solid-State Sensor and Actuator Workshop*, Hilton Head Island, 292–295, 1998.)

A capacitive cantilever beam electrostatic "zipper" switch (Fig. 6.61), can also be used as a variable capacitor.[28] A cantilever beam zipper switch is constructed similar to a normal cantilever beam switch except the bias electrode (bottom electrode) is shaped to enhance a zipper action on the beam. When an increasing voltage is applied between the beam and the bias electrode, the beam first bends downward, then collapses toward the substrate. At first, only the beam tip contacts the substrate, but as additional voltage is applied, the tip flattens and the beam zips along the substrate toward the cantilever beam anchor point. A specific capacitance-voltage characteristic can be tailored by lithographically "programming" the bias electrode geometry. This particular variable capacitor has demonstrated a capacitive tuning ratio of 1.7:1. A ratio of available gap sizes and large capacitances per unit area can be used to increase the capacitive tuning range.

Transmission Lines[29-31]

Obtaining transmission lines with low loss has always been a very challenging goal. Micromachining has enabled the fabrication of transmission lines on less than 2-µm thick dielectric layers. Examples of some transmission lines that have been fabricated are shown and described in Fig. 6.62. These transmission lines and resulting circuits have demonstrated zero dispersion, very low loss, and very small parasitics.

The circuitry is patterned on a gold film located on a stress-compensated 1.4-:m membrane layer consisting of SiO_2-Si_3N_4-SiO_2 layer on a high resistivity silicon substrate using thermal oxidation and low-pressure chemical vapor deposition. The silicon is completely etched away from under the circuitry until the circuitry is left on a thin dielectric membrane. When applicable, the mating wafers are patterned, etched, plated, and mounted to the main wafer. Besides low loss transmission lines, inductors,[32-34] resonators, and filters have been demonstrated based on this technology.

Resonators (Filters)

Resonators are a basic building block in frequency selective systems. Due to the diverse technologies involved and the low insertion loss associated with MEMS technology, several different resonator types exist. Three types of resonant structures, demonstrated over widely different frequency ranges, will be discussed: mechanical (300 KHz to 100 MHz), cavity (greater than 20 GHz), and piezoelectric film (1.5 to 7.5 GHz) resonators.

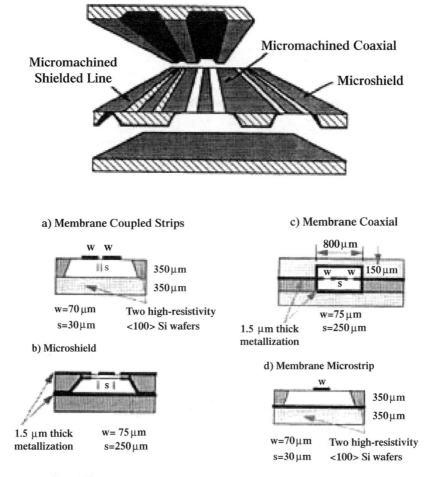

FIGURE 6.62 Different styles of demonstrated low loss transmission lines. (*Source:* Katehi, L.P.B. and Rebeiz, G.M., Novel micromachined approaches to MMICs using low-parasitic, high-performance transmission media and environments, in *1996 IEEE MTT-S International Microwave Symposium Digest*, 1145–1148, 1996.)

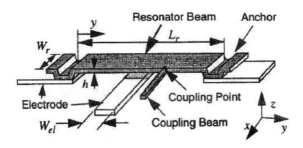

FIGURE 6.63 Mechanical MEMS resonator, a single pole. (*Source:* Nguyen, C.T.-C., Frequency-selective MEMS for miniaturized communication devices, in *1998 IEEE Aerospace Conference*, 1, 445–460, 1998.)

Mechanical resonators utilize the small feature sizes of the microstructures to create a mechanically resonant structure in the 300 KHz to 100 MHz range.[35-42] The mechanically resonant structure is a micromechanical "clamped-clamped" (clamped at both ends) beam resonator (Fig. 6.63). A bias electrode is located centrally under the beam. These structures are then electrostatically coupled to RF lines to create

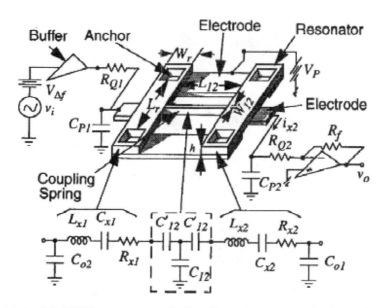

FIGURE 6.64 Mechanical MEMS resonator, 2-pole filter illustrating support circuitry and equivalent circuit. (*Source:* Bannon, F.D. III, Clark, J.R., and Nguyen, C.T.-C., High frequency microelectromechanical IF filters, in *1996 International Electron Devices Meeting*, 773–776, 1996.

electromechanical resonances. Bandpass filters can then be created by coupling two or more of these structures together.

A perspective view and a schematic of a mechanically resonant structure is shown in Fig. 6.64. In this figure key mechanical structures are indentified, along with the required bias and excitation scheme for proper operation.

A DC bias voltage, V_p, is applied to the resonator. The AC input signal, v_i, is applied to the bias electrode. When the applied AC signal frequency enters the passband of the beam resonator, the microstructure will vibrate in a direction perpendicular to the electrode, creating a DC-biased, time-varying capacitor. An AC current, i_{ac}, will thus flow

$$i_{ac} = V_P\left(\partial C/\partial t\right) \tag{6.18}$$

where C is the bias electrode to beam capacitance. The time-varying current is then AC coupled off of the beam resonator, (when more than one resonator is in the circuit, the resonators are coupled together and the signal is coupled off the second resonator with a sense electrode) and converted to an output voltage signal.

An example of a typical transmission spectrum for a second order (two resonator, Fig. 6.65) maximally flat bandpass filter centered at 14.54 MHz is shown in Fig. 6.66. The resulting filter Q is 1000 with a 13 dB insertion loss and 24 dB of stop band rejection.

Cavity resonators are created by using bulk micro-machining, plating, and epoxy bonding the wafers together to generate buried cavities. The cavity can be either a typical waveguide structure[43-45] (with and without obstructions) or a cavity containing transmission line resonating structures.[46-49] The structure shown in Fig. 6.67 is made by bonding three wafers together with epoxy. The central wafer has the desired circuitry while the other two wafers form a shield cavity around the circuitry. Micro-machined low-loss transmission lines with gold metallization are used to define the circuitry in the cavity. Silicon is completely etched away from under the circuitry, until the circuitry is left on a thin dielectric membrane. At the same time, via grooves are opened all around the circuitry to ensure complete shielding. This

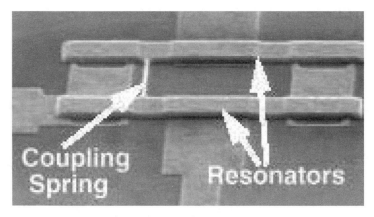

FIGURE 6.65 Two resonator, two-pole, MEMS bandpass filter. (*Source:* Nguyen, C.T.-C., Frequency-selective MEMS for miniaturized low-power communication devices, *IEEE Transactions on Microwave Theory and Techniques,* 47, 8, 1486–1503, August 1999.)

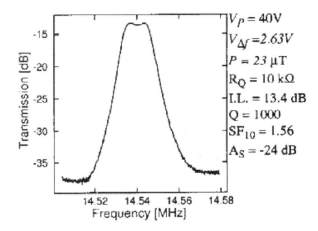

FIGURE 6.66 Transmission response of a two-resonator, two-pole, MEMS bandpass filter. (*Source:* Clark, J.R., Bannon, F.D. III, Wong, A.-C., and Nguyen, C.T.-C., Parallel-resonator HF micromechanical bandpass filters, in *Proceedings of 1997 International Conference on Solid-State Sensor and Actuators,* Chicago, 1161–1164, 1997.)

completes the processing of the central wafer. The upper and lower wafers are patterned, etched, and plated to form the required cavities. The resulting wafers are then joined together with silver epoxy as shown in Fig. 6.67. The four-pole, 60 GHz elliptic function filter demonstrated an insertion loss of 1.5 dB and rejection of better than -40 dB.

The final resonator structure to be considered is deposited piezoelectric films sandwiched between conductors in a similar fashion to quartz crystals.[50-52] Such resonators are constructed using aluminum nitride piezoelectric films with demonstrated Qs of over 1000 and resonance frequencies of 1.5 to 7.5 GHz. Although very promising, improvements in process and frequency trimming (tuning) are required.

Typical RF MEMS Applications

The utilization of RF MEMS devices and some implementation techniques will be discussed. Examples where chosen to illustrate the variety of applications in which RF MEMS can be used. Circuits included are phase shifters, tunable filters, oscillators (both fixed and voltage-controlled), and reconfigurable matching elements and antennas. Some of the examples include how the RF MEMS devices were packaged.

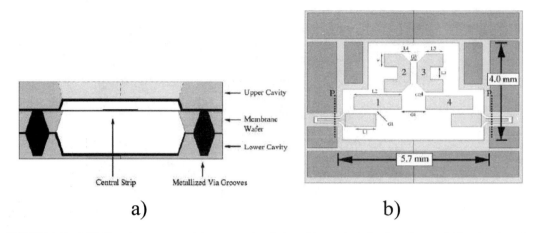

a) b)

FIGURE 6.67 MEMS cavity resonator, (a) cross-sectional view, (b) a typical circuitry that would be placed on the central strip, a 4-pole elliptic function filter. (*Source:* Blondy, P., Brown, A.R., Cros, D., and Rebeiz, G.M., Low-loss micromachined filters for millimeter-wave communication systems, *IEEE Transactions on Microwave Theory and Techniques*, 46, 12, 2283–2288, 1998.)

Phase Shifters (Time Delay)

Phase shifters and time delay networks are similar. A phase shifter adds a required phase shift to a signal relative to a reference path, while a time delay network shifts (delays) the signal in time relative to a reference path. At a fixed frequency, a time delay corresponds to a phase shift. As the frequency changes the time delay remains approximately the same whereas the phase shift changes linearly with frequency. An example of a time delay network is a transmission line of fixed length (distributed). An example of a phase shift circuit is a series capacitor or inductor (lumped elements). The following RF MEMS phase shifter circuits are a hybrid combination of lumped (RF MEMS switches and other discontinuities) and distributed (transmission lines) elements. Consequently, they are thought of more as phase shifters, although in a limited range they are also time delay networks.

Three circuits will be described: a switched path transmission line phase shifter, a reflection transmission line phase shifter, and a capacitively loaded transmission line phase shifter. Of these, the capacitive loaded transmission line phase shifter may be a true time delay phase shifter over the widest frequency range.

The phase shift in a switched path transmission line phase shifter occurs when an additional length of transmission line, corresponding to the required phase shift, is added to the signal path (see Fig. 6.68).[53,54] The switching is accomplished with RF MEMS capacitive membrane switches that are shunted to ground using shunt RF MEMS capacitively coupled membrane switches and quarter-wave length ($\lambda/4$) transmission lines. To turn off a section of line, the switch in that line is activated. Consequently, an open will appear at the tee junction for the activated line due to the two quarter-wave transformations, occurring from the tip of the resonant stub ($\lambda/4$ long) to the tee junction. In the other path (the desired path) the switches are not activated (off) and the RF signal follows the selected path. An RF MEMS Ka-band, four-bit switched path transmission line phase shifter is shown in Fig. 6.69. The circuit has an average insertion loss of 2.25 dB with better than a 15 dB return loss.

A 3-bit version of this same phase shifter with two phase shifters on a single die was fabricated. The die size was increased to permit a transparent glass lid (with a cavity over the circuitry) to be epoxy bonded onto the die (Fig. 6.70). The epoxy glass lid permitted the phase shifter to function in a non-controlled environment for a limited length of time.

A reflection transmission line phase shifter consists of a three port device, a variable length transmission line, and a short as shown in Fig. 6.71.[55] The signal enters through input port of the three port device (such as a Lange coupler or circulator). A variable length transmission line connects the three port device (for a Lange coupler, it would be both the direct and coupled arm) to a short, which reflects the signal

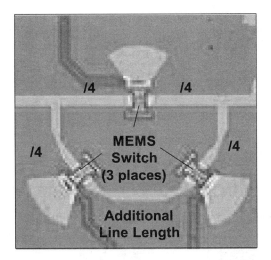

FIGURE 6.68 A single bit of a transmission line phase shifter.

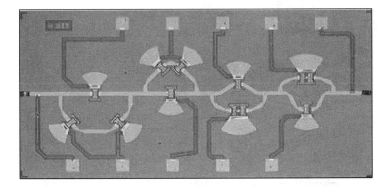

FIGURE 6.69 Ka-band 4-bit RF MEMS phase shifter. (*Source:* Pillans, B., Eshelman, S., Malczewski, A., Ehmke, J., and Goldsmith, C., Ka-band RF MEMS phase shifters, *IEEE Microwave and Guided Wave Letters,* 9, 12, 520–522, December 1999.)

FIGURE 6.70 Ka-bond 3-bit phase shifter with an epoxy bonded lid. (*Source:* Pillans, B., Eshelman, S., Malczewski, A., Ehmke, J., and Goldsmith, C., Ka-band RF MEMS phase shifters for phased array applications, in *2000 IEEE Radio Frequency Integrated Circuits Symposium Digest,* 195–198, 2000.)

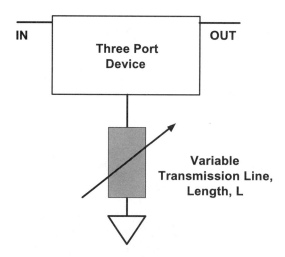

FIGURE 6.71   Basic reflection phase shifter.

FIGURE 6.72   10 GHz, 4-bit reflection phase shifter using Lange couplers as the nonreciprocal device. (*Source*: Malczewski, A., Eshelman, S., Pillans, B., Ehmke, J., and Goldsmith, C.L., X-band RF MEMS phase shifters for phased array applications, *IEEE Microwave and Guided Wave Letters*, 9, 12, 517–519, 1999.)

back to the three port device and out the exit port. The phase shift occurs when the length of the transmission line changes and the phase shift corresponds to twice the change in electrical length of the transmission line. A 10-GHz, 4-bit RF MEMS reflection phase shifter is shown in Fig. 6.72. The 4-bit phase shifter consists of two different cascaded 2-bit phase shifters. One phase shifter steps the phase in 90° steps from 0° to 270°, while the other phase shifter varies the phase between the 0°, 22.5°, 45°, and 67.5° states. A Lange coupler is used as the three port device and RF MEMS capacitive membrane switches are used to change the line length by changing the location of the short. After the Lange coupler, one of four different phase states can be selected from either phase shifter. The combined 4-bit phase shifter has an average insertion loss of 1.4 dB with less than 11 dB of return loss at 10 GHz.

A distributed phase shifter can be realized by capacitively loading a transmission line (see Fig. 6.73).[56,57] The capacitors can be either an analog or digital variable capacitor, as described earlier. Changing the shunt capacitance changes the line impedance and hence the phase shift of the line. This type of phase

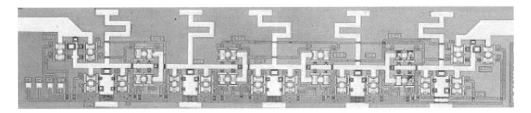

FIGURE 6.74   Five-pole, 1 GHz tunable bandpass filter. (*Source:* Goldsmith, C., RF MEMS devices and circuits for radar and receiver applications, Microwave and Photonic Applications of MEMs Workshop, 2000 IEEE MTT-S International Microwave Symposium, 2000.)

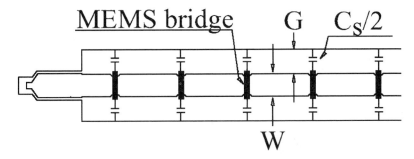

FIGURE 6.73   Capacitvely loaded transmission line phase shifter. MEMS bridge refers to a MEMS switch, G the gap between the transmission line and ground, W the width of the transmission line, and C_2 the shunt capacitor. (*Source:* Hayden, J.S. and Rebeiz, G.M., One- and two-bit low-loss cascadable MEMS distributed X-band phase shifters, in *2000 IEEE MTT-S International Microwave Symposium Digest,* 161–164, 2000.)

shifter has demonstrated, in a single bit configuration, insertion loss of 0.6 to 0.7 dB for roughly a 180° of phase shift.

Tunable Filters[18]

The RF MEMS variable capacitors (both analog and digital versions) have been used to fabricate RF tunable bandpass filters utilizing both discrete and distributed elements. The frequency range covered by the filters ranged from 100 MHz to 30 GHz, with tunability bandwidths of 2.5% to greater than 35%. An example of a 5-pole, 1 GHz discrete element bandpass filter tunable over a 20% band is shown in Fig. 6.74. The filter utilizes the digital variable capacitors and contains 44 capacitive switches.

This filter is part of a filter bank of seven filters covering the frequency range of 100 MHz to 2800 MHz. The filters were packaged into a 1.25 inch by 1.75 inch ball grid array hermetic package (Fig. 6.75). The package contained around 450 capacitive membrane switches, both discrete and integral (part of the RF MEMS integrated circuit) inductors, and a controller to convert an incoming data stream into the control signals required for band selection and filter tuning. Hermetic packaging permits assembly testing for an extended time in an uncontrolled environment.

In Figs. 6.76 and 6.77, a photograph and the corresponding responses of K and Ka band, bandpass filters are shown.[58] These filters used MEMS variable capacitors to tune either a lumped element filter center at 26 GHz or a distributed, half-wave resonator filter at 30 GHz. Respective minimum insertion losses were 4.9 dB and 3.8 dB were measured with corresponding tuning ranges of 4.2% and 2.5%.

Oscillators

A key oscillator parameter is phase noise. Phase noise is directly related to the circuit losses (inversely related to the square of the unloaded Q). RF MEMS provides a means of reducing circuit losses. The RF MEMS structures that have been used in oscillator design are low loss transmission lines, resonators, and variable capacitors.

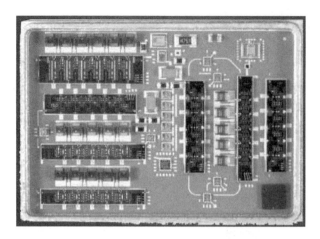

FIGURE 6.75 Seven-filter, filter bank in a hermetic package, plus control circuity. (*Source:* Goldsmith, C., RF MEMS devices and circuits for radar and receiver applications, Microwave and Photonic Applications of MEMs Workshop, 2000 IEEE MTT-S International Microwave Symposium, 2000.)

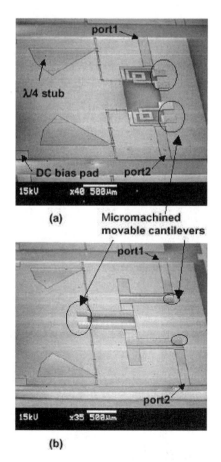

FIGURE 6.76 K and Ka band tunable MEMS filters (a) lumped element, (b) distributed elements. (*Source:* Kim, H.-T., Park, J.-H., Kim, Y.-K., and Kwon, Y., Millimeter-wave micromachined tunable filters, in *1999 IEEE MTT-S International Microwave Symposium Digest,* 3, 1235–1238, 1999.)

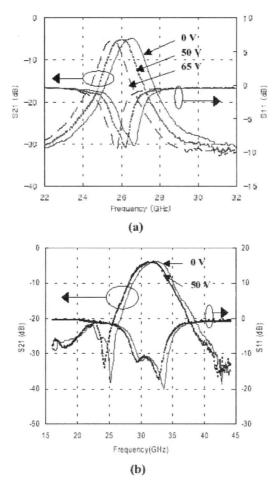

FIGURE 6.77 Response of the K and Ka band tunable MEMS filters in Fig. 6.70, (a) lumped element, (b) distributed elements. (*Source:* Kim, H.-T., Park, J.-H., Kim, Y.-K., and Kwon, Y., Millimeter-wave micromachined tunable filters, in *1999 IEEE MTT-S International Microwave Symposium Digest*, 3, 1235–1238, 1999.)

Low loss transmission lines have been used to provide stronger coupling into a dielectric resonator.[59,60] The stronger coupling permits the dielectric resonator to be decoupled more from the circuit, which in turn increases the unloaded Q of the resonator and lowers the phase noise of the oscillator. An unloaded Q of 1600 was obtained for a whispering gallery mode dielectric resonator oscillator at 35 GHz with micromachined coplanar transmission lines.

An additional use for a low loss transmission line has been to form a low loss resonating structure (this scheme has also been used for fixed tuned filters) (see Fig. 6.78).[61,62] This resonating structure is then coupled to an active device in such a way that the circuit will resonate. Unloaded Qs of 460 have been measured with oscillators at 28.65 GHz, with corresponding phase noise of −92 dBc/Hz at a 100 kHz offset frequency and −122 dBc/Hz at a 1 MHz offset frequency. The RF MEMS resonating structure resulted in a 10 dB improvement in phase noise.

RF MEMS variable capacitors have been used as the tunable element in a voltage controlled oscillator (VCO).[22,63-66] The frequency range of VCOs report to date has been less than 2.4 GHz and tuning ranges of 3.4% or less. A photograph of the RF MEMS variable capacitor used is shown in Fig. 6.79 and a schematic utilizing the RF MEMS variable capacitor is shown in Fig. 6.80. This oscillator had an output power of −14 dBm at 2.4 GHz and a measured phase noise of −93 dBc/Hz at a 100 kHz offset and −122 dBc/Hz at a 1 MHz offset.

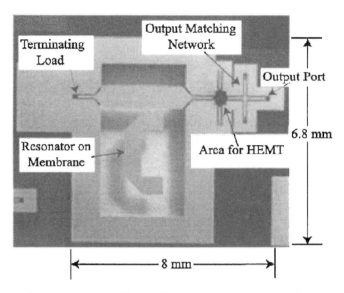

FIGURE 6.78 Photograph of a 28.65 GHz oscillator utilizing a low-loss transmission line resonator. (*Source:* Brown, A.R. and Rebeiz, G.M., A Ka-band micromachined low-phase-noise oscillator, *IEEE Transactions on Microwave Theory and Techniques*, 47, 8, 1504–1508, August 1999.)

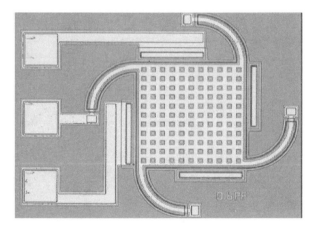

FIGURE 6.79 Photograph of MEMS variable capacitor used in VCO. (*Source:* Dec, A. and Suyama, K., A 2.4 GHz CMOS LC VCO using micromachined variable capacitors for frequency tuning, in *1999 IEEE MTT-S International Microwave Symposium Digest*, 1, 78–82, 1999.)

Reconfigurable Elements

RF MEMS devices have also been used to actively reconfigure matching elements and antennas. Three different examples will be discussed. Two are using RF MEMS switches to activate different portions of an antenna/array. The other example mechanically changes the shape of a radiating element.

The approach consisting on an array of MEMS switches is illustrated in Fig. 6.81.[17] The array of RF MEMS cantilever beam ohmic switches activates metal patches. By appropriately selecting the desired row and column actuator control lines, a specific metal patch will be selected. When this array concept was applied to a patch antenna (see Fig. 6.82), the patch demonstrated the capability of changing its radiating frequency from roughly 10 GHz to 20 GHz. This approach was also used to vary the input and output matching network of an amplifier from an operation center frequency of 10 GHz to 20 GHz.

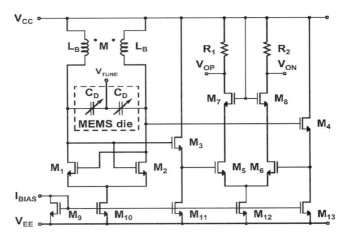

FIGURE 6.80 Schematic of a 2.4 GHz MEMS VCO. (*Source:* Dec, A. and Suyama, K., A 2.4 GHz CMOS LC VCO using micromachined variable capacitors for frequency tuning, in *1999 IEEE MTT-S International Microwave Symposium Digest,* 1, 79–82, 1999.)

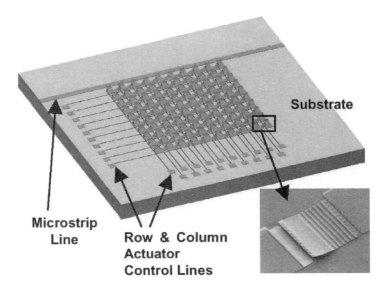

FIGURE 6.81 RF MEMS switches used in a reconfigurable array. (*Source:* Bozler, C., Drangmeister, R., Duffy, S., Gouker, M., Knecht, J., Kushner, L., Parr, R., Rabe, S., and Travis, L., MEMS Microswitch arrays for reconfigurable distributed microwave components, in *2000 IEEE MTT-S International Microwave Symposium Digest,* 153–156, 2000.)

Another, conceptually straightforward concept is shown in Fig. 6.83.[67,68] By using a MEMS cantilever beam ohmic switches, the length of a half-wave dipole antenna is changed. This reconfigurable antenna element was demonstrated with the antenna response changing from being centered at 12 GHz to above 18 GHz.

A MEMS reconfigurable Vee antenna (Fig. 6.84) has demonstrated the capability of beam steering and beam shaping.[69] The mechanical arrangement utilizes metal-coated polysilicon and metal-to-metal contacts to construct three-dimensional reconfigurable radiating and wave-guiding structures. The illustrated example contains fixed and moveable rotating hinges activated by scratch drive actuators.[70] The resulting structure demonstrated beam steering up to 48° off bore site along with the ability to reshape the beam.

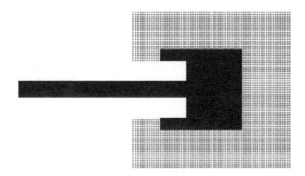

FIGURE 6.82 RF MEMS switch array layout for a patch antenna. (*Source:* Bozler, C., Drangmeister, R., Duffy, S., Gouker, M., Knecht, J., Kushner, L., Parr, R., Rabe, S., and Travis, L., MEMS Microswitch arrays for reconfigurable distributed microwave components, in *2000 IEEE MTT-S International Microwave Symposium Digest,* 153–156, 2000.)

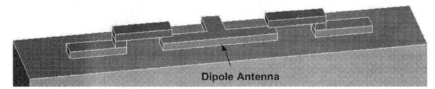

FIGURE 6.83 Reconfigurable dipole antenna using MEMS cantilever switches.

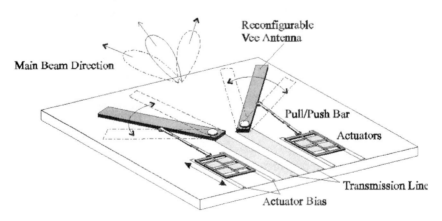

FIGURE 6.84 A MEMS reconfigurable vee antenna. (*Source:* Chiao, J.-C., Fu, Y., Chio, I.M., DeLisio, M., and Lin, L.-Y., MEMS reconfigurable Vee antenna, in *1999 IEEE MTT-S International Microwave Symposium Digest,* 4, 1515–1518, 1999.)

References

1. Goldsmith, C., Randall, J., Eshelman, S., Lin, T.H., Denniston, D., Chen, S., and Norvell, B., Characteristics of micromachined switches at microwave frequencies, in *1996 IEEE MTT-S International Microwave Digest,* 1141–1144, 1996.
2. Feng, Z., Zhang, H., Zhang, W., Su, B., Gupta, K.C., Bright, V. M., and Lee, Y.C., MEMS- based variable capacitor for millimeter-wave applications, in *2000 Solid-State Sensor and Actuator Workshop,* Hilton Head Island, 255–258, 2000.

3. Feng, Z., Zhang, W., Su, B., Harsh, K.F., Gupta, K.C., Bright, V., and Lee, Y.C., Design and modeling of RF MEMS tunable capacitors using electrothermal actuators, in *1999 IEEE MTT-S International Microwave Digest*, 1507–1510, 1999.

4. Wu, H.D., Harsh, K.F., Irwin, R.S., Zhang, W., Mickelson, A.R., and Lee, Y.C., MEMS designed for tunable capacitors, in *1998 IEEE MTT-S International Microwave Digest*, 127–129, 1999.

5. Hyman, D., Lam, J., Warneke, B., Schmitz, A., Hsu, T.Y., Brown, J., Schaffner, J., Walston, A., Loo, R.Y., Mehregany, M., and Lee, J., Surface-micromachineded RF MEMS switches on GaAs Substrates, in *International Journal of RF & Microwave Computer-Aided Engineering*, 348–361, 9, 4, 1999.

6. Madou, M., *Fundamentals of Microfabrication*, CRC Press, Boca Raton, FL, 1997.

7. Wolf, S. and Tauber, R.N., *Silicon Processing for the VLSI Era*, Lattice Press, Sunset Beach, CA, 1986.

8. Tang, W.C., Overview of MEMS Fabrication Technology in Microwave and Photonic Applications of MEMS at the 2000 IEEE MTT-S International Microwave Symposium, Boston, MA, 2000.

9. Willke, T.L. and Gearhart, S.S., LIGA micromachined planar transmission lines and filters, *IEEE Transactions on Microwave Theory and Techniques*, 45, 10, 1681–1688, Oct. 1997.

10. Becker, E.W., Ehrfeld, W., Hagmann, P., Maner, A., and Munchmeyer, D., Fabrication of microstructures with high-aspect ratios and great structural heights by synchrotron radiation lithography, galvanoforming, and plastic moulding (LIGA process), *Microelectronic Engineering*, 4, 35–36, 1986.

11. Gear picture from University of Wisconson — Madison web site, http://mems.engr.wisc.edu/MEMS.html.

12. Harendt, C., Appel, W., Graf, H.-G., Höfflinger, B., and Penteker, E., Wafer fusion bonding and its application to silicon-on-insulator fabrication, *Journal of Micromechanical Microengineering*, 1, 145–151, 1991.

13. Hyman, D., Schmitz, A., Warneke, B., Hsu, T.Y., Lam, J., Brown, J., Schaffner, J., Walston, A., Loo, R.Y., Tangonan, G.L., Mehregany, M., and Lee, J., GaAs-compatible surface-micromachined RF MEMS switches, *Electronics Letters*, 35, 3, 224–226, February 4, 1999.

14. Zavracky, P.M., McGruer, N.E., Morrison, R.H., and Potter, D., Microswitches and microrelays with a view toward microwave applications, *International Journal of RF & Microwave Computer-Aided Engineering*, 9, 4, 338–347, 1999.

15. McGruer, N.E., Zavracky, P.M., Majumder, S., Morrison, R., and Adams, G.G., Electrostatically actuated microswitches; scaling properties, in *Proceedings of 1998 Solid-State Sensor and Actuator Workshop*, Hilton Head Island, 132–135, 1998.

16. Micromachined-relay illustration courtesy of Analog Devices, http://www.analog.com/industry/umic/relay.html.

17. Bozler, C., Drangmeister, R., Duffy, S., Gouker, M., Knecht, J., Kushner, L., Parr, R., Rabe, S., and Travis, L., MEMS Microswitch arrays for reconfigurable distributed microwave components, in *2000 IEEE MTT-S International Microwave Symposium Digest*, 153–156, 2000.

18. Goldsmith, C., RF MEMS devices and circuits for radar and receiver applications, Microwave and photonic applications of MEMs Workshop, 2000 IEEE MTT-S International Microwave Symposium, 2000.

19. Goldsmith, C.L., Yao, Z., Eshelman, S., and Denniston, D., Performance of low-loss RF MEMS capacitive switches, *IEEE Microwave and Guided Wave Letters*, 8, 8, 517–519, August 1998.

20. Yao, Z.J., Chen, S., Eshelman, S., Denniston, D., and Goldsmith, C., Micromachined low-loss microwave switches, *IEEE Journal of Microelectromechanical Systems*, 8, 2, 129–134, June 1999.

21. Peroulis, D., Pacheco, S., Sarabandi, K., and Katehi, L.P.B., MEMS devices for high isolation switching and tunable filtering, in *2000 IEEE MTT-S International Microwave Symposium Digest*, 1217–1220, 2000.

22. Young, D.J. and Boser, B.E., A micromachine-based RF low-noise voltage-controlled oscillator, in *Proceedings of IEEE 1997 Custom Integrated Circuits Conference*, 431–434, 1997.

23. Pierce, B., Private Communication.

24. Malczewski, A., Private Communication.

25. Goldsmith, C.L., Malczewski, A., Yao, Z.J., Chen, S., Ehmke, J., and Hinzel, D.H., RF MEMS variable capacitors for tunable filters, in *International Journal of RF & Microwave Computer-Aided Engineering,* 9, 4, 362–374, 1999.

26. Yao, J.J., Park, S., Anderson, R., and DeNatale, J., A low power / low voltage electrostatic actuator for RF MEMS applications, in *Proceedings of 2000 Solid-State Sensor and Actuator Workshop,* Hilton Head Island, 246–249, 2000.

27. Yao, J.J., Park, S., and DeNatale, J., High tuning-ratio MEMS-based tunable capacitors for RF communications applications, in *Proceedings of 1998 Solid-State Sensor and Actuator Workshop,* Hilton Head Island, 124–127, 1998.

28. Hung, E. and Senturia, S., Tunable capacitors with programmable capacitance-voltage characteristics, in *Proceedings of 1998 Solid-State Sensor and Actuator Workshop,* Hilton Head Island, 292–295, 1998.

29. Katehi, L.P.B. and Rebeiz, G.M., Novel micromachined approaches to MMICs using low-parasitic, high-performance transmission media and environments, in *1996 IEEE MTT-S International Microwave Symposium Digest,* 1145–1148, 1996.

30. Ayon, A.A., Kolias, N.J., and MacDonald, N.C., Tunable, micromachined parrallel-plate transmission lines, in *IEEE/Cornell Conference Proceedings on Advanced Concepts in High Speed Semiconductor Devices and Circuits,* 201–208, 1995.

31. Kudrie, T.D., Neves, H.P., and MacDonald, N.C., Microfabricated single crystal silicon transmission lines, in *RAWCON'98 Proceedings,* 269–272, 1998.

32. Burghartz, J.N., Edelstein, D.C., Ainspan, H.A., and Jenkins, K.A., RF circuit design aspects of spiral inductors on silicon, in *IEEE J. Solid-State Circuits,* 33, 12, Dec. 1998.

33. Chang, J.Y.C., Abidi, A.A., and Gaitan M., Large suspended inductors on silicon and their use in a 2-mm CMOS RF Amplifier, in *IEEE Electron Device Letters,* 14, 5, 246–248, 1993.

34. Chi, C.Y. and Rebeiz, G.M., Planar microwave and milimeter-wave lumped elements and coupled-line filters using micro-machining techniques, in *IEEE Transactions on Microwave Theory and Techniques,* 43, 4, 730–738, 1995.

35. Clark, J.R., Bannon, F.D. III, Wong, A.-C., and Nguyen, C.T.-C., Parallel-resonator HF micromechanical bandpass filters, in *Proceedings of 1997 International Conference on Solid-State Sensor and Actuators,* Chicago, 1161–1164, 1997.

36. Bannon, F.D. III, Clark, J.R., and Nguyen, C.T.-C., High frequency microelectromenchanical IF filters, in *1996 International Electron Devices Meeting,* 773–776, 1996.

37. Bannon, F.D. III, Clark, J.R., and Nguyen, C.T.-C., High-Q HF microelectromechanical filters, in *IEEE Journal of Solid-State Circuits,* 35, 4, 512–526, April 2000.

38. Nguyen, C.T.-C., Frequency-selective MEMS for miniaturized communication devices, in *1998 IEEE Aerospace Conference,* 1, 445–460, 1998.

39. Nguyen, C.T.-C., Frequency-selective MEMS for miniaturized low-power communication devices, in *IEEE Transactions on Microwave Theory and Techniques,* 47, 8, 1486–1503, August 1999.

40. Nguyen, C.T.-C., Wong, A.-C., and Ding, H., Tunable, switchable, high-Q VHF microelectromechanical bandpass filters, in *1999 IEEE International Solid-State Circuits Conference Digest,* 78–79, 448, 1999.

41. Nguyen, C.T.-C. and Howe, R.T., An integrated CMOS micromechanical resonator high-Q oscillator, *IEEE Journal of Solid-State Circuits,* 34, 4, 440–454, April 1999.

42. Wang, K., Bannon, F.D. III, Clark, J.R., and Nguyen, C.T.-C., Q-enhancement of microelectromechanical filters via low-velocity spring coupling, in *1997 IEEE Ultrasonics Symposium Digest,* 323, 1997.

43. Becker, J.P. and Katehi, L.P.B., Toward a novel planar circuit compatible silicon micromachined waveguide, in *1999 Electrical Performance and Electronic Packaging Digest,* 221–224, 1999.

44. Katehi, L.P.B., Rebeiz, G.M., and Nguyen, C.T.-C., MEMS and Si-micromachined components for low-power, high-frequency communications systems, in *1998 IEEE MTT-S International Microwave Symposium Digest,* 1, 331–333, 1998.

45. Papapolymerou, J., Jui-Ching Cheng, East, J., and Katehi, L.P.B., A micromachined high-Q X-band resonator, *IEEE Microwave and Guided Wave Letters,* 7, 6, 168–170, June 1997.

46. Blondy, P., Brown, A.R., Cros, D., and Rebeiz, G.M., Low-loss micromachined filters for millimeter-wave communication systems, *IEEE Transactions on Microwave Theory and Techniques,* 46, 12, 2283–2288, 1998.

47. Brown, A.R. and Rebeiz, G.M., Micromachined micropackaged filter banks and tunable bandpass filters, in *1997 Wireless Communications Conference Digest,* 193–197, 1997.

48. Robertson, S.V., Katehi, L.P.B., and Rebeiz, G.M., Micromachined self-packaged W-band bandpass filters, in *1995 IEEE MTT-S International Microwave Symposium Digest,* 1543–1546, 1995.

49. Robertson, S.V., Katehi, L.P.B., and Rebeiz, G.M., Micromachined W-band filters, *IEEE Transactions on Microwave Theory and Techniques,* 44, 4, 598–606, April 1996.

50. Nguyen, C.T.-C., Katehi, L.P.B., and Rebeiz, G.M., Micromachined devices for wireless communications, in *Proceedings of the IEEE,* 86, 8, 1756–1768, 1998.

51. Krishnaswamy, S.V., Rosenbaum, J., Horwitz, S., Yale, C., and Moore, R.A., Compact FBAR filters offer low-loss performance, *Microwave & RF,* 127–136, Sept. 1991.

52. Rudy, R. and Merchant, P., Micromachined thin film bulk acoustic resonators in *Proceedings of the 1994 IEEE International Frequency Control Symposium,* Boston, MA, June 1–3, 1994, 135–138, 1994.

53. Pillans, B., Eshelman, S., Malczewski, A., Ehmke, J., and Goldsmith, C., Ka-band RF MEMS phase shifters, *IEEE Microwave and Guided Wave Letters,* 9, 12, 520–522, December 1999.

54. Pillans, B., Eshelman, S., Malczewski, A., Ehmke, J., and Goldsmith, C., Ka-band RF MEMS phase shifters for phased array applications, in *2000 IEEE Radio Frequency Integrated Circuits Symposium Digest,* 195–198, 2000.

55. Malczewski, A., Eshelman, S., Pillans, B., Ehmke, J., and Goldsmith, C.L., X-band RF MEMS phase shifters for phased array applications, in *IEEE Microwave and Guided Wave Letters,* 9, 12, 517–519, 1999.

56. Hayden, J.S. and Rebeiz, G.M., One and two-bit low-loss cascadable MEMS distributed X-band phase shifters, in *2000 IEEE MTT-S International Microwave Symposium Digest,* 161–164, 2000.

57. Barker, N.S. and Rebeiz, G.M., Optimization of distributed MEMS phase shifters, in *1999 IEEE MTT-S International Microwave Symposium Digest,* 299–302, Anaheim, CA, 1999.

58. Kim, H.-T., Park, J.-H., Kim, Y.-K., and Kwon, Y., Millimeter-wave micromachined tunable filters, in *1999 IEEE MTT-S International Microwave Symposium Digest,* 3, 1235–1238, 1999.

59. Guillon, B., Cros, D., Pons, P., Gazaux, J.L., Lalaurie, J.C., Plana, R., and Graffeuil, J., Ka band micromachined dielectric resonator oscillator, Electronics Letters, 35, 11, 909–910, May 1999.

60. Guillon, B., Cros, D., Pons, P., Grenier, K., Parra, T., Cazaux, J.L., Lalaurie, J.C., Graffeuil, J., and Plana, R., Design and realization of high Q millimeter-wave structures through micromachining techniques, in *1999 IEEE MTT-S International Microwave Symposium Digest,* 4 , 1519–1522, 1999.

61. Brown, A.R. and Rebeiz, G.M., Micromachined high-Q resonators, low-loss diplexers, and low phase-noise oscillators for a 28 GHz front-end, in *1999 IEEE Radio and Wireless Conference,* 247–253, 1999.

62. Brown, A.R. and Rebeiz, G.M., A Ka-band micromachined low-phase-noise oscillator, *IEEE Transactions on Microwave Theory and Techniques,* 47, 8, 1504–1508, August 1999.

63. Young, D.J., Malba, V., Ou, J.-J., Bernhardt, A.E., and Boser, B.E., A low-noise RF voltage-controlled oscillator using on-chip high-Q three-dimensional coil inductor and micromachined variable capacitor, in *Proceedings of 1998 Solid-State Sensor and Actuator Workshop,* Hilton Head Island, 128–131, 1998.

64. Young, D.J. and Boser, B.E., A micromachined variable capacitor for Monolithic low-noise VCOs, in *Proceedings of 1996 Solid-State Sensor and Actuator Workshop,* Hilton Head Island, 86–89, 1996.

65. Dec, A. and Suyama, K., A 2.4 GHz CMOS LC VCO using micromachined variable capacitors for frequency tuning, in *1999 IEEE MTT-S International Microwave Symposium Digest,* 1, 79–82, 1999.

66. Dec, A. and Suyama, K., A 1.9 GHz micromachined-based low-phase-noise CMOS VCO, in *1999 IEEE International Solid-State Circuits Conference Digest,* 80–81, 449, 1999.

67. Tangonan, G., Loo, R., Schaffner, J., and Lee, J.J., Microwave photonic applications of MEMS technology, in *1999 International Topical Meeting on Microwave Photonics,* 1, 109–112, 1999.

68. Izadpanah, H., Warneke, B., Loo, R., and Tangonan, G., Reconfigurable low power, light weight wireless system based on the RF MEM switches, in *1999 IEEE MTT-S Symposium on Technologies for Wireless Applications,* 175–180, 1999.

69. Chiao, J.-C., Fu, Y., Chio, I.M., DeLisio, M., and Lin, L.-Y., MEMS reconfigurable Vee antenna, in *1999 IEEE MTT-S International Microwave Symposium Digest,* 4, 1515–1518, 1999.

70. Akiyama, T. and Shono, K., Controlled step-wise motion in polysilicon microstructures, *Journal of MEMS,* 2, 3, 106, Sept. 1993.

6.5 Surface Acoustic Wave (SAW) Filters

Donald C. Malocha

A **surface acoustic wave (SAW),** also called a Rayleigh wave, is composed of a coupled compressional and shear wave in which the SAW energy is confined near the surface. There is also an associated electrostatic wave for a SAW on a piezoelectric substrate which allows electroacoustic coupling via a transducer. SAW technology's two key advantages are its ability to electroacoustically access and tap the wave at the crystal surface and that the wave velocity is approximately 100,000 times slower than an electromagnetic wave. Assuming an electromagnetic wave velocity of 3×10^8 m/s and an acoustic wave velocity of 3×10^3 m/s, Table 6.7 compares relative dimensions versus frequency and delay. The SAW wavelength is on the same order of magnitude as line dimensions that can be photolithographically produced and the lengths for both small and long delays are achievable on reasonable size substrates. The corresponding E&M transmission lines or waveguides would be impractical at these frequencies.

Because of SAWs' relatively high operating frequency, linear delay, and tap weight (or sampling) control, they are able to provide a broad range of signal processing capabilities. Some of these include linear and dispersive filtering, coding, frequency selection, convolution, delay line, time impulse response shaping, and others. There are very broad ranges of commercial and military system applications that include components for radars, front-end and IF filters, CATV and VCR components, cellular radio and pagers, synthesizers and analyzers, navigation, computer clocks, tags, and many, many others [Campbell, 1989; Matthews, 1977].

There are four principal SAW properties: transduction, reflection, regeneration, and nonlinearities. Nonlinear elastic properties are principally used for convolvers and will not be discussed. The other three properties are present, to some degree, in all SAW devices, and these properties must be understood and controlled to meet device specifications.

A finite-impulse response (FIR) or transversal filter is composed of a series of cascaded time delay elements that are sampled or "tapped" along the delay line path. The sampled and delayed signal is summed at a junction which yields the output signal. The output time signal is finite in length and has no feedback. A schematic of an FIR filter is shown in Fig. 6.85.

A SAW transducer is able to implement an FIR filter. The electrodes or fingers provide the ability to sample or "tap" the SAW and the distance between electrodes provides the relative delay. For a uniformly sampled SAW transducer, the delay between samples, Δt, is given by $\Delta t = \Delta L/v_a$, where ΔL is the electrode period and v_a is the acoustic velocity. The typical means for providing attenuation or weighting is to vary

TABLE 6.7 Comparison of SAW and E&M Dimensions versus Frequency and Delay, Where Assumed Velocities are $v_{SAW} = 3000$ m/s and $v_{EM} = 3 \times 10^8$ m/s

Parameter	SAW	E&M
$F_0 = 10$ MHz	$\lambda_{SAW} = 300$ μm	$\lambda_{EM} = 30$ m
$F_0 = 2$ GHz	$\lambda_{SAW} = 1.5$ μm	$\lambda_{EM} = 0.15$ m
Delay = 1 ns	$L_{SAW} = 3$ μm	$L_{EM} = 0.3$ m
Delay = 10 μs	$L_{SAW} = 30$ mm	$L_{EM} = 3000$ m

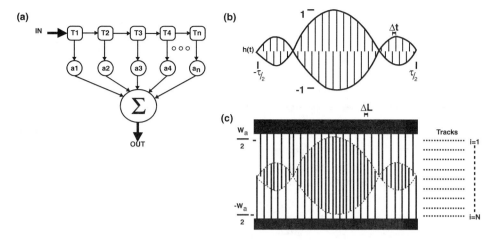

FIGURE 6.85 (a) Schematic of a finite-impulse response (FIR) filter. (b) An example of a sampled time function; the envelope is shown in the dotted lines. (c) A SAW transducer implementation of the time function $h(t)$.

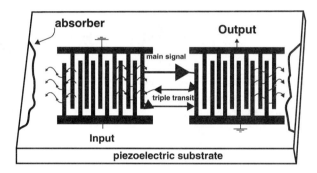

FIGURE 6.86 Schematic diagram of a typical SAW bidirectional filter consisting of two interdigital transducers. The transducers need not be identical. The input transducer launches waves in either direction and the output transducer converts the acoustic energy back to an electrical signal. The device exhibits a minimum 6-dB insertion loss. Acoustic absorber damps unwanted SAW energy to eliminate spurious reflections that could cause distortions.

the overlap between adjacent electrodes which provides a spatially weighted sampling of a uniform wave. Figure 6.85 shows a typical FIR time response and its equivalent SAW transducer implementation. A SAW filter is composed of a minimum of two transducers and possibly other SAW components. A schematic of a simple SAW bidirectional filter is shown in Fig. 6.86. A **bidirectional transducer** radiates energy equally from each side of the transducer (or port). Energy not being received is absorbed to eliminate spurious reflections.

SAW Material Properties

There are a large number of materials currently being used for SAW devices. The most popular single-crystal piezoelectric materials are quartz, lithium niobate ($LiNbO_3$), and lithium tantalate ($LiTa_2O_5$). The materials are anisotropic, which will yield different material properties versus the cut of the material and the direction of propagation. There are many parameters that must be considered when choosing a given material for a given device application. Table 6.8 shows some important material parameters for consideration for four of the most popular SAW materials [Datta, 1986; Morgan, 1985].

The coupling coefficient, k^2, determines the electroacoustic coupling efficiency. This determines the fractional bandwidth versus minimum insertion loss for a given material and filter. The static capacitance is a function of the transducer electrode structure and the dielectric properties of the substrate. The

TABLE 6.8 Common SAW Material Properties

Parameter/Material	ST-Quartz	YZ LiNbO$_3$	128° YX LiNbO$_3$	YZ LiTa$_2$O$_3$
k^2 (%)	0.16	4.8	5.6	0.72
C_s (pf/cm-pair)	0.05	4.6	5.4	4.5
v_0 (m/s)	3,159	3,488	3,992	3,230
Temp. coeff. of delay (ppm/°C)	0	94	76	35

values given in the table correspond to the capacitance per pair of electrodes having quarter wavelength width and one-half wavelength period. The free surface velocity, v_0, is a function of the material, cut angle, and propagation direction. The temperature coefficient of delay (TCD) is an indication of the frequency shift expected for a transducer due to a change of temperature and is also a function of cut angle and propagation direction.

The substrate is chosen based on the device design specifications and includes consideration of operating temperature, fractional bandwidth, and insertion loss. Second-order effects such as diffraction and beam steering are considered important on high-performance devices [Morgan, 1985]. Cost and manufacturing tolerances may also influence the choice of the substrate material.

Basic Filter Specifications

Figure 6.87 shows a typical time domain and frequency domain device performance specification. The basic frequency domain specification describes frequency bands and their desired level with respect to a given reference. Time domain specifications normally define the desired impulse response shape and any spurious time responses. The overall desired specification may be defined by combinations of both time and frequency domain specifications. Since time, $h(t)$, and frequency, $H(\omega)$, domain responses form unique Fourier transform pairs, given by

$$h(t) = 1/2\pi \int_{-\infty}^{\infty} H(\omega) e^{j\omega t} d\omega \tag{6.19}$$

$$H(\omega) = \int_{-\infty}^{\infty} h(t) e^{-j\omega t} dt \tag{6.20}$$

it is important that combinations of time and frequency domain specifications be self-consistent.

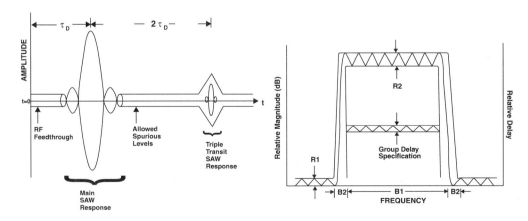

FIGURE 6.87 Typical time and frequency domain specification for a SAW filter. The filter bandwidth is B_1, the transition bandwidth is B_2, the inband ripple is R_2 and the out-of-band sidelobe level is R_1.

The electrodes of a SAW transducer act as sampling points for both transduction and reception. Given the desired modulated time response, it is necessary to sample the time waveform. For symmetrical frequency responses, sampling at twice the center frequency, $f_s = 2f_0$, is sufficient, while nonsymmetric frequency responses require sampling at twice the highest frequency of interest. A very popular approach is to sample at $f_s = 4f_0$. The SAW frequency response obtained is the convolution of the desired frequency response with a series of impulses, separated by f_s, in the frequency domain. The net effect of sampling is to produce a continuous set of harmonics in the frequency domain in addition to the desired response at f_0. This periodic, time-sampled function can be written as

$$g(t_n) = \sum_{-N/2}^{N/2} a_n \cdot \delta(t - t_n) \tag{6.21}$$

where a_n represents the sample values, $t_n = n\Delta t$, $n = n$th sample, and $\Delta t =$ time sample separation. The corresponding frequency response is given by

$$G(f) = \sum_{-N/2}^{N/2} g(t_n) e^{-j2\pi f t_n} = \sum_{-N/2}^{N/2} g(t_n) e^{-j2\pi f / f_s} \tag{6.22}$$

where $f_s = 1/\Delta t$. The effect of sampling in the time domain can be seen by letting $f = f + mf_s$, where m is an integer, which yields $G(f + mf_s) = G(f)$, which verifies the periodic harmonic frequency response.

Before leaving filter design, it is worth noting that a SAW filter is composed of two transducers that may have different center frequencies, bandwidth, and other filter specifications. This provides a great deal of flexibility in designing a filter by allowing the product of two frequency responses to achieve the total filter specification.

SAW Transducer Modeling

The four most popular and widely used models include the transmission line model, the coupling of modes model, the impulse response model, and the superposition model. The superposition model is an extension of the impulse response model and is the principal model used for the majority of SAW bidirectional and multiphase filter synthesis which do not have inband, interelectrode reflections. As is the case for most technologies, many models may be used in conjunction with each other for predicting device performance based on ease of synthesis, confidence in predicted parameters, and correlation with experimental device data.

The SAW Superposition Impulse Response Transducer Model

The impulse response model was first presented by Hartmann et al. [1973] to describe SAW filter design and synthesis. For a linear causal system, the Fourier transform of the device's frequency response is the device impulse time response. Hartmann showed that the time response of a SAW transducer is given by

$$h(t) = 4k \sqrt{C_s} f_i^{3/2}(t) \sin[\theta(t)] \text{ where } \theta(t) = 2\pi \int_0^t f_i(\tau) d\tau \tag{6.23}$$

where the following definitions are $k^2 =$ SAW coupling coefficient, $C_s =$ electrode pair capacitance per unit length (pf/cm-pair), and $f_i(t) =$ instantaneous frequency at a time, t. This is the general form for a uniform beam transducer with arbitrary electrode spacing. For a uniform beam transducer with periodic electrode spacing, $f_i(t) = f_0$ and $\sin \theta(t) = \sin \omega t$. This expression relates a time response to the physical device parameters of the material coupling coefficient and the electrode capacitance.

Given the form of the time response, energy arguments are used to determine the device equivalent circuit parameters. Assume a delta function voltage input, $v_{in}(t) = \delta(t)$, then $V_{in}(\omega) = 1$. Given $h(t), H(\omega)$ is known and the energy launched as a function of frequency is given by $E(\omega) = 2 \cdot |H(\omega)|^2$. Then

$$E\left(\omega\right) = V_{in}^2\left(\omega\right) \cdot G_a\left(\omega\right) = 1 \cdot G_a\left(\omega\right) \tag{6.24}$$

or

$$G_a\left(\omega\right) = 2 \cdot \left|H\left(\omega\right)\right|^2 \tag{6.25}$$

There is a direct relationship between the transducer frequency transfer function and the transducer conductance. Consider an **interdigital transducer (IDT)** with uniform overlap electrodes having N_p interaction pairs. Each gap between alternating polarity electrodes is considered a localized SAW source. The SAW impulse response at the fundamental frequency will be continuous and of duration τ, where $\tau = N \cdot \Delta t$, and $h(t)$ is given by

$$h\left(t\right) = \kappa \cdot \cos\left(\omega_0 t\right) \cdot rect\left(t/\tau\right) \tag{6.26}$$

where $\kappa = 4k\sqrt{C_s} \, f_0^{3/2}$ and f_0 is the carrier frequency. The corresponding frequency response is given by

$$H\left(\omega\right) = \frac{\kappa\tau}{2}\left\{\frac{\sin\left(x_1\right)}{x_1} + \frac{\sin\left(x_2\right)}{x_2}\right\} \tag{6.27}$$

where $x_1 = (\omega - \omega_0) \cdot \tau/2$ and $x_2 = (\omega + \omega_0) \cdot \tau/2$.

This represents the ideal SAW continuous response in both time and frequency. This can be related to the sampled response by a few substitutions of variables. Let

$$\Delta t = \frac{1}{2 \cdot f_0}, \quad t_n = n \cdot \Delta t, \quad N \cdot \Delta t = \tau, \quad N_p \cdot \Delta t = \tau/2 \tag{6.28}$$

Assuming a frequency bandlimited response, the negative frequency component centered around $-f_0$ can be ignored. Then the frequency response, using Eq. (6.27), is given by

$$H\left(\omega\right) = \kappa\left\{\frac{\pi N_p}{\omega_0}\right\} \cdot \frac{\sin\left(x_n\right)}{x_n} \tag{6.29}$$

where

$$x_n = \frac{\left(\omega - \omega_0\right)}{\omega_0}\pi N_p = \frac{\left(f - f_0\right)}{f_0}\pi N_p$$

The conductance, given using Eqs. (6.24) and (6.28), is

$$G_a\left(f\right) = 2\kappa^2\left\{\frac{\pi N_p}{2\pi f_0}\right\}^2 \frac{\sin^2\left(x_n\right)}{x_n^2} = 8k^2 f_0 C_s N_p^2 \cdot \frac{\sin^2\left(x_n\right)}{x_n^2} \tag{6.30}$$

This yields the frequency-dependent conductance per unit width of the transducer. Given a uniform transducer of width, W_a, the total transducer conductance is obtained by multiplying Eq. (6.30) by W_a. Defining the center frequency conductance as

$$G_a(f_0) = G_0 = 8k^2 f_0 C_s W_a N_p^2 \tag{6.31}$$

the transducer conductance is

$$G_a(f_0) = G_0 \cdot \frac{\sin^2(x_n)}{x_n^2} \tag{6.32}$$

The transducer electrode capacitance is given as

$$C_e = C_s W_a N_p \tag{6.33}$$

Finally, the last term of the SAW transducer's equivalent circuit is the frequency-dependent susceptance. Given any system where the frequency-dependent real part is known, there is an associated imaginary part that must exist for the system to be real and causal. This is given by the Hilbert transform susceptance, defined as B_a, where [Datta, 1986]

$$B_a(\omega) = \frac{1}{\pi} \int_{-\infty}^{\infty} \frac{G_a(u)}{(u - \omega)} du = G_a(\omega) * 1/\omega \tag{6.34}$$

where "*" indicates convolution.

These three elements compose a SAW transducer equivalent circuit. The equivalent circuit, shown in Fig. 6.88, is composed of one lumped element and two frequency-dependent terms that are related to the substrate material parameters, transducer electrode number, and the transducer configuration. Figure 6.89 shows the time and frequency response for a uniform transducer and the associated frequency-dependent

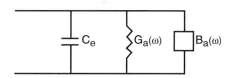

FIGURE 6.88 Electrical equivalent circuit model.

conductance and Hilbert transform susceptance. The simple impulse model treats each electrode as an ideal impulse; however, the electrodes have a finite width that distorts the ideal impulse response. The actual SAW potential has been shown to be closely related to the electrostatic charge induced on the transducer by the input voltage. The problem is solved assuming a quasi-static and electrostatic charge distribution, assuming a semi-infinite array of electrodes, solving for a single element, and then using superposition and convolution. The charge distribution solution for a single electrode with all others grounded is defined as the basic charge distribution function (BCDF). The result of a series of arbitrary voltages placed on a series of electrodes is the summation of scaled, time-shifted BCDFs. The identical result is obtained if an array factor, $a(x)$, defined as the ideal impulses localized at the center of the electrode or gap, is convolved with the BCDF, often called the element factor. This is very similar to the analysis of antenna arrays. Therefore, the ideal frequency transfer function and conductance given by the impulse response model need only be modified by multiplying the frequency-dependent element factor. The analytic solution to the BCDF is given in Datta [1986] and Morgan [1985], and is shown to place a small perturbation in the form of a slope or dip over the normal bandwidths of interest. The BCDF also predicts the expected harmonic frequency responses.

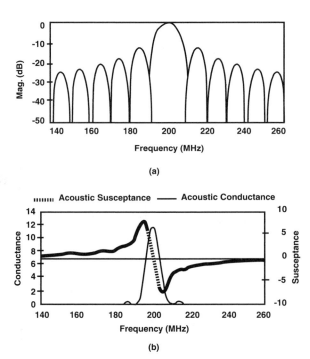

FIGURE 6.89 (a) Theoretical frequency response of a $rect(t/\tau)$ time function having a time length of 0.1 μs and a 200-MHz carrier frequency. (b) Theoretical conductance and susceptance for a SAW transducer implementing the frequency response. The conductance and susceptance are relative and are given in millisiemens.

Apodized SAW Transducers

Apodization is the most widely used method for weighting a SAW transducer. The desired time-sampled impulse response is implemented by assigning the overlap of opposite polarity electrodes at a given position to a normalized sample weight at a given time. A tap having a weight of unity has an overlap across the entire beamwidth while a small tap will have a small overlap of adjacent electrodes. The time impulse response can be broken into tracks which have uniform height but whose time length and impulse response may vary. Each of these time tracks is implemented spatially across the transducer's beamwidth by overlapped electrode sections at the proper positions. This is shown in Fig. 6.85. The smaller the width of the tracks, the more exact the approximation of uniform time samples. There are many different ways to implement the time-to-spatial transformation; Fig. 6.85 shows just one such implementation.

The impulse response can be represented, to any required accuracy, as the summation of uniform samples located at the proper positions in time in a given track. Mathematically this is given by

$$h(t) = \sum_{i=1}^{I} h_i(t) \tag{6.35}$$

and

$$H(\omega) = \sum_{i=1}^{I} H_i(\omega) = \sum_{i=1}^{I} \left\{ \int_{-\tau/2}^{\tau/2} h_i(t) e^{-j\omega t} dt \right\} \tag{6.36}$$

The frequency response is the summation of the individual frequency responses in each track, which may be widely varying depending on the required impulse response. This spatial weighting complicates the calculations of the equivalent circuit for the transducer. Each track must be evaluated separately for its acoustic conductance, acoustic capacitance, and acoustic susceptance. The transducer elements are then obtained by summing the individual track values yielding the final transducer equivalent circuit parameters. These parameters can be solved analytically for simple impulse response shapes (such as the rect, triangle, cosine, etc.) but are usually solved numerically on a computer [Richie et al., 1988].

There is also a secondary effect of apodization when attempting to extract energy. Not all of the power of a nonuniform SAW beam can be extracted by an a uniform transducer, and reciprocally, not all of the energy of a uniform SAW beam can be extracted by an apodized transducer. The transducer efficiency is calculated at center frequency as

$$E = \frac{\left| \sum_{i=1}^{I} H(\omega_0) \right|^2}{I \cdot \sum_{i=1}^{I} H^2(\omega_0)} \tag{6.37}$$

The apodization loss is defined as

$$\text{apodization loss} = 10 \cdot \log(E) \tag{6.38}$$

Typical apodization loss for common SAW transducers is 1 dB or less.

Finally, because an apodized transducer radiates a nonuniform beam profile, the response of two cascaded apodized transducers is not the product of each transducer's individual frequency responses, but rather is given by

$$H_{12}(\omega) = \sum_{i=1}^{I} H_{1i}(\omega) \cdot H_{2i}(\omega) \neq \sum_{i=1}^{I} H_{1i}(\omega) \cdot \sum_{i=1}^{I} H_{2i}(\omega) \tag{6.39}$$

In general, filters are normally designed with one apodized and one uniform transducer or with two apodized transducers coupled with a spatial-to-amplitude acoustic conversion component, such as a multistrip coupler [Datta, 1986].

Distortion and Second-Order Effects

In SAW devices there are a number of effects that can distort the desired response from the ideal response. The most significant distortion in SAW transducers is called the **triple transit echo (TTE)** which causes a delayed signal in time and an inband ripple in the amplitude and delay of the filter. The TTE is primarily due to an electrically regenerated SAW at the output transducer, which travels back to the input transducer, where it induces a voltage across the electrodes, which in turn regenerates another SAW which arrives back at the output transducer. This is illustrated schematically in Fig. 6.86. Properly designed and matched **unidirectional transducers** have acceptably low levels of TTE due to their design. Bidirectional transducers, however, must be mismatched in order to achieve acceptable TTE levels. To first order, the TTE for a bidirectional two-transducer filter is given as

$$\text{TTE} \approx 2 \cdot IL + 6 \text{ dB} \tag{6.40}$$

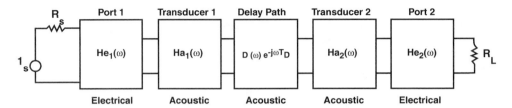

FIGURE 6.90 Complete transfer function of a SAW filter including the acoustic, electrical, and delay line transfer functions. The current generator is I_s, and R_s and R_L are the source and generator resistances, respectively.

where IL = filter insertion loss, in dB [Matthews, 1977]. As examples, the result of TTE is to cause a ghost in a video response and intersymbol interference in data transmission.

Another distortion effect is electromagnetic feedthrough which is due to direct coupling between the input and output ports of the device, bypassing any acoustic response. This effect is minimized by proper device design, mounting, bonding, and packaging.

In addition to generating a SAW, other spurious acoustic modes may be generated. Bulk acoustic waves (BAW) may be both generated and received, which causes passband distortion and loss of out-of-band rejection. BAW generation is minimized by proper choice of material, roughening of the crystal backside to scatter BAWs, and use of a SAW track changer, such as a multistrip coupler.

Any plane wave that is generated from a finite aperture will begin to diffract. This is exactly analogous to light diffracting through a slit. Diffraction's principal effect is to cause effective shifts in the filter's tap weights and phase which results in increased sidelobe levels in the measured frequency response. Diffraction is minimized by proper choice of substrate and filter design.

Transducer electrodes are fabricated from thin film metal, usually aluminum, and are finite in width. This metal can cause discontinuities to the surface wave which cause velocity shifts and frequency-dependent reflections. In addition, the films have a given sheet resistance which gives rise to a parasitic electrode resistance loss. The electrodes are designed to minimize these distortions in the device.

Bidirectional Filter Response

A SAW filter is composed of two cascaded transducers. In addition, the overall filter function is the product of two acoustic transfer functions, two electrical transfer functions, and a delay line function, as illustrated in Fig. 6.90. The acoustic filter functions are as designed by each SAW transducer. The delay line function is dependent on several parameters, the most important being frequency and transducer separation. The propagation path transfer function, $D(\omega)$, is normally assumed unity, although this may not be true for high frequencies ($f > 500$ MHz) or if there are films in the propagation path. The electrical networks may cause distortion of the acoustic response and are typically compensated in the initial SAW transducer's design.

The SAW electrical network is analyzed using the SAW equivalent circuit model plus the addition of packaging parasitics and any tuning or matching networks. Figure 6.91 shows a typical electrical network that is computer analyzed to yield the overall transfer function for one port of the two-port SAW filter [Morgan, 1985]. The second port is analyzed in a similar manner and the overall transfer function is obtained as the product of the electrical, acoustic, and propagation delay line effects.

Multiphase Unidirectional Transducers

The simplest SAW transducers are single-phase bidirectional transducers. Because of their symmetrical nature, SAW energy is launched equally in both directions from the transducer. In a two-transducer configuration, half the energy (3 dB) is lost at the transmitter, and reciprocally, only half the energy can be received at the receiver. This yields a net 6-dB loss in a filter. However, by adding nonsymmetry into the transducer, either by electrical multiphases or nonsymmetry in reflection and regeneration, energy can be unidirectionally directed yielding a theoretical minimum 0-dB loss.

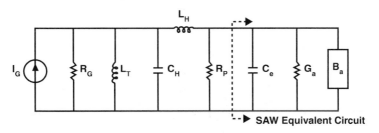

FIGURE 6.91 Electrical network analysis for a SAW transducer. I_G and R_G represent the generator source and impedance, L_T is a tuning inductor, C_H and L_H are due to the package capacitance and bond wire, respectively, and R_P represents a parasitic resistance due to the electrode transducer resistance. The entire network, including the frequency-dependent SAW network, is solved to yield the single-port transfer function.

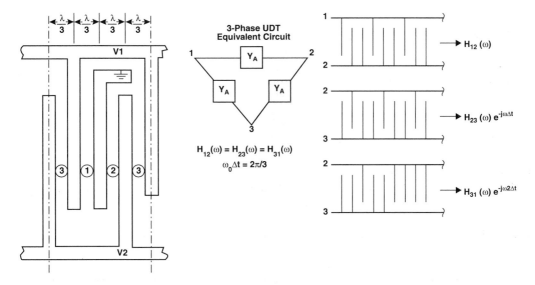

FIGURE 6.92 Schematic of a unit cell of a 3PUDT and the basic equivalent circuit. The 3PUDT can be analyzed as three collinear transducers with a spatial offset.

The most common SAW UDTs are called the three-phase UDT (3PUDT) and the group type UDT (GUDT). The 3PUDT has the broadest bandwidth and requires multilevel metal structures with crossovers. The GUDT uses a single-level metal but has a narrower unidirectional bandwidth due to its structure. In addition, there are other UDT or equivalent embodiments that can be implemented but will not be discussed [Morgan, 1985]. The basic structure of a 3PUDT is shown in Fig. 6.92. A unit cell consists of three electrodes, each connected to a separate bus bar, where the electrode period is $\lambda_0/3$. One bus bar is grounded and the other two bus bars will be driven by an electrical network where $V_1 = V_2 \angle 60°$. The transducer analysis can be accomplished similar to a simple IDT by considering the 3PUDT as three collinear IDTs with a spatial phase shift, as shown in Fig. 6.92. The electrical phasing network, typically consisting of one or two reactive elements, in conjunction with the spatial offset results in energy being launched in only one direction from the SAW transducer. The transducer can then be matched to the required load impedance with one or two additional reactive elements. The effective unidirectional bandwidth of the 3PUDT is typically 20% or less, beyond which the transducer behaves as a normal bidirectional transducer. Figure 6.93 shows a 3PUDT filter schematic consisting of two transducers and their associated matching and phasing networks. The overall filter must be analyzed with all external electrical components in place for accurate prediction of performance. The external components can be

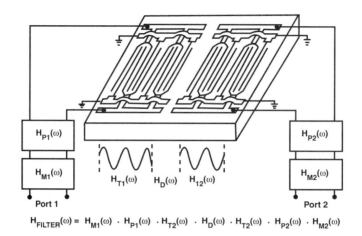

$$H_{FILTER}(\omega) = H_{M1}(\omega) \cdot H_{P1}(\omega) \cdot H_{T2}(\omega) \cdot H_D(\omega) \cdot H_{T2}(\omega) \cdot H_{P2}(\omega) \cdot H_{M2}(\omega)$$

FIGURE 6.93 Schematic diagram of a 3PUDT which requires the analysis of both the acoustic transducer responses as well as electrical phasing and matching networks.

miniaturized and may be fabricated using only printed circuit board material and area. This type of device has demonstrated as low as 2 dB insertion loss.

Single-Phase Unidirectional Transducers

Single-phase unidirectional transducers (SPUDT) use spatial offsets between mechanical electrode reflections and electrical regeneration to launch a SAW in one direction. A reflecting structure may be made of metal electrodes, dielectric strips, or grooved reflectors that are properly placed within a transduction structure. Under proper design and electrical matching conditions, the mechanical reflections can exactly cancel the electrical regeneration in one direction of the wave over a moderate band of frequencies. This is schematically illustrated in Fig. 6.94 which shows a reflector structure and a transduction structure merged to form a SPUDT. The transducer needs to be properly matched to the load for optimum operation. The mechanical reflections can be controlled by modifying the width, position, or height of the individual reflector. The regenerated SAW is primarily controlled by the electrical matching to the load of the transduction structure. SPUDT filters have exhibited as low as 3 dB loss over fractional bandwidths of 5% or less and have the advantage of not needing phasing networks when compared to the multiphase UDTs.

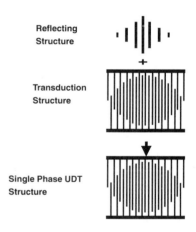

FIGURE 6.94 Schematic representation of a SPUDT which is a combination of transduction and reflecting structures to launch a SAW in one direction over moderate bandwidths.

Dispersive Filters

SAW filters can also be designed and fabricated using nonuniformly spaced electrodes in the transducer. The distance between adjacent electrodes determines the "local" generated frequency. As the spacing between the electrodes changes, the frequency is slowly changed either up (decreasing electrode spacing) or down (increasing electrode spacing) as the position progresses along the transducer. This slow frequency change with time is often called a "chirp." Figure 6.95 shows a typical dispersive filter consisting of a chirped

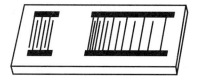

FIGURE 6.95 A SAW dispersive filter consisting of a uniform transducer and a "down chirp" dispersive transducer. The high frequencies have a shorter delay than the low frequencies in this example.

transducer in cascade with a uniform transducer. Filters can be designed with either one or two chirped transducers and the rate of the chirp is variable within the design. These devices have found wide application in radar systems due to their small size, reproducibility, and large time bandwidth product.

Coded SAW Filters

Because of the ability to control the amplitude and phase of the individual electrodes or taps, it is easy to implement coding in a SAW filter. Figure 6.96 shows an example of a coded SAW filter implementation. By changing the phase of the taps, it is possible to generate an arbitrary code sequence. These types of filters are used in secure communication systems, spread spectrum communications, and tagging, to name a few [Matthews, 1977].

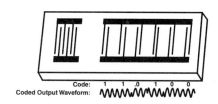

FIGURE 6.96 Example of a coded SAW tapped delay line.

SAW devices can also be used to produce time impulse response shapes for use in modulators, equalizers, and other applications. An example of a SAW modulator used for generating a cosine envelope for a minimum shift keyed (MSK) modulator is shown in Fig. 6.97 [Morgan, 1985].

Resonators

Another very important class of devices is SAW resonators. Resonators can be used as frequency control elements in oscillators, as notch filters, and as narrowband filters, to name a few. Resonators are typically fabricated on piezoelectric quartz substrates due to its low TCD which yields temperature-stable devices. A resonator uses

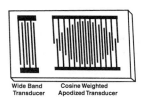

FIGURE 6.97 A SAW filter for implementing an MSK waveform using a wideband input transducer and a cosine envelope apodized transducer.

one or two transducers for coupling energy in/out of the device and one or more distributed reflector arrays to store energy in the device. This is analogous to an optical cavity with the distributed reflector arrays acting as the mirrors. A localized acoustic mirror, such as a cleaved edge, is not practical for SAW because of spurious mode coupling at edge discontinuities which causes significant losses.

A distributive reflective array is typically composed of a series of shorted metal electrodes, etched grooves in the substrate, or dielectric strips. There is a physical discontinuity on the substrate surface due to the individual reflectors. Each reflector is one-quarter wavelength wide and the periodicity of the array is one-half wavelength. This is shown schematically in Fig. 6.98. The net reflections from all the individual array elements add synchronously at center frequency, resulting in a very efficient reflector. The reflection from each array element is small and very little spurious mode coupling results.

Figure 6.98 shows a typical single-pole, single-cavity, two-port SAW resonator. Resonators can be made multipole by addition of multiple cavities, which can be accomplished by inline acoustic coupling, transverse acoustic coupling, and by electrical coupling. The equivalent circuit for SAW two-port and one-port resonators is shown in Fig. 6.99. SAW resonators have low insertion loss and high electrical Qs of several thousand [Campbell, 1989; Datta, 1986; Morgan, 1985].

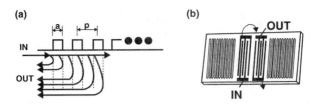

FIGURE 6.98   (a) SAW reflector array illustrating synchronous distributed reflections at center frequency. Individual electrode width (a) is 1/4 wavelength and the array period is 1/2 wavelength at center frequency. (b) A schematic of a simple single-pole, single-cavity two-port SAW resonator.

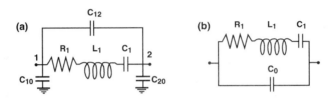

FIGURE 6.99   (a) Two-port resonator equivalent circuit and (b) one-port resonator equivalent circuit.

Defining Terms

Bidirectional transducer: A SAW transducer which launches energy from both acoustic ports which are located at either end of the transducer structure.

Interdigital transducer: A series of collinear electrodes placed on a piezoelectric substrate for the purpose of launching a surface acoustic wave.

Surface acoustic wave (SAW):   A surface acoustic wave (also known as a Rayleigh wave) is composed of a coupled compressional and shear wave. On a piezoelectric substrate there is also an electrostatic wave which allows electroacoustic coupling. The wave is confined at or near the surface and decays away rapidly from the surface.

Triple transit echo (TTE):   A multiple transit echo received at three times the main SAW signal delay time. This echo is caused due to the bidirectional nature of SAW transducers and the electrical and/or acoustic mismatch at the respective ports. This is a primary delayed signal distortion which can cause filter distortion, especially in bidirectional transducers and filters.

Unidirectional transducer (UDT):   A transducer that is capable of launching energy from primarily one acoustic port over a desired bandwidth of interest.

References

D.S. Ballintine, *Acoustic Wave Sensors,* San Diego, Calif.: Academic Press, 1995.

D.S. Ballantine, R.M. White, S.J. Martin, A.J. Ricco, E.T. Zllers, G.C. Frye, and H. Wohltjen, *Acoustic Wave Sensors, Theory, Design, Physico-Chemical Applications,* San Diego, CA: Academic Press, 1997.

C. Campbell, *Surface Acoustic Wave Devices and their Signal Processing Applications,* San Diego, Calif.: Academic Press, 1989.

S. Datta, *Surface Acoustic Wave Devices,* Englewood Cliffs, N.J.: Prentice-Hall, 1986.

A.J. DeVries and R. Adler, Case history of a surface wave TV IF filter for color television receivers, *IEEE Proceedings,* 64, 5, 671–676, 1976.

C.S. Hartmann, D.T. Bell, and R.C. Rosenfeld, Impulse model design of acoustic surface wave filters, *IEEE Transactions on Microwave Theory and Techniques,* 21, 162–175, 1973.

C.S. Hartmann, J.C. Andle, and M.B. King, SAW notch filters, *1987 IEEE Ultrasonics Symposium,* 131–138.

J. Machui, J. Baureger, G. Riha, and I. Schropp, SAW devices in cellular and cordless phones, *1995 Ultrasonics Symposium Proceedings,* 121–130.

D.C. Malocha, Surface Acoustic Wave Applications, *Wiley Encyclopedia of Electrcial and Electronics Engineering,* 21, 117–127, 1999.

H. Matthews, *Surface Wave Filters,* New York: Wiley Interscience, 1977.

D.P. Morgan, *Surface Wave Devices for Signal Processing,* New York: Elsevier, 1985.

D.P. Morgan, Surface Acoustic Wave, *Wiley Encyclopedia of Electrical and Electronics Engineering,* 21, 127–139, 1999.

S.M. Richie, B.P. Abbott, and D.C. Malocha, Description and development of a SAW filter CAD system, *IEEE Transactions on Microwave Theory and Techniques,* 36, 2, 1988.

C.C.W. Ruppel, R. Dill, A Fischerauer, W. Gawlik, J. Machui, F. Mueller, L. Reindl, G. Scholl, I. Schropp, and K. Ch. Wagner, SAW devices for consumer applications, *IEEE Trans. on Ultrasonics, Ferroelectrics, and Frequency Control,* 40, 5, 438–452, 1993.

Special Issue on Surface Acoustic Wave Devices and Applications, *IEEE Proceedings,* 64, 5, May 1976.

Special Issue on Applications, *IEEE Trans. on Ultrasonics, Ferroelectrics, and Frequency Control,* 40, 5, Sept. 1993.

K. Tsubouchi, H. Nakase, A. Namba, and K. Masu, Full duplex transmission operation of a 2.45 GHz asynchronous spread spectrum using a SAW convolver, *IEEE Trans. on Ultrasonics, Ferroelectrics, and Frequency Control,* 40, 5, 478–482, 1993.

Further Information

The *IEEE Transactions on Ultrasonics, Ferroelectrics, and Frequency Control* provides excellent information and detailed articles on SAW technology.

The *IEEE Ultrasonics Symposium Proceedings* provides information on ultrasonic devices, systems, and applications for that year. Articles present the latest research and developments and include invited articles from eminent engineers and scientists.

The *IEEE Frequency Control Symposium Proceedings* provides information on frequency control devices, systems, and applications (including SAW) for that year. Articles present the latest research and developments and include invited articles from eminent engineers and scientists.

For additional information, see the following references:

IEEE Transaction on Microwave Theory and Techniques, vol. 21, no. 4, 1973, special issue on SAW technology.

IEEE Proceedings, vol. 64, no. 5, special issue on SAW devices and applications.

Joint Special Issue of *IEEE Transaction on Microwave Theory and Techniques* and *IEEE Transactions on Sonics and Ultrasonics,* MTT-vol. 29, no. 5, 1981, on SAW device systems.

M. Feldmann and J. Henaff, *Surface Acoustic Waves for Signal Processing,* Norwood, Mass.: Artech House, 1989.

B.A. Auld, *Acoustic Fields and Waves in Solids,* New York: Wiley, 1973.

V.M. Ristic, *Principles of Acoustic Devices,* New York: Wiley, 1983.

A. Oliner, *Surface Acoustic Waves,* New York: Springer-Verlag, 1978.

6.6 RF Coaxial Cables

Michael E. Majerus

RF/microwave cables come in many different forms and types. The one thing they all have in common is that they are used to convey RF energy between locations. For RF/microwave frequencies (500 MHz to 50 GHz) the coaxial cable is the most widely used medium for information transfer. The following sections present a short abbreviated history of the coaxial cable. This is followed by sections on the basic characteristics of a coaxial cable and some of the materials used to make a modern coaxial cable.

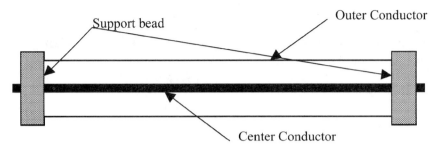

FIGURE 6.100 Air line with dielectric bead supports.

History of the Coaxial Cable

Heinrich Hertz used the coaxial cable in experiments to prove Faraday's and Maxwell's theories about the wave nature of RF energy. His use of the coaxial structure helped him demonstrate standing waves by producing minimum and maximum wave voltages at alternating quarter wavelength intervals along a coaxial line. Hertz published his investigations from 1887 to 1891.[1] Through out the 1920s many improvements in the design of coaxial lines were realized. The state-of-the-art in the mid 1920s was a coaxial cable that used a ridged outer conductor and an air dielectric with bead supports for the center conductor as illustrated in Fig. 6.100.[2-4] In the 1930s, coaxial cable was used almost exclusively in low frequency VHF and UHF radio applications. Primarily, it was used because its excellent shielding properties reduced static interference, which is a common problem at low VHF and UHF frequencies. For most high frequency applications, 1 GHz and up, a waveguide structure was used. The waveguide was the preferred option because it had lower losses, and reducing the required power directly reduces cost.

With the start of World War II, the use and development of higher frequency coaxial components and the production of coaxial cable increased at a rate that had never been seen before. This was due to the size reduction achievable with coaxial cable compared to waveguides. There was no real plan for miniaturization at the time; in the field, coaxial cable was easier to work with than waveguide. The development of radar during World War II was the driving force behind the need for higher frequency components and cable. To manage this development and procurement, the joint Army-Navy Cable Coordinating Committee was established. This committee published the specifications for cables and connectors for the Army and Navy.

Why 50 Ohms?

Why was 50 ohms selected as the reference impedance of radar and radio equipment? There are many misconceptions surrounding this bit of history. "The first coaxial dimensions just happened to define the 50 ohm reference impedance." "Fifty ohms matched up well with the antennas in use 60 years ago." In actuality, the 50 ohm reference impedance was selected from a trade-off between the lowest loss and maximum power-handling dimension for an air line coaxial cable. The optimum ratio of the outer conductor to inner conductor, for minimum attenuation in a coaxial structure with air as the dielectric, is 3.6. This corresponds to an impedance, Zo, of 77 ohms.[5] Although this yields the best performance from a loss standpoint, it does not provide the maximum peak power handling before dielectric breakdown occurs. The best power performance is achieved when the ratio of the outer conductor to inner conductor is 1.65. This corresponds to a Zo of 30 ohms.[5] The geometric mean of 77 ohms and 30 ohms is approximately 50 ohms [Eq. (6.41)]; thus, the 50 ohm standard is a compromise between best attenuation performance and maximum peak power handling in the coaxial cable.

$$50 \approx \sqrt{30 * 77} \qquad (6.41)$$

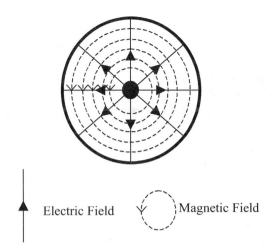

FIGURE 6.101 TEM Mode in coax.

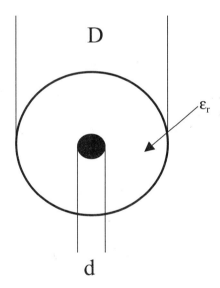

FIGURE 6.102 Cross-section of a coaxial cable showing outer diameter and inner diameter dimensions.

Characteristics of Coaxial Cable

TEM Mode of Propagation

A key feature of coaxial cable is that its characteristic impedance is very broadband. The fundamental mode setup in a coaxial cable is the TEM mode. This means that the electric and magnetic fields are transverse to the direction of propagation. Figure 6.101 shows a snapshot in time for a propagating wave.

Characteristic Impedance and Cutoff Frequency

In the TEM mode of propagation, below the cutoff frequency, the characteristic impedance of the coaxial cable is independent of frequency. The Zo of the cable is defined by the ratio of outer conductor "D" to the inner conductor "d" (see Fig. 6.102), and the relative dielectric constant (ε_r) of the dielectric material. The relationship is shown in Eq. (6.42):

$$Z_o = \frac{60}{\sqrt{\varepsilon_r}} * \ln\frac{D}{d} \qquad (6.42)$$

The approximate cutoff frequency for a cable, the frequency at which the first non-TEM mode of propagation begins, can be calculated using Eq. (6.43). At frequencies above F_{cutoff}, other propagation modes dominate and the characteristic impedance becomes frequency dependent.

$$F_{cutoff}\, GHz \approx \frac{7.51}{\sqrt{\varepsilon_r}} * \frac{1}{D+d} \qquad (6.43)$$

Electrical Length

The electrical length of a cable can be expressed in degrees of phase shift. These units are useful when using a Smith chart to synthesize a matching network. To find the electrical length, the dielectric material properties need to be known. The most common dielectric is Polytetrafluoroethylene (PTFE) with a ε_r of 2.10. Using Eq. (6.44) the velocity factor (υ_f) may be calculated: The value of υ_f represents the velocity of propagation of the

$$\upsilon_f = \frac{1}{\sqrt{\varepsilon_r}} = \frac{1}{\sqrt{2.1}} = 0.69 \qquad (6.44)$$

RF signal down the cable relative to the velocity in free space. The velocity of propagation may be calculated by multiplying the free space velocity "c" (300 million meters/sec) by the υ_f. This is shown in Eq. (6.45). Then Eq. (6.46) is used to calculate the electrical length. To get the desired cable length, take the ratio of the desired

$$\upsilon_f = \frac{1}{\sqrt{\varepsilon_r}} * c \qquad (6.45)$$

$$\frac{1}{Frequency} * \left(\upsilon_f * 3.0 \times 10^8\right) = \text{length in meters} \qquad (6.46)$$

angle or phase shift to 360° and multiply by the cable length found in Eq. (6.46). This will yield the length for the desired phase shift [(Eq. (6.47)]. As an example, a phase shift of 45° at 500 MHz, using a PTFE dielectric $\upsilon_f = 0.69$, is desired.

$$\frac{\text{Phase shift needed in deg.}}{360} * \text{wavelength} = \text{cable length} \qquad (6.47)$$

Using Eq. (6.46) a full wavelength or 360° of phase shift at 500 MHz is determined to be ~41.1 cm. Since we only need 45° of electrical length, we need to take the ratio of 45° to 360° and multiply that by 41.1 cm as shown in Eq. (6.48). This will give us the cable length of 5.137 cm, which is needed for an electrical length of phase shift of 45°.

$$\frac{45}{360} * 41.1\, cm = 5.137\, cm \qquad (6.48)$$

Cable Attenuation

The losses in a coaxial cable arise from two sources: one is from the resistance of the conductors and the currents flowing in the center and outer conductor; the second is from the conduction current in the dielectric. The conductor losses are ohmic and due to the skin effect in the conductor's increase with square root of the frequency. The skin effect loss increases proportional to the square root of the frequency.[6] Equations (6.49) and (6.50) can be used to calculate this loss. This represents the key trade-off in coaxial cable selection. The trade-off of cable cutoff frequency and the skin effect loss implies that the larger the diameter of the cable used for the frequency of operation, the better. The dielectric loss is the resistance of the dielectric material to the conduction currents, is linear with frequency, and is calculated with Eq. (6.51). The losses in the center conductor, outer conductor, and dielectric added together to give the total loss of the cable, as shown in Eq. (6.52). A third type of loss in a coaxial cable is

$$L_C\left(dB/100_{ft}\right) = \frac{.435 * \sqrt{F_{MHz}}}{Z_O * d} \qquad L_C = \text{Loss in center conductor} \qquad (6.49)$$

$$L_O\left(dB/100_{ft}\right) = \frac{.435 * \sqrt{F_{MHz}}}{Z_O * D} \qquad L_O = \text{Loss in outer conductor} \qquad (6.50)$$

$$L_D\left(dB/100_{ft}\right) = 2.78\rho * \sqrt{\varepsilon_r} * F_{MHZ} \qquad L_D = \text{Dielectric loss} \qquad (6.51)$$

$\varepsilon_r = 2.1$ for solid PTFE and $\qquad\qquad F_{MHz} = \text{Frequency in MHz}$
1.6 for expanded PTFE

$$(6.52)$$

$$\text{Loss}_{Total} = L_C + L_O + L_D$$

due to radiation. This loss is usually minimal in the coaxial cable because the outer conductor confines.[1] The losses from the major contributors add to give the approximate total losses in a coaxial cable.[7]

Maximum Peak Power

The maximum peak voltage that a cable can sustain is that voltage at which dielectric breakdown occurs. In an air filled coaxial cable, this happens when the maximum electric field, E_m, reaches about 2.9×10^4 volts per cm.[5] The maximum power (Pmax) under this condition is given by Eq. (6.53). The maximum power in an air dielectric coaxial cable is realized with a characteristic impedance of 30 ohms.

$$P_{max} = \frac{E_m^2}{480} * D^2 * \frac{\ln D/d}{\left(D/d\right)^2} \qquad (6.53)$$

Cable Types

Semirigid

The outer conductor as well as the center conductor can be made of different materials. The outer conductor will be discussed first. A type of cable called a semirigid cable has an outer conductor that is a solid sheath of extruded metal such as copper (see Fig. 6.103). This cable is the hardest to form into complex shapes and care needs to be taken when bending. The cable must be cut to size and then bent into the shape required. After shaping, heat is applied to the cable to expand the dielectric and relax the stress in the dielectric. After this step, the desired connector may be attached.

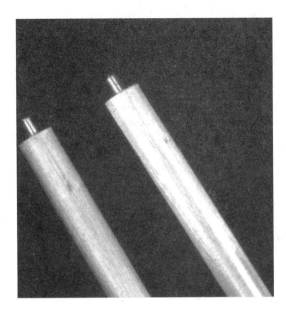

FIGURE 6.103   Photograph of copper- and tin-plated copper semi-rigid coaxial cable.

Pliable Semirigid

A variation to the semirigid cable is a type manufactured with a soft outer conductor, such as dead soft aluminum or an un-annealed copper. These types of cables are easier to form, and can often be formed without special tooling.

Flexible Cable

Another type of coaxial cable is the flexible cable, which uses a braided outer conductor, as shown in Fig. 6.104 (upper right). This cable has less phase stability with flexure than semirigid because of the limited dimensional rigidity around the dielectric material, but can be much easier to use. This type of cable typically uses a mechanical means of attaching the connector, such as a crimp-on fitting or a screw-on fitting.

A variation on the braided outer conductor cable has the exterior braid coated with solder as shown in the lower half of Fig. 6.104. This makes this cable similar to a semirigid cable. The use of the soft solder sheath makes the cable very easy to shape. The drawback is that only a limited number of bends are permitted before the cable fails.

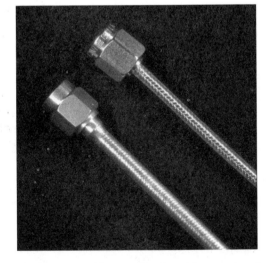

FIGURE 6.104   Photograph of a solder-coated and braided coaxial cable.

Center Conductors

The center conductor of a coaxial cable can take different forms. The most common are solid and stranded. The solid center conductor is the most prevalent. It commonly is made of copper, beryllium copper, and aluminum. Very often, the center conductor is plated with silver or tin. The stranded center conductor is not as prevalent because its performance advantage, reduced attenuation, is at the lower frequencies, 1 GHz and below. Above 1 GHz, the performance is the same as the solid center conductor.

TABLE 6.9 Common Cable Types

Cable RG #	Dielectric Type	Outer Conductor Dia.	Dielectric Dia.	Center Conductor Dia.	Attenuation per 100 ft	Average Power Max
.405/U	Solid PTFE	.0865	.0658	.0201	19 dB	.1 Kwatts
.402/U	Solid PTFE	.141	.1175	.036	11 dB	.3 Kwatts
.401/U	Solid PTFE	.250	.208	.0641	6.5 dB	.7 Kwatts
N/A	Air expanded PTFE	.141	.116	.043	9.5 dB	.55 Kwatts
N/A	Air expanded PTFE	.250	.210	.074	.55 dB	12 Kwatts

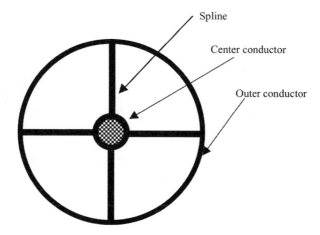

FIGURE 6.105 Cross-section of a spline coaxial cable.

Dielectric Materials

Solid PTFE

The dielectric used in most modern coaxial cable is polytetrafluoroethylene (PTFE), also known as TeflonS (a Dupont registered trademark). Solid PTFE is an extruded form, which is relatively sensitive to temperature changes. The solid PTFE has a negative phase shift with temperature,[6] i.e., as the temperature increases the electrical length decreases. Both the phase and characteristic impedance solid PTFE coaxial will change with changes in temperature.

Expanded PTFE

To improve the temperature performance of the coaxial cable, a more stable dielectric is PTFE that has been expanded with air. Another benefit due to a smaller dialectic constant of expanded PTFE is the reduction of the dielectric losses (ε_r is approximately 1.60). See Table 6.9 for examples.

Splined PTFE

To further reduce the dielectric losses, a unique cable configuration was developed: the spline cable. This dielectric sheath is made such that ridges or splines run the length of the cable, supporting the center conductor within the outer conductor, as shown in Fig. 6.105. The spline structure reduces the dielectric loss to nearly that of air. In addition, this configuration has a positive phase shift with temperature.[6]

Standard Available Coaxial Cable Types

Coaxial cable can be made in many sizes and impedances. Table 6.9 is a lists some of the standard commercially available cables.[8,9] The power maximums listed are average power ratings. This is the maximum average power that can be maintained without damaging the dielectric.

TABLE 6.10   Common Coaxial Connectors

Connector Type	Cutoff Frequency	Mating Torque
BNC	4.0 GHz	N/A
SMB	4 GHz	N/A
SMC	10 GHz	30–50 in-oz
TNC	15 GHz	12–15 in-lbs
Type-N	18 GHz	12–15 in-lbs
7 mm	18 GHz	12–15 in-lbs
SMA	18 GHz	7–10 in-lbs
3.5 mm	26.5 GHz	7–12 in-lbs
2.9 mm	46 GHz	8–10 in-lbs
2.4 mm	50 GHz	8–10 in-lbs

Connectors

The coaxial cable is of little use if the RF energy is not efficiently coupled into it, thus, high quality connectors are necessary. The connector usually defines the usable frequency range of a coaxial cable. In most, but not all cases, the connector will have a cutoff frequency lower than the cable itself. Table 6.10 is a list of common RF connectors and their cutoff frequencies.

Each type of connector can be attached to a cable in several ways: direct solder, crimp-on, and screw-on. Figure 6.106 indicates a direct solder connection where the connector body is soldered onto the outer conductor.[9] This makes the most reliable connection. The drawback to this method is that it is the most demanding on assembly techniques to prevent damage to the dielectric. A second method is "crimp-on" as shown in Fig. 6.107.[8,9] This type of connection is easy to make, but is not durable under hard usage. The last method, "screw-on" is used for lower frequency cables (under 1 GHz). This method uses a swage that tightens onto the outer conductor (typically a braided outer conductor), as shown in Fig. 6.108.[10] It is important to understand the proper methods of connector attachment to realize optimum performance. Proper assembly tools and techniques must be used.

Summary

The coaxial cable remains the backbone for signal interconnects between equipment and systems since the early 1920s due to its desirable features. The ease of use, relatively low cost, and broad operating

FIGURE 6.106   Photograph of a solder-on, connector-to-cable assembly.

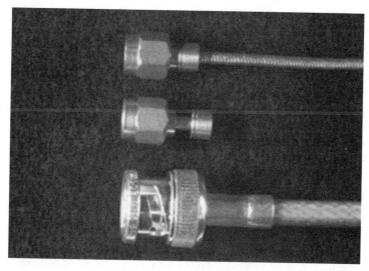

FIGURE 6.107   Photograph of a crimp-on, connector-to-cable assembly.

FIGURE 6.108   Photograph of a swage-on, connector-to-cable assembly.

bandwidth ensure its continued popularity. The wireless world is made possible through the lowly coaxial cable. This basic component brings the wireless technology to us.

References

1. J.H. Bryant, Coaxial Transmission Lines, Related Two-conductor Transmission lines, Connectors, and Components: A U.S. Historical Perspective, *IEEE Trans. Microwave Theory and Techniques*, MTT-32, 970–983, Sept. 1984.
2. R.S. Ohl, Means for transferring high frequency energy, U.S. Patent 1 619 882, Mar. 8 1927 (filed 1924).
3. H.A. Affel, E.I. Green, Concentric conducting system, U.S. Patent 1 781 092, Nov 11 1930 (filed 1929).

4. L. Espenscheid, H.A. Affel, Concentric conducting system,, U.S. Patent 1 835 031, Dec. 8 1931 (filed 1929).
5. G. Matthaei, L. Young, E.M.T. Jones, Properties of some common microwave filter elements, in *Microwave Filters, Impedance-Matching Networks, and Coupling Structures,* Artech House, Norwood, MA, 1985, 165–168.
6. P.H. Smith, *Electronic Applications of the Smith Chart,* 2nd edition, Noble, Atlanta, 1995, 38.
7. Astrolab Catalog.
8. Precision Tube Co. Coaxitube Division Catalog, 1993.
9. M/A-COM Coaxial Connectors, Adapters, Tools and Accessories. Catalog, 1995.
10. Pasternack Catalog.

6.7 Coaxial Connectors

David Anderson

Coaxial connectors are one of the fundamental tools of microwave technology and yet they are taken for granted in many instances. A good understanding of coaxial connectors, both electrically and mechanically, is required to utilize them fully and derive their full benefits.

History[1]

The only coaxial connector in general use in the early 1940s was the UHF connector, which is still manufactured and used today. E. C. Quackenbush of the American Phenolic Company (later Amphenol) in Chicago developed it. The UHF connector was not deemed suitable for higher frequency use, and the connector committee undertook to develop one in 1942. The result was the original type N (Navy) connector. This connector does not have a constant impedance from interface to interface. The three main coaxial parts, center conductor, dielectric, and outer conductor or shell were held together by a system of steps and shoulders. These mechanical discontinuities represent electrical discontinuities and limit performance. However, modified designs were made in an effort to improve microwave performance and still maintain interface-mating capability.

Several others, none of which appears to have received microwave design attention, followed the type N connector design. The type BNC, a "baby type N," is somewhat scaled down in diameter and provides a twist-lock coupling mechanism. The TNC followed the BNC in 1956.

Development of the SMA connector was driven by the need to provide compact, low-reflection connections for microwave components. By September 1962, the market for connectors as well as other microwave/RF components was emerging. Omni Spectra was manufacturing connectors for use in its components. Based on demand, the decision was made to offer connectors to help promote the use of their components. Thus, a microwave company was in the business of selling connectors. OSM, for Omni Spectra Miniature, was adopted as the trademark for this first microwave connector. The name SMA was adopted in 1968 as the military designation for this connector under specification MIL-C-39012.

Definitions[2]

There are two distinct categories of connectors, sexless and sexed. A sexless connector has two equal mating halves. A sexed connector has either a female or male configuration to form a mated pair. All sexed connectors are of the pin and socket type, with a male contact or pin and a female contact or socket. Two basic types of female sockets are used for precision connectors, either slotted or slotless.

---

[1] Bryant, John H., Coaxial Transmission Lines, Related Two-Conductor Transmission Lines, Connectors, and Components: A U.S. Historical Perspective, *IEEE MTT Transactions,* 9, 970, 1990.

[2] Maury Jr., Mario A., Microwave Coaxial Connector Technology: A Continuing Evolution, *Microwave Journal,* State of the Art Reference, 39, 1990.

To insure mechanical mating compatibility and electrical repeatability, the mechanical configuration, dimensions, and tolerances of a connector and its mating properties must be clearly defined. This set of data defines the connector's interface.

A connector's reference plane is defined as the outer conductor-mating surface of a coaxial connector. It is desirable to have both the outer and center conductors mating surface coplanar. When this occurs, the connectors are referred to as coplanar. Typical coplanar connectors are 14 mm, 7 mm, 3.5 mm, and SMA connectors. Typical noncoplanar connectors are Type-N, TNC, and BNC.

Connector coupling describes how the outer conductors are connected when the connector is mated. Coupling is either threaded, bayonet, or snap on. Precision connectors generally use threaded coupling. The best example of the bayonet or twist coupling is the BNC.

Dielectric styles in connectors are of two types, air or solid dielectric. Air dielectric simplifies connector construction and generally is used on precision connectors so accurate standards can be created. Solid dielectrics such as Teflon are used two configurations, flush and overlapping. Overlapping dielectrics are generally used for higher power applications and to prevent voltage breakdown.

Design[3]

Two equations define the general design parameters for coaxial connectors. One defines the coaxial characteristic impedance and the other defines the TE_{11} mode cutoff frequency.

Characteristic Impedance:

$$Z_o = 60 * \left(\sqrt{\mu_r} / \sqrt{\varepsilon_r} \right) * \ln b/a \; \Omega \tag{6.54}$$

Cutoff Frequency TE_{11} mode:

$$f_c = \frac{c}{\pi(a+b)\sqrt{\mu_r \varepsilon_r}} \; Hz \tag{6.55}$$

where: b is the inner radius of the outer conductor.
a is the radius of the center conductor.
ε_r is the relative permittivity of the transmission medium.
μ_r is the relative permeability of the transmission medium.
c is the velocity of light.

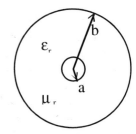

The frequency range of any connector is limited by the occurrence of the first circular waveguide mode in the coaxial structure. Decreasing the diameter of the outer conductor increases the highest usable frequency. Filling the air space with dielectric lowers the highest usable frequency and increases the system loss.[4]

Connector Care[5]

Dimensions of microwave connectors are small and some of the mechanical tolerances are very precise. Seemingly minor defects, damage and dirt, can significantly degrade repeatability and accuracy. In

[3] Rizzi, Peter A., *Microwave Engineering Passive Circuits*, Prentice-Hall, Englewood Cliffs, 1988, chap. 5.
[4] Microwave Test Accessories Catalog, Hewlett Packard, 1991, 80.
[5] Microwave Connector Care, Hewlett Packard, April 1986, Part One.

addition, the mating surfaces of most precision connectors are gold plated over a beryllium-copper alloy. This makes them very susceptible to mechanical damage due to the comparative softness of the metals.

Among the most important general care and usage recommendations for all types of microwave connectors are:

1. Connectors must be kept clean and the mating plane surfaces protected from harm during storage.
2. Connectors should be inspected visually before every connection and damaged connectors discarded immediately.
3. Connectors should be cleaned first with compressed air. Solvent should never be sprayed into a connector; use a cotton swab or lint-free cloth and the least amount of solvent possible. The plastic support beads should not come in contact with solvent.
4. Connectors should be inspected mechanically with a connector gage before being used for the first time and periodically thereafter.
5. Connectors should be aligned carefully when mated and the preliminary connection should only be made by lightly turning the connector nut to pull the connectors together. The final connection should be made by using a torque wrench.
6. Turning one connector body relative the other should never be done to make connections and disconnections. This is extremely harmful and can occur whenever the connector body rather than the connector nut alone are turned.

Handling

Microwave connectors must be handled carefully, inspected before each use, and stored for maximum protection.

The connector mating surface should never be touched and should never be placed in contact with a foreign surface, such as a table. Natural skin oils and microscopic particles of dirt are easily transferred to the connector interface and are very difficult to remove. In addition, damage to the plating and to the mating plane surfaces occurs readily when the interface comes into contact with any surface. Connectors should not be stored with the contact end exposed. Plastic end caps are provided with precision connectors and these should be retained after unpacking and placed over the connector ends before storage. Connectors should never be stored loose in a box or in a desk or bench drawer. Careless handling of this kind is the most common cause of connector damage during storage.

Visual Inspection

Visual inspection and, if necessary, cleaning should be done every time a connection is made. Metal and metal by-product particles from the connector threads often find their way onto the mating plane surfaces when a connection is disconnected, and even one connection made with a dirty or damaged connector can damage both connectors beyond repair.

The connectors should be examined prior to use for obvious defects or damage, e.g., badly worn plating, deformed threads, or bent, broken, or misaligned center conductors. Connector nuts should turn smoothly and be free of burrs or loose metal particles. Connectors with these types of problems should be discarded or repaired.

The mating plane surfaces of connectors should also be examined. Flat contact between the connectors at all points on the mating plane surfaces is essential for a good connection. Therefore, particular attention should be paid to deep scratches or dents and to dirt and metal or metal by-product particles on the connector mating plane surfaces. The mating plane surfaces of the center and outer conductors should be examined for bent or rounded edges and also for any signs of damage due to excessive or uneven wear or misalignment.

If a connector displays deep scratches or dents, or particles on the mating plane surfaces, clean it and inspect it again. Damage or defects of these kinds — dents or scratches deep enough to displace metal on the mating plane surface of the connector — may indicate the connector is damaged and should not be used.

Cleaning

Careful cleaning of all connectors is essential to assure long, reliable connector life, to prevent accidental damage to connectors, and to obtain maximum measurement accuracy and repeatability. However, this one step is most often neglected.

Loose particles on the connector mating plane surfaces can usually be removed with a quick blast of compressed, dry, air. Clean dry air cannot damage the connectors or leave particles or residues behind and should therefore be employed as the first attempt at connector cleaning. Dirt and other contaminants that cannot be removed with compressed air can often be removed with either a cotton swab or lint-free cleaning cloth and a solvent. The least possible amount of solvent should be used and neither the bead nor dielectric supports in the connectors should come in contact with the solvent. A very small amount of solvent should be applied to a cotton swab or a lint-free cleaning cloth. The connector should then be wiped as gently as possible. The precaution should be taken to always use solvents in a well-ventilated area, avoid prolonged breathing of solvent vapors, and avoid contact of solvents with the skin.

The threads of the connectors should be cleaned first since a small amount of metal wears off the threads every time a connection or a disconnection is made, and this metal often finds its way onto the mating plane surfaces of the connectors. Then, wet a clean swab, and clean the mating plane surfaces of the connector.

Mechanical Inspection

Even a perfectly clean unused connector can cause trouble if it is mechanically out of specification. Since the critical tolerances in precision microwave connectors are very small, using a connector gage is essential.

Before using any connector for the first time, inspect it for mechanical tolerance by using a connector gage. Connectors should be gaged after that depending upon their usage. In general, connectors should be gaged whenever visual inspection or electrical performance suggests that the connector interface may be out of specification, for example due to wear or damage. Connectors on calibration and verification devices should also be gaged whenever they have been used on another system or piece of equipment.

A different gage is required for each type of connector. See Fig. 6.109 for a typical connector gage and Fig. 6.110 for a master gage. Sexed connectors require two gages for female and male. Every connector gage requires a master gage for zeroing the gage. Care is necessary in selecting a connector gage to measure precision microwave connectors. Some have a very strong plunger springs, strong enough, in some cases, to push the center conductor back through the connector, damaging the connector itself.

The critical dimension to be measured is the position, recession or setback, of the center conductor relative to the outer conductor mating plane. Mechanical specifications for connectors specify a maximum and minimum distance the center conductor can be positioned with respect to the outer conductor-mating plane. Before gauging any connector, consult the mechanical specifications provided with the connector or the device itself.

Before using any connector gage it must be inspected, cleaned and zeroed. Inspect and clean the gage and master gage for dirt and particles exactly as the connector was cleaned and inspected. Dirt on the gage or master gage will make the gage measurements of the connectors incorrect and can transfer dirt to the connectors themselves, damaging the connectors during gauging or connection. Zero the gage using the correct connector gage master gage. Hold the gage by the plunger barrel only. Slip the gage master gage

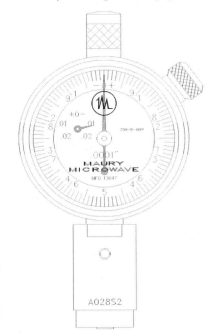

FIGURE 6.109 Typical connector gauge.

into the circular housing of the connector gage. Carefully bring the gage and the master gage together applying only enough pressure to the gage and the master gage to result in the dial indicator settling at a reading. Gently rock the two surfaces together to verify they have come together flatly. The gage indicator should now line up exactly with the zero mark of the gage. If it does not, inspect and clean both the gage and master gage again.

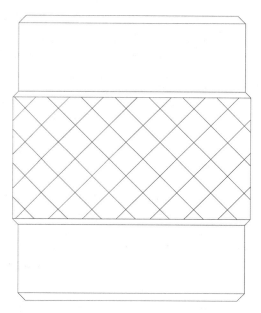

FIGURE 6.110 Typical connector master gauge.

Measuring the recession of the center conductor behind the outer conductor mating plane in a connector is performed in exactly the same manner as zeroing the gage. If the connector has a retractable sleeve or sliding connector nut, extend the sleeve or nut fully. This makes it easier to keep the gage centered in the connector. Hold the gage by the plunger barrel and slip the gage into the connector so that the gage plunger rests against the center conductor. Carefully bring the gage into firm contact with the outer conductor mating plane. Apply only enough pressure to the gage as results in the gage indicator settling at a reading. Gently rock the connector gage within the connector, to insure the gage and the outer conductor have come together flatly. Read the recession or protrusion from the gage dial. See Fig. 6.111 for a typical 7 mm example. For maximum accuracy, measure the connector several times and average the readings. Rotate the gage relative to the connector between each measurement.

Connections

Skill is essential in making good connections. The sensitivity of modern test instruments and the mechanical tolerances of precision microwave connectors are such that slight errors that once went unnoticed now have a significant effect on measurements.

Before making any connections, inspect all connectors visually, clean them if necessary, and use a connector gage to verify that all the center conductors are within specifications. If connections are made to static-sensitive devices, avoid electrostatic discharge by wearing a grounded wrist strap and grounding yourself and all devices before making any connections.

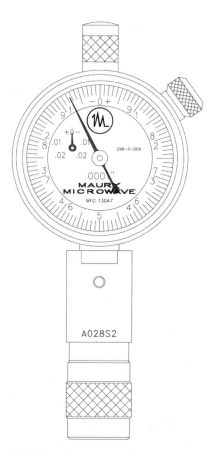

FIGURE 6.111 Typical 7 mm connector gauge inspection.

Careful alignment of the connectors is critical in making a good connection, both to avoid damaging connectors and devices and to assure accurate measurements. As one connector is connected to the other, care should be taken to notice any signs that the two connectors are not aligned perfectly. If misalignment has occurred, stop and begin again.

Alignment is especially important for sexed connectors to avoid bending or breaking the contact pins. The center pin on the male connector must slip concentrically into the contact fingers of the female connector and requires great care aligning the two connectors as they are mated.

When they have been aligned, the center conductors must be pushed straight together, not twisted or screwed together and only the connector nut, not the connector body, should be rotated to make the connection. Slight resistance is generally felt as the center conductors mate; very light finger pressure, 2 inch-ounces of torque, is enough.

When the preliminary connection has been made, use a torque wrench to make the final connection. The connection should be tightened only until the break point of the wrench is reached, when the handle gives way at its internal pivot point.

Disconnections

Disconnect connectors by grasping the body firmly to prevent it from rotating. Then loosen the connector nut that was tightened to make the connection. If necessary, use the torque wrench or an open-end wrench to start the process, but leave the connection finger tight. At all times support the connectors and the connection to avoid putting lateral or bending forces on the connector mating planes.

Complete the process by disconnecting the connector nut completely. As in making the connections, turn only the connector nut. Never allow the connectors to rotate relative to each other. Twisting the connection can damage the connector by damaging the center conductors or the interior components. It can also scrape the plating off the male contact pin or even unscrew the male or female contact pin slightly from its interior mounting.

Type N

Precision Type-N connectors are similar in size to 7 mm connectors but relatively inexpensive. They are more rugged, and were developed for severe operating environments or applications in which many connections and disconnections must be made. They are among the most popular general-purpose connectors used in the DC to 18 GHz frequency range.

Unlike precision 7 mm connectors, Type-N connectors are sexed connectors. The male contact pin slides into the female contact fingers and electrical contact is made by the inside surfaces of the tip of the female contact on the sides of the male contact pin. The position of the center conductor in the male connector is defined as the position of the shoulder of the male contact pin, not the position of the tip.

Type-N connectors differ from other connectors in that the outer conductor mating plane is offset from the mating plane of the center conductors. The outer conductor sleeve in the male connector extends in front of the shoulder of the male contact pin. When the connection is made, this outer conductor sleeve fits into a recess in the female outer conductor behind the tip of the female contact fingers.

Type-N connectors should not be used if there is a possibility of interference between the shoulder of the male contact pin and the tip of the female contact fingers when the connectors are mated. In practice, this means that no Type-N connector pair should be mated when the separation between the tip of the female contact fingers and the shoulder of the male contact pin could be less than zero when the connectors are mated. Care should be taken when gauging Type-N connectors to avoid damage.

As Type-N connectors wear, the protrusion of the female contact fingers generally increases, due to wear of the outer conductor mating plane inside the female connector. This decreases the total center conductor contact separation and must be monitored carefully. At lower frequencies the effects of wide contact separation are small; only at higher frequencies does the contact separation become important.

Type-N connectors are available in both 50 Ω and 75 Ω impedance. However 75 Ω Type-N connectors differ from 50 Ω Type-N connectors most significantly in that the center conductor, male contact pin, and female contact hole are smaller. Therefore, mating a male 50 Ω Type-N connector to a female 75 Ω Type-N connector will destroy the female 75 Ω connector by spreading the female contact fingers apart permanently or even breaking them. If both 50 Ω and 75 Ω Type-N connectors are among those on the devices you are using, mark the 75 Ω Type-N connectors to insure they are never mated with any 50 Ω Type-N connectors.

Type-N Specifications

Frequency Range: DC to 18 GHz
Impedance: 50 Ω and 75 Ω
Mating Torque: 12 in-lbs.
Female Socket: 0.207   +0.000   −0.010 inches
Male Pin: 0.207   +0.010   −0.000 inches

Type-N Dimensions

See Figs. 6.112 through 6.115 for details.

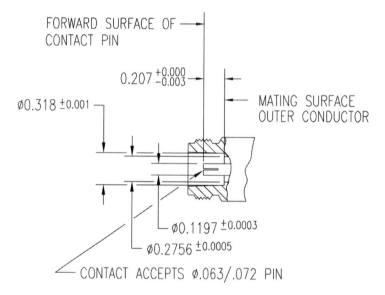

FIGURE 6.112   Type-N female 50 Ω interface.

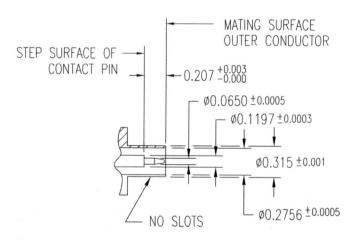

FIGURE 6.113   Type-N male 50 Ω interface.

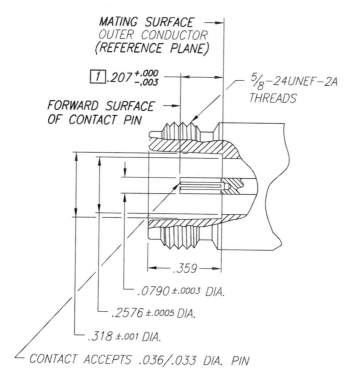

FIGURE 6.114 Type-N female 75 Ω interface.

BNC

The BNC is a general-purpose connector for low frequency uses. It is a dielectrically loaded, sexed connector. The male contact pin slides into the female contact fingers and electrical contact is made by the inside surfaces of the tip of the female contact fingers on the sides of the male contact pin. BNC connectors are available in both 50 Ω and 75 Ω versions, and the two versions will mate successfully with each other.

BNC Specifications

Frequency Range:	DC to 4 GHz		
Impedance:	50 Ω and 75 Ω		
Mating Torque:	None		
Female Socket:	0.206	+0.000	−0.003 inches
Female Dielectric Top:	0.208	+0.000	−0.008 inches
Female Dielectric Bottom:	0.000	+0.008	−0.000 inches
Male Pin:	0.209	+0.003	−0.000 inches
Male Dielectric Top:	0.008	+0.004	−0.000 inches
Male Dielectric Bottom:	0.212	+0.006	−0.000 inches

BNC Dimensions

See Figs. 6.116 through 6.119 for details.

TNC

The TNC is a dielectric loaded sexed connector with many different configurations. The male contact pin slides into the female contact fingers and electrical contact is made by the inside surfaces of the tip of the female contact fingers on the sides of the male contact pin.

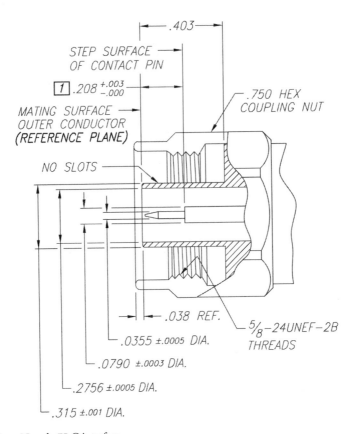

FIGURE 6.115 Type-N male 75 Ω interface.

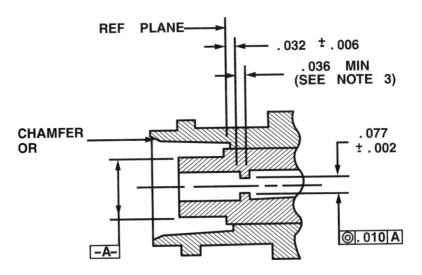

FIGURE 6.116 BNC female 50 Ω outer conductor.

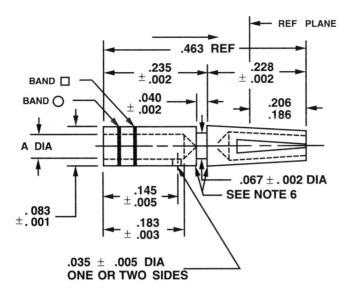

FIGURE 6.117   BNC female 50 Ω center conductor.

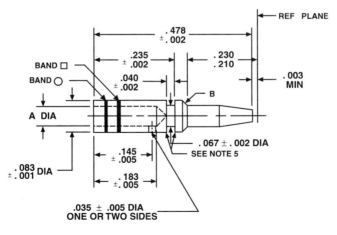

FIGURE 6.118   BNC male 50 Ω outer conductor.

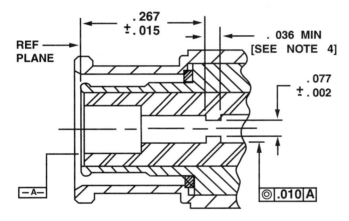

FIGURE 6.119   BNC male 50 Ω center conductor.

Table 6.11 TNC Connector Compatibility

A. Mating these TNC connectors together will result in non-contacting outer conductors.
B. These TNC connectors should not be mixed except in cases where one connector has been chosen as a test connector and it is characterized on a network analyzer for error corrected measurements.
C. The male contact pin interface of this TNC connector specification has not been fully defined.

Female and male TNC connectors of the same specification are designed to provide the best-matched condition when mated together. When female and male TNC connectors of different specifications are mated, less than optimum electrical performance can be experienced even though they are mechanically compatible. Table 6.11 displays TNC connector compatibility.[6]

TNC Specifications

Frequency Range:	DC to 18 GHz		
Impedance:	50 Ω		
Mating Torque:	12 in-lbs.		
Female Socket:	0.208	+0.000	−0.005 inches
Female Dielectric Top:	0.208	+0.000	−0.008 inches
Female Dielectric Bottom:	0.000	+0.000	−0.004 inches
Male Pin:	0.209	+0.005	−0.000 inches
Male Dielectric Top:	0.000	+0.004	−0.000 inches
Male Dielectric Bottom:	0.209	+0.005	−0.000 inches

TNC Dimensions

See Figs. 6.120 and 6.121 for details.

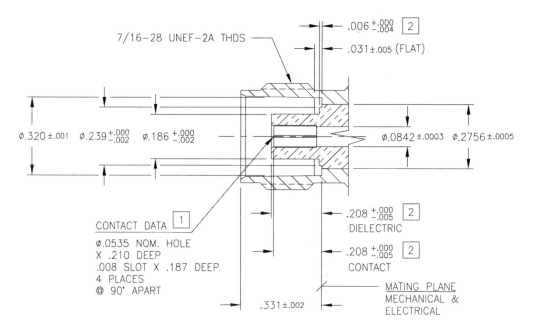

FIGURE 6.120 Typical TNC female interface.

---

[6] TNC Compatibility Chart, 5E-057, Maury Microwave, Dec. 1998.

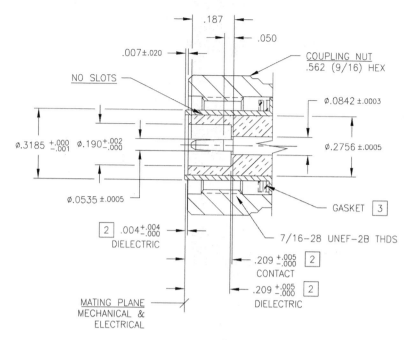

FIGURE 6.121 Typical TNC male interface.

SMA

SubMiniature Type-A or SMA is one of the most popular microwave connectors. Because of their smaller size, SMA connectors can be used at higher frequencies than Type-N connectors. The most common application is for semirigid cable and components that are connected only a few times because of the fragility of the outer connector wall.

SMA connectors are not precision devices. They are not designed for repeated connections and disconnections. SMA connectors wear out quickly and are often found to be out of specification even before they have been used. They are used most often as "one-time-only" connectors in internal component assemblies and in similar applications in which few connections or disconnections will be made.

The SMA is a dielectrically loaded, sexed connector. The male contact pin slides into the female contact fingers and electrical contact is made by the inside surfaces of the tip of the female contact fingers on the sides of the male contact pin. The mechanical specifications for the SMA connector gives a maximum and a minimum recession of the shoulder of the male contact pin and a maximum and minimum recession of the tip of the female contact fingers behind the outer conductor mating plane. An SMA connector will mate with 3.5 mm and 2.92 mm connectors without damage.

SMA Specifications

Frequency Range:	DC to 18 GHz
Impedance:	50 Ω
Mating Torque:	5 in-lbs.
Female Socket:	0.000 +0.005 −0.000 inches
Female Dielectric:	0.000 ±0.002 inches
Male Pin:	0.000 +0.005 −0.000 inches
Male Dielectric:	0.000 ±0.002 inches

SMA Dimensions

See Figs. 6.122 through 6.125 for details.

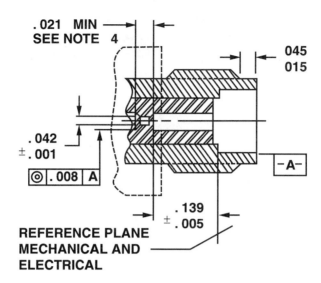

FIGURE 6.122   SMA female outer conductor.

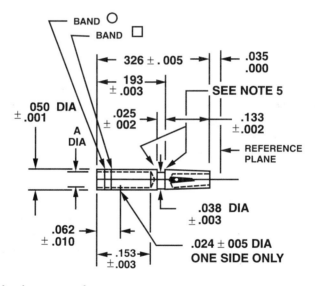

FIGURE 6.123   SMA female center conductor.

7-16

The 7-16 connector is a air dielectric loaded, sexed connector. The male contact pin slides into the female contact fingers and electrical contact is made by the inside surfaces of the tip of the female contact fingers on the sides of the male contact pin. The 7-16 is intended as a replacement connector for Type-N connectors in high power applications.

7-16 Specifications

Frequency Range:	DC to 7 GHz	
Impedance:	50 Ω	
Mating Torque:	20 in-lbs.	
Female Socket:	0.0697 +0.012	−0.000 inches
Male Pin:	0.0697 +0.000	−0.010 inches

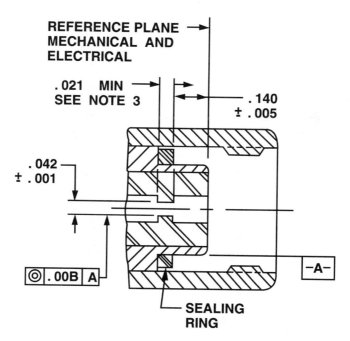

FIGURE 6.124 SMA male outer conductor.

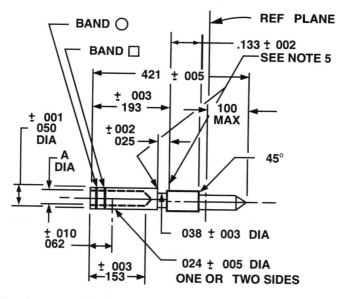

FIGURE 6.125 SMA male center conductor.

7-16 Dimensions

See Figs. 6.126 and 6.127 for details.

7 mm

Precision 7 mm connectors, among them APC-7® (Amphenol Precision Connector-7 mm) connectors, are used in the DC to 18 GHz frequency band and offer the lowest SWR and the most repeatable

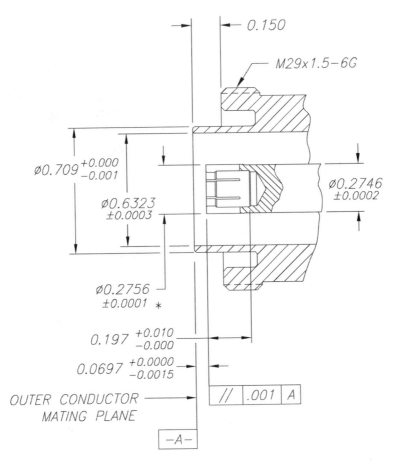

FIGURE 6.126   7-16 female interface.

connections of any 7 mm connector type. Development of these connectors was begun by Hewlett Packard (Agilent Technologies) in the mid-1960s and improved upon Amphenol Corporation.

Precision 7 mm connectors are air dielectric devices. Only a plastic support bead inside the connector body supports the center conductor. The conductors are generally made of beryllium copper alloy plated with gold.

Precision 7 mm connectors are durable and are suitable for many connections and disconnections. Therefore they are widely used in test and measurement applications requiring a high degree of accuracy and repeatability.

Precision 7 mm connectors are generally designed for use as sexless connectors, able to mate with all other precision 7 mm connectors. There is no male or female and contact between the center is made by replaceable inserts called collets designed to make spring-loaded butt contact when the connector is torqued. Small mechanical differences do sometimes exist between precision 7 mm connectors made by different manufacturers and occasionally these differences can cause difficulty in making connections. Always inspect all connectors mechanically, using a precision connector gage, to insure the connectors meet their critical interface specifications.

In precision 7 mm connectors, contact between the center conductors is made by spring-loaded contacts called collets. These protrude slightly in front of the outer conductor mating plane when the connectors are disconnected. When the connection is tightened, the collets are compressed into the same plane as the outer conductors. For this reason two mechanical specifications are generally given for precision 7 mm connectors:

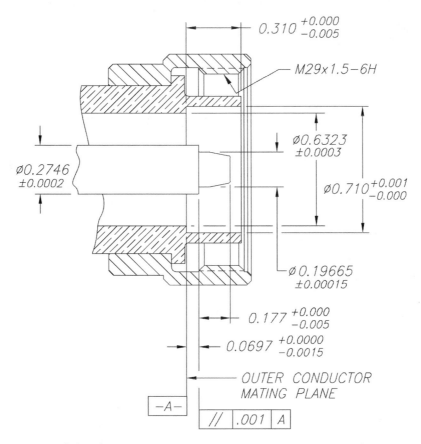

FIGURE 6.127 7-16 male interface.

1. The maximum and minimum allowable recession of the center conductor behind the outer conductor mating plane with the center conductor collet removed. The critical mechanical specification is the recession or setback of the center conductor relative to the outer conductor mating plane with the center conductor collet removed.
2. The maximum and minimum allowable protrusion of the center conductor collet in front of the outer conductor mating plane with collet in place. No protrusion of the center conductor in front of the outer conductor mating plane is ever allowable and sometimes a minimum recession is required.

The center conductor collet should also spring back immediately when pressed with a blunt plastic rod or with the rounded plastic handle of the collet removing tool.

Nominal specifications for precision 7 mm connectors exist, but the allowable tolerances differ from manufacturer to manufacturer and from connector to connector. Before gaging any precision 7 mm connector, consult the manufacturer's mechanical specifications provided with the connector.

7 mm Specifications

Frequency Range: DC to 18 GHz
Impedance: 50 Ω
Mating Torque: 12 in-lbs.
Contact: 0.0000 +0.0000 −0.0015 inches

7 mm Dimensions

See Figs. 6.128 and 6.129 for details.

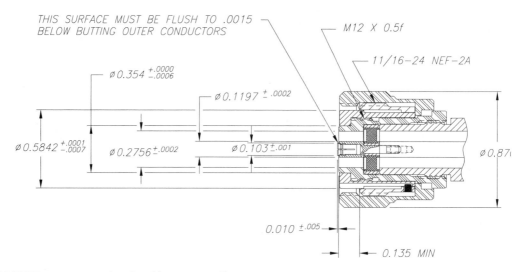

FIGURE 6.128   7 mm interface (sleeve retracted).

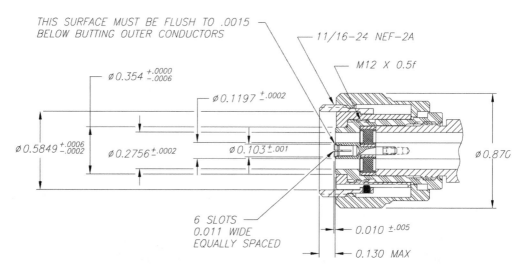

FIGURE 6.129   7 mm interface (sleeve extended).

3.5 mm

Precision 3.5 mm connectors, also known as APC-3.5® (Amphenol Precision Connector-3.5 mm) connectors, were developed during the early 1970s jointly by Hewlett Packard (Agilent Technologies) and Amphenol Corporation. The goal was to produce a durable high-frequency connector that could mate with SMA connectors, exhibit low SWR and insertion loss, and be mode free up to 34 GHz.

Unlike SMA connectors, precision 3.5 mm connectors are air dielectric devices. That is, air is the insulating dielectric between the center and outer conductors. A plastic support bead inside the connector body supports the center conductor. APC- 3.5 mm connectors are precision devices. Therefore, they are more expensive than SMA connectors, but are durable enough to permit repeated connections and disconnections. A 3.5 mm connector will mate with either a SMA or 2.92 mm connectors.

3.5 mm Specifications

Frequency Range: DC to 34 GHz
Impedance: 50 Ω

Mating Torque:	8 in-lbs.		
Female Socket:	0.000	+0.003	–0.000 inches
Male Pin:	0.000	+0.003	–0.000 inches

3.5 mm Dimensions

See Figs. 6.130 and 6.131 for details.

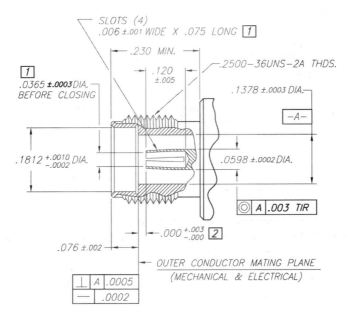

FIGURE 6.130   3.5 mm female interface.

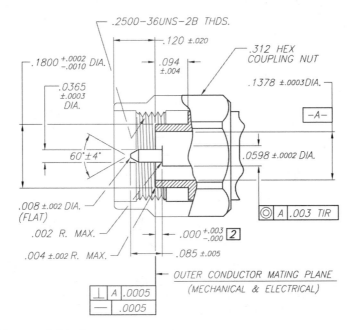

FIGURE 6.131   3.5 mm male interface.

2.92 mm

Precision 2.92 mm connectors are air dielectric loaded, sexed connectors. The male contact pin slides into the female contact fingers and electrical contact is made by the inside surfaces of the tip of the female contact fingers on the sides of the male contact pin. Air is the insulating dielectric between the center and outer conductors. A plastic support bead inside the connector body supports the center conductor. A 2.92 mm connector will mate with SMA, 3.5 mm, or 2.4 mm connectors.

2.92 mm Specifications

Frequency Range:	DC to 40 GHz		
Impedance:	50 Ω		
Mating Torque:	8 in-lbs.		
Female Socket:	0.000	+0.003	−0.000 inches
Male Pin:	0.000	+0.003	−0.000 inches

2.92 mm Dimensions

See Figs. 6.132 and 6.133 for details.

2.4 mm

Precision 2.4 mm connectors are air dielectric loaded, sexed connectors. The male contact pin slides into the female contact fingers and electrical contact is made by the inside surfaces of the tip of the female contact fingers on the sides of the male contact pin. Air is the insulating dielectric between the center and outer conductors. A plastic support bead inside the connector body supports the center conductor. A 2.4 mm connector will mate with SMA, 3.5 mm, or 2.92 mm connectors.

2.4 mm Specifications

Frequency Range:	DC to 50 GHz		
Impedance:	50 Ω		
Mating Torque:	8 in-lbs.		
Female Socket:	0.000	+0.002	−0.000 inches
Male Pin:	0.000	+0.002	−0.000 inches

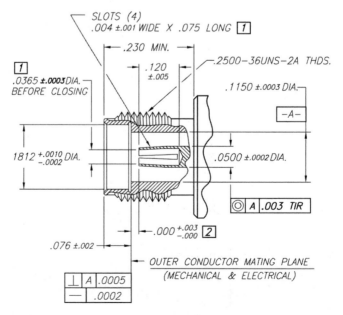

FIGURE 6.132   2.92 mm female interface.

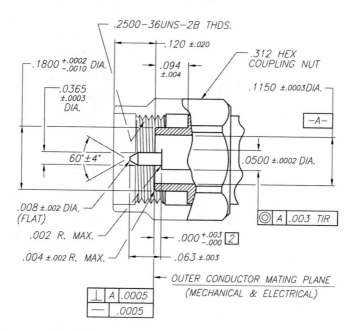

FIGURE 6.133   2.92 mm male interface.

2.4 mm Dimensions

See Figs. 6.134 and 6.135 for details.

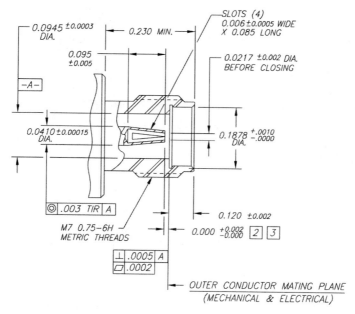

FIGURE 6.134   2.4 mm female interface.

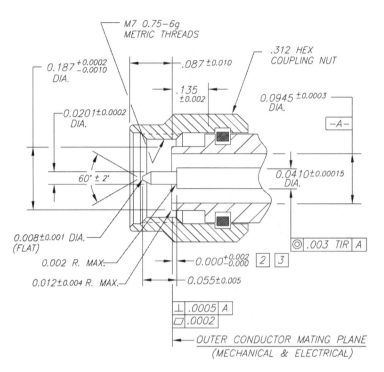

FIGURE 6.135   2.4 mm male interface.

References

1. Bryant, John H., Coaxial Transmission Lines, Related Two-Conductor Transmission Lines, Connectors, and Components: A U.S. Historical Perspective, *IEEE MTT Transactions*, 9, 970, 1990.
2. Maury Jr., Mario A., Microwave Coaxial Connector Technology: A Continuing Evolution, *Microwave Journal*, State of the Art Reference, 39, 1990.
3. Rizzi, Peter A., *Microwave Engineering Passive Circuits*, Prentice-Hall, Englewood Cliffs, NJ, 1988, chap. 5.
4. Microwave Test Accessories Catalog, Hewlett Packard, 1991, 80.
5. Microwave Connector Care, Hewlett Packard, April 1986, Part One.
6. TNC Compatibility Chart, 5E-057, Maury Microwave, Dec. 1998.

6.8   Antenna Technology

James B. West

This chapter is a brief overview of contemporary antenna types used in cellular, communication links, satellite communication, radar, and other microwave and millimeter wave systems. In this presentation microwave is presumed to cover the frequency spectrum from 800 MHz to 94 GHz. A discussion of Maxwell's equations of electromagnetic theory and the wave equation, which together mathematically describe nonionizing electrodynamic wave propagation and radiation are described in detail in Chapter 9. The reader is referred to the literature for detailed electromagnetic analysis of the multitude of antennas used in contemporary microwave and millimeter wave systems.

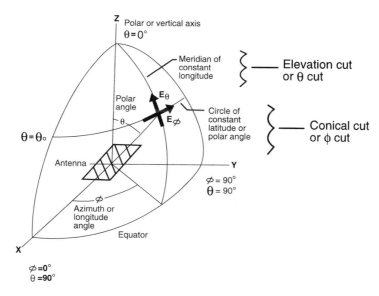

FIGURE 6.136 Coordinate system for radiation-pattern measurement.

Fundamental Antenna Parameter Definitions[1-4]

Electromagnetic radiation: The omission of energy from a device in the form of electromagnetic waves.

Radiation pattern: A graphical or mathematical description of the radiation properties of an antenna as a function of space coordinates. The standard (r, θ, φ) spherical coordinate system is typically used, as shown in Fig. 6.136. Radiation properties include **radiation intensity (Watt/solid angle), radiation density (Watt/m²), gain, directivity, radiated phase**, and **polarization parameters**, all of which are discussed in subsequent paragraphs. Radiated power, rather than electric field, is commonly used at microwave frequencies since radiated power is readily measured with contemporary antenna metrology equipment. As an example, a **power radiation pattern** is an expression of the variation of received or transmitted power density as quantified at a constant radius about the antenna. Power patterns are often graphically depicted as normalized to their main beam peak.

Radiation pattern cuts are two-dimensional cross-sections of the three-dimensional power radiation pattern. A cut in the plane of the theta unit vector is the **theta cut**, while a pattern cut in the phi unit vector direction is the **phi cut**, as shown in Fig. 6.136. The **principal plane cuts** are two dimensional, orthogonal antenna cuts taken in the E field and H field planes of the power radiation patterns, as illustrated in Fig. 6.138.

Isotropic radiator: A hypothetical, lossless antenna having equal radiation in all directions. The three-dimensional radiation pattern of an isotropic radiator is a sphere.

Directional antenna: An antenna that has a peak sensitivity that is a function of direction, as illustrated in Figs. 6.137 and 6.138.

Omnidirectional antenna: An antenna that has essentially a nondirectional radiation pattern in one plane and a directional pattern in any orthogonal plane.

Radiation pattern lobes: A radiation lobe is a portion of the radiation pattern that is bound by regions of lesser radiation intensity. Typical directional antennas have one major **main lobe** that contains the antenna's radiation peak, and several **minor lobes**, or **sidelobes**, which are any lobes other than the major lobe. A **back lobe** is a radiation lobe that is located approximately 180° from the main lobe. The **sidelobe level** is a measure of the power intensity of a minor lobe, usually referenced in dB to the main lobe peak of the power radiation pattern. The **front-to-back ratio** is measure of the power of the main beam to that of the back lobe. A **pattern null** is an angular position in the radiation pattern where the power radiation pattern is at a minimum. The parameters are depicted in Fig. 6.137b.

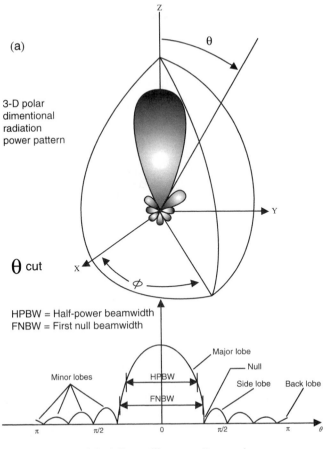

(a)

3-D polar
dimentional
radiation
power pattern

θ cut

HPBW = Half-power beamwidth
FNBW = First null beamwidth

(b) 2-D rectilinear pattern cut

FIGURE 6.137   Antenna parameters.

Field regions: The three-dimensional space around an antenna is divided into three regions. The region closest to the antenna is called the **reactive near field**, the intermediate region is the **radiating near field (Fresnel)** region, and the farthest region is the **far field (Fraunhofer)** region. The reactive near field region is where reactive fields dominate. The Fresnel region is where radiation fields dominate, but the angular radiation pattern variation about the antenna is a function of distance away from the antenna. The far field is where the angular variation of the radiation pattern is essentially independent of distance. Various criteria have been used to quantify the boundaries of these radiation regions. The commonly accepted boundaries between the field regions are:

Reactive Near field/Fresnel Boundary: $R_{RNF} = 0.62 \cdot \sqrt{\dfrac{D^3}{\lambda}}$ (6.56)

Fresnel/Far Field Boundary: $R_{FF} = \dfrac{2D^2}{\lambda}$,   (6.57)

where: D = the largest dimension of the antenna and
           λ = the wavelength in free space.

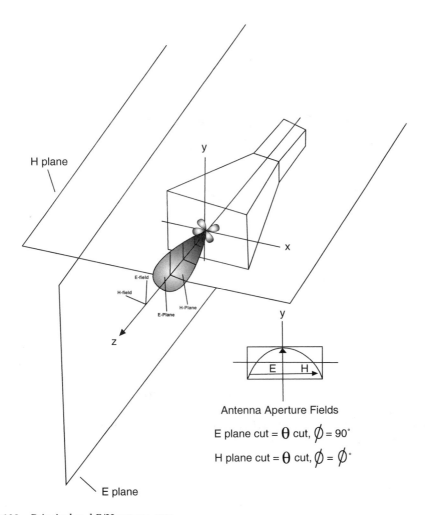

FIGURE 6.138 Principal and E/H pattern cuts.

The far field formulation assumes a 22.5° phase error across a circular aperture designed for a −25 dB sidelobe level, which creates a ±1.0 dB error at the 25 dB far field sidelobe level.[5] More stringent far field criterion is applicable to lower sidelobe levels.[6] Antennas operate in the far field for typical communication systems and radar applications.

Directivity: The directivity of an antenna is the ratio of the radiation intensity in a given direction to that of the radiation intensity of the antenna averaged over all directions. This ratio is usually expressed in decibels. The term directivity most often implicitly refers to the **maximum directivity**.

Directivity is a measure of an antenna's ability to focus power density (transmitting) or to preferentially receive an incoming electromagnetic wave's power density as a function of spatial coordinates. A passive, lossless antenna does not amplify its input signal due to conservation of energy considerations, but rather redistributes input energy as a function of spatial coordinates.

Consider an isotropic radiator, and a highly directive antenna such as a parabolic reflector. Assume that each antenna is lossless, impedance and polarization are matched, and both have the same power at their input terminals. The matched isotropic radiator couples the same total input power into free space as the parabolic reflector, but the power density as a function of spatial coordinates is drastically different. The parabola focuses the majority of the total input power into a very narrow spatial sector, and is commensurately more "sensitive" in this region. In contrast, the isotropic radiator uniformly

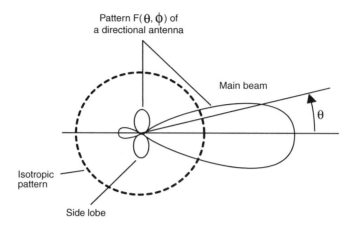

FIGURE 6.139   Comparison of 2-D isotropic and directional pattern cuts.

distributes the input power over all directions and lacks the directional sensitivity. These concepts are illustrated in Fig. 6.139.

Directivity does not account for resistive loss mechanisms, polarization mismatch, or impedance mismatch factors in an antenna. As previously noted, a high value of directivity means that a high percentage of antenna input power is focused in a small angular region. By way of example, parabolic reflector antennas used in radio astronomy can have directivities exceeding 80 dB. An isotropic radiator, in contrast, has a directivity of 0 dB (1.0 numeric) since it radiates equally in all directions. A half-wave resonant dipole radiating in free space has a theoretical directivity of 2.14 dB.

Half power beam width is the angle between two directions in the maximum lobe of the power radiation pattern where the directivity is one half the peak directivity, or 3 dB lower, as illustrated in Fig. 6.137b. The half power beam width is a measure of the spatial selectivity of an antenna. Low beam width antennas are very sensitive over a small angular region. Half power beam width and directivity are inversely proportional for antennas in which the major portion of the radiation resides within the main beam.

Power gain: The gain of an antenna is the ratio of the radiation intensity in a given direction to the total input power accepted by the antenna's input port. This ratio is usually expressed in decibels above an isotropic radiator (**dBi** for linear polarization and **dBic** for circular polarization).

Gain is related to directivity through the **radiation efficiency** parameter. Radiation efficiency is a measure of the dissipative power losses internal to the antenna:

$$G = \eta_{cd} * D \qquad (6.58)$$

where: G   = power gain,

η_{cd} = antenna efficiency due to conductor and dielectric dissipative losses, and

D   = the antenna's directivity.

Power gain usually implicitly refers to the peak power gain. The gain parameter includes the effects of dissipative losses, but does not include the effects of polarization and impedance mismatches.

Antenna polarization: The polarization of an antenna refers to the electric field polarization properties of the propagating wave received or transmitted by the antenna. Wave polarization is a description of the contour that the radiating electric field vector traces as the wave propagates through space away from the observer. The most general wave polarization is **elliptical**. **Circular** and **linear polarizations** are special cases of elliptical polarization. Examples of elliptical, circular, and linear polarized propagating electric field vectors are illustrated in Fig. 6.140. **Vertical** and **horizontal polarizations** are sometimes used as well, and loosely refer to the orientation of the linear polarized electric field vector at the plane of the antenna.

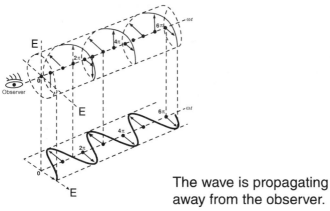

The wave is propagating away from the observer.

(a) Elliptically polarized radiating wave

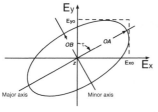

(b) Polarization ellipse

Rotation of a plane electromagnetic wave and its polarization ellipse at z = 0 as a function of time.

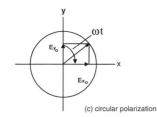

(c) circular polarization

Exo = Eyo, 90° phase shift between Exo and Eyo

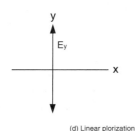

(d) Linear plorization

FIGURE 6.140   Antenna polarization.

The elliptically polarized wave is described by its **polarization ellipse**, which is the planar projection of the contour that its electric field vector sweeps out as the wave propagates through space. The orientation of the polarization ellipse is the **tilt angle**, and the ratio of the **major diameter** to the **minor diameter** of the ellipse is the **axial ratio**. The direction of rotation of the radiated field is expressed as either left-handed or right-handed as the wave propagates from a reference point. **Left-hand circular polarization (LHCP)** and **right-hand circular polarization (RHCP)** are commonly used in contemporary microwave systems.

In most applications it is desirable to have the receive antenna's polarization properties match those of the transmitting antennas, and vise versa. If an antenna is polarized differently than the wave it is

WAVE POLARIZATION						
	VERTICAL ↑	HORIZONTAL →	RIGHT HAND CIRCULAR ↻	LEFT HAND CIRCULAR ↺	45° RIGHT LINEAR ↗	45° LEFT LINEAR ↖
VERTICAL ↑	0 dB	∞	3 dB	3 dB	3 dB	3 dB
HORIZONTAL →	∞	0 dB	3 dB	3 dB	3 dB	3 dB
RIGHT HAND CIRCULAR ↻	3 dB	3 dB	0 dB	∞	3 dB	3 dB
LEFT HAND CIRCULAR ↺	3 dB	3 dB	∞	0 dB	3 dB	3 dB
45° RIGHT LINEAR ↗	3 dB	3 dB	3 dB	3 dB	0 dB	∞
45° LEFT LINEAR ↖	3 dB	3 dB	3 dB	3 dB	∞	0 dB

(Left axis label: WAVE POLARIZATION; Right axis label: POLARIZATION LOSS)

NOTE: Direction of propagation is into the page.

FIGURE 6.141   Polarization loss between receive and transmit antennas.

attempting to receive, then power received by the antenna will be less than maximum, and the effect is quantified by a **polarization mismatch loss**, which is defined by:

$$\text{PML} = \left| \rho_r^* \, \rho_t^* \right|^2 = \left| \cos \Psi_p \right|^2 \tag{6.59}$$

where: ρ_r = the receive polarization unit vector,
ρ_t = the transmit polarization unit vector, and
Ψ_p = the angle between the two unit vectors.

This formulation is applicable to elliptical, circular, and linear polarizations.

Some commonly encountered polarization mismatch losses are depicted in Fig. 6.141. It is apparent from the figure that a pure RHCP wave is completely isolated from a LHCP wave. This property is exploited in polarization diversity systems, e.g., satellite communications systems and cellular radio systems.

The **axial ratio**, usually expressed in dB, is a measure of the ellipticity of a polarized wave. A 0 dB axial ratio describes pure circular polarization. 6 dB axial ratio is typically used as a rule of thumb for the maximum axial ratio in which an elliptical wave is considered circular. Axial ratio does not describe the sense of the propagating wave. The axial ratio of an antenna is a three-dimensional function of spatial coordinates.

An arbitrary elliptically polarized wave can be mathematically decomposed into a LHCP component and a RHCP component. Consider two circular antennas of opposite polarization sense, one in the transmit mode and the other in the receiving mode. As each antenna deviates from pure circular polarization toward elliptical polarization, the orthogonal (undesired) polarization components of each antenna become increasingly influential, which in turn reduces the isolation between the antennas. Similarly, a deviation from pure circular polarizations between receive and transmitting antennas of the same polarization sense manifests itself as an increased loss. Figure 6.142 is quantifies the minimum polarization isolation given the axial ratio and polarization sense of each antenna.[7]

Radiation efficiency accounts for losses at the input terminal of the antenna and the losses within the structure. Efficiency due to conductor and dielectric losses is defined as:

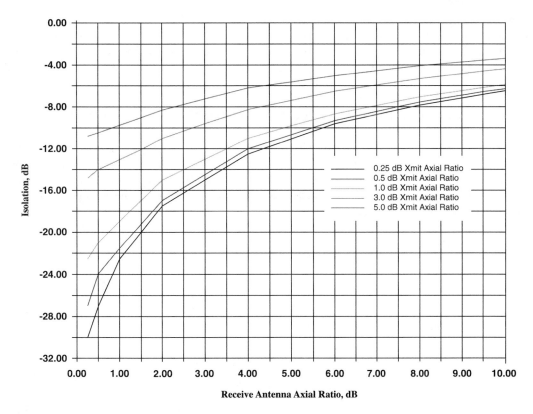

FIGURE 6.142   Receive antenna axial ratio, dB.

$$\eta_{cd} = \eta_c * \eta_d, \tag{6.60}$$

where: η_c = the efficiency due to conduction loss, and,
   η_d = the efficiency due to dielectric loss.

Conduction loss is a function of the conductivity of the metal used. The dielectric loss is typically specified in terms of the loss tangent parameter, and can be anisotropic. Tabulations of both parameters are available in the literature.[8] Conduction and dielectric loss efficiencies can also be determined experimentally.

Input impedance: The input impedance of an antenna is the impedance at the input terminal reference plane. **Impedance** is the ratio of the voltage to the current at the terminal plane, which is related to ratio of the electric and magnetic fields at the terminal plane. An antenna can be thought of as a **mode translator** of electromagnetic waves. Guided waves propagating through the input transmission line from the input generator to the antenna input terminal are transformed into unbounded (radiation) electromagnetic waves. Thus the input port of an antenna radiating into free space can be modeled as a frequency-dependent impedance with real and imaginary components.

The real or resistive impedance component is required for the transfer of real power from the input generator into the radiated electrical and magnetic fields to initiate an average power flow out of the antenna into free space. The **radiation resistance** is used to represent this power transfer for an antenna with no dissipative losses. The antenna's input resistance is the sum of the **conduction resistance**, the **dielectric loss resistance**, and the radiation resistance. The conduction resistance and dielectric loss resistance are related to the metallic and the dielectric material dissipative losses within the antenna, respectively. Both of these resistances are related to the radiation efficiency. The reactive near fields in the immediate vicinity of the antenna structure, i.e., the reactive near field region, are manifested as a reactance at the antenna input terminal. A **matched impedance condition** over the desired frequency range is required to optimally transfer input generator power into radiated waves. Both the real and imaginary components of the input impedance can vary with frequency, further complicating the impedance-matching problem. Standard transmission line and broadband circuit-matching techniques can be applied to ensure the optimal transfer of energy between the input transmission line and the antenna's input terminal.[9,10]

Impedance mismatch loss is a measure the amount of power from the input generator that passes through the antenna's input terminals. The input reflection coefficient is a measure of the amount of power that is reflected from the antenna input back into the generator. Mismatch loss is not a dissipative or resistive loss, but rather a loss due to reflections of guided electromagnetic waves back to the input generator.

The impedance mismatch loss is defined as:

$$\text{MML} = \left(1 - |\Gamma|^2\right) \tag{6.61}$$

where Γ = the voltage reflection coefficient at the antenna input terminal.

The voltage reflection coefficient is usually specified and measured as either a return loss or a Voltage Standing Wave Ratio (VSWR):

$$\text{Return Loss}_{dB} = 10 * \log\left(|\Gamma|^2\right), \text{ and} \tag{6.62}$$

$$vswr = \frac{1 + |\Gamma|}{1 - |\Gamma|}. \tag{6.63}$$

Overall antenna efficiency: The overall efficiency of an antenna is minimally related to the previously mentioned parameters of conductor/dielectric losses, polarization loss, and impedance mismatch factors. In addition, there are other efficiency parameters that are specific to certain classes of antennas. These additional terms include:

- Illumination efficiency (reflector and lens antennas)
- Spillover efficiency (reflector and lens antennas)
- Blockage efficiencies (reflector antennas)
- Random phase and amplitude errors (arrays, lens, apertures, reflector antennas)
- Aperture taper efficiencies (arrays, lens, apertures, and reflector antennas)
- Others

Effective area is the ratio of the available power at the antenna input terminals due to a polarization matched plane wave incident on the antenna from a given direction to the power density of that same plane wave incident on the antenna. Effective area is a measure of an antenna's power capturing properties under plane wave illumination. The effective area of an aperture antenna is related to its physical area by its **aperture efficiency** as shown in Eq. (6.64) below:

$$\eta_{ap} = \frac{A_{em}}{A_p} = \frac{maximum\_effective\_area}{physical\_area} \tag{6.64}$$

The effective area is related to its directivity by:

$$A_{em} = \frac{\lambda^2}{4\pi} * D_o \eta_{cd}\left(1 - \left|\Gamma^2\right|\right) \cdot \left|\overline{\rho}_r \cdot \overline{\rho}_t^{-*}\right|^2 \tag{6.65}$$

where D_o is the peak directivity.

The effective area parameter is useful in first-order array and aperture antenna directivity calculations.

Bandwidth is a frequency range in which a particular electrical parameter of an antenna conforms to a specified performance. Example parameters include gain, beamwidth, sidelobe level, efficiency, axial ratio, input impedance, and beam direction or squint. Since all of the above-mentioned parameters have different frequency dependencies, it is impossible to uniquely specify a single bandwidth parameter for all antenna properties.

Reciprocity is a theorem of electromagnetics that requires receive and transmit properties to be identical for a linear and reciprocal antenna structure operating in a linear, isotropic medium. This concept is very useful in antenna measurements since it is normally more convenient to measure the properties of an antenna under test in the receive mode. Also, it can be easier to understand some antenna concepts through a receive mode formulation, while others are easier through a transmit mode interpretation.

Friis transmission formula describes the coupling of electromagnetic energy between two antennas under far field radiation conditions. The power received by the receiving antenna is related to the power transmitted through the transmit antenna by the following expression:

$$\frac{P_r(\theta,\phi)}{P_t(\theta,\phi)} = \eta_{cdr} \cdot \eta_{cdt} \cdot \left(1 - \left|\Gamma_r\right|^2\right) \cdot \left(1 - \left|\Gamma_t\right|^2\right) \cdot \left(\frac{\lambda}{4\pi \cdot R}\right)^2 D_{or}(\theta,\phi) \cdot D_{ot}(\theta,\phi) \cdot \left|\overline{\rho}_r(\theta,\phi) \cdot \overline{\rho}_t(\theta,\phi)^*\right|^2 \tag{6.66}$$

where: θ and ϕ are the usual spherical coordinate angles, R = the distance between the two antennas, and the r and t subscripts refer to the receive and transmit modes, respectively.

Antenna noise temperature: All objects with a physical temperature greater than 0°K radiate energy. Antenna noise temperature is an important parameter in radio astronomy and satellite communications systems since the antenna, ground, and sky background noise contributes to the total system noise. This system noise ultimately sets a limit for the system signal-to-noise ratio. Satellite system link performance is typically specified in terms of the ratio of system gain to the system noise temperature, which is called the **G/T** figure of merit. The noise temperature of an antenna is defined as follows:

$$T_A = \frac{\int_0^{2\pi}\int_0^{\pi} T_B(\theta,\phi)G(\theta,\phi)\sin\theta d\theta d\phi}{\int_0^{2\pi}\int_0^{\pi} G(\theta,\phi)\sin\theta d\theta d\phi} \tag{6.67}$$

where:

$$T_B = \varepsilon(\theta,\phi)T_m = \left(1 - \left|\Gamma\right|^2\right) \cdot T_m = \text{the } \textbf{equivalent brightness temperature,} \tag{6.68}$$

where: $\varepsilon(\theta,\phi)$ = *emissivity*,
T_m = molecular (physical) temperature (K),
$G(\theta,\phi)$ = power gain of the antenna.

The brightness temperature is a function of frequency, polarization of the emitted radiation, and the atomic structure of the object. In the microwave frequency range, the ground has an equivalent temperature of about 300°K, and the sky temperature is about 5°K when observing toward the zenith (straight up, perpendicular to the ground), and between 100 and 150°K when looking toward the horizon.

Radiating Element Types

Several of the most prominent radiating elements used in contemporary microwave and millimeter wave systems are shown in table at the end of this section, which is an extension of the work of Salati,[1] but also draws on the information sited in the references contained herein. The intent is to catalog the salient features of these radiating elements in terms of broad classifications and to summarize the most important electrical parameters useful for system design. Detailed analysis and design methodologies are beyond the scope of this presentation and the reader is referred to the extensive literature for further information.

Wire Antennas[2-4]

Wire antennas are arguably the most common antenna type and they are used extensively in contemporary microwave systems. The classic **dipole, monopole,** and **whip** antennas are used extensively throughout the microwave frequency bands. Broadband variations of the classic wire antenna include the **electrically thick monopole and dipole, biconical dipole, bow tie, coaxial dipole and monopole, the folded dipole, discone, and conical skirt monopole.** Various techniques to reduce the size of this type of antenna for a given resonant frequency include foreshortening the antenna's electrical length and compensating with lumped-impedance loading, top hat capacitive loading, and dielectric material loading. Also, helical winding of a monopole element retains the proper electrical length, but with foreshortened effective height. An excellent discussion on electrically small antennas is found in Fujimoto, et al.[5]

Loop Antennas[2,7]

Loop antennas are used extensively in the HF through UHF band, and have application at L band and above as field probes. The circular loop is the most commonly used, but other geometric contours, such as rectangular, are used as well. Loops are typically classified as **electrically small**, where the overall wire length (circumference multiplied by the number of turns) is less than 0.1 wavelength, and electrically large, where the loop circumference is approximately one wavelength. Electrically small antennas suffer from low radiation efficiency, but are used in portable radio receivers and pagers and as field probes for electromagnetic field strength measurements. The ubiquitous AM portable radio antenna is a ferrite material loaded, multi-turn loop antenna. **Electrically large** loops are commonly used as array antenna radiating elements. The radiation pattern in this case is end-fire toward the axis of the loop.

Slot Antennas[6]

Slot antennas are used extensively in aircraft, missile, and other applications where physical low profile and ruggedness are required. Slot antennas are usually half-wave resonant and are fed by introducing an excitation electric field across their gap. **Cavity backed slots** have application in the UHF bands. The **annular slot antenna** is a very low profile structure that has a monopole wire-like radiation pattern. Circular polarization is possible with crossed slots fed in phase quadrature. Stripline-fed slots are frequently used in phased array applications for the microwave bands.[8]

Waveguide-fed slotted array antennas find extensive use in the microwave and millimeter wave bands.[9] Linear arrays of resonant slots are formed by machining slots along the length of standard waveguide transmission lines, and a collection of these waveguide linear array "sticks" are combined together to form two-dimensional planar phased arrays. The most common waveguide slots currently in use are the **edge slot**, and the **longitudinal broad wall slot.** This array type is a cost-effective way to build high efficiency, controlled sidelobe level arrays. One-dimensional electrically scanned phased arrays can be realized by introducing phase shifters in the feed manifold that excites each linear waveguide array stick within the two-dimensional aperture.

Helical Antennas[3]

The helical antenna has seen extensive use in the UHF through microwave frequencies, both for single radiating elements, and as phased array antenna elements. An excellent helical antenna discussion can be found in Kraus.[3] The helical antenna can operate either in the **axial (end fire) mode**, or in the **normal mode**. The axial mode results in a directional, circularly polarized pattern that can operate over a 2:1 frequency bandwidth.

The normal mode helix has found wide application for broad-beam, cardioid shaped radiation patterns with very good axial ratio performance. Kilgus developed bifilar and quadrafilar helical antennas for satellite signal reception applications.[10,11] Circular polarization is generated by exciting each quadrafilar element of the helix in phase quadrature.

Helical antennas are also used as radiating elements for phased arrays, and as feeds for parabolic reflector antennas.[12]

Yagi-Uda Array[z]

The Yagi-Uda array is a very common directional antenna for the VHF and UHF bands. End fire radiation is realized by a complex mutual coupling (surface wave) mechanism between the driven radiation element, a reflector element, and one or more director elements. Dipole elements are most commonly used, but loops and printed dipoles elements are used as well. Balanis describes a detailed step-by-step design procedure for this type of antenna.[2] Method-of-moment computer simulations are often used to verify and optimize the design.

Frequency-Independent Antennas[13]

Classic antenna radiator types, such as the wire dipole antenna radiate efficiently when their physical dimensions are a certain fraction of the operating wavelength. As an example, the half-wave dipole is strictly resonant only at one frequency, and acceptable performance can be realized for finite bandwidths. The frequency-independent antenna is based on designs that are specified in terms of geometric angles. Rumsey has shown that if an antenna shape can be completely specified by angles, then it could theoretically operate over an infinite bandwidth.[14] In practice, the upper operating frequency limit is dictated by the antenna feed structure, and the lower operating frequency is limited by the physical truncation of the antenna structure. The **planar spiral, planar slot spiral, conical spiral,** and the **cavity back planar spiral** antennas all exploit this concept.

A closely related and very useful antenna is the **log-periodic antenna**. This type of antenna has an electrical periodicity that is a logarithm of frequency. Because the antenna shape cannot be completely specified by angles, it is not truly frequency independent, but is nevertheless extremely broadband. The most common architecture is the log periodic dipole array. Balanis has documented a detailed design procedure for this structure.[2]

A key feature of frequency-independent antennas is that the electrically active region of the antenna is a subset of the entire antenna structure, and the active region migrates to different regions of the antenna as a function of frequency.[15]

Aperture Antennas[2,16,17]

Horn antennas are the most commonly used aperture antennas in the microwave and millimeter wave frequency bands. There is an abundance of horn antenna design information available in the literature. **E plane and H plane sectoral horns** generate fan beam radiation, i.e., a narrow beamwidth in one principal plane, and a broad beamwidth in the orthogonal principal plane. The **pyramidal horn** provides a narrow beam in both principal planes. Sector and pyramidal horns are typically fed with rectangular waveguide and are linearly polarized. The **ridged horn** is a broadband variation of the pyramidal horn.[18] A **dual ridge square aperture horn** in conjunction with an orthomode transducer (OMT) can be used to generate broadband, low axial, circularly polarized radiation patterns.

Conical horns are typically fed with circular waveguide. An important variation of the basic conical horn includes the **corrugated horn**, which is typically used as a feed for high efficiency, electrically large reflector antenna systems. The corrugations are used to extinguish field diffractions off the edge of the aperture plane that can lead to spurious radiation in the back lobe and sidelobe regions. The **aperture-matched horn** also minimizes aperture edge diffraction by blending the aperture edge to make a gradual transition into free space. **Dual mode conical horns** use a superposition of waveguide modes in the throat region to suppress beamwidth, control orthogonal beamwidths, and minimize cross-polarization. **Multimode horns** can be used in monopulse radar applications where sum and difference beams can be generated by means of mode generating/combining feed structures.

Reflector Antennas[2,19]

Reflector antennas are used extensively where very high gain, high radiation efficiency, and narrow beam widths are required. Applications include satellite communication systems, navigation systems, terrestrial communications systems, deep space communication systems, and radio astronomy. Gain is excess of 80 dBi has been reported in radio astronomy applications.[19] **Prime focus reflector** systems have the feed antenna at the focal point of the reflector, which is typically a paraboloid. A disadvantage of this approach is the blockage created by the feed and its mounting struts. **Multiple reflector** systems are used in some applications for improved performance over prime focus designs. A multiple reflector system has several advantages, including: (1) the feed can be mounted on the back of the main reflector, which removes the interconnecting transmission line between the feed and the receiver front end, (2) spillover radiation and sidelobe radiation can be reduced, (3) a large focal length can be realized with a shorter physical length, and (4) beamshaping can be realized by shaping the main reflector and sub-reflector surfaces. **Cassegrain** and **offset subreflector** systems are in common use. The offset reflector has the advantage of significantly reducing feed blockage and minimizing cross-polarization.

Variations of the reflector concepts include the planar reflector,[20] the microstrip reflect array,[21] and the use of clustered feeds[22] and phased arrays as reflector feeds to generate limited scan arrays.[23]

Microstrip Antennas[2,24,25]

Microstrip antennas are a fairly new and exciting antenna technology. They offer the advantage of being low profile, light weight, and are easily producible using contemporary printed circuit board materials and fabrication techniques. This technology has found extensive use in the Global Positioning System, wireless and satellite communication systems, and the cellular phone industry. They have the disadvantage of narrow bandwidth and high loss at upper microwave and millimeter wave frequencies.

The most common microstrip radiating elements used today are the **rectangular, square**, and **circular** geometries. Linear polarized designs are realized with single feeds, either by a **microstrip transmission line feed** on the same layer as the radiating element, or by **coaxial probe feeding**. The microstrip line feed has the advantage of ease of fabrication due to single layer printed circuit board construction, but disadvantage of spurious radiation caused by the feed lines, particularly in array applications. The coaxial probe feed method works well, but the feed probe inductance must be taken into account. **Aperture coupling** is a broader band excitation method, but it increases the complexity of fabrication since multilayer printed circuit boards are required.

Circular polarization can be generated with a single feed using rectangular patch geometry. The resultant circularly polarized radiation is very narrow band. Square and circular patches can create CP with dual feeds fed in phase quadrature. The resultant CP signal is much broader band, with 6.0 dB axial ratio bandwidths greater than 5 percent.

First order simple patch design can be accomplished with the **transmission line model**, and with the more accurate **cavity model**.[2] Method of Moment (MOM), Finite Element (FEM), and Finite Difference Time Domain (FDTD) techniques are used for detailed design.

Microstrip radiating elements are commonly used in contemporary phased array systems. Scan blindness (as discussed in the phased array antenna section) is problematic since a large dielectric slab can sustain a surface wave mode of guided wave propagation, which deteriorates the array radiation pattern. Mutual coupling further complicates the phased array design problem. The reader is referred to the literature for detailed treatises.

Millimeter Wave Antennas[26]

The most widely used millimeter wave antennas currently in use are the reflector antenna, the lens antenna, and the horn antenna. In addition to these structures, several lower frequency antennas can be adapted to millimeter wave frequencies, including the slot, dipole, biconical dipole, and monopole antenna.[27]

Slotted waveguide arrays have been demonstrated up to 94 GHz. The **Purcell slot array** has renewed interest in the millimeter wave regime due to its increased slot dimensions and relaxed mechanical tolerances.[28]

Printed microstrip patch arrays on low permitivity substrates have been demonstrated to 100 GHz. Feed loss of conventional microstrip lines is especially troublesome in the millimeter frequency bands. Several methods have been studied to reduce feed loss, including feeding microstrip radiating element with the fringing fields associated with dielectric image guide, and E field probe coupling of the microstrip patches to a feed waveguide. Special attention must be given to mechanical tolerances and assembly techniques in the millimeter wave implementation of these structures.

Open dielectric structure antennas are attractive since most millimeter wave transmission lines are open structures. Strategically placed discontinuities in these structures initiate a **surface wave** phenomenon, which in turn initiates radiation. Examples include the **tapered dielectric rod** and the **periodic dielectric antenna**.[29,30]

Leaky wave structures include the **long slot in waveguide, periodically loaded waveguide,** and **trough waveguide** structures.[31] Leaky waves are initiated when an open or closed waveguide is perturbed at periodic intervals, or by a continuous perturbation. Usually the amount of leakage per waveguide length is kept low to minimize impedance mismatch with the antenna feed. These antennas are designed to be electrically long so that most of the energy radiates away before the guided wave reaches the end of the antenna. Beam scanning as a function of input frequency is possible with this type of antenna.

The **periodically loaded waveguide antenna** utilizes a periodic surface perturbation, usually either a corrugation or a metal grating to initiate the surface wave mode. Other leaky wave antenna structures are discussed in the literature.

Dielectric resonator antennas are resonant volumes of dielectric material. The shape of the radiation pattern is a function of the resonator shape. This structure holds promise for the upper millimeter wave frequency bands.[32]

Lens antennas are very similar to reflector antennas since they are used to collimate spherical waves into plane waves. Reflector antennas are more versatile than lens antennas in general since lens technology suffers from the following phenomenon: surface loss from internal and external reflections, spillover loss, lens zone blockage, dielectric material dissapative losses, lens technology tends to be bulky and heavy, and lenses must be edge supported. However, lenses are superior in some applications, such as wide angle scanning.[33]

Dielectric lenses have very strong optical analogies. If the subtended feed angle between the feed and the extremities of the lens are large, the lenses can become very bulky and heavy. Placing steps in the lens to create separate zones can reduce bulk. The reader is referred to Elliot for a detailed exposition.[34] Lenses may also be constructed of artificial dielectrics, parallel metal plates, and waveguide technology.[35]

Fresnel zone plate antennas are lenses that collimate spherical electromagnetic waves impinging upon them by means of diffraction, rather than refraction.[36] The zone plate architecture offers several advantages over conventional lens technology, including a simpler, lighter, and planar construction with lower material induced insertion loss. Wiltse reviews the history of Fresnel zone plate antenna development in Reference 37. Hristov and Herben describe a Fresnel Zone plate lens with enhanced focusing capability.[38] Dual band Fresnel zone plate antennas are discussed in Reference 39. Gouker and Smith described an integrated circuit version of the Fresnel zone plate antenna operating at 230 GHz.[40]

Notch antennas find wide spread use in active phased array antenna applications since they are compatible with integrated circuit applications. Vertically polarized linear arrays are easily realized by stacking several notch antennas along a line. Close element spacing the H plane is possible due to the thin cross-section of these elements. Yngvesson et al. describe the salient features of the Vivaldi notch, the linear tapered slot, and the constant width slot.[41] These antennas are a subset of the general class of endfire traveling wave antennas and they offer moderately high directivity and are broadband.

Antenna Metrology [1,2,3]

Several excellent books are available on the subject of antenna metrology, so only a few of the basic concepts are briefly described herein. The reader is referred to the literature for a more detailed exposition.

Impedance, voltage standing wave ratio (VSWR), return loss, radiation patterns, cross-polarization, axial ratio, and the polarization ellipse tilt angle are the radiation parameters most often measured.

Impedance and **VSWR** are measured using normal network analyzer techniques with the antenna radiating in an anechoic environment.[4,5] Typically, swept frequency measurements are made in small anechoic chambers designed only for impedance and VSWR testing, in a full-scale anechoic chamber where radiation patterns are measured, or in an open, reflectionless environment.

Radiation pattern cuts are measured by precisely rotating the antenna under test (AUT) and recording the resultant phase and amplitude responses as a function of position, under the condition of constant source antenna illumination. The amplitude and phase response are recorded as a function of two orthogonal spherical coordinates. Figure 6.143 illustrates a conceptual block diagram of a typical far field data acquisition setup appropriate for both outdoor or anechoic chamber use.

Gain is most often measured by the substitution method. A signal level is recorded from a calibrated gain standard and the source antenna. The two antennas are typically axially aligned. Then the antenna under test is measured using the same source antenna and power levels. The difference in the receive level of both measurements is calculated and this difference is added to the gain standard's calibrated gain number to determine the antenna under test's gain above isotropic at the measurement frequency. Dipoles and dipole arrays are used as gain standards in the VHF and UHF frequency bands. Pyramidal and conical horns are used above 800 MHz and into the millimeter wave region. Axial mode helical

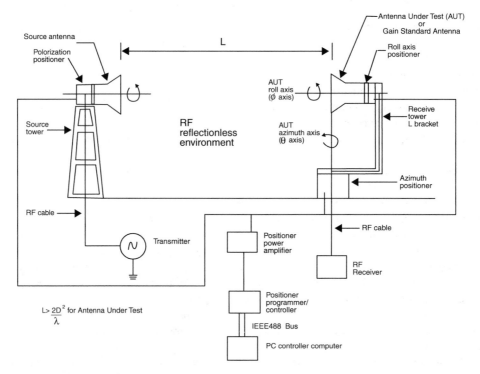

FIGURE 6.143   Basic roll/az for field test setup.

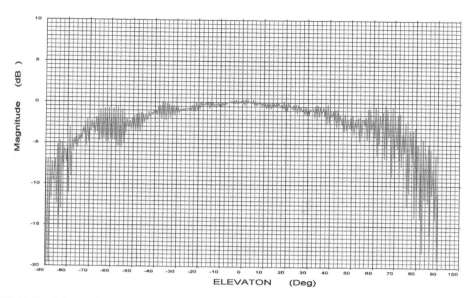

FIGURE 6.144 Spinning linear source axial ratio measurement.

antennas, dual mode ridged square aperture/Orthomode Transducer (OMT) fed waveguide horns, and OMT fed circular horns of various types can be used directly as circularly polarized gain standards.

Cross-polarization of a linearly polarized antenna is easily measured by rolling the source antenna so that its polarization is orientated 90° from the AUT's co-polarization orientation. Cross-polarization is typically referenced to the co-polarized, peak of beam signal. The cross-polarization of a circularly polarized antenna can be directly measured by comparing the received signal from the AUT that is co-polarized with the source antenna to its cross-polarized received signal using an identical source antenna of opposite polarization. In addition, cross-polarization rejection can be calculated knowing the source antenna and AUT's axial ratio and polarization tilt angles.

Axial ratio is usually measured with the spinning linear source antenna method. A linear source antenna with low cross-polarization is rapidly spun about its roll axis while the AUT's positioner is recording a pattern at a much slower rotation rate. The resulting pattern has a high frequency sinusoidal-like pattern superimposed on the basic antenna pattern. If the antenna under test is highly linear, its response will vary drastically as the source antenna is rolled. On the other hand, if the AUT is purely circular, there will be no undulation of the signal since the circular antenna is equally sensitive to all orientations of linear polarization. It can be shown that the difference in amplitude of adjacent maxima and minima in the spinning linear plot is equivalent to the axial ratio of the antenna under test (AUT) for that particular direction. The spinning linear measurement therefore provides an easy way to determine the axial ratio of the AUT across a pattern cut. An example axial ratio pattern cut is shown Fig. 6.144.

The **axial ratio** and **polarization tilt angle** for a single point on the antenna radiation pattern can be determined by a **polarization ellipse** measurement. In this test the AUT is parked in the position of interest and the source antenna is rolled one complete revolution. The recorded contour is the polarization ellipse and the ratio of the major to minor diameters is the axial ratio. The angle of the major diameter relative to a specified reference is related to the polarization tilt angle. A typical polarization ellipse of an axial mode helix is illustrated in Fig. 6.145.

Near field measurements are becoming increasingly popular for phased array antenna design and diagnostics.[6-8] The basic concept of near field metrology is related to the electromagnetic theorem called the **equivalence theorem**.[9] This theorem can be summarized as follows: if two orthogonal orientations of electric field are known everywhere on a closed surface, the electromagnetic field at any other observation point can be determined. In the near field measurement scenario, a field probe is used to sample two orthogonal polarizations of electrical field along the closed surface, a correction algorithm is applied

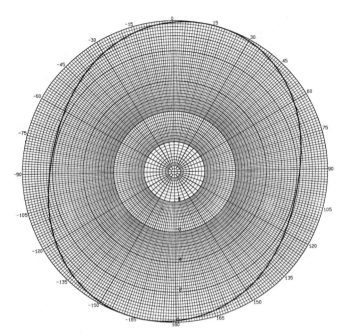

FIGURE 6.145 Bore site polarization ellipse for an axial mode helix fo ≈ 1.5 GHz.

to negate the perturbation of the measured field by the probe, and finally Fourier transform techniques are used to calculate the far field radiation pattern.

The three fundamental types of near field ranges are the **planar, cylindrical**, and **spherical near field**, named after the closed surfaces in which the data is sampled. In the planer near field setup, the enclosed measurement surface is approximated as a plane. This is appropriate for directional antennas such as planar arrays and reflector antennas. The cylindrical near field range is appropriate for fan beam antennas such as linear arrays. The spherical near field is appropriate for all antenna types, including broad beam antennas. The planar near field algorithm is the most computationally efficient, while the spherical near field is the least computationally and data acquisition efficient.

The near field measurement techniques offer several advantages over the traditional far field pattern cut procedures, even though the required data acquisition and data processing time is much longer. The advantages are listed below:

1. The complete data set is available to the analyst for future use.
2. In phased array measurements, the measurement frequency and the beam position can be time multiplexed such that multiple data sets from numerous beam positions and operating frequencies can be extracted from one data acquisition session.
3. Inverse Fourier transform techniques can be used to holographically image the aperture excitation.

The ability to directly produce an **image of the array aperture excitation** is a very powerful tool in the array design process. It can be used to diagnose element or phase shifter failures, etc., along the array face.

Emerging Antenna Technologies

Microwave and millimeter wave frequency material research presents several new possibilities for antenna technology advancement. **Ferroelectric materials, photonic band gap materials (PBG), frequency selective surfaces (FSS)**, and **radar absorbing materials (RAM)** are all being pursued for advanced antenna applications.

Frequency selective surfaces (FSS) are used in multiband, wideband phased array applications, and in applications where it is desirable to block signal transmission for one frequency band while passing another frequency band.[1] Diachroic reflector technology is another example of a FSS application.[2]

Radar Absorbing Material (RAM), traditionally used for Radar Cross Section (RCS) reduction for stealth military vehicles, is finding application in antenna design. RAM materials are used to alter surface currents of reflector antennas, to prevent diffraction off ground plane edges, and to extinguish surface currents to minimize parasitic mutual coupling, which causes co-site interference between antenna systems on vehicular platforms.[3,4]

MEMS (Micromachined ElectroMechanical Systems) switches are miniature semiconductor switches used in a variety of antenna applications. Brown provides an excellent overview of MEMS technology for RF (radio frequency) applications.[5] The MEMS switch is physically small; has very low on-state loss; high off-state isolation; and low average DC power consumption, as compared to ferrite waveguide and semiconductor switching elements such as PIN diode and FET transistor switches.

The two most common switch topologies are the cantilever, or type 1 switch[6] and the membrane, or type 2 switch.[7] Phase shifters and true time delay devices using MEMS type 2 switches have been reported at C Band, through W Band. The type 1 switch topology is superior for lower frequency (below X Band) applications due to its high off state isolation.

An intriguing application of MEMS device technology is for reconfigurable antenna technology.[8] The desire to use multi-band, multifunction shared asset apertures for advanced radar and communications systems has led researchers to consider radiating elements whose physical properties can be dynamically adjusted. This is already demonstrated, to a simpler degree, by the use of reactive loads distributed along the length of wire antennas, such as dipoles, to give multiband performance.[9] The MEMS switch offers the potential to generalize this concept in two dimensions using a switching matrix.

MEMS technology has been successfully demonstrated within the research community and efforts are underway to commercialize the technology, including lifetime reliability studies and RF compatible, environmental packaging techniques.[10,11] Device packaging issues are particularly challenging since the MEMS switch is an electromechanical device that requires unrestricted mechanical motion within the package.

Micromachined antenna technology can be used to synthesize a localized artificial dielectric material with low dielectric constant for microstrip an antenna substrate from a high dielectric substrate. The lower effective dielectric constant improves the radiation efficiency and reduces surface waves for a given substrate thickness. Gauthier et al. describe a method of drilling a densely spaced two-dimensional grid of holes into the substrate to reduce the local dielectric constant of the substrate.[12] A reduction of dielectric constant to 2.3 from the intrinsic substrate dielectric constant of 10.8 was reported. Nguyen et al. describe micromachining techniques that are used to suspend dipole or slot radiators on a very thin membrane to realize a local substrate dielectric constant of approximately 1.0, which is quite useful for millimeter wave applications.[13] Veidt et al. describe a micromachined diagonal horn at 110 GHz.[14]

Fractal antenna radiating elements have physical shapes based on fractal geometric curves popularized by Benoit Mandelbrot, and others.[15-17] An overview of fractal electromagnetic research is given by Werner in Reference 18. There are two basic types of fractal curves, random and deterministic. Deterministic fractals utilize a repetitive motif, or pattern, that is applied to successive size scales. Fractals are mathematically described in terms of "n" iterations of the motif, where n goes to infinity. Practically, fractal structures are based on a finite number of iterations.

The general concept for a fractal-based radiating element is to introduce fractal-based impedance loading into the radiating element structure to shrink the physical size of the element for a given operational frequency. Nonharmonically related multiband frequency response has been reported. Cohen has demonstrated fractal loop arrays and monofilar helical antennas.[19,20] Breden and Langley discuss printed fractal technology as applied to multiband radio and television service antennas.[21] Multiband and wideband printed fractal branched antennas are discussed by Sinduo and Sourdois.[22]

The relationship of fractal antennas to log periodic, frequency-independent, and genetic algorithm-based wire antenna designs is described in References 23 and 24.

Genetic Algorithm Based Antenna Design: Although the genetic algorithm-based optimization techniques strictly fall under the category of computational electromagnetic techniques, they are briefly mentioned here because of the nontraditional class of antennas that result from these design techniques. An introduction to genetic algorithm optimization techniques is given in Haupt and Haupt.[25] An excellent overview of genetic algorithms for electromagnetic applications, and comparisons to traditional local optimization techniques, is given by Haupt[26,27] and Johnson and Rahmat-Sami.[28] Traditional local-minimum based gradient optimization techniques are detailed by Cuthbert.[29]

A sampling of genetic-based, algorithm-based wire antenna designs are discussed in References 30 through 32.

Spatial Power Combining is a method for combining the output power of a large number of active microwave or millimeter wave semiconductor devices quasi-optically to achieve high radiated power levels. Spatial power combining is used as a means to alleviate the high insertion losses of millimeter wave distributed passive, planar, combining structures such as microstrip combining circuit elements. Rutledge et al. describe a 4-by-4 grid of Gunn diode driven microstrip patches to realize 22 watts of radiated power at 9.6 GHz.[33]

References

Antenna Parameters

1. *IEEE Standard Definitions of Terms for Antennas,* IEEE Standard 145-1983, Institute of Electrical and Electronics Engineers, 1983.
2. Balanis, C. A., *Antenna Theory, Analysis, and Design,* 2nd Ed., John Wiley and Sons, New York, 1997.
3. Macnamara, T., *Handbook of Antennas for EMC,* Artech House, Inc., Norwood, MA, 1995.
4. *Antennas and Antenna Systems,* Watkins-Johnson Company, Catalog Number 200, September 1990, chap. 12, pp. 101–127.
5. Hollis, J. S. et al., *Microwave Antenna Measurements,* Scientific-Atlanta, Inc., Atlanta Georgia, November 1985.
6. Hacker, P. S., and Schrank, H. E., Range Distance Requirements for Measuring Low and Ultralow Sidelobe Antenna Patterns, IEEE Transactions on Antennas and Propagation, AP-30, 5, 956–966, September 1982.
7. Offutt, W. B., and DeSize, L.K., *Antenna Engineering Handbook,* 2nd Ed., Johnson, R.C., and Jasik, H., Editors, McGraw-Hill, New York, 1984, Methods of Polarization Synthesis, 23-8–23-9.
8. Reference Data for Radio Engineers, 6th Edition, Howard K. Sams/ITT, New York, 1979, 4-21–4-23.
9. Bowman, D. F, *Antenna Engineering Handbook,* 2nd Ed., Johnson, R.C., and Jasik, H, Editors, McGraw-Hill, New York, 1984, chap 43, Impedance Matching and Broadbanding, 43-1–43-32.
10. Cuthbert, T.R., *Broad Band Direct-Coupled and Matching RF Networks,* TRCPEP Publication, Greenwood, Arkansas, 1999.

Radiating Elements

1. Salati, O.M., Antenna Chart for Systems Designers, *Electronic Engineer,* January, 1968.
2. Balanis, C. A., *Antenna Theory, Analysis, and Design,* 2nd Ed., John Wiley and Sons, New York, 1997.
3. Kraus. J. D., *Antennas, 2nd Edition,* McGraw-Hill, New York, 1988.
4. Tai. C. T, *Antenna Engineering Handbook,* 2nd Ed., Johnson, R.C., and Jasik, H., Editors, McGraw-Hill, New York, 1984, chap. 4, Dipoles and Monopoles, 4-1–4-34.
5. Fujimoto, K., Henderson, A., Hirasaw, K., and James, J. R., *Small Antennas,* John Wiley and Sons, Inc., New York, 1987.
6. Blass, J., *Antenna Engineering Handbook,* 2nd Ed., Johnson, R.C., and Jasik, H., Editors, McGraw-Hill, New York, 1984, Slot Antennas, 18-1–18-26.
7. Macnamara, T., *Handbook of Antennas for EMC,* Artech House, Inc., Norwood, MA, 1995.
8. Mailloux, R. J., On the Use of Metallized Cavities in Printed Slot Arrays with Dielectric Substrates, *IEEE Transactions on Antennas and Propagation,* AP-35, 55, May, 1987, 477–487.
9. Elliot, R.S., *Antenna Theory and Design,* Prentice-Hall, Inc., Englewood Cliffs, New Jersey, 1981.

10. Kilgus, C. C., Resonant Quadrafilar Helix, *IEEE Transactions on Antennas and Propagation,* AP-23, 392–397, May 1975.

11. Donn, C., Imbraie, W. A., Wong, G. G., An S Band Phased Array Design for Satellite Application, *IEEE International Symposium on Antennas and Propagation,* 1977, 60–63.

12. Holland, J. Multiple Feed Antenna Covers L, S, and C Band Segments,, *Microwave Journal,* October 1981, 82–85.

13. Mayes. P. E., Frequency Independent Antennas, *Antenna Handbook, Theory, Applications, and Design,* Lo, Y.T., and Lee, S.W., editors, Van Nostrand Reinhold Co., New York, 1988.

14. Rumsey, V.H., *Frequency Independent Antennas,* Academic Press, New York, 1966.

15. DuHamel, R. H., *Antenna Engineering Handbook, 2nd Ed.,* Johnson, R.C., and Jasik, H, Editors, McGraw-Hill, New York, 1984, Frequency Independent Antennas, 14-1–14-44.

16. Love, A. W., *Electromagnetic Horn Antennas,* IEEE Press, the Institute of Electrical and Electronics Engineers, Inc., New York, 1976.

17. Olver, A. D., Clarricoats, P.J. B., Kishk, A. A., Shafai, L., *Microwave Horns and Feeds,* IEEE Press, The Institute of Electrical and Electronics Engineers, Inc., New York, 1994.

18. Walton, K. L., and Sunberg, V. C., Broadband Ridge Horn Design, *Microwave Journal,* 96–101, March 1964.

19. Love, A. W., *Reflector Antennas,* IEEE Press, The Institute of Electrical and Electronics Engineers, Inc., New York, 1978.

20. Encinar, J. A., Design of Two-Layer printed Reflectarrays for Bandwidth Enhancement, *Proceedings of the IEEE Antennas and Propagation Society 1999 International Symposium,* 1999, 1164–1167.

21. Haung, J. A., Capabilities of Printed Reflectarray Antenna, *Proceedings from the 1996 IEEE International Symposium on Phased Array Systems and Technology,* Institute of Electrical and Electronics Engineers, Inc., 1996, 131–134.

22. Ford Aerospace and Communications Corporation, Design for Arabsat C Band Communication antenna System, Palo Alto, California.

23. Mailloux, R. J., *Phased Array Antenna Handbook,* Artech House, Inc., Norwood, MA, 1994, 480–504.

24. James, J. R., and Hall, P.S., *Handbook of Microstrip Antennas,* Volumes 1 and 2, Peter Peregrinus, London, U.K., 1989.

25. Pozar, D. M., and Schaubert, D. H., *Microstrip Antennas: the Analysis and Design of Microstrip Antennas and Arrays,* IEEE Press, The Institute of Electrical and Electronics Engineers, Inc., New York, 1995.

26. Schwering, F., and Oliner, Millimeter Wave Antennas, *Antenna Handbook, Theory, Applications, and Design,* Lo, Y.T., and Lee, S.W., editors, Van Nostrand Reinhold Co., New York, 1988, chap. 17.

27. A.F. Kay, Millimeter-Wave Antennas, *Proc. IEEE,* 54, 641–647, April 1966.

28. S. Silver, Ed. *Microwave Antenna Theory and Design,* New York: McGraw-Hill Book Co., 1949.

29. Zucker, F., J. *Antenna Engineering Handbook, 2nd Ed.,* Johnson, R.C., and Jasik, H, Editors, McGraw-Hill, New York, 1984, chap. 12, Surface Wave Antennas and Surface-Wave-Excited Arrays, 12-1–12-36.

30. Schwering, F., and Peng, S. T., Design of Dielectric Grating Antennas for Millimeter Wave Applications, *IEEE Transactions on Microwave Theory and Techniques,* MTT-31, February 1983, 199–209.

31. Mittra, R., *Antenna Engineering Handbook, 2nd Ed.,* Johnson, R.C., and Jasik, H., Editors, McGraw-Hill, New York, 1984, chap. 10, Leaky Wave Antennas, 10-1-1-021.

32. Long, S. A., McAllister, and Shen, L. C., The resonant Cylindrical Dielectric Cavity Antenna, *IEEE Transactions on Antennas and Propagation,* AP-31, 406–412, May 1983.

33. Peeler, G. D. M., *Antenna Engineering Handbook, 2nd Ed.,* Johnson, R.C., and Jasik, H, Editors, McGraw-Hill, New York, 1984, chap. 16, Lens Antennas, 16-1–16-32.

34. Elliot, R.S., *Antenna Theory and Design,* Prentice-Hall, Inc., Englewood, New Jersey, 1981, 529–532.

35. Elliot, R.S., *Antenna Theory and Design,* Prentice-Hall, Inc., Englewood, New Jersey, 1981, 538–545.

36. Wiltse, J. C., and Garret, J. E., The Fresnel Zone Plate Antenna, *Microwave Journal,* 101–114, January 1991.

37. Wiltse, J., History and Evolution of Fresnel Zone Plate Antennas for Microwaves and Millimeter Waves, IEEE Antennas and Propagation Society, 1999 IEEE International Symposium, 1999, 722–725.

38. Hristow, H. D., and Herbon, M. H. A. J., Millimeter-Wave Fresnel-Zone Plate and Lens Antenna, *IEEE Transactions on Microwave Theory and Techniques,* 43, 12, 2779–2785, December 1995.

39. Wiltse, J. C., Dual Band Fresnel Zone Plate Antennas, *Proceedings SPIE Aerospace Conference,* Orlando, FL, April 1997, 181–185.

40. Gouker, M. A., and Smith, G. S., A Millimeter-Wave Integrated-Circuit Antenna Based in the Fresnel Zone Plate, *IEEE Microwave Theory and Techniques Symposium Digest,* 1991, 157–160.

41. Yngvesson, K. S., Schaubert, D. H., Korzeniowski, T. L., Kollberg, E. L., Thungren, T., and Johansson, J. F., Endfire Tapered Slot Antennas on Dielectric Substrates, *IEEE Transactions on Antennas and Progagation,* AP-33. 12, 1392–1400, December 1985.

Antenna Metrology

1. *IEEE Standard Test Procedures for Antennas,* IEEE Std 149-1979, the Institute of Electrical and Electronics Engineers, 1979.

2. Hollis, J. S. et al., *Microwave Antenna Measurements,* Scientific-Atlanta, Inc., Atlanta Georgia, November 1985.

3. Evans, G., *Antenna Metrology,* Artech House, Norwood, MA, 1990.

4. Hewlett Packard, H. P. Understanding the Fundamental Principles of Vector Network Analysis, Application Note 1278-1, Hewlett Packard Company, No. 5965–770E, May 1997.

5. Hewlett Packard, H. P. Antenna Measurements Using the HP 8753C Network Analyzer, Product Note 8753-4, Hewlett Packard Company, No. 5952–2776, October 1990.

6. Rudge, A. W., Milne, K., Olver, A. D., Knight, P., *The Handbook of Antenna Design,* Volume 1, Peter Peregrinus, Ltd, London, U.K., 1982, 594–628.

7. Slater, D., *Near Field Antenna Measurements,* Artech House, Inc., Norwood. MA, 1991.

8. Kerns, D.M., *Plane-Wave Scattering-Matrix Theory of Antenna and Antenna-Antenna Interactions,* National Bureau of Standards Monograph 162, Washington, D.C., 1981.

9. Balanis, C.A., *Advanced Engineering Electromagnetics,* John Wiley and Sons, New York, 1989, 329–334.

Emerging Technologies

1. Munk, B. E., *Frequency Selective Surfaces,* John Wiley, and Sons, New York, 2000.

2. Schennum, G. H., Frequency Selective Surfaces for Multiple-Frequency Antennas, *Microwave Journal,* 55–76, May 1973.

3. Mahmoud, M. S., Lee, The-Hong, and Burnside, W. D., A New Approach of Edge Treatment For Compact Range Reflectors, *Proceedings of the 1995 17ᵗʰ Meeting and Symposium Antenna Measurement Techniques Association,* November 13–17, 1995, 415–418.

4. Bunside. W. D., and Smith, B., R Card Ground Planes, *Proceedings of the 1997 Nineteenth Meeting and Symposium, Antenna Measurement Techniques Association,* November 12–17, 1997, 159–163.

5. Brown, E. R., RF-MEMS Switches for Reconfigurable Integrated Circuits, *IEEE Transactions on Microwave Theory and Techniques,* 11, November 1998, 1868–1880.

6. Yao, J. J., and Chang, M. F., A Surface Micromachined Miniature Switch for Telecommunications with Signal Frequencies from DC to 4 GHz, *8ᵗʰ International Conference on Solid State Sensors and Actuators,* Stockholm, Sweden, June 25, 1995, 384–387.

7. Goldsmith, C. L., Yao, Z., Eshleman, S., and Denniston, D., Performance of Low-Loss RF MEMS Capacitive Switches, *IEEE Microwave and Guided Wave Letters,* 8, 8, 269–271, August, 1998.

8. Lee, J. J., Atkinson, D., Lam, J. J., Hackett, L., Lohr, R., Larson, L., Loo, R. Matloubian, M., Tangonon, G., De Los Santos, H., and Brunner, R. MEMS in Antenna Systems: Concepts, Design, and Systems Implications, National Radio Science Meeting, Boulder, Colorado, 1998.

9. *The ARRL Antenna Handbook,* The American Radio Relay League, Inc., Newington, CT, 1974, 183–186.

10. Lyle, J. C., Packaging Technologies for Space-Based Microsystems and Their Elements, Micoengineering Technologies for Space Systems, Helvajian, H., Ed., The Aerospace Corporation, El Segundo, CA, 1995, 441–504.

11. Connally, J., and Brown, S., Micromechanical Fatigue Testing, *Experimental Mechanics*, 81–90, June 1993.

12. Gauthier, G. P., Courtay, A., and Rebiez, G. M., Microstrip Antennas on Synthesized Low Dielectric-Constant Substrates, *IEEE Transactions on Antennas and Propagation*, 45, 8, 1310–1314, August 1997.

13. Nguyen, C. T. C., Katehi, L.P.B., and Rabeiz, G. M., Micromachined Devices for Wireless Communications, *Proceedings of the IEEE*, 86, 8, 1756–1768, August 1998.

14. Veidt, B., Kornelsen, K., Vaneldik, J. F., Routledge, D., and Brett, M. J., Diagonal Horn Integrated with Micromachined Waveguide for Sub-Millimetre Applications, *Electronics Letters*, 31, 16, 1307–1309, August 3, 1995.

15. Fractal Antenna Systems, Inc., Fractal Antenna White Paper, July 10, 1999. This document is available on the Fractal Antenna Systems, Inc. Web site.

16. Cohen, Nathan, Fractal Antennas, Part 1, *Communications Quarterly*, Summer, 7, 1995.

17. Mendelbrot, B. B., *The Fractal Geometry of Nature*, W.H. Freeman, New York, 1983.

18. Werner, D, H., An Overview of Fractal Electrodynamics Research, *Proceedings of the 11th Annual Review of Progress in Applied Computational Electromagnetics*, Monterey, California, March 20–24, 1995, 964–969.

19. Cohen, Nathan, Simple CP Fractal Loop Array With Parasitic, *14th Annual Review of Progress on Applied Computational Electromagnetics*, The Applied Computational Electromagnetics Society, March, 1998, 1047–1050.

20. Cohen, Nathan, NEC4 Analysis of a Fractalized Monofilar Helix in an Axial Mode, *14th Annual Review of Progress on the Applied Computational Electromagnetics*, Applied Computational Electromagnetics Society, March, 1998, 1051–1057.

21. Breden, R., and Langley, R. J., Printed Fractal Antennas, IEE National Conference on Antennas and Propagation, 1999.

22. Sindou, M. A., and Sourdois, C., Multiband and Wideband Properties of Printed Fractal Branched Antennas, *Electronics Letters*, 35, 3, 181–182, Feb. 4, 1999.

23. Werner, D. H., and Werner, P. L., Frequency Independent Features of Self-Similar Fractal Antennas, *Radio Science*, 31, 6, 1331–13343, 1996.

24. Werner, D. H., and Werner, P. L., Fractal Radiation Pattern Synthesis, National Radio Science Meeting, Boulder, CO, January 6–11, 1992, 66.

25. Haupt, R. L., and Haupt, S. E., *Practical Genetic Algorithms*, John Wiley & Sons, New York, 1998.

26. Haupt. R. L., Introduction to Genetic Algorithms for Electromagnetics, *IEEE Antennas and Propagation Magazine*, 37, 2, 7–15, April 1995.

27. Haupt, R. L., Comparison Between Genetic and Gradient-Based Optimizaiton Algorithms for Solving Electromagnetics Problems, *IEEE Transactions on Magnetics*, MAG-31(3), 1932–1935, May, 1995.

28. Johnson, J. M., and Rahmat-Samii, Y., Genetic Algorithms in Engineering Electromagnetics, *IEEE Antennas and Propagation Magazine*, 39(4), 7–25, August 1997.

29. Cuthbert, T. R., *Optimization Using Personal Computers with Applications to Electrical Networks*, John Wiley & Sons, New York, 1987.

30. Bahr, M., Boag, A., Michielson, E., and Mittra, R., Design of Ultra-Broadband Loaded Monopoles, IEEE Antenna and Propagation Society, International Symposium Digest, Seattle, WA, 2, 1290–1293, June, 1994.

31. Altshuler, E. E., and Linden, D. S., Wire-Antenna Design Using Genetic Algorithms, *IEEE Antennas and Propagation Magazine*, 39(2), 33–43, April 1997.

32. Werner, P. L., Altman, Z., Mittra, R., Werner, D. H., Ferraro, A. J., Genetic Algorithm Optimization of Stacked Vertical Dipoles Above a Ground Plane, *1997 International Symposium Digest*, Montreal, Canada, 3, 1976–1979, 1997.

33. Rutledge, D. B., Popovic', A. B., Weikle, R. M., Kim, M., Potter, K. A., Compton, R. C., and York, R. A., Quasi-Optical Power-Combining Arrays, *IEEE 1991 MTT S International Symposium Digest*, 1201–1204, 1990.

Antenna Class	Antenna Type	Configuration	Parameters	Typical Frequency Range (MHz)	Gain dB above isotropic	Beam Width at Center Frequency (Degrees)	3 dB Gain Reduction Bandwidth	Input Resistance at Center Frequency	2.0:1 VSWR Bandwidth %	Polarization	Side Lobe Levels
Ideal Isotropic	Isotropic	1		N/A	0	N/A	N/A	N/A	N/A	LP,CP	N/A
Wire Antennas	Small Dipole	2	$L<\lambda/2$		1.74	90				LP	N/A
	Thin $\lambda/2$ Dipole	3	$L=\lambda/2$, $L/D=276$	to X Band	2.14	78	34%	60 Ω		LP	N/A
	Thick $\lambda/2$ Dipole	4	$L=\lambda/2$, $L/D=51$	to X Band	2.14	78	55%	49 Ω		LP	N/A

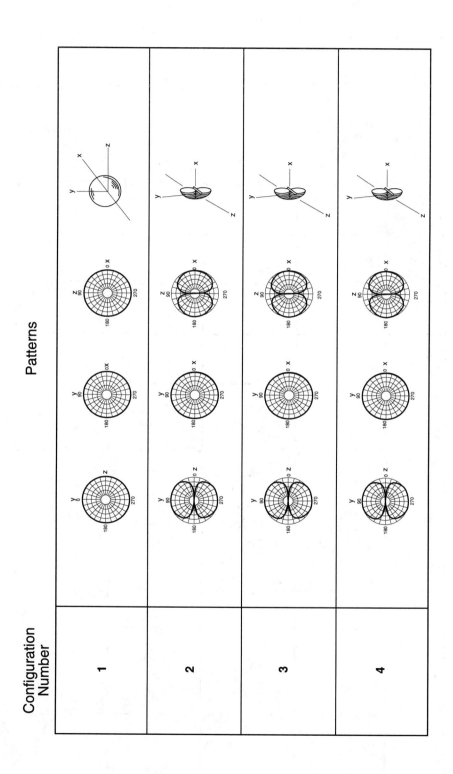

Antenna Class	Antenna Type	Configuration	Parameters	Typical Frequency Range (MHz)	Gain dB above isotropic	Beam Width at Center Frequency (Degrees)	3 dB Gain Reduction Bandwidth	Input Resistance at Center Frequency	2.0:1 VSWR Bandwidth %	Polarization	Side Lobe Levels
Wire Antennas (cont.)	Cylindrical Dipole	5	L=λ/2, L/D = 10	to X Band	2.14	78	100%	37 Ω		LP	N/A
	Biconical Dipole	6	L=λ/2 phi = 40 deg.	to X Band	2.14	78	100%	72 Ω		LP	N/A
	Folded Dipole	7	L=λ/2, L/D = 25.5	to S Band	2.14	78	45%	300 Ω		LP	N/A
	Folded Dipole λ/8 Above Perfect Ground	8	L=λ/2, L/D = 25.5	to S Band	7.14	89	20%	150 Ω		LP	N/A

Patterns

Configuration Number

5	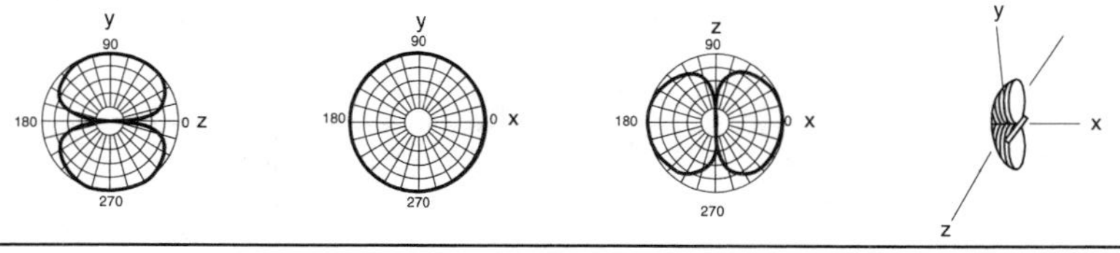
6	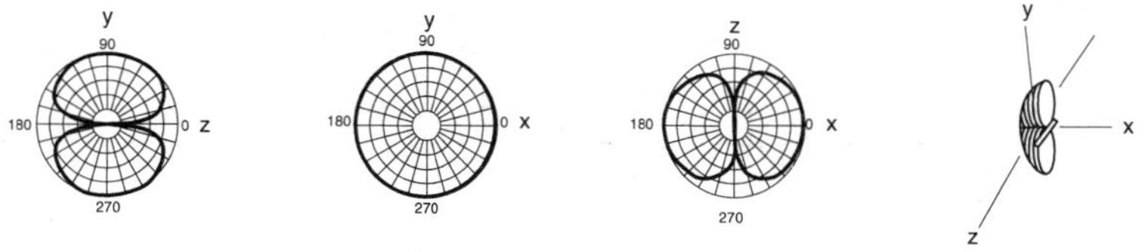
7	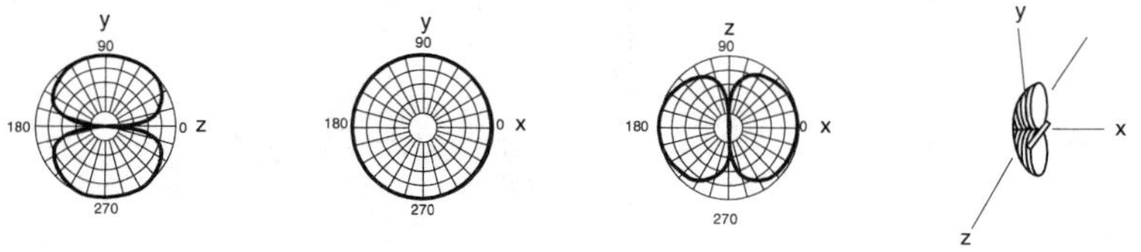
8	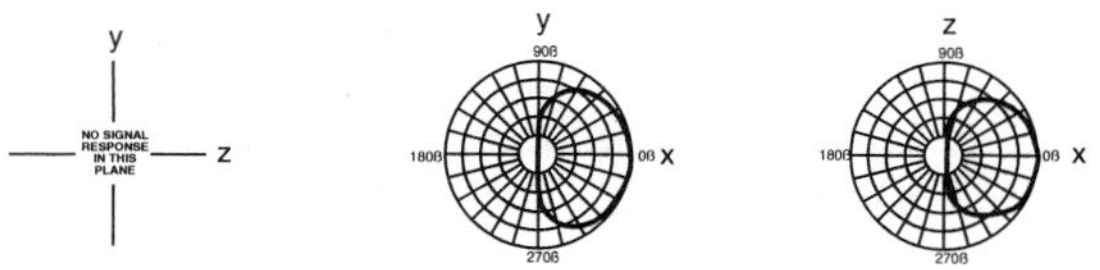

Antenna Class	Antenna Type	Configuration	Parameters	Typical Frequency Range (MHz)	Gain dB above isotropic	Beam Width at Center Frequency (Degrees)	3 dB Gain Reduction Bandwidth	Input Resistance at Center Frequency	2.0:1 VSWR Bandwidth %	Polarization	Side Lobe Levels
Wire Antennas (cont.)	Coaxial Dipole	9	$L = \lambda/4$, $L/D = 40$	to X Band	2.14	78	16%	50 Ω		LP	N/A
	Turnstile Dipole	10	$L/d = 25.5$	to L Band	-0.86		50%	150 Ω		LP	N/A
	Collinear Dipole Array	11	n, a, s	to L Band	2 to 4.5	elevation A_3=omni				LP	N/A
	Yagi - Uda Array	12	λn, sn, n, a	to X Band	2 to 16	variable		50 Ω with match	10 - 60%	LP	N/A

Configuration Number	Patterns
9	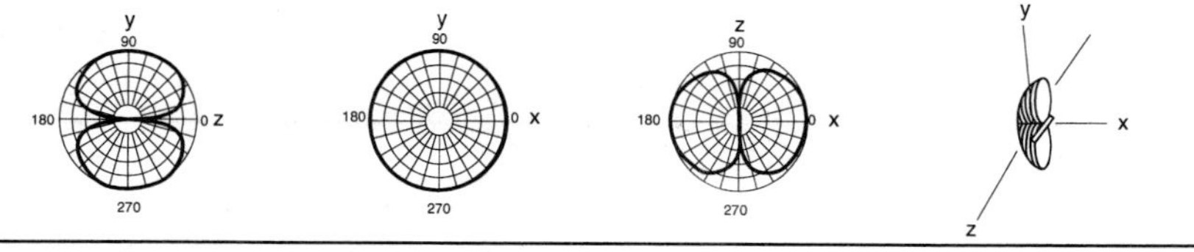
10	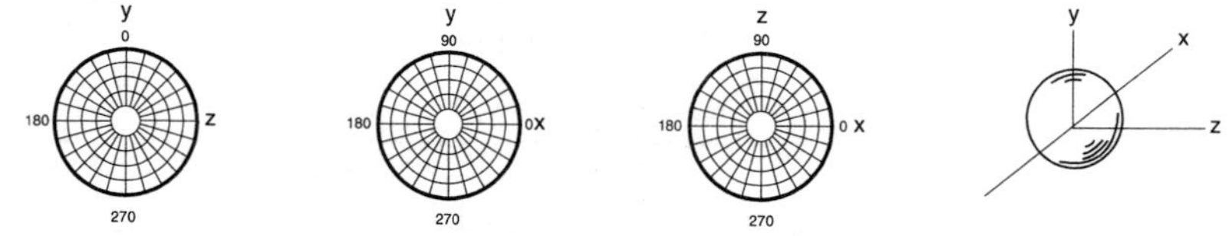
11	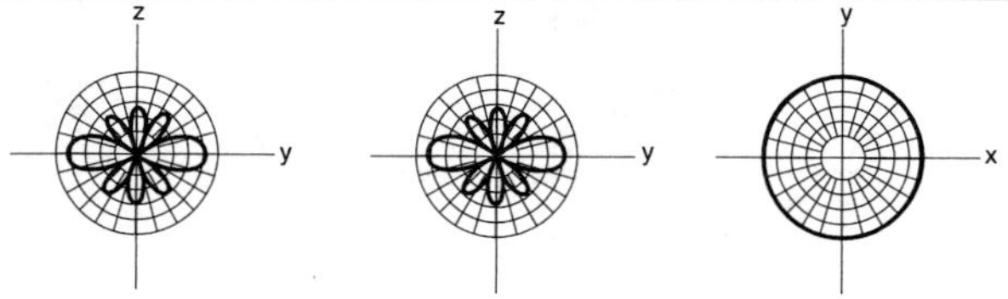
12	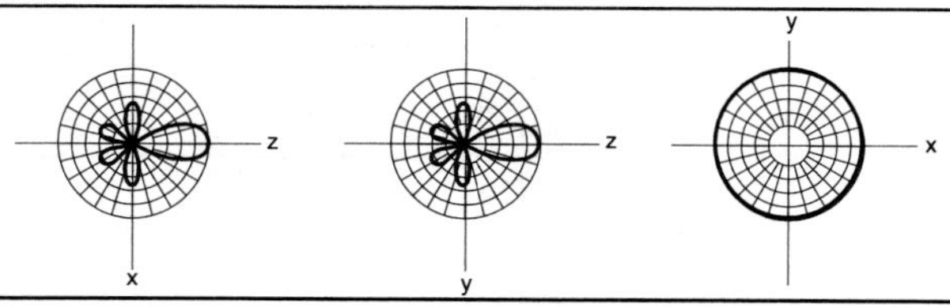

Antenna Class	Antenna Type	Configuration	Parameters	Typical Frequency Range (MHz)	Gain dB above isotropic	Beam Width at Center Frequency (Degrees)	3 dB Gain Reduction Bandwidth	Input Resistance at Center Frequency	2.0:1 VSWR Bandwidth %	Polarization	Side Lobe Levels
Wire Antennas (cont.)	λa/4 Monopole over Infinite Ground	**13**	L=λ/4, thin	to X Band	5.14			36.5 Ω		LP	N/A
	λ/4 Monopole over Finite Perfect Ground	**14**	L = λ/4, L/a = 53, D = λ	to X Band	2.14	78		28 Ω		LP	N/A
	Helical Monopole over Infinite Ground	**15**	L total = λ/2, h > 0.05*λ	to L Band	5.14	78		35 Ω	<4%	LP	N/A
	Center loaded λ/4 by λ/4 Whip over Infinite Ground	**16**	L/d = 0.05, d = dia.	to X Band	8.3			50 Ω	10 - 60%	LP	-10

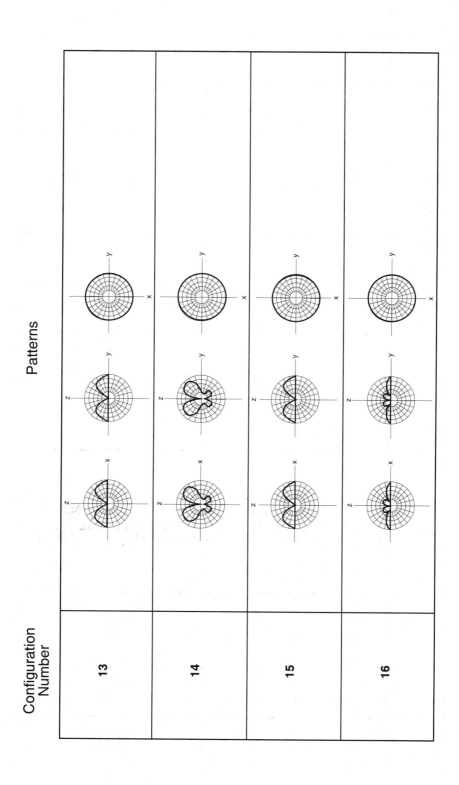

Antenna Class	Antenna Type	Configuration	Parameters	Typical Frequency Range (MHz)	Gain dB above isotropic	Beam Width at Center Frequency (Degrees)	3 dB Gain Reduction Bandwidth	Input Resistance at Center Frequency	2.0:1 VSWR Bandwidth %	Polarization	Side Lobe Levels
Wire Antennas (cont.)	Inverted F Monopole Over Infinite Ground		S, B, Px, Py, h	to X Band	5.14	78		50 Ω	2.50%	Linear:VP	N/A
	Discone Monopole		C, A, B	to S Band	2.14	78	4:01	50 Ω	Octaves	LP	N/A
	Helical Antenna		D, C, S, d, N, L, α, l	to S Band	6.0 to 18	Variable	<200%	approx. 140 Ω	<40%	CP	-13
Microstrip	Rectangular, Single Feed, Edge, or Probe Fed		ε_r, L, w, h	to mmWave	6	90		50 Ω	a few %	LP	N/A

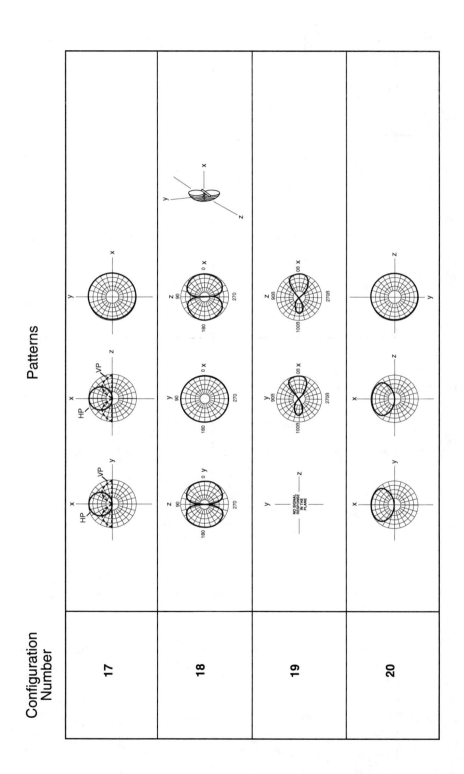

Antenna Class	Antenna Type	Configuration	Parameters	Typical Frequency Range (MHz)	Gain dB above isotropic	Beam Width at Center Frequency (Degrees)	3 dB Gain Reduction Bandwidth	Input Resistance at Center Frequency	2.0:1 VSWR Bandwidth %	Polarization	Side Lobe Levels
Microstrip (cont.)	Circular, single feed, edge or probe fed	21	ε_r, h, a, e	to mmWave	5.0 to 12	90		50 Ω	a few %	LP	N/A
	Circular Dual Fed	22	ε_r, h, r, a, 90 deg. feed	to mmWave	5 to 6	90		50 Ω	<5%	CP	N/A
	Circular, Higher Ordered Mode TM_{210}	23	ε_r, h, r, a, feed phase as shown		approx 5.14	78		50 Ω	a few %	LP, CP	N/A
Slot	$\lambda/2$ planer slot in infinite ground plane	24	$L=\lambda/2$, a	to mmWave	2.14	78		350 Ω		LP	N/A

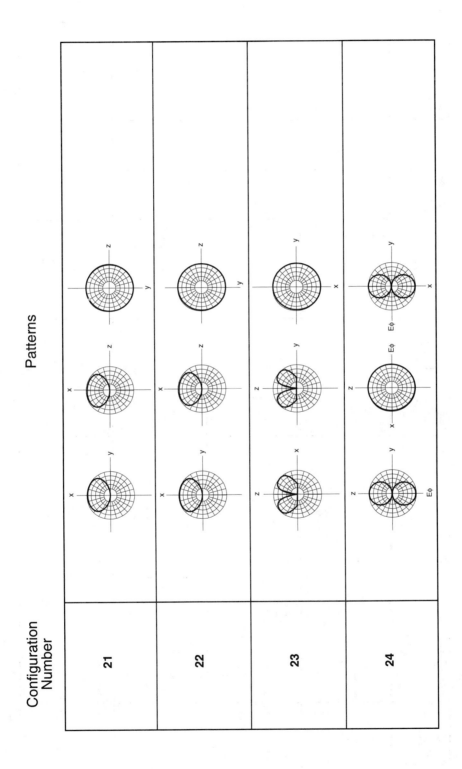

Antenna Class	Antenna Type	Configuration	Parameters	Typical Frequency Range (MHz)	Gain dB above isotropic	Beam Width at Center Frequency (Degrees)	3 dB Gain Reduction Bandwidth	Input Resistance at Center Frequency	2.0:1 VSWR Bandwidth %	Polarization	Side Lobe Levels
Slot (cont.)	Annular Slot	25	a & b < λa	to mmWave	1.5	70				LP	N/A
	Longitudinal Broadwall Waveguide Slot	26	ℓ, a, b, d, λg	to mmWave	approx. 5.0		feed dependent	high	N/A	LP	N/A
Aperture	Rectangular Aperture, Uniform Excitation	27	a by b	to mmWave	10 log(g), $g = 4\pi ab/(\lambda)^2$	yz plane: 50.6(λ)/a xz plane: 50.6(λ)/b		N/A	N/A	LP	-13.2 -13.2
	Rectangular Aperture, TE11 Excitation	28	a,b	to mmWave	10 log(g), $g = 3.24\pi ab/(\lambda)^2$	yz plane: 50.6(λ)/b xz plane: 68.8(λ)/a		N/A	N/A	LP	-13.2 -23

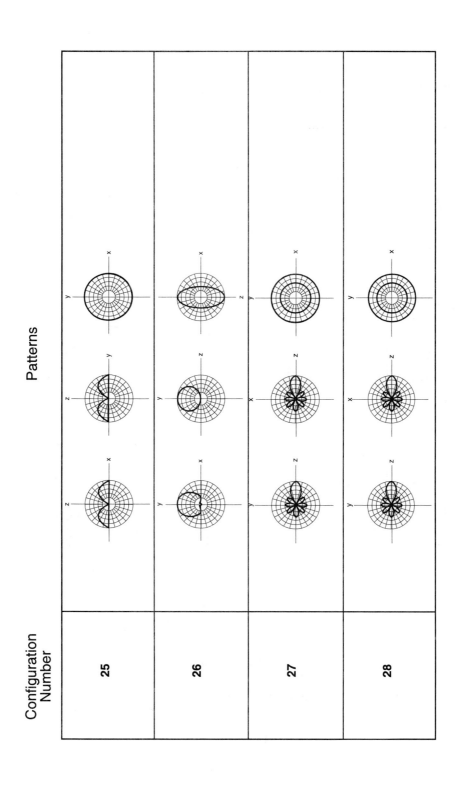

Antenna Class	Antenna Type	Configuration	Parameters	Typical Frequency Range (MHz)	Gain dB above isotropic	Beam Width at Center Frequency (Degrees)	3 dB Gain Reduction Bandwidth	Input Resistance at Center Frequency	2.0:1 VSWR Bandwidth %	Polarization	Side Lobe Levels
Aperture	Circular Aperture, Uniform Excitation	**29**	a	to mmWave	$G = 10 \log(g)$, $g = (2\pi a/\lambda)^2$	xz plane: $29.2(\lambda)/a$ yz plane: $29.2(\lambda)/a$		N/A	N/A	LP	-17.6 -17.6
	Circular Aperture, TM11 Excitation	**30**	a	to mmWave	$G = 10 \log(g)$, $g = 10.5\pi(a/\lambda)^2$	yz plane: $29.2(\lambda)/a$ xz plane: $37.0(\lambda)/a$		N/A	N/A	LP	-17.6 -26
	Circular Aperture, Tapered Distribution	**31**	a, $1-(r/a)^2$ distribution	to mmWave	$G = 10 \log(g)$, $g = 0.75*(2\pi*a/\lambda)^2$	$36.4\lambda/a$		N/A	N/A	LP	-24.6
	Prime Focus Reflector (dipole feed)	**32**	$D = \dfrac{5\lambda}{2}$	to mmWave	14.75			300 Ω	30		

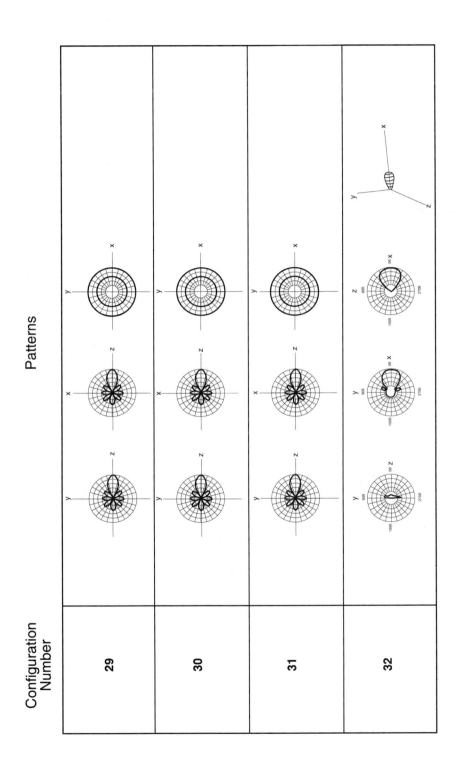

Antenna Class	Antenna Type	Configuration	Parameters	Typical Frequency Range (MHz)	Gain dB above isotropic	Beam Width at Center Frequency (Degrees)	3 dB Gain Reduction Bandwidth	Input Resistance at Center Frequency	2.0:1 VSWR Bandwidth %	Polarization	Side Lobe Levels
Aperture	Pyramidal Horn	33	L, l, b	to mmWave	< 35		35	50 Ω		LP	
Loop	Full wave loop	34	d, D	to X Band	3.14		13	45 Ω		LP	
Frequency Independent	Logperiodic Diapole Array	35	length dipole: length diam spacing	to X Band	7 to 12	> 50 deg.	N/A	50 - 100 Ω	approx. 200	LP & CP	< -13 dB
	Cavity Backed Planer Sprial	36	ρ, φ mode 2 has null	to mmWave	< +4.0	approx. 60 deg. Model 1	N/A	Balun required	> Decade	LP & CP	N/A

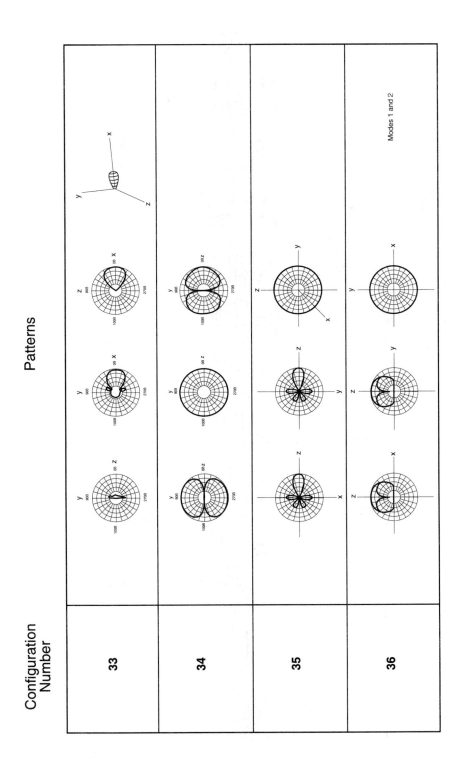

Antenna Class	Antenna Type	Configuration	Parameters	Typical Frequency Range (MHz)	Gain dB above Isotropic	Beam Width at Center Frequency (Degrees)	3 dB Gain Reduction Bandwidth	Input Resistance at Center Frequency	2.0:1 VSWR Bandwidth %	Polarization	Side Lobe Levels
Frequency Independent	Conical Spiral	37	α, ρ, φ mode 2 has null	to mmWave	1 to 15	140 to 40	N/A	50 to 300 Ω Balun req.		LP & CP	N/A
	Tapered Dielectric Rod	38	d1, d2, d3, d4, P, L, l1, l2, l3, l4	to mmWave	18 - 20	55 $(\lambda/L)^{0.5}$	+/- 15%	Zo matched feed	40	LP	<-20
Leaky Wave	Long Shot in Waveguide	39	h, w, d, L, θ L = slot length	mmWave	10 log (2L/λ)	$\Delta\theta = \frac{l}{\left(\frac{L}{\lambda o}\right)\cos\theta m}$ θm = Angle of max. radiation	Freq, beam scan	Zo matched to feed	Broad	LP & CP	-30
	Leaky Waveguide Trough	40	w, d1, d2,	mmWave	10 log (2L/λ)	$\Delta\theta = \frac{l}{\left(\frac{L}{\lambda o}\right)\cos\theta m}$ θm = Angle of max. radiation	Freq, beam scan	Zo matched to feed	Broad	LP & CP	-30

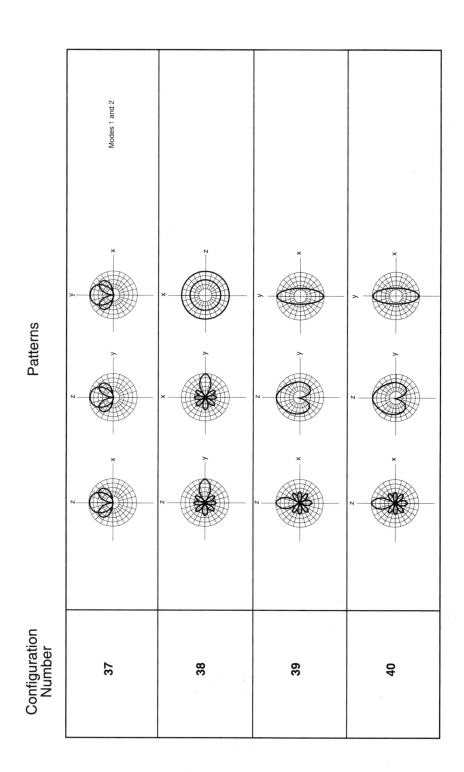

Antenna Class	Antenna Type	Configuration	Parameters	Typical Frequency Range (MHz)	Gain dB above isotropic	Beam Width at Center Frequency (Degrees)	Gain Reduction Bandwidth	Input Resistance at Center Frequency	2.0:1 VSWR Bandwidth %	Polarization	Side Lobe Levels
Lens	Dielectric	**41**	$N, f, c, o, C0, \varepsilon_r$ n = refractive index $n = \sqrt{\varepsilon_r}$ $k = \frac{2\pi}{\lambda}$ N = number of zones	x Band to mmWave	$<= 50$	prop. to gain	$25/(N-1)$ percent	N/A	N/A	LP & CP	< -20
	Waveguide	**42**	$a, c0, c1, s, D, dn, \ell$ n_o= refractive index	x Band to mmWave	$<= 50$	prop. to gain	$25\pi n_o/(1+Nno)$ percent	N/A	N/A	LP	< -20
	Fresnel Zone Plate Antennas	**43**	f, D, Z Z = number of transport zones fo = antenna freq.	mmWave	<40	prop. to gain	fo/Z	N/A	N/A	LP	< -10
Notch	Linear Tapered Slot	**44**	$t, W, L,$ teff, ε_r $\frac{teff}{\lambda o} = \frac{(\sqrt{\varepsilon_r}-1)}{\lambda o} \leq 0.01$ $L = 4\lambda$ to 10λ $\propto = 11.2°$	X Band to mmWave	$10 \log\left(\frac{4L}{\lambda o}\right) - 2$	$77/\sqrt{\frac{L}{\lambda o}}$ degrees	5:1 (3 dB)	$50 - 200\ \Omega$	N/A	LP	< -10

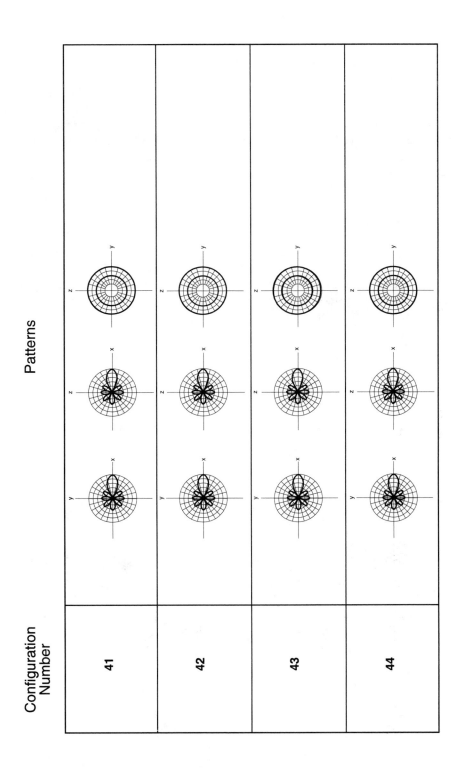

6.9   Phased Array Antenna Technology[1-3]

James B. West

An **array antenna** is a structure comprised of a collection of individual radiation elements, each excited with a specific amplitude and phase, whose electromagnetic fields are spatially combined to realize a specific radiation pattern, usually in the far field. An array antenna can also be thought of as a continuous aperture distribution that is sampled at discrete locations within the aperture. A **phased array** antenna is an array whose main beam can be electrically steered in one or more dimensions by changing the excitation phases on the individual radiation elements of the array. Beam scanning rates on the order of one microsecond are possible for certain array architectures, which is orders of magnitude more rapid than state-of-the-art mechanical scanning with motor technology. Rapid beam scanning is crucial in fire control and other radar systems where wide scan volumes must be rapidly interrogated and where multiple target tracking is a requirement.

Array antennas can take on one of four forms: (1) a **linear array,** where the array radiating elements are in a single line; (2) a **circular array,** where the radiating elements are laid out in a circular arrangement relative to a center reference location; (3) a **planar array,** where the radiating elements are laid out in a two-dimensional grid; and (4) a **conformal array,** where the radiating elements locations are described by a three-dimensional surface. Figure 6.146 illustrates the various array architectures.

The discussion herein will focus on linear phased array technology for brevity. The fundamental phased array concepts developed for linear arrays can be extended to two-dimensional and conformal arrays. The reader is referred to the literature for more detailed expositions.

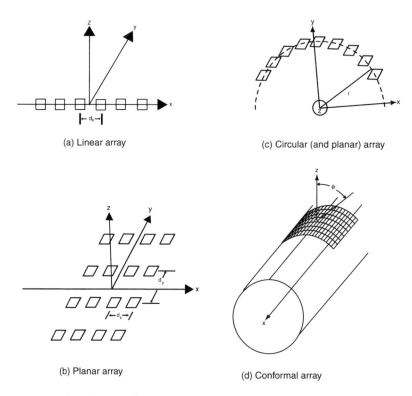

(a) Linear array

(c) Circular (and planar) array

(b) Planar array

(d) Conformal array

FIGURE 6.146   Various phased array architectures.

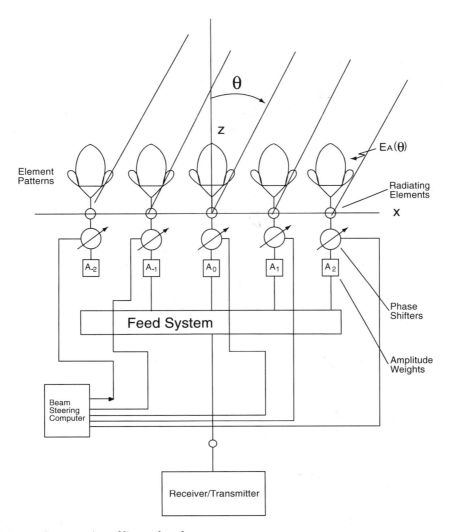

FIGURE 6.147   Center section of linear phased array.

Linear Array Technology

A linear array is conceptually illustrated in Fig. 6.147. This structure consists of a collection of **radiating elements** laid out in a straight line that form the transition from free space to the array feed system. Several common antennas, as summarized in the antenna catalog of the previous section, are appropriate choices for phased array design. Dipoles; monopoles; crossed dipoles; bow tie monopoles; open sleeve dipoles; edge and broad wall waveguide fed slots; printed notches; stripline or microstrip fed linear and crossed printed slots; circular, rectangular, and ridged open end waveguide; TEM horns; microstrip patches; and printed dipoles are commonly used. Each radiating element has a radiation pattern which, along with its element amplitude and phase excitation, contribute to the aggregate radiation pattern. The **feed system** is an apparatus that distributes the microwave energy between the radiating elements and the receiver/transmitter. The feed systems can either be constructed out of guided wave transmission lines, or be a radiation-based space feed. The **phase shifter** is a microwave/millimeter wave circuit device that outputs variable insertion phase, or time delay for a broadband array system, relative to its input. By adjusting the insertion phase of each phase shifter, the proper relative phase shift across the radiating elements is realized for beam scanning. **Amplitude weighting** is the nonuniform distribution of power across the array and is used to generate low sidelobe far field radiation patterns. The amplitude weighting is usually implemented in the feed system, but is shown conceptually as blocks A_{-N} through A_N in Fig. 6.147. The **beam steering computer (BSC)** computes the proper insertion phase value for a given beam pointing direction and sends the appropriate commands to each phase shifter.

The first order array analysis assumes isotropic radiating elements that are isolated from one another, i.e., there is no **mutual coupling**. Mutual coupling causes a given radiating element its impedance properties and its radiation pattern, to be effected by its proximity to neighboring radiating elements and their individual amplitude and phase excitations. This effect will be discussed in further detail in subsequent paragraphs.

Consider a linear array in the transmit mode. In the far field, the observation point is at infinity, which causes the line-of-sight rays from each radiating element and the observation point to be parallel. This approximation leads to the following expression for the far field electric field patterns as a function of θ:

$$E_A\left(\theta\right) \equiv \underbrace{\left[\frac{\exp\left(-j\dfrac{2\pi r_o}{\lambda}\right)}{r_o}\right]}_{\text{Isotropic Element Pattern}} \cdot \underbrace{\sum_N A_n \exp\left[j\frac{2\pi}{\lambda}n\Delta x\left(\sin\theta - \sin\theta_o\right)\right]}_{\text{Array Factor}} \qquad (6.69)$$

where: N = number of radiating elements,
r_o = distance for the center of the array to the far field observation point,
A_n = array element amplitude weighting coefficients,
θ = beam pointing, referenced to broadside ($\theta = 0$),
Δx = the array element spacing,
λ = wavelength,
θ_o = the desired beam scan angle, and
$\Phi_n \equiv \frac{2\pi}{\lambda} n\Delta x \sin\theta_o$ = The insertion phase for the nth phase shifter required to scan the beam to θ_o.

The isotropic pattern does not influence the far field amplitude pattern of the array and is typically dropped from the expression. One can see that the number of radiating elements, their relative spacing, the operating frequency (wavelength), and phase shifter settings of the array all influence the radiation pattern.

Only one main beam is desired in most system applications. Secondary, or false main beams are called **grating lobes**, and they are a function of: (1) the array element spacing, and (2) the scan volume of the phased array. Consider Eq. (6.69). All radiation elements add constructively in phase when:

$$\frac{1}{\lambda}n\Delta x\left(\sin\theta - \sin\theta_o\right) = m, \qquad (6.70)$$

where $m = 0, \pm1, 2, 3, \ldots$

The $m = 0$ is the principal beam and the others represent grating lobes. Grating lobes, as a function of element spacing and beam scanning, can be predicted by the following expression:

$$\sin\theta_{gl} \equiv \sin\theta_o + \frac{\lambda}{\Delta x} \qquad (6.71)$$

where: θ_{gl} is the first grating lobe in visible space.

If the array is required to scan over the $\pm\theta_m$ scan volume, then the required spacing is:

$$\Delta x \leq \frac{\lambda}{\sin\theta_m}. \qquad (6.72)$$

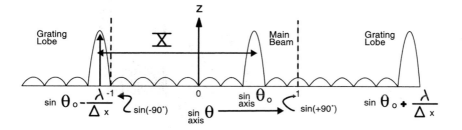

- **X** is a function of element spacing

- Grating lobes and the main Beam shift in concert along the sin θ axis as the array is scanned

- The left grating lobe is on the edge of visible space

FIGURE 6.148   Relationship of grating lobe to the main beam.

If the array inter-element spacing is one-half free wavelength, then the array can scan ±90° without grating lobes. Figure 6.148 illustrates the relationship between the main beam, grating lobes, and the region of visible space for a given array element spacing.

Amplitude weighting is used to create low sidelobe patterns. The theoretical sidelobe level for a uniformly excited linear array is −13.5 dB relative to the main beam peak. The uniformly illuminated array also has maximum gain and minimum beamwidth for a given number of array elements and inter-element spacing. Lower sidelobe level designs require a non-uniform amplitude illumination across the array. There is a trade-off between gain and beamwidth, and sidelobe level. Low sidelobe level arrays exhibit lower gain, and increased beamwidth for a given array relative to uniform illumination.

The **directivity** of a linear array of ideal isotropic architecture elements with spacings that are multiples of one half wave can be expressed as:

$$D_o \equiv \frac{\left| \sum_N A_n \right|^2}{\sum_N \left| A_n \right|^2}, \tag{6.73}$$

where A_n = the element excitation coefficients.

The reduction in gain due to amplitude weighting, relative to uniform aplitude weighting can be expressed as aperture efficiency by the following formula:

$$\eta_{ap} \equiv \frac{\left| \sum_N A_n \right|^2}{N \sum_N \left| A_n \right|^2}, \tag{6.74}$$

where A_n = the element excitation coefficients.

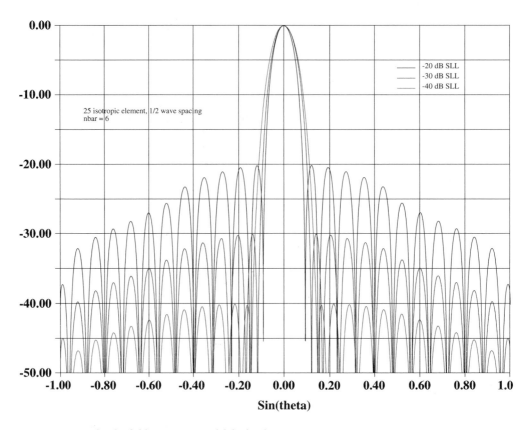

FIGURE 6.149   Taylor far field patterns vs. sidelobe level.

Several pattern synthesis methodologies have been created for low sidelobe level designs for continuous line sources and two-dimensional apertures. These procedures have been extended to array theory by continuous aperture sampling at the array element locations, or null matching techniques.[4] The most commonly used proceeding is the **Taylor weighting**. The Taylor design yields a far field radiation pattern that is an optimal compromise between beamwidth and sidelobe level. The details of the generation of the Taylor coefficients are beyond the scope of this article and the reader is referred to the literature.[5] Figure 6.149 compares uniform illumination, a −20, −25, and −30 dB Taylor sidelobe level designs.

Realistic radiation elements are not isotropic, but have an element pattern, and element pattern weights the array factor, as shown in Eq. (6.75) below:

$$E_A(\theta) = \underbrace{\left[E_A(\theta)\right]}_{\text{Element Pattern}} \cdot \underbrace{\sum_N A_n \exp\left[j\frac{2\pi}{\lambda}n\Delta x\left(\sin\theta - \sin\theta_o\right)\right]}_{\text{ArrayFactor}} \tag{6.75}$$

The isotropic element pattern is suppressed in this expression for clarity. A useful approximation for an ideal element is:

$$E_A(\theta) = \cos(\theta), \tag{6.76}$$

where θ is referenced off the Z axis.

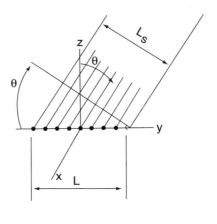

$$L_S = L \cos \theta$$

$$D_O \approx 2 \left(\frac{L}{\lambda} \right) \Bigg|_{L \gg \Delta X}$$

$$D_S \simeq \frac{2 L \cos \theta}{\lambda}$$

ΔX = element spacing

FIGURE 6.150   Aperture projection loss.

The ideal element pattern takes into account the aperture projection loss as the array scans off bore sight, as illustrated in Fig. 6.150, but does not account for the effects of mutual coupling between the radiating elements. All linear and planar phased arrays suffer at least a $\cos(\theta)$ scan loss. This loss can actually be more severe, i.e., $\cos^x(\theta)$, where x is a real number greater than 1.0, when mutual coupling is present. This effect will be described in subsequent paragraphs. Figure 6.151 illustrates the effect of the element pattern weighting on the main beam and sidelobes of the array space factor. Several points are evident from this figure:

1. The main beam gain falls off at least by $\cos(\theta)$ with scan angle.
2. Grating lobes just coming into visible scan manifest themselves as high sidelobe levels.
3. A broad side array cannot scan to end fire since $\cos(90°) = 0$. Arrays can be designed to operate under this "end fire" condition. The reader is referred to the literature.[6]

Array bandwidth is affected by several parameters, including impedance, polarization, radiation pattern bandwidth of the radiating elements, and mutual coupling between the radiation elements. The array factor experiences a beam squint as a function of frequency due to the frequency response of the phase shifters. For a linear array the frequency change that causes the beam position to shift $\pm^1/_2$ of is nominal 3 dB beam width can be expressed as a fractional bandwidth by the following expression:

$$\text{fractional band width} = \frac{\Delta f}{f} = \frac{BW_o}{\sin \theta_s} \tag{6.77}$$

where: θ_s = the scan angle off bore site, in radians, and
BW_o = the beam width of the array at bore site.

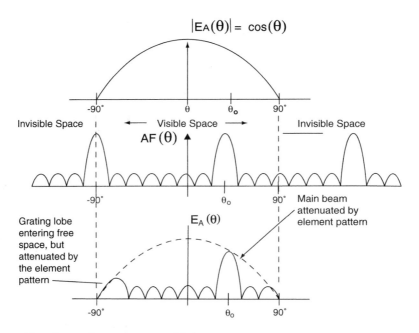

FIGURE 6.151 Phased array element pattern weighting.

Broad bandwidth can be realized when the array phase shifters are replaced with true time delay devices. Beam scanning is achieved by programming time delay into each radiating element to compensate for the difference in propagation time from each radiating element to an arbitrary observation point in the far field. In this case the bandwidth is not limited by the array factor because there is no beam squint with frequency. The array bandwidth is ultimately determined by the radiating element bandwidth within the array environment along with feed network bandwidth.

Array Error Sources

Random amplitude and **phase errors** cause increased sidelobe levels, pointing errors, and reduced directivity.[7] Random phase and amplitude errors can be treated statistically. The phase error ϕ_n is described by a Guassian probability distribution with zero mean and variance $\overline{\phi_n}$. The amplitude error δ_n has variance $\overline{\delta_n^2}$ and zero mean. The **average side lobe level due to random phase and amplitude errors** can be expressed as:

$$\overline{\sigma_1^2} = \left[\overline{\sigma_n^2 + \delta_n^2} \right] \cdot g_e, \tag{6.78}$$

where: $\overline{\sigma_1^2}$ = the average side lobe level relative to the isotropic side lobe level in the plane of the linear array axis, and
  g_e = the element directivity.

The **reduction of directivity for a linear array due to random phase and amplitude errors** is given by:

$$\frac{D}{D_o} = \frac{1}{1 + \overline{\phi_n^2} + \overline{\delta_n^2}}, \tag{6.79}$$

where: D = the directivity of the array with errors, and,
  D_o = the directivity of the error free array.

The **beam pointing error due to random phase errors** is given by:

$$\overline{\Delta^2} = \overline{\phi_n^2} \cdot \frac{\sum I_i^2 \dfrac{x_i}{\Delta x}}{\left(\sum I_i \left(\dfrac{x_i}{\Delta x}\right)\right)^2},$$

(6.80)

where: $\dfrac{x_i}{\Delta x}$ = the array element position from the origin relative to the element spacing, and

I_i = the element amplitude weighting.

Most contemporary phased arrays incorporate digital phase shifters. The development discussed thus far has considered only a perfect linear phase front across the elements to initiate beam scanning. This linear insertion phase required for beam scanning is typically implemented with finite insertion phase values, e.g., a 4-bit phase shifter has 180°, 90°, 45°, and 22.5° phase settings. In this case, the 180° bit is referred to as the most significant bit (MSB), while the 22.5° bit is referred to as the least significant bit (LSB). Any phase shift required by the radiating element for a given beam position is approximated by linear combinations of these phase bits. The higher the bit count, the better the approximation is for a given analog phase slope across the array. Three to 8 bits of insertion phase are typically used for practical phased array systems. These discrete phase approximations lead to errors in the beam pointing angle and can also deteriorate the array's far field sidelobe levels as a function of beam scanning.

In addition to random error efforts, **quantization errors due to digital phase shifters** add additional errors. The number of bits in the phase shifters determine the **minimum beam steering increment** of the array by the following expression:

$$\theta_{min} \cong \frac{1}{2^P} \cdot BW_o,$$

(6.81)

where: P = the number of bits in the phase shifter,

θ_{min} = the minimum steering increment of the array, and

BW_o = the bore site 3 dB beam width of the array.

It can readily be seen that a 4-bit phase shifter allows a minimum beam steering increment of 1/16 of the 3 dB beamwidth.

The following development is after Miller.[8] The digital approximation to a linear phase front required to scan the main beam creates a sawtooth phase error across the array as shown in Fig. 6.152. This is due to a P-bit phase shifter that has states separated by the least significant bit:

$$\phi_o = \frac{2\pi}{2^P}.$$

(6.82)

The **peak sidelobe level due to phase quantization** can be approximated as:

$$SLL_{Pk\_QL} = \frac{1}{2^{2P}}, \quad \text{or} \quad SLL_{Pk\_DL}(dB) = -6P.$$

(6.83)

This result is shown in Table 6.13.

Figure 6.153 shows how the peak sidelobe level is affected by the number of bits of the phase shifting device.

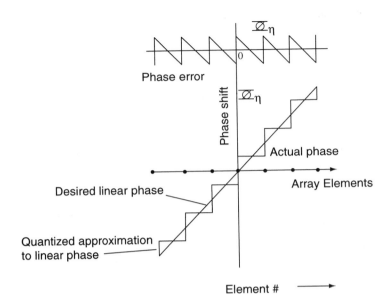

FIGURE 6.152   Quantization-induced phase error across a linear array.

TABLE 6.13   Peak Quantization Side Lobe Levels

Number of Bits	Peak SLL (dB), Relative to Main Beam
3	−18
4	−24
5	−30
6	−36
7	−42

The average sidelobe level due to the triangular phase error has also been analyzed. The mean squared error is given as:

$$\overline{\phi^2} = \frac{1}{3} \cdot \frac{\pi^2}{2^{2P}}. \tag{6.84}$$

The average sidelobe level due to quantization errors alone is:

$$\overline{\sigma_{QL}^2} = \frac{1}{3g_A} \cdot \frac{\pi^2}{2^{2P}} \tag{6.85}$$

where: g_A = the directivity of the linear array.

The **effects of phase quantization on pointing errors** can be described by:

$$\Delta_{QL} = \frac{\pi}{4} \cdot \frac{1}{2^P} \text{ beamwidths.} \tag{6.86}$$

The periodic nature of the sawtooth phase error function can be broken up by adding a random phase offset at each element.[9] The random, but known, offsets are included in the calculation of the phase shifter settings for a given beam position. The procedure tends to "whiten" the peak quantization sidelobe level without changing the average phase error level.[10]

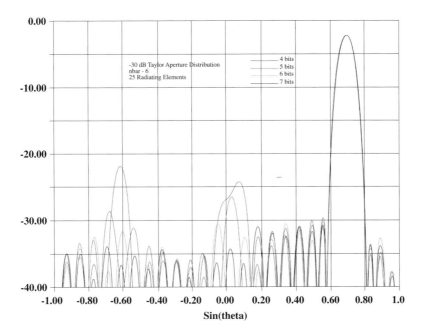

FIGURE 6.153 Taylor 45 deg, beam scan vs. phase shifter bit count.

True Time Delay Steering and Subarrays

Time delayed phased arrays replace the phase shifter devices in traditional devices with programmable time delay units at each radiating element. Scanning is realized by using time delay to compensate for the differences in propagation time from each radiating element to the target. The propagation delay across the phased array aperture becomes significant as the array is scanned off bore site. The instantaneous bandwidth is not limited by the antenna array factor, i.e., there is no beam squint as a function of frequency. True time delay devices have historically been lossy, bulky, heavy, and expensive, but recent advances in Micro-machined ElectroMechanical Systems (MEMS) hold promise for high performance at moderate input power levels in compact monolithic semiconductor architectures. A time delayed steered phased array is illustrated in Fig. 6.154. After a series of mathematical simplifications, the array factor for a time delayed steered phased array can be expressed as:

$$E(\theta) = \sum_{n=1}^{M} A_n \cdot \exp\left[\left(\left(j \cdot \frac{2\pi}{\lambda}\right) \cdot n \cdot \Delta x\right) \cdot \left(\sin\theta - \sin\theta_o\right)\right] \tag{6.87}$$

The maximum of the electric field occurs at θ_o independent of the wavelength (operating frequency), so no beam squint occurs. Two points should be noted: (1) the array beamwidth and gain are still a function of frequency, and (2) the ultimate bandwidth of an actual array is dependent on the actual bandwidth of the radiating elements and the true time delay devices used in the array.

A large phased array can be partitioned as a group of smaller arrays called **subarrays**.[11] Subarrays are used for the following: (1) to improve the broadband performance of an array while incorporating fewer true time delay devices, (2) to simplify the feed manifold structure, and (3) to provide convenient amplifier/feed network integration.

Grouping an array into subarrays will generally increase the sidelobe level of the array, as illustrated in Fig. 6.155. Random errors have a larger effect at the sub array level, as opposed to the individual element level, since there are typically much fewer subarrays than individual radiation elements, and these errors tend to be statistically correlated. Peak sidelobe errors are also larger because they are

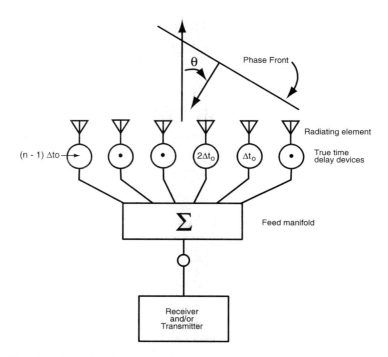

FIGURE 6.154   True time delay phased array architecture.

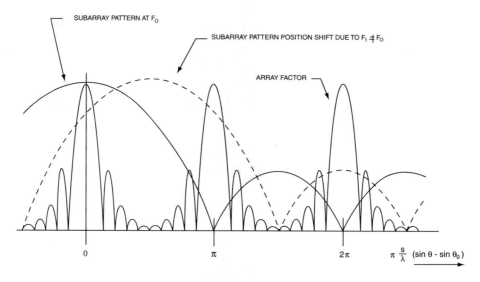

• Subarray pattern null cancels the array factor
  null only at the center frequency
• S = Adjacent subarray spacing

FIGURE 6.155   Subarray sidelobes as a function of frequency.

manifested as grating lobes since the subarray's physical implementation requires a large electrical spacing between adjacent subarrays.

The subarrayed phased array antenna, as shown in Fig. 6.156, can be treated as the product of the radiation element pattern, the subarray pattern, and the array factor for the array of subarrays. This

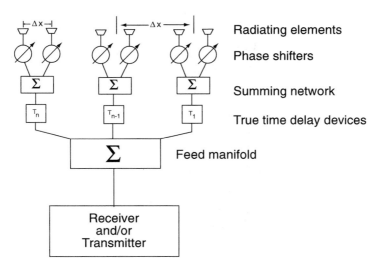

FIGURE 6.156 True time delay subarrayed architecture.

analysis neglects the effects of element pattern edge effects and mutual coupling between the radiation elements. Under these conditions the electric field expression can be written as:

$$E(\theta) = E_A(\theta) \cdot \sum_{m=1}^{M} W_m \cdot \exp\left[j\left(\frac{2\pi}{\lambda}\right) \cdot m \cdot \Delta s \cdot \left(\sin\theta - \sin\theta_o\right) \right]$$

$$\cdot \sum_{n=1}^{N} A_n \cdot \exp\left[j\left(\frac{2\pi}{\lambda}\right) \cdot n \cdot \Delta x \cdot \left(\frac{\sin\theta}{\lambda} - \frac{\sin\theta_o}{\lambda_o}\right) \right]$$

(6.88)

where: Δx = the element spacing within the subarray,
M = the number of radiating elements within each subarray,
A_n = the amplitude weighting within each subarray,
Δs = the subarray spacing within the array,
N = the number of subarrays in the array, and
W_m = the amplitude weighting of the subarrays across the array.

The bandwidth of this type of array is essentially the bandwidth of an individual subarray. Grating lobes generated off frequency due to the array factor ultimately set the array bandwidth. Unequal array spacing techniques can be used to reduce these grating lobes at the expense of higher overall sidelobe level for the top-level array.[12]

Subarray level time delay steering without the generation of grating lobes can be accomplished by means of overlapping subarrays. The technique involves pattern synthesis of an aperture size larger than the inter-subarray spacing. Flat topped, narrow, subarray patterns are possible that can suppress the array factor grating lobes, but the required feed networks become quite complex.[13]

Feed Manifold Technology

The **feed manifold** in a phased array is a guided wave electromagnetic system that distributes the RF energy from an input and/or output to each of the radiating elements in some prescribed manner. The introductory discussion herein will assume the antenna is in a transmission mode, but the principals are directly applicable to the receive case due to reciprocity. **Constrained feeds** utilize guided wave transmission lines to distribute the microwave energy to the radiating elements. A **space feed** utilizes a smaller

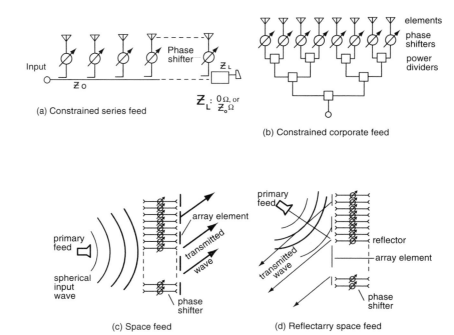

FIGURE 6.157   Phased array feed.

antenna system that excites the main phased array through radiation between the feed antenna and the receive elements of the phased array. The signals are then phase shifted for beam steering, perhaps amplified, and perhaps amplitude weighted for pattern control, and finally transmitted. Space feeds are much simpler to implement than constrained feeds for very large phased arrays, albeit with commensurate performance trades.

Constrained phased array feeds are generally classified as either **series**, where the radiating elements are in series with the feeding transmission line, or **shunt**, where the radiating elements are in parallel with the feeding transmission line. Series array feeds can be further subdivided into either **resonant**, or **traveling wave** feeds. Parallel feeds can be further subdivided into **corporate** and **distributed** feeds. Figure 6.157 illustrates the various feed architectures.

Series feeds are typically used for waveguide linear arrays. Microstrip patches can also be series fed via interconnecting series microstrip transmission lines, but potential problems with this approach include spurious radiation off the series transmission lines, particularly for low sidelobe level designs, and limitations of element amplitude tapering.

In resonant series feed design, each radiating element should be resonant (a pure real impedance) at the operating frequency. The sum of the radiating element's resistance (conductance) must equal the characteristic impedance (admittance) of the feed transmission line. Short-circuited quarter-wave or half-wave transmission line sections are used as the feed line loads to excite the radiating elements at voltage or current standing wave peaks. The resonant array exhibits high efficiency since all of the input power to the feed is radiated (ignoring dissipative losses in the system). The feed is inherently narrow band since the standing wave peaks must be precisely located at the radiating element, which strictly happens only at one frequency. It can be shown[14] that for a series resonant array the relationship between the number of array elements and VSWR is:

$$N * B = 66 \text{ for a uniformly excited array,} \tag{6.89}$$

and

$$N * B = 50, \text{ for a 25 dB SLL Taylor aperture distribution,} \qquad (6.90)$$

where: N = the number of array elements in the series fed array, and
B = the % Bandwidth of the array.

A trade between the number of radiating elements in a series feed and the percentage bandwidth is immediately apparent. Parallel feeding of smaller series-fed subarrays is sometimes used to increase bandwidth.

A **traveling wave feed** also has radiating elements distributed along a feed transmission line, but the input is at one end of the transmission line and a load is at the opposite end of the feed, as shown in Fig. 6.157a. As the input wave propagates from the input toward the load, a portion of the input power is coupled to free space by each radiation slot, and if the feed is properly designed, only a small fraction of the input power is absorbed by the load. Since this structure is nonresonant, wider bandwidths can be realized, but there is beam squint as a function of frequency. Also, this type of array is less efficient than the resonant array since approximately 15% of the power is typically delivered to the load.[15] The reader is referred to the literature for series feed design details.

Corporate feeds are typically used in dipole, waveguide aperture, microstrip patch, and printed slot radiating elements. In this architecture, there are successive power subdivisions until the all of the radiating elements of the array are excited. Figure 6.157b illustrates a binary corporate feed exciting an eight-element linear array. The division ratio at each junction is typically binary, but it can be from two to five, depending on the number of radiating elements and the type of power splitters used. The performance of a corporate feed is critically dependent on the power splitter architecture used. Hybrid couplers such as the Wilkinson, Lange, or branch line hybrids are typically used since they reduce the detrimental effects of radiating element impedance mismatch on the feed, which causes amplitude and phase imbalance. This imbalance can deteriorate the array's radiation pattern, particularly in low sidelobe level designs.

Distributed feeds have each radiating element, or a group of radiating elements, connected to their own receive/transmit (T/R) module. Contemporary examples of this approach are based on MMIC (Monolithic Microwave Integrated Circuit) T/R modules. A MMIC T/R module can contain transmit, low noise amplifier (LNA), phase shift, attenuation, low and higher power T/R switch, bias, and beam steering control logic functions. Collections of distributed feeds for a large array are typically fed with a passive corporate feed. Depending on the complexity of the T/R module, the passive corporate feed can operate at an intermediate frequency with the frequency translation to the actual radiated carrier frequencies occurring within the T/R module. The reader is referred to Chapters 5 and 7 of this text for a detailed description of MMIC-based T/R modules.

The actual hardware implementation of the feed manifold, phase shifter, and time delay device technology are covered in Chapter 5 of this handbook.

Space fed arrays typically take one of two basic forms, the in-line space fed-array, and the reflect array, as shown in Figs. 6.157c and 6.157d. The in-line space-fed phased array is analogous to exciting a lens structure. The array collects the energy radiated from the primary feed, compensates for its spherical nature, and adds the desired phase shifter for beam scanning, and the output set of radiating elements reradiates the beam into free space. This type of array has a natural amplitude pattern due to the spherical wave spreading of the primary feed's radiated energy. Low sidelobe level antennas can be designed, but careful attention must be applied to the feed antenna design to minimize spillover loss. Space-fed arrays are currently the only practical alternative for extremely large arrays, e.g., space-based systems.

A variation of the space-fed array is the **reflect array**, as shown in Fig. 6.157d. In this case, the array receives the energy from the primary feed, collimates and phase shifts it to form the desired beam position, and reradiates it from the same set of radiating elements used to receive the signal due to the reflection off the back plane of the array. This arrangement is convenient for the following reasons: (1) the phase shifters only have to realize one-half of the required phase shift since the signal passes through the phase shifter twice, and (2) the control and bias circuitry of for the phase shifters can be conveniently mounted on the back of the reflector plane. One disadvantage of this approach is the blocking of the aperture due to the primary feed of the antenna.

Element Mutual Coupling Effects and the Small Array

The mathematical array theory development to this point has assumed ideal radiating elements. The impedance properties and radiation pattern of each element is assumed to be independent of is environment, unaffected by its neighboring elements, and identical. Array pattern multiplication is valid under these conditions. However, this is rarely the case in actual arrays. **Mutual coupling** describes the interaction between one element and all others. Mutual coupling has been analyzed as a simple current sheet model by Wheeler, and also with infinite array techniques, finite array techniques, and with measurement-based techniques.[16,17] The infinite array exploits periodicity of the array structure to reduce the problem to a "unit cell," which allows significant mathematical simplification. This technique is commonly used. Several finite array techniques, such as direct impedance formulations (Method of Moment) and other semi-infinite array techniques can be applied to the finite array problem. Measurement-based techniques using measured scan impedance and scan element gain pattern data are commonly used. A variation of the measurement-based approach is to simulate measured data with an electromagnetic simulator on small test arrays as though the resulting analysis were measured data.

Wheeler has derived fundamental impedance and element radiation pattern properties of planar phased arrays from an ideal phased current sheet formulation. The scan resistance of a phased array increases in one plane, and decreases in the perpendicular plane, and the reflection coefficient is given as:

$$\Gamma = \pm \tan^2\left(\frac{\theta}{2}\right),$$
(6.91)

where θ is the scan angle off bore site.

This analysis does not account for radiating element reactance.

Another outcome of the Wheeler analysis the "ideal" element pattern:

$$E(\theta) = \sqrt{\cos(\theta)}$$
(6.92)

where θ is the scan angle off bore site.

Equations (6.91) and (6.92) illustrate that the gain of an ideal phased array falls off rapidly beyond 45° scan off bore site. This is manifested both as a roll off of element pattern gain, and resistive impedance mismatch. These equations suggest that you cannot scan a broad side array to end fire. Special conditions for end fire performance exist and are described in the literature.[18]

The above analysis is idealized in the sense that it does not account for the impedance and mutual coupling properties of practical radiating elements.

Array element mutual coupling can be modeled in terms of multi-element S parameter network theory. Consider Fig. 6.158, which illustrates linear array in the transmit mode. The development equally applies to the receive mode through reciprocity. Each element radiates a signal that is received by neighboring elements, and is subsequently reradiated. The impedance properties of a given element are due to itself, or **driving point impedance**, which is the impedance of an isolated element, along with the **mutual impedance** due to signals arriving from neighboring elements. Array impedance can be formulated as a multiport S parameter network. The analysis, termed **free excitation**, assumes that the feed network is replaced by constant available power sources connected to each radiating element. Under the conditions of an electrically large array, the scan reflection coefficient for the (m, n) th array element is given by:

$$\Gamma_{mn} = \sum_p \sum_q S_{mn,pq} \cdot \frac{A_{pq}}{A_{mn}}$$
(6.93)

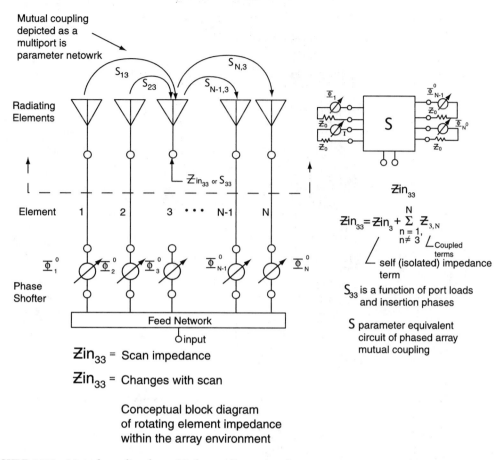

FIGURE 6.158   Mutual coupling for an N-element linear phased array.

where: $S_{mn,pq}$ is the coupling coefficient between the m,nth, and p,qth radiating elements, and,
        A_{pq}, A_{mn} are amplitude coefficients.

It is possible for the scan reflection coefficient to have a magnitude of 1, in which case total reflection exists at the element port. This condition is termed **scan blindness** and one electromagnetic interpretation is that it is due to the generation of surface waves, traveling on the array surface, that do not radiate into free space.

The **scan impedance**, which is closely related to the scan reflection coefficient, is defined as the impedance of a radiating element as a function of scan angle, when all of the radiating elements of the array are excited with the proper magnitude and phase. The scan impedance of a nonideal radiating element is a function of the phase front across the array aperture.

Mutual impedance also affects the element gain pattern. The scanned element pattern can be approximated as:

$$g_s(\theta) \cong \frac{4\pi A_{elem}}{\lambda^2} \cdot \cos(\theta) \cdot \left(1 - \left|\Gamma(\theta)\right|^2\right)$$

(6.94)

where, $g_s(\theta)$ = scanned element pattern,
        A_{elem} = a unit cell radiating element area,
        λ = the operating wavelength,
        θ = the scan angle off bore site.

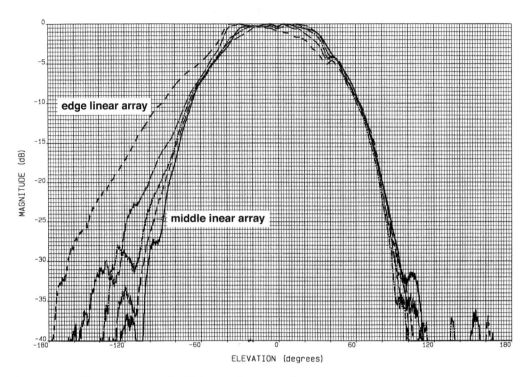

FIGURE 6.159 Element patterns of a 16-element linear array in an aperture.

   This expression is applicable to electrically large arrays where no grating lobes exist, and where higher modes can be neglected.

   Edge effects are present due to asymmetries at the edge of the array face. These effects are neglected in infinite array analysis, and often can be neglected in the design of electrically large arrays. These effects, however, can be significant in small array design. The radiation pattern of an edge element can be significantly different that that of an element in the center of the array. Figure 6.159 illustrates the asymmetries present in the H plane across an edge slotted waveguide aperture consisting of 16 edge slot linear arrays spaced at 0.764λ.

   Measurement-based array design uses experimental measurements of small test arrays, or electromagnetic simulations that predict the same test array parameters. The scan reflection coefficient formulation of Eq. (6.93) can predict reflection coefficient, which in turn is used to calculate an effect mismatch loss, and worse case, scan blindness as a function of scan. Pattern measurements of individual radiation elements, with all other elements properly terminated in a matched load, will give the scanned element pattern. Nulls in the scanned element patterns are related to scan blindness at specific scan angles, assuming that these nulls are not present on an isolated element.

   The scan performance of a linear array with mutual coupling can be predicted from the following expression:

$$E_A(\theta) \equiv \cdot \sum_N \underbrace{\left[E_{e_N}(\theta) \right]}_{\substack{Element \\ Pattern}} \cdot A_n \exp\left[j\frac{2\pi}{\lambda} n\Delta x \left(\sin\theta - \sin\theta_o\right) \right] \qquad (6.95)$$

$$\underbrace{}_{ArrayFactor}$$

   In this expression each element pattern can be different to account for the effects of mutual coupling, edge array effects, etc.

Thinned Arrays

Large arrays of regularly spaced radiation elements are often prohibitively large, expensive, and can exhibit excessive mutual coupling. Beamwidth specifications often dictate the length of the array. It is possible to retain the same array length, but thinning the number of elements to retain an equivalent beamwidth with a commensurate loss in gain. Periodic thinning of the array elements in a uniformly spaced array will create grating lobes, but randomizing the element positions (spacing) can offset this effect. A useful probabilistic approach to thinned array design was formulated by Aggarwal and Lo,[19] where the array sidelobe level threshold is described by a probability density function. The probability of power sidelobe level being below a threshold is given b:

$$P = \left[1 - \exp(-\alpha)\right] \exp\left(-\frac{2}{\lambda}\sqrt{\frac{\pi\alpha}{3}}\exp(-\alpha)\right),$$

(6.96)

where: α = N/SLR,
  N = the number of elements, and
  SLR = the power sidelobe level.

The N parameter sets an average element spacing for a given array length L. The use of directive radiation elements can reduce far out sidelobe levels. An analysis by Steinberg[20] reveals that the use of directive element can reduce the overall length of the linear array described in Eq. (6.96), by the following:

$$\frac{L_{eff}}{\lambda} = \frac{L/\lambda}{\sqrt{\alpha}\left(W/\lambda\right)}$$

(6.97)

where: L = the linear array length, and
  W/λ = the electrical size of the radiating element.

Emerging Phased Array Antenna Technologies

Monolithic microstrip antenna phased arrays are becoming increasingly important with advances in integrated circuit technology. The required dimensions of the microstrip radiating elements at millimeter wave frequencies are on the same order as the active circuit devices, thus making monolithic integration possible. The radiating elements are integrated with the Transmit/Receive (T/R) module sub-circuits, such as power amplifiers, low noise amplifies, phase shifters, attenuators, and T/R switches. The details of T/R module technology are described in Chapter 5 of this handbook. Packaging to protect the semiconductor device from mechanical damage and environmental harshness is a current research topic within the industry.[1] Additional challenges to the antenna designer include bandwidth, radiation efficiency, and surface wave/scan blindness concerns due to the electrical parameter constraints dictated by silicon (Si) and gallium arsenide (GaAs) semiconductor substrate technology.

Phased array applications of **micromachining** technology is currently an active research project. Parasitic mutual coupling due to grounded dielectric slab substrate surface wave modes is problematic in wide beam scanned microstrip phased array applications. This problem is particularly troublesome for substrates with a high relative dielectric constant, such as GaAs and Si, the typical semiconductor substrates used for integrated radiating element T/R module configurations. Micromachining techniques can be used to synthesize a localized artificial dielectric material with low dielectric constant for microstrip antenna substrate from a high dielectric substrate. The lower effective dielectric constant improves the radiation efficiency and reduces surface waves for a given substrate thickness. Gauthier et al. describe a method of drilling a densely spaced two-dimensional grid of holes into the substrate to reduce the local dielectric constant of the substrate.[2] A reduction of dielectric constant to 2.3 from the intrinsic substrate dielectric constant of 10.8 was reported. Nguyen et al. describe micromachining techniques that are used to suspend dipole or slot radiators on a very thin membrane to realize a local substrate dielectric constant of approximately 1.0, which is quite useful for millimeter wave applications.[3] Papapolymerou et al. have documented a 64 percent bandwidth improvement and 28 percent radiation efficiency improvement with the membrane technique[4] for substrate materials with dielectric constants similar to Si.

Micromachining techniques are also used to monolithically realize **microwave** and **millimeter wave distributed circuit devices** that can be integrated with antenna radiation elements. Examples include acoustic resonators, oscillators, tunable capacitors, high Q spiral inductors, and low loss filters.[3]

Microwave and millimeter wave frequency **material research** presents several new possibilities for phased array advancement. **Ferroelectric materials**, and **photonic band gap materials (PBG)** are all being pursued for advanced antenna applications.

A **ferroelectric material's** dielectric constant can be modulated by adjusting its static electric bias field. It is therefore possible to build antennas that are monolithically loaded with ferroelectric material and realize electronic scanning by modulating the biasing electric field.[5] This is similar in concept to the classic waveguide frequency scanned array. Both phased array and steerable, space-fed lense prototypes have been demonstrated. Progress in ferroelectric arrays has been slow to date due to the inability to fabricate ferroelectric materials with the appropriate tuning and dielectric constant parameters while simultaneously exhibiting low insertion loss at microwave frequencies.[6]

Frequency selective surfaces (FSS) are used in multiband, wide band phased array applications, and in applications where it is desirable to block signal transmission for one frequency band while passing another frequency band.[7]

Photonic band gap materials (PBG) provide a band stop function to electromagnetic wave propagation through a dielectric medium. This is useful for eliminating dielectric surface wave mode induced scan blindness in printed phased arrays.[8]

MEMS (Micromachined ElectroMechanical Systems) switches are miniature semiconductor switches used in a variety of antenna applications. Brown also provides an excellent overview of MEMS technology for RF (radio frequency) applications (see Section 6.40).[9] The MEMS switch is physically small, has very low on-state loss, high off-state isolation, and low average DC power consumption, as compared to ferrite waveguide and semiconductor switching elements such as PIN diode and FET transistor switches. These features have led some researchers to propose MEMS phase shifter based, passive electronically scanned array (PESA) architectures as a viable alternative to the monolithic T/R (transmit/receive) module based, active electronically scanned array (AESA) architectures for very large space-based and military radar and communications systems.[10] The MEMS architectures compare favorably in terms of weight, DC power consumption, and required system EIRP (effective isotropic radiated power).

The two most common switch topologies are the cantilever, or type 1 switch[11] and the membrane, or type 2 switch.[12] Phase shifters and true time delay devices using MEMS type 2 switches have been reported at C band, through W band. The type 1 switch topology is superior for lower frequency (below X band) applications due to its high off-state isolation.

An intriguing application of MEMS device technology is for reconfigurable antenna technology.[13,14] The desire to use multiband, multifunction, shared asset apertures for advanced radar and communications systems has led researchers to consider radiating elements whose physical properties can be dynamically adjusted. This is already demonstrated, to a simpler degree, by the use of reactive loads distributed along the length of wire antennas, such as dipoles, to give multi-band performance.[15] The MEMS switch offers the potential to generalize this concept to two dimensions using a switching matrix.

Another potential application for MEMS switches is to dynamically impedance match phased array radiating elements as a function of beam scan. Mutual coupling between radiating elements within the array environment causes the scan impedance of the radiating element to change with beam scan position. A dynamic impedance matching network consisting of switchable banks of reactive elements and planar transmission lines can be used to dynamically tune the radiating element.

MEMS technology has been successfully demonstrated within the research community and efforts are underway to commercialize the technology, including lifetime reliability studies, and RF compatible, environmental packaging techniques.[16,17] Device packaging issues are particularly challenging since the MEMS switch electromechanical device requires unrestricted mechanical motion within the package.

Fractal Antenna Array Technology: Fractal radiating elements have physical shapes based on fractal geometric curves popularized by Benoit Mandelbrot, and others.[18] An overview of fractal electromagnetic research is given by Werner.[19] There are two basic types of fractal curves, random and deterministic. Deterministic fractals utilize a repetitive motif, or pattern, that is applied to successive size scales. Fractals are mathematically described in terms of "n" iterations of the motif, where n goes to infinity. Practically, fractal structures are based on a finite number of iterations. **Fractal array antennas** can have the array element shapes, the array element spacing, or both, based on fractal curves.

Fractal-based arrays have been studied by many researchers.[20-26] A comprehensive overview of **fractal antenna array theory** is given by Werner, Haupt, and Werner.[24] Fractal arrays hold potential to combine the robustness of a random array and the efficiency of a periodically spaced array, but with a reduced element count.[25] Fractal design has also been applied to **sidelobe level microstrip arrays** and **multiband array designs.**[26]

Genetic Algorithm-Based Phased Array Antenna Design

Genetic algorithm-based optimization techniques are of current interest to the phased array antenna research community. **Optimal array thinning** is discussed in References 27 and 28. **Optimal subarray amplitude tapering** for low sidelobe level design is detailed by Haupt.[29] An introduction to genetic algorithm optimization techniques is given in Haupt and Haupt.[30] An excellent overview of genetic algorithms for electromagnetic applications, and comparisons to traditional local optimization techniques is given by Haupt,[31-32] and Johnson and Rahmat-Samii.[33] Traditional, local minimum, gradient-based optimization techniques are described by Cuthbert.[34]

Photonics Systems for the phased array application offer several advantages, including Electromagnetic Interference (EMI) immunity, greater than one octave signal bandwidths, low radar cross-section, and extremely low loss for digital control and RF signal interconnections.[35] The photonic system can perform one, or all of the following system functions within the phased array architectures: (1) the distribution of digital control signals, (2) the distribution of RF energy to and from the array radiating elements, and (3) optical signal processing of the array's RF signals.

The transmission of digital control signals are required to adjust the amplitude and phase weighting for each radiating element within the phased array antenna architecture. Traditional interconnect schemes such as cable bundles, printed circuit boards, and flexible circuit board technology are very cumbersome for phased arrays with hundreds or thousands of radiating elements. Kunath et al. describe an optically controlled Ka Band phased array antenna that incorporates an OptoElectronic Interface Circuit (OEIC) which allows the distribution of control signals to individual antenna elements through a single fiber-optic cable.[36]

Traditional microwave/millimeter wave feed manifold technology can be replaced by modulating the RF signal to optical frequencies for the distribution of RF energy to and from the phased array's radiating elements. Traditional microwave/millimeter wave guided wave transmission line technology is very bulky, heavy, and inherently narrow band, as compared to future optical cable technology. In this case, optical demodulation would most likely be resident in a T/R model at the radiating element level.

Optically based space-time signal processing is feasible once the RF signal is translated to the optical domain. The wide bandwidth of the optical channel would allow for the high processing burden that is required for large digital beamforming phased arrays.

Figure 6.160 illustrates one possible T/R module optical control scheme. Optoelectronic devices are now feasible with III-V compound semiconductor technology such as GaAs and InP (Indium Phosphide). This development offers promise for merging optoelectronic technology with microwave/millimeter wave circuitry on a single MMIC chip, as shown in the Fig. 6.160.[7,38] In the transmit mode, the RF signal's wave shape and phase information to be transmitted by the phased array is modulated to optical frequencies and delivered to the T/R module, where it is detected. The detected RF signal is then passed through the power amplifier (PA) module before being radiated. In the receive mode, the RF signal is

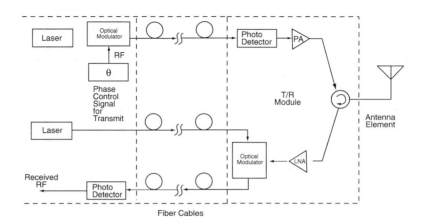

FIGURE 6.160 Optically controlled T/R module.

amplified through the RF LNA, and then is modulated to an optical carrier and is sent to the system receiver via fiber cable. In this example, the T/R module is connected optically, except for the bias signal for the PA and LNA circuitry. Digital beamforming is possible if the receive fiber output cable undergoes an analog-to-digital conversion scheme.

Phase shifting and true time delay required for array beam scanning can be accomplished in the optical domain. Parent[39] and Newberg[40] discuss true time delay schemes using optical technology. Beamforming based on differential–delay networks have been demonstrated and provide a means for frequency-independent beam steering. Beak et al. describe a two-dimensional, 480-element phased array with a true time delay photonic beamforming network that operates over a 1.5 octave bandwidth.[41] Along with beam steering, optoelectronic technology allows for photonic coherent beamforming techniques and signal processing for multiple beam generation, target identification, etc. True time delay devices can be realized by switching in different lengths of fiber-optic cable, but this can become mechanically burdensome for very large arrays with the current state-of-the-art technology.

References

Phased Arrays

1. Hansen, R. C., *Phased Array Antennas,* John Wiley and Sons, Inc., New York, 1997.
2. Mailloux, R. J., *Phased Array Antenna Handbook,* Artech House, Inc., Norwood, MA, 1994.
3. Corey, L. et al., *Phased Array Antennas: Theory, Design, and Technology,* Short Course Notes, Georgia Tech Research Institute/STL, The Georgia Institute of Technology Department of Continuing Education, Atlanta, GA, November 18–21, 1986.
4. Elliot, R. S., *Antenna Theory and Design,* Prentice-Hall, Inc., Englewood Cliffs, New Jersey, 1981, 172–180.
5. Elliot, R. S., *Antenna Theory and Design,* Prentice-Hall, Inc., Englewood Cliffs, New Jersey, 1981, 157–165.
6. Balanis, C. A, *Antenna Theory, Analysis, and Design, 2nd Ed.,* John Wiley and Sons, New York, 1997, 249–338.
7. Mailloux, R. J., *Phased Array Antenna Handbook,* Artech House, Inc., Norwood, MA, 1994, 394–403.
8. Mailloux, R. J., *Phased Array Antenna Handbook,* Artech House, Inc., Norwood, MA, 1994, 403–421.
9. Miller, C. J., Minimizing the Effects of Phase Quantization Errors in an Electronically Scanned Array, *Proc. 1964 Symposium Electronically Scanned Phase Arrays and Applications,* RADCTDR-64-225, RADC Griffis AFB, Vol. 1, 17–38.

10. Smith, M. S., and Guo, Y. C., A Comparison of Methods for Randomizing Phase Quantization Errors in Phased Arrays, *IEEE Transactions on Antennas and Propagation,* AP-31, 6, 821–827, Nov. 1983.

11. Mailloux, R. J., Periodic Arrays, *Antenna Handbook Theory, Applications, and Design,* Lo, Y. T., and Lee, S. W., editors, Van Nostrand Reinhold Co., New York, 1988, chap. 13, 13.30–13.32.

12. Tang. R., Survey of Time Delay Beam Steering Techniques, *Phased Array Antennas: Proceedings of the 1970 Phased Array Antenna Symposium,* Dedham, MA, Artech House, 1972, 254–260.

13. Tang, R., Survey of Time-Delayed Steering Techniques, *Phased-Array Antennas: Proceedings of the 1970 Phased Array Symposium,* Artech House, Dedham, MA, 254–260.

14. Hansen, R. C., *Phased Array Antennas,* John Wiley and Sons, Inc., New York, 1997, p. 166.

15. Rudge, A. W., Milne, K., Olver, A. D., Knight, P., *The Handbook of Antenna Design,* Volume 1, Peter Peregrinus, Ltd, London, U.K., 1982, 87–98.

16. Hansen, R. C., *Phased Array Antennas,* John Wiley and Sons, Inc., New York, 1997, 215–301.

17. Wheeler, H. A., Simple Relations for a Phased Array Antenna Made of an Infinite Current Sheet, *IEEE Transactions on Antennas and Propagation,* AP-13, 4, 506–514, July 1965.

18. Balanis, C. A., *Antenna Theory, Analysis, and Design,* 2nd Ed., John Wiley and Sons, New York, 1997, 264–276.

19. Aggarwal, V. D., and Lo, Y. T., Mutual Coupling in phased Arrays of Randomly Spaced Antennas, *IEEE Transactions on Antennas and Propagation,* AP-20, 288–295, May 1972.

20. Steinberg, B. D., Comparison Between the Peak Sidelobe Level of the Random Arrays and Algorithmically Designed Aperiodic Arrays, *IEEE Transactions on Antennas and Propagation,* AP-21, 366–369, May 1973.

Emerging Phased Array Antenna Technologies

1. Cohen, E. D., Trends in the Development of MMICS and Packages for Electronically Scanned Arrays (AESAs), *1996 Proceedings of the IEEE International Symposium on Phased Array Systems and Technology,* IEEE Press, Institute of Electrical and Electronic Engineers, Piscataway, New Jersey, 1996, 1–4.

2. Gauthier, G. P., Courtay, A., and Rebiez, G. M., Microstrip Antennas on Synthesized Low Dielectric-Constant Substrates, *IEEE Transactions on Antennas and Propagation,* 45, 8, 1310–1314, August, 1997.

3. Nguyen, C, T. C., Katehi, L. P. B., and Rabeiz, G. M., Micromachined Devices for Wireless Communications, *Proceedings of the IEEE,* Institute of Electrical and Electronics Engineers, 86, 8, 1756–1768, August, 1998.

4. Papapolymerou, I., Drayton, R. F., and Katehi, L .P .B., Micromachined Patch Antennas, *IEEE Transactions on Antennas and Propagation,* 46, 2, 275–283, February, 1998.

5. Rao, J. B. L., Trunk, G. V., and Patel, D. P., Two Low-cost Arrays, *1996 Proceedings of the IEEE International Symposium on Phased Array Systems and Technology,* IEEE Press, Institute of Electrical and Electronic Engineers, Piscatway, New Jersey, 1996, 119–124.

6. Sengupta, L. C. et al., Investigation of the Electronic Properties of Ceramic Phase Shifting Material, Proceedings of the Advanced Ceramic Technology for Electronic Applications Conference, U.S. Army, CECOM, Fort Monmouth, September 28, 1993.

7. Schennum, G. H., Frequency Selective Surfaces for Multiple-Frequency Antennas, *Microwave Journal,* 55–76, May 1973.

8. Sievenpiper, D., and Yablonovitch, E., Eliminating Surface Currents with Metallodieletric Photonic Crystals, *IEEE Antennas and Propagation International Symposium Proceedings,* 1998, 663–666.

9. Brown, E. R., RF-MEMS Switches for Reconfigurable Integrated Circuits, *IEEE Transactions on Microwave Theory and Techniques,* 11, 1868–1880, November 1998.

10. Norvell, B. R., Hancock, R. J., Smith, J. K., Pugh, M. L., Theis, S. W., and Kviatkofsky, J., Micro Electromechanical Switch (MEMS) Technology Applied to Electronically Scanned Arrays for Space Based Radar, *Proceedings of the IEEE Aerospace Conference,* 1999, 239–247.

11. Yao, J. J., and Chang, M. F., A Surface Micromachined Miniature Switch for Telecommunications with Signal Frequencies from DC to 4 GHz, *8th International Conference on Solid State Sensors and Actuators,* Stockholm, Sweden, June 25, 1995, 384–387.

12. Goldsmith, C. L., Yao, Z., Eshleman, S., and Denniston, D., Performance of Low-Loss RF MEMS Capacitive Switches, *IEEE Microwave and Guided Wave Letters,* 8, 8, 269–271, August, 1998.

13. Lee, J. J., Atkinson, D., Lam, J. J., Hackett, L., Lohr, R., Larson, L., Loo, R. Matloubian, M., Tangonon, G., De Los Santos, H., and Brunner, R. MEMS in Antenna Systems: Concepts, Design, and Systems Implications, National Radio Science Meeting, Boulder, Colorado, 1998.

14. Weedon, W. H., Payne, W. J., MES-Switched Reconfigurable Multi-Band antenna: Design and Modeling, Proceedings of the 1999 Antenna Applications Symposium, September 15–17, 1999, Robert Allerton Park, The University of Illinois, 1999.

15. *The ARRL Antenna Handbook,* The American Radio Relay League, Inc., 1974, Newington, CT, 183–186.

16. Connally, J., and Brown, S., Micromechanical Fatigue Testing, *Experimental Mechanics,* 81–90, June 1993.

17. Lyle, J. C., Packaging Technologies for Spaced Based Microsystems and Their Elements, in *Microengineering Technologies for Space Systems,* Helvajian, H., Ed., The Aerospace Corporation, El Segundo, CA, 1995, 441–504.

18. Mendelbrot, B. B., *The Fractal Geometry of Nature,* W.H. Freeman, New York, 1983.

19. Werner, D. H., An Overview of Fractal Electrodynamics Research, *Proceedings of the 11th Annual Review of Progress in Applied Computational Electromagnetics,* Monterey, California, March 20–24, 1995, 964–969.

20. Werner, D.H., and Werner, P.L., Fractal Radiation Pattern Synthesis, National Radio Science Meeting, Boulder Colorado, January 6–11, 1992, 66.

21. Werner, P., L., and Werner, D.H., Fractal Arrays and Fractal Radiation Patterns, Proceedings of the 11th Annual Review of Progress in Applied Computational Electromagnetics, Monterey, California, March 20–24, 1995, 970–978.

22. Werner, P. L., and Werner, D. H., On the Synthesis of Fractal Radiation Patterns, *Radio Science,* 30, 1, 29–45, 1995.

23. Werner, P., L., and Werner, D. H., Correction to: On the Synthesis of Fractal Radiation Patterns, *Radio Science,* 30, 3, 603, 1995.

24. Werner, D. H., Haupt, R. L., and Werner, P.L., Fractal Antenna Engineering: The Theory and Design of Fractal Antenna Arrays, *IEEE Antennas and Propagation Magazine,* 41, 5, 37–59, October 1999.

25. Technology and Business Section, *Scientific American,* July, 1999. A quotation of Dr. Dwight Jaggard of the University of Pennsylvania.

26. Pueente-Baliarda, C., Pous, R., Fractal Design of Multi-band and Low Side-Lobe Arrays, *IEEE Transactions on Antennas and Propagation,* 44, 5, 730, May 1996.

27. Haupt, R. L., Menozzi, J. J., and McCormack, C. J., Thinned Arrays Using Genetic Algorithms, *1993 International Symposium Digest,* 3, Ann Arbor, MI, 1993, 712–715.

28. O'Neill, D. J. Element Placement in Thinned Array Using Genetic Algorithms, *Proceedings OCEANS '94,* Brest, France, September 1994, II/301-3-6, 712–715.

29. Haupt, R. L., Optimization of Sub-Array Amplitude Tapers, *1995 International Symposium Digest,* 3, Newport Beach, California, June 1995, 1830–1833.

30. Haupt, R. L., and Haupt, S. E., *Practical Genetic Algorithms,* John Wiley and Sons. Inc. New York, 1998.

31. Haupt. R. L., Introduction to Genetic Algorithms for Electromagnetics, *IEEE Antennas and Propagation Magazine,* 37, 2, 7–15, April 1995.

32. Haupt, R. L., Comparison between Genetic and Gradient-Based Optimization Algorithms or Solving Electromagnetics Problems, *IEEE Transactions on Magnetics,* MAG-31, 3, 1932–1935, May 1995.

33. Johnson, J. M., and Rahmat-Samii, Y., Genetic Algorithms in Engineering Electromagnetics, *IEEE Antennas and Propagation Magazine,* 39, 4, 7–25, August 1997.

34. Cuthbert, T., R., *Optimization Using Personal Computers with Applications to Electrical Networks*, John Wiley and Sons, Inc., New York, 1987.

35. VanBlarium, M. L., Photonic Systems for Antenna Applications, *IEEE Antennas and Propagation Magazine*, 36, 5, 30–38, October 1994.

36. Kurath, R. R., Lee, R. Q., Martzaklis, K. S., Shalkhauser. K. A., Downey, A. N., and Simmons, R., An Optically Controlled Ka-Band Phased Array Antenna, Proceedings of the 1992 Antenna Applications Symposium, Robert Allerton Park, The University of Illinois, Monticello, Illinois, September 23–25, 1992.

37. Herezfeld, P., Monolithic Micowave-Photonic Integrated Circuits: A Possible Follow-up to MIMIC, *Microwave Journal*, 121–131, April 1985.

38. Simons, R., *Optical Control of Microwave Devices*, Artech House. Boston, 1990.

39. Parent, M. G., A Survey of Optical Beamforming Techniques, Proceedings of the 1995 Antenna Applications Symposium, Robert Allerton Park, The University of Illinois, Monticello, Illinois, September 20–22, 1995.

40. Newberg. I. L., Antenna True-Time-Delay Beamsteering Utilizing Fiber Optics, Proceedings of the 1992 Antenna Applications Symposium, September 23–25, 1992, Robert Allerton Park, The University of Illinois, 1995.

41. Beyak, P., Bobowics, D., and Collier, D., Performance Testing of a Wideband Phased Array for Shipboard Application, Proceedings of the 1999 Antenna Applications Symposium, September 15–17, 1999, Robert Allerton Park, The University of Illinois, 1999.

6.10 RF Package Design and Development

Jeanne S. Pavio

Successful RF and microwave package design involves adherence to a rigorous and systematic methodology in package development together with a multi-disciplined and comprehensive approach. This formal planning process and execution of the plan ultimately insures that the package and product will perform as expected, for the predicted lifetime duration in the customer's system, under the prescribed application conditions.

Probably the first concern is having a thorough and in-depth knowledge of the application and the system into which the microwave component or module will be placed. Once these are understood, then package design can begin. Elements that must be considered do not simply include proper electrical performance of the circuit within the proposed package. Mechanical aspects of the package design must be thoroughly analyzed to assure that the package will not come apart under the particular life conditions. Second, the substrates, components, or die within the package must not fracture or lose connection. Third, any solder, epoxy, or wire connections must be able to maintain their integrity throughout the thermal and mechanical excursions expected within the application. Once these elements are thoroughly investigated, the thermal aspects of the package must be simulated and analyzed to appropriately accommodate heat transfer to the system. Thermal management is probably one of the most critical aspects of the package design because it not only contributes to catastrophic circuit overload and failure in out-of-control conditions, but it could also contribute to reduced life of the product and fatigue failures over time. Thermal interactions with the various materials used for the package itself and within the package may augment mechanical stress of the entire package system, ultimately resulting in failure.

Once proper simulation and analysis have been completed from a mechanical and thermal point of view, the actual package design can be finalized. Material and electrical properties and parameters then become the primary concern. Circuit isolation and electromagnetic propagation paths within the package need to be thoroughly understood. In addition, impedance levels must be defined and designed for input and output to and from the package. New developments in package design systems have paved the way for rapid package prototyping through computer integrated manufacturing systems by tying the design itself to the machining equipment that will form the package. These systems can prototype a part in

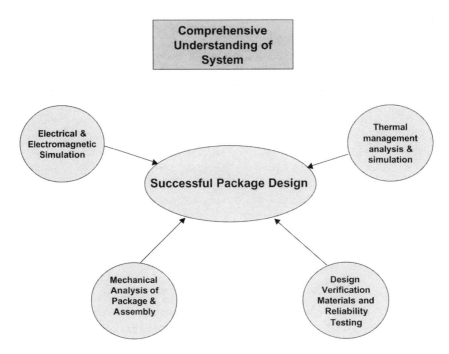

FIGURE 6.161 Elements of successful package design.

plastic for further study or can actually build the prototypes in metal for delivery of prototype samples. Finally, design verification must take place. The verification process typically includes the various long-term reliability tests that gives the designer, as well as his or her customer confidence that the package and its contents will live through the predicted lifetime and application conditions. Other testing may be more specific, such as fracture testing, material properties tests, or precise design tolerance testing. Much of the final testing may also include system-level integration tests. Usually specific power levels are defined and the packages, fully integrated into the system, are tested to these levels at particular environmental conditions.

These are some of the key elements in RF and microwave package development. Although this is not an all-encompassing list, these elements are critical to success in design implementation. These will be explored in the following discussion, hopefully defining a clear path to follow for RF package design and development. Figure 6.161 depicts these key elements leading to successful package design.

Thermal Management

From an MTBF (mean time before failure) point of view, the thermal aspects of the circuit/package interaction are one of the most important aspects of the package design itself. This can be specifically due to actual heat up of the circuit, reducing lifetime. It may also be due to thermal effects that degrade performance of the materials over time. A third effect may be a materials/heat interaction that causes severe thermal cycling of the materials resulting in stress concentrations and degradation over time.

It is clear that the package designer must have a fully encompassing knowledge of the performance objectives, duty cycles, and environmental conditions that the part will experience in the system environment. The engineer must also understand the thermal material properties within the entire thermal path. This includes the die, the solder or epoxy attachment of that die, the package or carrier base, package system attachment, and material connection to the chassis of the system. There are relatively good databases in the industry that provide the engineer with that information right at his or her fingertips. Among the many are the CINDAS [1] database and the materials' database developed at Georgia Institute of Technology. Other information may be gleaned from supplier datasheets or testing.

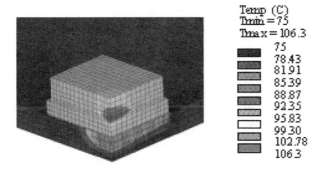

Temp (C)
Tmin = 75
Tmax = 106.3
75
78.43
81.91
85.39
88.87
92.35
95.83
99.30
102.78
106.3

FIGURE 6.162 Ansys output showing thermal gradient across silicon die.

Thermal density within the package, and in particular, at the die level becomes an all-important consideration in the thermal management equation. To insure proper heat transfer and to eliminate any potential thermal failure modes (such as materials breakdown or diffusion and migration), analysis of heat transfer within the die must be completed at the die layout level. This thermal analysis will ultimately be parametrically incorporated into an analysis at the next level up, which may be at the circuit substrate or at the package level itself. The analysis is usually completed with standard finite element simulation techniques present in various software packages available in the industry. Ansys, MSC Nastran, Mechanica, Flowtherm, and Computational Fluid Dynamics (CFD) are some of those available. Material properties that are critical to input into the model would be thermal conductivity and the change in conductivity with temperature. Figure 6.162 shows a typical output of one of these software tools. The FEM simulation uses $1/4$ model symmetry. In this particular figure, the analysis demonstrates the thermal gradient across a silicon die, which is an 8 W power amplifier transistor, solder attached to a via structure, with 75°C applied to the bottom of the heat sink. The die junction is at 106.3°C.

It is through such simulated analysis that the entire heat transfer methodology of component to system can be developed. Assuming that there is good correlation between simulated and verified results, the engineer can then gain confidence that the product will have a reasonable lifetime within the specific application. The correlation is typically achieved through the use of infrared microscopy techniques. A number of these infrared microscopes are available in the industry. Usually, the component or module is fixtured on a test station under the infrared camera. The camera is focused on the top surface of the die, which is the heat-generating element. As power is applied to the component or module, the die begins to heat up to a steady-state level. The heat can be measured under RF or DC power conditions. A measurement is done of the die surface temperature. At the same time, a thermocouple impinges on the bottom of the case or package and makes a temperature measurement there. With the maximum die junction temperature (Tjmax in °C), the case temperature (Tc in °C) and the dissipated power in watts, the packaged device junction to case thermal resistance (in °C/W) can be calculated from the following equation:

$$\theta jc = \left(Tjmax - Tc\right)/Pdis$$

In a fully correlated system, the agreement between simulated and measured results is usually within a few percentage points.

Mechanical Design

The mechanical design usually occurs concurrently with thermal analysis and heat transfer management. In order to adequately assess the robustness of the package system and its elements, an in-depth understanding of all the material properties must be achieved. In addition to the mechanical properties such as Young's modulus and stress and strain curves for materials, behavior of those materials under thermal loading conditions must be well understood. Once again, these material properties can be found in standard databases in the industry as mentioned above. Typically, the engineer will first insure that the packaging materials under consideration will not cause fracture of the semiconductor devices or of the substrates, solder joints, or other interconnects within the package. This involves knowing the Coefficient of Thermal Expansion (CTE) for each of these materials, and understanding the processing temperatures and the subsequent temperature ramp up that will be experienced under loading conditions in the application. The interaction of the CTEs of various materials may create a mismatched situation and create residual stresses that could result in fracture of any of the elements within the package. The engineer also assesses the structural requirements of the application and weight requirements in order to form appropriate decisions on what materials to use. For instance, a large microwave module may be housed within an iron-nickel (FeNi) package that sufficiently addresses all of the CTE concerns of the internal packaged elements. However, this large, heavy material might be inappropriate for an airborne application where a lightweight material such as AlSiC (aluminum silicon carbide) would be more suitable.

It is not sufficient to treat the packaged component or module as a closed structure without understanding and accounting for how this component or module will be mounted, attached, or enclosed within the actual system application. A number of different scenarios come to mind. For instance, in one situation a packaged component may be soldered onto a printed circuit board (PCB) of a wireless phone. Power levels would not be of concern in this situation, but the mechanical designer must develop confidence through simulation that the packaged RF component can be reliably attached to the PCB. He or she also must insure that the solder joints will not fracture over time due to the expansion coefficient of the PCB compared to the expansion of the leads of the package. Finally, the engineer must comprehend the expected lifetime in years of the product. Cost is obviously a major issue in this commercial application. A solution may be found that is perfectly acceptable from a thermo-mechanical perspective, but it may be cost prohibitive for a phone expected to live for three to five years and then be replaced.

Another scenario on the flip side of the same application is the power device or module that must be mounted into a base station. Here, obviously, the thermal aspects of the packaged device become all important. And great pain must be taken to insure that an acceptable heat transfer path is clearly delineated. With the additional heat from the power device and within the base station itself, heat degradation mechanisms are thoroughly investigated both with simulation techqniques and with rigorous testing. It is common for the RF power chains within base station circuits to dissipate 100 to 200 watts each. Since the expected lifetime of base stations may be over fifteen years, it would be a great temptation for a mechanical engineer to utilize optimum heat transfer materials for the package base, such as diamond for instance, with a thermal conductivity of 40.6 W/in°C. A high power device attached to diamond would operate much cooler than a device attached to FeNi or attached to ceramic. Since, over time, it is the heat degradation mechanisms that eventually cause failure of semiconductor devices, a high power die mounted over a diamond heat sink would be expected to have a much longer lifetime than one mounted over iron nickel or over ceramic. However, once again, the cost implications must enter into the equation. Within the multifunctioned team developing the package and the product, a cost trade-off analysis must be done to examine cost comparisons of materials vs. expected lifetimes. The mechanical package designer may develop several simulations with various materials to input into the cost-reliability matrix. It is necessary that such material substitutions can be done easily and effectively in the parametric model that was initially developed.

The mechanical analysis must encompass attachment of the RF or microwave component or module to the customer system. As we have discussed, in a base station, the thermal path is all important. In order to provide the best heat transfer path, engineers may inadvertently shortcut mechanical stress concerns, which then compromise package integrity. An example was a system mounting condition initially created for the eight-watt power device shown in Fig. 6.162. This semiconductor die was packaged on a copper lead frame to which plastic encapsulation was applied. The lead frame was exposed on the bottom side of the device to insure that there would be a good thermal path to the customer chassis. The copper leads were solder attached (using the typical lead-tin, PbSn, solder) to the printed circuit board. At the same time, the bottom of the device was solder attached to a brass heat sink, as shown in

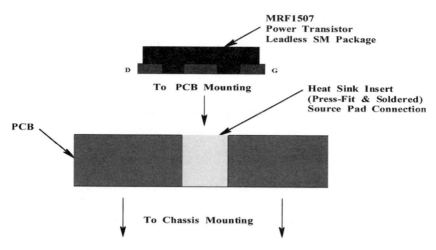

FIGURE 6.163 Eight watt power device attached to brass heat sink.

Stress In Mold Compound

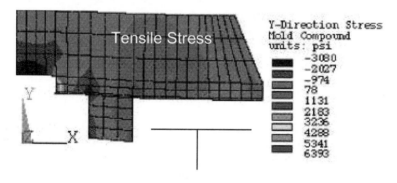

Interface Under Tension

FIGURE 6.164 Modeled stresses in plastic mold compound resulting in failure.

Fig. 6.163, which was then screw mounted to the aluminum chassis to provide thermal transfer to the chassis.

The CTE mismatch of materials resulting in residual stresses during thermal excursions, caused the plastic to rip away from the copper leads. It was the expansion of the aluminum chassis impacting the brass heat sink that created both tensile and shear forces on the leads of the device. The brass heat sink, in effect, became a piston pushing up at the center of the component. The stress levels in the plastic mold compound, which resulted in the failure of the mold/copper interface, can be seen in Fig. 6.164.

Through subsequent simulation, a solution was found that provided the proper heat transfer for the eight-watt device as well as mechanical stability over time and temperature. This was verified through thousands of hours of temperature cycle testing and device power conditioning over temperature excursions.

Package Electrical and Electromagnetic Modeling

Quite obviously, the electrical design cannot stop at the circuit model for the silicon or GaAs die itself. Particularly at higher frequencies, such as those in the RF or microwave arena, the electromagnetic propagation due to all circuit elements create interactions, interference, and possibly circuit oscillations

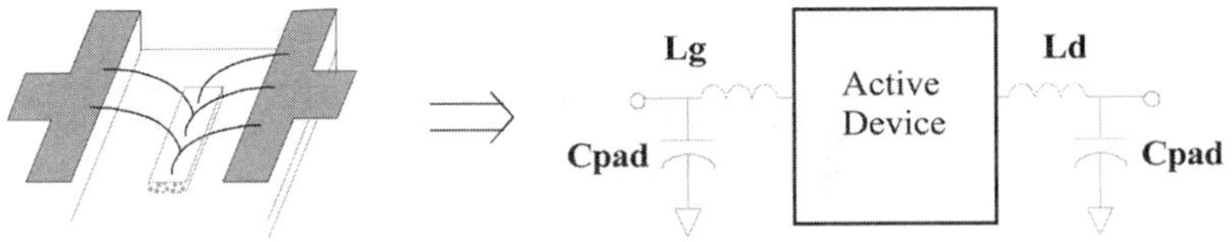

Parasitic effects are represented by:
- Leadframe capacitance, Cpad
- Total gate side wire inductance, Lg
- Total drain side wire inductance, Ld

FIGURE 6.165   Equivalent circuit representation of simple package.

if these electrical effects are not accounted for and managed. Of course the customer's initial requirement will be that a packaged device, component, or module have a specific impedance into and out of their system. Typically, this has been 50 ohms for many microwave systems. It can be achieved through properly dimensioned microstrip input and output leads, through coaxial feeds, or through stripline to microstrip connections that feed into the customer system. These are modeled using standard industry software such as that provided by Hewlett Packard or Ansoft. The next consideration for the package designer is that all of the circuit functions that require isolation are provided that isolation. This can be accomplished through the use of actual metal wall structures within the package. It can also be done by burying those circuit elements in cavities surrounded by ground planes or through the use of solid vias all around the functional elements. These are only some of the predictive means of providing isolation. The need for isolating circuit elements and functions is ascertained by using full wave electromagnetic solvers such as HFSS, Sonnet, or other full wave tools. The EM analysis of the packaged structure will output an S parameter block. From this block, an electrical equivalent circuit can then be extracted with circuit optimization software such as Libra, MDS, ADS, etc. An example of an equivalent circuit representation can be seen in Fig. 6.165.

After proper circuit isolation is achieved within the package, the designer must insure that there will not be inductive or capacitive effects due to such things as wire bonds, leads, or cavities. Wire bonds, if not controlled with respect to length in particular, could have serious inductive effects that result in poor RF performance with respect to things such as gain, efficiency, and intermodulation distortion, etc. In the worst case, uncontrolled wire bonds could result in circuit oscillation. In the same way, RF and microwave performance could be severely compromised if the capacitive effects of the leads and other capacitive elements are not accounted for. These are modeled with standard RF and microwave software tools, and then the materials or processes are controlled to maintain product performance within specifications. Most software tools have some type of "Monte Carlo" analysis capability in which one can alter the material or process conditions and predict the resulting circuit performance. This is especially useful if the processes have been fully characterized and process windows are fully defined and understood. The Monte Carlo analysis then can develop expected RF performance parameters for the characterized process within the defined process windows.

Design Verification, Materials, and Reliability Testing

After all of the required simulation and package design has been completed, the time has come to begin to build the first prototypes to verify the integrity of the design. During the simulation phase, various material property studies may have been undertaken in order to insure that the correct properties are input into the various models. These may be studies of dielectric constant or loss on a new material, fracture studies to determine when fracture will occur on a uniquely manufactured die or on a substrate, or thermal studies, such as laser flash, to determine the precise thermal conductivity of a material. Figure 6.166 shows one technique used to measure the fracture strength of a GaAs or silicon die. A load

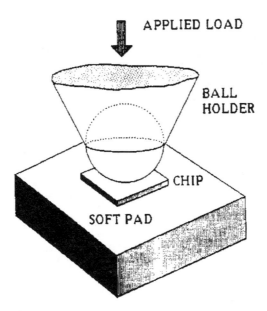

APPLIED LOAD

BALL HOLDER

CHIP

SOFT PAD

FIGURE 6.166 Technique used to measure fracture strength of semiconductor die.

is applied to a fixtured sphere, which then impacts the die at a precise force level. From the test, the critical value of the force to break the die is recorded. Then this force is converted to the maximum die stress via the following well-known [5] equation:

$$\sigma_t = \frac{3W}{2\pi m t^2}\left[(m+1)\ln\left(\frac{a}{r_o}\right)+(m-1)\left(1-\left(\frac{r_o}{a}\right)^2\right)\right]$$

After various material properties tests and all simulations and models have been completed, initial prototypes are built and tested. This next phase of tests typically assess long-term reliability of the product through thermal cycling, mechanical shock, variable random vibration, long-term storage, and high temperature and high humidity under biasing conditions. These, as well as other such tests, are the mainstay of common qualification programs. The levels of testing and cycles or hours experienced by the packaged device are often defined by the particular final application or system. For instance, a space-qualified product will require considerably more qualification assessment than a component or module going into a wireless handset that is expected to live 3 to 5 years. The temperature range of assessment for the space-qualified product may span from cryogenic temperatures to +150°C. The RF component for the wireless phone, on the other hand, may simply be tested from 0°C to 90°C.

In high power applications, often part of the reliability assessment involves powering up the device or module after it is mounted to a simulated customer board. The device is powered up and down at a specific duty cycle, through a number of cycles often as the ambient progresses through a series of thermal excursions. This represents what the RF packaged component would experience in the customer system, though usually at an accelerated power and/or temperature condition. Lifetime behavior can then be predicted, using standard prediction algorithms such as Black's equation, depending on the test results.

The RF product and packaging team submit a series of prototype lots through final, standard qualification/verification testing. If all results are positive, then samples are usually given to the customer at this time. These will undergo system accelerated life testing. The behavior of the system through this series of tests will be used to predict expected life cycle.

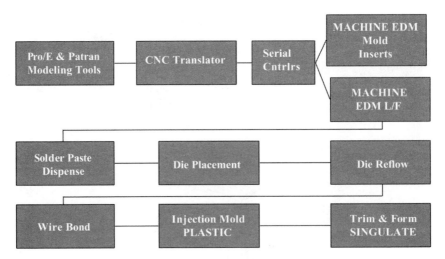

FIGURE 6.167 Rapid prototyping system.

Computer-Integrated Manufacturing

As mentioned above, there are tools available in the industry that can be used to rapidly develop prototypes directly from the package design files. These prototypes may be constructed of plastic or of various metals for examination and further assessment. Parametric Technologies offers such design and assembly software modules, although they are not by any means the only company with this type of software. The package design is done parameterically in Pro-E so that elements of the design can easily be changed and/or uploaded to form the next higher assembly. The package design elements then go through a series of algorithms to which processing conditions can be attached. These algorithms translate the information into CNC machine code which is used to operate equipment such as a wire EDM for the cutting of metals. Thus, a lead frame is fashioned automatically, in a construction that is a perfect match to the requirements of the die to be assembled. A process flow chart for this rapid prototyping scenario is shown in Fig. 6.167.

Computer-integrated manufacturing is also a highly effective tool utilized on the production floor, once the designed package has been accepted by the customer and is ready for production implementation. Here it is utilized for automated equipment operation, for statistical process control (SPC), for equipment shut down in out-of-control situations, etc. Coupled with neural networks, computer-integrated manufacturing can also be used for advanced automated process optimization techniques.

Conclusions

The development and design of packages for RF and microwave applications must involve a rigorous and systematic application of the proper tools and methodology to create a design that "works the first time" and every time for the predicted lifetime of the product. This encompasses an in-depth knowledge of the system requirements, the environmental conditions, and the mounting method and materials to be used for package assembly into the customer's system. Then modeling and simulation can take place. Often, in order to understand material properties and to use these more effectively in the models, material studies are done on specific parameters. These are then inserted into electrical, mechanical, and thermal models, which must be completed for effective package design. After a full set of models is completed, verification testing of the design can be done on the first prototypes. Rapid prototyping is made simple through techniques that automatically convert design parameters into machine code for operation of machining equipment. Computer-integrated manufacturing is a highly effective technique that can be utilized at various levels of the product introduction. In package design, it is often used for rapid prototyping and as a tool for better understanding the design. At the production level, it is often used for automated equipment operation and for statistical process control.

Verification testing of the prototypes may include IR scanning to assess thermal transfer. It may include instron testing to test the integrity of a solder interface or of a package construction. It may include power cycling under DC or RF conditions to insure that the packaged design will work in the customer application.

The final phase of assessment is the full qualification of the RF packaged device or module. This certifies to the engineer, and ultimately to the customer that the packaged product can live through a series of thermal cycles, through high temperature and high humidity conditions. It certifies that there will be no degradation under high temperature storage conditions. And it certifies that the product will still perform after appropriate mechanical shock or vibration have been applied. Typically, predictive lifetime assessment can be made using performance to accelerated test conditions during qualification and applying these results to standard reliability equations.

These package design elements, when integrated in a multidisciplined approach, provide the basis for successful package development at RF and microwave frequencies.

References

1. CINDAS = Center for Information and Data Analysis; Operated by Purdue University; Package Materials Database created under SRC (Semiconductor Research Corporation) funding.
2. G. Hawkins, H. Berg, M. Mahalingam, G. Lewis, and L. Lofgran "Measurement of silicon strength as affected by wafer back processing," International Reliability Physics Symposium, 1987.
3. T. Liang, J. Pla, and M. Mahalingam, Electrical Package Modeling for High Power RF Semiconductor Devices, Radio and Wireless Conference, IEEE, Aug. 9-12, 1998.
4. R.J. Roark, *Formulas for Stress and Strain*, 4th Edition, McGraw-Hill, New York, 219.

Jan Stake
Chalmers University of Technology

Jack East
University of Michigan

Robert J. Trew
U.S. Department of Defense

John C. Cowles
*Analog Devices — Northwest
Laboratories*

William Liu
Texas Instruments

Leonard MacEachern
Carleton University

Tajinder Manku
University of Waterloo

Michael S. Shur
Rensselaer Polytechnic Institute

Prashant Chavarkar
CREE Lighting Company

Umesh Mishra
University of California

Karen E. Moore
Motorola

Jerry C. Whitaker
Technical Press

Lawrence P. Dunleavy
University of South Florida

7

Active Device Technologies

7.1   Semiconductor Diodes

7.1.1   Varactors

Jan Stake

A varactor is a nonlinear reactive device used for harmonic generation, parametric amplification, mixing, detection, and voltage-variable tuning.[1] However, present applications of varactors are mostly for harmonic generation at millimeter and submillimeter wave frequencies, and as tuning elements in various microwave applications. Varactors normally exhibit a voltage-dependent capacitance and can be fabricated from a variety of semiconductor materials.[2] A common varactor is the reverse biased Schottky diode. Advantages of varactors are low loss and low noise. The maximum frequency of operation is mainly limited by a parasitic series resistance (see Fig. 7.1).

Basic Concepts

Many frequencies may interact in a varactor, and of those, some may be useful inputs or outputs, while the others are *idlers* that, although they are necessary for the operation of the device, are not part of any

0-8493-8592-X/01/$0.00+$.50
© 2001 by CRC Press LLC

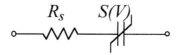

FIGURE 7.1   Equivalent circuit of a pure varactor.[4]

input or output. For instance, to generate high harmonics in a frequency multiplier it is more or less necessary to allow current at intermediate harmonics (idlers) to flow. Such idler circuits are usually realized as short-circuit resonators, which maximize the current at idler frequencies.

Manley-Rowe Formulas

The Manley-Rowe formulas[3] for *lossless* nonlinear reactances are useful for intuitive understanding of multipliers, frequency converters, and dividers. Consider a varactor excited at two frequencies f_p and f_s; the corresponding general Manley-Rowe formulas are

$$\sum_{m=1}^{\infty}\sum_{n=-\infty}^{\infty}\frac{mP_{m,n}}{nf_p+mf_s}=0$$

$$\sum_{m=-\infty}^{\infty}\sum_{n=1}^{\infty}\frac{nP_{m,n}}{nf_p+mf_s}=0$$

where m and n are integers representing different harmonics and $P_{m,n}$ is the average power flowing into the nonlinear reactance at the frequencies nf_p and mf_s.

- Frequency multiplier (m = 0): if the circuit is designed so that only real power can flow at the input frequency, f_p, and at the output frequency, nf_p, the above equations predict a theoretical efficiency of 100%. The Manley-Rowe equation is $P_1 + P_n = 0$.
- Parametric amplifier and frequency converter: assume that the RF-signal at the frequency f_s is small compared to the pump signal at the frequency f_p. Then, the powers exchanged at sidebands of the frequencies nf_p and mf_s for m different from 1 and 0 are negligible. Furthermore, one of the Manley-Rowe formulas only involves the small-signal power as

$$\sum_{n=-\infty}^{\infty}\frac{P_{1,n}}{nf_p+f_s}=0$$

Hence, the nonlinear reactance can act as an amplifying upconverter for the input signal at frequency f_s and output signal extracted at $f_u = f_s + f_p$ with a gain of

$$\frac{P_u}{P_s}=\frac{P_{1,1}}{P_{1,0}}=-\left(1+\frac{f_p}{f_s}\right)=-\frac{f_u}{f_s}$$

Varactor Model

The intrinsic varactor model in Fig. 7.1 has a constant series resistance, R_s, and a nonlinear differential elastance, $S(V) = dV/dQ = 1/C(V)$, where V is the voltage across the diode junction. This simple model is used to describe the basic properties of a varactor and is adequate as long as the displacement current is much larger than any conduction current across the junction. A rigorous analysis should also include the effect of a frequency- and voltage-dependent series resistance, and the equivalent circuit of parasitic elements due to packaging and contacting. The differential elastance is the slope of the voltage-charge

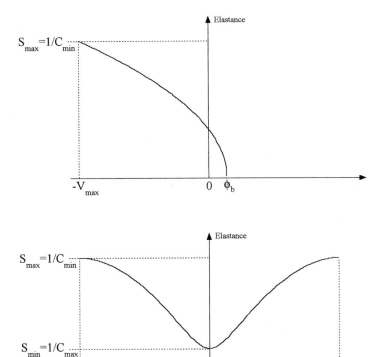

FIGURE 7.2 Elastance as a function of voltage for (top) one side junction diodes and (bottom) symmetric diodes.

relation of the diode and the reciprocal of the differential capacitance, $C(V)$. Since the standard varactor model consists of a resistance in series with a nonlinear capacitance, the elastance is used rather than the capacitance. This simplifies the analysis and gives, generally, a better understanding of the varactor. The differential elastance can be measured directly and is used rather than the ratio of voltage to charge.

The elastance versus voltage for a conventional varactor and a symmetric varactor are shown in Fig. 7.2. In both cases, the maximum available elastance is achieved at the breakdown voltage. The conventional varactor is reverse biased in order to allow maximum elastance swing and avoid any forward conduction current. A symmetric varactor will only produce odd harmonics when a sinusoidal signal is applied. This means that a varactor frequency tripler can be realized without any second harmonic idler circuit or DC bias. This simplifies the design of such circuits and, hence, many novel symmetric varactors have been proposed. Among these, the Heterostructure Barrier Varactor (HBV)[5] has so far shown the best performance.

Pumping

The pumping of a varactor is the process of passing a large current at frequency f_p through the varactor. The nonlinear varactor then behaves as a time-varying elastance, $S(t)$, and the series resistance dissipates power due to the large-signal current. The allowable swing in elastance is limited by the maximum elastance of the device used. Hence, the time domain equation describing the varactor model in Fig. 7.1 is given by:

$$V(t) = R_s i(t) + \int S(t) i(t) dt$$

The above equation, which describes the varactor in the time domain, must be solved together with equations describing the termination of the varactor. How the varactor is terminated at the input, output, and idler frequencies has a strong effect on the performance. The network has to terminate the varactor at some frequencies and couple the varactor to sources and loads at other frequencies. Since this embedding circuit is best described in the frequency domain, the above time domain equation is converted to the frequency domain. Moreover, the varactor is usually pumped strongly at one frequency, f_p, by a local oscillator. If there are, in addition, other small signals present, the varactor can be analyzed in two steps: (1) a large signal simulation of the pumped varactor at the frequency f_p, and (2) the varactor behaves like a time-varying linear elastance at the signal frequency, f_s. For the large signal analysis, the voltage, current, and differential elastance can be written in the forms

$$i(t) = \sum_{k=-\infty}^{\infty} I_k e^{jk\omega_p t}, \quad I_{-k} = I_k^*$$

$$v(t) = \sum_{k=-\infty}^{\infty} V_k e^{jk\omega_p t}, \quad V_{-k} = V_k^*$$

$$S(t) = \sum_{k=-\infty}^{\infty} S_k e^{jk\omega_p t}, \quad S_{-k} = S_k^*$$

Hence, the time domain equation that governs the above varactor model can be converted to the frequency domain and the relation between the Fourier coefficients, I_k, V_k, S_k, reads

$$V_k = R_s I_k + \frac{1}{jk\omega_p} = \sum_{l=-\infty}^{\infty} I_l S_{k-1}$$

The above equation is the general starting point for analyzing varactors. Since there is a relation between the Fourier coefficients S_k and I_k, the above equation is nonlinear and hard to solve for the general case. Today, the large signal response is usually calculated with a technique called harmonic balance.[6] This type of nonlinear circuit solver is available in most commercial microwave CAD tools.

Assume that the varactor is fully pumped and terminated so that the voltage across the diode junction becomes sinusoidal. The corresponding elastance waveforms for the conventional and symmetrical varactor in Fig. 7.2 are shown in Fig. 7.3. It is important to note that the fundamental frequency of the symmetric elastance device is twice the pump frequency.

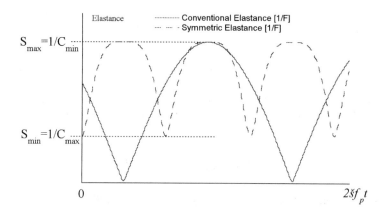

FIGURE 7.3   Elastance waveform $S(t)$ during full pumping with a sinusoidal voltage across the diode junction.

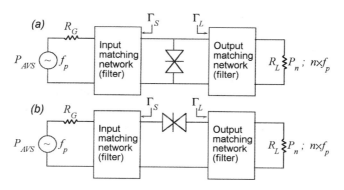

FIGURE 7.4 Block scheme of nth-order frequency multiplier circuit with (a) shunt-mounted and (b) series-mounted diodes.

The nonlinear part of the elastance, $S(t) - S_{min}$, creates harmonics and its impedance should be large compared to the series resistance, R_s, for a good varactor. The impedance ratio of the series resistance and the nonlinear part of the elastance at the fundamental frequency can be written as

$$\frac{R_s}{\dfrac{S_{max} - S_{min}}{2\pi f_p}} = \frac{f_p}{f_c}$$

and the dynamic cutoff frequency is introduced as

$$f_c = \frac{S_{max} - S_{min}}{2\pi R_s}$$

The dynamic cutoff frequency is an important figure of merit for many varactor applications and a typical value for a state-of-the-art varactor is more than 1 THz. The starting point for a varactor design is, hence, to maximize the elastance swing, $S_{max} - S_{min}$, and minimize any losses, R_s. For semiconductor varactors, the maximum elastance swing is limited by at least one of the following conditions:

- depletion layer punch-through,
- large electron conduction from impact ionization or forward conduction current,
- current saturation. The saturated electron velocity in the material determines the maximum length an electron can travel during a quarter of a pump cycle.[7]

Increasing the pump power beyond any of the above conditions will result in reduced performance and probably introduce extra noise.

Varactor Applications

Frequency Multipliers

Varactor frequency multipliers are extensively used to provide LO power to sensitive millimeter- and submillimeter-wavelength receivers. Today, frequency multipliers are the main application of varactors. Solid-state multipliers are relatively inexpensive, compact, lightweight, and reliable compared to vacuum tube technology, which makes them suitable for space applications at these frequencies. State-of-the-art balanced Schottky doublers can deliver 55 mW at 174 GHz[8] and state-of-the-art four-barrier HBV triplers deliver about 9 mW at 248 GHz.[9]

Frequency multiplication or harmonic generation in devices occur due to their nonlinearity. Based on whether the multiplication is due to a nonlinear resistance or a nonlinear reactance, one can differentiate between the varistor and varactor type of multipliers. Varactor type multipliers have a high potential conversion efficiency, but exhibit a narrow bandwidth and a high sensitivity to operating conditions. According to the Page-Pantell inequality, multipliers that depend upon a nonlinear resistance have at most an efficiency of $1/n^2$, where n is the order of multiplication.[10,11] The absence of reactive energy storage in varistor frequency multipliers ensures a large bandwidth. For the ideal varactor multiplier, i.e., a lossless nonlinear reactance, the theoretical limit is a conversion efficiency of 100% according to the Manley-Rowe formula. However, real devices exhibit properties and parameters that are a mixture of the ideal varistor and the ideal varactor multiplier. The following set of parameters is used to describe and compare properties of frequency multipliers:

- Conversion loss, L_n, is defined as the ratio of the available source power, P_{AVS}, to the output harmonic power, P_n, delivered to the load resistance. It is usually expressed in decibels. The inverted value of L_n, i.e., the conversion efficiency, η_n, is often expressed as a percent.
- In order to minimize the conversion loss, the optimum source and load embedding impedances, Z_S and Z_L, should be provided to the diode. Optimum source and load impedances are found from maximizing, e.g., the conversion efficiency, and they depend on each other and on the input signal level. In a nonlinear circuit, such as a multiplier, it is not possible to define a true impedance. However, a "quasi-impedance", Z_n, can be defined for periodic signals as

$$Z_n = \frac{V_n}{I_n}$$

where V_n and I_n are the voltage and the current, respectively, at the nth harmonic.

Basic Principles of Single Diode Frequency Multipliers — Single diode frequency multipliers can either be shunt or series mounted. In both cases the input and the output filter should provide optimum embedding impedances at the input and output frequencies, respectively. The output filter should also provide an open circuit for the shunt-mounted varactor and a short circuit for the series-mounted varactor at the pump frequency. The same arguments apply to the input filter at the output frequency. Analysis and design of conventional doublers and high order varactor multipliers are described well in the book by Penfield et al.[1] and in Reference 12.

In addition to the above conditions, the correct impedances must be provided at the idler frequencies for a high order multiplier (e.g., a quintupler). In general, it is hard to achieve optimum impedances at the different harmonics simultaneously. Therefore, a compromise has to be found.

Performance of Symmetric Varactor Frequency Multipliers — In Fig. 7.5 a calculation of the minimum conversion loss for a tripler and a quintupler is shown. To systematically investigate how the tripler and quintupler performance depends on the shape of the S-V characteristic, a fifth degree polynomial model was employed by Dillner et al.[13] The best efficiency is obtained for a S-V characteristic with a large nonlinearity at zero volts or a large average elastance during a pump cycle. The optimum idler circuit for the quintupler is an inductance in resonance with the diode capacitance (i.e., maximized third harmonic current).

Practical Multipliers — Since frequency multipliers find applications mostly as sources at higher millimeter and submillimeter wave frequencies, they are often realized in waveguide mounts[14,15] (see Fig. 7.6). A classic design is the arrangement of crossed rectangular waveguides of widths specific for the input and output frequency bands. The advantages are:

- The input signal does not excite the output waveguide, which is cut off at the input frequency.
- Low losses.
- The height of the waveguide in the diode mounting plane may be chosen to provide the electrical matching conditions. Assuming a thin planar probe, the output embedding impedance is given by analytical expressions.[16]
- Movable short circuits provide input/output tunability.

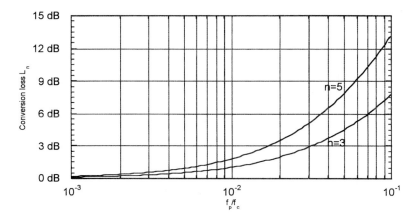

FIGURE 7.5 The minimum conversion loss for a tripler and a quintupler for the symmetric *S-V* characteristic shown in Fig. 7.2. The pump frequency is normalized to the dynamic cutoff frequency. For the quintupler case, the idler circuit is an inductance in resonance with the diode capacitance.

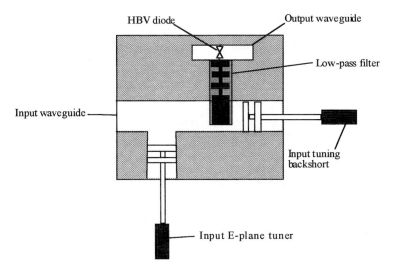

FIGURE 7.6 Schematic diagram of a crossed waveguide frequency multiplier. The output signal is isolated from the input waveguide with a low pass filter.

Today, whole waveguide mounts can be analyzed and designed using commercially available high frequency electromagnetic CAD tools. They either solve Maxwell's equations in the frequency domain or in the time domain using the FDTD method.

The inherently limited bandwidth of varactors can be improved by employing a transmission line periodically loaded with varactors.[17,18] These NonLinear Transmission Lines (NLTLs) can currently provide the largest bandwidth and still achieve a reasonable conversion efficiency as a frequency multiplier. Simultaneous effects of nonlinearity and dispersion may also be used for pulse compression (soliton propagation).

Frequency Converters

The varactor is useful as a frequency converter because of its good noise properties, and because gain can be achieved. The nonlinear reactance is pumped at a frequency f_p, and a small signal is introduced at a frequency f_s. Power will exchange at frequencies of the form $nf_p + f_s$ (n can be negative). If the output frequency is higher than the input frequency, the varactor acts as an upconverter, otherwise it is a downconverter. Furthermore, one differ between lower (n = −1) and upper sideband (n = 1) converters, and according to whether or not power is dissipated at any other sidebands (idlers).

Assume that the elastance is pumped sinusoidally (i.e., $S_k = 0$ for $|k| > 1$), the varactor is open-circuited at all frequencies except $f_s, f_p, f_u = f_p + f_s$, and that the varactor termination tunes out the average elastance. The source resistance is then adjusted to give optimum gain or optimum noise temperature. For the upper sideband upconverter, the minimal noise temperature is

$$T_{min} = T_d \frac{2f_s}{m_1 f_c} \left[\frac{f_s}{m_1 f_c} + \sqrt{1 + \left(\frac{f_s}{m_1 f_c} \right)^2} \right]$$

when the source resistance is

$$R_o^T = R_s \sqrt{1 + \left(\frac{m_1 f_c}{f_s} \right)^2}$$

where T_d is the diode temperature, f_c is the dynamic cutoff frequency, and m_1 is the modulation ratio defined as

$$m_1 = \frac{|S_1|}{S_{max} - S_{min}}$$

It can be shown that there is gain under optimum noise termination conditions only for signal frequencies smaller than $0.455 m_1 f_c$.[1] A different source resistance will result in maximum gain

$$R_o^G = R_s \sqrt{1 + \frac{m_1^2 f_c^2}{f_s f_u}}$$

The corresponding optimum gain is

$$G_{MAX} = \left(\frac{\dfrac{m_1 f_c}{f_s}}{1 + \sqrt{1 + \dfrac{m_1^2 f_c^2}{f_s f_u}}} \right)^2$$

As predicted by the Manley-Rowe formula for a lossless varactor, the gain increases as the output frequency, f_u, increases. The effect of an idler termination at $f_I = f_p − f_s$ can further increase the gain and reduce the noise temperature.

The above expressions for optimum noise and the corresponding source impedance are valid for the lower sideband upconverter as well. However, the lower sideband upconverter may have negative input and output resistances and an infinite gain causing stability problems and spurious oscillations. All pumped varactors may have such problems. With a proper choice of source impedance and pump frequency, it is possible to simultaneously minimize the noise and make the exchangeable gain infinite. This occurs for an "optimum" pump frequency of $f_p = \sqrt{m_1^2 f_c^2 + f_s^2}$ or approximately $m_1 f_c$ if the signal frequency is small. Further information on how to analyze, design, and optimize frequency converters can be found in the book by Penfield et al.[1]

Parametric Amplifiers

The parametric amplifier is a varactor pumped strongly at frequency f_p, with a signal introduced at frequency f_s. If the generated sidebands are terminated properly, the varactor can behave as a negative resistance at f_s. Especially the termination of the idler frequency, $f_p - f_s$, determines the real part of the impedance at the signal frequency. Hence, the varactor can operate as a negative resistance amplifier at the signal frequency, f_s. The series resistance limits the frequencies f_p and f_c for which amplification can be achieved and it also introduces noise.

The explanation of the effective negative resistance can be described as follows: The application of signal plus pump power to the nonlinear capacitance causes frequency mixing to occur. When current is allowed to flow at the idler frequency $f_p - f_s$, further frequency mixing occurs at the pump and idler frequencies. This latter mixing creates harmonics of f_p and $f_p - f_s$, and power at f_s is generated. When the power generated through mixing exceeds that being supplied at the signal frequency f_s, the varactor appears to have a negative resistance. If idler current is not allowed to flow, the negative resistance vanishes. Assuming that the elastance is pumped sinusoidally (i.e., $S_k = 0$ for $|k| > 1$), and the varactor is open circuited at all frequencies except f_s, f_p, $f_i = f_p - f_s$, and that the varactor termination tunes out the average elastance, gain can only be achieved if

$$f_s f_i \left(R_s + R_i \right) < R_s m_1^2 f_c^2$$

where R_i is the idler resistance. By terminating the varactor reactively at the idler frequency, it can be shown that a parametric amplifier attains a minimum noise temperature when pumped at the optimum pump frequency, which is exactly the same as for the simple frequency converter. This is true for nondegenerated amplifiers where the frequencies are well separated. The degenerate parametric amplifier operates with f_i close to f_s, and can use the same physical circuit for idler and signal frequencies. The degenerate amplifier is easier to build, but ordinary concepts of noise figure, noise temperature, and noise measure do not apply.

Voltage Tuning

One important application of varactors is voltage tuning. The variable capacitance is used to tune a resonant circuit with an externally applied voltage. This can be used to implement a Voltage Controlled Oscillator (VCO), since changing the varactor capacitance changes the frequency of oscillation within a certain range. As the bias is increased, the resonant frequency f_o increases from $f_{o,min}$ to $f_{o,max}$ as the elastance changes from S_{min} to S_{max}. If the present RF power is low, the main limitations are the finite tuning range implied by the minimum and maximum elastance and the fact that the series resistance degrades the quality factor, Q, of the tuned circuit. The ratio of the maximum and minimum resonant frequency gives a good indication of the tunability

$$\frac{f_{o,max}}{f_{o,min}} \leq \sqrt{\frac{S_{max}}{S_{min}}}$$

However, if the present RF power level is large, the average elastance, which determines the resonant frequency, depends upon drive level as well as bias. Second, the allowed variation of voltage is reduced for large RF power levels.

Since the varactor elastance is nonlinear, quite steep at low voltages, and almost flat at high voltages, the VCO tuning range is not naturally linear. However, an external bias circuit can improve the linearity of the VCO tuning range. It is also possible to optimize the doping profile of the varactor in terms of linearity, Q-value, or elastance ratio.

Varactor Devices

Conventional Diodes

Common conventional varactors at lower frequencies are reverse biased semiconductor abrupt p^+-n junction diodes made from GaAs or silicon.[2] However, metal-semiconductor junction diodes (Schottky diodes) are superior at high frequencies since the carrier transport only relies on electrons (unipolar device). The effective mass is lower and the mobility is higher for electrons compared to holes. Furthermore, the metal-semiconductor junction can be made very precisely even at a submicron level. A reverse biased Schottky diode exhibits a nonlinear capacitance with a very low leakage current. High frequency diodes are made from GaAs since the electron mobility is much higher than for silicon.

The hyperabrupt p^+-n junction varactor diode has a nonuniform n-doping profile and is often used for voltage tuning. The n-doping concentration is very high close to the junction and the doping profile is tailored to improve elastance ratio and sensitivity. Such doping profiles can be achieved with epitaxial growth or by ion implantation.

The Heterostructure Barrier Varactor Diode

The Heterostructure Barrier Varactor (HBV), first introduced in 1989 by Kollberg et al.,[5] is a symmetric varactor. The main advantage compared to the Schottky diode is that several barriers can be stacked epitaxially. Hence, an HBV diode can be tailored for a certain application in terms of both frequency and power handling capability. Moreover, the HBV operates unbiased and is a symmetric device, thus generating only odd harmonics. This greatly simplifies the design of high order and broadband multipliers.

The HBV diode is an unipolar device and consists of a symmetric layer structure. An undoped high band gap material (barrier) is sandwiched between two moderately n-doped, low band gap materials. The barrier prevents electron transport through the structure. Hence, the barrier should be undoped (no carriers), high and thick enough to minimize thermionic emission and tunnelling of carriers. When the diode is biased a depleted region builds up (Fig. 7.7), causing a nonlinear CV curve.

Contrary to the Schottky diode, where the barrier is formed at the interface between a metallic contact and a semiconductor, the HBV uses a heterojunction as the blocking element. A heterojunction, i.e., two adjacent epitaxial semiconductor layers with different band gaps, exhibits band discontinuities both in the valence and in the conduction band. Since the distance between the barriers (>1000 Å) is large compared to the de Broglie wavelength of the electron, it is possible to understand the stacked barrier structure as a series connection of N individual barriers. A generic layer structure of an HBV is shown in Table 7.1.

The HBV Capacitance — The parallel plate capacitor model, where the plate separation should be replaced with the sum of the barrier thickness, b, the spacer layer thickness, s, and the length of the depleted region, w, is normally an adequate description of the (differential) capacitance. The depletion

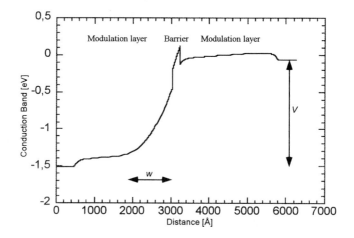

FIGURE 7.7 Conduction band of a biased GaAs/Al$_{0.7}$GaAs HBV.

TABLE 7.1 Generic Layer Structure of an HBV

Layer		Thickness [Å]	Doping Level [cm⁻³]	
7	Contact	~3000	n^{++}	
6	Modulation	$l = 3000$	$N_d \sim 10^{17}$	
5	Spacer	$s = 50$	undoped	
4	Barrier	$b \geq 100$	undoped	xN
3	Spacer	$s = 50$	undoped	
2	Modulation	$l = 3000$	$N_d \sim 10^{17}$	
1	Buffer	—	n^{++}	
0	Substrate		n^{++} or SI	

Note: For N epitaxially stacked barriers, the layer sequence 2–5 is repeated N times.

length is bias dependent and the layer structure is symmetric, therefore the elastance is an even function of applied voltage and is given by

$$S = \frac{1}{C} = \frac{N}{A}\left(\frac{b}{\varepsilon_b} + \frac{s}{\varepsilon_d} + \frac{w}{\varepsilon_d}\right)$$

$$w = \sqrt{\frac{2\varepsilon_d|V_d|}{qN_d}}$$

where V_d is the voltage across the depleted region, N_d is the doping concentration in the modulation layers, b is the barrier thickness, s is the undoped spacer layer thickness, A is the device area, and ε_b and ε_d are the dielectric constants in the barrier material and modulation layers, respectively. The maximum capacitance or the minimum elastance, S_{min}, occurs at zero bias. However, due to screening effects, the minimum elastance, S_{min}, must include the extrinsic Debye length, L_D, as:

$$S_{min} = \frac{1}{C_{max}} = \frac{N}{A}\left(\frac{b}{\varepsilon_b} + \frac{2s}{\varepsilon_d} + \frac{2L_D}{\varepsilon_d}\right)$$

$$L_D \equiv \sqrt{\frac{\varepsilon_d kT}{q^2 N_d}}$$

To achieve a high C_{max}/C_{min} ratio, the screening length can be minimized with a sheet doping, N_s, at the spacer/depletion layer interface. The minimum capacitance, C_{min}, is normally obtained for punch through condition, i.e., $w = l$, or when the breakdown voltage, V_{max}, is reached.

An accurate quasi-empirical expression for the *C-V* characteristic of homogeneously doped HBVs has been derived by Dillner et al.[19] The voltage across the nonlinear capacitor is expressed as a function of its charge as

$$V(Q) = N\left(\frac{bQ}{\varepsilon_b A} + 2\frac{sQ}{\varepsilon_d A} + Sign(Q)\left(\frac{Q^2}{2qN_d\varepsilon_d A^2} + \frac{4kT}{q}\left(1 - e^{\frac{|Q|}{2L_D AqN_d}}\right)\right)\right)$$

where T is the device temperature, q is the elementary charge, and Q is the charge stored in the HBV.

FIGURE 7.8   Planar four-barrier HBV (37 μm^2/CTH-NU2003J).

The substrate is either highly doped or semi-insulating (SI), depending on how the device is intended to be mounted. The contact layers (Nos. 1 and 7) should be optimized for low losses. Therefore, the buffer layer (No. 1) must be relatively thick ($\delta \sim 3\,\mu m$) and highly doped for planar HBVs (see Fig. 7.8). The barrier itself can consist of different layers to further improve the blocking characteristic. The spacer prevents diffusion of dopants into the barrier layer and increases the effective barrier height. The thickness of the barrier layer will not influence the cutoff frequency, but it has some influence on the optimum embedding impedances. Hence, the thickness is chosen to be thick enough to avoid tunneling of carriers. Several III-V semiconductor material systems have been employed for HBVs. The best choices to date for HBVs are the lattice matched $In_{0.53}Ga_{0.47}As/In_{0.52}Al_{0.48}As$ system grown on InP substrate and the lattice matched GaAs/AlGaAs system grown on GaAs substrate. High dynamic cutoff frequencies are achieved in both systems. However, the GaAs/AlGaAs system is well characterized and relatively easy to process, which increases the probability of reproducible results. The $In_{0.53}GaAs/In_{0.52}AlAs$ system exhibits a higher electron barrier and is therefore advantageous from a leakage current point of view. The thickness and doping concentration of the modulation layers should be optimized for maximal dynamic cutoff frequency.[20]

In the future, research on wide bandgap semiconductors (e.g., InGaN) could provide solutions for very high power HBVs, a combination of a II-VI barrier for low leakage current and a III-V modulation layer for high mobility and peak velocity. Today, narrow bandgap semiconductors from the III-V groups (e.g., $In_xGa_{1-x}As$) seem to be the most suitable for submillimeter wave applications.

The Si/SiO$_2$/Si Varactor

By bonding two thin silicon wafers, each with a thin layer of silicon dioxide, it is possible to form a structure similar to HBV diodes from III-V compounds. The SiO_2 layer blocks the conduction current very efficiently, but the drawback is the relatively low mobility of silicon. If a method to stack several barriers can be developed, this material system may be interesting for lower frequencies where the series resistance is less critical.

The Ferroelectric Varactor

Ferroelectrics are dielectric materials characterized by an electric field and temperature-dependent dielectric constant. Thin films of $Ba_xSr_{1-x}TiO_3$ have been proposed to be used for various microwave applications. Parallel plate capacitors made from such films can be used in varactor applications. However, the loss mechanisms at strong pump levels and high frequencies have not yet been fully investigated.

References

1. P. Penfield and R. P. Rafuse, *Varactor Applications.* Cambridge: M.I.T. Press, 1962.
2. S. M. Sze, *Physics of Semiconductor Devices,* 2nd ed. Singapore: John Wiley & Sons, 1981.

3. J. M. Manley and H. E. Rowe, Some General Properties of Nonlinear Elements, *IRE Proc.*, 44, 78, 904–913, 1956.
4. A. Uhlir, The potential of semiconductor diodes in high frequency communications, *Proc. IRE*, 46, 1099–1115, 1958.
5. E. L. Kollberg and A. Rydberg, Quantum-barrier-varactor diode for high efficiency millimeter-wave multipliers, *Electron. Lett.*, 25, 1696–1697, 1989.
6. S. A. Maas, Harmonic Balance and Large-Signal-Small-Signal Analysis, in *Nonlinear Microwave Circuits*. Artech House, Norwood, MA, 1988.
7. E. L. Kollberg, T. J. Tolmunen, M. A. Frerking, and J. R. East, Current saturation in submillimeter wave varactors, *IEEE Trans. Microwave Theory and Techniques*, 40, 5, 831–838, 1992.
8. B. J. Rizzi, T. W. Crowe, and N. R. Erickson, A high-power millimeter-wave frequency doubler using a planar diode array, *IEEE Microwave and Guided Wave Letters*, 3, 6, 188–190, 1993.
9. X. Mélique, A. Maestrini, E. Lheurette, P. Mounaix, M. Favreau, O. Vanbésien, J. M. Goutoule, G. Beaudin, T. Nähri, and D. Lippens, 12% Efficiency and 9.5 dBm Output Power from InP-based Heterostructure Barrier Varactor Triplers at 250 GHz, presented at IEEE-MTT Int. Microwave Symposium, Anaheim, CA, 1999.
10. R. H. Pantell, General power relationship for positive and negative nonlinear resistive elements, *Proceedings IRE*, 46, 1910–1913, December 1958.
11. C. H. Page, Harmonic generation with ideal rectifiers, *Proceedings IRE*, 46, 1738–1740, October 1958.
12. C. B. Burckhardt, Analysis of varactor frequency multipliers for arbitrary capacitance variation and drive level, *The Bell System Technical Journal*, 675–692, April 1965.
13. L. Dillner, J. Stake, and E. L. Kollberg, Analysis of symmetric varactor frequency multipliers, *Microwave Opt. Technol. Lett.*, 15, 1, 26–29, 1997.
14. J. W. Archer, Millimeter wavelength frequency multipliers, *IEEE Trans. on Microwave Theory and Techniques*, 29, 6, 552–557, 1981.
15. J. Thornton, C. M. Mann, and P. D. Maagt, Optimization of a 250-GHz Schottky tripler using novel fabrication and design techniques, *IEEE Trans. on Microwave Theory and Techniques*, 46, 8, 1055–1061, 1998.
16. E. L. Eisenhart and P. J. Khan, Theoretical and experimental analysis for a waveguide mounting structure, *IEEE Trans. on Microwave Theory and Techniques*, 19, 8, 1971.
17. S. Hollung, J. Stake, L. Dillner, M. Ingvarson, and E. L. Kollberg, A distributed heterostructure barrier varactor frequency tripler, *IEEE Microwave and Guided Wave Letters*, 10, 1, 24–26, 2000.
18. E. Carman, M. Case, M. Kamegawa, R. Yu, K. Giboney, and M. J. Rodwell, V-band and W-band broadband, monolithic distributed frequency multipliers, *IEEE Microwave and Guided Wave Letters*, 2, 6, 253–254, 1992.
19. L. Dillner, J. Stake, and E. L. Kollberg, Modeling of the Heterostructure Barrier Varactor Diode, presented at 1997 International Semiconductor Device Research Symposium, Charlottesville, SC, 1997.
20. J. Stake, S. H. Jones, L. Dillner, S. Hollung, and E. Kollberg, Heterostructure barrier varactor design, *IEEE Trans. on Microwave Theory and Techniques*, 48, 4, Part 2, 2000.

Further Information

M. T. Faber, J. Chramiec, and M. E. Adamski, *Microwave and Millimeter-Wave Diode Frequency Multipliers*. Artech House Publishers, Norwood, MA, 1995.
S. A. Maas, *Nonlinear Microwave Circuits*. Artech House, Norwood, MA, 1988.
P. Penfield and R. P. Rafuse, *Varactor Applications*. M.I.T. Press, Cambridge, 1962.
S. Yngvesson, *Microwave Semiconductor Devices*. Kluwer Academic Publishers, Boston, 1991.

7.1.2 Schottky Diode Frequency Multipliers

Jack East

Heterodyne receivers are an important component of most high frequency communications systems and other receivers. In its simplest form a receiver consists of a mixer being pumped by a local oscillator. At lower frequencies a variety of local oscillator sources are available, but as the desired frequency of operation increases, the local oscillator source options become more limited. The "lower frequencies" limit has increased with time. Early transistor oscillators were available in the MHz and low GHz range. Two terminal transit time devices such as IMPATT and Gunn diodes were developed for operation in X and Ka band in the early 1970s. However, higher frequency heterodyne receivers were needed for a variety of communications and science applications, so alternative local oscillator sources were needed. One option was vacuum tubes. A variety of vacuum tubes such as klystrons and backward wave oscillators grew out of the radar effort during the Second World War. These devices were able to produce large amounts of power over most of the desired frequency range. However, they were large, bulky, expensive, and suffered from modest lifetimes. They were also difficult to use in small science packages.

An alternative solid state source was needed and the technology of the diode frequency multiplier was developed beginning in the 1950s. These devices use the nonlinear reactance or resistance of a simple semiconductor or diode to produce high frequency signals by frequency multiplication. These multipliers have been a part of many high frequency communications and science applications since that time. As time passed the operating frequencies of both transistors and two-terminal devices increased. Silicon and GaAs transistors have been replaced by much higher frequency HFETs and HBTs with f_{max} values of hundreds of GHz. Two-terminal IMPATT and Gunn diodes can produce more than 100 milliwatts at frequencies above 100 GHz. However, the desired operating frequencies of communications and scientific applications have also increased. The most pressing needs are for a variety of science applications in the frequency range between several hundred GHz and several THz. Applications include space-based remote sensing of the earth's upper atmosphere to better understand the chemistry of ozone depletion and basic astrophysics to investigate the early history of the universe. Both missions will require space-based heterodyne receivers with near THz local oscillators. Size, weight, and prime power will be important parameters. Alternative approaches include mixing of infrared lasers to produce the desired local oscillator frequency from higher frequencies, and a multiplier chain to produce the desired frequency from lower frequencies. Laser-based systems with the desired output frequencies and powers are available, but not with the desired size and weight. Semiconductor diode-based frequency multipliers have the modest size and weight needed, but as of now cannot supply the required powers, on the order of hundreds of microwatts, needed for the missions. This is the subject of ongoing research.

The goal of this chapter is to briefly described the performance of diode frequency multipliers in order to better understand their performance and limitations. The chapter is organized as follows. The next section will describe the properties of Schottky barrier diodes, the most useful form of a varactor multiplier. The following section will describe the analytic tools developed to predict multiplier operation. Two limitations, the reactive multiplier described by Manley and Rowe and the resistive multiplier discussed by Page will be discussed. The results of these two descriptions can be used to understand the basic limits of multiplier operation. However, these analytic results do not provide enough information to design operating circuits. A more realistic computer-based design approach is needed. This will be discussed in the next section. Limitations on realistic devices and some alternative structures will then be described followed by a look at one future application and a brief summary.

Schottky Diode Characteristics

Multiplier operation depends on the nonlinear properties of Schottky diodes. The diode has both a capacitive and a resistive nonlinearity. Consider the simple representation of a uniformly Schottky diode shown in Fig. 7.9. Figure 7.11(a) shows a semiconductor with a Schottky barrier contact on the right and an ohmic contact on the left. The semiconductor has a depletion layer width w, with the remaining portion of the structure undepleted. The depletion layer can be represented as a capacitor and the

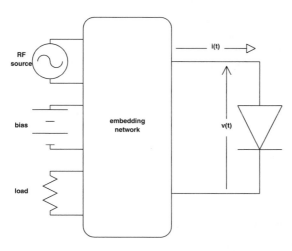

FIGURE 7.9 Schottky diodes (a) structure, (b) electric field, (c) capacitance vs. voltage, and (d) equivalent circuit.

undepleted portion can be represented as a resistor. The depletion layer will act as a parallel plate capacitor with a capacitance of

$$C = \frac{\epsilon A}{w}, \tag{7.1}$$

where C is the capacitance, ϵ is the dielectric constant, A is the area, and W is the depletion width w from Fig. 7.9(a). The electric field vs. distance for this structure is shown in Fig. 7.9(b). For reasonable conditions the electric field in the undepleted region is small. The field in the depletion regions extends over the width of the depletion region and depends linearly on x. The area under the electric field curve in Fig. 7.9(b) is the depletion layer voltage, the sum of the applied voltage V_b and the built-in potential ϕ_{bi}. The resulting depletion width vs. applied reverse voltage is

$$w = \sqrt{\frac{2\epsilon\left(\phi_{bi} + V_{bios}\right)}{qN_d}}, \tag{7.2}$$

where ϕ_{bi} is the built-in potential of the metal semiconductor or junction, V_{bios} is the applied reverse bias, q is the electronic charge, and N_d is the uniform semiconductor doping. This width vs. applied bias will result in the capacitance vs. applied voltage of the form

$$c\left(V\right) = \frac{C_{i0}}{\sqrt{\phi_{bi} + V_{bios}}} \tag{7.3}$$

where C_{i0} is the capacitance at zero applied bias. The resulting capacitance vs. bias voltage is shown in Fig. 7.9(c). This capacitance characteristic is the starting point for the analytic models in the next section. However, other effects are also present. Under realistic pumping conditions, the diode can also be forward biased allowing forward conduction current flow. This can be approximated with a nonlinear voltage-dependent resistance. The resulting equivalent circuit then becomes the right 2 nonlinear elements in Fig. 7.9(c). There are also parasitic elements. The undepleted region of the device and various contact and package resistance will appear in series with the nonlinear junction. Although the undepleted region width is voltage dependent, this series resistance is usually modeled with a constant value. However, at very high frequencies the current through the undepleted region can crowd to the outside edge of the material due to the skin effect, increasing the resistance. This frequency-dependent resistance is sometimes included in multiplier simulations [9]. The varactor diode must be connected to the external circuit. This resulting physical connection usually results in a parasitic shunt capacitance associated with the connection and a parasitic series inductance associated with the current flow through the wire connection. A major part of multiplier research over the past decade has involved attempts to reduce these parasitic effects.

This nonlinear capacitance vs. voltage characteristic can be used as a frequency multiplier. Consider the charge Q in the nonlinear capacitance as a function of voltage

$$Q(v) = \sqrt{a(\phi_{bi} - V)}, \tag{7.4}$$

where a is a constant. This function can be expanded in a series

$$Q(V) = a_0 + a_1 V + a_2 V^2 + a_3 V^3 + \dots, \tag{7.5}$$

The current $I(t)$ is the time derivative of the charge,

$$I(t) = \frac{dQ}{dt} = \left[a_1 + 2a_2 V + 3a_3 V^2 + \dots \right] \frac{dV}{dt} \tag{7.6}$$

If $V(t)$ is of the form $V_{rf} \sin \omega t$, then the higher order V terms will produce harmonics of the input pump frequency.

Analytic Descriptions of Diode Multipliers

Equation (7.6) shows that we can get higher order frequencies out of a nonlinear element. However, it is not clear what the output power or efficiency will be. The earliest research on frequency multipliers were based on closed form descriptions of multiplier operation to investigate this question. This section will discuss the ideal performance of reactive and resistive frequency multipliers.

A nonlinear resistor or capacitor, when driven by a pump source, can generate a series of harmonic frequencies. This is the basic form of a harmonic multiplier. The Manley-Rowe relations are a general description of power and frequency conversion relations in nonlinear reactive elements [1, 2]. They describe the properties of frequency conversion and general in nonlinear reactances. The earliest work on these devices sometimes used nonlinear inductances, but all present work involves the nonlinear capacitance vs. voltage characteristic of a reverse-biased Schottky barrier diode. Although the Manley Rowe equations describe frequency multiplication, mixer operation, and parametric amplification, they are also useful as an upper limit on multiplier operation. If an ideal nonlinear capacitance is a pump with a local oscillator at frequency f_0, and an embedding circuit allows power flow at harmonic frequencies, then the sum of the powers into and out of the capacitor is zero,

$$\sum_{m=0}^{\infty} P_m = 0. \tag{7.7}$$

This expression shows that we can have an ideal frequency multiplier at 100% efficiency converting input power to higher frequency output power if we properly terminate all the other frequencies. Nonlinear resistors can also be used as frequency multipliers [3, 4]. For a nonlinear resistor pumped with a local oscillator at frequency f_0, the sum of the power is

$$\sum_{m=0}^{\infty} m^2 P_m \geq 0. \tag{7.8}$$

For an mth order resistive harmonic generator with only an input and output signal, the efficiency is at best $\frac{1}{m^2}$, 25% for a doubler and 11% for a tripler.

Although Eqs. (7.7) and (7.8) give upper limits on the efficiency to be expected from a multiplier, they provide little design information for real multipliers. The next step in multiplier development was the development of closed form expressions for design based on varactor characteristics [5, 6]. Burckardt [6] gives design tables for linear and abrupt junction multipliers based on closed form expressions for the charge in the diode. These expressions form the starting point for waveform calculations. Computer simulations based on the Fourier components of these waveforms give efficiency and impedance information from 2nd to 8th order. However, these approximations limit the amount of design information available. A detailed set of information on multiplier design and optimization requires a computer-based analysis.

Computer-Based Design Approaches

The analytic tools discussed in the last section are useful to predict the ideal performance of various frequency multipliers. However, more exact techniques are needed for useful designs. Important information such as input and output impedances, the effects of series resistance, and the effect of harmonic terminations at other harmonic frequencies are all important in multiplier design. This information requires detailed knowledge of the current and voltage information at the nonlinear device. Computer-based simulations are needed to provide this information. The general problem can be described with the help of Fig. 7.10. The multiplier consists of a nonlinear microwave diode, an embedding network that provides coupling between the local oscillator source and the output load, provisions for DC bias and terminations for all other non-input or output frequencies. Looking toward the embedding network from the diode, the device sees embedding impedances at the fundamental frequency and each of the harmonic frequencies. The local oscillator power available is usually specified along with the load impedances and the other harmonic frequencies under consideration. The goal is to obtain the operating conditions, the output power and efficiency, and the input and output impedances of the overall circuit. The nonlinear nature of the problem makes the solution more difficult. Within the context of Fig. 7.10, the current as a function of time is a nonlinear function of the voltage. Device impedances can be obtained from ratios of the Fourier components of the voltage and current. The impedances of the diode must match the embedding impedances at the corresponding frequency. A "harmonic balance" is required for the correct situation. Many commercial software tools use nonlinear optimization techniques to solve

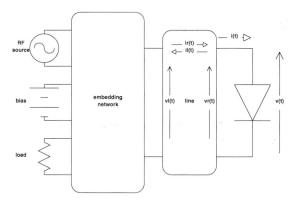

FIGURE 7.10 Generalized frequency multiplier.

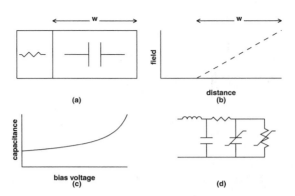

FIGURE 7.11 Multiple reflection circuit.

this "harmonic balance" optimization problem. An alternative approach is the multiple reflection algorithm. This solution has the advantage of a very physical representation of the actual transient response of the circuit. This technique is discussed in References 7, 8, and 9 and will be described here.

The basic problem is to calculate the diode current and voltage waveforms when the device is embedded in an external linear circuit. The diode waveforms are best represented in the time domain. However, the embedding circuit consists of linear elements and is best represented in the frequency or impedance domain. One approach is the multiple reflection technique [7, 8, 9]. This technique splits the simulation into two parts, a nonlinear time domain description of the diode multiplier and a linear frequency domain description of the embedding network. The solution goal is to match the frequency domain impedances of the embedding circuit with the time domain impedances of the nonlinear device. The circuit is shown in Fig. 7.11. The initial circuit is modified by including a long transmission line with a length l equal to an integral number of pump frequency wavelengths and an arbitrary characteristic impedance Z_0 between the diode and the embedding network. Waves will propagate in both directions on this transmission line depending on the conditions at the ends. When steady-state conditions are reached, the waveforms at the circuit and the diode will be the same, with or without the transmission line. This transmission line allows us to determine the waveforms across the diode as a series of reflections from the circuit. The signals on the transmission line are composed of left- and right-traveling current and voltage waves. The voltage at the diode is the sum of the right- and left-traveling wave voltages,

$$v\left(x\right)=v_r\left(x\right)+v_l\left(x\right) \tag{7.9}$$

and the current at the diode is the difference

$$i\left(x\right)=i_r\left(x\right)-i_l\left(x\right)=\frac{v_r\left(x\right)-v_l\left(x\right)}{Z_0} \tag{7.10}$$

Since the transmission line is an integral number of pump frequency wavelengths long, the conditions are the same at each end under steady-state conditions,

$$v\left(x=0\right)=v\left(x=l\right) \tag{7.11}$$

$$i\left(x=0\right)=i\left(x=l\right) \tag{7.12}$$

At the start of the simulation, we assume that there is a right-traveling wave associated with the DC bias at frequency 0 and the pump at frequency 1. This wave with a DC component and a component at the local oscillator frequency will arrive at the diode at $x = 0$ as a $V_r(t)$. The resulting voltage across the diode will produce a diode current. This current driving the transmission line will produce a first reflected voltage,

$$v^1_{reflected} = v^1_l = \left(v^1_{diode} - i^1_{diode}(t)Z_0\right)/2 \tag{7.13}$$

This time domain reflected or left-traveling wave will then propagate to the embedding network. Here is can be converted into the frequency domain with a Fourier transform. The resulting signal will contain harmonics of the local oscillator signal due to the nonlinear nature of the diode current vs. voltage and charge vs. voltage characteristic. The resulting frequency domain information can then be used to construct a new reflected voltage wave from the embedding network. This process of reflections from the diode and the circuit or "multiple reflections" continues until a steady-state solution is reached.

This computer-based solution has several advantages over the simpler analysis-based solutions. It can handle combinations of resistive and reactive nonlinearities. Most high-efficiency multipliers are pumped into forward conduction during a portion of the RF cycle so this is an important advantage for accurate performance predictions. In practice the varactor diode will also have series resistance, parasitic capacitances, and series inductances. These additional elements are easily included in the multiplier simulation. At very high frequencies the series resistance can be frequency dependent, due to current crowding in the semiconductor associated with the skin effect. This frequency dependence loss can also be included in simulations. Computer programs with these features are widely used to design high performance frequency multipliers.

An alternative solution technique is the fixed point method [10]. This technique uses a circuit similar to Fig. 7.11, with a nonlinear device and an embedding network connected with a transmission line. However, this approach uses a fixed point iteration to arrive at a converged solution. The solution starts with arbitrary voltages at the local oscillator frequency at the harmonics. With these starting conditions, a new voltage is obtained from the existing conditions using

$$V_{n,k+1} = \frac{Z_0 V_{rf}}{Z^L_n + Z_0} + \frac{Z^L_n\left(V_{n,k} - I_{n,k}Z_0\right)}{Z^L_n + Z_0} \tag{7.14}$$

for the driven local oscillator frequency and

$$V_{n,k+1} = \frac{Z_0}{Z^L_n + Z_0}\left(V_{n,k} - I_{n,k}Z_0\right) \tag{7.15}$$

for the remaining frequencies, where Z^L_n are the embedding impedances at frequency n, Z_0 is the same line characteristic impedance used in the multiple reflection simulation, $Z^{NL}_{n,k}$ is the nonlinear device impedance at frequency n and iteration number k, $V_{n,k}$, and $I_{n,k}$ are the frequency domain voltage and current at iteration k, and V^S_n is the RF voltage at the pump source. These two equations provide an iterative solution of the nonlinear problem. They are particularly useful when a simple equivalent circuit for the nonlinear device is not available. We now have the numerical tools to investigate the nonlinear operation of multipliers. However, there are some operating conditions where this equivalent circuit approach breaks down. Some of these limitations will be discussed in the next section.

Device Limitations and Alternative Device Structures

The simulation tools and simple device model do a good job of predicting the performance of low frequency diode multipliers. However, many multiplier applications require very high frequency output signals with reasonable output power levels. Under these conditions, the output powers and efficiencies predicted are always higher than the experimental results. There are several possible reasons. Circuit loss increases with frequency, so the loss between the diode and the external connection should be higher. Measurements are less accurate at these frequencies, so the differences between the desired designed circuit embedding impedances and the actual values may be different. Parasitic effects are also more important, degrading the performance. However, even when all these effects are taken into account, the experimental powers and efficiencies are still low. The problem is with the equivalent circuit of the diode in Fig. 7.9(d). It does not correctly represent the high frequency device physics [11]. The difficulty can be explained by referring back to Fig. 7.9(a). The device is a series connection of the depletion layer capacitance and a bulk series resistance. The displacement current flowing through the capacitor must be equal to the conduction current flowing through the undepleted resistive region. The capacitor displacement current is

$$I(t) = \frac{dC(V)V(t)}{dt} \tag{7.16}$$

If we approximate $V(t)$ with $V_{rf} \cos(\omega t)$, then the current becomes

$$I(t) = V_{rf}\omega C(V)\sin \omega t + V(t)\frac{dC(V)}{dt} \tag{7.17}$$

For a given device, the displacement current and the resulting current through the resistor increase with the frequency and the RF voltage. At modest drive levels and frequencies the undepleted region can support this current and the equivalent resistance remains constant. However, the current density through the undepleted region is

$$J_{dep} = qN_d v(E), \tag{7.18}$$

where J_{dep} is the conduction current density in the undepleted region and $v(E)$ is the carrier velocity vs. electric field. At low electric fields the slope of the velocity field curve is constant with a value equal to the carrier mobility μ. However at higher electric fields, the velocity begins to saturate at a constant value. Additional increases in the electric field do not increase the conduction current through the varactor. This current saturation is a fundamental limit on the multiplier performance. A more physical explanation also uses Fig. 7.9(a). A nonlinear capacitor implies a change in the depletion layer width with voltage. However, changing the depletion layer width involves moving electrons from the depletion layer edge. These electrons are limited by their saturated velocity, so the time rate of change of the capacitance is also unlimited.

This simple theory allows a modified device design. Equations (7.18) and (7.17) are the starting point. Our goal is usually to optimize the power or efficiency available from a multiplier circuit. Clearly one option is to increase the doping N_d in Eq. (7.18) to increase the available J_{dep}. However, increasing the doping decreases the breakdown voltage and thus the maximum RF voltage that can be present across the reverse biased depletion layer. The device doping design becomes a parameter in the overall multiplier design. The optimum efficiency and optimum power operating points for the same input and output frequency are usually different. Although this simple description provides useful information, a detailed physical model is usually needed for best results [12].

The discussion so far has been based on a uniformly doped abrupt junction varactor. However, other doping or material layer combinations are possible [13]. One option is to tailor the doping or material profile to obtain a capacitance vs. voltage that is more nonlinear than the $1/\sqrt{V_{bias}}$ dependence in Eq. (7.2). Two options are the hyperabrupt varactor and the BNN structure. These devices are shown in Fig. 7.12.

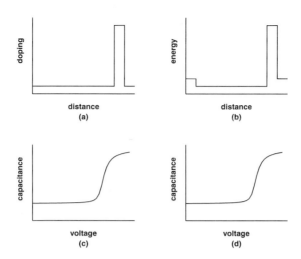

FIGURE 7.12 Alternative structures (a) hyperabrupt doping, (b) BNN energy, (c) hyperabrupt capacitance, and (d) BNN capacitance.

The doping profile of a hyperabrupt varactor is shown in Fig. 7.12(a). Instead of a uniform doping, this structure has a much smaller doping over most of the structure with a high doping or doping spike near the metal semiconductor junction. The corresponding capacitance vs. voltage characteristic is shown in Fig. 7.12(c). At modest reverse biases the depletion layer extends from the metal contact to the doping spike. The resulting narrow depletion layer produces a high capacitance. Higher applied voltages begin to deplete the charge in the doping spike. When the spike charge is depleted, there is a rapid increase in the depletion layer width through the second lightly doped region and a corresponding decrease in the capacitance. This structure can produce a more nonlinear capacitance variation than a uniformly doped device. However, the structure combines both lightly doped and heavily doped regions, so saturation effects can occur in the lightly doped portion. An alternative structure is the BNN or BIN device. This structure uses combinations of **B**arriers **I**ntrinsic and **N** doped regions formed with combinations of different epitaxial materials and doping to produce optimized capacitance structures. This structure can have either ohmic or Schottky contacts. A typical structure is shown in Fig. 7.12(b). Notice that conduction band energy rather than doping is being plotted. The structure consists of an *n*+ ohmic contact on the right followed by an *n* region that can be depleted, a wide bandgap barrier region, and a second *n*+ ohmic contact on the left. This structure can have a highly nonlinear capacitance characteristics as shown in Fig. 7.12(d), depending on the choice of layer doping and energy band offsets. These BNN structures have potential advantages in monolithic integration and can be fabricated in a stacked or series configuration for higher output powers.

Since the major application of frequency multipliers is for high frequency local oscillator sources, it would be reasonable to try to fabricate higher order multipliers, triplers for example, instead of doublers. Although based on Eq. (7.7), this is possible, there are some problems. Efficient higher order multiplication requires currents and voltages at intermediate frequencies, at the second harmonic in a tripler for example. This is an "idler" frequency or circuit. However, in order to avoid loss, this frequency must have the correct reactive termination. This adds to the complexity of the circuit design. Some details of doublers and triplers are given in Reference 14. An alternative that avoids the idlers is an even or symmetrical capacitance voltage characteristic. One possibility is the single barrier or quantum barrier varactor [15]. The structure and associated capacitance voltage characteristic are shown in Fig. 7.13. This structure, shown in Fig. 7.13(a), is similar to the BNN expect the barrier is the middle with lightly doped regions on either side. This will be a series connection of a depletion layer, a barrier and a depletion region, with ohmic contacts on each end. The capacitance is maximum at zero applied bias, with a built-in depletion layer on each side of the barrier. When a voltage is applied, one to the depletion layers will become forward biased and shrink and the other one will be reverse biased and expand. The series

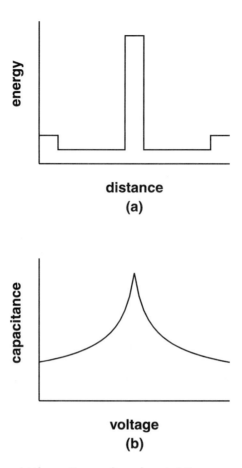

FIGURE 7.13   Structure and associated capacitance voltage characteristic.

combination capacitance will become smaller. Reversing the applied voltage will produce the same capacitance. The resulting symmetrical capacitance is shown in Fig. 7.13(b). This capacitance characteristic is a useful starting point for odd order multipliers.

Summary and Conclusions

This chapter has briefly discussed the properties of Schottky barrier diodes that are useful for frequency multipliers. Although a variety of other solid state sources are available for lower frequency sources, these devices are a critical component of future space-based applications.

References

1. J.M. Manley and H.E. Rowe, Some General Properties of Nonlinear Elements, I: General Energy Relations, *Proceedings of the IRE,* 44, 904, 1956.
2. H.A. Watson, *Microwave Semiconductor Devices and Their Circuit Applications,* McGraw-Hill, New York, 1969, chap. 8.
3. C.H. Page, Frequency Conversion With Positive Nonlinear Resistors, *Journal of Research of the National Bureau of Standards,* 56, 4, 179–182, April 1956.
4. R.H. Pantell, General Power Relationships for Positive and Negative Nonlinear Resistive Elements, *Proceedings of the IRE,* 46, 12, 1910–1913, December 1958.
5. J.A. Morrison, Maximization of the Fundamental Power in Nonlinear Capacitance Diodes, *Bell System Technical Journal,* 41, 677–721, 1962.

6. C.B. Burckardt, Analysis of Varactor Frequency Multipliers for Arbitrary Capacitance Variation and Drive Level, *Bell System Technical Journal*, 44, 675–692, 1965.

7. D.N. Held and A.R. Kerr, Conversion Loss and Noise of Microwave and Millimeter Wave Mixers: Part 1-Theory; Part 2 Experiment, *IEEE Transactions on Microwave Theory and Techniques*, MTT-26, 49, 1978.

8. S.A. Maas, *Microwave Mixers*, Artech House, Norwood, MA, 1986.

9. P. Siegel, A. Kerr and W. Hwang, Topics in the Optimization of Millimeter Wave Mixers, NASA Technical Paper 2287, March 1984.

10. G.B. Tait, Efficient Solution Method Unified Nonlinear Microwave Circuit and Numerical Solid-State Device Simulation, *IEEE Microwave and Guided Wave Letters*, 4, 12, 420–422, December 1994.

11. E.L. Kollberg, T.J. Tolmunen, M.A. Frerking, and J.R. East, Current Saturation in Submillimeter Wave Varactors, *IEEE Transactions on Microwave Theory and Techniques*, MTT-40, 5, 831–838, May 1992.

12. J.R. East, E.L. Kollberg and M.A. Frerking, Performance Limitations of Varactor Multipliers, Fourth International Conference on Space Terahertz Technology, Ann Arbor, MI, March 1993.

13. M. Frerking and J.R. East, Novel Heterojunction Varactors, *Proceedings of the IEEE Special Issue on Terahertz Technology*, 80, 11, 1853–1860, November 1992.

14. A. Raisanen, Frequency Multipliers for Millimeter and Submillimeter Wavelengths, *Proceedings of the IEEE Special Issue on Terahertz Technology*, 80, 11, 1842–1852, November 1992.

15. E. Kollberg and A. Rydberg, Quantum Barrier Varactor Diodes for High-Efficiency Millimeter-Wave Multipliers, *Electronics Letters*, 25, 1696–1697, December 1989.

7.1.3 Transit Time Microwave Devices

Robert J. Trew

There are several types of active two-terminal diodes that can oscillate or supply gain at microwave and millimeter-wave frequencies. These devices can be fabricated from a variety of semiconductor materials, but Si, GaAs, and InP are generally used. The most common types of active diodes are the IMPATT (an acronym for IMPact Avalanche Transit-Time) diode, and the Transferred Electron Device (generally called a Gunn diode). Tunnel diodes are also capable of producing active characteristics at microwave and millimeter-wave frequencies, but have been replaced in most applications by three-terminal transistors (such as GaAs MESFETs and AlGaAs/GaAs HEMTs), which have superior RF and noise performance, and are also much easier to use in systems applications. The IMPATT and Gunn diodes make use of a combination of internal feedback mechanisms and transit-time effects to create a phase delay between the RF current and voltage that is more than 90°, thereby generating active characteristics. These devices have high frequency capability since the saturated velocity of an electron in a semiconductor is high (generally on the order of ~10^7 cm/sec) and the transit time is short since the length of the region over which the electron transits can be made on the order of a micron (i.e., 10^{-4} cm) or less. The ability to fabricate devices with layer thicknesses on this scale permits these devices to operate at frequencies well into the millimeter-wave region. Oscillation frequency on the order of 400 GHz has been achieved with IMPATT diodes, and Gunn devices have produced oscillations up to about 150 GHz. These devices have been in practical use since the 1960s and their availability enabled a wide variety of solid-state system components to be designed and fabricated.

Semiconductor Material Properties

Active device operation is strongly dependent upon the charge transport, thermal, electronic breakdown, and mechanical characteristics of the semiconductor material from which the device is fabricated. The charge transport properties describe the ease with which free charge can flow through the material. This is described by the charge velocity-electric field characteristic, as shown in Fig. 7.14 for several commonly used semiconductors. At low values of electric field, the charge transport is ohmic and the charge velocity

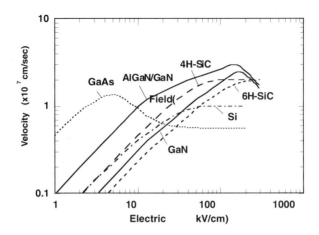

FIGURE 7.14   Electron velocity versus electric field characteristics for several semiconductors.

is directly proportional to the magnitude of the electric field. The proportionality constant is called the mobility and has units of cm²/V-sec. Above a critical value for the electric field, the charge velocity (units of cm/sec) saturates and either becomes constant (e.g., Si) or decreases with increasing field (e.g., GaAs). Both of these behaviors have implications for device fabrication, especially for devices intended for high frequency operation. Generally, for transit time devices, a high velocity is desired since current is directly proportional to velocity. The greatest saturated velocity is demonstrated for electrons in the wide bandgap semiconductors, SiC and GaN. Both of these materials have saturated electron velocities on the order of $v_s \sim 2 \times 10^7$ cm/sec. This is one of the main reasons these materials are being developed for high frequency electronic devices. Also, a low value for the magnitude of the electric field at which velocity saturation occurs is desirable since this implies high charge mobility. High mobility produces low resistivity, and therefore low values for parasitic and access resistances for semiconductor devices.

The decreasing electron velocity with electric field characteristic for compound semiconductors such as GaAs and InP makes active two-terminal devices called Transferred Electron Devices (TED's) or Gunn diodes possible. The negative slope of the velocity versus electric field characteristic implies a decreasing current with increasing voltage. That is, the device has a negative resistance. When a properly sized piece of these materials is biased in the region of decreasing current with voltage, and placed in a resonant cavity, the device will be unstable up to very high frequencies. By proper selection of embedding impedances, oscillators or amplifiers can be constructed.

Other semiconductor material parameters of interest include thermal conductivity, dielectric constant, energy bandgap, electric breakdown critical field, and minority carrier lifetime. The thermal conductivity of the material is important because it describes how easily heat can be extracted from the device. The thermal conductivity has units of W/cm-°K, and in general, high thermal conductivity is desirable. Compound semiconductors, such as GaAs and InP, have relatively poor thermal conductivity compared to elemental semiconductors such as Si. Materials such as SiC have excellent thermal conductivity and are used in high power electronic devices. The dielectric constant is important since it represents capacitive loading and, therefore, affects the size of the semiconductor device. Low values of dielectric constant are desirable since this permits larger device area, which in turn results in increased RF current and increased RF power that can be developed. Electric breakdown characteristics are important since electronic breakdown limits the magnitudes of the DC and RF voltages that can be applied to the device. A low magnitude for electric field breakdown limits the DC bias that can be applied to a device, and thereby limits the RF power that can be handled or generated by the device. The electric breakdown for the material is generally described by the critical value of electric field that produces avalanche ionization. Minority carrier lifetime is important for bipolar devices, such as pn junction diodes, rectifiers, and

TABLE 7.2 Material Parameters for Several Semiconductors

Material	E_g(eV)	ε_r	κ(W/K-cm) @ 300°K	E_c(V/cm)	$\tau_{minority}$ (sec)
Si	1.12	11.9	1.5	3×10^5	2.5×10^{-3}
GaAs	1.42	12.5	0.54	4×10^5	$\sim 10^{-8}$
InP	1.34	12.4	0.67	4.5×10^5	$\sim 10^{-8}$
6H-SiC	2.86	10.0	4.9	3.8×10^6	$\sim(10–100) \times 10^{-9}$
4H-SiC	3.2	10.0	4.9	3.5×10^6	$\sim(10–100) \times 10^{-9}$
3C-SiC	2.2	9.7	3.3	$(1–5) \times 10^6$	$\sim(10–100) \times 10^{-9}$
GaN	3.4	9.5	1.3	2×10^6	$\sim(1–100) \times 10^{-9}$

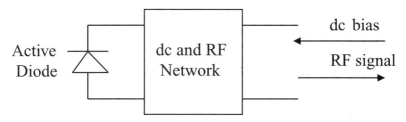

FIGURE 7.15 One-port network.

bipolar junction transistors (BJTs). A low value for minority carrier lifetime is desirable for devices such as diode temperature sensors and switches where low reverse bias leakage current is desirable. A long minority carrier lifetime is desirable for devices such as bipolar transistors. For materials such as Si and SiC the minority carrier lifetime can be varied by controlled impurity doping. A comparison of some of the important material parameters for several common semiconductors is presented in Table 7.2. The large variation for minority lifetime shown in Table 7.2 for SiC and GaN is due to relatively immature materials growth technology for these wide bandgap semiconductors.

Two-Terminal Active Microwave Devices

The IMPATT diode, transferred electron device (often called a Gunn diode), and tunnel diode are the most commonly used two-terminal active devices. These devices can operate from the low microwave through high mm-wave frequencies, extending to several hundred GHz. They were the first semiconductor devices that could provide useful RF power levels at microwave and mm-wave frequencies and were extensively used in early systems as solid-state replacements for vacuum tubes. The three devices are similar in that they are fabricated from diode or diode-like semiconductor structures. DC bias is applied through two metal contacts that form the anode and cathode electrodes. The same electrodes are used for both the DC and RF ports and since only two electrodes are available, the devices must be operated as a one-port RF network, as shown in Fig. 7.15. This causes little difficulty for oscillator circuits, but is problematic for amplifiers since a means of separating the input RF signal from the output RF signal must be devised. The use of a nonreciprocal device, such as a circulator can be used to accomplish the task. Circulators, however, are large, bulky, and their performance is sensitive to thermal variations. In general, circulators are difficult to use for integrated circuit applications. The one-port character of diodes has limited their use in modern microwave systems, particularly for amplifiers, since transistors, which have three terminals and are two-port networks, can be designed to operate with comparable RF performance, and are much easier to integrate. Diodes, however, are often used in oscillator circuits since these components are by nature one-port networks.

IMPATT and Gunn diodes require a combination of charge injection and transit time effects to generate active characteristics and they operate as negative immittance components (the term "immittance" is a

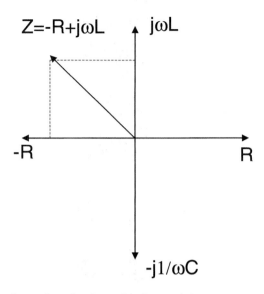

FIGURE 7.16 Complex impedance plane showing active characteristic.

general reference that includes both "impedance" and "admittance"). When properly designed and biased, the active characteristics of the diodes can be described as either a negative resistance or a negative conductance. Which description to use is determined by the physical operating principles of the particular device, and the two descriptions are, in general, not interchangeable. Bias and RF circuits for the two active characteristics must satisfy different stability and impedance matching criteria.

Transit time effects alone cannot generate active characteristics. This is illustrated in Fig. 7.16, which shows a general impedance plane. All passive circuits, no matter how complex or how many circuit elements are included, when arranged into a one-port network as shown in Fig. 7.15, and viewed from an external vantage point, will have an input impedance that lies in the right-hand plane of Fig. 7.16. The network resistance will be positive and real, and the reactance will be inductive or capacitive. This type of network is not capable of active performance and cannot add energy to a signal. Transit time effects can only produce terminal impedances with inductive or capacitive reactive effects, depending upon the magnitude of the delay relative to the RF period of the signal. In order to generate active characteristics it is necessary to develop an additional delay that will result in a phase delay between the terminal RF voltage and current that is greater than 90° and less than 270°. The additional delay can be generated by feedback that can be developed by physical phenomena internal to the device structure, or created by circuit design external to the device. The IMPATT and Gunn diodes make use of internal feedback resulting from electronic charge transfer within the semiconductor structure. The internal feedback generally produces a phase delay of ~90°, which when added to the transit time delay will produce a negative real component to the terminal immittance.

Tunnel Diodes

Tunnel diodes [1] generate active characteristics by an internal feedback mechanism involving the physical tunneling of electrons between energy bands in highly doped semiconductors, as illustrated in the energy band diagram shown in Fig. 7.17. The illustration shows a p$^+$n junction diode with heavily doped conduction and valence bands located in close proximity. When a bias is applied, charge carriers can tunnel through the electrostatic barrier separating the p-type and n-type regions, rather than be thermionically emitted over the barrier, as generally occurs in most diodes. When the diode is biased (either forward or reverse bias) current immediately flows and ohmic conduction characteristics are obtained. In the forward bias direction conduction occurs until the applied bias forces the conduction and valence

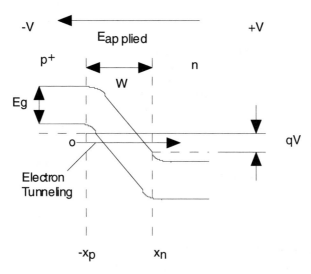

FIGURE 7.17 Energy-band diagram for a p⁺n semiconductor junction showing electron tunneling behavior.

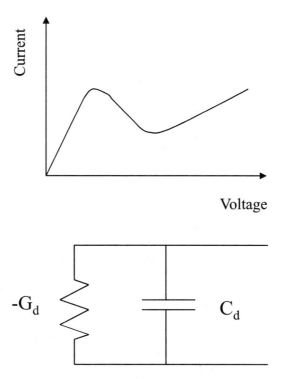

FIGURE 7.18 (a) Current-voltage characteristic for an N-type active device. (b) Small-signal RF equivalent circuit for an N-type active device.

bands to separate. The tunnel current then decreases and normal, thermionic junction conduction occurs. In the forward bias region where the tunnel current is decreasing with increasing bias voltage an N-type negative immittance characteristic is generated, as shown in Fig. 7.18a. The immittance is called "N-type" because the I-V characteristic looks like the letter N. This type of active element is current driven and is short-circuit stable. It is described by a negative conductance in shunt with a capacitance, as shown in

Fig. 7.18b. Tunnel diodes are limited in operation frequency by the time it takes for charge carriers to tunnel through the junction. Since this time is very short (on the order of 10^{-12} s) operation frequency can be very high, approaching 1000 GHz.

Tunnel diodes have been operated at 100s of GHz, and are primarily limited in frequency response by practical packaging and parasitic impedance considerations. The RF power available from a tunnel diode is limited (~100s of mW level) since the maximum RF voltage swing that can be developed across the junction is limited by the forward turn-on characteristics of the device (typically 0.6 to 0.9 v). Increased RF power can only be obtained by increasing device area to increase RF current. However, increases in diode area will limit operation frequency due to increased diode capacitance. Tunnel diodes have moderate DC-to-RF conversion efficiency (<10%) and very low noise figures and have been used in low noise systems applications, such as microwave and mm-wave receivers used for radioastronomy.

Transferred Electron Devices

Transferred electron devices (i.e., Gunn diodes) [2] also have N-type active characteristics and can be modeled as a negative conductance in parallel with a capacitance, as shown in Fig. 7.18b. Device operation, however, is based upon a fundamentally different principle. The negative conductance derives from the complex conduction band structure of certain compound semiconductors, such as GaAs and InP. In these direct bandgap materials the lower valley central (or Γ) conduction band is in close energy-momentum proximity to secondary, higher order conduction bands (i.e., the X and L) valleys (illustrated schematically as the L upper valley in Fig. 7.19). The electron effective mass is determined by the shape of the conduction bands and the effective mass is "light" in the Γ valley, but "heavy" in the higher order X and L upper valleys. When the crystal is biased, current flow is initially due to electrons in the light effective mass Γ valley and conduction is ohmic. However, as the bias field is increased, an increasing proportion of the free electrons are transferred into the X and L valleys where the electrons have heavier effective mass. The increased effective mass slows down the electrons, with a corresponding decrease in conduction current through the crystal. The net result is that the crystal displays a region of applied bias voltages where current decreases with increasing voltage. That is, a negative resistance is generated. A charge dipole domain is formed in the device, and this domain will travel through the device generating a transit time effect. The combination of the transferred electron effect and the transit time delay will produce a phase shift between the terminal RF current and voltage that is greater than 90°. The device is unstable and when placed in an RF circuit or resonant cavity, oscillators or amplifiers can be fabricated.

The Gunn device is not actually a diode since no pn or Schottky junction is used. The transferred electron phenomenon is a characteristic of the bulk material and the special structure of the conduction bands in certain compound semiconductors. In order to generate a transferred electron effect, a semiconductor must have Γ, X, and L valleys in the conduction bands in close proximity so that charge can be transferred from the lower Γ valley to the upper valleys at reasonable magnitude of applied electric field. It is desirable that the charge transfer occur at low values of applied bias voltage in order for the device to operate with good DC-to-RF conversion efficiency. Most semiconductors do not have the conduction band structure necessary for the transferred electron effect, and in practical use, Gunn diodes have only been fabricated from GaAs and InP. It should be noted that the name "Gunn diode" is actually a misnomer since the device is not actually a diode, but rather a piece of bulk semiconductor.

TEDs are widely used in oscillators from the microwave through high mm-wave frequency bands. They can be fabricated at low cost and provide an excellent price-to-performance ratio. They are, for example, the most common oscillator device used in police automotive radars. They have good RF output power capability (mW to W level), moderate efficiency (< 20%), and excellent noise and bandwidth capability. Octave or multi-octave band tunable oscillators are easily fabricated using devices such as YIG (yttrium iron garnet) resonators. High tuning speed can be achieved by using varactors as the tuning element. Many commercially available solid-state sources for 60 to 100 GHz operation (for example, automotive collision-avoidance radars) often use InP TEDs.

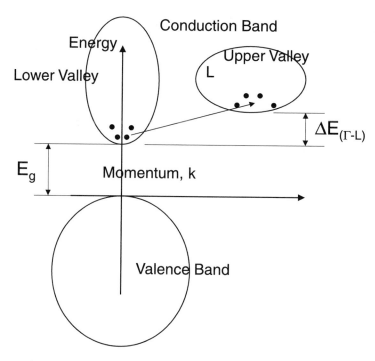

FIGURE 7.19   Energy band diagram for a semiconductor that demonstrates the transferred electron effect.

IMPATT Diodes

IMPATT (IMPact Avalanche Transit Time) diodes [3] are fabricated from pn or Schottky junctions. The doping profile in the device is generally tailored for optimum performance and a typical p^+nn^+ junction device structure as shown in Fig. 7.20. The diode is designed so that when it is reverse biased, the n-region is depleted of free electrons. The electric field at the p^+n junction exceeds the critical magnitude for avalanche breakdown, and the electric field exceeds the magnitude required to maintain electron velocity saturation throughout the n-region. Saturated charge carrier velocity must be maintained throughout the RF cycle in order for the device to operate with maximum efficiency. In operation, the high electric field region at the p^+n interface will generate charge when the sum of the RF and DC bias voltage produces an electric field that exceeds the critical value. A pulse of free charge (electrons and holes) will be generated. The holes will be swept into the p^+ region and the electrons will be injected into the depleted n-region, where they will drift through the diode, inducing a current in the external circuit as shown in Fig. 7.21. Due to the avalanche process, the RF current across the avalanche region lags the RF voltage by 90°. This inductive delay is not sufficient, by itself, to produce active characteristics. However, when the 90° phase shift is added to that arising from an additional inductive delay caused by the transit time of the carriers drifting through the remainder of the diode external to the avalanche region, a phase shift between the terminal RF voltage and current greater than 90° is obtained. A Fourier analysis of the resulting waveforms reveals a device impedance with a negative real part. That is, the device is active and can be used to generate or amplify RF signals. The device impedance has an "S-type" active i-v characteristic, as shown in Fig. 7.22a, and the device equivalent circuit consists of a negative resistance in series with an inductor, as shown in Fig. 7.22b. An S-Type active device is voltage driven and is open-circuit stable. For IMPATT diodes, the active characteristics only exist under RF conditions. That is, there is a lower frequency below which the diode does not generate a negative resistance. Also, the negative resistance is generally small in magnitude, and on the order of $-1\ \Omega$ to $-10\ \Omega$. Therefore, it is necessary

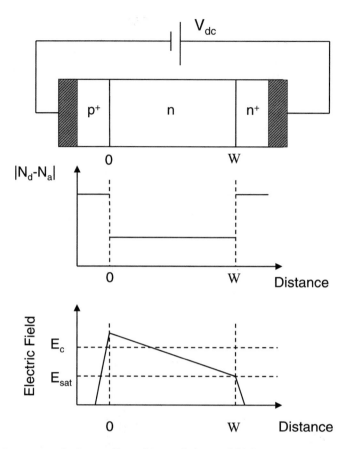

FIGURE 7.20   Diode structure, doping profile, and internal electric field for a p^+nn^+ IMPATT diode.

to reduce all parasitic resistances in external circuits to the maximum extent possible since parasitic series resistance will degrade the device's negative resistance, thereby limiting device performance. An IMPATT diode has significant pn junction capacitance that must be considered and a complete equivalent circuit includes the device capacitance in parallel with the series negative resistance-inductance elements, as shown in Fig. 7.22b.

For optimum performance the drift region is designed so that the electric field throughout the RF cycle is sufficiently high to produce velocity saturation for the charge carriers. In order to achieve this it is common to design complex structures consisting of alternating layers of highly doped and lightly doped semiconductor regions. These structures are called "high-low," "low-high-low," or "Read" diodes, after the man who first proposed their use [1]. They can also be fabricated in a back-to-back arrangement to form double-drift structures [4]. These devices are particularly attractive for mm-wave applications since the back-to-back arrangement permits the device to generate RF power from series-connected diodes, but each diode acts independently with regard to frequency response.

IMPATT diodes can be fabricated from most semiconductors, but are generally fabricated from Si or GaAs. The devices are capable of good RF output power (mW to tens of W) and good DC-to-RF conversion efficiency (~10 to 20%). They operate well into the mm-wave region and have been operated at frequencies in excess of 400 GHz. They have moderate bandwidth capability, but have relatively poor noise performance due to the impact ionization process associated with avalanche breakdown. The power-frequency performance of IMPATT diodes fabricated from Si, GaAs, and SiC is shown in Fig. 7.23. The Si and GaAs data are experimental, and the SiC data are predicted from simulation. The numbers associated with the data points are the conversion efficiencies for each point.

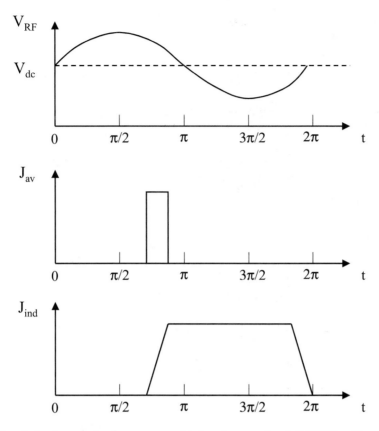

FIGURE 7.21 Terminal voltage, avalanche current, and induced current for an IMPATT oscillator.

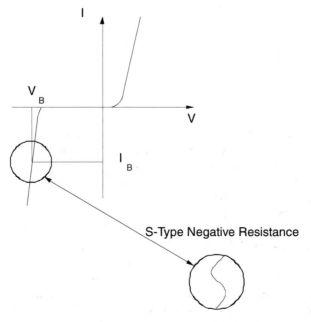

FIGURE 7.22 (a) Current-voltage characteristic for an S-type active device. The S-characteristic behavior does not exist under DC conditions for an IMPATT diode. The inset shows the dynamic i-v behavior that would exist about the DC operation point under RF conditions. (b) Small-signal RF equivalent circuit for an S-type active device.

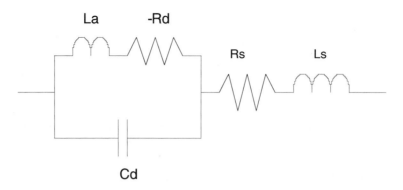

FIGURE 7.22   (continued)

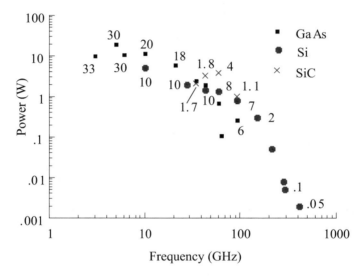

FIGURE 7.23   RF power-frequency performance for IMPATT oscillators (numbers associated with data points are DC-to-RF conversion efficiency).

Defining Terms

Immittance: A general term that refers to both impedance and admittance.

Active device: A device that can convert energy from a DC bias source to a signal at an RF frequency. Active devices are required in oscillators and amplifiers.

Two-terminal device: An electronic device, such as a diode, that has two contacts. The contacts are usually termed the cathode and anode.

One-port network: An electrical network that has only one RF port. This port must be used as both the input and output to the network. Two-terminal devices result in one-port networks.

Charge carriers: Units of electrical charge that produce current flow when moving. In a semiconductor two types of charge carriers exist: electrons and holes. Electrons carry unit negative charge and have an effective mass that is determined by the shape of the conduction band in energy-momentum space. The effective mass of an electron in a semiconductor is generally significantly less than an electron in free space. Holes have unit positive charge. Holes have an effective mass that is determined by the shape of the valence band in energy-momentum space. The effective mass of a hole is generally significantly larger than that for an electron. For this reason electrons generally move much faster than holes when an electric field is applied to the semiconductor.

References

1. Sze, S.M., 1981. *Physics of Semiconductor Devices*, 2nd Edition, Wiley-Interscience, New York.
2. Bosch, B.G., Engelmann, R.W. 1975. *Gunn-Effect Electronics*, Halsted Press, New York.
3. S.M. Sze, 1998. *Modern Semiconductor Device Physics*, Wiley-Interscience, New York.
4. P. Bhartia and I.J. Bahl, 1984. *Millimeter Wave Engineering and Applications*, Wiley-Interscience, New York.

7.2 Transistors

7.2.1 Bipolar Junction Transistors (BJTs)

John C. Cowles

The topic of bipolar junction transistors (BJTs) is obviously quite broad and a full treatment would consume volumes. This work focuses on basic principles to develop an intuitive feel for the transistor behavior and its application in contemporary high-speed integrated circuits (IC). The exponential growth in high bandwidth wired, wireless, and fiber communication systems coupled with advanced IC technologies has created an interesting convergence of two disparate worlds: the microwave and analog domains. The traditional microwave IC consists of a few transistors in discrete form or in low levels of integration surrounded by a sea of transmission lines and passive components. The modern high-speed analog IC usually involves 10s to 1000s of transistors with few passive components. The analog designer finds it a more cost-effective solution to use extra transistors rather than passive components to resolve performance issues. The microwave designer speaks in terms of noise figure, IP3, power gain, stability factor, VSWR, and s-parameters while the analog designer prefers noise voltages, harmonic distortion, voltage gain, phase margin, and impedance levels. Present BJT IC technologies fall exactly in this divide and thus force the two worlds together. Since parasitics within an IC are significantly lower than those associated with packages and external interconnects, analog techniques can be applied well into the microwave region and traditional microwave techniques such as impedance matching are only necessary when interfacing with the external world where reference impedances are the rule. This symbiosis has evolved into what is now termed radio frequency IC (RFIC) design. With this in mind, both analog and microwave aspects of bipolar transistors will be addressed. Throughout the discussion, the major differences between BJTs and field-effect transistors (FETs) will be mentioned.

A Brief History

The origin of the BJT was more a fortuitous discovery rather than an invention. The team of Shockley, Brattain, and Bardeen at Bell Labs had been pursuing a field effect device in the 1940s to replace vacuum tubes in telephony applications when they stumbled upon bipolar transistor action in their experimental point-contact structure. The point contact transistor consisted of a slab of n-type germanium sitting on a metal "base" with two metal contacts on either side. With proper applied bias, the "emitter" contact would inject current into the bulk and then the "collector" contact would sweep it up. They announced their discovery in 1948 for which they later received the Nobel Prize in 1956. At that point, the device was described as a current-controlled voltage source. This misinterpretation of the device as a trans-resistor led to coining of the universal term transistor that today applies to both BJTs and their field-effect brethren.

Within a few years of the discovery of the transistor, its theory of operation was developed and refined. The decade of the 1950s was a race toward the development of a practical technique to fabricate them. Research into various aspects of the material sciences led to the development of the planar process with silicon as the cornerstone element. Substrate and epitaxial growth, dopant diffusion, ion-implantation, metalization, lithography, and oxidation had to be perfected and integrated into a manufacturable process flow. These efforts culminated with Robert Noyce's patent application for the planar silicon BJT IC concept that he invented while at Fairchild. In the 1960s, the first commercial ICs appeared on the market, launching the modern age of electronics. Since then, transistor performance and integration levels have skyrocketed driven primarily by advances in materials, metrology, and process technology as well as competition, cost, and opportunity.

Basic Operation

The basic structure of the bipolar transistor given in Fig. 7.24a consists of two back-to-back intimately coupled p-n junctions. While the npn transistor will be chosen as the example, note that the pnp structure operates identically, albeit in complementary fashion. Bipolar transistor action is based on the injection of minority carrier electrons from the emitter into the base as dictated by the base-emitter voltage. These carriers diffuse across the thin base region and are swept by the collector, which is normally reverse biased. In high speed ICs, BJTs are nearly always biased in this way, which is known as the forward active mode. The net effect is that an input signal voltage presented across the base-emitter terminals causes a current to flow into the collector terminal. The relationship between collector current, I_c, and base-emitter voltage, V_{be}, originates from fundamental thermodynamic arguments via the statistical Boltzmann distribution of electrons and holes in solids. It is given by,

$$I_c = I_s \exp\left(V_{be}/V_t\right) \qquad (7.19)$$

where I_s is called the saturation current and V_t is Boltzmann's thermal voltage kT/q, which is 26 mV at room temperature (RT). This expression plotted in Fig. 7.24b is valid over many decades of current and

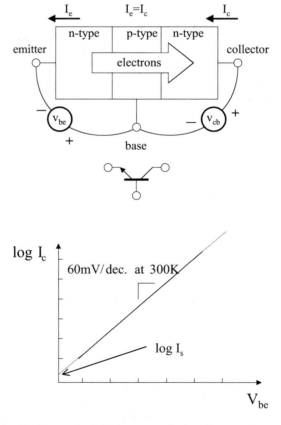

FIGURE 7.24 (a) BJT action involves carrier injection across the base from emitter to collector; (b) the ideal BJT exhibits a perfectly exponential input-output transfer function.

its implications are far reaching in BJT IC design. Consider the small-signal transfer characteristics relating the output current to input voltage at a given bias current, I_c, i.e., the transconductance, g_m. It is given by the remarkably simple expression,

$$\delta I_c / \delta V_{be} \big|_{I_c} = g_m = I_c / V_t. \tag{7.20}$$

The notion that transconductance is precisely and linearly proportional to the bias current, I_c and inversely proportional to temperature through V_t is known as translinearity. Notice that I_s, which has a strong dependence on temperature, base-width, and other physical device parameters, has dropped out of the picture. This view of bipolar transistors as a voltage-controlled current source suggests that it should have been named the transductor rather than the transistor. Although bipolar transistors are usually considered hopelessly nonlinear because of the exponential law in Eq. (7.19), extremely precise and robust linear and nonlinear functions can be realized at very high frequencies by proper circuit techniques that exploit the translinear property embodied in Eq. (7.20).

The intrinsic bandwidth of high frequency BJTs is ultimately limited by minority carrier storage primarily in the base region. Current is conveyed from the emitter to the collector by the relatively slow process of diffusion resulting from the gradient in the distribution of stored minority carriers in the base. In contrast, current in FETs is transported by the faster process of majority carrier drift in response to an electric field. By invoking the quasi-static approximation, which assumes that signals of interest change on a much longer time scale than the device time constants, carrier storage can be modeled as a lumped diffusion capacitance, C_d given by,

$$C_d = \delta Q_d / \delta V_{be} = \delta \left(I_c \tau_f \right) / \delta V_{be} = g_m \tau_f \tag{7.21}$$

where Q_d is the stored charge and τ_f is the forward transit time. The parameter τ_f can be viewed as the average time an electron spends diffusing across the base and can be expressed as a function of basic material and device parameters,

$$\tau_f = W_B^2 / 2\eta D_n \tag{7.22}$$

where W_B is the physical base width, D_n is the electron diffusion coefficient, and η is a dimensionless factor that accounts for any aiding fields in the base. Despite the limitations associated with carrier storage and diffusion, BJTs have demonstrated excellent high frequency performance. This is partly due to their vertical nature where dimensions such as W_B can be significantly less than 100 nm. In contrast, FETs are laterally arranged devices where the critical dimension through which carriers drift is determined by lithography and is typically larger than 100 nm even in the most advanced technologies. Furthermore, the fundamental limit contained in τ_f is never achieved in practice since other factors, both device and circuit-related, conspire to reduce the actual bandwidth.

A first order model of BJTs is illustrated in Fig. 7.25 consisting of a diode at the base-emitter junction whose current is conveyed entirely to the collector according to Eq. (7.19) and a diffusion capacitance that models the carrier storage. From this simplistic model, the transistor f_t, defined as the frequency at which the common-emitter short-circuit current gain, h_{21}, reaches unity, can be calculated from,

$$h_{21} = i_c / i_b = f_t / f. \tag{7.23a}$$

$$f_t = g_m / 2\pi C_d = 1 / 2\pi \tau_f. \tag{7.23b}$$

Note that f_t is independent of current and is solely a function of material parameters and device design details. Pushing this figure of merit to higher frequencies has dominated industrial and academic research

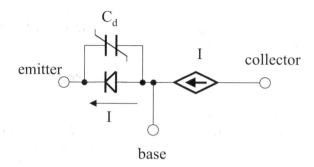

FIGURE 7.25   First order BJT model includes the nonlinear carrier injection and storage.

for years. Polysilicon emitter BJTs, Si/SiGe heterojunction BJTs (HBTs), and III-V-based HBTs are all attempts to craft the active layers in order to maximize f_t without significantly compromising other device aspects.

Parasitics and Refinements

The first order model captures the essential operation of the BJT. However, actual devices deviate from that ideal in many aspects. As electrons diffuse across the base, some are lost to the process of recombination with holes as illustrated in Fig. 7.26a. Excess holes are also parasitically injected into the emitter. This overall loss of holes must originate at the base terminal and represents a finite base current, I_b, which is related to I_c via the parameter, β, known as the common emitter current gain,

$$\beta = I_c / I_b. \tag{7.24}$$

The effect of finite β is to modify the emitter-base diode current by a correction factor, α, known as the common-base current gain, that accounts for the extra base current. Since $I_e = I_b + I_c$, it is easy to show that,

$$\alpha = \beta / \beta + 1 < 1 \tag{7.25}$$

so that the emitter-base diode current now becomes I_s/α. Many factors figure into determining β, including W_B and the highly variable recombination lifetime τ_r. Since β is a poorly controlled parameter with potential variations of ±50%, it is inadvisable to design circuits that depend on its exact value. Fortunately, it is usually possible to design ICs that are insensitive to β as long as β is large, say >50, thanks to matching of co-integrated devices. Figure 7.26b illustrates the well-known BJT Gummel plot, which shows the translinearity of I_c as well as the finite I_b in response to V_{be}. The separation between the two curves represents β. As shown in Fig. 7.27, β represents a low frequency asymptote of h_{21} that exhibits a single pole roll off toward f_t at a frequency, f_β given by f_t/β. Since RFICs most often operate well above f_β, the useful measure of current gain is actually f_t/f rather than β, although high DC β is still important for low noise and ease of biasing.

Traditionally, BJTs have been characterized as current-controlled devices where a forced I_b drives an I_c into an load impedance, consistent with Shockley's trans-resistor description. Now if I_b is considered a parasitic nuisance rather than a fundamental aspect, it becomes even more appropriate to view the BJT as a voltage-controlled device that behaves as a transconductor, albeit with exponential characteristics. In fact, contemporary analog IC design avoids operating the BJT as a current-controlled device due to the unpredictability of β. Instead, the designer takes advantage of device matching in an IC environment and translinearity to provide the appropriate voltage drive. This approach can be shown to be robust against device, process, temperature, and supply voltage variations.

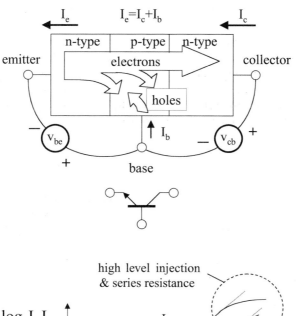

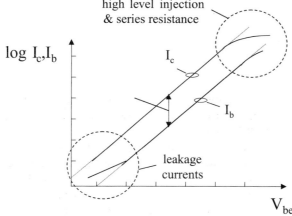

FIGURE 7.26 (a) Recombination of electrons and holes leads to finite current gain, β; (b) the Gummel plot illustrates the effects of finite β and other nonidealities.

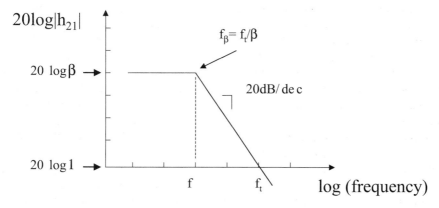

FIGURE 7.27 The frequency dependence of h_{21} captures finite β and f_t.

Superimposed on the basic model are parasitic ohmic resistances in series with each active terminal (R_b, R_e, R_c) and parasitic capacitances associated with all pn junction depletion regions (C_{jc}, C_{je}, C_{js}), including the collector-substrate junction present to some extent in all technologies. Since parasitics degrade the idealized DC and RF characteristics captured in Eqs. (7.19) and (7.23), a major effort is focused on minimizing them through aggressive scaling and process engineering. Their values and bias dependencies can be estimated from the physical device structure and layout.

In particular, the base and emitter resistance, R_b and R_e, soften the elegant exponential characteristics in Eq. (7.19) by essentially de-biasing the junction since

$$I_c = I_s \exp\left(\left[V_{be} - I_b R_b - I_e R_e\right]\middle/ V_t\right). \tag{7.26}$$

This effect is also illustrated in Fig. 7.26b at high values of V_{be} where both curves appear to saturate. This departure from ideal translinearity can introduce unwelcome distortion into many otherwise linear ICs. Furthermore, these resistances add unwelcome noise and degeneration (voltage drops) as will be discussed later.

The idealized formulation for f_t given in Eq. (7.23) also needs to be modified to account for parasitics. A more comprehensive expression for f_t based on a more complex equivalent circuit results in,

$$f_t = \left(2\pi^* \left[\tau_f + \left(C_{je} + C_{jc}\right)\middle/ g_m + C_{jc}\left(R_c + R_e\right)\right]\right)^{-1} \tag{7.27}$$

where now f_t is dependent on current through g_m charging of C_{je} and C_{jc}. To achieve peak f_t, high currents are required to overcome the capacitances. As the intrinsic device τ_f has been reduced, the parasitics have become a dominant part of the BJTs' high frequency performance requiring even higher currents to reach the lofty peak values of f_t. It should also be kept in mind that f_t only captures a snapshot of the device high frequency performance. In circuits, transistors are rarely current driven and short circuited at the output. The base resistance and substrate capacitance that do not appear in Eq. (7.27) can have significant impact on high frequency IC performance. While various other figures of merit such as f_{max} have been proposed, none can capture the complex effects of device interactions with source and load impedances in a compact form. The moral of the story is that ideal BJTs should have low inertia all around, i.e., not only high peak f_t but also low parasitic capacitances and resistances, so that time constants in general can be minimized at the lowest possible currents.

At the high currents required for high f_t, second order phenomena known in general as high level injection begin to corrupt the DC and RF characteristics. Essentially, the electron concentration responsible for carrying the current becomes comparable to the background doping levels in the device causing deviations from the basic theory that assumes low level injection. The dominant phenomenon known as the Kirk effect or base-pushout manifests itself as a sudden widening of the base width at the expense of the collector. This translates into a departure from translinearity with dramatic drops in both β and f_t. A number of other effects have been identified over the years such as Webster effect, base conductivity modulation, and current crowding. High level injection sets a practical maximum current at which peak f_t can be realized. Figure 7.28 illustrates the typical behavior of f_t with I_c. To counteract high level injection, doping levels throughout the device have increased at a cost of higher depletion capacitances and lower breakdown voltages. Since modeling of these effects is very complex and not necessarily included in many models, it is dangerous to design in this regime.

So far, the transistor output has been considered a perfect current source with infinite output resistance. In reality, as the output voltage swings, the base width is modulated, causing I_s and thus I_c to vary for a fixed V_{be}. This is exactly the effect of an output resistance and is modeled by a parameter V_a, the Early voltage,

$$r_o = \delta V_{ce}/\delta I_c = V_a/I_c. \tag{7.28}$$

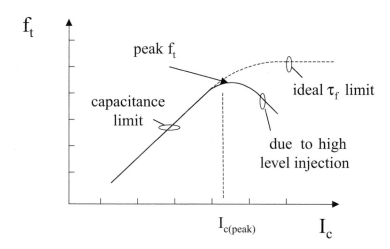

FIGURE 7.28 The f_t peaks at a particular current when high level injection occurs.

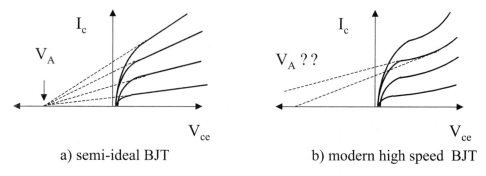

a) semi-ideal BJT   b) modern high speed BJT

FIGURE 7.29 (a) The traditional interpretation of V_A is not unique in modern BJTs as shown in (b).

As illustrated in the output common-emitter characteristics, I_c vs. V_{ce}, shown in Fig. 7.29a, V_a represents the common extrapolation point where all tangents intersect the V_{ce} axis. The effect of r_o is to set the maximum small-signal unloaded voltage gain since

$$A_v = \delta V_o / \delta V_i = \left(\delta I_c / \delta V_i\right)\left(\delta V_o / \delta I_c\right) = -g_m r_o = -V_a / V_t \qquad (7.29)$$

For typical values of $V_a = 50$ V, $A_v = 2000$ (66 dB) at room temperature. Note that this gain is independent of I_c and is quite high for a single device, representing one of the main advantages of BJTs over FETs where the maximum gain is usually limited to < 40 dB for reasonable gate lengths. In modern devices, however, V_a is not constant due to complex interactions between flowing electrons and internal electric fields. The net effect illustrated in Fig. 7.29b indicates a varying V_a depending on I_c and V_{ce}. This is often termed soft breakdown or weak avalanche in contrast to actual breakdown which will be discussed shortly. The effect of a varying V_a is to introduce another form of distortion since the gain will vary according to bias point. Figure 7.30 shows a more complete model that includes the fundamental parameters as well as the parasitics and refinements considered so far. It resembles a simplified version of the popular Gummel-Poon model found in many simulators.

Another form of pseudo-breakdown occurs in ultra-narrow base BJTs operating at high voltages. If the Early effect or base-width modulation is taken to an extreme, the base eventually becomes completely depleted. After this point, a further change in collector voltage directly modulates the emitter-base junction leading to an exponential increase in current flow. This phenomenon known as punchthrough

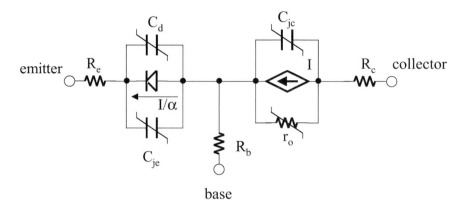

FIGURE 7.30 Augmented BJT model with parasitic resistances and finite base current.

fortunately has been mitigated by the fact that as base widths have narrowed, the base doping has been forced to increase so as to maintain a reasonable base resistance. Furthermore, since higher collector doping levels have been necessary to fight high level injection, true breakdown has become the voltage limiting mechanism rather than punchthrough.

Just as high level injection limits the maximum operating current, junction breakdown restricts the maximum operating voltage. When the collector voltage is raised, the collector base junction is reverse biased. The resulting electric field reaches a critical point where valence electrons are literally ripped out of their energy band and promoted to the conduction band while leaving holes in the valence band. The observed effect known as avalanche breakdown is a dramatic increase in current. The breakdown of the collector-base junction in isolation, i.e., with the emitter open circuited, is termed the BV_{cbo} where the "o" refers to the emitter open. Another limiting case of interest is when the base is open while a collector-emitter voltage is applied. In this case, an initial avalanche current acts as base current that induces more current flow, which further drives the avalanche process. The resulting positive feedback process causes the BV_{ceo} to be significantly lower than BV_{cbo}. The relationship clearly depends on β and is empirically modeled as,

$$BV_{ceo} = BV_{cbo}\big/\beta^{1/n} \tag{7.30}$$

where n is a fit parameter. A third limit occurs when the base is AC shorted to ground via a low impedance. In this scenario, avalanche currents are shunted to ground before becoming base current and BV_{ces} ("s" means shorted) would be expected to be $BV_{cbo} + V_{be}$. Therefore, BV_{ces} represents an absolute maximum value for V_{ce} while BV_{ceo} represents a pessimistic limit since the base is rarely open. Operating in the intermediate region requires care in setting the base impedance to ground and knowing its effect on breakdown. Figure 7.31 illustrates the transition from BV_{ceo} to BV_{ces}. The base-emitter junction is also

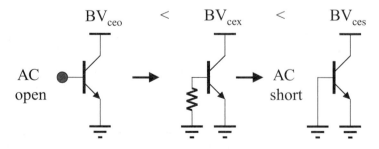

FIGURE 7.31 The AC base impedance helps determine the effective breakdown voltage.

sensitive to reverse bias. In this case, the breakdown is usually related to Zener tunneling and is represented by BV_{ebo}. Excessive excursions toward BV_{ebo} can cause injection of energetic carriers into surrounding dielectrics, leading to leakage currents, degraded reliability, and possibly device failure. Since these breakdown mechanisms in general are poorly modeled, it is recommended to operate well below BV_{ceo} and BV_{ebo}.

Even when these guidelines are followed, there are still situations where peak voltages might exceed breakdown limits. For instance, modern ICs in portable products often require a disable mode in which the part consumes virtually no current so as to conserve battery life. When disabled, it is common for transistors that are normally in a safe operating condition to find themselves with reverse-biased junctions in excess of BV_{ceo} and BV_{ebo}. The resulting leakage currents can lead to expectedly high disable currents, accidental activation of certain circuit blocks, and again to potential reliability failures. A second situation is when output stages drive variable reactive loads. It is possible (and likely) that certain conditions will cause 2 to 3 times the supply voltage to appear directly across a transistor leading to certain catastrophic breakdown.

As devices have been scaled down in size, isolated in dielectrics, and driven with higher current densities, self-heating has become increasingly problematic. The thermal resistance, R_{th}, and capacitance, C_{th}, model the change in junction temperature with power and the time constant of its response. The DC and AC variations in the junction temperature induce global variations in nearly all model parameters causing bias shifts, distortion, and even potentially catastrophic thermal runaway. Furthermore, local heating in one part of a circuit, say an ouput power stage, can start affecting neighboring areas such as bias circuits creating havoc. It is essential to selectively model thermal hotspots and ensure that they are within maximum temperature limits and that their effect on other circuit blocks is minimized by proper layout.

Dynamic Range

The limits on signal integrity are bounded on the low end by noise and on the high end by distortion. It is not appropriate to claim a certain dynamic range for BJTs since the circuit topology has a large impact on it. However, the fundamental noise sources and nonlinearities in BJTs can be quantified and then manipulated by circuit design. It is customary to refer the effects of noise and nonlinearity to either the input or the output of the device or circuit to allow fair comparisons and to facilitate cascaded system-level analyses. Here, all effects will be referred to the input, noting that the output quantity is simply scaled up by the appropriate gain.

There are three fundamental noise mechanisms that are manifested in BJTs. Figure 7.32 presents a linearized small-signal model that includes the noise sources. The objective will be to refer all noise sources to equivalent voltage and current noise sources, e_n and i_n, respectively, at the input. Shot noise is associated with fluctuations in current caused by the discrete nature of electronic charge overcoming the junction potential. The presence of the forward biased base-emitter junction results in collector and base shot noise, i_{cn} and i_{bn}, which are given by,

$$i_{cn}^2 = 2qI_c; \; i_{bn}^2 = 2qI_b \tag{7.31}$$

where q is the electronic charge. This is a major disadvantage of BJTs with respect to FETs, which are free of shot noise due to the absence of potential barriers that determine current flow. Each physical resistance carries an associated Johnson noise represented as

$$e_{Rx}^2 = 4kTR_x \tag{7.32}$$

where R_x is the appropriate resistance and k is Boltzmann's constant. Finally, a 1/f noise source, $i_{1/f}^2$ associated primarily with recombination at semiconductor surfaces appears at the input. The 1/f noise component decreases with increasing frequency, as expected. A corner frequency is often quoted as the

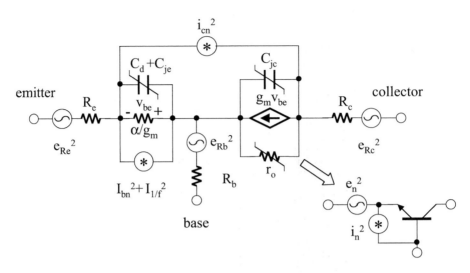

FIGURE 7.32   The fundamental noise sources can be transformed to equivalent input-referred sources.

point where the 1/f noise and the white thermal and shot noise cross over. It is an empirical value that varies from device to device and depends on bias conditions and source impedance. BJTs typically offer lower 1/f corner frequency than FETs since surfaces play a lesser role in BJTs. Except for broadband DC-coupled circuits and strongly nonlinear circuits such as oscillators and mixers that perform frequency conversion, 1/f noise is insignificant at high frequencies and will be ignored for simplicity. Note that treating noise in nonlinear systems is quite complex and remains an active research topic.

Once the noise sources are input referred, e_n and i_n are approximately given by

$$e_n^2 = 4kT\left(R_b + R_e + 1/2g_m\right) \tag{7.33a}$$

$$i_n^2 = 2q\left(I_b + I_c\Big/\left|h_{21}(f)\right|^2\right). \tag{7.33b}$$

An arbitrarily higher I_c reduces e_n through the larger g_m but has mixed effects on i_n depending on how $h_{21}(f)$ varies with current. An arbitrarily larger device reduces e_n by lowering R_b and R_e but the commensurate increase in capacitance might degrade $h_{21}(f)$ and thus i_n. The moral here is that minimization of noise requires a trade-off in I_c and device size given an f_t equation and parasitic scaling rules with device geometry.

Distortion originates in nonlinearities within the device. For simplicity, distortion will be described in terms of a power series to derive expressions for harmonic distortion, intermodulation distortion, and compression. It is assumed that linearities up to only third order are considered, limiting the accuracy of the results as compression is approached. While this approach is valid at low frequencies, capacitive effects at higher frequencies greatly complicate nonlinear analysis and more generalized Volterra series must be invoked. However, the general idea is still valid. Note that not all systems can be mathematically modeled by power or Volterra series; for example, a hard limiting amplifier that shows sudden changes in characteristics.

In BJTs, the fundamental transfer function given in Eq. (7.19) represents one of strongest nonlinearities. As shown in Fig. 7.33, by expanding Eq. (7.19) in its power series up to 3rd order, the following measures of input-referred figures of merit are derived in dBV (mV),

$$P_{1db} = 0.933\, V_t = 24.3\ \text{mV} = -32.3\ \text{dBV} \tag{7.34a}$$

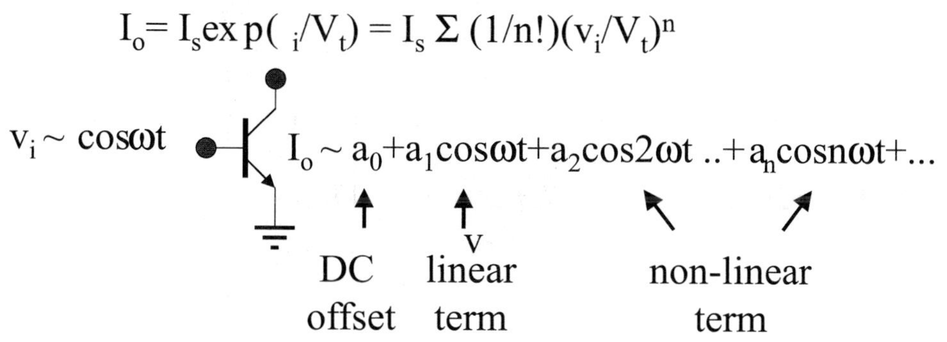

$$I_o = I_s \exp(v_i/V_t) = I_s \Sigma (1/n!)(v_i/V_t)^n$$

$v_i \sim \cos\omega t$

$$I_o \sim a_0 + a_1\cos\omega t + a_2\cos2\omega t .. + a_n\cos n\omega t + ...$$

DC linear non-linear
offset term   term

FIGURE 7.33 A power series expansion of the exponential junction nonlinearity can be used to quantify distortion.

$$IIP3 = 2.83\ V_t = 73.5\ mV = -22.7\ dBV \tag{7.34b}$$

$$P3OI = 4.89\ V_t = 127\ mV = -17.9\ dBV \tag{7.34c}$$

where P_{1db} is the 1-dB gain compression point, IIP3 is the third order intercept point, and P3OI is the third harmonic intercept point. Note that the conversion from dBV to dBm referenced to 50 Ω is + 10 dB. From the analysis, it can be inferred that P_{1db} occurs for peak input swings of about V_t while a third order intermodulation level of 40 dBc is achieved for an input swing of only 0.0283 V_t or 0.73 mV at room temperature. These values are quite low, indicating that circuit linearization techniques must be invoked to exceed these limits. FETs on the other hand possess a more benign input nonlinearity that ranges from square law in long channel devices to nearly linear in short channels. While this is not the sole source of nonlinear behavior, FETs are expected to have superior linearity as compared to BJTs from this perspective. It is interesting to observe that the commonly used rule of thumb stating that IIP3 is approximately 10 dB higher than P_{1db} is true in this case. As mentioned earlier, higher order terms, parasitics, and reactive effects modify the results in Eq. (7.34) but they do provide a useful order of magnitude.

A common technique for extending overall dynamic range is emitter degeneration, which as illustrated in Fig. 7.34, essentially amounts to series-series feedback. The impedance in series with the emitter reduces the signal that appears directly across the emitter base leading to lower input-referred distortion. Note that the gain is proportionately reduced so that output-referred distortion is not improved to the same

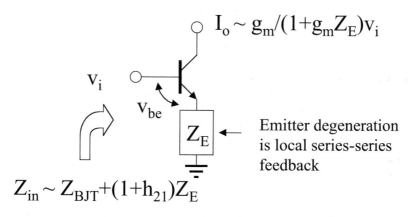

$$I_o \sim g_m/(1+g_m Z_E)v_i$$

V_i

V_{be}

Z_E

Emitter degeneration
is local series-series
feedback

$$Z_{in} \sim Z_{BJT} + (1+h_{21})Z_E$$

FIGURE 7.34 Emitter degeneration can improve dynamic range and change impedance levels.

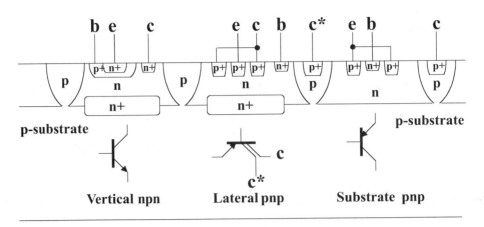

FIGURE 7.35   Both lateral and substrate pnp BJTs can be generated in a typical npn process.

degree. The potentially superior distortion performance of FETs is analogous to a strongly degenerated BJT with the accompanying lower gain. Since this degenerating impedance is directly in the signal path, it also affects noise. If the impedance is a resistor, it will contribute noise as if it were an extra R_e in Eq. (7.33a). If R_e is not a dominant noise contributor, this approach can extend dynamic range significantly. Another option is to use a degenerating inductor, which at high frequencies provides adequate degeneration while only contributing noise associated with its finite quality factor, Q. Other feedback techniques are also possible to improve dynamic range, but care must be taken to ensure stability particularly around longer feedback loops. Additional benefits of using local feedback are desensitization of input impedance and gain to device parameters.

Complementary Pnp

The availability of a pnp device to complement the npn offers numerous advantages, particularly at lower supply voltages. For example, pnp active loads can provide more voltage gain at a given supply voltage; signals can be "folded" up and down to avoid hitting the rails and balanced complementary push-pull topologies can operate near rail-to-rail. In general, the p-type device is slower than the n-type device across most technologies due to the fundamentally lower hole mobility with respect to electron mobility. Furthermore, in practice, the pnp transistor is often implemented as a parasitic device, i.e., the fabrication process is optimized for the npn and the pnp shares the available layers. Figure 7.35 illustrates how two different pnp's can be constructed in an npn process (p-type substrate). The so-called substrate pnp has the substrate double as its collector and thus only appears in a grounded collector arrangement. The lateral pnp frees the collector terminal, but carries along a parasitic substrate pnp of its own that injects large currents into the substrate and degrades f_t and β. In some cases, true complementary technologies are available in which both the npn's and pnp's are synthesized and optimized separately.

When only lateral and substrate pnp's are available, they are usually used outside of the high-speed signal path due to their compromised RF performance. They appear prominently in bias generation and distribution as well as in low and intermediate frequency blocks. When using either pnp, care must be taken to account for current injection into the substrate, which acts locally as a collector. This current can induce noise and signal coupling as well as unintentional voltage drops along the substrate. It is customary to add substrate contacts around the pnp's to bring this current to the surface before it disperses throughout the IC. If true high-quality vertical pnp's are available that are balanced in performance with respect to the npn's, they can be used for high-frequency signal processing, enabling a number of circuit concepts commonly used at lower frequencies to be applied at RF.

Topologies

Several basic single-transistor and composite transistor configurations are commonly encountered in RFICs. Figure 7.36 shows the three basic single-transistor connections known as common-emitter (CE),

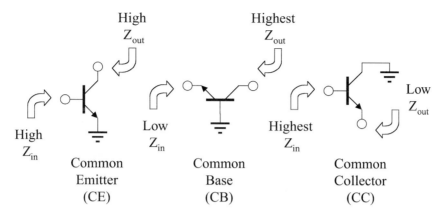

FIGURE 7.36 The three basic single BJT stages offer different features.

common-base (CB), and common-collector (CC). The name refers to the fact that the common terminal is AC grounded. Their properties are well covered in numerous texts and only a brief discussion on their general properties and applications will be presented here.

The CE stage is the most commonly considered configuration since it provides a fairly high input impedance, high output impedance, and both current gain, $h_{21}(f)$, and voltage gain, $g_m R_L$. The noise and distortion properties are the same as considered earlier for the BJT. It is commonly used in single-ended, low noise amplifiers (LNA) and power amplifiers (PA). A well-known drawback of the CE stage at high frequencies is the feedback C_{jc}, which degrades gain via Miller multiplication and couples input and output networks leading to detuning of matching networks and possible instability.

As noted earlier, emitter degeneration can be used not only to improve dynamic range but also to help set impedance levels. It is sometimes necessary for the IC to present a standard impedance level at its input and output terminals to provide proper terminations. The use of emitter inductive degeneration has the fortuitous property of generating a synthetic resistive input impedance at high frequencies without contributing noise. As illustrated in Fig. 7.34 earlier, the transformed value is given by

$$R_{eff} = h_{21}(f)Z_E = (f_t/f)(2\pi f L_E) = 2\pi\, f_t\, L_E \qquad (7.35)$$

which now appears in series with the input terminal.

The CB stage provides a well-predicted low input impedance given by $1/g_m$, high output impedance and voltage gain, $g_m R_L$, but has near unity current gain, α. It acts as an excellent current buffer and is often used to isolate circuit nodes. The CB appears most often in tight synergy with other transistors arranged in composite configurations. Although it lacks the capacitive feedback present in the CE that degrades gain and destabilizes operation, any series feedback at the grounded base node can lead to instability. The feedback element might be due to metallization, package parastics, or even the actual device base resistance. Extreme care must be exercised when grounding the CB stage. Another difference is that with a grounded base, the larger BV_{cbo} sets the voltage limit. The noise and distortion properties are identical to the CE.

The CC stage is often known as an emitter follower. It is predominantly used as a level shifter and as a voltage buffer to isolate stages and provide extra drive. It offers high input impedance, low output impedance, and near unity voltage gain. The impedance buffering factor is roughly $h_{21}(f)$ and at frequencies approaching f_t, its effectiveness is diminished as $h_{21}(f)$ nears unity. In this case it is common to see several stages of cascaded followers to provide adequate drive and buffering. The CC stage is a very wideband stage since direct capacitive feedthrough via C_d and C_{je} cancels the dominant pole to first order. However, this same capacitive coupling from input to output can cause destabilization, particularly with capacitive loads. In fact, this is the basis of operation of the Colpitts oscillator. From a noise point of

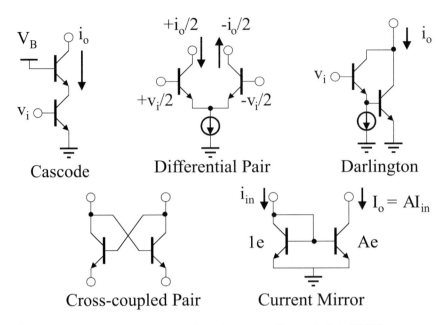

FIGURE 7.37   These common two-transistor configurations are used extensively in BJT ICs.

view, the CC simply transfers noise at its output directly back to its input and then adds on its own noise; therefore it is not used where low noise is essential. From a distortion perspective, the CC can be fairly linear as long as the impedance that it is driving is significantly higher than its own output impedance $1/g_m$. For minimal added shot noise and low distortion, CC stages must run at high bias currents.

The strength of IC technology is the ability to use transistors at will at practically no extra cost. Figure 7.37 illustrates five very common composite transistor configurations, namely the cascode, the differential pair, the Darlington, the cross-coupled pair, and the current mirror. These ubiquitous forms are common enough to be considered in more detail. Discussion of the current mirror appears later.

The cascode topology is a CE-CB connection. The CB provides a low impedance to the output of the CE stage, eliminating Miller multiplication of its C_{jc} and increasing the overall bandwidth. It also raises the output impedance of the single transistor by a factor of approximately β, which is useful in current sources. More importantly at RF, it improves input-output isolation, minimizing interactions that reduce the gain. The cascode configuration is common in LNAs since higher tuned power gain is achievable with minimal degradation in noise. Note that the CB transistor can be significantly smaller than the CE transistor, minimizing the parasitic capacitance at its output. The cascode requires a higher supply voltage than a stand-alone CE stage by at least a $V_{ce(sat)}$ to keep the lower BJT in its active region.

The differential pair (also known as a long-tailed pair) can be thought of as a CC-CB connection when driven in a single-ended fashion or as parallel CE stages when driven differentially. A tail current source or a current setting resistance is required to establish the operating point. This ubiquitous, cannonical form has a long, distinguished history in operational amplifiers, mixers, IF/RF amplifiers, digital ECL/CML gates, and even oscillators and latches when cross-coupled. The basic operation relies on the controlled steering of current from one branch to the other. Notice that the dynamic range is modified by the fact that noise sources from two transistors contribute to the total noise while the signal is divided across two junctions. The noise associated with the tail current source appears as common mode noise and does not affect the input noise if differential signaling is used. The structure can be enhanced with the addition of emitter degeneration, cascoding CB stages, and buffering CC stages.

The Darlington connection is a CE stage buffered at the input by CC stage. This configuration behaves essentially like a super-transistor with greater impedance buffering capability and higher $h_{21}(f)$. It typically appears at input and output interfaces where drive and buffering requirements are most severe. The

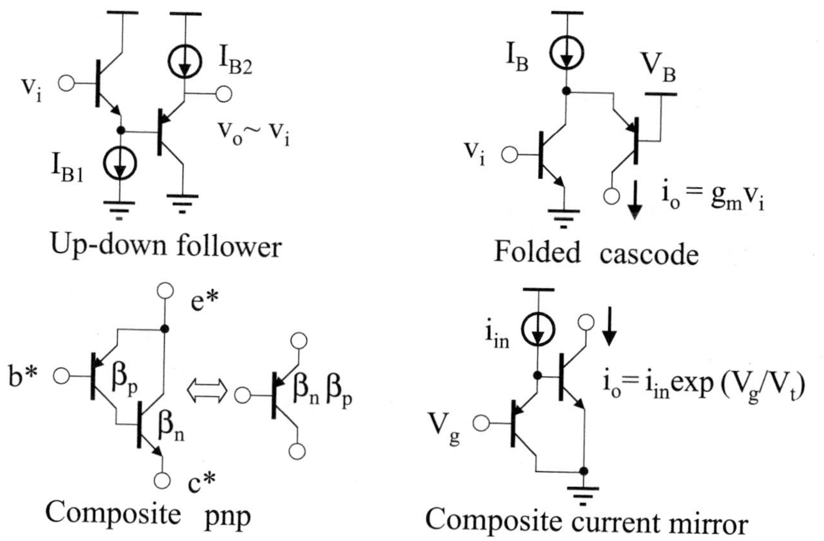

FIGURE 7.38 Composite npn-pnp combinations add functionality at lower supply voltages.

Darlington and the differential pair are sometimes called f_t multipliers since the input signal appears across two transistor junctions in series while the output signal is a parallel combination of currents. For a fixed input bias current per device, the effective input capacitance is $C_d/2$ while the effective transconductance is still g_m, giving from Eq. (7.23) an $f_t^{eff} = 2f_{tBJT}$. This of course ignores additional higher order poles and requires twice the current of a single BJT. In analogy to the differential pair, noise and distortion occur across two junctions in series.

The cross-coupled connection is derived from the differential pair. Following the signals around the loops leads to a positive feedback condition that can be used to induce regeneration for Schmidt triggers, multivibrators, and latches or negative resistance for tuned VCOs and gain peaking. It has also been used to synthesize voltage-to-current converters, bias cells, multipliers, and other functions. Since positive feedback is inevitably invoked in cross-coupled devices, care must be taken to avoid instability when it is undesired.

Note that npn-pnp composite structures are also possible creating many permutations that might offer certain advantages. These are illustrated in Fig. 7.38 in the form of the double up-down emitter-follower, the folded cascode, the composite pnp, and complementary mirror. The first two examples reduce the supply voltage requirements over their npn-only embodiment, the third transforms a poor quality pnp into a nearly ideal device, and the fourth provides a voltage controlled mirror ratio.

Translinear Circuits

The concept of translinearity was observed in deriving the basic properties of BJTs and its usefulness was alluded to in general terms. In particular, if the signals of interest are conveyed as currents rather than voltages, a large class of linear and nonlinear functions can be synthesized. When current-mode signals are used, extremely large dynamic ranges are possible since the junction essentially compresses the current signal logarithmically to a voltage; i.e., a 60 dB (1000x) signal current range corresponds to 180 mV of voltage swing at room temperature. The relatively small voltage swings represent a low impedance through which capacitances charge and discharge and thus the bandwidth of these circuits achieves broadband operation out to near the device limits represented by f_t. Furthermore, reduced internal voltage swings are consistent with lower supply voltages.

The simplest example of a translinear circuit is the well-known current mirror shown in Fig. 7.37. The input current, I_{in}, is mirrored to the output, I_{out} according to a scaling factor associated with the ratio of device sizes. Inherent in this process is a nonlinear conversion from I_{in} to the common V_{be} and then a second related nonlinear conversion from V_{be} to I_{out}. The impedance at the common base node is

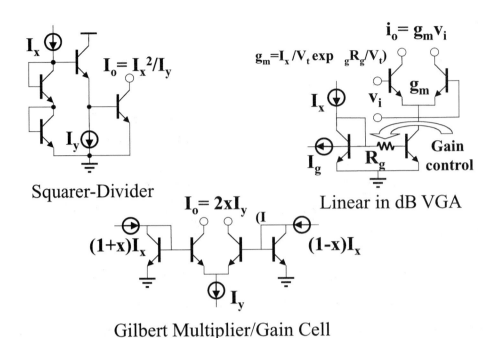

Squarer-Divider

Linear in dB VGA

Gilbert Multiplier/Gain Cell

FIGURE 7.39 Numerous functions can be implemented using translinear principles.

nominally a parallel combination of $1/g_m$ and $(1 + A)C_d$, which results in a low time constant on the order of τ_f.

A large family of translinear circuits has been synthesized by applying the translinear principle stated as: *In a closed loop containing only junctions, the product of current densities flowing in the clockwise direction is equal to the product of current densities flowing in the counterclockwise direction.*

The principle has a corollary when voltage sources are inserted into the loop, namely that the products are equal to within a factor $\exp(V_a/V_t)$, where V_a is the applied voltage. It is merely a restatement that logarithmic multiplication and division, i.e., addition and subtraction of V_{be}'s, is tantamount to products and quotients of currents. In this sense, the BJT can be thought of as the mathematical transistor. Note that since the signals are in current form, the distortion is not caused by the exponential characteristics but is actually due to *departure* from it, i.e., series resistances, high level injection, and Early voltage effects conspire to distort the signals.

Fig. 7.39 illustrates three examples of translinear IC design. The first example is an analog squarer/divider with input signals I_x, I_y and output, I_o. By following the translinear principle, it can be shown that the mathematical operation

$$I_o = I_x^2 \big/ I_y \tag{7.36a}$$

is performed. The second example is the well-known Gilbert-cell multiplier/gain cell with predistortion. Again, by following the rules, the differential output currents have the same form as the two inputs. The third example is a linear in dB variable gain amplifier. In this case a voltage $V_g = R_g I_g$ is inserted into the loop and Eq. (7.36a) is modified so that

$$i_o = I_x \big/ V_t \exp\!\big(V_g \big/ V_t\big) V_i = I_x \big/ V_t \, G_o \, 10^{Vg/Vt} \tag{7.36b}$$

Note that if V_g and I_x are engineered to be proportional to temperature, then the gain control becomes stable with temperature. In all cases, the operating frequency of these circuits approaches the technology f_t.

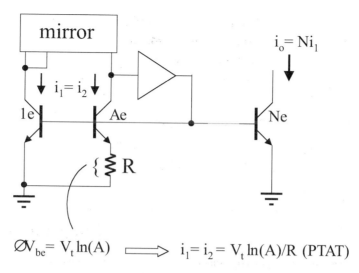

$$\varnothing V_{be} = V_t \ln(A) \implies i_1 = i_2 = V_t \ln(A)/R \ (PTAT)$$

FIGURE 7.40 The ΔV_{be} cell that generates a PTAT current is ubiquitous in BJT biasing.

Biasing Techniques

Microwave IC designers have historically focused on the signal path with biasing treated as an after-thought. Analog IC designers, on the other hand, have placed a strong emphasis on biasing since it usually determines the sensitivity and robustness of a circuit. In particular, as was already seen in the variable gain amplifier, it is often necessary to generate a current that is proportional to absolute temperature, PTAT. Other times it is necessary to generate a current that is complementary to absolute temperature, CTAT. Finally sometimes a zero temperature coefficient current, ZTAT, is desired. The choice on temperature shaping depends on what is needed. In differential amplifiers where the gain is given by $g_m R_L = I_o R_L / 2V_t$, it is appropriate to choose a PTAT bias current for a stable gain; however, if the amplifier is actually a limiter, then stable limited voltage swing requires a ZTAT shaping. By far the most common and fundamental bias current shape for BJTs turns out to be PTAT. From the PTAT form, other shapes can be derived.

The fundamental way to synthesize a bias sub-circuit is to start with a ΔV_{be} cell that generates a PTAT current. This cell, illustrated in Fig. 7.40, is a modified translinear loop. Two area ratioed BJTs each forced to carry the same current generate a difference in V_{be} given by

$$\Delta V_{be} = V_t \ln\left(A\right) \tag{7.37}$$

where A is the area ratio. This voltage develops a current in R that is PTAT and can now be mirrored throughout the IC by adding a driver that buffers the ΔV_{be} cell.

In some cases, a stable voltage reference is desired for scaling and it is beneficial to derive the various currents from this reference. To achieve a stable reference, the bandgap principle is invoked. As illustrated in Fig. 7.41, the idea is to add a CTAT voltage to a PTAT voltage so that the sum is a constant with temperature. It can be shown that the desired reference voltage V_g nearly corresponds to the bandgap energy, E_g of silicon, a very fundamental property. Intuitively, this is expected since at the extreme temperature of 0 K, it takes an energy equivalent to E_g to promote an electron into conduction. It so happens that a transistor V_{be} at a fixed I_c can be shown to be CTAT while a PTAT voltage can be generated from the ΔV_{be} cell. With this in mind, the PTAT component should be scaled such that it adds to the CTAT component to synthesize V_g. A physical realization of this principle known as a bandgap reference is illustrated in its simplest form in Fig. 7.42. It consists of a ΔV_{be} cell to generate a PTAT current and a resistor that converts it to the required PTAT voltage. Transistor Q1 plays a double role as part of the

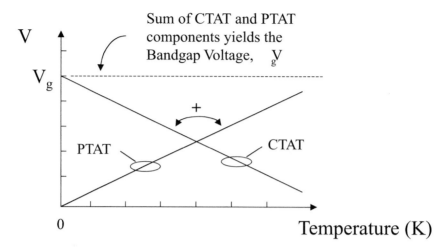

FIGURE 7.41 V_g is generated from properly proportioned PTAT and CTAT voltages.

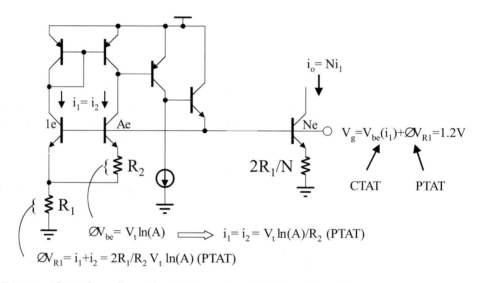

FIGURE 7.42 The Brokaw cell provides PTAT currents and a stable voltage reference.

ΔV_{be} cell and as the V_{be} responsible for the CTAT voltage. The current mirror above ensures that both currents are the same and the buffer provides drive to the reference output. This output can be used to generate PTAT currents elsewhere in the circuit as shown in Fig. 7.42. The target transistor and resistor scaling are necessary to preserve the PTAT shape. This topology can be made insensitive to supply voltage, temperature, and process variations by the use of more sophisticated cells and by the prudent choice of component sizes and layout.

Fabrication Technology

The technology for fabricating BJTs has been performance and cost driven toward smaller transistor geometries, lower device and interconnect parasitics, higher yield/integration, and greater functionality. Silicon-based BJT technology has benefited greatly from the synergy with CMOS processes targeting VLSI/ULSI digital applications. Several variants of the simple npn process have evolved over the years that feature better npn's and/or addition of other transistor types. Examples include BiCMOS, which integrates BJTs with CMOS; fully complementary bipolar with true vertical npn's and pnp's; SiGe/Si HBT,

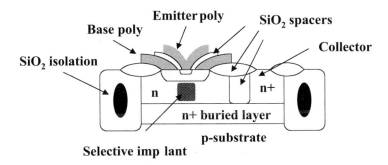

FIGURE 7.43 Modern processes feature double-poly, SiO_2 trenches and sub-μm lithography.

which offers higher f_t and lower R_b than traditional BJTs; and silicon-on-insulator (SOI) processes that rely on an insulating substrate to isolate devices and reduce substrate parasitics. III-V based HBTs have developed in a different direction since raw device speed has been more than adequate, but integration levels and process complexity have limited their availability. Process features can be grouped into two general categories: the active device, which determines the transistor performance, and the back-end process, which defines the interconnect metallization, dielectric isolation, passive components, and through-vias. The back-end process is particularly critical in RFICs since they set limits to IC parasitics and are responsible for the quality of resistors, capacitors, and inductors.

The silicon BJT process has evolved from the junction isolated buried collector process shown in Fig. 7.35 to the oxide trench isolated double-poly (emitter and base) process diagrammed in Fig. 7.43. The double poly structure with self-aligned spacers allows ultra-small emitter-base structures to be defined while simultaneously providing low extrinsic base resistance and low emitter charge storage. A selective collector implant under the active emitter helps delay the Kirk effect without significantly increasing the depletion capacitance. In general, the devices are defined by diffusions into the substrate with interconnects and passives constructed over dielectrics above the substrate. Standard aluminum metals have been silicided for improved reliability. Overall, this has led to significantly reduced device footprints and feature sizes with emitter geometries down to < 0.35 μm and f_t ranging from 30 GHz for standard BJTs to > 100 GHz for SiGe HBTs. Refinements such as silicided base handles and optimized epitaxially deposited bases promise even further improvements in device performance. On the back-end side, thin-film resistors and metal-metal capacitors are replacing their polysilicon and diffused versions bringing lower temperature coefficients, reduced parasitics and coupling to the substrate, trimming capability, and bias insensitivity. Furthermore, advanced multiple level interconnect modules and planarization techniques have been adopted from CMOS processes in the form of copper metalization, chemical-mechanical polishing, tungsten plugs, and low dielectric insulators. For RFICs, advanced interconnects are desirable not for high packing density, but for defining high quality passives away from the lossy substrate, particularly inductors, which are essential in certain applications.

The fabrication technology for the III-V HBTs is significantly different from that of silicon BJTs. The main difference stems from the fact that the starting material must be grown by advanced epitaxial techniques such as molecular beam epitaxy (MBE) or metallo-organic chemical vapor deposition (MOCVD). Each layer in the stack is individually optimized as needed. As a result of the predefined layers as well as thermal limits that disallow high temperature diffusion, the device is literally carved into the substrate leading to the wedding-cake triple mesa structure depicted in Fig. 7.44. The nature of the process limits minimum emitter geometries to about 1 μm, although self-aligned structures are used to minimize extrinsic areas. Values of f_t range from 30 GHz for power devices to 200 GHz for the highest speed structures. Interestingly, the improvements being made to silicon BJTs such as epitaxial bases, polysilicon emitters, and silicided bases, make use of deposited active layers, which is reminiscent of III-V technologies based on epitaxy. The back-end process inevitably consists of thin film resistors and metal-metal capacitors. Since the substrate is semi-insulating by definition, high-Q monolithic inductors are

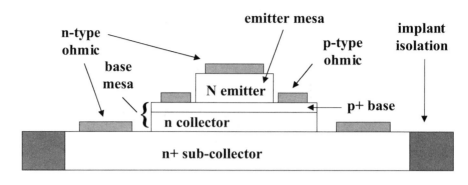

Semi-insulating substrate

FIGURE 7.44   III-V HBTs are fabricated with mesa structures and relaxed lithography.

already available. Some processes offer backside vias that literally provide a direct connection from the front-side to the back-side of the wafer enabling easy and effective grounding at any point on the circuit.

7.2.2   Heterostructure Bipolar Transistors (HBTs)

William Liu

Basic Device Principle

Heterojunction bipolar transistors (HBTs) differ from conventional bipolar junction transistors (BJTs) in their use of hetero-structures, particularly in the base-emitter junction. In a uniform material, an electric field exerts the same amount of force on an electron and a hole, producing movements in opposite directions as shown in Fig. 7.45a. With appropriate modifications in the semiconductor energy gap, the forces on an electron and a hole may differ, and at an extreme, drive the carriers along the same direction as shown in Fig. 7.45b.[1] The ability to modify the material composition to independently control the movement of carriers is the key advantage of adopting hetero-structures for transistor design.

Figure Fig. 7.46 illustrates the band diagram of a *npn* BJT under normal operation, wherein the base-emitter bias (V_{BE}) is positive and the base-collector bias (V_{BC}) is negative. The bipolar transistor was

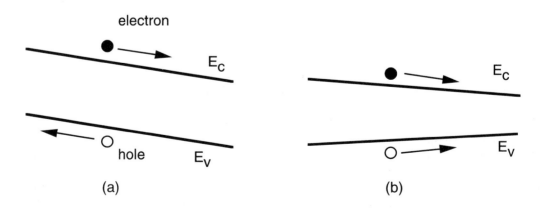

FIGURE 7.45   (a) An electric field exerts the same amount of force (but in opposite directions) on an electron and a hole. (b) Electron and hole can move in the same direction in a hetero-structure. (After Ref. [1], with permission.)

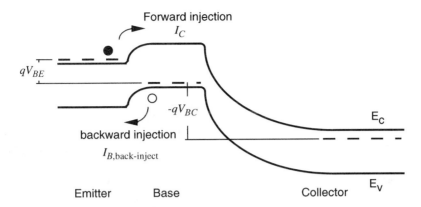

FIGURE 7.46 Band diagram of a *npn* bipolar junction transistor.

detailed in the previous section. Here we use Fig. 7.46 to emphasize that the energy gap is the same throughout the entire transistor structure. The Fermi levels and the depletion regions of the band diagram reflect the BJT design constraint that the doping level is highest in the emitter, and lowest in the collector. The Fermi level in the emitter is above the conduction band, signifying that the emitter is degenerately doped.

A BJT has two key current components. The first is the forward-injection current conducted by the electrons. These electrons are emitted at the emitter, crossing the base-emitter potential barrier with the help of the V_{BE} bias, diffusing through the thin base layer as minority carriers, and then are swept by the large electric field in the reverse-biased base-collector junction. The carriers finally leave the collector, forming the collector current (I_C). In their journey, only a few electrons are lost through recombinations with the majority holes in the base. We therefore denote the electron current as I_C in Fig. 7.46, even though the current magnitude in the emitter is slightly larger than that which finally exits the collector terminal. Of the three phases of the journey, the diffusion through the base layer is the rate-limiting step. Hence, a quantitative expression of I_C, to be developed shortly, relates intimately to the base layer parameters.

By varying V_{BE}, we control the amount of the electron injection from the emitter to the base, and subsequently, the amount of electrons collected at the collector. This current flow, however, is independent of V_{BC}, as long as the base-collector junction is reverse biased. The property that the desired signal (I_C) is modified by the input bias (V_{BE}) but unaffected by the output bias (V_{BC}) fulfills the requirement of a sound three-terminal device. As the input and output ports are decoupled from one another, complicated circuits based on the transistors can be easily designed.

The second current component is composed of holes that are back-injected from the base to the emitter. This current does not get collected at the collector terminal and does not contribute to the desired output signal. However, the moment a V_{BE} is applied so that the desired electrons are emitted from the emitter to the base, these holes are back-injected from the base to the emitter. The bipolar transistor is so named to emphasize that both electron and hole play significant roles in the device operation. A good design of a bipolar transistor maximizes the electron current transfer while minimizing the hole current. As indicated on Fig. 7.46, we refer the hole current as $I_{B,back-inject}$. It is a base current because the holes come from the base layer where they are the majority carriers.

We mentioned that I_C is limited by the diffusion process in the base, wherein the electrons are minority carriers. Likewise, $I_{B,back-inject}$ is limited by the diffusion in the emitter, wherein the holes are minority carriers. Fisk's law states that a diffusion current density across a layer is equal to the diffusion coefficient times the carrier concentration gradient.[2] (From the Einstein relationship, the diffusion coefficient can be taken to be kT/q times the carrier mobility.) The carrier concentrations at one end of the base layer (for the calculation of I_C) and the emitter layer (for the calculation of $I_{B,back-inject}$) are both proportional to $\exp(qV_{BE}/kT)$, and are 0 at the other end. An application of Fisk's Law leads to:[3,4]

$$I_C = \frac{q A_{\text{emit}} D_{n,\text{base}}}{X_{\text{base}}} \frac{n^2_{i,\text{base}}}{N_{\text{base}}} \exp\left(\frac{q V_{BE}}{kT}\right) \tag{7.38}$$

$$I_{B,\text{back-inject}} = \frac{q A_{\text{emit}} D_{p,\text{emit}}}{X_{\text{emit}}} \frac{n^2_{i,\text{emit}}}{N_{\text{emit}}} \exp\left(\frac{q V_{BE}}{kT}\right) \tag{7.39}$$

where, for example, A_{emit} is the emitter area; $D_{n,\text{base}}$ is the electron diffusion coefficient in the base layer; N_{base} is the base doping; X_{base} is the base thickness; and $n_{i,\text{base}}$ is the intrinsic carrier concentration in the base layer. I_C is proportional to the emitter area rather than the collector area because I_C is composed of the electrons injected from the emitter. For homojunction BJTs, $n_{i,\text{base}}$ is identical to $n_{i,\text{emit}}$. The ratio of the desired I_C to the undesired $I_{B,\text{back-inject}}$ is,

$$\frac{I_C}{I_{B,\text{back-inject}}} = \frac{X_{\text{emit}}}{X_{\text{base}}} \frac{D_{n,\text{base}}}{D_{p,\text{emit}}} \times \frac{N_{\text{emit}}}{N_{\text{base}}} \quad \left(\text{homojunction BJT}\right) \tag{7.40}$$

Because the diffusion coefficients and layer thicknesses are roughly equal on the first order, Eq. (7.40) demonstrates that the emitter doping of a BJT must exceed the base doping in order for I_C to exceed $I_{B,\text{back-inject}}$. For most applications, we would actually prefer the base doping be high and the emitter doping, low. A high base doping results in a small base resistance and a low emitter doping reduces the base-emitter junction capacitance, both factors leading to improved high frequency performance. Unfortunately, these advantages must be compromised in the interest of minimizing $I_{B,\text{back-inject}}$ in homojunction transistors.

A heterojunction bipolar transistor (HBT), formed by a replacement of the homojunction emitter by a larger energy gap material, enables the design freedom of independently optimizing the $I_C/I_{B,\text{back-inject}}$ ratio and the doping levels. Figure 7.47 illustrates the band diagrams of *Npn* HBTs. The capital *N* in "*Npn*" rather than a small letter *n* emphasizes that the emitter is made of a larger energy gap material than the rest. It is implicit that the base-collector junction of an HBT is a homojunction. An HBT whose base-collector junction is also a heterojunction is called a double heterojunction bipolar transistor (DHBT), typically found in the InP/InGaAs material system.

The energy gap difference in the base-emitter heterojunction of an HBT burdens the holes from the base to experience a much larger energy barrier than the electrons from the emitter. With the same application of V_{BE}, the forces acting on the electrons and holes differ, favoring the electron injection from

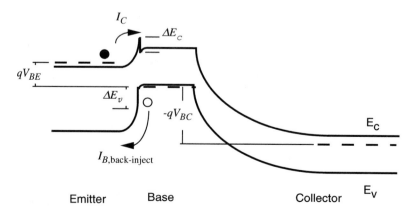

FIGURE 7.47 Band diagram of a *Npn* abrupt heterojunction bipolar transistor.

the emitter into the base to the hole back-injection from the base into the emitter. Figure 7.47 shows that the difference in the electron and hole barriers is ΔE_v, the valence band discontinuity. The $I_C/I_{B,\text{back-inject}}$ ratio of Eq. (7.40) can be extended to the HBT as:[1]

$$\frac{I_C}{I_{B,\text{back-inject}}} = \frac{X_{\text{emit}}}{X_{\text{base}}} \frac{D_{n,\text{base}}}{D_{p,\text{emit}}} \times \frac{N_{\text{emit}}}{N_{\text{base}}} \exp\left(\frac{\Delta E_v}{kT}\right) \quad \left(\text{abrupt HBT}\right) \tag{7.41}$$

The exponential factor is the key to the fact that base doping can be made larger than the emitter doping without adversely affecting the HBT performance.

We qualify the expression in Eq. (7.41): that is for an *abrupt* HBT. This means that an abrupt change of material composition exists between the emitter and the base. An example is an InP/In$_{0.53}$Ga$_{0.47}$As HBT. The indium and gallium compositions form a fixed ratio so that the resulting In$_{0.53}$Ga$_{0.47}$As layer has the same lattice constant as InP, the starting substrate material. Otherwise, the dislocations due to lattice mismatch prevent the device from being functional. In fact, although the idea of HBT is as old as the homojunction transistor itself,[5] HBTs have emerged as practical transistors only after molecular beam epitaxy (MBE) and metal-organic chemical vapor deposition (MOCVD) were developed to grow high-quality epitaxial layers in the 1980s.[6] Another example of an abrupt HBT is the Al$_{0.3}$Ga$_{0.7}$As/GaAs HBT, in which the entire emitter layer consists of Al$_{0.3}$Ga$_{0.7}$As, while the base and collector are GaAs. When the AlGaAs/GaAs material system is used, we can take advantage of the material property that AlAs and GaAs have nearly the same lattice constants. It is therefore possible to grow an Al$_x$Ga$_{1-x}$As layer with any composition x and still be lattice-matched to GaAs. When the aluminum composition of an intermediate AlGaAs layer is graded from 0 at the GaAs base to 30% at the Al$_{0.3}$Ga$_{0.7}$As emitter, then the resulting structure is called a *graded* HBT.

The band diagram of a graded HBT is shown in Fig. 7.48. A graded HBT has the advantage that the hole barrier is larger than the electron barrier by ΔE_g, the energy-gap discontinuity between the emitter and the base materials. Previously, in an abrupt HBT, the difference in the hole and electron barriers was only ΔE_v. The additional barrier brings forth an even larger $I_C/I_{B,\text{back-inject}}$ ratio for the graded HBT:

$$\frac{I_C}{I_{B,\text{back-inject}}} = \frac{X_{\text{emit}}}{X_{\text{base}}} \frac{D_{n,\text{base}}}{D_{p,\text{emit}}} \times \frac{N_{\text{emit}}}{N_{\text{base}}} \exp\left(\frac{\Delta E_g}{kT}\right) \quad \left(\text{Graded HBT}\right) \tag{7.42}$$

Equation (7.42) is obtained from the inspection of the band diagram. It is also derivable from Eqs. (7.38) and (7.39) by noting that $n_{i,\text{base}}^2 = n_{i,\text{emit}}^2 \times \exp(\Delta E_g/kT)$.

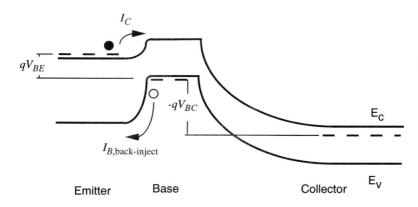

FIGURE 7.48 Band diagram of a *Npn* graded heterojunction bipolar transistor.

TABLE 7.3   Heterojunction Parameters at Room Temperature

	$Al_xGa_{1-x}As/GaAs$ $(x \leq 0.45)$	$Ga_{0.51}In_{0.49}P/GaAs$	$InP/In_{0.53}Ga_{0.47}As$	$In_{0.52}Al_{0.48}As/In_{0.53}Ga_{0.47}As$	$Si/Si_{0.8}Ge_{0.2}$
ΔE_g (eV)	$1.247 \cdot x$	Ordered: 0.43 Disordered: 0.46	0.60	0.71	0.165
ΔE_c (eV)	$0.697 \cdot x$	Ordered: 0.03 Disordered: 0.22	0.23	0.50	
ΔE_v (eV)	$0.55 \cdot x$	Ordered: 0.40 Disordered: 0.24	0.37	0.21	
E_g (eV)	GaAs: 1.424	Ordered: 1.424	InP: 1.35	InGaAs: 0.75	Si: 1.12

The amount of improvement from a homojunction to a heterojunction depends on the band alignment properties of the two hetero-materials. Table 7.3 lists the conduction band discontinuity (ΔE_c), valence band discontinuity (ΔE_v), and their sum which is the energy-gap discontinuity (ΔE_g), for several material systems used for HBTs. The energy gap of a semiconductor in the system is also given for convenience. The values shown in Table 7.3 are room temperature values.

The most popular III-V material system is $Al_xGa_{1-x}As/GaAs$. The table lists its band alignment parameters for $x \leq 0.45$. At these aluminum mole fractions, the AlGaAs layer is a direct energy-gap material, in which the electron wave function's wave vectors **k** at both the conduction band minimum and the valence band maximum are along the same crystal direction.[2] Since nearly all semiconductor valence band maximums take place when **k** is [000], the conduction band minimum in a direct energy-gap material is also at **k** = [000]. The conduction band structure surrounding this **k** direction is called the Γ valley. There are two other **k** directions of interest because the energy band structures in such directions can have either the minimum or a local minimum in the conduction band. When **k** is in the [100] direction, the band structure surrounding it is called the X valley, and when **k** is in the [111] direction, the L valley. As the aluminum mole fraction of $Al_xGa_{1-x}As$ exceeds 0.45, it becomes an indirect energy-gap material, with its conduction band minimum residing in the X valley. The band alignment parameters at $x > 0.45$ are found elsewhere.[3] SiGe is an indirect energy-gap material; its conduction band minimum is also located in the X valley. Electron-hole generation or recombination in indirect energy-gap materials requires a change of momentum ($\mathbf{p} = \hbar\mathbf{k}$), a condition that generally precludes them from being useful for laser or light-emitting applications.

Particularly due to the perceived advantage in improved reliability, HBTs made with the GaInP/GaAs system have gained considerable interest.[7] The band alignment of GaInP/GaAs depends on whether the grown GaInP layer is ordered or disordered, as noted in Table 7.3. The crystalline structure in an *ordered* GaInP layer is such that sheets of pure Ga, P, In, and P atoms alternate on the (001) planes of the basic unit cell, without the intermixing of the Ga and In atoms on the same lattice plane.[8] When the Ga, In and P atoms randomly distribute themselves on a plane, the GaInP layer is termed *disordered*.

The processing of AlGaAs/GaAs and GaInP/GaAs HBTs is fairly simple. Both wet and dry etching are used in production environments, and ion implantation is an effective technique to isolate the active devices from the rest. The processing of InP/InGaAs and InAlAs/InGaAs materials, in contrast, is not straightforward. Because of the narrow energy-gap of the InGaAs layer, achieving an effective device isolation often requires a complete removal of the inactive area surrounding the device, literally digging out trenches to form islands of devices. Further, the dry-etching, and its associated advantages such as directionality of etching, is not readily/easily available for InGaAs.[9] However, the material advantages intrinsic to InP/InGaAs and InAlAs/InGaAs make them the choice for applications above 40 GHz. In addition, the turn-on voltage, the applied V_{BE} giving rise to a certain collector current, is smaller for HBTs formed with InGaAs base compared to GaAs base (due to the energy-gap difference). The turn-on characteristics of various HBTs are shown in Fig. 7.49.

A calculation illustrates the advantage of an HBT. Consider an AlGaAs/GaAs HBT structure designed for power amplifier applications, as shown in Fig. 7.50.[10] The emitter and the base layers of the transistor are: $N_{emit} = 5 \times 10^{17}$ cm^{-3}; $N_{base} = 4 \times 10^{19}$ cm^{-3}; $X_{emit} \approx 1300$ Å; and $X_{base} = 1000$ Å. We shall use the

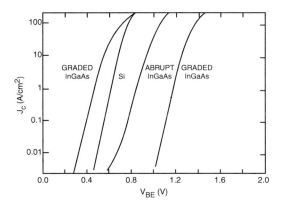

FIGURE 7.49 Turn-on characteristics of bipolar transistors based on various material systems. (After Ref. [6], with permission.)

Material	Thickness (Å)	Composition x	Doping (cm^{-3})	
n-In$_x$Ga$_{1-x}$As	800	$0 \to 0.6$	$> 3 \times 10^{19}$	
n-GaAs	2000		5×10^{18}	
N-Al$_x$Ga$_{1-x}$As	300	$0.3 \to 0$	1×10^{18}	
N-Al$_x$Ga$_{1-x}$As	1000	0.3	5×10^{17}	Emitter
N-Al$_x$Ga$_{1-x}$As	300	$0 \to 0.3$	5×10^{17}	
(The above grading layer is absent in abrupt HBT)				
p-GaAs	1000		4×10^{19}	Base
n-GaAs	7000		3×10^{16}	Collector
n-GaAs	6000		5×10^{18}	Subcollector
Semi-insulating GaAs Substrate				

FIGURE 7.50 Typical HBT epitaxial structure designed for power amplifier applications. (After Ref. [10], with permission.)

following diffusion coefficients for the calculation: $D_{n,\text{base}} = 20$ and $D_{p,\text{emit}} = 2.0$ cm^2/V-s. For a graded Al$_{0.3}$Ga$_{0.7}$As/GaAs heterojunction, ΔE_g is calculated from Table 7.3 to be $0.3 \times 1.247 = 0.374$ eV. The ratio for the graded HBT, according to Eq. (7.42), is,

$$\frac{I_C}{I_{B,\text{back-inject}}} = \frac{1300}{1000}\frac{20}{2.0} \times \frac{5 \times 10^{17}}{4 \times 10^{19}} \exp\left(\frac{0.374}{0.0258}\right) = 2.5 \times 10^5$$

For an abrupt Al$_{0.3}$Ga$_{0.7}$As/GaAs HBT, ΔE_v is the parameter of interest. According to the table, ΔE_v at $x = 0.3$ is $0.55 \times 0.3 = 0.165$ eV. Therefore, Eq. (7.41) leads to:

$$\frac{I_C}{I_{B,\text{back-inject}}} = \frac{1300}{1000}\frac{20}{2.0} \times \frac{5 \times 10^{17}}{4 \times 10^{19}} \exp\left(\frac{0.165}{0.0258}\right) = 97$$

Consider a Si BJT with identical doping levels, layer thicknesses, and diffusion coefficients, except that it is a homojunction transistor so that $\Delta E_g = 0$. Using Eq. (7.40), we find the $I_C/I_{B,\text{back-inject}}$ ratio to be:

$$\frac{I_C}{I_{B,\text{back-inject}}} = \frac{1300}{1000}\frac{20}{2.0} \times \frac{5 \times 10^{17}}{4 \times 10^{19}} = 0.16.$$

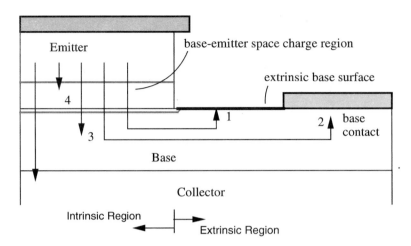

FIGURE 7.51 Locations of the four base recombination currents. The fifth base current component, not a recombination current, is $I_{B,\text{back-inject}}$ shown in Figure 7.46. (After Liu, W., Microwave and DC characterizations of Npn and Pnp HBTs, PhD. Dissertation, Stanford University, Stanford, CA, 1991.)

The useful collector current in the homojunction transistor is only 1/6 of the undesirable back-injection current. This means the device is useless. In contrast, both the graded and the abrupt HBTs remain functional, despite the higher base doping in comparison to the emitter.

Base Current Components

$I_{B,\text{back-inject}}$ is only one of the five dominant base current components in a bipolar transistor. We have thus far considered only $I_{B,\text{back-inject}}$ because it is the distinguishing component between a HBT and a BJT. Once $I_{B,\text{back-inject}}$ is made small in a HBT through the use of a heterojunction, the remaining four components become noteworthy. All of these components are recombination currents; they differ only in the locations where the recombinations take place, as shown in Fig. 7.51. They are: (1) extrinsic base surface recombination current, $I_{B,\text{surf}}$; (2) base contact surface recombination current, $I_{B,\text{cont}}$; (3) bulk recombination current in the base layer, $I_{B,\text{bulk}}$; and (4) space-charge recombination current in the base-emitter junction depletion region, $I_{B,\text{scr}}$. In the discussion of bipolar transistors, an easily measurable quantity of prime importance is the current gain (β), defined as the ratio of I_C to the total base current I_B:

$$\beta = \frac{I_C}{I_B} = \frac{I_C}{I_{B,\text{back-inject}} + I_{B,\text{surf}} + I_{B,\text{cont}} + I_{B,\text{bulk}} + I_{B,\text{scr}}} \tag{7.43}$$

Depending on the transistor geometrical layout, epitaxial layer design, and the processing details that shape each of the five base current components, the current gain can have various bias and temperature dependencies. In the following, the characteristics of each of the five base components are described, so that we can better interpret the current gain from measurement and establish some insight about the measured device.

Figure 7.52a illustrates a schematic cross-section of an HBT. Without special consideration, a conventional fabrication process results in an exposed base surface at the extrinsic base regions. (*Intrinsic region* is that underneath the emitter, and *extrinsic region* is outside the emitter mesa, as shown in Fig. 7.51.) Because the exposed surface is near the emitter mesa where the minority carrier concentration is large, and because the surface recombination velocity in GaAs is high (on the order of 10^6 cm/s), $I_{B,\text{surf}}$ is significant in these unpassivated devices. Various surface passivation techniques have been tested. The most effective method is *ledge passivation*,[11,12] formed with, for example, an AlGaAs layer on top of the GaAs base. The AlGaAs ledge must be thin enough so that it is fully depleted by a combination of the free surface Fermi level pinning above and the base-emitter junction below. If the passivation ledge is

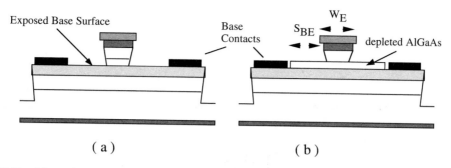

FIGURE 7.52 Schematic cross-sections of HBTs: (a) unpassivated HBTs, and (b) passivated HBTs. (After Liu, W., Microwave and DC characterizations of Npn and Pnp HBTs, PhD. Dissertation, Stanford University, Stanford, CA, 1991.)

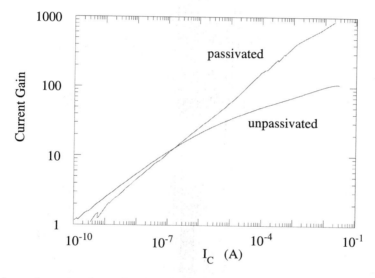

FIGURE 7.53 Measured current gain as a function of collector current for both passivated and unpassivated HBTs. (After Liu, W. et al., Diode ideality factor for surface recombination current in AlGaAs/GaAs heterojunction bipolar transistors, *IEEE Trans. Electron Devices*, 39, 2726–2732, 1992, with permission.)

not fully depleted, the active device area would be much larger than the designed emitter. The requirement for the AlGaAs layer to be fully depleted limits the AlGaAs thickness to the order of 1000 Å, and emitter doping to low to mid 10^{17} cm^{-3}. Although the ledge passivation was originally designed to minimize $I_{B,surf}$, it is also crucial to long-term reliability.[3,7]

Unlike I_C, $I_{B,surf}$ is proportional to the emitter periphery rather than the emitter area. For high frequency devices whose emitter is in a strip form (thus the perimeter-to-area ratio is large), $I_{B,surf}$ is a major component to the overall base current. The current gain is substantially reduced from that of a large squarish device whose perimeter-to-area ratio is small. The discrepancy in β due to emitter geometry is termed the *emitter-size effect*. Figure 7.53 displays β vs. I_C for both passivated and unpassivated devices with $A_{emit} = 4 \times 10 \ \mu m^2$. The emitter area is small enough to demonstrate the benefit of the surface passivation. A large device has negligible surface recombination current and the current gain does not depend on whether the surface is passivated or not. Because the two devices are fabricated simultaneously and physically adjacent to each other, the difference between the measured βs is attributed to the additional $I_{B,surf}$ of the unpassivated device.

The second base recombination current, $I_{B,cont}$, is in principle the same as $I_{B,surf}$. Both are surface recombination currents, except $I_{B,cont}$ takes place on the base contacts whereas $I_{B,surf}$, on the extrinsic base surfaces. Because the contacts are located further away from the intrinsic emitter than the extrinsic base surface, $I_{B,\text{cont}}$ is generally smaller than $I_{B,\text{surf}}$ when the surface is unpassivated. However, it may replace $I_{B,\text{surf}}$ in significance in passivated devices (or in Si BJTs whose silicon dioxide is famous in passivating silicon). There is a characteristic distance for the minority carrier concentration to decrease exponentially from the emitter edge toward the extrinsic base region.[3] As long as the base contact is placed at roughly 3 times this characteristic length away from the intrinsic emitter, $I_{B,\text{cont}}$ can be made small. The base contact cannot be placed too far from the emitter, however. An excessively wide separation increases the resistance in the extrinsic base region and degrades the transistor's high frequency performance.

The above two recombination currents occur in the extrinsic base region. Developing analytical expressions for them requires a solution of the two-dimensional carrier profile. Although this is possible without a full-blown numerical analysis,[3] the resulting analytical equations are quite complicated. The base bulk recombination current, in contrast, can be accurately determined from a one-dimensional analysis since most of the base minority carriers reside in the intrinsic region. It is convenient to express $I_{B,\text{bulk}}$ through its ratio with I_C:

$$\frac{I_C}{I_{B,\text{bulk}}} = \frac{\tau_n}{\tau_b} \tag{7.44}$$

where τ_n is the minority carrier lifetime in the base, and τ_b, the minority carrier transit time across the base. In a typical Si BJT design, an electron spends about 10 ns diffusing through the base layer while in every 1 µs an electron is lost through recombination. The transistor then has a current gain of 1 µs/10 ns = 100. The recombination lifetime in the GaAs base material is significantly shorter than in Si, at about 1 ns. However, the transit time through the GaAs is also much shorter than in Si due to the higher carrier mobility in GaAs. A well-designed HBT has a τ_b of 0.01 ns; therefore, a $\beta = 100$ is also routinely obtainable in III-V HBTs.

Equation (7.44) indicates that $I_{B,\text{bulk}}$ is large if τ_n is small. The recombination lifetime of a semiconductor, a material property, is found to be inversely proportional to the doping level. When the base doping in an AlGaAs/GaAs (or GaInP/GaAs) HBT is 5×10^{18} cm^{-3}, the current gain easily exceeds 1000 when proper device passivation is made and the base contacts are placed far from the emitter. As the base doping increases to 10^{20} cm^{-3}, $I_{B,\text{bulk}}$ dominates all other base current components, and the current gain decreases to only about 10, independent of whether the extrinsic surface is passivated or not. The base doping in III-V HBTs for power applications, as shown in Fig. 7.50, is around $3 - 5 \times 10^{19}$ cm^{-3}. It is a compromise between achieving a reasonable current gain (between 40 and 200) and minimizing the intrinsic base resistance to boost the high frequency performance.

Equation (7.44) also reveals that $I_{B,\text{bulk}}$ is large when the base transit time is long. This is the time that a base minority carrier injected from the emitter takes to diffuse through the base. Unlike the carrier lifetime, which is mostly a material constant, the base transit time is a strong function of the base layer design:

$$\tau_b = \frac{X_{\text{base}}^2}{2D_{n,\text{base}}} \tag{7.45}$$

Because τ_b is proportional to the square of X_{base}, the base thickness is designed to be thin. Making the base too thin, however, degrades high frequency performance due to increased base resistance. A compromise between these two considerations results in a X_{base} at around $800 - 1000$ Å, as shown in Fig. 7.50.

The derivation of Eq. (7.45) assumes that the minority carriers traverse through the base purely by diffusion. This is certainly the scenario in a bipolar transistor whose base layer is uniformly doped and of the same material composition. With energy-gap engineering, however, it is possible to shorten the base transit time (often by a factor of 3) by providing a drift field. In a Si/SiGe HBT for example, the Ge

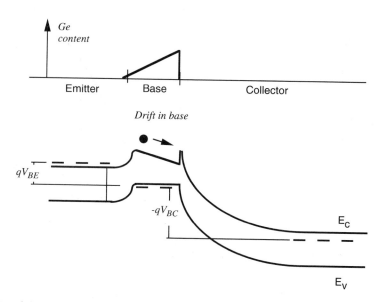

FIGURE 7.54 Band diagram of a SiGe HBT with a base quasi-electric field established by grading the germanium content.

content can be linearly graded from 0 to 8% across a 300 Å base to result a *quasi-electric field* on the order of 30 to 50 kV/cm.[13] We used the term quasi-electric field to describe the electric field generated by grading the energy gap, or more specifically, the gradient of the electron affinity (χ_e). In a conventional bipolar transistor, an electric field can be established only in the presence of space charges (such as the depletion region in a *p-n* junction). In an HBT with a graded base, the overall electric field in the base layer is nonzero even though the entire base region remains charge neutral. A SiGe HBT band diagram, shown in Fig. 7.54, illustrates how a minority carrier can speed up in the presence of the band grading. The figure shows that the energy gap becomes narrower as the Ge content increases. Because the base layer is heavily doped, the quasi-Fermi level in the base is pinned to be relatively flat with respect to position. The entire energy gap difference appears in the conduction band. The base quasi-electric field, being proportional to the slope of the band bending, propels the electrons to drift from the emitter toward the collector. As the carrier movement is enhanced by the drift motion, in addition to the diffusion, the base transit time decreases.

Figure 7.54 is characteristic of the SiGe HBT pioneered by a U.S. company,[14] in which the Ge content is placed nearly entirely in the base and graded in a way to create a base quasi-electric field. The base-emitter junction is practically a homojunction; therefore, it perhaps does not strictly fit the definition of being a heterojunction bipolar transistor and the base must be doped somewhat lighter than the emitter. This type of transistor resembles a drift homojunction transistor,[15] in which a base electric field is established by grading the base doping level. A drift transistor made with dopant grading suffers from the fact that part of the base must be lightly doped (at the collector side), thus bearing a large base resistance. An alternative school of SiGe HBT places a fixed Ge content in the base, mostly promoted by European companies.[16] Transistors fabricated with the latter approach do not have a base quasi-electric field. However, the existence of the base-emitter heterojunction allows the emitter to be more lightly doped than the base, just as in III-V HBTs. In either type of SiGe HBTs, there can be a conduction band discontinuity between the base and collector layers. This base-collector junction spike (Fig. 7.54), also notable in InP-based DHBT, has been known to cause current gain fall off.[3] The spike can be eliminated by grading the Ge content from the base to inside the collector layer.

Likely due to reliability concerns or for purely historical reasons, most commercial III-V HBTs have a uniformly doped base without a base quasi-electric field. If a base electric field is desired, in

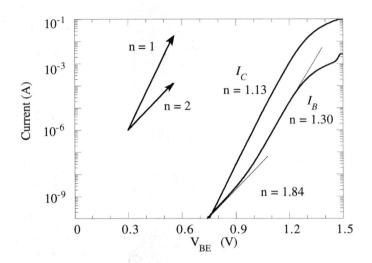

FIGURE 7.55 Measured Gummel plot of an abrupt AlGaAs/GaAs HBT. (After Liu, W., Experimental comparison of base recombination currents in abrupt and graded AlGaAs/GaAs heterojunction bipolar transistors, *Electronic Letters*, 27, 2115–2116, 1991, with permission.)

AlGaAs/"GaAs" HBTs in particular, the field can be established by grading of the aluminum concentration in the AlGaAs base layer.

The fourth recombination current is the space-charge recombination current in the base-emitter depletion region. $I_{B,\mathrm{scr}}$ differs from the other base current components in its bias dependency. Equations (7.38) and (7.39) show that I_C and $I_{B,\mathrm{back\text{-}inject}}$ are proportional to $\exp(qV_{BE}/nkT)$ with n, the *ideality factor*, being equal to 1. Equation (7.44) also shows that $I_{B,\mathrm{bulk}}$ is directly proportional to I_C. Hence, $I_{B,\mathrm{bulk}}$ has a unity ideality factor as well. Extensive measurement experiments and theoretical calculations indicate that $I_{B,\mathrm{surf}}$ and hence, $I_{B,\mathrm{cont}}$, have an ideality factor closer to 1 than 2.[3,17] The ideality factor of $I_{B,\mathrm{scr}}$, in contrast, is nearly 2 because the electron and hole concentrations in the forward-biased base-emitter junction are both proportional to $\exp(qV_{BE}/2kT)$.[2] This means $I_{B,\mathrm{scr}}$ is most significant when V_{BE} is small, at which operating region I_C is also small. A *Gummel plot*, I_B and I_C as a function of V_{BE} taken at $V_{BC} = 0$, illustrates dominance of $I_{B,\mathrm{scr}}$ in low-current regions, as shown in Fig. 7.55. There are times when $I_{B,\mathrm{scr}}$ dominates other base current components even at high I_C levels, particularly in graded HBTs.[3]

The previous four base current components are all recombination current. The fifth component is the back-injection current, $I_{B,\mathrm{back\text{-}inject}}$. This component is made small in a heterojunction transistor, at least at room temperature. However, as temperature increases, the extra energy barrier provided to the hole carriers becomes less effective in impeding the back injection. When the HBT is biased with a high I_C and a certain amount of V_{BC}, the power dissipated in the device can heat up the device itself (called *self-heating*). As the HBTs junction temperature rises, $I_{B,\mathrm{back\text{-}inject}}$ increases and the current gain decreases. This β's temperature dependency is to be contrasted with silicon BJTs current gain, which increases with temperature.[18]

Kirk Effects

Understanding the properties of the various base current components facilitates the description of the measured device characteristics, such as the β vs. I_C curve of Fig. 7.53. The previous analysis of the transistor currents implicitly assumes that the transistor operates in the normal bias condition, under which the transistor gain is seen to increase steadily with the collector current. However, I_C cannot increase indefinitely without adverse effects. Figure 7.53 reveals that, after a certain critical I_C is reached while V_{BC} is kept constant, the current gain plummets, rendering the device useless. For high-power applications, such as a power amplifier for wireless communications, HBTs are biased at a large current (with a current density on the order of 10^4 A/cm²), not only because the output power is directly proportional to I_C, but also because the high frequency performance is superior at large I_C (but before the current gain falls). Therefore, it is imperative to understand the factor setting the maximum I_C level, below which the current gain is maintained at some finite values.

The Poisson equation relates charges to the spatial variations of the electric field (ε).[2] It is a fundamental equation, applicable to any region at any time in a semiconductor:

$$\frac{d\varepsilon}{dx} = \frac{q}{\in_s}\left(p - n + N_d - N_a\right) \tag{7.46}$$

$\in_s$ is the dielectric constant of the semiconductor; N_d and N_a are the donor and acceptor doping levels, respectively; and n and p are the mobile electron and hole carrier concentrations, respectively. We apply this equation to the base-collector junction of HBTs, which is typically a homojunction. Since the base doping greatly exceeds the collector doping, most of the depletion region of the base-collector junction is at the collector side. In the depleted collector region where it is doped n-type, N_d in Eq. (7.46) is the collector doping level, N_{coll}, and N_a is 0. The collector current flowing through the junction consists of electrons. If the field inside the depletion region was small, then these electrons would move at a speed equal to the product of the mobility and the electric field: $\mu_n \cdot \varepsilon$. It is a fundamental semiconductor property that, once the electric field exceeds a certain critical value ($\varepsilon_{crit} \sim 10^3$ V/cm for GaAs), the carrier travels at a constant velocity called the saturation velocity (v_{sat}). Because the electric field inside most of the depletion region exceeds 10^4 V/cm, practically all of these electrons travel at a constant speed of v_{sat}. The electron carrier concentration inside the collector can be related to the collector current density (J_C; equal to I_C/A_{emit}) as:

$$n(x) = \frac{J_C}{qv_{sat}} = \text{constant inside the base-collector junction} \tag{7.47}$$

Lastly, because there is no hole current, p in Eq. (7.46) is zero.

Equation (7.46), when applied to the base-collector junction of a HBT, is simplified to:

$$\frac{d\varepsilon}{dx} = \frac{q}{\in_s}\left(-\frac{J_C}{qv_{sat}} + N_{coll}\right) \tag{7.48}$$

When J_C is small, the slope of the electric field is completely determined by the collector doping, N_{coll}. Because the doping is constant with position, solving Eq. (7.48) at negligible J_C gives rise to a field profile that varies linearly with position, as shown in Fig. 7.56a. As the current density increases, the mobile electron concentration constituting the current partially cancels the positive donor charge concentration N_{coll}. As the net charge concentration decreases, the slope of the field decreases, as shown in Fig. 7.56b. While the current density increases, the base-collector bias V_{BC} remains unchanged. Therefore, the enclosed area of the electric field profile, which is basically the junction voltage, is the same before and after the current increase. The simultaneous requirements of having a decreasing field slope and a constant enclosed area imply that the depletion region extends toward the subcollector layer and the maximum electric field decreases. The depletion thickness continues to increase until the collector is fully depleted, as shown in Fig. 7.56c. The depletion thickness does not extend beyond the collector layer because the subcollector is a heavily doped layer. Afterwards, further increase of current results in a quadrangle field profile, as shown in Fig. 7.56d, replacing the previous triangular profile. As the current density increases to a level such that $J_C = qN_{coll} \cdot v_{sat}$, the term inside the parentheses of Eq. (7.48) becomes zero. A field gradient of zero means that the field profile stays constant with the position inside the junction (slope = 0). This situation, depicted in Fig. 7.56e, marks the beginning the *field reversal*. When J_C increases further such that $J_C > qN_{coll} \cdot v_{sat}$, the mobile electrons brought about by the collector current more than

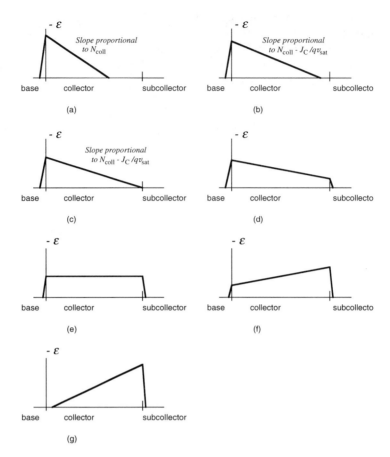

FIGURE 7.56 Electric field profile inside the base-collector junction of HBT during normal operation. (a) $J_C = 0$; (b) J_C is small; (c) J_C further increases so the entire collector is depleted; (d) further increase in J_C; (e) J_C reaches $qN_{coll} \cdot v_{sat}$; (f) $J_C > qN_{coll} \cdot v_{sat}$ and field reversal occurs; (g) further increase in J_C results in base push out as the electric field at the base-collector boundary decreases to zero.

compensates the fixed charges inside the collector. The net charge concentration for the first time becomes negative and the electric field takes on a negative slope (Fig. 7.56f), with a smaller magnitude at the base side of the junction than at the subcollector side. As the trend progresses, the magnitude of the field base-collector junction eventually diminishes to zero (Fig. 7.56g). When there is no more field to block the holes from "spilling" into the collector, the *base pushout* is said to occur and the current gain falls. The device characteristics as a result of the base pushout are referred to as *Kirk effects*.[19] The above description suggests that the threshold current due to Kirk effects increases if the collector doping increases. However, in many applications where the collector doping may not be increased arbitrarily (so the operating voltage is greater than a certain value), Kirk effects then become an important mechanism affecting the current gain falloff in HBTs. For an HBT with a collector doping of 3×10^{16} cm^{-3} (Fig. 7.50), the threshold current density is roughly $J_C = qN_{coll} \cdot v_{sat} = 3.9 \times 10^4$ A/cm^2 (v_{sat} is ~ 8×10^6 cm/s). Clearly, the value of such threshold current density depends on the magnitude of the saturation velocity. Since the saturation velocity decreases with the ambient temperature, the threshold density due to Kirk effects is lower at higher temperatures.

Kirk effects confine the operating collector current to some values. Similarly, the collector-to-emitter bias (V_{CE}) has its limit, set by two physical phenomena. The first one, well analyzed in elementary device physics, is the avalanche breakdown in the base-collector junction. The base-collector junction is a reverse-biased junction with a finite electric field. The mobile electrons comprised of J_C, while moving through the junction, quickly accelerate and pick up energy from the field. When V_{CE} is small, the magnitude of the field is not too large. The energy the carriers acquired is small and is quickly dissipated in the lattice as the carriers impact upon the lattice atoms. The distance within which a carrier travels between successive impacts with the lattice atoms is called a *mean free path*. As V_{CE} increases such that the electric field approaches $10^5 - 10^6$ V/cm, the energy gained by a carrier within one mean free path can exceed the energy gap of the collector material. As the highly energetic carrier impacts the lattice atoms, the atoms are ionized. The act of a carrier impacting the lattice and thereby creating electron-hole pairs is called *impact ionization*. One single impact ionization creates an electron-hole pair, which leads to further impact ionization as the recently generated carriers also pick up enough energy to ionize the lattice. The net result of this positive feedback mechanism is a rapid rise of I_C, which puts the transistor out of useful (or controllable) range of operation. The V_{CE} corresponding the rapid rise in I_C is called the *breakdown voltage*.

Collapse of Current Gain

The breakdown voltage represents the absolute maximum bias that can be applied to a bipolar transistor. There is, in addition, one more physical phenomenon that further restricts V_{CE} to values smaller than the breakdown voltage. This phenomenon occurs in multi-finger HBTs, having roots in the thermal-electrical interaction in the device. It is termed the *collapse of current gain* (or *gain collapse*) to emphasize the abrupt decrease of current gain observed in measured I-V characteristics when V_{CE} increases past certain values. Figure 7.57 is one such example, measured from a 2-finger HBT. The figure reveals two distinct operating regions, separated by a dotted curve. When V_{CE} is small, I_C decreases gradually with V_{CE}, exhibiting a negative differential resistance (NDR). We briefly describe the cause of NDR, as it relates to the understanding of the gain collapse. The band diagrams in Figs. 7.47 and 7.48 showed that the back-injected holes from the base into the emitter experience a larger energy barrier than the emitter electrons forward injected into the base. The ratio of the desirable I_C to the undesirable $I_{B,\text{back-inject}}$ is proportional to $\exp(\Delta E_g / kT)$ in a graded HBT, and $\exp(\Delta E_v / kT)$ in an abrupt HBT. At room temperature, this ratio is large in either HBT. However, as V_{CE} increases, the power dissipation in the HBT increases, gradually elevating the device temperature above the ambient temperature. $I_C / I_{B,\text{back-inject}}$, and hence the current gain, gradually decrease with increasing V_{CE}. Since Fig. 7.57 is measured I_C for several constant I_B, the decreasing β directly translates to the gradual decrease of I_C.

As V_{CE} crosses the dotted curve, NDR gives in to the collapse phenomenon, as marked by a dramatic lowering of I_C. The *collapse locus*, the dotted curve, is the collection of I_C as a function of V_{CE} at which the gain collapse occurs. When several identical transistors are connected together to common emitter, base, and collector electrodes, we tend to expect each transistor to conduct the same amount of collector current for any biases. Contrary to our intuition, equal conduction takes place only when the power dissipation is low to moderate, such that the junction temperature rise above the ambient temperature is small. At high V_{CE} and/or high I_C operation where the transistor is operated at elevated temperatures, one transistor spontaneously starts to conduct more current than the others (even if all the transistors are ideal and identical). Eventually, one transistor conducts all the current while the others become electrically inactive. This current imbalance originates from a universal bipolar transistor property that as the junction temperature increases, the bias required to turn on some arbitrary current level decreases. Quantitatively, this property is expressed by an empirical expression relating I_C, V_{BE}, and T:

$$I_C = I_{C,\text{sat}} \cdot \exp\left[\frac{q}{kT_0} \cdot \left(V_{BE,\text{junction}} - \phi \cdot (T - T_0) \right) \right] \qquad (7.49)$$

where $I_{C,\text{sat}}$ is the collector saturation current; $V_{BE,\text{junction}}$ is the bias across the base-emitter junction; T_0 is the ambient temperature; and T is the actual device temperature. The degree of the turn-on voltage change in response to a junction temperature change is characterized by ϕ, which is called the *thermal-electrical feedback coefficient*. ϕ decreases logarithmically with I_C^3, and can be approximately as 1.1 mV/°C.

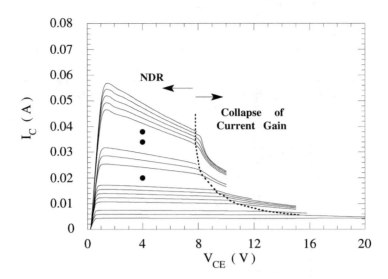

FIGURE 7.57 Measured I-V characteristics of a 2-finger AlGaAs/GaAs HBT showing two distinct regions of operation.

This means that when the junction temperature exceeds the ambient temperature by 1°C, turning on the same amount of I_C requires 1.1 mV less of $V_{BE,\text{junction}}$.

A multi-finger HBT can be viewed as consisting of several identical sub-HBTs, with their respective emitter, base, and collector leads connected together. If one finger (i.e., one sub-HBT) becomes slightly warmer than the others, its base-emitter junction turn-on voltage becomes slightly lower. Consequently, this particular finger conducts more current for a given fixed base-emitter voltage. The increased collector current, in turn, increases the power dissipation in the junction, raising the junction temperature even further. The gain collapse occurs when the junction temperature in one finger of the entire device becomes much hotter than the rest of the fingers, so that the feedback action of increased collector current with junction temperature quickly leads to the fact that just one particular finger conducts the entire device current. Since the transition from uniform current conduction to one finger domination occurs suddenly, the surge in junction temperature in the conducting finger quickly lowers the overall device current gain. The fundamental cause of both NDR and collapse is the current gain lowering at elevated temperatures. Their difference, however, lies in the degree of temperature increase as V_{CE} increases. In the NDR region, all fingers share relatively the same amount of current and the junction temperatures increase gradually with V_{CE}. In contrast, in the collapse region, as the device power is entirely dissipated in one finger and the junction temperature rises sharply, the current gain suddenly plummets.

The equation governing the collapse locus (per unit finger) is given by:[3,20]

$$I_{C,\text{collapse}} = \frac{kT_0}{q} \frac{1}{R_{th} \cdot \phi \cdot V_{CE} - R_E} \tag{7.50}$$

where R_{th} is the thermal resistance per finger and R_E is the emitter resistance per finger. When the individual finger current is below this critical current level (or when $I_{C,\text{collapse}}$ is negative), all fingers share the same amount of current. Above this critical current level, one finger conducts most of the current, whereas the rest of the fingers share the remaining current equally. Equation (7.50) shows that an effective method to increase $I_{C,\text{collapse}}$ is to increase R_E. The portion of the resistance that is intentionally introduced into device fabrication (such as by connecting a TaN thin-film resistor in series with the emitter electrode) is called the *ballasting resistance*. Alternatively, $I_{C,\text{collapse}}$ can be increased by reducing the thermal resistance, a goal often requiring more elaborate processes.[21]

Equation (7.50) neglects the contribution from the base resistance, R_B. For III-V HBTs, it is actually advantageous to use base ballasting; i.e., with the ballasting resistance placed in the base terminal.[22] The reason why the base ballasting approach is undesirable for Si BJT has been analyzed.[3]

The collapse of current gain occurring in III-V HBTs, closely relates to the thermal runaway in Si BJTs. HBTs suffering from gain collapse remain functional and can be biased out of the collapse region by reducing V_{CE}. Si BJTs suffering from thermal runaway, however, die instantly. The bias conditions triggering the thermal runaway in Si BJTs are identically given by the collapse locus equation [Eq. (7.50)]. The main cause of the difference between the collapse in HBTs and thermal runaway in Si BJTs is that the current gain increases with temperature in Si BJTs whereas it decreases with temperature in HBTs.[3]

High Frequency Performance

Current gain is the most important DC parameter characterizing a bipolar transistor. The high frequency properties are generally measured by two figures of merit: f_T, the cutoff frequency, and f_{max}, the maximum oscillation frequency. The cutoff frequency is the frequency at which the magnitude of the AC current gain (small-signal collector current divided by small-signal base current) decreases to unity. As far as analytical expression is concerned, it is easier to work with the related emitter-collector transit time (τ_{ec}), which is inversely proportional to f_T:

$$f_T = \frac{1}{2\pi \, \tau_{ec}}. \tag{7.51}$$

The emitter-collector transit time can be broken up into several components. The emitter charging time, τ_e, is the time required to change the base potential by charging up the capacitances through the differential base-emitter junction resistance:

$$\tau_e = \frac{kT}{q \, I_C} \cdot \left(C_{j,\mathrm{BE}} + C_{j,\mathrm{BC}} \right). \tag{7.52}$$

$C_{j,\mathrm{BE}}$ and $C_{j,\mathrm{BC}}$ denote the junction capacitances of the base-emitter and the base-collector junctions, respectively. The inverse dependence of τ_e on I_C is the primary reason why BJTs and HBTs are biased at high current levels. When the current density is below 10^4 A/cm², this term often dominates the overall collector-emitter transit time.

The second component is the base transit time, the time required for the forward-injected charges to diffuse/drift through base. It is given by,

$$\tau_b = \frac{X_{base}^2}{v \cdot D_{n,base}} \tag{7.53}$$

The value of v depends on the magnitude of the base quasi-electric field. In a uniform base without a base field, v is 2, as suggested by Eq. (7.45). Depending on the amount of energy-gap grading, v can easily increase to 6 to 10.[3]

The space-charge transit time, τ_{sc}, is the time required for the electrons to drift through the depletion region of the base-collector junction. It is given by,

$$\tau_{sc} = \frac{X_{dep}}{2 \, v_{sat}} \tag{7.54}$$

where X_{dep} is the depletion thickness of the base-collector junction. The factor of 2 results from averaging the sinusoidal of carriers current over a time period.[3] It is assumed in the derivation that, because the electric field is large throughout the entire reverse-biased base-collector junction, the carriers travel at a constant saturation velocity, v_{sat}. With a p^- collector layer placed adjacent to the p^+ base of an otherwise conventional *Npn* HBT,[23] the electric field near the base can be made smaller than ε_{crit}. Consequently, the electrons travel at the velocity determined completely by the Γ valley. Without scattering to the L valley, the electrons continue to travel at a velocity that is much larger than v_{sat}, and τ_{sc} is significantly reduced. When carriers travel faster than v_{sat} under an off-equilibrium condition, *velocity overshoot* is said to occur.

The last term, the collector charging time, τ_c, is given by,

$$\tau_c = \left(R_E + R_C \right) \cdot C_{j,\text{BC}}, \tag{7.55}$$

where R_E and R_C are the device emitter and collector resistances, respectively. The value of this charging time depends greatly on the parasitic resistances. This is the term that degrades the HBT's high frequency performance when the contacts are poorly fabricated.

The overall transit time is a sum of the four time constants:

$$\tau_{ec} = \frac{kT}{q\,I_C} \cdot \left(C_{j,\text{BE}} + C_{j,\text{BC}} \right) + \frac{X_{base}^2}{v \cdot D_{n,\text{base}}} + \frac{X_{dep}}{2\,v_{sat}} + \left(R_E + R_C \right) \cdot C_{j,\text{BC}} \tag{7.56}$$

In most HBTs, R_E and R_C are dominated by the electrode resistances; the epitaxial resistances in the emitter and collector layers are insignificant. The cutoff frequency relates to τ_{ec} through Eq. (7.51).

The maximum oscillation frequency is the frequency at which the unilateral power gain is equal to 1. The derivation is quite involved,[3] but the final result is elegant:

$$f_{\max} = \sqrt{\frac{f_T}{8\pi\,R_B C_{j,\text{BC}}}}. \tag{7.57}$$

The base resistance has three components that are roughly equal in magnitude. They are base electrode resistance ($R_{B,\text{eltd}}$); intrinsic base resistance ($R_{B,\text{intrinsic}}$); and extrinsic base resistance ($R_{B,\text{extrinsic}}$). $R_{B,\text{eltd}}$ is intimately related to processing, depending on the contact metal and the alloying recipe used to form the contact. The other two base resistance components, in contrast, depend purely on the designed base layer and the geometrical details. The HBT cross-sections illustrated in Fig. 7.52 show two base contacts placed symmetrically beside the central emitter mesa. For this popular transistor geometry, the intrinsic base resistance is given by,

$$R_{B,\text{intrinsic}} = \frac{1}{12} \times R_{SH,\text{base}} \frac{W_E}{L_E} \tag{7.58}$$

where W_E and L_E are the emitter width and length, respectively; and $R_{SH,\text{base}}$ is the base sheet resistance in Ω/square. Where does the 1/12 factor come from? If all of the base current that goes into one of the base contacts in Fig. 7.52 leaves from the other contact, the intrinsic base resistance would follow our intuitive formula of simple resistance, equal to $R_{SH,\text{base}} \times W_E/L_E$. However, during the actual transistor operation, the base current enters through both contacts, but no current flows out at the other end. The holes constituting the base current gradually decrease in number as they are recombined at the extrinsic base surface, in the base bulk layer, and in the base-emitter space-charge region. Some base current carriers remain in the base layer for a longer portion of the width before getting recombined. Other carriers get recombined sooner. The factor 1/12 accounts for the distributed nature of the current conduction, as derived elsewhere.[3] Because of the way the current flows in the base layer, the intrinsic base resistance is called a *distributed resistance*. If instead there is only one base contact, the factor 1/12 is replaced by 1/3 (not 1/6!).[24] The distributed resistance also exists in the gate of MOS transistors,[25] or III-V field-effect transistors.[4]

The extrinsic base resistance is the resistance associated with the base epitaxial layer between the emitter and the base contacts. It is given by,

$$R_{B,\text{extrinsic}} = \frac{1}{2} \times R_{SH,\text{base}} \frac{S_{BE}}{L_E} \tag{7.59}$$

where S_{BE} is the separation between the base and emitter contacts. The factor 1/2 appears in the transistor shown in Fig. 7.52, which has two base contacts. The presence of $R_{B,\text{extrinsic}}$ is the reason why most transistors are fabricated with self-aligned base-emitter contacts, and that the base contacts are deposited right next to the emitter contacts, regardless of the finite alignment tolerance between the base and emitter contact photolithographical steps. (A detailed fabrication process will be described shortly.) With self-alignment, the distance S_{BE} is minimized, to around 3000 Å.

A detailed calculation of f_T and $f_{\max}$ has been performed for a $W_E \times L_E = 2 \times 30~\mu\text{m}^2$ HBT,[4] with a S_{BE} of 0.2 μm, a base thickness of 800 Å, a base sheet resistance of 280 Ω/square, and a base diffusion coefficient of 25.5 cm²/s. Although device parameters will have different values in other HBT structures and geometries, the exemplar calculation gives a good estimation of the relative magnitudes of various terms that determine a HBTs high frequency performance. We briefly list the key results here. The HBT has: $R_E = 0.45~\Omega$; $R_{B,\text{eltd}} = 7~\Omega$; $R_{B,\text{extrinsic}} = 0.94~\Omega$; $R_{B,\text{intrinsic}} = 1.56~\Omega$; $C_{j,BE} = 0.224$ pF; $C_{j,BC} = 0.026$ pF. It was further determined at the bias condition of $J_C = 10^4$ A/cm² and $V_{BC} = -4$ V, that the collector depletion thickness $X_{dep} = 0.59$ μm. Therefore,

$$\tau_e = \frac{kT}{q\,I_C} \cdot \left(C_{j,BE} + C_{j,BC}\right) = \frac{0.0258\left(2.24 \times 10^{-13} + 2.59 \times 10^{-14}\right)}{1 \times 10^4 \cdot 2 \times 10^{-4} \cdot 3 \times 10^{-4}} = 1.08~\text{ps.}$$

$$\tau_b = \frac{X^2 \text{base}}{v \cdot D_{n,\text{base}}} = \frac{\left(800 \times 10^{-8}\right)^2}{2 \cdot 25.5} = 1.25~\text{ps.}$$

$$\tau_{sc} = \frac{X_{dep}}{2\,v_{sat}} = \frac{0.591 \times 10^{-4}}{2 \cdot 8 \times 10^6} = 3.69~\text{ps.}$$

$$\tau_c = \left(R_E + R_C\right) C_{j,BC} = \left(0.453 + 4.32\right) \cdot 2.59 \times 10^{-14} = 0.124~\text{ps.}$$

Summing up these four components, we find the emitter-collector transit time to be 6.14 ps. The cutoff frequency is therefore $1/(2\pi \cdot 6.14~\text{ps}) = 26$ GHz. As mentioned, although this calculation is specific to a particular power HBT design, the relative magnitudes of the four time constants are quite representative. Generally, the space-charge transit time is the major component of the overall transit time. This is unlike silicon BJTs, whose τ_b and τ_e usually dominate, because of the low base diffusion coefficient in the silicon material and the high base-emitter junction capacitance associated with high emitter doping. HBT design places a great emphasis on the collector design, while the Si BJT design concentrates on the base and emitter layers.

The total base resistance is $R_B = 7 + 0.96 + 1.95 = 9.91~\Omega$. The maximum oscillation frequency, calculated with Eq. (7.57), is 65 GHz.

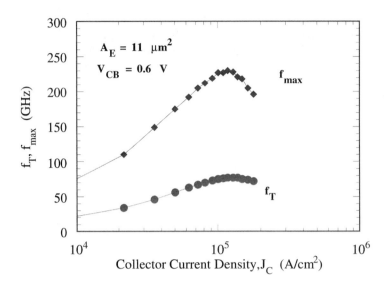

FIGURE 7.58   High-frequency performance of a state-of-the-art InAlAs/InGaAs HBT. (After Chau, H. and Kao, Y., High f_{max} InAlAs/InGaAs heterojunction bipolar transistors, IEEE IEDM, 783–786, 1993, with permission.)

Figure 7.58 illustrates the cutoff and maximum oscillation frequencies of a state-of-art InAlAs/InGaAs HBT. The frequency responses are plotted as a function of the collector current to facilitate the circuit design at the proper bias condition. When the current is small, the cutoff frequency is small because the emitter charging time is large. As the current increases, f_T increases because the emitter charging time decreases. When the current density exceeds roughly 10^5 A/cm^2, Kirk effects take place and the transistor performance degrades rapidly. f_{max}'s current dependence follows f_T's, as governed by the relationship of Eq. (7.57).

The high frequency transistor of the calculated example has a narrow width of 2 μm. What happens if W_E increases to a large value, such as 200 μm? A direct application of Eq. (7.56) would still lead to a cutoff frequency in the GHz range. In practice, such a 200 μm wide device will have a cutoff frequency much smaller than 1 GHz. The reason for the discrepancy is simple; the time constants appearing in Eq. (7.56), which all roughly scale with the emitter area, are based on the assumption that the current conduction is uniform within the entire emitter mesa. In HBTs, and more so in Si BJTs, the base resistance is significant. As the base current flows horizontally from the base contacts to the center of the emitter region, some finite voltage is developed, with a increasingly larger magnitude toward the center of the mesa. The effective base-emitter junction voltage is larger at the edge than at the center. Since the amount of carrier injection depends exponentially on the junction voltage at the given position, most of the injection takes place near the emitter edges, hence the term *emitter crowding*. Sometimes this phenomenon of nonuniform current conduction is referred to as the *base crowding*. Both terms are acceptable, depending on whether the focus is on the crowding of the emitter current or the base current. A figure of merit quantifying the severity of emitter crowding is the effective emitter width (W_{eff}). It is defined as the emitter width that would result in the same current level if current crowding were absent and the emitter current density were uniform at its edge value. Figure 7.59 illustrates the effective emitter width (normalized by the defined emitter width) as a function of base doping and collector current.

Because of the emitter crowding, all high-frequency III-V HBTs capable of GHz performance have a narrow emitter width on the order of 2 μm. For wireless applications in the 2 GHz range, 2 to 3 μm W_E is often used,[26] but a 6.4 μm W_E has also been reported.[27] The choice of the emitter width is a trade-off between the ease (cost) of device fabrication versus the transistor performance. We intuitively expected

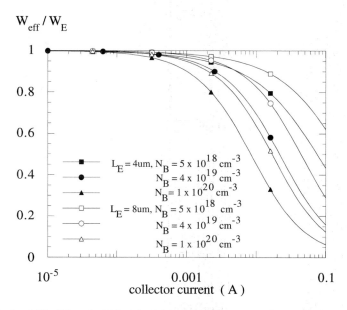

FIGURE 7.59 Calculated W_{eff}/W_E ratio. (After Liu, W. and Harris, J., Dependence of base crowding effect on base doping and thickness for Npn AlGaAs/GaAs heterojunction bipolar transistors, *Electronic Letters*, 27, 2048–2050, 1991, with permission.)

the W_{eff}/W_E ratio to increase as the doping increased (since the base resistance decreased). However, an increase in base doping was accompanied by shortened minority lifetime. The large increase in $I_{B,bulk}$ causes the emitter crowding to be more pronounced as N_{base} increases, as shown in Fig. 7.59. Due to differences in material properties, the emitter width in Si BJT or SiGe HBTs tends to be on the order of 0.5 μm.

Device Fabrication

An HBT mask layout suitable for high-frequency applications is shown in Fig. 7.60, and a corresponding fabrication process is delineated in Fig. 7.61. The following discussion of processing steps assumes the AlGaAs/GaAs HBT epitaxial structure shown in Fig. 7.50. The first step is to deposit ~7000 Å of silicon dioxide, onto which the first-mask pattern "ISOL" is defined, and 1.5 μm of aluminum, evaporated and lifted off. The aluminum protects the silicon dioxide underneath during the subsequent reactive ion etch (RIE), resulting in the profile shown in Fig. 7.61a. Oxygen atoms or protons are then implanted everywhere on the wafer. This implantation is designed to make the region outside of "ISOL" pattern electrically inactive. A shallow etching is applied right after the implantation so that a trail mark of the active device area is preserved. This facilitates the alignment of future mask levels to the first mask. Afterward, both aluminum and oxide are removed with wet etching solutions, reexposing the fresh InGaAs top surface onto which a refractory metal (such as W) is sputtered. The "EMIT" mask level is used to define the emitter mesas, and Ti/Al, a conventional contact metal used in III-V technologies, is evaporated and lifted off. If a refractory metal is not inserted between the InGaAs semiconductor and the Ti/Al metal, long-term reliability problems can arise as titanium reacts with indium. During the ensuing RIE, the refractory metal not protected by the emitter metal is removed.

The resulting transistor profile is shown in Fig. 7.61b. Wet or dry etching techniques are applied to remove the exposed GaAs cap and the AlGaAs active emitter layer, and eventually reach the top of the base layer. Silicon nitride is deposited by plasma enhanced chemical vapor deposition (PECVD). This deposition is conformal, forming a nitride layer everywhere, including the vertical edges. Immediately

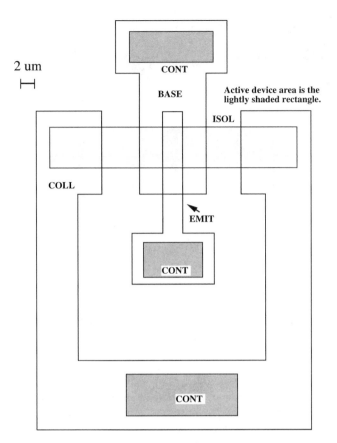

FIGURE 7.60 A HBT mask layout suitable for high-frequency applications. (After Liu, W., Microwave and DC Characterizations of Npn and Pnp HBTs, PhD. Dissertation, Stanford University, Stanford, CA, 1991.)

after the nitride deposition, the whole wafer is etched by RIE. Because a vertical electric field is set up in the RIE chamber to propel the chemical species in the vertical direction, the sidewall nitride covering the sides of the emitter mesas remains untouched while the nitride layer lying on a flat surface is etched away (Fig. 7.61c). The nitride sidewall improves the device yield by cutting down the possible electrical short between the emitter contact and the soon-to-be-deposited base contacts.

The step after the nitride sidewall formation is the "BASE" lithography. As shown in the layout of Fig. 7.60, there is no separation between the "BASE" and the "EMIT" levels. The base metal during the evaporation partly lands on the emitter mesa, and some of it lands on the base layer as desired (Fig. 7.61d). The process has the desired feature that, even if the "BASE" level is somewhat misaligned with the "EMIT" level, the distance S_{BE} between the emitter and base contacts is unchanged, and is at the minimum value determined by the thickness of the nitride sidewall. In this manner, the base contact is said to be *self-aligned* to the emitter. Self-alignment reduces $R_{B,\text{extrinsic}}$ [Eq. 7.59)] and $C_{j,BC}$, hence improving the high-frequency performance. Typically, Ti/Pt/Au is the choice for the base metal.

Following the base metal formation, the collector is defined and contact metal of Au/Ge/Ni/Au is deposited (Fig. 7.61e). After the contact is alloyed at ~450°C for ~1 minute, a polyimide layer is spun to planarize the device as shown in Fig. 7.61f. The contact holes are then defined and Ti/Au is evaporated to contact various electrodes. The final device cross-section after the entire process is shown in Fig. 7.61g.

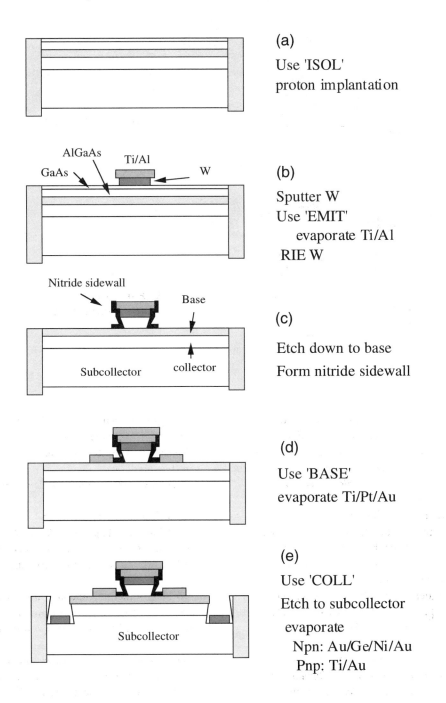

FIGURE 7.61 A high-frequency HBT process flow. (After Liu, W., Microwave and DC Characterizations of Npn and Pnp HBTs, PhD. Dissertation, Stanford University, Stanford, CA, 1991.)

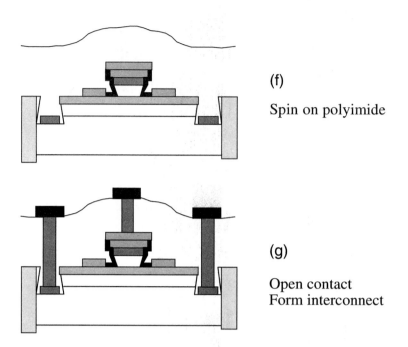

(f)

Spin on polyimide

(g)

Open contact
Form interconnect

FIGURE 7.61 (continued)

References

1. Kroemer, H, Heterostructure bipolar transistors and integrated circuits, *Proc. IEEE*, 70, 13, 1982.
2. Sah, C.T., *Fundamentals of Solid-State Electronics*, World Scientific, Singapore, 1991.
3. Liu, W., *Handbook of III-V Heterojunction Bipolar Transistors*, Wiley & Sons, New York, 1998. An in-depth discussion of several topics can be found in this handbook.
4. Liu, W., *Fundamentals of III-V Devices: HBTs, MESFETs, and HFETs/HEMTs*, Wiley & Sons, New York, 1999.
5. Kroemer, H., Theory of wide gap emitter for transistors, *Proc. IRE*, 45, 1535, 1957.
6. For pioneering HBT papers, see for example, Asbeck, P., Chang, M., Higgins, J., Sheng, N., Sullivan, G., and Wang, K., GaAlAs/GaAs heterojunction bipolar transistors: issues and prospects for application, *IEEE Trans. Electron Devices*, 36, 2032–2041, 1989. For introduction to MBE and MOCVD growth techniques, see References 3 and 4.
7. Henderson, T., Physics of degradation in GaAs-based heterojunction bipolar transistors, *Microelectronics Reliability*, 39, 1033–1042, 1999. See also, Low, T., et al., Migration from an AlGaAs to an InGaP emitter HBT IC process for improved reliability, *IEEE GaAs IC Symposium*, 153–156, 1998.
8. Liu, W. et al., Recent developments in GaInP/GaAs heterojunction bipolar transistors, in *Current Trends in Heterojunction Bipolar Transistors*, Chang, M.F., Ed., World Scientific, Singapore, 1996.
9. Chau, H. and Liu, W., Heterojunction bipolar transistors and circuit applications, in *InP-Based Material and Devices: Physics and Technology*, Wada, O. and Hasegawa, H., Eds., Wiley & Sons, New York, 1999.
10. Ali, F., Gupta, A. and Higgins, A., Advances in GaAs HBT power amplifiers for cellular phones and military applications, *IEEE Microwave and Millimeter-Wave Monolithic Circuits Symposium*, 61–66, 1996.
11. Lin, H. and Lee, S., Super-gain AlGaAs/GaAs heterojunction bipolar transistor using an emitter edge-thinning design, *Appl. Phys. Lett.*, 47, 839–841, 1985.

12. Lee, W., Ueda, D., Ma, T., Pao, Y. and Harris, J., Effect of emitter-base spacing on the current gain of AlGaAs/GaAs heterojunction bipolar transistors, *IEEE Electron Device Lett.*, 10, 200–202, 1989.

13. Patton, G., 75-GHz fr SiGe-base heterojunction bipolar transitors, *IEEE Electron Devices Lett.*, 11, 171–173, 1990.

14. Meyerson, B. et al., Silicon:Germanium heterojunction bipolar transistors; from experiment to technology, in *Current Trends in Heterojunction Bipolar Transistors*, Chang, M., Ed., World Scientific, Singapore, 1996.

15. Pritchard, R., *Electrical Characteristics of Transistors*, McGraw-Hill, New York, 1967.

16. Konig, U., SiGe & GaAs as competitive technologies for RF applications, *IEEE Bipolar Circuit Technology Meeting*, 87–92, 1998.

17. Tiwari, S., Frank, D. and Wright, S., Surface recombination current in GaAlAs/GaAs heterostructure bipolar transistors, *J. Appl. Phys.*, 64, 5009–5012, 1988.

18. Buhanan, D., Investigation of current-gain temperature dependence in silicon transistors, *IEEE Trans. Electron Devices*, 16, 117–124, 1969.

19. Kirk, C., A theory of transistor cutoff frequency falloff at high current densities, *IRE Trans. Electron Devices*, 9, 164–174, 1962.

20. Winkler, R., Thermal properties of high-power transistors, *IEEE Trans. Electron Devices*, 14, 1305–1306, 1958.

21. Hill, D., Katibzadeh, A., Liu, W., Kim, T. and Ikalainen, P., Novel HBT with reduced thermal impedance, *IEEE Microwave and Guided Wave Lett.*, 5, 373–375, 1995.

22. Khatibzadeh, A. and Liu, W., *Base Ballasting*, U.S. Patent Number 5,321,279, issued on June 14, 1994.

23. Maziar, C., Klausmeier-Brown, M. and Lundstrom, M., Proposed structure for collector transit time reduction in AlGaAs/GaAs bipolar transistors, *IEEE Electron Device Lett.*, 7, 483–385, 1986.

24. Hauser, J., The effects of distributed base potential on emitter-current injection density and effective base resistance for stripe transistor geometries, *IEEE Trans. Electron Devices*, 11, 238–242, 1964.

25. Liu, W., *MOSFET Models for SPICE Simulation, Including BSIM3v3 and BSIM4*, Wiley & Sons, New York, in press.

26. RF Microdevices Inc., A high efficiency HBT analog cellular power amplifier, *Microwave Journal*, 168–172, January 1996.

27. Yoshimasu, T., Tanba, N. and Hara, S., High-efficiency HBT MMIC linear power amplifier for L-band personal communication systems, *IEEE Microwave and Guided Wave Lett.*, 4, 65–67, 1994.

7.2.3 Metal-Oxide-Semiconductor Field-Effect Transistors (MOSFETs)

Leonard MacEachern and Tajinder Manku

The insulated-gate field-effect transistor was conceived in the 1930s by Lilienfeld and Heil. An insulated-gate transistor is distinguished by the presence of an insulator between the main control terminal and the remainder of the device. Ideally, the transistor draws no current through its gate (in practice a small leakage current on the order of 10^{-18} A to 10^{-16} A exists). This is in sharp contrast to bipolar junction transistors that require a significant base current to operate. Unfortunately, the Metal-Oxide-Semiconductor Field-Effect Transistor (MOSFET) had to wait nearly 30 years until the 1960s when manufacturing advances made the device a practical reality. Since then, the explosive growth of MOSFET utilization in every aspect of electronics has been phenomenal. The use of MOSFETs in electronics became ever more prevalent when "complementary" types of MOSFET devices were combined by Wanlass in the early 1960s to produce logic that required virtually no power except when changing state. MOSFET processes that offer complementary types of transistors are known as Complementary Metal Oxide Semiconductor (CMOS) processes, and are the foundation of the modern commodity electronics industry. The MOSFET's primary advantages over other types of integrated devices are its mature fabrication technology, its high integration levels, its mixed analog/digital compatibility, its capability for low voltage operation,

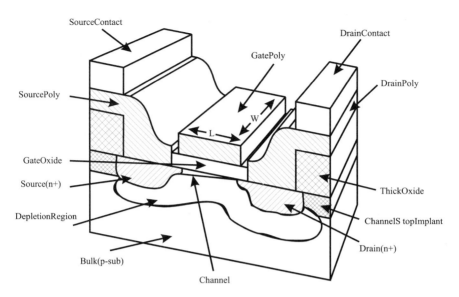

FIGURE 7.62   MOSFET physical structure.

its successful scaling characteristics, and the combination of complementary MOSFETs yielding low power CMOS circuits.

In this section, basic material concerning the MOSFET physical structure and operation is first presented. Nonideal effects are briefly discussed, as are important parasitics and distributed effects found in MOSFETs. Once the important parasitics and distributed effects are understood, the operation of MOSFETs at radio frequencies is examined, and important MOSFET operating parameters are given. Following this, MOSFET noise sources relevant to radio frequency designs are discussed. This section concludes with a discussion of MOSFET design and physical layout appropriate for radio frequency implementations.

MOSFET Fundamentals

Today, each of the tens of millions of MOSFETs that can occupy mere square centimetres of silicon area shares the same basic physical structure. The physical structure of the MOSFET is deceptively simple, as illustrated by the MOSFET cross-section appearing in Fig. 7.62. Visible in the figure are the various materials used to construct a MOSFET. These materials appear in layers when the MOSFET cross-section is viewed; this is a direct consequence of the processes of "doping," deposition, growth, and etching which are fundamental in conventional processing facilities.

The fabrication process of silicon MOSFET devices has evolved over the last 30 years into a reliable integrated circuit manufacturing technology. Silicon has emerged as the material of choice for MOSFETs, largely because of its stable oxide, SiO_2, which is used as a general insulator, as a surface passivation layer, and as an excellent gate dielectric.

Full appreciation of the MOSFET structure and operation requires some knowledge of silicon semiconductor properties and "doping." These topics are briefly reviewed next.

On the scale of conductivity that exists between pure insulators and perfect conductors, semiconductors fall between the extremes. The semiconductor material commonly used to make MOSFETs is silicon. Pure, or "intrinsic" silicon exists as an orderly three-dimensional array of atoms, arranged in a crystal lattice. The atoms of the lattice are bound together by covalent bonds containing silicon valence electrons. At absolute-zero temperature, all valence electrons are locked into these covalent bonds and are unavailable for current conduction, but as the temperature is increased, it is possible for an electron to gain enough thermal energy to escape its covalent bond, and in the process leave behind a covalent bond with a missing electron, or "hole." When that happens the electron that escaped is free to move about the crystal lattice. At the same time, another electron, which is still trapped in nearby covalent bonds because of its lower energy state, can move into the hole left by the escaping electron. The mechanism of current conduction in intrinsic silicon is therefore by hole-electron pair generation, and the subsequent motion of free electrons and holes throughout the lattice.

At normal temperatures intrinsic silicon behaves as an insulator because the number of free electron-hole pairs available for conducting current is very low, only about 14.5 hole-electron pairs per 1000 μm^3 of silicon. The conductivity of silicon can be adjusted by adding foreign atoms to the silicon crystal. This process is called "doping," and a "doped" semiconductor is referred to as an "extrinsic" semiconductor. Depending on what type of material is added to the pure silicon, the resulting crystal structure can either have more electrons than the normal number needed for perfect bonding within the silicon structure, or less electrons than needed for perfect bonding. When the dopant material increases the number of free electrons in the silicon crystal, the dopant is called a "donor." The donor materials commonly used to dope silicon are phosphorus, arsenic, and antimony. In a donor-doped semiconductor the number of free electrons is much larger than the number of holes, and so the free electrons are called the "majority carriers" and the holes are called the "minority carriers." Since electrons carry a negative charge and they are the majority carriers in a donor-doped silicon semiconductor; any semiconductor that is predominantly doped with donor impurities is known as "n-type." Semiconductors with extremely high donor doping concentrations are often denoted "n+ type."

Dopant atoms that accept electrons from the silicon lattice are also used to alter the electrical characteristics of silicon semiconductors. These types of dopants are known as "acceptors." The introduction of the acceptor impurity atoms creates the situation in which the dopant atoms have one less valence electron than necessary for complete bonding with neighboring silicon atoms. The number of holes in the lattice therefore increases. The holes are therefore the majority carriers and the electrons are the minority carriers. Semiconductors doped with acceptor impurities are known as "p-type," since the majority carriers effectively carry a positive charge. Semiconductors with extremely high acceptor doping concentrations are called "p+ type." Typical acceptor materials used to dope silicon are boron, gallium, and indium.

A general point that can be made concerning doping of semiconductor materials is that the greater the dopant concentration, the greater the conductivity of the doped semiconductor. A second general point that can be made about semiconductor doping is that n-type material exhibits a greater conductivity than p-type material of the same doping level. The reason for this is that electron mobility within the crystal lattice is greater than hole mobility, for the same doping concentration.

MOSFET Physical Structure and Operation

A cross-section through a typical n-type MOSFET, or "NFET," is shown in Fig. 7.63(a). The MOSFET consists of two highly conductive regions (the "source" and the "drain") separated by a semiconducting channel. The channel is typically rectangular, with an associated length (L) and width (W). The ratio of the channel width to the channel length, W/L, is an important determining factor for MOSFET performance.

As shown in Fig. 7.63(a), the MOSFET is considered a four-terminal device. These terminals are known as the gate (G), the bulk (B), the drain (D), and the source (S), and the voltages present at these terminals collectively control the current that flows within the device. For most circuit designs, the current flow from drain to source is the desired controlled quantity.

The operation of field-effect transistors (FETs) is based upon the principal of capacitively controlled conductivity in a channel. The MOSFET gate terminal sits on top of the channel, and is separated from the channel by an insulating layer of SiO_2. The controlling capacitance in a MOSFET device is therefore due to the insulating oxide layer between the gate and the semiconductor surface of the channel. The conductivity of the channel region is controlled by the voltage applied across the gate oxide and channel region to the bulk material under the channel. The resulting electric field causes the redistribution of holes and electrons within the channel. For example, when a positive voltage is applied to the gate, it is possible that enough electrons are attracted to the region under the gate oxide, that this region experiences a local inversion of the majority carrier type. Although the bulk material is p-type (the majority carriers

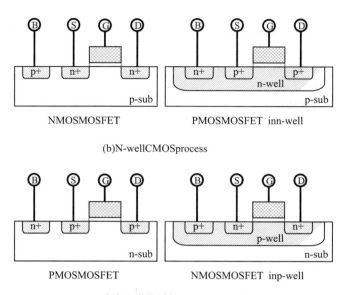

NMOSMOSFET PMOSMOSFET inn-well

(b)N-wellCMOSprocess

PMOSMOSFET NMOSMOSFET inp-well

(b)P-wellCMOSprocess

FIGURE 7.63 MOSFET terminals and active regions. (a) MOSFETs constructed on p-type substrate, (b) MOSFETs constructed on n-type substrate.

are holes and minority carriers are electrons), if enough electrons are attracted to the same region within the semiconductor material, the region becomes effectively n-type. Then the electrons are the majority carriers and the holes are the minority carriers. Under this condition electrons can flow from the n+ type drain to the n+ type source if a sufficient potential difference exists between the drain and source.

When the gate-to-source voltage exceeds a certain threshold, called V_T, the conductivity of the channel increases to the point where current may easily flow between the drain and the source. The value of V_T required for this to happen is determined largely by the dopant concentrations in the channel, but it also depends in part upon the voltage present on the bulk. This dependence of the threshold voltage upon the bulk voltage is known as the "body effect" and is discussed further in another section.

In MOSFETs, both holes and electrons can be used for conduction. As shown in Fig. 7.63, both n-type and p-type MOSFETs are possible. If, for an n-type MOSFET, all the n-type regions are replaced with p-type regions and all the p-type regions are replaced with n-type regions, the result is a p-type MOSFET. Since both the n-type and p-type MOSFETs require substrate material of the opposite type of doping, two distinct CMOS technologies exist, defined by whether the bulk is n-type or p-type. If the bulk material is p-type substrate, then n-type MOSFETs can be built directly on the substrate while p-type MOSFETs must be placed in an n-well. This type of process is illustrated in Fig. 7.63(a). Another possibility is that the bulk is composed of n-type substrate material, and in this case the p-type MOSFETs can be constructed directly on the bulk, while the n-type MOSFETs must be placed in an p-well, as in Fig. 7.63(b). A third type of process known as twin-well or twin-tub CMOS requires that both the p-type and n-type MOSFETs be placed in wells of the opposite type of doping. Other combinations of substrate doping and well types are in common use. For example, some processes offer a "deep-well" capability, which is useful for threshold adjustments and circuitry isolation.

Modern MOSFETs differ in an important respect from their counterparts developed in the 1960s. While the gate material used in the field effect transistors produced thirty years ago was made of metal, the use of this material was problematic for several reasons. At that time, the gate material was typically aluminium and was normally deposited by evaporating an aluminium wire by placing it in contact with a heated tungsten filament. Unfortunately, this method led to sodium ion contamination in the gate oxide, which caused the MOSFET's threshold voltage to be both high and unstable. A second problem with the earlier methods of gate deposition was that the gate was not necessarily correctly aligned with the source and drain regions. Matching between transistors was then problematic because of variability in the gate location with respect to source and drain for the various devices. Parasitics also varied greatly between devices because of this variability in gate location with respect to the source and drain regions. In the worst case, a nonfunctional device was produced because of the errors associated with the gate placement.

Devices manufactured today employ a different gate material, namely "polysilicon," and the processing stages used to produce a field-effect transistor with a poly gate are different than the processing stages required to produce a field-effect transistor with a metal gate. In particular, the drain and source wells are patterned using the gate and field oxide as a mask during the doping stage. Since the drain and source regions are defined in terms of the gate region, the source and drain are automatically aligned with the gate. CMOS manufacturing processes are referred to as self-aligning processes when this technique is used. Certain parasitics, such as the overlap parasitics considered later in this chapter, are minimized using this method. The use of a polysilicon gate tends to simplify the manufacturing process, reduces the variability in the threshold voltage, and has the additional benefit of automatically aligning the gate material with the edges of the source and drain regions.

The use of polysilicon for the gate material has one important drawback: the sheet resistance of polysilicon is much larger than that of aluminium and so the resistance of the gate is larger when polysilicon is used as the gate material. High-speed digital processes require fast switching time from the MOSFETs used in the digital circuitry, yet a large gate resistance hampers the switching speed of a MOSFET. One method commonly used to lower the gate resistance is to add a layer of silicide on top of the gate material. A silicide is a compound formed using silicon and a refractory metal, for example $TiSi_2$. Later in this chapter the importance of the gate resistance upon the radio frequency performance of the MOSFET is examined. In general, the use of silicided gates is required for reasonable radio frequency MOSFET performance.

Although the metal-oxide-semiconductor sandwich is no longer regularly used at the gate, the devices are still called MOSFETs. The term **Insulated-Gate Field-Effect Transistor (IGFET)** is also in common usage.

MOSFET Large Signal Current-Voltage Characteristics

When a bias voltage in excess of the threshold voltage is applied to the gate material, a sufficient number of charge carriers are concentrated under the gate oxide such that conduction between the source and drain is possible. Recall that the majority carrier component of the channel current is composed of charge carriers of the type opposite that of the substrate. If the substrate is p-type silicon then the majority carriers are electrons. For n-type silicon substrates, holes are the majority carriers.

The threshold voltage of a MOSFET depends on several transistor properties such as the gate material, the oxide thickness, and the silicon doping levels. The threshold voltage is also dependent upon any fixed charge present between the gate material and the gate oxides. MOSFETs used in most commodity products are normally the "enhancement mode" type. Enhancement mode n-type MOSFETs have a positive threshold voltage and do not conduct appreciable charge between the source and the drain unless the threshold voltage is exceeded. In contrast, "depletion mode" MOSFETs exhibit a negative threshold voltage and are normally conductive. Similarly, there exists enhancement mode and depletion mode p-type MOSFETs. For p-type MOSFETs the sign of the threshold voltage is reversed.

Equations that describe how MOSFET threshold voltage is affected by substrate doping, source and substrate biasing, oxide thickness, gate material, and surface charge density have been derived in the literature.[1-4] Of particular importance is the increase in the threshold voltage associated with a nonzero source-to-bulk voltage. This is known as the "body effect" and is quantified by the equation,

$$V_T = V_{T0} + \gamma\left(\sqrt{\left|2\phi_F\right| + V_{SB}} - \sqrt{\left|2\phi_F\right|}\right) \qquad (7.60)$$

where V_{T0} is the zero-bias threshold voltage, γ is the body effect coefficient or body factor, ϕ_F is the bulk surface potential, and V_{SB} is the bulk-to-source voltage. Details on the calculation of each of the terms in Eq. (7.60) is discussed extensively in the literature.[5]

Operating Regions

MOSFETs exhibit fairly distinct regions of operation depending upon their biasing conditions. In the simplest treatments of MOSFETs, the three operating regions considered are subthreshold, triode, and saturation.

Subthreshold — When the applied gate-to-source voltage is below the device's threshold voltage, the MOSFET is said to be operating in the subthreshold region. For gate voltages below the threshold voltage, the current decreases exponentially toward zero according to the equation

$$i_{DS} = I_{DS0} \frac{W}{L} \exp\left(\frac{v_{GS}}{nkT/q} \right) \tag{7.61}$$

where n is given by

$$n = 1 + \frac{\gamma}{2\sqrt{\phi_j - v_{BS}}} \tag{7.62}$$

in which γ is the body factor, ϕ_j is the channel junction built-in voltage, and v_{BS} is the source-to-bulk voltage.

For radio frequency applications, the MOSFET is normally operated in its saturation region. The reasons for this will become clear later in this section. It should be noted, however, that some researchers feel that it may be advantageous to operate deep-submicron radio frequency MOSFETs in the moderate inversion (subthreshold) region.[6]

Triode — A MOSFET operates in its triode, also called "linear" region, when bias conditions cause the induced channel to extend from the source to the drain. When $V_{GS} > V_T$ the surface under the oxide is inverted and if $V_{DS} > 0$ a drift current will flow from the drain to the source. The drain-to-source voltage is assumed small so that the depletion layer is approximately constant along the length of the channel. Under these conditions, the drain source current for an NMOS device is given by the relation,

$$I_D = \mu_n C'_{ox} \frac{W}{L} \left(\left(V_{GS} - V_T \right) V_{DS} - \frac{V_{DS}^2}{2} \right) \Bigg|_{\substack{V_{GS} \geq V_T \\ V_{DS} \leq V_{GS} - V_T}} \tag{7.63}$$

where μ_n is the electron mobility in the channel and C'_{ox} is the per unit area capacitance over the gate area. Similarly, for PMOS transistors the current relationship is given as,

$$I_D = \mu_p C'_{ox} \frac{W}{L} \left(\left(V_{SG} - V_T \right) V_{SD} - \frac{V_{SD}^2}{2} \right) \Bigg|_{\substack{V_{SG} \geq V_T \\ V_{SD} \leq V_{SG} - V_T}} \tag{7.64}$$

in which the threshold voltage of the p-type MOSFET is assumed to be positive and μ_p is the hole mobility.

Saturation — The conditions $V_{DS} \leq V_{GS} - V_T$ in Eq. (7.63) and $V_{SD} \leq V_{SG} - V_T$ in Eq. (7.64) ensure that the inversion charge is never zero for any point along the channel's length. However, when $V_{DS} = V_{GS} - V_T$ (or $V_{SD} = V_{SG} - V_T$ in PMOS devices) the inversion charge under the gate at the channel-drain junction is zero. The required drain-to-source voltage is called $V_{DS,sat}$ for NMOS and $V_{SD,sat}$ for PMOS.

For $V_{DS} > V_{DS,sat}$ ($V_{SD} > V_{SD,sat}$ for PMOS), the channel charge becomes "pinched off," and any increase in V_{DS} increases the drain current only slightly. The reason that the drain currents will increase for increasing V_{DS} is because the depletion layer width increases for increasing V_{DS}. This effect is called channel length modulation and is accounted for by λ, the channel length modulation parameter. The channel length modulation parameter ranges from approximately 0.1 for short channel devices to 0.01 for long channel devices. Since MOSFETs designed for radio frequency operation normally use minimum channel lengths, channel length modulation is an important concern for radio frequency implementations in CMOS.

When a MOSFET is operated with its channel pinched off, in other words $V_{DS} > V_{GS} - V_T$ and $V_{GS} \geq V_T$ for NMOS (or $V_{SD} > V_{SG} - V_T$ and $V_{SG} \geq V_T$ for PMOS), the device is said to be operating in the saturation region. The corresponding equations for the drain current are given by,

$$I_D = \frac{1}{2}\mu_n C'_{ox} \frac{W}{L}\left(V_{GS} - V_T\right)^2\left(1 + \lambda\left(V_{DS} - V_{DS,sat}\right)\right) \tag{7.65}$$

for long-channel NMOS devices and by,

$$I_D = \frac{1}{2}\mu_p C'_{ox} \frac{W}{L}\left(V_{SG} - V_T\right)^2\left(1 + \lambda\left(V_{SD} - V_{SD,sat}\right)\right) \tag{7.66}$$

for long-channel PMOS devices.

Figure 7.64 illustrates a family of curves typically used to visualize a MOSFET's drain current as a function of its terminal voltages. The drain-to-source voltage spans the operating region while the gate-to-source voltage is fixed at several values. The bulk-to-source voltage has been taken as zero. As shown in Fig. 7.64, when the MOSFET enters the saturation region the drain current is essentially independent of the drain-to-source voltage and so the curve is flat. The slope is not identically zero however, as the drain-to-source voltage does have some effect upon the channel current due to channel modulation effects.

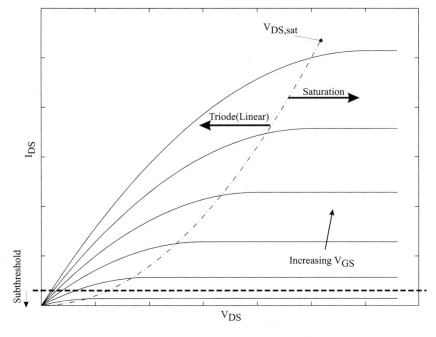

FIGURE 7.64 Characteristic drain-current versus terminal-voltage curves for a MOSFET.

Three simplifying assumptions are advantageous when one is trying to gain an intuitive understanding of the low-frequency operation of a circuit containing MOSFETs. When operated in the linear region, the MOSFET can be treated much like a resistor with terminal voltage V_{DS}. When operated in the saturation region, the MOSFET may be considered a voltage-controlled current source where the controlling voltage is present at the gate. Finally, when operated below the voltage threshold, the MOSFET can be considered an open circuit from the drain to the source.

Nonideal and Short Channel Effects

The equations presented for the subthreshold, triode, and saturation regions of the MOSFET operating characteristic curves do not include the many nonidealities exhibited by MOSFETs. Most of these nonideal behaviors are more pronounced in deep submicron devices such as those employed in radio frequency designs, and so it is important for a designer to be aware of them.

Velocity saturation — Electron and hole mobility are not constants; they are a function of the applied electric field. Above a certain critical electric field strength the mobility starts to decrease, and the drift velocity of carriers does not increase in proportion to the applied electric field. Under these conditions the device is said to be velocity saturated. Velocity saturation has important practical consequences in terms of the current-voltage characteristics of a MOSFET acting in the saturation region. In particular, the drain current of a velocity saturated MOSFET operating in the saturation region is a linear function of V_{GS}. This is in contrast to the results given in Eqs. (7.65) and (7.66). The drain current for a short channel device operating under velocity saturation conditions is given by

$$I_D = \mu_{crit} C'_{ox} W \left(V_{GS} - V_T \right) \tag{7.67}$$

where μ_{crit} is the carrier mobility at the critical electric field strength.

Drain-induced barrier lowering — A positive voltage applied to the drain terminal helps to attract electrons under the gate oxide region. This increases the surface potential and causes a threshold voltage reduction. Since the threshold decreases with increasing V_{DS}, the result is an increase in drain current and therefore an effective decrease in the MOSFET's output resistance. The effects of drain-induced barrier lowering are reduced in modern CMOS processes by using lightly doped drain (LDD) structures.

Hot carriers — Velocity saturated charge carriers are often called hot carriers. Hot carriers can potentially tunnel through the gate oxide and cause a gate current, or they may become trapped in the gate oxide. Hot carriers that become trapped in the gate oxide change the device threshold voltage. Over time, if enough hot carriers accumulate in the gate oxide, the threshold voltage is adjusted to the point that analog circuitry performance is severely degraded. Therefore, depending upon the application, it may be unwise to operate a device so that the carriers are velocity saturated since the reliability and lifespan of the circuit is degraded.

Small Signal Models

Small signal equivalent circuits are useful when the voltage and current waveforms in a circuit can be decomposed into a constant level plus a small time-varying offset. Under these conditions, a circuit can be linearized about its DC operating point. Nonlinear components are replaced with linear components that reflect the bias conditions.

The low-frequency small signal model for a MOSFET is shown in Fig. 7.65. Only the intrinsic portion of the transistor is considered for simplicity. As shown, the small signal model consists of three components: the small signal gate transconductance, g_m; the small signal substrate transconductance, g_{mb}; and the small signal drain conductance, g_d. Mathematically, these three components are defined by

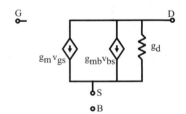

FIGURE 7.65 Low-frequency small signal MOSFET model.

$$g_m = \frac{\partial I_D}{\partial V_{GS}}\bigg|_{V_{BS}, V_{DS}\text{ constant}} \tag{7.68}$$

$$g_{mb} = \frac{\partial I_D}{\partial V_{BS}}\bigg|_{V_{GS}, V_{DS}\text{ constant}} \tag{7.69}$$

and

$$g_d = \frac{\partial I_D}{\partial V_{DS}}\bigg|_{V_{GS}, V_{BS}\text{ constant}} \tag{7.70}$$

Equations (7.68), (7.69), and (7.70) can be evaluated using the relationships given earlier. For the saturation region, the small signal transconductances and the drain conductance are given by,

$$g_m = \mu C'_{ox} \frac{W}{L}\left(V_{GS} - V_T\right) \tag{7.71}$$

$$g_{mb} = g_m \eta \tag{7.72}$$

and

$$g_d = \frac{1}{2}\mu C'_{ox} \frac{W}{L}\left(V_{GS} - V_T\right)^2 \lambda \tag{7.73}$$

where η in Eq. (7.72) is a factor that describes how the threshold voltages changes with reverse body bias. For small V_{BS}, $\eta \approx 0$.

The small signal model shown in Fig. 7.65 is only valid at very low frequencies. At higher frequencies capacitances present in the MOSFET must be included in the small signal model, and at radio frequencies distributed effects must be taken into account. In the next section these two factors are explored and the small signal model is revised.

CMOS at Radio Frequencies

Integrated radio frequency transceiver design is an evolving field, particularly in terms of understanding device performance and maximizing integration level. Typical commercial implementations of highly-integrated high-performance wireless transceivers use a mixture of technologies, including CMOS, BiC-MOS, BJTs, GaAs FETs, and HBTs. Regardless of the technology, all radio frequency integrated circuits contend with the same issues of noise, linearity, gain, and efficiency.

The best technology choice for an integrated radio frequency application must weigh the consequences of wafer cost, level of integration, performance, economics, and time to market. These requirements often lead designers into using several technologies within one transceiver system. Partitioning of transceiver functionality according to technology implies that the signal must go on-chip and off-chip at several locations. Bringing the signal off-chip and then on-chip again complicates the transceiver design because proper matching at the output and input terminals is required. Also, bringing the signal off-chip implies that power requirements are increased because it takes more power to drive an off-chip load than to keep the signal completely on the same integrated circuit. Generally, taking the signal off and then on-chip results in signal power loss accompanied by an undesirable increase in noise figure.

Recent trends apply CMOS to virtually the entire transceiver design, since CMOS excels in its level of integration. The level of integration offered by a particular technology determines the required die size, which in turn affects both the cost and the physical size of the final packaged circuitry. CMOS technology currently has the performance levels necessary to operate in the 900 MHz to 2.4 GHz frequency range, which is important for existing cellular and wireless network applications. Upcoming technologies should be able to operate in the 5 GHz ISM band, which is seen as the next important commodity frequency. CMOS devices manufactured with gate lengths of 0.18 µm will function at these frequencies, albeit with generous biasing currents. Future generations of CMOS scaled below the 100 nm gate length range are anticipated to provide the performance required to operate beyond the 5 GHz frequency range for receiver applications.

High Frequency Modeling

The majority of existing analog MOSFET models predate the use of CMOS in radio frequency designs and generally are unable to predict the performance of MOSFETs operating at microwave frequencies with the accuracy required and expected of modern simulators. These modeling shortcomings occur on two fronts. In the first case, existing models do not properly account for the distributed nature of the MOSFET, meaning that at high frequencies, geometry-related effects are ignored. In the second case, existing models do not properly model MOSFET noise at high frequencies. Typical problems with noise modeling include:

- not accounting for velocity-saturated carriers within the channel and the associated noise;
- discounting the significant thermal noise generated by the distributed gate resistance;
- ignoring the correlation between the induced gate noise and the MOSFET drain noise.

Accurate noise modeling is extremely important when low noise operation is essential, such as in a front-end low noise amplifier.

High Frequency Operation

MOSFET dimensions and physical layout are important determining factors for high frequency performance. As MOSFET operating frequencies approach several hundred MHz, the MOSFET can no longer be considered a lumped device. The intrinsic and extrinsic capacitance, conductance, and resistance are all distributed according to the geometry and physical layout of the MOSFET. The distributed nature of the MOSFET operating at high frequencies is particularly important for the front-end circuitry in a receiver, such as in the low noise amplifier and first stage mixer input MOSFETs. The devices used in these portions of the input circuitry are normally large, with high W/L ratios. Large W/L ratios are required because of the inherently low transconductance offered by CMOS, and in order to realize reasonable gain, the devices are therefore relatively wide compared to more conventional analog circuit layouts. Additionally, minimum gate lengths are preferred because the maximum operating frequency of the MOSFET scales as $1/L^2$. Shorter channels imply higher frequency because the time it takes the carriers to move from drain to source is inversely proportional to the length of the channel. Also, the mobility of the carriers is proportional to the electric field strength. Since the electric field strength along the length of the channel is inversely proportional to the distance between the source and the drain, the carrier mobility is inversely proportional to the length of the channel. Combined, these two effects have traditionally allowed the maximum operating frequency of the MOSFET to scale as $1/L^2$. It must be noted that in modern deep submicrometer MOSFETs experiencing velocity saturation, the maximum operating frequency no longer scales as $1/L^2$, but more closely to $1/L$. In any event, for maximum operating frequency, the device channel length should be the minimum allowable.

Since the gate width in RF front-end MOSFET devices is typically on the order of several hundred microns, the gate acts as a transmission line along its width. The gate acting as a transmission line is modeled similarly to a microstrip transmission line and can be analyzed by utilizing a distributed circuit model for transmission lines. Normally a transmission line is viewed as a two-port network in which the transmission line receives power from the source at the input port (source end) and delivers the power

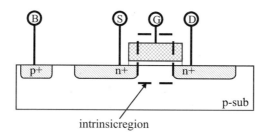

intrinsic region

FIGURE 7.66   MOSFET intrinsic and extrinsic capacitance boundary.

to the load of the output port (load end). In order to apply transmission line analysis to the gate of a MOSFET along its width, the width of the MOSFET gate is divided into many identical sections of incremental width Δx. Each portion of the transmission line with the width Δx is modeled by a resistance "R" per unit width, an inductance "L" per unit width, a capacitance "C" per unit width, and a conductance "G" per unit width. Normally the transmission line is assumed to be uniform, so that these parameters are constants along the transmission line's width. When analyzing signal propagation along the MOSFET gate width, it is important to note that there is no single output node. The transmission line cannot be treated as a two-port, since the gate couples to the channel in a distributed fashion.

Important Parasitics and Distributed Effects

Parasitic capacitances — At high operating frequencies the effects of parasitic capacitances on the operation of the MOSFET cannot be ignored. Transistor parasitic capacitances are subdivided into two general categories; extrinsic capacitances and intrinsic capacitances. The extrinsic capacitances are associated with regions of the transistor outside the dashed line shown in Fig. 7.66, while the intrinsic capacitances are all those capacitances located within the region illustrated in Fig. 7.66.

Extrinsic capacitances — Extrinsic capacitances are modeled by using small-signal lumped capacitances, each of which is associated with a region of the transistor's geometry. Seven small-signal capacitances are used, one capacitor between each pair of transistor terminals, plus an additional capacitor between the well and the bulk if the transistor is fabricated in a well. Figure 7.67(a) illustrates the seven extrinsic transistor capacitances added to an intrinsic small signal model, and Fig. 7.67(b) assigns a location to each capacitance within the transistor structure. In order of importance to high frequency performance, the extrinsic capacitances are as follows:

Gate overlap capacitances — Although MOSFETs are manufactured using a self-aligned process, there is still some overlap between the gate and the source and the gate and the drain. This overlapped area gives rise to the gate overlap capacitances denoted by C_{GSO} and C_{GDO} for the gate-to-source overlap capacitance and the gate-to-drain overlap capacitance, respectively. Both capacitances C_{GSO} and C_{GDO} are proportional to the width, W, of the device and the amount that the gate overlaps the source and the drain, typically denoted as "LD" in SPICE parameter files. The overlap capacitances of the source and the drain are often modeled as linear parallel plate capacitors, since the high dopant concentration in the source and drain regions and the gate material implies that the resulting capacitance is largely bias independent. However, for MOSFETs constructed with a lightly doped drain (LDD-MOSFET), the overlap capacitances can be highly bias dependent and therefore nonlinear. For a treatment of overlap capacitances in LDD-MOSFETs, refer to Klein.[7] For non-lightly doped drain MOSFETs, the gate-drain and gate-source overlap capacitances are given by the expression $C_{GSO} = C_{GDO} = W\,LD\,C_{ox}$, where C_{ox} is the thin oxide field capacitance per unit area under the gate region.

When the overlap distances are small, fringing field lines add significantly to the total capacitance. Since the exact calculation of the fringing capacitance requires an accurate knowledge of the drain and source region geometry, estimates of the fringing field capacitances based on measurements are normally used.

Extrinsic junction capacitances — The bias-dependent junction capacitances that must be considered when evaluating the extrinsic lumped-capacitance values are illustrated in Fig. 7.67(a) and summarized

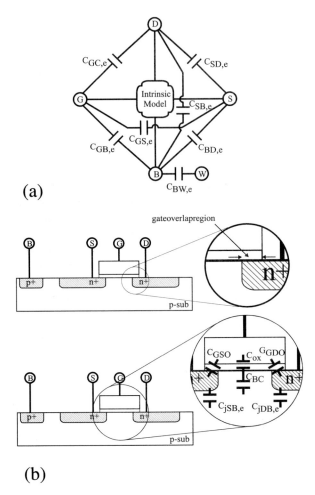

(a)

(b)

FIGURE 7.67 Extrinsic capacitances. (a) Lumped terminal capacitances added to the intrinsic model, (b) physical location of extrinsic capacitances.

in Table 7.4. At the source region there is a source-to-bulk junction capacitance, $C_{jBS,e}$, and at the drain region there is a drain-to-bulk junction capacitance, $C_{jBD,e}$. These capacitances can be calculated by splitting the drain and source regions into a "side wall" portion and a "bottom wall" portion. The capacitance associated with the side wall portion is found by multiplying the length of the side wall perimeter (excluding the side contacting the channel) by the effective side wall capacitance per unit length. Similarly, the capacitance for the bottom wall portion is found by multiplying the area of the bottom wall by the bottom wall capacitance per unit area. Additionally, if the MOSFET is in a well, a well-to-bulk junction capacitance, $C_{jBW,e}$, must be added. The well-bulk junction capacitance is calculated similar to the source and drain junction capacitances, by dividing the total well-bulk junction capacitance into side wall and bottom wall components. If more than one transistor is placed in a well, the well-bulk junction capacitance should only be included once in the total model.

Both the effective side wall capacitance and the effective bottom wall capacitance are bias dependent. Normally the per unit length zero-bias, side wall capacitance and the per unit area zero-bias, bottom wall capacitance are estimated from measured data. The values of these parameters for nonzero reverse-bias conditions are then calculated using the formulas given in Table 7.4.

Extrinsic source-drain capacitance — Accurate models of short channel devices may include the capacitance that exists between the source and drain region of the MOSFET. As shown in Fig. 7.67(a), the source-drain capacitance is denoted as $C_{sd,e}$. Although the source-drain capacitance originates in the

TABLE 7.4 MOSFET Extrinsic Junction Capacitances

Extrinsic Source Capacitance	$C_{jBS,e} = C'_{jBS}A_S + C''_{jswBS}P_S$	$C'_{jBS} = \dfrac{C'_j}{\left(1 - \dfrac{V_{BS}}{\phi_j}\right)^{m_j}}$	$C''_{jswBS} = \dfrac{C''_{jsw}}{\left(1 - \dfrac{V_{BS}}{\phi_{jsw}}\right)^{m_{jsw}}}$
Extrinsic Drain Capacitance	$C_{jBD,e} = C'_{jBD}A_D + C''_{jswBD}P_D$	$C'_{jBD} = \dfrac{C'_j}{\left(1 - \dfrac{V_{BD}}{\phi_j}\right)^{m_j}}$	$C''_{jswBD} = \dfrac{C''_{jsw}}{\left(1 - \dfrac{V_{BD}}{\phi_{jsw}}\right)^{m_{jsw}}}$
Extrinsic Well Capacitance	$C_{jBW,e} = C'_{jBW}A_W + C''_{jswBW}P_W$	$C'_{jBW} = \dfrac{C'_j}{\left(1 - \dfrac{V_{BW}}{\phi_j}\right)^{m_j}}$	$C''_{jswBW} = \dfrac{C''_{jsw}}{\left(1 - \dfrac{V_{BW}}{\phi_{jsw}}\right)^{m_{jsw}}}$

Notes: m_j and m_{jsw} are process dependent, typically 1/3 … 1/2.

$$C'_j = \sqrt{\frac{\varepsilon_{si}qN_B}{2\phi_j}}$$

Where: ϕ_j and ϕ_{jsw} are the built-in junction potential and side wall junction potentials, respectively. ε_{si} is the dielectric constant of silicon, q is the electronic charge constant, and N_B is the bulk dopant concentration.

$$C''_{jsw} = \sqrt{\frac{\varepsilon_{si}qN_B}{2\phi_{jsw}}}$$

A_S, A_D, and A_W are the source, drain, and well areas, respectively.
P_s, P_d, and P_w are the source, drain, and well perimeters, respectively. The source and drain perimeters do not include the channel boundary.

region within the dashed line in Fig. 7.66, it is still referred to as an extrinsic capacitance.[1] The value of this capacitance is difficult to calculate because its value is highly dependent upon the source and drain geometries. For longer channel devices, $C_{sd,e}$ is very small in comparison to the other extrinsic capacitances, and is therefore normally ignored.

Extrinsic gate-bulk capacitance — As with the gate-to-source and gate-to-drain overlap capacitances, there is a gate-to-bulk overlap capacitance caused by imperfect processing of the MOSFET. The parasitic gate-bulk capacitance, $C_{GB,e}$, is located in the overlap region between the gate and the substrate (or well) material outside the channel region. The parasitic extrinsic gate-bulk capacitance is extremely small in comparison to the other parasitic capacitances. In particular, it is negligible in comparison to the intrinsic gate-bulk capacitance. The parasitic extrinsic gate-bulk capacitance has little effect on the gate input impedance and is therefore generally ignored in most models.

Intrinsic capacitances — Intrinsic MOSFET capacitances are significantly more complicated than extrinsic capacitances because they are a strong function of the voltages at the terminals and the field distributions within the device. Although intrinsic MOSFET capacitances are distributed throughout the device, for the purposes of simpler modeling and simulation, the distributed capacitances are normally represented by lumped terminal capacitances. The terminal capacitances are derived by considering the change in charge associated with each terminal with respect to a change in voltage at another terminal, under the condition that the voltage at all other terminals is constant. The five intrinsic small signal capacitances are therefore expressed as,

$$C_{gd,i} = \left.\frac{\partial Q_G}{\partial V_D}\right|_{V_G, V_S, V_B} \tag{7.74}$$

$$C_{gs,i} = \left.\frac{\partial Q_G}{\partial V_S}\right|_{V_G, V_D, V_B} \tag{7.75}$$

TABLE 7.5 Intrinsic MOSFET Capacitances

Operating Region	$C_{gs,i}$	$C_{gd,i}$	$C_{gb,i}$	$C_{bs,i}$	$C_{bd,i}$
Triode	$\approx \frac{1}{2}C_{ox}$	$\approx \frac{1}{2}C_{ox}$	≈ 0	$k_0 C_{ox}$	$k_0 C_{ox}$
Saturation	$\approx \frac{2}{3}C_{ox}$	≈ 0	$K_1 C_{ox}$	$K_2 C_{ox}$	≈ 0

Notes: Triode region approximations are for $V_{DS} = 0$.
k_0, k_1, and k_2 are bias dependent. See [1].

$$C_{bd,i} = \left. \frac{\partial Q_B}{\partial V_D} \right|_{V_G, V_S, V_B} \tag{7.76}$$

$$C_{bs,i} = \left. \frac{\partial Q_B}{\partial V_D} \right|_{V_G, V_D, V_B} \tag{7.77}$$

and

$$C_{gb,i} = \left. \frac{\partial Q_G}{\partial V_B} \right|_{V_G, V_S, V_D} \tag{7.78}$$

These capacitances are evaluated in terms of the region of operation of the MOSFET, which is a function of the terminal voltages. Detailed models for each region of operation were investigated by Cobbold.[8] Simplified expressions are given in Table 7.5, for the triode and saturation operating regions. The total terminal capacitances are then given by combining the extrinsic capacitances and intrinsic capacitances according to,

$$C_{gs} = C_{gs,i} + C_{gs,e} = C_{gs,i} + C_{gso}$$

$$C_{gd} = C_{gd,i} + C_{gd,e} = C_{gd,i} + C_{gdo}$$

$$C_{gb} = C_{gb,i} + C_{gb,e} = C_{gb,i} + C_{gbo} \tag{7.79}$$

$$C_{sb} = C_{bs,i} + C_{sb,e} = C_{bs,i} + C_{jsb}$$

$$C_{db} = C_{bd,i} + C_{db,e} = C_{bd,i} + C_{jdb}$$

in which the small-signal form of each capacitance has been used.

The contribution of the total gate-to-channel capacitance, C_{GC}, to the gate-to-drain and gate-to-source capacitances is dependent upon the operating region of the MOSFET. The total value of the gate-to-channel capacitance is determined by the per unit area capacitance C_{ox} and the effective area over which the capacitance is taken. Since the extrinsic overlap capacitances include some of the region under the gate, this region must be removed when calculating the gate-to-channel capacitance. The effective channel length, L_{eff}, is given by $L_{eff} = L - 2LD$ so that the gate-to-channel capacitance can be calculated by the formula $C_{GC} = C_{ox} W L_{eff}$. The total value of the gate-to-channel capacitance is apportioned to both the drain and source terminals according to the operating region of the device. When the device is in the triode region, the capacitance exists solely between the gate and the channel and extends from the drain to the source. Its value is therefore evenly split between the terminal capacitances C_{gs} and C_{gd} as shown in Table 7.5. When the device operates in the saturation region, the channel does not extend all the way from the source to the drain. No portion of C_{GC} is added to the drain terminal capacitance under these circumstances. Again, as shown in Table 7.5, analytical calculations demonstrated that an appropriate amount of C_{GS} to include in the source terminal capacitance is 2/3 of the total.[1]

Finally, the channel to bulk junction capacitance, C_{BC}, should be considered. This particular capacitance is calculated in the same manner as the gate-to-channel capacitance. Also similar to the gate-to-channel capacitance proportioning between the drain in the source when calculating the terminal capacitances, the channel-to-bulk junction capacitance is also proportioned between the source-to-bulk and drain-to-bulk terminal capacitances, depending on the region of operation of the MOSFET.

Wiring capacitances — Referring to Fig. 7.62, one can see that the drain contact interconnect overlapping the field oxide and substrate body forms a capacitor. The value of this overlap capacitance is determined by the overlapping area, the fringing field, and the oxide thickness. Reduction of the overlapping area will decrease the capacitance to a point, but with an undesirable increase in the parasitic resistance at the interconnect to MOSFET drain juncture. The parasitic capacitance occurring at the drain is particularly troublesome due to the Miller effect, which effectively magnifies the parasitic capacitance value by the gain of the device. The interconnects between MOSFET devices also add parasitic capacitive loads to the each device. These interconnects may extend across the width of the IC in the worst case, and must be considered when determining the overall circuit performance.

Modern CMOS processes employ thick field-oxides that reduce the parasitic capacitance that exists at the drain and source contacts, and between interconnect wiring and the substrate. The thick field-oxide also aids in reducing the possibility of unintentional MOSFET operation in the field region.

Distributed gate resistance — Low-frequency MOSFET models treat the gate as purely capacitive. This assumption is invalid for frequencies beyond approximately 1 GHz, because the distributed gate resistance is typically larger than the capacitive reactance present at the gate input for frequencies beyond 1 GHz.

The impact of the distributed gate resistance upon the high frequency performance of MOSFETs has been investigated both experimentally and analytically by several researchers.[1,6,9–15] The distributed gate resistance affects the radio frequency performance of the MOSFET in three primary ways. In the first case, discounting the gate resistance causes nonoptimal power matching to off-chip source impedances. In the second case, discounting the distributed gate resistance in noise figure calculations causes an underestimation of the noise figure of the transistor. Finally, in the third case, since the power gain of the MOS transistor is strongly governed by the gate resistance, discounting the gate resistance causes an overestimation of the MOSFET's available power gain and maximum oscillation frequency. The gate resistance of MOSFET transistors therefore deserves important consideration during the design phase of integrated RF CMOS receivers.

Nonzero gate resistances have been factored into recent successful designs. Rofougaran et al.[16] noted that matching, input noise, and voltage gain are all ultimately limited by transistor imperfections such as gate resistance. The effects were most recently quantified by Enz.[6]

Channel charging resistance — Charge carriers located in the channel cannot instantaneously respond to changes in the MOSFET gate-to-source voltage. The channel charging resistance is used to account for this non-quasi-static behavior along the channel length. In Bagheri et al.,[15] the channel charging resistance was shown to be inversely proportional to the MOSFET transconductance, $r_i \approx (kg_m)^{-1}$. For long channel devices, with the distributed nature of the channel resistance between the source and drain taken into account, the constant of proportionality, k, was shown to equal five. Measurements of short channel devices indicate that the proportionality constant can go as low as one.

The channel charging resistance of a MOSFET is important because it strongly influences the input conductance and the forward transconductance parameters of the device. Both the input conductance and the forward transconductance are monotonically decreasing functions of the channel charging resistance. Since the transconductance of even a large MOSFET is small, on the order of 10 mS, the charging resistance of typical front-end transistors is large, potentially on the order of hundreds of ohms.

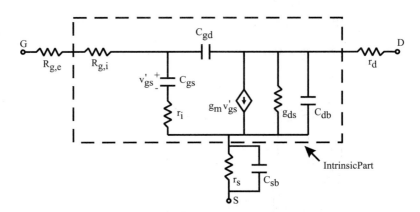

FIGURE 7.68   MOSFET small-signal high-frequency model.

Transconductance delay — MOSFET transconductance does not respond instantaneously to changes in gate voltage. The time it takes for the charge in the channel to be redistributed after an excitation of the gate voltage is dictated by the time constant due to the gate-to-source capacitance and the channel charging resistance. This time constant is denoted as τ, and is given by the expression $\tau \approx r_i C_{gs}$. The transconductance delay is generally ignored for frequencies less than $2\pi/\tau$.

Small Signal Models

Several high frequency small signal models incorporating the effects described in the previous sections have been proposed in the literature. The small signal model presented in Fig. 7.68 is useful for MOSFETs operating in saturation in a common-source configuration. Evident in the figure are the various lumped terminal capacitances, the drain conductance, and the output transconductance. Note that the transconductance g_{mb} is taken as zero because $V_{SB} = 0$ is assumed. Also evident in the model are the high-frequency related elements, namely the charging resistance, the transconductance delay, and the extrinsic and intrinsic gate resistances.

MOSFET Small Signal Y-parameters

Small-signal y-parameters are useful in radio frequency design work involving MOSFETs. As discussed later in this section, radio frequency MOSFETs are typically laid out in a "fingered" style, and if the y-parameters are found for a single finger, they are easily combined for the complete device.

The y-parameters corresponding to the small signal equivalent circuit shown in Fig. 7.68 have been evaluated and appear in Table 7.6. These y-parameters cannot accurately portray the distributed nature of the MOSFET at high frequencies because the model presented in Fig. 7.68 is composed of lumped elements. Recall that the gate resistance in MOS transistors operating at GHz frequencies in conjunction with intrinsic and extrinsic device capacitances acts as a distributed RC network. This is shown schematically in Fig. 7.69, where a MOSFET is represented as a network of smaller MOSFETs, interconnected by gate material.

Several models developed over the last three decades incorporate high frequency and distributed effects. Noise arising from the distributed gate was modeled by Jindal,[12] but high frequency effects were not incorporated in this model. Distributed geometry effects were recognized as important,[14,15] but the distributed nature of the gate was not fully explored analytically.[10] As the viability of CMOS in RF transceiver applications improved, significant progress in modeling wide devices, such as those required for RF applications, was made by Kim et al.[11] and Razavi et al.,[9] in which wide transistors were treated as arrays of smaller transistors interconnected by resistors representing the gate. Recently in a paper by Abou-Allam,[17,18] a closed-form, small signal model incorporating the distributed gate for wide transistors was derived, taking into account the distributed nature of the gate resistance and intrinsic capacitances. The y-parameters developed in Abou-Allam's paper[17] appear here in Table 7.6.

TABLE 7.6 MOSFET Small-Signal Y-Parameters

Parameter	Y-Parameters for the Intrinsic MOSFET Model of Fig. 7.68	Distributed MOSFET Y-Parameters*
y_{11}	$\dfrac{s\left(\kappa + C_{gd}\right)}{1 + s\left(\kappa + C_{gd}\right)R_{g,i}}$	$\dfrac{s\left(C_{gs} + C_{gd}\right) + s^2 C_{gs} C_{gd} r_i}{1 + s C_{gs} r_i} \dfrac{\tanh\left(\gamma W\right)}{\gamma W}$
y_{12}	$\dfrac{-s C_{gd}}{1 + s\left(\kappa + C_{gd}\right)R_{g,i}}$	$-s C_{gd} \dfrac{\tanh\left(\gamma W\right)}{\gamma W}$
y_{21}	$\dfrac{g_m - s C_{gd}}{1 + s\left(\kappa + C_{gd}\right)R_{g,i}}$	$\left(\dfrac{g_m}{1 + s C_{gs} r_i} - s C_{gd}\right) \dfrac{\tanh\left(\gamma W\right)}{\gamma W}$
y_{22}	$g_d + s C_{gd} + \dfrac{\left(g_m - s\left(\kappa - C_{gd}\right)\right) s C_{gd} R_{g,i}}{1 + s\left(\kappa + C_{gd}\right)R_{g,i}}$	$g_d + s C_{gd} + \dfrac{\left(\dfrac{g_m}{1 + s C_{gs} r_i} - s C_{gd}\right) C_{gd}}{C_{gd} + \dfrac{C_{gs}}{1 + s C_{gs} r_i}}\left(1 - \dfrac{\tanh\left(\gamma W\right)}{\gamma W}\right)$
Notes:	$\kappa = \dfrac{C_{gs}}{1 + s C_{gs} r_i}$	$\gamma = \dfrac{\sqrt{s C_{gd} r_i \dfrac{s C_{gs} r_i}{1 + s C_{gs} r_i}}}{W}$

*Source: Abou-Allam, E. and Manku, T., An Improved Transmission-Line Model for MOS Transistors, *IEEE Transactions on Circuits and Systems II: Analog and Digital Signal Processing*, 46, 1380–1387, Nov. 1999.

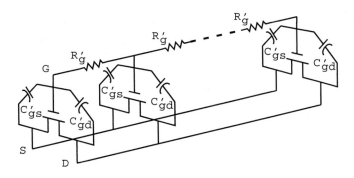

FIGURE 7.69 Illustration of distributed MOSFET network model.

A parallel between the results presented by Razavi et al.[9] and Abou-Allam[18] was drawn in a paper by Tin et al.,[19] in which a useful small signal lumped circuit model was presented and leads to the model here in Fig. 7.70. The lumped model shown in Fig. 7.70 incorporates the distributed effects represented by the $\tanh(\gamma W)/\gamma W$ factor within the expressions for the y-parameters presented in Table 7.6.

The distributed gate resistance appears as a lumped resistor of value $R_g/3$ and the distributed intrinsic capacitances appear as a lumped capacitor with value $C_g/5$. It is important to note that these expressions were derived for a gate connected from one end only. For example, when the gate is connected from both ends, the equivalent resistor changes to $R_g/12$.

The performance limitations imposed by distributed effects at radio frequencies was summarized by Manku.[20] Analysis of a two-port constructed from the y-parameters as given in Table 7.6 for a traditional MOSFET yields several important device performance metrics, as now discussed.

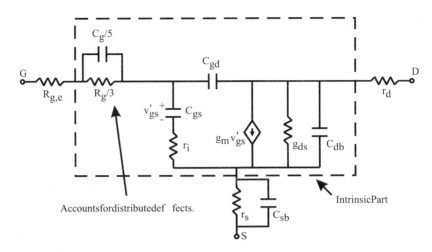

FIGURE 7.70   A lumped small-signal MOSFET model that accounts for distributed gate resistance effects.

Unity Current Gain Frequency: f_t

The unity current gain frequency is defined as the signal input frequency at which the extrapolated small-signal current gain of the MOSFET equals one. The small-signal current gain is defined as the amplitude of the small-signal drain current to the small-signal gate current. The symbol used in the literature to denote the unity current gain frequency is f_t and is read as the "transit frequency."

The unity current gain frequency, or transit frequency, is used as a benchmark to describe the speed of the intrinsic device. The performance of the complete device, which includes the additional effects of the extrinsic parasitics, is always lower. For linear amplifier configurations, the small-signal unity current gain is of primary concern since it determines the maximum achievable gain-bandwidth product of the amplifier. Small signal linear two-port models are useful for estimating the unity current gain frequency. The value of f_t is most easily found from the y-parameters using the relation,[13]

$$\left| \frac{y_{21}}{y_{11}} \right| = 1 \tag{7.80}$$

which holds when the current gain is unity. From y-parameters given in Table 7.6, the value of f_t is found as,

$$f_t = \frac{g_m}{2\pi\sqrt{C_g^2 - \left(g_m R_{gi} C_{gs} - C_{gs}\right)^2}} \approx \frac{g_m}{2\pi C_g}\bigg|_{C_g = C_{gs} + C_{gd}} \tag{7.81}$$

where $C_g^2 \gg (g_m R_{gi} C_{gs} - C_{gs})^2$ is assumed.

Note that the unity current gain frequency is independent of the distributed gate resistance. The most important determining factors of the unity current gain frequency are the device transconductance and the gate-to-source and gate-to-drain capacitances. Since the device f_t is directly proportional to g_m, the analog circuit designer can trade power for speed by increasing the device bias current and therefore g_m. Note that f_t cannot be arbitrarily increased by an increase in drain-source bias current; eventually g_m becomes independent of I_{DS}, and therefore f_t becomes independent of the bias current. MOSFETs used at radio frequencies are normally operated in saturation, because g_m is maximum for a given device in the saturation region. Recall that g_m can also be increased by increasing the width of the transistor, but this will not increase f_t because the parasitic capacitance C_g is proportional to the width of the device.

The transit frequency must not be confused with another often used performance metric, the "intrinsic cutoff frequency," f_τ. While both the "t" in f_t and the "τ" in f_τ refer to the carrier transit time along the length of the device channel, the two symbols have decidedly different meanings. The intrinsic cutoff frequency is given by $f_{\tau = (2\pi\tau)^{-1}}$ where τ is the mean transit time of the carriers along the channel length. Typically, f_τ is five or six times larger than f_t.[6]

Maximum Available Power Gain: G_{max}

Maximum port-to-port power gain within a device occurs when both the input and the output ports are matched to the impedance of the source and the load, respectively. The maximum available power gain of a device, G_{max}, provides a fundamental limit on how much power gain can be achieved with the device. The maximum available power gain achievable within a linear two-port network can be described in terms of the two-port y-parameters as,[13]

$$G_{max} = \frac{1}{4} \frac{\left| y_{21} - y_{12} \right|^2}{\text{Re}\left(y_{11}\right)\text{Re}\left(y_{22}\right) - \text{Re}\left(y_{21}\right)\text{Re}\left(y_{12}\right)} \tag{7.82}$$

where the two-port system is assumed to be unconditionally stable. Treating a common-source MOSFET as a two-port and using the y-parameters shown in Table 7.6 and the relation for f_t given in Eq. (7.81), the MOSFET G_{max} is derived in terms of the intrinsic device parameters as,

$$G_{max} \approx \frac{\left(f_t/f\right)^2}{4R_{g,i}\left(g_{ds} + g_m C_{gd}/C_g\right) + 4r_i g_{ds}} \tag{7.83}$$

$$\approx \frac{f_t}{8\pi R_{g,i} C_{gd} f^2}$$

in which the simplifying assumption $(R_{g,i} + r_i)g_{ds} \ll g_m R_{g,i} C_{gd}/C_g$ is made. To first order, the maximum achievable power gain of a MOSFET is proportional to the device's f_t, and the maximum achievable power gain is inversely proportional to the device's intrinsic gate resistance, $R_{g,i}$. Note that linear amplification is assumed in Eqs. (7.82) and (7.83). Equation (7.83) therefore applies to small-signal analyses such as for a receiver front-end, and would possibly be inadequate for describing the maximum power gain achievable using a nonlinear power amplifier in the transmitter portion of a transceiver.

Unity Power Gain Frequency: f_{max}

The third important figure of merit for MOSFET transistors operating at radio frequencies is the maximum frequency of oscillation, f_{max}. This is the frequency at which the maximum available power gain of the transistor is equal to one. An estimate of f_{max} for MOSFET transistors operating in saturation is found from Eq. (7.83) by setting $G_{max} = 1$ and is given by,

$$f_{max} = \sqrt{\frac{f_t}{8\pi R_{g,i} C_{gd}}} \tag{7.84}$$

From a designer's perspective, f_{max} can be optimized by a combination of proper device layout and careful choice of the operating bias point. From Eq. (7.81) $f_t \propto g_m$, therefore $f_{max} \propto \sqrt{g_m}$, and hence g_m should be maximized. Maximizing g_m requires some combination of increasing the gate-source overdrive, increasing device bias current I_{DS}, or increasing the device width. Note that increasing the device width will increase $R_{g,i}$ and reduce f_{max} unless a fingered layout is used.

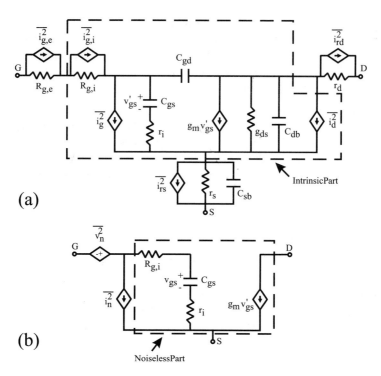

(a)

(b)

FIGURE 7.71 MOSFET small-signal noise model. (a) High-frequency small-signal model augmented with noise sources, (b) input-referred noise model.

MOSFET Noise Sources

The intrinsic and extrinsic noise sources in a MOSFET operating at microwave frequencies are predominantly thermal in origin. The extrinsic noise arises from the parasitic resistances found at the MOSFET terminal connections, and the metal to semiconductor junctions found at the contacts at these terminals. The intrinsic noise arises from three important sources:

- Drain channel noise, $\overline{i_d^2}$, which is the noise generated by the carriers in the channel region and appears as a noise current;
- Gate resistance noise, $\overline{v_{r_g}^2}$, which is the thermal noise due to the resistance of the distributed gate material;
- Induced gate noise, $\overline{i_g^2}$, which is a gate noise current that is capacitively coupled onto the gate from the distributed noise generated by the carriers in the channel and the distributed channel charging resistance. Induced gate noise is one of the main components of noise within the intrinsic portion of a MOSFET transistor.

MOSFET Noise Models

The noise sources discussed previously can be added to an appropriate lumped network model of a MOSFET transistor. The small-signal equivalent circuit shown in Fig. 7.71(a) incorporates the noise sources that are important at radio frequencies. The model is essentially the same model as presented in Fig. 7.68 except for the inclusion of the thermally generated noise. It is possible to refer all of the internal MOSFET noise sources to the gate input as shown in Fig. 7.71(b). This procedure is discussed in a later section for optimum noise matching. The various noise sources are now described.

Gate Resistance Noise

The resistance of the gate material, given by $R_g = (W \times R_\square)/(N \times L)$, where $R_\square$ is the sheet resistance of the gate material, contributes to the thermal noise present in the device. The lumped resistor shown at the gate of the MOSFET in Figs. 7.68 and 7.71(a) is intended to represent the thermal noise resistance of the distributed gate material. The value of this resistor is dependent upon whether the gate is connected from one end or both ends, and can be approximated analytically by treating the gate as a transmission line along its width, since for practical radio frequency MOSFET dimensions $W \gg L$. In the case of the gate connected from one end only, $R_{g,i} = R_g/3$. When the gate is connected from both ends, $R_{g,i} = R_g/12$. In terms of the equivalent gate resistance and the transistor's dimensions, the gate resistance noise is given by,

$$\overline{v_{g,i}^2} = 4kTR_{g,i}\Delta f \tag{7.85}$$

where k is Boltzmann's constant and T is the temperature in Kelvin. Note that the gate resistance noise is proportional to the width, inversely proportional to the length of the device, and scales in inverse proportion to the number of fingers used in the transistor layout.

Thermal Channel Noise

Thermal noise within the channel produces both drain channel noise and induced gate noise. Since both the channel drain noise and the induced gate noise are generated by the same physical noise sources, they exhibit a degree of correlation. The normalized correlation coefficient between the drain current noise and the gate current noise is in general a complex quantity given by the expression,[21]

$$c = \frac{\overline{i_g i_d^*}}{\sqrt{\overline{i_g i_g^*} \cdot \overline{i_d i_d^*}}} \tag{7.86}$$

Simulations and experimental measurements indicate that the real part of "c" in Eq. (7.86) is approximately equal to zero. Intuitively this makes sense because the gate noise is induced from the channel current noise capacitively. There is, therefore, a 90° phase shift between the induced gate current noise and the source of this noise, which is the channel current noise.

The value of "c" has been found for low frequencies and longer gate lengths as c = –j0.395.[21] For submicron MOSFETs, and as the frequency of operation increases, c approaches –j0.3.

Channel noise — The drain channel noise is a complicated function of the transistor bias conditions. For radio frequency applications, the MOSFET is assumed to operate in the saturation region and the drain channel noise is given approximately by

$$\overline{i_d^2} = 4kT\gamma g_{do}\Delta f \tag{7.87}$$

where γ is a bias dependent parameter and g_{do} is the zero drain voltage conductance of the channel. Usually $g_{do} = g_m$ is assumed. The factor γ is an increasing function of V_{DS}, but a value of 2/3 is often used for hand calculations and simple simulations. For quasi-static MOSFET operation, $\overline{i_d^2}$ is essentially independent of frequency.

Induced gate noise — Fluctuations in the channel are coupled to the transistor gate via the oxide capacitance. This produces a weak noise current at the gate terminal. The mean square value of this noise current was evaluated by van der Ziel[21] and is approximated by,

$$\overline{i_g^2} = 4kT\beta g_{do}\frac{(\omega C_{gs})^2}{k_{gs}g_{do}}\Delta f \tag{7.88}$$

where β is a bias-dependent parameter typically greater than or equal to 4/3. The factor $1/k_{gs}$ arises from a first-order expansion that gives $k_{gs} = 5$ for long channel devices. Interestingly, the induced gate noise is proportional to the square of the frequency. Clearly this expression cannot hold as the frequency becomes extremely large. The expression given in Eq. (7.88) is valid up to approximately $\frac{2}{3}f_t$.

1/f-Noise

Experimental measurements of the noise spectral density in MOSFETs demonstrate that the noise increases with decreasing frequency. The noise spectral density at very low frequencies exceeds the noise levels predicted for purely thermally generated noise. This excess noise is evident up to a corner frequency of approximately 100 kHz to 1 MHz for MOSFETs. The low-frequency excess noise is generally known as a flicker noise, and its spectral density is inversely proportional to the frequency raised to some power. Due to this inverse relationship to frequency, flicker noise is also called 1/f-noise or "pink" noise.

There are two dominant theories on the origins of 1/f-noise in MOSFETs. First, there is the carrier density fluctuation theory in which flicker noise is taken as a direct result of the random trapping and release of charges by the oxide traps near the Si-SiO$_2$ interface beneath the gate. The channel surface potential fluctuates because of this charge fluctuation, and the channel carrier density is in turn modulated by the channel surface potential fluctuation. The carrier density fluctuation theory predicts that the input-referred flicker noise is independent of the gate bias voltage and the noise power is proportional to the interface trap density. The carrier density fluctuation model is supported by experimental measurements that demonstrate the correlation between the flicker noise power and the interface trap density.

The second major theory on the origins of flicker noise is the mobility fluctuation theory. This theory treats flicker noise as arising from the fluctuation in bulk mobility based on Hooge's empirical relation for the spectral density of flicker noise in a homogenous medium. In contrast with the charge density fluctuation theory, the mobility fluctuation theory does predict that the power spectral density of 1/f-noise is dependent upon the gate bias voltage.

Neither of these two main theories satisfactorily accounts for the observed 1/f-noise power spectral density in MOSFETs under all conditions. Current thinking applies both models for an overall understanding of 1/f-noise in MOSFETs. Expressions for the MOSFET 1/f-noise have been derived by various researchers and normally some amount of "fitting" is required to agree with theory. Common expressions for the 1/f-noise of a MOSFET operating in saturation include

$$\overline{di_f^2} = K_f \frac{g_m^2}{C'_{ox} W L} \frac{df}{f^a} \tag{7.89}$$

for the flicker-noise current, and

$$\overline{dv_{in,f}^2} = K_f \frac{1}{C'_{ox} W L} \frac{df}{f^a} \tag{7.90}$$

for the equivalent input noise voltage. The value of α is typically close to unity, and K_f is in general a bias-dependent parameter, on the order of 10^{-14} C/m^2 for PMOS devices and 10^{-15} C/m^2 for NMOS devices.

Although 1/f-noise is negligible at radio frequencies, it is still an important consideration for transceiver design. For example, 1/f-noise can be a limiting factor in direct conversion receivers. Since the direct conversion receiver directly translates the signal channel of interest to base band, the 1/f-noise corrupts the desired information content. Modern modulation formats place important information signal content near the center of a channel. When such a channel is directly downconverted to DC, this important central region experiences the worst of the 1/f-noise contamination. If the 1/f-noise degrades the signal-to-noise ratio sufficiently, reception becomes impossible.

A second area of transceiver design in which 1/f-noise can create problems and cannot be ignored involves oscillator design. The close-in phase noise of an oscillator is normally dominated by 1/f-noise. At frequencies close to the oscillator center frequency, the phase noise fall off is typically −30 dB per decade, while further from the center frequency the phase noise fall off is normally −20 dB per decade. The additional phase noise close to the center frequency can be attributed to flicker noise present in the active devices comprising the oscillator circuitry.

Designers seeking to minimize 1/f-noise have four options. First, the choice of technology and process is important since the flicker noise exhibited by devices of equal sizes across different processes can vary significantly. Second, a designer can opt to use PMOS devices instead of NMOS devices since PMOS devices typically exhibit one tenth the amount of flicker noise a comparable NMOS device produces. Thirdly, devices operating in weak inversion exhibit markedly lower 1/f-noise than devices operated in saturation. However, operating in weak inversion is normally not an option for devices working at radio frequencies, since the device f_t is too low in the weak inversion region. Finally, as shown in Eqs. (7.89) and (7.90), the area of the device plays an important role in determining the flicker noise produced by the device, and so the area should be maximized. Unfortunately, for radio frequency designs there are constraints on the maximum practical width and length of the MOSFETs. Both the width and length determine the highest usable operating frequency of the device and so they are normally kept small. Small width and length implies that the flicker noise of devices designed for operation at radio frequencies is comparatively large.

Extrinsic Noise

The primary extrinsic noise sources of concern are due to the parasitic resistances at the MOSFETs terminals. The corresponding mean-square noise currents due to these parasitic resistances are given by,

$$\overline{i_{D,e}^2} = \frac{4kT}{R_{D,e}} \Delta f$$

$$\overline{i_{S,e}^2} = \frac{4kT}{R_{S,e}} \Delta f$$

$$\overline{i_{G,e}^2} = \frac{4kT}{R_{G,e}} \Delta f \tag{7.91}$$

$$\overline{i_{B,e}^2} = \frac{4kT}{R_{B,e}} \Delta f$$

where $R_{D,e}$, $R_{S,e}$, $R_{G,e}$ and $R_{B,e}$, are the extrinsic parasitic resistances at the drain, source, gate, and bulk, respectively.

MOSFET Design for RF Operation

Radio frequency integrated circuit design with CMOS requires numerous choices regarding device dimensions and bias voltages and currents. A combination of microwave theory and analog design fundamentals must be pooled with a firm knowledge of the overall system-level functional requirements. System-level calculations will normally dictate the allowable gain, linearity, and noise of the basic circuitry. Two important aspects of radio frequency design in CMOS involve MOSFET input impedance matching for power gain, and impedance matching for minimum noise figure. These two important topics are discussed next.

MOSFET Impedance Matching

Circuitry that accepts a signal from off-chip, such as the input to a receiver's low noise amplifier, must be conjugately matched to the source impedance. The concept of conjugate matching is often unfamiliar to digital designers of CMOS, but is a common theme in microwave design. When a source and load are conjugately matched, maximum power transfer from the source to the load occurs. Input impedance matching is also important because pre-select filters preceding the low noise amplifier may be sensitive to the quality of their terminating impedances.

A load is conjugately matched to its driving impedance when the input impedance of the load equals the complex conjugate of the driving impedance. Since the input impedance of any two-port network is given by,

$$Z_{in} = \frac{y_{22}}{y_{11}y_{22} - y_{21}y_{12}} \tag{7.92}$$

the y-parameters given in Table 7.6 can be used in conjunction with circuit analysis to give the total circuit input impedance. Note that a linear system has been tacitly assumed in Eq. (7.92). For small-signal applications, such as in the receiver portion of a transceiver, linearity is a reasonable assumption. However, for some applications, such as a nonlinear power amplifier, conjugate matching is not possible and so a resistive load line is used instead.

While matching can be augmented by a high-Q matching network constructed off-chip, this method adds expense to mass-market products, in which commodity radio frequency ICs may terminate an unknown length of transmission line. It is therefore desirable to provide as close to a stable 50 Ω input impedance as possible within the integrated circuit. The quality of the external matching network can then be relaxed.

As an example of input impedance matching, consider the common source low noise amplifier configuration shown in Fig. 7.72. The matching components include L_1, which is typically implemented as a bond wire having a value of 1 to 2 nH. Inductor L_1 is used for producing the real part of the input impedance for matching. For simplicity, the bonding pad capacitance is ignored. The LNA is assumed to be operating in a cascode configuration and so the Miller capacitance of the MOSFET is ignored. Conjugate matching to the source resistance R_S requires that the approximate matching condition $R_g + \frac{g_m}{C_{gs}} L_1 = R_S$ is satisfied. Proper choice of the device width, biasing, and number of fingers used for construction allow this relation to be met. Note that $2\pi f_t \approx g_m/C_{gs}$, and as discussed in the section on Maximum Available Power Gain, the maximum achievable power gain of a MOSFET is proportional to the device's f_t. Also, as shown in the next section, the device noise figure is inversely proportional to f_t. Hence the MOSFET biasing should be designed such that the resulting f_t is adequate for the application.

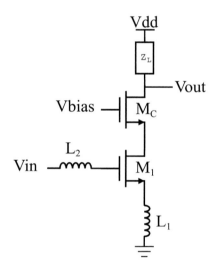

FIGURE 7.72   Common-source low-noise amplifier configuration.

Matching for Noise

Classical noise matching analysis is applicable to the MOSFET shown in Fig. 7.71(b). Classical methods are directly applicable when all MOSFET internal noise sources are referred to the input. For the common source MOSFET configuration shown in Fig. 7.71(b), the input referred noise sources v_n and i_n are given by,

$$v_n = v_{g,i} - \frac{i_d}{g_m} + \left(R_{g,i} + r_i\right)i_g - j\frac{f}{f_t}\left(R_{g,i} + r_i\right)i_d \tag{7.93}$$

and

$$i_n = i_g - j\frac{f}{f_t}i_d \tag{7.94}$$

Recall from the section on Thermal Channel Noise that the two noise sources in Eqs. (7.93) and (7.94) are correlated since i_d and i_g arise from the same physical process.

The noise performance of a transistor with driving impedance $Z_{SS} = R_{SS} + jX_{SS}$ is fully described in terms of four noise parameters: the minimum noise figure (F_{min}), the noise conductance (G_n), and the optimum driving impedance (Z_{opt}), which consists of a resistance (R_{opt}) and a reactance (X_{opt}). Following the procedure of Gonzalez[13] the noise parameters are found as,

$$G_n \approx \left(\frac{f}{f_t}\right)^2 \gamma g_{do} F_1 \tag{7.95}$$

$$R_{opt} \approx \sqrt{\left(R_{g,i} + r_i\right)^2 + \left(\frac{f_t}{f}\right)^2 \frac{R_{g,i} F_4}{\gamma g_{do}} + \left(\frac{f_t}{f}\right)^2 \frac{F_2^2 F_5}{\gamma g_{do}^2 F_1}} \tag{7.96}$$

$$X_{opt} \approx \frac{F_2}{2\pi f C_{gs}} \approx -F_2 X_{in} \tag{7.97}$$

and

$$F_{min} \approx 1 + 2G_n \left(R_{opt} + R_{g,i} + r_i\right)$$
$$\approx 1 + 2F_3 \frac{f}{f_t} \sqrt{\gamma g_{do} R_{g,i}} \tag{7.98}$$

The factors $\{F_1, F_2, F_3, F_4, F_5\}$ result from algebraic manipulations of Eqs. (7.93) and (7.94). If induced gate noise is ignored, then $\{F_1, F_2, F_3, F_4\}$ are all equal to one, otherwise they are less than one. The fifth factor, F_5, is equal to zero if induced gate noise is ignored, and equal to one if induced gate noise is included. From Eq. (7.98) the minimum noise figure is proportional to $\sqrt{R_{g,i}}$. Therefore to realize a low noise figure the intrinsic gate resistance must be made as small as practical. Noting that g_{do} in Eq. (7.98) is approximately equal to g_m, and using the results for f_t from Eq. (7.81), the minimum noise figure is shown to be inversely proportional to $\sqrt{g_m}$. Hence, increasing the g_m of the device decreases the minimum noise figure. Since both g_m and γ are functions of the transistor bias conditions, an optimum biasing point can be found that gives the lowest minimum noise figure.[20]

For optimum noise matching, the optimum noise resistance should equal the driving source resistance. For MOSFETs, near-optimum noise matching is possible by correctly choosing the number of gate fingers used to construct the MOSFET. In Fig. 7.71(b), the equivalent input noise current i_{eq} and noise voltage v_{eq}, yield a transistor noise figure given by

$$F = F_{min} + \frac{G_n}{R_S} \left| R_S - Z_{opt} \right|^2 \tag{7.99}$$

where R_S is the driving source resistance, G_n is the noise conductance of a single finger, and Z_{opt} is the optimum noise impedance. Neglecting interconnect parasitics, if N MOSFETs are placed in parallel, the equivalent noise current is $i_{eq}^{(N)} = i_{eq} \sqrt{N}$ and the equivalent noise voltage is $v_{eq}^{(N)} = v_{eq} / \sqrt{N}$. The noise figure for the complete MOSFET consisting of N fingers is then given by

$$F = F_{min} + \frac{NG_n}{R_S} \left| R_S - \frac{Z_{opt}}{N} \right|^2 \tag{7.100}$$

TABLE 7.7 MOSFET Scaling Rules for the Saturation Region

| | | Number of Fingers | Finger Width | Bias Current | |
	Parameter	N	W	$I_{DS} < I_{DS,sat}$	$I_{DS} > I_{DS,sat}$
Performance	g_m	N	W	Increases with I_{DS}	~constant
	f_t	Independent	Independent	Increases with I_{DS}	~constant
	f_{max}	Independent	~1/W	Increases with I_{DS}	~constant
Parasitics	C_{gs}, C_{gd}	N	W	~constant	~constant
	r_d, r_s, r_i	1/N	1/W	Independent	Independent
	$R_{g,i}$	1/N	W	Independent	Independent
Noise	F_{min}	Independent	~1/W	Decreases with I_{DS}	Increases with I_{DS}
	G_n	N	W	Increases with I_{DS}	~constant
	R_{opt}	1/N	~1/W	Decreases with I_{DS}	~constant
	X_{opt}	1/N	1/W	Independent	Independent

in which the minimum noise finger is independent of the number of fingers, but the noise conductance scales as N, and the optimum noise impedance scales as 1/N. Therefore, by using fingered MOSFETs, the transistor can be matched for noise using the number of fingers as a design parameter.

MOSFET Layout

Radio frequency MOSFET layout is a nontrivial task. A poor transistor layout adversely affects every aspect of the transistor's radio frequency performance. Recall that a MOSFET's f_{max}, NF_{min}, and input impedance-matching characteristics are all determined by the transistor's W/L ratio and the number of fingers used to construct the device. Additionally, both matching and flicker noise are affected by the area of the gate. Devices with a large gate area will match better and have lower flicker noise.

The various scaling rules for MOSFETs operating in saturation appear in Table 7.7. The MOSFET length is assumed to be the minimum allowed. Also shown in Table 7.7 is the effect of biasing upon the MOSFET. A typical design trade-off is between maximum gain and maximum f_{max}, assuming power requirements are a secondary concern. Using multiple MOSFET fingers is a convenient way of improving the device performance.

Fingered Layout

MOSFETs designed for radio frequency applications normally have large W/L ratios. This is primarily so that the MOSFET has a large transconductance and therefore has a large f_t so that its noise figure is minimal. As shown in Table 7.7, increasing the MOSFET width causes a decrease in f_{max}. To counteract this effect, the MOSFET is laid out as a parallel connection of narrower MOSFETs. The resulting fingered structure is shown in Fig. 7.73. Neglecting the interconnect resistance between fingers, the intrinsic gate resistance scales as 1/N, where N is the number of fingers.

Substrate Connections

Parasitic resistance to the substrate affects both the impedance matching and the noise figure of MOSFETs. The substrate resistance is in general bias dependent. Local variations in threshold voltage are therefore possible, and matching between devices is reduced. This may affect the performance of differential pairs and differential circuitry. Voltage fluctuations induced in the substrate, especially by digital clocks and switching circuitry, will raise the noise figure of radio frequency analog MOSFETs. Adequate substrate connections are therefore important for low-noise, precision MOSFET design. A substrate connection method is shown in Fig. 7.73.

The Future of CMOS

CMOS has definite potential for applications in commodity radio frequency electronics, and is expected to find increased use in radio frequency integrated circuits. Although this section has considered CMOS operating at radio frequencies mostly in terms of analog small-signal performance, it is important to note that existing implementations of digital logic are also operating at radio frequencies. Microprocessors

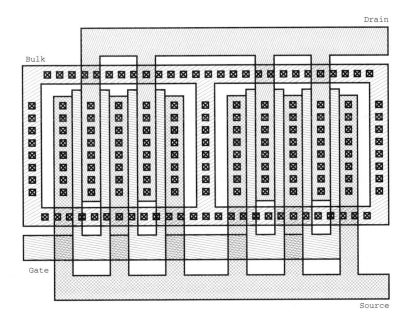

FIGURE 7.73 MOSFET finger layout style for high frequency applications.

operating above 1 GHz are now commonplace. The modeling of radio frequency logic (RFL) and the study of radio frequency digital design will be important contributing factors for future high-speed microprocessors, digital signal processors, and ASICs. Additionally, as the demand for extremely high-speed digital microprocessors increases, the base-level digital CMOS processes will improve to accommodate the performance expectations of designers. Exotic process features that were once considered purely for analog design use are now appearing in digital designs.

CMOS is not the ideal radio frequency technology, but it does compare favorably with bipolar and BiCMOS. For example, although not discussed in this section, communications circuits constructed with CMOS devices often exhibit higher linearity than their counterparts do when constructed with other technologies such as bipolar. Furthermore, in applications where the MOSFET's square-law characteristic applies, differential circuitry can effectively reduce second-order distortion. For an LNA constructed using deep submicron MOSFETs operating in velocity saturation, the device transconductance is essentially constant with v_{gs}, which implies an increased level of linearity.

CMOS may become the technology of choice in the near term for consumer radio frequency applications in the sub-10 GHz bands. The extent to which this becomes possible is dependent on the ability of the device performance to improve as the technology scales and improves. With this in mind, this section concludes with a brief examination of scaling issues in radio frequency CMOS, and recent advances in MOSFET manufacture.

The Impact of Technology Scaling and Process Improvements

A common measure for the gain efficiency of a MOSFET is its transconductance per drain current g_m/I_{DS}. MOSFETs require a large W/L ratio and a high DC biasing current to offer a transconductance comparable to that of other RF device technologies, such as bipolar. In bipolar transistors, the transconductance-to-bias current ratio equals kT/q, which was proved by Johnson[22] to be the maximum transconductance for both MOSFET and bipolar devices. As MOSFET processing improves and the minimum gate length decreases, the g_m/I_{DS} ratio has been found to tend toward the maximum value of kT/q.

Continual reduction in MOSFET minimum gate length has enabled commensurate increases in device f_t and f_{max}. Existing 0.18 μm devices have demonstrated f_t's in excess of 60 GHz, and experimental processes with gate lengths less than 0.1 μm have demonstrated f_t values in the order of 150 GHz. These increases in device f_t also mean that the overall device noise performance improves as the gate length scales down.

Although the radio frequency performance of MOSFETs has traditionally scaled in accordance with a decreasing minimum gate length, this trend will probably not continue along its present path. For example, the gate oxide thickness scales approximately linearly with the minimum gate length. The gate source capacitance therefore remains roughly constant per unit width of the gate for minimum channel length devices. As shown in Eq. (7.81), the MOSFET f_t is inversely proportional to the gate source capacitance, and if the gate source capacitance does not scale in inverse proportion to the gate length, it becomes a limiting factor. Other limitations for increased radio frequency performance of MOSFETs are found in the extrinsic parasitics, such as the gate overlap capacitances. For example, the gate-source overlap capacitance does not scale as 1/L for submicron devices, and C_{gso} becomes an increasingly large proportion of the total gate-to-source terminal capacitance, therefore limiting the high frequency gain of the device.

Certain process enhancements have a profound effect upon the performance of integrated MOSFET circuitry. The use of copper interconnects between MOSFETs is expected to increase the operating frequency of both digital and analog circuitry when used in place of aluminum. Copper has two primary advantages over aluminum when used as an interconnect material. First, the resistivity of copper is approximately 40% lower than that of aluminum. Second, electromigration effects are lower in copper interconnects implying that copper interconnects will exhibit higher reliability, longer life spans, and greater current-handling capability.

Since copper has a lower resistivity than aluminum, interconnect lines can be thinner and yet still allow for the same circuit performance. Alternatively, maintaining the same interconnect thickness gives a lower series resistance. Decreasing the series resistance of metal lines improves the quality factor of integrated inductors and capacitors.

Recent CMOS processes have given designers six or more interconnect metal layers. The increased number of metal layers has important consequences for radio frequency design. For example, several metal layers can be linked in parallel in order to decrease the series resistance of metal lines used to create inductors. Additionally, the top-level metal layers may be used so that the metal-to-substrate capacitance is reduced and therefore the quality factor of the integrated circuit is improved.

While not yet offered on any commodity CMOS processes, substrate etching has been used to increase the quality factor of integrated inductors by removing a section of substrate below the metal lines. Isolation between lower-level metal lines and substrate has also been achieved using patterned ground shields and spun-on thick dielectrics. Future process enhancements will increase the applicability of CMOS to radio frequency applications. Sub-micron mixed-signal processes featuring six metal layers, metal-insulator-metal capacitors, deep n-wells, and threshold voltage-adjust capability are available now, and hint at what will become available in the near future.

To conclude, CMOS is a mature, easily sourced, and reliable technology. If the performance of CMOS can be improved via scaling to within that of bipolar, BiCMOS, and GaAs, there is no reason why it will not supplant those technologies in commodity radio frequency integrated circuit applications.

References

1. Tsividis, Y. P., *Operation and Modeling of the MOS Transistor*, New York: McGraw-Hill, 1987.
2. Sze, S. *Physics of Semiconductor Devices*, New York: Wiley and Sons, 1981.
3. Grove, A. *Physics and Technology of Semiconductor Devices*, New York: Wiley and Sons, 1967.
4. Muller, R. and Kamins, T. *Device Electronics for Integrated Circuits*, New York: Wiley and Sons, 1986.
5. Laker, K. R. and Sansen, W. M. C. *Design of Analog Integrated Circuits and Systems*, New York: McGraw-Hill, 1994.
6. Enz, C. C. and Cheng, Y., MOS Transistor Modeling for RF IC Design, *IEEE Journal of Solid-State Circuits*, 35, 186–201, Feb. 2000.

7. Klein, P., A Compact-Charge LDD-MOSFET Model, *IEEE Transactions on Electron Devices*, 44, 1483–1490, Sept. 1997.

8. Cobbold, R. S. C. *Theory and Applications of Field-Effect Transistors*, New York: Wiley-Interscience, 1970.

9. Razavi, B., Yan, R.-H., and Lee, K. F., Impact of Distributed Gate Resistance on the Performance of MOS Devices, *IEEE Trans. Circuits and Systems I*, 41, 750–754, 1994.

10. Park, H. J., Ko, P. K., and Hu, C., A Non-Quasi-static MOSFET Model for SPICE-AC Analysis, *IEEE Trans. Computer Aided Design*, 11, 1247–1257, 1992.

11. Kim, L.-S. and Dutton, R. W., Modeling of the Distributed Gate RC Effect in MOSFETs, *IEEE Trans. Computer Aided Design*, 8, 1365–1367, 1989.

12. Jindal, R. P., Noise Associated with Distributed Resistance of MOSFET Gate Structure in Integrated Circuits, *IEEE Trans. Electron Devices*, ED-31, 1505–1509, 1984.

13. Gonzalez, G. *Microwave Transistor Amplifiers Analysis and Design*, Prentice-Hall, Englewood Cliffs, NJ, 1997.

14. Das, M. B., High Frequency Network Properties of MOS Transistors Including the Substrate Resistivity Effects, *IEEE Trans. Electron Devices*, ED-16, 1049–1069, 1969.

15. Bagheri, M. and Tsividis, Y., A Small-Signal DC-to-High-Frequency Non-Quasistatic Model for Four-Terminal MOSFETs Valid in All Regions of Operation, *IEEE Trans. Electron Devices*, ED-32, 2383–2391, 1985.

16. Rofougaran, A., Chang, J. Y. C., Rofougaran, M., and Abidi, A. A., A 1GHz CMOS RF Front-End IC for a Direct-Conversion Wireless Receiver, *IEEE Journal of Solid-State Circuits*, 31, 880–889, July 1996.

17. Abou-Allam, E. and Manku, T., An Improved Transmission-Line Model for MOS Transistors, *IEEE Transactions on Circuits and Systems II: Analog and Digital Signal Processing*, 46, 1380–1387, Nov. 1999.

18. Abou-Allam, E. and Manku, T., A Small-Signal MOSFET Model for Radio Frequency IC Applications, *IEEE Transactions on Computer-Aided Design of Integrated Circuits and Systems*, 16, 437–447, May 1997.

19. Tin, S. F., Osman, A. A., and Mayaram, K., Comments on A Small-Signal MOSFET Model for Radio Frequency IC Applications, *IEEE Trans. Computer-Aided Design*, 17, 373–374, 1998.

20. Manku, T., Microwave CMOS-Device Physics and Design, *IEEE Journal of Solid-State Circuits*, 34, 277–285, Mar. 1999.

21. Aldert van der Ziel. *Noise in Solid State Devices and Circuits*, John Wiley & Sons, New York, 1986.

22. Johnson, E. O., The Insulating-Gate Field Effect Transistor — a Bipolar Transistor in Disguise, *RCA Review*, 34, 80–94, 1973.

7.2.4 Metal Semiconductor Field Effect Transistors (MESFETs)

Michael S. Shur

Silicon Metal Oxide Semiconductor Field Effect Transistors (MOSFETs) dominate modern microelectronics. Gallium Arsenide Metal Semiconductor Field Effect Transistors (GaAs MESFETs) are "runners-up," and they find many important niche applications in high-speed or high frequency circuits. After the first successful fabrication of GaAs MESFETs by Mead in 1966[1] and after the demonstration of their performance at microwave frequencies in 1967 by Hooper and Lehrer,[2] these devices emerged as contenders with silicon MOSFETs and bipolar transistors. In the late 1970s and early 1980s, high quality semi-insulating substrates and ion-implantation processing techniques made it possible to fabricate GaAs MESFET VLSI circuits, such as 16×16 multipliers with a multiplication time of 10.5 ns and less than 1 W power dissipation.[3]

Today GaAs MESFETs play an important role in both analog and digital applications, such as for satellite and fiber-optic communication systems, in cellular phones, in automatic IC test equipment, and for other civilian and military uses. The microwave performance of GaAs MESFETs approaches that of

Heterostructure Field Effect Transistors (HFETs).[4] As discussed below, the record maximum frequency of oscillations and the record cutoff frequency, f_T, for GaAs MESFETs reached 190 and 168 GHz, respectively. Even though the record numbers of f_{max} and f_T for GaAs- and InP-based HFETs reach 400 GHz and 275 GHz, respectively, their more typical f_{max} and f_T are well within the reach of GaAs MESFET technology. The integration scale of GaAs MESFET integrated circuits approaches 1,000,000 transistors.

Emerging materials for MESFET applications are SiC and GaN wide bandgap semiconductors that have a much higher breakdown voltage, a higher thermal conductivity, and a higher electron velocity than GaAs. SiC MESFETs are predicted to reach breakdown voltages up to nearly 100 kV.[5]

However, SiGe and even advanced, deep submicron Si technologies emerge as a serious competitor to GaAs MESFET technology at relatively low frequencies (below 40 GHz or so). This trend toward SiGe and Si might be alleviated by a shift toward 150 mm GaAs substrates, which are now used by the leading GaAs IC manufacturers, such as Vitesse, Anadigics, Infinion, Motorola, Tektronix, and RFMD.

In this section, we first discuss the MESFET principles of operation. Then we review the material properties of semiconductors competing for applications in MESFETs and the properties of Schottky barrier contacts followed by a brief review of MESFET fabrication, and MESFET modeling. We also consider wide bandgap semiconductor MESFETs, new emerging hetero-dimensional MESFETs, and discuss applications of the MESFET technology.

Principle of Operation

Figure 7.74 shows a schematic MESFET structure. In *n*-channel MESFETs, an n-type channel connects n^+ drain and source regions. The depletion layer under the Schottky barrier gate contact constricts the current flow across the channel between the source and drain. The gate bias changes the depletion region thickness, and, hence, modulates the channel conductivity.

This device is very different from a silicon Metal Oxide Semiconductor Field Effect Transistor (MOSFET), where a silicon dioxide layer separates the gate from the channel. MOSFETs are mainstream devices in silicon technology, and silicon MESFETs are not common. Compound semiconductors, such as GaAs, do not have a stable oxide, and a Schottky gate allows one to avoid problems related to traps in the gate insulator, such as hot electron trapping in the gate insulator, threshold voltage shift due to charge trapped in the gate insulator, and so on.

In **normally-off** (**enhancement mode**) MESFETs, the channel is totally depleted by the gate built-in potential even at zero gate voltage (see Fig. 7.75). The threshold voltage of normally-off devices is positive. In **normally-on** (**depletion mode**) MESFETs, the conducting channel has a finite cross-section at zero gate voltage. The drawback of normally-off MESFET technology is a limited gate voltage swing due to the low turn-on voltage of the Schottky gate. This limitation is much less important in depletion mode FETs with a negative threshold voltage. Also, this limitation is less important in low power digital circuits operating with a low supply voltage.

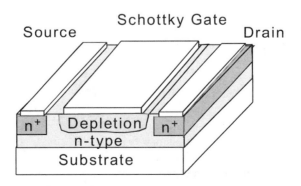

FIGURE 7.74   Schematic MESFET structure.

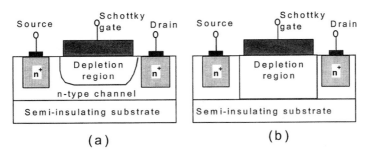

FIGURE 7.75 Normally-on and normally-off MESFETs at zero gate bias.

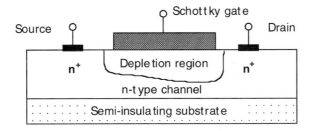

FIGURE 7.76 Depletion region in MESFET with positive drain bias.

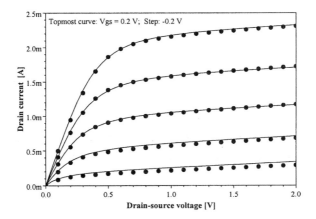

FIGURE 7.77 Measured (symbols) and simulated (using AIM-Spice, lines) drain current characteristics of MESFET operating at room temperature. (After Ytterdal et al.[6])

Usually, the source is grounded, and the drain is biased positively. A schematic diagram of the depletion region under the gate of a MESFET for a finite drain-to-source voltage is shown in Fig. 7.76.

The depletion region is wider closer to the drain because the positive drain voltage provides an additional reverse bias across the channel-to-gate junction. With an increase in the drain-to-source bias, the channel at the drain side of the gate becomes more and more constricted. Finally, the velocity of electrons saturates leading to the current saturation (see Fig. 7.77 that shows typical MESFET current-voltage characteristics).

MESFETs have been fabricated using many different semiconductor materials. However, GaAs MESFETs are mainstream MESFET devices. In many cases, GaAs MESFETs are fabricated by direct ion implantation into a GaAs semi-insulating substrate, making GaAs IC fabrication less complicated than silicon CMOS fabrication.

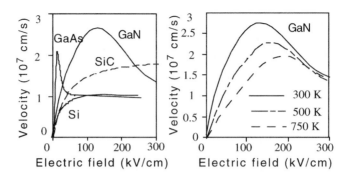

FIGURE 7.78 (a) Electron drift velocity at 300 K in GaN, SiC, and GaAs. (b) Electron drift velocity in GaN at 300 K, 500 K, and 750 K.[12]

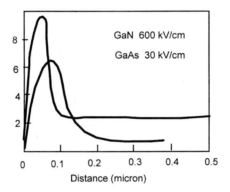

FIGURE 7.79 Computed velocity of electrons injected with low velocities into a constant electric field region into GaN and GaAs (after Foutz et al.[13]).

Properties of Semiconductor Materials Used in MESFET Technology

The effective mass of electrons in GaAs is very small (0.067 m_e in GaAs compared to 0.98 m_e longitudinal effective mass and 0.19 m_e transverse effective mass in Si, where m_e is the free electron mass). This leads to a much higher electron mobility in GaAs — approximately 8500 cm²/Vs in pure GaAs at room temperature compared to 1500 cm²/Vs in Si. As shown in Fig. 7.78, the electron velocity in GaAs exceeds that for the electrons in Si. This is an important advantage for modern day short channel devices, where the electric fields are higher than the peak velocity field under normal operating conditions. The light electrons in GaAs also experience so-called overshoot or even ballistic transport in short channel devices,[7–11] where the electron transit time becomes comparable to or even smaller than the electron energy or even momentum relaxation time. This boosts the electron velocity well above the expected steady-state values (see Fig. 7.79). GaN also has a high electron velocity and pronounced overshoot effects (see Figs. 7.78 and 7.79).

Another important advantage of GaAs and related compound semiconductors is the availability of semi-insulating material that could serve as a substrate for device and circuit fabrication. A typical resistivity of semi-insulating GaAs is 10^7 Ω-cm or larger, compared to 2.5×10^5 Ω-cm for intrinsic silicon at room temperature. The semi-insulating GaAs is used as a substrate for fabricating GaAs MESFETs and other devices. Passive elements can also be fabricated on the same substrate, which is a big advantage for fabricating Monolithic Microwave Integrated Circuits (MMICs). As mentioned above, an important advantage of the GaAs MESFET is the possibility of fabricating these devices and integrated circuits using a direct implantation into the semi-insulating GaAs substrate.

TABLE 7.8 Material Properties of Si, GaAs, α-SiC, and GaN

Property	Si	GaAs	α-SiC(6H)	GaN
Energy gap (eV)	1.12	1.42	2.9	3.4
Lattice constant(a) Å	5.43107	5.6533	3.081	3.189
Lattice constant (c), Å	—	—	15.117	5.185
Density (g/cm³)	2.329	5.3176	3.211	6.1
Dielectric constant	11.7	12.9	9.66($\perp$) 10.03($\parallel$)	9.5 (8.9)
Electron mobility (cm²/Vs)	1450	8500	330	1200
Hole mobility (cm²/Vs)	500	400	60	<30
Saturation velocity (m/s)	10^5	1.2×10^5	2–2.5×10^5	2–2.5×10^5
Electron effective mass ratio	0.92/0.19	0.067	0.25/1.5	0.22
Light hole mass ratio	0.16	0.076	0.33	0.7
Optical phonon energy (eV)	0.063	0.035	0.104	0.092
Thermal conductivity (W/cm°C)	1.31	0.46	4.9	1.5

Since GaAs, InP, and related semiconducting materials are direct gap materials, they are widely used in optoelectronic applications. Hence, electronic and photonic devices can be integrated on the same chip for use in optical interconnects or in optoelectronic circuits.

The direct band gap leads to a high recombination rate, which improves radiation hardness. GaAs-based devices can survive over 100 megarads of ionizing radiation.[14]

GaAs, and especially SiC and GaN MESFETs, are suitable for use in high-temperature electronics and as power devices (because of a small intrinsic carrier concentration and a high breakdown field).

Table 7.8 summarizes important material properties of semiconductors used in MESFET technology.

Schottky Barrier Contacts

Figure 7.80 shows the energy band diagram of a GaAs Schottky barrier metal-semiconductor contact at zero bias. The Fermi level will be constant throughout the entire metal-semiconductor system, and the energy band diagram in the semiconductor is similar to that for an *n*-type semiconductor in a p^+-n junction.

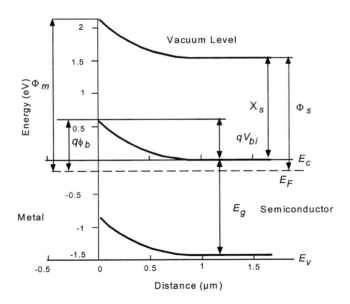

FIGURE 7.80 Simplified energy diagram of GaAs metal-semiconductor barrier $q\phi_b$ is the barrier height (0.75 eV), X_s is the electron affinity in the semiconductor, Φ_s and Φ_m are the semiconductor and the metal work functions, and V_{bi} (0.59 V) is the built-in voltage. Donor concentration in GaAs is 10^{15} cm^{-3} (From Shur[15]).

Energies Φ_m and Φ_s shown in Figure 7.80 are called the metal and the semiconductor **work functions**. The **work function** is equal to the difference between the **vacuum level** (which is defined as a free electron energy in vacuum) and the Fermi level. The **electron affinity** of the semiconductor, X_s (also shown in Fig. 7.80), corresponds to the energy separation between the vacuum level and the conduction band edge of the semiconductor.

In the idealized picture of the Schottky junction shown in Fig. 7.80, the energy barrier between the semiconductor and the metal is

$$q\phi_b = \Phi_m - X_s \qquad (7.101)$$

Here X_s is the electron affinity in the semiconductor, Φ_s and Φ_m are the metal work functions, and ϕ_b is the Schottky barrier height, q is the electronic charge [in Eq. (7.101) and in Fig. 7.80, ϕ_b is measured in eV, and Φ_s and Φ_m are measured in Joules). Since $\Phi_m > \Phi_s$ the metal is charged negatively. The positive net space charge in the semiconductor leads to a band bending

$$qV_{bi} = \Phi_m - \Phi_s, \qquad (7.102)$$

where V_{bi} is called the **built-in voltage**, in analogy with the corresponding quantity in a *p-n* junction.

Equation (7.101) and Fig. 7.80 are not quite correct. In reality, a change in the metal work function, Φ_m, is not equal to the corresponding change in the barrier height, ϕ_b, as predicted by Eq. (7.101). In actual Schottky diodes, ϕ_b increases with an increase in Φ_m, but only by 0.1 to 0.3 eV when Φ_m increases by 1 to 2 eV. This difference is caused by interface states and is determined by the properties of a thin interfacial layer. However, even though a detailed and accurate understanding of Schottky barrier formation remains a challenge, many properties of Schottky barriers may be understood independently of the exact mechanism determining the barrier height. In other words, we can simply determine the effective barrier height from experimental data.

A forward bias decreases the potential barrier for electrons moving from the semiconductor into the metal and leads to an exponential rise in current. At high forward biases (approaching the built-in voltage), the voltage drop across the series resistance (comprised of the contact resistance and the resistance of the neutral region between the ohmic contact and the depletion region) becomes important, and the overall current-voltage characteristic of a Schottky diode can be described by the following **diode equation**

$$I = I_S \left[\exp\left(\frac{V - IR_S}{\eta V_{th}} \right) - 1 \right], \qquad (7.103)$$

where I_s is the saturation current, R_s is the series resistance, $V_{th} = k_B T/q$ is the thermal voltage, η is the ideality factor (η typically varies from 1.02 to 1.6), q is the electronic charge, k_B is the Boltzmann constant, and T is temperature.

The diode saturation current, I_s, is typically much larger for Schottky barrier diodes than in *p-n* junction diodes since the Schottky barrier height is smaller than the barrier height in *p-n* junction diodes. For a *p-n* junction, the height of the barrier separating electrons in the conduction band of the *n*-type region from the bottom of the conduction band in the *p*-region is on the order of the energy gap.

The current mechanism in Schottky diodes depends on the doping level. In a relatively low-doped semiconductor, the depletion region between the semiconductor and the metal is very wide, and electrons can only reach the metal by going over the barrier. In higher doped samples, the barrier near the top is narrow enough for the electrons to tunnel through. Finally, in very highly doped structures, the barrier is thin enough for tunneling at the Fermi level. Figure 7.81 shows the band diagrams illustrating these three conduction mechanisms.

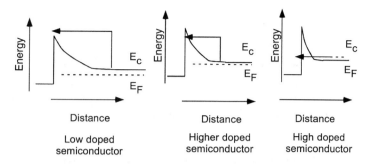

FIGURE 7.81 Current paths in low-doped, higher doped, and high-doped Schottky diodes.

For low-doped devices, the saturation current density, j_{ss}, in a Schottky diode is given by

$$j_{ss} = A * T^2 \exp\left(-\frac{q\phi_b}{k_B T}\right),$$

(7.104)

where A^* is called the Richardson constant, T is temperature (in degrees Kelvin), and k_B is the Boltzmann constant. For a conduction band minimum with the spherical surface of equal energy (such as the Γ minimum in GaAs),

$$A* = \alpha \frac{m_n q k_B^2}{2\pi^2 \hbar^3} \approx 120\, \alpha \frac{m_n}{m_e} \left(\frac{A}{cm^2 K^2}\right),$$

(7.105)

where m_n is the effective mass, m_e is the free electron mass, $\hbar$ is the Planck constant, and α is an empirical factor on the order of unity. The Schottky diode model described by Eqs. (7.104) and (7.105) is called the **thermionic emission model**. For Schottky barrier diodes fabricated on the {111} surfaces of Si, $A^* = 96\ A/(cm^2K^2)$. For GaAs, $A^* = 4.4\ A/(cm^2K^2)$.

As stated above, in higher doped semiconductors, the depletion region becomes so narrow that electrons can tunnel through the barrier near the top. This conduction mechanism is called thermionic-field emission.

The current-voltage characteristic of a Schottky diode in the case of thermionic-field emission (i.e., for higher doped semiconductors) under forward bias is given by:

$$j = j_{stf} \exp\left(\frac{qV}{E_o}\right)$$

(7.106)

where

$$E_o = E_{oo} \coth\left(\frac{E_{oo}}{k_B T}\right)$$

(7.107)

$$E_{oo} = \frac{qh}{4\pi}\sqrt{\frac{N_d}{m_n \varepsilon_s}} = 1.85 \times 10^{-11} \left[\frac{N_d\left(cm^{-3}\right)}{\left(m_n/m_e\right)\left(\varepsilon_s/\varepsilon_o\right)}\right]^{1/2} \left(eV\right)$$

(7.108)

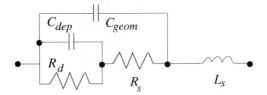

FIGURE 7.82   The small signal equivalent circuit of a Schottky diode.

$$j_{stf} = \frac{A*T\sqrt{\pi E_{oo}\left(\phi_b - qV - E_c + E_{Fn}\right)}}{k_B \cosh\left(E_{oo}/k_B T\right)} \exp\left[-\frac{E_c - E_{Fn}}{k_B T} - \frac{\left(\phi_b - E_c + E_{Fn}\right)}{E_o}\right] \qquad (7.109)$$

Here E_c is the bottom of the conduction band in a semiconductor (outside of the depletion region and E_{Fn} is the electron quasi-Fermi level. In GaAs Schottky diodes, the thermionic-field emission becomes important for $N_d > 10^{17}$ cm^{-3} at 300 K and for $N_d > 10^{16}$ cm^{-3} at 77 K. In silicon, the corresponding values of N_d are several times larger.

In degenerate semiconductors, especially in semiconductors with a small electron-effective mass, such as GaAs, electrons can tunnel through the barrier near or at the Fermi level, and the tunneling current is dominant. This mechanism is called **field emission**. The resistance of the Schottky barrier in the field emission regime is quite low. Metal-n^+ contacts operated in this regime are used as ohmic contacts.

Figure 7.82 shows a small signal equivalent circuit of a Schottky diode, which includes a parallel combination of the differential resistance of the Schottky barrier

$$R_d = \frac{dV}{dI} \qquad (7.110)$$

and the differential capacitance of the space charge region:

$$C_{dep} = S\sqrt{\frac{qN_d\varepsilon_s}{2\left(V_{bi} - V\right)}} \qquad (7.111)$$

Here V and I are the voltage drop across the Schottky diode and the current flowing through the Schottky diode, respectively, V_{bi} is the built-in voltage of the Schottky barrier, and N_d is the ionized donor concentration in the semiconductor. The equivalent circuit also includes the series resistance, R_s, which accounts for the contact resistance and the resistance of the neutral semiconductor region between the ohmic contact and the depletion region, the equivalent series inductance, L_s, and the device geometric capacitance:

$$C_{geom} = \varepsilon_s S/L \qquad (7.112)$$

where L is the device length and S is the device cross-section.

MESFET Technology

The most popular MESFETs (GaAs MESFETs) found in applications in both analog microwave circuits (including applications in Microwave Monolithic Integrated Circuits) and in digital integrated circuits. Ion implanted GaAs MESFETs represent the dominant technology for applications in digital integrated circuits. They also have microwave applications. Figure 7.83 shows a typical process sequence for ion implanted GaAs MESFETs (developed in late 1970s[16,17]).

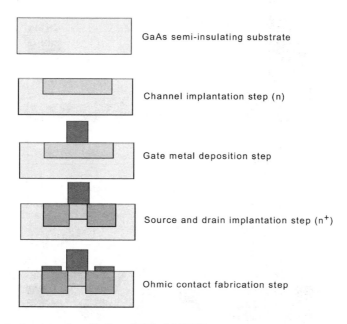

FIGURE 7.83   Fabrication steps for self-aligned GaAs MESFET.

In a typical fabrication process, a GaAs semi-insulating substrate is coated with a thin silicon nitride (Si_3N_4) film. Implantation steps shown in Fig. 7.83 are carried out through this layer. As shown in Fig. 7.83, the first implant defines the active layer including the MESFET channel. A deeper and a higher dose implant is used for ohmic contacts. This implant is often done as a self-aligned implant. In this case, a temperature-stable refractory metal-silicide gate (typically tungsten silicide) is used as a mask for implanting the n^+ source and drain contacts. This technique reduces parasitic resistances. Also, this fabrication process is planar. However, the n^+ implant straggle under the gate might increase gate leakage current and also cause carrier injection into the substrate.[18]

After the implants, an additional insulator is deposited in order to cap the GaAs surface for the subsequent annealing step. This annealing (at 800°C or above) activates the implants. For microwave applications, the devices are often grown by molecular beam epitaxy. In this design, a top of n^+ layer doping extending from the source and drain contacts helps minimize the series resistances.

Figure 7.84 shows the recessed gate MESFET structure, where the thickness of the active layer under the gate is reduced. A thick n-doped layer between the gate and the source and drain contacts leads to a

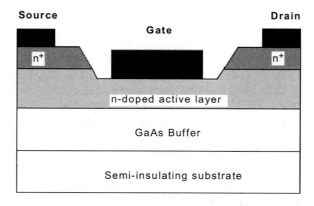

FIGURE 7.84   Recessed gate structure.

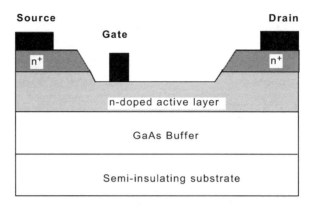

FIGURE 7.85   Recessed structure with offset gate for power devices.

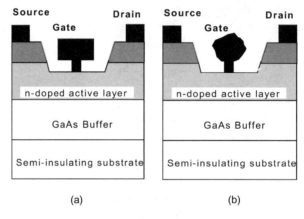

(a)   (b)

FIGURE 7.86   T-gate (a) and mushroom gate (b) for gate series resistance reduction.

relatively low series source and drain resistances. The position and the shape of the recess are very important, since they strongly affect the electric field distribution and the device breakdown voltage.

In power devices, the gate contact in the recess is usually closer to the source than to the drain (see Fig. 7.85). Such placement reduces the source parasitic resistance and enhances the drain-source breakdown voltage by allowing additional expansion space for the high-field region at the drain side of the gate.

Another important issue is the reduction of the gate series resistance. This can be achieved by using a T-shape gate or a so-called mushroom gate (which might be obtained by side etching the gate), see Fig. 7.86.

In this design, the gate series resistance is reduced without an increase in the gate length, which determines the device cutoff frequency.

MESFETs are usually passivated by a Si_3N_4 layer. This passivation affects the surface states and the surface depletion layer, and stress-related and piezoelectric effects can lead to shifts in the threshold voltage.[19]

A more detailed discussion of GaAs MESFET fabrication can be found in References 20 and 21.

Wide band gap semiconductors, such as SiC, GaN, and related materials, might potentially compete with GaAs for applications in MESFETs and other solid-state devices (see Fig. 7.87).

SiC exists in more than 170 polytypes. The three most important polytypes are hexagonal 6H (α-SiC) and 4H, and cubic 3C (β-SiC). As stated in Table 7.8, SiC has the electron saturation drift velocity of 2×10^7 cm/s (approximately twice that of silicon), a breakdown field larger than 2,500 to 5,000 kV/cm

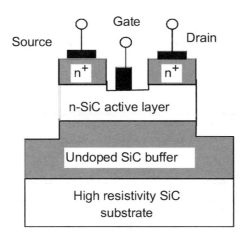

FIGURE 7.87 Cross-section of SiC MESFET (From Shur[22]).

(compared to 300 kV/cm for silicon), and a high thermal conductivity of 4.9 W/cm°C (compared to 1.3 W/cm°C for silicon and 0.5 W/cm°C for GaAs).

These properties make SiC important for potential applications in high-power, high frequency devices as well as in devices operating at high temperatures and/or in harsh environments.

Palmour et al.[23] reported operation of α-SiC MESFETs at a temperature of 773°K. In a 6H-SiC MESFET fabricated by CREE (gate length 24 μm, channel depth 600 nm, doping 6.5×10^{16} cm^{-3}), the room temperature transconductance was approximately 4 mS/mm. At elevated temperatures, the device transconductance decreases owing to the decrease in mobility. MESFETs did not exhibit breakdown even with drain voltages up to 100 V. Using the square law MESFET model (described below), Kelner and Shur[24] estimated that the field effect mobility in these MESFETs was approximately 300 cm²/Vs. β-SiC MESFETs have also been fabricated[25] but α-SiC MESFETs exhibit better performance because of better material quality.

α-SiC MESFETs have achieved microwave operation.[26] A cutoff frequency of 5 GHz, 12 dB gain at 2 GHz, and a breakdown voltage of 200 V was demonstrated in an α-SiC MESFET with a 0.4 μm gate length.[27]

GaN is another material that is potentially important for MESFET applications. For GaN at room temperature and with an n-type doping density of 10^{17} cm^{-3}, Monte Carlo simulations predict a high peak velocity (2.7×10^5 m/s), a high saturation velocity (1.5×10^5 m/s), and a high electron mobility (1000 cm²/Vs).[28–31] Khan et al.[32] and Binari et al.[33] reported on microwave performance of GaN MESFETs. However, most of the research on GaN-based FETs has concentrated on GaN-based Heterostructure Field Effect Transistors.[34,35]

MESFET Modeling

MESFET modeling has been done at several different levels. Most advanced numerical simulation techniques rely on self-consistent simulation based on the Monte Carlo approach. In this approach, random number generators are used to simulate random electron scattering processes. The motion of these electrons is simulated in the electric field that is calculated self-consistently by solving the Poisson equation iteratively. The particle movements between scattering events are described by the laws of classical mechanics, while the probabilities of the various scattering processes and the associated transition rates are derived from quantum mechanical calculations. Some of the results obtained by using this approach were reviewed in Reference 36. Table 7.9 from Reference 36 describes the Monte Carlo algorithm in more detail.

Self-consistent Monte Carlo simulations are very useful for revealing the device physics and verifying novel device concepts and ideas.[22,37–41]

TABLE 7.9 Monte Carlo Algorithm[36]

Generate random number r and determine the duration of the free flight.
↓
Record the time the particle spends in each cell of k-space during the free flight.
↓
Generate random numbers to determine which scattering process has occurred, and which is the final state. Repeat until the desired number of scattering events is reached.
↓
Calculate the distribution function, the drift velocity, the mean energy, etc.

A less rigorous, but also less numerically demanding approach is to use solving the balance equations. These partial differential equations describe conservation laws derived from the Boltzmann Transport Equation.[42,43] Two-dimensional device simulators based on the balance equations and on the drift-diffusion model can be used to optimize device design and link the device characteristics to the device fabrication process.[44-46]

A more simplistic and easier approach is to use conventional drift-diffusion equations implemented in commercial two-dimensional and three-dimensional device simulators, such as ATLAS or MEDICI. However, even this approach might be too complicated and too numerically involved for the simulation of MESFET-based digital VLSI and/or for the simulation of MESFET-based analog circuits.

The simplest model that relates the MESFET current-voltage characteristics to the electron mobility, the electron saturation velocity, the device dimensions, and applied voltages is called the square-law model. This model predicts the following equation for the drain saturation current:

$$I_{sat} = \beta \left(V_{GS} - V_T \right)^2, \tag{7.113}$$

where[47,48]

$$\beta = \frac{2\varepsilon_s \mu_n v_s W}{A\left(\mu_n V_{po} + 3v_s L\right)} \tag{7.114}$$

is the transconductance parameter,

$$V_T = V_{bi} - V_{po} \tag{7.115}$$

is the threshold voltage, V_{GS} is the intrinsic gate-to-source voltage, and

$$V_{po} = \frac{qN_d A^2}{2\varepsilon_s} \tag{7.116}$$

is the pinch-off voltage. Here A is the channel thickness, μ_n is a low field mobility, and v_s is the electron saturation velocity.

This "square law" model is fairly accurate for devices with relatively low pinch-off voltages ($V_{po} = V_{bi} - V_T \leq 1.5 \sim 2$ V). For devices with higher pinch-off voltages, the model called the **Raytheon model** (which is implemented in many versions of SPICE) yields a better agreement with experimental data:

$$I_{sat} = \frac{\beta\left(V_{GS} - V_T\right)^2}{1 + t_c\left(V_{GS} - V_T\right)} \tag{7.117}$$

Here t_c is an empirical parameter that depends on the doping profile in the MESFET channel. Another empirical model (called the **Sakurai-Newton model**) is also quite useful for MESFET modeling:

$$I_{sat} = \beta_{sn}\left(V_{GS} - V_T\right)^{m_{sn}} \tag{7.118}$$

The advantage of this model is simplicity. The disadvantage is that the empirical parameters β_{sn} and m_{sn} cannot be directly related to the device and material parameters. (The Sakurai-Newton model is implemented in several versions of SPICE. In AIM-Spice,[49,50] this model is implemented as Level 6 MOSFET model.)

The source and drain series resistances, R_S and R_d, may play an important role in determining the current-voltage characteristics of GaAs MESFETs. The intrinsic gate-to-source voltage, V_{GS}, is given by

$$V_{GS} = V_{gs} - I_{ds}R_S \tag{7.119}$$

where V_{gs} is the applied (extrinsic) gate-to-source voltage. Substituting Eq. (7.119) into Eq. (7.116) and solving for I_{sat} we obtain

$$I_{sat} = \frac{2\beta V_{gt}^2}{1 + 2\beta V_{gt}R_S + \sqrt{1 + 4\beta V_{gt}R_S}} \tag{7.120}$$

In device modeling suitable for computer-aided design, one has to model the current-voltage characteristics in the entire range of drain-to-source voltages, not only in the saturation regime. In 1980, Curtice proposed the use of a hyperbolic tangent function for the interpolation of MESFET current-voltage characteristics

$$I_d = I_{sat}\left(1 + \lambda V_{ds}\right)\tanh\left(\frac{g_{ch}}{I_{sat}}\right), \tag{7.121}$$

where

$$g_{ch} = \frac{g_i}{1 + g_i\left(R_S + R_d\right)} \tag{7.122}$$

is the MESFET conductance at low drain-to-source voltages, and

$$g_i = g_{cho}\left(1 - \sqrt{\frac{V_{bi} - V_{GS}}{V_{po}}}\right) \tag{7.123}$$

is the intrinsic channel conductance at low drain-to-source voltages predicted by the Shockley model.

The constant λ in Eq. (7.121) is an empirical constant that accounts for the output conductance in the saturation regime. This output conductance may be related to short channel effects and also to

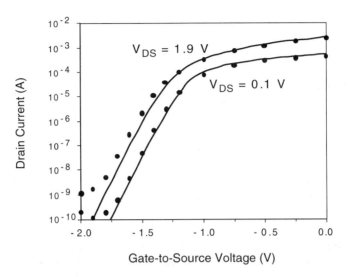

FIGURE 7.88 Subthreshold experimental (symbols) and calculated (solid lines) *I-V* characteristics for ion-implanted MESFET with nominal gate length $L = 1$ μm.[52]

parasitic leakage currents in the substrate. Hence, output conductance may be reduced by using a heterojunction buffer layer between the device channel and the substrate or by using a *p*-type buffer layer. Such a layer creates an additional barrier, which prevents carrier injection into the substrate.[18]

The **Curtice model** is implemented in PSpice™. The Curtice model and the Raytheon model [see Eq. (7.123)] have become the most popular models used for MESFET circuit modeling. A more sophisticated model, which describes both subthreshold and above-threshold regimes of MESFET operation, is implemented in AIM-Spice. This model accurately reproduces current-voltage characteristics over several decades of currents and is suitable for both analog and digital circuit simulations.[51] One of the simulation results obtained using this model is depicted in Fig. 7.88.

In order to simulate MESFET circuits, one also needs to have a model describing the MESFET capacitances. Meyer[53] proposed a simple charge-control model, in which capacitances ($C_{ij} = C_{ji}$) were obtained as derivatives of the gate charge with respect to the various terminal voltages. Fjeldly et al.[52,54] approximated a unified gate-channel capacitance C_{gc} of a MESFET at zero drain-source bias by the following combination of the above-threshold capacitance C_a and the below-threshold capacitance C_b:

$$C_{gc} = \frac{C_a C_b}{C_a + C_b}.$$ (7.124)

This approach in conjunction with Meyer's model, leads to the following expressions for the gate-to-source (C_{gs}) and gate-to-drain capacitance (C_{gd}) valid for the sub-threshold and the above-threshold regimes:

$$C_{gs} = \frac{2}{3} C_{gc} \left[1 - \left(\frac{V_{sat} - V_{dse}}{2V_{sat} - V_{dse}} \right)^2 \right],$$ (7.125)

$$C_{gs} = \frac{2}{3} C_{gc} \left[1 - \left(\frac{V_{sat}}{2V_{sat} - V_{dse}} \right)^2 \right].$$ (7.126)

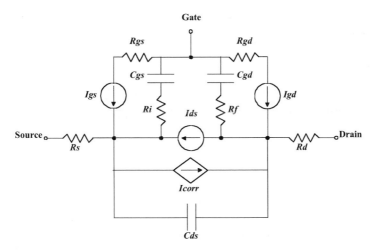

FIGURE 7.89 MESFET equivalent circuit. (After Shur et al.[52])

Here, V_{sat} is the extrinsic saturation voltage and V_{dse} is an effective extrinsic drain-source voltage that equals V_{ds} for $V_{ds} < V_{sat}$ and V_{sat} for $V_{ds} > V_{sat}$.

A more accurate model of the intrinsic capacitances requires an analysis of the variation of the charge distribution in the channel versus terminal bias voltages. For the MESFET, the depletion charge under the gate has to be partitioned between the source and drain terminals.[54]

Finally, the gate leakage current has to be modeled in order to accurately reproduce MESFET current voltage characteristics in the entire range of bias voltages including positive gate biases. To a first order approximation, the gate leakage current can be described in terms of simple diode equations assuming that each "diode" represents half of the gate area:

$$I_g = J_{ss} \frac{LW}{2} \left[\exp\left(\frac{V_{gs}}{m_{gs} V_{th}} \right) + \exp\left(\frac{V_{gd}}{m_{gd} V_{th}} \right) - 2 \right]. \tag{7.127}$$

Here L and W are the gate length and width, respectively, J_{ss} is the saturation current, V_{gs} and V_{gd} are gate-to-source and gate-to-drain voltages, V_{th} is the thermal voltage, and m_{gs} and m_{gd} are the ideality factors. Figure 7.89 shows a more accurate equivalent circuit, which accounts for the effect of the leakage current on the drain current. In the equivalent circuit shown in Fig. 7.89, this effect is accounted for by the current controlled current source, I_{corr}.

Here, J_{ss} is the reverse saturation current density, and m_{gs} and m_{gd} are the gate-source and gate-drain ideality factors, respectively.

A more accurate description proposed by Berroth et al.[55] introduced effective electron temperatures at the source side and the drain side of the channel. The electron temperature at the source side of the channel T_s is taken to be close to the lattice temperature, and the drain side electron temperature T_d is assumed to increase with the drain-source voltage to reflect the heating of the electrons in this part of the channel. The resulting gate leakage current can be written as

$$I_g = J_{gs} \frac{LW}{2} \left[\exp\left(\frac{V_{gs}}{m_{gs} V_{ths}} \right) - 1 \right] + \frac{LW}{2} \left[J_{gd} \exp\left(\frac{V_{gd}}{m_{gd} V_{thd}} \right) - J_{gs} \right], \tag{7.128}$$

where J_{gs} and J_{gd} are the reverse saturation current densities for the gate-source and the gate-drain diodes, respectively, and $V_{ths} = k_B T_s/q$ and $V_{thd} = k_B T_d/q$. The second term in Eq. (7.128) accounts for the gate-drain leakage current and for the fact that the effective temperature of the electrons in the metal is maintained at the ambient temperature.

In GaAs MESFETs, the reverse gate saturation current is usually also dependent on the reverse bias.[56] The following expression accounts for this dependence:[52]

$$
I_g = J_{gs} \frac{LW}{2} \left[\exp\left(\frac{V_{gs}}{m_{gs} V_{ths}} \right) - 1 \right] + \frac{LW}{2} g_{gs} V_{gs} \exp\left(-\frac{q V_{gs} \delta_g}{k_B T_s} \right) +
$$

$$
\frac{LW}{2} \left[J_{gd} \exp\left(\frac{V_{gd}}{m_{gd} V_{thd}} \right) - J_{gs} \right] + \frac{LW}{2} g_{gd} V_{gd} \exp\left(-\frac{q V_{gd} \delta_g}{k_B T_s} \right),
$$

(7.129)

where g_{gs} and g_{gd} are the reverse diode conductances and δ_g is the reverse bias conductance parameter. However, using the above expressions directly will cause a kink in the gate current and a discontinuity in its derivatives at zero applied voltage. Equation (7.129) is valid for both negative and positive values of V_{gs} and V_{gd}.

Figure 7.90 compares the measured gate leakage current with the model implemented in AIM-Spice and described above.[6]

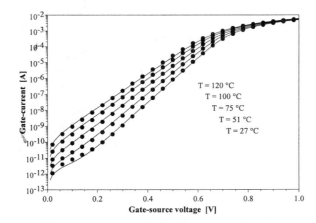

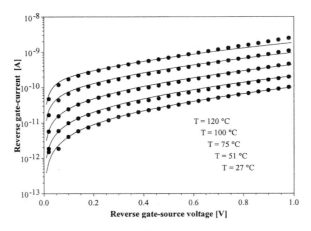

FIGURE 7.90 Measured (symbols) and simulated (lines) gate current versus gate bias for (a) positive and (b) negative gate-source voltages at different temperatures. Temperature parameters: $\Phi_{b1} = 0.96$ meV/K, $\xi = 0.033$ K^{-1}. (After Ytterdal et al.[6])

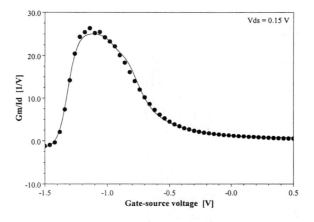

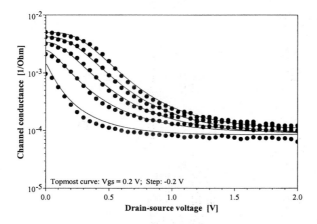

FIGURE 7.91 Measured (symbols) and simulated using AIM-Spice (lines) drain current characteristics of device A operating at room temperature; (a) ratio of transconductance and drain current, (b) channel conductance. (After Ytterdal et al.[6])

Figure 7.91 shows that the GaAs MESFET model implemented in AIM-Spice accurately reproduces the differential characteristics of the devices. Therefore, this model is suitable for the simulations of analog, microwave, and mixed-mode circuits.

Hetero-dimensional (2D MESFETs)

Hetero-dimensional MESFET technology utilizes the Schottky contact between a 3D metal and a 2D electron-gas in a semiconductor. This technology holds promise for the fabrication of high-speed devices with low power consumption.[57–62] However, this is still a very immature technology that has not found its way into production.

Figure 7.92 shows the 3D-2D Schottky barrier junction.

The depletion width d_{dep} of the semiconductor 2D electron gas for a reverse biased 3D-2D Schottky barrier shows a linear instead of a square root dependence of voltage:[58]

$$d_{dep} = \frac{\varepsilon}{qn_s}\left(V_{bi} - V\right) \tag{7.130}$$

Here n_s is the sheet density in the 2D electron gas (2-DEG), V_{bi} is the built-in voltage of the junction, and V is the voltage applied to the junction.

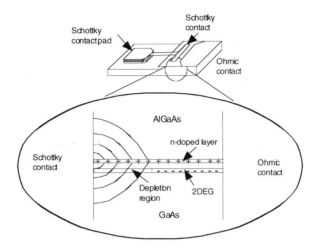

FIGURE 7.92   Schematic structure of a 3D-2D Schottky diode. (After Peatman et al.[63])

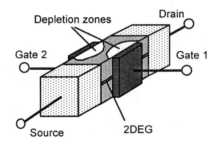

FIGURE 7.93   Schematic structure of a 2D MESFET. (After Peatman et al.[67])

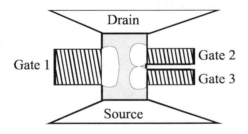

FIGURE 7.94   Structure of a three-gate 2D MESFET. (After Iñiguez et al.[69])

Figure 7.93 shows the two-dimensional metal-semiconductor field-effect transistor (2D MESFET). This transistor utilizes Schottky gates on both sides of a degenerate 2-DEG channel to laterally modulate the current between the drain and source.[65–66]

The novel geometry of this 2D MESFET eliminates or reduces parasitic effects associated with top planar contacts of conventional FETs, such as narrow-channel and short-channel effects. The output conductance in the saturation regime is quite small, and the junction capacitance of the 3D-2D Schottky diode is also small. This results in a low power-delay product.

The functionality of the 2D MESFET can be further enhanced by using multiple gates on both sides of the channel, as shown in Fig. 7.94. Two- and three-gate 2D MESFETs with excellent electrical performance have been fabricated.[68] The 2D MESFET also holds promise for microwave analog applications,

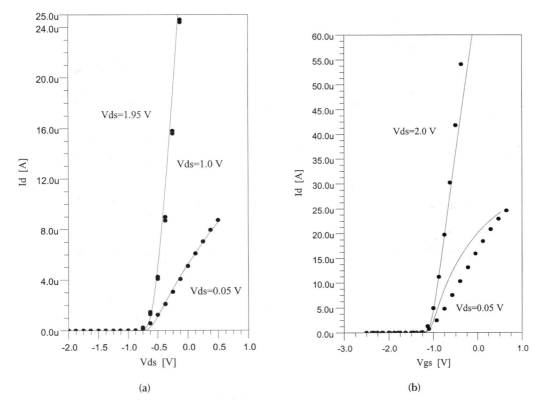

FIGURE 7.95 I-V characteristics for 2D MESFETs with the gates tied together for (a) L = 3 μm, (b) L = 0.5 μm. (From Iñiguez et al.[69])

where the small capacitance of the 3D-2D junction should lead to low channel and amplifier noise. Also, the high transconductance, and the fact that the transconductance and the output conductance do not vary much across the broad range of gate biases, are advantageous factors for linear amplification.

The 2D MESFET prototype devices were fabricated using a pseudomorphic $Al_{0.25}Ga_{0.75}As/In_{0.2}Ga_{0.8}As$ hetero-structure grown on a semi-insulating GaAs substrate.[68–70,72–74] Ni/Ge/Au ohmic contacts were formed using standard contact UV lithography and evaporation/lift-off techniques. The gate pattern was defined using electron beam lithography. The Pt/Au gates were deposited into the gate trench using capacitor discharge electroplating. Cr/Au contact pads were evaporated on the wafer and a wet etch was used to isolate the ohmic and Schottky pads (see Reference 64 for more details.)

A unified 2D MESFET is described in Reference 69. Figures 7.95 and 7.96 show the comparison between the measured and calculated 2D-MESFET I-V characteristics.

Applications

GaAs MESFETs play an important role in both analog and digital applications, such as in satellite and fiber-optic communication systems, in cellular phones and other wireless equipment, in automatic IC test equipment, and for other diverse civilian and military uses. GaAs MESFETs have been used in highly efficient microwave power amplifiers, since they combine low on resistance and high cutoff frequency. GaAs semi-insulating substrates also present a major advantage for microwave applications, since they decrease parasitic capacitance and allow for fabrication of passive elements with low parasitics for microwave monolithic integration. GaAs MESFETs have also found applications in linear low-noise amplifiers.

Figure 7.97 compares the cutoff frequencies and maximum frequencies of oscillations for different GaAs technologies.

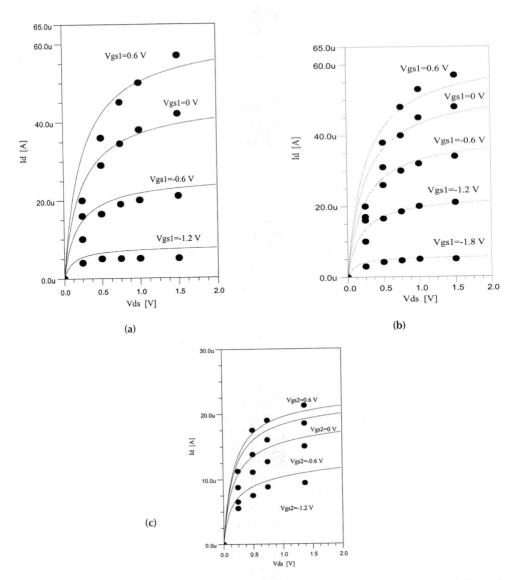

FIGURE 7.96   *I-V* characteristics for a three-gate 2D MESFET with $L = 2$ μm and $W = 1$ μm; (a) $V_{GS1} = V_{GS2} = V_{GS2} = V_{GS3}$, (b) $V_{GS2} = V_{GS3} = 0.6$ V. (c) $V_{GS1} = -1.2$ V, $V_{GS3} = 0.6$ V. Symbols: measurements. Solid lines: AIM-Spice simulations.

As can be seen from the figure, GaAs MESFETs exhibit quite respectable microwave performance and, given a lower cost of GaAs MESFETs compared to more advanced hetero-structure devices, they could capture a sizeable portion of the microwave market.

GaAs MESFET technology has also been used in efficient DC-to-DC converters that demonstrated a high switching speed.[71] These devices are capable of operating at a higher switching speeds and require a less complex circuitry.

Vitesse Semiconductor Corporation is one of the leaders in digital GaAs MESFET technology. VSC8141 and the VSC8144 SONET/SDH OC-48 multi-rate transceivers include multiplexer and demultiplexer with integrated clock generation capabilities for the physical layer.[72] Both ICs dissipate the lowest power available in the industry today — 1.2 W typically. These integrated circuits are suitable for transmission systems, optical networking equipment, networking and digital cross-connect systems, and they have 20% lower power dissipation than competing products.

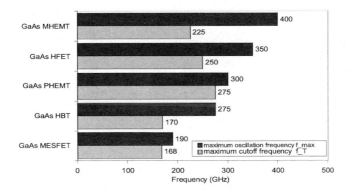

FIGURE 7.97 Maximum frequency of oscillation f_{max} and maximum cutoff frequency f_T of different GaAs technologies.[70]

Acknowledgments

The author is grateful to Mr. Tobias Werner for useful comments.

References

1. C. A. Mead, Schottky barrier gate field effect transistor, *Proc. IEEE*, 54, 2, 307–308, Feb. 1966.
2. W. W. Hooper and W. I. Lehrer, An epitaxial GaAs field-effect transistor, *Proc. IEEE*, 55, 7, 1237–1238, July 1967.
3. Y. Nakayama, K. Suyama, H. Shimizu, N. Yokoyama, H. Ohnishi, A. Shibatomi, and H. Ishikawa, A 16 × 16 bit parallel multiplier, *IEEE J. Solid-State Circuits*, SC-18, 599–603, 1983.
4. M. Feng, C. L. Lau, V. Eu, and C. Ito, Does the two-dimensional electron gas effect contribute to high-frequency and high speed performance of field-effect transistors?, *Appl. Phys. Lett.*, 57, 1233, 1990.
5. A. Mills, The GaAs IC business never so healthy! *GaAs IC Report, III-Vs Review*, 13, 1, 35–39, Jan. 2000.
6. T. Ytterdal, B-J. Moon, T. A. Fjeldly, and M. S. Shur, Enhanced GaAs MESFET CAD model for a wide range of temperatures, *IEEE Trans. Electron Devices*, 42, 10, 1724–1734, 1995.
7. M. S. Shur and L. F. Eastman, Ballistic transport in semiconductors at low-temperatures for low power high speed logic, *IEEE Trans. Electron Devices*, 26, 11, 1677–1683, Nov. 1979.
8. M. Heiblum, M. I. Nathan, D. C. Thomas, and C. M. Knoedler, Direct observation of ballistic transport in GaAs, *Phys. Rev. Lett.*, 55, 2200, 1985.
9. A. F. J. Levi, J. R. Hayes, P. M. Platzman, and W. Wiegmann, Injected hot electron transport in GaAs, *Phys. Rev. Lett.*, 55, 2071–2073, 1985.
10. G. Ruch, Electronics dynamics in short channel field-effect transistors, *IEEE Trans. Electron Devices*, ED-19, 652–654, 1972.
11. A. Cappy, B. Carnes, R. Fauquembergue, G. Salmer, and E. Constant, *IEEE Trans. Electron Devices*, ED-27, 2158–2168, 1980.
12. M. S. Shur and M. Asif Khan, Electronic and Optoelectronic AlGaN/GaN Heterostructure Field Effect Transistors, in *Proceedings of the Symposium on Wide Band Gap Semiconductors and the Twenty-Third State-of-the-Art Program on Compound Semiconductors* (SOTAPOCS XXIII), F. Ren, D. N. Buckley, S. J. Pearton, P. Van Daele, G. C. Chi, T. Kamijoh, and F. Schuermeyer, eds., Proceedings Volume 95-21, 128–135, The Electrochemical Society, Inc., New Jersey, 1995.
13. Foutz, B. E., L. F. Eastman, U. V. Bhapkar, and M. S. Shur, Comparison of high electron transport in GaN and GaAs, *Appl. Phys. Lett.*, 70, 21, 2849–2851, 1997.

14. S. Roosild, in *Microprocessor Design for GaAs Technology*, V. Milutinovic, Editor, Prentice Hall, Englewood Cliffs, NJ, 1990.

15. M. S. Shur, *Introduction to Electronic Devices*, John Wiley and Sons, New York, 1996.

16. J. A. Higgins, R. L. Kuvaas, F. H. Eisen, and D. R. Chen, *IEEE Trans. Electron Devices*, ED-25, 587–596, 1978.

17. B. M. Welch and R. C. Eden, *Int. Solid State Circuits Conf. Tech. Digest*, 205–208, 1977.

18. L. F. Eastman and M. S. Shur, Substrate Current in GaAs MESFET's, *IEEE Trans. Electron Devices*, ED-26, 9, 1359–61, Sept. 1979.

19. C. H. Chen, A. Peczalski, M. S. Shur, and H. K. Chung, Orientation and ion-implanted transverse effects in self-aligned GaAs MESFETs, *IEEE Trans. Electron Devices*, ED-34, 7, 1470–1481, July 1987.

20. M. S. Shur, *GaAs Devices and Circuits*, Plenum Publishing Corporation, New York, 1987.

21. I. Brodie and J. J. Muray, *The Physics of Microfabrication*, Plenum Press, New York and London, 1982.

22. M. S. Shur, SiC Transistors, in *SiC Materials and Devices*, Vol. 52, Y. S. Park, ed., Academic Press, New York, 161–193, 1998.

23. J. W. Palmour, H. S. Kong, D. G. Waltz, J. A. Edmond, and C. H. Carter, Jr., *Proc. of First Intern. High Temperature Electronics Conference*, Albuquerque, NM, 511–518, 1991.

24. G. Kelner and M. Shur, SiC Devices, in *Properties of Silicon Carbide*, G. Harris, ed., M. Faraday House, IEE, England, 1995.

25. G. Kelner, S. Binari, K. Sleger, and H. Kong, *IEEE Electr. Dev. Lett.*, 8, 428, 1987.

26. J. W. Palmour and J. A. Edmond, *Proc. 14th IEEE Cornell Conf.*, Ithaca, NY, 1991.

27. S. Sriram, R. C. Clarke, M. H. Hanes, P. G. McMullin, C. D. Brandt, T. J. Smith, A. A. Burk, Jr., H. M. Hobgood, D. L. Barrett, and R. H. Hopkins, SiC Microwave Power MESFETS, *Inst. Phys. Conf. Ser.*, 137, 491–494, 1993.

28. R. F. Davis, G. Kelner, M. Shur, J. W. Palmour, and J. A. Edmond, Thin Film Deposition and Microelectronic and Optoelectronic Device Fabrication and Characterization in Monocrystalline Alpha and Beta Silicon Carbide, *Proc. of IEEE*, 79, 5, 677–701, May 1991.

29. H. Morkoc, S. Strite, G. B. Gao, M. E. Lin, B. Sverdlov, and M. Burns, Large-band-gap SiC, III-V Nitride, and II-VI ZnSe-based semiconductor device technologies, *J. Appl. Phys.*, 76, 3, 1363–1398, Aug. 1994.

30. M. A. Littlejohn, J. R. Hauser, and T. H. Glisson, *Appl. Phys. Lett.*, 26, 625, 1975.

31. B. Gelmont, K. S. Kim, and M. Shur, *J. Appl. Phys.*, 74, 1818, 1993.

32. M. A. Khan, J. N. Kuznia, D. T. Olson, W. Schaff, G. Burm, M. S. Shur, and C. Eppers presented at Device Research Conference, Boulder, Colorado, 1994.

33. S. Binari, L. B. Rowland, W. Kruppa, G. Kelner, K. Doverspike, and D. K. Gatskill, *Electronics Letters*, 30, 15, 1248, July 1994.

34. R. Gaska, M. S. Shur, and A. Khan, GaN-based HEMTs, Gordon & Breach Science Publishers, O. Manasreh an Ed Yu, eds., in press.

35. M. S. Shur and M. A. Khan, *GaN and AlGaN Devices: Field Effect Transistors and Photodetectors*, S. Pearton, ed., Gordon and Breach Science Publishers, Amsterdam, 1999, 47–86.

36. G. U. Jensen, B. Lund, M. S. Shur, and T. A. Fjeldly, Monte Carlo simulation of semiconductor devices, *Computer Physics Communications*, 67, 1, 1–61, 1991.

37. K. Hess K and C. Kizilyalli, Scaling and transport properties of high electron mobility transistors, *IEDM Technical Digest*, Los Angeles, 556–558, 1986.

38. A. Afzalikushaa and G. Haddad, High-frequency characteristics of MESFETs, *Solid-State Electronics*, 38, 2, 401–406, Feb. 1995.

39. C. Jacoboni and P. Lugli, *The Monte Carlo Method for Semiconductor Simulation*, Springer Verlag, Vienna, 1989.

40. C. Moglestue, *Monte Carlo Simulation of Semiconductor Devices*, Chapman & Hall, London, 1993.

41. K. Tomizawa, *Numerical Simulation of Submicron Semiconductor Devices*, Artech House, Boston, 1993.

42. M. Lundstrom, *Fundamentals of Carrier Transport*, Addison-Wesley, Reading, MA, 1990.

43. K. Bløtekjær, Transport equations for two-valley semiconductors, *IEEE Trans. Electron Devices*, 17, 38–47, 1970.

44. J. Jyegal and T. A. Demassa, New nonstationary velocity overshoot phenomenon in submicron Gallium-Arsenide field-effect transistors, *J. of Appl. Physics*, 75, 6, 3169–3175, Mar. 1994.

45. S. H. Lo and C. P. Lee, Analysis of surface-state effect on gate lag phenomena in GaAs-MESFETs, *IEEE Trans. Electron Devices*, 41, 9, 1504–1512, Sept. 1994.

46. BLAZE, *Atlas II User's Manual*, Silvaco, 1993.

47. M. S. Shur, Analytical Models of GaAs FETs, *IEEE Trans. Electron Devices*, ED-32, 1, 70–72, Jan. 1985.

48. M. S. Shur, *Introduction to Electronic Devices*, John Wiley and Sons, New York, 1996.

49. K. Lee, M. S. Shur, T. A. Fjeldly, and T. Ytterdal, *Semiconductor Device Modeling for VLSI*, Prentice Hall, Englewood Cliffs, NJ, 1993.

50. T. Fjeldly, T. Ytterdal, and M. S. Shur, *Introduction to Device and Circuit Modeling for VLSI*, John Wiley and Sons, New York, 1998.

51. B. Iñiguez, T. A. Fjeldly, M. S. Shur, and T. Ytterdal, Spice modeling of compound semiconductor devices, in Special Issue Silicon and Beyond. Advanced Device Models and circuit simulators, *IJHSES*, M. S. Shur and T. A. Fjeldly, eds., in press.

52. M. S. Shur, T. Fjeldly, Y. Ytterdal, and K. Lee, Unified GaAs MESFET model for circuit simulations, *International Journal of High Speed Electronics*, 3, 2, 201–233, June 1992.

53. J. E. Meyer, MOS models and circuit simulation, *RCA Review*, 32, 42–63, 1971.

54. M. Nawaz and T. A. Fjeldly, A new charge conserving capacitance model for GaAs MESFETs, *IEEE Trans. Electron Devices*, 44, 11, 1813–1821, 1997.

55. M. Berroth, M. Shur, and W. Haydl, Experimental studies of hot electron effects in GaAs MESFETs, in *Extended Abstracts of the 20th International Conf. on Solid State Devices and Materials* (SSDM-88), Tokyo, 255–258, 1988.

56. C. Dunn, *Microwave Semiconductor Devices and Their Applications*, H. A. Watson, ed., McGraw Hill, New York, 1969.

57. B. Gelmont, M. Shur, and C. Moglestue, Theory of junction between two-dimensional electron gas and p-type semiconductor, *IEEE Trans. Electron Devices*, 39, 5, 1216–1222, 1992.

58. S. G. Petrosyan and Y. Shik, Contact phenomena in a two dimensional electron gas, *Sov. Phys.-Semicond.*, 23, 6, 696–697, 1989.

59. W. C. B. Peatman, T. W. Crowe, and M. S. Shur, A novel Schottky/2-DEG diode for millimeter and submillimeter wave multiplier applications, *IEEE Electron Device Lett.*, 13, 11–13, 1992.

60. W. C. B. Peatman, H. Park, B. Gelmont, M. S. Shur, P. Maki, E. R. Brown, and M. J. Rooks, Novel metal/2-DEG junction transistors, *Proc. 1993 IEEE/Cornell Conference*, Ithaca, NY, 314–319, 1993.

61. W. C. B. Peatman, H. Park, and M. Shur, Two-dimensional metal-semiconductor field effect transistor for ultra low power circuit applications, *IEEE Electron Device Lett.*, 15, 7, 245–247, 1994.

62. M. S. Shur, W. C. B. Peatman, H. Park, W. Grimm, and M. Hurt, Novel heterodimensional diodes and transistors, *Solid-State Electronics*, 38, 9, 1727–1730, 1995.

63. W. C. B. Peatman, T. W. Crowe, and M. S. Shur, A novel Schottky/2-DEG diode for millimeter and submillimeter wave multiplier applications, *IEEE Electron Device Lett.*, 13, 1, 1992.

64. W. C. B. Peatman, M. J. Hurt, H. Park, T. Ytterdal, R. Tsai and M. Shur, Narrow channel 2-D MESFET for low power electronics, *IEEE Trans. Electron Devices*, 42, 9, 1569–1573, 1995.

65. W. C. B. Peatman, R. Tsai, T. Ytterdal, M. Hurt, H. Park, J. Gonzales, and M. Shur, Sub-half-micrometer width 2-D MESFET, *IEEE Electron Device Lett.*, 17, 2, 40–42, 1996.

66. M. Hurt, W. C. B. Peatman, R. Tsai, T. Ytterdal, M. Shur, and B. J. Moon, An ion-implanted 0.4 μm wide 2-D MESFET for low-power electronics, *Electronics Lett.*, 32, 8, 772–773, 1996.

67. W. C. B. Peatman, H. Park, B. Gelmont, M. S. Shur, P. Maki, E. R. Brown, and M. J. Rooks, Novel metal/2-DEG junction transistors, in *Proc. 1993 IEEE/Cornell Conf.*, Ithaca, NY, 314, 1993.

68. J. Robertson, T. Ytterdal, W. C. B. Peatman, R. Tsai, E. Brown, and M. Shur, RTD/2-D MESFET/RTD logic elements for compact, ultra low-power electronics, *IEEE Trans. Electron Devices*, 44, 7, 1033–1039, 1997.

69. B. Iñiguez, J. –Q. Lü, M. Hurt, W. C. B. Peatman, and M. S. Shur, Modeling and simulation of single and multiple gate 2-D MESFETs, *IEEE Trans. Electron Devices*, 46, 8, 1999.

70. T. Werner and M. S. Shur, GaAs microwave transistors, unpublished.

71. http://www.anadigics.com/GaAsline/mesfets.html

72. http://www.vitesse.com/news/101199.htm

7.2.5 High Electron Mobility Transistors (HEMTs)

Prashant Chavarkar and Umesh Mishra

The concept of modulation doping was first introduced in 1978.[1] In this technique electrons from remote donors in a higher bandgap material transfer to an adjacent lower gap material. The electrostatics of the heterojunction results in the formation of a triangular well at the interface, which confines the electrons in a two-dimensional (2D) electron gas (2DEG). The separation of the 2DEG from the ionized donors significantly reduces ionized impurity scattering resulting in high electron mobility and saturation velocity.

Modulation-doped field effect transistors (MODFETs) or high electron mobility transistors (HEMTs), which use the 2DEG as the current conducting channel have proved to be excellent candidates for microwave and millimeter-wave analog applications and high-speed digital applications. This progress has been enabled by advances in crystal growth techniques such as molecular beam epitaxy (MBE) and metal-organic chemical vapor deposition (MOCVD) and advances in device processing techniques, most notably electron beam lithography, which has enabled the fabrication of HEMTs with gate lengths down to 0.05 μm.

However, using a high electron mobility channel alone does not guarantee superior high-frequency performance. It is crucial to understand the principles of device operation and to take into consideration the effect of scaling to design a microwave or millimeter-wave HEMT device. The advantages and limitations of the material system used to implement the device also need to be considered. This section therefore begins with a discussion on the device operation of a HEMT. This is followed by a discussion of scaling issues in HEMT, which are of prime importance, as the reduction of gate length is required to increase the operating frequency of the device.

The first HEMT was demonstrated in the AlGaAs/GaAs material system in 1981. It demonstrated significant performance improvements over the GaAs MESFET at microwave frequencies. However, the high-frequency performance was not sufficient for operation at millimeter-wave frequencies. In the past twenty years, the AlGaAs/InGaAs psuedomorphic HEMT on GaAs substrate (referred to as GaAs pHEMT) and the AlInAs/GaInAs HEMT on InP substrate (referred to as InP HEMT) have emerged as premier devices for microwave and millimeter-wave circuit applications. This highlights the importance of choosing the appropriate material system for device implementation. This will be discussed in the section on Material Systems for HEMT Devices.

The next two sections will discuss the major advances in the development of the GaAs pHEMT and InP HEMT. Traditionally these devices have been used in low-volume, high-performance and high-cost military and space-based electronic systems. Recently the phenomenal growth of commercial wireless and optical fiber-based communication systems has opened up new applications for these devices. This also means that new issues like manufacturability and operation at low bias voltage have to be addressed.

HEMT Device Operation and Design

Linear Charge Control Model

The current control mechanism in the HEMT is control of the 2DEG density at the heterojunction interface by the gate voltage. Figure 7.98 shows the band diagram along the direction perpendicular to the heterojunction interface using the AlGaAs/GaAs interface as an example.

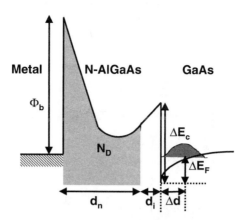

FIGURE 7.98 Schematic of conduction band diagram at the AlGaAs/GaAs interface.

The first HEMT charge control model was proposed by Delagebeaudeuf and Linh in 1982.[2] The potential well at the AlGaAs/GaAs interface is approximated by a triangular well. The energy levels in this triangular well and the maximum 2DEG density, n_{sm} can be calculated by solving the Schrödinger equation in the triangular well and Poisson equation in AlGaAs donor layers.[3] For $0 < n_s < n_{sm}$, the sheet charge density n_s as a function of gate voltage V_g can be expressed as

$$qn_s = C_s\left(V_g - V_{th}\right) \tag{7.131}$$

where C_s is the 2DEG capacitance per unit area and is given by the following expression:

$$C_s = \frac{\varepsilon}{d_n + d_i + \Delta d} \tag{7.132}$$

Here Δd is the distance of the centroid of the 2DEG distribution from the AlGaAs/GaAs interface and is typically or the order of 80 Å for $n_s \sim 10^{12}/\text{cm}^2$. Here V_{th} is the threshold voltage or pinch-off voltage and is given by,

$$V_{th} = \phi_b - \frac{qN_D}{2\varepsilon}d_n^2 - \Delta E_c + \Delta E_F \tag{7.133}$$

where ϕ_b, N_D, and d_n are the Schottky barrier height on the donor layer, doping density, and doped layer thickness as illustrated in Fig. 7.98. Here, ΔE_F is the Fermi potential of the 2DEG with respect to the bottom of the conduction band. It can be expressed as a function of 2DEG density as follows

$$\Delta E_F = \Delta E_{FO}\left(T\right) + an_s \tag{7.134}$$

where $\Delta E_{FO}(T) = 0$ at 300 K, $a = 0.125 \times 10^{-16}$ V/m^2.

This simplified version of charge control is accurate only at low temperature. At room temperature, apart from the 2DEG charge density n_s, the gate voltage also modulates the bound carrier density, n_{bound} in the donor layer and the free electrons, n_{free} in the donor layers. This results in premature saturation of the sheet charge and degradation of device performance. A more accurate model for charge control, which solves Poisson's and Schrödinger's equations in a self-consistent manner was proposed by Vinter.[4]

Modulation Efficiency

The parasitic modulation of charge in the higher bandgap donor layer reduces the efficiency of the gate voltage to modulate the drain current, as the carriers in the donor layers do not contribute the drain current. The modulation efficiency (η) of the FET is proportional to ratio between the change in drain current (δI_{ds}) and the change in total charge (δQ_{tot}) required to cause this change.[5] This ratio is defined as follows,

$$\eta \propto \frac{\delta I_{ds}}{\delta Q_{tot}} = \frac{\delta(q v_{sat} n_s)}{\delta q(n_s + n_{bound} + n_{free})} \tag{7.135}$$

Dividing the numerator and denominator by the change in gate voltage, δV_g that is required to cause this change, the following expression is obtained,

$$\frac{\delta I_{ds}}{\delta Q_{tot}} = v_{sat} \frac{\delta(n_s)/\delta V_g}{\delta(n_s + n_{bound} + n_{free})/\delta V_g} \tag{7.136}$$

The modulation efficiency is defined as the ratio of the rate of change of the useful charge, i.e., the 2DEG over that of the total charge,

$$\eta = \frac{\delta(n_s)/\delta V_g}{\delta(n_s + n_{bound} + n_{free})/\delta V_g} = \frac{\delta(n_s)/\delta V_g}{C_{TOT}} = \frac{C_s}{C_{TOT}} \tag{7.137}$$

The relation between the modulation efficiency and high frequency performance of the FET is evident in the expressions for transconductance (g_m) and current gain cutoff frequency (f_T).

$$g_m = \frac{\delta I_{ds}}{\delta V_g} = \frac{\delta(q v_{sat} n_s)}{\delta V_g} = q v_{sat}(\delta n_s/\delta V_g);$$

$$C_{gs} = C_{TOT} L_g; \tag{7.138}$$

$$f_T = \frac{g_m}{2\pi C_{gs}} = \frac{q v_{sat}(\delta n_s/\delta V_g)}{2\pi L_g C_{TOT}} = \frac{v_{sat}}{2\pi L_g}\eta$$

Hence, to improve the high frequency performance it is essential to improve the modulation efficiency.

Equation (7.138) must be used with caution in case of short gate length HEMTs. The saturation velocity v_{sat} may be replaced by the effective velocity v_{eff}. Usually v_{eff} is higher than v_{sat} due to high field and velocity overshoot effects. Using v_{sat} in this case may lead to values of modulation efficiency that are greater than 100%.

Current-Voltage (I-V) Models for HEMTs

By assuming linear charge control, gradual channel approximation, and a 2-piece linear velocity-field model, the expression for the saturated drain current I_{DSS} in a HEMT is given by,[2]

$$I_{DSS} = C_s v_{sat}\left(\sqrt{(E_c L_g)^2 + (V_g - V_c(0) - V_{th})^2} - E_c L_g\right) \tag{7.139}$$

Here E_c is defined as the critical electric field at which the electrons reach their saturation velocity v_{sat} and $V_c(0)$ is the channel potential at the source end of the gate. For a long gate length HEMT, Eq. (7.139) is valid until the onset of donor charge modulation, that is, $0 < n_s < n_{sm}$. The intrinsic transconductance of the device obtained by differentiating this expression with respect to the gate voltage and is expressed as follows:

$$g_{mo} = \frac{\delta I_{ds}}{\delta V_g} = C_s v_{sat} \frac{V_g - V_c(0) - V_{th}}{\sqrt{\left(V_g - V_c(0) - V_{th}\right)^2 + \left(E_c L_g\right)^2}} \tag{7.140}$$

For a short gate length HEMT, the electric field in the channel is much greater in magnitude than the critical electric field E_c. Assuming that the entire channel of the FET operates in saturated velocity mode, we can make the following assumption, that is, $V_g - V_c(0) - V_{th} \gg E_c L_g$. Then using Eqs. (7.131), (7.139), and (7.140) are reduced to the following:

$$I_{DSS} = q n_s v_{sat} \tag{7.141}$$

$$g_m = C_s v_{sat} \tag{7.142}$$

More insight can be obtained in terms of device parameters if the equation for charge control [Eq. (7.131)] is substituted in the expressions for I_{ds} and V_g as follows:[6]

$$I_{ds} = q v_{sat} n_s \left[\sqrt{1 + \left(\frac{n_c}{n_s}\right)^2} - \frac{n_c}{n_s} \right] \tag{7.143}$$

$$g_{mo} = C_s v_{sat} \frac{1}{\sqrt{1 + \left(n_c/n_s\right)^2}} \tag{7.144}$$

where $n_c = E_c C_s L_g / q$ and $0 < n_s < n_{sm}$. Dividing both sides of Eq. (7.144) by $C_s v_{sat}$ the following expression for modulation efficiency is obtained:

$$\eta = \frac{1}{\sqrt{1 + \left(n_c/n_s\right)^2}} \tag{7.145}$$

Hence it is necessary to maximize the 2DEG density n_s to maximize the current drive, transconductance, and modulation efficiency of the HEMT. Although this is in contrast with the saturated-velocity model, it agrees with the experimental results. The foregoing results can also be used to select the appropriate material system and layer structure for the fabrication of high-performance microwave and millimeter-wave HEMTs.

Although the analytical model of device operation as was described here provides great insight into the principles of device operation and performance optimization, it fails to predict some of the nonlinear phenomena such as reduction of g_m at high current levels (g_m compression) and soft pinch-off characteristics. A model has been developed to explain these phenomena.[5] The total charge in the HEMT is divided into three components. The first, Q_{SVM}, is the charge required to support a given I_{ds} under the saturated velocity model (SVM). This charge is uniformly distributed under the gate. In reality this is not the case as the electron velocity under the gate varies. To maintain the current continuity under the

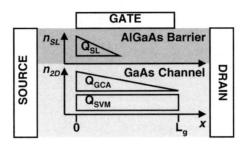

FIGURE 7.99 Schematic diagram showing the location and distribution of Q_{SVM}, Q_{GCA}, and Q_{SL} in a HEMT (Foisy et al. 1988).

gradual channel approximation (GCA), extra charge under the channel has to be introduced. This is defined as Q_{GCA} and is maximum at the source end of the gate and minimum at the drain end.

The excess charge in the wide bandgap electron supply layer is denoted by Q_{SL}. Figure 7.99 shows the location and distribution of these charges in the HEMT. Only Q_{SVM} supports current density and thus contributes to the transconductance of the HEMT. The other two components contribute only to the total capacitance of the device. Hence the modulation efficiency (ME) of the HEMT in terms of these charges is expressed as

$$\eta = \frac{\delta Q_{SVM}}{\delta\left(Q_{SVM} + Q_{GCA} + Q_{SL}\right)} \tag{7.146}$$

and the transconductance can be expressed as $g_m = C_s v_{sat} \eta$.

Figure 7.100 shows the variation of ME as a function of drain current density for an AlGaAs/GaAs HEMT and an AlGaAs/InGaAs pHEMT. At low current density, ME is low as most of the charge in the 2DEG channel has to satisfy the gradual channel approximation. This low value of ME results in low transconductance and soft pinch-off characteristics at low drain current densities. In the high current regime, modulation of Q_{SL} reduces the ME, resulting in gain compression. In the intermediate current regime the ME is maximum. However, if there exists a bias condition where both Q_{GCA} and Q_{SL} are modulated (as in the low band offset AlGaAs/GaAs system), it severely affects the ME.

For optimal high-power and high-frequency performance, it is necessary to maximize the range of current densities in which ME is high. The drop off in ME due to parasitic charge modulation in the donor layers can be pushed to higher current density by increasing the maximum 2DEG density n_{sm}. The 2DEG density can be maximized by using planar doping in the donor layer and by increasing the conduction band discontinuity at the barrier/2DEG interface. The drop off in ME due to operation in gradual channel mode can be pushed to lower current densities by reducing the saturation voltage V_{Dsat}. This is achieved by increasing the mobility of the electrons in the 2DEG channel and by reducing the gate length. As seen from Fig. 7.100 higher modulation efficiency is achieved over a larger range of current density for the AlGaAs/InGaAs pHEMT, which has higher sheet charge density, mobility, and band discontinuity at the interface than the AlGaAs/GaAs HEMT.

Small Signal Equivalent Circuit Model of HEMT

The small signal equivalent circuit model of the HEMT is essential for designing HEMT-based amplifiers. The model can also provide insights into the role of various parameters in the high-frequency performance of the device. Figure 7.101 shows the small signal equivalent circuit for a HEMT. The grey box highlights the intrinsic device. The circuit elements in the preceding model are determined using microwave S-parameter measurements.[7,8] The intrinsic circuit elements are a function of the DC bias, whereas the extrinsic circuit elements or parasitics are independent of it. The two measures of the high frequency performance of a FET can now be defined in terms of the small signal model of the device as follows. The current gain cutoff frequency, f_T can be defined as

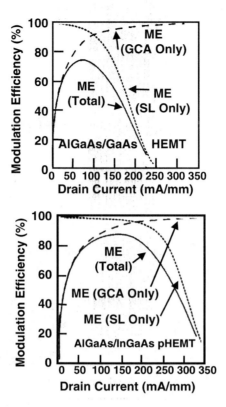

FIGURE 7.100   Modulation efficiency as a function of current density for GaAs HEMT and GaAs pHEMT.

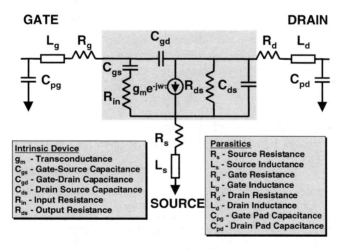

FIGURE 7.101   Small signal equivalent circuit of a HEMT.

$$f_T = \frac{g_m}{2\pi\left(C_{gs} + C_{gd}\right)} \qquad (7.147)$$

Hence, to increase the current gain cutoff frequency it is essential to increase the g_m and reduce C_{gs} and C_{gd}. Referring to Eq. (7.138), it is clear that this can be achieved by increasing electron velocity in the channel and reducing gate length. The current gain cutoff frequency is mainly a physical measure of device performance. A more practical measure of high-frequency device performance is f_{max}, the power gain cutoff frequency. This is the frequency at which the power gain of the FET is unity. It is defined as follows,[9]

$$f_{max} = \frac{f_T}{\sqrt{4g_{ds}\left(R_{in} + \frac{R_s + R_g}{1 + g_m R_s}\right) + \frac{4}{5}\frac{C_{gd}}{C_{gs}}\left(1 + \frac{2.5C_{gd}}{C_{gs}}\right)(1 + g_m R_s)^2}} \qquad (7.148)$$

A simple form of Eq. (7.148) is:

$$f_{max} = f_T \sqrt{\frac{R_{ds}}{4R_{in}}} = \frac{f_T}{\sqrt{4g_{ds}R_{in}}} \qquad (7.149)$$

To improve the f_{max} of the device it is necessary to minimize the quantities in the denominator of Eq. (7.148). The crucial parameters here are the output conductance of the device g_{ds}, the source and gate parasitic resistances R_s and R_g, and the gate-drain feedback capacitance C_{gd} that need to be minimized. Reduction of g_{ds} can be achieved by appropriate vertical scaling (to be discussed in the next section). Reduction of R_s and R_g depends mainly on the process technology. Reduction of C_{gd} can be achieved by proper design of the gate-drain region of the FET. The crucial parameter in the design of the gate drain depletion region is the gate-drain separation L_{gd}.[10] Increasing L_{gd} reduces C_{gd} but also increases the effective gate length of the device, reducing the short channel effects. The optimum value of L_{gd} is 2.3 times that of the gate length L_g. Thus it is clear that f_{max} is a better measure of the high-frequency performance of a FET as it is determined not only by the material system used but also by the process technology and device design parameters. Large signal models of HEMTs are essential for designing power amplifiers and are similar to those of MESFETs.

Scaling Issues in Ultra-High-Speed HEMTs

The frequency at which a HEMT operates is limited by the electron transit time from the source to the drain. Therefore to increase the frequency of operation it is necessary to reduce the gate length. However, as the gate length approaches 0.1 μm it is necessary to reduce the other parasitic delays in the device and take into account short channel effects to maintain the high-frequency performance of the HEMT.

Delay Time Analysis

The reduction of parasitic delays in a FET is essential to improve the high frequency performance as these delays can be as high as 45% of the intrinsic delay.[11] Considering the small-signal model of a FET, the total delay t_T in a FET can be expressed as follows:[12]

$$t_T = t_{pad} + t_{fringe} + t_{channel} + t_{transit} + t_{drain} = 1/(2\pi f_T) \qquad (7.150)$$

Here t_{pad} is the charging time for the parasitic pad capacitance and is given by

$$t_{pad} = C_{pad}/g_m \cdot W \qquad (7.151)$$

where C_{pad} is the pad capacitance and is typically 10 *fF* per 50 μm × 50 μm bonding pad, g_m is the extrinsic transconductance per unit gate width, and W is the width of the device. To minimize t_{pad} it is necessary to have a high gate width, high transconductance HEMT.

The gate fringe capacitance charging time (t_{fringe}) is given by

$$t_{fringe} = C_{fringe}/g_{mo} \qquad (7.152)$$

where g_{mo} is the intrinsic transconductance of the HEMT and is related to the extrinsic transconductance (g_m) and source resistance R_s by the following expression:

$$g_m = g_{mo} / \left(1 + g_{mo} \cdot R_s\right) \qquad (7.153)$$

The gate fringe capacitance C_{fringe} is typically 0.18 pF/mm, hence for a HEMT with an intrinsic transconductance of 1000 to 1500 mS/mm, t_{fringe} is approximately 0.1 to 0.2 ps.

Channel charging delay $t_{channel}$ is associated with RC delays and is proportional to channel resistance. The channel charging delay is minimum at high current densities. The channel charging delay can be considered as a measure of the effectiveness of a FET operating in the saturated velocity mode.

The transit delay of the FET, $t_{transit}$, can be expressed as the time required to traverse under the gate and is given by

$$t_{transit} = L_g / v_{sat} \qquad (7.154)$$

The drain delay (t_{drain}) is the time required by the electron to traverse the depletion region between the gate and the drain and is a function of bias conditions.[13] The drain delay increases with drain bias as the length of the depletion region beyond the gate increases. Drain delay is an important parameter for millimeter-wave power HEMTs. To increase the breakdown voltage of the device, the gate-to-drain spacing has to be increased. When the device is biased at a high drain voltage to maximize the power output, it creates a drain depletion region that is on the order of gate length of the device. Thus the drain delay becomes a major component of the total delay in the device, and can limit the maximum f_T and f_{max}.

Vertical Scaling

Aspect ratio (the ratio between the gate length L_g and the gate-to-channel separation $d_{Barrier}$) needs to be maintained when gate length is reduced. Aspect ratio is a critical factor affecting the operation of the field effect transistor and should be maintained above five. As the gate length is reduced, the distance between the gate and 2DEG (the distance $d_n + d_i$ as seen in Fig. 7.98) has to be reduced so that the aspect ratio of the device is maintained.

However, maintaining the aspect ratio alone does not guarantee improvement in device performance. This is clear if the variation of threshold voltage with the reduction in $d_{Barrier}$ is examined. It is clear that d_i cannot be reduced, as it will result in degradation of mobility in the 2DEG channel due to scattering from the donors in the barrier layers. Therefore, to maintain aspect ratio, the thickness of the doped barrier layer d_n has to be reduced. By examining Eq. (7.133) for threshold voltage, it is clear that this makes the threshold voltage more positive.

At first glance, this does not seem to affect device performance. The effect of the more positive threshold voltage is clear if the access regions of the device are considered. A more positive threshold voltage results in reduction of sheet charge in the access region of the device. This increases the source and drain resistance of the device, which reduces the extrinsic transconductance [see Eq. (7.153)] and also increases the channel charging time (due to increased RC delays). Thus the increased parasitic resistances nullify the improvements in speed in the intrinsic device.

The threshold voltage of the device must be kept constant with the reduction in d_n. From Eq. (7.133) it can be seen that the doping density in the high bandgap donor has to be increased. Since the threshold voltage varies as a square of the doped barrier thickness, a reduction in its thickness by a factor of 2 requires that the doping density be increased by a factor of 4. High doping densities can be difficult to achieve in wide bandgap materials such as AlGaAs due to the presence of DX centers. Increased doping also results in higher gate leakage current, higher output conductance, and a lower breakdown voltage. Utilizing planar or delta doping wherein all the dopants are located in a single plane can alleviate these problems. This leaves most of the higher bandgap layer undoped and enables reduction of its thickness.

The threshold voltage of a planar-doped HEMT is given as follows[14]

$$V_T = \phi_B - \frac{qN_{2D}d_n}{\varepsilon} - \Delta E_c + \Delta E_F \qquad (7.155)$$

where N_{2D} is the per unit area concentration of donors in the doping plane and d_n is the distance of the doping plane from the gate. In this case, the 2D doping density has to increase linearly with the reduction in barrier thickness. The transfer efficiency of electrons from the donors to the 2DEG channel also is increased, as all the dopant atoms are close to the 2DEG channel. Hence higher 2DEG sheet densities can be achieved in the channel and thus planar doping enables efficient vertical scaling of devices with reduction in gate length.[15] From a materials point of view, efficient vertical scaling of a HEMT requires a high bandgap donor/barrier semiconductor that can be doped efficiently.

The voltage gain of the device (g_m/g_{ds}) can be considered as a measure of short channel effects in the device. The reduction of gate length and the gate-to-channel separation results in an increase in the transconductance of the device. However, to reduce the output conductance g_{ds} of the device, it is also necessary to reduce the channel thickness, which then increases the carrier confinement in the channel. Enoki et al. have investigated the effect of the donor/barrier and channel layer thickness on the voltage gain of the device.[16] The gate-to-channel separation ($d_{Barrier}$) and the channel thickness ($d_{channel}$) were varied for a 0.08-μm gate length AlInAs/GaInAs HEMT. For a $d_{Barrier}$ of 170 Å and a $d_{channel}$ of 300 Å, the g_m was 790 mS/mm and g_{ds} was 99 mS/mm, resulting in a voltage gain of 8. When $d_{Barrier}$ was reduced to 100 Å and $d_{channel}$ was reduced to 150 Å, the g_m increased to 1100 mS/mm and g_{ds} reduced to 69 mS/mm; this doubled the voltage gain to 16. This illustrates the necessity to reduce the channel thickness to improve charge control in ultra-short gate length devices.

Subthreshold slope is an important parameter to evaluate short channel effects for digital devices. A high value of subthreshold slope is necessary to minimize the off-state power dissipation and to increase the device speed. Two-dimensional simulations performed by Enoki et al. indicate that reduction in channel thickness is more effective than the reduction in barrier thickness, for maintaining the subthreshold slope with reduction in gate length.[16]

The high-frequency performance of a device is a function of the electrical gate length $L_{g,eff}$ of the device, which larger than the metallurgical gate length L_g due to lateral depletion effects near the gate. The relation between $L_{g,eff}$ and L_g is given by,[14]

$$L_{g,eff} = L_g + \beta\left(d_{Barrier} + \Delta d\right) \qquad (7.156)$$

where $d_{Barrier}$ is the total thickness of the barrier layers, Δd is the distance of the centroid of the 2DEG from the channel barrier interface and is on the order of 80 Å. The value of parameter β is 2.

Consider a long gate length HEMT ($L_g = 1$ μm) with a barrier thickness of 300 Å. Using Eq. (7.156), the value of 1.076 μm is obtained for $L_{g,eff}$. Thus the effective gate length is only 7.6% higher than the metallurgical gate length. Now consider an ultra-short gate length HEMT ($L_g = 0.05$ μm) with an optimally scaled barrier thickness of 100 Å. Using the same analysis, a value of 0.086 μm is obtained for $L_{g,eff}$. In this case the effective gate length is 43% higher than the metallurgical gate length. Hence, to improve the high-frequency performance of a ultra-short gate length HEMT, effective gate length reduction along with vertical scaling is required.

Horizontal Scaling

Reduced gate length is required for the best high-frequency performance. However, it should be kept in mind that the gate series resistance increases with the reduction in gate length. This problem can be solved with a T-shaped gate. This configuration lowers the gate series resistance while maintaining a small footprint. Another advantage of the T-shaped gate is reduced susceptibility to electromigration under large signal RF drive as the large gate cross-section reduces current density. For a 0.1-μm gate length using a T-gate instead of a straight gate, reduces the gate resistance from 2000 Ω/mm to 200 Ω/mm.

The simplified expression for f_T as expressed in Eq. (7.147) does not include the effect of parasitics on the delay time in a FET. A more rigorous expression for f_T, which includes the effects of parasitics on f_T was derived by Tasker and Hughes and is given here,[17]

$$f_T = \frac{g_m/2\pi}{\left[C_{gs}+C_{gd}\right]\left[1+\left(R_s+R_d\right)/R_{ds}\right]+C_{gd}g_m\left(R_s+R_d\right)} \tag{7.157}$$

It is clear from Eq. (7.157) that it is necessary to reduce source and drain resistances R_s and R_d, respectively, to increase the f_T of a FET. Mishra et al. demonstrated a record f_T of 250 GHz for a 0.15-μm device with a self-aligned gate, which reduces the gate-source and gate-drain spacing and results in the reduction of R_s and R_d.[18] Equation (7.157) can be rearranged as follows,[17]

$$\frac{1}{2\pi f_T} = \frac{\left(C_{gs}+C_{gd}\right)}{g_m} + \frac{\left(C_{gs}+C_{gd}\right)\left(R_s+R_d\right)}{g_m R_{ds}} + C_{gd}\left(R_s+R_d\right) \tag{7.158}$$

where the first term on the right-hand side is the intrinsic delay of the device (τ_{int}) and the rest of the terms contribute to parasitic delay (τ_p). From this equation the ratio of parasitic delay to the total delay ($\tau_t = \tau_p + \tau_{int}$) is given as

$$\frac{\tau_p}{\tau_t} = g_m\left(R_s+R_d\right)\left[\frac{G_{ds}}{g_m} + \frac{1}{\left[1+C_{gs}/C_{gd}\right]}\right] \tag{7.159}$$

Hence to improve the f_T of the device, the parasitic source and drain resistance have to be reduced as the gate length of the device is reduced. This minimizes the contribution of the parasitic delays to the total delay of the device.

Material Systems for HEMT Devices

The previous portions of this section discussed the various device parameters crucial to high-frequency performance of HEMTs. In this section the relationship between material and device parameters will be discussed. This will enable the selection of the appropriate material system for a particular device application. Table 7.10 illustrates the relationship between the device parameters and material parameters

TABLE 7.10   Relationship between Device and Material Parameters

Device Type	Device Parameters	Material Parameters	
		2DEG Channel Layer	Barrier/Buffer Layer
Short Gate Length Devices	High Electron Velocity	High Electron Velocity High Electron Mobility	
	High Aspect Ratio		High Doping Efficiency
Power Devices	High Current Density	High 2DEG Density	
	Low Gate Leakage		High Schottky Barrier
	High Breakdown Voltage	High Breakdown Field	High Breakdown Field
	Low Output Conductance		High Quality Buffer
	Good Charge Control	High Modulation Efficiency	
	Low Frequency Dispersion		Low Trap Density
Low Noise Devices	Low Rs	High 2DEG density	
	High Electron Velocity	High Electron Velocity High Electron Mobility	
Digital Devices	Low Gate Leakage Current		High Schottky Barrier
	High Current Drive	High 2DEG Density	

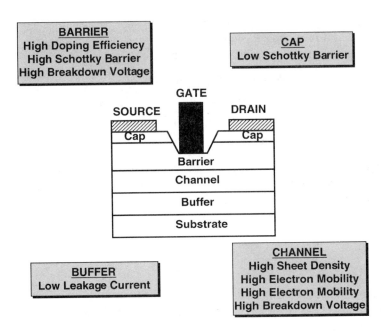

FIGURE 7.102   Material requirements for HEMT devices.

for the various constituent layers of the HEMT, namely the high bandgap donor and buffer layers, and the 2DEG channel. Figure 7.102 shows a schematic diagram of a HEMT, illustrating the material requirements from each component layer.

The first HEMT was implemented in the lattice-matched AlGaAs/GaAs system in 1981.[19] The AlGaAs/GaAs HEMT demonstrated significant improvement in low noise and power performance over GaAs MESFET due to superior electronic transport properties of the 2DEG at the AlGaAs/GaAs interface and better scaling properties. However, the limited band discontinuity at the AlGaAs/GaAs interface limits the 2DEG sheet charge density. Other undesirable effects, such as formation of a parasitic MESFET in the donor layer and real space transfer of electrons from the channel to donor, are prevalent, however.

One way to increase band discontinuity is to increase the Al composition in AlGaAs. However, the presence of deep level centers (DX centers) associated with Si donors in AlGaAs prevents the use of high Al composition AlGaAs donor layers to increase the band discontinuity and also limits doping efficiency. Problems relating to low band discontinuity can also be solved by reducing the bandgap of the channel, and by using a material that has higher electron mobility and electron saturation velocity. The first step in this direction was taken by the implementation of an AlGaAs/InGaAs pseudomorphic HEMT (GaAs pHEMT).[20] In an AlGaAs/InGaAs pHEMT the electron channel consists of a thin layer of narrow bandgap InGaAs that is lattice mismatched to GaAs by 1 to 2%. The thickness of the InGaAs channel is thin enough (~200 Å) so that the mismatch strain is accommodated coherently in the quantum well, resulting in a dislocation free "pseudomorphic" material. However the indium content in the InGaAs channel can be increased only up to 25%. Beyond this limit the introduction of dislocations due to high lattice mismatch degrades the electronic properties of the channel. The maximum Al composition that can be used in the barrier is 25% and the maximum indium composition that can be used in the channel is 25%.

Using the $Al_{0.48}In_{0.52}As/Ga_{0.47}In_{0.53}As$ material system lattice matched to InP can simultaneously solve the limitations of the high bandgap barrier material and the lower bandgap channel material. The AlInAs/GaInAs HEMT (InP HEMT) has demonstrated excellent low-noise and power performance that extends well into the millimeter-wave range; they currently hold all the high-frequency performance records for FETs. The GaInAs channel has high electron mobility (>10,000 cm^2/Vs at room temperature), high electron saturation velocity (2.6×10^7 cm/s) and higher intervalley (Γ-L) energy separation. The higher conduction band offset at the AlInAs/GaInAs interface ($\Delta E_c = 0.5$ eV) and the higher doping

TABLE 7.11 Material Parameters of AlGaAs/GaAs, AlGaAs/InGaAs, and AlInAs/GaInAs Material Systems

Material Parameter	AlGaAs/GaAs	AlGaAs/InGaAs	AlInAs/GaInAs
ΔE_c	0.22 eV	0.42 eV	0.51 eV
Maximum donor doping	$5 \times 10^{18}/cm^3$	$5 \times 10^{18}/cm^3$	$1 \times 10^{19}/cm^3$
Sheet charge density	$1 \times 10^{12}/cm^2$	$1.5 \times 10^{12}/cm^2$	$3 \times 10^{12}/cm^2$
Mobility	8000 cm²/Vs	6000 cm²/Vs	12,000 cm²/Vs
Peak Electron Velocity	2×10^7 cm/s		2.7×10^7 cm/s
Γ-L valley separation	0.33 eV		0.5 eV
Schottky Barrier	1.0 eV	1.0 eV	0.45 eV

efficiency of AlInAs (compared to AlGaAs) results in a sheet charge density that is twice that of the AlGaAs/InGaAs material system. Higher doping efficiency of AlInAs also enables efficient vertical scaling of short gate length HEMTs. The combination of high sheet charge and electron mobility in the channel results in low source resistance, which is necessary to achieve high transconductance. However, the low bandgap of the InGaAs channel results in low breakdown voltage due to high impact ionization rates.

Table 7.11 summarizes the material properties of the three main material systems used for the fabrication of HEMTs. The emergence of growth techniques like Metal Organic Chemical Vapor Deposition (MOCVD) and Gas Source Molecular Beam Epitaxy (GSMBE) and continuing improvement in the existing growth techniques like molecular beam epitaxy (MBE) have enabled a new class of phosphorus-based material systems for fabrication of HEMTs. On the GaAs substrate, the GaInP/InGaAs has emerged as an alternative to the AlGaAs/InGaAs material system. GaInP has a higher bandgap than AlGaAs and hence enables high 2DEG densities due to the increased conduction band discontinuity (ΔE_c) at the GaInP/InGaAs interface. As GaInP has no aluminum it is less susceptible to environmental oxidation. The availability of high selectivity etchants for GaAs and GaInP simplifies device processing. However, the high conduction band discontinuity is achieved only for disordered GaInP, which has a bandgap of 1.9 eV. Using graded GaInP barrier layers and an $In_{0.22}Ga_{0.78}As$ channel, 2DEG density as high as $5 \times 10^{12}/cm^2$ and a mobility of 6000 cm²/Vs was demonstrated.[21]

On InP substrates, the InP/InGaAs material system can be used in place of the AlInAs/GaInAs material system. The presence of deep levels and traps in AlInAs degrades the low frequency noise performance of AlInAs/GaInAs HEMT. Replacing the AlInAs barrier by InP or pseudomorphic InGaP can solve this problem. One disadvantage of using the InP-based barrier is the reduced band discontinuity (0.25 eV compared to 0.5 eV for AlInAs/GaInAs) at the InP/InGaAs interface. This reduces 2DEG density at the interface and modulation efficiency. Increasing the indium content up to 75% in the InGaAs channel can increase the band discontinuity at the InP/InGaAs interface. The poor Schottky characteristics on InP necessitate the use of higher bandgap InGaP barrier layers or depleted p-type InP layers. A sheet density of $3.5 \times 10^{12}/cm^2$ and mobility of 11,400 cm²/Vs was demonstrated in an $InP/In_{0.75}Ga_{0.25}As/InP$ double heterostructure.[22]

Despite the large number of material systems available for fabrication of HEMTs, the GaAs pHEMT implemented in the $Al_xGa_{1-x}As/In_yGa_{1-y}As$ (x ~ 0.25; y ~ 0.22) material system and the InP HEMT implemented in the $Al_{0.48}In_{0.52}As/Ga_{0.47}In_{0.53}As$ material system have emerged as industry vehicles for implementation of millimeter-wave analog and ultra high-speed digital circuits. The next two portions of this section will discuss the various performance aspects of GaAs pHEMT and InP HEMT.

AlGaAs/InGaAs/GaAs Pseudomorphic HEMT (GaAs pHEMT)

The first AlGaAs/InGaAs pseudomorphic HEMT was demonstrated in 1985.[23] Significant performance improvement over AlGaAs/GaAs HEMT was observed. Devices with a 1-µm gate length had peak transconductance of 270 mS/mm and maximum drain current density of 290 mA/mm.[20] The current gain cutoff frequency (f_T) was 24.5 GHz and the power gain cutoff frequency (f_{max}) was 40 GHz. An f_T of 120 GHz was reported for 0.2-µm gate length devices with $In_{0.25}Ga_{0.75}As$ channel.[24] Devices with a 0.1-µm gate length with an f_{max} of 270 GHz were demonstrated in 1989.[14]

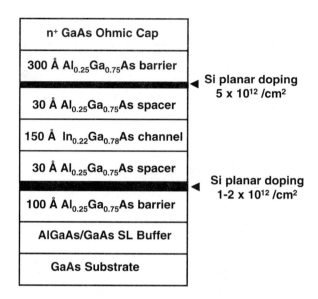

FIGURE 7.103   Layer structure of a GaAs power pHEMT.

Millimeter-Wave Power GaAs pHEMT

In the past few years, the GaAs pHEMT has emerged as a device of choice for implementing microwave and millimeter-wave power amplifiers. To achieve a high output power density, device structures with higher current density and consequently higher sheet charge are required. As the sheet charge density in a single heterojunction AlGaAs/InGaAs pHEMT is limited to $2.3 \times 10^{12}/cm^2$, a double heterojunction (DH) device structure must be used to increase the sheet charge. In a DH GaAs pHEMT, carriers are introduced in the InGaAs channel by doping the AlGaAs barriers on both sides of the InGaAs channel. The AlGaAs barriers are doped with silicon using atomic planar doping to increase the electron transfer efficiency. A typical charge density of $3.5 \times 10^{12}/cm^2$ and a mobility of 5000 cm^2/Vs is obtained for a double heterojunction GaAs pHEMT structure. The high sheet charge thus obtained enables higher current drive and power handling capability. Figure 7.103 shows the layer structure of a typical millimeter-wave power GaAs pHEMT. In some cases a doped InGaAs channel is also used to increase sheet charge density.[25,26]

Breakdown voltage is an important parameter for power devices. A device with high breakdown voltage can be biased at high drain voltages, which increases the drain efficiency, voltage gain, and power added efficiency (PAE). Typical breakdown voltages of GaAs pHEMTs range from 8 to 15 V. The breakdown mechanism of a GaAs pHEMT can be either at the surface in the gate-drain of the device or in the channel (due to impact ionization).

There are several approaches used to increase the breakdown voltage of a GaAs pHEMT. The planar doping of AlGaAs barriers (as already described) helps in maintaining a high breakdown voltage, as most of the AlGaAs barrier is undoped. Another approach to increase the breakdown voltage uses a low-temperature grown (LTG) GaAs buffer below the channel. Using this approach, a 45% increase in channel breakdown voltage with a 12% increase in output power was demonstrated.[27] Using a double recessed gate structure to tailor the electric field in the gate drain depletion region can also increase breakdown voltage. The increase in breakdown voltage is mainly due to reduction in the electric field at the gate edge by surface states in the exposed recess region.[28]

The output power obtained from a HEMT also depends on the biasing conditions. To achieve high efficiency devices (as in Class B operation), the device is biased near pinch off, and therefore, high gain is required near pinch off. The mode of operation is ideally suited for pHEMTs, which typically have high transconductance near pinch off due to their superior charge-control properties. The effect for gate bias on the power performance of HEMT has been investigated.[29] Higher gain is achieved under Class

TABLE 7.12 Summary of Power Performance of GaAs pHEMTs

Frequency	Gate Length (µm)	Gate Width	Power Density (W/mm)	Output Power[a]	Gain[b] (dB)	PAE[b] (%)	Device, Drain Bias (Reference)
12 GHz	0.45	1.05 mm	0.77	0.81 W	10.0	60	Double HJ[c] V_{ds} = 7 V[32]
(Ku-Band)	0.25	1.6 mm	1.37	2.2 W	14.0	39	Double HJ[33]
20 GHz	0.25	600 µm	0.51	306 mW	7.4	45	Prematched, V_{ds} = 7 V [34]
(K-Band)	0.15	400 µm	1.04	416 mW	10.5	63	LTG Buffer, V_{ds} = 5.9 V[27]
	0.15	600 µm	0.12	72 mW	8.6	68	V_{ds} = 2 V[35]
	0.15	600 µm	0.84	501 mW	11	60	V_{ds} = 8 V[35]
35 GHz	0.25	500 µm	0.62	310 mW	6.8	40	Double HJ, V_{ds} = 5 V[36]
(Ku-Band)	0.15	150 µm	0.63	95 mW	9.0	51	Double HJ[37]
			0.91	*137 mW*	7.6	40	
44 GHz	0.25		0.79		5.1	41	Doped Channel[d 38]
(Q-Band)	0.15	400 µm	0.5	200 mW	9.0	41	Double HJ, V_{ds} = 5 V[39]
	0.2	600 µm	0.53	318 mW	5.0	30	Double HJ, V_{ds} = 5 V[40]
	0.15	1.8 mm	*0.44*	*800 mW*	5.8	25	Double HJ, V_{ds} = 5 V[41]
55 GHz	0.25	400 µm	*0.46*	*184 mW*	4.6	25	Doped Channel, V_{ds} = 5.5 V[42]
(V-Band)	0.2	50 µm	*0.85*	*42 mW*	3.3	22	Doped Channel, V_{ds} = 4.3[43]
60 GHz	0.15	150 µm	0.83	125 mW	4.5	32	Double HJ[37]
(V-Band)			*0.55*	82 mW	4.7	38	
	0.15	400 µm	*0.55*	*225 mW*	4.5	25	Double HJ, V_{ds} = 5 V[44]
			0.44	174 mW	4.4	*28*	
94 GHz	0.25	75 µm	*0.43*	*32 mW*	3.0	15	Doped Channel, V_{ds} = 4.3 V[25]
W-Band				25 mW	4.0	14	
	0.15	150 µm	*0.38*	*57 mW*	2.0	16	Double HJ[37]
			0.30	45 mW	3.0	16	
	0.1	40 µm	*0.31*	*13 mW*	6.0	13	Doped Channel V_{ds} = 3.4 V[26]
	0.1	160 µm	0.39	63 mW	4.0	13	Doped Channel V_{ds} = 4.3 V[26]

[a] *Output Power in italics indicates device biased for maximum power output.*
[b] *PAE/Gain in italics indicates device biased for maximum PAE/Gain.*
[c] Double HJ — Double heterojunction GaAs pHEMT.
[d] Doped channel — Doped channel GaAs pHEMT.

A conditions. Biasing the device at higher drain voltages can increase the output power. Table 7.12 presents a summary of power performance of GaAs pHEMTs at various microwave and millimeter-wave frequencies.

From Table 7.12 it can be seen that at a given frequency, the device power output and gain increases with a decrease in gate length due to better high-frequency operation. Reduction in power gain is also observed for wider devices. This is due to the increase in source inductance, which increases with gate width and frequency and is due to the increase in gate resistance as a square of gate width. Low inductance via hole source grounding and proper gate layout is required to reduce these parasitics.

Reliability is important for space applications, typically a mean-time-to-failure (MTTF) of 10^7 h (1142 yr) is required for space applications. GaAs pHEMTs have demonstrated a MTTF of 1×10^7 h at a channel temperature of 125°C. The main failure mechanism is the atmospheric oxidation of the exposed AlGaAs barrier layers and interdiffusion of the gate metallization with the AlGaAs barrier layers (gate sinking).[30] Using dielectric passivation layers to reduce the oxidation of AlGaAs can solve these problems. Using refractory metal for gate contacts will minimize their interaction with the AlGaAs barrier layers. A MTTF of 1.5×10^7 hours at a channel temperature of 150°C was achieved using Molybdenum based gate contacts.[31]

Traveling wave tubes (TWT) have been traditionally used as multiwatt power sources for microwave applications up to K-band (20 GHz). Using GaAs pHEMT in place of TWT for these applications has many advantages, including lower cost, smaller size, smaller weight, and higher reliability. However, the typical power density of a GaAs pHEMT at 20 GHz is on the order of 1 W/mm. Hence, the output power from a large number of devices has to be combined. To minimize combining losses, it is desirable to maximize power output of a single device. When large devices (gate width on the order of mm) are used

TABLE 7.13　Summary of Power Performance of Multiwatt pHEMT Power Modules

Width (mm)	Frequency (GHz)	Power (W)	Gain (dB)	PAE (%)	Ref.
16.8	2.45	10.0	13.5	63.0	$V_{ds} = 7$ V[46]
24	2.45	11.7	14.0	58.2	$V_{ds} = 8$ V[47]
32.4	8.5–10.5	12.0	7.2	40.0	$V_{ds} = 7$ V[48]
16.8	12	12.0	10.1	48.0	[32]
25.2	12	15.8	9.6	36.0	$V_{ds} = 9$ V[49]
8	12	6	10.8	53.0	$V_{ds} = 9$ V[50]
9.72	18–21.2	4.7	7.5	41.4	$V_{ds} = 5.5$ V[45]

Note: V_{ds} = Drain bias for power measurements.

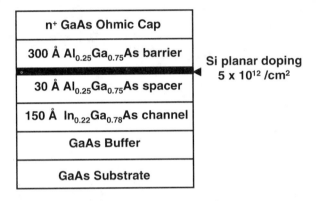

FIGURE 7.104　Layer structure of a low noise GaAs pHEMT.

to increase the total power output, other factors such as device layout, input signal distribution, output power combining networks, and substrate thickness are of critical importance. Several multiwatt GaAs pHEMT power modules have been demonstrated recently. A power module with 9.72-mm wide GaAs pHEMTs that delivered an output power of 4.7 W in the 18 to 21.2 GHz band with a PAE of 38% was demonstrated.[45] Table 7.13 summarizes the recent results of multiwatt GaAs pHEMT power modules.

Low Noise GaAs pHEMTs

Figure 7.104 shows the structure of a generic low-noise GaAs pHEMT. As the drain current requirement for a low-noise bias is low, single-side doped heterojunctions are sufficient for low-noise devices. The emphasis here is on achieving higher mobility to reduce the parasitic source resistance. As already discussed, the AlInAs/GaInAs material system is the ideal choice for fabrication of low-noise microwave and millimeter-wave devices. However, GaAs pHEMTs also find significant use in millimeter-wave low noise applications due to wafer size, cost, and process maturity related advantages.

Henderson et al. first reported on the low noise performance of GaAs pHEMTs in 1986. Devices with 0.25-μm gate length had a noise figure of 2.4 dB and an associated gain of 4.4 dB at 62 GHz.[51] A 0.15-μm gate length $Al_{0.25}Ga_{0.75}As/In_{0.28}Ga_{0.72}As$ pHEMT with a noise figure of 1.5 dB and an associated gain of 6.1 dB at 61.5 GHz was demonstrated in 1991.[52] The reduction in noise figure was a direct result of reducing the gate length, which increased the f_T of the device. Low noise operation of a GaAs pHEMT was also demonstrated at 94 GHz.[53] For a 0.1-μm gate length device a noise figure of 3.0 dB and associated gain of 5.1 dB was achieved. A noise figure of 2.1 dB and an associated gain of 6.3 dB was reported for a 0.1 μm gate length GaAs pHEMT at 94 GHz.[54] The improvement in noise figure is attributed to the use of a T-shaped gate with end-to-end resistance of 160 Ω/mm by Tan et al.,[52] compared to a trapezoidal gate with end-to-end resistance of 1700 Ω/mm used by Chao et al.[53] This further emphasizes the need to reduce parasitic resistances in low-noise devices.

TABLE 7.14 Summary of Low Noise Performance of GaAs pHEMTs

Frequency (GHz)	Gate Length (μm)	F_{min} (dB)	Ga (dB)	Ref.	Comments
12	0.25	0.6	11.3	55	
12	0.17	0.35	12.5	57	Packaged Device
18	0.25	0.9	10.4	51	
62	0.25	2.4	4.4	51	Improvement by
60	0.15	1.5	6.1	52	Reduction in L_g
94	0.10	3.0	5.1	53	Improvement by
94	0.10	2.1	6.3	54	Reduction in R_g

One of the main system applications of low-noise GaAs pHEMTs is satellite direct broadcasting receiver systems (DBS) that are in increasing demand worldwide. Low-noise amplifiers operating at 12 GHz are a critical component in these systems. The low-noise performance of GaAs pHEMTs is more than adequate for these applications. A 0.25-μm gate length GaAs pHEMT with a noise figure of 0.6 dB and an associated gain of 11.3 dB at 12 GHz was reported by Tokue et al.[55] Performance coupled with low cost packaging is one of the crucial factors in the high volume DBS market. Hwang et al. have demonstrated a 0.2-μm gate length GaAs pHEMT in plastic packaging with a 1.0 dB noise figure and 9.9 dB associated gain at 12 GHz.[56] A plastic packaged GaAs pHEMT device with a gate length of 0.17 μm demonstrated a noise figure of 0.35 dB, and 12.5 dB associated gain at 12 GHz.[57] Table 7.14 summarizes the low-noise performance of GaAs pHEMTs at various microwave and millimeter-wave frequencies.

GaAs pHEMT for Wireless Applications

The explosive growth of the wireless communication industry has opened up a new area of application for GaAs pHEMTs. Unlike the millimeter-wave military and space applications, the frequencies of operation of these applications are much lower. The frequencies used in typical cellular phones range from 850 MHz for the American Mobile Phone System (AMPS) to 1.9 GHz for the Japanese Personal Handy Phone System (PHS) and the Digital European Cordless Telephone (DECT). The device parameters of interest when considering device technologies for wireless applications are operating voltages, which must be positive, power density, output match and linearity requirements, and gate leakage current.[58] Using enhancement mode devices (threshold/pinch-off voltage > 0) eliminates the negative supply voltage generator and power-cutoff switch. The high gate turn-on voltage of an enhancement mode GaAs pHEMT, as compared to the enhancement mode GaAs MESFET, enables higher input voltage swing.

The power performance of MESFET and GaAs pHEMT for wireless applications has been compared.[59] For the same saturation drain current density, at a frequency of 950 MHz, the saturated power output from the pHEMT is 2.5 W, whereas it is 1.8 W from the MESFET. The power-added efficiency of the pHEMT is 68%, which is 8% higher than that of the MESFET. This difference is due to the transfer characteristics of the two devices. The pHEMT performs as a better power amplifier than the MESFET because the input power is effectively amplified with higher g_m near the pinch-off voltage. This is a direct consequence of better charge control properties of the HEMT when compared to the MESFET. The pHEMT also has a lower gate leakage current than the MESFET due to a higher Schottky barrier on AlGaAs.

The power performance of enhancement mode GaAs pHEMT with a threshold voltage of 0.05 V for wireless applications has also been investigated.[60] A device with a gate width of 3.2 mm delivered an output power of 22 dBm with power-added-efficiency of 41.7%. The standby current at a gate bias of 0 V was 150 μA. Enhancement mode GaAs HFET with a higher threshold voltage of 0.5 V has also been demonstrated.[61] A 1-μm gate length device with a gate width of 12 mm delivered an output power of 31.5 dBm with a PAE of 75% at 850 MHz and at a drain bias of 3.5 V. The standby current at a gate bias of 0 V was 1 μA. This eliminates the need for a switch in the drain current of the power amplifier. The device was manufactured using Motorola's CGaAs™ process, which is cost effective as it uses processes that are similar to standard silicon MOS and bipolar processes. Table 7.15 shows a summary of power performance of GaAs pHEMTs for cellular phones.

TABLE 7.15   Summary of Power Performance of GaAs pHEMTs for Cellular Phones

Frequency	Device Width (μm)	Drain Bias (V)	Power Output	PAE (%)	ACPL	Device/Ref.
850 MHz	5	1.2	19.6 dBm	65.2		AlGaInP/InGaAs pHEMT[62]
	10	1.3	21.5 dBm	57.4		InGaP/InGaAs pHEMT[63]
	12	3.5	31.5 dBm	75		Enhancement Mode HFET[61]
	12	3.5	33.1 dBm	84.8		GaAs pHEMT[64]
	30	3.7	31.0 dBm	59.0	−30.1 dBc @ 30 kHz	1 μm GaAs MESFET[65]
900 MHz	12	3.5	31.5 dBm	75		Enhancement Mode CGaAs™[66]
	21	2.3	31.3 dBm	68		0.8 μm MESFET[67]
	14	3.0	32.3 dBm	71		pHEMT[68]
	40	1.5	31.5 dBm	65		MESFET[69]
950 MHz	28	1.2	1.1 W	54		pHEMT[70]
	21	2.2	32.7 dBm	62.8	−50.5 dBc	PHEMT[71]
	12	3.0	1.4 W	60.0	—	pHEMT[72]
	7	3.4	30.9 dBm	56.3	−51.5 dBc @ 50 kHz	pHEMT[73]
	16	3.4	1.42 W	60.0	−48.2 dBc @ 50 kHz	pHEMT[74]
	12	4.7	2.5 W	68.0		pHEMT[59]
	12	4.7	1.8 W	60.0		MESFET[59]
1.9 GHz	1	2.0	20.2 dBm	45.3	−55.2 dBc @ 600 kHz	pHEMT[75]
	2.4	2.0	21.1 dBm	54.4	−55 dBc @ 600 kHz	MESFET[76]
	3.2	3.0	22.0 dBm	41.7	−58.2 dBc @ 600 kHz	Enhancement Mode pHEMT[60]
	12	3.5	30 dBm	50	−30 dBc	Enhancement Mode pHEMT[61]
	5	2.0	25.0 dBm	53.0		AlGaInP/InGaAs pHEMT[62]

AlInAs/GaInAs/InP (InP HEMT)

Future military and commercial electronic applications will require high-performance microwave and millimeter-wave devices. Important applications include low-noise amplifiers for receiver front ends, power amplifiers for phased-array radars, ultra high-speed digital circuits for prescalers, and MUX/DEMUX electronics for high-speed (> 40 Gb/s) optical links.

A HEMT device capable of operating at millimeter-wave frequency requires a channel with high electron velocity, high current density, and minimal parasitics. As discussed, the $Al_{0.48}In_{0.52}As/Ga_{0.47}In_{0.53}As$ material system lattice matched to InP satisfies these criteria. A 1-μm gate length AlInAs/GaInAs HEMT with extrinsic transconductance as high as 400 mS/mm was demonstrated.[77] The microwave performance of 1-μm gate-length devices showed an improvement of 20 to 30% over the AlGaAs/GaAs HEMT.[78] In 1988 Mishra et al. demonstrated a 0.1 μm InP HEMT with a f_T of 170 GHz.[79] Using a T-gate to self-align the source and drain contacts results in reduction of source-gate and source-drain spacing. This not only reduces the parasitic source and drain resistances, but also the drain delay. Using the preceding technique an f_T of 250 GHz was achieved in a 0.13-μm gate length self-aligned HEMT.[18] Recently, a 0.07-μm AlInAs/GaInAs HEMT with an f_T of 300 GHz and an f_{max} of 400 GHz was reported.[80]

The high-frequency performance of the InP HEMT can be further improved by using a pseudomorphic InGaAs channel with an indium content as high as 80%. The f_T of a 0.1-μm InP HEMT increased from 175 to 205 GHz when the indium content in the channel was increased from 53 to 62%.[81] Although devices with high indium content channels have low breakdown voltages, they are ideal for low noise applications and ultra high-speed digital applications. An f_T of 340 GHz was achieved in a 0.05-μm gate length psuedomorphic InP HEMT with a composite $In_{0.8}Ga_{0.2}As/In_{0.53}Ga_{0.47}As$ channel.[82] This is the highest reported f_T of any 3-terminal device.

Compared to the GaAs pHEMT, the AlInAs/GaInAs HEMT has a higher current density that makes it suitable for ultra high-speed digital applications. The high current gain cutoff frequency and low parasitics make the AlInAs/GaInAs HEMT the most suitable choice for low- noise applications extending well beyond 100 GHz. The high current density and superior high-frequency performance can be utilized for high-performance millimeter-wave power applications provided the breakdown voltage is improved.

TABLE 7.16 Summary of Low-Noise Performance of AlInAs/GaInAs HEMTs

Frequency (GHz)	Gate Length (μm)	F_{min} (dB)	G_a (dB)	Comments/Ref.
12	0.15	0.39	16.5	$In_{0.7}Ga_{0.3}As$ channel[87]
18	0.25	0.5	15.2	[85]
18	0.15	0.3	17.2	[86]
26	0.18	0.43	8.5	Passivated device [88]
57	0.25	1.2	8.5	[85]
60	0.1	0.8	8.9	[89]
63	0.1	0.8	7.6	Passivated[90]
		0.7	8.6	Unpassivated
94	0.15	1.4	6.6	[86]
94	0.1	1.2	7.2	[89]

Some of the state-of-the-art millimeter-wave analog circuits and ultra high-speed digital circuits have been implemented using InP HEMTs. A low-noise amplifier with 12 dB gain at a frequency of 155 GHz using a 0.1-μm InP HEMT with a $In_{0.65}Ga_{0.35}As$ psuedomorphic channel was demonstrated.[83] Pobanz et al. demonstrated an amplifier with 5 dB gain at 184 GHz using a 0.1 μm gate $In_{0.8}Ga_{0.2}As$/InP composite channel HEMT.[84]

Low-Noise AlInAs/GaInAs HEMT

The superior electronic properties of the GaInAs channel enable the fabrication of extremely high f_T and f_{max} devices. The superior carrier confinement at the AlInAs/GaInAs interface results in a highly linear transfer characteristic. High transconductance is also maintained very close to pinch off. This is essential because the noise contribution of the FET is minimized at low drain current levels. Hence high gain can be achieved at millimeter-wave frequencies under low-noise bias conditions. The high mobility at the AlInAs/GaInAs interface also results in reduced parasitic source resistance of the device. AlInAs/GaInAs HEMTs with 0.25-μm gate length exhibited a noise figure of 1.2 dB at 58 GHz.[85] At 95 GHz a noise figure of 1.4 dB with associated gain of 6.6 dB was achieved in a 0.15-μm gate length device.[86] Table 7.16 summarizes the low-noise performance of AlInAs/GaInAs HEMTs.

Millimeter-Wave AlInAs/GaInAs Power HEMT

The millimeter-wave power capability of single heterojunction AlInAs/GaInAs HEMTs has been demonstrated.[91,92] The requirements for power HEMT are high gain, high current density, high breakdown voltage, low access resistance, and low knee voltage to increase power output and power-added efficiency. The AlInAs/GaInAs HEMT satisfies all of these requirements with the exception of breakdown voltage. This limitation can be overcome by operating at a lower drain bias. In fact, the high gain and PAE characteristics of InP HEMTs at low drain bias voltages make them ideal candidates for battery-powered applications.[93] Another advantage is the use of InP substrate that has a 40% higher thermal conductivity than GaAs. This allows higher dissipated power per unit area of the device or lower operating temperature for the same power dissipation. As low breakdown voltage is a major factor that limits the power performance of InP HEMTs, this section will discuss in detail the various approaches used to increase breakdown voltage.

Breakdown in InP HEMT is a combination of electron injection from the gate contact and impact ionization in the channel.[94] The breakdown mechanism in the off-state (when the device is pinched off) is electron injection from the gate. The low Schottky barrier height of AlInAs results in increased electron injection from the gate and, consequently, higher gate leakage current compared to the GaAs pHEMT. These injected hot electrons cause impact ionization in the high-field drain end of the GaInAs channel. The high impact ionization rate in the low bandgap GaInAs channel is the main mechanism that determines the on-state breakdown. Some of the holes generated by impact ionization are collected by the negatively biased gate and result in increased gate leakage. The potential at the source end of the channel is modulated by holes collected by the source. This results in increased output conductance.

Lowering the electric field in the gate-drain region can reduce the impact ionization rate. This is achieved by using a double recess gate fabrication process that increases the breakdown voltage from 9 to 16 V.[95] A gate-drain breakdown voltage of 11.2 V was demonstrated for 0.15-μm gate length devices with a 0.6-μm recess width.[96] In addition, reduction in output conductance (g_{ds}) and gate-drain feedback capacitance (C_{gd}) was observed when compared to single recessed devices. The f_{max} of a double recessed device increased from 200 to 300 GHz.[97] Hence, it is desirable for power devices. Another approach to reduce electric field in the gate-drain region is to use an undoped GaInAs cap instead of a doped GaInAs cap.[98] The output conductance can be reduced from 50 to 20 mS/mm for a 0.15-μm gate length device by replacing the doped GaInAs cap by an undoped cap.[99] This also improved the breakdown voltage from 5 to 10 V. The reduction in C_{gd} and g_{ds} resulted in an f_{max} as high as 455 GHz. Redistribution of the dopants in the AlInAs barrier layers can also increase breakdown voltage. An increase in breakdown voltage from 4 to 9 V is achieved by reducing doping in the top AlInAs barrier layer and transferring it to the AlInAs barrier layers below the channel.[92]

The gate leakage current can be reduced and the breakdown voltage increased by using a higher bandgap strained AlInAs barrier.[100,101] By increasing the Al composition in the barrier layers from 48 to 70%, the gate-to-drain breakdown voltage was increased from 4 to 7 V. This also results in reduction of gate leakage as the Schottky barrier height increases from 0.5 to 0.8 eV. The use of $Al_{0.25}In_{0.75}P$ as an Schottky barrier improves the breakdown voltage from −6 to −12V.[102]

The on-state breakdown can be improved in two ways. The first is to reduce the gate leakage current by the impact ionization generated holes by increasing the barrier height for holes. This was achieved by increasing the valence band discontinuity at the channel-barrier interface. The use of a strained 25-Å $In_{0.5}Ga_{0.5}P$ spacer instead of AlInAs increases the valence band discontinuity at the interface from 0.2 to 0.37 eV. An on-state breakdown voltage of 8 V at a drain current density of 400 mA/mm for a 0.7-μm gate length InP HEMT was achieved by using a strained InGaP barrier.[103]

The various approaches to increase the breakdown voltage, as already discussed here, concentrate on reducing the electron injection from the Schottky gate and reducing the gate leakage current. These approaches also have an inherent disadvantage as Al-rich barriers result in high source resistance and are more susceptible to atmospheric oxidation. Additionally, these approaches do not address the problem of a high impact ionization rate in the GaInAs channel and carrier injection from contacts. In the recent past, various new approaches have been investigated to increase breakdown voltage without compromising the source resistance or atmospheric stability of the device. These include the junction-modulated AlInAs/GaInAs HEMT (JHEMT), the composite GaInAs/InP channel HEMT, and the use of regrown contacts. Table 7.17 summarizes the power performance of AlInAs/GaInAs HEMTs.

GaInAs/InP Composite Channel HEMT

The high speed and power performance of InP HEMT can be improved by the use of composite channels that are composed of two materials with complementing electronic properties. The high-speed performance of an InP HEMT can be improved by inserting InAs layers in the InGaAs channel. The current gain cutoff frequency of a 0.15-μm gate length device increased from 179 to 209 GHz due to improved electron properties.[111] An f_T as high as 264 GHz was achieved for a 0.08-μm gate length device.

The GaInAs channel has excellent electronic properties at a low electric field, but suffers from high impact ionization at high electric fields. On the other hand, InP has excellent electronic transport properties at high fields but has lower electron mobility. In a composite InGaAs/InP channel HEMT, the electrons are in the InGaAs channel at the low field source end of the channel and are in the InP channel at the high field drain end of the channel. This improves the device characteristics at high drain bias while still maintaining the advantages of the GaInAs channel at low bias voltages.[112] A typical submicron gatelength AlInAs/GaInAs HEMT has an off-state breakdown voltage (BV_{dsoff}) of 7 V, and on-state breakdown voltage (BV_{dson}) of 3.5 V. Using a composite channel, (30 Å GaInAs/50 Å InP/100 Å n^+ InP), Matloubian et al. demonstrated a BV_{dsoff} of 10 V and BV_{dson} of 8 V for a 0.15-μm gate length device.[113] A 0.25-μm GaInAs/InP composite channel HEMT with a two-terminal gate drain voltage of 18 V was also demonstrated.[114]

TABLE 7.17 Summary of Power Performance of InP HEMTs

Frequency (GHz)	Gate Length (μm)	Gate Width	Power Density (W/mm)	Power Output (mW)	Gain (dB)	PAE (%)	Device/Drain Bias (Ref.)
4	0.5	2 mm	0.13	269	18	66	V_{ds} = 2.5 V[93]
	0.15	0.8mm	0.4	320	18	57	V_{ds} = 3 V[93]
12	0.22	150 μm	0.78	117	8.4	47	V_{ds} = 4 V[92]
			0.47	70	11.3	59	V_{ds} = 3 V, Double HJ
18	0.15	600 μm	0.74	446	13	59	Double recessed V_{ds} = 7 V[104]
20	0.15	50 μm	0.78	39	10.2	44	V_{ds} = 4.9 V[91]
(K-Band)			0.41	21	10.5	52	V_{ds} = 2.5 V Single heterojunction
	0.15	50 μm	0.61	30	12.2	44	Single heterojunction $In_{0.69}Ga_{0.31}As$ channel V_{ds} = 4.1 V[91]
	0.15	800 μm	0.65	516	7.1	47	70% AllnAs, V_{ds} = 4 V[105]
44	0.15	450 μm	0.55	251	8.5	33	$Al_{0.6}In_{0.4}As$ barrier
(Q-Band)			0.88	398	6.7	30	Doped Channel[100]
	0.2	600 μm	0.37	225	5	39	$Al_{0.6}In_{0.4}As$ barrier Single heterojunction, V_{ds} = 4 V[106]
57	0.22	450 μm	0.33	150	3.6	20	$Al_{0.6}In_{0.4}As$ barrier
(V-Band)			0.44	200		17	Doped channel, V_{ds} = 3.5 V[101]
60	0.15	50 μm	0.35		7.2	41	V_{ds} = 2.6 V[91]
(V-Band)			0.52	26	5.9	33	V_{ds} = 3.6 V Single heterojunction
	0.1	50 μm	0.30	15	8.6	49	V_{ds} = 3.35 V[107]
			0.41	21	8.0	45	
	0.1	400 μm	0.48	192	4.0	30	V_{ds} = 4.12 V, 67% In[107]
94	0.15	50 μm	0.30	15	4.6	21	Single HJ, Passivated device V_{ds} = 2.6 V[108]
(W-Band)	0.15	640 μm	0.20	130	4.0	13	Double HJ, V_{ds} = 2.7[109]
	0.1	200 μm	0.29	58		33	$In_{0.68}Ga_{0.32}As$ channel[110]

The increased breakdown voltage of a composite channel HEMT enables operation at a higher drain bias. This increases the drain efficiency and the PAE of the device. An output power of 0.9 W/mm with a PAE of 76% at 7 GHz was demonstrated for a 0.15-μm GaInAs/InP composite channel HEMT at a drain bias of 5 V.[115] At 20 GHz, an output power density of 0.62 W/mm (280 mW) and a PAE 46% was achieved for a 0.15 μm gate length device at a drain bias of 6 V.[113] At 60 GHz, a 0.15-μm GaInAs/InP composite channel HEMT demonstrated an output power of 0.35 W/mm, and a power gain of 6.2 dB with a PAE of 12% at a drain bias of 2.5 V.[116]

Technology Comparisons

Comparison between FETs and HBTs for RF/Microwave Applications

In recent years hetero-structure bipolar transistors (HBTs) have emerged as strong contenders for wireless, microwave, and millimeter-wave applications. HBTs have similar high frequency performance with modest lateral dimensions as transport is along the vertical direction. This also offers higher current drivability per unit chip area. On the other hand, submicron gate lengths are required to achieve microwave and millimeter-wave operation in FETs. In addition, HBTs have high transconductance due to an exponential relationship between the input (base) voltage and output (collector) current. Device uniformity over a wafer can be easily achieved in a HBT as the turn-on voltage depends on the built-in voltage of a pn junction (and is governed by the uniformity of the epitaxial growth technique). The threshold voltage of a HEMT (which is a measure of device uniformity) is mainly determined by uniformity in gate recess etching. The FET is mainly a surface device and therefore has a higher 1/f noise due generation-recombination processes in the surface traps. On the other hand, current transport in a

HBT is in the bulk, hence HBTs have lower 1/f noise which makes them ideal for low-phase-noise microwave and millimeter-wave oscillators.

Semiconductor processing of HBTs is more complex than FETs owing to the vertical structure. Another advantage of FETs when compared to HBTs is lower operating voltage. The turn-on voltage of GaAs HBTs is significantly higher (1.4 V) than that of HEMTs (0.5 V). This makes the HEMT an ideal candidate for low-voltage battery-operated applications. Owing to low parasitics, HEMTs have a lower noise figure than HBTs at any frequency and therefore are preferred for low-noise receivers. For microwave and millimeter-wave power applications gain, HEMTs are again the preferred choice. This is because at millimeter-wave frequency (> 50 GHz) the gain in a HBT is mainly limited by the high collector base feedback capacitance. Another disadvantage of large area HBTs is thermal runaway due to nonuniform junction temperatures. This can be solved by the use of emitter ballast resistors, but this reduces device gain and processing complexity. In the following sections, the various HEMT technologies are compared for specific applications.

Power Amplifiers for Wireless Phones

The two main aspects that govern the choice of technology for wireless power amplifiers are the operating voltage and the cost. As far as operating voltages are concerned, the use of GaAs HBTs is limited to 3.6 V supply voltages. This is due to the high turn-on voltage (1.4 V) for the emitter-base junction in the GaAs HBT. On the other hand, GaAs MESFETs and pHEMT can be used in 1.2 V systems due to their low knee voltages. As far as cost is concerned, the major contenders are ion-implanted GaAs MESFETs and GaAs HBTs. However, GaAs MESFETs require a negative gate voltage, which potentially increases circuit complexity and cost. The enhancement mode GaAs pHEMT, which has a low knee voltage (like a GaAs MESFET) and positive voltage operation (like a GaAs HBT) is therefore a viable technology option for high performance, low voltage wireless phones.

Microwave Power Amplifiers

The GaAs pHEMT and the InP HEMT are the two major competing technologies for microwave and millimeter-wave power amplifiers. The relation between output power density (P_{out}) and power-added efficiency (PAE) and device parameters is given by the following expressions,

$$P_{out} = \frac{1}{8}\left(I_{max}\right)\left(BV_{gd} - V_{knee}\right) \tag{7.160}$$

$$PAE = \alpha\left(\frac{V_{DD} - V_{knee}}{V_{DD}}\right)\left(1 - \frac{1}{G_a}\right) \tag{7.161}$$

In Eq. (7.161), α is ½ for Class A operation and $\pi/4$ for Class B operation. Figure 7.105 compares the P_{out} and PAE of GaAs pHEMTs and InP HEMTs as a function of operating frequency. It can be seen that GaAs pHEMTs have a higher power density than InP HEMTs at all frequencies than InP HEMT. This is due to higher breakdown voltages (BV_{gd}) in GaAs pHEMTs that enables higher operating voltage (V_{DD}). The InP HEMT operating voltage is limited by the low breakdown voltage, hence the power output is low. However, InP HEMTs have comparable power output at millimeter-wave frequencies. This is enabled by the low knee voltage (V_{knee}) and high current drive (I_{max}). On the other hand, due to their low knee voltage and high gain (G_a), InP HEMTs have higher gain and PAE at frequencies exceeding 60 GHz. The potential of InP HEMTs as millimeter-wave power devices is evident in the fact that comparable power performance is achieved at drain biases 2 to 3 V lower than those for GaAs pHEMTs. Hence a high voltage InP HEMT technology should be able to outperform the GaAs pHEMT for power applications at all frequencies.

Low Noise Amplifiers

Figure 7.106 compares the gain and noise figure of GaAs pHEMTs and InP HEMTs as a function of frequency. At any given frequency, the InP HEMT has a noise figure about 1 dB lower than the GaAs

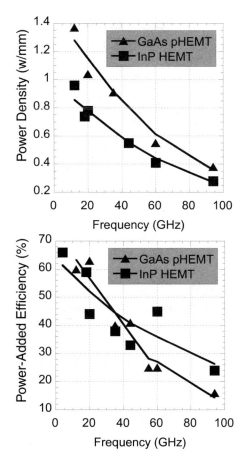

FIGURE 7.105 Comparison of output power density and power-added-efficiency of GaAs pHEMTs and InP HEMTs as a function of frequency.

pHEMT and gain 2 dB higher than the GaAs pHEMT. This is mainly due to the excellent transport properties of the AlInAs/GaInAs material system.

Digital applications — High current driving capability with low voltage operation is essential for high speed, low power digital circuits. High f_T devices are also required for increasing the frequency of operation. In this area, GaAs MESFETs and GaAs pHEMTs are sufficient for 10 Gb/s digital circuits. However, for digital circuits operating at bit rates exceeding 40 Gb/s, the InP HEMT is the preferred technology option.

Table 7.18 lists the various frequency bands and military and space applications in each band with the appropriate HEMT device technology for each application.

Conclusion

GaAs- and InP-based high electron mobility transistors have emerged as premier devices for the implementation of millimeter-wave analog circuits and ultra high-speed digital circuits. In this section, principles of HEMT operation were discussed. The design aspects of HEMT for both low-noise and high-power applications were discussed. Reduction in gate length is essential for improved performance at high frequencies. Appropriate device scaling with gate length reduction is necessary to minimize the effect of parasitics on device performance.

Millimeter-wave power modules have been demonstrated using GaAs pHEMT devices. The superior device performance of GaAs pHEMTs is being used to improve the performance of power amplifiers for

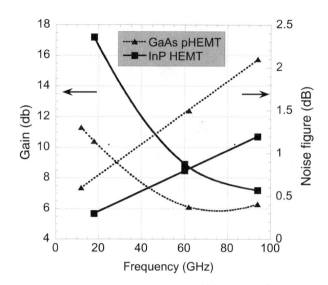

FIGURE 7.106   Comparison of noise figure and gain of GaAs pHEMTs and InP HEMTs as a function of frequency.

TABLE 7.18   Frequency Bands and Military and Commercial Applications

Frequency	Military/Space	Commercial	Device Technology
850 MHz–1.9 GHz		Wireless	Low noise — GaAs pHEMT
			Power — GaAs pHEMT
12 GHz (Ku-Band)	Phased array radar	Direct broadcast satellite	Low noise — GaAs pHEMT
			Power — GaAs pHEMT
20 GHz (K-Band)	Satellite downlinks		
27–35 GHz (Ka-Band)	Missile seekers	LMDS — Local multipoint	Low noise — GaAs pHEMT
		distribution system	Power — GaAs pHEMT
44 GHz (Q-Band)	SATCOM ground terminals	MVDS — Multipoint video	Low noise — InP HEMT
		distribution system	Power — GaAs pHEMT
60 GHz (V-Band)	Satellite crosslinks	Wireless LAN	Low noise — InP HEMT
			Power — GaAs pHEMT/InP HEMT
77 GHz		Collision avoidance radar	Low noise — InP HEMT
			Power — GaAs pHEMT/InP HEMT
94 GHz (W-Band)	FMCW radar		Low noise — InP HEMT
			Power — InP HEMT
100–140 GHz	Radio astronomy		Low noise — InP HEMT
Digital 10 Gb/s		Fiber-optic communication	GaAs pHEMT
Digital 40 Gb/s		Fiber-optic communication	InP HEMT

wireless phone systems. The superior material characteristics of the AlInAs/GaInAs material system have been used to achieve record low-noise performance at millimeter-wave frequencies using InP HEMTs. Despite their low breakdown voltage, InP HEMTs have demonstrated superior power performance at millimeter-wave frequency. Improving the breakdown voltage using approaches that include composite channel GaInAs/InP HEMT and junction-modulated HEMT will further improve power performance. The development of the GaAs pHEMT and InP HEMT technology was traditionally supported by low-volume, high-cost military and space applications. The recent emergence of high-volume commercial applications such as wireless and optical communications systems has new constraints that include manufacturability and low-voltage operation.

References

1. Dingle, R., Stromer, H. L., Gossard, A. C., and Wiemann, W., Electron mobility in modulation doped semiconductor superlattices, *Applied Physics Letters*, 33, 665, 1978.
2. Delagebeaudeuf, D. and Linh, N. T., Metal-(n) AlGaAs-GaAs two-dimensional electron gas FET, *IEEE Transactions on Electron Devices*, 29, 955, 1982.
3. Drummond, T. J., Masselink, W. T., and Morkoc, H., Modulation Doped GaAs/(Al,Ga)As hetero-junction field-effect transistors: MODFETs, *Proceedings of the IEEE*, 74, 773, 1986.
4. Vinter, B., Subbands and charge control in two-dimensional electron gas field effect transistor, *Applied Physics Letters*, 44, 307, 1984.
5. Foisy, M. C., Tasker, P. J., Hughes, B., and Eastman, L. F., The role of insufficient charge modulation in limiting the current gain cutoff frequency of the MODFET, *IEEE Transactions of Electron Devices*, 35, 871, 1988.
6. Nguyen, L. D., Larson, L. E., and Mishra, U. K., Ultra-high-speed modulation-doped field-effect transistors: A Tutorial Review, *Proceedings of the IEEE*, 80, 494, 1992.
7. Dambrine, G., Cappy, A., Heliodore, F., and Playez, E., A new method for determining the FET small-signal equivalent circuit, *IEEE Transactions on Microwave Theory and Techniques*, 36, 1151, 1988.
8. Berroth, M. and Bosch, R., Broad-band determination of the FET small-signal equivalent circuit, *IEEE Transactions on Microwave Theory and Techniques*, 38, 891, 1990.
9. Das, M. B., A high aspect ratio design approach to millimeter-wave HEMT structures, *IEEE Transactions on Electron Devices*, 32, 11, 1985.
10. Lester, L. F., Smith, P. M., Ho, P., Chao, P. C., Tiberio, R. C., Duh, K. H. G., and Wolf, E. D., 0.15 μm gate length double recess psuedomorphic HEMT with f_{max} of 350 GHz, *IEDM Technical Digest*, 172, 1988.
11. Nguyen, L. D., Tasker, P. J., Radulescu, D. C., and Eastman, L. F., Characterization of ultra-high speed pseudomorphic AlGaAs/InGaAs (on GaAs) MODFET's, *IEEE Transactions on Electron Devices*, 36, 2243, 1989.
12. Nguyen, L. D. and Tasker, P. J., Scaling issue of ultra-high speed HEMTs, *SPIE Conf. on High Speed Electronics and Device Scaling*, 1288, 251, 1990.
13. Moll, N., Hueschen, M. R., and Fischer-Colbrie, A., Pulse doped AlGaAs/InGaAs pseudomorphic MODFETs, *IEEE Transactions on Electron Devices*, 35, 879, 1988.
14. Chao, P. C., Shur, M. S., Tiberio, R. C., Duh, K. H. G., Smith, P. M., Ballingall, J. M., Ho, P., and Jabra, A. A., DC and microwave characteristics of Sub 0.1 μm gate length planar doped pseudomorphic HEMTs, *IEEE Transactions on Electron Devices*, 36, 461, 1989.
15. Nguyen, L. D., Brown, A., Delaney, M., Mishra, U., Larson, L., Jelloian, L., Melendes, M., Hooper, C., and Thompson, M., Vertical scaling of ultra-high-speed AlInAs-GaInAs HEMTs, *IEDM Technical Digest*, 105, 1989.
16. Enoki, T., Tomizawa, M., Umeda, Y., and Ishii, Y., 0.05 μm Gate InAlAs/InGaAs High Electron mobility transistor and reduction of its short channel effects, *Japanese Journal of Applied Physics*, 33, 798, 1994.
17. Tasker, P. J. and Hughes, B., Importance of source and drain resistance to the maximum f_T of millimeter-wave MODFETs, *IEEE Electron Device Letters*, 10, 291, 1989.
18. Mishra, U. K., Brown, A. S., Jelloian, L. M., Thompson, M., Nguyen, L. D., and Rosenbaum, S. E., Novel high performance self aligned 0.15 μm long T-gate AlInAs-GaInAs HEMTs, *IEDM Technical Digest*, 101, 1989.
19. Mimura, T., Hiyamizu, S., and Hikosak, S., Enhancement mode high electron mobility transistors for logic applications, *Japanese Journal of Applied Physics*, L317, 1981.
20. Ketterson, A. A., Masselink, W. T., Gedymin, J. S., Klem, J., Peng, C.-K., Kopp, W. F., Morkoç, H., and Gleason, K. R., Characterization of InGaAs/AlGaAs pseudomorphic modulation-doped field-effect transistors, *IEEE Transactions on Electron Devices*, 33, 564, 1986.

21. Pereiaslavets, B., Bachem, K. H., Braunstein, J., and Eastman, L. F., GaInP/InGaAs/GaAs graded barrier MODFET grown by OMVPE: design, fabrication, and device results, *IEEE Transactions on Electron Devices*, 43, 1659, 1996.

22. Mesquida-Kusters, A. and Heime, K., Al-Free InP-based high electron mobility transistors: design, fabrication and performance, *Solid State Electronics*, 41, 1159, 1997.

23. Ketterson, A., Moloney, M., Masselink, W. T., Klem, J., Fischer, R., Kopp, W., and Morkoc, H., High transconductance InGaAs/AlGaAs pseudomorphic modulation doped field effect transistors, *IEEE Electron Device Letters*, 6, 628, 1985.

24. Nguyen, L. D., Radulescu, D. C., Tasker, P. J., Schaff, W. J., and Eastman, L. F., 0.2 μm gate-length atomic-planar doped pseudomorphic $Al_{0.3}Ga_{0.7}As/In_{0.25}Ga_{0.75}As$ MODFET's with f_t over 120 GHz, *IEEE Electron Device Letters*, 9, 374, 1988.

25. Smith, P. M., Lester, L. F., Chao, P.-C., Ho, P., Smith, R. P., Ballingall, J. M., and Kao, M.-Y., A 0.25 μm gate length pseudomorphic HFET with 32 mW Output power at 94 GHz, *IEEE Electron Device Letters*, 10, 437, 1989.

26. Streit, D. C., Tan, K. L., Dia, R. M., Liu, J. K., Han, A. C., Velebir, J. R., Wang, S. K., Trinh, T. Q., Chow, P. D., Liu, P. H., and Yen, H. C., High gain W-band pseudomorphic InGaAs power HEMTs, *IEEE Electron Device Letters*, 12, 149, 1991.

27. Actis, R., Nichols, K. B., Kopp, W. F., Rogers, T. J., and Smith, F. W., High performance 0.15 μm gate length pHEMTs enhanced with a low temperature grown GaAs buffer, *IEEE MTT-S Int. Microwave Symp. Dig.*, 445, 1995.

28. Huang, J. C., Boulais, W., Platzker, A., Kazior, T., Aucoin, L., Shanfield, S., Bertrand, A., Vafiades, M., and Niedzwiecki, M., The effect of channel dimensions on the millimeter wave performance of pseudomorphic HEMT, *Proceedings of GaAs IC Symposium*, 177, 1993.

29. Danzilio, D., White, P., Hanes, L. K., Lauterwasser, B., Ostrowski, B., and Rose, F., A high efficiency 0.25 μm pseudomorphic HEMT power process, *Proceedings of GaAs IC Symposium*, 255, 1992.

30. Chen, C. H., Saito, Y., Yen, H. C., Tan, K., Onak, G., and Mancini, J., Reliability study on pseudomorphic InGaAs power HEMT devices at 60 GHz, *IEEE MTT-S Int. Microwave Symp. Dig.*, 817, 1994.

31. Hori, Y., Onda, K., Funabashi, M., Mizutani, H., Maruhashi, K., Fujihara, A., Hosoya, K., Inoue, T., and Kuzuhara, M., Manufacturable and reliable millimeter wave HJFET MMIC technology using novel 0.15 μm MoTiPtAu gates, *IEEE MTT-S Int. Microwave Symp. Dig.*, 431, 1995.

32. Matsunaga, K., Okamoto, Y., and M. Kuzuhara, A 12-GHz, 12-W HJFET amplifier with 48% peak power added-efficiency, *IEEE Microwave and Guided Wave Letters*, 5, 402, 1995.

33. Helms, D., Komiak, J. J., Kopp, W. F., Ho, P., Smith, P. M., Smith, R. P., and Hogue, D., Ku band power amplifier using pseudomorphic HEMT devices for improved efficiency, *IEEE MTT-S Int. Microwave Symp. Dig.*, 819, 1991.

34. Yarborough, R., Heston, D., Saunier, P., Tserng, H. Q., Salzman, K., and B. Smith, Four watt, Kt-band MMIC amplifier, *IEEE MTT-S Int. Microwave Symp. Dig.*, 797, 1994.

35. Kao, M.-Y., Saunier, P., Ketterson, A. A., Yarborough, R., and Tserng, H. Q., 20 GHz power PHEMTs with power-added efficiency of 68% at 2 volts, *IEDM Technical Digest*, 931, 1996.

36. Dow, G. S., Tan, K., Ton, N., Abell, J., Siddiqui, M., Gorospe, B., Streit, D., Liu, P., and Sholley, M., Ka-band high efficiency 1 watt power amplifier, *IEEE MTT-S Int. Microwave Symp. Dig.*, 579, 1992.

37. Kao, M.-Y., Smith, P. M., Ho, P., Chao, P.-C., Duh, K. H. G., Abra, A. A. J., and Ballingall, J. M., Very high power-added efficiency and low noise 0.15 μm gate-length pseudomorphic HEMTs, *IEEE Electron Device Letters*, 10, 580, 1989.

38. Ferguson, D. W., Smith, P. M., Chao, P. C., Lester, L. F., Smith, R. P., Ho, P., Jabra, A., and Ballingall, J. M., 44 GHz hybrid HEMT power amplifiers, *IEEE MTT-S Int. Microwave Symp. Dig.*, 987, 1989.

39. Kasody, R., Wang, H., Biedenbender, M., Callejo, L., Dow, G. S., and Allen, B. R., Q Band high efficiency monolithic HEMT power prematch structures, *Electronics Letters*, 31, 505, 1995.

40. Boulais, W., Donahue, R. S., Platzker, A., Huang, J., Aucoin, L., Shanfield, S., and Vafiades, M., A high power Q-band GaAs pseudomorphic HEMT monolithic amplifier, *IEEE MTT-S Int. Microwave Symp. Dig.*, 649, 1994.

41. Smith, P. M., Creamer, C. T., Kopp, W. F., Ferguson, D. W., Ho, P., and Willhite, J. R., A high power Q-band HEMT for communication terminal applications, *IEEE MTT-S Int. Microwave Symp. Dig.*, 809, 1994.

42. Tan, K. L., Streit, D. C., Dia, R. M., Wang, S. K., Han, A. C., Chow, P. D., Trinh, T. Q., Liu, P. H., Velebir, J. R., and Yen, H. C., High power V-band pseudomorphic InGaAs HEMT, *IEEE Electron Device Letters*, 12, 213, 1991.

43. Saunier, P., Matyi, R. J., and Bradshaw, K., A double-heterojunction doped-channel pseudomorphic power HEMT with a power density of 0.85 W/mm at 55 GHz, *IEEE Electron Device Letters*, 9, 397, 1988.

44. Lai, R., Wojtowicz, M., Chen, C. H., Biedenbender, M., Yen, H. C., Streit, D. C., Tan, K. L., and Liu, P. H., High power 0.15 μm V-band pseudomorphic InGaAs-AlGaAs-GaAs HEMT, *IEEE Microwave and Guided Wave Letters*, 3, 363, 1993.

45. Kraemer, B., Basset, R., Baughman, C., Chye, P., Day, D., and Wei, J., Power PHEMT module delivers 4 Watts, 38% PAE over the 18.0 to 21.2 GHz band, *IEEE MTT-S Int. Microwave Symp. Dig.*, 801, 1994.

46. Aucoin, L., Bouthillette, S., Platzker, A., Shanfield, S., Bertrand, A., Hoke, W., and Lyman, P., Large periphery, high power pseudomorphic HEMTs, *Proceedings of GaAs IC Symposium*, 351, 1993.

47. Bouthillette, S., Platzker, A., and Aucoin, L., High efficiency 40 Watt PsHEMT S-band MIC power amplifiers, *IEEE MTT-S Int. Microwave Symp. Dig.*, 667, 1994.

48. Kraemer, B., Basset, R., Chye, P., Day, D., and Wei, J., Power PHEMT module delivers 12 watts, 40% PAE over the 8.5 to 10.5 GHz band, *IEEE MTT-S Int. Microwave Symp. Dig.*, 683, 1994.

49. Matsunaga, K., Okamoto, Y., Miura, I., and Kuzuhara, M., Ku-band 15 W single chip HJFET power amplifier, *IEEE MTT-S Int. Microwave Symp. Dig.*, 697, 1996.

50. Fu, S. T., Lester, L. F., and Rogers, T., Ku band high power high efficiency pseudomorphic HEMT, *IEEE MTT-S Int. Microwave Symp. Dig.*, 793, 1994.

51. Henderson, T., Aksun, M. I., Peng, C. K., Morkoc, H., Chao, P. C., Smith, P. M., Duh, K.-H. G., and Lester, L. F., Microwave performance of quarter micrometer gate low noise pseudomorphic InGaAs/AlGaAs modulation doped field effect transistor, *IEEE Electron Device Letters*, 7, 649, 1986.

52. Tan, K. L., Dia, R. M., Streit, D. C., Shaw, L. K., Han, A. C., Sholley, M. D., Liu, P. H., Trinh, T. Q., Lin, T., and Yen, H. C., 60 GHz pseudomorphic $Al_{0.25}Ga_{0.75}As/In_{0.28}Ga_{0.72}As$ low- noise HEMTs, *IEEE Electron Device Letters*, 12, 23, 1991.

53. Chao, P. C., Duh, K. H. G., Ho, P., Smith, P. M., Ballingall, J. M., Jabra, A. A., and Tiberio, R. C., 94 GHz low noise HEMT, *Electronics Letters*, 25, 504, 1989.

54. Tan, K. L., Dia, R. M., Streit, D. C., Lin, T., Trinh, T. Q., Han, A. C., Liu, P. H., Chow, P. D., and Yen, H. C., 94 GHz 0.1 μm T-gate low noise pseudomorphic InGaAs HEMTs, *IEEE Electron Device Letters*, 11, 585, 1990.

55. Tokue, T., Nashimoto, Y., Hirokawa, T., Mese, A., Ichikawa, S., Negishi, H., Toda, T., Kimura, T., Fujita, M., Nagasako, I., and Itoh, T., Ku band super low noise pseudomorphic heterojunction field effect transistor (HJFET) with high producibility and high reliability, *IEEE MTT-S Int. Microwave Symp. Dig.*, 705, 1991.

56. Hwang, T., Kao, T. M., Glajchen, D., and Chye, P., Pseudomorphic AlGaAs/InGaAs/GaAs HEMTs in low-cost plastic packaging for DBS application, *Electronics Letters*, 32, 141, 1996.

57. Hirokawa, T., Negishi, H., Nishimura, Y., Ichikawa, S., Tanaka, J., Kimura, T., Watanbe, K., and Nashimoto, Y., A Ku-band ultra super low-noise pseudomorphic heterojunction FET in a hollow plastic PKG, *IEEE MTT-S Int. Microwave Symp. Dig.*, 1603, 1996.

58. Halchin, D. and Golio, M., Trends for portable wireless applications, *Microwave Journal*, 40, 62, 1997.

59. Ota, Y., Adachi, C., Takehara, H., Yangihara, M., Fujimoto, H., Masato, H., and Inoue, K., Application of heterojunction FET to power amplifier for cellular telephone, *Electronics Letters*, 30, 906, 1994.

60. Kunihisa, T., Yokoyama, T., Nishijima, M., Yamamoto, S., Nishitsuji, M., K. Nishii, Nakayama, and Ishikawa, O., A high-efficiency normally-off MODFET power MMIC for PHS operating under 3.0V single-supply condition, *Proceedings of GaAs IC Symposium*, 37, 1997.

61. Glass, E., Huang, J.-H., Abrokwah, J., Bernhradt, B., Majerus, M., Spears, E., Droopad, R., and Ooms, B., A true enhancement mode single supply power HFET for portable applications, *IEEE MTT-S Int. Microwave Symp. Dig.*, 1399, 1997.

62. Wang, Y. C. and J.M. Kuo, J. R. L., F. Ren, H.S. Tsai, J.S. Weiner, J. Lin, A. Tate, Y.K. Chen, W.E. Mayo, An $In_{0.5}(Al_{0.3}Ga_{0.7})_{0.5}P/In_{0.2}Ga_{0.8}As$ power HEMT with 65.2% power-added efficiency under 1.2 V operation, *Electronics Letters*, 34, 594, 1998.

63. Ren, F., Lothian, J. R., Tsai, H. S., Kuo, J. M., Lin, J., Weiner, J. S., Ryan, R. W., Tate, A., and Chen, Y. K., High performance pseudomorphic InGaP/InGaAs power HEMTs, *Solid State Electronics*, 41, 1913, 1997.

64. Martinez, M. J., Schirmann, E., Durlam, M., Halchin, D., Burton, R., Huang, J.-H., Tehrani, S., Reyes, A., Green, D., and Cody, N., P-HEMTs for low-voltage portable applications using filled gate fabrication process, *Proceedings of GaAs IC Symposium*, 241, 1996.

65. Masato, H., Maeda, M., Fujimoto, H., Morimoto, S., Nakamura, M., Yoshikawa, Y., Ikeda, H., Kosugi, H., and Ota, Y., Analogue/digital dual power module using ion-implanted GaAs MESFETs, *IEEE MTT-S Int. Microwave Symp. Dig.*, 567, 1995.

66. Huang, J. H. and E. Glass, J. A., B. Bernhardt, M. Majerus, E. Spears, J.M. Parsey Jr., D. Scheitlin, R. Droopad, L.A. Mills, K. Hawthorne, J. Blaugh, Device and process optimization for a low voltage enhancement mode power heterojunction FET for portable applications, *Proceedings of GaAs IC Symposium*, 55, 1997.

67. Lee, J.-L., Mun, J. K., Kim, H., Lee, J.-J., and Park, H.-M., A 68% PAE, GaAs power MESFET operating at 2.3 V drain bias for low distortion power applications, *IEEE Transactions on Electron Devices*, 43, 519, 1996.

68. Inosako, K., Matsunaga, K., Okamoto, Y., and Kuzuhara, M., Highly efficient double doped heterojunction FET's for battery operated portable power applications, *IEEE Electron Device Letters*, 15, 248, 1994.

69. Tanaka, T., Furukawa, H., Takenaka, H., Ueda, T., Noma, A., Fukui, T., Tateoka, K., and Ueda, D., 1.5 V Operation GaAs spike-gate power FET with 65% power-added efficiency, *IEDM Technical Digest*, 181, 1995.

70. Inosako, K., Iwata, N., and Kuzuhara, M., 1.2 V operation 1.1 W heterojunction FET for portable radio applications, *IEEE Electron Devices Meeting*, 185, 1995.

71. Iwata, N., Inosako, K., and Kuzuhara, M., 2.2 V operation power heterojunction FET for personal digital cellular telephones, *Electronics Letters*, 31, 2213, 1995.

72. Iwata, N., Inosako, K., and Kuzuhara, M., 3V operation L-band power double-doped heterojunction FETs, *IEEE MTT-S Int. Microwave Symp. Dig.*, 1465, 1993.

73. Iwata, N., Tomita, M., Yamaguchi, K., Oikawa, H., and Kuzuhara, M., 7 mm gate width power heterojunction FETs for Li-ion battery operated personal digital cellular phones, *Proceedings of GaAs IC Symposium*, 119, 1996.

74. Bito, Y., Iwata, N., and Tomita, M., Single 3.4 V operation power heterojunction FET with 60% efficiency for personal digital cellular phones, *Electronics Letters*, 34, 600, 1998.

75. Lai, Y.-L., Chang, E. Y., Chang, C.-Y., Liu, T. H., Wang, S. P., and Hsu, H. T., 2-V-operation δ-doped power HEMT's for personal handy-phone systems, *IEEE Microwave and Guided Wave Letters*, 7, 219, 1997.

76. Choumei, K., Yamamoto, K., Kasai, N., Moriwaki, T., Y, Y., Fujii, T., Otsuji, J., Miyazaki, Y., Tanino, N., and Sato, K., A high efficiency, 2 V single-supply voltage operation RF front-end MMIC for 1.9 GHz personal hand phone systems, *Proceedings of GaAs IC Symposium*, 73, 1998.

77. Hirose, K., Ohata, K., Mizutani, T., Itoh, T., and Ogawa, M., 700 mS/mm 2DEGFETs fabricated from high electron mobility MBE-grown n-AlInAs/GaInAs heterostructures, *GaAs and Related Compounds*, 529, 1985.

78. Palamateer, L. F., Tasker, P. J., Itoh, T., Brown, A. S., Wicks, G. W., and Eastman, L. F., Microwave characterization of 1 μm gate $Al_{0.48}In_{0.52}As/Ga_{0.47}In_{0.53}As/InP$ MODFETs, *Electronics Letters*, 23, 53, 1987.

79. Mishra, U. K., Brown, A. S., Rosenbaum, S. E., Hooper, C. E., Pierce, M. W., Delaney, M. J., Vaughn, S., and White, K., Microwave performance of AlInAs-GaInAs HEMT's with 0.2 μm and 0.1 μm gate length, *IEEE Electron Device Letters*, 9, 647, 1988.

80. Suemitsu, T., Enoki, T., Yokoyama, H., Umeda, Y., and Ishii, Y., Impact of two-step-recessed gate structure on RF performance of InP-based HEMTs, *Electronics Letters*, 34, 220, 1998.

81. Mishra, U. K., Brown, A. S., and Rosenbaum, S. E., DC and RF performance of 0.1 μm gate length $Al_{0.48}In_{0.52}As-Ga_{0.38}In_{0.62}As$ pseudomorphic HEMTs, *IEDM Technical Digest*, 180, 1988.

82. Nguyen, L. D., Brown, A. S., Thompson, M. A., and Jelloian, L. M., 50-nm self-aligned-gate pseudomorphic AlInAs/GaInAs high electron mobility transistors, *IEEE Transactions on Electron Devices*, 39, 2007, 1992.

83. Lai, R., Wang, H., Chen, Y. C., Block, T., Liu, P. H., Streit, D. C., Tran, D., Barsky, M., Jones, W., Siegel, P., and Gaier, T., 155 GHz MMIC LNAs with 12 dB gain fabricated using a high yield InP HEMT MMIC process, *Microwave Journal*, 40, 166, 1997.

84. Pobanz, C., Matloubian, M., Lui, M., Sun, H.-C., Case, M., Ngo, C., Janke, P., Gaier, T., and Samoska, L., A high-gain monolithic D-band InP HEMT amplifier, *Proceedings of GaAs IC Symposium*, 41, 1998.

85. Ho, P., Chao, P. C., Du, K. H. G., Jabra, A. A., Ballingall, J. M., and Smith, P. M., Extremely high gain, low noise InAlAs/InGaAs HEMTs grown by molecular beam epitaxy, *IEDM Technical Digest*, 184, 1988.

86. Chao, P. C., Tessmer, A. J., Duh, K. H. G., Ho, P., Kao, M.-Y., Smith, P. M., Ballingall, J. M., Liu, S.-M. J., and Jabra, A. A., W-band low-noise InAlAs/InGaAs lattice matched HEMTs, *IEEE Electron Device Letters*, 11, 59, 1990.

87. Onda, K., Fujihara, A., Miyamoto, H., Nakayama, T., Mizuki, E., Samoto, N., and Kuzuhara, M., Low noise and high gain InAlAs/InGaAs heterojunction FETs with high indium composition channels, *GaAs and Related Compounds*, 139, 1993.

88. Umeda, Y., Enoki, T., Arai, K., and Ishii, Y., Silicon nitride passivated ultra low noise InAlAs/InGaAs HEMT's with n^+ InGaAs/n^+-InAlAs cap layer, *IEICE Transactions on Electronics*, E75-C, 649, 1992.

89. Duh, K. H. G., Chao, P. C., Liu, S. M. J., Ho, P., Kao, M. Y., and Ballingall, J. M., A super low noise 0.1 μm T-gate InAlAs-InGaAs-InP HEMT, *IEEE Microwave and Guided Wave Letters*, 1, 114, 1991.

90. Kao, M.-Y., Duh, K. H. G., Ho, P., and Chao, P.-C., An extremely low noise InP-based HEMT with silicon nitride passivation, *IEDM Technical Digest*, 907, 1994.

91. Kao, M. Y., Smith, P. M., Chao, P. C., and Ho, P., Millimeter wave power performance of InAlAs/InGaAs/InP HEMTs, *IEEE/Cornell Conference on Advanced Concepts in High Speed Semiconductor Devices and Circuits*, 469, 1991.

92. Matloubian, M., Nguyen, L. D., Brown, A. S., Larson, L. E., Melendes, M. A., and Thompson, M. A., High power and high efficiency AlInAs/GaInAs on InP HEMTs, *IEEE MTT-S Int. Microwave Symp. Dig.*, 721, 1991.

93. Larson, L. E., Matloubian, M., Brown, J. J., Brown, A. S., Rhodes, R., Crampton, D., and Thompson, M., AlInAs/GaInAs on InP HEMTs for low power supply voltage operation of high power-added efficiency microwave amplifiers, *Electronics Letters*, 29, 1324, 1993.

94. Bahl, S. R., Azzam, W. J., delAlamo, J. A., Dickmann, J., and Schildberg, S., Off-state breakdown in InAlAs/InGaAs MODFET's, *IEEE Transactions on Electron Devices*, 42, 15, 1995.

95. Boos, J. B. and Kruppa, W., InAlAs/InGaAs/InP HEMTs with High Breakdown voltages using double recess gate process, *Electronics Letters*, 27, 1909, 1991.

96. Hur, K. Y., McTaggart, R. A., LeBlanc, B. W., Hoke, W. E., Lemonias, P. J., Miller, A. B., Kazior, T. E., and Aucoin, L. M., Double recessed AlInAs/GaInAs/InP HEMTs with high breakdown voltages, *Proceedings of GaAs IC Symposium*, 101, 1995.

97. Hur, K. Y., Mctaggart, R. A., Miller, A. B., Hoke, W. E., Lemonias, P. J., and Aucoin, L. M., DC and RF characteristics of double recessed and double pulse doped AlInAs/GaInAs/InP HEMTs, *Electronics Letters*, 31, 135, 1995.

98. Pao, Y. C., Nishimoto, C. K., Majidi-Ahy, R., Archer, J., Betchel, N. G., and Harris, J. S., Characterization of surface undoped $In_{0.52}Al_{0.48}As/In_{0.53}Ga_{0.47}As$ high electron mobility transistors, *IEEE Transactions on Electron Devices*, 37, 2165, 1990.

99. Ho, P., Kao, M. Y., Chao, P. C., Duh, K. H. G., Ballingall, J. M., Allen, S. T., Tessmer, A. J., and Smith, P. M., Extremely high gain 0.15 μm gate-length InAlAs/InGaAs/InP HEMTs, *Electronics Letters*, 27, 325, 1991.

100. Matloubian, M., Larson, L., Brown, A., Jelloian, L., Nguyen, L., Lui, M., Liu, T., Brown, J., Thompson, M., Lam, W., Kurdoghlian, A., Rhodes, R., Delaney, M., and Pence, J., InP based HEMTs for the realization of ultra high efficiency millimeter wave power amplifiers, *IEEE/Cornell Conference on Advanced Concepts in High Speed Semiconductor Devices and Circuits*, 520, 1993.

101. Matloubian, M., Brown, A. S., Nguyen, L. D., Melendes, M. A., Larson, L. E., Delaney, M. J., Pence, J. E., Rhodes, R. A., Thompson, M. A., and Henige, J. A., High power V-band AlInAs/GaInAs on InP HEMT's, *IEEE Electron Device Letters*, 14, 188, 1993.

102. Brown, J. J., Matloubian, M., Liu, T. K., Jelloian, L. M., Schmitz, A. E., Wilson, R. G., Lui, M., Larson, L., Melendes, M. A., and Thompson, M. A., InP Based HEMTs with $Al_xIn_{1-x}P$ Schottky layers grown by gas source MBE, *Proceedings of International Conference on InP and Related Materials*, 419, 1994.

103. Scheffer, F., Heedt, C., Reuter, R., Lindner, A., Liu, Q., Prost, W., and Tegude, F. J., High breakdown voltage InGaAs/InAlAs HFET using $In_{0.5}Ga_{0.5}P$ spacer layer, *Electronics Letters*, 30, 169, 1994.

104. Hur, K. Y., McTaggart, R. A., Lemonias, P. J., and Hoke, W. E., Development of double recessed AlInAs/GaInAs/InP HEMTs for millimeter wave power applications, *Solid State Electronics*, 41, 1581, 1997.

105. Matloubian, M., Brown, A. S., Nguyen, L. D., Melendes, M. A., Larson, L. E., Delaney, M. J., Thompson, M. A., Rhodes, R. A., and Pence, J. A., 20 GHz high efficiency AlInAs-GaInAs on InP power HEMT, *IEEE Microwave and Guided Wave Letters*, 3, 142, 1993.

106. Hur, K. Y., McTaggart, R. A., Ventresca, M. P., Wohlert, R., Hoke, W. E., Lemonias, P. J., Kazior, T. E., and Aucoin, L. M., High efficiency single pulse doped $Al_{0.60}In_{0.40}As/GaInAs$ HEMTs for Q band power applications, *Electronics Letters*, 31, 585, 1995.

107. Ho, P., Smith, P. M., Hwang, K. C., Wang, S. C., Kao, M. Y., Chao, P. C., and Liu, S. M. J., 60 GHz power performance of 0.1 μm gate length InAlAs/GaInAs HEMTs, *Proceedings of International Conference on InP and Related Materials*, 411, 1994.

108. Hwang, K. C., Ho, P., Kao, M. Y., Fu, S. T., Liu, J., Chao, P. C., Smith, P. M., and Swanson, A. W., W Band high power passivated 0.15 μm InAlAs/InGaAs HEMT device, *Proceedings of International Conference on InP and Related Materials*, 18, 1994.

109. Chen, Y. C., Lai, R., Lin, E., Wang, H., Block, T., Yen, H. C., Streit, D., Jones, W., Liu, P. H., Dia, R. M., Huang, T.-W., Huang, P.-P., and Stamper, K., A 94-GHz 130-mW InGaAs/InAlAs/InP HEMT High-Power MMIC Amplifier, *IEEE Microwave and Guided Wave Letters*, 7, 133, 1997.

110. Smith, P. M., Liu, S. M. J., Kao, M. Y., Ho, P., Wang, S. C., Duh, K. H. G., Fu, S. T., and Chao, P. C., W-Band high efficiency InP-based power HEMT with 600 GHz f_{max}, *IEEE Microwave and Guided Wave Letters*, 5, 230, 1995.

111. Akazaki, T., Enoki, T., Arai, K., Umeda, Y., and Ishii, Y., High frequency performance for sub 0.1 μm Gate InAs-inserted-channel InAlAs/InGaAs HEMT, *Electronics Letters*, 28, 1230, 1992.

112. Enoki, T., Arai, K., Kohzen, A., and Ishii, Y., InGaAs/InP Double channel HEMT on InP, *Proceedings of International Conference on InP and Related Materials*, 14, 1992.

113. Matloubian, M., Liu, T., Jelloian, L. M., Thompson, M. A., and Rhodes, R. A., K-Band GaInAs/InP channel power HEMTs, *Electronics Letters*, 31, 761, 1995.

114. Shealy, J. B., Matloubian, M., Liu, T. Y., Thompson, M. A., Hashemi, M. M., DenBaars, S. P., and Mishra, U. K., High-performance submicrometer gatelength GaInAs/InP composite channel HEMT's with regrown contacts, *IEEE Electron Device Letters*, 16, 540, 1996.

115. Shealy, J. B., Matloubian, M., Liu, T. Y., Lam, W., and Ngo, C., 0.9 W/mm, 76% PAE (7 GHz) GaInAs/InP composite channel HEMTs, *Proceedings of International Conference on InP and Related Materials*, 20, 1997.

116. Chevalier, P., Wallart, X., Bonte, B., and Fauquembergue, R., V-band high-power/low-voltage InGaAs/InP composite channel HEMTs, *Electronics Letters*, 34, 409, 1998.

7.2.6 RF Power Transistors from Wide Bandgap Materials

Karen E. Moore

Wide bandgap materials such as silicon carbide (SiC) and gallium nitride (GaN) are known as such because their bandgaps are much larger than those of more conventional semiconductors such as silicon, germanium, or gallium arsenide. 4H-SiC, for example, has a bandgap of 3.2 eV, and GaN has a bandgap of 3.4 eV, as compared to 1.11 eV for silicon. These materials have been studied in theory for over 30 years; however, it has only been in the past decade that device development from wide bandgap semiconductors has occurred at any level, with significant breakthroughs occurring in SiC substrate and epi technology in the late 1980s, and in GaN epi technology in the mid-1990s. The first commercial applications for wide bandgap materials were blue LEDs fabricated from SiC, later followed by blue LEDs with greatly increased brightness, from GaN-related materials. While these applications are not RF applications, they have helped drive development of these very new, experimental materials.

In recent years, wide bandgap semiconductors have received a great deal of attention as a nearly ideal material for the fabrication of high speed, high power transistors [1–5], particularly for cellular base station, broadcast, and high frequency radar applications. The large bandgaps should allow SiC- and GaN-based transistors to have stable DC and RF operation at very high temperatures [6]. Several other material properties of wide bandgap materials also make them attractive for fabrication of high voltage, high power, and high frequency transistors. These include a high electric breakdown field of 3 to 4 × 10⁶ V/cm, high saturated electron drift velocities of 1.5 to 2.2 × 10⁷ cm/sec, and high thermal conductivity (for SiC substrates) of 4.9 W/cm-K.

This article will present a basic description of figures of merit for microwave devices fabricated from wide bandgap materials, a very basic description of the operation of some of the commonly studied wide bandgap RF power transistors, a discussion of the material properties needed for RF power generation and how those properties translate into improved performance of microwave systems, and a summary of state-of-the-art wide bandgap high frequency device performance. The article will focus primarily on AlGaN/GaN HEMTs, SiC MESFETs, and SiC SITs, since they are currently the most mature and most widely studied wide bandgap RF device technologies.

Figures of Merit for RF Power Transistors

Virtually all RF systems require active circuit elements for use as oscillators, amplifiers, etc. These elements permit conversion of energy from DC bias sources to RF bands where the energy can be used to provide useful gain at specified frequencies. The ideal RF power transistor has high current, high breakdown voltage, and a low "knee" voltage (the voltage at which the transistor current saturates), as illustrated in Fig. 7.107. The device is given DC bias at one-half its maximum operating voltage and one-half or less of its maximum operating current, and any RF signal superimposed over the device is amplified over the I-V curve as shown in Fig. 7.107. The maximum possible RF output power of a transistor is [7]

$$P_{out,max} = (1/8)(V_{DS} - V_{knee})^2 / R_L \qquad (7.162)$$

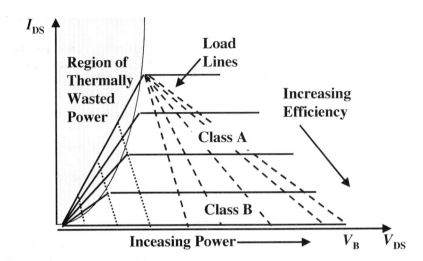

FIGURE 7.107 I-V curves and load lines for a power transistor.

where P_{out} is the RF output power, V_{DS} is the drain bias, V_{Knee} is the knee voltage, and R_L is the load resistance, which is determined by the bias current and voltage in the device. The Power Added Efficency (PAE), η, of the device is another important figure of merit, and quantifies the amount of DC bias that is converted to RF power:

$$\eta = \left(P_{out} - P_{in}\right)\big/P_{dc} \qquad (7.163)$$

The maximum efficiency for an RF transistor is 50% under class A operation (transistor is biased at 50% of its open channel current), or 78.5% under class B operation (transistor is biased at pinch off or $I_{dq} = 0A$); however, the maximum possible output power for the device remains unchanged with bias as long as the device is operated at 50% or less of its open channel current. Qualitatively, any RF drive in the region below the knee voltage will result only in resistive loss, which leads to lower output power, decreased gain, and lower efficiency. The increased resistive loss also leads to device heating that will further degrade device performance. Thus, the ideal RF transistor would have a knee voltage of 0V.

Two other important figures of merit for a microwave transistor are gain and intermodulation distortion (IMD). Gain is generally quantified under both large-signal and small-signal conditions, and is the ratio of output power to input power for a device [8]. Intermodulation distortion, in its simplest sense, is a measure of how constant gain is in a device over a wide range of instantaneous drive conditions as induced by a large RF signal. Any variation of gain from its linear, or constant value will cause generation of new signals at harmonic frequencies. These harmonics can cause system level problems. For example, a harmonic signal could be in a channel that is adjacent to its carrier, and be mistaken for a signal in that adjacent channel. IMD becomes increasingly more important, particularly in cellular systems, as more and more carriers are used in a given bandwith with increasing technology capabilities.

The unity current gain cutoff frequency of a device, f_t, and the maximum frequency of oscillation, f_{max}, are both very important parameters. These two figures of merit will determine the highest frequency at which a device is useful as an amplifier.

Finally, the power density of an RF transistor is very important for several reasons. Power density is most often expressed in units of watts/mm of gate periphery or watts/square cm of die area. For a given power rating, a high power density results in a smaller device, which will mean higher output impedance, easier matching, and possibly more power per die and/or fewer combining networks.

Common RF Power Devices from Wide Bandgap Materials

The three most commonly studied wide bandgap RF power devices are the SiC MESFET (MEtal-Semiconductor Field Effect Transistor), the SiC SIT (Static Induction Transistor), and the AlGaN/GaN

HEMT (High Electron Mobility Transistor), sometimes called an AlGaN/GaN MODFET (Modulation Doped Field Effect Transistor). A brief description of operation of these three devices follows.

SiC MESFETs

The MESFET was first proposed by Mead in 1966 [9], and fabricated in GaAs by Hooper and Lehrer in 1967 [10]. A basic schematic of a MESFET is shown in Fig. 7.108a [11]. The MESFET is a planar device fabricated by growth of a thin, doped epitaxial layer located on either a semi-insulating substrate or a low-doped layer of conductivity type opposite to that of the channel material. MESFETs in SiC can be fabricated on substrates of the same conductivity as the device channel (typically, n-type), but with a 1 to 5 micron buffer layer of opposite conductivity between the channel and the substrate. High resistivity substrates are most desirable for high frequency devices and result in improved DC and RF performance for the transistor by better confining electrons to the conducting channel and reducing microwave losses. This will be described in more detail in the next section.

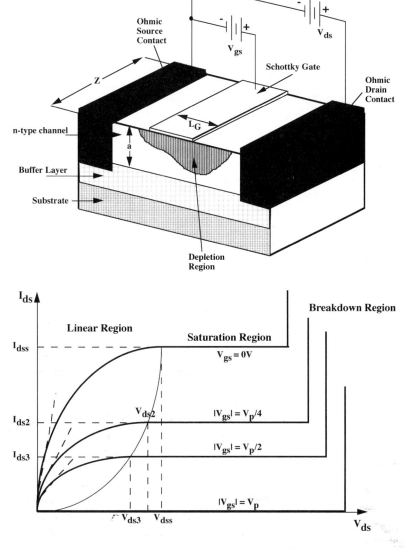

FIGURE 7.108 (a) Cross-section of a typical SiC metal-semiconductor field effect transistor (MESFET); (b) I-V curves of a typical MESFET.

In the MESFET, current is passed through the conducting channel by means of two ohmic contacts (the source and drain), which are typically separated by a distance of 3 to 10 μm. The actual dimensions are dependent upon the operating frequency, and smaller dimensions are used as operating frequency is increased. A rectifying Schottky contact (the gate) is located between the two ohmic contacts, and is typically 0.1 to 2 μm in length for modern microwave devices.

Fig. 7.108b shows the DC I-V curves for a MESFET [11]. During operation, the drain contact is biased at a specified potential (positive drain potential for an n-channel device) and the source is grounded. The flow of current through the conducting channel is controlled by a combination of a negative DC (for n-channel devices) gate bias with a superimposed RF signal also at the gate. The DC bias sets the quiescent operating current for the device, while the RF signal modulates the channel current, thereby providing RF gain. The operation of the transistor is determined by the ability of the gate signal to effectively modulate and control the current in the conducting channel. For this reason, any electrons that leak into the substrate or through the gate electrode will lead to performance degradation. The availability of both high resistivity, low leakage substrates and high quality, low leakage Schottky barrier gates in SiC makes high performance SiC MESFETs possible for high voltage RF power applications.

SiC SITs

Static Induction Transistors (SITs) are useful as high power RF sources and amplifiers at UHF and microwave frequencies. This device was originally proposed by Nishizawa [12] as a power amplifier for applications such as audio amplifiers. A cross-section of the transistor is shown in Fig. 7.109a. The device has a structure very similar to a vacuum triode with source and drain contacts separated by a certain distance. Electrons are emitted from the source, which is generally at ground potential, and are accelerated to the drain, which is biased at a positive potential, where they are collected. A grid structure is located in the space between the source and drain electrodes so the charge carriers can be externally modulated. The RF gain of the device is determined by the efficiency with which the modulation is accomplished. The grid structure is generally fabricated using pn or Schottky junctions. SIT devices are capable of high voltage gain, but limited current gain. However, since the device has good impedance characteristics, high power gain is available and the device can be effectively used at frequencies significantly in excess of the f_t for the device.

SITs have four possible modes of operation, as illustrated in the I-V curves shown in Fig. 7.109b. These modes are ohmic, thermionic emission, space-charge limited current flow (SCLC), and space-charge limited current flow under saturated velocity conditions (SCLC with V_{sat}). Static induction transistors perform best when designed so they operate under either of the space-charge-limited current flow conditions. This occurs when the conducting channel is lightly doped so that the region between the grid bars is completely depleted of free charge under typical bias conditions. Under these conditions a saddle point potential is created in the conducting channel between the grid bars, and modulation of this potential controls the flow of current from drain to source. In this mode of operation the current that can flow through the device is limited by the number of electrons that can be forced to flow between the grid bars and through the saddle point potential. Since this region is depleted of free charge carriers, the injected charge flowing in the channel generates a space-charge that reduces the electric field in front of the moving charge, and increases the electric field behind it. The net result is that current flow is self-limited by the space charge of the moving electrons. This limits the density of charge and RF currents that can flow through the device and establishes a limit to the current drive capability of the device.

The use of SiC offers improved RF performance for SIT transistors since the current density that flows through the semiconductor is a function of both the injected charge density and the velocity at which the electrons move. Since the saturated velocity of electrons in SiC is very high (i.e., $v_s = 2.2 \times 10^7$ cm/sec [13]), a high channel current is possible by designing the device so that the magnitude of the electric field is above that necessary to maintain electron velocity saturation. Also, the use of highly doped n+ source and drain contact regions permit low resistance contacts to be fabricated, thereby minimizing the degrading effects of low electron mobility. The net result is that higher channel current can be achieved, and this results in good RF performance.

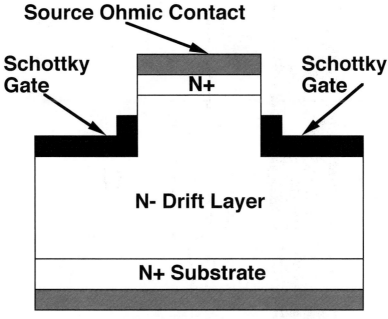

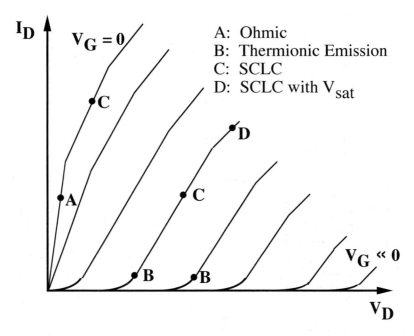

FIGURE 7.109 (a) Cross-section of a single unit cell of a typical SiC static induction transistor (SIT); (b) I-V curves of a typical SIT.

AlGaN/GaN HEMTs

AlGaN/GaN HEMTs (High Electron Mobility Transistors) are somewhat similar in structure and operation to SiC MESFETs, with the notable difference that they are heterojunction devices. A cross-section of an AlGaN/GaN HEMT is shown in Fig. 7.110. In this device, all of the conduction is in a channel

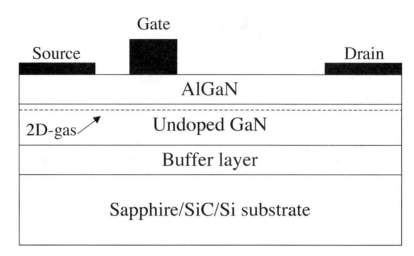

FIGURE 7.110   Cross-section of a typical AlGaN/GaN HEMT. I-V curves are similar to those of the SiC MESFET.

formed by a sheet of charge immediately under the AlGaN/GaN heterojunction, with the channel usually being about 50 to 250Å total thickness. Typical sheet charge densities in the AlGaN/GaN materials system are 1 to 1.5×10^{13} cm^{-2}, which is 3 to 10 times that normally seen in GaAs-based HEMTs. The charge in this device can either come from doping in the AlGaN layer, which spills over into the lower energy GaN layer or can be induced by the piezoelectric field found in strained GaN-based materials grown on heteroexpitaxial substrates like sapphire or SiC [14]. The thick GaN layer and the buffer layer, most often AlN, provide isolation from the substrate and carrier confinement. These devices have been fabricated most frequently on sapphire substrates, but have also been extensively studied on SiC substrates [15–16], and have recently been fabricated on p-type ⟨111⟩ silicon wafers [17]. To date, the best AlGaN/GaN HEMT performance has been from devices utilizing semi-insulating SiC substrates.

Like in the MESFET, current flows in the AlGaN/GaN HEMT from source to drain, induced by a bias at the drain, and the gate modulates charge through the channel through the same combination of DC bias and RF signal. The high sheet charge density in these structures creates a very high current density to complement the high breakdown voltage already inherent in the GaN channel material. The high sheet charge will also allow very low on-resistance, which will improve the microwave performance of the devices, and the combination of high current density with high voltage operation makes the AlGaN/GaN HEMT the transistor with the highest power density of the wide bandgap devices.

Other Wide Bandgap Devices

There are several other wide bandgap power devices that are beginning to be studied. These include AlGaN/GaN HBTs (Heterojunction Bipolar Transistors), which should have excellent efficiency and breakdown characteristics [18], MOS-HFETs (Metal-Oxide-Semiconductor Heterojunction Field Effect Transistors), which are expected to have improved linearity as compared to AlGaN/GaN HEMTs [19], JFETs (Junction Field Effect Transistors), which should have very good high temperature performance [20], and IMPATT diodes, which are expected to generate power at very high frequencies with much greater efficiency and lower access resistances than their GaAs and InP-based counterparts. All of these devices are in their technological infancy, and have yet to demonstrate the remarkable RF results of the more conventional wide bandgap devices. Nevertheless, the potential for these devices to become important in the future should not be overlooked. In the meantime, the rest of this article will focus only on results from SiC MESFETs, SiC SITs, and AlGaN/GaN HEMTs.

Desirable Material Properties for RF Power Transistors

There are a number of semiconductor material properties that affect the performance of a high speed, high power transistor [1–2]. These include the bandgap, critical (breakdown) electric field, and thermal

TABLE 7.19 Material Properties of Common Semiconductors

Property	Units	Silicon	GaAs	4H-SiC	GaN
Bandgap	EV	1.11	1.43	3.2	3.4
Breakdown field	V/cm	7×10^5	7×10^5	35×10^5	35×10^5
Saturation velocity	cm/sec	1×10^7	1×10^7	2×10^7	1.5×10^7
Saturation field	V/cm	8×10^3	3×10^3	25×10^3	15×10^3
Thermal conductivity	W/cm-K	1.5	0.46	4.9	1.7/substrate
Electron mobility	Cm^2/V-sec	1350	6000	800	1000
Hole mobility	Cm^2/V-sec	450	330	120	300

conductivity. Electron and hole transport properties, the saturated electron velocity, and the electric field at which electron velocity saturates will also strongly influence the DC and RF characteristics of a high frequency power transistor. Several of these properties are summarized in Table 7.19, which compares the material properties of Si, GaAs, 4H-SiC, and GaN. In addition to the properties summarized in Table 7.19, the electrical conductivity of the substrate material can strongly affect RF losses in a transistor. Finally, the ability to make both low resistance ohmic contacts and good rectifying Schottky contacts are critical to transistor fabrication and performance. Figure 7.108 attempts to capture some of the many relationships between material properties and power device and system performance. The system level advantage is particularly important as a metric for evaluating devices from wide bandgap materials, as that is where their real importance and impact for a customer or consumer will be evident.

Critical or Breakdown Field

SiC and GaN have critical fields of 3.5 MV/cm, five times that of Si or GaAs. This critical, or breakdown field of a material is possibly the most important material parameter for design of a high power density device, as it determines the highest operating voltage of a transistor for a given device design and channel doping, and thus limits the RF power swing in the device. Analagously, for a given voltage requirement, a higher breakdown field will allow a device designer to use a higher doping level in the device than for a lower breakdown field material, and, with the higher doping, tighter device dimensions. Higher operating voltage, as shown in Eq. (7.162), results in both higher power and higher power density in the device. Higher doping and reduced dimensions in an FET will also enable a device with increased transconductance, lower parasitic resistances, increased power gain, higher f_t and f_{max}, and improved efficiency due to the decreased access resistances.

Furthermore, large output power levels can be achieved through either high current or high voltage operation, but a device that obtains its power from high voltage rather than high current will be much smaller due to the fact that voltage scaling in a device only involves design changes in a device channel on the order of 1 to 2 μm, whereas high current devices require complete scaling of the device periphery. This becomes even more important for a very high power device as increased current in the linear region of device operation introduces more resistive heating, and further degrades device performance. In addition, a smaller geometry, high voltage part will have output impedance levels that are much larger than for a larger geometry, high current device of the same output power level, making the high voltage FET easier to incorporate into a circuit design. At a system level, the higher power density of a wide bandgap transistor will lead to more power per die, and thus a smaller die count per system, and greater bandwidth due to improved output impedance characteristics. Higher efficiency will translate into lower total energy usage for the microwave system in question, and smaller die sizes will use smaller and therefore cheaper packages.

Thermal Conductivity

The thermal conductivity of a material determines the ease with which heat generated from unconverted DC power can be removed from the device. Any temperature rise from undissipated heat will further degrade device performance by causing a drop in the mobility and saturated electron velocity in the transistor, which in turn causes the device to become progressively less efficient and to generate more heat. Thermal conductivity can also influence whether or not special packaging and/or system cooling become necessary for successful device operation. The thermal conductivity of SiC of 4.9 W/cm-K is three times that of Si and ten times that of GaAs or sapphire and, as such, is a tremendous advantage for SiC-based devices. Since GaN-based devices are grown on both sapphire and SiC substrates at the present time, the limiting thermal conductivity for these devices depends on the choice of substrate.

Wide Bandgap

A large bandgap is considered desirable for high voltage power devices for two reasons. First, the bandgap determines the upper temperature limit of device operation. This is particularly important when transistor scaling and power density are examined, as higher temperature operation makes it possible to design a smaller, denser device that will withstand the heat it generates under bias. Second, wide bandgap materials have been shown to be more resistant to radiation, such as α particles, which are known to be very destructive in Si MOS devices. The large bandgaps of 3.4 eV for GaN and 3.2 eV for 4H-SiC give these materials a high temperature performance advantage over Si and GaAs. In the system limit, a device that can withstand higher channel and ambient temperatures may be suitable for a cheaper package and relaxed system cooling requirements.

Saturated Electron Velocity

The saturated electron velocity of a material is very important for sub-micron gate length devices, which typically operate at very high electric fields. In this regime, the frequency performance (in particular, f_t) of a device is largely determined by electron velocity. SiC has a measured saturated electron velocity of 2.2×10^7 cm/second, twice that of Si or GaAs, and GaN has an electron velocity of 1.5×10^7 cm/second, still significantly higher than that of Si or GaAs. The electric field at which the electron velocity saturates is also important as it determines how quickly the charge carriers can be accelerated to their saturated values. The saturation fields for GaN and SiC are 1.5 to 2.5×10^4 V/cm, or 2 to 8 times higher than the 3 to 8×10^3 V/cm values for GaAs and Si. The high saturation fields combined with the low mobilities of wide bandgap semiconductors result in devices that will have higher knee voltages and therefore will have to be operated at considerably higher supply voltages before they operate in a saturated electron velocity regime. This will move their optimal performance to much higher voltage levels than for more conventional semiconductor technologies. Overall, the high electron velocities, at whatever voltage is necessary, will contribute to a very high speed device and, as a result, as very high system frequency.

Electron and Hole Mobilities

A primary disadvantage of fabricating transistors from wide bandgap semiconductors is the relatively low values for electron and hole mobilities in these materials. Electron and hole transport properties are also critical to successful device performance, and play a dominant role in determining the on-resistance and knee voltage of a device in its low-field region of operation. Low mobility results in increased parasitic resistance, increased losses, and reduced gain. These problems are worsened both as operating frequency is increased, where series resistances play an increasingly stronger role, and at elevated temperatures, where mobility will rapidly decrease.

SiC and GaN have electron mobilities of 800 to 1000 cm²/V-sec in nominally undoped material. These values are much less than the 1350 cm²/V-sec electron mobility of silicon, let alone the 6000 cm²/V-sec electron mobility of GaAs. At the typical 10^{17} cm^{-3} doping levels commonly seen in high voltage MESFETs, the electron mobility in 4H-SiC drops to around $\mu_e \sim 500$ cm²/V-sec. Fortunately, the electron mobility in a 2D-gas in an AlGaN/GaN heteostructure typically remains close to 1000 cm²/V-sec as the electrons in the AlGaN/GaN heterojunction are physically separated from their donor atoms and thus are not affected by ionized impurity scattering as in the case of SiC. The hole mobility in both SiC and GaN is very low, on the order of $\mu_p \sim 120$ to 300 cm²/V-sec. The low value for hole mobility severely limits the use of p-type SiC in RF transistors that are intended for operation above about 1 GHz.

Substrate Conductivity

In addition to the material properties detailed in Table 7.19, there are several other material parameters than can affect RF power device performance. The electrical conductivity of the device substrate is a very important material property for higher frequency operation of RF transistors. The SIT, which is a vertical device, must have a conducting substrate in order to accomodate its backside source contact. The effects of substrate conductivity are much more complicated in MESFET and HEMTs. Substrate conductivity is strongly related to losses at high frequency in microwave MESFETs and HEMTs, since they are lateral devices. At microwave frequencies, the contact pads of the transistor act as lossy transmission lines, and as such create both loss and dispersion in the small-signal characteristics of the device, including small-signal gain, which is the upper limit of power gain in a transistor. Frequency dispersion of various device elements makes device model extrapolation, wideband device operation, and circuit design all very difficult. Development of semi-insulating SiC substrates of both 6H and 4H polytypes has clearly demonstrated the importance of this material improvement. The device gain in MESFETs on semi-insulating SiC is higher, the substrate losses are smaller, and the dispersion seen in parasitic device elements is decreased as compared to devices on highly doped SiC substrates. Similarly, AlGaN/GaN HFETs fabricated on sapphire (an insulator) or semi-insulating silicon carbide have much better microwave performance than those fabricated on highly doped silicon substrates. This will be discussed in more detail in the device results in the next section.

Electrical Contacts

Electrical contacts, which are critical to successful device operation, are also strongly influenced by material properties. One of the disadvantages of wide bandgap materials is that they tend to form ohmic contacts with higher contact resistance than ohmic contacts fabricated on smaller bandgap semiconductors. Unfortunately, ohmic contact resitance can strongly influence both DC and RF operation of a device by contributing to the resistive region of the I-V curves shown in Fig. 7.107. As discussed earlier, this increase in the on-resistance adversely affects the extrinsic transconductance, small-signal gain, frequency performance, and power-added efficiency of a device. With optimization, reasonably low contact resistances have been realized in both SiC and GaN n-type materials.

High quality Schottky contacts are also critical to successful transistor performance. The quality of its Schottky contact determines the ability of the gate to completely pinch off the device channel, and controls the gate leakage current in the device under both DC and RF conditions. High gate leakage will decrease RF gain and power-added efficiency, and, in turn, the maximum output power of a device. Since good Schottky contacts are typically easy to fabricate on wide bandgap materials, good Schottky gate operation is usually possible in SiC technology.

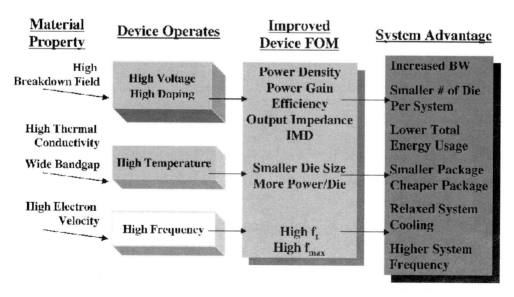

FIGURE 7.111 Relationships between material properties and device and system level performance for transistors and systems using wide bandgap materials.

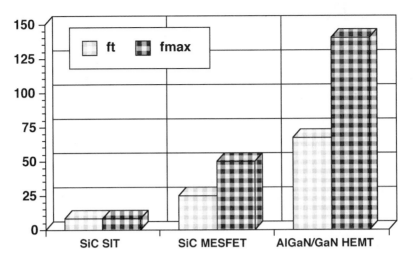

FIGURE 7.112   Comparison of highest reported f_t and f_{max} for SiC SITs, SiC MESFETs, and AlGaN/GaN HEMTs.

State-of-the-Art Wide Bandgap Microwave Transistor Data

The effects of the material properties of wide bandgap semiconductors are best illustrated through the phenomenal device performance being reported today. Despite the immaturity of wide bandgap technology, the DC, small-signal, and large signal results from wide bandgap transistors are already challenging or exceeding the best results reported in many semiconductor material systems.

High Frequency Performance

f_t and f_{max} are commonly cited for SiC MESFETs, SITs, and AlGaN/GaN HEMTs because they provide an upper limit to the frequencies at which these devices are useful in system applications. Figure 7.112 illustrates both the best performance reported to date for these devices and the performance differences from the device types. The best performance is from the AlGaN/GaN HEMT, with $f_t = 67$ GHz and $f_{max} = 140$ GHz [21]. The SiC MESFET is also very fast, with peak $f_t = 25$ GHz and $f_{max} = 50$ GHz [22]. The SiC SIT is the weakest performer, with an f_t of only 7 GHz [23] and f_{max} of 8 GHz.

The reasons for the differences in performance can be explained through simple device physics. f_t is linear with transconductance, which is very high (~240 to 250 mS/mm) for the AlGaN/GaN HEMT. Transconductance in the SiC MESFET is typically on the order of 50 mS/mm, leaving the MESFET with much lower f_t. The strong effect of gate capacitance on f_t can also be seen with the SiC MESFET. The MESFETs cited with 25 GHz f_t are fabricated on semi-insulating substrates. When the same device is fabricated on a conducting SiC substrate, which adds a very large parasitic pad capacitance to both the gate and the drain of the device, the best reported f_t drops to about 8 GHz and f_{max} falls to 16 GHz [24]. Similarly, when an AlGaN/GaN HEMT was fabricated on a highly doped p-type silicon substrate, f_t and f_{max} were both 25 GHz [17]. The SiC SIT has a very large parasitic gate-drain capacitance due to its vertical design, and low transconductance due to the lower doping levels used in this device, so even with careful design and processing, the upper limit of f_t for the SIT is expected to be around 20 GHz.

f_{max} is linear with f_t, but also depends strongly on gate, source, drain, and channel resistances, output conductance, and gate-drain capacitance. The large differences in f_{max} among the wide bandgap devices are probably primarily due to the dramatic differences in transconductances in the different devices. This is validated by the fact that the f_{max}/f_t ratio is nearly the same for the AlGaN/GaN HEMT and the SiC MESFET.

Power Density

Power density, shown in Fig. 7.113, is also an important parameter for evaluation as it is instrumental in determining the minimum device, die, and package size of a microwave transistor. The trend for power density is similar to that of frequency performance. AlGaN/GaN HEMTs have the highest reported power density, 9.1 W/mm at Vds = 30 V [25], due to both the high current density and high breakdown voltage

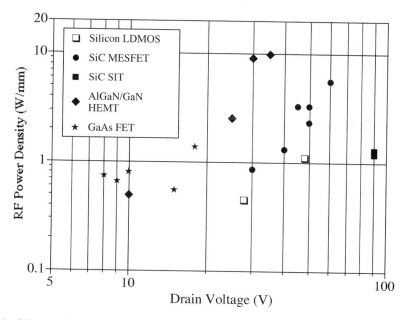

FIGURE 7.113   RF power density vs. drain voltage for different RF power technologies.

of these devices. SiC MESFETs follow with a significant power density of 5.6 W/mm at Vds = 60 V [26]. While the drain voltage of the MESFET is higher, its current carrying capability (drain current density) is significantly lower than that of the AlGAN/GaN HEMT, and so the power density of the SiC MESFET ends up being lower than that of the AlGaN/GaN HEMT. SiC SITs have lower power density at higher drain voltage, due to their lower current carrying capability. The peak power density for a SiC SIT is 1.3 W/mm at Vds = 90 V [27]. Overall, as seen in Fig. 7.113, the wide bandgap power devices significantly outperform silicon LDMOS and GaAs devices at higher supply voltages. Also worth noting is the fact that the highest power density for an AlGaN/GaN HEMT on a sapphire substrate is only 3.1 W/mm, much less than the 9.1 W/mm reported for a very similar device design on a SiC substrate. This further underscores the importance of substrate thermal conductivity.

Total Power

Total RF output power, shown in Fig. 7.114, is very important. In the final analysis, it will not matter what power density a transistor achieves if it cannot be scaled up into a total power that is useful for a system-level application. In the case of wide bandgap devices, the power requirements for their intended uses (basestations, radar, broadcast, etc.) range from approximately 30 to 1500 W. For total power, the SiC SIT is the device with superior performance. Some of the best reported SIT power data includes a 34.5 cm gate periphery module with 470 W total output power at 600 MHz, and a higher frequency, 3 GHz, 3 cm gate periphery module with 36 W output power and 42% power-added efficiency [27]. The SIT has also been used to make a UHF high power amplifier module designed for HDTV transmission [28]. This module, designed for operation in the 470 to 806 MHz frequency range, had 200 W average output power and 1 kW peak output power. SiC MESFETs have been fabricated with total CW output power of 80 W, and total pulsed output power of 120 W, both at 3.1 GHz [26]. The AlGaN/GaN HEMTs are most limited in total output power at the present time, with the highest reported power of 9.86 W at 8.2 GHz [25]. Also represented in Fig. 7.114 are currently available Si LDMOS parts with output powers as high as 120 W, and GaAs HFETs with 200 W total output power at 2.16 GHz [29] and 15.8 W at 12 GHz [30].

Challenges to Production

While there is little doubt that wide bandgap materials have tremendous potential in the high power RF arena, there are still several challenges to production that have yet to be answered. The challenges include

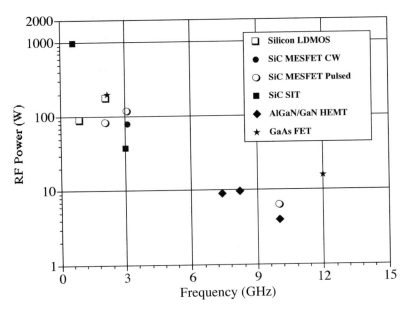

FIGURE 7.114 Total output power vs. frequency for different RF power technologies.

the size, defectivity, and choice (SiC/sapphire/silicon) of the device substrate along with the quality and defect density of the epilayers, elimination of dispersion phenomena currently seen in many of these devices, and improved understanding of system packaging and thermal limitations. In addition, market drivers will be needed to provide the impetus for continuing material and device development. It will be very difficult to implement any type of large-scale production of wide bandgap power transistors until these issues are resolved.

SiC Substrates

SiC substrates have come a long way since they were first introduced over a decade ago. However, they still have a long way to go. The most common "killer" defect in SiC wafers, the micropipe (a screw dislocation that can run the length of a boule of material), has been reduced in density from over $100/cm^2$ in 1993 to less than $1/cm^2$ in 1998 [31]. Those numbers, however, are best research results, and commercially available wafers will have higher micropipe densities, are significantly more expensive than silicon wafers, and are only available in 2″ to 3″ diameters, depending on polytype and doping. The high defect densities in SiC substrates make it difficult to yield large, high voltage, high power devices, while the small wafer sizes increase cost through decreased economy of scale, and can actually be a limiting factor in that it is now difficult to even obtain used semiconductor process equipment for 2″ to 3″ wafers.

Substrates for GaN Devices

As mentioned earlier, GaN devices have been grown on a number of different substrates. This is because larger diameter (2″) GaN substrates will become available, in prototype quantities only, for the first time in 2000. In the absence of a homoepitaxial GaN wafer, substrate choice for GaN is something of a conundrum. Sapphire is relatively cheap, is offered in large diameter (4″ to 6″) wafers, and provides an excellent low-loss microwave substrate. However, the thermal conductivity of sapphire is extremely poor (0.46 W/cmK) and will severely limit the power density and total power performance of devices fabricated on it. On the other hand, semi-insulating SiC is also an excellent microwave substrate, but has the cost, size, and defectivity issues described above. Silicon is yet another substrate possibility, but there is very limited data as to its performance at this time.

Dispersion and Instability Due to Material Defects

There have been many reports on both SiC [32] and GaN [33–35] devices of different types of device instabilities and dispersions. Drain current reduction under RF drive and surface instabilities have both been documented for SiC MESFETs. These phenomena have been attributed to traps at both the surface of the device and the buffer/substrate interface. Drain current compression has been widely reported for AlGaN/GaN HEMTs, along with frequency dispersion of transconductance and capacitances. These phenomena in GaN devices have been attributed to hole traps in the buffer layer, traps in the AlGaN barrier, and have even been modeled as a lossy dielectric layer under the gate of the device. In the final analysis, any type of DC-to-RF dispersion is extremely undesirable, not only because of device degradation, but also because is makes circuit design very difficult. Any dispersion effects in devices will have to be eliminated before serious production of wide bandgap devices can occur.

System Level Issues

There are two system level issues that will need much more study in order to make wide bandgap devices a commercial reality. First, significantly improved packaging will be needed because devices from wide bandgap materials will have much greater power per die and thus power per package than currently available silicon or GaAs devices. Thus, the thermal capabilities of the package for the wide bandgap device will have to be significantly improved over the currently available packaging technology used for silicon power devices. Second, the overall temperature handling and cooling capabilities of the systems using RF power devices will have to be improved. Put simply, there cannot be any advancement in technology if a very high temperature tolerant device operating at a very high temperature and power level causes thermal runaway in the non wide bandgap, temperature-sensitive devices that surround it.

Market Drivers

One bit of good news for wide bandgap technology is that it appears to be in high demand despite its immature status. Cree Inc. has already announced release of a SiC MESFET product. The U.S. government is providing very large amounts of funding to drive development of wide bandgap technology for military applications [36]. Most importantly, development and sales of blue LEDs from GaN are growing very fast in both demand and level of technology development [37]. The large LED market will drive nitride-based technology to improved materials, which will in turn feed back into microwave device technology, which should in turn mitigate many of the issues described above.

Conclusions

This article has discussed the properties of wide bandgap, high frequency, high power transistors, and how the materials from which they are fabricated affect device performance. SiC and GaN are shown to be outstanding materials for use in microwave device applications. This is clearly exemplified by the outstanding device performance seen from these relatively immature technologies. Device results including f_{max} = 139 GHz and power densities of 9.1 W/mm for AlGaN/GaN HEMTs, power densities of 5.6 W/mm and power-added efficiency of 65.7% for SiC MESFETs, device modules of over 470 W at power densities of 175 kW/in², and the design and execution of an HDTV transmitter module in SiC SIT technology all show the tremendous potential of wide bangap materials for use in microwave transistors. With the improvements of larger diameter, low defect density substrates, more perfect epitaxial films, and improved package and system thermal tolerances, wide bandgap materials should prove an excellent high power, high temperature technology for microwave applications.

References

1. R. Trew, J. Yan, and P. Mock, The Potential of Diamond and SiC Electronic Devices for Microwave and Millimeter-Wave Power Applications, *Proceedings of the IEEE*, 79, 5, May 1991, 598–620.
2. R. Davis, Thin Films and Devices of Diamond, Silicon Carbide, and Gallium Nitride, *Physica B*, 185, 1993, 1–15.
3. C.E. Weitzel et al., Silicon Carbide High-Power Devices, *IEEE Transactions on Electron Devices*, 43, 10, 1996, 1732–1741.
4. C.E. Weitzel and K.E. Moore, Performance Comparison of Wide Bandgap Semiconductor RF Power Devices, *Journal of Electronic Materials*, 27, 4, 1998, 365–369.

5. J. Zolper, Wide Bandgap Semiconductor Microwave Technologies: From Promise to Practice, IEEE International Electron Devices Meeting, 1999.

6. R.J. Trew, Wide Bandgap Semiconductor Transistors for Microwave Power Amplifiers, *Microwave,* March 2000, 46–54.

7. I. Bahl and P. Bhartia, eds., *Microwave Solid State Circuit Design,* John Wiley & Sons, Inc., New York, 1988, 483–536.

8. S. Liao, *Microwave Circuit Analysis and Amplifier Design,* Prentice-Hall, Englewood Cliffs, NJ, 1987, 78–122 and 236–274.

9. C.A. Mead, Schottky Barrier Gate Field-Effect Transistor, *Proceedings of the IEEE,* 54, 1966, 307.

10. W. W. Hooper and W. I. Lehrer, An Epitaxial GaAs Field-Effect Transistor, *Proceedings of the IEEE,* 55, 1967, 1237.

11. S. Sze, *Physics of Semiconductor Devices,* John Wiley & Sons, New York, 1981, 312–361.

12. J. Nishizawa, T. Terasaki, and J. Shbata, Field-Effect Transistor versus Analog Transistor (Static Induction Transistor), *IEEE Transactions on Electron Devices,* ED-22, 1975, 185–197.

13. I. Khan and J. Cooper, Measurement of High-Field Electron Transport in Silicon Carbide, *IEEE Transactions on Electron Devices,* ED-47, 2, 2000, 269–273.

14. M.S. Shur, A.D. Bykhovski, and R. Gaska, Pyroelectric and Piezoelectric Propeties of GaN-Based Materials, *MRS Internet Journal of Nitride Semiconductor Research,* Res. 4S1, G1.6, 1999.

15. M.A. Khan et al., GaN Based Heterostructure for High Power Devices, *Solid State Electronics,* 41, 10, 1997, 1555–1559.

16. U.K. Mishra, Y.-F. Wu, B.P. Keller, S. Keller, and S. Denbaars, GaN Microwave Electronics, *IEEE Transactions on Electron Devices,* ED-46, 6, 1998, 756–761.

17. E.M. Chumbes, A.T. Schremer, J.A. Smart, D. Hogue, J. Komiak, and J. Shealy, Microwave Performance of AlGaN/GaN High Electron Mobility Transistors on Si(111) Substrates, *IEEE IEDM Digest,* 1999, 397–400.

18. L.S. McCarthy, P. Kozodoy, M. Rodwell, S. Denbaars, and U.K. Mishra, AlGaN/GaN Heterojunction Bipolar Transistor, *IEEE Electron Device Letters,* 20, 6, 1999, 277–279.

19. M.A. Khan, X. Hu, G. Sumin, A. Lunev, J. Yang, R. Gaska, and M. Shur, AlGaN/GaN Metal Oxide Semiconductor Heterostructure Field Effect Transistor, *IEEE Electron Device Letters,* 21, 2, 2000, 63–65.

20. L. Zhang et al., Epitaxially-Grown GaN Junction Field Effect Transistors, *IEEE Transactions on Electron Devices,* ED-47, 3, 2000, 507–511.

21. L. Eastman, K. Chu, J. Smart, and R. Shealy, GaN Materials for High Power Microwave Amplifiers, *Proceedings of the Materials Research Society Symposium,* 512, 1998, 3–7.

22. S. Allen, R. Sadler, T. Alcorn, J. Palmour, and C. Carter, Jr., Silicon Carbide MESFETs for High-Power S-Band Applications, *1997 MTT-S Digest,* 57–60.

23. J. Henning, A. Przadka, M. Melloch, and J. Cooper, Design and Demonstration of C-Band Static Induction Transistors in Silicon Carbide, *1999 57ᵗʰ Annual Device Research Conference Digest,* 48–49, 1999.

24. C. Weitzel, J. Palmour, C. Carter Jr., K. Nordquist, K. Moore and S. Allen, SiC Microwave Power MESFETs and JFETs, *Compound Semiconductors 1994,* U. Mishra and H. Goronkin, eds., 389–392, 1994.

25. Y.-F. Wu, D. Kapolnek, J. Ibbetson, N.-Q. Zhang, P. Parikh, B. Keller, and U. Mishra, High Al-Content AlGaN/GaN HEMTs on SiC Substrates with Very High Power Performance, *IEEE IEDM Digest,* 1999, 925–927.

26. J. Palmour, S. Allen, S. Sheppard, W. Pribble, R. Sadler, T. Alcorn, Z. Ring, and C. Carter, Jr., Progress in SiC and GaN Microwave Devices Fabricated on Semi-Insulating 4H-SiC Substrates, *1999 57ᵗʰ Annual Device Research Conference Digest,* 38–41, 1999.

27. R. Siergiej, R. Clarke, A. Agarwal, C. Brandt, A. Burke, A. Morse, and P. Orphanoa, High Power 4H-SiC Static Induction Transistors, *IEEE IEDM Digest,* 1995, 353–356.

28. A First for Silicon Carbide, *Compound Semiconductor,* July/August 1996, 4.

29. H. Ishida, T. Yokoyama, H. Furukawa, T. Tanaka, M. Maeda, S. Morimoto, Y. Ota, D. Ueda, and C. Hamaguchi, 200W GaAs-Based MODFET Power Amplifier for W-CDMA Base Stations, *IEEE IEDM Digest*, 1999, 393–396.

30. K. Matsunaga, Y. Okamoto, I. Miura, and M. Kuzuhara, Ku-Band 15W Single-Ship HJFET Power Amplifier, *1996 MTT-S Digest*, 697–700.

31. B. Foutz, Industry Shows Support for Wide Bandgap Semiconductors at the Spring MRS Meeting, *Compound Semiconductor,* June 1999, 21–28.

32. K. Hilton, M. Uren, P. Wilding, H. Johnson, J. Guest, and B. Smith, Surface Induced Instabilities in 4H-SiC Microwave MESFETs, International Conference on Silicon Carbide and Related Materials, 1999, paper 416.

33. C. Nguyen, N. Nguyen, and D. Grider, Drain Current Compression in GaN MODFET's Under Large-Signal Modulation at Microwave Frequencies, *Electronics Letters*, 35, 16, 1380–1382.

34. E. Kohn, I. Daumiller, P. Schmid, N. Nguyen, and C. Nguyen, Large Signal Frequency Dispersion of AlGaN/GaN Heterostructure Field Effect Transistors, *Electronics Letters*, 35, 12, 1022–1024.

35. S. Trassaert, B. Boudart, C. Gaquiere, D. Theron, Y. Crosnier, F. Huet, and M. Poisson, Trap Effects Studies in GaN MESFETs by Pulsed Measurements, *Electronics Letters*, 35, 16, 1386–1388.

36. J. Zolper, C. Wood, and M. Yoder, Nitride Semiconductor Transistors Poised to Revolutionize DOD Systems, *Compound Semiconductor*, June 1999, 29–30.

37. R. Dixon, Who's Who in Blue and Green LEDs, *Compound Semiconductor*, June 1999, 15–20.

7.3 Tubes

Jerry C. Whitaker

Microwave Power Tubes

Microwave power tubes span a wide range of applications, operating at frequencies from 300 MHz to 300 GHz with output powers from a few hundred watts to more than 10 MW. Applications range from the familiar to the exotic. The following devices are included under the general description of microwave power tubes:

- Klystron, including the *reflex* and *multicavity* klystron
- *Multistage depressed collector* (MSDC) klystron
- *Klystrode* (IOT) tube
- Traveling-wave tube (TWT)
- **Crossed-field tube**
- *Coaxial magnetron*
- *Gyrotron*
- **Planar triode**
- High frequency tetrode

This wide variety of microwave devices has been developed to meet a wide range of applications. Some common uses include:

- UHF-TV transmission
- Shipboard and ground-based radar
- Weapons guidance systems
- Electronic counter-measure (ECM) systems
- Satellite communications
- Tropospheric scatter communications
- Fusion research

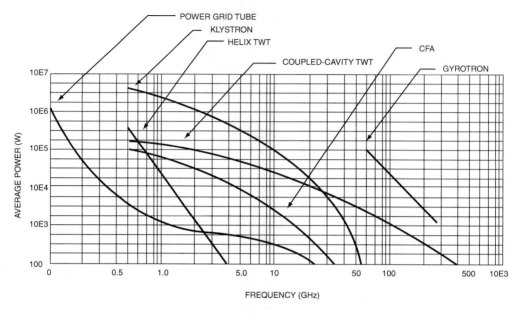

FIGURE 7.115 Microwave power tube type as a function of frequency and output power.

As new applications are identified, improved devices are designed to meet the need. Microwave power tubes manufacturers continue to push the limits of frequency, operating power, and efficiency. Microwave technology is an evolving science. Figure 7.115 charts device type as a function of operating frequency and power output.

Two principal classes of microwave vacuum devices are in common use today:

- **Linear beam tubes**
- Crossed-field tubes

Each class serves a specific range of applications. In addition to these primary classes, some power grid tubes are also used at microwave frequencies.

Linear Beam Tubes

As the name implies, in a linear beam tube the electron beam and the circuit elements with which it interact are arranged linearly. The major classifications of linear beam tubes are shown in Fig. 7.116. In

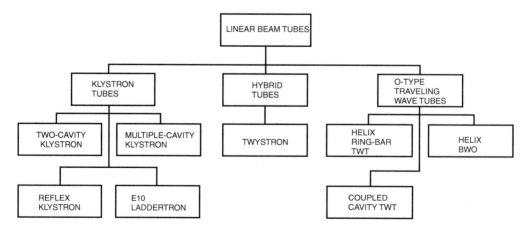

FIGURE 7.116 Types of linear beam microwave tubes.

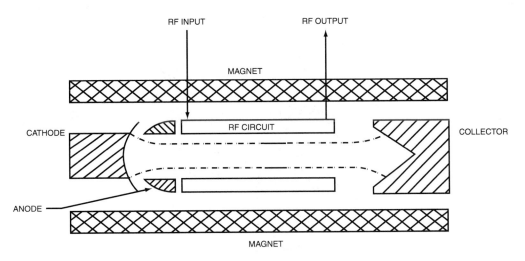

FIGURE 7.117 Schematic diagram of a linear beam tube.

such a device, a voltage applied to an anode accelerates electrons drawn from a cathode, creating a beam of kinetic energy. Power supply potential energy is converted to kinetic energy in the electron beam as it travels toward the microwave circuit. A portion of this kinetic energy is transferred to microwave energy as RF waves slow down the electrons. The remaining beam energy is dissipated either as heat or returned to the power supply at the collector. Because electrons will respell one another, there is usually an applied magnetic focusing field to maintain the beam during the interaction process. The magnetic field is supplied either by a solenoid or permanent magnets. Figure 7.117 shows a simplified schematic of a linear beam tube.

Crossed-Field Tubes

The magnetron is the pioneering device of the family of crossed-field tubes. The family tree of this class of devices is shown in Fig. 7.118. Although the physical appearance differs from that of linear beam tubes, which are usually circular in format, the major difference is in the interaction physics that requires a magnetic field at right angles to the applied electric field. Whereas the linear beam tube sometimes requires a magnetic field to maintain the beam, the crossed-field tube always requires a magnetic focusing field.

Figure 7.119 shows a cross-section of the magnetron, including the magnetic field applied perpendicular to the cathode-anode plane. The device is basically a diode with the anode composed of a pluarlity of resonant cavities. The interaction between the electrons emitted from the cathode and the crossed electric and magnetic fields produces a series of space charge spokes that travel around the anode-cathode space in a manner that transfers energy to the RF signal supported by the multicavity circuit. The mechanism is highly efficient.

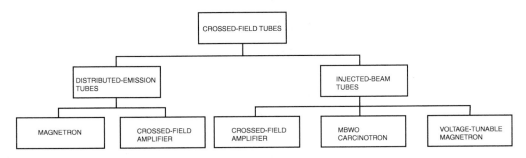

FIGURE 7.118 Types of crossed-field microwave tubes.

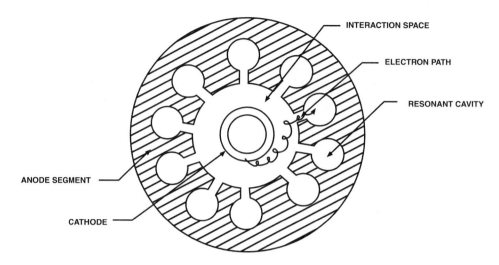

FIGURE 7.119 Magnetron electron path looking down into the cavity the with magnetic field applied.

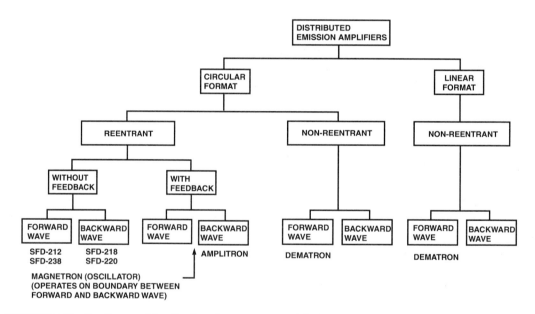

FIGURE 7.120 Family tree of the distributed emission crossed-field amplifier (CFA).

Crossed-Field Amplifiers — Figure 7.120 shows the family tree of the crossed-field amplifier (CFA). The configuration of a typical present day distributed emission amplifier is similar to that of the magnetron except that the device has an input for the introduction of RF into the circuit. Current is obtained primarily by secondary emission from the negative electrode that serves as a cathode throughout all or most of the interaction space. The earliest versions of this tube type were called **amplitrons**.

The CFA is deployed in radar systems operating from UHF to the Ku band, and at power levels up to several megawatts. In general, bandwidth ranges from a few percent to as much as 25% of the center frequency.

Grid Vacuum Tubes

The physical construction of a vacuum tube causes the output power and available gain to decrease with increasing frequency. The principal limitations faced by grid-based devices include the following:

- Physical size. Ideally, the RF voltages between electrodes should be uniform, however, this condition cannot be realized unless the major electrode dimensions are significantly less than 1/4-wavelength at the operating frequency. This restriction presents no problems at VHF frequencies, however, as the operating frequency increases into the microwave range, severe restrictions are placed on the physical size of individual tube elements.
- Electron transit-time. Interelectrode spacing, principally between the grid and the cathode, must be scaled inversely with frequency to avoid problems associated with electron transit-time. Possible adverse conditions include: (1) excessive loading of the drive source, (2) reduction in power gain, (3) back-heating of the cathode as a result of electron bombardment, and (4) reduced conversion efficiency.
- Voltage standoff. High power tubes operate at high voltages. This presents significant problems for microwave vacuum tubes. For example, at 1 GHz the grid-cathode spacing must not exceed a few mils. This places restrictions on the operating voltages that may be applied to individual elements.
- Circulating currents. Substantial RF currents may develop as a result of the inherent interelectrode capacitances and stray inductances/capacitances of the device. Significant heating of the grid, connecting leads and vacuum seals may result.
- Heat dissipation. Because the elements of a microwave grid tube must be kept small, power dissipation is limited.

Still, some grid vacuum tubes find applications at high frequencies. Planar triodes are available that operate at several gigahertz, with output powers of 1 to 2 kW in pulsed service. Efficiency (again for pulsed applications) ranges from 30 to 60 percent, depending on the frequency.

Planar Triode

A cross-sectional diagram of a planar triode is shown in Fig. 7.121. The envelope is made of ceramic, with metal members penetrating the ceramic to provide for connection points. The metal members are shaped either as disks or as disks with cylindrical projections.

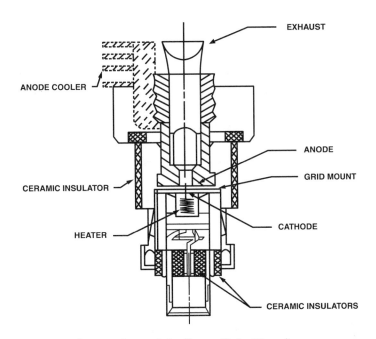

FIGURE 7.121 Cross-section of a 7289 planar triode. (*Source:* Varian/Eimac.)

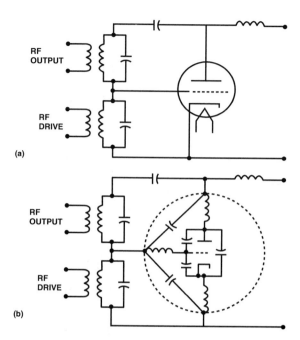

FIGURE 7.122 Grounded-grid equivalent circuits: (a) low frequency operation, (b) microwave frequency operation. The cathode-heating and grid-bias circuits are not shown.

The cathode is typically oxide-coated and indirectly-heated. The key design objective for a cathode is high emission density and long tube life. Low temperature emitters are preferred because high cathode temperatures typically result in more evaporation and shorter life.

The grid of the planar triode represents perhaps the greatest design challenge for tube manufacturers. Close spacing of small-sized elements is needed, at tight tolerances. Good thermal stability is also required, because the grid is subjected to heating from currents in the element itself, plus heating from the cathode and bombardment of electrons from the cathode.

The anode, usually made of copper, conducts the heat of electron bombardment to an external heat sink. Most planar triodes are air cooled.

Planar triodes designed for operation at 1 GHz and above are used in a variety of circuits. The grounded grid configuration is most common. The plate resonant circuit is cavity-based, using waveguide, coaxial line, or stripline. Electrically, the operation of planar triode is much more complicated at microwave frequencies than at low frequencies. Figure 7.122 compares the elements at work for a grounded-grid amplifier operating at low frequencies (*a*) and at microwave frequencies (*b*). The equivalent circuit is made more complicated by:

- Stray inductance and capacitance of the tube elements
- Effects of the tube contact rings and socket elements
- Distributed reactance of cavity resonators and the device itself
- Electron transit-time effects, which result in resistive loading and phase shifts

Reasonable gains of 5 to 10 dB may be achieved with a planar triode. Increased gain is available by cascading stages. Interstage coupling may consist of waveguide or coaxial-line elements. Tuning is accomplished by varying the cavity inductance or capacitance. Additional bandwidth is possible by stagger-tuning cascaded stages.

High Power UHF Tetrode

New advancements in vacuum tube technology have permitted the construction of high power UHF transmitters based on tetrodes. Such devices are attractive because they inherently operate in an efficient class A-B mode. UHF tetrodes operating at high power levels provide essentially the same specifications, gain, and efficiency as tubes operating at lower powers. The anode power supply is much lower in voltage than the collector potential of a klystron- or Klystrode™ tube-based system (8 kV is common). The tetrode also does not require focusing magnets.

Efficient removal of heat is the key to making a tetrode practical at high power levels. Such devices typically use water or vapor-phase cooling. Air cooling at such levels is impractical because of the fin size that would be required. Also, the blower for the tube would have to be quite large, reducing the overall transmitter AC-to-RF efficiency.

The expected lifetime of a tetrode in UHF service is usually shorter than a klystron of the same power level. Typical lifetimes of 8,000 to 15,000 hours have been reported. Work is underway as of this writing to extend the operating limits of the tetrode, while retaining the benefits of its inherent class A-B operation.

Defining Terms

Amplitron: A classic crossed-field amplifier in which output current is obtained primarily by secondary emission from the negative electrode that serves as a cathode throughout all or most of the interaction space.

Linear beam tube: A class of microwave devices where the electron beam and the circuit elements with which it interact are arranged linearly.

Crossed-filed tube: A class of microwave devices where the electron beam and the circuit elements with which it interacts are arranged in a nonlinear, usually circular, format. The interaction physics of a crossed-field tube requires a magnetic field at right angles to the applied electric field.

Planar triode: A grid power tube whose physical design make it practical for use at microwave frequencies.

References

Badger, G., The Klystrode: A New High-Efficiency UHF-TV Power Amplifier, *Proceedings of the NAB Engineering Conference*, National Association of Broadcasters, Washington, D.C., 1986.

Clayworth, G. T., H. P. Bohlen, and R. Heppinstall, Some Exciting Adventures in the IOT Business, *NAB 1992 Broadcast Engineering Conference Proceedings*, National Association of Broadcasters, Washington, D.C., pp. 200–208, 1992.

Collins, G. B. 1947. *Radar System Engineering*, McGraw-Hill, New York.

Crutchfield, E. B. 1992. *NAB Engineering Handbook*, 8th ed., National Association of Broadcasters, Washington D.C.

Fink, D., and Christensen, D., eds. 1989. *Electronics Engineer's Handbook*, 3rd Ed., McGraw-Hill Book Company, New York.

IEEE, 1984. *IEEE Standard Dictionary of Electrical and Electronics Terms*, Institute of Electrical and Electronics Engineers, Inc., New York,.

McCune, E. 1988. Final Report: The Multi-Stage Depressed Collector Project, *Proceedings of the NAB Engineering Conference*, National Association of Broadcasters, Washington, D.C.

Ostroff, N., Kiesel, A., Whiteside, A., and See, A. 1989. Klystrode-Equipped UHF-TV Transmitters: Report on the Initial Full Service Station Installations. In *Proceedings of the NAB Engineering Conference*. National Association of Broadcasters, Washington, D.C.

Pierce, J. A. 1945. Reflex Oscillators, *Proc. IRE* 33 (Feb.):112.

Pierce, J. R. 1947. Theory of the Beam-type Traveling Wave Tube, *Proc. IRE* 35 (Feb.):111.

Pond, N. H., and Lob, C. G. 1988. Fifty Years Ago Today or On Choosing a Microwave Tube. *Microwave Journal* (Sept.):226-238.

Spangenberg, K. 1947. *Vacuum Tubes*, McGraw-Hill, New York.

Terman, F. E. 1947. *Radio Engineering*, McGraw-Hill, New York.

Varian, R., and Varian, S. 1939. A high-frequency oscillator and amplifier. *J. Applied Phys.* 10 (May):321.

Whitaker, J. C. 1991. *Radio Frequency Transmission Systems: Design and Operation*, McGraw-Hill, New York.

Whitaker, J. C. 1994. *Power Vacuum Tubes Handbook*, Van Nostrand Reinhold, New York.

Further Information

Specific information on the application of microwave tubes can be obtained from the manufacturers of those devices. More general application information can be found in the following publications:

Radio Frequency Transmission Systems: Design and Operation, written by Jerry C. Whitaker, McGraw-Hill, New York, 1991.
Power Vacuum Tubes Handbook, written by Jerry C. Whitaker, Van Nostrand Reinhold, New York, 1994.

The following classic texts are also recommended:

Vacuum Tubes, written by Karl R. Spangenberg, McGraw-Hill, New York, 1948.
Radio Engineering, 2nd Ed., written by Frederick E. Terman, McGraw-Hill, New York, 1947.

Operational Considerations for Microwave Tubes

Long-term reliability of a microwave power tube requires detailed attention to the operating environment of the device, including supply voltages, load point, and cooling. Optimum performance of the system can only be achieved when all elements are functioning within specified parameters.

Microwave Tube Life

Any analysis of microwave tube life must first identify the parameters that define *life*. The primary *wear out* mechanism in a microwave power tube is the electron gun at the cathode. In principal, the cathode will eventually evaporate the activating material and cease to produce the required output power. Tubes, however, rarely fail because of low emission, but for a variety of other reasons that are usually external to the device.

Power tubes designed for microwave applications provide long life when operated within their designed parameters. The point at which the device fails to produce the required output power can be predicted with some accuracy, based on design data and in-service experience. Most power tubes, however, fail because of mechanisms other than predictable chemical reactions inside the device itself. External forces, such as transient overvoltages caused by lightning, cooling system faults, and improper tuning more often than not lead to the failure of a microwave tube.

Life-Support System

Transmitter control logic is usually configured for two states of operation:

- An *operational level*, which requires all of the "life-support" systems to be present before the high voltage (HV) command is enabled
- An *overload level*, which removes HV when one or more fault conditions occur

The cooling system is the primary life-support element in most RF generators. The cooling system should be fully operational before the application of voltages to the tube. Likewise, a cool-down period is usually recommended between the removal of beam and filament voltages and shut-down of the cooling system.

Most microwave power tubes require a high voltage removal time of less than 100 ms from the occurrence of an overload. If the trip time is longer, damage may result to the device. **Arc detectors** are often installed in the cavities of high power tubes to sense fault conditions, and shut down the high voltage power supply before damage can be done to the tube. Depending on the circuit parameters, arcs can be sustaining, requiring removal of high voltage to squelch the arc. A number of factors can cause RF arcing, including:

- Overdrive condition
- Mistuning of one or more cavities
- Poor cavity fit (applies to external types only)

- Under coupling of the output to the load
- Lightning strike at the antenna
- High VSWR

Regardless of the cause, arcing can destroy internal elements or the vacuum seal if drive and/or high voltage are not removed quickly. A lamp is usually included with each arc detector photocell for test purposes.

Protection Measures

A microwave power tube must be protected by control devices in the amplifier system. Such devices offer either visual indications, aural alarm warnings, or actuate interlocks within the system. Figure 7.123 shows a klystron amplifier and the basic components associated with its operation, including metering for each of the power supplies. Other types of microwave power devices use similar protection schemes. Sections of coaxial transmission line, representing essential components, are shown in the figure attached to the RF input and RF output ports of the tube. A single magnet coil is shown to represent any coil configuration that may exist; its position in the drawing is for convenience only and does not represent the true position in the system.

    Heater Supply — The heater power supply can be either AC or DC. If DC, the positive terminal must be connected to the common heater-cathode terminal and the negative terminal to the heater terminal. The amount of power supplied to the heater is important because it establishes the cathode operating temperature. The temperature must be high enough to provide ample electron emission but not so high that emission life is jeopardized. The test performance sheet accompanying each tube lists the proper operating values of heater voltage and current for that tube. Heater voltage should be measured at the terminals of the device using a true-RMS-reading voltmeter.

    Because the cathode and heater are connected to the negative side of the beam supply, they must be insulated to withstand the full beam potential.

    Beam Supply — The high voltage **beam supply** furnishes the DC input power to the klystron. The positive side of the beam supply is connected to the body and collector of the klystron. The negative terminal is connected to the common heater-cathode terminal. Never connect the negative terminal of the beam supply to the heater-only terminal because the beam current will then flow through the heater

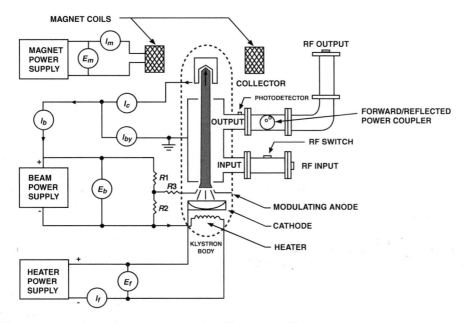

FIGURE 7.123 Protection and metering system for a klystron amplifier.

to the cathode and cause premature heater failure. The voltmeter, E_b in Fig. 7.123, measures the beam voltage applied between the cathode and the body of the klystron.

Current meter I_c measures collector current, typically 95 percent or more of the total device current. Current meter I_{by} measures **body current**. An interlock should interrupt the beam supply if the body current exceeds a specified maximum value.

The sum of the body current I_{by} and collector current I_c is equal to the beam current I_b, which should stay constant as long as the beam voltage and modulating anode voltage are held constant.

Magnet Supply — Electrical connections to the DC magnet supply typically include two meters, one for measuring current through the circuit (I_m in Fig. 7.123), and one for measuring voltage (E_m). When the device is first installed in its magnet assembly, both parameters should be measured and recorded for future reference. If excessive body current or other unusual symptoms occur, this data will be valuable for system analysis.

Undercurrent protection should be provided to remove beam voltage if the magnetic circuit current falls below a preset value. The interlock should also prevent the beam voltage from being applied if the magnetic circuit is not energized. This scheme will not, however, provide protection if the coils are shorted. Shorted conditions can be determined by measuring the normal values of voltage and current and recording them for future reference.

The body-current overload protection should actuate if the magnetic field is reduced for any reason.

RF Circuits — In Fig. 7.123, monitoring devices are shown on the RF input and output of the device. These monitors protect the tube should a failure occur in the RF output circuit. Two directional couplers and a photodetector are attached to the output. These components and an RF switching device on the input form a protective network against output transmission line mismatch. The RF switch is activated by the photodetector or the reflected power monitor and must be capable of removing RF drive power from the klystron in less than 10 ms (typically).

In the RF output circuit, the forward power coupler is used to measure the relative power output of the klystron. The reflected power coupler is used to measure the RF power reflected by the output circuit components, or antenna. Damaged components or foreign material in the RF line will increase the RF reflected power. The amount of reflected power should be no more than 5 percent of the actual forward RF output power of the device in most applications. An interlock monitors the reflected power and removes RF drive to the klystron if the reflected energy reaches an unsafe level. To protect against arcs occurring between the monitor and the output window, a photodetector is placed between the monitor and the window. Light from an arc will trigger the photodetector, which actuates the interlock system and removes RF drive before the window is damaged.

Filament Voltage Control

Extending the life of a microwave tube begins with accurate adjustment of filament voltage. The filament should not be operated at a reduced voltage in an effort to extend tube life. Unlike a thoriated tungsten grid tube, reduced filament voltage may cause uneven emission from the surface of the cathode with little or no improvement in cathode life.

Voltage should be applied to the filament for a specified warm-up period before the application of beam current to minimize thermal stress on the cathode/gun structure. However, voltage normally should not be applied to the filaments for extended periods (2 hours or more) if no beam voltage is present. The net rate of evaporation of emissive material from the cathode surface is greater without beam voltage. Subsequent condensation of material on gun components may lead to voltage standoff problems.

Cooling System

The cooling system is vital to any RF generator. In a high power microwave transmitter, the cooling system may need to dissipate as much as 70% of the input AC power in the form of waste heat. For vapor phase-cooled devices, pure (distilled or demineralized) water must be used. Because the collector is usually only several volts above ground potential, it is generally not necessary to use de-ionized water.

TABLE 7.20 Variation of Electrical and Thermal Properties of Common Insulators as a Function of Temperature

	20°C	120°C	260°C	400°C	538°C
Thermal Conductivity[a]					
99.5% BeO	140	120	65	50	40
99.5% Al_2O_3	20	17	12	7.5	6
95.0% Al_2O_3	13.5				
Glass	0.3				
Power Dissipation[b]					
BeO	2.4	2.1	1.1	0.9	0.7
Electrical Resistivity[c]					
BeO	10^{16}	10^{14}	5×10^{12}	10^{12}	10^{11}
Al_2O_3	10^{14}	10^{14}	10^{12}	10^{12}	10^{11}
Glass	10^{12}	10^{10}	10^8	10^6	
Dielectric Constant[d]					
BeO	6.57	6.64	6.75	6.90	7.05
Al_2O_3	9.4	9.5	9.6	9.7	9.8
Loss Tangent[d]					
BeO	0.00044	0.00040	0.00040	0.00049	0.00080

[a] Heat transfer in Btu/ft^2/hr/°F.
[b] Dissipation in W/cm/°C.
[c] Resistivity in Ω·cm.
[d] At 8.5 GHz.

Excessive dissipation is perhaps the single greatest cause of catastrophic failure in a power tube. The critical points of almost every tube type are the metal-to-ceramic junctions or seals. At temperatures below 250°C these seals remain secure, but above that temperature, the bonding in the seal may begin to disintegrate. Warping of internal structures also may occur at temperatures above the maximum operating level of the device. The result of prolonged overheating is shortened tube life or catastrophic failure. Several precautions are usually taken to prevent damage to tube seals under normal operating conditions. Air directors or sections of tubing may be used to provide spot-cooling at critical surface areas of the device. Airflow and waterflow sensors typically prevent operation of the RF system in the event of a cooling system failure.

Temperature control is important for microwave tube operation because the properties of many of the materials used to build a device change with increasing temperature. In some applications, these changes are insignificant. In others, however, such changes can result in detrimental effects, leading to — in the worst case — catastrophic failure. Table 7.20 details the variation of electrical and thermal properties with temperature for various substances.

Water Cooling Systems

A water-cooled tube depends upon an adequate flow of fluid to remove heat from the device and transport it to an external heat sink. The recommended flow as specified in the technical data sheet should be maintained at all times when the tube is in operation. Inadequate water flow at high temperature may cause the formation of steam bubbles at the collector surface where the water is in direct contact with the collector. This condition can contribute to premature tube failure.

Circulating water can remove about 1.0 kW/cm^2 of effective internal collector area. In practice, the temperature of water leaving the tube is limited to 70°C to preclude the possibility of spot boiling. The water is then passed through a heat exchanger where it is cooled to 30 to 40°C before being pumped back to the device.

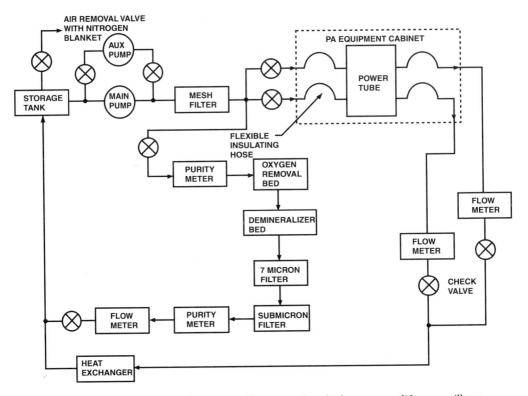

FIGURE 7.124 Functional schematic of a water-cooling system for a high power amplifier or oscillator.

Cooling System Design — A liquid cooling system consists of the following principal components:

- A source of coolant
- Circulation pump
- Heat exchanger
- Coolant purification loop
- Various connection pipes, valves, and gauges
- Flow interlocking devices (required to insure coolant flow anytime the equipment is energized)

Such a system is shown schematically in Fig. 7.124. In most cases the liquid coolant will be water, however, if there is a danger of freezing it will be necessary to use an anti-freeze solution such as ethylene glycol. In these cases, coolant flow must be increased or plate dissipation reduced to compensate for the poorer heat capacity of the ethylene glycol solution. A mixture of 60% ethylene glycol to 40% water by weight will be about 75% as efficient as pure water as a coolant at 25°C. Regardless of the choice of liquid, the system volume must be maintained above the minimum required to insure proper cooling of the vacuum tube(s).

The main circulation pump must be of sufficient size to ensure necessary flow and pressure as specified on the tube data sheet. A filter screen of at least 60 mesh is usually installed in the pump outlet line to trap any circulating debris that might clog coolant passages within the tube.

The heat exchanger system is sized to maintain the outlet temperature such that the outlet water from the tube at full dissipation does not exceed 70°C. Supplementary coolant courses may be connected in parallel or series with the main supply as long as the maximum outlet temperature is not exceeded.

Valves and pressure meters are installed on the inlet lines to the tube to permit adjustment of flow and measurement of pressure drop, respectively. A pressure meter and check valve is employed in the

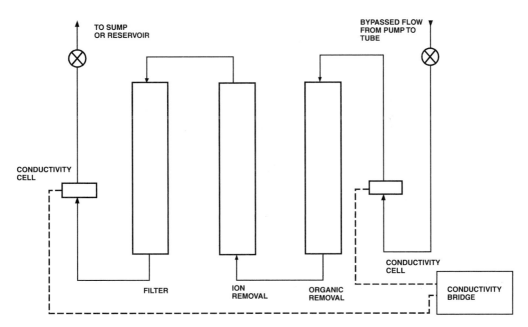

FIGURE 7.125   Typical configuration of a water purification loop.

outlet line. In addition, a flow meter, sized in accordance with the tube data sheet, and a thermometer, are included in the outlet line of each coolant course. These flow meters are equipped with automatic interlock switches, wired into the system electrical controls, so that the tube will be completely deenergized in the event of a loss of coolant flow in any one of the coolant passages. In some tubes, filament power alone is sufficient to damage the tube structure in the absence of proper water flow.

The lines connecting the plumbing system to the inlet and outlet ports of tube are made of a flexible insulating material configured so as to avoid excessive strain on the tube flanges. The coolant lines must be of sufficient length to keep electrical leakage below approximately 4 mA at full operating power.

The hoses are coiled or otherwise supported so they do not contact each other or any conducting surface between the high voltage end and ground. Conducting hose *barbs*, connected to ground, are provided at the low potential end so that the insulating column of water is broken and grounded at the point it exits the equipment cabinet.

Even if the cooling system is constructed using the recommended materials, and is filled with distilled or deionized water, the solubility of the metals, carbon dioxide, and free oxygen in the water make the use of a coolant purification **regeneration loop** essential. The integration of the purification loop into the overall cooling system is shown in Fig. 7.125. The regeneration loop typically taps 5 to 10% of the total cooling system capacity, circulating it through oxygen scavenging and de-ionization beds, and submicron filters before returning it into the main system. The purification loop can theoretically process water to 18 MΩ resistivity at 25°C. In practice resistivity will be somewhat below this value.

Figure 7.125 shows a typical purification loop configuration. Packaged systems such as the one illustrated are available from a number of manufacturers. In general, such systems consist of replaceable cartridges that perform the filtering, ion exchange, and organic solid removal functions. The system will usually include flow and pressure gauges and valves, and conductivity cells for continuous evaluation of the condition of the water and filters.

Water Purity and Resistivity — The purity of the cooling water is an important operating parameter for any water-cooled amplifier or oscillator. The specific **resistivity** typically must be maintained at 1 MΩ · cm minimum at 25°C. Distilled or deionized water should be used and the purity and flow protection periodically checked to insure against degradation.

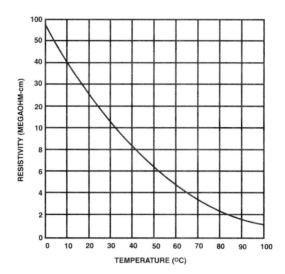

FIGURE 7.126   The effect of temperature on the resistivity of ultrapure water.

Oxygen and carbon dioxide in the coolant will form copper oxide, reducing cooling efficiency, and electrolysis may destroy the coolant passages. In addition, a filter screen should be installed in the tube inlet line to trap any circulating debris that might clog coolant passages within the tube.

After normal operation for an extended period, the cooling system should be capable of holding 3 to 4 $M\Omega \cdot cm$ until the filter beds become contaminated and must be replaced. The need to replace the filter bed resins is indicated when the purification loop output water falls below 5 $M\Omega \cdot cm$. Although the resistivity measurement is not a test for free oxygen in the coolant, the oxygen filter bed should always be replaced when replacing the de-ionizing bed.

The resistivity of the coolant is also affected by the temperature of the water. The temperature dependence of the resistivity of pure water is charted in Fig. 7.126.

Generally speaking, it is recommended that the coolant water be circulated at all times. This procedure provides the following benefits:

- It maintains high resistivity
- Reduces bacteria growth
- Minimizes oxidation resulting from coolant stagnation

If it is undesirable to circulate the coolant at the regular rate when the tube is deenergized, a secondary circulating pump can be used to move the coolant at a lower rate to purge any air that might enter the system and to prevent stagnation. Recommended minimum circulation rates within the coolant lines are as follows:

- 2 m/s during normal operation
- 30 cm/s during standby

The regeneration loop is typically capable of maintaining the cooling system such that the following maximum contaminant levels are not exceeded:

- Copper: 0.05 PPM by weight
- Oxygen: 0.5 PPM by weight
- CO_2: 0.5 PPM by weight
- Total solids: 3 PPM by weight

These parameters represent maximum levels; if the precautions outlined in this section are taken, actual levels will be considerably lower.

If the cooling system water temperature is allowed to reach 50°C, it will be necessary to use cartridges in the coolant regeneration loop that are designed to operate at elevated temperatures. Ordinary cartridges will decompose in high temperature service.

Defining Terms

Arc detector: A device placed within a microwave power tube or within one or more of the external cavities of a microwave power tube whose purpose is to sense the presence of an overvoltage arc.

Beam supply: The high voltage power supply that delivers operating potential to a microwave power tube.

Body-current: A parameter of operation for a microwave power tube that describes the amount of current passing through leakage paths within or around the device, principally through the body of the tube. Some small body-current is typical during normal operation. Excessive current, however, can indicate a problem condition within the device or its support systems.

Regeneration loop: A water purification system used to maintain proper conditions of the cooling liquid for a power vacuum tube.

Water resistivity: A measure of the purity of cooling liquid for a power tube, typically measured in $M\Omega \cdot cm$.

References

Crutchfield, E. B. 1992. *NAB Engineering Handbook*, 8th ed., National Association of Broadcasters, Washington D.C.

Harper, C. A. 1991. *Electronic Packaging and Interconnection Handbook*, McGraw-Hill, New York.

Kimmel, E. 1983. Temperature sensitive indicators: how they work, when to use them, *Heat Treating* (March).

Spangenberg, K. 1947. *Vacuum Tubes*. McGraw-Hill, New York.

Terman, F. E. 1947. *Radio Engineering*. McGraw-Hill, New York.

Varian. n.d. *Integral Cavity Klystrons for UHF-TV Transmitters*. Varian Associates, Palo Alto, CA.

Varian. 1971. Foaming test for water purity. Application Engineering Bulletin No. AEB-26. Varian Associates, Palo Alto, CA, Sept.

Varian. 1977. Water purity requirements in liquid cooling systems. Application Bulletin No. 16. Varian Associates, San Carlos, CA, July.

Varian. 1978. Protecting integral-cavity klytrons against water leakage. Technical Bulletin No. 3834. Varian Associates, Palo Alto, CA, July.

Varian. 1982. Cleaning and flushing klystron water and vapor cooling systems. Application Engineering Bulletin No. AEB-32. Varian Associates, Palo Alto, CA, Feb.

Varian. 1984. *Care and Feeding of Power Grid Tubes*. Varian Associates, San Carlos, CA.

Varian. 1984. Temperature measurements with eimac power tubes. Application Bulletin No. 20. Varian Associates, San Carlos, CA, Jan.

Varian. 1991. Water purity requirements in water and vapor cooling systems. Application Engineering Bulletin No. AEB-31. Varian Associates, Palo Alto, CA, Sept.

Whitaker, J. C. 1991. *Radio Frequency Transmission Systems: Design and Operation*, McGraw-Hill, New York.

Whitaker, J. C. 1992. *Maintaining Electronic Systems*, CRC Press, Boca Raton, FL.

Whitaker, J. C. 1994. *Power Vacuum Tubes Handbook*, Van Nostrand Reinhold, New York.

Further Information

Specific information on the application of microwave tubes can be obtained from the manufacturers of those devices. More general application information can be found in the following publications:

Maintaining Electronic Systems, written by Jerry C. Whitaker, CRC Press, Boca Raton, FL, 1990.

Radio Frequency Transmission Systems: Design and Operation, written by Jerry C. Whitaker, McGraw-Hill, New York, 1991.

Power Vacuum Tubes Handbook, written by Jerry C. Whitaker, Van Nostrand Reinhold, New York, 1994.

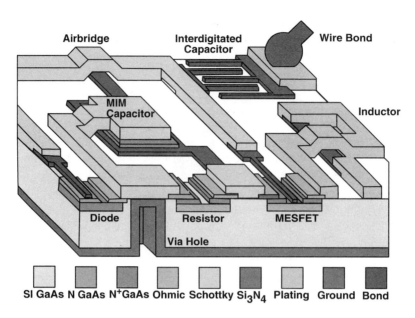

FIGURE 7.127   Three-dimensional conceptual illustration of MMIC technology (from Ladbrooke[15]).

7.4   Monolithic Microwave IC Technology

Lawrence P. Dunleavy

Monolithic Microwave Integrated Circuit Technology

MMIC Definition and Concepts

Pucel gives an excellent review of Monolithic Microwave Integrated Circuit (MMIC) technology in a 1981 paper.[1] Pucel went on to assemble a collection of papers on the subject in which he states in the introduction:[2]

> the monolithic approach is an approach wherein all active and passive circuit elements and interconnections are formed, in situ on or within a semi-insulating semi-conductor substrate by a combination of deposition schemes such as epitaxy, ion implantation, sputtering, and evaporation.

Figure 7.127 is a conceptual MMIC chip illustrating most of the major components. These include field effect transistor (FET) active devices, metal-insulator-metal (MIM) capacitors, thin film resistors, spiral strip inductors, via hole grounding, and air bridges.

As implied by the above quote and Fig. 7.127, in a MMIC all of the circuit components, including transistors, resistors, capacitors, and interconnecting transmission lines are integrated onto a single semi-insulating/semiconducting (usually GaAs) substrate. Use of a mask set and a corresponding series of processing steps achieves the integrated circuit fabrication. The mask set can be thought of as a mold. Once the mold has been cast, the process can be repeated in a "turn-the-crank" fashion to batch process tens, hundreds, or thousands of essentially identical circuits on each wafer.

A Brief History of GaAs MMICs

As noted by Pucel,[3] the origin of MMICs may be traced to a 1964 government program at Texas Instruments. A few key milestones are summarized in the following:

- 1964 — U.S. government funded a research program at TI based on silicon integrated circuit technology:[4] The objective was a transmit/receive module for a phased array radar antenna. The results were disappointing due to the poor semi-insulating properties of silicon.

- 1968 — Mehal and Wacker[5] used semi-insulating gallium arsenide (GaAs) as the substrate with Schottky diodes and Gunn devices as active devices to fabricate an integrated circuit comprising a 94 GHz receiver "front-end."
- 1976 — Pengelly and Turner[6] used MESFET devices on GaAs to fabricate an X-band (~10 GHz) amplifier and sparked an intense activity in GaAs MMICs.
- 1988 (approximately) — U.S. government's Defense Advanced Research Projects Agency (DARPA, today called ARPA) launched a massive research and development program called the MIMIC program (included Phase I, Phase II, and Phase III efforts) that involved most of the major MMIC manufacturing companies.

In the early 1980s a good deal of excitement was generated and several optimistic projections were made predicting the rapid adoption of GaAs MMIC technology by microwave system designers, with correspondingly large profits for MMIC manufacturers. The reality is that there was a much slower rate of progress to widespread use of MMIC technology, with the majority of the early thrust being provided by the government for defense applications. Still, steady progress was made through the 1980s and the government's MIMIC program was very successful in allowing companies to develop lower cost design and fabrication techniques to make commercial application of the technology viable. The 1990s have seen good progress toward commercial use with applications ranging from direct broadcast satellite (DBS) TV receivers, to automotive collision avoidance radar, and the many wireless communication applications (cell phones, WLANs, etc.).

Hybrid vs. MMIC Microwave IC Technologies

The conventional approach to microwave circuit design that MMIC technology competes with, or is used in combination with, is called "hybrid microwave integrated circuit," "discrete microwave integrated circuit" technology, or simply MIC technology. In a hybrid MIC, the circuit pattern is formed using photolithography. Discrete components are then assembled onto the substrate (e.g., using solder or silver epoxy) and connected using bondwires. In contrast to the batch processing afforded the MMIC approach, MICs have to be assembled with discrete components attached using relatively labor-intensive manufacturing methods. Table 7.21 summarizes some of the contrasting features of hybrid and monolithic approaches.

The choice of MMICs vs. the hybrid approach is mainly a matter of volume requirements. The batch processing of MMICs gives this approach advantages for high volume applications. Significant cost savings can be reaped in reduced assembly labor, however, for MMIC the initial design and mask preparation costs are considerable. The cost of maintaining a MMIC manufacturing facility is also extremely high and this has forced several companies out of the business. A couple of examples are Harris, which sold its GaAs operation to Samsung and put its resources into silicon. Another is AT&T, which is also relying on silicon for is anticipated microwave IC needs.

The high cost of maintaining a facility can only be offset by high volume production of MMICs. Still this does not prevent companies without MMIC foundries from using the technology, as there are several "commercial foundries" who offset the costs of maintaining their facilities by manufacturing MMIC chip products for third party companies through a foundry design working relationship.

TABLE 7.21   Features of Hybrid and Monolithic Approaches

Feature	Hybrid	Monolithic
Type of substrate	Insulator	Semiconductor
Passive components	Discrete/Deposited	Deposited
Active components	Discrete	Deposited
Interconnects	Deposited and wire-bonded	Deposited
Batch processed?	No	Yes
Labor intensive/chip	Yes	No

[a] After Pucel[7] with permission from IEEE.

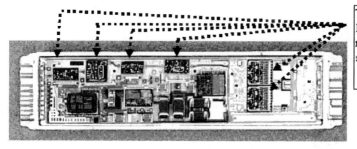

There are five GaAs MMIC Chips that can be recognized as black substrates with gold traces.

FIGURE 7.128 An example of combined MMIC and hybrid MIC technologies, this radar module includes several MMIC chips interconnected using microstrip lines. The hybrid substrate (white areas) is an alumina insulating substrate; the traces on the hybrid substrate are microstrip lines (courtesy Raytheon Systems Company).

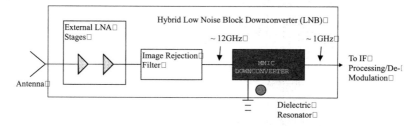

FIGURE 7.129 Application of MMICs in Ku-Band Direct Broadcast Satellite downconverters. The antenna is typically a small parabolic dish.

A key advantage of MMICs is small size. To give an example, a hybrid MIC the size of a business card can easily be reduced to a small chip one or two millimeters on a side. An associated advantage is the ease of integration that allows several functions to be integrated onto a single chip. For example, Anadigics and Raytheon have both manufactured DBS-related MMICs, wherein the functions of amplification (LNA and IF amplifiers), signal generation (VCO), and signal conversion (mixer) and filtering are all accomplished on a 1 mm × 2 mm, or smaller, chip.

In contrast to MMIC, MIC lithography is quite inexpensive and a much smaller scale investment is required to maintain a MIC manufacturing capability. There are also some performance advantages of the hybrid approach. For example, it is much easier to tune or repair a hybrid circuit after fabrication than it is for a MMIC. For this reason, for applications where the lowest noise figure is required, such as in a satellite TV receiver, an individually tuned hybrid LNA may be preferred as the first stage.

Ultimately, there is no such thing as a truly all MMIC system. Monolithic technology can be used to integrate single functions, or several system functions, but cannot sustain a system function in isolation. Usually, a MMIC is packaged along with other circuitry to make a practically useful component, or system. Figure 7.128, a radar module made by Raytheon, exemplifies how the advantages of both MMIC and hybrid approaches are realized in a hybrid connection of MMIC chips. Another example of combined hybrid/MMIC technology is shown in Fig. 7.129, in the form of a low noise block downconverter for Direct Broadcast Satellite applications. An example of combined MMIC and hybrid MIC technologies, this radar module includes several MMIC chips interconnected using microstrip lines. The hybrid substrates (white areas) is an alumina insulating substrate; the traces on the hybrid substrate are microstrip lines (courtesy Raytheon Systems Company).

GaAs MMICs in Comparison to Silicon VLSI Computer Chips

Everyone is familiar with silicon digital IC chips, or at least the enormous impact silicon-based Very Large Scale Integrated (VLSI) circuits have had on the computer industry. Silicon computer chips are digital circuits that contain hundreds to thousands of transistors on each chip. In a digital circuit the transistors are used as switches that are in one of two possible states depending on the "logic" voltage across a pair of terminals. The information processed by a digital circuit consists of a sequence of "1s" and "0s," which translate into logic voltages as the signal passes through the digital IC. Noise distorts the logic waveform in much the same way that it distorts a sinusoidal signal, however, as long as the signal distortion due to noise is not severe, the digital circuitry can assign the correct (discrete) logic levels to the signal as it is processed. Signal interpretation errors that occur due to noise are measured in terms of a bit error rate (BER).

The speed of the digital processing is related to how fast the transistors can switch between one state and another, among other factors. Because of certain material factors, such as electron mobility, digital circuits made on GaAs have been demonstrated to have speed advantages over silicon digital ICs, however, the speed advantages have not been considered by the majority of companies to outweigh the significant economic advantages of well-established, lower cost, silicon processing technology.

Because of the large volume of silicon chips that have been produced over the last twenty years, silicon processing techniques are significantly more established and in many cases standardized as compared to GaAs processing techniques, which still vary widely from foundry to foundry. The digital nature of the signals and operation modes of transistors in digital ICs makes uniformity between digital ICs, and even similar ICs made by different manufacturers, much easier to achieve than achieving uniformity with analog GaAs MMICs.

In contrast, GaAs MMICs are analog circuits that usually contain less than 10 transistors on a typical chip. The analog signals processed, which can take on any value between certain limits, may generally be thought of as combinations of noisy sinusoidal signals. Bias voltages are applied to the transistors in such a way that each transistor will respond in one of several predetermined ways to an applied input signal. One common use for microwave transistors is amplification, whereby the result of a signal passing through the transistor is for it to be boosted by an amount determined by the gain of the transistor.

A complication that arises is that no two transistors are identical in the analog sense. Taking gain for example, while there will be a statistical distribution of gain for a set of amplifier chips measured on the same GaAs MMIC wafer, a different (wider) set of statistics applies to variations in gain from wafer-to-wafer for the same design. These variations are caused primarily by variations in transistors, but also by variations in other components that make up the MMIC, including MIM capacitors, spiral inductors, film resistors, and transmission line interconnects. Successful foundries are able to control the variations within acceptable limits in order to achieve a satisfactory yield of chips meeting a customer's requirements. However, translating a MMIC design mask set to a different manufacturing foundry is a different story altogether.

This is not to say that foundry translation of MMIC designs cannot be accomplished. Under the federally funded MIMIC program, mentioned earlier, several pairs of foundries were tasked to translate designs from one to the other to demonstrate a "second-sourcing" capability. These efforts met with varying degrees of success, but not without considerable effort on the part of the participating GaAs MMIC foundries. Two pairs of companies involved in this second-sourcing demonstration effort for the MIMIC program are Raytheon and Texas Instruments, and Hughes Aircraft Company (GaAs foundry since bought and closed by Raytheon Company) and General Electric Company (now part of Lockheed-Martin).

MMIC Yield and Cost Considerations

Yield is an important concept for MMICs and refers to the percentage of circuits on a given wafer with acceptable performance relative to the total number of circuits fabricated. Since yield may be defined at several points in the MMIC process, it must be interpreted carefully.

- DC yield is the number of circuits whose voltages and currents measured at DC are within acceptable limits.
- RF yield is generally defined as the number of circuits that have acceptable RF/microwave performance when measured "on-wafer," before circuit dicing.
- Packaged RF yield is the final determination of the number of acceptable MMIC products that have been assembled using the fabricated MMIC chips.

If measured in terms of the total number of circuits fabricated, each of these yields will be successively lower numbers. For typical foundries, DC yields exceed 90%, while packaged yields may be around 50%. Final packaged RF yield depends heavily on the difficulty of the RF specifications, the uniformity of the process (achieved by statistical process control), as well as how sensitive the RF performance of the circuit design is to fabrication variations.

The costs involved with MMICs include:

- Material
- Design
- Mask set preparation
- Wafer processing
- Capital equipment
- Testing
- Packaging
- Inspection

A typical wafer run may cost $20,000 to $50,000, with $5,000 to $10,000 attributable to the mask set alone. These figures do not include design costs. Per-chip MMIC costs are determined by:

- Difficulty of design specifications
- Yield
- Material (wafer size and quality)
- Production volume
- Degree of automation

Some 1989–90 example prices for MMIC chips are as follows:[8]

1 to 5 GHz Wideband Amplifier — $30.00
2 to 8 GHz Wideband Amplifier — $45.00
6 to 18 GHz Wideband Amplifier — $100.00
DC to 12 GHz Attenuator — $60.00
DBS downconverter chip — $10.00.

In comparison, example prices for 1999 MMIC chips are as follows:[9]

DBS Downconverter chip — <$2.50
DC to 8 GHz HBT MMIC amp. — ~$3
Packaged power FET MMIC — $1 for ~2 GHz
4 to 7 GHz and 8 to 11 GHz high power/high efficiency power amps. — $85 (2 Watt) to $330 (12 Watt)

Prices for low volume specials at high frequency have not moved much since 1989.

Si vs. GaAs for Microwave Integrated Circuits

The subject of Silicon versus GaAs has been a hotly debated subject since the beginning of the MMIC concept in around 1965 (see "A brief history of GaAs MMICs" above). Two of the main discriminating issues between the technologies are microwave transistor performance and the loss of the semiconductor when used as a semi-insulating substrate for passive components.

A comparison of relevant physical parameters for silicon and GaAs materials is given in Table 7.22. The dielectric constants of the materials mainly affects the velocity of propagation down transmission line interconnects, and for the materials compared, this parameter is on the same order. For the other factors considered, significant differences are observed. The thermal conductivity, a measure of how efficiently the substrate conducts heat (generated by DC currents) away from the transistors, is best for silicon and is one key advantage of the silicon approach. This advantage is offset by silicon's lower mobility

TABLE 7.22 Properties of GaAs, Silicon, and Common Insulating Substrates

Property	GaAs	Silicon	Semi-Insulating GaAs	Semi-Insulating Silicon	Sapphire	Alumina
Dielectric constant	12.9	11.7	12.9	11.7	11.6	9.7
Thermal cond. (watts/cm-K)	0.46 (fair)	1.45 (good)	0.46 (fair)	1.45 (good)	0.46 (fair)	0.37 (fair)
Resistivity (ohm-cm)	—	—	10^7–10^9 (fair)	10^3–10^5 (poor)	$>10^{14}$ (good)	10^{11}–10^{14} (good)
Elec. mobility	4300^a (best)	700^a (good)	—	—	—	—
Sat. elec. velocity	1.3×10^7 (best)	9×10^6 (good)	—	—	—	—

Adapted from Pucel,[13] with permission from IEEE.

[a] At a doping concentration of $10^{17}/cm^3$.

and lower resistivity. Mobility in a semiconductor is a measure of how easily electrons can move through the "doped" region of the semiconductor (see discussion of FET operation in the following section). The mobility, as well as the saturated velocity, have a strong influence on the maximum frequency at which a microwave transistor can have useful gain.

Turning our attention to passive component operation, GaAs has much better properties for lower loss passive circuit realization. With the exception of a resistor, the ideal passive component is a transmission line, inductor, or capacitor that causes no signal loss. Resistivity is a measure of how "resistant" the substrate is to leakage currents that could flow, for example, from the top conductor of a microstrip line and the ground plane below. Looking first at the properties of the insulators sapphire and alumina, the resistivity is seen to be quite high. Semi-insulating GaAs is almost as high, and silicon has the lowest resistivity (highest leakage currents for a given voltage).

These considerations have led many companies to invest heavily in GaAs technology for microwave applications over the last several years. However, silicon remains a strong contender. In fact there has been a very strong renewal in development of silicon MMICs[10] with the advent of numerous markets for microwave "wireless" products. The front lines of the battlefield between silicon and GaAs are at frequencies below 6 GHz, where potential commercial opportunities are numerous. Silicon-based microwave ICs are also beginning to appear in higher frequency applications, such as Ku-band DBS satellite receivers.[11] Silicon-germanium heterojunction bipolar transistor technology is paving the way for increasing the applicability of silicon technology to even higher frequencies.[12]

Basic Principles of GaAs MESFETs and HEMTs

Basic MESFET Structure

The primary active device in a GaAs MMIC is the metal electrode semiconductor field-effect transistor, or MESFET. The basic construction of a MESFET is shown in Figs. 7.130 and 7.131. An "active layer" is first formed on top of a semi-insulating GaAs substrate by intentionally introducing an n-type impurity onto the surface of the GaAs, and isolating specific channel regions. These channel regions are semiconducting in that they contain free electrons that are available for current flow. When a metal is placed in direct contact with a semiconductor, as in the case of the gate, a "Schottky diode" is formed. One of the consequences of this is that a natural "depletion region," a region depleted of available electrons is formed under the gate. A diode allows current flow easily in one direction, while impeding current flow in the other direction. In the case of a MESFET gate, a positive bias voltage between the gate and the source "turns on" the diode and allows current to flow between the gate and the source through the substrate. A negative bias between the gate and the drain "turns off" the diode and blocks current flow, it also increases the depth of the depletion region under the gate.

In contrast to the gate contact, the drain and source contacts are made using what are called "ohmic contacts." In an ohmic contact, current can flow freely in both directions. Whether an ohmic contact or

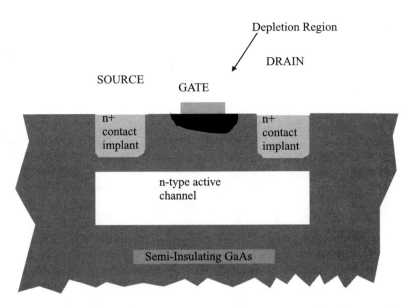

FIGURE 7.130 Cross-section of a MESFET transistor. In operation, current I_{ds} flows between the gate and drain terminals through the doped n-type active channel. An AC voltage applied to the gate modulates the size of the depletion region causing the I_{ds} current to be modulated as well. Notice that in a MMIC the doped n-type region is restricted to the transistor region leaving semi-insulating GaAs outside to serve as a passive device substrate.

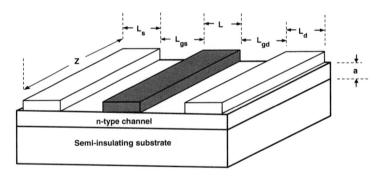

FIGURE 7.131 Aspect view of MESFET. Important dimensions to note are the gate length (Lg) and the gate width (Z). (From Golio.[16])

Schottky diode is formed at the metal-semiconductor interface is determined by the composition of the metal placed on the interface and the doping of the semiconductor region directly under the metal. The introduction of "pocket n+ implants" help form the ohmic contacts in the FET structure illustrated in Fig. 7.130. In the absence of the gate, the structure formed by the active channel in combination with the drain and source contacts essentially behaves as a resistor obeying ohms law. In fact this is exactly how one type of GaAs-based resistor commonly used in MMICs is made.

FETs in Microwave Applications

The most common way to operate a MESFET, for example in an amplifier application, is to ground the source (also called "common source" mode), introduce a positive bias voltage between the drain and source, and a negative bias voltage between the gate and source. The positive voltage between the drain and source V_{ds} causes current I_{ds} to flow in the channel.

As negative bias is applied between the gate and source V_{gs} the current I_{ds} is reduced as the depletion region extends farther and farther into the channel region. The value of current that flows with zero

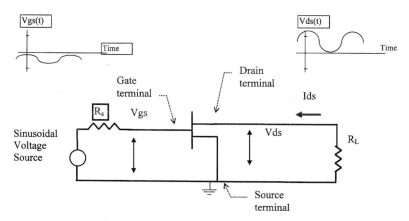

FIGURE 7.132 Simplified schematic of a FET in an amplifier application. Not shown are matching networks needed to match between the source resistance R_s and the FET input impedance, and the load resistance R_L and the FET output impedance. Also omitted are the networks needed to properly apply the DC bias voltages to the device and provide isolation between the RF and DC networks.

gate-to-source voltage is called the saturation current I_{dss}. Eventually, at a sufficiently large negative voltage, the channel is completely depleted of free electrons and the current I_{ds} will be reduced to essentially zero. This condition is called "pinch off." In most amplifier applications, the negative gate voltage is set to a "bias condition" between 0 volts and the pinch-off voltage V_{po}.

Figure 7.132 gives a simplified view of a FET configured in an amplifier application. An input sinusoidal signal $V_{gs}(t)$ is shown offset by a negative DC bias voltage. The sinusoidal variation in V_{gs} causes a likewise sinusoidal variation in depth of the depletion region that in turn creates a sinusoidal variation (or modulation) in the output current. Amplification occurs because small variations in the V_{gs} voltage cause relatively large variations in the output current. By passing the output current through a resistance RL the voltage waveform $V_{ds}(t)$ is formed. The V_{ds} waveform is shown to have higher amplitude than V_{gs} to illustrate the amplification process.

Other common uses, which involve different configurations and biasing arrangements, include use of FETs as the basis for mixers and oscillators (or VCOs).

FET Fabrication Variations and Layout Approaches

Figure 7.133 illustrates a MESFET fabricated with a recessed gate, along with a related type of FET device called a high electron mobility transistor (HEMT). A recessed gate is used for a number of reasons. First in processing it can aid in assuring that the gate stripe is placed in the proper position between the drain and source, and it can also result in better control and uniformity in I_{dss} and V_{po}. A HEMT is a variation of the MESFET structure that generally produces a higher performing device. This translates, for example, into a higher gain and a lower noise figure at a given frequency.

In light of the above cursory understanding of microwave FET structures, some qualitative comments can be made about some of the main factors that cause intended and unintended variations in FET performance. The first factor is the doping profile in the active layer. The doping profile refers to the density of the charge carriers (i.e., electrons) as a function of depth into the substrate. For the simplest type, uniform doping, the density of dopants (intended impurities introduced in the active region) is the same throughout the active region. In practice, there is a natural "tail" of the doping profile that refers to a gradual decrease in doping density as the interface between the active layer and the semi-insulating substrate is approached. One approach to create a more abrupt junction is the so-called "buried P-layer" technique. The buried p-layer influences the distribution of electrons versus depth from the surface of the chip in the "active area" of the chip where the FET devices are made. The idea is create better definition between where the conducting channel stops and where the nonconducting substrate begins. (More specifically the buried p-layer counteracts the n-type dopants in the tail of the doping

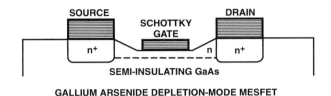

GALLIUM ARSENIDE DEPLETION-MODE MESFET

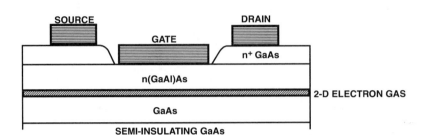

FIGURE 7.133   MESFET (top) and HEMT (bottom) structures showing "gate recess" structure whose advantages include better control of drain-to-source saturation current I_{dss} and pinch-off voltage V_{po}. (From Goyal.[17])

profile.) The doping profile and density determine the number of charge carriers available for current flow in a given cross-section of the active channel. This has a strong influence on the saturation current I_{dss} and pinch-off voltage V_{po}. The depth of the active layer (dimension "a" of Fig. 7.131) also plays a critical role in determining the current characteristics.

Other variables that influence MESFET performance are the gate length and gate width ("L" and "Z" of Fig. 7.131). The names for these two parameters are counterintuitive since the gate length refers to the shorter of the two dimensions. The gate width and channel depth determine the cross-sectional area available for current flow. An increase in gate width increases the value of the saturation current, which translates into the ability to operate the device at higher RF power levels (or AC voltage amplitudes). Typical values for gate widths are in the range of 100 microns for low noise devices to over 10 millimeters for high power devices. The gate length is usually the minimum feature size of a device and is the most significant factor in determining the maximum frequency where useful gain can be obtained from a FET; generally, the smaller the gate length the higher the gain for a given frequency. However, the fabrication difficulty increases, and processing yield decreases, as the gate length is reduced. The difficulty arises from the intricacy in controlling the exact position and length (small dimension) of the gate.

The geometrical layout of the FET also influences performance. Figure 7.134 shows common FET layouts. The layout affects what are called "external parasitics," which are undesired effects that can be modeled as a combination of capacitors, inductors, and resistors added to the basic FET electrical model.

In MMIC fabrication, variations in the most of the above-mentioned parameters are a natural consequence of a real process. These variations cause variations in observed FET performance even for identical microwave FETs made using the same layout geometry on the same wafer. Certainly there are many more subtle factors that influence performance, but the factors considered here should give some intuitive understanding of how unavoidable variations in the physical structure of fabricated FETs cause variations in microwave performance. As previously mentioned, successful GaAs MMIC foundries use statistical process control methods to produce FET devices within acceptable limits of uniformity between devices.

MMIC Lumped Elements: Resistors, Capacitors, and Inductors

MMIC Resistors

Figure 7.135 shows three common resistor types used in GaAs MMICs. For MMIC resistors the type of resistor material, and the length and width of the resistor determine the value of the resistance. In practice,

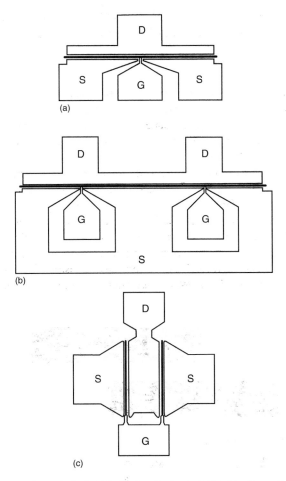

FIGURE 7.134 Three common layout approaches for MESFETs and HEMTs. (From Ladbrooke.[18])

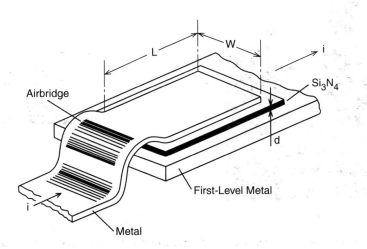

FIGURE 7.135 Common MMIC resistor fabrication approaches: (a) implanted GaAs, (b) mesa-etched/epitaxialy grown GaAs, (c) thin film (e.g., TaN). (From Goyal.[19])

there are also unwanted "parasitic" effects associated with MMIC resistors that can be modeled generally as a combination of series inductance, and capacitance to ground in addition to the basic resistance of the component.

MMIC Capacitors

The most commonly used type of MMIC capacitor is the metal-insulator-metal capacitor shown in Fig. 7.136. In a MIM capacitor the value of capacitance is determined from the area of the overlapping metal (smaller dimension of two overlapping plates), the dielectric constant ε_r of the insulator material, typically silicon nitride, and the thickness of the insulator. For values less than about 0.2 pF, series connected MIM capacitors can be used.

Smaller values of capacitance can be achieved with one of the various arrangements of coupled lines illustrated in Fig. 7.137. For these capacitors, the capacitance is determined from the width and spacing of strips on the surface of the wafer.

At microwave frequencies, "parasitic" effects limit the performance of all these capacitors. The two main effects are signal loss due to leakage currents, as measured by the quality factor Q of the capacitor, and a self-resonance frequency, beyond which the component no longer behaves as a capacitor. The final wafer or chip thickness can have a strong influence on these parasitic effects and the associated performance of the capacitors in the circuit. Parasitic effects must be accurately modeled for successful MMIC design usage.

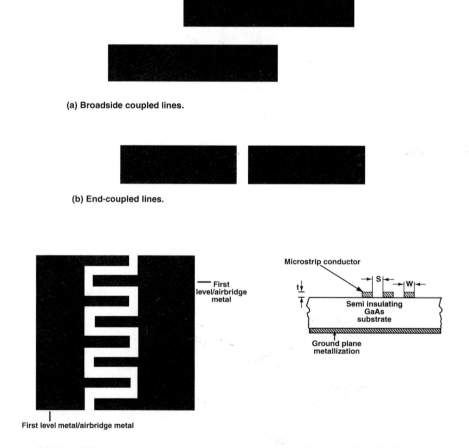

(a) Broadside coupled lines.

(b) End-coupled lines.

(c) 'Interdigitated capacitor layout and vertical cross section of an interdigitated capacitor.

FIGURE 7.136 Metal-insulator-metal conceptual diagram. (From Ladbrooke.[20])

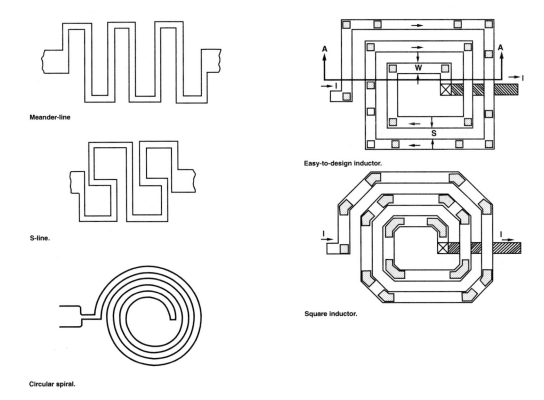

Meander-line

S-line.

Circular spiral.

Easy-to-design inductor.

Square inductor.

FIGURE 7.137 MMIC approaches for small valued capacitors. (From Goyal.[21])

MMIC Inductors

MMIC inductors are realized with narrow strips of metal on the surface of the chip. Various layout geometries are possible as illustrated in Fig. 7.138(a). The choice of layout is dictated mainly by the available space and the amount of inductance L that is required in the circuit application, with the spiral inductors providing the highest values. The nominal value of inductance achievable from strip inductors is determined from the total length, for the simpler layouts, and by the number of turns, spacing, and line width for the spiral inductors.

At microwave frequencies, "parasitic" effects limit the performance of these inductors. The two main effects are signal loss due to leakage currents, as measured by the quality factor Q of the capacitor, and a self-resonance frequency, beyond which the component no longer behaves as an inductor. The final thickness of the substrate influences not only the nominal value of inductance, but also the quality factor and self-resonance frequency. Inductor parasitic effects must be accurately modeled for successful MMIC design usage.

Air Bridge Spiral Inductors

Air bridge spiral inductors are distinguished from conventional spiral inductors by having the metal traces that make up the inductor suspended from the top of the substrate using MMIC air bridge technology. MMIC air bridges are generally used to allow crossing lines to jump over one another without touching and are almost invariably used in conventional spiral inductors to allow the center of the spiral to be brought through the turns of the spiral inductor for connection to the circuit outside of the spiral. In an air bridge spiral inductor, all of the turns are suspended off the substrate using a series of air bridges supported by metalized posts. The reason for doing this is to improve inductor performance by reducing loss as well as the effective dielectric constant of the lines that make up the spiral. The latter can have the effect of reducing inter-turn capacitance and increasing the resonant frequency of the inductor. Whether or not air bridge inductors are "worth the effort" is a debatable subject as the air bridge process

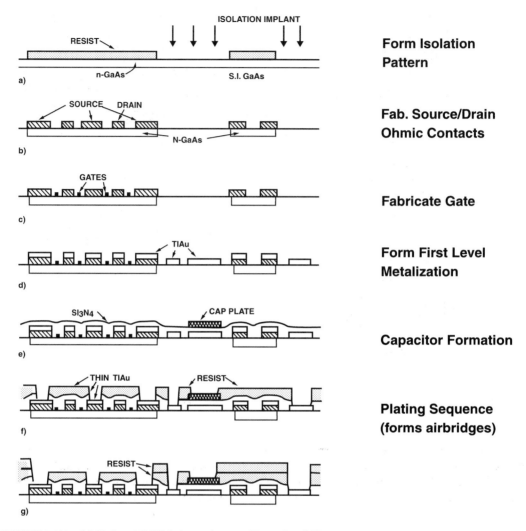

ISOLATION IMPLANT

RESIST

n-GaAs   S.I. GaAs

a)   **Form Isolation Pattern**

SOURCE   DRAIN

N-GaAs

b)   **Fab. Source/Drain Ohmic Contacts**

GATES

c)   **Fabricate Gate**

TlAu

d)   **Form First Level Metalization**

Sl3N4   CAP PLATE

e)   **Capacitor Formation**

THIN TlAu   RESIST

f)   **Plating Sequence (forms airbridges)**

RESIST

g)

FIGURE 7.138   (a) Various MMIC inductor layouts. (From Goyal.[22])

is an important yield-limiting factor. This means circuit failure due to collapsed air bridges, for example, occur at an increasing rate, the more air bridges that are used.

Typical Values for MMIC Lumped Elements

Each MMIC fabrication foundry sets its limits on the geometrical dimensions and range of materials available to the designer in constructing the MMIC lumped elements discussed in the previous section. Accordingly, the range of different resistor, capacitor, and inductor values available for design will vary from foundry to foundry. With this understanding, a "typical" set of element values associated with MMIC lumped elements are presented in Table 7.23.

MMIC Processing and Mask Sets

The most common MMIC process approach in the industry can be characterized as having 0.5 micron gate length MESFETs fabricated on a GaAs wafer whose final thickness is 100 microns, or 4 mils. The back side of the wafer has plated gold; via holes are used to connect from the back side of the wafer to the topside of the wafer. Although specific procedures and steps vary from foundry to foundry, Fig. 7.138b illustrates a typical process.

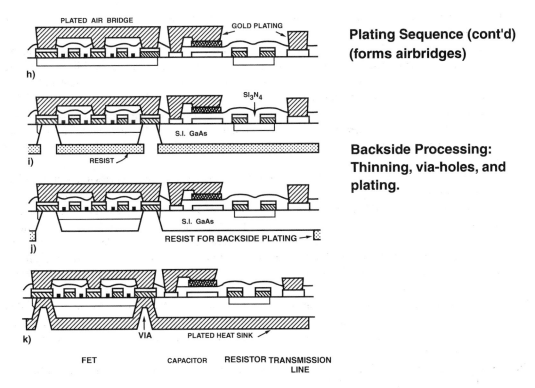

**Plating Sequence (cont'd)
(forms airbridges)**

**Backside Processing:
Thinning, via-holes, and
plating.**

FIGURE 7.138　(b) Conceptual diagrams illustrating flow for typical MMIC process. (From Williams.[23,25])

TABLE 7.23　Ranges of MMIC Lumped Element Values Available to the Designer for a "Typical" Foundry Process

Type	Value	Dielectric or Metal	Application
Inductor: Single loop, meander line, etc.	0.01–0.5 nH	Plated gold	Matching
Inductor: Spiral	0.5–10 nH	Plated gold	Matching, DC power (bias) supply choke
Capacitor: MIM	0.1–20 pF	Plated and unplated gold, silicon nitride	Matching, RF/DC signal separation
Capacitor: Coupled lines and interdigital	0.001–0.1 pF	Plated or unplated gold	Matching, RF/DC signal separation
Resistor: Thin film	5 Ω–1 kΩ	NiCr, TaN	DC bias ckts., feedback, matching, stabilization
Resistor: GaAs Monolithic	10 Ω–10 kΩ	Implanted or epitaxial GaAs	DC biasing, feedback, matching, stabilization

Adapted from Goyal.[14]

Defining Terms

Active load:　A transistor connected in a way as to replace a function that would conventionally be performed by a passive component such as a resistor, capacitor, or inductor.

Air bridge:　A bridge made of metal strip suspended in air that can connect components on an integrated circuit in such a way as to cross over another strip. Air bridges are also used to suspend metallization in spiral inductors off of the semiconducting substrate, in a way that can lead to improved performance in some cases.

Attenuation: Ratio of output signal to input signal, usually expressed in dB (see below), for a device that reduces the signal level.

Bias voltage or current: The DC power applied to a transistor allowing it to operate as an active amplifying or signal-generating device. Typical voltage levels in GaAs FETs used in receivers are 1 to 7 volts between the drain and source terminals, and 0 to –5 volts between the gate and source terminals. For microwave systems, DC voltages and currents, provided by batteries or AC/DC converters required to bias transistors to a region of operation where they will either amplify, mix or frequency translate, or generate (oscillators) microwave energy. Since energy can be neither created or destroyed, microwave energy amplification or creation is accomplished at the expense of DC energy.

Bias network: A key aspect of microwave circuit design is to apply the proper DC bias to the appropriate terminals of transistors (e.g., FETs) without disturbing the AC microwave operation of the circuit. In some cases, on-chip DC circuitry needs to be designed in such a way as to provide stable bias voltage/current conditions for the device even when the chip DC supply voltages vary (due to weakening batteries, etc.). The other aspect of bias network design is to isolate the DC network form interfering with the AC or RF/microwave operation of the circuit and vice versa. In a lumped element design, this is generally accomplished by a combination of spiral inductors and MIM capacitors.

Characteristic impedance: Inherent property of a transmission line that defines the impedance that would be seen by a signal if the transmission line were infinitely long. If a signal source with a source or reference impedance equal to the characteristic impedance is connected to the line there will be zero reflections.

Chip or die: An individual MMIC circuit or subsystem that is one of several identical chips that are produced after dicing up an MMIC wafer.

DBS Receiver: Electronic assembly that accepts as input a microwave signal from a satellite containing transmitted TV signals modulated onto the signal. The receiver first amplifies the low level signal, then processes the signal by first converting it to a lower IF frequency and then demodulating the signal to separate the TV signals from the microwave carrier signal. A basic way of looking at the relationship of the microwave carrier to the TV signal is to think of the carrier signal as an envelope with a message inside. The message is the TV signal. Demodulation is the process of carefully removing the message from the envelope (carrier). The noise figure of receiver is a measure of the amount of noise that will be added to the signal (carrier and TV signal) by the receiver. If the receiver adds too much noise, the result will be a snowy picture on the TV screen.

DBS: Direct Broadcast Satellite: Refers to TV signal transmission and distribution from a base station up to a satellite, and then down to consumers who have suitable satellite receiving antennas and downconverter receivers.

$$F(dB) = 10 Log_{10} \left(SNR_{in} / SNR_{out} \right)$$

dB: Decibel: A unit of measure that describes the ratio between two quantities in terms of a base 10 logarithm. For example, the ratio between the power level at the input and output of an amplifier is called the power gain and may be expressed in decibels as follows:

$$G(dB) = 10 Log_{10} \left(P_{out} / P_{in} \right)$$

dBm: Decibel referenced to milliwatts: A common unit of power in decibels referenced to 1 milliwatt.

$$P(dBm) = 10 Log_{10} \left(P \text{ in } mW / 1 \, mW \right)$$

DC: Direct current/voltage: Constant voltage or current with no variation over time. This can be considered in general terms as an alternating current/voltage (AC) with a frequency of variation of zero, or a zero frequency signal. For microwave systems, DC voltages and currents, provided by batteries or AC/DC converters required to bias transistors to a region of operation where they will either amplify, mix or frequency translate, or generate (oscillators) microwave energy.

DRO: Dielectric resonator (stabled) oscillator: A dielectric resonator is a cylindrically shaped piece of material, or "puck," that has low loss resonant frequencies that are determined primarily by the size of the cylinder. Placing a dielectric resonator near a microstrip line can form a resonant circuit that will frequency stabilize a voltage controlled oscillator.

Dielectric constant (e_r): The dielectric constant is an electric property of an insulator or semiconducting material that describes how differently electric fields will behave inside the material as compared to air. As an example, $e_r = 12.9$ for GaAs as compared to $e_r = 1$ for air. In integrated circuits, an effective dielectric constant (e_{eff}) is used since the electric fields supported by the signals traveling through the conductors on the circuit flow through both air and the insulator or semiconductor simultaneously.

FET: Field Effect Transistor: The MESFET (Metal-Electrode-Semiconductor-Field-Effect-Transistor) is a specific type of FET that is the dominant active (amplifying) device in GaAs MMICs. A FET is composed of three terminals called the gate, drain, and source, and a conducting "channel." In an amplifier application, the source is connected to ground, and DC bias is applied between the drain and source causing a current to flow in the channel. The current flow is controlled and modulated by the AC or DC voltage applied to the gate.

Filter: A network, usually composed of inductors and capacitors (for lumped circuit), or transmission lines of varying length and characteristic impedance (for distributed circuit), that passes AC signals over a certain frequency range while blocking signals at other frequencies. A bandpass filter passes signals over a specified range (f_{low} to f_{hi}), and rejects frequencies outside this range. For example, for a DBS receiver that is to receive satellite transmitted microwave signals in a frequency range of 11 GHz to 12 GHz, a bandpass filter (BPF) would allow signals in this frequency range to pass through with minimum signal loss, while blocking all other frequencies. A low pass filter (LPF) would allow signals to pass with minimum signal loss as long as their frequency was less than a certain cutoff frequency above which significant signal blocking occurs.

Frequency: The repetition rate of a periodic signal used to represent or process a communication signal. Frequency is expressed in units of Hertz (Hz). One Hz represents one cycle per second, 1 MHz represents one million cycles per second, and 1 GHz represents one billion cycles per second.

Gain: Ratio of the output signal over the input signal of a component. See example given with dB definition. For an amplifying device the gain will be greater than 1 when expressed as a ratio, and greater than 0 dB when expressed in decibels.

Image rejection filter: A filter usually placed before a mixer to preselect the frequency that when mixed with the local oscillator frequency, will downconvert to the desired intermediate frequency. In the absence of an image rejection filter, there are two RF frequencies which, when mixed with the local oscillator frequency f_{lo}, will down convert to the intermediate frequency f_{if}. Specifically these are given by $f_{lo} + f_{if}$ and $f_{lo} - f_{if}$. For example, if the local oscillator frequency is 10 GHz and the local oscillator frequency is 1 GHz, then signals at both 11 GHz and 9 GHz will produce an intermediate frequency signal of 1 GHz without an image rejection filter.

Impedance (Z): Electrical property of a network that measures its ability to conduct electrical AC current for a given AC voltage. Impedance is defined as the ratio of the AC Voltage divided by the AC current at a given point in the network. In general, impedance has two parts, a real (resistive) part, and an imaginary (inductive or capacitive reactive) part. Unless the circuit is purely resistive (made up of resistors only), the value of impedance will change with frequency.

Impedance matching: One of the main design activities in microwave circuit design. An impedance matching network is made up of a combination of lumped elements (resistors, capacitors, and inductors), or distributed elements (transmission lines of varying characteristic impedance and length). Impedance matching networks transform network impedance from one value to another. For example, on the input to a low noise transistor, the impedance of an incoming 75 ohm transmission line would be transformed by the input matching network to the impedance Z_{opt}, required to achieve the minimum noise figure of the transistor. The Smith Chart is a tool commonly used by microwave engineers to aid with impedance matching.

Insulator: A nonconducting material, also called a dielectric. An example is Teflon. Insulators are often used as substrates onto which hybrid electrical circuits are constructed.

Ku-Band: Frequency band of approximately 11 to 12 GHz.

L-Band: Frequency band of approximately 1 to 2 GHz.

LNA: Low noise amplifier: Boosts low level radio/microwave signal received without adding substantial distortions to the signal.

Loss: See **attenuation**.

Mask or mask set: A mask defines the geometrical pattern to be used for a single step in the fabrication process of a MMIC. A mask set consists of the dozen or so (varies with process and company) individual masks that are required to complete a MMIC wafer fabrication from start to finish. Examples of masks or mask levels are first level metal (defines all the primary metal structure on the circuit), capacitor top plate (defines the pattern for the metal used to form the top plate of MIM capacitors), and dielectric etch (defines areas where dielectric (insulator) material will be removed after coating the entire wafer with it).

Microstrip: A transmission line commonly used in MMICs, a microstrip consists of a conducting strip suspended over a ground plane on a slab of insulating or semiconducting material. The characteristic impedance of a microstrip line is determined by the width of the line and the height or thickness of the insulating or semiconducting slab. Conceptually, microstrip can be visualized as a transmission line as follows: envision a section of coaxial cable that is sliced lengthwise down to the center conductor. Now, uncurl the line so that the outer shield is laying flat and the center conductor is suspended over the shield by the insulator (now a flat slab). If we now flatten out the center conductor into a strip, a microstrip will be the result. Microstrip is a convenient way to route signals between components in a MMIC.

Microwave: Term used to refer to a radio signal at a very high frequency. One broad definition gives the microwave frequency range as that from 300 MHz to 300 GHz.

MIM Capacitor: Metal-Insulator-Metal Capacitor: (also called a Thin Film Capacitor) Integrated circuit implementation of a common electrical element that stores electric energy (a car battery can be thought of as a big capacitor). Two extreme behaviors of a capacitor are that it will act as an open circuit to low frequencies or DC (zero frequency), and as a short frequency at sufficiently high frequency (how high is determined by the capacitor value).

Mixer: A mixer is a nonlinear device containing either diodes or transistors, the function of which is to combine signals of two different frequencies in such a way as to produce energy at other frequencies. In a typical downconverter application, a mixer has two inputs and one output. One of the inputs is the modulated carrier RF or microwave signal at a frequency f_{rf}, the other is a well-controlled signal from a local oscillator or VCO at a frequency f_{lo}. The result of downconversion is a signal at the difference frequency $f_{rf} - f_{lo}$ which is also called the intermediate frequency f_{if}. A filter is usually connected to the output of the mixer to allow only the desired IF frequency signal to be passed on for further processing. For example, for an RF frequency of 10.95 GHz (=10,950 MHz) and an LO frequency of 10 GHz (=10,000 MHz), the IF frequency would be 950 MHz.

MMIC: Monolithic Microwave Integrated Circuit: The word monolith refers to a single block of stone that does not (in general) permit individual variations. MMICs are made of gallium arsenide (GaAs), silicon, or other semiconducting materials. In a MMIC, all of the components needed to make a circuit (resistors, inductors, capacitors, transistors, diodes, transmission lines) are formed onto a single wafer of material using a series of process steps. Attractive features of MMICs over competing hybrid (combination of two or more technologies) circuits are that a multitude of nearly identical circuits can be processed simultaneously with no assembly (soldering) using batch processing manufacturing techniques. A disadvantage is that circuit adjustment after manufacture is difficult or impossible. As a consequence, significantly more effort is required to use accurate computer-aided design (CAD) techniques to design MMICs that will perform as desired without adjustment. Of course, eventually assembly and packaging of MMICs is performed in order to connect them into a system such as a DBS receiver. MMICs are only cost effective for very high volume applications because the cost of the initial design is very high, as is the cost of wafer manufacture. These costs can only be recovered through high volume manufacture.

Noise: Random perturbations/distortions in signal voltage, current, or power.

Noise figure: Property of a microwave component that describes the amount of noise added to a signal passing through the component. Technically defined as the signal-to-noise ratio at the component input to the signal-to-noise ratio at the component output. For a transistor, the noise figure is highly dependent on the impedance the transistor sees when it looks back at the input matching network from its input (gate for FET) terminal. The minimum noise figure F_{min} is the lowest noise figure that a FET can exhibit under optimum input impedance matching conditions (Z_{opt} or in terms of reflection coefficient G_{opt}). Noise figure is usually specified in decibels.

Package: In MMIC technology, die or chips have to ultimately be packaged to be useful. An example of a package is the T07 "can." The MMIC chip is connected within the can with bond wires connecting from pads on the chip to lead pins on the package. The package protects the chip from the environment and allows easy connection of the chip with other components needed to assemble an entire system, such as a DBS TV receiver.

P1dB: 1 dB Compression power: Like TOI, this gives a measure of the maximum signal power level that can be processed without causing significant signal distortion or saturation effects. Technically, this refers to the power level at the input or the output of a component or system at which the saturation of active devices like transistors causes the gain to be compressed by 1dB from the linear gain.

Reflection coefficient (G): Another way of expressing the impedance. The reflection coefficient is defined as how much signal energy would be reflected at a given frequency. Like impedance, the reflection coefficient will vary with frequency if inductors or capacitors are in the circuit. The reflection coefficient is always defined with respect to a reference or characteristic impedance ($=(Z - Z0)/(Z + Z0)$). For example, the characteristic impedance of one typical TV transmission line is 75 ohms, whereas another type of TV transmission line has a characteristic impedance of 300 ohms. Hooking up a 75 ohm transmission line to a 300 ohm transmission line will result in a reflection coefficient of value $(300 - 75)/(300 + 75) = 0.6$, which means 60% of the energy received from the antenna.

RF: Radio frequency: A general term used to refer to radio signals in the general frequency range from thousands of cycles per second (kHz) to millions of cycles per second (MHz). It is also is often used generically and interchangeably with the term microwave to distinguish the high frequency AC portion of a circuit or signal from the DC bias signal or the downconverted intermediate frequency (IF) signal.

Self bias: A technique employed whereby a transistor only needs a single bias supply voltage between the drain terminal and ground. This is commonly accomplished by placing a parallel combination of a resistor and capacitor between the source terminal and ground.

Semiconductor (or semi-insulator): A material that is partially conducting (can support electrical current flow), but also has properties of an insulator. Common examples are silicon and GaAs. The amount of current conduction that can be supported can be varied by doping the material with appropriate materials that result in the increased presence of free electrons for current flow.

Smith chart: A complicated looking two-dimensional chart used by RF/microwave engineers that allows for impedance to be plotted.

Spiral inductor: Integrated circuit implementation of a common electrical element that stores magnetic energy. Two extreme behaviors of an inductor are that it will act as a short circuit to low frequency or DC energy, and as an open circuit to energy at a sufficiently high frequency (how high is determined by the inductor value). In a MMIC, a spiral inductor is realized by a rectangular or circular spiral layout of a narrow strip of metal. The value of the inductance increases as the number of turns and total length of the spiral is increased. Large spiral inductors are very commonly used as bias chokes to isolate the DC input connection from the RF circuit. Since a large-valued inductor essentially looks like an open circuit to high frequency RF/microwave energy, negligible RF/micro-wave energy will leak through and interact with the DC bias circuitry.

Third Order Intercept (TOI) Point: This gives a measure of the power level where significant undesired nonlinear distortion of a communication signal will occur. It is related to the maximum signal that can be processed without causing significant problems to the accurate reproduction of the desired information (e.g., TV signal). Technically, the TOI is the hypothetical power in dBm at which the power of the third order intermodulation nonlinear distortion product between two signals input to a component would be equal to the linear extrapolation of the fundamental power.

VCO: Voltage Controlled Oscillator: A device that produces microwave energy at a frequency that is adjustable over a certain range depending on an input DC voltage. An oscillator contains an active device, such as a FET that is connected in such a way as to be susceptible to breaking into oscillation at a frequency that is controlled by a resonant circuit. A voltage-controlled oscillator typically contains a diode that allows the resonant frequency of the resonator to be varied according to the voltage placed across its terminals. Without external stabilization with a high quality factor (low loss) resonant circuit, for example a dielectric resonator, the frequency of a VCO will not be very steady or stable. This will cause unacceptable noise and instability in the received signal.

Via holes: Holes chemically etched from the back of a MMIC wafer and filled with metal in such a way as to allow an electrical connection between the back side of a wafer and the topside of the wafer.

VSWR: Voltage Standing Wave Ratio: Another way of expressing impedance mismatch resulting in signal reflection. With respect to reflection coefficient G (see **Reflection Coefficient**) the VSWR may be expressed mathematically as:

$$\text{VSWR} = \left(1 + /G/\right)\big/\left(1 - /G/\right)$$

Yield: Percentage of acceptably good chips to the total chips considered at a certain level of a MMIC process. High yield is one of the most important parameters of a cost-efficient process. DC yield refers to the percentage of chips that behave appropriately to the application of DC biasing voltages and currents (see **Bias Voltage or Current**). RF yield refers the percentage of chips that properly process RF/microwave signals.

References

1. Pucel, R. A., Design considerations for monolithic microwave circuits, *IEEE Trans. Microwave Theory and Techniques*, MTT-29, 513–534, June 1981.
2. Pucel, R. A., ed., *Monolithic Microwave Integrated Circuits*, IEEE Press, 1985, 1.
3. Pucel, R.A., ed., *Monolithic Microwave Integrated Circuits*, IEEE Press, 1985, 1–2.
4. Hyltin, T.M., Microstrip transmission on semiconductor substrates, *IEEE Trans. Microwave Theory and Techniques*, MTT-13, 777–781, Nov. 1965.
5. Mehal, E., and Wacker, R., GaAs integrated microwave circuits, *IEEE Trans. Microwave Theory and Tech.*, MTT-16, 451–454, July 1968.
6. Pengelly, R. S., and Turner, J.A., Monolithic broadband GaAs FET amplifiers, *Electron Lett.*, 12, 251–252, May 13, 1976.
7. Pucel, R.A., ed., *Monolithic Microwave Integrated Circuits*, IEEE Press, 1985, 1.
8. Goyal, R., *Monolithic Microwave Integrated Circuits: Technology and Design*, Artech House, Norwood, MA, 1989, 17.

9. Pengelly R., Cree, and Schmitz, Bert, M/A-Com personal communication, August 1999.

10. GaAs fights back against Silicon assault, *Military and Aerospace Electronics*, 27–30, October 18, 1993.

11. Mike Frank, Hewlett Packard Company, personal communication, December 22, 1995.

12. Cressler, J.D. Re-Engineering Silicion: Si-Ge heterojunction bipolar technology. *IEEE Spectrum*, 49–55, March 1995.

13. Pucel, R., Ed., *Monolithic Microwave Integrated Circuits*, IEEE Press, 1985, 2.

14. Reprinted with permission from *Monolithic Microwave Integrated Circuits: Technology and Design*, by R. Goyal, Artech House, Norwood, MA, (www.artech-house.com), 1989, 320.

15. Reprinted with permission from, *MMIC Design GaAs FETs and HEMTs*, by P. Ladbrooke Artech House, Norwood, MA, (www.artech-house.com), 1989, 29.

16. Reprinted with permission from, *Microwave MESFETs and HEMTs*, Edited by J. M. Golio, Ed., Artech House, Norwood, MA, (www.artech-house.com), 1991, 28.

17. Reprinted with permission from *Monolithic Microwave Integrated Circuits: Technology and Design*, by R. Goyal, Artech House, Norwood, MA, (www.artech-house.com), 1989, 113.

18. Reprinted with permission from, *MMIC Design GaAs FETs and HEMTs*, by P. Ladbrooke Artech House, Norwood, MA, (www.artech-house.com), 1989, 92.

19. Reprinted with permission from *Monolithic Microwave Integrated Circuits: Technology and Design*, by R. Goyal, Artech House, Norwood, MA (www.artech-house.com), 1989, 342.

20. Reprinted with permission from, *MMIC Design GaAs FETs and HEMTs*, by P. Ladbrooke Artech House, Norwood, MA, (www.artech-house.com), 1989.

21. Reprinted with permission from *Monolithic Microwave Integrated Circuits: Technology and Design*, by R. Goyal, Artech House, Norwood, MA, (www.artech-house.com), 1989, 331.

22. Reprinted with permission from *Monolithic Microwave Integrated Circuits: Technology and Design*, by R. Goyal, Artech House, Norwood, MA, (www.artech-house.com), 1989, 320–325.

23. Reprinted with permission from *Modern GaAs Processing Methods*, by R. Williams, Artech House, Norwood, MA, (www.artech-house.com), 11–15.

8

CAD, Simulation, and Modeling

Joseph Staudinger
Motorola

Manos M. Tentzeris
Georgia Institute of Technology

Daniel G. Swanson, Jr.
Bartley R.F. Systems

Michael B. Steer
North Carolina State University

John F. Sevic
Ultra RF, Inc.

Michael Lightner
University of Colorado

Ron Kielmeyer
Motorola

Walter R. Curtice
W. R. Curtice Consulting

Peter A. Blakey
Motorola

0-8493-8592-X/01/$0.00+$.50
© 2001 by CRC Press LLC

8.1   System Simulation

Joseph Staudinger

The concept of system simulation is an exceptionally broad topic. The term itself, system, does not have a rigid definition and in practice the term is applied to represent dramatically differing levels of circuit integration, complexity, and interaction. In a simple sense, the term is applied to represent the interconnection and interaction of several electrical circuits. In a broader sense, such as in communication systems, the term may be applied to represent a much higher level of complexity including part of, or the composite mobile radio, base unit, and the transmission medium (channel). Regardless of complexity of the level, the issue of simulation is of critical importance in the area of design and optimization.

As one might expect, the techniques and methods available in the engineering environment to simulate system level performance are quite diverse in technique and complexity [1–3]. Techniques include mathematically simple formula-based methods based upon simplified models of electrical elements. Such methods tend to be useful in the early design phase and are applied with the intent of providing insight into performance level and trade-off issues. While such methods tend to be computationally efficient allowing simulations to be performed rapidly, accuracy is limited in large part due to the use of simplified models representing electrical elements. Other techniques tend to be computationally intensive CAD-based where waveforms are generated and calculated throughout the system. The waveform level technique is more versatile and can accommodate describing electrical elements to almost any level of detail required, thereby exploring the system design and performance space in fine detail. The models may be simple or complex in mathematical form. In addition, it is possible to use measured component data (e.g., scattering parameters) or results from other simulators (e.g., small- and large-signal circuit simulations using harmonic balance where the active device is represented by a large-signal electrical model). The price for the improvement in accuracy is significantly longer, perhaps much longer simulation times and the requirement to very accurately describe the characteristics of the electrical components.

The intent of this section is to examine fundamental issues relative to simulating and evaluating performance of microwave/RF related system components. A number of terms describing system level performance characteristics will be examined and defined. In addition, first order methods for calculating the performance of systems consisting of cascaded electrical circuits will be examined. To begin, consider three parameters of interest in nearly all systems: gain, noise figure (NF), and intermodulation distortion (IMD).

Gain

A usual parameter of interest in all systems is the small signal (linear) gain relationship describing signal characteristics at the output port relative to the input port and/or source for a series of cascaded circuits. Numerous definitions of gain have been defined in relationship to voltage, current, and power (e.g., power gain can be defined in terms of transducer, available, maximum stable, etc.) [1]. In general for system analysis, the concept of transducer power gain (G_T) is often applied to approximate the small signal gain response of a series of cascaded elements. Transducer power gain (G_T) is defined as the magnitude of the forward scattering parameter (S_{21}) squared (i.e., $G_T = |S_{21}|^2$, the ratio of power delivered to the load to that available from the source). This assumes the source (Γ_S) and load (Γ_L) voltage reflection coefficient are equal to zero, or alternatively defined as a terminating impedance equal to characteristic impedance Z_0 (typically 50 ohms).

Consider several two-port networks cascaded together as illustrated in Fig. 8.1 where the transducer power gain of the ith element is represented as G_{Ti}. The transducer power gain of the cascaded network is:

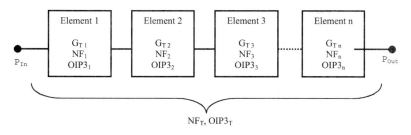

FIGURE 8.1 System formed by cascading three electrical elements.

$$G_{T_T} = G_{T1}(dB) + G_{T2}(dB) + G_{T3}(dB) + \dots G_{Tn}(dB) \tag{8.1}$$

The accuracy of Eq. (8.1) relies on the assumption that that i^{th} two port network is terminated by characteristic impedance Z_0 per the above definition. In practice, the source and load termination provided to the ith element is defined by the $i^{th} - 1$ and $i^{th} + 1$ elements, respectively. Even though in a well-designed subsystem, each two-port network is designed such that the input (S_{11}) and output (S_{22}) scattering parameters are near zero in magnitude, they cannot be exactly zero resulting in impedance mismatch effects. Hence, Eq. (8.1) is approximate and its accuracy dependent on each element satisfying the above criteria. A more thorough analysis accounting for impedance mismatches can be performed at the expense of more complexity. In general this requires a more precise description of each element using perhaps some form of network parameters. For example, the T and scattering parameters (T_T and S_T, respectively) for two networks, A and B, cascaded together are given by [3,7]

$$T_T = \begin{bmatrix} T_{11}^A & T_{12}^A \\ T_{21}^A & T_{22}^A \end{bmatrix} \begin{bmatrix} T_{11}^B & T_{12}^B \\ T_{21}^B & T_{22}^B \end{bmatrix} \tag{8.2}$$

$$S_T = \begin{bmatrix} S_{11}^A & 0 \\ 0 & S_{22}^B \end{bmatrix} + \begin{bmatrix} S_{12}^A & 0 \\ 0 & S_{21}^B \end{bmatrix} \begin{bmatrix} -S_{22}^A & 1 \\ 1 & -S_{11}^B \end{bmatrix}^{-1} \begin{bmatrix} S_{21}^A & 0 \\ 0 & S_{12}^B \end{bmatrix} \tag{8.3}$$

While the above methods allow an exact analysis for cascaded linear circuits, it is often difficult to apply them to practical systems since the network parameters for each of the elements comprising the system are usually not known precisely. For example, in systems consisting of interconnected circuits, board layout effects (e.g., coupling between elements) and interconnecting structures (board traces) must also be included in applying the network analysis techniques shown in Eqs. (8.2) and (8.3). This, of course, requires accurate knowledge of the electrical nature of these structures, which is often unknown. In critical situations, the network parameters for these elements can be determined by measurement or through the use of electromagnetic simulations assuming the geometrical and physical nature of these structures are known.

Noise

A second parameter of interest important to all systems is noise. In receivers, noise performance is often specified by noise figure, defined as

$$NF(dB) = \frac{S_i/N_i}{S_o/N_o} \tag{8.4}$$

where S_i/N_i and S_o/N_o are the signal-to-noise ratio at the input and output ports, respectively. Note that *NF* is always greater than or equal to unity (0 dB). When several circuits are cascaded together as illustrated in Fig. 8.1, the cascaded noise figure (NF_T) is given by

$$NF_T = NF_1 + \frac{NF_2 - 1}{G_1} + \frac{NF_3 - 1}{G_1 G_2} + \frac{NF_4 - 1}{G_1 G_2 G_3} + \cdots \frac{NF_n - 1}{G_1 G_2 G_3 \dots G_n} \tag{8.5}$$

where G_i and NF_i are the gain and noise figure of the ith element, respectively. Note the importance of the contribution of the first element's noise figure to the total cascaded noise figure. Hence, the noise figure of the low noise amplifier contained in a receiver is a major contributor in setting the noise performance of the receiver.

Intermodulation Distortion

Intermodulation distortion (IMD) has been a traditional spectral measure (frequency domain) of linearity applied to both receiver and transmitter elements. The basis of *IMD* is formed around the concept that the input-output signal relationship of an electrical circuit can be expressed in terms of a series expansion taking the form:

$$E_o = a_1 E_{in} + a_2 E_{in}^2 + a_3 E_{in}^3 + a_4 E_{in}^4 + a_5 E_{in}^5 + \cdots \tag{8.6}$$

where E_{in} and E_o are instantaneous signal levels at the input and output ports of the electrical circuit, respectively. If the circuit is exactly linear, all terms in the expansion are zero except for a_1 (i.e., gain). In practice, all circuits exhibit some nonlinear behavior and hence higher order coefficients are nonzero. Of particular interest is the spectral content of the output signal when the circuit is driven by an input consisting of two sinusoids separated slightly in frequency taking the form

$$E_{in}(t) = \cos(\omega_1 t) + \cos(\omega_2 t) \tag{8.7}$$

where ω_1 and ω_2 are the angular frequencies of the input stimuli and where ω_2 is slightly greater than ω_1. The output signal will exhibit spectral components at frequencies $m\omega_1 \pm n\omega_2$, m = 0,1,2, ... and n = 0,1,2, ... as illustrated in Fig. 8.2. Notice that the third series term ($a_3 Ein^3$) generates spectral tones at $2\omega_1 - \omega_2$ and $2\omega_2 - \omega_1$. These spectral components are termed third order intermodulation distortion (*IM3*). It can also be shown that series terms greater than three also produce spectral components at these frequencies and hence the total *IM3* is the vector sum from Ein^3, Ein^4.... In a similar manner, higher order IMD products exist (e.g., *IM5* @ $3\omega_1 - 2\omega_2$ and $3\omega_2 - 2\omega_1$) due to the higher order series terms. Notice however, that all IMD products are close in frequency to ω_1 and ω_2, fall within the desired frequency

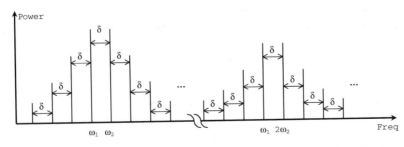

FIGURE 8.2 Resultant spectrum representing the nonlinear amplification of two equal amplitude sinusoidal stimuli at frequencies ω_1 and ω_2 ($\delta = \omega_2 - \omega_1$).

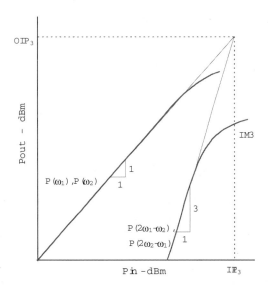

FIGURE 8.3 Expected relationship between fundamental frequency components and third order intermodulation distortion products neglecting effects due to higher order series coefficients.

band of the circuit, and hence cannot be removed by external filtering. In practice, third order products are often the highest in magnitude and thus of greatest concern, although in some cases 5th and higher order products may also be of interest. Spectral analysis of the circuit can be greatly simplified if the input signal is assumed small in magnitude such that the dominant contributor to *IM3* is from E_{in}^3.

Intermodulation distortion products (*IM3, IM5,…*) can be expressed in terms of power, either absolute (dBm) or relative to the carrier (dBc), or by a fictitious intercept point. For certain circuits where the input stimulus is small in magnitude (e.g., low noise amplifier and certain receiver components), an intercept point distortion specification is useful. Consider the nonlinear response of a circuit represented by Eq. (8.6) and driven with an equal amplitude two-tone sinusoidal stimuli as given in Eq. (8.7). Assume further than only a_1 and a_3 in Eq. (8.6) are nonzero. A plot of the circuit's output power spectrum as a function of input power is illustrated in Fig. 8.3. The output spectral tones at ω_1 and ω_2 increase on a 1:1 (dB) basis with input power. The *IM3* products ($2\omega_1 - \omega_2$, $2\omega_2 - \omega_1$) increase on a 3:1 (dB) basis with input power (due to E_{in}^3). The intersection of the fundamental tones with the third order products is defined as the third order intercept point. Note that the intercept point can be specified relative to input or output power of each tone, IIP_3 and OIP_3, respectively. Given this linear relationship, the output intercept point can easily be calculated based on the power of the fundamental and third order terms present at the output port [2,6]

$$OIP3\left(dBm\right) = P_{out}\left(\omega_1\right) + \frac{P_{out}\left(\omega_1\right) - P_{out}\left(2\omega_1 - \omega_2\right)}{2} \tag{8.8}$$

where $P_{out}(\omega_1)$ and $P_{out}(2\omega_2 - \omega_1)$ is the power (dBm) in the fundamental and third order products referenced to the output port. Notice that since the input stimuli is an equal amplitude two-tone sinusoidal stimulus, like order spectra products are assumed equal in magnitude (i.e., $P_{out}(\omega_1) = P_{out}(\omega_2)$, $P_{out}(2\omega_2 - \omega_1) = P_{out}(2\omega_1 - \omega_2)$, …).

The relationship between input and output intercept points is given by

$$IIP3\left(dBm\right) = OIP3\left(dBm\right) - G\left(dBm\right) \tag{8.9}$$

In a similar manner, the fifth order output intercept point can be defined as [5]

$$OIP5\left(dBm\right) = P_{out}\left(\omega_1\right) + \frac{P_{out}\left(\omega_1\right) - P_{out}\left(3\omega_1 - 2\omega_2\right)}{4} \tag{8.10}$$

where $P_{out}(3\omega_2 - 2\omega_1)$ is power (dBm) of fifth order products referenced to the output port.
Similarly,

$$IIP5\left(dBm\right) = OIP5\left(dBm\right) - G\left(dBm\right) \tag{8.11}$$

In system analysis, it is often desirable to consider the linearity performance for a series of cascaded two-port circuits and the contribution of each circuit's nonlinearity to the total. The IM3 distortion of the complete cascaded network can be approximated based on the third order intercept point of each element. Consider the two-tone sinusoidal stimulus [Eq. (8.7)] applied to the input of the cascaded circuits shown in Fig. 8.1. The magnitude of the fundamental and IM3 products (i.e., ω_1, ω_2, $2\omega_1 - \omega_2$, and $2\omega_2 - \omega$) at the output port of the 1st element can be calculated based on knowledge of the input power level of tones ω_1 and ω_2, transducer gain, and the third order intercept point of element one using Eq. (8.8). Next, consider the second element where the input stimulus now consists of spectral components at $2\omega_1 - \omega_2$ and $2\omega_2 - \omega_1$ in addition to those at ω_1 and ω_2. The IM3 spectral products at the second element's output port will be the result of contributions from two sources, (1) those due to intermodulation distortion of ω_1 and ω_2 in element 2, and (2) those due to amplifying spectral products $2\omega_1 - \omega_2$, and $2\omega_2 - \omega$ present at the input port of element 2. The IM3 products due to the former are, again, calculated from Eq. (8.8). The IM3 products at the output of element 2 due to the latter will be the IM3 products at the input amplified by G_2. Hence, the total *IM3* spectral products are the vector sum from each. Both a minimum and maximum value is possible depending on the vector relationship between the various signals. A worst case (lowest *OIP3*) results when they combine in phase and are given by [2]

$$\frac{1}{IIP3_T} = \frac{1}{IIP3_1} + \frac{G_1}{IIP3_2} + \frac{G_1 G_2}{IIP3_3} \cdots \tag{8.12}$$

with *IIP3* expressed in Watts.
Or alternatively from [6]

$$OIP3_{T\,min} = \left(\sum_{i=1}^{n} \frac{1}{OIP3_i g_i} \right)^{-1} \tag{8.13}$$

with *OIP3* expressed in Watts and where g_i is the cascaded gain from the output of the ith element to the system output, including impedance mismatch effects. A best case scenario (highest *OIP3*) results when they combine out of phase with the results given by [6]:

$$OIP3_{T\,max} = \left(\sum_{i=1}^{n} \frac{1}{OIP3_i^2 g_i^2} - 2 \sum_{i=2}^{n} \sum_{j=1}^{n} \frac{1}{OIP3_i OIP3_j g_i g_j} \right)^{-1/2} \tag{8.14}$$

$$i > j$$

Hence, Eqs. (8.13) and (8.14) specify bounds for intercept performance of cascaded networks.

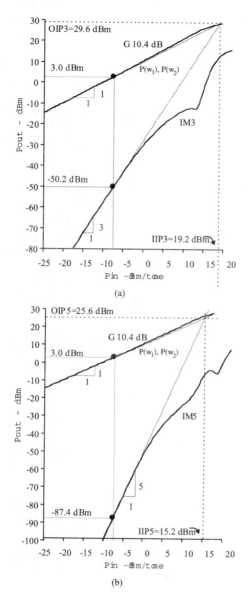

FIGURE 8.4 Typical relationship between fundamental frequency components and 3rd and 5th order intermodulation distortion products including effects due to higher order series coefficients. a) Third order IMD, and b) Fifth order IMD.

An illustration of the measured spectral content of an amplifier driven with a two-tone sinusoidal stimuli is shown in Fig. 8.4. At low power levels, third and fifth order IM products closely follow a 3:1 and 5:1 (dB) relationship with input power. Hence, per the previous discussion, $OIP3$ and $OIP5$ can be calculated based on measurements of the output spectral products at a given input power level. In this example, the spectral content is $P_{out}(3\omega_2 - 2\omega_1) = -87.4$ dBm, $P_{out}(2\omega_2 - \omega_1) = -50.2$ dBm, $P_{out}(\omega_1) = +3.0$ dBm and $G = 10.4$ dB for the input level shown. Applying Eqs. (8.8) and (8.10) yield $OIP3 = 29.6$ dBm and $OIP5 = 25.6$ dBm, respectively. The input intercept points are determined from Eqs. (8.9) and (8.11).

Limitations rooted in the approximations in deriving intercept points become more apparent at higher power levels where the relationship between input power and spectral products deviates dramatically from their assumed values. At some increased power level, the effects due to the higher order series coefficients in Eq. (8.6) become significant and cannot be ignored. Hence, for certain circuits, such as power amplifiers for example, the concept of intercept point is meaningless. A more meaningful measure of nonlinearity is the relative power in the IMD products (dBc) referenced to the fundamental tones, with the reference generally made to output rather than input power.

System Simulation with a Digitally Modulated RF Stimuli

Many modern communications systems, including 2nd and 3rd generation cellular, satellite communications, and wireless local area networks (W-LAN), to name but a few, utilize some form of digital modulation to encode information onto an RF carrier. As discussed earlier in this chapter, these signals are complex in that the RF carrier's phase, amplitude, or both are modulated in some manner to represent digital information. An extensive examination of the mathematical techniques and available methods to simulate the system response of passing such signals through various RF circuits is well beyond the scope of this section. The reader is referred to [1] for a more detailed discussion of simulation techniques. Nevertheless, some of the fundamental RF-related aspects of system simulation will be examined in the context of a mobile wireless radio. Consider the architecture illustrated in Fig. 8.5 which is intended to represent major elements of wireless radio such as presently utilized in 2G and 3G cellular systems. The radio utilizes frequency division multiplexing whereby a diplexer confines RX and TX signals to the respective receiver and transmitter paths.

To begin, consider the TX path where digital information is first generated by a DSP. This data is modulated onto an RF carrier whereby the information is encoded and modulated onto a carrier conforming to a particular modulation format. From this point, the signal is injected into a mixer where it is raised in frequency to coincide with the TX frequency band. The signal is then amplified by a power amplifier and passed to the antenna via a diplexer.

Simulation of the TX signal path begins by considering the digital information present at the modulator. In a cellular system, this information corresponds to a digital sequence representing voice information and/or data. The precise structure of the sequence (i.e., patterns of zeros and ones) is important in that it is a major contributor in defining the envelope characteristics of the RF signal to be transmitted. Also note for simulation purposes, the RF stimulus is generally formed by repeating this sequence and modulating it onto an RF carrier. Hence, the resultant RF signal is periodic per the digital bit sequence. The effect of a particular digital bit sequence in defining the RF signal is illustrated by considering the two randomly generated NRZ bit patterns shown in Fig. 8.6. The amplitude modulated RF envelope voltage developed by utilizing these sequences in a π/4 DQPSK modulator is also shown in Fig. 8.6. While the two envelope signals are nearly identical in their average power levels, they are substantially different, especially in their peak voltage excursions. Hence, the digital bit sequence and the resultant modulated

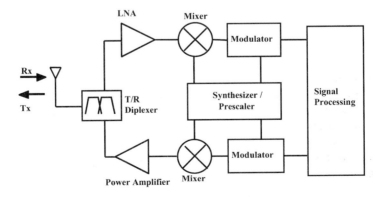

FIGURE 8.5 Block diagram representing major elements in mobile cellular radio.

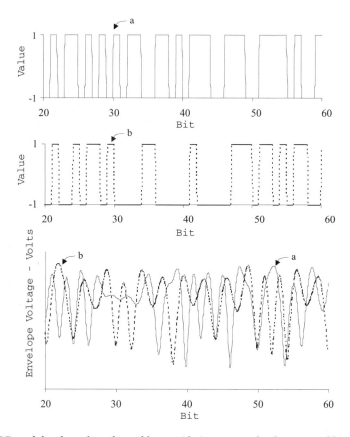

FIGURE 8.6 The RF modulated envelope formed by considering two randomly generated bit sequences.

RF waveform can be particularly important when evaluating the performance of nonlinear circuits such as power amplifiers in that the bit sequence can affect spectral distortion. Nevertheless for simulation purposes, it is necessary to choose a suitable sequence to represent the digital information to be transmitted. In general, either a predefined binary NRZ sequence, a randomly generated one, or a pseudo-noise (PN) sequence is generally chosen. Often the latter is considered a more desirable choice due to its statistical properties. For example, a maximal length PN sequence of length 2^m-1 contains all but one m bit combinations of 1s and 0s. This property is particularly important in that it allows all possible bit patterns (except for the all zeros pattern) to be utilized in generating the RF modulated waveform. In contrast, a much longer random sequence would be needed with no guarantee of this property. Further, the autocorrelation function of a PN sequence is similar to a random one [1]. A potentially significant disadvantage of applying a random sequence in evaluating nonlinear circuit blocks is that the simulation results will change from evaluation to evaluation as the randomly generated sequence is not identical for each simulation.

Once a sequence is chosen, the performance of the modulator can be evaluated. The modulator can be modeled and simulated at the component level using time-based methods such as SPICE. However, in the early system design phase, such detail and the time required to perform a full circuit level simulation may be unattractive. On the other hand, it may be more appropriate to model the modulator at a higher level (e.g., behavioral model) and only consider selected first order effects. For example, the simplified diagram in Fig. 8.7 depicts the functionality of a quadrature modulator (in this case to represent a QPSK modulator). Starting with a data steam, every two bits are grouped into separate binary streams representing even and odd bits as indicated by X_K and Y_K. These signals (X_K and Y_K) are encoded in some manner (e.g., as relative changes in phase for IS-136 cellular) and are now represented as I_K and Q_K with

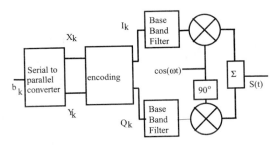

FIGURE 8.7   The process of generating modulated signal $S(t)$ in a QPSK system.

each symbol extending over two bit time intervals. These signals now pass though a baseband filter, often implemented as a finite impulse response filter with impulse response h(t). The filtered signal can be calculated based upon the convolution theorem for time sampled signals. These signals are then modulated onto a carrier (IF) with the output modulated signal taking the form

$$S(t) = \sum_n g(t-nT)\cos(\Phi_n)\cos(\omega_c t) - \sum_n g(t-nT)\sin(\Phi_n)\sin(\omega_c t) \qquad (8.15)$$

where ω_c is the radian carrier frequency, Φ_n represents phase, $g(t)$ is a pulse shaping factor, and n = 0, 1, 2, … are discrete time samples.

At this point, some first order effects can be evaluated by considering the mathematical nature of Eq. (8.15). For example, phase imbalance (Φ_{imb}) within the modulator can be modeled as:

$$S(t) = \sum_n g(t-nT)\cos(\Phi_n)\cos(\omega_c t) - \sum_n g(t-nT)\sin(\Phi_n)\sin(\omega_c t + \Phi_{imb}) \qquad (8.16)$$

Given a higher level model, the modulated envelope can be simulated using time-based methods to determine $S(t)$.

The power amplifier represents an element in the transmitter chain where linearity is of concern, especially in those RF systems employing modulation methods resulting in a nonconstant amplitude envelope. These cases, which incidentally include a number of cellular and PCS systems, require a linear power amplifier to preserve the amplitude/phase characteristics of the signal. The nonlinear characteristics of the amplifier can be simulated at the circuit level using time- and/or frequency-based methods. However, circuit-based simulations require accurate and detailed knowledge of all circuit components within the amplifier, including an appropriate large signal model for all active devices as well as highly accurate models for all passive structures. Such knowledge is often unavailable at the system level with sufficient detail and accuracy to perform such a simulation. In addition, circuit level nonlinear simulations of the amplifier driven by digitally modulated RF stimulus are generally quite computationally intensive resulting in long simulation times, making this approach even more unattractive.

A more common approach to modeling the nonlinearity of a power amplifier at the system level is through the use of behavioral models [1,7,8]. While a number of behavioral models have been proposed with varying levels of complexity and sophistication, all of them rely to some extent on certain approximations regarding the circuit nonlinearity. A common assumption in many of the behavioral models is that the nonlinear circuit/RF-modulated stimulus can be represented in terms of a memoryless and bandpass nonlinearity [1]. Although the active device within the amplifier generally exhibits some memory behavior, and additional memory-like effects can be caused by bias networks, these assumptions are generally not too limiting for power amplifiers utilized in cellular communications systems. Further, when the above-noted assumptions are met or nearly met, the simulation results are very accurate and the needed simulation time is very short.

In general, an input-output complex envelope voltage relationship is assumed with the complex output envelope voltage $v_{out}(t)$ taking the form

$$v_{out}(t) = RE\left\{ G(V(t)) e^{j\left\{ \Phi(t) + \phi(V(t)) + \omega_c t \right\}} \right\}$$ (8.17)

where $G(V(t))$ and $\phi(V(t))$ describe the instantaneous input-output envelope voltage gain and phase. Note that functions $G(V)$ and $\phi(V)$ represent the amplifier's am-am and am-pm response, respectively. The term ω_c represents the carrier frequency.

An inspection of Eq. (8.17) suggests the output envelope voltage can be calculated in a time-based method by selecting time samples with respect to the modulation rate rather than at the RF carrier frequency. This feature is particularly advantageous in digitally modulated systems where the bit rate and modulation bandwidth are small in comparison to the carrier frequency. Significantly long and arbitrary bit sequences can be simulated very quickly since the time steps are at the envelope rate. For example, consider an NRZ bit sequence on the order of several mS which is filtered at baseband and modulated onto an RF carrier with a 1nS period (i.e., 1 GHz). Time-based simulation at the RF frequency would likely require time samples significantly less than 0.1 nS and the overall number of sample points would easily exceed 10^7. Alternatively, simulating the output envelope voltage by choosing time steps relative to the modulation rate would result in several orders of magnitude fewer time samples.

A particular advantage of the above model is that the entire power amplifier nonlinearity is described in terms of its am-am (gain) and am-pm (insertion phase) response. The am-am response is equivalent to RF gain and the am-pm characteristics are equivalent to the insertion phase — both measured from input to output ports with each expressed as a function of input RF power. For simplicity, both characteristics are often determined using a single tone CW stimulus, although modulated RF signals can be used at the expense of more complexity. The nature of the behavioral model allows it to be characterized based on the results from another simulator (e.g., harmonic balance simulations of the amplifier circuit), an idealized or assumed response represented in terms of mathematical functions describing am-am and am-pm characteristics, or from measurements made on an actual amplifier.

Computation of the output envelope voltage from Eq. (8.17) allows evaluation of many critical system characteristics, including ACPR (Adjacent and/or Alternate Channel Power Ratio), Error Vector Magnitude (EVM), and Bit Error Rate (BER) to name but a few. For example, using Fourier methods, the spectral properties of the output signal expressed in Eq. (8.17) can be calculated. Nonlinear distortion mechanisms are manifested in spectral regrowth or equivalently a widening of the spectral properties. This regrowth is quantified by figure of merit ACPR, which is defined as a ratio of power in the adjacent (or in some cases alternative) transmit channel to that in the main channel. In some systems (e.g., IS-136), the measurement is reference to the receiver (i.e., the transmitted signal must be downconverted in frequency and filtered at base band). An illustration of spectral distortion for a $\pi/4$ DQPSK stimulus compliant to the IS-136 cellular standard is shown in Fig. 8.8. In this system the channel bandwidth is 30 KHz. Hence, using the specified root raised cosine baseband filter with excess bandwidth ratio $\alpha = 0.35$, a 24.3 KS/s data sequence will result in a non-distorted modulated RF stimulus band limited to 32.805 KHz (i.e., $1.35 * 24.3$ KS/s) as illustrated in Fig. 8.8. Nonlinear distortion mechanisms cause regrowth as illustrated. Note the substantial increase in power in adjacent and alternate channels.

Another measure of nonlinear distortions involves examining the envelope characteristics of $S(t)$ in time as measured by the receiver. Consider an input $\pi/4$ DPSK modulated RF signal which is developed from a 256 length randomly selected NRZ symbol sequence and passed through both linear and nonlinear amplifiers. Figure 8.9 illustrates the constellation diagram for both signals as measured at the receiver. Note that both signals represent the I and Q components of the envelope where $S(t)$ has been passed through a root-raised-cosine receiver filter. The constellation plot only shows the values of the I and Q components at the appropriate time sample intervals. The non-distorted signal exhibits expected values of ± 1, $\pm\sqrt{1/2} \pm i\sqrt{1/2}$, and $\pm i$. The distorted signal is in error from these expected values in terms of both amplitude and phase.

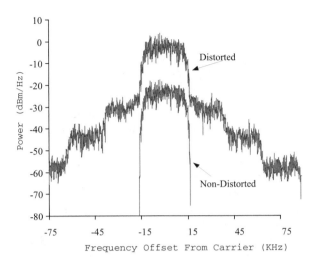

FIGURE 8.8   Nonlinear distortions due to a power amplifier result in a widening of the spectral properties of the RF signal.

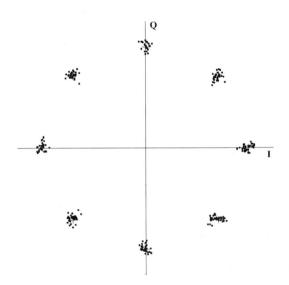

FIGURE 8.9   Constellation plot for both a non-distorted and distorted $\pi/4$ DPSK stimulus.

References

1. Michel C. Jeruchim, Philip Balaban, and K. Sam Shanmugan, *Simulation of Communication Systems*, Plenum Press, New York, 1992.
2. J. Tsui, *Microwave Receivers and Related Components*, Air Force Avionics Lab., Wright-Patterson AFB, OH, 1983.
3. G. Gonzalez, *Microwave Transistor Amplifiers*, Prentice-Hall, Inc., Englewood Cliffs, N.J., 1984.
4. J. Sevice, Nonlinear Analysis Methods for the Simulation of Digital Wireless Communication Systems, *Int. Journal of Microwave and Millimeter-Wave Computer-Aided Eng.*, 6, 1997, 197–216.
5. H. Xiao, Q. Wu, F. Li, Measure a Power Amplifier's Fifth-Order Intercept Point, *RF Design*, April 1999, 54–56.
6. N. Kanaglekar, R. Mcintosh, and W. Bryant, Analysis of Two-Tone, Third-Order Distortion in Cascaded Two-Ports, *IEEE Transactions on Microwave Theory and Techniques*, April 1988, 701–705.

7. System Theory of Operation, OmniSys Manual, Hewlett-Packard Co.
8. J. Staudinger, Applying the Quadrature Modeling Technique to Wireless Power Amplifiers, *Microwave Journal*, Nov. 1997, 66–86.

8.2 Numerical Techniques for the Analysis and Design of RF/Microwave Structures

Manos M. Tentzeris

Recent advances in wireless and microwave communication systems (higher operation frequency bands, more compact topologies containing MMICs and MEMS) have increased the necessity of fast and accurate numerical simulation techniques [1–20]. Unlike hybrid microwave integrated circuits at low frequencies, it is extremely difficult and essentially impossible to adjust the circuit and radiation characteristics of communication modules once they are fabricated. The starting point for the development of efficient numerical algorithms is an accurate characterization of the passive and active structures involved in the topologies. Although most commercial CAD programs are based on curve-fitting formulas and lookup tables and not on accurate numerical characterization, the latter can be used if it is fast enough. In addition, it can be used to generate lookup tables and to check the accuracy of empirical formulas.

Any numerical method for characterization needs to be as efficient and economical as possible in both CPU time and temporary storage requirement, although recent rapid advances in computers impose less severe restrictions on the efficiency and economy of the method. Another important aspect in the development of numerical methods has been the versatility of the method. In reality, however, numerical methods are chosen on the basis of trade-offs between accuracy, speed, storage requirements, versatility, etc. and are often structure dependent. Among these techniques, the most popular ones include the Moment Method (MoM), Integral Equation Based Techniques, Mode-Matching (MM), the Finite-Difference Time-Domain (FDTD), the Transmission Line Matrix method (TLM) and the Finite Element (FEM) method.

Integral Equation Based Techniques

Method of Moments (MoM) — Integral Equation

The term "Moment Method" was introduced in electromagnetics by R. Harrington [3] in 1968 in order to specify a certain general method for reducing linear operator equations to finite matrix solutions [4–7]. MoM computations invariably reduce the physical problem, specified by the Maxwell's equations and the boundary conditions, into integral equations having finite, and preferably small, domains. In this small domain, the discretization is performed through the expansion of unknowns as a series of basis functions. An example is the Magnetic Field Integral Equation (MFIE) for the scattering of a perfectly conducting body illuminated by an incident field H^i [21],

$$\hat{n} \times \mathbf{H}(\mathbf{r}) = 2\hat{n} \times \mathbf{H}^i(\mathbf{r}) + 2\hat{n} \times \int_s \left[\hat{n}' \times \mathbf{H}(\mathbf{r}')\right] \times \nabla' G(r, r') ds' \quad on \ S \qquad (8.18)$$

where $\mathbf{H}^i$ is defined as the field due to the source in the absence of the scattering body S, and

$$G(r, r') = \frac{e^{-jk|r-r'|}}{4\pi|r - r'|} \qquad (8.19)$$

where r and r' are the position vectors for the field and source positions, respectively. A continuous integral, such as the one above, can be written in an abbreviated form as

$$L(f) = g \qquad (8.20)$$

where f denotes the unknown, which is **H** above, and g denotes the given excitation, which is **H**^{i}. Also, L is a linear operator. Let f be expanded in a series of functions $f_1, f_2, f_3, \ldots$ in the domain S of L, as

$$f = \sum_n a_n f_n \qquad (8.21)$$

where a_n are constants and f_n are called expansion (basis) functions. For exact solutions, the above summation is infinite and f_n forms a complete set of basis functions. For approximate solutions, this solution is truncated to

$$\sum_n a_n L(f_n) = g \qquad (8.22)$$

Assume that a suitable inner product $\langle f, g \rangle = \int f(x) g(x) dx$ has been determined for the problem. Defining a set of weighting (testing) functions, $w_1, w_2, w_3, \ldots$ in the range of L, and taking the inner product of the above summation with each w_m leads to the result

$$\sum_n a_n \langle w_m, L(f_n) \rangle = \langle w_m, g \rangle, \quad m = 1, 2, 3, \ldots \qquad (8.23)$$

which can be written in matrix form as

$$\left[l_{mn} \right] \left[a_n \right] = \left[g_m \right], \qquad (8.24)$$

where

$$\left[l_{mn} \right] = \begin{bmatrix} \langle w_1, L(f_1) \rangle & \langle w_1, L(f_2) \rangle & \cdots \\ \langle w_2, L(f_1) \rangle & \langle w_2, L(f_2) \rangle & \cdots \\ \cdots & \cdots & \cdots \end{bmatrix}, \quad \left[a_n \right] = \begin{bmatrix} a_1 \\ a_2 \\ \vdots \end{bmatrix}, \quad \left[g_m \right] = \begin{bmatrix} \langle w_1, g \rangle \\ \langle w_2, g \rangle \\ \vdots \end{bmatrix}. \qquad (8.25)$$

If the matrix $[l]$ is nonsingular, its inverse $[l]^{-1}$ exists and the a_n are given by

$$\left[a_n \right] = \left[l_{mn} \right]^{-1} \left[g_m \right] \qquad (8.26)$$

and the unknown f is given from the weighted summation [Eq. (8.21)]. Assuming that the finite expansion basis is defined by $[\tilde{f}_n] = [f_1 \ f_2 \ f_3 \ \ldots]$, the approximate solution for f is

$$f = \left[\tilde{f}_n \right] \left[a_n \right] = \left[\tilde{f}_n \right] \left[l_{mn} \right]^{-1} \left[g_m \right] \qquad (8.27)$$

Depending upon the choice of f_n and w_n this solution could be exact or approximate [22]. The most important aspect of MoM is the choice of expansion and testing functions. The f_n should be linearly independent and chosen such that a finite-term superposition approximates f quite well. The w_n should also be linearly independent. In addition, this choice is affected by the size of the matrix that has to be

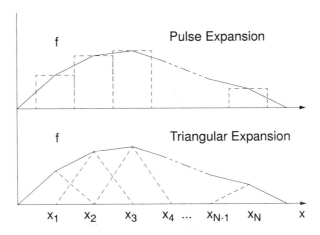

FIGURE 8.10 Moment method expansion in pulse and triangular basis.

inverted (should be minimal), the ease of evaluation of the matrix elements, the accuracy of the desired solution, and the realization of a well-conditioned matrix [1]. The special choice $w_n = f_n$ gives Galerkin's method. The two most popular subsectional bases are the pulse function (step approximation) and the triangle function (piecewise linear approximation), as shown in Fig. 8.10. The numerical Gaussian Quadrature rule [7] is used when the integrations involved in the evaluation of $l_{mn} = \langle w_m, Lf_n \rangle$ are difficult to perform for common w_m's for a specific problem or when a more complex expansion basis is used. An alternative approach makes use of the Dirac delta functions for testing. This technique is called point-matching and effectively satisfies Eq. (8.19) at discrete points in the region of interest. When the basis functions f_n exist only over subsections of the domain of f, each a_n affects the approximation of f only over a subsection of the region of interest (method of subsections). In this way, the evaluation of l_{mn} is simplified and the form of the matrix [l] is stripped and easier to invert.

The Method of Moments (MoM) involves setting up and solving dense, full, complex-valued systems. For a one-dimensional structure (e.g., a wire) of length L wavelengths, the size of the matrix is typically on the value of 10 L; for a 3D structure with surface area S square wavelengths, the size is typically on the order of 100 S. Consequently, there are applications such as radar scattering from aircraft when the system of equations has order in excess of one million. Though MoM has been proven to be a very robust technique, it is plagued by the significant computational burden of having to solve a dense matrix equation of large order. In addition, modeling of a new structure requires the reformulation of the integral equation, a task that may require the very difficult derivation of a geometry-specific Green's function. MoM is used for the solution of various integral equations such as the Electric Field Integral Equation (EFIE) and the Magnetic Field Integral Equation (MFIE). The integral equation approach has the advantage that it rigorously takes into account the radiation conditions of any open structure and therefore it is not necessary to implement absorbing boundary conditions. The kernel of the integral equation is the Green's function that accurately describes all possible wave propagation effects, such as radiation, leakage, anisotropy, stratification, etc.

The Method of Moments has been used in many scattering problems [23–25], microstrip lines on multilayered dielectrics [26], microstrip antennas [27, 28], integrated waveguides in microwave [29, 30], and optical frequencies [31] even on anisotropical substrates [32]. In addition, results have been derived for the characterization of junctions [33], high-speed interconnects [34], viaholes [35], couplers [36], and infinite aperture arrays [37]. It should be emphasized that the discretization may be nonuniform, something that has been demonstrated in the application of MoM to planar circuits in [38].

Spectral Domain Approach

Electromagnetic analysis in the spectral (Fourier transform) domain is preferable to many spatial domain numerical techniques, especially for planar transmission lines, microstrip antennas, and other planar layered structures. If applied to an equation describing EM fields in planar media, a Fourier transform reduces a fully coupled, three-dimensional equation to a one-dimensional equation that depends on two parameters (the transform variables), but is uncoupled and can be solved independently at any values of those parameters. After solving the one-dimensional equation, the inverse Fourier transform performs a superposition of the uncoupled one-dimensional solutions to obtain the three-dimensional solution. Thus, the process offers substantial savings since it effectively converts a 3D problem into a series of 1D problems.

Yamashita and Mittra [39] introduced the Fourier domain analysis for the calculation of the phase velocity and the characteristic impedance of a microstrip line. Using the quasi-TEM approximation (negligible longitudinal E- and H-fields), the line capacitance is determined by the assumed charge density through the application of the Fourier domain variational method. Denlinger [40] extended this approach to the full wave analysis of the same line. However, his solution depends strongly to the assumed current strip distributions. Itoh and Mittra [41] introduced a new technique, the Spectral Domain Approach (SDA) that allows for the systematic improvement of the solution accuracy to a desired degree. In SDA, the transformed equations are discretized using MoM, yielding a homogeneous system of equations to determine the propagation constant and the amplitude of current distributions from which the characteristic impedance is derived.

For metallic strip problems, the Fourier transform is performed along the direction parallel to the substrate interface and perpendicular to the strip. The first step is the formulation of the integral equation that correlates the E-field and the current distribution J along the strip and the application of the boundary conditions for E- and H-fields. Then, the Fourier transform is applied over E and J and the MoM technique produces a system of algebraic equations that can be solved. Different choices of expansion and testing functions have been discussed in [42]. SDA is applicable to most planar transmission lines (microstrips, finlines, CPWs) [43–48], microstrip antennas and arrays [49, 50], interconnects [51], junctions [52], dielectric waveguides [53], resonators of planar configurations [54], and micromachined devices [55] on single or multilayered dielectric substrates. This method requires significant analytical preprocessing, something that improves its numerical efficiency, but also restricts its applicability especially for structures with finite conductivity strips and arbitrary dielectric discontinuities.

Mode Matching Technique

This technique is usually applied to the analysis of waveguiding/packaging discontinuities that involve numerous field modes. The fields on both sides of the discontinuity are expanded in summations of the modes in the respective regions with unknown coefficients [56] and appropriate continuity conditions are imposed at the interface to yield a system of equations. As an example, to analyze a waveguide step discontinuity with TE_{n0} excitation, E_y and H_x fields are written as the superposition of the modal functions $\phi_{an}(x)$ and $\phi_{bn}(x)$ for $n = 1, 2, \ldots$, respectively for the left waveguide (waveguide A) and the right waveguide (waveguide B), as it is displayed in Fig. 8.11. Both of these fields should be continuous at the interface $z = 0$. Thus:

$$\sum_{n=1}^{\infty}\left(A_n^+ + A_n^-\right)\phi_{an} = \begin{cases} \sum_{n=1}^{\infty}\left(B_n^+ + B_n^-\right)\phi_{bn}, & 0 < x < b \\ 0, & b < x < a \end{cases} \tag{8.28}$$

$$\sum_{n=1}^{\infty}\left(A_n^+ - A_n^-\right)Y_{an}\phi_{an} = \begin{cases} \sum_{n=1}^{\infty}\left(B_n^+ - B_n^-\right)Y_{bn}\phi_{bn}, & 0 < x < b \\ 0, & b < x < a \end{cases} \tag{8.29}$$

where (+) and (–) indicate the modal waves propagating to the positive and negative z direction and Y_{an}, Y_{bn} are the mode impedances. Sampling these equations with ϕ_{bm} (m-mode in waveguide B) and making use of the mode orthogonality in waveguide B,

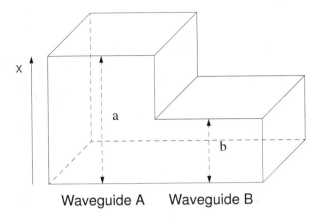

FIGURE 8.11 Rectangular waveguide discontinuity.

$$\sum_{n=1}^{\infty} H_{nm}\left(A_n^+ + A_n^-\right) = B_m^+ + B_m^- \tag{8.30}$$

$$\sum_{n=1}^{\infty} H_{nm} Y_{an}\left(A_n^+ - A_n^-\right) = Y_{bm}\left(B_m^+ - B_m^-\right) \tag{8.31}$$

with: $H_{nm} = \int_0^b \phi_{an}(x)\phi_{bm}(x)dx$. Similarly, sampling Eqs. (8.28) and (8.29) with ϕ_{am} (m-mode in waveguide A) and making use of mode orthogonality in waveguide A, we have

$$A_m^+ + A_m^- = \sum_{n=1}^{\infty} H_{mn}\left(B_n^+ + B_n^-\right) \tag{8.32}$$

$$Y_{am}\left(A_m^+ - A_m^-\right) = \sum_{n=1}^{\infty} H_{mn} Y_{an}\left(B_n^+ - B_n^-\right) \tag{8.33}$$

Assuming that the structure is excited through A_n^+ terms, the calculation of A_n^- (reflected modes in A) and B_m^+ (transmitted modes in B) will provide the scattering parameters for the analyzed structure through a procedure that involves matrix inversions.

The foundation of this technique is the expansion of an electromagnetic field in terms of an infinite series of normal modes. Because a computer's capacity for numerical calculation is finite, these summations have to be truncated, something that could lead to incorrect solutions if not performed efficiently. The main criterion for this truncation is the convergence of the summation. A natural way to check it is to plot the numerical values of some desired parameters versus the number of retained terms. The truncation is considered sufficient when the change in the parameters is smaller than prespecified criteria. The procedure becomes more complicated where there is a need for the truncation of two or more infinite series (bifurcated waveguide, step junction). The numerical results appear to converge to different values depending on the manner of the truncation, a phenomenon that is called relative convergence. It has been found that relative convergence is related to the violation of field distributions at the edge of a conductor at the boundary [9] and to the ill-conditioned situation of the linear system of the computation process [57]. Thus, either the edge condition or the condition number of the linear system can be used as a criterion to ensure the validity of modal analyses. Another common criterion is to plot the field distributions on both sides of the boundary and observe their matching conditions. The mode matching method has been applied to analyze various discontinuities in waveguides [58, 59], finlines [60, 61], microstrip lines [62, 63], and coplanar waveguides [64, 65]. In addition, it is used for closed-region scattering geometries involving a discrete set of modes, such as E-plane filters [66, 67], waveguide impedance transformers [68], power dividers [69], and microstrip filters [70]. Moreover, this technique has been implemented for the solution of eigenvalue problems, such as the resonant frequency of a cavity [71] and the performance of evanescent mode filters [72], since it can efficiently model both evanescent and propagating modes.

PDE-Based Techniques FDTD, MRTD, FEM

In contrast to the previous techniques, numerical methods based on the partial differential equation (PDE) solutions of Maxwell's equations yield either sparse matrices (frequency domain, finite-element methods) or no matrices at all (time domain, finite-difference or finite-volume methods). In addition, specifying a new geometry is reduced to a problem of mesh generation only. Thus, PDE solvers could provide a framework for a space/time (frequency) microscope permitting the EM designer to visualize with submicron/subpicosecond resolution the dynamics of electromagnetic wave phenomena propagating at light speed within proposed geometries.

Finite-Difference Time Domain (FDTD) Technique

The Finite-Difference Time-Domain (FDTD) [10–13] is an explicit solution method for Maxwell's time-dependent curl equations. It is based upon volumetric sampling of the electric and magnetic field distribution within and around the structure of interest over a period of time. The sampling is set below the Nyquist limit and typically more than 10 samples per wavelength are required. The time step has to satisfy a stability condition. For simulations of open geometries, absorbing boundary conditions (ABC) are employed at the outer grid truncation planes in order to reduce spurious numerical reflections from the grid termination.

In 1966, Yee [73] suggested the solution of the first-order Maxwell equations in time and space instead of solving the second order wave equation. In this way, the solution is more robust and more accurate for a wider class of structures. In Yee's discretization cell, E- and H-fields are interlaced by half space and time gridding steps, as shown in Fig. 8.12. The spatial displacement is very useful in specifying field boundary conditions and singularities and leads to finite-difference expressions for the space derivatives that are central in nature and second-order accurate. The time displacement (leapfrog) is fully explicit, completely avoiding the problems involved with simultaneous equations and matrix inversion. The

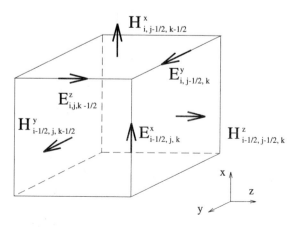

FIGURE 8.12 FDTD Yee cell.

resulting scheme is non-dissipative; numerical wave modes propagating in the mesh do not spuriously decay due to a nonphysical artifact of the time-stepping algorithm. Denoting any function u of space and time evaluated at a discrete point in the grid and at a discrete point in time as $u(i\Delta x, j\Delta y, k\Delta z, l\Delta t) = {}_l u_{i,j,k}$, where Δt is the time step and Δx, Δy, Δz the cell size along the x-, y- and z-direction, the first partial space derivative of u in the x-direction and the first time derivative of u are approximated with the following central differences, respectively

$$\frac{\partial u}{\partial x}\left(i\Delta x, j\Delta y, k\Delta z, l\Delta t\right) = \frac{{}_l u_{i+1/2,j,k} - {}_l u_{i-1/2,j,k}}{\Delta x} + O\left[\left(\Delta x\right)^2\right]$$

$$\frac{\partial u}{\partial t}\left(i\Delta x, j\Delta y, k\Delta z, l\Delta t\right) = \frac{{}_{l+1/2} u_{i,j,k} - {}_{l-1/2} u_{i,j,k}}{\Delta t} + O\left[\left(\Delta t\right)^2\right]$$

(8.34)

Applying the above notation, the FDTD equations are derived for all field components. For example,

$$\begin{aligned}
{}_{l+0.5} H^x_{i,j-0.5,k-0.5} &= \left(\frac{1 - \dfrac{\rho'_{i,j,k}\Delta t}{2\mu_{i,j,k}}}{1 + \dfrac{\rho'_{i,j,k}\Delta t}{2\mu_{i,j,k}}}\right) {}_{l-0.5} H^x_{i,j-0.5,k-0.5} \\[2em]
&+ \left(\frac{\dfrac{\Delta t}{\mu_{i,j,k}}}{1 + \dfrac{\rho'_{i,j,k}\Delta t}{2\mu_{i,j,k}}}\right)\left(\frac{{}_l E^y_{i,j-0.5,k} - {}_l E^y_{i,j-0.5,k-1}}{\Delta z} - \frac{{}_l E^z_{i,j,k-0.5} - {}_l E^z_{i,j-1,k-0.5}}{\Delta y}\right)
\end{aligned}$$

(8.35)

where $\rho'_{i,j,k}$ is the magnetic loss coefficient for the (i, j, k) cell. It can be observed that a new value of a field vector component at any space lattice point depends only on its previous value and the previous values of the components of the other field vectors at adjacent points. Therefore, at any given time step, the value of a field vector component at p different lattice points can be calculated simultaneously if p parallel processors are employed, demonstrating that the FDTD algorithm is highly parallelizable. Holland [74] suggested an exponential time stepping to model the exponential decay of propagating waves in certain highly lossy media that the standard Yee time-stepping algorithm fails to describe. Stability analysis [75] has shown that the upper bound for the FDTD time step for a homogeneous region of space (ϵ_r, μ_r) is given by

$$\Delta t \leq \frac{\sqrt{\epsilon_r \mu_r}}{c\sqrt{\dfrac{1}{\left(\Delta x\right)^2} + \dfrac{1}{\left(\Delta y\right)^2} + \dfrac{1}{\left(\Delta z\right)^2}}} \left(\textit{3D simulations}\right), \quad \Delta t \leq \frac{\sqrt{\epsilon_r \mu_r}}{c\sqrt{\dfrac{1}{\left(\Delta x\right)^2} + \dfrac{1}{\left(\Delta y\right)^2}}} \left(\textit{2D simulations}\right)$$

Lower values of upper bounds are used in case a highly lossy material or a variable grid is employed. Discretization with at least 10 to 20 cells/wavelength almost guarantees that the FDTD algorithm will have satisfactory dispersion characteristics (phase error smaller than $5°/\lambda$ for time step close to the upper bound value).

The computational domain must be large enough to enclose the structure of interest, and a suitable absorbing boundary condition (ABC) on the outer perimeter of the domain must be used to simulate its extension to infinity to minimize numerical reflections for a wide range of incidence angles and frequencies. Central differences cannot be implemented at the outermost lattice planes, since by definition no information exists concerning the fields at points one-half space cell outside of these planes. The perfectly matched layer (PML) ABC, introduced in 2D by Berenger in 1994 [76] and extended to 3D by Katz et al. [77], provides numerical reflection comparable to the reflection of anechoic chambers with values −40 dB lower than the previous absorbers. A new ABC based on Green's functions that absorbs efficiently propagating and evanescent modes has also been demonstrated [78, 79] for waveguide and RF packaging structures.

An incoming plane wave source [73] is very useful in modeling radar scattering problems, since in most cases of this type the target of interest is in the near field of the radiating antenna, and the incident illumination can be considered to be a plane wave. The hard source [80] is another common FDTD source implementation. It is set up simply by superimposing a desired time function onto specific electric or magnetic field components in the FDTD space lattice that are regularly updated by the FDTD equations. Collinear arrays of hard-source field vector components in 3D can be useful for exciting waveguides and strip lines. In the FDTD simulations of microstrip and stripline structures, the Gaussian pulse (nonzero DC content) is used as the excitation of the microstrip and stripline structures. The Gabor function

$$s(t) = e^{-\left((t-t_o)/(pw)\right)^2} \sin(wt) \tag{8.36}$$

where $pw = 2 \cdot \frac{\sqrt{6}}{\pi(f_{max} - f_{min})}$, $t_o = 2pw$, $w = \pi(f_{min} + f_{max})$, is used as the excitation of the waveguide structures, since it has zero DC content. By modifying the parameters pw and w, the frequency spectrum of the Gabor function can be practically restricted to the interval $[f_{min}, f_{max}]$. As a result, the envelope of the Gabor function represents a Gaussian function in both time and frequency domain. Monochromatic simulations are performed through the use of continuous-wave (sinusoidal) excitations.

It is very common, especially for high-speed circuit structures, to use a cell size Δ that is dictated by the very fine dimensions of the circuit and is almost always much finer than needed to resolve the smallest spectral wavelength propagating in the circuit. As a result, with the time step Δt bound to Δ by numerical stability considerations, FDTD simulations have to run for tens of thousands of time steps in order to fully evolve the impulse responses needed for calculating impedances, S-parameters, or resonant frequencies. One popular way to avoid virtually prohibitive execution time has been to apply contemporary analysis techniques from the discipline of digital signal processing and spectrum estimation. The strategy is to extrapolate the electromagnetic field time waveform by 10:1 or more beyond the actual FDTD time window, allowing a very good estimate of the complete system response with 90% or greater reduction in computation time. This extrapolation can be performed using forward-backward predictors [81] or autoregressive (AR) models [82].

The FDTD technique has found numerous applications in modeling microwave devices such as waveguides, resonators, transmissions lines, vias, antennas, and active and passive elements. In 1985, DePourcq [83] used FDTD to analyze various 3D waveguide devices. Navarro et al. [84] investigated rectangular, circular, and T junctions in square coaxial waveguides and narrow-wall, multiple-slot couplers. Wang et al. [85] studied the Q factors of resonators using FDTD. Liang et al. [86] used FDTD to analyze coplanar waveguides and slotlines and Sheen et al. [87] presented FDTD results for various microstrip structures including a rectangular patch antenna, a low-pass filter, and a branch-line coupler. Cangellaris estimated the effect of the staircasing approximation of conductors of arbitrary orientation in [88].

The characterization of interconnect transitions in multichip and microwave circuit modules has also been investigated using FDTD. Lam et al. [89] used a nonuniform mesh to model microstrip-to-via-to-stripline connections. Piket-May et al. [90] studied pulse propagation and crosstalk in a computer module with more than ten metal-dielectric-metal layers and numerous vias. Luebbers et al. [91] and Shlager and Smith [92] developed and described in detail efficient three-dimensional, time domain, near-to-far-field transformations. In 1990, Maloney et al. [93] presented accurate results for the radiation from rotationally symmetric simple antennas such as cylindrical and conical monopoles, while Luebbers and coworkers [94, 95] presented mutual coupling and gain computations for a pair of wire dipoles and Tirkas and Balanis [96] modeled three-dimensional horn antennas. Uehara and Kagoshima [97] analyzed microstrip phased-array antennas and Jensen and Rahmat-Samii [98] presented results for the input impedance and gain of monopoles, planar inverted-F antennas (PIFAs) and loop antennas on handheld transceivers. In addition, Taflove [99] used FDTD to model scattering and compute near/far fields and radar cross-section (RCS) for 2D and 3D structures. Britt [100] calculated the RCS of both two- and three-dimensional perfectly conducting and dielectric scatterers.

Another area of FDTD applications is active and passive device modeling. Two different approaches are used. In the first, analytical device models are coupled directly with FDTD. In the second, lumped-element subgrid models are used with the device behavior being determined by other software, something that may be preferable in the modeling of active devices with complicated equivalent circuits. In 1992, Sui et al. [101] reported a 2D FDTD model with lumped circuit elements, including nonlinear devices, such as diodes and transistors and this approach was extended to 3D by Piket-May et al. [90] and Ciampollini et al. [102]. Kuo et al. [103] presented a large-signal analysis of packaged nonlinear active microwave circuits. Alsunaidi et al. [104] developed an active device model that couples the Yee update equations with the solution of the current continuity equation, the energy-conservation equation, and the momentum-conservation equations. Thomas et al. [105] developed an approach for coupling SPICE lumped elements into the FDTD method.

Transmission Line Matrix Method (TLM)

The TLM method [106–108] is similar to FDTD. The main difference is that the electromagnetic problem is analyzed through the use of a three-dimensional equivalent network problem [109]. It is a very versatile time domain technique and discretizes the computational domain using cubic cells with a period Δl. Boundaries corresponding to perfect electric (magnetic) conductors are represented by short-circuited (open-circuited) parallel nodes on the boundary. Variations of dielectric and diamagnetic constant [110] are introduced by adding short-circuited series stubs of length $\Delta l/2$ at the series (H-field) nodes and open-circuited $\Delta l/2$ stubs at the shunt (E-field) nodes. Losses can be introduced by resistively loading the shunt nodes. After the time domain response is obtained, the frequency response is calculated using the Fourier transform. Due to the introduction of periodic lattice structures, a typical passband-stopband phenomenon appears in the frequency domain data. The frequency range must be below the upper bound of the lowest passband and is determined by the mesh size Δl. The TLM technique has been used in the analysis of waveguiding structures [111] making use of the Diakoptics Theorem [112] and has been extensively compared to the FDTD technique [113]. Effective numerical absorbers [114, 115] including the FDTD-popular PML absorber have been derived and implemented in the modeling of radiating structures [116]. In addition, TLM has been employed in the analysis of Bondwire Packaging [117] and MEMS switches [118].

Finite Element Method (FEM)

In the finite element method [14–20], instead of partial differential equations with boundary conditions, corresponding functionals (e.g., power) are set up and variational expressions are applied to each cell (element) of the area of interest. Most of the time the elements are rectangles or triangles for two-dimensional problems and parallelepiped (bricks) or tetrahedra for three-dimensional problems, something that allows for the efficient representation of most arbitrary shapes.

Assume that the two-dimensional (*x-y*) Laplace equation is to be solved

$$\frac{\partial^2 \phi}{\partial x^2} + \frac{\partial^2 \phi}{\partial y^2} = 0 \tag{8.37}$$

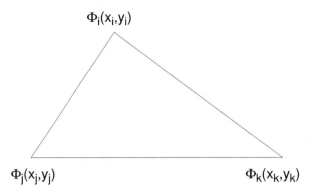

FIGURE 8.13   FEM triangular (2D) element.

The solution is equivalent to the minimization of the functional

$$I(\phi) = \langle \phi, \nabla^2 \phi \rangle = \iint_S \phi \left(\frac{\partial^2 \phi}{\partial x^2} + \frac{\partial^2 \phi}{\partial y^2} \right) dx\, dy = -\iint_S \left[\left(\frac{\partial \phi}{\partial x} \right)^2 + \left(\frac{\partial \phi}{\partial y} \right)^2 \right] dx\, dy. \qquad (8.38)$$

The last integral is the superposition of each element's contribution. In each 2D element ϕ can be approximated as polynomial of variables x and y. For example, for a triangular element p (Fig. 8.13),

$$\phi = a + a_x x + a_y y, \qquad (8.39)$$

where the constants a, a_x, a_y depend on the ϕ values at the three vertices of the triangle

$$\phi_p = a + a_x x_p + a_y y_p, \qquad (8.40)$$

where $p = i, j, k$ are the three triangle vertices. Due to the first derivatives of Eq. (8.39), only a_x, a_y are needed for the calculation of $I(\phi)$, thus

$$\begin{bmatrix} a_x \\ a_y \end{bmatrix} = A \begin{bmatrix} \phi_i \\ \phi_j \\ \phi_k \end{bmatrix} \qquad (8.41)$$

For each (i, j, k) element, $I(\phi)$ gets the value

$$I_{i,j,k}(\phi) = [\phi_i, \phi_j, \phi_k] A^T A \begin{bmatrix} \phi_i \\ \phi_j \\ \phi_k \end{bmatrix} \cdot |\Delta S|, \qquad (8.42)$$

where A^T is the transpose of A and ΔS is the area of the triangle (element) calculated by

$$\Delta S = \frac{1}{2} \begin{vmatrix} 1 & x_i & y_i \\ 1 & x_j & y_j \\ 1 & x_k & y_k \end{vmatrix}. \qquad (8.43)$$

The Rayleigh-Ritz technique is used for the minimization of $I_{i,j,k}$

$$\frac{\partial I_{i,j,k}}{\partial \phi_i} = \frac{\partial I_{i,j,k}}{\partial \phi_j} = \frac{\partial I_{i,j,k}}{\partial \phi_k} = 0 \tag{8.44}$$

As a result, for each (i, j, k) element

$$A^T A \begin{bmatrix} \phi_i \\ \phi_j \\ \phi_k \end{bmatrix} = 0 \tag{8.45}$$

After iterating this procedure to all elements of the computational domain S and using a connection matrix to account for points that are vertices common to more than one element, the following matrix equation is derived

$$\mathbf{B} \begin{bmatrix} \phi_1 \\ \phi_2 \\ \vdots \\ \phi_N \end{bmatrix} = 0 \tag{8.46}$$

After plugging in the known values of the ϕ_i's that are located on the boundaries, the rest of the ϕ's are calculated through the inversion of matrix $\mathbf{B}$. In this way, the potentials of all interior points can be given by Eq. (8.39).

Various element forms [119–123] have been used in order to minimize the memory requirements and facilitate the gridding procedure and the modeling of boundary conditions (PECs, dielectric interfaces). The effect of discretization error on the numerical dispersion has been extensively studied and gridding guidelines have been derived in References 124 and 125. In addition, the analysis of radiating (antennas) and scattering problems have led to the development of numerical absorbers [126–128] with very low numerical reflection coefficients. Due to the shape of the finite elements, the FEM technique can accurately represent very complex geometries and is one of the most popular techniques for scattering problems [129], discontinuities and transitions [130], packaging [131], interconnects [132], and MMIC modeling [133].

The boundary element technique (BE) [134, 135] is a combination of the boundary integral equation and a finite-element discretization applied to the boundary. In essence, it is a form of the integral equation — MoM approach discussed previously. The wave equation for the volume is converted to the surface integral equation through Green's identity. The surface integrals are discretized into N elements, and the evaluation is performed for each element after E- and H-fields are approximated by polynomials. Due to the reduction of the number of dimensions, there is a significant reduction in memory and CPU time requirements. The BE technique has been utilized in the analysis of cavities [136], of planar layered media [137], and in the incorporation of lumped elements in the FEM analysis [138].

Hybrid Techniques

As has become clear from the previous discussion, all numerical techniques have advantages and disadvantages depending on the geometry to be modeled. Integral equation techniques allow for the quick and efficient modeling of radiation phenomena, but derivation of Green's function for complex structures is tedious. The MM method is more appropriate for waveguiding structures where modes are easily determined. The FDTD (and TLM) technique is quite general and requires no preprocessing, though it must often to be run for medium to large execution times. The FEM technique is adaptive due to the shape of the elements, but gridding and functional optimization demands significant computational effort. Thus, there have been numerous efforts for the development of hybrid simulation approaches that use different techniques for different subgeometries and utilize connection relationships for the areas of numerical interfaces. Jin [139] and Gedney [140] proposed a hybrid FEM/MoM method for the modeling of wave scattering by 3D apertures and wave diffraction in gratings. Wu [142] and Monorchio [143] suggested an FEM/FDTD approach for the multifrequency modeling of complex geometries. The MM technique has been coupled with Integral Equation [144], spectral domain [145], and FEM [146] to analyze complicated waveguiding problems including inductive loading and wave scattering. Lindenmeier [147] introduced a hybrid TLM/MIE for thin wire modeling and Pierantoni [148] analyzed numerical aspects of MoM-FDTD and TLM-Integral Equation used in EMC Modeling.

Wavelets: A Memory-Efficient Adaptive Approach?

The term "wavelet" [149–153] has a very broad meaning, ranging from singular integral operators in harmonic approach to sub-band coding algorithms in signal processing, from coherent states in quantum analysis to spline analysis in approximation theory, from multi-resolution transform in computer vision to a multilevel approach in the numerical solution of partial differential equations, and so on. Most of the time wavelets could be considered mathematical tools for waveform representations and segmentations, time-frequency analysis, and fast and efficient algorithms for easy implementation in both time and frequency domains. One of the most important characteristics of expansion to scaling and wavelet functions is the time (domain)-frequency (Fourier-transformed domain) localization. Another very salient feature of these new expansions is that the entire set of basis functions is generated by the dilation and shifting of a single function, called the mother wavelet. The standard approach in ideal lowpass ("scaling," Fig. 8.14) and bandpass ("wavelet," Fig. 8.15) filtering for separating an analog signal into different frequency bands emphasizes the importance of time localization. Multiresolution Analysis (MRA), introduced by Mallat and Meyer [154, 155], provides a very powerful tool for the construction of wavelets and implementation of the wavelet decomposition/reconstruction algorithms. In the case of cardinal B-splines [156], an orthonormalization process is used to produce an orthonormal scaling function and, hence, its corresponding orthonormal wavelet by a suitable modification of the two-scale

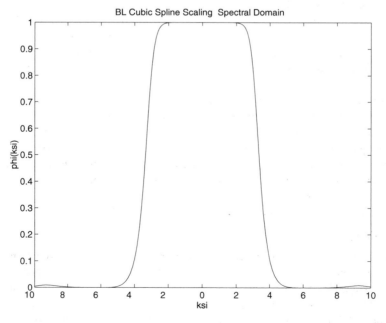

FIGURE 8.14   Battle-Lemarie scaling-spectral ("lowpass") domain.

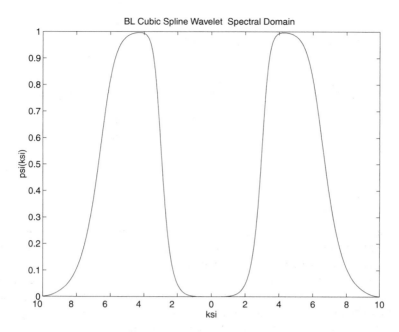

FIGURE 8.15 Battle-Lemarie wavelet-spectral ("bandpass") domain.

sequence. The orthonormalization process was introduced by Schweinler and Wigner [157] and the resulting wavelets are the Battle-Lemarie wavelets, obtained independently by Battle [158] and Lemarie [159] using different methods. The only orthonormal wavelet that is symmetric or antisymmetric and has compact support (to give finite decomposition and reconstruction series) is the Haar [160] wavelet. Nevertheless, these wavelets exhibit poor time-frequency localization. Another set of orthonormal basis is the Daubechies wavelets [161]. At present, MRA has been applied to alleviate the numerical disadvantages mainly of Integral Equation and FDTD methods, though preliminary efforts for FEM are currently being investigated.

It is well known that the Integral Equation method described in previous sections offers a straightforward and efficient numerical solution when applied to small- to medium-scale problems. Difficulties arise when the complexity of the geometry and subsequently the number of the unknowns increases, resulting in very large matrices. All conventional basis functions traditionally used in MoM generate full moment matrices. The computational problems associated with the storage and manipulation of large, densely populated matrices easily rules out the practicality of the integral equation techniques. The potential application of wavelet theory in the numerical solution of integral equations led to the finding that wavelet expansion of certain types of integral operators generates highly sparse linear systems [162]. This proposition was used [163–165] in the moment method formulation of one-dimensional electromagnetic scattering problems and by in the analysis of integrated millimeter-wave and submillimeter-wave waveguides with a Battle-Lemarie orthonormal basis [166]. Sparsity results above 90% allowed for the accurate modeling of structures that could not be analyzed with IE using conventional expansions.

As far as it concerns FDTD, despite its numerous applications, many practical geometries, especially in microwave and millimeter-wave integrated circuits (MMIC), packaging, interconnects, subnanosecond digital electronic circuits [such as multichip modules (MCM)], and antennas used in wireless and microwave communication systems, have been left untreated due to their complexity and the inability of the existing techniques to deal with requirements for large size and high resolution. Krumpholz has shown that Yee's FDTD scheme can be derived by applying the method of moments for the discretization of Maxwell's equations using pulse basis functions for the expansion of the unknown fields. The use of scaling and wavelet functions as a complete expansion basis of the fields demonstrates that MultiResolution Time Domain (MRTD) [167] schemes are generalizations of Yee's FDTD and can extend this

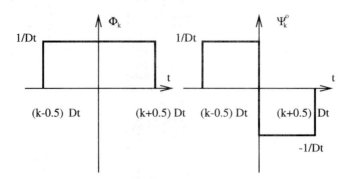

FIGURE 8.16 Haar expansion basis.

technique's capabilities by improving computational efficiency and substantially reducing computer resources. In an MRTD scheme the fields are represented by a twofold expansion in scaling and wavelet functions with respect to time/space. Scaling functions ("lowpass") guarantee a correct modeling of smoothly varying fields. In regions characterized by strong field variations or field singularities, higher resolution is enhanced by incorporating wavelets in the field expansions. The major advantage of the use of MRTD in the time domain is the capability to develop time and space adaptive grids through the thresholding of the wavelet coefficients for each time step throughout the grid. MRTD schemes based on cubic spline Battle-Lemarie scaling and wavelet functions have been used for the derivation of time/space-adaptive schemes [168, 169] in real time. They have been successfully applied to the simulation of 2.5D transmission lines [170] and 3D dielectric cavities [171], filters [172], and mixers [173] offering economies in memory and execution time of order(s) of magnitude with respect to FDTD. Dispersion analysis [174] shows the capability of excellent accuracy with up to 2 points/wavelength (Nyquist limit)! Nevertheless, the functions of this family do not have compact support; thus the MRTD schemes have to be truncated with respect to space, something that requires the use of image theory for the modeling of hard boundaries (e.g., PECs). Thus, specific problems may require the use of functions with compact support (e.g., Haar, Fig. 8.16) especially for the approximation of time derivatives. In this way, the modeling of boundary conditions is most straightforward, though the computational economies are not so dramatic [175–177].

Conclusions

This chapter has briefly presented various numerical techniques that are commonly used for the analysis and design of RF and microwave circuits. Their fundamental features as well as their advantages and disadvantages have been discussed and representative references have been reported. It must be emphasized that there is no numerical scheme that can achieve optimal performance for all types of structures. Thus, the hybridization of these techniques or the implementation of novel approaches (e.g., wavelets), successfully applied in other research areas, would be better candidates for the generalized, efficient, and accurate modeling of modern complex devices used in telecommunication, radar, and computing applications.

References

1. T. Itoh, *Numerical Techniques for Microwave and Millimeter-Wave Passive Structures*, John Wiley & Sons, New York, 1989.
2. R.C. Booton, *Computational Methods for Electromagnetics and Microwaves*, John Wiley & Sons, New York, 1992.
3. R.F. Harrington, *Field Computation by Moment Methods*, The Macmillan Company, New York, 1968.
4. R.C. Hansen, *Moment Methods in Antennas and Scattering*, Artech House, Norwood, MA, 1990.

5. J.H. Wang, *Generalized Moment Methods in Electromagnetics,* John Wiley & Sons, New York, 1991.
6. R. Bancroft, *Understanding Electromagnetic Scattering Using the Moment Method: A Practical Approach,* Artech House, Norwood, MA, 1996.
7. A.F. Peterson, S.L. Ray, and R. Mittra, *Computational Methods for EM,* IEEE Press, 1998.
8. C.R. Scott, *The Spectral Domain Method in Electromagnetics,* Artech House, Norwood, MA, 1989.
9. R. Mittra and W.W. Lee, *Analytical Techniques in the Theory of Guided Waves,* Macmillan, New York, 1971.
10. K.S. Kunz and R.J. Luebbers, *The Finite Difference Time-Domain Method for Electromagnetics,* CRC Press, Boca Raton, FL, 1993.
11. A. Taflove, *Computational Electrodynamics: The Finite-Difference Time-Domain Technique,* Artech House, Norwood, MA, 1995.
12. A. Taflove, *Advances in Computational Electrodynamics,* Artech House, Norwood, MA, 1998.
13. T. Itoh and B. Housmand, *Time-Domain Methods for Microwave Structures: Analysis and Design,* IEEE Press, 1998.
14. G. Strang and G.J. Fix, *An Analysis of the Finite Element Method,* Prentice-Hall, Englewood Cliffs, NJ, 1973.
15. P.P. Silvester and R.L. Ferrari, *Finite Elements for Electrical Engineers,* Cambridge University Press, Cambridge, 1983.
16. J.N. Reddy, *An Introduction to the Finite Element Method,* McGraw Hill, New York, 1984.
17. J.-C. Sabonnadiere and J.-L. Coulomb, *Finite Element Methods in CAD: Electric and Magnetic Fields,* Springer-Verlag, New York, 1987.
18. O.C. Zienkiewicz and R.L. Taylor, *The Finite Element Method,* McGraw Hill, London, 1988.
19. T. Itoh, G. Pelosi, and P.P. Silvester, *Finite Element Software for Microwave Engineering,* John Wiley and Sons, New York, 1996.
20. J.L. Volakis, A Chaterjee, and L.C. Kempel, *Finite Elements Method for Electromagnetics,* IEEE Press, 1998.
21. A.W. Glisson and D.R. Wilton, Simple and efficient methods for problems of radiation and scattering from surfaces, *IEEE Trans. Antennas and Propagation,* 28, 593–603, 1980.
22. T.K. Sarkar, A.R. Djordjevic, and E. Arvas, On the choice of expansion and weighting functions in the numerical solution of operator equations, *IEEE Trans. Antennas and Propagation,* 33, 988–996, 1985.
23. J.R. Mossig and F.E. Gardiol, Analytical and numerical techniques in the Green's function treatment of microstrip antennas and scatterers, *IEE Proc.,* Part H, 130, 175–182, 1983.
24. A.J. Poggio and E.K. Miller, Integral equation solutions of 3D scattering problems, in *Computer Techniques for Electromagnetics,* R. Mittra, Ed., Hemisphere, New York, 1987.
25. A.F. Peterson and P.W. Klock, An improved MFIE formulation for TE-wave scattering from lossy, inhomogeneous dielectric cylinders, *IEEE Trans. Antennas and Propagation,* 25, 518–524, 1987.
26. F. Ling, D. Jiao, and J.M. Jin, Efficient EM modeling of microstrip structures in multilayer media, *IEEE Trans. Microwave Theory Tech.,* 47, 1810–1818, 1999.
27. D.M. Pozar, Input impedance and mutual coupling of rectangular microstrip antennas, *IEEE Trans. Antennas and Propagation,* 30, 1191–1196, 1982.
28. W.-T. Chen and H.-R. Chuang, Numerical computation of the EM coupling between a circular loop antenna and a full-scale human-body model, *IEEE Trans. Microwave Theory Tech.,* 1516–1520, 1998.
29. G. Athanasoulias and N.K. Uzunoglu, An accurate and efficient entire-domain basis Galerkin's method for the integral equation analysis of integrated rectangular dielectric waveguides, *IEEE Trans. Microwave Theory Tech.,* 43, 2794–2804, 1995.
30. M. Swaminathan, T.K. Sarkar, and A.T. Adams, Computation of TM and TE modes in waveguides based on surface integral formulation, *IEEE Trans. Microwave Theory Tech.,* 40, 285–297, 1992.
31. S.J. Polychronopoulos and N.K. Uzunoglu, Propagation and coupling properties of integrated optical waveguides — an integral equation formulation, *IEEE Trans. Microwave Theory Tech.,* 44, 641–650, 1996.

32. P. Cottis and N. Uzunoglu, Integral equation approach for the analysis of anisotropic channel waveguides, *J. Opt. Soc. A. A,* 8, 4, 608–614, 1991.

33. P. Guillot, P. Couffignal, H. Baudrand, and B. Theron, Improvement in calculation of some surface integrals: application to junction characterization in cavity filter design, *IEEE Trans. Microwave Theory Tech.,* 41, 2156–2160, 1993.

34. T.E. van Deventer, L.P.B. Katehi, and A.C. Cangellaris, Analysis of conductor losses in high-speed interconnects, *IEEE Trans. Microwave Theory Tech.,* 42, 78–83, 1994.

35. A.W. Mathis and A.F. Peterson, Efficient electromagnetic analysis of a doubly infinite array of rectangular apertures, *IEEE Trans. Microwave Theory Tech.,* 46–54, 1998.

36. S-G. Hsu and R.-B. Wu, Full-wave characterization of a through hole via in multi-layered packaging, *IEEE Trans. Microwave Theory Tech.,* 1073–1081, 1995.

37. A.M. Rajeek and A. Chakraborty, Analysis of a wide compound slot-coupled parallel waveguide coupler and radiator, *IEEE Trans. Microwave Theory Tech.,* 43, 802–809, 1995.

38. G.V. Eleftheriades, J.R. Mosig, and M. Guglielmi, A fast integral equation technique for shielded planar circuits defined on nonuniform meshes, *IEEE Trans. Microwave Theory, Tech.,* 44, 2293–2296, 1996.

39. E. Yamashita and R. Mittra, Variational method for the analysis of microstrip line, *IEEE Trans. Microwave Theory, Tech.,* 16, 251–256, 1968.

40. E.J. Denlinger, A frequency dependent solution for microstrip transmission lines, *IEEE Trans. Microwave Theory, Tech.,* 19, 30–39, 1971.

41. T. Itoh and R. Mittra, Spectral-domain approach for calculating the dispersion characteristics of microstrip lines, *IEEE Trans. Microwave Theory, Tech.,* 21, 496–499, 1973.

42. M.I. Aksun and R. Mittra, Choices of expansion and testing functions for MoM applied to a class of EM problems, *IEEE Trans. Microwave Theory, Tech.,* 41, 503–509, 1993.

43. R.H. Jansen, Unified user-oriented computation of shielded, covered and open planar microwave and millimeter wave transmission-line characteristics, *IEE J. Microwave Opt. Acoust.,* 3, 14–22, 1979.

44. T. Itoh, Spectral domain immitance approach for dispersion characteristics of generalized printed transmission lines, *IEEE Trans. Microwave Theory Tech.,* 28, 733–736, 1980.

45. J. Sercu, N. Fache, F. Libbrecht, and D. De Zutter, Full-wave space-domain analysis of open microstrip discontinuities including the singular current-edge behavior, *IEEE Trans. Microwave Theory Tech.,* 41, 1581–1588, 1993.

46. F. Olyslager, D. De Zutter, and K. Blomme, Rigorous full-wave analysis of propagation characteristics of general lossless and lossy multiconductor transmission lines in multilayered media, *IEEE Trans. Microwave Theory Tech.,* 41, 79–88, 1993.

47. K.K.M. Cheng and J.K.A. Everard, A new technique for the quasi-TEM analysis of conductor-backed coplanar waveguide structures, *IEEE Trans. Microwave Theory Tech.,* 41, 1589–1592, 1993.

48. N. Gupta and M. Singh, Investigation of periodic structures in a fin line: A space-spectral domain approach, *IEEE Trans. Microwave Theory Tech.,* 43, 2708–2710, 1995.

49. Y. Imaizumi, M. Shinagawa, and H. Ogawa, Electric field distribution measurement of microstrip antennas and arrays using electro-optic sampling, *IEEE Trans. Microwave Theory Tech.,* 43, 2402–2407, 1995.

50. Y.-D. Lin, J.-W. Sheen, and C.-K.C. Tzuang, Analysis and design of feeding structures for microstrip leaky wave antenna, *IEEE Trans. Microwave Theory Tech.,* 44, 1540–1547, 1996.

51. P. Petre and M. Swaminathan, Spectral domain technique using surface wave excitation for the analysis of interconnects, *IEEE Trans. Microwave Theory Tech.,* 42, 1744–1749, 1994.

52. B.L. Ooi, M.S. Leong, P.S. Kooi, and T.S. Yeo, Enhancements of the spectral-domain approach for analysis of microstrip Y-junction, *IEEE Trans. Microwave Theory Tech.,* 45, 1800–1805, 1997.

53. K. Sabetfakhri and L.P.B. Katehi, Analysis of general class of open dielectric waveguides by spectral-domain technique, *Proc. IEEE-MTT Symposium,* 3, 1523–1526, 1993.

54. T. Itoh, Analysis of microstrip resonators, *IEEE Trans. Microwave Theory Tech.,* MTT-22, 946–952, 1974.

55. T.M. Weller, K.J. Herrick, and L.P.B. Katehi, Quasi-static design technique for mm-wave micromachined filters with lumped elements and series stubs, *IEEE Trans. Microwave Theory Tech.,* 45, 931–938, 1997.

56. Y.C. Shih and K.G. Gray, Convergence of numerical solutions of step-type waveguide discontinuity problems by modal analysis, *Proc. IEEE-MTT Symposium,* 233–235, 1983.

57. M. Leroy, On the convergence of numerical results in modal analysis, *IEEE Trans. Antennas and Propagation,* 31, 655–659, 1983.

58. A. Wexler, Solution of waveguide discontinuities by modal analysis, *IEEE Trans. Microwave Theory Tech.,* 15, 508–517, 1967.

59. W. Wessel, T. Sieverding, and F. Arndt, Mode-matching analysis of general waveguide multiport junctions, *Proc. IEEE-MTT Symposium,* 1273–1276, 1999.

60. A.S. Omar and K. Schunemann, Transmission matrix representation of finite discontinuity, *IEEE Trans. Microwave Theory Tech.,* 33, 830–835, 1985.

61. R. Vahldieck and W.J.R. Hoefer, Finline and metal insert filters with improved passband separation and increased stopband attenuation, *IEEE Trans. Microwave Theory Tech.,* 33, 1333–1339, 1985.

62. W. Menzel and I. Wolff, A method for calculating the frequency-dependent properties of microstrip discontinuities, *IEEE Trans. Microwave Theory Tech.,* 25, 107–112, 1977.

63. T.S. Chu, T. Itoh, and Y.-C. Shih, Comparative study of mode-matching formulations for microstrip discontinuity problems, *IEEE Trans. Microwave Theory Tech.,* 33, 1018–1023, 1985.

64. F. Alessandri, G. Baini, M. Mongiardo, and R. Sorrentino, A 3D mode matching technique for the efficient analysis of coplanar MMIC discontinuities with finite metallization thickness, *IEEE Trans. Microwave Theory Tech.,* 41, 1625–1629, 1993.

65. R. Schmidt and P. Russer, Modeling of cascaded coplanar waveguide discontinuities by the mode-matching approach, *IEEE Trans. Microwave Theory Tech.,* 43, 2910–2917, 1995.

66. Y.C. Shih, Design of waveguide E-plane filters with all metal insert, *IEEE Trans. Microwave Theory Tech.,* 32, 695–704, 1984.

67. F. Arndt et al., E-plane integrated filters with improved stopband attenuation, *IEEE Trans. Microwave Theory Tech.,* 32, 1391–1394, 1984.

68. F. Arndt et al., Computer-optimized multisection transforms between rectangular waveguides of adjacent frequency bands, *IEEE Trans. Microwave Theory Tech.,* 32, 1479–1484, 1984.

69. F. Arndt et al., Optimized E-plane T-junction series power divider, *IEEE Trans. Microwave Theory Tech.,* 35, 1052–1059, 1987.

70. R. Mehran, Computer-aided design of microstrip filters considering dispersion, loss and discontinuity effects, *IEEE Trans. Microwave Theory Tech.,* 27, 239–245, 1979.

71. L. Accatino, G. Bertin, and M. Mongiardo, Elliptical cavity resonators for dual-mode narrow-band filters, *IEEE Trans. Microwave Theory Tech.,* 45, 2393–2401, 1997.

72. J. Bornemann and F. Arndt, Rigorous design of evanescent-mode E-plane finned waveguide bandpass filters, *Proc. IEEE-MTT Symposium,* 603–606, 1989.

73. K.S. Yee, Numerical solution of initial boundary value problems involving Maxwell's equations in isotropic media, *IEEE Trans. Antennas and Propagation,* 14, 302–307, 1966.

74. R. Holland, THREDE: A free-field EMP coupling and scattering code, *IEEE Trans. Nuclear Science,* 24, 2416–2421, 1977.

75. A. Taflove and M.E. Brodwin, Numerical solution of steady-state electromagnetic scattering problems using the time-dependent Maxwell's equations, *IEEE Trans. Microwave Theory Tech.,* 23, 623–630, 1975.

76. J.-P. Berenger, A perfectly matched layer for the absorption of electromagnetic waves, *Computational Physics,* 114, 185–200, 1994.

77. D.S. Katz, E.T. Thiele, and A. Taflove, Validation and extension to three dimensions of the Berenger PML absorbing boundary conditions for FDTD meshes, *IEEE Microwave and Guided Wave Letters,* 4, 344–346, 1994.

78. E. Tentzeris, M. Krumpholz, N. Dib, J.-G. Yook, and L.P.B. Katehi, FDTD characterization of waveguide probe structures, *IEEE Trans. Microwave Theory Tech.*, 46, 10, 1452–1460, 1998.

79. M. Werthen, M. Rittweger, and I. Wolff, FDTD simulation of waveguide junctions using a new boundary condition for rectangular waveguides, *Proc. 24th European Microwave Conf.*, Cannes, France, 1715–1719, 1994.

80. A. Taflove, Computation of the electromagnetic fields and induced temperatures within a model of the microwave-irradiated human eye, Ph.D. Dissertation, Department of Electrical Engineering, Northwestern University, Evanston, IL, June 1975.

81. J. Chen, C. Wu, T.K. Lo, K.-L. Wu, and J. Litva, Using linear and nonlinear predictors to improve the computational efficiency of the FDTD algorithm, *IEEE Trans. Microwave Theory Tech.*, 42, 1992–1997, 1994.

82. V. Jandhyala, E. Michielssen, and R. Mittra, On the performance of different AR methods in the spectral estimation of FDTD waveforms, *Microwave and Optical Technology Letters*, 7, 690–692, 1994.

83. M. DePourcq, Field and power density calculations in closed microwave systems by 3D finite differences, *IEE Proc. H: Microwaves, Antennas and Propagation*, 132, 360–368, 1985.

84. E.A. Navarro, V. Such, B. Gimeno, and J.L. Cruz, T-junctions in square coaxial waveguide: an FDTD approach, *IEEE Trans. Microwave Theory Tech.*, 42, 347–350, 1994.

85. C. Wang, B.-Q. Gao, and C.-P. Ding, Q factor of a resonator by the FDTD method incorporating perturbation techniques, *Electronics Letters*, 29, 1866–1867, 1993.

86. G.-C. Liang, Y.-W. Liu, and K.K. Mei, Full-wave analysis of coplanar waveguide and slotline using FDTD, *IEEE Trans. Microwave Theory Tech.*, 37, 1949–1957, 1989.

87. D.M. Sheen, S.M. Ali, M.D. Abouzahra, and J.A. Kong, Application of the 3D FDTD method to the analysis of planar microstrip circuits, *IEEE Trans. Microwave Theory Tech.*, 38, 849–857, 1990.

88. A.C. Cangellaris and D.B. Wright, Analysis of the numerical error caused by the stair-stepped approximation of a conducting boundary in FDTD simulations of electromagnetic phenomena, *IEEE Trans. Antennas and Propagation*, 39, 1518–1525, 1991.

89. C.-W. Lam, S.M. Ali, and P. Nuytkens, Three-dimensional modeling of multichip module interconnects, *IEEE Trans. Components, Hybrids and Manufacturing Technology*, 16, 699–704, 1993.

90. M. Piket-May, A. Taflove, and J. Baron, FDTD modeling of digital signal propagation in 3D circuits with passive and active loads, *IEEE Trans. Microwave Theory Tech.*, 42, 1514–1523, 1994.

91. R.J. Luebbers, K.S. Kunz, M. Schneider, and F. Hunsberger, An FDTD near zone to far zone transformation, *IEEE Trans. Antennas and Propagation*, 39, 429–433, 1991.

92. K.L. Shlager and G.S. Smith, Comparison of two FDTD near-field to near-field transformations applied to pulsed antenna problems, *Electronics Letters*, 31, 936–938, 1995.

93. J.G. Maloney, G.S. Smith, and W.R. Scott, Jr., Accurate computation of the radiation from simple antennas using the FDTD method, *IEEE Trans. Antennas and Propagation*, 38, 1059–1069, 1990.

94. R. Luebbers and K. Kunz, FDTD calculations of antenna mutual coupling, *IEEE Trans. Electromagnetic Compatibility*, 34, 357–359, 1992.

95. R.J. Luebbers and J. Beggs, FDTD calculation of wideband antenna gain and efficiency, *IEEE Trans. Antennas and Propagation*, 40, 1403–1407, 1992.

96. P.A. Tirkas and C.A. Balanis, FDTD method for antenna radiation, *IEEE Trans. Antenna and Propagation*, 40, 334–340, 1992.

97. K. Uehara and K. Kagoshima, Rigorous analysis of microstrip phased array antennas using a new FDTD method, *Electronics Letters*, 30, 100–101, 1994.

98. M.A. Jensen and Y. Rahmat-Samii, Performance analysis of antennas for hand-held transceivers using FDTD, *IEEE Trans. Antennas and Propagation*, 42, 1106–1113, 1994.

99. A Taflove and K.R. Umashankar, Radar cross section of general three-dimensional scatterers, *IEEE Trans. Electromagnetic Compatibility*, 25, 433–440, 1983.

100. C.L. Britt, Solution of electromagnetic scattering problems using time-domain techniques, *IEEE Trans. Antennas and Propagation*, 37, 1181–1192, 1989.

101. W. Sui, D.A. Christensen, and C.H. Durney, Extending the 2D FDTD method to hybrid electromagnetic systems with active and passive lumped elements, *IEEE Trans. Microwave Theory Tech.*, 40, 724–730, 1992.

102. P. Ciampolini, P. Mezzanotte, L. Roselli, and R. Sorrentino, Accurate and efficient circuit simulation with lumped-element FDTD, *IEEE Trans. Microwave Theory Tech.*, 44, 2207–2215, 1996.

103. C.-N. Kuo, B. Housmand, and T. Itoh, Full-wave analysis of packaged microwave circuits with active and nonlinear devices: an FDTD approach, *IEEE Trans. Microwave Theory Tech.*, 45, 819–826, 1997.

104. M.A. Alsunaidi, S.M. Sohel Imtiaz, and S.M. El-Ghazaly, Electromagnetic wave effects on microwave transistors using a full-wave time-domain model, *IEEE Trans. Microwave Theory Tech.*, 44, 799–808, 1996.

105. V.A. Thomas, M.E. Jones, M. Piket-May, A. Taflove, and E. Harrigan, The use of SPICE lumped circuits as sub-grid models for FDTD analysis, *IEEE Microwave and Guided Wave Letters*, 4, 141–143, 1994.

106. S. Akhtarzad and P.B. Johns, Three-dimensional transmission-line matrix computer analysis of microstrip resonators, *IEEE Trans. Microwave Theory Tech.*, MTT-23, 990–997, 1975.

107. W.J.R. Hoefer, The transmission-line matrix method-theory and applications, *IEEE Trans. Microwave Theory Tech.*, MTT-33, 882–893, 1985.

108. C. Christopoulos, *The Transmission-Line Modeling Method*, IEEE/OUP, 1995.

109. P.B. Johns, A symmetrical condensed node for the TLM-method, *IEEE Trans. Microwave Theory Tech.*, 35, 370–377, 1987.

110. L.R.A.X. de Menezes and W.J.R. Hoefer, Modeling of general (nonlinear) constitutive relationships in SCN TLM, *IEEE Trans. Microwave Theory Tech.*, MTT-44, 854–861, 1996.

111. M. Krumpholz and P. Russer, Discrete time-domain Green's functions for three-dimensional TLM modeling of the radiating boundary conditions, *9th Annual Review of Progress in ACES Digest*, 458–466, 1993.

112. M. Righi and W.J.R. Hoefer, Efficient 3D-SCN-TLM diakoptics for waveguide components, *IEEE Trans. Microwave Theory Tech.*, 42, 2381–2384, 1994.

113. M. Krumpholz, C. Huber, and P. Russer, A field theoretical comparison of FDTD and TLM, *IEEE Trans. Microwave Theory Tech.*, 43, 1935–1950, 1995.

114. C. Eswarappa and W.J.R. Hoefer, One-way equation absorbing boundary condition for 3D TLM analysis of planar and quasi-planar structures, *IEEE Trans. Microwave Theory Tech.*, 42, 1669–1677, 1994.

115. J.L. Dubard and D. Pompei, A modified 3D-TLM variable node for the Berenger's perfectly matched layer implementation, *13th Annual Review of Progress in ACES Digest*, 661–665, 1997.

116. M.I. Sobhy, M.W.R. Ng, R.J. Langley, and J.C. Batchelor, TLM simulation of patch antenna on magnetized ferrite substrate, *16th Annual Review of Progress in ACES Digest*, 562–569, 2000.

117. A.P. Duffy, J.L. Herring, T.M. Benson, and C. Christopoulos, Improved wire modeling in TLM, *IEEE Trans. Microwave Theory Tech.*, 42, 1978–1983, 1994.

118. F. Coccetti, L. Vietzorreck, V. Chtchekatourov, and P. Russer, A numerical study of MEMS capacitive switches using TLM, *16th Annual Review of Progress in ACES Digest*, 580–587, 2000.

119. D.H. Schaubert, D.R. Wilton, and A.W. Glisson, A tetrahedral method for electromagnetic scattering by arbitrarily shaped inhomogeneous dielectric bodies, *IEEE Trans. Antennas and Propagation*, 32, 77–85, 1984.

120. Z.J. Cendes, Vector finite elements for electromagnetic field computation, *IEEE Trans. Magnetics*, 27, 3958–3966, 1991.

121. A. Chatterjee, J.M. Jin, and J.L. Volakis, Edge-based finite elements and vector ABC's applied to 3D scattering, *IEEE Trans Antennas and Propagation*, 41, 221–226, 1993.

122. A.F. Peterson and D.R. Wilton, Curl-conforming mixed-order edge elements for discretizing the 2D and 3D vector Helmholtz equation, in *Finite Element Software for Microwave Engineering*, T. Itoh, G. Pelosi, and P.P. Silvester, Eds., Wiley, New York, 1996.

123. Z. Pantic-Tanner, J.S. Savage, D.R. Tanner, and A.F. Peterson, Two-dimensional singular vector elements for finite-element analysis, *IEEE Trans. Microwave Theory Tech.*, 46, 178–184, 1998.

124. R. Lee and A.C. Cangellaris, A study of discretization error in the finite element approximation of wave solutions, *IEEE Trans. Antennas and Propagation*, 40, 542–549, 1992.

125. G.S. Warren and W.R. Scott, An investigation of numerical dispersion in the vector finite element method, *IEEE Trans. Antennas and Propagation*, 42, 1502–1508, 1994.

126. L.W. Pearson, R.A. Whitaker, and L.J. Bahrmasel, An exact radiation boundary condition for the finite-element solution of electromagnetic scattering on an open domain, *IEEE Trans. Magnet.*, 25, 3046–3048, 1989.

127. J.P. Webb and V.N. Kanellopoulos, A numerical study of vector absorbing boundary conditions for the finite element solution of Maxwell's equations, *IEEE Microwave Guided Wave Letters*, 1, 325–327, 1991.

128. W. Sun and C.A. Balanis, Vector one-way absorbing boundary conditions for FEM applications, *IEEE Trans. Antennas and Propagation*, 42, 872–878, 1994.

129. J.M. Jin and V.V. Liepa, A note on hybrid finite element for solving scattering problems, *IEEE Trans. Antennas and Propagation*, 36, 1486–1490, 1988.

130. J.S. Wang, Analysis of microstrip discontinuities based on edge-element 3D FEM formulation and absorbing boundary conditions, *Proc. IEEE-MTT Symposium*, 2, 745–748, 1993.

131. J.-S. Wang and R. Mittra, Finite element analysis of MMIC structures and electronic packages using absorbing boundary conditions, *IEEE Trans. Microwave Theory Tech.*, 42, 441–449, 1994.

132. J.-G. Yook, N.I. Dib, and L.P.B. Katehi, Characterization of high frequency interconnects using finite difference time-domain and finite element methods, *IEEE Trans. Microwave Theory Tech.*, 42, 1727–1736, 1994.

133. A.C. Polycarpou, M.R. Lyons, and C.A. Balanis, Finite element analysis of MMIC waveguide structures with anisotropic substrates, *IEEE Trans. Microwave Theory Tech.*, 44, 1650–1663, 1996.

134. C.A. Brebbia, *The Boundary Element Method for Engineers*, Pentech Press, London, 1978.

135. S. Kagami and I. Fukai, Application of boundary element method to EM field problems, *IEEE Trans. Microwave Theory Tech.*, 32, 455–461, 1984.

136. H. Cam, S. Toutain, P. Gelin, and G. Landrac, Study of Fabry-Perot cavity in the microwave frequency range by the boundary element method, *IEEE Trans. Microwave Theory Tech.*, 40, 298–304, 1992.

137. T.F. Eibert and V. Hansen, 3-D FEM/BEM-hybrid approach based on a general formulation of Huygen's principle for planar layered media, *IEEE Trans. Microwave Theory Tech.*, 45, 1105–1112, 1997.

138. K. Guillouard, M.F. Wong, and V. Fouad Hanna, A new global time domain electromagnetic simulator of microwave circuits including lumped elements based on finite element method, *Proc. IEEE-MTT Symposium*, 1239–1242, 1997.

139. J.M. Jin, J.L. Volakis, A FE-boundary integral formulation for scattering by 3D cavity-blocked apertures, *IEEE Trans. Antennas and Propagation*, 39, 97–104, 1991.

140. S.D. Gedney, J.F. Lee, and R. Mittra, A combined FEM/MoM approach to analyze the plane wave diffraction by arbitrary gratings, *IEEE Trans. Antenna and Propagation*, 40, 363–370, 1992.

141. R.D. Graglia, D.R. Wilton, and A.F. Peterson, Higher order interpolatory vector bases for computational EM, *IEEE Trans. Antennas and Propagation*, 45, 329–342, 1997.

142. R.-B. Wu and T. Itoh, Hybridizing FDTD analysis with unconditionally stable FEM for objects of curved boundary, *Proc. IEEE-MTT Symposium*, 833–836, 1995.

143. A. Monorchio and R. Mittra, A hybrid FE/FDTD technique for solving complex electromagnetic problems, *IEEE Microwave Guided Wire Letters*, 8, 93–95, 1998.

144. A.G. Engel and L.P.B. Katehi, Frequency and time domain characterization of microstrip-ridge structures, *IEEE Trans. Microwave Theory Tech.*, 1251–1262, 1993.

145. H. Esteban et al., A new hybrid mode-matching method for the analysis of inductive obstacles and discontinuities, *Proc. IEEE-AP Symposium*, 966–969, 1999.

146. D.C. Ross, J.L. Volakis, and H. Anastasiu, Hybrid FE-modal analysis of jet engine inlet scattering, *IEEE Trans. Antennas and Propagation*, 43, 1995.

147. S. Lindenmeier, C. Christopoulos, and P. Russer, Thin wire modeling with TLM/MIE method, *16th Annual Review of Progress in ACES Digest*, 587–594, 2000.

148. L. Pierantoni, G. Cerri, S. Lindemeier, and P. Russer, Theoretical and numerical aspects of the hybrid MoM-FDTD, TLM-IE and ARB methods for the efficient modeling of EMC problems, *Proc. 29 EuMC*, 2, 313–316, 1999.

149. I. Daubechies, *Ten Lectures on Wavelets*, SIAM, 1992.

150. G. Kaiser, *A Friendly Guide to Wavelets*, Springer Verlag, Berlin, 1994.

151. C. Burrus, R.A. Gopinath, and H. Guo, *Introduction to Wavelets and Wavelet Transforms: A Primer*, Prentice Hall, Englewood Cliffs, NJ, 1997.

152. J. Goswami and A. Chan, *Fundamentals on Wavelets: Theory, Algorithms and Applications*, John Wiley & Sons, New York, 1999.

153. W. Dahmen, A.J. Kurdila, and P. Oswald, *Multiscale Wavelet Methods for Partial Differential Equations*, Academic Press, New York, 1997.

154. S. Mallat, Multiresolution representation and wavelets, Ph.D. Thesis, University of Pennsylvania, Philadelphia, 1988.

155. Y. Meyer, Ondelettes et Fonctions Splines, Seminaire EDP, Ecole Polytechnique, Paris, December 1986.

156. L.L. Schumaker, *Spline Functions: Basic Theory*, Wiley-Interscience, New York, 1981.

157. H.C. Schweinler and E.P. Wigner, Orthogonalization methods, *J. Math. Phys.*, 11, 1693–1694, 1970.

158. G. Battle, A block spin construction of ondelettes, Part I: Lemarie functions, *Comm. Math. Phys.*, 110, 601–615, 1987.

159. P.G. Lemarie, Ondelettes a localization exponentielles, *J. Math. Pures Appl.*, 67, 227–236, 1988.

160. A. Haar, Zur theorie der orthogonalen funktionsysteme, *Math. Ann.*, 69, 331–371, 1910.

161. I. Daubechies, Orthonormal bases of compactly supported wavelets, *Comm. Pure Appl. Math.*, 41, 909–996, 1988.

162. G. Beylkin, R. Coifman, and V. Rokhlin, Fast wavelet transforms and numerical algorithms I, *Commun. Pure Appl. Math.*, *44, 141–183, 1991*.

163. B.Z. Steinberg and Y. Leviatan, On the use of wavelet expansions in the method of moments, *IEEE Trans. Antennas and Propagation*, 41, 610–619, 1993.

164. G. Wang, On the utilization of periodic wavelet expansions in the moment methods, *IEEE Trans. Microwave Theory Tech.*, 43, 2495–2498, 1995.

165. J.C. Goswami, A.K. Chan, and C.K. Chui, On solving first-kind integral equations using wavelets on a bounded interval, *IEEE Trans. Antennas and Propagation*, 43, 614–622, 1995.

166. K. Sabetfakhri and L.P.B. Katehi, Analysis of integrated millimeter-wave and submillimeter-wave waveguides using orthonormal wavelet expansions, *IEEE Trans. Microwave Theory Tech.*, 42, 2412–2422, 1994.

167. M. Krumpholz and L.P.B. Katehi, MRTD: New time domain schemes based on multiresolution analysis, *IEEE Trans. Microwave Theory Tech.*, 44, 4, 555–561, April 1996.

168. M.M. Tentzeris, J. Harvey, and L.P.B. Katehi, Time adaptive time-domain techniques for the design of microwave circuits, *IEEE Microwave Guided Wave Letters*, 9, 96–98, 1999.

169. M. Tentzeris, R. Robertson, A. Cangellaris, and L.P.B. Katehi, Space- and time-adaptive gridding using MRTD, *Proc. IEEE-MTT Symposium*, 337–340, 1997.

170. M. Tentzeris, M. Krumpholz, and L.P.B. Katehi, Applications of MRTD to printed transmission lines, *Proc. IEEE-MTT Symposium*, 573–576, 1996.

171. R.L. Robertson, M. Tentzeris, and L.P.B. Katehi, Modelling of dielectric-loaded cavities using MRTD, *Int. Journal of Numerical Modeling, Special Issue on Wavelets in Electromagnetics*, 11, 55–68, 1998.

172. M. Tentzeris and L.P.B. Katehi, Space adaptive analysis of evanescent waveguide filters, *Proc. IEEE-MTT Symposium*, 481–484, 1998.

173. L. Roselli, M. Tentzeris, and L.P.B. Katehi, Nonlinear circuit characterization using a multiresolution time domain technique, *Proc. IEEE-MTT Symposium*, 1387–1390, 1998.

174. M. Tentzeris, R. Robertson, J. Harvey, and L.P.B. Katehi, Stability and dispersion analysis of Battle-Lemarie based MRTD schemes, *IEEE Trans. Microwave Theory Tech.*, 47, 7, 1004–1013, 1999.

175. K. Goverdhanam, M. Tentzeris, M. Krumpholz, and L.P.B. Katehi, An FDTD multigrid based on multiresolution analysis, *Proc. IEEE-AP Symposium*, 352–355, 1996.

176. M. Fujii and W.J.R. Hoefer, Formulation of a Haar-wavelet based multiresolution analysis similar to 3D FDTD method, *Proc. IEEE-MTT Symposium*, 1393–1396, 1998.

177. C. Sarris and L.P.B. Katehi, Multiresolution time-domain (MRTD) schemes with space-time Haar wavelets, *Proc. IEEE-MTT Symposium*, 1459–1462, 1999.

8.3 Computer Aided Design of Passive Components

Daniel G. Swanson, Jr.

Computer-aided design (CAD) of passive RF and microwave components has advanced slowly but steadily over the past four decades. The 1960s and 1970s were the decades of the mainframe computer. In the early years, CAD tools were proprietary, in-house efforts running on text-only terminals. The few graphics terminals available were large, expensive, and required a short, direct connection to the mainframe. Later in this period, commercial tools became available for use on in-house machines or through time sharing services. A simulation of a RF or microwave network was based on a combination of lumped and distributed elements. The elements were connected in cascade using ABCD parameters or in a nodal network using admittance- or Y-parameters. The connection between elements and the control parameters for the simulation were stored in a text file called a netlist. The netlist syntax was similar but unique for each software tool. The mathematical foundations for a more sophisticated analysis based on Maxwell's equations were being laid down in this same time period.[1-4] However, the computer technology of the day could not support effective commercial implementation of these more advanced codes.

The 1980s brought the development of the microprocessor and UNIX workstations. The UNIX workstation played a large role in the development of more sophisticated CAD tools. For the first time there was a common operating system and computer language (the C language) to support the development of cross-platform applications. UNIX workstations also featured large, bit mapped graphics displays for interaction with the user. The same microprocessor technology that launched the workstation also made the personal computer possible. Although the workstation architecture was initially more sophisticated, personal computer hardware and software has grown steadily more elaborate. Today, the choice between a workstation and a personal computer is largely a personal one. CAD tools in this time period were still based on lumped and distributed concepts. The innovations brought about by the cheaper, graphics-based hardware had largely to do with schematic capture and layout. Schematic capture replaced the netlist on the input side of the analysis and automatic or semi-automatic layout provided a quicker path to the finished circuit after analysis and optimization.

The greatest innovation in the 1990s was the emergence of CAD tools based on the direct solution of Maxwell's equations. Finally, there was enough computer horsepower to support commercial versions of the codes that had been in development since the late 1960s and early 1970s. These codes are in general labeled electromagnetic field-solvers although any one code may be based on one of several different numerical methods. Sonnet *em*,[5] based on the Method of Moments (MoM), was the first commercially viable tool designed for RF and microwave engineers. Only a few months later, Hewlett-Packard HFSS,[6] a Finite Element Method (FEM) code co-developed with Ansoft Corp., was released to the design community. All of these tools approximate the true fields or currents in the problem space by subdividing the problem into basic "cells" or "elements" that are roughly one tenth to one twentieth of a guide wavelength in size. For any guided electromagnetic wave, the guide wavelength is the distance spanned by one full cycle of the electric or magnetic field. The problem is to find the magnitude of the assumed current on each cell or the field at the junction of elements. The final solution is then just the sum of

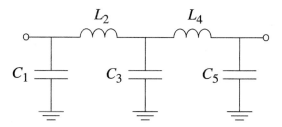

FIGURE 8.17   Lumped element lowpass filter or matching network.

each small contribution from each basic unit. Most of these codes first appeared on UNIX workstations and then migrated to the personal computer, as that hardware became more powerful. In the later years of the 1990s, field-solver codes appeared that were developed on and for the personal computer. In the early years, the typical field-solver problem was a single discontinuity or some other structure that was small in terms of wavelengths. Today, groups of discontinuities, complete matching networks, or small parts of a multilayer printed circuit (PC) board are all suitable problems for a field-solver. Field-solver data in S-parameter form is typically imported into a circuit simulator and combined with lumped and distributed models to complete the analysis of the structure.

Circuit Theory Based CAD

CAD of low frequency circuits is at least 30 years old and microwave circuits have been analyzed by computer for at least 20 years. At very low frequencies, we can connect inductors, capacitors, resistors, and active devices in a very arbitrary way. The lumped lowpass filter shown in Fig. 8.17 is a simple example. This very simple circuit has only three nodes. Most network analysis programs will form an admittance matrix (Y-matrix) internally and invert the matrix to find a solution. The Y-matrix is filled using some fairly simple rules. A shunt element connected to node two generates an entry at Y_{22}. A series element connected between nodes two and three generates entries at Y_{22}, Y_{23}, Y_{32}, and Y_{33}. A large ladder network with sequential node numbering results in a large tri-diagonal matrix with many zeros off axis.

$$
\mathbf{Y} = \begin{bmatrix}
j\omega C_1 - j\dfrac{1}{\omega L_2} & j\dfrac{1}{\omega L_2} & 0 \\[2ex]
j\dfrac{1}{\omega L_2} & j\omega C_3 - j\dfrac{1}{\omega L_2} - j\dfrac{1}{\omega L_4} & j\dfrac{1}{\omega L_4} \\[2ex]
0 & j\dfrac{1}{\omega L_4} & j\omega C_5 - j\dfrac{1}{\omega L_4}
\end{bmatrix}
$$

The Y-matrix links the known source currents to the unknown node voltages. **I** is a vector of source currents. Typically the input node is excited with a one amp source and the rest of the nodes are set to zero. **V** is the vector of unknown node voltages. To find **V**, we invert the matrix **Y** and multiply by the known source currents.

$$\mathbf{I} = \mathbf{YV}$$

$$\mathbf{V} = \mathbf{Y}^{-1}\mathbf{I}$$

The time needed to invert an N × N matrix is roughly proportional to N^3. Filling and inverting the Y-matrix for each frequency of interest will be very fast, in this case, so fast it will be difficult to measure the computation time unless we specify a very large number of frequencies. This very simple approach might be good up to 1 MHz or so.

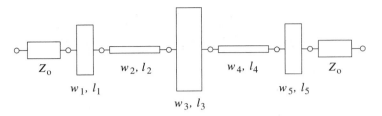

FIGURE 8.18   Distributed lowpass filter circuit. Step discontinuities are ignored.

In our low-frequency model there is no concept of wavelength or even physical size. Any phase shift we compute is strictly due to the reactance of the component, not its physical size. There is also no concept of radiation; power can only be dissipated in resistive components. As we move into the HF frequency range (1 to 30 MHz) the real components we buy will have significant parasitics. Lead lengths and proximity to the ground plane become very important and our physical construction techniques will have a big impact on the results achieved.

By the time we reach VHF frequencies (50 to 150 MHz) we are forced to adopt distributed concepts in the physical construction and analysis of our circuits. The connections between components become transmission lines and many components themselves are based on transmission line models. Our simple lowpass circuit might become a cascade of low and high impedance transmission lines, as seen in Fig. 8.18.

If this was a microstrip circuit, we would typically specify the substrate parameters and the width and length of each transmission line. We have ignored the step discontinuities due to changes in line width in this simplified example. Internally, the software would use analytical equations to convert our physical dimensions to impedances and electrical lengths. The software might use a Y-matrix, a cascade of ABCD parameter blocks, or a cascade of scattering-parameter (S-parameter) blocks for the actual analysis. At the ports, we typically ask for S-parameters referenced to the system impedance.

Notice that we still have a small number of nodes to consider. Our circuit is clearly distributed but the solution time does not depend on its size in terms of wavelengths. Any phase shift we compute is directly related to the physical size of the network. Although we can include conductor and substrate losses, there is still no radiation loss mechanism. It is also difficult to include enclosure effects; there may be box resonances or waveguide modes in our physical implementation. There is also no mechanism for parasitic coupling between our various circuit models.

The boundary between a lumped circuit point of view and a distributed point of view can be somewhat fuzzy. A quick review of some rules of thumb and terminology might be helpful. One common rule of thumb says that the boundary between lumped and distributed behavior is somewhere between a tenth and an eighth of a guide wavelength. Remember that wavelength in inches is defined by

$$\lambda = \frac{11.803}{\sqrt{\varepsilon_{eff} \cdot f}}$$

where ε_{eff} is the effective dielectric constant of the medium and f is in GHz. At 1 GHz, $\lambda = 11.803$ inches in air and $\lambda = 6.465$ inches for a 50 ohm line on 0.014-inch thick FR4. FR4 is a common, low cost printed circuit board material for digital and RF circuits. In Fig. 8.19 we can relate the physical size of our structure to the concept of wavelength and to some common terminology. Again, the boundary between purely lumped and purely distributed behavior is not always distinct.

Wavelength

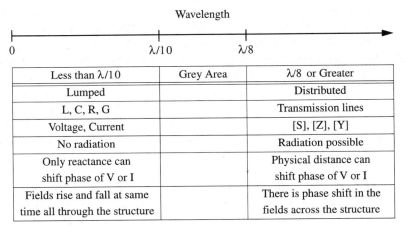

Less than λ/10	Grey Area	λ/8 or Greater
Lumped		Distributed
L, C, R, G		Transmission lines
Voltage, Current		[S], [Z], [Y]
No radiation		Radiation possible
Only reactance can shift phase of V or I		Physical distance can shift phase of V or I
Fields rise and fall at same time all through the structure		There is phase shift in the fields across the structure

FIGURE 8.19 The transition between lumped and distributed behavior and some common terminology.

Field Theory Based CAD

A field-solver based solution is an alternative to the previous distributed, circuit theory based approach. The field-solver takes a more microscopic view of any distributed geometry. Any field-solver we might employ must subdivide the geometry based on guide wavelength. Typically we need 10 to 30 elements or cells per guide wavelength to capture the fields or currents in our structure. Figure 8.20 shows a typical mesh generated by Agilent Momentum[7] for our microstrip lowpass filter example. Narrow cells are used on the edges of the strip to capture the spatial wavelength, or highly nonuniform current distribution across the width of the strips. This Method of Moments code has subdivided the microstrip metal and will solve for the current on each small rectangular or triangular patch. The default settings for mesh generation were used.

For this type of field-solver there is a strong analogy between the Y-matrix description we discussed for our lumped element circuit and what the field-solver must do internally. Imagine a lumped capacitor to ground at the center of each "cell" in our field-solver description. Series inductors connect these capacitors to each other. Coupling between non-adjacent cells can be represented by mutual inductances. So we have to fill and invert a matrix, but this matrix is now large and dense compared to our simple, lumped element circuit Y-matrix. For the mesh in Figure 8.20, N = 474 and we must fill and invert an N × N matrix.

One reason we turn to the field-solver is because it can potentially include all electromagnetic effects from first principles. We can include all loss mechanisms including surface waves and

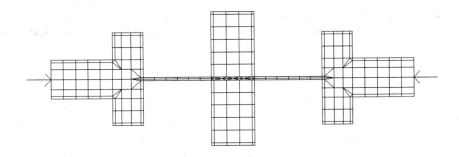

FIGURE 8.20 A typical model mesh fo rthe distributed lowpass filter circuit. The number of unknowns, N is 474. *Agilent Momentum, ADS 1.3.*

radiation. We can also include parasitic coupling between elements and the effects of compacting a circuit into a small space. The effects of the package or housing on our circuit performance can also be included in the field-solver analysis. However, the size of the numerical problem is now proportional to the structure size in wavelengths. The details of how enclosures are included in our analysis will vary from solver to solver. In some tools an enclosure is part of the basic formulation. In other tools, the analysis environment is "laterally open"; there are no sidewalls, although there may be a cover. One of the exciting aspects of field-solvers is the ability to observe fields and currents in the circuit, which sometimes leads to a deeper understanding of how the circuit actually operates. However, the size of the numerical problem will also be greater using a field-solver versus circuit theory, so we must carefully choose which pieces of global problem we will attack with the field-solver.

Although our discussion so far has focused on planar, distributed circuits there are actually three broad classes of field-solver codes. The 2D cross-section codes solve for the modal impedance and phase velocity of 1 to N strips with a uniform cross-section. This class of problem includes coupled microstrips, coupled slots, and conductors of arbitrary cross-section buried in a multilayer PC board. These tools use a variety of numerical methods including Method of Moments, the Finite Element Method, and the Spectral Domain Method. Field-solver engines that solve for multiple strips in a layered environment are built into several linear and nonlinear simulators. A multistrip model of this type is a building block for more complicated geometries like Lange couplers, spiral inductors, baluns, and many distributed filters. The advantage of this approach is speed; only the 2D cross-section must be discretized and solved.

The second general class of codes mesh or subdivide the surfaces of planar metals. The assumed environment for these surface meshing codes is a set of homogeneous dielectric layers with patterned metal conductors at the layer interfaces. Vertical vias are available to form connections between metal layers. There are two fundamental formulations for these codes, closed box and laterally open. In the closed box formulation the boundaries of the problem space are perfectly conducting walls. In the laterally open formulation, the dielectric layers extend to infinity. The numerical method for this class of tool is generally Method of Moments (MoM). Surface meshing codes can solve a broad range of strip- and slot-based planar circuits and antennas. Compared to the 2D cross-section solvers, the numerical effort is considerably higher.

The third general class of codes meshes or subdivides a 3D volume. These volume meshing codes can handle virtually any three-dimensional object, with some restrictions on where ports can be located. Typical problems are waveguide discontinuities, various coaxial junctions, and transitions between different guiding systems, such as transitions from coax to waveguide. These codes can also be quite efficient for computing transitions between layers in multilayer PC boards and connector transitions between boards or off the board. The more popular volume meshing codes employ the Finite Element Method, the Finite Difference Time Domain (FDTD) method, and the Transmission Line Matrix (TLM) method. Although the volume meshing codes can solve a very broad range of problems, the penalty for this generality is total solution time. It typically takes longer to set up and run a 3D problem compared to a surface meshing or cross-section problem. Sadiku[8] has compiled a very thorough introduction to many of these numerical methods.

Solution Time for Circuit Theory and Field Theory

When we use circuit theory to analyze a RF or microwave network, we are building a Y-matrix of dimension N, where N is the number of nodes. A typical amplifier or oscillator design may have only a couple of dozen nodes. Depending on the solution method, the solution time is proportional to a factor between N^2 and N^3. When we talk about a "solution" we really mean matrix inversion. In Fig. 8.21 we have plotted solution time as a function of matrix size N. The vertical time scale is somewhat arbitrary but should be typical of workstations and personal computers today.

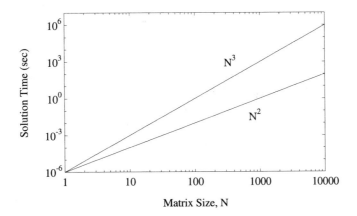

FIGURE 8.21   Solution time as a function of matrix size, N. Solution time for circuit simulators, MoM field-solvers, and FEM field-solvers is roughly proportional to N^3.

When we use a MoM field-solver, a "small" problem has a matrix dimension of N = 300–600. Medium size problems may be around N = 1500 and large problems quickly get into the N = 2000–3000 range. Because of the N^2/N^3 effect, the solution time is impacted dramatically as the problem size grows. In this case we can identify two processes, filling the matrix with all the couplings between cells and inverting or solving that matrix. So we are motivated to keep our problem size as small as possible. The FEM codes also must fill and invert a matrix. Compared to MoM, the matrix tends to be larger but more sparse.

The time domain solvers using FDTD or TLM are exceptions to the N^2/N^3 rule. The solution process for these codes is iterative; there is no matrix to fill or invert with these solvers. Thus the memory required and the solution time grow more linearly with problem size in terms of wavelengths. This is one reason these tools have been very popular for radar cross-section (RCS) analysis of ships and airplanes. However, because these are time stepping codes, we must perform a Fast Fourier Transform (FFT) on the time domain solution to get the frequency domain solution. Closely spaced resonances in the frequency domain require a large number of time samples in the time domain. Therefore, time stepping codes may not be the most efficient choice for structures like filters, although there are techniques available to speed up convergence. Veidt[9] presents a good summary of how solution time scales for various numerical methods.

A Hybrid Approach to Circuit Analysis

If long solution times prevent us from analyzing complete circuits with a field-solver, what is the best strategy for integrating these tools into the design process? I believe the best approach is to identify the key pieces of the problem that need the field-solver, and to do the rest with circuit theory. Thus the final result is a "hybrid solution" using different techniques, and even different tools from different vendors. As computer power grows and software techniques improve, we can do larger and larger pieces of the problem with a field-solver. A simple example will help to demonstrate this approach. The circuit in Fig. 8.22 is part of a larger RF printed circuit board. In one corner of the board we have a branchline coupler, a resistive termination, and several mitered bends.

Using the library of elements in our favorite linear simulator, there are several possible ways to subdivide this network for analysis (see Fig. 8.23). In this case we get about 21 nodes in our circuit. Solution time is roughly proportional to N^3, so if we ignore the overhead of computing any of the individual models, we would expect the solution to come back very quickly. But we have clearly neglected several things in our analysis. Parasitic coupling between the arms of the coupler, interaction between the discontinuities, and any potential interaction with the package have all been ignored. Some of our analytical models may not be as accurate as we would like, and in some cases a combination of models

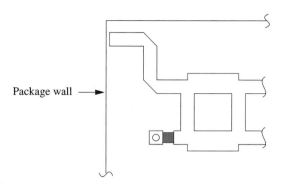

Package wall →

FIGURE 8.22 Part of an RF printed circuit board which includes a branchline coupler, a resistive termination to ground, and several mitered bends.

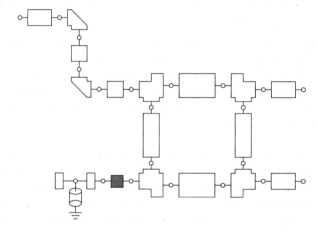

FIGURE 8.23 The layout in Fig. 8.22 has been subdivided for analysis using the standard library elements found in many circuit-theory-based simulations.

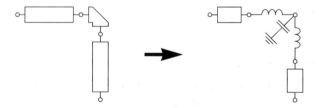

FIGURE 8.24 The equivalent circuit of a microstrip mitered bend. The physical dimensions are mapped to an equivalent lumped element circuit.

may not accurately describe our actual circuit. If this circuit were compacted into a much denser layout, all of the effects mentioned above would become more pronounced.

Each one of the circuit elements in our schematic has some kind of analytic model inside the software. For a transmission line, the model would relate physical width and length to impedance and electrical length through a set of closed form equations. For a discontinuity like the mitered bend, the physical parameters might be mapped to an equivalent lumped element circuit (Fig. 8.24), again through a set of closed form equations. The field-solver will take a more microscopic view of the same mitered bend discontinuity. Any tool we use will subdivide the metal pattern using 10 to 30 elements per guide wavelength. The sharp inside corner where current changes direction rapidly will force an even finer subdivision. If we want to solve the

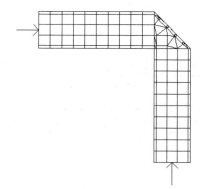

FIGURE 8.25 A typical MoM mesh for the microstrip mitered bend. The number of unknowns, N, is 221. *Agilent Momentum, ADS 1.3.*

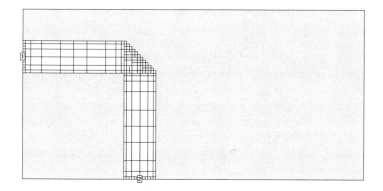

FIGURE 8.26 An analysis of the input line and the mitered bend in the presence of the package walls. The number of unknowns, N, is 360. *Sonnet em 6.0.*

bend discontinuity individually, we must also connect a short length of series line to each port. Agilent Momentum generated the mesh in Fig. 8.25. The number of unknowns is 221. If the line widths are not variable in our design, we could compute this bend once, and use it over and over again in our circuit design.

Another potential field-solver problem is in the corner of the package near the input trace. You might be able to include the box wall effect on the series line, but wall effects are generally not included in discontinuity models. However, it is quite easy to set up a field-solver problem that would include the microstrip line, the mitered bend, and the influence of the walls. The project in Fig. 8.26 was drawn using Sonnet *em.* The box walls to the left and top in the electromagnetic simulation mimic the true location of the package walls in the real hardware. There are 360 unknowns in this simulation.

One of the more interesting ways to use a field-solver is to analyze groups of discontinuities rather than single discontinuities. A good example of this is the termination resistor and via[10,11] in our example circuit. A field-solver analysis of this group may be much more accurate than a combination of individual analytical models. We could also optimize the termination, then use the analysis data and the optimized geometry over and over again in this project or other projects. The mesh for the resistor via combination (Fig. 8.27) was generated using Sonnet *em* and represents a problem with 452 unknowns.

Our original analysis scheme based on circuit theory models alone is shown in Fig. 8.23. Although this will give us the fastest analysis, there may be room for improvement. We can substitute results in our field-solver for the elements near the package walls and for the resistor/via combination (Fig. 8.28).

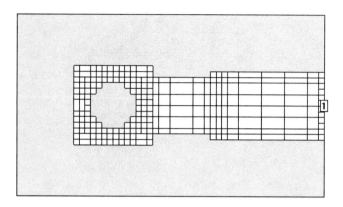

FIGURE 8.27 A MoM analysis of a group of discontinuities including a thin-film resistor, two steps in width, and a via hole to ground. The number of unknowns, N, is 452. *Sonnet em 6.0.*

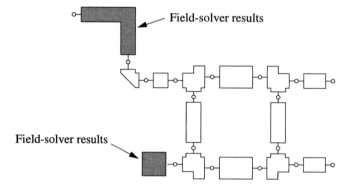

FIGURE 8.28 Substituting field-solver results into the original solution scheme mixes field-theory and circuit theory in a cost effective way.

The data from the field-solver would typically be S-parameter files. This "hybrid" solution mixes field theory and circuit theory in a cost-effective way.[12] The challenge for the design engineer is to identify the critical components that should be addressed using the field-solver.

The hybrid solution philosophy is not limited to planar components; three-dimensional problems can be solved and cascaded as well. The right angle coax bend shown in Fig. 8.29 is one example of a 3D component that was analyzed and optimized using Ansoft HFSS.[13] In this case we have taken advantage of a symmetry plane down the center of the problem in order to reduce solution time. This component includes a large step in inner conductor diameter and a Teflon sleeve to support the larger inner conductor. After optimizing two dimensions, the computed return loss is greater than −30 dB. The coax bend is only one of several problems taken from a larger assembly that included a lowpass filter, coupler, amplifier, and bandpass filter.

Optimization

Optimization is a key component of modern linear and nonlinear circuit design. Many optimization schemes require gradient information, which is often computed by taking simple forward or central differences. The extra computations required to find gradients become very costly if there is a field-solver inside the optimization loop. So it is important to minimize the number of field-solver analysis runs. It is also necessary to capture the desired changes in the geometry and pass this information to the field-solver. Bandler et al.[14,15] developed an elegant solution to both of these problems in 1993. The key concept was a "data pipe" program sitting between the simulator and the field-solver (see Fig. 8.30). When the

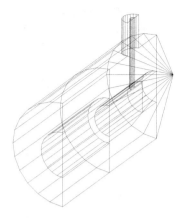

FIGURE 8.29   A right angle coax-to-coax transition that was optimized for return loss. The number of unknowns, N, is 8172. *An soft HFSS 7.0.*

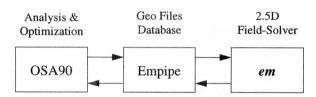

FIGURE 8.30   The first commercially successful optimization scheme which included a field-solver inside the optimization loop.

linear simulator calls for a field-solver analysis, the data pipe generates a new geometry file and passes it to the field-solver. In the reverse direction, the data pipe stores the analysis results and interpolates between data sets if possible. The final iterations of the optimization operate entirely on interpolated data without requiring any new field-solver runs. This concept was applied quite successfully to both surface meshing[16] and volume meshing solvers. The same basic rules that lead to successful circuit theory based optimization apply when a field-solver is in the loop as well. First, a good starting point leads to more rapid and consistent convergence. Second, it is important to limit the number of variables.

The Next Decade

The need for inexpensive wireless systems has forced the RF community to rapidly adopt low cost, multilayer PC board technology. In the simpler examples, most circuitry and components are mounted on the top layer while inner layers are used for routing of RF signals and DC bias. However, more complex examples can be found where printed passive components and discontinuities are located in one or more buried layers. Given the large number of variables in PC board construction it will be difficult for vendors of linear and nonlinear circuit simulators to support large libraries of passive models that cover all possible scenarios. However, a field-solver can be used to generate new models as needed for any novel layer stack up. Of course the user is also free to use the field-solver data to develop custom, proprietary models for his or her particular technology.

The traditional hierarchy of construction for RF systems has been a chip device, mounted to leaded package, mounted to printed circuit board located in system cabinet or housing. Today however, the "package" may be a multilayer Low Temperature Co-fired Ceramic (LTCC) substrate or a multilayer PC board using Ball Grid Array (BGA) interconnects. Thus the boundary between package and PC board has blurred somewhat. No matter what the technology details, the problem remains to transfer a signal from the outside world into the system, onto the main system board, through the package, and into the chip. And of course there is an analogous connection from the chip back to the outside world. From this point of view, the problem becomes a complex, multilevel passive interconnect that must support not only the signal currents but also the ground currents in the return path. It is often the ground return path that limits package isolation or causes unexpected oscillations in active circuits.[17] The high-speed digital community is faced with very similar passive interconnect challenges at similar, if not higher frequencies and typically much higher signal densities. Again, there is ample opportunity to apply field-solver technology to these problems, although practical problem size is still somewhat limited. The challenge to the practitioner is to identify and correct problems at multiple points in the signal path.

Conclusion

At very low frequencies we can use lumped element models to describe our circuits. Connection lengths and device parasitics are not issues. At higher frequencies we use distributed models to capture the effects of guide wavelength, but spurious couplings between elements and other effects due to circuit compaction are typically not captured. A field-solver can potentially capture all the macro and micro aspects of our circuit. It should capture spatial wavelength effects, guide wavelength, spurious couplings among elements, and interference among elements due to dense packing. Although the size of a practical field-solver problem is still somewhat small, there are many useful and cost effective problems that can be identified and solved using a combination of circuit theory based and field theory based CAD.

References

1. K. S. Yee, Numerical solution of initial boundary-value problems involving Maxwell's equations in isotropic media, *IEEE Trans. Ant. Prop.*, AP-14, 302–207, May 1966.
2. R. F. Harrington, *Field Computation by Moment Methods*, Macmillan, New York, 1968.
3. P. B. Johns and R. L. Beurle, Numerical solution of 2-dimensional scattering problems using a transmission-line matrix, *Proc. Inst. Electr. Eng.*, 118, 1203–1208, Sept. 1971.
4. P. Silvester, Finite element analysis of planar microwave networks, *IEEE Trans. Microwave Theory Tech.*, MTT-21, 104–108, Feb. 1973.
5. *em*™, Sonnet Software, Liverpool, NY.
6. HFSS, Hewlett-Packard, Santa Rosa, CA and Ansoft, Pittsburgh, PA.
7. Momentum, Agilent EEsof EDA, Santa Rosa, CA.
8. M. Sadiku, *Numerical Techniques in Electromagnetics*, CRC Press, Boca Raton, 1992.
9. B. Veidt, Selecting 3D electromagnetic software, *Microwave Journal*, 126–137, Sept. 1998.
10. M. Goldfarb and R. Pucel, Modeling via hole grounds in microstrip, *IEEE Microwave and Guided Wave Letters*, 1, 135–137, June 1991.
11. D. Swanson, Grounding microstrip lines with via holes, *IEEE Trans. Microwave Theory Tech.*, MTT-40, 1719–1721, Aug. 1992.
12. D. Swanson, Using a microstrip bandpass filter to compare different circuit analysis techniques, *Int. J. MIMICAE*, 5, 4–12, Jan. 1995.
13. HFSS, Ansoft Corp., Pittsburgh, PA.
14. J. W. Bandler, S. Ye, R. M. Biernacki, S. H. Chen, and D. G. Swanson, Jr., Minimax microstrip filter design using direct em field simulation, *IEEE MTT-S Int. Microwave Symposium Digest*, 889–892, 1993.
15. J. W. Bandler, R. M. Biernacki, S. H. Chen, D. G. Swanson, Jr., and S. Ye, Microstrip filter design using direct em field simulation, *IEEE Trans. Microwave Theory Tech.*, MTT-42, 1353–1359, July 1994.
16. D. Swanson, Optimizing a microstrip bandpass filter using electromagnetics, *Int. J. MIMICAE*, 5, 344–351, Sept. 1995.
17. D. Swanson, D. Baker, and M. O'Mahoney, Connecting MMIC chips to ground in a microstrip environment, *Microwave Journal*, 58–64, Dec. 1993.

8.4 Nonlinear RF and Microwave Circuit Analysis

Michael B. Steer and John F. Sevic

The two most popular circuit-level simulation technologies are embodied in SPICE-like simulators, operating entirely in the time domain, and in Harmonic Balance (HB) simulators, which are hybrid time and frequency domain simulators. Neither is ideal for modeling RF and microwave circuits and in this chapter their concepts and bases of operation will be explored with the aim of illuminating the limitations and advantages of each. All of the technologies considered here have been implemented in commercial microwave simulators. An effort is made to provide sufficient background for these to be used to full advantage.

Simulation of digital and low frequency analog circuits at the component level is performed using SPICE, or commercial equivalents, and this has proved to be very robust. The operation of SPICE will be considered in detail later, but in essence SPICE solves for the state of the circuit at a time point and then uses this state to estimate the state of the circuit at the next time point (and so is referred to as a time-marching technique). The state of the circuit at the new time point is iterated to minimize error. This process captures the transient response of a circuit and the algorithm obtains the best waveform estimate. That is, the best estimate of the current and voltages in the circuit at each time point are obtained. The accurate calculation of the waveform in a circuit is what we want in low pass circuits such as digital and low frequency analog circuits. However with RF and microwave circuits, especially in communications, it is more critical to accurately determine the spectrum of a signal (i.e., the frequency components and their amplitudes) than the precise waveform. In part this is because regulations require strict control of spurious spectral emissions so as not to interfere with other wireless systems, and also because the generation of extraneous emissions compromises the demodulation and detection of communication signals by other radios in the same system. The primary distortion concern in radio is spectrum spreading or more specifically, adjacent channel interference. In-band distortion is also important especially with base station amplifiers where filtering can be used to eliminate spectral components outside the main channel. Distortion is largely the result of the nonlinear behavior of transmitters and so characterization of this phenomenon is important in RF design. In addition, provided that the designer has confidence in the stability and well-behaved transient response of a circuit, it is only necessary to determine its steady-state response. In order to determine the steady-state response using a time-marching approach, it is necessary to determine the RF waveform for perhaps millions of RF cycles, including the full transient interval, so as to extract the superimposed modulated signal. The essential feature of HB is that a solution form is assumed, in particular, a sum of sinusoids and the unknowns to be solved for are the amplitudes and phases of these sinusoids. The form of the solution then allows simplification of the equations and determination of the unknown coefficients. HB procedures work well when the signal can be described by a simple spectrum. However, it does not enable the transient response to be determined exactly.

In the following sections we will first look at the types of signals that must be characterized and identify the information that must be extracted from a circuit simulation. We will then look at transient SPICE-like simulation and HB simulation. Both types of analyses have restrictions and neither provides a complete characterization of an RF or microwave circuit. However, there are extensions to each that improve their basic capabilities and increase applicability. We will also review frequency domain analysis techniques as this is also an important technique and forms the basis of behavioral modeling approaches.

Modeling RF and Microwave Signals

The way nonlinear effects are modeled and characterized depends on the properties of the input signal. Signals having frequency components above a few hundred megahertz are generally regarded as RF or microwave signals. However, the distinguishing features that identify RF and microwave circuits are the design methodologies used with them. Communication systems generally have a a small operating fractional bandwidth — rarely is it much higher than 10%. Generated or monitored signals in sensing systems (including radar and imaging systems) generally have small bandwidths. Even broadband systems including instrumentation circuits and octave (and more) bandwidth amplifiers have passband characteristics. Thus RF and microwave design and modeling technology has developed specifically for narrowband systems.

The signals to be characterized in RF and microwave circuits are either correlated, in the case of communication and radar systems, or uncorrelated noise in the case of many imaging systems. We are principally interested in handling correlated signals as uncorrelated noise is nearly always very small and can be handled using relatively straightforward linear circuit analysis techniques. There are two families of correlated signals, one being discrete tone and the other being digitally modulated. In the following, three types of signals will be examined and their response to nonlinearities described.

Discrete Tone Signals

Single tone signals, i.e., a single sinewave, are found in frequency sources but such tones do not transmit information and must be modulated. Until recently, communication and radar systems used amplitude, phase, or frequency modulation (AM, PM, and FM, respectively) to put information on a carrier and transmission of the carrier was usually suppressed. These modulation formats are called analog modulation and the resulting frequency components can be considered as being sums of sinusoids. The signal and its response are then deterministic and a well-defined design methodology has been developed to characterize nonlinear effects. With multifrequency sinusoidal excitation consisting possibly of nonharmonically related (or non-commensurable) frequency components, the waveforms in the circuit are not periodic yet the nonlinear circuit does have a steady-state response. Even considering a single-tone signal (a single sinewave) yields directly usable design information. However, being able to model the response of a circuit to a multitone stimulus increases the likelihood that the fabricated circuit will have the desired performance.

In an FM modulated scheme the transmitted signal can be represented as

$$x(t) = \cos\left\{\left[\omega_c + \omega_i(t)\right]t\right\} + \sin\left\{\left[\omega_c + \omega_q(t)\right]t\right\} \qquad (8.47)$$

where the signal information is contained in $\omega_i(t)$ and can be adequately represented as a sum of sinewaves. The term $\omega_q(t)$ is the quadrature of $\omega_i(t)$, meaning that it is 90° out of phase. The net result is that $x(t)$ can also be represented as a sum of sinusoids. Other forms of analog modulation can be represented in a similar way. The consequence of this is that all signals in a circuit with analog modulation can be adequately represented as comprising discrete tones.

With discrete tones input to a nonlinear circuit, the output will also consist of discrete tones but will have components at frequencies that were not part of the input signal. Power series expansion analysis of a nonlinear subsystem illustrates the nonlinear process involved. When a single frequency sinusoidal signal excites a nonlinear circuit, the response "usually" includes the original signal and harmonics of the input sinewave. We say "usually," because if the circuit contains nonlinear reactive elements, subharmonics and autonomous oscillation could also be present. The process is more complicated when the excitation includes more than one sinusoid, as the circuit response may then include all sum and difference frequencies of the original signals. The term *intermodulation* is generally used to describe this process, in which power at one frequency, or group of frequencies, is transferred to power at other frequencies. The term intermodulation is also used to describe the production of sum and difference frequency components, or intermodulation frequencies, in the output of a system with multiple input sinewaves. This is a macroscopic definition of intermodulation as the generation of each intermodulation frequency component derives from many separate intermodulation processes. Here a treatment of intermodulation is developed at the microscopic level.

To begin with, consider a nonlinear system with output $y(t)$ described by the power series

$$y(t) = \sum_{l=1}^{\infty} a_l x(t)^l \tag{8.48}$$

where $x(t)$ is the input and is the sum of three sinusoids:

$$x(t) = c_1 \cos(\omega_1 t) + c_2 \cos(\omega_2 t) + c_3 \cos(\omega_3 t). \tag{8.49}$$

Thus

$$x(t)^l = \left[c_1 \cos(\omega_1 t) + c_2 \cos(\omega_2 t) + c_3 \cos(\omega_3 t) \right]^l. \tag{8.50}$$

This equation includes a large number of components the radian frequencies of which are the sum and differences of ω_1, ω_2, and ω_3. These result from multiplying out the term $[\cos(\omega_1 t)]^k [\cos(\omega_3 t)]^{l-p}$. For example

$$\cos(\omega_1 t)\cos(\omega_2 t)\cos(\omega_3 t) = \left[\cos(\omega_1 + \omega_2 + \omega_3)t + \cos(\omega_1 + \omega_2 - \omega_3)t + \cos(\omega_1 - \omega_2 + \omega_3)t \right.$$
$$\left. + \cos(\omega_1 - \omega_2 - \omega_3)t \right] / 4 \tag{8.51}$$

where the (radian) frequencies of the components are, in order, $(\omega_1 + \omega_2 + \omega_3)$, $(\omega_1 + \omega_2 - \omega_3)$, $(\omega_1 - \omega_2 + \omega_3)$, and $(\omega_1 - \omega_2 - \omega_3)$. This mixing process is called intermodulation and the additional tones are called intermodulation frequencies with each separate component of the intermodulation process called an intermodulation product or IP. Thus when a sum of sinusoids is input to a nonlinear element additional frequency components are generated. In order to make the analysis tractable, the number of frequency components considered must be limited. With a two-tone input, the frequencies generated are integer combinations of the two inputs, e.g., $f = mf_1 + nf_2$. One way of limiting the number of frequencies is to consider only the combinations of m and p such that

$$|m| + |n| \leq p_{MAX} \tag{8.52}$$

assuming that all products of order greater than p_{MAX} are negligible. This is called a triangular truncation scheme and is depicted as shown in Fig. 8.31. The alternative rectangular truncation scheme is shown in Fig. 8.32 and is defined by

$$|m| \leq m_{MAX} \quad \text{and} \quad |n| \leq n_{MAX}. \tag{8.53}$$

With one-tone excitation, the spectra of the input and output of a nonlinear circuit consists of a single tone at the input and the original, fundamental tone, and its harmonics. Here intermodulation converts power at f_1 to power at DC (this intermodulation is commonly referred to as rectification), and to power at the harmonics ($2f_1, 3f_1, \ldots$), as well as to power at f_1. Simply squaring a sinusoidal signal will give rise to a second harmonic component. The measured and simulated responses of a class A amplifier operating at 2 GHz are shown in Fig. 8.33. This exhibits classic responses. At low signal levels the fundamental response has a slope of 1:1 with respect to the input signal level — corresponding to the linear response.

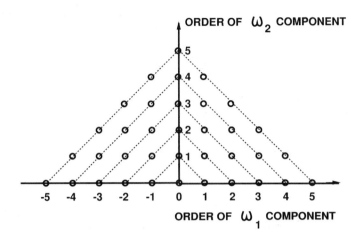

FIGURE 8.31   A triangular scheme for truncating higher order tones.

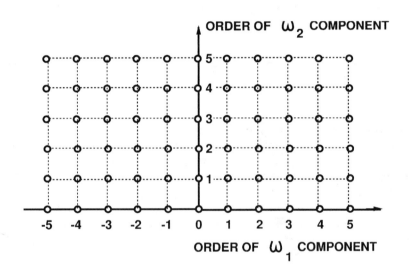

FIGURE 8.32   A rectangular scheme for truncating higher order tones. Here $m_{max} = 5 = n_{max}$.

Initially the second harmonic varies as the square of the input fundamental level and so has a 2:1 slope on the log-log plot. This is because the dominant IP contributing to the second harmonic level at low input powers is second order. Similarly the third harmonic response has a 3:1 slope because the dominant IP here is third order. As the input power increases, the second harmonic exhibits classic nonlinear behavior which is observed with many intermodulation tones and results from the production of a second, or more significant IP tone, which is due to higher order intermodulation than the dominant IP. In this situation, the dominant and additional IPs vectorially combine, with the result that the tone almost cancels out.

It is much more complicated to describe the nonlinear response to multifrequency sinusoidal excitation. If the excitation of an analog circuit is sinusoidal, then specifications of circuit performance are generally in terms of frequency domain phenomena, e.g., intermodulation levels, gain, and the 1 dB gain compression point. However, with multi-frequency excitation by signals that are not harmonically related, the waveforms in the circuit are not periodic, although there is a steady-state response often called quasi-periodic.

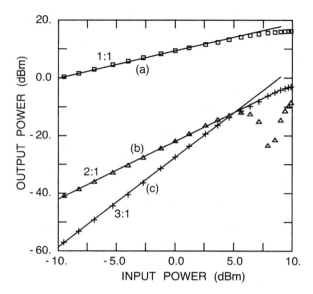

FIGURE 8.33 Measured (markers) and simulated (lines) response of a class A MESFET amplifier to a single tone input: (a) is the fundamental output; (b) is the second harmonic response; and (c) is the third harmonic response.

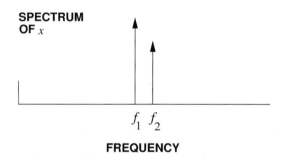

FIGURE 8.34 The spectrum of a two-tone signal.

Consider the nonlinear response of a system to the two-tone excitation shown in Fig. 8.34. The frequencies f_1 and f_2 are, in general, nonharmonically related and components at all sum and difference frequencies $mf_1 + nf_2$, $(m, n = -\infty,\ldots,-1,0,1,\ldots,\infty)$ of f_1 and f_2 will appear at the output of the system. If the nonlinear system has a quadratic nonlinearity, the spectrum of the output of the system is that of Fig. 8.35. With a general nonlinearity, the spectrum of the output will contain a very large number of components. An approximate output spectrum is given in Fig. 8.36. Also shown is a truncated spectrum that will be used in the following discussion. Most of the frequency components in the truncated spectrum of Fig. 8.36 have names: DC (f_6) results from rectification; $f_3, f_4, f_5, f_8, f_9, f_{10}$, and f_{11} are called intermodulation frequency components; f_4, f_5 are commonly called image frequencies, or "third order" intermods, as well; f_1, f_2 are the input frequencies; and f_7, f_8 are harmonics.

All of the frequencies in the steady-state output of the nonlinear system result from intermodulation — the process of frequency mixing. A classification of nonlinear behavior that closely parallels the way in which nonlinear responses are observed and specified is given below.

Gain Compression/Enhancement: Gain compression can be conveniently described in the time domain or in the frequency domain. Time domain descriptions refer to limited power availability or to limitations on voltage or current swings. At low signal levels, moderately nonlinear devices such as class

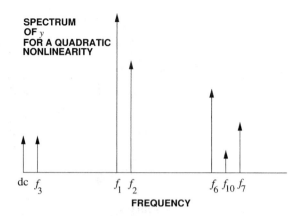

FIGURE 8.35   The spectrum at the output of a quadratic nonlinear system with a two-tone input.

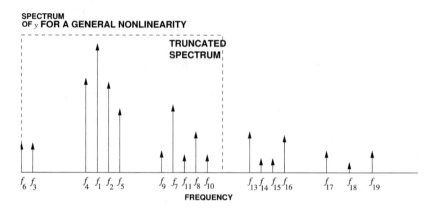

FIGURE 8.36   The approximate spectrum for a general nonlinearity with a two-tone input.

A amplifiers behave linearly so that there is one dominant IP with a zero saturation term. As signal levels increase, other IPs become important as harmonic levels increase. Depending on the harmonic loading condition, these IPs could be in phase with the original IP contributing to gain enhancement or out of phase contributing to gain compression.

Desensitization: Desensitization is the variation of the amplitude of one of the desired components due to the presence of another noncommensurable signal. This is an over-riding saturation effect affecting all output tones and comes out of the power series expansion.

Harmonic Generation: Harmonic generation is the most obvious result of nonlinear distortion and is identical to the process with a single-tone input.

Intermodulation: Intermodulation is the generation of spurious frequency components at the sum and difference frequencies of the input frequencies. In the truncated spectrum $f_3, f_4, f_5, f_9, f_{10},$ and f_{11} are intermodulation frequencies. Numerically $f_4 = 2f_1 - f_2$ and so this intermodulation tone is commonly called the lower third order intermod. There are other IPs that can contribute to the "third order" intermod that are not due to third order intermodulation. A particularly important intermodulation process begins with the generation of the difference frequency component $f_3 = f_2 - f_1$ as a second order IP. This is also referred to as the baseband component, envelope frequency, intermediate frequency, or difference frequency. This component then mixes with one of the original tones to contribute to the level of the "third order" intermod, e.g., $f_4 = f_1 - f_3$, again a second order process. The corresponding contribution to the upper third order intermod f_5, i.e., $f_5 = f_2 + f_3$, can (depending on the baseband impedance) have a phase that differs from the phase of the f_4 contribution and, in general, the result is that there can be asymmetry in the lower and upper third order intermod levels as the various IPs, at their respective frequencies, add vectorially.

Cross-modulation: Cross-modulation is modulation of one component by another noncommensurable component. Here it would be modulation of f_1 by f_2 or modulation of f_2 by f_1. However, with cross-modulation, information contained in the sidebands of one non-commensurable tone can be transferred to the other non-commensurable tone.

Detuning: Detuning is the generation of DC charge or DC current resulting in change of an active device's operating point. The generation of DC current with a large signal is commonly referred to as rectification. The effect of rectification can often be reduced by biasing using voltage and current sources. However, DC charge generation in nonlinear reactances is more troublesome as it can neither be detected nor effectively reduced.

AM-PM Conversion: The conversion of amplitude modulation to phase modulation (AM-PM conversion) is a troublesome nonlinear phenomenon in high frequency analog circuits and results from the amplitude of a signal affecting the delay through a system. Alternatively, the process can be understood by considering that at higher input levels, additional IPs are generated at the fundamental frequency and when these vectorially contribute to the fundamental response, phase rotation occurs.

Subharmonic Generation and Chaos: In systems with memory effects, i.e., with reactive elements, subharmonic generation is possible. The intermodulation products for subharmonics cannot be expressed in terms of the input non-commensurable components. (Components are non-commensurable if they cannot be expressed as integer multiples of each other.) Subharmonics are initiated by noise, possibly a turn-on transient, and so in a steady-state simulation must be explicitly incorporated into the assumed set of steady-state frequency components. The lowest common denominator of the subharmonic frequencies then becomes the basis non-commensurable component. Chaotic behavior can only be simulated in the time domain. The nonlinear frequency domain methods as well as the conventional harmonic balance methods simplify a nonlinear problem by imposing an assumed steady-state on the nonlinear circuit solution problem. Chaotic behavior is not periodic and so the simplification is not valid in this case. Together with the ability to simulate transient behavior, the capability to simulate chaotic behavior is the unrivaled realm of time domain methods.

Except for chaotic behavior, all nonlinear behavior with discrete tones can be viewed as an intermodulation process with IPs (the number of significant ones increasing with increasing signal level) adding vectorially. Understanding this process provides valuable design insight and is also the basis of frequency domain nonlinear analysis.

Digitally Modulated Signals

A digitally modulated signal cannot be represented by discrete tones and so nonlinear behavior cannot be adequately characterized by considering the response to a sum of sinusoids. Nonlinear effects with digital are difficult to describe as the signals themselves appear to be random, but there is an underlying correlation. It is more appropriate to characterize a digitally modulated signal by its statistics, such as power spectral density, than by its component tones. Most current (and future) wireless communication systems use digital modulation, in contrast to first-generation radio systems, which were based on analog modulation. Digital modulation offers increased channel capacity, improved transmission quality, secure channels, and the ability to provide other value-added services. These systems present significant challenges to the RF and microwave engineer with respect to representation and characterization of digitally modulated signals, and also with respect to nonlinear analysis of digital wireless communication systems.

Amplifier linearity in the context of digital modulation is therefore most suitably characterized by measuring the degree of spectrum regeneration. This is done by comparing the power in the upper and lower adjacent channels to the power in the main channel: the adjacent-channel power ratio (ACPR). The spectrum of a digitally modulated signal is shown in Fig. 8.37. This is the spectrum of a finite bit length digitally modulated signal and not the smooth spectrum of an infinitely long sequence often depicted.

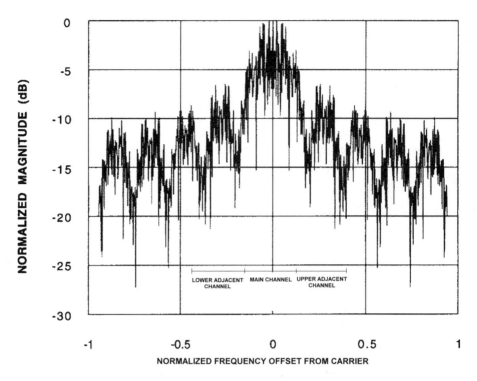

FIGURE 8.37   Spectrum of a digitally modulated signal.

Basics of Circuit Modeling

The solution, or simulation, of a circuit is obtained by solving a number of network equations developed by applying Kirchoff's current law (KCL) and Kirchoff's voltage law (KVL). There are two basic methods for developing the network equations for DC analysis, or steady-state analysis of linear circuits with sinusoidal excitation, based on Kirchoff's laws. These are the nodal formulation and mesh formulation of the network equations. The nodal formulation is best for electronic circuits as there are many fewer nodes than there are elements connecting the nodes. The nodal formulation, specifically node-voltage analysis, requires that the current in an element be expressed as a function of voltage. Some elements cannot be so described and so there is not a node-voltage description for them. Then the modified nodal approach is most commonly used wherein every element that can be described by an equation for current in terms of voltages is described in this way, and only for the exceptional elements are other constitutive relations considered. However, the general formulation approach can be illustrated by considering node-voltage analysis.

The nodal formulation of the network equations is based on the application of KCL, which in its general form states that if a circuit is partitioned, then the total instantaneous current flowing into a partition is zero. This is an instantaneous requirement — physically it is only necessary that the net current flow be zero on average to ensure charge conservation. So this is an artificial constraint imposed by circuit analysis technology. The approach used in overcoming this restriction is to cast this issue as a modeling problem: it is the responsibility of the device modeler to ensure that a model satisfies KCL instantaneously. This results in many of the modeling limitations that are encountered. A general network is shown in Fig. 8.38. The concept here is that every node of the circuit is pulled to the outside of the main body of the network. The main body contains only the constitutive relations and the required external nodes have the connectivity information to implement Kirchoff's laws. The result is that the

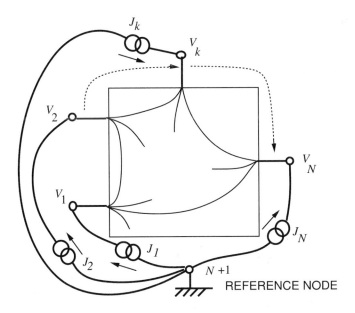

FIGURE 8.38   General network.

constitutive relations are contained in the main body, but the variables, the node-voltages, and the external currents are clearly separated. This representation of a network enables as uniform a treatment as possible. It makes it very easy to add one element at a time to the network as variables are already defined. Indeed this is how all general purpose simulators work and the network equations are built up by inspection. Initially the network is defined with nothing in the main body and only the variables defined. Then each element is considered in turn and the describing relations added to the evolving network equation matrix. This representation serves us well when it comes to harmonic balance. Applying KCL to each of the nodes of the network the following matrix network equation is obtained:

$$\mathbf{YV} = \mathbf{J}. \tag{8.54}$$

Here $\mathbf{Y}$ is the nodal admittance matrix of the network, $\mathbf{V}$ is the vector of node voltages (i.e., voltages at the nodes each referred to the reference node), and $\mathbf{J}$ is the vector of external current sources at each node. Expanding the matrix equation:

$$
\begin{bmatrix}
y_{11} & y_{12} & \cdots & y_{1N} \\
y_{21} & y_{22} & \cdots & y_{2N} \\
\vdots & \vdots & \ddots & \vdots \\
y_{N1} & y_{N2} & \cdots & y_{NN}
\end{bmatrix}
\begin{bmatrix}
V_1 \\ V_2 \\ \vdots \\ V_N
\end{bmatrix}
=
\begin{bmatrix}
J_1 \\ J_2 \\ \vdots \\ J_N
\end{bmatrix}. \tag{8.55}
$$

We will see this utilized in the formulation of SPICE and HB analyses.

Time-Domain Circuit Simulation

The principal advantage of simulating circuits in the time domain is that it most closely resembles the real world. Phenomena such as chaos, instability, subharmonic generation, and parametric effects can be accurately simulated without the *a priori* knowledge of the spectral components of the signals in a circuit.

Direct Integration of the State Equations

The most direct method for analyzing nonlinear circuits is numerical integration of the differential equations describing the network. By applying Kirchoff's voltage and current laws and using the characteristic equations for the circuit elements (generally using the modified nodal formulation), the state equations can be written as a set of coupled first-order differential equations:

$$\dot{\mathbf{X}} = f(\mathbf{X}, t) \tag{8.56}$$

where, for example, the time derivative of a quantity such as voltage or current is a function of time and of the voltages and currents in the circuit. More generally the state equations are rearranged and written in the implicit form

$$g(\dot{\mathbf{X}}, \mathbf{X}, t) = 0 \tag{8.57}$$

where $\mathbf{X} = [X_1, X_2, \ldots, X_N]^{\mathrm{T}}$ is a set of voltages and currents, typically at different nodes and different time instants. The general formulation of Eq. (8.57) is discretized in time and solved using a numerical integration procedure. This modeling approach can be used with many systems as well as circuits and was the only approach considered in the early days of circuit simulation (in the 1960s). Unfortunately, it was not robust except for the simplest of circuits. SPICE-like analysis, considered next, solves the same problem but in a much more robust way.

SPICE: Associated Discrete Circuit Modeling

SPICE is the most common of the time domain methods used for nonlinear circuit analysis. This method is fundamentally the same as that just described in that the state equations are integrated numerically, however the order of operations is changed. The time discretization step is applied directly to the equations describing the circuit element characteristics. The nonlinear differential equations are thereby converted to nonlinear algebraic equations. Kirchoff's voltage and current laws are then applied to form a set of algebraic equations solved iteratively at each time point.

Converting the differential equations describing the element characteristics into algebraic equations changes the network from a nonlinear dynamic circuit to a nonlinear resistive circuit. In effect, the differential equations describing the capacitors and inductors, for example, are approximated by resistive circuits associated with the numerical integration algorithm. This modeling approach is called associated discrete modeling or just companion modeling. The term "associated" refers to the model's dependence upon the integration method while "discrete" refers to the model's dependence on the discrete time value.

The numerical integration algorithm is the means by which the element characteristics are turned into difference equations. Three low order numerical integration formulas are commonly used: the Forward Euler formula, the Backward Euler formula, and the Trapezoidal Rule. A generalization of these to higher order is called the weighted integration formula from which the Gear Two method, available in some SPICE simulators, is derived. In all methods the aim is to estimate the state of a circuit at the next time instant from the current state of the circuit and derivative information. In one dimension and denoting the current state by x_0 and the next state by x_1, the basic integration step is

$$x_1 = x_0 + hx'. \tag{8.58}$$

The formulas differ by the method used to estimate x'.

In the Forward Euler Formula, $x' = x'_0$ is used and the basic numerical integration step [Eq. (8.58)] becomes

$$x_1 = x_0 + hx'_0. \tag{8.59}$$

Numerical integration using the forward Euler formula is called a predictor method as information about the behavior of the waveform at time t_0, x_0', is used to predict the waveform at t_1.

In the Backward Euler Formula, $x' = x_1'$ is used and the discretized numerical integration equation becomes

$$x_1 = x_0 + h x_1'. \qquad (8.60)$$

The obvious problem here is how to determine x_1' when x_1 is not known. The solution is to iterate as follows: (1) assume some initial value for x_1 (e.g., using the Forward Euler formula); and (2) iterate to satisfy the requirement $x_1' = f(x_1, t)$. Discretization using the Backward Euler formula is therefore called a predictor-corrector method.

In the Trapezoidal Rule, $x' = (x_0' + x_1')/2$ is used and the discretized numerical integration equation becomes

$$x_1 = x_0 + h\left(x_0' + x_1'\right)\big/2. \qquad (8.61)$$

So the essence of the trapezoidal rule is that the slope of the waveform is taken as the average of the slope at the beginning of the time step and the slope at the end of the time step determined using the Backward Euler formula.

There is a significant difference in the numerical stability, accuracy, and run times of these methods, although all will be stable with a small enough step size. Note that stability is a different issue than whether or not the correct answer is obtained. The Backward Euler and Trapezoidal Rules will always be stable and these are the preferred integration methods. The Forward Euler method of discretization does not always result in a numerically stable method. This can be understood by considering that the Forward Euler method always predicts the response into the future and does not improve on the guess using other information that can be obtained. The Backward Euler and Trapezoidal Rule approaches use a prediction of the future state of a waveform, but then require iteration to correct any error and use derivative information as well as instantaneous information to achieve this. Generally, when any simulation strategy is first developed, predictor methods are used. However, in the long run, predictor-corrector methods are always adopted as these have much better overall performance in terms of stability and accuracy but do require much more development effort. Except for the Forward Euler method, none of the other methods are clearly the best choice in all circumstances, and experimentation should occur. Generally, we can say that for RF and microwave circuits that have resonant bandpass-pass characteristics, the Trapezoidal Rule tends to result in an over-damped response and the Backward Euler method results in an under-damped response. The effect of this on accuracy, the prime requirement, is not consistent and must be investigated for a specific circuit.

Associated Discrete Model of a Linear Element

The development of the associated discrete model (ADM) of an element begins with a time discretization of the constitutive relation of the element. The development for a linear capacitor is presented here as an example. The simplest algorithm to use in developing this discretization is the Backward Euler integration formula. The Backward Euler algorithm for solving the differential equation

$$\dot{x} = f\left(x\right) \qquad (8.62)$$

with step size $h = t_{n+1} - t_n$ is

$$x_{n+1} = x_n + h f\left(x_{n+1}\right) = x_n + h \dot{x}_{n+1} \qquad (8.63)$$

where the subscript n refers to the nth time sample. The discretization is performed for each and every element independently by replacing the differential equation of Eq. (8.62) by the constitutive relation of the particular elements.

For a linear capacitor, the charge on the capacitor is linearly proportional to the voltage across it so that $q = Cv$. Thus

$$i(t) = \frac{dq}{dt} = C\frac{dv}{dt} = C\dot{v}$$

or

$$v_{n+1} = \frac{1}{C}i_{n+1} \tag{8.64}$$

where the reference convention for the circuit quantities are defined in Fig. 8.39.

Substituting Eq. (8.64) into Eq. (8.65) and rearranging leads to the discretized Backward Euler model of the linear capacitor:

$$i_{n+1} = \frac{C}{h}v_{n+1} - \frac{C}{h}v_n. \tag{8.65}$$

This equation has the form

$$i_{n+1} = g_{eq}v_{n+1} - i_{eq} \tag{8.66}$$

and so is modeled by a constant conductance $g_{eq} = C/h$ in parallel with a current source $i_{eq} = -C/hv_n$ that depends on the previous time step, as shown in Fig. 8.40. The associated discrete circuit models for all other elements are developed in the same way, but of course the development is usually much more complicated, especially for nonlinear and multiterminal elements, but the approach is the same. The final circuit combining the ADM of all of the elements is linear with resistors and current sources, as well as a few special elements such as voltage sources. This circuit

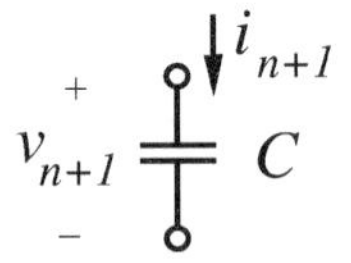

FIGURE 8.39 Reference direction for the circuit quantities of a capacitor.

is especially compatible with the nodal-formulation described by Eq. (8.55). The linear circuit is then solved repeatedly with the circuit elements updated at each step and, if the circuit voltage and current quantities change by less than a specified tolerance, the time step advanced.

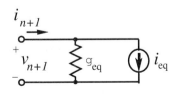

FIGURE 8.40 The associated discrete model of a two-terminal element.

The feature that distinguishes the associated discrete modeling approach from integration of the state equations for the system is that the discretization and particularly the Newton iteration is performed at the individual element level rather than at the top system level. The most important aspect of this is that special convergence treatments can be applied locally. For example, a diode has an exponential relationship between current and voltage and is the most difficult characteristic to handle. With the top-level systems-of-equations approach, any convergence scheme developed would need to be applied to all elements in a circuit, not just to the problem elements. In the associated discrete modeling, many local steps can be taken to improve convergence properties. This can include limiting the voltage and current changes from one iteration step to another. The scheme adopted depends on the characteristics of a particular element and heuristics developed in using it. It is this focus on local convergence control and embedding specific element knowledge in the element code that makes the SPICE approach so successful.

The Shooting Method

As has been mentioned, time-marching simulation has problems in determining the steady-state response because of the long simulation times that are involved. There is one elegant solution when the excitation is a sinusoid so that the response is known to be periodic. For strictly periodic excitation, shooting methods are often used to bypass the transient response altogether. This is advantageous in situations that would require many iterations for the transient components to die out. It is assumed that the nonlinear circuit has a periodic solution and that the solution can be determined by finding an initial state such that transients are not excited. If $\mathbf{x}(t)$ is the set of state variables obtained by a time-domain analysis, the boundary value constraint for periodicity is that $\mathbf{x}(t) = \mathbf{x}(t + T)$, where T is the known period. A series of iterations at time points between t and $t + T$ can be performed for a given set of initial conditions, and the condition for periodicity checked. Thus, in the shooting method, the problem of solving the state equations is converted into the two-point boundary value problem

$$\mathbf{x}(0) = \mathbf{x}(T)$$

$$\mathbf{x}(T) = \int_0^T \mathbf{f}(\mathbf{x}, \tau) d\tau + \mathbf{x}(0).$$

(8.67)

If $\mathbf{x}(t) \neq \mathbf{x}(t + T)$ then a new set of initial conditions can then be determined using a gradient method based upon the error in achieving a periodic solution. Once the sensitivity of the circuit to the choice of initial conditions is established in this way, a set of initial conditions that establishes steady-state operation can be determined; this set is, of course, the desired solution. This iterative procedure can be implemented using the Newton's method iteration

$$\mathbf{x}^{k+1} = \mathbf{x}^k - \left[\mathbf{I} - \frac{\partial \mathbf{x}^k(T)}{\partial \mathbf{x}^k(0)} \right]^{-1} \left[\mathbf{x}^k(0) - \mathbf{x}^k(T) \right]$$

(8.68)

where the superscripts refer to iteration numbers and $\mathbf{x}^k(T)$ is found by integrating the circuit equations over one period from the initial state $\mathbf{x}^k(0)$.

To begin the analysis, the period (T) is determined and the initial state $(\mathbf{x}^k(0))$ is estimated. Using these values, the circuit equations are numerically integrated from $t = 0$ to $t = T$ and the necessary derivatives calculated. Then, the estimate of the initial state is updated using the Newton iteration [Eq. (8.68)]. This process is repeated until $\mathbf{x}(0) = \mathbf{x}(T)$ is satisfied within a reasonable tolerance.

Shooting methods are attractive for problems that have small periods. Unlike the direct integration methods, the circuit equations are only integrated over one period (per iteration). They are therefore more efficient, provided that the initial state can be found in a number of iterations that is smaller than the number of periods that must be simulated before steady-state is reached in the direct methods. Unfortunately, shooting methods can only be applied to find periodic solutions. Also, shooting methods become less attractive for cases where the circuit has a large approximate period, for example, when several nonharmonic signals are present. The computation becomes further complicated when transmission lines are present, because functional initial conditions are then required to establish the initial conditions at every point along the line (corresponding to the delayed instants in time seen at the ports of the line).

In multitone situations when only one signal is large and when operating frequencies are not so high that distributed effects are important, the large tone response can be captured using the shooting method and then the frequency conversion method described in the next section can be used to determine the response with the additional small signals present.

Frequency Conversion Matrix Methods

In many multitone situations, one of two or more impressed non-commensurate tones is large while the others are much smaller. In a mixer, a large local oscillator, LO, (which is generally 20 dB or more larger than the other signals) pumps a nonlinearity, while the effect of the other signals on the waveforms at the nonlinearities is negligible. The pumped time-invariant nonlinearity can be replaced by a linear time-varying circuit without an LO signal. The electrical properties of the time-varying circuit are described by a frequency domain conversion matrix. This conversion matrix relates the current and voltage phasors of the first order sidebands with each other. In other words, by performing a fast, single-tone shooting method or harmonic balance analysis with only the LO impressed upon it, the AC operating point of the mixer may be determined and linearized with respect to small-signal perturbations about this point. This information is already available in the Jacobian, which is essentially a gradient matrix relating the sensitivity of one dependent variable to another independent variable. A two-tone signal can be rewritten to group the LO waveform, $x_{LO}(t)$ terms and the first order sidebands as

$$x(t,j) = x_{LO}(t) + \mathrm{Re}\left\{\sum_{p=0}^{N_A} \mathbf{X}_{p,1} e^{j(p\omega_{LO}+\omega_{RF})} + \sum_{p=0}^{N_A} \mathbf{X}_{p,-1} e^{j(p\omega_{LO}-\omega_{RF})}\right\} \tag{8.69}$$

where $\mathbf{X}_{p,1}$ and $\mathbf{X}_{p,-1}$ are vectors of the spectral components at the first order sidebands of the pth harmonic of the LO. For voltage controlled nonlinearities, the output quantities (the $\mathbf{X}$'s) are current phasors so that the expression relating the IF current to the RF voltage is

$$\left[\mathbf{I}_{p,1}, \mathbf{I}_{p,-1}\right]^{\mathrm{T}} = \mathbf{Y}_C\left[\mathbf{V}_{0,1}, \mathbf{V}_{0,-1}\right]^{\mathrm{T}}. \tag{8.70}$$

Here $\mathbf{Y}_C$ is the admittance conversion matrix and can be used in much the same manner as a nodal admittance matrix. Alternatively, for current-controlled nonlinearities the following could be used:

$$\left[\mathbf{V}_{0,1}, \mathbf{V}_{0,-1}\right]^{\mathrm{T}} = \mathbf{Z}_C\left[\mathbf{I}_{0,1}, \mathbf{I}_{0,-1}\right]^{\mathrm{T}} \tag{8.71}$$

where $\mathbf{Z}_C$ is the impedance conversion matrix. Nonlinearities with state variable descriptions or mixed voltage-controlled and current-controlled descriptions require a combination of Eqs. (8.70) and (8.71) to derive a modified nodal admittance formulation.

Convolution Techniques

The fundamental difficulty encountered in integrating RF and microwave circuits in a transient circuit simulator arises because circuits containing nonlinear devices or time dependent characteristics must be characterized in the time domain while distributed elements such as transmission lines with loss, dispersion, and interconnect discontinuities are best simulated in the frequency domain. Convolution techniques are directed at the simulation of these circuits.

The procedure begins by partitioning the network into linear and nonlinear subcircuits as shown in Fig. 8.41. In a typical approach the frequency domain admittance (y) parameter description of the distributed network is converted to a time domain description using a Fourier transform. This time domain description is then the Dirac delta impulse response of the distributed system. Using the method of Green's function, the system response is found by convolving the impulse response with the transient response of the terminating nonlinear load. Normally this requires that the impulse response be extended in time to include many reflections. While this technique can handle arbitrary distributed networks, a difficulty arises as the y parameters of a typical multiconductor array have a wide dynamic range. For a low loss, closely matched, strongly coupled system, the y parameters describing the coupling mechanism

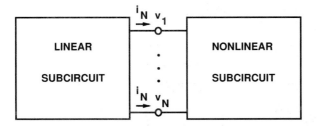

FIGURE 8.41 Circuit partitioned into linear and nonlinear sub-circuits.

approach zero at low frequencies and become very large at high frequencies. Conversely, the transmission and self-admittance *y* parameters approach infinity at DC and zero at resonance frequencies. Both numerical extremes are important in describing the physical process of reflections and crosstalk. The dynamic range of the time domain solution is similarly large and values close to zero are significant in determining reflections and crosstalk. Consequently, aliasing in the frequency domain to time domain transformation can cause appreciable errors in the simulated transient response. The problem is considerably reduced by using resistive padding at the linear-nonlinear circuit interface to reduce the dynamic range of the variables being transformed. The effect of the padding can then be removed in subsequent iteration.

Harmonic Balance: Mixed Frequency and Time Domain Simulation

The Harmonic Balance (HB) procedure has emerged as a practical and efficient tool for the design and analysis of nonlinear circuits in steady-state with sinusoidal excitation. The harmonic balance method is a technique that allows efficient solution of the steady-state response of a nonlinear circuit. For example, the steady-state response of a circuit driven by one or more sinusoidal signals is also a sum of sinusoids and includes tones at frequencies other than those of the input sinusoids (e.g., harmonics and difference frequencies). The response does not need to be periodic to be steady-state and with narrowband systems it is common to call the response to a complicated narrowband input as being quasi-periodic. Usually we are not interested in the transient response of the circuit such as when the power supply is turned on or when a signal is first applied. Thus much of the behavior of the circuit is not of interest. The harmonic balance procedure is a technique to extract just the information that is required to describe the steady-state response. The method may also be compared to the solution of a homogeneous, ordinary differential equation. A solution that is the sum of sinusoids of unknown amplitudes is substituted into the differential equation. Using the orthogonality of the sinusoids, the resulting problem simplifies to solving a set of nonlinear algebraic equations for the amplitudes of the sinusoids. There are several methods of solving for (complex) amplitudes, which will be discussed later in this section.

The HB method formulates the system of nonlinear equations in either domain (although more typically the time domain), with the linear contributions calculated in the frequency domain and the nonlinear contributions in the time domain. This is a distinct advantage for microwave circuits, in that distributed and dispersive elements are then much more readily modeled analytically or using alternative electromagnetic techniques based in the frequency domain.

While it is common to refer to the nonlinear calculations as being in the time domain, the most usual HB implementations require that the nonlinear elements be described algebraically, that is without derivatives or other memory effects. Thus a nonlinear resistor is described, for example, as a current as a nonlinear function of instantaneous voltage. So given the voltage across the nonlinear resistor at a particular time, the current that flows at the same instant can be calculated. A nonlinear capacitive element must be expressed as a charge which is a nonlinear function of instantaneous voltage. Then a sequence of charge values in time is Fourier transformed so that phasors of charge are obtained. Each phasor of charge is then multiplied by the appropriate $j\omega$ to yield current phasors.

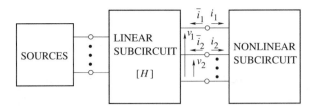

FIGURE 8.42   Circuit partitioned into linear, nonlinear, and source sub-circuits.

Problem Formulation

The harmonic balance method seeks to match the frequency components (harmonics) of current at the interface of two sub-circuits — one linear and one nonlinear. The sub-circuits are chosen in such a way that nonlinear elements are partitioned into one sub-circuit, linear elements into another, and (in some approaches) sources into a third (see Fig. 8.42). The edges at the linear/nonlinear interface connect the two circuits and define corresponding nodes; current flowing out of one circuit must equal that flowing into the other. Every node in the nonlinear circuit is "pulled out" of the nonlinear sub-circuit so that it is at the interface and becomes part of the error function formulation. Matching the frequency components in each edge satisfies the continuity equation for current. The current at each edge is obtained by a process of iteration so that dependencies are satisfied for both the linear and nonlinear sides of the circuit.

The unknowns are found by forming an error function — typically the Kirchoff's Current Law (KCL) error at the linear/nonlinear interface. This error function is minimized by adjusting the voltages at the interface. Every node in the nonlinear sub-circuit is therefore considered to be connected to the linear sub-circuit. If the total circuit has N nodes, and if $\mathbf{v}$ is the vector of node voltage waveforms, then applying KCL to each node yields a system of equations

$$f(\mathbf{v},t) = \mathbf{i}\big[\mathbf{v}(t)\big] + \frac{d}{dt}\mathbf{q}\big[\mathbf{v}(t)\big] + \int_{-\infty}^{t} y(t-\tau)\mathbf{v}(\tau)d\tau + \mathbf{i}_s(t) = 0 \tag{8.72}$$

where the nonlinear circuit is chosen to contain only voltage-controlled resistors and capacitors for representational ease. The quantities $\mathbf{i}$ and $\mathbf{q}$ are the sum of the currents and charges entering the nodes from the nonlinearities, $\mathbf{y}$ is the matrix impulse response of the linear circuit with all the nonlinear devices removed, and $\mathbf{i}_s$ are the external source currents.

In the frequency domain, the convolution integral maps into $\mathbf{YV}$, where $\mathbf{V}$ contains the Fourier coefficients of the voltages at each node and at each harmonic, and $\mathbf{Y}$ is a block node admittance matrix for the linear portion of the circuit. The system of Eq. (8.72) then becomes, on transforming into the frequency domain

$$F(\mathbf{V}) = \mathbf{IV} + \Omega\mathbf{QV} + \mathbf{YV} + \mathbf{I}_S = 0 \tag{8.73}$$

where Ω is a matrix with frequency coefficients (terms such as $j\Omega_k$) representing the differentiation step. The notation here uses small letters to represent the time domain waveforms and capital letters for the frequency domain spectra. This equation is, then, just KCL in the frequency domain for a nonlinear circuit. HB seeks a solution to Eq. (8.73) by matching harmonic quantities at the linear-nonlinear interface. The first two terms are spectra of waveforms calculated in the time domain via the nonlinear model, i.e.,

$$F(\mathbf{V}) = \Im i\big(\Im^{-1}\mathbf{V}\big) + \Omega\Im q\big(\Im^{-1}\mathbf{V}\big) + \mathbf{YV} + \mathbf{I}_S = 0 \tag{8.74}$$

where $\Im$ is the Fourier transform and $\Im^{-1}$ is the inverse Fourier transform.

The solution of Eq. (8.74) can be obtained by several methods. One method, known as relaxation, uses no derivative information and is relatively simple and fast, but is not robust. In a relaxation method the error function is taken to zero by adjusting current phasors or voltage phasors on successive iterations using what is in effect very limited derivative information. Alternatively, gradient methods can be used to solve either a system of equations (e.g., using a quasi-Newton method) or to minimize an objective function using a quasi-Newton or search method.

The Newton and quasi-Newton methods require derivative information to guide the error minimization process. Calculation of these derivatives is computationally intensive and generally the equations for these require considerable development. As with all the harmonic balance methods, the number of nodes used can be reduced by "burying" internal nodes within the linear network, which then becomes a single multiterminal sub-circuit as far as harmonic balance is concerned. The system of equations is then reduced accordingly. Once the "interfacial" node voltages are known, any internal node voltage can be found by using simple linear analysis and the full y matrix for the linear circuit.

Multitone Analysis

The problem with multitone analysis reduces to implementing a method to perform the multifrequency Fourier transform operations required in solving Eq. (8.74). This also requires developing the multifrequency Jacobian required in a Newton-like procedure. Time-frequency conversion for multitone signals can be achieved using nested Fourier transform operations. This is implemented using the multidimensional Fast Fourier Transform, or MFFT. Application of the multidimensional Fourier transform (MFFT), in which the Fourier coefficients are themselves periodic in the other dimensions, requires that the multiple tones (in each dimension) be truly orthogonal, i.e., not integer multiples of each other. If the two tones are frequency degenerate, then the method fails because orthogonality of the bases is a requirement for determining the Fourier coefficients in that basis. In such a case, one of the tones is slightly shifted to ensure that the technique can be applied.

The most general and easily programmed of the Fourier transform techniques applied to the HB method is the Almost-Periodic Discrete Fourier Transform (APDFT) algorithm. After truncation, consider the K arbitrarily spaced frequencies $0, \omega_1, \omega_2, \ldots, \omega_{K-1}$ generated by the nonlinearity. Then

$$\sum_{k=0}^{K-1} X_k^C \cos\omega_1 t_1 + X_k^2 \sin \omega_1 t_1 = x(t)$$ may be sampled at S time points, resulting in a set of S equations and $2K-1$ unknowns:

$$
\begin{bmatrix}
1 & \cos\omega_1 t_1 & \sin\omega_1 t_1 & \cdots & \cos\omega_{K-1} t_1 & \sin\omega_{K-1} t_1 \\
1 & \cos\omega_2 t_2 & \sin\omega_2 t_2 & \cdots & \cos\omega_{K-1} t_2 & \sin\omega_{K-1} t_2 \\
\vdots & \vdots & \vdots & \ddots & \vdots & \vdots \\
1 & \cos\omega_1 t_S & \sin\omega_2 t_S & \cdots & \cos\omega_{K-1} t_S & \sin\omega_{K-1} t_S
\end{bmatrix}
\begin{bmatrix}
X_0 \\
X_1^C \\
X_1^S \\
\vdots \\
X_{K-1}^C \\
X_{K-1}^S
\end{bmatrix}
=
\begin{bmatrix}
x(t_1) \\
x(t_1) \\
\vdots \\
x(t_S)
\end{bmatrix}.
\tag{8.75}
$$

The number of samples S must be at least $2K-1$ to uniquely determine the coefficients. This equation may compactly be written as

$$\Gamma^{-1}\mathbf{X} = x \text{ or } \Gamma x = \mathbf{X} \tag{8.76}$$

where Γ and Γ^{-1} are known as an almost-periodic Fourier transform pair. Thus the multifrequency transform can be performed as a matrix operation but spectrum mapping and Fast Fourier transformation is much faster.

Combining the above procedures yields the time invariant form of Harmonic Balance. This is referred to as just Harmonic Balance. This technique is very efficient for simulating circuits with just a few active devices and a few tones, as then there are only a few unknowns. Problems arise as the number of active devices increases or the number of tones becomes large as the size of the problem increases significantly. Still, digitally modulated signals can be reasonably modeled by considering a very large number of tones.

Method of Time-varying Phasors

Harmonic balance using time-variant phasors is ideally suited to the representation and characterization of circuits with digitally modulated signals. In contrast to time-variant harmonic balance, where the assumed phasor solution was time invariant, we instead assume a solution of the form

$$V_k(j\omega) = \text{real}\left\{\sum_{m=0}^{n} V_m(t)\exp\left[jm\omega(t) + \phi_m(t)\right]\right\}\tag{8.77}$$

where in general the amplitude, frequency, and phase of each term are allowed to vary with respect to time. If $V_m(t)$ varies slowly with respect to the carrier frequency, we are in essence solving for the envelope of the signal at each node without the requisite memory requirements of time-invariant harmonic balance, or the frequency domain dynamic range and resolution problems of time domain methods. Taking the Fourier transform of each summation term in Eq. (8.77) results in a highly resolved power spectral density distribution approximation of the digitally modulated signal, not an ill-conditioned approximation, as with time-invariant harmonic balance.

Frequency Domain Analysis of Nonlinear Circuits

Frequency domain characterization of RF and microwave circuits directly provides the types of performance parameters required in communication systems as well as many other applications of RF and microwave circuits. The Harmonic Balance (HB) method uses Fourier transformation to relate sequence of instantaneous current, voltage, and charge to their (frequency domain) phasor forms. In frequency domain nonlinear analysis techniques alternative mappings are used. There are many types of mappings for arriving at a set of (say, current) phasors as a nonlinear function of another set of (say, voltage) phasors.

The common underlying principle of frequency domain nonlinear analysis techniques is that the spectrum of the output of a broad class of nonlinear circuits and systems can be calculated directly given the input spectrum input to the nonlinear system. The mapping operation is depicted in Fig. 8.43 and is the concept behind most RF and microwave behavioral modeling approaches. Some techniques determine

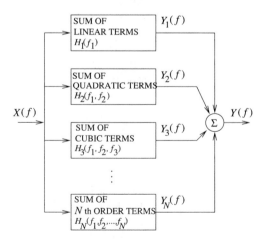

FIGURE 8.43   Mapping concept of frequency domain analysis.

an output frequency component by summing calculations of individual intermodulation products. For example, the product of two tones is, in the time domain, the product of two sinusoids. The trigonometric expansion of this yields two intermodulation products with frequencies that are the sum and difference, respectively, of the frequencies of the tones. Power series techniques use trigonometric identities to expand the power series and calculate each intermodulation product individually. Algorithms sum these by frequency to yield the output spectrum. At the coarse end of the scale are Volterra series-based techniques that evaluate groups of intermodulation products at a single frequency. Some frequency-domain non-linear analysis techniques are noniterative, although these are restricted to unilateral systems. Others, known as frequency domain spectral balance techniques, are iterative, being the frequency domain equivalent of the harmonic balance techniques. The term spectral balance is used to distinguish the frequency domain techniques from the harmonic balance techniques as the latter term has come to be solely applied to mixed time and frequency domain methods in which Fourier transformation is used. Intermediate between these extremes are techniques that operate by converting a nonlinear element into a linear element shunted by a number of controlled current sources. This process is iterative and at each iteration a residual nonlinear element is left that reduces from one iteration to another. This is the basis of one of the Volterra series analysis techniques called the method of nonlinear currents, which is also discussed in the next section.

Volterra Analysis

Expanding on Volterra analysis illustrates the concepts behind functional analysis of circuits in the frequency domain. Volterra series have the form

$$G(x) = \sum_{n=0}^{\infty} F_n(x) \tag{8.78}$$

where $F_n(x)$ is a regular homogeneous functional such that

$$F_n(x) = \int_a^b \cdots \int_a^b h_n(\chi_1, \chi_2, \ldots, \chi_n) x(\chi_1) x(\chi_2) \cdots x(\chi_n) d\chi_1 d\chi_2 \cdots d\chi_n \tag{8.79}$$

and the functions $h_n(\chi_1, \chi_2, \ldots, \chi_n)$ are known as the nth order Volterra kernels. It can be used as a time domain description (with the χ's replaced by t) of many nonlinear systems including nonlinear microwave circuits that do not exhibit hysteresis. In this case, the nth order kernel, h_n, is called the nonlinear impulse response of the circuit of order n. Equation (8.79) is then interpreted as an n dimensional convolution of an nth order impulse response (h_n) and the input signal (x). The total response $G(x)$ is the summation of the different order responses $F_n(x)$. Note that for a linear system there is only a first order response so that the total response of the system is the conventional convolution integral

$$G(x) = F_0 + \int_a^b h_1(\tau) d\tau \tag{8.80}$$

where F_0 is just a DC offset.

The important concept here is that the total response of a signal is the summation of a number of responses of different order. This scheme only works if, as the order n increases, the contribution to the response gets smaller and eventually insignificant. The reason this works for many RF and microwave circuits is that the response is close to linear and nonlinear behavior is a departure from linearity. A weak nonlinearity could be represented with just the first few terms of such a series.

In analyzing nonlinear circuits it is not necessary to deal with the Volterra series which, here, is in the time domain. Mostly the frequency domain form is used, which is expressed in terms of Volterra nonlinear transfer functions. Mathematically these are obtained by taking the n-fold Fourier transform of h_n:

$$H_n\left(f_1, f_2, \ldots, f_n\right) = \int_{-\infty}^{\infty} \cdots \int_{-\infty}^{\infty} h_n\left(\tau_1, \tau_2, \ldots, \tau_n\right) e^{-j2\left(f_1\tau_1 + \ldots \tau_n\right)} d\tau_1 d\tau_2 \cdots d\tau_n \tag{8.81}$$

where H_n is called the nonlinear transfer function of order n. The time-domain input-output relation $y(t) = f[x(t)]$ can be put in the form

$$y\left(t\right) = \sum_{n=1}^{\infty} y_n\left(t\right) \tag{8.82}$$

where

$$y_n\left(t\right) = \int_{-\infty}^{\infty} \cdots \int_{-\infty}^{\infty} h_n\left(\tau_1, \ldots, \tau_n\right) x\left(t - \tau_1\right) \cdots x\left(t - \tau_n\right) d\tau_1 \cdots d\tau_n \tag{8.83}$$

and $x(t)$ is the input. Taking the n-fold Fourier transform of both sides we have an expression for the spectrum of the nth order component of the output

$$\mathbf{Y}_n = \int_{-\infty}^{\infty} \cdots \int_{-\infty}^{\infty} H_n\left(f_1, \ldots, f_n\right) \delta\left(f - f_1 - \cdots - f_n\right) \prod_{i=1}^{n} X\left(f_i\right) df_i \tag{8.84}$$

where $\mathbf{X}_n(f)$ is the Fourier transform of $x(t)$, $\mathbf{Y}_n(f)$ is the Fourier transform of $y_n(t)$, and $\delta(\cdot)$ is the delta function. This expresses the nth order terms of the output as a function of the input spectrum. The order of the terms refers to the fact that multiplication of the input by a constant A results in multiplication of the nth order terms by A^n. Then a frequency domain series for the output can be written as

$$Y\left(f\right) = \sum_{n=1}^{\infty} Y_n\left(f\right) \tag{8.85}$$

in terms of the input spectrum and the nonlinear transfer functions. $Y_n(f)$ is the nth order response and corresponds to the response of the nth order term in the power series description of the nonlinearity.

The method of nonlinear currents enables the direct calculation of the response of a circuit with nonlinear elements that are described by a power series. Here a circuit is first solved for its linearized response described by zero and first order Volterra nonlinear transfer functions. Considering only the linearized response allows standard linear circuit nodal admittance matrix techniques to be used. The second order response, described by the second order Volterra nonlinear transfer functions can then be represented by controlled current sources. Thus the second order sources are used as excitations again enabling linear nodal admittance techniques to be used. The process is repeated for the third- and higher-order node voltages and is easily automated in a general purpose microwave simulator. The process is terminated at some specified order of the Volterra nonlinear transfer functions. This is a noniterative technique, but relies on rapid convergence of the Volterra series restricting its use to moderately strong nonlinear circuits.

Summary

SPICE is at its best when simulating large circuits as memory and computation time increase a little more than linearly after a circuit reaches a certain size. However, to determine the response to sinusoidal excitation requires simulation over a great many cycles until the transient response has died down. A major problem in itself is determining when the steady state has been achieved. A similar problem occurs with narrowband modulated signals, which can have many millions of RF cycles before the response appears to be steady state. For example, a typical sequence length for the (digitally modulated) DAMPS format is 10 ms, although the time step would be on the order of 100 ps to capture the fundamental and harmonics of the 850 MHz carrier. This results in 10^8 time-points and hence the same order discrete Fourier transforms. Fourier transformation, e.g., using a fast Fourier transform, of the simulated wave-form is required to determine its spectral content. This is not too complex a task if the exciting signal is a single frequency, but if the signal driving the nonlinear circuit has non-commensurable frequency components or is digitally modulated, then the procedure is more difficult and the effect of numerical noise is exaggerated. Even low-level numerical noise may make it impossible to extract a low-level tone in the presence of a large tone. The ability to detect a small tone defines the dynamic range of a simulator in RF and microwave applications and SPICE-analysis has poor performance in this case.

There is also a fundamental approximation error present in the SPICE algorithm due to what amounts to a *z*-domain approximation to the frequency domain characteristics of the circuit. The consequence is that time steps must be short for reasonable dynamic range. This also makes it particularly difficult to represent circuits with strongly varying narrowband frequency response. Recent extensions to SPICE — the shooting method with the frequency conversion method and convolution techniques — increase the applicability of SPICE to RF and microwave circuits. In spite of the difficulties, SPICE remains the only method of determining the transient response of a circuit.

Harmonic Balance analysis of circuits achieves significant computation savings by assuming that the signals in a circuit are steady state, described by a sum of sinusoids. The coefficients and phases of these sinusoids are solved for and not the transient response. Harmonic balance has a significant computation time advantage over SPICE for small to medium RF and microwave circuits. However, the time increases rapidly as circuit size increases. HB lends itself well to optimization and to analysis of multifunction circuits including amplifiers, oscillators, mixers, frequency converters, and numerous types of control circuits such as limiters and switches, if transient effects are not of concern. Another major advantage of the harmonic balance method is that linear circuits can be of practically any size, with no significant decrease in speed if additional internal nodes are added, or if elements of widely varying time constants are used (such is not the case with time domain simulators). Two extensions, separately implemented, also increase the usefulness of Harmonic Balance. The method of time-variant phasors enables digitally modulated signals to be handled. The second extension using matrix-free methods enables Harmonic Balance to handle very rich spectra and thus also approximately treat digitally modulated signals.

All of the techniques discussed here have been implemented in circuit simulators developed for RF and microwave circuit modeling. Many other simulator technologies exist, but these are within the overall framework of the discussion here. The reader is directed to the Further Information list for exploration of other technologies and for greater detail on those treated here.

Further Information

The bases of circuit simulation are described in J. Vlach and K. Singhal, *Computer Methods for Circuit Analysis and Design*, Van Nostrand Reinhold, 1983, ISBN 0442281080; and L. T. Pillage, R. A. Rohrer, and C. Visweswariah, *Electronic Circuit and System Simulation Methods*, McGraw-Hill, 1995, ISBN 0070501696. These two books are oriented toward SPICE-like analysis. Details on the algorithms used in SPICE are given in A. Vladimirescu, *The SPICE Book*, J. Wiley, 1994, ISBN 0471609269, and the techniques used in developing the associated discrete models used in SPICE in P. Antognetti and G. Massobrio, *Semiconductor Device Modeling with SPICE*, McGraw-Hill, 1988, ISBN 0070021538. In addition to the above, a short discussion of SPICE errors relevant to modeling RF and microwave circuits is contained in A. Brambilla and D. D'Amore, The simulation errors introduced by the SPICE transient analysis, *IEEE Trans. on Circuits and Systems-I: Fundamental Theory and Application*, 40, 57–60, January 1993. Circuit simulations oriented toward microwave circuit simulation are described in J. Dobrowolski, *Computer-Aided Analysis, Modeling, and Design of Microwave Networks: the Wave Approach*, Artech House, 1996, ISBN 0890066698; P. J. C. Rodrigues, *Computer-Aided Analysis of Nonlinear Microwave Circuits*, Artech House, 1998, ISBN 0890066906; and G. D. Vendelin, A. M. Pavio, and U L. Rohde, *Microwave Circuit Design Using Linear and Nonlinear Techniques*, Wiley, 1990, ISBN 0471602760. As well as providing a treatment of microwave circuit simulation, the following book provides a good treatment of Volterra analysis: S. A. Maas, *Nonlinear Microwave Circuits*, IEEE Press, 1997, ISBN 0780334035. Simulation of microwave circuits with digitally modulated signals is given in J. F. Sevic, M. B. Steer and A. M. Pavio, Nonlinear analysis methods for the simulation of digital wireless communication systems, *Int. J. on Microwave and Millimeter Wave Computer Aided Engineering*, 197–216, May 1996. A review of frequency domain techniques for microwave circuit simulation is given in M. B. Steer, C. R. Chang and G. W. Rhyne, Computer aided analysis of nonlinear microwave circuits using frequency domain spectral balance techniques: the state of the art, *Int. J. on Microwave and Millimeter Wave Computer Aided Engineering*, 1, 181–200, April 1991.

8.5   Time Domain Computer-Aided Circuit Simulation

Michael Lightner

The simulation of circuits at the electrical level has been the primary CAD activity of many designers. Electrical level simulation is used in both digital and analog system design. Its main use for RF and microwave circuits is to study waveform behavior and distortion characteristics of circuits under various environmental and parameter variations. Electrical level simulation is also one of the most costly CAD activities involved in the design of circuits. Because of the importance and expense of electrical simulation it has been widely studied and a huge literature exists of exact and approximate techniques. In this section we will consider the transient solution of nonlinear circuits, as this is the most important area of electrical simulation.

Before we begin the discussion of the details it is useful to see the big picture of how a program such as SPICE operates when solving for the transient response of a circuit. First the program must read in the description of the circuit and form the equations that will be solved. Actually, what SPICE does is read in the data associated with the circuit, the elements and values, the transistor models, and the element interconnections. In addition, controls such as the time range over which to solve the equations, the accuracy requirements, initial conditions, etc. are obtained from the user. Since we assume that the program will be solving nonlinear differential equations there is a set procedure (Table 8.2). These steps are common to solving for the transient response of any circuit. It is these steps that are built into the SPICE program. In fact, any program that produces exact solutions will use a procedure basically similar to this one. It is these steps which we will study in this chapter. Notice that the equations that are written and solved are linear equations. Since this forms the heart of the program we will begin with equation formulation and linear equation solution and follow that with the approximations used to deal with nonlinear and differential elements.

TABLE 8.2   Standard Steps in Solving Circuit Equations

Find the initial operating point (DC solution).
Choose a first time step (use the default).
   Discretize the derivatives (C's and L's) forming approximate models.
     Linearize the nonlinear elements forming approximate models.
      Form the *linear* equations associated with the approximate circuit.
      Solve the linear equations.
     Iterate until have solution or reduce the time step and try again.
   Check for accuracy.
If accurate continue forward in time else reduce time step and try again.

Hierarchy of Modeling and Simulation

Modern RF integrated circuits are becoming increasingly complex. Radio-on-a-chip and system-on-a-chip applications lead to extremely large circuits. In some cases. it is not possible to perform circuit simulation on the entire chip; indeed, besides taking much too much time and memory the data generated would overload any engineer's ability to assimilate information. In order to deal with the complexity of modern designs a hierarchy of models and associated simulators is used.

At the highest level, a behavioral model of the entire circuit is written, often in a standard programming language, and the large-scale input/output behavior of the system is studied. This level of modeling is done for both analog and digital circuits. Depending on the type of circuit, i.e., analog or digital, other more complex levels of modeling and simulation are used in the design process. At the lowest level typically dealt with by designers we find interconnections of individual transistors.

The transistor level is what concerns us in this section. We will assume that the higher levels of modeling and simulation have already been done. In addition, we will assume that the integrated circuit process that will be used in building the chip is well characterized and that models for the individual transistors and the interconnect are available. Thus the task that we face is the simulation of a collection of transistors in order to study the details of the voltage and current waveforms and their timing.

Typically microwave and RF transistor level simulation is used to characterize key circuit performance characteristics such as gain, phase, power saturation, and harmonic distortion. Section 8.6 of this handbook describes the application of time domain simulations in greater detail. Circuit size can range from a single transistor with input and output matching elements to dozens or hundreds of transistors for full RF radio-on-a-chip applications. For some mixed-mode circuit applications, transistor counts can get into the thousands. Each of these situations can be addressed using the methods in this section.

Modeling and Basic Circuit Equations

It is clear that no matter how accurate and efficient a simulation program is, the most important factor is the models that are used to represent the circuit. These include models for transistors, interconnects, passive components, and process and temperature dependencies. We will concentrate on the solution of network equations given the basic models. For detailed discussion of modeling the reader is referred to Section 8.7 in this handbook.

Equation Formulation

There are two common methods that are used to write circuit equations for simulators, the tableau approach [Hachtel et al., 1971] and the modified nodal method [Ho et al., 1975]. Both methods will be discussed briefly in this section.

The Tableau Approach

The simplest way to write the equations associated with a circuit is to use the tableau approach. In this technique the Kirchhoff voltage and current laws (**KVL** and **KCL**) and the individual branch/element equations are written in one large matrix. Although there are several ways of writing the KVL and KCL equations we will concentrate on one.

To begin we will assume that we are using the **associated reference direction** to align current and voltage for each element. This reference scheme is shown in Fig. 8.44.

FIGURE 8.44 Associated reference directions.

We will further assume that each element has an orientation of + and − associated with it. In SPICE the first node is the + node and the second the −, for a two-terminal element. Finally, we assume that each node in the network has a unique identifier (we will use numbers) and that the ground node has been identified and has reference number 0. For convenience, let n be the number of nodes including the ground node and b be the number of branches (assuming two terminal branches). With these assumptions we can now form the node-branch incidence matrix A. This matrix, A, is $(n-1) \times b$ with entries:

$$A_{ij} = 1 \text{ if the } + \text{ of element } j \text{ is connected to node } i$$

$$= -1 \text{ if the } - \text{ of element } j \text{ is connected to node } i$$

$$= 0 \text{ otherwise}$$

Now if we let i_b be the vector of branch currents, then KCL can be expressed as:

$$A\, i_b = 0$$

Further, if v_b is the vector of branch voltages and v_n is the vector of node voltages, not including the reference node, then KVL can be expressed:

$$v_b = A^T v_n$$

Finally, we need to express the **branch relationships.** For complete generality in dealing with two-terminal elements, including independent and dependent sources, we need a form of the branch equations such as:

$$R\, i_b + G\, v_b = s$$

where R is a matrix of resistances, G is a matrix of conductances, and s is a vector of source values. For example, if the kth element was a resistor, R_2, then $R_{kk} = R_2$, $G_{kk} = -1$, and $s_k = 0$. Further, if the nth component was a voltage source of value V_n, then $R_{nn} = 0$, $G_{nn} = 1$, and s_n and V_n. If we arrange these equations in one large matrix, we have the **tableau formulation** which has the form:

$$\begin{bmatrix} A & 0 & 0 \\ 0 & -I & A^T \\ R & G & 0 \end{bmatrix} \begin{bmatrix} i_b \\ v_b \\ v_n \end{bmatrix} = \begin{bmatrix} 0 \\ 0 \\ s \end{bmatrix}$$

A simple example is given in Fig. 8.45. For Fig. 8.45 the entries to the tableau are:

$$A = \begin{bmatrix} 1 & 1 & 0 & 0 \\ 0 & -1 & 1 & 1 \end{bmatrix}$$

$$R = \begin{bmatrix} 0 & & & \\ & R_1 & & \\ & & -1 & \\ & & & R_2 \end{bmatrix} \quad G = \begin{bmatrix} 1 & & & \\ & -1 & & \\ & & G_1 & \\ & & & -1 \end{bmatrix} \quad s = \begin{bmatrix} V \\ 0 \\ 0 \\ 0 \end{bmatrix}$$

This method of formulating equations is quite easy to implement. The one drawback is that the set of equations tends to be quite large. For example, if a small chip had 1000 nodes, not including ground, and there were 2500 branches, and an average of 2.5 branches per node, then the tableau equations would be a system of equations 6000 by 6000. In addition, most of the entries would be zero; thus sparse matrix techniques would be required to solve the equations. However, note that any element can be modeled

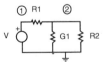

FIGURE 8.45 Example circuit for tableau equations.

using this approach and there are no limitations as there are with the standard nodal and loop analysis methods.

The Modified Nodal Approach

Another method of writing circuit equations that is less demanding of space and almost as flexible is called the **modified nodal approach.** In this approach, the typical nodal equations are written. However, for elements that are a function of current, e.g., inductors and certain controlled sources, extra variables and equations are added. This approach allows a broad range of elements to be modeled and yet keeps the size of the matrix relatively small. In order to describe, in a simple way, the modified nodal method we will consider networks with two terminal elements only (this is not a restriction of the method, but simply a convenience). We will label the elements as either being current controlled or voltage controlled. For example, a resistor, inductor, and independent voltage source will be considered current controlled, while conductors, capacitors, and independent current sources will be considered voltage controlled. Specifically, if the branch relation is written

$$i = f(v)$$

the element is considered voltage controlled. If the branch relation is written

$$v = g(i)$$

the element is considered current controlled. Note that the standard nodal equation approach only deals with voltage controlled elements. Consider S_I to be the set of current controlled elements and S_V to be the set of voltage controlled elements. Now we label each current controlled element, $l \in S_I$ with a new unknown I_l, the current through the element and we write the KCL at node k in the following way:

$$\underset{j\in S_V,\, j\,\text{connected to node }k}{\sum f_j(v)} \quad + \quad \underset{j\in S_I,\, j\,\text{connected to node }k}{\sum I_j} \quad = \quad \underset{j\in S_V,\, \text{current source currents connected to }k}{-\sum I_j}$$

The summations are all algebraic sums taking the reference directions of the current into account with current flow leaving the node considered positive. This is, of course, the normal way to write a nodal equation except that the unknown currents through current controlled elements have been added as unknowns. This generates a set of equations with more unknowns than equations. In order to provide the necessary additional equations we add the branch relations for each element in S_I to the previous set of equations. There are methods for easily generating these sets of equations. For our purposes a simple example will illustrate the important principles.

For the circuit in Fig. 8.46 the modified nodal equations are given by

$$\begin{bmatrix} 0 & 0 & 0 & 1 & 1 & 0 \\ 0 & G_1 & -G_1 & 0 & -1 & 1 \\ 0 & -G_1 & G_1 & 0 & 0 & 0 \\ 1 & 0 & 0 & 0 & 0 & 0 \\ 1 & -1 & 0 & 0 & -R_1 & 0 \\ 0 & 1 & 0 & 0 & 0 & -L_1\dfrac{d}{dt} \end{bmatrix} \begin{bmatrix} V_1 \\ V_2 \\ V_3 \\ I_V \\ I_R \\ I_L \end{bmatrix} = \begin{bmatrix} 0 \\ 0 \\ I \\ V \\ 0 \\ 0 \end{bmatrix}$$

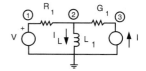

FIGURE 8.46 Example circuit for modified nodal equations.

Solving Linear Equations

The general form of the equations we generated in the previous section is

$$Ax = b$$

This is a set of linear algebraic equations and for our applications the system is completely specified with a unique solution. The goal of our discussion in this section is to explore some of the methods for solving this type of system of equations.

In general, for a system of linear equations we distinguish between direct and iterative methods of solution. Direct methods, the ones we shall examine in this section, are concerned with the calculation of the exact solution of the equation in a single computational step. In contrast, the iterative techniques, sometimes called relaxation techniques, use a sequence of steps to solve the equations and have to specify the conditions under which the computational sequence will converge to the correct answer (or indeed, converge at all).

We will not be concerned to any great extent with accuracy, however, the reader should know that there is a tremendous amount of literature on all facets of these methods.

Direct Methods for Full Systems

We will now consider the classical **Gaussian elimination** scheme for solving

$$Ax = b$$

We will assume that the matrix is relatively full and that whenever we choose a matrix location to use as a divisor that the entry in that location will be nonzero. Methods to deal with relaxing these assumptions are discussed later and in detail in the references.

For an n by n system of equations the standard Gaussian elimination procedure is given in Table 8.3.

A simple example of this procedure is given below. Forward elimination produces the sequence of matrices given below.

$$\begin{bmatrix} 2 & 2 & 1 \\ 3 & 1 & 0 \\ 0 & 1 & 2 \end{bmatrix} \begin{bmatrix} x_1 \\ x_2 \\ x_3 \end{bmatrix} = \begin{bmatrix} 1 \\ 2 \\ 1 \end{bmatrix}$$

$$\begin{bmatrix} 2 & 2 & 1 \\ 0 & -2 & -3/2 \\ 0 & 1 & 2 \end{bmatrix} \begin{bmatrix} x_1 \\ x_2 \\ x_3 \end{bmatrix} = \begin{bmatrix} 1 \\ 1/2 \\ 1 \end{bmatrix}$$

$$\begin{bmatrix} 2 & 2 & 1 \\ 0 & -2 & -3/2 \\ 0 & 0 & 5/4 \end{bmatrix} \begin{bmatrix} x_1 \\ x_2 \\ x_3 \end{bmatrix} = \begin{bmatrix} 1 \\ 1/2 \\ 5/4 \end{bmatrix}$$

Backward substitution yields:

$$x_3 = 1$$

$$x_2 = \left(\tfrac{1}{2} + \tfrac{3}{2} 1 \right) / 2 = -1$$

$$x_1 = \left(1 - 1 - 2(-1) \right) / 2 = 1$$

TABLE 8.3 Gaussan Elimination

Forward Elimination	Backward Substitution
For $i = 1$ to $n - 1$	For $i = n$ to 1
For $j = i + 1$ to n	
For $k = i$ to n	$x_i = b_i - \sum\limits_{j=i+1}^{n} a_{ij} x_j$
$a_{jk} = a_{jk} - \dfrac{a_{ji}}{a_{ii}} a_{ik}$	
End for	where
	$\sum\limits_{k=n+1}^{n} = 0$
$b_j = b_j - \dfrac{b_i}{a_{ii}} a_{ji}$	End for
End for	
End for	

This is the simple procedure that you may have seen before. The element that is used as the divisor in the forward elimination is called the *pivot element*. The procedure obviously depends on the pivot elements being nonzero. Furthermore, there is unnecessary work that is done. In particular, there is no need to zero out the column underneath the pivot element. Modifications based upon these observations are common. We note that the complexity of Gaussian elimination is $O(n^3)$ and that this complexity comes from the three nested loops in the forward elimination portion of the algorithm.

In solving linear systems it is often the case that the same equation will be solved with several right-hand sides. In this case, there is a procedure that can result in considerable savings. This procedure is known as *LU factorization* and also incorporates the savings from not zeroing out the column below the pivot element.

In LU factorization, the A matrix is decomposed into the product of a lower triangular matrix, I, and an upper triangular matrix, U. Thus

$$Ax = b$$

becomes

$$LUx = b$$

With this decomposition we can perform a forward substitution and a backward substitution to solve the system of equations. Specifically, we solve

$$Ly = b$$

using a forward substitution and

$$Ux = y$$

using a backward substitution. Both of these systems are triangular and are easy to solve requiring $O(n^2)$ operations each. Furthermore, if the right-hand-side vector is changed it is a simple task to perform the substitutions to find the new answer.

Thus the question becomes: How do we decompose the A matrix into LU? Although we will not analyze this carefully it will turn out that the steps required for the decomposition are the same as those for the forward elimination portion of Gaussian elimination except that we will not zero out the elements below the diagonal. In fact, the procedure given in Table 8.4 will write the LU matrices on top of the A matrix. Also, because the substitution process is straightforward we will only give the procedure for generating the LU decomposition.

For illustration purposes we present a matrix A and its LU factors. If

$$A = \begin{bmatrix} 3 & 5 & 1 \\ 1 & 4 & 3 \\ 2 & 2 & 3 \end{bmatrix}$$

then

$$L = \begin{bmatrix} 3 & 0 & 0 \\ 1/3 & 1 & 0 \\ 2/3 & -4/7 & 1 \end{bmatrix}, \quad U = \begin{bmatrix} 3 & 0 & 1 \\ 0 & 7/3 & 5/3 \\ 0 & 0 & 69/21 \end{bmatrix}$$

TABLE 8.4 LU Decomposition

For $i = 1$ to n
  For $j = i$ to n

$$u_{ij} = a_{ij} - \sum_{p=1}^{i-1} l_{ip} u_{pj}$$

$$l_{ji} = a_{ji} - \sum_{p=1}^{j-1} l_{jp} u_{pi} / u_{ii}$$

  End for
End for
where

$$\sum_{p=1}^{0} = 0$$

Pivoting for Accuracy

One subject that we have ignored in our discussion of Gaussian elimination and LU factorization is what to do when the element on the diagonal is zero or very small. In that case we cannot simply divide by the diagonal element without incurring significant numerical error. The choice of which element to use as the divisor in each row is known as *pivot selection* and the element chosen is known as a *pivot element*.

Significant research has gone into the subject of pivot selection, which we will not discuss in this section. However, we do wish to point out that the discussion of **pivoting** in most references will leave a suggestion that after the pivot element is selected there will be a row and column permutation of the matrix to bring that element to the diagonal. In point of fact this is not necessary. In an efficient implementation there would be no data movement, merely a keeping track of the proper order in which to perform the operations. This extra bookkeeping is much cheaper than repeated data motion.

Finally, note that moving columns causes the order of the variables to change and moving rows causes the right-hand side to move. This information should be part of the bookkeeping involved in the solution process.

Solving Sparse Linear Systems

As pointed out earlier, most of the entries in the matrices associated with circuit equations are zero. That is, the matrices are very **sparse**. There are two issues of importance in dealing with sparse matrices: the storage scheme and the pivot strategy.

If proper care is taken in solving sparse systems, the computational cost can, in practice, be reduced from $O(n^3)$ to $O(n^{1.3})$. This is a significant savings and well worth the programming investment when we are solving the systems repeatedly as is the case in circuit simulation.

Storage Scheme for Sparse Matrices

If the matrix is very sparse, it is wasteful of space to store using a n^2 array. In fact, for circuit equations there tends to be a small constant number of nonzeros per row independent of the size of the system. Typically, this is between 3 and 10. Thus in a medium-sized circuit having 1000 nodes there will be about 1% nonzeros.

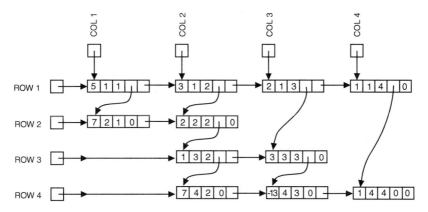

FIGURE 8.47 Bidirectional doubly linked list pivoting for sparsity.

Many storage schemes have been proposed for sparse matrices. The general trade-off is space versus ease of operation. Remember that we still have to search the rows and columns of the matrix during the elimination process. A bidirectional doubly linked list is a very easy data structure to traverse and modify and is a common one for use in storing sparse matrices. Note that this data structure has a relatively high degree of overhead and thus requires a very sparse system to make viable.

Figure 8.47 illustrates the bidirectional doubly linked list structure used to store a sparse matrix.

It is quite possible in the course of the Gaussian elimination process for an entry in the matrix that was zero to become nonzero. This is known as **fill-in.** The example shown below is an extreme case. For an array with the zero/nonzero structure given by

$$A = \begin{bmatrix} x & x & x & x \\ x & x & & \\ x & & x & \\ x & & & x \end{bmatrix}$$

after the first step of Gaussian elimination the zero/nonzero structure becomes

$$\begin{bmatrix} x & x & x & x \\ x & x & x & x \\ x & x & x & x \\ x & x & x & x \end{bmatrix}$$

All of the zeros have become nonzero. In this case there will be $O(n^3)$ operations. This was not necessary. If the matrix had been pivoted to the following form:

$$\begin{bmatrix} x & & & x \\ & x & & x \\ & & x & x \\ x & x & x & x \end{bmatrix}$$

there would have been no fill-in during the elimination process. Clearly there is a requirement that the solution order be chosen to preserve the sparse structure of the equations to the extent possible.

Significant research has been done to find the best pivoting strategy to preserve sparseness. However, of all the methods available, one of the earliest is still robust and used widely. The *Markowitz criteria* is a simple method used to determine the pivot order.

In describing the Markowitz criteria, remember that we are always working on a portion of the matrix, the whole matrix at the first step and reduced portions thereafter. Markowitz looks at each nonzero, a_{ij} in the reduced matrix and counts the number of nonzeros in the associated row, r_i, and column, c_j, and forms the quantity, $(r_i - 1)(c_j - 1)$ (known as the Markowitz count). Then the pivot element is chosen as that nonzero with the smallest Markowitz count (ties are broken arbitrarily).

To illustrate this process we take the arrow head matrix shown above and attach the Markowitz count as superscripts to each of the nonzeros.

$$A = \begin{bmatrix} x^9 & x^3 & x^3 & x^3 \\ x^3 & x^1 & & \\ x^3 & & x^1 & \\ x^3 & & & x^1 \end{bmatrix}$$

Clearly, the (1,1) position would be the last chosen or, in other words, is the worst possible choice. Note that the count has to be recalculated at every step of the elimination process, which emphasizes the need for easy traversal of the data structure.

Solving Nonlinear Equations

Obviously we must deal with nonlinear elements when simulating integrated circuits. These elements arise from the nonlinearities in transistor device models. For the sake of simplicity we will first consider two terminal nonlinearities.

For a simple nonlinear equation

$$f(x) = 0$$

the usual method of solution is some variant on the **Newton-Raphson** (NR) procedure.

In order to understand the NR procedure we must realize that the entire procedure is based upon a linear approximation to the function at the current point, x_c. Thus, if we expand the function in a Taylor series about the current point we have

$$f(x_+) = f(x_c) + \frac{\partial f}{\partial x}\bigg|_{x=x_c} (x_+ - x_c)$$

Of course, we are ignoring the error involved in the approximation. Now, if we further assume that the x_+ is the solution, i.e.,

$$f(x_+) = 0$$

then we have

$$x_+ = x_c - f(x_c) / (\partial f / \partial x)\big|_{x=x_c}$$

which is the basic NR step.

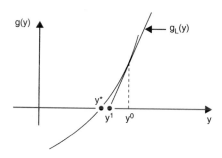 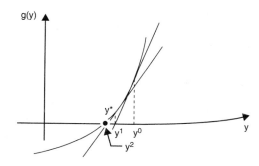

FIGURE 8.48 Newton-Raphson example.

The process of repeated calculation of linear approximations and their solution is illustrated in Fig. 8.48.

The difficulties associated with this scheme are numerous. For example, if the partial derivative (the normal derivative in the single variable case) is zero, then the step fails. Further, we have no guarantee of convergence for the general case. In addition, with the applications to electrical circuits with exponential nonlinearities, numerical accuracy can be a tremendous problem. Significant research has been done to deal with all of these cases, and more. Yet the nonlinear solution problem, especially the initial DC solution, continues to be the least robust portion of the electrical level simulation package.

The question of convergence of the NR algorithm is answered by the following theorem.

Theorem: If f is twice continuously differentiable, $[\partial f(x)/\partial x]|_{x=x^*} \neq 0$, i.e., the derivative is nonzero at the solution, x^*, and the initial guess x_0 is close enough to the solution, then the NR method always converges to the solution. Furthermore, letting

$$\varepsilon_k = |x_k - x^*|$$

be the error at the kth iteration, then

$$\varepsilon_{k+1} \leq C\varepsilon_k^2$$

where C is a positive constant, i.e., the NR methods have quadratic convergence.

Notice that there is no mechanism for determining how close is close enough in the previous theorem. This type of convergence is known as *local convergence*; if the iteration converges from *any* starting point, the algorithm is said to possess *global convergence*. In addition, the quadratic convergence property is

TABLE 8.5 Newton-Raphson

Error = ∞
Accuracy = ε (user supplied)
X_c = current estimate of the solution, initially provided by user
While (Error ≥ Accuracy)
        calculate an update of the estimated solution
$$x_+ = x_c - f(x_c) / (\partial f / \partial x)\Big|_{x=x_c}$$
        calculate the new error
            Error = $|x_+ - x_c|$
        update the estimate of the solution
            $x_c = x_+$
End While

very important because it shows that if the error is ever less than one that it will be driven to zero very quickly. In fact, once the error is below one, the number of significant decimal places effectively doubles at each iteration.

Newton-Raphson Applied to Circuit Simulation

Given the general picture of the NR scheme, how is this applied to circuits? The first difficulty is that nonlinear circuit equations are not just a function of a single variable. Further, there is a set of equations and not just a single one. This problem is easily handled by extending the one-dimensional NR scheme to multidimensions using vector calculus. The update formula now becomes

$$x_+ = x_c + \left[\frac{\partial f}{\partial x}\right]_{x=x_c}^{-1} f(x_c)$$

were $\partial f/\partial x$ is known is the *Jacobian* of the set of circuit equations and is the matrix of partial derivatives.

$$\frac{\partial f}{\partial x} = \begin{bmatrix} \dfrac{\partial f_1}{\partial x_1} & \cdots & \dfrac{\partial f_1}{\partial x_n} \\ \vdots & \ddots & \vdots \\ \dfrac{\partial f_n}{\partial x_1} & \cdots & \dfrac{\partial f_n}{\partial x_n} \end{bmatrix}$$

Rewriting this set of equations we have

$$\left.\frac{\partial f}{\partial x}\right|_{x=x_c} (x_+ - x_c) = f(x_c)$$

This is precisely in the form

$$Ax = b$$

which we studied in the previous section. A similar convergence condition applies and the reader is referred to the literature. Most importantly, we note that local convergence and a quadratic rate of convergence apply in the multidimensional case as well as the single-dimensional case.

The final difficulty we face is how to generate the equations in the first place and then how to take the partial derivative of this large set of equations. What can be shown without too much difficulty, although we will skip the demonstration, is that the individual elements can be linearized and the equations written for this linear network and we still have the same set of equations required for NR.

To illustrate this we will consider the network in Fig. 8.49 with two diodes. We will assume a simple form for the diode equation

$$i = f(v)$$

The linearized form of this becomes

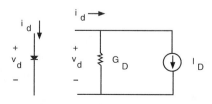

FIGURE 8.49 Diode linearized model.

$$i_+ = i_c + \left.\frac{\partial f}{\partial v}\right|_{v=v_c} \left(v_+ - v_c\right)$$

If we rewrite this grouping the known and unknown quantities we have

$$i_+ = \left[i_c - \left.\frac{\partial f}{\partial v}\right|_{v=v_c} v_c\right] + \left.\frac{\partial f}{\partial v}\right|_{v=v_c} v_+$$

This can be interpreted as an electrical model of the linearized diode that has a current source of value

$$I_D = \left[i_c - \left.\frac{\partial f}{\partial v}\right|_{v=v_c} v_c\right]$$

in parallel with a conductance of value

$$G_D = \left.\frac{\partial f}{\partial v}\right|_{v=v_c}$$

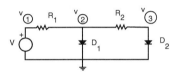

FIGURE 8.50 Nonlinear network used for NR example.

This model is illustrated in Fig. 8.50.

With this model the original circuit can be transformed into a linearized model as shown in Fig. 8.51.

Now we can use the normal equation formulation techniques to generate the set of linear circuit equations that represent the Jacobian of the nonlinear network equations. This is the process used in most circuit simulation programs.

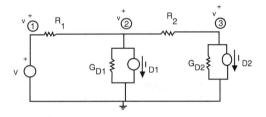

FIGURE 8.51 Linearized network.

Accuracy Control in Circuit Simulation with NR

Two major areas of concern in using NR for solving circuit equations are

- Numerical overflow during the iterations
- Nonconvergence of the iterations

The first problem is primarily due to the nature of the nonlinearities in the circuit itself. This is especially a problem with the exponential nonlinearity associated with pn junction and Schottky barrier models. For example, in the circuit shown in Fig. 8.52 if the initial guess is v_0, then the result of the first iteration is v_1 and to continue the iteration we have to evaluate the diode current for a voltage of 10 V; this requires evaluating e^{400} (these iterations are shown in Fig. 8.53). Clearly overflow can be a problem.

FIGURE 8.52 Circuit example with numerical overflow problems.

Two classes of methods have been proposed to alleviate this problem. Both limit the step in voltage that is actually taken and are known as *limiting methods*. The first group of methods, which we will not discuss in detail, require some arbitrary limitation of the NR step. Although some of these can work well, they are quite heuristic.

The second class of limiting methods, which are used more often in solving circuit equations, involve the changing of the variable of iteration. We have been writing the diode equation as

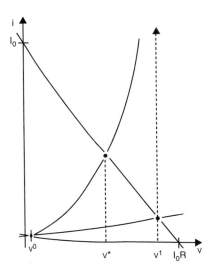 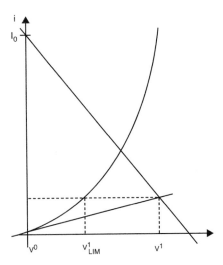

FIGURE 8.53 Alternating V and I as unknowns in NR.

$$i = f(v)$$

but we could have just as well written

$$v = g(i)$$

As we know, the function g is a logarithmic function and so is much less prone to numerical difficulties than the exponential function f. The decision on which form to represent the function can be made during the iterations and can change from step to step. The difference between these two steps is illustrated in Fig. 8.53.

The second class of difficulties, nonconvergence, can have several causes:

- The equations may not have a solution due to error in their formulation.
- The first derivatives of the elements' branch relationships lead to a singular Jacobian.
- The initial guess is not sufficiently close to the solution.
- The branch relationships may not have continuous derivatives.

Difficulties with model equations can only be solved with a refinement of the equations themselves and so this is out of our range of discussion.

There are two broad classes of methods to attempt to force convergence of the NR equations:

- Optimization-based methods
- Continuation methods

We will not discuss the first class of methods. The second class of methods is best characterized in circuit simulators as *source stepping methods*. Source stepping methods are generally used to help find the initial DC solution of the circuit. First we note that if all the DC sources are set to zero, then we know the solution to the network — zero. So if we increase the source values slightly from zero to their final value, we have a very good guess at the solution — zero. This process is continued until the sources reach their final value.

There is much known on how to improve the numerical behavior of NR in circuit simulation. Unfortunately, it is mostly known to practitioners and is not in the public domain. This is primarily because good nonlinear solution techniques (especially for the DC solution) are valuable and are not lightly shared from one company to the next.

Solving Differential Equations

The last task we face in constructing and understanding electrical level simulation is how to deal with elements with memory such as capacitors and inductors. For these problems we could write the tableau equations in the following general form

$$F\left(\dot{x}, x, T\right)=0 \quad x\left(0\right)=X_0 \text{ for } 0\leq t\leq T$$

This is known as an initial value problem and has no closed form solution in general.

In attempting to solve this problem we will find approximate solutions at a finite set of points over the time interval of interest

$$t_0=0, t_N=T, t_{n+1}=t_n+h_{n+1} \quad n=0,1,\ldots,N$$

The h_{n+1} are called *time steps* and their values are called **step sizes.** At each point, t_n, we are going to compute an approximation, x_n, to the exact solution, $x(t_n)$.

A large number of methods are available for solving this problem. The class of methods known as **linear multistep methods** are the ones most widely used in circuit simulation programs and we shall concentrate on this class. For sake of simplicity we will first examine these methods in terms of general, first-order, ordinary scalar differential equations of the form

$$\dot{y}=f\left(y, t\right)$$

A linear multistep method is a numerical method that computes y_{n+1} based on the values of y and $\dot{y}$ at the previous $p+1$ time points. Specifically,

$$y_{n+1}=\sum_{i=0}^{p}a_i y_{n-i}+\sum_{i=-1}^{p}h_{n-1}b_i \dot{y}_{n-1}$$

or written using the definition of $\dot{y}$ we have

$$y_{n+1}=\sum_{i=0}^{p}a_i y_{n-1}+\sum_{i=-1}^{p}h_{n-i}b_i f\left(y_{n-i}, t_{n-i}\right)$$

A simple example of a multistep method is the forward Euler method (FE) which is given by

$$y_{n+1}=y_n+h_{n+1}\dot{y}_n$$

Where $p=0$, $a_0=1$, $b_0=1$ and all other coefficients are zero. This method can also be considered as a Taylor series expansion about t_n truncated at its first term.

Other well-known multistep methods with $p=0$, also called single-step methods, are

$$\text{Backward Euler}\left(\text{BE}\right) \quad y_{n+1}=y_n+h_{n+1}\dot{y}_{n+1}$$

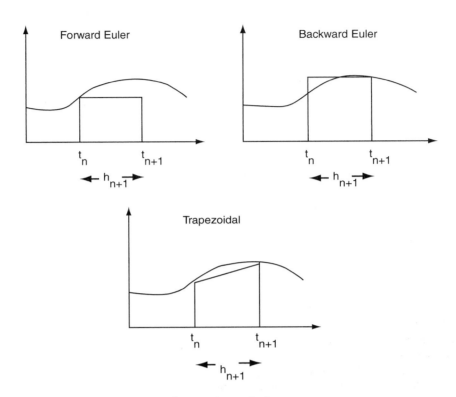

FIGURE 8.54   Geometrical interpretation of integration methods.

and

$$\text{Trapezoidal } \left(\text{TR}\right) \quad y_{n+1} = y_n = \left(\frac{h_{n+1}}{2}\right)\left(\dot{y}_{n+1} + \dot{y}_n\right)$$

Each of these three methods has a simple interpretation. First, note that the solution we are calculating is really the integral of $\dot{y}$. So that if we plot $f(y, t)$ and then look at estimating the area under the curve between t_n and t_{n+1} we generate a simple graphical interpretation of the methods given in Fig. 8.54.

Truncation Error for Linear Multistep Methods

A major concern in solving systems of differential equations is the accuracy of the computed solution. In particular, what is the difference between the computed solution at t_n, y_n, and the exact solution, $y(t_n)$? This particular type of error is known as the **truncation error.**

There are two types of truncation error, local and global. Local truncation error (LTE) of a method at t_{n+1} is the difference between the computed value y_{n+1} and the exact value, $y(t_{n+1})$ *assuming that no previous error has been made,* that is, that

$$y\left(t_i\right) = y_i, \quad i = 0, 1, 2, \ldots, n$$

Stability of Linear Multistep Methods

In order to study the global behavior of multistep methods we must examine how any errors made accumulate. This allows us to gain an understanding of the global truncation error. This type of study addresses the **stability** of the method.

To illustrate that there might be a problem consider the differential equation

$$\dot{y} = -y, \quad y(0) = 1$$

If we apply FE to this equation with step size $h = 1$, we generate the sequence

$$0, \ 0, \ 0, \ ...$$

Although not a very accurate answer for the beginning, at least it does not blow up and, indeed, the difference between this sequence and the correct answer goes to zero as time increases. However, if we apply FE with a step size of $h = 3$, we generate the sequence

$$-2, \ 4, \ -8, \ 16, \ -32, \ ...$$

which shows a growing oscillation. So whatever error is being made in the calculation accumulates in a disastrous manner.

Applying BE to the same equation with $h = 3$ yields a well-behaved sequence.

Stiff Systems

Especially for microwave and millimeter wave circuits, it is important to include parasitic elements for an accurate model. These parasitics are modeled as very small capacitors and inductors. This results in a system of equations that have very widely separated time constants. This type of system is said to be **stiff.**

The solution of stiff systems poses some severe problems. For example, consider

$$\dot{y} = -\lambda_1 \left(y - s(t) \right) + ds/dt, \quad y(0) = y_0$$

where $s(t) = 1 - e^{-\lambda_2 t}$. The exact solution is

$$y(t) = y_0 e^{-\lambda_1 t} + \left(1 - e^{-\lambda_2 t} \right)$$

and is shown in Fig. 8.55.

Suppose that λ_1 and λ_2 are widely separated, with, for example, $\lambda_1 = 10^6$ and $\lambda_2 = 1$ yielding a stiff system. For a good picture of the waveform we have to integrate at least 5 s. But note that the first exponential waveform dies out in 5 μs.

We now are faced with a dilemma: to get an accurate solution of the fast waveform we need a small time step and to have an efficient simulation of the slow waveform we need a large time step. The obvious solution is to use a variable time step, small at the beginning of the simulation and large at the end.

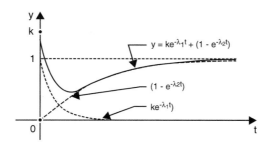

FIGURE 8.55 Exact solution to stiff system.

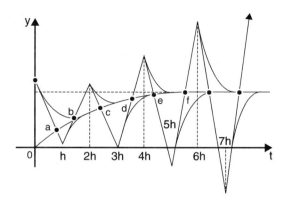

FIGURE 8.56   Numerical instability in FE applied to stiff system.

However, varying the time step changes the analysis of the regions of stability of the linear multistep methods (recall that we assume a constant step size in our analysis). If we apply the FE method to the previous example with an initial step size of 10^{-6} and after five steps we change to a step size of 10^{-4}, the behavior shown in Fig. 8.56 is obtained.

It turns out that the region of absolute stability for FE can be found by looking at the fastest time constant. In Fig. 8.56 y_{n+1} computed by FE is based upon the first derivative of the solution *passing through* y_n. If y_n is not very close to the exact solution, then FE picks up a spurious fast transient which causes instability if h is not chosen small enough.

Thus we see that

- The analysis of stiff systems requires variable time step.
- Not all linear multistep methods can be efficiently used to integrate stiff equations.

Clearly, to be useful for stiff systems the integration methods have to have a region of absolute stability which allows for a large step size for large time constants without being constrained by the small time constants. Thus A-stable methods are fine (remember that FE is not A-stable).

Time Step Control for Multistep Methods

Now that we know that the time step must be changed for efficiency and we are aware of the importance of different orders of integration schemes, how can we control these in a circuit simulator?

If we let E_{n+1} be a bound on the absolute value of the LTE at t_{n+1}, we know for a multistep method of order k that

$$\left|LTE_{n+1}\right| = \left|\left[C_{k+1}h_{n+1}^{k+1}\Big/(k+1)!\right]y^{(k+1)}\left(t_{n+1}\right)\right|$$

Thus we must have

$$h_{n+1} \le \left|\frac{(k+1)!\,E_{n+1}}{C_{k+1}h^{(k+1)}\left(t_{n+1}\right)}\right|^{\frac{1}{k+1}}$$

in order to meet the error requirement. Since the $(k+1)^{st}$ derivative of y is not available we will have to form some approximation of it.

The scheme used in SPICE is the method of *divided differences* (DD). The DD are defined recursively as

$$DD_1\left(t_{n+1}\right) = \frac{y_{n+1} - y_n}{h_{n+1}}$$

$$DD_2\left(t_{n+1}\right) = \frac{DD_1\left(t_{n+1}\right) - DD_1\left(t_n\right)}{h_{n+1} - h_n}$$

$$\vdots$$

$$DD_{k+1}\left(t_{n+1}\right) = \frac{DD_k\left(t_{n+1}\right) - DD_k\left(t_n\right)}{\displaystyle\sum_{i=-1}^{k-1} h_{n-1}}$$

It is relatively easy to show that

$$y^{(k+1)}\left(t_{n+1}\right) \cong (k+1)! \, DD_{k+1}\left(t_{n+1}\right)$$

Thus we have an algorithm for time step control.

Given a step h_{n+1}, y_{n+1} is computed according to the method chosen for integration. Then, y_{n+1} and h_{n+1} are used to compute $DD_{k+1}(t_{n+1})$ and h_{n+1} is checked using

$$h_{n+1} \le \left[\frac{E_{n+1}}{C_{k+1} DD_{k+1}\left(t_{n+1}\right)}\right]^{\frac{1}{k+1}}$$

If h_{n+1} satisfies the test, it is accepted: otherwise it is rejected and a new h_{n+1} given by the right-hand side is used.

After h_{n+1} is accepted h_{n+2} must be chosen. A common strategy is to use the right-hand side of the error checking equation as the new step size.

An important factor in the selection of the step size is the choice of E_{n+1}. In general, a circuit designer has a rough idea of the precision that is required in the simulation. This is generally related to the GTE and not the LTE. Thus a mechanism to transfer the GTE estimate to the LTE is required.

It is often assumed that the LTE accumulates linearly thus giving the following estimate

$$\left|GTE_{n+1}\right| \le \sum_{i=1}^{n+1} \left|LTE_i\right|$$

This is quite conservative in practice and so E_{n+1} is often obtained from

$$E_{n+1} = \varepsilon_u h_{n+1}$$

where ε_u is the maximum allowed error per unit step.

There is a large practice, theory, and lore about the choice of step size and integration method. To construct a useful, efficient, and robust circuit simulation package this information must be tapped and utilized. The interested reader is referred to the literature for further information.

Techniques for Large Circuits

The techniques that we have been discussing are used in most standard time-domain circuit simulation, e.g., SPICE in all its various forms. Although very common, these techniques have limitations when solving very large circuits. We have indicated that circuit size may vary from one transistor with matching elements to several hundred or even thousands for high levels of radio and system integration. When very large circuits must be simulated special techniques must be used.

There are three types of specialized techniques that have been used: techniques for partitioning large circuits into smaller ones and combining the solutions, techniques that exploit special characteristics of certain classes of circuits (see the next section), and techniques that utilize parallel and vector high-speed computers.

All of these techniques are beyond the scope of this chapter. However, the reader is referred to the references for further information.

Defining Terms

Associated reference directions: A method of assigning the current and voltage directions to an electrical element so that a positive current-voltage product always means that the element is absorbing power from the network and a negative product always means that the element is delivering power to the network. This method of assigning directions is used in most circuit simulation programs.

Branch relation: The relationship between voltage and current for electrical components. Common branch relations are Ohm's law and the lumped equations for capacitors and inductors. More complex branch relationships would be transistor models.

CAD: Computer-aided design for the electronics industry is concerned with producing new algorithms/programs that aid the designer in the complex tasks associated with designing and building an integrated circuit. There are many subfields of electrical CAD: simulation, synthesis, physical design, testing, packaging, and semiconductor process support.

Circuit simulation: Constructing a mathematical model of an electrical circuit and then solving that model to find the behavior of the model. Since the model usually cannot be solved in closed form, the equations are evaluated for a specific set of input conditions to produce a sequence of voltages and currents that approximates the true solution.

Fill-in: When solving a set of sparse linear equations using Gaussian elimination it is possible for a zero location to become nonzero. This new nonzero is termed a fill-in.

Gaussian elimination: The standard direct method for solving a set of linear equations. It is termed direct because it does not involve iterative solutions. Variations of this scheme are used in most circuit simulators.

KVL, KCL: Kirchhoff voltage and current laws provide the basic physical constraints on voltage and current in an electrical circuit.

Linear multistep method: This is a class of techniques for solving ordinary differential equations that is widely used in circuit simulators.

Modified nodal formulation: A modification of the classical nodal formulation that allows any network to be described. The modification consists of adding extra equations and unknowns when an element not normally modeled in classical nodal analysis is encountered.

Newton-Raphson: A numerical method for finding the solution to a set of simultaneous nonlinear equations. Variations of this method are commonly used in circuit simulation programs.

Pivoting: When applying Gaussian elimination to solve a set of simultaneous linear equations the natural solution order is sometimes varied. The process of varying the natural solution order is termed pivoting. Pivoting is used to avoid fill-in and to maintain the accuracy of a solution.

Sparse equations: When a set of linear simultaneous equations has very few nonzeros in any row the system is said to be sparse. Normally for a system to be considered sparse less than 10% of the possible entries should be nonzero. For large integrated circuits less than 1% of the possible entries are nonzero.

SPICE: The most widely used circuit simulation program. Originally developed at the University of California, Berkeley, this program is available from several commercial sources. SPICE is an acronym for Simulation Program with Integrated Circuit Emphasis.

Stability: In the context of circuit simulation the main stability concern is in the solution of differential equations. Specifically, if small errors made by the integration methods — such as a linear multistep method — are amplified as the solution progresses, the integration method is exhibiting unstable behavior. Controlling the step size is one of the primary mechanisms for maintaining stability.

Step size: When solving for the transient behavior of an electrical circuit the associated differential equations are solved at specific points in time. The difference between two adjacent solution time points is known as the step size.

Stiff system: When an electrical circuit has widely separated time constants the circuit is said to be stiff. The system of equations associated with the circuit is known as a stiff system and special numerical methods must be used to maintain stability and accuracy when simulating a stiff system.

Tableau formulation: A method for formulating the equations governing the behavior of electrical networks. The tableau method simply groups the KVL, KCL, and branch relationships into one huge set of equations.

Truncation error: When numerically solving the differential equations associated with electrical circuits, approximation techniques are used. The errors associated with the use of these methods are termed truncation error. Controlling the local and global truncation error is an important part of a circuit simulator's task. Limits on these errors are often given by the user of the program.

References

W. Banzhaf, *Computer-Aided Circuit Analysis Using PSpice*, Englewood Cliffs, N.J.: Prentice-Hall, 1992.

R. E. Bryant, A survey of switch-level algorithms, *IEEE Design and Test of Computers*, 4, 4, 26–40, 1987.

L. O. Chua and P.-M. Lin, *Computer-Aided Analysis of Electronic Circuits: Algorithms and Computational Techniques*, Englewood Cliffs, N.J.: Prentice-Hall, 1975.

S. W. Director, *Circuit Theory: A Computational Approach*, New York: John Wiley & Sons, 1975.

G. F. Forsythe, M. A. Malcom, and C. B. Moler, *Computer Methods for Mathematical Computations*, Englewood Cliffs, N.J.: Prentice-Hall, 1977.

G. D. Hachtel, R. K. Brayton, and F. G. Gustavson, The sparse tableau approach to network analysis and design, *IEEE Trans. Circuit Theory*, CT-18, 1, 101–113, 1971.

C. H. Ho, A. E. Ruehli, and P. A. Brennan, The modified nodal approach to network analysis, *IEEE Trans. Circuits and Systems*, CAS-22, 6, 504–509, 1975.

A. Jennings, *Matrix Computation for Engineers and Scientists*, New York: John Wiley & Sons, 1977.

J. D. Lambert, *Computational Methods in Ordinary Differential Equations*, New York: John Wiley & Sons, 1973.

J. M. Ortega and W. C. Rheinboldt, *Iterative Solutions of Nonlinear Equations in Several Variables*, New York: Academic Press, 1970.

V. B. Rao, D. V. Overhauser, T. N. Trick, and I. B. Hajj, *Switch-Level Timing Simulation of MOS VLSI Circuits*, Dordrecht, Netherlands: Kluwer Academic Publishers, 1989.

R. A. Saleh and A. R. Newton, *Mixed-Mode Simulation*, Dordrecht, Netherlands: Kluwer Academic Publishers, 1990.

A. F. Schwarz, *Computer-Aided Design of Microelectronic Circuits and Systems* (two volumes), New York: Academic Press, 1987.

R. Spence and J. P. Burgess, *Circuit Analysis by Computer — From Algorithms to Package*, Englewood Cliffs, N.J.: Prentice-Hall, 1986.

J. Vlach and K. Singhal, *Computer Methods for Circuit Analysis and Design*, New York: Van Nostrand Reinhold, 1983.

Further Information

Simulation of circuits at all levels of abstraction is an active research area. The interested reader is encouraged to examine the *IEEE Transactions on Computer-Aided Design,* the *IEEE Transactions on Circuits and Systems,* and the *IEEE Transactions on Computers.* In addition to journal articles there are a number of conferences held each year at which the latest results on circuit simulation are presented. The proceedings of the following conferences have a wealth of information on circuit simulation developments: *Proceedings of the Design Automation Conference (DAC), Proceedings of the International Conference on Computer-Aided Design (ICCAD), Proceeding of the International Conference on Computer Design (ICCD), Proceedings of the European Design Automation Conference (EDAC),* and the *Proceedings of Euro-DAC.* Finally, there are a large number of numerical analysis, computer science, and applied mathematics conferences and journals in which results related to circuit simulation are published.

8.6 Computer Aided Design of Microwave Circuitry

Ron Kielmeyer

The growth of personal communication and Internet industries along with the need for portability has resulted in an ever-increasing demand for low cost, high volume microwave circuitry. The commercialization of GaAs wafer processing and the simultaneous reduction in the physical size of silicon devices has enable the development of complex microwave circuitry which can no longer be designed without the aid of sophisticated CAD circuit simulators. This article will discuss the typical steps involved in a design cycle, some basic requirements for a CAD program, a look at the theory behind the most popular CAD programs in use today, and some emerging CAD technologies.

Initial Design

The design cycle shown in Fig. 8.57 starts with a circuit and/or system function such as an amplifier, a mixer, or a whole receiver along with appropriate specifications for that function. Then active devices such as transistors or diodes, if required to achieve the function, are chosen. Circuit topologies may be explored simultaneously with device selection. With these active devices will come a computer representation for the device, usually from the device vendor. This computer representation is in the form of a mathematical model or measured S-parameters.

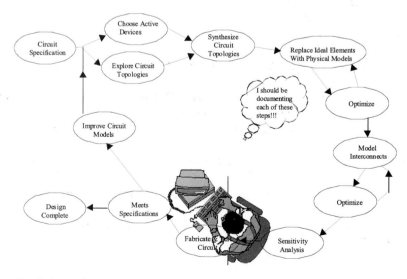

FIGURE 8.57   The design cycle.

Synthesis programs, if available, are used to determine the best possible performance. Many different topologies can be explored or identified as possible candidates for realizing the function. The ideal topologies generated by the synthesis program must exceed the design specifications since performance will only deteriorate from the idealized case.

Physical Element Models

Ideal elements must be replaced by models of the physical devices. These models are typically sub-circuits made up of ideal elements. For example, an ideal capacitor, physically realized by a chip capacitor, must have a model that can account for the finite inductance of the terminals and the finite resistive loss inherent in the physical device. For example, Fig. 8.58 shows a model for a physical capacitor mounted on a printed circuit board. In this model, C is the desired or ideal capacitance while Cpad represents the shunt capacitance of the metal pads on the circuit board to which the capacitor is soldered. Rs and Ls take into account the inductance and conductivity of the metal plates that form the capacitance. Rpar and Cpar account for the parallel resonant

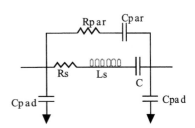

FIGURE 8.58 Physical model of a chip capacitor.

characteristics of the capacitor. It is common to refer to the extra elements Rs, Rpar, Cpar, Cpad, and Ls as "parasitic" elements. By replacing the ideal elements with nonideal models one at a time, the designer can accomplish two important practicalities. First, the sensitivity of the circuit to the nonideal element can be evaluated. Second, an optimization step can be performed on the remaining ideal elements in order to bring the circuit back within the design specifications. The actual value of the nonideal elements can also be optimized as long as the resulting optimized values do not change significantly from the pre-optimized values.

Layout Effects

As the process of replacing the idealized elements continues, the physical layout of the circuit must be introduced into the analysis. Two physical elements cannot be connected with zero length metal patterns. How close the elements can be placed is usually determined by how close the manufacturing process can place them. The metal patterns used to interconnect the physical devices must be introduced in the form of transmission lines and/or transmission line junctions. Figure 8.59 illustrates a simple PI resistive attenuator as might be realized using MMIC or MIC technology. The metal interconnects are modeled as a microstrip transmission line, followed by a Tee junction model. The resistors in this model are modeled as ideal resistors. The ground vias are represented as inductors. As each of these physical effects are introduced, an optimization step is performed in an attempt to meet or exceed the design specifications.

Sensitivity to Process Variation

After transforming many possible design topologies, sensitivity of the circuit to the nonideal or parasitic elements of the circuits as well as to the actual element values must be determined. One measure of the sensitivity of a circuit is the percent yield. To determine the percent yield of a circuit, the element values, parasitic elements, and/or known physical tolerances are treated as independent random variables. A range for each random variable is given and the computer analysis program iterates through random samples of the variables with their given ranges. This process is called a Monte Carlo analysis. The ideal outcome of the Monte Carlo analysis would be that the circuit passes all specifications under any combination of the random variable values. The percent yield can be improved by performing an optimization on the circuit. The optimization includes the ranges assigned to the random variables in the analysis. At each step in the optimization process, the mean values of the random variables are varied. The end result is an overall increase in the percent yield.

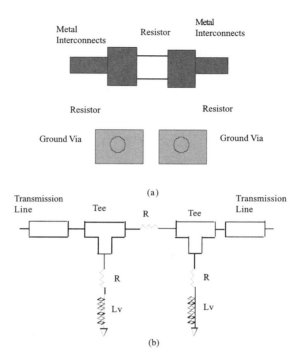

FIGURE 8.59 Simple PI pad resistive attenuator; (a) physical layout, (b) CAD model.

Design of Experiments (DOE) is another way to measure the sensitivity of the circuit to the random variables. In performing the DOE, each of the random variable values are independently incremented from their mean values. The computer keeps track of the analyzed responses and does the tedious bookkeeping task of incrementing all of the variables. Once this process is complete, statistical techniques are used to determine the sensitivity of each of the random variables to each of the specifications.

Design Tool Requirements

There are certain features that a microwave CAD tool must have in order to improve the productivity of the microwave designer. The most important is that it must be accurate. Once a circuit has been simulated, it must be fabricated and tested against the specifications for that circuit. If the computer simulation is inaccurate, the reasons for this must be determined. The fabrication of the circuit and the debugging of the circuit are by far the most time consuming part of the design cycle. Therefore, the goal of any CAD design is to model the circuit so that only one fabrication cycle needs to be implemented.

Inaccurate simulations that are tool dependent can result from numerical difficulties, model inaccuracies, or a lack of acceptable models. Inaccurate simulations that are directly caused by the user usually result from failure to model the circuit correctly or the misuse of models by using them outside the range for which they are valid. The designer has little control over the numerical problems. In some cases, the designer can overcome model accuracy by knowing the limitations of the models being used and if possible compensating for them. However, since model accuracy and availability usually are the features that distinguish one program from the other, companies tend to treat the models as proprietary and therefore do not want to release information about the models and how they are implemented. This can be a great disadvantage since the designer has no way of knowing how the model is implemented and when the range of validity for the model has been exceeded.

Programs usually provide a way to implement custom models. For example, many programs will allow the user to import S-parameter data files to represent a part of the circuit that cannot be accurately represented by standard transmission line discontinuity models. These S-parameter files can be generated from a measurement of the actual discontinuity. Likewise, active device S-parameters can be measured

TABLE 8.6 Summary of Desirable CAD Features

Analysis Features	Synthesis Features	User Interface Features
Accurate models	Filter/impedance matching	Schematic input
Optimization	Constant noise circles	Text and graphical output
Import/export data	Constant gain circles	Easy documentation
Robust model library	Stability circles	Drawing complex shapes

at a specific bias point and imported for use in the design process. Measured data is usually limited to two-port networks since multiport network analyzers are not typically available to the designer. For multiport passive networks, electromagnetic simulators are often used to create data file representations of the passive network. In some cases, programs will allow the designer access to the code or hooks into the code that can be used to implement custom models.

Ease of use is a program feature that can greatly increase the productivity of a designer. Early implementations of circuit analysis programs used a descriptive language called a netlist to describe the circuit to be analyzed. Most modern programs however, use a graphical interface to create a schematic representation of the circuit. Schematics are much easier to create and are less prone to error than netlists. For electromagnetic simulators, a drawing package is used to draw the geometric shapes. These drawing packages, at a minimum, must provide a means to implement complex, difficult-to-draw structures from basic shapes that are easy for a designer to draw. In addition to this, the ability to change dimensions without the need to redraw the shape can greatly decrease the time it takes to perform tolerance studies or to modify the structure in order to meet the desired performance criteria.

Because microwave circuit performance specifications can be both time and frequency domain quantities, the circuit analysis program must provide an easy means to display the resulting data in both the time and frequency domains. The program must also be able to export the results into standard text or graphic formats for importation into word processors or view cell generating programs for documentation and presentation purposes.

Electromagnetic simulators need to be able to display field quantities such as current density on the conductors, E and H field intensity and direction, as well as outputting S, Z, or Y parameters for importation into circuit simulators. Data must be displayed in either graphical or text format.

Synthesis programs are available that provide impedance matching network topologies for amplifier/mixer design and filter response functions. These programs could be simple spreadsheet implementations of well-known design equations, an electronic Smith chart, or sophisticated implementations of impedance matching theory or filter design. The most sophisticated programs can provide networks based on both ideal and nonideal elements. The ability to display constant gain, noise circles, and stability circles can also be considered part of the synthesis capability since these are often used to determine the matching network impedances [1].

Time Domain vs. Frequency Domain Simulation

A circuit can be analyzed in either the frequency domain or the time domain. SPICE is a program developed at the University of California at Berkeley that analyzes a circuit in the time domain. Harmonic balance programs solve the circuit equations partially in the frequency domain and partially in the time domain. The choice of which program is used depends on the parameters that are specified. Since SPICE does the computations in the time domain, it is quite naturally used when the design parameters involve time-dependent quantities. SPICE is used for transient parameters such as the turn-on time of an oscillator or amplifier, the switching time of a switch, or perhaps the impulse response of a circuit. SPICE can also solve for the DC bias point of the circuit and then perform a small-signal, frequency domain analysis of the circuit about this bias point. By adding some special circuit elements to the file description, S-parameters can be computed from the results of this small-signal analysis. However, many SPICE based programs have added direct S-parameter output capability in order to accommodate the needs of microwave designers.

SPICE can be used to predict the effects of noise and distortion within the circuit. Using small-signal, frequency domain analysis, the linear noise parameters of a circuit can be predicted. However, the noise due to mixing effects of nonlinearities within the circuit can not be predicted. Likewise the up- or downconversion noise and phase noise of an oscillator are not analyzed. Distortion analysis using SPICE is usually performed through the use of a transient analysis followed by conversion of a part of the time waveform into the frequency domain using a Discrete Fourier Transform (DFT). The microwave designer is typically interested in two types of distortions — the harmonic content of the time waveform and the intermodulation products caused by the excitation of the circuit by two signals typically close to each other in frequency.

Harmonic balance is used when the circuit is driven by periodic sources and when the design parameters, input and output, are specified in the frequency domain. The assumed periodicity of the circuit response avoids the need to compute the circuit response from time zero until the steady-state response is obtained. Therefore, much less computer time is required to predict the circuit response. Since the harmonic balance techniques were developed specifically to aid the microwave designer in the design of nonlinear circuits, the available programs are custom tailored to provide the results in a format familiar to the designer. For example, the input source for an amplifier can be swept in both frequency and power, and nonlinear parameters such as gain, 1 db compression point, saturated power output, power-added efficiency, and harmonic levels can all be displayed in a graphical or text form. Indeed these parameters are all natural artifacts of the computations. Other parameters that can easily be computed are intermodulation products, third order intercept point, noise side bands, and mixer conversion.

The harmonic balance method is a hybrid of the small-signal, frequency domain analysis and a nonlinear time domain analysis. The circuit is divided into two subcircuits. One sub-circuit contains only those circuit elements that can be modeled in the small signal frequency domain. This sub-circuit results in a Y matrix representation that relates the frequency domain currents to the frequency domain voltages. At this point, the matrix can be reduced in size by eliminating voltages and currents for the nodes that are not connected to the nonlinear devices or the input and output nodes. These matrix computations need only be performed once for each frequency and harmonic frequency of interest. The second subcircuit includes all of the active or nonlinear elements that are modeled in the time domain. These circuit models relate the instantaneous branch currents to the instantaneous voltages across the device nodes. These models are the same models used in SPICE programs.

The two sub-circuits make up two systems of equations having equal node voltages whose branch currents must obey Kirchhoff's current law. The system of equations corresponding to the linear sub-circuit is now solved by making an initial guess at the frequency spectrum of the node voltages. The node voltage frequency spectrum is then used to solve for the frequency spectrum of the branch currents of the linear sub-circuit. In addition to this, the node voltage frequency spectrum is transformed into the time domain using a FFT algorithm. The result of this operation is a sampling of the periodic time voltage waveform. The sample voltages are applied to the time domain sub-circuit resulting in time domain current waveforms, which are then transformed into the frequency domain again using a FFT algorithm. The two frequency domain current spectrums are compared, and based on the error between them, the voltage spectrums are updated. This process is repeated until the error is sufficiently small.

Early implementations of harmonic balance programs used either an optimization routine to solve for the node voltage spectrums [2] or Newton's method [3]. The main advantage of Newton's method is that it uses the derivatives of the nonlinear device currents with respect to the node voltages to predict the next increment in the node voltages. By taking advantage of these derivatives, convergence can be achieved for a relatively large number of nonlinear devices. This method appears to work well as long as the nonlinearity of the system is not too severe.

Emerging Simulation Developments

Current research directed toward improving the implementation of harmonic balance programs is concentrating on techniques that can handle the large number of nonlinear devices typically found in integrated circuits. Currently, Krylov-subspace solutions have been implemented [4]. When Krylov subspace techniques are used, the harmonic balance method can be used to solve circuit problems containing hundreds of transistors.

When the excitation of a circuit consists of multiple sinusoids, closed spaced in frequency, both SPICE and conventional harmonic balance methods tax computer hardware resources as they require large amounts of memory and computer time. A program must be able to efficiently handle this type of excitation in order to be able to predict the effects of spectral reqrowth in digitally modulated circuits, as well as noise-power ratio simulations for these circuits. For these types of circuit analysis, the excitation consists of multiple sinusoids, closely spaced in frequency. Borich [5] has proposed a means of overcoming these problems for harmonic balance programs by adjusting the sampling rate and the spacing between excitation carriers in order to reduce the computations of the multitone distorted spectra to an efficient one-dimensional FFT operation.

Envelope following methods [6] have been implemented to solve for circuits in which the excitation consists of a high frequency carrier modulated by a much slower information signal. The method performs a transient analysis consistent with the time scales of the information signal. At each time step, a harmonic balance analysis is performed at the harmonic frequencies of the carrier. This method can be used to study PPL phase noise, oscillator turn-on time, and mixer spectral regrowth due to digital modulation on the RF carrier [7].

References

1. George D. Vendelin, *Design of Amplifiers & Oscillators by the S-Parameter Method*. John Wiley & Sons, New York, 1982.
2. M. Nakhla and J. Vlach, A Piecewise Harmonic Balance Technique for Determination of Periodic Response of Nonlinear Systems, *IEEE Transactions on Circuits and Systems*, CAS-23, 2, February 1976.
3. K. Kundert and A. Sangiovanni-Vincentelli, Simulation of Nonlinear Circuits in the Frequency Domain, *IEEE Transactions Computer-Aided Design*, CAD-5, 4, October 1986.
4. R. Telichevesky, K. Kundert, I. Elfadel, and J. White, Fast Simulation Algorithms for RF Circuits, IEEE 1996 Custom Integrated Circuits Conference.
5. V. Borich, J. East, and G. Haddad. An Efficient Fourier Transform Algorithm for Multitone Harmonic Balance, *IEEE Transactions Microwave Theory and Techniques*, 47, 2, February 1999.
6. P. Feldmann and J. Roychowdhury, Computation of Circuit Waveform Envelopes Using an Efficient, Matrix-Decomposed Harmonic Balance Algorithm, in *Proc. IC-CAD*, November 1996.
7. K. Mayaram, D.C. Lee, S. Moinian, D. Rich, and J. Roychowdhury, Overview of Computer-Aided Analysis Tools for RFIC Simulation: Algorithms, Features, and Limitations, IEEE 1997 Custom Integrated Circuits Conference.

8.7 Nonlinear Transistor Modeling for Circuit Simulation

Walter R. Curtice

Modeling in General

By definition, a transistor model is a simplified representation of the physical entity, constructed to enable analysis to be made in a relatively simple manner. It follows that models, although useful, may be wrong or inaccurate for some application. Designers must learn the useful range of application for each model.

It is interesting to note that all transistors are fundamentally nonlinear. That is, under any bias condition, one can always measure harmonic output power or intermodulation products at any input RF power level, as long as the power is above the noise threshold of the measurement equipment. In that sense, the nonlinear model is more physical than the linear model.

The purpose of this work is give a tutorial presentation of nonlinear transistor modeling. After reviewing the types of models, we will concentrate on equivalent circuit models, of the type used in

SPICE.[1] Recent improvements in models will be described and the modeling of temperature effects and the effects of traps will be discussed. Finally, parameter extraction and model verification is described.

Two-Dimensional Models

The models constructed for describing the non-linear behavior of transistors fall into several distinctly different categories, as depicted in Table 8.7. The most complex is the "physics-based" model. Here electron and hole transport is described by fundamental transport and current continuity relationships and the physical geometry may be described in one-, two-, or

TABLE 8.7   Types of Large-Signal Transistor Models

I. Physical or "Physics-Based" Device Models
II. Measurement-Based Models
    1 - Anaytical Models, such as SPICE Models
    2 - Black Box Models
      Table-Based Models
      Artificial Neural Network (ANN) Models

even three-dimensional space. The electric field is found by solution of Poisson's equation consistent with the distribution of charge and boundary conditions. Such a model may use macro-physics, such as drift-diffusion equations,[2-4] or more detailed descriptions, such as a particle-mesh model with scattering implemented using Monte-Carlo methods.[5,6] Two- and three-dimensional models must be used if geometrical effects are to be included. The matter of how much detail to put into the model is often decided by the time it takes for the available computer to run a useful simulation using the model. In fact, as computers have increased their speed, modelers have increased the complexity of the model simulated.

The lengthy execution time required for Monte-Carlo analysis can be reduced by using electron temperature[7] as a measure of electron energy. Electron temperature is determined by the standard deviation of the energy distribution function and is well defined in the case of the displaced Maxwellian distribution function. Electron and hole transport coefficients are developed as a function of electron temperature, and nonequilibrium effects, such as velocity overshoot in GaAs, may be simulated in a more efficient manner. However, the solution of Poisson's equation still require appreciable computational time.

BLAZE[8] is a good example of a commercial, physically-based device simulator that uses electron temperature models. BLAZE is efficient enough to model interaction of a device with simple circuits.

The physics-based model would be constructed with all known parameters and simulations of current control for DC and transient or RF operation, then compared with measured data. Using the data, some transport coefficients or physical parameters would be fine-tuned for best agreement between the model and the data. This is the process of calibration of the model.[9] After calibration, simulations can be trusted to be of good accuracy as long as the model is not asked to produce effects that are not part of its construction. That is, if trapping effects[10] have not been incorporated into the model, the model will disagree with data when such effects are important. With the physics-based model, as with all others, a range of validity must be established.

Present physics-based models still require too much computational time to be used to any extent in circuit design work. Optimization of a circuit design will involve invoking the device model frequently enough to be impractical with physics-based models. These models can be used if only one or two nonlinear transistors are used in a specific circuit, but typically, the circuit designer has a larger number of nonlinear devices.

Several quasi two-dimensional models[11,12] have been developed that execute more efficiently. Initial results look good, but accuracy may depend upon the simplifications made in the development of the code and will vary with the application.

Measurement-Based Models

The next general category is that of "measurements-based" models. These are empirical models either constructed using analytical equations and called "analytical models" or based upon a lookup table developed from the measured data. The latter are called "table-based" models. Multidimensional spline functions are used to fit the data in some of these models[13] and only the coefficients need be stored.

In the case of analytical models, the coefficients of the equations serve as fitting parameters to permit the equations to approximate the measured data. Functions are usually chosen with functional behavior similar to measured data so that the number of fitting parameters is reduced.

The advantages of analytical models are: computational efficiency, automatic data smoothing, accommodation of device statistics, physical insight, and the ability to be modified in a systematic manner. Disadvantages are: restriction of behavior often due to use of over-simplified expressions, difficulty in parameter extraction, and guaranteed nonphysical behavior in some operating condition. The nonphysical behavior is often associated with the use of a function, such as a polynomial, to fit data over a specific range of voltages and subsequent application of the model to voltages outside this range. The function may not behave well outside the fitting range. The best example of analytical models is the set of transistor models used in the various forms of the SPICE program. A major advantage of analytical models is that all the microwave nonlinear simulators provide some sort of user-defined model interface for analytical model insertion.

Table-based models have some properties of black-box models. The equations used result from fitting to the data, using splines, or other such functions. These models can therefore "learn" the behavior of the nonlinear device and are ideal for applications where the functional form of the behavior is unknown. Table-based models are efficient but do not provide the user with any insight, since there is a minimal "circuit model." They have difficulty incorporating dispersive effects, such as "parasitic gating" due to traps (see the section on Modeling the Effects Due to Traps) and do not accommodate self-heating effects.

The model cannot be accurately extrapolated into regions where data was not taken, and the models are often limited in their application due to the particular coding used by their author. This means that the model cannot be tailored by users other than the author. Customization of models is important to improve the "performance" of a model. The first table-based model that has been widely used is the Root model.[13]

Physical Parameter Models

One may argue that there is a class of models between physics-based and analytic models, namely, physical-parameter models. A good example would be the Gummel-Poon model.[14] Here, analytical equations are used but the fitting parameters or equation coefficients have physical significance. For example, NF, the ideality factor of the emitter-base junction, is one model parameter. This model is an analytical model but more useful device information may be gleaned from the value of the coefficients. This is often the case and the model is widely used for various forms of bipolar devices.

A useful physical parameter model for the AlGaAs/InGaAs/GaAs PHEMT has been published and verified by Daniel and Tayrani.[15] No information has been given on the range of validity of the analytical model, and it is expected that such a simple model will have inaccuracy when two-dimensional effects or nonequilibrium effects are important. It is interesting that the HEMT structure has less two-dimensional effects than the MESFET because of the sheet current layer produced in the HEMT.

There are many analytical models for which the coefficient has the name and dimensions of a physical parameter but the coefficient is only a fitting parameter and is not strongly related to the physical parameter. Khatibzadeh and Trew[16] have presented one commonly called the Trew model and Ladbrooke[17] has presented a second model commonly called the Ladbrooke model. Ladbrooke's model is an extension of the much earlier Lehovec and Zuleeg[18] model and it is more empirical than physical, as Bandler has shown.[19] Such models must be tested to see how strong the relationship is between the model parameter and the physical parameter. It is a matter of the degree of correlation between the two quantities. It is dangerous to attach too much physical significance to these coefficients. One should verify the relationship before doing so.

There is a unique case where physical parameters have been installed in a previously developed analytical model. The Statz-Pucel (analytical) GaAs MESFET model[20] has been converted to a physical parameter model by D'Agostino et al.[21]

Neural Network Modeling

Rather recently, a new approach has been developed for the modeling of nonlinear devices and networks. It utilizes artificial neural networks, or ANN. ANN models are similar to table-based or black-box models in that there is no assumption of particular analytical functions. As with table-based models, the ANN model "learns" the relationship between current and voltage from the data, and model currents are efficiently calculated after application of voltages. ANN analysis can treat linear or nonlinear operation of devices or complex circuits. ANN models have many of the advantages and deficiencies of table-based models. An excellent special issue of *RF and Microwave Computer-Aided Engineering*[22] has been devoted to this modeling method. Unfortunately, discussion of this approach is beyond the scope of this work.

Scope of This Work

The purpose of this work is to present a tutorial on the modeling of the nonlinear behavior of transistors. The scope is limited to nonlinear models useful for the development of circuit designs. It would not be possible to cover all the important material on other types of models, such as physical models, in this article. This paper will deal primarily with nonlinear analytical models for MESFETs, (P)HEMTs and HBTs. The emphasis is on GaAs device models, although many of these models are also used for transistors fabricated in InP, silicon, and other materials. The RF LDMOS power device is also discussed because it can be treated as a three-terminal device, much like a MESFET.

We address the concerns of analog and digital circuit designers who must choose between a wide variety of nonlinear models for transistors. Of particular concern here is the MMIC designer who must select the proper model to use for his GaAs microwave transistor. Even with very complex models presently supplied in circuit simulators, some specific behaviors are not modeled and new model features are required. We will spend much time on SPICE models and SPICE-type models. We will inspect some of the recent models that incorporate important device effects, omitted in previous models. We will discuss the modeling of gate charge as a function of local and remote voltages, the modeling of self-heating effects, the modeling of trapping effects, and model verification.

Unfortunately, the references presented will only be representative of prior work because there is a wealth of papers in each area of transistor modeling. I apologize to any author not included, as there are now and have been many people working in this area. I do recommend some modeling tutorial articles, previously published. Trew[23] and Snowden[24] and Dortu et al.[25] have presented excellent reviews of SPICE type transistor models and are recommended reading.

Equivalent Circuit Models

We concern ourselves here with equivalent circuit models because they are formulated to be efficiently exercised in a circuit simulator, and thus are efficient for circuit design and optimization. This is because the simulator is accustomed to dealing with resistors, capacitors, inductors, and voltage or current-controlled sources.

One problem is that "de facto" standard models have evolved in the industry, and often these models are inadequate to describe the device behavior. This is even true for small-signal equivalent-circuit models. Still, all circuit simulators utilize standard model topologies for small-signal and large-signal MESFET and PHEMT models. These models represent a minimum number of elements and are efficient for evaluation of transistor characteristics. Figures 8.60 and 8.61 show the conventional topologies for small-signal and large-signal simulation, respectively. It is conventional to separate the extrinsic parameters from the intrinsic device parameters, as shown in Fig. 8.60. The intrinsic parameters are assumed to contain all the bias-dependent behavior and the extrinsic parameters are assumed to be of constant values.

Curtice and Camisa[26] and Viakus[27] have discussed the trade-offs that exist between simple models with a small number of parameters, and complex models with a large number of parameters. A primary concern is the significant increase in the uncertainty for each model parameter in a complex model.

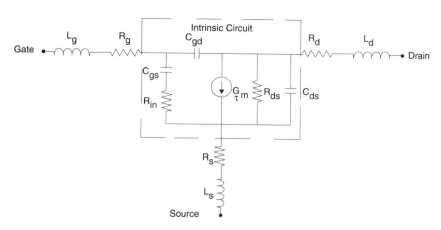

FIGURE 8.60 The conventional small-signal model for a MESFET, showing intrinsic and extrinsic elements.

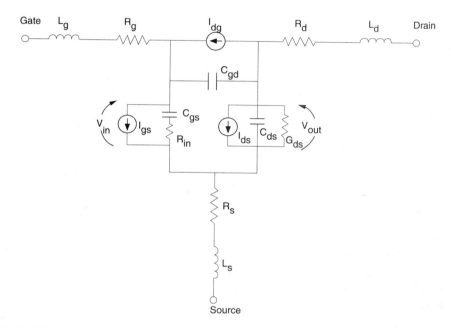

FIGURE 8.61 The conventional large-signal model for MESFET.

Viakus showed that a small increase in the number of elements in a small-signal model could easily increase the uncertainty of critical elements beyond the standard deviation of the element value resulting from the fabrication process.

Byun[28] and others assert that source resistance should be taken as bias dependent. This decision is actually a choice made by the modeler. If source and drain resistance are taken as constant, then the reference planes defining these resistances are taken as being close to the metal ohmic contacts and not too close to the Schottky contact. That is, no region that may become depleted of change is included. All bias-dependent behavior is then lumped into the intrinsic elements. This is the convention followed by most modelers. It results in a simpler model, with fewer parameters.

Unfortunately, many of the published nonlinear device models have inconsistencies with the conventional small-signal model. Some of the inconsistencies may go unnoticed but can cause design errors. A good example that will be described later is the modeling of capacitance as a function of two independent voltages. The small-signal model must contain "transcapacitance" elements to be consistent with the large-signal model.

The large-signal model should agree with the small-signal model, but is not expected to be as efficient. Because transconductance and some resistances and capacitances must be evaluated from functions of voltages, numerical evaluation will take more computational time. However, the behavior of the large-signal model will gracefully go from small-signal to large-signal in a good model.

SPICE Models and Application-Specific Models

SPICE was developed to help in the design of switching circuits. Thus, the SPICE transistor models were developed to model the time domain behavior of devices in such circuits. However, transistors operating in RF analog circuits have a different locus of operation. For example, a simple class A amplifier will have locus of operation around its quiescent bias point. This should be compared with a logic circuit where the transistors go from a biased-off condition (high voltage, low current) to a strongly turned on condition (low voltage, high current). If the same transistor is used in these two applications, one would expect the SPICE model would approximate both behaviors, but not be optimum for either. In fact, if the model is fine-tuned to be more accurate for one of these applications, it will by default, be less accurate for the other.

For these reasons, accurate nonlinear transistor models will be application specific. In order to make the model more general, the model can be made more complex and more model parameters will be added. This may result in poorer execution efficiency.

Clearly one goal of transistor modeling should be to keep the model simple and to keep the number of model parameters small so that the extraction of these parameters is more efficient. My experience is that designers like simple models for initial work and are willing to work with more complicated models for difficult design specifications. The SPICE transistor models serve the function of the initial, simple models. These models are also universally known by name and have history and familiarity associated with them. A designer attempting to use a new GaAs foundry would not be intimidated by obscure nonlinear models if SPICE transistor models are used in that foundry.

Many modelers have attempted to extend or enhance the SPICE models so that their accuracy is improved, particularly in microwave analog applications. Usually, the default model is the original SPICE model and the designer will feel comfortable with this approach. There are many examples. The Gummel-Poon BJT model has been extended by Samelis and Pavlidis[29] and others[30] for application to heterojunction bipolar devices. The JFET SPICE model was extended by Curtice[31] in 1980 for better application to GaAs MESFET logic circuits.

Because of the increasing use of harmonic-balance simulators for microwave applications, many new equivalent circuit models have been developed specifically for these simulators. Nevertheless, these models are "SPICE-type" models, and can also be executed in a time domain simulation. The requirements of a model for SPICE are the same as for harmonic balance since the device is operated in the time domain in both simulators.

The producers of commercial harmonic balance (HB) software recognized the need of users to customize their transistor models. All commercial packages now contain user-defined modeling interfaces that permit the installation of customized models into the transistor model library. The process of installing or customizing a model in SPICE is much more difficult and not available to the average user. However, the ease of installation of models into HB software has produced a rash of new models for many transistor types.

Table 8.8 shows the typical array of SPICE equivalent circuit models available as part of a commercial simulator software package. The models are categorized, in general, as diode models, GaAs MESFET or (P)HEMT models, MOS models, and bipolar device models. The list is not complete for any specific product but representative of the models available. The models listed in Table 8.8 are in most commercial simulator products whether a version of SPICE or a harmonic balance simulator.

TABLE 8.8   SPICE Models

GaAs MESFET/HEMT	MOS Models
Curtice (Cubic and quadratic)	BSIM 1,2,3 3v3
STATZ (Raytheon)	UC Berkeley 2 and 3
JFET (N & P)	HSPICE
TOM (TriQuint's Own Model)	MOSFET (various levels)
Materka	
Diode Models	BJT Models
P/N diode	Gummel-Poon
PIN diode	METRAM
	VBIC

Improved Transistor Models for Circuit Simulation

Early SPICE models have shown a number of deficiencies. One problem is that the models developed before 1980 were developed for silicon devices and they do not reflect the behavior of GaAs devices. Most SPICE models need to be customized to be accurate enough for present design requirements. With regard to GaAs MESFET and PHEMT modeling, the strong dependency of gate-source capacitance upon drain-source voltage as well as gate-source voltage is not modeled in the early SPICE models. None of the standard SPICE models accommodate self-heating effects. These effects are more important in GaAs applications due to the poorer thermal conductivity of GaAs compared to silicon. Some GaAs transistors exhibit important dispersion effects in transconductance as well as in drain admittance. All the GaAs models in Table 8.8 were added during the 1980s. These models represented major improvements; however, deficiencies remain. These deficiencies are summarized below for SPICE large-signal models:

- Insufficient accuracy for GaAs applications.
- Poor modeling of nonlinear capacitance.
- Poor modeling of self-heating effects.
- No modeling of dispersion of transconductance.
- Model parameter extraction not defined.
- Poor modeling of nonlinear effects dependent upon higher order derivatives.

The improved large-signal models of the 1990s exhibit some common features. It is quite popular to utilize analytical functions that have an infinite number of derivatives. For example, in the modeling of PHEMTs, Angelov et al.[32] have relied heavily on the hyperbolic tangent function for current because its derivative with respect to gate-source voltage is a bell-shaped curve, much like transconductance in PHEMTs. All further derivatives also exit. The Parker[33] model also utilizes higher order continuity in the drain current description and its derivatives.

Some SPICE models do have continuous derivatives but may not be accurate. The differences between the COBRA[34] model and the previous Materka[35] model are more evident when the derivatives of current (first through third) are compared. Since the COBRA model has derivatives closer to the data, Cojocaru and Brazil[34] show that the model predicts intermodulation products more accurately.

Many SPICE models use a simple expression for junction capacitance in MESFETs and PHEMTs. However, capacitance values extracted from data show that the gate-source and gate-drain capacitance depends strongly on the remote voltage as well as the local voltage (or capacitance terminal voltage). The Statz (Raytheon), the TOM, and the EEFET3 SPICE models[36] all have detailed equations for the gate, drain, and source charge as a function of local and remote voltages. Extraction of coefficients for these expressions is not simple and this will be discussed in the next section.

Modeling Gate Charge as a Function of Local and Remote Voltages in MESFETS and PHEMTS

Small-signal modeling of GaAs and InP MESFETs and PHEMT show that both C_{gd}, the gate-drain capacitance, and C_{gs}, the gate-source capacitance, vary with change of both V_{gs}, the gate-source voltage, and V_{ds}, the drain-source voltage. Thus, these capacitances are dependent upon the local, or terminal voltage, and a remote voltage. The dependency upon the local voltage is expected for capacitances, but the dependency upon the remote voltage leads to a term called "transcapacitance." The modeling of these capacitances can lead to nonphysical effects if not handled properly, as Calvo et al.[37] have shown.

Harmonic balance simulators work with charge functions whose derivatives are capacitive terms. The conventional approach is to find some charge function for total gate charge, such as $Q_g (V_{gs}, V_{ds})$. Then,

$$C_{11} = \text{Partial Derivative of } Q_g \text{ with respect to } V_{gs}$$

$$C_{12} = \text{Partial Derivative of } Q_g \text{ with respect to } V_{ds}$$

For consistency with small-signal models:

$$C_{11} = C_{gs} + C_{gd}$$

$$C_{12} = -C_{gd}$$

The problem remaining is to partition the total gate charge Q_g into charge associated with the gate-source region, Q_{gs}, and charge associated with the gate-drain region, Q_{gd}. Then, for large signal modeling, the gate node has charge Q_g, the source node has charge $-Q_{gs}$, and the drain node has charge $-Q_{gd}$, and

$$Q_g = Q_{gs} + Q_{gd}.$$

One scheme for partitioning the charge is used in the EEFET model and described in the HP-EEsof Manual.[36] One advantage of this approach is that the model becomes symmetrical, meaning that drain and source may be interchanged and the expressions are still valid.

A simpler approach is presented by Jansen et al.[38] Jansen assumes that all of the gate charge is associated with the gate source region and:

$$Q_{gd} = 0$$

$$Q_g = Q_{gs}.$$

The topology for this approach is different than the conventional model. There is no drain-gate capacitance element. Instead, the transcapacitance term accounts for the conventional drain-to-gate capacitive effects. That is, a change in V_{ds} produces current in the gate source region through the transcapacitance term. Figure 8.62 shows the new topology for the intrinsic circuit.

The procedure for modeling the capacitive effects is the same for both approaches and is the following:

1. Measure small-signal values of C_{gs} and C_{gd} as a function of V_{gs} and V_{ds}
2. Choose a Q_g function and optimize the coefficients of the Q_g expression for best fit of

$$C_{11} = C_{gs} + C_{gd}$$

$$C_{12} = -C_{gd}$$

A good example of the fitting functions and the type of fit obtained is given by Mallavarpu, Teeter, and Snow[39] for a PHEMT.

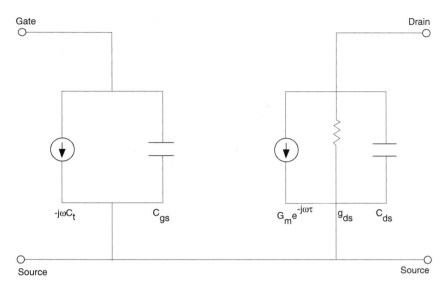

FIGURE 8.62 Jansen's topology for the intrinsic circuit.

In summary, large-signal capacitive effects are modeled by constructing a total gate charge function, $Q_g(V_{gs}, V_{ds})$, whose partial derivative approximates the measured capacitance functions. If device symmetry is important, the EEFET charge partitioning scheme may be used. For amplifier applications, the Jansen model is simplest to code and implement because there is no charge partitioning expression. However, the Jansen model uses a topology that is not conventional.

Modeling the Effects Due to Traps

Electron and hole traps exist in GaAs materials and cause numerous effects during operation of a GaAs MESFET or PHEMT transistor. The following is a brief listing of these effects:

- Dispersion in transconductance and output admittance
- Backgating
- Parasitic bipolar effects
- "Kinks" in the I/V relationship
- Surface gating
- Gate and drain lag effects during switching
- Light sensitivity
- Substrate current or lack of current pinchoff

These effects have been studied and circuit-level models developed to simulate the effects. In many cases, the details of the behavior have been made clear using two-dimensional simulation modeling. For example, Li and Dutton[10] used PISCES-IIB to show that the common EL2 trap causes dispersion in the output conductance of a GaAs MESFET up to several hundred Hz.

The circuit-level modeling of dispersion is described by Cojocaru and Brazil.[34] They extend the previous conventional modeling of dispersion of the output conductance to include dispersion of the transconductance. The circuit is very simple. A second voltage-control current source in parallel with a resistance is capacitively coupled to the internal drain-source terminals. This enables the model's transconductance and drain-source conductance to be tailed for high frequencies using these new elements.

Horio and Usarni[40] use two-dimensional simulation to show that a small amount of avalanche breakdown in the presence of traps causes excess hole charge in the substrate that produces "kinks" in the low-frequency I/V data. Since the traps cannot be easily eliminated, the kinks may be removed by removing the conditions initiating avalanche breakdown.

Upon switching the gate voltage, the drain current of a MESFET will have lag effects in the microsecond and millisecond regions that are produced by traps. Curtice et al.[41] have shown the circuit-level modeling of such gate lag effects as well as drain lag effects. Others, such as Kunihiro and Ohno[42] have also presented circuits for the modeling of drain lag effects.

The transistor model with such circuits may then be used to determine if the lag effects interfere with proper operation of the circuit. The work of Curtice et al. was directed toward GaAs digital circuits where the switching waveform must be of high quality. Using the new transistor model, one may determine not only if the circuit will perform, but also circuit changes that will permit operation in the presence of strong lag effects.

Light sensitivity has been described and modeled by Chakrubarti et al.[43] and by Madjar et al.[44] Some circuit-level models are presented in their discussions.

In the microwave application arena, the principal difficulty with traps is that they cause difficulty in determining an accurate microwave model for the transistor. An excellent experimental study of surface gating effects is given by Teyssier et al.[45] Surface gating means that the charge stored in surface states and traps influences the I/V behavior by acting as a second gate. The amount of charge stored in the traps will vary, depending upon the applied voltages, the ambient light, the temperature, and trapping time constants. Teyssier et al. show that measured trap capture time constants are quite different than trap emission time constants. They show how they are able to accurately characterize the I/V behavior for RF operation by using 150 ns bias pulse width. They also describe how they characterize the thermal behavior using longer pulse width.

Many others have published data showing surface gating effects (see, for example, Platzker et al.[46]). The approach a modeler should use is to first determine how important such trapping effects are in the transistor operation. In many transistor designs, the active region is shielded sufficiently from surface charge so that negligible surface gating occurs. In that case, low-frequency I/V data and the resulting transconductance may be very predictive of microwave-frequency behavior. If the trapping effects are found to be of importance, then short-pulsed characterization is required and low-frequency I/V data will not suffice.

In any case, device characterization must include the behavior changes due to self-heating and ambient temperature effects. The modeling of heating effects will be discussed next.

Modeling Temperature Effects and Self-Heating

Anholt and Swirhun[47] and others have documented the changes of GaAs MESFETs and HEMTs at elevated temperatures. However, the modeling of a device over a temperature range is often accomplished using temperature coefficients. With regard to drain current and DC transconductance, the effects of elevated temperature is quite different at large channel current as compared to operation near pinch off. At large channel current, the electron mobility decrease with temperature increase is most important, whereas at low current, the decrease in the pinch-off voltage with temperature is most important. This produces the interesting effect that transconductance decreases with temperature at large current but increases with temperature at very low currents. This behavior is best modeled using a temperature analog circuit that will be described later.

Figure 8.63 shows the behavior of DC drain current for a 0.25-μm PHEMT at four different ambient temperatures. Heating effects are obvious in Fig. 8.63 where current (and transconductance) near positive V_{gs} is reduced and current (and transconductance) near pinch off is increased.

The effects of temperature upon drain current in MESFETs and PHEMTs may be considered second order but they are of first order in bipolar devices. The reason is due to the current exponential dependence upon temperature in a bipolar device. The gate current effects due to temperature in MESFETs and PHEMTs are of first order for the same reason. First order and even some second order changes with temperature may be modeled using linear temperature coefficients if the changes are reasonably linear.

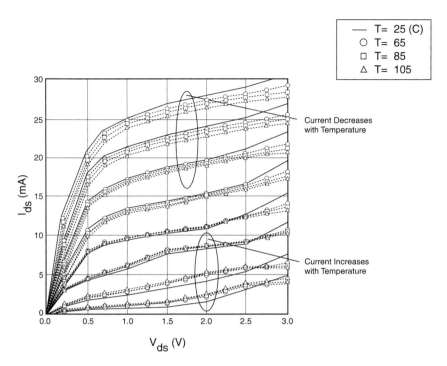

FIGURE 8.63 The behavior of DC drain current for a 0.25-μm PHEMT at four different temperatures.

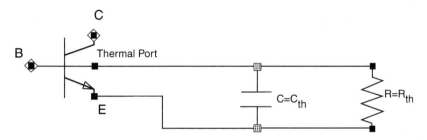

FIGURE 8.64 The HBT transistor model with the thermal analog circuit.

Many devices are operated such that their self-heating effects are quite important. This can occur, for example, in a power amplifier design where the standby biasing current is low, but the advent of input RF power turns on the drain current and output RF power. The ambient temperature may not change but the device will operate with more elevated temperature with the application of the RF power. The use of temperature coefficients may not be accurate for such simulations.

More accurate modeling of self-heating effects in transistors circuits simulators has been done for about ten years using a thermal analog circuit, first used in SPICE applications. It has been found to work well for bipolar simulation as well as for MESFET and PHEMT simulations. The early studies were reported by Grossman and Oki,[48] F. Q. Ye,[49] and others.[50-52] The main differences in these early studies relate to the description of temperature effects in the transistor and not to the CAD model used for simulation in SPICE.

Figure 8.64 shows an HBT transistor model with the additional thermal analog circuit. The device can be of any type but must have well-defined descriptions of the model coefficients as a function of temperature. The analog circuit consists of a current source, a resistor, and a capacitor. R_{th} is the value of thermal resistance and the $R_{th}*C_{th}$ time constant is the thermal time constant of the device. The current source to the thermal circuit, I_{th}, is equal in magnitude to the instantaneous internal dissipated power

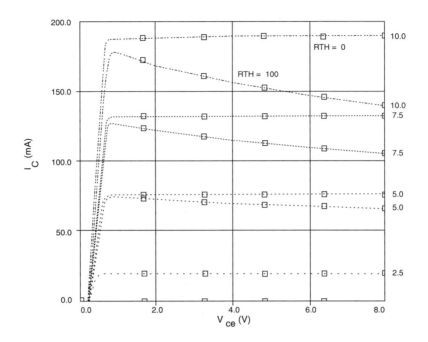

FIGURE 8.65   The collector I-V relationship for an HBT with and without self-heating effects.

to the device. For DC biasing, I_{th} to the thermal circuit would be equal to the total biasing power to the device and the temperature rise would be numerically equal to I_{th} times R_{th}. For RF or transient conditions, the average temperature rise would be that evaluated over some period of time including the effects of the thermal time constant. Thus, for an RF amplifier application, I_{th} would be equal to the DC biasing power plus the RF heating effects less the net RF power leaving the device. Convergence in a simulator is not assured since the value of the model coefficients must be consistent for the temperature of the device.

Harmonic balance simulators usually find the steady-state RF condition efficiently. Using this method, the simulator must find the solution with temperature rise consistent with the device parameters producing the temperature rise. Surprisingly, the harmonic balance simulators do not seem to be much less efficient when the thermal circuit is used, unless a thermal runaway condition exists. In that case, no solution will be found.

Figure 8.65 shows the collector I-V relationship for an HBT exhibiting self-heating effects. The current curves without heating effects are flat in the saturation region. Heating of the lattice reduces the electron mobility, and thus reduces the collector current.

Enhancing the Gummel-Poon Model for Use with GaAs and InP HBTs

The Gummel-Poon model, or the GP model,[14] is a complex, physical parameter model with 55 parameters, and is widely used. It was developed early for SPICE, and all colleges and universities teach their electrical engineering students to use this model. Although the GP model has many parameters, the current expressions are relatively simple. In addition, the current parameters are more closely tied to material parameters rather than manufacturing tolerances, so that there is less variation in current control characteristics than with MESFETs and PHEMTs. The standard bipolar device has less two-dimensional effects than do MESFETs and PHEMTs.

Much effort has been expended to improve the accuracy of compact BJT circuit models for silicon devices. Fossum[53] has reviewed the effort to 1989 and it continues to this day.

Whether the bipolar device is all silicon or a heterojunction device made with SiGe on Si, AlGaAs on GaAs, GaInP on GaAs, or InP on InGaAs, the bipolar action with current gain is the same physical process. So one expects some similarities in the analysis and modeling of the device. However, the heterojunction with the wide bandgap emitter causes the details of the analysis and model to have significant differences from a homojunction device.

The standard SPICE GP model has a number of major deficiencies that must be addressed before it can be used to accurately model a GaAs-based HBT in a large-signal microwave application. First of all, the SPICE code has silicon bandgap parameters hard coded into it and this must be changed to produce the correct temperature effects upon the bandgap. Next, collector-to-base avalanche breakdown must be added because it is important to most GaAs applications. The GP model uses PTF, a phase function, to accommodate the time delay associated with transconductance. It is more convenient for microwave engineers to use the time delay term TAU, as used in MESFET models.

The parameter "Early Voltage" is not as important in GaAs modeling, as it is usually very large. This is because the base doping can be made an order of magnitude larger than for silicon devices, because of the wide bandgap emitter. The large base doping reduces the importance of collector biasing upon the base region, and thus upon the collector current.

There may be dispersion in the collector admittance in HBTs, so some RF conductive element may be needed between collector and emitter.

The manner in which F_t, the frequency for unity current gain, changes with voltage and current is quite different in GaAs devices than with silicon devices. Therefore, a new functional form is need here. Most of the behavior of F_t with respect to collector voltage is related to the electron velocity-electric field (v-E) curve for the material. In the case of silicon devices, F_t generally increases with collector voltage and saturates until heating effects cause a decrease. Figure 8.66 shows such behavior for a SiGe HBT. The I/V characteristic of the device is given in Fig. 8.67.

In the case of GaAs HBTs, F_t peaks at a low voltage and monotonically decrease with further increase of collector voltage. Figure 8.68 shows such data and the devices I/V characteristics are given in Fig. 8.69. This difference in behavior reflects the striking differences between silicon and GaAs v-E curves in the high field region.

There are a multitude of new CAD models formulated to model the behavior of GaAs and InP HBTs. Two examples are the VBIC model[54] and the MEXTRAM model[55] and these are installed on a number

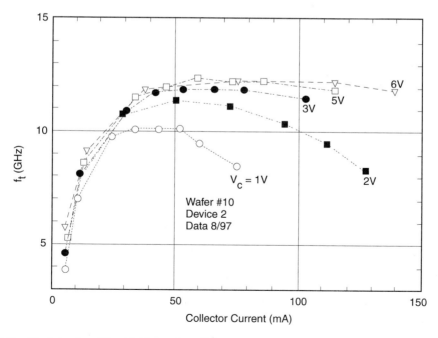

FIGURE 8.66 The behavior of F_t with biasing for a SiGe HBT.

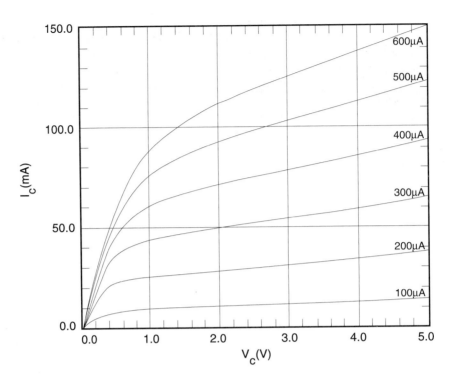

FIGURE 8.67   The I-V relationship for the device of Fig. 8.66.

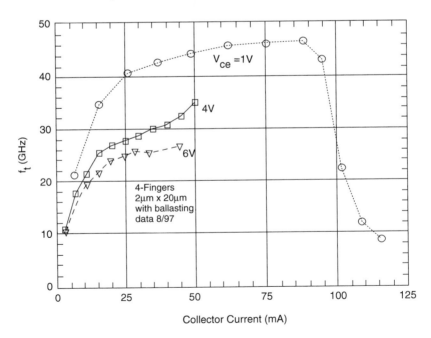

FIGURE 8.68   The behavior of F_t with biasing for GaAs HBT.

of circuit simulators. Most of the new models include self-heating effects because of their importance to device operation and accurate modeling. Because of the significantly poorer thermal conductivity in GaAs compared to silicon and because of the higher power density for best operation in GaAs, self-heating effects are usually important to the operation of the GaAs-based HBTs.

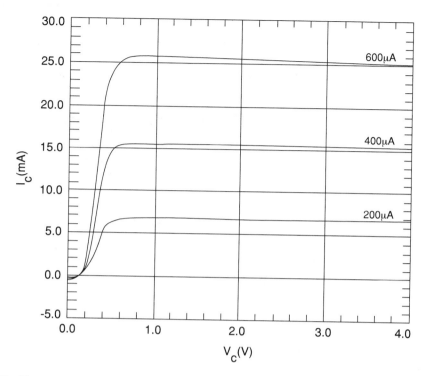

FIGURE 8.69   The I-V relationship for the device of Fig. 8.68.

Modeling the RF LDMOS Power Transistor

There are numerous silicon MOS models available for DC and RF modeling of silicon transistors. Because the silicon LDMOS transistor has become important for cost-effective consumer applications, many companies have developed nonlinear models specifically for this device. The device incorporates a p-type sinker diffusion to ground the source to the substrate, and thus can be treated as a three-terminal device. This makes it possible to construct a much simpler model, one very similar to the SPICE models developed for the GaAs MESFET.

Perugupalli et al.[56] have used a SPICE circuit network incorporating the standard NMOS SPICE element. Motorola uses the Root model developed for GaAs MESFETs for characterizing the device. However, self-heating effects, important for power applications, cannot be incorporated into this model. The Ho, Green, and Culbertson model[57] based upon the SPICE BSIM3v3 model suffers from the same problem.

Miller, Dinh, and Shumate[58] developed analytical current equations for the device, which led to the development of a new, simpler SPICE model by Curtice, Pla, Bridges, Liang, and Shumate.[59] The model includes self-heating effects, is accurate for both small and large-signal simulations, and operates in transient or harmonic balance simulators. Figure 8.70 shows the model predicts power-added efficiency in excellent agreement with the data for an RF power sweep. A similar model based upon the same equations has been developed and verified by Heo et al.[60]

Parameter Extraction for Analytical Models

The extraction of parameters for a device model has become less laborious since the advent of new extraction and equipment control programs, such as IC-CAP by Hewlett Parkard and UTMOST by Sylvaco International. These programs provide data acquisition and parameter extraction. Such software first enables the engineer to collect I/V and RF data in a systematic fashion on each device tested. This

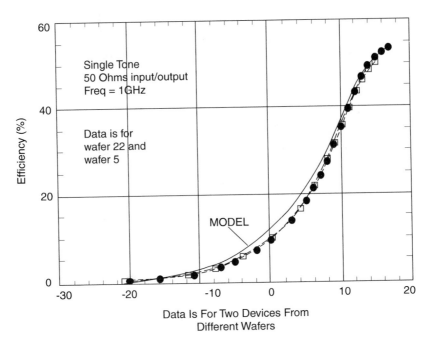

FIGURE 8.70   Efficiency prediction by the model and data for two RF LDMOS devices.

provides consistency between I/V and RF data. The parameter extraction routines then permit the extraction to specific, standard (SPICE) models, such as the Gunnel-Poon, or to new models for which equations may be user defined. Optimizers are used to provide the best fit between the data and the model. Testing can be with pulsed biasing or DC. Heating effects can be separately studied using thermal chucks during testing. Teyssier et al.[45] have discussed the merits of long- and short-pulse testing.

It is usually necessary to use devices of small sizes for characterization and then scale the model to devices actually used in the circuit design. Most SPICE models provide scaling with device area. However, the scaling laws for devices should be verified. It is usually possible to scale MESFETs and HEMTs accurately for a larger number of fingers of the same size. Golio[61] shows the scaling rules if the finger width is different. Because the biasing power may not be uniform on large devices and because the interelectrode capacitance does not scale simply, more complicated scale rules may be found for relatively large devices at high frequencies, i.e., above 5 GHz.

The Vector Nonlinear Network Analyzer

A large-signal, waveform measurement system has been used by many researchers to measure device characteristics dynamically. The equipment provides time domain voltage and current waveforms during RF excitation of the transistor. The equipment is often called the Vectorial Nonlinear Network Analyzer, or VNNA.

Demmler and Tasker[62] have shown that it is possible to accurately determine the drain current relationship to gate voltage for RF excitation at 2 GHz of a MODFET. The characteristic time delay is found by adding delay until "looping" is minimized. Furthermore, the drain-source I-V relationship can also be evaluated at 2 GHz. There is some difference from that obtained from the DC data. Thus the VNNA provides large-signal transfer characteristics from which more accurate model extraction can be done. Wei et al.[63] have also utilized this techniques to provide the data used for device model parameter extraction for a GaAs HBT.

Model Verification

The usual approach to verification of a large-signal model is to compare measured device performance with simulations under the same conditions. Initial verification should be comparison of a power sweep of the transistor at the application frequency and with no matching at the input and the output. For this test, we know that all harmonics at the input and output see 50-ohm impedance.

One should measure not only the output power at the fundamental, but also the power at second and third harmonics. If the model is not fully optimized, it will usually agree well with fundamental output power, gain, and efficiency, but not agree with the harmonic power production. The usual cause is due to poor modeling of the I/V relationship; however, in some cases, the nonlinear capacitive modeling may be the problem. After this problem is fixed, testing of third- and fifth-order IMD (intermodulation distortion) should be made, again in a 50-ohm system. If the harmonics now agree, the third-order IMD should agree and some further work may be required to get agreement for the fifth-order IMD.

In the previous test, it is important that the large-signal model be reasonably accurate at small-signal levels. It need not be as accurate as the best small-signal model for most applications.

Further verification work would involve power sweeps under tuned conditions. That is, the transistor may be tuned for best efficiency and the tuner impedance measured. It is important to measure the tuner impedance for fundamental, second, and third harmonics and to use these values in the simulation. The effects of the second harmonic voltage at either input[64] or output can be extremely important.

Testing of the load-pull characteristics should be made and compared with the model's behavior. Here, again, one has to be careful about the effects of harmonics. There are load-pull systems that operate separately on fundamentals and harmonics.

Further tests that may be important to the application may be testing with various ambient temperatures, noise testing, switch-on testing, and others. The specific application of the transistor will dictate the importance of the agreement for each test as well as the RF frequencies, power, and modulation to use for testing.

Foundry Models and Statistics

GaAs chip foundries provide design manuals that utilize small-signal as well as large-signal models. These are developed from measurements and statistical analysis of the data. However, these are guidelines for the designer, and often the best procedure is to obtain foundry test devices and develop more accurate models based upon new data. Device uniformity has improved greatly and yield prediction is becoming more accurate.

In addition, software programs such as IC-CAP and others provide statistical analysis of models extracted from test wafers. Both corner models and standard statistical patterns are available.

There is the fundamental problem that most foundries continue to tune their processes. It is often the case that the process has been changed and previous statistics are no longer valid. However, the design engineer is better off with approximate guidelines as to statistical patterns or corner models, than none at all.

Future Nonlinear Transistor Models

One can expect that with the ever-increasing speed of computers, circuit simulators will be able to utilize more physics-based models. This will aid in determining the effect of device design parameters upon chip yield and performance.

Improvements will be made in nonlinear model extraction software. The extraction parameters will be much less dependent upon the expertise of the tester. There will be improved collection schemes for transistor model statistics.

Finally, one expects that the nonlinear models will be made to be more easily tailored for adaptation to specific device behaviors. One often would like to start with a template for one of the standard nonlinear models and then tailor its behavior. Future simulators should make this procedure simpler than present procedures.

References

1. Nagle, L. W., SPICE 2: A computer program to simulate semiconductor circuits, Electronics Research Laboratory, College of Engineering, University of California, Berkeley, Memo, ERL-M520, 1975.

2. Pinto, M. R., Conor, R. S., and Dutton, R. W., PIECES2 - Poisson and continuity equation solver, Stanford Electronics Laboratory, Technical Report, Stanford University, 1984.

3. Wada, T. and Frey, J., Physical basis of short-channel MESFET operation, *IEEE Trans. on Electron Devices*, ED-26, 476, 1979.

4. Curtice, W. R., Analysis of the properties of three-terminal transferred electron logic gates, *IEEE Trans. on Electron. Devices*, ED-24, 1553, 1977.

5. Warriner, R. A., Computer simulation of gallium arsenide field-effect transistors using Monte-Carlo methods, *Solid-State Electron Devices*, 1, 105, 1977.

6. Moglestue, C., A self-consistent Monte Carlo particle model to analyze semiconductor microcomponents of any geometry, *IEEE Trans. on CAD*, CAD-5, 326, 1986.

7. Curtice, W. R. and Yun, Y-H, A temperature model for the GaAs MESFET, *IEEE Trans. on Electron Devices*, ED-28, 954, 1981.

8. *ATLAS User's Manual*, Silvaco International, Santa Clara, CA, Version 4.0, 1995.

9. Curtice, W. R., Direct comparison of the electron-temperature model with the particle-mesh (Monte-Carlo) model for the GaAs MESFET, *IEEE Trans. on Electron Devices*, ED-29, 1942, Dec. 1982.

10. Li, Q. and Dutton, R. W., Numerical small-signal AC modeling of deep-level-trap related frequency-dependent output conductance and capacitance for GaAs MESFETs on semi-insulating substrates, *IEEE Trans. on Electron Devices*, 38, 1285, 1991.

11. Snowden, D. M. and Pantoia, R. R., Quasi-two-dimensional MESFET simulation for CAD, *IEEE Trans. on Electron Devices*, 36, 1989.

12. Morton, C. G., Atherton, J. S., Snowden, C. M., Pollard, R. D., and Howes. M. J., A large-signal physical HEMT model, 1996 *International Micrwave Symposium Digest*, 1759, 1996.

13. Root, D. E. et al., Technology independent large-signal non quasi-static FET models by direct construction from automatically characterized device data, *21st European Microwave Conference Proceedings*, 927, 1991.

14. Gummel and Poon, An integral charge-control relationship for bipolar transistors, *Bell System Tech. Journal*, 49, 115, 1970.

15. Daniel, T. T. and Tayrani, R., Fast bias dependent device models for CAD of MMICs, *Microwave Journal*, 74, 1995.

16. Khatibzadeh, M. A. and Trew, R. J., A large-signal analytical model for the GaAs MESFET, *IEEE Trans. on Microwave Theory and Tech.*, 36, 231, 1988.

17. Ladbrooke, P. H., *MMIC Design: GaAs FETs and HEMTs*, Artech House, Inc., Boston, 1989, chap. 6.

18. Lehovec, K. and Zuleeg, R., Voltage-current characteristics of GaAs JFETs in the hot electron range, *Solid State Electron.*, 13, 1415, 1970.

19. Bandler, J. W. et al., Statistical modeling of GaAs MESFETs, 1991 *IEEE MTT-S International Microwave Symposium Digest*, 1, 87, 1991.

20. Statz, H, Newman, P., Smith, I. W., Pucel, R. A., and Haus, H. A., GaAs FET device and circuit simulation in SPICE, *IEEE Trans. Electron Devices*, 34, 160, 1987.

21. D'Agostino, S. et al., Analytic physics-based expressions for the empirical parameters of the Statz-Pucel MESFET model, *IEEE Trans. on MTT*, MTT-40, 1576, 1992.

22. *International Journal of Microwave and Millimeter-Wave CAE*, 9, No. 3, 1999.

23. Trew, R. J., MESFET models for microwave CAD applications, *Internation Journal of Microwave and Millimeter-Wave CAE*, 1, 143, 1991.

24. Snowden, C. M., Nonlinear modeling of power FETs and HBTs, *International Journal of Microwave and Millimeter-Wave CAE*, 6, 219, 1996.

25. Dortu, J-M, Muller, J-E, Pirola, M., and Ghione, G. Accurate large-signal GaAs MESFET and HEMT modeling for power MMIC amplifier design, *International Journal of Microwave and Millimeter-Wave CAE*, 5, 195, 1995.

26. W. R. Curtice and R. L. Camisa, Self-consistent GaAs FET models for amplifier design and device diagnostics, *IEEE Trans. on Microwave Theory and Tech.*, MTT-32, 1573, 1984.

27. R. L. Vaitkus, Uncertainty in the Values of GaAS MESFET Equivalent Circuit Elements Extracted from Measured Two-Port Scattering Parameters, Presented at 1983 IEEE Cornell Conference on High Speed Semiconductor Devices and Circuits, Cornell University, Ithaca, NY, 1983.

28. Byun, Y. H., Shur, M. S., Peczalski, A., and Schuermeyer, F. L., Gate voltage dependence of source and drain resistances, *IEEE Trans. on Electron Devices*, 35, 1241, 1998.

29. Samelis, A. and Pavlidis, D., Modeling HBT self-heating, *Applied Microwave & Wireless*, Summer Issue, 56, 1995.

30. Teeter, D. A. and Curtice, W. R., Comparison of hybrid pi and tee HBT circuit topologies and their relationship to large-signal modeling, 1997 *IEEE MTT-S International Microwave Symposium Digest*, 2, 375, 1997.

31. Curtice, W. R., A MESFET model for use in the design of GaAs integrated circuits, *IEEE Trans. on Microwave Theory and Techniques*, 23, 448, 1980.

32. Angelov, I., Zirath, H., and Rorsman, N., New empirical nonlinear model for HEMT and MESFET and devices, *IEEE Trans. on Microwave Theory and Techniques*, 40, 2258, 1992.

33. Qu, G. and Parker, A. E., Continuous HEMT model for SPICE, *IEE Electronic Letters*, 32, 1321, 1996.

34. Cojocaru, V. I. and Brazil, T. J., A scalable general-purpose model for microwave FETs including the DC/AC dispersion effects, *IEEE Trans. on Microwave Theory and Techniques*, 12, 2248, 1997.

35. Materka, A. and Kacprzak, T., Computer calculation of large-signal GaAs FET amplifier characteristics, *IEEE Trans. on Microwave Theory and Tech.*, 33, 129, 1985.

36. Circuit Network Items, Series IV, Hewlett Packard, HP Part. No. E4605-90038, 1161, 1995.

37. Calvo, M. V., Snider, A. D., and Dunleavy, L. P., Resolving Capacitor Discrepancies Between Large and Small Signal FET Models, 1995 IEEE MTT-S International Microwave Symposium, 1251, 1995.

38. Jansen, P. et al., Consistent small-signal and large-signal extraction techniques for heterojunction FET's, *IEEE Transaction on Microwave Theory and Tech.*, 43, 1, 87, 1995.

39. Mallavarpu, R., Teeter. D., and Snow, M., The importance of gate charge formulation in large-signal PHEMT modeling, *GaAs IC Symposium Technical Digest*, 87, 1998.

40. Horio, K. and Usarni, K., Analysis of kink-related backgating effect in GaAs MESFETs, *IEEE Electron Devices Letters*, 537, 16, 1995.

41. Curtice, W. R., Bennett, J. H., Suda, D., and Syrett, B. A., Modeling of current lag effects in GaAs IC's, 1998 *IEEE MTT-S International Microwave Symposium Digest*, 2, 603, 1998.

42. K. Kunihiro and Y. Ohno, An equivalent circuit model for deep trap induced drain current transient behavior in HJFETs, *1994 GaAs IC Symposium Digest*, 267, 1994.

43. Chakrabarti, P., Shrestha, S. K., Srivastava, A., and Skxena, D., Switching characteristics of an optically controlled GaAs-MESFET, *IEEE Trans. on Microwave Theory and Techniques*, 42, 365, 1994.

44. Madjar, K., Paolella, A., and Herczfeld, P. R., Modeling the optical switching of MESFET's considering the external and internal photovoltaic effects, *IEEE Trans. on Microwave Theory and Tech.*, 42, 62, 1994.

45. Teyssier, J-P, Bouysse, P., Ouarch, A., Barataud, D., Peyretaillade, T., and Quere, R., 40-GHz/150-ns versatile pulsed measurement system for microwave transistor isothermal characterization, *IEEE Trans. on Microwave Theory and Tech.*, 46, 2043, 1998.

46. A. Platzker et al., Characterization of GaAs devices by a versatile pulsed I-V measurement system, *1990 IEEE MTT Symposium Digest*, 1137.

47. Anlholt and Swirhun, Experimental characterization of the temperature dependence of GaAs FET equivalent circuits, *IEEE Trans. on Electron Devices*, 39, 2029, 1992.

48. Grossman, P. C. and Oki, A., A large signal DC model for GaAs/GaAlAs heterojunction bipolar transistors, *Proc. IEEE BCTM*, 258, 1989.

49. Ye, F. Q., A BJT model with self-heating for WATAND computer simuation, M. S. Thesis, Youngstown State University, Youngstown, OH, 1990.

50. McAndrew, C. C., A complete and consistent electrical/thermal HBT model, *Proc. IEEE BCTM*, 200, 1992.

51. Corcoran, J., Poulton, K. and Knudsen, K., GaAs HBTs: an analog circuit design perspective, *Proc. IEEE BCTM*, 245, 1991.

52. Fox, R. M. and Lee, S-G, Predictive modeling of thermal effects in BJTs, *Proc. IEEE BCTM*, 89, 1991.

53. Fossum, J. G., Modeling issues for advanced bipolar device/circuit simulation, *Proc. 1989 IEEE BCTM*, 234, 1989.

54. McAndrew, C. C., Seitchik, J., Bowers, D., Dunn, M., Foisy, M., Getreu, I., Moinian, S., Parker, J., van Wijnen, P., and Wagner, L., VBIC95: An improved vertical IC bipolar transistor model, *Proc. 1995 BCTM*, 170, 1995.

55. de Graaff, H. C. and Kloosterman, W. J., New formulation of the current and charge relations in bipolar transistor modeling for CACD purposes, *IEEE Trans. on Electron Devices*, ED-32, 2415, 1986.

56. Perugupalli, P., Trivedi, M., Shenai, K., and Leong, S. K., Modeling and characterization of 80v LDMOSFET for RF communications, *1997 IEEE BCTM*, 92, 1997.

57. Ho, M. C., Green, K., Culbertson, R., Yang, J. Y., Ladwig, D., and Ehnis, P., A physical large signal Si model for RF circuit design, *1997 MTT-S International Microwave Symposium Digest*, 1997.

58. Miller, M., Dinh, T., and Shumate, E., A new empirical large signal model for silicon RF LDMOS FET's, *1997 IEEE MTT Symposium on Technologies for Wireless Applications Digest*, Vancouver, Canada, 19, 1997.

59. Curtice, W. R., Pla, J. A., Bridges, D., Liang, T., and Shumate, E., A new dynamic electro-thermal model for silicon RF LDMOS FET's, *1999 IEEE MTT-S International Microwave Symposium Digest*, 1999.

60. Heo, D., Chen, E., Gebara, E., Yoo, S., Lasker, J., and Anderson, T., Temperature dependent MOSFET RF large signal model incorporating self-heating effects, *1999 MTT-S International Microwave Symposium Digest*, 1999.

61. Golio, J. M., *Microwave MESFETs and HEMTs*, Artech House, Boston, 1991.

62. Demmler, M. and Tasker, P. J., A vector corrected on-wafer large-signal waveform system for novel characterization and onlinear modeling techniques for transistors, Presented at the workshop on New Direction in Nonlinear RF and Microwave Characterization, 1996 International Microwave Symposium, 1996.

63. Wei, C.-J., Lan, Y. E., Hwang, J. C. M., Ho, W.-J., and Higgins, J. A. Waveform-based modeling and characterization of microwave power heterojunction transistors, *IEEE Trans. on Microwave Theory and Techniques*, 43, 2898, 1995.

64. Watanabe, S, Takatuka, S., Takagi, K., Kukoda, H., and Oda, Y., Simulation and experimental results of source harmonic tuning on linearity of power GaAs FET, 1996 MTT-S International Microwave Symposium, 1996.

8.8 Technology Computer Aided Design

Peter A. Blakey

Computer-aided design (CAD) is used in response to complexity in engineering problems. The complexity may be associated with the manipulation and storage of large amounts of relatively simple information; or it may be associated with the interactions of complicated nonlinear physical phenomena. Both types of complexity are encountered in the field of microelectronics. Electronics CAD (ECAD) is used to design new products using an *existing* semiconductor technology, and technology CAD (TCAD) is used to accelerate the development of *new* semiconductor technologies. ECAD is able to manage large amounts of relatively simple information, such as the number, size, and locations of the polygons used to define lithographic masks. TCAD predicts the way in which nonlinear process and device physics would impact the performance of proposed technologies. The subject matter of this chapter is TCAD and its application to the development of RF, microwave, and millimeter-wave semiconductor technologies.

An Overview of TCAD

The core tools of technology CAD (TCAD) are process simulators and device simulators. Process simulators predict the structures that result from applying a sequence of processing steps to an initial structure. Device simulators predict the electrical characteristics of specified structures. Process and device simulators are often used in combination to predict the impact of process variations on electrical behavior. The basic flow of information is indicated in Fig. 8.71.

Process and device simulators use numerical techniques to solve mathematical equations that describe the underlying physics of an experiment. They are able to predict the result of experiments, without requiring the experiments to be run. Predictive simulation is useful whenever it is time consuming, expensive, dangerous, or otherwise difficult to run real experiments. In the case of microelectronics development, it is both time consuming and expensive to run real experiments in a semiconductor fab. As a result, it can be cost effective to substitute simulated experiments for some of the real experiments that would otherwise be required by purely empirical development procedures.

Process simulators incorporate physical models for phenomena such as diffusion, oxidation, ion implantation, deposition, and etching. Device simulators solve equations that express charge continuity and the dependency of the electrostatic potential on the distribution of charge. In order to solve the charge continuity equations, device simulators need models for charge transport and for the generation and recombination of charge carriers. Implementing process and device simulators is a challenging task that requires knowledge of physics, numerical methods, visualization techniques, user interfaces, and software engineering. Research groups in universities and in industry were the first to develop this type of software. However, these groups were not well suited to the task of maintaining the software and providing long-term support to users. General-purpose process and device simulation software is now available from several commercial vendors who provide maintenance and support. Most microelectronics companies use commercial software as the foundation of their TCAD capabilities.

Commercial TCAD systems also include tool integration utilities that make it convenient to use the core tools in combination, and task integration environments that automate large-scale simulation-based experimentation. The tool integration utilities include run time environments, visualization tools, structure editors, layout editors and interfaces, and optimization capabilities. Task integration usually supports a "Virtual Wafer Fab" paradigm that makes it convenient to define and run simulated split lot experiments. After a simulated experiment is defined, the system automatically generates all of the associated jobs, farms out the jobs across a computer network, and collects the results. It may also provide modeling capabilities that make it easy to analyze the simulated results.

Predictive, physics-based simulation is different from data modeling. Simulation predicts the data that an experiment would yield if it were run. Modeling provides a compact description of existing data that

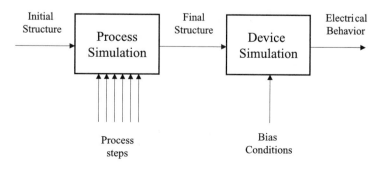

FIGURE 8.71 The basic flow of TCAD information.

may come from either real or simulated experiments. Data models are used to interpolate, but cannot be used to extrapolate. Expectations regarding "accuracy" are quite different for data models and for physics-based simulators. The agreement between modeled and experimental data is normally very good. The agreement between simulated data and experimental data is subject to several sources of uncertainty that can lead to significant discrepancies between simulated data and experimental data. These sources of uncertainty are discussed in a later section.

Process and device simulators were implemented initially to assist the development of digital Si VLSI technologies. In recent years the capabilities of TCAD software have been extended to cover a range of different material systems, and to provide convenient RF-oriented analysis capabilities. The widespread use of epitaxially grown materials and low thermal budget processing means that process simulation for III-V hetero-structure technologies can often be reduced to the definition of material layers and doping densities, with geometric specification of deposition and etch steps. This avoids difficult problems associated with process simulation of silicon technologies, which must account for moving boundaries during oxidation, complicated diffusion mechanisms, and so forth. Device simulation for advanced material systems often requires advanced models for charge transport, and it is usually necessary to account for the impact of deep level traps. The appropriate models have been incorporated into some of the commercially available device simulators.

In principle, device simulators can predict any electrical behavior that can be measured. Simulators targeted toward the development of digital silicon VLSI technologies are primarily concerned with the calculation of DC I-V characteristics, but they also provide basic capabilities for calculating small-signal AC behavior and large-signal transient behavior. Mixed-mode device-and-circuit simulation capabilities are also available. Mixed-mode simulation is basically SPICE-like circuit simulation that uses numerical device simulation to model the behavior of one or more of the active devices in a circuit. Other types of analysis are feasible, but are not available in commercial simulators. One example is the prediction of noise behavior. Another example is harmonic balance analysis of device operation that is implemented at the level of the underlying physics, rather than at the circuit level. Computational load-pull systems seek to mirror the functionality of experimental load-pull systems. These systems predict RF output power, power-added efficiency, gain, linearity, and tunability, and present the results as a function of load (or source) impedance plotted on a Smith chart. Such systems have been implemented in industry, but are not yet offered by commercial vendors.

Benefits of TCAD

Simulated experiments can replace a significant fraction of the real experiments that are otherwise required by traditional, empirical development procedures. Simulated experiments are almost always performed more rapidly and more cheaply than real experiments. The use of TCAD can therefore yield significant reductions in both cycle times and development costs. Reduced development times lead in turn to a faster time to market, and this normally translates into higher profits. The magnitude of the financial benefits depends both on the skill with which TCAD is deployed and on the characteristics of the market for a given technology. TCAD is most useful when the market is high volume, and both performance and cost driven. Until recently, this meant that TCAD was applied primarily to the development of silicon VLSI technologies. As a result of the wireless revolution, TCAD is now being used to accelerate the development of analog, RF, microwave, and millimeter-wave technologies.

The high costs of semiconductor fabs, material, and personnel mean that reducing development costs by even a few percent is enough to justify significant investments in TCAD capabilities. When TCAD is used effectively it can yield direct savings in development times and development costs of around 50%, and can result in enhanced profits (due to faster time to market) that are much greater than the direct cost savings. Savings of this magnitude can, in principle, justify investments of tens or even hundreds of millions of dollars in TCAD capabilities. The investments that companies make in TCAD have traditionally been much smaller than this.

Simulation can predict all of the quantities that are measured routinely. It can also predict the values of quantities that are either very difficult or impossible to measure. Examples of such quantities are the doping profiles, carrier profiles, and electrostatic potential distributions within a semiconductor device. By examining these quantities, engineers gain insight into the subtleties of device operation. This insight can guide the direction of future development, and it can lead to the development of simple analytic theories that capture the essence of device operation. Although the enhanced physical insight is very useful, it is difficult to quantify the value in monetary terms.

Process and device simulators encapsulate and integrate the knowledge of experts in a way that makes detailed knowledge and experience accessible on demand to less expert process and device engineers. For example, it is possible for a device engineer to account for the impact of complicated transport phenomena on device performance, without having any knowledge of the associated physics. This function of encapsulation and integration helps to ameliorate both the shortage of experienced technical personnel and the comparatively narrow technical span of most experts. Once again, it is difficult to quantify the value in monetary terms.

Limitations of TCAD

For certain widely used types of simulation, the underlying physics is well established and, for practical purposes, "exact." For example, electromagnetic simulation is based on Maxwell's equations, thermal simulation is based on the heat flow equation, and circuit simulation is based on Kirchoff's Current and Voltage Laws. The underlying physics of process and device simulation is less well defined. For each physical phenomenon (e.g., ion implantation, or electron transport) there is a hierarchy of models of varying complexity. All of these models introduce approximations. In general, the more complicated models provide a more complete description, but are more difficult to implement and involve longer (often much longer) computation times. The need to select appropriate accuracy-efficiency trade-offs can make it difficult for nonexperts to utilize TCAD effectively.

After a physical model has been selected, it is necessary to define values for the parameters associated with the model. These parameters must usually be defined over a range of different conditions (e.g., for different temperatures, doping concentrations, or electric fields.) The values of the parameters may be obtained from experiments or from theory. In either case, there is often considerable uncertainty in the accuracy of the model parameters.

Process and device simulators solve systems of coupled, nonlinear equations that do not have general analytic solutions. The equations are therefore solved using numerical techniques that are implemented in computer software. Numerical techniques provide an approximate solution that is defined on a discrete "mesh" of points. The choice of mesh and numerical techniques has a major impact on the accuracy and efficiency with which solutions are obtained. Using a finer mesh provides solutions that have higher accuracy but are calculated more slowly. Numerical techniques for solving nonlinear equations employ sequences of approximations that are expected to converge to a solution. In some cases the sequence of approximations fails to converge. Because of these issues, users of TCAD need some expertise in defining meshes and numerical techniques that provide acceptable trade-offs between accuracy, efficiency, and stability.

Discrepancies between measured and simulated results arise because of difficulties in specifying equivalent experiments. For example, a real diffusion step may specify a thermal ramp in terms of the times and temperatures that are dialed in as input to a controller. Process simulation will normally use the same times and temperatures. However, in the real experiment the temperature of the wafer as a function of time is different from the nominally specified ramp. The temperature as a function of time may also vary from wafer to wafer within a lot, and from point to point on a given wafer. Another example is provided by device simulation. The lattice temperature used in device simulation is often assumed to be 300 K, but the real lattice temperature is a function of both the true ambient temperature and any self-heating that occurs. Other sources of uncertainty are systematic measurement and repeatability.

In order to apply TCAD effectively, users need technical skills that encompass physics, numerical techniques, and the application domain. They need to understand and employ effective application methodologies, and they need good judgment to assess the validity and usefulness of TCAD predictions that are impacted by several sources of uncertainty. They also need to be able to "sell" their insights to process and device engineers who do not understand the role of simulation and are skeptical of its value, and they need to be comfortable with having a secondary support role within a technology development organization.

The Role of Calibration

By calibrating simulated results to experimental results, it is possible to ameliorate some of the issues that were outlined in the previous section. The existence of high-quality calibration is a prerequisite for using TCAD for certain applications. Achieving a good calibration involves a considerable amount of effort, and poor calibration can exacerbate the problems outlined previously. In view of the importance of calibration to the effective use of TCAD, a review of the associated issues is presented in this section.

Calibration attempts to reduce some of the errors associated with simulation. The measure of the errors is taken as the differences between simulated and measured results. Calibration seeks to reduce these errors by adjusting the values of the model parameters used by the simulators. In essence, it is assumed that the various sources of errors can be partially compensated for by using appropriate "effective" values of model parameters. A good calibration must use a sufficiently complete set of output variables, i.e., quantities for which the predicted and measured results are to be brought into better agreement. It must also use an appropriate set of input variables, i.e., model parameters whose values are adjusted to improve the agreement between predicted and measured values. A good calibration must not allow the adjusted values of the model parameters to assume implausible values.

Calibration is often performed using ad hoc procedures that lead to unsatisfactory results. A very common error is to adjust too few input variables, or the wrong input variables. A related error is to allow input variables to assume values that are implausible or nonphysical. Another common problem is that some key output variables are neglected, and the associated errors diverge as a result of the calibration attempt. Calibration should be set up as a clearly defined optimization problem and the techniques used for calibration should draw on known methods for nonlinear optimization involving multiple constrained input variables. Black-box optimization can be performed, using small-change parameter sensitivity matrices calculated using TCAD. However, it is usually more efficient to use a sequential, step-by-step calibration procedure in which only a subset of the input and output variables are varied at each step. For example, in the case of a very simple calibration of a GaAs MESFET simulation, the threshold voltage could be matched by adjusting the channel charge. The low-field mobility could then be adjusted to fit the on-resistance in the linear region, and the saturated carrier velocity could be adjusted to fit the saturated current.

Performing a good calibration takes a lot of effort, but certain applications of TCAD only become viable after a good calibration has been achieved. Regardless of the amount of effort expended on calibration, there will still be residual errors. This is because calibration does not remove approximations in the underlying physics; it simply allows a portion of their impact to be absorbed into the values of the model parameters. No set of model parameters can correct for an underlying physical model that is inadequate. The converse of this observation is that the degree to which a calibration is successful provides insight into the adequacy of the underlying physics.

Applications of TCAD

The block diagram shown in Fig. 8.72 provides a high-level view of the activities associated with traditional, empirical development of semiconductor technologies. The development process involves a make-and-measure cycle that is repeated until acceptable results are obtained. The process recipe is then transferred to a fab; and layout rules and electrical models for active and passive components are

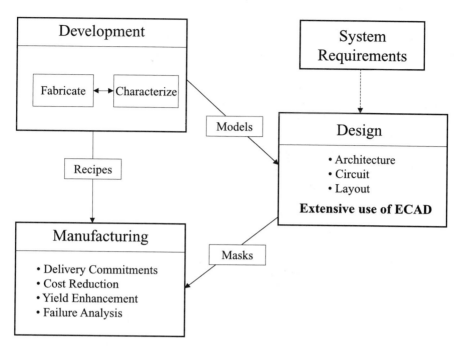

FIGURE 8.72 A high-level view of the traditional development methodology.

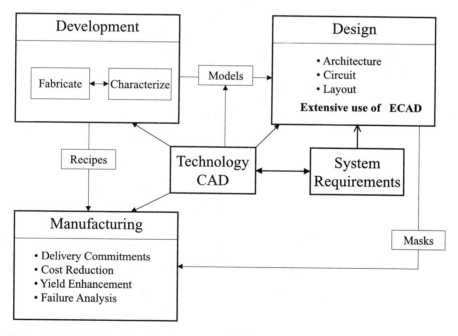

FIGURE 8.73 A high-level of modern development methodologies.

transferred to the circuit design community. Circuits are designed, masks are made, and the circuits are fabricated.

The block diagram shown in Fig. 8.73 provides a high-level view of the activities that are associated with more modern simulation-oriented development strategies. Depending on the ways in which TCAD is used, modern development methodologies may be viewed as being "TCAD influenced," "TCAD driven," or "system driven via TCAD." TCAD-influenced development uses process and device simulation to accelerate the make-and-measure phase of technology development. TCAD-driven development makes more extensive use of process and device simulation in the areas of modeling, characterization, and manufacturing. System-driven development supplements TCAD-driven development by adding explicit bidirectional links to system requirements and basic architectural choices. Much of modern development is TCAD influenced. Some companies use TCAD-driven methodologies, and a few companies are implementing system-driven methodologies.

It is useful to order individual TCAD activities by the amount of calibration required to achieve useful results. One such list, ordered by increasing calibration requirements, is as follows:

- Concept evaluation
- Technology discrimination
- Technology development
- Simulation-based model development
- Simulation-assisted characterization
- Statistical process design
- Technology transfer
- Process control
- Yield enhancement and failure analysis

Prior to the availability of TCAD, there was considerable reliance on simplified analytic models. Gross oversimplification of these models would lead to unrealistic performance projections, e.g., to the prediction of useful transistor performance at THz frequencies. TCAD is now used to perform rapid evaluations of proposed new concepts. Accounting for Poisson's equation, one or more carrier continuity equations, the coupling between these equations, and the need for device contacts ensures more realistic estimates of device performance. Projections of wildly overoptimistic device performance have been substantially eradicated by the widespread availability of TCAD. Concept evaluation usually requires little calibration effort, since all that is required is a go/no-go assessment.

At the initial stages of technology development, it is sometimes necessary to select between two or more competing approaches. TCAD is well suited to this purpose. Calibration is usually not very critical because the errors that are introduced by simulation tend to affect both technologies approximately equally, and are unlikely to cause a simulation-based ranking to be different from an experimentally determined ranking.

Once an overall approach has been selected, TCAD is used to explore different sub-options, and to assist process optimization. This is presently the single most important application of TCAD. Simulation-based design and optimization can and should be done systematically, but in practice it is often done using inefficient ad hoc techniques. Application methodologies associated with simulation-based technology development are discussed in the next section. As the results of real experiments become available they should be fed back to refine the simulator calibrations, in order to improve the usefulness of simulation at each stage of design and optimization. A general goal is for the calibration achieved for one generation of a technology to provide a solid starting point for the simulation-based design and optimization of the next generation of the technology.

TCAD can also be used to assist modeling and characterization activities. In the area of device modeling, TCAD can be used as a reference standard that enables careful testing of assumptions made in the course of developing device models, and supports systematic assessments of the global accuracy of a device model. TCAD is especially useful for assessing the accuracy of large-signal models that are constructed from DC and small-signal AC data. The large-signal behavior predicted by the model is compared to the large-signal behavior predicted directly using TCAD. The development and characterization of charge-based models is also facilitated by the use of TCAD, since the calculated charge distributions within a device can easily be integrated and partitioned in accordance with charge-based modeling approaches. TCAD can also be used to assist in device characterization. For example, it is quite straightforward to predict the values of bias- and temperature-dependent source and drain access resistances. Simulation can also assist in the interpretation of pulsed I-V measurements that are used to characterize thermal effects and low-frequency dispersive effects associated with deep-level traps.

When there is sufficiently high confidence in the quality of an underling calibration, the uses of TCAD can be expanded further. Calculation of process sensitivities provides a foundation for statistical process design, design centering, and yield modeling. Information concerning process sensitivity is also a useful input for process control. Simulation of the impact of large-change variations (e.g., the impact on electrical behavior of skipping a process step) can provide "signature" information that assists failure analysis. The ideal situation is when a comprehensive set of simulated data, together with an associated set of input data, forms an integral part of the technology transfer package that is supplied by a development organization to a manufacturing organization.

Application Protocols

TCAD is often deployed using inefficient ad hoc techniques. In order to use TCAD efficiently, it is helpful to develop, refine, and document sets of useful simulation-based procedures. The details of these procedures often vary considerably from one application (e.g., technology discrimination) to another (e.g., failure analysis.) A description of a simulation-based procedure that is used for a particular application is referred to as an application protocol. From a high-level perspective, the development of simulation-based application protocols is often a two-stage procedure. In the first stage, the simulation-based procedures are set up to mirror the existing experimental procedures associated with traditional, empirical technology development. In the second stage, there is explicit consideration of the additional possibilities that can be exploited due to the use of simulated rather than real experiments.

Since technology development is presently the most important application of TCAD, an example of an application protocol for technology development will be outlined. The first step is to mirror traditional, empirical development procedures that use split-lot experiments. This is the underlying idea of Virtual Wafer Fab (VWF) methodology, variants of which have been implemented by several commercial vendors of TCAD software. VWF methodology makes it convenient to define and run large-scale simulated split-lot experiments. Users select from a broad range of built-in experimental designs, define split variables among input quantities, and specify the output quantities to be extracted. Once these variables have been specified, all of the input descriptions associated with the simulated experiment are generated automatically, and farmed out automatically within a networked computing environment. The results are collected automatically, stored in a database, and are conveniently viewed and analyzed using built-in tools.

The next step is to define ways in which a simulation-based approach can go beyond traditional development procedures. When using TCAD it is possible to use sophisticated "black box" optimization techniques that require the ability to calculate accurate derivatives in a multidimensional search space. Due to long cycle times, high costs, and noisy experimental data, this approach is not feasible in the real world. However, it is quite suitable for use in the simulation world, where results can be obtained quickly and reproducibly. The use of black-box optimization focuses explicit attention on the need to define an objective function, and also on the need to define all of the constraints on the input variables. Split-lot experimentation also involves objective functions and constraints, but they are usually defined and implemented less formally.

A focus on objective functions and constraints leads to a useful high-level view of technology development in general and the role of simulation in particular. Technology development can be viewed as a search, within a multidimensional design space, for the point corresponding to an optimum design. Various techniques are available for narrowing down the regions of the design space that need to be searched. These techniques include analytic theory and scaling arguments, numerical simulation, and real experiments. Each of these techniques has its advantages and limitations. Analytic theory and scaling arguments are quick and easy to apply, and can eliminate large regions of the design space, but their absolute accuracy is poor. Simulation can provide much better accuracy, and is relatively fast and

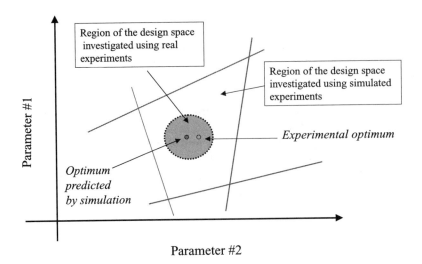

FIGURE 8.74   A simplified view of the search process.

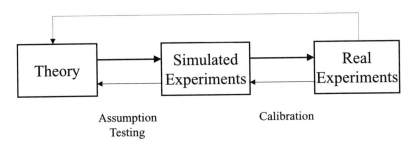

FIGURE 8.75   A three-level hierarchical approach to technology development.

inexpensive, but is subject to model errors and numerical error. Experiments provide "real" data, but are expensive, time consuming, and subject to experimental error. Analytic theory, simulation, and experiments can be viewed as a hierarchy of search techniques that can be applied sequentially and iteratively to maximize the efficiency of the search process.

A simplified view of the search for an optimum design in a two-dimensional space is presented in Fig. 8.74. Analytic theories and scaling arguments are used to establish the boundaries of the region of the design space that needs to be searched. Simulation is then used to predict the optimum point in the design space. Due to modeling and numerical errors, this predicted optimum (indicated by the closed circle) is not exactly coincident with the true optimum point. The region around the predicted optimum is therefore searched experimentally, to determine the location of the true optimum (the open circle). According to this view, the goal of simulation is to predict an optimum design with sufficient accuracy that the true optimum can be found using only a small number of real experiments. Thus, simulation does not eliminate the need for experiments, but it is realistic to expect a reduction of 50 to 80% in the number of real experiments that would be required if simulation is not used.

Figure 8.75 provides another view that indicates how theory, simulated experiments, and real experiments are linked in modern development procedures. The forward arrows indicate the basic sequence of development activities as outlined in the previous paragraph. The reverse arrows indicate that information is fed back to allow simulator calibration and assumption testing. Subsets of this view correspond to other development methodologies. For example, eliminating simulation results in a "traditional" development methodology, in which analytic theory and scaling arguments are used to bracket a design space that is then searched experimentally. Eliminating theory results in a black-box-simulation-reliant development methodology that will usually be inefficient in its search of an unconstrained design space.

Similar approaches can be used to develop application protocols for the other uses of TCAD.

Conclusions

Technology CAD provides benefits associated with faster development times, lower development costs, enhanced physical insight, and the encapsulation and integration of expert knowledge. These benefits are especially important in the context of high-performance, high-volume, consumer-oriented applications. Technology CAD is very well established as an integral part of the procedures used to develop silicon VLSI technologies, and is assuming an important role in the development of semiconductor technologies for RF, microwave, and millimeter-wave applications.

Although the benefits of using TCAD can be very great, achieving the benefits is not easy. Some of the challenges, such as the definition of physical models, model parameters, calibration procedures, and application protocols, are primarily technical in nature. Other challenges, such as a shortage of trained manpower able to apply TCAD effectively, are associated with economic and social factors. Companies that are able to overcome these challenges can obtain a competitive advantage. Conversely, companies that do not deploy TCAD effectively can expect to find themselves at a competitive disadvantage.

9
Underlying Physics

Nicholas E. Buris
Motorola Labs

Matthew Sadiku
Avaya Inc.

W.R. Deal
Malibu Networks, Inc.

Vesna Radisic
HRL Laboratories, LLC

Y. Qian
University of California

T. Itoh
University of California

Wayne E. Stark
University of Michigan

Alfy Riddle
Macallan Consulting

Mike Golio
Rockwell Collins

K.F. Etzold
IBM Thomas J. Watson Research Center

Mike Harris
Georgia Tech Research Institute

9.1 Maxwell's Equations

Nicholas E. Buris

Microwaves and RF is a branch of electrical engineering that deals ultimately with special cases of the physics of electrically charged particles and their interactions via electromagnetic waves. The fundamental branch of science describing the physics of electrically charged particles is electromagnetism. Electromagnetism deals with the electromagnetic force and is based on the concept of electric and magnetic vector fields, $E(r, t)$ and $H(r, t)$, respectively. The fields $E(r, t)$ and $H(r, t)$ were first introduced to resolve the issues of the "action at a distance" experienced between charges. Maxwell's equations are four coupled partial differential equations describing the electromagnetic field in terms of its sources, the charges, and their associated currents (charges in motion). Electromagnetic waves are one special solution of Maxwell's

0-8493-8592-X/01/$0.00+$.50
© 2001 by CRC Press LLC

equations that microwave engineering is built upon. In engineering, of course, depending on the technology of interest, we deal with a full range of special circumstances of electromagnetism. At one end of the spectrum are applications such as solid state devices where electromagnetics is applied to just a few charges, albeit in a phenomenological sense and in conjunction with quantum mechanics. In this realm the forces on individual charges are important. At the other end we have applications where the wavelength of the electromagnetic waves is much smaller than the dimensions of the problem and electromagnetics is reduced to optics where only simple, plane wave phenomena are at play. In the middle of the spectrum, we deal with structures whose size is comparable to the wavelength and electromagnetics is treated as a rigorous mathematical boundary value problem. The majority of microwave applications is somewhat in the middle of this spectrum with some having connections to either end. When studying time harmonic events in microwaves, the frequency domain version of Maxwell's equations is very convenient. In the frequency domain, we have developed a number of high level descriptions of electromagnetic phenomena and several specialized disciplines such as circuits, filtering, antennas, and others have been created to efficiently address the engineering problems at hand. This section of the handbook will describe Maxwell's equations and their solution in order to establish the connection between the various microwave and RF topics and their basic physics, electromagnetism.

Time Domain Differential Form of Maxwell's Equations

As implied earlier, the fundamental description of the physics involved in the study of microwaves and RF is based on the concept of electric and magnetic vector fields, $E(r, t)$ and $H(r, t)$, respectively. According to the well-known Helmholtz's theorem, any vector field can be uniquely specified in terms of its rotation (curl) and divergence components [3]. Maxwell's equations in vacuum essentially define the sources of the curl and divergence of E and H. in the International System of units (SI) these equations take the following form:

$$\nabla \times \mathbf{E} = -\mathbf{J}_m - \mu_o \frac{\partial \mathbf{H}}{\partial t} \tag{9.1}$$

$$\nabla \cdot \mathbf{E} = \frac{\rho_e}{\varepsilon_o} \tag{9.2}$$

$$\nabla \times \mathbf{H} = \mathbf{J}_e + \varepsilon_o \frac{\partial \mathbf{E}}{\partial t} \tag{9.3}$$

$$\nabla \cdot \mathbf{H} = \frac{\rho_m}{\mu_o} \tag{9.4}$$

where the constants ε_o and μ_o are the permittivity and permeability of vacuum, ρ_e and J_e are the electric charge and current densities, and ρ_m and J_m are the magnetic charge and current densities.

Equations (9.1) through (9.4) indicate that, in addition to the charges and the currents, time variation in one field serves as a source to the other. In that sense, in microwaves, dealing with high frequency harmonic time variations, the electric and magnetic fields are always coupled and we refer to them combined as the electromagnetic field. It is interesting to note that Maxwell developed his equations by abstraction and generalization from a number of experimental laws that had been discovered previously. Up to Maxwell's time, electromagnetism was a collection of interesting experimental and theoretical laws from Coulomb, Gauss, Faraday, and others. Maxwell, in 1864, combined and extended all these into a remarkably complete system of equations thus founding the science of electromagnetism. Maxwell's generalizations helped start and propel work in electromagnetic waves, and also facilitated the introduced of the special theory of relativity. Interestingly, Maxwell's equations were not covariant under a Galilean transformation (observer moving with respect to the environment). However, after the postulates of the special theory of relativity, no modification of any kind was needed to Maxwell's equations. The speed of light, derived from the wave solutions to Maxwell's equations, is a constant for all inertial frames of reference.

As mentioned above, the electromagnetic fields are a conceptual contraption, the result of an effort to systematically describe how electrically charged particles move. Maxwell's equations describe the field in terms of its sources but they do not describe how the charges move. The motion of charges is governed by Coulomb's law. The force, *F*, on an electric charge, q, moving with velocity v inside an electromagnetic field is

$$\mathbf{F} = q\mathbf{E} + q\mu_o \mathbf{v} \times \mathbf{H}$$

A very important assumption made here is that the charge q is a test charge. That is, small enough that it does not alter the field in which it exists. The problem of the self fields of charges cannot be solved by classical means, but only through quantum electrodynamics [12]. The fields themselves cannot be measured directly. It is through their effects on charges particles that they are experienced. In most microwave applications we do not see Coulomb's force because we seldom deal with just a few particles. Instead, we devised complicated quantities such as voltage, impedance, and others to arrive at efficient engineering designs.

Some Comments on Maxwell's Equations

Because the divergence of the curl of any vector is identically equal to zero, combining Eqs. (9.2) and (9.3) results in the current continuity equation (charges conservation), i.e.,

$$\nabla \cdot \mathbf{J}_e + \frac{\partial \rho_e}{\partial t} = 0 \tag{9.5a}$$

Similarly,

$$\nabla \cdot \mathbf{J}_m + \frac{\partial \rho_m}{\partial t} = 0 \tag{9.5b}$$

This is a peculiar fact, as the continuity equation is somewhat of a statement on the nature of the field sources, and as such, one would not expect it to be an intrinsic property of Maxwell's equations.

Another peculiar property of Maxwell's equations is that they are covariant under a duality transformation (see part on duality later in this section). Critical quantities quadratic in the fields, remain invariant under such a transformation. Under the proper choice of a duality transformation, Eqs. (9.1) through (9.4) can be made to have only electric, or only magnetic sources. Therefore, the question of whether magnetic sources exist is equivalent to whether the ratio of electric to magnetic sources is the same for all charged particles [6]. Again, this is another statement related to the structure of the field sources and it is rather peculiar that it should be contained in Maxwell's equations. In microwaves, we frequently analyze problems where we consider both electric and magnetic sources to be present. The concept of duality is used extensively to simplify certain problems where apertures are significant parts of the geometry at hand.

Frequency Domain Differential Form of Maxwell's Equations

It is customary to restrict investigation of Maxwell equations to the case where the time variations are harmonic, adopting the phasor convention $e^{j\omega t}$ where ω represents the angular frequency. According to this convention, $E(r, t) = \text{Re}\{\mathbf{E}(\mathbf{r},\omega)\, e^{j\omega t}\}$. Similar expressions hold for all scalar and vector quantities. We say that $E(r, t)$ is the time domain representation of the field while $\mathbf{E}(\mathbf{r},\omega)$ is the phasor, or frequency domain representation. This convention simplifies the mathematics of the partial differential equations with respect to time, reducing time derivatives to simple multiplication by $j\omega$. It should be noted, however, that the product of two harmonic signals in the time domain does not correspond to the product of their phasors.

To effectively treat complicated materials such as dielectrics, ferrites, and others, the electric and magnetic flux density vector fields are introduced, $\mathbf{D}(\mathbf{r},\omega)$ and $\mathbf{B}(\mathbf{r},\omega)$, respectively. These fields essentially account for complicated material mechanisms such as losses and memory (dispersion), in a phenomenological way. They represent average field quantities when large quantities of particles are present. In fact, in the Gaussian system of units, $\mathbf{E}$ and $\mathbf{D}$ have the same units and so do $\mathbf{H}$ and $\mathbf{B}$. To 1st order approximation, for example, $\mathbf{D}(\mathbf{r},\omega)$, the electric flux density field in a dielectric equals the externally applied electric field plus the field due to the dipole moment, created by the atoms being "stretched" by the external field. Similar arguments could be made for the magnetic field flux density, although to be correct, quantum mechanical considerations are needed for a more satisfying explanation. Additional discussions on materials will be made in the sections to follow. Maxwell's equations in complicated media and in the frequency domain become

$$\nabla \times \mathbf{E} = -\mathbf{J}_m - j\omega \mathbf{B} \tag{9.6a}$$

$$\nabla \cdot \mathbf{D} = \rho_e \tag{9.6b}$$

$$\nabla \times \mathbf{H} = \mathbf{J}_e + j\omega \mathbf{D} \tag{9.6c}$$

$$\nabla \cdot \mathbf{B} = \rho_m \tag{9.6d}$$

with the constitutive relations

$$\mathbf{D} = \bar{\bar{\varepsilon}} \cdot \mathbf{E} \tag{9.7}$$

and

$$\mathbf{B} = \bar{\bar{\mu}} \cdot \mathbf{H} \tag{9.8}$$

and with the continuity equations

$$\nabla \cdot \mathbf{J}_e + j\omega \rho_e = 0 \tag{9.9}$$

$$\nabla \cdot \mathbf{J}_m + j\omega \rho_m = 0 \tag{9.10}$$

In this formulation, $\bar{\bar{\mu}}$ and $\bar{\bar{\varepsilon}}$ are the generalizations of the parameters ε_o and μ_o and they can be complex, tensorial functions of frequency. There are also some rare materials for which the constitutive relations are even more general. For a simplified, molecular level derivation of ε for dielectrics, see the text by Jackson [5].

To study inhomogeneous materials, we need proper boundary conditions to govern the behavior of the fields across the interface between two media. Consider such an interface as depicted in Fig. 9.1.

Equations (9.1) through (9.4) are associated with the following boundary conditions (derivable from Maxwell's equations themselves).

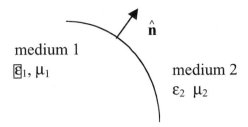

FIGURE 9.1 Geometry for the boundary conditions at the interface between two media.

$$\hat{n} \cdot \left(\mathbf{D}_2 - \mathbf{D}_1\right) = \rho_{es} \tag{9.11a}$$

$$\hat{n} \cdot \left(\mathbf{B}_2 - \mathbf{B}_1\right) = \rho_{ms} \tag{9.11b}$$

$$-\hat{n} \times \left(\mathbf{E}_2 - \mathbf{E}_1\right) = \mathbf{J}_{ms} \tag{9.11c}$$

$$-\hat{n} \times \left(\mathbf{H}_2 - \mathbf{H}_1\right) = \mathbf{J}_{es} \tag{9.11d}$$

where $\hat{n}$ is the unit vector normal to the interface pointing from medium 1 to medium 2. The quantities ρ_{es}, $\mathbf{J}_{es}$, ρ_{ms}, and $\mathbf{J}_{ms}$, represent the free electric and magnetic charge and current surface densities on the interface, respectively. Note that these are free sources, and as such, only exist on conductors and are equal to zero on the interface between two dielectric media. There are bound (polarization) charges on dielectric interfaces, but they are accounted for by Eq. (9.11a) and the constitutive relation in Eq. (9.7).

General Solution to Maxwell's Equations (the Stratton–Chu Formulation)

One of the most comprehensive solutions to Maxwell's equations in a general homogeneous and isotropic domain is given by Stratton and Chu [1] and is also further discussed by Silver [2]. Consider a volume V with a boundary consisting of a collection of closed surfaces, S_1, S_2, ..., S_n. Next consider the unit vectors, $\hat{n}$, normal to the boundary surface with direction pointing inside the volume of interest as depicted in Fig. 9.2. The volume V is occupied uniformly by a material with dielectric constant ε, and magnetic permeability μ. Inside V there exist electric and magnetic charge and current density distributions, ρ_e, $\mathbf{J}_e$, ρ_m, and $\mathbf{J}_m$, respectively.

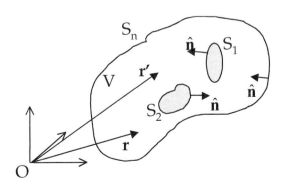

FIGURE 9.2 The volume of interest and its boundary surface consisting of $S_1 + S_2 + \ldots + S_n$.

The electric and magnetic fields at an arbitrary point, **r**, inside V are then given in terms of the sources and the values of the fields at the boundary by the following equations.

$$\mathbf{E}(\mathbf{r}) = -\int_V \left[j\omega\mu G(\mathbf{r},\mathbf{r}')\mathbf{J}_e(\mathbf{r}') + \mathbf{J}_m(\mathbf{r}') \times \nabla'G - \frac{\rho_e(\mathbf{r}')}{\varepsilon}\nabla'G \right] dV'$$

$$- \int_{S_1+S_2+\cdots+S_n} \left[j\omega\mu G(\mathbf{r},\mathbf{r}')\left(\hat{n}' \times \mathbf{H}(\mathbf{r}')\right) + \left(-\hat{n}' \times \mathbf{E}(\mathbf{r}')\right) \times \nabla'G - \left(\hat{n}' \cdot \mathbf{E}(\mathbf{r}')\right)\nabla'G \right] dS' \qquad (9.12)$$

and

$$\mathbf{H}(\mathbf{r}) = -\int_V \left[j\omega\varepsilon G(\mathbf{r},\mathbf{r}')\mathbf{J}_m(\mathbf{r}') - \mathbf{J}_e(\mathbf{r}') \times \nabla'G - \frac{\rho_m(\mathbf{r}')}{\mu}\nabla'G \right] dV'$$

$$- \int_{S_1+S_2+\cdots+S_n} \left[j\omega\varepsilon G(\mathbf{r},\mathbf{r}')\left(-\hat{n}' \times \mathbf{E}(\mathbf{r}')\right) - \left(\hat{n}' \times \mathbf{H}(\mathbf{r}')\right) \times \nabla'G - \left(\hat{n}' \cdot \mathbf{H}(\mathbf{r}')\right)\nabla'G \right] dS' \qquad (9.13)$$

where

$$G(\mathbf{r},\mathbf{r}') = \frac{e^{-jk|\mathbf{r}-\mathbf{r}'|}}{4\pi|\mathbf{r}-\mathbf{r}'|} \qquad (9.14)$$

is called the free space Green's function and

$$k = \sqrt{\omega^2\mu\varepsilon} \qquad (9.15)$$

is called the wavenumber with ω representing the angular frequency.

When the surface S_n is at infinity, it can be shown that the fields there satisfy the radiation conditions (**E** and **H** attenuate at least as fast as $1/r$, they are perpendicular to each other, and to **r**, and their magnitudes are related by the wave impedance, $\sqrt{\mu/\varepsilon}$). Moreover, the continuity equations can be employed to eliminate ρ_e and ρ_m from the field solution. When the volume V is unbounded ($S_n \to$ infinity) and the charges are substituted by their current density expressions, the solution of Maxwell's equations becomes

$$\mathbf{E}(\mathbf{r}) = -\frac{j}{\omega\varepsilon}\int_V \left[\left(\mathbf{J}_e(\mathbf{r}') \cdot \nabla'\right)\nabla' + k^2\mathbf{J}_e(\mathbf{r}') - j\omega\varepsilon\,\mathbf{J}_m(\mathbf{r}') \times \nabla' \right] G(\mathbf{r},\mathbf{r}')dV' \qquad (9.16)$$

and

$$\mathbf{H}(\mathbf{r}) = -\frac{j}{\omega\mu}\int_V \left[\left(\mathbf{J}_m(\mathbf{r}') \cdot \nabla'\right)\nabla' + k^2\mathbf{J}_m(\mathbf{r}') + j\omega\mu\,\mathbf{J}_e(\mathbf{r}') \times \nabla' \right] G(\mathbf{r},\mathbf{r}')dV'. \qquad (9.17)$$

These forms of the fields are used routinely for the numerical solution to Maxwell's equations, particularly using the method of moments.

It should be noted here that in the special case where the boundary surfaces S_1, S_2, ..., S_{n-1} are perfect electric conductors, the field solution can be expressed by Eqs. (9.12) and (9.13), or their unbounded counterparts, Eqs. (9.16) and (9.17) provided one recognizes that the free currents on an electric conductor are $\hat{n} \times \mathbf{H}$ and the charges are $\hat{n} \cdot \mathbf{D}$. From the materials point of view (also discussed later in this chapter) a perfect conductor has unlimited capacity to provide free charges that distribute themselves on its surface in such a way as to effectively eliminate their internal fields regardless of the external field they are in. Therefore, since the fields vanish inside, perfect conductors have no energy in them. They simple shape and guide the energy in the space around them. Thus, a coil has no energy inside its metal, it simply stores energy inside and outside its windings. In this light, electromagnetic design of especially small, densely populated electronic structures is very difficult and suffers from potential ElectroMagnetic Interference (EMI) problems.

Far Field Approximation

At the limit where the observation point is at infinity ($r \to \infty$, far field) it can be shown that Eqs. (9.16) and (9.17) simplify to

$$\mathbf{E}(\mathbf{r}) = -j\omega\mu \frac{e^{-jkr}}{4\pi r} \int_V \left[\mathbf{J}_e(\mathbf{r}') - \left(\hat{\mathbf{r}} \cdot \mathbf{J}_e(\mathbf{r}') \right)\hat{\mathbf{r}} + \sqrt{\frac{\varepsilon}{\mu}} \mathbf{J}_m(\mathbf{r}') \times \hat{\mathbf{r}} \right] e^{jk\hat{\mathbf{r}}\cdot\mathbf{r}'} \, d \tag{9.18}$$

with

$$\mathbf{H}(\mathbf{r}) = \sqrt{\frac{\varepsilon}{\mu}} \, \hat{\mathbf{r}} \times \mathbf{E}(\mathbf{r}). \tag{9.19}$$

$\mathbf{E}(\mathbf{r})$ and, consequently, $\mathbf{H}(\mathbf{r})$ in Eqs. (9.18) and (9.19) have zero radial components and can be found often in the following, more practical forms.

$$E_\theta(\mathbf{r}) = -j\omega\mu \frac{e^{-jkr}}{4\pi r} \int_V \left[J_\theta(\hat{\mathbf{r}}') + \sqrt{\frac{\varepsilon}{\mu}} J_{m_\phi}(\mathbf{r}') \right] e^{jk\hat{\mathbf{r}}\cdot\mathbf{r}'} \, dV' \tag{9.20}$$

and

$$E_\phi(\mathbf{r}) = -j\omega\mu \frac{e^{-jkr}}{4\pi r} \int_V \left[J_\phi(\mathbf{r}') - \sqrt{\frac{\varepsilon}{\mu}} J_{m_\theta}(\mathbf{r}') \right] e^{jk\hat{\mathbf{r}}\cdot\mathbf{r}'} \, dV' \tag{9.21}$$

or, equivalently,

$$E_\theta(\mathbf{r}) = -j\omega\mu \frac{e^{-jkr}}{4\pi r} \int_V \left[J_x \cos\theta\cos\phi + J_y \cos\theta\sin\phi - J_z \sin\theta + \sqrt{\frac{\varepsilon}{\mu}} \left(-J_{m_x} \sin\phi + J_{m_y} \cos\phi \right) \right] e^{jk\hat{\mathbf{r}}\cdot\mathbf{r}'} \, dV' \tag{9.22}$$

and

$$E_\phi(\mathbf{r}) = -j\omega\mu \frac{e^{-jkr}}{4\pi r} \int_V \left[-J_x \sin\phi + J_y \cos\phi - \sqrt{\frac{\varepsilon}{\mu}} \left(J_{m_x} \cos\theta\cos\phi + J_{m_y} \cos\theta\sin\phi - J_{m_z} \sin\theta \right) \right] e^{jk\hat{\mathbf{r}}\cdot\mathbf{r}'} \, dV' \tag{9.23}$$

where it also holds that

$$H_\phi = \sqrt{\frac{\varepsilon}{\mu}}\, E_\theta \qquad\qquad (9.24)$$

and

$$H_\theta = \sqrt{\frac{\varepsilon}{\mu}}\, E_\phi. \qquad\qquad (9.25)$$

These solutions to Maxwell's equations can also be derived via a potential field formulation. The interested reader can refer to the treatment in Harrington [10] and Balanis [7].

General Theorems in Electromagnetics

Uniqueness of Solution

It can be shown that the field in the volume V, depicted in Fig. 9.2, is uniquely defined by the source distributions in it and the values of the tangential electric field, or tangential magnetic field on all the boundary surfaces, $S_1, S_2, \ldots, S_n$. Uniqueness is also guaranteed when the tangential **E** is specified for part of the boundary while tangential **H** is specified for the remaining part. A detailed discussion of the uniqueness of the solution can be found in Stratton [9].

Duality

As briefly mentioned earlier, Maxwell's equations are covariant under duality transformations. Consider the transformation,

$$\begin{pmatrix} \mathbf{E} \\ \mathbf{H} \end{pmatrix} = \begin{bmatrix} \cos\xi & \sin\xi \\ -\sin\xi & \cos\xi \end{bmatrix} \cdot \begin{pmatrix} \mathbf{E}' \\ \mathbf{H}' \end{pmatrix} \qquad\qquad (9.26)$$

$$\begin{pmatrix} \mathbf{D} \\ \mathbf{B} \end{pmatrix} = \begin{bmatrix} \cos\xi & \sin\xi \\ -\sin\xi & \cos\xi \end{bmatrix} \cdot \begin{pmatrix} \mathbf{D}' \\ \mathbf{B}' \end{pmatrix} \qquad\qquad (9.27)$$

$$\begin{pmatrix} \mathbf{J}_e \\ \mathbf{J}_m \end{pmatrix} = \begin{bmatrix} \cos\xi & \sin\xi \\ -\sin\xi & \cos\xi \end{bmatrix} \cdot \begin{pmatrix} \mathbf{J}'_e \\ \mathbf{J}'_m \end{pmatrix} \qquad\qquad (9.28)$$

and

$$\begin{pmatrix} \rho_e \\ \rho_m \end{pmatrix} = \begin{bmatrix} \cos\xi & \sin\xi \\ -\sin\xi & \cos\xi \end{bmatrix} \cdot \begin{pmatrix} \rho'_e \\ \rho'_m \end{pmatrix} \qquad\qquad (9.29)$$

It can be easily shown that under this duality transformation, Maxwell's equations are covariant. That is, if the primed fields satisfy Maxwell's equations with the primed sources, the non-primed fields satisfy Maxwell's equations with the non-primed sources. Moreover, critical quantities quadratic in the fields, such as $\mathbf{E} \times \mathbf{H}$ and $\mathbf{E} \cdot \mathbf{D} + \mathbf{B} \cdot \mathbf{H}$ remain invariant under this duality transformation (for real ξ). That is, for example, $\mathbf{E} \times \mathbf{H} = \mathbf{E}' \times \mathbf{H}'$ [6]. As a special case of this property, $\xi = -\pi/2$, we have the following correspondence.

$$\mathbf{E} \leftrightarrow -\mathbf{H}$$

$$\mathbf{H} \leftrightarrow \mathbf{E}$$

$$\mathbf{J}_m \leftrightarrow \mathbf{J}_e$$

$$\rho_m \leftrightarrow \rho_e$$

$$\mu \leftrightarrow \varepsilon$$

$$\varepsilon \leftrightarrow \mu$$

It is worth mentioning here that, since the fields themselves cannot be measured, and since all the important quantities (such as the power and energy terms, $\mathbf{E} \times \mathbf{H}$ and $\mathbf{E} \cdot \mathbf{D} + \mathbf{B} \cdot \mathbf{H}$) remain invariant under duality, it is just a matter of convenience which of several possible dual solutions to an electromagnetic problem we consider. This is precisely the reason why in some antenna problems, for example, one can formulate them with electric currents on the conductors, or with magnetic currents on the apertures and obtain the same detectable (measurable) results.

Lorentz Reciprocity Theorem

Consider the fields $\mathbf{E}_1$ and $\mathbf{H}_1$, generated by the system of sources $\mathbf{J}_{e1}$ and $\mathbf{J}_{m1}$ and the fields $\mathbf{E}_2$ and $\mathbf{H}_2$, generated by the system of sources $\mathbf{J}_{e2}$ and $\mathbf{J}_{m2}$. Then it can be shown that [8]

$$\oiint_S \left(\mathbf{E}_1 \times \mathbf{H}_2 - \mathbf{E}_2 \times \mathbf{H}_1\right) \cdot d\mathbf{S} = \int_V \left(\mathbf{H}_1 \cdot \mathbf{J}_{m2} - \mathbf{E}_1 \cdot \mathbf{J}_{m2} - \mathbf{H}_2 \cdot \mathbf{J}_{m1} + \mathbf{E}_2 \cdot \mathbf{J}_{e1}\right) dV \qquad (9.30)$$

As a special case of this theorem, consider the open volume of interest V and two sources in it, $\mathbf{J}_{e1}$ and $\mathbf{J}_{e2}$. When $\mathbf{J}_{e1}$ is present alone, it generates the field $\mathbf{E}_1$. When $\mathbf{J}_{e2}$ is present alone, it generates the field $\mathbf{E}_2$. Lorentz' reciprocity theorem as stated by Eq. (9.30) implies that

$$\mathbf{E}_1 \cdot \mathbf{J}_{e2} = \mathbf{E}_2 \cdot \mathbf{J}_{e1} \qquad (9.31)$$

The reciprocity theorem is used frequently in electromagnetic problem studies to facilitate solution of complicated problems. One well-known application of reciprocity has been in the expressions of the mutual and self-impedance of simple radiators. Reciprocity also manifests itself in some linear circuits so that Eq. (9.13) above holds when the electric fields and the current densities are replaced by voltages and currents, respectively. For details of this property see the text by Valkenburg [13].

Equivalent Principles (Theory of Images)

Equivalent principles are used to describe the electromagnetic field problem in a volume of interest in more than one configuration of sources. While the volume of interest is the same in all the equivalent configurations (materials, geometry, and source distributions), the geometry outside the volume of interest is different for each equivalent configuration. A detailed discussion of field equivalence principles can be found in Collin [8]. Here only the method of images is mentioned. The method of images is based on the uniqueness of the field. Namely, as long as the sources inside the volume of interest, V, and the boundary conditions on S_1, S_2, ... remain the same, the fields inside V are unique regardless of what sources exist outside V.

The method of images is based on this principle. Its applicability is limited to just a few canonical geometries. Consider for example, a current-carrying wire over an infinite perfect electric conductor, as shown in Fig. 9.3(a). The tangential electric field is zero on the pec ground plane. Next consider the configuration (b) where the ground plane has been removed and the volume below it has a second current-carrying wire, shaped in the mirror image of the original and with an opposite direction for the

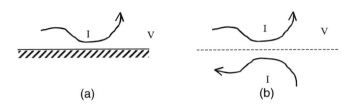

FIGURE 9.3   The fields in V are identical for both configurations, (a) and (b). The fields outside V vanish in (a) while they are certainly nonzero in (b).

current. It can be shown with simple symmetry arguments that the tangential electric field on the dashed line (where the ground plane used to be) in configuration (b) vanishes. Therefore, since (a) and (b) are identical within V and they have the same boundary conditions (zero tangential electric field) on the boundary, the fields inside V are the same.

Clearly, it is much easier to numerically evaluate the solution in configuration (b) than that of configuration (a). Images are very often used in electromagnetic problems when appropriate. It is often easy to extend the theory of images to current-carrying wires in the presence of corners of ground planes. Moreover, there have been several studies to extend the theory of images to more complex geometries [11]. However, the complexity of the images quickly escalates diminishing the benefits of removing the ground planes.

Simple Solution to Maxwell's Equations I (Unbounded Plane Waves)

In electromagnetics, a homogeneous and isotropic region of space with no sources in it is called free space. In free space, $\overline{\overline{\mu}} = \mu_o$, $\overline{\overline{\varepsilon}} = \varepsilon_o$ and Eqs. (9.6a) through (9.6d) decouple giving rise to the following second order wave equations.

$$\nabla^2 \mathbf{E} + \omega^2 \mu_o \varepsilon_o \mathbf{E} = 0 \tag{9.32}$$

$$\nabla^2 \mathbf{H} + \omega^2 \mu_o \varepsilon_o \mathbf{H} = 0 \tag{9.33}$$

Applying separation of variables, Eqs. (9.32) and (9.33) are found to have plane wave solutions of the form

$$\mathbf{E} = \mathbf{E}_o^+ e^{-j\mathbf{k}\cdot\mathbf{r}} + \mathbf{E}_o^- e^{+j\mathbf{k}\cdot\mathbf{r}} \tag{9.34}$$

$$\mathbf{H} = \mathbf{H}_o^+ e^{-j\mathbf{k}\cdot\mathbf{r}} + \mathbf{H}_o^- e^{+j\mathbf{k}\cdot\mathbf{r}} \tag{9.35}$$

The electromagnetic wave propagating in the $+\mathbf{k}$ direction ($e^{-j\mathbf{k}\cdot\mathbf{r}}$) travels with the speed of light, $1/(\mu_o\varepsilon_o)^{1/2}$ and has a wave impedance,

$$\left|\frac{\mathbf{E}_o^+}{\mathbf{H}_o^+}\right| = \sqrt{\frac{\mu_o}{\varepsilon_o}} . \tag{9.36}$$

The wave number, k, obeys the relation

$$k = \frac{\omega}{c} = \omega\sqrt{\mu_o \varepsilon_o} \tag{9.37}$$

where c is the speed of light. It can be shown that the power density carried by the plane wave is along the direction of propagation, k, and its average value over one period is given by the Poynting vector

$$\mathbf{S}^+ = \frac{1}{2}\mathbf{E}_o^+ \times \mathbf{H}_o^+ \tag{9.38}$$

where the fields **E** and **H** are represented by their amplitude values in the phasor convention. It is worth noting that the Poynting vector itself is not a phasor. Being the product of two harmonic signals, the power density represented by the Poynting vector has a steady (DC) component and a harmonic component of twice the frequency of the electromagnetic field. The DC component of the Poynting vector is exactly given by Eq. (9.38). Similar discussions hold for the wave propagating in the –**k** direction. These wave solutions obey all the usual wave phenomena of reflection and refraction when incident on interfaces between two different media. For detailed discussions on these phenomena and more general cases of wave solutions to Maxwell's equations, the reader is referred to the following sections of this chapter and the text by Collin [8].

Simple Solution to Maxwell's Equations II (Guided Plane Waves)

It is possible to have electromagnetic waves inside cavities and waveguides. A waveguide is any structure that supports waves traveling in one direction and confined in the transverse plane by its boundaries. A rectangular duct made out of a conductor and a coaxial cable are waveguides with closed boundaries. A twin lead transmission line, or a dielectric plated conductor are among the many open boundary waveguides. Waves supported by waveguides are special solutions to Maxwell's equations.

Let us consider a waveguide along the z axis. We are looking for solutions to Maxwell's equations that represent waves traveling along z. That is, we are looking for solutions in the form:

$$\mathbf{E} = \mathbf{E}(x,y)e^{-j\beta z} = \left(\mathbf{E}_t(x,y) + \hat{z}E_z(x,y)\right)e^{-j\beta z} \tag{9.39}$$

and

$$\mathbf{H} = \mathbf{H}(x,y)e^{-j\beta z} = \left(\mathbf{H}_t(x,y) + \hat{z}H_z(x,y)\right)e^{-j\beta z} \tag{9.40}$$

where $\mathbf{E}_t$ and $\mathbf{H}_t$ are vectors in the transverse, (x,y) plane. Solving Eq. (9.6) in a region of space free from any sources and looking specifically for solutions in the form of Eqs. (9.39) and (9.40) reduces the problem to the following wave equations:

$$\nabla_t^2 E_z + \left(\omega^2\mu\varepsilon - \beta^2\right)E_z = 0 \tag{9.41}$$

and

$$\nabla_t^2 H_z + \left(\omega^2\mu\varepsilon - \beta^2\right)H_z = 0 \tag{9.42}$$

The transverse fields are given in terms of the axial components of the waves as

$$\mathbf{E}_t(x,y) = \frac{1}{k^2 - \beta^2}\left(-j\beta\nabla_t E_z(x,y) + j\omega\mu\hat{z} \times \nabla H_z(x,y)\right) \tag{9.43}$$

and

$$\mathbf{H}_t\left(x,y\right) = \frac{1}{k^2 - \beta^2}\left(-j\beta\nabla_t H_z\left(x,y\right) - j\omega\varepsilon\,\hat{\mathbf{z}} \times \nabla E_z\left(x,y\right)\right) \qquad (9.44)$$

∇_t in the above equations stands for the transverse del operator, i.e., $\nabla_t = \hat{\mathbf{x}}\,\partial/\partial x + \hat{\mathbf{y}}\,\partial/\partial y$.

The wave number of the guided wave, β, is, in general, different from the wave number in the medium ($k^2 = \omega^2\mu\varepsilon$). The wave number is determined when the boundary condition specific to the waveguide are applied. There are some special cases of these guided wave solutions that are examined at a later section in this chapter in much more detail for practical microwave waveguides.

TEM Modes

$$E_z = H_z = 0$$

$$\nabla_t \times \mathbf{E}_t = 0 \quad \beta^2 = k^2 = \omega^2\mu\varepsilon \quad \text{and} \quad \mathbf{H} = \sqrt{\frac{\varepsilon}{\mu}}\left(\hat{\mathbf{z}} \times \mathbf{E}_t\right)$$

TM Modes

$$E_z \neq 0; \quad H_z = 0$$

$$\nabla_t^2 E_z + \left(\omega^2\mu\varepsilon - \beta^2\right)E_z = 0 \quad \mathbf{E}_t\left(x,y\right) = \frac{-j\beta}{k^2 - \beta^2}\nabla_t E_z\left(x,y\right) \quad \text{and} \quad \mathbf{H}_t = -\frac{\omega\varepsilon}{\beta}\mathbf{E}_t \times \hat{\mathbf{z}}$$

TE Modes

$$H_z \neq 0; \quad E_z = 0$$

$$\nabla_t^2 H_z + \left(\omega^2\mu\varepsilon - \beta^2\right)H_z = 0 \quad \mathbf{H}_t\left(x,y\right) = \frac{-j\beta}{k^2 - \beta^2}\nabla_t H_z\left(x,y\right) \quad \text{and} \quad \mathbf{E}_t = \frac{\omega\mu}{\beta}\mathbf{H}_t \times \hat{\mathbf{z}}$$

References

1. J.A. Stratton and L.J. Chu, *Phys. Res.*, 56, 99, 1939.
2. S. Silver, Ed., *Microwave Antenna Theory and Design*, McGraw-Hill, New York, 1949, chap. 3.
3. G. Arfken, *Mathematical Methods for Physicists*, 2nd ed., Academic Press, 1970, 66–70.
4. W. Pauli, *Theory of Relativity*, Dover, New York, 1981.
5. J.D. Jackson, *Classical Electrodynamics*, 2nd ed., John Wiley & Sons, New York, 1975, 226–235.
6. J.D. Jackson, *Classical Electrodynamics*, 2nd ed., John Wiley & Sons, New York, 1975, 251–260.
7. C. Balanis, *Advanced Engineering Electromagnetics*, John Wiley & Sons, New York, 1989.
8. R.E. Collin, *Field Theory of Guided Waves*, 2nd ed., IEEE Press, 1991.
9. J.A. Stratton, *Electromagnetic Theory*, McGraw-Hill, New York, 1941.
10. R.F. Harrington, *Time-Harmonic Electromagnetic Fields*, McGraw-Hill, New York, 1961.
11. I.V. Lindel, Image Theory for the Soft and Hard Surface, *IEEE Trans. Antenn. Propagat.*, 43, 1, January 1995.
12. R.P. Feynman, R.B. Leighton, and M. Sands, Electromagnetic Mass, in *The Feynman Lectures on Physics*, Addison Wesley, New York, 1964.
13. M.E. Valkenburg, *Network Analysis*, 3rd ed., Prentice-Hall, Englewood Cliffs, NJ, 1974, 255–259.

9.2 Wave Propagation in Free Space

Matthew Sadiku

The concept of propagation refers to the various ways by which an electromagnetic (EM) wave travels from the transmitting antenna to the receiving antenna. Propagation of EM wave may also be regarded as a means of transferring energy or information from one point (a transmitter) to another (a receiver). The transmission of analog or digital information from one point to another is the largest application of microwave frequencies. Therefore, understanding the principles of wave propagation is of practical interest to microwave engineers. Engineers cannot completely apply formulas or models for microwave system design without an adequate knowledge of the propagation issue.

Wave propagation at microwave frequencies has a number of advantages (Veley, 1987). First, microwaves can accommodate very wide bandwidth without causing interference problems because microwave frequencies are so high. Consequently, a huge amount of information can be handled by a single microwave carrier. Second, microwaves propagate along a straight line like light rays and are not bent by the ionosphere as are lower frequency signals. This straight-line propagation makes communication satellites possible. In essence, a communication satellite is a microwave relay station that is used in linking two or more grounded-based transmitters and receivers. Third, it is feasible to design highly directive antenna systems of a reasonable size at microwave frequencies. Fourth, compared with low-frequency electromagnetic waves, microwave energy is more easily controlled, concentrated, and directed. This makes it useful for cooking, drying, and physical diathermy. Moreover, the microwave spectrum provides more communication channels than the radio and TV bands. With the ever-increasing demand for channel allocation, microwave communication has become more common.

EM wave propagation is achieved through guided structures such as transmission lines and waveguides or through space. In this chapter, our major focus is on EM wave propagation in free space and the power resident in the wave.

EM wave propagation can be described by two complimentary models. The physicist attempts a theoretical model based on universal laws, which extends the field of application more widely than currently known. The engineer prefers an empirical model based on measurements, which can be used immediately. This chapter presents the complimentary standpoints by discussing theoretical factors affecting wave propagation and the semiempirical rules allowing handy engineering calculations. First, we consider wave propagation in idealistic simple media, with no obstacles. We later consider the more realistic case of wave propagation around the earth, as influenced by its curvature and by atmospheric conditions.

Wave Equation

The conventional propagation models, on which the basic calculation of microwave links is based, result directly from Maxwell's equations (Sadiku, 2001):

$$\nabla \cdot \mathbf{D} = \rho_v \tag{9.45}$$

$$\nabla \cdot \mathbf{B} = 0 \tag{9.46}$$

$$\nabla \times \mathbf{E} = -\frac{\partial \mathbf{B}}{\partial t} \tag{9.47}$$

$$\nabla \times \mathbf{H} = \mathbf{J} + \frac{\partial \mathbf{D}}{\partial t} \tag{9.48}$$

In these equations, **E** is electric field strength in volts per meter, **H** is magnetic field strength in amperes per meter, **D** is electric flux density in coulombs per square meter, **B** is magnetic flux density in webers per square meter, **J** is conduction current density in ampere per square meter, and ρ_v is electric charge density in coulombs per cubic meter. These equations go hand in hand with the constitutive equations for the medium:

$$\mathbf{D} = \epsilon\mathbf{E} \tag{9.49}$$

$$\mathbf{B} = \mu\mathbf{H} \tag{9.50}$$

$$\mathbf{J} = \sigma\mathbf{E} \tag{9.51}$$

where $\epsilon = \epsilon_o\epsilon_r$, $\mu = \mu_o\mu_r$, and σ are the permittivity, the permeability, and the conductivity of the medium, respectively.

Consider the general case of a lossy medium that is charge-free ($\rho_v = 0$). Assuming time-harmonic fields and suppressing the time factor $e^{j\omega t}$, Eqs. (9.45) to (9.51) can be manipulated to yield Helmholtz's wave equations

$$\nabla^2\mathbf{E} - \gamma^2\mathbf{E} = 0 \tag{9.52}$$

$$\nabla^2\mathbf{H} - \gamma^2\mathbf{H} = 0 \tag{9.53}$$

where $\gamma = \alpha + j\beta$ is the *propagation constant,* α is the *attenuation constant* in nepers per meter or decibels per meter, and β is the *phase constant* in radians per meter. Constants α and β are given by

$$\alpha = \omega\sqrt{\frac{\mu\epsilon}{2}\left[\sqrt{1+\left(\frac{\sigma}{\omega\epsilon}\right)^2}-1\right]} \tag{9.54}$$

$$\beta = \omega\sqrt{\frac{\mu\epsilon}{2}\left[\sqrt{1+\left(\frac{\sigma}{\omega\epsilon}\right)^2}+1\right]} \tag{9.55}$$

where $\omega = 2\pi f$ is the frequency of the wave. The wavelength λ and wave velocity u are given in terms of β as

$$\lambda = \frac{2\pi}{\beta} \tag{9.56}$$

$$u = \frac{\omega}{\beta} = f\lambda \tag{9.57}$$

Without loss of generality, if we assume that wave propagates in the z direction and the wave is polarized in the x direction, solving the wave equations (9.52) and (9.53) results in

$$\mathbf{E} = E_o e^{-\alpha z}\cos\left(\omega t - \beta z\right)\mathbf{a}_x \tag{9.58}$$

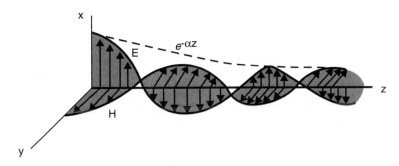

FIGURE 9.4 The magnetic and electric field components of a plane wave in a lossy medium.

$$\mathbf{H} = \frac{E_o}{|\eta|} e^{-\alpha z} \cos\left(\omega t - \beta z - \theta_\eta\right) \mathbf{a}_y \qquad (9.59)$$

where $\eta = |\eta| / \theta_\eta$ is the *intrinsic impedance* of the medium and is given by

$$|\eta| = \frac{\sqrt{\dfrac{\mu}{\epsilon}}}{\sqrt[4]{\left[1 + \left(\dfrac{\sigma}{\omega\epsilon}\right)^2\right]}}, \qquad \tan 2\theta_\eta = \frac{\sigma}{\omega\epsilon}, \qquad 0 \le \theta_\eta \le 45° \qquad (9.60)$$

Equations (9.58) and (9.59) show that as the EM wave propagates in the medium, its amplitude is attenuated according to $e^{-\alpha z}$, as illustrated in Fig. 9.4. The distance δ through which the wave amplitude is reduced by a factor of e^{-1} (about 37%) is called the *skin depth* or *penetration depth* of the medium, i.e.,

$$\delta = \frac{1}{\alpha} \qquad (9.61)$$

The power density of the EM wave is obtained from the Poynting vector

$$\mathbf{P} = \mathbf{E} \times \mathbf{H} \qquad (9.62)$$

with the time-average value of

$$P_{ave} = \mathrm{Re}\left(\mathbf{E} \times \mathbf{H}^*\right)$$
$$= \frac{E_o^2}{2|\eta|} e^{-2\alpha z} \cos\theta_\eta \mathbf{a}_z \qquad (9.63)$$

It should be noted from Eqs. (9.58) and (9.59) that **E** and **H** are everywhere perpendicular to each other and also to the direction of wave propagation. Thus, the wave described by Eqs. (9.58) and (9.59) is said to be *plane polarized,* implying that the electric field is always parallel to the same plane (the *xz* plane in this case) and is perpendicular to the direction of propagation. Also, as mentioned earlier, the wave decays as it travels in the *z* direction because of loss. This loss is expressed in the *complex relative permittivity* of the medium.

$$\epsilon_c = \epsilon_r' - \epsilon_r'' = \epsilon_r \left(1 - \frac{\sigma}{\omega\epsilon} \right) \tag{9.64}$$

and measured by the *loss tangent,* defined by

$$\tan \delta = \frac{\epsilon_r''}{\epsilon_r'} = \frac{\sigma}{\omega\epsilon} \tag{9.65}$$

The imaginary part $\epsilon_r'' = \sigma/\omega\epsilon_o$ corresponds to the losses in the medium. The refractive index of the medium n is given by

$$n = \sqrt{\epsilon_r} \tag{9.66}$$

Having considered the general case of wave propagation through a lossy medium, we now consider wave propagation in other types of media. A medium is said to be a good conductor if the loss tangent is large ($\sigma \gg \omega\epsilon$) or a lossless or good dielectric if the loss tangent is very small ($\sigma \ll \omega\epsilon$). Thus, the characteristics of wave propagation through other types of media can be obtained as special cases of wave propagation in a lossy medium as follows:

1. Good conductors: $\sigma \gg \omega\epsilon$, $\epsilon = \epsilon_o$, $\mu = \mu_o\mu_r$
2. Good dielectrics: $\sigma \ll \omega\epsilon$, $\epsilon = \epsilon_o$, $\mu = \mu_o\mu_r$
3. Free space: $\sigma = 0$, $\epsilon = \epsilon_o$, $\mu = \mu_o$

where $\epsilon_o = 8.854 \times 10^{-11}$ F/m is the free-space permittivity, and $\mu_o = 4\pi \times 10^{-7}$ H/m is the free-space permeability. The conditions for each medium type are merely substituted in Eqs. (9.54) to (9.65) to obtain the wave properties for that medium.

The classical model of wave propagation presented in this section helps us understand some basic concepts of EM wave propagation and the various parameters that play a part in determining the progress of the wave from the transmitter to the receiver. We will apply the ideas to the particular case of wave propagation in free space or the atmosphere in the section on Propagation in the Atmosphere. Before then, we digress a little and consider the important issue of wave polarization.

Wave Polarization

The concept of polarization is an important property of an EM wave that has been developed to described the various types of electric field variation and orientation. It is therefore a common practice to describe an EM wave by its polarization. The polarization of an EM wave depends on the transmitting antenna or source. It is determined by the direction of the electric field. It is regarded as the locus of the tip of the electric field (in a plane perpendicular to the direction of propagation) at a given point in space as a function of time. For this reason, there are four types of polarization: linear or plane, circular, elliptic, and random.

In linear or plane polarized waves, the orientation of the field is constant in space and time. For a plane traveling in the $+z$ direction, the electric field may be written as

$$\mathbf{E}(z,t) = E_x(z,t)\mathbf{a}_x + E_y(z,t)\mathbf{a}_y \tag{9.67}$$

where

$$E_x = \mathrm{Re}\left[E_{ox} e^{j(\omega t - kz + \phi_x)} \right] = E_{ox}\cos(\omega t - kz + \phi_x) \tag{9.68a}$$

$$E_y = \text{Re}\left[E_{oy} e^{j\left(\omega t - kz + \phi_x\right)} \right] = E_{oy} \cos\left(\omega t - kz + \phi_y\right) \tag{9.68b}$$

For linear polarization, the phase difference between the x and y components must be

$$\Delta\phi = \phi_y - \phi_x = n\pi, \quad n = 0, 1, 2, \ldots \tag{9.69}$$

This allows the two components to maintain the same ratio at all times, which implies that the electric field always lies along a straight line in a constant $-z$ plane. In other words, if we observe the wave in the direction of propagation (z in this case), we will notice that the tip of the electric field follows a line. Hence, the term *linear polarization*. Linearly polarized plane waves can be generated by simple antennas (such as dipole antennas) or lasers.

Circular polarized waves are characterized by an electric field with constant magnitude and orientation rotating in a plane transverse to the direction of propagation. Circular polarization takes place when the x and y components are the same in magnitude ($E_x = E_y$) and the phase difference between them is an odd multiple of $\pi/2$, i.e.,

$$\Delta\phi = \phi_y - \phi_x = \pm\left(\frac{1}{2} + 2n\right)\pi, \quad n = 0, 1, 2, \ldots \tag{9.70}$$

The two components of the electric field rotate around the axis of propagation as a function of time and space. Circularly polarized waves can be generated by a helically wound wire antenna or by two linear sources that are oriented perpendicular to each other and fed with currents that are out of phase by 90°.

Linear and circular polarizations are special cases of the more general case of the elliptical polarization. An elliptically polarized wave is one in which the tip of the field traces an elliptic locus in a fixed transverse plane as the field changes with time. Elliptical polarization is achieved when the x and y components are not equal in magnitude ($E_x \neq E_y$) and the phase difference between them is an odd multiple of $\pi/2$, i.e.,

$$\Delta\phi = \phi_y - \phi_x = \pm\left(\frac{1}{2} + 2n\right)\pi, \quad n = 0, 1, 2, \ldots \tag{9.71}$$

This allows the tip of the electric field to trace an ellipse in the $x - y$ plane.

The polarized waves described so far are illustrated in Fig. 9.5. They are deterministic meaning that the field is a predictable function of space and time. If the field is completely random, the wave is said to be randomly polarized. Typical examples of such waves are radiation from the sun and radio stars.

Propagation in the Atmosphere

Wave propagation hardly occurs under the idealized conditions assumed in previously. For most communication links, the previous analysis must be modified to account for the presence of the earth, the ionosphere, and atmospheric precipitates such as fog, raindrops, snow, and hail. This is done in this section.

The major regions of the earth's atmosphere that are of importance in radio wave propagation are the troposphere and the ionosphere. At radar frequencies (approximately 100 MHz to 300 GHz), the troposphere is by far the most important. It is the lower atmosphere comprised of a nonionized region extending from the earth's surface up to about 15 km. The ionosphere is the earth's upper atmosphere in the altitude region from 50 km to one earth radius (6370 km). Sufficient ionization exists in this region to influence wave propagation.

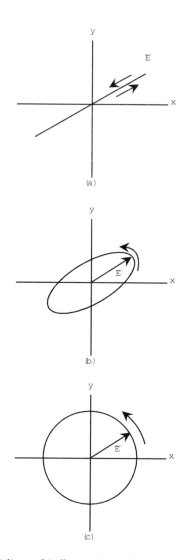

FIGURE 9.5 Wave polarizations: (a) linear, (b) elliptic, (c) circular.

Wave propagation over the surface of the earth may assume any of the following three principal modes:

- Surface wave propagation along the surface of the earth;
- Space wave propagation through the lower atmosphere;
- Sky wave propagation by reflection from the upper atmosphere.

These modes are portrayed in Fig. 9.6. The sky wave is directed toward the ionosphere, which bends the propagation path back toward the earth under certain conditions in a limited frequency range (0 to 50 MHz approximately). This is highly dependent on the condition of the ionosphere (its level of ionization) and the signal frequency. The surface (or ground) wave takes effect at the low-frequency end of the spectrum (2 to 5 MHz approximately) and is directed along the surface over which the wave is propagated. Since the propagation of the ground wave depends on the conductivity of the earth's surface, the wave is attenuated more than if it were propagation through free space. The space wave consists of the direct wave and the reflected wave. The direct wave travels from the transmitter to the receiver in nearly a straight path while the reflected wave is due to ground reflection. The space wave obeys the optical laws in that direct and reflected wave components contribute to the total wave component.

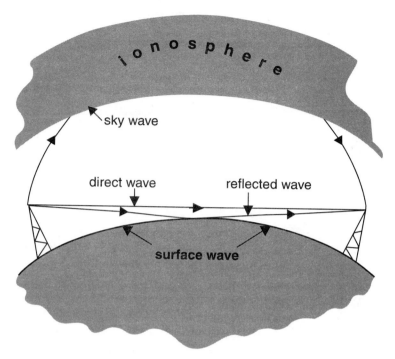

FIGURE 9.6   Modes of wave propagation.

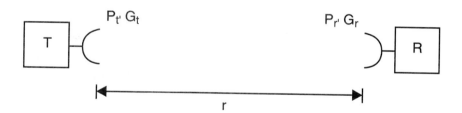

FIGURE 9.7   Transmitting and receiving antennas in free space.

Although the sky and surface waves are important in many applications, we will only consider the space wave in this chapter.

Figure 9.7 depicts the electromagnetic energy transmission between two antennas in space. As a wave radiates from the transmitting antenna and propagates in space, its power density decreases, as expressed ideally in Eq. (9.63). Assuming that the antennas are in a lossless medium or free space, the power received by the receiving antenna is given by the *Friis transmission equation* (Liu and Fang, 1988):

$$P_r = G_r G_t \left(\frac{\lambda}{4\pi r} \right)^2 P_t \qquad (9.72)$$

where the subscripts t and r respectively refer to transmitting and receiving antennas. In Eq. (9.72), $P =$ power in watts, $G =$ antenna gain (dimensionless), $r =$ distance between the antennas in meters, and $\lambda =$ wavelength in meters. The Friis equation relates the power received by one antenna to the power transmitted by the other, provided that the two antennae are separated by $r > 2D^2/\lambda$, and D is the largest dimension of either antenna. Thus, the Friis equation applies only when the two antennas are in the far field of each other. It shows that the received power decays at a rate of 20 dB/decade with distance. In case the propagation path is not in free space, a correction factor F is included to account for the effect of the medium. This factor, known as the *propagation factor*, is simply the ratio of the electric field intensity E_m in the medium to the electric field intensity E_o in free space, i.e.,

$$F = \frac{E_m}{E_o} \tag{9.73}$$

The magnitude of F is always less than unity since E_m is always less than E_o. Thus, for a lossy medium, Eq. (9.72) becomes

$$P_r = G_r G_t \left(\frac{\lambda}{4\pi r}\right)^2 P_t |F|^2 \tag{9.74}$$

For practical reasons, Eqs. (9.72) or (9.73) are commonly expressed in logarithmic form. If the all the terms are expressed in decibels (dB), Eq. (9.74) can be written in logarithmic form as

$$P_r = P_t + G_r + G_t - L_o - L_m \tag{9.75}$$

where P = power in dB referred to 1 W (or simply dBW), G = gain in dB, L_o = free-space loss in dB, and L_m = loss in dB due to the medium. The free-space loss is obtained from standard nomograph or directly from

$$L_o = 20 \log\left(\frac{4\pi r}{\lambda}\right) \tag{9.76}$$

while the loss due to the medium is given by

$$L_m = -20 \log|F| \tag{9.77}$$

Our major concern in the rest of this section is to determine L_o and L_m for two important cases of space propagation that differ considerably from the free-space conditions.

Effect on the Earth

The phenomenon of multipath propagation causes significant departures from free-space conditions. The term *multipath* denotes the possibility of an EM wave propagating along various paths from the transmitter to the receiver. In multipath propagation of an EM wave over the earth's surface, two such paths exist: a direct path and a path via reflection and diffractions from the interface between the atmosphere and the earth. A simplified geometry of the multipath situation is shown in Fig. 9.8. The reflected and diffracted component is commonly separated into two parts: one *specular* (or coherent) and the other *diffuse* (or incoherent), that can be separately analyzed. The specular component is well defined in terms of its amplitude, phase, and incident direction. Its main characteristic is its conformance to Snell's law for reflection, which requires that the angles of incidence and reflection be equal and coplanar. It is a plane wave, and as such, is uniquely specified by its direction. The diffuse component, however, arises out of the random nature of the scattering surface, and as such, is nondeterministic. It is not a plane wave and does not obey Snell's law for reflection. It does not come from a given direction but from a continuum.

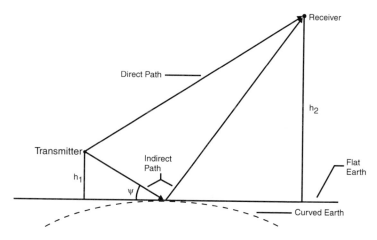

FIGURE 9.8 Multipath geometry.

The loss factor F that accounts for the departures from free-space conditions is given by

$$F = 1 + \Gamma \, \rho_s \, D \, S(\theta) \, e^{-j\Delta} \tag{9.78}$$

where Γ = Fresnel reflection coefficient,
  ρ_s = roughness coefficient,
  D = divergence factor,
  $S(\theta)$ = shadowing function,
  Δ = is the phase angle corresponding to the path difference

The Fresnel reflection coefficient Γ accounts for the electrical properties of the earth's surface. Since the earth is a lossy medium, the value of the reflection coefficient depends on the complex relative permittivity ϵ_c of the surface, the grazing angle ψ, and the wave polarization. It is given by

$$\Gamma = \frac{\sin \psi - z}{\sin \psi + z} \tag{9.79}$$

where

$$z = \sqrt{\epsilon_c - \cos^2 \psi} \quad \text{for horizontal polarization,} \tag{9.80}$$

$$z = \frac{\sqrt{\epsilon_c - \cos^2 \psi}}{\epsilon_c} \quad \text{for vertical polarization,} \tag{9.81}$$

$$\epsilon_c = \epsilon_r - j \frac{\sigma}{\omega \epsilon_o} = \epsilon_r - j60\sigma\lambda \tag{9.82}$$

ϵ_r and σ are the dielectric constant and conductivity of the surface; ω and λ are the frequency and wavelength of the incident wave; and ψ is the grazing angle. It is apparent that $0 < |\Gamma| < 1$.

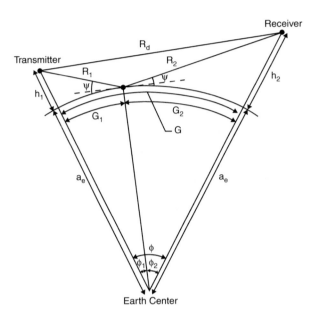

FIGURE 9.9 Geometry of spherical earth reflection.

To account for the spreading (or divergence) of the reflected rays due to earth curvature, we introduce the divergence factor D. The curvature has a tendency to spread out the reflected energy more than a corresponding flat surface. The divergence factor is defined as the ratio of the reflected field from a curved surface to the reflected field from flat surface (Kerr, 1951). Using the geometry of Fig. 9.9, D is given by

$$D \simeq \left(1 + \frac{2G_1 G_2}{a_e G \sin \psi}\right)^{-1/2} \tag{9.83}$$

where $G = G_1 + G_2$ is the total ground range and $a_e = 6370$ km is the effective earth radius. Given the transmitter height h_1, the receiver height h_2, and the total ground range G, we can determine G_1, G_2, and ψ. If we define

$$p = \frac{2}{\sqrt{3}} \left[a_e (h_1 + h_2) + \frac{G^2}{4} \right]^{1/2} \tag{9.84}$$

$$\alpha = \cos^{-1} \left[\frac{2a_e (h_1 - h_2) G}{p^3} \right] \tag{9.85}$$

and assume $h_1 \le h_2$, $G_1 \le G_2$, using small angle approximation yields

$$G_1 = \frac{G}{2} + p \cos\left(\frac{\pi + \alpha}{3}\right) \tag{9.86}$$

$$G_2 = G - G_1 \tag{9.87}$$

$$\phi_i = \frac{G_i}{a_e}, \quad i = 1, 2 \tag{9.88}$$

$$R_i = \left[h_i^2 + 4a_e \left(a_e + h_i \right) \sin^2 \left(\phi_i / 2 \right) \right]^{1/2} \quad i = 1, 2 \tag{9.89}$$

The grazing angle is given by

$$\psi = \sin^{-1} \left[\frac{2a_e h_1 + h_1^2 - R_1^2}{2a_e R_1} \right] \tag{9.90}$$

or

$$\psi = \sin^{-1} \left[\frac{2a_e h_1 + h_1^2 + R_1^2}{2\left(a_e + h_1 \right) R_1} \right] - \phi_1 \tag{9.91}$$

Although D varies from 0 to 1, in practice D is a significant factor at low grazing angle ψ (less than 0.1%). The phase angle corresponding to the path difference between direct and reflected waves is given by

$$\Delta = \frac{2\pi}{\lambda} \left(R_1 + R_2 - R_d \right) \tag{9.92}$$

The roughness coefficient ρ_s takes care of the fact that the earth's surface is not sufficiently smooth to produce specular (mirror-like) reflection except at a very low grazing angle. The earth's surface has a height distribution that is random in nature. The randomness arises out of the hills, structures, vegetation, and ocean waves. It is found that the distribution of the different heights on the earth's surface is usually the Gaussian or normal distribution of probability theory. If σ_h is the standard deviation of the normal distribution of heights, we define the roughness parameters as

$$g = \frac{\sigma_h \sin \psi}{\lambda} \tag{9.93}$$

If $g < 1/8$, specular reflection is dominant; if $g > 1/8$, diffuse scattering results. This criterion, known as the *Rayleigh criterion,* should only be used as a guideline since the dividing line between a specular and a diffuse reflection or between a smooth and a rough surface is not well defined. The roughness is taken into account by the roughness coefficient ($0 < \rho_s < 1$), which is the ratio of the field strength after reflection with roughness taken into account to that which would be received if the surface were smooth. The roughness coefficient is given by

$$\rho_s = \exp \left[-2 \left(2\pi g \right)^2 \right] \tag{9.94}$$

The shadowing function $S(\theta)$ is important at a low grazing angle. It considers the effect of geometric shadowing — the fact that the incident wave cannot illuminate parts of the earth's surface shadowed by higher parts. In a geometric approach, where diffraction and multiple scattering effects are neglected, the reflecting surface will consist of well-defined zones of illumination and shadow. As there will be no

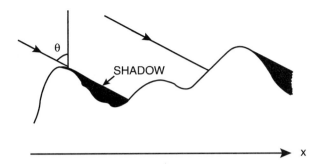

FIGURE 9.10   Rough surface illuminated at an angle of incidence θ.

field on a shadowed portion of the surface, the analysis should include only the illuminated portions of the surface. The phenomenon of shadowing of a stationary surface was first investigated by Beckman in 1965 and subsequently refined by Smith (1967) and others. A pictorial representation of rough surfaces illuminated at the angle of incidence θ (= 90° − ψ) is shown in Fig. 9.10. It is evident from the figure that the shadowing function $S(\theta)$ equals unity when θ = 0 and zero when θ = π/2. According to Smith (1967),

$$S\!\left(\theta\right) \simeq \frac{\left[1 - \tfrac{1}{2}\,\mathrm{erfc}\!\left(a\right)\right]}{1 + 2B} \tag{9.95}$$

where erfc(x) is the complementary error function,

$$\mathrm{erfc}\!\left(x\right) = 1 - \mathrm{erf}\!\left(x\right) = \frac{2}{\sqrt{\pi}} \int_{x}^{\infty} e^{-t^{2}}\,dt \tag{9.96}$$

and

$$B = \frac{1}{4a}\left[\frac{1}{\sqrt{\pi}}\,e^{a^{2}} - a\,\mathrm{erfc}\!\left(a\right)\right], \tag{9.97}$$

$$a = \frac{\cot\theta}{2s}, \tag{9.98}$$

$$s = \frac{\sigma_{h}}{\sigma_{l}} = \text{rms surface slope} \tag{9.99}$$

In Eq. (9.99), σ_{h} is the rms roughness height and σ_{l} is the correlation length. Alternative models for $S(\theta)$ are available in the literature. Using Eqs. (9.79) to (9.99), the loss factor in Eq. (9.78) can be calculated. Thus

$$L_{o} = 20\log\!\left(\frac{4\pi R_{d}}{\lambda}\right) \tag{9.100}$$

$$L_{m} = -20\log\!\left(1 + \Gamma\,\rho_{s}\,D\,S\!\left(\theta\right)e^{-j\Delta}\right) \tag{9.101}$$

Effect of Atmospheric Hydrometeors

The effect of atmospheric hydrometeors on satellite-earth propagation is of major concern at microwave frequencies. The problem of scattering of electromagnetic waves by atmospheric hydrometeors has attracted much interest since the late 1940s. The main hydrometeors that exist for long duration, and have the greatest interaction with microwaves are rain, snow, and dust particles. At frequencies above 10 GHz, rain has been recognized as the most fundamental obstacle in the earth-space path. Rain has been known to cause attenuation, phase difference, and depolarization of radio waves. For analog signals, the effect of rain is more significant above 10 GHz while for digital signals, rain effects can be significant down to 3 GHz. Attenuation of microwaves due to precipitation becomes severe owing to increased scattering and beam energy absorption by raindrops thus impairing terrestrial as well as earth-satellite communication links. Cross-polarization distortion due to rain has also been engaging the attention of researchers. This is particularly of interest when frequency reuse employing signals with orthogonal polarizations are used for doubling the capacity of a communication system. Thorough reviews on the interaction of microwaves with hydrometeors have been done by Oguchi (1983).

The loss due to rain-filled medium is given by

$$L_m = \gamma\left(R\right) \ell_e\left(R\right) p\left(R\right) \tag{9.102}$$

where γ = attenuation per unit length at rain rate R,
ℓ_e = equivalent path length at rain rate R, and
$p(R)$ = probability in percentage of rainfall rate R.

The attenuation is a function of the cumulative rain-rate distribution, drop-size distribution, refractive index of water, temperature, and other variables. A rigorous calculation of $\gamma(R)$ using various numerical modeling tools and incorporating raindrop size distribution, velocity of raindrops, and the refractive index of water can be found in Sadiku (2001). For practical engineering purposes, what is needed is a simple formula relating attenuation to rain parameters. Such is found in the aR^b empirical relationship, which has been employed to calculate rain attenuation directly (Collin, 1985), i.e.,

$$\gamma\left(R\right) = aR^b \quad \text{dB}/\text{km} \tag{9.103}$$

where R is the rain rate and a and b are constants. At 0° C, the values of a and b are related to frequency f in gigahertz as follows:

$$a = G_a \, f^{E_a} \tag{9.104}$$

where

$$G_a = 6.39 \times 10^{-5}, \quad E_a = 2.03, \quad \text{for } f < 2.9 \text{ GHz}$$

$$G_a = 4.21 \times 10^{-5}, \quad E_a = 2.42, \quad \text{for } 2.9 \text{ GHz} \leq f \leq 54 \text{ GHz}$$

$$G_a = 4.09 \times 10^{-2}, \quad E_a = 0.699, \quad \text{for } 54 \text{ GHz} \leq f < 100 \text{ GHz}$$

$$G_a = 3.38, \quad E_a = -0.151, \quad \text{for } 180 \text{ GHz} < f$$

and

$$b = G_b \, f^{E_b} \tag{9.105}$$

where

$$G_b = 0.851, \quad E_b = 0.158, \qquad \text{for } f < 8.5 \text{ GHz}$$

$$G_b = 1.41, \quad E_b = -0.0779, \quad \text{for } 8.5 \text{ GHz} \le f < 25 \text{ GHz}$$

$$G_b = 2.63, \quad E_b = -0.272, \quad \text{for } 25 \text{ GHz} \le f < 164 \text{ GHz}$$

$$G_b = 0.616, \quad E_b = 0.0126, \quad \text{for } 164 \text{ GHz} \le f.$$

The effective length $\ell_e(R)$ through the medium is needed since rain intensity is not uniform over the path. Its actual value depends on the particular area of interest and therefore has a number of representations (Liu and Fang, 1988). Based on data collected in western Europe and eastern North America, the effective path length has been approximated as (Hyde, 1984)

$$\ell_e(R) = \left[0.00741 R^{0.766} + \left(0.232 - 0.00018 R \right) \sin \theta \right]^{-1} \qquad (9.106)$$

where θ is the elevation angle.

The cumulative probability in percentage of rainfall rate R is given by (Hyde, 1984)

$$p(R) = \frac{M}{87.66} \left[0.03 \beta e^{-0.03R} + 0.2 \left(1 - \beta \right) \left(e^{-0.258R} + 1.86 e^{-1.63R} \right) \right] \qquad (9.107)$$

where M is mean the annual rainfall accumulation in mm and β is the Rice–Holmberg thunderstorm ratio.

The effect of other hydrometeors such as vapor, fog, hail, snow, and ice is governed by fundamental principles similar to the effect of rain (Collin, 1985). However, their effects are at least an order of magnitude less than the effect of rain in most cases.

Other Effects

Besides hydrometeors, the atmosphere has the composition given in Table 9.1. While attenuation of EM waves by hydrometeors may result from both absorption and scattering, gases act only as absorbers. Although some of these gases do not absorb microwaves, some possess permanent electric and/or magnetic dipole moment and play some part in microwave absorption. For example, nitrogen molecules do not possess permanent electric or magnetic dipole moment and therefore play no part in microwave absorption. Oxygen has a small magnetic moment that enables it to display weak absorption lines in the

TABLE 9.1 Composition of Dry Atmosphere from Sea Level to about 90 km (Livingston, 1970)

Constituent	Percent by Volume	Percent by Weight
Nitrogen	78.088	75.527
Oxygen	20.949	23.143
Argon	0.93	1.282
Carbon dioxide	0.03	0.0456
Neon	1.8×10^{-3}	1.25×10^{-3}
Helium	5.24×10^{-4}	7.24×10^{-5}
Methane	1.4×10^{-4}	7.75×10^{-5}
Krypton	1.14×10^{-4}	3.30×10^{-4}
Nitrous oxide	5×10^{-5}	7.6×10^{-5}
Xenon	8.6×10^{-6}	3.90×10^{-5}
Hydrogen	5×10^{-5}	3.48×10^{-6}

centimeter- and millimeter-wave regions. Water vapor is a molecular gas with a permanent electric dipole moment. It is more responsive to excitation by an EM field than is oxygen.

Other mechanisms that can affect EM wave propagation in free space, not discussed in this chapter, include clouds, dust, and the ionosphere. The effect of the ionosphere is discussed in detail in standard texts.

References

Collin, R. E., 1985. *Antennas and Radiowave Propagation,* McGraw-Hill, New York, 339–456.

Freeman, R. L., 1994. *Reference Manual for Telecommunications Engineering,* 2nd ed., John Wiley & Sons, New York, 711–768.

Hyde, G., 1984. Microwave propagation, in R. C. Johnson and H. Jasik (eds.), *Antenna Engineering Handbook,* 2nd ed., McGraw-Hill, New York, 45.1–45.17.

Kerr, D. E., 1951. *Propagation of Short Radio Waves,* McGraw-Hill, New York, (Republished by Peter Peregrinus, London, U.K., 1987), 396–444.

Liu, C. H. and D. J. Fang, 1988. Propagation, in Y. T. Lo and S. W. Lee, *Antenna Handbook: Theory, Applications, and Design,* Van Nostrand Reinhold, New York, 29.1–29.56.

Livingston, D. C., 1970. *The Physics of Microwave Propagation,* Prentice-Hall, Englewood Cliffs, NJ, 11.

Oguchi, T., 1983. Electromagnetic Wave Propagation and Scattering in Rain and Other Hydrometeors, *Proc. IEEE,* 71, 1029–1078.

Sadiku, M. N. O., 2001. *Numerical Techniques in Electromagnetics,* 2nd ed., CRC Press, Boca Raton, FL, 95–105.

Sadiku, M. N. O., 2001. *Elements of Electromagnetics,* 3rd ed. Oxford University Press, New York, 410–472.

Smith, B. G., 1967. Geometrical shadowing of a random rough surface, *IEEE Trans. Ant. Prog.,* 15, 668–671.

Veley, V. F., *Modern Microwave Technology,* Prentice-Hall, Englewood Cliffs, NJ, 1987, 23–33.

Further Information

The subject of wave propagation could easily fill many chapters, and here it has only been possible to point out some of the main points of concern to microwave systems engineer. There are several sources of information dealing with the theory and practice of wave propagation in space. Some of these are in the references section. Journals such as *Radio Science, IEE Proceedings Part H, IEEE Transactions on Antenna and Propagation* are devoted to EM wave propagation. *Radio Science* is available at American Geophysical Union, 2000 Florida Avenue, NW, Washington DC 20009; *IEE Proceedings Part H* at IEE Publishing Department, Michael Faraday House, 6 Hills Way, Stevenage, Herts SG1 2AY, U.K.; and *IEEE Transactions on Antenna and Propagation* at IEEE, 445 Hoes Lane, P. O. Box 1331, Piscataway, NJ 08855-1331.

9.3 Guided Wave Propagation and Transmission Lines

W.R. Deal, V. Radisic, Y. Qian, and T. Itoh

At higher frequencies where wavelength becomes small with respect to feature size, it is often necessary to consider an electronic signal as an electromagnetic wave and the structure where this signal exists as a waveguide. A variety of different concepts can be used to examine this wave behavior. The most simplistic view is transmission line theory, where propagation is considered in a simplistic 1-D manner and the cross-sectional variation of the guided wave is entirely represented in terms of distributed transmission parameters in an equivalent circuit. This is the starting point for transmission line theory that is commonly used to design microwave circuits. In other guided wave structures, such as enclosed waveguides, it is more appropriate to examine the concepts of wave propagation from the perspective of Maxwell's equations, the solutions of which will explicitly demonstrate the cross-sectional dependence of the guided wave structure.

Most practical wave guiding structures rely on single-mode propagation, which is restricted to a single direction. This allows the propagating wave to be categorized according to its polarization properties. A convenient method is classifying the modes as TEM, TE, or TM. TEM modes have both the electric and magnetic field transverse in the direction of propagation. Only the magnetic field transverses in the direction of propagation in TM modes, and only the electric field transverses in the direction of propagation in TE modes.

In this chapter, we first briefly examine the telegrapher's equation, which is the starting point for transmission line theory. The simple transmission line model accurately describes a number of guided wave structures and is the starting point for transmission line theory. In the next section, enclosed waveguides including rectangular and circular waveguides will be discussed. Relevant concepts such as cutoff frequency and modes will be given. In the final section, four common planar guided wave structures will be discussed. These inexpensive and compact structures are the foundation for the modern commercial RF front end.

TEM Transmission Lines, Telegrapher's Equations, and Transmission Line Theory

In this section, the concept of guided waves in simple TEM-guiding structures will be explored in terms of the simple model provided by Telegrapher's Equations, also referred to as transmission line equations. Telegrapher's equations demonstrate guided wave properties in terms of lumped equivalent circuit parameters available for many types of simple two-conductor transmission lines, and are valid for all types of TEM waveguides if their corresponding equivalent circuit parameters are known. These parameters must be found from Maxwell's equations in their fundamental form. Finally, properties and parameters for several types of two-wire TEM transmission line structures are introduced.

A transmission line or waveguide is used to transmit power and information from one point to another in an efficient manner. Three common types of transmission lines that support TEM guided waves are shown in Figure 9.11(a–c), including the parallel-plate transmission line, two-wire line, and coaxial transmission line. The parallel-plate transmission line consists of a dielectric slab sandwiched between two parallel conducting plates of width w. More practical, commonly used variations of this structure at microwave and millimeter-wave frequencies include microstrip and stripline, which will be briefly discussed in the final part of Section 9.3. A two-wire transmission line, consisting of two parallel conducting lines separated by a distance d is shown in Fig. 9.11b. This is commonly used for power distribution at low frequencies. Finally, the coaxial transmission line consists of two concentric conductors separated by a dielectric layer. This structure is well shielded and commonly used at high frequencies well into the microwave range.

The telegrapher's equations form a simple and intuitive starting point for the physics of guided wave propagation in these structures. An equivalent circuit model is shown in Fig. 9.12 for a two-conductor transmission line of differential length Δz in terms of the following four parameters:

R, resistance per unit length of both conductors (Ω/m).
L, inductance per unit length of both conductors (H/m).
G, conductance per unit length (S/m).
C, capacitance per unit length of both conductors (F/m).

These parameters represent physical quantities for each of the relevant transmission lines. For each of the structures shown in Fig. 9.11(a–c), R represents conductor losses, L represents inductance, G represents dielectric losses, and C represents the capacitance between the two lines.

Returning to Fig. 9.12, the quantities $v(z,t)$ and $v(z + \Delta z,t)$ represent change in voltage along the differential length of transmission line, while $i(z,t)$ and $i(z + \Delta z,t)$ represent the change in current. Writing Kirchoff's voltage law and current laws for the structure, dividing by Δz, and applying the fundamental

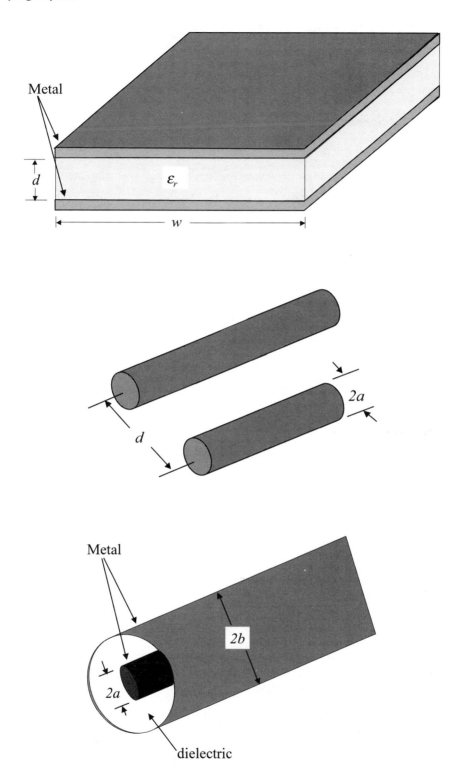

FIGURE 9.11 Three simple TEM-type transmission line geometries including (a) parallel-plate transmission line, (b) two-wire line, and (c) coaxial line.

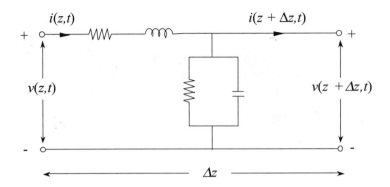

FIGURE 9.12   Distributed equivalent circuit model for a transmission line.

theorem of calculus as $\Delta z \to 0$, two coupled differential equations known as the telegrapher's equations are obtained:

$$-\frac{\partial v(z,t)}{\partial z} = Ri(z,t) + L\frac{\partial i(z,t)}{\partial t} \tag{9.108}$$

$$-\frac{\partial i(z,t)}{\partial z} = Gi(z,t) + D\frac{\partial v(z,t)}{\partial t} \tag{9.109}$$

However, typically we are interested in signals with harmonic time dependence ($e^{j\omega t}$). In this case, the time harmonic forms of the telegrapher's equations are given by

$$-\frac{dV(z)}{dz} = (R + j\omega L)I(z) \tag{9.110}$$

$$-\frac{dI(z)}{dz} = (G + j\omega C)V(z) \tag{9.111}$$

The constant γ is defined as the propagation constant with real and imaginary parts, α and β, corresponding to the attenuation constant (Np/m) and phase constant (rad/m) in the following manner

$$\gamma = \alpha + j\beta = \sqrt{(R + j\omega L)(G + j\omega C)} \tag{9.112}$$

This may then be substituted into the telegrapher's equations, which may then be solved for $V(z)$ and $I(z)$ to yield the following one-dimensional wave equations:

$$\frac{d^2V(z)}{dz^2} - \gamma^2 V(z) = 0 \tag{9.113}$$

$$\frac{d^2I(z)}{dz^2} - \gamma^2 I(z) = 0 \tag{9.114}$$

The form of this equation is the well-known wave equation. This indicates that the transmission line will support a guided electromagnetic wave traveling in the z-direction. The telegrapher's equations use a physical equivalent circuit and basic circuit theory to demonstrate the wave behavior of an electromagnetic signal on a transmission line. Alternatively, the same result can be obtained by starting directly with Maxwell's equations in their fundamental form, which may be used to derive the wave equation for a propagating electromagnetic wave. In this case, the solution of the wave equation will be governed by the boundary conditions. Similarly, the parameters R, L, G, and C are determined by the geometry of the transmission line structures.

Returning to the telegrapher's equations, several important facts may be noted. First, the characteristic impedance of the transmission line may be found by taking the ratio of the forward traveling voltage and current wave amplitudes, and is given in terms of the equivalent circuit parameters as

$$Z_0 = \sqrt{\frac{R + j\omega L}{G + j\omega C}} \qquad (9.115)$$

In the case of a lossless transmission line, this reduces to $Z_0 = \sqrt{L/C}$. The phase velocity, also known as the propagation velocity, is the velocity of the wave as it moves along the waveguide. It is defined as

$$v_p = \frac{\omega}{\beta} \qquad (9.116)$$

In the lossless case, this reduces to:

$$v_p = \frac{1}{\sqrt{LC}} = \frac{1}{\sqrt{\mu\varepsilon}} \qquad (9.117)$$

This shows that the velocity of the signal is directly related to the medium. In the case of an air-filled, purely TEM mode, the wave will propagate at the familiar value $c = 3 \times 10^8$ m/s. Additionally, it provides the relationship between L, C and the medium in which the wave is guided. Therefore, if the properties of the medium are known, it is only necessary to determine either L or C. Once C is known, G may be determined by the following relationship:

$$\frac{G}{C} = \frac{\sigma}{\varepsilon} \qquad (9.118)$$

Note that σ is the conductivity of the medium, not of the metal conductors. The final parameter, the series resistance R, is determined by the power loss in the conductors. Simple approximations for the transmission line parameters R, L, G, and C for the three types of transmission lines shown in Figs. 9.11(a–c) are well known and are shown in Table 9.2. Note that μ, ε, and σ relate to the medium separating the conductors, and σ_c refers to the conductor. Once the equivalent circuit parameters are determined, the characteristic impedance and propagation constant of the transmission line may be determined. Note that R_s represents the surface resistance of the conductors, given as

$$R_s = \sqrt{\frac{\pi f \mu_c}{\sigma_c}} \qquad (9.119)$$

TABLE 9.2 Transmission Line Parameters for Parallel-Plate,
Two-Wire Line and Coaxial Transmission Lines

	Parallel-Plate Waveguide	Two-Wire Line	Coaxial Line
R (Ω/m)	$\dfrac{2}{w}R_s$	$\dfrac{R_s}{\pi a}$	$\dfrac{R_s}{2\pi}\left(\dfrac{1}{a}+\dfrac{1}{b}\right)$
L (H/m)	$\mu\dfrac{d}{w}$	$\dfrac{\mu}{\pi}\cosh^{-1}\left(\dfrac{D}{2a}\right)$	$\dfrac{\mu}{2\pi}\ln\left(\dfrac{b}{a}\right)$
G (S/m)	$\sigma\dfrac{w}{d}$	$\dfrac{\pi\sigma}{\cosh^{-1}\left(D/2a\right)}$	$\dfrac{2\pi\sigma}{\ln\left(b/a\right)}$
C (F/m)	$\varepsilon\dfrac{w}{d}$	$\dfrac{\pi\varepsilon}{\cosh^{-1}\left(D/2a\right)}$	$\dfrac{2\pi\varepsilon}{\ln\left(b/a\right)}$

Guided Wave Solution from Maxwell's Equations, Rectangular Waveguide, and Circular Waveguide

A waveguide is any structure that guides an electromagnetic wave. In the preceding section, several simple TEM transmission structures were discussed. While these structures do support a guided wave, the term waveguide more commonly refers to a closed metallic structure with a fixed cross-section within which a guided wave propagates, as shown for the arbitrary cross-section in Fig. 9.13. The guide is filled with a material of permittivity ε and permeability μ, and is defined by its metallic wall parallel to the z-axis. These structures demonstrate lower losses than the simple transmission line structures of the first section, and are used to transport power in the microwave and millimeter-wave frequency range. Ohmic losses are low and the waveguide is capable of carrying large power levels. Disadvantages are bulk, weight, and limited bandwidth, which cause planar transmission lines to be used wherever possible in modern communications circuits. However, a wide variety of components are available in this technology, including high performance filters, couplers, isolators, attenuators, and detectors.

Inside this type of enclosed waveguide, an infinite number of distinct solutions exist, each of which is referred to as a *waveguide mode*. At a given operating frequency, the cross-section of the waveguide

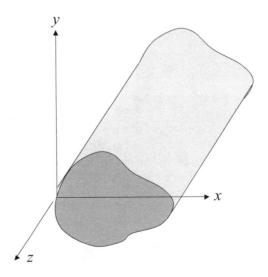

FIGURE 9.13 Geometry of enclosed waveguide with arbitrary cross-section. Propagation is in the z direction.

and the type of material in the waveguide determine the characteristics of these modes. These modes are usually classified by the longitudinal components of the electric and magnetic fields, E_z and H_z, where propagation is in the z direction. The most common classifications are TE (Transverse Electric), TM (Transverse Magnetic), EH, and HE modes. The basic characteristics are described in the next two paragraphs. The TEM modes that were discussed in the previous section do not propagate in this type of metallic enclosed waveguide. This is because a TEM mode requires two conductors to propagate, where a conventional enclosed waveguide has only a single enclosing conductor.

The two most common waveguide modes are the TE and TM modes. TE modes have no component of E in the z direction, which means that E is completely transverse to the direction of propagation. Similarly, TM modes have no component of H in the z direction.

EH and HE modes are hybrid modes that may be present under certain conditions, such as a waveguide partially filled with dielectric. In this case, pure TE and TM are unable to satisfy all of the necessary boundary conditions and a more complex type of modal solution is required. With both EH and HE, neither E nor H are zero in the direction of propagation. In EH modes, the characteristics of the transverse fields are controlled more by H_z than by E_z. HE modes are controlled more by E_z than by H_z. These types of hybrid modes may also be referred to as LSE (Longitudinal Section Electric) and LSM (Longitudinal Section Magnetic). It should be noted that most commonly used waveguides are homogenous, being entirely filled with material of a single permittivity (which may of course be air) and these types of modes will not be present.

Inside a homogenous waveguide, E_z and H_z satisfy the scalar wave equation inside the waveguide:

$$\left(\frac{\partial^2}{\partial x^2} + \frac{\partial^2}{\partial y^2}\right)E_z + h^2 E_z = 0 \tag{9.120}$$

$$\left(\frac{\partial^2}{\partial x^2} + \frac{\partial^2}{\partial y^2}\right)H_z + h^2 H_z = 0 \tag{9.121}$$

Note that h is given as:

$$h^2 = \omega^2 \mu \varepsilon + \gamma^2 = k^2 + \gamma^2 \tag{9.122}$$

The wavenumber, k, is for the material filling the waveguide. For several simple homogenous waveguides with commonly used waveguide geometries, applying boundary equations on the walls of the waveguide may be used to solve these equations to obtain closed form solutions. The resulting modal solution will possess distinct eigenvalues determined by the cross-section of the waveguide. One important result obtained from this procedure is that waveguide modes, unlike the fundamental TEM mode that propagates in two-wire structures at any frequency, will have a distinct cutoff frequency. It may be shown that the propagation constant varies with frequency as

$$\gamma = \alpha + j\beta = h\sqrt{1 - \left(\frac{f}{f_c}\right)^2} \tag{9.123}$$

where the cutoff frequency, f_c is given by:

$$f_c = \frac{h}{2\pi\sqrt{\mu\varepsilon}} \tag{9.124}$$

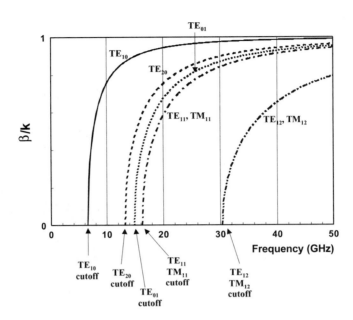

FIGURE 9.14 β/k diagram for WR-90 waveguide illustrating the concept of higher mode propagation and cutoff frequency.

By inspection of Eq. (9.133), and recalling the $\exp(j\omega t - \gamma z)$ dependence of the wave propagating in the $+z$ direction (for propagation in the $-z$ direction, replace z with $-z$), the physical significance of the cutoff frequency is clear. For a given mode, when $f > f_c$, the propagation constant γ is imaginary and the wave is propagating. Alternatively, when $f < f_c$, the propagation constant γ is real and the wave decays exponentially. In this case, modes operated below the cutoff frequency attenuate rapidly and are therefore referred to as evanescent modes. In practice, a given waveguide geometry is seldom operated at a frequency where more than one mode will propagate. This fixes the bandwidth of the waveguide to operate at some point above the cutoff frequency of the fundamental mode and below the cutoff frequency of the second order mode, although in some rare instances higher order modes may be used for specialized applications.

The guided wavelength is also a function of the cross-section geometry of the waveguide structure. The guided wavelength is given as:

$$\lambda_g = \frac{\lambda_0}{\sqrt{1 - \left(\dfrac{f_c}{f}\right)^2}} \tag{9.125}$$

Note that λ_0 is the wavelength of a plane wave propagating in an infinite medium of the same material as the waveguide. Two important facts may be noted about this expression. First, at frequencies well above the cutoff frequency, $\lambda_g \approx \lambda$. Secondly, as $f \to f_c$, $\lambda \to \infty$, further illustrating that the mode does not propagate. This is another reason that the operating frequency is always chosen above the cutoff frequency. This concept is graphically depicted in Fig. 9.14, a β/k diagram for a standard WR-90 waveguide. At the cutoff frequency, the phase constant goes to zero, indicating that the wave does not propagate. At high frequencies, β approaches the phase constant in an infinite region of the same medium. Therefore, β/k approaches one.

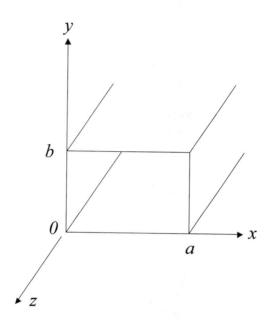

FIGURE 9.15 Geometry of a rectangular waveguide.

The wave impedance of the waveguide is given by the ratio of the magnitudes of the transverse electric and magnetic field components, which will be constant across the cross-section of the waveguide. For a given mode, the wave impedance for the TE and TM modes are given as:

$$Z_{TE} = \frac{E_T}{H_T} = \frac{j\omega\mu}{\mu} \tag{9.126}$$

$$Z_{TM} = \frac{E_T}{H_T} = \frac{\gamma}{j\omega\varepsilon} \tag{9.127}$$

E_T and H_T represent the transverse electric and magnetic fields. Note that at frequencies well above cutoff, the wave impedance for both the TE and TM modes approaches $\sqrt{\mu/\varepsilon}$, the characteristic impedance of a plane wave propagating in an infinite medium of the same material as the waveguide. Further, as $f \to f_c$, then $Z_{TE} \to \infty$ and $Z_{TM} \to 0$, again demonstrating the necessity of choosing an operating point well above cutoff.

A variety of geometries are used for waveguides, the most common being the rectangular waveguide, which is used in the microwave and well into the millimeter-wave frequency regime. Shown in Fig. 9.15, it is a rectangular metallic guide of width a and height b. Rectangular waveguide propagate both TE and TM modes. For conciseness, the field components of the TE_{mn} and TM_{mn} modes are presented in Table 9.3. From the basic form of the equations, we see that the effect of the rectangular cross-section is a standing wave dependence determined by the dimensions of the cross-section, a and b. Further, h (and therefore the propagation constant, γ) are determined by a and b. The dimensions of the waveguide are chosen so that only a single mode propagates at the desired frequency, with all other modes cut off. By convention, $a > b$ and a ratio of $a/b = 2.1$ is typical for commercial waveguide types.

The dominant mode in rectangular waveguide is the TE_{10} mode, which has a cutoff frequency of:

$$f_{c_{10}} = \frac{1}{2a\sqrt{\mu\varepsilon}} = \frac{c}{2a} \tag{9.128}$$

TABLE 9.3   Field Components for Rectangular Waveguide

	TE	TM
E_z	0	$E_0 \sin\left(\dfrac{m\pi x}{a}\right)\sin\left(\dfrac{n\pi y}{b}\right)e^{-\gamma_{mn}z}$
H_z	$H_0 \cos\left(\dfrac{m\pi x}{a}\right)\cos\left(\dfrac{n\pi y}{b}\right)e^{-\gamma_{mn}z}$	0
E_x	$H_0 \dfrac{j\omega\mu n\pi}{h_{mn}^2 b}\cos\left(\dfrac{m\pi x}{a}\right)\sin\left(\dfrac{n\pi y}{b}\right)e^{-\gamma_{mn}z}$	$-E_0 \dfrac{\gamma_{mn} m\pi}{h_{mn}^2 a}\cos\left(\dfrac{m\pi x}{a}\right)\sin\left(\dfrac{n\pi y}{b}\right)e^{-\gamma_{mn}z}$
H_x	$H_0 \dfrac{\gamma_{mn} m\pi}{h_{mn}^2 a}\sin\left(\dfrac{m\pi x}{a}\right)\cos\left(\dfrac{n\pi y}{b}\right)e^{-\gamma_{mn}z}$	$H_0 \dfrac{j\omega\varepsilon n\pi}{h_{mn}^2 b}\sin\left(\dfrac{m\pi x}{a}\right)\cos\left(\dfrac{n\pi y}{b}\right)e^{-\gamma_{mn}z}$
E_y	$-H_0 \dfrac{j\omega\mu m\pi}{h_{mn}^2 a}\sin\left(\dfrac{m\pi x}{a}\right)\cos\left(\dfrac{n\pi y}{b}\right)e^{-\gamma_{mn}z}$	$-E_0 \dfrac{\gamma_{mn} n\pi}{h_{mn}^2 b}\sin\left(\dfrac{m\pi x}{a}\right)\cos\left(\dfrac{n\pi y}{b}\right)e^{-\gamma_{mn}z}$
H_y	$H_0 \dfrac{\gamma_{mn} n\pi}{h_{mn}^2 b}\cos\left(\dfrac{m\pi x}{a}\right)\sin\left(\dfrac{n\pi y}{b}\right)e^{-\gamma_{mn}z}$	$-E_0 \dfrac{j\omega\varepsilon m\pi}{h_{mn}^2 a}\cos\left(\dfrac{m\pi x}{a}\right)\sin\left(\dfrac{n\pi y}{b}\right)e^{-\gamma_{mn}z}$
h_{mn}	$\sqrt{\left(\dfrac{m\pi x}{a}\right)^2 + \left(\dfrac{n\pi y}{b}\right)^2} = 2\pi f_c\sqrt{\mu\varepsilon}$	$\sqrt{\left(\dfrac{m\pi x}{a}\right)^2 + \left(\dfrac{n\pi y}{b}\right)^2} = 2\pi f_c\sqrt{\mu\varepsilon}$

The concept of cutoff frequency is further illustrated in Fig. 9.14, a β/k diagram for a lossless WR-90 waveguide (note that in the lossless case, the propagation constant will be equal to $j\beta$). It is apparent that higher order modes may propagate as the operating frequency increases. At the cutoff frequency, β is zero because the guided wavelength is infinity. At high frequencies, the ratio β/k approaches one.

A number of variations of the rectangular waveguide are available, including single and double-ridged waveguides, which are desirable because of increased bandwidth. However, closed solutions for the fields in these structures do not exist and numerical techniques must be used to solve for the field distributions, as well as essential design information such as guided wavelength and characteristic impedance. Additionally, losses are typically higher than standard waveguides.

The circular waveguide is also used in some applications, although not nearly as often as rectangular geometry guides. Closed form solutions for the fields in a circular geometry, perfectly conducting waveguide with an inside diameter of $2a$ are given in Table 9.4. Note that these equations use a standard cylindrical coordinate system with ρ the radial distance from the z-axis, and ϕ is the angular distance measured from the y-axis. The axis of the waveguide is aligned along the z-axis. For both the TE_{mn} and TM_{mn} modes, any integer value of $n \geq 0$ is allowed, and $J_n(x)$ and $J_n'(x)$ are Bessel functions of order n and its first derivative. As with the rectangular waveguide, only certain values of h are allowed. For the TE_{mn} modes, the allowed values of the modal eigenvalues must satisfy the roots of $J_n'(h_{mn}a) = 0$, where m signifies the root number and may range from one to infinity with $m = 1$ the smallest root. Similarly, for the TM_{mn} modes, the values of the modal eigenvalues are the solutions of $J_n(h_{mn}a) = 0$. The dominant mode in the circular waveguide is the TE_{11} mode, with a cutoff frequency given by:

$$f_{c_{11}} = \frac{0.293}{a\sqrt{\mu\varepsilon}} \tag{9.129}$$

The cutoff frequencies for several of the lowest order modes are given in Table 9.5, referenced to the cutoff frequency of the dominant mode.

TABLE 9.4   Field Components for Circular Waveguide

	TE	TM
E_z	0	$E_0 J_n\left(h_{nm}\rho\right)\cos\left(n\phi\right)e^{-\gamma_{nm}z}$
H_z	$H_0 J_n\left(h_{nm}\rho\right)\cos\left(n\phi\right)e^{-\gamma_{nm}z}$	0
E_ρ	$H_0 \dfrac{j\omega\mu n}{h_{nm}^2 \rho} J_n\left(h_{nm}\rho\right)\sin\left(n\phi\right)e^{-\gamma_{nm}z}$	$-E_0 \dfrac{\gamma_{nm}}{h_{nm}} J_n'\left(h_{nm}\rho\right)\cos\left(n\phi\right)e^{-\gamma_{nm}z}$
H_ρ	$-H_0 \dfrac{\gamma_{nm}}{h_{nm}} J_n'\left(h_{nm}\rho\right)\cos\left(n\phi\right)e^{-\gamma_{nm}z}$	$-E_0 \dfrac{j\omega\varepsilon n}{h_{nm}^2 \rho} J_n'\left(h_{nm}\rho\right)\sin\left(n\phi\right)e^{-\gamma_{nm}z}$
E_φ	$-H_0 \dfrac{j\omega\mu}{h_{nm}} J_n'\left(h_{nm}\rho\right)\cos\left(n\phi\right)e^{-\gamma_{nm}z}$	$E_0 \dfrac{\gamma_{nm}}{h_{nm}^2 \rho} J_n'\left(h_{nm}\rho\right)\sin\left(n\phi\right)e^{-\gamma_{nm}z}$
H_φ	$H_0 \dfrac{\gamma_{nm}}{h_{nm}^2} J_n\left(h_{nm}\rho\right)\sin\left(n\phi\right)e^{-\gamma_{nm}z}$	$-E_0 \dfrac{j\omega\varepsilon}{h_{nm}} J_n'\left(h_{nm}\rho\right)\cos\left(n\phi\right)e^{-\gamma_{nm}z}$

TABLE 9.5   Cutoff Frequencies
for Several Lower Order Waveguide
Modes for Circular Waveguide

$f_c/f_{c_{10}}$	Modes
1.0	TE_{11}
1.307	TM_{01}
1.66	TE_{21}
2.083	TE_{01}, TM_{11}
2.283	TE_{31}
2.791	TE_{21}
2.89	TE_{41}
3.0	TE_{12}

Note: Frequencies have been normalized to the cutoff frequency of the TE_{10} mode.

Planar Guiding Structures

Planar guiding structures are composed of a comparatively thin dielectric substrate with metallization on one or both planes. By controlling the dimensions of the metallization, a variety of passive components, transmission lines, and matching circuits can be constructed using photolithography and photoetching. Further, active devices are readily integrated into planar guiding structures. This provides a low-cost and compact way of realizing complicated microwave and millimeter-wave circuits. Microwave integrated circuits (MICs) and monolithic microwave integrated circuits (MMICs) based on this concept are commonly available.

A variety of planar transmission lines have been demonstrated, including microstrip, coplanar waveguide (CPW), slotline, and coplanar stripline. The cross-section of each of these planar transmission lines is shown in Figs. 9.16(a–d). Once the dielectric substrate is chosen, characteristics of these transmission lines are controlled by the width of the conductors and/or gaps on the top planes of the geometry. Of these, the microstrip is by far the most commonly used planar transmission line. CPW is also often used, with slotlines and coplanar striplines being the least common at microwave frequencies, for a variety of reasons that will briefly be discussed later. In this section, we will describe the basic properties of planar transmission lines. Because of its prevalence, the microstrip will be described in detail and closed form expressions for the design of the microstrip will be given.

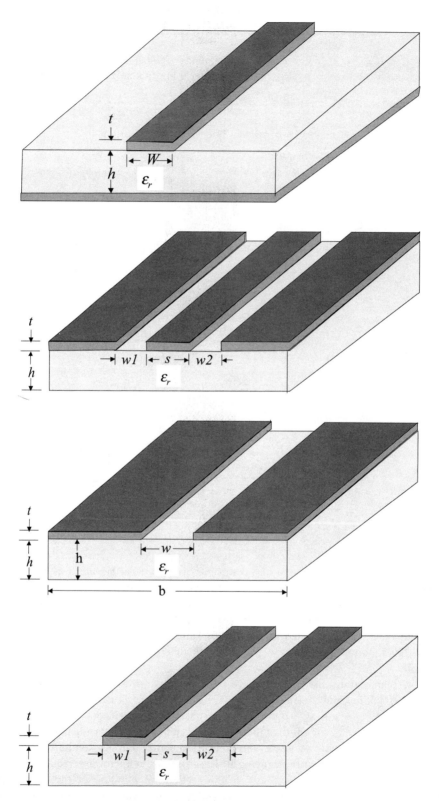

FIGURE 9.16 Cross-section of four of the most popular types of planar guiding structures, including (a) microstrip, (b) coplanar waveguide, (c) slotline, and (d) coplanar stripline.

Microstrip

As seen in Fig. 9.16(a), the simplest form of microstrip consists of a single conductor on a grounded dielectric slab. Microstrip is the most common type of planar transmission line used in microwave and millimeter-wave circuits, with a great deal of design data freely available. A broad range of passive components may be designed with the microstrip, including filters, resonators, diplexers, distribution networks, and matching components. Additionally, three terminal active components can be integrated by using vias to ground. However, this may introduce considerable inductances at high frequencies.

The fundamental mode of propagation for this type of planar waveguide is often referred to as quasi-TEM, because of its close resemblance to pure TEM modes. In fact, noting that the majority of the power is confined in the region bounded by the width of the microstrip, the basic characteristics of microstrip are quite similar to the parallel-strip transmission line of Fig. 9.11(a). Because of the presence of the air-dielectric interface, it is not a true TEM mode. The use of the dielectric between the ground and top conductor confines the majority of the fields in this region, but some energy may radiate from the structures. Using a high permittivity substrate and shielding the structure helps to minimize this factor. Microstrip is capable of carrying moderate power levels (a 50 Ω microstrip line on 25 mil alumina can handle several kW of power), is broadband, and enables realization of a variety of circuit topologies, both active and passive.

To design the basic microstrip line, it is necessary to be able to determine characteristic impedance and effective permittivity, preferably as a function of frequency. A wide variety of approximations have been presented in the literature, with most techniques using a quasi-static approximation for the characteristic impedance, Z_0, at low frequencies, and a dispersion model for the characteristic impedance as a function of frequency, $Z_0(f)$ in terms of Z_0. One fairly accurate and simple model commonly used to obtain Z_0 and the effective permittivity, ε_{re}, neglecting the effect of conductor thickness is given as[1]:

$$Z_0 = \frac{\eta}{2\pi\sqrt{\varepsilon_{re}}} \ln\left(\frac{8h}{W} + 0.25\frac{W}{h}\right) \quad \text{for} \left(\frac{W}{h} \leq 1\right) \tag{9.130}$$

$$Z_0 = \frac{\eta}{\sqrt{\varepsilon_{re}}} \left\{\frac{W}{h} + 1.393 + 0.667 \ln\left(\frac{W}{h} + 1.444\right)\right\}^{-1} \quad \text{for} \left(\frac{W}{h} \geq 1\right) \tag{9.131}$$

Note that η is 120π-Ω, by definition. The effective permittivity is given as:

$$\varepsilon_{re} = \frac{\varepsilon_r + 1}{2} + \frac{\varepsilon_r - 1}{2} F\left(W/h\right) \tag{9.132}$$

$$F\left(W/h\right) = \left(1 + 12h/W\right)^{-1/2} + 0.04\left(1 - W/h\right)^2 \quad \text{for} \left(\frac{W}{h} \leq 1\right)$$

$$F\left(W/h\right) = \left(1 + 12h/W\right)^{-1/2} \quad \text{for} \left(\frac{W}{h} \geq 1\right)$$

With these equations, one can determine the characteristic impedance in terms of the geometry. For a desired characteristic impedance, the line width can be determined from:

$$W/h = \frac{8\exp\left(A\right)}{\exp\left(2A\right) - 2} \quad \text{for } A > 1.52 \tag{9.133}$$

$$W/h = \frac{2}{\pi} \left\{ B - 1 - \ln(2B-1) + \frac{\varepsilon_r - 1}{2\varepsilon_r} \left[\ln(B-1) + 0.39 - \frac{0.61}{\varepsilon_r} \right] \right\} \quad \text{for } A > 1.52 \quad (9.134)$$

where

$$A = \frac{Z_0}{60} \left\{ \frac{\varepsilon_r + 1}{2} \right\}^{1/2} + \frac{\varepsilon_r - 1}{\varepsilon_r + 1} \left\{ 0.23 + \frac{0.11}{\varepsilon_r} \right\}$$

$$B = \frac{60\pi^2}{Z_0 \sqrt{\varepsilon_r}}$$

Once Z_0 and ε_{re} have been determined, effects of dispersion may also be determined using expressions from Hammerstad[2] and Jensen for $Z_0(f)$ and Kobayashi[3] for $\varepsilon_{re}(f)$. To illustrate the effects of dispersion, the characteristic impedance and effective permittivity of several microstrip lines on various substrates are plotted in Figs. 9.17(a–b) using the formulas from the previously mentioned papers. The substrates indicated by the solid ($\varepsilon_r = 2.33$, h = 31 mils, W = 90 mils) and dashed ($\varepsilon_r = 10.2$, h = 25 mils, W = 23 mils) lines in these figures are typical for those that might be used in a hybrid circuit at microwave frequencies. We can see in Fig. 9.17(a) that the characteristic impedance is fairly flat until X-band, above which it may be necessary to consider the effects of dispersion for accurate design. The third line in the figure is an alumina substrate ($\varepsilon_r = 9$, h = 2.464 mils, W = 2.5 mils) on a thin substrate. The characteristic impedance is flat until about 70 GHz, indicating that this thin substrate is useful at higher frequency operation. The effective permittivity as a function of frequency is shown in Fig. 9.17a. Frequency variation for this parameter is more dramatic. However, it must be remembered that guided wavelength is inversely proportional to the square root of the effective permittivity. Therefore, variation in electrical length will be less pronounced than the plot suggests.

In addition to dispersion, higher frequency operation is complicated by a number of issues, including decreased Q-factor, radiation losses, surface wave losses, and higher order mode propagation. The designer must be aware of the limitations of both the substrate on which he is designing and the characteristic impedance of the lines he is working with. In terms of the substrate, a considerable amount of energy can couple between the desired quasi-TEM mode of the microstrip and the lowest order surface wave mode of the substrate. In terms of the substrate thickness and permittivity, an approximation for determining the frequency where this coupling becomes significant is given by the following expression.[4]

$$f_T = \frac{150}{\pi h} \sqrt{\frac{2}{\varepsilon_r - 1}} \arctan(\varepsilon_r) \quad (9.135)$$

Note that f_T is in gigahertz and h is in millimeters. In addition to the quasi-TEM mode, microstrip will propagate undesired higher order TE and TM-type modes with cutoff frequency roughly determined by the cross-section of the microstrip. The excitation of the first mode is approximately given by the following expression.[4]

$$f_c = \frac{300}{\sqrt{\varepsilon_r} \left(2W + 0.8h\right)} \quad (9.136)$$

Again, note that f_c is in gigahertz, and h and W are both in millimeters. This expression is useful in determining the lowest impedance that may be reliably used for a given substrate and operating frequency. As a rule of thumb, the maximum operating frequency should be chosen somewhat lower. A good choice for maximum frequency may be 90% of this value or lower.

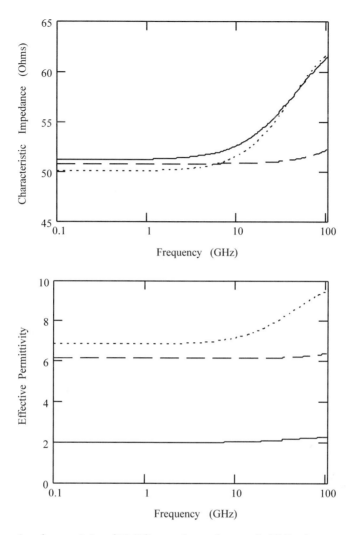

FIGURE 9.17 Dispersion characteristics of 50 Ω line on three substrates (solid line is $\varepsilon_r = 2.33$, $h = 31$ mils, $W = 90$ mils, dotted line is $\varepsilon_r = 10.2$, $h = 25$ mils, $W = 23$ mils and the dashed line is $\varepsilon_r = 9$, h = 2.464 mils, W = 2.5 mils). Shown in (a), the impedance changes significantly at high frequencies for the thicker substrates as does the effective permittivity shown in (b).

A variety of techniques have also been developed to minimize or characterize the effects of discontinuities in microstrip circuits, a variety of which are shown in Figs. 9.18(a,b) including a microstrip bend and a T-junction. Another common effect is the fringing capacitance found at impedance steps or open-circuited microstrip stubs.

The microstrip bend allows flexibility in microstrip circuit layouts and may be at an arbitrary angle with different line widths at either end. However, by far the most common is the 90° bend with equal widths at either end, shown on the left of Fig. 9.18a. Due to the geometry of the bend, excess capacitance is formed causing a discontinuity. A variety of techniques have been used to reduce the discontinuity by eliminating a sufficient amount of capacitance, including the mitered bend shown on the right. Note that another way of reducing this effect is to use a curved microstrip line with sufficiently large radius to minimize the effect. A second type of discontinuity commonly encountered by necessity in layouts is the T-junction, shown in Fig. 9.18b, which is formed at a junction of two lines. As with the bend, excess capacitance is formed, degrading performance. The mitered T-junction below is used to reduce this problem. Again, a variety of other simple techniques have also been developed.

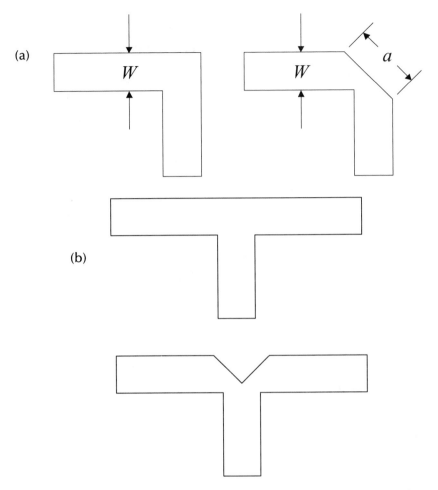

FIGURE 9.18 Two common microstrip discontinuities encountered in layout, including (a) the microstrip bend and (b) the T-junction.

Fringing capacitance will be present with microstrip open-circuited stubs and at impedance steps. With the open-circuited stub, this causes the electrical length of the structure to be somewhat longer. For an impedance step, the lower impedance line will also appear to be electrically longer. The simplest way of compensating for this problem is by modeling the capacitance and effective length of the fringing fields. Again, a variety of simple models have been developed to perform this task, most based on quasi-static approximations. A commonly used expression for the length extension of an open end based on empirical data is given by the following expression.[5]

$$\frac{\Delta l_{oc}}{h} = 0.412 \frac{\varepsilon_{re} + 0.3}{\varepsilon_{re} - 0.258} \left[\frac{W/h + 0.264}{W/h + 0.8} \right] \tag{9.137}$$

This expression is reported to yield relatively accurate results for substrates with permittivity in the range of 2 to 50, but is not as accurate for wide microstrip lines. For the impedance step, a first order approximation for determining the excess length of the impedance step is to multiply the open-end extension, $\Delta l_{oc}/h$ by an appropriate factor to obtain a useful value, i.e., $\Delta l_{step}/h \approx \Delta l_{oc} (w_1/w_2 - 1)/h$.

Because of the prevalence of microstrip, modern microwave CAD tools typically have extensive libraries for microstrip components, including discontinuities effects.

Coplanar Waveguide (CPW)

Coplanar Waveguide (CPW), shown in Fig. 9.16b, consists of a signal line and two ground planes on a dielectric slab with metallization on one side. For a given substrate, characteristic impedance is determined by the signal line width, s, and the two gaps, w_1 and w_2. This structure often demonstrates better dispersion characteristics than microstrip. Additionally, three terminal devices are easily integrated into this uniplanar transmission line that requires no vias for grounding. For this reason, parasitics are lower than microstrip making CPW a good choice for high frequency operation where this is a primary design concern.

The three-conductor line shown in Fig. 9.16b supports two fundamental modes, including the desired CPW-mode and an undesired coupled slotline mode if the two ground planes separating the signal line are not kept at the same potential. For this reason, wires or metal strips referred to as *air bridges* are placed at discontinuities where mode conversion may occur.

Packaging may be a problem for this type of structure, because the bottom plane of the dielectric may come in close proximity with other materials, causing perturbations of the transmission line characteristics. In practice, this is remedied by using *grounded* or *conductor-backed* CPW (CB-CPW) where a ground plane is placed on the backside for electrical isolation. At high frequencies, this may present a problem with additional losses through coupling to the parallel-plate waveguide mode. These losses can be minimized using vias in the region around the transmission line to suppress this problem.

Although CPW was first proposed by Wen[6] in 1969, acceptance of CPW has been much slower than microstrip. For this reason, simple and reliable models for CPW are not as readily available as for microstrip. A compilation of some of the more useful data can be found in Reference 6.

Slotline and Coplanar Stripline

Two other types of planar transmission lines are slotline and coplanar stripline (CPS). These structures are used less often than either microstrip or CPW, but do find some applications. Both of these structures consist of a dielectric slab with metallization on one side. Slotline has a slot of width w etched into the ground plane. CPS consists of two metal strips of width w_1 and w_2 separated by a distance s on the dielectric slab. Due to their geometry, both of these structures are balanced transmission line structures, and are useful in balanced circuits such as mixers and modulators. Only limited design information is available for these types of transmission lines.

The slotline mode is non-TEM and is almost entirely TE. However, no cutoff frequency exists as with the waveguide TE modes discussed previously in this section. Microwave circuits designed solely in slotline are seldom used. However, slotline is sometimes used in conjunction with other transmission line types such as microstrip or CPW for increased versatility. Examples of these include filters, hybrids, and resonators. Additionally, slotline is sometimes used in planar antennas, such as the slot antenna or some kinds of multilayer patch antennas.

The CPS transmission line has two conductors on the top plane of the circuit, allowing series or shunt elements to be readily integrated into CPS circuits. CPS is often used in electro-optic circuits such as optic traveling wave modulators, as well as in high-speed digital circuits. Due to its balanced nature, CPS also makes an ideal feed for printed dipoles. Difficulties (or benefits, depending on the application) with CPS include high characteristic impedances.

References

1. E. Hammerstad, Equations for microstrip circuit design, *Proc. European Microwave Conf.,* 1975, 268–272.
2. E. Hammerstad and O. Jensen, Accurate models for microstrip computer-aided design, *IEEE MTT-S Int. Microwave Symp. Dig.,* 1980, 407–409.
3. M. Kobayashi, A dispersion formula satisfying recent requirements in microstrip CAD, *IEEE Trans.,* MTT-36, August 1988, 1246–1250.
4. G.D. Vendelin, Limitations on stripline Q, *Microwave J.,* 13, May 1970, 63–69.

5. R. Garg and I.J. Bahl, Microstrip discontinuities, *Int. J. Electron.*, 45, July 1978, 81–87.
6. C.P. Wen, Coplanar waveguide: A surface strip transmission line suitable for non-reciprocal gyro-magnetic device applications, *IEEE Trans.*, MTT-23, 1975, 541–548.
7. K.C. Gupta, R. Garg, I. Bahl, and R. Bhartia, *Microstrip Lines and Slotlines*, Artech House, Inc., Norwood MA, 1996.

9.4 Effects of Multipath Fading in Wireless Communication Systems

Wayne E. Stark

The performance of a wireless communication system is heavily dependent on the channel over which the transmitted signal propagates. Typically in a wireless communication system the channel consists of multiple paths between the transmitter and receiver with different attenuation and delay. The paths have different attenuation and delays because of the different distances between transmitter and receiver along different paths. For certain transmitted signals the entire transmitted signal may experience a deep fade (large attenuation) due to destructive multipath cancellation. The multipath signals may also add constructively giving a larger amplitude. In addition to the multipath effect of the channel on the transmitted signal there are other effects on the transmitted signal due to the channel. One of these is distance related and is called the propagation loss. The larger the distance between the transmitter and receiver the smaller the received power. Another effect is known as shadowing. Shadowing occurs due to buildings and other obstacles obstructing the line-of-sight path between the transmitter and receiver. This causes the received signal amplitude to vary as the receiver moves out from behind buildings or moves behind buildings.

In this chapter we examine models of fading channels and methods of mitigating the degradation in performance due to fading. We first discuss in detail models for the multipath fading effects of wireless channels. We then briefly discuss the models for propagation loss as a function of distance and shadowing. Next we show an example of how diversity in receiving over multiple, independent faded paths can significantly improve performance. We conclude by discussing the fundamental limits on reliable communication in the presence of fading.

Multipath Fading

In this section we discuss the effects of multiple paths between the transmitter and receiver. These effects depend not only on the delays and amplitudes of the paths but also on the transmitted signal. We give examples of frequency selective fading and time selective fading.

Consider a sinusoidal signal transmitted over a multipath channel. If there are two paths between the transmitter and receiver with the same delay and amplitude but opposite phase (180 degree phase shift), the channel will cause the received signal amplitude to be zero. This can be viewed as destructive interference. However, if there is no phase shift, the received signal amplitude will be twice as large as the signal amplitude on each of the individual paths. This is constructive interference.

In a digital communication system data is modulated onto a carrier. The data modulation causes variations in the amplitude and phase of the carrier. These variations occur at a rate proportional to the bandwidth of the modulating signal. As an example, in a cellular system the data rates are on the order of 25 kbps and carrier frequency is about 900 MHz. So the amplitude and phase of the carrier is changing at a rate on the order of 25 kHz. Equivalently, the envelope and phase of the carrier might change significantly every 1/25 KHz = 0.04 ms = 40 μs. If this signal is transmitted over a channel with two paths with differential delay of 1 μs, the modulation part of the signal would not differ significantly. However, if this signal was received on two paths with a differential delay of 40 μs, then there would be a significant difference in the modulated part of the signal. If the data rate of the signal was increased to 250 kbps then the modulated signal would change significantly in a 4 μs time frame and thus the effect of multipath would be different.

Thus the type of fading depends on various parameters of the channel and the transmitted signal. Fading can be considered as a filtering operation on the transmitted signal. The filter characteristics are time varying due to the motion of the transmitter/receiver. The faster the motion the faster the change in the filter characteristics operation. Fading channels are typically characterized in the following ways.

1. *Frequency Selective Fading:* If the transfer function of the filter has significant variations within the frequency band of the transmitted signal, the fading is called frequency selective.
2. *Time Selective Fading:* If the fading changes relatively quickly (compared to the duration of a data bit), the fading is said to be time selective.

If the channel is both time and frequency selective, it is said to be doubly selective.

To illustrate these types of fading we consider some special cases. Consider a simple model for fading where there are a finite number, k, of paths from the transmitter to the receiver. The transmitted signal is denoted by $s(t)$. The signal can be represented as a baseband signal modulated onto a carrier as

$$s(t) = \text{Re}\left[s_0(t)\exp\left\{j2\pi f_c t\right\}\right]$$

where f_c is the carrier frequency and $s_0(t)$ is the baseband signal or the envelope of the signal $s(t)$. The paths between the transmitter and receiver have delays τ_k and amplitudes α_k. The received signal can thus be expressed as

$$r(t) = \text{Re}\left[\sum_k \alpha_k s_0\left(t - \tau_k\right)\exp\left\{j2\pi f_c\left(t - \tau_k\right) + j\phi_k\right\}\right]$$

where ϕ_k is a phase term added due the kth path that might be due to a reflection off an object. The baseband received signal is given by

$$r_0(t) = \sum_k \alpha_k s_0\left(t - \tau_k\right)\exp\left\{j\phi_k - j2\pi f_c \tau_k\right\}.$$

To understand the effects of multipath we will consider a couple different examples.

Frequency/Time Nonselective (Flat) Fading

First we consider a frequency and time nonselective fading model. In this case the multipath components are assumed to have independent phases. If we let W denote the bandwidth of the transmitted signal, the envelope of the signal does not change significantly in time smaller than $1/W$. Thus if the maximum delay satisfies $\tau_{\max} \ll 1/W$, that is

$$\frac{1}{f_c} \ll \tau_k \ll T = W^{-1}$$

Then $s_0(t - \tau_k) \approx s_0(t)$. In this case

$$r_0(t) = s_0(t)\left(\sum_k \alpha_k \exp\left\{j\theta_k\right\}\right)$$

$$= X s_0(t)$$

where $\theta_k = \phi_k - 2\pi f_c \tau_k$. The factor $X = \sum_k \alpha_k \exp\{j\theta_k\}$ by which the signal is attenuated/phase shifted is usually modeled by a complex Gaussian distributed random variable. The magnitude of X is a Rayleigh distributed random variable. The phase of X is uniformly distributed. The fading occurs because the random phases sometimes add destructively and sometimes add constructively. Thus for narrow enough signal bandwidths ($\tau_k \ll W^{-1}$) the multipath results in an amplitude attenuation by a Rayleigh distributed random variable. It is important to note that the transmitted signal in this example has not been distorted. The only effect on the transmitted signal is an amplitude and phase change. This will not be true in a frequency selective channel.

Usually the path lengths change with time due to motion of the transmitter or receiver. Here we have assumed that the motion is slow enough relative to the symbol duration so that $\alpha_k(t)$ and $\phi_k(t)$ are constants. In this model the transmitted signal is simply attenuated by a slowly varying random variable. This is called a flat fading model or frequency and time nonselective fading.

Frequency Selective/Time Nonselective Fading

Now consider the case where the bandwidth of the modulating signal $s_0(t)$ is W and the delays satisfy

$$\tau_k \gg T = W^{-1}.$$

In this case we say the channel exhibits frequency selective fading. For example, consider a discrete multipath model. That is,

$$r_0(t) = \alpha_1 e^{j\theta_1} s_0(t - \tau_1) + \cdots \alpha_M s_0(t - \tau_M) e^{j\theta_M}.$$

The impulse response of this channel is

$$h(t) = \sum_{k=1}^{M} \alpha_k e^{j\theta_k} \delta(t - \tau_k)$$

and the transfer function is

$$H(f) = \sum_{k=1}^{M} \alpha_k \exp\{j\theta_k - j2\pi f \tau_k\}.$$

More specifically, assume $M = 2$ and that the receiver is synchronized to the first path (so that we assume $\tau_1 = \phi_1 = \theta_1 = 0$). Then

$$H(f) = 1 + \alpha_2 \exp\{j\theta_k - j2\pi f \tau_k\}$$

At frequencies where $2\pi f \tau_2 = \theta_2 + 2n\pi$ or $f = (\theta_2 + 2n\pi)/2\pi\tau_2$ the transfer function will be $H(f) = 1 + \alpha_2$. If $\alpha_2 > 0$, the amplitude of the received signal will be larger because of the second path. This is called constructive interference. At frequencies where $2\pi f \tau_2 = \theta_2 + (2n + 1)\pi$ or $f = (\theta_2 + (2n + 1)\pi)/2\pi\tau_2$, the transfer will be $H(f) = 1 - \alpha_2$. Again, for $\alpha_2 > 0$ the amplitude of the received signal will be smaller due to the second path. This is called destructive interference. The frequency range between successive nulls (destructive interference) is $1/\tau$. Thus if $\tau \gg \frac{1}{W}, \frac{1}{\tau} \ll W$ there will be multiple nulls in the spectrum of the received signal. In Fig. 9.19 we show the transfer function of a multipath channel with two equal strength paths with differential delay of 1 microsecond. In Fig. 9.20 we show the transfer function of a channel with eight equal strength paths with delays from 0 to 7 μs. The frequency selectivity of the channel is seen in the fact that the transfer function varies as a function of frequency. Narrowband systems

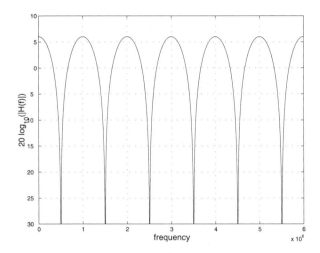

FIGURE 9.19 Transfer function of multipath channel with two equal strength paths and relative delay of 1 μs.

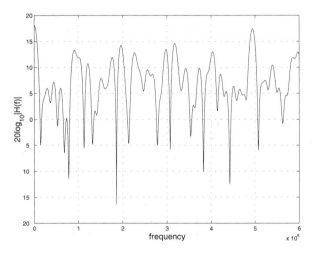

FIGURE 9.20 Transfer function of multipath channel with eight equal strength paths.

have the potential for the whole signal band to experience a deep fade while in wideband systems the transfer function of the channel varies within the band of the transmitted signal and thus the channel causes distortion of the transmitted signal.

Time Selectivity

The dual concept to frequency selectivity is time selectivity. In this case the path strength is changing as a function of time (e.g., due to vehicle motion) and the envelope of the received signal (as the vehicle moves) undergoes time-dependent fading. A model for this would be that of a time-varying impulse response (without frequency selectivity):

$$h(t; t - \beta) = \alpha_k(t) e^{j\theta(t)} \delta(t - \beta - \tau(t))$$

where $\tau(t)$ is the time-varying delay between the transmitter and receiver. The output $r_0(t)$ of the channel is related to the input $s_0(t)$ via

$$r_0(t) = \int_{-\infty}^{\infty} h(t; t - \beta) s_0(\beta) d\beta$$

$$= \int_{-\infty}^{\infty} \alpha_k(t) e^{j\theta(t)} \delta(t - \beta - \tau(t)) s_0(\beta) d\beta$$

$$= \alpha_k(t) e^{j\theta(t)} s_0(t - \tau(t))$$

Because the impulse response is time varying, the fading at different time instances is correlated if the time instances are very close and uncorrelated if they are very far apart. Consider the simple case of a sinusoidal at frequency f_c as the signal transmitted. In this case, the baseband component of the transmitted signal, $s_0(t)$, is a constant DC term. However the output of the channel is given by

$$r_0(t) = \alpha_k(t) e^{j\theta(t)} s_0.$$

The frequency content of the baseband representation of the received signal is no longer just a DC component, but has components at other frequencies due to the time-varying nature of α and θ. If we consider just a single direct path between the transmitter and receiver and assume that the receiver is moving away from the transmitter, then because of the motion there will be a Doppler shift in the received spectrum. That is the received frequency will be shifted down in frequency. Similarly if the receiver is moving toward the transmitter, there will be a shift up in the frequency of the received signal. Because there can be paths between the transmitter and receiver that are direct and paths that are reflected, some paths will be shifted up in frequency and some paths will be shifted down in frequency. The overall received signal will be spread out in the frequency domain due to these different frequency shifts on different paths. The spread in the spectrum of the transmitted signal is known as the Doppler spread. If the data duration is much shorter than the time variation of the fading process, the fading can be considered a constant or a slowly changing random process. In Fig. 9.21 we plot the fading amplitude for a single path as a function of time for a vehicle traveling at 10 miles per hour. In Fig. 9.22 a similar plot is done for a vehicle at 30 mph. It is clear that the faster a vehicle is moving the more quickly the fading amplitude varies. Fading amplitude variations with time can be compensated for by power control at low vehicle velocities. At high velocities the changes in amplitude can be averaged out by proper use of error control coding.

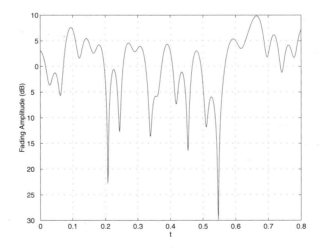

FIGURE 9.21 Received signal strength as a function of time for vehicle velocity 10 mph.

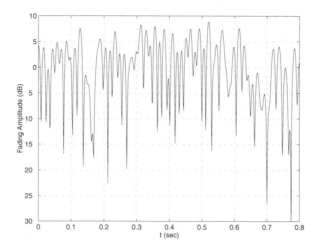

FIGURE 9.22 Received signal strength as a function of time for vehicle velocity 30 mph.

General Model

In this section we describe a general mode for fading channels and discuss the relevant parameters that characterize a fading channel. The most widely used general model for fading channels is the wide-sense stationary, uncorrelated scattering (WSSUS) fading model. In this model the received signal is modeled as a time-varying filter operation on the transmitted signal. That is

$$r_0(t) = \int_{-\infty}^{\infty} h(t; t - \alpha) s_0(\alpha) d\alpha$$

where $h(t; t - \tau)$ is the response due to an impulse at time τ and is modeled as a zero mean complex Gaussian random process. Note that it depends not only on the time difference between the output and the input, but also on the time directly. The first variable in h accounts for the time-varying nature of the channel while the second variable accounts for the delay between the input and output. This is the result of the assumption that there are a large number of (possibly time-varying) paths at a given delay with independent phases. If there is no direct (unfaded) path, then the impulse response will have zero mean. In this case the channel is known as a Rayleigh faded channel. If there is a (strong) direct path between the transmitter and receiver, then the filter $h(t, \tau)$ will have nonzero mean. This case is called a Rican faded channel. In the following we will assume the mean of the channel is zero.

The assumption for WSSUS is that the impulse response, $h(t, \tau)$, is uncorrelated for different delays and the correlation at different times depends only on the time difference. Mathematically we write the correlation of the impulse response at different delays and times as an expectation;

$$E\left[h(t; \tau_1) h^*(t + \Delta t; \tau_2)\right] = \phi(\tau_1; \Delta t) \delta(\tau_2 - \tau_1)$$

where $E[h(t; \tau_1) h^*(t + \Delta t; \tau_2)]$ denotes the expected (average) value of the impulse response at two different delays and times. The function $\phi(\tau; \Delta t)$ is the intensity delay profile and $\delta(\tau)$ is the usual Dirac delta function.

For a wide-sense stationary uncorrelated scattering (WSSUS) model the correlation between the responses at two different times depends only on the difference between times. This is indicated by the Dirac delta function. Also, the response at two different delays are uncorrelated. The amount of power received at a given delay τ is $\gamma(\tau; 0)$. This is called the intensity delay profile or the delay power spectrum. The mean excess delay, μ_m is defined to be the average excess delay above the delay of the first path

$$\mu = \frac{\int_{\tau_{min}}^{\tau_{max}} \tau \phi(\tau;0) d\tau}{\int_{\tau_{min}}^{\tau_{max}} \phi(\tau;0) d\tau} - \tau_{min}$$

The rms delay spread is defined as

$$s = \left[\frac{\int_{\tau_{min}}^{\tau_{max}} (\tau - \mu - \tau_{min})^2 \phi(\tau;0) d\tau}{\int_{\tau_{min}}^{\tau_{max}} \phi(\tau;0) d\tau} \right]^{1/2}$$

The largest value τ_{max} of τ such that $\phi(\tau; 0)$ is nonzero is called the multipath spread of the channel. The importance of the rms delay spread is that it is a good indicator of the performance of a communication system with frequency selective fading. The larger the rms delay spread the more inter-symbol interference. In the general model the delays cause distortion in the received signal.

Now consider the frequency domain representation of the channel response. The time-varying function of the channel $H(f; t)$ is given by the Fourier transform of the impulse response with respect to the delay variable. That is,

$$H(f;t) = \int_{-\infty}^{\infty} h(t;\tau) e^{-j2\pi f \tau} d\tau.$$

Since $h(t; \tau)$ is assumed to be a complex Gaussian random variable, $H(f; t)$ is also a complex Gaussian random process. The correlation $\Phi(f_1, f_2; \Delta t)$ between the transfer function at two different frequencies and two different times as defined as

$$\Phi(f_1, f_2; \Delta t) = E\left[H(f_1;t) H^*(f_2; t + \Delta t) \right]$$

$$= \int_{-\infty}^{\infty} \phi(\tau; \Delta t) e^{-j2\pi(f_2 - f_1)\tau} d\tau$$

Thus the correlation between two frequencies for the WSSUS model (and at two times) depends only on the frequency difference. If we let $\Delta t = 0$ then we obtain

$$\Phi(\Delta f; 0) = \int_{-\infty}^{\infty} \phi(\tau; 0) e^{-j2\pi(\Delta f)\tau} d\tau$$

As the frequency separation becomes larger the correlation in the response between those two frequencies generally decreases. The smallest frequency separation, B_c, such that the correlation of the response at two frequencies separated by B_c is zero is called the coherence bandwidth of the channel. It is related to the delay spread by

$$B_c \approx \frac{1}{\tau_{max}}$$

The rms delay spread and coherence bandwidth are important measures for narrowband channels. The performance of an equalizer for narrowband channels often does not depend on the exact delay power profile, but simply on the rms delay spread.

Now consider the time-varying nature of the channel. In particular, consider $\Phi(\Delta f; \Delta t)$, which is the correlation between the responses of the channel at two frequencies separated by Δf and at times separated by Δt. For $\Delta f = 0$ $\Phi(0; \Delta t)$ measures the correlation between two responses (at the same frequency) but separated in time by Δt. The Fourier transform gives the Doppler power spectral density

$$S(\lambda) = \int_{-\infty}^{\infty} \phi(0; \gamma) e^{-j2\pi\lambda\gamma} d\gamma.$$

The Doppler power spectral density gives the distribution of received power as a function of frequency shift. Since there are many paths coming from different directions and the receiver is moving, these paths will experience different frequency shifts. Consider a situation where a vehicle is moving toward a base station with velocity v.

Example

If we assume that there are many multipath components that arrive with an angle uniformly distributed over $[0, 2\pi]$, then the Doppler spectral density is given by

$$S(\lambda) = \frac{1}{2\pi f_m} \left[1 - \left(\lambda / f_m\right)^2\right]^{-1/2}, \quad 0 \leq |\lambda| \leq f_m$$

where $f_m = vf_c/c$, f_c is the center frequency and c is the speed of light (3×10^8 m/s). For example, a vehicle moving at 100 m/s with 1 GHz center frequency has maximum Doppler shift of 33.3 Hz. A vehicle moving at 30 m/s would have a maximum Doppler shift of 10 Hz. Thus most of the power is either at the carrier frequency plus 10 Hz or at the carrier frequency minus 10 Hz. The corresponding autocorrelation function is the inverse Fourier transform and is given by

$$\Phi(0, \gamma) = \int_{-\infty}^{\infty} S(\lambda) e^{j2\pi\lambda\gamma} d\lambda$$

$$= J_0\left(2\pi f_m \gamma\right)$$

The channel correlation and Doppler spread are illustrated in Figs. 9.23 and 9.24 for vehicle velocities of 10 km/hour and 100 km/hour. From these figures it is clear that a lower vehicle velocity implies a small spread in the spectrum of the received signal and a larger correlation between the fading at different times. It is often useful for the receiver in a digital communication system to estimate the fading level. The faster the fading level changes the harder it is to estimate. The product of maximum Doppler spread f_m times the data symbol duration T is a useful tool for determining the difficulty in estimating the channel response. For $f_m T$ products much smaller than 1, the channel is easy to estimate while for $f_m T$ much larger than 1, the channel is hard to estimate. Channel estimation can improve the performance of coded systems as shown in the last section of this chapter. The availability of channel information is sometimes called "side information."

If the channel is not time varying (i.e., time invariant), then the responses at two different times are perfectly correlated so that $\Phi(0; \Delta t) = 1$. This implies that $S(\lambda) = \delta(f)$.

The largest value of λ for which $S(\lambda)$ is nonzero is called the Doppler spread of the channel. It is related to the coherence time T_c, the largest time difference for which the responses are correlated by

$$B_d = \frac{1}{T_c}$$

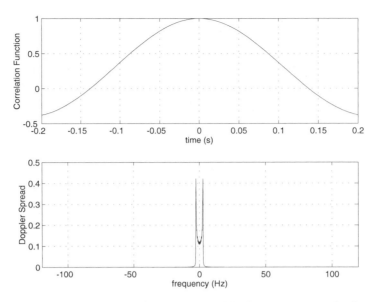

FIGURE 9.23   Channel correlation function and Doppler spread for $f_c = 1$ GHz, $v = 10$ km/hour.

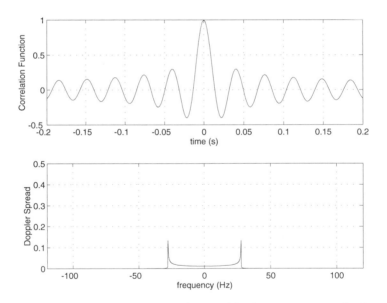

FIGURE 9.24   Channel correlation function and Doppler spread for $f_c = 1$ GHz, $v = 100$ km/hour.

GSM Model

The GSM (Global System for Mobile Communications) model was developed in order to compare different coding and modulation techniques. The GSM model is a special case of the general WSSUS model described in the previous section. The model consists of N_p paths, each time varying with different power levels. In Fig. 9.25 one example of the delay power profile for a GSM model of an urban environment is shown. In the model each path's time variation is modeled according to a Doppler spread for a uniform angle of arrival spread for the multipath. Thus the vehicle velocity determines the time selectivity for each path. The power delay profile shown below determines the frequency selectivity of the channel.

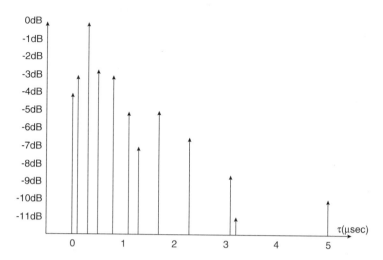

FIGURE 9.25 Power delay profile for the GSM model for typical urban channel.

TABLE 9.6 Parameters for Power Delay Profile GSM Model of Urban Area

Path	Delay (μ sec)	Average Power (dB)
1	0.0	−4.0
2	0.1	−3.0
3	0.3	0.0
4	0.5	−2.6
5	0.8	−3.0
6	1.1	−5.0
7	1.3	−7.0
8	1.7	−5.0
9	2.3	−6.5
10	3.1	−8.6
11	3.2	−11.0
12	5.0	−10.0

In Table 9.6 the parameters for the GSM model are given. The usefulness of this model is that it gives communication engineers a common channel to compare the performance of different designs.

Propagation Loss

The fading discussed above is referred to as short-term fading as opposed to long-term fading. Long-term fading refers to shadowing of the receiver from the transmitter due to terrain and buildings. The time scale for long-term fading is much longer (on the order of seconds or minutes) than the time scale for short-term fading. It is generally modeled as lognormal. That is the received power (in dB) has a normal (or Gaussian) distribution.

In this section we discuss the received power as a function of distance from the receiver. Suppose we have a transmitter and receiver separated by a distance d. The transmitter and receiver have antennas with gain G_t and G_r respectively. If the transmitted power is P_t the received power is

$$P_r = P_t G_r G_t \left(\frac{\lambda}{4\pi d} \right)^2$$

where $\lambda = c/f$ is the wavelength of the signal. The above equation holds in free space without any reflections or multipath of any sort.

Now consider the case where there is an additional path due to a single reflection from the ground. The multipath has a different phase from the direct path. If we assume the reflection from the ground causes a 180 degree phase change, then for large distances relative to the heights of the antennas the relation between the transmitted power and the received power changes to

$$P_r = P_t G_r G_t \frac{h_1^2 h_2^2}{d^4}$$

where h_1 and h_2 are the heights of the transmitting and receiving antenna. Thus the relation of received power to distance becomes an inverse fourth power law, or equivalently, the power decreases 40 dB per decade of distance. Experimental evidence for a wireless channel shows that the decrease in power with distance is 20 dB per decade near the base station, but as the receiver moves away, the rate of decrease increases. There are other models based on experimental measurements in different cities that give more complicated expressions for the path loss as a function of distance, antenna height, and carrier frequency. See [1] for further details.

Shadowing

In addition to the multipath effect on the channel and the propagation loss there is an effect due to shadowing. If a power measurement at a fixed distance from the transmitter was made there would be local variations due to constructive and destructive interference (discussed previously). At a fixed distance from the transmitter we would also have fluctuations in the received power because of the location of the receiver relative to various obstacles (e.g., buildings). If we measured the power over many locations separated by a distance of a wavelength or more from a given point we would see that this average would vary depending on the location of measurement. Measurements with an obstacle blocking the direct line-of-sight path would have much smaller averages than measurements without the obstacle. These fluctuations due to obstacles are called "shadowing." The fluctuation in amplitude changes much slower than that due to multipath fading. Multipath fading changes as the receiver moves about a wavelength (30 cm for a carrier frequency of 1 GHz) in distance while shadowing causes fluctuations as the receiver moves about 10 m or more in distance.

The model for these fluctuations is typically that of a log-normal distributed random variable for the received power. Equivalently, the power received expressed in dB is a Gaussian distributed random variable with the mean being the value determined by the propagation loss. The variance is dependent on the type of structures where the vehicle is located and varies from about 3 to 6 dB. The fluctuations, however, are correlated. If $v(d)$ is a Gaussian random process modeling the shadowing process (in dB) at some location, then the model for the correlation between the shadowing at distance d_1 and the shadowing at distance d_2 is

$$E\left[v(d_1)v(d_2)\right] = \sigma^2 \exp\left\{-\left|d_1 - d_2\right|/d_0\right\}$$

where d_0 is a parameter that determines how fast the correlation decays with distance. If the velocity is known, then the correlation with time can be determined from the correlation in space. A typical value for d_0 is 10 m. Because shadowing is relatively slow, it can be compensated for by power control algorithms.

Performance with (Time and Frequency) Nonselective Fading

In this section we derive the performance of different modulation techniques with nonselective fading. We will ignore the propagation loss and shadowing effect and concentrate on the effects due only to multipath fading. First the error probability conditioned on a particular fading level is determined. Then the conditional error probability is averaged with respect to the distribution of the fading level.

Coherent Reception, Binary Phase Shift Keying (BPSK)

First consider a modulator transmitting a BPSK signal received with a faded amplitude. The transmitted signal is

$$s(t) = \sqrt{2P} b(t) \cos(2\pi f_c t)$$

where $b(t)$ is a data bit signal consisting of a sequence of rectangular pulses of amplitude +1 or −1. The received signal is

$$r(t) = R\sqrt{2P} b(t) \cos(2\pi f_c t + \phi) + n(t)$$

where $n(t)$ is additive white Gaussian noise with two-side power spectral density $N_0/2$. Assuming the receiver can accurately estimate the phase, the demodulator (matched filter) output at time kT is

$$z_k = R\sqrt{E} b_{k-1} + \eta_k$$

where $E = PT$ is the transmitted energy, b_{k-1} is the data bit transmitted during the time interval $[(k-1)T, kT]$, and η_k is a Gaussian random variable with mean 0 and variance $N_0/2$. The random variable R represents the attenuation due to fading ($R = |X|$) or fading level and has probability density

$$p_R(r) = \begin{cases} 0, & r < 0 \\ \dfrac{r}{\sigma^2} e^{-r^2/2\sigma^2} & r \geq 0 \end{cases}$$

The density function determines the probability that the fading is between any two levels as

$$P\{a < R \leq b\} = \int_a^b p_R(r)\, dr.$$

The error probability for a given fading level R is

$$P_e(R) = Q\left(\sqrt{\frac{2ER^2}{N_0}} \right).$$

The unconditional error probability is the average of the conditional error probability for a given fade level with respect to the density of the fading level.

$$P_e = \int_{r=0}^{\infty} p_R(r) Q\left(\sqrt{\frac{2Er^2}{N_0}} \right) dr$$

$$= \frac{1}{2} - \frac{1}{2} \sqrt{\frac{\overline{E}/N_0}{1 + \overline{E}/N_0}}.$$

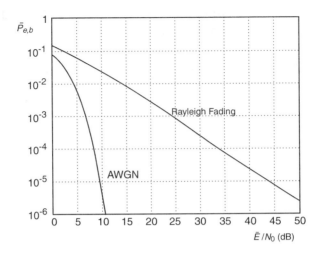

FIGURE 9.26   Bit error probability for BPSK with Rayleigh fading.

The error probability is shown in Fig. 9.26 for the case of no fading (additive white Gaussian noise) and Rayleigh fading. For the additive white Gaussian noise channel the error probability decreases exponentially with signal-to-noise ratio, E/N_0. However, with fading the decrease in error probability is much slower. In fact, for large E/N_0 the error probability is

$$P_e \simeq \frac{1}{4E/N_0}.$$

Thus for high E/N_0, the error probability decreases inverse linearly with signal-to-noise ratio.

To achieve an error probability of 10^{-5} requires a signal-to-noise ratio of 44.0 dB, whereas in additive white Gaussian noise the required signal-to-noise ratio for the same error probability is 9.6 dB. Thus fading causes a loss in signal-to-noise ratio of 34.4 dB. This loss in performance is at the same average received power. The cause of this loss is the fact that the signal amplitude sometimes is very small and causes the error probability to be close to 1/2. Of course, sometimes the signal amplitude is large and results in a very small error probability (say 0). However, when we average the error probability the result is going to be much larger than the error probability at the average signal-to-noise ratio because of the highly nonlinear nature of the error probability as a function of signal amplitude without fading.

While the specific error probabilities change when the modulation changes the general nature of the error probabilities remain the same. That is, without fading the error probability decreases exponentially with signal-to-noise ratio, while with fading the error probability decreases inverse linearly with signal-to-noise ratio. This typically causes a loss in performance of between 30 and 40 db and forces a designer to consider mitigation techniques as will be discussed subsequently.

BPSK with Diversity

To overcome this loss in performance (without just increasing power) a number of techniques are applied. Many of the techniques attempt to receive the same information with independent fading statistics. This is generally called diversity. The diversity could be the form of L different antennas suitably separated so that the fading on different paths from the transmitter are independent. The diversity could be in the form of transmitting the same data L times suitably separated in time so that the fading is independent.

In any case, consider a system with L independent paths. The receiver demodulates each path coherently. Assume that the receiver also knows exactly the faded amplitude on each path. The decision statistics are then given by

$$z_l = r_l \sqrt{E} b + \eta_l, \ \ l = 1, 2, \ldots, L$$

where r_l are Rayleigh η_l is Gaussian, and b represents the data bit transmitted, which is either $+1$ or -1. The optimal method to combine the demodulator outputs can be derived as follows. Let $p_1(z_1 \ldots, z_L | r_1, \ldots, r_L)$ be the conditional density function of $z_1, \ldots, z_L$ given the transmitted bit is $+1$ and the fading amplitude is $r_1, \ldots, r_L$. The unconditional density is

$$p_1\left(z_1, \ldots, z_L, r_1, \ldots, r_L\right) = p_1\left(z_1, \ldots, z_L | r_1, \ldots, r_L\right) p\left(r_1, \ldots, r_L\right)$$

The conditional density of z_1 given $b = 1$ and r_1, is Gaussian with mean $r_1\sqrt{E}$ and variance $N_0/2$. The joint distribution of $z_1, \ldots, z_L$ is the product of the marginal density functions. The optimal combining rule is derived from the ratio

$$\Lambda = \frac{p_1\left(z_1, \ldots, z_L, r_1, \ldots, r_L\right)}{p_{-1}\left(z_1, \ldots, z_L, r_1, \ldots, r_L\right)}$$

$$= \exp\left\{\frac{4}{N_0} \sum_{l=1}^{L} z_l r_l \sqrt{E}\right\}.$$

The optimum decision rule is to compare Λ with 1 to make a decision. Thus the optimal rule is

$$\sum_{l=1}^{L} r_l z_l \mathop{\gtrless}_{b=-1}^{b=+1} 0$$

The error probability with diversity L can be determined using the same technique as used without diversity. The expression for error probability is

$$P_e\left(L\right) = P_e\left(1\right) - \frac{1}{2}\sum_{k=1}^{L-1} \frac{(2k)!}{k!k!}\left(1 - 2P_e\left(1\right)\right)\left(P_e\left(1\right)\right)^k\left(1 - P_e\left(1\right)\right)^k.$$

The error probability as a function of the signal-to-noise ratio is shown in Fig. 9.27. The signal-to-noise ratio in this case is defined as $E_b/N_0 = E * L/N_0$ where E is the energy transmitted per transmitting antenna or time diversity. Thus we assume that LE_b is the energy needed to have the signal received with L independent fading amplitudes. If we had L receiving antennas, then the performance as a function of the transmitting energy would be L times better. In any case, we plot the error probability as a function of the total received energy. In the case of diversity transmission, the energy transmitted per bit E_b is LE. For a fixed E_b, as L increases, each transmission contains less and less energy, but there are more transmissions over independent faded paths. In the limit, as L becomes large using the weak law of large numbers it can be shown that

$$\lim_{L\to\infty} P_e\left(L\right) = Q\left(\sqrt{\frac{2E_b}{N_0}}\right)$$

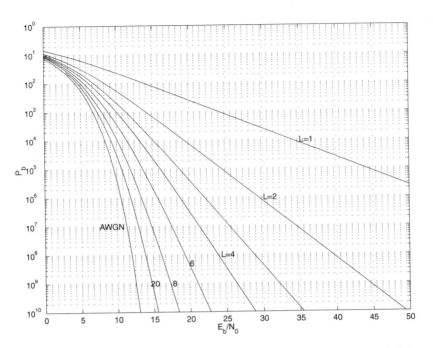

FIGURE 9.27   Error probability for BPSK (coherent demodulation) with and without Rayleigh fading.

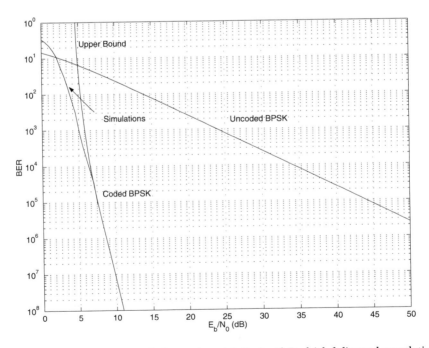

FIGURE 9.28   Error probability for BPSK (coherent demodulation) with Rayleigh fading and convolutional coding.

For large signal-to-noise ratio the error probability with diversity L is decreasing as $1/(E_b/N_0)^L$. While these curves show it is possible to get back to the performance with additive white Gaussian noise by using sufficient resources (diversity), it is possible to do even better with the right coding. In Fig. 9.28 we show the performance of a rate 1/2 constraint length 7 convolutional code on a Rayleigh faded channel (independent fading on each bit) where the receiver knows the fading level (side information) for each bit and can appropriately weight the metric in the decoder. Notice that the required E_b/N_0 for 10^{-5} bit error probability is about 7.5 dB, which is less than that required for uncoded BPSK without fading. The gain compared to uncoded performance is more than 36 dB.

Fundamental Limits

The fundamental limits on performance can be determined for a variety of circumstances. Here we assume that the transmitter has no knowledge of the fading amplitude and assume the modulation in binary phase shift keying. When the receiver knows exactly the amplitude (and phase) of the fading process we say that side information is available. The maximum rate of transmission (in bits/symbol) is called the capacity of the channel C. If an error control code of rate r information bits/channel use is used, then reliable (arbitrarily small error probability) is possible provided the rate is less than the capacity. For the case of side information available this condition is

$$r < C = 1 - \int_{r=0}^{\infty} \int_{y=-\infty}^{\infty} f(r)g(y)\log_2\left(1+e^{-2y\beta}\right)dydr$$

where $f(r) = 2r\exp\{-r^2\}$, $\beta = \sqrt{2\bar{E}/N_0}$ and

$$g(y) = \frac{1}{\sqrt{2\pi}}\exp\left\{-\left(y-\sqrt{2\bar{E}r^2/N_0}\right)^2\Big/2\right\}.$$

If the receiver does not know the fading amplitude (but still does coherent demodulation) then we say no side information is available. The rate at which reliable communication is possible in this case satisfies

$$r < C = 1 - \int_{r=0}^{\infty} \int_{y=-\infty}^{\infty} p(y|1)\log_2\left(1+\frac{p(y|0)}{p(y|1)}\right)dy$$

where

$$p(y|0) = \int_0^{\infty} f(r)\frac{1}{\sqrt{2\pi N_0}}e^{-\left(y-\sqrt{Er}\right)^2/N_0}dr$$

and

$$p(y|1) = \int_0^{\infty} f(r)\frac{1}{\sqrt{2\pi N_0}}e^{-\left(y+\sqrt{Er}\right)^2/N_0}dr$$

If the receiver makes a hard decision about each modulated symbol and the receiver knows the fading amplitude, then the capacity is

$$C = \int_0^{\infty} f(r)\left[1+p(r)\log_2\left(p(r)\right)+\left(1-p(r)\right)\log_2\left(1-p(r)\right)\right]dr$$

where $p(r) = Q\left(\sqrt{\frac{2\bar{E}r^2}{N_0}}\right)$. For a receiver that does not know the fading amplitude and makes hard decisions on each coded bit, the capacity is given by

$$C = 1 + \bar{p}\log_2\left(\bar{p}\right) + \left(1 - \bar{p}\right)\log_2\left(1 - \bar{p}\right)$$

where

$$\bar{p} = \frac{1}{2} - \frac{1}{2}\sqrt{\frac{\bar{E}/N_0}{\bar{E}/N_0 + 1}}.$$

Finally if the transmitter is not restricted to binary phase shift keying but can use any type of modulation, then the capacity when the receiver knows the fading level is

$$C = \int_0^\infty f(r)\frac{1}{2}\log_2\left(1 + 2\bar{E}r^2/N_0\right)dr$$

In Fig. 9.29 we show the minimum signal-to-noise ratio $E_b/N_0 = E/N_0/C$ per information bit required for arbitrarily reliable communication as a function of the code rate ($r = C$) being used. In this figure the top curve (a) is the minimum signal-to-noise ratio necessary for reliable communication with hard decisions and no side information. The second curve (b) is the case of hard decisions with side information. The third curve (c) is the case of soft decisions with side information and binary modulation (BPSK). The bottom curve (d) is the case of unrestricted modulation and side information available at the receiver. There is about a 2 dB gap between hard decisions and soft decisions when side information is available. There is an extra one dB degradation in hard decisions if the receiver does not know the amplitude. A roughly similar degradation in performance is also true for soft decisions with and without side information. The model shown here assumes that the fading is constant over one symbol duration,

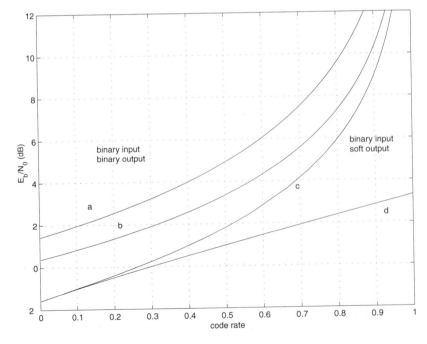

FIGURE 9.29 Capacity of Rayleigh faded channel with coherent detection.

but independent from one symbol to the next. However, for the case of the receiver knowing the fading level (side information) the capacity actually does not depend on the time selectivity as long as the fading is constant for at least one symbol duration. When there is no side information, the capacity gets larger when there is less selectivity. In this case the receiver can better estimate the channel. In fact, as the channel coherence time becomes large, the capacity without side information approaches the capacity with side information.

As can be seen in Fig. 9.29 it is extremely important that some form of encoding be used with fading. The required signal-to-noise ratio for small error probabilities has decreased from on the order of 45 dB for an uncoded system to a mere 2 dB for a coded system. When a repetition code is used, the error probability can be made to decrease exponentially with signal-to-noise ratio provided that we use a large number of antennas or repeat the same symbol a large number of times, which results in a small rate of transmission (information bits/modulated symbol). However, with error control coding such as a convolutional code and independent fading we can greatly improve the performance. The minimum required signal-to-noise ratio is no different than an unfaded channel when very low rate coding is used. For rate 1/2 coding, the loss in performance is less than 2 dB compared to an unfaded channel.

In conclusion, multipath fading causes the signal amplitude to vary and the performance of typical modulation techniques to degrade by tens of dB. However, with the right amount of error control coding, the required signal-to-noise ratio can be decreased to less than 2 dB of the required signal-to-noise ratio for a additive white Gaussian channel when the code rate is 0.5.

Reference

1. Pahlavan, K. and A. H. Levesque, *Wireless Information Networks,* John Wiley & Sons, New York, 1995.

9.5   Electromagnetic Interference (EMI)

Alfy Riddle

Fundamentals of EMI

Electromagnetic interference (EMI) is a potential hazard to all wireless and wired products. Most EMI concerns are due to one piece of equipment unintentionally affecting another piece of equipment, but EMI problems can arise within an instrument as well. Often the term electromagnetic compatibility (EMC) is used to denote the study of EMI effects. The following sections on generation of EMI, shielding of EMI, and probing for EMI will be helpful in both internal product EMI reduction and external product EMI compliance.

EMI compliance is regulated in the U.S. through the Federal Communications Commission (FCC). Specifically, Parts 15 and 18 of the Code of Federal Regulations (CFR) govern radiation standards and standards for industrial, scientific, and medical equipment. In Europe, Publication 22 from the Comite International Special des Perturbations Radioelectriques (CISPR) governs equipment radiation. Although the primary concern is compliance with radiation standards, conduction of unwanted signals onto power lines causes radiation from the long power lines, so conducted EMI specifications are also included in FCC Part 15 and CISPR 22 [1,2].

Figure 9.30 shows the allowed conducted EMI. FCC and CISPR specifications do not set any limits above 30 MHz. All of the conducted measurements are to be done with a line impedance stabilization network (LISN) connected in the line. The LISN converts current-based EMI to a measurable voltage. The LISN uses series inductors of 50 μH to build up a voltage from line current interference, and 0.1 μF capacitors couple the noise voltage to 50 ohm resistors for measurement [1]. Capacitors of 1 μF also bridge the output so the inductors see an AC short. The measurements in Fig. 9.30 are reported in dBμV, which is dB with respect to 1 μV. Both FCC and CISPR measurements specify an RF bandwidth of at least 100 kHz. The CISPR limitations given in Fig. 9.30 denote that a quasi-peak (QP) detector should

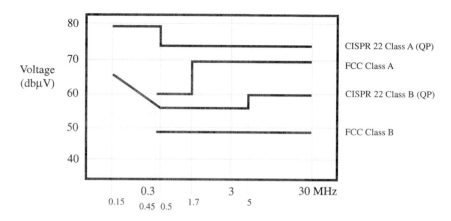

FIGURE 9.30 Conducted EMI specifications, measured with LISN.

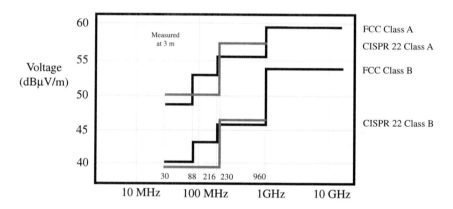

FIGURE 9.31 Radiated EMI specifications referred to 3 m.

be used. The QP detector is more indicative of human responses to interference. CISPR specifications for an averaging detector are 10 dB below that of the QP detector. FCC specifications require a QP detector. Both FCC and CISPR limitations have two classes. Class A is basically for industrial use, while Class B is for residential use.

Both CISPR and FCC radiated EMI specifications begin at 30 MHz. Fig. 9.31 shows the radiation limits for CISPR and FCC Classes A and B [1,2]. Because these measurements are made with an antenna, they are specified as a field strength in dB referenced to 1 µV/m. The measurement distances for radiation limits varies in the specifications, but all of the limits shown in Fig. 9.31 are referred to 3 m. Other distances can be derived by reducing the limits by 20 dB for every factor of ten increase in distance.

Generation of EMI

Almost any component can generate EMI. Oscillators, digital switching circuits, switching regulators, and fiber-optic transmitters can radiate through PCB traces, inductors, gaps in metal boxes, ground loops, and gaps in ground planes [1].

Switching Regulators

Because switching regulators have a fundamental frequency below 30 MHz and switch large currents at high speed, they can contribute to both conducted and radiated emissions. Very careful filtering is required so switching regulators do not contaminate ground planes with noise. The input filters on switching

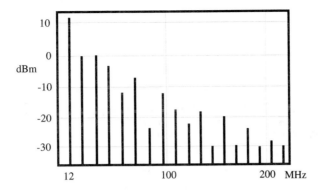

FIGURE 9.32 Harmonic clock spectra.

regulators are just as important as the output filters because the noise can travel to other power supplies and out to the line [3].

Digital Switching

Digital networks have several characteristics that increase EMI. Digital networks tend to have many lines switching simultaneously. The currents on the lines add in phase and increase radiation. Also, as CMOS digital circuits increase in clock frequency they require more current to drive their loads. Increasing both the current and the frequency creates more di/dt noise through the inductive connection from ICs to ground [4]. The fast switching of digital waveforms produces harmonics decades beyond the oscillator fundamental frequency. Fig. 9.32 shows the spectra from a typical 12 MHz crystal square-wave clock oscillator. Note that even though the output appears as a square wave, both even and odd harmonics are present. The harmonics of this oscillator fall off at roughly 20 dB/decade. As will be seen in the next section, most sources of coupling increase at about 20 dB/decade, which causes not only a relatively flat coupling spectrum, but significant EMI at ten or even 100 times the oscillator frequency. While this 12 MHz clock oscillator is at a very low frequency, similar phenomena happens with the laser drivers for high speed fiber-optic networks that operate with clocks of 2.5 GHz and higher.

Coupling

Any inductor is a potential source of coupling and radiation. Ribbon cables act like very long coupling loops and can spread signals or power supply noise all over an instrument and out to the outside world. Fundamentally, lengths of wire radiate an electric field and loops of wire radiate a magnetic field. The electric field, E_{Far}, far from a short radiator, is given by Eq. (9.138) [1]. Eq. (9.138) is in volts per meter where I is the element current, l is the element length, λ is the radiation wavelength, and r is the distance from the radiating element to the measured field. Because the field strength is inversely proportional to frequency, EMI coupling tends to increase with frequency. The far magnetic field, H_{Far}, due to a current loop is given in A/m by Eq. (9.139). In Eq. (9.139) a is the radius of the loop, c is the speed of light, I is the current in the loop, and r is the distance from the loop to the measured radiation.

$$E_{Far} = 377 \, I \, l / (2\lambda r) \qquad (9.138)$$

$$H_{Far} = \omega^2 \, a^2 \, \mu \, I / (1508 \, c \, r) \qquad (9.139)$$

Cabling

Cables form a source of radiation and susceptibility. In general it is the signal on the shield of a coaxial cable or the common-mode signal on twisted pairs that generates most of the radiation [1]. However, even high-quality coaxial cables will leak a finite amount of signal through their shields. Multiple braids, solid outer conductors, and even solder-filled braid coaxial cables are used to increase shielding. Twisted pair cables rely on twisting to cancel far field radiation from the differential mode. The effectiveness of the twisting reduces as frequency increases [1].

Shielding

Shielding is the basic tool for EMC work. Shielding keeps unwanted signals out and potential EMI sources contained. For the most part, a heavy metal box with no seams or apertures is the most effective shield. While thin aluminum enclosures with copper tape over the seams appear to enclose the RF currents, in fact they are a poor substitute for a heavy gauge cast box with an EMI gasket. It has been said that if spectrum analyzer manufacturers could make their instruments any lighter, primarily by leaving out some of the expensive casting, they would. The basic equation for shielding is given in Eq. (9.140) [5].

$$S = A + R + B \qquad\qquad (9.140)$$

In Eq. (9.140), A is the shield absorption in dB, R is the shield reflection in dB, and B is a correction factor for multiple reflections within the shield [5]. Shield effectiveness depends on the nature of the field. Purely electric fields are well isolated by thin conductive layers while purely magnetic fields are barely attenuated by such layers. Magnetic fields require thick layers of high permeability material for effective shielding at low frequencies. Plane waves contain a fairly high impedance mix of electric and magnetic fields that are both reflected and absorbed by thin metal layers provided the frequency is high enough. One of the subtle points in EMI shielding is that any slot can destroy shielding effectiveness. It is the length of a slot in comparison to a wavelength that determines how easily a wave can pass through the slot [5].

Measurement of EMI

EMI compliance must be verified. Unfortunately, EMI measurements are time consuming, tedious, plagued by local interference sources, and often of frustrating variability. The FCC requires measurements to be verified at an Open Area Test Site (OATS) [2]. Many manufactures use a local site or a shielded TEM cell to estimate FCC compliance during product development. With care, OATS and TEM cell measurements can be correlated [6]. In any case, even careful EMI measurements can wander by several dB so most manufacturers design for a healthy margin in their products.

Open Area Test Site (OATS)

A sketch of an OATS site is given in Fig. 9.33a. OATS testing involves setting the antenna at specified distances from the device under test (DUT). FCC and CISPR regulations use distances of 3,10, and

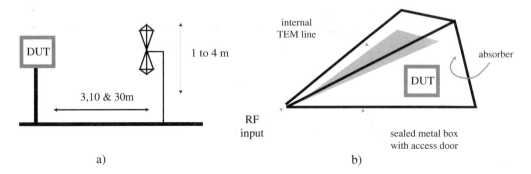

FIGURE 9.33   EMI measurements methods: a) OATS; and b) TEM cell.

30 meters depending on the verification class [1,2]. The antenna height must also be varied to account for ground reflections. Finally, the DUT must be rotated about all axes and the antenna must be utilized to test the DUT under radiation by both horizontally and vertically polarized fields.

TEM Cell

TEM cells are a convenient and relatively low cost method for making accurate and well-isolated EMI measurements [2,7]. A sketch of one configuration of TEM cell is shown in Fig. 9.33b [2]. The TEM cell uses a transmission line in a box to create a TEM field for testing devices. The box is driven from a narrow end and usually expands into an area where the DUT can be placed. The box terminates in a resistor surrounded by RF absorber. For EMI measurements the DUT must be placed away from the box walls and rotated about each axis so that all possible radiated waves can be measured.

Probes

EMI probes can be a very effective way of solving EMI problems. Articles have been written on building probes and commercial probes are available that provide a flat frequency response [8]. These probes can be used to "sniff" around a device until signals with the same spectral spacing as the EMI problem can be found. For clocks and switching power supplies, a close examination of the spectral spacing will indicate the fundamental frequency, which can be traced to a component on a schematic. In the time domain the suspected EMI source can be used to trigger an oscilloscope with the probed signal as the oscilloscope input. If the trace is stable, then the suspected EMI source has been found. Electric field probes are based on Eq. (9.138) and can be as simple as a wire extending from a connector. Magnetic field probes can be as simple as a loop of wire completing the path from a connector's center pin to its flange. A rectangular loop of length l with the near side a distance a from a current source I, and having the far side a distance b from the current source will yield the voltage given in Eq. (9.141). With all small probes it is useful to have at least a 6 dB pad after the probe to establish a load and minimize reflections.

$$V = j \, \omega \, l \, \mu \, I \ln(b/a) / (2 \, \pi) \tag{9.141}$$

Summary

EMI problems create an inexhaustible supply of work for those in the field. The EMC field has well documented requirements and a long history of measurement. Many excellent sources of information exist for those working in this area [1,2,5,9].

References

1. Paul, C.R., *Introduction to Electromagnetic Compatibility*, John Wiley & Sons, New York, 1992.
2. Morgan, D., *A Handbook for EMC Testing and Measurement*, Peter Peregrinus Ltd., London, 1994.
3. Lee, F.C. and Yu, Y., Input-Filter Design for Switching Regulators, *IEEE Trans. Aerosp. and Electron. Sys.*, 627–634, September 1979.
4. Dolle, M., Analysis of Simultaneous Switching Noise, *IEEE ISCAS*, 904–907, 1995.
5. Ott, H.W., *Noise Reduction Techniques in Electronic Systems*, John Wiley & Sons, New York, 1976.
6. Wilson, P., On Correlating TEM Cell and OATS Emission Measurements, *IEEE Trans. EMC*, 1–16, February 1995.
7. Konigstein, D., and Hansen, D., A New Family of TEM-Cells with Enlarged bandwidth and Optimized Working Volume, *Proc. 7th Int'l. Zurich Symp. on EMC*, 127–132, March 1987.
8. Johnson, F., Simple "Homemade" Sensors Solve Tough EMI Problems, *Electron. Design*, 109–114, November 8, 1999.
9. Kodali, V.P. and Kanda, M., *EMC/EMI Selected Readings*, IEEE Press, Piscataway, NJ, 1996.

9.6   Material Properties

9.6.1   Metals

Mike Golio

Metals serve several different functions in the realization of RF and microwave products. These functions include:

- The wire or guided wave boundary material for circuits and transmission media.
- The carrier or structural support for dielectric substrates or semiconductor chips.
- The heat sink for devices or circuits that exhibit high power density.
- The reflector element for antennas or screen room applications.

Each of these functions imposes different electrical, thermal, chemical, and mechanical requirements on the metal material selection. Thus the optimum metal for each application will vary. Consideration of a wide range of material properties for each metal is needed to choose an appropriate metal for most applications.

Resistance, Resistivity, and Conductivity

A first-order consideration in the choice of metals for many electrical applications is the electrical resistance of the metal conductor. DC resistance of a metal rod is given by

$$R = \frac{\rho L}{A} \qquad (9.142)$$

where R is the resistance of the rod, ρ is the resistivity of the metal, L is the length of the rod, and A is the cross-sectional area of the rod. The DC electrical properties of metals are also sometimes discussed in terms of conductivity. Conductivity is the inverse of resistivity given by

$$\sigma = \frac{1}{\rho} \qquad (9.143)$$

where σ is the conductivity of the material. For most applications, high conductivity, or conversely, low resistivity, is desirable. The resistivity of a number of metals is listed in Table 9.7.

TABLE 9.7   Electrical Resistivity at Room
Temperature of Several Metals in 10^{-8} Ω m

Metal	Electrical Resistivity
Aluminum	2.7
Beryllium	34.0
Chromium	12.6
Copper	1.7
Gold	2.3
Lead	21.1
Magnesium	4.5
Manganese	144
Molybdenum	5.5
Nickel	7.1
Palladium	10.7
Platinum	10.7
Silver	1.6

TABLE 9.8 Temperature Coefficient of Metal
Resistivity at Room Temperature in (1/K)

Metal	Temp. Coefficient of Resistivity
Aluminum	4.38×10^{-3}
Copper	3.92×10^{-3}
Silver	3.71×10^{-3}
Gold	3.61×10^{-3}

The resistivity of metal is also a function of temperature. For small perturbations in temperature, this temperature dependence may be characterized by the equation

$$\alpha = \frac{1}{\rho}\frac{\partial\rho}{\partial T} \tag{9.144}$$

where α is the temperature coefficient of resistivity. The measured temperature dependence of resistivity changes as a function of the nominal temperature and can vary significantly at temperatures near 0 K, or well above room temperature. For most applications, a low value for the temperature coefficient of resistivity is desirable. Table 9.8 presents the temperature coefficient of resistivity at room temperature for several metals that might be chosen for their low resistivity values.

Skin Depth

An electromagnetic field can penetrate into a conductor only a minute distance at microwave frequencies. The field amplitude decays exponentially from its surface value according to

$$A = e^{-x/\delta_s} \tag{9.145}$$

where x is the normal distance into the conductor measured from the surface, and δ_s is the skin depth. The skin depth or depth of penetration into a metal is defined as the distance the wave must travel in order to decay by an amount equal to $e^{-1} = 0.368$ or 8.686 dB. The skin depth δ_s is given by

$$\delta_s = \frac{1}{\sqrt{\pi f \mu \sigma}} \tag{9.146}$$

where f is the frequency, σ is the metal conductivity, and μ is the permeability of the metal given as

$$\mu = \mu_o \mu_r \tag{9.147}$$

with μ_o equal to the permeability of free space and μ_r the relative permeability of the metal. For most metals used as conductors for microwave and RF applications, the relative permeability, $\mu_r = 1$. The relative permeability of ferroelectric materials such as iron and steel are typically on the order of several hundred.

Skin depth is closely related to the shielding effectiveness of a metal since the attenuation of electric field strength into a metal can be expressed as in Eq. (9.145). For static or low frequency fields, the only method of shielding a space is by surrounding it with a high-permeability material. For RF and microwave frequencies, however, a thin sheet or screen of metal serves as an effective shield from electric fields.

Skin depth can be an important consideration in the development of guided wave and reflecting structures for high frequency work. For the best conductors, skin depth is on the order of microns for 1 GHz fields. Since electric fields cannot penetrate very deeply into a conductor, all current is concentrated

TABLE 9.9 Thermal Conductivities of Typical Metals (W/m K) at Room Temperature

Metal	Thermal Conductivity
Silver	419
Copper	395
Gold	298
Aluminum	156
Brass	101
Lead	32
Kovar	17

near the surface. As conductivity or frequency approach infinity, skin depth approaches zero and the current is contained in a narrower and narrower region. For this reason, only the properties of the surface metal affect RF or microwave resistance. A poor conductor with a thin layer of high conductivity metal will exhibit the same RF conduction properties as a solid, high conductivity structure.

Heat Conduction

One of the uses of metal in the development of RF and microwave parts and modules is as a heat spreader. For many applications that involve high power density electronic components, the efficient removal of heat is of great importance in order to preserve component reliability. Section 3.1 discusses heat transfer fundamentals and Section 3.8 discusses hardware reliability in greater detail.

The one-dimensional heat flow equation that applies to metals (as well as other media) is given as

$$q = kA \frac{\partial T}{\partial x} \tag{9.148}$$

where q is the heat flow, k is the thermal conductivity of the metal, A is the cross-sectional area for heat flow, and $\frac{\partial T}{\partial x}$ the temperature gradient across the metal. For applications where good heat sinking characteristics are desired, high thermal conductivity, k, is desirable. Table 9.9 lists thermal conductivity values for several metals.

Temperature Expansion

Because RF and microwave components often must operate over a wide range of temperatures, consideration of the coefficient of linear expansion of the metals must be included in making a metal selection for many applications. When temperature is increased, the average distance between atoms increases. This leads to an expansion of the whole solid body with increasing temperature. The changes to the linear dimension of the metal can be characterized by

$$\beta = \frac{1}{l} \frac{\Delta l}{\Delta T} \tag{9.149}$$

where β is called the coefficient of linear expansion, l is the linear dimension of the material, ΔT is the change in temperature, and Δl is the change in linear dimension arising from the change in temperature.

Linear expansion properties of metals are important whenever metallic structures are bonded to other materials in an electronic assembly. When two materials with dissimilar thermal expansion characteristics are bonded together, significant stress is experienced during temperature excursions. This stress can result in one or both of the materials breaking or cracking and this can result in degraded electrical performance or catastrophic failure. The best choice of metals to match thermal linear expansion properties is, therefore, determined by the thermal coefficient of linear expansion of the material that is used with the metal. Kovar, for example, is often chosen as the metal material of preference for use as a carrier when

TABLE 9.10 Thermal Coefficient of Linear Expansion of Some of the Materials Used in Microwave and RF Packaging Applications (at Room Temperature, in 10^{-6}/K)

Material	Thermal Coefficient of Expansion
Dielectrics	
Aluminum nitride	4
Alumina 96%	6
Beryllia	6.5
Diamond	1
Glass-ceramic	4–8
Quartz (fused)	0.54
Metals	
Aluminum	23
Beryllium	12
Copper	16.5
Gold	14.2
Kovar	5.2
Molybdenum	5.2
Nickel	13.3
Platinum	9
Silver	18.9
Semiconductors	
GaAs	5.9
Silicon	2.6
Silicon Carbide	2.2

alumina dielectric substrates are used to fabricate RF or microwave guided wave elements. Although Kovar is neither a superior electrical conductor nor a superior thermal conductor, its coefficient of linear expansion is a close match to that of the dielectric material, alumina. Table 9.10 presents the coefficients of linear expansion for several metals as well as other materials that are often used for RF and microwave circuits.

Chemical Properties

The chemical properties of metals can be especially important in the selection of metals to be used for semiconductor device contacts and in integrated circuits. Metal is used extensively in the development of transistors and ICs. Uses include:

- the contact material to establish ohmic and rectifying junctions,
- the interconnect layers, and
- the material used to fabricate passive components such as inductors and transmission line segments.

For these applications, the chemical properties of the metal when exposed to heat and in contact with the semiconductor material play a significant role in the metal selection criteria. The process of fabricating a semiconductor device often involves hundreds of individual process steps and exposure to significant thermal cycling. The temperature ranges associated with device fabrication will far exceed the environment the final device will be exposed to.

For silicon processes, for example, aluminum (or its alloys) is often the metal chosen for contacts and interconnects. Aluminum has high conductivity, but it is also chosen because it adheres well to silicon and silicon dioxide and because it does not interact significantly with silicon during the thermal cycling associated with processing. In contrast, gold also has high conductivity, but its use is typically avoided in silicon fabrication facilities because gold forms deep levels (traps) in silicon that dramatically degrade device performance. Other metals of specific interest in silicon processing include gallium and antiminide, which are often used in the formation of ohmic contacts.

Because interconnects continue to shrink in size as fabricated devices continue to be scaled down, interconnect resistance has begun to pose significant limitations on the levels of integration that can be achieved. One solution to extend these limits is to use copper rather than aluminum for interconnect metal. Copper's higher conductivity translates directly into improved interconnect performance. Rapid progress is being made in this area.

The formation of good ohmic contacts and Schottky barriers is critical to the fabrication of most GaAs devices. Different metals react chemically in distinct ways when exposed to a GaAs surface and high temperature. Metals such as gold, tin, and zinc tend to form ohmic contacts when placed in contact with a GaAs surface. In contrast, aluminum, titanium, and nickel normally form Schottky barriers on GaAs. In order to obtain optimum electrical conductivity and still produce good contacts, sandwiched layers of different metals are sometimes used. For example, a thick layer of gold is often utilized over layers of titanium and platinum to produce Schottky barrier contacts. When this technique is employed, the titanium resting on the surface of the GaAs forms the Schottky barrier, the platinum serves as a diffusion barrier to keep the gold and titanium from diffusing together, and the gold is used to produce a low resistance connection to the contact pads or remaining IC circuitry.

Certain metals can also react with GaAs to produce undesirable effects. Chromium, for example, produces undesirable deep levels in GaAs that can degrade device performance.

TABLE 9.11 Density of Several Metals in g/cm³

Metal	Density
Aluminum	2.7
Beryllium	1.85
Copper	8.93
Gold	19.4
Kovar	7.7
Molybdenum	10.2
Nickel	8.9
Platinum	21.45
Silver	10.5

Weight

Over the past several decades a dominant trend in the development of electronic circuits has been the continued reduction of size and weight. Although much of this progress has been made possible by the continued scaling of semiconductor devices and ICs, metal portions of many electronic assemblies still dominate the weight of the system. Metal density can be an important factor in choosing metals for certain applications. Table 9.11 presents the density in g/cm³ for several metals of interest.

References

Halliday, D. and Resnick, R., *Fundamentals of Physics*, John Wiley & Sons, New York, 1970.
Schroder, D. K., *Semiconductor Material and Device Characterization*, John Wiley & Sons, New York, 1990.
Plonus, M. A., *Applied Electromagnetics*, McGraw-Hill, New York, 1978.
Elliott, D. J., *Integrated Circuit Fabrication Technology*, McGraw-Hill, New York, 1982.
Collin, R. E., *Foundations for Microwave Engineering*, McGraw-Hill, New York, 1966.
Smith, A. A., *Radio Frequency Principles and Applications*, IEEE Press, New York, 1998.

9.6.2 Dielectrics

K.F. Etzold

Basic Properties

When electrical circuits are built, the physical environment consists of conductors separated from each other by insulating materials. Depending on the application, the properties of the insulating materials are chosen to satisfy requirements of low conductivity, a high or low dielectric permittivity, and desirable loss properties. In most cases the high field properties of the materials do not play a role (certainly not in the usual signal processing applications). This is also true with respect to electrical breakdown. However, there are exceptions. For instance, in transmitters the fields can be very large and the choice of insulators has to reflect those conditions. In semiconductor devices dielectric breakdown can be a problem that must be considered, but in many cases it is actually tunneling through the dielectric that is responsible for charge transport. In DRAM the leakage currents in the memory storage capacitor must be small enough to give a charge half-life of about one second over the full temperature range. In Flash memories the charge is transferred via tunneling to the controlling site in the FET where it must then remain resident for years. This requires a very good insulator. In most applications it is desirable that the material properties be independent of the strength of the applied field. This is not always true and there are now materials becoming available that can be used as transmission line phase shifters for steerable antennas, taking advantage of the field dependent dielectric permittivity.

Probably the most important property of dielectrics in circuit applications is the dielectric permittivity, often colloquially called the dielectric constant. This quantity relates the electric field in the material to the free charge on the surfaces. Superficially the connection between this property and the application of dielectrics in a circuit seems rather tenuous. Consider now, however, the capacity C of a device

$$C = \frac{q}{V} \tag{9.150}$$

where q is the charge on the plates and V the voltage across the device. Usually in a discrete capacitor (understood here as a circuit element), we would like store as much charge as possible in a given volume. Therefore, if we can somehow double the stored charge, the capacity C will double. What allows us to do this is the choice of the material that fills the device. Thus the material in this example has twice the charge storage capability or dielectric permittivity. This quantity is typically labeled by the Greek letter epsilon (ε).

Consider, as an example, a parallel plate capacitor. This simple geometry is easily analyzed, and in general we can calculate the capacity of a device exactly if it is possible to calculate the electric field. For the parallel plate device the capacity is

$$C = \varepsilon^* \frac{A}{d}, \tag{9.151}$$

where A is the area, and d the separation between the plates. If the material between the plates is vacuum, $\varepsilon = 8.85*10^{-12}$ F/m, or in units that are a little easier to remember, 8.85 pF/m. The unit of capacity is the Farad. Using Eq. (9.151) a capacitor with 1m x 1m plates and with a separation of 1m has a capacity of 8.85 pF, neglecting edge effects, i.e., assuming parallel field lines between the plates, which clearly will not be the case here unless a special, guarded geometry is chosen (to guard a circuit, conductors at the same potential are placed nearby). There are no field lines between equipotentials and thus there are no contributions to the capacity. We can consider Eq. (9.151) for the capacity as a definition of dielectric permittivity. In general, the capacity is proportional to ε and the general expression for the capacity of a device is

$$C = \varepsilon^* \text{ geometry factor.} \tag{9.152}$$

Also, for practical applications the dielectric permittivity is usually specified relative to that of a vacuum. Thus,

$$\varepsilon = \varepsilon_0^* k. \tag{9.153}$$

Here ε_0 is redefined as the dielectric permittivity of vacuum (8.85 pF/m) and k is the relative dielectric permittivity, most often referred to as simply, but somewhat inaccurately, the dielectric permittivity or the dielectric constant; k of course is dimensionless. As an example, the dielectric permittivity of typical Glass-Epoxy circuit board is 4.5, where this is understood to be the relative dielectric permittivity. On physical grounds all materials have a dielectric permittivity greater than 1. Also, for the same reason, the dielectric permittivity of a vacuum has no frequency dependence. In Table 9.12 the important properties of some typical engineering insulating materials are given.

TABLE 9.12   Properties of Some Typical Engineering Insulating Materials

Material	k	Loss	Frequency	Resistivity
Vacuum	1.00	0	All	Zero
Air	1.0006	0		
Glass Vycor 7910	3.8	$9.1*10^{-4}$		
Glass Corning 0080	6.75	$5.8*10^{-2}$		
Al_2O_3	8.5	10^{-3}	1 MHz	
Teflon (PTFE)	2.0	$2*10^{-4}$	1 MHz	10^{17}
Arlon 25N circuit board	3.28	$2.5*10^{-3}$	1 MHz	
Epoxy-glass circuit board	4.5			
Beryllium oxide	7.35			
Diamond	5.58			10^{16}
PZT (lead zirconium oxide)	~1000			
Undoped silicon	11.8			
TaO_5	28			
Quartz (SiO_2)	3.75–4.1	$2*10^{-4}$		
Mica (Ruby)	6.5–8.7	$3.5*10^{-4}$		
Water	78.2	0.04	1 MHz	

Note that for crystalline materials the dielectric permittivity is not isotropic, i.e., the permittivity depends on the orientation of the crystal axes relative to the electric field direction. The angular variation is directly related to the spatial symmetry of the crystal.

So far, we have treated dielectrics as essentially perfect and lossless insulators that allow energy storage greater than that in a vacuum. However, as can be seen from the fact that the table has a loss factor column, frequency-dependent energy losses will occur in all materials (except, again, in a vacuum). As could be anticipated, there are a number of loss mechanisms. Perhaps the easiest loss mechanism to deal with conceptually is the low frequency conductivity of the material. If there is a current flow as a result of an applied field, energy will be dissipated. The DC conductivity can be due to the basic properties of the materials, such as the presence of impurities and their associated energy levels. It can also be due to crystalline defects, and conduction can take place along the surfaces of crystallites that constitute many materials (for example, ceramics). Most often, suitable materials can be selected based on desired properties. In Table 9.12 the conductivity for selected materials is presented. It is seen that some natural and man-made materials have an extremely low conductivity. Indeed the time constant for loss of charge for a free-floating (disconnected) device can be weeks. The time constant τ (in seconds) of a RC network is defined as

$$\tau = RC \qquad (9.154)$$

where R is the resistance in parallel with the capacitor C (R is in Ohms and C in Farads). Consider Teflon (PTFE), for example. The DC resistivity at room temperature (25°C) is 10^{17} ohm cm.

Therefore, for this material, the time constant for the charge to decay to 1/e of its original value is 23 days using the dielectric permittivity of 2.0 from Table 9.12. This demonstrates that insulating materials exist for the cabling and supporting structures that are suitable for applications where only the smallest leakage can be tolerated. An example would be systems that measure very small charges, such as quartz-based force transducers and accelerometers.

Frequency Dependence of the Properties

We need now to consider the behavior at all user frequencies. These extend from DC into the microwave and optical range. In this range a number of different loss mechanisms have to be considered. But before this can be done, the definition of the dielectric permittivity needs to be extended so that AC effects can be properly considered. The losses are accounted for if an imaginary component is added to the definition of the dielectric permittivity. Thus

$$\varepsilon = \varepsilon' - j^*\varepsilon'' \tag{9.155}$$

where ε' is the real part of the dielectric permittivity and j is the square root of -1. The existence of losses is now explicitly included by the imaginary quantity ε''. It is also obvious that the term "permittivity" as a property of dielectrics is somewhat of a misnomer and is partly historical. In fact both the real and imaginary parts are functions of frequency. As is often the case in describing physical parameters that are complex (in the mathematical sense) the two quantities in Eq. (9.155) are actually related. This is a consequence of the fact that the electromagnetic behavior is governed by Maxwell's equations, the absence of discontinuities, and also that causality holds, a statement that relates to the time evolution of electric fields. Thus the so-called Kramers-Kronig equations relate the two quantities [1]

$$\varepsilon'(\omega) = \frac{1}{\pi^*} \int \frac{\varepsilon''}{x - \omega} dx;$$

$$\varepsilon''(\omega) = \frac{1}{\pi^*} \int \frac{\varepsilon'}{x - \omega} dx; \tag{9.156}$$

$$-\infty < \lim its < +\infty$$

$$\omega = 2\pi \text{frequency}$$

where the dependence of the real and imaginary parts of ε on frequency is now explicitly stated and where the integrals are the Cauchy principal values. Thus we only need to know the real or imaginary parts and the corresponding complimentary part can be calculated. For instance, if we know the real part of the dielectric response over the full frequency range we can calculate the losses by integration. Even if our knowledge of the real part (capacity) is only over a limited frequency range, the complementary part (loss) can often still be estimated, albeit with reduced accuracy.

Let us now discuss the behavior of dielectrics between the frequency extremes of DC and the optical range. At optical frequencies there is a relation from Maxwell's equations that relates the optical index of refraction to the dielectric and magnetic properties. Most dielectrics are nonmagnetic so the magnetic susceptibility is just that of free space, μ_0. The optical index of refraction n is given by

$$n = \sqrt{\frac{\varepsilon}{\mu_0}}. \tag{9.157}$$

This is of interest because the optical index for all materials varies only over a fairly narrow range, which also relates to the dielectric properties at high (optical) frequencies. Vacuum has an index of one. Many insulators have an index near two, which means that the dielectric permittivity is around four. High optical indices are rare and one of the highest ones known is that of diamond, which has an index of 2.42 (hence the sparkle) and therefore a calculated dielectric permittivity of 5.81. The measured value is 5.58 at low frequencies. Thus there is a slight inconsistency between these values and one would expect a value equal to or slightly higher than 5.58.

Consider now the other end of the frequency range. At low frequencies (Fig. 9.34, up to about 1 GHz) the dielectric permittivity for different materials has a much wider range. Here a small diversion is necessary. We need to address the question of what gives rise to dielectric constants greater than one. The attribute that allows additional charge to be placed on the plates of a capacitor is the (bulk) polarizability of the material or the ability to deposit charge on interfaces associated with the contacts or grain boundaries. Note that except for amorphous ones, most materials are made up of aggregates of small crystallites. The solid then consists of adjacent crystallites separated by a "grain boundary." The properties of the grain boundary can be vastly different from the properties within a crystallite.

The polarizability can take various forms. It can be in the form of permanent dipoles that are observed, for example, in water or PZT. For many materials there is no permanent polarization but there is an

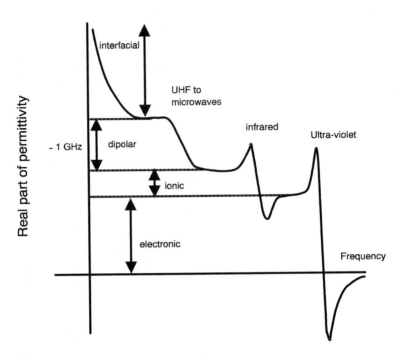

FIGURE 9.34 Adapted from Kittel, C., *Introduction to Solid State Physics,* 3rd ed., John Wiley & Sons, New York, 1967.

induced polarization caused by an externally applied field. The latter materials typically have relatively low dielectric constants, up to about 20. Materials that have permanent dipoles can have very high dielectric constants (up to about 20,000 and higher) especially in the presence of cooperative effects. These same cooperative effects imply the presence of a structural phase transition typically in a temperature range of somewhat below room temperature up to several hundred ºC depending on the material. The inherent (molecular) polarization does not disappear above the phase transition where the materials change from ferroelectric to paraelectric, but the long-range order (domains) that prevails at temperatures below the phase transition temperature (the Curie temperature) disappears. This kind of behavior is typical of PZT and many other perovskites. (A perovskite is a crystal with a specific crystal geometry. Many of the materials with this kind of geometry have recently become technologically important; examples are PZT and $BaTiO_3$). Water is an example of a material whose molecules are polarized permanently but which does not have a ferroelectric phase, a phase in which the dipolar vector for a large number of molecules point in the same direction.

Let us now describe the trend of the dielectric permittivity over the entire frequency range. As discussed earlier, at very high (optical) frequencies the dielectric permittivity tends to have its lowest value (up to ~4) and increases as the frequency decreases toward DC. What connects these end points? Where do the changes take place and do they need to be considered in the RF and microwave regime? The general trend of the dielectric permittivity as a function of frequency is shown in Fig. 9.34. As can be seen, there are several regimes in which the dielectric permittivity is changing. At each transition the interaction mechanism between the field and the material is different.

Consider the low frequency extreme in Fig. 9.34. The low frequency rise in the permittivity is attributable to interface effects. These take place at the contacts or in the bulk of the material at the grain boundaries. Charge is able to accumulate at these interfaces and thereby able to contribute to the total capacity or dielectric permittivity. As the frequency is raised, eventually the system can no longer follow these time-dependent charge fluctuations and the dielectric permittivity settles to a value equal to the one without the interface charges. As a widely discussed phenomenon (papers on ferroelectrics or high permittivity thin films), surface and interface effects can take place in ferroelectric (for example PZT)

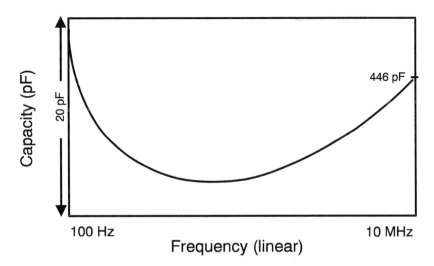

FIGURE 9.35

and non-ferroelectric materials such as $BaTiO_3$ or glassy amorphous materials. Barium titanates are the materials for chip capacitors.

A manifestation of this low frequency effect is memory or dielectric relaxation that is attributable to the same interface behavior. This relaxation can be readily observed if a capacitor is charged or discharged over long periods of time. Consider a discharged capacitor that is suddenly connected to a steady-state source such as a DC power supply or battery. If we measure the current in this circuit after a long time we will not find a zero current, as we would expect after the capacitor is fully charged. Rather, a slowly decreasing current is observed in the circuit. This current is not due to DC current leakage or a current variation associated with the RC time permittivity of the capacitor and the charging circuit. A leakage current would eventually become unvarying and the time constant of the charging RC circuit is much shorter than the observed decay. It is also observed that the capacitor returns some of this low charging current. If the power supply is suddenly removed and the leads of the capacitor are connected to the current meter we see the initial ordinary discharge from the capacitor. After this discharge is completed, a small, slowly decaying current in the opposite direction from the charging current is observed. It is also, of course, opposite to the residual charging current described above. These relaxation currents are due to charge that slowly accumulates on the various interior interfaces in the device. An example of areas where charge often becomes resident is the interface between the contacts and the dielectric. Of course the charge is returned when the capacitor is discharged. By way of contrast if there is leakage in a device there is no stored charge.

As a concrete example, consider the variation of the capacity of a commercial 470 pF chip capacitor as a function of frequency, shown in Fig. 9.35. We observe the increase of the capacity due to interface effects as the frequency decreases. However, somewhat unexpectedly, the response does not flatten or drop at high frequencies, as Fig. 9.34 implies. Instead the capacity goes through a minimum and then rises again. This rise is due to a parasitic effect, a resonance in the capacitor and the test jig. The unavoidable resonant elements are the lead inductance and the inherent capacity of the device. All capacitors exhibit this phenomenon. Indeed above the resonance the impedance can become positive and the device behaves like an inductor! In bypass applications (i.e., larger value capacitors) this usually requires paralleling capacitors of different values (and therefore different resonance frequencies) to obtain a composite capacitor that has a negative reactance over a wider frequency range. Manufacturers of chip capacitors have also developed devices with a particularly low inductance. In any case, the rise in the response toward high frequencies in Fig. 9.35 is not a material property, but rather is a consequence of the physical construction of the device. Therefore, the low frequency behavior of this chip capacitor is comparable to that shown in Fig. 9.34, modified for the resonance. As can be seen for the 470 pF capacitor, the interface effect can be neglected at frequencies above 2 to 3 MHz.

Returning now to the material dependent dielectric behavior as a function of frequency. As the frequency is increased, the next polarizability is associated with the material rather than interfaces. Perovskites have oxygen vacancies that will exhibit a dipolar polarization. This effect can have associated frequencies as low as 100 MHz. It can therefore play a role in chip capacitors, which are typically made from mixed or doped perovskites. Therefore, frequency-dependent effects can appear in these devices at relatively low frequencies. However, it is possible to control the number of vacancies in the manufacturing process so that in most cases this effect will be small.

As the frequency is increased again, into the range above 1 GHz, there will be contributions due to the ionic separation of the components of a material. For instance, in PZT the electronic charge cloud associated with a unit cell is slightly distorted. Therefore, the octahedron associated with the unit cell has a dipole moment, which will contribute to the polarization. Again, as the frequency of the applied field or voltage is raised, eventually these charges will no longer be able to follow the external field and the polarization, and therefore the dielectric permittivity will decrease to a new plateau.

The only polarization left at this point is the electronic contribution, i.e., the distortion induced in the electronic cloud of the material by the external field. Note that this is an induced polarization unlike the permanent polarization in water molecules or PZT. This happens at optical frequencies, hence the diamond example earlier in the section.

The above view of the dielectric behavior of many materials is somewhat idealized. In fact, the transition from one regime to another is usually not characterized by a single, well-located (in frequency) transition, but is presented to highlight the various mechanisms that enhance the dielectric permittivity. Most often there are multiple mechanisms resulting in a distribution of relaxation times. This in turn causes the transitions in Fig. 9.34 to be smeared out. In fact, the more typical behavior is that of the "Universal Law." Here the dielectric permittivity follows a power law typically with an exponent less than one. In many materials this behavior is observed over many decades of frequency. Thus the observed response follows

$$\varepsilon\left(\omega\right) - \varepsilon_\infty = \varepsilon_0 \omega^{\left(n-1\right)}, \tag{9.158}$$

with $0 < n < 1$. Interestingly this relation holds (at least for some perovskites) over as many as 12 orders of magnitude.

Measurements

The measurement of devices was initially done with a bridge circuit that is an AC modification of the Wheatstone bridge. The method essentially involves manually finding a pair of values representing the capacitance and dissipation factor. This is done by adjusting calibrated resistors to balance a bridge circuit and observing a minimum voltage across the diagonal terminals of the bridge. Because this is a manual method it is slow and tedious, particularly because the two values are interacting, necessitating an adjustment of the first value after the minimum was found for the second value. It is therefore a slowly converging cyclic adjustment. Classical AC bridges are also limited to low frequencies (below approximately 1 MHz).

In recent years the measurement of capacity and the loss of devices has become significantly easier with the development of fixed frequency or swept impedance meters by commercial vendors. At moderate frequencies, up to about 100 MHz, the measurement is typically done as a 4-terminal measurement. Two terminals deliver a known AC current (I). The complex voltage (V) with its corresponding phase angle relative to the phase of the exciting current across the device is measured with another pair of terminals. The complex impedance (Z) is then given by the AC equivalent of Ohm's law

$$Z = V/I. \tag{9.159}$$

TABLE 9.13

Measured Quantity	Series Equivalent	Parallel Equivalent
$|Z|$	$\sqrt{R^2 + X^2}$	
$|Y|$		$\sqrt{G^2 + B^2}$
Θ	$\text{Tan}^{-1}(X/R)$	$\text{Tan}^{-1}(B/G)$
C	$-1/(\omega X)$	B/ω
Q	$|X|/R$	$|B|/G$
D	$R/|X|$	$G/|B|$

The transformations between the measured quantity $|Z|$ and the associated phase angle Θ, and the engineering values of the capacity and loss for a series equivalent circuit and parallel equivalent are given in Table 9.13.

It should be remembered that a lossy two terminal capacitor necessarily has to be represented either as an ideal capacitor connected in series with its loss equivalent resistor or as the two equivalent elements connected in parallel. In the series column the resistance R and the capacity C are in series. In the parallel column the (equivalent) conductance G and inverse reactance B are connected in parallel. For the parallel connection the measured quantity is typically the admittance Y. The quality factor Q and the dissipation factor D are also included in the table. These transformations allow us to find any of the desired values.

It is possible to force an adjustable DC voltage across the current terminals. This allows a measurement of the capacity of the device as a function of an applied steady-state bias field. This application is particularly important in the analysis of semiconductor devices where the junction properties have strong field dependencies. An example is the acquisition of capacity vs. voltage or CV curves. Another example is the characterization of phase shifter materials for phased array antennas, mentioned at the beginning of this article.

At higher frequencies, i.e., above about 100 MHz, it not possible to accurately measure the current or voltage, primarily because of the aberrations introduced by the stray capacities and stray inductances in the connections. It is similarly difficult to measure the current and voltages without interactions between them. Because of this, the impedance measurement is instead made indirectly and is based on reflections in a transmission line using directional couplers that are capable of separating traveling waves going in the forward or reverse direction. Network analyzers exist that are designed to measure the transmission and reflection on a 50 ohm transmission line over a very wide range of frequencies. Thus if a complex impedance terminates such a line, the impedance of the terminating device can be determined from the reflected signals. In fact, the complex reflection coefficient Gamma (Γ) is given by

$$\Gamma = \frac{Z - Z_0}{Z + Z_0} \qquad (9.160)$$

where Z is the reflecting impedance and Z_0 is the impedance of the instrument and transmission line (usually 50 ohms). This can be inverted to solve for the unknown impedance

$$Z = Z_0 * \frac{\Gamma + 1}{\Gamma - 1}. \qquad (9.161)$$

Both the 4-terminal and the network analyzer impedance measurement methods place the sample at the terminals of the respective analyzer. In fact, it is often desirable to make an *in situ* measurement requiring cables. An example is the characterization of devices that are being probed on a silicon wafer. To accommodate this requirement many instruments have the capability to move the reference plane from the terminals to the measurement site. An accurate, purely resistive 50 ohm termination is placed at the actual test site. The resulting data is used to calibrate the analyzer. An internal algorithm does a little more than simply remeasure the calibration resistor. Two other measurements are also necessary to compensate for imperfections in the network analyzer and the cabling. The compensation is made for an imperfect directional coupler and direct reflection or imperfect impedance match at the sample site. This is accomplished with additional measurements with an open circuit and a short circuit in addition to the 50 ohm standard at the sample site. Using this data one can recalculate the value of the unknown at the end of the cable connections to remove the stray impedances. Similar compensations can be made for the four-terminal measurement at lower frequencies.

References

1. A. K. Jonscher, *Dielectric Relaxation in Solids*, Chelsea Dielectric Press, London, 1983.
2. C. Kittel, *Introduction to Solid State Physics*, 3rd Ed., John Wiley & Sons, New York, 1967.
3. J. D. Baniecki, *Dielectric Relaxation Of Barium Strontium Titanate and Application to Thin Films For DRAM Capacitors*, Ph.D. Dissertation, Columbia University, New York, 1999.
4. Hewlett Packard, Instruction Manuals for the HP4194A Impedance and Gain/Phase Analyzer and HP8753C Network Analyzer with the 85047A S-Parameter Test Set.

9.6.3 Ferroelectric and Piezoelectric Materials

K.F. Etzold

Piezoelectric materials have been used extensively in actuator and ultrasonic receiver applications, while **ferroelectric** materials (in thin film form) have recently received much attention for their potential use in nonvolatile (NV) memory applications. We will discuss the basic concepts in the use of these materials, highlight their applications, and describe the constraints limiting their uses. This section emphasizes properties that need to be understood for the effective use of these materials but are often very difficult to research. Among the properties discussed are **hysteresis** and **domains**.

Ferroelectric and piezoelectric materials derive their properties from a combination of structural and electrical properties. As the name implies, both types of materials have electric attributes. A large number of materials that are ferroelectric are also piezoelectric. However, the converse is not true. Pyroelectricity is closely related to ferroelectric and piezoelectric properties via the symmetry properties of the crystals.

Examples of the classes of materials that are technologically important are given in Table 9.14. It is apparent that many materials exhibit electric phenomena that can be attributed to ferroelectric, piezoelectric, and **electret** materials. It is also clear that vastly different materials (organic and inorganic) can exhibit ferroelectricity or piezoelectricity, and many have actually been commercially exploited for these properties.

As shown in Table 9.14, there are two dominant classes of ferroelectric materials, ceramics and organics. Both classes have important applications of their piezoelectric properties. To exploit the ferroelectric

TABLE 9.14   Ferroelectric, Piezoelectric, and Electrostrictive Materials

Type	Material Class	Example	Applications
Electret	Organic	Waxes	No recent
Electret	Organic	Fluorine based	Microphones
Ferroelectric	Organic	PVF2	No known
Ferroelectric	Organic	Liquid crystals	Displays
Ferroelectric	Ceramic	PZT thin film	NV-memory
Piezoelectric	Organic	PVF2	Transducer
Piezoelectric	Ceramic	PZT	Transducer
Piezoelectric	Ceramic	PLZT	Optical
Piezoelectric	Single crystal	Quartz	Freq. control
Piezoelectric	Single crystal	LiNbO3	SAW devices
Electrostrictive	Ceramic	PMN	Actuators

property, recently a large effort has been devoted to producing thin films of **PZT** (lead [Pb] Zirconate Titanate) on various substrates for silicon-based memory chips for nonvolatile storage. In these devices, data is retained in the absence of external power as positive and negative **polarization**. Organic materials have not been used for their ferroelectric properties but have seen extensive application as electrets (the ability to retain a permanent charge). Liquid crystals in display applications are used for their ability to rotate the plane of polarization of light but not their ferroelectric attribute.

It should be noted that the prefix *ferro* refers to the permanent nature of the electric polarization in analogy with the magnetization in the magnetic case. It does not imply the presence of iron, even though the root of the word means iron. The root of the word piezo means pressure; hence the original meaning of the word piezoelectric implied "pressure electricity" — the generation of electric field from applied pressure. This early definition ignores the fact that these materials are reversible, allowing the generation of mechanical motion by applying a field.

Mechanical Characteristics

Materials are acted on by forces (stresses) and the resulting deformations are called strains. An example of a strain due to a force to the material is the change of dimension parallel and perpendicular to the applied force. It is useful to introduce the coordinate system and the numbering conventions that are used when discussing these materials. Subscripts 1, 2, and 3 refer to the *x*, *y*, and *z* directions, respectively. Displacements have single indices associated with their directions. If the material has a preferred axis, such as the poling direction in PZT, the axis is designated the *z* or 3 axis. Stresses and strains require double indices such as *xx* or *xy*. To make the notation less cluttered and confusing, contracted notation has been defined. The following mnemonic rule is used to reduce the double index to a single index:

$$
\begin{array}{ccc}
1 & 6 & 5 \\
xx & xy & xz \\
 & 2 & 4 \\
 & yy & yz \\
 & & 3 \\
 & & zz
\end{array}
$$

This rule can be thought of as a matrix with the diagonal elements having repeated indices in the expected order, then continuing the count in a counterclockwise direction. Note that $xy = yx$, etc. so that subscript 6 applies equally to *xy* and *yx*.

Any mechanical object is governed by the well-known relationship between stress and strain,

$$\mathbf{S} = \mathbf{sT} \tag{9.162}$$

where **S** is the strain (relative elongation), **T** is the stress (force per unit area), and **s** contains the coefficients connecting the two. All quantities are tensors; **S** and **T** are second rank, and **s** is fourth rank. Note, however, that usually contracted notation is used so that the full complement of subscripts is not visible. PZT converts electrical fields into mechanical displacements and vice versa. The connection between the two is via the *d* and *g* coefficients. The *d* coefficients give the displacement when a field is applied (motion transmitter), while the *g* coefficients give the field across the device when a stress is applied (motion receiver). The electrical effects are added to the basic Eq. (9.162) such that

$$\mathbf{S} = \mathbf{sT} + \mathbf{dE} \tag{9.163}$$

where **E** is the electric field and **d** is the tensor that contains the electromechanical coupling coefficients. The latter parameters are reported in Table 9.15 for representative materials. One can write the matrix equation [Eq. (9.163)],

TABLE 9.15 Properties of Well-Known PZT Formulations (Based on the Original Navy Designations)

	Units	PZT4	PZT5A	PZT5H	PZT8
ε_{33}	—	1300	1700	3400	1000
d_{33}	10^{-2} Å/V	289	374	593	225
d_{13}	10^{-2} Å/V	−123	−171	−274	−97
d_{15}	10^{-2} Å/V	496	584	741	330
g_{33}	10^{-3} Vm/N	26.1	24.8	19.7	25.4
k_{33}	—	70	0.705	0.752	0.64
T_Θ	°C	328	365	193	300
Q	—	500	75	65	1000
ρ	g/cm³	7.5	7.75	7.5	7.6
Application	—	High signal	Medium signal	Receiver	Highest signal

$$
\begin{bmatrix} S_1 \\ S_2 \\ S_3 \\ S_4 \\ S_5 \\ S_6 \end{bmatrix} = \begin{bmatrix} s_{11} & s_{12} & s_{13} & & & \\ s_{12} & s_{11} & s_{13} & & 0 & \\ s_{13} & s_{13} & s_{33} & & & \\ & & & s_{44} & & \\ & 0 & & & s_{44} & \\ & & & & & 2(s_{11}-s_{12}) \end{bmatrix} \begin{bmatrix} T_1 \\ T_2 \\ T_3 \\ T_4 \\ T_5 \\ T_6 \end{bmatrix} + \begin{bmatrix} 0 & 0 & d_{13} \\ 0 & 0 & d_{13} \\ 0 & 0 & d_{33} \\ 0 & d_{15} & 0 \\ d_{15} & 0 & 0 \\ 0 & 0 & 0 \end{bmatrix} \begin{bmatrix} E_1 \\ E_2 \\ E_3 \end{bmatrix} \qquad (9.164)
$$

Note that **T** and **E** are shown as column vectors for typographical reasons; they are in fact row vectors. This equation shows explicitly the stress-strain relation and the effect of the electromechanical conversion.

A similar equation applies when the material is used as a receiver:

$$
\mathbf{E} = -\mathbf{gT} + \left(\varepsilon^\mathrm{T}\right)^{-1} \mathbf{D} \qquad (9.165)
$$

where **T** is the transpose and **D** the electric displacement. The matrices are not fully populated for all materials. Whether a coefficient is nonzero depends on the symmetry. For PZT, a ceramic given a preferred direction by the poling operation (the z-axis), only d_{33}, d_{13}, and d_{15} are nonzero. Also, again by symmetry, $d_{13} = d_{23}$ and $d_{15} = d_{25}$.

Applications

Historically the material that was used earliest for its piezoelectric properties was single-crystal quartz. Crude sonar devices were built by Langevin using quartz transducers, but the most important application was, and still is, frequency control. Crystal oscillators are today at the heart of every clock that does not derive its frequency reference from the AC power line. They are also used in every color television set and personal computer. In these applications at least one (or more) "quartz crystal" controls frequency or time. This explains the label "quartz," which appears on many clocks and watches. The use of quartz resonators for frequency control relies on another unique property. Not only is the material piezoelectric (which allows one to excite mechanical vibrations), but the material has also a very high mechanical "Q" or quality factor ($Q > 100,000$). The actual value depends on the mounting details, whether the crystal is in a vacuum, and other details. Compare this value to a Q for PZT between 75 and 1000. The Q factor is a measure of the rate of decay and thus the mechanical losses of an excitation with no external drive. A high Q leads to a very sharp resonance and thus tight frequency control. For frequency control it has been possible to find orientations of cuts of quartz that reduce the influence of temperature on the vibration frequency.

Ceramic materials of the PZT family have also found increasingly important applications. The piezo-electric but not the ferroelectric property of these materials is made use of in transducer applications. PZT has a very high efficiency (electric energy to mechanical energy coupling factor k) and can generate high-amplitude ultrasonic waves in water or solids. The coupling factor is defined by

$$k^2 = \frac{\text{energy stored mechanically}}{\text{total energy stored electrically}} \tag{9.166}$$

Typical values of k_{33} are 0.7 for PZT 4 and 0.09 for quartz, showing that PZT is a much more efficient transducer material than quartz. Note that the energy is a scalar; the subscripts are assigned by finding the energy conversion coefficient for a specific vibrational mode and field direction and selecting the subscripts accordingly. Thus k_{33} refers to the coupling factor for a longitudinal mode driven by a longi-tudinal field.

Probably the most important applications of PZT today are based on ultrasonic echo ranging. Sonar uses the conversion of electrical signals to mechanical displacement as well as the reverse transducer property, which is to generate electrical signals in response to a stress wave. Medical diagnostic ultrasound and nondestructive testing systems devices rely on the same properties. Actuators have also been built but a major obstacle is the small displacement which can conveniently be generated. Even then, the required drive voltages are typically hundreds of volts and the displacements are only a few hundred angstroms. For PZT the strain in the z-direction due to an applied field in the z-direction is (no stress, $\mathbf{T} = 0$)

$$s_3 = d_{33}E_3 \tag{9.167}$$

or

$$s_3 = \frac{\Delta d}{d} = d_{33}\frac{V}{d} \tag{9.168}$$

where s is the strain, E the electric field, and V the potential; d_{33} is the coupling coefficient which connects the two. Thus

$$\Delta d = d_{33}V \tag{9.169}$$

Note that this expression is independent of the thickness d of the material but this is true only when the applied field is parallel to the displacement. Let the applied voltage be 100 V and let us use PZT8 for which d_{33} is 225 (from Table 9.15). Hence $\Delta d = 225$ Å or 2.25 Å/V, a small displacement indeed. We also note that Eq. (9.167) is a special case of Eq. (9.163) with the stress equal to zero. This is the situation when an actuator is used in a force-free environment, for example, as a mirror driver. This arrangement results in the maximum displacement. Any forces that tend to oppose the free motion of the PZT will subtract from the available displacement with the reduction given by the normal stress-strain relation, Eq. (9.162).

It is possible to obtain larger displacements with mechanisms that exhibit mechanical gain, such as laminated strips (similar to bimetallic strips). The motion then is typically up to about 1 millimeter but at a cost of a reduced available force. An example of such an application is the video head translating device to provide tracking in VCRs.

There is another class of ceramic materials that recently has become important. **PMN** (lead [Pb], Magnesium Niobate), typically doped with $\approx$10% lead titanate) is an **electrostrictive** material that has seen applications where the absence of hysteresis is important. For example, deformable mirrors require

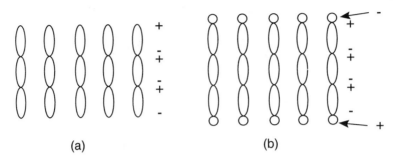

FIGURE 9.36   Charge configurations in ferroelectric model materials: (a) uncompensated and (b) compensated dipole arrays.

repositioning of the reflecting surface to a defined location regardless of whether the old position was above or below the original position.

Electrostrictive materials exhibit a strain that is quadratic as a function of the applied field. Producing a displacement requires an internal polarization. Because the latter polarization is induced by the applied field and is not permanent, as it is in the ferroelectric materials, electrostrictive materials have essentially no hysteresis. Unlike PZT, electrostrictive materials are not reversible; PZT will change shape on application of a field and generate a field when a strain is induced. Electrostrictive materials only change shape on application of a field and, therefore, cannot be used as receivers. PZT has inherently large hysteresis because of the domain nature of the polarization.

Organic electrets have important applications in self-polarized condenser (or capacitor) microphones where the required electric bias field in the gap is generated by the diaphragm material rather than by an external power supply.

Structure of Ferroelectric and Piezoelectric Materials

Ferroelectric materials have, as their basic building block, atomic groups that have an associated electric field, either as a result of their structure or as result of distortion of the charge clouds that make up the groups. In the first case, the field arises from an asymmetric placement of the individual ions in the group (these groupings are called unit cells). In the second case, the electronic cloud is moved with respect to the ionic core. If the group is distorted permanently, then a permanent electric field can be associated with each group. We can think of these distorted groups as represented by electric dipoles, defined as two equal but opposite charges separated by a small distance. Electric dipoles are similar to magnetic dipoles which have the familiar north and south poles. The external manifestation of a magnetic dipole is a magnetic field and that of an electric dipole an electric field.

Figure 9.36(a) represents a hypothetical slab of material in which the dipoles are perfectly arranged. In actual materials the atoms are not as uniformly arranged, but, nevertheless, from this model there would be a very strong field emanating from the surface of the crystal. The common observation, however, is that the fields are either absent or weak. This effective charge neutrality arises from the fact that there are free, mobile charges available that can be attracted to the surfaces. The polarity of the mobile charges is opposite to the charge of the free dipole end. The added charges on the two surfaces generate their own field, equal and opposite to the field due to the internal dipoles. Thus the effect of the internal field is canceled and the external field is zero, as if no charges were present at all [Fig. 9.36(b)].

In ferroelectric materials a crystalline asymmetry exists that allows electric dipoles to form. If the dipoles are absent the internal field disappears. Consider an imaginary horizontal line drawn through the middle of a dipole. We can see readily that the dipole is not symmetric about that line. The asymmetry thus requires that there be no center of inversion when a material is in the ferroelectric state.

All ferroelectric and piezoelectric materials have phase transitions at which the material changes crystalline symmetry. For example, in PZT there is a change from tetragonal or rhombohedral symmetry to cubic as the temperature is increased. The temperature at which the material changes **crystalline**

phases is called the **Curie temperature**, T_Θ. For typical PZT compositions the Curie temperature is between 250 and 450°C.

A consequence of a phase transition is that a rearrangement of the lattice takes place when the material is cooled through the transition. Intuitively we would expect that the entire crystal assumes the same orientation throughout as the temperature passes through the transition. By orientation we mean the direction of the preferred crystalline axis (say the tetragonal axis). Experimentally it is found, however, that the material breaks up into smaller regions in which the preferred direction and thus the polarization is uniform. Note that cubic materials have no preferred direction. In tetragonal crystals the polarization points along the c-axis (the longer axis) whereas in rhombohedral lattices the polarization is along the body diagonal. The volume in which the preferred axis is pointing in the same direction is called a domain and the border between the regions is called a domain wall. The energy of the multidomain state is slightly lower than the single-domain state and is thus the preferred configuration. The direction of the polarization changes by either 90° or 180° as we pass from one uniform region to another. Thus the domains are called 90° and 180° domains. Whether an individual crystallite or grain consists of a single domain depends on the size of the crystallite and external parameters such as strain gradients, impurities, etc. It is also possible that the domain extends beyond the grain boundary and encompasses two or more grains of the crystal.

Real materials consist of large numbers of unit cells, and the manifestation of the individual charged groups is an internal and an external electric field when the material is stressed. Internal and external refer to inside and outside of the material. The interaction of an external electric field with a charged group causes a displacement of certain atoms in the group. The macroscopic manifestation of this is a displacement of the surfaces of the material. This motion is called the piezoelectric effect, the conversion of an applied field into a corresponding displacement.

Ferroelectric Materials

PZT ($PbZr_xTi_{(1-x)}O_3$) is an example of a ceramic material that is ferroelectric. We will use PZT as a prototype system for many of the ferroelectric attributes to be discussed. The concepts, of course, have general validity. The structure of this material is ABO_3 where A is lead and B is one or the other atoms, Ti or Zr. This material consists of many randomly oriented crystallites that vary in size between approximately 10 nm and several microns depending on the preparation method. The crystalline symmetry of the material is determined by the magnitude of the parameter x. The material changes from rhombohedral to tetragonal symmetry when $x > 0.48$. This transition is almost independent of temperature. The line that divides the two phases is called a **morphotropic phase boundary** (change of symmetry as a function of composition only). Commercial materials are made with $x \approx 0.48$, where the d and g sensitivity of the material is maximum. It is clear from Table 9.15 that there are other parameters that can be influenced as well. Doping the material with donors or acceptors often changes the properties dramatically. Thus niobium is important to obtain higher sensitivity and resistivity and to lower the Curie temperature. PZT typically is a p-type conductor and niobium will significantly decrease the conductivity because of the electron Nb^{5+} contributes to the lattice. The Nb ion substitutes for the **B-site** ion Ti^{4+} or Zr^{4+}. The resistance to depolarization (the hardness of the material) is affected by iron doping. Hardness is a definition giving the relative resistance to depolarization. It should not be confused with mechanical hardness. Many other dopants and admixtures have been used, often in very exotic combinations to affect aging, sensitivity, etc.

The designations used in Table 9.15 reflect very few of the many combinations that have been developed. The PZT designation types were originated by the U.S. Navy to reflect certain property combinations. These can also be obtained with different combinations of compositions and dopants. The examples given in the table are representative of typical PZT materials, but today essentially all applications have their own custom formulation. The name PZT has become generic for the lead zirconate titanates and does not reflect Navy or proprietary designations.

When PZT ceramic material is prepared, the crystallites and domains are randomly oriented, and therefore the material does not exhibit any piezoelectric behavior [Fig. 9.37(a)]. The random nature of

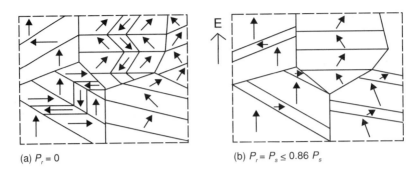

(a) $P_r = 0$ (b) $P_r = P_s \leq 0.86\, P_s$

FIGURE 9.37 Domains in PZT, as prepared (a) and poled (b).

the displacements for the individual crystallites causes the net displacement to average to zero when an external field is applied. The tetragonal axis has three equivalent directions 90° apart and the material can be poled by reorienting the polarization of the domains into a direction nearest the applied field. When a sufficiently high field is applied, some but not all of the domains will be rotated toward the electric field through the allowed angle 90° or 180°. If the field is raised further, eventually all domains will be oriented as close as possible to the direction of the field. Note however, that the polarization will not point exactly in the direction of the field [Fig. 9.37(b)]. At this point, no further domain motion is possible and the material is saturated. As the field is reduced, the majority of domains retain the orientation they had with the field on leaving the material in an oriented state which now has a net polarization. This operation is called poling and is accomplished by raising the temperature to about 150°C, usually in oil to avoid breakdown. Raising the temperature lowers the **coercive field**, E_c and the poling field of about 30–60 kV/cm is applied for several minutes. The temperature is then lowered but it is not necessary to keep the applied field turned on during cooling because the domains will not spontaneously re-randomize.

Electrical Characteristics

Before considering the dielectric properties, consider the equivalent circuit for a slab of ferroelectric material. In Fig. 9.38 the circuit shows a mechanical (acoustic) component and the static or clamped capacity C_o (and the dielectric loss R_d) which are connected in parallel. The acoustic components are due to their motional or mechanical equivalents, the compliance (capacity, C) and the mass (inductance, L). There will be mechanical losses, which are indicated in the mechanical branch by R. The electrical branch has the clamped capacity C_o and a dielectric loss (R_d), distinct from the mechanical losses. This configuration will have a resonance that is usually assumed to correspond to the mechanical thickness mode but can represent other modes as well. This simple model does not show the many other modes a slab (or rod) of material will have. Thus transverse, plate, and flexural modes are present. Each can be represented by its own combination of L, C, and R. The presence of a large number of modes often causes difficulties in characterizing the material since some parameters must be measured either away from the

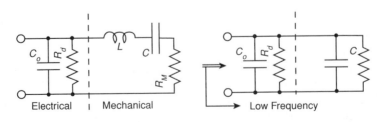

Electrical │ Mechanical   Low Frequency

FIGURE 9.38 Equivalent circuit for a piezoelectric resonator. The reduction of the equivalent circuit at low frequencies is shown on the right.

resonances or from clean, non-overlapping resonances. For instance, the clamped capacity (or clamped dielectric constant) of a material is measured at high frequencies where there are usually a large number of modes present. For an accurate measurement these must be avoided and often a low-frequency measurement is made in which the material is physically clamped to prevent motion. This yields the static, non-mechanical capacity, C_o. The circuit can be approximated at low frequencies by ignoring the inductor and redefining R and C. Thus, the coupling constant can be extracted from the value of C and C_o. From the previous definition of k we find

$$k^2 = \frac{\text{energy stored mechanically}}{\text{total energy stored electrically}} = \frac{CV^2/2}{\left(C + C_o\right)V^2/2} = \frac{1}{\dfrac{C_o}{C}} + 1 \tag{9.170}$$

It requires charge to rotate or flip a domain. Thus, there is charge flow associated with the rearrangement of the polarization in the ferroelectric material. If a bipolar, repetitive signal is applied to a ferroelectric material, its hysteresis loop is traced out and the charge in the circuit can be measured using the Sawyer Tower circuit (Fig. 9.39). In some cases the drive signal to the material is not repetitive and only a single cycle is used. In that case the starting point and the end point do not have the same polarization value and the hysteresis curve will not close on itself.

The charge flow through the sample is due to the rearrangement of the polarization vectors in the domains (the polarization) and contributions from the static capacity and losses (C_o and R_d in Fig. 9.38). The charge is integrated by the measuring capacitor which is in series with the sample. The measuring capacitor is sufficiently large to avoid a significant voltage loss. The polarization is plotted on a X-Y oscilloscope or plotter against the applied voltage and therefore the applied field.

Ferroelectric and piezoelectric materials are lossy. This will distort the shape of the hysteresis loop and can even lead to incorrect identification of materials as ferroelectric when they merely have nonlinear conduction characteristics. A resistive component (from R_d in Fig. 9.38) will introduce a phase shift in the polarization signal. Thus the display has an elliptical component, which looks like the beginnings of the opening of a hysteresis loop. However, if the horizontal signal has the same phase shift, the influence of this lossy component is eliminated, because it is in effect subtracted. Obtaining the exact match is the function of the optional phase shifter, and in the original circuits a bridge was constructed with a second measuring capacitor in the comparison arm (identical to the one in series with the sample). The phase was then matched with adjustable high-voltage components that match C_o and R_d.

This design is inconvenient to implement and modern Sawyer Tower circuits have the capability to shift the reference phase either electronically or digitally to compensate for the loss and static components. A contemporary version, which has compensation and no voltage loss across the integrating capacitor,

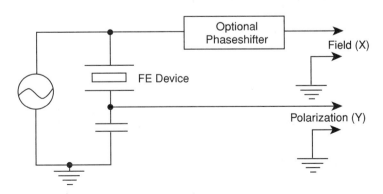

FIGURE 9.39 Sawyer Tower circuit.

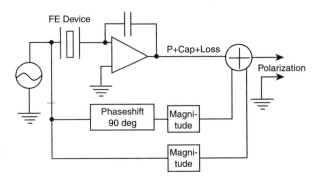

FIGURE 9.40 Modern hysteresis circuit. An op amp is used to integrate the charge; loss and static capacitance compensation are included.

is shown in Fig. 9.40. The op-amp integrator provides a virtual ground at the input, reducing the voltage loss to negligible values. The output from this circuit is the sum of the polarization and the capacitive and loss components. These contributions can be canceled using a purely real (resistive) and a purely imaginary (capacitive, 90° phase shift) compensation component proportional to the drive across the sample. Both need to be scaled (magnitude adjustments) to match them to the device being measured and then have to be subtracted (adding negatively) from the output of the op amp. The remainder is the polarization. The hysteresis response for typical ferroelectrics is frequency dependent and traditionally the reported values of the polarization are measured at 50 or 60 Hz.

The improved version of the Sawyer Tower (Fig. 9.41) circuit allows us to cancel C_o and R_d and the losses, thus determining the active component. This is important in the development of materials for ferroelectric memory applications. It is far easier to judge the squareness of the loop when the inactive components are canceled. Also, by calibrating the "magnitude controls" the values of the inactive components can be read off directly. In typical measurements the resonance is far above the frequencies used, so ignoring the inductance in the equivalent circuit is justified.

The measurement of the dielectric constant and the losses is usually very straightforward. A slab with a circular or other well-defined cross-section is prepared, electrodes are applied, and the capacity and loss are measured (usually as a function of frequency). The relative dielectric permittivity is found from

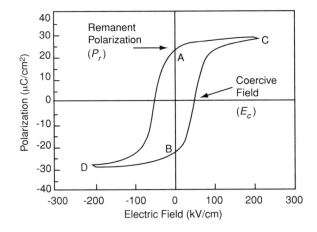

FIGURE 9.41 Idealized hysteresis curve for typical PZT materials. Many PZT materials display offsets from the origin and have asymmetries with respect to the origin. The curve shows how the remanent polarization $(\vec{P_r})$ and the coercive field $(\vec{E_c})$ are defined. While the loop is idealized, the values given for the polarization and field are realistic for typical PZT materials.

$$C = \varepsilon_o k \frac{A}{t}$$ (9.171)

where A is the area of the device and t the thickness. In this definition (also used in Table 9.15) k is the relative dielectric permittivity and ε_o is the permittivity of vacuum. Until recently, the relative dielectric permittivity, like the polarization, was measured at 50 or 60 Hz (typical powerline frequencies). Today the dielectric parameters are typically specified at 1 kHz or higher, which is possible because impedance analyzers with high-frequency capability are readily available. To avoid low-frequency anomalies, even higher frequencies such as 1 MHz are often selected. This is especially the case when evaluating PZT or non-ferroelectric high k thin films. Low frequency anomalies can be present but are not modeled in the equivalent circuit (Fig. 9.37). These are typically due to interface layers. These layers will cause both the resistive and reactive components to rise at low frequencies producing readings that are not representative of the dielectric material properties (see Section 9.6.2).

A piezoelectric component often has a very simple geometric shape, especially when it is prepared for measurement purposes. There will be mechanical resonances associated with the major dimensions of a sample piece. The resonance spectrum will be more or less complicated, depending on the shape of a sample piece. If the object has a simple shape, then some of the resonances will be well separated from each other and can be associated with specific vibrations and dimensions (modes). Each of these resonances has an electrical equivalent, and inspection of the equivalent circuit shows that there will be a resonance (minimum impedance) and an antiresonance (maximum impedance). Thus an impedance plot can be used to determine the frequencies and also the coupling constants and mechanical parameters for the various modes.

Ferroelectric and High Epsilon Thin Films

While PZT and other ferroelectric (FE) bulk materials have had major commercial importance, thin films prepared from these materials have only recently been the focus of significant research efforts. In this section the material properties and process issues will be discussed. Because of the potentially large payoff, major efforts have been directed at developing the technologies for depositing thin films of ferroelectric and non-ferroelectric but high epsilon (high dielectric permittivity) thin films.

A recent trend has been the ever increasing density of dynamic random access memory (DRAM). The data storage capacitor in these devices is becoming a major limiting factor because its dielectric has to be very thin in order to achieve the desired capacitance values to yield, in turn, a sufficient signal for the storage cell. It is often also desirable to have nonvolatile operation (no data loss on power loss). These two desires have, probably more than anything else, driven the development of high epsilon and FE thin films. Of course, these are not the only applications of FE films. Table 9.16 lists the applications of FE (nonvolatile, NV) and high epsilon films (volatile) and highlights which of the properties are important for their use. It is seen that NV memory application is very demanding. Satisfying all these requirements simultaneously has produced significant challenges in the manufacture of these films.

Perhaps the least understood and to some extent still unsolved problem is that of fatigue. In nonvolatile memory applications the polarization represents the memory state of the material (up $\equiv$ bit 1; down $\equiv$ bit 0). In use the device can be switched at the clock rate, say 100 MHz. Thus for a lifetime of 5 years

TABLE 9.16 Material Properties and Applications Areas

	Ferroelectric	Epsilon	Polarization	Coercive Field	Leakage	Aging	Electro-Optical	Electro-Mechanical
NV RAM	X		X	X	X	X		
DRAM		X			X	X		
Actuator				X				X
Display	X				X	X	X	
Optical Modulator	X				X	X	X	

the material must withstand $\simeq 10^{16}$ polarization reversals or large field reversals. Typical candidate materials for ferroelectric applications are PZTs with the ratio of zirconium to titanium adjusted to yield the maximum dielectric constant and polarization. This maximum will be near the morphotropic phase boundary for PZT. Small quantities of other materials can be added, such as lanthanum or niobium to modify optical or switching characteristics. The Sol-Gel method discussed below is particularly suitable for incorporating these additives. Devices made from materials at the current state of the art lose a significant portion of their polarization after 10^{10} to 10^{12} cycles, rendering them useless for their intended memory use because of the associated signal loss. This is a topic of intensive investigation and only one proprietary material has emerged that might be suitable for memory use (Symetric Corporation).

High epsilon non-ferroelectric materials are of great interest for DRAM applications. As an example, major efforts are extant to produce thin films of mixtures of barium and strontium titanate (BST). Dielectric constants of 600 and above have been achieved (compared to 4 to 6 for silicon oxides and nitrides).

In applications for FE films, significant opportunities also exist for electro-optical modulators for fiber-optic devices and light valves for displays. Another large scale application is actuators and sensors. For the latter the electromechanical conversion property is used and large values of d_{33} (the conversion coefficient) are desirable. However, economically the importance of all other applications are, and probably will be in the foreseeable future, less significant than that of memory devices.

Integration of ferroelectric or nonferroelectric materials with silicon devices and substrates has proved to be very challenging. Contacts and control of the crystallinity and crystal size and the stack structure of the capacitor device are the principal issues. In both volatile and nonvolatile memory cells the dielectric material interacts with the silicon substrate. Thus an appropriate barrier layer must be incorporated while which at the same time is a suitable substrate on which to grow the dielectric films. A typical device structure starts with an oxide layer (SiO_x) on the silicon substrate followed by a thin titanium layer which prevents diffusion of the final substrate layer, platinum (the actual growth substrate). If electrical contact to the silicon is desired, ruthenium based barrier layer materials can be used. These are structurally similar to the FE or high epsilon films.

While the major effort in developing thin films for data storage has been in the films described above, a significant effort has recently been made to produce low epsilon films. In contemporary logic chips the device speed is not only determined by the speed of the active devices but also by the propagation delay in the interconnect wiring. To reduce the delay the conductivity of the metal must be increased (hence the use of copper as conductors) and the dielectric separating the conductors must have as low as possible dielectric permittivity. Materials for this purpose are now also being developed in this film form.

Significant differences have been observed in the quality of the films depending on the nature of the substrate. The quality can be described by intrinsic parameters such as the crystallinity (i.e., the degree to which non-crystalline phases are present). The uniformity of the orientation of the crystallites also seems to play a role in determining the electrical properties of the films. In the extreme case of perfect alignment of the crystallites of the film with the single crystal substrate an epitaxial film is obtained. These films tend to have the best electrical properties. In addition to amorphous material, other crystalline but non-ferroelectric phases can be present. An example is the pyrochlore phase in PZT. These phases often form incidentally to the growth process of the desired film and usually degrade one or more of the desired properties of the film (for instance the dielectric constant). The pyrochlore and other oxide materials can accumulate between the Pt electrode and the desired PZT or BST layer. The interface layer is then electrically in series with the desired dielectric layer and degrades its properties. The apparent reduction of the dielectric constant which is often observed in these films as the thickness is reduced can be attributed to the presence of these low dielectric constant layers.

There are many growth methods for these films. Table 9.17 lists the most important techniques along with some of the critical parameters. Wet methods use metal organic compounds in liquid form. In the Sol-Gel process the liquid is spun onto the substrate. The wafer is then heated, typically to a lower, intermediate temperature (around 300°C). This spin-on and heat process is repeated until the desired thickness is reached. At this temperature only an amorphous film forms. The wafer is then heated to

TABLE 9.17   Deposition Methods for PZT and Perovskites

	Process Type	Rate nm/min	Substrate Temperature	Anneal Temperature	Target/Source
Wet	Sol-Gel	100 nm/coat	RT	450–750	Metal organic
Wet	MOD	300 nm/coat	RT	500–750	Metal organic
Dry	RF sputter	.5–5	RT–700	500–700	Metals and oxides
Dry	Magnetron sputter	5–30	RT–700	500–700	Metals and oxides
Dry	Ion beam sputter	2–10	RT–700	500–700	Metals and oxides
Dry	Laser sputter	5–100	RT–700	500–700	Oxide
Dry	MOCVD	5–100	400–800	500–700	MO vapor and carrier gas

between 500 and 700°C usually in oxygen and the actual crystal growth takes place. Instead of simple long-term heating (order of hours), rapid thermal annealing (RTA) is often used. In this process the sample is only briefly exposed to the elevated temperature, usually by a scanning infrared beam. It is in the transition between the low decomposition temperature and the firing temperature that the pyrochlore tends to form. At the higher temperatures the more volatile components have a tendency to evaporate, thus producing a chemically unbalanced compound that also has a great propensity to form one or more of the pyrochlore phases. In the case of PZT, 5 to 10% excess lead is usually incorporated which helps to form the desired perovskite material and compensates for the loss. In preparing Sol–Gel films it is generally easy to prepare the compatible liquid compounds of the major constituents and the dopants. The composition is then readily adjusted by appropriately changing the ratio of the constituents. Very fine quality films have been prepared by this method, including epitaxial films.

The current semiconductor technology is tending toward dry processing. Thus, in spite of the advantages of the Sol-Gel method, other methods using physical vapor deposition (PVD) are being investigated. These methods use energetic beams or plasma to move the constituent materials from the target to the heated substrate. The compound then forms *in situ* on the heated wafer ($\simeq 500$°C). Even then, however, a subsequent anneal is often required. With PVD methods it is much more difficult to change the composition since now the oxide or metal ratios of the target have to be changed or dopants have to be added. This involves the fabrication of a new target for each composition ratio. MOCVD is an exception here; the ratio is adjusted by regulating the carrier gas flow. However, the equipment is very expensive and the substrate temperatures tend to be high (up to 800°, uncomfortably high for semiconductor device processing). The laser sputtering method is very attractive and it has produced very fine films. The disadvantage is that the films are defined by the plume that forms when the laser beam is directed at the source. This produces only small areas of good films and scanning methods need to be developed to cover full size silicon wafers. Debris is also a significant issue in laser deposition. However, it is a convenient method to produce films quickly and with a small investment. In the long run MOCVD or Sol-Gel will probably evolve as the method of choice for realistic DRAM devices with state of the art densities.

Defining Terms

A-site: Many ferroelectric materials are oxides with a chemical formula ABO_3. The A-site is the crystalline location of the A atom.

B-site: Analogous to the definition of the A-site.

Coercive field: When a ferroelectric material is cycled through the hysteresis loop the coercive field is the electric field value at which the polarization is zero. A material has a negative and a positive coercive field and these are usually, but not always, equal in magnitude to each other.

Crystalline phase: In crystalline materials the constituent atoms are arranged in regular geometric ways; for instance in the cubic phase the atoms occupy the corners of a cube (edge dimensions ≈ 2–15 Å for typical oxides).

Curie temperature: The temperature at which a material spontaneously changes its crystalline phase or symmetry. Ferroelectric materials are often cubic above the Curie temperature and tetragonal or rhombohedral below.

Domain: Domains are portions of a material in which the polarization is uniform in magnitude and direction. A domain can be smaller, larger, or equal in size to a crystalline grain.

Electret: A material that is similar to ferroelectrics but charges are macroscopically separated and thus are not structural. In some cases the net charge in the electrets is not zero, for instance when an implantation process was used to embed the charge.

Electrostriction: The change in size of a non-polarized, dielectric material when it is placed in an electric field.

Ferroelectric: A material with permanent charge dipoles which arise from asymmetries in the crystal structure. The electric field due to these dipoles can be observed external to the material when certain conditions are satisfied (ordered material and no charge on the surfaces).

Hysteresis: When the electric field is raised across a ferroelectric material the polarization lags behind. When the field is cycled across the material the hysteresis loop is traced out by the polarization.

Morphotropic phase boundary (MPB):   Materials that have a MPB assume a different crystalline phase depending on the composition of the material. The MPB is sharp (a few percent in composition) and separates the phases of a material. It is approximately independent of temperature in PZT.

Piezoelectric: A material that exhibits an external electric field when a stress is applied to the material and a charge flow proportional to the strain is observed when a closed circuit is attached to electrodes on the surface of the material.

PLZT: A PZT material with a lanthanum doping or admixture (up to approximately 15% concentration). The lanthanum occupies the A-site.

PMN: Generic name for electrostrictive materials of the lead (Pb) magnesium niobate family.

Polarization: The polarization is the amount of charge associated with the dipolar or free charge in a ferroelectric or an electret, respectively. For dipoles the direction of the polarization is the direction of the dipole. The polarization is equal to the external charge that must be supplied to the material to produce a polarized state from a random state (twice that amount is necessary to reverse the polarization). The statement is rigorously true if all movable charges in the material are reoriented (i.e., saturation can be achieved).

PVF2:   An organic polymer that can be ferroelectric. The name is an abbreviation for polyvinyledene difluoride.

PZT: Generic name for piezoelectric materials of the lead (Pb) zirconate titanate family.

Remanent polarization: The residual or remanent polarization of a material after an applied field is reduced to zero. If the material was saturated, the remanent value is usually referred to as the polarization, although even at smaller fields a (smaller) polarization remains.

References

J. C. Burfoot and G. W. Taylor, *Polar Dielectrics and Their Applications*, Berkeley: University of California Press, 1979.

H. Diamant, K. Drenck, and R. Pepinsky, *Rev. Sci. Instrum.*, 28, 30, 1957.

T. Hueter and R. Bolt, *Sonics*, New York: John Wiley and Sons, 1954.

B. Jaffe, W. Cook, and H. Jaffe, *Piezoelectric Ceramics*, London: Academic Press, 1971.

M. E. Lines and A. M. Glass, *Principles and Applications of Ferroelectric Materials*, Oxford: Clarendon Press, 1977.

R. A. Roy and K. F. Etzold, Ferroelectric film synthesis, past and present: a select review, *Mater. Res. Soc. Symp. Proc.*, 200, 141, 1990.

C. B. Sawyer and C. H. Tower, *Phys. Rev.*, 35, 269, 1930.

Z. Surowiak, J. Brodacki, and H. Zajosz, *Rev. Sci. Instrum.*, 49, 1351, 1978.

Further Information

IEEE Transactions on Ultrasonics, Ferroelectrics, and Frequency Control (UFFC).

IEEE Proceedings of International Symposium on the Application of Ferroelectrics (ISAF) (these symposia are held at irregular intervals).

Materials Research Society, Symposium Proceedings, vols. 191, 200, and 243 (this society holds symposia on ferroelectric materials at irregular intervals).

K.-H. Hellwege, Ed., *Landolt-Bornstein: Numerical Data and Functional Relationships in Science and Technology,* New Series, Gruppe III, vols. 11 and 18, Berlin: Springer-Verlag, 1979 and 1984 (these volumes have elastic and other data on piezoelectric materials).

American Institute of Physics Handbook, 3rd ed., New York: McGraw-Hill, 1972.

D. E. Kotecki et al., $(BaSr)TiO_3$ dielectrics for future stacked-capacitor DRAM, *IBM J. Res. Dev.,* 43, 367, 1999.

Grill, V. Patel, Novel low-k Dual Phase Materials Prepared by PECVD, *Mat. Res. Soc. Proc.,* Spring Meeting 2000 (San Francisco).

9.6.4   Semiconductors

Mike Harris

Semiconductor is a class of materials that can generally be defined as having an electrical resistivity in the range of 10^{-2} to 10^9 ohm-cm.[1] Addition of a very small amount of impurity atoms can make a large change in the conductivity of the semiconductor material. This unique materials property makes all semiconductor devices and circuits possible. The amount of free charge in the semiconductor and the transport characteristics of the charge within the crystalline lattice determine the conductivity. Device operation is governed by the ability to generate, move, and remove free charge in a controlled manner. Material characteristics vary widely in semiconductors and only certain materials are suitable for use in the fabrication of microwave and RF devices.

Bulk semiconductors, for microwave and RF applications, include germanium (Ge), silicon (Si), silicon carbide (SiC), gallium arsenide (GaAs), and indium phosphide (InP). Electronic properties of these materials determine the appropriate frequency range for a particular material. Epitaxial layers of other important materials are grown on these host substrates to produce higher performance devices that overcome basic materials limitations of homogeneous semiconductors. These specialized semiconductors include silicon germanium, gallium nitride, aluminum gallium arsenide, and indium gallium arsenide among others. Many of the advanced devices described in Chapter 7 are made possible by the optimized properties of these materials. Through the use of "bandgap engineering," many of the material compromises that limit electronic performance can be overcome with these hetero-structures.

Electron transport properties determine, to a large extent, the frequency at which the various semiconductors are used. On the basis of maturity and cost, silicon will dominate all applications in which it can satisfy the performance requirements. Figure 9.42 is a plot showing the general range of frequencies over which semiconductor materials are being used for integrated circuit applications. It should be noted that the boundary tends to shift to the right with time as new device and fabrication technologies emerge. Discrete devices may be used outside these ranges for specific applications.

This section provides information on important materials properties of semiconductors used for microwave and RF applications. Basic information about each semiconductor is presented followed by tables of electronic properties, thermal properties, and mechanical properties. In order to use these materials in microwave and RF applications, devices and circuits must be fabricated. Device fabrication requires etching and deposition of metal contacts and this section provides a partial list of etchants for these semiconductors along with a list of metallization systems that have been used to successfully produce components.

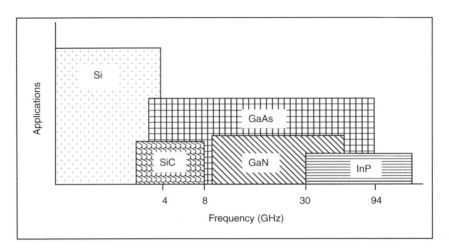

FIGURE 9.42 Frequency range for semiconductor materials.

Silicon

Silicon is an elemental semiconductor that is by far the best known and most mature. A measure of the maturity of silicon is the fact that it is available in 300-mm (12″) diameter wafers. Single crystal silicon wafers are made from electronic grade silicon, which is one of the most refined materials in the world, having an impurity level of no more than one part per billion. Silicon and germanium have the same crystal structure as diamond. In this structure, each atom is surrounded by four nearest neighbor atoms forming a tetrahedron as shown in Figure 9.43. All of the atoms in the diamond lattice are silicon. Silicon is a "workhorse" material at lower frequencies; however, its electron transport properties and low bulk resistivity limit its application in integrated circuit form to frequencies typically below 4 GHz. Silicon-based discrete devices such as PIN diodes find application at higher frequencies.

Si wafers are available with dopant atoms that make them conductive. Wafers having phosphorus impurities are n-type containing excess electrons. Boron-doped wafers are p type and have an excess of holes. Flats are cut on the wafers to distinguish between conductivity types and crystal orientation as shown in Fig. 9.44.[2]

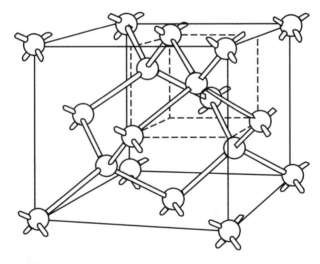

FIGURE 9.43 Cyrstalline structure of Si.

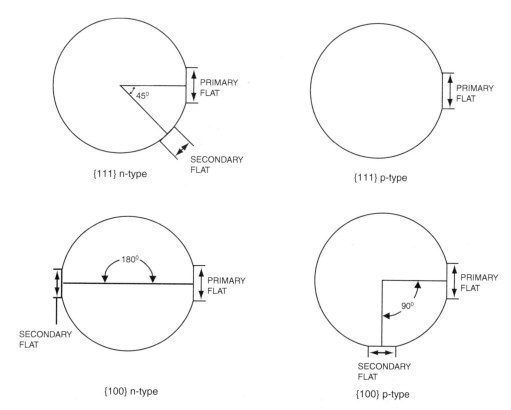

FIGURE 9.44 Silicon wafer type identification (Sze, VLSI Technology, Ref. 2).

Silicon is an indirect band gap material meaning that when an electron and hole recombine, the energy produced is dissipated as a lattice vibration. This should be compared to material like gallium arsenide that is direct gap. When an electron and hole recombine in GaAs, energy is released in the form of light.

Gallium Arsenide

Silicon is the most widely used semiconductor to make electronic devices, however, there are compounds that perform functions beyond the physical limits of the electronic properties of silicon. There are many different kinds of compound semiconductor materials, but the most common material combinations used in microwave and RF applications come from the group III and group V elements.

Gallium arsenide has a zincblende crystalline structure and is one of the most important compound semiconductors. Zincblende consists of two interpenetrating, face-centered cubic (fcc) sublattices as seen in Fig. 9.45. One sublattice is displaced by 1/4 of a lattice parameter in each direction from the other sublattice, so that each site of one sublattice is tetrahedrally coordinated with sites from the other sublattice. That is, each atom is at the center of a regular tetrahedron formed by four atoms of the opposite type. When the two sublattices have the same type of atom, the zincblende lattice becomes the diamond lattice as shown above for silicon. Other examples of compound semiconductors with the zincblende lattice include indium phosphide and silicon carbide.

GaAs wafers, for microwave and RF applications, are available in 4″ and 6″ diameters. Six-inch wafers are currently used for high speed digital and wireless applications and will be used for higher frequency applications as demand increases. Figure 9.46 shows the standard wafer orientation for semi-insulating GaAs. The front face is the (100) direction or 2 degrees off (100) toward the [110] direction. Figure 9.47 shows the different edge profiles that occur when etching GaAs. These profiles are a function of the crystal orientation as seen in the diagram.[3]

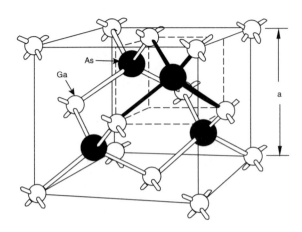

FIGURE 9.45 Zinc blende crystalline structure of gallium arsenide (GaAs).

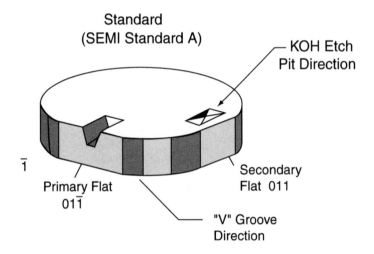

FIGURE 9.46 Standard wafer orientation for semi-insulating GaAs.

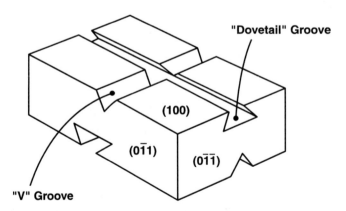

FIGURE 9.47 Orientation-dependent etching profiles of GaAs[3].

Figure 9.48 is a plot of the bandgap energy of various semiconductors as a function of temperature.[4] GaAs has a bandgap energy at room temperature of 1.43 eV compared to 1.12 eV for silicon. This means that the intrinsic carrier concentration of GaAs can be very low compared to silicon. Since the intrinsic

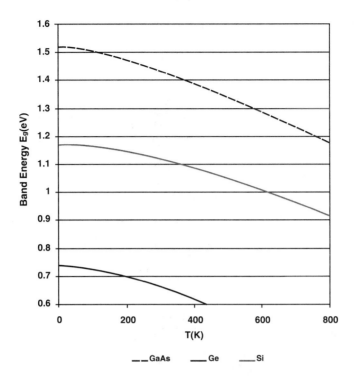

FIGURE 9.48 Energy bandgaps of GaAs, Si, and Ge as a function of temperature (after Sze, Ref. 3).

carrier concentration in gallium arsenide is less than that in silicon, higher resistivity substrates are available in gallium arsenide. High resistivity substrates are desirable since they allow the active region of a device to be isolated using a simple ion implantation or mesa etching. Availability of high resistivity gallium arsenide substrates is one reason that this material has found such widespread use for microwave and wireless applications.

The ability to move charge is determined by the transport characteristics of the material. This information is most often presented in the charge carrier velocity electric field characteristic as shown in Figure 9.49.[5] For low values of electric field, the carrier velocity is linearly related to the electric field

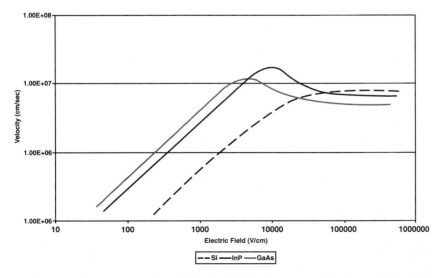

FIGURE 9.49 Electron velocity as a function of electric field for common semiconductors (after Bahl, Ref. 5).

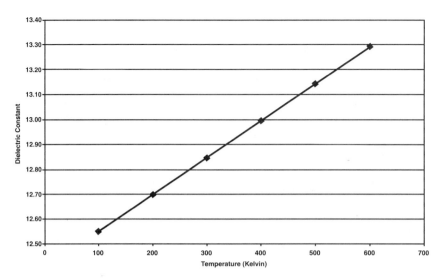

FIGURE 9.50   Relative dielectric constant of GaAs as a function of temperature.

strength. The proportionality constant is the mobility and this parameter is important in determining the low field operation of a device. Generally a high value of mobility is desired for optimum device performance. Since the mobility of electrons in gallium arsenide is about six times that of silicon, gallium arsenide is a more attractive material for high frequency RF and high-speed digital applications. This mobility advantage along with the availability of high resistivity substrates makes gallium arsenide the preferred and most widely used semiconductor material for these applications.[5]

The relative dielectric constant of GaAs is of critical importance in the design of monolithic microwave integrated circuits (MMICs). This parameter is used to determine the width of transmission lines used to interconnect transistors. The characteristic impedance of the transmission line is a function of the relative dielectric constant, the substrate thickness, and the width of the line. Figure 9.50 shows how the relative dielectric constant behaves with temperature. Microwave and RF designs must accommodate this variation over the expected temperature range of operation.

Reliable design and use of GaAs-based MMICs depends on keeping the device channel temperature below a certain absolute maximum level. Channel temperature is defined by the equation,

$$\textbf{Tch} = \textbf{Tsink} + \theta \times \textbf{P}$$

where **Tch** is the channel temperature in (K), and **Tsink** is the heat sink temperature in (K). θ is the thermal resistance defined as L/kA, where L is the thickness of the GaAs substrate, k is the thermal conductivity of GaAs, and A is the area of the channel. **P** is the power dissipated (W). Thermal conductivity is a bulk material property that varies with temperature. Figure 9.51 shows the variation in thermal conductivity as a function of temperature for various semiconductors. At temperatures on the order of 100 K, the thermal conductivity of GaAs approaches that of copper.

III-V Heterostructures

A new class of high performance materials, based on GaAs and InP, has been developed using advanced epitaxial processes such as molecule beam epitaxy (MBE). These materials have heterojunctions that are formed between semiconductors having different compositions and bandgaps. The bandgap discontinuity can provide significant improvement in the transport properties of the material and allows optimization of device characteristics not possible with homojunctions. In FET devices, the current density can be high while retaining high electron mobility. This is possible because the free electrons are separated from their donor atoms and are confined to a region where the lattice scattering is minimal. Table 9.18 lists

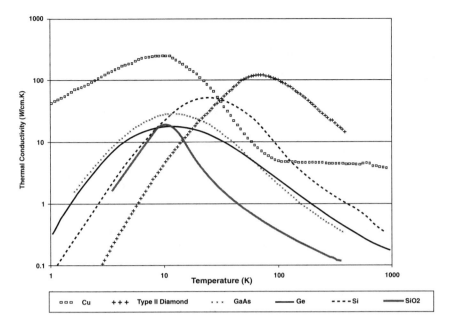

FIGURE 9.51 Thermal conductivity of various materials as a function of temperature (after Sze, Ref. 4).

TABLE 9.18 Common Heterostructures
Used for Microwave and RF Applications

$Al_xGa_{1-x}As/GaAs$
$Al_xIn_{x-1}As/InGaAs$
$Al_xGa_{1-x}As/In_yGa_{1-y}As/GaAs$
InGaAs/InP
$Al_xIn_{x-1}As/InP$

the most common hetero-structures for microwave and RF applications. Material compositions vary and are designated by mole fraction using subscripts. In order to match the lattice spacing of GaAs, $Al_{0.25}Ga_{0.75}As$ is routinely used as shown in the diagram of Fig. 9.52. This diagram also shows the compounds that are lattice matched to InP. When a compound is formed with materials that have different lattice spacing, there is strain in the material and under certain conditions, improved performance is possible. This materials system is called pseudomorphic. Figure 9.53 is a cross-section diagram of a double pulsed doped pseudomorphic layer structure used for microwave power transistors.

Indium Phosphide

Indium phosphide (InP) is an important compound semiconductor for microwave and RF devices due to its physical and electronic properties. Some of the most important properties of InP are high peak electron velocity, high electric field breakdown, and relatively high thermal conductivity. Three-inch diameter bulk, semi-insulating InP wafers are available and four-inch diameter wafers are being validated. Indium phosphide has a zincblende crystalline structure like gallium arsenide and its lattice constant is 5.8687 angstroms compared to 5.6532 angstroms for GaAs. This materials property is important in the growth of hetero-structures as discussed below. Bulk InP is used in the fabrication of opto-electronic devices but is not used directly for microwave and RF applications. In microwave and RF devices, InP is used as a substrate for epitaxial growth to support GaInAs/AlInAs pseudomorphic high electron mobility transistors. This material system has proved to be a more desirable choice for microwave power amplifiers and millimeter-wave low noise amplifiers.

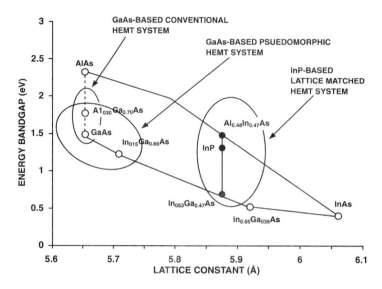

FIGURE 9.52 Energy bandgap and associated lattice constants for II-V hetero-structures.

	Thickness	Dopant	Doping
N⁺ GaAs Cap	500 Å	Si	$5.0 \times 10^{18} cm^{-3}$
i AlGaAs Donor	300Å	None	....
Si Planar Doping	-------	Si	$5.0 \times 10^{12} cm^{-2}$
i AlGaAs Spacer	20 Å	None	....
i In GaAs Channel	120Å	None	....
i GaAs Spacer			
Si Planar Doping	-------	Si	$1.2 \times 10^{12} cm^{-2}$
S/L Buffer	1000Å	None	....
i GaAs Buffer	5000Å	None	....
GaAs Substrate			

FIGURE 9.53 Double pulsed doped pseudomorphic HEMT layer structure. (From Quantum Epitaxial Designs. With permission.)

A fundamental design goal for microwave and RF device engineers, is to achieve the best electron transport properties possible within the reliability, breakdown, and leakage requirements of the application. Transition from GaAs/AlGaAs high electron mobility transistors to pseudomorphic GaAs/InGaAs/AlGaAs HEMTs resulted in significant improvements in device capability for both low noise and power applications.[6] This was due primarily to the increased electron velocity associated with the smaller electron effective mass in InGaAs compared to GaAs. However, the InAs lattice parameter of about 0.606 nm is considerably larger than the GaAs lattice constant of 0.565 nm, and due to strain effects the compositional limit of psuedomorphic InGaAs on GaAs substrates is limited to about x = 0.30. One method to achieve higher InAs content in the InGaAs channel is to use an indium phosphide substrate with a lattice constant of 0.587 nm. This newer generation of HEMTs devices uses $In_{0.53}Ga_{0.47}As$ channel lattice matched to InP substrates. InP-based pseudomorphic HEMTs with $In_xGa_{1-x}As$ channel compositions of up to about x = 0.80 have achieved improvements in performance capability for both low noise and power amplifier applications compared to the best GaAs-based devices.

InP-based MMICs require semi-insulating substrates with resistivities from 10^{-6} to 10^{-8} ohm cm. To achieve such resistivities, in nominally undoped crystals would require the residual donor concentration to be reduced by a factor of at least 10^6 from current values. This is not practical and an alternate method is required that employs acceptor doping to compensate the residual donors. In principal, any acceptor can compensate the donors. However, because bulk indium phosphide crystals commonly have a short range variation of at least 5% in donor and acceptor concentration, the maximum resistivity that can be obtained by compensation with shallow acceptors in the absence of p-n junction formation, is only about 15 ohm cm. Resistivities in the semi-insulating range are usually obtained by doping with a deep acceptor such as iron (Fe). Fe is by far the most widely used deep acceptor to produce semi-insulating substrates. As a substitutional impurity, Fe is generally stable under normal device operating temperatures. At high temperatures there is concern of possible diffusion of Fe into the epitaxial layer leading to adverse effects on the devices. Diffusion studies of Fe in InP at temperature of 650°C for four hours indicated virtually no change from the control sample.[7]

Use of InP materials is sometimes restricted by its cost. InP substrates are significantly more costly than those made from GaAs, and even more so when compared to silicon. In addition, the technology of indium phosphide substrate manufacturing is much more difficult than for GaAs or silicon. This situation is not simply the result of lower market demand but is linked fundamentally to the high vapor pressure of phosphorus that creates an obstacle to the synthesis of single crystal boules of larger diameters. While 8-inch silicon substrates and 6-inch GaAs substrates are the rule in commercial fabrication, indium phosphide substrates are still primarily 2 inch. Three-inch diameter wafers are becoming available and 4-inch wafers are being validated. New concepts of single crystal growth may provide larger diameter wafers at low cost leading to wider acceptance of this compound semiconductor for microwave and RF applications.

Silicon Carbide

Silicon carbide possesses many intrinsic material properties that make it ideal for a wide variety of high power, high temperature, and high frequency electronic device applications. In comparison with silicon and gallium arsenide, SiC has greater than 2.5x larger bandgap, 3x greater thermal conductivity, and a factor of 10 larger breakdown electric field. These characteristics enable SiC devices to operate at higher temperatures and power levels, with lower on-resistances, and in harsh environments inaccessible to other semiconductors. However, electron transport properties may limit SiC applications to frequencies less than 10 GHz. While excellent prototype SiC devices have been demonstrated, SiC devices will not be widely available until material quality improves and material cost drops.[8]

SiC belongs to a class of semiconductors commonly known as "wide bandgap," which makes it, among other things, less sensitive to increased temperatures. Properly designed and fabricated SiC devices should operate at 500°C or higher, a realm that Si does not even approach. Furthermore, the thermal conductivity of SiC exceeds even that of copper; any heat produced by a device is therefore quickly dissipated. The inertness of SiC to chemical reaction implies that devices have the potential to operate even in the most caustic of environments. SiC is extremely hard and is best known as the grit coating on sandpaper. This hardness again implies that SiC devices can operate under conditions of extreme pressure. SiC is extremely radiation hard and can be used close to reactors or for electronic hardware in space. Properties of particular importance to the microwave and RF device design engineer are high electric field strength and relatively high saturation drift velocity.

SiC exists not as a single crystal type, but as a whole family of crystals known as polytypes. Each crystal structure has its own unique electrical and optical properties.[9,10] Polytypes differ not in the relative numbers of Si and C atoms, but in the arrangement of these atoms in layers as illustrated in Fig. 9.54. The polytypes are named according to the periodicity of these layers. For example, one of the most common polytypes is called 6H, which means a hexagonal type lattice with an arrangement of 6 different Si+C layers before the pattern repeats itself. In total, more than 200 different polytypes of SiC have been shown to exist, some with patterns that do not repeat for hundreds of layers. The exact physical properties of SiC depend on the crystal structure adopted. Some of the most common structures used are 6H, 4H, and 3C, the last being the one cubic form of SiC.

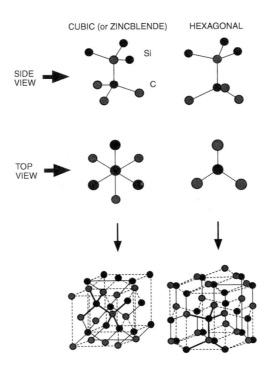

FIGURE 9.54   Difference between the cubic and hexagonal polytypes of SiC. (From Virgil B. Shields, 1994. With permission.)

Silicon carbide (SiC) is also emerging as a substrate material that may meet the challenging requirements for GaN growth.[11,12] Type 6H-SiC is lattice matched to within 3.5% of GaN compared to 16% for sapphire.[13] SiC has a thermal conductivity that is over 10 times higher than sapphire.

Figure 9.55 compares the energy band and lattice constants of the conventional III-V semiconductor material systems shown in the circle and the wide gap semiconductors. As shown in the figure, SiC is reasonably lattice matched to GaN and makes a good candidate for a substrate material.

Gallium Nitride

Gallium nitride in the wurtzite form (2H polytype) has a bandgap of 3.45 eV (near UV region) at room temperature. It also forms a continuous range of solid solutions with AlN (6.28 eV) and a discontinuous range of solid solutions with InN (1.95 eV). The wide bandgap, the heterojunction capability, and the strong atomic bonding of these materials make them good candidates for RF and microwave devices. Other pertinent device-related parameters include a good thermal conductivity of 1.5 W/cm-K, a type I heterojunction with AlN and AlGaN alloys, large band discontinuities with resultant large interface carrier concentrations, and a large breakdown voltage.

Intrinsic material properties of gallium nitride combined with hetero-structure designs are producing revolutionary improvement in output power for microwave devices. GaN-based FET devices have demonstrated power densities of near 10 W/mm of gate width compared to 1 W/mm for the best GaAs-based devices.[14] GaN has a bandgap of 3.45 eV compared to 1.43 eV for GaAs. This property leads to orders of magnitude lower thermal leakage. GaN has a thermal conductivity almost 3 times higher than GaAs, a parameter that is critically important for power amplifier applications. The dielectric breakdown strength of GaN is a factor of 2 greater than GaAs, further supporting the consideration of this material for microwave power amplifier applications. A key limitation in the development of GaN devices is the fact that there is no native GaN substrate. Currently, GaN must be deposited on host substrates such as sapphire or SiC.

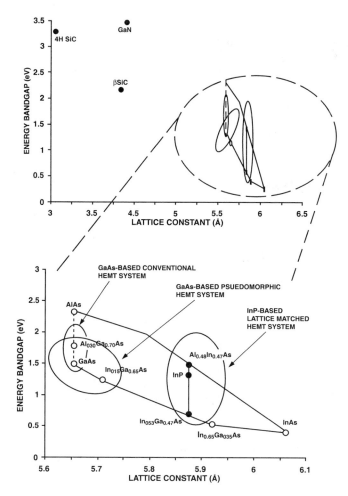

FIGURE 9.55 Comparison of conventional and wide bandgap materials.

GaN, AlN, and InGaN have a polar wurtzite structure as shown in Fig. 9.56 and epitaxial films of these materials typically grow along the polar axis. Although the polarity of these nitrides has been studied by a number of techniques, many results in the literature are in conflict.[15] The wurtzite lattice can be considered as two interpenetrating hexagonal close-packed lattices. The wurtzite structure has a tetrahedral arrangement of four equidistant nearest neighbors, similar to the GaAs zincblende structure.

Electron transport in GaN is governed by the electron mobility. Low field electron mobility, in bulk wurtzite GaN, is limited by charged states that are dislocation related, as well as by isolated donor ions and phonons. With no dislocations or donor ions, the 300°K phonon-limited mobility is near 2000 cm^2/V-s. Combined scattering effects reduce the mobility for both highly-doped and lightly-doped material, for a given dislocation density. Dislocation density depends on the nucleation methods used at growth initiation, usually ranging from 3×10^9/cm^2 to 3×10^{10}/cm^2.[16]

The average drift velocity at high fields in bulk GaN, or in GaN MODFETs, is expected to be above 2×10^7 cm/s. These high fields extend increasingly toward the drain with higher applied drain-source voltage. The average transit velocity is $>1.25 \times 10^7$ cm/s when only the effective gate length is considered, without the extended high-field region. This result is consistent with an average drift velocity of $\sim 2 \times 10^7$ cm/s over an extended region.

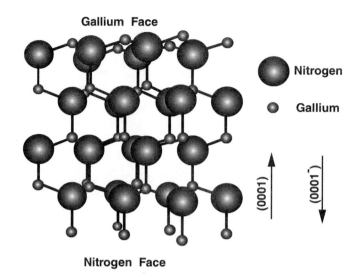

FIGURE 9.56   Wurtzite structure of GaN (Hellman[15] used with permission).

Growth of GaN and its related compounds is dominated by metal-organic vapor phase epitaxy (MOVPE) or MOCVD metal-organic chemical vapor deposition, techniques with obvious advantages in fields where high throughput is required. Molecular beam epitaxy (MBE) is also used to grow GaN films. The III-nitride community grows films of GaN and related nitride materials using hetero-epitaxial growth routes because of the lack of bulk substrates of these materials. This results in films containing dislocations because of the mismatches in the lattice parameters and the coefficients of thermal expansion between the buffer layer and the film and/or the buffer layer and the substrate. These high concentrations of dislocations may also limit the performance of devices.

Advanced growth procedures, including selective area growth (SAG) and lateral epitaxial overgrowth (LEO) techniques for GaN deposition, are being used specifically to significantly reduce the dislocation density. The latter technique involves the initial vertical growth of a configuration and material composition through windows etched in an SiO_2 or Si_3N_4 mask previously deposited on an underlying GaN seed layer and the subsequent lateral and vertical growth of a film over and adjacent to the mask. Reduction in the number of defects in the material and growth of GaN on larger diameter wafers will lead to the production of GaN-based devices and circuits for microwave and higher frequency applications.

Selected Material Properties

Table 9.19 contains the most common material properties on semiconductors used for microwave and RF applications. Sources of information in this table include textbooks, articles, and World Wide Web sites. Web sites are a convenient way to find materials data and several of the sites used to prepare Table 9.19 are listed in Table 9.20. Most of these sites include detailed references that will provide the user with more extensive information. When using these material properties in the design of components, it is recommended that the user examine the temperature and frequency behavior of the specific property before finalizing the design.

Etching Processes for Semiconductors

Table 9.21 lists typical etching processes for the various semiconductors described in this section.

Ohmic and Schottky Contacts

Table 9.22 is a list of metallizations that have been used to successfully fabricate low resistance ohmic contacts for drain and source contacts in FET devices and low leakage Schottky barrier contacts for metal semiconductor diode gates.

TABLE 9.19 Selected Material Properties of Semiconductors for Microwave and RF Applications

Property	Si	SiC	InP	GaAs	GaN
Atoms/cm^3	5.0×10^{22}	3.95×10^{22}	4.43×10^{22}	4.96×10^{22}	
Atomic weight	28.09	40.1	72.90	72.32	41.87
Breakdown Field (V/cm)	3×10^5 [32]	20×10^4 3C-SiC[27]	5×10^5 [32]	6×10^5	$>10 \times 10^5$
		30×10^5 4H-SiC[27]			
Crystal structure	Diamond	Zincblende	Zincblende	Zincblende	Wurtzite
Density (g/cm^3)	2.3283[23]	4.787[23]	5.316[33]	6.1[23]	
Dielectric constant	11.8[23]	9.75[17]	12.4[23]	12.5[23]	9[35]
		9.66[18]			
Effective mass m*/m$_0$	1.1	0.37 3C-SiC[19]	0.067[23]	0.068[23]	0.22[35,36]
Electron		0.45 6H-SiC[20]			
Electron Affinity, eV	4.05[31]	—	4.38[31]	4.07[31]	3.4[34]
Energy Gap (eV) at 300 K	1.107[23]	2.403 3C-SiC[23]	1.29[31]	1.35[31]	3.34[37]
		3.101 6H-SiC[23]			
Intrinsic carrier concentration (cm^{-3})	1.45×10^{10} [23]	3×10^6 3C-SiC[21]	1.6×10^7 [23]	1.8×10^6 [23]	$3\text{-}6 \times 10^9$ [23]
		$10^{15}\text{--}10^{16}$ 6H-SiC[22]			
Lattice constant (Angstroms)	5.431[31]	4.3596[27]	5.860[31]	5.651[31]	3.190[38]
Linear Coeff. of thermal expansion (10^{-6} K^{-1})	2.49[23]	5.48[22]	4.6[23]	5.4[23]	5.6[27]
Melting point (K)	1685[23]	3070[23]	1335[31]	1511[31]	—
Electron mobility (cm^2/V-S) μ_m	1900[23]	1000 3C-SiC[24]	4600[23]	8800[23]	1000[39]
		600 6H-SiC[24]			
Holes mobility μ_p (cm^2/V-S)	500[23]	40 3C-SiC[24]	150[23]	400[23]	30[39]
		40 6H-SiC[24]			
Optical phonon energy (eV)	.063 eV[31]	—	0.43[31]	0.35[31]	.912[40]
Refractive index	3.42[23]	2.65 3C-SiC[25]	3.1[31]	3.66[31]	2.7[41]
		2.72 6H-SiC[26]			(at band edge)
Resistivity, intrinsic (Ω-cm)	1000[31]	150 3C-SiC[27]	8.2×10^7 [31]	3.8×10^8 [31]	$>10^{13}$ [27]
		$>10^{12}$ 4H-SiC[27]			
Specific heat (J/kg°K)	702[23]	640[28]	310[31]	325[32]	847.39[42]
Thermal conductivity at 300°K (Watt/cm°K)	1.24[23]	3.2 3C-SiC[29]	.77[32]	.56[31]	1.3[43]
		4.9 6H-SiC[30]			

TABLE 9.20 World Wide Web Sites for Semiconductor Material Properties

http://www.ioffe.rssi.ru/SVA/NSM/Nano/index.html
http://mems.isi.edu/mems/materials/
http://www.webelements.com/webelements/elements/
http://www.sensors-research.com/senres/materials.htm
http://nsr.mij.mrs.org/
http://nina.ecse.rpi.edu/shur/nitride.htm

TABLE 9.21 Etching Processes for Various Semiconductors

Etchant	Substrate and Conditions	Applications	Etch Rate	Ref.
15 HNO_3, 5 CH_3COOH, 2 HF (planar)	Si	For general etching	Variant	44
110 ml CH_3COOH, 100 ml HNO_3, 50 ml HF, 3 g I_2 (iodine etch)	Si	For general etching	Variant	45
KOH solutions; hydrazine (a) KOH (3–50%) (b) 60 hydrazine, 40 H_2O	(100) Si 70–90 C, SiO_2 unmasked 110 C, 10 min, unmasked	For texturing and V grooving solar etch (no texturizing) for diffuse-reflectivity texturizing	Variant	46
28% by weight KOH at 90–95°C	[100] Si, with oxide mask	V-shaped channel etching for optical waveguides	~2 microns/min	47
3CH3OH, 1H3PO4, 1H2O2	[110], [100], Ga [111]GaAs	Preferential structural etching	~2 microns/min, except Ga[111] reduced twofold	48
3 ml HPO_4, 1 ml H_2O_2, 50 ml H_2O	GaAs, 300 K with Shipley 1813 mask	Wet etch	~.1 microns/min	
BCL_3 Gas, reactive ion etch	GaN, BCL_3, 150 W, 10 mTorr	Slow etching, used to obtain a vertical profile for mesa isolation	~.02 microns/min	

TABLE 9.22 Metallization Systems for Ohmic and Schottky Barrier Contacts

Contact	SiC	GaAs	GaN	InP
Ohmic contact mettalization	Ta, Al/Ti, Ni, Ti, Mo, Ta[50]	Au/Ge/Ni, In, Sn, Ag and Ag alloys, Au, Ni[52]	Ti/Al/Ni/Aun-GaN Ni/Aup-GaN[53]	Au and Au alloys, near noble transition metals (Co, Ni, Ti, Pd, Pt)[49]
Schottky gate mettalization	Au and Au alloys[50]	Ti/Pt/Au multilayer structures, Au alloys[51]	Ni/Au/Ti n-GaN Aup-GaN[53]	For AlInP/InP 15% Al superlattice, 20% Al quantum well[52]

References

1. Yu, P.Y. and Cardona M., *Fundamentals of Semiconductors,* New York, Springer-Verlag, 1999, 1.
2. Sze, S.M., Ed., *VLSI Technology,* New York, McGraw-Hill, 1983, 35.
3. Sumitomo Electric III-V Semiconductors Specifications, 32.
4. Sze, S.M., Ed., *Physics of Semiconductors,* New York, John-Wiley & Sons, 1981, 15, 43.
5. Bahl, I. and Bhartia P., *Microwave Solid State Circuit Design*, New York, John Wiley & Sons, 1988, 307.
6. Pearsall, T.P., *Properties, Processing and Applications of Indium Phosphide,* Eisele, H. and Haddad, G.I., Eds., INSPEC, London, 2000, 40.
7. Katz, A., *Induim Phosphide and Related Materials: Processing, Technology and Devices,* Byrne, E.K. and Katz, A., Eds., Boston, Artech House, 1992, 169.
8. Brandes G.R., Growth of SiC Boules and Epitaxial Films for Next Generation Electronic Devices, American Physical Society Centennial Meeting, March 20–26, Atlanta, GA, 1999.
9. von-munch, W., Silicon Carbide, in *Physik der Elemete der* IV. Gruppe und der III-V Verbindungen, K.-H. Hellwege, Ed. Berlin, Heidelberg, Springer-Verlag, 1982, 132–142.
10. Powell, J. A., Pirouz, P., and Choyke, W.J., Growth and Characterization of Silicon Carbide Polytypes for Electronic Applications, in *Semiconductor Interfaces, Microstructures, and Devices: Properties and Applications*, Feng Z.C., Ed. Bristol, United Kingdom, Institute of Physics Publishing, 1993, 257–293.

11. Mohammad, S.N., Salvador A.A and Morkoc Hardel, Emerging Gallium Nitride Based Devices, *Proceedings of the IEEE*, 83, 10, 1306–1355, Oct. 1995.

12. Lin, M.E., Sverdlov, B., Zhou, G.L., and Morkoc, H., A Comparative Study of GaN Epilayers Grown on Sapphire and SiC Substrates by Plasma-assisted Molecular-beam Epitaxy, *Appl. Phys. Lett.*, 62, 26, 3479–3481, 1993.

13. Lin, M.E. et al. Low Resistance Ohmic Contacts on Wide Band-gap GaN, *Appl. Phys. Lett.*, 64, 1003–1005, 1994.

14. Sheppard, S., Doverspike, K., Leonard, M., Pribble, W., Allen, S., and Palmour, J., Improved Operation of GaN/AlGaN HEMTS on Silicon Carbide, International Conference on Silicon Carbide and Related Materials 1999, Paper 349.

15. Hellman, E.S., The Polarity of GaN: a Critical Review. *Materials Research Society Internet Journal*, Vol 3, Res 3, May 19, 1998.

16. Eastman, L., Chu, K., Schaff, W., Murphy, M.,Weimann, M., and Eustis, T., High Frequency AlGaN/GaN MODFET's, *Materials Research Society Internet Journal*, Vol. 2, Res 2, August 12, 1997.

17. Patrick, L., Choyke, W.J., *Physics Review Journal*, 2, 2255, 1977.

18. Choyke, W.J., *NATO ASI Ser. E, Applied Sciences*, 185, and references therin, 1990.

19. Suzuki, A., Ogura A., Furukawa, K., Fujii, Y., Shigeta, M., and Nakajima, S., *Journal of Applied Physics*, 64, 2818, 1988.

20. Wessels, B.W. and Gatos, J.C., *Journal of Physics Chemical Solids*, 15, 83, 1977.

21. Anikin, M.M., Zubrilov, A.S., Lebedev, A.A., Sterlchuck, A.M., and Cherenkov, A.E., *Fitz. Tekh. Polurpovdn.*, 24, 467, 1992.

22. Schackelford, J., Alexander, W., and Parker, J., Eds., *CRC Materials Science and Engineering Handbook*, 2nd ed., Boca Raton, FL, CRC Press, 1994, 304.

23. Lide, D.R., Properties of Semiconductors, *CRC Handbook of Chemistry and Physics*, 12–98, 1999–2000 ed. Boca Raton, FL, CRC Press, 2000.

24. http://vshields.jpl.nasa.gov/windchime.html.

25. Schaffer, P.T.B. and Naum, R.G., *Journal of Optical Society of America*, 59, 1498, 1969.

26. Scaffer, P.T.B., *Applied Optics*, 10, 1034, 1977.

27. Yoder, M.N.,*Wide Bandgap Semiconductor Materials and Devices* , 43, 10, 1634, 1966.

28. Kern, E.L., Hamill, H.W., Deem, H.W., and Sheets, H.D., *Mater Res. Bull.*, 4, 107, 1969.

29. Morelli, D., Hermans, J., Bettz, C., Woo, W.S., Harris, G.L., and Taylor, C., *Inst. Physics Conf. Ser.*, 137, 313–6, 1990.

30. Parafenova, I.I., Tairov, Y.M., and Tsvetkov, V.F., *Sov. Physics-Semiconductors*, 24, 2, 158–61, 1990.

31. http://www.ioffe.rssi.ru/SVA/NSM/Nano/index.html.

32. Garland, C.W. and Parks, K.C., *Journal Applied Physics*, 33, 759, 1962.

33. http://mems.isi.edu/mems/materials/measurements.cgi?MATTAG=galliumarsenidegaas-bulk&PAGE_SIZE=20.

34. Benjamin, M.C., Wang, C., Davis, R.F., and Nemanich, R.J., *Applied Phys. Letter*, 64, 3288, 1994.

35. Mohammad, S.N. and Morkoc, H., *Prog. Quant. Electron.*, 20, 361, 1996.

36. Barker, A.S. and Ilegems, M., *Phys. Rev. B*, 7, 743, 1973.

37. Maruska, H.P. and Tietjen, J.J., *Appl. Phys. Lett.*, 15, 327, 1969.

38. Shur, M.S. and Khan, A.M., *Mat. Res. Bull.*, 22, 2, 44, 1977.

39. Bhapkar, U.V. and Shur, M. S. , *Journal of Applied Physics*, 82, 4, 1649, 1997.

40. Chin, V. W. L., Tansley, T. L., andOsotchan, T., *Journal of Applied Physics*, 75, 7365, 1994.

41. Billeb, A., Grieshhaber, W., Stocker, D., Schubert, E.F., and Karlicek, R.F., *Applied. Physics. Letter*, 69, 2953, 1996.

42. Koshchenko, V.I., Grinberg, Ya, K.H., and Demidienko, A.F., *Inorganic Matter*, 11, 1550–3, 1984.

43. Sichel, E.K. and Pankove, J.I., *J. Phys. Chem. Solids*, 38, 330, 1977.

44. Ruynun, W.R., *Semiconductor Measurements and Instrumentation*, McGraw-Hill, New York, 1975, 129, Table 7.3, Chaps. 1, 2, 7, 9.

45. Integrated Circuit Silicon Device Technology; X-Chemical Metallurgical Properties of Silicon, ASD-TDR-63-316, Vol. X, AD 625, 985. Research Triangle Inst., Research Triangle Park, North Carolina.

46. Baroana, C.R. and Brandhorst, H.W., *IEEE Photovoltaic Spec. Conf. Proc.*, Scottsdale, AZ, 44–48, 1975.

47. http://peta.ee.cornell.edu/~jay/res/vgroove/

48. Merz, J.L. and Logan, R.A., *J. Appl. Physics*, 47, 3503, 1976.

49. Pearshall, T.P., Processing Technologies, in *Properties, Processing, and Applications of Induim Phosphide,* Katz A. and Pearshall, T.P., Eds., Inspec, London, 2000, 246.

50. Harris, G.L., SiC Devices and Ohmic Contacts, in *Properties of SiC,* Harris, G.L, Kelner, G., and Shur, M., Eds., Inspec, London, 1995, 233, 243.

51. Misssous, M., Interfaces and Contacts, in *Properties of Gallium Arsenide,* Morgan, D.V. and Wood, J., Eds., Inspec, London, 1990, 386–387.

52. Lammasniemi, J., Tappura K., and Smekalin K., *Applied Physics Letter,* 65, 20, 2574–5, 1998.

53. Edgar, J. H., Strite, S., Akasaki, I., Amano, H., and Wetzel, C., Specifications, characterisation and applications of GaN based devices, in *Properties, Processing and Applications of Gallium Nitride and Related Semiconductors,* Mohney, S.E., Ed., Inspec, London, 1999, 491–96.

APPENDIX A

Mathematics, Symbols, and Physical Constants

Greek Alphabet

	Greek letter		Greek name	English equivalent		Greek letter		Greek name	English equivalent
A	α		Alpha	a	N	ν		Nu	n
B	β		Beta	b	Ξ	ξ		Xi	x
Γ	γ		Gamma	g	O	o		Omicron	ŏ
Δ	δ		Delta	d	Π	π		Pi	p
E	ε		Epsilon	ĕ	P	ρ		Rho	r
Z	ζ		Zeta	z	Σ	σ		Sigma	s
H	η		Eta	ē	T	τ		Tau	t
Θ	θ	ϑ	Theta	th	Y	υ		Upsilon	u
I	ι		Iota	i	Φ	ϕ	φ	Phi	ph
K	κ		Kappa	k	X	χ		Chi	ch
Λ	λ		Lambda	l	Ψ	ψ		Psi	ps
M	μ		Mu	m	Ω	ω		Omega	ō

International System of Units (SI)

The International System of units (SI) was adopted by the 11th General Conference on Weights and Measures (CGPM) in 1960. It is a coherent system of units built form seven *SI base units,* one for each of the seven dimensionally independent base quantities: they are the meter, kilogram, second, ampere, kelvin, mole, and candela, for the dimensions length, mass, time, electric current, thermodynamic temperature, amount of substance, and luminous intensity, respectively. The definitions of the SI base units are given below. The *SI derived units* are expressed as products of powers of the base units, analogous to the corresponding relations between physical quantities but with numerical factors equal to unity.

In the International System there is only one SI unit for each physical quantity. This is either the appropriate SI base unit itself or the appropriate SI derived unit. However, any of the approved decimal prefixes, called *SI prefixes,* may be used to construct decimal multiples or submultiples of SI units.

It is recommended that only SI units be used in science and technology (with SI prefixes where appropriate). Where there are special reasons for making an exception to this rule, it is recommended always to define the units used in terms of SI units. This section is based on information supplied by IUPAC.

Definitions of SI Base Units

Meter—The meter is the length of path traveled by light in vacuum during a time interval of 1/299 792 458 of a second (17th CGPM, 1983).

Kilogram—The kilogram is the unit of mass; it is equal to the mass of the international prototype of the kilogram (3rd CGPM, 1901).

Second—The second is the duration of 9 192 631 770 periods of the radiation corresponding to the transition between the two hyperfine levels of the ground state of the cesium-133 atom (13th CGPM, 1967).

Ampere—The ampere is that constant current which, if maintained in two straight parallel conductors of infinite length, of negligible circular cross-section, and placed 1 meter apart in vacuum, would produce between these conductors a force equal to 2×10^{-7} newton per meter of length (9th CGPM, 1948).

Kelvin—The kelvin, unit of thermodynamic temperature, is the fraction 1/273.16 of the thermodynamic temperature of the triple point of water (13th CGPM, 1967).

Mole—The mole is the amount of substance of a system which contains as many elementary entities as there are atoms in 0.012 kilogram of carbon-12. When the mole is used, the elementary entities must be specified and may be atoms, molecules, ions, electrons, or other particles, or specified groups of such particles (14th CGPM, 1971).

Examples of the use of the mole:

   1 mol of H_2 contains about 6.022×10^{23} H_2 molecules, or 12.044×10^{23} H atoms

   1 mol of HgCl has a mass of 236.04 g

   1 mol of Hg_2Cl_2 has a mass of 472.08 g

   1 mol of Hg_2^{2+} has a mass of 401.18 g and a charge of 192.97 kC

   1 mol of $Fe_{0.91}S$ has a mass of 82.88 g

   1 mol of e^- has a mass of 548.60 µg and a charge of −96.49 kC

   1 mol of photons whose frequency is 10^{14} Hz has energy of about 39.90 kJ

Candela—The candela is the luminous intensity, in a given direction, of a source that emits monochromatic radiation of frequency 540×10^{12} hertz and that has a radiant intensity in that direction of (1/683) watt per steradian (16th CGPM, 1979).

Names and Symbols for the SI Base Units

Physical quantity	Name of SI unit	Symbol for SI unit
length	meter	m
mass	kilogram	kg
time	second	s
electric current	ampere	A
thermodynamic temperature	kelvin	K
amount of substance	mole	mol
luminous intensity	candela	cd

SI Derived Units with Special Names and Symbols

Physical quantity	Name of SI unit	Symbol for SI unit	Expression in terms of SI base units	
frequency[1]	hertz	Hz	s^{-1}	
force	newton	N	$m \ kg \ s^{-2}$	
pressure, stress	pascal	Pa	$N \ m^{-2}$	$= m^{-1} \ kg \ s^{-2}$
energy, work, heat	joule	J	$N \ m$	$= m^2 \ kg \ s^{-2}$
power, radiant flux	watt	W	$J \ s^{-1}$	$= m^2 \ kg \ s^{-3}$
electric charge	coulomb	C	$A \ s$	
electric potential, electromotive force	volt	V	$J \ C^{-1}$	$= m^2 \ kg \ s^{-3} \ A^{-1}$

Physical quantity	Name of SI unit	Symbol for SI unit	Expression in terms of SI base units	
electric resistance	ohm	Ω	$V\ A^{-1}$	$= m^2\ kg\ s^{-3}\ A^{-2}$
electric conductance	siemens	S	Ω^{-1}	$= m^{-2}\ kg^{-1}\ s^3\ A^2$
electric capacitance	farad	F	$C\ V^{-1}$	$= m^{-2}\ kg^{-1}\ s^4\ A^2$
magnetic flux density	tesla	T	$V\ s\ m^{-2}$	$= kg\ s^{-2}\ A^{-1}$
magnetic flux	weber	Wb	$V\ s$	$= m^2\ kg\ s^{-2}\ A^{-1}$
inductance	henry	H	$V\ A^{-1}\ s$	$= m^2\ kg\ s^{-2}\ A^{-2}$
Celsius temperature[2]	degree Celsius	°C	K	
luminous flux	lumen	lm	cd sr	
illuminance	lux	lx	$cd\ sr\ m^{-2}$	
activity (radioactive)	becquerel	Bq	s^{-1}	
absorbed dose (of radiation)	gray	Gy	$J\ kg^{-1}$	$= m^2\ s^{-2}$
dose equivalent (dose equivalent index)	sievert	Sv	$J\ kg^{-1}$	$= m^2\ s^{-2}$
plane angle	radian	rad	1	$= m\ m^{-1}$
solid angle	steradian	sr	1	$= m^2\ m^{-2}$

[1]For radial (circular) frequency and for angular velocity the unit rad s^{-1}, or simply s^{-1}, should be used, and this may not be simplified to Hz. The unit Hz should be used only for frequency in the sense of cycles per second.

[2]The Celsius temperature θ is defined by the equation:
$$\theta/°C = T/K - 273.15$$
The SI unit of Celsius temperature interval is the degree Celsius, °C, which is equal to the kelvin, K. °C should be treated as a single symbol, with no space between the ° sign and the letter C. (The symbol °K, and the symbol °, should no longer be used.)

Units in Use Together with the SI

These units are not part of the SI, but it is recognized that they will continue to be used in appropriate contexts. SI prefixes may be attached to some of these units, such as milliliter, ml; millibar, mbar; megaelectronvolt, MeV; kilotonne, ktonne.

Physical quantity	Name of unit	Symbol for unit	Value in SI units	
time	minute	min	60 s	
time	hour	h	3600 s	
time	day	d	86 400 s	
plane angle	degree	°	$(\pi/180)$ rad	
plane angle	minute	'	$(\pi/10\ 800)$ rad	
plane angle	second	"	$(\pi/648\ 000)$ rad	
length	ångstrom[1]	Å	10^{-10} m	
area	barn	b	10^{-28} m^2	
volume	litre	l, L	dm^3	$= 10^{-3}\ m^3$
mass	tonne	t	Mg	$= 10^3$ kg
pressure	bar[1]	bar	10^5 Pa	$= 10^5\ N\ m^{-2}$
energy	electronvolt[2]	eV ($= e \times V$)	$\approx 1.60218 \times 10^{-19}$ J	
mass	unified atomic mass unit[2,3]	u ($= m_a(^{12}C)/12$)	$\approx 1.66054 \times 10^{-27}$ kg	

[1]The ångstrom and the bar are approved by CIPM for "temporary use with SI units," until CIPM makes a further recommendation. However, they should not be introduced where they are not used at present.

[2]The values of these units in terms of the corresponding SI units are not exact, since they depend on the values of the physical constants e (for the electronvolt) and N_a (for the unified atomic mass unit), which are determined by experiment.

[3]The unified atomic mass unit is also sometimes called the dalton, with symbol Da, although the name and symbol have not been approved by CGPM.

Physical Constants

General

Equatorial radius of the earth = 6378.388 km = 3963.34 miles (statute).
Polar radius of the earth, 6356.912 km = 3949.99 miles (statute).
1 degree of latitude at 40° = 69 miles.
1 international nautical mile = 1.15078 miles (statute) = 1852 m = 6076.115 ft.
Mean density of the earth = 5.522 g/cm³ = 344.7 lb/ft³
Constant of gravitation $(6.673 \pm 0.003) \times 10^{-8}$ cm³ gm⁻¹ s⁻².
Acceleration due to gravity at sea level, latitude 45° = 980.6194 cm/s² = 32.1726 ft/s².
Length of seconds pendulum at sea level, latitude 45° = 99.3575 cm = 39.1171 in.
1 knot (international) = 101.269 ft/min = 1.6878 ft/s = 1.1508 miles (statute)/h.
1 micron = 10^{-4} cm.
1 ångstrom = 10^{-8} cm.
Mass of hydrogen atom = $(1.67339 \pm 0.0031) \times 10^{-24}$ g.
Density of mercury at 0°C = 13.5955 g/ml.
Density of water at 3.98°C = 1.000000 g/ml.
Density, maximum, of water, at 3.98°C = 0.999973 g/cm³.
Density of dry air at 0°C, 760 mm = 1.2929 g/l.
Velocity of sound in dry air at 0°C = 331.36 m/s – 1087.1 ft/s.
Velocity of light in vacuum = $(2.997925 \pm 0.000002) \times 10^{10}$ cm/s.
Heat of fusion of water 0°C = 79.71 cal/g.
Heat of vaporization of water 100°C = 539.55 cal/g.
Electrochemical equivalent of silver 0.001118 g/s international amp.
Absolute wavelength of red cadmium light in air at 15°C, 760 mm pressure = 6438.4696 Å.
Wavelength of orange-red line of krypton 86 = 6057.802 Å.

π Constants

π = 3.14159 26535 89793 23846 26433 83279 50288 41971 69399 37511
$1/\pi$ = 0.31830 98861 83790 67153 77675 26745 02872 40689 19291 48091
π^2 = 9.8690 44010 89358 61883 44909 99876 15113 53136 99407 24079
$\log_e\pi$ = 1.14472 98858 49400 17414 34273 51353 05871 16472 94812 91531
$\log_{10}\pi$ = 0.49714 98726 94133 85435 12682 88290 89887 36516 78324 38044
$\log_{10}\sqrt{2\pi}$ = 0.39908 99341 79057 52478 25035 91507 69595 02099 34102 92128

Constants Involving *e*

e = 2.71828 18284 59045 23536 02874 71352 66249 77572 47093 69996
$1/e$ = 0.36787 94411 71442 32159 55237 70161 46086 74458 11131 03177
e^2 = 7.38905 60989 30650 22723 04274 60575 00781 31803 15570 55185
$M = \log_{10}e$ = 0.43429 44819 03251 82765 11289 18916 60508 22943 97005 80367
$1/M = \log_e10$ = 2.30258 50929 94045 68401 79914 54684 36420 67011 01488 62877
$\log_{10}M$ = 9.63778 43113 00536 78912 29674 98645 -10

Numerical Constants

$\sqrt{2}$ = 1.41421 35623 73095 04880 16887 24209 69807 85696 71875 37695
$\sqrt[3]{2}$ = 1.25992 10498 94873 16476 72106 07278 22835 05702 51464 70151
$\log_e2$ = 0.69314 71805 59945 30941 72321 21458 17656 80755 00134 36026
$\log_{10}2$ = 0.30102 99956 63981 19521 37388 94724 49302 67881 89881 46211
$\sqrt{3}$ = 1.73205 08075 68877 29352 74463 41505 87236 69428 05253 81039
$\sqrt[3]{3}$ = 1.44224 95703 07408 38232 16383 10780 10958 83918 69253 49935
$\log_e3$ = 1.09861 22886 68109 69139 52452 36922 52570 46474 90557 82275
$\log_{10}3$ = 0.47712 12547 19662 43729 50279 03255 11530 92001 28864 19070

Symbols and Terminology for Physical and Chemical Quantities

Name	Symbol	Definition	SI unit
		Classical Mechanics	
mass	m		kg
reduced mass	μ	$\mu = m_1 m_2/(m_1 + m_2)$	kg
density, mass density	ρ	$\rho = M/V$	kg m^{-3}
relative density	d	$d = \rho/\rho^\theta$	1
surface density	ρ_A, ρ_S	$\rho_A = m/A$	kg m^{-2}
specific volume	v	$v = V/m = 1/\rho$	m^3 kg^{-1}
momentum	p	$\boldsymbol{p} = mv$	kg m s^{-1}
angular momentum, action	L	$l = r \times p$	J s
moment of inertia	I, J	$I = \Sigma m_i r_i^2$	kg m^2
force	F	$\boldsymbol{F} = d\boldsymbol{p}/dt = ma$	N
torque, moment of a force	$\boldsymbol{T}, (\boldsymbol{M})$	$T = r \times F$	N m
energy	E		J
potential energy	E_p, V, Φ	$E_p = -\int \boldsymbol{F} \cdot ds$	J
kinetic energy	E_k, T, K	$e_k = (1/2)mv^2$	J
work	W, w	$w = \int \boldsymbol{F} \cdot ds$	J
Hamilton function	H	$H(q, p)$ $= T(q, p) + V(q)$	J
Lagrange function	L	$L(q, \dot{q})$ $T(q, \dot{q}) - V(q)$	J
pressure	p, P	$p = F/A$	Pa, N m^{-2}
surface tension	γ, σ	$\gamma = dW/dA$	N m^{-1}, J m^{-2}
weight	$G, (W, P)$	$G = mg$	N
gravitational constant	G	$F = Gm_1 m_2/r^2$	N m^2 kg^{-2}
normal stress	σ	$\sigma = F/A$	Pa
shear stress	τ	$\tau = F/A$	Pa
linear strain, relative elongation	ε, e	$\varepsilon = \Delta l/l$	1
modulus of elasticity, Young's modulus	E	$E = \sigma/\varepsilon$	Pa
shear strain	γ	$\gamma = \Delta x/d$	1
shear modulus	G	$G = \tau/\gamma$	Pa
volume strain, bulk strain	θ	$\theta = \Delta V/V_0$	1
bulk modulus, compression modulus	K	$K = -V_0(dp/dV)$	Pa
viscosity, dynamic viscosity	η, μ	$\tau_{x,z} = \eta(dv_x/dz)$	Pa s
fluidity	ϕ	$\phi = 1/\eta$	m kg^{-1} s
kinematic viscosity	v	$v = \eta/\rho$	m^2 s^{-1}
friction coefficient	$\mu, (f)$	$F_{frict} = \mu F_{norm}$	1
power	P	$P = dW/dt$	W
sound energy flux	P, P_a	$P = dE/dt$	W
acoustic factors			
reflection factor	ρ	$\rho = P_r/P_0$	1
acoustic absorption factor	$\alpha_a, (\alpha)$	$\alpha_a = 1 - \rho$	1
transmission factor	τ	$\tau = P_{tr}/P_0$	1
dissipation factor	δ	$\delta = \alpha_a - \tau$	1

Fundamental Physical Constants

Summary of the 1986 Recommended Values
of the Fundamental Physical Constants

Quantity	Symbol	Value	Units	Relative Uncertainty (ppm)
Speed of light in vacuum	c	299 792 458	ms^{-1}	(exact)
Permeability of vacuum	μ_o	$4\pi \times 10^{-7}$	N A^{-2}	
		$= 12.566\ 370614\ ...$	10^{-7} N A^{-2}	(exact)
Permittivity of vacuum	ϵ_o	$1/\mu_o c^2$		
		$= 8.854\ 187\ 817\ ...$	10^{-12} F m^{-1}	(exact)
Newtonian constant of gravitation	G	6.672 59(85)	10^{-11} m^3 kg^{-1} s^{-2}	128
Planck constant	h	6.626 0755(40)	10^{-34} J s	0.60
$h/2\pi$	$\hbar$	1.054 572 66(63)	10^{-34} J s	0.60
Elementary charge	e	1.602 177 33(49)	10^{-19} C	0.30
Magnetic flux quantum, $h/2e$	Φ_o	2.067 834 61(61)	10^{-15} Wb	0.30
Electron mass	m_e	9.109 3897(54)	10^{-31} kg	0.59
Proton mass	m_p	1.672 6231(10)	10^{-27} kg	0.59
Proton-electron mass ratio	$m_p jm_e$	1836.152701(37)		0.020
Fine-structure constant, $\mu_o ce^2/2h$	α	7.297 353 08(33)	10^{-3}	0.045
Inverse fine-structure constant	α^{-1}	137 035 9895(61)		0.045
Rydberg constant, $m_e c\alpha^2/2h$	R_∞	10 973 731.534(13)	m^{-1}	0.0012
Avogadro constant	N_A, L	6.022 1367(36)	10^{23} mol^{-1}	0.59
Faraday constant, $N_A e$	F	96 485.309(29)	C mol^{-1}	0.30
Molar gas constant	R	8.314 510(70)	J mol^{-1} K^{-1}	8.4
Boltzmann constant, R/N_A	k	1.380 658(12)	10^{-23} J K^{-1}	8.5
Stafan-Boltzmann constant, $(\pi^2/60)k^4/\hbar^3 c^2$	σ	5.670 51(19)	10^{-8} W m^{-2} K^{-4}	34
Non-SI units used with SI				
Electronvolt, $(e/C)J = \{e\}J$	eV	1.602 17733(40)	10^{-19} J	0.30
(Unified) atomic mass unit, 1 u $= m_u = 1/12m(^{12}C)$	u	1.660 5402(10)	10^{-27} kg	0.59

Note: An abbreviated list of the fundamental constants of physics and chemistry based on a least-squares adjustment with 17 degrees of freedom. The digits in parentheses are the one-standard-deviation uncertainty in the last digits of the given value. Since the uncertainties of many entries are correlated, the full covariance matrix must be used in evaluating the uncertainties of quantities computed from them.

PERIODIC TABLE OF THE ELEMENTS

New Notation / Previous IUPAC Form / CAS Version

Key to Chart

Atomic Number →	50
Symbol →	Sn
1995 Atomic Weight →	118.710
Oxidation States →	+2 +4
Electron Configuration →	-18-18-4

Group header cross-reference (New Notation — Previous IUPAC Form — CAS Version):

New	Prev. IUPAC	CAS
1	IA	IA
2	IIA	IIA
3	IIIA	IIIB
4	IVA	IVB
5	VA	VB
6	VIA	VIB
7	VIIA	VIIB
8	VIIIA	VIII
9	VIIIA	VIII
10	VIIIA	VIII
11	IB	IB
12	IIB	IIB
13	IIIB	IIIA
14	IVB	IVA
15	VB	VA
16	VIB	VIA
17	VIIB	VIIA
18	VIIIA	0

Element data (Atomic Number, Symbol, Atomic Weight, Oxidation States, Electron Configuration, Shell):

Z	Symbol	Atomic Weight	Oxidation States	Electron Config.	Shell
1	H	1.00794	+1 -1	1	K
2	He	4.002602	0	2	K
3	Li	6.941	+1	2-1	K-L
4	Be	9.012182	+2	2-2	K-L
5	B	10.811	+3	2-3	K-L
6	C	12.0107	+2 +4 -4	2-4	K-L
7	N	14.00674	+1 +2 +3 +4 +5 -1 -2 -3	2-5	K-L
8	O	15.9994	-2	2-6	K-L
9	F	18.9984032	-1	2-7	K-L
10	Ne	20.1797	0	2-8	K-L
11	Na	22.989770	+1	2-8-1	K-L-M
12	Mg	24.3050	+2	2-8-2	K-L-M
13	Al	26.981538	+3	2-8-3	K-L-M
14	Si	28.0855	+2 +4 -4	2-8-4	K-L-M
15	P	30.973761	+3 +5 -3	2-8-5	K-L-M
16	S	32.066	+4 +6 -2	2-8-6	K-L-M
17	Cl	35.4527	+1 +5 +7 -1	2-8-7	K-L-M
18	Ar	39.948	0	2-8-8	K-L-M
19	K	39.0983	+1	2-8-8-1	-L-M-N
20	Ca	40.078	+2	2-8-8-2	-L-M-N
21	Sc	44.955910	+3	-8-9-2	-L-M-N
22	Ti	47.867	+2 +3 +4	-8-10-2	-L-M-N
23	V	50.9415	+2 +3 +4 +5	-8-11-2	-L-M-N
24	Cr	51.9961	+2 +3 +6	-8-13-1	-L-M-N
25	Mn	54.938049	+2 +3 +4 +7	-8-13-2	-L-M-N
26	Fe	55.845	+2 +3	-8-13-2	-L-M-N
27	Co	58.933200	+2 +3	-8-15-2	-L-M-N
28	Ni	58.6934	+2 +3	-8-16-2	-L-M-N
29	Cu	63.546	+1 +2	-8-18-1	-L-M-N
30	Zn	65.39	+2	-8-18-2	-L-M-N
31	Ga	69.723	+3	-8-18-3	-L-M-N
32	Ge	72.61	+2 +4	-8-18-4	-L-M-N
33	As	74.92160	+3 +5 -3	-8-18-5	-L-M-N
34	Se	78.96	+4 +6 -2	-8-18-6	-L-M-N
35	Br	79.904	+1 +5 -1	-8-18-7	-L-M-N
36	Kr	83.80	0	-8-18-8	-L-M-N
37	Rb	85.4678	+1	-18-8-1	-M-N-O
38	Sr	87.62	+2	-18-8-2	-M-N-O
39	Y	88.90585	+3	-18-9-2	-M-N-O
40	Zr	91.224	+4	-18-10-2	-M-N-O
41	Nb	92.90638	+3 +5	-18-12-1	-M-N-O
42	Mo	95.94	+6	-18-13-1	-M-N-O
43	Tc	(98)	+4 +6 +7	-18-13-2	-M-N-O
44	Ru	101.07	+3	-18-15-1	-M-N-O
45	Rh	102.90550	+3	-18-16-1	-M-N-O
46	Pd	106.42	+2 +4	-18-18-0	-M-N-O
47	Ag	107.8682	+1	-18-18-1	-M-N-O
48	Cd	112.411	+2	-18-18-2	-M-N-O
49	In	114.818	+3	-18-18-3	-M-N-O
50	Sn	118.710	+2 +4	-18-18-4	-M-N-O
51	Sb	121.760	+3 +5 -3	-18-18-5	-M-N-O
52	Te	127.60	+4 +6 -2	-18-18-6	-M-N-O
53	I	126.90447	+1 +5 +7 -1	-18-18-7	-M-N-O
54	Xe	131.29	0	-18-18-8	-M-N-O
55	Cs	132.90545	+1	-18-8-1	-N-O-P
56	Ba	137.327	+2	-18-8-2	-N-O-P
57*	La	138.9055	+3	-18-9-2	-N-O-P
72	Hf	178.49	+4	-32-10-2	-N-O-P
73	Ta	180.9479	+5	-32-11-2	-N-O-P
74	W	183.84	+6	-32-12-2	-N-O-P
75	Re	186.207	+4 +6 +7	-32-13-2	-N-O-P
76	Os	190.23	+3 +4 +6	-32-14-2	-N-O-P
77	Ir	192.217	+3 +4	-32-15-2	-N-O-P
78	Pt	195.078	+2 +4	-32-17-1	-N-O-P
79	Au	196.96655	+1 +3	-32-18-1	-N-O-P
80	Hg	200.59	+1 +2	-32-18-2	-N-O-P
81	Tl	204.3833	+1 +3	-32-18-3	-N-O-P
82	Pb	207.2	+2 +4	-32-18-4	-N-O-P
83	Bi	208.98038	+3 +5	-32-18-5	-N-O-P
84	Po	(209)	+2 +4	-32-18-6	-N-O-P
85	At	(210)		-32-18-7	-N-O-P
86	Rn	(222)	0	-32-18-8	-N-O-P
87	Fr	(223)	+1	-18-8-1	-O-P-Q
88	Ra	(226)	+2	-18-8-2	-O-P-Q
89**	Ac	(227)	+3	-18-9-2	-O-P-Q
104	Rf	(261)	+4	-32-10-2	-O-P-Q
105	Db	(262)		-32-11-2	-O-P-Q
106	Sg	(266)		-32-12-2	-O-P-Q
107	Bh	(264)		-32-13-2	-O-P-Q
108	Hs	(269)		-32-14-2	-O-P-Q
109	Mt	(268)		-32-15-2	-O-P-Q
110	Uun	(271)		-32-16-2	-O-P-Q
111	Uuu	(272)		-32-17-1	-O-P-Q
112	Uub				

* Lanthanides (Shell -N-O-P):

Z	Symbol	Atomic Weight	Oxidation States	Electron Config.
58	Ce	140.116	+3 +4	-19-9-2
59	Pr	140.90765	+3 +4	-21-8-2
60	Nd	144.24	+3	-22-8-2
61	Pm	(145)	+3	-23-8-2
62	Sm	150.36	+2 +3	-24-8-2
63	Eu	151.964	+2 +3	-25-8-2
64	Gd	157.25	+3	-25-9-2
65	Tb	158.92534	+3	-27-8-2
66	Dy	162.50	+3	-28-8-2
67	Ho	164.93032	+3	-29-8-2
68	Er	167.26	+3	-30-8-2
69	Tm	168.93421	+3	-31-8-2
70	Yb	173.04	+2 +3	-32-8-2
71	Lu	174.967	+3	-32-9-2

** Actinides (Shell -O-P-Q):

Z	Symbol	Atomic Weight	Oxidation States	Electron Config.
90	Th	232.0381	+4	-18-10-2
91	Pa	231.03588	+4 +5	-20-9-2
92	U	238.0289	+3 +4 +5 +6	-21-9-2
93	Np	(237)	+3 +4 +5 +6	-22-9-2
94	Pu	(244)	+3 +4 +5 +6	-24-8-2
95	Am	(243)	+3 +4 +5 +6	-25-8-2
96	Cm	(247)	+3	-25-9-2
97	Bk	(247)	+3 +4	-27-8-2
98	Cf	(251)	+3	-28-8-2
99	Es	(252)		-29-8-2
100	Fm	(257)		-30-8-2
101	Md	(258)		-31-8-2
102	No	(259)		-32-8-2
103	Lr	(262)	+3	-32-9-2

The new IUPAC format numbers the groups from 1 to 18. The previous IUPAC numbering system and the system used by Chemical Abstracts Service (CAS) are also shown. For radioactive elements that do not occur in nature, the mass number of the most stable isotope is given in parentheses.

References
1. G. J. Leigh, Editor, *Nomenclature of Inorganic Chemistry*, Blackwell Scientific Publications, Oxford, 1990.
2. *Chemical and Engineering News*, 63(5), 27, 1985.
3. Atomic Weights of the Elements, 1995, *Pure & Appl. Chem.*, 68, 2339, 1996.

Electrical Resistivity

Electrical Resistivity of Pure Metals

The first part of this table gives the electrical resistivity, in units of 10^{-8} Ω m, for 28 common metallic elements as a function of temperature. The data refer to polycrystalline samples. The number of significant figures indicates the accuracy of the values. However, at low temperatures (especially below 50 K) the electrical resistivity is extremely sensitive to sample purity. Thus the low-temperature values refer to samples of specified purity and treatment.

The second part of the table gives resistivity values in the neighborhood of room temperature for other metallic elements that have not been studied over an extended temperature range.

Electrical Resistivity in 10^{-8} Ω m

T/K	Aluminum	Barium	Beryllium	Calcium	Cesium	Chromium	Copper
1	0.000100	0.081	0.0332	0.045	0.0026		0.00200
10	0.000193	0.189	0.0332	0.047	0.243		0.00202
20	0.000755	0.94	0.0336	0.060	0.86	0.00280	
40	0.0181	2.91	0.0367	0.175	1.99		0.0239
60	0.0959	4.86	0.067	0.40	3.07		0.0971
80	0.245	6.83	0.075	0.65	4.16		0.215
100	0.442	8.85	0.133	0.91	5.28	1.6	0.348
150	1.006	14.3	0.510	1.56	8.43	4.5	0.699
200	1.587	20.2	1.29	2.19	12.2	7.7	1.046
273	2.417	30.2	3.02	3.11	18.7	11.8	1.543
293	2.650	33.2	3.56	3.36	20.5	12.5	1.678
298	2.709	34.0	3.70	3.42	20.8	12.6	1.712
300	2.733	34.3	3.76	3.45	21.0	12.7	1.725
400	3.87	51.4	6.76	4.7		15.8	2.402
500	4.99	72.4	9.9	6.0		20.1	3.090
600	6.13	98.2	13.2	7.3		24.7	3.792
700	7.35	130	16.5	8.7		29.5	4.514
800	8.70	168	20.0	10.0	34.6	5.262	
900	10.18	216	23.7	11.4		39.9	6.041

T/K	Gold	Hafnium	Iron	Lead	Lithium	Magnesium	Manganese
1	0.0220	1.00	0.0225		0.007	0.0062	7.02
10	0.0226	1.00	0.0238		0.008	0.0069	18.9
20	0.035	1.11	0.0287		0.012	0.0123	54
40	0.141	2.52	0.0758		0.074	0.074	116
60	0.308	4.53	0.271		0.345	0.261	131
80	0.481	6.75	0.693	4.9	1.00	0.557	132
100	0.650	9.12	1.28	6.4	1.73	0.91	132
150	1.061	15.0	3.15	9.9	3.72	1.84	136
200	1.462	21.0	5.20	13.6	5.71	2.75	139
273	2.051	30.4	8.57	19.2	8.53	4.05	143
293	2.214	33.1	9.61	20.8	9.28	4.39	144
298	2.255	33.7	9.87	21.1	9.47	4.48	144
300	2.271	34.0	9.98	21.3	9.55	4.51	144
400	3.107	48.1	16.1	29.6	13.4	6.19	147
500	3.97	63.1	23.7	38.3		7.86	149
600	4.87	78.5	32.9			9.52	151
700	5.82		44.0			11.2	152
800	6.81		57.1			12.8	
900	7.86					14.4	

Electrical Resistivity in 10^{-8} Ω m (continued)

T/K	Molybdenum	Nickel	Palladium	Platinum	Potassium	Rubidium	Silver
1	0.00070	0.0032	0.0200	0.002	0.0008	0.0131	0.00100
10	0.00089	0.0057	0.0242	0.0154	0.0160	0.109	0.00115
20	0.00261	0.0140	0.0563	0.0484	0.117	0.444	0.0042
40	0.0457	0.068	0.334	0.409	0.480	1.21	0.0539
60	0.206	0.242	0.938	1.107	0.90	1.94	0.162
80	0.482	0.545	1.75	1.922	1.34	2.65	0.289
100	0.858	0.96	2.62	2.755	1.79	3.36	0.418
150	1.99	2.21	4.80	4.76	2.99	5.27	0.726
200	3.13	3.67	6.88	6.77	4.26	7.49	1.029
273	4.85	6.16	9.78	9.6	6.49	11.5	1.467
293	5.34	6.93	10.54	10.5	7.20	12.8	1.587
298	5.47	7.12	10.73	10.7	7.39	13.1	1.617
300	5.52	7.20	10.80	10.8	7.47	13.3	1.629
400	8.02	11.8	14.48	14.6			2.241
500	10.6	17.7	17.94	18.3			2.87
600	13.1	25.5	21.2	21.9			3.53
700	15.8	32.1	24.2	25.4			4.21
800	18.4	35.5	27.1	28.7			4.91
900	21.2	38.6	29.4	32.0			5.64

T/K	Sodium	Strontium	Tantalum	Tungsten	Vanadium	Zinc	Zirconium
1	0.0009	0.80	0.10	0.00016		0.0100	0.250
10	0.0015	0.80	0.102	0.000137	0.0145	0.0112	0.253
20	0.016	0.92	0.146	0.00196	0.039	0.0387	0.357
40	0.172	1.70	0.751	0.0544	0.304	0.306	1.44
60	0.447	2.68	1.65	0.266	1.11	0.715	3.75
80	0.80	3.64	2.62	0.606	2.41	1.15	6.64
100	1.16	4.58	3.64	1.02	4.01	1.60	9.79
150	2.03	6.84	6.19	2.09	8.2	2.71	17.8
200	2.89	9.04	8.66	3.18	12.4	3.83	26.3
273	4.33	12.3	12.2	4.82	18.1	5.46	38.8
293	4.77	13.2	13.1	5.28	19.7	5.90	42.1
298	4.88	13.4	13.4	5.39	20.1	6.01	42.9
300	4.93	13.5	13.5	5.44	20.2	6.06	43.3
400		17.8	18.2	7.83	28.0	8.37	60.3
500		22.2	22.9	10.3	34.8	10.82	76.5
600		26.7	27.4	13.0	41.1	13.49	91.5
700		31.2	31.8	15.7	47.2		104.2
800		35.6	35.9	18.6	53.1		114.9
900			40.1	21.5	58.7		123.1

Electrical Resistivity of Pure Metals (continued)

Element	T/K	Electrical Resistivity $10^{-8}\ \Omega$ m
Antimony	273	39
Bismuth	273	107
Cadmium	273	6.8
Cerium	290–300	82.8
Cobalt	273	5.6
Dysprosium	290–300	92.6
Erbium	290–300	86.0
Europium	290–300	90.0
Gadolinium	290–300	131
Gallium	273	13.6
Holmium	290–300	81.4
Indium	273	8.0
Iridium	273	4.7
Lanthanum	290–300	61.5
Lutetium	290–300	58.2
Mercury	273	94.1
Neodymium	290–300	64.3
Niobium	273	15.2
Osmium	273	8.1
Polonium	273	40
Praseodymium	290–300	70.0
Promethium	290–300	75
Protactinium	273	17.7
Rhenium	273	17.2
Rhodium	273	4.3
Ruthenium	273	7.1
Samrium	290–300	94.0
Scandium	290–300	56.2
Terbium	290–300	115
Thallium	273	15
Thorium	273	14.7
Thulium	290–300	67.6
Tin	273	11.5
Titanium	273	39
Uranium	273	28
Ytterbium	290–300	25.0
Yttrium	290–300	59.6

Electrical Resistivity of Selected Alloys

Values of the resistivity are given in units of 10^{-8} Ω m. General comments in the preceding table for pure metals also apply here.

	273 K	293 K	300 K	350 K	400 K		273 K	293 K	300 K	350 K	400 K
Alloy Aluminum-Copper						**Alloy—Copper-Nickel**					
Wt % Al						**Wt % Cu**					
99[a]	2.51	2.74	2.82	3.38	3.95	99[c]	2.71	2.85	2.91	3.27	3.62
95[a]	2.88	3.10	3.18	3.75	4.33	95[c]	7.60	7.71	7.82	8.22	8.62
90[b]	3.36	3.59	3.67	4.25	4.86	90[c]	13.69	13.89	13.96	14.40	14.81
85[b]	3.87	4.10	4.19	4.79	5.42	85[c]	19.63	19.83	19.90	2032	20.70
80[b]	4.33	4.58	4.67	5.31	5.99	80[c]	25.46	25.66	25.72	26.12[aa]	26.44[aa]
70[b]	5.03	5.31	5.41	6.16	6.94	70[i]	36.67	36.72	36.76	36.85	36.89
60[b]	5.56	5.88	5.99	6.77	7.63	60[i]	45.43	45.38	45.35	45.20	45.01
50[b]	6.22	6.55	6.67	7.55	8.52	50[i]	50.19	50.05	50.01	49.73	49.50
40[c]	7.57	7.96	8.10	9.12	10.2	40[c]	47.42	47.73	47.82	48.28	48.49
30[c]	11.2	11.8	12.0	13.5	15.2	30[i]	40.19	41.79	42.34	44.51	45.40
25[f]	16.3[aa]	17.2	17.6	19.8	22.2	25[c]	33.46	35.11	35.69	39.67[aa]	42.81[aa]
15[h]	—	12.3	—	—	—	15[c]	22.00	23.35	23.85	27.60	31.38
19[g]	10.8[aa]	11.0	11.1	11.7	12.3	10[c]	16.65	17.82	18.26	21.51	25.19
5[e]	9.43	9.61	9.68	10.2	10.7	5[c]	11.49	12.50	12.90	15.69	18.78
1[b]	4.46	4.60	4.65	5.00	5.37	1[c]	7.23	8.08	8.37	10.63[aa]	13.18[aa]
Alloy—Aluminum-Magnesium						**Alloy—Copper-Palladium**					
Wt % Al						**Wt % Cu**					
99[c]	2.96	3.18	3.26	3.82	4.39	99[c]	2.10	2.23	2.27	2.59	2.92
95[c]	5.05	5.28	5.36	5.93	6.51	95[c]	4.21	4.35	4.40	4.74	5.08
90[c]	7.52	7.76	7.85	8.43	9.02	90[c]	6.89	7.03	7.08	7.41	7.74
85	—	—	—	—	—	85[c]	9.48	9.61	9.66	10.01	10.36
80	—	—	—	—	—	80[c]	11.99	12.12	12.16	12.51[aa]	12.87
70	—	—	—	—	—	70[c]	16.87	17.01	17.06	17.41	17.78
60	—	—	—	—	—	60[c]	21.73	21.87	21.92	22.30	22.69
50	—	—	—	—	—	50[c]	27.62	27.79	27.86	28.25	28.64
40	—	—	—	—	—	40[c]	35.31	35.51	35.57	36.03	36.47
30	—	—	—	—	—	30[c]	46.50	46.66	46.71	47.11	47.47
25	—	—	—	—	—	25[c]	46.25	46.45	46.52	46.99[aa]	47.43[aa]
15	—	—	—	—	—	15[c]	36.52	36.99	37.16	38.28	39.35
10[b]	17.1	17.4	17.6	18.4	19.2	10[c]	28.90	29.51	29.73	31.19[aa]	32.56[aa]
5[b]	13.1	13.4	13.5	14.3	15.2	5[c]	20.00	20.75	21.02	22.84[aa]	24.54[aa]
1[a]	5.92	6.25	6.37	7.20	8.03	1[c]	11.90	12.67	12.93[aa]	14.82[aa]	16.68[aa]
Alloy—Copper-Gold						**Alloy—Copper-Zinc**					
Wt % Cu						**Wt % Cu**					
99[c]	1.73	1.86[aa]	1.91[aa]	2.24[aa]	2.58[aa]	99[b]	1.84	1.97	2.02	2.36	2.71
95[c]	2.41	2.54[aa]	2.59[aa]	2.92[aa]	3.26[aa]	95[b]	2.78	2.92	2.97	3.33	3.69
90[c]	3.29	4.42[aa]	3.46[aa]	3.79[aa]	4.12[aa]	90[b]	3.66	3.81	3.86	4.25	4.63
85[c]	4.20	4.33	4.38[aa]	4.71[aa]	5.05[aa]	85[b]	4.37	4.54	4.60	5.02	5.44
80[c]	5.15	5.28	5.32	5.65	5.99	80[b]	5.01	5.19	5.26	5.71	6.17
70[c]	7.12	7.25	7.30	7.64	7.99	70[b]	5.87	6.08	6.15	6.67	7.19
60[c]	9.18	9.13	9.36	9.70	10.05	60	—	—	—	—	—
50[c]	11.07	11.20	11.25	11.60	11.94	50	—	—	—	—	—
40[c]	12.70	12.85	12.90[aa]	13.27[aa]	13.65[aa]	40	—	—	—	—	—
30[c]	13.77	13.93	13.99[aa]	14.38[aa]	14.78[aa]	30	—	—	—	—	—
25[c]	13.93	14.09	14.14	14.54	14.94	25	—	—	—	—	—
15[c]	12.75	12.91	12.96[aa]	13.36[aa]	13.77	15	—	—	—	—	—
10[c]	10.70	10.86	10.91	11.31	11.72	10	—	—	—	—	—
5[c]	7.25	7.41[aa]	7.46	7.87	8.28	5	—	—	—	—	—
1[c]	3.40	3.57	3.62	4.03	4.45	1	—	—	—	—	—

Alloy—Gold-Palladium

Wt % Au	273 K	293 K	300 K	350 K	400 K
99[c]	2.69	2.86	2.91	3.32	3.73
95[c]	5.21	5.35	5.41	5.79	6.17
90[i]	8.01	8.17	8.22	8.56	8.93
85[b]	10.50[aa]	10.66	10.72[aa]	11.100[aa]	11.48[aa]
80[b]	12.75	12.93	12.99	13.45	13.93
70[c]	18.23	18.46	18.54	19.10	19.67
60[b]	26.70	26.94	27.01	27.63[aa]	28.23[aa]
50[a]	27.23	27.63	27.76	28.64[aa]	29.42[aa]
40[a]	24.65	25.23	25.42	26.74	27.95
30[b]	20.82	21.49	21.72	23.35	24.92
25[b]	18.86	19.53	19.77	21.51	23.19
15[a]	15.08	15.77	16.01	17.80	19.61
10[a]	13.25	13.95	14.20[aa]	16.00[aa]	17.81[aa]
5[a]	11.49[aa]	12.21	12.46[aa]	14.26[aa]	16.07[aa]
1[a]	10.07	10.85[aa]	11.12[aa]	12.99[aa]	14.80[aa]

Alloy—Iron-Nickel

Wt % Fe	273 K	293 K	300 K	350 K	400 K
99[a]	10.9	12.0	12.4	—	18.7
95[c]	18.7	19.9	20.2	—	26.8
90[c]	24.2	25.5	25.9	—	33.2
85[c]	27.8	29.2	29.7	—	37.3
80[c]	30.1	31.6	32.2	—	40.0
70[b]	32.3	33.9	34.4	—	42.4
60[c]	53.8	57.1	58.2	—	73.9
50[d]	28.4	30.6	31.4	—	43.7
40[d]	19.6	21.6	22.5	—	34.0
30[c]	15.3	17.1	17.7	—	27.4
25[b]	14.3	15.9	16.4	—	25.1
15[c]	12.6	13.8	14.2	—	21.1
10[c]	11.4	12.5	12.9	—	18.9
5[c]	9.66	10.6	10.9	—	16.1[aa]
1[b]	7.17	7.94	8.12	—	12.8

Alloy—Gold-Silver

Wt % Au	273 K	293 K	300 K	350 K	400 K
99[b]	2.58	2.75	2.80[aa]	3.22[aa]	3.63[aa]
95[a]	4.58	4.74	4.79	5.19	5.59
90[j]	6.57	6.73	6.78	7.19	7.58
85[j]	8.14	8.30	8.36[aa]	8.75	9.15
80[j]	9.34	9.50	9.55	9.94	10.33
70[j]	10.70	10.86	10.91	11.29	11.68[aa]
60[j]	10.92	11.07	11.12	11.50	11.87
50[j]	10.23	10.37	10.42	10.78	11.14
40[j]	8.92	9.06	9.11	9.46[aa]	9.81
30[a]	7.34	7.47	7.52	7.85	8.19
25[a]	6.46	6.59	6.63	6.96	7.30[aa]
15[a]	4.55	4.67	4.72	5.03	5.34
10[a]	3.54	3.66	3.71	4.00	4.31
5[i]	2.52	2.64[aa]	2.68[aa]	2.96[aa]	3.25[aa]
1[b]	1.69	1.80	1.84[aa]	2.12[aa]	2.42[aa]

Alloy—Silver-Palladium

Wt % Ag	273 K	293 K	300 K	350 K	400 K
99[b]	1.891	2.007	2.049	2.35	2.66
95[b]	3.58	3.70	3.74	4.04	4.34
90[b]	5.82	5.94	5.98	6.28	6.59
85[k]	7.92[aa]	8.04[aa]	8.08	8.38[aa]	8.68[aa]
80[k]	10.01	10.13	10.17	10.47	10.78
70[k]	14.53	14.65	14.69	14.99	15.30
60[i]	20.9	21.1	21.2	21.6	22.0
50[k]	31.2	31.4	31.5	32.0	32.4
40[m]	42.2	42.2	42.2	42.3	42.3
30[b]	40.4	40.6	40.7	41.3	41.7
25[k]	36.67[aa]	37.06	37.19	38.1[aa]	38.8[aa]
15[i]	27.08[aa]	26.68[aa]	27.89[aa]	29.3[aa]	30.6[aa]
10[i]	21.69	22.39	22.63	24.3	25.9
5[b]	15.98	16.72	16.98	18.8[aa]	20.5[aa]
1[a]	11.06	11.82	12.08[aa]	13.92[aa]	15.70[aa]

[a]　Uncertainty in resistivity is ± 2%.
[b]　Uncertainty in resistivity is ± 3%.
[c]　Uncertainty in resistivity is ± 5%.
[d]　Uncertainty in resistivity is ± 7% below 300 K and ± 5% at 300 and 400 K.
[e]　Uncertainty in resistivity is ± 7%.
[f]　Uncertainty in resistivity is ± 8%.
[g]　Uncertainty in resistivity is ± 10%.
[h]　Uncertainty in resistivity is ± 12%.
[i]　Uncertainty in resistivity is ± 4%.
[j]　Uncertainty in resistivity is ± 1%.
[k]　Uncertainty in resistivity is ± 3% up to 300 K and ± 4% above 300 K.
[m]　Uncertainty in resistivity is ± 2% up to 300 K and ± 4% above 300 K.
[a]　Crystal usually a mixture of α-hep and fcc lattice.
[aa]　In temperature range where no experimental data are available.

Resistivity of Selected Ceramics (Listed by Ceramic)

Ceramic	Resistivity (Ω-cm)
Borides	
Chromium diboride (CrB_2)	21×10^{-6}
Hafnium diboride (HfB_2)	$10–12 \times 10^{-6}$ at room temp.
Tantalum diboride (TaB_2)	68×10^{-6}
Titanium diboride (TiB_2) (polycrystalline)	
85% dense	$26.5–28.4 \times 10^{-6}$ at room temp.
85% dense	9.0×10^{-6} at room temp.
100% dense, extrapolated values	$8.7–14.1 \times 10^{-6}$ at room temp.
	3.7×10^{-6} at liquid air temp.
Titanium diboride (TiB_2) (monocrystalline)	
Crystal length 5 cm, 39 deg. and 59 deg. orientation with respect to growth axis	$6.6 \pm 0.2 \times 10^{-6}$ at room temp.
Crystal length 1.5 cm, 16.5 deg. and 90 deg. orientation with respect to growth axis	$6.7 \pm 0.2 \times 10^{-6}$ at room temp.
Zirconium diboride (ZrB_2)	9.2×10^{-6} at 20°C
	1.8×10^{-6} at liquid air temp.
Carbides: boron carbide (B_4C)	0.3–0.8

Dielectric Constants

Dielectric Constants of Solids

These data refer to temperatures in the range 17–22°C.

Material	Freq. (Hz)	Dielectric constant	Material	Freq. (Hz)	Dielectric constant
Acetamide	4×10^8	4.0	Diphenylmethane	4×10^8	2.7
Acetanilide	—	2.9	Dolomite $\perp$ optic axis	10^8	8.0
Acetic acid (2°C)	4×10^8	4.1	Dolomite $\parallel$	10^8	6.8
Aluminum oleate	4×10^8	2.40	Ferrous oxide (15°C)	10^8	14.2
Ammonium bromide	10^8	7.1	Iodine	10^8	4
Ammonium chloride	10^8	7.0	Lead acetate	10^8	2.6
Antimony trichloride	10^8	5.34	Lead carbonate (15°C)	10^8	18.6
Apatite $\perp$ optic axis	3×10^8	9.50	Lead chloride	10^8	4.2
Apatite $\parallel$ optic axis	3×10^8	7.41	Lead monoxide (15°C)	10^8	25.9
Asphalt	$<3 \times 10^4$	2.68	Lead nitrate	6×10^7	37.7
Barium chloride (anhyd.)	6×10^7	11.4	Lead oleate	4×10^8	3.27
Barium chloride ($2H_3O$)	6×10^7	9.4	Lead sulfate	10^4	14.3
Barium nitrate	6×10^7	5.9	Lead sulfide (15°C)	16^8	17.9
Barium sulfate (15°C)	10^8	11.40	Malachite (mean)	10^{12}	7.2
Beryl $\perp$ optic axis	10^4	7.02	Mercuric chloride	10^8	3.2
Beryl $\parallel$ optic axis	10^4	6.08	Mercurous chloride	10^8	9.4
Calcite $\perp$ optic axis	10^4	8.5	Naphthalene	4×10^8	2.52
Calcite $\parallel$ optic axis	10^4	8.0	Phenanthrene	4×10^8	2.80
Calcium carbonate	10^4	6.14	Phenol (10°C)	4×10^8	4.3
Calcium fluoride	10^4	7.36	Phosphorus, red	10^8	4.1
Calcium sulfate ($2H_2O$)	10^4	5.66	Phosphorus, yellow	10^8	3.6
Cassiterite $\perp$ optic axis	10^{12}	23.4	Potassium aluminum sulfate	10^8	3.8
Cassiterite $\parallel$ optic axis	10^{12}	24	Potassium carbonate (15°C)	10^8	5.6
d-Cocaine	5×10^8	3.10	Potassium chlorate	6×10^7	5.1
Cupric oleate	4×10^8	2.80	Potassium chloride	10^4	5.03
Cupric oxide (15°C)	10^8	18.1	Potassium chromate	6×10^7	7.3
Cupric sulfate (anhyd.)	6×10^7	10.3	Potassium iodide	6×10^7	5.6
Cupric sulfate ($5H_2O$)	6×10^7	7.8	Potassium nitrate	6×10^7	5.0
Diamond	10^8	5.5	Potassium sulfate	6×10^7	5.9

Material	Freq. (Hz)	Dielectric constant	Material	Freq. (Hz)	Dielectric constant
Quartz $\perp$ optic axis	3×10^7	4.34	Sodium carbonate ($10H_2O$)	6×10^7	5.3
Quartz $\parallel$ optic axis	3×10^7	4.27	Sodium chloride	10^4	6.12
Resorcinol	4×10^8	3.2	Sodium nitrate	—	5.2
Ruby $\perp$ optic axis	10^4	13.27	Sodium oleate	4×10^8	2.75
Ruby $\parallel$ optic axis	10^4	11.28	Sodium perchlorate	6×10^7	5.4
Rutile $\perp$ optic axis	10^8	86	Sucrose (mean)	3×10^8	3.32
Rutile $\parallel$ optic axis	10^8	170	Sulfur (mean)	—	4.0
Selenium	10^8	6.6	Thallium chloride	10^4	46.9
Silver bromide	10^4	12.2	*p*-Toluidine	4×10^8	3.0
Silver chloride	10^4	11.2	Tourmaline $\perp$ optic axis	10^4	7.10
Silver cyanide	10^4	5.6	Tourmaline $\parallel$ optic axis	10^4	6.3
Smithsonite $\perp$ optic axis	10^{12}	9.3	Urea	4×10^8	3.5
Smithsonite $\parallel$ optic axis	10^{10}	9.4	Zircon $\perp, \parallel$	10^8	12
Sodium carbonate (anhyd.)	6×10^7	8.4			

Dielectric Constants of Ceramics

Material	Dielectric constant 10^4 Hz	Dielectric strength Volts/mil	Volume resistivity Ohm-cm (23°C)	Loss Factor[a]
Alumina	4.5–8.4	40–160	10^{11}–10^{14}	0.0002–0.01
Corderite	4.5–5.4	40–250	10^{12}–10^{14}	0.004–0.012
Forsterite	6.2	240	10^{14}	0.0004
Porcelain (dry process)	6.0–8.0	40–240	10^{12}–10^{14}	0.0003–0.02
Porcelain (wet process)	6.0–7.0	90–400	10^{12}–10^{14}	0.006–0.01
Porcelain, zircon	7.1–10.5	250–400	10^{13}–10^{15}	0.0002–0.008
Steatite	5.5–7.5	200–400	10^{13}–10^{15}	0.0002–0.004
Titanates (Ba, Sr, Ca, Mg, and Pb)	15–12.000	50–300	10^8–10^{13}	0.0001–0.02
Titanium dioxide	14–110	100–210	10^{13}–10^{18}	0.0002–0.005

Dielectric Constants of Glasses

Type	Dielectric constant At 100 MHz (20°C)	Volume resistivity (350°C megohm-cm)	Loss factor[a]
Corning 0010	6.32	10	0.015
Corning 0080	6.75	0.13	0.058
Corning 0120	6.65	100	0.012
Pyrex 1710	6.00	2,500	0.025
Pyrex 3320	4.71	—	0.019
Pyrex 7040	4.65	80	0.013
Pyrex 7050	4.77	16	0.017
Pyrex 7052	5.07	25	0.019
Pyrex 7060	4.70	13	0.018
Pyrex 7070	4.00	1,300	0.0048
Vycor 7230	3.83	—	0.0061
Pyrex 7720	4.50	16	0.014
Pyrex 7740	5.00	4	0.040
Pyrex 7750	4.28	50	0.011
Pyrex 7760	4.50	50	0.0081
Vycor 7900	3.9	130	0.0023
Vycor 7910	3.8	1,600	0.00091
Vycor 7911	3.8	4,000	0.00072
Corning 8870	9.5	5,000	0.0085
G. E. Clear (silica glass)	3.81	4,000–30,000	0.00038
Quartz (fused)	3.75 4.1 (1 MHz)	—	0.0002 (1 MHz)

[a] Power factor $\times$ dielectric constant equals loss factor.

Properties of Semiconductors

Semiconducting Properties of Selected Materials

Substance	Minimum energy gap (eV)		$\frac{dE_g}{dT}$ $\times 10^4$ eV/°C	$\frac{dE_g}{dP}$ $\times 10^6$ eV·cm²/kg	Density of states electron effective mass m_{d_n} (m_o)	Electron mobility and temperature dependence		Density of states hole effective mass m_{d_p} (m_o)	Hole mobility and temperature dependence	
	R.T.	0 K				μ_n (cm²/V·s)	$-x$		μ_p (cm²/V·s)	$-x$
Si	1.107	1.153	−2.3	−2.0	1.1	1,900	2.6	0.56	500	2.3
Ge	0.67	0.744	−3.7	±7.3	0.55	3,80	1.66	0.3	1,820	2.33
αSn	0.08	0.094	−0.5		0.02	2,500	1.65	0.3	2,400	2.0
Te	0.33				0.68	1,100		0.19	560	
III–V Compounds										
AlAs	2.2	2.3				1,200			420	
AlSb	1.6	1.7	−3.5	−1.6	0.09	2..	1.5	0.4	500	1.8
GaP	2.24	2.40	−5.4	−1.7	0.35	300	1.5	0.5	150	1.5
GaAs	1.35	1.53	−5.0	+9.4	0.068	9,000	1.0	0.5	500	2.1
GaSb	0.67	0.78	−3.5	+12	0.050	5,000	2.0	0.23	1,400	0.9
InP	1.27	1.41	−4.6	+4.6	0.067	5,000	2.0		200	2.4
InAs	0.36	0.43	−2.8	+8	0.022	33,000	1.2	0.41	460	2.3
InSb	0.165	0.23	−2.8	+15	0.014	78,000	1.6	0.4	750	2.1
II–VI Compounds										
ZnO	3.2		−9.5	+0.6	0.38	180	1.5			
ZnS	3.54		−5.3	+5.7		180			5 (400°C)	
ZnSe	2.58	2.80	−7.2	+6		540			28	
ZnTe	2.26			+6		340			100	
CdO	2.5 ± 0.1		−6		0.1	120				
CdS	2.42		−5	+3.3	0.165	400		0.8		
CdSe	1.74	1.85	−4.6		0.13	650	1.0	0.6		
CdTe	1.44	1.56	−4.1	+8	0.14	1,200		0.35	50	
HgSe	0.30				0.030	20,000	2.0			
HgTe	0.15		−1		0.017	25,000		0.5	350	
Halite Structure Compounds										
PbS	0.37	0.28	+4		0.16	800		0.1	1,000	2.2
PbSe	0.26	0.16	+4		0.3	1,500		0.34	1,500	2.2
PbTe	0.25	0.19	+4	−7	0.21	1,600		0.14	750	2.2
Others										
ZnSb	0.50	0.56			0.15	10				1.5
CdSb	0.45	0.57	−5.4		0.15	300			2,000	1.5
Bi_2S_3	1.3					200			1,100	
Bi_2Se_3	0.27					600			675	
Bi_2Te_3	0.13		−0.95		0.58	1,200	1.68	1.07	510	1.95
Mg_2Si		0.77	−6.4		0.46	400	2.5		70	
Mg_2Ge		0.74	−9			280	2		110	
Mg_2Sn	0.21	0.33	−3.5		0.37	320			260	
Mg_3Sb_2		0.32				20			82	
Zn_3As_2	0.93					10	1.1		10	
Cd_3As_2	0.55				0.046	100,000	0.88			
GaSe	2.05		3.8						20	
GaTe	1.66	1.80	−3.6			14	−5			
InSe	1.8					9000				
TlSe	0.57		−3.9		0.3	30		0.6	20	1.5
$CdSnAs_2$	0.23				0.05	25,000	1.7			
Ga_2Te_3	1.1	1.55	−4.8							
$α\text{-}In_2Te_3$	1.1	1.2			0.7				50	1.1
$β\text{-}In_2Te_3$	1.0								5	
$Hg_5In_2Te_8$	0.5								11,000	
SnO_2									78	

Band Properties of Semiconductors

Part A. Data on Valence Bands of Semiconductors (Room Temperature)

	Band curvature effective mass (expressed as fraction of free electron mass)				Measured light
Substance	Heavy holes	Light holes	"Split-off" band holes	Energy separation of "split-off" band (eV)	hole mobility (cm²/V·s)
Semiconductors with Valence Bands Maximum at the Center of the Brillouin Zone ("F")					
Si	0.52	0.16	0.25	0.044	500
Ge	0.34	0.043	0.08	0.3	1,820
Sn	0.3				2,400
AlAs					
AlSb	0.4			0.7	550
GaP				0.13	100
GaAs	0.8	0.12	0.20	0.34	400
GaSb	0.23	0.06		0.7	1,400
InP				0.21	150
InAs	0.41	0.025	0.083	0.43	460
InSb	0.4	0.015		0.85	750
CdTe	0.35				50
HgTe	0.5				350

Semiconductors with Multiple Valence Band Maxima

		Band curvature effective masses			Measured (light)
Substance	Number of equivalent valleys and directions	Longitudinal m_L	Transverse m_T	Anisotropy $K = m_L/m_T$	hole mobility cm²/V·s
PbSe	4 "L" [111]	0.095	0.047	2.0	1,500
PbTe	4 "L" [111]	0.27	0.02	10	750
Bi_2Te_3	6	0.207	~0.045	4.5	515

Part B. Data on Conduction Bands of Semiconductors (Room Temperature Data)

Single Valley Semiconductors

Substance	Energy gap (eV)	Effective mass (m_o)	Mobility (cm²/V·s)	Comments
GaAs	1.35	0.067	8,500	3 (or 6?) equivalent [100] valleys 0.36 eV above this maximum with a mobility of ~50
InP	1.27	0.067	5,000	3 (or 6?) equivalent [100] valleys 0.4 eV above this minimum
InAs	0.36	0.022	33,000	Equivalent valleys ~1.0 eV above this minimum
InSb	0.165	0.014	78,000	
CdTe	1.44	0.11	1,000	4 (or 8?) equivalent [111] valleys 0.51 eV above this minimum

Multivalley Semiconductors

			Band curvature effective mass			
Substance	Energy Gap	Number of equivalent valleys and direction	Longitudinal m_L	Transverse m_T	Anisotropy $K = m_L/m_T$	Comments
Si	1.107	6 in [100] "Δ"	0.90	0.192	4.7	
Ge	0.67	4 in [111] at "L"	1.588	~0.0815	19.5	
GaSb	0.67	as Ge (?)	~1.0	~0.2	~5	
PbSe	0.26	4 in [111] at "L"	0.085	0.05	1.7	
PbTe	0.25	4 in [111] at "L"	0.21	0.029	5.5	
Bi_2Te_3	0.13	6			~0.05	

Resistance of Wires

The following table gives the approximate resistance of various metallic conductors. The values have been computed from the resistivities at 20°C, except as otherwise stated, and for the dimensions of wire indicated. Owing to differences in purity in the case of elements and of composition in alloys, the values can be considered only as approximations.

B. & S. Gauge	Diameter mm	Diameter mills 1 mil = .001 in	B. & S. gauge	Diameter mm	Diameter mills 1 mil = .001 in
10	2.588	101.9	26	0.4049	15.94
12	2.053	80.81	27	0.3606	14.20
14	1.628	64.08	28	0.3211	12.64
16	1.291	50.82	30	0.2546	10.03
18	1.024	40.30	32	0.2019	7.950
20	0.8118	31.96	34	0.1601	6.305
22	0.6438	25.35	36	0.1270	5.000
24	0.5106	20.10	40	0.07987	3.145

B. & S. No.	Ohms per cm	Ohms per ft	B. & S. No.	Ohms per cm	Ohms per ft
Advance (0°C) $Q = 48. \times 10^{-6}$ ohm cm			Brass $Q = 7.00 \times 10^{-6}$ ohm cm		
10	.000912	.0278	10	.000133	.00406
12	.00145	.0442	12	.000212	.00645
14	.00231	.0703	14	.000336	.0103
16	.00367	.112	16	.000535	.0163
18	.00583	.178	18	.000850	.0259
20	.00927	.283	20	.00135	.0412
22	.0147	.449	22	.00215	.0655
24	.0234	.715	24	.00342	.104
26	.0373	1.14	26	.00543	.166
27	.0470	1.43	27	.00686	.209
28	.0593	1.81	28	.00864	.263
30	.0942	2.87	30	.0137	.419
32	.150	4.57	32	.0219	.666
34	.238	7.26	34	.0348	1.06
36	.379	11.5	36	.0552	1.68
40	.958	29.2	40	.140	4.26
Aluminum $Q = 2.828 \times 10^{-6}$ ohm cm			Climax $Q = 87. \times 10^{-6}$ ohm cm		
10	.0000538	.00164	10	.00165	.0504
12	.0000855	.00260	12	.00263	.0801
14	.000136	.00414	14	.00418	.127
16	.000216	.00658	16	.00665	.203
18	.000344	.0105	18	.0106	.322
20	.000546	.0167	20	.0168	.512
22	.000869	.0265	22	.0267	.815
24	.00138	.0421	24	.0425	1.30
26	.00220	.0669	26	.0675	2.06
27	.00277	.0844	27	.0852	2.60
28	.00349	.106	28	.107	3.27
30	.00555	.169	30	.171	5.21
32	.00883	.269	32	.272	8.28
34	.0140	.428	34	.432	13.2
36	.0223	.680	36	.687	20.9
40	.0564	1.72	40	1.74	52.9

B. & S. No.	Ohms per cm	Ohms per ft	B. & S. No.	Ohms per cm	Ohms per ft
Constantan (0°C) $Q = 44.1 \times 10^{-6}$ ohm cm			Excello $Q = 92. \times 10^{-6}$ ohm cm		
10	.000838	.0255	10	.00175	.0533
12	.00133	.0406	12	.00278	.0847
14	.00212	.0646	14	.00442	.135
16	.00337	.103	16	.00703	.214
18	.00536	.163	18	.0112	.341
20	.00852	.260	20	.0178	.542
22	.0135	.413	22	.0283	.861
24	.0215	.657	24	.0449	1.37
26	.0342	1.04	26	.0714	2.18
27	.0432	1.32	27	.0901	2.75
28	.0545	1.66	28	.114	3.46
30	.0866	2.64	30	.181	5.51
32	.138	4.20	32	.287	8.75
34	.219	6.67	34	.457	13.9
36	.348	10.6	36	.726	22.1
40	.880	26.8	40	1.84	56.0
Copper, annealed $Q = 1.724 \times 10^{-6}$ ohm cm			German silver $Q = 33. \times 10^{-6}$ ohm cm		
10	.0000328	.000999	10	.000627	.0191
12	.0000521	.00159	12	.000997	.0304
14	.0000828	.00253	14	.00159	.0483
16	.000132	.00401	16	.00252	.0768
18	.000209	.00638	18	.00401	.122
20	.000333	.0102	20	.00638	.194
22	.000530	.0161	22	.0101	.309
24	.000842	.0257	24	.0161	.491
26	.00134	.0408	26	.0256	.781
27	.00169	.0515	27	.0323	.985
28	.00213	.0649	28	.0408	1.24
30	.00339	.103	30	.0648	1.97
32	.00538	.164	32	.103	3.14
34	.00856	.261	34	.164	4.99
36	.0136	.415	36	.260	.794
40	.0344	1.05	40	.659	20.1
Eureka (0°C) $Q = 47. \times 10^{-6}$ ohm cm			Gold $Q = 2.44 \times 10^{-6}$ ohm cm		
10	.000893	.0272	10	.0000464	.00141
12	.00142	.0433	12	.0000737	.00225
14	.00226	0.688	14	.000117	.00357
16	.00359	.109	16	.000186	.00568
18	.00571	.174	18	.000296	.00904
20	.00908	.277	20	.000471	.0144
22	.0144	.440	22	.000750	.0228
24	.0230	.700	24	.00119	.0363
26	.0365	1.11	26	.00189	.0577
27	.0460	1.40	27	.00239	.0728
28	.0580	1.77	28	.00301	.0918
30	.0923	2.81	30	.00479	.146
32	.147	4.47	32	.00762	.232
34	.233	7.11	34	.0121	.369
36	.371	11.3	36	.0193	.587
40	.938	28.6	40	.0487	1.48

B. & S. No.	Ohms per cm	Ohms per ft	B. & S. No.	Ohms per cm	Ohms per ft
Iron $Q = 10. \times 10^{-6}$ ohm cm			Manganin $Q = 44. \times 10^{-6}$ ohm cm		
10	.000190	.00579	10	.000836	.0255
12	.000302	.00921	12	.00133	.0405
14	.000481	.0146	14	.00211	.0644
16	.000764	.0233	16	.00336	.102
18	.00121	.0370	18	.00535	.163
20	.00193	.0589	20	.00850	.259
22	.00307	.0936	22	.0135	.412
24	.00489	.149	24	.0215	.655
26	.00776	.237	26	.0342	1.04
27	.00979	.299	27	.0431	1.31
28	.0123	.376	28	.0543	1.66
30	.0196	.598	30	.0864	2.63
32	.0312	.952	32	.137	4.19
34	.0497	1.51	34	.218	6.66
36	0.789	2.41	36	.347	10.6
40	.200	6.08	40	.878	26.8
Lead $Q = 22. \times 10^{-6}$ ohm cm			Molybdenum $Q = 5.7 \times 10^{-6}$ ohm cm		
10	.000418	.0127	10	.000108	.00330
12	.000665	.0203	12	.000172	.00525
14	.00106	.0322	14	.000274	.00835
16	.00168	.0512	16	.000435	.0133
18	.00267	.0815	18	.000693	.0211
20	.00425	.130	20	.00110	.0336
22	.00676	.206	22	.00175	.0534
24	.0107	.328	24	.00278	.0849
26	.0171	.521	26	.00443	.135
27	.0215	.657	27	.00558	.170
28	.0272	.828	28	.00704	.215
30	.0432	1.32	30	.0112	.341
32	.0687	2.09	32	.0178	.542
34	.109	3.33	34	.0283	.863
36	.174	5.29	36	.0450	1.37
40	.439	13.4	40	.114	3.47
Magnesium $Q = 4.6 \times 10^{-6}$ ohm cm			Monel Metal $Q = 42. \times 10^{-6}$ ohm cm		
10	.0000874	.00267	10	.000798	.0243
12	.000139	.00424	12	.00127	.0387
14	.000221	.00674	14	.00202	.0615
16	.000351	.0107	16	.00321	.0978
18	.000559	.0170	18	.00510	.156
20	.000889	.0271	20	.00811	.247
22	.00141	.0431	22	.0129	.393
24	.00225	.0685	24	.0205	.625
26	.00357	.109	26	.0326	.994
27	.00451	.137	27	.0411	1.25
28	.00568	.173	28	.0519	1.58
30	.00903	.275	30	.0825	2.51
32	.0144	.438	32	.131	4.00
34	.0228	.696	34	.209	6.36
36	.0363	1.11	36	.331	10.1
40	.0918	2.80	40	.838	25.6

B. & S. No.	Ohms per cm	Ohms per ft	B. & S. No.	Ohms per cm	Ohms per ft
*Nichrome $Q = 150. \times 10^{-6}$ ohm cm			Silver (18°C) $Q = 1.629 \times 10^{-6}$ ohm cm		
10	.0021281	.06488	10	.0000310	.000944
12	.0033751	.1029	12	.0000492	.00150
14	.0054054	.1648	14	.0000783	.00239
16	.0085116	.2595	16	.000124	.00379
18	.0138383	.4219	18	.000198	.00603
20	.0216218	.6592	20	.000315	.00959
22	.0346040	1.055	22	.000500	.0153
24	.0548088	1.671	24	.000796	.0243
26	.0875760	2.670	26	.00126	.0386
28	.1394328	4.251	27	.00160	.0486
30	.2214000	6.750	28	.00201	.0613
32	.346040	10.55	30	.00320	.0975
34	.557600	17.00	32	.00509	.155
36	.885600	27.00	34	.00809	.247
38	1.383832	42.19	36	.0129	.392
40	2.303872	70.24	40	.0325	.991
Nickel $Q = 7.8 \times 10^{-6}$ ohm cm			Steel, piano wire (0°C) $Q = 11.8 \times 10^{-6}$ ohm cm		
10	.000148	.00452	10	.000224	.00684
12	.000236	.00718	12	.000357	.0109
14	.000375	.0114	14	.000567	.0173
16	.000596	.0182	16	.000901	.0275
18	.000948	.0289	18	.00143	.0437
20	.00151	.0459	20	.00228	.0695
22	.00240	.0730	22	.00363	.110
24	.00381	.116	24	.00576	.176
26	.00606	.185	26	.00916	.279
27	.00764	.233	27	.0116	.352
28	.00963	.294	28	.0146	.444
30	.0153	.467	30	.0232	.706
32	.0244	.742	32	.0368	1.12
34	.0387	1.18	34	.0586	1.79
36	.0616	1.88	36	.0931	2.84
40	.156	4.75	40	.236	7.18
Platinum $Q = 10. \times 10^{-6}$ ohm cm			Steel, invar (35% Ni) $Q = 81. \times 10^{-6}$ ohm cm		
10	.000190	.00579	10	.00154	.0469
12	.000302	.00921	12	.00245	.0746
14	.000481	.0146	14	.00389	.119
16	.000764	.0233	16	.00619	.189
18	.00121	.0370	18	.00984	.300
20	.00193	.0589	20	.0156	.477
22	.00307	.0936	22	.0249	.758
24	.00489	.149	24	.0396	1.21
26	.00776	.237	26	.0629	1.92
27	.00979	.299	27	.0793	2.42
28	.0123	.376	28	.100	3.05
30	.0196	.598	30	.159	4.85
32	.0312	.952	32	.253	7.71
34	.0497	1.51	34	.402	12.3
36	.0789	2.41	36	.639	19.5
40	.200	6.08	40	1.62	49.3

B. & S. No.	Ohms per cm	Ohms per ft	B. & S. No.	Ohms per cm	Ohms per ft
Tantalum $Q = 15.5 \times 10^{-6}$ ohm cm			Tungsten $Q = 5.51 \times 10^{-6}$ ohm cm		
10	.000295	.00898	10	.000105	.00319
12	.000468	.0143	12	.000167	.00508
14	.000745	.0227	14	.000265	.00807
16	.00118	.0361	16	.000421	.0128
18	.00188	.0574	18	.000669	.0204
20	.00299	.0913	20	.00106	.0324
22	.00476	.145	22	.00169	.0516
24	.00757	.231	24	.00269	.0820
26	.0120	.367	26	.00428	.130
27	.0152	.463	27	.00540	.164
28	.0191	.583	28	.00680	.207
30	.0304	.928	30	.0108	.330
32	.0484	1.47	32	.0172	.524
34	.0770	2.35	34	.0274	.834
36	.122	3.73	36	.0435	1.33
40	.309	9.43	40	.110	3.35
Tin $Q = 11.5 \times 10^{-6}$ ohm cm			Zinc (0°C) $Q = 5.75 \times 10^{-6}$ ohm cm		
10	.000219	.00666	10	.000109	.00333
12	.000348	.0106	12	.000174	.00530
14	.000553	.0168	14	.000276	.00842
16	.000879	.0268	16	.000439	.0134
18	.00140	.0426	18	.000699	.0213
20	.00222	.0677	20	.00111	.0339
22	.00353	.108	22	.00177	.0538
24	.00562	.171	24	.00281	.0856
26	.00893	.272	26	.00446	.136
27	.0113	.343	27	.00563	.172
28	.0142	.433	28	.00710	.216
30	.0226	.688	30	.0113	.344
32	.0359	1.09	32	.0180	.547
34	.0571	1.74	34	.0286	.870
36	.0908	2.77	36	.0454	1.38
40	.230	7.00	40	.115	3.50

Credits

Material in Section XII was reprinted from the following sources:

D. R. Lide, Ed., *CRC Handbook of Chemistry and Physics*, 76th ed., Boca Raton, Fla.: CRC Press, 1992: International System of Units (SI), conversion constants and multipliers (conversion of temperatures), symbols and terminology for physical and chemical quantities, fundamental physical constants, classification of electromagnetic radiation.

W. H. Beyer, Ed., *CRC Standard Mathematical Tables and Formulae*, 29th ed., Boca Raton, Fla.: CRC Press, 1991: Greek alphabet, conversion constants and multipliers (recommended decimal multiples and submultiples, metric to English, English to metric, general, temperature factors), physical constants, series expansion, integrals, the Fourier transforms, numerical methods, probability, positional notation.

R. J. Tallarida, *Pocket Book of Integrals and Mathematical Formulas,* 2nd ed., Boca Raotn, Fla.: CRC Press, 1991: Elementary algebra and geometry; determinants, matrices, and linear systems of equations; trigonometry; analytic geometry; series; differential calculus; integral calculus; vector analysis; special functions; statistics; tables of probability and statistics; table of derivatives.

J. F. Pankow, *Aquatic Chemistry Concepts,* Chelsea, Mich.: Lewis Publishers, 1991: Periodic table of the elements.

J. Shackelford and W. Alexander, Eds., *CRC Materials Science and Engineering Handbook,* Boca Raton, Fla.: CRC Press, 1992: Electrical resistivity of selected alloy cast irons, resistivity of selected ceramics.

APPENDIX B
Microwave Engineering Appendix

John P. Wendler
*M/A — Com Components
Business Unit*

Attenuator Design Values

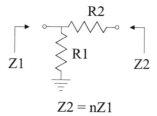

FIGURE 1 Equivalent circuit for a minimum loss pad.

TABLE 1 Minimum Loss Matching Pad Resistance Values as a Function of Transformation Ratio For Z1 = 1 Ohm, 50 Ohms, and 75 Ohms

n Z2/Z1	r1 Z1 = 1	r2 Z1 = 1	R1 Z1 = 50	R2 Z1 = 50	R1 Z1 = 75	R2 Z1 = 75	Loss [dB]
1.1	3.3166	0.3317	165.8	16.6	248.7	24.9	2.7
1.2	2.4495	0.4899	122.5	24.5	183.7	36.7	3.8
1.3	2.0817	0.6245	104.1	31.2	156.1	46.8	4.5
1.4	1.8708	0.7483	93.5	37.4	140.3	56.1	5.2
1.5	1.7321	0.8660	86.6	43.3	129.9	65.0	5.7
1.6	1.6330	0.9798	81.6	49.0	122.5	73.5	6.2
1.7	1.5584	1.0909	77.9	54.5	116.9	81.8	6.6
1.8	1.5000	1.2000	75.0	60.0	112.5	90.0	7.0
1.9	1.4530	1.3077	72.6	65.4	109.0	98.1	7.3
2.0	1.4142	1.4142	70.7	70.7	106.1	106.1	7.7
2.1	1.3817	1.5199	69.1	76.0	103.6	114.0	8.0
2.2	1.3540	1.6248	67.7	81.2	101.6	121.9	8.2
2.3	1.3301	1.7292	66.5	86.5	99.8	129.7	8.5
2.4	1.3093	1.8330	65.5	91.7	98.2	137.5	8.7
2.5	1.2910	1.9365	64.5	96.8	96.8	145.2	9.0
2.6	1.2748	2.0396	63.7	102.0	95.6	153.0	9.2
2.7	1.2603	2.1424	63.0	107.1	94.5	160.7	9.4
2.8	1.2472	2.2450	62.4	112.2	93.5	168.4	9.6
2.9	1.2354	2.3473	61.8	117.4	92.7	176.1	9.8
3.0	1.2247	2.4495	61.2	122.5	91.9	183.7	10.0
3.1	1.2150	2.5515	60.7	127.6	91.1	191.4	10.1

0-8493-8592-X/01/$0.00+$.50
© 2001 by CRC Press LLC

TABLE 1 (continued)   Minimum Loss Matching Pad Resistance Values as a
Function of Transformation Ratio For Z1 = 1 Ohm, 50 Ohms, and 75 Ohms

n Z2/Z1	r1 Z1 = 1	r2 Z1 = 1	R1 Z1 = 50	R2 Z1 = 50	R1 Z1 = 75	R2 Z1 = 75	Loss [dB]
3.2	1.2060	2.6533	60.3	132.7	90.5	199.0	10.3
3.3	1.1978	2.7550	59.9	137.7	89.8	206.6	10.5
3.4	1.1902	2.8566	59.5	142.8	89.3	214.2	10.6
3.5	1.1832	2.9580	59.2	147.9	88.7	221.9	10.8
3.6	1.1767	3.0594	58.8	153.0	88.3	229.5	10.9
3.7	1.1706	3.1607	58.5	158.0	87.8	237.1	11.0
3.8	1.1650	3.2619	58.2	163.1	87.4	244.6	11.2
3.9	1.1597	3.3630	58.0	168.2	87.0	252.2	11.3
4.0	1.1547	3.4641	57.7	173.2	86.6	259.8	11.4
4.1	1.1500	3.5651	57.5	178.3	86.3	267.4	11.6
4.2	1.1456	3.6661	57.3	183.3	85.9	275.0	11.7
4.3	1.1415	3.7670	57.1	188.3	85.6	282.5	11.8
4.4	1.1376	3.8678	56.9	193.4	85.3	290.1	11.9
4.5	1.1339	3.9686	56.7	198.4	85.0	297.6	12.0
4.6	1.1304	4.0694	56.5	203.5	84.8	305.2	12.1
4.7	1.1271	4.1701	56.4	208.5	84.5	312.8	12.2
4.8	1.1239	4.2708	56.2	213.5	84.3	320.3	12.3
4.9	1.1209	4.3715	56.0	218.6	84.1	327.9	12.4
5.0	1.1180	4.4721	55.9	223.6	83.9	335.4	12.5
5.1	1.1153	4.5727	55.8	228.6	83.6	343.0	12.6
5.2	1.1127	4.6733	55.6	233.7	83.5	350.5	12.7
5.3	1.1102	4.7739	55.5	238.7	83.3	358.0	12.8
5.4	1.1078	4.8744	55.4	243.7	83.1	365.6	12.9
5.5	1.1055	4.9749	55.3	248.7	82.9	373.1	13.0
5.6	1.1034	5.0754	55.2	253.8	82.8	380.7	13.1
5.7	1.1013	5.1759	55.1	258.8	82.6	388.2	13.2
5.8	1.0992	5.2764	55.0	263.8	82.4	395.7	13.3
5.9	1.0973	5.3768	54.9	268.8	82.3	403.3	13.3
6.0	1.0954	5.4772	54.8	273.9	82.2	410.8	13.4
6.1	1.0937	5.5776	54.7	278.9	82.0	418.3	13.5
6.2	1.0919	5.6780	54.6	283.9	81.9	425.9	13.6
6.3	1.0903	5.7784	54.5	288.9	81.8	433.4	13.6
6.4	1.0887	5.8788	54.4	293.9	81.6	440.9	13.7
6.5	1.0871	5.9791	54.4	299.0	81.5	448.4	13.8
6.6	1.0856	6.0795	54.3	304.0	81.4	456.0	13.9
6.7	1.0842	6.1798	54.2	309.0	81.3	463.5	13.9
6.8	1.0828	6.2801	54.1	314.0	81.2	471.0	14.0
6.9	1.0814	6.3804	54.1	319.0	81.1	478.5	14.1
7.0	1.0801	6.4807	54.0	324.0	81.0	486.1	14.1
7.1	1.0789	6.5810	53.9	329.1	80.9	493.6	14.2
7.2	1.0776	6.6813	53.9	334.1	80.8	501.1	14.3
7.3	1.0764	6.7816	53.8	339.1	80.7	508.6	14.3
7.4	1.0753	6.8819	53.8	344.1	80.6	516.1	14.4
7.5	1.0742	6.9821	53.7	349.1	80.6	523.7	14.5
7.6	1.0731	7.0824	53.7	354.1	80.5	531.2	14.5
7.7	1.0720	7.1826	53.6	359.1	80.4	538.7	14.6
7.8	1.0710	7.2829	53.6	364.1	80.3	546.2	14.6
7.9	1.0700	7.3831	53.5	369.2	80.3	553.7	14.7
8.0	1.0690	7.4833	53.5	374.2	80.2	561.2	14.8
8.1	1.0681	7.5835	53.4	379.2	80.1	568.8	14.8
8.2	1.0672	7.6837	53.4	384.2	80.0	576.3	14.9
8.3	1.0663	7.7840	53.3	389.2	80.0	583.8	14.9
8.4	1.0654	7.8842	53.3	394.2	79.9	591.3	15.0
8.5	1.0646	7.9844	53.2	399.2	79.8	598.8	15.0
8.6	1.0638	8.0846	53.2	404.2	79.8	606.3	15.1

TABLE 1 (continued) Minimum Loss Matching Pad Resistance Values as a Function of Transformation Ratio For Z1 = 1 Ohm, 50 Ohms, and 75 Ohms

| n | r1 | r2 | R1 | R2 | R1 | R2 | |
Z2/Z1	Z1 = 1	Z1 = 1	Z1 = 50	Z1 = 50	Z1 = 75	Z1 = 75	Loss [dB]
8.7	1.0630	8.1847	53.1	409.2	79.7	613.9	15.2
8.8	1.0622	8.2849	53.1	414.2	79.7	621.4	15.2
8.9	1.0614	8.3851	53.1	419.3	79.6	628.9	15.3
9.0	1.0607	8.4853	53.0	424.3	79.5	636.4	15.3
9.1	1.0599	8.5855	53.0	429.3	79.5	643.9	15.4
9.2	1.0592	8.6856	53.0	434.3	79.4	651.4	15.4
9.3	1.0585	8.7858	52.9	439.3	79.4	658.9	15.5
9.4	1.0579	8.8859	52.9	444.3	79.3	666.4	15.5
9.5	1.0572	8.9861	52.9	449.3	79.3	674.0	15.6
9.6	1.0565	9.0863	52.8	454.3	79.2	681.5	15.6
9.7	1.0559	9.1864	52.8	459.3	79.2	689.0	15.7
9.8	1.0553	9.2865	52.8	464.3	79.1	696.5	15.7
9.9	1.0547	9.3867	52.7	469.3	79.1	704.0	15.7
10.0	1.0541	9.4868	52.7	474.3	79.1	711.5	15.8

$$R_1 = \frac{Z_1\left(n + \sqrt{n^2 - n}\right)}{n - 1 + \sqrt{n^2 - n}} \qquad R_2 = Z_1\sqrt{n^2 - n} \qquad \frac{P_O}{P_A} = \frac{1}{n}\left(\frac{1}{1 + \sqrt{\frac{1}{n}}}\right)^2$$

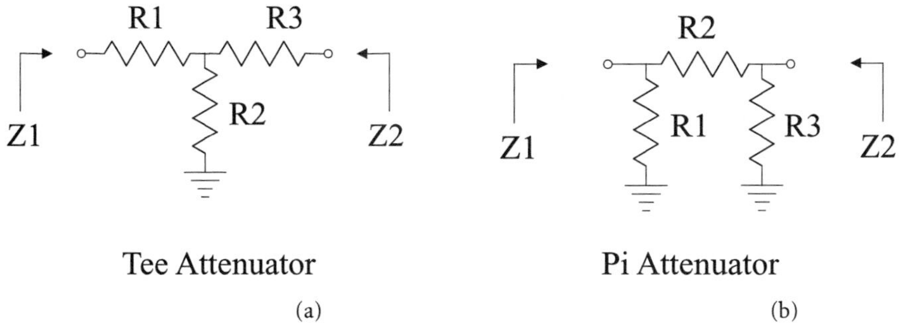

Tee Attenuator

(a)

Pi Attenuator

(b)

FIGURE 2   (a) Equivalent circuit for a Tee attenuator; (b) Equivalent circuit for a Pi attenuator.

TABLE 2   Tee- and Pi-Pad Resistor Values for Zo = 1 Ohm and Zo = 50 Ohms

Loss [dB]	Voltage Atten	r1, r3, g1, g3 Z1 = Z2 = 1	r2, g2 Z1 = Z2 = 1	Tee R1, R3 Z1 = Z2 = 50	Tee R2 Z1 = Z2 = 50	Pi R1, R3 Z1 = Z2 = 50	Pi R2 Z1 = Z2 = 50
0.1	0.98855	0.0058	86.8570	0.3	4342.8	8686.0	0.6
0.2	0.97724	0.0115	43.4256	0.6	2171.3	4343.1	1.2
0.3	0.96605	0.0173	28.9472	0.9	1447.4	2895.6	1.7
0.4	0.95499	0.0230	21.7071	1.2	1085.4	2171.9	2.3
0.5	0.94406	0.0288	17.3622	1.4	868.1	1737.7	2.9
0.6	0.93325	0.0345	14.4650	1.7	723.2	1448.2	3.5
0.7	0.92257	0.0403	12.3950	2.0	619.7	1241.5	4.0
0.8	0.91201	0.0460	10.8420	2.3	542.1	1086.5	4.6
0.9	0.90157	0.0518	9.6337	2.6	481.7	966.0	5.2
1.0	0.89125	0.0575	8.6667	2.9	433.3	869.5	5.8
1.2	0.87096	0.0690	7.2153	3.4	360.8	725.0	6.9
1.4	0.85114	0.0804	6.1774	4.0	308.9	621.8	8.1
1.6	0.83176	0.0918	5.3981	4.6	269.9	544.4	9.3
1.8	0.81283	0.1032	4.7911	5.2	239.6	484.3	10.4
2.0	0.79433	0.1146	4.3048	5.7	215.2	436.2	11.6

TABLE 2 (continued)   Tee- and Pi-Pad Resistor Values for Zo = 1 Ohm and Zo = 50 Ohms

Loss [dB]	Voltage Atten	r1, r3, g1, g3 Z1 = Z2 = 1	r2, g2 Z1 = Z2 = 1	Tee R1, R3 Z1 = Z2 = 50	Tee R2 Z1 = Z2 = 50	Pi R1, R3 Z1 = Z2 = 50	Pi R2 Z1 = Z2 = 50
2.2	0.77625	0.1260	3.9062	6.3	195.3	396.9	12.8
2.4	0.75858	0.1373	3.5735	6.9	178.7	364.2	14.0
2.6	0.74131	0.1486	3.2914	7.4	164.6	336.6	15.2
2.8	0.72444	0.1598	3.0490	8.0	152.5	312.9	16.4
3.0	0.70795	0.1710	2.8385	8.5	141.9	292.4	17.6
3.2	0.69183	0.1822	2.6539	9.1	132.7	274.5	18.8
3.4	0.67608	0.1933	2.4906	9.7	124.5	258.7	20.1
3.6	0.66069	0.2043	2.3450	10.2	117.3	244.7	21.3
3.8	0.64565	0.2153	2.2144	10.8	110.7	232.2	22.6
4.0	0.63096	0.2263	2.0966	11.3	104.8	221.0	23.8
4.2	0.61660	0.2372	1.9896	11.9	99.5	210.8	25.1
4.4	0.60256	0.2480	1.8921	12.4	94.6	201.6	26.4
4.6	0.58884	0.2588	1.8028	12.9	90.1	193.2	27.7
4.8	0.57544	0.2695	1.7206	13.5	86.0	185.5	29.1
5.0	0.56234	0.2801	1.6448	14.0	82.2	178.5	30.4
5.5	0.53088	0.3064	1.4785	15.3	73.9	163.2	33.8
6.0	0.50119	0.3323	1.3386	16.6	66.9	150.5	37.4
6.5	0.47315	0.3576	1.2193	17.9	61.0	139.8	41.0
7.0	0.44668	0.3825	1.1160	19.1	55.8	130.7	44.8
7.5	0.42170	0.4068	1.0258	20.3	51.3	122.9	48.7
8.0	0.39811	0.4305	0.9462	21.5	47.3	116.1	52.8
8.5	0.37584	0.4537	0.8753	22.7	43.8	110.2	57.1
9.0	0.35481	0.4762	0.8118	23.8	40.6	105.0	61.6
9.5	0.33497	0.4982	0.7546	24.9	37.7	100.4	66.3
10.0	0.31623	0.5195	0.7027	26.0	35.1	96.2	71.2
10.5	0.29854	0.5402	0.6555	27.0	32.8	92.6	76.3
11.0	0.28184	0.5603	0.6123	28.0	30.6	89.2	81.7
11.5	0.26607	0.5797	0.5727	29.0	28.6	86.3	87.3
12.0	0.25119	0.5985	0.5362	29.9	26.8	83.5	93.2
12.5	0.23714	0.6166	0.5025	30.8	25.1	81.1	99.5
13.0	0.22387	0.6342	0.4714	31.7	23.6	78.8	106.1
13.5	0.21135	0.6511	0.4425	32.6	22.1	76.8	113.0
14.0	0.19953	0.6673	0.4156	33.4	20.8	74.9	120.3
14.5	0.18836	0.6830	0.3906	34.1	19.5	73.2	128.0
15.0	0.17783	0.6980	0.3673	34.9	18.4	71.6	136.1
15.5	0.16788	0.7125	0.3455	35.6	17.3	70.2	144.7
16.0	0.15849	0.7264	0.3251	36.3	16.3	68.8	153.8
16.5	0.14962	0.7397	0.3061	37.0	15.3	67.6	163.3
17.0	0.14125	0.7525	0.2883	37.6	14.4	66.4	173.5
17.5	0.13335	0.7647	0.2715	38.2	13.6	65.4	184.1
18.0	0.12589	0.7764	0.2558	38.8	12.8	64.4	195.4
18.5	0.11885	0.7875	0.2411	39.4	12.1	63.5	207.4
19.0	0.11220	0.7982	0.2273	39.9	11.4	62.6	220.0
19.5	0.10593	0.8084	0.2143	40.4	10.7	61.8	233.4
20.0	0.10000	0.8182	0.2020	40.9	10.1	61.1	247.5
20.5	0.09441	0.8275	0.1905	41.4	9.5	60.4	262.5
21.0	0.08913	0.8363	0.1797	41.8	9.0	59.8	278.3
21.5	0.08414	0.8448	0.1695	42.2	8.5	59.2	295.0
22.0	0.07943	0.8528	0.1599	42.6	8.0	58.6	312.7
22.5	0.07499	0.8605	0.1508	43.0	7.5	58.1	331.5
23.0	0.07079	0.8678	0.1423	43.4	7.1	57.6	351.4
23.5	0.06683	0.8747	0.1343	43.7	6.7	57.2	372.4
24.0	0.06310	0.8813	0.1267	44.1	6.3	56.7	394.6
24.5	0.05957	0.8876	0.1196	44.4	6.0	56.3	418.2
25.0	0.05623	0.8935	0.1128	44.7	5.6	56.0	443.2

TABLE 2 (continued)   Tee- and Pi-Pad Resistor Values for Zo = 1 Ohm and Zo = 50 Ohms

Loss [dB]	Voltage Atten	r1, r3, g1, g3 Z1 = Z2 = 1	r2, g2 Z1 = Z2 = 1	Tee R1, R3 Z1 = Z2 = 50	Tee R2 Z1 = Z2 = 50	Pi R1, R3 Z1 = Z2 = 50	Pi R2 Z1 = Z2 = 50
26.0	0.05012	0.9045	0.1005	45.2	5.0	55.3	497.6
27.0	0.04467	0.9145	0.0895	45.7	4.5	54.7	558.6
28.0	0.03981	0.9234	0.0797	46.2	4.0	54.1	627.0
29.0	0.03548	0.9315	0.0711	46.6	3.6	53.7	703.7
30.0	0.03162	0.9387	0.0633	46.9	3.2	53.3	789.8
31.0	0.02818	0.9452	0.0564	47.3	2.8	52.9	886.3
32.0	0.02512	0.9510	0.0503	47.5	2.5	52.6	994.6
33.0	0.02239	0.9562	0.0448	47.8	2.2	52.3	1116.1
34.0	0.01995	0.9609	0.0399	48.0	2.0	52.0	1252.5
35.0	0.01778	0.9651	0.0356	48.3	1.8	51.8	1405.4
36.0	0.01585	0.9688	0.0317	48.4	1.6	51.6	1577.0
37.0	0.01413	0.9721	0.0283	48.6	1.4	51.4	1769.5
38.0	0.01259	0.9751	0.0252	48.8	1.3	51.3	1985.5
39.0	0.01122	0.9778	0.0224	48.9	1.1	51.1	2227.8
40.0	0.01000	0.9802	0.0200	49.0	1.0	51.0	2499.8
41.0	0.00891	0.9823	0.0178	49.1	0.9	50.9	2804.8
42.0	0.00794	0.9842	0.0159	49.2	0.8	50.8	3147.1
43.0	0.00708	0.9859	0.0142	49.3	0.7	50.7	3531.2
44.0	0.00631	0.9875	0.0126	49.4	0.6	50.6	3962.1
45.0	0.00562	0.9888	0.0112	49.4	0.6	50.6	4445.6

Note: P_i values are duals of Tee values.

$$a = \sqrt{\frac{P_{Z_2}}{P_{Z_1}}} \qquad R_{1T} = \left(\frac{2}{(1-a^2)} - 1\right)Z_1 - \frac{2a}{(1-a^2)}\sqrt{Z_1 Z_2}$$

$$R_{2T} = 2\sqrt{Z_1 Z_2}\,\frac{a}{(1-a^2)}$$

$$R_{3T} = \left(\frac{2}{(1-a^2)} - 1\right)Z_2 - \frac{2a}{(1-a^2)}\sqrt{Z_1 Z_2}$$

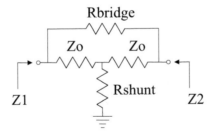

Bridged-Tee   Attenuator
Z1=Z2=Zo

FIGURE 3   Equivalent circuit for a Bridged-T attenuator.

TABLE 3   Bridged-T Attenuator Resistance Values for Zo = 1 Ohm, 50 Ohms, 75 Ohms

Loss [dB]	Voltage Atten	Bridge Arm Z1 = Z2 = 1	Shunt Arm Z1 = Z2 = 1	Bridge Arm Z1 = Z2 = 50	Shunt Arm Z1 = Z2 = 50	Bridge Arm Z1 = Z2 = 75	Shunt Arm Z1 = Z2 = 75
0.1	0.98855	0.0116	86.3599	0.6	4318.0	6477.0	0.9
0.2	0.97724	0.0233	42.9314	1.2	2146.6	3219.9	1.7
0.3	0.96605	0.0351	28.4558	1.8	1422.8	2134.2	2.6
0.4	0.95499	0.0471	21.2186	2.4	1060.9	1591.4	3.5
0.5	0.94406	0.0593	16.8766	3.0	843.8	1265.7	4.4
0.6	0.93325	0.0715	13.9822	3.6	699.1	1048.7	5.4
0.7	0.92257	0.0839	11.9151	4.2	595.8	893.6	6.3

TABLE 3 (continued)   Bridged-T Attenuator Resistance Values for Zo = 1 Ohm, 50 Ohms, 75 Ohms

Loss [dB]	Voltage Atten	Bridge Arm Z1 = Z2 = 1	Shunt Arm Z1 = Z2 = 1	Bridge Arm Z1 = Z2 = 50	Shunt Arm Z1 = Z2 = 50	Bridge Arm Z1 = Z2 = 75	Shunt Arm Z1 = Z2 = 75
0.8	0.91201	0.0965	10.3650	4.8	518.3	777.4	7.2
0.9	0.90157	0.1092	9.1596	5.5	458.0	687.0	8.2
1.0	0.89125	0.1220	8.1955	6.1	409.8	614.7	9.2
1.2	0.87096	0.1482	6.7498	7.4	337.5	506.2	11.1
1.4	0.85114	0.1749	5.7176	8.7	285.9	428.8	13.1
1.6	0.83176	0.2023	4.9440	10.1	247.2	370.8	15.2
1.8	0.81283	0.2303	4.3428	11.5	217.1	325.7	17.3
2.0	0.79433	0.2589	3.8621	12.9	193.1	289.7	19.4
2.2	0.77625	0.2882	3.4692	14.4	173.5	260.2	21.6
2.4	0.75858	0.3183	3.1421	15.9	157.1	235.7	23.9
2.6	0.74131	0.3490	2.8656	17.4	143.3	214.9	26.2
2.8	0.72444	0.3804	2.6289	19.0	131.4	197.2	28.5
3.0	0.70795	0.4125	2.4240	20.6	121.2	181.8	30.9
3.2	0.69183	0.4454	2.2450	22.3	112.2	168.4	33.4
3.4	0.67608	0.4791	2.0872	24.0	104.4	156.5	35.9
3.6	0.66069	0.5136	1.9472	25.7	97.4	146.0	38.5
3.8	0.64565	0.5488	1.8221	27.4	91.1	136.7	41.2
4.0	0.63096	0.5849	1.7097	29.2	85.5	128.2	43.9
4.2	0.61660	0.6218	1.6082	31.1	80.4	120.6	46.6
4.4	0.60256	0.6596	1.5161	33.0	75.8	113.7	49.5
4.6	0.58884	0.6982	1.4322	34.9	71.6	107.4	52.4
4.8	0.57544	0.7378	1.3554	36.9	67.8	101.7	55.3
5.0	0.56234	0.7783	1.2849	38.9	64.2	96.4	58.4
5.5	0.53088	0.8836	1.1317	44.2	56.6	84.9	66.3
6.0	0.50119	0.9953	1.0048	49.8	50.2	75.4	74.6
6.5	0.47315	1.1135	0.8981	55.7	44.9	67.4	83.5
7.0	0.44668	1.2387	0.8073	61.9	40.4	60.5	92.9
7.5	0.42170	1.3714	0.7292	68.6	36.5	54.7	102.9
8.0	0.39811	1.5119	0.6614	75.6	33.1	49.6	113.4
8.5	0.37584	1.6607	0.6021	83.0	30.1	45.2	124.6
9.0	0.35481	1.8184	0.5499	90.9	27.5	41.2	136.4
9.5	0.33497	1.9854	0.5037	99.3	25.2	37.8	148.9
10.0	0.31623	2.1623	0.4625	108.1	23.1	34.7	162.2
10.5	0.29854	2.3497	0.4256	117.5	21.3	31.9	176.2
11.0	0.28184	2.5481	0.3924	127.4	19.6	29.4	191.1
11.5	0.26607	2.7584	0.3625	137.9	18.1	27.2	206.9
12.0	0.25119	2.9811	0.3354	149.1	16.8	25.2	223.6
12.5	0.23714	3.2170	0.3109	160.8	15.5	23.3	241.3
13.0	0.22387	3.4668	0.2884	173.3	14.4	21.6	260.0
13.5	0.21135	3.7315	0.2680	186.6	13.4	20.1	279.9
14.0	0.19953	4.0119	0.2493	200.6	12.5	18.7	300.9
14.5	0.18836	4.3088	0.2321	215.4	11.6	17.4	323.2
15.0	0.17783	4.6234	0.2163	231.2	10.8	16.2	346.8
15.5	0.16788	4.9566	0.2018	247.8	10.1	15.1	371.7
16.0	0.15849	5.3096	0.1883	265.5	9.4	14.1	398.2
16.5	0.14962	5.6834	0.1759	284.2	8.8	13.2	426.3
17.0	0.14125	6.0795	0.1645	304.0	8.2	12.3	456.0
17.5	0.13335	6.4989	0.1539	324.9	7.7	11.5	487.4
18.0	0.12589	6.9433	0.1440	347.2	7.2	10.8	520.7
18.5	0.11885	7.4140	0.1349	370.7	6.7	10.1	556.0
19.0	0.11220	7.9125	0.1264	395.6	6.3	9.5	593.4
19.5	0.10593	8.4406	0.1185	422.0	5.9	8.9	633.0
20.0	0.10000	9.0000	0.1111	450.0	5.6	8.3	675.0
20.5	0.09441	9.5925	0.1042	479.6	5.2	7.8	719.4
21.0	0.08913	10.2202	0.0978	511.0	4.9	7.3	766.5
21.5	0.08414	10.8850	0.0919	544.3	4.6	6.9	816.4

TABLE 3 (continued)   Bridged-T Attenuator Resistance Values for Zo = 1 Ohm, 50 Ohms, 75 Ohms

Loss [dB]	Voltage Atten	Bridge Arm Z1 = Z2 = 1	Shunt Arm Z1 = Z2 = 1	Bridge Arm Z1 = Z2 = 50	Shunt Arm Z1 = Z2 = 50	Bridge Arm Z1 = Z2 = 75	Shunt Arm Z1 = Z2 = 75
22.0	0.07943	11.5893	0.0863	579.5	4.3	6.5	869.2
22.5	0.07499	12.3352	0.0811	616.8	4.1	6.1	925.1
23.0	0.07079	13.1254	0.0762	656.3	3.8	5.7	984.4
23.5	0.06683	13.9624	0.0716	698.1	3.6	5.4	1047.2
24.0	0.06310	14.8489	0.0673	742.4	3.4	5.1	1113.7
24.5	0.05957	15.7880	0.0633	789.4	3.2	4.8	1184.1
25.0	0.05623	16.7828	0.0596	839.1	3.0	4.5	1258.7
26.0	0.05012	18.9526	0.0528	947.6	2.6	4.0	1421.4
27.0	0.04467	21.3872	0.0468	1069.4	2.3	3.5	1604.0
28.0	0.03981	24.1189	0.0415	1205.9	2.1	3.1	1808.9
29.0	0.03548	27.1838	0.0368	1359.2	1.8	2.8	2038.8
30.0	0.03162	30.6228	0.0327	1531.1	1.6	2.4	2296.7
31.0	0.02818	34.4813	0.0290	1724.1	1.5	2.2	2586.1
32.0	0.02512	38.8107	0.0258	1940.5	1.3	1.9	2910.8
33.0	0.02239	43.6684	0.0229	2183.4	1.1	1.7	3275.1
34.0	0.01995	49.1187	0.0204	2455.9	1.0	1.5	3683.9
35.0	0.01778	55.2341	0.0181	2761.7	0.9	1.4	4142.6
36.0	0.01585	62.0957	0.0161	3104.8	0.8	1.2	4657.2
37.0	0.01413	69.7946	0.0143	3489.7	0.7	1.1	5234.6
38.0	0.01259	78.4328	0.0127	3921.6	0.6	1.0	5882.5
39.0	0.01122	88.1251	0.0113	4406.3	0.6	0.9	6609.4
40.0	0.01000	99.0000	0.0101	4950.0	0.5	0.8	7425.0
41.0	0.00891	111.2018	0.0090	5560.1	0.4	0.7	8340.1
42.0	0.00794	124.8925	0.0080	6244.6	0.4	0.6	9366.9
43.0	0.00708	140.2538	0.0071	7012.7	0.4	0.5	10519.0
44.0	0.00631	157.4893	0.0063	7874.5	0.3	0.5	11811.7
45.0	0.00562	176.8279	0.0057	8841.4	0.3	0.4	13262.1

Return Loss, Reflection Coefficient, VSWR, and Mismatch Loss

TABLE 4   Conversion Between Return Loss, Reflection Coefficient, VSWR, and Mismatch Loss

Return Loss [dB]	Reflection Coefficient (Rho)	VSWR ( ):1	Mismatch Loss [dB]	Return Loss [dB]	Reflection Coefficient (Rho)	VSWR ( ):1	Mismatch Loss [dB]
Infinite	0.0000	1.00	0.00	33.00	0.0224	1.05	0.00
50.00	0.0032	1.01	0.00	32.00	0.0251	1.05	0.00
49.00	0.0035	1.01	0.00	31.00	0.0282	1.06	0.00
48.00	0.0040	1.01	0.00	30.00	0.0316	1.07	0.00
47.00	0.0045	1.01	0.00	29.00	0.0355	1.07	0.01
46.00	0.0050	1.01	0.00	28.00	0.0398	1.08	0.01
45.00	0.0056	1.01	0.00	27.00	0.0447	1.09	0.01
44.00	0.0063	1.01	0.00	26.00	0.0501	1.11	0.01
43.00	0.0071	1.01	0.00	25.00	0.0562	1.12	0.01
42.00	0.0079	1.02	0.00	24.00	0.0631	1.13	0.02
41.00	0.0089	1.02	0.00	23.00	0.0708	1.15	0.02
40.00	0.0100	1.02	0.00	22.00	0.0794	1.17	0.03
39.00	0.0112	1.02	0.00	21.00	0.0891	1.20	0.03
38.00	0.0126	1.03	0.00	20.00	0.1000	1.22	0.04
37.00	0.0141	1.03	0.00	19.50	0.1059	1.24	0.05
36.00	0.0158	1.03	0.00	19.00	0.1122	1.25	0.06
35.00	0.0178	1.04	0.00	18.50	0.1189	1.27	0.06
34.00	0.0200	1.04	0.00	18.00	0.1259	1.29	0.07

TABLE 4 (continued)   Conversion Between Return Loss, Reflection Coefficient, VSWR, and Mismatch Loss

Return Loss [dB]	Reflection Coefficient (Rho)	VSWR ( ):1	Mismatch Loss [dB]	Return Loss [dB]	Reflection Coefficient (Rho)	VSWR ( ):1	Mismatch Loss [dB]
17.50	0.1334	1.31	0.08	6.02	0.5000	3.00	1.25
17.00	0.1413	1.33	0.09	6.00	0.5012	3.01	1.26
16.50	0.1496	1.35	0.10	5.80	0.5129	3.11	1.33
16.00	0.1585	1.38	0.11	5.60	0.5248	3.21	1.40
15.50	0.1679	1.40	0.12	5.40	0.5370	3.32	1.48
15.00	0.1778	1.43	0.14	5.20	0.5495	3.44	1.56
14.50	0.1884	1.46	0.16	5.11	0.5556	3.50	1.60
14.00	0.1995	1.50	0.18	5.00	0.5623	3.57	1.65
13.50	0.2113	1.54	0.20	4.80	0.5754	3.71	1.75
13.00	0.2239	1.58	0.22	4.60	0.5888	3.86	1.85
12.50	0.2371	1.62	0.25	4.44	0.6000	4.00	1.94
12.00	0.2512	1.67	0.28	4.40	0.6026	4.03	1.96
11.50	0.2661	1.73	0.32	4.20	0.6166	4.22	2.08
11.00	0.2818	1.78	0.36	4.00	0.6310	4.42	2.20
10.50	0.2985	1.85	0.41	3.93	0.6364	4.50	2.25
10.00	0.3162	1.92	0.46	3.80	0.6457	4.64	2.34
9.80	0.3236	1.96	0.48	3.60	0.6607	4.89	2.49
9.60	0.3311	1.99	0.50	3.52	0.6667	5.00	2.55
9.54	0.3333	2.00	0.51	3.40	0.6761	5.17	2.65
9.40	0.3388	2.03	0.53	3.20	0.6918	5.49	2.83
9.20	0.3467	2.06	0.56	3.00	0.7079	5.85	3.02
9.00	0.3548	2.10	0.58	2.80	0.7244	6.26	3.23
8.80	0.3631	2.14	0.61	2.60	0.7413	6.73	3.46
8.60	0.3715	2.18	0.65	2.40	0.7586	7.28	3.72
8.40	0.3802	2.23	0.68	2.20	0.7762	7.94	4.01
8.20	0.3890	2.27	0.71	2.00	0.7943	8.72	4.33
8.00	0.3981	2.32	0.75	1.80	0.8128	9.69	4.69
7.80	0.4074	2.37	0.79	1.74	0.8182	10.00	4.81
7.60	0.4169	2.43	0.83	1.60	0.8318	10.89	5.11
7.40	0.4266	2.49	0.87	1.40	0.8511	12.44	5.60
7.36	0.4286	2.50	0.88	1.20	0.8710	14.50	6.17
7.20	0.4365	2.55	0.92	1.00	0.8913	17.39	6.87
7.00	0.4467	2.61	0.97	0.80	0.9120	21.73	7.74
6.80	0.4571	2.68	1.02	0.60	0.9333	28.96	8.89
6.60	0.4677	2.76	1.07	0.40	0.9550	43.44	10.56
6.40	0.4786	2.84	1.13	0.20	0.9772	86.86	13.47
6.20	0.4898	2.92	1.19	0.00	1.0000	Infinite	Infinite

Notes:

1. Return Loss = -20*log(|Rho|)
2. Mismatch Loss = -10*log(1-|Rho|^2)
3. VSWR = (1+|Rho|)/(1-|Rho|)

Waveguide Components

TABLE 5 Waveguide Performance and Dimensions

EIA WR-	Mil-W-85E RG( )/U	TE10 Cutoff Frequency [GHz]	Recommended Frequency Range Min [GHz]	Max [GHz]	Theoretical Attenuation Fmin [dB/100']	Fmax [dB/100']	Inside Dimensions a [inches]	b [Inches]	Tolerance ± [Inches]	Outside Dimensions [Inches]	[Inches]	Tolerance ± [Inches]	Wall Thickness [Inches]	Material	Contact Flange	Choke Flange	Cover Flange	Hole Pattern Figure
3		173.5726	220.00	325.00	503.90	352.59	0.0340	0.0170	0.00020	4.156	(Diameter)	0.001		Silver				
4		137.2434	140.00	260.00	371.25	246.94	0.0430	0.0215	0.00020	3.156	(Diameter)	0.001		Silver				
5	135	115.7151	140.00	220.00	303.47	190.96	0.0510	0.0255	0.00025	2.156	(Diameter)	0.001		Silver				
7	136	90.7918	110.00	170.00	210.19	133.39	0.0650	0.0325	0.00025	1.156	(Diameter)	0.001		Silver				
8	138	73.7683	90.00	140.00	151.38	97.26	0.080	0.040	0.0003	0.156	(Diameter)	0.001		Silver				1
10		59.0147	75.00	110.00			0.100	0.050	0.0005	0.180	0.130	0.002	0.040					
12		48.3727	60.00	90.00	122.71	82.37	0.122	0.061	0.0005	0.202	0.141	0.002	0.040	Brass				1
12	99	48.3727	60.00	90.00	77.46	51.99	0.122	0.061	0.0005	0.202	0.141	0.002	0.040	Silver			387	1
15		39.8748	50.00	75.00	89.78	61.41	0.148	0.074	0.0010	0.228	0.154	0.002	0.040	Brass				1
15	98	39.8748	50.00	75.00	56.67	38.76	0.148	0.074	0.0010	0.228	0.154	0.002	0.040	Silver			385	1
19		31.3908	40.00	60.00			0.188	0.094	0.0010	0.268	0.174	0.002	0.040					
22		26.3458	33.00	50.00	48.33	32.89	0.224	0.112	0.0010	0.304	0.192	0.002	0.040	Brass				1
22	97	26.3458	33.00	50.00	30.50	20.76	0.224	0.112	0.0010	0.304	0.192	0.002	0.040	Silver			383	1
28		21.0767	26.50	40.00	28.00	19.20	0.280	0.140	0.0015	0.360	0.220	0.002	0.040	Aluminum				1,2
28	96	21.0767	26.50	40.00	34.32	23.53	0.280	0.140	0.0015	0.360	0.220	0.002	0.040	Brass		600	599	1,2
28		21.0767	26.50	40.00	21.66	14.85	0.280	0.140	0.0015	0.360	0.220	0.002	0.040	Silver				1,2
34		17.3573	22.00	33.00			0.340	0.170	0.0020	0.420	0.250	0.003	0.040					
42	121	14.0511	18.00	26.50	16.86	12.40	0.420	0.170	0.0020	0.500	0.250	0.003	0.040	Aluminum	425	598	597	2
42	53	14.0511	18.00	26.50	20.66	15.19	0.420	0.170	0.0020	0.500	0.250	0.003	0.040	Brass	425	596	595	2
42	66	14.0511	18.00	26.50	13.04	9.59	0.420	0.170	0.0020	0.500	0.250	0.003	0.040	Silver	425	596A	595	2
51		11.5715	15.00	22.00			0.510	0.255	0.0025	0.590	0.335	0.003	0.040					
62	91	9.4879	12.40	18.00	7.88	5.80	0.622	0.311	0.0025	0.702	0.391	0.003	0.040	Aluminum			419	2
62		9.4879	12.40	18.00	9.66	7.11	0.622	0.311	0.0025	0.702	0.391	0.003	0.040	Brass		541		2
62	107	9.4879	12.40	18.00	6.10	4.49	0.622	0.311	0.0025	0.702	0.391	0.003	0.040	Silver				2
75		7.8686	10.00	15.00			0.750	0.375	0.003	0.850	0.475	0.003	0.050					
90	67	6.5572	8.20	12.40	5.29	3.66	0.900	0.400	0.003	1.000	0.500	0.003	0.050	Aluminum		136A	135	2
90	52	6.5572	8.20	12.40	6.48	4.49	0.900	0.400	0.003	1.000	0.500	0.003	0.050	Brass		40A	39	2
112	68	5.2598	7.05	10.00	3.39	2.63	1.122	0.497	0.004	1.250	0.625	0.004	0.064	Aluminum		137A	138	2
112	51	5.2598	7.05	10.00	4.15	3.23	1.122	0.497	0.004	1.250	0.625	0.004	0.064	Brass		52A	51	2
137	106	4.3014	5.85	8.20	2.42	1.91	1.372	0.622	0.004	1.500	0.750	0.004	0.064	Aluminum		440A	441	2
137	50	4.3014	5.85	8.20	2.96	2.34	1.372	0.622	0.004	1.500	0.750	0.004	0.064	Brass		343A	344	2
159		3.7116	4.90	7.05			1.590	0.795	0.004	1.718	0.923	0.004	0.064					

TABLE 5 (continued)　Waveguide Performance and Dimensions

EIA WR-	Mil-W-85E RG( )/U	TE10 Cutoff Frequency [GHz]	Recommended Frequency Range Min [GHz]	Max [GHz]	Theoretical Attenuation Fmin [dB/100']	Fmax [dB/100']	Inside Dimensions a [inches]	b [Inches]	Tolerance ± [Inches]	Outside Dimensions [Inches]	[Inches]	Tolerance ± [Inches]	Wall Thickness [Inches]	Material	Contact Flange	Choke Flange	Cover Flange	Hole Pattern Figure
187	95	3.1525	3.95	5.85	1.70	1.18	1.872	0.872	0.005	2.000	1.000	0.005	0.064	Aluminum		406A	407	
187	49	3.1525	3.95	5.85	2.09	1.45	1.872	0.872	0.005	2.000	1.000	0.005	0.064	Brass		148B	149A	
229		2.5771	3.30	4.90			2.290	1.145	0.005	2.418	1.273	0.005	0.064					
284	75	2.0780	2.60	3.95	0.91	0.62	2.840	1.340	0.005	3.000	1.500	0.005	0.080	Aluminum		585	584	
284	48	2.0780	2.60	3.95	1.11	0.76	2.840	1.340	0.005	3.000	1.500	0.005	0.080	Brass		54B	53	
340	113	1.7357	2.20	3.30	0.65	0.45	3.400	1.700	0.005	3.560	1.860	0.005	0.080	Aluminum			554	
340	112	1.7357	2.20	3.30	0.80	0.56	3.400	1.700	0.005	3.560	1.860	0.005	0.080	Brass			553	
430	105	1.3724	1.70	2.60	0.48	0.32	4.300	2.150	0.005	4.460	2.310	0.005	0.080	Aluminum			437A	
430	104	1.3724	1.70	2.60	0.59	0.39	4.300	2.150	0.005	4.460	2.310	0.005	0.080	Brass			435A	
510		1.1572	1.45	2.20			5.100	2.550	0.005	5.260	2.710	0.005	0.080					
650	103	0.9079	1.12	1.70	0.26	0.17	6.500	3.250	0.005	6.660	3.410	0.005	0.080	Aluminum			418A	
650	69	0.9079	1.12	1.70	0.32	0.21	6.500	3.250	0.005	6.660	3.410	0.005	0.080	Brass			417A	
770	205	0.7664	0.96	1.45	0.20	0.13	7.700	3.850	0.005	7.950	4.100	0.005	0.125	Aluminum				
975	204	0.6053	0.75	1.12	0.14	0.09	9.750	4.875	0.010	10.000	5.125	0.010	0.125	Aluminum				
1150	203	0.5132	0.64	0.96	0.11	0.07	11.500	5.750	0.015	11.750	6.000	0.015	0.125	Aluminum				
1500	202	0.3934	0.49	0.75	0.07	0.05	15.000	7.500	0.015	15.250	7.750	0.015	0.125	Aluminum				
1800	201	0.3279	0.41	0.63	0.05	0.04	18.000	9.000	0.020	18.250	9.250	0.020	0.125	Aluminum				
2100		0.2810	0.35	0.53	0.04	0.03	21.000	10.500	0.020	21.250	10.750	0.020	0.125	Aluminum				
2300		0.2566	0.32	0.49	0.04	0.03	23.000	11.500	0.020	23.250	11.750	0.020	0.125	Aluminum				

Notes:

1. Conductivity of 63.0e6 Mhos/Meter used for Silver.
2. Conductivity of 37.7e6 Mhos/Meter used for Aluminum.
3. Conductivity of 25.1e6 Mhos/Meter used for Brass.
4. Loss is inversely proportional to the square root of conductivity.

Source:

1. Balanis, C.A., *Advanced Engineering Electromagnetics*, John Wiley & Sons, New York, 1990.
2. Catalog, Microwave Development Labs, Natick, MA.
3. Catalog, Aerowave Inc, Medford, MA.
4. Catalog, Formcraft Tool Co, Chicago, IL.
5. Catalog, Penn Engineering Components, No. Hollywood, CA.

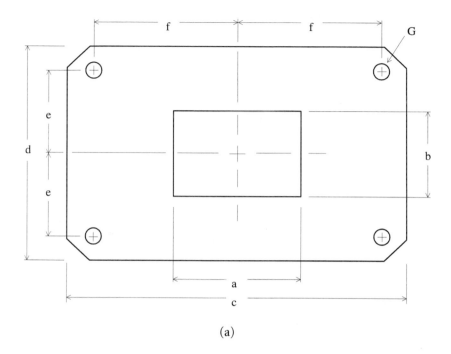

(a)

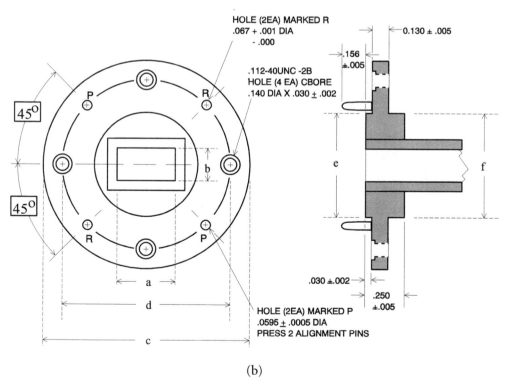

(b)

FIGURE 4 (a) Standard waveguide flange dimensions (rectangular flange); (b) Standard waveguide flange dimensions (circular flange).

TABLE 6 Waveguide Flange Dimensions

WR-()	Rectangular Flange Dimensions						
	a [inches] (Ref.)	b [inches] (Ref.)	c [inches] ±0.015	d [inches] ±0.015	e [inches] ±0.005	f [inches] ±0.005	G [inches-dia] ±0.0015
28	0.280	0.140	0.750	1.750	0.265	0.250	0.1175
42	0.420	0.170	0.875	0.875	0.335	0.320	0.1175
51	0.510	0.255	1.313	1.313	0.497	0.478	0.1445
62	0.620	0.311	1.313	1.313	0.478	0.497	0.1445
75	0.750	0.375	1.500	2.500	0.560	0.520	0.1445
90	0.900	0.400	1.625	1.625	0.610	0.640	0.1705
112	1.122	0.497	1.875	1.875	0.676	0.737	0.1705

WR-()	Circular Flange Dimensions					
	a [inches] (Ref.)	b [inches] (Ref.)	c [inches] +0.000 −0.002	d [inches] BSC.	e [inches] ±0.005	f [inches] ±0.005
10	0.1	0.05	0.75	0.5625	0.375	0.312
12	0.122	0.061	0.75	0.5625	0.375	0.312
15	0.148	0.074	0.75	0.5625	0.375	0.312
19	0.188	0.094	1.125	0.9375	0.5	0.468
22	0.224	0.112	1.125	0.9375	0.5	0.468
28	0.28	0.14	1.125	0.9375	0.5	0.468
42	0.42	0.17	1.125	0.9375	0.625	0.625

Coaxial Cables

TABLE 7 Flexible Coax Specifications

RG-()/U	Mil-C-17/()	Zo	Loss dB/100' @1 MHz	Loss dB/100' @10 MHz	Loss dB/100' @100 MHz	Loss dB/100' @1000 MHz	Center Conductor	Outer Conductor	Jacket	Outside Diameter [Inches]	Dielectric	Velocity Factor [%]	Capacitance pF/foot	Dielectric Core Diameter [inches]	Maximum Voltage [RMS]
8		52	0.16	0.56	1.9	7.4	#13 Stranded Bare Copper .058 Dia	Bare Copper Braid, 97%	Black PVC	0.405	Polyethylene	66	29.5	0.285	3700
8A	[3] 163	52	0.16	0.56	1.9	7.4	#13 Stranded Bare Copper .058 Dia	Bare Copper Braid, 97%	Black PVC, Noncontaminating	0.405	Polyethylene	66	29.5	0.285	3700
9		51	0.18	0.62	2.1	8.2	#13 Stranded Silver Coated Copper .086 Dia	Double Braid, Silver Coated Inner, Bare Copper Outer, 97%	Gray PVC, Noncontaminating	0.42	Polyethylene	66	30	0.28	3700
9B	[4] 164	50					#13 Stranded Silver Coated Copper .089 Dia	Double Braid, Silver Coated Inner, Bare Copper Outer, 97%	PVC	0.36	Polyethylene	66		0.285	3701
11	[6]	75	0.19	0.66	2	7.1	#18 Stranded Tinned Copper .048	Bare Copper Braid, 97%	Black PVC	0.405	Flame Retardant Semi-Foam Polyethylene	66	20.5	0.285	300
11A		75	0.19	0.66	2	8.5	#18 Stranded Tinned Copper .048	Bare Copper Braid, 97%	Black PVC, Noncontaminating	0.405	Polyethylene	66	20.5	0.285	3700
58	[28]	50	0.42	1.5	5.4	22.8	#20 Tinned Copper .035 Dia	Tinned Copper Braid, 95%	Black PVC, Noncontaminating	0.193	Polyethylene	66	30.8	0.116	1400
58A		50	0.42	1.5	5.4	22.8	#20 Solid Bare Copper .035 Dia	Tinned Copper Braid, 95%	Black PVC	0.193	Polyethylene	66	30.8	0.116	1400
58C	155	50	0.42	1.5	5.4	22.8	#20 Tinned Copper .035 Dia	Tinned Copper Braid, 95%	Black PVC, Noncontaminating	0.193	Polyethylene	66	30.8	0.116	1400
59	[29]	75	0.6	1.1	3.4	12	#23 Solid Bare Copper Covered Steel .023 Dia	Bare Copper Braid, 95%	Black PVC, Noncontaminating	0.241	Polyethylene	66	20.5	0.146	1700
59B		75	0.6	1.1	3.4	12	#23 Solid Bare Copper Covered Steel .023 Dia	Bare Copper Braid, 95%	Black PVC, Noncontaminating	0.241	Polyethylene	66	20.5	0.146	1700
62A	[30]	93	0.25	0.85	2.7	8.7	#22 Solid Bare Copper Covered Steel .023 Dia	Bare Copper Braid, 95%	Black PVC, Noncontaminating	0.242	Semi-solid Polyethylene	84	13.5	0.146	750

TABLE 7 (continued) Flexible Coax Specifications

RG-()/U	Mil-C-17/()	Zo	Loss dB/100' @1 MHz	Loss dB/100' @10 MHz	Loss dB/100' @100 MHz	Loss dB/100' @1000 MHz	Center Conductor	Outer Conductor	Jacket	Outside Diameter [Inches]	Dielectric	Velocity Factor [%]	Capacitance pF/foot	Dielectric Core Diameter [inches]	Maximum Voltage [RMS]
62B	[91] 97	93	0.31	0.9	2.9	11	#24 Solid Bare Copper Covered Steel .025 Dia	Bare Copper Braid, 95%	Black PVC, Noncontaminating	0.242	Semi-solid Polyethylene	84	13.5	0.146	750
63	[31]	125	0.19	0.52	1.5	5.8	#22 Solid Bare Copper Covered Steel .025 Dia	Bare Copper Braid, 97%	Black PVC, Noncontaminating	0.405	Semi-solid Polyethylene	84	9.7	0.285	750
71	[90]	93	0.25	0.85	2.7	8.7	#22 Solid Bare Copper Covered Steel .025 Dia	Double Braid, Tinned Copper Outer, Bare Copper Inner, 98%	Black Polyethylene	0.245	Semi-solid Polyethylene	84	13.5	0.146	750
122	[54] 157	50	0.4	1.7	7	29	#22 Stranded Tinned Copper .030 Dia	Tinned Copper Braid, 95%	Black PVC, Noncontaminating	0.16	Polyethylene	66	30.8	0.096	1400
141	[59] 170	50					#18 Solid Silver Coated Copper Covered Steel .037 Dia	Silver Coated Copper Braid, 94%	Fluorinated Ethylene-Propylene	0.17	TFE Teflon	69.5		0.116	1400
141A		50	0.34	1.1	3.9	13.5	#18 Solid Silver Coated Copper Covered Steel .037 Dia	Silver Coated Copper Braid, 94%	Tinted Brown Fiberglass	0.187	TFE Teflon	69.5	29.2	0.116	1400
142	[60] 158	50	0.34	1.1	3.9	13.5	#18 Solid Silver Coated Copper Covered Steel .037 Dia	Double Silver Coated Copper Braid, 94%	Tinted Brown Fluorinated Ethylene-Propylene	0.187	TFE Teflon	69.5	29.2	0.116	1400
174	[119] 173	50	1.9	3.3	8.4	34	#26 Stranded Bare Copper Covered Steel .019 Dia	Tinned Copper Braid, 90%	Black PVC Jacket	0.11	Polyethylene	66	30.8	0.059	1100
178	[93] 169	50					#30 Stranded Silver Coated Copper Covered Steel .012 Dia	Silver Coated Copper Braid, 96%	Fluorinated Ethylene-Propylene	0.071	TFE Teflon	69.5		0.033	750
178B		50	2.6	5.6	14	46	#30 Solid Silver Coated Copper Covered Steel .012 Dia	Silver Coated Copper Braid, 96%	White Fluorinated Ethylene-Propylene	0.071	TFE Teflon	69.5	29.2	0.033	750

Part No.	Spec.	Z				Conductor	Shield	Jacket		Dielectric				
179	[94]	75	3	5.3	10	24	#30 Solid Silver Coated Copper Covered Steel .012 Dia	Silver Coated Copper Braid, 95%	Tinted Brown Fluorinated Ethylene-Propylene	0.1	TFE Teflon	69.5	19.5	0.062	900
180	[95]	95	2.4	3.3	5.7	17	#30 Solid Silver Coated Copper Covered Steel .012 Dia	Silver Coated Copper Braid, 95%	Tinted Brown Fluorinated Ethylene-Propylene	0.141	TFE Teflon	69.5	15.4	0.102	1100
187	[68] 94	75					#30 Solid Silver Coated Copper Covered Steel .012 Dia	Silver Coated Copper Braid, 92.3%	Fluorinated Ethylene-Propylene	0.1	TFE Teflon	69.5		0.063	900
212	[73] 162	50	0.26	0.83	2.7	9.8	#15.5 Solid Silver Coated Copper .0556 Dia	Double Silver Coated Copper Braid, 95%	Black PVC Noncontaminating	0.332	Polyethylene	66	30.8	0.185	2200
213	[74] 163	50	0.18	0.62	2.1	8.2	#13 Stranded Bare Copper .089 Dia	Bare Copper Braid 97%	Black PVC Noncontaminating	0.405	Polyethylene	66	30.8	0.285	3700
214	[75] 164	50	0.17	0.55	1.9	8	#13 Stranded Silver Coated Copper .089 Dia	Double Silver Coated Copper Braid, 97%	Black PVC Noncontaminating	0.425	Polyethylene	66	30.8	0.285	3700
216	[77]	75	0.19	0.66	2	7.1	#18 Stranded Tinned Copper .048	Double Bare Copper Braid 95%	Black PVC Noncontaminating	0.425	Polyethylene	66	20.5	0.285	3700
223	[84]	50	0.35	1.2	4.1	14.5	#19 Solid Silver Coated Copper .034 Dia	Double Silver Coated Copper Braid, 95%	Black PVC Noncontaminating	0.212	Polyethylene	66	30.8	0.117	1700
303	[111] 170	50	0.34	1.1	3.9	13.5	#18 Solid Silver Coated Copper Covered Steel .037 Dia	Silver Coated Copper Braid, 95%	Tinted Brown Fluorinated Ethylene-Propylene	0.17	TFE Teflon	69.5	29.2	0.116	1400
316	[113] 172	50	1.2	2.7	8.3	29	#26 Stranded Silver Coated Copper Covered Steel .020 Dia	Silver Coated Copper Braid, 95%	White Fluorinated Ethylene-Propylene	0.098	TFE Teflon	69.5	29.2	0.06	900

Note: Mil-C-17/() part numbers were revised. Initial specification numbers are shown in brackets, current specification numbers are unbracketed.

Source:
Mil-C-17G
Mil-C-17G Supplement 1
Belden Master Catalog, Belden Wire & Cable Co, Richmond, IN.

TABLE 8 Semirigid Coax Specification

RG-()/U	Mil-C-17 Part Number M17/	Zo	Loss [dB/100 ft] Power [W] @500 MHz	Loss [dB/100 ft] Power [W] @1 GHz	Loss [dB/100 ft] Power [W] @3 GHz	Loss [dB/100 ft] Power [W] @5 GHz	Loss [dB/100 ft] Power [W] @10 GHz	Loss [dB/100 ft] Power [W] @18 GHz	Loss [dB/100 ft] Power [W] @20 GHz	Center Conductor Material	Center Conductor Diameter	Outer Conductor	Outside Diameter [Inches]	Dielectric	Dielectric Constant (1 GHz)	Max Capacitance pF/foot	Dielectric Diameter [inches]	Maximum Voltage (60Hz) [RMS]
	154-00001	50 ± 3.0	42 / 14	60 / 10	100 / 6	140 / 4.5	190 / 3.1		280 / 2	Silver Plated Copper Coated Steel	0.008 ±0.0005	Copper	0.034 ±0.001	Solid PTFE	2.03	29.9	0.026 ±0.001	750
	154-00002	50 ± 3.0	42 / 14	60 / 10	100 / 6	140 / 4.5	190 / 3.1		280 / 2	Silver Plated Copper Coated Steel	0.008 ±0.0005	Tin-Plated Copper	0.034 ±0.002	Solid PTFE	2.03	29.9	0.026 ±0.001	750
	151-00001	50 ± 2.5	28 / 45	40 / 32	70 / 18	90 / 13	130 / 9		190 / 65	Silver Plated Copper Coated Steel	0.0113 ±0.0005	Copper	0.047 ±0.001	Solid PTFE	2.03	29.9	0.037 ±0.001	1000
	151-00002	50 ± 2.5	28 / 45	40 / 32	70 / 18	90 / 13	130 / 9		190 / 6.5	Silver Plated Copper Coated Steel	0.0113 ±0.0005	Tin-Plated Copper	0.047 +0.002 -0.001	Solid PTFE	2.03	29.9	0.037 ±0.001	1000
405	133-RG-405	50 ± 1.5	15 / 180	22 / 130		50 / 54	80 / 35		130 / 20	Silver Plated Copper Coated Steel	0.0201 ±0.0005	Copper	0.0865 ±0.001	Solid PTFE	2.03	29.9	0.066 ±0.002	5000
	133-00001	50 ± 1.5	15 / 180	22 / 130		50 / 54	80 / 35		130 / 20	Silver Plated Copper Coated Steel	0.0201 ±0.0005	Tin-Plated Copper	0.0865 +0.002 -0.001	Solid PTFE	2.03	29.9	0.066 ±0.002	5000
	133-00002	50 ± 1.5	15 / 180	22 / 130		50 / 54	80 / 35		130 / 20	Silver Plated Copper	0.0201 ±0.0005	Copper	0.0865 ±0.001	Solid PTFE	2.03	29.9	0.066 ±0.002	5000
	133-00003	50 ± 1.5	15 / 180	22 / 130		50 / 54	80 / 35		130 / 20	Silver Plated Copper	0.0201 ±0.0005	Tin-Plated Copper	0.0865 +0.002 -0.001	Solid PTFE	2.03	29.9	0.066 ±0.002	5000
	133-00004	50 ± 1.5	15 / 180	22 / 130		50 / 54	80 / 35		130 / 20	Silver Plated Nickel Copper Coated Steel	0.0201 ±0.0005	Copper	0.0865 ±0.001	Solid PTFE	2.03	29.9	0.066 ±0.002	5000
	133-00005	50 ± 1.5	15 / 180	22 / 130		50 / 54	80 / 35		130 / 20	Silver Plated Nickel Copper Coated Steel	0.0201 ±0.0005	Tin-Plated Copper	0.0865 +0.002 -0.001	Solid PTFE	2.03	29.9	0.066 ±0.002	5000
	133-0006	50 ± 1.5	15 / 180	22 / 130		50 / 54	80 / 35		130 / 20	Silver Plated Copper Coated Steel	0.0201 ±0.0005	Copper	0.0865 ±0.001	Solid PTFE	2.03	29.9	0.066 ±0.002	5000

Part No.	Impedance						Material								
133-00007	50 ± 1.5	15 / 180	22 / 130		50 / 54	80 / 35	Silver Plated Copper Coated Steel	0.0201 ±0.0005	Tin-Plated Copper	0.086 +0.0021 -0.001	Solid PTFE	2.03	29.9	0.066 ±0.002	5000
133-00008	50 ± 1.5	15 / 180	22 / 130		50 / 54	80 / 35	Silver Plated Copper	0.0201 ±0.0005	Copper	0.0865 ±0.001	Solid PTFE	2.03	29.9	0.066 ±0.002	5000
133-00009	50 ± 1.5	15 / 180	22 / 130		50 / 54	80 / 35	Silver Plated Copper	0.0201 ±0.0005	Tin-Plated Copper	0.0865 +0.002 -0.001	Solid PTFE	2.03	29.9	0.066 ±0.002	5000
133-00010	50 ± 1.5	15 / 180	22 / 130		50 / 54	80 / 35	Silver Plated Nickel Copper Coated Steel	0.0201 ±0.0005	Copper	0.086 ±0.001	Solid PTFE	2.03	29.9	0.066 ±0.002	5000
133-00011	50 ± 1.5	15 / 180	22 / 130		50 / 54	80 / 35	Silver Plated Copper Coated Steel	0.0201 ±0.0005	Tin-Plated Copper	0.086 +0.002 -0.001	Solid PTFE	2.03	29.9	0.066 ±0.002	5000
130-RG-402	50 ± 1.0	8 / 600	12 / 450	21 / 250	29 / 180	45 / 120	Silver Plated Copper Coated Steel	0.0362 ±0.0007	Copper	0.141 ±0.001	Solid PTFE	2.03	29.9	0.1175 ±0.001	5000
130-00001	50 ± 1.0	8 / 600	12 / 450	21 / 250	29 / 180	45 / 120	Silver Plated Copper Coated Steel	0.0362 ±0.0007	Tin-Plated Copper	0.141 +0.002 -0.001	Solid PTFE	2.03	29.9	0.1175 ±0.001	5000
130-00002	50 ± 1.0	8 / 601	12 / 451	21 / 251	29 / 181	45 / 121	Silver Plated Nickel Copper Coated Steel	0.0362 ±0.0007	Copper	0.141 ±0.001	Solid PTFE	2.03	29.9	0.1175 ±0.001	5000
130-00003	50 ± 1.0	8 / 602	12 / 452	21 / 252	29 / 182	45 / 122	Silver Plated Copper Coated Steel	0.0362 ±0.0007	Tin-Plated Copper	0.141 +0.002 -0.001	Solid PTFE	2.03	29.9	0.1175 ±0.001	5000
130-00004	50 ± 1.0	8 / 603	12 / 453	21 / 253	29 / 183	45 / 123	Silver Plated Copper Coated Steel	0.0362 ±0.0007	Copper	0.141 ±0.001	Solid PTFE	2.03	29.9	0.1175 ±0.001	5000
130-00005	50 ± 1.0	8 / 604	12 / 454	21 / 254	29 / 184	45 / 124	Silver Plated Copper Coated Steel	0.0362 ±0.0007	Tin-Plated Copper	0.141 +0.002 -0.001	Solid PTFE	2.03	29.9	0.1175 ±0.001	5000
130-00006	50 ± 1.0	8 / 605	12 / 455	21 / 255	29 / 185	45 / 125	Silver Plated Nickel Copper Coated Steel	0.0362 ±0.0007	Copper	0.141 ±0.001	Solid PTFE	2.03	29.9	0.1175 ±0.001	5000

402

TABLE 8 (continued) Semirigid Coax Specification

RG-()/U	Mil-C-17 Part Number M17/	Zo	Loss [dB/100 ft] Power [W] @500 MHz	Loss [dB/100 ft] Power [W] @1 GHz	Loss [dB/100 ft] Power [W] @3 GHz	Loss [dB/100 ft] Power [W] @5 GHz	Loss [dB/100 ft] Power [W] @10 GHz	Loss [dB/100 ft] Power [W] @18 GHz	Loss [dB/100 ft] Power [W] @20 GHz	Center Conductor Material	Center Conductor Diameter	Outer Conductor	Outside Diameter [Inches]	Dielectric	Dielectric Constant (1 GHz)	Max Capacitance pF/foot	Dielectric Diameter [inches]	Maximum Voltage (60Hz) [RMS]
	130-00007	50 ± 1.0	8 / 606	12 / 456	21 / 256	29 / 186	45 / 126		70 / 76	Silver Plated Nickel Copper Coated Steel	0.0362 ±0.0007	Tin-Plated Copper	0.141 +0.002 -0.001	Solid PTFE	2.03	29.9	0.1175 ±0.001	5000
	130-00008	50 ± 1.0	8 / 607	12 / 457	21 / 257	29 / 187	45 / 127		70 / 77	Silver Plated Copper Coated Steel	0.0362 ±0.0007	Aluminum	0.141 ±0.001	Solid PTFE	2.03	29.9	0.1175 ±0.001	5000
	130-00009	50 ± 1.0	8 / 608	12 / 458	21 / 258	29 / 188	45 / 128		70 / 78	Silver Plated Copper Coated Steel	0.0362 ±0.0007	Aluminum	0.141 ±0.001	Solid PTFE	2.03	29.9	0.1175 ±0.001	5000
	130-00010	50 ± 1.0	8 / 609	12 / 459	21 / 259	29 / 189	45 / 129		70 / 79	Silver Plated Nickel Copper Coated Steel	0.0362 ±0.0007	Aluminum	0.141 ±0.001	Solid PTFE	2.03	29.9	0.1175 ±0.001	5000
	130-00011	50 ± 1.0	8 / 610	12 / 460	21 / 260	29 / 190	45 / 130		70 / 80	Silver Plated Nickel Copper Coated Steel	0.0362 ±0.0007	Aluminum	0.141 ±0.001	Solid PTFE	2.03	29.9	0.1175 ±0.001	5000
	130-00012	50 ± 1.0	8 / 611	12 / 461	21 / 261	29 / 191	45 / 131		70 / 81	Silver Plated Copper Coated Steel	0.0362 ±0.0007	Silver Plated Copper	0.141 +0.002 -0.001	Solid PTFE	2.03	29.9	0.1175 ±0.001	5000
	130-00013	50 ± 1.0	8 / 612	12 / 462	21 / 262	29 / 192	45 / 132		70 / 82	Silver Plated Nickel Copper Coated Steel	0.0362 ±0.0007	Silver Plated Copper	0.141 +0.002 -0.001	Solid PTFE	2.03	29.9	0.1175 ±0.001	5000
401	129-RG-401	50 ± 0.5	4.5 / 1900	7.5 / 1400	11 / 750		33 / 350	48 / 200	—	Silver Plated Copper	0.0641 ±0.001	Copper	0.250 ±0.001	Solid PTFE	2.03	29.9	0.209 ±0.002	7500
	129-00001	50 ± 0.5	4.5 / 1900	7.5 / 1400	11 / 750		33 / 350	48 / 200	—	Silver Plated Copper	0.0641 ±0.001	Tin-Plated Copper	0.250 +0.002 -0.001	Solid PTFE	2.03	29.9	0.209 ±0.002	7500

Notes: Attenuation/Power Ratings are maximum values for families.
Sources: Mil-C-17/130E
Semi-Rigid Coaxial Cable Catalog, Micro-Coax Components, Inc, Collegeville, PA.

Metal Compatibility (See MIL-STD-889B)

Single Sideband and Image Reject Mixers

TABLE 10 Maximum Tolerable Phase Error (Degrees) for Single Sideband and Image Reject Mixers as a Function of Suppression and Amplitude Imbalance

Amplitude Imbalance [dB]	Suppression												
	−10 dBc	−13 dBc	−15 dBc	−17 dBc	−20 dBc	−23 dBc	−25 dBc	−27 dBc	−30 dBc	−33 dBc	−35 dBc	−37 dBc	−40 dBc
0.00	35.10	25.24	20.17	16.08	11.42	8.10	6.44	5.12	3.62	2.56	2.04	1.62	1.15
0.05	35.10	25.24	20.16	16.08	11.42	8.09	6.43	5.10	3.61	2.54	2.01	1.58	1.10
0.10	35.09	25.23	20.16	16.07	11.40	8.07	6.40	5.07	3.56	2.48	1.93	1.48	0.94
0.15	35.08	25.22	20.14	16.05	11.38	8.04	6.36	5.02	3.48	2.37	1.78	1.28	0.58
0.20	35.08	25.21	20.13	16.03	11.35	7.99	6.30	4.94	3.37	2.20	1.55	0.94	
0.25	35.06	25.19	20.10	16.00	11.30	7.93	6.22	4.84	3.23	1.97	1.20		
0.30	35.05	25.17	20.07	15.96	11.25	7.86	6.13	4.72	3.04	1.63	0.49		
0.35	35.03	25.14	20.04	15.92	11.19	7.77	6.01	4.57	2.79	1.12			
0.40	35.01	25.11	20.00	15.87	11.12	7.66	5.87	4.38	2.48				
0.45	34.99	25.07	19.96	15.81	11.03	7.54	5.72	4.17	2.08				
0.50	34.96	25.04	19.91	15.75	10.94	7.40	5.53	3.91	1.50				
0.55	34.93	24.99	19.85	15.68	10.84	7.25	5.32	3.61					
0.60	34.90	24.95	19.79	15.60	10.72	7.07	5.08	3.25					
0.65	34.87	24.90	19.73	15.51	10.60	6.88	4.81	2.80					
0.70	34.83	24.84	19.65	15.42	10.46	6.66	4.50	2.21					
0.75	34.79	24.78	19.58	15.32	10.31	6.42	4.13	1.32					
0.80	34.75	24.72	19.49	15.21	10.15	6.16	3.70						
0.85	34.70	24.65	19.41	15.10	9.97	5.86	3.18						
0.90	34.66	24.58	19.31	14.98	9.78	5.53	2.52						
0.95	34.61	24.50	19.21	14.84	9.58	5.16	1.53						
1.00	34.55	24.42	19.10	14.70	9.35	4.73							
1.10	34.44	24.24	18.87	14.40	8.86	3.65							
1.20	34.31	24.05	18.62	14.06	8.28	1.83							
1.30	34.17	23.84	18.34	13.67	7.61								
1.40	34.02	23.61	18.03	13.25	6.80								
1.50	33.86	23.35	17.69	12.78	5.82								
1.60	33.68	23.08	17.32	12.25	4.53								
1.70	33.50	22.79	16.92	11.67	2.52								
1.80	33.30	22.48	16.48	11.01									
1.90	33.09	22.14	16.00	10.28									
2.00	32.86	21.78	15.49	9.44									
2.10	32.63	21.39	14.93	8.47									
2.20	32.37	20.98	14.31	7.31									
2.30	32.11	20.54	13.64	5.87									
2.40	31.83	20.07	12.90	3.81									
2.50	31.54	19.56	12.08										
2.60	31.23	19.03	11.17										
2.70	30.90	18.45	10.13										
2.80	30.56	17.83	8.93										
2.90	30.21	17.17	7.48										
3.00	29.83	16.46	5.59										
3.10	29.44	15.68	2.41										
3.20	29.03	14.84											
3.30	28.60	13.92											
3.40	28.16	12.91											
3.50	27.69	11.77											
3.60	27.19	10.47											
3.70	26.68	8.94											
3.80	26.14	7.02											
3.90	25.57	4.24											
4.00	24.98												
4.10	24.35												
4.20	23.69												
4.30	23.00												
4.40	22.27												
4.50	21.49												
4.60	20.67												
4.70	19.79												
4.80	18.86												
4.90	17.85												
5.00	16.76												

Notes: Example: An image reject mixer requires 25 dB of unwanted image suppression, and a maximum amplitude imbalance of 1 dB is expected between the channels. Looking at the intersection of the 1 dB row and −25 dBc column shows that a maximum interchannel phase imbalance of 9.35 degrees is tolerable.
Suppression [dBc] = $10 \log ((1 - 2 a \cos(p) + a^2) / (1 + 2 a \cos(p) + a^2))$ where a is the voltage ratio amplitude imbalance and p is the phase imbalance.

Power, Voltage and Decibels

TABLE 11 Power, Voltage, dB Conversion Table

dB	Power Factor	Voltage Factor	dB	Power Factor	Voltage Factor	dB	Power Factor	Voltage Factor	dB	Power Factor	Voltage Factor
0	1	1	9	7.943282347	2.818382931	34	2511.886432	50.11872336	60	1000000	1000
0.1	1.023292992	1.011579454	9.5	8.912509381	2.985382619	35	3162.27766	56.23413252	61	1258925.412	1122.018454
0.2	1.047128548	1.023292992	10	10	3.16227766	36	3981.071706	63.09573445	62	1584893.192	1258.925412
0.3	1.071519305	1.035142167	11	12.58925412	3.548133892	37	5011.872336	70.79457844	63	1995262.315	1412.537545
0.4	1.096478196	1.047128548	12	15.84893192	3.981071706	38	6309.573445	79.43282347	64	2511886.432	1584.893192
0.5	1.122018454	1.059253725	13	19.95262315	4.466835922	39	7943.282347	89.12509381	65	3162277.66	1778.27941
0.6	1.148153621	1.071519305	14	25.11886432	5.011872336	40	10000	100	66	3981071.706	1995.262315
0.7	1.174897555	1.083926914	15	31.6227766	5.623413252	41	12589.25412	112.2018454	67	5011872.336	2238.721139
0.8	1.202264435	1.096478196	16	39.81071706	6.309573445	42	15848.93192	125.8925412	68	6309573.445	2511.886432
0.9	1.230268771	1.109174815	17	50.11872336	7.079457844	43	19952.62315	141.2537545	69	7943282.347	2818.382931
1	1.258925412	1.122018454	18	63.09573445	7.943282347	44	25118.86432	158.4893192	70	10000000	3162.27766
1.5	1.412537545	1.188502227	19	79.43282347	8.912509381	45	31622.7766	177.827941	71	12589254.12	3548.133892
2	1.584893192	1.258925412	20	100	10	46	39810.71706	199.5262315	72	15848931.92	3981.071706
2.5	1.77827941	1.333521432	21	125.8925412	11.22018454	47	50118.72336	223.8721139	73	19952623.15	4466.835922
3	1.995262315	1.412537545	22	158.4893192	12.58925412	48	63095.73445	251.1886432	74	25118864.32	5011.872336
3.5	2.238721139	1.496235656	23	199.5262315	14.12537545	49	79432.82347	281.8382931	75	31622776.6	5623.413252
4	2.511886432	1.584893192	24	251.1886432	15.84893192	50	100000	316.227766	76	39810717.06	6309.573445
4.5	2.818382931	1.678804018	25	316.227766	17.7827941	51	125892.5412	354.8133892	77	50118723.36	7079.457844
5	3.16227766	1.77827941	26	398.1071706	19.95262315	52	158489.3192	398.1071706	78	63095734.45	7943.282347
5.5	3.548133892	1.883649089	27	501.1872336	22.38721139	53	199526.2315	446.6835922	79	79432823.47	8912.509381
6	3.981071706	1.995262315	28	630.9573445	25.11886432	54	251188.6432	501.1872336	80	100000000	10000
6.5	4.466835922	2.11348904	29	794.3282347	28.18382931	55	316227.766	562.3413252	81	125892541.2	11220.18454
7	5.011872336	2.238721139	30	1000	31.6227766	56	398107.1706	630.9573445	82	158489319.2	12589.25412
7.5	5.623413252	2.371373706	31	1258.925412	35.48133892	57	501187.2336	707.9457844	83	199526231.5	14125.37545
8	6.309573445	2.511886432	32	1584.893192	39.81071706	58	630957.3445	794.3282347	84	251188643.2	15848.93192
8.5	7.079457844	2.66072506	33	1995.262315	44.66835922	59	794328.2347	891.2509381	85	316227766	17782.7941
86	398107170.6	19952.62315	-5.5	0.281838293	0.530884444	-35	0.000316228	0.017782794	-69	1.25893E-07	0.000354813
87	501187233.6	22387.21139	-6	0.251188643	0.501187234	-36	0.000251189	0.015848932	-70	0.0000001	0.000316228
88	630957344.5	25118.86432	-6.5	0.223872114	0.473151259	-37	0.000199526	0.014125375	-71	7.94328E-08	0.000281838
89	794328234.7	28183.82931	-7	0.199526231	0.446683592	-38	0.000158489	0.012589254	-72	6.30957E-08	0.000251189
90	1000000000	31622.7766	-7.5	0.177827941	0.421696503	-39	0.000125893	0.011220185	-73	5.01187E-08	0.000223872
91	1258925412	35481.33892	-8	0.158489319	0.398107171	-40	0.0001	0.01	-74	3.98107E-08	0.000199526
92	1584893192	39810.71706	-8.5	0.141253754	0.375837404	-41	7.94328E-05	0.008912509	-75	3.16228E-08	0.000177828
93	1995262315	44668.35922	-9	0.125892541	0.354813389	-42	6.30957E-05	0.007943282	-76	2.51189E-08	0.000158489
94	2511886432	50118.72336	-9.5	0.112201845	0.334965439	-43	5.01187E-05	0.007079458	-77	1.99526E-08	0.000141254
95	3162277660	56234.13252	-10	0.1	0.316227766	-44	3.98107E-05	0.006309573	-78	1.58489E-08	0.000125893
96	3981071706	63095.73445	-11	0.079432823	0.281838293	-45	3.16228E-05	0.005623413	-79	1.25893E-08	0.000112202
97	5011872336	70794.57844	-12	0.063095734	0.251188643	-46	2.51189E-05	0.005011872	-80	0.00000001	0.0001
98	6309573445	79432.82347	-13	0.050118723	0.223872114	-47	1.99526E-05	0.004466836	-81	7.94328E-09	8.91251E-05
99	7943282347	89125.09381	-14	0.039810717	0.199526231	-48	1.58489E-05	0.003981072	-82	6.30957E-09	7.94328E-05
100	10000000000	100000	-15	0.031622777	0.177827941	-49	1.25893E-05	0.003548134	-83	5.01187E-09	7.07946E-05
0	1	1	-16	0.025118864	0.158489319	-50	0.00001	0.003162278	-84	3.98107E-09	6.30957E-05
-0.1	0.977237221	0.988553095	-17	0.019952623	0.141253754	-51	7.94328E-06	0.002818383	-85	3.16228E-09	5.62341E-05
-0.2	0.954992586	0.977237221	-18	0.015848932	0.125892541	-52	6.30957E-06	0.002511886	-86	2.51189E-09	5.01187E-05
-0.3	0.933254301	0.966050879	-19	0.012589254	0.112201845	-53	5.01187E-06	0.002238721	-87	1.99526E-09	4.46684E-05
-0.4	0.912010839	0.954992586	-20	0.01	0.1	-54	3.98107E-06	0.001995262	-88	1.58489E-09	3.98107E-05
-0.5	0.891250938	0.944060656	-21	0.007943282	0.089125094	-55	3.16228E-06	0.001778279	-89	1.25893E-09	3.54813E-05
-0.6	0.87096359	0.933254301	-22	0.006309573	0.079432823	-56	2.51189E-06	0.001584893	-90	0.000000001	3.16228E-05
-0.7	0.851138038	0.922571427	-23	0.005011872	0.070794578	-57	1.99526E-06	0.001412538	-91	7.94328E-10	2.81838E-05
-0.8	0.831763771	0.912010839	-24	0.003981072	0.063095734	-58	1.58489E-06	0.001258925	-92	6.30957E-10	2.51189E-05
-0.9	0.812830516	0.901571138	-25	0.003162278	0.056234133	-59	1.25893E-06	0.001122018	-93	5.01187E-10	2.23872E-05
-1	0.794328235	0.891250938	-26	0.002511886	0.050118723	-60	0.000001	0.001	-94	3.98107E-10	1.99526E-05
-1.5	0.707945784	0.841395142	-27	0.001995262	0.044668359	-61	7.94328E-07	0.000891251	-95	3.16228E-10	1.77828E-05
-2	0.630957344	0.794328235	-28	0.001584893	0.039810717	-62	6.30957E-07	0.000794328	-96	2.51189E-10	1.58489E-05
-2.5	0.562341325	0.749894209	-29	0.001258925	0.035481339	-63	5.01187E-07	0.000707946	-97	1.99526E-10	1.41254E-05
-3	0.501187234	0.707945784	-30	0.001	0.031622777	-64	3.98107E-07	0.000630957	-98	1.58489E-10	1.25893E-05
-3.5	0.446683592	0.668343918	-31	0.000794328	0.028183829	-65	3.16228E-07	0.000562341	-99	1.25893E-10	1.12202E-05
-4	0.398107171	0.630957344	-32	0.000630957	0.025118864	-66	2.51189E-07	0.000501187	-100	1E-10	0.00001
-4.5	0.354813389	0.595662144	-33	0.000501187	0.022387211	-67	1.99526E-07	0.000446684			
-5	0.316227766	0.562341325	-34	0.000398107	0.019952623	-68	1.58489E-07	0.000398107			

Notes:
1. Multiply by appropriate factor.
2. Use voltage factor to convert S-parameters.

Microstrip Impedances

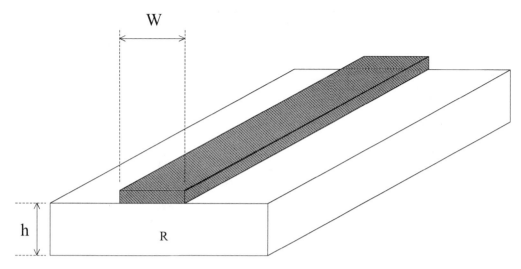

FIGURE 12

TABLE 12a Zero Thickness Microstrip Dimensions, Effective Dielectric Constant, and PUL Capacitance and Inductance for Er = 2.2

Zo [Ohms]	W/h	Keff	pF/cm	nH/mm	Zo [Ohms]	W/h	Keff	pF/cm	nH/mm
1	250.3363	2.1861	49.3192	0.0493	57	2.5199	1.8500	0.7959	2.5860
2	123.5739	2.1728	24.5846	0.0983	58	2.4514	1.8471	0.7816	2.6294
3	81.4007	2.1601	16.3417	0.1471	59	2.3854	1.8443	0.7678	2.6727
4	60.3537	2.1480	12.2218	0.1955	60	2.3218	1.8416	0.7544	2.7160
5	47.7490	2.1364	9.7510	0.2438	61	2.2603	1.8389	0.7415	2.7592
6	39.3615	2.1253	8.1046	0.2918	62	2.2010	1.8362	0.7290	2.8024
7	33.3816	2.1146	6.9294	0.3395	63	2.1437	1.8336	0.7170	2.8456
8	28.9051	2.1044	6.0485	0.3871	64	2.0883	1.8310	0.7053	2.8887
9	25.4299	2.0946	5.3639	0.4345	65	2.0347	1.8285	0.6939	2.9318
10	22.6550	2.0851	4.8166	0.4817	66	1.9829	1.8259	0.6829	2.9749
11	20.3889	2.0760	4.3692	0.5287	67	1.9327	1.8235	0.6723	3.0179
12	18.5040	2.0673	3.9967	0.5755	68	1.8919	1.8214	0.6620	3.0612
13	16.9121	2.0589	3.6817	0.6222	69	1.8447	1.8190	0.6520	3.1042
14	15.5503	2.0508	3.4120	0.6688	70	1.7990	1.8166	0.6423	3.1471
15	14.3723	2.0429	3.1784	0.7152	71	1.7548	1.8143	0.6328	3.1900
16	13.3435	2.0354	2.9743	0.7614	72	1.7119	1.8120	0.6236	3.2329
17	12.4375	2.0280	2.7943	0.8075	73	1.6704	1.8097	0.6147	3.2757
18	11.6337	2.0210	2.6344	0.8536	74	1.6301	1.8075	0.6060	3.3186
19	10.9159	2.0141	2.4915	0.8994	75	1.5910	1.8053	0.5976	3.3614
20	10.2711	2.0075	2.3631	0.9452	76	1.5531	1.8031	0.5894	3.4041
21	9.6889	2.0010	2.2469	0.9909	77	1.5163	1.8010	0.5814	3.4469
22	9.1607	1.9948	2.1414	1.0365	78	1.4806	1.7988	0.5736	3.4896
23	8.6793	1.9887	2.0452	1.0819	79	1.4458	1.7968	0.5660	3.5322
24	8.2389	1.9828	1.9571	1.1273	80	1.4121	1.7947	0.5586	3.5749
25	7.8346	1.9771	1.8761	1.1726	81	1.3793	1.7926	0.5514	3.6175
26	7.4621	1.9715	1.8014	1.2177	82	1.3474	1.7906	0.5443	3.6601
27	7.1178	1.9661	1.7323	1.2628	83	1.3164	1.7886	0.5375	3.7027
28	6.7988	1.9608	1.6682	1.3078	84	1.2862	1.7867	0.5308	3.7453
29	6.5024	1.9557	1.6085	1.3528	85	1.2568	1.7847	0.5243	3.7878

TABLE 12a (continued) Zero Thickness Microstrip Dimensions, Effective Dielectric Constant, and PUL Capacitance and Inductance for Er = 2.2

Zo [Ohms]	W/h	Keff	pF/cm	nH/mm	Zo [Ohms]	W/h	Keff	pF/cm	nH/mm
30	6.2263	1.9507	1.5529	1.3976	86	1.2282	1.7828	0.5179	3.8303
31	5.9686	1.9458	1.5010	1.4424	87	1.2003	1.7809	0.5117	3.8728
32	5.7274	1.9410	1.4523	1.4871	88	1.1732	1.7791	0.5056	3.9152
33	5.5013	1.9364	1.4066	1.5318	89	1.1467	1.7772	0.4996	3.9577
34	5.2890	1.9319	1.3636	1.5763	90	1.1210	1.7754	0.4938	4.0001
35	5.0892	1.9274	1.3231	1.6208	91	1.0959	1.7736	0.4882	4.0424
36	4.9009	1.9231	1.2849	1.6653	92	1.0714	1.7718	0.4826	4.0848
37	4.7231	1.9189	1.2488	1.7096	93	1.0476	1.7700	0.4772	4.1272
38	4.5550	1.9147	1.2146	1.7539	94	1.0243	1.7683	0.4719	4.1695
39	4.3959	1.9107	1.1822	1.7982	95	1.0016	1.7665	0.4667	4.2118
40	4.2450	1.9067	1.1515	1.8424	96	0.9795	1.7658	0.4617	4.2552
41	4.1018	1.9028	1.1223	1.8865	97	0.9579	1.7651	0.4569	4.2987
42	3.9658	1.8990	1.0945	1.9306	98	0.9369	1.7644	0.4521	4.3421
43	3.8363	1.8953	1.0680	1.9746	99	0.9163	1.7637	0.4475	4.3855
44	3.7129	1.8917	1.0427	2.0186	100	0.8962	1.7629	0.4429	4.4289
45	3.5953	1.8881	1.0185	2.0625	101	0.8767	1.7621	0.4384	4.4722
46	3.4831	1.8846	0.9955	2.1064	102	0.8575	1.7613	0.4340	4.5154
47	3.3758	1.8811	0.9734	2.1502	103	0.8389	1.7605	0.4297	4.5586
48	3.2733	1.8778	0.9523	2.1940	104	0.8206	1.7596	0.4255	4.6018
49	3.1751	1.8745	0.9320	2.2378	105	0.8028	1.7588	0.4213	4.6449
50	3.0811	1.8712	0.9126	2.2814	106	0.7855	1.7579	0.4172	4.6880
51	2.9909	1.8680	0.8939	2.3251	107	0.7685	1.7570	0.4132	4.7310
52	2.9044	1.8649	0.8760	2.3687	108	0.7519	1.7561	0.4093	4.7740
53	2.8214	1.8618	0.8588	2.4122	109	0.7357	1.7552	0.4054	4.8169
54	2.7416	1.8587	0.8422	2.4557	110	0.7198	1.7543	0.4016	4.8599
55	2.6648	1.8558	0.8262	2.4992	111	0.7043	1.7534	0.3979	4.9027
56	2.5910	1.8528	0.8108	2.5426	112	0.6892	1.7524	0.3943	4.9456
113	0.6744	1.7515	0.3907	4.9884	132	0.4484	1.7330	0.3327	5.7964
114	0.6600	1.7505	0.3871	5.0312	133	0.4389	1.7321	0.3301	5.8387
115	0.6459	1.7496	0.3837	5.0739	134	0.4296	1.7311	0.3275	5.8810
116	0.6320	1.7486	0.3803	5.1166	135	0.4206	1.7302	0.3250	5.9232
117	0.6185	1.7477	0.3769	5.1593	136	0.4117	1.7292	0.3225	5.9654
118	0.6053	1.7467	0.3736	5.2020	137	0.4030	1.7283	0.3201	6.0076
119	0.5924	1.7457	0.3704	5.2446	138	0.3945	1.7273	0.3177	6.0498
120	0.5798	1.7447	0.3672	5.2872	139	0.3862	1.7264	0.3153	6.0920
121	0.5675	1.7438	0.3640	5.3298	140	0.3781	1.7254	0.3130	6.1342
122	0.5554	1.7428	0.3609	5.3723	141	0.3702	1.7245	0.3107	6.1763
123	0.5436	1.7418	0.3579	5.4148	142	0.3624	1.7236	0.3084	6.2185
124	0.5321	1.7408	0.3549	5.4573	143	0.3547	1.7226	0.3062	6.2606
125	0.5208	1.7399	0.3520	5.4998	144	0.3473	1.7217	0.3039	6.3027
126	0.5097	1.7389	0.3491	5.5422	145	0.3400	1.7208	0.3018	6.3448
127	0.4989	1.7379	0.3462	5.5846	146	0.3328	1.7199	0.2996	6.3868
128	0.4884	1.7369	0.3434	5.6270	147	0.3259	1.7190	0.2975	6.4289
129	0.4780	1.7360	0.3407	5.6694	148	0.3190	1.7181	0.2954	6.4709
130	0.4679	1.7350	0.3380	5.7118	149	0.3123	1.7172	0.2934	6.5130
131	0.4580	1.7340	0.3353	5.7541	150	0.3058	1.7163	0.2913	6.5550

Notes: Calculation of W/H has an error of less than 1%.

Source: Gupta, K.C., Garg, R., Bahl, I., Bhartia, P., *Microstrip Lines and Slotlines,* 2nd Ed., Artech House, Norwood, MA, 1996, 103.

TABLE 12b   Zero Thickness Microstrip Dimensions, Effective Dielectric Constant, and PUL Capacitance and Inductance for Er =3.78

Zo [Ohms]	W/h	Keff	pF/cm	nH/mm	Zo [Ohms]	W/h	Keff	pF/cm	nH/mm
1	190.5772	3.7382	64.4927	0.0645	26	5.4498	3.1668	2.2831	1.5433
2	93.9049	3.6989	32.0763	0.1283	27	5.1890	3.1537	2.1940	1.5994
3	61.7511	3.6619	21.2770	0.1915	28	4.9474	3.1410	2.1113	1.6553
4	45.7086	3.6271	15.8817	0.2541	29	4.7230	3.1287	2.0345	1.7110
5	36.1035	3.5942	12.6477	0.3162	30	4.5140	3.1167	1.9629	1.7666
6	29.7137	3.5632	10.4941	0.3778	31	4.3190	3.1051	1.8961	1.8221
7	25.1593	3.5337	8.9578	0.4389	32	4.1365	3.0938	1.8335	1.8775
8	21.7508	3.5059	7.8070	0.4997	33	3.9655	3.0827	1.7747	1.9327
9	19.1053	3.4794	6.9133	0.5600	34	3.8050	3.0720	1.7195	1.9878
10	16.9935	3.4542	6.1994	0.6199	35	3.6539	3.0616	1.6676	2.0428
11	15.2694	3.4301	5.6162	0.6796	36	3.5116	3.0514	1.6185	2.0976
12	13.8357	3.4072	5.1309	0.7389	37	3.3773	3.0414	1.5722	2.1524
13	12.6252	3.3853	4.7210	0.7978	38	3.2504	3.0317	1.5284	2.2070
14	11.5899	3.3643	4.3702	0.8566	39	3.1302	3.0222	1.4869	2.2616
15	10.6946	3.3442	4.0666	0.9150	40	3.0164	3.0130	1.4475	2.3160
16	9.9129	3.3249	3.8014	0.9732	41	2.9083	3.0039	1.4101	2.3703
17	9.2247	3.3064	3.5678	1.0311	42	2.8056	2.9951	1.3745	2.4246
18	8.6143	3.2885	3.3605	1.0888	43	2.7080	2.9864	1.3406	2.4787
19	8.0694	3.2714	3.1754	1.1463	44	2.6150	2.9780	1.3082	2.5327
20	7.5800	3.2549	3.0090	1.2036	45	2.5263	2.9697	1.2774	2.5867
21	7.1382	3.2389	2.8586	1.2607	46	2.4417	2.9615	1.2479	2.6406
22	6.7375	3.2235	2.7222	1.3175	47	2.3609	2.9536	1.2197	2.6943
23	6.3724	3.2086	2.5978	1.3743	48	2.2836	2.9458	1.1927	2.7480
24	6.0386	3.1942	2.4840	1.4308	49	2.2097	2.9381	1.1669	2.8016
25	5.7321	3.1803	2.3794	1.4871	50	2.1389	2.9306	1.1421	2.8552
51	2.0711	2.9233	1.1183	2.9086	101	0.5140	2.7126	0.5439	5.5488
52	2.0060	2.9160	1.0954	2.9620	102	0.5008	2.7099	0.5383	5.6008
53	1.9435	2.9089	1.0734	3.0152	103	0.4878	2.7071	0.5328	5.6529
54	1.8840	2.9020	1.0523	3.0685	104	0.4752	2.7043	0.5274	5.7048
55	1.8268	2.8952	1.0320	3.1217	105	0.4630	2.7016	0.5222	5.7568
56	1.7719	2.8886	1.0124	3.1748	106	0.4511	2.6989	0.5170	5.8086
57	1.7190	2.8820	0.9935	3.2278	107	0.4394	2.6961	0.5119	5.8605
58	1.6682	2.8756	0.9752	3.2807	108	0.4281	2.6934	0.5069	5.9123
59	1.6192	2.8693	0.9577	3.3336	109	0.4171	2.6907	0.5020	5.9640
60	1.5720	2.8631	0.9407	3.3865	110	0.4064	2.6880	0.4972	6.0157
61	1.5265	2.8569	0.9243	3.4392	111	0.3960	2.6853	0.4924	6.0674
62	1.4826	2.8509	0.9084	3.4919	112	0.3858	2.6826	0.4878	6.1190
63	1.4402	2.8450	0.8931	3.5446	113	0.3759	2.6800	0.4832	6.1706
64	1.3993	2.8392	0.8782	3.5971	114	0.3663	2.6774	0.4788	6.2221
65	1.3597	2.8334	0.8638	3.6496	115	0.3569	2.6747	0.4744	6.2736
66	1.3215	2.8278	0.8499	3.7021	116	0.3477	2.6721	0.4701	6.3251
67	1.2845	2.8222	0.8364	3.7545	117	0.3388	2.6696	0.4658	6.3765
68	1.2488	2.8167	0.8233	3.8068	118	0.3301	2.6670	0.4616	6.4279
69	1.2142	2.8113	0.8106	3.8591	119	0.3217	2.6644	0.4575	6.4793
70	1.1807	2.8060	0.7982	3.9113	120	0.3134	2.6619	0.4535	6.5307
71	1.1482	2.8008	0.7862	3.9635	121	0.3054	2.6594	0.4496	6.5820
72	1.1168	2.7956	0.7746	4.0156	122	0.2976	2.6569	0.4457	6.6333
73	1.0863	2.7905	0.7633	4.0676	123	0.2900	2.6544	0.4418	6.6845
74	1.0568	2.7855	0.7523	4.1196	124	0.2826	2.6520	0.4381	6.7358
75	1.0282	2.7805	0.7416	4.1716	125	0.2754	2.6496	0.4344	6.7870
76	1.0004	2.7756	0.7312	4.2235	126	0.2683	2.6472	0.4307	6.8382
77	0.9735	2.7737	0.7215	4.2776	127	0.2615	2.6448	0.4271	6.8893
78	0.9474	2.7717	0.7120	4.3316	128	0.2548	2.6424	0.4236	6.9405
79	0.9220	2.7696	0.7027	4.3855	129	0.2483	2.6401	0.4201	6.9916

TABLE 12b (continued) Zero Thickness Microstrip Dimensions, Effective Dielectric Constant, and PUL Capacitance and Inductance for Er =3.78

Zo [Ohms]	W/h	Keff	pF/cm	nH/mm	Zo [Ohms]	W/h	Keff	pF/cm	nH/mm
80	0.8974	2.7675	0.6936	4.4393	130	0.2420	2.6378	0.4167	7.0427
81	0.8735	2.7653	0.6848	4.4929	131	0.2358	2.6355	0.4134	7.0938
82	0.8503	2.7630	0.6762	4.5465	132	0.2298	2.6332	0.4101	7.1449
83	0.8277	2.7606	0.6677	4.6000	133	0.2239	2.6309	0.4068	7.1959
84	0.8058	2.7582	0.6595	4.6534	134	0.2182	2.6287	0.4036	7.2469
85	0.7845	2.7557	0.6514	4.7067	135	0.2126	2.6265	0.4004	7.2980
86	0.7639	2.7532	0.6436	4.7599	136	0.2072	2.6243	0.3973	7.3490
87	0.7438	2.7506	0.6359	4.8130	137	0.2019	2.6221	0.3943	7.3999
88	0.7242	2.7481	0.6284	4.8660	138	0.1968	2.6200	0.3912	7.4509
89	0.7052	2.7454	0.6210	4.9190	139	0.1917	2.6179	0.3883	7.5019
90	0.6868	2.7428	0.6138	4.9718	140	0.1869	2.6158	0.3853	7.5528
91	0.6688	2.7401	0.6068	5.0246	141	0.1821	2.6137	0.3825	7.6037
92	0.6513	2.7374	0.5999	5.0773	142	0.1775	2.6116	0.3796	7.6546
93	0.6344	2.7347	0.5931	5.1300	143	0.1729	2.6096	0.3768	7.7055
94	0.6178	2.7320	0.5865	5.1826	144	0.1685	2.6076	0.3741	7.7564
95	0.6018	2.7292	0.5801	5.2351	145	0.1642	2.6056	0.3713	7.8073
96	0.5861	2.7265	0.5737	5.2875	146	0.1600	2.6036	0.3687	7.8582
97	0.5709	2.7237	0.5675	5.3399	147	0.1560	2.6017	0.3660	7.9091
98	0.5561	2.7209	0.5615	5.3922	148	0.1520	2.5998	0.3634	7.9599
99	0.5417	2.7182	0.5555	5.4444	149	0.1481	2.5979	0.3608	8.0108
100	0.5277	2.7154	0.5497	5.4966	150	0.1444	2.5960	0.3583	8.0616

Notes: Calculation of W/H has an error of less than 1%.

Source: Gupta, K.C., Garg, R., Bahl, I., Bhartia, P., *Microstrip Lines and Slotlines*, 2nd Ed., Artech House, Norwood, MA, 1996, 103.

TABLE 12c Zero Thickness Microstrip Dimensions, Effective Dielectric Constant, and PUL Capacitance and Inductance for Er =5.75

Zo [Ohms]	W/h	Keff	pF/cm	nH/mm	Zo [Ohms]	W/h	Keff	pF/cm	nH/mm
1	154.1545	5.6626	79.3758	0.0794	55	1.3098	4.1200	1.2310	3.7239
2	75.8057	5.5818	39.4035	0.1576	56	1.2664	4.1088	1.2074	3.7864
3	49.7547	5.5068	26.0920	0.2348	57	1.2248	4.0978	1.1846	3.8488
4	36.7611	5.4372	19.4449	0.3111	58	1.1847	4.0869	1.1627	3.9112
5	28.9839	5.3723	15.4628	0.3866	59	1.1461	4.0763	1.1415	3.9734
6	23.8117	5.3116	12.8127	0.4613	60	1.1089	4.0658	1.1210	4.0355
7	20.1263	5.2548	10.9235	0.5352	61	1.0731	4.0555	1.1012	4.0976
8	17.3689	5.2014	9.5094	0.6086	62	1.0386	4.0453	1.0821	4.1596
9	15.2296	5.1512	8.4118	0.6814	63	1.0054	4.0353	1.0636	4.2214
10	13.5223	5.1037	7.5357	0.7536	64	0.9733	4.0305	1.0464	4.2859
11	12.1289	5.0589	6.8204	0.8253	65	0.9423	4.0265	1.0297	4.3507
12	10.9706	5.0163	6.2257	0.8965	66	0.9124	4.0222	1.0136	4.4153
13	9.9929	4.9759	5.7236	0.9673	67	0.8836	4.0178	0.9979	4.4797
14	9.1570	4.9375	5.2942	1.0377	68	0.8557	4.0132	0.9827	4.5439
15	8.4343	4.9008	4.9229	1.1077	69	0.8288	4.0084	0.9679	4.6080
16	7.8035	4.8659	4.5987	1.1773	70	0.8028	4.0035	0.9535	4.6719
17	7.2484	4.8324	4.3133	1.2466	71	0.7777	3.9985	0.9394	4.7357
18	6.7562	4.8004	4.0602	1.3155	72	0.7534	3.9933	0.9258	4.7993
19	6.3169	4.7697	3.8342	1.3841	73	0.7299	3.9881	0.9125	4.8628
20	5.9225	4.7403	3.6312	1.4525	74	0.7072	3.9828	0.8996	4.9261
21	5.5666	4.7119	3.4479	1.5205	75	0.6853	3.9774	0.8870	4.9893
22	5.2438	4.6847	3.2817	1.5883	76	0.6641	3.9720	0.8747	5.0524
23	4.9499	4.6584	3.1302	1.6559	77	0.6435	3.9665	0.8628	5.1153
24	4.6812	4.6331	2.9916	1.7232	78	0.6236	3.9609	0.8511	5.1781
25	4.4346	4.6087	2.8644	1.7902	79	0.6044	3.9554	0.8397	5.2408

TABLE 12c (continued)   Zero Thickness Microstrip Dimensions, Effective Dielectric Constant, and PUL Capacitance and Inductance for Er =5.75

Zo [Ohms]	W/h	Keff	pF/cm	nH/mm	Zo [Ohms]	W/h	Keff	pF/cm	nH/mm
26	4.2075	4.5851	2.7471	1.8571	80	0.5858	3.9498	0.8287	5.3034
27	3.9979	4.5623	2.6388	1.9237	81	0.5678	3.9442	0.8178	5.3659
28	3.8037	4.5402	2.5384	1.9901	82	0.5503	3.9386	0.8073	5.4283
29	3.6233	4.5187	2.4451	2.0563	83	0.5334	3.9329	0.7970	5.4905
30	3.4554	4.4980	2.3581	2.1223	84	0.5171	3.9273	0.7870	5.5527
31	3.2988	4.4778	2.2769	2.1881	85	0.5012	3.9217	0.7771	5.6148
32	3.1523	4.4583	2.2010	2.2538	86	0.4859	3.9161	0.7676	5.6768
33	3.0151	4.4393	2.1297	2.3193	87	0.4710	3.9105	0.7582	5.7387
34	2.8863	4.4208	2.0628	2.3846	88	0.4566	3.9049	0.7490	5.8005
35	2.7652	4.4028	1.9997	2.4497	89	0.4427	3.8994	0.7401	5.8623
36	2.6511	4.3853	1.9403	2.5147	90	0.4292	3.8938	0.7314	5.9239
37	2.5434	4.3682	1.8842	2.5795	91	0.4161	3.8883	0.7228	5.9855
38	2.4417	4.3516	1.8311	2.6441	92	0.4034	3.8829	0.7144	6.0470
39	2.3455	4.3353	1.7808	2.7087	93	0.3912	3.8774	0.7063	6.1085
40	2.2543	4.3195	1.7331	2.7730	94	0.3793	3.8720	0.6983	6.1699
41	2.1678	4.3040	1.6878	2.8373	95	0.3677	3.8667	0.6904	6.2312
42	2.0856	4.2889	1.6448	2.9014	96	0.3565	3.8614	0.6828	6.2925
43	2.0075	4.2741	1.6037	2.9653	97	0.3457	3.8561	0.6753	6.3537
44	1.9289	4.2588	1.5645	3.0288	98	0.3352	3.8508	0.6679	6.4148
45	1.8591	4.2449	1.5272	3.0926	99	0.3250	3.8456	0.6607	6.4759
46	1.7926	4.2312	1.4916	3.1562	100	0.3152	3.8405	0.6537	6.5369
47	1.7291	4.2178	1.4576	3.2198	101	0.3056	3.8354	0.6468	6.5979
48	1.6684	4.2048	1.4250	3.2832	102	0.2963	3.8304	0.6400	6.6588
49	1.6104	4.1919	1.3938	3.3464	103	0.2874	3.8254	0.6334	6.7197
50	1.5549	4.1794	1.3638	3.4096	104	0.2787	3.8204	0.6269	6.7806
51	1.5017	4.1671	1.3351	3.4727	105	0.2702	3.8155	0.6205	6.8414
52	1.4507	4.1550	1.3076	3.5356	106	0.2620	3.8107	0.6143	6.9022
53	1.4018	4.1431	1.2811	3.5985	107	0.2541	3.8059	0.6082	6.9629
54	1.3549	4.1315	1.2556	3.6612	108	0.2464	3.8011	0.6022	7.0236
109	0.2389	3.7964	0.5963	7.0842	130	0.1255	3.7101	0.4942	8.3525
110	0.2317	3.7918	0.5905	7.1449	131	0.1217	3.7065	0.4902	8.4127
111	0.2247	3.7872	0.5848	7.2055	132	0.1180	3.7030	0.4863	8.4729
112	0.2179	3.7827	0.5792	7.2660	133	0.1144	3.6996	0.4824	8.5331
113	0.2113	3.7782	0.5738	7.3265	134	0.1110	3.6962	0.4786	8.5933
114	0.2049	3.7738	0.5684	7.3871	135	0.1076	3.6928	0.4748	8.6535
115	0.1987	3.7694	0.5631	7.4475	136	0.1044	3.6895	0.4711	8.7137
116	0.1927	3.7651	0.5580	7.5080	137	0.1012	3.6863	0.4675	8.7739
117	0.1869	3.7608	0.5529	7.5684	138	0.0982	3.6830	0.4639	8.8341
118	0.1813	3.7566	0.5479	7.6288	139	0.0952	3.6799	0.4603	8.8943
119	0.1758	3.7524	0.5430	7.6892	140	0.0923	3.6767	0.4569	8.9544
120	0.1705	3.7483	0.5382	7.7496	141	0.0896	3.6737	0.4534	9.0146
121	0.1653	3.7443	0.5334	7.8099	142	0.0869	3.6706	0.4500	9.0748
122	0.1603	3.7403	0.5288	7.8703	143	0.0842	3.6676	0.4467	9.1350
123	0.1555	3.7363	0.5242	7.9306	144	0.0817	3.6647	0.4434	9.1952
124	0.1508	3.7324	0.5197	7.9909	145	0.0792	3.6618	0.4402	9.2553
125	0.1462	3.7286	0.5153	8.0512	146	0.0768	3.6589	0.4370	9.3155
126	0.1418	3.7248	0.5109	8.1115	147	0.0745	3.6561	0.4339	9.3757
127	0.1375	3.7210	0.5066	8.1717	148	0.0723	3.6533	0.4308	9.4359
128	0.1334	3.7173	0.5024	8.2320	149	0.0701	3.6505	0.4277	9.4961
129	0.1294	3.7137	0.4983	8.2922	150	0.0680	3.6478	0.4247	9.5562

Notes: Calculation of W/H has an error of less than 1%.

Source: Gupta, K.C., Garg, R., Bahl, I., Bhartia, P., *Microstrip Lines and Slotlines,* 2nd Ed., Artech House, Norwood, MA, 1996, 103.

TABLE 12d Zero Thickness Microstrip Dimensions, Effective Dielectric Constant, and PUL Capacitance and Inductance for Er = 9.4

Zo [Ohms]	W/h	Keff	pF/cm	nH/mm	Zo [Ohms]	W/h	Keff	pF/cm	nH/mm
1	120.1221	9.2047	101.2010	0.1012	24	3.4066	7.1750	3.7229	2.1444
2	58.8859	9.0280	50.1125	0.2004	25	3.2161	7.1309	3.5630	2.2269
3	38.5353	8.8676	33.1101	0.2980	26	3.0409	7.0885	3.4157	2.3090
4	28.3901	8.7212	24.6268	0.3940	27	2.8791	7.0475	3.2797	2.3909
5	22.3208	8.5871	19.5493	0.4887	28	2.7293	7.0079	3.1537	2.4725
6	18.2864	8.4635	16.1735	0.5822	29	2.5902	6.9697	3.0366	2.5538
7	15.4132	8.3493	13.7691	0.6747	30	2.4609	6.9326	2.9276	2.6348
8	13.2647	8.2433	11.9712	0.7662	31	2.3403	6.8967	2.8258	2.7156
9	11.5985	8.1445	10.5771	0.8567	32	2.2275	6.8619	2.7306	2.7961
10	10.2695	8.0521	9.4653	0.9465	33	2.1220	6.8281	2.6413	2.8764
11	9.1854	7.9655	8.5584	1.0356	34	2.0229	6.7952	2.5574	2.9564
12	8.2846	7.8841	7.8050	1.1239	35	1.9221	6.7606	2.4780	3.0356
13	7.5248	7.8074	7.1695	1.2116	36	1.8364	6.7301	2.4037	3.1152
14	6.8754	7.7348	6.6264	1.2988	37	1.7556	6.7004	2.3336	3.1947
15	6.3143	7.6661	6.1571	1.3853	38	1.6792	6.6715	2.2673	3.2740
16	5.8248	7.6009	5.7477	1.4714	39	1.6070	6.6434	2.2045	3.3530
17	5.3942	7.5389	5.3875	1.5570	40	1.5386	6.6159	2.1449	3.4319
18	5.0126	7.4798	5.0682	1.6421	41	1.4738	6.5891	2.0884	3.5105
19	4.6722	7.4234	4.7833	1.7268	42	1.4122	6.5629	2.0346	3.5890
20	4.3668	7.3694	4.5276	1.8110	43	1.3537	6.5373	1.9834	3.6673
21	4.0913	7.3178	4.2968	1.8949	44	1.2981	6.5122	1.9346	3.7454
22	3.8416	7.2683	4.0876	1.9784	45	1.2451	6.4877	1.8880	3.8233
23	3.6143	7.2207	3.8971	2.0616	46	1.1946	6.4638	1.8436	3.9010
47	1.1465	6.4403	1.8011	3.9786	99	0.1530	5.8352	0.8139	7.9771
48	1.1005	6.4173	1.7604	4.0560	100	0.1472	5.8268	0.8052	8.0518
49	1.0567	6.3948	1.7215	4.1332	101	0.1417	5.8184	0.7966	8.1265
50	1.0148	6.3728	1.6841	4.2103	102	0.1365	5.8102	0.7883	8.2012
51	0.9747	6.3596	1.6494	4.2901	103	0.1314	5.8021	0.7801	8.2758
52	0.9364	6.3507	1.6165	4.3711	104	0.1265	5.7942	0.7720	8.3504
53	0.8997	6.3412	1.5849	4.4519	105	0.1217	5.7864	0.7642	8.4251
54	0.8646	6.3312	1.5543	4.5323	106	0.1172	5.7787	0.7565	8.4997
55	0.8309	6.3208	1.5248	4.6124	107	0.1128	5.7712	0.7489	8.5742
56	0.7987	6.3100	1.4963	4.6923	108	0.1086	5.7638	0.7415	8.6488
57	0.7678	6.2989	1.4687	4.7719	109	0.1046	5.7565	0.7342	8.7234
58	0.7382	6.2875	1.4421	4.8512	110	0.1006	5.7493	0.7271	8.7979
59	0.7098	6.2759	1.4163	4.9303	111	0.0969	5.7423	0.7201	8.8725
60	0.6826	6.2641	1.3914	5.0091	112	0.0933	5.7354	0.7133	8.9470
61	0.6564	6.2521	1.3673	5.0877	113	0.0898	5.7286	0.7065	9.0216
62	0.6313	6.2400	1.3439	5.1661	114	0.0864	5.7219	0.6999	9.0961
63	0.6072	6.2278	1.3213	5.2443	115	0.0832	5.7154	0.6934	9.1707
64	0.5841	6.2156	1.2994	5.3223	116	0.0801	5.7090	0.6871	9.2452
65	0.5619	6.2032	1.2781	5.4001	117	0.0771	5.7026	0.6808	9.3197
66	0.5406	6.1909	1.2575	5.4777	118	0.0743	5.6964	0.6747	9.3943
67	0.5201	6.1786	1.2375	5.5552	119	0.0715	5.6903	0.6687	9.4688
68	0.5004	6.1662	1.2181	5.6325	120	0.0688	5.6843	0.6627	9.5434
69	0.4814	6.1539	1.1992	5.7096	121	0.0662	5.6785	0.6569	9.6179
70	0.4632	6.1417	1.1809	5.7866	122	0.0638	5.6727	0.6512	9.6925
71	0.4457	6.1295	1.1631	5.8634	123	0.0614	5.6670	0.6456	9.7670
72	0.4289	6.1173	1.1458	5.9401	124	0.0591	5.6615	0.6401	9.8416
73	0.4127	6.1053	1.1290	6.0166	125	0.0569	5.6560	0.6346	9.9162
74	0.3972	6.0933	1.1127	6.0931	126	0.0548	5.6506	0.6293	9.9907
75	0.3822	6.0814	1.0968	6.1694	127	0.0527	5.6454	0.6241	10.0653
76	0.3679	6.0696	1.0813	6.2456	128	0.0508	5.6402	0.6189	10.1399
77	0.3540	6.0579	1.0662	6.3217	129	0.0489	5.6351	0.6138	10.2146
78	0.3407	6.0464	1.0516	6.3977	130	0.0471	5.6301	0.6088	10.2892
79	0.3279	6.0350	1.0373	6.4736	131	0.0453	5.6252	0.6039	10.3638

TABLE 12d (continued)   Zero Thickness Microstrip Dimensions, Effective Dielectric Constant, and PUL Capacitance and Inductance for Er = 9.4

Zo [Ohms]	W/h	Keff	pF/cm	nH/mm	Zo [Ohms]	W/h	Keff	pF/cm	nH/mm
80	0.3156	6.0236	1.0233	6.5494	132	0.0436	5.6204	0.5991	10.4385
81	0.3038	6.0125	1.0098	6.6251	133	0.0420	5.6157	0.5943	10.5131
82	0.2924	6.0014	0.9965	6.7007	134	0.0404	5.6111	0.5897	10.5878
83	0.2815	5.9905	0.9836	6.7762	135	0.0389	5.6065	0.5850	10.6625
84	0.2709	5.9797	0.9710	6.8517	136	0.0375	5.6021	0.5805	10.7372
85	0.2608	5.9691	0.9588	6.9271	137	0.0361	5.5977	0.5761	10.8119
86	0.2510	5.9586	0.9468	7.0024	138	0.0347	5.5934	0.5717	10.8867
87	0.2416	5.9482	0.9351	7.0777	139	0.0334	5.5892	0.5673	10.9614
88	0.2326	5.9380	0.9237	7.1529	140	0.0322	5.5850	0.5631	11.0362
89	0.2239	5.9280	0.9125	7.2281	141	0.0310	5.5809	0.5589	11.1110
90	0.2155	5.9180	0.9016	7.3032	142	0.0298	5.5770	0.5547	11.1858
91	0.2074	5.9083	0.8910	7.3782	143	0.0287	5.5730	0.5507	11.2606
92	0.1997	5.8986	0.8806	7.4532	144	0.0276	5.5692	0.5467	11.3354
93	0.1922	5.8891	0.8704	7.5281	145	0.0266	5.5654	0.5427	11.4103
94	0.1850	5.8798	0.8605	7.6031	146	0.0256	5.5617	0.5388	11.4852
95	0.1781	5.8706	0.8507	7.6779	147	0.0247	5.5581	0.5350	11.5600
96	0.1715	5.8616	0.8412	7.7528	148	0.0237	5.5545	0.5312	11.6350
97	0.1651	5.8526	0.8319	7.8276	149	0.0229	5.5510	0.5274	11.7099
98	0.1589	5.8439	0.8228	7.9023	150	0.0220	5.5476	0.5238	11.7848

Notes: Calculation of W/H has an error of less than 1%.

Source: Gupta, K.C., Garg, R., Bahl, I., Bhartia, P., *Microstrip Lines and Slotlines*, 2nd Ed., Artech House, Norwood, MA, 1996, 103.

TABLE 12e   Zero Thickness Microstrip Dimensions, Effective Dielectric Constant, and PUL Capacitance and Inductance for Er = 9.8

Zo [Ohms]	W/h	Keff	pF/cm	nH/mm	Zo [Ohms]	W/h	Keff	pF/cm	nH/mm
1	117.6023	9.5914	103.3045	0.1033	57	0.7347	6.5379	1.4963	4.8615
2	57.6329	9.4030	51.1424	0.2046	58	0.7059	6.5254	1.4691	4.9421
3	37.7044	9.2322	33.7840	0.3041	59	0.6783	6.5128	1.4428	5.0224
4	27.7700	9.0767	25.1237	0.4020	60	0.6519	6.5000	1.4174	5.1025
5	21.8272	8.9344	19.9408	0.4985	61	0.6265	6.4870	1.3927	5.1824
6	17.8771	8.8035	16.4952	0.5938	62	0.6022	6.4740	1.3689	5.2621
7	15.0641	8.6827	14.0413	0.6880	63	0.5788	6.4609	1.3458	5.3415
8	12.9606	8.5706	12.2066	0.7812	64	0.5564	6.4477	1.3234	5.4208
9	11.3295	8.4662	10.7841	0.8735	65	0.5349	6.4346	1.3017	5.4999
10	10.0285	8.3688	9.6496	0.9650	66	0.5142	6.4214	1.2807	5.5788
11	8.9673	8.2775	8.7244	1.0557	67	0.4944	6.4082	1.2603	5.6575
12	8.0857	8.1917	7.9558	1.1456	68	0.4753	6.3951	1.2405	5.7360
13	7.3419	8.1109	7.3075	1.2350	69	0.4570	6.3820	1.2213	5.8144
14	6.7063	8.0345	6.7535	1.3237	70	0.4395	6.3690	1.2026	5.8927
15	6.1572	7.9622	6.2749	1.4118	71	0.4226	6.3561	1.1844	5.9708
16	5.6782	7.8937	5.8573	1.4995	72	0.4064	6.3432	1.1668	6.0488
17	5.2568	7.8285	5.4900	1.5866	73	0.3908	6.3305	1.1497	6.1266
18	4.8834	7.7664	5.1644	1.6733	74	0.3758	6.3179	1.1330	6.2043
19	4.5503	7.7071	4.8738	1.7595	75	0.3614	6.3053	1.1168	6.2820
20	4.2515	7.6505	4.6131	1.8452	76	0.3476	6.2929	1.1010	6.3595
21	3.9820	7.5963	4.3778	1.9306	77	0.3343	6.2807	1.0857	6.4368
22	3.7377	7.5443	4.1645	2.0156	78	0.3215	6.2685	1.0707	6.5141
23	3.5154	7.4944	3.9703	2.1003	79	0.3092	6.2565	1.0561	6.5913
24	3.3122	7.4464	3.7926	2.1846	80	0.2974	6.2447	1.0419	6.6684
25	3.1259	7.4002	3.6296	2.2685	81	0.2860	6.2329	1.0281	6.7455

TABLE 12e (continued)   Zero Thickness Microstrip Dimensions, Effective Dielectric Constant, and PUL
Capacitance and Inductance for Er = 9.8

Zo [Ohms]	W/h	Keff	pF/cm	nH/mm	Zo [Ohms]	W/h	Keff	pF/cm	nH/mm
26	2.9545	7.3557	3.4795	2.3522	82	0.2751	6.2214	1.0146	6.8224
27	2.7962	7.3128	3.3408	2.4355	83	0.2646	6.2100	1.0015	6.8993
28	2.6497	7.2713	3.2124	2.5185	84	0.2545	6.1987	0.9887	6.9760
29	2.5138	7.2312	3.0930	2.6012	85	0.2448	6.1876	0.9762	7.0528
30	2.3873	7.1923	2.9819	2.6837	86	0.2355	6.1766	0.9640	7.1294
31	2.2693	7.1547	2.8781	2.7659	87	0.2265	6.1658	0.9520	7.2060
32	2.1591	7.1182	2.7811	2.8478	88	0.2179	6.1552	0.9404	7.2825
33	2.0558	7.0827	2.6901	2.9295	89	0.2096	6.1447	0.9291	7.3590
34	1.9590	7.0483	2.6046	3.0109	90	0.2016	6.1344	0.9180	7.4355
35	1.8614	7.0124	2.5237	3.0916	91	0.1939	6.1242	0.9071	7.5118
36	1.7777	6.9805	2.4480	3.1727	92	0.1865	6.1142	0.8965	7.5882
37	1.6987	6.9494	2.3766	3.2535	93	0.1794	6.1044	0.8862	7.6645
38	1.6241	6.9191	2.3090	3.3342	94	0.1726	6.0947	0.8760	7.7408
39	1.5535	6.8896	2.2450	3.4146	95	0.1660	6.0852	0.8661	7.8170
40	1.4867	6.8609	2.1843	3.4948	96	0.1597	6.0758	0.8565	7.8932
41	1.4233	6.8328	2.1266	3.5749	97	0.1536	6.0666	0.8470	7.9693
42	1.3632	6.8053	2.0718	3.6547	98	0.1478	6.0575	0.8377	8.0455
43	1.3060	6.7785	2.0197	3.7343	99	0.1422	6.0486	0.8286	8.1216
44	1.2517	6.7523	1.9699	3.8138	100	0.1368	6.0398	0.8198	8.1977
45	1.1999	6.7266	1.9225	3.8931	101	0.1316	6.0312	0.8111	8.2737
46	1.1507	6.7015	1.8772	3.9721	102	0.1266	6.0227	0.8026	8.3498
47	1.1037	6.6770	1.8339	4.0510	103	0.1217	6.0143	0.7942	8.4258
48	1.0588	6.6529	1.7924	4.1298	104	0.1171	6.0062	0.7860	8.5018
49	1.0161	6.6293	1.7527	4.2083	105	0.1127	5.9981	0.7780	8.5778
50	0.9752	6.6149	1.7158	4.2895	106	0.1084	5.9902	0.7702	8.6538
51	0.9361	6.6054	1.6810	4.3722	107	0.1043	5.9824	0.7625	8.7298
52	0.8988	6.5953	1.6474	4.4545	108	0.1003	5.9748	0.7549	8.8057
53	0.8630	6.5846	1.6150	4.5365	109	0.0965	5.9673	0.7476	8.8817
54	0.8289	6.5735	1.5837	4.6182	110	0.0928	5.9600	0.7403	8.9576
55	0.7961	6.5619	1.5536	4.6996	111	0.0893	5.9527	0.7332	9.0336
56	0.7648	6.5501	1.5245	4.7807	112	0.0859	5.9456	0.7262	9.1095
113	0.0826	5.9387	0.7194	9.1855	132	0.0396	5.8280	0.6101	10.6295
114	0.0795	5.9318	0.7126	9.2614	133	0.0381	5.8232	0.6052	10.7056
115	0.0765	5.9251	0.7060	9.3374	134	0.0366	5.8185	0.6005	10.7818
116	0.0736	5.9185	0.6996	9.4133	135	0.0352	5.8139	0.5958	10.8579
117	0.0708	5.9120	0.6932	9.4893	136	0.0339	5.8093	0.5912	10.9341
118	0.0681	5.9057	0.6870	9.5652	137	0.0326	5.8049	0.5866	11.0102
119	0.0655	5.8994	0.6808	9.6412	138	0.0314	5.8005	0.5821	11.0864
120	0.0630	5.8933	0.6748	9.7172	139	0.0302	5.7962	0.5777	11.1626
121	0.0606	5.8873	0.6689	9.7931	140	0.0290	5.7920	0.5734	11.2389
122	0.0583	5.8814	0.6631	9.8691	141	0.0279	5.7879	0.5691	11.3151
123	0.0561	5.8756	0.6574	9.9451	142	0.0269	5.7838	0.5649	11.3914
124	0.0540	5.8699	0.6517	10.0211	143	0.0258	5.7799	0.5608	11.4677
125	0.0519	5.8643	0.6462	10.0971	144	0.0249	5.7760	0.5567	11.5440
126	0.0499	5.8588	0.6408	10.1731	145	0.0239	5.7722	0.5527	11.6203
127	0.0480	5.8534	0.6354	10.2492	146	0.0230	5.7684	0.5487	11.6966
128	0.0462	5.8482	0.6302	10.3252	147	0.0221	5.7647	0.5448	11.7730
129	0.0445	5.8430	0.6250	10.4013	148	0.0213	5.7611	0.5410	11.8494
130	0.0428	5.8379	0.6200	10.4773	149	0.0205	5.7576	0.5372	11.9258
131	0.0411	5.8329	0.6150	10.5534	150	0.0197	5.7541	0.5334	12.0022

Notes: Calculation of W/H has an error of less than 1%.

Source: Gupta, K.C., Garg, R., Bahl, I., Bhartia, P., *Microstrip Lines and Slotlines*, 2nd Ed., Artech House, Norwood, MA, 1996, 103.

TABLE 12f   Zero Thickness Microstrip Dimensions, Effective Dielectric Constant, and PUL Capacitance and Inductance for Er = 11.6

Zo [Ohms]	W/h	Keff	pF/cm	nH/mm	Zo [Ohms]	W/h	Keff	pF/cm	nH/mm
1	107.9253	11.3278	112.2672	0.1123	26	2.6229	8.5446	3.7502	2.5351
2	52.8209	11.0843	55.5270	0.2221	27	2.4782	8.4928	3.6003	2.6246
3	34.5132	10.8654	36.6506	0.3299	28	2.3444	8.4426	3.4615	2.7138
4	25.3888	10.6674	27.2364	0.4358	29	2.2202	8.3942	3.3325	2.8026
5	19.9316	10.4873	21.6044	0.5401	30	2.1047	8.3473	3.2124	2.8912
6	16.3052	10.3226	17.8617	0.6430	31	1.9970	8.3019	3.1003	2.9794
7	13.7232	10.1712	15.1973	0.7447	32	1.8884	8.2543	2.9948	3.0667
8	11.7929	10.0313	13.2059	0.8452	33	1.7964	8.2125	2.8967	3.1545
9	10.2964	9.9016	11.6625	0.9447	34	1.7101	8.1718	2.8045	3.2420
10	9.1031	9.7809	10.4321	1.0432	35	1.6291	8.1324	2.7178	3.3293
11	8.1299	9.6682	9.4289	1.1409	36	1.5527	8.0939	2.6361	3.4163
12	7.3215	9.5625	8.5958	1.2378	37	1.4807	8.0565	2.5589	3.5031
13	6.6397	9.4632	7.8932	1.3340	38	1.4128	8.0201	2.4859	3.5897
14	6.0573	9.3696	7.2931	1.4294	39	1.3485	7.9846	2.4168	3.6759
15	5.5541	9.2812	6.7747	1.5243	40	1.2877	7.9499	2.3513	3.7620
16	5.1153	9.1975	6.3226	1.6186	41	1.2301	7.9161	2.2890	3.8478
17	4.7294	9.1180	5.9249	1.7123	42	1.1754	7.8830	2.2299	3.9335
18	4.3875	9.0424	5.5725	1.8055	43	1.1235	7.8507	2.1735	4.0189
19	4.0826	8.9703	5.2581	1.8982	44	1.0742	7.8192	2.1199	4.1041
20	3.8091	8.9016	4.9760	1.9904	45	1.0273	7.7883	2.0687	4.1890
21	3.5625	8.8358	4.7215	2.0822	46	0.9827	7.7654	2.0207	4.2758
22	3.3390	8.7728	4.4908	2.1736	47	0.9402	7.7532	1.9762	4.3653
23	3.1357	8.7124	4.2807	2.2645	48	0.8998	7.7401	1.9334	4.4545
24	2.9499	8.6543	4.0887	2.3551	49	0.8612	7.7262	1.8922	4.5432
25	2.7796	8.5984	3.9125	2.4453	50	0.8244	7.7117	1.8526	4.6315
51	0.7893	7.6966	1.8145	4.7195	101	0.0957	6.9814	0.8726	8.9017
52	0.7558	7.6810	1.7778	4.8072	102	0.0917	6.9719	0.8635	8.9837
53	0.7238	7.6650	1.7425	4.8945	103	0.0880	6.9625	0.8545	9.0657
54	0.6933	7.6487	1.7084	4.9816	104	0.0844	6.9534	0.8458	9.1477
55	0.6641	7.6322	1.6755	5.0683	105	0.0809	6.9444	0.8372	9.2297
56	0.6362	7.6154	1.6438	5.1548	106	0.0776	6.9356	0.8287	9.3116
57	0.6095	7.5985	1.6131	5.2410	107	0.0744	6.9269	0.8205	9.3936
58	0.5840	7.5815	1.5835	5.3270	108	0.0714	6.9185	0.8124	9.4756
59	0.5596	7.5643	1.5549	5.4127	109	0.0684	6.9102	0.8044	9.5576
60	0.5363	7.5472	1.5273	5.4982	110	0.0656	6.9020	0.7967	9.6396
61	0.5139	7.5301	1.5005	5.5835	111	0.0630	6.8940	0.7890	9.7216
62	0.4925	7.5129	1.4747	5.6686	112	0.0604	6.8862	0.7815	9.8036
63	0.4721	7.4959	1.4496	5.7535	113	0.0579	6.8786	0.7742	9.8857
64	0.4525	7.4789	1.4253	5.8382	114	0.0555	6.8710	0.7670	9.9677
65	0.4337	7.4620	1.4018	5.9227	115	0.0533	6.8637	0.7599	10.0498
66	0.4158	7.4452	1.3790	6.0071	116	0.0511	6.8565	0.7530	10.1318
67	0.3986	7.4286	1.3569	6.0913	117	0.0490	6.8494	0.7461	10.2139
68	0.3821	7.4121	1.3355	6.1753	118	0.0470	6.8425	0.7394	10.2960
69	0.3663	7.3957	1.3147	6.2592	119	0.0450	6.8357	0.7329	10.3781
70	0.3512	7.3795	1.2945	6.3430	120	0.0432	6.8290	0.7264	10.4602
71	0.3367	7.3635	1.2749	6.4266	121	0.0414	6.8225	0.7201	10.5423
72	0.3228	7.3477	1.2558	6.5101	122	0.0397	6.8161	0.7138	10.6245
73	0.3095	7.3321	1.2373	6.5935	123	0.0381	6.8099	0.7077	10.7067
74	0.2967	7.3166	1.2193	6.6768	124	0.0365	6.8037	0.7017	10.7888
75	0.2845	7.3014	1.2018	6.7600	125	0.0350	6.7977	0.6957	10.8711
76	0.2728	7.2864	1.1847	6.8430	126	0.0336	6.7919	0.6899	10.9533
77	0.2616	7.2716	1.1682	6.9260	127	0.0322	6.7861	0.6842	11.0355
78	0.2508	7.2570	1.1520	7.0089	128	0.0309	6.7804	0.6786	11.1178
79	0.2405	7.2426	1.1363	7.0918	129	0.0296	6.7749	0.6730	11.2001
80	0.2306	7.2285	1.1210	7.1745	130	0.0284	6.7695	0.6676	11.2824

TABLE 12f Zero Thickness Microstrip Dimensions, Effective Dielectric Constant, and PUL Capacitance and Inductance for Er = 11.6

Zo [Ohms]	W/h	Keff	pF/cm	nH/mm	Zo [Ohms]	W/h	Keff	pF/cm	nH/mm
81	0.2211	7.2146	1.1061	7.2572	131	0.0273	6.7642	0.6622	11.3647
82	0.2121	7.2009	1.0916	7.3398	132	0.0261	6.7590	0.6570	11.4471
83	0.2033	7.1874	1.0774	7.4224	133	0.0251	6.7539	0.6518	11.5294
84	0.1950	7.1741	1.0636	7.5049	134	0.0240	6.7489	0.6467	11.6118
85	0.1870	7.1611	1.0501	7.5873	135	0.0231	6.7440	0.6417	11.6943
86	0.1793	7.1483	1.0370	7.6697	136	0.0221	6.7392	0.6367	11.7767
87	0.1719	7.1357	1.0242	7.7520	137	0.0212	6.7345	0.6318	11.8592
88	0.1649	7.1233	1.0117	7.8343	138	0.0203	6.7300	0.6271	11.9417
89	0.1581	7.1111	0.9994	7.9166	139	0.0195	6.7255	0.6223	12.0242
90	0.1516	7.0992	0.9875	7.9988	140	0.0187	6.7210	0.6177	12.1067
91	0.1454	7.0874	0.9758	8.0810	141	0.0179	6.7167	0.6131	12.1893
92	0.1394	7.0759	0.9645	8.1632	142	0.0172	6.7125	0.6086	12.2718
93	0.1337	7.0646	0.9533	8.2453	143	0.0165	6.7084	0.6042	12.3545
94	0.1282	7.0535	0.9424	8.3274	144	0.0158	6.7043	0.5998	12.4371
95	0.1230	7.0426	0.9318	8.4095	145	0.0152	6.7003	0.5955	12.5197
96	0.1179	7.0319	0.9214	8.4916	146	0.0146	6.6964	0.5912	12.6024
97	0.1131	7.0214	0.9112	8.5736	147	0.0140	6.6926	0.5870	12.6851
98	0.1085	7.0111	0.9013	8.6556	148	0.0134	6.6889	0.5829	12.7679
99	0.1040	7.0010	0.8915	8.7377	149	0.0128	6.6852	0.5788	12.8506
100	0.0998	6.9911	0.8820	8.8197	150	0.0123	6.6817	0.5748	12.9334

Notes: Calculation of W/H has an error of less than 1%.

Source: Gupta, K.C., Garg, R., Bahl, I., Bhartia, P., *Microstrip Lines and Slotlines,* 2nd Ed., Artech House, Norwood, MA, 1996, 103.

TABLE 12g Zero Thickness Microstrip Dimensions, Effective Dielectric Constant, and PUL Capacitance and Inductance for Er =11.9

Zo [Ohms]	W/h	Keff	pF/cm	nH/mm	Zo [Ohms]	W/h	Keff	pF/cm	nH/mm
1	106.5298	11.6168	113.6899	0.1137	55	0.6453	7.8084	1.6947	5.1265
2	52.1270	11.3637	56.2223	0.2249	56	0.6179	7.7908	1.6626	5.2139
3	34.0530	11.1365	37.1049	0.3339	57	0.5918	7.7731	1.6316	5.3009
4	25.0454	10.9312	27.5710	0.4411	58	0.5667	7.7554	1.6016	5.3878
5	19.6583	10.7446	21.8678	0.5467	59	0.5428	7.7375	1.5726	5.4744
6	16.0785	10.5741	18.0780	0.6508	60	0.5199	7.7197	1.5446	5.5607
7	13.5298	10.4175	15.3802	0.7536	61	0.4980	7.7019	1.5176	5.6469
8	11.6245	10.2730	13.3640	0.8553	62	0.4771	7.6841	1.4914	5.7328
9	10.1474	10.1390	11.8014	0.9559	63	0.4571	7.6664	1.4660	5.8186
10	8.9696	10.0144	10.5558	1.0556	64	0.4379	7.6488	1.4414	5.9041
11	8.0091	9.8981	9.5403	1.1544	65	0.4195	7.6313	1.4176	5.9895
12	7.2113	9.7891	8.6970	1.2524	66	0.4020	7.6140	1.3946	6.0748
13	6.5385	9.6867	7.9859	1.3496	67	0.3852	7.5968	1.3722	6.1598
14	5.9637	9.5902	7.3784	1.4462	68	0.3691	7.5797	1.3505	6.2447
15	5.4672	9.4991	6.8538	1.5421	69	0.3536	7.5628	1.3295	6.3295
16	5.0342	9.4128	6.3961	1.6374	70	0.3389	7.5462	1.3090	6.4142
17	4.6534	9.3309	5.9937	1.7322	71	0.3247	7.5297	1.2892	6.4987
18	4.3160	9.2531	5.6370	1.8264	72	0.3112	7.5134	1.2699	6.5831
19	4.0152	9.1789	5.3189	1.9201	73	0.2982	7.4973	1.2511	6.6674
20	3.7454	9.1081	5.0334	2.0134	74	0.2858	7.4814	1.2329	6.7516
21	3.5020	9.0404	4.7759	2.1062	75	0.2739	7.4658	1.2152	6.8356
22	3.2816	8.9755	4.5424	2.1985	76	0.2625	7.4504	1.1980	6.9196
23	3.0810	8.9134	4.3298	2.2905	77	0.2516	7.4352	1.1812	7.0035
24	2.8977	8.8536	4.1355	2.3820	78	0.2411	7.4202	1.1649	7.0873
25	2.7297	8.7961	3.9572	2.4732	79	0.2311	7.4055	1.1490	7.1711

TABLE 12g (continued) Zero Thickness Microstrip Dimensions, Effective Dielectric Constant, and PUL Capacitance and Inductance for Er =11.9

Zo [Ohms]	W/h	Keff	pF/cm	nH/mm	Zo [Ohms]	W/h	Keff	pF/cm	nH/mm
26	2.5751	8.7408	3.7930	2.5641	80	0.2215	7.3910	1.1335	7.2547
27	2.4324	8.6874	3.6413	2.6545	81	0.2123	7.3767	1.1185	7.3383
28	2.3004	8.6359	3.5009	2.7447	82	0.2035	7.3627	1.1038	7.4218
29	2.1779	8.5861	3.3704	2.8345	83	0.1950	7.3489	1.0895	7.5053
30	2.0640	8.5378	3.2489	2.9240	84	0.1869	7.3353	1.0755	7.5887
31	1.9578	8.4911	3.1355	3.0132	85	0.1791	7.3220	1.0619	7.6721
32	1.8513	8.4425	3.0288	3.1014	86	0.1717	7.3089	1.0486	7.7554
33	1.7607	8.3995	2.9295	3.1902	87	0.1646	7.2960	1.0356	7.8386
34	1.6756	8.3577	2.8362	3.2787	88	0.1577	7.2834	1.0230	7.9219
35	1.5956	8.3171	2.7485	3.3669	89	0.1512	7.2709	1.0106	8.0051
36	1.5204	8.2776	2.6658	3.4549	90	0.1449	7.2588	0.9985	8.0882
37	1.4494	8.2391	2.5877	3.5426	91	0.1389	7.2468	0.9868	8.1713
38	1.3824	8.2016	2.5139	3.6301	92	0.1331	7.2350	0.9752	8.2544
39	1.3190	8.1651	2.4440	3.7173	93	0.1276	7.2235	0.9640	8.3375
40	1.2591	8.1295	2.3777	3.8043	94	0.1223	7.2122	0.9530	8.4206
41	1.2023	8.0947	2.3147	3.8910	95	0.1172	7.2011	0.9422	8.5036
42	1.1484	8.0607	2.2548	3.9775	96	0.1124	7.1902	0.9317	8.5866
43	1.0973	8.0275	2.1979	4.0639	97	0.1077	7.1795	0.9214	8.6696
44	1.0487	7.9951	2.1436	4.1499	98	0.1033	7.1691	0.9113	8.7526
45	1.0026	7.9633	2.0918	4.2358	99	0.0990	7.1588	0.9015	8.8356
46	0.9586	7.9500	2.0446	4.3263	100	0.0949	7.1487	0.8919	8.9185
47	0.9168	7.9367	1.9994	4.4167	101	0.0909	7.1388	0.8824	9.0015
48	0.8770	7.9226	1.9560	4.5067	102	0.0872	7.1292	0.8732	9.0845
49	0.8391	7.9078	1.9143	4.5962	103	0.0836	7.1197	0.8641	9.1674
50	0.8029	7.8923	1.8742	4.6854	104	0.0801	7.1104	0.8552	9.2504
51	0.7684	7.8762	1.8356	4.7743	105	0.0768	7.1012	0.8466	9.3333
52	0.7354	7.8598	1.7984	4.8628	106	0.0736	7.0923	0.8380	9.4163
53	0.7040	7.8429	1.7626	4.9510	107	0.0705	7.0835	0.8297	9.4992
54	0.6740	7.8258	1.7280	5.0389	108	0.0676	7.0749	0.8215	9.5822
109	0.0648	7.0665	0.8135	9.6652	130	0.0266	6.9243	0.6752	11.4107
110	0.0621	7.0583	0.8056	9.7481	131	0.0255	6.9190	0.6698	11.4940
111	0.0595	7.0502	0.7979	9.8311	132	0.0245	6.9138	0.6645	11.5774
112	0.0571	7.0423	0.7903	9.9141	133	0.0235	6.9086	0.6592	11.6608
113	0.0547	7.0345	0.7829	9.9971	134	0.0225	6.9036	0.6541	11.7442
114	0.0524	7.0269	0.7756	10.0801	135	0.0216	6.8987	0.6490	11.8276
115	0.0503	7.0195	0.7685	10.1632	136	0.0207	6.8939	0.6440	11.9110
116	0.0482	7.0122	0.7615	10.2462	137	0.0198	6.8892	0.6391	11.9945
117	0.0462	7.0050	0.7546	10.3293	138	0.0190	6.8845	0.6342	12.0780
118	0.0443	6.9980	0.7478	10.4123	139	0.0182	6.8800	0.6294	12.1615
119	0.0424	6.9911	0.7412	10.4954	140	0.0174	6.8756	0.6247	12.2451
120	0.0407	6.9844	0.7346	10.5785	141	0.0167	6.8713	0.6201	12.3287
121	0.0390	6.9778	0.7282	10.6617	142	0.0160	6.8670	0.6156	12.4123
122	0.0374	6.9714	0.7219	10.7448	143	0.0154	6.8628	0.6111	12.4959
123	0.0358	6.9651	0.7157	10.8280	144	0.0147	6.8588	0.6067	12.5795
124	0.0343	6.9589	0.7096	10.9112	145	0.0141	6.8548	0.6023	12.6632
125	0.0329	6.9528	0.7036	10.9944	146	0.0135	6.8509	0.5980	12.7469
126	0.0316	6.9469	0.6978	11.0776	147	0.0130	6.8471	0.5938	12.8306
127	0.0302	6.9411	0.6920	11.1608	148	0.0124	6.8433	0.5896	12.9144
128	0.0290	6.9354	0.6863	11.2441	149	0.0119	6.8396	0.5855	12.9982
129	0.0278	6.9298	0.6807	11.3274	150	0.0114	6.8361	0.5814	13.0820

Notes: Calculation of W/H has an error of less than 1%.

Source: Gupta, K.C., Garg, R., Bahl, I., Bhartia, P., *Microstrip Lines and Slotlines,* 2nd Ed., Artech House, Norwood, MA, 1996, 103.

TABLE 12h   Zero Thickness Microstrip Dimensions, Effective Dielectric Constant, and PUL Capacitance and Inductance for Er =12.88

Zo [Ohms]	W/h	Keff	pF/cm	nH/mm	Zo [Ohms]	W/h	Keff	pF/cm	nH/mm
1	102.3159	12.5596	118.2135	0.1182	24	2.7402	9.5011	4.2841	2.4676
2	50.0315	12.2746	58.4323	0.2337	25	2.5791	9.4384	4.0991	2.5619
3	32.6633	12.0197	38.5483	0.3469	26	2.4309	9.3779	3.9288	2.6559
4	24.0084	11.7903	28.6339	0.4581	27	2.2942	9.3197	3.7715	2.7494
5	18.8329	11.5823	22.7043	0.5676	28	2.1677	9.2635	3.6258	2.8427
6	15.3941	11.3928	18.7648	0.6755	29	2.0503	9.2091	3.4905	2.9355
7	12.9460	11.2191	15.9610	0.7821	30	1.9316	9.1518	3.3637	3.0273
8	11.1161	11.0591	13.8660	0.8874	31	1.8324	9.1020	3.2463	3.1197
9	9.6977	10.9111	12.2425	0.9916	32	1.7397	9.0536	3.1365	3.2117
10	8.5668	10.7736	10.9487	1.0949	33	1.6528	9.0068	3.0335	3.3035
11	7.6446	10.6455	9.8939	1.1972	34	1.5714	8.9612	2.9369	3.3950
12	6.8787	10.5255	9.0182	1.2986	35	1.4948	8.9170	2.8459	3.4862
13	6.2329	10.4130	8.2799	1.3993	36	1.4228	8.8739	2.7602	3.5772
14	5.6812	10.3071	7.6492	1.4993	37	1.3549	8.8320	2.6792	3.6678
15	5.2048	10.2071	7.1046	1.5985	38	1.2908	8.7911	2.6027	3.7582
16	4.7893	10.1125	6.6296	1.6972	39	1.2302	8.7513	2.5302	3.8484
17	4.4239	10.0228	6.2119	1.7952	40	1.1729	8.7125	2.4614	3.9383
18	4.1003	9.9376	5.8418	1.8927	41	1.1186	8.6745	2.3962	4.0280
19	3.8117	9.8565	5.5117	1.9897	42	1.0672	8.6375	2.3341	4.1174
20	3.5529	9.7791	5.2155	2.0862	43	1.0184	8.6014	2.2751	4.2066
21	3.3196	9.7051	4.9483	2.1822	44	0.9721	8.5791	2.2205	4.2989
22	3.1082	9.6342	4.7061	2.2778	45	0.9281	8.5645	2.1693	4.3928
23	2.9159	9.5663	4.4856	2.3729	46	0.8863	8.5487	2.1202	4.4863
47	0.8465	8.5321	2.0731	4.5794	99	0.0845	7.6726	0.9333	9.1471
48	0.8086	8.5147	2.0278	4.6720	100	0.0808	7.6620	0.9233	9.2331
49	0.7726	8.4967	1.9843	4.7643	101	0.0774	7.6516	0.9136	9.3192
50	0.7383	8.4781	1.9425	4.8562	102	0.0740	7.6415	0.9040	9.4052
51	0.7056	8.4591	1.9023	4.9478	103	0.0709	7.6315	0.8946	9.4912
52	0.6744	8.4397	1.8635	5.0390	104	0.0678	7.6218	0.8855	9.5773
53	0.6447	8.4201	1.8263	5.1299	105	0.0649	7.6123	0.8765	9.6633
54	0.6163	8.4002	1.7903	5.2206	106	0.0621	7.6030	0.8677	9.7494
55	0.5893	8.3802	1.7557	5.3109	107	0.0595	7.5938	0.8591	9.8354
56	0.5634	8.3601	1.7223	5.4010	108	0.0569	7.5849	0.8506	9.9215
57	0.5388	8.3399	1.6900	5.4908	109	0.0545	7.5761	0.8423	10.0076
58	0.5152	8.3198	1.6588	5.5804	110	0.0521	7.5675	0.8342	10.0937
59	0.4928	8.2996	1.6288	5.6697	111	0.0499	7.5592	0.8262	10.1798
60	0.4713	8.2796	1.5997	5.7588	112	0.0477	7.5509	0.8184	10.2659
61	0.4508	8.2596	1.5716	5.8477	113	0.0457	7.5429	0.8107	10.3521
62	0.4312	8.2397	1.5443	5.9365	114	0.0437	7.5350	0.8032	10.4382
63	0.4125	8.2200	1.5180	6.0250	115	0.0418	7.5273	0.7958	10.5244
64	0.3946	8.2005	1.4925	6.1133	116	0.0400	7.5198	0.7885	10.6106
65	0.3775	8.1811	1.4678	6.2015	117	0.0383	7.5124	0.7814	10.6968
66	0.3611	8.1619	1.4439	6.2895	118	0.0367	7.5052	0.7744	10.7830
67	0.3455	8.1429	1.4207	6.3774	119	0.0351	7.4981	0.7676	10.8693
68	0.3306	8.1242	1.3982	6.4652	120	0.0336	7.4912	0.7608	10.9556
69	0.3163	8.1057	1.3763	6.5528	121	0.0321	7.4844	0.7542	11.0419
70	0.3026	8.0874	1.3551	6.6402	122	0.0308	7.4778	0.7477	11.1282
71	0.2895	8.0694	1.3346	6.7276	123	0.0294	7.4713	0.7413	11.2146
72	0.2770	8.0517	1.3146	6.8148	124	0.0282	7.4649	0.7350	11.3009
73	0.2651	8.0342	1.2952	6.9020	125	0.0270	7.4587	0.7288	11.3873
74	0.2536	8.0169	1.2763	6.9890	126	0.0258	7.4526	0.7227	11.4737
75	0.2427	8.0000	1.2579	7.0759	127	0.0247	7.4467	0.7167	11.5602
76	0.2323	7.9833	1.2401	7.1628	128	0.0236	7.4409	0.7109	11.6466
77	0.2222	7.9669	1.2227	7.2496	129	0.0226	7.4352	0.7051	11.7331
78	0.2127	7.9507	1.2058	7.3363	130	0.0217	7.4296	0.6994	11.8197
79	0.2035	7.9348	1.1894	7.4229	131	0.0207	7.4241	0.6938	11.9062

TABLE 12h (continued) Zero Thickness Microstrip Dimensions, Effective Dielectric Constant, and PUL Capacitance and Inductance for Er =12.88

Zo [Ohms]	W/h	Keff	pF/cm	nH/mm	Zo [Ohms]	W/h	Keff	pF/cm	nH/mm
80	0.1947	7.9192	1.1734	7.5095	132	0.0198	7.4188	0.6883	11.9928
81	0.1864	7.9039	1.1578	7.5960	133	0.0190	7.4136	0.6829	12.0794
82	0.1783	7.8889	1.1425	7.6825	134	0.0182	7.4085	0.6775	12.1660
83	0.1707	7.8741	1.1277	7.7689	135	0.0174	7.4035	0.6723	12.2527
84	0.1633	7.8596	1.1133	7.8552	136	0.0166	7.3986	0.6671	12.3393
85	0.1563	7.8453	1.0992	7.9415	137	0.0159	7.3938	0.6621	12.4261
86	0.1496	7.8314	1.0854	8.0278	138	0.0152	7.3891	0.6570	12.5128
87	0.1431	7.8176	1.0720	8.1140	139	0.0146	7.3845	0.6521	12.5996
88	0.1370	7.8042	1.0589	8.2002	140	0.0140	7.3800	0.6473	12.6863
89	0.1311	7.7910	1.0461	8.2864	141	0.0134	7.3756	0.6425	12.7732
90	0.1255	7.7781	1.0336	8.3725	142	0.0128	7.3713	0.6378	12.8600
91	0.1201	7.7654	1.0215	8.4587	143	0.0122	7.3671	0.6331	12.9469
92	0.1149	7.7529	1.0095	8.5448	144	0.0117	7.3630	0.6286	13.0338
93	0.1100	7.7407	0.9979	8.6309	145	0.0112	7.3590	0.6241	13.1207
94	0.1052	7.7288	0.9865	8.7169	146	0.0107	7.3551	0.6196	13.2077
95	0.1007	7.7171	0.9754	8.8030	147	0.0103	7.3512	0.6152	13.2946
96	0.0964	7.7056	0.9645	8.8890	148	0.0098	7.3475	0.6109	13.3817
97	0.0922	7.6944	0.9539	8.9751	149	0.0094	7.3438	0.6067	13.4687
98	0.0883	7.6833	0.9435	9.0611	150	0.0090	7.3402	0.6025	13.5558

Notes: Calculation of W/H has an error of less than 1%.

Source: Gupta, K.C., Garg, R., Bahl, I., Bhartia, P., *Microstrip Lines and Slotlines,* 2nd Ed., Artech House, Norwood, MA, 1996, 103.

TABLE 12i Zero Thickness Microstrip Dimensions, Effective Dielectric Constant, and PUL Capacitance and Inductance for Er =35

Zo [Ohms]	W/h	Keff	pF/cm	nH/mm	Zo [Ohms]	W/h	Keff	pF/cm	nH/mm
1	61.2525	33.5453	193.1947	0.1932	57	0.1141	20.3209	2.6380	8.5709
2	29.6179	32.3412	94.8478	0.3794	58	0.1063	20.2652	2.5890	8.7093
3	19.1302	31.3265	62.2321	0.5601	59	0.0990	20.2113	2.5417	8.8477
4	13.9141	30.4569	46.0216	0.7363	60	0.0923	20.1591	2.4961	8.9860
5	10.8011	29.7005	36.3573	0.9089	61	0.0860	20.1087	2.4521	9.1243
6	8.7367	29.0345	29.9561	1.0784	62	0.0801	20.0598	2.4096	9.2626
7	7.2699	28.4417	25.4132	1.2452	63	0.0746	20.0126	2.3686	9.4010
8	6.1757	27.9094	22.0275	1.4098	64	0.0695	19.9670	2.3289	9.5393
9	5.3291	27.4273	19.4101	1.5722	65	0.0648	19.9228	2.2906	9.6776
10	4.6555	26.9878	17.3286	1.7329	66	0.0604	19.8801	2.2534	9.8160
11	4.1073	26.5845	15.6351	1.8918	67	0.0562	19.8389	2.2175	9.9543
12	3.6529	26.2124	14.2315	2.0493	68	0.0524	19.7990	2.1827	10.0928
13	3.2705	25.8674	13.0501	2.2055	69	0.0488	19.7604	2.1490	10.2312
14	2.9445	25.5460	12.0424	2.3603	70	0.0455	19.7232	2.1163	10.3697
15	2.6635	25.2453	11.1732	2.5140	71	0.0424	19.6872	2.0846	10.5082
16	2.4189	24.9629	10.4161	2.6665	72	0.0395	19.6525	2.0538	10.6468
17	2.2042	24.6968	9.7510	2.8180	73	0.0368	19.6189	2.0239	10.7855
18	2.0144	24.4452	9.1623	2.9686	74	0.0343	19.5865	1.9949	10.9242
19	1.8358	24.1924	8.6351	3.1173	75	0.0319	19.5551	1.9667	11.0630
20	1.6891	23.9716	8.1658	3.2663	76	0.0298	19.5249	1.9394	11.2018
21	1.5569	23.7610	7.7427	3.4145	77	0.0277	19.4957	1.9127	11.3407
22	1.4372	23.5597	7.3594	3.5619	78	0.0258	19.4674	1.8869	11.4796
23	1.3284	23.3669	7.0106	3.7086	79	0.0241	19.4402	1.8617	11.6187
24	1.2292	23.1820	6.6918	3.8545	80	0.0224	19.4139	1.8372	11.7578
25	1.1385	23.0043	6.3995	3.9997	81	0.0209	19.3884	1.8133	11.8970

TABLE 12i (continued)   Zero Thickness Microstrip Dimensions, Effective Dielectric Constant, and PUL
Capacitance and Inductance for Er =35

Zo [Ohms]	W/h	Keff	pF/cm	nH/mm	Zo [Ohms]	W/h	Keff	pF/cm	nH/mm
26	1.0554	22.8334	6.1304	4.1442	82	0.0195	19.3639	1.7900	12.0362
27	0.9790	22.6972	5.8857	4.2907	83	0.0181	19.3402	1.7674	12.1755
28	0.9087	22.6289	5.6670	4.4429	84	0.0169	19.3173	1.7453	12.3149
29	0.8438	22.5532	5.4624	4.5939	85	0.0157	19.2952	1.7238	12.4544
30	0.7840	22.4719	5.2708	4.7437	86	0.0147	19.2739	1.7028	12.5940
31	0.7286	22.3863	5.0911	4.8925	87	0.0137	19.2533	1.6823	12.7336
32	0.6774	22.2977	4.9222	5.0403	88	0.0127	19.2334	1.6624	12.8733
33	0.6300	22.2069	4.7633	5.1873	89	0.0119	19.2142	1.6429	13.0131
34	0.5861	22.1148	4.6136	5.3333	90	0.0111	19.1957	1.6238	13.1530
35	0.5453	22.0220	4.4724	5.4787	91	0.0103	19.1778	1.6052	13.2929
36	0.5074	21.9291	4.3390	5.6233	92	0.0096	19.1605	1.5871	13.4329
37	0.4723	21.8365	4.2128	5.7673	93	0.0089	19.1439	1.5693	13.5730
38	0.4397	21.7445	4.0933	5.9107	94	0.0083	19.1278	1.5520	13.7132
39	0.4093	21.6536	3.9800	6.0535	95	0.0078	19.1122	1.5350	13.8535
40	0.3811	21.5638	3.8724	6.1959	96	0.0072	19.0972	1.5184	13.9938
41	0.3549	21.4755	3.7702	6.3378	97	0.0067	19.0828	1.5022	14.1342
42	0.3305	21.3889	3.6730	6.4792	98	0.0063	19.0688	1.4863	14.2747
43	0.3078	21.3039	3.5805	6.6203	99	0.0059	19.0553	1.4708	14.4152
44	0.2867	21.2208	3.4923	6.7610	100	0.0055	19.0422	1.4556	14.5559
45	0.2670	21.1396	3.4081	6.9014	101	0.0051	19.0297	1.4407	14.6966
46	0.2487	21.0603	3.3278	7.0416	102	0.0047	19.0175	1.4261	14.8374
47	0.2317	20.9831	3.2510	7.1814	103	0.0044	19.0058	1.4118	14.9782
48	0.2158	20.9078	3.1776	7.3211	104	0.0041	18.9945	1.3978	15.1191
49	0.2010	20.8346	3.1073	7.4605	105	0.0038	18.9836	1.3841	15.2601
50	0.1873	20.7635	3.0399	7.5998	106	0.0036	18.9730	1.3707	15.4012
51	0.1745	20.6944	2.9753	7.7388	107	0.0033	18.9629	1.3575	15.5423
52	0.1625	20.6272	2.9134	7.8778	108	0.0031	18.9530	1.3446	15.6835
53	0.1514	20.5621	2.8539	8.0166	109	0.0029	18.9436	1.3319	15.8247
54	0.1411	20.4989	2.7967	8.1553	110	0.0027	18.9344	1.3195	15.9661
55	0.1314	20.4377	2.7418	8.2939	111	0.0025	18.9256	1.3073	16.1074
56	0.1224	20.3784	2.6889	8.4324	112	0.0023	18.9170	1.2954	16.2489
113	0.0022	18.9088	1.2836	16.3904	132	0.0006	18.7969	1.0956	19.0896
114	0.0020	18.9009	1.2721	16.5320	133	0.0005	18.7928	1.0872	19.2321
115	0.0019	18.8932	1.2608	16.6736	134	0.0005	18.7889	1.0790	19.3747
116	0.0018	18.8858	1.2497	16.8153	135	0.0005	18.7851	1.0709	19.5173
117	0.0016	18.8786	1.2387	16.9571	136	0.0004	18.7815	1.0629	19.6600
118	0.0015	18.8717	1.2280	17.0989	137	0.0004	18.7779	1.0551	19.8027
119	0.0014	18.8651	1.2175	17.2407	138	0.0004	18.7745	1.0473	19.9454
120	0.0013	18.8587	1.2071	17.3826	139	0.0003	18.7713	1.0397	20.0882
121	0.0012	18.8524	1.1970	17.5246	140	0.0003	18.7681	1.0322	20.2310
122	0.0012	18.8465	1.1870	17.6666	141	0.0003	18.7650	1.0248	20.3738
123	0.0011	18.8407	1.1771	17.8087	142	0.0003	18.7621	1.0175	20.5167
124	0.0010	18.8351	1.1675	17.9508	143	0.0003	18.7592	1.0103	20.6596
125	0.0009	18.8297	1.1580	18.0930	144	0.0002	18.7565	1.0032	20.8026
126	0.0009	18.8245	1.1486	18.2352	145	0.0002	18.7538	0.9962	20.9456
127	0.0008	18.8195	1.1394	18.3775	146	0.0002	18.7513	0.9893	21.0886
128	0.0008	18.8146	1.1304	18.5198	147	0.0002	18.7488	0.9825	21.2316
129	0.0007	18.8100	1.1215	18.6622	148	0.0002	18.7464	0.9758	21.3747
130	0.0007	18.8055	1.1127	18.8046	149	0.0002	18.7441	0.9692	21.5178
131	0.0006	18.8011	1.1041	18.9471	150	0.0002	18.7419	0.9627	21.6609

Notes: Calculation of W/H has an error of less than 1%.
Source: Gupta, K.C., Garg, R., Bahl, I., Bhartia, P., *Microstrip Lines and Slotlines,* 2nd Ed., Artech House, Norwood, MA, 1996, 103.

TABLE 12j Zero Thickness Microstrip Dimensions, Effective Dielectric Constant, and PUL Capacitance and Inductance for Er =85

Zo [Ohms]	W/h	Keff	pF/cm	nH/mm	Zo [Ohms]	W/h	Keff	pF/cm	nH/mm
1	38.5925	79.6824	297.7560	0.2978	26	0.3739	51.7462	9.2288	6.2387
2	18.3705	75.6651	145.0765	0.5803	27	0.3349	51.4124	8.8583	6.4577
3	11.6859	72.5009	94.6738	0.8521	28	0.3000	51.0886	8.5150	6.6757
4	8.3710	69.9235	69.7318	1.1157	29	0.2688	50.7757	8.1961	6.8930
5	6.3982	67.7680	54.9189	1.3730	30	0.2409	50.4745	7.8994	7.1095
6	5.0937	65.9270	45.1398	1.6250	31	0.2159	50.1850	7.6226	7.3253
7	4.1695	64.3277	38.2191	1.8727	32	0.1935	49.9076	7.3640	7.5407
8	3.4821	62.9184	33.0733	2.1167	33	0.1734	49.6422	7.1218	7.7557
9	2.9517	61.6613	29.1034	2.3574	34	0.1554	49.3886	6.8947	7.9702
10	2.5309	60.5285	25.9513	2.5951	35	0.1393	49.1467	6.6813	8.1845
11	2.1895	59.4981	23.3904	2.8302	36	0.1249	48.9161	6.4804	8.3986
12	1.8946	58.5089	21.2622	3.0618	37	0.1119	48.6966	6.2911	8.6125
13	1.6651	57.6611	19.4840	3.2928	38	0.1003	48.4877	6.1124	8.8263
14	1.4696	56.8730	17.9682	3.5218	39	0.0899	48.2891	5.9435	9.0400
15	1.3012	56.1365	16.6614	3.7488	40	0.0806	48.1003	5.7835	9.2537
16	1.1551	55.4455	15.5236	3.9740	41	0.0723	47.9210	5.6319	9.4673
17	1.0274	54.7949	14.5245	4.1976	42	0.0648	47.7509	5.4881	9.6810
18	0.9153	54.4535	13.6748	4.4306	43	0.0581	47.5893	5.3514	9.8947
19	0.8165	54.1607	12.9202	4.6642	44	0.0521	47.4361	5.2213	10.1085
20	0.7290	53.8383	12.2376	4.8950	45	0.0467	47.2907	5.0975	10.3224
21	0.6514	53.4976	11.6179	5.1235	46	0.0418	47.1529	4.9794	10.5364
22	0.5825	53.1467	11.0534	5.3498	47	0.0375	47.0222	4.8667	10.7505
23	0.5211	52.7921	10.5375	5.5743	48	0.0336	46.8984	4.7590	10.9648
24	0.4664	52.4383	10.0645	5.7972	49	0.0301	46.7810	4.6561	11.1792
25	0.4176	52.0888	9.6297	6.0185	50	0.0270	46.6697	4.5575	11.3938
51	0.0242	46.5643	4.4631	11.6085	101	0.0001	44.8028	2.2106	22.5503
52	0.0217	46.4644	4.3726	11.8234	102	0.0001	44.7963	2.1888	22.7720
53	0.0195	46.3698	4.2857	12.0385	103	0.0001	44.7901	2.1674	22.9936
54	0.0175	46.2801	4.2023	12.2538	104	0.0001	44.7842	2.1464	23.2153
55	0.0156	46.1952	4.1221	12.4693	105	0.0001	44.7787	2.1258	23.4371
56	0.0140	46.1148	4.0449	12.6849	106	0.0001	44.7734	2.1056	23.6589
57	0.0126	46.0386	3.9707	12.9008	107	0.0001	44.7685	2.0858	23.8808
58	0.0113	45.9665	3.8992	13.1168	108	0.0000	44.7638	2.0664	24.1027
59	0.0101	45.8982	3.8302	13.3330	109	0.0000	44.7593	2.0474	24.3247
60	0.0091	45.8335	3.7637	13.5495	110	0.0000	44.7551	2.0287	24.5467
61	0.0081	45.7722	3.6996	13.7661	111	0.0000	44.7511	2.0103	24.7688
62	0.0073	45.7142	3.6376	13.9829	112	0.0000	44.7473	1.9923	24.9908
63	0.0065	45.6592	3.5777	14.1999	113	0.0000	44.7437	1.9745	25.2130
64	0.0059	45.6072	3.5198	14.4170	114	0.0000	44.7404	1.9572	25.4351
65	0.0052	45.5579	3.4638	14.6344	115	0.0000	44.7371	1.9401	25.6573
66	0.0047	45.5112	3.4095	14.8519	116	0.0000	44.7341	1.9233	25.8795
67	0.0042	45.4670	3.3570	15.0696	117	0.0000	44.7312	1.9068	26.1018
68	0.0038	45.4252	3.3061	15.2875	118	0.0000	44.7285	1.8906	26.3241
69	0.0034	45.3856	3.2568	15.5056	119	0.0000	44.7259	1.8746	26.5464
70	0.0030	45.3481	3.2089	15.7238	120	0.0000	44.7235	1.8589	26.7688
71	0.0027	45.3126	3.1625	15.9422	121	0.0000	44.7212	1.8435	26.9911
72	0.0024	45.2789	3.1174	16.1607	122	0.0000	44.7190	1.8284	27.2135
73	0.0022	45.2471	3.0736	16.3794	123	0.0000	44.7169	1.8135	27.4360
74	0.0020	45.2169	3.0311	16.5982	124	0.0000	44.7149	1.7988	27.6584
75	0.0018	45.1884	2.9897	16.8172	125	0.0000	44.7131	1.7844	27.8809
76	0.0016	45.1614	2.9495	17.0363	126	0.0000	44.7113	1.7702	28.1034
77	0.0014	45.1358	2.9104	17.2556	127	0.0000	44.7097	1.7562	28.3259
78	0.0013	45.1115	2.8723	17.4750	128	0.0000	44.7081	1.7425	28.5484
79	0.0011	45.0886	2.8352	17.6946	129	0.0000	44.7066	1.7289	28.7710
80	0.0010	45.0669	2.7991	17.9142	130	0.0000	44.7052	1.7156	28.9936

TABLE 12j (continued)   Zero Thickness Microstrip Dimensions, Effective Dielectric Constant, and PUL
Capacitance and Inductance for Er =85

Zo [Ohms]	W/h	Keff	pF/cm	nH/mm	Zo [Ohms]	W/h	Keff	pF/cm	nH/mm
81	0.0009	45.0463	2.7639	18.1340	131	0.0000	44.7038	1.7025	29.2162
82	0.0008	45.0268	2.7296	18.3539	132	0.0000	44.7026	1.6896	29.4388
83	0.0007	45.0084	2.6962	18.5739	133	0.0000	44.7014	1.6768	29.6614
84	0.0007	44.9909	2.6636	18.7941	134	0.0000	44.7002	1.6643	29.8840
85	0.0006	44.9744	2.6317	19.0143	135	0.0000	44.6992	1.6519	30.1067
86	0.0005	44.9587	2.6007	19.2347	136	0.0000	44.6981	1.6398	30.3293
87	0.0005	44.9439	2.5704	19.4551	137	0.0000	44.6972	1.6278	30.5520
88	0.0004	44.9299	2.5408	19.6757	138	0.0000	44.6963	1.6160	30.7747
89	0.0004	44.9166	2.5118	19.8963	139	0.0000	44.6954	1.6043	30.9974
90	0.0003	44.9040	2.4836	20.1171	140	0.0000	44.6946	1.5929	31.2201
91	0.0003	44.8921	2.4560	20.3379	141	0.0000	44.6938	1.5816	31.4429
92	0.0003	44.8808	2.4290	20.5588	142	0.0000	44.6931	1.5704	31.6656
93	0.0002	44.8701	2.4026	20.7798	143	0.0000	44.6924	1.5594	31.8883
94	0.0002	44.8600	2.3767	21.0009	144	0.0000	44.6917	1.5486	32.1111
95	0.0002	44.8504	2.3515	21.2220	145	0.0000	44.6911	1.5379	32.3339
96	0.0002	44.8414	2.3267	21.4432	146	0.0000	44.6905	1.5273	32.5567
97	0.0002	44.8328	2.3025	21.6645	147	0.0000	44.6899	1.5169	32.7794
98	0.0001	44.8247	2.2788	21.8859	148	0.0000	44.6894	1.5067	33.0022
99	0.0001	44.8170	2.2556	22.1073	149	0.0000	44.6889	1.4966	33.2250
100	0.0001	44.8097	2.2329	22.3288	150	0.0000	44.6884	1.4866	33.4478

Notes: Calculation of W/H has an error of less than 1%.

Source: Gupta, K.C., Garg, R., Bahl, I., Bhartia, P., *Microstrip Lines and Slotlines,* 2nd Ed., Artech House, Norwood, MA, 1996, 103.

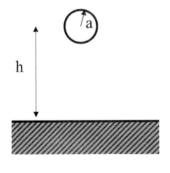

$$L_{WIREoverGND} = \frac{\mu_0 \mu_r}{2\pi} \cosh^{-1} \frac{h}{a} \ (nH/m) \qquad \text{for } h \gg a$$

FIGURE 5   Inductance of a wire over a ground plane. When consistent units are used for the wire radius, a, and the height over the ground plane, h, the formula provides an estimate of the inductance per unit length in (nH/m).

Bondwires, Ribbons, Mesh

The assembly of semiconductor and hybrid integrated circuits for microwave and millimeter-wave frequencies generally requires the use of gold bondwires, ribbons, or mesh. The impedance of the interconnection must be accounted for in a good design. Unfortunately, there is no single accepted electrical model. The complexity required of the model will depend on the frequency of operation and the general impedance levels of the circuits being connected. At low frequencies and moderate to high impedances, the connection is frequently modeled as an inductor (sometimes in series with a resistor); at high frequencies, a full 3-D electromagnetic simulation may be required for accurate results. At intermediate points it may be modeled as a high impedance transmission line or as a lumped LC circuit. Note that the resistances of the RF interconnects should be included in the design of extremely low-noise circuits as they will affect the noise figure. In connecting a semiconductor die to package leads, it may also be necessary to model the mutual inductances and interlead capacitances in addition to the usual self-inductances and shunt capacitance. Figure 5 illustrates one method of modeling bond wire inductance that has been shown adequate for many microwave applications. More sophisticated methods of modeling bond wires, ribbon or mesh are described in the references.

References

1. Grover, F.W., *Inductance Calculations*, Dover Publications, New York, Available through Instrument Society of America, Research Triangle Park, NC.
2. Caulton, M., Lumped Elements in Microwave Circuits, in *Advances in Microwaves 1974*, Academic Press, New York, 1974, 143–202.
3. Wadell, B. C., *Transmission Line Design Handbook*, Artech House, Boston, MA, 1991, 151–155.
4. Terman, F.E., *Radio Engineer's Handbook*, McGraw-Hill, New York, 1943, 48–51.
5. Caverly, R. H., Characteristic impedance of integrated circuit bond wires, *IEEE Transactions on Microwave Theory and Techniques*, MTT-34, 9, September 1986, 982–984.
6. Kuester, E. F., Chang, D. C., Propagating modes along a thin wire located above a grounded dielectric slab, *IEEE Transactions on Microwave Theory and Techniques*, MTT-25, 12, December 1977, 1065–1069.
7. Mondal, J. P., Octagonal spiral inductor measurement and modelling for MMIC applications, *Int. J. Electronics*, 1990, 68, 1, 113–125.
8. MIL-STD-883E, Test Method Standard, Microcircuits U.S. Department of Defense, Washington, D.C.

Index